10th Textile Bioengineering and Informatics Symposium Proceedings (TBIS 2017)

Wu Han, China
16 – 19 May 2017

Volume 1 of 2

Editors:

Yi Li
Wei Lin Xu

ISBN: 978-1-5108-4951-8

Printed from e-media with permission by:

Curran Associates, Inc.
57 Morehouse Lane
Red Hook, NY 12571

Some format issues inherent in the e-media version may also appear in this print version.

Printed by Curran Associates, Inc. (2018)

For permission requests, please contact Textile Bioengineering and Informatics Society (TBIS) at the address below.

Textile Bioengineering and Informatics Society (TBIS)
Unit 503, 5/F. Silvercord
Tower 2, 30 Canton Roa
Tsim Sha Tsui
Hong Kong

Phone: (852) 66382781

tbis@tbisociety.org

Additional copies of this publication are available from:

Curran Associates, Inc.
57 Morehouse Lane
Red Hook, NY 12571 USA
Phone: 845-758-0400
Fax: 845-758-2633
Email: curran@proceedings.com
Web: www.proceedings.com

TBIS

Textile Bioengineering and Informatics Symposium Proceedings

2017

ISSN:19423438

Edited by
Yi Li
Wei Lin Xu

The 10th Textile Bioengineering and Informatics Symposium

May 16-19, 2017 , Wuhan Textile University, Wu Han, China

Welcoming Message

Professor Wei Lin Xu
Chairman, TBIS 2017 Organizing Committee
Wuhan Textile University
Vice President of China Textile Engineering Society
Changjiang Scholar
China

Welcome to Wuhan!

TBIS is an international platform that nurtures the scientific textile community in the exchange and development of innovative and revolutionary ideas in the field of textiles and apparel, and it is a our great pleasure to host the incoming 2017 TBIS.

With the aim of cultivating textile specialists and serving China's textile industry, Wuhan Textile University (WTU) was found in 1958 in central China. Hubei's capital, Wuhan is a metropolis full of universities and talents. WTU is the only college in Central and Southwest China that features studies in textile and apparel, and there are more than 10,000 students studying in the field of textile and apparel. WTU adheres to upgrade its competence by implementing high-level featured strategies for development, promoting its innovative capacity by implementing the strategy for high level multidisciplinary development, achieving breakthroughs in its faculty team building by introducing and fostering high-end talents. WTU is now extending to different textile universities and forming many collaborations, we look forward to cooperating with you in the near future!

Welcoming Message from TBIS

Professor Yi Li
Chair of Textile Science and Engineering
School of Materials, The University of Manchester
Chairman, TBIS 2017 International Scientific Committee
Chairman, Textile Bioengineering and Informatics Society
Editor-in-Chief, Journal of Fiber Bioengineering and Informatics

With great pleasure, we are now collaborating with Wuhan Textile University to hold TBIS 2017, the 10th Textile Bioengineering and Informatics Symposium in May 16-19, 2017 in Wu Han China.

With continuous success and enthusiastic support from authors and experts around the world, Textile Bioengineering and Informatics Symposiums (TBIS) have been recognized internationally as a brand of high quality conference with vision, creativity, excitement and global network. TBIS proceedings and JFBI (Journal of Fiber Bioengineering and Informatics) have been indexed by worldwide scientific and academic reputable databases such as EI Compendex, Scopus, CPCI/ISI Web of Knowledge. Since 2008, TBIS has published 1686 research papers indexed in the CPCI/ISI Web of Knowledge database, becoming a leading academic conference organizer and publisher.

TBIS has established an effective platform for academic institutions and individual scholars, researchers and students to publish their research works and communicate with peers and experts in the field to get recognitions and increase impact such as citations and H-indexes. The TBIS symposium is a peer-reviewed conference aiming to set up a unique global stage and platform for people working in the whole supply chain from materials, fibers, textiles and apparel and technical textiles to relevant service industries such as biomedical industry, fashion industry and sport industry with common interests in providing high quality of products and services for protection, health, performance and comfort of human populations living in various environments and exposing to various natural and human disasters. To further enhance global networking, this year we have the pleasure to collaborate with Wuhan Textile University to organize TBIS in Wuhan.

TBIS 2017 welcomes you and is looking forward to your active participation and contribution.

TBIS 2017 Organizing and International Scientific Committees

Organizing Committee

Chairman:	Prof. Weilin Xu		
Vice Chairmen:	Prof. Keqin Zhang	Prof. Ruomei Wang	Prof. Xin Zhang
Secretariat:	Ms. Mai Wang Li	Dr. Xin Liu	Dr. Genyang Cao
Members:	Prof. Xue-Mei Ding	Prof. Jun Feng	Prof. Jianxun Gao
	Dr. Junyan Hu	Prof. Jianqiang Li	Prof. Xiangpeng Li
	Prof. Li Lin	Prof. Xiaonan Luo	Dr. Subramanian S.K. Muthu
	Prof. Xiaojun Tong	Prof. Dong Wang	Dr. Zhong Wang
	Prof. Xiongying Wu	Ms. Hui Wu	Prof. Zhaofei Xiong
	Prof. Zhaoyang Xu	Dr. Lei Yao	Prof. Jinbo Yao
	Dr. Boan Ying	Prof. Shangyong Zhang	

International Scientific Committee

Honorary Chairman:	Academician Mu Yao	China		
Chairman:	Prof. Henry Yi Li	UK		
Co-Chairmen:	Prof. Weilin Xu (China)	Prof. Frank Ko (Canada)	Prof. Kanji Kajiwara (Japan)	
Members:	Prof. Seung Kook An (Korea)	Prof. Senem K. Bahadir (Turkey)	Prof. Aizheng Chen (China)	Prof. Guoqiang Chen (China)
	Prof. Weiguo Chen (China)	Dr. Stewart R. Collie (New Zealand)	Prof. ZhaoQun Du (China)	Prof. Emiel Den Hartog (USA)
	Prof. Lubos Hes (Czech Republic)	Dr. Yang Hu (USA)	Prof. Tanveer Hussain (Pakistan)	Prof. Kap Jin Kim (Korea)
	Prof. Gang Li (China)	Dr. Luximon Ameersing (Hong Kong, China)	Prof. Jiri Militky (Czech Republic)	Prof. Qingqing Ni (Japan)
	Prof. Rajiv Padhye (Australia)	Prof. Mishra Rajesh (Czech Republic)	Prof. Tetsuya Sato (Japan)	Prof. Guowen Song (USA)
	Prof. Masayuki Takatera (Japan)	Prof. Kaoru Wakatsuki (Japan)	Prof. Qiang Wang (China)	Prof. Lijing Wang (Australia)
	Dr. Xin Wang (Australia)	Prof. Xungai Wang (Australia)	Prof. Qufu Wei (China)	Prof. Calvin Wong (Hong Kong, China)

Organizing and Supporting Institutions

Hosted by:

Wuhan Textile University
State Key Lab for New Textile
Materials and Advanced Technology

Co-organized by:

Textile Bioengineering and Informatics Society

Supporting Institutions

Aimer Human Engineering Research Centre (HEC)

Soochow University

Shaanxi Province's Research Center of Apparel Engineering and Technology

The University of Manchester

Xian Polytechnic University

Exclusive Sponsor

纤维谷
FIBRE VALLEY
R&D / COMMUNICATION
INCUBATION / CONSULTING
INVESTMENT / COOPERATION / MARKETING
Covering the whole industry chain of Fibre

PROCEEDINGS OF TBIS 2017

TABLE OF CONTENT

Track 1 Textile Biomedical Engineering

Track 2 Textile Biomechanical Engineering

Track 3 Clothing Bioengineering

Track 4 Textile Supply Chain Informatics

Track 5 Textile Thermal Bioengineering

Track 6 Textile Science and Engineering Chemistry

Track 7 Textile Science and Engineering--Physics

Track 8 Textile Material Bioengineering-Chemistry

Track 9 Textile Material Bioengineering-Physics

Fabrication and Water Permeability of Bifurcated Woven Endovascular Prosthesis

Ze-Kun Liu[1], Gang Li[1*], Yi Li[2], Yu-Ling Li[3], Xiao-Qin Wang[1]

[1]*Soochow University, Ren Ai Road, Suzhou, Jiangsu, 215123, China*
[2]*The University of Manchester, Oxford Road, Manchester, M13 9PL, UK*
[3]*Donghua University, North Renmin Road, Shanghai, 201620, China.*

[*]Corresponding author's email: tcligang@suda.edu.cn

Abstract

Blood vessel, as one of the important organs in the human body, playing the function of blood transportation, transfusion and nutrition exchanges. Proper treatment such as repair or replacement is needed when blood vessel tissue are injured. Therefore, it is crucial to develop ideal artificial vascular prosthesis. Using non-invasive surgery of endovascular prosthesis has the advantage of causing minimal injury to patients and returns them to health faster than traditional larger-scale surgeries, which can be one of the best methods for the treatment of vascular diseases. Substantial diseases in tissue bifurcations such as abdominal aorta bifurcations and the aortoiliac vessels have been found in the morphological structure of the cardiovascular system in the human body. Hence, endovascular prosthesis should not only have small wall thickness, but also low permeability. In our study, there is an orthogonal design of three factors at three different levels which contains weave, warp × weft density and warp × weft materials. Nine kinds of bifurcated woven endovascular prosthesis (BWEP) were prepared using the modified rigid rapier weaving loom and the weaving process was optimized, subsequently. Finally, the most important factors, water permeability and wall thickness were characterized according to ISO 7198:1998. The results showed that the BWEP made from pure silk possesses better performance than other groups in terms of wall thickness and water permeability. In conclusion, the silk BWEP may be more suitable as endovascular prosthesis in human body for transplant.

Keywords: Artificial Endovascular Prosthesis; Bifurcation; Woven; Wall Thickness; Water Permeability

1. Introduction

Endovascular graft exclusion has appeared for arteriosclerosis, thromboembolism, intracranial aneurysm and other artery expansion diseases. It has the advantages of causing small trauma, less bleeding, faster recovery and less complication compared with traditional remedy [1, 2]. Diagrams of endovascular graft exclusion and traditional remedy were shown in Fig.1. Laparotomy or thoracoscopic is not obligatory for the surgeon, it just opens a small mouth at the patients' arteries of the limbs, and then guide the stent-graft (SG) into the lesion area with a catheter [3]. SG expands through a balloon assisted technique to get back into shape, so as to isolate lesion blood vessel, ensuring the pathway. Endovascular prosthesis is made up of metal stent and SG. SG is important because it plays an important role in determining success of the endovascular graft exclusion. The application of mature artificial blood vessel in commercial field has been always made by using a sutural method, especially for the bifurcated woven endovascular prosthesis (BWEP). Consequently, it may cause exudation for the branched part.. Endovascular grafts should have high performance requirements when compared with ordinary artificial blood vessels. It should not only meet the general performance of the artificial blood vessels such as biocompatibility and long term stability, but also should have good seepage resistance under the condition of thinner tube wall. Based on previous studies [4, 5], the water permeability should be lower than 300ml/(cm^2 × min) and the wall thickness should be less than 0.1 mm, otherwise, exudation would happen after transplantation and guiding is hard during surgery procedures. However, it is different to balance the relationship between wall thickness and water permeability, because if the thickness of endovascular prosthesis is too thin, good anti-seepage function can't be guaranteed. So developing endovascular prosthesis with thinner thickness and better water permeability is of great significance. In fact, because of human's own individual differences, ideally the wall thickness should be less than 0.1 mm to prevent against seepage.

Biomedical polyester (PET) has been widely used in the preparation of artificial blood vessel for many years as it has good mechanical properties and chemical stability, good social and economic benefits [6, 7]. Studies have shown that a layer of coagulation would arise after PET SG transplanted. It is advantageous to the vascular endothelial cells attachment and growth [8]. However, the blood compatibility of polyester is unsatisfied, so it has defects as a biological material [9]. Silk contains large amounts of silk fibroin which is widely used in the field of biomedical materials [10-12], which is used as sutures for a long time. It is good for cell proliferation as it has good stability, blood compatibility, biocompatibility and appropriate mechanical properties [13-15]. Therefore silk is also very suitable as the material of artificial blood vessels. Studies have shown that artificial blood vessel made from silk has a moderate amount of pores, less tissue reaction and a more complete tissue lining [16, 17]. Therefore, polyester and silk are chosen to develop BWEP, in order to make up the shortage of single material performance.

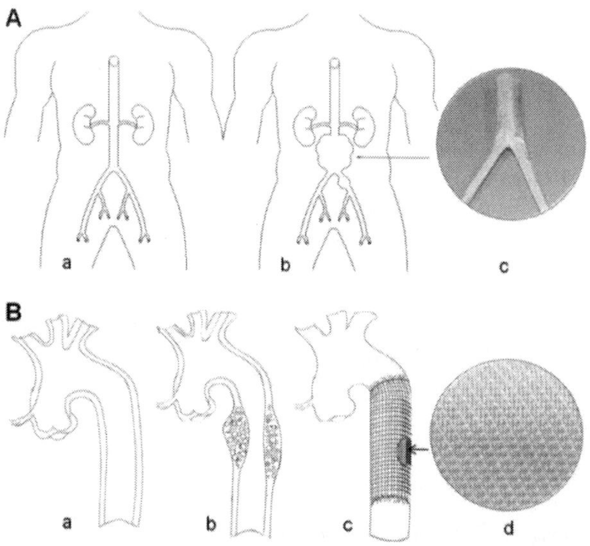

Fig.1 Schematic diagram of endovascular stent (A) and replacement artificial vascular prosthesis (B): (a) normal blood vessel, (b) abnormal vessel, (c) and (d) artificial blood vessels prototyped in our team.

In this paper, a novel seamless BWEP with special woven design was developed to avoid blood leakage. An orthogonal design of three factors was conducted including basic weave, warp × weft density and warp × weft materials. Then nine kinds of BWEP samples were prepared using a modified rigid rapier loom. After a serious of post-treatment process for instance cleaning, degumming and design finalization, the most important performance index, wall thickness and water penetration were characterized according to ISO 7198:1998. The results of wall thickness and water penetration were analyzed from range analysis in order to detect the relationship between factors and performance of BWEP, which provided reference for woven vascular prosthesis preparation, especially for the development of BWEP which is used in endovascular graft exclusion.

2. Materials & Methods

2.1 Materials

100% raw silk (2.4 tex) (RS) and partly degummed silk (2.4 tex) (DS) filament (Bombyx mori) were supplied by Xiehe Silk Co., Ltd, (Zhejiang China). Monofilament and multifilament PET were purchased from New Material Technology Co., Ltd, (Jiaxing China). The Na_2CO_3 was purchased from Sinopharm Chemical Reagent Co., Ltd, (Shanghai China).

2.2 Design of Experiment

The main factors which affect the wall thickness and water permeability of BWEP are fabric weave, warp × weft density and warp × weft materials. As a result, they were chosen as the factors of orthogonal design. In order to develop an ideal BWEP that can meet transplant requirements, there are three levels for each factor. The artificial blood vessels with plain weave have been applied to vascular disease therapy for many years, so it can be one of fabric weaves. The choice of the other two factors, 2/2 twill and 3/1 twill are not only to contrast differences, but also for their similar structure to plain weave. Warp × weft density is of great importance for it is the influential factor of water permeability. Based some studies, the warp × weft density of 1100/10 cm × 800/10 cm, 1100/10 cm × 1400/10 cm and 1100/10 cm × 2000/10 cm were selected. We all know that the fabric thickness is two to three times thicker than the yarn. For this reason, 2.4 tex yarn was chosen as both warp and weft of BWEP. It must be mentioned that warp and weft materials selecting is not only to achieve good compatibility and permeability but also ensure weaving process successfully. Warping machine was used for the PET monofilament and RS before weaving. Considering the low strength of the material used, the warping rate should be used was in a low range. The warp density and warping speed were set at 75 yarns/inches and 20 r/minute. A modified rigid rapier weaving loom (China Patent: CN101215749A), suitable for BWEP weaving, was used for the preparation. The Factors and levels for orthogonal design can be seen in Table 1.

Table.1 Factors and levels for orthogonal design

Variable	Levels		
(A) Weft density inserts/10cm	800	1400	2000
(B) Fabric weave	plain	3/1twill	2/2twill
(C) Warp×weft materials	PET×PET	PET×silk	silk×silk

2.3 Post-treatment of BWEP

There were three kinds of BWEP, pure 100% PET, PET × DS and RS × DS. Firstly, ultrasonic cleaning rinsed samples in distilled water for 30 minutes to remove rust, dirt and other impurities which were attached on the fabric surface due to reed's friction and static adsorption during weaving process. Secondly, BWEP made from PET × DS and RS × DS were degummed by treating 2 and 4 times, respectively, with 0.05% (w/w) Na_2CO_3 solution at 98°C in order to remove surface sericin, each time was for 20 min. Then they were washed in distilled water and naturally air-dried at room temperature for 24h. The effectiveness and accomplishment of the degumming course were confirmed by measuring weight loss. After the cleaning process, all samples would be placed under the condition of 120 °C for 30 minutes to finalize the design.

2.4 Characterization of wall Thickness and Water Permeability

According to international standard ISO7198:1998, the wall thickness of BWEP was tested under the pressure of 981Pa and the sampling area is 0.5 cm^2. Five parallel samples were characterized to get reasonable results by fabric thickness tester (YG-B 141D, Guoliang Company, China). The water permeability device was set up according to describe in ISO7198:1998, and it have applied Chinese patent (CN1460847A). It can meet the requirements of the transplantation when the water permeability is lower than 300 ml/ (cm^2 × min) under the 120 mmHg hydrostatic pressure.

2.5 Statistical Analysis

The range analysis method described in SPSS (16.0) was used for the orthogonal experiment to identify the significant factor influencing the performance of BWEP. The size of range "R" reflected the influence of the corresponding factors. The bigger the value of R is, the more powerful the factor influencing the target in different levels, it is usually the major factor of all. The smaller the value of R is, the less powerful the factor influencing the target in different levels, it is usually the minor factor of all.

3. Results & Discussion

3.1 Design of BWEP

According to the design principle of three factors and three levels orthogonal experiment, the orthogonal design specifications of experiment can be got, the results were shown in Table 2. There are three parts for BWEP fabric: straight tube part, transitional part and bifurcation part according to the design principle of BWEP fabric we have done before [18]. Weaving diagrams of three basic weaves was showed in Fig.2. The main trunk and transition can be weaved by choosing one of shuttles on the machine. In order to achieve a good water permeability resistance in the transition part, the integrated structure can be used. Diagram of sample view and weaving process using the modified loom were shown in Fig.3.

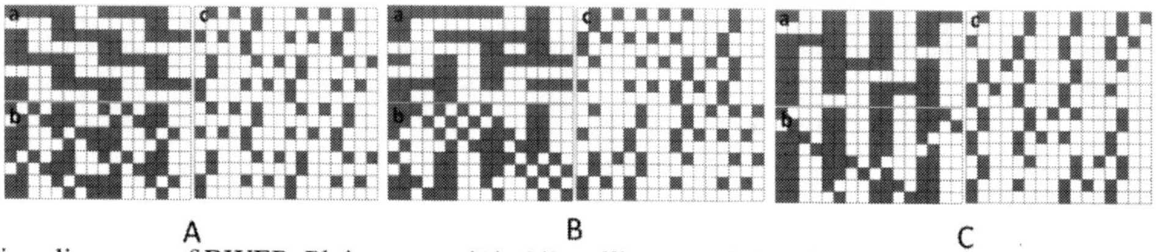

Fig.2 Weaving diagrams of BWEP: Plain weave (A), 2/2 twill weave (B) and 3/1 twill weave (C); weave designs of different BWEP: (a) main trunk, (b) transition, (c) branch.

Table.2 The specifications of BWEPs

Samples	Weaves	Materials (warp × weft)	Warp× weft counts (D/filament count)	Fabric counts (ends/10cm × picks/10cm)
a	plain	PET × PET	22D/1f × 22D/12f	1100×800
b	3/1twill	PET × DS	22D/1f × 22D/12f	1100×800
c	2/2twill	RS × DS	22D/12f × 22D/12f	1100×800
d	plain	PET × DS	22D/1f × 22D/12f	1100×1400
e	3/1twill	RS × DS	22D/12f × 22D/12f	1100×1400
f	2/2twill	PET × PET	22D/1f × 22D/12f	1100×1400
g	plain	RS × DS	22D/12f × 22D/12f	1100×2000
h	3/1twill	PET × PET	22D/1f × 22D/12f	1100×2000
i	2/2twill	PET × DS	22D/1f × 22D/12f	1100×2000

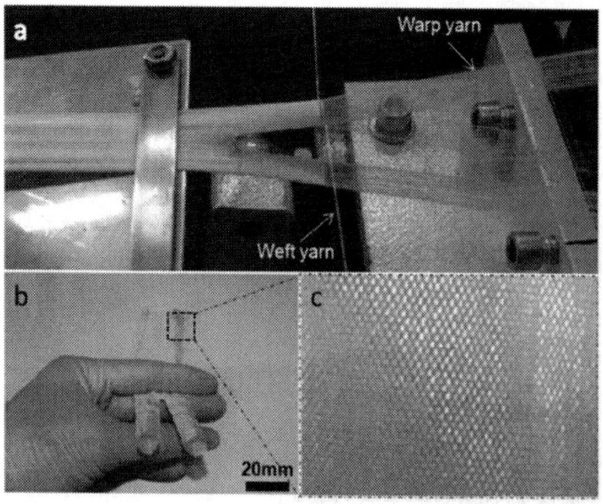

Fig.3 Sample design and weaving process using the modified loom: (a) diagrammatic figure, (b) full view of the sample, (c) surface of the sample.

3.2 Wall Thickness

BWEP thickness is of great importance because it would be hard when guided into human body if that is too thick. The wall thickness of different kinds of BWEP was shown in Fig.4. As seen from it, the BWEP we have prepared can almost meet the thickness requirements of lumen isolation technique as thickness of more than half of BWEP was less than 0.1mm. What is more, the thickness of all samples is around 0.1mm, which proved the rationality of orthogonal experiment. Because the higher thickness can provide better mechanical property, that ensure a long term stability, but it must be under the condition of being less than 0.1m. The range analysis results of wall thickness and diagram of factor and level effect on the thickness were shown in table 3 and figure 5, respectively. It can be concluded that the thickness increased when the density is from 800/cm to 1400/cm while it declined when the density is from 1400/10cm to 2000/10cm. The decreasing trend from 800/cm to 1400/cm can be attributed to cross wave highness. In other words, the cross wave is higher when weft density is 800/10cm. Nonetheless, the trend has changed from 1400/10cm to 2000/10cm. It indicated that the cross wave highness increases firstly and then decreases when weft density changed from 800/10cm to 2000/10cm. As expected, the fabric materials were with less effect when compared them with fabric density in term of BWEP thickness, in that the materials were all with the same diameter. The insignificant difference can be attributed to the discrepancy of their different yield stress, which affected the cross wave during weaving process. The factor (B) fabric weave was the most significant which affected the fabric floating, so as to determine the thickness of BWEP.

Range analysis results of BWEP thickness showed that factors (B) fabric weave, (A) weft density and (C) warp × weft materials were with a gradual decline influence on BWEP thickness. What is more, factor B processed a more significant effect by comparing factors A and C as the R of it was several times larger than the other two. BWEP with 1100×1400 density, plain weave and PET×PET materials processed better performance in terms of just taking wall thickness into consideration.

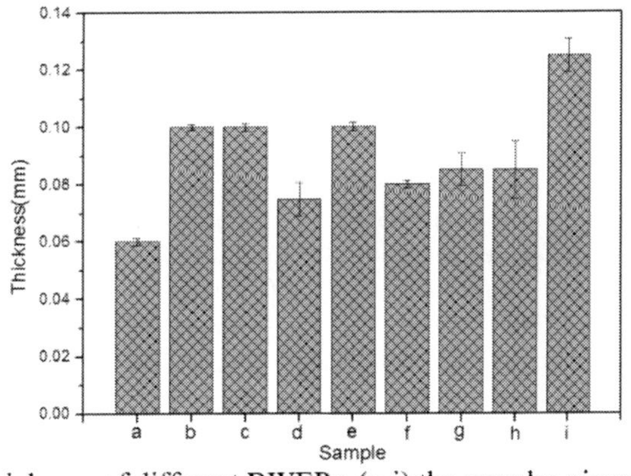

Fig.4 Wall thickness of different BWEPs: (a-i) the samples given into Table 2

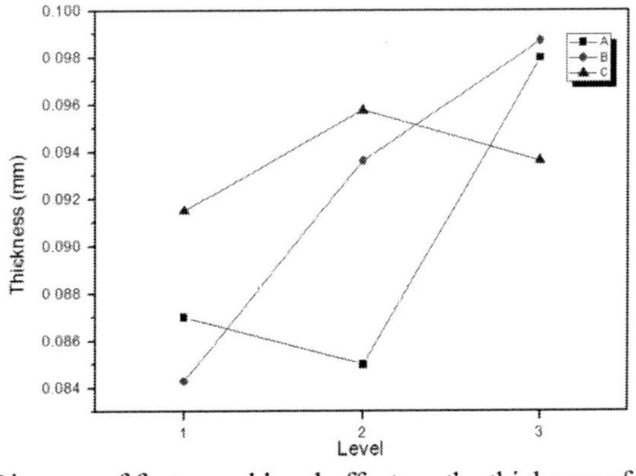

Fig.5 Diagram of factor and level effect on the thickness of BWEP

Table.3 Results of wall thickness from range analysis

$L9(3^3)$			
Test number and rang analysis	Factor		
	A	B	C
a	1	1	1
b	1	2	2
c	1	3	3
d	2	1	2
e	2	2	3
f	2	3	1
g	3	1	3
h	3	2	1
i	3	3	2
K_1	0.26	0.22	0.27
K_2	0.255	0.285	0.3
K_3	0.295	0.305	0.285
k_1 (K_1/8)	0.087	0.073	0.09
k_2 (K_2/8)	0.085	0.095	0.1
k_3 (K_3/8)	0.098	0.107	0.095
R	0.013	0.034	0.01

3.3 Water Permeability

There is no denying that blood leakage would not happen after transplant if vascular graft is with an ideal permeability performance, so it pays vital role in transplantation. As can be seen from Fig. 6, water permeability of BWEP made from pure PET was far above 300 ml/ (cm^2 × min), which cannot satisfy application requirement. However, BWEP containing silk possessed a better water permeability. It must also be mentioned that pure silk BWEPs have ideal water permeability, especially for sample g with just only 5.19 ml/ (cm^2 × min). They fully manifested that silk fibroin based endovascular prosthesis can easily meet permeability requirements while pure PET based endovascular prosthesis cannot. Moreover, permeability result confirmed rationality of the orthogonal experimental design again. Results of water permeability from range analysis in Table 4 showed that there were big gaps among factors (A) weft density, (B) fabric weave and (C) warp × weft materials. Furthermore, Range analysis results in Table 4 showed that factor C is more critical than factor B and factor A in water permeability determining.

Water permeability is not only related to fabric porosity but also surface energy. Fabric porosity determined the water permeability from proportion of blank area while surface energy influenced it from the different ability to combine with the water among BWEP. As can be seen from Fig.7, the water permeability would be lower with fabric density increasing, which resulted from higher density providing less porosity. The factor (C) warp × weft materials was the most significant as there is essential difference of surface energy between silk and PET. In addition, the better water permeability of silk BWEP indicated the ability of silk to combine with water is much better than that of PET. The factor (B) fabric weave was not as significant as warp × weft materials and fabric density, in that there were less influence on water permeability of fabric weave due to no distinct differences between porosity and surface energy for each BWEP.

Range analysis results of BWEP water permeability showed that factors (C) warp × weft materials, (A) weft density and (B) fabric weave were with a gradual decline influence on the water permeability. What is more, factor A and C processed a more significant effect by comparing to factor B. It can be concluded that BWEP with 1100×2000 density, 2/2twill weave and silk × silk materials processed better performance in terms of just taking water permeability into consideration.

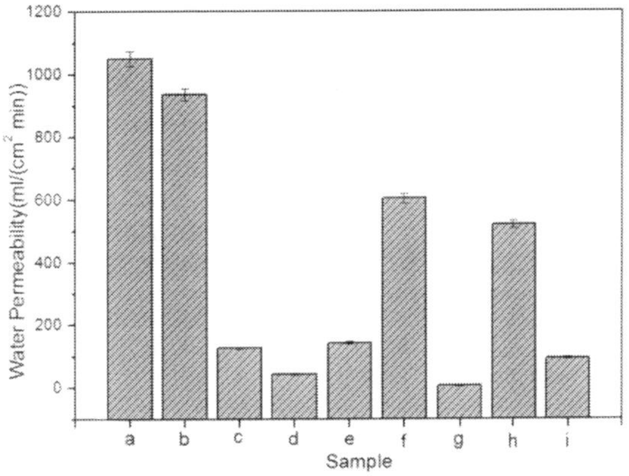

Fig.6 Water permeability of different BWEPs: (a-i) the samples given into Table 2.

Table.4 Results of water permeability from range analysis

L9(3^3)			
Test number and rang analysis	Factor		
	A	B	C
a	1	1	1
b	1	2	2
c	1	3	3
d	2	1	2
e	2	2	3
f	2	3	1
g	3	1	3
h	3	2	1
i	3	3	2
K_1	2112	1098	2173
K_2	789	1599	1073
K_3	619	823	274
k_1 (K_1/8)	704	366	724
k_2 (K_2/8)	263	533	358
k_3 (K_3/8)	206	274	91
R	498	259	633

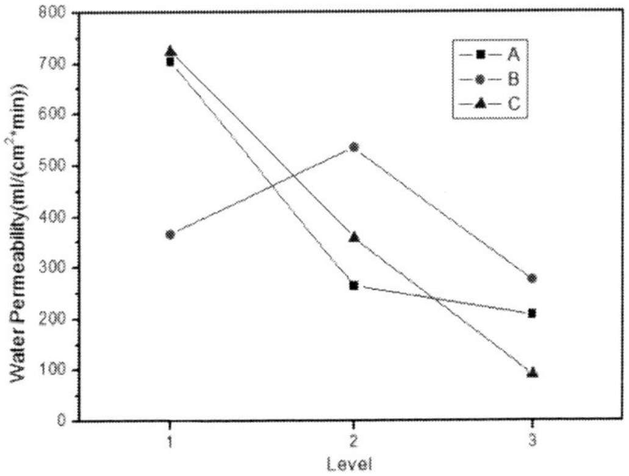

Fig.7 Diagram of factor and level effect on the water permeability of BWEP

4. Conclusion

In the present study, a novel seamless bifurcated woven endovascular prosthesis (BWEP) with special woven design was successfully developed using a modified rigid rapier loom to avoid blood leakage. Using the method of orthogonal design, it was found that the factors influenced the thickness of BWEP in order of importance were: (B) fabric weave > (A) weft density> (C) warp × weft materials. Moreover, Range analysis results of BWEP water permeability showed that factors (C) warp × weft materials, (A) weft density and (B) fabric weave were with a gradual decline influence on the water permeability of BWEP. It can be concluded that BWEP with 1100×1400 density, plain weave and PET×PET materials processed better performance in terms of just taking wall thickness into consideration while it with 1100×2000 density, 2/2twill weave and silk × silk materials processed better performance when as to water permeability. The study we have done provided references for the development of woven artificial blood vessels, especially for BWEP. Further study will be focused on others properties of the BWEP such as mechanical property, cell and blood compatibility.

5. Acknowledgement

Thanks for the support of Natural Science Foundation of China (51603140), Sino-German Center Project (GZ1094), Natural Science Foundation of Jiangsu Province (BK20150372), University Science Research Project of Jiangsu Province (16KJB540003), Key industry technology innovation, Science and technology project of Suzhou (SYG201638).

References

[1] Chuter TAM, Ivancev K, Lindblad B, Brunkwall J, Arén C, Bo R. Endovascular Stent-Graft Exclusion of an Aortobronchial Fistula. Journal of Vascular & Interventional Radiology: 1996; 7: 357-369.
[2] Marty B, Sanchez LA, Ohki T, Wain RA, Faries PL, Cynamon J, et al. Endoleak after endovascular graft repair of experimental aortic aneurysms: Does coil embolization with angiographic "seal" lower intraaneurysmal pressure. Journal of Vascular Surgery: 1998; 27: 454-462.
[3] Bray PJ, Robson WJ, Bray AE. Percutaneous treatment of long superficial femoral artery occlusive disease: efficacy of the Hemobahn stent-graft. Journal of Endovascular Therapy: 2003; 10: 619-628.
[4] Rousseau H, Puel J, Joffre F, Sigwart U, Duboucher C, Imbert C, et al. Self-expanding endovascular prosthesis: an experimental study. Radiology: 1987; 164: 709-714.
[5] J Z, ZP J, ZT W, H Y. Value of digital subtraction angiography in endovascular graft exclusion for abdominal aortic aneurysms. Journal of Medical Colleges of PLA: 2000; 13-26.
[6] Cheng ST, Chen ZF, Chen GQ. The expression of cross-linked elastin by rabbit blood vessel smooth muscle cells cultured in polyhydroxyalkanoate scaffolds. Biomaterials: 2008; 29: 4187-4194.
[7] Blanchemain N, Haulon S, Martel B, Traisnel M, Morcellet M, Hildebrand HF. Vascular PET Prostheses Surface Modification with Cyclodextrin Coating: Development of a New Drug Delivery System. European Journal of Vascular & Endovascular Surgery: 2005; 29: 628-632.
[8] Wang SG. Investigation of Relationship between Surface Properties and Blood Compatibility of Blended Polyether Polyester Block Copolymer. Polymer Journal: 1989; 21: 179-183.
[9] Kadono H, Furuzono T, Masuda M, Okada M, Ueki M, Takamizawa K, et al. In vivo evaluation of hydroxyapatite nanocoating on polyester artificial vascular grafts and possibility as soft-tissue compatible material. Asaio Journal: 2010; 56: 61-66.
[10] Li G, Li F, Zheng Z, Luo T, Liu J, Wu J, et al. Silk microfiber-reinforced silk composite scaffolds: fabrication, mechanical properties, and cytocompatibility. Journal of Materials Science: 2015; 51: 3025-3035.
[11] Gang L, Jian L, Zheng Z, Wang X, Kaplan DL. Structural Mimetic Silk Fiber-Reinforced Composite Scaffolds Using Multi-Angle Fibers. Macromolecular Bioscience: 2015; 15: 1125-1133.
[12] Zhao Z, Li Y, Xie MB. Silk fibroin-based nanoparticles for drug delivery. International Journal of Molecular Sciences: 2015; 16: 4880-4903.
[13] Lovett ML, Cannizzaro CM, Vunjak-Novakovic G, Kaplan DL. Gel spinning of silk tubes for tissue engineering. Biomaterials: 2008; 29: 4650-4657.

[14] Wang HY, Zhang YQ. Processing silk hydrogel and its applications in biomedical materials. Biotechnology Progress: 2015; 31: 630-640.
[15] Li G, Chen Y, Hu J, Wu X, Hu J, He X, et al. A 5-fluorouracil-loaded polydioxanone weft-knitted stent for the treatment of colorectal cancer. Biomaterials: 2013; 34: 9451-9461.
[16] Ding X, Zou T, Gong X, Ren C, Kang H, Xu P, et al. Trilayered sulfated silk fibroin vascular grafts enhanced with braided silk tube. Journal of Bioactive and Compatible Polymers: Biomedical Applications: 2016; 31: 886-898
[17] Li G, Li Y, Chen G, He J, Han Y, Wang X, et al. Silk-Based Biomaterials in Biomedical Textiles and Fiber-Based Implants. Advanced Healthcare Materials: 2015; 4: 1134-1151.
[18] Li G, Liu Y, Lan P, Li Y, Li Y. A prospective bifurcated biomedical stent with seamless woven structure. Journal of the Textile Institute: 2013; 104: 1017-1023.

N-carboxyethyl Chitosan Hydrofiber Prepared as a Potential Wound Dressing

Kai-Li Liang[1], Can Zhang[1], Jun Li[2], Ying-Shan Zhou[1,2], Xin Liu[1], Wei-Lin Xu[1,2]

[1]*College of Materials Science and Engineering, Wuhan Textile University, Fangzhi Road, Wuhan, 430073, People's Republic of China*
[2]*Key Laboratory of Green Processing and Functional Textiles of New Textile Materials, Ministry of Education, Wuhan Textile University, Fangzhi Road, Wuhan, 430073, People's Republic of China*

[*]Corresponding author's email: zyssyz@126.com

Abstract

To improve liquid-absorption capacity of wound dressing based on chitosan fiber, N-carboxyethyl chitosan hydrofiber was successfully synthesized by a Michael addition reaction of chitosan fiber with acrylic acid. A series of properties of N-carboxyethyl chitosan hydrofibers including swelling behaviours, hydrophilicity, antibacterial activity, as well as in vitro degradation were investigated. The degree of N-substitution was varied from 0.10 to 0.51. Main results showed that, N-carboxyethyl chitosan hydrofibers had obviously higher liquid-absorption capacity and higher level of hydrophilicity than chitosan fiber, and had antibacterial activity towards E. coli and S. aureus. N-carboxyethyl chitosan hydrofibers had nearly complete degradation in 3 weeks. Cytotoxicity tests showed that N-carboxyethyl chitosan hydrofibers were biocompatible to L929 cells. The N-carboxyethyl chitosan hydrofibers could be a good candidate to be used as wound dressing for skin regeneration.

Keywords: Chitosan; Carboxyethylation; Hydrophilicity; Hydrofiber; Wound Dressing

1. Introduction

Wound dressings are often used to protect the wound from environmental irritants, caused by fire, heat, electricity, and chemicals [1]. An ideal wound dressing should keep the moisture in healing environment, allow gaseous exchange and have antibacterial activity [2, 3]. Most importantly, it should have wound exudates absorbing ability because excessive exudates production can have a very detrimental effect on the wound healing [4] and excessive exudates accumulated between the wound and the dressing would increase the risk of infection leading to wound deterioration [5]. Additionally, it should be biocompatible, biologically interactive and no adhesion without damaging newly regenerated tissue when it is ripped from the wound surface [6]. Recent advances in the development of more efficacious wound dressings reveal that natural polymers have greater potential in these applications [7, 8].

Among natural polymers, chitosan, the second most abundant polysaccharide after cellulose [9], is one of the most promising biomacromolecules for wound dressing, due to its excellent biological contributions to the wound, such as biodegradability, biocompatibility, and antibacterial activity [10-12]. Furthermore, chitosan can heal the wound faster and scar prevention [13], because N-acetyl-β-D-glucosamine from degradation of chitosan initiates fibroblast proliferation, which helps in ordered collagen deposition and stimulates increased level of natural hyaluronic acid synthesis at the wound site [5]. Chitosan fiber obtained by wet-spun process [14, 15] are paid more attention to as wound dressing, due to its various merits including gentleness to the body and moisture retain [16], as well as allowing adequate drainage of exudates [17]. However due to chitosan fiber's poor liquid-absorbing capacity and weak hydrophilicity [18] it makes it not appropriate to use in current form, especially when dealing with moderate to high exuding wounds, such as partial thickness burns and diabetic foot ulcers.

To overcome these problems, in this research, N-carboxyethyl chitosan hydrofiber was synthesized by a Michael addition reaction of chitosan fiber with acrylic acid. Series of properties of N-carboxyethyl chitosan hydrofibers including swelling behaviours, hydrophilicity, antibacterial activity, as well as in vitro degradation were investigated. The potential use of N-carboxyethyl chitosan hydrofibers as wound dressings were assessed by cytotoxicity in vitro with mouse fibroblasts.

2. Experimental

2.1 Materials

Chitosan fiber (CS fiber) with viscosity of 505 mPa•s was purchased from Shandong Huaxing Haici New Materials Co., Ltd., China. The degree of deacetylation of CS fiber was calculated as 78.2 % by ^{1}H NMR. Mouse fibroblasts (L929) were obtained from Wuhan Beinglay Biological Technology Co., Ltd., China. Dulbecco's modified eagle medium (DMEM), 1 % Penicillin-streptomycin, trypsin, 10 % fetal bovine serum (FBS), MTT powder was supplied by Shanghaikayon Biological Technology Co., Ltd., China.

2.2 Synthesis of *N*-carboxyethyl Chitosan Hydrofibers (CECFs)

CECFs were synthesized according to the method reported by us [19], and the degree of substitution (DS) of CECFs was calculated by ^{1}H NMR (Bruker AV400). The results were listed in Table 1.

Table.1 The CECFs with DS.

Sample	DS
CECF1	0.10
CECF2	0.16
CECF3	0.20
CECF4	0.23
CECF5	0.25
CECF6	0.32
CECF7	0.51

2.3 Swelling Behaviours

The swelling characteristics of chitosan fiber and CECFs with different reaction conditions were investigated by immersing dried chitosan fiber and CECFs in water or solution A (2.298g sodium chloride, 0.368g calcium chloride dehydrate were added to 1 liter of de-ionized water). Briefly, each fiber was wrapped in a piece of polyester cloth and was then placed in water or solution A with the volume of 40 times than its own weight at 37 ℃ for 30 min. After centrifuged for 5 min and weighed as m_1, the fiber was dried to constant weight at 105 ℃ for 8 h and the dry weight was recorded (m_0). The swelling ratio of the fiber was quantified as the following expression: $(m_1-m_0)/m_0$, which represented fluid absorption amount per gram of dry fiber.

The swelling processes of chitosan fiber and CECFs in water were recorded by Fluorescence Microscopy (XSZ-HY, Chongqing Mic Optical & Electrical Instrument Co., Ltd., China).

2.4 Antibacterial Study

Gram-negative E. coli and Gram-positive S. aureus were used to assess the antibacterial activity of CECFs. Firstly, E. coli or S. aureus grown overnight in meat-peptone broth at 37 ℃ under orbital agitation. The culture was diluted with fresh medium to about 1.0×10^6 cells/mL. 0.1 mL above bacteria culture was taken to inoculate in nutrient agar. The non-woven fabrics made from CECFs were cut into circle with the diameter of 2 cm and placed on agar plates. The plates were incubated at 37 °C for 48 h. And zone under non-woven fabrics were examined as an inhibition against E. coli or S. aureus.

2.5 In Vitro Degradation

CECFs were wrapped with nylon net and incubated in a 1.5 μg/mL lysozyme PBS solution at pH 7.4 which simulated as tissue fluids. The degradation medium was replaced with fresh buffer containing lysozyme every seven days. The degradation was monitored for 3 weeks at 37 ℃. Samples were removed at set time points, vacuum-dried for 24 h and weighed to obtain the final weight after degradation. Also, Samples were dry samples were lyophilized and sputtered with gold for observation of morphology of fibers by SEM.

2.6 Cytotoxicity Assays

Indirect cytotoxicity test. Extracts were prepared from chitosan fiber or CECFs by adding fragments of the sterile samples to the culture medium at a concentration of 1.25 cm^2/mL and inculating at 37 ℃ for 24 h without shaking. After this period, the medium was obtained and the chitosan fiber or CECFs were removed. This procedure aimed to assess potential deleterious effects of substances released by chitosan fiber or CECFs into the culture medium.

According to ISO-10993, MTT [3- (4, 5 – dimethylthiazol – 2 - yl) - 2, 5 - diphenyltetrazolium bromide] was used to determine the extracts toxicity. Initially, 200μL of L929 cell suspension at density of 10^5 cells per well were seeded in wells of 96–well plate and cultured for 24 h at 37 ℃. After this period, the culture medium was removed and replaced with 100μL of as-prepared extraction medium. After 24 h at 37 ℃, the extracts were removed and 10 μL of MTT solution was added to each well. After 4 h incubation at 37 ℃, 150 μL of dimethyl sulfoxide was added to dissolve the formazan crystals. The dissolved solution was swirled homogeneously for about 10 min by the shaker. The optical density of the formazan solution was detected by an ELISA reader (Multiscan MK3, Labsystem Co. Finland) at 490 nm.

For reference purposes, cells were seeded to a fresh culture medium (negative control) under the same seeding conditions, respectively.

Results are depicted as mean ± standard deviation. Significance between the mean values was calculated using ANOVA one-way analysis (Origin 7.0 SRO, Northampton, MA, USA). Probability values $P<0.05$ were considered significant (n= 8).

Direct cytotoxicity test. The direct cytotoxicity test consisted of assessing the possible toxic effects of CECFs in direct contact with L929 cells. Briefly, the sterilized samples were placed on a 96-well plate and covered with 200 μL of DMEM with 10 % FBS at 37 ℃ for 24 h. After this period, 100 μL of cell suspension at density of 106 cells per well were seeded in wells with samples. After incubated for 48 h, 72 h at 37 ℃, state of cells growth were observed by optical microscope.

3. Results and Discussions

3.1 Swelling Behaviours

The high levers of exudate combined with the pro-inflammatory mediators have a detrimental effect on wound healing, including wound enlargement and damage to the periwound skin such as maceration and excoriation [20]. So, it is necessary to absorb excess fluid from wound skin to prevent maceration or excoriation of the wounds for good wound dressing, which can be characterized by swelling behaviours of fibers. Fig.1 depicts swelling ratio of CS fiber and CECFs with different DS value in water or solution A.

Swelling ratio of CECFs in water increases significantly from 2.7 to 5.2 with DS value rising from 0.10 to 0.51, obviously higher than CS fiber of 1.4 (Fig.1a). Statistically significant differences ($p < 0.05$) were observed in the swelling ratio in water of CECFs, in comparison with CS fiber. The results of improved swelling ratio suggested that, introduction of carboxylethyl group into polysaccharide structure by Michael addition reaction not only would form new more hydrophilic centers (carboxyl group) which could bind more water molecules [21], but also disrupt the crystalline structure of chitosan leading to reduced interaction forces among molecular chains so as to form a loose structure to favor more free water molecules to diffuse into the molecular chains of CECFs [18], which resulted in improvement of its liquid absorbing ability. As shown in Fig.1b, swelling ratio of CECFs in solution A increases significantly from 1.5 to 2.0, obviously higher than CS fiber of 0.3. Statistically significant differences (p

< 0.05) were observed in the swelling ratio in solution A of CECFs, in comparison with CS fiber. However, swelling ratio of CECFs in solution A was lower than that in water, which probably attributed to chelation between Ca^{2+} ions in solution A and carboxyl group of CECFs resulting in reduce of carboxyl group.

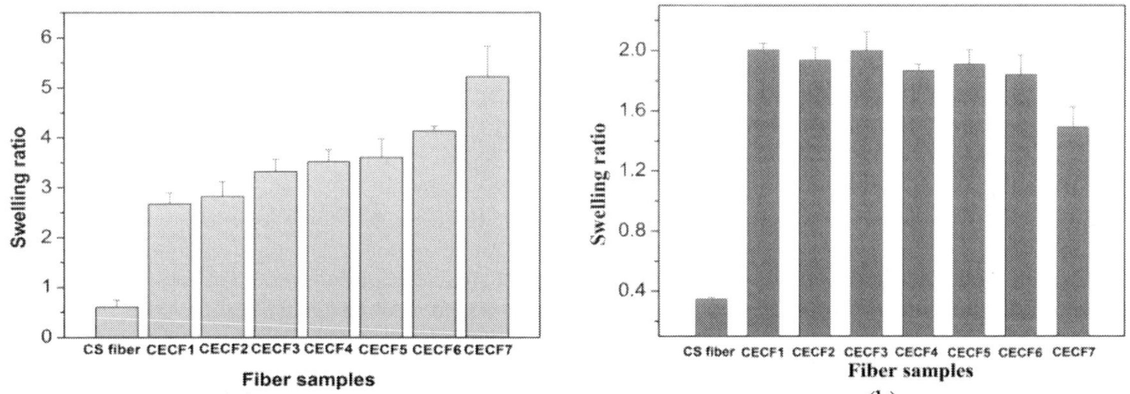

(a) (b)

Fig.1 Swelling ratio of CS fiber and CECFs in liquid. (a) Water; (b) Solution A.

Fig.2 shows the photomicrographs of the dry and wet structures of CS fiber and CECFs. It could be seen that the CS fiber showed a slight degree of wetting when placed in contact with water, which were good agreement with the results of swelling ratio. However, when CECFs was placed in contact with water, a large number of water was sucked into the fiber structure, which made the fiber expand with a high degree to form clear, soft gel, while keeping its fibrous structure, as shown in Fig.2 b2-h2.

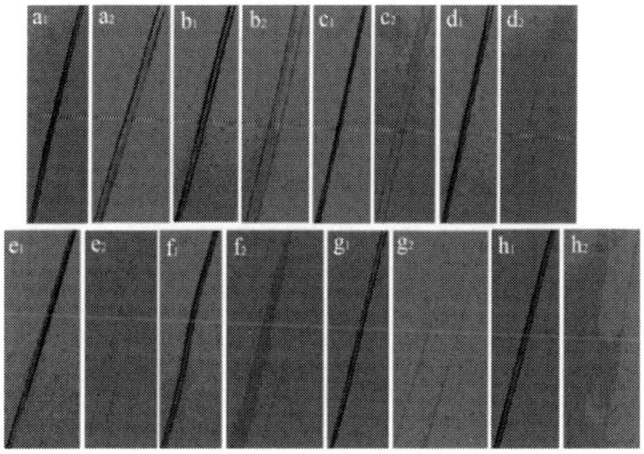

Fig.2 Photomicrographs of the dry and wet structures of CS fiber and CECFs. (×100) (a) CS fiber; (b) CECF1; (c) CECF2; (d) CECF3; (e) CECF4; (f) CECF5; (g) CECF6; (h) CECF7. Subscripts 1 and 2 are for dry and wet state of fibers, respectively.

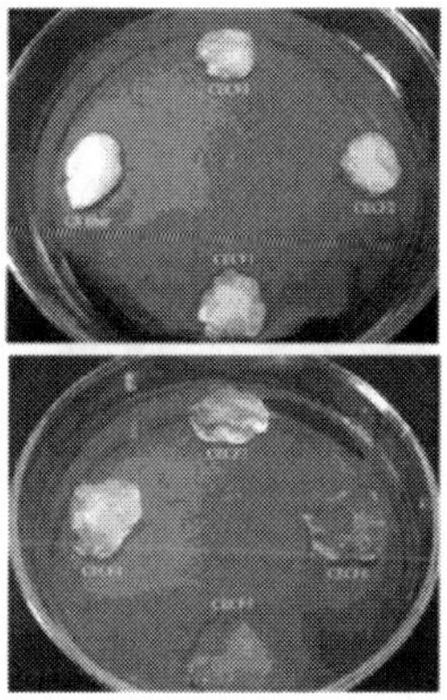

Fig.3 Photomicrographs of non-woven fabrics from CS fiber and CECFs in water.

The morphological change after the fibers absorbed water also investigated by morphology of non-woven fabrics from the fibers in water. Fig.3 shows that non-woven fabric from CS fiber has not shown obvious change in morphology, while non-woven fabrics from CECFs have morphological change from fiber to gel with increase of DS. When DS goes to 0.25, CECFs transform into a clear gel, while retaining overall structural integrity.

The carboxyl groups on the surface of CECFs allow the fibers to swell just enough for gel formation [4]. The gel formed can promote a moist wound healing environment and it allows the vertical wicking and retention of wound exudates which ensures that periwound tissue is not macerated [22]. At the same time, the gelling promoted non-traumatic removal of wound dressings without destroying the new granulation tissue. Additionally, these nonwoven fabrics can bind bacteria within their structure thus reducing the chance of infection [4]. They were therefore suitable for management of a wide range of moderately to highly exuding wounds, such as partial thickness burns, diabetic foot ulcers and traumatic wounds [23].

3.2 Antibacterial Study

Gram-negative E. coli and Gram-positive S. aureus which was commonly seen in burn wounds, was considered as the model bacteria. Fig. 4 depicts the antibacterial activity of CS fiber and CECFs against E. coli or S. aureus. It could be seen that, there is obviously not bacterium seen under the CECFs zone while a large number of colonies are found around the CECFs zone, which showed antibacterial activity of CECFs against E. coli and S. aureus, which probably attributed to the damaging interaction polycations of CECFs with negatively charged surface of bacteria resulting in loss of membrane permeability, cytoplasm leakage and cell death, as the literatures indicated [24, 25]. Once the non-woven fabrics from CECFs had transformed into a gel, it was able to contour closely to the wound, which meant that there were no gaps at the fabrics-wound interface, limiting the amount of space for bacteria to proliferate [26]. From the clinical perspective, CECFs could contribute to the prevention of secondary infections in wounds meanwhile CECFs absorb exudates, which accelerated the wound healing [27].

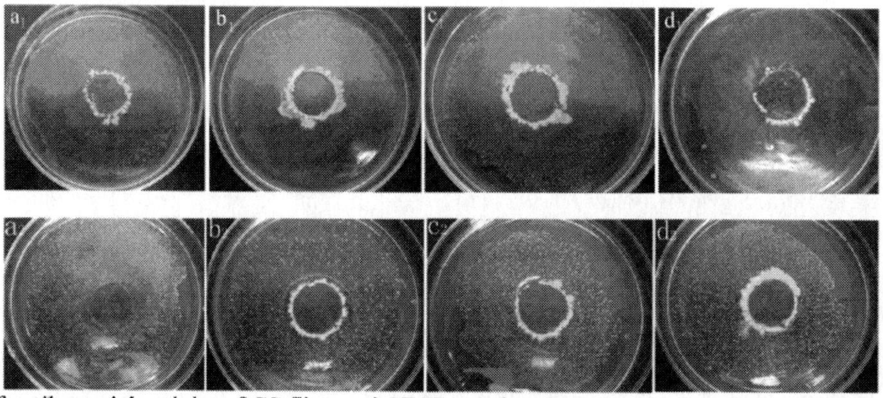

Fig.4 Image of antibacterial activity of CS fiber and CECFs against E. coli or S. aureus. (a) CECF1; (b) CECF3; (c) CECF5; (d) CECF7. Subscripts 1 and 2 are for E. coli and S. aureus, respectively.

3.3 In Vitro Degradation

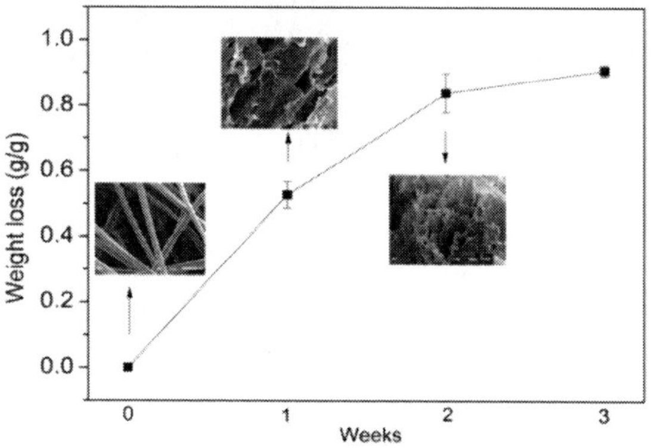

Fig.5 Degradation behaviors of CECF7.

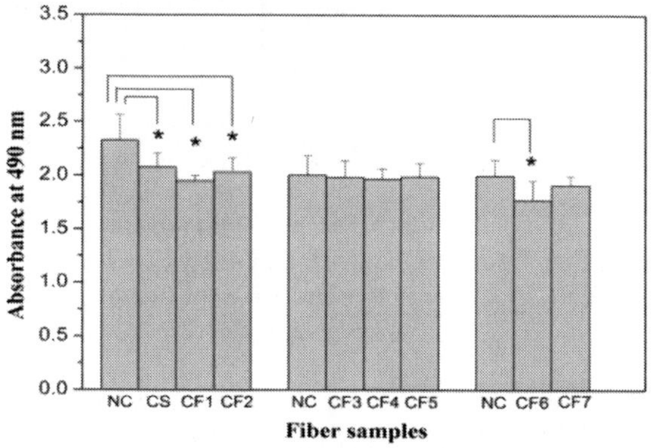

Fig.6 Indirect cytotoxicity test of CS fiber and CECFs with negative controls ($p < 0.05$) *$p < 0.05$ when compared to the negative control.

The controllable biodegradability are the important requirements for the development of any medical implant materials [28]. Fig.5 shows the degradation behaviors of CECF7. CECF7 loses 91% of its weight after 27 days, even has 84% weight loss at 2 weeks. SEM micrographs shows that honeycomb structure instead of fibrous structure was formed with the degradation of the fiber, which is possibly favorable for new tissue ingrowth. The results indicated, introduction of more carboxylethyl groups disrupted intra/intermolecular hydrogen bonding, allowing for more ready access of the enzyme to the chitosan backbone, as the active binding site of lysozyme bound six sugar rings [29]. From the clinical perspective, CECFs are fitted well to wound dressing for hemostasis in vivo, due to their rapid biodegradability.

3.4 Cytotoxicity Assays

An ideal wound dressing should not release toxic products or produce adverse reactions, which could be evaluated through in vitro cytotoxic tests [30]. In the evaluation, mouse fibroblast cells (L929) were used as reference. Fig.6 shows the absorbance obtained from an MTT assay of L929 cells which were cultured with the extraction medium from various fiber samples in comparison with control. It could be seen that, when CS fiber extract was used, statistically significant differences ($p<0.05$) were observed in the cell activity in comparison with control within 24 h, but the viability of the cell still reached 89 % of that of the negative control, which meant that CS fiber was less toxic to L929 cell. Also, statistically significant differences ($p<0.05$) were observed in the cell activity of CECF1 or CECF2 or CECF6 in comparison with control, but viability of the cell still reached 84 % of that of the negative control, which meant that CECF1 or CECF2 or CECF6 was less toxic to L929 cell. However, no statistically significant differences ($p<0.05$) were observed in the cell activity of L929 cell within 24h in the presence of other CECFs extracts in comparison with control, which meant that other CECFs was nontoxic to L929 cell.

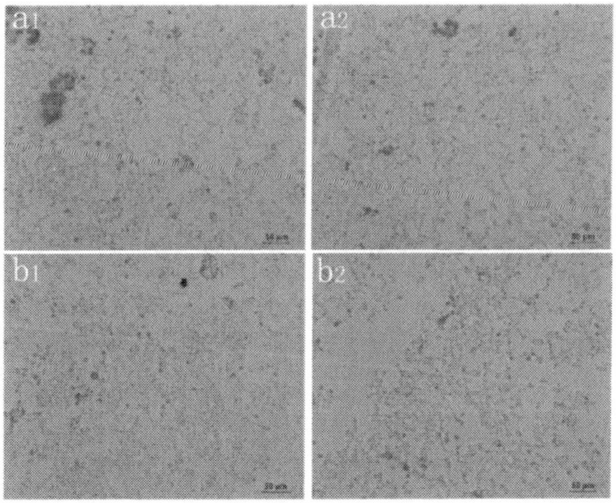

Fig.7 Direct cytotoxicity test of CECFs with negative controls in different culture time. (a) Negative controls; (b) CECF7. Subscripts 1 and 2 are for 48 h and 72 h, respectively.

Fig.7 shows direct cytotoxicity test of CECFs after 48 h, 72 h cell culture, respectively. No obvious differences were observed in the cell activity of CECF7 in comparison with negative control. Cells that settled on the bottom of the well plate proliferated well and displayed in spindle-like shapes in the presence of the CECF7, which attributed to biological property of chitosan to stimulate cell proliferation [31]. The obtained results clearly suggest that CECFs are biocompatible and are good candidates to be as wound dressings.

4. Conclusion

N-carboxyethyl chitosan hydrofiber was successfully synthesized by a Michael addition reaction of chitosan fiber with acrylic acid. Swelling behaviours, water contact angle and antibacterial activity, as well as in vitro degradation were investigated. The results indicated that, CECFs had high liquid absorption capacity, high level of hydrophilicity and antibacterial activity towards E. coli and S. aureus with DS increased, as well as rapid biodegradability. Cytotoxicity tests showed that CECFs were biocompatible to L929 cells. The CECFs could be good candidate to be used as wound dressing for skin regeneration.

5. Acknowledgment

This study was supported by National Natural Science Foundation of China (No.51203123) and Educational Commission of Hubei Province of China (No. D20151605).

References

[1] Zhou Y, Yang D, Chen X, Xu Q, Lu F, Nie J. Electrospun water-soluble carboxyethyl chitosan/poly (vinyl alcohol) nanofibrous menbrance as potential wound dressing for skin regeneration. Biomacromolecules: 2008; 9: 349-354.
[2] Liao N, Unnithan AR, Joshi MK, Tiwari AP, Hong ST, Park CH, Kim CS. Electrospun bioactive poly (ε-caprolactone)-cellulose acetate-dextran antibacterial composite mats for wound dressing applications. Colloids Surf A Physicochem Eng Asp: 2015; 469: 194-201.
[3] Xu F, Weng B, Gilkerson R, Materon LA, Lozano K. Development of tannic acid/chitosan/pullulan composite nanofibers from aqueous solution for potential applications as wound dressing. Carbohydr Polym: 2015; 115: 16-24.
[4] Sweeney IR, Miraftab M, Collyer G. A critical review of modern and emerging absorbent dressings used to treat exuding wounds. Int Wound J: 2012; 9: 601-612.
[5] Jayakumar R, Prabaharan M, Kumar PTS, Nair SV, Tamura H. Biomaterials based on chitin and chitosan in wound dressing applications. Biotechnol Adv: 2011; 29: 322-337.
[6] Liu BS, Yao CH, Fang SS. Evaluation of a non-woven fabric coated with a chitosan bilayer composite for wound dressing. Macromol Biosci: 2008; 8: 432-440.
[7] Chandika P, Ko SC, Jung WK. Marine-derived biological mocromolecule-based biomaterials for wound healing and skin tissue regeneration, Int J Biol Macromol: 2015; 77: 24-35.
[8] Dreifke MB, Jayasuriya AA, Jayasuriya AC. Current wound healing procedures and potential care. Mater Sci Eng C: 2015; 48: 651-662.
[9] Tang R, Yu Z, Zhang Y, Qi C. Synthesis, characterization, and properties of antibacterial dye based on chitosan. Cellulose: 2016; 23: 1741-1749.
[10] Khor E, Lim LY. Implantable applications of chitin and chitosan. Biomaterials: 2003; 24: 2339-2349.
[11] No HK, Park NY, Lee,SH, Meyers SP. Antibacterial activity of chitosans and chitosan oligomers with different molecular weights. Int J Food Microbiol: 2002; 74: 65-72.
[12] Ueno H, Mori T, Fujinaga T. Topical formulations and wound healing applications of chitosan. Adv Drug Deliv Rev: 2001; 52: 105-115.
[13] Anjum S, Arora A, Alam MS, Gupta B. Development of antimicrobial and scar preventive chitosan hydrogel wound dressings. Int J Pharm: 2016; 508: 92-101.
[14] Agboh OC, Qin Y. Chitin and chitosan fibers. Polym Adv Technol: 1997; 8: 355-365.
[15] Knaul JZ, Hudson SM, Creber KAM. Improved mechanical properties of chitosan fibers. J Appl Polym Sci: 1999; 72: 1721-1732.
[16] Liu L, Zhang J, Tang R. Adsorption and functional properties of natural lac dye on chitosan fiber. React Funct Polym: 2013; 73: 1559-1566.
[17] Zhou Y, Yang H, Liu X, Mao J, Gu S, Xu W. Potential of quaternization-functionalized chitosan fiber for wound dressing. Int. J Biol Macromol: 2013; 52: 327-332.
[18] Yang SS, Dong Q, Yang HJ, Liu,X Gu SJ, Zhou Y, Xu WL. N-carboxyethyl chitosan fibers prepared as potential use in tissue engineering. Int J Biol Macromol: 2016; 82: 1018-1022.
[19] Zhou Y, Xu W, Shi L, Li F. Preparation of N-carboxyethyl chitosan fibers. CN Patent No. 201310201041.2: 2015.
[20] Uzun M, Anand SC, Shah T. A novel approach for designing nonwoven hybrid wound dressings: processing

and characterization. J Ind Text: 2016; 45: 1383-1398.
[21] Zhou YS, Yang DZ, Gao XY, Chen XM, Xu Q, Lu FM, J. Nie. Semi-interpenetrating polymer network hydrogels based on water-souluble N-carboxylethyl chitosan and photopolymerized poly (2-hydroxyethyl methacrylate). Carbohydr Polym: 2009; 75: 293-298.
[22] Barnea Y, Weiss J, Gur E. A review of the applications of the hydrofiber dressing with silver (Aquacel Ag○R) wound care. Ther Clin Risk Manag: 2010; 6: 21-27.
[23] Robinson BJ. The use of a hydrofibre dressing in wound management. J Wound Care: 2000; 9: 32-34.
[24] Jia ZS, Shen D, Xu WL. Synthesis and antibacterial activities of quaternary ammonium salt of chitosan. Carbohydr. Res.: 2001; 333: 1-6.
[25] Kim JY, Lee JK, Lee TS, Park WH. Synthesis of chitooligosaccharide derivative with quaternary ammonium group and its antimicrobial activity against Streptococcus mutans. Int J Biol Macromol: 2003; 32: 23-27.
[26] Hoekstra MJ, Hermans MH, Richters CD, Dutrieux RP. A histological comparison of acute inflammatory responses with a hydrofibre or tulle gauze dressing. J Wound Care: 2002; 11: 113-117.
[27] Demir A, Arik B, Ozdogan E, Seventekin N. The comparison of the effect of enzyme, peroxide plasma and chitosan processes on wool fabrics and evaluation for antimicrobial activity. Fiber Polym: 2010; 11: 989-995.
[28] Shuangguan Y, Wang Y, Wu Q, Chen G. The mechanical properties and in vitro biodegradation and biocompatibility of UV-treated poly (3-hydroxybutyrate-co-3- hydroxyhexanoate). Biomaterials: 2006; 27: 2349-2357.
[29] Nordtveit R, Varum KM, Smidsrod O. Degradation of fully water-soluble, partially N-acetylated chitosans with lysozyme. Carbohydr Polym: 1994; 23: 253-260.
[30] Marreco PR, Moreira PL, Genari SC, Moraes AM. Effects of different sterilization methods on the morphology, mechanical properties, and cytotoxicity of chitosan membranes used as wound dressings. J Biomed Mater Res Part B Appl Biomater: 2004; 71B: 268-277.
[31] Yusof NLBM, Lim LY, Khor E. Preparation and characterization of chitin beads as a wound dressing precursor. J Biomed Mater Res: 2001; 54: 59-68.

A Review on Skin Regeneration and Silk Fibroin

Zhong-Da Chen, Yi Li*, Olga Tsigkou, Xu-Qing Liu

School of Material, The University of Manchester, Oxford Street, M13 9PL

*Corresponding author's email: henry.yili@manchester.ac.uk

Abstract

Skin is one of the most important organs surrounding the human body, which is responsible for protecting people from infection, injury and temperature. The healing of a skin wound is a complex process requiring the collaborative effects of various tissues, cell lineages and cytokines. However, the healing process can be improved by the use of biomaterials such as silk fibroin. Here, a brief discussion for cell and material choice for wound healing has been given. Currently, for instance, the attempts for skin regeneration mainly using rat skin fibroblasts (3T3 or L929), human keratinocytes and human dermal fibroblasts by seeding on collagen, gelatin, aginate, chitosan, fibronectin and fibrin to provide an excellent Excetral Cellular Matrix (ECM) environment have been reported. Otherwise Bioprinting technology has been reported for fabrication of 3D artificial ECM instead of freeze drying or electrospinning technologies. Furthermore, a discussion of the recent achievements and their limitations has been stated and some expectations have been listed.

Keywords: Skin Regeneration; Wound Healing; Burn; Tissue Engineering; Silk Fibroin

Introduction

A skin wound is one of the most common injuries to the human body. Various reasons could cause a wound to form, including virus and fungal infection, physical and chemical injury and autoimmunity effect [1]. Once a wound is formed, complications may occur, such as infection, due to viruses or bacteria intruding into the body, they can cause necrosis of tissues or organs and haemorrhagic shock. The traditional treatments for wound healing are based on body repair itself, with the main purpose being to avoid infection complications. However, the self-regeneration process of the adult human body does not always heal the wound perfectly, this is due to the varying abilities of different cells to regenerate. The generation of granulation tissue and the subsequent formation of the scar cannot achieve the same strength as the original skin and the shrinking of the scar may cause malformation. Nowadays, tissue engineering and regenerative medicine (TERM) has been developed for several decades, in which cells and scaffolds have shown their potential ability to proliferate and differentiate into all kinds of tissues [2]. In this review, a brief introduction of skin structure and the healing process of the body itself will be summarised; meanwhile, some attempts for cut and burn wound repair will be listed and analysed. Based on this review, the perspectives and challenges of wound healing using TERM methods are concluded.

1. The Skin Structure

Fig.1 The structure of skin [1]

Skin coats all the surface of human body, it is the biggest organ, contributing 16% weight of the whole body. Skin consists of epidermis layer, dermis layer, subcutaneous tissue (hypodermis) and cutaneous appendages including glandula sebacea, sweat gland, hair and fingernail (Fig.1) [1].

1.1 Epidermis

Epidermis, located on the surface of skin, which has several stratums: stratum basale, stratum spinosum, stratum granulosum, stratum lucidum and stratum corneum. This structure is shown by Fig.2. All of stratums are consisted by keratinocytes with different morphologies and take part in keratinisation and produce keratin, stratum corneum consist of dead keratinocytes and its thickness depends on the part of skin, such as usually 40-50 layer on palm and 5-10 layers on face [3]. Usually, it will spend 14 days for corneum move from stratum basale to stratum granulosum and another 14 days from stratum granulosum to stratum corneum [4].

Otherwise, melanocyte, Langerhan cell and Merkel cell exist in epidermis. Melanocyte usually locate in stratum basale, it affects the colour of skin and has the function of ultraviolet ray resist and protect deeper tissues. Melanocyte is a kind of transparent cell with plenty of dendritic apophysis and contains many melansome which is produced by Golgi complex and could compound melaningeanule by using tyrosine (Tyr). Melaningeanule migrates from melansome to these dendritic apophysis [1].

Langerhan cell is a type of immunocompetence cell and originates from macrophagocyte. It usually locates in stratum basale and stratum spinosum. It occupies about 10% of all cells in stratum basale [5]. It has similar function to T lymphocyte which focus on recognising IgG, IgE, FcR and C3b receptors and MHC III, CD4, CD45 and S-100 antigens, thus it could play an important role in contact allergy, antiviral infection and rejection of allograft.

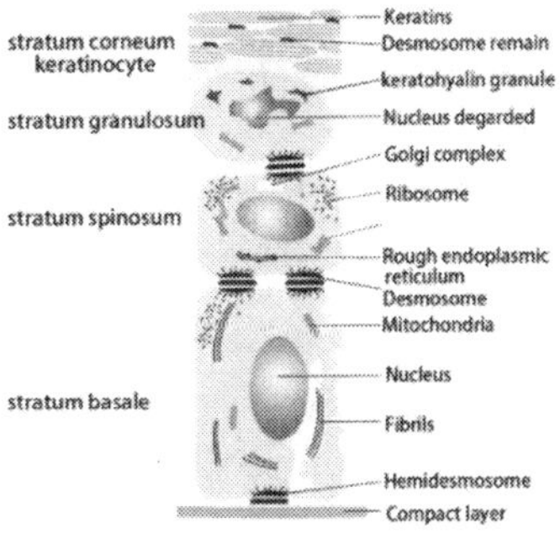

Fig.2 Structure of epidermis [1]

Merkel cells only locate in stratum basale and never migrate [4]. The functions of Merkel cells are not very clear, but usually they are considered to have close connection with sense, because the density of Merkel cells is relative higher in these sensitive area such as fingertip or apex nail; In these areas, nerve fibre loses its myelin sheath and connect with Merkel cells by chemical synapse, which are called Merkel cells-neurite complex [1]. Also, other researches shows Merkel cells arew a kind of paracrine and involved in the regulation of keratinocytes, cutaneous appendages and nerve fibre in skin [5].

1.2 Dermis

Plenty of blood vessels, lymph-vessels and nerves pass through dermis which has two layers: papillary layer and reticular layer but there is no obvious boundary between them. Dermis is a kind of tight connective tissue and consists of fibre, matrix and cells [6]. Papillary layer has many embossments and could connect well with stratum basale of epidermis and could help epidermis obtain nutrition from blood vessel in dermis. Reticular layer is under papillary layer and is the main part of dermis [7].

Fibres in dermis are fabricated from collagen fibres, reticular fibres and elastic fibres. The diameter of collagen fibres is between 70-140nm and increase along with the depth of dermis, these fibres have higher toughness and strength, but are less elastic; reticular fibre is immature collagen fibres III, has fibre diameter around 40-65nm, and usually, distributed in papillary layer; elastic fibres consist by elasticin and microfibril and the diameter is between 1-3nm which distribute in reticular layer twine around collagen fibres thus reticular layer has higher elasticity [8].

The matrix in dermis is used to filling the space between fibre and cells, mainly is proteoglycan [8]. The structure of proteoglycan has many micro pores which allow water, electrolyte, metabolite and nutrition pass, but large stuff such as bacteria will deposit and facilitate macrophage to clear it [1]. Immune response within skin mainly happens in dermis [9]. This is because most of cells such as macrophage, T cell, endotheliocyte, fibroblast and mastocyte who join in skin immune response distribute capillaries in papillary layer. The interaction of these cells and cytokines produced by these cells play an important role in repair of inflammatory injury [9].

1.3 Hypodermis, Cutaneous Appendages, Nerve, Vessel and Muscle

Hypodermis is a part of skin which consist of loosened connective tissue and fat, blood vessels, lymph-vessels, nerves and sweat gland through the hypodermis. Besides this, the hypodermis can retain the human heat and buffer mechanical stress [1, 8]. Cutaneous appendages, include eccrine gland, apocrine gland, sebaceous gland, hair and nail. They are both accessory of skin and could cooling skin (sweat gland), protecting skin (hair and nail) and lubricate skin (sebaceous gland).

There are two kinds of nerves in skin, one is for motion and another is for sense. Motor nerve including cholinergic nerve and adrenergic nerve, which is for secretory cell in eccrine gland and epithelial cell of arrector pili muscle, apocrine gland and blood vessel respectively [10]. Sensory nerves include free nerve endings and sensory corpuscles, mainly distributed under pileous skin and glabrous skin respectively [1].

The blood vessel in skin could regulate body temperature by contraction and diastole. They combine a network which is parallel to epidermis by continuous endothelial cells [8]. Blood vessels have 3 layers, and thus have a poor permeability. In contrast with blood vessel, lymph-vessel has one layer which consists of endothelial cells and fibre, so that immune cell could go out of lymph-vessel easier. Arrector pili muscle is the most common muscle in skin which is located near hair follicles and can stand the hair when people feel cold [5].

2. Burn, Wound Healing and Regeneration of Skin

2.1 Burn

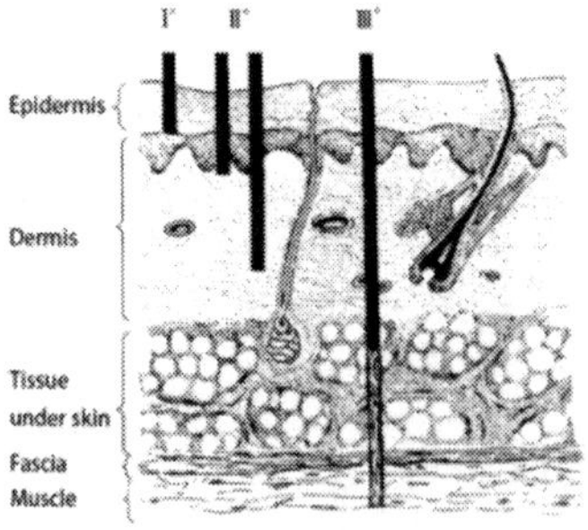

Fig.3 Classification of a burn [10]

Burn can be caused by various reasons which include flame, hot liquid, gas and solid, steam, intense radiation, electricity, and chemical medicine, the way to classify the degree of burn has shown on Fig.3, Superficial (1st-degree) means the only layer involved is only epidermis; superficial partial thickness (2nd-degree) means injury extends into superficial papillary dermis; deep partial thickness (2nd-degree) means injury extends into deep reticular dermis; Full thickness (3rd-degree) means injury extends through entire dermis; 4th-degree means injury extends through entire skin, and into underlying fat, muscle and bone [5, 11-13]. This classification states the degree of injury of skin clearly, not only for the burn. A large area of burn will also cause swelling because of an increase of capillaries' permeability. Further problems are encephaledema, kidney failure, heart function failure and gastric ulcer [8]. Moreover, loss of a large amount of skin will make people very weak in defending bacteria and virus, so that generalized infection may happen. In this period, pulmonary complications may occur [9]. If the degree of burn is higher than 2nd-degree, melanin deposition, scar and contracture will occur on the injured skin. This is because with the rise of burn degree, the remaining epithelial cells will be increasingly lost, and causing body stress response which can lead to pathologic pigmentation and hyperplasia of connective tissue [11].

The injury mechanism in burn is high temperature and ionizing radiation, during these situation, proteins begin losing their 3-dimensional shape and start breaking down [14], which will cause cell death and tissue damage.

2.2 Regeneration of Skin and Wound Healing

In cell lever, the proliferation of cell, to large extend, affects by the stimulation of cell factor. The most important thing is to promote motionless cell into the cell cycle [6]. When injury happens, growth factors will release and stimulate cell from the same germinal layer. Skin cells are all from ectoderm. For skin regenerate, plenty of cell factors such as epidermal growth factor (EGF), fibroblast growth factor (FGF), transforming growth factor (TGF) and vascular endothelial growth factor (VEGF) take part in the promotion of cell growth [5, 8, 23]. Chalone such as angiostatin and endostatin could inhibit the growth of capillaries. The inhibition of epidermis is mainly through contract inhibition: when epidermis contact with each other, proliferation stops [4, 14].

When the burn happens, the regeneration of skin starts. Firstly, the blood coagulation happens in the injury area, the inflammation appears which cause congestion and exudation of neutrophils [1]. Sometimes, incrustation made by the collagen and blood clot cover the surface of wound to protect it. The blood clot is caused by a cross-linked fibrin fibres embedded platelets plasma fibronectin, vitronectin, and thrombospondin [15]. Then inflammatory cells are recruited toward to the site, mainly neutrophils and monocytes. During the recruitment of inflammatory cells, molecular changes in the surface of endothelial cells lining capillaries and selectin and β2 class of integrin express to allow adhesion, leukocytes slow down and diapedesis, which causes neutrophils crawl out between endothelial cells into the extravascular space and arrive at wound site in several minutes [16-17]. For the injuries, monocyte enters into the injured tissue, and grows and transfers to macrophagocyte which is response for the clearing of dead cells and releases some chemokines which can attract fibroblast and epithelial cell in the injury area [8]. Monocytes also release a chain of growth factors and cytokines [17]. Hübner et al found that neutrophils could also provide some of the earliest signals to activate local fibroblasts and keratinocytes [18].

In the wound area, large amount of fibres and matrixes are produced by fibroblast. During this period, various capillaries are made by epithelial cell, so that granulation tissue forms [19]. Granulation tissue can be thought of as a substrate that provides nutrition for epidermis formation. However, cutaneous appendages cannot regenerate completely after burning [20]. Under the granulation tissue, reconstituting of epidermal cells happens. During the process of reepithelialisation, hemidesmosome attached keratinocyte with dermis are dissolved by the new expressed integrins, which caused keratinocyte to crawl over wound matrix [21-22]. A large amount of researches have revealed that keratins play a significant role for crawling motility of adult keratinocytes and purse-string closure of embryonic epidermal wounds [23]. Once keratinocytes migrated to the boundary of wound, a proliferative burst of epidermis cells occurs [24-25]. The reepithelialisation will finish when a monolayer of keratinocytes covers the wound site, epidermal migration will be stopped and new anchoring fibrils and hemidesmosome links to basal lamina, integrins and matrix metalloproteinase stop expression [5]. General opinion of this phenomenon is contact inhibition [26]. Meanwhile, regeneration of hair and sweat glands start at dermis. However, if wound is deeper than dermis and there are no hair follicles remain, hair regeneration will not happen [5]. In a mice mould research [27], lymphoid enhancer factor-1 (LEF-1), sonic hedgehog (Shh), bone morphogenetic protein-2 (BMP-2) and FGF-4 has significate effect for hair growth. Thus the main reason for being unable to regenerate hair is because adult wound epidermis does not receive such signals from the underlying wound dermis [23]. Hair follicle could improve reepithelialisation, if the remaining hair follicles could provide a space for keratinocytes proliferation and they spread out like growing islands [25]. The proliferative scar forms a barrier that prevents sweat gland regeneration, which is because the stem cells in the injured sweat glands cannot achieve the regeneration of sweat glands, and the new epidermal stem cells also fail to differentiate into sweat gland cells after scar healing [28]. Regeneration of sweat gland cells is also regulated by the same factors because the sweat gland cells are homologous with the basal epidermal stem cells [29]. Human umbilical cord- derived mesenchymal stem cells (hUC-MSCs) could differentiate into sweat gland-like cells (SGCs) by inducing of Keratinocyte Growth Factor (KGF) [30]. During this process, blood vessels and nerves grow as well. For angiogenesis, VEGF and FGF-2 release for blood vessel growth [23]. FGF is indispensable and produced from damaged endothelial cells and macrophages at the wound site, VEGF is induced keratinocytes and macrophages in wound-edge when response to KGF and TGF-α [31-32]. Otherwise capillary morphogenesis has some relative evidence of regulated proteolysis in wound site matrix [33]. For the neural regeneration, nerve growth factor (NGF) is the directly signal for nerve overgrowth [34], and it is up-regulated by the effect of TGF-α isoforms [35]. Oppositely, sprouting nerves could deliver some neuropeptides and other growth factors to the wound site so that there may some stimulatory effect for the healing process [36-37].

In a deep burn (higher than 2nd degree), the necrotic tissue cannot be dissolved or absorbed by the body, and granulation tissue occurs [9]. About 3-4 days after wound, resident dermal fibroblasts proliferate, platelet-derived growth factor (PDGF) and TGF-β acting as mitogens and chemotactic factors for them [23]. Granulation tissue is new connective tissue which is full of microscopic blood vessels that form on the surfaces of a wound during the healing process. When granulation tissue take shape, organisation will start and during this period, granulation tissue matures and replaces necrosis tissue [5, 15]. A week after wound, wound clot is replaced by a new collagen-rich matrix which produced by fibroblast stimulated by TGF-β1 [23]. The wound contraction starts and fibroblasts reduce expressing collagen receptors and raise the amount of integrins in order to crawl into the clot [38]. At this stage, some of fibroblasts transform into myofibroblasts, which is the source of contractile forces because of the expression of α-smooth muscle actin [15]. Finally, a scar forms. Granulation tissue plays a significate role in formation of extracellular matrix, operation of the immune system and vascularisation [1]. In serious burn or deep cut wound, scars usually remain. This is because there are large areas to regenerate and the repair rate cannot satisfy [8, 15]. In order to ensure the completeness of skin tissue, the body fills these areas by granulation tissue which grow into scar [13]. A scar has stronger mechanical strength than granulation tissue. This process is led by the structure change of extracellular matrix (ECM) [5]. Many researches have showed that a balance among the TGF-β isoforms and neutralise TGF-β1 and -β2 at the time of wounding reduces scarring [23]. Mannose-6-phosphate (M-6-P) directly and indirectly applied to wounds will prevent scarring [39].

2.3 Current Treatment

In clinical treatment, except for burns higher than 2nd-degree, the skin repair depends on the body itself. Some slight treatments like the extracting vesicular fluid and the clear pustule may be applied [11]. The most of treatment, especially for large area burn, is to sterilize and avoid infection, such as application of silver sulfadiazine which is used for inhibition of bacteria growth [11]. However, for deeper burn, skin-grafting may necessary [5].

Table.1 Some SF fabricated wound dressing [40-44, 53, 65].

Material	Morpholog y	Method	Antimicrobial way	Cell	Reference
SF nanofiber	membranes	Thiol-maleimide coupling	immobilized antimicrobial peptide (Cys-KR12)	Human keratinocytes (HaCaT) Human dermal fibroblasts (NHDF) Murine monocytes (Raw264.7)	[40]
SF/gelatin nanofibers	Mat	Electrospinning	Ceftazidime	Human skin fibroblast cells	[41]
SF fiber	Mat	Electrospinning	Coated ciprofloxacin hydrochloride	Full thickness cut rabbit model HaCaT and NHDF cells	[42]
SF nanoparticles Sodium-carboxymethyl	hydrocolloi d dressing	Membrane grind Ball milling	Exchange dressing	2nd burn rat model 3T3 fibroblast cells	[43]
Isabgol (psyllium husk) SF	3D scaffold	Freeze drying	Not mentioned	Full thickness cut wound rat model	[44]
SF fiber	Nano matrix	Electrospinning	Not mentioned	2nd burn rat model	[53]
Amorphous SF Transgenic worms Wild-type worm	Film	Not mentioned	Not mentioned	Full-thickness skin wounds on mice NHDF cells	[65]

3. Biomaterial Application for Skin Regeneration

Biomaterial has been widely used for tissue regeneration, such as for wound healing. Currently, the attempts for the skin regeneration mainly uses rat skin fibroblasts (3T3 or L929), human keratinocytes and human dermal fibroblasts [40-44], which is similar to the skin regeneration in natural environment. For this situation, it is important to provide an excellent ECM environment. Collagen, gelatin, aginate, chitosan, fibronectin and fibrin has been reported for wound healing [45].

Table.2 The achievements and Limitations SF fabricated wound dressing [40-44, 53, 65].

Achievements/Limitations	Reference
Significantly inhibited TNF-α expression Only 1-week antimicrobial activity retained on body temperature (37°C)	[40]
Good antibacterial effect Drug release activity very depends on water	[41]
Functionalized with EGF and Scar-less healing tending The rate of drug release not even, up to 60% released within the first 10 hours	[42]
Addition of SF NPs enhances tensile strength and increase the density of collagen fibers and PCNA expression Cannot closure wound within 3 weeks	[43]
Enhanced the fluid uptake ability with enough fibroblast attachment and good viability, cause dense collagen fibers, and neovascularization The ability of cells migration may not great	[44]
Accelerates re-epithelialization and wound closure in burn wound and scarless wound healing by regulating TGF-β1 expression No anti-microbiology method refers	[53]
Wound closure accelerates (transgenic worms), granulation tissue was thicker and larger and neovascularization was promoted significantly (wild-type silkworm)	[65]

Silk fibroin composite material has been reported in skin regeneration, silk fibroin-alginate scaffolds sufficiently enhance reepithelisation in a full-thickness rat wound model [46]. Otherwise, nanofibrous silk-chitin [47], silk-collagen [48], silk-elastin [49] and intermolecular cross-linked recombinant human-like collagen (RHLC) with fibroin [50] have also been applied successfully in skin repair. These natural materials are used for as the cell scaffold and perform a good property of biocompatibility. Shellac wax, beeswax, or carnauba wax has been reported to coat on silk fibroin material successfully which could provide the wound dressing a hydrophobic surface [51] and protect the wound from infection. Polyethylenimine has been reported that could be antibacterial when it was blended with electrospinning fibroin scaffold [52].

Skin fibroin (SF) material has been wildly used for wound healing. It can be fabricated in various forms including matrix [53], microsphere, nanoparticle [54] and film or membranes [55], this is because SF material supports well the human keratinocytes and fibroblasts [56]. In full-thickness burns, gelatin microspheres were impregnated with the antibiotic gentamycin sulfate, and the microspheres were then embedded in a silk fibroin matrix to fabricate GS/GM/SF scaffolds, these scaffolds could effectively inhibit injection during dermal regeneration [57]. SF coated with polypyrrole (pPy) by chemical polymerization has been published and this method could improve mechanical resistance [58]. SF electrospinning scaffold was also reported to improve vascular growth by using human aortic endothelial and human coronary artery smooth muscle cell [59]. SF electrospinning scaffold is a kind of degradable material; it can be degraded 65% by protease within 24 days [60] and full degraded within 21 days for aqueous-derived scaffold [61].

Bioprinting technology has been applied for fabrication of 3D artificial ECM, for skin repair. Sha Huang et al prove a 3D printed artificial ECM made by mixture of cell, EGF and homogenates could lead sweat gland regeneration in a rat mould [62]. Dextran hydrogel scaffolds made by group of Guoming Sun in a mice mould show an improvement of angiogenic formation [63]. Reducing the scar is one of the most significant task to do, Chouhan *et al* [42] indicated that their SF mat functionalized with EGF could acquire a scar-less healing treatment. This may be caused by EFG lead the degradation of collagen I [64] and SF nanofiber could significantly inhibit TNF-α expression [40, 53]. Otherwise, Ponrasu *et al* reported their Isabgol (psyllium husk) and SF 3D composite scaffolds could accelerate the cellular activity to heal the wounds faster with minimal scar formation [44]. SF has been used for research, it can come from various kinds of worm, Baba *et al* documented that SF accelerates cutaneous wound healing and this effect is enhanced by over-express the RG sequence SF (TG-SF) [65]. Table 1 summarised some weighted outcomes during 2016 and each of their achievements and limitations have been listed on Table 2 [40-44, 53, 65].

4. Conclusion

In conclusion, to fabricate a biomaterial wound dressing, the material must have the following functions: firstly, encapsulate the wound, inhibit bacterium growth and eradicate infection; secondly, have low toughness and fit wound very well; thirdly, have the similar structure with ECM in skin, lastly, be biodegradable, which could disappearing when new skin forms. Meanwhile, some cytokines could be loaded on the material, in order to enhance and accelerate the process of wound closure and reduce scar scale. From the previous research, SF fibres were made from various types of silkworm cocoons, and each of them has their own properties [65], a mixed electrospinning SF fibre scaffold may be considered. Otherwise, collagen, fibrin, fibronectin, elastin, chitin and chitosan has been reported applied on the wound healing and skin regeneration respectively [45, 47-49], which should be commendable choices for wound healing. Collagen fibrin, fibronectin and elastin are important part of human skin [2] and could blend in the skin while the healing process, chitosan and chitin are famous material which has been used in TERM aspect for decades, including bone, cartilage, tendon, ligament, skin, nerve and liver regeneration, also applied for drug and growth factor deliver [66]. These materials could be considered for scaffold fabrication.

In the future, our aim is to build a composite system for a perfect wound repair, a wound dressing (which is also a scaffold) could culture fibroblast, keratinocytes and monocytes in three-dimensional scale and steerable release of different growth factors in various layers. This could simulate a ECM environment like a wound site, and also, to avoid scar formation by regulating the elements in this artificial ECM.

5. Acknowledgment

We would like to thank the support of the EU Horizon 2020 and University of Manchester through projects with project codes 644268 - ETEXWELD - H2020-MSCA-RISE-2014, AA14512 (UMRI project "Graphene-Smart Textiles E-Healthcare Network") and AA01906.

References

[1] Zhang XJ. Dermatovenereology 8th edition. People's Medical Publishing House: 2013; 2: 5-13.
[2] Li Z, Xie MB, Li Y, Ma Y, Li JS, Dai FY. Recent Progress in Tissue Engineering and Regenerative Medicine. Journal of Biomaterials and Tissue Engineering: 2016; 10: 755-766.
[3] Gao YM. Histology and embryology. People's Medical Publishing House: 2001; 11: 165-178.
[4] Yang T. Cell biology 2nd edition. People's Medical Publishing House: 2010; 6: 143-170.
[5] Zhang XJ. Dermatovenereology 8th edition. People's Medical Publishing House: 2013; 3: 14-18.
[6] Yang T. Cell biology 2nd edition. People's Medical Publishing House: 2010; 3: 32-47.
[7] He W. Medical immunology 2nd edition. People's Medical Publishing House: 2010; 6: 93-101.
[8] Zhao B. Clinical Dermatology 2nd edition. Jiangsu science Publishing House: 1992: 2; 15-34.
[9] He W. Medical immunology 2nd edition, People's Medical Publishing House: 2010; 4: 57-75.
[10] Wu J. Neurology 2nd edition. People's Medical Publishing House: 2010; 2: 3-48.
[11] Chen XP. Surgery, 8th edition. People's Medical Publishing House: 2013; 13: 138-152.
[12] Granger, Joyce. An Evidence-Based Approach to Pediatric Burns. Pediatric Emergency Medicine Practice. 2009: 6; 1.
[13] Marx, John. Chapter 60: Thermal Burns. Rosen's emergency medicine: concepts and clinical practice, 7th ed. Philadelphia: Mosby/Elsevier: 2010; ISBN 978-0-323-05472-0.
[14] Zhou AR. Biological chemistry 5th edition. People's Medical Publishing House: 2001; 2: 6-31.

[15] Clark RAF. The Molecular and Cellular Biology of Wound Repair (Plenum, New York): 1996 2: 62-131.
[16] Springer TA. Traffic signals for lymphocyte recirculation and leukocyte emigration: the multistep paradigm. Cell: 1994; 76; 2: 301-314.
[17] Mayadas TN, Johnson RC, Rayburn H., Hynes RO, Wagner DD. Leukocyte rolling and extravasation are severely compromised in P selectin-deficient mice. Cell: 1993; 74; 3: 541-554.
[18] Hübner G, Brauchle M, Smola H, Madlener M, Fässler R, Werner S. Differential regulation of pro-inflammatory cytokines during wound healing in normal and glucocorticoid-treated mice. Cytokine: 1996; 8; 7: 548-556.
[19] Li GD. Pathology. People's Medical Publishing House: 2001; 2: 34-56
[20] Chen ZC. Pathophysiology. People's Medical Publishing House: 2001; 5: 52-82
[21] Paglia P, Girolo, moni G, Robbiati F, Granucci F, Ricciardi-Castagnoli P. Immortalized dendritic cell line fully competent in antigen presentation initiates primary T cell responses in vivo. The Journal of experimental medicine: 1993; 178; 6: 1893-1901.
[22] Cavani A, Zambruno G, Marconi A, Manca V, Marchetti M, Giannetti A. Distinctive integrin expression in the newly forming epidermis during wound healing in humans. Journal of Investigative Dermatology: 1993: 101; 4: 600-604.
[23] Martin P. Wound healing--aiming for perfect skin regeneration. Science: 1997; 276; 5309: 75-81.
[24] Garlick JA, Taichman LB. Fate of human keratinocytes during reepithelialization in an organotypic culture model. Laboratory investigation; a journal of technical methods and pathology: 1994; 70; 6: 916-924.
[25] Matoltsy AG, Viziam CB. Further observations on epithelialization of small wounds. Journal of Investigative Dermatology: 1970; 55; 1: 20-25.
[26] Compton CC, Gill JM, Bradford DA, Regauer S, Gallico GG, O'connor NE. Skin regenerated from cultured epithelial autografts on full-thickness burn wounds from 6 days to 5 years after grafting. A light, electron microscopic and immunohistochemical study. Laboratory investigation; a journal of technical methods and pathology: 1989; 60; 5: 600-612.
[27] Zhou P, Byrne C, Jacobs J, Fuchs E. Lymphoid enhancer factor 1 directs hair follicle patterning and epithelial cell fate. Genes & development: 1995; 9; 6: 700-713.
[28] Ma K, Tan Z, Zhang C, Fu X. Mesenchymal stem cells for sweat gland regeneration after burns: From possibility to reality. Burns: 2016; 42; 3: 492-499.
[29] Biedermann T, Pontiggia L, Böttcher-Haberzeth S, Tharakan S, Braziulis E, Schiestl C, Reichmann E. Human eccrine sweat gland cells can reconstitute a stratified epidermis. Journal of Investigative Dermatology: 2010; 130; 8: 1996-2009.
[30] Xu Y, Hong Y, Xu M, Ma K, Fu X, Zhang M, Wang G. Role of keratinocyte growth factor in the differentiation of sweat gland-like cells from human umbilical cord-derived mesenchymal stem cells. Stem cells translational medicine: 2016; 5; 1: 106-116.
[31] Broadley KN, Aquino AM, Woodward SC, Buckley-Sturrock A, Sato Y, Rifkin DB, Davidson JM. Monospecific antibodies implicate basic fibroblast growth factor in normal wound repair. Laboratory investigation; a journal of technical methods and pathology: 1989; 61; 5: 571-575.
[32] Brown LF, Yeo K, Berse B, Yeo TK, Senger DR, Dvorak HF, Van De Water L. Expression of vascular permeability factor (vascular endothelial growth factor) by epidermal keratinocytes during wound healing. The Journal of experimental medicine: 1992; 176; 5: 1375-1379.
[33] Fisher C, Gilbertson-Beadling S, Powers EA, Petzold G, Poorman R, Mitchell MA. Interstitial collagenase is required for angiogenesis in vitro. Developmental biology: 1994; 162; 2: 499-510.
[34] Constantinou J, Reynolds ML, Woolf CJ, Safieh-Garabedian B, Fitzgerald M. Nerve growth factor levels in developing rat skin: unregulation following skin wounding. Neuroreport: 1994; 5; 17: 2281-2284.
[35] Buchman VL, Sporn M, Davies AM. Role of transforming growth factor-beta isoforms in regulating the expression of nerve growth factor and neurotrophin-3 mRNA levels in embryonic cutaneous cells at different stages of development. Development: 1994; 120; 6: 1621-1629.
[36] Nilsson J, von Euler AM, Dalsgaard CJ. Stimulation of connective tissue cell growth by substance P and substance K. Nature: 1984; 315; 6014: 61-63.
[37] Lee KF, Li E, Huber LJ, Landis SC, Sharpe AH, Chao MV, Jaenisch R. Targeted mutation of the gene encoding the low affinity NGF receptor p75 leads to deficits in the peripheral sensory nervous system. Cell: 1992: 69; 5: 737-749.

[38] Xu J, Clark RA. Extracellular matrix alters PDGF regulation of fibroblast integrins. The Journal of cell biology: 1996: 132; 1: 239-249.
[39] Shah M, Foreman DM, Ferguson MW. Neutralising antibody to TGF-beta 1, 2 reduces cutaneous scarring in adult rodents. Journal of cell science: 1994; 107; 5: 1137-1157.
[40] Song DW, Kim SH, Kim HH, Lee KH, Ki CS, Park YH. Multi-biofunction of antimicrobial peptide-immobilized silk fibroin nanofiber membrane: Implications for wound healing. Acta biomaterialia: 2016
[41] Safdari M, Shakiba E, Kiaie SH, Fattahi A. Preparation and characterization of Ceftazidime loaded electrospun silk fibroin/gelatin mat for wound dressing. Fibers and Polymers: 2016; 17; 5: 744-750.
[42] Chouhan D, Chakraborty B, Nandi SK, Mandal B. Role of non-mulberry silk fibroin in deposition and regulation of extracellular matrix towards accelerated wound healing. Acta Biomaterialia: 2017; 48: 157-174.
[43] Lee OJ, Kim JH, Moon BM, Chao JR, Yoon J, Ju HW, Park HS. Fabrication and characterization of hydrocolloid dressing with silk fibroin nanoparticles for wound healing. Tissue Engineering and Regenerative Medicine: 2016; 13; 3: 218-226.
[44] Ponrasu T, Vishal P, Kannan R, Suguna L, Muthuvijayan V. Isabgol–silk fibroin 3D composite scaffolds as an effective dermal substitute for cutaneous wound healing in rats. RSC Advances: 2016; 6; 77: 73617-73626.
[45] Huang S, Fu X. Naturally derived materials-based cell and drug delivery systems in skin regeneration. Journal of Controlled Release: 2010; 142; 2: 149-159.
[46] Roh DH, Kang SY, Kim JY, Kwon YB, Young Kweon H, Lee KG, Park YH, Baek RM, Heo CY, Choe J, Lee JH. Wound healing effect of silk fibroin/ alginate-blended sponge in full thickness skin defect of rat, J. Mater. Sci. Mater. Med: 2006; 17: 547-552.
[47] Yoo CR, Yeo IS, Park KE, Park JH, Lee SJ, Park WH, Min BM. Effect of chitin/silk fibroin nanofibrous bicomponent structures on interaction with human epidermal keratinocytes, Int. J. Biol. Macromol: 2008; 42: 324-334.
[48] Yeo IS, Oh JE, Jeong L, Lee TS, Lee SJ, Park WH, Min BM. Collagen-based biomimetic nanofibrous scaffolds: preparation and characterization of collagen/silk fibroin bicomponent nanofibrous structures, Biomacromolecules: 2008; 9: 1106-1116.
[49] Vasconcelos A, Gomes AC, Cavaco-Paulo A. Novel silk fibroin/elastin wound dressings. Acta Biomaterialia: 2010; 8; 8: 3049-3060.
[50] Khor HL, Ng KW, Htay AS, Schantz JT, Teoh SH, Hutmacher DW. Prelimi- nary study of a polycaprolactone membrane utilized as epidermal substrate, J. Mater. Sci. Mater. Med: 2003; 14: 113-120.
[51] Sorada Kanokpanont, Siriporn Damrongsakkul, Juthamas Ratanavaraporn, Pornanong Aramwit. Physico-chemical properties and efficacy of silk fibroin fabric coated with different waxes as wound dressing. International Journal of Biological Macromolecules: 2013; 55: 88-97
[52] Çalamak S, Erdoğdu C, Özalp M, Ulubayram K. Silk fibroin based antibacterial bionanotextiles as wound dressing materials. Materials Science and Engineering: C: 2014; 43: 11-20.
[53] Ju HW, Lee OJ, Lee JM, Moon BM, Park HJ, Park YR, Park CH. Wound healing effect of electrospun silk fibroin nanomatrix in burn-model. International journal of biological macromolecules: 2016; 85: 29-39.
[54] Zhou Y, Yang H, Liu X, Mao J, Gu S, Xu W. Electrospinning of carboxyethyl chitosan/poly (vinyl alcohol) /silk fibroin nanoparticles for wound dressings. International journal of biological macromolecules: 2013; 53: 88-92.
[55] Gu Z, Xie H, Huang C, Li L, Yu X. Preparation of chitosan/silk fibroin blending membrane fixed with alginate dialdehyde for wound dressing. International journal of biological macromolecules: 2013; 58: 121-126.
[56] Zhang X, Reagan MR, Kaplan DL. Electrospun silk biomaterial scaffolds for regenerative medicine, Adv. Drug Deliv. Rev: 2009; 61: 988-1006.
[57] Lan Y, Li W, Jiao Y, Guo R, Zhang Y, Xue W, Zhang Y. Therapeutic efficacy of antibiotic-loaded gelatin microsphere/silk fibroin scaffolds in infected full-thickness burns. Acta biomaterialia: 2014; 10; 7: 3167-3176.
[58] Aznar-Cervantes S, Roca MI, Martinez JG, Meseguer-Olmo L, Cenis JL, Moraleda JM, Otero TF. Fabrication of conductive electrospun silk fibroin scaffolds by coating with polypyrrole for biomedical applications. Bioelectrochemistry: 2012; 85: 36-43.
[59] Zhang X, Baughman CB, Kaplan DL. In vitro evaluation of electrospun silk fibroin scaffolds for vascular cell growth. Biomaterials: 2008; 29; 14: 2217-2227.

[60] Zhou J, Cao C, Ma X, Hu L, Chen L, Wang C. In vitro and in vivo degradation behavior of aqueous-derived electrospun silk fibroin scaffolds. Polymer Degradation and Stability: 2010; 95; 9: 1679-1685.
[61] Kim UJ, Park J, Kim HJ, Wada M, Kaplan DL. Three-dimensional aqueous-derived biomaterial scaffolds from silk fibroin. Biomaterials: 2005; 26; 15: 2775-2785.
[62] Huang S, Yao B, Xie J, Fu X. 3D bioprinted extracellular matrix mimics facilitate directed differentiation of epithelial progenitors for sweat gland regeneration. Acta biomaterialia: 2016; 32: 170-177.
[63] Sun G, Zhang X, Shen YI, Sebastian R, Dickinson LE, Fox-Talbot K, Gerecht S. Dextran hydrogel scaffolds enhance angiogenic responses and promote complete skin regeneration during burn wound healing. Proceedings of the National Academy of Sciences: 2011; 108; 52: 20976-20981.
[64] Mimura Y, Ihn H, Jinnin M, Asano Y, Yamane K, Tamaki K. Epidermal growth factor affects the synthesis and degradation of type I collagen in cultured human dermal fibroblasts. Matrix biology: 2006; 25; 4: 202-212.
[65] Baba A., Kitayama K, Asakura T, Sezutsu H, Tanimoto A, Kanekura T. Silk fibroin produced by transgenic silkworms overexpressing the Arg-Gly-Asp motif accelerates cutaneous wound healing in mice. Journal of Dermatological Science: 21016; 84; 1: e150.
[66] Anitha A, Sowmya S, Kumar PS, Deepthi S, Chennazhi KP, Ehrlich H, Jayakumar R. Chitin and chitosan in selected biomedical applications. Progress in Polymer Science: 2014; 39; 9: 1644-1667.

Montmorillonite-doped Electrospun Polyurethane Nanofibers for Drug Encapsulation and Sustained Release

Qing-Yan Peng , Hao Wang, Shi-Li Xiao*

School of Textile Science and Engineering, Wuhan Textile University, Wuhan, Hubei 430000, China

*Corresponding author's email: xshili@yahoo.com

Abstract

A novel electrospun composite nanofiber-based drug delivery system was developed. In this study, montmorillonite (MMT) nanoparticles were first used to load a model drug, aspirin (ASP). The ASP-loaded MMT nanohybrids were mixed with polyurethane (PU) for subsequent electrospinning to form drug-loaded composite nanofibrous mats. The PU/MMT/ASP composite nanofibrous mats were systematically characterized using scanning electron microscopy (SEM), Fourier transform infrared spectroscopy, and mechanical testing. In vitro drug release showed that this composite nanofibrous drug delivery system can effectively mitigate the burst release of the drug and the introduction of MMT can improve the tensile stress property.

Keywords: Electrospinning; Polyurethane; Montmorillonite; Aspirin

1. Introduction

It is known that tissue engineering includes cells, materials and biochemical elements for the construction of a new tissue to replace the impaired one [1, 2]. The electrospinning technique is considered to be a versatile approach to generate sequential nanofibers, because of its superior properties of electrospun nanofibers such as large surface area, volume ratio and high porosity. In particular, the three-dimensional network structure of electrospun nanofibrous mats that can mimic human extracellular matrix (ECM) has aroused great interest from scientific research workers [3, 4]. Many biodegradable, non-degradable, and natural materials have been used to produce electrospun nanofibers, the characteristics like diameter, appearance and mechanical properties can be regulated to fit cells adhesion and proliferation [5-7].

Recently, nanofibers have been widely studied as drug delivery system on account of their advanced properties [8]. A majority of drug-loaded electrospun nanofibers were fabricated from blends [9, 10] using core-shell method [11] or separate-spinneret manner [12]. However, although this method enables easy incorporation of drug molecules into the nanofibers, a burst release always occurs which is undesirable in many cases [13, 14]. For instance, Verreck and co-worker incorporated poorly water-soluble drug, itraconazole and ketanserin drugs into electrospun segmented polyurethane nanofibers for potential drug administration and wound healing. Drug release study showed that in the first 4 h, the content of model drug ketanserin released from PU nanofibrous mats reached 5 $\mu g/cm^2$, 50% of loaded drug has been released out compared to 10$\mu g/cm^2$ within 24 h, suggesting an obvious burst release phenomenon [15]. Thus, developing nanofibrous drug delivery system that can mitigate the burst release of the model drugs is still a challenge.

Montmorillonite (MMT), a kind of non-toxic and layered silicate material has been frequently used in biomedical field as a drug carrier because of its high internal surface area, excellent dispersive performance and biocompatibility [16]. Moreover, negatively charged MMT has good swelling property in the presence of water and hydrophilic solvents because of which the positively charged bioactive compounds can be intercalated into the interlayer spaces by electrostatic interaction [17, 18]. In view of the advantages of MMT, we hypothesize that incorporating drug-loaded MMT into electrospun nanofbers might be effective to reduce the bust release of model drug. In this study, we choose PU, a widely used polymer in the biomedical field owing to its well physical, chemical and biological properties, especially excellent antimicrobial activity and cytocompatibility [19] to produce electrospun nanofibers. The anti-inflammatory drug aspirin (ASP) was chose as a model to be incorporated within MMT firstly, then ASP-loaded MMT was blended with PU solution to form a PU/MMT/ASP nanofibers by electrospinning technology. The developed PU/MMT/ASP nanofibrous mats were systematically characterized using scanning electron microscopy (SEM), Fourier transform infrared (FTIR) spectroscopy and mechanical testing. The drug release kinetics of the PU/MMT/ASP nanofibrous mat was monitored using ultraviolet-visible (UV-Vis) spectrometer. The biocompatibility of nanofibrous mats with different compositions was investigated using fibrous blast cell L929 as a model.

2. Materials and Methods

Polyurethane (PU) was purchased from Dow Chemical Co., Ltd. Acetone and dimethyl formamide (DMF) were bought from Sinopharm Chemical Reagent Co., Ltd. Montmorillonite (MMT) was purchased from Zhejiang Feng Hong New Material Co., Ltd. Aspirin(ASP) (purity≥99%) was from Aladdin Industrial Corporation (Shanghai).

2.1 Preparation of the Drug-loaded MMT Nanohybrids

ASP was dissolved into phosphate buffered solution (PBS) (pH=7.4) to acquire ASP aqueous solution (0.25 mg/mL) at room temperature. After that, MMT powder was dispersed into ASP solution with different concentrations, namely 0.25, 0.5, 1, 2 and 4 mg/mL, respectively, to get serious of MMT/ASP suspensions. Subsequently, the MMT/ASP suspensions were vigorously stirred for 12 h to allow sufficient interaction between ASP and MMT. The mixture were then centrifuged under 6000 rpm for 10 min to remove the free ASP and then rinsed with PBS for 3 times. The supernatants after 4 times centrifugation were collected together. The ASP encapsulation and loading efficiencies were determined by measuring the absorbance at 274 nm using a UV-756PC UV-visible Spectrometer and a standard curve for ASP (prepared using ASP solution of different known concentration). The prepared MMT/ASP nanohybrids were lyophilized and stored in desiccator before use. The drug loading efficiency can be calculated from the following equation [20]:

$$\text{Loading efficiency} = (M_1 - M_0) / M \times 100\% \quad (1)$$

Where M_1 and M_0 stand for the mass of initial ASP used for encapsulation and the free ASP, respectively, and M stands for the total mass of ASP and MMT.

2.2 Preparation of the PU/MMT/ASP Composite Nanofibers

PU was dissolved in a mixture solvent of acetone/DMF (v/v = 1:1) with an optimized concentration of 15% (w/v). Subsequently, MMT/ASP nanohybrids (with the optimized proportion, 2% w/v) were dispersed in PU solution under magnetic stirring for 12 h for preparing PU/MMT/ASP nanofibrous mats. Pure PU and PU with MMT nanofibrous mats were also prepared as a control. The electrospinning process was carried out under ambient condition with a fixed electrical potential of 20 kV, a collector distance of 20 cm, and a feeding rate of 0.5 mL/h. The prepared electrospun nanofibrous mats with different compositions were taken off from the collector and vacuum-dried for 12 h to remove the residual organic solvent before use.

2.3 Characterization

The PU and PU/MMT/ASP nanofibrous mats were characterized using FTIR spectroscopy. The spectra were recorded using a Nicolet Nexus 470 FTIR spectrometer (Thermo Nicolet Corporation, United States) in a wavenumber range of 4000-400 cm^{-1} at ambient conditions. The surface morphology of PU, PU/MMT composites nanofibers was detected with scanning electron microscope (SEM) (HITACHI UHR FE-SEM SU1510) at an accelerating voltage of 15 kV. Before detection, each specimen was spray-coated with a 15 nm thick gold film. Fiber diameter was measured using ImageJ1.49 (http://rsb.info.nih.gov/ij/download.html), at least 200 nanofibers from SEM images were randomly selected and analyzed. Tensile test of the electrospun nanofibrous mats were performed by electronic material testing machine (INSTRON3300, UK) at constant temperature (20 ^{0}C) and humidity (63%) with a cross-head speed of 1cm/min. Specimens were cut into rectangular pieces with width × length= 1cm × 5cm and were gripped in the top and bottom collets 2 cm apart. The tensile testing was performed with a loading of 10 N. According to the collected data, the stress-strain curves of the samples can be calculated through mathematical conversions [21]:

$$\delta(MPa) = \frac{f(N)}{w(mm) \times h(mm)} \tag{2}$$

$$\varepsilon(\%) = \frac{L(mm)}{g(mm)} \times 100\% \tag{3}$$

Where f is the breaking force of sample; w and h are the width and thickness of the sample, respectively; l is the breaking length and g is the distance between top and bottom collets.

2.4 In Vitro Drug Release Analysis

In vitro release of ASP from PU/ASP and PU/MMT/ASP (the content of ASP in nanofibrous mat was 2%) nanofibrous mats was evaluated using ultraviolet-visible (UV-Vis) spectroscopy (UV-756PC). 60 mg of each specimen was put into 20-mL vial containing 5 mL PBS (pH = 7.4) solution with gently shaking in a oscillator (SKY-2102C) at 37 ^{0}C. At different time intervals, 2 mL of solution was taken out and replenished with an equal volume of corresponding buffer solution. The optical density value was measured using an UV-Vis spectrometer at 274 nm. Thus, the concentration of released ASP at different time interval can be calculated using standard curve of the concentration vs. absorption. Simultaneously, the accumulated release concentration can also be obtained. All the samples were analyzed in triplicate.

2.5 Cytotoxicity Evaluation

Cytotoxicity of electrospun PU, PU/ASP, and PU/MMT/ASP nanofibrous mats were evaluated using L929 cells directly cultured on the samples. Glass coverslips were covered with nanofibrous mats and placed in the bottom of the wells of 24-well culture plates. For sterilization, the mats were initially left under UV radiation overnight, followed by soaking with 75% ethanol solution for 1.5 h and ultimately washed three times with PBS solution. The electrospun nanofibrous mats were then soaked in culture medium overnight prior to cell seeding. Cells were seeded on the top of the mats at a density of 2×10^4 cells per well, and cultured in 1 mL DMEM cell culture medium (Gibco) containing 1% Antibiotic-Antimycotic solution (AA, Gibco) and 10% fetal bovine serum (FBS, Gibco) in a humidified atmosphere with 5% CO_2 at 37 ^{0}C. After 24 h, the medium was changed by fresh medium containing 10% (v/v) of a rezasurin solution, followed by incubation for another 3 h. Subsequently, 100 Ul of resultant medium was transferred to an opaque 96-well plate and the resorufin fluorescence (λ_{ex}= 530 nm, λ_{em}= 590 nm) was measured in a microplate reader (Perkin Elmer, Victor^3 1420). The cell morphology was observed by SEM (JEOL JSM-5600LV, Japan) with an acceleration voltage of 10 kV. Before this, cell cultured on each sample were rinsed 3 times with PBS solution to remove non-adherent cells. Then each sample was fixed with 2.5% glutaraldehyde at 4 ^{0}C for 2 h, followed by dehydrating through a series of gradient ethanol, and air dried overnight. Before SEM observation, samples were sputter coated with a 10 nm thick gold film.

3. Results and Discussion

3.1 Preparation of ASP/MMT Nanohybrid

Montmorillonite (MMT) is a layered aluminum silicate with exchangeable cations and reactive-OH groups on the surface. It is one of the most widely used layered silicates because its lamellar elements display high in-plane strength, stiffness, high aspect ratio, and high surface areas in particular [22, 23]. Recent studies have shown that incorporating bioactive cargos or drugs into silicate clay-doped polymers could be a promising way to develop functional tissue engineering materials [24]. Herein, we load anti-inflammatory drug ASP onto MMT nanoparticles to produce nanohybrids for controlled drug release materials. Fig.1 shows the loading efficiency of ASP with a function of MMT concentration, where the concentration of ASP solution was fixed at 0.25 mg/mL. As we can see, when the concentration of MMT was lower than 2 mg/mL, the loading efficiency of ASP on MMT nanoparticles increased with the concentration of MMT solution. The loading efficiency of ASP on MMT nanoparticles reached up to 51.7% when the concentration of MMT was 2 mg/mL. However, further increasing the concentration of MMT solution resulted in a decreased loading efficiency of ASP on MMT nanoparticles, which is likely due to the easier assembling of the MMT nanodisks at higher concentration, impeding the ASP molecules into the interlayer space of MMT. According to the results, 2 mg/mL was selected for forming the MMT/ASP nanohybrids.

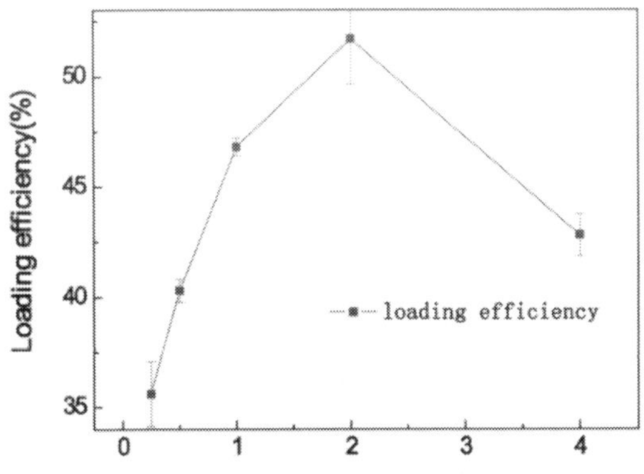

Fig.1 ASP loading efficiency as a function of MMT concentration.

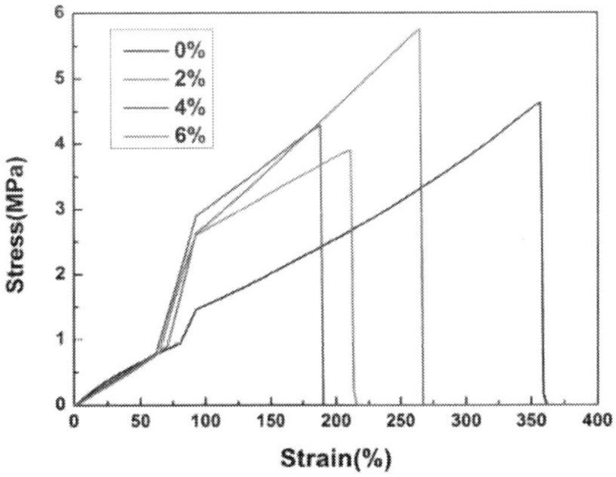

Fig.2 Typical strain-stress curves of PU nanofibrous mats with 0%, 2%, 4% and 6% of MMT nanoparticles.

3.2 Formation of ASP/MMT/PU Nanofibrous Mats

It is known that mechanical property is important for tissue engineering scaffolds, therefore before producing PU/MMT/ASP nanofibrous mat, the effect of MMT content on the mechanical property of PU nanofibrous mats was investigated. Fig.2 shows the typical stress-strain profiles of PU/MMT nanofibrous mat with different content of MMT nanoparticles. We can see that incorporating MMT into PU nanofibers impacted the mechanical properties of PU nanofibrous mats (see Table 1). For PU nanofibrous mat with MMT, the tensile strength was significantly enhanced, whereas the strain to failure decreased. When the content of MMT was 2%, the tensile strength of PU/MMT nanofibrous mat was 9.55 MPa, 4.24 MPa higher than that of pure PU nanofibrous mats, which might be ascribed to the presence of MMT uniformly dispersed in the PU nanofibers. However, further increasing the content of MMT in PU solution resulted in a gradual decreased tensile strength, although still higher than that of pure PU nanofibrous mats. This might be due to the agglomeration of MMT nanoparticles in the PU solution, which made MMT nanoparticles not disperse in nanofibers uniformly. Therefore, 2% MMT was chose to produce PU/MMT/ASP nanofibers.

Table.1 Mechanical properties of PU/MMT nanofibrous

Sample	MMT content(%)	tensile stress (MPa)	Stress to failure (%)
PU/MMT nanofibrous mat	0	5.31±0.32	383.50±43.35
	2	9.55±1.96	333.23±36.33
	4	6.73±2.80	232.54±33.84
	6	6.60±2.85	182.68±35.29

The MMT/ASP nanohybrid was then incorporated within PU nanofibers (2% ASP relative to PU) via electrospining to form PU/MMT/ASP nanofibers. Meanwhile, PU, PU/ASP (2% ASP relative to PU) and PU/MMT were prepared as controls. Here, SEM was used to characterize the morphology of the formed electrospun nanofibers with different compositions (Fig.3). It is clear that PU/MMT nanofibers with smooth and uniform fibrous morphology can be formed, even after loading ASP drug, similar to the pure PU and PU/ASP nanofibers. The diameters of PU/MMT (580 nm), PU/MMT/ASP (591 nm) and PU/ASP nanofibers (517 nm) are smaller than that of pure PU nanofibers (792 nm), which might be due to the increase of solution conductivity caused by the introduction of charged species in the electrospinning solution [14].

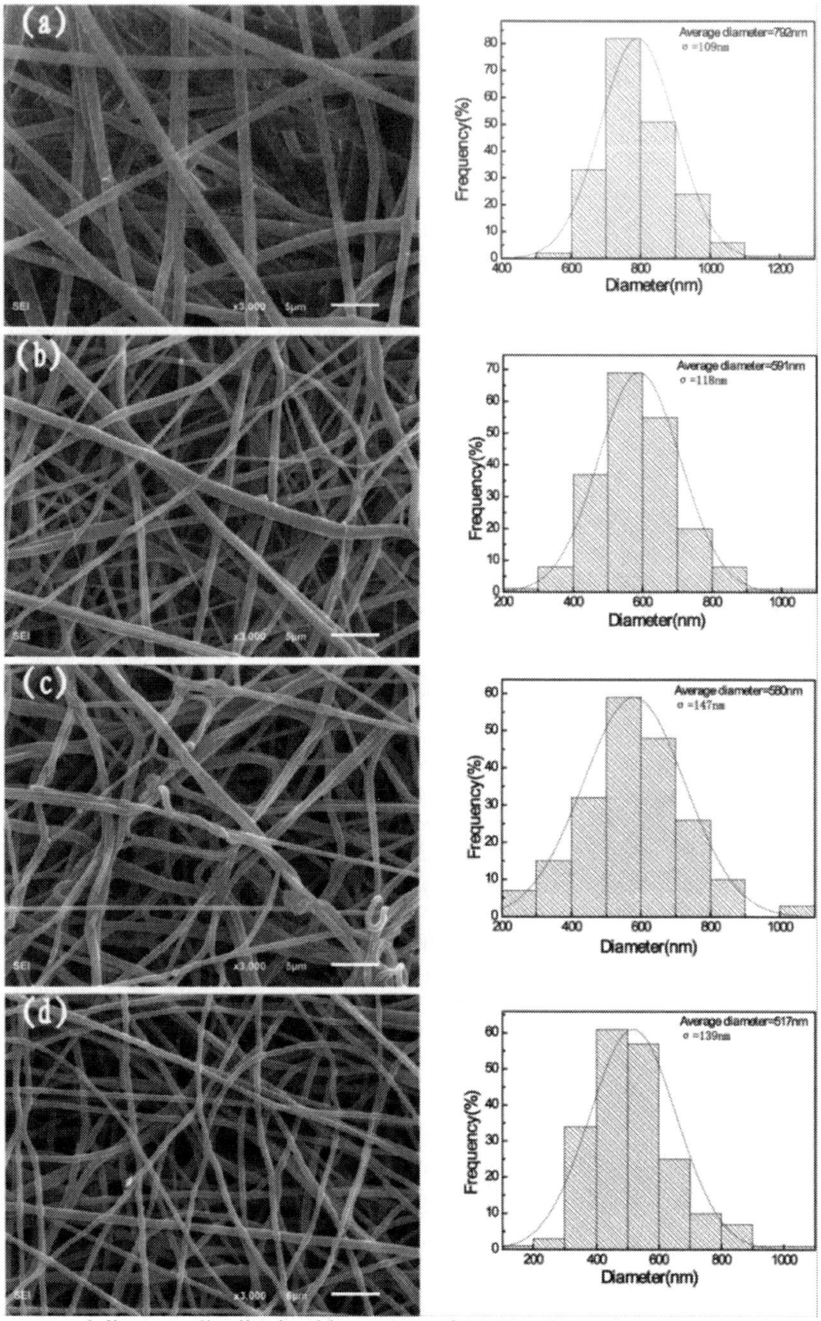

Fig.3 SEM images and diameter distribution histograms of (a) PU, (b) PU/MMT/ASP, (c) PU/MMT and (d) PU/MMT nanofibers.

The successful loading of MMT/ASP nanohybrids within PU nanofibers was confirmed using FTIR spectroscopy, which was shown in Fig.4 Comparing the spectrum of PU/MMT/ASP with that of PU nanofibers, significant difference could be observed. The new peaks emerged at 3640 cm^{-1} ascribed to the −Si−O−stretching vibration of MMT was detected. Except that, the peaks at 1690 cm^{-1} and 1610 cm^{-1} due to vibration of CH2=CH-COOH and benzene skeleton from ASP were obtained, proving the existence of MMT/ASP nanohybrid in PU nanofibers.

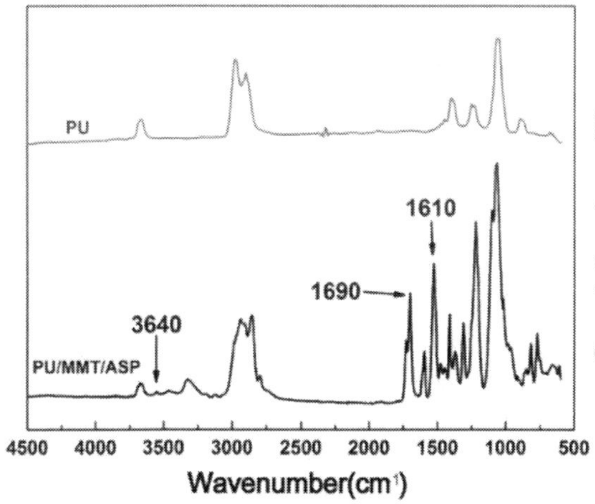

Fig.4 FTIR spectra of pure PU and PU/MMT/ASP nanofibrous mats.

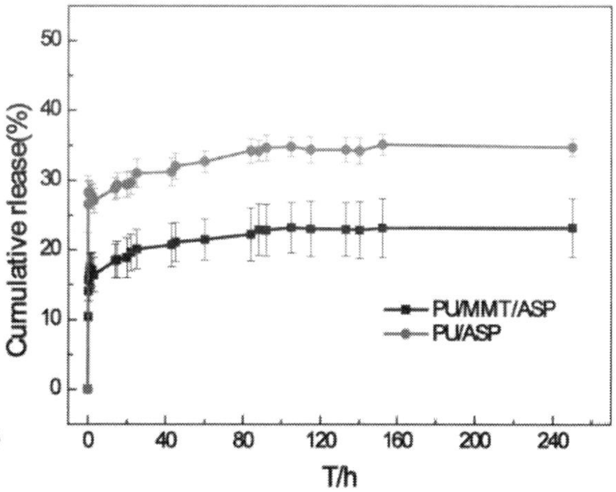

Fig.5 Drug release profiles of PU/ASP and PU/MMT/ASP nanofibrous mats at pH= 7.4.

3.3 In Vitro Drug Release

The main idea behind this study is that loading ASP onto MMT nanoparticle can endow them controllable drug release properties. Thus, the study of the drug release kinetics is of vital importance to evaluate the developed new drug delivery systems. The ASP release behaviors of PU/MMT/ASP and PU/ASP nanofibrous mats were investigated in PBS at pH=7.4. It can be seen from the ASP release profiles of PU/MMT/ASP nanofibrous mats that ASP was released in a sustained manner during 10 days of incubation, the cumulative drug release was 23%. In addition, the initiative drug release was 10.5% of the loaded drug in the first 5 min, much lower than that from PU/ASP nanofibrous mat (26%), proving that loading ASP onto MMT nanoparticles could significantly mediate the bust drug release (Fig.5). This is because that drug should first run out from the reservoir of MMT and then from the polymer matrix, which forms an additional obstacle for the drug release.

3.4 Cytotoxicity of the ASP/MMT/PU Nanofibrous Mats.

For tissue engineering applications of the PU/MMT/ASP nanofibrous mats, it is important to testify the cytotoxicity of fabricated fibrous scaffolding material. The viabilities of fibroblasts L929 cells seeded onto TCPs, cover slips, and all nanofibrous scaffolds were shown in Fig.6a. Cell proliferations on C, D and E were much better than the cells seeded onto A and B control groups. The incorporation of MMT and ASP did not have severe influence on the cell viability when comparing to the pure PU nanofibers. The better cell proliferation onto nanofibrous scaffolds than that onto TCPs suggests that these PU-based nanofibrous mats have an excellent biocompatibility to promote cell growth. The morphology of L929 cells cultured on different electrospun nanofibrous scaffolds was observed via SEM imaging. Cells cultured on all nanofibrous scaffolds displayed a phenotypic shape, indicating that the cells can penetrate and migrate within the scaffolds in a manner similar to native ECM. There is no significant difference among PU/MMT/ASP, PU and PU/ASP, suggesting that PU/MMT/ASP has a good cytocompatibility after loading ASP and MMT nanoparticles.

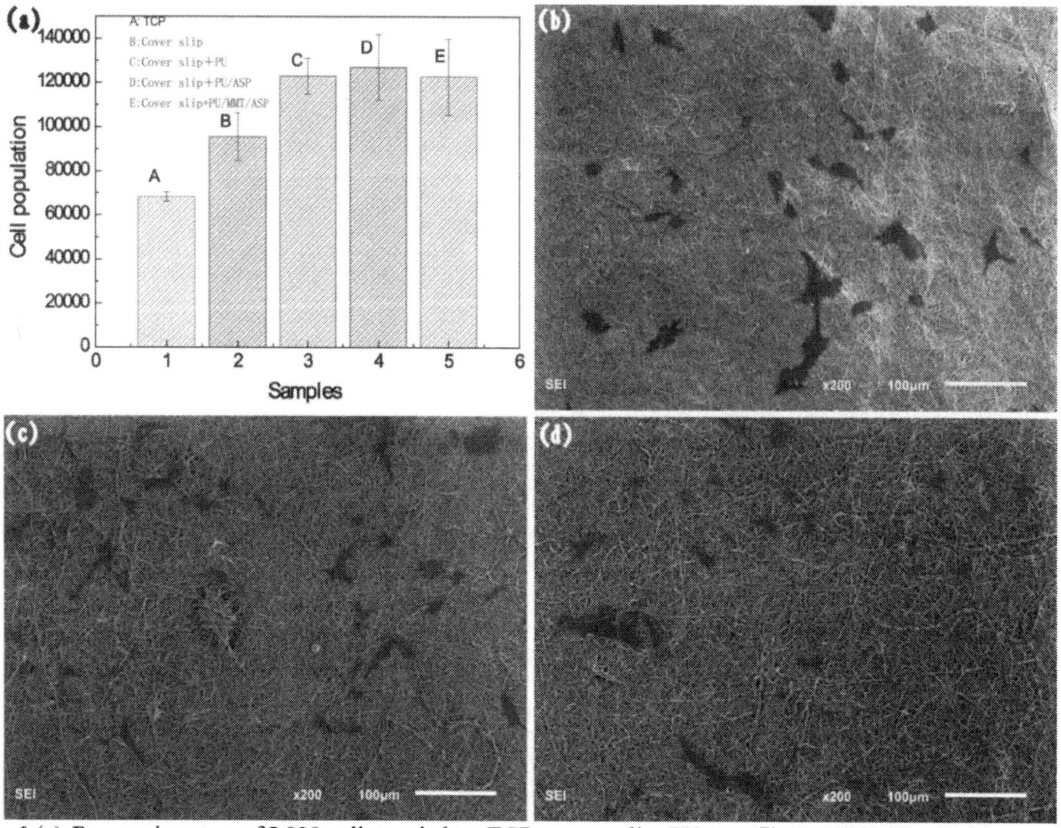

Fig.6 (a) Rezasurin assay of L929 cells seeded on TCPs, cover slip, PU nanofibers, PU/ASP nanofibers, and PU/MMT/ASP nanofibers. SEM images of L929 cells cultured onto (b) PU, (c) PU/ASP and (d)PU/MMT/ASP composite fibrous mats.

4. Conclusion

In summary, we have developed a facile approach to produce composite nanofibers for controlled, sustained drug delivery. The MMT/ASP nanohybrids with the ASP loading efficiency of 51.7% were successfully incorporated into PU electrospun nanofibers. The incorporation of 2% MMT nanoparticles not only improved the mechanical properties of the PU nanofibrous mats, but also weakened the initial burst release characteristics of conventional electrospun polymer fiber-based drug delivery system. The PU/MMT/ASP nanofibrous mats showed excellent biocompatibility confirmed via resazurin assay and SEM observation of cell morphology, suggesting that the developed composite system would be amendable for the applications in tissue engineering and wound dress.

5. Acknowledgment

This research was financially supported by the National Natural Science Foundation of China (51403165), the Education Commission of Hubei Province (D20161602) and the Science Foundation of Wuhan Textile University (2014CFB161).

References

[1] Ramalingam K, Castro R, Pires P, et al. Gene delivery using dendrimer/pDNA complexes immobilized in electrospun fibers using the Layer-by-Layer technique. RSC Adv: 2016; 6: 97116-97128.

[2] Marx V. Tissue engineering: Organs from the lab. Nature: 2015; 522: 373-377.

[3] Ramakrishna S, Fujihara K, Teo WE, et al. Electrospun nanofibers: solving global issues. Mater Today: 2006; 9: 40-50.
[4] Barnes CP, Sell SA, Boland ED, et al. Nanofiber technology: designing the next generation of tissue engineering scaffolds. Adv Drug Deliv Rev: 2007; 59: 1413-1433.
[5] Lannutti J, Reneker D, Ma T, et al. Electrospinning for tissue engineering scaffolds. Mater. Sci. Eng., C: 2007; 27: 504-509.
[6] Sill TJ, von Recum HA. Electrospinning: applications in drug delivery and tissue engineering. Biomaterials: 2008; 29: 989-2006.
[7] Jayasinghe SN. Cell electrospinning: a novel tool for functionalising fibres, scaffolds and membranes with living cells and other advanced materials for regenerative biology and medicine. Analyst: 2013; 138: 2215-2223.
[8] Tan L, Hu J, Zhao H. Design of bilayered nanofibrous mats for wound dressing using an electrospinning technique. Mater Lett: 2015; 156: 46-49.
[9] Stone SA, Gosavi P, Athauda TJ, et al. In situ citric acid crosslinking of alginate/polyvinyl alcohol electrospun nanofibers. Mater Lett: 2013; 112: 32-35.
[10] Kim SE, Heo DN, Lee JB, et al. Electrospun gelatin/polyurethane blended nanofibers for wound healing. Biomed Mater: 2009; 4: 044106.
[11] Tong HW, Zhang X, Wang M. A new nanofiber fabrication technique based on coaxial electrospinning. Mater Lett: 2012; 66: 257-260.
[12] Detta N, Errico C, Dinucci D, et al. Novel electrospun polyurethane/gelatin composite meshes for vascular grafts. J Mater Sci: Mater Med: 2010; 21: 1761-1769.
[13] Kenawy ER, Bowlin GL, Mansfield K, et al. Release of tetracycline hydrochloride from electrospun poly (ethylene-co-vinylacetate), poly (lactic acid), and a blend. J Control Release: 2002; 81: 57-64.
[14] Wang S, Zheng F, Huang Y, et al. Encapsulation of amoxicillin within laponite-doped poly (lactic-co-glycolic acid) nanofibers: preparation, characterization, and antibacterial activity. ACS Appl Mater Inter: 2012; 4: 6393-6401.
[15] Verreck G, Chun I, Rosenblatt J, et al. Incorporation of drugs in an amorphous state into electrospun nanofibers composed of a water-insoluble, nonbiodegradable polymer. J Control Release: 2003; 92: 349-360.
[16] Wenhua Lü, Guangjie Zhao. Design of wood/montmorillonite (MMT) intercalation nanocomposite. Forest Sci Practice: 2004; 6: 54-62.
[17] Kevadiya BD, Bajaj HC. The layered silicate, montmorillonite (MMT) as a drug delivery carrier. Key Eng. Mater: 2013; 571: 111-132.
[18] Fudala Á, Pálinkó I, Kiricsi I. Preparation and characterization of hybrid organic− inorganic composite materials using the amphoteric property of amino acids: amino acid intercalated layered double hydroxide and montmorillonite. Inorg Chem: 1999; 38: 4653-4658.
[19] Kimmer D, Slobodian P, Petráš D, et al. Polyurethane/multiwalled carbon nanotube nanowebs prepared by an electrospinning process. J Appl Polym Sci: 2009; 111; 2711-2714.
[20] Liu H, Zhang Y, Han Y, et al. Characterization and cytotoxicity studies of DPPC: M 2+ novel delivery system for cisplatin thermosensitivity liposome with improving loading efficiency. Colloid Surface B: 2015; 131: 12-20.
[21] Xiao S, Shen M, Guo R, et al. Immobilization of zerovalent iron nanoparticles into electrospun polymer nanofibers: synthesis, characterization, and potential environmental applications. J Phys Chem C: 2009; 113: 18062-18068.
[22] Bao Y, Ma J, Li N. Synthesis and swelling behaviors of sodium carboxymethyl cellulose-g-poly(AA-co-AM-co-AMPS)/MMT superabsorbent hydrogel. Carbohydr Polym: 2011; 84: 76-82.
[23] Mansoori Y, Atghia SV, Zamanloo MR, et al. Polymer–clay nanocomposites: free-radical grafting of polyacrylamide onto organophilic montmorillonite. Eur Polym J: 2010; 46: 1844-1853.
[24] Nair BP, Sindhu M, Nair PD. Polycaprolactone-laponite composite scaffold releasing strontium ranelate for bone tissue engineering applications. Colloid Surface B: 2016; 143: 423-430.

The Study on Enhancing Smoothness of the Surface of Alcohol/ Oil/ Blood Repellent Finishing PP SMS Nonwoven for Medical Applications

Ya Wang[1], Rong-Fang Xiao[2], Ru-Quan Zhang[1*]

[1]*Ya Wang, Wuhan Textile University, Wuhan, Hubei, 430073, China*
[2]*Rong-Fang Xiao, Xiantao, CHTC Jiahua Nonwoven CO. , LTD, XianTao, Hubei, 433000, China*

[*]Corresponding author's email: zry2010@yahoo.com

Abstract

During the industrial manufacturing progress, after the processing of the jet convention drying pit, the surface of the alcohol/ oil/ blood repellent SMS Nonwoven will no longer be smooth. This paper analyzes and designs the experiment to examine the influence of the heating machine temperature on the variable heating air volume, the heating time and cooling time. The results show that increasing nozzles and decreasing the temperature and drying wind can improve the condition of smoothness.

Keywords: Jet Convention Drying Machine; Alcohol/ Oil/ Blood Repellent Finishing; SMS Nonwoven; Out of Smoothness

1. Introduction

The stenter frame range Monfongs 328 TwinAir from Monforts Fong's Textile Machinery CO., Ltd separately blows air from two directions and forms a jet convection to dry fabric. It is designed to match the requirements of all kinds of fabric finishing [1]. Besides the finishing of the traditional knitting and tatting fabric, it also can be used for the PP SMS's alcohol/ oil/ blood repellent finishing. The finished products can be used in medical applications such as surgery gown and surgery drape [2].

Through the spun-bond and melt-blown process, the PP SMS has already equipped with some hydrostatic head and tensile strength [3]. During the finishing process, the nonwoven would undergo rolling, saturating, drying and ranging. After these processes, not only will the hydrostatic head and the tensile strength decrease the surface of the fabric would also be wrinkled (Fig.1) [4, 5]. That issue could affect the appearance of the products and the post processing [6, 7].

Fig.1 The fabric out of smoothness sample.

1.1 Manufacture Process

Semi-finished products uploaded on the roll→delivered by the storage shelf→passing through the oil sink→circled by the chain of finishing paddler→Ranging→drying→cooling→rolling.

1.2 The Principle of the Technics

The SMS nonwoven which is used for the alcohol/ oil/ blood repellent finishing is manufactured by the technics of spun-bond and melt-blown. First spun-bond nonwoven equipped with high tensile strength and minimal difference between the across strength [8]. And the hydrophobicity, ventilation property and shielding property of melt-blown nonwoven are excellent. This SMS nonwoven combines the advantages both of spun-bond and the melt-blown technics. Also the dropping off phenomena of the SMS is not very serious. The purpose of the finishing is to equip the nonwoven with alcohol/ oil/ blood repellent and antistatic functions. The gaps between the superfine fibre which come from the M part of SMS nonwoven are minimum. So the SMS nonwoven could partly prevent the blood, ashes and the germs. But it's not antistatic. In order to match the higher requirements of medical protection, by taking the further dip method technic to handle, the products can enhance its resistance value and surface. At this moment, the antistatic performance and the function of the blood repellent would be better. When the products are used in the medical applications, they will be safer [9, 10].

After going through the oil sink, the SMS nonwoven must be heated to dry. When the water is evaporated from the SMS nonwoven, the molecules of the complex oil would be attached to the surface of the SMS nonwoven. After the above procedure, the fabric can be rolled to translate to the next process. The stenter frame range Monfongs 328 TwinAir separately controls the heating air blow from top and bottom. The efficiency of the evaporation is decided by the speed and the temperature of the heating air blow, the distance between nozzles and fibre web and the shape of the nozzles [11, 12].

1.3 The Analysis of the Issues

Under the manufacturing, during the first part, the SMS nonwoven would go through the storage shelf which is to bring support for the next procedures. The mechanical stretching of the storage shelf could affect the fabric surface. The second part is delivering the SMS nonwoven in the oil sink. In this process, as the molecules of the complex oil would be attached to the surface of the SMS nonwoven, the physical properties of the SMS nonwoven would be changed. That third part is the main part of the whole finishing line is drying and ranging. The whole drying equipment has ten sections. Each section's length is 3 meters. In the regular industrial manufacturing, the speed of the chain should be controlled nearly 60 m/min. The temperature of the drying equipment should be keep as 135 to 140 degree. In the drying and ranging process, as the high speed jet-like heating air blow striking the fibre web badly, the fibre web would be vibrated violently [13]. This can cause the wrinkle of the products. Considering that the good thermos ability and the character of distortion temperature that is about 100 degree [14], So that some special experiments must be designed to adjust the equipment by change the index of the machine. The last part of the finishing line is cooling system. The fibre line which just comes from the drying equipment has a high temperature. This part is to cool the fabric and roll. This whole cooling system only has one section. The length of the cooling zone is about 3 meters. The cooling zone also can be a major part on ranging the fabric [15].

2. The Experiments Part

2.1 The Materials and the Laboratory Instruments

45gsm PP SMS nonwoven, wetting agent 6112, anstatic agent AMD, protective agent E600 Stenter frame range Monfongs 328 TwinAir.

2.2 The Methods for the Experiments

In the experiments, the main indexes of the machine shown in table 1.

Table.1 The indexes of machine

Index	Unit	Usual setup
Speed of main chain	m/min	60
Heating of zone 1	℃	144
Heating of zone 2~10	℃	139
Power of top air blow	%	50
Power of bottom air blow	%	80
Cooling air1#~2#	%	ON

The influences of the temperature of the heating machine, variable heating air volume, the heating time and cooling time. By taking the method of comparative study, we can separately watch out the corresponding appearance of the products. We choose randomly 5 pieces of fabric which is 3m in length from the outer ring and inner ring. Then we should lay them on the ground flatness. By observing the picture 3, the appearance of the examples could be judged as Excellent, Good, out of Bad and Very bad (Fig.2).

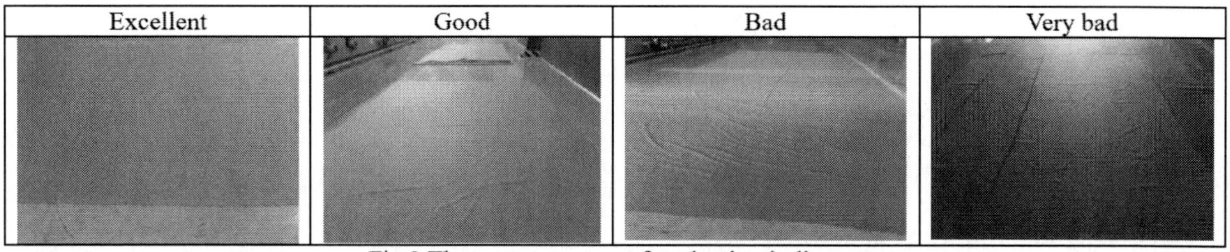

Fig.2 The contrast group of evaluating indicator

2.3 The Analysis of the Experiment Results

The fabric roll we use for the experiment is nearly 4000 meters. In taking examples process, we took one piece from outer ring to inner ring by the distance of 500m. We named the examples as example 1 to example 5. From the appearance of those examples, the manufacturing factors could be adjusted to reach the most appropriate index combination.

2.3.1 The Drying Temperature's Influence to the Surface of the Fabric

In order to consider the influence of Heating temperature on it. We design 5 group experiment shown in the Table 2.

Table.2 The technics parameter of T1 to T5

Type	Temp.	Air volume	Speed and time	cooling
T1	130℃	10 Zones of 50|80	60m/min|30s	On
T2	135℃	10 Zones of 50|80	60m/min|30s	On
T3	140℃	10 Zones of 50|80	60m/min|30s	On
T4	145℃	10 Zones of 50|80	60m/min|30s	On
T5	150℃	10 Zones of 50|80	60m/min|30s	On

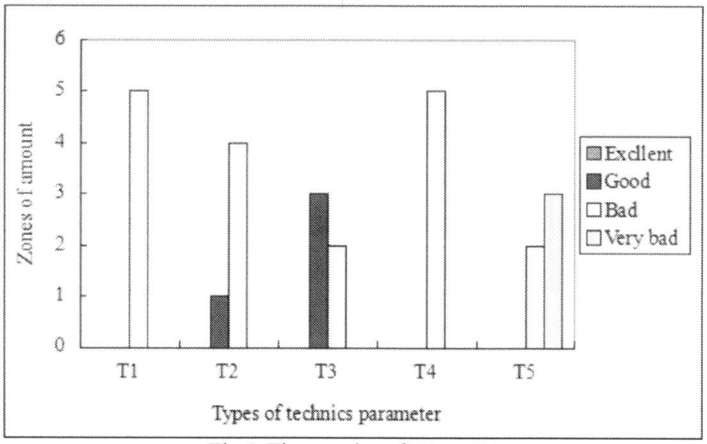

Fig.3 The results of T1 to T5

From the experimental results shown in the picture (Fig.3), 150℃of the heating temperature could ruin the fabric. However when the temperature decreased to the 130℃, the appearance of the fabric is not satisfied or even still wet. From the results, we could know that the 135℃ to 140℃ is appropriate.

2.3.2 The Air Volume's Influence to the Surface of the Fabric

In order to consider the influence of Air volume on it. We design 5 group experiment shown in the Table 3.

Table.3 The technics parameter of T6 to T10

Type	Temp.	Air volume	Speed and time	Cooling air
T6	135℃~140℃	10 Zones of 40|70	60m/min|30s	On
T7	135℃~140℃	10 Zones of 50|70	60m/min|30s	On
T8	135℃~140℃	6 Zones of 50|80 and 4 zones of 40|70	60m/min|30s	On
T9	135℃~140℃	10 Zones of 50|80	60m/min|30s	On
T10	135℃~140℃	10 Zones of 65|85	60m/min|30s	On

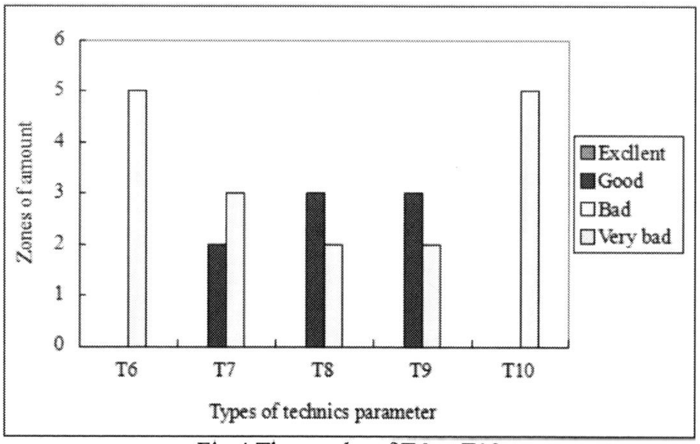

Fig.4 The results of T6 to T10

The results show that both low and high air volume are not suitable (Fig.4). Low air volume could not dry the fabric enough. Blowing too hard, the high air volume would cause the wrinkle of the surface of the fabric.

2.3.3 The Cooling Time's Influence to the Surface of the Fabric

In order to consider the influence of Cooling air on it. we design 5 group experiment shown in the Table 4.

Table.4 The technics parameter of T11 to T15

Type	Temp.	Air volume	Speed and time	Cooling air
T11	135℃~140℃	10 Zones of 40|70	60m/min|30s	off
T12	135℃~140℃	6 Zones of 50|80 and 4 zones of 40|70	60m/min|30s	off
T13	135℃~140℃	6 Zones of 50|80 and 2 zones of 40|70	60m/min|24s	On and add 2 zones
T14	135℃~140℃	6 Zones of 50|80 and 2 zones of 40|70	48m/min|30s	On and add 2 zones
T15	135℃~140℃	6 Zones of 65|85	36m/min|30s	On and add 2 zones

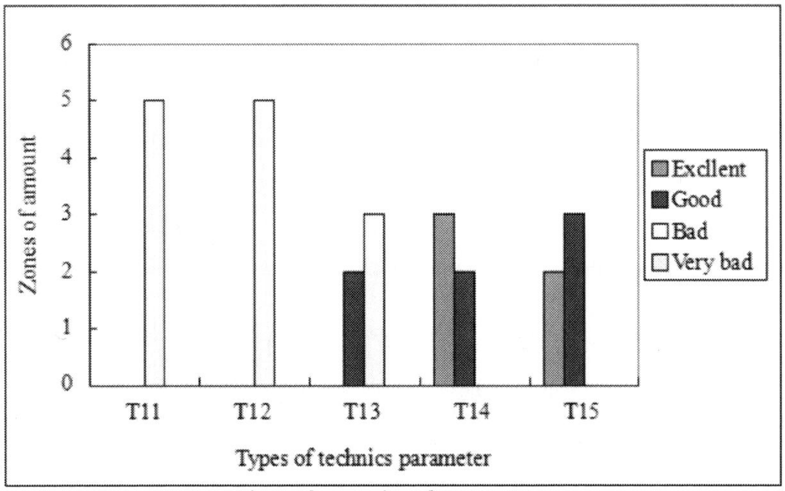

Fig.5 The results of T11 to T15

From the results we could see the surface of the fabric was worse when the cooling zone was closed. After spreading the cooling area, the condition of the fabric was becoming much better. In the daily manufacturing, changing the heating area to the cooling area could waste the equipment's productivity and cause the major decreasing of productivity. So it's not practical in the industrial manufacturing. We could draw a conclude that increasing the cooling time do better to the fabric condition.

2.3.4 The Distance Between Nozzles and Fibre Influence to the Surface of the Fabric

To slow down the impact force form hot air does better to the fabric condition. So we reformed the machine that increasing the distance between nozzles and fibre. In order to know the results of reform we design 5 group experiment shown in the Table 5.

Table.5 The technics parameter of T15 to T20

Type	Temp.	Air volume	Speed and time	Cooling air
T16	135℃~140℃	10 Zones of 50|55	60m/min|30s	On
T17	135℃~140℃	10 Zones of 65|55	60m/min|30s	On
T18	135℃~140℃	10 Zones of 65|65	60m/min|30s	On
T19	135℃~140℃	10 Zones of 65|70	60m/min|30s	On
T20	135℃~140℃	10 Zones of 70|70	60m/min|30s	On

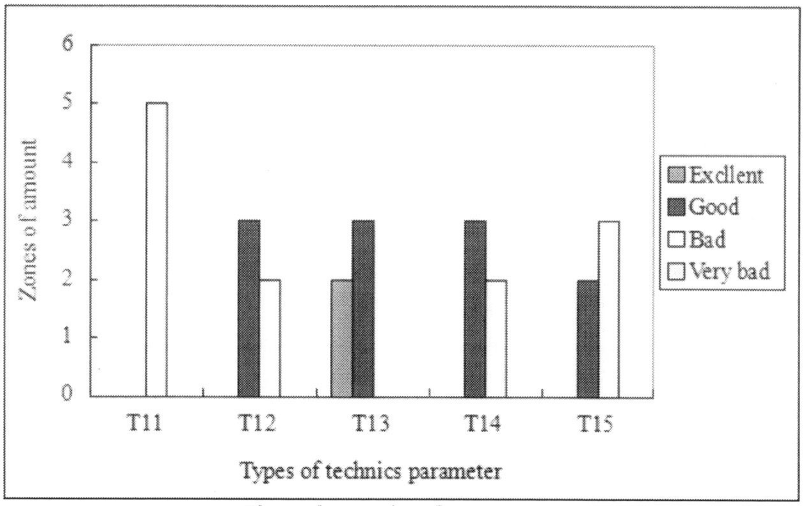

Fig.5 The results of T16 to T20

After increasing the distance between nozzles and fibre, the appearance of the fabric could be the best by controlling the air top and bottom volume as 65. From the experiments we could see that products were becoming better before the adjustment. There were so many difficulties on adding the cold rolls. So this plan fetched up in the adjustment program.

3. Conclusions

From the above experiments, the heating and drying technics were proved to be the main reasons that cause the wrinkle phenomena in the alcohol/ oil/ blood repellent finishing. From the experiments results, we could have the following conclusions.

1) The heating and drying temperature should be keep as 135 degree to 140 degree. Using the low heating temperature is not enough to dry the fabric. Over 150 degree heating temperature would ruin the products.

2) Low air volumes could also be not enough to dry the fabric. High air volumes would strike the surface of the fabric badly and cause the more serious wrinkle problems. From the data we collected, the best combinations were 50 of the top air volume and 80 of the bottom air volume before the adjustment and 65 of the top and bottom air volume.

3) Increasing the cooling time can do much better for the fabric's surface. Adding the cold rolls could eliminate this problem forever.

4) Increasing the distance between the nozzles and fibre web could decrease air blow strike to the fabric. When the adjustment was done, the surface of the products was becoming so much better.

References

[1] Information on http://www.monfongs.com/MON/display.

[2] Cao YX, Liu XL, Zhang YM. The property and applicantion of Polypropylene for medical nonwoven spunbound layer. Synthetic Fiber in China: 2015; 44 (1): 1-3.

[3] Zhao B. Study on and testing analyzes the performances of SMS compound nonwoven fabric. Issue Name: 2017; 30 (1): 22-25.

[4] Ma RY, Qu FY, Pang SS. Research on finishing process for impermeable and directional water-transfer nonwoven operation coat. Technical Textiles: 2015; 33 (1): 38-43.

[5] Zhang TT, Wu HB. Study on the process of coating dyeing for PP spunbouned nonwovens. Technical Textiles: 2007; 25 (1): 27-30.

[6] Xu J, Chen HW, Zhou YJ. Analysis of the flow field in the air duct of a Stentor's drying room. Development & Innovation of Machinery & Electrical Products: 2013; 26 (6): 87-89.

[7] Wei HL. Causes and Preventive Measures of pure cotton elastic fabric surface wrinkling. Cotton Textile Technology: 2010; 38 (10): 6-9.
[8] Craig IH, White JR, Kin PC. Crystallization and chemi-crystallization of recycled photo-degraded polypropylene. Polymer: 2005; 46 (2): 505-512.
[9] Qiao XL. Study on water and repellent finish of SMS nonwovens [D]. Donghua University: 2007.
[10] Yuan H, Xu HM. The introduce of the alcohol/oil/blood repellent finishing SMS nonwoven for medical application. The 17th of spun-bonded/melt-blown nonwoven industry conference proceedings of China: 2010.
[11] Liu S, Mao JM. A new open-width airflow for drying and softing. Textile Dyeing and Finishing Journal: 2014 (12); 52-53.
[12] Ke QF, Jin XY. Nonwovens. Donghua University Press: 2010.
[13] Wang XQ, Qi DY, Zhu RX. Study on the light aging of polypropylene nonwoven fabric. China Fiber Inspection: 2015 (7); 70-72.
[14] Yuan H, Zhou X, Zhang YG. Storage conditions of sensitized fabrics for post cure durable press finish. Dyeing & Finishing: 2016 (20); 16-20.
[15] Zhang WH. The practicality manual of injection mold design. Chemical Industry Press: 2011.

Current Situation and Prospects of Antibacterial Healthcare Underpants

Si-Fan Wang[1*], Yong-Mei Deng[2]

[1]*School of Fashion Technology, Zhongyuan University of Technology, Zhengzhou, 450007, China*
[2]*School of Apparel & Art Design, Xi'an Polytechnic University, Xi'an, 710048, China*

[*]Corresponding author's email: 411711757@qq.com

Abstract

With the development of antibacterial underpants, the safety and the practical effects have been increasingly emphasized. By the method of literature research, the research status of antibacterial underpants on fabrics and structures were studied in this paper. The evaluation of practical effects of underpants were also studied. The study found that antibacterial fabrics is the core of antibacterial underpants. The natural green, safe, broad spectrum and efficient of antibacterial fabrics should be taken into attention. For the structures of underpants, the development is slow and more innovations are needed. The relevant microbial environment, application fields and practical effects of antibacterial underpants were discussed. The paper underlines the idea that making relevant evaluation standard of antibacterial underpants' wearing effect will help the research and development of underpants as well as the choosing antibacterial underpants.

Keywords: Antibacterial; Underpants; Wear; Evaluation

1. Introduction

In daily life, the excretions of the human body such as sweat, sebum and scurfy accumulated on the underpants may result in bacterial growth. As such, there is a demand in antibacterial textiles, especially for close-fitting underpants. According to a survey conducted by Textile Intelligence, consumers' demand for antibacterial functional clothing is growing. However, unlike the people's demand for antibacterial healthcare underpants, little research has been done about the underpants in recent three years. In order to understand the present situation of underpants development, and provide references for manufacturer and consumer, this paper have expatiated the current situation of antibacterial healthcare underpants.

2. Research Status of Antimicrobial Underpants Fabrics

The function of Antibacterial underpants lies in antibacterial deodorization, preventing bacteria growth and skin infections etc. It has potential for application and development in the fields of the military, medicine, athletics, daily healthcare and others. Underpants is an essential part of people's daily lives as the human body's waist, abdomen, perineum, and hips all come in contact with it. It is one of the most intimately contacting garments.

In daily life, due to a lack of personal hygiene and environmental factors, a great deal of bacteria can accumulate on underpants, and any clothing worn by humans will become contaminated with microorganisms [1-2], which can cause eczema, itching, and other symptoms. Candida albicans vaginitis caused by unclean underpants is relatively common, and whether underpants are of good quality or not is closely related to health [3]. In order to keep the skin healthy and avoid bacterial growth in the long term, which can cause skin disease, underpants should be designed with better air permeability and moisture permeability knitting fabric. As the main purpose of garment, its antibacterial function has become the core of anti-bacterial underpants. At present, the two main methods to develop anti-bacterial underpants are antibacterial fibres and antibacterial finish.

2.1 Antibacterial Fibres

Antibacterial fibres used on underpants refers to the antibacterial functional fibre being obtained first, then made into various types of antibacterial fabric which is finally used for underpants processing.

2.1.1 Plant Source Antibacterial Fibre

With the increasing demand for natural products, underpants made using plant source antibacterial fibre was developed.

There exists a large number of viscose fibre underpants on the market, however ordinary viscose fibre has no antibacterial function [4]. In fact, it easily accumulates dirt and stains when worn, which results in bacterial growth. One of the plant sources, viscose acetal fibre, known as aloe glue, can be used to make into a knitted fabric with different structures, moisture transported and quick drying properties. This kind of knitted fabric can also be used to make into underpants or other intimate garments. This kind of fibre can slow Staphylococcus aureus and Candida albicans growth, however the antibacterial ability to Escherichia Coli is unknown [5].

The application of antimicrobial substances containing tea is also a trend. Fibre finished with antibacterial extract, then made into intimate garments, has an antibacterial effect and is comfortable to wear [6]. Tea viscose fibre have antibacterial activity on three typical bacteria (S.aureus, E.Coli and Candida albicans). However, the antibacterial washability was not mentioned [7].

The application of Chinese herbal medicine has also been developed in recent years. Jia-Qian Wu et al. invented a Chinese herbal medicine antibacterial fibre, which improved the strength and heat retention of fabric. At the same time, the Chinese herbal medicine fabric also has skin care properties and antibacterial function [8].

Certainly, nature plant fibre such as cotton fibre, Linum usitatissimum and bamboo fibre are with good antibacterial effect and have obtained better performance by compositing with other materials for underpants. Antibacterial fabrics made with bamboo charcoal fibre as the main component are widely used in the underpants market. What's more, underpants used in the army also contains bamboo charcoal fibre in the fabric content, and with mainly nature fibre [9-10], it provides both comfort and antibacterial properties when wearing. However, studies show that there were limitations in the antibacterial species [11-12].

2.1.2 Silver Antibacterial Fibre

Silver antibacterial fibre used the excellent antibacterial properties of silver, but for the safety of the human body and for the comfort and appearance of clothing, they have always been the direction of the research [13]. Using nanotechnology, nano-silver antibacterial fibre became more widely studied. However, the safety of nano-silver antibacterial fibre has always been in question by researchers [14]. The safety of silver ion is better than nano silver, and its use on underpants is increasing, however the safe and effective amount of silver ion needs furthermore confirm [15-16].

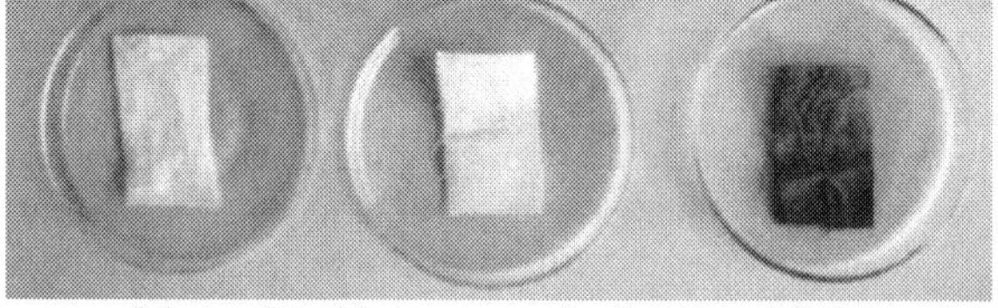

Fig.1 Photograph of (a) plain cotton fabric, (b) chitosan-attached cotton fabric, and (c) nanosilver-loaded chitosan-attached fabric.

Another widely studied silver antibacterial fibre is called silver-loaded antibacterial fibre. The antibacterial fibre combines the effect of silver with other substances, to achieve a synergistic antibacterial effect. The same as nano-silver and silver iron, the problems occur when the silver-loaded antibacterial agent is used for light color fabric. Fig.1 shows, according to Varsha Thomas's research [17], the fabric color after being treated by the nano-silver antibacterial agent. Silver antibacterial agent can be easily oxidized and discolored. Thus, reducing the amount of silver and releasing silver slowly are two ways to ensure the safety of the fabric as well as maintain its appearance. At the same time, achieving an efficient broad-spectrum antibacterial effect is the research direction of the future.

2.1.3 Chitosan Antibacterial Fibre

Chitosan comes from natural shells, shrimp shells, fish bones, shell of insects and other animals. The advantages of chitosan, such as being antibacterial, natural and providing skin care, made it ideal for use in underpants. Lipatova et al. argue that chitosan is healthy for the skin and is non-toxic, and its film-formation and other excellent features make it very suitable for a fabric finishing agent [18]. Peng Nie invented a blended fabric mainly using chitosan antibacterial fibre, which can be made into antibacterial underpants [19].

The antibacterial function of chitosan was influenced by various factors, such as its molecular weight, relative molecular mass, and deacetylation degree. And chitosan was often the carrier, combined with other antimicrobial substances such as silver, copper, or oxide [20]. However, fabrics made by these composite fibres encountered problems with the handle, appearance and antibacterial endurance. The amount of chitosan and antibacterial compounds should be studied further.

In general, antibacterial fibre applied to underpants has been developing for a long time [21]. Natural-occurring substances' using onto the fibre is increasing day by day. The plant-source antibacterial fibre and animal-source antibacterial fibre are with healthcare function. However, their broad-spectrum antibacterial functions and antibacterial durability are not enough. The inorganic antimicrobial fibre such as silver, copper and oxide fibres are with excellent broad-spectrum antibacterial functions and antibacterial durability. However, the safety using amount was not immediately known and discoloration problems should be taken into consideration when making underpants.

2.2 Antibacterial Finish

The antibacterial finishing method is divided into antibacterial finishing of fabrics and garments, while fibre antibacterial finishing belongs to antibacterial fibre in section 2.1.

2.2.1 Antibacterial Finishing of Fabrics

The technology of finishing methods giving fabric an antibacterial function has been very widely applied. At present, it is one of the main methods of producing antibacterial fabric. For instance, it can improve the hygroscopicity of fabric to finish dacron by chitin, which can imbue fabric with both antibacterial and antistatic functions; natural fibre fabrics finished with Chinese herbal medicine can be antibacterial; fabric finished with the SCJ series of antibacterial agents has obtained good security, broad spectrum efficiency, and excellent antibacterial properties reaching almost 100%.

The fabric finishing process is relatively simple, and easy to adjust. The types of fabrics that can be finished is increasing, the range of antibacterial finishing agents to choose from is wider, and the antibacterial agents' combining to the fabric surface can produce a positive antibacterial effect. The finishing methods contain mainly surface coating method, resin finishing method and microcapsule [22].

In the process of constant development and improvement by adjusting the antibacterial agents and finishing process, safer and more durable methods are also in the process of development. These include the cotton fabric dipping-steaming method, which aims to develop nano-silver antibacterial finishing process for cotton fabrics, which has a uniform distribution of antibacterial agent on the fabric and has little effect on the performance of the fabric itself. It has made new progress in improving appearance and antibacterial durability of the antibacterial finished fabric [23].

2.2.2 Antibacterial Finishing of Ggarments

An additional means of antibacterial finishing is garment finishing methods, which is to say, antibacterial finishing after the fabric has been made into a garment. One antibacterial finishing method is the garments dipping method. For example, Japan Megi Company developed odor-proof technology for garments, finishing the garments after sewing. Antibacterial garments were obtained after dipping and drying. According to domestic research in recent years, garments dipped in Chinese herbal medicine have received a certain antibacterial effect and can be applied to actual production [24]. Another garment finishing method is the spray ironing method, which is used for underpants. After finishing, the finished garment has an antibacterial function, wearing comfort and a certain degree of antibacterial washing resistance [25].

In general, antibacterial finishing is still the main method to develop anti-bacterial underpants, but further studies are needed to improve the process.

3. Research Status of Antimicrobial Underpants Structures

The groin, base of the thighs and perineum are all in touch with underpants. These body parts produce stagnate sweat easily and are not easy to dry, which provides an excellent environment for microbial proliferation. Over time, if underpants and the skin are not clean, the body will develop an odor and can even suffer from skin disease such as eczema or ringworm. There are two kinds of antibacterial underpants in terms of the structure.

3.1 Opening Design

Underpants, contacting with the human body, is the innermost garment. The environment of underpants and body is relatively closed and doesn't allow for air or water vapor permeability, thus the environment facilitates bacteria reproduction. Research has found that bacteria breeding can be reduced by reducing the covering of underpants which can improve its ventilation and air permeability. For instance, the leg opening was designed to be the shape of a lotus leaf, as in Fig.2 (a) [26]. With this design underpants are relatively loose, with good permeability, and sweat can be evaporated and spread out over time, which can help keep the relative area cleaner [27]. Some underpants use a variety of shapes of hollow design, which can also increase the permeability of underpants [28]. However, the acceptable degree of appearance, practicability and comfort of hollow pants need further verification.

Some medical underpants need to prevent external pollution and infection caused by the invasion of bacteria. So, the waist and leg opening need an enclosed design (usually using a lacing type or pull chain type). At the same time, the degree of tightness should be ensured. It should reduce the discomfort caused by pressure [29].

3.2 Partial use of Antibacterial Materials

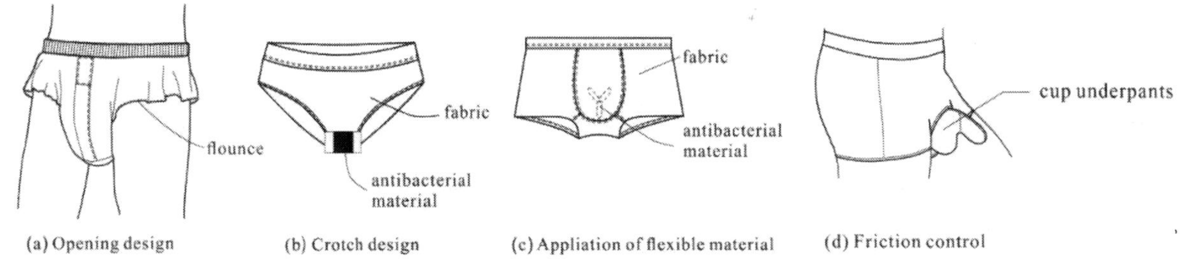

Fig.2 Research status of antimicrobial underwear structures.

The pantyhose, a collaboration of the British company Flude and the American company AEGIS, combined trousers and socks, use antibacterial fabrics in the crotch area, conform to the body, and prevent vaginitis. In clinical research, antimicrobials were produced in bags, loaded in the part of underpants where anti-bacterial in needed, and should be changed in the period of validity. In addition, other research found a flexible material treated by natural antimicrobial agent loaded in the crotch (women) and in front of the crotch (men), as shown in Fig.2 (b), (c) [30-31]. Partial application of antibacterial material, can adopt the method of stitching antibacterial fabric to the corresponding parts of underpants.

3.3 Friction Control

Excessive friction between skin or between skin and underpants can cause the body (especially in high intensity activity or training) not only to be invaded by bacteria breeding in sweat, but can also easily cause erosion of the wound. The wound will become inflamed and cankered due to bacterial infection.

The Seamless knitting technology can weaken the effects of the seam of underpants on the human body, reduce skin infections caused by friction, and keep the skin from suffering worse aggression by bacteria [32-33]. Sports close-fitting shorts made using this method can reduce the athletes' tinea cruris and "festered hip" which often appears in athletes or soldiers [34]. What's more, a kind of cup underpants for men, can also reduce skin infections caused by friction, as shown in Fig.2 (d) [35].

To sum up, for daily wear antibacterial underpants plays an important antibacterial healthcare role, and it combines the use of antibacterial fabrics as the main application methods. For those who engage in intense exercise and training, it is important that the underpants have ventilation and is moisture permeable. At the same time, it must produce low friction to keep skin healthy. For medical antimicrobial underpants, the underpants were asked to have the function of preventing outside pollution as well as bacteria. The underpants also should have barrier properties and flexible characteristics in order to better protect the patient.

4. Research Status of Underpants Wearing Evaluation

Antimicrobial underpants belong to the field of healthcare textiles. They are used in daily life, army, sports and medicine. Daily antibacterial underpants only need to meet common sanitary requirements. The effect is not strict disinfection and sterilization, nor can they provide a therapeutic effect. Daily antibacterial underpants should not be expected to provide a complete, high curative effect or disinfection function.

Nowadays, people around the world are paying more attention and caution to the use of antibiotics. Because applying a large amount of antibacterial or disinfectant will damage the originally healthy micro-ecological environment of the skin. For example, some lotions or sanitary napkin use excess antimicrobial agent and disinfectant for efficient sterilization. Similarly, wearing underpants incorrectly will lead to the damage of normal bacteria of women's perineum and an imbalance of the micro-ecological environment. The news reported that some antibacterial deodorant insoles which were finished with aldehydes antimicrobial stimulated the skin and then damaged the skin and sweat glands with the excessive use of the aldehydes substance.

At present, there is no independent evaluation standard for antibacterial underpants. The evaluation standard based on antibacterial underpants wearing does not exist. With no wearing evaluation standard and only the use of fabric washability and antibacterial effects as the evaluation index, there is a large gap between the physical truth after wearing and the developed antibacterial fabric. There exists a gap between antibacterial fabrics and its practical antibacterial wearing effect. Establishing a wearing antibacterial evaluation method can make developers working on antibacterial formulas to meet practical use more accurately. At the same time, it will offer a reference for establishing safer, more rigorous and effective antibacterial underpants evaluation standards. In addition, it can offer consumers an easy guide to understand and buy antibacterial garments, especially underpants.

4.1 The Function Category and Health Standards of Antibacterial Underpants

Daily antimicrobial underpants should restrain the multitude of microbes adherent on the fabric, which plays a role in healthcare and disease prevention. Medical antimicrobial underpants have higher requirements of disinfection or treatment effects for humans. Antibacterial underpants are not able to protect against all microbes, only for some bacteria. If beyond the actual request, it may cause high dose of antibacterial, which may lead to the normal balance of human skin being broken by wearing the antibacterial underpants. This in turn may lead to the skin glands being damaged or another bacterial infection. At the same time, antibacterial underpants are not required to kill the bacteria completely. Especially for children and women, the absolutely disinfected underpants may lower the body's resistance to disease, ultimately leading to poorer health.

Antibacterial underpants belong to the category of antibacterial clothing products in contact with human skin, and the main health evaluation standards are as follows [36]:

1) General health requirements. No odor, no stimulation and damage to the skin or mucous membrane. The adding of banned substances or toxic substances shall not exceed the prescribed limits. Good appearance.

2) Safety requirements. Anti-bacterial underpants need to be safe; their antimicrobial component dosage should be in the range of safety.

The function category and sanitary standards of antibacterial underpants should be divided and classified according to the application object and environment. It is understood that the World Health Organization has attached great importance to the global antibiotics phenomenon, and has put forward the warning accordingly. Antibacterial clothes, especially underpants, has begun to be affected and the subsequent research needs to be carefully studied.

4.2 The Microbial Environment of Underpants

There are numerous superficial microbes in the areas of the body covered by underpants which are closely related to human health. To evaluate the wearing of underpants, the microbial environment of these body parts must be understood first. Then the designer can make pointed references to design underpants.

4.2.1 Bacteria on the Parts of the Body Covered by Underpants

On the parts of the body covered by underpants, the superficial microbes normally maintained a precise balance. The normal superficial microbes play important roles in rejecting alien microbes, and are immunocompetence owned by the skin. But some bacteria which inhabit the skin surface are susceptible to the environment, which may lead to the destruction of the balance of bacterial growth leading to problems regarding the health of the skin. For example, for Gram-negative bacterium (G^-) in connection with our body's urinary tract, the main disease-causing representative bacteria are escherichia coli and pseudomonas aeruginosa. Similarly, for the gram-positive bacteria (G^+) the main disease-causing representative bacteria are staphylococcus aureus and micrococcus. In addition, there is candida albicans on behalf of fungus [37].

Researchers separated the microorganisms from fibrous material and found that there is a great variety of bacterial genus on these parts of the skin. At the same time, they also found that disease of skin and underpants mildew were mainly because sweat, sebum cutaneum and other secretions were absorbed on underpants.

Table.1 The amount of bacteria absorbed by underpants unit: cfu /cm^2

Season	Abdomen	Waist	Hip	Mean	Std.
Spring	200.0	280.0	300.0	260.0	52.9
Summer	400.0	1100.0	260.0	586.7	450.0
Autumn	280.0	270.0	240.0	263.3	20.8
Winter	58.0	80.0	90.0	76.0	16.4

As can be seen in Table 1 [2], underpants absorbed the most bacteria in the summer, about eight times as much as in winter, and about four times as much as in spring and autumn. The absorbing amount of the waist is greater than the abdomen, and the abdomen is greater the hip. In clinical studies, the body parts likely to cause illness include the abdomen, groin, the base of the thighs, perineum and some other areas which become unclean easily. If underpants are to have an antibacterial effect, it needs to have contact with the body so the antibacterial can effect the skin and cover the areas likely to be infected [38].

4.2.2 Pathogenic Bacterium on the Areas of the Body Covered by Underpants

Multiplying bacteria will produce toxins that cause skin disease. For instance, bacteria G^+ are parasitic on the surface and mucosa of human skin, which may cause pyogenic infection (such as pneumonia, papule). The spread of fungi may lead to skin tinea (such as tinea pedis, tinea unguium, tinea capitis). Bacteria G^-(e. coli) thriving may cause scytitis (red, itchy skin). Fabrics antibacterial testing will not evaluate each kind of bacteria for textiles, but to select the representative bacteria. So, the representative bacteria should be analyzed if the antibacterial properties of the underpants will be evaluated. If the targeted pathogenic bacteria is found, the evaluation should use the targeted bacteria. Table 2 lists representative bacteria for evaluating antibacterial underpants [37-39].

Table.2 Representative bacteria for evaluating underpants [37-39]

Microorganism species		The harm for underpants & Related diseases or state
Gram-negative bacterium (G^-)	Escherichia coli	Urinary tract infection; Pruritus; Dermatitis; Eczema; Jock itch
	Proteusbacillus vulgaris	Diapers of textile; Scytitis caused by ammonia decomposed by urea;
	Aeruginosa	Wound & Burn infection
Gram-positive bacterium (G^+)	Escherichia coli	Rancidness & Bromidrosis
	Proteusbacillus vulgaris	Body odor
	Corynebacterium diphtheroides	Body odor
	Brevibacterium ammoniagenes	Diaper rash
Fungus	Candida albicans	Diaper rash; Vaginal infection;
	Aspergillus niger	The spore is black brown, widely distributed in the cotton, wool and chemical fiber
	Penicillium citrinum (one of Penicillin)	Producing uranidin

4.3 Evaluation of Antibacterial Underpants Wearing

4.3.1 Assessment Index of Antibacterial Underpants Wearing Evaluation

Wearing antibacterial underpants has an effect which is in close relationship with the human body, so in order to put forward the reliability index for evaluation, analyzing the influencing factors is important. Because of the complexity of the environment and human body, the putting forward of related factors and the extraction of key factors will determine the evaluation index. Currently, antibacterial wearing evaluations still mainly rely on the evaluation of the fabric, and the main effect of wearing is given priority over comfort. The antibacterial effect of wearing is more common, so the confirming of its factors needs to follow the following method principles [40]:

1) Evaluation method of antibacterial wearing

A: Ensure the wearing subjects;

B: Ensure the focused bacteria;

C: Collecting the existing relevant research materials and available and reliable products;

D: Consulting, site inspection, screening, and finally determining factors reasonable and effective.

2) Evaluation principles of antibacterial wearing

A: Consistency of the evaluation targets and elimination of inconsistent factors;

B: Factors of evaluation should be observable quantitative index. For objective testing results, the factors should be based on regular data, while subjective testing results should be quantified by appraisal standard;

C: All these factors finally can form an independent evaluation index;

D: Influence factors in the research are comprehensive within a reasonable range.

4.3.2 Safety Evaluation of Antibacterial Underpants Wearing

Due to antibacterial underpants involving the safety of the body, the safety evaluation of the antibacterial is critical. In recent years, toxicity studies of antibacterial clothing products have undergone more research [41]. At present, the latest methods of safety evaluation of antibacterial fibres are urgent toxicity test through the mouth, variation of the original test, skin irritation test, and skin sensitivity test. The classification of antibacterial textiles in China are designated A, AA, AAA level, referring to the antibacterial textiles standards "FZ/T73023-2006 antibacterial knitwear". Standards about antibacterial underpants need to be put forward. And current antibacterial textiles standards still need to be updated.

4.3.3 Wearing Effect of Antibacterial Underpants

Current antibacterial testing standards are only concentrated on evaluating the fibres and fabrics, and the standards for wearing antibacterial underpants is rare. The difference between antibacterial fabric and antibacterial underpants is unclear. And the fabric meeting the antibacterial standards implies that the antibacterial underpants after wearing are with the same antibacterial effect. These problems directly relate to the actual utility of the antibacterial products, and the products need to be tested. For instance, a kind of reliably antibacterial underpants bought from the market, which truthfulness claim that its antibacterial effect can reach almost 100% after washing 30 times. That is about the antibacterial effect of fabric, and it is true after verify (antibacterial rate is 99.97%). However, we worn the underpants for 30 days and washed 30 times, the antibacterial rate went down to 36.60% which is not qualified according to the antibacterial textiles standards "FZ/T73023-2006 antibacterial knitwear" [42]. Wearing do greatly affect the antibacterial effect of underpants.

What's more, the common antimicrobial activity tests are focused on the antibacterial effects of representative bacteria, with no simulation experiment or clinic which are required in medicine. The antibacterial effect of wearing is related to many factors, of which the antibacterial efficiency is the main factors, but this is just a preliminary evaluation. In subsequent studies, researchers need to take more factors into consideration, cross analyzation and perform correlation analysis as well.

Due to the role of both the internal and external environment, antibacterial agent under the condition of wearing may produce more utility. For the two sides of protecting human body and fabric, that will consume more antibacterial agent.

5. Conclusion

1) At present, antibacterial effects are obtained through antibacterial fabrics and structures design of underpants. Natural fibre fabrics and natural antibacterial agent have received attention and development for their safety and healthcare to make antibacterial underpants. However, broad-spectrum antibacterial property and antibacterial endurance of the natural fibre fabrics and natural antibacterial agent were not good enough and need to be studied further. Metal ions antibacterial agent such as Ag^{+} and Gu^{2+} are very popular in Europe and the United States in recent years, however the safety is still controversial and the recognized of safety is much-needed. In addition, the structure studies on antibacterial underpants develops slowly and we would need to innovate more.

2) Not only the antibacterial fabrics but also the microbiology and clinical medicine have relations with antibacterial underpants. The target pathogen, environment both inside and outside the underpants and the medical conditions caused by antibacterial growth are all the relative evaluation indexes of antibacterial underpants when wearing. These relative indexes should be noted and considered by researchers as well as consumers. Because the antibacterial effect of underpants after wearing is worse than not wearing, the evaluation standard of antibacterial wearing effect is necessary to put forward. And the evaluation standard can provide a reference for the safety research of antibacterial underpants.

References

[1] Doctors' and nurses' clothing found to harbor potentially dangerous bacteria. Medical Economics [serial online]. October 25, 2011; 88 (20): 24. Available from: Business Source Premier, Ipswich, MA. Accessed April 20: 2017.

[2] Cheng XG. Clothing environment. In: Jin YS, Gao XS, editors. Beijing: China Textile Press: 1999; 7.

[3] Liang LP. Why to wear linen underwear. J China textile & apparel: 2014; 10: 156.

[4] Liu W, Zhao YY, Yu HS. Comparison study of several kinds of plant source antibacterial viscose fiber. J Shan Dong Sci & Tech: 2016; 57 (1): 19-21.

[5] Liu L, Chen Y. Aloe vera viscose fiber antibacterial knitted sports leisure fabric development. J Shanghai textile science & technology: 2016; 4: 33-35.

[6] Ren R, Huang YH. Processing technology research of new type antibacterial fiber containing tea. J Guangdong tea: 2016; 2: 17-19.

[7] Yu HS, Wu J, Liu YX, et al. Property study of plant origin functional viscose fiber. J Knitting industries: 2016;

8: 7-8.
[8] Wu JQ, Wu YR. A Chinese herbal medicine antibacterial fiber fabric and its producing method. CN Patent No. 104562260A: 2015.
[9] Zhang H, Gao Y, Zhang J, et al. Development and performance testing of the new military knitted underwear fabrics. J Knitting Industries: 2012; 10: 12-14.
[10] Bu E. Bamboo fiber ecological textile stores - breathing ecological textile. J Public investment guide: 2013; 10: 45.
[11] Huang SQ. Bamboo charcoal antibacterial thermal underwear. CN Patent No. 202774138U: 2013.
[12] Mo BR, Mo RM, Zhong KX. Natural anti-bacterial skin care plain knitted underwear fabrics. CN Patent No. CN202340829U: 2012.
[13] Wang SF, Deng YM, Wang SJ, et al. Study on the whiteness of cotton fabric finished with chitosan-silver. J Shanghai textile science & technology: 2015; 7: 56-58+72.
[14] Li ZH, Yu FB, Wang SJ, et al. Application of nanometer silver antibacterial fiber in textile industry. J Knitting Industries: 2013; 7: 65-66.
[15] Yan YJ, Song Y, Ji YM, et al. Advances in antimicrobial finishing for textiles based on silver. J Textile Research: 2008; 29 (4): 12-13.
[16] Chu JY. Current research status of nano-silver anti-bacterial textiles. J Chengdu Textile College: 2016; 33 (4): 194-196.
[17] Thomas V, Bajpai M, Bajpai SK. In situ formation of silver nanoparticles within chitosan-attached cotton fabric for antibacterial property. J Industrial Textiles: 2011; 40 (3): 229-245.
[18] Lipatova IM. Mechanoacoustic method for production of composite Chitosan finishing agents for textile materials. Russ J Gen Chem: 2013; 83: 205-213.
[19] Nie P. Chitosan fiber health care underwear. CN Patent No. 202476441U: 2012.
[20] Wang SF, Deng YM, Wang SJ, et al. Research of chitosan-based antibacterial agent in textiles and apparel. J China Textile Leader: 2013; 12: 55-57.
[21] Ma N. Application of new fiber in knitted underwear. J China Fiber Inspection: 2016; 5: 143-144.
[22] Dai FW. Antibacterial mechanism, processing and ecological safety analysis of antibacterial fiber. J China Fiber Inspection: 2015; (1): 68-69.
[23] Zhu PF. Research on antibacterial finishing processes of cotton fabric with the in-situ growthed nano-silver. D Zhejiang Sci-Tech University: 2015.
[24] Huang YQ. Bamboo charcoal antibacterial thermal underwear. CN Patent No. 101612258: 2009.
[25] Li KJ, Wang XC, Liu Z. Development of antibacterial and fragrant cotton underwear on the microcapsule technology. J Shanghai textile science & technology: 2006; 8: 48-50.
[26] Xu SY. A male health care underwear. CN Patent No. 88206301.4: 1989.
[27] Zhou ZP. Antibacterial underwear with ruffles at neckline. CN Patent No. 103082421A : 2013.
[28] Sun JL. Hollowed-out antibacterial underwear. CN Patent No. 103120366A: 2013.
[29] Yao L. The development of nanometer Ag ion antibacterial clothing. J Luohe Vocational College: 2011; 5: 18-20.
[30] Pan L, He LM, Zhu DP. Mint viscose antibacterial health underpants. CN Patent No. 205197009U: 2016
[31] Ding WL, Wu HL, Lin FL. Ventilate and antibacterial underpants. CN Patent No. 201388539 Y: 2010
[32] Chen W. Performance testing and comprehensive evaluation of germanium fiber seamless fabrics. D Zhejiang Sci-Tech University: 2016.
[33] Jin GJ, Tang QD, Zhao H. A kind of production method process for antibacterial recombination seamless underwear. CN, Patent No. 104287106A: 2015.
[34] Ren YZ, Wang KL, Yu WD, et al. Application of nanotechnology to sport costumes. J Sport Science Research: 2006; 3: 36-38.
[35] Alligator S. Men's anatomic underwear/swimwear: US, Patent No. 6061840: 2000.
[36] Shi H, Wang JP, Wang YZ, et al. Knitted underwear functional theory research. J Knitting Industries: 2011; 12: 48-49.
[37] Shen P. Microbiology. M Beijing: China Education Press: 2015; 312.
[38] Li DM, Huang C. Bacterial skin infection and its diagnosis countermeasures. Southern China J of dermato-venereology: 2013; 3: 221-223.
[39] Gu JL, Zhu YJ, Chen JH. The prevention and control measures of troops` skin disease under special

environmental conditions. J People's military surgeon: 2009; 3: 145-14.
[40] Tu JW. Comprehensive evaluation on wearability of health protection functional socks. D Donghua University: 2015.
[41] Maryan AS, Montazer M. Natural and organo-montmorillonite as antibacterial nanoclays for cotton garment. J IND & Engineering Chemistry: 2015; 22: 164-170.
[42] Wang SF. Research on Antibacterial Underwear Finished with Chitosan-silver. D Xi'an Polytechnic University: 2015.

Controlled Release and Antibacterial Properties of Berberine-loaded Silk Fibroin Film

Chao-Heng He[1], Zhao-Zhu Zheng[1], Jue Zhang[1], Gang Li[1*], Xiao-Qin Wang[1, 2**]

[1]*National Engineering Laboratory for Modern Silk, College of Textile and Clothing Engineering, Soochow University, Suzhou 215123, P.R.China*
[2]*Department of Biomedical Engineering, Tufts University, 4 Colby St, Medford, MA 02155, USA*

[*]Corresponding authors' email: tcligang@suda.edu.cn, wangxiaoqin@suda.edu.cn

Abstract

A significant challenge still remains for the development of drug delivery system (DDS) using herbal medicine for clinical applications. The present study describes a novel DDS consisting of a natural antibacterial drug berberine and silk fibroin (SF) for controlled drug delivery application. Composite films were synthesized using varied ratios of SF and berberine mixtures. The unique structural features of SF molecules were analyzed using Fourier transform infrared spectroscopy (FTIR) during various film processing and post-processing conditions. Results revealed that water-annealed and glycerol modified method accurately controlled crystallization via, modification of the secondary structure of the SF matrix and the interaction with the non-crystalline domains resulting in control of film degradation as well as drug diffusion rate. A linear relationship between crystallinity content (beta-sheet dominated silk II structure) and the release of entrapped berberine was achieved. The result indicated that berberine retaining its initial antibacterial activity. The stability of the drug in SF film was likely due to intermolecular interactions between SF and berberine molecules. In conclusion, the composite materials consisting of SF and berberine have a great potential application as a DDS providing antimicrobial activities, such as antimicrobial suture and biomedical textiles.

Keywords: Silk Fibroin; Berberine; Film; Antibacterial; Drug Delivery System

1. Introduction

Microbial infection is a kind of commonly and frequently occurring infective disease worldwide. So far, a great number of antibiotics and synthetic drugs have been extensively used in clinics and have played a significant role in the treatment of microbial infections [1]. The wide nonmedical use of antibiotics may contribute to antibiotics-adapted cross resistance. In order to control drug dosage, reduce drug toxicity and side effects, improve drug stability and efficacy, and reduce frequency and cost of the treatment, physical or chemical agents and other means can be used to maintain the effective concentration of the drug for a long time, and to release the drug at a predetermined time in the corresponding organization [2-7]. Among various means for controlled release, using polymeric materials as delivering carriers is attractive due to their high efficiency, low toxicity and resistance, as well as high structural stability and capability controlling drug release rate.

Silk fibroin (SF) is a polymer protein biomaterial with desirable biocompatibility and mechanical properties [8]. It has excellent film-forming capabilities and is also compatible for use in the human body [9]. The processed SF can be processed into a variety of forms, such as tube, film, gel, scaffold, fiber and microspheres [10-18]. SF fibers consist of a surface layer with amorphous regions and an inner layer with crystalline regions [19]. The crystalline regions of the inner layer contain two types of structures: Silk I and Silk II [20]. Silk I structure is not stable, and easy to turn into silk II structure after certain treatments, thus making materials aqueous insoluble, and providing options for the use of the material in a range of biomedical and other applications [21]. SF can be extracted with chemicals from the beta-sheet dominated fibers, existing in a random coil-rich structure in solution. The random coil structure can be transformed into beta-sheet (silk II) form by heating and water vapor treatment, immersion in polar organic solvents (such as methanol, glycerol), acidic solution and mechanical stretching [22-25].

Silk film is proposed as an excellent material for coating drug carrying matrices; the porous structure of the film itself has very good adsorption and slow release functions [26]. Hofmann et al. reported that silk films with different crystallinities were obtained by water vapor and methanol treatment. When dextran with different molecular weights were used as model compounds to test the release of the power curve of the two kinds of films, it was found that the release of dextran was delayed from the SF film with high crystallinity as compared to that with low crystallinity [27]. Jian et al. found that the surface coating of vancomycin / polycaprolactone with SF effectively prolonged the release of vancomycin and improved its anti-infection effect [28].

Herbal medicine is attracting attention because of its lower cost, better compatibility with the human body and fewer side effects as compared to the synthetic medicine [29]. Berberine, one of the main alkaloids isolated from Rhizoma coptidis, has a long history of use in the Chinese and Ayurvedic medicinal system. It exhibits many pharmacological effects, including antibacterial, anti-inflammatory, reducing blood pressure and blood glucose, etc. [30-34]. Researches showed that berberine has a significant inhibitory effect, on bacteria gram negative Escherichia coli and gram positive bacteria Staphylococcus aureus [35]. It has been widely used in clinical treatment of various diseases, such as bacterial dysentery, acute gastroenteritis, cholera and other related diseases [36]. Lin et al. with collagen as a carrier, the load of berberine electrospun into nanofibers. He found that the dressing had a good antibacterial activity against bacteria, resulting in an improvement for the healing of wound [37]. So far there was no report yet with a focus on using berberine-based SF as an antimicrobial material.

The present study is about the preparation of berberine loaded SF films, with a focus on elucidating the relationships between beta sheet content of SF, water content, berberine release and antibacterial activity of the SF films. The new berberine-loaded SF films with a sustained-release feature and antibacterial function can be used, as a coating material, biological dressing and will be useful for the basic studies in the future.

2. Materials and Methods

2.1 Materials

Cocoons of Bombyx mori silkworm were purchased from Zhejiang Shengzhou Xiehe Silk Textile Ltd. Co., Berberine (98 %+), was purchased from Sigma-Aldrich (St. Louis, MO). All other chemicals used in the study were from Sigma-Aldrich (St. Louis, MO).

2.2 Purification of SF

SF aqueous stock solution was prepared as previous described [38]. Briefly, cocoons of Bombyx mori were boiled for 20 min in an aqueous solution of 0.02 M sodium carbonate and then rinsed thoroughly with distilled water. After drying, the extracted SF was dissolved in 9.3 M LiBr solution at 60°C for 4h, yielding a 20% w/v solution. This solution was dialyzed against distilled water using Slide-a-Lyzer dialysis cassettes (MWCO 3500, Pierce) for 3 days to remove salt. The solution was optically clear after dialysis and was then centrifuged to remove the small amount of silk aggregates that formed during the process. The final concentration of SF aqueous solution was around 7.5% w/v. This concentration was determined by weighing the residual solid from a known volume of solution after drying.

2.3 Preparation of SF/berberine Films

To prepare SF/berberine films without glycerol, berberine was dissolved in distilled water at a concentration of 4 $mg \cdot ml^{-1}$ and this solution were mixed with 7% silk solution at a volume ratio of 1:10 and 1:20. The mixed solution (2 ml) was transferred into a Petri dish (diameter 20 mm) and dried at 60°C with 3h. Dried films were treated by water vapor for 6h; in order to induce SF beta-sheet structure and thus water insolubility. The untreated and water-annealed films were termed as S-B-NO and S-B-WA, respectively.

To prepare glycerol-modified SF films, berberine was dissolved in distilled water at a concentration of 4 mg·ml-1, and the purified SF solution was mixed with glycerol at a weight ratio of SF/glycerol=3/1, and the two solutions were mixed at a volume ratio of 1:10 and 1:20. The mixed solution (2ml) was transferred into a Petri dish (diameter 20 mm) and dried at 60°C for 3h. Based on our previous study, water insoluble silk films were prepared directly after drying overnight when the content of glycerol retain 30%. Because the glycerol-modified SF films with 30% of glycerol were prepared and stretched to control secondary structure. The glycerol-modified SF films were flexible in the dry state and the secondary structure changed from silk I to silk II after stretching [39]. The glycerol-modified films were termed as S-B-G.

2.4 Characterization

2.4.1 Fourier Transforms Infrared (FTIR) Spectroscopy

The structure of the various films was analyzed by Fourier transform infrared (FTIR) spectroscopy (Nicolet 5700) in reflection mode. Background measurements were taken with an empty cell and subtracted from the sample reading. For each measurement, 64 scans were recorded with a resolution of 4 cm^{-1}, and the wavenumber ranged from 400 to 4000 cm^{-1}. The secondary structures present in the film, including random coil, α-helix, beta-sheets, and beta-turn were evaluated using Fourier self-deconvolution (FSD) of the infrared absorbance spectra. Glycerol did not show peaks at the amide I band region so its influence was negligible. FSD of the infrared spectra coving the amide I region (1595-1705 cm^{-1}) was performed by Opus 5.0 software, as previously described [40]. Absorption bands in the frequency ranges of 1616-1637 cm^{-1} and 1695-1705 cm^{-1} represented enriched beta-sheet structure (silk II); those in the range of 1638-1655 cm^{-1} were ascribed to random coil structure, those in the range of 1656-1663 cm^{-1} were ascribed to α-helix, and those in the range of 1663-1695 cm^{-1} were ascribed to turns [40]. The contribution of these bands (silk I or silk II content) to the amide I band was assessed by integrating the area under the band and normalizing to the total area under the amide I band region (1595-1705 cm^{-1}).

2.5 Antibacterial Activity Measurement

2.5.1 Qualitative Assay-agar Diffusion Plate Test

The sensitivity of the organism to the antibacterial agents was estimated by the zone of inhibition [41]. This assay was carried out on round films with 10 mg in weight, and 15 mm in diameter for each film. One piece of film sample was inoculated with approximately 1ml of bacterial suspension (1×10^8 CFU/ml) per tryptic soya agar plate. The plate was incubated at 37°C for 24h and then measured by a digital vernier caliper to determine the zone of inhibition around the film (Fig.1A). Gram-positive S. aureus and Gram-negative E. coli were selected as the testing organisms. The assessments were performed according to antibacterial activity evaluation standard (ISO 20645:2004). In this assay, the film samples without encapsulation berberine were used as control. The zone of inhibition was calculated according to the formula (1):

$$H = (D-d)/2 \quad (1)$$

Where: H is the zone of inhibition (mm), D represents the total diameter of the specimen and the zone of inhibition (mm), and d represents the diameter of the specimen (mm).

2.5.2 Sustained Efficacy Assay-serial Plate Transfer test

The purpose of this assay was to determine the duration of antibacterial activity of the SF films. The samples were transferred onto new Petri dishes inoculated with similar number of bacteria. The inhibition zones were determined as described above after 24h. The test was continued until the films ceased antibacterial functions.

2.6 In vitro Berberine Release

In vitro release kinetics of berberine from SF films was determined by incubating 10 mg of the dry film in 3 ml PBS buffer, pH 7.4, in an Eppendorf tube at 37°C. Three types of films (as-cast, water-annealed, glycerol-modified) were prepared using the same methods as described in the stability tests. To start the release experiment, at specific time points (1, 3, 6, 12, 24, 48, 72, 96, 120, 144, and168 h), 1 mL of release medium was moved to an empty tube and 1 mL of fresh medium was added into the tube. Berberine concentration in the release medium was determined by measuring absorbance at 345 nm. Berberine concentration was calculated based on the standard curves measured under the same condition. The percentage release was obtained by comparing the amount of drug in the release medium with the initial loading in the film. Each test group contained four repeat samples.

2.7 Statistical Analysis

All experiments were performed with a minimum of N=3 for each data point. Statistical analysis was performed by one-way analysis of variance (ANOVA) and Student -Newman-Keuls multiple comparisons test. Difference were considered significant when $p \le 0.05$ and very significant $p \le 0.01$.

3. Results and Discussion

3.1 Preparation of SF/berberine Films

Silk/berberine films were fabricated by mixing berberine and fibroin solutions at weight ratios of 10/90, 20/80, 30/70, 40/60, 50/50, and 0/100. During the mixing process of 20/80, 30/70, 40/60, and 50/50 solutions crystal precipitation of berberine formed, while the 10/90 solution had the minimal amount of crystals formed SF/berberine films were then fabricated by mixing berberine and fibroin solutions at weight ratios of 1/10, 1/20, 1/30, 1/40, 1/50, followed by water-annealing treatment. Anti-microbial efficacy of the films was determined by Agar plate diffusion tests using Gram-positive S. aureus and Gram-negative E. coli. As shown in the Fig.1B, the results of berberine loaded films at weight ratios of 1/10 (a, g) and 1/20 (b, h) had a better antibacterial effect than 1/30 (c, i). In addition, 1/40 (d, j) and 1/50 (e, k) showed almost no inhibition effect. As-cast films (f, m) without berberine encapsulation and treated with water annealing showed no zones of inhibition.

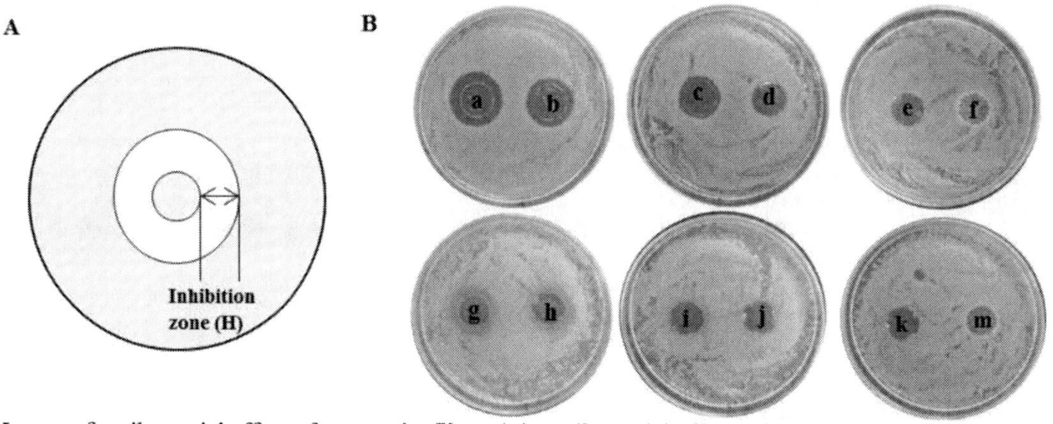

Fig.1 Image of antibacterial effect of composite films. (A) Antibacterial effect schematic diagram; (B) Images of different proportion of BBH/SF antibacterial effect, (a, g) BBH/SF:1/10, (b, h) BBH/SF:1/20, (c, i) BBH/SF:1/30, (d, j) BBH/SF:1/40, (e, k) BBH/SF:1/50 and (f, m) As-cast silk film, the empty films treated with water annealing without berberine loading.

3.2 Antibacterial Activity

Berberine-loaded SF films showed the following E. coli inhibition zones after 24h incubation of 7.06 mm (SB-WA:1/10), 6.04 mm (SB-WA:1/20), 5.23 mm (SBG:1/10), and 4.51 mm (SBG:1/20). The inhibition zones decreased rapidly during the first 2 days, and then decreased more slowly (Fig.2). The inhibition zones were maintained for 3-4 days, with that of the glycerol-blended film (SBG: 1/20) ending after 3 days, whereas that of the water-annealed (SB-WA: 1/20) and glycerol-blended (SBG: 1/10 and 1/20) films ending after 4 days. Thus, water-annealed SF film showed a more sustained antibacterial effect than the glycerol-blended SF film. Water-annealed (SB-WA: 1/20) and glycerol-blended (SBG: 1/10) SF films showed a similar size of inhibition zones and duration of antibacterial effect. Especially, the water annealed (SB-WA: 1/10) have the best sustained antibacterial efficacy, the inhibition zones of 0.63 mm until for 4 days. Thus, water-annealed SF film showed a more sustained antibacterial effect than the glycerol-blended SF film. However, berberine antibacterial effect from the samples, it was significantly delayed compared with that from the difference samples ($p<0.01$, for all time points).

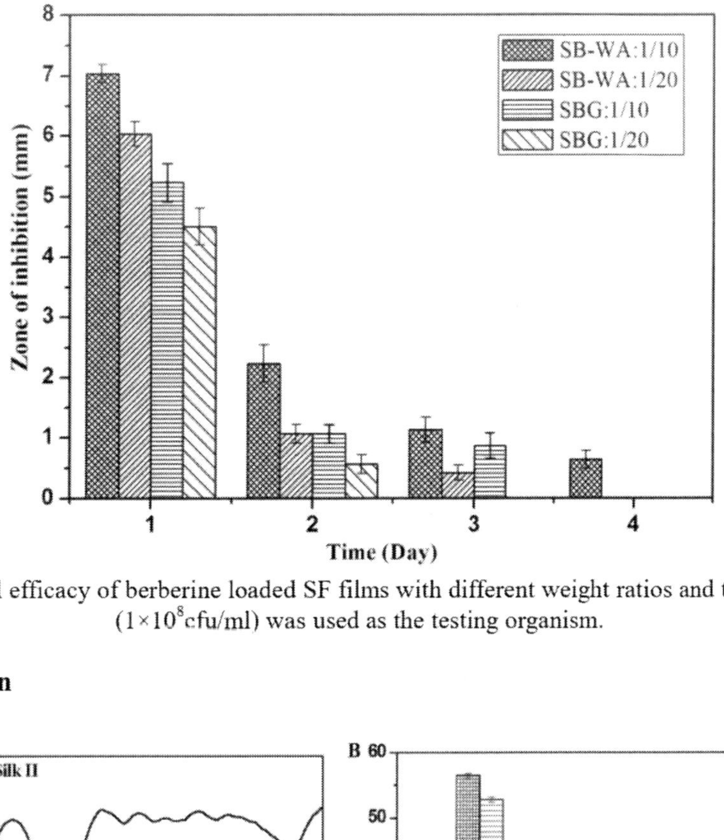

Fig.2 Antibacterial efficacy of berberine loaded SF films with different weight ratios and treatments. E. coli (1×10^8cfu/ml) was used as the testing organism.

3.3 Characterization

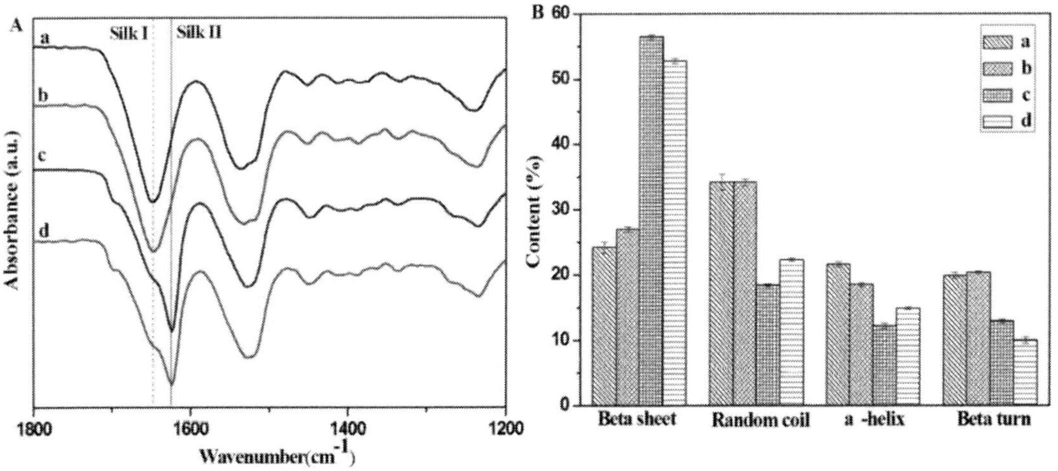

Fig.3 Structural analyses the silk-berberine film, (A) ATR-FTIR spectra, (B) conformational fraction s after deconvolution of amide I and II peaks, (a) SF is As-cast film, (b) S-B-NO is berberine-loaded silk film, (c) S-B-G means glycerol-modified BBH-loaded silk film, (d) S-B-WA represents berberine-loaded silk film after water annealed.

The structure of dried SF films with and without berberine loading was determined by FTIR. Characteristic absorbance peaks in the amide I region (1200-1800 cm^{-1}) are shown in Fig.3A. As-cast films and Berberine-loaded SF films without treatment showed peaks around 1645 cm^{-1}, indicating the films were dominated by random coil structure (Fig.3A). Berberine did not show peaks in the amide I and II band region and thus had a negligible influence on silk structure analysis. Absorbance peaks of water-annealed treatment and glycerol blended films were around 1622 cm^{-1} and 1627 cm^{-1}, respectively, indicating both types of films were dominated by beta-sheet structures (Silk II). Amide I peaks were deconvoluted to quantify secondary structures (Fig.3B). Beta-sheet content was approximately 56.5±0.4% in glycerol blended films and 52.8±0.5 % in water-annealed treatment films, higher than that in the as cast film (24.2±0.8%). As-cast films contained more random coil structure (34.2±1.2%) than glycerol blended (18.4±0.3%) and water-annealed treatment (22.3±0.2%) films.

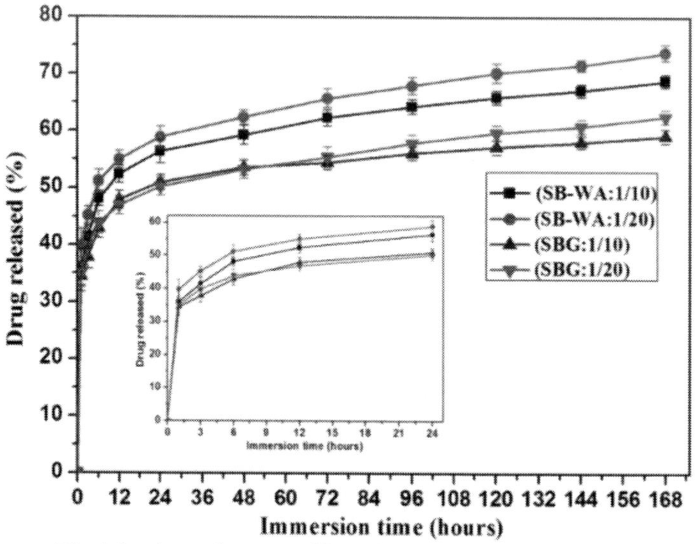

Fig.4 In vitro release profiles of berberine loading SF films.

3.4 In Vitro Berberine Release

Berberine was loaded in silk film at two different weight ratios (1/10 and 1/20) and two different types of treatment (water-annealed and glycerol-blended). As shown in Fig.4, all samples demonstrated continuous drug release within 6-7 days of experiments (Fig.4). The initial burst release was observed in the first 12h, and the amount of the drug released remained stable after 96h. The berberine-loaded and water-annealed SF films showed a more rapid release than the glycerol-blended films, with 68.15±1.1% of total loaded berberine being released for SB-WA: 1/10 and 74.11±1.5% for SB-WA: 1/20 (Fig.4), while 59.1.9±1.2% for SBG: 1/10 and 62.94±1.3% for SBG: 1/20 (Fig.4). As compared to water annealed treatment films, the glycerol blended films showed significant delayed berberine release from 3 to 12 h ($p<0.01$) but no difference at the early and late stage of release ($p>0.05$)(Fig.4).

4. Conclusion

In this study, a novel antibacterial material containing berberine and SF was prepared using two different processing methods with a berberine and SF antibacterial system. All prepared films showed effective antibacterial function against S. aureus and E. coli and could effectively prevent bacteria adherence. The water annealed (BBH/SF: 1/10) SF film samples showed, the largest antibacterial zone which lasted for 4 days. Based on this research, a promising herbal medicine delivery system with high stability, controllable burst release of drug, and sustained antibacterial functionality was achieved by fine-tuning the structure of berberine-loaded SF films via various post-processing strategies. However, more work is required to confirm the clinical safety and biocompatibility of the materials in vivo.

5. Acknowledgement

This work was supported by the Natural Science Foundation of China (51603140), Natural Science Foundation of Jiangsu Province (BK20150372), State Key Laboratory for Modification of Chemical Fibers and Polymer Materials, Donghua University (LK1421), and University Science Research Project of Jiangsu Province (16KJB540003) and Natural Science Foundation for Young Scholars of Soochow University (SDY2014A06). We would like to thank the support of Key industry technology innovation, Science and technology project of Suzhou (SYG201638).

References

[1] Wen S, Jeyakkumar P, Avula SR, et al. Discovery of novel berberine imidazoles as safe antimicrobial agents by down regulating ROS generation. Bioorganic & medicinal chemistry letters: 2016; 26: 2768-2773.
[2] Endo T, Nakagawa T, Kita T, Kim TS, et al. A novel drug delivery system: A biodegradable gel for the treatment of inner ear diseases. Otolaryngology-Head and Neck Surgery: 2004; 131: 260-260.
[3] Li D, Zhong X, Zeng Z, et al. Application of targeted drug delivery system in Chinese medicine. Journal of Controlled Release: 2009; 138: 103-112.
[4] Basit AW. Advances in colonic drug delivery. Drugs: 2005; 65: 1991-2007.
[5] Braun K, Pipkorn R, Waldeck W et al. Development and characterization of drug delivery systems for targeting mammalian cells and tissues: a review. Current Medicinal Chemistry: 2005; 12: 1841-1858.
[6] Phillips KA, Veenstra DL, Sadee W et al. Implications of the genetics revolution for health services research: pharmacogenomics and improvements in drug therapy. Health services research: 2000; 35: 128.
[7] Allen TM, Cullis PR. Drug delivery systems: Entering the mainstream. Science: 2004; 303: 1818-22.
[8] Vepari C, Kaplan DL. Silk as a biomaterial. Progress in polymer science: 2007; 32: 991-1007.
[9] Altman GH, Diaz F, Jakuba C, et al. Silk-based biomaterials. Biomaterials: 2003; 24: 401-416.
[10] Lovett ML, Cannizzaro CM, Vunjak-Novakovic G, et al. Gel spinning of silk tubes for tissue engineering. Biomaterials: 2008; 29: 4650-4657.
[11] Zhao Z, Li Y, Xie M. Silk fibroin-based nanoparticles for drug delivery. International Journal of Molecular Sciences: 2015; 16 (3): 4880-4903.
[12] Seib FP, Pritchard EM, Kaplan DL. Self-assembling doxorubicin silk hydrogels for the focal treatment of primary breast cancer. Advanced Functional Materials: 2013; 23: 58-65.
[13] Rnjakkovacina J, Wray LS, Burke KA, et al. Lyophilized Silk Sponges: A Versatile Biomaterial Platform for Soft Tissue Engineering. Acs Biomater.sci.eng: 2015; 1: 260-270.
[14] Chen J, Altman GH, Karageorgiou V, et al. Human bone marrow stromal cell and ligament fibroblast responses on RGD-modified silk fibers. Journal of Biomedical Materials Research Part A: 2003; 67: 559-70.
[15] Breslauer DN, Muller SJ, Lee LP. Generation of Monodisperse Silk Microspheres Prepared with Microfluidics. Biomacromolecules: 2010; 11: 643-7.
[16] Li G, Li Y, Chen G, et al. Silk-Based Biomaterials in Biomedical Textiles and Fiber-Based Implants. Advanced Healthcare Materials: 2015; 4: 1134-1151.
[17] Li G, Li F, Zheng Z, et al. Silk microfiber-reinforced silk composite scaffolds: fabrication, mechanical properties, and cytocompatibility. Journal of Materials Science: 2016; 51: 3025-3035.
[18] Li G, Liu J, Zheng Z, et al. Structural mimetic silk fiber-reinforced composite scaffolds using multi-angle fibers. Macromolecular Bioscience: 2015; 15: 1125-1133.
[19] Lu Q, Hu X, Wang X, and et al. Water-insoluble silk films with silk I structure. Acta Biomaterialia: 2010; 6: 1380-1387
[20] Lawrence BD, Wharram S, Kluge JA, et al. Effect of hydration on silk film material properties. Macromolecular Bioscience: 2010; 10: 393-403.
[21] Zhang Y. Natural silk fibroin as a support for enzyme immobilization. Biotechnology Advances: 1998; 16: 961-971.
[22] Park SH, Gil ES, Shi H, et al. Relationships between degradability of silk scaffolds and osteogenesis.

Biomaterials: 2010; 31: 6162-6172.
[23] Rajkhowa R, Levin B, Redmond SL, et al. Structure and properties of biomedical films prepared from aqueous and acidic silk fibroin solutions. Journal of Biomedical Materials Research Part A: 2011; 97: 37-45.
[24] Li X, Wu L, Huang M, et al. Conformational transition and liquid crystalline state of regenerated silk fibroin in water. Biopolymers: 2008; 89: 497-505.
[25] Hu X, Shmelev K, Sun L, et al. Regulation of silk material structure by temperature-controlled water vapor annealing. Biomacromolecules: 2011;12: 157-182
[26] Rockwood DN, Preda RC, Yücel T, et al. Materials fabrication from Bombyx mori silk fibroin. Nature protocols: 2011; 6: 1612-1631.
[27] Hofmann S, Foo CTWP, Rossetti F, et al. Silk fibroin as an organic polymer for controlled drug delivery. Journal of Controlled Release: 2006; 111: 219-227.
[28] Zhou J, Fang T, Wen J, et al. Silk coating on poly (ε-caprolactone) microspheres for the delayed release of vancomycin. Journal of Microencapsulation: 2011; 28: 99-107.
[29] El-Wahab AEA, Ghareeb DA, Sarhan EEM, et al. In vitro biological assessment of berberins vulgaris and its active constituent, berberine: antioxidants, anti-acetylcholinesterase, anti-diabetic and anticancer effects. BMC complementary and alternative medicine: 2013; 13: 1.
[30] Yi Z, Yu Y, Liang Y, et al. Evaluation of the antimicrobial mode of berberine by LC/ESI-MS combined with principal component analysis. Journal of Pharmaceutical & Biomedical Analysis: 2007; 44: 301-4.
[31] Huang WP, Chan P, Cheng JT. The antihypertensive effect of the berberine derivative 6-protoberberine in spontaneously hypertensive rats. Pharmacology: 1999; 12: 283-9.
[32] Zhang W, Xu Y, Guo F, et al. Anti-diabetic effects of cinnamaldehyde and berberine and their impacts on retinol-binding protein 4 expression in rats with type 2 diabetes mellitus. Chinese Medical Journal: 2008; 121: 2124-8.
[33] Kuo C, Chi C, Liu T. The anti-inflammatory potential of berberine in vitro and in vivo. Cancer Letters: 2004; 203: 127-37.
[34] Kong W, Wei J, Abidi P, et al. Berberine is a Novel Cholesterol-Lowering Drug Working Through a Unique Mechanism Distinct from Stains. Nature Medicine: 2005; 10: 1344-51.
[35] Chen L, Bu Q, Xu H, et al. The effect of berberine hydrochloride on Enterococcus faecalis biofilm formation and dispersion in vitro. Microbiological Research: 2016; 186: 44-51.
[36] Tang L, Ni W, Cai M, et al. Renoprotective effects of berberine and its potential effect on the expression of β -arrestins and intercellular adhesion molecule-1 and vascular cell adhesion molecule-1 in streptozocin diabetic nephropathy rats. Journal of Diabetes: 2015.
[37] Lin J, Li C, Zhao Y, et al. Co-electrospun Nanofibrous Membranes of Collagen and Zein for Wound Healing. ACS Applied Materials & Interfaces: 2012; 4: 1050-7.
[38] Sofia S, McCarthy MB, Gronowicz G, et al. Functionalized silk-based biomaterials for bone formation. Journal of biomedical materials research: 2001; 54: 139-148.
[39] Lu S, Wang X, Lu Q, et al. Insoluble and flexible silk films containing glycerol. Biomacromolecules: 2009; 11: 143-150.
[40] Hu X, Kaplan D, Cebe P. Determining beta-sheet crystallinity in fibrous proteins by thermal analysis and infrared spectroscopy. Macromolecules: 2006; 39: 6161-6170.
[41] Chen X, Hou D, Wang L, et al. Antibacterial surgical silk sutures using a high-performance slow-release carrier coating system. ACS applied materials & interfaces: 2015; 7: 22394-22403.

An Insight into Development of Electrospun Cells Scaffolds

Budimir Mijovic[1], Emilija Zdraveva[1], Emi Govorcin Bajsic[2], Mirna Tominac Trcin[3], Tamara Holjevac Grguric[4], Iva Dekaris[5], Tamara Dolenec[3], Ivana Vrgoc[3]

[1]*University of Zagreb, Faculty of Textile Technology, Prilaz baruna Filipovica 28a, Zagreb, 10000, Croatia*
[2]*University of Zagreb, Faculty of Chemical Engineering and Technology, Trg Marka Marulica 19, Zagreb, 10000, Croatia*
[3]*University Hospital Centre Sestre Milosrdnice, Tissue bank at University Department of Traumatology, Zagreb, 10000, Croatia*
[4]*University of Zagreb, Faculty of Metallurgy, Aleja narodnih heroja 3, Sisak, 44000, Croatia*
[5]*Specialty Eye Hospital Svjetlost, Zagreb and Department of Ophthalmology Medical Faculty, University of Rijeka, Rijeka, 51000, Croatia*

*Corresponding author's email: budimir.mijovic@ttf.hr

Abstract

This paper work is based on a brief discussion of a recent project proposal, approved by the Croatian Science Foundation. The project proposal focuses on the application of the technique of electrospinning in the development of scaffold prototype to be used for in vitro culture of ocular and skin cells. A brief review was also conducted on recent studies focusing on the usage of electrospinning combined with other advanced techniques for the fabrication of materials used for biomedical applications. The project itself has proposed scaffolds fabrication combining both solution and melt electrospinning.

Keywords: Scaffolds; Solution; Melt; Electrospinning; Skin Cells; Ocular Cells

1. Introduction

Among the many fields of applications of electrospun materials, the field of biomedicine has become very important, since these materials can primarily mimic the natural structure of the extracellular matrix (ECM). Secondly, their porous interconnected structure and ease of functionalization have proven their role in the design of scaffolds for both soft and hard tissues regeneration. From a tissue engineering point of view, the continuous shortage of tissue/organ donors has intensified the research in the fabrication of manmade scaffolds to (potentially) fully replace the natural biological materials.

Among the many techniques reported [1, 2] for the production of scaffolds, including: solvent casting, porogen (particle) leaching, gas foaming, phase separation, melt moulding, freeze drying, rapid prototyping, electrospinning [3, 4] was selected as a cost-effective one, versatile in polymers and fillers selection, fibers morphology and diameter control, and functionalization [5].

Requirements to be fulfilled in the process of scaffold design is concerned with both the construction properties and *in vitro* properties. The first group includes biocompatibility, biological or chemical mediated degradation, adjustable porosity (fibre and pore size) according to cell type, 3D structure and simulation of the native environment stiffness. The second group includes scaffold configuration favourable for cells adhesion, migration and inner penetration, simulation of the native cell/scaffold interface and permeability to oxygen and nutrients important for cells vitality.

When choosing the technique of electrospinning for the fabrication of nanofibrous scaffolds it is important to consider advantages including ECM mimic but also the large number of cell receptors binding sites, which can prevent (e.g. in wound healing) desiccation by liquid exudation, further controlled evaporation, excellent oxygen permeability and promoted fluid drainage capacity, as well as microorganism invasion inhibition [6, 7].

2. The Challenge for the Design of an Ideal Electrospun Scaffold

Apart from the advantageous characteristics of electrospun scaffolds there are some major challenges that researchers are still facing with. The intrinsically two-dimensional scaffold structure inhibits cellular migration and inhibits in-growth penetration. It also results in pores close up due to fiber swelling when cells media is added [8, 9].

To overcome the forth-mentioned drawbacks researches have offered different solutions concerning both chemical and physical scaffold structure modifications. Scaffolds chemical alterations may include: different polymer compositions and polymers combinations for the production of hybrid scaffolds, inclusion of functional species, from both synthetic and biological nature, and surface modifications (e.g. application of coatings). The second approach is focusing on the modification of the physical scaffolds structure. This is achieved by electrospinning set up modifications, e.g. collectors, or by simultaneous usage of the electrospinning with other advanced techniques.

Some of the chemical modifications reported involved: usage of co-polymers, blending of the polymers with natural compounds, or physical blending, based on layer by layer electrospinning of two different polymers.
For example, skin and bone tissue scaffolds for periodontal repair were prepared from polyhydroxybutyrate-co-hydroxyvalerate (PHBV), polylactic acid (PLA) and poly ε-caprolactone (PCL) electrospun into PHBV–PLA or PHBV–PCL bilayers and PLA–PHBV–PLA trilayers providing cell attachment and proliferation for one type of cells and barrier to cell penetration for second type of cells [10].

Similarly, studies reported on the electrospinning of co-polymers such as PLA and PGA to support the growth of limbal epithelial cells [11], blended chitosan with PCL providing more hydrophilic sites suitable for cells growth and proliferation [12]. Further, plant derived proteins, such as soy protein isolate solution (with a small addition of polyethylene oxide) and zein from corn have yield cytocompatibility and supported human dermal fibroblasts adhesion and proliferation [13]. Combinations were also made using silk fibroin [14], collagen [15, 16] or layered combinations of PCL, silk fibroin, collagen as well as bovine neck ligament elastin [17]. Examples of coating procedures after electrospinning included: graphene oxide and laminin electrospun PCL coating via plasma [18], or heparin coating applied on electrospun PLLA [19].

Metal and metal oxides nanoparticles [20, 21] or antibiotics [22] were added to polymers to functionalize the scaffolds against bacteria and similar contaminants. the scaffolds have the ability to release these compounds with different paces in compliance with the neo-tissue formation. Controlled release of the therapy is achieved by physical blending, core-shell fiber production, through chemical bonding, encapsulation or posttreatment [23].

Some examples reported concerned fibrocartilage regeneration in mini pigs with the release of fibroblast growth factor (bFGF) and insulin from polyvinyl alcohol containing liposomes [24]. Similarly, sustained release of pDNA, improved the viability of endothelial and smooth muscle cells. Also, little inflammation in rats was obtained from electrospun poly (DL-lactide)–poly(ethylene glycol) (PELA) and calcium phosphate nanoparticles loaded with the pDNA [25].

2.1 Recent Solutions - Electrospinning and other Advanced Techniques

One of recently used advanced techniques reported, was photopatterning which involves photopolymerization with the aid of reactive macromer solution with a photoinitiator through a mask [26]. Thus, multi-scale porous scaffolds with macro-channels to enhance cellular infiltration and vascularization were prepared by photopatterning and electrospinning of hyaluronic acid (HA) [27]. For the fabrication of precisely sized and spaced pores into PLA electrospun scaffolds, authors reported on the usage of excimer laser micro-machining. Scaffolds with the biggest micro-scale pores showed highest values of human adipose-derived stem cell (hASC) DNA values, after 7 days of seeding [28]. Similarly, micropatterned scaffolds from a copolymer of lactide and PCL were fabricated using direct laser writing technique and electrospinning. The scaffold construct comprised an electrospun nanofibrous layer, laser cut into different patterns, on which a new electrospun layer was deposited (to stand as a base), thus forming a multiscale structure. As prepared scaffolds were reported to not only support bone marrow derived human mesenchymal stem cells growth, but also support alignment and differentiation into cardiac myogenic lineage [29].

Another approach reported was the combination of the technique of 3D printing and electrospinning. Thus, a 3D composite scaffold was prepared by filling the mesh of a 3D printed PCL construct with electrospun PCL/gelatin, previously cut and dispersed to discontinuous nanofibers. Apart from good mechanical integrity, the scaffolds showed good cells migration, proliferation and were infiltrated into the micro pores of the electrospun scaffold.These scaffolds have the potential to be used in bone tissues repair [30].

Generally, rapid prototyping is combined with electrospinning as to provide: controlled fiber deposition or patterned structure design and small scaled fibers production. As so, more advanced approach was reported with the development of an electrospinning-based rapid prototyping machine. The machine could fabricate stacks of patterned layers (both woven and non-woven configurations) with control of the fiber deposition to a certain extent. The lay-down patterns reported were 0°/90°, 45°/135° and 45° twists with max of 45 layers. The authors also reported on the problem concerning thickness control which is related to fibers solidification prior deposition [31]. In a study of stable jet electrospinning, which provides jet bending instability elimination, combined with X-Y translation stage, fibers deposition was performed into predetermined patterns. 3D-patterned PLLA fibrous scaffolds were fabricated with patterns of 0°/90°and 0°/45°. The study confirmed advantageous scaffolds architecture, in regard to improved human umbilical arterial smooth muscle cells penetration and proliferation, when compared to conventional electrospun scaffold [32].

Although many studies have offered a solution to fabricate 3D scaffold architectures in order to eliminate the problem with the intrinsically two-dimensional scaffold structure, there is still a lack of systematic studies analysing cells in-depth penetration within the developed 3D structures. Also, no attention is given to the possible differences, in the pace of migration to the lower layers of the scaffolds, between cultured animal and human cells.

3. Project Proposal - COMBOELECTROSPUN

To fulfil both physical and chemical aspect requirements in the design of the scaffold, an innovative solution is suggested based on the combination of both solution and melt electrospinning, with additional functionalization of the target electrospun scaffold layers.

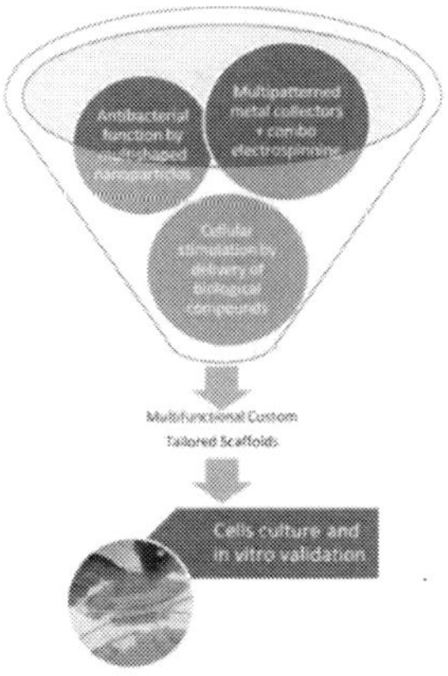

Fig.1 Project objectives scheme

Thus, from a physical point of view, the proposed project objectives cover the design of target scaffold topography based on custom made collectors to enhance cells adhesion and spreading.

The chemical aspect covers two objectives: 1) addition of multi shaped (sized) metal oxide nanoparticles fillers to provide antibacterial properties; 2) addition of cells growth factors to improve cellular vitality processes. Both fillers will be released in a controlled manner, thus in a compliance with tissues repair. The final goal is to fabricate a multifunctional scaffold with optimal architecture, antibacterial feature and cellular processes stimulating property which is based on the alternating melt and solution electrospun nanofibrous layers. The summary of the project proposed objectives is given in Fig.1.

As designed scaffolds, are used for the cultivation of human, skin and ocular cells including: fibroblasts and keratinocytes, as well as limbal stem cells, respectively. The choice of cells is based on the need for these types of tissue repair. Each year more than 300 000 patients die due to high level burns among 6 millions of them being hospitalized [33]. Further, global health issues reported, include visually impaired people estimated to 285 million, of whom 39 million are blind, WHO [34]. Additionally limbal stem cells, situated in the limbus are able to renew themselves and give new cells that differentiate progressively, thus can provide corneal epithelial tissue repair [35, 36].

4. Conclusion

Although there are many new contributions in recent years concerning the problem of electrospun scaffolds 2D configuration, the challenge to further deal with cells in-depth penetration and total tissue growth still remains. On one hand, there are new advanced techniques which can fabricate scaffolds constructs in a fully controlled manner and according to desired design. On the other hand, the advantageous side of electrospinning, that is the nanoscale fibers, is still somehow irreplaceable. This still keeps this technique on the thrown among other nanofibrous fabrication techniques and in combination with e.g. rapid prototyping techniques. Finally, this project proposes the usage of sole electrospinning but, combining the solution electrospinning with the not as explored one, the melt electrospinning technique. Melt electrospinning can provide micro scaled fibers fabrication, while solution electrospinning nano scaled fibers. Thus, multi-scaled architectures can be designed by layer stacking. In this way cells will be provided with a 3D scaffold architecture comprising macro pores for in depth penetration and nanofibers important for permeability to oxygen and nutrients. Additionally, each layer can be functionalized with the introduction of a filler, or agent in separate layers, depending on their nature (hydrophobic or hydrophilic).

5. Acknowledgement

This work has been fully supported by Croatian Science Foundation under the project IP-2016-06-6878, Custom Tailored Fibrous Scaffold Prototype for Tissue Cells Culture via Combined Electrospinning, COMBOELECTROSPUN.

References

[1] Traversa E, Mecheri B, Mandoli C, Solimanet S, et al. Tuning hierarchical architecture of 3D polymeric scaffolds for cardiac tissue engineering. J Exp Nanosci: 2008; 3: 97-110.
[2] Zhu N, Chen X. Biofabrication of Tissue Scaffolds. In: Pignatello R, editor. Advances in Biomaterials Science and Biomedical Applications. InTech: 2013; p. 568.
[3] Mijovic B, Tominac Trcin M, Agic A, Zdraveva E, et al. Study on Cell Adhesion Detection onto Biodegradable Electrospun PCL Scaffolds. JFBI 2012; 5: 33-40.
[4] Tominac Trcin M, Dekaris I, Mijovic B, Bujic M, et al. Viability of Human Limbal Epithelial Cell Cultured on Different Types of Scaffolds. XXV Annual Meeting of the European Eye Bank Association 2013. Zagreb, Croatia.
[5] Zdraveva E, Fang J, Mijovic B, Lin T. Electrospun nanofibers. In: Gajanan B, editor. Structure and properties of high performance fibers. UK: Woodhead Publishing: 2016; p. 267-300.
[6] Stevens MM, George JH. Exploring and Engineering the Cell Surface Interface. Mater Biol: 2005; 310: 1135-1138.
[7] He JH, Liu Y, Mo LF, Wan YQ et al. Electrospun Nanofibres and Their Applications, UK: iSmithers: 2008; p. 6-7.
[8] Khanam N, Mikoryak C, Draper RK, Balkus KJJr, et al. Electrospun linear polyethyleneimine scaffolds for cell growth. Acta Biomater: 2007; 3: 1050-1059.
[9] Shabani I, Haddadi-Asl V, Seyedjafari E, Soleimani M. Cellular infiltration on nanofibrous scaffolds using a modified electrospinning technique. Biochem Biophys Res Commun: 2012; 423: 50-4.
[10] Bye FJ, Bissoli J, Black L, Bullock JA, et al. Development of bilayer and trilayer nanofibrous/microfibrous scaffolds for regenerative medicine. Biomater Sci: 2013; 1: 942-951.
[11] Sefataet F, Deshpande P, Ramachandran C, et al. Production, sterilisation and storage of biodegradable electrospun PLGA membranes for delivery of limbal stem cells to the cornea, 3rd International Conference on Tissue Engineering, ICTE 2013, Procedia Engineering: 2013; 59: 101-116.

[12] Chen H, Fan X, Xia J, Chen P, et al. Electrospun chitosan-graft-poly (ε-caprolactone)/poly (ε-caprolactone) nanofibrous scaffolds for retinal tissue engineering. Int J Nanomedicine: 2011; 6: 453-461.
[13] Lin L, Perets A, Har-el YE, Varma D, et al. Alimentary 'green' proteins as electrospun scaffolds for skin regenerative engineering. J Tissue Eng Regen Med: 2012; 7: 994-1008.
[14] Jose RR, Elia R, Firpo MA, Kaplan DL, et al. Seamless, axially aligned, fiber tubes, meshes, microbundles and gradient biomaterial constructs. J Mater Sci Mater Med: 2012; 23: 2679-2695.
[15] Wingate K, Bonani W, Tan Y, Bryant SJ, et al. Compressive elasticity of three-dimensional nanofiber matrix directs mesenchymal stem cell differentiation to vascular cells with endothelial or smooth muscle cell markers. Acta Biomater: 2012; 8: 1440-1449.
[16] Meng L, Arnoult O, Smithc M, Wnek GE. Electrospinning of in situ crosslinked collagen nanofibers. J Mater Chem: 2012; 22: 19412-19417.
[17] McClure MJ, Simpson DG, Bowlin GL. Tri-layered vascular grafts composed of polycaprolactone, elastin, collagen, and silk: Optimization of graft properties. J Mech Behav Biomed Mater: 2012; 10: 48-61.
[18] Shah S, Fan X, Xia J, Chen P et al. Guiding Stem Cell Differentiation into Oligodendrocytes Using Graphene-Nanofiber Hybrid Scaffolds. Adv Mater: 2014; 26: 3673-3680.
[19] Kurpinski KT, Stephenson JT, Janairo RR, Lee H, et al. The effect of fiber alignment and heparin coating on cell infiltration into nanofibrous PLLA scaffolds. Biomater: 2010; 31: 3536-3542.
[20] Jeong L, Kim MH, Jung JY, Min BM, et al. Effect of silk fibroin nanofibers containing silver sulfadiazine on wound healing. Int J Nanomed: 2014; 14: 5277-5287.
[21] Tijing LD, Amarjargal A, Jiang Z. Antibacterial tourmaline nanoparticles/polyurethane hybrid mat decorated with silver nanoparticles prepared by electrospinning and UV photoreduction. Curr Appl Phys: 2013; 13: 205-210.
[22] Baranowska-Korczyc A, Warowicka A, Jasiurkowska-Delaporte M, Grześkowiak B, et al. Antimicrobial electrospun poly(ε-caprolactone) scaffolds for gingival fibroblast growth. RSC Adv: 2016; 6: 19647-19656.
[23] Gao Y, Truong JB, Zhu Y, Kyratzis IL, et al. Electrospun antibacterial nanofibers: production, activity, and in vivo applications. J Appl Polym Sci: 2014; 131: 40797.
[24] Filová E, Rampichová M, Litvinec A, Držík M, et al. A cell-free nanofiber composite scaffold regenerated osteochondral defects in miniature pigs. Int J Pharm: 2013; 44: 139-149.
[25] Chen F, Wan H, Xia T, Guo X, et al. Promoted regeneration of mature blood vessels by electrospun fibers with loaded multiple pDNA-calcium phosphate nanoparticles. Eur J Pharm Biopharm: 2013; 85: 699-710.
[26] Bryant SJ, Cuy JL, Hauch KD, Ratner BD. Photo-patterning of Porous Hydrogels for Tissue Engineering. Biomater: 2007; 28: 2978.
[27] Sundararaghavan HG, Metter RB, Burdick JA. Electrospun Fibrous Scaffolds with Multiscale and Photopatterned Porosity. Macromol Biosci: 2010; 10: 265-270.
[28] Mccullen SD, Gittard SD, Miller PR, Pourdeyhimi B, et al. Laser Ablation Imparts Controlled Micro-Scale Pores in Electrospun Scaffolds for Tissue Engineering Applications. Ann Biomed Eng: 2011; 39: 3021-3030.
[29] Huaqiong Li, Wong YS, Wen F, Ng KW, et al. Human Mesenchymal Stem-Cell Behaviour On Direct Laser Micropatterned Electrospun Scaffolds with Hierarchical Structures. Macromol Biosci: 2013; 13: 299-310.
[30] Yu Y, Hua S, Yang M, Fu Z, et al. Fabrication and Characterization of Electrospinning/3D Printing Bone Tissue Engineering Scaffold. RSC Adv: 2016 DOI: 10.1039/C6RA17718B.
[31] Chanthakulchan A, Koomsap P, Auyson K. Development of an electrospinning-based rapid prototyping for scaffold fabrication. Rapid Prototyping J: 2015; 21: 329-339.
[32] Yuan H, Zhou Q, Li B, Bao M, et al. Direct printing of patterned three-dimensional ultra fine fibrous scaffolds by stable jet electrospinning for cellular ingrowth. Biofabrication: 2015; 4: 5004.
[33] Torpy JM, Lynm C, Glass RM. Burn injuries. J Am Med Assoc: 2009; 302: 1828.
[34] Metcalfe AD, Ferguson MWJ. Bioengineering skin using mechanisms of regeneration and repair. Biomater: 2007; 28: 5100-5113.
[35] Information at: http://www.who.int/mediacentre/factsheets/fs282/en/
[36] Meek KM, The Cornea and Sclera in Collagen, In: Fratzl P, editor. Structure and Mechanics, US: Springer: 2008; p. 359-396.

Ion Exchange Materials with Antimicrobial Properties

Jaromír Marek[1]

[1]*Technical University of Liberec, Studentská 2/1402, Liberec, 461 17, Czech Republic*

*Corresponding author's email: jaromir.marek@tul.cz

Abstract

Polymers, biopolymers and resins containing quarternary ammonium groups exhibit antibacterial and antifungal properties. Strong anion exchange resins in the granular form are mostly functionalized with the quarternary ammonium salts. It was approved that such materials also exhibit antimicrobial properties especially when doped with iodide salts. Heterogeneous ion exchange membranes tend to degrade the quarternary ammonium groups (by Hoffman elimination) while being prepared by milling strong base ion exchange resin and mixing with inert binder by high temperature. Thus, it creates a weak cation exchange material. Anion exchange membranes doped with silver are also described. This work compares antimicrobial properties of anion exchange membranes doped and non-doped as well as the durability of the antimicrobial properties along with resistance against fouling in the water environment and resistance against bacterial growth in the air conditioning systems ("enthalpy exchangers").

Keywords: Ion Exchange; Antibacterial; Antimicrobial; Membranes; Strong Base Exchangers; Fouling

1. Introduction

Current water treatment technologies and so called "enthalpy exchangers" representing state-of-the-art in the air conditioning systems utilizes ion exchange based membranes [1]. The antimicrobial properties can have a very high impact on their effectivity, durability and industrial application. The main goal of this study is to avoid the growth of bacteria and fungi of ion exchange membranes used for enthalpy exchangers. This contamination can be easily carried by the air stream into the indoor environment which can cause breathing and allergy problems (including asthma etc.). The membranes should also exhibit antimicrobial properties because of their durability in the system. Destroyed membrane (i.e. fouled and holey - porous membranes) will not have any selectivity for the dissociated ions in the water treatment and dissolved gasses in the air conditioning systems. In the air conditioning systems, these holes will result in the mixing of the exhaled gasses (i.e. discharged air stream) with inhaled streams – resulting in nonfunctional air conditioning. Furthermore, fouled membranes in the water treatment technologies provides rapidly lower efficiency or no efficiency at all.

2. Microbiological Characterization of Ion Exchange Materials

The ion exchange membranes suffer from fouling similar to other types of water treatment membranes (e.g. reverse osmosis membranes etc.). The fouling includes inorganic and organic matter including micro-organisms. The growth of micro-organisms on the surface of membranes is the objective of this work. Ion exchange membranes include both heterogeneous and homogeneous membranes. Due to the charge of the membranes we can recognize cation and anion exchange membranes. This work is focused on heterogeneous anion exchange membranes. The reason is that heterogeneous anion exchange membranes are being prepared from the strong anion exchange granular resin. This resin contains quarternary ammonium groups which are widely known for their antimicrobial properties.

Regarding preparation of membranes from the material with antimicrobial properties it can be expected that anion exchange type of membranes should exhibit these features. Despite these facts it does not happen – neither cation nor anion exchange membranes escape from the fouling. This work makes basic overview of the behavior of the ion exchange materials and their treatment by preparation of the heterogeneous membranes according to their antimicrobial properties and it should reveal the weak side of technological processes and describe possible solutions.

2.1 Ion Exchange Materials Doped with Antimicrobial Agents

In the past it was observed that materials consisting of ion exchange materials doped with iodide [2] or silver [3] clearly exhibited antimicrobial properties. However results regarding our research shows there is not any necessity for additional functionalization of ion exchange material to enhance their antimicrobial properties.

2.2 Anion Exchange Material Itself for Disinfection

In this work it is analysed whether the ion exchange material itself exhibits antibacterial properties. Preparation of anion exchange heterogeneous membranes and microbial characterization of used material reveals many important facts not related only with the anti-fouling process.

First step of preparation of heterogeneous anion exchange membranes is milling the anion exchange granular resins. Commonly strong base resin containing quarternary ammonium groups is being used. As mentioned before, these groups exhibit antimicrobial properties, so it is obvious that the powder should exhibit such properties also.

The above mentioned idea was characterized by the Cornell test. During the first 120 minutes the anion exchange milled material exhibited antimicrobial properties resulting in decreasing number of colonies of St.Gallinarum.

Table.1 Cornell test of anion exchange materials

Material	Contact time	Concentration of test bacteria	Aditional dilution	CPU Number
Control$_0$	T_0	10^4 S.g./ml		394
	T30	10^4 S.g./ml		362
	T60	10^4 S.g./ml		275
	T120	10^4 S.g./ml		335
	T240	10^4 S.g./ml		301
Control$_1$	T30	10^4 S.g./ml		> 2.10^3
	T30	10^4 S.g./ml	1:20	308
	T60	10^4 S.g./ml		> 2.10^3
	T60	10^4 S.g./ml	1:20	282
	T120	10^4 S.g./ml		811
	T120	10^4 S.g./ml	1:20	49
	T240	10^4 S.g./ml		732
	T240	10^4 S.g./ml	1:20	35
Sample "a"	T30	10^4 S.g./ml		Contamination
	T30	10^4 S.g./ml	1:20	
	T60	10^4 S.g./ml		
	T60	10^4 S.g./ml	1:20	
	T120	10^4 S.g./ml		
	T120	10^4 S.g./ml	1:20	
	T240	10^4 S.g./ml		
	T240	10^4 S.g./ml	1:20	
Sample "b"	T30	10^4 S.g./ml		Contamination
	T30	10^4 S.g./ml	1:20	
	T60	10^4 S.g./ml		
	T60	10^4 S.g./ml	1:20	
	T120	10^4 S.g./ml		
	T120	10^4 S.g./ml	1:20	
	T240	10^4 S.g./ml		
	T240	10^4 S.g./ml	1:20	

Legend:

Control$_0$	bacterial suspension only
Control$_1$	Cation exchange membrane
Sample "a"	Anion exchange membrane
Sample "b"	duplicate of sample "a"
Contact time	contact time of material with bacterial suspencion
CPU	Colony producing Units / ml
Contamination	very high proportion of contaminating bacteria
Test bacteria	Staphylococcus gallinarum CCM 3572

On the other hand anion exchange membrane did not show any antimicrobial properties during first 2 hrs. This result would be explained as degradation of quarternary group during melting anion exchange powder with inert binder and supporting polyester textile at 130°C. Due to this fact the membranes were analysed by infrared spectroscopy. It should have approved the lack of the quarternary ammonium groups in the membrane. The results were surprising, as in both – the anion exchange powder and ion exchange membrane, quarternary ammonium group was detected. The stretching vibration of N-C at 3000 cm^{-1}, and then deformation vibration of $N\text{-}(CH_3)_3^+$ at 1490, 1419 and 975 cm^{-1} proves the presence of this strong basic functional group.

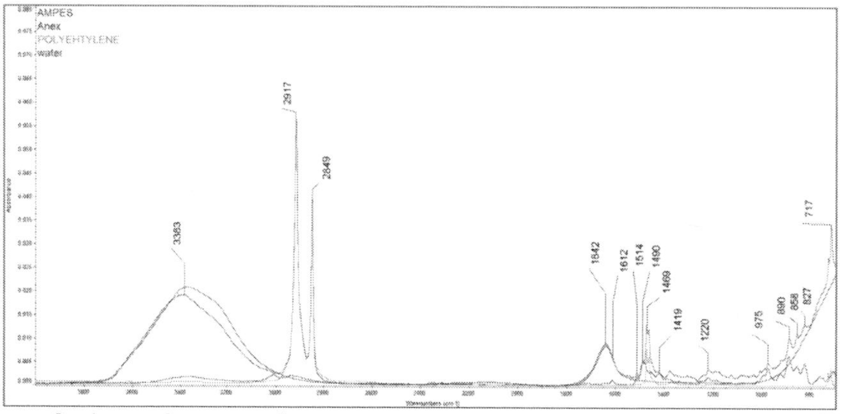

Fig.1 IR spectrum of anion exchange membrane (AMPES = anion exchange membrane, Anex = anion exchange powder, polyethylene = inert binder in membrane)

Nevertheless, during the next few hours of Cornell test the anion exchange powder was also contaminated as the membrane. This was next unexpected fact showing that presence of this quarternary ammonium salt e.g. strong base group does not necessarily mean the antimicrobial properties.

2.3 Cation Exchange Material

Fig.2 Cornell test of anion and cation exchange membranes. Legend: T0 – T240 = time 0 min – time 240 min; Blank 0 = bacterial suspension only, Blank 1 = cation exchange membrane; Vz.A,B = anion and duplicate of anion exchange membrane; 10^4 = number of colony of Staphilococus gallinarium; 10^4 zr= dilluted 1:20.

From the Cornell test it is also clearly visible that the cation exchange powder and cation exchange membrane both exhibit antimicrobial properties (Fig.2). This next surprising fact could be easily explained by the presence of strongly acidic sulphonic groups. After 120 minutes, there were 49 colonies and after 240 minutes 35 colonies covering the surface of membrane. This result shows strong antibacterial features of strongly acidic cation exchange membranes.

3. Conclusion

The simplicity of this idea promises fast and effective application of the results in industry – the latest technology of air conditioning so called "enthalpy exchanger" which transfers the heat together with humidity for healthy and comfortable living [4] already utilize ion exchange membranes. This way avoids drying of the air – which is important because drying of the air is supporting allergies and breathing problems. The enthalpy exchanger utilizes ion exchange membranes based on pretty complicated pentablock copolymer. The results of this test show the possibility and effectivity of using standard cation exchange membranes providing also antimicrobial properties – so important for the air conditioning to avoid bacterial and mold and fungus growth in the system and breathing their spores.

Despite of the expectations that strongly basic anion exchange materials should exhibit the antibacterial features due to presence of quarternary ammonium functional group it was observed that the bacterial colony grows easily on them. It was also expected that heterogeneous anion exchange membranes prepared from the strongly alkaline anion exchange granules will suffer from the heat treatment followed by the loss of the desirable group (quarternary ammonium) by the Hoffman elimination. In contrast, the membranes analyzed by infrared spectroscopy proved the presence of such group while also did not exhibit the antimicrobial properties similarly as the raw material.

Surprisingly the most efficient antimicrobial properties are exhibited by strongly acidic cation exchange materials – both the granules and the membranes.

This fact can effectively be used in the future technologies including air conditioning and water treatment when utilizing only one type of membranes – i.e. enthalpy exchangers and shock electrodialysis [5].

4. Acknowledgement

The results of this project LO1201 were obtained with co-funding from the Ministry of Education, Youth and Sports as part of targeted support from the "Národní program udržitelnosti I" programme.

References

[1] Kruzer B, De Oude C. Process of the preparation of an aqueous emulsion of a midblock sulfonated copolymer. World patent WO2015/153750.

[2] Polievkova E, Hubalek J, Krivcik J, Drbohlavova J. Ion exchange membrane with antimicrobial effect. In: EEA Grants: https://www.ceitec.cz/e/f32955

[3] Taylor SL, Fina LR, Lambert JL. New Water Disinfectant: an Insoluble Quaternary Ammonium Resin-Triiodide Combination that Releases Bactericide on Demand: App. And Environm. Microbiology, American Society for microbiology: November 1970; vol. 20 no. 5: 720-722

[4] Vu NV, Dvorak V. Evaluating the moisture transfer of membranes for total heat exchange. EPJ Web of conferences: 02080 (2017) 714302080 143.

[5] Nam S, Cho I, Heo J, Lim G, Bazant MZ, Sung G, Kim SJ. Experimental Verification of Overlimiting Current by Surface Conduction and Electro-osmotic Flow in Microchannels: 2014.

Zwitterionic Poly (Cysteine Methacrylate) Functional Polysulfone Via ATRP and Their Anti-protein Fouling Property

Ling Yang, Si-Tian Ma, Huan Xia, Yi-Ming Bu, Jing-Jing Huang, Shao-Jin Gu*

State Key Laboratory Breeding Base of New Textile Materials and Advanced Processing Technologies, School of Materials Science and Engineering, Wuhan Textile University, Wuhan, 430200, P. R. China

*Corresponding author's email: gushaojin@hotmail.com

Abstract

Zwitterionic polymer brushes based on cysteine methacrylate (CysMA) were successfully grafted to polysulfone substrates by surface-initiated atomic-transfer radical polymerization (SI-ATRP), and exhibited excellent anti-fouling activities. Various characterization techniques, including attenuated total reflectance Fourier transform infrared spectra (ATR-FTIR), atomic force microscopy (AFM), and X-ray photoelectron spectroscopy (XPS) were used to characterize the polymer brush-modified polysulfone membrane. Subsequently, the hydrophilicity and anti-biofouling activities of the polymer brush substrates were evaluated. The experimental results showed that the pCysMA displays better hydrophilicity and effectively resisted the adsorption of bovine serum albumin (BSA) protein.

Keywords: Polysulfone; Cysteine; ATRP; Protein Adsorption

1. Introduction

Polysulfone (PSU) has been used in biomedical field for artificial organs, medical devices, and disposable clinical apparatus [1, 2], due to its excellent mechanical strength, chemical resistance and thermal stability. However, the blood compatibility of PSU membrane is inadequate and needs to be modified before use.

Polymers containing zwitterionic structural units have been used for a wide range of biomedical and engineering applications. Such surface coatings are highly resistant to nonspecific protein adsorption, bacterial adhesion, and biofilm formation. Recently, amino acids, as a natural zwitterion, have shown good anti-biofouling performance [3-5]. Rosen et al. functionalized silica nanoparticles with cysteine in order to create a low-fouling surface for nano medicine applications [6]. Shen et al. used self-polymerized dopamine to form a thin and surface-adherent polydopamine layer onto poly (ethylene terephthalate) (PET) sheet, followed by covalent grafting cysteine (Cys) to improve hemocompatibility and anti-biofouling property [7]. Shi et al. immobilized lysine onto the surface of the polyacrylonitirile porous filtration membrane and showed reduced bovine serum albuminand lysozyme adsorption. Liu et al. prepared zwitterionic poly (serine methacrylate) (PSerMA) brushes on a planar gold substrate using surface-initiated photo iniferter-mediated polymerization (SI-PIMP) for evaluation as a potential antibiofouling material [8]. Very recently, Armes et al. grafted the Poly (cysteine methacrylate) (PCysMA) brushes into the surface of silicon wafers by atom-transfer radical polymerization [9]. The obtained the film strongly resist cell adhesion when immersed in culture media in the presence of HDF cells.

To the best of our knowledge, there are no reports that grow poly (amino acid) brush on polysulfone membrane surface. Our interest in developing synthetic methods for the preparation and modification of polymer materials [10-14], is part of our continuous research. In the present study, we modified the PSU membrane with a cysteine-based zwitterionic poly (cysteine methacrylate) via ATRP method as shown in Scheme 1. The anti-fouling performances of modified membranes were tested by bovine serum albumin (BSA) protein absorption.

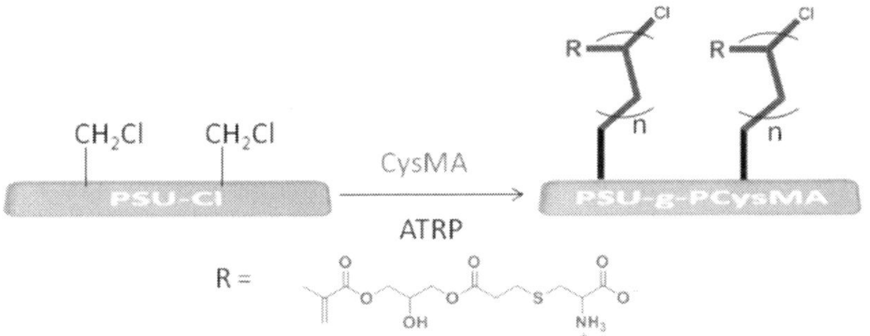

Scheme.1 Grafted zwitterionic poly (amino acid) brushes with anti-fouling properties.

2. Experimental Procedure

2.1 Materials

Polysulfone (PSU Udel P1700LCD) was purchased from Solvay Advanced Polymers.2, 2'-dipyridyl, and CuBr were purchased from Sinopharm (Shanghai, China).Cysteine and 3-(acryloyloxy)-2-hydroxypropyl methacrylate were purchased from Aladdin Reagent (Shanghai) Co., Ltd. and used without purification. Chloromethylation of PSU was performed according to the reported methods [2]. Cysteine Methacrylate Monomer (CysMA) was synthesized according to the published work [9].

2.2 Grafting of pCysMA Brushes from PSU Film

Cysteine methacrylate (CysMA) monomer (1.5g) was dissolved with 12 mL deionized water and stirred for 15 minutes, then 2,2'-dipyridyl (0.070 g) was added into the mixture. After being stirred for 15 min under a nitrogen atmosphere, the mixture was degassed by three freeze–pump–thaw cycles. In the frozen state, PSU-Cl membrane (2*2 cm) and CuBr (0.032 g) were placed into the glass tube. The tube was then subjected to three additional freeze–pump–thaw cycles before being kept at 60 ^{o}C. After a predetermined period of time, the modified membrane was removed from the reaction system. After being thoroughly washed with methanol and water, the membrane was dried to constant weight under vacuum.

2.3 Characterization

The chemical structures of PSU-Cl, and PSU-g-PCysMA were characterized by Fourier Transform Infrared-Attenuated Total Reflectance (FTIR-ATR, Tensor 27, Bruker). X-ray photoelectron spectrometer (XPS) with the Aluminum Kα as X-ray optical source (15 kV, 10 mA) and an initiation angle of 90°. Roughness Analysis of membranes was measured by atomic force microscopy (AFM). Contact angle measurements of a drop of water (3 uL) on the membranes were carried out using the sessile drop method with a CAM 100 KSV (KSV, Finland). Recording the drop profile with a CCD camera allowed monitoring changes in contact angle.SEM measurements were carried out on a SIRION 200 field emission scanning electron microscope.

2.4 Protein Adhesion Study

Protein solutions with concentrations of 0.2, 0.4, 0.6, 0.8, and 1.0 mg/ml were freshly prepared by dissolving bovine serum albumin (BSA) into 0.1M phosphate buffer solution (PBS, pH = 7.4). The membrane samples with the size of 1 cm×1 cm were placed in PH buffer solution (pH= 7.4) for 12h. After then, the membranes were put into vials filled with 5 mL protein solution respectively. The vials were maintained at 37 ^{o}C for 13h with interval waving to reach an adsorption–desorption equilibrium. Next, equilibrium concentrations of BSA (C) were measured by UV-vis spectrophotometer at 280 nm, and the adsorbed mass of protein per unit volume of membrane

(Q μg/cm^2) was calculated by the equation: $Q=(C_0-C)/S$, where C and C_0 are the equilibrium and initial concentrations, respectively, S is the area of the membrane.

3. Results and Discussion

In this work, the surface grafting polymerization of CysMA from the PSU-Cl membrane was carried out by atom transfer radical polymerization method. The PSU-g-pCysMA were characterized by ATR-FTIR, as shown in Fig.1. The presence of the pCysMA could be ascertained from the ester groups observed from the band of the O-C=O stretch at 1729 cm^{-1}.

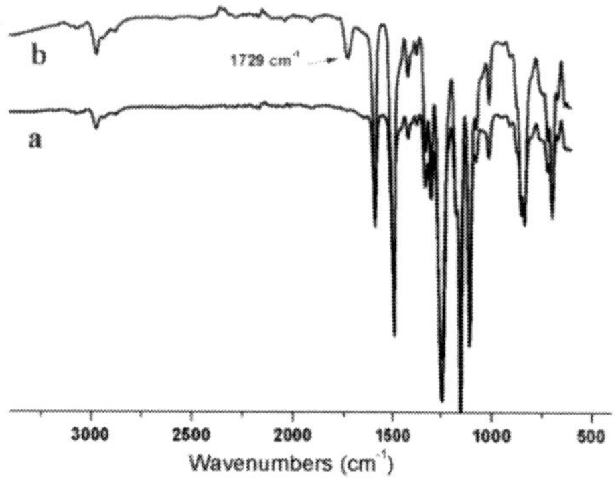

Fig.1 ATR-FTIR spectrum of the reaction process of SI-ATRP reaction.

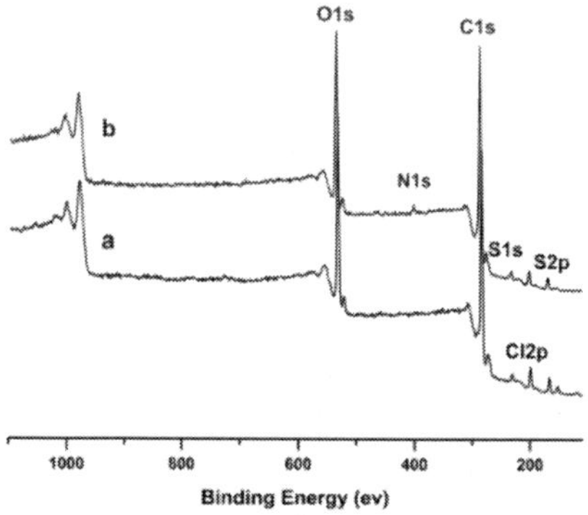

Fig.2 The full XPS survey spectra of PSU-Cl membrane (a), and PSU-g-pCysMA (b)

3.1 XPS Analysis of Membrane Surface

The chemical compositions of the PSU film surfaces at various stages of surface modification were determined by X-ray photoelectron spectroscopy (XPS). Fig.2 illustrates the full XPS survey spectra of PSU-Cl membrane, and modified PSU membrane. As shown in Fig.2(b), after 5 h of SI-ATRP of the monomer cysteine methacrylate monomer, the PSU-g-pCysMA exhibited an N1s signal at 404 eV, which was not detected with PSU-Cl membrane, confirming the successful grafting of the Poly (cysteine methacrylate) (PCysMA) brushes.

3.2 AFM Analysis

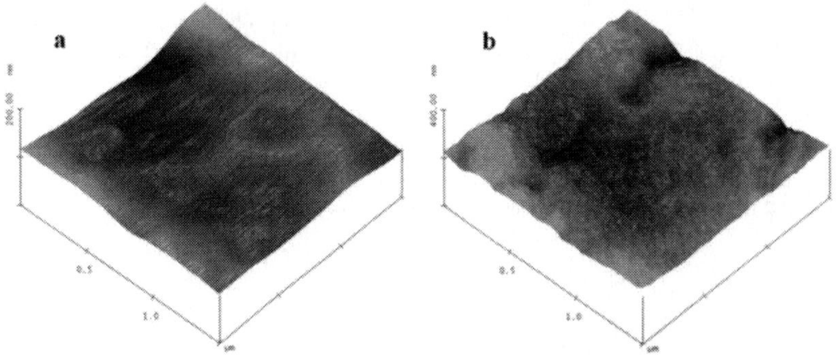

Fig.3 AFM three-dimensional images of PSU-Cl(a) and PSU-g-PDEPAS(b) membranes

The topography of surfaces modified by Poly (cysteine methacrylate) was examined using tapping AFM. The AFM topographic images in Fig.3 show the morphology of the surfaces modified by the initiators, poly (cysteine methacrylate) brush. Compared to the initiator-modified PSU-Cl membrane, bare silicon wafers appeared very smooth (Fig.3(a)), and after polymer modification (Fig.3(b)), the root mean square roughness values of PCysMA-PSU increased to 1.486 nm and the entire surface was uniformly covered by polymer brushes.

3.3 Hydrophilicity

Hydrophilicity of the modified membranes was characterized by examining the water contact angle of the membranes. After the successful functionalization of the PSU-Cl membrane the static water contact angles of the different surfaces were measured. In Fig.4 the static water contact angle of pCysMA modified PSU-Cl membrane is seen to be 56°, which is well below that of PSU membrane (106 °) or PSU-Cl membrane (101°). The zwitterionic polymer brushes of pCysMA clearly exhibited good water affinity because of the introduction of zwitterionic cysteine groups.

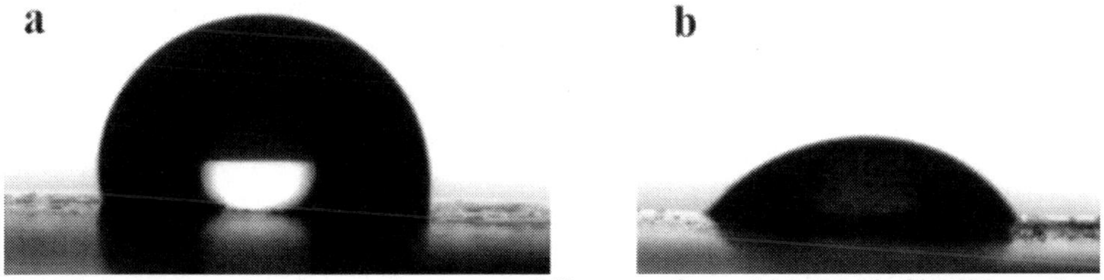

Fig.4 Digital photographs of water droplets on different surfaces: (a) PSU-Cl, (b) PSU-g- pCysMA

3.4 Protein Adsorption

Protein adsorption and fouling are closely related to surface wettability and charge. The protein resistance of different surfaces was tested by exposure to BSA in PBS buffer (pH 7.4). As shown in Figure 7, the BSA adsorption amount for the PSU-Cl membrane was higher (about 78μg/cm^2) than that for the sulfobetaine functionalization membranes PSU-g-pCysMA (about 25μg/cm^2), which indicates the sulfobetaine functionalization PSU membrane display excellent nonspecific protein antifouling property.

4. Conclusion

Zwitterionic cysteine surface was successfully fabricated through surface-initiated atomic-transfer radical polymerization on a polysulfone substrate. The obtained zwitterionic poly (cysteine methacrylate) functional polysulfone displayed increased hydrophillicity properties and nonspecific protein antifouling property.

5. Acknowledgement

This work was financially supported by National Natural Science Foundation of China (21202127), Educational Commission of Hubei Province (Q20151606), and Foundation of Wuhan Textile University.

References

[1] Ishihara K, Fukumoto K, Iwasaki Y, Nakabayashi N. Modification of polysulfone with phospholipid polymer for improvement of the blood compatibility. Part 1. Surface characterization. Biomaterials: 1999; 20: 1545-51.

[2] Yilmaz G, Toiserkani H, Demirkol DO, Sakarya S, Timur S, Yagci Y, Torun L. Modification of Polysulfones by Click Chemistry: Amphiphilic Graft Copolymers and Their Protein Adsorption and Cell Adhesion Properties. J Polym Sci Pol Chem: 2011; 49: 110-7.

[3] Liu Q, Li W, Singh A, Cheng G, Liu L. Two amino acid-based superlow fouling polymers: Poly (lysine methacrylamide) and poly(ornithine methacrylamide). Acta Biomater: 2014; 10: 2956-64.

[4] Zhi X, Li P, Gan X, Zhang W, Shen T, Yuan J, Shen J. Hemocompatibility and anti-biofouling property

improvement of poly (ethylene terephthalate) via self-polymerization of dopamine and covalent graft of lysine. J Biomat Sci-Polym E: 2014; 25: 1619-28.
[5] Cui J, Ju Y, Liang K, Ejima H, Loercher S, Gause KT, Richardson JJ, Caruso F. Nanoscale engineering of low-fouling surfaces through polydopamine immobilisation of zwitterionic peptides. Soft Matter: 2014; 10: 2656-63.
[6] Rosen JE, Gu FX. Surface functionalization of silica nanoparticles with cysteine: a low-fouling zwitterionic surface. Langmuir: 2011; 27: 10507-13.
[7] Li P, Cai X, Wang D, Chen S, Yuan J, Li L, Shen J. Hemocompatibility and anti-biofouling property improvement of poly (ethylene terephthalate) via self-polymerization of dopamine and covalent graft of zwitterionic cysteine. Colloid Surfaces B: 2013; 110: 327-332.
[8] Liu Q, Singh A, Liu L. Amino Acid-Based Zwitterionic Poly (serine methacrylate) as an Antifouling Material. Biomacromolecules: 2013; 14: 226-231.
[9] Alswieleh AM, Cheng N, Canton I, Ustbas B, Xue X, Ladmiral V, Xia S, Ducker RE, El Zubir O, Cartron ML, Hunter CN, Leggett GJ, Armes SP. Zwitterionic Poly(amino acid methacrylate) Brushes. J Am Chem Soc: 2014; 136: 9404-9413.
[10] Huang W, Huang J, Xu C, Gu S, Xu W. Surface functionalization of cellulose membrane via heterogeneous "click" grafting of zwitterionic sulfobetaine. Poly Bull: 2014; 71: 2559-2569.
[11] Huang J, Xu W. Zwitterionic monomer graft copolymerization onto polyurethane surface through a PEG spacer. Appl Surf Sci: 2010; 256: 3921-3927.
[12] Huang J, Gu S, Zhang R, Xia Z, Xu W. Synthesis, spectroscopic, and thermal properties of polyurethanes containing zwitterionic sulfobetaine groups. J Therm Anal Calorim: 2013; 112: 1289-95.
[13] Huang J, Wang D, Lu Y, Li M, Xu W. Surface zwitterionically functionalized PVA-co-PE nanofiber materials by click chemistry. Rsc Advances: 2013; 3: 20922-9.
[14] Xu C, Gu S, Huang J, Xu W, Xia H, Du J, Liu H, Yang H, Zhou Y, Bai Z. Modification of Polysulfones by Click Chemistry: Zwitterionic Graft Complex and Their Antiprotein Fouling Property. J Appl Polym Sci: 2015; 132: 41327.

Application of Knitting Structure in Implantable Medical Textiles

Xiao-Hui Zhang[1], Pi-Bo Ma[2*]

[1,2]*Name of institution of the first author, Wuxi, 214122, China*

[*]Corresponding author's email: mapibo@jiangnan.edu.cn

Abstract

Knitting structures, including weft knitting structure and warp knitting structure, are popular for the looser structure, higher porosity, more flexibility and better forming technology. They are greatly applied in high-tech medical textiles, containing artificial blood vessel, hernia patch, and cardiac support device. This present article will introduce some knitting structures and materials applied in medical textiles in accordance to implantable textiles.

Keywords: Implantable Medical Textiles; Knitting Structure; Warp-knitted Fabric; Weft-knitted Fabric

1. Introduction

Medical textiles are high-added value textile structures and products, used in mass care and patient treatment in medical institutions. This is a new field, combining textile technology and medical science. Medical textiles contain knitted fabric, woven fabric, braided fabric and non-woven fabric. Compared with other fabrics, knitted fabrics have loose structures, good flexibility, high porosity, flexible and changeable structure design, etc. Therefore, they are more suitable for high-tech medical textiles, especially implantable medical textiles.

2. Knitting Structure

Knitting is done by a set of connected loops from a series of yarn in warp or weft direction. Therefore, knitted fabrics include weft knitted structures and warp knitted structures. In a weft knitted fabric, a horizontal row of loops are made using one thread running in the horizontal direction. While in a warp knitted fabric, each loop in the horizontal direction is made from different threads. Different types of machines, structures, stitch types, yarns and needle sizes may be used to create knitted fabrics with different properties and applications.

3. Implantable Medical Textiles

Implantable textiles are a series of materials implanted into human bodies. It is flexible, biocompatible and easy to weave. The materials refer to natural fibers (e.g. cotton and silk) and chemical fibers (e.g. polyester, polypropylene and polytetrafluoroethylene). Implantable textiles mainly include artificial blood vessel, hernia patch, artificial ligaments, etc. The structures varies for different parts andfunction.

3.1 Artificial Blood Vessel

Artificial blood vessel is knitted on double needle-bar warp knitting machine with more patterns and ground bars. The materials include polyester, polypropylene, polyethylene, polyurethane, polytetrafluoroethylene and some other synthetic fibers [1]. Silk fibroin is also used for its high strength and toughness [2]. The specifications are changeable as they are applied in different occasions. Reverse locknit and sharkskin are frequently-used to knit artificial blood vessel. Warp-knitted artificial blood vessel has stable structure and high compliance. Excessive expansion, edge roll and ladder rarely exist.

3.2 Hernia Patch

Hernia patch is an important implant for hernia repair. Polymer meshes are frequently used in hernia surgery. Warp-knitted pillar stitch, tricot stitch and atlas stitch are generally structures for hernia patch. Marlex® adopts tricot stitch as the basic structure and Prolene® adopts atlas stitch. They all have small porous [3]. Surgipro® PP hernia patch and Dexon® PGA hernia patch are created by tricot stitch and pillar stitch. The structures have higher strength and greater stability. The materials include non-absorbable polymers like polyester, polypropylene, expanded polytetrafluoroethylene, and polyvinylidene fluorideand absorbable polymers polyglycolic acid.

3.3 Artificial Ligament

Artificial ligaments can be used to replace the damaged knee joint. Ligament advance reinforcement system (LARS) studied by Laboureau is widely used in anterior cruciate ligament reconstruction. The structure is warp-knitted made by medical polyester. As is shown in Fig.1, the longitudinal fibers are bound together with a transverse knitted structure. And the longitudinal fibers are parallel and pre-twisted at 90°.

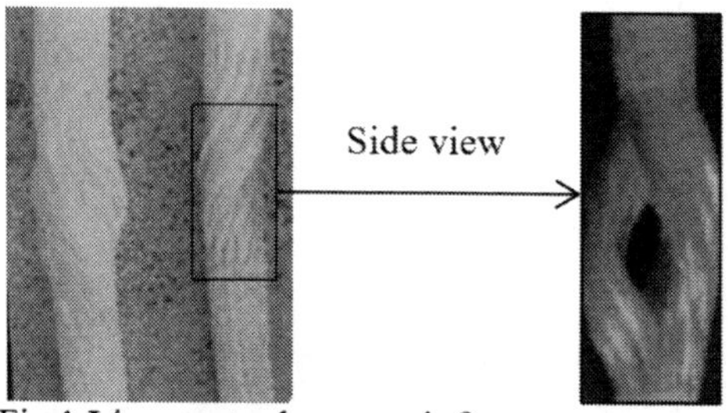

Fig.1 Ligament advance reinforcement system

3.4 Cardiac Support Device

Cardiac Support Device (CSD) is a warp-knitted elastic mesh bag, which is cut and sewn on the basis of the shape of the heart. The material is multi-filamentous yarn to provide high-strength and fatigue-resistant characteristics while maintaining flexibility [4]. The fabric with warp-knitted atlas stitch is glossy and it is easy to slide on the surface of heart to provide acute wall support.

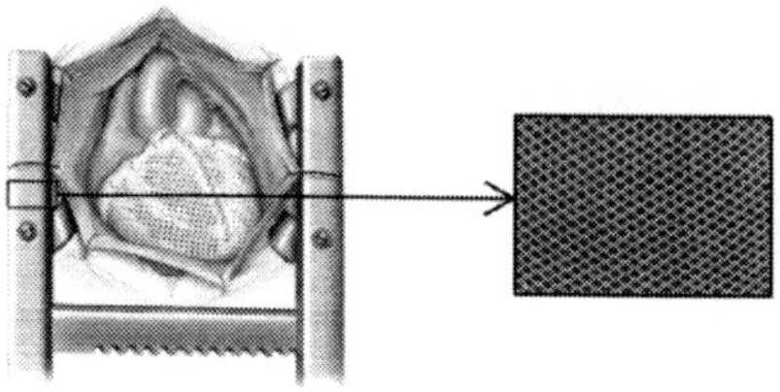

Fig.2 Positioning and securing the CSD[4]

3.5 Knitted Medical Expandable Metallic Stent

The use of metallic expandable tracheal stents can provide a palliative relief for severe airway obstruction. The Ultraflex™ stent (Micro-invasive, Boston Scientific, and Watertown, MA, USA) is meshes knitted by a single strand of nitinol wire [5, 6]. Nitinol is an alloy of 55% nickel and 45% titanium, and it has thermal memory. The stent is compressed along a supple guide by a coiled thread. If the thread is pulled, the stent will self-expand to the final diameter (Fig.3a). The Ultraflex™ stent may be uncovered form or covered by a sheath of silastic. The covered one is often used to prevent growth of tumor tissue (Fig.3b). And according to the difference of genders, the stent diameter for patients differs.

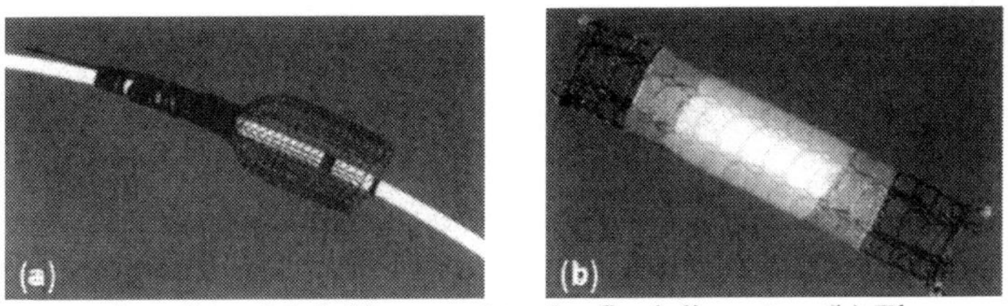

Fig.3 Urethral suspension sling. (a) Self-expand to the final diameter; (b) The covered one to prevent growth of tumor tissue [5, 6]

3.6 Artificial Tracheal Stent

Artificial tracheal stent is biodegradable and can be absorbed by human body. Artificial tracheal stent can temporarily replace the removed tracheal and maintain the airway travelable for easy regeneration of the tracheal. Plain plating stitch can be used to knit artificial tracheal stent. The plating yarn is polyglycolidelactide (PLGA) and the ground is polypropylene (PP). The fabric is knitted on small diameter weft-knitting machine. Before implanted in human body, the tracheal stent need to be coated and sterilized.

3.7 Periodontal Guided Tissue Regeneration

Periodontal patch is a kind of micro porous membrane for the treatment of periodontal disease and guide the regeneration of periodontal tissue. The basic structure is plain stitch knitted on single circular knitting machine. Biodegradable fibers, PLGA and PGA, are applied as the materials. According to the requirements, different shapes of patch can be sewn.

3.8 Tendon Scaffold

Compared with braided fabrics, knitted scaffolds are proven to favor deposition of collagenous connective tissue matrix and crucial for tendon/ligament reconstruction. Weft plain stitch is greatly applied in tendon scaffold. PLGA and PGA are the usual materials for tendon scaffold. PLGA is used to knit plain stitch as the external sheath and PGA fibers keep straight as the core. And warp-knitting technology can be applied in the field of tendon scaffold.

3.9 Artificial Chest Wall

Artificial chest wall is greatly used in chest wall reconstruction. Tricot stitch based on rib and tricot stitch based on interlock are two warp-knitted structures chosen for its good properties to resist ladder. The material is poly-p-dioxanone (PDS).

3.10 Radiotherapy Stent

Axial fabric has enough strength to be applied in radiotherapy stent. Generally, weft and warp yarns are inlayed in plain stitch to form axial fabric. Catheter with diamond structure is woven first on braiding machine. After coating and heat setting, the catheter is imported to the special weft- knitting machine to knit the composite structure. The materials are all biodegradable biomaterials, including collagen, polylactic acid (PLA), polyglycolic acid (PGA) and polyglycolidelactide (PGLA).

3.11 Department of Orthopedics Splint

Weft yarn and warp yarn can be inlayed in rib stitch to form multi-axial fabric. This kind of fabric is multi-layered biaxial weft knitted fabrics (MBWK Fabrics). It can be used as department of orthopedics splint, such as the foot splint. Weft and warp yarns can be glass fiber or some other high-performance fibers and the yarn knitting rib stitch can be PET or other chemical fibers. The fabric is often compounded with photosensitive resin to form the department of orthopedics splint with the same shapes with the injured part.

4. Conclusions

There are many studies about knitting structures greatly applied in medical field. Knitted medical textiles are one of the popular trends in technical textiles with high-added value and high technology content. It is supported by many subjects. As the development of fibers and knitting technology, there may be more high-tech products appeared and applied.

References

[1] Nemeno-Guanzon JG, et al. Trends in Tissue Engineering for Blood Vessels. Journal of Biomedicine and Biotechnology: 2012; p. 1-14.

[2] Yagi T, et al. Preparation of double-raschel knitted silk vascular grafts and evaluation of short-term function in a rat abdominal aorta. Journal of Artificial Organs: 2011; 14 (2): p. 89-99.

[3] Klosterhalfen BK, Junge, Klinge U. The lightweight and large porous mesh concept for hernia repair. Expert Review of Medical Devices: 2005; 2 (1): p. 103-117.

[4] Oz MC, et al. Global surgical experience with the Acorn cardiac support device. The Journal of Thoracic and Cardiovascular Surgery: 2003; 126 (4): p. 983-991.

[5] Madden BP, Datta S, Charokopos N. Experience with ultraflex expandable metallic stents in the management of endobronchial pathology. The Annals of Thoracic Surgery: 2002; 73 (3): p. 938-944.

[6] Gonfiotti A, et al. The first tissue-engineered airway transplantation: 5-year follow-up results. The Lancet: 383 (9913); p. 238-244.

Multifunctional Polysaccharide Fibers for Wound Healing

Muhammad Umar*, Yi Li, Olga Tsigkou, Xu-Qing Liu

School of Materials, University of Manchester, Oxford Road, Manchester, M13 9PL, UK

Corresponding author's email: muhammad.umar@postgrad.manchester.ac.uk

Abstract

Wound healing techniques have rapidly evolved over the last sixty years. It all started with the better understanding of skin and wound healing process. Any type of injury can cause pain, scars and results in stress but skin protects body from external environment. Wounded skin is healed through a lot of phases i.e. inflammation, migration, proliferation and maturation phase. Moist wound healing along with antimicrobial properties and controlled targeted drug delivery are main focus in the current wound care market. A plethora of highly absorbent fibers with multifunctional properties is centre of focus in current wound care market. Different types of polysaccharide fibers has been developed which claims to have high absorbence properties to manage exudates and provides antibacterial properties to encourage wound healing. However there is still a need for more research to develop fibers with high mechanical propertiesn so that they can be converted into wound dressing and scaffolds without loosing their integrity. These fibers must provide high drug loading capacity with controlled and targeted drug delivery to further enhance wound healing. This article presents a review of skin structure, wound healing processes and requirements, different type of materials, their structures and fibers develop from those materials for medical applications especially wound healing.

Keywords: Skin; Wound Healing; Fibers; Dressings; Polysaccharide

1. Introduction

Polysaccharide based fibers and polymers have amazing applications including nutrition, therapeutics and wound dressings. These fibers possess biologically active compounds for the design of biocompatible, biodegradable and environment friendly materials owing to their remarkable molecular structure. In last twenty years, a lot of research has been done to explore the biological properties of polysaccharides for their utilization in biomedical field. Polysaccharides exist in form of plants and animals like psyllium and alginate which can be extracted.

The earliest of occlusive wound therapy was discovered by Smith papyrus in 1615 BC and the basis of occlusive dressing was to provide wound with moisture to heal faster. No doubt, until the 19th century, exudate from wound was encouraged in case of both chronic and acute wounds. A frequently used phrase "laudable puss" (1).

In middle ages, it was customary for wound to rot a little. In 19th century Dr. Joseph Lister, demonstrated that pus is cause of an infection. He used antiseptic in surgery and his treatment of wounds with germ theory and infection were then understood. He concluded that pus formation in wound is the result of fight of body against the Staphylococcus and gangrene (2).

It was George D. Winter in 1962, who demonstrated that if wound is artificially dried or dried on exposure to air than more of the dermis is incorporated in the scab. He suggested that occlusive moist treatment must be given to the wound so that no microorganism can enter the wound and if the surface of wound is kept moist, the epidermis will migrate over the surface of the dermis. The migration of the epidermis is twice as rapid as when it is forced to pass through the fibrous tissues [3] [4].

1.1 Wound and its Types

Wound is a break down in the skin or other body tissues: a state where the normal structure of body tissue is damaged by an externally applied force or mechanical abuse (5). Wound usually happen due to an accident, they can be in form of cuts, scrapes, scratches and punctured skin. Other causes of wound can be surgery, sutures and stitches (6). Wound can be categorized according to wound healing speed (7). Acute wounds have healing time 8-12 weeks e.g. sword, gun, or surgery as physical cause, radiation, electricity, or chemical substances as chemical cause (8). Chronic wounds have healing time of more than 12 weeks e.g. ulcer of diabetes or tumor (9).

1.2 History and use of Dressings for Wound Care

In earlier days, wound dressings such as gauze, bandages and lint were made of natural fibers; the gauze or other dressings absorb exudate and make the wound dry and clean. The Egyptians had been certainly the first people who use adhesive bandages and honey on the wound (10). Ancient Greece started the use of vinegar and wine for washing the wounds in 460-370 BC, further treated with oil, honey, wine and used boiled wool as a bandage (11) (12).

1.3 Skin Structure

Skin is the largest organ of the body and it constitutes 16% of human body mass. It performs various functions. One of its most critical functions is to work as physical hindrance to the environment [13] (14), permitting and restraining the inward and outward passage of water, electrolytes and numerous substances. At the same time it provides protection against microbes, ultraviolet radiation [6] [7], noxious substances and mechanical abuses (15) [16]. Skin constitutes of three layers; the epidermis, the dermis and subcutis as shown in Fig.1. Hairs, nails, sebaceous, sweat and apocrine glands are regarded as subsidiaries of skin [8] [9].

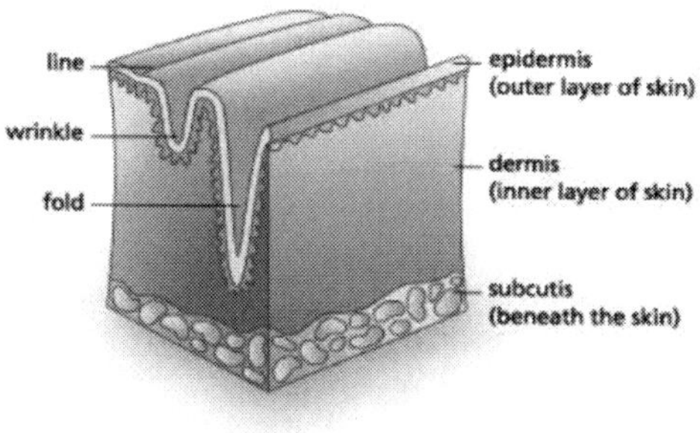

Fig.1 General scheme of skin structure - Modified from (16)

The epidermis is the external layer, acting as the physical and chemical barrier between the inner body and outer environment; the dermis is the deeper inner layer which provides the structural nourishment of the skin, there is a loose connective tissue layer beneath this layer, the subcutis (also known as hypodermis), which is an important storehouse of fat. Brief description of the layers of the skin is given in Table 1.

Table.1 Layers of the human skin

Skin layer	Description
Epidermis	It is the outer layer which is basically consist of layers of keratinocytes but also includes melanocytes, Langerhans cells and Merkel cells.
Dermis	It is the central area consists of supportive connective tissues between the external epidermis layer and the lower subcutis layer: contains sweat glands, hair roots, nervous cells and fibers, blood and lymph vessels.
Subcutis	The layer of loose connective tissue and fat below the dermis.

Since skin plays an important role to protect our body from the outer environment, therefore any type of damage to skin can cause disturbance in life and results in post trauma sequel and stress.

1.4 Significance of Moist Wound Healing

The concept and application of moist wound healing is not a new one. In modern world, this concept demands enhancement to increase the level of life, ease of application and fasten the healing process (17) (18). Moist wound healing process was introduced by George Winter in 1962; in his theory moist wound heals faster than dry wounds (19) [20]. Many other studies have been also documented which describes that the moist dressings are beneficial in quality of healing and reducing the time of healing (20) [22]. Maintaining moist environment of wound facilitates the healing procedure. The beneficial effects of moist environment is to prevent tissues or cell from death by dehydration and accelerated angiogenesis [23] [24]. The fundamental idea in moist wound healing is that the suitable quantity exudate will provide an atmosphere concept behind moist wound healing is that the presence of exudate in a wound will provide an environment that stimulates healing by delivering a range of cells and cytokines necessary for wound repair [23]. The dermal tissues exposed in air become dehydrated and epidermis migrates under the dehydrated tissue where it can get sufficient moisture to live. If the surface of the wound is kept moist by using moist wound dressings, the epidermis will migrate towards the surface of dermis. Later on migration of the epidermis becomes rapid than dry wound [3] [4].

1.5 Wound Healing Process

It is biological process linked to regeneration of tissues and normal process of growth, and it is also an uninterrupted process for substituting impaired tissues and completely restorative the wound [14]. The wound healing process consists of four phases; haemostasis and inflammation phase, migration phase, proliferation phase, and the maturation phase. Brief description of all the healing phases is given in this section [20].

1.5.1 Haemostasis and Inflammation Phase

Haemostasis starts within a few minutes of injury as platelets starts aggregating and plays a major role by discharging cytokines, chemokines, and hormones. In this phase, vasoactive mediators (epinephrine, norepinephrine, prostaglandins, serotonin, and thromboxane) help to reduce loss of blood which results in temporary blanching of the wound. Collagen and tissue factor along with exposed sub endothelium facilitates stimulation of platelet aggregation and results in formation of clot [22]. Moreover, to stimulate fibrin, thrombin trigger by the coagulation cascades helps migration of inflammatory cells to the wound site by increasing vascular permeability [23].

1.5.2 Migration Phase

This phase is characterized by the movement of the epidermis and fibroblasts to the site of injury to regenerate the damaged and lost tissues. Regeneration starts from the edges of the wound and quickly spread across the entire wound via thickening of the epidermis below the crust [14].

1.5.3 Proliferative Phase

In this phase, 'angiogenesis' occurs to reconstruct a wound with new granulation tissues composed of ECM and collagen and builds a new network of blood vessels [20] [22]. Healthier granulation tissues are granular and irregular in texture, pinkish/reddish in colour and do not bleed easily. 'Re-epithelialization' process restores the wound site by epithelial cells [24] [25].

1.5.4 Maturation Phase

During the maturation phase, continued collagen synthesis in the presence of collagen destruction is responsible for collagen remodelling. Excessive collagen is removed by the help of collagenases and matrix metalloproteinase in the wound but generation of new collagen continues. Collagen becomes more organized during remodelling. Proteoglycans takes place of hyaluronic acid and glycosaminoglycan and fibronectin slowly disappears. Water is absorbed from the scar and collagen III is replaced by collagen I [20] [23] [25].

1.6 Wound Dressings

Wound dressings play an effective role in the management of the warm moist wound environment and these dressings also protect the wound from the environmental microorganisms hence they facilitate the wound healing process. Fig.2 shows the functionality mechanism of wound dressings. Different types of fibers such as cotton fibers, polyester fibers, rayon fibers, nylon fibers, polyolefin fibers, acrylic fibers, elastomeric fibers, chitosan fibers, alginate fibers etc. are used to produce the dressings for the wound care. Absorption, antibacterial activity, tensile strength, biocompatibility and toxicity are some general properties of the wound care materials (23) (24).

Wound dressings are classified into two main categories; Primary dressings and Secondary dressings. The dressings which are applied directly to the wound base are known as primary dressings. Alginate dressing, Vaseline impregnated gauze, foam dressings are some primary dressings. The dressings, used to secure the primary dressings in place are known as secondary dressings such as gauze dressing, transparent film etc. Wound care products are classified by the wound bed color, wound moisture level and wound depth.

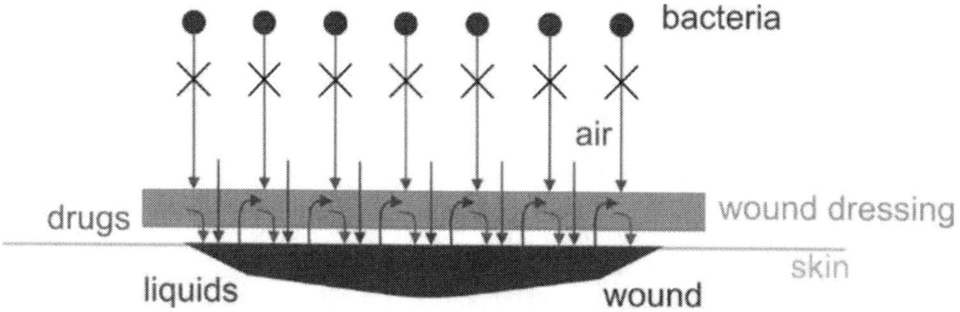

Fig.2 A wound dressing functionality - Modified from (25)

1.6.1 Requirements for an Ideal Wound Dressing

The dressing is said to be an ideal dressing if it maintains the warm moist wound environment, protect the wounds from the microbes, be non-toxic and biocompatible, eliminate the dead space, facilitate the autolytic debridement etc. Ideal wound dressings should be cheap and easily available and they can be converted into desired shape. The application, removal and maintenance of an ideal wound dressing must be easy (26). The dressing should be easily replaceable or removable without causing trauma and damaging any body tissues.

Characteristics of an Ideal dressing are as follows [13] [14]:

1) Dressing must help to reduce the flow of blood by increasing the rate of clotting process.

2) Dressings should keep the wound and dressing interface moist to assist hemostasis, debridement and faster healing rate.

3) Dressings must protect the wound from germs and external forces i.e. mechanical damage along with protection from cold by providing thermal insulation.

4) Dressings need to provide an efficient immunity to micro-organisms along with 'reducing the increase in number of bacteria by strengthening the body's defense system.'

5) Dressing should soak up and remove the excess exudate, blood and plasma. It should also facilitate absorption and removal of toxic fluids and contaminants.

6) Dressings should accomplish the elimination or consumption of exudate and ingredients harmful to tissues, so that chance of secondary infections must be reduced.

7) Dressings need to reduce discomfort and pain for the individual during elimination, application and use. Dressings must be hypoallergenic and biodegradable.

8) Dressings must be convenient to use and may assist in monitoring of the wound.

9) Dressings must be conformable to non-uniform wound shapes along with entire body curves and may not necessarily limit activities of daily routine.

10) Dressings must be easily changeable without causing trauma during this process.

11) Dressings need to reduce expense per use regarding health and fitness improved.

2. Materials

In the development of wound dressings having drug delivery systems, attention has been increasingly focused on naturally occurring herbs and materials due to their broad range of favorable biological functions including biocompatibility, anti-bacterial, anti-oxidant and anti-inflammatory activities while having minimum toxicity.

2.1 Alginate

Alginate belongs to a family of polysaccharides obtained from brown algae and bacteria. Alginic acid was found, produced and patented by Stanford in 1881 (27). It has been used in a variety of sectors, such as food, textile printing, medication, paper industry and for many other novel end uses. Alginate is a remarkable gel developing content blessed of having a lot of water. In recent times, alginate continues to be traditionally used in the wound management industry as a novel content for the development of 'moist healing' items for example skin gels, ointments, foams and fibrous non-woven dressings which can be utilized to protect chronic wounds.

Alginate materials are beneficial as raw materials for the development of most absorbing wound dressings. In this unique area, the characteristics of alginate fibers are remarkable in many aspects. Some important characteristics of alginate are:

1) It is non-toxic and safe material to use on wound surfaces and in cavities due to its natural origin.

2) Calcium alginate fiber is put on the outer face of wound and it soaks up extra fluids and exudate. In this process calcium ions are exchanged with sodium ions from body fluids, which often help in hemostasis.

3) As Calcium alginate fiber changes to sodium alginate fiber, it soaks up huge amount of fluids and exudates and becomes a gel, which helps to maintain moist environment for faster wound recovery

4) It is organic and evergreen material with never ending source in nature.

Alginate can be produced from fresh seaweeds by treating with aqueous alkali solutions, usually sodium hydroxide. In this process, organic alginate in many salt types is converted into water soluble sodium alginate. After filtration, the sodium alginate in solution can be precipitated by the addition of calcium chloride. After further purification and conversion, water soluble sodium alginate powder is produced [28].

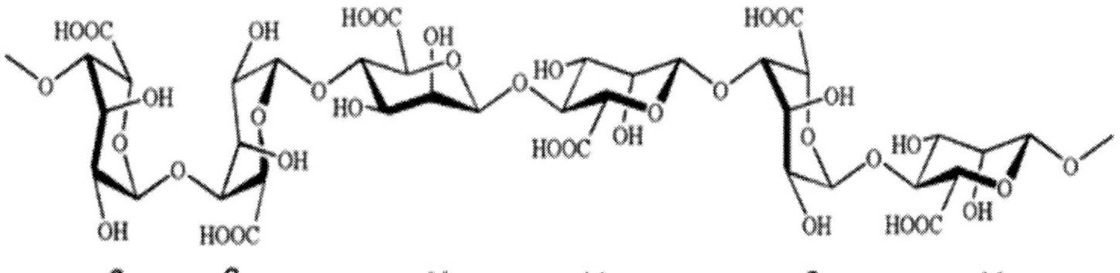

Fig.3 Chemical structure of alginate (29)

Alginate exhibits a chain of (1,4) linked β-D-mannuronic acid (M) and α-L-guluronic acid (G) monomers (29). Fig.3. shows the chemical structures of these two monomers. These sugar acids are distributed in blocks and have extensively varying proportions and sequence. Alginate consists of homopolymeric regions of M and G, known as 'M' and 'G' blocks, which combine as an alternating structure. Typically, the blocks are composed of three different forms of polymer segments: consecutive G residues, consecutive M residues and alternating MG residues resultant hydrogels [29].

Calcium alginate has emerged as a popular wound dressing material. Calcium alginate fibers are hemostatic, biocompatible and resorb able by human body, as the calcium ion from the fiber is exchanged from sodium ion of the body fluids. They form gels which absorb exudates from the wounds. Sodium alginate is soluble in water while calcium and magnesium salts are insoluble (30). Alginate exhibits great biocompatibility and biodegradability, and is widely used in field of biomedicine. Alginate fibers forms gel to prevent wound from becoming dry that enhance the healing process and prevent from scar formation. When Calcium ions released into wounds, stimulate platelet activation and blood coagulation (31). Alginate dressings have good permeability of gases and water content, which imitate natural skin, resulting in wound remaining moist throughout the healing process (30) (32). Table 2 summarizes manufacturing and properties of some alginate fibers.

Table.2 Manufacturing and properties of alginate fibers

Fiber Materials	Coagulation Bath	Spinning Technique	Post Processing	Limitations/ Gaps	Features	Reference
Calcium Alginate/ PVA	N/A	Electrospinning	Nanofiber matrixes	Excessive exudate cause maceration Poor mechanical performance Nanofiber bandages cannot be developed for commercial purposes due to low integrity	Better antibacterial properties Adequate absorbency for wound healing Use in biomedical applications	(33)
Zinc Alginate/ TCP	10 wt. % $CaCl_2$ 2 wt. % HCl	Wet Spinning	Bone grafting Implantable materials	Optimization of Conditions for Nanocomposites manufacturing	Antibacterial properties Enhanced wound healing	(34)
Sodium Alginate/ Soy Protein	$ZnCl_2$ 0.03 wt. % HCl	Wet Spinning	Fibers	Need to develop eco-friendly matrices Need to develop Green chemistry & sustainable development Strategy	Best Strength & Elongation at 10% Soy Protein	(35)
Sodium Alginate/ Gelatin	10% $CaCl_2$ + Ethanol	Wet Extrusion	Fibers	Less absorption and strength for further processing	More water retention properties Best Strength & Elongation at 30% Gelatin	(36)

2.1.1 Manufacturing and Properties of Alginate Fibres

K. Tarun, et al. developed calcium alginate/PVA blended nano fiber matrixes for wound healing. Electro-spinning was used for the development of the calcium alginate/PVA blended nano fibers. These fibers have high surface area to volume ratio and have many other biomedical applications. Calcium alginate has good biomedical properties and poly vinyl alcohol has high mechanical properties so calcium alginate/PVA blended nano fibers have good biomedical properties along with good mechanical properties. Concentration of the alginate and PVA in the blended nano fibers also influences the properties of the final product. Increase in the concentration of the calcium alginate results in reduction in diameter of fiber, improved antimicrobial activity and reduction in moisture vapor transmission rate etc (33).

T. Mikołajczyk, et al. developed zinc alginate fibers with tricalcium phosphate (TCP) nano additives by using the wet spinning technique. Zinc alginate fibers have antibacterial activity and kill the germs and promote the wound healing activity. The tricalcium phosphate nano additives have osteoconductive activity. The zinc alginate fibers containing the TCP nano additives have antibacterial activities and they can be used as implantable materials (34).

Q. Wang, et al. developed alginate/soy protein isolate blended fibers by the combined extrusion of the alginate solution and soy protein isolate solution into the coagulation bath containing the calcium chloride, hydrochloric (HCL) acid and ethanol. Content of the soy protein isolate in the alginate/soy protein isolate blended fibers influence the water retention properties, tensile strength and elongation of the fiber. Best results are achieved when content of the soy protein isolate is 10 % by weight in the blended fiber, as its content increases it results in decrease in the water retention value (35).

L. Fan, et al. described that water retention properties of the alginate/gelatin blend fibers are superior to the pure alginate fiber. Gelatin content in the blend fibers influence the mechanical and water retention properties. Gelatin and alginate solutions are mixed together and mixed solution is extruded through the spinneret into the coagulation bath containing the calcium chloride and ethanol. Strong bond is developed between the gelatin and alginate. As the content of the gelatin is increased it improves the water retention properties of the blend fibers. If the content of the gelatin is 30 wt. % in the blend fiber then it gives the optimum values of the tensile strength and breaking elongation (36).

2.2 Chitosan

Chitosan was found by Rouget in 1859. Chitosan is a deacetylated derivate of chitin, which is an organic polysaccharide mainly obtained from the actual covers and seashells of shrimps and other sea crustaceans. It can be chemically specified as poly-β-(1→4)-D-glucosamine or poly-(1, 4) - 2-amido-deoxy-β-D-glucose (37). Fig.4. shows the structure of (a) chitin and (b) chitosan polymer. In addition to its anti-microbial action, chitosan features several significant essential benefits such as non-toxicity, biocompatibility and bio-degradability. Chitosan can be used as an additive for spinning antimicrobial fibers and as a finishing agent for surface treatments primarily of cellulose, cellulose/poly- ester wool fibers (38).

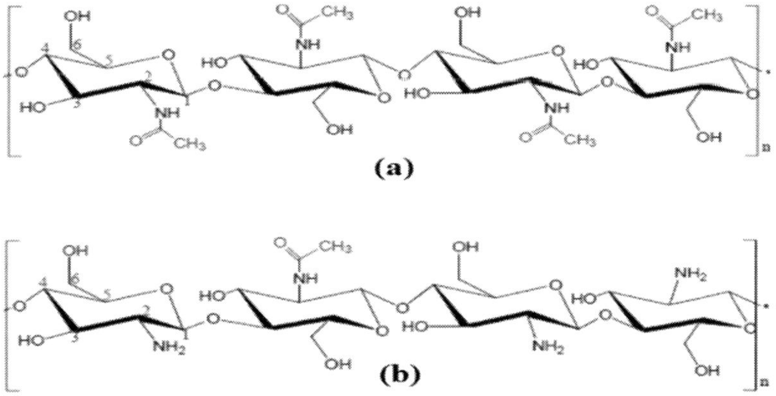

Fig.4 Chemical structures of (a) chitin and (b)chitosan (37)

Chitosan is prepared by deacetylation of chitin in a hot alkali solution (Fig.2.2). Chemically, Chitin is poly-β-(1,4) linked-N-acetyl-D-glucosamine). When the C-2s of chitin's monomers substitute all or partial vinyl amine with amine groups, chitin becomes Chitosan (38).

Chitosan is soluble in an aqueous solution of acids. Chitosan possesses neutral molecules which form inter-molecular hydrogen bonds that hinder its dissolution in water. But, the presence of amino groups in chitosan chain makes it dissolvable in acidic aqueous solution (pH<6.5) and acquires properties of a cationic polyelectrolyte. Hence in solutions, the protonization of glucosamine amino group into $R\text{-}NH_3^+$ makes the chitosan salts positive (39).

Chitosan is semi-crystalline polymer with somewhat organized polymer chains (40) and hence exhibits some rigidity and stiffness in feel. It is also nontoxic, biodegradable, and biocompatible (41). Various studies have shown that Chitosan possesses anti-microbial effects against bacteria, fungi and viruses (42).

Biological activity of chitosan is of interest, especially its susceptibility to degradation under the influence of enzymes present in the body fluid such as lysozyme and N-acetylglucosoamidase. The degradation results in stimulation of macrophages and positively influence collagen sedimentation, thus accelerating the wound healing process [31] [32].

Binding on cells and biodegradability of chitosan may lead to a variety of biomedical applications such as wound dressing, carriers for drug delivery systems and space filling implants (43). Table 3 summarizes manufacturing and properties of some chitosan fibers.

Table.3 Manufacturing and properties of chitosan fibers

Fiber Materials	Coagulation Bath	Spinning Technique	Post Processing	Limitations/ Gaps	Features	Reference
Sodium Alginate/ Hydrolyzed Chitosan	2% $CaCl_2$ + 0.15% – 0.6% of chitosan 2% $CaCl_2$ + 0.3 % chitosan	Wet Spinning + Coating	Two Routes	Development of Alginate/Chitosan Fibers Viscosity of Chitosan for wet spinning Need to determine parameters required for best wet spinning conditions	Very high water retention values Good mechanical properties Antimicrobial properties	(44)
Hydroxypropyl Chitosan/ Sodium Alginate	5% $CaCl_2$ + Ethanol	Wet Spinning + Dip Coating	Treatment with $AgNO_3$	Need to improve tensile, absorption and antimicrobial properties Low interaction of chitin and chitosan with wound sites	Good antibacterial activity to Staphylococcus aureus Best tensile strength and breaking elongation when HPCS content was 30 wt. %. Increased water retention properties	(45)
Sodium Alginate/ N-Succinyl-chitosan (SCS)	5 wt. % $CaCl_2$	Wet Spinning	Treatment with AgNO3	Low interaction of healing agents with wound Less Compatibility Chitosan inability to dissolve in H_2O	Good antibacterial properties Improved water retention Maximum tensile strength and elongation at 30% SCS	(46)
Hyaluronic Acid/Chitosan	N/A	Electrospinning	Nanofibers mats	Process ability of chitosan and HA due to high viscosity Controlled drug delivery	Useful for wound healing Relatively better controlled drug delivery	(47)
Hydrolyzed Chitosan/ Sodium Alginate	Hydrolyzed Chitosan	Wet Spinning	Fibers	Development of wet extruder for commercial setup Use of hydrolysed Chitosan in Coagulation Bath	Improved tensile properties Better absorption enhances wound healing rate	(48)

2.2.1 Manufacturing and Properties of Chitosan Fibres

A research on the process of manufacturing alginate-chitosan fiber was made by Steplewski et al. In this work two different methods were used to produce fibers, in first method coagulation bath contain both chitosan and $CaCl_2$, while in second method only $CaCl_2$ was used in coagulation bath. The developed fiber was then dip coated into another bath containing chitosan. It was concluded that the method in which chitosan was applied at the stage of finishing processing was more effective, and enable to obtain alginate chitosan fibers with higher chitosan content (44).

L. Fan, et al. developed alginate/hydroxypropyl chitosan blend fibers with wet spinning technique. Chitosan was treated with propylene oxide under alkaline conditions to develop the hydroxypropyl chitosan fiber. Alginate solution and hydroxypropyl chitosan solutions were prepared and then both were mixed and stirred. This solution was extruded through the spinneret into the coagulation bath containing 50% calcium chloride and 50% ethanol. Strong forces were developed between the alginate and the hydroxypropyl chitosan due to the development of hydrogen bond between them. The quantity of the hydroxypropyl chitosan directly influences the tensile and water retention properties of the blend fiber and best results were achieved when its content in the blend fiber was 30 % by weight. Alginate/hydroxypropyl chitosan blend fibers treated with silver nitrate exhibit good antibacterial properties against the Staphylococcus aureus. These alginate/hydroxypropyl chitosan blend fibers treated with silver nitrate were used to treat wounds to promote wound healing by maintaining the moist wound environment and providing protection against microbes (45).

L.Fan, et al. developed N-succinyl-chitosan/alginate fibers by wet spinning technique. In this method, solutions of the N-succinyl-chitosan and alginate were mixed together and then extruded through the spinneret into a coagulation bath containing calcium chloride solution and ethanol. Strong intermolecular forces were developed between the blended fibers due to hydrogen bonding. The blended fibers were then treated with the silver nitrate solution and dried. Water retention of the blended fibers and tensile strength of the dry fibers were superior to pure alginate fiber. Increase in the concentration of the N-succinyl-chitosan results in decrease in the wet tensile strength of the blended fibers. Blended fibers also have antimicrobial properties and can be used for the treatment of the infected wounds (46).

N. Maeda, et al manufactured chitosan-hyaluronic acid hybrid polysaccharide fibers by electrospinning of poly cationic chitosan with PEO and then coated them by poly anionic hyaluronic acid. Scanning Electron Microscope was used to check the fiber size distribution. It was found that more than 50% humidity affects the spinnability of fibers and beads started to appear on small diameter fibers. Chitosan-hyaluronic acid fibers remained stable in water although they show controlled release of hyaluronic acid in phosphate buffered saline solution (PBS). Swelling ratio of newly developed was increased upto 30% on coating of 3-wt% hyaluronic acid. Newly developed chitosan-hyaluronic acid fibers have significant contribution in faster wound healing and controlled drug delivery along with other biomedical applications (47).

Sweeney, et al. manufactured alginate-chitosan fibers by using hydrolyzed chitosan as a coagulant in one step direct wet spinning extrusion process. These newly developed alginate-chitosan fibers contain about 4.50% and 5.10% chitosan. The inclusion of hydrolysed chitosan improves the tensile properties i.e. elongation and tenacity of core alginate fibers. The gelling and absorption properties were also enhanced in both distilled and saline water with increasing quantity of hydrolysed chitosan within the fibers. These fibers with improved properties can be converted into nonwoven dressings and have useful application in wound care (48).

2.3 Hyaluronic Acid

Hyaluronic acid is a linear polysaccharide sourced from nature. It is composed of alternating disaccharide units of a-1,4-D-glucuronic acid and b-1,3-N-acetyl-D-glucosamine. Fig.5 shows the chemical structure of hyaluronic acid. Its molecular weight varies from 103 to 107. This polyanionic polymer is in connective tissues including umbilical cord, synovial fluid, vitreous, etc. Hyaluronic acid has been largely applied in the fields of controlled medication delivery, cellular encapsulation and muscle regeneration due to its exclusive viscoelastic properties and great biocompatibility (49).

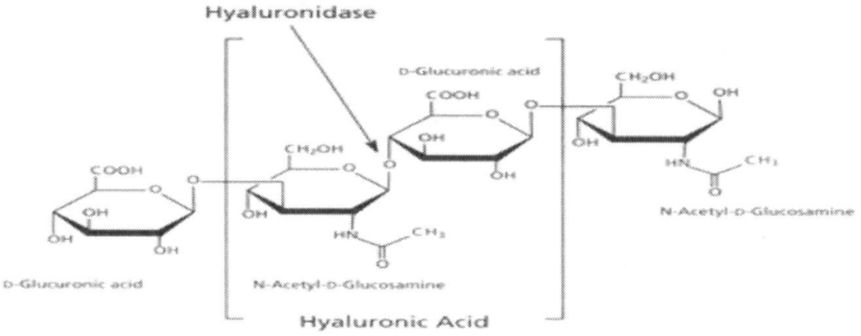

Fig.5 Chemical structures of hyaluronic acid (50)

Table.4 Manufacturing and properties of hyaluronic acid fibers

Fiber Materials	Coagulation Bath	Spinning Technique	Post Processing	Limitations/ Gaps	Features	Reference
Hyaluronic Acid/Chitosan	$CaCl_2$	Wet Spinning	Fibers Yarns Scaffolds	Need to determine adequate shape & pore size that enhances chondrogenesis	Enhanced mechanical properties Improved extracellular matrix activity	(52)
Alginate/ Chitosan/ Hyaluronic acid	$CaCl_2$	Wet Spinning	Fibers	Need of controlled drug delivery mechanisms Need to check effect of different processing parameters on fibers	Improved absorption and tensile properties Controlled release of hyaluronic acid is useful for faster wound healing	(53)
Hyaluronic Acid + Chitosan	64% $CaCl_2$ + 50% Methanol 50% Methanol 0.04-0.07% HA + 50% Methanol	Wet Spinning	Scaffolds Manufacturing	No suitable materials & procedures for cartilage repair Mechanical weakness Need to check biocompatibility & immunological reactions Process of degradation or absorption of fibrous material is unclear Need research to find adequate shape, pore size and mechanical properties of 3D fabrication for cartilage tissue regeneration	Desirable biomaterials for cartilaginous tissue scaffolds and issue Engineering Enhanced Cell adhesivity and cell proliferation	(54)
Hyaluronic Acid + Chitosan	N/A	Plastic tubes + Freeze Drying	Scaffolds	Need to find shape, size & 3D fabrication for cartilage repair	Non-cytotoxicity Cell binding Better pore network, Extra cellular matrix generation & cartilage fix	(55)

Hyaluronic acid is a linear polysaccharide and it performs significant natural and biological tasks. Hyaluronic acid is a glycosaminoglycan discovered in conjunctive cells of almost any vertebrate. The macromolecule consists of N-acetyl glucosamine as well as glucuronic acid solution. Being a part of extracellular matrix, Hyaluronic acid may impact several cellular characteristics such as attachment, migration, as well as growth. The latest biomedical applications of hyaluronic acid include things like scaffolds pertaining to injure therapeutic as well as cells technological innovation, and surgery therapy, osteo-arthritis cure, in addition to being a part of implant materials (50).

Hyaluronic acid solution was discovered in excessive levels throughout embryonic skin advancement. Hyaluronic acid solution can also be cross-linked. It can be used as short-term prostheses for synovial liquid to cure arthrosis. Hyaluronic acid solution can be quickly biodegraded within a body (51). Table 4 summarizes manufacturing and properties of some hyaluronic acid fibers.

2.3.1 Manufacturing and Properties of Hyaluronic Acid Fibres

S. Yamane, et al developed three dimensional scaffolds designed from the composite chitosan-based hyaluronic acid polymer fibers. Wet spinning technique was used to manufacture chitosan based hyaluronic acid composite polymer fibers followed by coating of hyaluronic acid (0.07%). The tensile strength, diameter and tension at fracture of the composite polymer fiber were 144.4 N/mm^2, 0.03mm and 9.7%, respectively. Composite polymer fibers were used to make yarns, these yarns were woven to develop 3-D scaffolds of varying pores (100, 200, and 400 mm). After testing, it was discovered that scaffold with 400mm pore size substantially improves the actual extracellular matrix activity through chondrocytes. There seemed to be no main difference in these scaffolds in concern with the cell growth. Moreover, most of these scaffolds demonstrated high technical properties in contrast to liquid and gel components. According to this study, the future of a novel designed material with comparatively large pore size scaffold for fibrous regeneration was guaranteed (52).

T. Hussain, et al. developed alginate/chitosan/hyaluronic acid (ACH) composite fibers by wet spinning and subsequent coating procedures. Sodium alginate was extruded into calcium chloride bath to produce calcium alginate fibers and then passed through hydrolysed chitosan bath and dip coated in hyaluronic acid to develop new fibers. The inclusion of hyaluronic acid increases the swelling and absorption properties of fibers which are useful for faster wound healing. The inclusion of hydrolysed chitosan and hyaluronic acid also improves tensile properties of fibers. Newly developed fibers have also shown controlled release of hyaluronic acid and found useful applications in wound care (53).

S. Yamane, et al. developed a chitosan polymer fibers and chitosan based hyaluronic acid hybrid polymer fibers by wet spinning technique. The novel fiber of chitosan based hyaluronic acid comprised of chitosan coated hyaluronic acid (0.04% and 0.07%). These can be useful as a scaffold biomaterial for cartilage tissue engineering. It has been found that composite fibers has more cell adhesivity, proliferation and the synthesis of cartilage-specific proteoglycan core protein (aggrecan) than in the chitosan fiber. The chitosan-based hyaluronic acid hybrid polymer fibers demonstrate excellent prospects as a suitable biomaterial for cartilaginous tissue scaffolds (54).

L. S. Moreira-teixeira et al prepared hybrid scaffolds of chitosan and hyaluronic acid of varying weight ratios containing no hyaluronic acid (control) or 1%, 5%, or 10% of hyaluronic acid. Biological and structural features of chitosan scaffolds were improved by hyaluronic acid. Results demonstrate that chitosan/hyaluronic acid hybrid scaffolds are non-cytotoxic and improves cell bonding. The incorporation involving hyaluronic acid increased cartilage extra cell matrix generation. Chitosan with 5% hyaluronic acid got an even better pore network configuration in addition to increased extra cell matrix cartilage development. It has been found that chitosan and hyaluronic acid hybrid matrixes have got significant use in cartilage fix (55).

2.4 Psyllium

Psyllium is the seed extract of plants from the Plantago genus (56), which has 200 species widely distributed throughout the temperate regions of the world. Psyllium is locally known as Isabgol meaning "horse ear" in hindi, which describes the shape of the seed (57) (58). The main thing isolated from psyllium seed is psyllium husk while remaining part of the seed is used as animal feed. Psyllium contains about 78% soluble fibre and 13% insoluble fibre and is a good dietary source of both soluble and insoluble fibre (59). Psyllium intake has been proved useful in reduction of plasma and LDL cholesterol levels (60), reduces the probability of colon cancer, and hyperglycemia and may have potential in body weight control and treatment of constipation and irritable bowel syndrome (59).

It is a complicated polysaccharide which consists of a highly branched non-starch polysaccharide with mainly a basic chain comprises of (1→4) and (1→3) linkages, while the side chains are associated with basic chain via O-3 and/or O-2 linkage. Xylopyranose residues are primarily a part of basic chain of this polysaccharide. In same way, arabinofuranose and xylopyranose residues are part of side chains (56).

Fig.6. shows the polymer chain of the psyllium. Health benefits of psyllium are largely associated with its gelling ability. The wound cleansing and wound healing of this polysaccharide are originated from Mucopolysaccharides which is derived from the husk of psyllium (Plantago ovata). Mucopolysaccharides is proven to be useful in reduction of scar formation according to latest research (60). Table 5 summarizes manufacturing and properties of some psyllium fibers.

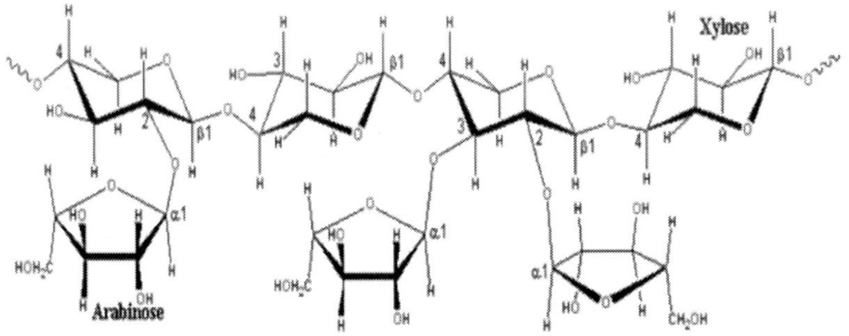

Fig.6 Polymer chain of Psyllium (57)

Table.5 Manufacturing and properties of psyllium fibers

Fiber Materials	Coagulation Bath	Spinning Technique	Post Processing	Limitations/ Gaps	Features	Reference
Psyllium/ Sodium Alginate	$CaCl_2$	Wet Spinning	Fibers	No reports available on minimum quantity of silver/control release agents for antimicrobial properties	Long lasting highly antimicrobial properties, Inbuilt ability to gel	(61)
Psyllium/Sodium Alginate /Silver Carbonate	1% $CaCl_2$	Wet Extrusion	Fibers	Absorption deficiencies Need to optimize silver quantity	Long lasting antimicrobial & gelling properties	(62)
Psyllium/ Sodium Alginate	$CaCl_2$	Wet Spinning	Fibers in cold, hot and hydrolysed routes	Psyllium was not used in fibers before To check effect of different media on fibers production and properties	Cold media helps to achieve high tensile strength, absorption and gelling properties for faster healing rates	(63)
Psyllium/ Sodium Alginate/ Silver	$CaCl_2$	Wet Spinning	Fibers	Poor antimicrobial properties Need of targeted and controlled drug delivery mechanisms	Good Tensile Properties Superior Liquid absorption improves healing rates	(64)

2.4.1 Manufacturing and Properties of Psyllium Fibres

M. Miraftab, et al. developed psyllginate fibers for biomedical applications such as wound management. It is made as a bicomponent fiber with alginate and psyllium as polymers and silver salt was introduced in it for antimicrobial activity. The alginate psyllium dope can be prepared by using alginate (4 – 4.5%) concentration and psyllium (0.3 – 1.5%) concentration. Dopes can be extruded in aqueous mixture of alkaline (0.1 - 0.5% NaOH) solubilized psyllium and sodium alginate into a calcium chloride coagulation bath containing 1-2% Calcium Chloride. Then fibers were dried by passing through baths of aqueous solutions containing (50, 70, 100%) acetone. The fibers produced in this invention have good tensile properties and superior liquid absorption (g/g) compared to pure alginate fibers (61).

M. Miraftab, et al. targeted highly exudation wounds that can had infection and unnecessary trauma if proper management is ignored. So, three fibers were developed, the controlled fiber having only alginate and the other two were psyllginate fibers made from natural polymers alginate and psyllium. The psyllium alginate dope was prepared with concentration of 4% alginate and 0.75% psyllium and 0.05% silver carbonate. The dope was extruded on a pilot size wet extruder in coagulation bath having 1% aqueous solution of Calcium chloride. The developed psyllginate Ag fiber has long lasting antimicrobial, carries up to six times less silver than that in conventionally available silver dressings and gelling properties better than alginate and commercially available dressings (62).

Masood, et al. studied different routes of producing high absorbent and gel forming fibers from alginate and psyllium. They used three routes i.e. hot, cold and hydrolysed method for extraction of psyllium gel. In the extracted gel, dry powder of alginate was dissolved to form the dope solution. This research confirms that composite fibers can be developed by various method of psyllium gel extraction. However, it was discovered that cold water extraction results in better composite fibers with improving tensile strength properties, liquid absorbance and gelling. It is also a cost effective method in terms of maintaining room temperature and efficient working performance (63).

In another research, Masood et al. coextrude psyllium with Na/Ca alginate along with silver compound to develop controlled releasing fibers for wound care application. The study reveals that the produced natural fiber has an inherent capability to form gel and absorb a high quantity of exudate. The produced fiber encapsulates lesser amount of silver than conventional silver dressings. It also exhibits slow release while providing full potency all the time against microorganism including Methicillin Resistant Staphylococcus Aureus. This fiber producing method is least expensive and is highly effecient against gram positive and negative microbes and also work effective against silver staining of the skin (64).

2.5 Mechanism of Antimicrobial Activity of Metals

Table.6 Manufacturing and properties of antimicrobial metals incorporated fibers

Fiber Materials	Coagulation Bath	Spinning Technique	Post Processing	Limitations/ Gaps	Features	Reference
Silver sulphadiazine + Sodium Alginate	$CaCl_2$ + $AgNO_3$	Wet Spinning	Wound Dressing	Overuse of silver cause toxicity Research required to find optimize quantity	Enhanced absorbency and antibacterial properties	(70)
Chitosan + $AgNO_3$	5% NaOH	Wet Spinning	Fibers	Need to control release of silver	Zinc containing dressings are useful for wound healing but silver containing dressings are highly antimicrobial and absorbent	(71)
Copper/Sodium Alginate/ Cellulose	N/A	Dip Method	Fibers	Lack of ion exchange capability of cellulose fibers	Improved antimicrobial and mechanical properties	(72)

It is reported that metals and their salts have comprehensively toxic effects on bacterial cells, mainly as a result of their ability to denature protein molecules, but many of them are also essential for growth of human cells (65). The silver ions are highly reactive and when bind to protein, it causes structural changes in the bacterial cell wall, intracellular and nuclear membranes. Furthermore the silver ion exhibits its bacteriostatic properties by denaturing bacterial RNA and DNA, and that's why bacterial replication stops (66) (67). It has been reported that the copper on contact targets the DNA of bacteria that leads to rapid fragmentation of DNA, causes damage to cell membrane and results in cell death (68). The antibacterial mechanism of zinc oxide nanoparticles is very different as they increase levels of oxidative stress in bacterial cells. The possibilities of membrane damage due to direct or electrostatic interaction between zinc oxide and bacterial cell surfaces also increases (69).

2.5.1 Manufacturing and Properties of Antimicrobial Metals Incorporated Fibres

In a research Y. Qin explained about bandages having alginate and silver which can work as antimicrobial agent. In this study, highly absorbent and antimicrobial alginate fibers were developed by incorporating silver in ionic form. Silver was incorporated in alginate fiber by reaction of solutions of calcium chloride and silver nitrate in coagulation bath to produce silver and calcium alginate fiber. The incorporation of silver sulphadiazine (SSD) in alginate fiber was achieved by mixing the water soluble SSD with sodium alginate to make dope solution, which was extruded in 2% calcium chloride solution containing silver nitrate. Silver was added in alginate fibers without reaction by mixing silver sodium hydrogen zirconium phosphate particles with sodium alginate dope solution and then extruded. The alginate and host cell can be prevented from oxidization by using sodium hydrogen zirconium phosphate framework. The developed silver alginate fibers maintain their white color and have sustained silver ions release when came in contact with wound exudate (70).

Y. Qin et al. introduced silver and zinc in chitosan fiber with an ability to gel and absorb large quantities of wound exudates. The dope contains 5% chitosan dissolved in 2% acetic acid aqueous solution. The dope was extruded on extruder in coagulation bath having 5% aqueous solution of sodium hydroxide. The silver was incorporated by treating the chitosan fibers with aqueous solutions of $AgNO_3$ having concentrations (0.001, 0.002, 0.003, 0.005 and 0.05%) in aqueous solutions for 24 hours at room temperature. The results concluded that zinc and silver content in chitosan fibers can be controlled by changing the concentrations of treatment solution and treatment time. The chitosan fibers having silver are highly antimicrobial than chitosan fibers containing zinc (71). The copper alginate cotton cellulose (CACC) fibers were synthesized and characterized by M. Grace et al. CACC fibers were developed by dipping cotton fibers in 5 % sodium alginate solution for 4 hours at room temperature, then these fibers were immersed in cuprous sulphate solution for 2 hours. Copper containing cotton fiber was also developed by immersing cotton fiber in cuprous sulphate solution for 2 hours. This study concluded that CACC fibers had both, good antibacterial action and mechanical strength. The antimicrobial properties depend upon copper loaded on fiber while it also depends upon alginate loaded on cellulose fibers. The fibers released more copper ion in physiological fluid as compared to distilled water (72).

2.5.2 Effect and Mechanism of Controlled Release of Metals

The controlled release of strong antibacterial reagent leads to better antimicrobial properties with the help of polymeric materials (73). The silver loaded alginate fibers when encountered wound exudates releases silver into exudate by the ion exchange between the silver ions present in fibers and the sodium or calcium ions in the wound fluid. Silver ions can also be chelated by protein molecules in the wound fluid or particles detachment from surface of fibers (74). However, this release should be controlled because slow release of the silver improves the life of bandage and also reduces the possibility of toxicity and side effects due to higher silver concentration (75). In addition, biocidal properties are provided by an improved and sustained release of silver ions (76). Table 6 summarizes manufacturing and properties of some antimicrobial metals incorporated fibers.

3. Conclusion

Wound dressings are used to cover wounds and improve healing rates for many years. In earlier times, traditional cotton dressings were used which cause pain while removal and sometimes produce infection. In recent years, a lot of research and innovation has been done in both textiles and medical sciences which help to enhance wound healing rate. Many type of biocompatible materials having high absorbency and antimicrobial properties, are being used to develop fibers for wound management. They provide moist wound healing environment and prevention against microbes to reduce chance of infections and increase wound healing time. Different techniques have been used to control drug delivery, but there is still need of further research to increase drug loading capacity of these fibers and to improve targeted and controlled drug delivery at wound site. Further work is required to bring these fibers from lab to clinical and commercial level.

4. Future Perspectives

- Hyaluronic acid fibers and dressings have low mechanical strength and can cause pain while removal.
- Chitosan fibers also have low mechanical properties to be processed for nonwoven wound dressings.
- These fibers have limited drug loading capacity, issues relating to high targeted concentration and efficiency at tumor sites.
- There is need to develop fibers which have both absorbency and strong antimicrobial properties along with mechanical strength to produce nonwoven wound dressings.
- There is also need to improve drug loading capacity and targeted drug delivery systems in fibers.
- No type of wound dressings has been developed by using combination of these materials.
- There is need to bring newly developed fibers and dressings from lab to clinical and commercial level.

5. Acknowledgment

We would like to thank the support of the EU Horizon 2020 and University of Manchester through projects with project codes 644268 - ETEXWELD - H2020-MSCA-RISE-2014, AA14512 (UMRI project "Graphene-Smart Textiles E-Healthcare Network") and R119938 ('Protective Efficiency of Respiratory Protective Equipment (RPE) against Byssinosis for Cotton Workers'.

References

[1] Henry G, Garner W. Inflammatory mediator in wound healing. Surg Clin North Am: 2003; 83 (3): 483-507.
[2] Broughton G, Janis JE, Attinger CE. A brief history of wound care. Plast Reconstr Surg: 2006; 117: 6-11.
[3] Winter GD, Scales JT. Effect of Air drying and Dressings on the Surface of a Wound. Nature: 1963; 4862: 91-92.
[4] Hinman C, Maibach H. Effect of air exposure and occlusion on experimental human skin wounds. Nature: 1963; 200: 377-379.
[5] Kim GH, Kang YM, Kang KN, Kim DY, Kim HJ, Min BH, et al. Wound dressings for wound healing and drug delivery. Tissue Eng Regen Med: 2011; 8: 1-7.
[6] Plus M. Wounds and injuries [Internet]. Medline Plus: 2014; p. 1-7. Available from: https://www.nlm.nih.gov/medlineplus/woundsandinjuries.html
[7] Walburn J, Vedhara K, Hankins M, Rixon L, Weinman J. Psychological stress and wound healing in humans: a systematic review and meta-analysis. J Psychosom Res: 2009; 67: 253-271.
[8] Percival NJ. Classification of wounds and their management. Surgery: 2002; 20: 114-117.
[9] Moore K, McCallion R, Searle RJ, Stacey MC, Harding KG. Prediction and monitoring the therapeutic response of chronic dermal wounds. Int Wound J: 2006; 3: 89-98.
[10] Sipos P, Győry H, Hagymási K, Ondrejka P, Blázovics A. Special wound healing methods used in ancient egypt and the mythological background. World J Surg: 2004; 28: 211-216.
[11] Shah JB. The history of wound care. J Am Col Certif Wound Spec: 2011; 3: 65-66.
[12] Daunton C, Kothari S, Smith L, Steele D. A history of materials and practices for wound management. Wound Pract Res: 2012; 20: 174-186.
[13] Bouwstra JA, Honeywell-Nguyen PL. Skin structure and mode of action of vesicles. Adv Drug Deliv Rev: 2002; 54: 41-55.
[14] Structure and Function of the Skin. Clinimed: 2014; p. 1-2.

[15] Structure and functions of the skin Skin structure. Health and safety Executive: 2014; p. 1-2.
[16] Bensouilah J, Buck P. Skin structure and function. In: Aromadermatology. Radcliffe Publishing Ltd, UK: 2006; p. 1-11.
[17] Kerstein, D M. Introduction : Moist Wound Healing. In: The American Journal of Surgery: 1994.
[18] Field CK, Kerstein MD. Overview of wound healing in a moist environment. Am J Surg: 1994; 167: 2-6.
[19] Eaglstein WH. Moist Wound Healing with Occlusive Dressings: A Clinical Focus. Dermatologic Surg: 2001; 27: 175-182.
[20] Blume PA, Walters J, Payne W, Ayala J, Lantis J. Comparison of negative pressure wound therapy using vacuum-assisted closure with advanced moist wound therapy in the treatment of diabetic foot ulcers: a multicenter randomized controlled trial. Diabetes Care: 2008; 31: 631-636.
[21] Garner L, Rodriguez JL, Aii I. Infection and Healing. Am J Surg: 1994; 167: 46-48.
[22] Winterswijk PJ Van, Nout E. Tissue Engineering and Wound Healing : An Overview of the Past , Present , and Future. wounds: 2007; 19: 277-284.
[23] Gupta B, Agarwal R, Alam M. Textile-based smart wound dressings. Indian J Fibre Text Res: 2010; 35: 174-187.
[24] Mortimer D. Moist wound dressings and pressure relieving surfaces mechanisms , materials and a review of some cost- effectiveness findings. 104th ed. Centre for Health Program Evaluation: 2001; p.75.
[25] Horrocks AR, Anand AC. Handbook of technical textiles. Cambridge: Woodhead Publishing Limited: 2000; p. 439-447.
[26] Sciences B, Building JA. Wound healing dressings and drug delivery systems : A review. J Pharm Sci: 2008; 97: 2892-2923.
[27] Qin Y. Alginate fibres : an overview of the production processes and applications. Polym Int: 2008; 180: 171-180.
[28] Qin Y. Alginate fibres : an overview of the production processes and applications in wound management. Polym Int: 2007; 57: 171-180.
[29] Qin Y. The characterization of alginate wound dressings with different fiber and textile structures. J Appl Polym Sci: 2006; 100: 2516-2520.
[30] Yong K, Mooney DJ. Alginate : properties and biomedical applications. Prog Polym Sci: 2012; 37: 106-26.
[31] Draget KI, Taylor C. Chemical, physical and biological properties of alginates and their biomedical implications. Food Hydrocoll: 2011; 25: 251-256.
[32] Donati I, Paoletti S. Material properties of alginates. In: Rehm BHA, editor. Alginates: Biology and Applications. 1st ed. Springer: 2008; p. 2-54.
[33] Tarun K, Gobi N. Calcium alginate / PVA blended nano fibre matrix for wound dressing. Indian J Fibre Text Res: 2012; 37: 127-132.
[34] Mikołajczyk T, Boguń M, Kurzak A. Zinc alginate fibres with a tricalcium phosphate (TCP) nanoadditive. Fibers Text East Eur: 2009; 17: 12-18.
[35] Wang Q, Du Y, Hu X, Yang J, Fan L, Feng T. Preparation of alginate/soy protein isolate blend fibers through a novel coagulating bath. J Appl Polym Sci: 2006; 101: 425-431.
[36] Fan L, Du Y, Huang R, Wang Q, Wang X, Zhang L. Preparation and characterization of alginate/gelatin blend fibers. J Appl Polym Sci: 2005; 96: 1625-1629.
[37] Raafat D, Bargen K Von, Haas A, Sahl H. Insights into the mode of action of chitosan as an antibacterial compound. Appl Environ Microbiol: 2008; 74: 3746-3773.
[38] Elsabee MZ, Naguib HF, Morsi RE. Chitosan based nanofibers, review. Mater Sci Eng C: 2012; 32: 1711-1726.
[39] Viarsagh MS, Janmaleki M, Falahatpisheh HR. Chitosan preparation from persian gulf shrimp shells and investigating the effect of time on the degree of deacetylation. J Paramed Sci: 2008; 1: 2-7.
[40] Bagheri-Khoulenjani S, Taghizadeh SM, Mirzadeh H. An investigation on the short-term biodegradability of chitosan with various molecular weights and degrees of deacetylation. Carbohydr Polym: 2009; 78: 773-778.
[41] Sashiwa H, Aiba S. Chemically modified chitin and chitosan as biomaterials. Prog Polym Sci: 2004; 29: 887-908.
[42] Vandevord PJ, Matthew HWT, Desilva SP, Mayton L, Wu B, Wooley PH. Evaluation of the biocompatibility of a chitosan scaffold in mice. John Wiley Sons, Inc: 2001; 59: 585-590.
[43] Niekraszewicz A. Chitosan medical dressings. FIBRES Text East Eur: 2005; 13: 16-18.

[44] Stęplewski W, Wawro D, Niekraszewicz A, Ciechanska D. Research into the process of manufacturing alginate-chitosan fibres. Fibres Text Easten Eur: 2006; 14: 25-31.
[45] Fan L, Li M, Gong Y, Peng K, Xie W. Preparation and characterization of alginate / hydroxypropyl chitosan blend fibers. J Appl Polym Sci: 2011; 125: 829-835.
[46] Fan L, Yu L, Xu Y, Yi C, Cai J, Li M, et al. The novel alginate / N- succinyl-chitosan antibacterial blend fibers. J Appl Polym Sci: 2010; 116: 2151-2156.
[47] Maeda N, Miao J, Simmons TJ, Dordick JS, Linhardt RJ. Composite polysaccharide fibers prepared by electrospinning and coating. Carbohydr Polym: 2013; 102: 950-955.
[48] Sweeney IR, Miraftab M, Collyer G. Absorbent alginate fibres modified with hydrolysed chitosan for wound care dressings-II. Pilot scale development. Carbohydr Polym: 2014; 102: 920-927.
[49] Necas J, Bartosikova L, Brauner P, Kolar J. Hyaluronic acid (hyaluronan): a review. Vet Med (Praha): 2008; 53: 397-411.
[50] Ji Y, Ghosh K, Zheng X, Li B, Sokolov JC, Prestwich GD, et al. Electrospun three-dimensional hyaluronic acid nanofibrous scaffolds. Biomaterials: 2006; 27: 3782-3792.
[51] Doillon CJ, Silver FH. Collagen-based wound dressing: Effects of hyaluronic acid and fibronectin on wound healing. Biomaterials: 1986; 7: 3-8.
[52] Yamane S, Iwasaki N, Kasahara Y, Harada K, Majima T, Monde K, et al. Effect of pore size on in vitro cartilage formation using chitosan-based hyaluronic acid hybrid polymer fibers. J Biomed Mater Res Part A: 2006; 81: 586-593.
[53] Hussain T, Masood R, Umar M, Areeb T, Ullah A. Development and characterization of alginate-chitosan-hyaluronic acid (ACH) composite fibers for medical applications. Fibers Polym: 2016; 17: 1749-1756.
[54] Yamane S, Iwasaki N, Majima T, Funakoshi T. Feasibility of chitosan-based hyaluronic acid hybrid biomaterial for a novel scaffold in cartilage tissue engineering. Biomaterials: 2005; 26: 611-619.
[55] Moreira-teixeira LS, Sc M, Moroni L, Ph D, Reis RL, Mano F. Chitosan scaffolds containing hyaluronic acid. Tissue Eng Part C: 2011; 17: 717-730.
[56] Ahmadi R, Kalbasi-Ashtari A, Oromiehie A, Yarmand MS, Jahandideh F. Development and characterization of a novel biodegradable edible film obtained from psyllium seed (Plantago ovata Forsk). J Food Eng: 2012; 109: 745-751.
[57] Guo Q, Cui SW, Wang Q, Goff HD, Smith A. Microstructure and rheological properties of psyllium polysaccharide gel. Food Hydrocoll: 2009; 23: 1542-1547.
[58] Patil BS, Mastiholimath VS, Kulkarni AR. Development and evaluation of psyllium seed husk polysaccharide based wound dressing films. Orient Pharma Exp Med: 2011; 11: 123-129.
[59] Cheng Z, Blackford J, Wang Q, Yu L (Lucy). Acid treatment to improve psyllium functionality. J Funct Foods: 2009; 1: 44-49.
[60] Westerhof W, Das P, Middelkoop E, Verschoor J, Storey J, Rehnier C. Mucopolysaccharides from psyllium involved in wound healing. Drugs Exp Clin Res: 2001; 27: 165-175.
[61] Miraftab M, Masood R. Polysaccharide fibres for wound dressings. US Patent No. 20140227370 A1: 2014.
[62] Miraftab M, Masood R, Edward-Jones V. A new carbohydrate-based wound dressing fibre with superior absorption and antimicrobial potency. Carbohydr Polym: 2014; 101: 1184-1190.
[63] Masood R, Miraftab M. Novel materials for moist wound management: Alginate-psyllium hybrid fibres. J Wound Care: 2014; 23: 153-159.
[64] Masood R, Miraftab M, Hussain T, Edward-jones V. Development of slow release silver-containing biomaterial for wound care applications. J Ind Text: 2013; 44: 1-10.
[65] Gadd GM, Griffiths AJ. Microorganisms and Heavy Metal Toxicity. Microb Ecol: 1978; 4: 303-317.
[66] Castellano JJ, Shafii SM, Ko F, Donate G, Wright TE, Mannari RJ, et al. Comparative evaluation of silver-containing antimicrobial dressings and drugs. Int Wound Journa: 2007; 4: 114-122.
[67] Ruparelia JP, Chatterjee AK, Duttagupta SP, Mukherji S. Strain specificity in antimicrobial activity of silver and copper nanoparticles. Acta Biomater: 2008; 4: 707-716.
[68] Grass G, Rensing C, Solioz M. Metallic copper as an antimicrobial surface. Appl Environ Microbiol: 2011; 77: 1541-1547.
[69] Xie Y, He Y, Irwin PL, Jin T, Shi X. Antibacterial activity and mechanism of action of zinc oxide nanoparticles against Campylobacter jejuni. Appl Environ Microbiol: 2011; 77: 2325-2331.
[70] Qin Y. Silver-containing alginate fibres and dressings. Int Wound J: 2005; 2: 172-176.

[71] Qin Y, Zhu C, Chen J, Chen Y, Zhang C. The absorption and release of silver and zinc ions by chitosan fibers. J Appl Polym Sci: 2006; 101: 766-771.
[72] Grace M, Chand N, Bajpai, Kumar S. Copper Alginate-Cotton Cellulose (CACC) Fibers with Excellent Antibacterial Properties. J Eng Fiber Fabr: 2009; 4: 24-35.
[73] Zhang W, Zhang Y, Ji J, Yan Q, Huang A, Chu PK. Antimicrobial polyethylene with controlled copper release. J Biomed Mater Res Part A: 2007; 838-844.
[74] Ureyen ME, Dogan A, Koparal AS. Antibacterial functionalization of cotton and polyester fabrics with a finishing agent based on silver-doped calcium phosphate powders. Text Res J: 2012; 82: 1731-1742.
[75] Parikh DV., Fink T, DeLucca AJ, Parikh AD. Absorption and swelling characteristics of silver (I) antimicrobial wound dressings. Text Res J: 2010; 81: 494-503.
[76] Tessier D, Filteaus M. Antimicrobial Fabrics Coated with Nano-Sized Silver Salt Crystals. NSTI-Nanotech: 2005; 1: 762-764.

Melt Spinning of Medical Grade Polypropylene

Yi-Ke Wang, Hong-Tao Liu*

School of Material Science & Engineering, Wuhan Textile University, Wuhan, Hubei 430200, China

*Corresponding author's email: Lht@wtu.edu.cn

Abstract

Charpy impact strength and melt flow rate of a commercially available medical grade polypropylene for injectors and infusion bottles were measured, then the melt-spun filament of the medical polypropylene (MPP) was produced for potential hernia repair mesh by a laboratory mixing extruder with a take-up system. The results showed that the average Charpy impact strength for the notched MPP plate specimen is 4.4 kJ/m^2. Melt flow rates for MPP and control PP increased with the increasing temperature. Melt flow rate of MPP is lower than that of control general PP samples. The Young's modulus and sonic velocity orientation angle of the as-spun MPP filament are calculated as 1.94 GPa and 54.2 °, respectively. From the result of this study, the relatively proper condition for the spinning of MPP is at the extrusion temperature of 210 °C, the winding speed of 2.15 m/min and the extrusion rate of 16 r/min.

Keywords: Melt Spinning; Medical Polypropylene; Thermal Property; Mechanical Property

1. Introduction

Polypropylene (PP) has been extensively used in the manufacture of medical devices such as tubing, catheters, dialysis units and syringes [1, 2]. With the increasing popularity of mesh use, a wide variety of synthetic meshes (including PP) are now commercially available, making the selection of the most appropriate mesh difficult [3]. However, it is necessary and interesting to make more new product and find the corresponding detailed process. In this preliminary study, we aimed to gain a better understanding of the mechanical properties of a commercially available medical grade polypropylene and its as-spun filament.

2. Experimental

2.1 Starting Materials

Commercially available medical grade polypropylene (MPP, random copolymer) named GM1600E was derived from Sinopec Shanghai Petrochemical Company Limited. Control general PP samples (model Jin'ao and Luoyang) were also obtained from Sinopec.

2.2 Preparation of MPP Plate Specimens

To form the notched impact specimens, dried MPP granules were melt at 230 °C and injected directly into the predetermined geometry 80 mm (length) × 10 mm (width) × 4 mm (thickness) by using an injection molding machine (Model HBL-1300, Ningbo Haibo Machinery Manufacture Co. Ltd., China).

2.3 Melt Spinning Process and As-spun Filament of MPP

To obtain the as-spun filament of medical polypropylene (MPP), a laboratory mixing extruder with a take-up system (Dynisco, Franklin MA, USA) was employed. Drawing the MPP melt from the laboratory mixing extruder into fibers, the take-up system pulls the fibers into smaller diameters, wrapping them around a spindle. The winding speed was controlled at 2.15 m/min to match the extrusion rate (15-20 r/min) and produce the desired fiber diameter (0.1-0.2 mm) at the varying extrusion temperatures of 180 °C, 190 °C, 200 °C, and 210 °C, respectively. Fig.1. shows the starting medical polypropylene (left) and its as-spun filament (right).

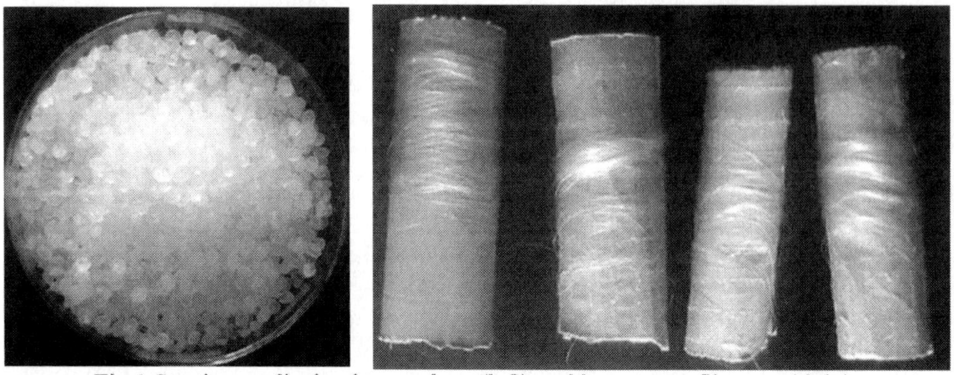

Fig.1 Starting medical polypropylene (left) and its as-spun filament (right).

2.4 Property Measurements

2.4.1 Charpy Impact Strength of MPP Plate

For notched MPP plate specimens, Charpy impact strengths of MPP were determined according to the Chinese standard GB/T 1043-93 (idt. DIN EN ISO 179 standard) in the edgewise notched (notch type A) modes, using an XJJ-50 pendulum-type impact tester (Chengde Jinjian Testing Instrument Co. Ltd., China) [4]. The test samples were conditioned at 23 °C and 50% relative humidity for 3 days before testing and all the tests were performed under the same conditions. For each specimen, 12 measurements were carried out and the average values were calculated. Standard deviation was within 1% as well.

2.4.2 Melt Flow Rate of MPP

With a steady load of 2160 g, the melt flow rate (MFR) of MPP and control PP samples (model jinao and luoyang, also from Sinopec) was measured using a melt indexer (model XNR-400AM, Chengde Dahua Testing Instrument Co. Ltd., China), according to Chinese standard GB/T 3682-2000 at 210 °C, 220 °C, 230 °C, and 240 °C, respectively. The inner diameter of hole of the melt flow is 2.095 ± 0.005 mm. For each specimen, 5 measurements were carried out and the average values were calculated. Standard deviation was within 1% as well.

2.4.3 Young's Modulus and Sonic Velocity Orientation Angle of as-spun MPP Filament

A Young's modulus and sonic velocity orientation angle tester for fibers (model SCY-III, Shanghai Donghua Kaili Chemical Fiber High-Tech Co. Ltd., China) was employed.

3. Results and Discussion

3.1 Charpy Impact Strength of MPP Plate

A pendulum-type single-blow impact test in which the specimen is supported at both ends as a simple beam and broken by a falling pendulum was used. The absorbed energy per unit area (kJ/m^2), as determined by the subsequent rise of the pendulum, is a measure of Charpy impact strength. The results show that the average Charpy impact strength for the notched MPP plate specimen is 4.4 kJ/m^2.

3.2 Melt Flow Rate of MPP

Melt flow rates for MPP and control PP samples are shown in Tab. 1 and Fig. 2, indicating that the melt flow rates of all PP samples increased with the increasing temperature. Melt flow rate of MPP is lower than that of control general PP samples. All data are the averages from less than 5 samples.

Table.1 Melt flow rates for MPP and control PP samples (g/10min)

PP samples	210 °C	220 °C	230 °C	240 °C
MPP	10.560	12.774	15.030	18.318
PP (Jin'ao)	12.834	14.631	19.843	22.584
PP (Luoyang)	18.498	24.516	30.120	34.728

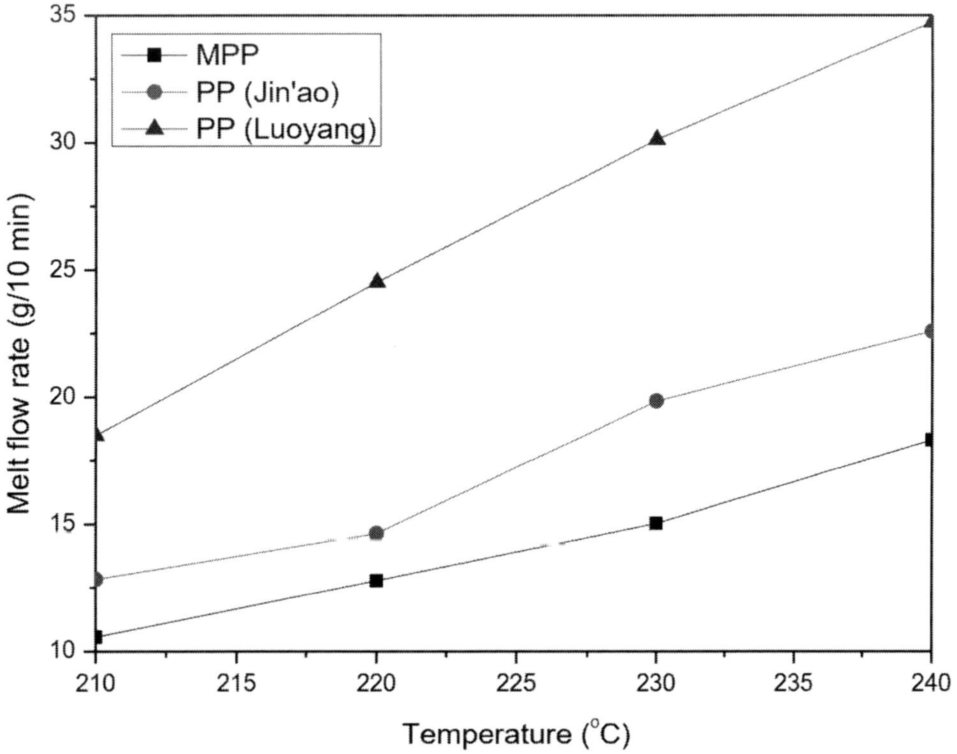

Fig.2 Melt flow rates for MPP and control PP samples with a steady load of 2160 g.

3.3 Young's Modulus and Ssonic Velocity Orientation Angle of As-spun MPP Filament

The Young's modulus and sonic velocity orientation angle tester for MPP fiber were calculated by the following formula (1) and (2):

$$\overline{COS^2\theta} = 1 - \frac{2}{3}(\frac{C_u}{C})^2 \quad (1)$$

Where θ is the sonic velocity orientation angle (°) for MPP fiber, the sonic velocity in completely amorphous PP, i.e. Cu (PP) = 1.45 km/s, and C is the sonic velocity (km/s) tested in MPP fiber.

$$E = \rho \cdot C^2 \tag{2}$$

Where E is the Young's modulus (MPa) for MPP fiber, ρ is the density (910 kg/m3) for MPP fiber, and C is the sonic velocity (km/s) tested in MPP fiber.

The sonic velocity tested in MPP fiber, i.e. C = 1.46 km/s, and the Young's modulus (E) and sonic velocity orientation angle (θ) of the as-spun MPP filament are calculated as 1.94 GPa and 54.2 °, respectively. The value of Young's modulus (E) is consistent with the data of polypropylene reported [5-6], and the θ value is very close to that of completely amorphous PP (54.7 °). This may be due to the as-spun MPP filament was produced without enough high-speed extension. From the result of this study, the relatively proper condition for the spinning of MPP is at the extrusion temperature of 210 °C, the winding speed of 2.15 m/min and the extrusion rate of 16 r/min.

4. Conclusions

In this paper, Charpy impact strength and melt flow rate of a commercially available medical grade polypropylene for injectors and infusion bottles were measured. The melt-spun filament of the medical polypropylene (MPP) was produced for potential hernia repair material by a laboratory mixing extruder with a take-up system. The results show that the average Charpy impact strength for the notched MPP plate specimen is 4.4 kJ/m^2. Melt flow rates for MPP and control PP increased with the increasing temperature. Melt flow rate of MPP was lower than that of control general PP samples. The Young's modulus and sonic velocity orientation angle of the as-spun MPP filament are calculated as 1.94 GPa and 54.2 °, respectively. From the result of this study, the relatively proper condition for the spinning of MPP is at the extrusion temperature of 210 °C, the winding speed of 2.15 m/min and the extrusion rate of 16 r/min.

5. Acknowledgements

This work was financially supported by the projects of Hubei Provincial Department of Education (XD2012220 and 201210495002) and the project of Wuhan Textile University (2013CXXL005).

References

[1] Fuzail M, Hill DJT, Anwar J, Jahan MS, Rintoul L. Effectiveness of DOP mobilizer on the radiolysis of a semi-crystalline ethylene-propylene copolymer. Nulear Instru Math Phys Res B: 2007; 265: 285-289.

[2] Aguayo-Albasini JL, Moreno-Egea A, Torralba-Martínez JA. The labyrinth of composite prostheses in ventral hernias. Cirugía Española (Eng. Ed.): 2009; 86: 139-146.

[3] Li X, Kruger JA, Jor JWY, Wong V, Dietz HP, Nash MP, Nielsen PMF. Characterizing the ex vivo mechanical properties of synthetic polypropylene surgical mesh. J Mech Behav Biomed Mater: 2014; 37: 48-55.

[4] Liu H, Zuo D, Liu H, Li L, Li J, Xu W. Enhanced impact strength and deviated thermal decomposition of PP and PS toughened with graft-modified rubbers. e-Polym: 2010; 134: 1-5.

[5] Mareri P, Bastide S, Binda N, Crespy A. Mechanical behaviour of polypropylene composites containing fine mineral filler: Effect of filler surface treatment. Compos Sci Technol: 1998; 58: 747-752.

[6] Iroh JO, Berry JP. Mechanical properties of nucleated polypropylene and short glass fiber-polypropylene composites. Eur Polym J: 1996; 32: 1425-1429.

An Exploration on the Influence of Shapewear on Skin Blood Flow

Ao-Xue Wang, Jie Zhou*, Sha-Sha Liu, Hui -Hua Wen

Apparel & Art Design College, Xian Polytechnic University, No.19 South Jinhua Road, Xian, Shaanxi, 710048, China

*Corresponding author's email: 446961933@qq.com

Abstract

Aim: To explore the impact of garment pressure on skin blood flow. Method: Three healthy adult women were recruited to be subjects for this study. A clothing contact pressure testing system was used to measure the skin blood flow under bare and dressed conditions, the findings of which were subject to a comparative analysis. Result: After comparing and analysing the findings, it was found that the skin blood flow obviously changed after dressing. Conclusion: Research shows that wearing shapewear reduces the skin blood flow, while skin blood flow velocity and skin temperature are increased.

Keywords: Shapewear; Garment Pressure; Skin Blood Flow; Skin Temperature

1. Introduction

In recent years, shapewear has become favoured by many consumers because it can reshape the body curve. Shapewear is made by special tailoring, increasing the density of fabric produced by the mechanical force to localize the fat shift, thereby improving the local shape. Today, shapewear is not only a simple piece of tight underwear, but includes a variety of corrective underwear, such as bras, one-piece and split-type shapewear, staylace and pantyhose. Most shapewear contains spandex or other elastic fibres, while the size of the finished product is smaller than the size of the body. So, when wearing shapewear, the body will experience pressure. Garment pressure is mainly generated in two respects [1, 2]: the weight of the garment itself and the pressure generated by the garment as it wraps the body. Garment pressure may have a positive or negative influence on the body shape and physiological system. The safety and comfort of clothing are also increasingly valued by practitioners and consumers.

Different degrees of oppression can affect the body's skin temperature, blood pressure, breathing, heart rate, skin blood flow and even autonomic nervous system [3, 4]. Meanwhile, soft tissues, such as skin, fat and muscle, are easy to deform. If the microvasculature contained in the soft tissue is squeezed by external mechanical compression, it may be forced to deform or crush, which may affect blood microcirculation. Normally, the microcirculation of blood flow and organ metabolic levels adapt to each other in order to ensure the organ's blood perfusion and adjust the blood accordingly. If there are obstacles to microcirculation, the physiological functions of various organs will be directly affected. One Japanese gynaecologist has long pointed out that shapewear, when "tied" in the body, will affect normal blood circulation and sweat excretion. In addition, it will make the vulva humid, cause bacteria to multiply and lead to gynaecological diseases. Some studies, such as [5], show that, when the abdomen is without bone support, even a small amount of pressure can change the location and shape of internal organs. Shapewear will also affect the normal activities of the digestive and reproductive system, as well as easily induce genital inflammation, breast hyperplasia, cysts and other diseases, which can cause infertility [6].

Currently, there is a great deal of interest, particularly among Japanese scholars, in research that brings together the medical field of physiology and the field of textile and clothing. The main method used in such research is to test the physiological response to textile and clothing in order to understand the stressful effects on human physiology. However, to date, studies have mainly focused on pressure socks [7-11], pants [12-13], the waist [14] and the corset [15]. Despite the fact that domestic and foreign experts on the impact of oppression on human physiology have reported many valuable results, this study mainly focuses on the individual parts of the compression effect on skin blood flow, given that there has been very little research on the impact of comprehensive shapewear on skin blood flow. Therefore, this paper investigates the compression effect of comprehensive shapewear on skin blood flow in order to measure the impact of shapewear on skin blood flow.

2. Research Approach

2.1 Experimental Preparation

2.1.1 Laboratory Equipment

This experiment seeks to measure the pressure, temperature and blood flow in undressed and dressed states. The experimental test equipment, as shown in Fig.1, is a system for collecting pressure and blood flow signals. The integrated sensor can measure pressure, temperature and blood flow. Methods for measuring blood flow include the isotope method, an infrared camera, and the laser Doppler approach (laser Doppler technology is widely used because it is non-interventionist, non-invasive and easy to operate, among other advantages). In this study, the instrument used to measure skin blood flow is a clothing contact pressure test system (AMI-3037) manufactured in Japan. The blood flow sensor uses a laser probe, which can detect the subcutaneous radius of about 1mm in the hemisphere within the skin (mainly capillaries), as well as continuous changes in blood flow every 0.1 s, in order to collect data. The three parameters of "Flow", "Mass" and "Velocity" can be obtained simultaneously. Among them, "Flow" refers to the total amount of blood flow in 100 g of soft skin tissue within 1 min. "Mass" means the amount of blood measured at a given point in time. "Velocity" means the equivalent velocity of blood flow in the blood vessel.

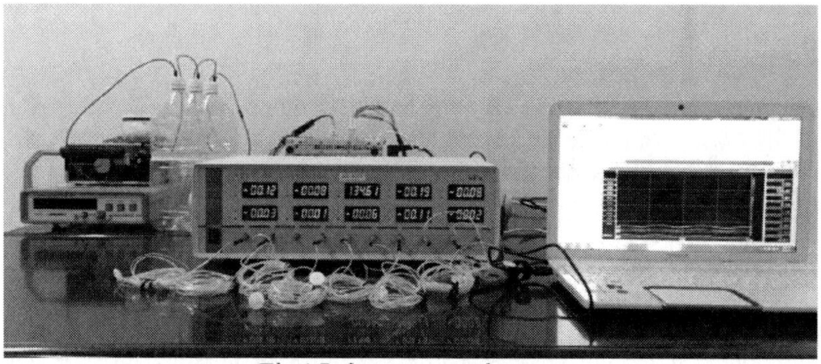

Fig.1 Laboratory equipment

2.1.2 Experimental Environment

The room was kept quiet, and had no direct sunlight nor infrared radiation. The laboratory temperature was controlled at 27± 3℃, while the relative temperature was 60±10% and the wind speed was ≤0.1 m/s. Subjects entered the laboratory at the same time every day (14:00-16:00). All experiments were completed in November and December 2016.

2.1.3 Subjects

Three undergraduates were selected as subjects and labelled as S1-S3. The age of the three subjects was 21, 22 and 24 years. Their body mass index was 23.18, 22.86 and 19.83, respectively. All subjects were asked to avoid strenuous exercise for 24 hours before the experiment, and to not consume spicy food, alcoholic or caffeine drinks, or other stimulating edible items 17 hours beforehand. In order to eliminate the influence of the external environment, personal emotions and exercise on skin blood flow, each subject was required to maintain an emotional stability at the set temperature on entering the laboratory for 15 min, during time which subjects were informed about the test methods and purpose. The subjects were allowed to sit for half an hour before the experiment started. In the course of the experiment, the subjects were asked to remain calm and relax.

2.2 Determining the Measurement Point

The theoretical basis of the test points is as follows:

1) Capillary-rich parts of skin blood flow changes significantly and can be easily observed, which should be considered as the key point. As hair and sweat glands typically have more capillaries, armpits were chosen as a test point.

2) Tanaka Xi Mi's findings suggest that changes in blood flow are related to the size of the dressing pressure. According to the method for calculating clothing pressure, by using the radius of curvature, it can be seen that the garment pressure applied to the contact part increases in line with the increase in the curvature of a part of the human body. In addition, the lack of adipose tissue protection or muscle wrap means that muscle or thin bone flaps are more easily damaged by pressure. Therefore, some of the salient points of the human body and the protruding part of the human body are considered as key areas of interest during an experimental test.

3) According to medical knowledge, some parts of the body are more sensitive to pressure, causing discomfort in the stomach and liver, for example. As their capacity to tolerate pressure is small, in the course of the experiment, it is necessary to take such parts of the body into account.

4) Some parts are neither body bumps, nor sensitive parts, but are needed for shaping functions, such as hips or legs.

Based on the above analysis, and after the pre-experiment, the study selected nine test points as the research objects: P1 (armpit), P2 (areola), P3 (shoulder tip), P4 (ventral point), P5 (stomach), P6 (under the bust line near the right chest), P7 (side waist point), P8 (about 3 cm under the thigh), P9 (peak of buttock).

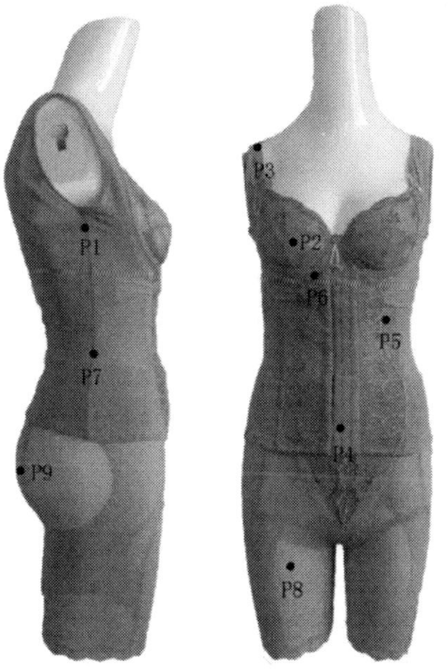

Fig.2 Measurement points

2.3 Experimental Scheme

In this study, the measurement of each measurement point of skin blood flow was completed in both undressed and dressed states. Each subject was asked to adopt the standing posture, with her upper body exposed, while the lower half wore pressure-free underpants. The laser beam centre was aligned with the test point, and each point was measured for 2 min. After wearing the shapewear, the subject had to rest for 30 min, after which the skin blood flow of each test point was measured. Finally, the measurement results were saved for analysis.

2.4 Statistical Processing

SPSS 17.0 statistical software was used for statistical processing. All data were expressed by X±SD. The average and standard deviation of each test point were calculated, along with the rate of change for each parameter. Finally, a comparative analysis was made.

3. Experimental Results and Analysis

3.1 Pressure Distribution of Shapewear

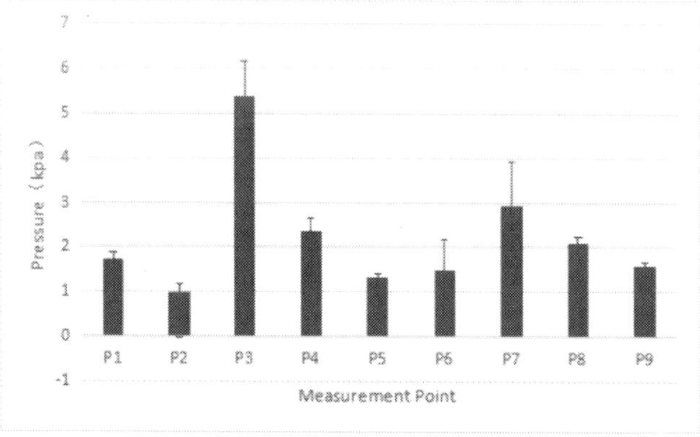

Fig.3 Pressure distribution histogram

According to Fig.3, the pressure from high to low was in the following order: P3, P7, P4, P8, P1, P9, P6, P5, P2. The largest pressure was experienced by the shoulder (5.39 kPa). This was caused by the shoulder strap applying a downward force through the entire garment to the point of support, resulting in greater pressure compared to other aspects. The next largest pressure was located at the waist, abdomen and legs, where the pressure is relatively greater given the garment's function in shaping the waist and legs. Pressure on the chest and hips was relatively small, which may be explained by the thickness of the fat at these locations, as fat can offer a buffer effect. Usually, shapewear supports and holds up the entire breast, which should mean that the pressure is large. But, in this experiment, the area under the bra was wide, such that the contact stress area was large, resulting in a small amount of pressure.

3.2 Changes in Skin Blood Flow

3.2.1 Comparison of Skin Blood Flow Before and After Dressing

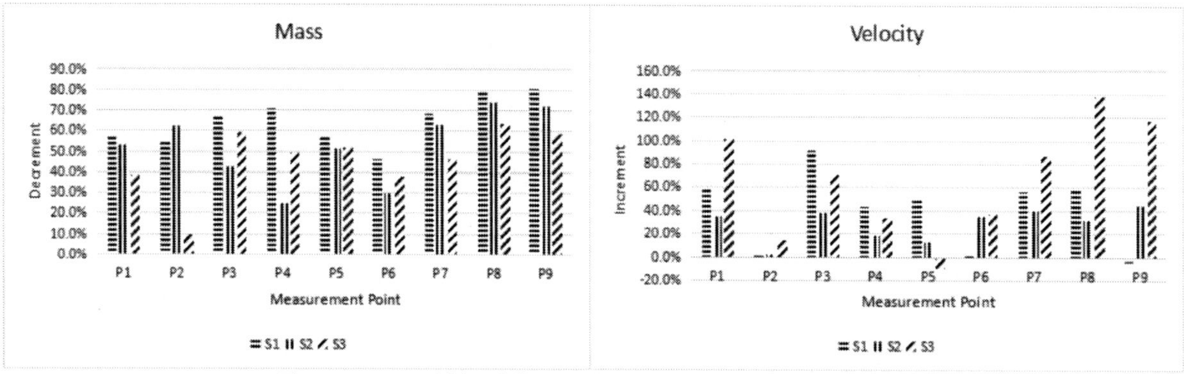

Fig.4 Decreased blood flow and increased blood flow velocity

As is shown in Fig.4, after dressing, the overall trend of blood flow was reduced and the blood flow velocity increased. By using Excel to calculate the rate of change in blood flow velocity at each measuring point for each subject, we found two subjects who had a blood flow rate measuring point that was comparable to the reduction in the non-dress state. This may be explained by the fact that the pressure at the measuring point was greater than the pressure of the other two. Studies have shown that very small pressure (<2 kPa) applied to the skin's surface can significantly reduce the skin's blood flow. Therefore, research should be carried out to further explore the pressure range of the blood flow for each test point.

3.2.2 Comparison of Skin Temperature Before and After Dressing

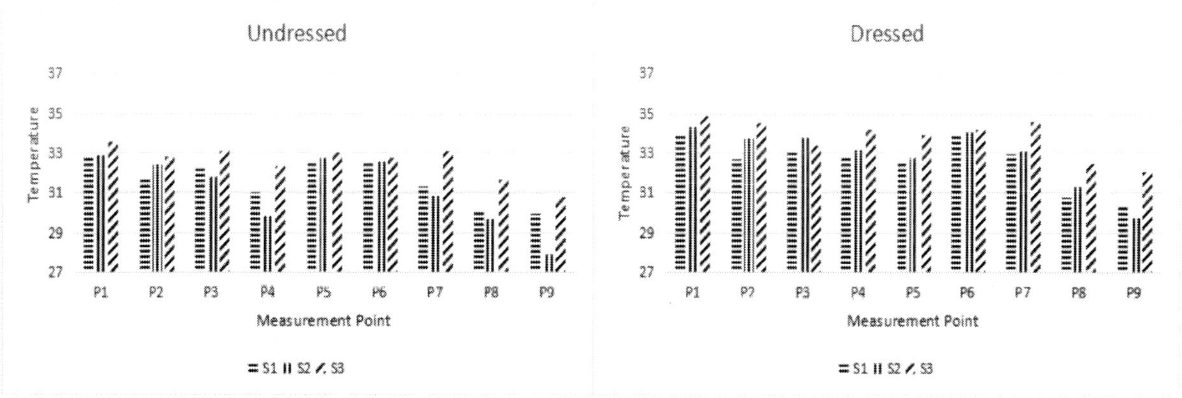

Fig.5 Comparison of skin blood temperature before and after dressing

Generally speaking, for normal human skin, upper limb temperature is higher than lower limb temperature by about 2~3℃. Indeed, as can be seen in Fig.5, our study found that the upper body temperature of the human body was higher than that of the lower body temperature. During the experiment, we also found that, in addition to controlling the experimental environment and factors relevant to the subjects, the measurement sequence should be arranged reasonably. In addition to a lower extremity temperature, the temperature of the subcutaneous fat and other parts of the skin is also low. The experimental results show that temperature of the hip bumps, both in the lower limbs and the subcutaneous fat, at a certain time, is more likely to fall compared to other parts. As such, the order of the hip bumps should be measured beforehand in order to avoid experimental errors. After dressing, the skin temperature of the subjects was generally increased, with the most obvious changes seen in the temperature of the ventral point. The most significant change in the temperature of the ventral point was due to the thickness of the fabric in the shapewear. This experiment used shapewear, which comprised a bra, waist clip and bundle pants. The waist clip and the bundle pants all covered the abdomen, so the temperature change in this regard was the most obvious.

4. Conclusion

1) The pressure distribution of shapewear was P3>P7>P4>P8>P1>P9>P6>P5>P2. As the largest pressure was found on the shoulder, the appropriate adjustment of the shoulder strap length is recommended to increase comfort.

2) Wearing shapewear was found to reduce blood flow, while skin blood flow velocity showed an increasing trend. Subsequent work should explore the relationship between pressure, skin temperature and blood flow velocity.

3) When undressed, the upper extremity temperature of the skin was found to be higher than the lower extremity temperature; in other words, the skin temperature increased after dressing.

5. Acknowledgement

We would like to thank Nanjing Joymain Science & Technology Development Co. Ltd for funding this research.

We would also like to thank the teachers of Shanxi Province Garment Engineering Technology Research Centre for providing experimental sites and equipment for our study.

References

[1] Wang J, Dai XQ, Liu GL. Clothing pressure and its influence on human body. Silk Text Technol Overseas: 2007; 5: 20-22.
[2] Xu J, Tao KS. Introduction to Ergonomics. 2002, Beijing: Chin Text & Apparel Press: 147-152.
[3] Liu L, Zhang WY, Wang YJ. The effect of adjusting underwear pressure on human physiological indexes. J Text Research: 2008 (11): 97-100.
[4] Morooka H, Nakahashi M, Kitamura K. Effects of clothing pressure exerted on a trunk on heart rate, blood pressure, skin blood flow and respiratory function. J Text Mach Soc Jpn: 2001; 54 (2): T37-T42.
[5] Watanabe MT. Effect of clothing pressure on human. J Home Econ: 1976; 27 (1): 44-50.
[6] Pan JH. Human Ergonomics and Costume Design. Beijing: Chin Light Ind Press: 2000.
[7] Lu YH. Effects of Lower Limb Dressing on Skin Blood Flow and Its Mechanism. Jiangsu: Suzhou University: 2010.
[8] Wang J. Study on Dressing Stress and Human Physiological Comfort: Study on Skin Blood Flow Changes of Lower Extremities. Jiangsu: Suzhou University: 2008.
[9] Sajid MS, Tai N, Goli G et al. Knee versus thigh length graduated compression stockings for prevention of deep venous thrombosis: a systematic review. Eur J Vasc Endovasc Surg: 2006; 32: 730-736.
[10] Nakahashi M, Morooka H, Nakajima C et al. Effects of clothing pressure of pantyhose with controlled loop length on the human circulatory system. Sen'i Gakkaishi: 2003; 59 (10): 407-413.
[11] Ni HY. Study on pressure and human physiological comfort of pressure socks. Light Text Ind Technol: 2014; 06: 12-14, 80.
[12] Watanuki S, Murata H. Effects of wearing compression stockings on cardiovascular responses. Ann Physiol Anthropol: 1994; 13 (3): 121-7.
[13] Wu LY. Effect of shorts on skin blood flow. Melliand China: 2012; 08: 71-72 + 74-76 +78.
[14] Nakahashi M et al. An analysis of waist-nipper factors that affect subjective feeling and physiological response: for the design of comfortable women's foundation garments. Sen'iGakkaishi: 2005; 61 (1): 6-12.
[15] Morooka H, Fukuda R, Sasaki H et al. The effects of the tensile properties of cup-stand materials on clothing pressure and comfort of push-up type brassieres. Sen'iGakkaishi: 2006; 62 (12): 287-292.

Effects of Women's Socks Opening Specification on Tensile Properties of Socks' Mouths

Lan Zhang, Xiao-Wen Jiang*

Apparel & Art Design College, Xi'an Polytechnic University, Beilin District, Xi'an, Shaan xi, 710048,P. R. China

*Corresponding author's email: xwenj20134@163.com

Abstract

The purpose of this paper is to explore the factors that affect the tensile properties under low stress of female sox mouth. The experimental samples include eighteen kinds of woven female socks with three types. They have the same raw material but with different specifications and organizational structures. We used the YG026PC-500 type electronic fabric strength machine to carry out the test regarding the fixed elongation of the sox mouth. The experimental data was analyzed by two pairs of paired T test. Results show that sox mouth height, sox mouth width, and sox mouth weaving structure are significant differences in tensile properties under low stress.

Keywords: Women's Socks; Socks Opening Specifications; Specified Elongation; Tensile Properties

1. Introduction

With the continuous development of variety of tights, elastic fabrics, the research on the clothing pressure comfort is becoming more and more important. At the same time, clothing pressure comfort has become one of the hot topics at home and abroad [1-6]. The circumference of sock openings is the key part of its pressure. Appropriate pressure of sock openings play an important role on the clothing wearing comfort and the health of the lower limb. From the physical mechanism of producing clothing pressure, to the relation of clothing pressure to its material elastic performance and tensile length in the same body parts. As we study the situation of the sock mouth's pressure on the human body, we must first consider the elasticity and tensile properties of socks' mouths. Therefore, the research of sox mouth tensile properties has practical significance for the optimization design of socks' mouths.

There are some foreign scholars such as Homata and Toshiyuki who studied variances of socks' mouths and gave different comfort pressure values respectively [7-8]. In China, Daqi Shen and Sudan Zhu from Xi'an Textile and Engineering College studied the relationship between elastic socks and its transverse tensile elasticity. They also analyzed the relationship between the length of the coil and the elongation tensile force and tensile elastic recovery rate [9]. Aifeng Wang and SuzhenPang analyzed the relationship between the structure and performance of socks, and suggested that the structure and performance of socks are one of the important factors affecting the appearance and performance of socks [10]. Lei Tan studied the performance of the rubber hose comprehensively and analyzed the effects of various parameters on the performance [11]. Guiyuan Liao analyzed the deformation of the rubber socks and discussed the factors caused the deformation of elastic stockings' mouths, and proceeded to put forward the first problem to solve the deformation of elastic stockings is to overcome horizontal stroke pulls [12]. These abnormal sock shapes are shown in Fig.1.

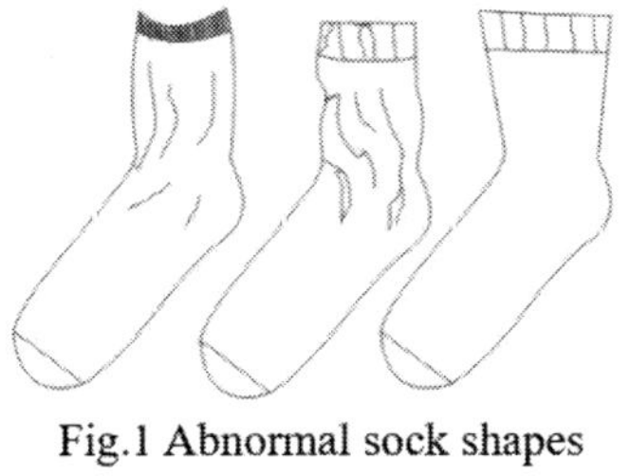

Fig.1 Abnormal sock shapes

2. Experiment

2.1 Experimental Method

Using the YG026PC-500 type electronic fabric strength machine to carry out the experiments revolving around the recovery of socks' mouths constant stretch elongation. Refer to the standard of FZ-T5007 to set experimental parameters. Each test sample was stretched twice. The final experimental data comprised of arithmetic means of the data obtained by the two tensile experiments. The specific parameter settings are shown in Table 1, and the experimental instrument and test sample style are shown in Fig.2.

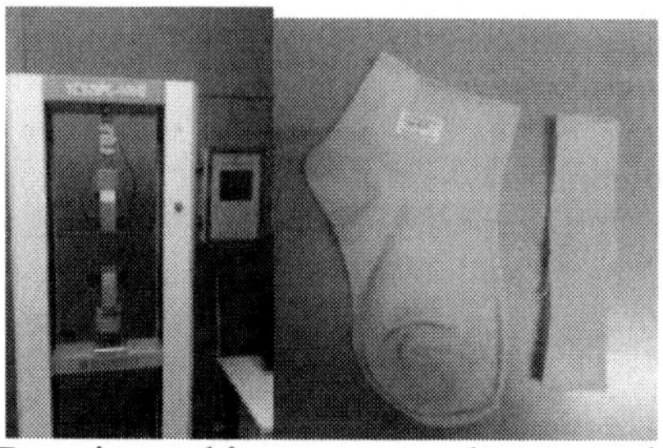

Fig.2 Experimental instrument and test sample style

Table.1 Specific parameter settings

Number of stretches	2	Initial clamping length	100mm
Predetermined elongation length	150mm	Pre-tensioning	0.2N
Driving speed	50mm/min	Down speed	100mm/min

2.2 Experimental Principle

In the national standard "FZ/T73001-2008", the horizontal elongation of the socks is used to reflect the sock extension performance, and the experience value of adult female socks horizontal extension is 21cm [13]. However, when the value of sock horizontal extension meets the standard requirements, the phenomenon of elastic discomfort also appears when worn. Socks are knitted products. The flexibility of the knitted fabrics is measured by means of elongation at specified load test or load at specified elongation test. Taking into account the shape of the sock on the foot, we can use the method of fixed length load test.

The purpose of elongation of tensile recovery experiment is to study the extension performance under low stress and tensile elastic recovery performance of socks' mouths [9]. Extracting load force (tensile force) and tensile recovery rate from this reflects the tensile deformation and elastic resilience of sox mouth under low stress. The extension performance affects the fit of the socks when worn, and the tensile recovery performance can control the phenomena of torsion and slippage in the process of wearing socks. The tensile curve is shown in Fig.3.

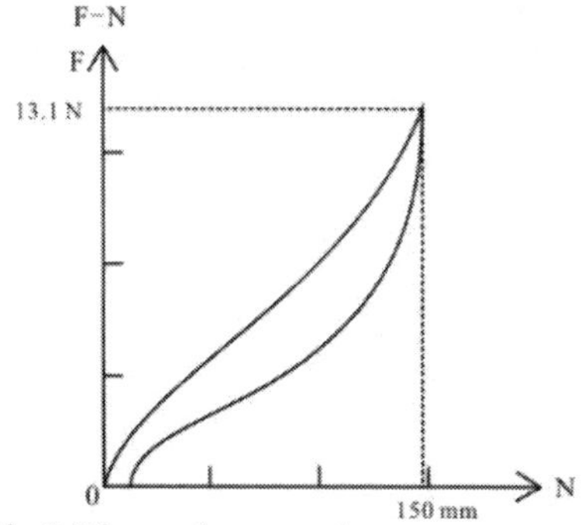

Fig.3 Elongation stretch recovery curve

2.3 Experimental Sample

Our experimental samples consists of eighteen woven female socks with different specifications, different organizational structures, but made out of the same raw materials, as shown in Table 2 and Fig.4. The raw material is 32s twist combed cotton, number 100 for rubber, nylon wrap yarn of 3070. Socks' mouths' organizational specification are 1×1 rib top and 3×1 rib top. The height of sox mouth are 2.5cm, 3.0cm and 3.5 cm. Their widths are 6.5 cm, 7.0 cm and 7.5 cm respectively.

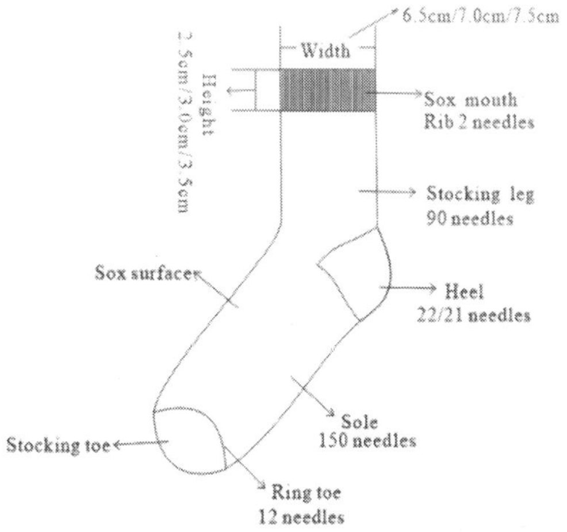

Fig.4 The specification parameters of experimental sample

Table.2 The specification parameters of experimental sample

Sample number	Knitting parameters	Raw material	Organizational structures	Width (cm)	Height (cm)	Grams per square meter (g)
1#	In the mouth: two needles; Elastic: sixty needles; Sock tube: ninety needles; Socks with: twtenty-two /twtenty-one needles; Sole of stocking: one hundred and fifty needles.	Cotton yarn: 32 S twist; Sackage yarn: nylon Wrap yarn of 3070; elastic: number 100.	1×1rib top	6.5	2.5	10.2
2#			1×1rib top	6.5	3.0	11.5
3#			1×1rib top	6.5	3.5	11.5
4#			1×1rib top	7.0	2.5	10.1
5#			1×1rib top	7.0	3.0	11.4
6#			1×1rib top	7.0	3.5	11.6
7#			1×1rib top	7.5	2.5	10.5
8#			1×1rib top	7.5	3.0	11.3
9#			1×1rib top	7.5	3.5	11.6
10#			3×1rib top	6.5	2.5	10.4
11#			3×1rib top	6.5	3.0	11.1
12#			3×1rib top	6.5	3.5	11.6
13#			3×1rib top	7.0	2.5	10.2
14#			3×1rib top	7.0	3.0	11.4
15#			3×1rib top	7.0	3.5	12.0
16#			3×1rib top	7.5	2.5	10.2
17#			3×1rib top	7.5	3.0	11.6
18#			3×1rib top	7.5	3.5	12.0

3. Experimental Results and Analysis

Stretch recovery experiment was carried out for all kinds of socks' mouths. The initial clamping length is 150mm. Pre-tension is 0.2N. The instrument calculates the tensile force, recovery rate, retraction force, residual elongation and elongation of the two tensile experiments automatically. Then we calculated the arithmetic mean of the two stretches for the above measurement indicators, which are regarded as final experimental data. In order to better analyze the data, we calculated the average and standard deviation of the data. The specific data is shown in Table3.

Table.3 Specific data for different indicators

Sample number	Load force (N)	Reversion rate (%)	Deformati on rate (%)	Retractio n rate (N)	Residual elongation (cm)	Elongati on (%)
1#	10.4	91.5	12.6	10.4	12.99	12.9
2#	6.1	93.8	9.1	6.1	9.68	9.6
3#	13.1	94.5	8.1	13.1	8.5	8.5
4#	10.1	93.8	9	10.1	9.7	9.7
5#	6.4	93.6	9.4	6.4	9.96	9.9
6#	18.2	93.6	9.5	18.2	9.63	9.6
7#	8.7	94.3	8.3	8.7	8.82	8.8
8#	6.5	93.1	10.1	6.5	10.33	10.3
9#	14.9	94.2	8.5	14.9	8.87	8.8
10#	6.9	94.4	8.2	6.9	8.66	8.6
11#	6.3	95.3	6.9	6.3	7.2	7.2
12#	11.0	95.1	7.2	11.0	7.59	7.5
13#	8.4	93.8	9.1	8.5N	9.41	9.4
14#	5.8	94.4	8.2	5.8	9.24	9.2
15#	10.7	94.2	8.5	10.8	8.93	8.9
16#	8.6	93.5	9.6	8.6	9.96	9.9
17#	6.0	93.9	8.9	6.1	9.55	9.5
18#	13.2	94.0	8.8	13.2	9.07	9.0
Average	9.51	93.94	8.88	9.59	9.33	9.29
Stdev	3.53	0.81	1.22	3.62	1.21	1.20

3.1 Effect of Sox Mouth Height on the Properties of Extension Under Low Stress

Two paired T test was did for two different height of socks mouth. The test result is shown in table 4.

Table.4 Paired samples test results of sox mouth high

Sox mouth high	Paired differences					T	Df	Sig (2-tailed)
	Mean	Std deviation	Std error mean	95% confidence interval of the difference				
				lower	upper			
2.5cm-3.0cm	2.6667	1.28323	.52387	1.32000	4.01333	5.090	5	.004
2.5cm-3.5cm	-4.6667	2.18785	.89318	-6.96267	-2.37066	-5.225	5	.003
3.0cm-3.5cm	-7.33333	2.61049	.06573	-10.07288	-4.59379	-6.881	5	.001

For this example, we will take the height of 3.0cm and 3.5cm of sox mouth as subjects for a two paired T test. Table 4 illustrated that the average tensile force difference of two kinds of socks mouth high is 7.3333N, the standard deviation is 2.6104, the standard error is 0.06573. The average difference of 95% confidence interval is (10.07288,-4.59379), because do not contain 0. Namely the probability of no difference is not zero, sig=0.001<0.05, which is a significant difference.

3.2 Effect of Sox Mouth Width on the Properties of Extension Under Low Stress

Two paired T test was did was for two different width of socks' mouths. The test result is shown in Table 5.

Table.5 Paired samples test results of sox mouth wide

Sox mouth wide	Paired differences					T	Df	Sig (2-tailed)
	Mean	Std deviation	Std error mean	95% confidence interval of the difference				
				lower	upper			
6.5cm-7.5cm	-1.15000	1.07098	.43723	-2.27323	-.02607	-2.630	5	.0474

Table 5 illustrated that the average tensile force difference of two kinds of socks mouth width is -1.15000N, the standard deviation is 1.07098, the standard error is 0.43723. The average difference of 95% confidence interval is (-2.27323,-0.02607), because do not contain 0. Namely the probability of no difference is not zero, sig=0.0474<0.05, significant difference. So the variation of the socks' mouths' width would have significant influence on tensile properties under low stress. From tensile force of 1#, 4#, 7#, 2#, 5#, 8# and 3#, 6#, 9# to know that the larger the sox mouth width the smaller the tensile force. By adjusting the screw which controls the feeding speed of the elastic yarn, in the process of weaving hosiery socks' mouth. As the elastic yarn feeding speed slows, socks' mouth's width grows bigger. The low rubber content of finished products lead to the low degree of bending of wire coil. On the same premise, the smaller force was required when socks's mouths were stretched up to the same length.

3.3 Effect of Socks' Mouths' Organizational Structures on the Properties of Extension Under Low Stress

Two paired T test was did for socks mouth of different rib top. The test result in shown in Table 6.

Table.6 Paired samples test results of different rib top

Organization	Paired differences					T	Df	Sig (2-tailed)
	Mean	Std deviation	Std error mean	95% confidence interval of the difference				
				lower	upper			
1×1-3×1	3.23333	3.55352	1.18451	.50186	5.96481	2.730	8	.026

Table 6 illustrated that the average tensile force difference of two kinds of different rib top is 3.23333N, the standard deviation is 3.55352, the standard error is 1.18451. The average difference of 95% confidence interval is (0.50186, 5.96481), because do not contain 0. Namely the probability of no difference is not zero, sig=0.026<0.05, significant difference. Table 1 illustrates the average tensile force of 1×1 smaller than 3×1. The extension of these screw-type organization performance depends on the coil shape of the structure. The coil bending degree of 1×1 structure larger than 3×1 structure. When were stretched up to the same length the former one need small tensile force because of the straightening coil.

4. Conclusion

We can get the following conclusions through the research and analysis of this paper.

1) The width has a great influence on the tensile of socks' mouths under low stress. Different socks' mouths have significant differences on the level of 0.05. When designing socks' mouths' widths, we must meet the requirement of the stretched mouth, but to consider the size of the exerted force as well.

2) Different socks' mouths' height corresponds to different tensile forces. They have significant difference on the

level of 0.05. The height of socks' mouths was decided by the factors such as the customer requirements and appearance in the practical production. Through the test its known that design of mouth high affects the size of socks' mouths' tensile force. The factors of fitness and comfort should also be taken into account.

3) The organization has a great influence on the tensile of socks' mouths under low stress. When they were stretched up to the same length, the average tensile force of 1×1 smaller than 3×1. Usually used different screw-type structure to meet the requirements of socks' inlet pressure during the design of functional socks.

4) With further study, we can change the raw materials of the socks' mouths, and use the orthogonal experiments to optimize the design of socks.

5. Acknowledgement

The authors wishes to acknowledge the support of research project 2016KJ-084 of Xi'an Polytechnic University and Datang Town People's Government.

References

[1] Niwaya H. Evaluation technology of clothing comfortableness. J Natl Inst Mater Chem Res: 1999; 7: 269-282.

[2] Toyonori N, Akaenichi O, Sadmu S, et al. Comfort pressure evaluation of man's socks using elastic optical fiber. Text Res J: 1998; 68: 435-440.

[3] Kazuya S, Kazuhiro M, Massayoshi E, et al. Evaluation of foundation comfort based on the sensory evaluation and dynamic clothing pressure measurement. Jpn Res Assn Text: 1997; 38: 53-58.

[4] Kirk W, Ibrahim SM. Fundamental relationship of extensibility to anthropomortric requirements and garment performance. Text Res J: 1996; 57: 37-47.

[5] You F. Research of elastic trousers of the pressure comfort. D Xi'an: Xi'an textile institute of technology: 2000.

[6] Song XX. Relationship between the clothing pressure of the knitting underwear and human thermal comfort. Journal of Knitting industry: 2007; 4: 33-37.

[7] Momota H, Makebe H, et.al. A Study of clothing pressure caused by Japanese men's socks. Journal of the Japan Research Association for Textile End-uses: 1993; 34: 175-186.

[8] Tsujisaka T, Azuma Y, etc. Comfort pressure of the Top Part of Men's Socks. T R J: 2004; 74: 598-602.

[9] Shen DQ, Zhu SD. Research of relationship between the structure of elastic legging and horizontal elastic. Journal of Knitting industry: 1996; 1-3.

[10] Wang AF, Pang SZ. Analysis for sox mouth of organization structure and performance.Journal of Knitting industry: 1990; 2-5.

[11] Tan L, Pan BR Tang Zhenmin. Discuss for clothing pressure of rubber hose mouth. Journal of China textile university: 1992; 6: 270-279.

[12] Liao GY. Analysis of rubber hose deformation. Journal of Knitting industry: 1994; 15-16.

[13] China Textile Association. FZ/T 73001-2008. Socks. Beijing: Standards press of China: 2008.

Test and Analysis of Electromagnetic Protective Clothing for Pregnant Women

Dan Wang, Rong Zheng*, Xiao-Ya Feng, Hai-Jing Liu, Jing Guo

Aimer Human Engineering Research Centre, Beijing Institute of Fashion Technology, Beijing, 100029, China

*Corresponding author's email: yuki.zheng@gmail.com

Abstract

Electromagnetic radiation is harmful to human reproductive, nervous and immune systems and has a significant impact on pregnancy. Increasing numbers of expectant mothers therefore wear electromagnetic field (EMF) shielding clothes to protect themselves and their unborn babies. This study selected and tested nine models popular on the market and analyzed their electromagnetic shielding effectiveness via the electromagnetic shielding enclosure method following national standards. Results show that their effectiveness is largely the same, regardless of brands, materials and designs.

Keywords: Protection; Maternity Clothes; Shielding Effectiveness

1. Introduction

Widespread usage of the Internet has increased instances of electromagnetic radiation, which is now the fourth largest source of pollution following air, water and noise pollution. Studies have proven that electromagnetic radiation impairs people's reproductive, nervous and immune systems, and also affects pregnancy [1]. Han, Cao and Liu's research found that more time spent watching TV and using mobile phones could significantly increase the risk of arrested intrauterine pregnancies for women in early pregnancy [2]. In their test on the impact of electromagnetic radiation on animals' reproductive system, Wang, Peng and Gao observed that even a very short term of exposure to 2-11 mW/cm^2 microwave could damage the structure and function of male rats' reproductive system [3]. Electromagnetic radiation poses tangible health risks to pregnant women and fetuses, and protective devices are hence necessary. More and more expectant mothers now wear electromagnetic field (EMF) shielding clothes.

There are often several brands of such clothes on display in large shopping centres, and a search of any major online-shopping websites may yield tens of thousands of results about dozens of brands. Such clothes use different EMF shielding materials; the most commonly used are coated shielding fabric, metal plating shielding fabric, composite shielding fabric, multi-ion shielding fabric and nanometer fabric [4]. Shielding performances vary due to the clothings' distribution and concentration of conductive fibres, their density and structure as well as fabric thickness [5, 6, 7].

The EMF shielding clothes for pregnant women available on the market include camisole, belly band, and overcoat among many others. Their overall and sleeve lengths, collar designs, as well as sizes of any holes/slits are different, as are the production techniques. What's more, decorations such as ruffles, lace and ribbons add more variants to EMF shielding garments.

There are relatively few academic studies to date on EMF shielding clothes (both in China and internationally), and still fewer about such clothes for pregnant women. Existing studies focus mostly on the shielding effectiveness of the materials used for such clothes, which cannot be equated however to the effectiveness of the finished product. The scarcity of test and analysis results for EMF shielding clothes on the market makes it difficult for consumers to compare products and make informed decisions when shopping for such garments.

This study is based on test and analysis of nine models of EMF shielding maternity clothes popular in the marketplace, which were selected according to a questionnaire survey. The purpose was to gauge the disparity, if any, in their shielding effectiveness, and therefore provide more information for consumers.

2. Problem Formulation

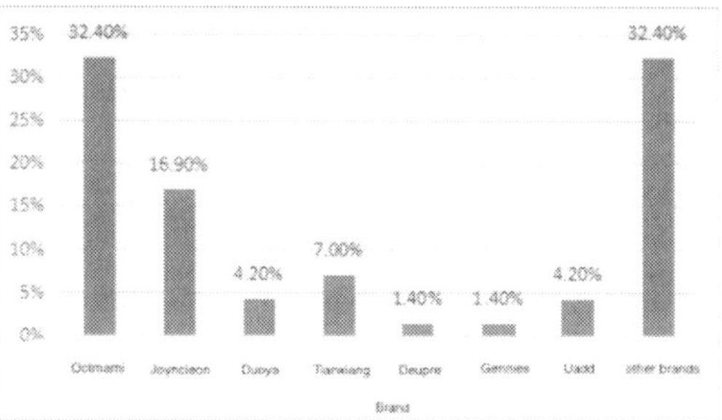

Fig.1 Brand preference.

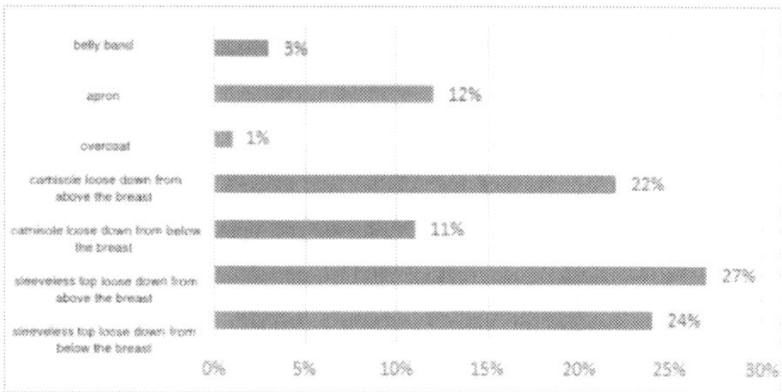

Fig.2 Design preference.

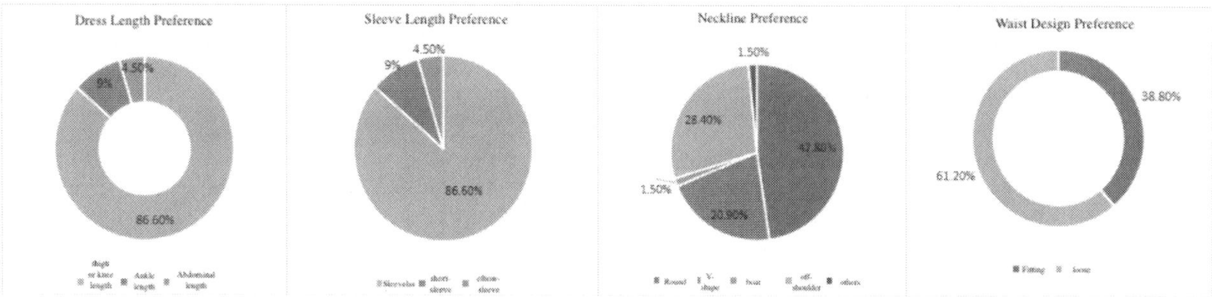

Fig.3 Survey on design preference.

Table.1 Samples' design and material specifications

Code:	1	2	3	4	5
Brand:	Octmami			Joyncleon	
Material:	57.8% silver-plated fibre	100% silver-plated fibre	39% metal fibre	100% silver-plated fibre	
Design:					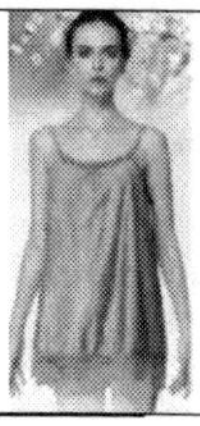
Code:	6	7	8	9	
Brand:	Joyncleon				
Material:	75% silver plated fibre	55% silver plated fibre, 45% polyester fibre		30% metal fibre	
Design:					

A questionnaire survey was conducted among pregnant women visiting Aviation General Hospital of China Medical University and Beijing Chaoyang District Maternal and Child Health Care Hospital. A total 109 valid questionnaires were retrieved. Of these respondents, 69 (63.3%) had worn EMF shielding clothes. The two most commonly named brands are Octmami (32.4%) and Joyncleon (16.9%) (see Fig.1). As for the type of clothes, the most popular are sleeveless top (51%) and camisole (33%) (see Fig.2). When asked about design, 86.6% of respondents prefer high- or knee-length garments; the same number of respondents prefer sleeveless garments; and 61.2% prefer baggy garments (see Fig.3).

We selected nine Octmami and Joyncleon brands of camisoles and sleeveless tops whose lengths range from the buttock to above the knee. We then tested their shielding effectiveness. Sample details are listed in Table 1.

3. Problem Solution

3.1 Electromagnetic Shielding Enclosure Test

3.1.1 Testing method

According to requirements of the GB/T 23463-2009 national standard on microwave protective garments, we conducted the electromagnetic shielding enclosure test in the electromagnetic semi-anechoic chamber of Beijing Municipal Institute of Labour Protection. The devices we used include N5183B signal generator, AR 50WD1000 power amplifier, ETS-3117 and VULB9163 antenna, and EP603 electric field probe. As the radiation most people are exposed to and should be protected from is that from mobile phones, Wifi, computers and microwave ovens among other electronic devices, we set the test frequencies at 800MHz, 915MHz, 1200MHz, 1800MHz, 2400MHz, 2450MHz, 3000MHz, 5500MHz, 5850MHz and 6000MHz (see Fig.4), and conducted both horizontal and vertical polarization tests [8, 9].

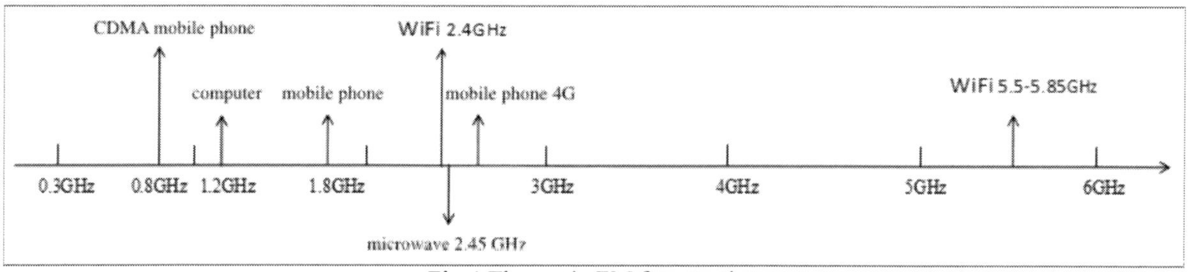

Fig.4 The test's EM frequencies.

3.1.2 Testing Procedure

In light of the data collected by the Aimer Human Engineering Research Centre, Beijing Institute of Fashion Technology, in studying the body shape of 26 pregnant women, we designed test mannequins resembling women's bodies in mid to late-stage pregnancy; specifications are shown in Figs.5 and 6 and Table 2.

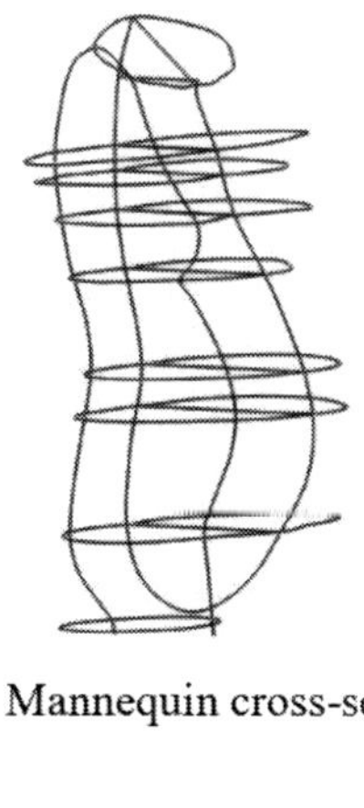

Fig.5 Mannequin cross-section.

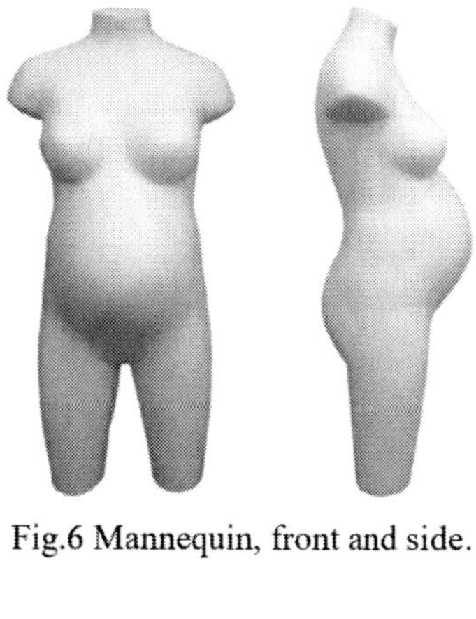

Fig.6 Mannequin, front and side.

To measure the shielding effectiveness, we tested the radiation intensity at four points of the mannequin-front of bust, side of bust, front of abdomen, and side of abdomen-at different frequencies and polarization directions, and with the mannequin both clad and not clad in EMF shielding clothes. An average was produced after three readings at the same spot, frequency and polarization direction for the same clothes. We then calculated according to its definition the shielding effectiveness of the clothes at the said frequency, polarization direction and spot.

Table.2 Sizes of different parts of the mannequin

Description	Size (cm)
Upper bust	89
Arm hole girth	41
Bust line	92
Waist line	94
Abdominal circumference	99
Hip girth	100
Widest point	117
Back length	41
Shoulder breadth	39
Height	159
Hip height	77
Knee height	41
Knee girth	36
Back length	41

3.2 Analysis of Test Results

Fig.7 shows the shielding performances of different sample clothes at the breast. As can been seen from the approximation of curves, the data are similar for the same electromagnetic frequency, with the disparity within -5 dB to 5 dB. This indicates that these samples, of different brands, materials and designs, are largely on par in terms of EMF shielding effectiveness at the part of the breast.

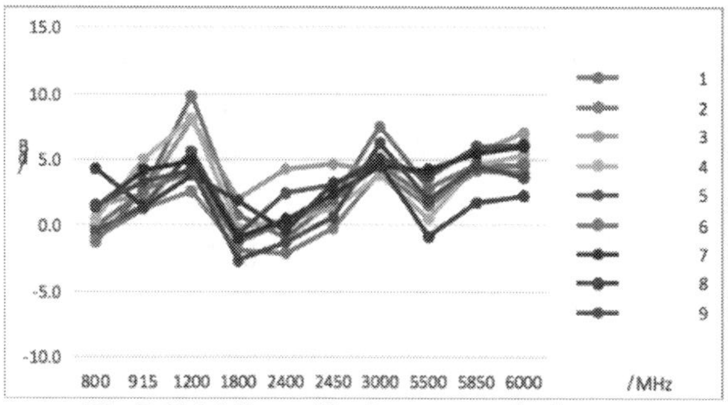

Fig.7 Shielding effectiveness at the breast.

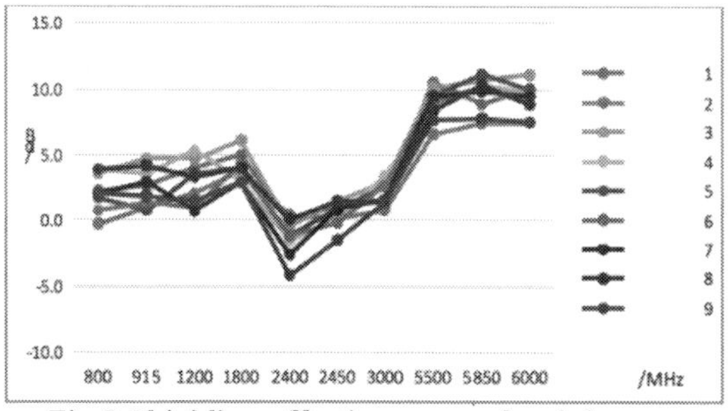

Fig.8 Shielding effectiveness at the abdomen.

Fig.8 shows test results of the nine samples at the abdomen. The pattern of change across various frequencies is noticeably different from that shown in Fig.7, but the data are still similar for the same frequency in this chart. There is a marked increase in shielding effectiveness in the range of 3000MHz to 6000MHz. Again the samples, of different brands, materials and designs, are largely on par in terms of EMF shielding effectiveness at the part of abdomen.

In the real world the polarization of electromagnetic waves rotates 360 degrees. To give each sample an overall evaluation of its shielding effectiveness, we tested the shielding results at the same part of the clothes, at the same frequency, and under horizontal and vertical polarizations, respectively. We then used the average in our analysis.

As Fig.9 indicates, the overall shielding effectiveness of the nine samples is similar at any given frequency, with the data narrowly fluctuating between -5dB~10dB. In sum, these sample garments of different brands, materials and designs are neck and neck in terms of overall shielding effectiveness.

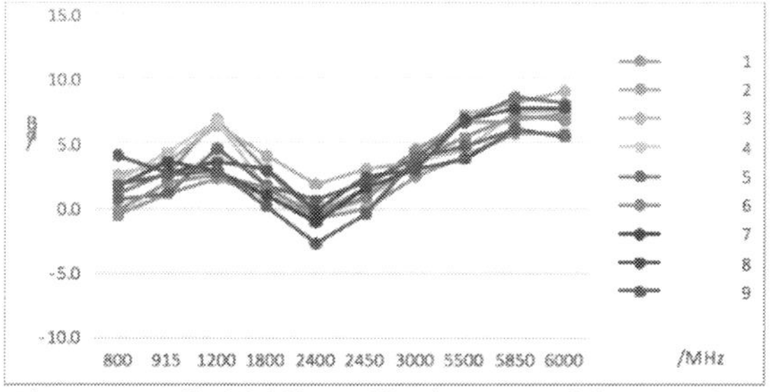

Fig.9 Overall shielding effectiveness of the samples.

4. Conclusion

This research tested the shielding effectiveness of nine samples of maternity dresses that according to our survey are popular in the marketplace. The test shows that the pattern of change in their shielding effectiveness is similar, as is their shielding effectiveness at the same EM frequency. This reveals that the EMF shielding clothes for pregnant women on the market are generally at the same level of protective effectiveness regardless of such factors as brand, material and design.

References

[1] Zhu QL, Li JT, Gao CF. Progress about biological damage effect of electromagnetic radiation. Pract J Med Pharm: 2015; 2: 100-102.

[2] Han JX, Cao ZJ, Liu XY. Effect of early pregnancy electromagnetic field exposure on embryo growth ceasing. J Hyg Res: 2010; 3: 349-352.

[3] Wang YM, Peng DY, Gao YB. Effect of extreme short-term microwave radiation on male reproduction in rats. Mil Med Sci: 2011; 35: 375-378.

[4] Zhang BT, Li GX, Zhai JY. Prepapation and application of electromagnetic shielding fabrics. Environ Eng: 1995; 10: 38-40.

[5] Wang X. Gray predominant analysis of influencing factors of serviceability of electromagnetic shielding fabrics. J Text Res: 2009; 8: 49-53.

[6] Cheng KB, Cheng TW, Nadaraj RN et al. Electromagnetic shielding effectiveness of the twill copper woven fabrics. J Reinf Plas Comp: 2006; 25: 699-709.

[7] Yang YL. Research on the influence factors of blended type electromagnetic shielding fabrics (master's thesis). Zhengzhou: Zhongyuan University of Technology: 2014.

[8] Cheng MJ. Study on the test methods of fabric electromagnetic shielding effectiveness (master's thesis). Shanghai: Donghua University: 2004.

[9] Zhang LL, Chen Y. Testing and simulation of electromagnetic shielding effects of electromagnetic protective clothing for pregnant women. J Text Res: 2011; 32: 108-112.

Evaluation and Research on the Comfortability of Corrective Brassieres for Outward Breasts

Min Lian

Fashion & Art Engineering Deparment, MinJiang University, Wenxian Road, Minhou, Fuzhou, Fujian, China

*Corresponding author's email: 6844264@qq.com

Abstract

Corrective brassieres for outward breasts are able to correct the shape of breasts and make them more beautiful through their structural designs. In order to research the corrective effect of corrective brassieres for outward breasts as well as the evaluation on their comfortability, 5 young women were selected as the testees to try on 6 sets of corrective brassieres in different types one by one. By conducting the direct measurement method, the clothing pressures at multiple points of the brassieres at stationary state were measured. Through the correlation between the objective pressure values from the measurement results and the subjective feeling, an evaluation method for the corrective effect and comfortability of corrective brassieres for outward breasts was developed.

Keywords: Corrective Brassieres for Outward Breasts; Clothing Pressure; Subjective Evaluation; Objective Evaluation

With the improvement of livelihood, people are paying more and more attention to the shape of their body curves, while for women, the silhouette of breasts will directly affect the overall beauty of their curves.

Outward breasts are mainly manifested as breasts in the shape of round plates, while some are also accompanied with mamma accessoria. (As shown in Fig.1) For the two BP points, m:n<4:5, and the distance length of n is relatively big. As the two bust points are not concentrated, the visual effect of the breasts is flat and the external circumferences of the breasts are relatively big. If mamma accessoria exists, part of the fat will be diverted to the mamma accessoria, making the distance between the breasts relatively larger. [1]

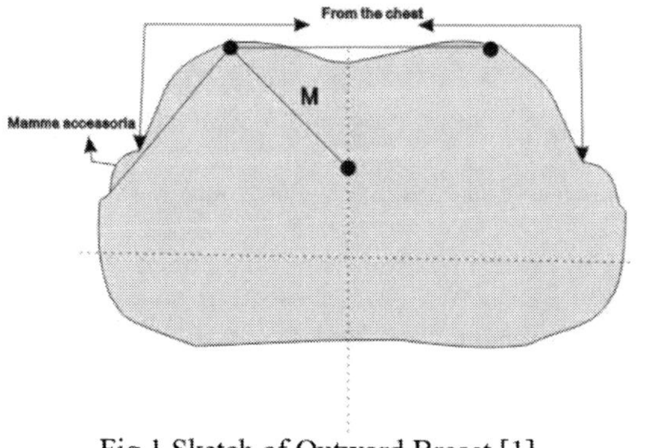

Fig.1 Sketch of Outward Breast [1]

The working principle of corrective brassieres for outward breasts ar as follows: the process in which the underwear gives shape to the contours of the body can be divided into three stages: the return of fat, the back-flow of fat and the fixture of fat. Corrective brassieres, also called as "heavily functioned" underwear, have adopted the principle of fat transposition in the design. Relying on the transposition of fat, the brassieres are able to wrap the fat from armpit, the back and the lower abdomen together at the breasts. After putting the brassieres on following the correct steps, the flabby and pendulous muscles will be supported and the body curves will be shaped instantly, so as to create a graceful and balanced posture.

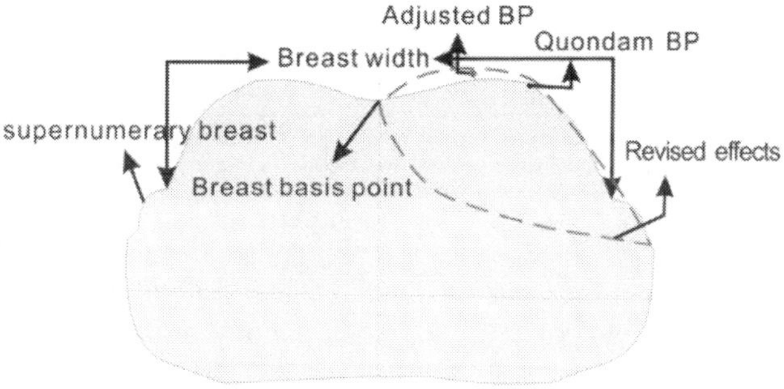

Fig.2 Contrast Before and After Wearing [1]

Comfortability, a complicated and vague concept and evaluative term, is wearers' determination of their physiological feelings for their clothes under specific circumstances. It involves both the physical effects and the physiological feelings, and will be influenced by people's psychological feelings. [2] The evaluation on the comfortability of clothes can be divided into the subjective evaluation and the objective evaluation.

In the literature [3], Professor Chen Dongsheng et al. have researched and developed the micro air sac, a pressure sensing element which is based on the micro sensor technology and is suitable for the flexible pressure measurement of human body (the diagram is as shown in Fig.3). The element has avoided the changes of clothing pressure test conditions on the part being tested due to the thickness of the sensing element and realized the collection and recording of clothing pressure value data at multiple points. The test system meets the requirements for the pressure tests of clothing pressure on most parts of human body and provides the technical support for the clothing pressure comfortability test. [5]

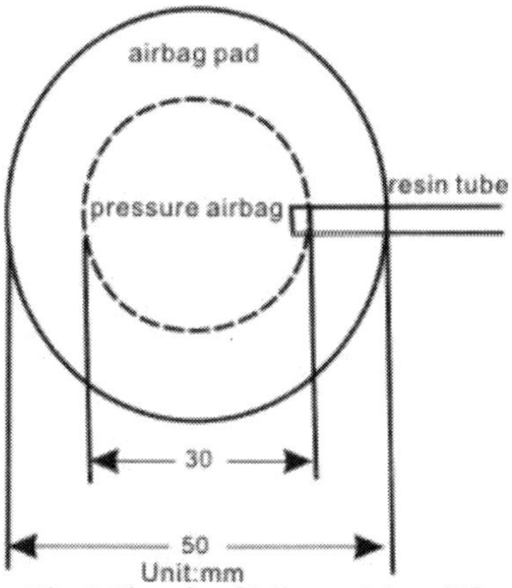

Fig.3 Sketch of Micro-airbag [3]

In the literature [4], Professor Xu Jun et al. have adopted the psychological scale to make subjective evaluations on the senses of pressure of sports underwear under different conditions and carried out the variance analysis on the values obtained, which have provided a method to conduct the subjective quantitative evaluation on the clothing pressure.

1. Test Section

1.1 Samples of Corrective Brassieres for Outward Breasts

Six corrective brassieres for outward breasts were purchased from the market and were numbered from S1 to S6 respectively, which were steel-wire-free knitted brassieres (S1, S2 and S3), and knitted brassieres with steel-wire

(S4, S5 and S6). The sizes of the brassieres were as shown in Table 1

Table.1 Sizes of Corrective Brassieres for Outward Breasts

Sample No.	Bust	Cup	Steel Wire
S1, S2 and S3	88	75/D	No
S4, S5 and S6	88	75/D	Yes

1.2 Test Subjects

A total of 5 young women, who are at the size of 75D and have outward breasts, were selected. See Table 2 for the basic situation of the test subjects. All the participants to the test had no strenuous exercise within the 24 hours before the test. The time of the test was from 14:30 to 16:30. At the same time, it should be avoided to carry out the test during the menstrual cycles of the women to reduce the fluctuation of test data.

Table.2 Basic Information of the Experimental Subjects

Item	Age	Bust	Under Bust	Weight	Height	BMI
Maximum	38	90	77	114	161	73.1
Minimum	21	86	74	107	155	68.2
Mean	29.2	87.8	75.2	110	157.8	69.7
Std.	6.8	1.6	1.3	3.1	2.6	2.0

1.3 Test Instrument

The AMI-3037-5s Air Sac Contact Pressure Tester produced by Japan AMI Company was adopted as the clothing pressure measuring instrument; before the test, the instrument had been reset to zero.

1.4 Test Environment

In order to reduce the interference from environmental factors, the test was carried out under constant temperature and humidity. The temperature was controlled at (22±2)℃, the relative humidity was 65%±3%, and the wind speed was ≤0.1m/s.

1.5 Test Scheme

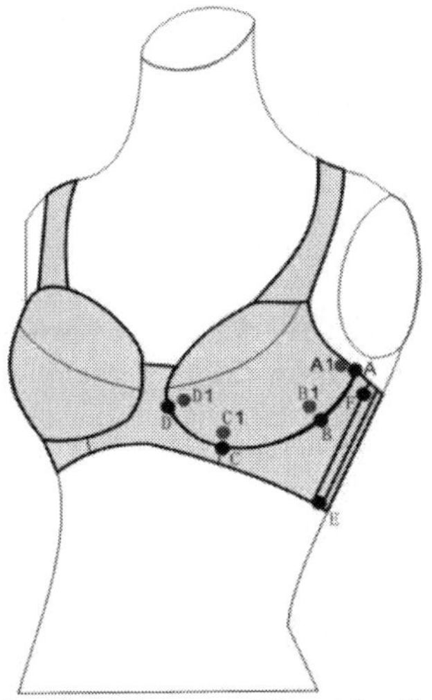

Fig.4 Sketch of Pressure Measuring Points

First of all, the testees simulated the brassieres with their hands to lift and adjust the outward breasts to the push-up effect. At the same time, the pressure values of the air sac placed at the four points on the breasts, from Point A to Point D, were recorded as the reference values for evaluating the corrective effect and comfortability.

Then, the testees put one the sample brassieres from S1 to S6 one by one and the pressure values at the four points on the breasts, from Point A to Point D were recorded. It was discovered in the test that when the air sac of the pressure tester was exactly under the wire, the pressure values measured were relatively high with large fluctuations. In order to obtain stable data, on the one hand, the center of the circle of the air sac should be placed right below the wire when the air sacs were arranged; on the other hand, four extra points, from A1 to D1 were added inside the wire. At the same time, point E and point F were added to measure the clothing pressures of the brassieres. The pressure testing points were as shown in Fig.4.

While recording the pressure values at these points, the subjective evaluations of all the corrective brassieres being worn were inquired from the testees, including the corrective effect and comfortability, which were rated from 1 to 10. Due to the limitation of space, the data records of only one testee as listed as in below table:

Table.3 The Data Records of One Testee

	Objective Evaluation (Pressure Value) kPa										Subjective Evaluation (1-10)	
	A	A1	B	B1	C	C1	D	D1	E	F	Corrective Effect	Comfortability
Hands Simulation	1.7		1.6		3.0		1.1				10	10
S1	2.2	0.9	2.4	1.1	3.3	1.5	2.8	2.8	2.5	1.8	8	9
S2	1.6	2.3	2.7	1.9	3.0	2.9	2.5	1.6	4.5	1.7	9	8
S3	2.1	1.0	2.4	1.2	3.4	1.8	2.8	2.0	2.6	1.9	8	9
S4	4.2	2.1	3.4	1.1	2.7	1.0	1.8	0.7	1.9	2.0	8	7
S5	2.9	1.3	4.0	1.3	4.6	1.3	3.0	0.6	2.8	2.2	9	6
S6	3.0	1.4	3.9	1.3	3.8	1.2	2.8	0.7	2.7	2.3	8	7

2. Data Analysis

2.1 Analysis of Data Error

On the one hand, the errors of pressure values in the objective evaluation were from the errors of the measuring instrument. On the other hand, since the measurement method had recorded the static data of pressures, while the actual clothing pressures would change with the respiratory rhythms, blood pressures and heart rates of the testees, it would be appropriate to record the middle values of the fluctuations. However, the accuracy was already enough for the qualitative analysis.

The errors of the subjective evaluations were from the fact that when the subjective feelings of the individuals toward the corrective effect and comfortability of the corrective brassieres were converted into the ten-point rating, there would be individual differences. In this test, the hands simulation was specially added, in which, the outward breasts were lifted and adjusted by hands to the push-up effect. At the same time, the pressure values of the air sac when placed at multiple points of the breasts were recorded as the reference for the status when both the corrective effect and the comfortability were rated at 10 points.

2.2 The Correlation between Pressure Value and Comfortability

For outward breasts, corrective brassieres have to provide proper pressure so as to achieve the effect of slimming and shaping. However, as the clothing pressure increases and exceeds certain value, comfortability will then decrease. Especially when the threshold of comfort is exceeded, the brassieres will bring people discomfort. According to document [6] literature, the comfort pressure threshold of women's brassieres straps is 3.2kPa and that for the under bust and the band is 1.47~2.13kPa; since the average blood pressure of the capillary vessels on the surface of human skin is 7.85kPa, when the clothing pressure exceeds this value, it will be very difficult for the blood to flow, and will cause harm to health.

According to the test data, when the outward breasts were lifted and adjusted to the push-up effect with the hands simulating the brassieres, the average value (kPa) of the pressure measured were:

Point A: 1.74; Point B: 1.71; Point C: 2.85; Point D: 1.02

These data basically meet the record of the comfort pressure threshold in documents [6]; therefore, it is reasonable to adopt the subjective evaluation of comfortability on the touch of the two hands as the reference value of 10 points.

According to test data, for the brassieres with steel-wire numbered from S4~S6, the average value (kPa) of the pressure measured were:

Point A: 3.44; Point B: 3.80; Point C: 3.71; Point D: 2.43

Among these brassieres, for No.S5, the brassieres with steel-wire, the maximum partial pressure occurred at Point C: 5.1kPa, which was far beyond the comfort pressure threshold. Therefore, these brassieres would be bad for health after long-term wearing.

And for the wire-free brassieres numbered from S1~S3, the average value (kPa) of the pressure measured were:

Point A: 1.94; Point B: 2.40; Point C: 3.12; Point D: 2.73

The test data have shown that because of the existence of wires, the brassieres with wires will cause the "pressure concentration" effect on parts of the human body. In the design of underwear, attention should be paid to reduce its adverse effect.

The data in Fig.5 have shown that as the main bearing point of the corrective brassieres, Point C is the point that bears the largest pressure among all measured points. The reasonable pressure value for it is 3.0~3.4kPa. As the pressure value continues to increase after exceeding 3.4kPa, the subjective evaluation index on the comfortability will decline rapidly.

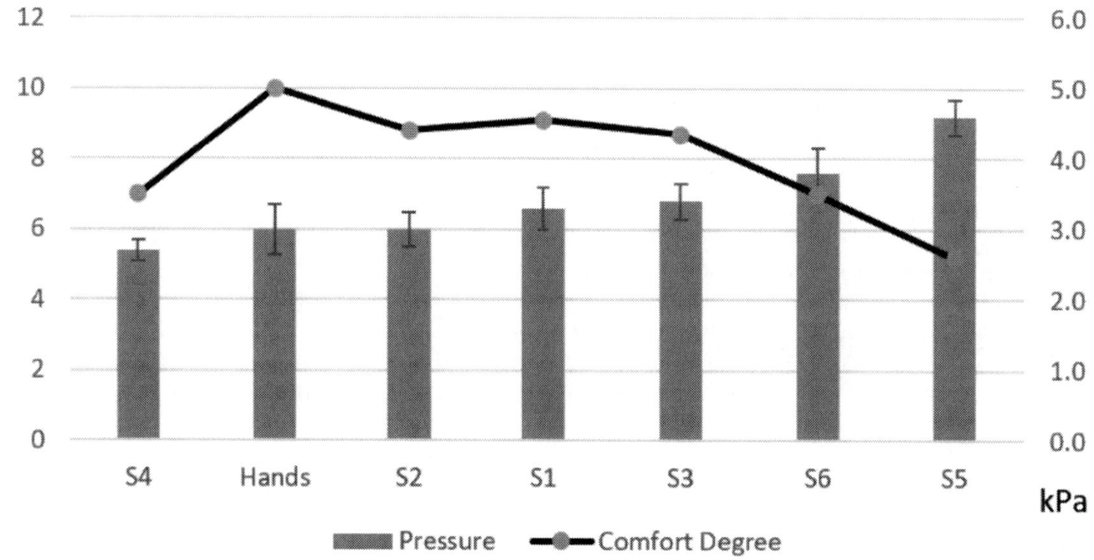

Fig.5 The Correlation between the Pressure Value and Comfortability at Point C

The data in Fig.6 have shown that as the important corrective bearing point of the corrective brassieres, the pressure value at Point B will influence the corrective effect of the brassieres directly. The increase of the pressure value at Point B will lead to the improvement of the corrective effect. However, it is accompanied with the decline of subjective evaluation on comfortability. How to strike a reasonable balance between the two is an important segment which is worth of considering in the design of corrective brassieres for outward breasts.

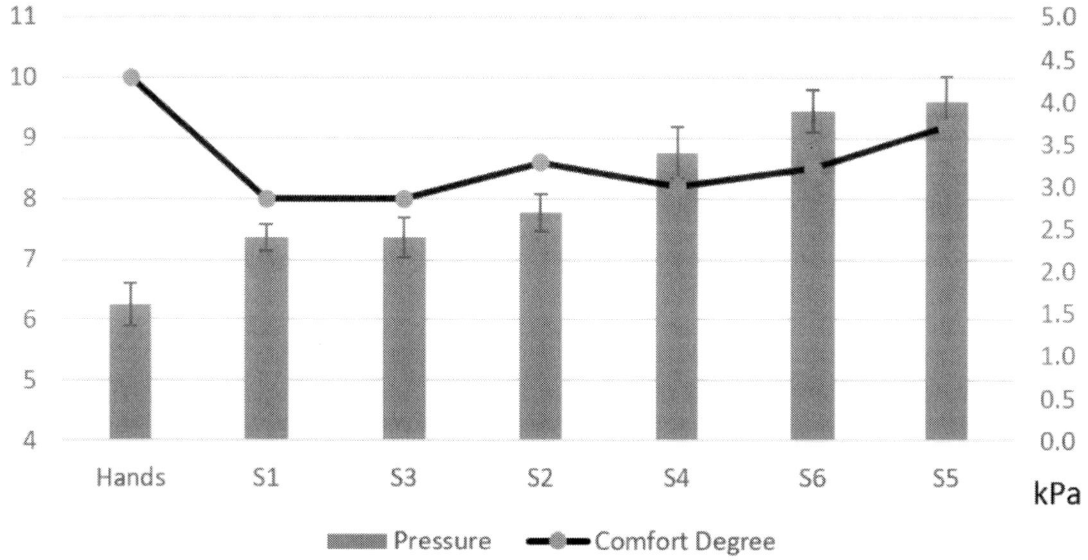

Fig.6 The Correlation between the Pressure Value and Corrective Effect at Point B

3. Conclusion

From above analysis, this study can be drawn in following 3 major outcomes:

1) Both subjective evaluation and objective evaluation can be adopted in the evaluation on the comfortability of corrective brassieres for outward breasts. In the objective evaluation, the clothing pressure test can be conducted with the air sac contact pressure tester to obtain the data of clothing pressure.

2) Through the combination of subjective evaluation and objective evaluation, the data basis can be provided for the improvement of the design of corrective brassieres for outward breasts.

3) In this test, the measurement method had recorded the static data of the clothing pressure and the rating system was adopted for the subjective evaluation. Therefore, there were certain errors in the data obtained. If other data collecting instruments could be adopted to acquire the physiological data of the testees, such as their heart rates, blood pressures, respirations and EEG [7], as well as the dynamic changes of the clothing pressure, richer and more accurate data support can be provided for the design of corrective brassieres for outward breasts.

References

[1] Lian M. Design and Study of the Expasion-adjusting Brassiere. Journal of Zhongyuan Institute of Technology: 2009; 20 (6): 50-53.

[2] Duan XY, Yu WD. Research Progress in Pressure Comfort of Foundation Garment. Journal of Textile Research: 2007; 11-0144: 05.

[3] Chen DS, Chui LM. Development of Clothing Pressure Measuring System: 2008; 29 (3): 72-75.

[4] Xu J, Zhou Q. Subjective Assessment of Pressure Distribution for Sports Underwear. Journal of Textile Research: 2005; 26 (2): 77-81.

[5] Liu H, Wang HG, Chen DS. Analysis of Objective Measurement Method for Clothing Pressure. Journal of Clothing Research: 2016; 03-0267: 04.

[6] Yin L, Zhang WB, Xia L. Review and analysis of pressure comfort of garment. Journal of Textile Research: 2008; 03-0137: 06.

[7] Chen DS, Liu YJ. Application of EEG Technology to Evaluation of Clothing Comfort. Journal of Clothing Research: 2016; 1: 1.

Versatile Silk/Graphene Composites for Biomedical Engineering

Yan Ma[1], Zu-Lan Liu[1], Si-Hao Chen[1], Zhi Li[1*], Fang-Yin Dai[1,2]

[1]*Chongqing Engineering Research Center of Biomaterial Fiber and Modern Textile, College of Textiles and Garments, Southwest University, Chongqing, 400716, P.R.China*
[2]*State Key Laboratory of Silkworm Genome Biology,Key Laboratory of Sericultural Biology and Genetic Breeding, Ministry of Agriculture,College of Biotechnology, Southwest University, Chongqing, 400716, P.R.China*

[*]Corresponding author's email: tclizhi@swu.edu.cn

Abstract

Biomedical engineering is an interdisciplinary subject that aims to promote disease prevention, diagnosis and therapy, patient rehabilitation and other healthcare purposes for the benefit of human beings. The urgent demand of society encourages the rapid development of biomedical engineering in recent years. Material filed is one of important areas which strongly support this development. Graphene is a relatively new material that attracted a lot of attention when it was discovered due to its excellent properties. Silk is a traditional textile fiber which is applied in a variety of areas. The present article reviews the fabricated silk/graphene composites through different ways and discusses the properties of these silk/graphene composites. Moreover, the versatile applications in BME area of these silk/graphene composites are summarized and provided. On the basis of this comprehensive review, the future demand and direction of related research are presented.

Keywords: Biomedical Engineering; Silk Fibroin; Graphene; Electrode; Biosensor; Drug Delivery

1. Introduction

Biomedical engineering (BME) develops materials, implants, devices and methods of processing for promoting disease prevention, diagnosis and therapy, patient rehabilitation and other healthcare purposes for the benefit of human beings [1, 2]. Modern BME dates back to two centuries ago; however, BME may have existed for more than thousands of years. Heavily depended on principles of physics, chemistry, mathematics, computer and engineering, BME have grown and has become interdisciplinary, overlapping significantly with medicine, biology and engineering. The scope of BME embraces some well-established areas involving genetic engineering, biomaterials, tissue engineering, bioinformatics, rehabilitation engineering, biomechanics, medical imaging and medical devices. Thanks to the rapid development of the material field, medical devices and tissue engineered scaffolds for BME have made a great improvements in the past decades. A great many of natural and synthetic materials have been applied for nanocapsules [3], fibers, films [4], nanosheets, gels, biosensors [5], scaffold [6, 7]. In this paper, versatile fabricated silk/graphene composites in recent years are reviewed and their applications in BME are summarized.

2. Graphene and its Derivatives

Graphene is a unique two-dimensional (2-D) crystal sheet constructed with sp^2 hybridized carbon atoms [8, 9]. Owing to its excellent mechanical properties (Young's modulus, ~1100 Gpa), high surface area (2630 m^2/g), remarkable electrical conductivity (mobility of charge carriers, 200,000 $cm^2\ V^{-1}\ s^{-1}$), high thermal conductivity (~5000 W/m/K) [10], good biocompatibility [11], antibacterial [12] and anticancer cell activity [13], graphene has attracted much attention and the studies on graphene has increased exponentially. With suitable processing, the graphene are able to transform to a variety of structures, such as 0-D fullerenes like a football, rolled 1-D carbon nanotubes and compressed 3-D graphite [14]. Graphene oxide (GO) has poor electrical conductivity since the conjugated structures of graphite are destroyed when abundant epoxide, hydroxyl groups and carboxyl groups are introduced on GO sheet during oxidation [15]. Furthermore, reducing GO by chemical reducing agents [16], flash [17] or thermal treatment [18] can easily produce reduced GO (rGO). Graphene, GO and rGO has different contents of oxygen and this apparently leads to the distinct structures and characteristics [13]. GO has highest density of oxygen and best hydrophilicity and reactivity; however, its electrical and optical conductivity is worst among them. Graphene with rare oxygen has superb electrical and optical conductivity while its water solubility and reactivity is relatively poor. The properties of rGO is between GO and graphene due to the middle content of oxygen. Intensive research have been conducted for the application of graphene and its derivatives in biomedical engineering, involving biosensors [19, 20], molecular imaging [21, 22], drug/gene delivery [23, 24], phototherapy [25, 26], and scaffolds [27, 28].

3. Silk

The domestication of silkworm can date back to five thousand years ago when silk was realized as a favorable natural fiber for textile application. Silk is made of double fibroin filaments embedded by superficial sericins for the purpose of protection and aggregation [29]. Structurally, fibroin filament, with a diameter of around 10-25 μm, is gradually assembled by hydrophobic blocks and hydrophilic blocks, fibroin micelle (100-200 nm) and fibroin globule (0.8-1.5 μm) [30]. In term of the constitution, silk fibroin (SF) consists of three protein subunits: heavy chain, light chain and P_{25} [31]. Since both heavy chain and light chain are dominant components in SF, their promoters are identified separately for transgenic silkworm study in which the silk gland is used as a 'bioreactor' [32, 33]. SF, with fine toughness, biocompatibility, and biodegradability, has an enormous value on biomedical application for scaffolds [34-36], enzyme immobilization [37], biosensors [38, 39], drug delivery [40-42] etc. Sericins, the glue-like glycoproteins, account for 20-30% of silk. Generally, sericins have to be removed to generate degummed silk for textile and biomedical application; however, the increasing research in recent years prove their great potential in various areas, such as skincare [43], food industry [44], drug delivery [45], constituent of culture media [46] and grafts [47, 48]. In addition to SF and sericin, the analysis of silk proteomics reveals some other silk proteins although a little part of these proteins and their functions are still unknown [49].

Although the history for the discovery and application of graphene and silk has been diverse for thousands of years, the increased research on fabricating the composites based on both materials are performed only in recent years and these results are published in the following related journals (Fig.1). Herein, the present review is to provide a comprehensive overview of fabricated composites for biomedical application based on graphene and silk, as well as the discussion for future demand and development.

4. Silk/graphene-based Composites

4.1 Scaffold

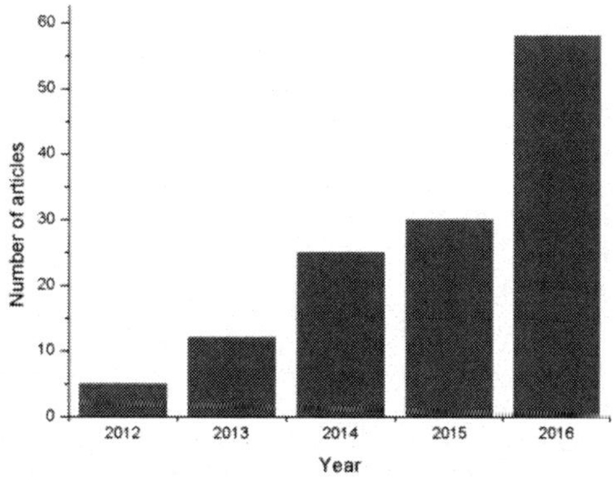

Fig.1 The number of articles related to the word 'silk' and 'graphene' together in recent five years.

SF is rich of glycine, alanine and serine in content and these amino acids mainly contribute to β-sheet crystallites which are known as hydrophobic repetitive sequences. These hydrophobic domains and hydrophilic domains (non-repetitive sequences) are alternating in SF protein, facilitating various interactions through hydrogen bond, polar bond and hydrophobic force. On the basis of these interactions, various silk composites are allowed to be fabricated [50, 51]. Graphene and its derivatives have strong abilities to interact with biomolecules by covalent binding and such interactions possibly influence the secondary structure of biomolecules [52, 53]. A recent study of large-scale molecular dynamics simulations reveals the interactions between graphene and the secondary structure of SF, indicating that graphene contends with the intramolecular interactions and weaken the stable crystal structure formed by β-sheet, while organizing the domains of random or less ordered secondary structures to more stable structures comprised by β-sheet or helical so that the strength and resilience of SF are improved [54].

Although these simulations need to be approved in further works, the appropriate use of GO as fillers for reinforcing mechanical properties of SF/GO composites are effectively verified and reported [55]. Through the spin-assisted layer by layer method (SA-LbL), the alternative single/double-layered GO flakes (0.95 nm) and SF layers (5 nm) are comprised of composite membrane in which the hydrogen bonding, polar-polar and hydrophobic interactions are reinforced, accordingly, the ultra-robust properties of these GO/SF nanomembranes are performed including a good toughness vale (above 2.2 MJ m^{-3}), an excellent stress value (more than 300 MPa), and the highest tensile modulus by far (145 GPa) [55]. However, a further process of water vapor annealing after SA-LbL assembly can increase the ultimate stress to around 450 MPa, which is the highest stress of GO/SF composites reported till now [56]. In addition, Young's modulus and toughness rise by 75% and 45% respectively after annealing. The significant improvement of these mechanical properties is possible because when water vapor anneals, it contributes to the recrystallization of SF followed by SF nanofibrils assembly, as a result, the reinforced hydrophobic–hydrophobic interactions between GO layers and assembled SF nanofibrils lead to the augment of shear force. Through the hydrophobic interaction in alkaline condition, the presence of graphene nanosheets promotes in situ growth of SF nanofibrils which are rich in β-sheet, as such it gas an average modulus of 2.5 GPa. Moreover, these colloidal hybrids are made to macroscopic films with the assistant of vacuum filtration process. On the basis of SF ratio, these macroscopic films have the electric conductivity of 1.4×10^{-3} - 1.1×10^{-2} S cm^{-1}, the tensile modulus of 6-10 GPa, and the toughness of 3×10^5 - 6×10^5 J/m^3. Most importantly, they have improved cell adhesion and proliferation compared that of pure graphene [57].

4.2 Electrode

The facile immersion of SF fiber in GO suspension can fabricate GO/SF fibers. GO is tightly coated on SF fibers through polar interactions. On the basis of this rapid progress, abundant short SF fibers are mixed in GO suspension and filtrated to create a GO/SF film [58]. The post-chemical reduction GO by ascorbic acid solution is enough to prepare a GR/SF film with two distinct sides, a rough and porous top surface and a compact and smooth bottom side. With a hierarchical structure, this fabricated GR/SF film is flexible, mechanically robust and highly effective in electrical conducting and after being decorated by spiky Pt nanosphere, it can directly be used as a H_2O_2 electrode with a sensitivity of 0.56 mA mM^{-1} cm^{-2}. This fabricated GR/SF can also be used to fabricate glucose biosensor electrode by enzyme immobilization.

To fabricate electrodes with superior electrochemical performance, SF blended with rGO is used as templates to control formation of porous Fe_2O_3 nanorod structure [59]. The induction of SF nanofiber on rGO sheet by hydrazine is crucial for developing nanorod structure. The prepared porous nanorod composites has a specific capacity of up to 1495 mAh g-1, with a maximum discharge-charge rate of 0.2C and 2C respectively, and this performance can be maintained up to 300 cycles. With these characteristics, these porous nanorod complexes have the potential for application in fields of energy storage and biomedical material.

4.3 Biosensor

Membranes prepared by SF have good elasticity and toughness and they can be in contact with human skin without causing immunogenicity, therefore they can be used as a base for a biosensor film [60]. The amino acid sequence of SF contains a large number of repeated hydrophobic sequences, contributing to β-sheet or helical structures through the backbone hydrogen bonds. The β-sheet or helical structures are further stack to a hydrophobic region, interlaced with the hydrophilic segments constituted with hydrophilic amino acids [30, 61]. This structure can provide sufficient space for the immobilization of the enzyme. Using this special structure of SF, glucose oxidase can be fixed. Further, it is used to fabricate a SF/graphene field effect transistor enzymatic biosensor together with a gate, a source, a drain and a graphene tablet which is prepared by chemical vapor deposition method [62]. The fabricated SF/graphene biosensor has a very sensitive reaction to glucose (less than 10s). It can detect a minimum glucose concentration of 0.1mM, and its detection ranges from 0.1 to 10mM. Through this hydrophobic immobilization, the enzymatic activity of the glucose oxidase in the SF membrane can be maintained for at least ten months at room temperature. With good strength, flexibility and biodegradability, the SF/graphene biosensor has potential to be applied for clinical monitoring of glucose concentration in type I diabetes, assisting patients to timely control the concentration of glucose in the body. To prepare a sensitive amperometric biosensor for the detection of phenolic compounds, SF/graphene nanosheets are first synthesized from the homogeneous mixture solution of SF and graphene. Then the tyrosinase is covalently bonded to the nanosheets through the crosslinking agent of glutaraldehyde [63]. This novel biosensor is highly sensitive, exhibiting a linear response on the concentration ranges of 0.001-16.91 mM for catechol with the sensitivity of 7634 mAM^{-1} cm^{-2}, 0.0015-21.12 mM for phenol with the sensitivity of 4082 mAM^{-1} cm^{-2}, and 0.002–5.48 mM for bisphenol A with the sensitivity of 2511 mAM^{-1} cm^{-2}. Tis SF/graphene nanosheets can also be used to immobilize different enzymes to produce the corresponding biosensor for detection of different compounds.

4.4 Drug/gene Delivery

Graphene material not only has good mechanical properties, it also exhibits antimicrobial and anticancer activity [64, 65]. The graphene is introduced into biomedicine after its discovery for application in drug/gene delivery [66, 67], molecular imaging [14] and tissue engineering [68, 69]. In the use of 'green' approach, porous silk fibroin/graphene scaffolds have been fabricated for drug delivery, successfully avoiding any toxic and hazardous chemical reagents [70]. The glycerol is used as a crosslinking agent in SF/graphene mixture followed by a freeze-drying method. The higher the graphene content (up to 1%), the better the mechanical properties of SF/graphene scaffolds while the the biodegradability decreases. The porous SF/graphene composite scaffolds with 0.5% graphene content are suitable for drug delivery of Simvastatin, a drug used to promote the formation of osteoblasts. These drug loaded scaffolds can effectively improve the proliferation of mouse osteoblasts MC3T3-E1, demonstrating their application for treating bone tissue diseases and bone tissue engineering.

5. Conclusion and Future Perspectives

It is proved that graphene and its derivatives have many excellent characteristics which enable them to be used in many physical, chemical and biomedical fields. A variety of composite materials based on graphene have been fabricated and characterized. However, seldom research has focused on modifying silk with graphene. As such there is a future research direction for the development of synthesized composite fiber with high performance. The research of silk/graphene complexes in the field of electrode and biosensor has made some progress. However, less focus has been put on the field of biomedicine, especially in tissue engineering. How to fabricated scaffolds by using graphene and silk/SF with ideal mechanical properties and biodegradability for different purposes, is an urgent problem to be solved. It is also an important direction for silk/graphene research in the future.

6. Acknowledgement

This work was supported by grants from the National Hi-Tech Research and Development 863 Program of China (Grant No. 2013AA102507), and Fundamental Research Funds for the Central Universities (SWU115041; XDJK2016B004; XDJK2016C019).

References

[1] Erogul O. Future trends in biomedical engineering. J Biotechnol: 2015; 208: 9.
[2] Ibrahim F, Thio TH, Faisal T, Neuman M. The application of biomedical engineering techniques to the

diagnosis and management of tropical diseases: a review. Sensors: 2015; 15 (3): 6947-6995.
[3] Chen Y, XU P, Wu M, Meng Q, Chen H, Shu Z, Wang J, Zhang L, Li Y, Shi J. Colloidal RBC-shaped, hydrophilic, and hollow mesoporous carbon nanocapsules for highly efficient biomedical engineering. Adv Mater: 2014; 26 (25): 4294-4301.
[4] Ailincai D, Farcau C, Paslaru E, Marin L. PDLC composites based on polyvinyl boric acid matrix - a promising pathway towards biomedical engineering. Liquid Crystals: 2016; 1-13.
[5] Xiao F, Li Y, Gao H, Ge S, Duan H. Growth of coral-like PtAu-MnO2 binary nanocomposites on free-standing graphene paper for flexible nonenzymatic glucose sensors. Biosens Bioelectron: 2012; 41 (1): 417-423.
[6] Li Z, Xie M B, Li Y, Ma Y, LiI JS, Dai FY. Recent Progress in Tissue Engineering and Regenerative Medicine. J Biomater Tiss Eng: 2016; 6 (10): 755-766.
[7] Lalwani G, D'agati M, Gopalan A, Rao M, Schneller J, Sitharaman B. Three-dimensional macroporous graphene scaffolds for tissue engineering. J Biomed Mater Res A: 2017; 105 (1): 73-83.
[8] Geim AK. Graphene: status and prospects. Science: 2009; 324 (5934): 1530-1534.
[9] Novoselov KS, Geim AK, Morozov SV, Jiang D, Zhang Y, Dubonos SV, Grigorieva IV, Firsov AA. Electric field effect in atomically thin carbon films. Science: 2004; 306 (5696): 666-669.
[10] Shen H, Zhang L, Liu M, Zhang Z. Biomedical applications of graphene. Theranostics: 2012; 2 (3): 283-294.
[11] Wu X, Ding S, Lin K, Su J. A Review on the Biocompatibility and Potential Applications of Graphene in Inducing Cell Differentiation and Tissue Regeneration. J Mater Chem B: 2017.
[12] Zou X, Zhang L, Wang Z, Luo Y. Mechanisms of the Antimicrobial Activities of Graphene Materials. J Am Chem Soc: 2016; 138 (7): 2064-2077.
[13] Zhang B, Wang Y, Zhai G. Biomedical applications of the graphene-based materials. Mater Sci Eng C-Mater: 2016; 61: 953-964.
[14] Zhang Y, Nayak TR, Hong H, Cai W. Graphene: a versatile nanoplatform for biomedical applications. Nanoscale: 2012; 4 (13): 3833-3842.
[15] Huang YW, Zeng M, Feng ZJ, Yin D, Xu QY, Fan LR. Graphene oxide-based composite hydrogels with self-assembled macroporous structures. Rsc Advances: 2016; 6 (5): 3561-3570.
[16] Stankovich S, Dikin DA, Piner RD, Kohlhaas KA, Kleinhammes A, Jia Y, Wu Y, Nguyen ST, Ruoff RS. Synthesis of graphene-based nanosheets via chemical reduction of exfoliated graphite oxide. Carbon: 2007; 45 (7): 1558-1565.
[17] Cote LJ, Cruz-silva R, Huang JX. Flash Reduction and Patterning of Graphite Oxide and Its Polymer Composite. J Am Chem Soc: 2009; 131 (31): 11027-11032.
[18] Ju HM, Huh SH, Choi SH, Lee HL. Structures of thermally and chemically reduced graphene. Mater Lett: 2010; 64 (3): 357-360.
[19] Kailashiya J, Singh N, Singh SK, Agrawal V, Dash D. Graphene oxide-based biosensor for detection of platelet-derived microparticles: A potential tool for thrombus risk identification. Biosens Bioelectron: 2015; 65 (274-280).
[20] Zhao H, JI X, Wang B, Wang N, Li X, Ni R, Ren J. An ultra-sensitive acetylcholinesterase biosensor based on reduced graphene oxide-Au nanoparticles-beta-cyclodextrin/Prussian blue-chitosan nanocomposites for organophosphorus pesticides detection. Biosens Bioelectron: 2015; 65: 23-30.
[21] Zhu S, Zhang J, Qiao C, Tang S, Li Y, Yuan W, Li B, Tian L, Liu F, Hu R, Gao H, Wei H, Zhang H, Sun H, Yang B. Strongly green-photoluminescent graphene quantum dots for bioimaging applications. Chem Commun: 2011; 47 (24): 6858-6860.
[22] Roy P, Periasamy AP, Lin CY, Her GM, Chiu WJ, Li CL, Shu CL, Huang CC, Liang CT, Chang HT. Photoluminescent graphene quantum dots for in vivo imaging of apoptotic cells. Nanoscale: 2015; 7 (6): 2504-2510.
[23] Kim H, Kim J, Lee M, Choi HC, Kim WJ. Photothermal Gene Delivery: Stimuli-Regulated Enzymatically Degradable Smart Graphene-Oxide-Polymer Nanocarrier Facilitating Photothermal Gene Delivery. Adv Healthc Mater: 2016; 5 (15): 1917.
[24] Jiang T, Sun W, Zhu Q, Burns NA, Khan SA, Mo R, Gu Z. Furin-mediated sequential delivery of anticancer cytokine and small-molecule drug shuttled by graphene. Adv Mater: 2015; 27 (6): 1021-1028.
[25] Yan X, Niu G, Lin J, Jin AJ, Hu H, Tang Y, Zhang Y, Wu A, Lu J, Zhang S, Huang P, Shen B, Chen X. Enhanced fluorescence imaging guided photodynamic therapy of sinoporphyrin sodium loaded graphene oxide. Biomaterials: 2015; 42: 94-102.

[26] Taratula O, Patel M, Schumann C, Naleway MA, Pang AJ, He H, Taratula O. Phthalocyanine-loaded graphene nanoplatform for imaging-guided combinatorial phototherapy. Int J Nanomed: 2015; 10: 2347-2362.
[27] Akhavan O. Graphene scaffolds in progressive nanotechnology/stem cell-based tissue engineering of the nervous system. J Mater Chem B: 2016; 4 (19): 3169-3190.
[28] Xie XY, Hu KW, Fang DD, Shang LH, Tran SD, Cerruti M. Graphene and hydroxyapatite self-assemble into homogeneous, free standing nanocomposite hydrogels for bone tissue engineering. Nanoscale: 2015; 7 (17): 7992-8002.
[29] Li G, Li Y, Chen GQ, He JH, Han YF, Wang XQ, Kaplan DL. Silk-Based Biomaterials in Biomedical Textiles and Fiber-Based Implants. Adv Healthc Mater: 2015; 4 (8): 1134-1151.
[30] Jin HJ, Kaplan DL. Mechanism of silk processing in insects and spiders. Nature: 2003; 424 (6952): 1057-1061.
[31] Kasoju N, Bora U. Silk Fibroin in Tissue Engineering. Adv Healthc Mater: 2012; 1 (4): 393-412.
[32] Ma SY, Shi R, Wang XG, Liu YY, Chang JS, Gao J, Lu W, Zhang JD, Zhao P, Xia QY. Genome editing of BmFib-H gene provides an empty Bombyx mori silk gland for a highly efficient bioreactor. Sci Rep-Uk: 2014; 4.
[33] Kuwana Y, Sezutsu H, Nakajima K, Tamada Y, Kojima K. High-Toughness Silk Produced by a Transgenic Silkworm Expressing Spider (Araneus ventricosus) Dragline Silk Protein. PloS one: 2014; 9.
[34] Kapoor S, Kundu SC. Silk protein-based hydrogels: Promising advanced materials for biomedical applications. Acta Biomater: 2016; 31: 17-32.
[35] Elahi MF, Guan GP, Wang L, Zhao XZ, Wang FJ, King MW. Surface Modification of Silk Fibroin Fabric Using Layer-by-Layer Polyelectrolyte Deposition and Heparin Immobilization for Small-Diameter Vascular Prostheses. Langmuir: 2015; 31 (8): 2517-2526.
[36] Chen BQ, Kankala RK, Chen AZ, Yang DZ, Cheng XX, Jiang NN, Zhu K, Wang SB. Investigation of silk fibroin nanoparticle-decorated poly(L-lactic acid) composite scaffolds for osteoblast growth and differentiation. International Int J Nanomed: 2017; 12: 1877-1890.
[37] Bezerra CS, Lemos CMGD, Desousa M, Goncalves LRB. Enzyme immobilization onto renewable polymeric matrixes: Past, present, and future trends. J Appl Polym Sci: 2015; 132.
[38] Delezuka JAM, Pavinatto A, Moraes ML, Shimizu FM, Rodrigues VC, Campana SP, Ribeiro SJL, Oliveira ON. Silk fibroin organization induced by chitosan in layer-by-layer films: Application as a matrix in a biosensor. Carbohyd Polym: 2017; 155: 146-151.
[39] Yin HS, Zhou YL, Xu J, AI SY, Cui L, Zhu LS. Amperometric biosensor based on tyrosinase immobilized onto multiwalled carbon nanotubes-cobalt phthalocyanine-silk fibroin film and its application to determine bisphenol A. Anal Chim Acta: 2010; 659 (1-2): 144-150.
[40] Zhao Z, Li Y, Xie MB. Silk fibroin-based nanoparticles for drug delivery. Int J Mol Sci: 2015; 16 (3): 4880-4903.
[41] Xie MB, Fan DJ, Chen YF, Zhao Z, He XW, Li G, Chen AZ, Wu XJ, Li JS, Li Z, Hunt JA, Li Y, Lan P. An implantable and controlled drug-release silk fibroin nanofibrous matrix to advance the treatment of solid tumour cancers. Biomaterials: 2016; 103: 33-43.
[42] Chen AZ, Chen LQ, Wang SB, Wang YQ, Zha JZ. Study of magnetic silk fibroin nanoparticles for massage-like transdermal drug delivery. International journal of nanomedicine: 2015; 10: 4639-4651.
[43] Kundu SC, Dash BC, Dash R, Kaplan DL. Natural protective glue protein, sericin bioengineered by silkworms: Potential for biomedical and biotechnological applications. Prog Polym Sci: 2008; 33 (10): 998-1012.
[44] Wu JH, Rong YZ, Wang ZW, Zhou YF, Wang SY, Zhao B. Isolation and characterisation of sericin antifreeze peptides and molecular dynamics modelling of their ice-binding interaction. Food Chem: 2015; 174:621-629.
[45] Lamboni L, Gauthier M, Yang G, Wang Q. Silk sericin: A versatile material for tissue engineering and drug delivery. Biotechnol Adv: 2015; 33 (8): 1855-1867.
[46] Aghaz F, Hajarian H, Shabankareh HK, Abdolmohammadi A. Effect of sericin supplementation in maturation medium on cumulus cell expansion, oocyte nuclear maturation, and subsequent embryo development in Sanjabi ewes during the breeding season. Theriogenology: 2015; 84 (9): 1631-1635.
[47] Vulpe R, Popa M, Picton L, Balan V, Dulong V, Butnaru M, Verestiuc L. Crosslinked hydrogels based on biological macromolecules with potential use in skin tissue engineering. Int J Biol Macromol: 2016; 84: 174-181.
[48] Worthington KS, Kaalberg E, Mullins RF, Stone EM, Tucker B. Fabrication and characterization of

tissue-engineered multilayer outer retinal grafts. Invest Ophth Vis Sci: 2015; 56.
[49] Dong ZM, Zhao P, Wang C, Zhang Y, Chen JP, Wang X, Lin Y, Xia QY. Comparative Proteomics Reveal Diverse Functions and Dynamic Changes of Bombyx mori Silk Proteins Spun from Different Development Stages. J Proteom Res: 2013; 12 (11): 5213-5222.
[50] Manchineella S, Thrivikraman G, Khanum KK, Ramamurthyp C, Basu B, Govindaraju T. Pigmented Silk Nanofibrous Composite for Skeletal Muscle Tissue Engineering. Adv Healthc Mater: 2016; 5 (10): 1222-1232.
[51] Lin NB, Meng ZH, Toh GW, Zhen Y, Diao YY, Xu HY, LIU XY. Engineering of Fluorescent Emission of Silk Fibroin Composite Materials by Material Assembly. Small: 2015; 11 (9-10): 1205-1214.
[52] Zhang Y, Wu CY, Guo SW, Zhang JY. Interactions of graphene and graphene oxide with proteins and peptides. Nanotechnol Rev: 2013; 2 (1): 27-45.
[53] Katoch J, Kim SN, Kuang ZF, Farmer BL, Nalk RR, Tatulian SA, Ishigami M. Structure of a Peptide Adsorbed on Graphene and Graphite. Nano Lett: 2012; 12 (5): 2342-2346.
[54] Cheng Y, Koh LD, LiI D, JiI B, Zhang Y, Yeo J, Guan G, Han MY, Zhang YW. Peptide-Graphene Interactions Enhance the Mechanical Properties of Silk Fibroin. ACS Appl Mater Inter: 2015; 7 (39): 21787-21796.
[55] Hu KS, Gupta MK, Kulkarni DD, Tsukruk VV. Ultra-Robust Graphene Oxide-Silk Fibroin Nanocomposite Membranes. Adv Mater: 2013; 25 (16): 2301-2307.
[56] Wang YX, Ma RL, Hu KS, Kim SH, Fang GQ, Shao ZZ, Tsukruk VV. Dramatic Enhancement of Graphene Oxide/Silk Nanocomposite Membranes: Increasing Toughness, Strength, and Young's modulus via Annealing of Interfacial Structures. ACS Appl Mater Inter: 2016; 8 (37): 24962-24973.
[57] Ling SJ, Li CX, Adamcik J, Wang SH, Shao ZZ, Chen X, Mezzenga R. Directed Growth of Silk Nanofibrils on Graphene and Their Hybrid Nanocomposites. Acs Macro Lett: 2014; 3 (2): 146-152.
[58] Liang B, Fang L, Hu Y, Yang G, Zhu Q, Ye X. Fabrication and application of flexible graphene silk composite film electrodes decorated with spiky Pt nanospheres. Nanoscale: 2014; 6 (8): 4264-4274.
[59] Sheng WQ, Zhu GB, Lu Q, Kaplan DL. Metal Oxide Nanomaterials with Nitrogen-Doped Graphene-Silk Nanofi ber Complexes as Templates. Part Part Syst Char: 2016; 33 (5): 286-292.
[60] Jiang CY, Wang XY, Gunawidjaja R, Lin YH, Gupta MK, Kaplan DL, Naik RR, Tsukruk VV. Mechanical properties of robust ultrathin silk fibroin films. Adv Funct Mater: 2007; 17 (13): 2229-2237.
[61] Lu Q, Hu X, Wang XQ, Kluge JA, Lu SZ, Cebe P, Kaplan DL. Water-insoluble silk films with silk I structure. Acta Biomater: 2010; 6 (4): 1380-1387.
[62] You X, Pak JJ. Graphene-based field effect transistor enzymatic glucose biosensor using silk protein for enzyme immobilization and device substrate. Sensor Actuat B-Chem: 2014; 202 (4): 1357-1365.
[63] Qu Y, Ma M, Wang ZG, Zhan GQ, Li BH, Wang X, Fang HF, Zhang HJ, Li CY. Sensitive amperometric biosensor for phenolic compounds based on graphene-silk peptide/tyrosinase composite nanointerface. Biosens Bioelectron: 2013; 44:85-88.
[64] Zou XF, Zhang L, Wang ZJ, Luo Y. Mechanisms of the Antimicrobial Activities of Graphene Materials. J Am Chem Soc: 2016; 138 (7): 2064-2077.
[65] Jarosz A, Skoda M, Dudek I, Szukiewicz D. Oxidative Stress and Mitochondrial Activation as the Main Mechanisms Underlying Graphene Toxicity against Human Cancer Cells. Oxid Med Cell Longev: 2016.
[66] Bitounis D, Ali-boucetta H, Hong BH, Min DH, KOSTARELOS K. Prospects and Challenges of Graphene in Biomedical Applications. Adv Mater: 2013; 25 (16): 2258-2268.
[67] Zhang BM, Wang Y, Zhai GX. Biomedical applications of the graphene-based materials. Mat Sci Eng C-Mater: 2016; 61: 953-964.
[68] Shadjou N, Hasanzadeh M. Graphene and its nanostructure derivatives for use in bone tissue engineering: Recent advances. J Biomed Mater Res A: 2016; 104 (5): 1250-1275.
[69] Xie YT, Li HQ, Zhang C, Gu X, Zheng XB, Huang LP. Graphene-reinforced calcium silicate coatings for load-bearing implants. Biomed Mater: 2014; 9.
[70] Wang L, Lu CX, Li YH, Wu F, Zhao B, Dong XZ. Green fabrication of porous silk fibroin/graphene oxide hybrid scaffolds for bone tissue engineering. Rsc Advances: 2015; 5 (96): 78660-78668.

A Review on Wearable NFC Antenna Devices

Lu-Lu Xu[1], Yi Li[2*], Yu-Tong Jang[3], Zhui-Run Hu[4], Yu-Xi Shi[5]

[1,2,5] *School of Materials, University of Manchester, Oxford Road, Manchester, M13 9PL, UK*
[3, 4] *School of Electrical and Electronic Engineering, University of Manchester, Oxford Road, Manchester, M13 9PL, UK*

[*] Corresponding author's email: henry.yili@manchester.ac.uk

Abstract

In the Internet of Things (IoT), wireless communication technologies are growing in interest through means of collecting and transferring data without interruption of human activities. The battery-free near field communication (NFC) has the potential to establish a smart wearable system based on textiles by the advances in materials, fabrication techniques and electronics design. In this paper, an overview of techniques and materials in antenna fabrication are presented. The design citations of NFC antenna on electromagnetic and mechanical points of view are presented. Their applications, outline challenges and opportunities of wearable NFC in the future are also discussed.

Keywords: Embroidery; Printing; Wearable Antenna; Near Field Communication; Textiles

1. Introduction

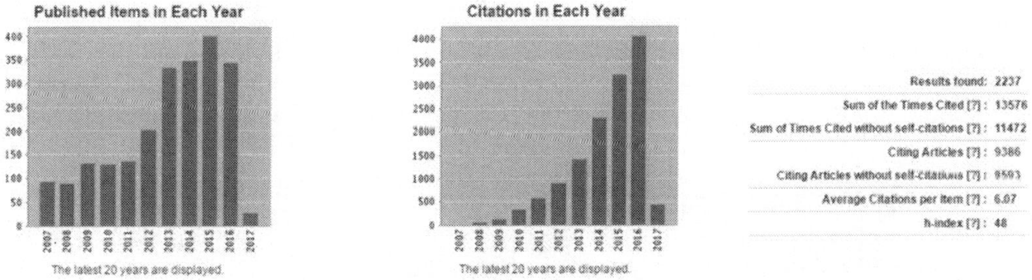

Fig.1 Citation report of NFC in last 10 years

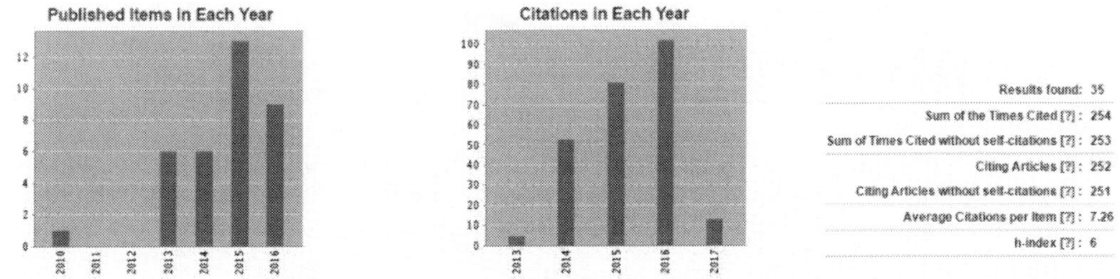

Fig.2 Citation report of wearable NFC in last 10 years

Smart wearable electronic textiles (e-textiles) refer to various electronic elements including sensors, antennas, batteries, harvesting devices, and circuit components (inductor, capacitor, amplifier, etc.) embedded into fabrics. It has increased rapidly with the development of advanced materials, fabrication machines, and electronic technologies [1, 2]. The wearable electronics market generated over $14 billion in 2014, and this figure is predicted to increase to over $100 billion by 2020 [3]. The battery-free and centimeter radiation capacities in NFC (near field communication or short range wireless communication) form the foundation for wireless communications in wearable e-textile system [4]. The citation reports of these topics from the web of science are showing in Fig.1 and in 2. The NFC becomes a heat topic in recent decade but the research of wearable NFC is still on infant period.

In recent years, the wearable NFC devices, a short range wireless communication technology operated in high-frequency (13.56 MHz), have a rapidly grown in commercial and research areas. Similar with passive Radio Frequency Identification (RFID) antennas, data and information could be shared by contactless matching. NFC devices become promising candidates in system of the Internet of Things by transferring measurement and sensing data to smart phone or other reading devices. NFC flexibility embedded into textiles in optimized shapes (like logo or letters) as a part of the garments worn by humans [5], and they emerge in various applications including: access tokens, personal authentication cards, transaction, social networking and biometric sensing [6-10].

This paper will be structured as following. Section 2 will review and compare two main technologies of embroidery and printing technologies in fabrication of wearable NFC. Section 3 will address the design citations from both electrical and mechanical points of view. Section 4 will illustrate the commercial models and applications. Section 5 will consider the opportunities and challenges from technical and commercial areas. Section 6 will summarize the key points in this paper.

2. Materials and Fabrication Techniques

Advances in materials, fabrication techniques and device architecture, create opportunities for realizing wearable properties, expanding the places of use, enhancing reliability of the system, minimizing the size in lightweight shape, and lower the cost of manufacturing. The techniques for the production of wearable antennas are various, including embroidered and sewed antenna loops [11]; woven and knitted conductive sheets [12, 13]; printed devices combining with coating and lamination techniques [14-17]. In the following section, we will present the previous main fabrication techniques of printing and embroidery, including materials and relative research.

2.1 Embroidery

The computerized embroidery machine (Fig.3(a) and (b)) is a promising technique in antenna fabrication, as conductive yarns can be integrated onto textile based materials in aesthetically shapes without glue, masks, or etching, which improve the washability and durability of the devices [18]. Mostly the embroidery machine could recognize the images from the software CAD/CAM, which provide repeatable, fast and efficient productivity in lower manufacturing costs [19].

(a) (b) (c)

Fig.3 Embroidery machines and samples: (a) Barudan embroidery machine [20] © 2014 by the authors; licensee MDPI AG, Basel, Switzerland. (b) ZSK embroidery machine with 2 different heads [21]. (c) Embroidered samples from ZSK using special W-head [22].

2.1.1 Conductive Yarns

By using the specialist conductive yarns, various shapes of antennas can be embroidered onto the textile substrates based on the similar principles with traditional embroidery process [23]. However, the high tension and speed of the embroidery machine require conductive yarns to be equipped with suitable flexibility and stretchability [19]. Adversely, the yarns effective conductivity is loss during the high tension process, varying between 1×10^4 and 5×10^4 S/m [24, 25]. Mallet [26] presented a finite element simulation model for sewing processes and demonstrated that sewing consisted of the needle tip force, the needle cutting force and the friction force, and reported that the speed of 1000 rpm and 2000 rpm created 5.6 Nat and 6.1 Nat respectively. Thus, to be used for embroidery, conductive yarns should have high mechanical properties to reduce the breaks, and also should remain effective conductivity after fabrication. It means the yarns are preferential with high modulus and tensile strength, as well as in smooth surface and uniform diameter as to avoid breakage and shrinkage during manufacturing. Commercially, conductive yarns are fabricated using composite metalized textile yarns, metal coated textile yarns or metallic wires [27]. Under this principle, Roh [1] divided the conductive yarns in three types: bundle of metal filaments, metal filament composite yarn and plied yarns of metal coated polymer filament (Table 1). The different amount of metal used on yarns resulting resistivity range from about 0.1 Ω/m to 1000 Ω/m. Fig. 4 shows various types of conductive yarns and their construction details.

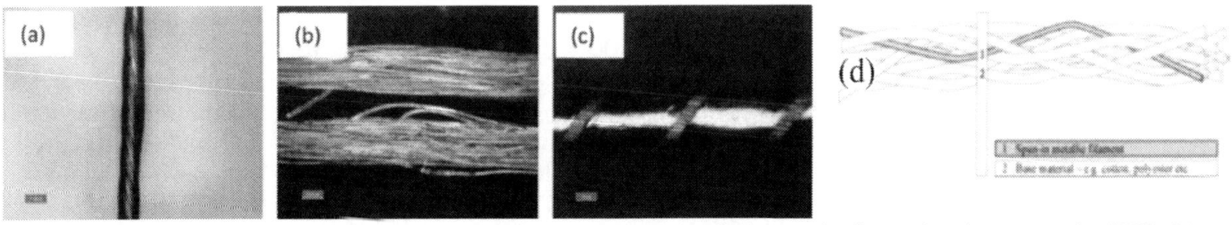

Fig.4 Various types of conductive yarns and their structures: (a) a fully metal, silver-plated copper wire [27]; (b) a silver-coated polyamide multifilament yarn; (c) a Kevlar multifilament yarn wrapped with metal foil; (d) Structure of conductive fiber twisted with the textile fibers [28]. Rights managed by AIP Publishing LLC.

Table.1 Characteristics of currently developed conductive embroidery yarns Roh [1]

Type	Research group	Organization and person	Thickness (linear density)	Resistivity (Ω/m)
Bundle of metal filaments	Plied yarn of stainless steel muti-filaments	Bekinox® VN (12/ 3×275/175S/316L) (BEKAERT 2006)	Three plies of stainless steel filaments (12 μm×275 f)	9.13
	Stainless steel yarns with a copper core	Coosemans (2006)	Stainless steel yarns (ø: 19 μm) with a copper wire core yarn (ø: 79 μm), limited circuit designs	0.2
Metal filament composite yarn	Stainless steel/nylon	IBM, Post et al. (Bekinox® VN 140 nyl/35×3) (2000)	Nylon core wrapped with three crossing stainless steel filaments, suitable as lower yarn for CNC embroidery	100
	Ag-copper/polyester	Ohmatex, Bi-component yarn (2004)	Polyester (240 dtex/48 f × 2) air textured yarn with silver-plated copper filament (40 μm: 123 dtex×2, 63 μm: 305 dtex×2)	7.75
	Ag-copper/polyester	Roh (2009). [26].	Three plies of Ag-copper filament (40 μm) and polyester (83 dtex/36 f)	3.89
Plied yarn of metal coated polymer filament	Silver plated nylon	Fraunhofer IZM (2009)	Two plies of Shieldex® 117/17dtex	350
	Silver-coated polyester	Mitsufuji Textile Ind. Co., Ltd. (AGposs®)	Three plies of Ag-coated polyester filament	26.2
	Metal plated aramid	DuPont (Aracon®) (2000)	Plied yarn of silver, nickel, copper, gold or tin clad aramid	0.1
	Silver-coated PBO	AmberStrand™ (Syscom Technology Inc. 2005)	Three plies of Ag-coated PBO	7.87

The targeted applications play crucial roles in yarns selection, for instant sensing and heating devices prefer yarns with higher resistance (lower conductivity), while antenna requires yarns with lower resistance (higher conductivity) as to achieve a considerable radiating power to send the data to a reading device. Therefore, the trade-off between conductivity and suitability are crucial. Such as, embroidered by metal wires will produce lower resistance but tend to break easily during embroidery. However, for ribbons, cords and fancy yarns, it is hard to sewn or embroidery onto the fabric by the traditional stitches. However, new embroidery technologies have been developed for fabrication these yarns. Specifically, the upper yarns are used to fix the conductive yarns to ensure them lie onto the textiles, move smoothly and place next to the needle (Fig.3(c)).

2.1.2 Substrates for Embroidery

The material of substrate plays a key role in antenna design and manufacturing. Selection of the flexible fabrics instead of circuit boards could improve the wearable capability and flexible placement of the targeted antennas [29]. However, the textile substrate should be equipped with low loss to increase the power efficiency when embedded onto body as wearing clothing or textiles. Also, the substrate dielectric permittivity varies from the materials and fabrication technologies. The materials' dielectric properties measurement method is based on capacitance extraction, using Agilent E4991A RF Impedance/Material Analyzer and 16453A calibration kit [30, 31]. Based on this principle, Koulouridis showed materials dielectric properties at different volume ratios [30] (Fig.5). Later, Osman and Rahim used the free space method to measure the permittivity of a certain fannel fabric (made of 100% cotton). After test of bandwidth, return loss (S11), gain and efficiency, they pointed that fannel fabric has satisfied properties in providing of antenna fabrication [32]. Therefore, the accurate information of substrate's dielectric material properties (i.e., loss tangents and permittivity) and thickness are fundamental to realize antenna's performance. A summary of fabrics used in papers are illustrated in Table 2.

Table.2 Type of substrate materials and their characters

Types	Permittivity	Thickness (mm)	Loss tangent
Cotton [19]	0.5-1.54	1.60	0.06
Denim [33]	1.0-2.0	1.50	0.90-5.00
Felt [19]	1.38	1.38	0.02
Hypalon coated Darcon fabric [34]	3.00	1.52	-
Woven fiberglass fabric [34]	2.20	1.524	-
Fleece [35]	1.17	2.56	2.6-3.95
Silk [35]	1.75	1.00	2.6-3.95
Jeans [34]	1.70	1.00	0.03

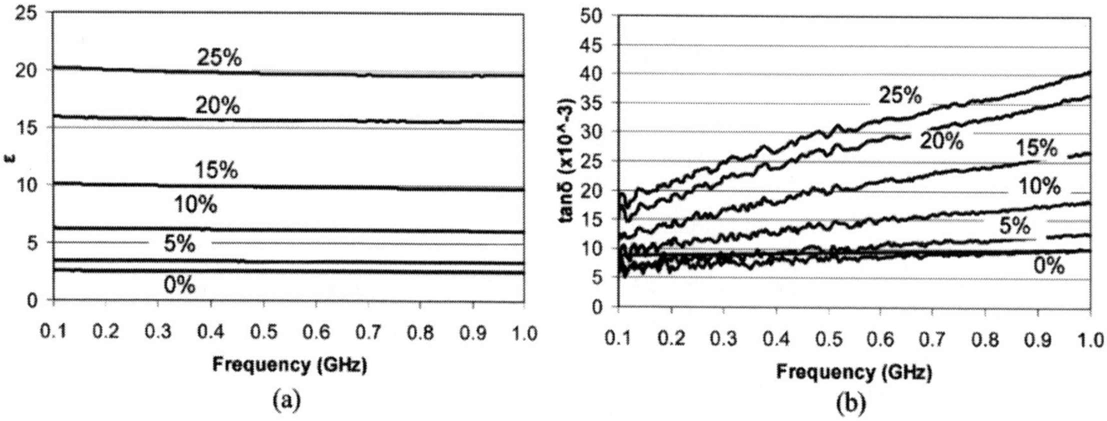

Fig.5 Materials dielectric properties test: (a) substrate dielectric permittivity. (b) Loss tangent for samples at different volume ratios [30]. Copyright © 2006, IEEE.

The antenna substrates fabricated either by a signal fabric layer or multiple layers with different materials. Kim, G., et al [36] designed alphabet images RFID, and sewed on the fabric directly. After test, its radiation range and efficiency were about 2.5m and 33% respectively (Fig.6 (a)). While, Wang, Z., et al. embroidered the conductive yarns on a regular fabric, followed by assembly with PDMS (Sylgard 184, Dow Corning Corp) (Fig.6(b)). In this novel fabrication, the antenna provided tunable permittivity (range from 3 to 12) and low Loss tangent ($tan\delta < 10^{-2}$), but also with flexibility, stretchability, and washability capacities [31]. Thus, the layers of wearable antenna should be designed by the target application and its electrical properties, which will be discussed in section 3.

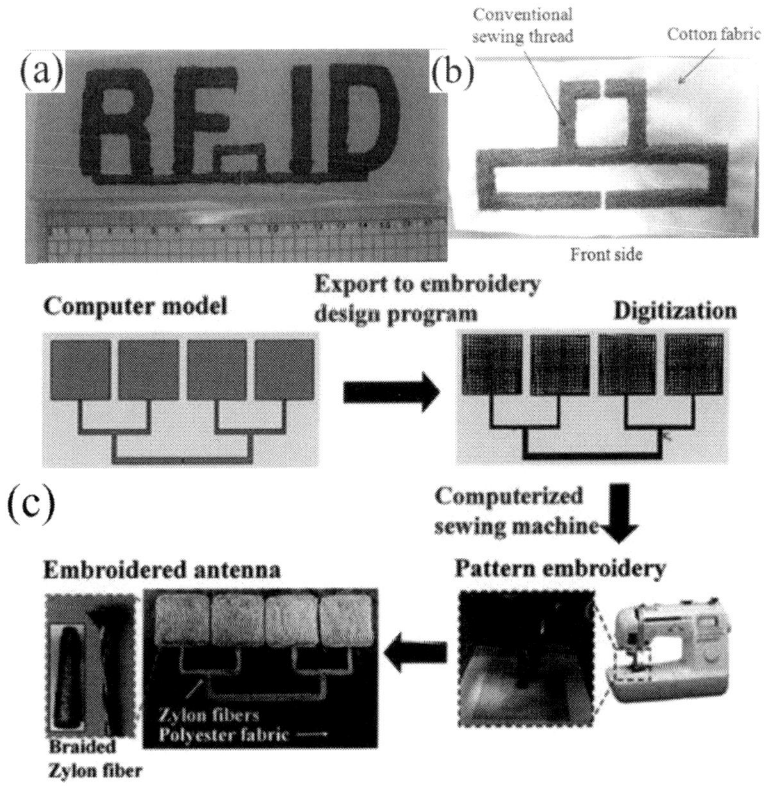

Fig.6 Embroidered antenna with different layers of substrate: (a) RFID on a single-layer fabric [36]; Copyright © 2008, IEEE. (b) RFID on nonwoven fabric [11]. Copyright © 2012, IEEE; (c) Embroidered antenna on polyester fabrics [31] Copyright © 2012, IEEE.

2.1.3 Related Work

Since 1997, wearable antenna has been a heated topic in smart wearable e-textiles system [32]. However, papers mainly focus on antenna's electrical properties or mechanical abilities (see section 3), fewer researches about fabrication technique, especially lack of addressing in embroidery process. The effects of the stitch type, stitch direction, stitch density and stitch spacing on the performance of antennas were discussed in recent papers. Acti, T., et al. [20] used the lock stich (Fig.7(a)), most commonly used in embroidery, to investigated effects of the stitch directions and densities on antenna inside connectivity. They found that the conductivity was better in parallel direction than the in-between one, as the current flow along the thread rather than interval to each other, and also addressed the higher density could improve the conductivity.

Later, Seager, R., et al. [37] used the commercial yarns (silver-coated Amberstrand yarn) embroidering three antennas in different stitch directions (vertical, horizontal and diagonal) and spacing (0.4, 0.6, 0.8 and1 mm). The structures are showed in Fig.7(b). After simulation and measurement, they reported that in patch antennas, vertical stitch showed had the best performance, and the diagonal stitch gave the worst results. Wang, Z., et al. [38]

embroidered microstrip antenna on fabric substrate, and found that electrical properties could be improved by tighter stitch spacing. Embroidery technique has widely used in antenna fabrication, but it hasn't applied in NFC fabrication.

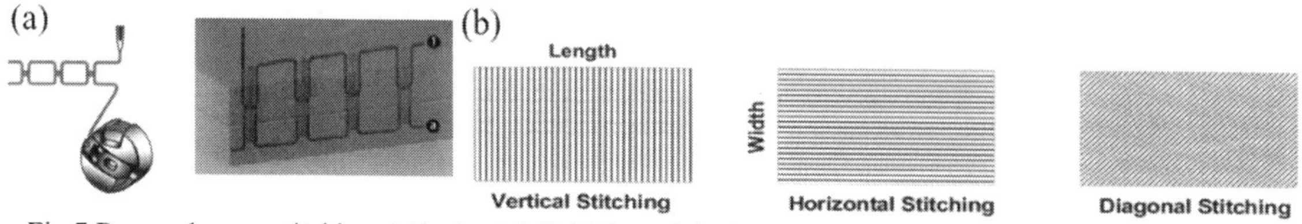

Fig.7 Researches on stitching: (a)lock stich [20] Copyright © 2011, IEEE; (b) stitch directions [37] © The Institution of Engineering and Technology.

2.2 Printing Techniques

Printing techniques has attracted high attention in antenna fabrication due to its easy process, low-cost and eco-friendly capabilities. Printed NFC and RFID antennas were discussed in papers [16, 39, 40]. Screen printing and inkjet printing are two main techniques in printing wearable electronics. The conductive sections work by depositing inks accurately on a certain substrate [39, 41]. Then, plasma and UV treatment will be applied to keep a good contact with ink and substrate, which keep direct printed antenna in good conducting quality. Therefore, for wearable antennas, inks and substrates with low-cost and high transparent abilities are essential in printing process.

Screen printing [42] has been developed since 1000 years ago and now is widely used in printing electronics. The working principle is showed in Fig.8(a). The flood blade moved across the screen to mesh area with the ink, and then the ink pull into the substrate by moving the squeegee. In operation, a special mask is required for each antenna and cannot be reused with different designs [43, 44]. However, no mask is required in inkjet printing as the shape of antenna can be designed in software directly and hence creates repeatable fabrications. Three basic elements of inkjet printing system including printer, ink and substrate are showed in Fig.8(b). Each of these should be compatible during the operation, such as the ink keep wet the substrate but not run off before drying, and the substrate must absorb the ink uniformly.

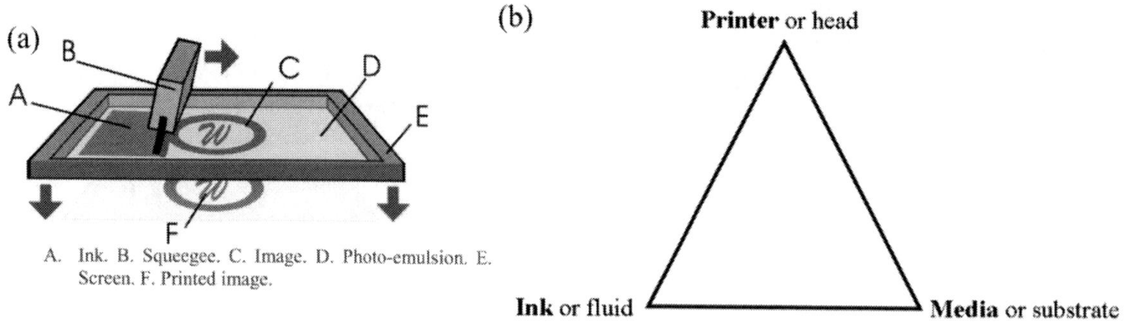

Fig.8 Illustration of screen printing and inject printing: (a) screen printing [42]; (b) inject printing system [45]. Copyright © 2016, IEEE

2.2.1 Substrates for Printing

Generally, wearable antennas were printing on paper based materials [45] and Kapton [40]. However, these materials are not robust enough, which become vulnerable and easy damaged in harsh environment. Later, the lower cost flexible polyimide-film [17, 46, 47], PET [16] and PMDS substrates emerged to overcome these problems. Table 3 illustrates the substrates and their characteristics in research papers.

Table.3 Different substrates for printing wearable antennas

Type	Loss tangent	Permittivity	Tensile	deformation
Paper based materials [48-50]	0.065	3.4	30 Low	High
Kapton [51, 52]	0.002	3.4	165 High	Low
Polyimide-film [46, 47]				High
PET [16, 53, 54]	0.003	2.8	266 High	High
PMDS [55, 56]	0.37	2.67	3.9 Low	High
Cotton textiles	0.06	0.5-1.54	-	High

2.2.2 Related Work

Paper, an organic-based substrate in low-cost and mass production, was first attempted in printing antenna. Rida, A., et al. [50] discussed the paper substrate electrical and dielectric performance and found that the papers dielectric slightly increase in RF modules, which was suitable in printing a low-cost passive RFID tag. However, the paper has limitations in printing process, as with the lower tensile, easy deformation and shrink thus restrict their application, especially in wearing system. Flexible substrates used in antennas printing have been developed fast in recent years. Babar, A.A., et al [55] designed a RFID antenna with a composite polymer-ceramic as a substrate. The ceramic powder was added into a stretchable, low loss and resistant PMDS material to achieve a high permittivity thus minimizing the size of antenna (Fig.9(a)). They yielded reasonable radiation performance as well as resistant to moisture and tension damages.

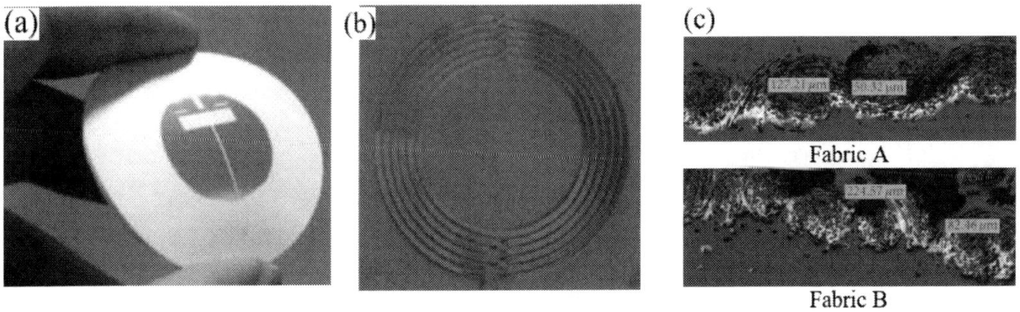

Fig.9 Printed antenna: (a) antenna on flexible ceramic-polymer substrate [55] Copyright © 2012, IEEE; (b) inject print NFC [40] Copyright © 2010, IEEE; (c) Cross-sections of antennas on Fabrics (magnification is ×20. The light part is the ink film, and the dark part is the fibre structure of the fabric) [43]. Copyright © 2012 Tiiti Kellomäki et al.

Research on printing wearable antenna was focusing on RFID system as demonstrated above. Fewer characterized the process on NFC printing. Mujal, J., et al. [40] explored the potential of the inkjet printing in manufacturing NFC antennas (Fig.9(b)) and found that inkjet printing was a promising technique in fabricating of passive NFC, which decreased the cost of fabrication and generated a reasonable radiation distance (2.0-3.7cm).

However, printing on textile surface is still challenging, especially for the roughness fabric. Kellomäki, T., et al.[43] explored screen printing for the wearable RFID by two different fabrics, 100% cotton woven fabric (A), and 35% cotton, 65% polyester knitted fabric (B). With the comparison of the cross-section of different fabrics (Fig.9(c)) the results emerged, uneven fabric surface caused discontinuities and breaks in screen printing progress. Chauraya, A. [19] point that the roughness of the fabric was reduced by multilayers of printing, while the long term conductivity, robustness to stretching and reliable in high temperatures still need to improve.

In comparison with printing (Table 4), embroidery allows fabricating repeatable via the computerized embroidery machines thus increasing higher interests with low-cost and providing the potential for mass production.

Table.4 Different technology for manufacturing wearable antennas

Type	advantage	disadvantage
Embroidery	(1) Potential in mass production (2) Repeatable and flexible geometries. (3) Without the use of glue, improved the washability	High tension and friction of the machine probably cause fiber breakage and will decrease the antenna's conductivity
Printing	Lightweight, small and flexible	Paper and Kapton are normally used, while textiles are not ideal for printing.

3. NFC Antenna Design and Performance Evaluation

NFC antenna as a short range device in wireless communication, the distance for exchange of information is not more than 10cm. It works at 13.5 MHz frequency and its data rate is less than 424 kbps [57, 58]. Its working principle is similar to RFID in a short un-contacted area, energy created via the inductive coupling between the platform of reader (cellular phones, cash dispenser, wrist devices, etc.) and the antennas through a magnetic field. It provides a security centimeter ranges to overcome eavesdrop information in transactions.

3.1 Antenna Design

The design of a wearable NFC antenna device is critical important in terms of its electrical and mechanical performance. The design process of a wearable NFC antenna should guarantee good radiation performance with the simulation result and should work reasonable under deformation conditions by wearers. In this section, the electromagnetic and mechanical design citations are addressed.

3.1.1 Electromagnetic Design

In the 13.56MHz area, the inductively coupled exists in the near field, where the reader (mobile phone or other devices) and the antenna loop would generate necessary energy by the magnetic field. Fig.10 illustrates the basic work principles of an inductive coupled NFC antenna:

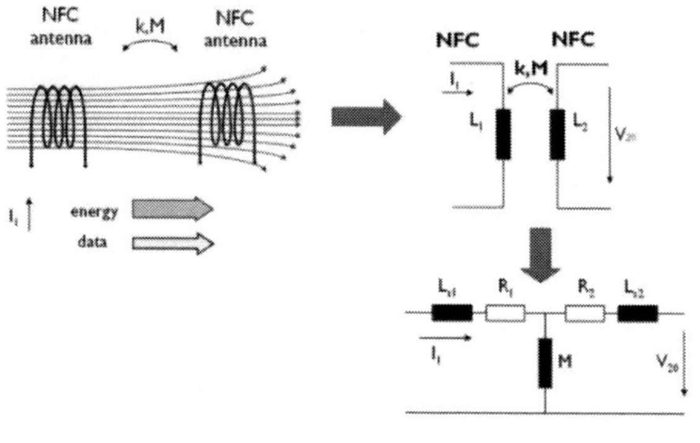

Fig.10 NFC principle for 13.56 MHZ[58]. Copyright © 2008, IEEE.

Loop antenna, an integrated circuit chips (IC) and readers are basic components in NFC system, illustrated in Fig.11(a) (H antenna for example). With a capacitor placed across in parallel, the two ports (start and the end port) of the antenna are linked to a radio frequency (RF) transceiver. The input DC resistance can be assumed to be zero, it will create an inductance with the increased frequency [59]. Fig.11(b) shows the circuit model for such antenna, where antenna is represented as an inductor. Typically, the commercial inductor values of NFC antenna should be in the range from 0.3 to 3μH. In order to obtain a self-resonance frequency higher than 35 MHz, the parasitic capacitance should be kept low enough [40] [12]. Due to the short range operation characteristic, the NFC antenna presents smaller quality factor (Q) than the RFID tags. In NFC loop antenna, Q is a measure of the bandwidth of it relative to the center frequency of the bandwidth. If an NFC operates brand between $f_1$andf_2, then the center frequency is given by equation (1), and the Q is by equation (2).

$$f_c = \frac{f_1+f_2}{2} \tag{1}$$

$$Q = \frac{f_c}{f_2-f_1} \tag{2}$$

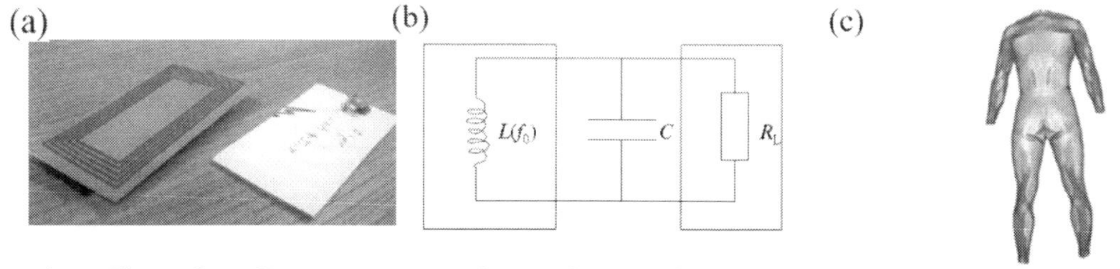

Fig.11 Illustration of NFC antenna. (a) shapes of antenna [60]. Copyright © 2011, IEEE. (b) Circuit model of a antenna [59]; (c) Curvature radii along the human body [26]. Copyright © 2011, IEEE.

In the equation (1) and (2), Q factor is inversely proportional to bandwidth. For a fixed resonance frequency (13.56 MHz), a higher Q (limited the bandwidth) results a lower data transfer rate. While with a lower Q factor, it implies a shorter transmission range in the system, because much energy is lost in transferring. Therefore, the NFC design should have a proper trade-off between maximum interactive power (minimum heat losses) and maximum amount of transmitted information [40].

3.1.2 Mechanical Design

In wireless communication system, a wearable NFC antenna device is always contacting naturally with the human body, can aware of users' situation but not restrict their activities. It features high stretchable quality, easy to use and supports unobtrusively in their routine wearing. Bending, friction, compression and stretchable deformation usually occur in garments along with the movements of the body, especially around joints. Fig.11(c) qualitatively shows the curvature distribution along the human body [61]. The darker regions mean the smaller the radius is. The bending of the human body is in arbitrary directions, that only textiles can satisfy these exposures. In contrast, a flexible PCB substrate such as polyamides can be bended only in a single direction at a time. Therefore, textile NFC antenna devices should have a proper ability to withstand deformation also maintaining their radiation specifications in design process.

3.2 Performance Evaluation and Related Work

After antenna design and fabrication, the performance evaluation plays a key role in realizing the functional properties of NFC devices. Kim, J., et al. [47] developed a thin and stretchable NFC monitoring device that were compatible with commercial reading devices, such as smart phones, wrist, computers and other NFC-read electronics (Fig.12(a-c)). For this antenna worked at 13.56 MHz, the Q factor was 9.8, and the inductance and resistance were 5.9 mH and 51 respectively. The layout/geometry of NFC died, chips and shapes of the serpentine interconnects were ascertained by finite element analysis (FEA) to reduce their strains and avoid entanglement under mechanical deformation (Fig.12(e-f)). Finally, the results from sensors demonstrated that the devices could work well under deformation thus have a potential in diagnostic of heart rate, temporal dynamic arterial blood flow, tissue oxygenation and ultraviolet dosimetry.

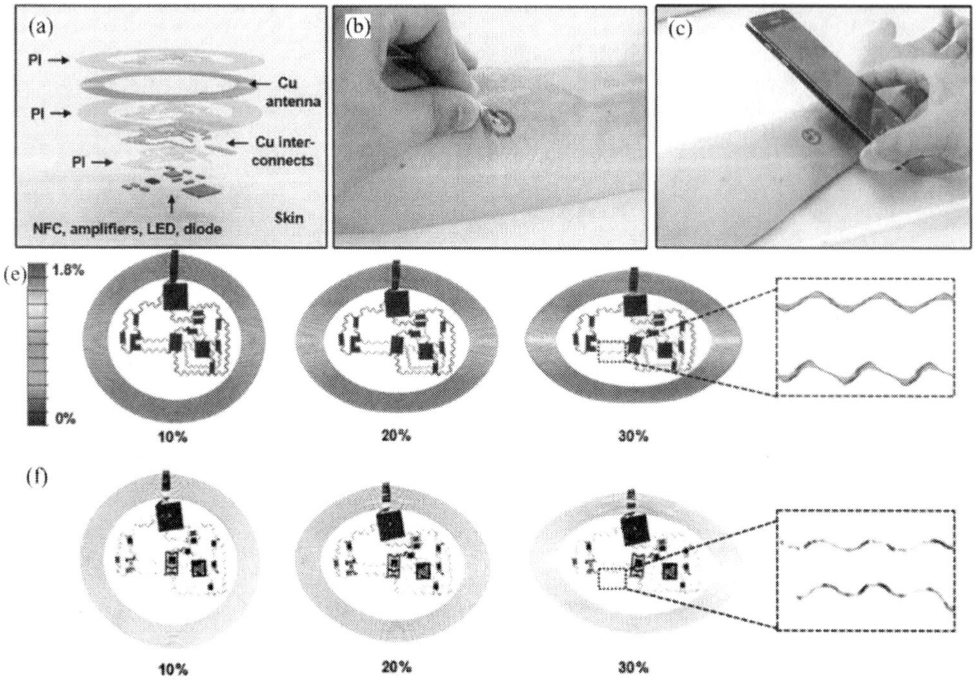

Fig.12 Wireless epidermal optoelectron NFC devices designed for heart rate and MAP tracking: (a-c) design and test; (e-f) FEA results of strains and mechanical deformation [47]. © 2016 WILEY-VCH Verlag GmbH & Co. KGaA, Weinheim.

John A. Rogers [17, 46, 62, 63] continued focusing on construction of NFC devices on epidermis with ultraminiaturized, thin, lightweight and flexible format. These devices could maintain their electromagnetic properties (Fig.13(a)) under mechanical strength and liquid corrosion, which could be applied as biosensors for hospital and home health monitoring.

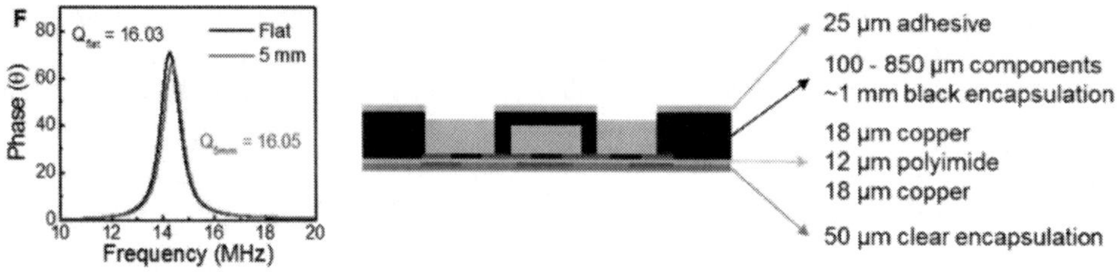

Fig.13 NFC electromagnetic properties and manufacturing details [17]: (a) Phase responsesand Q factors associated with the flat and curved devices; (b) Layered structure © 2016 WILEY-VCH Verlag GmbH & Co. KGaA, Weinheim

4. Applications

Previously, the NFC antenna devices was mainly focused in three different modes, reader with write mode, peer to peer mode and card emulation mode. For the reader and write model, the active device can not only read data from a passive tag but can also write data over it. For the peer to peer one, both are active devices, and data can be shared in a special format by close touching. For the card emulation mode, the active antenna works as a passive tag and the other antenna device reads or writes data over this card. The Table 5 shows the three working models and makes comparison between them.

Commercially, the NFC devices have been used via the existing mobile phone programs and services. Specifically, The NFC antenna combined with data stored chip, then transfer and exchange information by touching a mobile phone in a centimeter range [8] [64]. But now, the NFC devices have widened their versatile applications in different areas, such as reorganization devices, e-ticket, cash dispenser and access cards, etc. (Table 5).

In the Internet of Thing, the NFC device has changed the way of usage and setup a bridge between objects and electronics. By just touching their smart devices over the NFC antennas, users can share information conveniently. These antennas can be put or stick onto ordinary objects. Fig.14(a) shows a book with pasted NFC interacting with a smart phone, and Fig.14(b) shows a ring with NFC tag interacting with the phone [21].

Currently, fewer NFC devices have been used in wearable communication. The Rogers research group tried several thin and stretchable substrates to fabricate NFC and pasted them on skin, after that information can be transferred to the electronic devices. They applied these NFC devices in hospital and at-home diagnostics [47]. However, their applications were mainly fabricated by soldering coils in polyimide, laminated an encapsulation layer (silicone elastomer) to protect the device as shown Fig.13(b). The sweat secreted by body cannot transfer to the outside by adhesive these devices onto the skin interface that will cause irritation of the skin, also the skin-contact devices couldn't be durable as the human needs to shower, swim and wipe regularly.

Fig.14 Commercial NFC applications: (a) NFC tagged book interacting with a smart phone [57]; (b) NFC ring communicating with a smart phone [57]. Copyright © 2014, IEEE.

Table.5 Details of three modes of NFC

Mode	Function	Application	Advantage	Challenges
Reader/Writer	Reading or writing data	Wearable devices[57] advertisements; ticks[40]	(1) Rapid action in a centimeter range. (2) Battery free. (3) High security, reliability and quality levels without a complex pairing process. (4) Available bandwidth for multiple users in the same area and time[65].	(1) High volume and thickness restrictions in a miniature size. (2) Replace the sintered ferrite on a mobile phone battery cell to reduce the whole cost. (3) Improve the wearable-ability and wear-ability. (4) Explore thinner substrates and protecting encapsulating layers[63].
Peer-to-peer	Sharing data	Smart devices[57, 66]		
Card	Contactless identification	Contactless payment; premises access; transportation; electronic toll collection; car parking; school and campus cards; employee cards; loyalty cards[6, 60].		

5. Challenges and Opportunities

From the costs savings and secure protecting viewpoint, the NFC antenna devices with lower price, battery free and a better immunity to eavesdropping have advantages over the existing wireless communication technologies such as Bluetooth. The price of Bluetooth per node is several dollars, which is too much for a mass of production. For information exchange, tens of mill watts are used for Bluetooth that is too much for the continuous long-term operation and cannot work without battery. Typically, matching two Bluetooth devices together waste several tens of seconds, and manual intervention are required in re-connecting previously paired devices. Since NFC does not require a pairing process, just placing the NFC reader or mobile phone next to the tag without manual intervention. The short range of distances (about 0-20 centimeters) between the NFC and readers enable it can only read one tag at a time, which leads to a secure communication in card mode [64]. Further advantages, the NFC tags have the possibility to write data back to the tag without a direct line of sight. This allows the tags to be covered and increases the usage convenience. These advantages make it easy to place NFC in small and wearable devices, such as ring, wristbands, watch, etc. But currently, most of NFC antennas have been packed with a mobile phone battery cell in the industry, with limited applications on wearables because of their rigid format.

However, the wearable NFC faces several challenges, especially embroider it on textiles. The textile antennas are fabricated with several layers that are roughness and thick. But the volume of the antenna is limited in order to without interface the wearers. The deformation, compression, bending and friction degradation will interrupt the performance of reading and tagging range when an NFC wearing on human body.

Therefore, an overarching goal is to engineer the physical properties of wearable NFC antenna devices, and in particular of the elastic modulus and elastic stretchability, to match the curvature radio along the human body and to reduce irritation and discomfort at the skin.

6. Conclusion

Smart wearable NFC devices based on textiles have prominent advantages that can be integrated directly into clothing or wearable devices to contact with human body and do not affect the body's work which allow wireless communication. Combining a mediation design from mechanical and electromagnetic view of points with advanced materials, NFC would definitely change our way of living. However, current research is mainly focus on manufacturing of RFID antennas and test of their electromagnetic properties, NFC fabrication process and wearable properties evolution are still at infant stage. In the future, we would explore the potential applications of NFC on textiles or clothing to ensure it joins into the IoT system spontaneously.

7. Acknowledgment

Also, we would like to thank the support of the EU Horizon 2020 and University of Manchester through projects with project codes 644268 - ETEXWELD - H2020-MSCA-RISE-2014, AA14512 (UMRI project "Graphene-Smart Textiles E-Healthcare Network") and AA01906

Reference

[1] Roh JS, Chi YS, Kang TJ. Wearable textile antennas. International Journal of Fashion Design, Technology and Education: 2010; 3 (3): p. 135-153.

[2] Shi YX, et al. Advanced in Smart E-Textile Apparel Technology, in Textile Bioengineering and Informatics Symposium Proceedings, 2016, Vols 1 and 2, Y. Li and R. Padhye, Editors: 2016, Textile Bioengineering & Informatics Society Ltd: Hong Kong Sar: p. 53-60.

[3] Raupp G. (Invited) Flexible Thin Film Transistor Arrays as an Enabling Platform Technology: Opportunities and Challenges. ECS Transactions: 2011; 37 (1): p. 229-240.

[4] Marrocco G. Pervasive electromagnetics: sensing paradigms by passive RFID technology. IEEE Wireless Communications: 2010; 17 (6): p. 10-17.

[5] JaeHan C, et al. Various wearable embroidery RFID tag antenna using electro-thread. in 2008 IEEE Antennas and Propagation Society International Symposium: 2008.

[6] Al-Chalabi AE, et al. A wearable and ubiquitous NFC wallet. in 2015 IEEE 28th Canadian Conference on Electrical and Computer Engineering (CCECE): 2015.

[7] Köstinger H, et al. Developing a NFC based patient identification and ward round system for mobile devices using the android platform. in 2013 IEEE Point-of-Care Healthcare Technologies (PHT): 2013.

[8] López-Soriano S, Parrón J. Wearable RFID tag antenna for healthcare applications. in 2015 IEEE-APS Topical Conference on Antennas and Propagation in Wireless Communications (APWC): 2015.

[9] Pham NN, Chien Dao, Chung JY. A text-meandered RFID tag implemented with conductive threads. Microwave and Optical Technology Letters: 2016; 58 (8): p. 1978-1984.

[10] Bansal R. Coming soon to a Wal-Mart near you. IEEE Antennas and Propagation Magazine: 2003; 45 (6): p. 105-106.

[11] Ukkonen L, Sydänheimo L, Rahmat-Samii Y. Sewed textile RFID tag and sensor antennas for on-body use. in 2012 6th European Conference on Antennas and Propagation (EUCAP): 2012.
[12] Nguyen TM, Chung JY, Lee B. Radiation Characteristics of Woven Patch Antennas Composed of Conductive Threads. IEEE Transactions on Antennas and Propagation: 2015; 63 (6): p. 2796-2801.
[13] Salonen P, Hurme L. A novel fabric WLAN antenna for wearable applications. in IEEE Antennas and Propagation Society International Symposium. Digest. Held in conjunction with: USNC/CNC/URSI North American Radio Sci. Meeting (Cat. No.03CH37450): 2003.
[14] Wang J, et al. Brush-painting and photonical sintering of copper and silver inks on cotton fabric to form antennas for wearable ultra-high-frequency radio-frequency identification tags. Textile Research Journal: 2015; p. 0040517515596969.
[15] Chen ZN, See TSP, Qing X. Small Printed Ultrawideband Antenna With Reduced Ground Plane Effect. IEEE Transactions on Antennas and Propagation: 2007; 55 (2): p. 383-388.
[16] Hassan A, et al. All printed antenna based on silver nanoparticles for 1.8 GHz applications. Applied Physics A: 2016; 122 (8): p. 768.
[17] Kim J, et al. Miniaturized Battery-Free Wireless Systems for Wearable Pulse Oximetry. Advanced Functional Materials: 2017; 27 (1).
[18] Acti T, et al. Embroidered wire dipole antennas using novel copper yarns. IEEE Antennas and Wireless Propagation Letters: 2015; 14: p. 638-641.
[19] Tsolis A, et al. Embroidery and related manufacturing techniques for wearable antennas: challenges and opportunities. Electronics: 2014; 3 (2): p. 314-338.
[20] Acti T, et al. High performance flexible fabric electronics for megahertz frequency communications. in 2011 Loughborough Antennas & Propagation Conference: 2011.
[21] http://www.stocks.co.uk/zsk-embroidery-machine/2-head-embroidery/zsk-racer-2w.html. ZSk Website: 2017.
[22] http://www.stocks.co.uk/zsk-embroidery-machine/special-embroidery.html. ZSK Website: 2017.
[23] Jevšnik S, et al. Thermal-Mechanical Sensory Properties of Hot-Air Welded Textile Transmission Lines. Tekstilec: 2016; 59 (2).
[24] Chauraya A, et al. Addressing the challenges of fabricating microwave antennas using conductive threads: 2012.
[25] Fan YY, Li Y, Liu XQ. Granphene Fibre Review: from Spinning Methodologies to Applications, in Textile Bioengineering and Informatics Symposium Proceedings: 2016, Vols 1 and 2, Y. Li and R. Padhye, Editors: 2016, Textile Bioengineering & Informatics Society Ltd: Hong Kong Sar: p. 26-40.
[26] Mallet E, Du R. Finite element analysis of sewing process. International Journal of Clothing Science and Technology: 1999; 11 (1): p. 19-36.
[27] Cherenack K, L. van Pieterson, Smart textiles: challenges and opportunities. Journal of Applied Physics: 2012; 112 (9): p. 091301.
[28] http://www.swiss-shield.ch/Garne-Textilien.61.0.html: 2017.
[29] Van Langenhove L. Smart textiles for medicine and healthcare: materials, systems and applications: 2007; Elsevier.
[30] Koulouridis S, et al. Polymer–Ceramic Composites for Microwave Applications: Fabrication and Performance Assessment. IEEE Transactions on Microwave Theory and Techniques: 2006; 54 (12): p. 4202-4208.
[31] Wang Z, et al. Embroidered conductive fibers on polymer composite for conformal antennas. IEEE Transactions on Antennas and Propagation: 2012; 60 (9): p. 4141-4147.
[32] Osman MAR, et al. Embroidered fully textile wearable antenna for medical monitoring applications. Progress In Electromagnetics Research: 2011; 117: p. 321-337.
[33] Sanz-Izquierdo B, et al. Textile integrated waveguide slot antenna. in 2010 IEEE Antennas and Propagation Society International Symposium: 2010.
[34] Shakhirul M, et al. Embroidered wearable textile antenna on bending and wet performances for UWB reception. Microwave and Optical Technology Letters: 2014; 56 (9): p. 2158-2163.
[35] Zhu S, Langley R. Dual-Band Wearable Textile Antenna on an EBG Substrate. IEEE Transactions on Antennas and Propagation: 2009; 57 (4): p. 926-935.
[36] Kim G, et al. Design of a UHF RFID fiber tag antenna with electric-thread using a sewing machine. in 2008 Asia-Pacific Microwave Conference: 2008.

[37] Seager R, et al. Effect of the fabrication parameters on the performance of embroidered antennas. IET Microwaves, Antennas & Propagation: 2013; 7 (14): p. 1174-1181.
[38] Zhang S, et al. Embroidered wearable antennas using conductive threads with different stitch spacings. in Antennas and Propagation Conference (LAPC), 2012 Loughborough: 2012. IEEE.
[39] Park H, et al. Fully roll-to-roll gravure printed rectenna on plastic foils for wireless power transmission at 13.56 MHz. Nanotechnology: 2012; 23 (34): p. 344006.
[40] Mujal J, et al. Inkjet printed antennas for NFC systems. in 2010 17th IEEE International Conference on Electronics, Circuits and Systems: 2010.
[41] Zichner R, Sowade E, Baumann RR. Inkjet printed WLAN antenna for an application in smartphones. Japanese Journal of Applied Physics: 2014; 53 (5S3): p. 05HB06.
[42] Xiao GG, et al. Screen printing RF antennas. in 2016 17th International Symposium on Antenna Technology and Applied Electromagnetics (ANTEM): 2016.
[43] Kellomäki T, et al. Towards washable wearable antennas: a comparison of coating materials for screen-printed textile-based UHF RFID tags. International Journal of Antennas and Propagation: 2012; 2012.
[44] Kazani I, et al. Washable screen printed textile antennas. in Advances in Science and Technology: 2013. Trans Tech Publ.
[45] Hodgson A. The role of paper in the future of printed electronics. CoPADD: 2007; 2007: p. 39.
[46] Kim J, et al. Miniaturized Flexible Electronic Systems with Wireless Power and Near-Field Communication Capabilities. Advanced Functional Materials: 2015; 25 (30): p. 4761-4767.
[47] Kim J, et al. Battery-free, stretchable optoelectronic systems for wireless optical characterization of the skin. Science Advances: 2016; 2 (8): p. e1600418.
[48] Choi BO, Kim CH, Kim DS. Manufacturing ultra-high-frequency radio frequency identification tag antennas by multilayer printings. Proceedings of the Institution of Mechanical Engineers, Part C: Journal of Mechanical Engineering Science: 2010; 224 (1): p. 149-156.
[49] Amin R, et al. Design and integration of inkjet-printed paper-based UHF components for RFID and ubiquitous sensing applications. in 2007 European Microwave Conference: 2007.
[50] Rida A, et al. Conductive inkjet-printed antennas on flexible low-cost paper-based substrates for RFID and WSN applications. IEEE Antennas and Propagation Magazine: 2009; 51 (3).
[51] Virtanen J, et al. The effect of conductor thickness in passive inkjet printed RFID tags. in 2010 IEEE Antennas and Propagation Society International Symposium: 2010.
[52] Kanso A, et al. Inkjet printing of Coplanar Wire-Patch antenna on a flexible substrate. in 2012 15 International Symposium on Antenna Technology and Applied Electromagnetics: 2012.
[53] Mäntysalo M, Mansikkamäki P. An inkjet-deposited antenna for 2.4 GHz applications. AEU - International Journal of Electronics and Communications: 2009; 63 (1): p. 31-35.
[54] Palukuru VK, et al. Inkjet-Printed RF Structures on BST–Polymer Composites: An Application of a Monopole Antenna for 2.4 GHz Wireless Local Area Network Operation. International Journal of Applied Ceramic Technology: 2011; 8 (4): p. 940-946.
[55] Babar A.A, et al. Inkjet-printable UHF RFID tag antenna on a flexible ceramic- polymer composite substrate. in 2012 IEEE/MTT-S International Microwave Symposium Digest: 2012.
[56] Zichner R., et al. Optimized antenna design for gravure printed RFID Applications. Proc. LOPE-C: 2010; p. 13-16.
[57] Iqbal R, Saeed MD, Ahmad A. Wearable Near Field Communication ring antenna for mobile communication. in 2014 International Conference on Open Source Systems & Technologies: 2014.
[58] Rinner D, Witschnig H, Merlin E. Broadband NFC - A system analysis for the uplink. in 2008 6th International Symposium on Communication Systems, Networks and Digital Signal Processing: 2008.
[59] Chen ZN. Antennas for portable devices: 2007: John Wiley & Sons.
[60] Brown TWC, Diakos T. On the design of NFC antennas for contactless payment applications. in Proceedings of the 5th European Conference on Antennas and Propagation (EUCAP): 2011.
[61] Locher I, et al. Design and Characterization of Purely Textile Patch Antennas. IEEE Transactions on Advanced Packaging: 2006; 29 (4): p. 777-788.
[62] Araki H, et al. Materials and Device Designs for an Epidermal UV Colorimetric Dosimeter with Near Field Communication Capabilities. Advanced Functional Materials: 2017; 27 (2).

[63] Kim J, et al. Epidermal Electronics with Advanced Capabilities in Near - Field Communication. small: 2015; 11 (8): p. 906-912.
[64] Strommer E, et al. Application of Near Field Communication for Health Monitoring in Daily Life. in 2006 International Conference of the IEEE Engineering in Medicine and Biology Society 2006.
[65] Bansal R. Near-field magnetic communication. IEEE Antennas and Propagation Magazine: 2004; 46 (2): p. 114-115.
[66] Lee B, et al. NFC Antenna Design for Low-Permeability Ferromagnetic Material. IEEE Antennas and Wireless Propagation Letters: 2014; 13: p. 59-62.

Mechanical Similarity for Three-dimensional Fabric Drape Considering Bending Rigidity and Shear Stiffness

Liu Yang[1], KyoungOk Kim[2], Masayuki Takatera[2*]

[1] *Graduate School of Science and Technology, Shinshu University, 3-15-1 Tokida, Ueda, Nagano 386-8567, Japan*
[2] *Division of Kansei and Fashion Engineering, Institute for Fiber Engineering (IFES), Interdisciplinary Cluster for Cutting Edge Research (ICCER), Shinshu University, 3-15-1 Tokida, Ueda, Nagano 386-8567, Japan*

[*]Corresponding author's email: takatera@shinshu-u.ac.jp

Abstract

To determine a similarity rule that can be applied to fabric drape, experiments were conducted to explore the relationship among drape coefficient, mechanical properties of fabrics, and geometrical parameters in drape tests. Three sample fabrics were investigated: broadcloth, gabardine, and herringbone tweed. Mechanical properties of these fabrics were measured and drape tests were carried out under various geometrical conditions. Multiple regression analysis was conducted to find which parameters affect fabric drape coefficient. The results show that shear stiffness is related to drape coefficient in addition to the bending length of the fabric. Thus, parameters related to a similarity rule on fabric drape were found.

Keywords: Drapability; Similarity; Bending Rigidity; Shear Stiffness; Bending Length

1. Introduction

Garment simulation has recently been applied in the apparel industry. However, garment simulation lacks high accuracy, and evaluation of available garment simulators remains a problem. The simulation accuracy of a garment simulator could be evaluated if a similarity rule for fabric drapes could be found and applied.

The similarity rule for a fabric cantilever uses a non-dimensional parameter $K = L(w/B)^{1/3}$, where L is sample length, B is bending rigidity per unit width, and w is fabric weight per unit area [1]. When K values for any fabrics are the same, the fabrics display similar bending shapes although they have different values for other parameters. However, in three-dimensional fabric deformation such as a fabric drape, similarity rules cannot be established because the deformations have not yet been described in an analytical formulae.

Drape coefficient (DC) is the ratio of a projected fold area formed by a piece of fabric draping under its own weight to the original undraped area of the piece of fabric [2]. It was defined in Fabric Research Laboratories (F. R. L.) drape test and is used for evaluating fabric drapability [3]. Since the 1930s, many studies on fabric drape have been conducted [3-8]. According to these studies, drapability depends on the following fabric mechanical properties: bending rigidity, bending hysteresis, fabric own weight, shear stiffness, and shear hysteresis. Models and equations to describe DCs with drape parameters related to the Kawabata evaluation system (KES) [9] have also been proposed. However, the proposed models did not take into account the geometrical parameters; drape similarity is still unknown.

In the present study, we explore the basic parameters for establishing a similarity rule for fabric drape. To establish such a similarity rule, experiments were conducted to explore the relationships among DC, mechanical properties of fabrics, and geometrical parameters in drape tests.

2. Experimental

Drape tests were performed on three different fabrics: broadcloth, gabardine, and herringbone tweed. Bending rigidity and shear stiffness of the samples were measured by KES system. A summary of the samples is given in Table 1.

Table.1 Sample specifications

Sample	Material	Area density ($g\cdot cm^{-2}$)	Bending rigidity ($gf\cdot cm^2cm^{-1}$) Warp	Weft	45°	Mean	Shear stiffness ($gf\cdot cm^{-1}degree^{-1}$) Warp	Weft	Mean
Broadcloth	Cotton 100%	0.011	0.053	0.041	0.035	0.043	0.93	0.70	0.82
Gabardine	Wool 100%	0.019	0.091	0.054	0.069	0.071	0.77	0.56	0.67
Herringbone tweed	Wool 80%, others 20%	0.031	0.237	0.184	0.135	0.185	0.71	0.59	0.65

F. R. L. drape tests were performed to obtain DC, as shown in Fig.1, following the principles described by Chu et al [3]. Equation (1) is used for DC calculation,

$$DC\ (\%) = 100 \times (A_d - S_1)/\ (S_2 - S_1), \tag{1}$$

where A_d is the area of the sample draping fabric, S_1 is the area of the circular disk and S_2 is the area of the sample fabric without draping.

DC calculations are affected by the sizes of sample fabrics and disks, thus, the radius r of the circular disk was set from 1.5 cm to 8 cm at intervals of 0.5 cm. For each disk radius, sample radius R was set as twice the disk radius r, $R=2r$. Thus, the disk radius r is equal to the sample length L. The node number n of the drapes was manually set as 3, 4, 5, and 6. For the DC calculation, photos were taken from above the drape tester at a height of 150 cm. The pixels of each area in these photos were measured using Adobe Photoshop CC 2015 (Adobe Systems Incorporated, CA, USA).

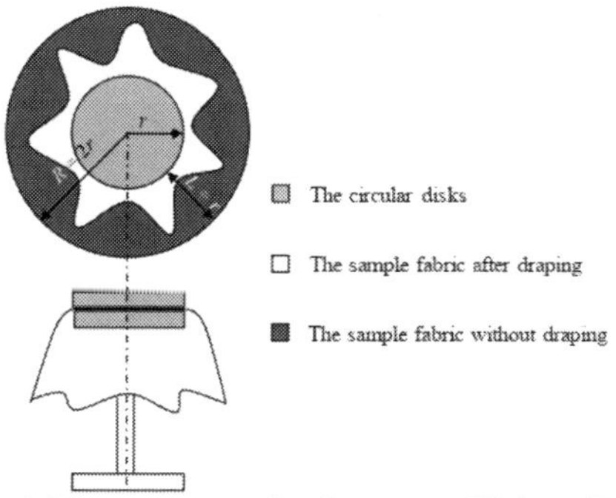

Fig.1 Measurement for drape coefficient (DC).

3. Results and Discussion

No sample fabrics showed drape for L of 1.5 cm or 2 cm. Fig.2 shows photos of the drape of Broadcloth for increasing values of r. Herringbone tweed formed nodes more easily, making it easier to increase the node number compared with other fabrics.

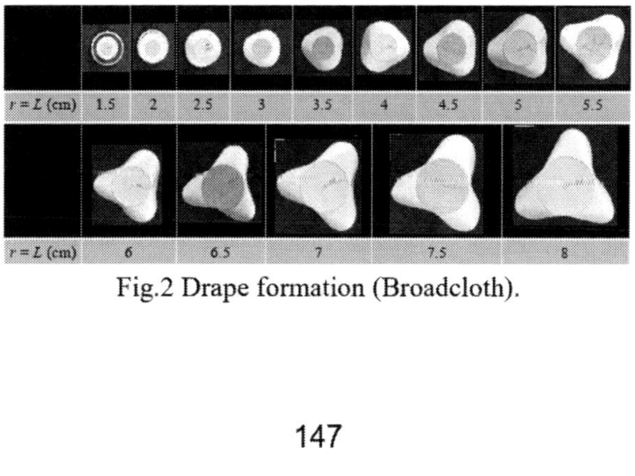

Fig.2 Drape formation (Broadcloth).

Fig.3 shows the relationship between K and DC for herringbone tweed. DC decreased linearly as K increased. DC did not depend on n.

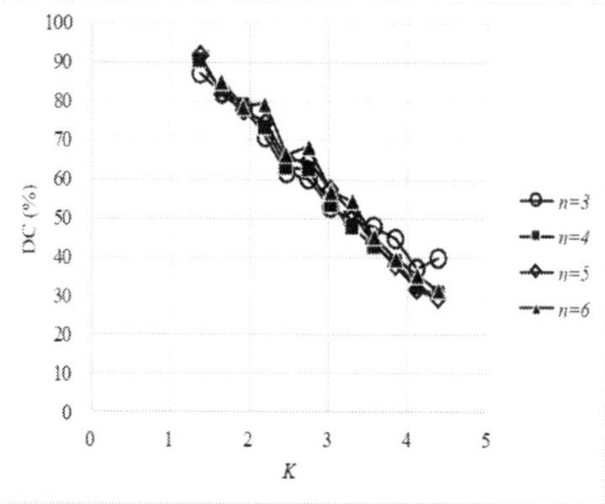

Fig.3 Relationship between the non-dimensional parameter K and the drape coefficient DC at different node numbers for herringbone tweed.

Fig.4 shows the relationship between K and DC for all samples when $n = 5$. The K–DC lines of different fabrics are almost parallel to each other. Broadcloth with higher shear stiffness and lower weight per unit area showed higher DC at the same K. Thus, shear stiffness and weight per unit area should be taken into account for establishing the similarity rule for fabrics.

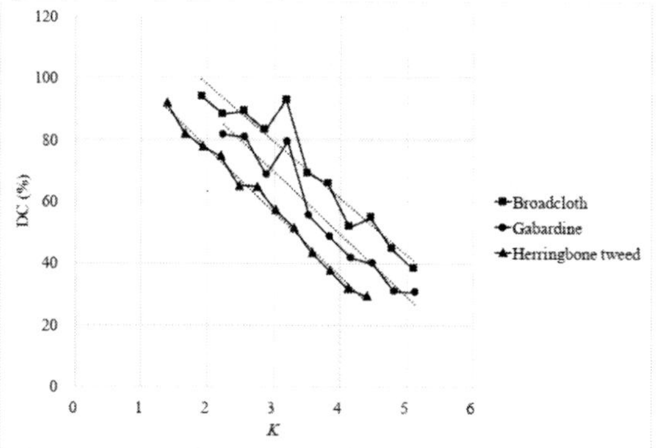

Fig.4 Relationship between the non-dimensional parameter K and the drape coefficient DC of the samples when n=5.

Multiple regression analysis for DC was conducted using K and $L\sqrt[3]{w/G}$. Equation (2) shows the regression equation for estimated DC (DCe).

$$\begin{aligned} \mathrm{DCe} &= -8.54K - 20.21\,L\sqrt[3]{w/G} + 122.42 \\ &= -(8.54\sqrt[3]{w/B} + 20.21\sqrt[3]{w/G})L + 122.42. \end{aligned} \quad (2)$$

Fig.5 shows the relationship between DCe and measured DC($\mathrm{DC_m}$). There was a good agreement, with a high correlation coefficient of 0.9626. The p-values of all parameters are below 0.01. Thus, DC can be estimated using K and $L\sqrt[3]{w/G}$ for different R when r/R is a constant.

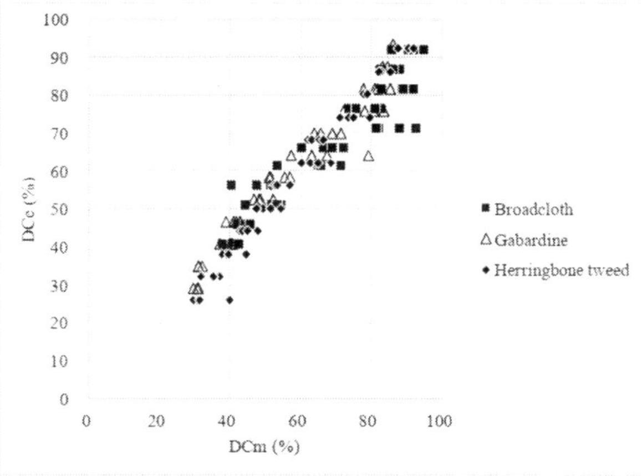

Fig.5 Relationship between estimated drape coefficient (DCe) and measured drape coefficient (DCm).

4. Conclusion

We investigated the relationship between drape coefficient and mechanical properties of fabrics considering geometrical parameters. For the same fabric, with various node numbers, the differences in DC are small enough that the effect of node number can be ignored in F. R. L. drape tests. Correlations between the non-dimensional parameter, K and DC for different fabrics are almost linear and parallel to each other. The differences between DCs for different fabrics are explained by the differences in shear stiffness and weight per unit area. DC can be estimated using K and $L\sqrt[3]{w/G}$ for different R when r/R is a constant. Therefore, similar drape shapes for different fabrics can be obtained by changing L.

5. Acknowledgment

This work was supported by a Grant-in-Aid for the Shinshu University Advanced Leading Graduate Program by the Ministry of Education, Culture, Sports, Science and Technology (MEXT), Japan. This work was also supported by the Japan Society for the Promotion of Science (JSPS) fellows and JSPS KAKENHI under grant number JP24220012.

References

[1] Takatera M, Shinohara A. A unified analysis to compare conventional methods for estimating bending rigidity of fabrics. J Text Machin Soc Jpn: 1994; 47: T250-T258.
[2] Chen Y, Rong GH, Chen J. Fabric database and fuzzy logic models for evaluating fabric performance. In: Galindo J, editor. Handbook of Research on Fuzzy Information Processing in Databases. Hershey: PA: 2006; p. 562.
[3] Chu CC, Cummings CL, Teixeira NA. Mechanics of elastic performance of textile materials: Part V: A study of the factors affecting the drape of fabrics—the development of a drape meter. Text Res J: 1950; 20: 539-548.
[4] Cusick GE. The resistance of fabrics to shearing forces. J Text Inst: 1961; 52: 395-406.
[5] Hearle JWS, Amirbayat J. Analysis of drape by means of dimensionless groups, Text Res J: 1986; 56: 727-733.
[6] Niwa M, Seto F. Relationship between drapability and mechanical properties of fabrics, J Text Machin Soc Jpn: 1986; 39: T161-T168.
[7] Lo WM, Hu JL, Li LK. Modeling a fabric drape profile, Text Res J: 2002; 72: 454-463.
[8] Al-Gaadi B, Göktepe F, Halász M. A new method in fabric drape measurement and analysis of the drape formation process. Text Res J: 2012; 82: 502-512.
[9] Kawabata S. The standardization and analysis of hand evaluation (2nd edition), Nishi-ku: Osaka: 1980.

Influences of Filling Materials on Pillows' Dynamic Pressure Comfort

Xuan Gao[1], Li-Ya Zhou[1,2*]

[1]*The College of Fashion and Design, Donghua University. Room 412, Yifu Building, West Yan'an Road No.1882, Shanghai, 200051, China*
[2]*Tongji University Shanghai Institute of Design and Innovation*

*Corresponding author's email: zly@dhu.edu.cn

Abstract

The type and its corresponding characteristics of a filling material have great influences on pillows' pressure comfort. This paper compared 3 groups of 7 typical filling materials, which were natural down, polyester, and down/polyester mixture, in terms of filling power and dynamic pillow pressure comfort. The dynamic pressure distribution on the pillows with each filling material were tested and recorded respectively based on self-built multipoint dynamic pressure testing system simulating the stress cause on the head when a person lies down and raise up his/her head on the pillow in a supine position. The experiment showed that down fiber had the highest filling power, followed by down/polyester, and polyester fiber. Down pillow had the highest maximum pressure value while insufficient support for the neck, leading to large pressure differences among the 7 testing points indicating uneven pressure distribution. Down/polyester mixture pillows had high maximum pressure values with even-distributed pressure, which contributed to high pressure comfort. 2 types of polyester pillow both had high maximum pressure values, and the pressure on the head was larger than that on the neck indicating unevenly distributed stress. Polyester wadding pillows showed better comfort with low maximum pressure value and even-distributed pressure. The correlation analysis between filling power and dynamic pressure values showed that filling power had significant influences on both maximum pressure value and the difference between maximum and minimum pressure value. A pillow with high degree of pressure comfort, should have suitable filling power under the prerequisite of reasonable head pressure and sufficient neck pressure.

Keywords: Filling Materials; Filling Power; Dynamic Pressure; Pressure Comfort

1. Introduction

The type and its corresponding characteristics of a filling material have great influences on pillows' pressure comfort [1]. As an essential component of a pillow, filling materials are classified into traditional Chinese medical herb, tea leaf, latex, down, memory foam, polyester and down/polyester mixture etc. found in the current market [2]. Different types of filling materials influence the pillow comfort diversely due to their characteristics and performances [3, 4].

Some researchers carried out experiments to explore the relationship between filling materials and pillow comfort. Lavin et al. compared 3 pillows (the subjects' usual pillow, a roll pillow and a water-based pillow) by subjective evaluations (visual analog scale, sleep questionnaire, sickness impact profile, and a satisfaction scale). The results showed that water-based pillow performed better for pain relief and sleep quality improvement due to its ability to spontaneously adapt to the position and shape of the head and neck [5]. Persson and Moritz tested 6 pillows in different shape and materials (polyester, polyurethane and inelastic foam) and founded that an optimal pillow should be soft, not-too-high and provide enough support to the cervical lordosis [6]. Gordon et al. recruited 106 subjects to evaluate pillow comfort by recording their sleep quality, cervical pain and pillow comfort ratings using their usual pillow, polyester, foam, feather, and rubber pillows. The study showed that rubber pillow performed better in the management of cervical pain, improvement of sleep quality and pillow comfort. Feather pillow are not recommended [7]. Jian-Jun Hou et al. examined the influences of filling materials on sleeping comfort in lateral position by using body pressure measuring system and subjective evaluation on ergonomics. The result demonstrated that latex pillow had higher comfort degree, with more uniform pressure distribution (on head, neck, shoulder and back), lower overall pressure, and fatigue relief of neck muscles; the polyester pillow had medium comfort degree, with low pressure on head and uneven distribution of pressure; the medicine/tea pillow had low comfort degree, with high pressure on head, neck and shoulder. The authors concluded there was remarkable negative correlation between overall pressure comfort and maximum pressure, and high negative correlation between hardness and pressure values [8]. In view of the above-mentioned studies, most of them used subjective comfort evaluation, such as questionnaires and sensory evaluation, to test filling materials affecting pillow comfort, and only a minority combined with objective experiment, which usually result in lower accuracy [9, 10]. This present paper contrasted the differences of materials' filling power, and analyzed the dynamic pressure change trend and pressure distributions on pillows.

2. Filling Power

2.1 Material Samples

The detailed information of 3 groups of 7 filling materials selected for the experiment are shown in Table 1. The material samples were placed in constant temperature and constant humidity environment for 24 hours, with temperature (20±2) ℃ and relative humidity (65±2) %. Before the experiment, the samples were put in a box and given scatter treatment by electrical drier. No. 6 and No. 7 (polyester mixture wadding), with differences of gram weight and physical appearance, were not given scatter treatment, due to their stable shape.

Table.1 The detailed information of filling material samples

No.	Component	Code	Type	Shape
1	Down	YR	Natural down	Down fiber
2	Down60%+PTT-PET 40%	YR/PTT-PET	Down/polyester	Mixture fiber
3	Down60%+polyester40%	YR/PET	Down/polyester	Mixture fiber
4	PTT-PET	PTT-PET	Polyester	Short-staple fiber pellet
5	Polyester	PET	Polyester	Short-staple fiber pellet
6	PTT-PET	X1	Polyester	Wadding
7	PTT PET	X2	Polyester	Wadding

2.2 Methods

According to modified GB/T 10288-2003 Testing methods for down and feather, the test-cylinder's height was 60 cm and 24.5 cm in internal diameter. The platen's weight was 68.4 g and 24 cm in diameter (Fig.1). The electronic balance's precision was 0.0001. Take 28.4 g specimen from each sample into horizontally arranged glass cylinder smoothly and evenly, then placed the platen and waited to a steady-state. Put weight on platen in proper order of 0g, 100g, 150g, 200g, 150g, 100g, and 0g. Record the variation of platen's height. Computational formula (Eq.1):

$$FP = V/G \quad (1)$$

FP is filling power, cm^3/g; V is fiber assemblies' volume, cm^3; G is fiber assemblies' weight, g.

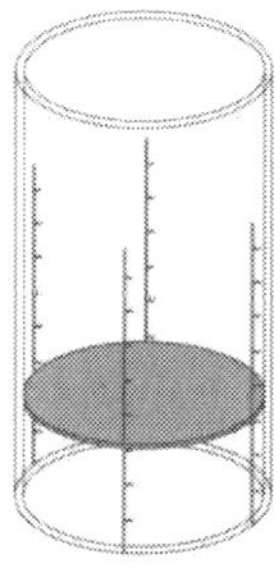

Fig.1 Test-cylinder.

2.3 Filling Power Results

Maximum and average value of filling power of fiber assemblies are shown in Fig.2. Volume ratio under different pressure of fiber assembly samples is shown in Fig.3.

According to Fig.2, down fiber had the highest filling power, followed by down/polyester, and polyester fiber. In the same quality situation, down fiber had the maximum initial volume. Down fiber's unique natural structure contributed to the characteristics of light and fluffy [11-13].

According to Fig.3, after variations of compression and decompression, the volume ratios of X1 and X2 were both more than 80%, meaning a higher elastic recovery property. Secondly, PET and PTT-PET were almost unanimous in variation trend. Down fiber had the lowest value of 58%. It was noteworthy in practical production that PTT-PET, with similar elastic recovery rate as PET, performed higher than PET after mix with equal proportion of down.

There was a negative correlation between elastic recovery property and filling power.

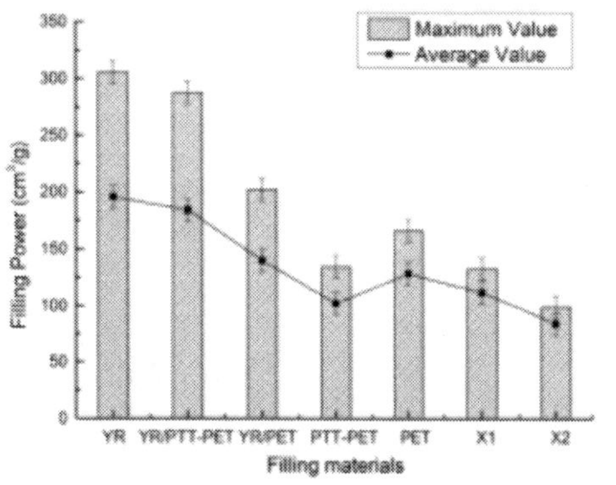

Fig.2 Filling power of fiber assemblies.

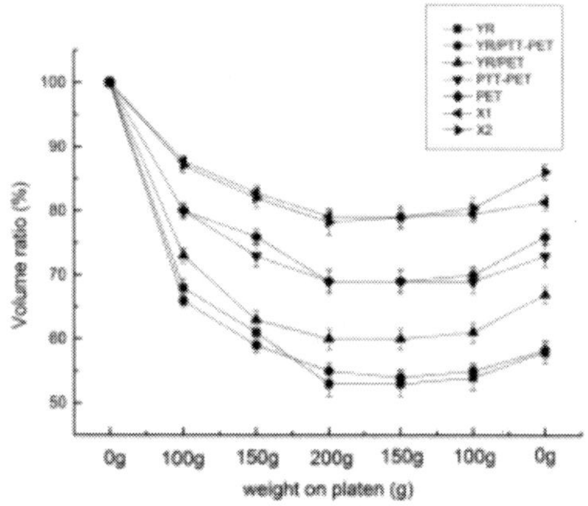

Fig.3 Volume ratio variation of fiber assemblies.

3. Multipoint Dynamic Pressure Test

3.1 Pillow Samples

Pillow samples were filled with above-mentioned materials and guaranteed to have the same height [14, 15]. The pillowcase were cotton twill fabrics in same specifications [16]. Pillow samples are shown in Table 2.

Table.2 The samples of pillows

No.	Filling material	Code	Fill weight (g)	Pillowcase size (cm*cm)	Pillowcase fabric
1	Down	YR	340	50*80	cotton twill
2	Down60%+PTT-PET 40%	YR/PTT-PET	355	50*80	cotton twill
3	Down60%+polyester40%	YR/PET	455	50*80	cotton twill
4	PTT-PET	PTT-PET	600	50*80	cotton twill
5	Polyester	PET	485	50*80	cotton twill
6	PTT-PET	X1	500	50*80	cotton twill
7	PTT-PET	X2	500	50*80	cotton twill

3.2 Methods

3.2.1 Experimental

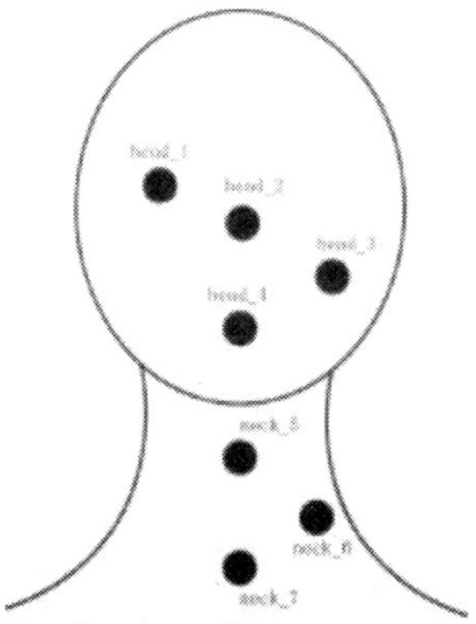

Fig.4 Distribution of pressure points.

Fig.5 Test posture.

The experiment used a dummy to imitate the situation between human body and pillow. There were 7 test points with sensors on dummy's head and neck (Fig.4). The dummy was placed supine on a pillow, so the head and neck contact to the pillow and generate pressure (Fig.5). During the test, weight was placed on dummy's head in turns of 0kg, 1kg, 2kg, 3kg, 4kg, and then decreased to 3kg, 2kg, 1kg, and 0kg. Record the pressure values of 7 test points.

3.2.2 Multipoint Dynamic Pressure System

The multipoint dynamic pressure system was based on LabVIEW software and NI data acquisition software, using Flexiforce sensor and self-built amplified circuit. Flexiforce sensor is an ultrathin flexible printed circuit used to measure the pressure between two surfaces. The effective area is a circular spot about 9.53 mm in diameter at the end of senor, which can output the pressure as voltage signals. The system can be used after calibrating the correspondence between output signals and pressure values. There are different ranges of Flexiforce sensor as options, and 0-4.4 N was selected in this experiment.

3.3 Multipoint Dynamic Pressure Results

Pressure value variation of test points on pillow samples are showed in Fig.6, 7, 8, and 9.

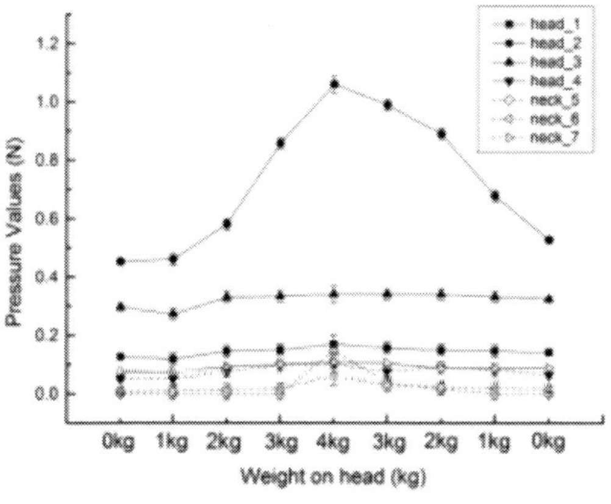

Fig.6 Pressure value variation of test points on YR.

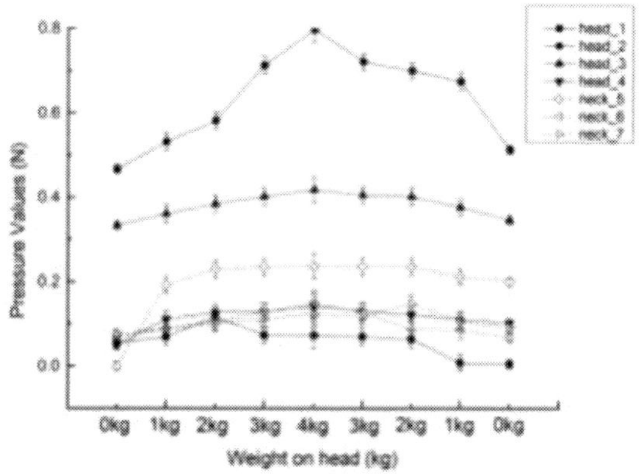

Fig.7 Pressure value variation of test points on YR/PET.

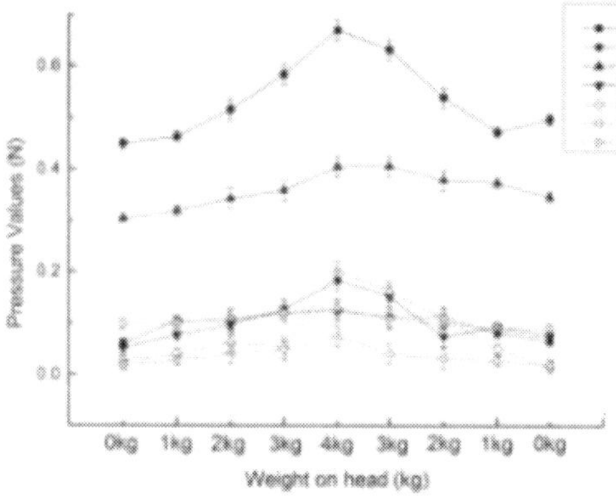

Fig.8 Pressure value variation of test points on PTT-PET

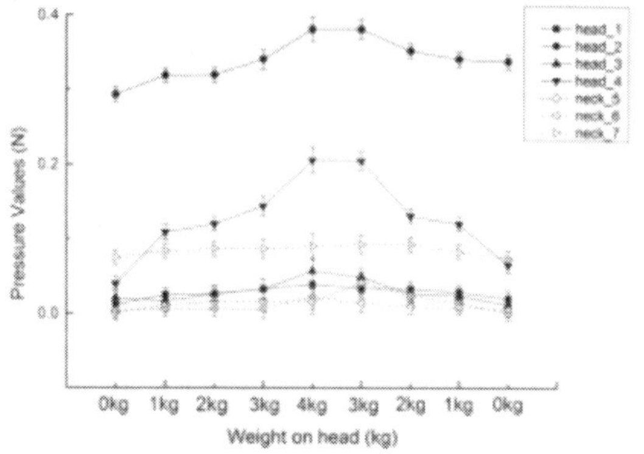

Fig.9 Pressure value variation of test points on X1.

From Fig.6, during dynamic pressure testing, down pillow had the highest maximum pressure value (1.06 N) at test point 2, and its pressure mainly distributed on head, but provided insufficient support to neck, so the large pressure differences among test points led to uneven pressure distribution.

From Fig.7, down/polyester pillows had high maximum pressure values and provided support to neck, so the even-distributed pressure contributed to a higher comfort degree.

From Fig.8, PTT-PET pillow and down-like polyester pillow also had high maximum pressure values, and the pressure mainly distributed on head while less on neck, so the uneven-distributed pressure led to medium comfort.

From Fig.9, polyester wadding pillows showed better comfort with low maximum pressure values (0.38 N), smaller pressure differences among test points, and even-distributed pressure.

4. Conclusion

Under the circumstances of same pillowcase and same pillow height, there were significant differences in both pillow samples' maximum pressure and pressure distribution. It showed that filling material was an important factor which influenced pillow dynamic pressure comfort [7].

Filling power: down> down/PTT-PET> down/polyester> polyester> PTT-PET> polyester wadding 1> polyester wadding 2 (YR> YR/PTT-PET> YR/PET> PET> PTT-PET> X1> X2). Down pillow with high filling power and polyester pillow with low filling power, both had poor performance in dynamic pressure test, due to high maximum pressure and major pressure on head while less support on neck. Down/polyester pillows, with high maximum pressure, showed evenly distributed pressure. Polyester wadding pillows, with low maximum pressure, showed even-distributed pressure. Among 7 pillow samples, X1 had the lowest maximum pressure and smallest pressure difference among test points, and the pressure between X1 and X2 differed due to the different physical properties of the filling materials in terms of weight, hardness, layer thickness, and physical appearance etc.

Through descriptive analysis and pearson correlation analysis, there were significant differences between filling power and pressure values. The sig. values were less than 0.05; pearson correlation coefficients of maximum pressure value and pressure extremum difference were 0.796 and 0.786. It could be seen that filling power had positive correlation with maximum pressure value and pressure extremum difference. A pillow with high degree of pressure comfort, should have suitable filling power under the prerequisite of reasonable head pressure and sufficient neck pressure.

5. Acknowledgments

This work was financially supported by National Natural Science Funds of China (Grant No. 51206021) and Shanghai Committee of Science and Technology, China (Grant No. 12ZR1400700) in experiment part. Thanks very much.

References

[1] Li J. Analysis of the structure and function development of pillows. Art & Design Research: 2012; S1:107-108.

[2] Xiao J. Choose the right pillow and have a good sleep. TCM Healthy Life-Nurturing: 2015; 10: 78-79.

[3] Li KJ, Ma Z. On functional innovation for the design of pillows. Shandong Textile Economy: 2015; 4: 18-20.

[4] Wang CM, Wamg R, Xiao H, et al. Research progress in structure of PET and PTT and PTT/PET composite fiber. China Synthetic Fiber Industry: 2009; 32: 48-51.

[5] Lavin RA, Pappagallo M, Kuhlemeier KV. Cervical pain: a comparison of three pillows. Archives of Physical Medicine and Rehabilitation: 1997; 78: 193-198.

[6] Persson L, Moritz U. Neck support pillows: a comparative study. Journal of Manipulative & Physiological Therapeutics: 1998; 21: 237-240.

[7] Gordon SJ, Grimmer-Somers K, Trott P. Pillow use: The behaviour of cervical pain, sleep quality and pillow comfort in side sleepers. Manual Therapy: 2009; 14: 671-678.

[8] Hou JJ, Shen LM, Song J, et al. Influence of the materials of pillow core on sleeping comfort in lateral position. Journal of Northwest Forestry University: 2012; 27: 217-221.

[9] Hou JJ, Shen LM. The design of ergonomics pillow suitable for sleeping comfort. Packaging Engineering: 2014; 35: 21-25.

[10] Cui MM. Research status of the comfort of sleeping pillow. Furniture: 2015; 36: 47-50.

[11] Lu Y, Wang LM, Zhang ZH, Yang C. The research progress of feather fiber structure and its physical and chemical properties. Shanghai Textile Science & Technology: 2015; 11: 6-8.

[12] Jin Y, Jin YF, Li YW. Study on surface feature and fluffiness of natural down. China Surfactant Detergent & Cosmetics: 2000; 30: 17-19.

[13] He DY, Guan XY, Wu QL. Probe into Linear Relation Between Down Content and Filling Power. China Fiber Inspection: 2011; 3: 76-78.

[14] Hou JJ, Shen LM. Influence of Pillow Height on Sleeping Comfort in Supine Position. Furniture: 2013; 1: 44-46.

[15] He YM, Chen Y, Tang SJ, Li N. Influence of pillow's structure on the evaluation of sleeping comfortableness. Journal of Beijing Institute of Clothing Technology: 2005; 25: 40-47.

[16] Tong DD, Li QJ, Zheng XQ. Fabric material selection for functional pillows. Journal of Southwest University: 2007; 29: 116-118.

Ballistic Performance of Angle-interlock Woven Fabrics in Comparison to Other Woven Structures

Dan Yang[1], Xiao-Gang Chen[2*]

[1]*College of Textile and Material Engineering, Wuhan Textile University, No. 1 Yangguang Avenue, Wuhan, Hubei 430200, P.R. China*
[2]*School of Materials, University of Manchester, Sackville Street, PO BOX 88, Manchester M60 1QD, UK*

[*]Corresponding author's email: xiaogang.chen@manchester.ac.uk

Abstract

This paper focuses on the ballistic performance of angle-interlock woven fabrics and the result shows that this kind of fabric has relatively lower capabilities to absorb impact energy when compared with fabrics of other structures.

This is because the angle-interlock woven fabric has less interlacements, which is determined by its unique structure consisting of layers of weft yarns laid straight and bound by a single layer of warp yarns. The interlacement can help transfer the energy to the adjacent yarns through propagating the stress wave, and therefore the projectile impact energy can be absorbed. Less interlacements in the angle-interlock woven fabric limitedly assist in this impact energy absorption.

Keywords: Angle-interlock; Interlacement; Ballistic Performance

1. Introduction

The national safety against terrorism has raised concerns about the importance of body armour, which could improve the potential to save lives of police or soldiers during conflicts. Body armour is an integral part of a fighting kit and plays a key role in protecting police or the military personnel from enemy threats. In the early 21st century, the National Institute of Justice reported that the risk of fatality from firearms assaults was 14 times higher for officers who were not wearing body armour than those who were [1]. Moreover, without a good customised fit, body armour can become an obstruction. Officers may feel uncomfortable or even pain, the armour may restrict their mobility thereby endangering them.

The traditional cutting and stitching method can help create a three-dimensional shape that accommodates the contour of a human's torso. Nevertheless, seams employed in the garment production are easily penetrated by small projectiles or lose fragments at close range. Overlapping seams are much stronger in fabric stitching due to more friction between the fabrics and threads; however, the small ballistic missiles can still break them directly by penetrating and severing the loop of threads among the seams. Fabric folding is another alternative but the downside is that there are too many layers folded under the armpits, which leads to the discomfort. In addition, there is still the problem on the weakness of stitches against the projectile impact as many stitches are required during the folding procedure. Therefore, it is imperative to explore an improved method for making the three-dimensional body armour with good wearability.

The angle-interlock woven fabric is an alternative to the traditional woven fabric in the shaping of a three-dimensional domed-pattern; it can form the 3D area without cutting or sewing as it has an advantage in good mouldability. A good wearability can be expected for the wearer if this kind of fabric is applied to make body armour. The mouldability of the angle-interlock woven fabric is demonstrated with the first attempt in creating a garment panel of three-dimensional body armour. The mathematical models were established for both single-piece front panel and multi-piece front panel of female body armour [2-3]. As the candidate of body armour material, the angle-interlock woven fabric was also tested in the ballistic range and the result showed that it has a certain level of ballistic resistance as the NIJ standard level II was achieved [4]. However, it is not very clear about the ballistic performance of angle-interlock woven fabric on the condition of comparison with other ballistic fabrics, especially the conventional plain weave fabric. This paper is attempted to explore the overall ballistic performance of fabrics with different woven structures in order to benchmark the angle-interlock woven fabric under the energy loss test.

2. Fabrics Preparation

Twenty-one different woven fabrics including angle-interlock woven fabrics were designed and manufactured using the Saurer 100W dobby loom with the maximum heald frames 22 and the maximum width 1.2 m. A negative let-off mechanism was used in controlling the weaver's beam and balancing the warp yarn tension. The structures of these woven fabrics are listed as follows.

2.1 Structures

2.1.1 Angle-interlock Woven Fabric

The angle-interlock woven fabric consists of layers of weft yarns which are laid straight and bound by a single layer of warp yarns to lock the layers of weft yarns together [5]. The stuffers (warp yarns) are oriented along the longitudinal direction. The fillers (weft yarns) are oriented transverse to the loom feed direction, and are inserted between layers of stuffers. The stuffers and fillers form an orthogonal array. The cross-section of a four-layer angle-interlock woven fabric is shown in Fig.1.

Fig.1 Cross-section of a four-layer angle-interlock woven fabric.

2.1.2 Plain Weave

Plain weave is most widely used in the ballistic application. It is the simplest weave, characterised by the tightest woven structure.

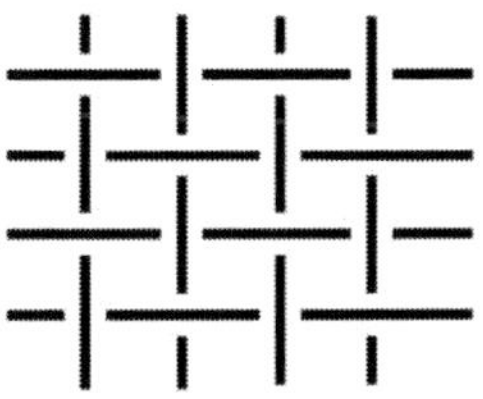

Fig.2 Plain weave.

2.1.3 Plain Weave with Leno Insertions

Leno weave is a type of structure that a pair of warp yarns is intertwined by embracing a pick of weft yarn in each of the interstices [6]. It has firmer gripping on the weft yarns and can be produced easily in broad fabrics. The fabrics with this weave could form the stable dimensions as such structure is to limit the yarn movement. The plain weave fabric with leno insertions is supposed to have better dimensional stability than that without leno insertions, and should have better restraining on the weft yarns. Fig.3 shows the leno weave structure.

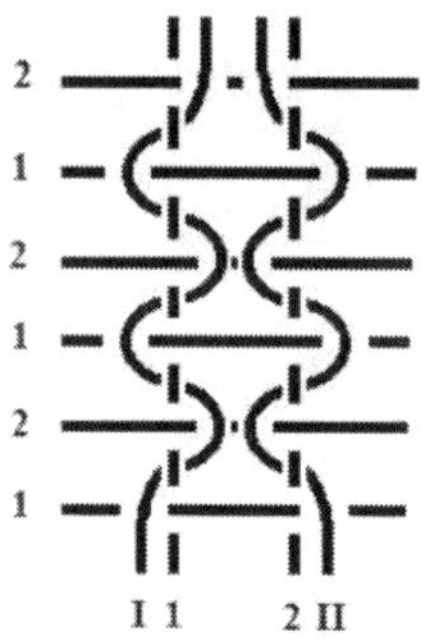

Fig.3 Leno weave.

2.1.4 Plain Weave with Weft Yarn Cramming

The plain weave could be added with weft yarn cramming. Weft yarn cramming comes out by stopping take-up, at the same time all other actions are unchanged as usual during weaving. Fabric weave structure keeps the same. Weft yarn density increases when weft yarn cramming takes place.

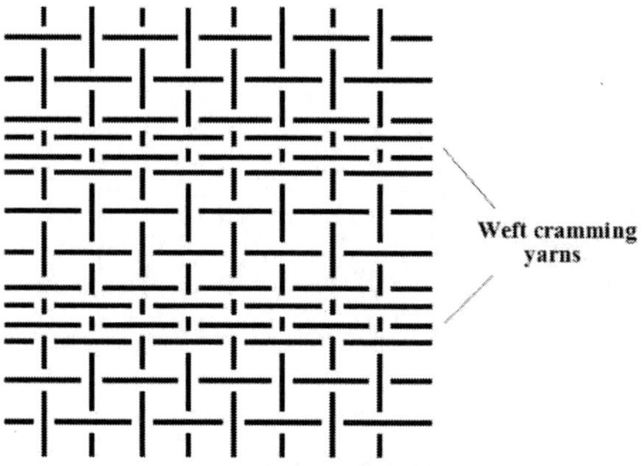

Fig.4 Weft yarn cramming.

2.1.5 Four-layer Cellular Fabric

The four-layer cellular fabric is formed by connecting the adjacent two layers together in the evenly spaced intervals. In this experiment, 8 warp yarns and 104 weft yarns composed one repeat unit of the weave structure. The conjunction area was characterised by 8 weft yarns and twice the yarn density in both warp and weft directions. The cross-section of four-layer cellular fabric with plain weave structure is shown in Fig.5.

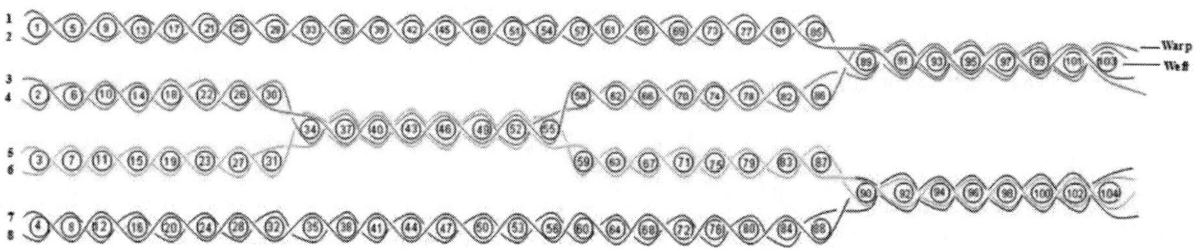

Fig.5 The cross-section of four-layered cellular fabric with plain weave.

2.2 Specifications

The detailed specifications of various fabrics are listed in Table 1.

Table.1 Specifications

Fabrics	Explanation	Fibre	Yarn densities		Yarn counts	
			Warp (ends/cm)	Weft (picks/cm)	Warp (tex)	Weft (tex)
4LAI12 x26	4L – four layer, AI – angle interlock fabric, 12×26 – 12 ends/cm and 26 picks/cm.		12	26		
4LAI12 x28	4L – four layer, AI – angle interlock fabric, 12×28 – 12 ends/cm and 28 picks/cm.			28		
4LAI12 x30	4L – four layer, AI – angle interlock fabric, 12×30 – 12 ends/cm and 30 picks/cm.			30		
BPlain	B – broad, Plain – plain woven fabric.	Kevlar® 49	7.5	7.5	158	158
BPL02	B – broad, P – plain woven fabric, L – leno insertions, 02 – every 2 cm intervals.					
BPL04	B – broad, P – plain woven fabric, L – leno insertions, 04 – every 4 cm intervals.					
BPL06	B – broad, P – plain woven fabric, L – leno insertions, 06 – every 6 cm intervals.					
BPL08	B – broad, P – plain woven fabric, L – leno insertions, 08 – every 8 cm intervals.					
BPL10	B – broad, P – plain woven fabric, L – leno insertions, 10 – every 10 cm intervals.					
BPL02 WC	B – broad, P – plain woven fabric, L – leno insertions, 02 – every 2 cm intervals, WC – weft yarn cramming.					
BPL04 WC	B – broad, P – plain woven fabric, L – leno insertions, 04 – every 4 cm intervals, WC – weft yarn cramming.					
BPL06 WC	B – broad, P – plain woven fabric, L – leno insertions, 06 – every 6 cm intervals, WC – weft yarn cramming.					
BPL08 WC	B – broad, P – plain woven fabric, L – leno insertions, 08 – every 8 cm intervals, WC – weft yarn cramming.					
BPL10 WC	B – broad, P – plain woven fabric, L – leno insertions, 10 – every 10 cm intervals, WC – weft yarn cramming.					
B2LRe	B – broad plain, 2LRe – regular two-layer.					
B2LIn	B – broad plain, 2LIn – interchange two-layer.					
B4LRe	B – broad plain, 4LRe – regular four -layer.					
HPWW 06	H – hand loom made, P – plain woven fabric, WW – weft winding, 06 – every 6 cm intervals.					
BDP	BD – broad, P – plain woven fabric.	Dyneema®	6.73	6.73	176	176
BDPL04	BD – broad, P – plain woven, L – leno insertions, 04 – every 4 cm intervals.					
BDUD	BD – broad, UD – unidirectional.					

The fabrics listed in Table 1 were cut into the size of 24×24 cm^2, clamped by a pair of metal clamps for the energy loss test. The target area via penetrating the fabric was locked by the needles, which were planted onto one part of the clamp around an aperture with the diameter of 15.2 cm, as shown in Fig.6. The target area was further locked by screwing the two clamps together in the four corners. The same condition of this full-clamped state was applied during the whole test procedure.

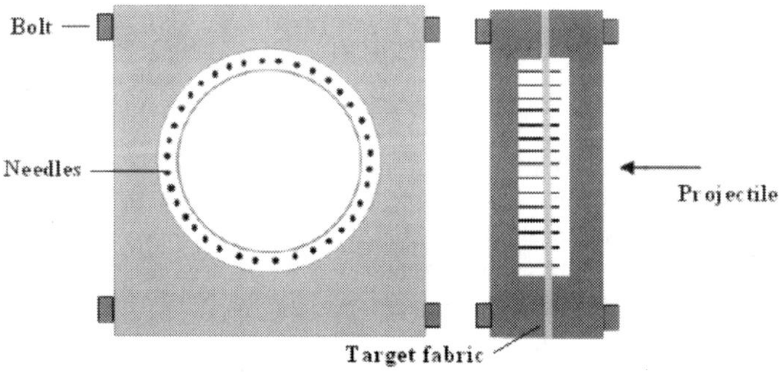

Fig.6 Fabric clamp.

3. Energy Loss Test

3.1 Principle of Fabric Energy Absorption

The longitudinal wave and the transverse wave are two typical waves coming out when the projectile hits the fabric. The longitudinal wave is propagating in the plane of the target fabric. This wave travels through the primary yarns which are known as the yarns directly hit by the projectile. The yarns not directly hit by the projectile are secondary yarns, which are also propagated through by the longitudinal wave because they interact with the primary yarns. Another wave, namely, the transverse wave, travels outward from the impact zone to make deformation in the perpendicular direction to the fabric plane. Because of the transverse wave, the fabric deformation is shaped like a cone with the impact point at its vertex. Fig.7 depicts the propagation of the two waves [7].

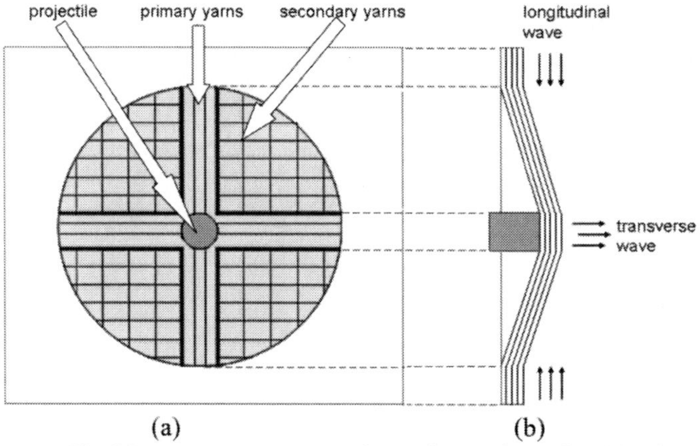

Fig.7 Impact wave propagation: (a) top view; (b) side view [7].

Impact energy could be converted into various forms, e.g. yarn failure, fibre pull-out, fibre fracture, and cone formation on the back face of the target. The impact energy conversion is a complex issue, influenced by several factors, including the material properties of the constituent fibres, boundary conditions of the target fabric, and the projectile geometry. Among these, friction between yarns is an important major factor. The increase of friction between yarns may consume more impact energy and therefore could lead to the increase in the ability to absorb the impact energy [8]. However, it also have been noticed that if the friction is so high, the corresponding movent of yarns could be blocked easily. Consequently, the reduction could happen in the fabrics' capability of absorbing impact energy. This is because the yarn movement is also an important way of absorbing the impact energy [9-10].

3.2 Calculation of Projectile Energy Loss

The fabric energy absorption on penetration could be measured by the loss in projectile kinetic energy, which is used to assess the ballistic performance of fabric. This projectile energy loss is determined by both the projectile impact velocity and the exit velocity. The overall loss of kinetic energy of the projectile could be calculated using the following formula [11]:

$$\Delta E = 1/2m(V_1^2 - V_2^2) \quad (1)$$

in which, ΔE is the loss of kinetic energy, m is the projectile mass, V_1 is the impact velocity, and V_2 is the exit velocity. The impact velocity V_1 can be calculated from the flying time of the projectile t_1 between two detectors placed at d_1 distance apart. The impact velocity V_1 can be then calculated from:

$$V_1 = d_1/t_1 \quad (2)$$

The exit velocity V_2 can be obtained using the same method, namely, measuring projectile flying time t_2 between the two detectors with a distance d_2 between them behind the target fabric. After that, the overall loss of kinetic energy carried by the projectile can be calculated according to Eq.1.

3. 3 Procedure of Test

The energy loss test was an effective method to investigate the fabrics' ballistic performance. The firing range for energy loss test was composed of a firing device with a 7.6 mm rifle barrel, target holder, and velocity detectors, which were positioned in an enclosed environment. The firing apparatus is shown in Fig.8.

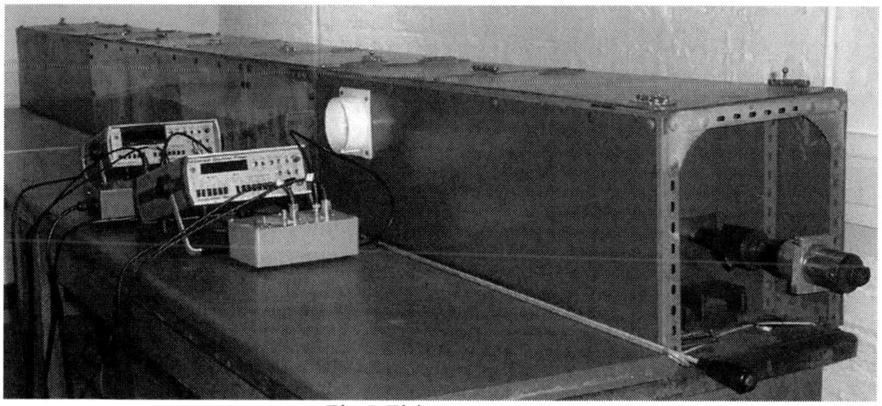

Fig.8 Firing apparatus.

In this experiment, the projectile was a steel cylinder with a diameter and a length of being both 5.56 mm, weighing 1.06 g, as shown in Fig.9(a). A blank cartridge was used to propel the projectile held in a sabot, as shown in Fig.9(b). After pulling the trigger, the projectile flied out from the barrel towards the target fabric at a high velocity, typically 480m/s. The time detectors then picked up the travelling time of the projectile before and after the target fabric. The loss in kinetic energy carried by the projectile after going through the fabric could be calculated using Eq.1. In this set-up, the distances between the two pairs of time detectors were 47 cm (front) and 36.2 cm (rear) respectively. Fig.9(c) shows the construction of this firing range.

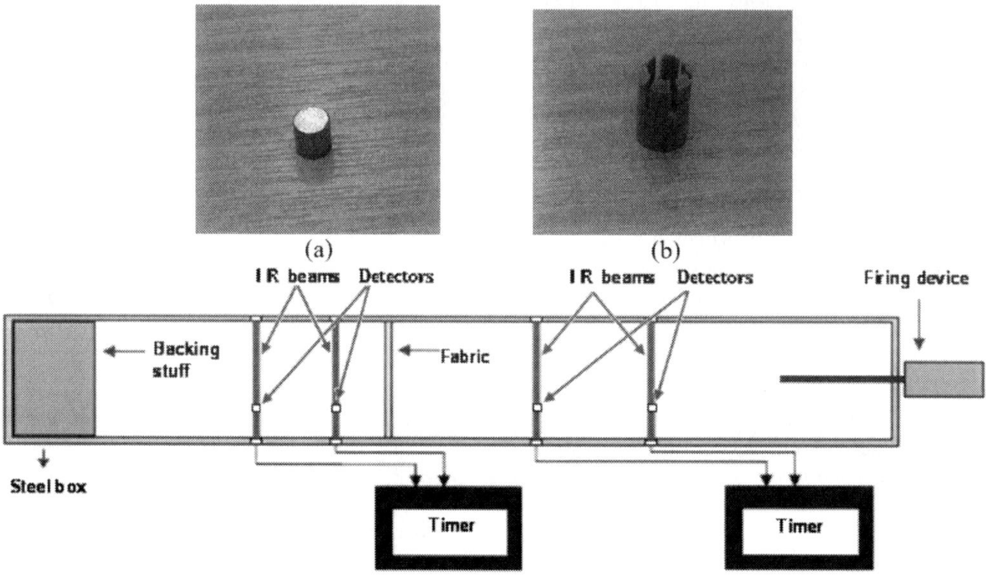

Fig.9 Construction of the testing apparatus:
(a) steel projectile; (b) plastic sabot; (c) schematic diagram.

4. Energy Loss Test Results

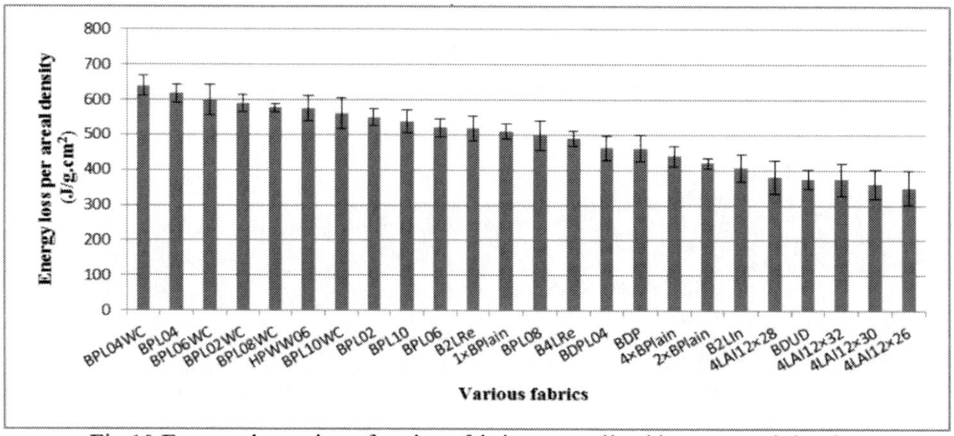

Fig.10 Energy absorption of various fabrics normalized by per areal density.

The energy loss test results for all fabrics are shown in Fig.10. The projectile energy loss was normalized by dividing the fabric areal density (in g/cm^2), in order to process the energy loss under the condition of the same amount of materials. It has been found that the ballistic performance of angle-interlock woven fabrics in general is relatively quite low as 4LAI12×26, 4LAI12×30, 4LAI12×32 and 4LAI12×28 are ranked as the 1st, 2nd ,3rd and 5th respectively from the bottom. More energy has been absorbed by almost all other woven fabrics than the angle-interlock woven fabrics in Fig.10. This is because that the number of interlacements is different in the fabrics, which directly influences the yarn gripping effect. The more interlacements, the tighter yarn gripping it could be expect. The interlacements assist in propagating energy to the adjacent yarns. The angle-interlock woven fabric have the least yarn interlacements, compared to other fabrics in Fig.10; therefore, the stress wave in angle-interlock woven fabrics is inclined to restrict in a smaller region of woven fabric and difficult to travel to a larger area.

Consequently, the angle-interlock woven fabric would absorb less projectile impact energy than other woven fabrics.

It is of necessity to compare the angle-interlock woven fabrics with the plain woven fabrics as the latter is considered as the most common ballistic material, via which way the ballistic performance of the angle-interlock woven fabrics could be benchmarked clearly. As shown in Fig.10, the plain woven fabrics absorb more kinetic energy than the angle-interlock woven fabrics. The top view of structures of both plain and 4-layer angle-interlock weaves are sketched in Fig.11. It could be noticed that there are more interlacements in the plain weave than in the 4-layer angle-interlock weave in the same unit area. Additionally, the former also has more interlacements than the latter in the cross-section. Interlacements play an important role in absorbing the impact energy in the fabric. Working as pivot points, the interlacements can stabilize the structure in order to enhance the yarn gripping by increasing friction among yarns. When the projectile hits the fabric, the impact energy propagates among the warp and weft yarns, and then is distributed on interlacements for further dissipation. More interlacements may make the energy distribution and dissipation more effectively because stronger gripping of yarns in the fabric helps the impact energy to propagate to a larger region of fabric. However, the energy absorption principle is a complicated issue affected not only by friction but also by other factors. It is of necessity to notice that if the friction among yarns is too high to block the yarn movement in the fabric, the impact energy absorption by yarns displacement could be minimised, and therefore the mechanism for the energy could be absorbed and left to the yarn breakage only. This can be well explained that the unidirectional fabric (BDUD) shows the lower energy absorption than the single layer woven fabrics (namely, BDPL04, 1×BPlain, BDP and etc.) and even than a certain type of angle-interlock woven fabrics (4LAI12×28), as demonstrated in Fig.10.

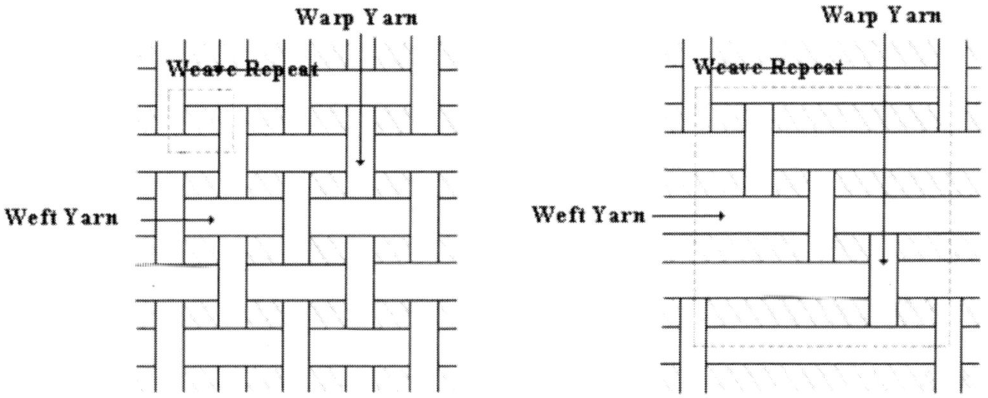

(a) (b)

Fig.11 Weave structure: (a) plain; (b) 4-layer angle-interlock.

Moreover, comparisons are also considered on the capability in absorbing impact energy for other various fabrics. Important findings are listed as follows according to the designed and manufactured fabrics in this experiment:

1) Fabrics with leno insertions in warp and cramming yarns in weft have shown stronger capability in impact energy absorption than their counterparts without weft yarn cramming. BPL04WC (Broad plain woven fabric with leno insertions in warp and cramming yarns in weft both in every 4cm intervals) demonstrates the best impact energy absorption. This is due to the reason that the weft yarn cramming could increase the yarn gripping effect to strengthen the friction among weft yarns by increasing the weft density. More kinetic energy could be consumed against this additional friction. However, for BPLWC fabric itself, optimal setting exits; in this paper, fabrics with 4cm insertion intervals are better than other settings. Further research may be required regarding this specific region.

2) 3D cellular fabrics and the same layer 2D plain fabric: the two and four layered cellular fabrics have demonstrated better impact energy absorption than the 2D plain woven fabric with the same layer of similar yarn densities.

3) HPWW06 (Hand loom made plain woven fabric with weft winding in every 6cm intervals) is better than BDPL06 (the fabric with same insertion intervals of leno weave) in respect of the impact energy absorption. The

reason is that in the former structure, much tighter warp yarns have gripped the winding weft yarns, which could positively influence on the impact energy absorption.

4) Kevlar® plain woven fabrics absorb more impact energy compared to their Dyneema® counterparts. These results may verify that the higher surface friction of Kevlar® fibre affects its ballistic-resistant performance.

5) The unidirectional Dyneema® fabrics have demonstrated lower energy absorption capability than the woven Dyeema® fabrics, but may have better trauma proof ability.

5. Conclusion

The angle-interlock woven fabrics has been compared with different woven structures to benchmark the ballistic performance of angle-interlock woven fabrics under the energy loss test. The test result has shown that in comparison to other woven structures, angle-interlock woven fabric demonstrates low ballistic resistance as absorbing less impact energy. It is due to less interlacements owned by the angle-interlock woven fabric than its counterparts. The interlacement assists in transferring the impact energy to the adjacent yarns: more area the stress wave could propagate if more interlacements exist, and therefore more impact energy could be absorbed.

References

[1] Ashcroft J, Daniels DJ, Hart SV. Selection and application guide to personal body armor NIJ guide 100-01 (update to NIJ guide 100-98). In: the National Institute of Justice's National Law Enforcement and Corrections Technology Center, editors. Lance Miller. Testing Manager. Maryland: MD 2001; p. 7.

[2] Chen X, Yang D. Use of 3D angle-interlock woven fabric for seamless female body armor: part II: mathematical modeling. Text. Res. J.: 2010; 80: 1589-1601.

[3] Yang D, Chen X. Multi-layer pattern creation for seamless front body armor panel using angle-interlock woven fabrics. Text. Res. J.: 2016; 0: 1-6.

[4] Chen X, Yang D. Use of 3D angle-interlock woven fabric for seamless female body armor: part I: ballistic evaluation. Text. Res. J.: 2010; 80: 1581-1588.

[5] Chen X, Lo WY, Tayyar AE, Day RJ. Mouldablity of angle-interlock woven fabrics for technical applications. Text. Res. J.: 2002; 72: 195-200.

[6] Grosicki Z. Watson's advanced textile design: compound woven structures, In: Butterworth & Co., editors: 1977.

[7] Naik NK, Shrirao P, Reddy BCK. Ballistic impact behaviour of woven fabric composites: formulation. Int. J. Impact Eng.: 2006; 32: 1521-1552.

[8] Backman ME, Goldsmith W. The mechanics of penetration of projectiles into targets. Int. J. Eng. Sci.: 1978; 16: 1-99.

[9] Cantwell WJ, Morton, J. The impact resistance of composite materials - a review. Composites: 1991; 22: 347-362.

[10] Kirkwood JE, Kirkwood KM, Lee YS, Ronald G, Egres RG, Wagner NJ, Wetzel ED. Yarn pull-out as a mechanism for dissipating ballistic impact energy in kevlar KM-2 fabric part II predicting ballistic performance. Text. Res. J.: 2004; 74: 939-948.

[11] Cork CR, Foster PW. The ballistic performance of narrow fabrics. Int. J. Impact Eng.: 2007; 34: 495-508.

Finite Element Analysis on the Effect of Yarn Reinforcement Distribution on the Tensile Behavior of Unidirectional Composite

Ke Wang[1†], Qian Ma[1†], Shu-Dong Wang[1], Ju-Mei Zhao[1], Li-Min Jin[2,3,4*], Pi-Bo Ma[5*],Sheng-Zhong Zhang[1]

[1]*Jiangsu Research and Development Center of the Ecological Textile Engineering and Technology, College of Textile and Clothing, Yancheng Vocational Institute of Industry Technology, 285 South Jiefang Road, Yancheng, Jiangsu, 224005, P. R. China.*

[2]*Chinese Academy of Sciences, Shanghai Institute of Applied Physics, 239 Zhangheng Road, Shanghai, 201203, P.R. China.*

[3]*Hubei Key Laboratory of Advanced Textile Materials & Application, Wuhan Textile University, Wuhan 430200, P.R. China*

[4]*College of Textiles, Donghua University, 2999 North Renmin Road, Shanghai 201620, P.R. China*

[5]*Key Laboratory of Eco-textiles, Ministry of Education, Jiangnan University,1800 Lihu Road, Wuxi, Jiangsu, 214122, P. R. China.*

[*] Corresponding author's email: lmjin@mail.dhu.edu.cn, mapibo@jiangnan.edu.cn

Abstract

To study the influence of yarn reinforcement distribution on the tensile behavior of the unidirectional composite, 4 types of micro-structural finite element models of the unidirectional composites with different styles of yarns distribution were established by using finite element software ABAQUS on the basis of constant fiber volume fraction, and the deflection-time curves, maximum von Mises stress-time curves, distribution characteristic of deformation and structural stress distribution were calculated respectively. By analyzing the obtained results, it was found that the "4 yarns" type of composite had shown the best tension-resistance behavior. This is due to the uniformly distribution state of the yarn reinforcement, i.e., the improved uniformity which is part of the mechanical property of the composite induced the progressive optimization on the structural strength. In addition, from looking at the stress concentration located at specific regions of the yarn-matrix interface, it was recommended to avoid or reduce such regions when designing the composite structures.

Keywords: Unidirectional Composite; Tensile Behavior; Distribution of Yarns; Finite Element Analysis (FEA)

1. Introduction

Unidirectional composites of yarn/matrix which use parallel yarns as reinforcement are, soaked and infiltrated by the matrix in a liquid form, and finally cured and consolidated. It has been widely used in many industries. Apart from having advantages such as being light-weight and high-strength, unidirectional composites also present other distinct advantages over conventional fiber reinforced composites, such as design ability, abundance of raw materials and high efficiency energy dissipation performance.

Recently, to maximize the mechanical properties of the fiber reinforced composites to meet the growing demand for high strength and light-weight materials has become a pretty hot research subject. So far, the mechanical properties of the fiber reinforced composites can be improved by the following 5 methods.

1) Selection of raw materials [1-3]

Select high performance fibers (carbon fiber, glass fiber, aramid fiber et.al) and resins (epoxy resin, unsaturated polyester resin et.al), i.e., directly select the high quality raw materials of reinforcement and matrix to manufacture the composites with good mechanical properties.

2) Proper volume fractions of constituents [4-6]

In a certain range, the mechanical properties of the composites can be improved with the higher fiber volume fraction. Therefore, appropriately increase the fiber content can significantly improve the mechanical performances.

3) Enhancement of interfacial bonding [7]

The efficiency of fiber-matrix interface bonding of the composite greatly affects its mechanical performance. Hence, enhance the fiber-matrix interfacial properties by some physical or chemical ways is also an effective method to improve the mechanical properties of composites.

4) Structural optimization [8-10]

The optimization of the composite structures can decrease stress concentration effects and hence improve the strength and structural stability, thus obtain the required composites with good mechanical properties.

5) Processing technique optimization [11-15]

Suitable processing techniques can reduce the occurrence of structural weak links (voids, micro-cracks et.al) in the composites. Therefore, the performance of the fiber reinforced composites can be improved by optimizing the processing techniques and parameters.

The mechanical properties of composites can partly be improved by the above summarized methods, a comprehensive and fundamental understanding of the mechanisms of the above methods for the purpose of exploring the best method or the combination of several methods remains a big challenge from the academic point of view.

In this paper, the tensile behavior of the unidirectional composites with 4 different styles of yarn distribution were investigated by using finite element analysis (FEA) on the basis of constant fiber volume fraction. The mechanisms of the yarn distribution on the tension-resistance properties of the composites was discussed by analyzing deflection-time curves, maximum von Mises stress-time curves, distribution characteristics of deformation and structural stress, thus to guide the designs of unidirectional composites with better mechanical performance.

2. Experimental

2.1 Material Specifications

The specifications of the component materials are listed in Table 1. The micro-structure models of 4 types of unidirectional composites with constant fiber volume fraction are shown in Fig.1.

Table.1 Material specifications of fiber and matrix

Components	Materials	Modulus/GPa	Poisson's ratio	Density/g·cm^{-3}
Reinforcement	Glass fiber	70.00	0.20	2.50±0.1
Matrix	Unsaturated polyester resin	3.65	0.35	1.36±0.1

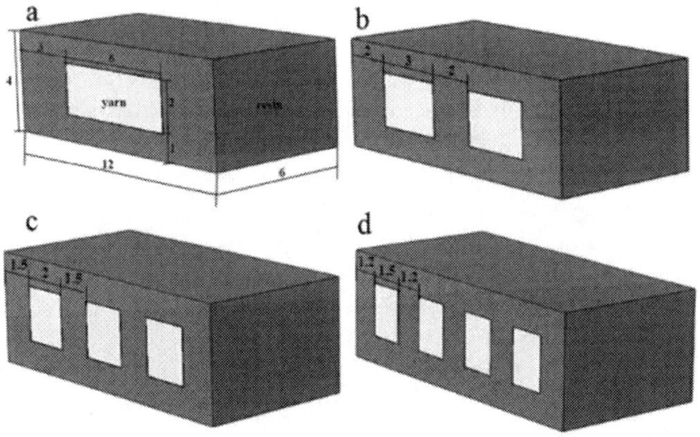

Fig.1 Models of 4 types of unidirectional composites (unit: mm): (a) "Single yarn" type, (b) "double yarns" type, (c) "three yarns" type, (d) "four yarns" type

2.2 FEA Model

FEA was performed using the finite element software ABAQUS and windows XP platform.

1) Micro-structure models

The FEA models of 4 types of unidirectional composites were established on the basis of constant fiber volume fraction and uniformly distributed structure as shown in Fig.1.

2) Interaction condition

The interaction property between the resin and yarns was set as "Surface-to-surface contact (Standard)". Furthermore, the resin was set as the master surface and the yarns were set as the slave surfaces. All yarns were defined as a set with the type of "Node". The discretization method was "Node to surface". Slave nodes were adjusted to the set of the yarns. And the bonding was limited to the nodes set of yarns in the contact condition.

3) Boundary and loading conditions

In the FEA model, as shown in Fig.2, the tensile load on one end of the composite was applied along the axial direction, and the freedom degrees of the other end surface of the composite was fully restrained in all directions. A constant load of 1000MPa was applied for 0.5s.

4) Mesh scheme

As shown in Fig.2, in the models, all parts were assigned to global seeds and meshed with the technique of nodes of coincidence. The element type was quad-dominated.

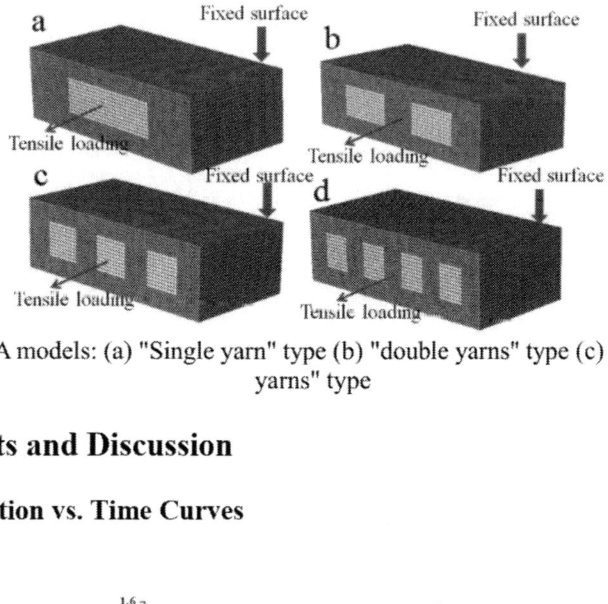

Fig.2 Mesh sheme for FEA models: (a) "Single yarn" type (b) "double yarns" type (c) "three yarns" type (d) "four yarns" type

3. Numerical Results and Discussion

3.1 Maximum Deformation vs. Time Curves

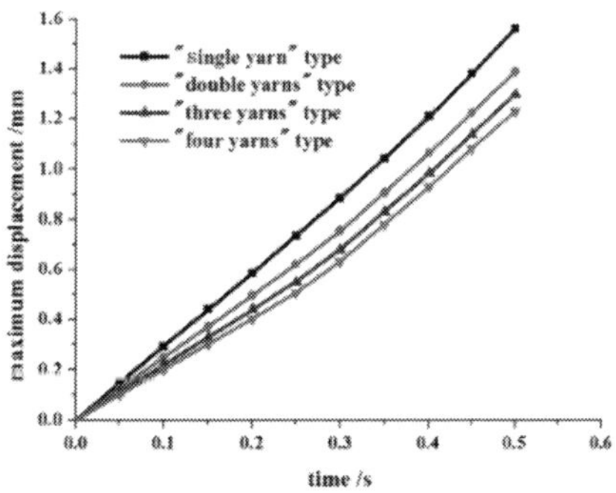

Fig.3 Maximum Deformation vs. Time curves

As one of the basic mechanical indexes of composites, the deformation can directly reveal the tension-resistance performance of composites. In order to illustrate the influence of yarn distribution on the tensile properties of the unidirectional composites, the maximum deformation vs. time curves of 4 types of unidirectional composites are shown in Fig.3. Herein, the maximum deformation is parallel to the loading direction.

It can be seen that the maximum deformation of 4 types of unidirectional composites increased as the time increased, the different tension-resistance performances are also presented. It can be clearly found that the "single yarn" type composite shows the largest deformation, which followed by the "double yarns" type and the "three yarns" type composites, whereas the "four yarns" type composite indicated the smallest deformation. The result was obviously related to the distribution status of yarns in the composites. As shown in Fig.1, it can be found that the yarn distribution gradually turned from "simply concentration" into "uniform distribution" from the "single yarn" type, "double yarns" type, "three yarns" type to the "four yarns" type, which also made the distribution of yarns more uniform. In other words, the non-reinforcement area or low level reinforcement area in the composites reduced gradually, i.e., the content of the areas with weak mechanical properties in the composites decreased. For the entire composite, the uniformity of mechanical properties increased. Therefore, the tensile strength of the composite was also progressively optimized while the deformation decreased.

In addition, Fig.4 also shows the difference of tension-resistance performance for 4 types of unidirectional composites. It can be seen that the maximum deformation occurred at some specific regions of the matrix, such regions were directly affected by the tensile loading. This is mainly due to the difference in the mechanical properties between the yarns and the matrix, as well as the non-uniform distribution of the fiber volume fraction. Due to the mechanical properties (elastic modulus, yield strength et.al) of the yarn are obviously greater than those of the matrix, which induced the deformation of matrix apparently larger than that of yarn. Furthermore, the distribution of the fiber volume fraction in the unidirectional composites decreased from the central region to the surrounding area, and the fiber reinforcement acted as the main load bearing part, which made the elongation at the surrounding area of the composites much larger than that of the central area in the composites. Both of the above mentioned reasons resulted in such phenomenon.

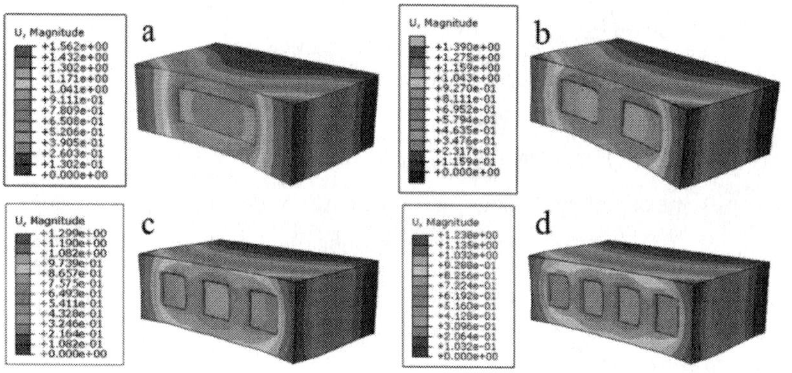

Fig.4 Contours of deformation distribution: (a) "Single yarn" type, (b) "double yarns" type, (c) "three yarns" type, (d) "four yarns" type

Moreover, it was found that the "four yarns" type unidirectional composites shown the best tensile properties, whereas the "single yarn" type composites shown the worst tensile properties on the basis of constant fiber volume fraction. The tensile property of the composites can be improved after the uniformly distributed state of the yarn reinforcement has been increased. Hence, the influence of distribution of yarns on the mechanical performance optimization of the composites needs to be considered while designing the composites with high tensile strength.

3.2 Maximum Stress vs. Time Curves

The structural stress variation in the composites during the loading process can be used to analyze the performance of energy dissipation and absorption. In order to further study the effect of yarn distribution on the tensile properties of the unidirectional composites, the Maximum Von Mises Stress vs. Time curves of 4 types of composites are shown in Fig.5.

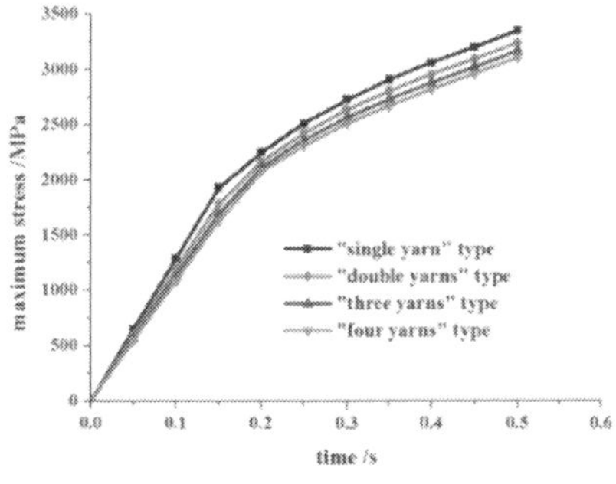

Fig.5 Maximum Von Mises Stress vs. Time curves

It was observed that the maximum stress in 4 types of composites increased as the time increased and show a similar trend. The maximum structural stress can be found in the "single yarn" type composite, followed by the "double yarns" type and "three yarns" type composites, while the minimum structural stress can be found at the "four yarns" type composite. According to the relationship between structural stress and the performance of energy absorption of composites, i.e., the lower the structural stress, the better energy absorption property for the composites under the same conditions (material selection, fiber content, loading condition, external environment, et. al). It can be concluded that the "four yarns" type composite presented the best tension-resistance performance. This was also due to distinct yarn distribution in the "four yarns" type composite. As described above, the mechanical properties of the composite can be improved as the increase of uniformly distributed state of the yarn reinforcement, which will significantly decrease the stress concentration to a certain extent and thus show lower stress in the composite structure.

3.3 Contours of Stress Distribution

In order to further clarify the structural advantages and energy absorption mechanisms of the "four yarns" type composite, the contours of stress distribution of 4 types of composites are shown in Fig.6.

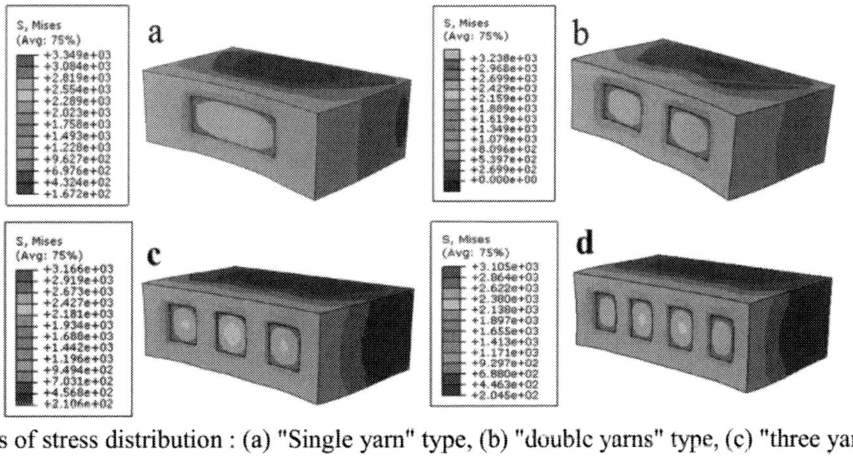

Fig.6 Contours of stress distribution : (a) "Single yarn" type, (b) "double yarns" type, (c) "three yarns" type, (d) "four yarns" type

The stress distribution of 4 types of composites under constant tensile loading can be clearly observed in Fig.6. Stress got the maximum value at the regions where the load was directly applied, then gradually decreased towards the surrounding area. In particular, stress in the yarn reinforcement was pretty higher than that in the matrix. This was also due to the significant difference of the mechanical properties, considering the majority of the energy was absorbed by the yarns, the yarn reinforcement was the main load-carrying part in the composite.

Furthermore, the similar phenomenon of stress distribution can be observed in the yarn reinforcement of 4 types of composites. The maximum stress can be found at the specific regions of the fiber-matrix interface. All of these areas were also the stress concentration regions of composites under tensile loading. The cracks or interfacial de-bonding were also easily occur and propagate at these regions. Therefore, such phenomena should be avoided or reduced while designing the required composite structures.

4. Conclusions

The tensile behavior of the unidirectional composites with 4 different styles of yarns distribution were investigated and analyzed by using FEA on the basis of constant fiber volume fraction. The following contents are concluded.

1) Under the condition of constant tensile loading, the "single yarn" type composite shown the largest deformation, whereas the "four yarns" type composite indicated the smallest deformation. The result was related to the distribution of yarns in the composites, i.e., the stress concentration was weakened with the increasing uniformity of mechanical properties in the composites, which also progressively optimized the structural strength of the composite.

2) The maximum deformation occurred at the resin, where the tensile loading was directly applied. This was owing to the difference in mechanical properties between the yarn reinforcement and the matrix, as well as the non-uniformity distribution of the fiber volume fraction within the composite structure.

3) Stress got the maximum value at the regions where the tensile load was directly applied, then gradually decreased towards the surrounding area. The maximum stress can be found at the specific regions of the fiber-matrix interface. All of these areas were the stress concentration regions of composites under tensile loading. The cracks or interfacial de-bonding were also easily occur and propagate at these regions. Therefore, such phenomena should be avoided or reduced while designing the required composite structures.

5. Acknowledgement

This research was financially supported by a grant from the financial support by Jiangsu Overseas Research & Training Program for University Prominent Young & Middle-aged Teachers and Presidents (2014) and the Open Project Program of Key Laboratory of Eco-textiles, Ministry of Education, Jiangnan University (No. KLET1505). The authors are also grateful for the project supported by Top-notch Academic Programs Project of Jiangsu Higher Education Institutions (PPZY2015C254), 2015 Program for Excellent Scientific and Technological Innovation Team of Jiangsu Higher Education [No. Education Department of Jiangsu Province (2015)4], Scientific Research Fund of Yancheng Institute of Industry Technology, and the Open Project Program of Hubei Key Laboratory of Advanced Textile Materials & Application, Wuhan Textile University (Fzxcl2017001).

References

[1] Ling LT, Li YJ, Frank KK. Fabrication and properties of lignin based carbon nanofiber. J. Fiber Bioeng. Informatics: 2013; 6 (4): 335-347.

[2] Deng SS, Li G. Structural features and microwave absorbing properties of polyaniline-montmorillonite composites prepared by in-situ polymerization. J. Fiber Bioeng. Informatics: 2013; 6 (1): 33-40.

[3] Wang X, Bi SM. Investigation on damage mechanisms of PE self-reinforced composites by acoustic emission technology. J. Fiber Bioeng. Informatics: 2012; 5 (3): 281-287.

[4] He HW, Gao F. Effect of fiber volume fraction on the flexural properties of unidirectional carbonfiber/epoxy composites. Int J Polym Anal Ch: 2015; 20: 180-189.

[5] Mini KM, Lakshmanan M, Mathew L, Mukundan M. Effect of fibre volume fraction on fatigue behaviour of glass fibre reinforced composite. Fatigue Fract Eng M: 2012; 35: 1160-1166.

[6] Feret V, Ghiasi H, Hubert P. Effect of fibre volume fraction on mixed-mode fracture of a fabric carbon/epoxy composite. Appl Compos Mater: 2013; 20: 415-429.

[7] Song JJ, Zhang YS, Fan HZ, Hu TC, Hu LT et. al. Design of interfaces for optimal mechanical properties in Al2O3/Mo laminated composites. J Eur Ceram Soc: 2015; 35: 1123-1127.
[8] Song JJ, Zhang YS, Fan HZ, Fang Y, Hu LT. Design of structure parameters and corrugated interfaces for optimal mechanical properties in alumina/graphite laminated nanocomposites. Mater Design: 2015; 65: 1205-1213.
[9] Hao MY, Luo RY, Hou ZH, Yang W, Zhang Y et. al. Effect of structure of pyrocarbon on the static and dynamic mechanical properties of carbon/carbon composites. Mater Sci Eng: 2014; 614: 156-161.
[10] Heo GY, Park SJ. Effect of structural change of pitch on the thermal conductivity of epoxy-based composites filled with heat-treated pitch. Polym Int: 2013; 62: 1524-1527.
[11] KOTOV VA. Effect of structural and processing parameters of the reinforcement on the strength of unidirectional metal-fiber composites under different types of loading. Mech Compos Mater: 1986; 22: 716-721.
[12] Kuo CFJ, Huang CC, Su TL, Lin MY, Lan WL. Optimization process parameters of multiple quality characteristics for polyoxymethylene/glass fiber composite material. Fiber Polym: 2014; 15: 2597-2606.
[13] Kravaev P, Stolyarov O, Seide G, Gries T. Influence of process parameters on filament distribution and blending quality in commingled yarns used for thermoplastic composites. J Thermoplast Compos: 2014; 27: 350-363.
[14] Robati H, Haghparast A, Shishesaz M, Attarroshan P. The effect of void shape and volume fraction of fibers on the stress distribution in a laminated composite plate with triangular fibers. Mech Compos Mater: 2014; 50: 155-164.
[15] Frank KK, Li YJ, Ling LT, Yang HJ. Multifunctional composite Nanofibers. J. Fiber Bioeng. Informatics: 2013; 6 (2): 129-138.

Evaluation of Basic Pattern Fitness on Men's Upper Body by Finite Element Modeling

Xue-Lian Kang, Bo-An Ying*, Xin Zhang, Long Wu, Jin Duan

Apparel & Art Design College, Xi'an Polytechnic University, Xi'an, Shaanxi, 710048, P.R. China

*Corresponding author's email: yingba2006@163.com

Abstract

It has become increasingly important to research the digital design and means of evaluation for basic patterns during the clothing design process. In this paper, the finite element model is established to detect the fitness of basic patterns. The pressure distribution of basic patterns for men's clothing is analyzed for fitness evaluation.

The main work of this paper is described in a chronological sequence. Firstly, the point cloud data of basic patterns for men's clothing and measurements of their upper body are acquired by the non-contact three-dimensional scanning equipment. Secondly, the reverse software aims to build, improve and perfect the geometric model, followed by meshing the upper body pattern model for finite element, and subsequently building a contact model. When the model is built, the upper body is assumed to be elastomer, and material of basic pattern is set as continuous, homogeneous and isotropic elastic in the computer. Deformation of the upper body and the basic patterns for men's clothing is large strain and nonlinear. The finite element model of the upper body and the basic pattern for men's clothing is used to simulate the pattern's gravity when the basic pattern is on the surface of the upper body. Through analyzing the pressure distribution, a conclusion can be reached on whether or not the model is available and the suitable for men's clothing. The results of this study provide a technique and method for the digital design and evaluation of basic patterns for men's clothing.

Keywords: Finite Element Model; Basic Pattern for Men's Clothing; Fitness Evaluation; Digital Design

1. Introduction

The fitness and comfort of clothing became one of the increasingly important considerations in modern life, which also makes the evaluation of basic patterns for men's clothing more significant. In the computer software, changing the materials or the structures of the basic patterns can be time-consuming, as it is timely to calculate the change in pressure distribution between the body and the basic pattern simultaneously. The digital simulation and evaluation method, which aims to test the comfortableness through pressure distribution, is an effective method for the apparel industry. Digital modeling technology and finite element analysis method can give designers a better understanding of their design's fitness for men's clothing, which can then improve the overall design.

At present, studies on the evaluation of basic patterns usually transform the basic patterns of clothing into a piece of garment. Almost all existing methods directly arrive conclusions using the complete garment. In the paper titled "Research on Knowledge-drive Female Jacket Model Design Method", Jing Chen used the camera to take photos of the completed design to evaluate the fitness of pattern in dressing state, and surveyed the amount and quality of space to evaluate the fitness of the garment for the human body [1]. In the paper titled "Research and Application of Garment Fitness Assessments", Yi-chao Xu and Yong-sheng Ding used subjective methods such as fuzzy mathematics, statistical method and wavelet analysis to study the physical fitness of the clothing, and the measurement data is vital for the analysis and evaluation [2]. Xing-xiang Qi meshed the plane of the clothing, then used space geometric transformation to set the initial three-dimensional position of the clothing with the characteristics lines of the human body matching the clothing's lines, which eventually evaluates clothing's fitness to the human body [3]. On the other hand, Wei Dai and Wei-yuan Zhang established a theoretical membership function using fuzzy mathematics methodology to evaluate fitness [4], while Zheng-yan Xu and Xiao-xia Song analyzed and summarized the application of finite element method and intelligent algorithm to study pressure on clothing [5]. In the paper titled "Digital Simulation of Bra Pressure Distribution", Lu Lu, during the design stage of a bra, put forward the body and bra contact model, and used the finite element method to analyze bra pressure distribution [6].

In this paper, we began with constructing a geometric model of the human body and clothing. Then, we selected the element type, gridding, which is vital to establishing the finite element contact model, and calculating and analyzing the results. Finally, according to the result, conclusions were made and suggestions were proposed regarding fitness of basic patterns for men's clothing. The finite element method is a creative means to study the subject here, as it is difficult to simulate real conditions when people dress the clothing, in which the method is used to predict the effect of the pattern and improve on it to generate a functional piece of garment. Also it can save time for customized production, as well as predict the fitness of the dresser. This paper provides an evaluation method that can be used to check the quality of the basic patterns and evaluate their fitness on men's upper body on a computer.

2. Establishment of Geometric Models for Men's Upper Body and the Basic Patterns

2.1 Establishment of the Upper body Geometric Model

In a realistic scenario, the uneven surface and three-dimensional structural features of the human body has to be taken into account when designing clothing patterns and fitting them onto a model. A geometrical model has to be established when describing a three-dimensional upper body and a pattern designed accordingly to the computer model in the form of numerical data to reconstruct an object prototype. The key technologies involved include acquisition of three-dimensional solid geometry data, handling of discrete point cloud data, and reconstruction of three-dimensional solid model [7, 8].

The three-dimensional body data was acquired by a non-contact three-dimensional scanner. The optical triangulation method lies the foundation of the data acquisition process, in which points are illuminated onto the surface of the object, accurately measuring the dimensions. In this paper, a simple, quick and smart non-contact three-dimensional scanner developed by the Germany company Vitus, which adopts more reliable calibration methods and more updated hardware technology and laser beam that is harmless to the human body with a bigger range of measurements and higher accuracy, was used. A 360-degree all-rounded vertical and uniform infrared laser scans the surface of the object in around 20 seconds, which is a simple and quick way of obtaining data.

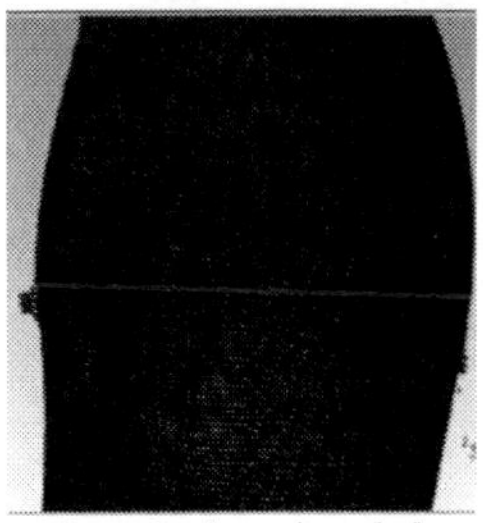
Fig.1 Point cloud data.

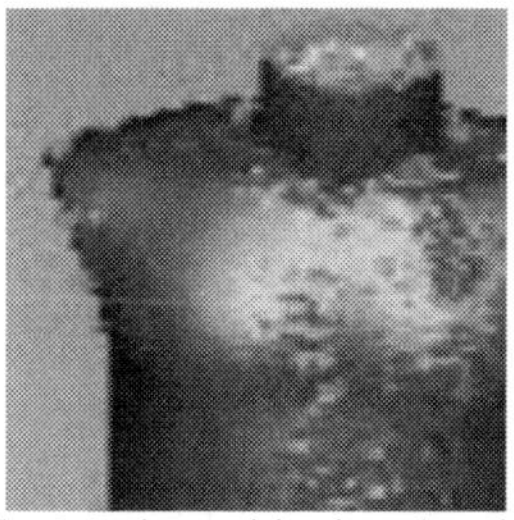
Fig.2 Selected isolated point.

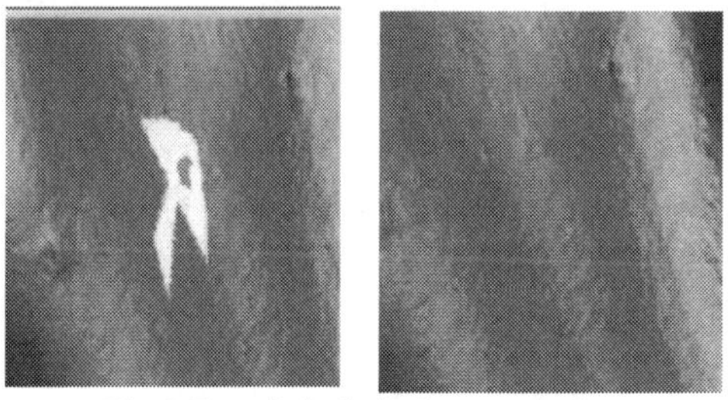
Fig.3 Repair hole before and after.

Point cloud data of the upper body is acquired by a three-dimensional non-contact scanner, which then displays the shape of the body on the computer as shown in Fig.1. Despite the convenience of the technology, unnecessary data points and noise points are produced, due to limitations of the scanning technology and influence from the scanning environment. The isolated points should be selected as shown in Fig.2 and be deleted to smooth the surface. Using the "wrapping the point model" function, a polygon model can be produced. The method of nail removal can reduces the number of irregular triangles and polygons to ensure simplified calculation and smooth surface while keeping the design realistic and original. Effects of nail removal is shown in Fig. 4. To make the model look more appealing, holes generated by short-distance scanning can be repaired using curvatures of planes, as shown in Fig.3. The whole model is a facet model, and the facet body is sutured as shown in Fig.5. Finally, a complete three-dimensional geometric solid model is obtained.

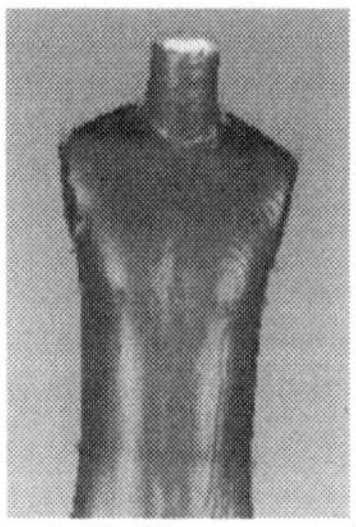

Fig.4 Nails removal.

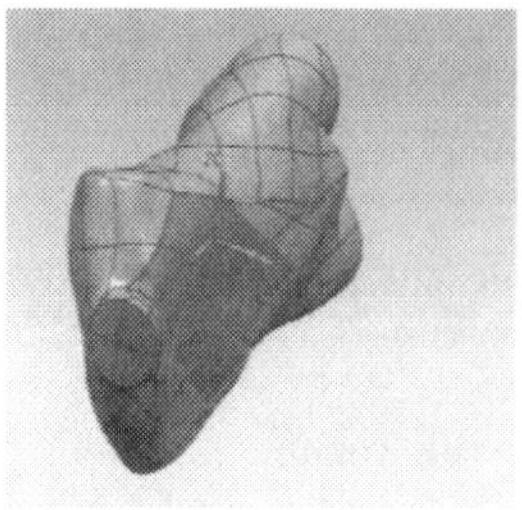

Fig.5 Suture facet.

2.2 Establishment of a Geometric Model for Basic Pattern for Men

The body has an uneven surface, and due to the outstanding shape of the chest and scapula, there should be a certain amount of floating margin in these areas when designing basic patterns. To ensure the pattern does not hinder human physiological characteristics or cause any discomfort, a set split line should be added to the shoulders and other areas that stand out from the flat surface. When designing near the split line, the two following principles should be followed: the dividing sharp points should point to the bumps; the split lines should be easily available. According to the two principles, it is necessary to draw split lines on the fit surface which has the lines on the pattern [9]. Firstly, according to the bust line, waist line, neck line, shoulder line, front center line, back center line, and other auxiliary lines involved in the basic pattern, the contour of characteristic curve is extracted from the 3D digital body model, which must be a closed curve. Fig.6 shows the lines of the body contours of the front and back of the basic pattern. Secondly, according to the trend of the point cloud in the front and rear contours, the surface of the pattern is reconstructed. Finally, adding the split line in the construction design of the curve attends to adjust the final basic pattern as shown in Fig.7.

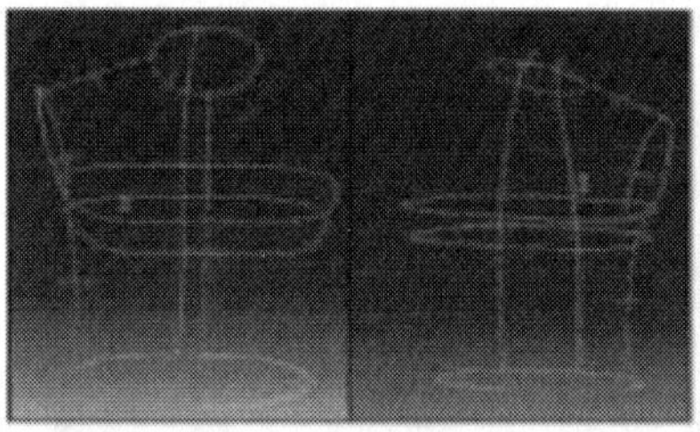

Fig.6 Contour lines of upper body.

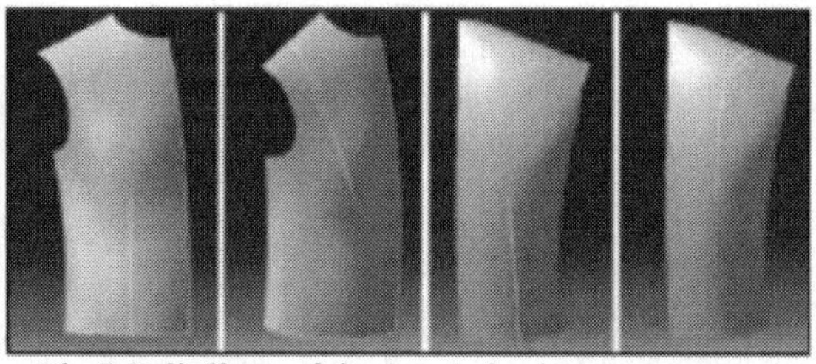

Fig.7 Split lines of the men's basic pattern.

3. Establishment of the Finite Element Model

3.1 Element Type

Before meshing the upper body and the basic pattern geometric model for men, the element type of both components have to be defined. The upper body is considered as a volume element because of the irregular surface on the solid model, and thus the ten nodes tetrahedron grid is chosen. The ten nodes tetrahedron grid allows for accurate spatial orientation and calculation [10, 11]. More attributes to the uneven surface of the body are the underlying bones, ligaments, muscles, and other irregularly shaped organs. In order to facilitate the division of the grid, the human body is assumed to be evenly distributed, and divided by a single grid. The basic pattern can then be meshed onto a two-dimensional plane triangular grid or a three-dimensional quadrilateral surface grid. A quadrilateral grid, whether in terms of calculation accuracy or convergence rate, is better than the triangular mesh [12].

3.2 Meshing

Since the quadrilateral grid and tetrahedral grid have good convergence and adaptability, and the human body and clothing have obvious irregular surface, the upper body uses a tetrahedral element while the basic pattern adopts a quadrilateral three-dimensional shell element. The size and type of the grid will affect the distribution of pressure and the accuracy of the results [13]. The size of body's element is set at 15mm, and the size of the basic pattern's element is set at 25mm. These are conducive to speed up calculation.

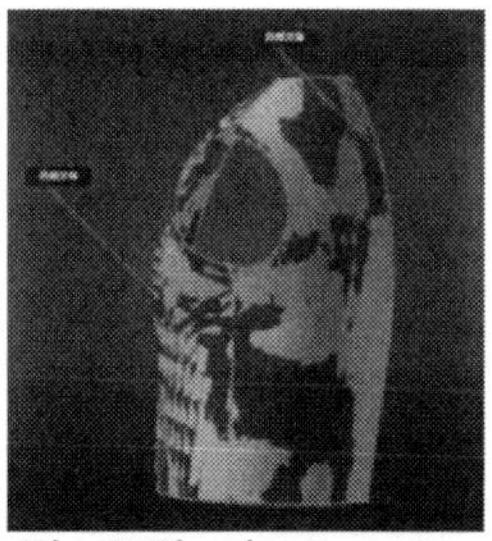

Fig.8 Check geometry of men's basic pattern.

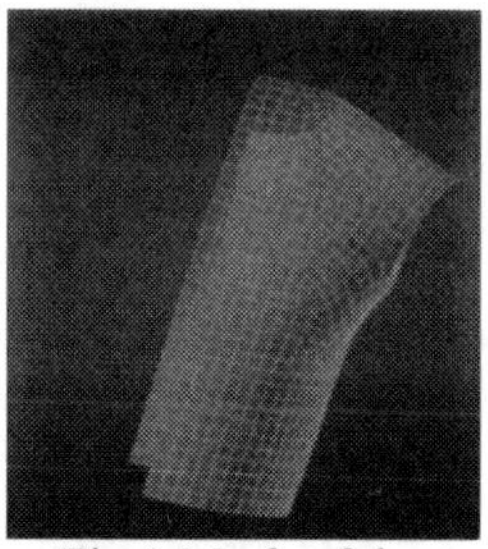

Fig.9 Mesh of the men's basic pattern.

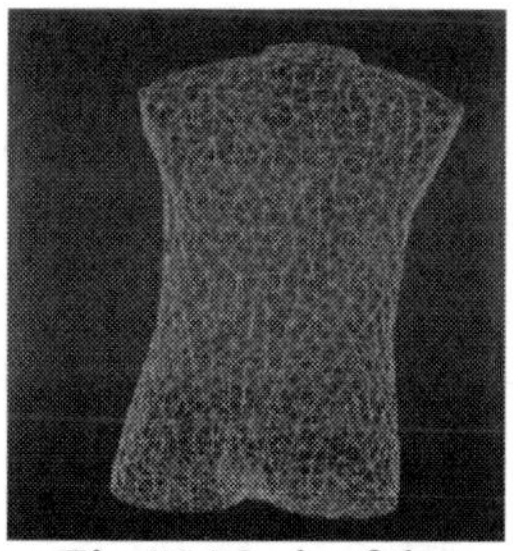

Fig.10 Mesh of the upper body.

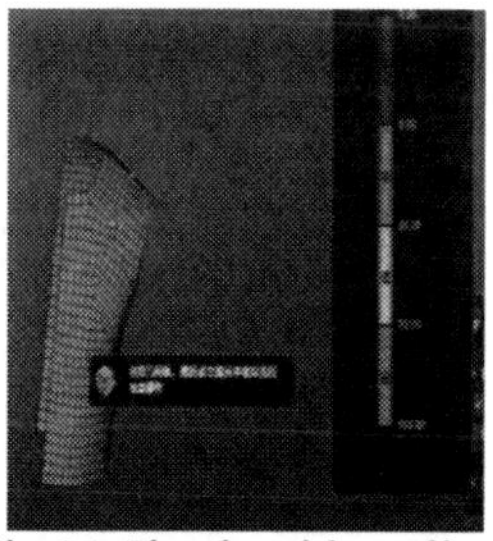

Fig.11 Check grid quality.

To achieve the requirements of meshing as shown in Fig.8, the geometric model is checked. Then the men's basic pattern is meshed by a quadrilateral grid and the upper body is meshed by a tetrahedral grid as shown in Fig.9 and 10. Finally, the quality of the grid is checked as shown in Fig.11. If the quality of the geometric model is not appropriate, then the geometric model should be adjusted. The model can be saved as a BDF file until the quality is improved, which can be identified by the finite element software.

3.3 Material Properties

In real life, the upper body is flexible, and has non-linear and elastic properties. As the finite element solver has constraints on the material, the material used for the upper body is simplified to reduce calculation. The upper body is assumed to be an elastomer made of the same material and has the same uniform density. The experimental data shows that the material properties of the basic pattern is anisotropic. Since the tensile coefficient of different directions in this paper is not obvious, it is assumed that the material of the basic pattern is isotropic. Also, a thickness value has to be assigned, as the clothing has a certain thickness. The human skin has viscoelastic properties, which is the same as Mooney [14]. The body chooses a Mooney elastic material, and the value of C10 and C01 represent the hardness and elasticity of Mooney elastic material, respectively (C10 = -12, C01= 17). The fabric material properties of the basic pattern for men is shown in Table 1.

Table.1 Fabric material parameters of the men's basic pattern

Type	Results
Name	The men's basic pattern
Numbers of element	959
Element type	Quadrilateral three-dimensional shell element
Material properties	Isotropic elastic material
Elastic modulus(N/mm^2)	220
Poisson ratio	0.3
Density(g/mm^2)	0.0002
Thickness(mm)	2

3.4 Contact Model between the Basic Pattern for Men and the Upper Body

After meshing, the basic pattern and the upper body are assembled in Marc, as shown in Fig.12. The contact between the basic pattern and the human body is realized by the gravity of the pattern. Therefore, the upper body and the basic pattern are defined as two contact bodies. Establishing the contact model between the upper body and the basic pattern requires the following steps: defining the contact point, defining the boundary condition, and defining the working condition.

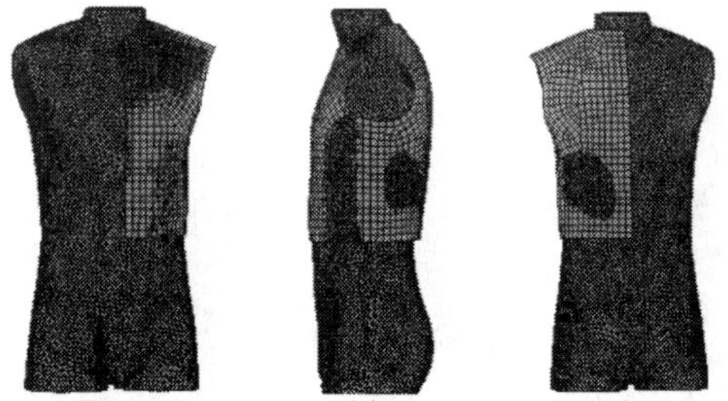

Front Side Back

Fig.12 Assembled model of the basic pattern for men on upper body.

When the basic pattern on the upper body remains in a standing state, due to the gravity of the basic pattern, the upper body exerts greater pressure on the shoulders. The shoulders of the pattern and the shoulders of body remain static; thus, the three degrees of freedom in the shoulders of the pattern should be fixed. The basic pattern is symmetrical in terms of total pattern, so selecting half of it is sufficient for a simple calculation, with the top and the bottom of the upper body defined as stationary as well.

The corresponding boundary conditions are as shown in Fig.13: a) front and back of the basic pattern for men is defined no displacement in the width direction; b) the shoulder line of the basic pattern for men in the height, width and thickness direction is defined as no displacement; c) gravity of the basic pattern for men in height is defined by acceleration, which is set at -0.0098; d) the neck and waist of upper body in the height are defined as no displacement, and the x direction refers to height; y refers to width; and z refers to thickness.

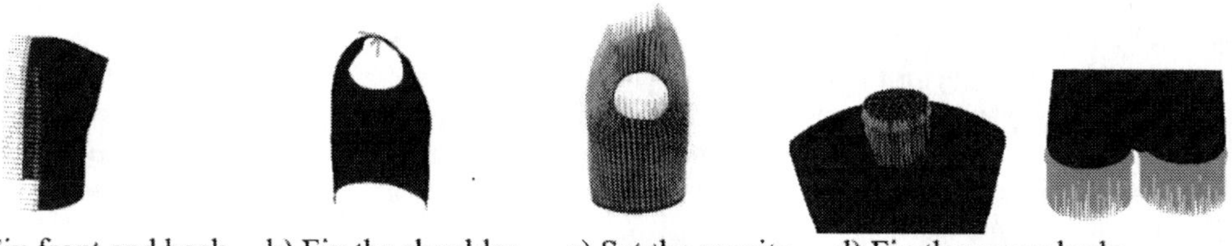

a) Fix front and back b) Fix the shoulder c) Set the gravity d) Fix the upper body

Fig.13 Boundary conditions.

In simulation the men's basic pattern covers the surface of the upper body, which indicates that it is only a surface contact. The working condition should be defined, according to the real dressing conditions, which includes contact, contact table, initial contact and so on.

3.5 Working Conditions and Tasks

The purpose of analyzing the working conditions and tasks attends to simulate the gravity effect of the basic pattern for men on the upper body.

1) Analysis of working conditions and load

The load is defined by conditions in a single analysis; all the boundary conditions in the listed load conditions are selected to compose the required conditions; the contact table in the analysis contact is used [15]; the maximum step increased is defined as 9999, the maximum cycle step is set at 10, the minimum cycle step is 0, and the analysis time is 0.5.

In the finite element analysis, the pressure distribution is affected by gravity, which is consistent with real life conditions. When analyzing pressure between the upper body and the basic pattern, the weight of the external and the internal weight must equal to zero. The Newton - Raphson method of algorithm is usually selected.

2) submit the task

The task is selected and submitted before analyzing, and at the same time, defining the initial contact and selecting the options of result must be done. Note that selecting the big strain and the follow force are vital to complete the calculations. Some options are selected, such as stress and Von Mises.

In addition, the fitness between the upper body and the basic pattern is evaluated by the distribution of the force and the displacement.

4. Result and Analysis

After calculation of the whole model is completed, the results are as follows. By analyzing the distribution of the contact normal force and the displacement, a conclusion can be derived. According to the boundary conditions and contact conditions as described above, where the displacement x is shown in Fig.14, the displacement y is shown in Fig.15, and the displacement z is shown in Fig.16. The different colors represent the different values of displacement. From the displacement cloud of the basic pattern, the largest displacement occurs in the shoulders. From the stress cloud, the force of the shoulders reaches a maximum, and the basic pattern has an obvious displacement towards the upper body, which is consistent with the real life scenario.

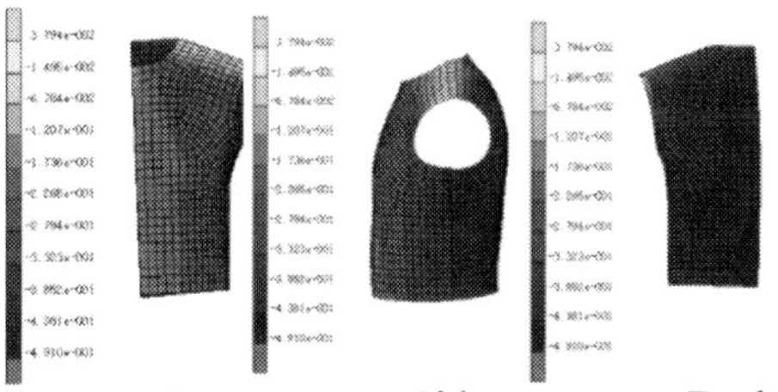

Front Side Back

Fig.14 Displacement x of basic pattern.

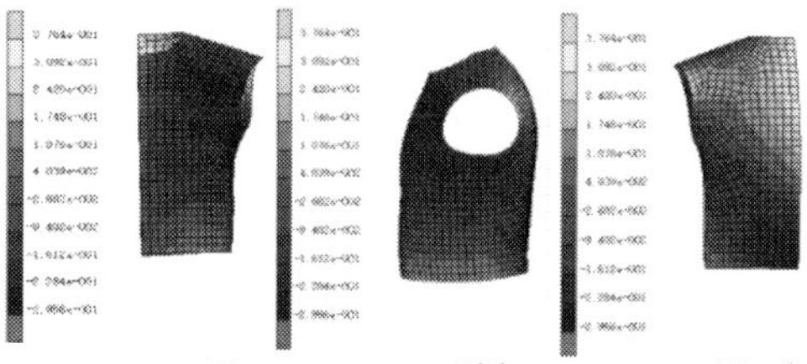

Front Side Back

Fig.15 Displacement y of basic pattern.

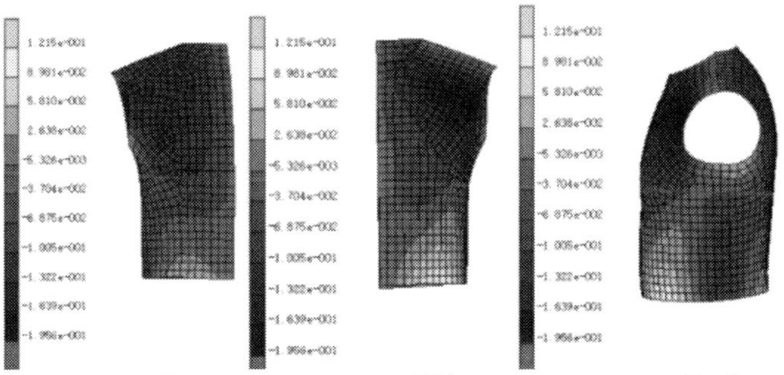

Front Side Back

Fig.16 Displacement z of basic pattern.

The distribution of contact normal force should be analyzed as shown in Fig.17 and 18. According to the contact normal force, the force is greater and more obvious on the shoulders, chest and back of the basic pattern for men. Also, the density of the fabric is small and the patterns have a certain displacement towards the direction of the upper body due to its rigidity, which is consistent with the real life scenarios. According to the results, it can be concluded that the basic pattern for men are basically in line with the fitness and comfort of the upper body.

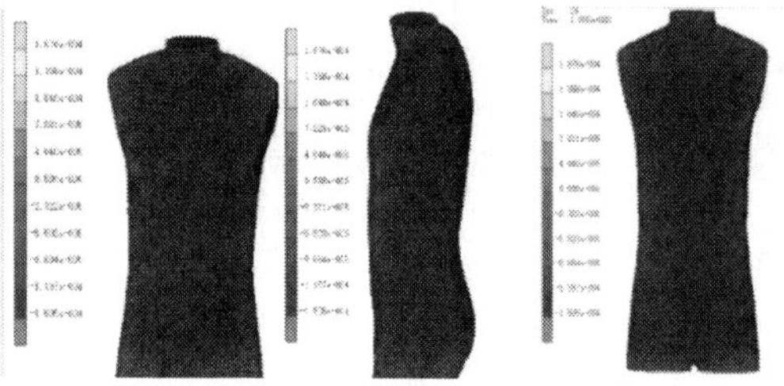

Fig.17 Contact normal force of the upper body.

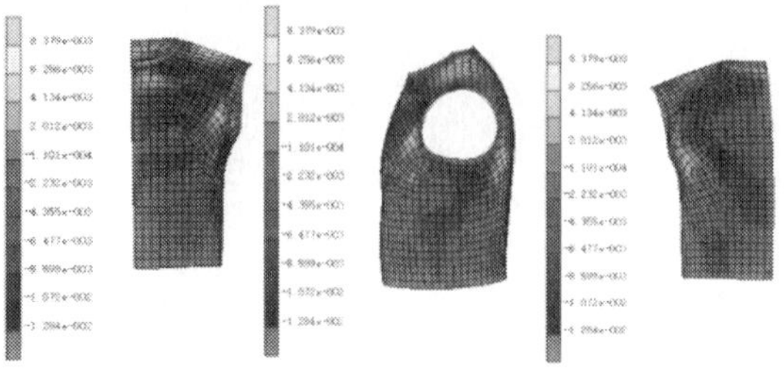

Fig.18 Contact normal force of the basic pattern for men.

5. Conclusion

The main achievements and conclusions of this research project are as follows:

1) In this paper, a feasible digital geometric model is put forward as a method to evaluate the fitness of pattern. The main works include processing the point cloud, reconstructing the surface model and the solid model, choosing the element type, meshing and establishing a contact model between basic pattern for men and the upper body.

2) In this paper, the finite element model of the upper body and the basic pattern for men simulates the gravity of the basic pattern on the upper body. By establishing the finite element model and analyzing the distribution of contact normal force provide effective references for the digital evaluation of the fitness of basic pattern for men. Since the simulation regards both the clothing and the human body are made of the same material, further studies should be conducted to improve the system.

3) The contact normal force, obtained by the finite element method, is bigger on the shoulders and back of the basic pattern for men, which is consistent with the real scenario. It shows that the digital evaluation model is more reasonable, and it can guide the design process.

Acknowledge

The study was supported by grants from Natural Science Foundation of Shaanxi Provincial Department of Education, China (Project No. 16JK1321), and Doctoral Scientific Research Start-up Foundation of Xi'an Polytechnic University (310-0400608).

References

[1] Chen J, Zhang X. Research on female jacket model design method based on knowledge driven: Xi'an Polytechnic University: 2014; 5.

[2] Xu YC, Ding YS. Research and application of garment fit assessments: TEXT RES J: 2007; 10: 127-130.

[3] Qi XX, Zhang WB. Research on evaluation model of clothes fit based on individual virtual mannequin: Donghua University: 2011; 9.

[4] Wei D, Zhang WY. Fuzzy mathematics methodology on clothing fitness evaluation: Journal of Donghua University: 2003; 29: 33-35.

[5] Xu ZY, Song XX. Research progress in relationship between human body and clothing pressure: Journal of Shanghai University of engineer Science: 2013; 27: 170-173.

[6] Lu L, Zhang X. Study on Digital Simulation of Bra's Pressure Distribution: Xi'an Polytechnic University: 2013; 2.

[7] Chen SY, Wu WT, Yang XR. Rapid surface reconstruction based on Geomagic Studio Software: Modern Manufacturing Engineering: 2011; 1: 8-12.

[8] Hu YF. The application of Geomagic Studio Software in reverse engineering post-processing: Manufacturing Automation: 2009; 9: 135-137.

[9] Duan J. Research on relational model for body block of menswear based on digitizing 3D human model: Proceedings of TBIS: 2016.
[10] Zhang R. Comparison of tetrahedral and hexahedral mesh features: Technological Development of Enterprise: 2012; 31: 101-102.
[11] Ji W, Li JL, Yang J. Research on mechanical damage of apple in robot harvesting using finite element analysis: Proceedings of TBIS: 2014; 430-435.
[12] Yang S. Quadrilateral mesh generation based on features: Dalian University of Technology: 2006; 6.
[13] Yan YX. Effects of mesh density on finite element analysis for TBR tire: Tire Industry: 2008; 10: 579-582.
[14] Wang X. Analysis and application of human-body finite element model: Bus Technology and Research: 2006; 2: 30-33.
[15] Liu YX. Marc-based FEA (Finite Element Analysis) of Hybrid Ceramic Ball Bearings and Its' Experiment Confirmation: Tianjin Polytechnic University: 2005; 1.

Development of FEA Contact Mechanical Model for Simulating Pressure Distribution between Female Breast and Moulded Bra Cup

Liang Ma, Long Wu*, Xin Zhang, Si-Chen Pan

Xi'an Polytechnic University, Xi'an, Shanxi Province, China

*Corresponding author's email: 83618518@qq.com

Abstract

With the development of people's living quality, the bra plays a more and more important role in a women's daily life. Clothing pressure distribution has become an important factor for clothing comfort. Researchers need to predict the function and property of the product through the establishment of the model to improve the level of design. Establishing an appropriate model is significant in digital clothing technology and evaluation field. This paper takes a 3D digital moulded bra cup geometric model and a female breast geometric as the foundation, and uses finite element method to build a digital contact mechanical model between female breast and the moulded bra cup. This paper aims to provide a new method and technology to evaluate the result of digital design of a moulded bra cup. The fundamental geometric female breast model [1] is established in a software through program. After that the FEA (Finite Element Analysis) contact mechanics model of female breast and a moulded bra cup is established. In this paper, female breasts are defined as elasticity, and the moulded bra cup is defined as anisotropic material. The hypothesis of the mechanical deformation between female breast and moulded bra cup is a large strain when simulated the pressure distribution of a moulded bra cup. The results of simulation prove that the situation of pressure distribution on moulded bra cup is clear and uneven, and the moulded bra cup model is suitable for female breast model. This paper presents a contact mechanical model for numerical simulations of 3D moulded bra cup pressure distribution during wearing with a finite element method. We could evaluate the clothing comfort based on the results of simulation, and the results of simulation can provide a numerical evaluation method for moulded bra cup comfort.

Keywords: Moulded Bra Cup; FEA; Pressure Distribution

1. Introduction

With the improvement of living condition, healthy dress concept is deeply rooted in people's heart. People put forward higher and higher requirements for the comfort of clothing. Except for heat-moisture, clothing pressure distribution has become an important indicator for clothing comfort. There are two major factors that can affect clothing pressure distribution; one is the human body, the other is clothing. Firstly, different shapes of female breast, curvature and modulus of elasticity have different impact on a moulded bra cup. Secondly, the different motion states of human have different impact on clothing pressure distribution as well. In this field, the study of the distribution of the moulded bra cup has become the focus of attention. Zhang Xin and Li Yi put forward the effects of fabric as an isotropy, and the friction between fabric and a rigid ball on fabric bagging are investigated by the use of finite element method [2]. Zhang Xin presents a mechanical model for numerical simulations of 3D dynamic garment pressure during wearing with a finite element method [3]. They finished a computational model that can simulate and predict the dynamic mechanical behavior of garments during wearing, such as garment deformation, pressure, and internal stresses, without actually producing the garment [4]. Gu Lin states that bodily form is classified by the difference between circumference and waistline according to the state standards. The method is cited in female somatotype, and the basic parts of female somatotype and the classification indexes are also given. Finally, the key data of influencing the circumference shape of basic parts are determined [5]. However, the existing problems in this classification are analyzed. In recent years, Tan Rui selected 50 Chinese college students aged between 20 and 25 with standard breast shape as subjects. Through three-dimensional body scanning, they presented a model of the cross-section of the lower leg where the top part of a man's sock is located [6]. They established a finite element simulation model of the lower leg cross-section, the model assumes that skin, soft tissue and bones are all liner elastic materials. They simulated the pressure and displacement distribution at the top part of socks. Subdivision of body shape is the foundation for digital processing and marketing of garments. Breast shape was proved as the critical factor influencing the fitness and comfort of women's wear [7]. They also measured them by means of the 3-D Interactive Body Scanning Instrument (TecMath), and provided technological parameter for designing the bras as intimate and made-to-measure (MTM) garments. Xu Yaoyao uses the finite element method to do the virtual fitting of the moulded bra cup, and the material properties will be set based on the actual situation of the moulded bra cup and the human body condition. They found the shortage of cup in hand and revised it on reverse engineering software to reduce the wasting of resource caused by repeated experimental in industrial production [8]. Through consulting references and practices and adopting 3D digital simulation technology, Lu Lu gets the point cloud data of both body and cup by different 3D scanners to reconstruct the body and the moulded bra based on the 3D scanning data in different ways due to their different shape and surface characters, and then, makes the two digital geometry models into suitable finite element models based on the knowledge and application of FEA software [9].

The FEA (Finite Element Analysis) is a numerical calculation that is widely used in engineering analysis field. From the perspective of functional analysis, finite element analysis method can be applied to solve all kinds of flow field including the material structure, mechanics research, electrical and magnetic field, fully integrated solution thermodynamics, dynamics, acoustics, super unit and fluid-structure coupling[8]. From the perspective of application, the finite element analysis is mainly applied in early high-tech fields of astronautics, but now it is widely used in machinery, chemical, textile materials [10], and other fields. Finite element analysis is applied in textile field and mainly divided into two aspects; one is to analyze mechanical properties of the textile materials, and the other is to analyze relationship between clothing and the human body. Now many scholars began to use the finite element method to analyze heat transformation between human environments and clothing environment [8] .

In this paper, FEA is a method used to deal with the problems of clothing pressure distribution. The basic thought is separating continuous medium into finite element. This element passes on the message through nodes. The FEA method has superiority in biomechanical filed [11,12,13]. In recent years, researchers in the field of textile and garment pay more attention to the FEA method, which is used in clothing pressure distribution [14,15]. However, scholars usually scan clothing and human's body to get the point cloud file, and then use reverse engineering software to deal with it, and finally get the useful geometric model. This paper adopts a 3D digital model including female breast and moulded bra cup, which is established in software through program. The surface of moulded bra cup depends on the human curvature of female breast geometric model and is then assembled together. Finally, the FEA elastic contact model for female breast and moulded bra cup is established. This method leaves out complicated steps of reverse engineering and achieves the digital simulation of a moulded bra cup from the basic model to FEA pressure distribution.

2. Development of FEA Contact Model

2.1 The Geometric Model of Female Breast and Moulded Bra Cup

Fig.1 Geometri model of female breast

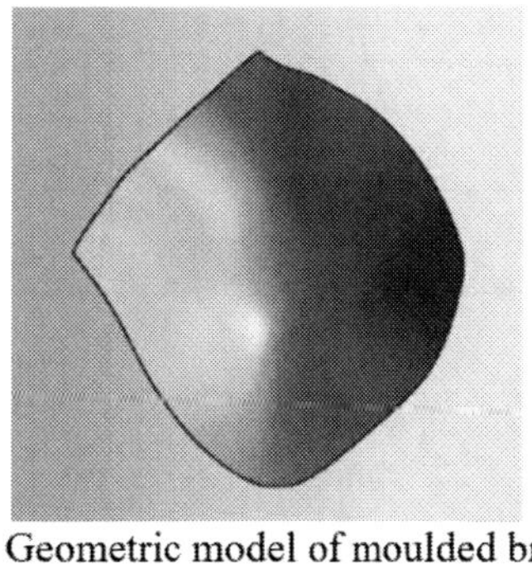

Fig.2 Geometric model of moulded bra cup

This paper adopts a 3D digital model including female body and moulded bra cup, which are established in software through program. As for geometric model of moulded bra cup, it must determine the first point on breast model, and then curve of bra is drawn according to the curvature of breast. The moulded bra's surface depends on the human body's curvature. Suitable geometric model is the foundation of finite element analysis. The geometric model of female breast and moulded bra cup are shown in Fig.1 and Fig.2.

2.2 Element Type and Meshing

This paper's FEA model is built based on a geometric model step by step. This is a FEA contact elastic model. There are 20214 elements for female breast model, and 62 elements for moulded bra cup. The female breast adopts entity element and eight-node hexahedron element. The moulded bra cup adopts shell elements and rectangular and triangular mixed element. Considering the symmetry of the human body, this paper just simulates the right part of bra pressure distribution. The result of meshing is shown in Fig.3 and Fig.4. The FEA model is shown in Fig.5 and Fig.6.

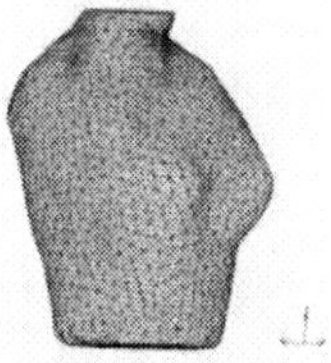

Fig.3 Meshing of female breast

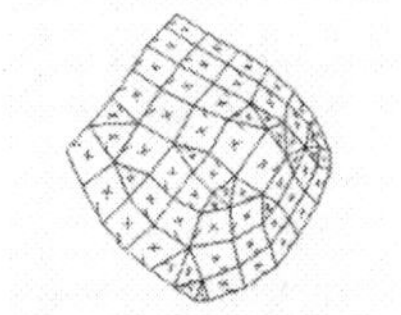

Fig.4 Meshing of moulded bra cup

Fig.5 FEA model of moulded bra cup

Fig.6 FEA model of female breast

2.3 Material Definition

Finite element meshing is an important part of FEA. It is the foundation of finite element analysis during the process of simulation. In order to simulate the real properties of female breast and the moulded bra cup, this paper simplifies the structure of model by simplifying the model in the process of simulation and considers the complicated structure of muscle and soft tissue. Because the human breast is elastic, the body is taken as an elastic entity. Since moulded bra cup is closed-fit with human breast during wear, and it has good elasticity and reshaping function, the bra is defined as isotropic material. Parameter of bra and human breasts is shown in Table 1 and Table 2.

Table.1 Parameter of breast

Type	Element number	Element type	Material property	C10	C 01
Human breast	20214	eight-node hexahedronentity	Mooney	-12	17

Table.2 Parameter of bra

Type	Element number	Element type	Material property	Elastic modulus	Poisson's ratio
bra mold cup	62	shell	Isotropic material	180	0.3

2.4 Model Assembly

This paper defines human breast and bra as the elastic model, and the female breast has a complicated structure that makes the FEA contact model become complicated, and therefore, it must be simplified, so that the element number could be reduced. The contact model includes the female breast model and moulded bra cup model. The coordinates of the moulded bra cup depend on the location of female breast. It can ensure the right position of the contact model of breast and bra. The rationality of the contact model can be ensured as well. There are two hypotheses before the simulation: first, human's medium is even-distributed, in order to ensure the continuity of calculation. Second, before the contact between human breast and bra, no initial stress exists. The assembled model is shown in Fig.7.

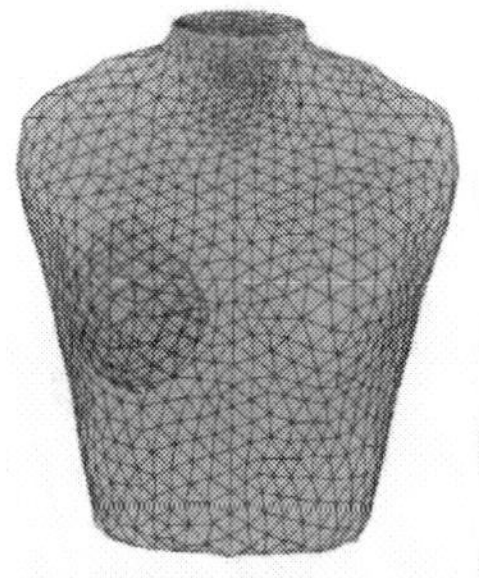

Fig.7 Assembled model

2.5 Contact Conditions

Three-dimensional digital human body and clothing contact problem is a non-linear behavior. Therefore, when the contact mechanics model is established, the human body and the garment are set as deformable contact model, and the contact relation is set as the contact between the deformable body and the deformable body. The contact table is established to determine the relationship between female breast model and moulded bra cup. The contact mechanics model is used to analyze the pressure distribution between the female breast and the moulded bra cup under the condition of large-strain simulation. This paper uses normal contact to determine the three-dimensional contact between the human body and clothing. Parameter of interference is set at 20%. Line chart of expansion time is shown in Fig.8.

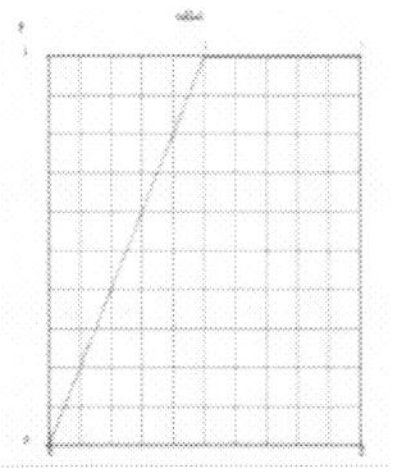

Fig.8 Line chart of expansion time

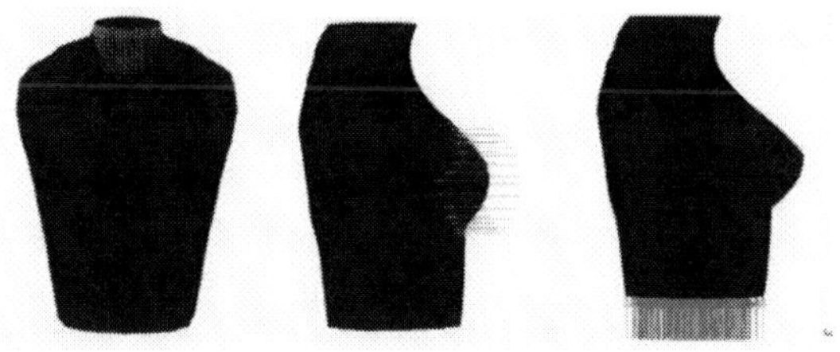

Fig.9 The boundary conditions

2.6 Boundary Conditions

As for boundary conditions, there are four conditions in the model to avoid the body's movement. First, human neck and back have a condition to ensure that the human body will not move up in the simulation process. In addition, the lower part of the body is also fixed to ensure the stability of the location during wearing. As for moulded bra cup, the boundaries around the moulded bra cup are fixed. The boundary conditions are shown in Fig.9.

3. Simulation Results and Analysis

The finite element analysis method could simulate the mold process of wearing, and the pressure distribution of the moulded bra cup. According to the results of the output to evaluate the fit of moulded bra cup, the simulation is not equivalent to the real situation. Virtual evaluation needs to be verified by a real situation. The simulation results of pressure distribution of moulded bra cup are shown in Fig.10. The FEA contact mechanics model of the female breast and moulded bra cup are analyzed by the method of static analysis under the condition of large strain simulation, and the pressure between the three-dimensional female breast and the moulded bra cup is analyzed. Fig.10 shows pressure distribution of moulded bra cup, different colors represent different pressure values, with the biggest pressure on the top of the largest, and then reduces accordingly in the colors chart. The results show that the moulded bra cup in the process of wearing has an uneven pressure distribution. The top of moulded bra cup has greater pressure, and the lower part has less pressure. This result simulates the effect of the moulded bra cup gathering upward. The pressure in boundary of moulded bra cup is slightly larger than the medial side, indicating that the moulded bra cup can gather the breast to the center.

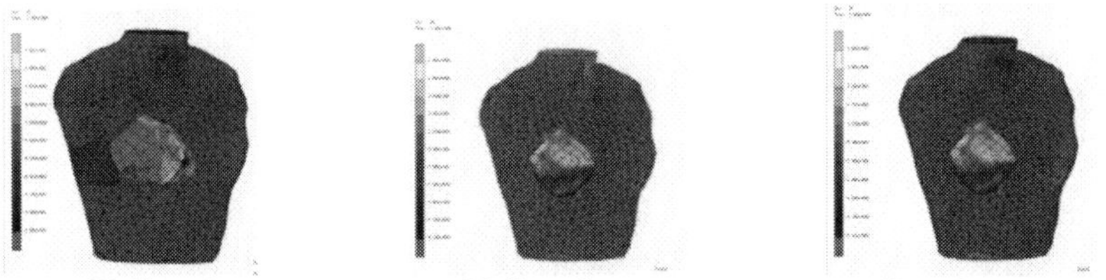

X direction Y direction Z direction

Fig.10 Pressure distribution in moulded bra cup

During the process of simulation, the distribution of human pressure and displacement can be placed in three directions. By analyzing the pressure and displacement distribution of the breast and the human body, a conclusion can be arrived. According to the cloud map distribution, the human's displacement of x is shown in Fig.11, the displacement of y is shown in Fig.12 and the displacement of z as Fig.13. The different colors represent the difference in displacement and pressure.

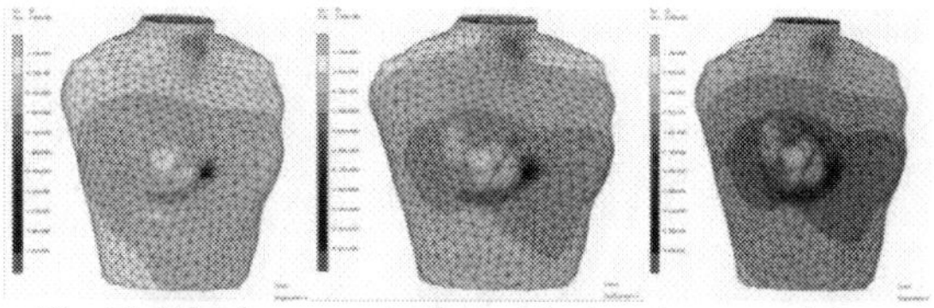

Fig.11 Displacement of x with the change of time

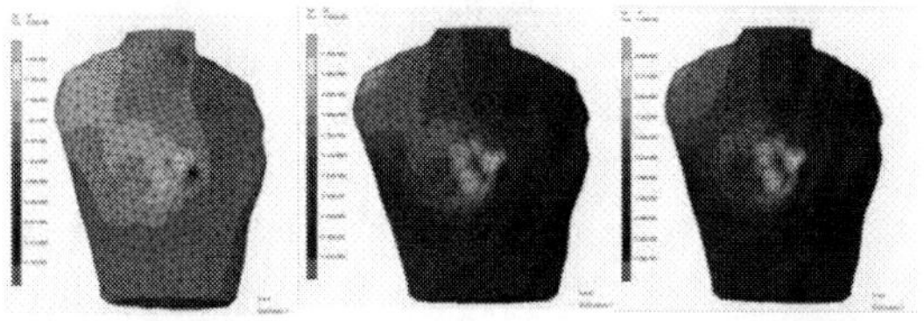

Fig.12 Displacement of y with the change of time

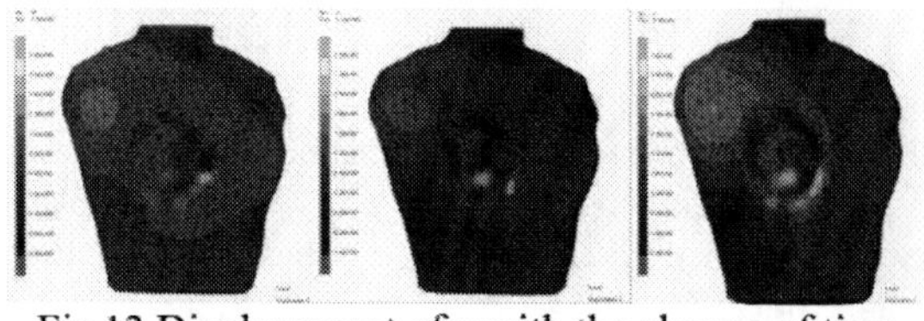

Fig.13 Displacement of z with the change of time

4. Conclusion

This paper directly adopts geometric model female breast model that is programmed in software platform, and uses finite element method to build a digital contact mechanical model between female breast and a moulded bra cup. The finite element simulation in this paper belongs to strain analysis, in the process of the moulded bra cup contact with breast deformation situation, its own gravity is not taken into consideration. The modeling method is simple and rapid. Through the simulation results, the situation of pressure distribution is obvious and the model of breast is suitable for moulded bra cup. This paper uses the finite element method to simulate the pressure distribution of the moulded bra cup and female breast and set the physical properties to improve the truth of simulation. The results of the simulation can provide a numerical evaluation method for moulded bra cup comfort.

5. Acknowledgement

This work was supported by grants from the Natural Science Foundation of Shaanxi Provincial Department of Education, China (Project No. 16JK1321) and the Science and Technology Guiding Project Supported by China National Textile and Apparel Council (2016124).

References

[1] Pan SC. The research and design of brassiere and 3D body measurement based on the surface modeling technology. Xi'an Polytechnic University: 2017; 5: 124-129.
[2] Zhang X, Li Y, Yeung KW, Yao MA. finite element study of stress distribution in textiles with bagging. Commputational Mechanics Techniques and Developments: 2000; 4: 235-242.
[3] Zhang X, Yeung KW, Yi Li. Numerical simulation of 3D dynamic garment pressure. Textile Research Journal: 2002: 72 (3): 245-252.
[4] Yeung KW, Li Y, Zhang X. A 3D biomechanical human model for numerical simulation of garment-body dynamic mechanical interactions during wearing. Textile Research Journal: 2004; 95: 59-79.
[5] Gu L, Zhang X. A method of somatotype for 3D parametric modeling of human body. Journal of Xi'an University of Engineering Science and Technology: 2007; Vol.21: No.3 (Sum No.85) .
[6] Tan R. Study on the contact pressure between the lower leg and the top part of socks using finite element method. Jiang Nan university: 2011; 12.
[7] Chang LX, Zhang X, Qi J. Research on subdividing of female breast shapes based on 3-D body measurement .Journal of Textile Research: 2016; 12.
[8] Xu YY. The Research of Bra Cup Fitness Based on Finite Element Analysis. Zhejiang Sci-Tech university: 2015; 12.
[9] Lu L. Study on Digital Simulation of Bra's Pressure Distribution. Xi'an Polytechnic University: 2013; 12.
[10] Wu XN. Study on Clothing Pressure Distribution Based On the Relationship between Elastic Fabrics and Clothing Pressure. Shanghai University of Engineering and Science: 2015; 11.
[11] Sha SX. Study on Mechanical impact Resistance of Fabrics for Functional clothing. Soochow university: 2010; 5.
[12] Li Y, Zhang X, Yeung KW. A 3D Biomechanical Model for Numerical Simulation of Dynamic Mechanical Interactions of Bra and Breast during Wearing. Sen'I Gakkaishi: 2003; 59 (1): 46-55.
[13] Roberts JC, Merkle AC, Biermann PJ. Computational and experimental models of the human torso for non-penetrating ballistic impact [J]. Journal of Biomechanics: 2007; 40: 125-136.
[14] Chen XY. An Investigation into Clothing Pressure Comfort on Human Neck based on Cardiovascular Physiology. Jiang Nan university: 2010; 6.
[15] Zhou J, Yu W, Sun P. Identifying effective design features of commercial sports bras. Textile Research Journal: 2013; 14: 1500-1513.

Research on the Method of Establishing the FEM Contact Mechanics Model for the Human Body and Clothing

Fang-Fang Qiao, Bu-Gao Xu, Bo-An Ying *, Xin Zhang, Qian Ma

Apparel & Art Design College, Xi'an Polytechnic University, Xi'an, Shaanxi, 710048, China

*Corresponding author's email: yingba2006@163.com

Abstract

Digital simulation of pressure distribution between human body and clothing is a key technology to achieve digital design and evaluation for clothing. In this paper, the finite element method (FEM) is used to establish the contact mechanics model between clothing and the human body, as well as analyze the pressure distribution between the human body and clothing.

Firstly, the non-contact three-dimensional body scanner is used to obtain clothing and three-dimensional human body point cloud data. Both clothing and human body geometric models are established through processing the point cloud data, reconstructing surface and so on. Then, the finite element contact mechanics model of clothing and the human body is established. The model is used to simulate the static pressure of the human body after wearing the garment. In this model, the human body model is assumed to be an elastic object, while the clothing is regarded as an isotropic flexible material. In addition, the contact method between the human body model and the clothing is set to the normal surface of the human body. By defining the material properties, geometric properties, contact bodies and boundary conditions of the garment and the 3D human body, the model can be used to calculate the pressure distribution and variation in both the garment and the human body. The results can provide technical methods for the realization of digital clothing design and evaluation.

Keywords: Digitization; Contact Mechanics Model; FEM; Pressure Distribution

1. Introduction

Digital and information technologies are key to intelligent production and intelligent design. The application of information and digital technologies for garment simulation has become the main area of interest in garment stress comfort research. In recent years, domestic and foreign experts have increasingly been concerned about the application of the FEM in understanding garment pressure in the field of textiles and garments. The basic idea of the FEM is to divide the complex continuum into a finite simple unit body. The unit body is connected only by the node, while the complex problem is analysed by means of converting the whole into parts and gathering the parts into the whole [1]. The FEM has a great advantage over other biomechanical research methods and been widely used in various engineering fields in recent decades [2].

Zhang et al. [3] used a FEM to establish a mechanical model that was used to simulate the dynamic pressure generated by a garment during the wearing process. In this paper, the garment was treated as a non-linear elastic shell, while the body was regarded as rigid body. The mode of contact between the human body and clothing was achieved by dynamic sliding. This model can also be used to calculate the pressure distribution in the body during the wearing process, as well as the dynamic distribution and stress-strain variation in the garment, soft tissue and skin.

Yu et al. [4] created a software dummy model for the study of clothing pressure, in which silica gel was used to simulate the human skin, and spandex foam was used to simulate human soft tissue. The enhanced glass fiber was used to simulate human bones, while nine different types of body tights, as the object of study, were used to measure the pressure of the tights, both on the real body and on the dummy, after which the two sets of data were compared and correlated.

Yeung et al. [5] established a 3D mechanical model for describing the dynamic balance of the human and clothing. The study proposed a geometric non-linear mathematical model to simulate the dynamic clothing pressure distribution during the wearing process of the human body. The model simulated the wearing process of the pants from the waist to the foot. The results showed that the model can describe a series of mechanical contact modes with regard to the dynamic contact surfaces, such as garment deformation and body deformation. However, as a powerful computer must be used to complete the calculation due to the complexity of the calculation process of the model, the promotion of the practical application of this model is still challenging.

Zhang et al. [6] established a foot-and-sock contact model using the FEM. In the model, the foot was made of bones and soft tissue, while the socks were considered to be elastomers. The study simulated the process of wearing socks and predicted the pressure and stress distribution of the socks with this model. By comparing the predicted values of the pressure at several points with the measured values, it was confirmed that the model could simulate the wearing state of the socks and predict the pressure values applied by the socks to the feet.

Research on the pressure distribution with regard to the degree of fit, style and pressure of 3D clothing may be achieved if the FEM is applied to the contact model of a 3D human body and clothing. As for the human FEM, which was established in the past, there remain some gaps, compared with the actual situation under the constraints of many boundary conditions. Based on the understanding of the contact mechanics model, we establish the contact mechanics model between the 3D human body and the clothing, and analyse the pressure distribution of the 3D human body under large strain conditions. This paper can provide a digital evaluation method for the study of garment pressure comfort.

2. Establishing Geometric Models of the Human Body and Clothing

The object modelling technique has always posed a difficult problem in the process of the finite element structural mechanics simulation of organisms. The organization or topology of the organism is very complex, so the accuracy of the geometric model directly affects the calculation process of finite element analysis and the accuracy of the results [7]. Reverse engineering is an effective way to solve such complex modelling. This paper will introduce the construction method of human and garment geometric models in detail.

2.1 The Digital Measurement of Human Body and Clothing

3D human surface reconstruction is an important part of 3D human body measurement. This article refers to the basic version of clothing in the literature [8] as a scanning object. As for the human body as a scanning object, the data in the paper come from the non-contact 3D human body scanning system, known as Vitus Smart, from Germany. It can be used to scan the human body with multiple angles and orientations in 20 s through the laser method, while moving the laser vertically and uniformly to sweep the surface of the target object. The cloud data obtained by scanning have the characteristics of uniform horizontal distribution and high precision [9]. But, during the scanning process, there may be some bad points due to the slight vibration of the scanning equipment, light changes and other reasons. In addition, the scanning system can only sweep the surface of the human body data, resulting in some notches appearing on some parts of the body, such as the armpit and crotch. The data need to be processed further. The part of the point cloud data on the 3D human body is as shown in Fig.1. The part of the point cloud data in clothing is as shown in Fig.2. In the picture, you can clearly see some bad points in the chest and back.

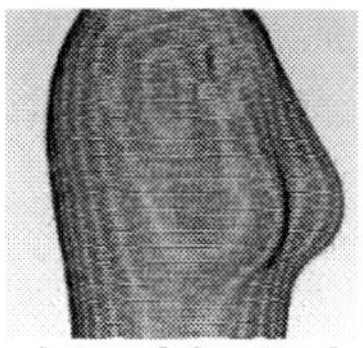

Fig.1 Scanning data of the 3D human body

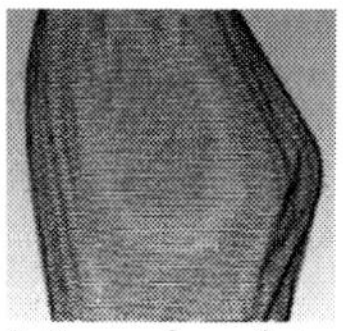

Fig.2 Scanning data of 3D clothing

2.2 Establishing Geometry Model of the 3D Human Body

The establishment of a 3D human body geometry model involve three stages: point cloud stage, polygon stage and surface forming stage.

The optimized point cloud data can more accurately reflect the actual human body than the original point cloud data. There are three main steps to process the data in the point cloud stage: reducing bad points, disconnecting the components, and filtering out the solitary points. The system will automatically deal with the point cloud by reducing the noise function. The fewer noise points obtained, the more accurate the cloud data. When disconnecting the components, the number of non-connected items should be less than 5% of the model points to ensure the integrity of the point cloud data and to remove the unconnected point cloud. When filtering out the solitary point, the sensitivity has to be set at 100. Sensitivity means setting a movement limit for the system in order to constrain the number of selected points. The higher the sensitivity, the more accurately the system removes the solitary points. As the optimized and coloured chest point cloud data in Fig.3 show, they are no more homogeneous than the original data, while there are no external interference data. The result of the optimization can be helpful for the post-processing of the point cloud.

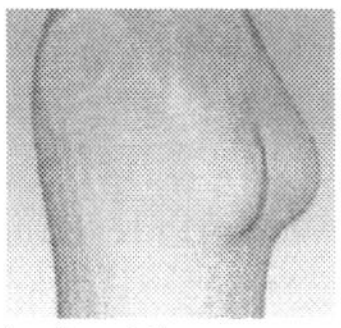

Fig.3 Optimized human point cloud

A human body surface composed of triangular meshes is obtained by encapsulating point cloud data after processing the original point cloud data. In the polygon stage, there are many commands to adjust the triangular mesh, which includes removing triangles and filling holes. In addition, other tools, such as redefining, filling holes, sharpening, plane sections, offset surfaces, thickened surfaces, and slack and fit boundaries, can also improve the surface and make the surface more complete. Polygon stage processing is very important because the higher the quality of the treated triangular mesh, the more accurate the human body surface is in the forming stage.

In this paper, the processing of the polygon stage mainly involves repairing the triangular mesh holes and simplifying the model. This can allow the missing point cloud data of the body to be completed by repairing the triangular mesh holes, based on curvature and single hole filling. The simplified model will reduce the number of polygons based on maintaining the main surface structure.

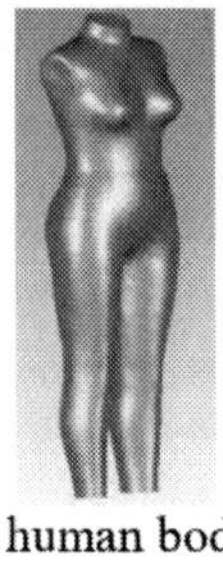

Fig.4 3D human body surface

It is not suitable for finite element calculation due to the large amount of triangular mesh data formed in the polygon stage, while it is necessary to obtain the human body surface with better quality through the shape stage. The processing of the body surface mainly includes making the surface fitting accurate, which means that the optimized triangular mesh is needed to automatically convert, using a computer, the optimized surface known as the human body surface. The 3D human body surface is as shown in Fig.4.

2.3 Establishing the Geometry Model of the Clothing

The clothing surface reconstruction reconstructs the surface of the point cloud data on the basis of the optimized point cloud. In addition, the same processes and parameters exist when the point cloud data are obtained for both the clothing and the 3D human body. The part of the optimized clothing's point cloud data is as shown in Fig.5.

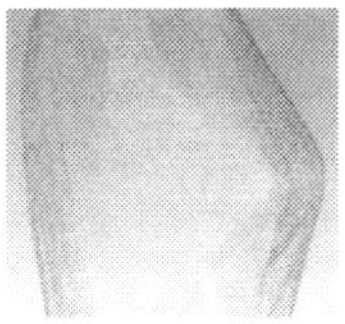

Fig.5 The optimized clothing's point cloud data

As the clothing surface is more complex, the way to reconstruct the surface, based on the optimized point cloud, is to rebuild the clothing surface. The process of surface reconstruction includes constructing 3D curves, constructing wireframes and reconstructing surfaces. According to the structure of the garment modelling, the key structural lines are established in the model, which contact the shoulder line, front centre line, rear centre line, collar arc, armhole arc, chest line, side suture, hip line and so on. The surface reconstruction process of the clothing's waist is as shown in Fig.6.

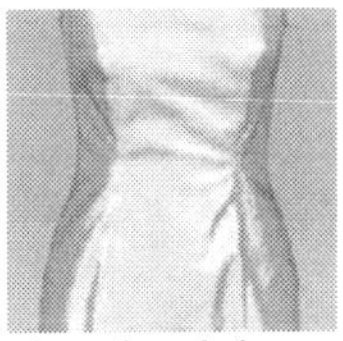

(1) Point cloud data of the waist

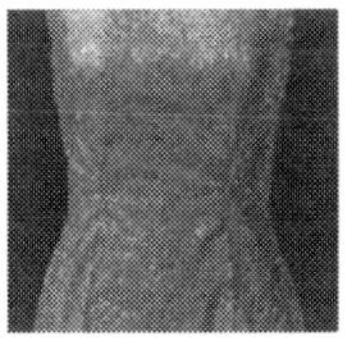

(2) Waist surface

(3) Horizontal line of chest and waist

(4) Curve line of the waist

(5) Waist surface reconstruction

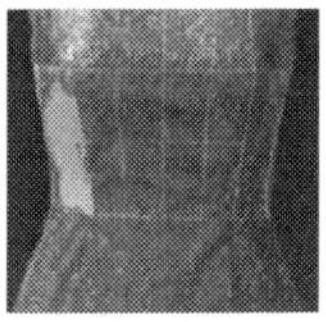

(6) Waist auxiliary lines

(7) Auxiliary wireframe of the waist

(8) Waist surface

Fig.6 The surface reconstruction process of the clothing's waist

In the process of surface reconstruction, it is necessary to extend the structure line based on the key structure line in order to reduce the error between the reconstructed surface and the scanning data. The principle of segmentation is based on horizontal lines and vertical lines, which can reflect the real clothing surface accurately. The reconstructed surface of the clothing is as shown in Fig.7.

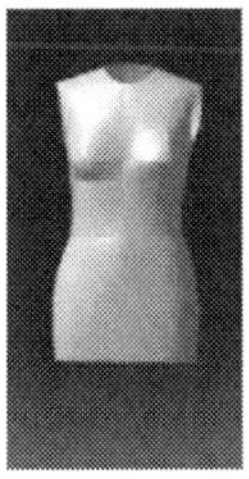

Fig.7 The reconstructed surface of the clothing

3. Establishing the FEM Contact Mechanics Mode for the Human Body and Clothing

3.1 The Assumption of Establishing the FEM Contact Mechanics Mode

In the process of finite element analysis, the accuracy and simplification of the model have great significance. The accuracy of the model determines the effectiveness of the analysis directly, while the simplification of the model can reduce the complexity of the model and save time for modelling and computing [10].

As the human organizational structure contacts bone, muscle, viscera etc., this will greatly increase the computational complexity, provided the human body structure is taken into account in simulation analysis. In this paper, the human body surface is used as a research object to simulate the human body in order to increase the speed of calculate. According to the literature [11], the linear elastic material and viscoplastic material model can express the human body property well, while rubber and linear elastic material have similar characteristics to the human body's surface. Thus, Mooney rubber is used to simulate the human body material in this paper. As the mechanical properties of clothing materials also have a certain impact on the simulation results, in this paper, the clothing materials are assumed to be elastic, uniform and orthogonally isotropic.

In summary, the contact simulation model of the 3D human body and clothing is based on the following assumptions:

1) The human body is assumed to be a rubber elastic and deformable body, which is considered to be of a shell structure with a certain thickness.

2) The garment material is assumed to be an isotropic elastic material, while the garment itself is considered to be of a shell structure with a thickness of 1 mm.

3) Human body surface pressure distribution is simulated by assuming that the garment is sagged by gravity.

3.2 The Method of Establishing the FEM Contact Mechanics Model for the Human Body and Clothing

The FEM contact mechanics model is divided into five stages: namely, defining geometric properties, defining material properties, assembling the human body and garment, setting contact conditions, and setting boundary conditions.

1) Defining geometric properties

Fig.8 3D human grid

Fig.9 Clothing grid

Fig.10 Quality check results for the human grid

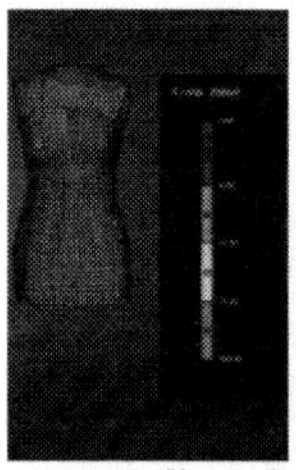

Fig.11 Quality check results for the clothing grid

In this paper, the 3D human body surface and the reconfigured clothing surface are set to a four-node shell element structure; the grid cell is 15 mm. The number of 3D human grid cells is 5,014, while the number of nodes is 5048. The clothing has 2,988 grid cells and 3,046 nodes. The 3D human grid is as shown in Fig.8, while the clothing grid is as shown in Fig.9. The results of the grid quality check for the human body and clothing are shown in Fig.10 and Fig.11, respectively.

2) Defining material properties

Biological tissues, such as long bones and ligaments, are always non-homogeneous, non-linear and viscoelastic [12]. Mooney material [13] has good flexibility and resilience, with similar properties to the human body. Therefore, the human body material is defined as Mooney-Rivlin material in this study in order to simulate the pressure distribution of the human body after contact with the clothing. The parameters "C10" and "C01" in the Mooney-Rivlin material indicate that the Rivlin coefficients are positive definite constants. The initial shear modulus, tensile modulus and strain energy density of the material can be calculated by these coefficients. For most rubbers, a reasonable approximation is obtained when the strain is within 150% [14]. The cheongsam is defined as an isotropic elastic material. The parameters of the human body material are as shown in Table 1, while the parameters of the cheongsam material are as shown in Table 2.

Table.1 The parameters of the human body material

Type	Number of elements	Element properties	Material properties	C10	C01
Human body	5,014	Shell	Mooney	-12	17

Table.2 The parameters of the clothing material

Type	Number of elements	Element properties	Material properties	Elastic Modulus (N/mm^2)	Poisson ratio
Cheongsam	3,297	Shell	Isotropic elastic material	175	0.3

3) Assembling the human body and garment

After establishing the finite element geometry of the human body and the clothing, it is necessary to assemble the human body and the clothing, which requires an assembly geometry model to conform to the state of the human body wearing the clothes, thereby ensuring the accuracy of the calculation results. The assembly drawings between the human body and clothing are as shown in Fig.12.

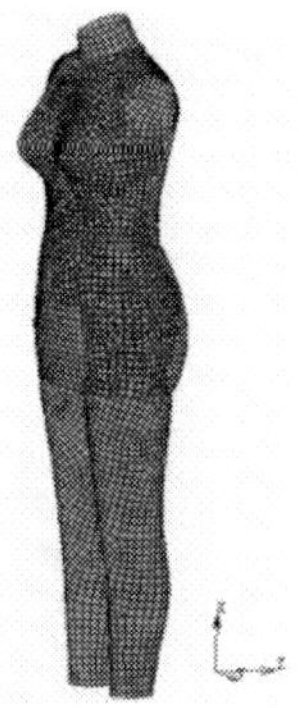

Fig.12 The assembly drawings between the human body and clothing

4) Setting contact conditions

The contact between the human body and the clothing is of a high degree of non-linear behaviour, while different modes of contact may lead to differing results [15]. Therefore, it is very important to take into account the flexible deformation of clothing and the contact surface uncertainty between the human body and clothing in the establishment of a contact mechanics model. In this paper, both the human body and clothing are set with regard to a deformable contact body, while the contact relationship applies gravity to the clothing in order to achieve contact with the human body.

5) Setting boundary conditions

The boundary condition is the condition that the solution of the equations for the motion boundary should be satisfied. In the finite element analysis, the correctness of the boundary condition is directly related to the correctness and rationality of the results [16]. During the setting of the boundary conditions in this article, the human body is fixed in the vertical direction to prevent displacement. It is necessary to apply gravity to the garment in order to simulate the contact and fix the clothing openings to prevent penetration. All boundary conditions are set as follows: 1) fixing the neck and feet of the human body in the x direction, 2) fixing the clothing collar in the x direction, 3) fixing the cheongsam shoulder in the x, y, z direction, and 4) setting the gravity of clothing. The boundary conditions are shown in Fig.13.

Based on the theoretical basis of elastic mechanics and FEM, this paper establishes a 3D human body and garment contact mechanics model in order to analyse the materials, geometric properties and contact condition of the 3D human body and clothing. The modelling method described in this paper is widely applicable to all kinds of finite element pressure analysis between a 3D human body and clothing.

Fig.13 The boundary conditions

4. Contact Mechanics Model Simulation and Calculation

The contact mechanics model established in this paper analysed the pressure distribution between the 3D body and clothing under the condition of a large strain simulation. Clothing displacement deformation, under the action of gravity, allowed us to analyse the clothing fit and the pressure on the human body. The clothing displacement deformation diagram is as shown in Fig.14, in which the ribbon shows both the size and the direction of displacement. The results show that the clothing's left shoulder droops to a greater extent than the right one, the chest has a smaller offset in the Y direction, and body parts below the waist swing within a larger range. Meanwhile, the digital evaluation for the basic version of clothing design allows us to analyse the displacement deformation of clothing under the action of gravity. The layout of the human body pressure distribution is as shown in Fig.15, with the ribbon showing both the size and the direction of the pressure distribution. It is clear that the human shoulder and neck are subject to greater pressure, while, in the chest, there is a small gathering effect. The model's analysis of human pressure reflects the comfort of the body after wearing clothing.

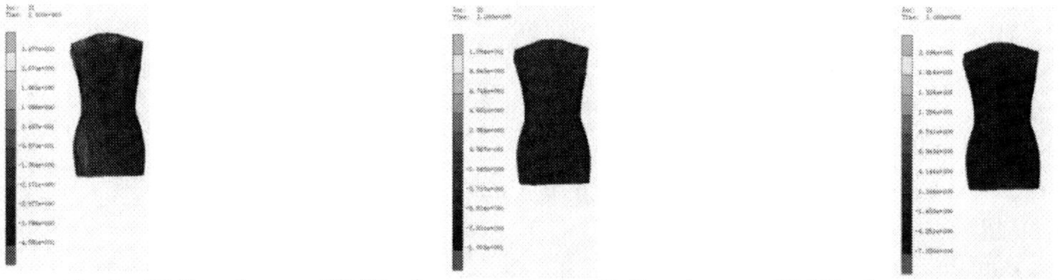

(1) Displacement in the X direction (2) Displacement in the Y direction (3) Displacement in the Z direction

Fig.14 Clothing displacement deformation diagram

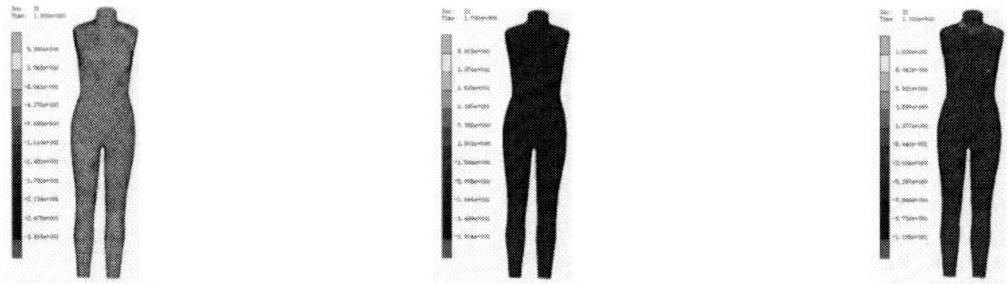

(1) Pressure in the X direction (2) Pressure in the Y direction (3) Pressure in the Z direction

Fig.15 Pressure distribution of the human body

5. Conclusion

1) This paper establishes a geometric model of the human body and clothing, based on the results measured by a non-contact 3D measuring instrument. In the process of garment geometric modelling, it is necessary to consider the way in which the structural line of clothing divides, based on the clothing plate. In addition, it is important to

reduce the error between the reconstructed surface and the measurement data by adding horizontal and vertical lines on the basis of the key line of clothing. In this paper, the 3D human body and clothing geometric model can be used for garment simulation, 3D human body measurement and so on.

2) This paper explores the establishment method of a finite element contact mechanics model for a 3D human body and clothing. The model is based on data obtained from human and real clothing scanning. During the course of the study, the human body and clothing were regarded as deformable bodies. In order to analyse the pressure distribution on the human body when wearing clothes, the clothing was assumed to be gravitational.

3) The FEM contact mechanics model of the human body and clothing established in this paper can provide a technical route for garment pressure comfort and garment version digital evaluation. The type of modelling method proposed in this paper can link reality with simulation, which in turn offers a technical approach to simulated pressure evaluation prior to the mass customization of clothing, as well as provide a digital evaluation method for costume design and pressure comfort.

References

[1] Cheng XS, Zhang SB. A course in the theory of elastic mechanics and the finite element [M]. Beijing: China Metrology Publishing House: 2008.

[2] Roberts JC, Merkle AC, Biermann PJ. Computational and experimental models of the human torso for non-penetrating ballistic impact [J]. Journal of Biomechanics: 2007 (40); 125-136.

[3] Zhang X, Yeung KW, Li Y. Numerical simulation of 3D dynamic garment pressure [J]. Textile Research Journal: 2002; 72 (3): 245-252.

[4] Yu W, Fan JT, Qian XM et al. A soft mannequin for the evaluation of pressure garments on the human body [J]. Sen Gakkaishi: 2004; 60 (2): 57-64.

[5] Yeung KW, Li Y, Zhang X. A 3D biomechanical human model for numerical simulation of garment-body dynamic mechanical interactions during wear [J]. Textile Research Journal: 2004; 95 (1-6): 59-79.

[6] Li Y, Zhang X, Yeung KW. A 3D biomechanical model for numerical simulation of dynamic mechanical interactions of bra and breast during wear [J]. Sen Gakkaishi: 2003; 59 (1): 46-55.

[7] Zhang MC, Zhao WD, Li JT. Reconstruction of the finite element simulation model by a three-dimensional laser scanning system. Laboratory of Medical Biomechanics [C]. First Military Medical University: 2003.

[8] Zhang X, Ma Q, Ying BA. Research on the relationship model of a cheongsam based on personalized customization. Textile Bioengineering and Informatics Symposium: 2017.

[9] Zhu S. Research on the comfort of 3D human waist garment based on finite element analysis [D]. Wuhan Textile University: 2012.

[10] Tan R. Finite element study on the contact pressure between foot and socks. Jiangnan University [D]. 2011.

[11] Wang X. Computer simulation analysis of vehicle frontal collision with human body model [D]. Chongqing University: 2003.

[12] Untaroiu C, Darvish K, Crandall J. A finite element model of the lower limb for simulating pedestrian impacts [J]. Stapp Car Crash Journal, Vol. 49: (November 2005); 157-181.

[13] Huang JL, Xie GJ, Liu ZW. Finite element analysis of superelastic rubber based on Mooney-Rivlin and Yeoh models [J]. Rubber & Plastics Technology and Equipment: 2008; 34 (12): 22-26.

[14] Wang W, Deng T, Zhao SG. Determination of material constants in rubber Mooney-Rivlin model [J]. Special Rubber Products: 2004; 25(4): 8-10.

[15] Ji S. Research and simulation realization of 3D human body garment contact mechanics model [D]. Sun Yat-sen University: 2008.

[16] Shang XJ, Qiu F, Zhao HF. Advanced analysis method and example application of ANSYS structural finite element [M]. Beijing: China Water Resources and Hydropower Press: 2006.

Studies of Breast Movement Under Shapewear in Dynamic Conditions

Sha-Sha Liu, Jie Zhou*, Ao-Xue Wang, Hui-Hua Wen

Apparel & Art Design college, Xi'an Polytechnic University, No. 19 South Jinhua Road, Xian, Shaanxi, 710048, China

*Corresponding author's email: 446961933@qq.com

Abstract

Shapewear not only has the function of beautifying the body, it also has a certain impact on some parts of the body's movement stability. In order to study the motor function of shapewear, this study uses a three-dimensional motion capture system to explore the female body in dynamic conditions with different frequencies under the shapewear and when bare-breasted. By analysing the movement of the breast, a certain control effect is observed, which affects the different regions of the breast in different dynamic conditions. This provides a theoretical reference for the follow-up study of shapewear and also has some reference value for the design of ordinary underwear and sports underwear.

Keywords: Shapewear; Breast Movement; Function; Relative Displacement

Introduction

First of all, with improvements in living standards over time, human clothing is not just about covering and protecting the body, as it now involves more aesthetic pursuits related to shapeliness and beauty needs. Second, due to gravity, age-related growth in females and the demands of sedentary work, especially after breastfeeding, women experience changes, such as a drooping chest, waist and abdomen fat accumulation, and hip sagging. At present, scholarly research on body sculpting underwear is limited to the study of its fabric [1-5], auxiliary material [6, 7], pattern [8, 9], structure [10, 11], technology [12] and comfort [13-15]. As there is a lack of functional research on sports body underwear sports, this study is interested in the motor function of body sculpting underwear.

Gait is the basic movement of people in normal life [16]; therefore, this study is carried out with regard to different frequency gait movements. There are two kinds of breast movement: one is relative to the ground of the three-dimensional movement, which we call absolute movement; the other is relative to the body of the three-dimensional movement of the trunk, which we call relative movement [17]. Therefore, this study selected two points to mark: one is able to represent the relative movement in the marked area, while the other is used as a reference coordinate point. It is important to change the position of the markers on the surface of the body caused by soft tissue relative to bone movement in the mark on the bones [18]. The coordinate system determined by the reference points is the reaction of the rigid body, so the mark point should not be placed in the soft tissue of the body relative to the place where the bone will move; it also needs to be as consistent with skeletal movement as possible and involving the same bone [19]. Marking points reflect the movement of the soft tissue, such as the breast without bones, so that it is marked on the corresponding soft tissue. In this study, the human body template is divided into sub-templates with relative rigidity [20]. In the process of motion capture, the mark points are tracked and matched. When the tracking process encounters an error, it can be better to deal with the exchange problem of the markers.

Previously, researchers have studied the breast movement characteristics when nude and when wearing a sports bra in order to evaluate the protective effect of a sports bra on the breast [21-24]. Lorentzen and Lawson [25] studied female movement in jogging, indicating that 56% of women feel pain in the breast. This phenomenon is better illustrated in the course of the movement of women's breasts, which should receive professional protection. Yan Ying [26] studied the relative displacement and relative velocity of breasts in female patients in three stages: a bare breast, with an ordinary bra and with a sports bra. When an ordinary bra and a sports bra were worn, the breast's relative displacement and relative velocity curve shape did not change significantly; only the size of the value changed. This study explores the motor function of body sculpting underwear on the basis of previous research, while using a new coordinate system that describes the movement of the breast relative to the torso, given that the acquisition of mammal movement data from a motion capture system is relative to the Earth and cannot describe the movement of the breast relative to the torso [27].

1.Human Motion Capture Device

The motion capture system consists of two parts: one is hardware equipment, including a display, a camera (in the application, the camera is fixed relative to the scene and objects), software dongles, a reflective ball (for marking the corresponding parts of the human body) and a controller; the other is the Cortex software system. In this study, the motion analysis device was used to identify the movement of the ball on the subject's body. The capture principle involves the use of an infrared high-speed camera on the movement of the human body for continuous data collection (the capture frequency is 60 frames per second). Once a series of continuous data is obtained, it is transferred to Excel for processing analysis.

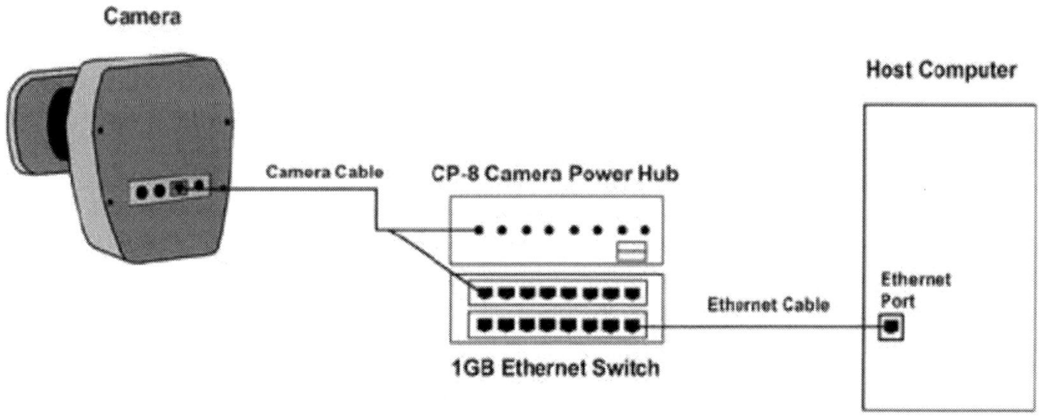

Fig.1 Part of the standard digital camera configuration

2.Research methods

2.1 Subjects

In order to make the results of this experiment more representative, this study chose five healthy women of different ages and in different stages of breast fullness. In turn, women aged from 24 to 48 years, with body mass indices ranging from 19.55 to 24.44 kg/m^2, volunteered to participate in this study. None of the breasts had received any operations affecting normal movement, while the bra sizes of participants were 90B, 85B, 85C and 80C. None of the women had been involved in breastfeeding or was pregnant. Before signing the consent form, the participants were fully informed of the experimental process to ensure that they knew their participation was voluntary. The women were also selected on the basis that their body shape was relative to the sizes of the shapewear available from the manufacturer (including essays, beam pants in the chest, and waist and back clips).

2.2 Research Environment

The experiment was conducted in winter, according to the most comfortable body temperature in winter, in a quiet laboratory room with the following conditions: temperature 23±0.5℃, relative humidity 65±3%, and wind speed less than 0.1m/s.

2.3 Selection of Marked Points

Table.1 The selection of the markers

Number	Definition
C7	Spinous process of the 7th cervical vertebra
IJ	Deepest point of the incisura jugularis (suprasternal notch)
T8	Spinous process of the 8th thoracic vertebra
PX	Processus xiphoideus, the most caudal point on the sternum
B1	Nipple
B2	Point of nipple lateral
B3	Point of nipple underside
B4	Point of nipple inside
B5	Point of nipple upper

Note: The nipple lateral point and the nipple inside point represent the points on the right and left, with the nipple level at 4 cm; the nipple upper point and the nipple underside point represent the points of the nipples pointing straight up and down by 4 cm.

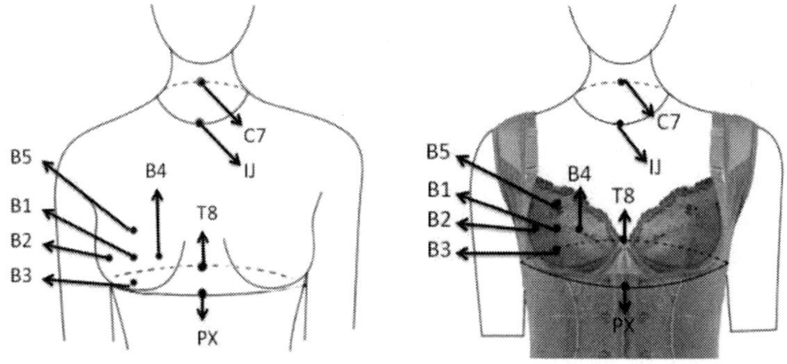

Fig.2-1 The position of the markers on the human body

As the coordinate system determined by the reference points is the movement of the bone, the mark point should not be placed on the soft part of the body as opposed to where the bone will move. There needs to be as much consistency as possible with the bones, and on the same bone. The breast has no skeletal support; during movement, when there is obvious shaking, the most representative breast movement involves the nipple. To explore the different aspects of breast movement differences, different quadrants of the breast were marked.

2.4 Experimental Procedure

a. Preparatory stage

Check that the equipment is functioning, including the device's online status, that each camera is intact. Subjects should be tested under two conditions: bare-breasted (dressed only in body-fitting briefs); and under the shapewear (equipped with firm shapewear technology). The experimenter readies the clothing for experiment. As the experiment requires a certain amount of exercise, a meal needs to be eaten by the subjects an hour before the test to avoid any physical discomfort. Subjects cannot wear any reflective jewellery, so as to avoid noise affecting the experimental results.

b. Template construct

Participants dress as required, while markers are attached to corresponding parts of the body. They are then stationed within the capture area as required in order to conduct static collection. The Cortex software is used to make connections with the markers in order to complete the template.

c. Static measurement

In order to avoid confusion about the markers of the experimental equipment at the time of capture, this study makes a rough manual measurement of the distance between the test points and the reference points for easy comparison and correction.

d. Motion capture

This study is mainly to study shapewear in the daily wear process on the breast stability, and therefore involves mild and moderate exercise. In order to unify the exercise intensity of different subjects during the course of the experiment, this study specifies the motion requirements. Subjects are required to step or run in three states: the first state of the running frequency is 120 hz/s; the second state of the running frequency is 140 hz/s; and the third state is the frequency is 140 hz/s while jumping with two legs. The purpose of selecting these three states is to compare the movement of breasts in both scenarios at the same frequency and the same movement at different exercise frequencies. The subjects are asked to exercise according to the set frequency beat using data obtained online. According to a predetermined frequency motion after motion stability for data collection, each participant's state was collected three times to reduce the error, with each acquisition take every minute. Everybody takes a rest for 5 min before each data collection to ensure that error caused by fatigue is reduced.

3.Experimental Results and Discussion

3.1 Static Capture

This is done by capturing the position of the markers on the subjects' body under shapewear and when bare-breasted. After dressing, the distance from B1, B2, B3, B4 and B5 to the reference point IJ is reduced, so the shapewear on the breast position has a promoting role.

3.2 Dynamic Capture

The degree of breast movement in different frequencies is represented by a change in DRD (DRD: dynamic RD, RD: as per formula (1)) and the distance from markers on the breast to the reference point IJ (X, Y, Z are the coordinates of the reference point IJ).

$$RD = \sqrt{(x_i - X)^2 + (y_i - Y)^2 + (z_i - Z)^2}, (i = B1, B2.B3, B4, B5) \quad (1)$$

(RD: relative displacement)

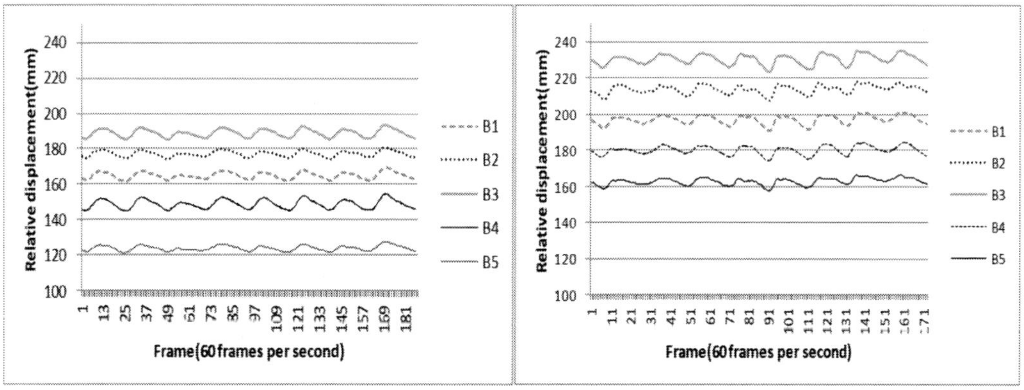

Fig.3-1 DRD change trends of the markers on the breast of the subject under shapewear (left) and when bare-breasted (right) in the first state

As we can see from the contrast between Fig.3-1, Fig.3-2 and Fig.3-3, B1, B2, B3, B4 and B5 under shapewear and when bare-breasted are consistent with the overall trend of relative motion. After dressing, the relative displacement trend of B1, B2, B3, B4 and B5 is smoother and the amplitude is reduced; thus, under shapewear, when compared with the bare-breasted state, there is a decrease in the frequency of slight wobbles of the breast.

The data from Tables 3-1 to 3-5 were from subjects with bra type 90B.

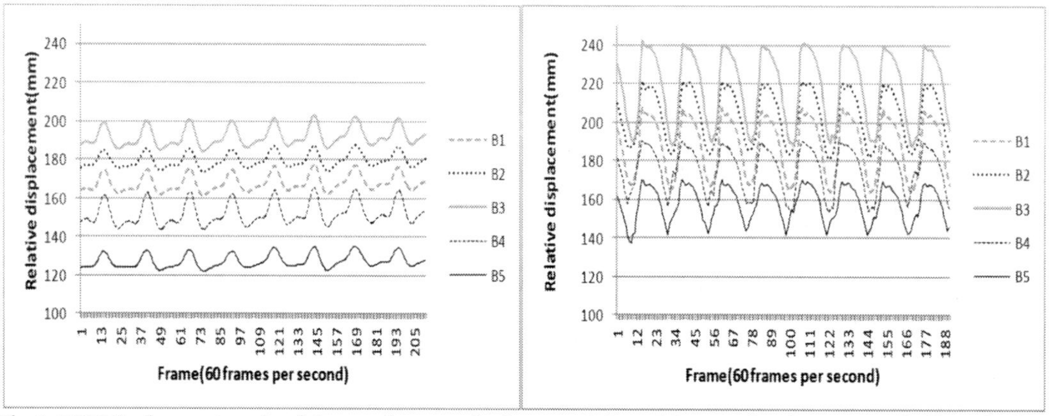

Fig.3-2 DRD change trends of the markers on the breast of the subject under shapewear (left) and when bare-breasted (right) in the second state

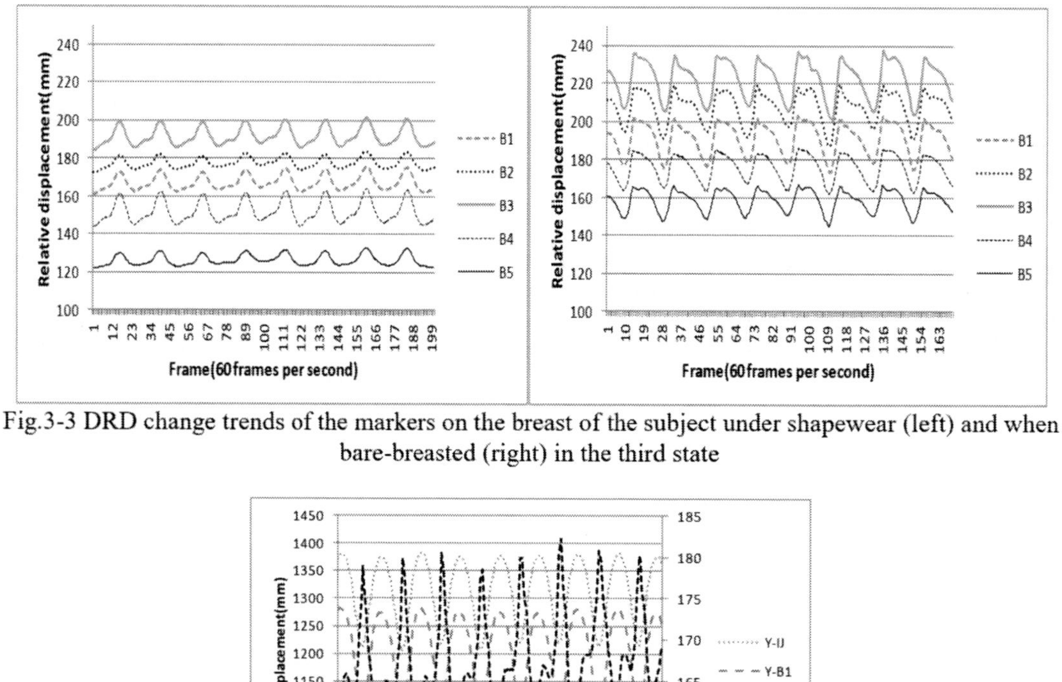

Fig.3-3 DRD change trends of the markers on the breast of the subject under shapewear (left) and when bare-breasted (right) in the third state

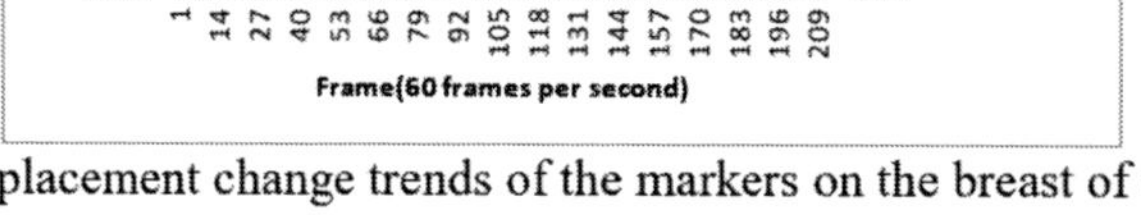

Fig.3-4 Displacement change trends of the markers on the breast of the subject

Fig.3-4 shows that, in the Y-axis direction of the breast with the IJ point of the movement of the trend, there is no significant difference, indicating that the movement of the breast with the trunk movement trend is consistent. The change trend of the displacement from B1 to IJ is opposite to the absolute movement of B1in the Y-axis direction. When the breast in the Y-axis direction of movement reaches the highest point, the displacement from B1 to IJ is the smallest; when B1 in the Y-axis movement is touching the bottom, the displacement from B1 to IJ is the largest, which can be seen in the impact with the ground, where the displacement from B1 to IJ is the largest. (Y-IJ: the displacement of IJ in the direction of the Y-axis direction, Y-B1: the displacement of B1 in the direction of the Y-axis direction, B1-IJ: the displacement from B1 to IJ.)

$$\overline{DRD_{\max}} = \frac{\sum_{k=1}^{n} DRD_{\max k}}{n}, (n = 1.,2,3,\ldots,8) \tag{2}$$

$$\overline{DRD_{\min}} = \frac{\sum_{k=1}^{n} DRD_{\min k}}{n}, (n = 1,2,3,\ldots,8) \tag{3}$$

$$VORD = \frac{\overline{DRD_{\max}} - \overline{DED_{\min}}}{SRD} \times 100\% \tag{4}$$

The VORD under the shapewear (VORD: variation of rate of dynamic relative displacement) is calculated using formula (4), while SRD (static RD) and RD are calculated using formula (1).

When the subjects under shapewear exercise as required, the VORD of the breast changes as the beat frequency increases. It can be seen that, with the change in movement intensity, the degree of breast movement relative to the trunk changes. As such, we learn that there is a different impact on the breast when there is the same exercise frequency of the different motion state of breast movement.

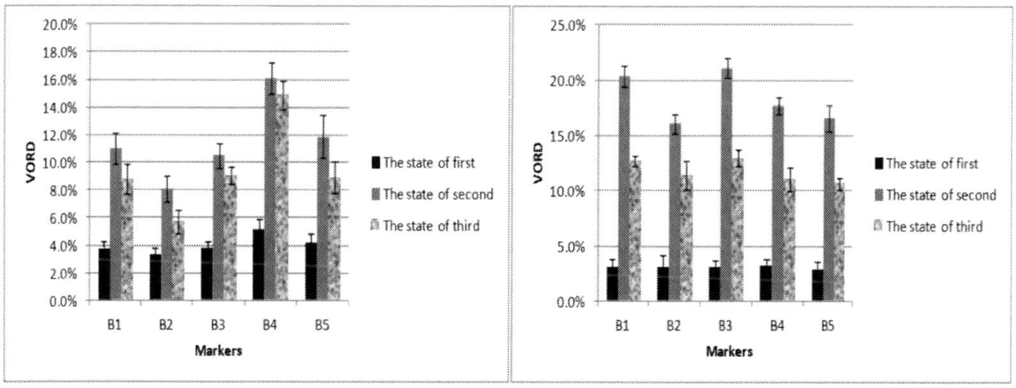

Fig.3-5 The VORD of each marker when the subject is under shapewear (left) and bare-breasted (right) in three states

We can learn from Fig.3-5 that the VORD of B1 and B3 is the largest in the three motion states when the subject is bare-breasted, indicating that the stability of B1 and B3 is the lowest in this state. The VORD of B2 and B5 is the lowest, while the stability of B2 and B5 is the highest in this state. The VORD of B4 and B5 is the highest among the three kinds of motion conditions under shapewear, indicating that the stability of B4 and B5 is the lowest in this state. The VORD of B2 and B3 is the lowest, which shows that B2 and B3 have the highest stability in the three motion states. In summary, the VORD of each region of the breast is different in the same state of motion. The same area in a different motion state is also different, while the shapewear on the stability of all markers has different effects. Previously, researchers have studied the chest, finding that the movement of the nipple (B1) and the lower nipple (B3) is greater than the other points when the subject is wearing a general bra and sports bra.

Combined with the findings of this study, the effect of shapewear on the control of breast movement is different between the ordinary bra and the sports bra.

When comparing the first state with the second state, the trend of breast movement is smaller. When under shapewear, breasts are more stable under low exercise intensity; as exercise intensity increases, the range of breast movement increases, while the rate of control breast movement (ROCM, calculated by formula (5)) under shapewear is excessive in the bare-breasted state at the same exercise intensity. In the second state, shapewear has the highest ROCM.

$$ROCM = \frac{VORD - VORD_0}{VORD_0} \times 100\% \quad (5)$$

($VORD_0$: the variation of rate of dynamic relative displacement on the bare-breasted subject.)

For the same size cup and different size of circumference under the breast, the VORD of the breast reduces when the size of circumference under the breast decreases. In the case of the same circumference under the breast, the VORD of the breast increases along with the cup size (as in Fig.3-6). With the increase in waist circumference, the VORD of W1 increases; and with the increase in exercise intensity, the control rate for the abdominal movement of a subject with a larger waistline increases. Meanwhile, the control rate of the VORD of the buttocks does not change much.

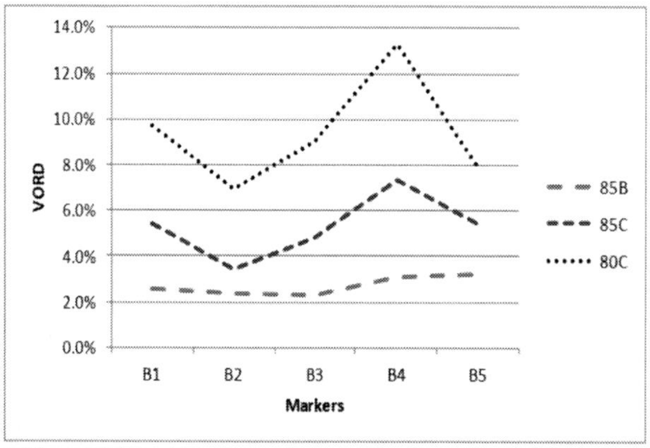

Fig.3-6 The VORD of each marker on the breasts of different subjects

4.Conclusion

This paper has studied the movement of female breasts in the state of exercise, and draws the following conclusions when comparing shapewear and the bare-breasted state:

1) Wearing shapewear can improve the female's breast. The impact before and after dressing of the breast on the overall movement of the trend is the same, while the breast avoids any movement in the form of frequent slight shaking. On wearing shapewear, the DRD of the breast is reduced, which shows that shapewear has a control effect on the movement of the breast.

2) With increases in the frequency of motion, the stability of the breast is reduced. The same motion frequency when running and bouncing has a different impact on breasts. When wearing shapewear before and after in each state of exercise, B2 is the highest point of stability of the chest. Body sculpting underwear's impact on the stability of different areas of breast control is different from that of a sports bra and an ordinary bra.

3) For the same size of cup and different sizes of circumference under the breast, as the size of circumference under the breast decreases, the stability of the breast under shapewear reduces; for the same size of circumference under the breast, the stability of the breast is reduced as the cup increases.

Acknowledgements

We would like to thank Nanjing Joymain Science & Technology Development Co. Ltd for funding this research.

References

[1] Lin L. Lycra Soft Raschel fabric in popular body underwear. J Knitt Ind: 2002; 3: 53-54.
[2] Wang MZ. Lycra - comfortable, easy, new experience. Wool Text Technol: 1999; (4): 32.
[3] Xing BA. Shirt underwear of China. Beijing: China Text Press: 1998; 177.
[4] Feng Y. Study on design of raschel lace fabric for underwear. Jiangnan University: 2014.
[5] Shen X, Feng Y. Study on pattern design of raschel lace fabric for body-moulding lingerie. J Nantong Text Vocational Technical College: 2014; 14 (4): 47-50.
[6] Li ZW. A plastic body with plastic blocks. China; 01215648.5.2002-01-23.
[7] Chang LX, Qi J, Zhang X. Sports corset both at home and abroad. Chin Text: 2004; 11: 164-171.
[8] Wang CC. Study of the solid cutting block pattern of adjustable shaping bra. J Knitt Ind: 2016; 02: 69-72.
[9] Wang CC, Pan HY. Study and application of the basic pattern of an adjustable shaping bra. J Knitt Ind: 2015; 5: 59-63.
[10] Yao YJ, Chen Y. Construction analysis of foundation garments. Int Text Guide: 2011; 8: 68, 70-71.
[11] Wang YQ. A study on the adjusted performance garment's structure innovation. Jiangnan University: 2010.
[12] Duan XY. Research on the function modification of re-shaping underwear. J Knitt Ind: 2007; 12: 28-30.
[13] Yin L. Study on pressure comfort in wearing a waist-nipper on the basis of a human's heart rate variability and electroencephalogram. Donghua University: 2012.
[14] Wang S. Research on the comfort of female body shaping underwear. Guangdong University Technol: 2012.
[15] Liu YP, Yan YX. Research of the relationship between stress comfort and shaping figure of seamless shapewear lingerie. J Zhejiang Inst Sci Technol: 2012; 29 (1): 48-52.
[16] Wu XX, Qian JG, Kang YL. Research overview on gait analysis technology and its application in hemiplegia rehabilitation. J Nanjing Inst Phys Educ (Nat Sci): 2016; 15 (3): 37-41.
[17] Yan Y, Zhou XL, Ren JP. Study on breast relative coordinate system in exercise. Chin Society Sports Biomech: 2016; 03: 37-41, 81.
[18] Hoschek J, Weber U, Ladstatter P, Schelske HJ. Mathematical kinematics in engineering of endoprostheses. Evaluation of the results of gait analysis. Arch Orthop Trauma Surg: 1984; 103 (5): 342.
[19] Su HT. Sports capture mark point placement program basic principles and application. Film Television Technol: 2005; 10: 13-16, 61.
[20] Wei XP, Liu Ri, Zhang Q, Xiao BX. Approach of human motion capture data processing based on template matching. J Syst Simul: 2010; 10: 2368-2372, 2391.
[21] Wang FY. Study on the comforts of sports bra by the combination of displacement and pressure. Donghua University: 2013.
[22] Li SJ, Ren JP, Yan Y, Zhou XL. The influence of a bra on the breast's kinematic parameters during walking. J Beijing Sport University: 2014; 7: 75-79.
[23] Zhou J, Yu W, Ng S P. Identifying effective design features of commercial sports bras. Text Res J: 2013; 83 (14): 1500-1513.
[24] Ren JP, Yan Y, Zhou XL. Research on the impact of sports bras on breast movement in walking and running with different speeds. J Cap Inst Phys Educ Sports: 2015; 27 (2): 172-177.
[25] Lorentzen D, Lawson L. Selected sports bras: a biomechanical analysis of breast motion while jogging [J]. Physician Sports Med: 1987; 15 (5): 128-139.
[26] Yan Y, Ren JP, Zhou XL. Study on the influence of Chinese chests on breast kinematic parameters in two-step jumping. Chin Society Sports Biomech: 2012; 2.
[27] Zhou J, Yu W, Ng SP. Studies of three-dimensional trajectories of breast movement for better bra design. Text Res J: 2012; 82 (3): 242-254.

Influencing Factors of Garments on Elbow Pressure

Liang-Su Zhen

Fashion and Art Engineering College, Min Jiang University, Fuzhou Fujian, China

*Corresponding author's email: 14679705@qq.com

Abstract

This paper adopted 4 kinds of common fabrics to produce garments with fit and body-fitted sleeves for testing. According to the test and data analysis, the main factors of garments that affect elbow pressure are fabric characteristics, body-fitted degree of garments at the elbow and the different elbow-joint motion. The greater the elasticity of fabric the smaller the elbow pressure exerted by the garment. For fabrics with smaller fiber-strength, the poorer the elasticity, the easier it is for the textile fabric to deform and will result in less elbow pressure from garments. For chemical fabrics such as polyester with larger fiber-strength and better shape-retention, the greater the elasticity, the smaller elbow pressure caused by garments. The thinner the fabric, the smaller the pressure on the elbow exerted from garments and the easier for garments to be deformed. Generally, the higher degree of body fitted garments at the elbow, the greater the pressure exerted at the elbow during elbow-joint motion. The findings will be of practical significance for the garment design and production by enterprises and purchase by consumers.

Keywords: Sleeves; Elbow Pressure; Pressure Test; Influencing Factors

1. Introduction

Clothing pressure is one of three factors to affect clothing comfort, which has become a primary research focus of many institutions and scholars at home and abroad. Denton and Harumi Morooka et al. analyzed the effects of the physical characteristics of the human body on clothing pressure, such as the body curvature, human body's elastic module, soft tissue structure, fat content and motion amplitude [1]. Noriko Ito, Harumi Morooka and Ayako Inamura et al. analyzed the effect of a fabrics' physical properties on clothing pressure and comfort from the aspect of fabric performance [2-4]. You Fang and Zhang Xin discussed the relationship between the garment fitness, elastic module of body-fitted garments and clothing pressure [5]. In recent years, scholars mainly focused on the effect of functional clothing pressure on human's health and wearing comfort [6-9]. There however has been a limited amount of research conducted to understand the influencing factors of local pressure exerted by the garment or the effect of local pressure on appearance characteristics of the garment. In particular, there is a lack of research on the influencing factors of local pressure on the knee and elbow parts exerted by garment and the bridging & deformation of garments caused by the pressure.

The pressure exerted on the elbow from a garment is an important factor to examine the wearing comfort of the garment, durability and shape retention [10, 11]. It is also critical for examining a garment's bulging at the elbow. When the arm moves, the garment will cause huge pressure at the elbow and resulting in the bulging. It will then directly affect the durability and beauty of the garment. Based upon the author's previous research, this paper adopted 4 kinds of common fabrics for test. This paper has measured the elbow's pressure when wearing garments with different fabrics. It has also analyzed influencing factors of the pressure. The findings will be of practical significance for the garment design and production by enterprises and the purchase by consumers, as well as offering the mechanical foundation for the further study of garment's bulging at the elbow.

2. Test

2.1 Garment Materials

This research adopts 4 kinds of common fabrics (cotton, linen, wool and polyester) to make garments with two kinds of sleeves, namely fit and body-fitted sleeves (the section girth of the subject's elbow is about 26cm, thus it chooses 28cm for the body-fitted sleeves at the elbow, while 30cm for the fit one. Because of limited pressure caused by the loosing garment at the elbow, the fit one will not be discussed in this paper.) The main specifications and parameters of fabrics are as in the following table.

Table.1 Main specifications and parameters of fabrics for test

Numbers of fabrics	Fabric content	Fabric structure	Thickness/ (mm)	Elastic elongation / (%)
1#	Cotton 100%	plain	0.401	16.28
2#	Cotton 100%	plain	0.384	10.31
3#	Linen 100%	plain	0.720	2.67
4#	Linen 100%	plain	0.521	2.36
5#	Wool 70%; Polyester 30%	plain	0.435	6.59
6#	Wool 70%; Polyester 30%	plain	0.722	2.51
7#	Polyester 100%	plain	0.718	12.17
8#	Polyester 100%	plain	0.713	2.38

Note: The elastic extension is 10N constant load.

2.2 Main Instruments

1) Pressure Tester. It adopts Japanese AM13037S-5 Clothing Contact Pressure Tester as shown in Fig.1. The instrument takes the test according to the principle of flow (air) pressure. The researcher will put the air-bag between the garment and the skin to be measured and transfer the pressure caused by the clothing weight or clothing deformation to the host machine through the air-bag. After being converted, the value can be shown on the display of the host machine, as the red figure in Fig.1.

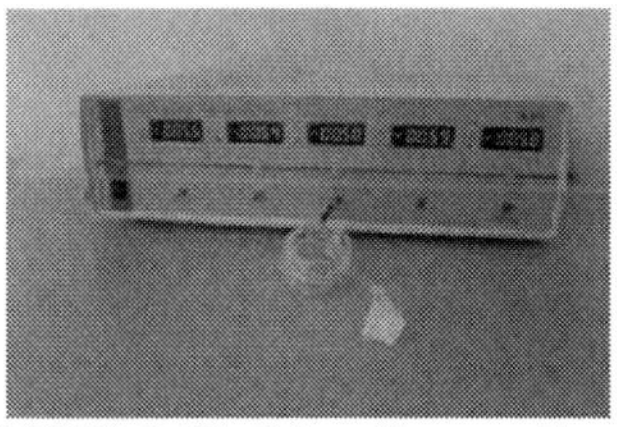

Fig.1 AM13037S-5 Clothing Contact Pressure Tester

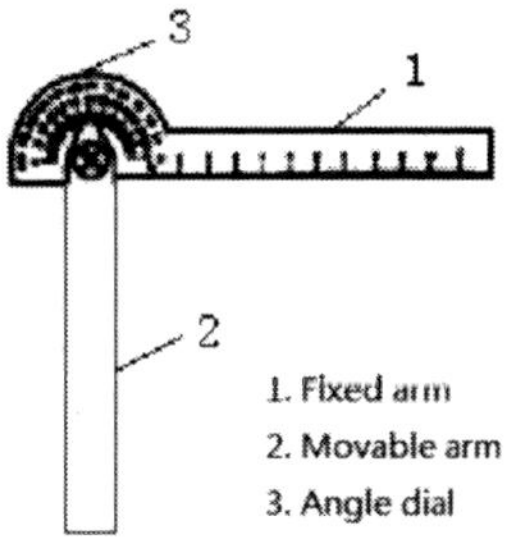

Fig.2 JKY/GS-102 Joint Angular Measuring Instrument

2) Joint Angular Measuring Instrument. The researcher will adopt the JKY/GS-102 Joint Angular Measuring Instrument which is also named as double-hinged joint angular instrument as shown in Fig.2. As the specialized instrument to measure the amplitude of joint motion, it consists of fixed arm, movable arm and angular dial. The angle on the angular dial where the tip of movable arm points is means the angle of joint motion.

2.3 Test Conditions

This test is done in the environment with standard atmosphere, temperature of 25℃ and relative humidity of 65%. Subjects are relaxed.

2.4 Test Procedures

This research is to test the static pressure, namely measure the pressure at the elbow caused by the garments made of different fabrics under different conditions of elbow-joint bending. The researcher will take the endpoint of ulna of elbow-joint (the projecting point of elbow-joint when the arm bends) as the measured point, and measure the pressure of 5 motion states of the arm from naturally hanging to bending (these 5 motion states are frequent for human's elbow-joint activity and cause the significant pressure to the elbow part of the garment). During the motion of elbow-joint, the bending angle cannot be randomly changed within 0-180°. Besides, according to the observation of test, the maximal bending angle of elbow-joint is about 170° (namely the included angle between the upper arm and forearm is about 10°, same as follows) and the frequent activity range is 0-120°. Thus the researcher sets the min measuring angle as 0° and the max one as 120°. Accordingly, the researcher will define 5 different motions of the arm from naturally hanging to bending as the bending angles of elbow-joint, namely 0°, 30°, 60°, 90° and 120°. Detailed steps are as follows.

1) To guarantee the accuracy of measurement, the researcher would maintain the measured point every time and make the subject bend his arm naturally. The researcher will mark the endpoint of ulna of elbow-joint with the marker.

2) The researcher will fix the air-bag on the marked point from step 1 using the adhesive tape and will not apply any pressure on the air-bag when fixing it, because it is the precondition to guarantee the accuracy of measurement data.

3) The researcher will ask the subject to wear the test garments and pay attention to keep from falling or damaging of air-bag when he wearing it.

4) The researcher will ask the subject to bend the arm and then use the Joint Angular Measuring Instrument to measure the angle of elbow-joint's bending. When the angle is 0°, 30°, 60°, 90° and 120°, the subject would maintain the posture of elbow until the data displayed on the pressure tester is stable and then the researcher will record the test data. To reduce the error, the test will repeat 3 times.

2.5 Data Treatment

Due to the accidental error and systematic error during the measurement, the researcher should preliminarily examine the measurement data at first and check the test data whether there is any error and then exclude the singular value or wrong data that may affect the results of analyses. SPSS, a Statistical Analysis Software will be adopted in this paper to analyze the measurement data of each group, and the singular values will be excluded and then the mean value obtained.

3. Results and Discussion

3.1 Effect of Garment's Fabrics on Elbow's Pressure

3.1.1 Fiber Composition of Fabric

Choose 3 samples (3#, 6# and 8#) with the similar elasticity and thickness for the comparison, as shown in Fig.3. According to the Fig.3, when the bending angle of elbow-joint is 0°-90°, there is little effect of different fiber compositions on the elbow's pressure from garments. But when the bending angle of elbow-joint is greater than 90°, there will be different effect of different fiber compositions on the elbow's pressure, with the order of polyester, linen and wool respectively.

The main reason is that: as for the elasticity of fibers, polyester has the poorest elasticity, followed by linen and wool; but for the surface smoothness of fabric, polyester is the smoothest and then linen and wool. When the bending angle of elbow-joint is small (within 90°) and the elbow's pressure caused by the garments is small, though the elasticity of polyester and linen is poor, the fabric surface is relatively smooth, and then during the elbow-joint motion the fabric will produce the certain slippage and thus the change of curved surface will be relieved to the certain extent to reduce the pressure accordingly.

While when the bending angle of elbow-joint is large and the elbow's pressure caused by the garments is large, the slippage of fabric stops and the fabric surface is applied by the frontal pressure of its surface deformation. Meanwhile, as the bending angle increases, the pressure will increase, and the elasticity of fabric fiber plays the critical role, namely the larger the elasticity is, the smaller elbow's pressure caused by the garment will be.

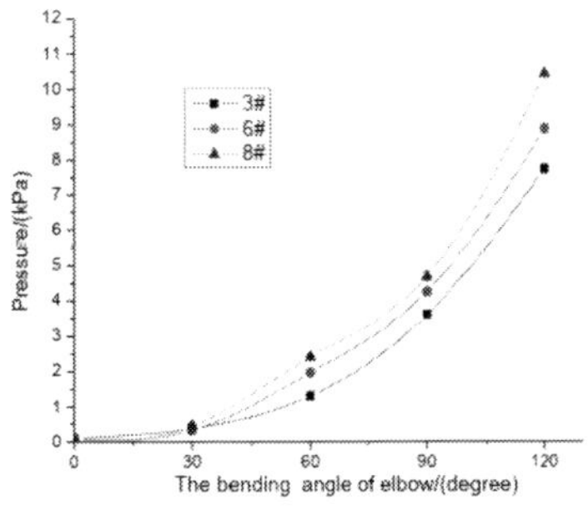

(a) Fit sleeves

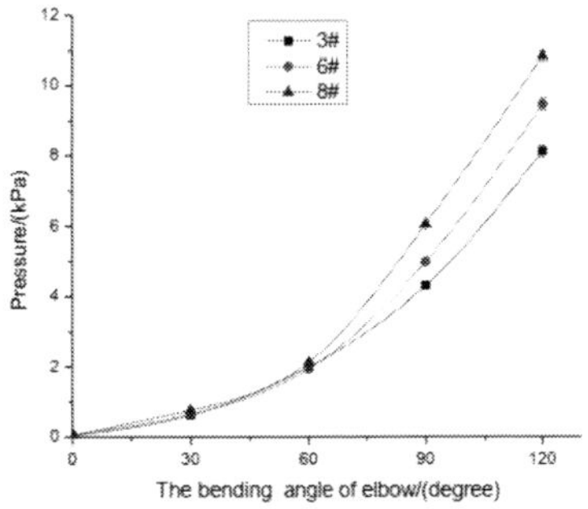

(b) Body-fitted sleeves

Fig.3 Diagram of elbow's pressure from garments with different fibers

3.1.2 Elasticity of Fabric

The researcher would choose samples with different elasticity, but same fiber composition and similar thickness (sample 1# and 2#, 7# and 8#) for the analysis.

1) Cotton Fabric

Fig.4 shows the elbow's pressure diagram of cotton fabrics with different elasticity. The elasticity of sample 2# is smaller than that of sample 1#. According to Fig. 4, there is a positive correlation between the elasticity of cotton fabric and the elbow's pressure. The reason is that the natural fibers, such as cotton, possess the small strength, as well as the small tensile strength and elastic resilience, and these fabrics are easily deformed.

As the elbow-joint bending angle is within 30° and the elbow's pressure is small, there is limited effect of elasticity on elbow's pressure, with the same pressure to both of them basically.

As the elbow-joint bending angle increases and the elbow's pressure from garments increases, due to larger elasticity, sample 1# garments can not only offset part of the pressure, but also maintain the shape, and the elbow receives the normal pressure; while sample 2# garments with smaller elasticity have the unrecovered bridging and deformation, so the elbow's pressure from the garment is reduced.

Therefore, when the garment enterprises design and produce, and the consumers purchase cotton clothes, they should realize that the proper elasticity will be of critical significance to maintain the shape of garments under the comfortable pressure.

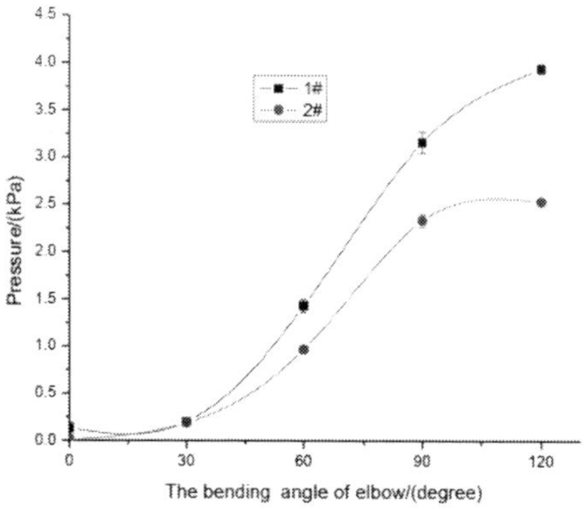

(a) Fit sleeves

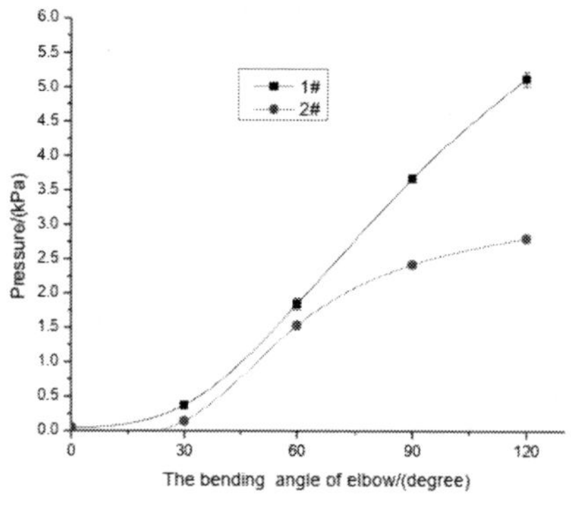

(b) Body-fitted sleeves

Fig.4 Diagram of elbow's pressure from cotton garments with different elasticity

2) Polyester Fabric

Fig.5 shows the elbow's pressure diagram of polyester fabrics with different elasticity. Compared with Fig.4, there is a negative correlation between the elasticity of polyester fabric and the elbow's pressure, the larger the elasticity is, the smaller the elbow's pressure caused by garments will be, and vice versa. The main reason is that the polyester fiber has large strength and its fabric has large tensile strength, making the fabric uneasy to be deformed. Thus some sports-wears prefer to the material of chemical fiber such as polyester. After the treatment of breathing and vapor permeability, as well as the blending of elastic materials, it can overcome the shortages of chemical fiber and also achieve the purpose of shape-retention and comfortable pressure.

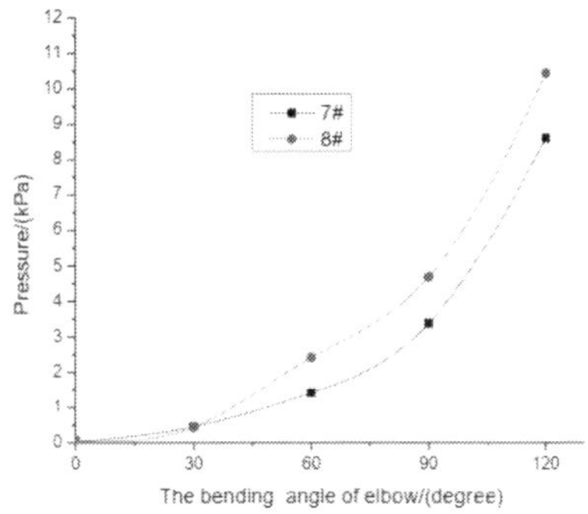

(a) Fit sleeves

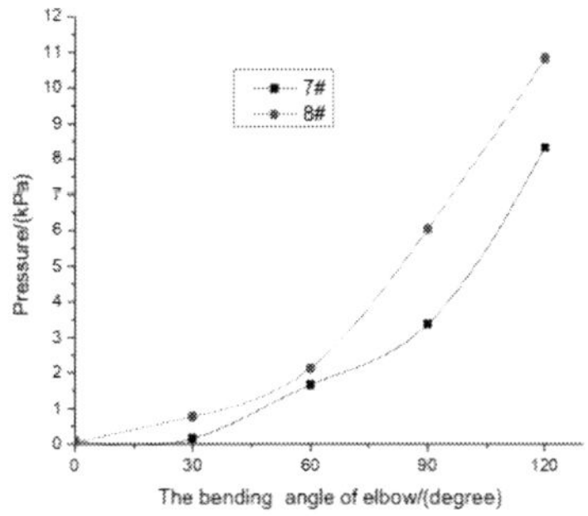

(b) Body-fitted sleeves

Fig.5 Diagram of elbow's pressure from polyester garments with different elasticity

3.1.3 Thickness of Fabric

The researcher would choose samples with the similar elasticity and same composition, but different thickness (sample 3# and 4#) for the comparison. There are two reasons to choose linen fabrics to analyze the influencing factor of fabric thickness: one is the limitation of test condition and materials, so it's impossible for us to test every kind of fabric; the other is the performance of linen fabric which is strongly representative. To be specific, its strength is comparable to the chemical fiber, while some other properties are similar with the natural fiber, such as cotton. As shown in the Fig.6, the thicker fabric will lead to the larger elbow's pressure from garments during the motion of elbow-joint.

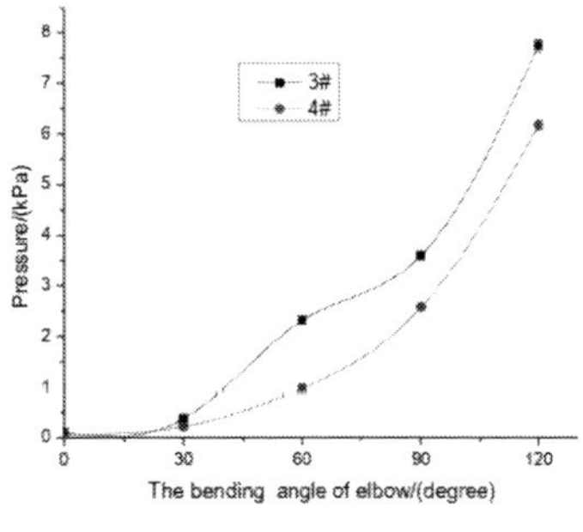

(a) Fit sleeves

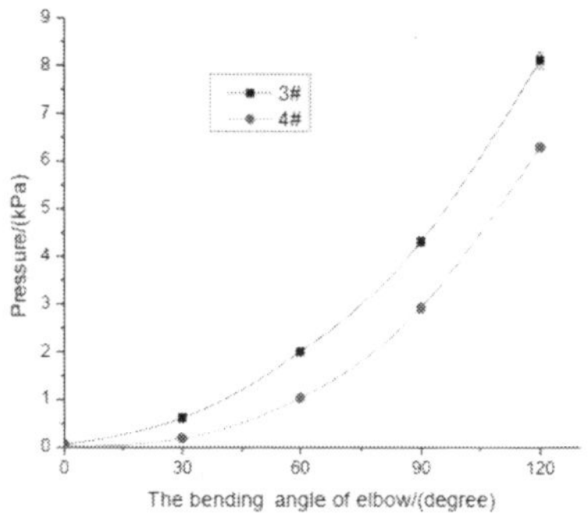

(b) Body-fitted sleeves

Fig.6 Diagram of elbow's pressure from linen garments with different thickness

3.2 Sleeve's Fitness

Taking sample 7# garments for an example, the researcher would analyze the elbow's pressure from garments with different sleeve's fitness. As shown in the Fig.7, during different states of elbow-joint motion, as long as the fabric is not deformed permanently, the higher body-fitted degree of sleeves and the larger pressure from garments will be applied to the elbow. Due to the limited space of this paper, garments with other fabrics will not be detailed.

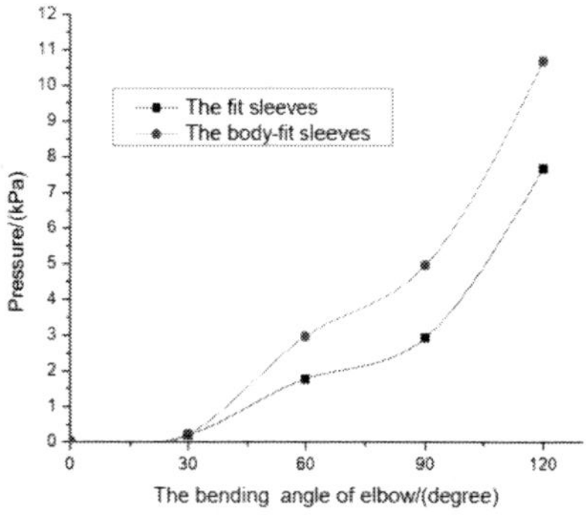

Fig.7 Diagram of elbow's pressure from wool-polyester garments with different fitness

Generally speaking, within the pressure scope that the textile has not been deformed permanently, there is a positive correlation between the body-fitted degree of garment and the pressure acting on the elbow, which is with the higher body-fitted degree of garment at the elbow, the larger pressure will act on the elbow during the different states of elbow-joint motion. While the pressure that makes the textile deformed permanently is closely related to the fiber strength and garment's body-fitted degree.

3.3 Elbow's Motion

According to pressure diagrams of Fig.3 to 7, within the pressure scope that the textile has not been deformed permanently, the larger bending angle of elbow-joint during the motion will lead to larger pressure applied to the elbow from garments regardless of the different fabrics and body-fitted degree.

4. Conclusion

According to the test and data analysis, the elbow's pressure from garments is 0-10.84kPa; the main factors that affect the pressure include:

1) Fabric characteristics, including the fiber composition, elasticity and thickness. The better elasticity will lead to smaller clothing pressure acting on the elbow and vice versa. For the fabrics with smaller fiber- strength, the poorer elasticity, the textile will be easier to be deformed and the clothing pressure acting on the elbow will be smaller, and vice versa; for some chemical fabrics, such as polyester, with larger fiber-strength and better shape-retention, the better the elasticity is, the smaller the clothing pressure acting on the elbow will be. The thinner the fabric is, the smaller the clothing pressure acting on the elbow will be and easier for the garment to be deformed.

2) Body-fitted degree. Generally, within the pressure scope that the textile has not been deformed, the higher the body-fitted degree of garment at the elbow is, the larger clothing pressure acting on the elbow during elbow-joint motion will be.

3) Elbow motion. The larger bending angle of elbow-joint during the motion will lead to larger clothing pressure acting on the elbow.

Furthermore, the surface smoothness of fabric also has the certain impact on the clothing pressure at the elbow. Within the certain scope of clothing pressure, the smoother the fabric surface is the easier for the garment at the elbow to be slipped and the smaller the clothing pressure acting on the elbow will be.

5. Acknowledgement

The author wishes to acknowledge the research project "Research on the Women's Bra Design and Pressure Comfort Based on 3D Human Body Measurement (JAT160384)" from Fujian Education Department.

References

[1] Morooka H, Fukuda R, Nakahashi M, et al. Clothing pressure and wear feeling at under-bust part on a push-up type brassiere [J]. Sen'i Gakkaishi: 2005; 61 (2): 53-58.

[2] Ito N, Inoue M, Nakanishi M, et al. The relation among the biaxial extension properties of girdle cloths and wearing comfort and clothing pressure of girdles [J]. Jpn Res Assn Text End Uses: 1995; 36: 102-108.

[3] Morooka H, Nakahashi M, Morooka H. Compressive property of legs and clothing pressure of pantyhose from the view point of difference in age [J]. Jpn Res Assn Text End Uses: 199; 38: 44-52.

[4] Ayako I, Masae N, Masako N. Relationship between wearing comfort and physical properties of girdles [J]. Jpn Res AssnText EndUses: 1995; 36: 109-118.

[5] You F, Zhang X. Predictability of pressure sensation by the garment's fitness and fabric's extensibility for tight-fitness [J]. Journal of Northwest Institute of Textile Science and Technology: 2000; 14 (2): 133-137.

[6] Yin L, Zhang WB, Xia L. Review and analysis of pressure comfort of garment [J]. Journal of Textile Research: 2008; 29 (3): 137-142.

[7] Chan AP, Fan J. Effect of clothing pressure on the tightness sensation of girdles [J]. International Journal of Clothing Science and Technology: 2003; 15 (5): 320-334.

[8] Yu W, Fan J, Qian XM, et al. A soft mannequin for predicting girdle pressure on human body [J]. Sen'i Gakkaishi: 2004; 60 (2): 57-64.

[9] Duan XY, Yu WD. Research progress in pressure comfort of foundation garment [J]. Journal of Textile Research: 2007; 28 (11): 144-148.

[10] Institute of Textile and Clothing of Hong Kong Polytechnic University. Dress bulging and mechanical engineering design [M]. Beijing: China Textile & Apparel Press: 2002; 10: 16-19.

[11] Li Y, Zhang X. Mechanical sensory engineering design of textile and apparel products [J]. Journal of Textile and Institute: Part 2; 2005; 93 (2): 54-67.

Review on Prediction of Physiological Comfort in Clothing

Inese Šroma[1*], Inese Ziemele[2]

[1, 2] *Institute of Design Technologies, Faculty of Material Science and Applied Chemistry, Riga Technical University, Kipsalas Str.6, Riga, LV-1048, Latvia*

[*]Corresponding author's email: inese.sroma@edu.rtu.lv

Abstract

It has been determined that the comfort in clothing is evaluated subjectively and depends on the consumer's both positive and negative senses which the consumer receives through his/her sensory system. It has been established that the consumer's subjective comfort in clothing is influenced by physical, physiological, neurophysiological, and psychological processes. Factors that make an impact on wearer comfort in clothing have been studied. The factors interact, form and influence the senses and perception of the individual. Indicators that have an influence on the physiological subjective comfort in clothing have been determined and hypothetically grouped. The prediction algorithm for physiological comfort has been created with the aim to predict the wearer's level of physiological comfort in clothing in its guaranteed period of use while performing specific work tasks in a certain ambient environment. Based on the developed algorithm and factors that influence comfort in clothing, it is envisaged to develop a classification of bipolar descriptors that depict a consumer's physiological comfort.

Keywords: Comfort in Clothing; Physiological Comfort; Senses; Influencing Factors

1. Introduction

From theoretical research, it has been determined that it is difficult to give a qualitative definition of the concept "comfort in clothing", because it is a complex combination of interrelated subjective senses. The combination of such senses is influenced by certain environmental conditions, different factors and their mutual combinations. Keith Slater defined comfort as "a pleasant state of physiological, psychological [1, 2], neuro-physiological [2] and physical harmony between a human being and the environment" [1, 2].

Upon reviewing and analysing existing explanations of comfort in clothing, it has been concluded that the comfort in clothing is also:

1) subjectively evaluated and depends on the consumer's senses that arise wearing a definite clothing under certain ambient conditions [3];

2) both positive and negative subjective senses, which the consumer receives through his/her sensory system and can be described by bipolar sensory descriptors [3], such as lightweight - heavy, smooth - rough, stiff - soft, hot - cool, etc. [1, 2];

3) both thermal and mechanical interaction between the body and clothing [1, 2];

4) a wide range of physical, ergonomic, bio-sensory, and other indicators, some of which can be expressed in physically measurable values [3].

Therefore, in using standardized research methods and equipment, it is possible to predict the wearer's comfort level in clothing which can be altered depending on definite fabrics and their layers under certain ambient conditions [3].

According to Yi Li, the consumer's subjective comfort in clothing is influenced by simultaneous physical (in the clothing and the surrounding environment), physiological (in the body), neurophysiological (neurophysiological mechanisms of the sensory reception system of the body), and psychological (the processes of the brain, subjective perceptions) processes [1].

Pontrelli wearer's subjective comfort in clothing can be considered as the process which is influenced by physical (environment and the clothing variables), psycho-physiological (wearer parameters) and psychological (brain filters) variables values and their mutual interaction [1, 2].

Keith Slater identified that the ambient environment was important for comfort and defined physiological (the human body's ability to maintain life), psychological (the mind's ability to keep itself functioning satisfactorily with external help) and physical (the external environment's effect on the body) comfort [1, 2].

From theoretical research, it has been concluded that the subjective comfort in clothing of the consumer can be divided into:

1) physical (wearer's ease of body movement in statics and dynamics, garment's cut, weight, fit, fabric-skin contact including tactile properties of fabrics);

2) physiological (wearer's body physiology, thermoregulation, activity level, clothing fabrics thermal resistance and moisture vapour permeability/transmission) and

3) psychological (garment's price, cultural and social factors, aesthetic parameters – design, fashion, brand, exclusivity, colours harmony, wearer's social affiliation, culture, psychological status) comfort [1, 2, 4], the characteristics of which are interacting and forming definite senses [3].

According to Yi Li, thermophysiological comfort is "attainment of a comfortable thermal and wetness state", which includes "transport of heat and moisture through a fabric" [1, 3].

In turn, physiological comfort is defined "as the achievement of thermal equilibrium at normal body temperature with the minimum amount of bodily regulation" [2].

Thermal comfort is "that condition of mind which expresses satisfaction with the thermal environment" [2, 5].

It has been concluded that the wearer's physiological comfort in clothing is essential during both physically active and passive work under different ambient conditions.

It has been determined that each subjective sense arises when a definite receptor (differs by its sensitivity) receives a particular irritation [6, 7]. Based on irritation properties there are: mechanical (kinaesthetic, touch, and other senses), acoustic (rigid body vibrations), chemical (smell and taste), thermal and optical sensitivity [6, 7].

From theoretical research, it has been determined that not every irritation produces sensations. The irritation may be so weak that it does not cause any sensations. The minimum intensity of irritation, which produces slight noticeable senses, is the lowest absolute sensitivity threshold [3, 6, 7]. The maximum intensity, which causes adequate senses to the existing irritation, is the highest absolute sensitivity threshold [3, 6, 7]. The determination of the senses threshold is important for recognition and naming the bipolar senses descriptors [3].

In turn, the original classification of senses is based on the receptor. There are: skin's senses – tactile (touch, pressure, pain) and thermal (heat, cold); touch; taste (sweet, salty, sour, bitter); smell; vision; hearing (volume, pitch, timbre, duration) senses; position and movement senses (static and kinaesthetic) and organic senses (hunger, thirst, sexual desire, visceral pain) [6, 7].

Each sense has its intensity, emotional tone, speed (or time threshold), differentiation (sense subtlety), adequacy (sense accuracy), quality (given modality sense) and stability of sensitivity level and sense length [7].

It has been determined that there are two major types of discomfort sensation – psychological discomfort (when the cloth is inappropriate for the human who is using it) and physiological discomfort (when the body feels uncomfortable) [2].

2. Problem Formulation

From theoretical research, it has been determined that a consumer's subjective comfort in clothing is influenced by the following factors which can be divided into three large groups:

1) Environment:
   a) certain ambient environmental conditions [1, 2, 8, 9];
   b) possible environmental threats [2, 8];
   c) human social and cultural environment [1, 2, 8];
   d) human economic situation [8];
2) Clothing:
   a) clothing, its structure/construction, functions [1, 2, 8];
   b) features and physical-mechanical properties of clothing fabrics [1, 2, 8];
3) Human:
   a) human body physical state [2, 8];

b) human body physiology and temperature [1, 2, 8];

c) metabolic rate and physical activity (intensity of movement, work) [1, 2, 8, 10, 11];

d) human psychological state [2];

e) heat losses [1, 2, 8, 9];

f) possible human-caused threats [2].

It has been concluded that consumer's comfort in clothing is also influenced by:

a) human vital needs;

b) surrounding hard surfaces and objects, their mean radiant temperature;

c) environmental ergonomics.

It has been deduced that subjective senses of a consumer's comfort (also physiological comfort) in clothing are formed and can be influenced by all of the above-mentioned factors in interaction and in various combinations.

Each descriptive subjective sense, whether positive or negative, of a consumer's comfort in clothing can be likened to measurable physical values of different fabrics and their layers [12]. Each sense can be depicted by bipolar descriptors that can be systematized in a way that is understandable for clothes designers and consumers [12].

A consumer assesses the physiological comfort in clothing subjectively by the feeling of warmth, cold, dryness or dampness that depends on the capacity of fabrics and their layers to absorb, transfer and discharge moisture. When designing a piece of clothing it is essential to forecast its physiological comfort level in use.

It has been concluded that there are several information sources, which forecast a consumer's physiological factors in clothing (Dlim, IREQ, WBGT, PMV, PPD etc.). It has been found that there are a lot of various pieces of information and methods on how to detect one or another factor or indicator that influences comfort. However, currently there is no common physiological comfort forecasting method described in a manner that would be clear to all clothes designers. Consumer's physiological comfort should be taken into consideration while designing a definite piece of clothing for a certain ambient environment, range of consumers and with the required functions (requirements).

The aim of the research – to develop a prediction algorithm for consumers' physiological comfort in clothing considering the factors and indicators that influence physiological comfort.

2.1 Environmental Factors Affecting Comfort in Clothing

A human body is constantly exposed to ambient environmental conditions and various physical activities [1]. A consumer's thermal comfort depends on clothing, climate and physical activities combinations [1].

It has been determined that a consumer's physiological comfort in clothing is influenced by the following factors of ambient environment – air temperature t_a; mean radiant temperature t_r; air velocity v_a [9, 13] and relative humidity RH [9] (in ISO 7933 – partial vapour pressure Pa) [13].

It has been stated that ergonomically the ambient environment can be hot, moderate and cold. According to the ergonomics standard ISO 7730, moderate ambient environment is an environment where the air temperature is below 26 °C [5]. The air temperature of hot ambient environment respectively is above 26 °C, and the air temperature of cold ambient environment is below +10 °C [5, 9].

The air temperature of moderate environment is between 10 °C and 20 °C [14]. In turn, most people have thermal neutrality feeling working in an office at operative temperature from 20 °C to 26 °C doing light, sedentary work (M = 70 W/m^2) in appropriate clothing (insulation values 0.6-1.0 clo) [14]. According to standard ISO 7730, for a person working in an office in summer the most comfortable air temperature is from 23.5 °C to 25.5 °C, and local air temperature is from 20 °C to 26 °C [5].

Absolute humidity of the environment (water vapour pressure in the air) leaves an impact on the evaporative heat loss from the body [5]. Consequently, it influences the total thermal comfort of the body (heat balance) [5]. In moderate temperature (< 26 °C) and at the level of moderate activity (< 2 met) this influence is rather limited [5]. In moderate environment air humidity has an insignificant impact on thermal feeling [5]. Usually, if relative humidity increases by 10 %, it feels like operative temperature increase of 0.3 °C [5].

There are environmental threats and human-caused threats. Not just an official but also a civilian, while carrying out their duties or being in a certain ambient environment, is subject either to environmental threats (heat, cold, wind, rain, snow, dust, ultraviolet radiation, etc.) or human-caused threats (explosives, poisons, fire, stab/incised wounds, high speed, biological attacks, etc.), or to both [12].

To assess a certain thermal ambient environment impact on human, on his/her work ability, comfort level and the clothing performance during physical activities, the following ergonomics standards must be used: ISO 7730 (PMV/PPD) [5], ISO 7243 (WBGT) [11], ISO 7933 (Sw_{req}) [13], ISO 11079 (IREQ) [9], ISO 9886 (physiology) [15], ISO 10551 (subjective) [12, 16, 17, 18]. Ergonomics standards ISO 7726 (instruments) [19], 8996 (metabolic rate) [10], 9920 (clothing) [20], ISO 15265 (risk assessment) [21] for assessment of the ambient environment can be used alone or in combination with the above-mentioned standards [12, 17, 18].

Assessing the thermal ambient environment, a human, while carrying out his/her work duties or during physical activities, may come into contact with various hard surfaces and objects that have their peculiar surface (smooth, rough, textured, combined, etc.) made of different raw materials and their mean radiant temperature. A person's thermal vulnerability level, while in contact with a hard surface [5], is evaluated in a hot environment by using the standard ISO 13732-1 [22], in a moderate environment – ISO 13732-2 [23], and in a cold environment – ISO 13732-3 [18, 24].

A factor that significantly influences a consumer's subjective comfort is the environment ergonomics (work or everyday environment and organising of human). To provide for an effective, easily performed and safe work, the work process and environment must be adjusted to the mental and physical abilities of a human [25]. Working environment for a physically and mentally healthy person differs from that of a disabled person.

2.2 Clothing and Fabrics

When the temperature of the certain ambient environment changes, clothing and body help a human to acclimatize oneself by using temperature-regulating mechanisms [2]. Nowadays clothing has a decoration, social status, protection and modesty function for people [2].

Construction of clothing, its fit, tightness and weight influence comfort, consumer' movements and the comfort of use of clothing elements. The colours of clothes, the used raw materials, type of technological processing, etc., are also of the essence. To decrease or prevent the ambient environmental threat or that caused by humans, clothing suitable for the type of threat must be used [8].

It has been determined that a body's redundant moisture can be discharged through textile material in the form of vapour (undetectable sweat) and in the form of liquid (detectable sweat) [26]. Processes involved in the moisture transfer are: water vapour transfer, liquid water transfer, combined vapour and liquid transfer, and combined heat and mass transfer [8, 26] (see Fig.1). The transfer of moisture is influenced by evaporation and condensation, which in turn depend on temperature and moisture distribution in a porous textile material [26].

Water wapour transfer processes are: diffusion through the layers; absorption, transmission and desorption by the fibres; adsorption and migration along the fibre surface and transmission by forced convection [8, 26] (see Fig.1). Liquid transfer through a porous structure involves two sequential processes – wetting and wicking [8, 26]. A simultaneous transfer of vapour and liquid is important during sweating and is carried out in three stages: at the beginning, water vapour diffusion and liquid water diffusion, then absorption of humidity and finally – stabilisation of moisture and heat transfer [26]. The heat transfer process is coupled with four different forms of moisture transfer due to the heat released or absorbed during sorption/desorption and evaporation/condensation [8, 26].

To decrease a consumer's body temperature and to sustain its thermal balance, the clothing used must transfer the sweat to the ambient environment [26]. In addition, if the transfer of moisture in the clothes is slow during sweating of the body, then the clothing's microclimate relative and absolute humidity levels increase causing overheating of the body and discomfort in hot environment [26]. Collection of moisture in the fabrics system inner layer reduces the heat insulation of the clothing as well and causes undesirable loss of body heat in cold environment [26].

It has been stated that fabrics moisture control and underwear microclimate, which has also a function of heat insulation, leave a strong impact on a consumer's physiological comfort. Sense of dryness is provided by fabrics that let moisture evaporate easily into the external environment.

Subjective psychological or physiological senses can arise in thermal, sensorial or body movement comfort [2]. In everyday life a consumer's comfort is characterised by clothing fabrics and their layers subjective tactile and touch senses (soft, rough, silky, unpleasant) that a consumer evaluates by being in contact on his/her subjective experience.

From theoretical research, it has been determined that fabric tactile properties (stiffness, softness, etc.), fabric thermal and moisture transfer properties and other physical mechanical properties (air permeability, drape ability, resistance to abrasion, etc.) are affected by the fiber type, structure, surface density, linear density, porosity, finishing and other characteristics of fabrics [2].

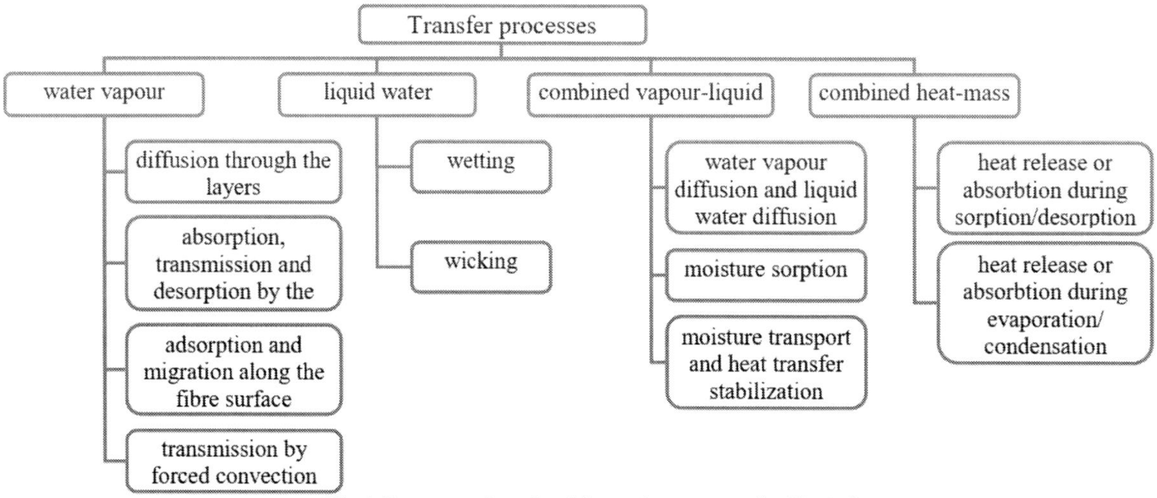

Fig.1 Processes involved in moisture transfer [8, 26].

2.3 Human Physical, Physiological and Psychological State

A consumer's physical state is influenced by age, health condition, body structure, physiological response, activity level, etc. [2, 8]. In turn, a consumer's health condition and body build are determined by body weight, fitness, gender, lifestyle, nutrition, etc. Human physiological comfort is affected also by the state and activity of the body physiological functions, namely of the heart and blood vessels, nervous and respiratory systems [2].

The human body core temperature is 37 °C (98,6 °F) ± 0,5 °C [1, 2] and its changes cause body hypothermia (< 26.5 °C) or hyperthermia (> 37.8 °C) [2]. Warm and cold receptors are located in the body and in the skin [1, 2, 8]. Cold receptors are excited by reduction in temperature and have a peak sensitivity of around 25-30 °C [1, 8]. Warm receptors are sensitive to increases in skin temperature and have a peak sensitivity of around 39-40 °C [1, 8]. When body skin temperature is higher than 37 °C, heat sensors in the hypothalamus, and cold sensors in the skin, when skin temperature is below 37 °C, send signals to the brain [2].

"The body temperature is a result of heat production and heat loss and it is mainly regulated through the nervous feedback mechanism" [2]. The hypothalamus (the body's thermostat) triggers heat controlling mechanisms to increase or decrease heat loss by controlling the flow of blood to the skin [2].

The human body metabolic rate (1 met = 58.2 $W \cdot m^{-2}$) is influenced by age, physical exercise, body weight and surface area, hormones, food consumption, sympathetic stimulation, climate, sleep, fever, malnutrition, physiological condition and amount of clothing [2]. Levels of metabolic rate and physical activity at each level are prescribed by ergonomics standards ISO 8996 [10] and ISO 7243 [10, 11]. It has been determined that consumer's subjective comfort in clothing (heat balance) is influenced by: metabolic rate M ($W \cdot m^{-2}$), effective mechanical power W ($W \cdot m^{-2}$) and respiratory convective C_{res} and evaporative E_{res} heat loss ($W \cdot m^{-2}$), convective C, evaporative E at the skin, conductive K and radiative R heat flow (exchange) ($W \cdot m^{-2}$) [1, 2, 8, 9, 13]. Heat balance is influenced also by body heat storage rate S [1, 9, 13] and heat storage for increase of core temperature associated with the metabolic rate dSeq [13]. According to ISO 7933 body heat balance is influenced also by clothing thermal characteristics [13].

It has been concluded that the wearer's subjective senses of physiological comfort in clothing are affected by human's fatigue, hunger, thirst, and other vital needs. In turn, consumer's social, cultural and ambient environment and economic situation affect the wearer's physical and psychological state. The psychological state of human is affected by psychological conditions, past experiences and future desires [2].

3. Problem Solution

From theoretical research, has been gathered and a hypothetical classification of descriptive and impacting factors which interact, form and influence an individual's senses and perception regarding a clothing wearer's psychological, physical and physiological subjective comfort in clothing has been performed (see Fig.2). A consumer's subjective comfort in clothing is influenced by the following factors, which can be divided into groups of environment, clothing and human.

It has been concluded that the wearer's subjective physiological comfort in clothing is mostly affected by (see Fig.2):

Environment – certain ambient environmental conditions (air temperature, mean radiant temperature, air velocity and relative humidity);

Clothing – clothing, its fabrics and their layers' hygienic parameters (water vapour permeability/resistance, water (im-) permeability and air (im-) permeability) and thermal resistance;

Human – human body physiology (state of physiological functions and physiological reactions), body core temperature, metabolic rate (intensity of body movement, heat losses during exhalation, evaporation, thermal emissivity, heat discharge and heat exchange processes), also basic human needs and possible threats.

Physical-mechanical properties of clothing fabrics and their layers can be determined instrumentally in accordance with standardized research methods by means of suitable equipments.

Standardized research methods and equipment have been studied for determination of a consumer's physiological comfort in clothing. For instance, heat resistance R_{ct} ($m^2 \cdot K \cdot W^{-1}$) or vapour resistance R_{et} ($m^2 \cdot Pa \cdot W^{-1}$) of clothing fabrics and their layers can be detected on a sweating guarded hot plate instrument in accordance with the standardized method ISO 11092 (Textiles - Physiological effects - Measurement of thermal and water - vapour resistance under steady - state conditions (sweating guarded – hotplate test)), LVS EN 11092:2014 [27] also.

As a result of the research the prediction algorithm for consumer's physiological comfort in clothing is created. The algorithm is foreseen for physiological comfort level estimation and prediction of definite fabrics layers in clothing during its guaranteed time of use while the wearer carries out the metabolic activity at a certain level in a certain ambient environment.

An example of physiological comfort prediction in clothing.

If a definite piece of clothing is being designed out of a specific fabric and one must predict the performance of the designed clothing and the wearer's comfort in that clothing, the following steps shall be taken:

1) First, from the database ENVIRONMENT it is necessary to identify factors of the ambient environment: the type of ambient envinroment (hot, moderate, cold), climatic conditions, mean radiant temperature, possible environment threats, if any.

2) From the database CLOTHING it is necessary to identify the functions of clothing which can be altered, its particularities, construction and fit, physical-mechanical properties of fabric layers used in the clothing. Thermal insulation of clothing fabrics is one of the most important features of physiological comfort and clothing performance. The basic clothing insulation I_{cl} and resultant clothing insulation $I_{cl,r}$ ($m^2 \cdot K \cdot W^{-1}$ or in clo) values are calculated if the value of total insulation I_T ($m^2 \cdot K \cdot W^{-1}$ or in clo) is known [9, 20]. The total insulation I_T is also the total thermal insulation of the garment ($m^2 \cdot K \cdot W^{-1}$ or in clo) [20]. In accordance with the ergonomics standard ISO 9920 or ISO 11079 method the total clothing evaporative resistance of clothing and boundary air layer $R_{e,T}$ is determined [9, 20]. If the clothing fabric thermal resistance or water vapour resistance is not known, the parameters are determined on a standardized instrument in accordance with the standard LVS EN 31092:2002 and the water permeability index i_{mt} is calculated [27]. To determine several interrelated thermal effects of the garment wearer's comfort simultaneously a thermal manikin is used in accordance with the ISO 15831 standard [28].

3) From the database HUMAN, in accordance with the ergonomics standard ISO 8996, the metabolic rate of a consumer is determined [9, 10]. The metabolic rate is influenced by the intensity of body movement, heat losses, body core temperature, physiological state and other factors. One must also identify possible human-caused threats and vital needs.

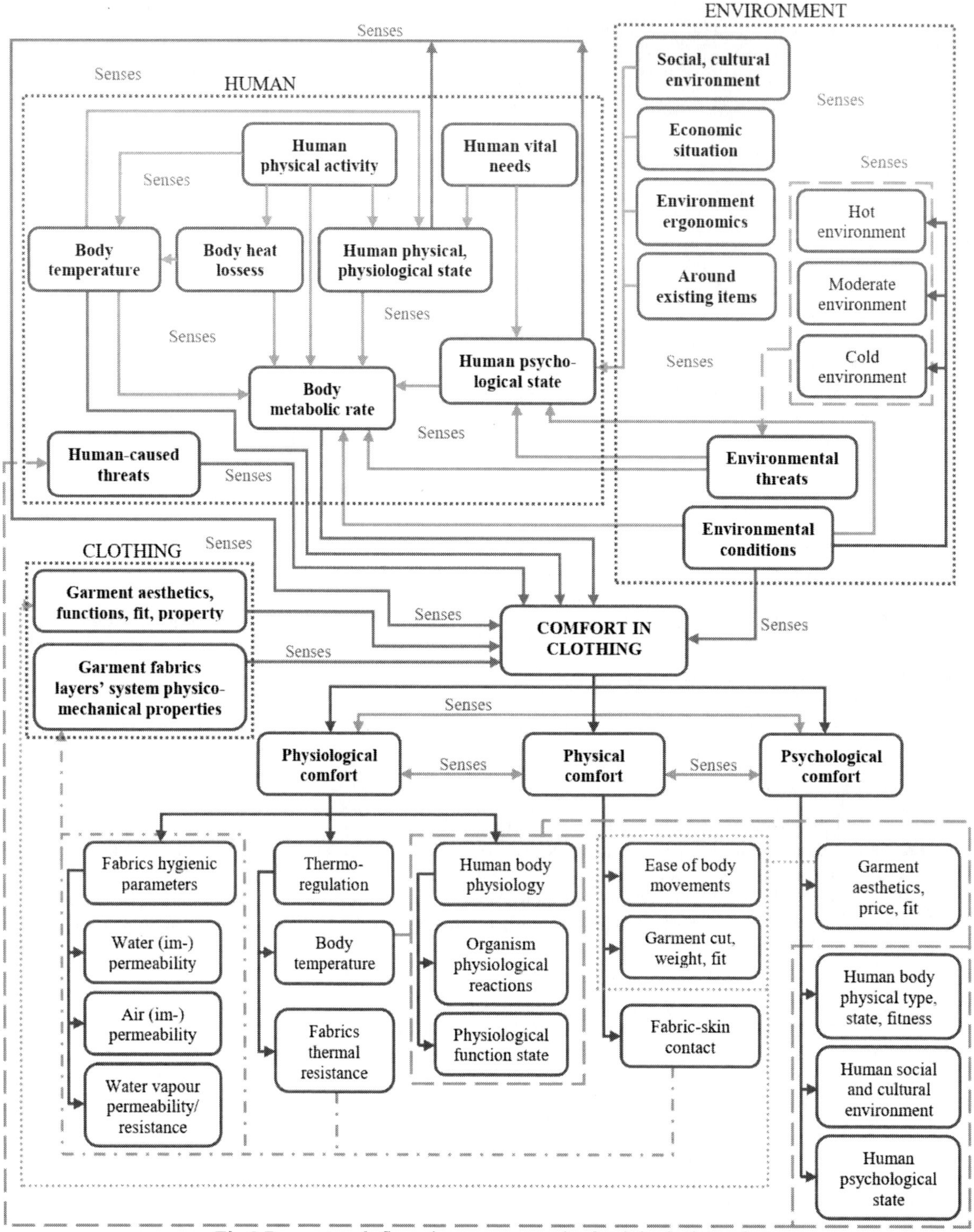

Fig.2 Parameters influencing a consumer's physiological comfort.

4) Having knowledge of the characteristics and values of factors that influence physiological comfort in clothing, it is possible, in accordance with the ergonomics standards, to assess impact of a specific thermal environment on

the work ability of a human and the comfort level in clothing which can be altered as well as the performance length of clothing during its wearing. Hot environment is assessed in accordance with the ISO 7243 (WBGT) [17] or ISO 7933 (S_{wreq}), moderate environment – ISO 7730 (PMV and PPD) and cold environment – ISO 11079 (IREQ and D_{lim}) standard method [17, 18, 20]. The length of comfort in clothing depends on the person's metabolic rate – the work that is carried out and its intensity. Carrying out moderate or passive work, a person will feel comfortable longer than doing active, physical work in hot or moderate ambient environment when there is strong body sweating. Besides, it must be taken into account that in cold ambient environment there are more layers in the clothing.

5) Having knowledge of a certain thermal ambient environmental impact on a human's work ability and the level of comfort in clothing which can be altered, it is possible to predict bipolar senses describing a consumer's physiological comfort that can be classified in a manner understandable for consumers and clothing designers.

4. Conclusion

It has been concluded that a wearer's physiological comfort in clothing is essential during any physical activity and that clothing must maintain a steady human body temperature and provide for heat exchange between the human body and the ambient environment in cold, moderate or hot ambient environmental conditions.

From theoretical research, knowledge has been acquired and a hypothetical classification of descriptive and impacting factors which interact, create and influence an individual's senses and perception regarding a clothing wearer's psychological, physical and physiological subjective comfort in clothing has been performed. Each subjective sense is formed when a particular sensitive receptor detects an irritation. All factors that leave an impact on a wearer's comfort in clothing are divided into groups of Environment, Clothing and Human.

It has been concluded that a wearer's physiological comfort in clothing is mostly affected by: certain ambient environmental conditions, clothing fabrics and their layers' hygienic indicators and thermal resistance, a consumer's body physiology, thermal regulation, metabolic rate (movement intensity and heat losses), as well as organic needs and possible threats.

The prediction algorithm for physiological comfort in clothing has been created in order to predict a wearer's physiological comfort level in definite clothing during its guaranteed time of use while the wearer carries out the particular duties in a certain ambient environment.

It is possible to predict a consumer's physiological comfort in clothing and the performance of the clothing. For this purpose, it is necessary to have information about the ambient environmental conditions, physical-mechanical properties of clothing fabrics layers, a person's metabolic rate and to calculate indicators of physiological comfort in accordance with ergonomics standards (ISO 7730 [5], ISO 7243 [11], ISO 11079 [9], ISO 9886 [15], ISO 10551 [16], ISO 15265 [21]).

It has been concluded that indicators describing and influencing the sense of a consumer's comfort in clothing can be used as starting point information for designing comfortable clothing.

Based on the worked-out algorithm and factors that leave an impact on comfort in clothing, it is envisaged to develop a classification of bipolar descriptors that depict a consumer's physiological comfort.

To assess and predict the actual consumer's level of physiological comfort in clothing, taking into consideration the factors that influence it, additional testing of physical-mechanical properties of definite clothing fabrics is necessary.

References

[1] Li Y. The Science of Clothing Comfort. In: Textile Progress: 2001; 31:1-2: p. 1-135.

[2] Fundamentals of comfort and assessment. Part I. In: Guowen Song, editors. Improving comfort in Clothing. Woodhead Publishing Limited: 2011; p. 3-162.

[3] Šroma I. Physiological Comfort in Clothing. In: The 13th International Conference on Global Research and Education "Inter Academia 2014", Digest: Riga, Latvia, September 10-12, 2014. Riga: RTU Press: 2014; p. 202-203.

[4] Martins SB. The OIKOS methodology for usability and comfort evaluation in clothing and fashion. Ergonomics and Fashion: Work 41: 2012; p. 6059-6067.

[5] Ergonomics of the thermal environment - Analytical determination and interpretation of thermal comfort using calculation of the PMV and PPD indices and local thermal comfort criteria. ISO 7730: 2005.
[6] Rubinshtejn SL. OSNOVY OBSHHEJ PSIXOLOGII: Glava VII OSHHUSHHENIE I VOSPRIYATIE. Izdatel'stvo "Piter", 2000. Information on http://psylib.org.ua/books/rubin01/index.htm
[7] Psixologiya oshhushhenij. Information on http://libsib.ru/obschaya-psichologiya/psichologiya-oschuscheniy/vse-stranitsi
[8] Das A, Alagirusamy R. Science in Clothing Comfort. Woodhead Publishing India Pvt. Ltd: 2010; 175: p.
[9] Ergonomics of the thermal environment - Determination and interpretation of cold stress when using required clothing insulation (IREQ) and local cooling effects. ISO 11079: 2007.
[10] Ergonomics of the thermal environment - Determination of metabolic rate. ISO 8996: 2004.
[11] Hot Environments - Estimation of the heat stress on working man, based on the WBGT-index (wet bulb globe temperature). ISO 7243: 1989.
[12] Šroma I, Ābele I, Ziemele I, Viļumsone A. Requirements for State Administrative Institution Employees' Functional Clothing and their Systemization. In: Textile Bioengineering and Informatics Symposium Proceedings: Paper of the 8th Textile Bioengineering and Informatics Symposium, Croatia, Zadar, June 14-17, 2015. The Hong Kong Polytechnic University: Binary Information Press, Textile Bioengineering and Informatics Society: 2015; p. 168-177.
[13] Ergonomics of the thermal environment - Analytical determination and interpretation of heat stress using calculation of the predicted heat strain. ISO 7933: 2004.
[14] Holmér I, Granberg PO, Dahlstrom G. Cold environments and cold work. In: Chapter 42 - Heat and Cold: Encyclopaedia of Occupational Health and Safety: p. 35.
[15] Ergonomics - Evaluation of thermal strain by physiological measurements. ISO 9886: 2004.
[16] Ergonomics of the thermal environment - Assessment of the influence of the thermal environment using subjective judgement scales. ISO 10551: 1995.
[17] Ergonomics of the thermal environment - Principles and application of relevant International Standards. ISO 11399: 1995.
[18] Parsons K. Heat Stress Standard ISO 7243 and its Global Application. Industrial Health: 2006; 44: p. 368-379.
[19] Ergonomics of the thermal environment - Instruments for measuring physical quantities. ISO 7726: 1998.
[20] Ergonomics of the thermal environment - Estimation of thermal insulation and water vapour resistance of a clothing ensemble. ISO 9920: 2007.
[21] Ergonomics of the thermal environment - Risk assessment strategy for the prevention of stress or discomfort in thermal working conditions. ISO 15265: 2004.
[22] Ergonomics of the thermal environment - Methods for the assessment of human responses to contact with surfaces - Part 1: Hot surfaces. ISO 13732-1: 2006.
[23] Ergonomics of the thermal environment - Methods for the assessment of human responses to contact with surfaces - Part 2: Human contact with surfaces at moderate temperature. ISO/TS 13732-2: 2001.
[24] Ergonomics of the thermal environment - Methods for the assessment of human responses to contact with surfaces - Part 3: Cold surfaces. ISO 13732-3: 2005.
[25] Information on http://www.ergonomika.lv/par-ergonomiku //www.weld.labs.gov.cn/
[26] Brojeswari D, Das A, Kothari VK, Fanguiero R, M. de Araújo. Moisture transmission through textiles. Part I: Processes involved in moisture transmission and the factors at play. AUTEX Research Journal, Vol. 7, No 2, June: 2007; p. 100-110.
[27] Textiles - Physiological effects - Measurement of thermal and water-vapour resistance under steady-state conditions (sweating guarded-hotplate test) (ISO 11092:2014). LVS EN ISO 11092: 2014.
[28] Clothing - Physiological effects - Measurement of thermal insulation by means of a thermal manikin. ISO 15831: 2004.

Effect of Wearing Sports Bras on Breast Motion During Jogging

Yang Zhang, Jin Li, Xiao-Qun Dai

Soochow University, College of Textile and Clothing Engineering, Suzhou 215006, China

*Corresponding author's email: daixqsz@suda.edu.cn

Abstract

It is well known that wearing sports bra helps to reduce breast movement during exercise. However, how much a sports bra affects breast motion during jogging is still not very clear. Four female undergraduates with breast size 75B, wearing a sports bra, the same bra with cup pads removed, and bared breast respectively, completed standing upright for 5s and a 7.5km/h treadmill running for 3 mins. The breast movement was recorded by using an 8-camera Vicon motion capturing system. It was found that during jogging wearing the sports bra can reduce the displacement amplitude of nipple by 51% in anteroposterior direction, 76% in vertical direction, and 55% in mediolateral direction. However, there was no significant difference found between the bras with and without cup pads.

Keywords: Sports Bra; Breast Motion; Jogging

1. Introduction

Breast tissue is relatively free to move over the chest wall during physical activities. Excessive body displacement may hurt the breast's anatomical structure and cause sagging or injury [1, 2]. Clinical experiments have shown that sports bras do help in reducing breast displacement, thereby relieving breast pain [3]. Sport bra design has significantly progressed in recent years with the aim to minimize breast motion during exercise [4]. In addition to reducing breast motion, reducing exercise-related breast pain and improving overall comfort have been regarded as important considerations for sports bra design. In recent years the structural components and material properties of sports bras have evolved dramatically from the first sports bra, with Lycra®, Elastane and Cool Max® materials incorporated into the more complex sports bras [5]. These materials are known to be lightweight, have high levels of both elastic and recovery properties, and are suggested to provide more support to the breast than a bra made out of Cotton [6]. Currently, there are three major sports bra designs: encapsulation, compression, and combination [7-8]. Compression sports bras have been designed to restrict the amount of movement of the breast by compressing and flattening the breast tissue against the body, and redistributing the mass evenly. On the other hand, an encapsulation bra provides more structured support, harnessing each of the breasts individually. This design is thought to be more effective for larger breast sizes (C-cup upwards) [4].

Motion capture systems have been widely used to measure kinematics and kinetics of subjects. However, the challenge in measuring breast motion is that its movement must be measured relative to the thorax, and the position and orientation of the breast is affected by both thoracic rotation and translation. To measure the breast movement relative to the thorax, it is necessary to measure the movement of the thorax as a reference. Various reference systems using one point [2, 9], three points [3] or four body points [10-14] were proposed to describe the rigid body movement. With these reference systems, the motion of breast under daily bras, various sports bras, and braless conditions during walking, running, jogging, and aerobics were investigated. Most of these studies examined the breast displacement only at the nipple, which was regarded as the best indicator of vertical breast displacement. However, as the breast tissues are soft and viscoelastic, the nipple movements may be insufficient to describe the whole breast movements. Therefore, Okabe and Kurokawa and Zhou et al. added four points or five points surrounding the nipple respectively [13, 14].

In our previous study, the breast-bra mechanical interaction was analysed while the subjects were doing vertical jumping, and the material effect on the interaction was also investigated [15]. As jogging becomes a popular recreational activity among people of all ages, the aim of this study is to investigate how the breast motion is affected by sports bras during jogging.

2. Methods

Four healthy, active and young female undergraduates (age 23 ± 1 years, height 163 ± 5 cm, body mass 50 ± 5 kg) denoted as S1 through to S4, volunteered to participate in this study. They were not pregnant, had not given birth or breast-fed in the last year, had no previous surgery to the breasts and were a 75B breast size. In addition, they usually had a good exercise habits, trained to run on a treadmill in a standard position.

Previous studies on breast kinematics have found that there is no significant difference in any direction or support condition between the left and right breast [16]. Therefore in this study, the movement of the left breast was measured. Retro-reflective markers were placed at five points shown in Fig.1, the left nipple (BL), 5cm right and left from left nipple (L1, L3), 5cm below and above the left nipple (L2, L4). For bare conditions, the markers were attached to the breast while for bra conditions, the markers had to be attached to the bra surface. The participants stood upright still for 5s, then kept jogging at speed of 7.5km/h on a treadmill running for three minutes. The treadmill test was completed in three randomized breast support conditions: (1) bare breast, (2) a sports bra with the cup pad removed (bra-1), and (3) a sports bra with cup pads (bra-2). The sports bra and cup pad are shown in Fig.2. The breast motion was measured using an 8-camera Vicon motion capturing system (Oxford, Metrics Limited, UK) at the frequency of 200 Hz, the breast displacement relative to the thorax was obtained by transforming the obtained breast trajectory from a global coordinate system (GCS) to a local breast coordinate system (BCS) proposed by Zhou et al. [14].

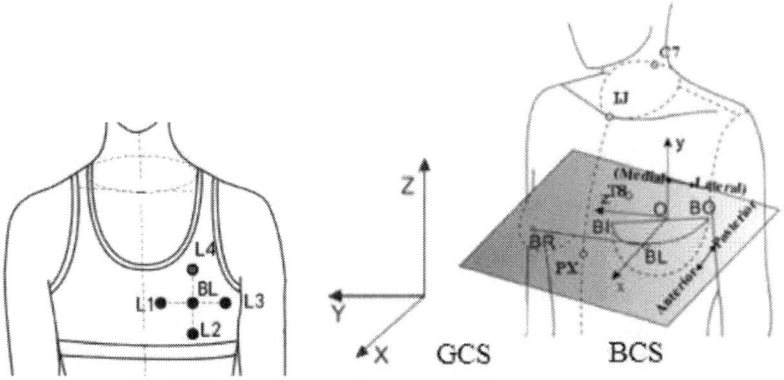

Fig.1 Mark points and breast coordinate system

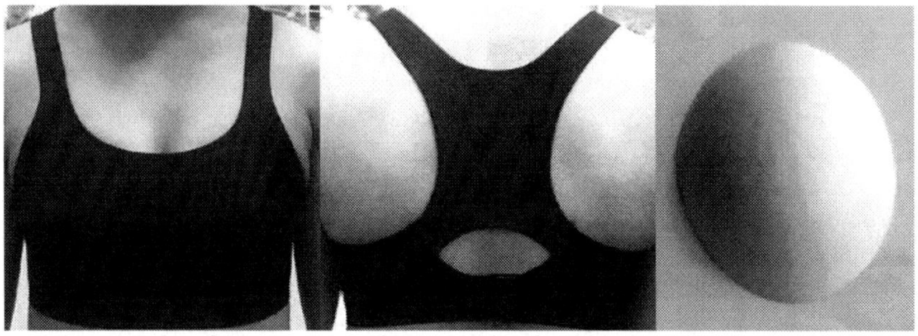
(a) Sports bra front (b) Sports bra back (c) Cup pad
Fig.2 Sports bra and cup pad

As illustrated in Fig.1, the X-Y plain of Global Coordinate System(GCS) is on the floor where the subjects stand, and Z is the upward vertical direction. Four reference markers at IJ (deepest point of the incisura jugularis (suprasternal notch)), PX (processus xiphoideus, the most caudal point on the sternum), C7 (spinous process of the

7th cervical vertebra) and T8 (spinous process of the 8th thoracic vertebra) were used to define the thorax, and four breast boundary markers, BR and BL (right and left nipples), BI and BO (most medial point and lateral point on the left breast intersecting with the breast root and the transverse plane through BR and BL), were used to define the BCS. In the BCS, the origin O is defined by the intersection point of line BI-BO and the line perpendicular to BR-BL. The z-axis is parallel to the line BL-BR pointing to the right, representing medial and lateral (ML) direction; the x-axis is perpendicular to the z-axis on the plane which is formed by four breast boundary markers, representing anterior and posterior (AP) direction; and the y-axis is the common line perpendicular to both x-axis and z-axis.

Each subject was instructed to wear a daily bra and adjust the two breasts to two semisphere shapes. The four breast boundary markers and four reference markers were captured at the same time while the subject was standing upright. The eight points data was used to establish the transformation from GCS to BCS. Then the boundary markers were removed, and the four reference markers and the five breast markers were captured while the same subject was jogging with three conditions. Paired-sample t-tests were carried out among three conditions. All data were analyzed using SPSS Version 17.0 software (IBM Inc., Armonk, USA, Version 17.0), with results deemed significant at an alpha level of $p \leq 0.05$, extremely significant at an alpha level of $p \leq 0.01$.

3. Results and Discussion

3.1 Static Position of Five Points in BCS

Fig.3 shows the static position of five breast points in BCS averaged among four subjects under three conditions. In Fig.2(a), X denotes the anteroposterior distance from the thorax, three conditions show similar trend but are slightly different. Except L2, Bare is in between of Bra-1 and Bra-2, indicating that the breast was squeezed by the compression of Bra-1, while the thick cup pads of Bra-2 increased this distance a little because the marker were placed on bra surface. In BCS, if the breast does not deform, BL, L1 and L3 are supposed to be on the plain formed by x-axis and z-axis. As shown in Fig.2(b), in vertical direction, the bare breast sagged due to the gravity, except the up points L3, the coordinates of the other four points were negative; wearing sports bra elevated all five points and BL, L1 and L3 were even elevated above the X-Z plain. As shown in Fig.2(c), BL, L2 and L4 lay in the center while L1 and L3 were in the both sides of the center line. There was no big difference among conditions, indicating that wearing sports bra didn't cause big breast displacement in mediolateral direction.

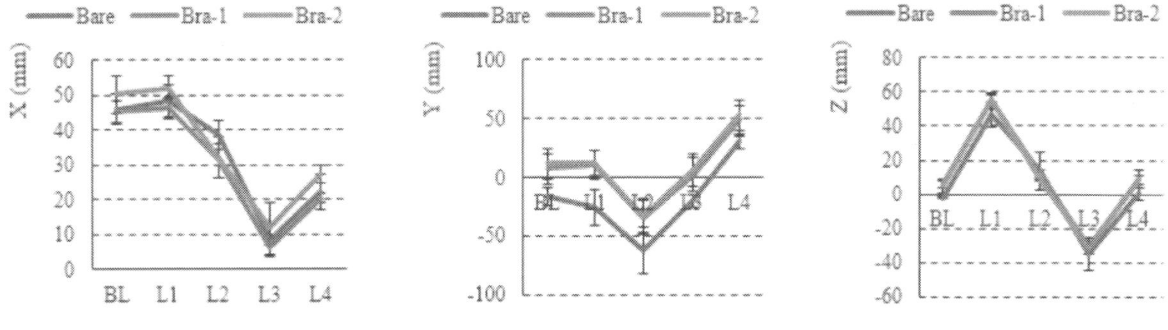

(a) X- Anteroposterior (b) Y- Vertical (c) Z- Mediolateral

Fig.3 Static position of five breast points in BCS under three conditions

3.2 Breast Motion in three Conditions During Jogging

Fig.4 shows the breast motion at the nipple (BL) of S2 during jogging, the displacements of three directions shows clear cyclic fluctuation with the jogging pace. Here, two complete gait cycles (double steps) were shown for three conditions. In Fig.3(b), the four peaks for bare condition corresponded to two double steps, (two left foot steps and two right foot steps). The first peak was supposed to be caused by the impact of the left heel touched the treadmill, and the second peak should be caused by the impact of the right heel touched the treadmill. The first and second valleys occurred when the left toe and the right toe departed from the treadmill respectively. When the left foot stepped forward, the left arm waved to the back and left, the left breast would be waved to the opposite due to the inertia effect. Therefore as shown in Fig.3(a) and (c), as the bare breast waved to the front and right side of the torso, the X and Z coordinates increased to peaks. On the contrary, when the right foot stepped forward, the left arm waved to the right and front, and the left breast would be waved to the left and back, the X and Z coordinates decreased to valleys. Also wearing sports bra greatly reduced the fluctuation in all three directions, the cyclic fluctuation along with the gait cycles still can be seen clearly.

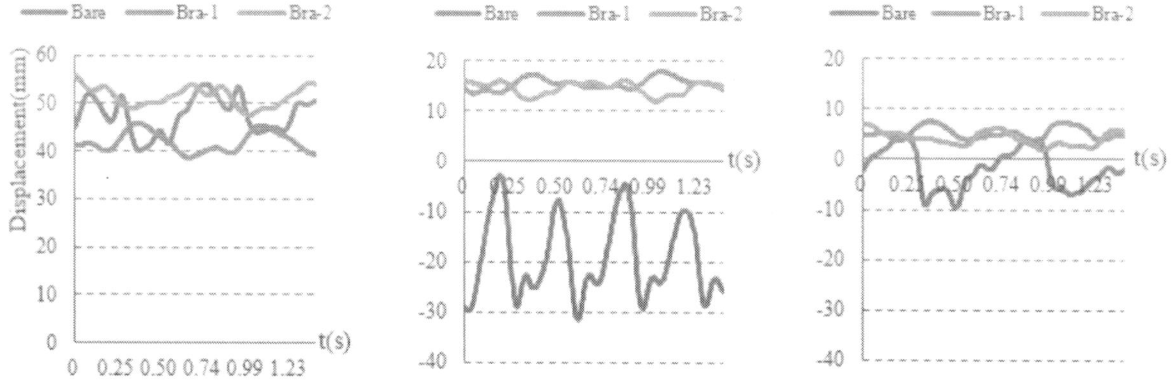

(a) X- Anteroposterior (b) Y- Vertical (c) Z- Mediolateral

Fig.4 Breast motion at the nipple (BL) in three conditions during jogging(S2)

3.3 Effect on Amplitude of Breast Displacement

The amplitude of displacement was obtained by calculating the difference between a peak and a valley within a gait cycle from the dynamic displacement change as shown in Fig.4. Fig.5 showed the mean amplitudes of displacements in three directions averaged among 4 subjects, together with the respective standard variation. The amplitude of wearing a sports bra decreased greatly in all three directions. Compared to the amplitude of bare breast, the amplitude of nipple when wearing Bra-1reduced by 51% in anteroposterior direction, 76% in vertical direction, 55% in mediolateral direction. At most points, the difference was significant. The amplitudes of Bra-2 were slightly smaller than the respective ones for Bra-1. However, the difference was not significant. The possible reason might be the softness of breast tissues of the four subjects varied greatly.

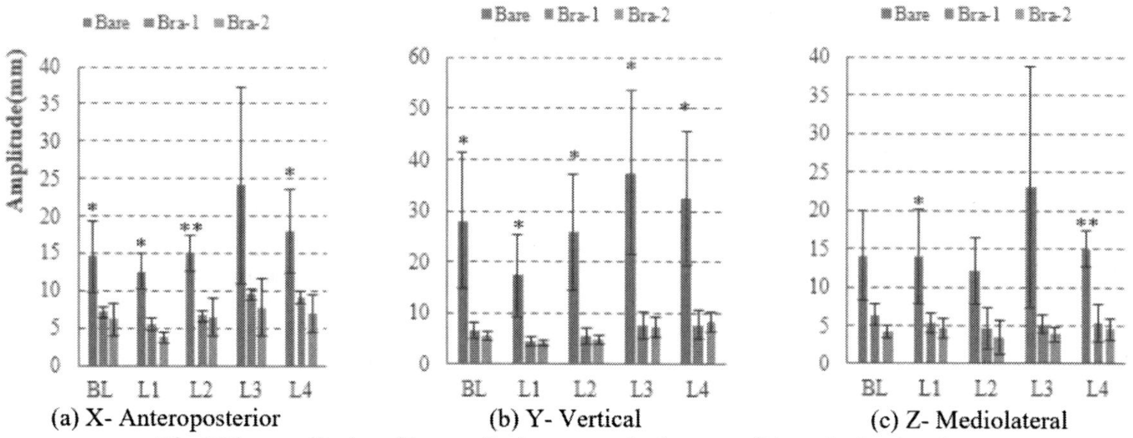

(a) X- Anteroposterior (b) Y- Vertical (c) Z- Mediolateral

Fig.5 The amplitudes of breast displacements in three conditions during jogging
The *p* value between bare and bra-1 *(p<0.05) **(p<0.01)

4. Conclusion

In this study, it was found that during jogging wearing the sports bra reduced the displacement amplitude of nipple by 51% in anteroposterior direction, 76% in vertical direction, and 55% in mediolateral direction. Although some subjects showed even reduced amplitude of displacement wearing bra with cup pads, the difference between the bras with and without cup pads was not significant. Further study need to be carried out with the tissue softness of subjects' breasts taken into account.

5. Acknowledgement

This is a project funded by the priority academic program Development (PAPD) of Jiangsu Higher Education Institutions.

References

[1] Steele JR. The biomechanics of better bras: improving support and comfort during exercise. ISBS Conference Proceedings: 2013.

[2] Mason BR, Page K, Fallon K. An analysis of the movement and discomfort of the female breast during exercise and the effects of the breast support in three case studies. Journal of Science and Medicine in Sport: 1999; 2: 134-144.

[3] Zhou J, Yu W. Evaluation of shock absorbing performance of sports bras. Journal of Fiber Bioengineering and Informatics: 2009; 2 (2): 108-113

[4] Starr C, Branson D, Shehab R, Farr C, Ownbey S, Swinney J. Biomechanical analysis of a prototype sports bra. Journal of Textiles and Apparel Technology and Management: 2005; 4: 1-14.

[5] Bowles KA, Steele J, Munroe B. Features of sports bras that deter Australian women. Journal of Science Medicine in Sport: 2011; 15: 195-200.

[6] Zhou J, Yu W, Ng S. Studies of three-dimensional trajectories of breast movement for better bra design. Textile Research Journal: 2012; 82 (3): 242-254.

[7] Krenzer G, Starr C, Branson D. Excellence in design: Development of a sports bra prototype. International Textile and Apparel Association: 2005; 23 (2): 131-134.

[8] Yip J, Yu W. Intimate apparel with special functions. (Chapter 8) Innovation and Technology of Women's Intimate Apparel. Woodhead Publishing Limited: Cambridge: 151-170.

[9] Okabe K, Kurokawa T. Characteristics of Three-Dimensional Displacement of the Breasts Wearing Different Kind of Brassieres in Young Japanese Women. j.jpn.res.assn.text.end-uses: 2004; 45: 416-424.

[10] Scurr JC, White JL, Hedger W. The effect of breast support on the kinematics of the breast during the running gait cycle. Journal of Sport Sciences: 2010; 28 (10): 1103-1109.

[11] Scurr JC, White JL, Hedger W. Supported and unsupported breast displacement in three dimensions across treadmill activity levels. Journal of Sports Science: 2011; 29: 55-61.

[12] McGhee DE, Steele JR. Breast elevation and compression decrease exercise-induced breast discomfort. Medicine and Science in Sports and Exercise: 2010; 42 (7): 1333-1338.

[13] McGhee DE, Steele JR, Zealey WJ. Bra-breast forces generated in women with large breasts while standing and during treadmill running: Implications for sports bra design. Applied Ergonomics: 2013; 44 (1): 112-118.

[14] Zhou J, Yu W, Ng S. Studies of three-dimensional trajectories of breast movement for better bra design. Textile Research Journal: 2012; 82 (3): 242-254.

[15] Lu MY, Qiu JY, Wang GD, Dai XQ. Mechanical analysis of breast–bra interaction for sports bra design. Materials Today Communications: 2016; 6: 28-36.

[16] Chris M, Risius D, Scurr JC. Breast motion asymmetry during running. Journal of Sports Science: 2015; 33(7): 746-753.

Research on the Elongation Properties of Breast Skin in Young Female

Li Zhao[1*], Jian-Gang Wang[1], Dong-Sheng Chen[2], Yun-Juan Liu[1]

[1]*Faculty of Clothing and Design, Minjiang University, Fuzhou 350000, P.R. China*
[2]*Textiles and Clothing Technology Innovation Center, Fuzhou 350000, P.R. China*

[*]Corresponding author's email: 81840611@qq.com

Abstract

The anthropometer and other measuring instruments are used to measure the data of 10 female undergraduate students (body type of A, height of 160cm and breast girth of 84cm). The measured data include the armpit girth, breast girth, under breast girth, cup spacing, inframammary distance, vertical line between the under arm point and the under breast girth, chest circumference line by standing naturally and passing the acromion and supraclavicular fossa, limb lateral raising, limb raising, limb front raising, spinal column bend down 80°, spinal column bend down 130°, standing straight and arms folded, spinal column lateral bending 30°, spinal column rear protection 15°. SPSS is used for data analysis to obtain the feature data including static skin length of different parts, dynamic skin stretching circumference, length and direction. The findings provide scientific and theoretical insight into the design of women's plastron and the choice of materials.

Keywords: Skin Stretching; Anthropometry; Clustering Analysis; Plastron Design; Body Comfort

1. Introduction

With the social progress and economic development there are changes in market demand and consumer awareness. Body sculpting underwear, tights, fitness clothes are gradually favored by consumers. In the new era of fashion, health factor is very important hence women demand that tights not only fit their body figure but also guarantee its' health, safety and comfort.

As early as the 19th century, a woman in Europe died because of her rib squeezing her heart while wearing a plastron. This incident brought researcher's and people's attention to the plastron. In 1978, Tomoko Miyoshi from Japan studied the structure and elastic material of the garment [1]. In 1991, Miyatsuji et al. studied the pressure of shaping pants and plastrons [2]. In 1997, Sasaki et al. worked out the relationship between pressure and comfort [3]. In 2009, Ishimaru proposed a high-elasticity shell model [4]. Compared with overseas researches, the domestic researches began relatively late. Relevant achievements were made by the pressure comfort team led by Professor Chen Dongsheng, from Mingjiang College. In 1999, they investigated the tricot's elastic modulus, the relationship between commodious quantity and the sense of pressure [5]. In 2002, they subjectively and objectively evaluated the garment pressure comfort through practical measurement from dynamic and static gestures, and analyzed the relationship between pressure and comfort of the men's suit [6]. In 2009, they comprehensively analyzed the pressure on different parts of the body after wearing the bra under different gestures [7]. From analysis of the plastron stress, they concluded that the pressure of plastron was influenced by the material, structure and the way of dressing [8]. In this paper, we focus on skin stretching under static and dynamic gestures of young women. The aim is to provide scientific and theoretical support for the design of women's plastron and the choice of materials.

2. Experimental Condition and Method

2.1 Experimental Subject

The subjects of our experiment consist of 10 female undergraduate students of Mingjiang University who are in the 20-24 age group and have standard body shapes, i.e. height of 160±1cm, body type A, and BMI 20±1.

2.2 Experimental Instrument

The instruments include the OMRON fat determination system (HBF-306), Martin anthropometer, band tape, straight ruler, angulometer, ink pen, scissor, and cardboard.

2.3 Measured Gesture

There are 9 sorts of gestures, including natural upright standing, limb lateral raising, limb front raising, spinal column bend down 80°, spinal column bend down 130°, standing straight and arms folded, spiral column lateral bending 30°, spinal column rear protraction 15°.

2.4 Determination of Baseline

Based on relevant studies [9], the baselines are as follows.

Circumference: armpit girth, breast girth and under breast girth.

Width: upper breast spacing, back spacing, cup spacing and inframammary distance.

Length: vertical line between the under arm point and the under breast girth (L1), chest circumference line by standing naturally and passing the acromion and supraclavicular fossa (L2).

2.5 Experimental Method

First, the subjects stand still and then the baselines are drawn from the breast in terms of circumference, spacing and length. The Martin anthropometer and the plastic band tape are used to measure the vertical and lateral dimensions of the statically standing body, as well as the vertical and lateral dimensions of body skin. The difference between the static and dynamic dimensions is computed to study the variation of breast skin.

During the measurement, we first measured the circumference and spacing and then measured the length in a top-down manner from left to right. Under different body gestures, we measured and recorded the data on armpit, breast and under breast.

3. Analysis of Experimental Data

3.1 Statistical Analysis of Static Data

Table.1 Statistical analysis of static data

Part	N	Minimal value	Maximal value	Mean	Standard deviation	Variance
Height	10	158.5	161.0	159.66	0.926	0.858
Weight	10	50.6	53.8	52.04	1.261	1.590
BMI	10	19.8	21.4	20.39	0.556	0.309
Armpit girth	10	84.0	86.3	85.06	0.742	0.550
Breast girth	10	82.8	84.8	83.80	0.716	0.512
Under breast girth	10	71.2	73.5	72.64	0.821	0.674
Back width	10	32.8	34.8	34.04	0.796	0.634
Front breast width	10	29.4	32.0	30.60	0.849	0.720
Cup spacing	10	0.70	1.60	1.28	0.319	0.102
Inframammary distance	10	17.3	18.5	17.88	0.431	0.186
L_1	10	11	12.1	11.74	0.408	0.166
L_2	10	47.2	48.9	48.22	0.571	0.326
Valid N	10					

In the case of natural and static standing, we measured 9 breast metrics, including height, weight, BMI, armpit girth, breast girth, under breast girth, cup spacing, inframammary distance, L_1 and L_2. The measured results are shown in Table 1.

The data in this table are obtained through SPSS analysis and it provides the raw data for the analysis of the body skin's original length.

3.2 Statistical Analysis of Dynamic Breast Data

From the above discussions, it can be learned that the experimental Data are measured in terms of circumference, spacing and length. In the case of circumference measurement, it is divided into the front and back parts with L_1 as the boundary. Let ε denote the skin's total percentage elongation. Hence, the elongation of the front and back parts can be denoted with εq and εh, respectively. Moreover, ε can be computed as follows:

$$\text{Rate of skin change}(\varepsilon)=\frac{\text{Dnamic length - Static length}}{\text{Static length}}\times 100\% \quad (1)$$

3.2.1 Collection and Analysis of Circumference Data

The rate of skin change along the circumference direction of the upper limb under different gestures is shown in Table 2.

Table.2 The rate of skin change along the circumference direction, ε(%)

Parts	Rate of skin change	Limb lateral raising	Limb raising	Limb front raising	Spinal column bend down 80°	Spinal column bend down 130°	Standing straight and arms folded	Spinal column lateral bending 30°	Spinal column rear protectio n 15°
Armpit girth	εq	-1.064	-14.792	-20.871	-22.746	-24.265	-20.973	0.709	5.522
	εh	2.574	22.469	19.808	31.021	35.166	17.190	0.393	-2.967
	ε	0.893	5.596	0.823	6.537	7.853	-0.658	0.917	1.340
Breast girth	εq	-3.397	-10.796	-10.191	-12.704	-14.518	-8.795	0.186	3.118
	εh	-0.049	7.876	6.660	19.981	22.606	6.514	2.868	-2.431
	ε	-1.599	-1.026	-1.456	3.962	3.580	-0.191	1.814	1.146
Under breast girth	εq	-2.220	-5.407	-3.699	-8.822	-7.285	-3.130	-0.114	2.903
	εh	1.758	5.381	3.463	15.237	17.155	3.250	1.119	-1.012
	ε	-0.798	-0.330	-0.551	3.442	4.598	-0.028	0.138	-0.220

(Note: a positive rate of change denotes skin elongation, and a negative rate of change denotes skin contraction.)

While designing and wearing the plastron, we regard the armpit girth, breast girth and under breast girth as a whole, and the total rate of skin change in the armpit girth is in the range of -0.658%-7.853%. The total rate of skin change in the breast girth is small and in the range of -1.599%-3.962%. The rate for the under breast girth is in the range of -0.798%-4.598%. The skin is mostly in a contracted state in the front skin of armpit girth, breast girth and under breast girth, and it is elongated in the back skin. It is particularly obvious when the spinal column bends down, while the rate of skin change being up to 35.166%.

3.2.2 Collection and Analysis of Spacing Data

The rate of skin change along the spacing direction of the upper limb under different gestures is shown in Table 3.

The rate of skin change is very evident for the back width in the dynamic case, and it ranges from -3.685% - 70.418%. Except the case of spinal column rear protraction 15°, the skin is always elongated. And it is elongated more than other parts, with the rate being 70.418%. The rate of skin change contracts most in the breast and cup spacing in the case of standing straight and arms folded. The skin is also elongated considerably for limb lateral raising, and the rate is in the range of -25.229%-15.3595 and -39.0635-17.188%. The rate of change is small for the inframammary distance, being in the range of 19.463-5.705%.

Table.3 The rate of skin change along the spacing direction,ε(%)

Parts	Rate of skin change	Limb lateral raising	Limb raising	Limb front raising	Spinal column bend down 80°	Spinal column bend down 130°	Standing straight and arms folded	Spinal column lateral bending 30°	Spinal column rear protectio n 15°
Back width	ε	14.343	50.797	48.506	61.653	70.418	34.463	6.375	-3.685
Front breast width	ε	15.359	5.163	-11.569	-14.510	-12.418	-25.229	-6.078	4.118
Cup spacing	ε	17.188	4.688	-7.813	-23.438	-23.438	-39.063	-12.500	12.500
Inframa -mmary distance	ε	3.132	-3.915	-7.159	-17.450	-19.463	-9.172	0.783	5.705

3.2.3 Collection and Analysis of Length Data

The rate of skin change along the length direction of the upper limb under different gestures is shown in Table 4.

Table.4 The rate of skin change along the length direction,ε(%)

Parts	Rate of skin change	Limb lateral raising	Limb raising	Limb front raising	Spinal column bend down 80°	Spinal column bend down 130°	Standing straight and arms folded	Spinal column lateral bending 30°	Spinal column rear protection 15°
L_1	ε	29.983	62.010	34.072	31.857	40.886	21.635	-12.095	-2.385
L_2	ε	1.286	1.286	1.078	-3.443	-6.719	1.908	1.825	-1.825

Elongation is the major form of skin change for L_1 and the change is the most obvious in limb raising at 62.010%. In the case of spinal column lateral bending 30° and spinal column rear protection 15°, the skin is contracted and the rate is in the range of-12.095%-2.385%. When it comes to L_2, the rate of change is small in the range of -6.719-1.908.

3.3 Clustering Analysis of the Rate Under Dynamic Scenarios

Table.5 Final clustering centers (%)

	1	2	3	4	5	6
VAR00001	1.18	37.37	-14.61	-39.06	64.69	18.48

Table.6 Observed values of each cluster

Clusters	1	140.000
	2	18.000
	3	52.000
	4	2.000
	5	6.000
	6	24.000
Valid		242.000
Omission		.000

The rate of skin change fluctuates radically under dynamic scenarios. In order to obtain the feature value of skin elongation, SPSS is used to perform K-means clustering analysis of the rate in armpit girth, breast girth, under breast girth, cup spacing, inframammary distance, L_1 and L_2. The skin elongation and contraction Data are shown in Tables 5 and 6.

From the tables, it can be seen that the rate of skin change among the young women can be clustered into six groups of data, as shown in Table 5. The rate is mostly concentrated in 1.18%, followed by skin contraction of -14.61%. Skin elongation is very obvious at 37.37% and 18.48%, with many data samples concentrating in these ranges. The rate of skin change peaks at -39.06% and 64.69%, and there are few data samples clustered around.

4. Conclusions

Female healthy undergraduate students (age group of 20-24, body type of A, height of 160±1cm, BMI of 20±1 and breast girth of 84±2 cm) are chosen to measure the breast skin, producing 9 groups of data. Analysis of the data reveals the skin elongation features of the young women.

1) The rate of skin change along the circumference exhibits the following features: the armpit girth skin changes radically at -0.658%-7.853%, and the breast girth skin changes slightly at -1.599%-3.962%. In the case of skin change along the spacing direction, the back skin provides the greatest elongation. Particularly for spinal column bend down, the rate is 70.418%. When it comes to the feature of change along the length direction, the skin change of L_1 has the greatest rate of 62.010%.

2) From analysis of the skin elongation and contraction data, it is learned that the lateral (spacing and circumference) skin change is larger than the rate along the length direction. K-means clustering analysis shows that the feature values of skin elongation or contraction are 1.18%, 37.37%, -14.61%, -39.06%, 64.69% and 18.48%. In detail, skin elongation is mostly clustered around 1.18%, and skin contraction is clustered around 37.37% and 18.48%. Theoretically, when the elongation of shell fabric is consistent with skin elongation, wearing the plastron made of this material produces a high level of comfort. Hence, 37.37% and 18.48% can be used as the reference value for the choice of shell fabric elasticity while making the female's plastron.

5. Acknowledgement

The authors wish to acknowledge the young and middle-aged education scientific research projects (JA15432) of Fujian Education Department, National university students innovation project (201510395007).

References

[1] Li Y. Clothing Comfort and Product Development. Beijing: China's textile press: 2002; p. 35-37.

[2] Miyatsuji A1, Matsumoto T, Matarai S, Kotabe T, Watanuki S. Effects of clothing pressure caused by different types of brassieres on autonomic nervous system activity evaluated by heart rate variability power spectral analysis. J Physiol Anthropol Appl Human Sci: 2002 Jan; 21 (1): 67-74

[3] Sasakik, Miyashitak. Evaluation of foundation comfort based on the sensory evaluation and dynamic clothing pressure measurement. Journal-Japan Research Association for Textile End Uses: 1997; 38 (3): 53-58.

[4] Ishimaru S, Isogia Y. Prediction of clothing pressure distribution by using finite element method. Journal of Textile Engineering: 2009; 55 (6): 179-186.

[5] Dan R, Chen DS, Fan XR, Wang Q. Advances and research on distribution and prediction of clothing pressure. Journal of Textile Research: 2010; (4):139-144.

[6] Chen DS Cui LM. Development of clothing measuring system. Journal of Textile Research: 2008; (3): 72-75.

[7] Chen HY, Chen DS, Yuan XH. Objective measurement and analysis of bra stress Distribution under Static and Dynamic Status.Textile science and technology progress: 2009; (5): 84-86.

[8] Gan YJ. Analysis of Pressure on the comfort of Bra. International textile guides: 2009 (2): 80-82.

[9] Zhao L, Chen DS, Wang JG. Study on relationship between ease and fabric elasticity of men's trousers based on skin expansion. Journal of Wuhan Textile University: 2016; 29 (4): 3-7.

Parametric Representation on the Curve of the Chest and the Design of the Basic Cup

Bo-An Ying*, Si-Chen Pan, Xin Zhang

Apparel & Art Design College, Xi'an Polytechnic University, Xi'an, Shaanxi, 710048, P.R. China

*Corresponding author's email: yingba2006@163.com

Abstract

Under the influence of customization, clothing consumers have a high demand for a proper fitting. Proper fitting and customization are the most important factors that affect appearance and clothing sales performance. The same is true for the lingerie market, where consumers are demanding more fit and tailored underwear. Cup bra when compared to the traditional bra is becoming popular in recent years in the market by consumers and bra products.

Fit cup bra requires chest cup design specifications which are completely based on the size of consumers. This paper puts forward a detailed study to form a feature point based on the human body feature points. The research content includes the extraction of human chest points and the required module design data points in the environment of computer data points to represent the bra mold cup design using chest surface finish, it starts from point to linė and finally to the cup surface design foundation. A complete model of consumer cup based on the human body is made, from the body to relation model cup is established to create a personalized brassiere.

Keywords: Moulded Bra-cup; Personalized Customization; Surface Modelling; Parameterization; Digital Design

1. Introduction

The most important part of a bra cup are the cups. Correction bra shaping is done by bra cups as it is directly in contact with the chest and as such it is a three-dimensional model [1]. Because of this, the shape and the fit of the human body directly affects the bra after wearing aesthetics and comfort. It is very important for individually tailored mold cup design.

Jia Li from Xi'an University of Engineering [2] has studied the computer model of the mold cup bra for western female college students, and summed up the key human characteristics data. Wang Yue [3] from Xi'an University of Engineering carried out an in-depth study based on the bra model design method, the feature points and characteristic curves of the shape of the brackets were clearly characterized, and the parametric characteristic relationship between them was established. Moreover, Han Jing from the Xi'an University of Engineering [4], through the detailed analysis of the relationship between the parameters of female breast shape and mold design, deduced that the main human body parameters can be used in mold cup design and established the digital mold cup model.

In this paper, the design idea of personalized brassiere based on the extraction of human detail size combined with 3D body measurement technology was put forward. The value point extraction of human chest type and the corresponding basic brassiere design value in the environment of computer data points was used to represent the bra mold cup design using the chest surface. This was done from from the point to the line to complete the basic model surface design.

2. The Relationship between Female Breasts Details Characteristics and Design Parameters of the Moulded Bra-cup

In anthropometry, the human feature point is a clear point used as a marker location on the human body. In general, the location of the human body points in the human body are either more prominent or are joint joints. In this paper, the female chest feature points are defined as to clearly reflect the female chest shape point. From these points on the human body, the relationship between the female chest feature points and the base model of the moulded bra-cup is established, and these points are used as the basis geometric elements for the model. According to the summary of the basic model design of the moulded bra-cup, it is required to obtain the human characteristics of the female chest feature points as shown in Fig.1.

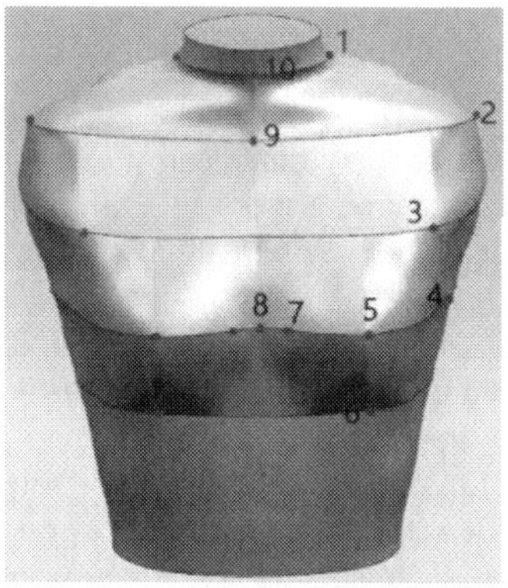

1.Neck side point
2.Shoulder end point
3.The upper point of the breast
4.The outer point of the breast
5.BP
6.The lower point of breast
7.The inner point of the breast
8.Front center point

Fig.1 The female chest feature points for the design of the basic moulded bra-cup

The human chest feature points are the reference points of the human body feature curve. The feature curve is the line formed by connecting two or more feature points on the human body surface. These curves, which characterize the size and position of the human body, are shaped according to the design of the model of moulded bra-cup, which not only provides an important reference for the design, but also serves as the main object for the later adjusting of the feature. Fig.2 shows the feature curves that can be obtained directly from the human body. These feature curves can be used to indicate the human body shape.

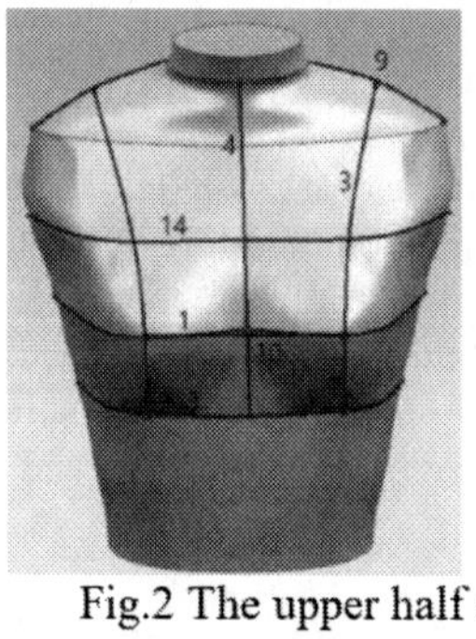

1.Chest line
3.Princess line
4.Front centerline
9.Shoulder line
13.Under the chest line
14.On the chest line

Fig.2 The upper half of the body feature curves

The goal of this research is to achieve the complete fit of the moulded bra-cup. Therefore, the design of the model of the moulded bra-cup needs to be based on the size and shape of the human body, so that the model of the moulded bra-cup can meet the requirement of the size ratio and body shape of the target body. Through the above summary of the female chest detail feature points and feature curves, the required geometric elements are determined for the corresponding design of the model of the moulded bra-cup. The corresponding relationship between the chest detail feature points and the design feature points of moulded bre-cup are shown in Table 1. The corresponding relationship between chest feature curves and the model design curves of moulded bra-cup are shown in Table 2.

Table.1 The correspondence between the feature points of the chest detail and the design feature points of moulded bra-cup

Serial number	Body chest	Moulded bra-cup
1	Neck side point	Neck side point
2	Shoulder end point	Ear position reference point
3	The upper point of the breast	loop position
4	The outer point of the breast	Lateral position
5	BP	The highest point of the moulded bra-cup
6	The lower point of the breast	Under the chop site
7	The inner point of the breast	Moulded bra-cup inner arc reference point
8	Front center point	BP

Table.2 The correspondence between the chest characteristic curves and the design curves of moulded bra-cup

Serial number	Body chest	Moulded bra-cup
1	Breast external arc length	Cup bowl side arc length
2	Breast inner arc length	Cup front bowl arc length
3	Upper and lower milk cup characteristic curve	Bowl cup on the characteristic curve
4	Breast root arc	Grilled and leading characteristic curve under the cup
5	Front cup curve	The front cup bowl curve
6	BP point to the chest under the vertical distance	Cup height
7	Breast circumference	Cup width

3. Definition and Parameterization of the Data Points of Human Chest

In this paper, the upper body shape of human body is built in UG (UG (Unigraphics NX) It is a product engineering solution produced by the Siemens PLM Software company. It provides digital modeling and verification methods for user's product design and processing) environment. According to the body's own feature points, through BP point, the upper point of the breast and the lower point of the breast, three horizontal datum planes were established in the part of human chest, as shown in Fig.3. Through the front center point, and points on the both right and left side including BP point, the outer point of the breast and the inner point of the breast, seven vertical datum planes were established as well, as shown in Fig.4.

Fig.3 Horizontal datum planes

Fig.4 Vertical datum planes

However, the established datum planes as shown in Fig.3 and Fig.4 are too simple to form enough feature points on the part of human chest for subsequent design operations. Therefore, some additional datum planes in the vertical and horizontal direction needs to be added on the basis of the established datum planes. In the horizontal direction, the number of n1 layers of horizontal datum plane were established equally between the BP point and the upper point of the breast, where n1 is equal to 3; the number of n2 layers of horizontal datum plane were established equally between the BP and the lower point of the breast, where n2 is equal to 3. As a result, the total number of 3+n1+n2=9 layers of horizontal datum plane were established in the part of human chest, as shown in Fig.5. As the symmetry of the human chest lies in the vertical direction, the same number of vertical datum planes can be built both on the both right and left side of the human chest. The number of n4 layers of vertical datum plane were established between the BP and the inner point of the breast on the both side, where n4 is equal to 2. The number of n5 layers of vertical datum plane were established between the BP and the outer point of the breast on the both side, where n5 is equal to 2. As the result, the total number of 7+2*n4+2*n5=15 layers of vertical datum plane were established in the part of human chest, as shown in Fig.5.

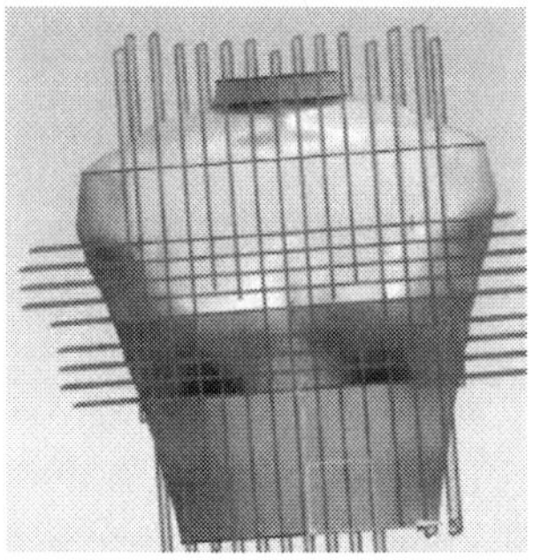

Fig.5 The system of horizontal and vertical datum planes

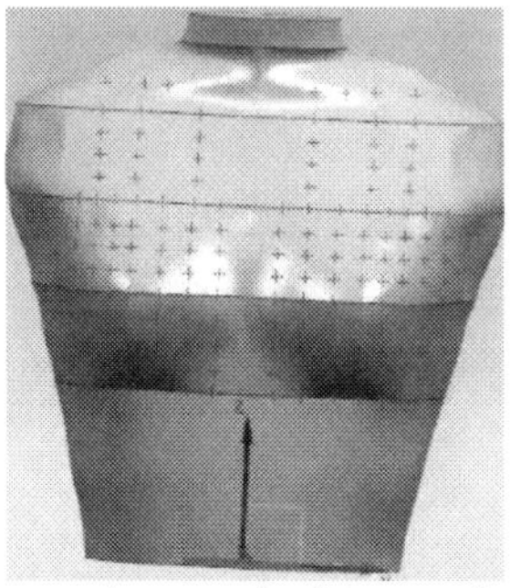

Fig.6 Data points on the human chest

Based on the 9 layers of horizontal datum plane and 15 layers of vertical datum plane established above, a total of 9*15 intersection points on the surface of human chest can be obtained, as shown in Fig.6. These intersection points are defined as the data points on the human chest, which can be expressed with the matrix parameterization relationship. The matrix of the data points on the left side of the human chest are expressed by equation (2-1), which is a matrix formed with 9*8 elements. This matrix can be used to represent all the values of the left chest (2-1):

$$B\_L=(b\_l_k)_{i,j}=\begin{bmatrix} (b\_l_k)_{1,1} & (b\_l_k)_{1,2} & (b\_l_k)_{1,3} & (b\_l_k)_{1,4} & (b\_l_k)_{1,5} & (b\_l_k)_{1,6} & (b\_l_k)_{1,7} & (b\_l_k)_{1,8} \\ (b\_l_k)_{2,1} & (b\_l_k)_{2,2} & (b\_l_k)_{2,3} & (b\_l_k)_{2,4} & (b\_l_k)_{2,5} & (b\_l_k)_{2,6} & (b\_l_k)_{2,7} & (b\_l_k)_{2,8} \\ (b\_l_k)_{3,1} & (b\_l_k)_{3,2} & (b\_l_k)_{3,3} & (b\_l_k)_{3,4} & (b\_l_k)_{3,5} & (b\_l_k)_{3,6} & (b\_l_k)_{3,7} & (b\_l_k)_{3,8} \\ (b\_l_k)_{4,1} & (b\_l_k)_{4,2} & (b\_l_k)_{4,3} & (b\_l_k)_{4,4} & (b\_l_k)_{4,5} & (b\_l_k)_{4,6} & (b\_l_k)_{4,7} & (b\_l_k)_{4,8} \\ (b\_l_k)_{5,1} & (b\_l_k)_{5,2} & (b\_l_k)_{5,3} & (b\_l_k)_{5,4} & (b\_l_k)_{5,5} & (b\_l_k)_{5,6} & (b\_l_k)_{5,7} & (b\_l_k)_{5,8} \\ (b\_l_k)_{6,1} & (b\_l_k)_{6,2} & (b\_l_k)_{6,3} & (b\_l_k)_{6,4} & (b\_l_k)_{6,5} & (b\_l_k)_{6,6} & (b\_l_k)_{6,7} & (b\_l_k)_{6,8} \\ (b\_l_k)_{7,1} & (b\_l_k)_{7,2} & (b\_l_k)_{7,3} & (b\_l_k)_{7,4} & (b\_l_k)_{7,5} & (b\_l_k)_{7,6} & (b\_l_k)_{7,7} & (b\_l_k)_{7,8} \\ (b\_l_k)_{8,1} & (b\_l_k)_{8,2} & (b\_l_k)_{8,3} & (b\_l_k)_{8,4} & (b\_l_k)_{8,5} & (b\_l_k)_{8,6} & (b\_l_k)_{8,7} & (b\_l_k)_{8,8} \\ (b\_l_k)_{9,1} & (b\_l_k)_{9,2} & (b\_l_k)_{9,3} & (b\_l_k)_{9,4} & (b\_l_k)_{9,5} & (b\_l_k)_{9,6} & (b\_l_k)_{9,7} & (b\_l_k)_{9,8} \end{bmatrix} \tag{2-1}$$

The above matrix representation denoted by *B_L*, refers to the data points on the left side of the chest surface, where *B* denotes the human body, *L* denotes the left chest. The elements of the matrix are denoted by *b_l*, where *k* denotes the axis in the x, y, z coordinates, k = (1, 2, 3), when k = 1, the data point is the x-axis coordinate value. When k = 2, the data point is the y-axis coordinate value. Wwhen k = 3, the point is the z-axis coordinate value; i and j denote the number of row and column in the matrix. As such, the matrix of data points on the left chest can be expressed with the following equation (2-2).

$$B\_L=(b\_l_k)_{i,j}$$
$$k=(1,2,3)_{,}\ i\in\{1,2,\ldots,3+n1+n2=9\}_{,}\ j\in\{1,2,\ldots,4+n4+n5=8\} \tag{2-2}$$

Similarly, the matrix of data points on the right chest denoted by B_L can be expressed as the equation (2-3), where R denotes the right chest and b_l denotes the elements of the matrix.

$$B\_R = (b\_r_k)_{i,j}$$
$$k = (1,2,3)\,,\; i \in \{1,2,\ldots 3 + n1 + n2 = 9\}\,,\; j \in \{1,2,\ldots,4 + n4 + n5 = 8\} \tag{2-3}$$

4. He Establishment of the Basic Model of Moulded Bra-cup

4.1 The Definition of the Basic Model of Moulded Bra-cup

The surface of the model based on the body's chest size information and the relationship between the size of the chest and the moulded bra-cap are called "basic model of moulded bra-cup"

4.2 Design and Representation of the Relationship Model of the Basic Model of Moulded Bra-cup

The following points explain the relation of the basic bra cup. (2-4):

$$F\_R = (f\_r_k)_{i,j}\; B\_R' = (b\_r_k)'_{i,j} = (B\_R) + (\Delta P\_B)$$
$$= (b\_r_k)_{i,j} + (\Delta P_{lj} - \Delta P_{rj})$$
$$F\_L = (f\_l_k)_{i,j} = B\_L\_v = (b\_l\_v_k)_{i,j} \tag{2-4}$$

It can be seen on the above relation that the point of the right level of the base mold represents the left level of the base mold cup. The left and right chest of the human body from the point of the human sagittal plane on the coordinates of the difference, the relationship is expressed as follows (2-5):

$$\Delta P_{lj} = (B\_L\_v) - (B\_A) = \begin{bmatrix} (b\_l\_v_k)_{i,j} \\ (b\_l\_v_k)_{i,j} \\ (b\_l\_v_k)_{i,j} \\ (b\_l\_v_k)_{i,j} \\ (b\_l\_v_k)_{i,j} \\ (b\_l\_v_k)_{i,j} \\ (b\_l\_v_k)_{i,j} \\ (b\_l\_v_k)_{i,j} \\ (b\_l\_v_k)_{i,j} \end{bmatrix} - \begin{bmatrix} (b\_a_k)_{i,8} \\ (b\_a_k)_{i,8} \\ (b\_a_k)_{i,8} \\ (b\_a_k)_{i,8} \\ (b\_a_k)_{i,8} \\ (b\_a_k)_{i,8} \\ (b\_a_k)_{i,8} \\ (b\_a_k)_{i,8} \\ (b\_a_k)_{i,8} \end{bmatrix} \quad k = \{1,2\}\; i \in \{1,2,\ldots 9\}\; j \in \{1,2,\ldots 7\}$$

$$\Delta P_{rj} = (B\_R\_v) - (B\_A) = \begin{bmatrix} (b\_r\_v_k)_{i,j} \\ (b\_r\_v_k)_{i,j} \\ (b\_r\_v_k)_{i,j} \\ (b\_r\_v_k)_{i,j} \\ (b\_r\_v_k)_{i,j} \\ (b\_r\_v_k)_{i,j} \\ (b\_r\_v_k)_{i,j} \\ (b\_r\_v_k)_{i,j} \\ (b\_r\_v_k)_{i,j} \end{bmatrix} - \begin{bmatrix} (b\_a_k)_{i,8} \\ (b\_a_k)_{i,8} \\ (b\_a_k)_{i,8} \\ (b\_a_k)_{i,8} \\ (b\_a_k)_{i,8} \\ (b\_a_k)_{i,8} \\ (b\_a_k)_{i,8} \\ (b\_a_k)_{i,8} \\ (b\_a_k)_{i,8} \end{bmatrix} \quad k = \{1,2\}\; i \in \{1,2,\ldots 9\}\; j \in \{1,2,\ldots 7\}$$

(2-5)

This difference is the adjustment reference value for the right chest type point. At present, the left side of the chest corresponding to the base mold cup value and the right side of the chest corresponding to the base mold cup point of the human sagittal section is set on the basis of a completely symmetrical state. As such, the formation of the type of point to facilitate the late mold cup design can be used directly in the mold-cup model design link.

In this study, according to the three generation curve spline method to generate, according to the definition of three spline curve, for this series of points will be generated by the curve, each two parameters which are connected to form curve segments, then the curve segment connected. That is formed by three spline curves of this series of points. The expression level of the parameter curve segment based on module (2-6) is shown as follows:

$$f_{a,b}(t)_l = B_m + B_{m+1}t + B_{m+2}t^2 + B_{m+3}t^3 \qquad 0 \le t \le t_m \tag{2-6}$$

In the UG environment, according to the relationship between the expression of this type of spline curves, points connecting the list of all types, form value point of spline curve of this type, other aspects of spline curve in accordance with the value calculated point coordinates and three spline curve to form. As shown in Fig.7, the horizontal and vertical spline curves are formed for the left chest. After the partition of the base level of all kind of cup curve and vertical spline curve, a total of 9*8* level vertical cup is formed based on the spline curve in the chest position, as shown in Fig.8.

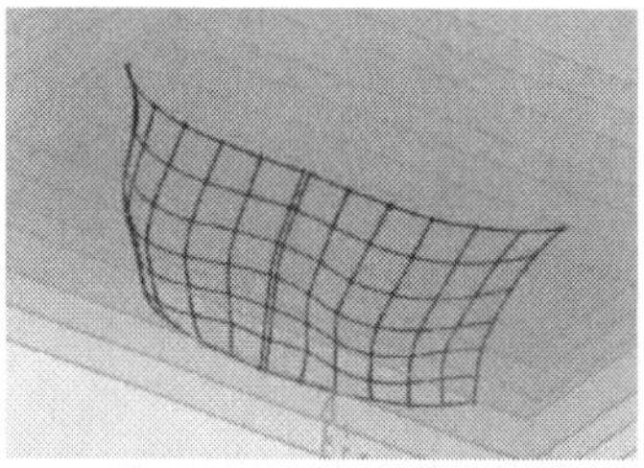

Fig.7 Body chest level, vertical spline curve

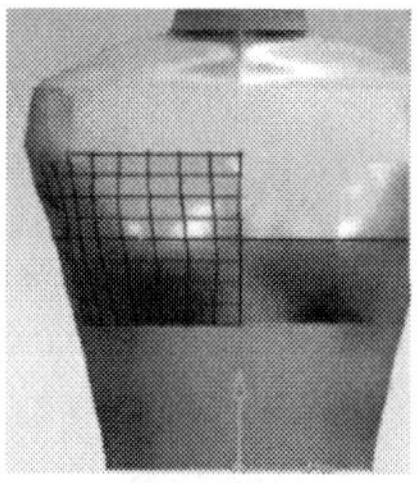

Fig.8 Basic mold cup spline curve

Based on the horizontal characteristic curve of the base mold cup and the cross grid formed by the vertical characteristic curve, the bimodal surface is used to form the matrix of the base mold cup. For the left base mold cup, the number of horizontal characteristic curve * vertical curve of the total number of curves for a total of 9 * 8, these characteristics of the grid cross the grid formed 8 * 7 = 56, as shown in Fig.9. Thus, the formation of the base mold surface is made up of the 56 grid surfaces. The 56 grids are connected smoothly to form the grid surface of the left chest base mold cup as shown in Fig.9

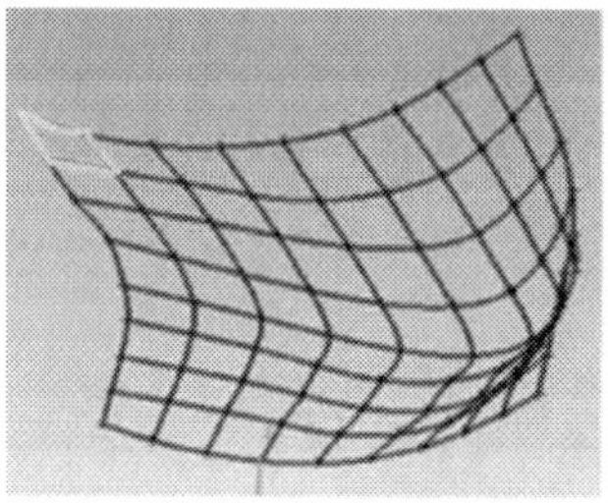

Fig.9 The vertical curve segment and the horizontal curve segment are generated by the grid

5. Conclusion

In this paper, the base mold cup which is the first step of the design process of the bracelet mold model, and that is based on the body's chest size information and the relationship between the size of the chest and the size of the bas e mold cup (called "Base mold cup relationship model") to design. The 3D scan is used to target individuals. After the informed data is formed in the computer parametric model and a model of the human body is generated, the relevant feature points is obtained and analyzed. The formation of the corresponding relationship between human body characteristics and mould design is established based on the relationship between the model parameters and the model of chest based mould design.. This form is established based on the model of consumer cup body.The body's relationship to the relational model cup is established, and a true personalized brassiere is achieved.

6. Acknowledgement

The authors wish to acknowledge the National Natural Science Foundation of China and the Youth Science Foundation (No. 61303120).

References

[1] Lv FF. Design of Human Body Based on Human Body Model [D], XUEM: 2012.
[2] Jia L. Underwear business model cup bra existing production process simplified study [J], VE: 2012 (30); 19-21.
[3] Wang Y. Study on Design of Brake Cup Model Based on Feature [D], XUEM: 2010.
[4] Han J. The design parameters of bra cups [D], Xi'an Polytechnic University, master's degree thesis: 2010.
[5] Zheng Y. Study on the somatotype of female college students based on the definition [D]. XXCEST: 2005.

Deconstruction Of 3D Printing Technology For Modern Clothing Design

Xiao-Hua Lin*, Jian-Gang Wang

Xi Yuan Road 200#,ShangJie Town, Ming Hou Country, the School of Fashion and Art Engineering, Min jiang University, Fu Zhou, Fu Jian, 350108, China

*Corresponding author's email: 120540961@qq.com

Abstract

Modern clothing design has been constantly in development due to 3D printing technology. Consequently, not only have three-dimensional shapes that are unattainable in traditional clothing design become achievable, but optimization and iteration of the production line for clothing has become faster, making customization of modern clothing feasible.

Keywords: 3D Printing Technology; Modern Clothing Design; Shoe Industry; Updating

In recent years, there has been continuous development of three-dimensional (3D) printing fabrication processes and materials with ever-increasing scope. This has included applications such as "3D printed skin", "3D printed cartilage" and "3D printed self-healing materials", *etc*., demonstrating that it is a disruptive technology which is shocks traditional manufacturing industries. Recently, Ray Kurzweil, a machine learning expert working for Google, has suggested that advanced technologies such as 3D printing will fev the fashion industry. There is no doubt that, with growing maturity and expansion of the scope of the technology, significant changes to manufacturing and will occur in the future [1].

The archetype of 3D printing technology is a 3D digital model which is translated using a hot-melt nozzle and laser beam to stack and overlap, layer-by-layer, special materials such as plastics, metal powder, cellular tissue and ceramic powder, *etc*. The procedure includes: 1) design of the digital model, generation of the 3D model file, selection of suitable printing materials and the printing process, and setting the resolution and size of the printer's output, via software; 2) outputting printed layers using specialized layer-control software. In this way, an object will be printed out that is almost the same as the initial model [2].

1. Advantages of 3D Printing

3D printing technology perfectly integrates the advantages of mechanization including high efficiency, low cost and high accuracy with the diversification and individualization of artificial design, accelerating the design lifecycle and improving fabrication precision and rate. The potential advantages that 3D printing technology have which traditional mass-production techniques do not, include: 1) short production period (continuous improvements of the technology will reduce the time required for 3D printing compared with current production times); 2) environmental protection (the 3D printing inks and materials currently used are non-toxic, recyclable and environmentally-friendly thus protecting the environment); 3) low manufacturing cost (3D printing technology eliminates expensive molding and tooling, allowing customers to immediately print objects they wish to purchase removing the need for storage of stock and reducing waste and unsalable products, thus lowering cost. In addition, 3D printing technology that has high output with short fabrication time has the potential to be cheaper and more precise than artificial manufacturing techniques, and it may provide a breakthrough at a technological level. This reduces previous dependence on craftsmanship, effectively improving product workmanship, and greatly reducing the reject ratio and so reducing production cost; 4) greater precision and safety (due to use of the mechanical arm and robotics of 3D printing technology for production, which provides automatic control and detection). Existing representative technology is SiSpis (Siemens Spiders), a 3D printer with laser scanner and camera which can output PLA printed materials. In the context of large-scale 3D printing, robots can possibly work together to complete printing tasks based on procedures input in advance. 3D printing technology might precisely perform true "customization" according to the scanning data [3].

Fig.1 Spire dress

Fig.2 3D printed petal dress

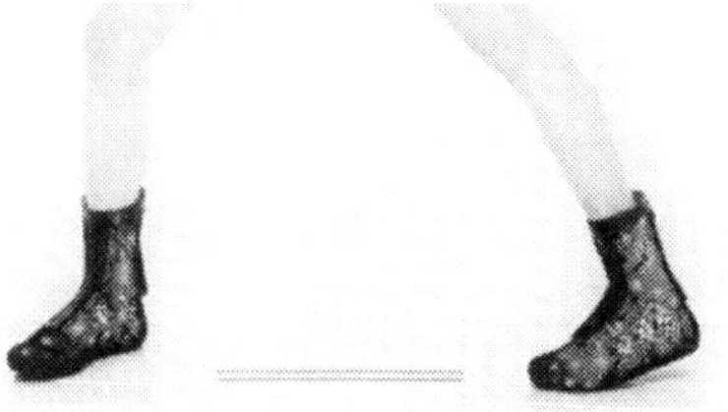

Fig.3 3D printed shoes

2. Updating 3D Printing Technology in the Clothing Industry

At present, the reproducibility of 3D printing is the greatest advantage of the technology. Even compared with mechanized mass-production, it has certain advantages [3]. Looking several years in the future, with continuous development and introduction of increasing numbers of materials for 3D printing and user-friendly CAD software, 3D printed clothing will be increasingly popular among consumers, and application of 3D printing technology in the field of fashion will be comprehensively updated [4].

During the New York Fashion Week of 2016, Alexis Walsh's LYSIS collection and the Spire formal dress were presented, all attractive works with unique style that were made using 3D printing technology. The LYSIS collection was produced using a combination of 3D printing and traditional clothing technologies, with a design that simulates a growing virus structure that connects organic shapes with emotional elements [5]. Firstly, the structure was fabricated with white nylon using 3D printing technology followed by polishing, coloration and surface treatments by hand. Finally, various printed 3D pieces of the clothing, and formation of the unique structure was achieved using manual tailoring. The Spire formal dress (Fig.1) was made from more than 400 individual parts printed using SLS technology then manually spliced using tiny metal linkers. Various geometrical shapes similar to the top of a cathedral on the formal dress were made using Grasshopper Software. The archetype and printing of the items were performed by the designer's personal 3D printer [6].

Customization of the 3D printing of clothing has been put more into practice during 2016. For customization, the designer requires a customer to enter the changing room wearing only their underwear and equipped with an ultra red sensor. The scanning process lasts less than 30s and provides 120 pieces of information on the customer's body type. With these detailed data, the designer can precisely judge the customer's body type and posture so as to customize the design of suitable, personalized clothing.

Another company, Nervous System, presented a 3D printed dress (Fig.2) decorated with 1600 petals in 2016, and this dress was tailored using measurements from each part of the body [6]. Their process of creating custom-made clothing using 3D printing concepts combines the results of the three-dimensional scanning of the customer's body, selecting the clothing style and adjusting the intensity of nylon petals. Selective laser sintering technology is used to print the clothes which are then finally cleaned of redundant chippings. At the same time, they have researched and developed software based on dynamics, simplifying the design and printing of collapsible clothing. Such technology not only avoids the cumbersome process of traditional clothing manufacture, but can also make the clothing fashionable.

A limitation of 3D printing materials means that existing 3D printed clothing is mostly relatively stiff and uncomfortable. However, Monika Januszkiewicz, a student from the Textile and Fashion Department of the University of Huddersfield in the UK has designed a comfortable 3D printed skirt (Fig.3) consisting of 2000 interlocked grids [6]. The inspiration for the skirt came from dragon scales. It was finished using three software programs: Marvelous Designer, 3ds Max and Zbrush. The material was TPU 92A-1, a flexible, elastic and durable thermoplastic polyurethane, and very suitable for fashion design.

3D printing not only brings innovation to the clothing industry, it is also changing the footwear industry. By comprehensively scanning a customer's feet in several short seconds, detailed data can be obtained, including toe and tarsal bone girth, heelpiece height and dozens of other parameters. In addition, based on the precise shape and demands of support for the user's feet, combined with video material and data, selection of the correct size suitable for the feet realize "shoemaking under feet measurement" for customers. This allows optimal iteration of product lines of the shoe industry and makes customization workable. Many brands now depend on 3D printing to achieve a competitive advantage [7].

The Nike brand, using 3D printing technology, has begun to popularize a "socks-type" design focusing on improving fit and support, which combines an avant-garde shape with improved performance to attract more buyers.

New Balance has generated an algorithm by researching a mass of data from the company Nervous System and has designed unique and fully customized shoes by selective laser sintering (SLS) 3D printing technology. This type of shoe is entirely suitable for the user's feet shape and their way of exercising by increasing the organizational density of shoe's sole to achieve an ideal pressure buffer.

The 3D printed shoes from SOLS (Fig.3), named Adaptiv, have been constructed with motion sensors in the body of the shoe and a silicone air pocket in the ankle area to monitor the motion of wearer. The pressure distribution is changed accordingly based on the demands and support required in the feet, ankles and other parts through inflation and deflation [6]. Meanwhile, they installed LED lamps and a color sensing camera on the shoes so that the colors changed according to demand. High-performance shoes help sportsmen better adapt to physical activity and improve their sports grade through excellent adaptability. These 3D printed shoes have the possibility of selecting from various materials.

Shamees Aden, a designer From London, has developed with researchers "Amoeba", a self-healing biological running shoe by taking advantage of 3D printing technology and integrating reproducible compound archaeocyte particles. The shoes fit the user's feet perfectly just like a second skin and react to pressure, movement, light and heat. Special areas of the shoe will expand when the user is running to provide an extra buffer making the person wearing the shoes feel more comfortable. The shoes also have a self-healing function.

For designers, 3D printing technology allows them to maximize their originality of design, making reality things you could previously only think of, no matter how complex the structure or how strange the form. 3D printing has become almost universal in Euromerican developed countries. Many countries have created various policies and raised investment for development, bringing convenience and intelligence to production [8].

3. Easily Occurring Problems in the Process of Translation of 3D Models Into Printed Objects

It is necessary to design a good 3D model for perfect unification between the 3D printing and 3D model. However, the following problems easily occur in the process of translation of the 3D model into a printed object:

3.1 Mismatching of Model File with the Selected Printing Consumables

The properties of printing consumables vary. They can be fragile, tough, flexible, stiff, smooth or rough, and can be either high or low density, etc. Therefore, before designing the model, the designer should assess the features of suitable materials and select appropriate consumables for printing [9].

3.2 Mismatching of Model File with Printing Process

As the properties of consumable materials vary, so do printing processes. Different printing processes are required for printing different objects. For example, 3D models with an interlocking structure requires ABS, PLA, aluminium alloy or rubber. The printing of precious metals must adopt a lost-wax process, and the printing of resin and ABS with high accuracy and toughness should adopt SLA printing technology which is based on liquid. Thus a suitable printing process needs to be selected.

3.3 Thickness of the Model File does not meet Printing Requirements

At present, the most common problem encountered during 3D printing is that the thickness of the 3D model does not meet the printing requirements. When the model is too thin, it is impossible to print out the small parts of the model, or it easily damages the model due to high fragility. When the model is too thick, the printout may suffer from cracks or damage. Hence, appropriate thickness is the most important factor for successful printing.

3.4 Resolution of the Model File does not meet Printing Requirements

A critical factor for printing a high-quality model is adoption of an appropriate resolution. If the resolution is too high or too low printing problems may arise. Poor resolution may lead to a rough surface, while too high a resolution may make it impossible for the 3D printer to process the file.

3.5 The Software for 3D Model File Generation is Unmatched with the Printing Process

There is a great variety of 3D modeling software, and different printing demands may lead to different software design and applications. Most modeling software requires the file to be further edited, automatically or manually, before printing. Since the settings and application should be designed according to the thickness and dimensions of the object, the corresponding settings and adjustments should be performed before printing depending on the software type.

4. Summary

With continuous development of 3D printing technology and a broader range of printing materials, there are an increasing variation of 3D printing process available. As of now, 3D printing technology need not merely be used for printing models, but more widely in the design of jewelry, clothing, medical devices and for space fight, aviation and automobile making etc. Products so far unachievable due to the limitations of the technology and materials are gradually being realized, bringing convenience and added innovation to the production of merchandise [10].

References

[1] Liu TT. 3D printing technology in the application of the custom-made gift packaging design. China Packaging Industry: VCH: 2015; 02: p. 103-105.

[2] Jia XL. From the perspective of design about 3 d printing to remake of various industries. Science and Technology&Innovation: 2015; 04: p. 03-04.

[3] Wu CG. The application of 3 d printing technology in the modern clothing. Art Science and Technology: 2014; 02: p. 9-10.

[4] Zhang MZ. Based on the analysis of 3 d printing clothing design development trend. Art research: 2015; 02: p. 95-97.

[5] Information on http:// mp.weixin.qq.com.cn

[6] Information on http:// mp.weixin.com.cn

[7] Zhang RH, Xiong Wei, Zhang Nan. 3D Printing technology in the application of traditional process performance research. Gems and precious stones: 2015; 01: p. 45-47.

[8] Zhou Li, Xin Yue, Zhang Li. Exploring the Fashion Design Based on Three-Dimensional Printing Technology. Decoration: 2014; 05: p. 87-89.

[9] Zhao YX, Liu SZ. Children's orthopedic insole design application and development. The creative front: 2014; 04: p. 10-12.

[10] Zhang YT. 3D printing technology in the application of the product modeling design. Design practice: 2015; 02: p. 17-18.

Study of Development on Apparel Additive Manufacturing Technology

Li-Wen Gu[1,2], Yan-Wen Ruan[1*], Cynthia Istook[2], Xiao-Gang Liu[1]

[1] *Donghua University, College of Fashion. Art Design, West Yan'an Road, Shanghai, 200051, P.R. China*
[2] *North Carolina State University, College of Textiles, 1020 Main Campus Drive, Raleigh, North Carolina, 27695, USA*

[*]Corresponding author's email: ruanyanwen.elsa@hotmail.com

Abstract

Additive Manufacturing (AM) is a process of building a 3D object by joining materials layer upon layer from a digital 3D model data. As a branch of AM technology, Apparel Additive Manufacturing (AAM) technologies can be defined as a technical system which is used to produce wearable apparel product directly from a 3D digital model through a dispersed material aggregation manufacturing method. Nowadays, more and more attempts have been made to transfer AAM from an experimental technology into a practical one, but for various reasons, AAM technology is not as developed as much as its equivalent in other fields, such as the industrial manufacturing field. Additionally, there is few literature research dissecting AAM's development status from multiple perspectives. In this paper, the literature related to the AAM technology development status is reviewed systematically and hierarchically. The research result will present a clear picture of developments in AAM technology, which is of great significance and value in technological progress.

Keywords: Additive Manufacturing; Apparel; Garment; 3D Printing; CAD

1. Introduction

1.1 AM Background

According to the ASTM standard ISO/ASTM 52900:2015(E), Additive Manufacturing (AM) can be defined as a process of building 3D objects by joining materials layer upon layer from digital 3D model data [1]. It is nowa generic term used for all similar and related manufacturing technologies [4]. There are also some other terms which are similar to AM. The most common ones are Rapid Prototyping (RP) [1], Rapid Manufacturing (RM) [4], Rapid Tooling (RT), Layer-Based Manufacturing (LBM) or Additive Layer Manufacturing (ALM) [2], Stereolithography (SL) [3] and 3D Printing (3DP). Compared to all the terms above, 3DP is becoming the most commonly used term to describe AM technologies since printing technology is quite common in our daily life and it is easy for people to understand [2].

From the perspective of the manufacturing process, the AM system can be divided into four basic types, Liquid Polymer Systems, Discrete Particle Systems, Molten Material Systems and Solid Sheet Systems [2]. The earliest technologies of these types are SL [4], Selective Laser Sintering (SLS) [10], Fused Deposition Modelling (FDM) [4] and Laminated Object Manufacturing (LOM) [2] respectively. Besides all the above, Poly jet [4] and Direct Metal Laser Sintering (DMLS) are also two important technologies which play an essential role in the process that turns RP into RM [4]. All of these technologies were developed from the 1980s to the beginning of the 1990s.

Nowadays, AM has been introduced in a variety of fields. In the industrial manufacturing field, AM is widely used in aerospace and automobile sectors. Compared with the traditional manufacturing process, AM has advantages in making aerospace components of advanced materials [6] with complex and high-performance geometry [7], light weight [7], reduced product lead-time and non-recurring product introduction costs [11]. In the construction field, AM was used to make building models in the early stage. Beyond that, full-size building had been realized entirely by AM in 2014 [12]. In the biomedical field, along with scanning technology, AM is not only used to fabricate complex anatomical parts for assisting in precise pre-surgical planning and to manufacture custom shaped implants and medical devices [7], but also to fabricate artificial organ parts [13]. In the food field, compared with mass production manufacturing, AM has obvious advantages to achieve customized food design with personalized nutrition, as a prototyping tool that can save time and enable a viable food supply by utilizing ingredients extracted from the sustainable resources [9]. In the energy field, AM has great potential to develop fuel cells, a "green" energy device [6]. In the fashion field, AM has been widely used in the jewelry industry [7] and shoe industry commercially [8]. The applications of AM in the apparel industry have made great progress, but there are still many limitations from variety of perspectives.

1.2 AAM Background

As a branch of AM technology, Apparel Additive Manufacturing (AAM) technologies can be defined as a technical system which is used to produce wearable apparel products directly from a 3D digital model through a dispersed material aggregation manufacturing method. The development direction of AAM can be generally summed up as moving from rigid to flexible material, from small to full size, from experimental to commercial application and so on. Until now, people have been able to produce apparel products for daily wearing through seamless knitting technology, a mature traditional knitting technology which is one approach to AAM [16]. AAM technology has great potential, but future development remains uncertain, which will be discussed in detail later.

AAM has not matured as a technology system, but as a branch of AM, AAM will inherit advantages from AM as it matures, which can be concluded in three perspectives. First, AAM will enable a significant reduction in the costs related to customized clothing from the perspectives of materials, equipment, labour, time, inventory and the environment [2-8, 19]. Second, through 3D digital modelling, AAM can expand the creative space of apparel products by simplifying much of the traditional garment making process. This will lead to one-step formation of garment products, enabling shapes which could not have been realized in the past [17]. Finally, garment fit can be improved because that AAM can construct 3D garment directly on a subject's avatar, without relying on 2D patterns and traditional cutting and sewing processes [17].

One of the disadvantage of AAM is related to the basic shortcoming of AM with its low scale effect, offering no economics of scale [14]. Compared with the traditional mass production garment making process, AAM cannot take advantage of material costs reduction and production efficiency improvement when it is used to manufacture a single style of garment in large quantities. Another important disadvantage is that AAM is still testing technologies at this moment.

2. Problem Formulation

The main purpose of this paper is to dissect the development status of AAM from multiple perspectives.

2.1 Problems Identified in Existing Literature

2.1.1 Lack of Academic Research

Table.1 Comparison of filtered keyword results [18]

AF Words	#	IMF Words	#	BIF Words	#	COF Words	#	ENF Words	#	FOF Words	#
textile	13	metal	661	bio*	413	building	189	"power"	114	food	15
fashion	9	engine	518	bone*	212	construction	109	energy	103	chocolate	6
"fabric"	5	alloy	225	medic*	182	house	2	electric*	85	agriculture	4
apparel	5	titan*	133	cell	149	bridge	2	fuel	32		
yarn	2	aero*	83	dental	50			battery	12		
knit*	2	vehicle	59	organ	38						
cloth*	1	automotive	46	health*	18						
garment	1	aircraft	31	heart*	9						
		plane	18	ear*	7						
		car	5								

* is used to search for words that begin with the same letters; "" is used to search for an exact phrase or word.

Compared with the fields of industrial manufacturing and biomedical engineering, there has been less academic research carried out in the field of AAM, which can be seen by the literature retrieval results through Engineering Village, the broadest and most complete engineering literature database available in the world [15]. Literature retrieval was divided into two steps; the first was to search literature related to AM, using terms such as additive manufacturing, 3D printing, 3D print, 3D printer, rapid prototyping, rapid manufacturing, rapid tooling, additive layer manufacturing, layer based manufacturing, stereolithography, or freeform fabrication. There were 14,285 articles found between 1884 and January, 2017. The second step was to filter the results of the search further by focusing on keywords related to the fields of apparel (AF), industrial manufacturing (IMF), biomedical (BIF), construction (COF), energy (ENF), and food (FOF). These results are shown in Table 1.

The number of articles specifically related to the apparel field were significantly less than those in other fields in AM. Most of the information related to AAM was released in non-academic articles online.

2.1.2 Drawback of Existing Literature

Based on the study of existing literature, most researchers analyzed the development of AAM from the perspectives of the differences between design process, production method of AAM and the traditional ways, the positive effect of AAM on product development within the apparel industry, the limitation of AAM [17], the changes caused by AAM on business operations within the fashion industry related to the supply chain, employment, tariffs, etc. [19], the applications of AAM in different fashion business models such as haute couture and mass-market, the applications of AAM realized by existing types of AM such as SL, FDM etc. [8], and the applications of AAM in conceptual fashion design [20]. The drawback of these research discussions can be concluded as two points. First, AAM was treated as a whole in the process to analyze its advantages, disadvantages and its influence on apparel industry, which lacked the internal analysis of AAM. Second, AAM was analyzed according to the classification of the mainstream AM technologies such as SL, SLS or FDM, which neglected the creative technologies beyond the mainstream AM technologies but demonstrated potential to be used as AAM. The analysis of AAM development in this paper will be improved to avoid these two problems.

2.2 Research Method of AAM Development Status

The research of AAM development status will be analyzed synthetically and hierarchically from four perspectives of key fields, five generation of technology and three development paths.

3. Problem Solution

3.1 Key Fields of AAM Technology

According to the review of literature, the maturity of AAM technologies require the support from at least four key fields of AM which are Material Technology (MT), Shaping Technology (ST), Texture Organizing Method (TOM) and CAD technology (CADT). MT and ST can be classified as hardware technologies while the TOM and CADT can be classified as software technologies.

3.1.1 Material Technology

MT of AM refers to the technology related to the productive material of AM for constructing 3D objects, which is the fundamental technology of AM and the precondition for transferring a digital model into a physical product. Excellent AAM materials should possess the three features: (1) it can be broken up into a small volume or basic units; (2) it can be joined together to form a larger object; and (3) the material properties must simulate properties of textile as closely as possible.

3.1.2 Shaping Technology

ST of AM refers to the technology through which the product material is shaped into desired stereo shape, which requires particular equipment to achieve. There is a close relationship between ST and MT. For example, FDM is used with the material of PLA filament, SL is used with liquid photosensitive polymer and SLS is used with metal or thermoplastic powder.

3.1.3 Texture Organizing Method

TOM of AM refers to the technology through which the product material is joined, or the units are assembled into particular form of texture or organized structure in the shaping process. There is no definite relationship between TOM and MT or ST. In the application of AAM, excellent TOM can make up the shortage of achieving flexibility through MT and ST.

3.1.4 CAD Technology

CADT of AM refers to the technology which is used to build a 3D digital model and to drive the product manufacturing process. CADT of AAM can be divided into two kinds: (1) is a 2D-based 3D apparel modelling which is based on traditional clothing technics such as 2D patterns for sewing or knitting; and (2) is a 3D apparel direct modelling process without the steps of the 2D operation. 3D apparel direct modelling is based on the same process used in the industrial engineering design field. The representative software for 2D-based 3D apparel modelling are CLO 3D, GERBER 3D, V-Stitcher and Stoll M1 Plus etc. The ones with direct 3D apparel modelling are Lookstailor X, Kinematics Cloth etc. Others of 3D direct modelling are 3DS Max, Maya, Blender, CATIA, UG NX, Pro Engineer and CREO. In the apparel industry at present, 2D-based 3D apparel modelling is a more matured technology than 3D apparel direct modelling.

3.2 Development Paths of AAM Technology

In each key field which was mentioned previously, the development path (DP) of AAM could be divided into three kinds:

1) DP1: Inventing brand-new AAM technologies or methods.
2) DP2: Utilizing existing matured mainstream non-textile AM technologies or methods.
3) DP3: Utilizing existing matured textile technologies or methods

3.3 Generations of AAM Technology

In this paper, AAM is divided into five generations in this paper. The products of the later generation possess features which are more closely related to the ones of textile

3.3.1 Generation 1st (G 1st)

G 1st produced one-piece rigid wearable product by AM technology. This generation of AAM technology introduced mainstream AM technologies used in the industrial manufacturing field into the apparel industry directly without any improvement. Therefore, the products of this type of technology were wearable industrial products rather than strictly garments. The common features of these products were rigid bodies: these products could not be bent or deformed on a large scale (some rigid materials can be bent on a small scale because they have a weak elasticity or flexibility). Additionally, they were wearable only because they were designed ergonomically. Finally, they could be manufactured integrally or be assembled by more than one pieces with unmovable joints. Iris Van Herpen's early designs such as the "Crystallization Collection" in 2010 were the representative works of the 1st generation AAM technology [21]. The "Parametric Sculpture Dress" wore by Lady Gaga in 2013 was also a famous works of this generation [22]. Though AAM technology of this generation couldn't produce wearable products that possess any feature of textiles, the value of the concept was much greater than the practical value. It established the connection between AM and wear-ability, which drew more attention to the potential for the apparel industry. Since then, AAM entered into a rapid period of development.

3.3.2 Generation 2nd (G 2nd)

The most important advantage of the 2nd generation AAM technology was that it enabled a flexible garment by upgrading TOM, rather than using flexible material. First, instead of being produced in one-piece, the garment was broken up into basic components for layer-based manufacturing. Second, the garment was constructed by assembling the components together with movable joints through which the textile-like flexibility could be enabled. Third, the garment could be manufactured one-step at a time leaving the components to be joined together or assembled manually after they were completed. One of the most representative examples of this technology was a 3D printed dress called the "Kinematics Dress" presented by Nervous System, a design studio in Boston, through its self-developed 3D apparel direct modelling software named Kinematics Cloth [23]. It was a full-sized integrally flexible dress composed by 2,279 unique and rigid triangular panels interlocked through 3,316 movable hinges. All the pieces were 3D printed in nylon [23]. There are also some other famous examples such as the N12 Bikini designed by Jenna Fizel, Mary Huang, made by the company of Shapways in 2011 [24], and the formal dress of Dita Von Teese created by designer Francis Bitonti and the company of Shapways in 2013 [25].

3.3.3 Generation 3rd (G 3rd)

The technology of this generation made a breakthrough in achieving flexibility of garment by improvement in MT, due to the progress of flexible AM material such as rubber or silica gel. The new material made it much easier to create flexible garments rather than relying on improvement to TOM, which ensured flexibility along with light weight and a comfortable wearing experience. One of the most representative examples were the 3D printed suits in the CHANEL 2015 F-W haute couture collection. These suits were made of flexible AM material by SLS. It was hard to distinguish them from normal textile suits visually [26]. The use of this technology by a well-known haute couture designer demonstrated that the 3rd generation AAM technology was moving from an experimental technology to a practical one. Another representative example is the silicon printed underwear created by Jess Haughton in 2016 [27].

3.3.4 Generation 4th (G 4th)

The technology of this generation made breakthroughs both on MT and ST. From the perspective of MT, liquid textile fiber solution was used as the printing material for garment manufacturing. From the perspective of ST, it required a rigid mold upon which the textile fiber solution was sprayed. Until now, most successful cases used 2D molds rather than 3D ones because it is hard to spray the solution evenly on the surface of a stereo molds [28]. The mold could be removed after the liquid solution turning into a membrane which possessed the features of non-woven fabric. One of the most representative cases was Electroloom, the world's first 3D fabric printer, which was developed in 2014 [28]. It printed garments by guiding the liquid nano-fibers solution onto the 2D mold by an electric field [28]. So far, it can create fabrics with a hand feeling similar to polyester/cotton, silk and acrylic, some of which were also elastic [28]. As early as 2012, a similar technology was presented by Manel Torres [29]. He invented a method to making fabrics by spraying liquid fiber solution from a portable sprinkler irrigation on to a support member, which could be seen as the embryonic form of Electroloom. The 4th generation technology realized textile AM in a real sense, which proved the feasibility of the application of AAM in the field of real textile manufacturing.

3.3.5 Generation 5th (G 5th)

The technology of this generation focused more on the development of existing technology in the apparel industry rather than creating a novel AAM technology. The primary focus was on seamless knitting technology which has been used in the apparel industry for many years [30], also called 3D machine knitting [31]. This generation focus on developing more intelligent or user-friendly CAD/ CAM technologies. One of best representations of this generation was the 3D machine knitting compiler developed by Disney Research, together with University of

California, Santa Cruz (UCSZ), Massachusetts Institute of Technology (MIT) and Carnegie Mellon University (CMU) in 2016 [31]. This compiler could translate the high-level shaping and scheduling operations into needle-level operations, which made it much easier to create 3D seamless kitting through a friendly visual design interface. Meanwhile, the technology of this generation has been used in apparel industry commercially. Ministry of Supply, a Boston based menswear brand, released its additively manufactured seamless jacket in 2016 [16]. In 2017, UNIQLO, a famous Japanese apparel brand, announced that they will present a new collection of seamless knitting garment called 3D U-Kints [32].

3.4 Composite Analysis

Table.2 Composite analysis AAM development

Key Field	Development Path	Generation of Technology	Advantages	Disadvantages
MT	DP1	G 4th	Textile properties	non-woven fabric only
	DP2	G 1st, G 2nd, G 3rd	Marketable	1) Poor tactile properties for daily wearing; 2) Difficult garment creation with thin substrate
	DP3	G 5th	1) Textile properties 2) Marketable	Knit fabric only
ST	DP1	G 4th	Automatic, one-step manufacturing	1) Only 2D molds 2) No multi-layer garment parts 3) Low economy of scale
	DP2	G 1st, G 2nd, G 3rd	1) Automatic manufacturing; 2) One-step manufacturing; 3) Almost unlimited 3D shaping	Weak strength in the Z-axis due to the lay-base manufacturing method [33]
	DP3	G 5th	1) Automatic, one-step manufacturing 2) High level of industrialization	Complicated knitting process compared to material joining process
TOM	DP1	G 2nd	1) Flexibility with rigid materials 2) Producible on small size AM equipment	1) Difficult garment creation with thin substrate 2) Low comfort due to rigid material
	DP2	G 1st, G 3rd	1) Limited flexibility by rigid materials or high flexibility by flexible materials; 2) Light weight garments	Little or no comfort due to rigid or non-textile material
	DP3	G 5th	Highly matured technology	Knitted fabric only
CADT	DP1	G 1st, G 2nd, G 3rd	1) What you see is what you get 2) Incorporation of apparel design tools;	Poor garment detail editing function
	DP2	G 1st, G 2nd, G 3rd	1) What you see is what you get 2) Almost unlimited shaping	Significant learning curve for fashion designers due to the lack of semantic information in the apparel industry
	DP3	G 4th, G 5th	1) Modelling process is similar to traditional garment making process 2) Easy to create regular geometric hem on 2D patterns	Hard to use for most fashion designers due to its requirement of 2D pattern making skill.

Based on the classification from different perspectives previously described, the composite analysis of AAM development status was derived as shown in Table 2.

4. Conclusion

According to the review of literature, AAM technologies were started around 2010. From Irs Van Herpen's presentation of her work "Crystallization Collection" in the 1st generation of AAM technology in 2010, to the release of Disney's user friendly compiler for seamless knitting as the 5th generation of AAM technology in 2016, it only took 6 years for AAM technologies to grow from being able to produce only rigid wearable objects to a level where real garments could be manufactured for daily wearing. Though there are still many problems regarding material diversity, material quality, apparel styling, etc., at the present rate of development, it is quite feasible that industrialized AAM technologies will impact the whole apparel category in the foreseeable future.

5. Acknowledgement

The authors wish to acknowledge the Shanghai Zhangjiang National Innovation Demonstration Zone Special Development Fund Major Project (No.ZX2014050 00012), China Scholarship Council.

References

[1] Standard terminology for additive manufacturing: general principles terminology 1 and 2. ISO/ASTM No. 52900: 2015.
[2] Gibson I, Rosen D, Stucker B. Additive manufacturing technologies: 3D printing, rapid prototyping, and direct digital manufacturing. Springer: 2014; p. 8-34.
[3] Melchels FP, Feijen J, Grijpma DW. A review on stereolithography and its applications in biomedical engineering. Biomaterials: 2010; 31 (24): 6121-30.
[4] Gebhardt A, Hötter JS. Additive manufacturing: 3D printing for prototyping and manufacturing. Carl Hanser Verlag GmbH Co KG: 2016; p. 3-9.
[5] Bandyopadhyay A, Bose S, editors. Additive manufacturing. CRC Press: 2015; p. 2-4.
[6] Guo N, Leu MC. Additive manufacturing: technology, applications and research needs. Frontiers of Mechanical Engineering: 2013; 8 (3): 215-43.
[7] Negi S, Dhiman S, Sharma RK. Basics, applications and future of additive manufacturing technologies: a review. Journal of Manufacturing Technology Research: 2013; 5 (1/2): 75.
[8] Vanderploeg A, Lee SE, Mamp M. The application of 3D printing technology in the fashion industry. International Journal of Fashion Design, Technology and Education: 2016; 25: 1-0.
[9] Sun J, Zhou W, Huang D, Fuh JY, Hong GS. An overview of 3D printing technologies for food fabrication. Food and bioprocess technology: 2015; 8(8): 1605-15.
[10] Mazzoli A. Selective laser sintering in biomedical engineering. Medical & biological engineering & computing: 2013; 51 (3): 245-56.
[11] Raja V, Zhang S, Garside J, Ryall C, Wimpenny D. Rapid and cost-effective manufacturing of high-integrity aerospace components. The International Journal of Advanced Manufacturing Technology: 2006; 27 (7-8): 759-73.
[12] Information on https://www.cnet.com/news/worlds-first-3d-printed-apartment-building-constructed-in-china/
[13] Mannoor MS, Jiang Z, James T, Kong YL, Malatesta KA, Soboyejo WO, Verma N, Gracias DH, McAlpine MC. 3D printed bionic ears. Nano letters: 2013 May 6; 13 (6): 2634-9.
[14] Lipson H, Kurman M. Fabricated: The new world of 3D printing. John Wiley & Sons: 2013; p. 27.
[15] Information on https://www.elsevier.com/solutions/engineering-village/content#compendex
[16] Infromation on http://www.3ders.org/articles/20160415-ministry-of-supplys-3d-knitted-seamless-jacket-brings-additive-manufacturing-to-menswear.html
[17] Wang BZ, Chen Y. The Effect of 3D printing technology on the future fashion design and manufacturing. Applied Mechanics and Materials: 2014; (496): 2687-2691.
[18] Information on https://www-engineeringvillage-com.prox.lib.ncsu.edu/search/quick.url
[19] Sedhom RV. 3D Printing and its Effect on the Fashion Industry: It's More Than Just About Intellectual Property. Santa Clara L. Rev: 2015; 55: 865.

[20] Yan Y, MZ Z, GZ Z, TZ C. Textile Bioengineering and Informatics Symposium Proceedings: The study on the approach of conceptual fashion design based on 3D printing technology: 2016.
[21] Information on http://www.irisvanherpen.com/home
[22] Information on http://www.materialise.com/en/cases/3d-printed-parametric-sculpture-dress-by-studio-xo-for-lady-gaga
[23] Information on http://n-e-r-v-o-u-s.com/blog/?p=6280
[24] Information on http://www.continuumfashion.com/N12.php
[25] Information on https://www.shapeways.com/blog/archives/30318-remembering-revolution-francis-bitonti-ditas-gown.html
[26] Information on http://chanel-news.chanel.com/en/home/2015/07/haute-couture-3d-technology-fall-winter-2015-16-collection.html.
[27] Information on https://3dprint.com/138006/stretched-silicone-lingerie/
[28] Information on https://www.kickstarter.com/projects/electroloom/electroloom-the-worlds-first-3d-fabric-printer#project faq 134401
[29] Manel T, Paul L. Non-woven fabric. US Patent: 8088315; 2012.
[30] Mahbub R, Wang L, Arnold L. Design of knitted three-dimensional seamless female body armour vests. International Journal of Fashion Design, Technology and Education: 2014; 7 (3): 198-207.
[31] McCann J, Albaugh L, Narayanan V, Grow A, Matusik W, Mankoff J, Hodgins J. A compiler for 3D machine knitting. ACM Transactions on Graphics (TOG): 2016; 35 (4): 49.
[32] Information on https://www.uniqlo.com/us/en/lifewear-fw17.html
[33] Winnan CD. Adventures in 3D printing: limitless possibilities and profit using 3D printers. Create Space Independent Publishing Platform: 2013; p. 17.

Study on the Influence Factors of the Seat Seam Gradient for Women's Trousers

Jia-Jia Yu, Jian Li*

Beijing Institute Of Fashion Technology, NO.2 Yinghua Road Chaoyang District, Beijing, 100029, China.

*Corresponding author's email: jean.xiaoyu@163.com

Abstract

The seat seam gradient is one of the significant factors in the structural design of trousers, as it has a significant influence on the hip shape and the mobility of the trousers. This paper compares the different dressing effect of 21 different pairs of trousers to find the relationship among the hip relaxed, the slack bottom, the back crease line and the seat seam gradient, and tries to establish the ideal gradient for women's trousers in general. The approach taken to solve the problem includes the fabric fitting experiment, the paper observation experiment and the virtual experiment. This analysis shows that the seat seam gradient of women's trousers is influenced by the hip relaxed, the slack bottom and the back crease line; the hip relaxed and the seat seam gradient are positively correlated; the slack bottom and the seat seam gradient are negatively correlated; the back crease line and the seat seam gradient are negatively correlated; the slack bottom and the back crease line can both affect the seat seam gradient; and the ideal gradient for the general women's trousers is 7.5°.

Keywords: Seat Seam Gradient; Hip Relaxed; Slack Bottom; Back Crease Line; Ideal Gradient

1. Introduction

The seat seam gradient is the angle between the seat seam and the plumb line, and is an important part of the design of the trousers' structure. This angle can influence the hip shape and the mobility of the trousers. In recent years, a large number of scholars have made considerable research effort on the design of the crotch structure and trouser silhouette [1-13]. Shu-Juan Li attempted to apply a mathematical method to find the numerical value of the seat seam gradient, but the results obtained were not tested by the fitting experiment [3]. Juan-Juan Cui measured the human body and used a mannequin as an experimental body to find the relationship between the seat seam gradient and the dressing effect, but the shape of the mannequin could not accurately reflect the shape of the human body, so the result did not represent the actual situation [4]. Shu-Fen Qiu divided trousers into two categories and set the angle of the seat seam gradient according to the different categories, but the author did not consider the influence factors of the seat seam gradient, so the result has low credibility [5]. This paper therefore attempts to figure out the relationship among the hip relaxed, the slack bottom, the back crease line and the seat seam gradient, and to analyze the influence mechanism by doing a series of experiments, and then explore the ideal seat seam gradient for women's trousers.

2. Experimental Design

The single woman body (Table 1) was selected as the experimental body [14, 15].

Muslin with 70% Cotton was selected as the experimental material. The width of the muslin is 90cm.

The experiment controls the design of the trousers' structure, the difference between the front slack bottom and back slack bottom, the difference in leg width and the difference in hipline.

The basic shape of the experimental trousers was defined as tapered, and six straight trousers were used in the experiment for confirmation.

Table.1 Body size and specifications unit/cm

Body parts		Body size	Specifications
Height		160	-
Waistline		67	Waistline + 1
Hipline		85	Hipline + Relaxed
Hip relaxed		-	According to the experiment
Length		-	0.6*Height+20
Seat		-	0.13*Height+0.1*Hipline-8
Slack Bottom		-	Hipline/5+Relaxed/2-20
Leg width	Tapered	-	Slack Bottom+3
	Straight	-	Slack Bottom

"Ha"(where "a" is a constant) represents the hip relaxed, "F±m"(where "m" is a constant, + indicates mains outward, - indicates mains inward) defines the position of the front crease line, "B±n"(where "n" is a constant, " +" indicates mains outward, - indicates mains inward) defines the position of the back crease line.

The experiment was divided into five parts, with a total of 21 pairs of trousers (Table 2). The comparative experiments were used to draw the final conclusion.

Table.2 Experimental design

Item	Hip relaxed 2cm (H2)				Hip relaxed 4cm (H4)		Hip relaxed 6cm (H6)		
Bottom 16cm	7.5°		9.5°		7.5°	9.5°	6.5°	7.5°	9.5°
Bottom 17cm	7.5°		9.5°		7.5°			7.5°	
Bottom 18cm	7.5°		9.5°		7.5°			7.5°	
Bottom 20cm	7°		10°		/			/	
Bottom 25cm	5°	6.5°	8°	9.5°	/			/	

Notes: The angles shown in the table represents the seat seam gradient.

AutoCAD is used to draw the pattern faster and more efficiently. Thc following figures show the different patterns drawn by AutoCAD for the "Slack Bottom 16cm experiment" as an example with the hip relaxed was changed (Fig.1-1, Fig.1-2, Fig.1-3, Fig.1-4, Fig.1-5, Fig.1-6).

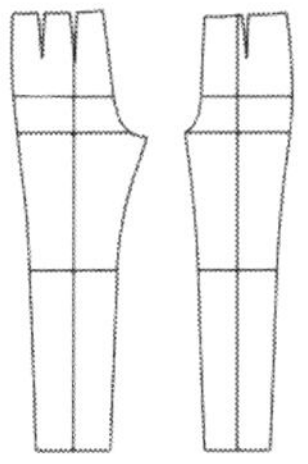

Fig.1-1 H2/7.5°

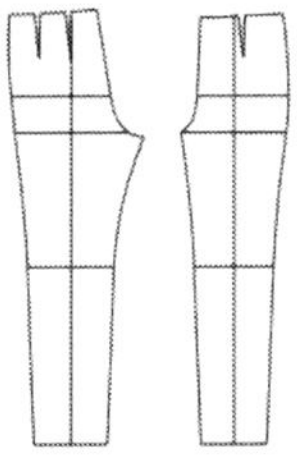

Fig.1-2 H2/9.5°

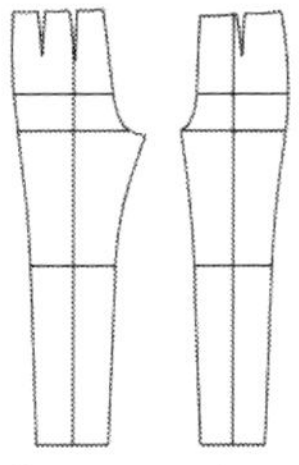

Fig.1-3 H4/7.5°

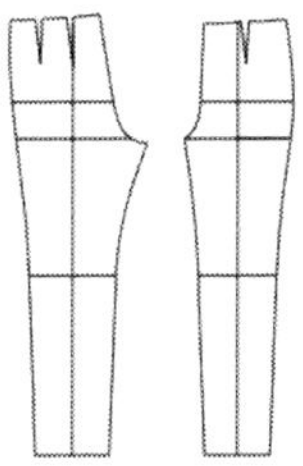

Fig.1-4 H4/9.5°

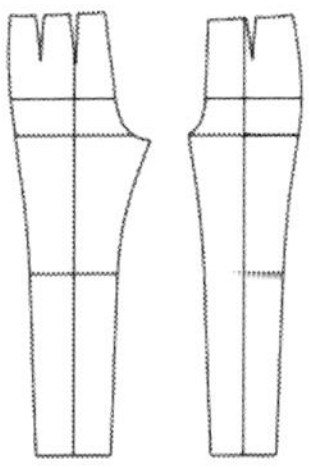

Fig.1-5 H6/7.5°

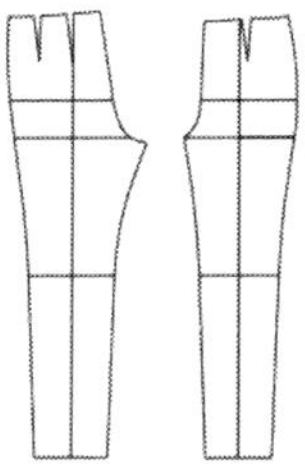

Fig.1-6 H6/9.5°

3. Paper Observation Experiment

The 1:4 patterns are drawn by AutoCAD, and the pattern are then printed and cut out carefully. Transparent tape is used to join the paper pattern together to obtain the trouser model. The models are then grouped together and fixed side by side on horizontal plane. Finally the shape of the models is observed and evaluated. The experiment shows that the gradient of the hip changes with changes in the seat seam gradient changed, and the condition of the hip fit and wrinkles change with the changes in the gradient.

4. Fabric Fitting Experiment

As a results of the application of the variable control method, the fabric fitting experiment is divided into four parts, the hip relaxed, the slack bottom, the back crease line, the interaction between the hip relaxed and the back crease line. This experiment explores the impact of various factors on the seat seam gradient.

4.1 Effect of the Hip Relaxed on the Seat Seam Gradient

The experiment is divided into three experimental groups according to the size of the slack bottom, group 16cm, group 17cm, and group 18cm. First of all, in the 16cm group, the slack bottom, the hip relaxed and the position of the crease line are held constant, and the seat seam gradient is set to be 7.5°and 9.5°. The dressing effect of two kinds of trousers with different angles of seat seam gradient is compared to find the better seat seam gradient when the hip relaxed is fixed. The seat seam gradient is then set as a benchmark and the bottom and this seat seam gradient are kept constant in each group, while the size of the hip relaxed changed, allowing the dressing effect to be compared (Fig.2-1, Fig.2-2, Fig.2-3, Fig.2-4, Fig.2-5, Fig.2-6) .

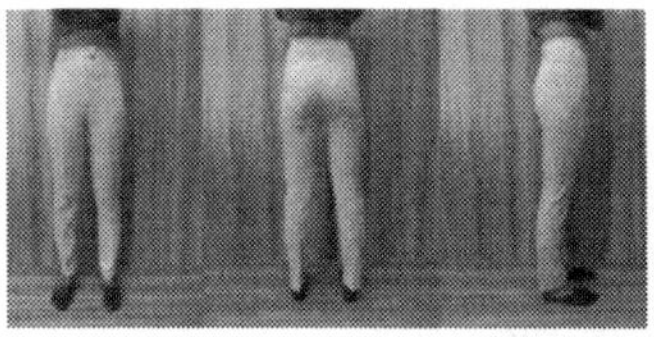

Fig.2-1 H2/SlackBottom16cm/ 7.5°

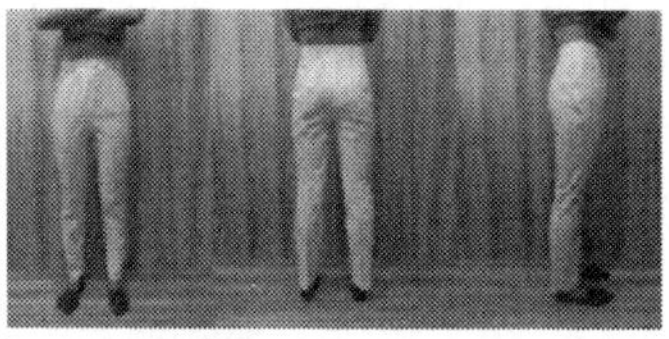

Fig.2-2 H4/SlackBottom16cm/ 7.5°

Fig.2-3 H6/SlackBottom16cm/ 7.5°

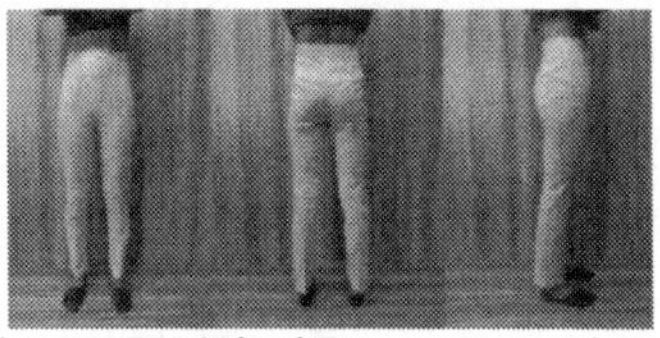

Fig.2-4 H2/SlackBottom16cm/ 9.5°

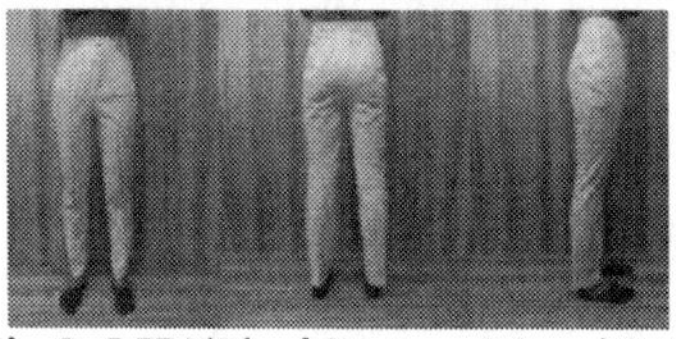

Fig.2-5 H4/SlackBottom16cm/ 9.5°

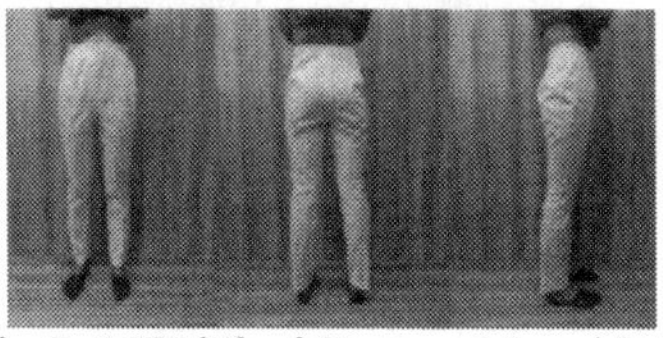

Fig.2-6 H6/SlackBottom16cm/ 9.5°

The SPSS is used to analysis the data (Table 3), the results show that for a better dressing effect, when the slack bottom and the position of the crease line are kept constant, the seat seam gradient changes with changes in the hip relaxed, and when the hip relaxed is increased, the seat seam gradient increases and the dressing effect improves.

Table.3 The Relationship Between the Seat Seam Gradient and the Hip Relaxed

		Seat seam gradient	Hip relaxed
Seat seam gradient	Pearson	1	.820*
	Sig.		.024
	N	7	7
Hip relaxed	Pearson	.820*	1
	Sig.	.024	
	N	7	7

4.2 Effect of the Slack Bottom on the Seat Seam Gradient

The tapered trousers experiment is set as the basis. In this experiment, the dressing effect of trousers with different slack bottom is compared, when the hip relaxed and seat seam gradient are held constant. After that, the straight trousers experiment is taken to be the confirmatory experiment.

4.2.1 Fabric Fitting Experiment of Tapered Trousers

The experiment is divided into three experimental groups according to the size of the hip relaxed, group H2, group H4, and group H6. First of all, the H2 group is taken as an example, the hip relaxed, the slack bottom and the position of the crease line are held constant, and the seat seam gradient is set to be 7.5°and 9.5°. The dressing effect of two kinds of trousers with different angles of seat seam gradient is compared to find the better seat seam gradient when the size of the slack bottom is fixed. The seat seam gradient is then set as a benchmark and the hip relaxed and this seat seam gradient are kept constant in each group, while the size of the slack bottom changed, allowing the dressing effect to be compared (Fig.2-7, Fig.2-8, Fig.2-9, Fig.2-10, Fig.2-11, Fig.2-12) .

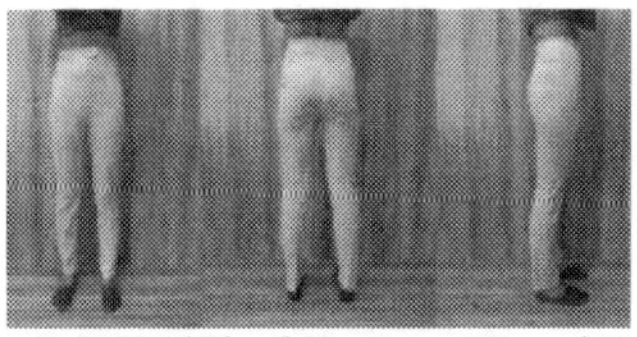
Fig.2-7 H2/SlackBottom16cm/ 7.5°

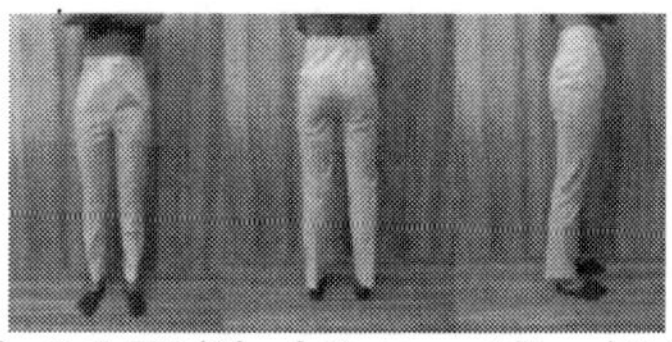
Fig.2-8 H2/SlackBottom 17cm/ 7.5°

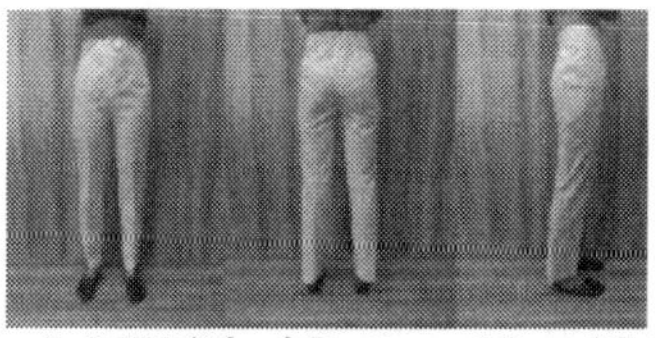
Fig.2-9 H2/SlackBottom 18cm/ 7.5°

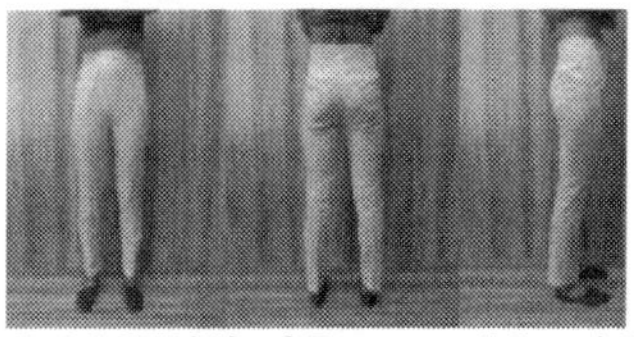
Fig.2-10 H2/SlackBottom 16cm/ 9.5°

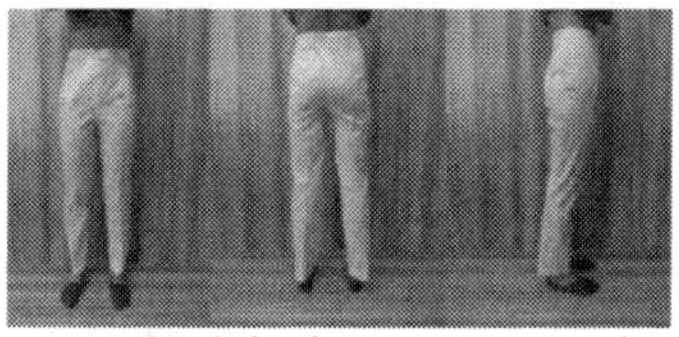
Fig.2-11 H2/SlackBottom 17cm/ 9.5°

Fig.2-12 H2/SlackBottom 18cm/ 9.5°

The SPSS is used to analysis the data (Table 4), the results show that when the slack bottom of the tapered trousers is in the range of 16cm~18cm, the seat seam gradient in 7.5° will lead to a better dressing effect while the hip relaxed and crease line are kept constant and only the size of the slack bottom of the tapered trousers are changed. And for a better dressing effect, when the hip relaxed and the position of the crease line are kept constant, the seat seam gradient changes with changes in the bottom of the tapered trousers, and when the slack bottom is increased, the seat seam gradient decreases and the dressing effect improves.

Table.4 The Relationship Between the Seat Seam Gradient and the Slack Bottom

		Seat seam gradient	Slack bottom
	Pearson	1	-.717*
Seat seam gradient	Sig.		.045
	N	8	8
	Pearson	-.717*	1
Slack bottom	Sig.	.045	
	N	8	8

4.2.2 Fabric Fitting Experiment of Straight Trousers

The experiment is divided into two experimental groups according to the size of the slack bottom, group 20cm and group 25cm.First of all, the dressing effect of trousers with different angles of seat seam gradient is compared when the size of the slack bottom is fixed, then the size of the slack bottom is changed and the dressing effect is compared again in order to get the general rule when it comes to the straight trousers. Group 20cm and group 25cm are taken as examples, the size of slack bottom and the hip relaxed are held constant, the seat seam gradient is set to be 7°and 10° in group 20cm, and the seat seam gradient is set to be 5°, 6.5°, 8°and 9.5° in group 25cm. The dressing effect of trousers with different angles of seat seam gradient is compared to find the better seat seam gradient(Fig.2-13, Fig.2-14, Fig.2-15, Fig.2-16, Fig.2-17, Fig.2-18) .

Fig.2-13 H2/SlackBottom20cm/ 7°

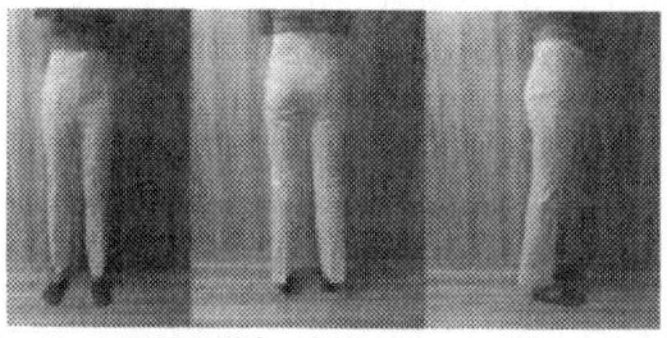

Fig.2-14 H2/SlackBottom20cm/ 10°

Fig.2-15 H2/SlackBottom25cm/ 5°

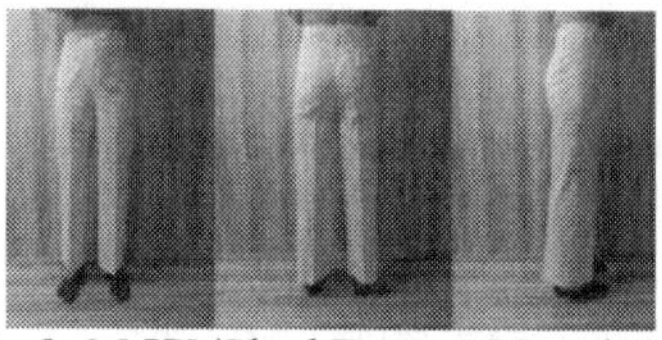

Fig.2-16 H2/SlackBottom25cm/ 6.5°

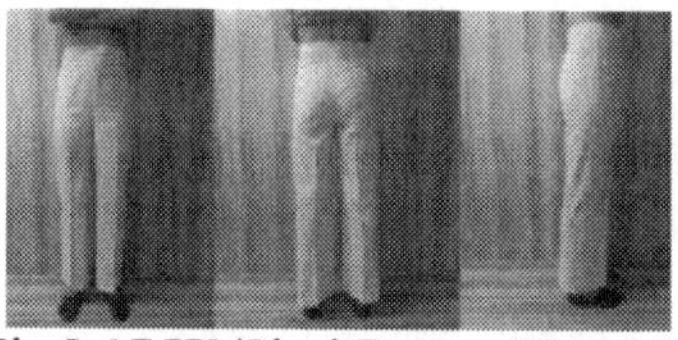

Fig.2-17 H2/SlackBottom25cm/ 8°

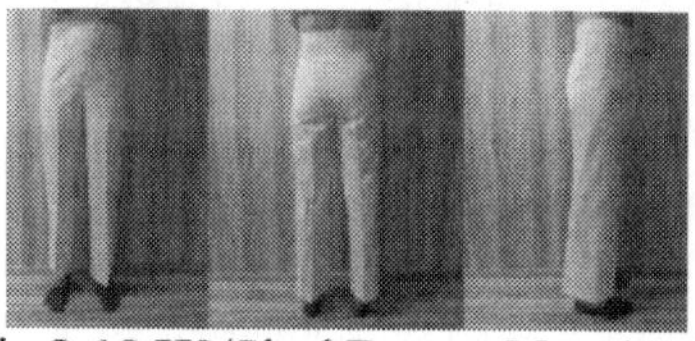

Fig.2-18 H2/SlackBottom25cm/ 9.5°

The results show that the change of the seat seam gradient has a very small influence to the dressing effect when the value of the seat seam gradient is excessive increased, but when the size of the seat seam gradient is smaller, it can still hold a good shape of the hip. So, when it comes to the big slack bottom trousers, the shape of the trousers should be taken into consideration.

So for a better dressing effect, when the hip relaxed and the position of the crease line are kept constant, the seat seam gradient changes with changes in the slack bottom of the straight trousers. And when the size of the slack bottom increased, the seat seam gradient decreases and the dressing effect improves.

4.3 Effect of the Back Crease Line on the Seat Seam Gradient

The experiment is divided into three experimental groups according to the size of the position of the back crease line, group B+1, group B+2, and group B+3. First of all, the dressing effect of the trousers with different angles of seat seam gradient is compared when the hip relaxed and the slack bottom are held constant, then the seat seam gradient is kept constant in each group, while the position of the back crease line changed, allowing the dressing effect to be compared. (Fig.2-19, Fig.2-20, Fig.2-21, Fig.2-22, Fig.2-23, Fig.2-24).

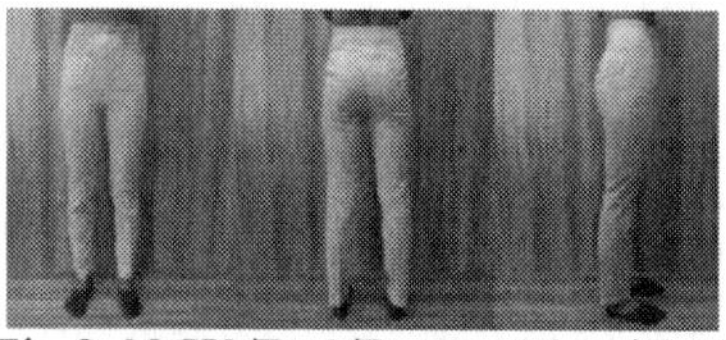

Fig.2-19 H2/B+1/Bottom16cm/ 7.5°

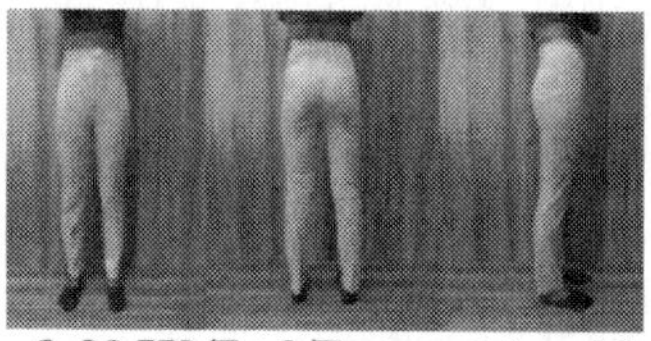

Fig.2-20 H2/B+2/Bottom16cm/ 7.5°

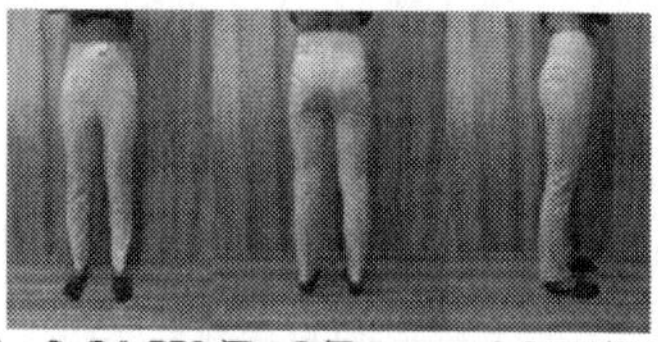

Fig.2-21 H2/B+3/Bottom16cm/7.5°

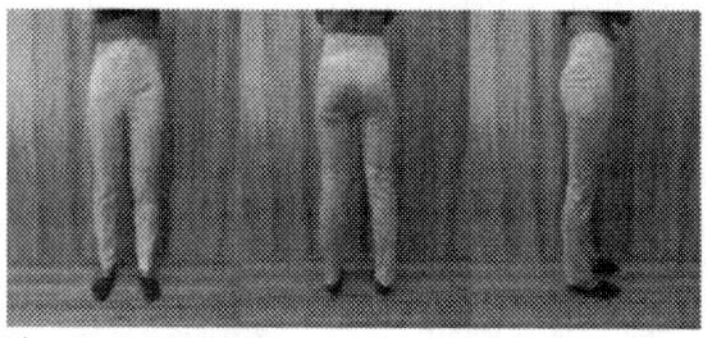

Fig.2-22 H2/B+1/Bottom16cm/9.5°

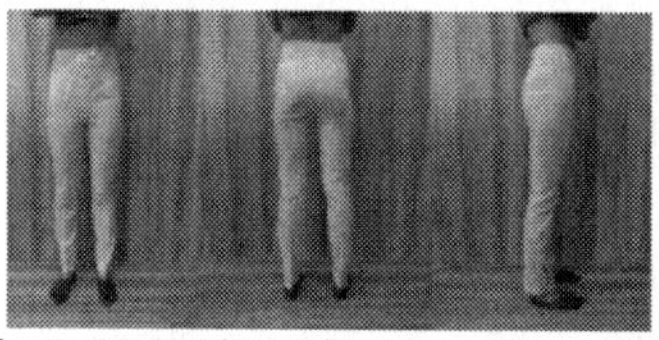

Fig.2-23 H2/B+3/Bottom16cm/6.5°

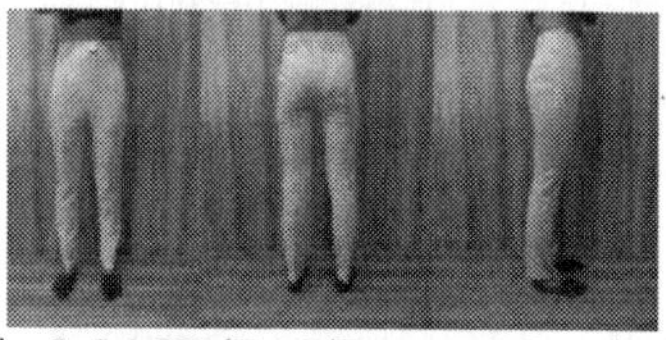

Fig.2-24 H2/B+3/Bottom16cm/9.5°

The SPSS is used to analysis the data (Table 5), the results show that for a better dressing effect, when the hip relaxed and the size of the slack bottom are kept constant, the seat seam gradient changes with changes in the position of the back crease line, and when the back crease line moves more outward, the seat seam gradient decreases and the dressing effect improves.

Table.5 The Relationship Between the Seat Seam Gradient and the Back Crease Line

		Seat seam gradient	Back crease line
Seat seam gradient	Pearson	1	-.764*
	Sig.		.046
	N	7	7
Back crease line	Pearson	-.764*	1
	Sig.	.046	
	N	7	7

4.4 Effect of the Hip Relaxed and the Slack Bottom on the Seat Seam Gradient

The experiment is divided three parts according to the different change degree of the hip relaxed and the slack bottom, group equal, group of hip relaxed-more and group of slack bottom-more. The dressing effect of trousers with different size of the hip relaxed and the slack bottom is compared (Fig.2-25, Fig.2-26, Fig.2-27, Fig.2-28, Fig.2-29, Fig.2-30) .

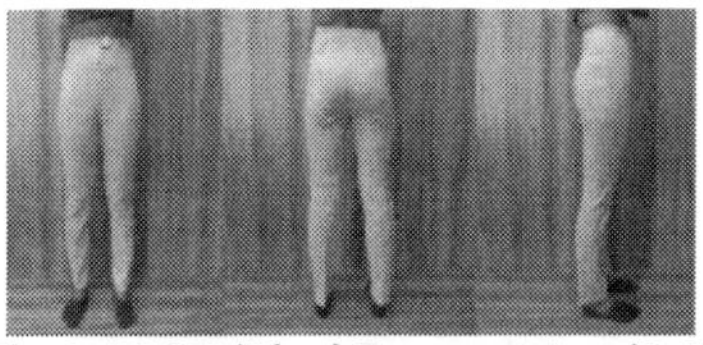

Fig.2-25 H2/SlackBottom16cm/7.5°

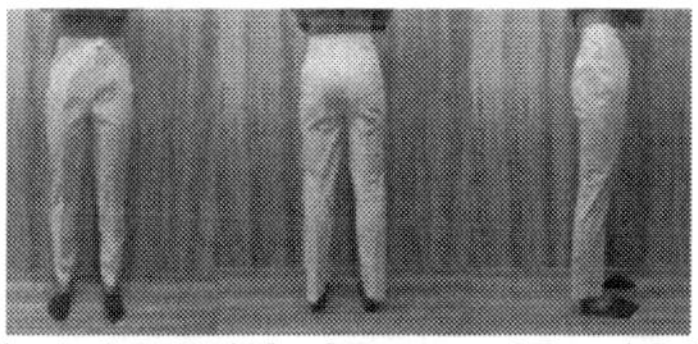

Fig.2-26 H4/SlackBottom17cm/ 7.5°

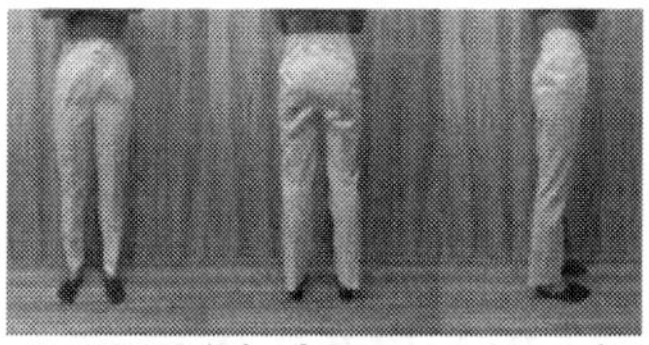

Fig.2-27 H6/SlackBottom18cm/ 7.5°

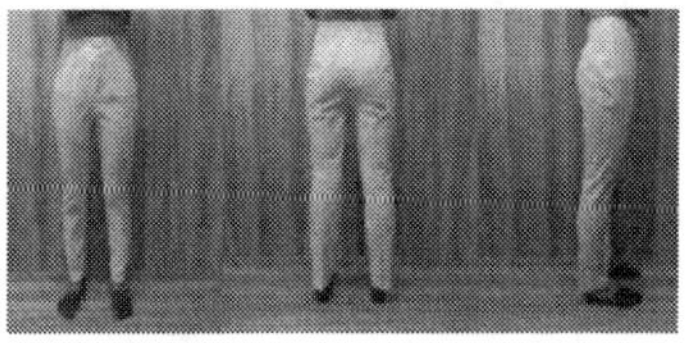

Fig.2-28 H4/SlackBottom16cm/ 7.5°

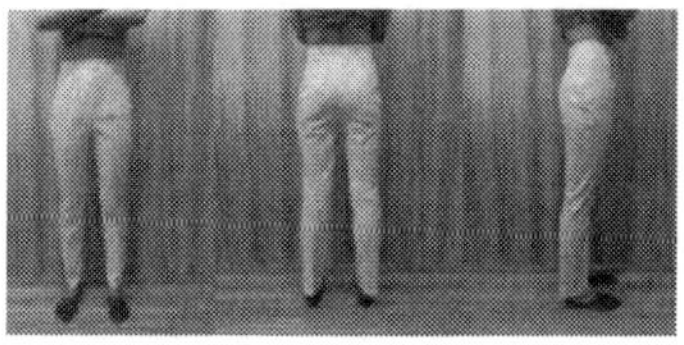

Fig.2-29 H4/SlackBottom16cm/ 9.5°

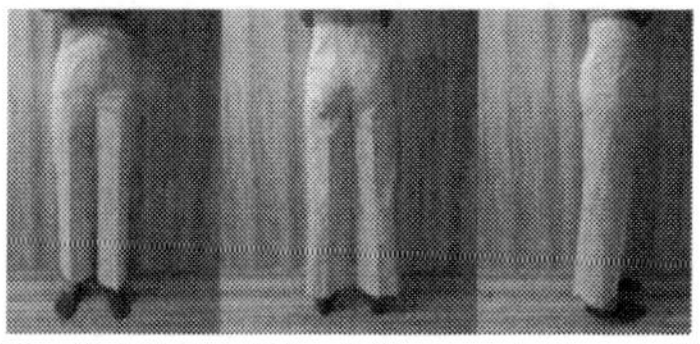

Fig.2-30 H2/SlackBottom25cm/ 5°

The results show that when the change degree of the hip relaxed is equal to the change degree of in the slack bottom, the hip relaxed and the slack bottom have little influence to the seat seam gradient; when the change degree of the hip relaxed is greater than the change degree of the slack bottom, the seat seam gradient increases and the dressing effect improves; when the change degree of the hip relaxed is less than the change degree of in the slack bottom, the seat seam gradient decreases and the dressing effect improves.

5. Virtual Evaluation of Clo 3D Modelist

Clo 3D Modelist is a virtual fitting software. It can simulate the real body posture to show the condition of the virtual fitting. In Clo 3D Modelist, the finished pattern is imported into the software and the virtual patterns are then arrayed and sewn together to become three-dimensional trouser models so that the virtual dressing effect can be seen. Fig.3-1 and Fig.3-2 show the virtual experiment demonstrating the effect of the hip relaxed on the seat seam gradient. Fig.3-3 and Fig.3-4 show the virtual experiment displaying the effect of the slack bottom on the seat seam gradient. Fig.3-5 and Fig.3-6 show the virtual experiment illustrating the effect of the back crease line on the seat seam gradient. It is clear that the results from the virtual software simulation experiment and the fabric fitting experiment are consistent.

In conclusion, three different experiments explore the influence factors of the seat seam gradient of women's trousers in different ways. The Paper Experiment is a qualitative experiment which provides the general direction for the Fabric Fitting Experiment. The Fabric Fitting Experiment is an intuitive display of the dressing effect, which provides the material evidence for the observation and analysis of the influence factors of seat seam gradient and provides reliable experimental data for the study. The Virtual Experiment is a verification of the Fabric Fitting Experiment. The conclusions of the fabric fitting experiment are verified through the representation of the virtual dressing effect by the software. The Virtual Experiment is additional experiment to the Fabric Fitting Experiment.

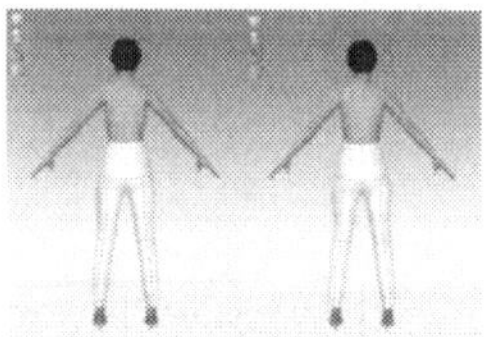

Fig.3-1 The back dressing effect of the basic pattern in H2 and H4

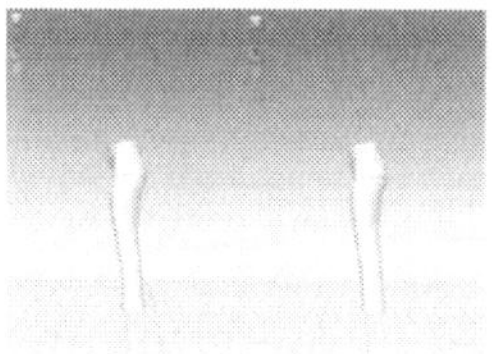

Fig.3-2 The side dressing effect of the basic pattern in H2 and H4

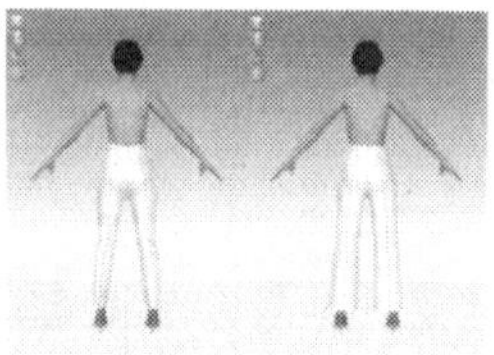

Fig.3-3 The back dressing effect of the basic pattern in bottom 16cm and 25cm

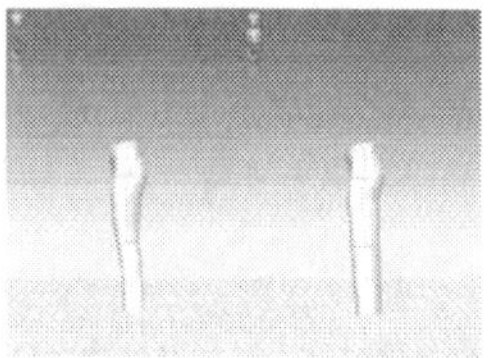

Fig.3-4 The side dressing effect of the basic pattern in bottom 16cm and 25cm

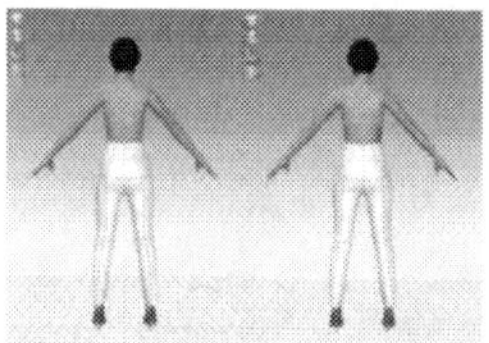

Fig.5-5 The back dressing effect of the basic pattern in B+2 and B+3

Fig.5-6 The side dressing effect of the basic pattern in B+2 and B+3

6. Conclusion

Through the Paper Experiment, the Fabric Fitting Experiment and the Virtual Experiment, the following conclusions are drawn.

Firstly, when the size of slack bottom and the position of the crease line are kept constant, as the hip width relaxed increase, the seat seam gradient increases and the dressing effect improves. Secondly, when the size of hip relaxed and the position of the crease line are kept constant, as the size of the slack bottom increase, the seat seam gradient decreases and the dressing effect improves. Thirdly, when the size of hip relaxed and the slack bottom are kept constant and the back crease line moves more outward, the seat seam gradient decreases and the dressing effect improves. Fourthly, when the change degree of the hip relaxed is equal to the change degree of the slack bottom, the hip relaxed and the slack bottom have little influence on the seat seam gradient; when the change degree of in the hip relaxed is greater than the change degree of the slack bottom, the seat seam gradient increases and the dressing effect improves; when the change degree of the hip relaxed is less than the change degree of in the slack bottom, the seat seam gradient decreases and the dressing effect improves. Finally, the smaller the seat seam gradient is, the better the dressing effect.

7. Acknowledgements

The paper is financially supported by the research projects (grant number KYTG02170202) of the Beijing Municipal Commission of Science and Technology.

References

[1] Zhang WB. Pattern making for fashion design. Beijing: 2006; 88-96.
[2] Zhang XH, Yu XK. Women's wear pattern making. Shanghai: 2009; 135-145.
[3] Li SJ. Quantitative Analysis of Seam Gradient of Back Trouser Crotch. Journal of Jiyuan Vocational and

Technical College: 2002; 1: 65-66.
[4] Cui JJ. Elimination the Standard-Fitted Front and Rear Crotch Pants Fold Quantitative Research. Master Dissertation of Dong-Hua University: 2010; 45-50.
[5] Qiu SF. Analysis of the Pants Structure Design and the Local Deformation. Packaging World: 2015; 6: 118-119.
[6] He YD, Lu LL. Study on the relationship of pants back crotch angle and back crotch length increment. Journal of Henan Institute of Engineering: 2013; 3: 21-24.
[7] He H. Optimization on the Butt Lifting Effect of Fit Jeans Pattern. Master Dissertation of Dong-Hua University: 2014; 32-38.
[8] Xia Y, Shi XQ, Wang HF. Analysis on young female's hip convex curve and trousers back part structure. Journal of Textiles Research: 2015; 11: 115-120.
[9] Tu Y. Research about Structure Design of Carry Buttock Suitable Pants. Journal of Zhejiang Sci-Tech University: 2011; 3: 373-377.
[10] Zhang YH. The Design of Trousers Crotch. Journal of Henan Institute of Engineering: 2010; 3: 9-12.
[11] Rui D. Research on the Fitted Pants Based on Surface Stripping Method. Master Dissertation of Zhejiang Sci-Tech University: 2013; 13-19.
[12] Wang S. Study on Standard Basic Pattern of Women's Fit-Trousers. Master Dissertation of Dong-Hua University: 2011; 44-65.
[13] Zhang R, Dai H. Research on the seat structure of trousers on prominent hip and scooped hip. Journal of Xi'an University of Engineering Science and Technology: 2005; 73: 41-45.
[14] Jiyoung Choi, Kyunghi Hong. 3D skin length deformation of lower body during knee joint flexion for the practical application of functional sportswear. Applied Ergonomics: 2015; 48: 186-201.
[15] Hyewon Seo, See-jo Kim, Frederic Cordier, Jiyoung Choi, Kyunghi Hong. Estimating dynamic skin tension lines in vivo using 3D scans. Computer-Aided Design: 2013; 45: 551-555.

Study on Matching of the Male Suit and the Corresponding Interlining Under Knowledge Engineering using Browser/Server mode

Jun Xu[1*], Tiao-Tiao Li[1], Yan-Chen Tao[2], Hong-Yan Liu[1]

[1]*Apparel and Art Design College, Xi'an Polytechnic University, Xi'an 710048, China*
[2]*College of Letters and Science, University of California, Berkeley, CA 94720, USA*

[*]Corresponding author's email: xujun0127@sina.com

Abstract

Rapid response is a critical developing mode of the modern textile and apparel industry. The aim of the paper is based on the rapid response of apparel manufacturing process. Using the semantic web method of knowledge engineering, knowledge representation, knowledge reasoning and knowledge acquisition, the matching of male suit and its fusible interlining are realized. The results showed that it is possible to use knowledge engineering to deduct the match system of face fabric and fusible interlining for a male suit. To get the rapid match of face fabric and its interlining on a male suit, the database and knowledge base were set up, the reasoning mechanism was also defined for the system of matching interlining on male suit in the article. Finally, the knowledge system based on PHP (language) and MYSQL (database) under the environment of Apache was developed successfully for the compatibility of the male suit and its fusible interlining. Knowledge system studied in the paper will help to improve the match efficiency of the textile product and its interlining in clothing factory, and also to make the match effect well to meet high-quality suit.

Keywords: Knowledge Engineering; Suits; Compatibility

The earliest knowledge engineering is proposed by artificial intelligence expert EA Feigenbaum from the United States [1]. A new task of "knowledge management" is created by combining computer technology, which has become one of the main ways of using computers to simulate the human brain in solving various problems and to answer all questions or new knowledge from existing knowledge, etc. Up to now, Knowledge engineering has been widely used in different fields [2]. However, few cases are found that applies the idea of Knowledge engineering in modern textile and apparel industry. One of the reasons for this is due to high individual experience domination of the apparel industry which restricts the developing pace of the apparel industry [3]. In this article knowledge engineering was applied on the matching of male suit and its fusible interlining. The reasoning logic was studied to question computing in system of compatibility for male suit and its interlining and the technical method is discovered for rapidly realizing the match of face fabric and its interlining on suit. These all provide a basis for the theory and technique on the rapid realization of the compatibility of textile and its interlining in a studied system so that the computer can simulate the human brain to solve the problem. Generally, knowledge engineering mainly consists of knowledge acquisition, knowledge representation and reasoning mechanism [4].

1. Knowledge Representation of Matching Interlining on Male Suit

In the system of matching interlining on male suit, knowledge includes the specification and performance of fabric, specifications of interlining, technology conditions of match and match position on male suit [5]. There is a relationship of genera or pattern match among the knowledge. The methods of knowledge indication include predicate logic, semantic network, production rules and framework rules etc. Due to the each method of knowledge representation reflects different relationship among knowledge [6], thus the method of semantic network is used to represent the knowledge in this article.

The event on the compatibility of the textile and the interlining on male suit is described as follows: under the temperature c1, pressure p1 and time t1, the fabric X matches interlining Y on the part ξ of suit. Its semantic network graph of knowledge system is shown in Fig.1.

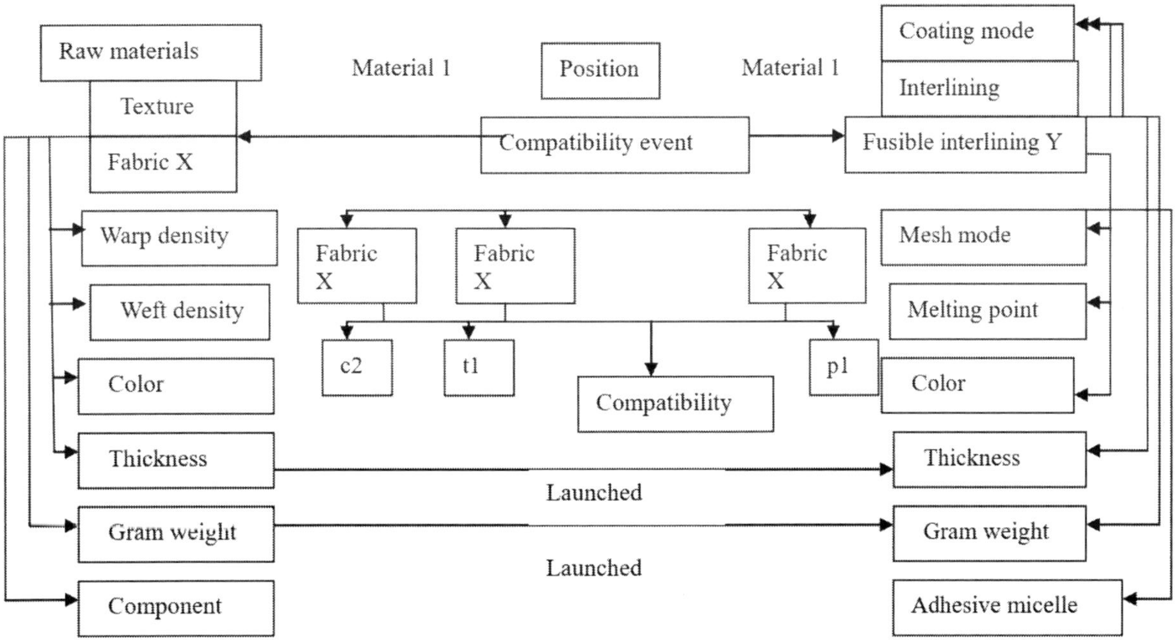

Fig.1 Knowledge semantic network of compatibility system for suit

At the temperature 125 ℃, pressure 2.5N and time 4.5S, the fabric X matches the interlining Y on the part ∮ of suit. The part ∮ on suit indicates the collar, shoulder, armhole, armpit, chest and pockets on suit. The specification parameters of fabric X includes the composition of raw material, weight, count, warp density, weft density and thickness. The specification parameters of interlining Y include adhesive micelle type, composition of base fabric, adhesive micelle distribution, type of base fabric, thickness, weight, melting point, weave and rubber density of adhesive micelle.

2. Knowledge Acquisition of Matching Interlining on Male Suit

Knowledge is an important symbol of an expert system that differs from other computer software system. The key elements to determine the system performance are the quality and quantity of knowledge.

The basic task of knowledge acquisition is to obtain knowledge for an expert system to establish a sound and effective knowledge base to resolve the solution of an issue in a related domain. Several needed works are knowledge extraction, knowledge conversion, knowledge input and knowledge detection. The modes of knowledge acquisition can be divided into the non-automatic knowledge acquisition and automatic knowledge acquisition based on the degree of automation knowledge acquisition.

The method of knowledge acquisition for the compatibility of textile and interlining on male suit is the non-automatic knowledge acquisition. These knowledges achieved through experiment, experience and technology documents of related domain experts, was inputted in the repository using knowledge editing software by knowledge engineers.

3. Construction of Knowledge and Database System

The quality of the system function is closely related with the affiliated knowledge in this system for match between face fabric and its fusible interlining on male suit. Here, the database and knowledge base used for date storage, were established to complete the validity, order and intelligence of knowledge management. In which, the database is used to store the original data of compatibility of the face fabric and its interlining on male suit. The knowledge base is used to store massive experience and knowledge that experts have in the field of male suit, the known facts and the rules are used in the knowledge.

3.1 Database

Database is a warehouse that organizes, stores and manages data with a data structure. It contains a number of whole data with format and structure, and the data storage is independent of the programs that use it. Usually database can be shown to add, modify and retrieve by a common and controlled manner. A database often contains several different files where each file contains several records with a number of sections per record, and the sections are directly put in the data item in the system.

Taking into account a little amount of data and simple data relationships, PHP language, mysql database and apache server are used in the system. The established databases for the match of system between face and its fusible interlining on male suit are shown in the graph from Fig.2 to Fig.7.

Data Number	Type	X-Value	Y-Value	Data Value
1	collar	0.35	0.4	0.3333
2	collar	0.35	0.45	0
3	collar	0.35	0.5	0.3333
4	collar	0.35	0.65	0.3333
6	collar	0.4	0.4	0.3333
7	collar	0.4	0.45	0.3333
8	collar	0.4	0.5	0.6667
9	collar	0.4	0.65	0
10	collar	0.45	0.4	0.6
11	collar	0.45	0.45	0
12	collar	0.45	0.5	0.6
13	collar	0.45	0.65	0
14	collar	235	30	1
15	collar	235	45	0.333
16	collar	235	70	0
17	collar	235	95	0
18	collar	235	100	1
19	collar	270	30	0
20	collar	270	45	0.333
21	collar	270	70	0
22	collar	270	95	0
23	collar	270	100	0
24	collar	275	30	1
25	collar	275	45	0
26	collar	275	70	1
27	collar	275	95	1
28	collar	275	100	0
29	collar	315	30	0
30	collar	315	45	0.333
31	collar	315	70	0
32	collar	315	95	0
33	collar	315	100	0

Fig.2 Collar parts database

Data Number	Type	X-Value	Y-Value	Data Value
47	shoulder	270	35	0
48	shoulder	270	50	1
49	shoulder	270	65	0
50	shoulder	270	100	0
51	shoulder	275	35	0.5
52	shoulder	275	50	0
53	shoulder	275	65	1
54	shoulder	275	100	0.5
55	shoulder	310	35	1
56	shoulder	310	50	0
57	shoulder	310	65	0
58	shoulder	310	100	0
39	shoulder	0.3	0.4	0
40	shoulder	0.3	0.5	0
41	shoulder	0.35	0.4	1
42	shoulder	0.35	0.5	0.667
43	shoulder	0.4	0.4	0
44	shoulder	0.4	0.5	0.333
45	shoulder	0.45	0.4	0
46	shoulder	0.45	0.5	0

Fig.3 Shoulder database

Data Number	Type	X-Value	Y-Value	Data Value
88	armpit	0.35	0.4	0
89	armpit	0.35	0.4	0.6
90	armpit	0.4	0.5	1
91	armpit	0.4	0.5	0.4
92	armpit	235	50	1
93	armpit	235	65	0.333
94	armpit	235	70	0
95	armpit	235	100	0
96	armpit	270	50	0
97	armpit	270	65	0.333
98	armpit	270	70	0
99	armpit	270	100	0
100	armpit	275	50	0
101	armpit	275	65	0
102	armpit	275	70	1
103	armpit	275	100	1
104	armpit	310	50	0
105	armpit	310	65	0.333
106	armpit	310	70	0
107	armpit	310	100	0

Fig.4 Armhole parts database

Data Number	Type	X-Value	Y-Value	Data Value
88	armpit	0.35	0.4	0
89	armpit	0.35	0.4	0.6
90	armpit	0.4	0.5	1
91	armpit	0.4	0.5	0.4
92	armpit	235	50	1
93	armpit	235	65	0.333
94	armpit	235	70	0
95	armpit	235	100	0
96	armpit	270	50	0
97	armpit	270	65	0.333
98	armpit	270	70	0
99	armpit	270	100	0
100	armpit	275	50	0
101	armpit	275	65	0
102	armpit	275	70	1
103	armpit	275	100	1
104	armpit	310	50	0
105	armpit	310	65	0.333
106	armpit	310	70	0
107	armpit	310	100	0

Fig.5 Armpit database

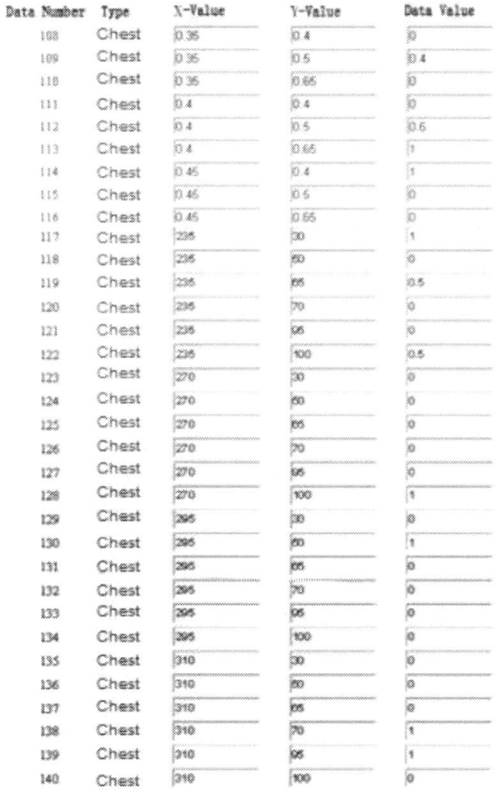

Data Number	Type	X-Value	Y-Value	Data Value
108	Chest	0.35	0.4	0
109	Chest	0.35	0.5	0.4
110	Chest	0.35	0.65	0
111	Chest	0.4	0.4	0
112	Chest	0.4	0.5	0.6
113	Chest	0.4	0.65	1
114	Chest	0.45	0.4	1
115	Chest	0.45	0.5	0
116	Chest	0.45	0.65	0
117	Chest	235	30	1
118	Chest	235	60	0
119	Chest	235	65	0.5
120	Chest	235	70	0
121	Chest	235	95	0
122	Chest	235	100	0.5
123	Chest	270	30	0
124	Chest	270	60	0
125	Chest	270	65	0
126	Chest	270	70	0
127	Chest	270	95	0
128	Chest	270	100	1
129	Chest	295	30	0
130	Chest	295	60	1
131	Chest	295	65	0
132	Chest	295	70	0
133	Chest	295	95	0
134	Chest	295	100	0
135	Chest	310	30	0
136	Chest	310	60	0
137	Chest	310	65	0
138	Chest	310	70	1
139	Chest	310	95	1
140	Chest	310	100	0

Fig.6 Chest parts database

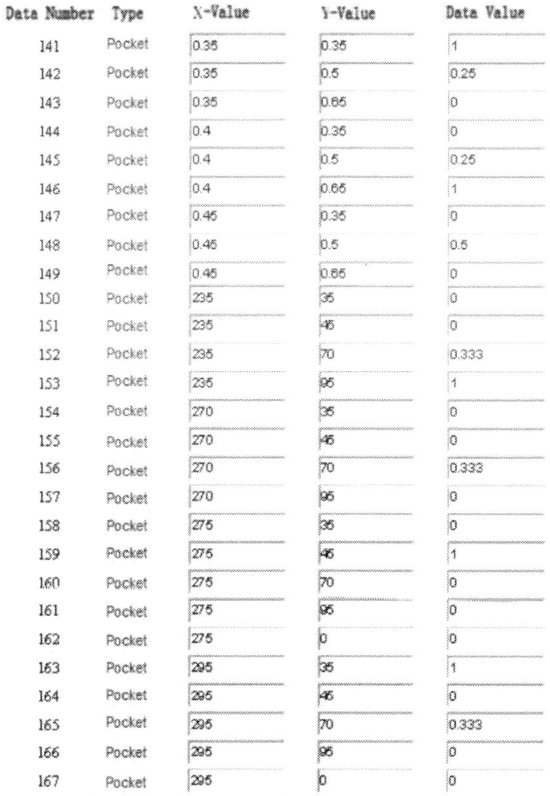

Data Number	Type	X-Value	Y-Value	Data Value
141	Pocket	0.35	0.35	1
142	Pocket	0.35	0.5	0.25
143	Pocket	0.35	0.65	0
144	Pocket	0.4	0.35	0
145	Pocket	0.4	0.5	0.25
146	Pocket	0.4	0.65	1
147	Pocket	0.45	0.35	0
148	Pocket	0.45	0.5	0.5
149	Pocket	0.45	0.65	0
150	Pocket	235	35	0
151	Pocket	235	45	0
152	Pocket	235	70	0.333
153	Pocket	235	95	1
154	Pocket	270	35	0
155	Pocket	270	45	0
156	Pocket	270	70	0.333
157	Pocket	270	95	0
158	Pocket	275	35	0
159	Pocket	275	45	1
160	Pocket	275	70	0
161	Pocket	275	95	0
162	Pocket	275	0	0
163	Pocket	295	35	1
164	Pocket	295	45	0
165	Pocket	295	70	0.333
166	Pocket	295	95	0
167	Pocket	295	0	0

Fig.7 Pocket parts database

3.2 Knowledge Base

Knowledge base is structured and can be easily operated, used and organized into clusters in knowledge engineering. It can also collect knowledge pieces required to resolve problems in computing. Knowledge pieces include knowledge in related areas, experimental data, or inspiration from the experience of experts such as some definitions, theorems and algorithms etc. [7].

System knowledge is used to store the experience of experts in the related field, the knowledge of experts, known facts and rules adopted in knowledge. The data that are available through certain channels can be adopted by inference and also managed, maintained and modified. Different database structure are used in different application to easily maintain and build appropriate reasoning mechanisms. In the system of match between face fabric and its fusible interlining on male suit, the established partial knowledge bases are separately shown in tables from 1 to 4 and figures 8.

Table.1 Knowledge base of face fabric on suit

Fabric sample numbers	Composition	Gram weight (g/m^2)	Count number	Warp density (/10cm)	Weft density (/10cm)	Thickness (mm)
A2	70%W: 30%T	275	80/2*50/1	366	374	0.34
A3	90%W: 10%S	270	80/2*74/2	378	356	0.34
A4	98%W: 2%U	295	64/2*64/2	378/	254	0.46
A5	50%W: 50%T	235	90/2*55/1	372	336	0.38

Table.2 Knowledge base of fusible interlining on suit

Interlinin gnumbers	Bottom-cloth material	Lining cloth type	Thickness (mm)	Gram weight (g/m^2)	Melting point (°c)	Weave	Colloidal particle ($/cm^2$)
B1	Polypro-pylene fiber	nonwoven	0.42	45.156	120~160	/	79
B2	Polypro-pylene fiber	nonwoven	0.48	32.912	120~160	/	52
B3	polyester	woven	0.48	63.54	120~160	plain	60
B4	polyester	woven	0.42	51.024	120~160	twill	76
B5	polyester	woven	0.51	69.624	120~160	Satin	46
B6	Polyester contain cotton	woven	0.51	98.344	120~160	Satin	36
B7	polyester	woven	0.67	96.88	120~160	Satin	47

Table.3 Knowledge base of fusing process condition of interlining on suit

Temperature (℃)	Pressure (N)	Time (S)
125	2.5	4.5

Table.4 Testing knowledge of relation model on fabrics and interlinings

Position on male suit	Relation model on fabrics and interlinings
Collar	Collar detection mode=3.18*X1+0.006*X2+0.009*X3-0.022*X4-1.143*X5+5.193
Shoulder	Shoulder detection mode=6.877*X1-0.009*X2+0.017*X3+0.018*X4-0.746*X5+0.439*X6-2.904*X7-0.576
Armhole	Armhole detection mode=11.306*X1+0.016*X2+0.015*X3-0.042*X4-2.143*X5+0.176*X6-0.635*X7+12.835
Armpit	Armpit detection mode=-1.156*X1+0.008*X2-0.011*X3+0.005*X4-0.218*X5+0.058*X6-0.26*X7+1.231
Chest	Chest detection mode=2.431*X1-0.005*X2-0.786*X3-0.059*X4-5.477*X5+21.987
Pocket	Pocket detection mode=4.168*X1+0.015*X2+0.031*X3+0.047*X4-0.854*X5-0.169*X6+1.998*X7-12.906

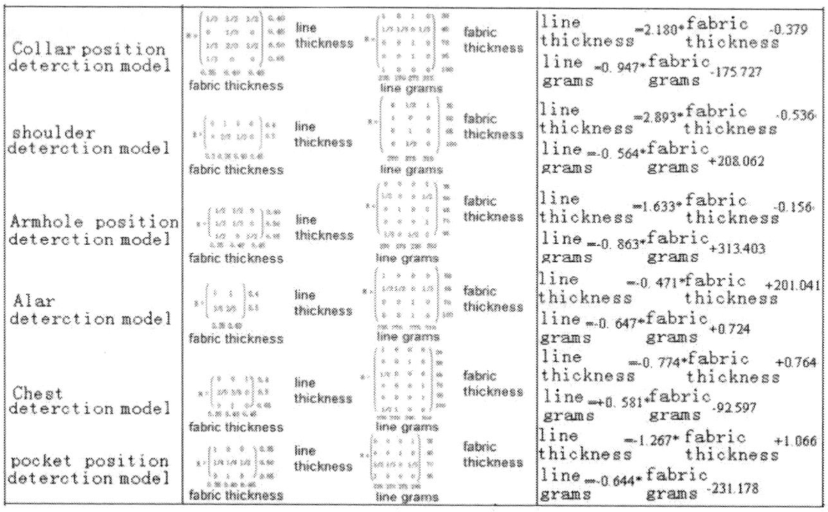

Fig.8 Knowledge of relation model on fabrics and interlinings

4. Inference Engine of Matching Interlining on Male Suit

Inference Engine is the use of computers for knowledge reasoning, which is an indispensable component for expert system. It mainly includes two aspects of reasoning and controlling with the actuator, dispatcher and consistency coordinator. Dispatcher's primary role is to choose an action for the next step in the system, which is based on the control strategy and the board records. The main function of the actuator is to carry out the action chosen by dispatcher using the knowledge in knowledge base and information recorded on the blackboard. Consistency Coordinator's main feature is to make update for assumptions and data timely. Reasoning, based on the existing axioms, rules and principles, is a process that leads to new facts or judge by a series of steps via the former fact or judge. The former fact or judge is called prerequisite, the new fact or judge are conclusions, and the existing axioms, rules and principles are taken as reasoning control strategy.

Reasoning strategies is currently divided into forward reasoning, backward reasoning and two-way reasoning based on the reasoning direction from control strategy.

4.1 Method of Knowledge Reasoning for Matching Interlining on Male Suit

Reasoning process of knowledge system can be described in four steps: first, according to the experiment data for different parts of the male suit, various matches combined between fabric and fusible interlining are analyzed. Then matching models are established. Following this, the database, knowledge base, and algorithm library are constructed. Finally, the system of matching interlining on male suit is completed. This process will achieve target state and remove unanswered questions after a series of steps through known information and data. Thus, the system uses the method of forward reasoning.

4.2 Reasoning Process of Reasoning Machine for Matching Interlining on Male Suit

Problem computing process of reasoning machine primarily contains the following steps: To solve problems, the semantic networks clips are constructed that contains the known and unknown knowledge, along with comparing the overall semantic networks of the system. The relationship between known knowledge and calculated knowledge is established. Then the previous corresponding formulas are applied to solve the knowledge or retrieve the subordinate knowledge. According to the local knowledge clips that is already resolved, further specific problems that requires solutions are retrieved.

The steps are shown for each individual case.

4.2.1 Reasoning of Specifications of Interlining Matching with Fabric Sample

The question is to get the specification of interlining matching with fabric 1. Thus, the reasoning process is as following:

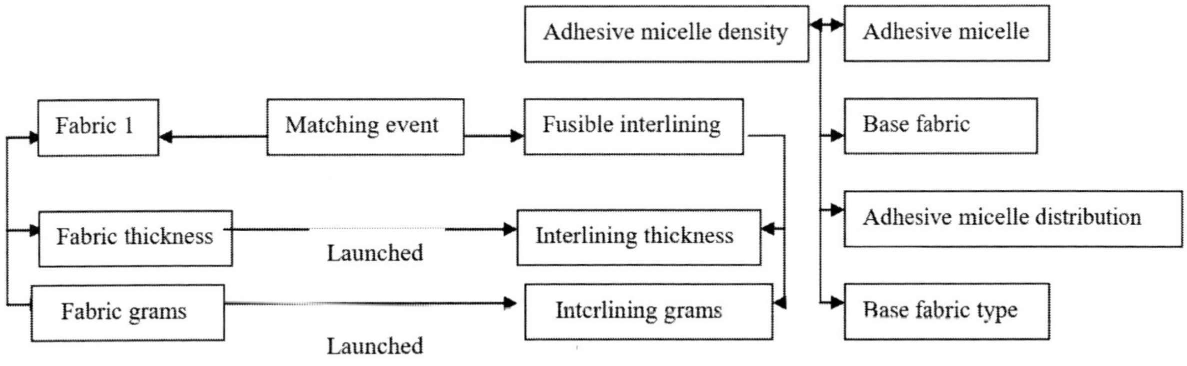

Fig.9 Semantic network segment for solving the problem

The specifications of interlining matching with fabric 1 are logically obtained as follows

1) The semantic network fragments are constructed for the compatibility of the fabric and its interlining (shown in Fig.9). Then the semantic networks of system knowledge map that is used for matching interlining on male suit is compared (shown in Fig.1).

2) According to the match model of compatibility for fabric and agglutinate line shown in Fig.10, the gram weight and thickness of corresponding agglutinate line 1 is marched out.

3) According to the above gram weight and thickness of interlining 1, the other specification data related to the corresponding thickness and grams of interlining 1 is retrieved from a database.

4.2.2 Reasoning of Adhesive Process and Applied Part on Suit between Fabric1 and Interlining 1

The challenge is to get agglutinate conditions and applicable parts on suit between fabrics 1 and corresponding interlining 1. Then the reasoning process is as following:

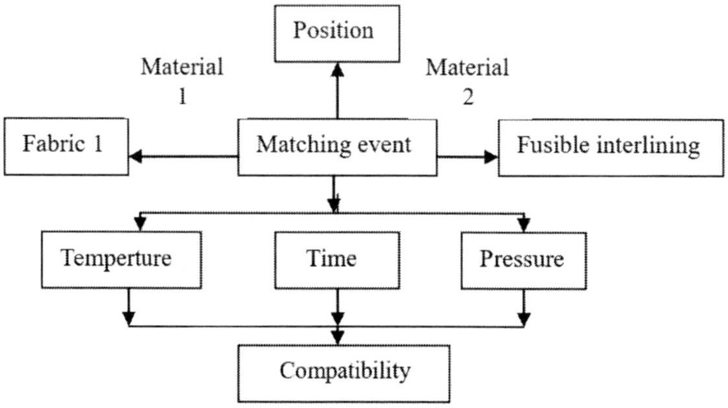

Fig.10 Semantic network segment for solving the problem

1) The semantic web segment is constructed and shown in Fig. 10 which demonstrates the adhesive process and the applicable part on suit for the compatibility of face fabric 1 and interlining 1 on male suit. The semantic networks of system knowledge map that is used for the matching interlining on male suit is compared (shown in Fig.1). Here fabrics 1 is matched with interlining 1, the aim is to get adhesive process conditions interlining 1 on fabric 1 and their applicable part on suit.

2) According to the specifications of fabric 1 and interlining 1, the data of whole matching events for fabric 1 and interlining 1 is retrieved from a database.

3) Through the whole matching event, the applicable part on male suit and the corresponding fusing pressure, temperature and time for matching between fabric 1 and interlining 1 is obtained.

5. The Realization of Knowledge System for Matching Interlining on Male Suit

5.1 Requirements Analysis of System used for Compatibility of Fabric and Interlining on Male Suit

The system used for compatibility of fabric and interlining on male suit have the following functions: when the user selects the part on suit and inputs the thickness and gram of used fabric, the system would automatically calculate the thickness R and gram G of interlining matching with the above face fabric according to the match model of the face fabric and interlining on male suit. Meanwhile the database looks for the sample of agglutinate line which has a range of thickness and gram that separately varies in $\pm$ 0.05 when compared to standard values. Then the retrieved interlining together with other corresponding specifications will be displayed on the computer interface for reference. Lastly the user can select the needed interlining according to the specifications shown on the interface of the computer.

5.2 The Structure of Knowledge System for Matching Interlining on Male Suit

Based on the request above, in the designed system of matching interlining on male suit, the logic diagram formed among all procedures are related to the system and their respective modules as shown in Fig.11. From this logic diagram these procedures' internal links, as well as the design ideas of these procedures based on these internal connections [8] are dispalyed visually. In this chart, the logic order of functional module also reflects basically the required steps of user operating.

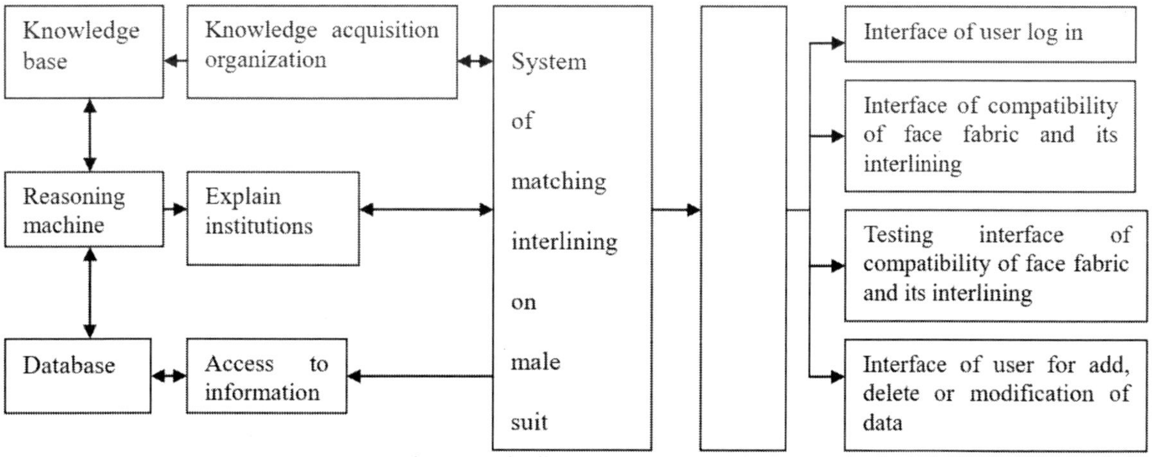

Fig.11 Logic diagram among modules

5.3 Flowchart of Knowledge for Matching Interlining on Male Suit

The system flowchart for matching interlining on male suit is shown in Fig.12.

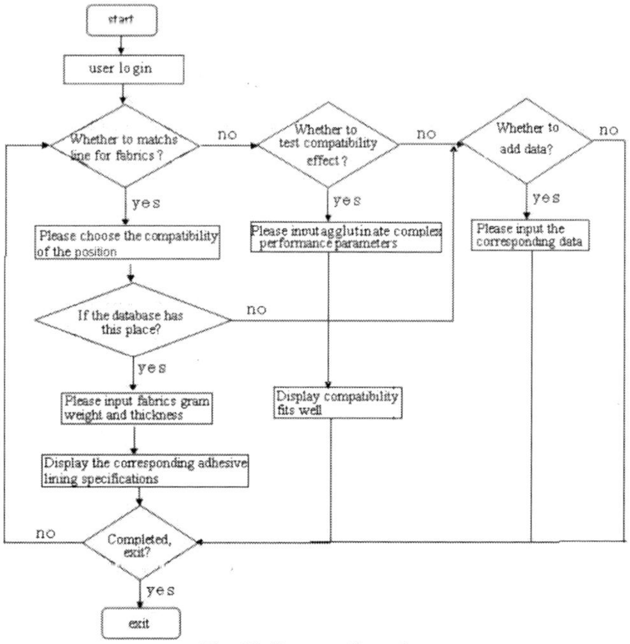

Fig.12 System flowchart

6. Conclusions

Based on the constructed relation model and the detection model of compatibility of interlining at different parts of the male suit, the Mysql database is used to obtain a knowledge base and database for the compatibility of face fabric and interlining on male suit. Knowledge engineering is used to construct the knowledge acquisition mechanism and the knowledge inference mechanism in the knowledge system of compatibility of textile and interlining on male suit. PHP is applied as a developing tool in backstage, finally, the B/S mode system of matching interlining on male suit is put forward with the applying procedure from HTML, PHP and ActiveX.

Acknowledgement

The Authors wish to acknowledge the Scientific Research Programs funded by Shaanxi Provincial Education Department (Program No.14JF008) and China National Textile and Apparel Council (Program No. 2014064)

References

[1] Li YX, Hu J. Basic and applied knowledge engineering case [M]. Sci. Press: 2006.

[2] Zhao H, Liu DF. Realiaty and Development of research of knowledge engineering in China [J]. Inf. of Chin. Manu. Ind.: 2018; 3: 27-30

[3] Ye M, Xu SW. Optimization of clothing fabric and its fusible Interlining [J]. Text. Technol. Overseas: 2003;

[4] Cao LM, Chen YL, Zhou Q. The knowledge of engineering principles [M]. Chin. Univ. of Mining Press: 1995.

[5] Fan JT, Leeuwner W, Hunter L. Compatibility of outer and fusible interlining fabrics in tailored garments. Part I.: Desirable range of mechanical properties of fused composites [J]. Text. Res. J.: 2005; 3: 33-36.

[6] Chen LH. Ten classified methods of knowledge [J]. Sci. Res.: 2007; vol.33; 2: 26-31.

[7] Wu JW. Ontology-based Knowledge Base System [J]. Innovation: 2009; 1.

[8] Wang J, Sun HX, Wu FH. Study on the Construction method of knowledge base for smart design system based on knowledge [J]. Mech. Design: 2007; 2: 24-27.

Design Research on Casual Wear for People with Physical Disabilities

Meng-Jie Hou[1], Hong Dai[2*], Cun-Xin Liang[1]

[1] School of Apparel & Art Design, Xi'an Polytechnic University, Xi'an, 710048, China
[2] Apparel & Art Design, Xi'an Polytechnic University, Xi'an, 710048, China

[*]Corresponding author's email: 406634917@qq.com

Abstract

China is the world's most populated country with the greatest amount of the disabled and those with physical disability. The physically disabled due to their impairment often lack self-confidence. In order to improve self-confidence, not only is emotional enrichment important, the appearance of the individual is also critical. For the healthy development of the physically handicapped, fashion design is necessary. Their clothing is different from a healthy person's clothing whose dress is primarily focused on appearance not functionality. This research, through conducting questionnaires towards people with physical disability, seeks to identify their needs and demands for clothing. It summarizes the factors influencing their clothing design. It put forward suggestions for improvement for designing appropriate clothing for the disabled, and provide references for the clothing design of physically disabled people.

Keywords: People With Physical Disability; Clothing; Survey; Comfort

1. Introduction

1.1 The Concept of the Disabled

If a country wants to thrive, the respect for the disabled cannot be neglected. According to the 2008 revised protection of the disabled persons of the People's Republic of China: "the disabled refer to the psychological or physiological impairment of the human body where some organization or function is lost or abnormal which results in either the partial or full inability to engage in everyday activities." The term "disabled persons" includes those with impaired vision, hearing, speech or those with physical disabilities, mental retardation, mental disorder etc. [1]

1.2 The Significance of the Research

Based on the current market situation, there are approximately 82.96 million disabled people in China, 24.12 million of which have physical disabilities. The clothing market for the physically disabled in China at the moment is still in an undeveloped state, and the market is still very broad. [2] Clothing for those with a disability from a limb in our country is still in theoretical discussion. The clothing designed for them is to conceal their body impairment so that they can blend in more with ordinary people. The purpose of the clothing is to help them better fit into the society, and to make their life more convenient at the same time to reduce their trouble of taking off clothes.

1.3 The Status Quo of Domestic and International

Domestic study of limb disabled clothing is still at an infancy stage. [3] In general, domestic clothing is still under theoretical research with little practice. The physically handicapped live with inconvenience; they also generally wear normal clothes that is an inconvenience to them, thereby making their life difficult.

The foreign clothing market for disabled people is very matured for there are many different kinds of clothes and large enterprises that specializes in the production of these kind of clothes. They are full of structural and functional innovation. The clothes are designed so that they do not hinder the individual's mobility and enables them the full use of the chest and the superior limbs. They must also be durable. [4] There are even foreign design institutes and courses in clothing design for the disabled. [5]

1.4 The Purpose of the Study

The level of social security for the disabled reflects a country's economic, political, and cultural development level and their degree of civilisation. [6] Clothing design for the limb disabled is a demonstration of the people's care towards the needy and the weak and helps promote social integration. Limb disabled clothing can satisfy the psychological and physiological needs of the disabled and make their lives more convenient while helping them build up their self-esteem and self-confidence.

1.5 The Research Methods

The research methods employed in this paper mainly include:

Literature retrieval method: by means of network access and retrieving relevant papers, documents and books, both at home and abroad to understand the current research status.

Survey method: carry out questionnaires on the limb disabled. Include information on color, profile, clothing categories and other issues. Analyze and summarize the clothing demand of limb disabled people.

2. Investigation and Analysis on the Present Situation of Body Clothing for the Disabled

The survey's coverage is mainly in a district of Xi'an dedicated to people with disabilities and caregivers. 100 copies of questionnaires were given away; 160 copies of questionnaires were sent online, with a combined total of 260 copies of questionnaires of which 216 are effective questionnaires. 200 questionnaires are analyzed to illustrate the problems that exist in the body clothing for the disabled.

Based on the questionnaire results completed by 112 people with disability (56%) and 88 caregivers (44%), it can be observed that caregivers are important when considering the design of clothes for the disabled.. It also strictly analyzes the collected questionnaire and confirms its credibility. The sex ratio in 1995 released by the China disabled persons' federation accounted for about 60% of limb disabled people. Men and women accounted for about 40% of similar data. According to figures released by the China disabled persons' federation in 1995, handicapped persons with physical disability over the age of 40 accounts for 75.22%. In this investigation, the elderly with physical disability over the age of 40 accounts for 79% of the physical disability totally. Compared to young disabled people, elderly disabled people requires clothing that are more practical and functional (although aesthetics are still important). [7]

2.1 Color Respect

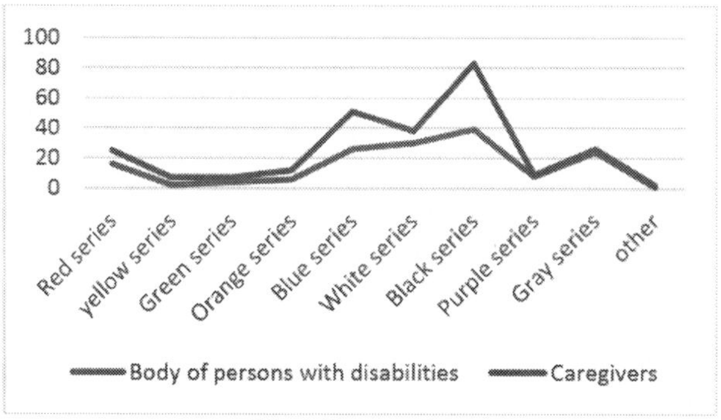

Fig.1 Survey data

Through the survey, it can be observed that limb disabled people prefer the colors black, white and gray. Yellow and orange are the least popular as they do not want to draw attention to themselves. There are however a few who like red because red represents enthusiasm and hope. Caregivers prefer black as black clothes are easy to clean. As Fig.1.

2.2 In Profile

From the survey, it can be observed that limb disabled people prefer A and H clothing profile more. This is because most disabled people are sensitive to their image. The profiles help to conceal their physical defect thereby giving them more confidence. X profile of clothing is higher in the requirement of body shape. The caretakers also prefer A profile suit. As Fig.2:

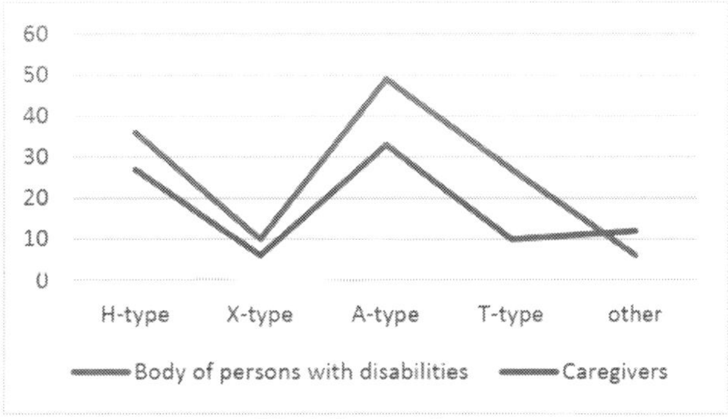

Fig.2 Survey data

2.3 Clothing Categories

According to the survey, limb disabled people prefer casual sportswear, loose and comfortable fabrics that are soft and can be taken off easily. There are limitations of a dress. A suit helps cultivate moral. Trench coat is especially more popular with body of persons with disabilities. As with the disabled people, caregivers are likely to opt for casual clothing. As Fig.3:

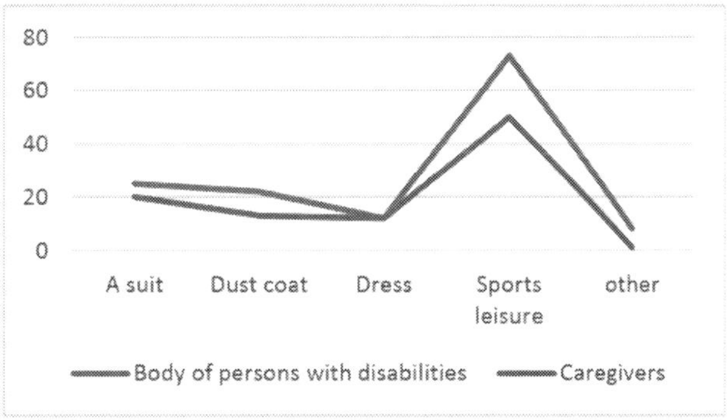

Fig.3 Survey data

3. Research on other Influencing Factors

3.1 Appearance Respect

Clothes that are beautiful not only makes the wearer more presentable, it helps with their self-confidence and esteem. Limb disabled people also seek to have beautiful clothes. Fashionable clothing helps to improve the psychological state of people with disabilities. It makes them more determined to integrate with the society. With the rapid development of the social economy, the aesthetics of clothing has become an essential part to help people with disabilities to better fit into society.

3.2 Fabric Comfort

Body for fabrics are comfortable for the disabled who are very seriously, because the uncomfortable fabric will bring a great burden to the disabled people. [8] Some people with disabilities have to be in a wheelchair. This limits the physical activity that they can do which may slow down their blood circulation resulting in all types of skin diseases. As such the comfortability of the fabrics are critically important for the disabled.

Flexibility refers to the degree of softness of a fabric. Fabric softness is affected by the twist of yarn, different fiber varieties, as well as its organization and finishing. The softness of the fabric has an effect on the human body comfort. Woven fibers are softer than any other type of fibers including cotton. Clothing made of soft fabric is not only comfortable it also gives a sense of security to those who are limb disabled. [9] Due to the lack of exercise, limb disabled people have fragile skin and weak muscles. Fabrics that are hard or rough when comes into direct contact with the skin can cause skin disease which can affect the health of the disabled people.

Permeability also affects the comfort of fabrics. From a clothing hygiene perspective, fabric with good permeability helps with air circulation and enhances body metabolism. It is conducive to people's health as it helps regulate the body's comfort performance. Suitable fabric for clothing is very important. The permeability of fabric is affected by material organization, density, surface shape and may other factors. Permeability is able to regulate the body's comfort performance through fabric softness. On a hot summer, movement hindered by the disabled limb can result in a lot of sweat. [8] The clothes therefore will easily stick to the body which makes the disabled person more uncomfortable. Furthermore, disabled people who wear prosthetic limbs can lead to injury when there is friction between the prosthetic limb and skin. As such fabrics with good permeability is essential to prevent secondary damage and in reducing the heat and humidity of the human body. In the cold winter, limb disabled people due to physical inconvenience often do not exercise. As such they are easy to feel cold and need to wear layers of clothing to maintain body heat.

The warmth retention property of clothes is affected not only by the structure of the clothing model, it is also affected by clothing material. Limb disabled people due to limited mobility during the cold winter are unable to generate sufficient heat and are prone to be colder than normal. As such the fabric warmth retention property of the fabrics are important for people with disabilities.

When designing clothes for people with disabilities, fabric softness, air permeability, warmth retention property and many other factors needs to be considered. For the clothing design of the limb disabled, if the outer clothing is used as the fabric, the middle layer garment can use the wool fabric as the clothing material; the inner layer clothes should use cotton as the fabric. The clothing comfort is the most important, which also needs considering. The clothing's air permeability, moisture absorption and flexibility are also.

3.3 Convenience for Taking On and Off

People without disabilities wear sleeves and trousers that are tightly sewn to keep them warm. These become restriction fo people with disabilities. [10] Disabled people find it difficult to take off clothes. Those with disabled upper limbs find it difficult to take off coats with fasteners. It takes too much effort to wear the clothes and some of the clothes fasteners are tto complex. As a result the clothes may even be bad for the health of the disabled. Disabled people with impaired lower limb find it difficult to take off bottom wear, especially those who use prosthetic limb. They are unable to bend their legs. This is also true for those who has to use wheelchair. They rely on caregivers to take off their bottom wear which requires a lot of time and energy. There is a higher clothing requirement for people with disabilities, as such the functionality of the clothing must be further considered during the design process.

3.4 Safety and Health

As a result of long term inconvenience and lack of exercise, Limb disabled's body immunity is low, and physiology is poor, which are susceptible to skin and respiratory disease.

Wearing convenience, safety and health are critical factors to consider when designing clothing for disabled people. Security is the foundation of clothing design which must incorporate the physiological and psychological wellbeing of the disabled. Health is the guarantee for clothing design which has requirements on health and environmental protection. In accordance with the physiological and psychological demands of the limb disabled people, their need is different to those without disability. For instance the coordination of disabled people is inferior to those abled people, as such the clothes designed must be short as to not cause harm to the body when taking off. [1] For the disabled with limb impairment, their movement is slow and to avoid burns flame retardant fabric is required.

3.5 Scientific and Technological Content

Limb disabled people's clothing needs to be more advanced than ordinary clothing, as more consideration must be taken which requires the further study in the science and technology department. The design of clothing for the disabled is complex and different to traditional clothing design. It needs a lot more time and a nd practical experience which will inevitably result in greater cost. The technology needed for creating limb disabled friendly clothing is slow and difficult, since it is also very expensive the profit margin is quite low. Moreover the cost of high tech products is often higher and not suitable for the general public, and this is especially true for disabled people as they often have more complicated health problems and their income is not high. To cost to create clothing suitable for limb disabled has not been solved, the graduate student postpartum will have direct effect on the purchasing power and quantity. [11]

3.6 Design Research

According to the surveys, people with disabilities favor dark gray clothes. A and H profile clothing and sport casual wear are convenient and safe.

Previous limb disabled clothing is inconvenient for the disabled to take off and they are not aesthetically pleasing, as such disabled people were not inclined to buy them. Ordinary products cannot meet their basic needs, [12] therefore the disabled basically has no special clothing to wear. [13] In regards to the situation of the disability on clothing, the main problem is aesthetics and stigma. [14]

Limb disabled women's clothing design is shaped like a cloak, because the shape like a cloak can block the mutilated parts of the physically disabled. In the case of upper limb deformity or disability, single limb activities will increase, A single limb needs to complete a lot of difficult movements. In cloak of women's clothing design, clothing left and right side of the each has a cardigan, convenient to the activities of the arm, buttons in lower position, lower the following don't do the stitching. Right upper limb disabilities only pass on the right of a front line close to complete shirt open, usually on the left side of the front line has been closed. Left upper limb disabled people wear method by the same token, the opposite position.

Body disabled men's clothing is to a man often wearing sweater design, limb disabled when wearing a jumper's biggest problem is that because of his physical cannot quickly come from the jumper inside his head. In the design of jumper, in view of the left upper limb disabled does on the one hand the shoulders open design. [15] this design in the left upper limb disabled, for example, did open the shoulders of the design, so in the case of open mouth, the size of the collar, do set of head movements, can be drilled from the inside. As Fig.4 and 5:

Fig.4 women's clothing

Fig.5 Men's clothing

4. Conclusion

Detailed questionnaires on people with disabilities were conducted. They are based on the real life situation and summarizes the needs of the limb diabled people. It analyzes factors that need to be considered when designing clothes for the disabled.

With the development of social economy and the improvement of people's living standard, there is a higher demand for advocating services to disadvantaged groups and therefore limb disabled clothing designs have becoming of greater value. When designing clothing for the disabled many factors must be considered including the aesthetics, the fabric comfort, health and safety. In regards to the health aspect, moisture wicking antibracterial fabrics are needed to ensure clothing hygiene for instance in the summer, a hundred grass fiber knitted fabrics can be used to make limb disabled clothing. As for comfort, we can select some soft andcomfortable fabric. Due to their disabilities and the affect it has on their everyday life, the amount of effort taken to take off the clothing must be taken into account. Limb disabled clothing design is a systematic project which cannot be completed overnight. It needs continuous exploration and discovery. Body clothing for the disabled has very different requirements to ordinary clothing. The research on clothing for disabilities is still at an infancy stage due to the lack of theory and practice. It does not resolve actual problems and has a long way to go to be effective in helping the disabled integrate into the society.

References

[1] Chen S. Research on sustainability in fashion design for the disabled [J]. Ecological Economy: 2014; 9: 261-264.

[2] Cui XJ. Our body functional clothing market demand survey for people with disabilities [J]. Modern Business Trade Industry: 2015; 28: 68-69.

[3] Sun SY. Men's clothing design problems research [D]. Northeast Normal University: 2014.

[4] Antonela C, Viorica C, Laura M, Marian P. Designing Functional Clothes for Persons with Locomotor Disabilities [J]. De Gruyter: 2014; 281-289.

[5] Wu J, Shi LM, Zhao YX, Guo RP, Zhou YK. The development present situation analysis of physically disabled persons garments [J]. China Personal Protective Equipment: 2009; 16: 26-30.

[6] Wu XM. Limb disabled clothing wear to take off the study [D]. Jiangnan University: 2009.

[7] Li M. The body the disabled garment research and design [D]. Donghua University: 2007.

[8] Tang H, Shen Y. Physical disabilities clothing structure parsing [J]. Journal of Textile Research: 2004; 26: 69-70.

[9] Wu XM. Body fabric comfort for the disabled [J]. Tianjin Textile Science & Technology: 2011; 32: 10-12.

[10] Qi Y, Du J. In view of the limb disabled costume design analysis [J]. Humor and the Humorist: 2016; 61: 137-138.

[11] Liang HE, Wu XM. The research status and prospects of our country's body clothing for the disabled [J]. Journal of Wuhan University of Science and Technology: 2008; 21: 45-48.

[12] J. Issues and techniques in the inclusive design of apparel for the active ageing population [J]. Elsevier: 2015.

[13] Wu XM. Body clothing coat cardigan design for the disabled [J]. Tianjin Textile Science & Technology: 2010; 31: 42-44.

[14] Wang N. The psychological factors and design thought of the clothing of disabled young people [D]. Southwest University: 2011.

[15] Zhang ZH, Dai XJ. Study on the comfort of trousers structure in the sitting posture of disabled persons with lower limb [J]. Tianjin Textile Science & Technology: 2016; 37: 55-60.

The Influence of Infant Clothing Structure Design on Safety

Xiu-E Bai[1*], Zi-Lan Li[1], Ling Chen[2]

College of Textile and Clothing Engineering, Soochow University, GAN JIANG East Road No.178, Suzhou, 215021, China

*Corresponding author's email: baixiue@suda.edu.cn

Abstract

In light of the domestic household, several accidents have taken place as a result of infants clothing. As such people are paying more attention to the security of infants clothing. This paper combines research results, analyzes the overall structural design and detail structural design while examining the fundamentals of clothing structural design on security. The aim is to provide a theoretical reference for the infant clothing industry to establish and meet a set of safety standards.

Keywords: Infant Clothing; Structural Designing; Security

1. Introduction

The term infant refers to babies, toddlers and young children between the ages of zero to three-years [1]. During this stage of development, the physical body of infants are not matured, as such they can be easily harmed or injured and are in need of more caution and protection [2]. Clothing is a critical element in an infant's daily life to ensure their safety. With the development of the society and the economy, the standard of infant clothing will be higher, and there is more expectation and demand for the structural design of infant clothing. This paper studies the structural design of infant clothing from two aspects, the first a general overview and then a more detailed examination.

2. The Influence of the Whole Structure Design on the Safety of Infants

Integral structure design is key to clothing design and encompasses garment body outline, garment body proportion, clothing prototype, province, pleated, dividing line changes and application [3]. The garment body outline and clothing prototype is key to the whole structure design of infant clothing. This section will discuss separately from them.

2.1 The Influence of Body Outline Design on Safety

Garment body outline is the foundation of the clothing structure design. Garment body profile type encompasses linear type and curve type. Linear model includes type H, A, T, X; Type curve includes type O, hourglass, bell-shaped, mermaid, bitter fleabane [3]. Infant clothing is more choice to type H, type A and type O due to the particularity of infants' bodily form [4]. Type H which features is simple and neat. It is convenient to dress. Type O is loose, be tighten and both ends are conducive to infant motor, so it can reduce accidents. Type A is presented as tighten above and below loose, which can protect infants' abdominal. To sum up, the proper type is conductive to wear off and play for infant.

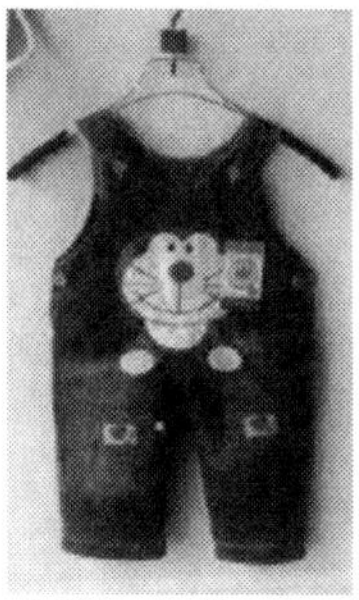

Fig.1 H type braces [9]

Fig.2 O jumpsuits [9]

Fig.3 Type A small coat [9]

2.2 The Clothing Prototype Design's Impact on Security

Infant clothing prototype is particular which designed for infants aged 0 to 3 years old. The prototype mainly consists of planar upper body prototype, which encompasses basic prototype, planar coat prototype and prototype kimono. As shown in Fig.4 [5], (a) prototype can be used in knitting clothing. In the quarter bust (3 - 12 horizontal line) increases the amount of 3 cm ~ 5 cm, increasing 2cm to 3cm loose in 1 / 2 back width (3 - 8 horizontal line). The design conforms to the physiological characteristics of infants' abdominal bulge and back sag. (b) prototype is modified on the basis of (a), which is suitable for woven clothing. As shown in Fig.5 [4],increasing the amount of losing 6 cm in the quarter bust (0 - 3 level line), the armhole depth (0 - 7 horizontal line) increased 4cm to make the coat more relaxed, in order to giving infants' body more space. The kimono structure is used frequently in coat design. As shown in Fig.6 [5], connecting sleeve designed the underarm point to fall to the waist. This design makes the armhole more lenient to wear off and protect infants against restrictions.

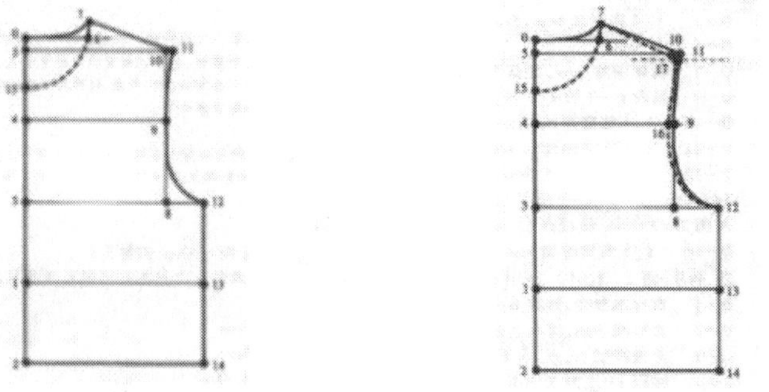

(a) (b)

Fig.4 Basic prototype of plane upper body [5]

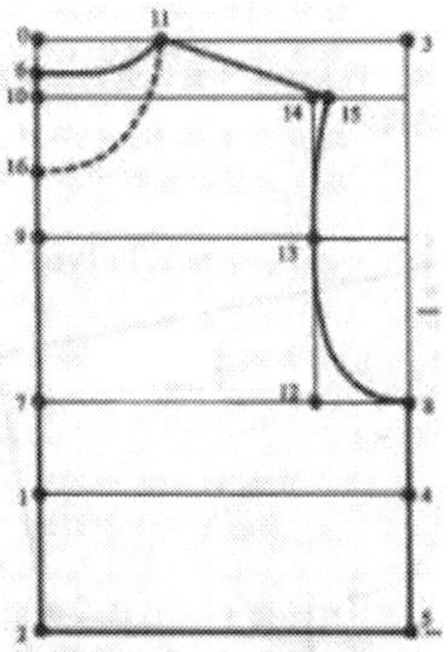

Fig.5 Infant planar garment prototype [5]

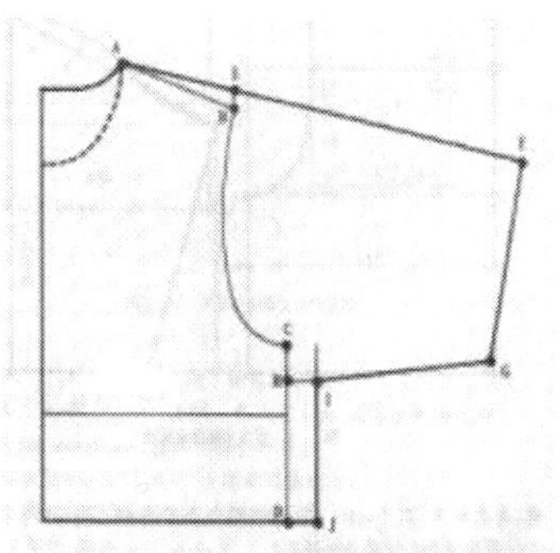

Fig.6 The kimono prototype [5]

3. Detail Structure Design's Impact on the Safety of Infants' Clothing

Different structure and shape expresses different styles. Detail design includes collar, sleeves, crotch, waist and hips, and decoration design [6]. There is a certain difference between infants and adults' clothing design because the special physiological characteristics. This section will discuss the influence of detail structure design on the safety of infants from these aspects.

3.1 Collar Design on the Impact of Security

The collar encompasses without collar, the collar, lapel collar and shawl collar. Collar design should be based on the characteristics of the wearer's face and neck and be coordinated with the appearance of clothing. Also you can use different decorative techniques [6]. Because infants' neck is very short and their skin is delicate, so the collar design of the fashion sense weakened and mainly focus on its comfort and safety. There are three kinds of methods used in collar design of infants' clothing at present. As shown in Fig.7 [7], the assembling type used for winter, The design protects the child from cold and sore throat by preventing air into easily. When in summer it can reduce the amount of assembling, while it needless to take off the shoulder. As shown in Fig.8 [6], Sleeveless garment in summer can be designed for opening in the shoulder, thus it is more convenient to wear off and make the child feels cool. As shown in Fig.9 [7], the oblique open collar type is suitable for infants dressed in autumn and winter. Oblique open designing increases the neckline space, so it can protect child's neck. These three types designed by increasing or decreasing the amount of opening on the shoulder or sleeve curve, which plays an important role in safety protection for infants' neck.

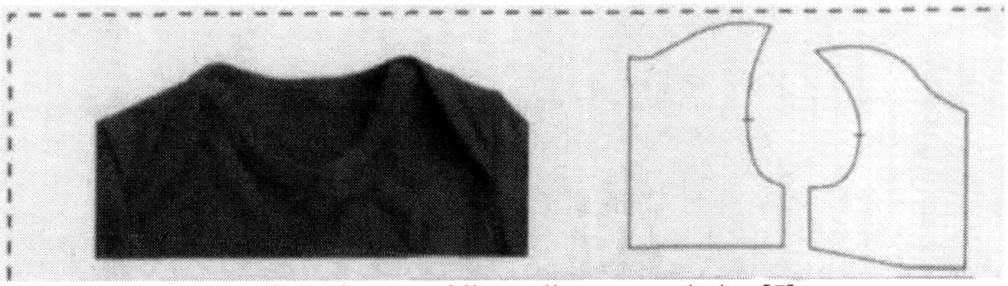

Fig.7 The assembling collar pattern design [7]

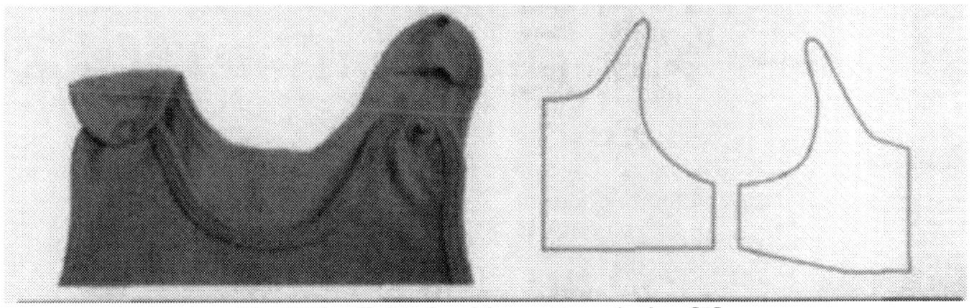

Fig.8 The fight shoulder pattern design [7]

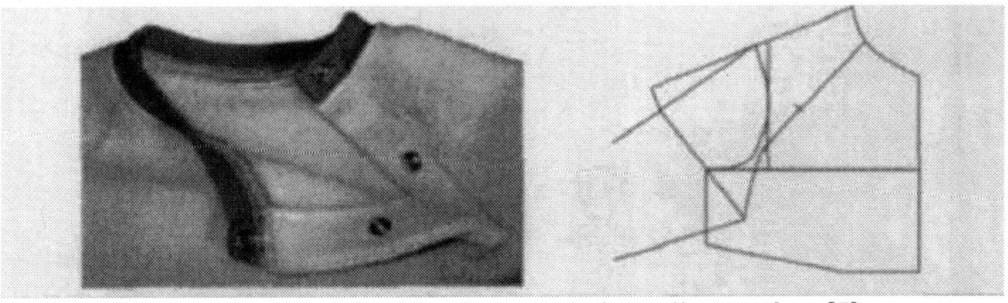

Fig.9 The use of shoulder sleeve to design collar openings [7]

3.2 The Influence of Sleeve on Security

According to the relationship between sleeve and garment, sleeves generally includes the sleeve, sleeve, raglan sleeves, without sleeves. Because infants' head is big and have short feet, so it is best to wear clothes with wide sleeves. The shoulder sleeve or attached sleeve is very suitable for infants [8]. Decorative shouldn't be considered too much. There are two common types sleeves in the market, the one is one-piece sleeve and another is raglan sleeve. Because the security function of the connected sleeve is analyzed in the prototype of the previous text, this section mainly analyzes the influence of the raglan sleeve. As shown in Fig.10 [9], it presented a long sleeved T-shirt with raglan sleeve structure. The sleeves' cut pieces are connected with the shoulder parts, which eliminates the dividing line of shoulder. It enable shoulders be relaxed and be easy to movement.

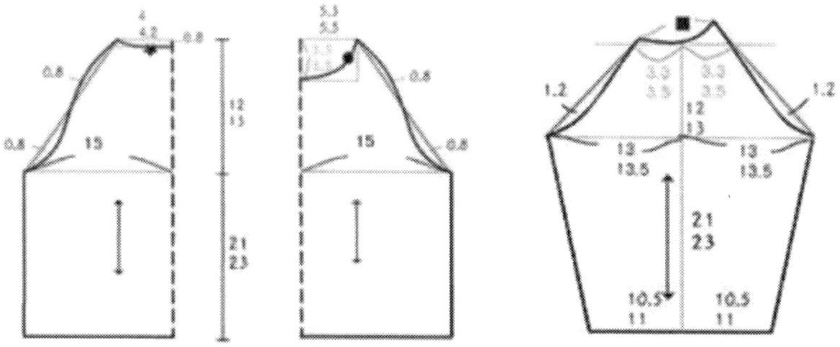

Fig.10 Insert the shoulder sleeve structure [9]

3.3 The Crotch Design's Impact on Security

It is found that crotch is the most complex part on pants design structure because of the complex relationship between pants and human hips. The crotch treatment is appropriate or not will directly affect the appearance of the trousers or their motor function [10]. It was used to designing open crotch to changing a diaper easily. But it cannot be open too large in case that infant will shrink their legs from the open to cause fracture. At the same time when infants have a fall the dirty things will stick to bottom, which causing some bacteria easily into the urethra, led to urinary tract infection especially for small girls [11]. Therefore the structural design of pants have a great influence on the safety of infants' dressing .There are most common three open-seat pants' open file in the market: the one is that opening the archives line edge. The structure design is convenient for adults to take care of babies, but the infant hip will be exposed. Second is that one side with zipper shut, the baby wear is relatively comfortable. But it inconvenient to urine. Third is that opening the medial suture of pants and buckled in the form of a snap fastener. The design led infant's mother to inconvenience and the end of button will rub pudendal of infants. The researchers on the basis of three pants above concluded that the design of the bottom should be added to a piece of cloth in addition. As shown in Fig.11 [12], structure diagram (a) shows that the central piece of gear for an expressed as a dotted line and a pocket file base cloth at the bottom .The original trousers was changed to no end of file structure and it shaped like a skirt. The bottom cloth is combined with the letter button on the upper side seam. It is conducive to avoid bacterial infection.

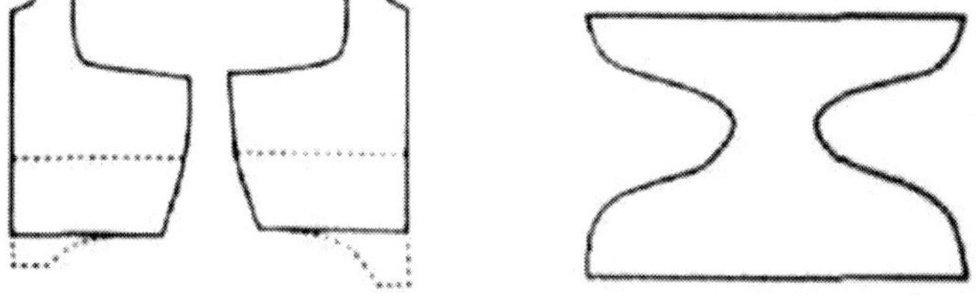

(a) (b)

Fig.11 Dress shorts structure and graphic [12]

3.4 The Waist and Hip Design's Impact on Security

Because the infants is characterized by a short neck, abdominal bulge, short limbs, the development of the waist is not perfect. So the structure should not be designed waist line [13]. The waist design should be loose so as not to squeeze the abdomen and chest of babies, to affect the baby's growth and development. The design of infants' hip is directly related to the trousers, it's necessary to consider the diaper space or facilitate activities. So the hip design mostly is more relaxed, such as hip pants and overalls which have sufficient amount of relaxation. Due to take the safety of baby activities into account the trousers cannot be too loose. As shown in Fig.12 [14], the (a), (b) two pants can be opened and closed easily in front gate, which designed ligature or elastic at the waist be convenient for infants to wear off. The front door uses the overlapping design, which connected with claw button may adjust the waist size. This design can effectively protect the abdomen of infants and ensure that the waist circumference can be adjusted.

Fig.12 Front door that opens fission pants [14]

3.5 Decorative Design on the Impact of Security

Decorative designing includes clothing logos, buttons, metal accessories, patch, bows and other. In order to increasing the consumption and showing clothing's aesthetics, so designers will try to minimize decoration design in infants clothing. It is need to pay attention to the safety design of adornment belt and ornaments. It's suggested that children under the age of 3 the band length should be less than 14 cm, bowknot tail should be shorter than 5 cm long, the exposed length on the finished rope is not more than 14 cm (about the length of the hand) [15, 16]. Clothing identification need detail, especially ingredients to be accurate. Trademarks, washing, size, those which should not be sewn at the back of collar arc, otherwise it will injure infants' delicate. Buttons are advised to use small one and must be nailed. At the back of the embroidery parts need adhesive a layer lining so that infants will not be injured by formaldehyde.

4. Conclusion

1) Infant clothing profile should be based on the H type, O type, A type, these three kinds of profile in line with the physical and sports characteristics of infants with safety protection function.

2) Infant plane body prototype design points are that increasing the amount of the back width and chest..Deeping the armhole, so as to ensure styles loose and easy to put on and take off.

3) The collar design can be used fight the collar, fight the shoulder, oblique open collar design, these three types played a role on infants' neck development.

4) Choosing the connecting sleeves and raglan sleeves are appropriate. They can protect the infant arm movement. At the same time increasing the sleeve length can prevent nails to scratching infants' skin.

5) Infant clothing waist and hip design focus on protecting the abdomen and it should not be designed with waistline. The hip circumference should be loose and the best design for the front door can be retractable styles, which is more convenient for infants to wear off.

6) Infant clothing should try to reduce the decoration design and band. It should not to use buttons, zippers, Velcro and other small accessories or hard objects, to avoid the baby swallowing.

7) Increased an extra bottom cloth in crotch and be connected with garment piece by snap fastener at the side

seam. Such a design will be conducive to the safety of infants and avoid bacterial infection.

References

[1] Shen H, Wang XY. Children's clothing structure design. Chemical Industry Press: 2014.
[2] Men M, Wang F. The safety and development trends of baby clothes [J]. Tianjin Textile Technology: 2012; 02: 34-36.
[3] Zhang WB. Structure design of garment. China Textile Press: 2006
[4] Chen SY. Study on the structure design of infant clothing [J]. Guangxi light industry: 2011; 06: 95-96.
[5] Fred A. 《The children's clothing, baby clothing pattern design (0~14 years old) international fashion series》.
[6] Liu RP. The design principle and application of garment pattern. Chinese Textile Press.
[7] Yang XM. Structural details in the design of infant and young children's clothing [J]. China apparel: 2007; 03: 140-141.
[8] Wang YJ. Analysis of the influence of human characteristics on the clothing design of infants [J].2007 (7); 86 -90.
[9] On http://www.baike.baidu.com/link information
[10] Zhang YH. Trousers crotch structure design [J]. Journal of Henan Institute of Engineering (NATURAL SCIENCE EDITION): 2010; 03: 9-12.
[11] Zhou YQ. Children should not wear pants [J]. China health of urban and rural enterprise: 1989; 04: 43.
[12] Wu ZM, Wang WW. The comfort and safety of baby clothing design [J]. Tianjin Textile Science and technology: 2006; (3).
[13] Lu Jun. Baby clothes to investigate the safety of functional [J]. (Education): 2013; 09: 286-287.
[14] Guan WL. Based on "humanized" baby improved style design [J]. Shanghai Textile Science and technology: 2014; 05: 33-34+49.
[15] FZ/T81014-2008 "infant clothing".
[16] There will be a standard for baby and baby clothes [J]. Textile clothing weekly: 2008; 31: 49.

The Application and Research of New Paper Cutting Art in Women's Clothing Design

Ying Wang [1], Yan-Bo Ji[2*]

[1] School of Apparel & Art Design, Xi'an Polytechnic University, Xi'an, 710048, China
[2] Apparel & Art Design, Xi'an Polytechnic University, Xi'an, 710048, China

*Corresponding author's email: 651577266@qq.com

Abstract

In the light of the historical development of the paper-cut art and the fashion and change of the clothing industry, the paper takes the ladies' clothing as an example, and adopts the methods of literature collection, physical analysis, and graphic cross examination. Therefore, this paper intends to combine the historical development of paper cut art with the fashion and garment industry to provide a practical and theoretical understanding to college students who are about to enter the clothing apparel industry.

Keywords: Paper-cut Art; Style Type; Fabric Shear; Practical Dress

1. Introduction

The emerging of fashion trends in society means that we should look at traditional paper cut art through an international lens so that it can flourish in the market. The purpose of this article is to combine paper-cut art with fashion through theoretical academic research. At the same time we will be testing its feasibility through experiments and providing a theoretical basis and practical scheme for the solid market [6].

In 2007, the book The history of Chinese folk paper-cut history was published on the theoretical position of paper cut art, which is part of a more comprehensive exposition. The author Chen Jing systematically studies from folklore, aesthetics, psychology and sociology and explains the emergence, application, development and prosperity of paper-cut art, as well as the different manifestations in folk custom [3]. Although there have been lots of extensive studies on paper cut art, there are only a few done on its application to clothing [7].

In this paper, the historical origin and development prospects of paper cutting were studied. After the comparative analysis of the new and traditional paper-cut art, we attempted to learn from the culture and methodology of traditional paper cut art to apply it to clothing. The new paper cut elements are better suited for dress design in terms of, order reduction, decoration, collage, fabric shear and other techniques. The target market is aimed at females between the ages of 20-30.

This paper is based on the extensive reading of ancient Chinese clothing, paper cutting, art books, the collection of relevant graphic and studying the history development process of paper-cut while also reading the combination of Chinese auspicious beliefs, Chinese folk art history and clothing magazines. The aim is to keep up with the fashion trend in the current modern society and provide a theoretical basis for applying paper cut-art to female clothing.

Firstly, the literature and pictures are arranged and the basic rules of paper cutting art are summed up through analysis and comparison. The paper cut art pictures used in women's clothing design are analyzed and then the paper-cut art expressions are studied to see how it can be applied to women's clothing.

An analysis was done using the paper-cut patterns of College gate in Shaanxi and Xi'an Dongjiao kangfulu clothing wholesale clothing market, which provided favorable information for the paper-cut application of women's dress.

2. The General Situation and Classification of New Paper Cutting Culture

2.1 The General Situation and Classification of New Paper Cutting Culture

2.1.1 New Definition of the Art of Paper Cutting

The generations of paper-cut art stems from the practical application of folk culture. The modern version of the "Cihai" explains paper cutting in this way: one of the Chinese folk craft decorations [1]. Jin Kaicheng's definition of"paper-cut art" is: paper cutting is using scissors or knife to carve patterns [2].

2.1.2 The General Situation of New Paper-cut Culture

Diversification is a major feature of paper-cut forms which can be viewed from four perspectives: the decoration can be pasted on glass, jewelry, daily necessities, such as bars and other traditional folk custom display; the ornament can be used for packaging of products during marriages, the ritual items used for embroidery and finally clothing and furniture etc. [8].

2.2 New types of paper cutting culture

The paper-cut culture is meaningful and its artistic style is not simple but can be summarized. It does not exist in isolation as it cannot be separated from its form and content. The style of the paper-cut art originates from four characteristics of the surface form created by the artist [5].

Hollow paper cutting art is belongs to the perspective style. Carved intaglio technique is used to produce the hollow effect. Transparent and hollow effect and exposed paper-cut incised, carved the manipulation of hollow effect.

Peoples heritage has formed concise and focused paper cut art features. The works are composed of different materials and have continued the belief that "the plan must be intentional and the meaning must be auspicious", which is another form of culture.

3. The Development of the Art of Paper Cutting

3.1 Spring and Autumn and Warring States Period - Qin and Han Dynasties Hollow Silhouette Art

Paper cutting art is dated back to the spring and autumn period in the Qin and Han Dynasty. (Fig.3.1.1, Fig.3.1.2.) With the arrival of various sheet cutting materials, paper cutting was able to flourish inspired by paintings, pottery, bronze, sculpture and other art that advanced and evolved into the art of hollow silhouette.

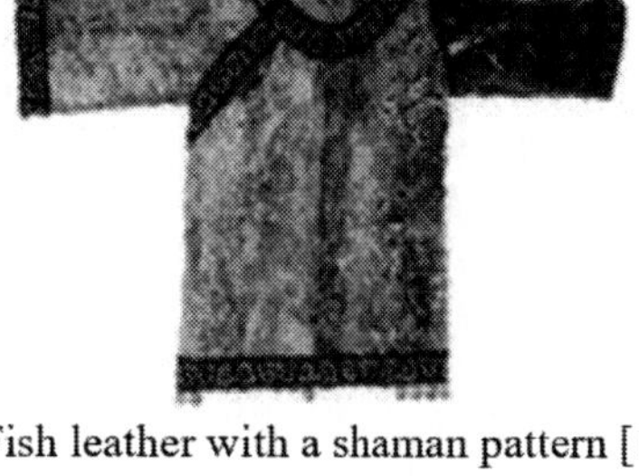

Fig.3.1.1 Fish leather with a shaman pattern [1]

Fig.3.1.2 Women's clothing in Chu state during the Warring States Period [1]

3.2 Paper Cutting Art in Wei Jin Southern and Northern Dynasties

During the historical development of this period, paper cutting was popularly called "paste and victory". Xinjiang Turpan Astana tomb unearthed in the Jin Dynasty had, paper hats, paper shoes, paper belt and the six flower paper cutting [9]. From the unearthed paper-cut objects, historical documents and research paper has documented the printing fabric process. Chinese folk paper-cut paper produced in the Wei and Jin Dynasties gad spread to the Yellow River basin [4].

3.3 The Application of Paper Cutting in Tang and Song Dynasties

During the heyday of Tang Dynasty, people lived in peace and the society and economy developed at an unprecedented height. Paper cutting was also widely applied to various fields. (Fig.3.3.1, Fig.3.3.2, Fig.3.3.3)

Fig.3.3.1 Women in Tang Dynasty [1]

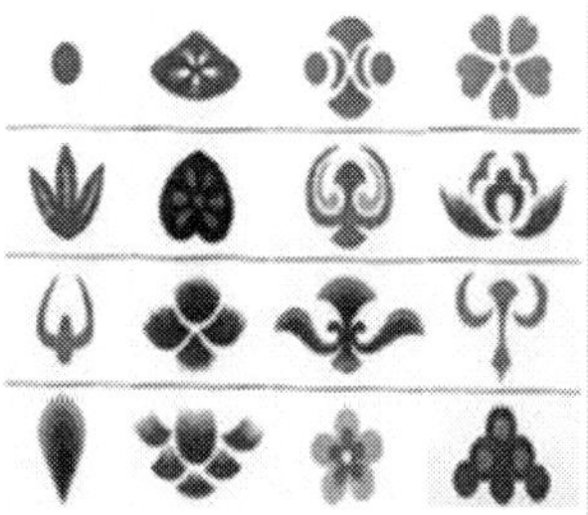

Fig.3.3.2 All kinds of floral [1]

Fig.3.3.3 The Tang Dynasty palace Sculptures [1]

3.4 The Development of Paper Cutting Art in Ming and Qing Dynasties

The classification of paper-cut art has also been unprecedentedly developed in this period. Monochrome paper-cut, multi-color paper-cut, point color paper-cut also came into being. Paper cutting masters diversified industry, which makes paper-cut in this period of development is also a big step forward quickly. (Fig.3.4.1, Fig.3.4.2).

Fig.3.4.1 Miao paper-cut [1]

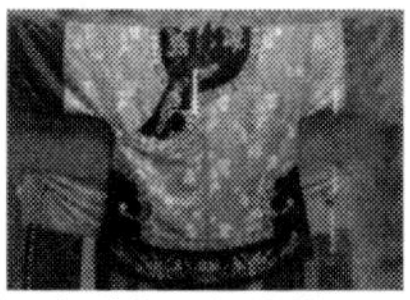

Fig.3.4.2 Women's clothing in Ming and Qing Dynasties [1]

4. Analysis and Extraction of New Paper Cutting Elements

4.1 An Analysis of the Elements of the New Paper Cutting Elements

In traditional Chinese culture, paper cutting is one form of art that displayed local characteristics. It has concise symmetrical features that are carved out.

4.2 Extraction of New Paper Cutting Elements

Dior's 2016 spring / summer haute couture week combines paper cutting with modern technology. It is crafted through hollow, laser cutting techniques and repeated geometric patterns. (Fig.4.2-a)

The Shiatzy Chen 2017 series launched a design based on nature's flora and fauna, and embodies the meaning of paper cutting on clothing. (Fig.4.2-b)

Fig.4.2-a Dior's 2016 summer haute couture week [7]

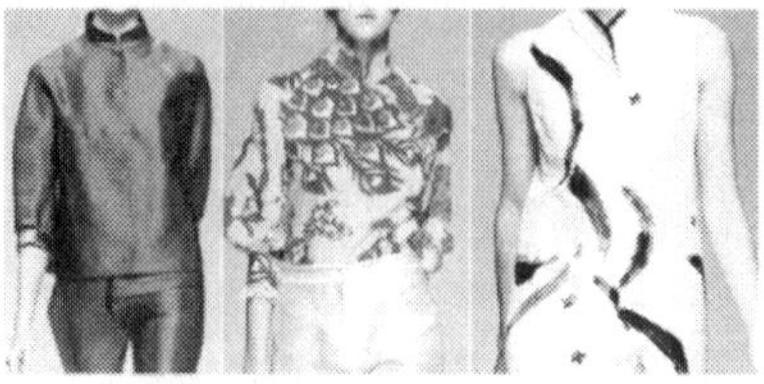

Fig.4.2-b The Shiatzy Chen 2017 series [7]

5. The Innovative Practice of the New Paper-cut Art in Women's Clothing Design

5.1 Comfortable Appearance and Ruler Design of Women's Wear

5.1.1 The Sensory Characteristics of a Woman's Color; the Design of a Psychological Scale

The color perception and psychological characteristics are designed from a psychological scale evaluation governed from five points, the gradient of the color, liveliness, whether it is cool or warm, simple or complex, monochrom correspondence is rich. Following table:

Talbe.1 The sensory characteristics of a woman's color; the design of a psychological scale

Sensory characteristics	Like	Some like	Neither	Some like	Like	Sensory characteristics
Dark	25	11	3	5	6	Light colour
Serious	10	30	2	5	13	Lively
Cool color	40	1	1	1	7	Warm
simple	9	28	4	3	6	Gorgeous
Single	10	25	8	2	5	Abundant

The source of information: from actual investigation

5.1.2 The Sensory Characteristics of Women's Fabrics are Designed based on Psychological Scale Evaluation

The fabric sensory characteristics of the psychological scale evaluation are designed based on how natural it is, heaviness, pendulous corresponding to quite broad, supple corresponding to rough, gorgeous corresponding to elegant, itching corresponding to skin friendly, creamy corresponding to hygroscopic as shown on the following table:

Table.2 The sensory characteristics of women's fabrics are designed of psychological scale evaluation

Sensory characteristics	completely	some	moderate	some	completely	Sensory characteristics
Natural	9	25	6	7	3	Artifical
Light	8	10	25	11	7	Heavy
Pendulous	28	12	8	1	1	Very broad
Supple	2	3	32	9	4	Rough
Gorgeous	2	10	6	7	25	Elegant
Itching	0	1	4	15	30	Skin friendly
Creamy	0	1	2	10	25	Hygroscopic

The source of information: from actual investigation

5.1.3 Data Statistics and Analysis

As we can see from above, most people prefer light colored clothes with a slightly livelier appeal. 80% of them prefer warm colors and simple design.

Furthermore, people prefer natural textured fabrics with moderate weight. 68% of the participants prefer soft, lively, skin friendly and hygroscopic fabrics and 57% of the participants prefer very broad and medium fabrics.

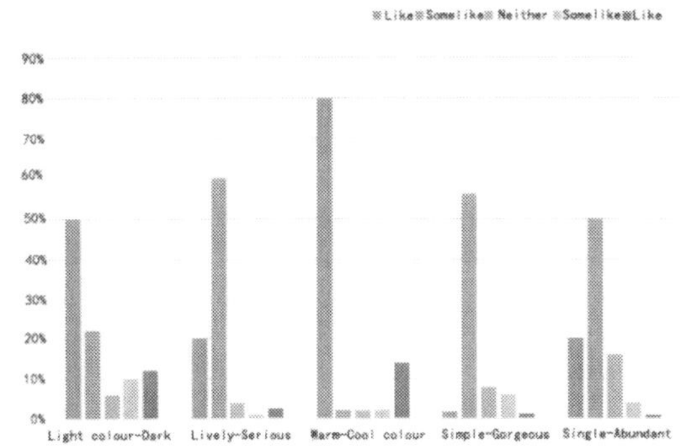

Fig.5.1.3a The sensory characteristics of women's dresses color

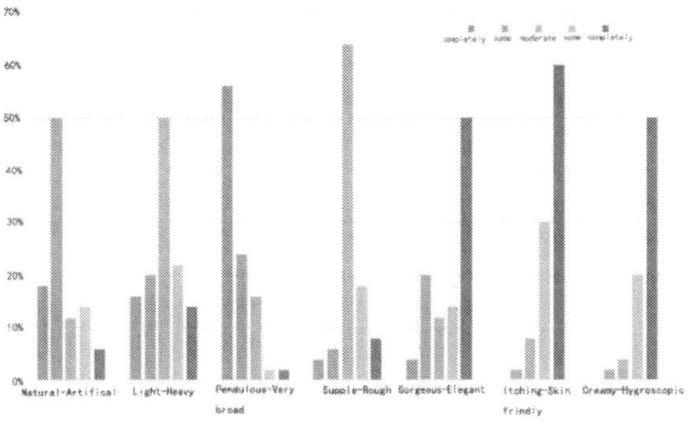

Fig.5.1.3b The sensory characteristics of women's fabrics

The source of information: from actual investigation.

5.2 Inspiration Source

The inspiration of this design practice comes from the study of paper-cut culture. By looking at a large number of pictures and materials, we find that there are lots of representation of flowers and animals, abstract reproduction, and the reappearance of life.

5.3 Design Subject and Description

Tibet flower, gives off a water moon hazy beauty. There is an unexpected charm added to the clothing design series. What people see is perhaps her hollowed out color, the glamour of color, or the ingenuity of the design. Paper cutting clothing has attracted attention for its unique beauty.

5.4 Production Process and Display

5.4.1 Order style Design and Practice

There is a new application for paper cut art when applied on a dress: It can be applied to leather garments or cloth. The hollowed out pattern of plants and animals are chosen by the wearer before stitching the suture in the required parts.

1) Experiment Name: Using the same process but different color contrasts

The purpose of this experiment is to study the importance of the color of the dress pattern and how it can be coordinated with the whole garment.

Two different color order implementation pattern is used when selecting the technique used.

Conclusion: Using the same color two kinds of background was used. The use of contrasting colors, decorated with yellow is nifty and lovely providing a harmony of colors.

Color has a huge impact on the clothing pattern image. The relationship between the dress pattern and the color of the clothing is roughly divided into three parts: unity, harmony and contrast. Similar colors refer to different colors of the same color system, which are combined with differences in brightness and chromaticity. These are more used for clothing, collar, foot and other parts [10]. (As shown in Figure 5.4.1 (I))

2) Experiment Name: The same part of the same fabric

The purpose of the experiment is to compare the clothing with different color and material.

Experimental procedure: First the decorated linen fabric dress will be wrapped around the waist, in order for the cotton fabric material to be compared.

Conclusion: Red stitching and decorated linen fabric material are paired to form a strong contrast. The flax crisp fabric, which is easy to fold with the cotton that is elastic and soft, brings out the beauty of women. The combination of different materials and color fabrics will greatly enhance the visual impact of the design, and form a rich appearance. (As shown in figure 5.4.1 (II))

Fig.5.4.1 (I)Physical comparison

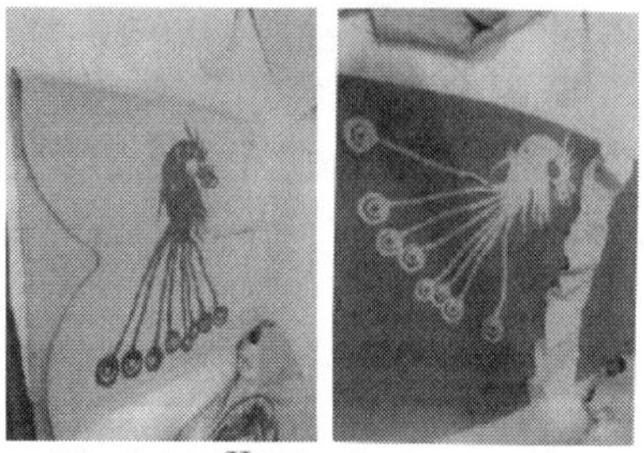

Fig.5.4.1(II)Physical comparison

The source of information: from experiment

5.4.2 The Design and Practice of the Reduction Method

With the help of modern high-tech means, there are many techniques used for paper cutting and costume design. The fabric texture is damaged to produce a broken yarn or hollow layered effect. Textured fabrics includes elements such as holes, burnt, flaw, embroidery and so on. This contrived, deliberate, casual, accidental, decorative texture effect gives an unlimited imagination (As Fig.5.4.2-a, b).

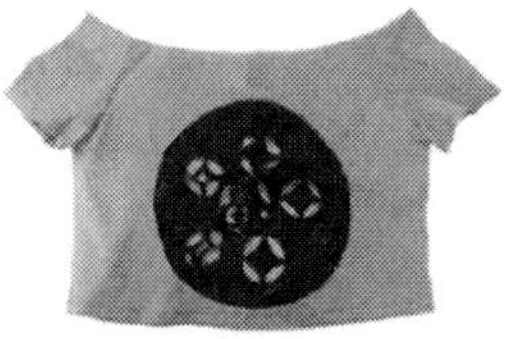

Fig.5.4.2-a Material object

Fig.5.4.2-b Material object

The source of information: from experiment

5.4.3 Practice of Shear Technique on the cutting art fabric design

The cloth sticking technology is a new process based on ancient exquisite craftsmanship, which is mainly cloth sticking. Shear refers to the weave motion in the fur and leather machine. Animal fabric is hollowed out using the above paper-cut method done by manual cutting and cutting machine two. Collage is a collection of techniques with different shapes and different color patterns which can be used as interior, clothing or graphic design. This is a new process based on ancient craftsmanship and the evolution of the main method. The material is made out of cloth. (As shown in Fig.5.4.3-a, b.)

Fig.5.4.3-a Physical display of ready-made clothes

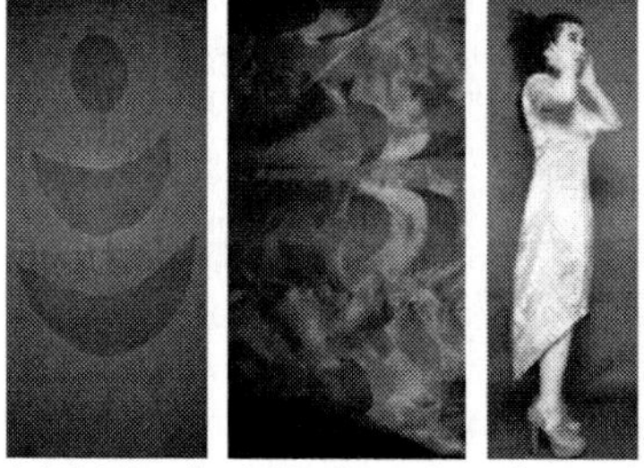

Fig.5.4.3-b Physical display of ready-made clothes

The source of information: from experiment

Manual cutting tool set includes blades, scissors, pencil and ruler. However, the standard and accuracy of the pattern is relatively highthus the end product is well done and delicate.

6. Conclusion

The paper is entitled the design and application of paper-cut art which is based on surveys, paper-cut cultural classification, historical development, analysis and research on the design of paper-cut art and its application to dresses. The conclusion is:

1) From the classification of paper-cut culture the new and traditional paper cut art were compared according to four characteristics: transparency, conciseness, decorative exaggeration and totem meaning. The traditional and new paper cutting art are compared through the characteristics of style and region.

2) We analyzed the extraction of paper cutting elements so that it can be applied to clothing, through the use of printing, embroidery, laser cutting and other paper cutting derivative technology. The application of new paper cutting art on clothes, not only take into account the cultural connotations, but also consider the rationalization of production costs.

Although market oriented, the design and application of the new paper-cut art is suitable for dresses. However the theoretical study of bringing paper cut art to the rest of the clothing industry and the formation of a brand is still far off.

References

[1] Wang WJ, Li YX. The aesthetic principle of Chinese paper cutting art and its application in fashion design [J]. Silk: 2016; 5: 50-54.

[2] Zhang P. The application of traditional paper cutting art in Shaanxi [J]. Northern Literature: 2016; 10: 109.

[3] Li LH. The inheritance and development of paper cutting art in the changing environment of folk belief [J]. Journal of Xichang College (Natural Science Edition): 2016; 3: 67-70.

[4] Wen W. Under the new economy era China magnetic paper-cut art of harmony [J]. Contemporary educational practice and teaching research: 2015; 5: 26.

[5] Wang Y. The protection and development of folk paper-cut art in Shandong [J]. Folklore. Studies: 2015; 3: 135-138.

[6] Xu JC. The deep thinking of the inheritance and value of paper cutting art [J]. Heilongjiang National Series: 2015; 6: 112-116.

[7] Lu N. The application of paper cutting art in fashion design [J].Textile Leader: 2015; 1: 72-74.

[8] Bao SD, Ruan XY. Folk paper cutting art modeling concept and expression method and Its Inspiration to modern design [J]. Shandong Social Science: 2015; S2: 533-534.

[9] Xu YH, Ren XJ, Li YC, Zhao YT. A new design patterns of flower type application in the design of sweater [J]. Textile Leader: 2015; 8: 99-102.

[10]Bao XL, Liu YR, Lu YF. Analysis and comparison of the traditional Chinese blue calico patterns and folk paper-cut patterns [J]. Silk: 2015; 1: 66-70.

Design And Research On The Variability Of School Age Girls' Clothing

Ning Guo[1], Hong Dai[2*]

[1] *School of Apparel & Art Design, Xi'an Polytechnic University, Xi'an, 710048, China*
[2] *Apparel & Art Design, Xi'an Polytechnic University, Xi'an, 710048, China*

[*]Corresponding author's email: 350307252@qq.com

Abstract

In reference to the design of changeable wear, this paper puts forward multiple innovative methods including application of component thinking and interaction design. Through analyzing similarities of the structural cutting and clothing structure, reusability of the component is introduced to the design of changeable clothing. We developed the exchanging, reassembling and synthetic design methods to give the clothing more styles and functions. At the same time, changeable clothing = should be combined with interaction design so that consumers can wind twine and buckle through space reservation and elastic fabric. This way, interaction effect between clothing and consumers can be generated with infinite possibilities. `According to the research, component thinking and interaction design are very valuable in the design of changeable clothing [11], make yourself your own designer.

Keywords: School-age Children's Clothes; Clothing of Changeable Wearing; Component Thinking; Interactive Design; Design Method

1. Introduction

1.1 The Significance of Topic Selection

In T=the constant progress of the society in the 21st century, people have gotten a lot richer and increased their demand for material possessions. Clothing which is an indispensable part of our life is constantly growing and developing. Therefore products that satisfy both the people's survival and spiritual needs are highly sought out.

1.2 Related Concept

In recent years, the contradiction between advocating low carbon and pursuing novelty is particularly prominent in the field of clothing. Variability of clothing is a new design feature that can be introduced. It gives clothes =a range of styles and looks, which can meet a variety of psychological needs and reduce the rate of elimination [21]. Instead of wearing more clothes, we attempt to create a garment that can be worn in many different ways. [7], This can enhance the awareness of environmental protection while slowing down the diversification of clothing design, and slow down the ecological problems of the postindustrial society [8].

1.3 Current Problems and Status Quo of School-age Children's Clothing in Our Country

Foreign countries have made remarkable achievement in the variability of clothing [3], where as China is a lot behind.

Donna Karan, Lanvin, and others international brands design clothes that can be worn in many different ways. Donna Karen launched the DKNY Cozy in 2003. From first impression it's just an ordinary clothe, however customers can freely transform the costume style in different way. Its official website points out twelve dressing methods, and many fans of Cozy are constantly finding innovative ways of dressing [5].

The famous Chinese clothing brand "JNBY" has been involved in the variability of clothing and has increased its sales in their New York Store. In the internet, this design is also a hot topic. Many tide web designers, young fashion bloggers have posted related video courses one after another, to teach methods and share their experiences. They are unconventional and unorthodox. The development of clothing variability is one of the direction of the clothing design industry..

1.4 The research Content and Meaning

By analyzing the structural characteristics and component [14], it is an attempt to bring clothing variability into the design of girl's clothing. By expanding the interchange =, dismantling and the synthetic me~ thod we gave one garment to many models so that we are able to discover its variability and other functions. At the same time, changeable clothing should be combined with interaction design. The preserved space and elastic fabric is able to make consumers entangle and buckle, producing the interaction effect, exerting possibilities of multiple-wear. (preserved space: providing consumers with actionable space to allow consumers to participate in a variety of attempts.)Studies suggest that component thinking and changeable design in school-age girls has an important value. ,.

1.5 The Research Methods

In the thesis, the methods we studied mainly include:

1) Literature research method: Literature research method is done through bookstores, libraries, network and other ways.

2) Design practice method: combining an analysis of girls' physical and mental characteristics, style of dress, color and fabrics to determine innovative clothing styles. [10].

3) Questionnaire method: Questionnaire was distributed among three types of people: We gathered the opinion of teachers, parents and students on changeable design of girl's clothing, style, fabric, color and function design. A total of 2, 000 questionnaires were issued, and we received 1800 pieces. tTe recovery rate was 90 percent. To ensure the accuracy and truth of the survey data, the respondents completed the questionnaire by themselves.

2. The Development History and Present Situation of Girl's Wear in School-age

The children's clothing industry is a part of the 20th century. Before that [12], children were dressed like miniature adults. At the end of the 19th century, children's clothing finally differentiated from adult's. They began to wear uniforms. The most successful example in the 19th century was the company named WILLIAM CARTER, which was founded in 1865 and became the largest children's underwear company that exists until today.. It was not until after the First World War that children's clothing started to be produced and sold. The television got introduced to American families and soon advertisers found that children are the largest audience. In the 1950, they became the primary target market.TV programs that accommodate different age of audience can help = understand the popular clothing styles of each age group [6]. Interestingly, the clothing manufacturers and retailers in Los Angeles have been in the lead when it comes to youth clothing, but they rarely developed girls' clothing. [2].

There are more babies' clothes in the children's market. The children's clothing counter is mainly occupied by infants, young children and preschoolers. Children's clothing market accounts for 44% of the children's population. There is not enough focus on such groups to create a market gap [1]. There are also problems such as fussy decoration, inappropriate size ratio, adult style, lack of practicality and unqualified fabric quality and so on [4].

3. Investigation and Analysis of the Current Situation of School Girls' Children's Wear in China

We studied the opinion's and wishes of the student, parents and teachers through analyzing the questionnaire survey and interviews which focused on aspects regarding style, color and fabric. A total of 200 international primary, secondary and general schools were selected which had 1500 students, 300 parents and 200 teachers. This paper analyzes the existing problems of children's clothing who are attending school, and provides reference for clothing design and production. A total of 2000 questionnaires were issued and 1800 valid questionnaires were recovered. The recovery rate of the questionnaire was 90%. There were 400 international high school questionnaires distributed, occupied 20% of the total individualnumber.There were 500 international primary school questionnaires distributed, sample 27%. There were an average of 500 school questionnaires and the sample was 27%. There were 290 parents and 16% of the sample and 110 teachers,ccupied 10% of the total individualnumber.

3.1 Pattern Color Choice of School-age Girls' Children's Clothes

In the survey 55% of the parents believe that the overall clothes for school aged children are not aesthetically pleasing. The teacher's dissatisfaction accounted for 30% while 60% of the students were not satisfied with their uniforms. Fig.1:

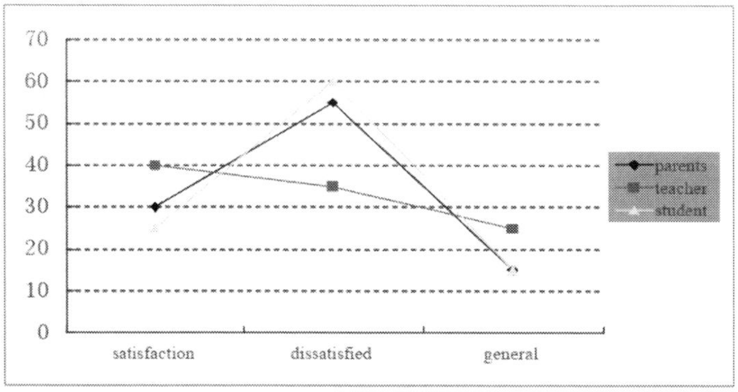

Fig.1 Survey data

3.2 Style Design of School-age Girls' Children's Clothes

This is a set of questions for students and teachers. As shown in the figure below, 55% of the students prefer formal attire. 30% of the students likes sportswear, and 15% of them like casual clothes. So by combining the three conditions, we will design a variety of changeable garments to meet their demand. Fig.2:

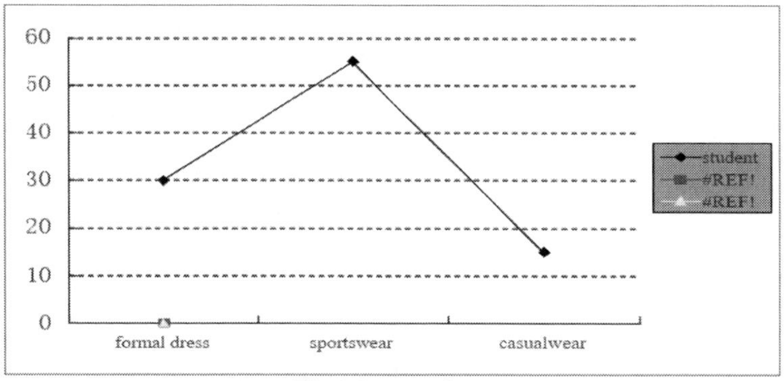

Fig.2 Survey data

3.3 Fabric Selection of School-age Children's Clothes

In the analysis, 85% of the parents think that the choice of medium fabric such as cotton blended fabric, which is cheaper is more suitable. A small number of children prefer to buy high-grade pure cotton fabric clothing. Of the 10% of parents only 5% of them believe that buying high-grade pure cotton fabric clothing is ok. In the field survey, many school-age children wear synthetic items rather than environmental protection material,not to mention comfortable. As shown in Fig.3:

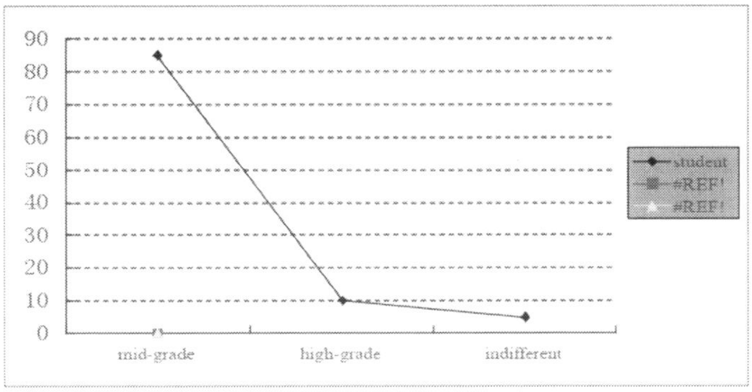

Fig.3 Survey data

The analysis shows that the present situation is not ideal as school age children, teachers and parents have their own opinions on colors, styles and fabrics. It is obvious that few of the original garments have met their demands. Therefore, it is imperative to design and develop multi-functional variants of clothing.

4. Variability Design is Applied in Clothing Design

The consumer crowd for existing variability of clothing designs are mainly female groups. This is because women are more into fashion. The types of women's clothing are always changing, so variability design in women's clothing is promising [15]. As shown in Fig.4 and 5:

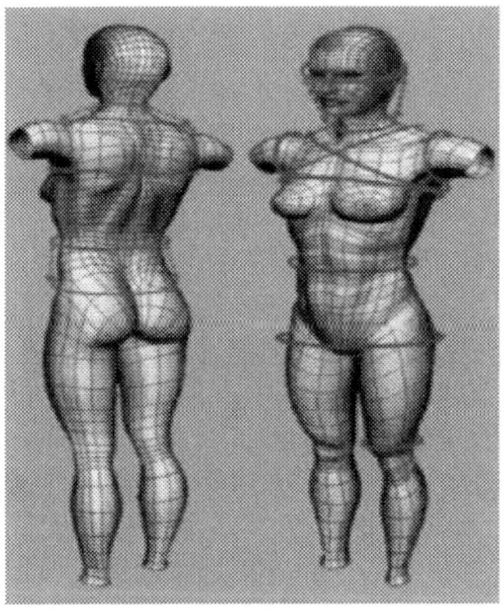

Fig.4 Human 3D

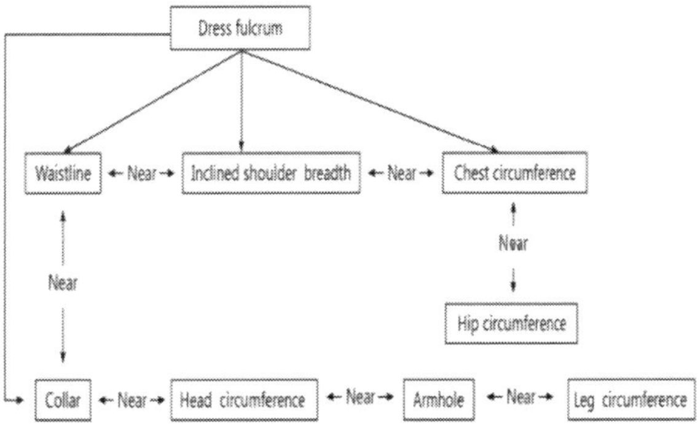

Fig.5 Relationship between of the elements and size

4.1 The Variability Classification in Garment Design

Transgender clothing puts emphasis on variability, including the color, fabric, shape, structure, style and function. People in modern society do not simply ask for clothing that only covers basic functions such as warmth and cover. People are now looking forward for clothing that expresses emotions. as After analyzing clothing characteristic and form, the variability of clothing design includes: twisting method, fastening method, dismounting method, combination method and conversion method, etc [9].

4.2 Variability is Designed for Innovative Applications in Girls' Clothing

4.2.1 The Innovative Design of the Girl's Dress

In this paper the combination of variable design and girls' clothing can be used for research and applied to real life. The variability of design can contribute to aesthetic and practicality [17]. According to the research three sets of costumes were designed [19]. One is a garment with changeable length, the other is a garment which can be disassembled and assembled arbitrarily, and the third is a loose garment which can adjust the tightness clothes [13].

4.2.2 School-age Girl's Wear Patterns

The profile of the A type style dress, upper body and skirt splicing and splice length adjustable [20]. As shown in the following Fig.6:

Fig.6 Style chart 1

The shape of the pattern belongs to type A. the collar is flat, the sleeves have little bubble sleeves and has an overall loose style. The collar and sleeve can be disassembled and reassembled [16]. This is shown in Fig.7and 8.

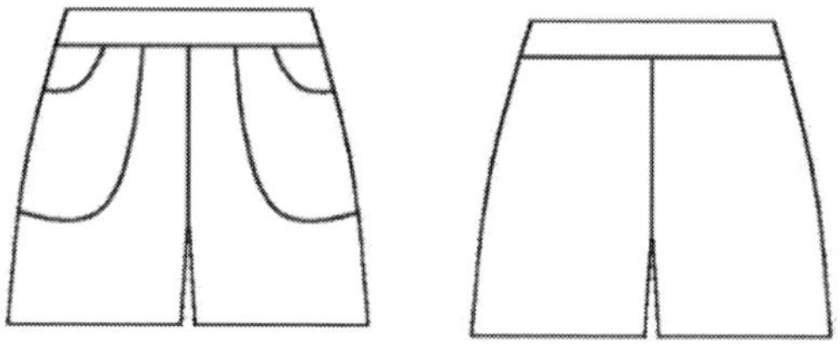

Fig.7 Style chart 2

Fig.8 Style chart 3

The shape of the coat in Fig.3 is H type. The skirt of the condole belt belongs to type A. The skirt can be adjusted to a loose degree. This is shown in Fig.9 and 10:

Fig.9 Style chart 4

Fig.10 Style chart 5

4.2.3 Series school-age Girl's Rendering

Color is an integral part of the three elements of clothing. This series of children's clothing is based on a bright blue color supplemented by yellow. Blue has a cool hue while yellow provides warmth. As shown in the following Fig.11:

Fig.11 Application of variability in girl's wear

5. Conclusion

In this paper, through the combination of a large number of theoretical analysis and physical experiments, we studied the practical application of the variability design in girls' clothing, and drew the following conclusions:

In the 21st century, everyone wants to become a designer and to dress uniquely. [6], Clothing variability can meet this very demand. Therefore is an integral part of the future development of garment design and industry. With the help of component reuse, the design method of changeable clothing has greatly improved its structure. [22]. The designer needs to be familiar with component thinking and interaction design which requires constant practice, communication and guidance.

Considering the psychological and physiological characteristics of children as well as analysis of consumer trends, it is important to optimize the structure and function of clothing styles. There are three kinds of function and aesthetics with the variability of girls' clothing which are function, protection, safety and warning. [18]. Finally, the design of girls' clothing is not only convenient to wear, beautiful and fashionable, but also has certain functionality and protection.

References

[1] Jill L. Elfenbein, Ruth A. Bentler, Juliam. Davis, Diane P. Status of School Children's Hearing Aids Relative to Monitoring Practices [J]. Niebuhr, Ear and Hearing: 1988 (4).

[2] Beth J. Macpherson, Jill L. Elfenbein, Robert L. Schum, Ruth A. Bentler. Thresholds of Discomfort in Young Children [J]. Ear and Hearing: 1991 (3).

[3] Roisen FJ, Klueber KM, Lu CL, Hatcher LM, Dozier A, Shields CB, Maguire S. Adult human olfactory stem cells. [J]. Brain Research: 2001 (1).

[4] Hediger ML, Overpeck MD, Kuczmarski RJ, Ruan WJ. Association between infant breastfeeding and overweight in young Children. [J]. JAMA: The Journal of the American Medical Association: 2001 (19).

[5] Genevieve Marie Johnson. The Internet Vocabulary Test for Children: Preliminary development [J]. Internet Research: 2007 (3).

[6] Ruthl F. Deformation fashion [J]. The Bund: 2011 (424); 8-10.

[7] Chen DY. Application of darts design in clothing of variety wearing [J]. Market Modernization: 2012 (9); 88-89.

[8] XU QJ. Creative espression of the silk apparel design under the influence of post-industrial society [J]. Journal of Silk: 2011; 48 (5): 35-39.

[9] Liu CX, Lu P. Basic methods of deconstruction application in fashion design [J]. Journal of Textile Research:

2011 (11); 96-99.
[10] Wu Q. The field boundary of interaction design [J]. ZHUANGSHI: 2010 (1); 34-37.
[11] Tang XY, Yan HM, Yan GD. Devoloping trend of industrial design [J]. Packaging Engineering: 2006 (8); 235-237.
[12] Feng ZM, Qi ZJ. The History of Develop of Clothing Course [M]. Beijing: China Textile & Apparel Press: 2003; 137-140.
[13] Xu GQ, Zhou B. Multifunctional detachable jacket [P]. China patent: CN204245219U: 2014-10-08.
[14] Yu F. Component and interactive design of multi clothing [J]. silk: 2014 (11); 51-56.
[15] Li D, Xia F. A study on the application of "More than a dress to wear" in clothing design [J]. silk: 2014 (1); 50-54.
[16] Liu YY. A removable hooded sun protection suit [P]. China Patent: CN104473347A: 2014-12-12.
[17] Li D. A study of "NWAY" garment design method based on the changing demands of consumers [D]. Zhejiang: Zhejiang Sci-Tech University: 2014.
[18] Yu F, Ma AD. Multi garment design based on knitting characteristics [J]. Shandong textile economics: 2015 (10).
[19] He YJ. The design of multifunctional knitting children's clothing based on the concept of "More than a dress to wear" [J]. knitting industry: 2014 (2); 56-58.
[20] Zhong S. Research on costume decoration for 4-6 year old children [D]. Beijing: Beijing Institute Of Fashion Technology: 2010.
[21] Ruan ND. Study on Design of children's clothes under ecological vision [D]. Zhejiang: The China Academy of Art: 2013.
[22] Wu RH. A brief discussion on modern children's clothing [J]. preschool education research: 2005 (9); 12-14.

Improvement and Designing Daily Animation Based Costumes

Xin-Yi Lyu[1], Jun Ren[2*]

[1] *School of Apparel & Art Design, Xi'an Polytechnic University, Xi'an, 710048, China*

*Corresponding author's email: 66980826@qq.com

Abstract

In the 21st century, the animation industry is developing rapidly with the boom of information technology. The number of followers in anime and animation apparels also has a significant growth. Due to the impractical feature of animation costumes, it is not feasible to wear them in the normal life. This study is going to have a deep research, fine design and wide application of wearable costumes based on animation, and seek a better combination of animation costumes with people's normal clothing and achieve the idea of wearing them in the daily life finally.

The first part of this paper analyses the basic theoretical knowledge of animation based clothing; bringing out animation elements and major outstanding images from animation based costumes, and provides a referential method of real life animation based costumes. As well as using the designed preference questionnaires to select acceptable animation based style clothing, and apply these costumes in people's daily life.

Keywords: Animation Clothing; Daily Clothing; Improve Design; Style; Color

1. Introduction

1.1 Research Background and Current Conditions

With the development of digital communication and information technology since the 21st century, animation has become a critical part of the cultural industry. An animation industry chain is formed by movies, publications, videos, the Internet, computer games, and other derivative products, thus showing that the industry is popular and prosperous. Animation products significantly affect people's daily lives and aesthetic standards due to their dreamlike themes and entertainment, which are rapidly becoming popular cultural products among most people, especially teenagers. [1]

At present, some companies manufacturing animation costumes in the world have gradually been on the right track and marketing operations. In Japan, there are official animation costume stores. In mainland China, only Cosplay costumes are available, which are usually used for stage performance instead of daily lives. Furthermore, most of these Cosplay costumes are bought from online stores. Currently, there are not any companies that specially sell animation costumes, however, the demand for these costumes are increasing steadily. [2]

1.2 Goals and Significance of Research

1) Being clear about the general conception of animation costumes.

2) Understanding the types and styles of animation access. Summarizing affecting factors and key points for each about how to make animation costumes more daily.

3) Having a clearly understand of costume stylings, colors, accessories, and design the daily animation costumes.

4) Combining animation factors together with clothes, making costumes available and wearable in daily lives.

1.3 Methodology

1) Literature review method: Access to information related to the subject, including a typical representative of the animation works, magazines, papers, as well as other relevant information of animation and its clothing.

2) Field research: Field research is in Xi'an. Attending the Xi'an animation festival, school animation exhibition and other activities about the modern animation which can bring fun and ideas to people, which can also provide first-hand information for this research. With this research, we can investigate these groups and tries to understand the popular type of animation and animation clothing styles.

3) Comprehensive analysis: Arranging many outstanding representatives of the animation works, which will be summarized in the animation elements, combined with fashion design, materials science, color science and other related knowledge such as modern design analysis. Finally, designing a theory of the animation system to improve the design from a subjective point of view.

2. Conception of Animation Costumes, Marketing Trend and Analysis

2.1 Conception of Animation Costumes

Animation costumes refer to costumes in animations, cartoons and games, playing a decisive role in shaping characters in animation works. [3] Except from attractive stories, costumes of characters could not be neglected when it comes to vivid Japanese animation, promoting the representation of characters, and giving the audience the feeling of their beauty.

2.2 Markets of Costumes and Analysis of Developing Trends

With the support of national policies since the 21st century, the Chinese animation industry has developed rapidly. The rise of animation culture is gradually changing lives, especially for the generations after 80s and 90s, who are main consumers in the field of animation. As important parts of animation culture, animation costumes have considerable influence on the style of dressing, life style, and the way they think and behave, which is the inevitable trend of this new industry. [4]

3. Costume Style Classification and Analysis of Elements

3.1 Animation Clothing Style Classification

Style is the results of features from things, which could be described as representation of thoughts and artistic of individuals or groups. [5] With the prevalence and emergence of animation culture, quite a few diverse products have emerged from daily lives, the main clothing styles are: Firstly, the traditional style; Secondly, the European style; Thirdly, the daily style; and Fourthly, the futuristic style.

3.2 Analysis of Cartoon Elements

Defined by length, width and thickness, costume style is an external form of representing and has a sense of space [6]. In general, the shape of clothing, which include the overall shape of clothing and local modeling. The overall shape of clothing refers to its outer contours; the local modeling shape, such as collar type, sleeve type, pocket shape [7]. Before we consider the clothing of anime characters or real life characters, clothing modeling will be a particularly important step. The costumes of Anime work to reflect the role of race, status, character, identity, and other characteristics, through the different animation theme and its style, clothing can be found in the shape of a variety of clothing in the animation.

Coloring is one of three elements in clothing design, primarily affecting people's visual sense and aesthetic standards. [8] When we watch an anime, the first thing we would see is the color layout, followed by the focus on the role of the characters, the story and other contents. The rich and varied color scheme in anime is not only for the viewer, but also to show the virtual role of the image and the plot to bring spiritual collision. Clothing in as an integral part of the animation, its color and animation works has a very close contact.

Animation clothes is an important part in this field, the colors of clothes are closely linked to the theme of animation product. And accessories have been used frequently to perfectly match with clothes. Many styles are still used currently and developing. [9] The reason is obvious, one is its practicality, the use of simple; Second, it is easy to model, and as a whole harmonious unity with clothing; Third, it has a visual beauty, a strong decorative effect. In the animation, the accessories have a wide variety, and each piece of accessories have their own unique significance. This section is mainly concerned with the animation and daily life accessories.

3.3 Summary of this Chapter

This section has a more detailed study of the animation works in the extraction of animation elements, animation clothing modeling, color, and accessories. In the study of the animation costumes shapes, its analysis different themes of the costumes, as well as different styles of animation costumes. Color as a part of the animation from different countries in the animation costumes and different styles of animation clothing color were analyzed. Accessories part of the analysis is focused on of the animation accessories and daily life-related accessories. Thus, this chapter analysis is for the back to improve the design and provide a more detailed point.

4. Design of Making Animation Costumes more Daily

4.1 Points of Improvement and Design

Daily animation clothes are popular products with most people. When we improve the design, some trivial details could be avoided. And special details and silhouette of a costume which could highlight the characteristics are encouraged to be kept. [10] Therefore, daily costumes include simplified animation element, that is to say, some elements would be filtered after simplification [11]. If these remaining elements are not represented in series, the final products may lack the feeling of animation. Only daily animation costumes which could convey features and reasonable design could make costumes attractive for consumers.

4.2 Research and Surveys

As for any design process, research and exploration are two important essential parts, which is the original train of thought and creative idea of the design. Research is an experimental process that supports and helps to find the particular design topic. In the creative process, exploration is an indispensable tool which will provide inspiration, information, and creative direction for the design as well as an indication a systematic design [12].

4.2.1 Detailed Rules of Research

Research is a very personalized and creative activity. According to its form of expression, it can provide a sense of the designer's ideas, pursue, interest and observation of the creative imagination. At the same time, according to the depth and breadth of exploration, there can start the relevant systematic design. [13] The animation clothing's' style, modelling, color and accessories, which all closely related to the daily animation costumes, will have its own position in the process of design. Through the exploration, they will all be found and applied creatively.

The web questionnaires and field questionnaires were distributed based on the proportion of 1:3 and distributed randomly in a particular location. The field questionnaires can timely recovery, and the time can be controlled. However, the number of recycling questionnaire is the most difficult part, such as some uncertain factors outside during the research have the influence [14] there, the questionnaire will exist a certain amount of loss. Because the time of network questionnaire is not controllable, the recovery time is relatively long. But ultimately all questionnaires can be recycled back, so the recovery rate will be relatively high compared with the field survey questionnaire.

1) Respondents: animation lovers, cosers

2) Method: Surveys (50 online, 50 paper-printed)

3) Location: Qujiang International Conference and Exhibition Center, Xi'an Jiaotong University, Xi'an Dahua 1935 Art District, Xi'an Greenland Pico International Convention and Exhibition Center (refer with: Table1, Table2)

Table.1 Ways of Conducting Surveys

Researching Ways	Printed Surveys	Printed Surveys	Total
Amount	150	50	200
Percentage	75%	25%	100%

Table.2 Response Rates of Surveys

	Amount(Out)	Amount(Back)	Response Rate
Qujiang International Conference and Exhibition Center	40	36	90%
Xi%terJiaotong University	30	29	96.7%
Dahua 1935 Art District	40	38	95%
Xi%35 Art Districttrictce and Exhibition Center Center Convention a	40	36	90%
Online	50	50	100%
Total	200	189	94.5%

4.2.2 Results Analysis

1) Basic situations

As Fig.1 shows in the following, the ratio of male respondents and female respondents is 1. Over half of them are students in the university because the survey was conducted in the animation show or animation festival in the campus. Therefore, college students make up a considerable part of respondents, indicating that students are main consumers of animation costumes. (refer with: Fig.1)

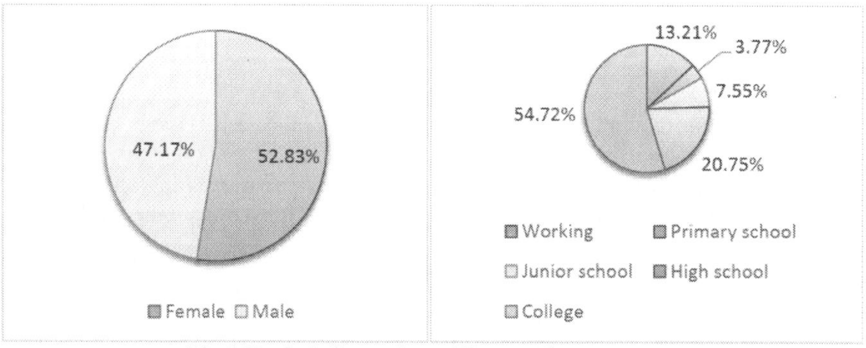

Fig.1 Basic Information of Respondents

2) Preferences for costume styles

From Fig.2, it could be concluded that most people prefer Lolita and Shota style, followed by royal luxury style. The amounts of people who are fond of traditional and daily ones are nearly the same. Only a small part of its audience is interested in futuristic styles. This result shows that the design is inclined to be adorable to meet more consumer groups.

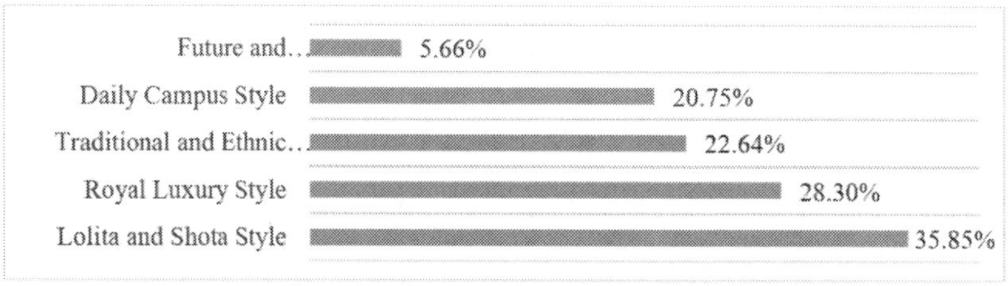

Fig.2 Preferences for various styles

3) Demands for costume styles

As Fig.3 shows that around half of people prefer X-Shape costumes, indicating that they are willing to represent the line of waist by this kind of design. In the next, A-Shape and H-Shape costumes are popular as well, which illustrates that people like animation costumes whose outline shape is consistent with daily clothes.

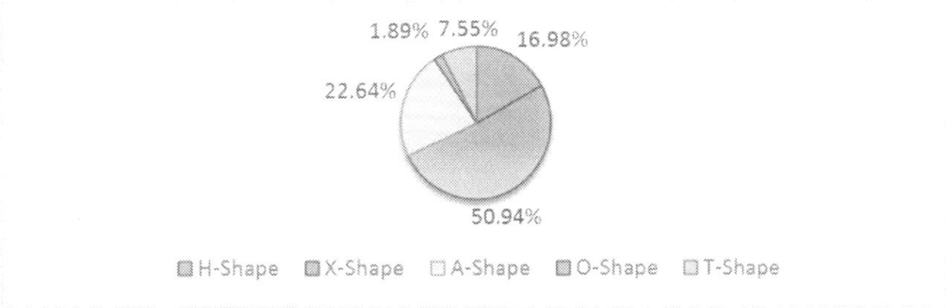

Fig.3 Preference for outline shapes of costume

Fig.4 explains that 40% of respondents like puff sleeves, followed by straight sleeves and flare sleeves. A minor group of people would be interested in tapering and bat-wing designs. So it is not hard to conclude that animation lovers like creatively-designed sleeves, and these styles should be taken into consideration when designing the clothes.

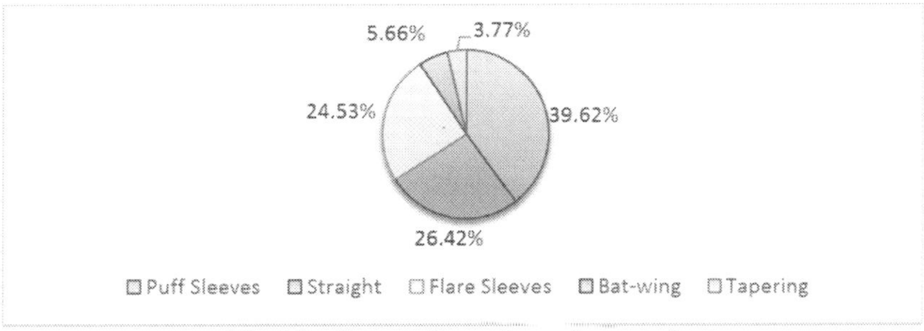

Fig.4 Preference for various sleeve styles

Fig.5 shows that majority of people are likely to pair clothes with matching socks or skirts, dresses and pants. Few of them would choose jumpsuits. And Fig.6 indicates that over 60% of people hope that the dress would be better above the knees and of a modest length. Fig.7 explains that pants which are used to match with clothes should be long pants. And from Figure8, we can draw a conclusion that length of dress could stop around the knees, but miniskirts are not favorable.

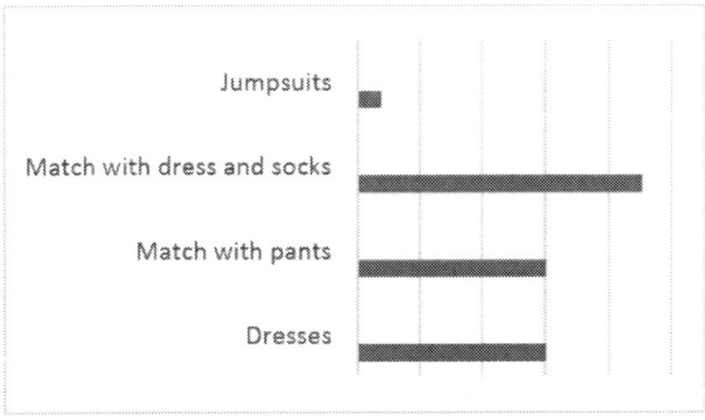

Fig.5 Preference for Matching Styles

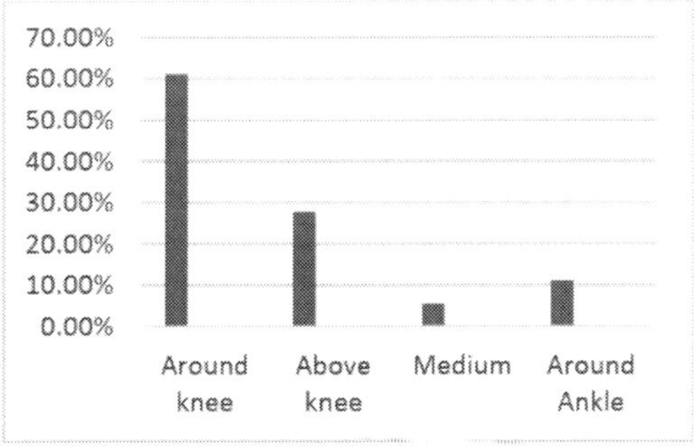

Fig.6 Preference for Length of Dresses

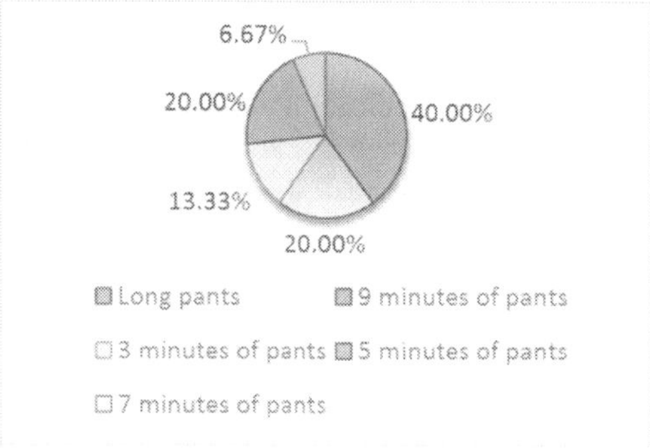

Fig.7 Preference for Length of Pant

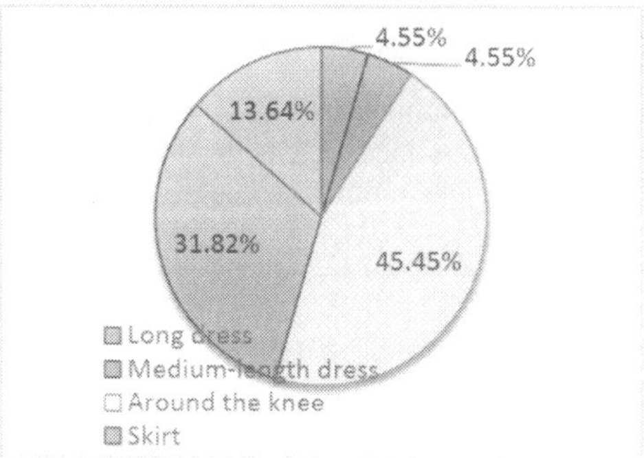

Fig.8 Preference for Length of Dresses

4) Demand for Colors

Fig.9 shows that nearly 34% of respondents are fond of warm colors, followed by neutral colors. And nearly no one likes cold colors.

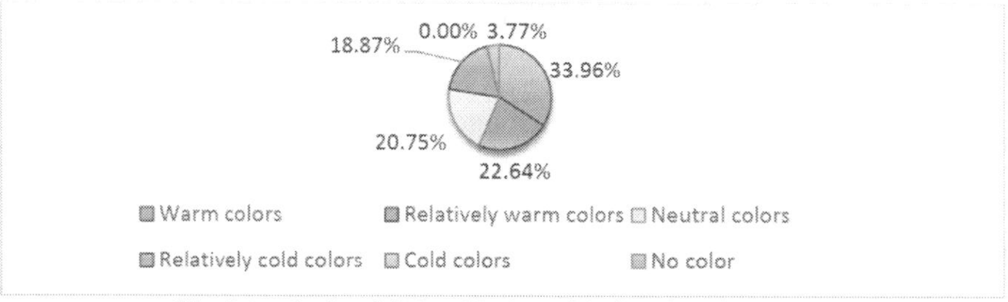

Fig.9 Analysis of Color Choice of Animation Costumes

5) Demands for Accessories

Fig.10 and Fig.11 show that people prefer accessories are put in certain parts of clothes, such as collar, sleeve and hemlines.

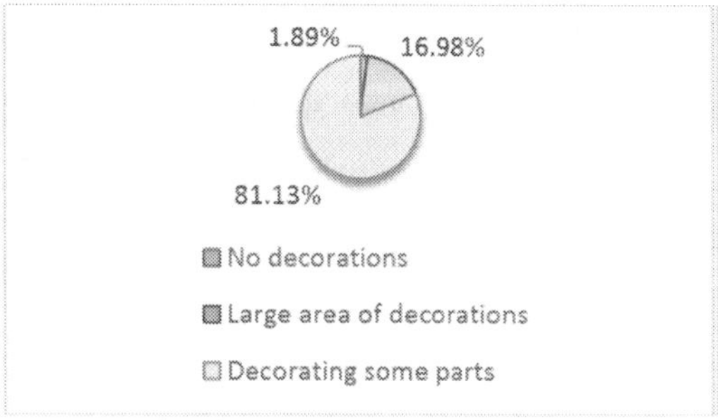

Fig.10 Areas for Accessories

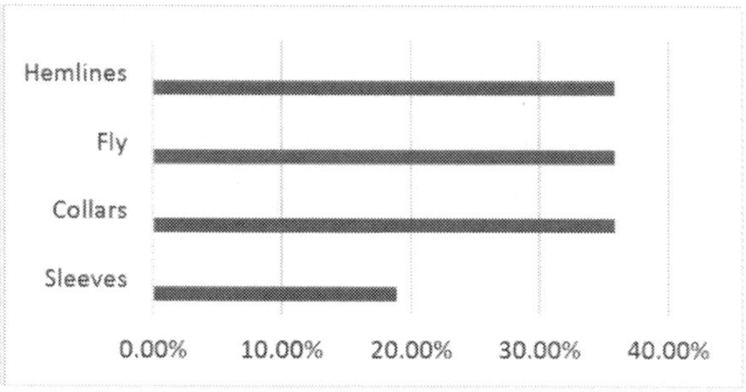

Fig.11 Preferences for Decorating Locations

From the survey, it could be concluded that the majority of respondents prefer a Lolita style in terms of costume style. And an increasing number of animation lovers are looking forward to a simple and daily style, with some accessories in certain parts of clothes. As for colors, costumes in cold colors seem not to be that popular among these lovers, on the contrary, warm colors are favored. And in terms of decorations, it would be better if some items is put on the collar, sleeves, and hemline.

4.3 Improving Design Drawings

Fig.12 is designed in order to meet the requirements and preferences of animation lovers, according to our aforementioned surveys. In this design, the style of shirt is modified into styles that would be suitable for daily use, and the coat collar is designed to be resembling a western-suit style. Middle-length sleeves and lace in the end of the dress make the overall style suitable in daily life, as shown below. [15]

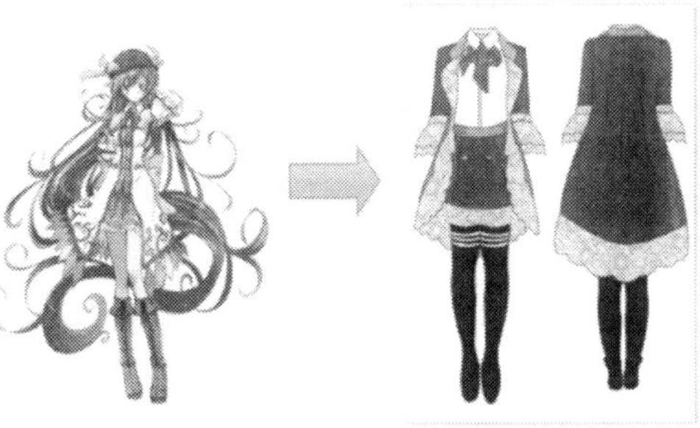

Fig.12 Improving Design 1

Fig.13 Improving Design 2

In Fig.13, the design is updated by adding a circuit lace and bow, and increasing the length of the sleeves. Through this redesign, the costume is represented in form of a dress to be in accordance with the results of survey. As a final addition, a hat is added as an accessory for flair.

5. Conclusion

The concept of animation clothing is defined, the basic theory of anime and animation costumes is explored and analyzed, as well as the relevant markets and developing trend of animation costumes. The style, color and accessories of the animation costumes are well studied. This paper analyzes the styles of animation costumes in different types of anime, and has successfully expressed the feeling and significance of these unique styles. The colors of anime costumes are analyzed in the context of different countries and styles of anime. This paper has also analyzed the accessories that are associated with daily life in different genres.

When analyzing the results of the questionnaire, most people like the lovely Lolita style, but they also want to simplify it; a preference for warmer colors. The best local decorations are in the collar, sleeve, front, and hem; a preference for tops with skirts and socks; Skirts tend to be above the knee; Like the collared collar and the medium length of the sleeves. According to the research results of daily work of the improvement to the original animation costumes design, make it more suitable for daily wear, have more audience, enhance its practicability.

The gate has been opened for a animation-based costume market. In this paper, the paper didn't involve too much, just slightly studied about it. Therefore, a later research can involve an in-depth study on Chinese animation clothing market, such as carrying out a series of inquiries into online to offline store experiences, and increase consumers' convenience in buying their favorite animation-based costumes and related accessories. The construction of Chinese animation clothing brand. products of animation, animation manga and anime characters are combined with animation clothing brands; its development of animation clothing brands amongst the world remains a deep and meaningful subject.

References

[1] Xie XF. Establishment of animation industry development, big data and the Internet technology [J]. Academic Forum: 2014; 3: 46-50.

[2] Wang X. Discovery and development of animation costumes derivation [J]. Cultural and Educational Materials: 2012; 4: 95-96.

[3] Yang Y. Meaning of exploring animation culture [J]. Chinese Education Innovation Herald: 2008; 34: 146-147.

[4] Zhou Y, Huang YF. Influence of animation clothes popular trends on current culture [N]. Journal of Zhejiang University of Technology: Social and Scientific Edition: 2010-03; 9.

[5] Liu XG. Practical operation of clothing design [M]. Shanghai Journal of Donghua University: 2008.

[6] Yu LH, Fang YM. Comfort level research into casual clothes [J]. Artistic Educational Research: 2013; 5: 71-72.

[7] Ding XJ. Outline shape and style research into modern clothes [J]. Artistic Educational Research: 2012; 13.
[8] Zhang HL. The design of animation character [M]. Beijing; Beijing Jiaotong University Press: 2006; 149-165.
[9] Jiang L. Analysis of Chinese elements of kimono [N]. Journal of Chongqing University of Science and Technology: 2010-02-03; 11.
[10] Liu XQ. Application of bow in popular women clothes [J]. Rural Education in China: 2012; 1: 46-48.
[11] Yu XF. Discussion of animation costumes brands [J]. Chinese Education Innovation Herald: 2009; 3: 213-214
[12] Wang YJ. Research on the design of animation derivative clothing [D]. Jiangnan University: 2013
[13] Jin YP. Introducyion to the creative animation industry [M]. Beijing Education: 2012; 190
[14] Xiao H. Clothing hygiene comfort and application [M]. Shanghai: Donghua University Press: 2009-09.
[15] Qin ML. Animation styles and design art [M]. Beijing: China Renmin University Press: 2005; 109-120.

Intelligent Guarding Clothing System for Children

Xuan Liu*, Zhi-Xin Zhang, Lu Wang

School of Apparel & Art Design, ShaanXi Institute of International Trade & Commerce, Xi'an, Shaanxi, 712046, China

*Corresponding author's email: 285797323@qq.com

Abstract

Children's safety has always been an important social issue. Children are unable to effectively protect themselves or accurately identify threats. Their mental and physical functioning is also relatively weak. Any hidden danger may threaten a child's safety. This paper analyzes the necessity of intelligent guarding clothing for children, and proposes two kinds of intelligent guarding clothing system for different groups of children. This can help parents to remotely monitor their children's heart rate, body temperature, location and safety status through mobile-phone technology to minimize the likelihood of children's disappearance or accidental injury.

Keywords: Intelligent Clothing; Child Care; Child Safety; Positioning

1. The Necessity of Intelligent Guarding Clothing for Children

1.1 Reducing Accidental Injury

An accidental injury is an injury with an unexpected cause. In most countries, accidents are the most common cause of injury and disability among children and adolescents. According to statistics on child accidents gathered by China's health-care workers, accidents have surpassed infectious diseases and malnutrition as the primary cause of children's death in China. Disability and death caused by accidental injury to children inflict great pain and place a heavy burden on Chinese families and society [1-4].

The findings of research in Xi'an offer a useful example, Zhao Yu and Li Ying of Xi'an Central Hospital [5] analyzed the causes of accidental death of children in Xi'an from 2008 to 2012. Of 2,616,885 people aged between 0-14, 2,562 died during this period. The average annual mortality rate was thus 97.90 per 10 million. Accidents were responsible for 493 of these cases. The top five causes of death in this period were traffic accidents, drowning, accidental asphyxia, accidental fall and accidental poisoning (Table 1).

Table.1 Relative importance of causes of child death in Xi'an (top 5) [5]

Order	Cause of death	Number of cases	Percentage
1	Traffic accident	189	38.34
2	Drowning	75	15.21
3	Accidental asphyxia	70	14.20
4	Accidental fall	44	8.92
5	Accidental poisoning	32	6.49
Total		410	83.16

The majority of child accidents are caused by improper parental care, and parents are often criticized as irresponsible in cases of accidental injury involving children. However, this reflects a social problem indirectly resulting from China's rapid economic development. Due to extremely intense work pressure, parents are rarely able to personally look after their children all of the time. Intelligent guarding clothing thus offers a very useful child-care tool for parents, with the potential to reduce the risk of child accidents. Such clothing allows children to be located while at school. When a child visits an accident-prone location, reminders can be given to ensure that he or she pays attention to vehicles and does not cross the road at random. During the summer, the positioning function can be used to monitor children's location in real time to prevent them from secretly entering the water or swimming in accident-prone waters, reducing the risk of drowning accidents. Physiological heart rate monitoring and real-time positioning can be used to promptly notify parents when children are at risk, ensuring that children are not left in cars or other spaces with a high risk of suffocation. When a child falls, his or her parents are able to rapidly check the child's injury and offer timely treatment and comfort.

1.2 Reducing the Risk of Child Disappearance

Statistics suggest that about 200,000 Chinese children go missing each year, but only about 0.1% are found. Many countries elsewhere in the world have missing child warning systems, such as the American Amber system [6]. China lags behind in the development of a warning system for missing children. In 2015, China's first missing child micro-warning platform went online, but many parents were concerned about the disclosure of personal information [7]. As this system is insufficiently mature, help for missing children is also limited. With the development of science and technology, geopositioning is increasingly used to prevent the disappearance of children at an initial stage [8-9]. Many smart watches have been marketed with a geopositioning function to reduce the risk of lost children. But this approach does not work in cases of child kidnapping, trafficking and similar crimes, as the watch is too conspicuous: once criminals have removed the watch, the child's positioning information is no longer accurate. This may even mislead the police and parents, distracting them from timely opportunities for rescue. Intelligent guarding clothing is far better concealed: the child's positioning information can be transferred to parents without the criminals' knowledge, helping parents and the police to promptly locate the missing children.

2. Function of Intelligent Guarding Clothing for Children

Children in early childhood and preschool children leave their parents to play alone or to go to kindergarten, but their self-protection ability is poor; it is easy for them to become lost or sustain an accidental injury, and their self-help ability is also weak. If parents can monitor children's positioning and health status through intelligent guarding clothing, many unexpected risks will be reduced. Children's shoes with a positioning function are currently on the market, but they rely on Bluetooth communication, so the scope of protection is very limited.

Therefore, the aim of this paper was to design an intelligent guarding clothing system for early-childhood and pre-school children to establish real-time links between children and parents by Bluetooth and mobile communication without affecting children's play, school or other activities to enable parents to constantly monitor the location and safety of their children. The ultimate goal was to significantly reduce the disappearance of children and accidental child injury. The proposed system has the following main functions.

2.1 Temperature Monitoring (Temperature Sensor)

When children are at kindergarten or school, parents can remotely observe changes in their body temperature and take timely measures in response to any abnormalities. Many kinds of human body temperature detection sensor are available; the most common is an infrared body temperature detection sensor. The human body emits infrared light that is detected by an infrared temperature sensor and converted directly into a temperature-value output. The temperature values are sent to the microcontroller, yielding the child's current temperature.

2.2 Fall Monitoring (Posture Sensor)

Young children are often slow moving and have poor physical coordination, making it easy for them to fall while playing. If a child has fallen over, the device will inform his or her parents, allowing them to keep abreast of the child's injury to deal with the problem and offer timely comfort. Fall monitoring works via a posture sensor, also known as an accelerometer, which is based on measurements of the Earth's gravity. The sensor can detect its own tilt angle on three axes, x, y and z, offering a simulation of three-dimensional movement. If the accelerometer sensor is placed horizontally, its x-axis = 0, its y-axis = 0 and its z-axis = 90.

2.3 Heart-rate Monitoring (Heart-rate Sensor)

When children are at risk, they experience tension, anxiety and fear, causing their heartbeat to accelerate. When a child's heartbeat exceeds a normal range, a heart rate monitoring function offers parents timely feedback; this can be combined with a positioning function to facilitate rescue, preventing tragedy. A heart-rate sensor detects changes to the heartbeat by measuring blood oxygen. When the heart beats, the oxygen concentration in the blood changes. The heart-rate sensor outputs voltage when detecting the human heart rate; the sensor outputs a relatively high voltage when the heartbeat increases due to exercise, for example, and outputs a lower voltage when the heartbeat is lower.

2.4 Positioning (GPRS Module)

The system's positioning function can be used to locate a child's position in real time, monitor the child's activities and provide instant feedback on the child's location. If children become lost, positioning information can be used by parents and the police to find them, greatly reducing the risk of disappearance. A positioning function can be achieved via a GPRS module.

3. Work System Design

In this study, two programs were designed according to whether children used mobile phones. This ensured that the proposed system most accurately met the needs of children and parents by accommodating children's age and parents' decision as to whether to allow their children to use a mobile phone.

3.1 First Option

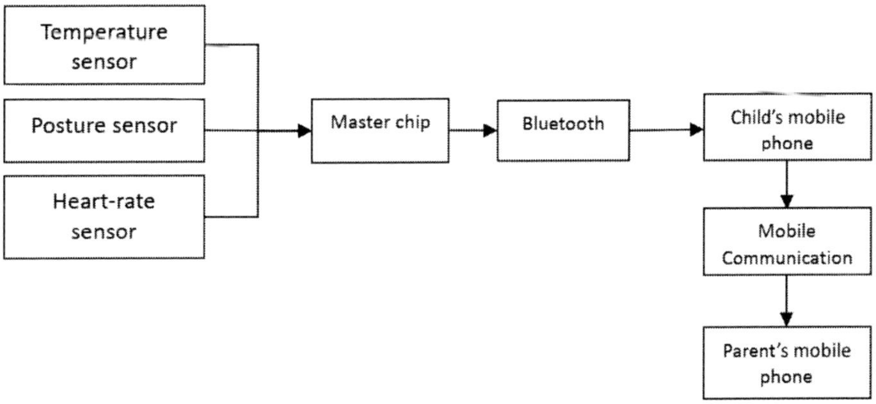

The master chip collects data on the child's temperature, posture and heart rate and then transfers these data to the child's mobile phone via Bluetooth. The child's mobile phone in turn transfers this location information to the parent phone via a 3G or 2G network. Parents can then access up-to-date information on the child's location through the phone software. The advantage of this program is that the energy required for the positioning function is provided by the mobile phone, greatly reducing the energy consumption of the monitoring clothing. A long battery life is advantageous, as users need not replace the battery frequently. However, this method relies on mobile-phone positioning and communication, and is thus only suitable for older children who can use a phone.

3.2 Second Option

The master chip collects data on the child's temperature, posture and heart rate, and sends this information to the parent phone via a GPRS module with a positioning function. Parents can then use their mobile-phone software to access their children's current information. The advantage of this program is that all of the features are provided by the clothing, removing the need for a mobile phone; therefore, this is a good option for young children. Although the energy consumption of the GPRS module is high, reducing battery life, this problem can be solved by increasing charging power.

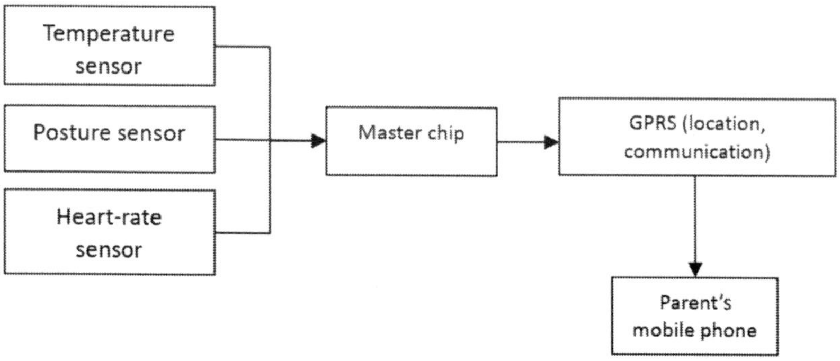

4. Conclusion

The disappearance or accidental injury of children is usually attributed to parents' neglect; in China, however, parents' care burden is expected to become more and more heavy with the recent implementation of the universal two-child policy. The development of intelligent technology offers potential tools enabling parents to care for their children, establishing real-time contact between children and parents, reducing parents' burden in taking care of their children, and enhancing the protection of children. The design has the following advantages: 1. it is embedded in children's clothing and thus better concealed, reducing the chance of its being removed by criminals; 2. monitoring range is not limited, due to the use of mobile phones for information transmission; and 3. monitoring is comprehensive: parents can monitor not only their children's location in real time, but their body temperature, heartbeat, other health conditions, falls during sports activities, etc.

References

[1] Zhang H, Li Y, Cui YX, et al. Correlation between child injury and parental injury cognition in Daqing City. Chinese Journal of School Health: 2015; 3: 427-429.

[2] Zhu P, Cai YH, Xu XY. Children of accidental injury and prevention. National Pediatric Nursing Academic Exchange and Symposium Conference Papers: 2006.

[3] Zhao XF, Liu LK, Tu GF, et al. Investigation on accidental injury in children in Chengdu. Nursing Research: Zhongguo Ban: 2016; 3: 952-953.

[4] Tian LY, Liu MH, Li L, et al. Knowledge and risk behaviors of school-age children with unintentional injuries in Changsha City . Journal of Central South University (Medical Scienc): 2016; 7: 741-749.

[5] Zhao Y, Li Y. Death cause analysis of accidental injuries among children in Xi'an from 2008 to 2012. Chinese Journal of Woman and Child Health Research: 2016; 3: 295-297.

[6] Li ZD. Search for missing children's "Amber Alert". Prosecution clouds: 2014; 1: 52-53.

[7] Luo P. China's first "early warning platform for missing children" on-line parents worried about information disclosure. Innovation era: 2016; 1: 34-36.

[8] Wang CY, Zhu Y, Zhang F, et al. Study on industrial design of children's anti-lost products. Continental Bridge Vision: 2016; 4.

[9] Li J, Zhong XL. Study on children's anti-lost system based on communication network and GPS. Information and Communications: 2015; 1: 74-75.

Two Color Jacquard Design for Warp Knitted Spacer Fabric and its Use for Sport Shoes

Md. Abdul Mueeid Alam, Wei Kei, Deng Zhongmin*, K. M. Faridul Hasan, Junaid Khalid, Md. Saiful Islam Bhuiyan`

Wuhan Textile University, No.1 Sunshine Avenue, Jiangxia District, Wuhan, Hubei, 430200 P.R. China

*Corresponding author's email: hzcad@163.com

Abstract

This research mainly introduces the principle of a new type of jacquard three-stitch technique. It uses an ingenious combination of sixteen kinds of stitches which enhances the ability of jacquard knitting that makes the fabric more stylish. This break through of traditional pattern in the form of single three-stitch technology introduces its application in warp knitted spacer fabric especially on shoes. The development of the production process of shoes material technology and two-color jacquard shoe design is the main focus of this article. New three stitches technology application in shoes materials, makes the mesh stitch transformation more versatile, which expands the scope of jacquard pattern, making the fabric more comfortable. They play a diverse role when using the material for different parts of the shoe such as the heel, tongue and side.. The main function required for the shoe head and side is air permeability and moisture permeability, so the footwear trademark or brand logo can be easily placed.

Keywords: Jacquard; Warp Knitted; Spacer Fabric; Shoe Material.

1. Introduction

The design is continuously changing for the development of sport shoes. There is a lot of knowledge not only on leather, fiber and warp knitting technology, but also plastic surgery and partial knowledge on physical shape and comfort [1, 2]. People normally expect comfort, wear resistance, dimensional stability, lighter weight, air permeability, moisture absorbency, etc. for sports shoes fabrics. At present, in the field of sporting items, function and fashion are the two important themes. Sneaker fabric design should be made based on high-technology, with a multi-functional, fashionable and a developing trend. Current research focuses on sports shoes fabrics qualities especially for lighter weight, decorative pattern, dimensional stability and fastness properties [3, 4].

2. Problem Formulation

The current demand changes with the improvement of people's living standard in developing countries. Now, people look for higher requirements of textile products in sport shoes especially for the design, quality, color and style. In order to adapt to these requirements, the warp knitted products have started from simplification to diversified development especially for the society and customer [5]. Warp knitted product variety and color has become the determinants of how well it sells to the consumer. For facing these challenges, artificial warp knitted design is going to be more difficult which is needed to bring up a new design approach [6, 7].

In two-color jacquard shoe design, a new type three stitches technology is used for the production of a fabric with double color effect. Interval of fabric, yarn surface the fabric would come together independently, and maintain a certain distance between unique three-dimensional materials, which belongs to the technical textiles. It has been widely used in shoes materials, especially to the sneakers item [8, 9]. Two-color jacquard would break the traditional single structure appearance, using a new version of Piezo jacquard system; where the structure of the jacquard stitch constantly enriches. Jacquard pattern scope expands to a greater extend for making the fabric most significant and attractive with psychological satisfaction and aesthetic appearance along with the diversification of fabric style [3, 10].

3. Problem Solution

3.1 Materials and Method

HZCAD software was used to design the warp knitted spacer fabric for sports shoes. Firstly, the sample is designed. The shoe-like cardboard or design samples must be analyzed first. Secondly the raw materials were analyzed, to determine the type of yarn used. Then the fabric width, GSM, color, machine number, fabric density are recorded and determined. The procedure of a shoe design is shown as below:

$$W = w \times E/2.54 \quad ; \quad H = h \times C/2$$

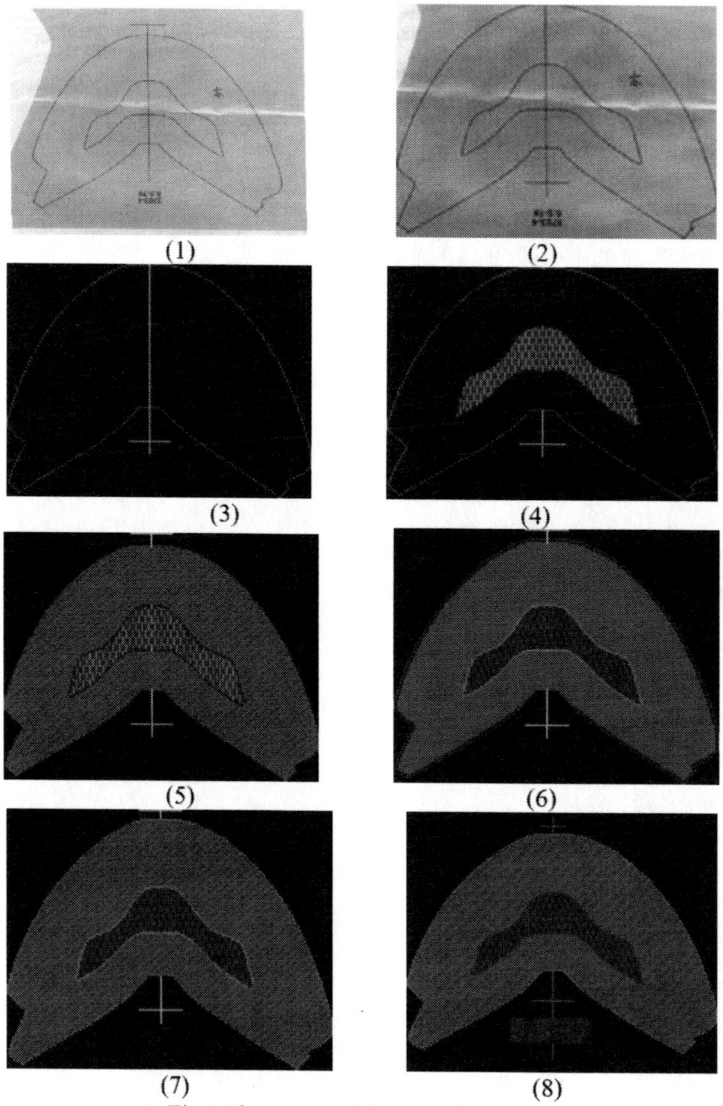

Fig.1 Shoes material processing steps.

1) The actual width (w) and height (h) in cm of shoe-like board is measured. It was printed on a transparent film

and scanned into the computer, as shown in Fig.1(1).

2) Coral drawing software was used to correct the scanned design, so that the cross-point positioning and vertical and horizontal coordinates respectively. Then the outer part is cut. The density of C and machine number E was used to calculate the width and Height (number of rows). They wereentered to adjust the size of the scan. It is then saved as a BMP format image, as shown in Fig.1(2).

3) The processed BMP format image was inserted into the HZCAD drawing software to check the parameters in Gy, select the appropriate model, fill in the density, and then draw the outline of the shoe design, Fig.1(3).

4) The 16 effects of the new Jacquard three-pin technique and the appropriate Jacquard stitches were combined. The tissue was filled into the corresponding area, as shown in Fig.1(4) and Fig.1(5).then to the height was then inreased to 20, from the toe with the top cross-point positioning to facilitate the control after dyeing machine size.

5) Then the process parameters in Gy was modified, the flower width was increased by 4-pin, flower height increased by 8 rows, the overall outline of the toe was expanded. The purpose was to ensure that the latter part of the machine when cutting the integrity of the toe, as shown in Fig.1(6). The inner contour line was deleted, while the toe was filled with stitches, the outer contour of the toe which is outside the other point, so that when knitting out the outline is obvious and convenient version and is cut, as shown in Fig.1(7).

6) The shoe size code number is at the center line, as shown in Fig.1. (1) and Fig.1(2) are for the local enlargement. A shoe stitch is designed, the Gy in the width of the flower width (W) the machine is used to change the number of needles, and then typesetting (Fig.3) and the size is adjusted if needed, to avoid waste, as shown in Fig.1.

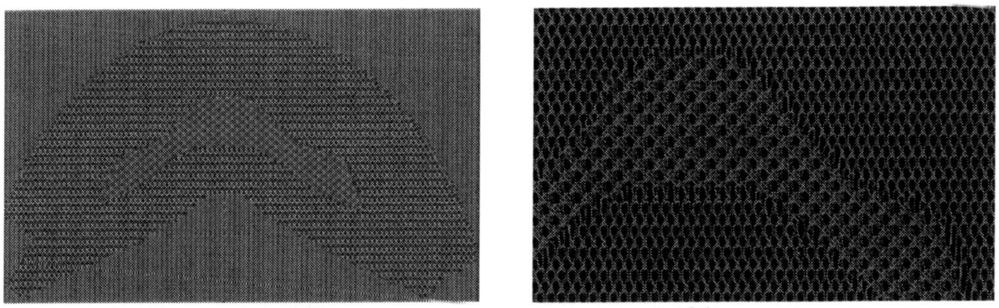

(1) (2)

Fig.2 Double-color shoes material process simulation diagram.

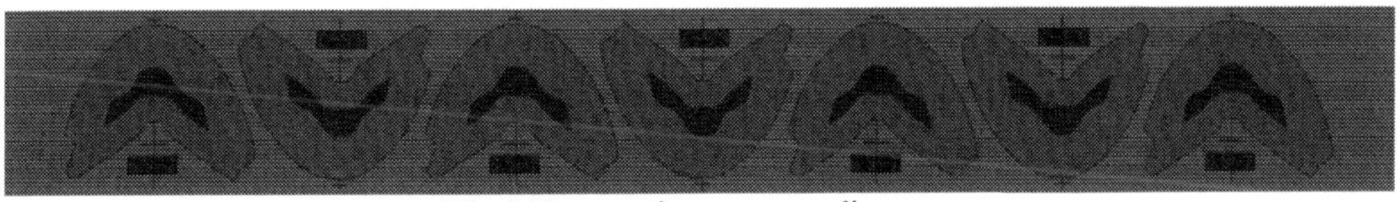

Fig.3 Typesetting process diagram.

Jacquard yarn isis divided into odd and even number row, which represents the different configuration of colors Fig.2) or different kinds of yarn, namely, two and a half the number of jacquard comb through the different color or different kinds of yarn. Fabric is knitted after using different dyes, creating a contrast effect n the vamp. The best contrast is created using two kinds of yarn color. After layout the file is saved and copied. Then a single process of comb bar arrangement was developed as follows:

GB1 (to comb): 2-2, 1-0 / 2-2, 3-4 / / T100DFB / 36F (480/6) full;

GB2 (to comb): 1-1, 1-0 / 0-0, 0-1 / / T100DFB / 36F (480/6) full;

GB3 (to comb): 1-0, 1-0 / 2-3, 2-3 / / T40D / 1F monofilament (480/6) full;

JB4 (Jacquard): 1-0, 0-0 / 1-2, 2 / / 150D polyester / 48F (240/6) full;

JB5 (jacquard): 1-0, 0-0 / 1-2, 2-2 / / T150D low stretch / 48F (240/6) full;

GB6 (to comb): 1-0, 0-0 / 0-1, 1-1 / / T100DFB / 36F (480/6) full;

According to a single process, the yarn is separated into odd and even columns which are arranged with different colors or types of yarns.

3.2 Results and Discussion

In the new Jacquard design process, each unit consists of two rows which is controlled by two pieces of guide bars. The offset Jacquard comb is controlled by four parts. aA color point is controlled by the guide. In contrast to the traditional Jacquard processes, these four signals are controlled by the pitch of the odd-numbered rows and the offset of the previous needle and the even numbered rows, and the offset of the pre-stitch yarns in the Piezo Jacquard control system. There are two control messages for each row. The first information controls the offset of the needle back padding, and the second information controls the offset of the needle before the stitch. Therefore, it is possible to form 16 kinds of mat yarns. The rational combination of these 16 kinds of gauze yarns can be used to design jacquard fabrics with different styles. As shown in Fig.4, H indicates that the guide bars are not offset, and T indicates that the guide bars are offset.

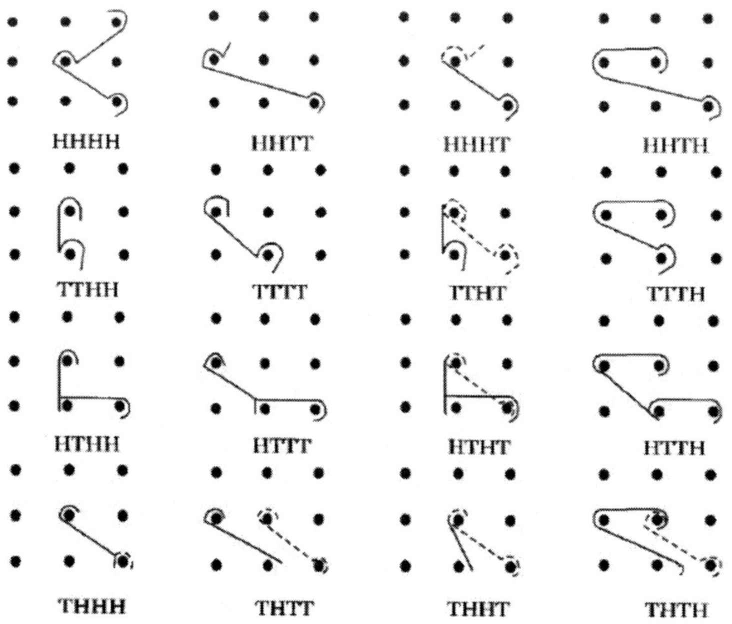

Fig.4 16 kinds of basic stitch way of mat yarn.

Thus canceling the mat before the needle forms on the floating line. The combination of these 16 kinds of stitches enhances the Jacquard's ability to make the fabric style more abundant. The application of this new type of jacquard needle selection technology has broken through the limitations of the traditional three-pin needle selection technology. Therefore, the application field of the Jacquard warp knitted fabric is further expanded.

Traditional Jacquard three-pin technology could only be applied for different raw materials, different structures, colors and gloss to make flower style richer. Therefore, the new three-pin technology would be a great breakthrough. The use of new three-pin technology is designed for shoe fabric, through setting the process parameters. The odd and even columns are arranged with different colors or different types of yarn. The appropriate stitch can produce the fabric with the same color giving a unique but simple design. Fig.5 for the new three stitches of jacquard technology creates a variation of 16 kinds of basic stitches, numbered from 0 to 15, respectively, and the pad in Fig.5 corresponds to the yarn. 1, 3, 8, 9, 11, 13 stitches are in single row. A needle mat yarn, can form a mesh; 0, 2, 5, 7, 10, 15 stitches in a double row, namely on the two needle mat yarn.

THTT	TTTH	THHH	TTTT
THHT	HHHH	TTHH	HTHT
THTH	HHHT	HTHH	HHTT
HHTH	HTTH	HTTT	TTHT

Fig.5: Assorted pieces of mat yarn.

A different combination of basic yarn can form single color effects or color mixing effect. As shown in Fig.6 the left side of the simulation design, the right side of the color design for the Jacquard. Here, different colors represent different ways of yarn. A wide range of styles such as monochrome bar effect, small mesh effect, as well as the color mixture of large mesh effect are able to be created using a combination of different basic stitches etc. which makes the fabric style rich and innovative.

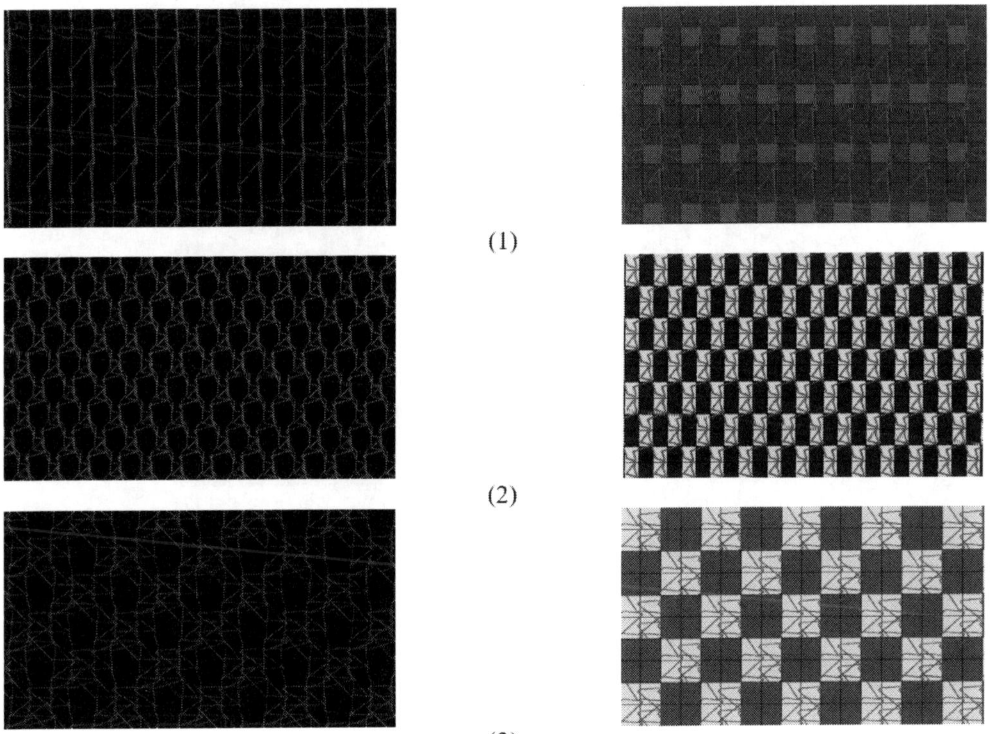

(3)

Fig.6 Combinations renderings of the organization.

In Fig.6(1), the no. 1, no. 6 and no. 9 color, constitute a double color stripes and mesh effect. Fig.6(2) is composed of no. 3 and 10 which constitutes a rule of double color small mesh style. Fig.6(3) is composed of 5 and 10which constitutes to a mixed color net hole style. It is worth noting: in process design, stitches should take into account the process of practical feasibility as each pad yarn stitch structure and excursion guide are not identitcal.

In the design of flower pattern, the formation of flower effects requires a variety of stitches. For example, while several stitches exist mutually, these are the most basic form of flower effect. Therefore, organizational design, where stitches need to be considered in order to form better patterns, it must be take into account the link between the various stitches. As shown in the left image in Fig.7, the combination of these two yarns can create a gradual effect. But only when these two stitches converge, there can be two different results. The left side of the transition is more natural, but in the right side there is a clear fault. Again, the 12-color and 4-color designs are thicker, but the 12-color structure is more left-leaning and the 4-color is right-shifted.

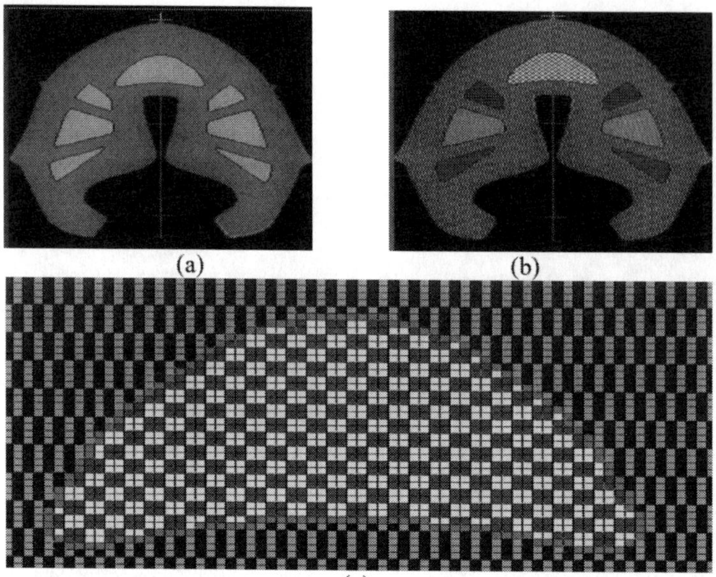
(a) (b)

(c)

Fig.7 The top of the shoe part of jacquard organization chart

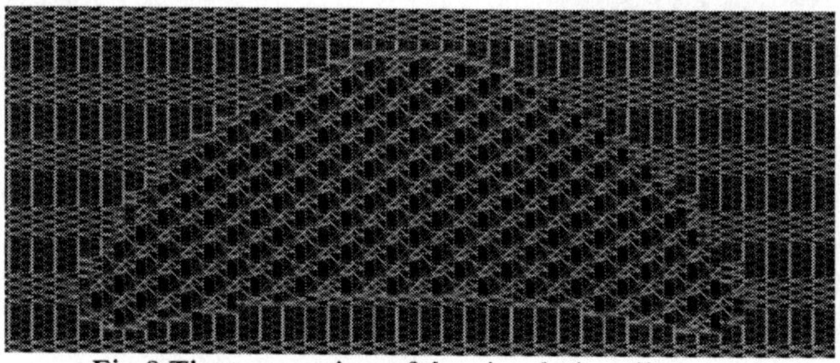
Fig.8 The toe portion of the simulation diagram.

Toe portion of the organization is composed of 5 and 10 color which constitutes to a kind of double lubricious net hole style, making the shoes more breathable and beautiful.

4. Conclusion

The subject analyzed the principle of the new type of jacquard three stitches technique. Spacer fabric for shoe fabric produced using the related technology is compared with the traditional technique of three stitches to highlight its advantages and innovation.

Traditional double color fabrics production efficiency is low compared with traditional multiple comb double color fabric. Double color fabric changes in stitch design are more flexible and more diverse with higher production efficiency. Three stitches technology reduces pattern design limitation, design variety and convenient design. Combined with the HZCAD design software, it can greatly shorten the production process. However, the design need to consider the connection problem. Attention should be paid to the collocation of different stitch pattern, otherwise easy leakage of needle design may create quality problem to the fabric. Spacer fabric attempts to find wider applications in shoes materials, and the new three stitches technology will make it more stylish and fashionable which can greatly increase its unique appearance drawing in more customers.

References

[1] Kadolph SJ, Marcketti SB. Textiles: Pearson: 2016.

[2] Li QZ, Zhou J. Research on Color Model of Digital Multi-warp Jacquard Fabric Design. Sil Mont: 2005; 5: 006.

[3] Dua B, Huffa B, Shaffer BA. Article of footwear having an upper incorporating a knitted component. Google Patents: 2013.

[4] Davies A, McCarthy J. Use of knitted spacer fabrics for hygiene applications. B J McCarthy, Editors. Textiles for hygiene and infection control. Woodhead Publishing: 2011; 27-47.

[5] Jiang GM, Li DJ. The Structure and Properties of the Warp-Knitted Spacer Fabrics. J. S. Yangtze Uni: 2003; 4: 017.

[6] Rock M, Lohmueller K. Three-dimensional sport medical support spacer fabric. Google Patents: 2000.

[7] Orei T, Matsuzaki T, Yamamoto T. Light weight shoes. Google Patents: 2006.

[8] Dua B, Thomas EN. Article of footwear having a textile upper. Google Patents: 2008.

[9] Little C. Seamless upper for footwear and method for making the same. Google Patents: 2012.

[10] Kondou T, Yoshida T, Furuya K. Elastic knitting fabric having multilayer structure. Google Patents: 2007.

Laser Engraving for Tie-dye Pattern Design

Guo-Xiang Yuan[1,2*], Zhuo-Ming Chen[3], Kinor Jiang[4]

[1] *Key Laboratory of Clothing Design and Technology, Donghua University, Ministry of Education, Shanghai, 200051, China*
[2] *College of Fashion and Design, Donghua University, Shanghai, 200051, China*
[3] *College of Fashion, Shanghai University of Engineering Science, Shanghai, China*
[4] *Institute of Textiles and Clothing, The Hong Kong Polytechnic University, Hong Kong*

[*] Corresponding author's email: vogue@126.com

Abstract

In this paper, an eco-friendly tie-dye textile pattern design method by using laser engraving was developed. After the irradiation of carbon dioxide laser on textile, some surface fibers were melted, evaporated and engraved with dyestuff. By combining the traditional tie-dye methods with computer aided design processes, unique visual appearances of textiles with shade changing tie-dye effects were achieved after laser engraving without the application of water and chemicals. The results revealed the potential of the laser engraving process for textile surface embellishment with tie-dye effect. With lower resolutions, engraved vague patterns with small laser beam dots can be achieved. While the treatment resolution increased, clear patterns could be performed. Based on the same resolution, the higher pixel time can export more energy, which makes the pattern look more clear. This paper indicates that the laser engraving technique is an environmentally friendly and feasible tie-dye pattern design approach which can create different pattern, color and texture for textile application.

Keywords: Tie-dye; Pattern Design; Textile; Laser Engraving; Eco-friendly

1. Introduction

There are different approaches to conduct the dyeing process for textile and garment. In textile and apparel industry, tie-dye technique was increasingly applied for design innovation. Tie-dye is a traditional resist dyeing technique which is achieved when certain parts of the textile product are treated or prevented from accepting the dyes that the rest of the fabric receives. Traditional tie-dye textiles are dyed by using natural dye or chemical dye for different colors. During the tie-dye process, attractive and intricate patterns can be made by hand processes via knotting, binding, folding, rolling or sewing certain parts of the fabrics in such a way that the dye cannot penetrate into these areas when the fabric is dyed. Originally, the dyestuffs used for tie-dye were natural dyes like madder, saffron, weld, Persian berries and Indigo. Nowadays, synthetic dyes are applied with tie-dye technique in textile design for visual excitement and tactile appeal. However the synthetic dyeing process and finishing process using water could cause pollution problems with dye-containing wastewaters in the textile industry. 10-25% of textile dyes are lost during the dyeing process, and 2-20% are directly discharged as aqueous effluents in different environmental components [1-3]. Based on the pollution problem, a laser fading process for textile could be applied for surface embellishment without the application of water and dyestuff.

Laser is used in the industry for a variety of manufacturing processes and applications, such as material removal, surface heating, bending, melting, alloying, cladding, texturing, roughening, marking, cleaning, shock hardening, stereo lithography and layered manufacturing processes. Laser has various advantages, such as no physical contact, a high degree of automation and fast and precise cutting and engraving [4]. The application of laser cutting and engraving in the market contains not only metals but also wood, plastics, textiles and compounds. Nowadays, laser has been increasingly applied in the industry to generate graphics onto material surface with laser beam. The CO_2 laser is a gas laser that produces an infrared light and is easily absorbed by organic material, making it the first choice when it comes to engraving fabrics [5]. Nd and Nd-YAG lasers are solid-state lasers that rely on a crystal to create the light beam which are suited for engraving, welding, cutting and drilling metals. With a computer control system, laser surface treatment is a relatively economical and flexible technology to treat and embellish fabric surface towards unique images [6]. This process applies laser beams to melt, burn or vaporize materials, such as material cutting for car airbags, lace fabrics and edging of embroideries [7]. Different materials can be engraved by using laser for different textile surface effects.

Laser engraving can be used to engrave the fabric surface without cutting through. It is applied to remove certain fibers as the laser beam vaporizes the surface [8]. Nowadays, laser has been applied in denim industry as a bleaching tool to create patterns on the denim fabric. After engraving, a color fading, worn appearance or printing effect can be created on fabrics. It demonstrates that the laser treatment is one of time and cost-saving alternatives and has the advantages of design and treatment flexibility without causing any pollution [9, 10]. In this paper, some experimental laser treated tie-dye pattern designs for textile design were presented to demonstrate the application of laser technique to produce tie-dye patterns on textiles with unique appearances.

2. Experimental

2.1 Laser Engraving

In this study, laser engraving is carried out by using a GFK Marcatex FLEXI-150, a commercially designed pulsed CO_2 laser machine (Eurotrend Group, Spain) coupled to a computer-controlled table under controlled atmospheric conditions. The main characteristics and parameters of the machine are shown in Table 1.

Table.1 Main characteristics and parameters of laser engraving machine

Items	Parameters
Machine model & manufacture	GFK Marcatex FLEXI-150, Spain
Lens size	80 cm (1.5 mm beam diameter)
Excitation frequency	81 MHz
Laser medium	CO_2
Wavelength of laser beam	10.6 μm (10600 nm)
Wave model	Pulse
Power (W)/energy (mJ/p)	150 W Pulse energy: 5–230 mJ Pulse activation time: <45 μs
Polarisation	Linear polarisation perpendicular to the bottom of the laser head

2.2 Process

The process of laser engraving on textile with tie-dye patterns is shown as follows: (1) create desired tie-dye graphics in graphics program or software, (2) convert the graphics to gray scale, (3) send designed graphics and patterns to laser system, (4) set engraving parameters of pixel time and resolution, (5) tie and place the textile on a honeycomb cutting table in a cabinet, (6) location identification of the laser engraving area, (7) conduct laser engraving.

2.3 Fabrics

Three kinds of commercially produced plain weave fabrics were used for laser engraving and design application. Table 2 shows the fabric specifications for the experiments.

Table.2 Fabric Specifications

Fabrics	Composition	Weave	Fabric sett, Ends/inch	Fabric sett, Picks/inch	Fabric unit weight (g/m^2)	Thickness* (mm)
F-a	100% Polyester	Plain	80	96	43.2	0.46
F-b	45% Wool/ 55% Polyester blends	Plain	48	40	80.35	1.35
F-c	45% Rayon/ 55% Polyester blends	Plain	80	88	48.3	0.65

Note. * Fabric thickness under 5g/cm^2 pressure.

2.4 Surface Observation

Figs.1, 2 and 3 show surface micrograph images of the three types of fabrics before and after laser engraving with different parameters of resolutions and pixel times.

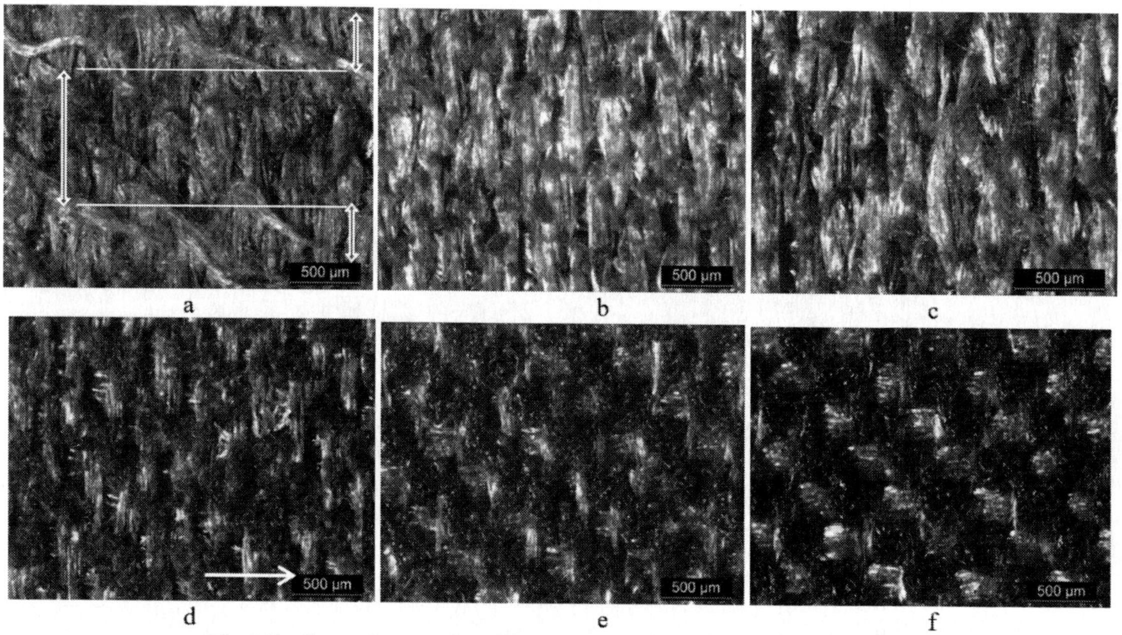

Fig.1 Surface micrographs of laser engraved polyester fabric samples.
a. Engraving at 20 dpi/120 μs; b. Engraving at 20 dpi/270 μs; c. Engraving at 30 dpi/120 μs; d. Engraving at 30 dpi/270 μs; e. Engraving at 40 dpi/120 μs; f. Engraving at 40dpi/270μs.

Fig.1 illustrates the micrograph images of polyester fabrics before and after laser engraving with different resolutions and pixel times. The resolutions were set at 20 dpi, 30 dpi and 40 dpi with increasing pixel times from 120 μs to 150 μs. Fig.1-a shows that the untreated fabric is covered by some downy polyester fibers. After laser engraving as shown in Fig.1-a to Fig.1-f, the fabric surface gradually changes. The results indicate that the application of parameters with lower values will give the least laser power density, and only a few fibers on the fabric surface are engraved, but some will have degraded and formed into beads as shown in Fig.1-a and Fig.1-b. With an increase in the resolution (from 30 dpi to 40 dpi) and pixel time (from 120 μs to 270 μs) as shown in Fig.1-c to Fig.1-f, different degrees of higher laser power density were achieved. With the application of thermal energy, more polyester fibers melted, gasified and vaporized due to the high temperature of the laser beams. As shown in Fig.1-e and Fig.1-f, a basic fabric texture can be observed after the engraving of the surface fibers, and some of the polyester fibers have melted, condensed and adhered onto the fabric surface accordingly.

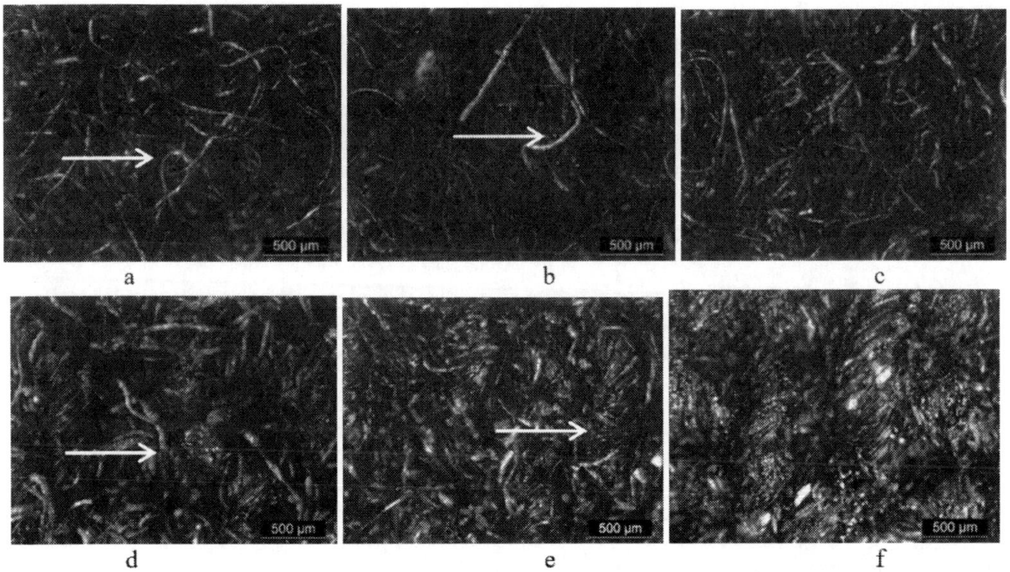

Fig.2 Surface micrographs of laser engraved wool/polyester blended fabric samples.
a. Engraving at 20 dpi/120 µs; b. Engraving at 20 dpi/270 µs; c. Engraving at 30 dpi/120 µs; d. Engraving at 30 dpi/270 µs; e. Engraving at 40 dpi/120 µs; f. Engraving at 40 dpi/270 µs.

With reference to the micrograph images of the original and laser engraved wool/polyester blended fabric samples as shown in Fig.2, a series of surface morphology images of wool fabric is observed. With the application of thermal laser energy, the wool fabrics show different engraved effects. Through laser engraving with a lower resolution and shorter pixel time, a few fibers become burned as shown in Fig.2-a. When the pixel time is prolonged, the fibers turn yellow as shown in Fig.2-b. With higher resolutions (e.g. 30 dpi and 40 dpi) and longer pixel times, some of the surface fibers form burnt, bubbled and etched effects on the fabrics as a result of the increased laser energy as shown in Fig.2-c to Fig.2-f.

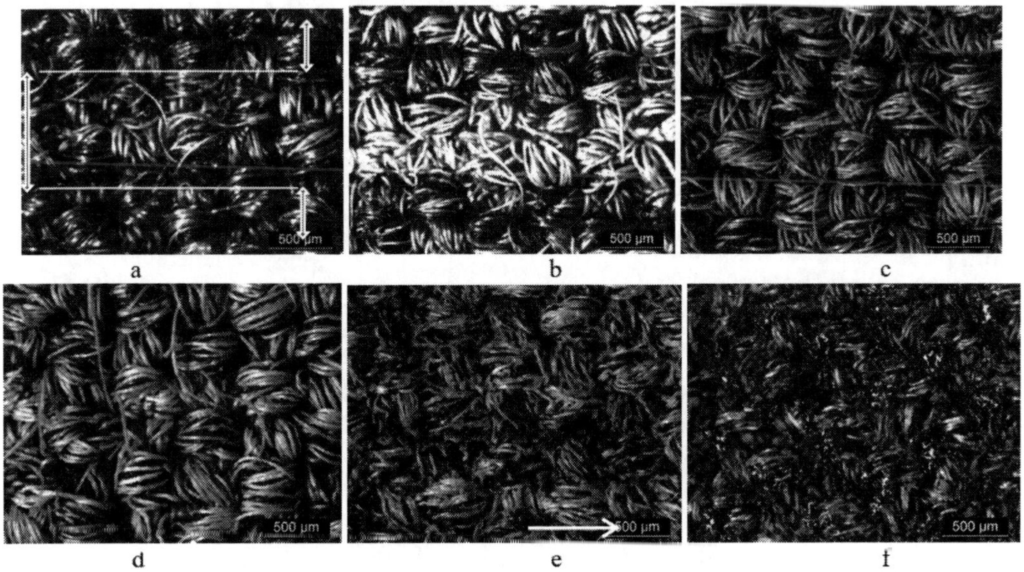

Fig.3 Surface micrographs of laser engraved rayon/polyester blended fabric samples.
a. Engraving at 20 dpi/120 µs; b. Engraving at 20 dpi/270 µs; c. Engraving at 30 dpi/120 µs; d. Engraving at 30 dpi/270 µs; e. Engraving at 40 dpi/120 µs; f. Engraving at 40 dpi/270 µs.

Fig.3 shows the differences of laser engraving treatment at resolutions of 20 dpi, 30 dpi and 40 dpi and pixel times of 120 μs and 270 μs on untreated and treated rayon/polyester blended fabric samples. During the laser engraving process, thermal energy was absorbed by the different fibers. Fig.3-a and 3-b are the samples treated by a lower resolution (20 dpi); the results reveal that at the same resolution, only parts of the fabric surface are treated, but with a higher pixel time at the same resolution, more laser beam energy is applied onto the sample. With high thermal treatment, the rayon fibers become engraved, and some bubbles were emerged in the polyester fibers due to the high temperature of laser beams. There is also a light gray color residue.

3. Design Method

This design method permits the visualization of diverse forms of design inspiration in textile design, resulting in new ways to design patterns and the visual effects similar to the tie-dye effect without the use of chemicals or dyeing. Different textiles are arranged into different shapes using resist methods such as crumpling, pleating and tying for laser engraving. After treatment, the original and engraved areas showed uneven engraving effects amidst the folds, creating various photo-realistic three dimensional patterns for textile. Some experimental designs of tie-dye patterns on textile treated by using laser beams with different parameters are shown below.

3.1 Crumpling

Crumpling results in spiral effects on laser engraved textiles. Textile samples were first laid on a laser engraving table and twisted to achieve a crumpled shape. After laser engraving, the fabric were flattened and crumpled design effects could be seen. Fig.4 shows the result of the engraving process achieved using this method with different parameters.

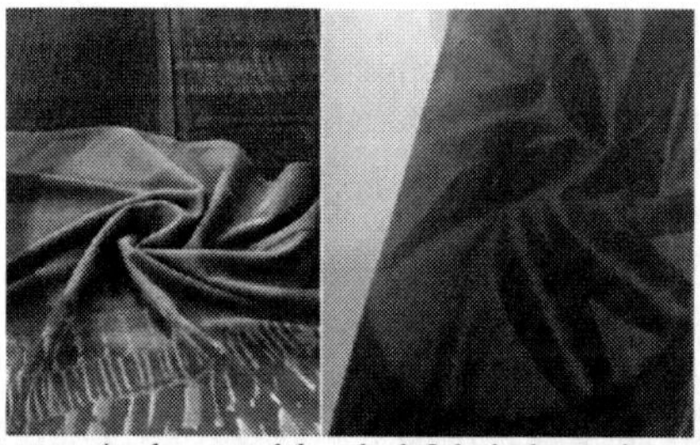

Fig.4 Laser engraved rayon/polyester blended fabric by using the crumpling method.
a. engraving process; b. laser engraving result

This figure shows the results of laser engraving on rayon/polyester blended fabric by using the crumpling method. Spiral and floral patterns appear on textile surface. During laser engraving, the laser beams engraved the surface of the crumpled fabric, and the colors of the treated areas changed and showed diverse gradient effects.

3.2 Pleating

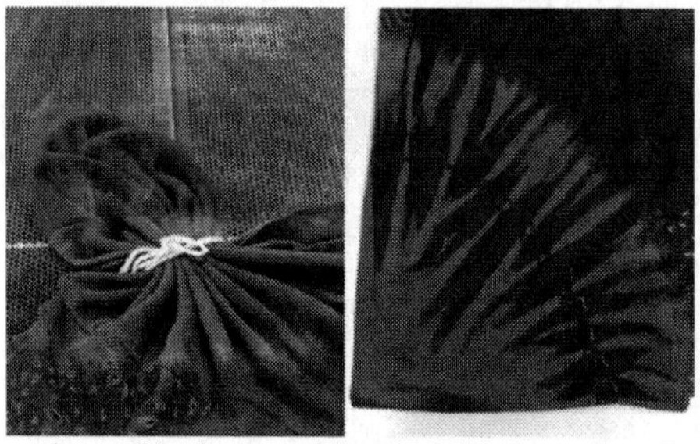

Fig.5 Laser engraved wool/polyester blended fabric using the pleating method.
a. engraving process; b. laser engraving result

With the pleating method, the textile were laid onto a laser treatment table and pinched to raise folds until the pleats met the requirements of the design, and then engraved by laser beams with different treatment parameters. As illustrated in Fig.5, some clearly delineated pleated engraving patterns are achieved.

The results show some indistinct as well as clear pattern edges are engraved onto the wool/polyester blended fabric with correspondingly different shade change effects.

3.3 Tying

This method utilizes thread to wrap and tie the fabric for laser engraving. It is similar to tie-dyeing, but the wrapping is done relatively loosely, because the laser beams engrave the surfaces of the tied textiles, and no chemicals are used, so there is no saturation of the dyestuff. After engraving, some vivid changes in the color shade and imaginative patterns are achieved. The polyester fabric designs based on the tying method are shown in Fig.6.

Fig.6 Laser engraved polyester fabric textile by using the tying method.
a. engraving process; b. laser engraving result.

As shown in Fig.6, the fabric was tied by threads to achieve the requisite appearance. The areas which were tied and folded by threads blocked the laser beams and preserved the original color of the fabric. The results indicate that this method could be used to make tie patterns more easily on textile which are similar to those made with the traditional tie-dye method.

The laser engraving for textile design use Adobe Photoshop software to design, edit and process patterns in computer. As shown in Fig.4, Fig.5 and Fig.6, the textiles are tied by threads. The areas which were tied, crumpled, pleated and folded by threads on the textiles blocked the laser beams and preserved the original colour of the fabric. The results of the engraving show that the resultant process has been succeeded on the tied textiles made of different kinds of fabric materials with sophisticated patterns. After laser engraving, the engraved areas show different shades in colour with diverse patterns which due to the area were burnt by the laser beam. All the design practices present that the design process from textile tying to laser surface engraving is a flexible design approach and it can mimic tie-dye patterns with integrity which differs with the traditional tie-dye method that need to splice patterns at the seams. The results indicate that this method could be used to make tie-dye designs on textiles more easily and achieve effects similar to those with the chemical tie-dye method.

4. Conclusion

Surface laser engraving on textile can mark patterns by engraving without touching the fabric. When processing textile, the laser can produce patterns with flexibility. The laser beam melts the material while engraving results in clean, perfectly patterns. In this paper, laser engraving creates unique patterns on textiles with tie-dye effect by using different methods. The textiles made of polyester, wool/polyester blends, and rayon/polyester blends were carried out on the basis of the proposed tying and design process. With the resist method, textiles with unique tie-dye patterns and colors were created. This design process indicates as alternatives to improve tie-dye textile design with laser engraving for future use by designers and researchers for surface embellishment as an area of academic study and design practice. It also emphasizes the integration of design, technology and material is applicable to textile design. By achieving ideal tie-dye appearances on textile without the application of dyestuff and water, appropriate laser power can be applied by changing the two parameters including pixel time and resolution individually to conduct laser engraving on certain areas of textiles. Under the conditions of pixel time unchanged, more energy of laser can be applied on textile to engrave some fibers and dyestuff with the increase of the resolution of laser treatment. When with set resolution, more energy of the laser can be applied on material with the extension of pixel time. It also indicates that relevant resist methods were developed and integrated with laser engraving for pattern creation. With these methods, textiles with unique patterns and color shades were created with the integration of inspiration and related design elements.

The results of textile design with tie-dye effects induced by laser engraving indicated that similar tie-dye patterns can be achieved with the parameters of resolutions and pixel time. In accordance with the design processes, the benefits are evident, i.e., (i) a design model developed by integrating traditional resist dyeing method and laser engraving; (ii) a design process from textile redesign to surface laser engraving could be applied for textile embellishment; and, (iii) non-water and chemical applied tie-dye textiles could be achieved by using eco-friendly laser engraving technique with the selected parameters. To conclude, the laser technique has a high potential to achieve surface treatment with tie-dye patterns and different shades of color effects for textile design as it is an eco-friendly process and can make special tie-dye surface appearances without involving chemicals and water.

5. Acknowledgement

This research was supported by Key Laboratory of Clothing Design and Technology, Donghua University, Ministry of Education, under grant number 279-10-0001076, and the Initial Research Funds for Young Teachers of Donghua University under grant number 107-07-0053106.

References

[1] Baban A, Yediler A, Ciliz NK. Integrated water management and CP implementation for wool and textile blend processes. Clean: 2010; 1: 84-90.

[2] Robinson T, McMullan G, Marchant R, Nigam P. Remediation of dyes in textile effluent: a critical review on current treatment technologies with a proposed alternative. Bioresource Technology: 2001; 77: 247-255.

[3] Soloman PA, Basha CA, Ramamurthi V, Koteeswaran K, Balasubramanian N. Electrochemical degradation of Remazol Black B dye effluent. Clean: 2009; 37 (11): 889-900.

[4] Yam KL. The wiley encyclopedia of packaging technology. New York: John Wiley & Sons: 2010.

[5] Yuan GX, Jiang SX, Newton E et al. Application of laser engraving for sustainable fashion design. Res J Textil Apparel: 2013; 17 (2): 21-27.

[6] Ready JF. Industrial applications of lasers. New York: Academic Press: 1997.

[7] Steele V. Encyclopedia of clothing and fashion. New York: Charles Scribner's Sons: 2005; p. 3, 12, 54.

[8] Ghorannevissa M, Shahidia S, Moazzenchia B et al. Comparison between decolourization of denim fabrics with oxygen and Argon Glow Discharge. Surf Coat Tech: 2007; 201: 4926-4930.

[9] Dascalu T, Acosta-Ortiz SE, Ortiz-Morales M et al. Removal of the indigo colour by laser beam-denim interaction. Opt Laser Eng: 2000; 34: 179-189.

[10] Jiang SX, Yuan GX, Huang JJ. The effect of laser engraving on aluminum foil-laminated denim fabric. Text Res J. 2015; doi: 10.1177/0040517515599738.

Innovation Design on Multi-bar Raschel Lace Fabrics

Md. Saiful Islam Bhuiyan[1], Kei Wei[1], Zhong-Min Deng[1], Soumya Chowdhury[1], Junaid Khalid[1], Md. Abdul Mueeid Alam[1]

[1] *Wuhan Textile University, Sunshine Road, Wuhan, Hubei, 430200, China*

*Corresponding author's email: hzcad@163.com

Abstract

Multi-bar Raschel lace designing process defines actual fabric lace technology analysis, popularly known as lace sample analysis or lace rendering process design. With the urging of product innovation regarding lace market requirements to continue to increase but the domestic R&D which is dealing with traditional lace spline is not that competitive in the global context. The domestic multi-bar Raschel lace pattern design needs to improve the status quo. The focus on lace design includes the, introduction of critical design rules/paths, raw material selection, process implementation on effect and finishing which summarizes the design method of the lace through modeling with enough characterization of deformation, recreation, and patterning rich design style. This paper aims to analyze the above-mentioned issues through pointing out the solution, and investigating the profound influence of different raw materials on lace effect. In addition, here the study of Raschel lace pattern rendering technology implementation also has a great contribution towards which will help lace flower designers to guide to expand their expertise on critical through this paper evaluating.

Keywords: Raschel Lace; Innovation Design; Multi-bar; Pattern Design

1. Introduction

Warp Raschel lace originated from the West. The 18th-century European men's lapels and socks also used lace decoration. In the twentieth century, the demand for lace in Europe and North America has increased, and other countries have produced handmade lace for many years which also promoted the development of lace industry especially Piezo, jacquard control system and application [1]. Previously the Jacquard warp knitting machine can only produce varieties of design with more than one raw material or different yarn colors. But using Piezo technology it can expand the scope of the identical Jacquard design through single color yarn and single material [2]. Raschel lace has rich and diversified organizational structures where the effect is prominent. Moreover and the appearance is attractive to all age groups.

However, the production of lace products in the market is mainly generated through sample processing whereas it has a lack of innovation and creativity which doesn't leave designers much independence to customize a design. As time progress market is demanding, ostentatious designs which leads to more requirements for consumer satisfaction. This is the latest challenges for designing warp lace products. In the manufacturing process of lace products, it should not be confined to using traditional natural fibers and chemical fibers. Moreover other materials such as metal wires and fancy yarns can also be used for manufacturing warp lace to produce special texture and color effect product. Secondly the design of flowers is a breakthrough of the traditional thinking where the domestic warp lace market should be bold & innovative [3]. Based on this analysis of the lack of innovation consciousness, single product and lack of market competitiveness slows down the pace of grabbing market share for the Chinese manufacturers and marketers. This paper introduces the texture of the lace fabric from the ground bar and the Jacquard bar, and the main body pattern.

2. Lace Patterns and the Use of Features

Raschel lace pattern design is the most common lace design off with basic factors. Through the re-creation of the general features of the pattern which presents a rich pattern style that's divided into two categories: the concrete pattern and the abstract pattern.

There are many patterns in the Raschel lace, including floral pattern, leaf pattern, insect pattern, etc. The more common patterns are Raschel lace [4]. Flower patterns in the lace looking more attractive because of its beautiful shape and moral texture always has been the theme of the designers to express happiness as the main creative object.

The generalization and sublimation figurative forms of Abstract patterns and the basic elements are modeling point, line, surface combination, to fully reflect the pure form of shape, such as geometric patterns, stripes, dot patterns, texture patterns [5]. Texture pattern with natural primitive, rhythmic, random and harmonious beauty usually pay attention to the performance of material and texture [6] combination of flowers and geometric patterns of design and transition.

2.1 Geometric Pattern Transition

The transition of the geometric pattern is mainly through the creation of the point, line, and surface design elements which have the characteristic of modeling. This simple and colorful pictorial language form completes the connection between the unit lines and the direction of the bar. Points are usually jumping, lively, flashing. The role of balance in the lace is usually made into the actual weft insertion or virtual network transition.

Fig.1 Abstract arc transition

Fig.2 Triangular geometry transition

2.2 Floral Pattern Transition

Flower pattern transition mainly using plant mosaic and other concrete patterns. For the transition pattern commonly used in the main patterns of flowers with four-leaf clover, five leaf flower plum and six flowers leaf and so on. In small pattern and dots patterns are usually used 1 to 3 bars and can formed patterns, which don't need to add the extra bar. It can play a good role as the main body of the background [7]. The same type of small flower-shaped transition connection and bar can be maintained consistency where the overall sense of lace more strongly (Fig.3, 4) shows transition of the same pattern.

Fig.3 Transition of a specific flower

Fig.4 the transition between leaves

2.3 Abstract Pattern Transition

Abstract pattern is also a unit pattern transition between common elements which are Persian lines, honeysuckle pattern, volume grass grain, grape pattern, and so on. These patterns are more integrated with a variety of flower characteristics of the imagination pattern as like leaves, vines, etc. can be a good foil against the main pattern and the role of alignment. These patterns are usually used when the unit pattern far away and sometimes on the edge of the transition.

Fig.5 Arc transition

Fig.6 General Arc transition

2.4 Raschel Lace Transition Arrangement

For arrangement of lace transition elements combination method is also very important. There is a relationship between combination of transition elements in the form of layout and the beautiful pattern which is directly affect the bar to achieve process problem.

2.4.1 Scatter Point as the Main Form of Organization

Scatter type is the unit pattern between the area configuration scatter patterns so that scattered flowers more evenly distributed into that unit pattern, the scatter arrangement should pay attention to pattern density, symmetry to avoid pattern organization too spacious or too dense and other ills, arranged to flexible layout, and connect to the natural pattern as shown in Fig.7.

Fig.7 Organization of scatter points as the main body

2.4.2 Curved as the Main Form of Organization

Curve type includes wave type (single wave type and double wave type) and broken line type. The transition and main pattern is the arrangement which forming a certain curve-shaped or linear shape, etc. and constitutes a rhythm of the lace as shown in Fig.8, 9. This type of lace is not only easy to distribute bar but also more natural and beautiful and full of lively and dynamic movement [8]. The abstract pattern is easy to change whereas the arrangement is flexible and rhythmical. Though the combination of the arrangement and the specific pattern [9], with the use of the Jacquard mesh, the line segmentation is more distinct. To meet the different design needs, the yarns should be reasonably organized to form different layers so that the special visual effects and tactile effects of the yarns can be matched with each other. It will enable the lace to form a changeable appearance effect [10].

Fig.8 Curve as the main form of organization

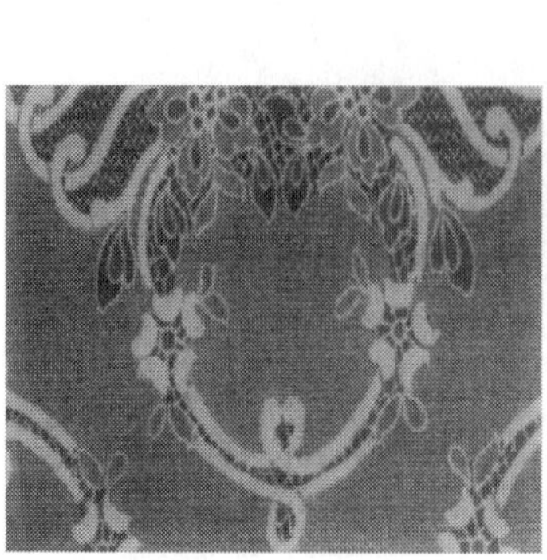

Fig.9 Curve as the main form of organization

2.4.3 Raschel Lace Material

Nylon with high strength, good wear resistance, and high elasticity taking into consideration for the production of lace because of the resultant of yarn such as tensile strength and strength requirements are relatively high compared to other fibers. Nylon has also become one of the important raw materials for production Raschel lace. Nylon can be used for Raschel lace flower bar organizations, such as high compressed yarn into the circle ornamentation. Spandex is usually used because of its' resiliency capacity, good elastic recovery, commonly used bare silk, core yarn wrapped in the form of yarn and other knitted production. Spandex has a good affinity for dyes [11], easy to color, light and soft, wearing from.

3. Lace Effect Diagram of the Process

The process realization of multi-bar Raschel Lace Drawing is to design the walking route of each yarn according to the shape of the pattern in the multi-bar Raschel Lace Design Software. Then design the flower Bar and the bottom of the organizational structure, and select the appropriate raw materials

And actual lace fabric process analysis is different, the lace effect of process design without physical reference, can only rely on the experience of process designers, through the analysis of the effect of pattern structure, shape and lace texture, self-setting fabric.

3.1 Definition and Basic Principles of Carding

Multi-bar Raschel lace to rely on the number of bars with each other, co-woven into a variety of patterns. In order to clarify the weaving range of each bar and to prevent the bar on the same gathering line from striking during the knitting process. It is necessary to make a clear description of the walking route of each bar before drawing the carded yarn. The same gathering line of the bar yarn often use the same color rendering, so that by observing the same color of the yarn is intersecting or overlapping can determine the same concentration line needle collision.

Fig.10 complete carding diagram

3.2 Determination of the Model

From the lace design style point of view, if you need to get a strong three-dimensional effect of the fabric flower, you can choose the machine lace weaving system, such as XGHM63 / 1/36; If you want to fabric pattern flat, Choose to weave in the pure weft insertion machine, for example XGHM59 / 1; If you want the lace effect in the figure without Jacquard, can be directly used without Jacquard machines, such as XGHM45 / 1.

3.3 The Calculation Method of Flower Height and Width

According to the height and width of the flower pattern, multiply the longitudinal and transverse densities of the finished product to calculate the flower width and flower height.

Flower width = pattern width * finished cross-cutting
Flower Height = Pattern Height * Product Thickness

Elastic fabric of the finished product is generally in the longitudinal row 38 / cm or so, no bombs lace finished product generally longitudinal density of 24 to 28 rows / cm; elastic and inelastic lace finished crosswise 10 to 12 columns / cm, design pattern are shown in Table 1.

Table1. Density of the pattern

	On the machine density	Finished product density
24E (pressed yarn type)	20.8	38
24E (Inserted Pattern)	19.2	35
28E	24	44
No Comb lace	24	30
No Comb (commonly used)	18-24	24-30

Note: The table has a bar lace using 140D spandex calculation densities.

In order to make the flower height and flower width to meet the technological requirements, the final flower width and flower height should be adjusted in accordance with the following principles: Flower width is the most common needle width, such as 152,172,184,192,220,236 needles. The following specifications for the high 14.5 cm, width 18.8 cm pattern, for example, to calculate the width and height of flowers, finished 38 longitudinal row / c m, finished cross-10 column / c m. Flower width = flower width * finished crosswise = 18. 8 * 10 = 188 needles, flower height = pattern height * finished longitudinal = 145 * 38 = 551 needles. If the machine is 188 with the nearest needle position is 192, then the flower width is adjusted to 192 pins. If there are two types of Jacquard tissue with a height of 4 and 6 in the unit pattern, the numbers of stitches that are closest to 551 and divisible by 4 and 6 are 552.

3.4 Determination of Number of Cycles

For the machine width of 132 inches multi-bar warp knitting machine is generally weave two, for 200-inch machine, consider three, because the width is too wide.

(I.e. the number of cycles per pattern) = (total number of stitches - number of stitches) / width of flowers

Width of the machine set: the number of cloth edge + flower width * the number of cloth + cloth edge * 2 + flower width * the number of cloth + the number of cloth edge = total number of stitches

Here, in the 132-inch machine weaving, the number of = (3168-96) / 192 = 16

The machine width is set to: 24 + 192 * 8 + 48 + 192 * 8 + 24 = 3168 stitches

Flower width 192 pins, the number of repeat 16, cloth edge of 24 needles. The number of commonly used stitches and the number of finished products as shown in the relationship between the widths of Table 3 (to 24 machine, 132 inches, for example):

Table2. Lace effective needle number and the number of finished width relationship

The number	10	11	12	13	14	15	16	17	18	19
The number of pins	306	270	256	236	220	204	192	180	170	162
No. of stitches	298	270	248	228	212	196	184	172	162	154
Conventional cm	26	25	24	21	20	19	18	17	16	15

3.5 Jacquard Drawing

According to the selected Jacquard drafting of the organization to draw the corresponding Jacquard. Jacquard basic organization as shown in Fig.11,

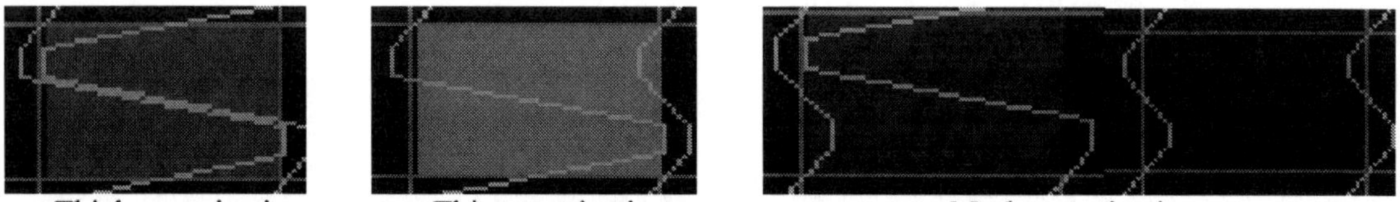

Thick organization Thin organization Mesh organization

Fig.11 basic organization of Jacquard drawing

Rule of the cell area just select a corresponding Jacquard organization can be covered; irregular cells have to draw one by one. Ground network in the formation of thin or thick Jacquard shadows, very vulnerable to the influence of the surrounding Jacquard organization, resulting in changes in pattern shape, contour is not clear. At this time you need to make the appropriate structure to adjust the pattern. In the shadow of the edge of the package to take the edge line, to strengthen the contours of the pattern to avoid being stretched around the Jacquard organization. Jacquard can be used when filling the central CAD software to fill the organization's library, quickly fill Jacquard, and then for intensive.

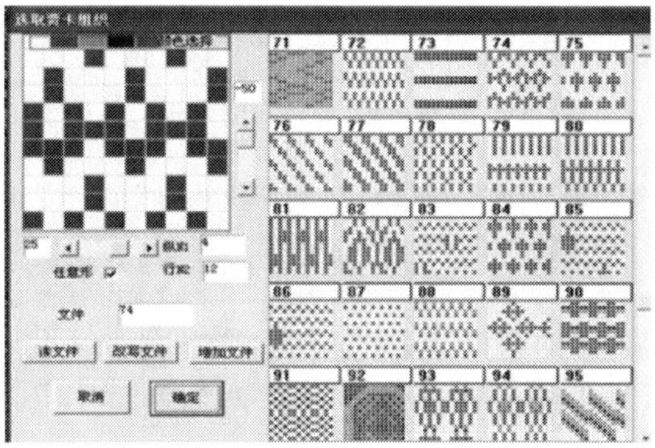

Fig.12 Jacquard filling stitch library

Fig.13 multi-bar Jacquard complete process map

3.6 Process Inspection

Check whether the bar in the same traverse line is a striker. Check whether the striker can be used in Central China CAD system analysis of the use of aggregation, as shown in Fig.14. Select the existing machine cluster relationship, such as manual input XGHM63 / 1 machine cluster relationship, and then click Save to facilitate the next Use, this function is currently the most advanced lace software. If there is a striker situation, which requires the striker bar and the other bar to the exchange, until the bar is not adjusted to the bar so far.

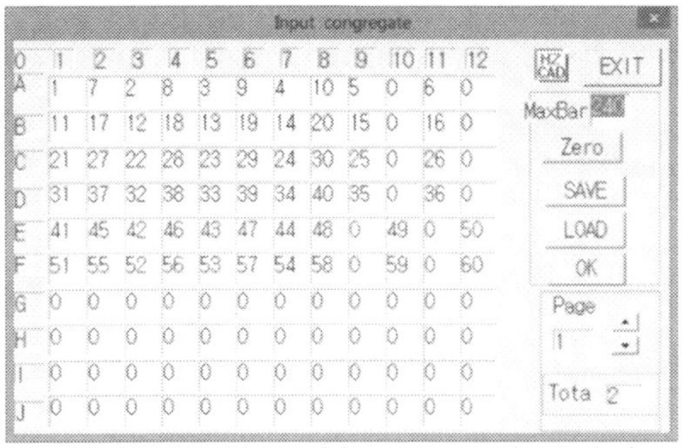

Fig.14 XGHM63 / 1 machine aggregation relationship

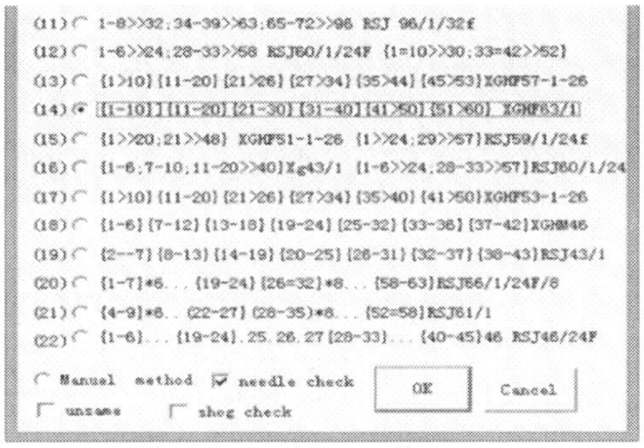

Fig.15 checking striker

Fig.14 shows the machine aggregation relationship, that means the arrangement of bar setting according to machine model. It depends on total number of bar in the machine, depends on total number of bar in the machine each group contains 6 or 7 bar.

Fig.15 shows the automatic bar arrangement, that means according to machine model the bar will arrange automatically and maintain the minimum distance between two bars of each group to minimizing the needle collusion. Needle collusion means the overlapping of two needles. When the minimum distance between two bar of each group not maintain that time the needle collusion will occurs two bar of needle. The minimum distance between two bars of each group is 3 or 4 and it must be maintain during making design to remove the needle collusion.

Each bar has a gauze range, and if the bar moves beyond its padding range, an error occurs in the knitting. Also check whether the location of the package line is row in front of the adjacent bar, so as not to be covered by other yarns.

To the ground bar, spandex bar, Jacquard bar after painting in the process, the last thing to do is to export the chain block file and pattern file, thus the entire lace design and process design process is over.

3.7 Lace Three-dimensional Effect of the Embodiment

The feature of the press yarn is that the yarn is connected with the ground tissue only at the two ends of the extension yarn. The extension line of the coil is not pressed by the chain organization but floats on the fabric surface, so the three-dimensional effect is very good as shown in Fig.16. In addition, since the press yarn is not knitted by the hook, thicker yarns can be used, movement using open or closed mat yarns, various patterns, and a strong three-dimensional feeling [12], which can be distinguished from other tissues. Three-dimensional pattern yarn specifications for the 280D-480D. Shadow patterns are usually used to show a thin fine silk, from the foil role, so that the main pattern of visual three-dimensional sense of a stronger [13].

Fig.16 Stereo effect of the press yarn pattern

4. Conclusion

Today's lace market needs product innovation. Lace effect map process design will inevitably become a multi-bar Raschel lace process design trends. Unique design and perfect technology to achieve and advanced finishing technology combined to make lace with more refined and perfect. Faster catch up with today's international lace fashion for the development of China's lace industry to provide world-class standards. In summary, we can consider some issues. Lace effect diagram design methods, this paper introduces the Raschel lace renderings of the specific patterns and abstract patterns, and the transition between them, according to the different lace style select the most appropriate design method to complete the yarn direction. Lace effect of process design without physical reference, this paper focuses on the effect of the carding of the general rules, models, flower width and width of the process parameters such as the determination of flower-type structure modification, Jacquard drawing and the process of machine inspection. In addition it also discussed the different models, raw materials, structure, and the overall style of the lace.

References

[1] Liu JJ, Wu ZM, Chu YM. RJPC4F Jacquard warp machine new jacquard process principle. Shan Text Sci T: 2006; (11): 1-2.

[2] Jiang GM, Xun WF. Has been developed in the Raschel lace machine. Text R: 2005; (8): 2-3.

[3] Liu L, Zhao C. The development trend of domestic warp lace production. Text Appar W: 2008; (12): 1-1.

[4] Liu JF. Vigorously develop eco-textiles to actively deal with green barriers. CIQ: 2006; (12): 2-3.

[5] Yan JJ, Wu WW. Discussion on green design of children's clothing. Text Text ST: 2010; (1): 4-5.

[6] Yu J, Jiang GM. Raschel lace flowers and geometric patterns combined design. STT: 2011; (9): 4-4.

[7] Lei JX, Liu JM. Application of Design Elements of Midpoint, Line and Surface of Decorative Pattern Design. Art Des: 2007; (6): 6-6.

[8] Jiang GM, Miao XH. Transitional design of multi-bar Raschel lace. CC: 2010; (1): 3-5.

[9] Zhao Y, Jiang GM. Lace design "permutation and combination of" magical [J]. CC: 2008; (9): 2-4.

[10] Wu ZM, Fan JJ. Rachel lace traditional auspicious peony pattern design [J]. Silk: 2014; (12): 5-5.

[11] Zhong MW, Wang LP. Lace fabric production process research [J]. Knitting industry: 2014; (8): 2-3.

[12] Wang CL, Xia FL, Miao XH. Method of realizing multi-bar Raschel lace effect drawing. Shan Text ST: 2010; (11): 2-5.

[13] Zhang T, He KJ. Pattern Design and Process Realization of Multi-bar Raschel Lace. J Zhe Text GTC: 2010; (2): 4-5.

Traditional Chinese Auspicious Patterns in Costume Design

Wei Yin*, Jian-Gang Wang, Yan Wang

Faculty of Clothing and Design, Minjiang University, Fuzhou, Fujian, China, 350121

*Corresponding author's email: wei_e80@126.com

Abstract

The traditional Chinese auspicious patterns are valuable to Chinese national culture. With the growth of Chinese style costume design and fashion worldwide, many of these patterns are used by designers from different countries. It is important to discuss and compare the pattern on traditional and modern Chinese costumes and also whether it is inherited or innovated.

Keywords: Auspicious Patterns; Costume Design; Chinese Style; Application

As a native Chinese art, traditional Chinese auspicious patterns are well-known for their profound culture and are widely used in almost every aspect of social life especially in the domains of costume design. It is worthwhile to understand the art dicipline and its development before it is used to promote Chinese fashion culture.

1. The Origin and Characteristics of Chinese Traditional Auspicious Patterns

Research show that traditional Chinese auspicious patters started from witchcraft and is the totem of a primitive society. In order to pursue good fortune and avoid disaster, sorcerers created unique symbols for sacrifice and prayer. These symbols evolved into auspicious patterns like Tai Ji and gossip. Simultaneously, the clans wish their patron saint to protect their people from enemies and natural disasters. Therefore, the original ancestors usually used animals or plants as their clan's titile of honor, and attached many nice and auspicious tales to them which we call totems. The auspicious patterns like the well known dragon, phoenix, and Kylin are all used as clan totems. Fig.1 shows the pheonix totem pattern found in ancient Chu kingdom relics.

Fig.1 Phoenix is the main totem of ancient Chu.

After much development, traditional Chinese auspicious patterns became an important part of Chinese culture and art because of their numerus varieties, deep ideological content and unique expressive force. These patterns show distinct individuality and characteristcs for their special usage and extensive feasibility.

The patterns can be divided into two typical series. is the first are patterns with implied meaning [1]. Different from simple paintings, each figure in traditional auspicious patterns has specific symbolic meaning. For example, pomegranates have many seeds inside and calabashes grow continously, so they are often used to symbolize family's flourish. Pine and cypress trees are symbols for longevity for their long lifetime. Peony flowers usually represent riches and honor for their beauty and grandeur. Plum blossom stands for pure and lofty character.

Some patterns combine flowers, fruits, birds and other animals together and then give them a nice meaning. For instance, plum, orchid, bamboo and chrysanthemum are called the "Four Junzi Flowers". Pine, bamboo and plum as drawn in Fig.2 are given another name the "Three Cold-weather Friends". Some glorious creatures are combined into one new image. The Chinese gods Fu His and Nue Wa's images are of human's head with snake body. These images enhance the artistic charm and aesthetics of the auspicious patterns, broadening their range of application.

Fig.2 The famous auspicious pattern "Three Cold-weather Friends"which stands for noble character and loyal friendship

The other type of patterns are using Chinese homonym characters when desgning a picture. Using a logo or symbol with a particular meaning, the symbol can take advantage of Chinese partial tone. For example, "deer" is a homophone for "emolument" in Chinese, while "elephant" sounds like "luck". "Fish" and "surplus" are homonymic in Chinese, so are "lotus" and "continuously" [2]. A pattern with a carp swimming in a lotus lake means rich and surplus in the successive years. In addition to these homophonic, a lot of patterns also utilize metonymy, puns and many other forms of rhetorical methods of Chinese language.

2. Auspicious Patterns and Chinese Ancient Costume

Compared with western ancient costume, Chinese ancient costume pays more attention to decorative patterns. This difference is due to their different culture in aesthetics and history between eastern and western.

First, the patterns are for the aesthetic reasons. In view of the west, the human body is the most beautiful thing, all clothing styles are to show and highlight the beauty of the human body. Western style clothing designers place emphasis on modeling. They consider fashion to be closely linked with the human form, and create the three-dimensional cutting method of clothing design. On the contrary, the aesthetic concept of the Oriental, believe clothing is not used to show the beauty of the human body, but to cover the body. Because of the influence and restriction of Confucian culture, fashion in ancient China is kind of implicative beauty. In order to show the inherent spirit, identity, and culture, patterns are considered to be important. In addition, the typical Chinese cutting method is plane cutting which made clothing less fit than the three-dimensional method. Therefore, using the patterns will make up for this deficiency.

Secondly, in the historical culture, publicity and self-expression Westerners focus on individuality, relatively less linkages between clothing, political, and social hierarchy. In contrast, in Chinese history, clothing has not only its natural function but also political function. For instance, the clan system had different grades to maintain the ruling class patriarchal. Since the Zhou Dynasty the officials had started to use clothing to distinguish different levels in the etiquette system. At the meantime, clothing decorative patterns have become a distinctive mark of social status and personal rating.

For example, in the Ming and Qing dynasties, the mandarin square of an official's clothing was used to distinguish the identity grade. The mandarin square is dressed on the chest and the back of a robe, with gold, color embroidery, rare fowls, propitious animals, and auspicious patterns. The patterns are usually in square shape with fowls patterns for civil officers and beasts for military officers. This is because fowls usually mean civilization while beasts express might. Each pattern corresponds to an official rank and cannot be used casually. For instance, in the Ming Dynasty, the highest official rank in imperial China uses the crane pattern, the second grade uses caragana, the third grade uses peacock, and the fourth uses wild goose. For the military officers' robes, patterns with lion, tiger, leopard, and bear are used for different ranks in a descending order [3]. Fig.3 shows the crane pattern on a mandarin square.

Fig.3 The mandarin square for the first rank minister of Qing Dynasty with a crane pattern which is exhibited in Metropolitan Museum of Art.

In thousands of years of Chinese history, all kinds of auspicious patterns were widely applied to various folk costumes under the influence of the ruling class and the culture of Confucianism. For example, "Five Fu Offering Birthday Congratulations", "Pine Trees and Cranes, Symbol of Longevity", "God of Longevity", "Fu Charactors" and such other patterns were generally used in old people clothing. "Goldfish Swimming" and other lucky and beatific patterns were used in children's clothing. The contents and forms of the auspicious patterns progressed well because of their extensive usage in costumes.

3. The Application of Traditional Auspicious Patterns in Modern Costume

Chinese traditional auspicious patterns become one of the most typical Chinese elements in costume design, for its artful conception, profound connotation, rich content, beautiful designs, national colors, and unique charm. The combination of auspicious patterns and fashion shows us not only the beautiful patterns but also a profound history and culture of the Chinese nation for the presentation of the optimistic and unyielding spirit.

Auspicious patterns are highly decorative. A ordinary clothing can become beautiful and charming with just one single suitable auspicious pattern. Patterns in Fig.4 are of typical Chinese style. For example, a variety of Medallion Flower patterns were produced in the Chinese Northern and Southern Dynasties, which have over one thousand years history. Some domestic apparel enterprises design and sell Tang series products with Medallion Flower patterns. They are well received by the market. In 2001, in the Asia Pacific Economic Cooperation of Shanghai APEC summit, the leaders of various countries appeared wearing such Tang suits together, which was a highlight of this important worldwide meeting and an instructive anecdote. After this, wearing Tang suit became a kind of fashion. Every decorative design or auspicious pattern on women skirts of Miao and Yao nationality in Yunan and Guizhou province represents a certain meaning and records the ethnic origin, migration, social life, and history. These national costumes are sold on the market as tourist souvenirs. It not only created good economic benefits, but also greatly promoted local tourism.

Fig.4 Wu's "What is Chinese" uses typical Chinese embroidery and auspicious patterns.

Chinese traditional auspicious patterns attract more and more attention from international garment industry. Utilizing Chinese traditional auspicious pattern as a partial decoration has already become one of the Western designers' commonly used gimmick. Some typical examples are "Chinese style" designed by Kenzo, the top hat, mandarin jacket, long gown of the Qing Dynasty designed by Yves Saint Laurent, the cornice and rake shoulder sculpt of Pierre Cardin, and Valentino's "Impression of China", etc. Additionally, embroidered shoes, bags, jewelry, and other products with auspicious patterns like mandarin duck, butterfly, magpie, plum, dragon, and phoenix are popular in China and other countries for the use of Chinese traditional auspicious patterns, which also introduce the Chinese culture to the world.

4. What Should Be Noticed While Applying the Patterns

In order to properly integrate the auspicious patterns into the fashion tide and make them become the "special language" of Chinese fashion, there are a few things need to be considered.

Fig.5 "Fu" character is the mostly used traditional auspicious pattern often seen on elderly clothing.

First, designers should absorb the quintessence of the traditional auspicious patterns and discard the dross. The auspicious patterns have both the decorative function and educational purpose [4]. The latter exerts a subtle infuence on people's character. Because the traditional auspicious patterns are the historical culture for thousands of years, there are some bad elements, such as "the More sons the More Happyness" which advocates gender discrimination. Therefore, when selecting a pattern, the designer should understand the internal content the pattern expresses, and then choose the pattern which best reflects positive Chinese national culture, and eventually transform it to a new one that best fits the new ages.

Second, a proper pattern selection must be based on both the type of the clothing and the customer. There are a variety of contents and forms of the auspicious patterns, which should be used in different conditions. For example, "Five Fu Offering Birthday Congratulations", "Pine Trees and Cranes, Symbol of Longevity", "God of Longevity" are appropriate patterns when used in elderly clothing and will be obviously odd when used in infant clothing. Similarly, " Friendlies" and "Tiger Head" are suitable for children's clothing and never should be used in elderly clothing.

Third, the style of the pattern should properly fit the style of the costume. The latter always affects and limits the former. For instance, looser clothing has more space for ornament, so the patterns may be larger in size and can be arranged more freely to fit the relexation style that the whole clothing wants to express. The slim-fitting clothing shows the beauty of human body, so the decorative patterns must not be so big as to steal the show, and their images should suit the feature of the human body.

The last but not the least, innovation should be encouraged. Adding auspicious patterns to a clothing mechanically will not create a fresh innovative design. Learning from traditional costume does not mean indiscriminately imitating either. The invention spirit of art inside Chinese nation is the essense that we should learn [5]. Technique like deconstruction and recombination are often used in design masters' works, which can bring an old pattern a new life.

5. Conclusion

The traditional Chinese culture and folkways contained in auspicious patterns bring profound enlightenment and profitable use of reference to modern costume design. The pattern of manifestation and artistic charm are very well appreciated by people of different countries.. Inheritance is needed for innovation. While discovery is critical for inheritance. The in-depth understanding of the essense of the auspicious patterns can improve our perception of creation and beauty-appreciation.

Today, Chinese style has ever increasing influence on the fashion tide. Designers should learn more from Chinese tradition art and emulate the creative spirit of Chinese ancients. The traditional skills of art and modern design methods can be combined together and improved. More and more works with Chinese characteristics can be created in this way, which will definitely influence the trend of costume design and fashion.

Author Introduction

Wei Yin (1980-), female, Master of Design Arts, Born in Hunan, China. Vice professor of Fashion and Art Engineering College of Minjiang University.

This manuscript is a research result for these projects: Pattern Design and Process Research Based on New EL Multi-bar Raschel Warp Knitting Machine, Project of Fuzhou Science Plans (Science and Technology Bureau of Fuzhou 2015-G-66); Research and Construction of Costume Design Major Courses System for Practical Undergraduates of Minjiang University (No. JAS151320); The Academician Workstation for Textile Research Institute of Minjiang University (No. 3140420402).

References

[1] Su J. The Chinese Traditional Auspicious Patterns' Artistic Expression in Modern Costume Design. J Design and Product: 2006; 6: 15-16.

[2] Li SS. Discussion on the Usage of Auspicious Patterns in Modern Costume Design. J Popular Art: 2008; 175-176.

[3] Ni JC, Qian X. Influence of Chinese Traditional Costume Culture on European Fashion Brands. Journal of Textile Research: 2006; 7: 27-28.

[4] Teng HJ. The Application of Auspicious Patterns in Traditional Costumes. J Liaoning Silk: 2005; 1: 32-33.

[5] Zhou F. On the Value of Traditional Chinese Auspicious Pattern for Modern Design. J Decoration: 2007; 3: 105-106.

Analysis of the Application of Transformable Elements in Fashion Design

Jia-Ai Zhu*, Hui Tao

School of Fashion, Wuhan Textile University, NO.1Textile Road, Hongshan District, Wuhan, Hubei, 430073, China

*Corresponding author's email: zja19931116@126.com

Abstract

Transformable clothing is able to change itself with the help of special fabrics or methods after being affected by the surroundings or the wearer. It is a new concept of clothing that have appeared in recent years. This paper mainly discussed the application forms and methods of transformable elements in fashion design. The concept of transformable clothing, the inspiration and the application of transformable elements are introduced in the present paper. Major part of the study encompasses literature review, web surfing, blogs and article readings. The analysis of the transformable elements and the research of transformable clothing design all over the world are used to study and figure out the relationship between innovative design and the development of science and technology. The paper preliminarily discusses how technology provides fashion design with new thoughts, methods and approaches by learning from the examples of transformable clothing design based on technology and innovative ideas. It can be demonstrated that innovative design has a mutually reinforcing relationship with the development of science and technology.

Keywords: Fashion Design; Transformable Clothing; Intelligent Clothing; Technology

1. What Is Transformable Clothing?

Transformable clothing, as its name implies, is a kind of clothing that can change itself with the help of special fabrics or methods after being affected by the environment. It is a new concept of clothing that has appeared in recent years. It is designed to find a better way to convey human emotions through fashion design, satisfying people's profound pursuit of aesthetics and provide fashion design with new thoughts and methods by combining the innovative design and science and technology together. This paper mainly discusses the application forms and methods of transformable elements in fashion design. It assimilates transformable elements from external feature or internal structure of natural ecological environment, innovation of science and technology, social culture and reflection of the changes in human emotions. Meanwhile, this paper classifies these elements and applies them to the ideology and visual style of the integral design and the style of the clothing.

2. The Inspiration of Transformable Elements in Fashion Design

2.1 Inspired by Natural Ecological Environment

In the design of transformable clothing, some of the transformable elements come from the original form of natural creatures and environment. Others rely on the expansion of design according to the dynamics of the natural creature and environment, in order to attach a new definition to the original. These elements will be integrated with the concept of fashion design and style and will enforce the expression of clothing.

2.2 Inspired by New Science and Technology

With the continuous innovation of science and technology, some new forms and methods of fashion design are emerging. The development of this high technology provides more possibilities for creative design of transformable clothing. The integration of technology and design concept translates more new design works into reality. The application of intelligent devices, special fabric and fiber, LED light-emitting diodes and other scientific and technological means enable the clothing to change its color or silhouette after being affected by the environment. In this way, it is possible to convey the wearer's emotions better, enrich the clothing styles, realize the new forms of artistic expression and find other innovative conception of fashion design.

2.3 Inspired by Human Emotions and Feelings

The design concept based on human emotions in the design of transformable clothing shows the concern and care for ourselves and others, which has been continuously developing in the field of fashion design. In this increasingly competitive society, the life of people is getting more and more stressful. It is important for humane design to improve interpersonal relationship, promote the transmission and expression of emotion, satisfy the emotional needs of people, and ease all kinds of psychological stress of modern people. Many designs of transformable clothing based on changes in human emotions are created by means of detection of sound, temperature, heart rate and other data of human body, enabling clothes to change accordingly to their visual properties such as form, color and pattern.

2.4 Inspired by Society and Culture

In the design of transformable clothing, there are often some innovative works based on the current social and cultural hotspots. As a mirror of social life, the design and style of clothing reflect the social and cultural trends of a certain period. Culture is not only the essence of an era, but also the source of the power for fashion design. New ideological trend of social culture, new trend of social movement and others will pass fashion information and affect all segments of society to a certain extent, at the same time provide designers with creative stimulation.

3. Application of Transformable Elements in Fashion Design

Since the appearance of the transformable clothing concept, there have been a lot of creative transformable clothing designs all over the world. Based on the analysis of the existing design cases, the following passage will analyze the transformable elements from two aspects: the analysis of transformable design elements and its application in fashion design.

3.1 Analysis of Transformable Design Elements

3.1.1 The Transformation of Clothing Silhouette

In the design of transformable clothing, some fashion designers directly imitate the shape of changing things to design and create. The silhouette of the clothing is changed after transformation, which may convey some unique design ideas and enrich the style of clothing. For instance, Hussein Chalayan combined clothing and furniture, then the transformations is displayed in his autumn and winter After Words series in 2000 (Fig.1). A table can be transformed into a half length skirt, the sofa cover turned into an elegant dress and the sofa changed into a suitcase [1]. Four models in simple gray dresses approached to stage near the chairs and started removing the covers of the chairs. After removing the covers, they started putting them on their bodies and eventually the covers became dresses. These covers were perfectly adjusted to their bodies. Finally the last model stepped inside the simple looking coffee table, she lifted it up and the table turned into a skirt. The inspiration of this collection was the refugees of war who are forced to flee their homes and leave their assets behind during the area of 60's and 70's. The outcome of this collection is the portable and wearable furniture.

Fig.1 Hussein Chalayan After Words A/W 2000
(From http://www.beautimode.com/article/content/10758/)

Another typical case in the transformation of clothing silhouette refers to the design of Issey Miyake. [2] In his most famous autumn winter 2012 collection entitled "132 5", Issey Miyake exhibited a range of transformable shirts, skirts, dresses and pants made of single piece fabric without any cut and stitching. These transformable dresses are flat origami style folded and turn to wearable when unfolded. It's a combination of mathematical techniques and permanent folding as well as origami and green fashion.

3.1.2 The Transformation of Clothing Color

The transformation of the clothing color in transformable clothing design is often associated with the wearer's emotions and feelings. It conveys the physical changes of the person through data such as the wearer's heart rate, voice, body temperature and other physiological data detected by the intelligent device in clothes. Through collecting these data, the clothes change accordingly to a specific color. This transformation of clothing color is often achieved by light fiber, intelligent foil and other special materials. For example, Holland designer Daan Rooseqaarde designed the Intimacy 2.0 intelligent clothing (Fig.2), which used leather, transparent intelligent foil and the light emitted by the LED to achieve the masking effect. [3] The clothing is internally connected with a sensor capable of detecting heart rate. When the feelings of wearer change, the color of clothing will transform simultaneously. When the heart rate of BPM reached a certain point, LED and other devices will shut down automatically, clothing will instantly become transparent. The purpose of this design is to convey the human emotion and express the intimate relationship through the change of the external color. At the same time, this design method and technology are also used for polygraph work.

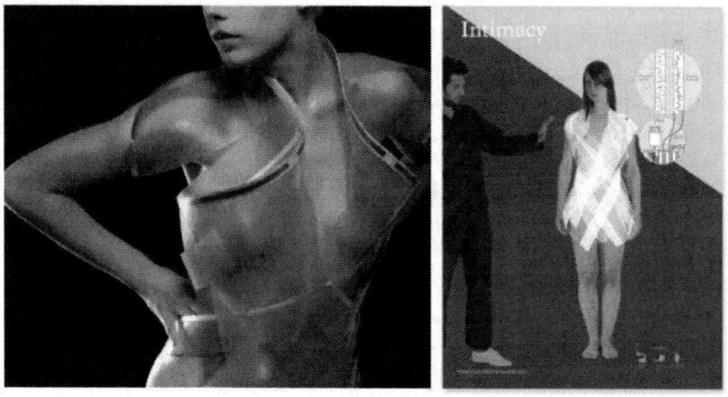

Fig.2 Daan Rooseqaarde "Intimacy 2.0" intelligent clothing
(From http://man.vsuch.com/2015/happy_0812/12928.html)

Another example for the transformation of clothing color is a London-based experimental design studio "The

Unseen" founded by Lauren Browker. [4] Unseen refers to the climate around us like air, heat, temperature and pollution etc. Lauren invented some inks that can react to the climate and transforms from one color to another accordingly. Her invented inks or dyes can react up to seven stimuli in the climate such as friction, sound, pollution, moisture, heat, chemicals and UV. Every of these stimuli have different color effect on the same surface. By using materials "the Unseen" studio integrates chemistry, biology, and electronic tech into amazing fashion design products. (Fig.3)

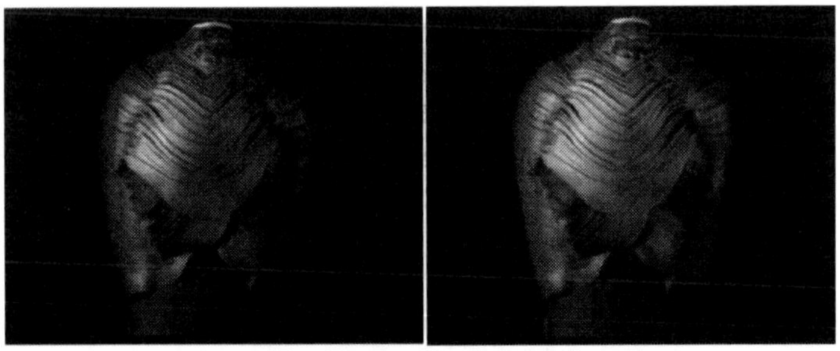

Fig.3 "The Unseen"by Lauren Browker
(From http://henan.163.com/14/0815/09/A3MAE69E02270J7F.html)

A number of designers are working on sustainability and environmental issues. Dahea Sun, a London based designer, combined fashion, science and sustainability in a different way. She created color-changing pH-sensitive dyes and the collection Rain Palette. [5] It aims to provide an easy and poetic approach to visualizing air quality through rainwater. The garments she has developed are dyed with natural dye that will change color from light gray to pink in reaction to the pH levels of rainwater. (Fig.4) This project provides an at-a-glance indication of atmospheric air quality, with the potential for wearers to record and upload rain pH readings online to create a global database of real-time environmental data.

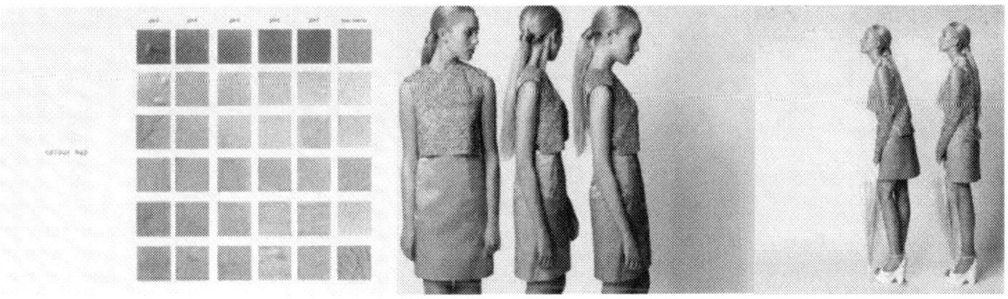

Fig.4 Rain Palette by Dahea Sun
(From http://www.morethangreen.es/en/rain-palette-by-dahea-sun/)

3.1.3 The Transformation of Pattern and Texture

In the design of transformable clothing, the change of pattern, texture and printing is very common. Everything in the world is composed of different patterns and textures, which can be used in fashion design to bring about more artistic effects for clothing. By the means of using transformable elements, the texture is no longer a monotonous pattern, which adds to the sense of art and aesthetics as well as a lingering charm on the clothing which can stimulate people's unconstrained imagination. (No)where And (Now)here series(Fig.5)designed by Canadian artist Ying Gao combined light emitting fiber, eye tracking and voice technology, which will transform in the flow of light or pattern when other people look at the dress or speaks.

She tries to express that clothes can also be animate. In another one of her design series Incertitudes (Fig.5), the dress is covered with pins and sensors. These sensors can be activated by the voice of other people just like they can communicate with human beings. The dress can transform based on the intensity of the sound received, this can lead to the shrinkage or expansion of the clothing. [6] This thoughtful dress is designed to explore the uncertainty of the world and future.

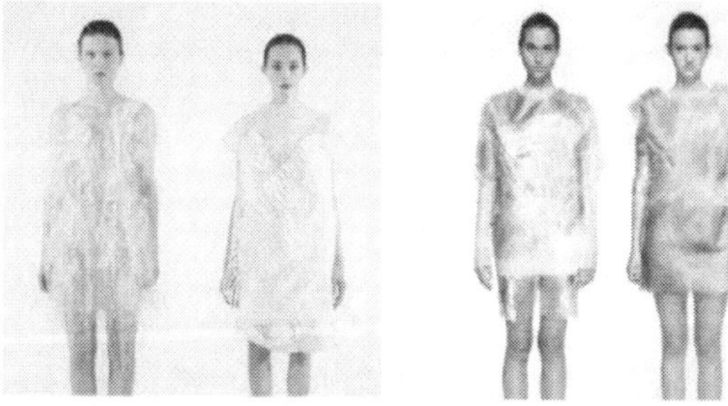

Fig.5 Ying Gao(No) where And (Now) here and Incertitudes series
(From http://radarcollective.com/fashion/nowhere-nowhere/)

3.1.4 The Transformation of Style and Structure

Attaching transformable elements to the style and structure of clothes is one of the common practices of transformable clothing, which greatly enriches the diversity of clothes, providing a piece of clothing two or more different styles and bringing new ways of thinking for fashion design. The transformation of style and structure is usually realized by the exquisite cutting and structural design, through different segmentation, three-dimensional modeling to achieve the effect of innovative design. Hussein Chalayan's Two-in-one dresses series (Fig.6) applied transformable design elements to achieve dress changes in a second [7]. As the model pulls down the collar of the transformable clothing, the romantic elegant long skirt instantly replaces the vest dress and achieves the gorgeous transfiguration of the wearer.

Fig.6. Hussein Chalayan Two-in-one dresses Autumn 2013
(From https://www.dezeen.com/2013/03/06/rise-by-hussein-chalayan/)

There are also a lot of examples for the transformation of garment style and structure. Such as the black and silver sheath dress designed by Sebastian Errazuriz, that can be converted as many ways as the wearer want. [8] This dress is made up of recycled zippers that can be attached and detached to change the look from a classic to sexy looking dress.

3.2 The Application Methods of Transformable Elements in Modern Fashion Design

3.2.1 Total or Partial Application

Transformable elements can be applied in the whole clothing. Transformable elements can also be used in a part of the clothing. This means that transformable elements are used in a certain part of the design. The butterfly dress designed by designer Ezra and Tuba has imitated fabric butterflies on its shoulder, which can preen their wings when the people around is detected. [9]

3.2.2 Direct or Indirect Application

Direct method refers to the repeated use of transformable elements in fashion design so that people can directly come in contact with the artistic visual affect, it displays the changes in clothing style strongly and the transformation can be realized many times. By using the indirect method of transformable elements in fashion design, clothing can transform imperceptibly. The change of the clothes style can be achieved through one transformation and does not restore to its original shape.

3.2.3 The Application of Color and Fabric

The application of color and fabric in the design of transformable clothing refers to the use of some new fabrics. The fabrics have special properties developed by science and technology to achieve the changes in color of the clothing, the structure and other aspects. By using them, designers can achieve the desired effect that conventional fabrics cannot achieve. For instance Hussein Chalayan's design of water ablation clothing, which used the soluble paper fabric and the shower device arranged on the runway. The paper windbreaker gradually disintegrated into an Indian palm group cocktail dress through water erosion. The transformation in styles and patterns of the clothing is realized by using the special fabric.

4. Conclusion

The application of transformable elements in fashion design reflects people's thoughts, methods and approaches in innovative fashion design. It also gives new ideas on how to create clothing that have more vitality and help people to communicate their emotions in the pursuit of combining technology and human nature [10]. Few dresses among the above-mentioned examples are not wearable, but they still give effect of transformability and convertibility. The diversified design of transformable clothing continuously puts forward new requirements for science and technology. A powerful and unconstrained design requires the continuous development of science and technology. Although modern science and technology reflects only a certain type of new high technology methods, it has a great potential for further innovation and breakthrough in changing thoughts notions and behavioral styles. Therefore, it is worth thinking on how to combine fashion design with science and technology together properly.

References

[1] Zhang H. Walking Space: An Interpretation of the Fashion Design Art by Hussein Chalayan [J]. Decoration: 2013 (7); 121-122.

[2] We LN. "132 5" by Issey Miyake [D]. Beijing Institute of Fashion Technology: 2015.

[3] A Garment can change its Transparency based on Heart Rate Change [J]. Melliand China: 2016 (7); 68-69.

[4] Perfect Combination of Technology and Fashion: Lauren Bowker created an Amazing Fabric by using Inks that can react to Climate [EB/OL]. http://www.sohu.com/a/114812976_500095, 2016-09-21/2017-03-25.

[5] Mora V. Rain Palette by Dahea Sun [EB/OL]. http://www.morethangreen.es/en/rain-palette-by-dahea-sun/, 2014-05-14/2017-03-21.

[6] Nettelbladt S. Fashion: nowhere-nowhere [EB/OL]. http://radarcollective.com/fashion/nowhere-nowhere/: 2017-03-15.

[7] Lau C K. Exhibition Review: Hussein Chalayan: From Fashion and Back [J]. Fashion Theory: 2011; 15 (1): 93-98.

[8] Various Zipper Dress Skirt [EB/OL]. http://houshidai.com/creative/sebastian-dress-design.html: 2012-04-8/ 2017-03-21.

[9] Kimani N. Ezra + Tuba took on Tech and gave Fashion Butterflies [EB/OL]. http://tdsblog.com/ezra-tuba-took-tech-gave-fashion-butterflies/: 2016-10-10/2017-03-25

[10] Li MY. Clothing Science and Technology and Fashion design [J]. Beauty and Times: 2012 (5); 106-108.

Evaluation of Functional Clothing Design for Posture Improvement in Young Females

Chao Shi, Rong Zheng*, Kasia Lee, Ying Wang

Aimer Human Engineering Research Centre, Beijing Institute of Fashion Technology, Beijing, 100029, China

*Corresponding author's email: yuki.zheng@gmail.com

Abstract

Prolonged deskwork-related posture with a forward head and rounded back causes muscle strain, neck and shoulder pain and other musculoskeletal disorders. Therefore, it is important to prevent and alleviate shoulder and neck pain through proper alignment of body posture. The study conducted a literature review to identify and analyze medical concepts and theories on posture improvement. Based on the analysis of merits and shortcomings of existing products in the market, this study designed a clothing prototype with posture correction function. A comparison experiment was undertaken in which 24 female volunteers were selected and divided into groups to try on both the prototype design and a market-based apparel commodity. The experiment included three parts: posture evaluation and investigation of shoulder and neck function. Study results show that over a 4-week period, the functional clothing prototype design helped improve rounded shoulder posture and alleviated neck and shoulder pain and dysfunction more so than the market-based apparel.

Keywords: Posture Evaluation; Functional Clothing Design; Anthropometry; Ergonomics

1. Introduction

The digital era has brought about the so-called iPosture—forward head, hunched back and slumped shoulders [1]. The human mechanics studies the bones, ligaments, muscles work and function in relation to one another. Good posture can help maintain the smooth functioning of these interrelated parts. Improper posture decreases blood flow, compresses spinal bones, creates pain due to heightened muscle strain, increases risk of headaches and mood swings, and ultimately may damage the back and spine [2, 3]. Griegel-Morris found that compared to people with standard posture, people with improper posture tend to have relatively higher rate of ailments. A correlation was found between scapular pain and hunched back/rounded shoulder, with a more forward head angle resulting in a greater occurrence of neck and scapular pain and headache [4]. Studies show that as the forward head angle increases, the load on the cervical spine increases [5]. In addition, college students were found to suffer greater arm discomfort than people in the workforce due to prolonged use of computers. Moreover, a study determined that college students' poor posture, prolonged slouching over a computer, bad sleep quality and ignorance of cervical vertebra care were major factors that impacted their health [6]. Some researchers also found that bending over a desk is a cause of various musculoskeletal problems [7], and lack of proper treatment can make such health problems last a lifetime [8].

Treatment for neck, shoulder and back ailments caused by poor postures includes exercise, posture supports [9] and surgery. However, due to patients' lack of time [10] and discomfort during treatment [11, 12] among other reasons, they often abandon the process, thus lessening the treatment's effect [13]. Despite an array of health care and rehabilitation products in the market, there is a lack of clothing products designed to correct posture. Therefore, it is important to develop a posture correction clothing that is functional, more readily available and better suited for daily life. The study carried out an experimental evaluation of posture improvement clothing.

2. Literature Review

This section discusses evaluation methods related to functional clothing design and posture improvement to provide a basis for the study's experiment on functional clothing evaluation. Posture correction products in the market can be divided into three types: rigid, semi-rigid and soft-type products.

Rigid products mainly include medical braces, such as scoliosis braces or orthoses. These products are more restrictive and impede movement when worn, and thus are not suitable for everyday wear.

Semi-rigid products include SpineCor's Dynamic Corrective Brace. Other adjustable products include orthopedic belts and waist protection belts. Among domestic products, Babaka (a back and posture correction support) is a superior example. Babaka claims to treat and prevent slumped shoulders, forwardhead, hunched back, myopia, uneven shoulders and scoliosis. However, no study has proven the product's corrective effect.

Soft-type products have the function of improving the posture and shape of the bodyExisting clothing with such function include shaping underwear and tights, which are usually made from flexible knitted fabrics that are more comfortable compared with rigid and semi-rigid products. The underlying principle of such posture correction clothing items is that by providing a combination of pressure, torque and angle through tension bands or structural design, the apparel can provide neuromuscular stimulation and, alter the wearer's posture by re-training muscle memory[14]. The disadvantages of this type of apparel reflected in evaluations mainly include tight fit, prickly or itchy feeling, being too tight under the armpit, and having a long torsal part. After structural analysis of the specific apparel, it was found that the pricky and itchy feeling was caused by seams.

3. Material and Methods

Current studies disagree on the effectiveness ofposture correction clothing, and an investigation on both postural evaluation and head and neck function has yet to be conducted Therefore, this study sought to fill such a gap. This section discusses the experiment design, participant selection and data analysis. The study selected 24 female volunteers for a 4-week-long comparison experiment divided into three parts: posture evaluation, investigation of shoulder and neck function, and subjective comfort evaluation. At the end of the experiment, a questionnaire was administered to volunteers to get the wearers' subjective evaluation of the tested apparels. The subjects are female students and working females aged 18-28 who had never had any spinal surgery and who experience suffer neck and shoulder discomfort and/or pain. They were equally divided into two groups; 12 in the prototype (PT) group, and 12 in the commercial product (COM) group.

3.1 Test Garments

The two apparels selected for the experiment were both soft products as shown in Fig.1. Fig.1(A) is the prototype design with a main body fabric made of 79% polyester fibre and 21% spandex, and side panels using a mesh fabric made of 82% polyester fibre and 18% spandex Fig.1(B) is the Intelliskin Women's Foundation 3/4 Zip Tee with the following fabric contents: for the main body, 37.5% Cool Jade, 37.5% Nylon and 25% Spandex; for the grey contrast panels, 91% Polyester and 9% Spandex. Participants were instructed to wear the test garments for at least 2 hours in the first week, and at least 8 hours in the second, third and fourth weeks.

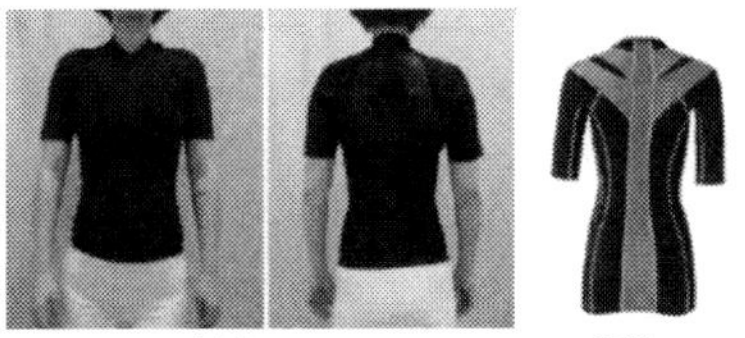

(A) (B)

Fig 1 Test garments: (A) prototype design; (B) Intelliskin Women's Foundation 3/4 Zip Tee.

3.2 Posture Analysis

There are two types of evaluations for standing posture: static and dynamic. Static posture evaluation methods include visual observation, plumbline, goniometry, photographic, radiographic, photogrammetric, flexiruler and electromagnetic tracking device [15]. Due to its high accuracy, quick and ease of operation, and being non-intrusive to the human body compared with radiography, the repeatable photogrammetric method is the most popular evaluation approach [16].

Thigpen et al. collected photographs of the neck and shoulder postures of 310 student volunteers and grouped them into two groups. The group with ideal posture had a forward head angle (FHA)≤36° and forward shoulder angle (FSA)≤22°. The forward head and rounded shoulder posture (FHRSP) group was defined as FHA≥46° and FSA≥52° [17].

Before the test, reflective markers were placed overvolunteers' right tragus, acromion and C7 spinous process. The volunteers, in bare feet, were instructed to bend and stretch three times, with the finger touching the toe while bending and arms extending up forward when stretching. During photographing, the volunteers were asked to look forward at a natural resting posture with hands placed naturally at both sides of the body. They would be photographed from the front, back and sides, respectively.

A digital camera was placed on a tripod 1m high and5m from the subject, with the camera focus set on the front of the neck. The digital images were uploaded to a computer forprocessing. Using Adobe Photoshop, the forward head angle (FHA) and forward shoulder angle (FSA) were measured. FHA is the angle between the vertical line passing the point C7 and the straight line connecting the tragus and C7. FSA is the angle between a vertical line passing through C7's and the straight line connecting C7 and acromion (Fig.2).

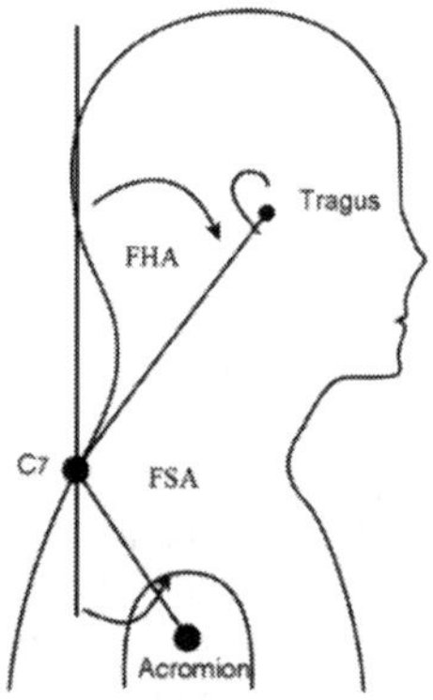

Fig.2 Measurement of FHA and FSA posture.

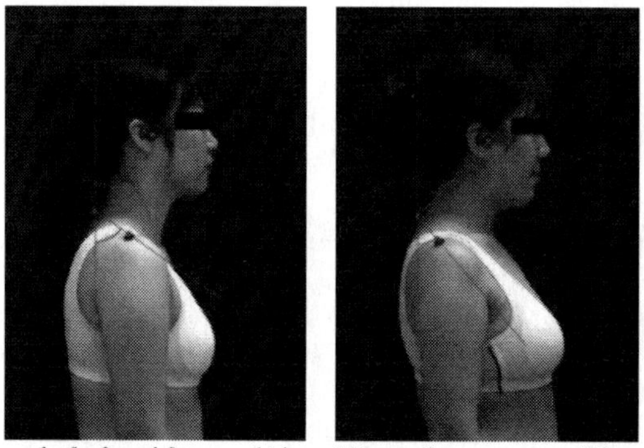
Rounded-shoulder (A) Normal posture (B)

Fig.3 Comparison of (A) rounded-shoulder and (B) normal posture.

3.3 Evaluation Test of Shoulder and Neck Functions

In order to learn about the volunteers' shoulder and neck health status, the study administered a questionnaire which included Neck Disability Index Table (NDI) and Shoulder Pain and Disability Index (SPADI).

NDI is a questionnaire widely used in the international medicine field for measuring neck pain. It includes10 items on different aspects of daily life, and the overall score reflects the extent of neck pain and functional disability. SPADI is 13 item self-report questionnaire divided into 2 subscales: pain and disability. The overall score is obtained based on the average of the pain score and disability score; the higher the score, the more serious the pain or dysfunction.

In designing the prototype, the researchers considered that the corrective posture apparel would be used for everyday activities. In order to learn about the effects of long-term wear on personal comfort, the study used a subjective questionnaire in the fourth (i.e., the last) week of the experiment to determine the 24 subjects' comfort evaluation of the test apparels. The questionnaire content includes wearer acceptability, thermal comfort, fit and social comfort.

4. Results and Discussion

The study recruited a total of 68 volunteers with an average age of 23.8 years, including 65 students and three working females, from which 24 volunteers were selected for the wear trial. The volunteers met the following conditions: FSA≥52° and the chest circumference in the size M category. The test results based on the collected posture photographs consist of two parts: 1) the instant effect of the tested apparel in correcting posture; 2) the long-term effect of the tested apparel in correcting posture.

4.1 Instant Effect

During the first day of the experiment, photos were collected of volunteers before wearing the test garment (nude condition) and 10 minutes after wearing the test garment. Table 1 shows that for the PT group, in the nude condition, the average FHA was 35.68° and the average FSA was 52.75°; 10 minutes after putting on the test garment, the average FHA was 37.8° and the average FSA was 43.23°. As for the COM group, in the nude condition the average FHA was 37. 8° and the average FSA was 55.5°; 10 minutes after putting on the test garment, the average FHA was 36.99° and the average FSA was 55.38°. For the PT group, after putting on the apparel, the average FSA was significantly reduced ($p<0.05$) by 9.52°, an improvement in rounded-shoulder posture. The COM group saw a 0.120 ° drop in FSA; no significant effect was found.

Table.1 Instant effect on FHA and FSA (unit: degree)

	FHA			FSA		
Group	Nude	Test garment	Difference	Nude	Test garment	Difference
PT	35.68	36.06	0.38	52.75	43.23	(9.52)
COM	37.77	36.99	(0.78)	55.51	55.38	(0.12)

4.2 Long-term Effect

The study collected FHA and FSA data of the 24 volunteers' data for before and 10 minutes after they wore the test garments. condition over four weeks.

Over the four-week period, the average FHA of the PT group dropped by 0.32±4.11°, the average FSA decreased by 4.13±10.46°; the average FHA of the COM group fell by 1.71±3.50°, the average FSA decreased by 5.20±7.97°. A significant effect was shown in the COM group ($p<0.05$); no significant effect was shown in the PT group.

Table.2 FHA changes during the four-week experiment (unit: degree)

		WK0	WK1		WK2		WK3		WK4	
Angle	Group	Nude	Nude	Test garment	Nude	Test garment	Nude	Test garment	Nude	Test garment
	PT	35.68	35.82	36.06	35.38	35.14	35.68	34.93	35.31	35.36
FHA	COM	37.77	37.16	36.21	35.70	36.99	36.39	36.10	35.57	36.06

Table.3 FSA changes during the four-week experiment (unit: degree)

		WK0	WK1		WK2		WK3		WK4	
Angle	Group	Nude	Nude	Test garment	Nude	Test garment	Nude	Test garment	Nude	Test garment
	PT	52.75	54.18	38.78	58.66	48.48	59.80	50.05	58.85	48.62
FSA	COM	55.51	54.45	50.54	54.40	54.83	52.28	51.01	54.74	50.31

4.3 NDI and SPADI Scores and Subjective Questionaire

The study also tracked participants' changes in shoulder and neck pain and disability via NDI and SPADI questionnaires during the four-week experiments. To eliminate errors of subjective scoring, a normalizing algorithm was used to obtain the normalized average score of NDI and SPADI.

NDI scores of both groups show an overall decrease. The overall average NDI score of the PT group dropped by 4.54±8.23%, and that of the COM group decreased by 1.46±7.47%. No significant effect was shown in the two groups.

Results show that in the PT group, the pain score in SPADI dropped from 33.64±17.83% to 31.15±23.75%; the disability score fell from 26.86±20.01% to 18.05±19.18%; the overall SPADI score decreased from 25.52±14.41% to 23.94±18.61%. In the COM group, the pain score in SPADI dropped from 30.85±37.03% to 30.85±37.03%; the disability score increased from 13.14±11.73% to 24.45±33.07%; the overall SPADI score rose from 26.83±32.84% to 26.83±32.84%. The prototype outperformed the commercial apparel in alleviating shoulder discomfort and dysfunction. However, there was no significant difference in pain, disability and overall scores for both groups during the four-week experiment.

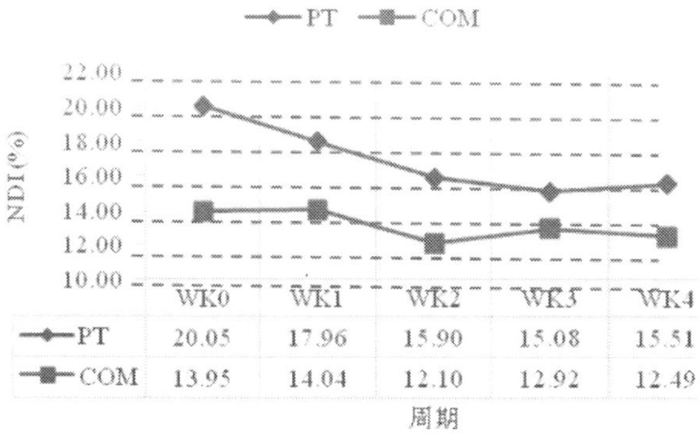

Fig.4 NDI changes during the four-week experiment.

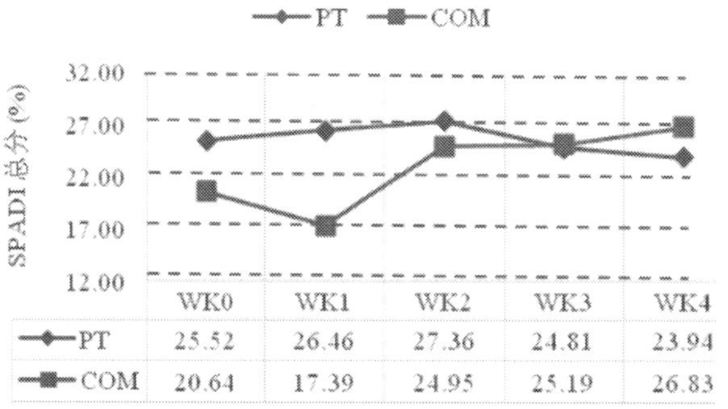

Fig.5 Changes of overall SPADI scores during the four-week experiment.

The final part of the questionnaire consisted of open-ended questions to acquire the subjects' psychological comfort evaluation on the tested apparels, covering such aspects as aesthetics, style, fashion, decency, function, design, colour, texture and fit effect. After data processing and analysis, merits and shortcomings of the two apparels were summarized. The 24 volunteers regarded the tested apparels as being comfortable to wear.

5. Conclusion

By combining the test methods used in both medical and clothing fields, the study investigated the function and comfort aspects of aposture-improvement apparel, Posture improvement assessment was conducted among the targeted consumer group via questionnaire investigation, photographing subjects' postures and asking them to try on functional clothing items. The study concluded that compared with existing products in the market, the prototype design showed more effect in improving posture.

References

[1] Fejer R, Kyvik KO, Hartvigsen J. The prevalence of neck pain in the world population: a systematic critical review of the literature. J Eur Spine: 2006; 15 (6): 834-848.

[2] Lawrence R, Helmick C, Arnett F, Deyo R, Felson D, Giannini E, Heyse S, Hirsch R, Hochberg M, Hochberg M, Hunder G, Liang M, Pillemer S, Steen V, Wolfe F. Estimates of the prevalence of arthritis and selected musculoskeletal disorders in the United States. Arthritis Rheum: 1998; 41: 778-799.
[3] Womersley L, Stephen M. Sitting posture of subjects with postural backache. J Manipulative Physiol Ther: 2006; 29: 213-218.
[4] Haughie LJ, Fiebert IM, Roach KE. Relationship of forward head posture and cervical backward bending to neck pain. J Manual Manipulative Ther: 1995; 3: 91-96.
[5] Lewis JS, Green A, Wright C. Subacromial impingement syndrome: the role of posture and muscle imbalance. J Shoulder Elbow Surg: 2005; 14 (4): 385-392.
[6] Enwemeka CS, Bonet IM, Ingle JA, Prudhithumrong S, Ogbahon FE, Gbenedio NA. Postural correction in persons with neck pain II. Integrated electromyography of the upper trapezius in three simulated neck positions. J Orthop Sports Phys Ther: 1986; 8 (5): 240-242.
[7] Dunn KR, Brace CL, Masud T, Haslam RA, Morris RO. Prevention of fall-induced hip fractures: usability evaluation of hip protectors. Contemp Ergo: 2005; 464-468.
[8] Huck J, Kim Y. Coveralls for grass fire fighting. Int J Cloth Sci Tech: 1997; 9 (5): 346-359.
[9] Lou E, Hill D, Hedden D, Mahood J, Moreau M, Raso J. An objective measurement of brace usage for the treatment of adolescent idiopathic scoliosis. Med Eng Phys: 2011; 33: 290-294.
[10] Pham V, Houlliez A, Carpentier A, Herbaux B, Schill A, Thevenon A. Determination of the influence of the Chêneau brace on quality of life for adolescent with idiopathic scoliosis. Ann Réadapt Méd Phys 2008;15:3-8.
[11] Ugwonali O, Lomas G, Choe J, Hyman J, Lee F, Vitale M, Roye D. Effect of bracing on the quality of life of adolescents with idiopathic scoliosis. J Spine: 2004; 4: 254-260.
[12] DeJonge JO (1984) Foreword: the design process. In Watkins SM, editor. Clothing: the portable environment. Ames, Iowa: Iowa State UP: 1984; p. vii-xi.
[13] Watkins SM. Clothing: the portable environment (2nd ed.). Ames, Iowa: Iowa State UP: 1995.
[14] Brown TW. Posture improvement device and method of use. United States Patent: 7871388 B2; 2011-01-18.
[15] Veqar Z. Methods of postural assessment used for sports persons. J Clin Diag Res: 2014; 8 (4): LE01-LE04.
[16] Fortin C, Feldman DE, Cheriet F, Labelle H. Clinical methods for quantifying body segment posture: a literature review. Dis Rehab: 2011; 33 (5): 367-383.
[17] Thigpen CA, Padua DA, Michener LA, Guskiewicz K, Giuliani C, Keener JD, Stergiou N. Head and shoulder posture affect scapular mechanics and muscle activity in overhead tasks. J Electromyogr Kines: 2010; 20 (4): 701-709.

Research on Humanistic Driving Factors of Fabric Recreation

Xiu-Qin Hu*, Injoo Kim, Yi-Fang Zhang

[1,3]*School of Art Design, Jiangsu University of Technology, 1801 Zhongwu Road, Changzhou City, Jiangsu province, 213001, China*
[2]*School of Design, College of Design, Art, Architecture, and Planning, University of Cincinnati, 2600 Clifton Avenue, Cincinnati, Ohio, 45221, U.S*

*Corresponding author's email: 929146522@qq.com

Abstract

In recent years, the use and performance of clothing fabric recreation have continuously been expanding, and playing an increasingly important role in fashion design. Through the development of fabric reconstruction, it is not difficult to discover the role of humanistic factors in the process. This paper will discuss and research the humanistic factors which influence the developments of fabric recreation, from the aspect of regional culture, popular aesthetics, and consumption concept, so as to provide reference for the study on the design theory of clothing as well as guidance for the practice of fabric recreation.

Keywords: Research; Humanistic; Driving Factors; Fabric; Recreation

In modern fashion design, depending on the properties of the materials, a variety of technological means are used to further process, combine and transform fabric, so as to change its original appearance. Fabric Recreation is the process by which there is a change in the form, texture and tactile sensation of the fabric, giving it a new look, quality and style [1].

Since the 21st century, clothing designers have placed a lot of design emphasis on clothing fabric recreation. The design theme and inspiration are realized by the perfect combination of creative fabric design and the whole production process; especially in creation of Haute Couture. The application of fabric reconstruction has become a standard to measure a fashion designer's innovation ability [2].

Since demand drives development, there is always a need for innovation in garment design. Fabric recreation methods, scope of application and performance space are also in constant development. Comparative studies of the history of human clothing in different regions and in different periods showed evidence that the clothing style was completely different, and the direction of the fabric innovation was also correspondingly different. The reason is closely related to the regional culture, popular aesthetics and the concept of consumption. It is these humanistic factors that directly or indirectly affected the development and change of fabric recreation.

1. Different Regional Culture Cause Different Styles of Fabric Recreation

During the process of human social development and the influence of different regional cultures, such as natural conditions, national beliefs and social thoughts, the style of human clothing has gradually developed with different characteristics. Different places have different ideas on what the form apparel fabrics should be, so the treatment of fabric recreation was recorded into a system.

People incorporate the national cultures into fabric recreation, eventually make it produce completely different visual effects. On a world scale, the eastern and western regions have their own distinctive style features: the Oriental people pay attention to the texture of their fabric, in both colour and decorative pattern, while Westerners emphasize the three-dimensional visual and spatial effect of their fabrics [3].

1.1 The Plane Style of Oriental Clothing Fabric Recreation

Oriental countries such as ancient China, who followed Confucianism and Taoism, studied the "Ethical philosophy" between man and man, and paid attention to etiquette and norms. So the Oriental Chinese focused on how to fade out "people" and placed the emphasis on "spirit" [4]. The Chinese were able to identify the status of a person with apparel fabrics reconstruction. They accomplished this by using fabrics made of different materials, processes and means of fabric recreation. The Chinese art of fabric recreation mainly used a kind of flat fabric decoration style, reflected in their embroidery, printing, dyeing, splicing and more, to obtain the plane design variations.

Plane embroidery is one of the earliest forms used in clothing fabric recreation in ancient China. During the Tang Dynasty, which was in the heyday of China's feudal society, printing, dyeing and weaving techniques were developed [5]. Fabric decorative techniques using plane embroidery had also been widely developed, and been used to decorate skirts, chests, backs, cuffs, etc. The most representative was the popular "Flower Cage Skirt" in the Tang dynasty, and the "Mandarin Square" pattern of robes in the Ming and Qing dynasties. The "Flower Cage Skirt" was a kind of short skirt covered with an outside skirt, woven with a soft, thin, translucent silk called "monofilament", which included embroidered flowers and bird patterns that had hair made of gold and silver thread as well as various colors of thread [5]. There was a hint of flowers overlapping and color compounding (refer to Fig.1).

The "Mandarin Square" pattern was used in the robes of the Ming and Qing dynasties. These robes were similar to the embroidered gowns used in the Tang dynasty, but were changed the decorative pattern of embroidery on the robes to the "Mandarin Square," Which could be patched. The embroidered cloth was first woven on a loom, then embroidered by hand. The different birds and animals used in the patterns symbolized different ranks (refer to Fig.2).

Fig.1 "Flower Cage Skirt"

Fig.2 Pattern of "Mandarin Square"

Fig.3 "Paddy Clothes"

In the Ming dynasty, the "Paddy Clothes" represented another kind of plane decoration style. This kind of clothing was made of all sorts of colored pieces of brocade cloth, using a splicing technique. The name Paddy came from the geometric shapes of the print, which resembled a paddy field, with all kinds of color cloths interlacing with each other repeated all over the whole garment. The pattern was simple, but had an extremely strong visual effect, which cannot be achieved with traditional embroidery (refer to Fig.3).

In addition to the embroidery and fabric splicing, other common forms of fabric recreation with Oriental characteristics are ribbon embroidery, bead embroidery, patch embroidery, and hand dyeing, just to name a few. All of these embellishments are focused on plane pattern processing. Plane clothing fabric recreation development has become one of the most attractive trends of modern costume design, and offers much opportunity for clothing design development (refer to Fig.3, Fig.4, Fig.5).

Fig.3 Ribbon embroidery

Fig.4 Bead embroidery

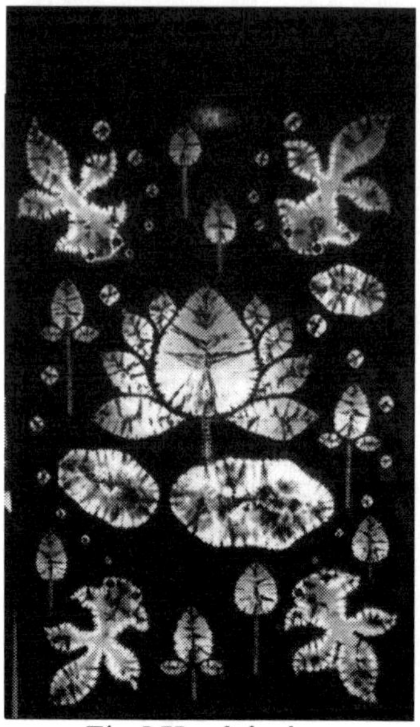

Fig.5 Hand dyeing

1.2 Three-Dimensional Style of Western Clothing Fabric Recreation

Ancient western cultures followed the "people-oriented" social thinking, so the clothing of ancient Greek and Roman garments usually reflected the curves of the human body [3, 4].The fabric changed with human body, and wrapped around the body naturally. "Fold," was a kind of fabric reconstruction that strives to express the visual and spatial effect of apparel, and was born in western civilization. Up until now, the ancient Egyptian clothing has changed a lot, but pleating is still a common element (refer to Fig.7).

During the Renaissance, a particular cutting technique was widely applied to fabrics, which presented a unique style of creative fabric recreation during that period. "Incision", the name of this solid modeling technique, added artistic charm to fabrics (refer to Fig.8) and could be seen on 15th-century fur cuffs. Longitudinal folding method used in fabrics resulted in clothes that reflected a clear sense of hierarchy. Most representative of this was the independent use of ruffled collars named "Ruff" (refer to Fig.9). This fabric reconstruction technique is often borrowed today by modern day fashion designers.

It is apparent that eastern and western clothing fabric recreation, clearly developed in different lands, show completely different style characteristics. Because of the influence of different regional cultures on dress design, there is a huge distinction between Chinese and Western clothing culture. Each culture's clothing continued to evolve along its own trajectory, and formed clothing with a unique regional style.

Fig.7 Ancient Greek "Fold" clothes

Fig.8 "Incision" sleeves

Fig.9 "Ruff" Collar

2. Trends and Aesthetics Lead the Direction of Fabric Recreation Development

In the history of clothing development, trend is a common psychological phenomenon in fashion. Trends are when popular attire is worn by a person or persons in a culture. A comprehensive trend embraced by a society over a given period of time, is generally a reflection of the people's aesthetic demand [6]. The popular trends and aesthetic choices in various periods seep into people's life, and deeply affect the form of wearing and fabric art during the period.

Take the Tang Dynasty, for example. Inclusive and open thinking prevailed during this time. The aesthetics on fashion were keen on gracious and gorgeous styles. Fabric artistic recreation focused on pursuing more elegant and refined embroidery, beautiful dye and bird feather decoration, etc. The flowers dyed decorative pattern of fabrics were popular, covered with translucent veils. During the Tang dynasty, the typical clothing form can be observed in the "Portrait of Flower-wearing Maids". The picture vividly recreates women's clothing worn at that time, and reflects the impact of fashion on the women's dress (refer to Fig.10).

Fig.10 Portrait of Flower-wearing Maids

In the 1950s, many chemical fibers were created, such as polyester, polypropylene, nylon, and etc., and expanded the types of fabric textures on the market. Knitwear fashion fabrics were attractive and grew quickly in popularity, because it offered a variety of different textures and pattern styles (refer to Fig.11).

The modern art movement in the 1960s led to the formation of different schools of art. The avant-garde and hippie style gradually rose, and personalized fabric gained popularity. Artificial leather, imitation metal, plastic, aluminium foil, folded fabrics, of some unexpected materials were used as fabrics for fashion [7], and as a result the wearers experienced the charm of fabric recreation many times over (refer to Fig.12).

At the same time, it became the goal for many designers to emphasize the artistic effect of the fabric itself, and enrich its visual effect on its surface, using techniques such as drilling (on the fur), printing, dyeing, decorative rivets and burrs on denim, organdie piping with embroidery, etc. The prevailing "POP" art style, also affected the fashion industry, as demonstrated by the clothing fabric and design innovation, which was deeply imprinted in the history of European clothing. It also brought a new style to fabric reconstruction (refer to Fig.13).

In today's society, with the development of science and technology, and the frequent communication between nations, vogue changes constantly. Once the fashion appears, it will immediately spread all over the world [7]. There was a trend of internationalization, diversification, and personality on fabric recreation art. The wide application of high and new technology made the forms of fabric recreation an unprecedented diversity. New technologies such as digital printing, 3D printing and laser engraving had greatly enhanced the effect of surface visual of fabric recreation [8, 9] (refer to Fig.14, refer to Fig.15, refer to Fig.16).

Fig.11 Knitting fashion

Fig.12 Clothing made of Aluminium foils

Fig.13 Pop art style

Fig.14 Digital printing clothing [15]

Fig.15 3D printed dress [15]

Fig.16 Laser engraving cloth [15]

3. The Concept of Consumption Drives the Development of Fabric Recreation

Throughout the history of fashion, social and economic changes, to varying degrees, drives consumers to purchase clothing, which presumes the idea that consumption may be psychological. The luxurious consumption concept, or conversely, environmentally frugal concept, suggests that every kind of consumer demand will lead the fabric art along the direction of the corresponding development.

The performance of luxury consumption concept was particularly prominent in the period of the 17th century "Baroque" and the 18th century "Rococo" in Europe. The surge of material wealth and greed led to the unlimited pursuit of a life of luxury. The fabrics usually chosen were luxurious silks, leather and wools, and not only to be printed and embroidered with gorgeous patterns, but also be decorated with a lot of lace, ribbons, bows and jewels, which made dressing up an extreme luxury [10]. Fabric art recreation showed a very strong decorative developmental trend. It can be seen in the remarkable portrait of Mrs. Ponpadu, which was painted by the famous painter Boucher, in the 18th century (refer to Fig.17).

Fig.17 Portrait of Mrs. Ponpadu [15]

Fig.18 McQueen's Bird feather clothing [15]

The rapid development of industry promoted the development and progress of human society, but also accelerated the consumption of resources, energy and led to environmental deterioration. The contradiction between people and nature forced people to begin to pay attention to environmental protection. Environmental consciousness also penetrated the field of clothing consumption. Influenced by the consumption concept of ecological environmental protection, the development of fabric recreation embodied two aspects: on the one hand, it is shown as the mining of natural material: designers get inspiration from the material pattern, color and texture, so as to inspire the desire of people to love life and harmoniously coexist with nature. British designer Alexander McQueen, for example, who has been described as a "Fashion urchin," designed a set of clothing on the subject of feathers (refer to Fig.18). The eye-catching colour and modelling called people to maintain the ecological balance, to love the natural environment [11].

On the other hand, regarding the subject of resource recycling, fashion consumers stopped buying new clothes, and chose to recycle or reconstruct their old clothes. For example, some people cut defective old clothes in stock into strips, then hand knit them into a coat with thick needles. Some people make fabrics old deliberately to reinforce the design concept of reuse, distressing denim, leather and corduroy fabrics by washing, wearing, tearing, and bleaching, etc [12]. These clothes, processed by fabric artistic reconstruction, have rich and wonderful fabric texture and an aesthetic with a unique personality.

With the passing of time and social inclusion, the aesthetic trends and pursuits of the consumer have also changed. People have become more receptive to new art styles and new things. The desires for personal style and fashion sense have become more intense. Personalized consumption concept drives the design idea of designers infinite [13].

When the style change is affected by requirements of clothing function, the recreation of the fabric can be enriched and expanded into new ideas in clothing design. The hand-made parts of fabric recreation such as embroidery, hand-painted, batik, weaving and other traditional handicraft application make the product arbitrary, which can fully serve society's pursuit of individuality, self-taste and other aspects of consumer demand [14]. In fact, it is the pursuit of personalized clothing that makes the development and design of clothing pursuit of endless possibilities.

4. Conclusion

From the abovementioned analysis, we conclude that modern design of fabrics has become a type of artistic expression, especially in regards to Haute Couture performance. The many works of master fashion designers fully illustrate the great influence of fabric recreation. At the same time, they reflect the current situation and development trend of the contemporary fashion design innovation.

The innovation of clothing fabrics must conform to the development direction of human factors in contemporary society. There is no theoretical basis, apart from humanistic factors, to explain the development of fabric recreation.

The conclusion of this paper can provide references for the study of the theory of clothing design and guidance for the practice of fabric recreation.

5. Acknowledgment

This study was financially supported by a project from the 2013 Jiangsu Province College of Philosophy and Social Science Fund Project (2013SJD760007) - Research on Technology and Humanistic Driving Factors of Fabric Recreation.

References

[1] Hu XQ, Duan XY, Zhang Y. Research on Technology Driving Factors of Fabric Reconstruction. Textile Bioengineering and Informatics Symposium Proceedings: 2015; 343-349.

[2] Shi LL. Study on Fashion Fabric Re-design [J]. Journal of Xi'an Polytechnic University: 2010; 24 (5): 574-575.

[3] Zhang S. Introduction the History and Development of Fabric Recreation in the Eastern and Western. China After School Education: 2014; (5):163.

[4] Zhang DZ, Li SP. The Influence of Eastern and Western Aesthetic Thought on the Fabrics Recreation. Journal of Hainan Radio & TV University: 2010; (4): 66-68.

[5] Yang J. Study on Tang dynasty costumes aesthetic cultural. Journal of Kaifeng Institute of Education: 2004; (7): 260-261.

[6] Yao HH. Vogue and Aesthetics of Design Products. Journal of Henan Science and Technology: 2010; (03): 31-32

[7] Hu XQ. Redesign for the Appearance of Costume Materials. Soochow University: 2004; 5-7.

[8] Liu ZY, Xu BJ. Application of Textile Digital Printing in Garment Face Fabrics Creation. Shanghai Textile Science & Technology: 2013; 41 (4): 46-50.

[9] Du Q. Application of Laser Engraving in Fabric Reconstruction and Pattern Design [J]. Advanced Textile Technology: 2014; 22 (2): 49-52.

[10] Che WD, Jin CY. Artistic Performance of Fabric Recreation in Fashion Design [J]. China Textile Leader, 2011; (6): 96-97.

[11] Wei YL. Applications and Representations of Low Carbon Concept in Modern Fashion Design. Journal of Zhengzhou University of Light Industry (Social Science Edition): 2014; 15 (4): 104-107.

[12] Wang F. "Make Old" Design of Fabric Graphics. New Arts: 2013; (12): 126-127.

[13] Ma XM. Diversifications of Fabric Recreation. China Textile Leader: 2012; (4): 92-93.

[14] Li H, Liang HE, Gu M. From the Modern Tie-dye Techniques to See the Innovative Fabric Design [J]. Silk Monthly: 2010; (9): 37-41.

[15] Images courtesy of http://www.pop-fashion.com and http://baike.baidu.com.

Research on the Relations between the Designs of Lingerie Straps and Collars

Jian-Fei Dong[1], Jie Zhou[2*]

Xian Polytechnic University, Apparel & Art Design College No.19 South Jinhua Road, Xian, Shanxi, 710048, China

*Corresponding author's email: 446961933@qq.com

Abstract

In this paper, the authors use the literature query method and the experimental method to explore the applied shoulder and collar structure of clothing to the structural design of lingerie shoulder straps, as well as analyse the pressure caused by the lifting of straps on the breast and collar stress on the shoulders in underwear. The purpose here is to find out how to reduce the pressure of current lingerie straps on the human shoulder. Finally, the authors propose underwear design addressing the aforementioned issues regarding a garment's shoulder and the collar structure on the basis of the theoretical research and data analysis, which in turn may effectively reduce pressure on the human shoulder.

Keywords: Lingerie Straps; Garment Collar; Design; Aesthetics

1. Introduction

The female breast consists of mostly adipose tissue, with some ligaments and skin providing internal support. However, sometimes, they are incapable of supporting or limiting pressure on women' breast, especially among the aged population, thereby causing pain and discomfort [1]. The main function of straps is to minimize the movement of the breast by fixing and lifting the breast. The straps can also can emphasize the woman's good figure by lifting the breast. From the perspective of fixing and shaping functions, keeping the breast upright is the basic requirement of a bra. In order to achieve this effect, bras and straps should bear and lift the breast [2].

Ensuring comfort is integral to the production of clothing, as the human body's physiological and psychological state are affected by it [3-4]. Nowadays, the bra is essential for all women, with many domestic and foreign scholars carrying out research on this topic. However, this research mainly includes the analysis of chest shape characteristics and the extraction characteristic index, the type and cup model of the bra, and the design of a virtual three-dimensional structure [5-6]. Little research exists on the structural design of the garment's shoulder and collar.

In clothing, the shoulder has the function of providing support to a garment and embodying its design. It does not matter whether the neck moves or not, it influences the design of the clothes. Besides the functions of supporting and fixing clothes and comfort, the garment collar attaches more importance to the design, aesthetics and innovation.

In this study, the author applies the structure of the shoulder and collar of a garment to the design of a shoulder strap for underwear. We hope to design a lingerie with garment collar feature and improve the local stress by straps.

2. Structural Features of the Female Breast, Shoulder and Neck

2.1 Outward Shape and Internal Organization of the Female Breast

The outward shapes of the female breast can be categorized into petite, busty, sagging, extended, asymmetric and dove-shaped. In terms of size, the breast can be categorized into cup A, B, C, D etc. Cup A features a 10 cm circumference difference between the top and upper breast, while 12.5 cm for cup B.

The breast constitutes a gland, duct, adipose tissue and fibrous tissue. The adipose tissues surround the breast and form a hemispherical shape called the adipose capsule. The thickness of the adipose capsule differs greatly according to age and fertility. The amount of adipose tissue is the key element deciding the breast size. As shown in the cross-section in Fig.1, the breast comprises 1) chest wall, 2) pectoralis muscles, 3) lobules, 4) nipple, 5) areola, 6) milk duct, 7) fatty tissue and 8) skin [5].

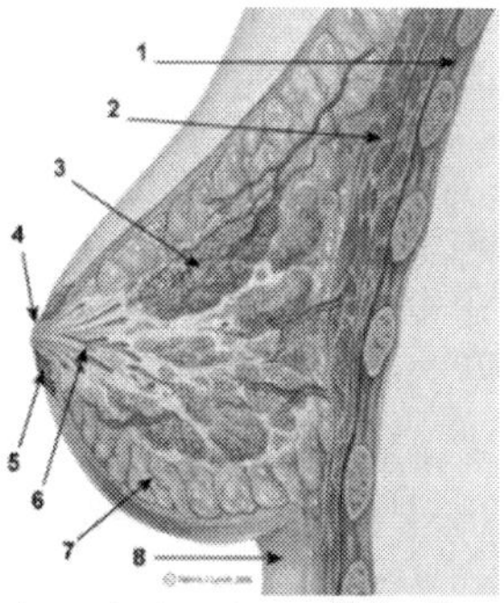

Fig.1 Internal structure of the female breast

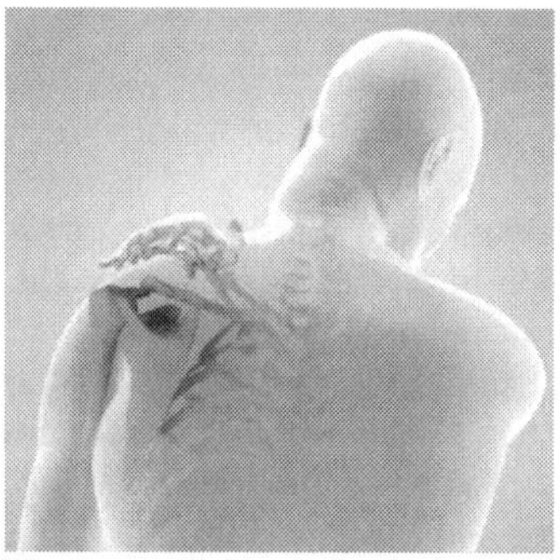

Fig.2 Plaster model - vertical view [5]

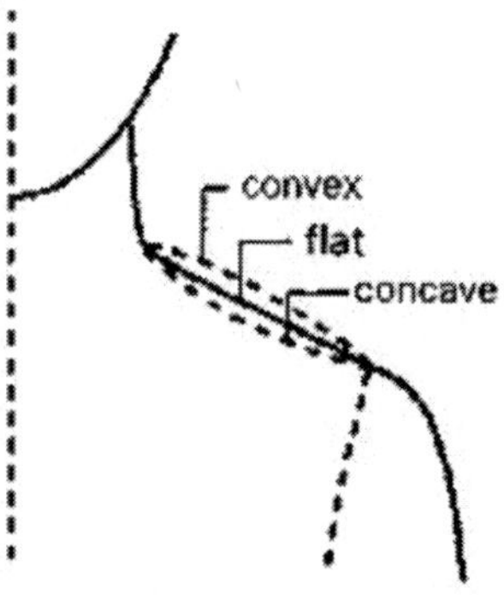

Fig.3 Three types of shoulder shape [6]

2.2 The Structural Features of the Female Shoulder and Neck

The end of a human shoulder features a spherical shape with front shoulder hyperboloid and forward shoulders. The back shoulder is arch-shaped (as shown in Fig.2). Due to the differences in the extent to which shoulder muscles bulge, the body consists of three different types of shoulder: flat shoulder, convex shoulder, and concave shoulder (as shown in Fig.3).

The upper bound of the neck is the ligature of the inferior border of mandible, with the chin resting on the trailing edge, along with the mastoid process and external occipital protuberance. The lower bound of the neck is the ligature of the superior border of sternum, clavicle, acromion and the seventh cervical spinous process. The local area is bounded by the anterior border of trapezius and divided into the inherent front neck and back neck. The inherent neck is bounded by the sternocleidomastoid, which divides into the anterior region of neck, the lateral region of neck and the sternocleidomastoid region [7].

3. Analysis of the Influence of a Shoulder Strap on the Breast

3.1 The Constitution and Function of Straps

A strap consists of a loop, the strap body and a belt buckle (as shown in Fig.4). The loop is a connection between the strap body and the bra. The belt buckle is a connection between the strap and the underwear, with which to adjust the shoulder girdle. Based on its functions, the belt buckle can be categorized into three types: the "0" shape buckle combing the strap body and the bra; the "8" shape buckle, which adjusts the length; the "9" shape, which is situated on both sides of the detachable straps.

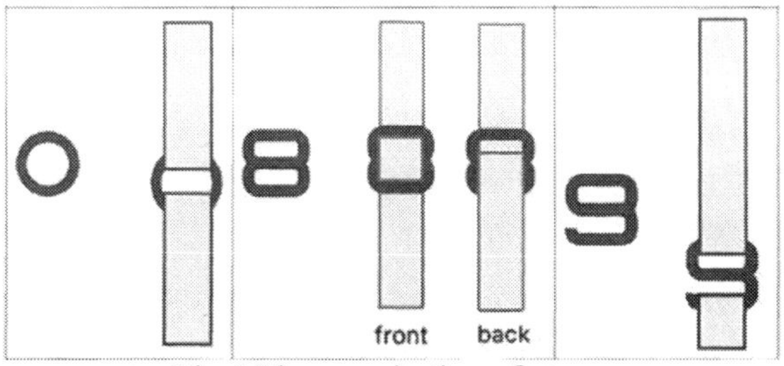

Fig.4 The constitution of straps

The function of shoulder straps is to fix and lift the breast, while minimizing vibration when the body moves. It can also enhance the profile of the wearer's figure, particularly the breast. No matter whether the function of the strap is to fix or shape, making the breast "busty" is the most basic requirement. In order to realize the effect, the bra should have a holding effect, a reinforced side and lifting straps [8].

3.2 The Force Analysis of Different Straps on the Same Woman

The bra straps' pressure distribution reveals that the rule of pressure distribution is that the highest point features the highest pressure, while it reduces forward and backward (as shown in Fig.5). At the junction between the end and the upfront area, the pressure will increase under the influence of the upper side. When women with a bra make different moves, the pressure point of the bra changes all the time. The influential point is the junction between the strap and clavicle, which is the intersection of the muscle and shoulder straps' endpoint [9]. Different widths of shoulder strap apply different pressure to the shoulder. A relatively broad strap with a broad area of thrust surface has more lifting force. On the contrary, a strap with a relatively smaller area of thrust surface has less lifting force (as shown in Fig.6). A big breasted woman wearing a bra with a narrow strap is prone to redness and swelling on the shoulder. However, while a broad strap is comfortable, it does not produce a good look [10].

The tensile property of the strap can reflect the basic features of its usage. The bound pressure caused by extensional deformation will directly affect clothing pressure. The degree of the tightness of the strap and the tensile elasticity will exert unignorable influence on the needs. The fatigue of shoulder straps and related issues are also concerns for women [11].

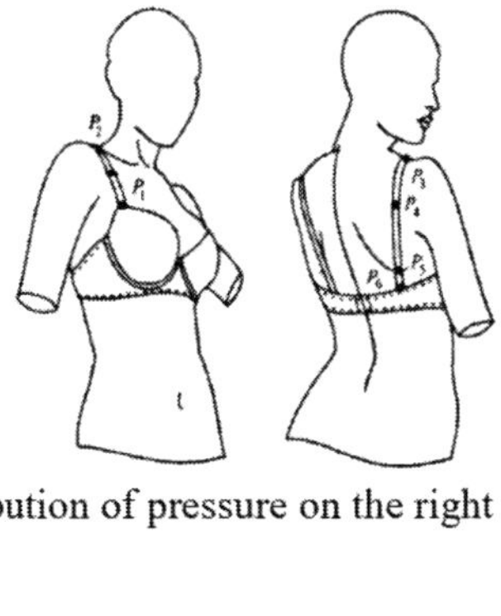

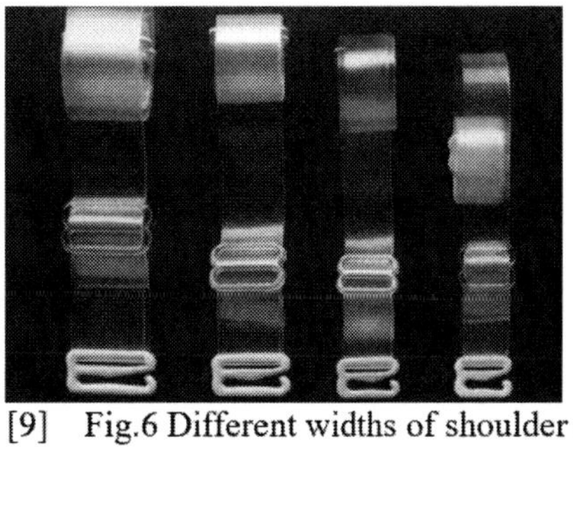

Fig.5 The distribution of pressure on the right shoulder of the bra [9] Fig.6 Different widths of shoulder strap

3.3 Force Analysis of the Same Straps on Different Women

The chest size and age of different women are not exactly the same. The need for a shoulder strap for women of a bigger size is greater than for those of a smaller size. The strap should have the function of lifting, shaping and providing aesthetics, while minimizing pressure on the shoulder. Generally, after lactation, because of reduced hormones, a woman's mammary gland bubble tube, gland and adipose tissue will atrophy, while the skin and support tissue will expand, causing mastoptosis (as shown in Fig.7). As women get older, the full extent and the shapeliness of the breasts will decrease. At this stage in life, women have greater need for comfortable and shape-enhancing lingerie. In other words, high-quality, well-fitting underwear embodies elegance too [12].

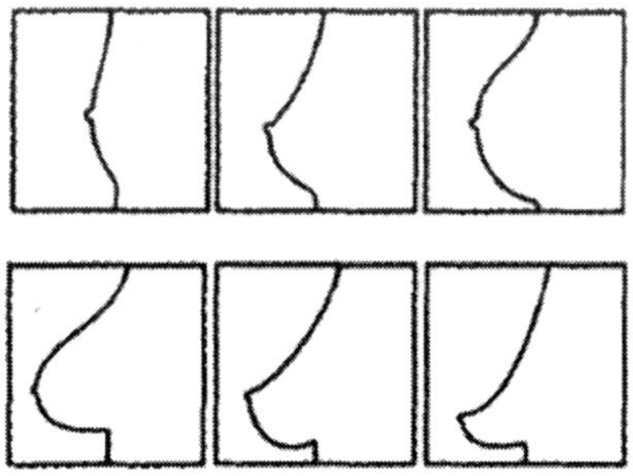

Fig.7 The process of breast sagging

The lifting effects of the same straps on different women differ in terms of breast size and the degree of sagging. Once the length, width and elastic force of the straps are set, a sagging breast reflects a poor effect; otherwise, an upright one means the straps' lifting effect is good.

4. Experiment and Evaluation

4.1 Experimental Theory and Framework of Contents

Based on the analysis of the straps' lifting force, with regard to the breast, and the analysis of a garment collar's stress on shoulders, the authors applied structural design features of the shoulder and collar to the design of underwear. In the process, they created a new style of underwear with a shoulder and collar structure, which can reduce the local stress on the shoulder by using currently available straps. The empirical research process was as follows. First, the straps from a 75B bra with a 1/2 cup were removed, while the stress area was analysed. Next, we designed a garment collar matching the bra, choosing certain fabrics and accessories to tailor a new one, while taking comfort, function and aesthetics into consideration. We then examined the effect of the bra on different bodies, as well as carried out a dressing evaluation involving try-on users and a sensory evaluation from indirect bystanders [13].

4.2 Experimental Procedures

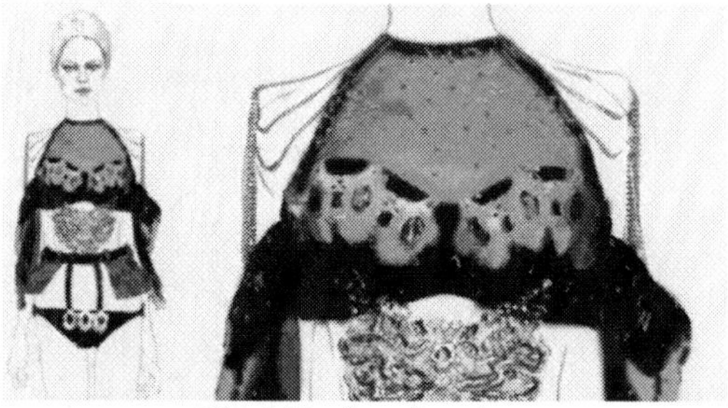

Fig.8 Design drawing

a) Designing a garment collar model combined with a bra (as shown in Fig.8). In the experiment, the bra size was 75B with a 1/2 cup. Originally, the bra comprised hanging neck shoulder straps (as shown in Fig.9). We then designed a garment collar with a rounded shoulder back buckle. The structure embodied the curves of a female shoulder, neck and collarbone (as shown in Fig.10).

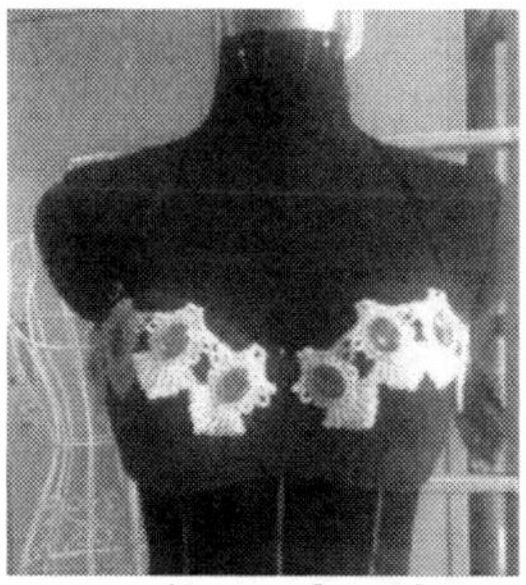

Fig.9 75B, 1/2 cup, hanging neck bra

Fig.10 Splitting design on the backline

The back side with the bra buckle was convenient to put on and off. Based on the fabric, the backline was integrated with the bra's flower design (as shown in Fig.11).

b) Choice of fabrics and accessories. The back side of the bra was covered by elastic pleuche in a red wine colour with the elastic force fitting the surface modelling (as shown in Fig.6). Part of collar was made with black gauze, fixed with micro elastic gold embroidery, which complemented the neck and fixed the shape (as shown in Fig.12). The collar overall was made from elastic black lace, whose edge was connected to the front and back sides of the collar by a back buckle (as shown in Fig.13) [14].

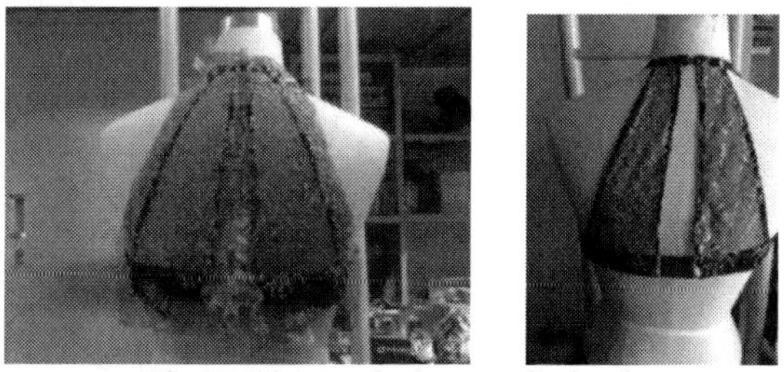

Fig.11 Flower design on the back

Fig.12 Black gauze fixed with micro elastic gold embroidery

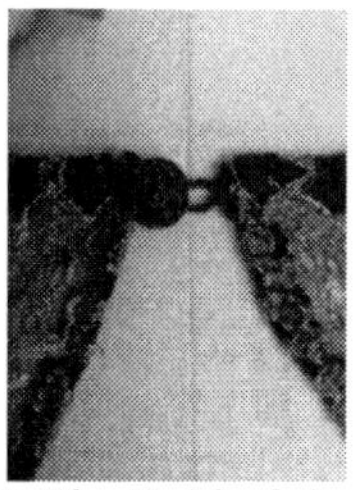

Fig.13 Elastic black lace edge

c) Adding the garment collar structure to the lingerie. Originally, the straps were for a hanging neck design, with the main force bearing point located 12 cm below the strap and the back of the neck. This design can cause creased scars and obvious pain if worn for too long a time. Given these drawbacks, we properly attached the garment collar structure to the lingerie. The experiment involved a 75B special lingerie model, for the front and back pieces were fixed and the bra was tailored by removing the straps. In the tailoring process, we analysed the original garment collar structure to keep the collar integrated and divide the back side. We changed the hanging neck structure into an upper collar and connected the edge and the bottom of the lingerie (as shown in Fig.14)

4.3 Analysis of the Experiment Outcome

After the completion of the underwear production, the authors selected 30 women whose bra size was 75B and invited them to try on the garment. The purpose here was to verify whether the design of the underwear shoulder strap was able to ease the wearer's shoulder pressure (as shown in Fig.15). The subjects' feelings about the experimental underwear after carrying out some simple exercises were then evaluated. The method of evaluation was the Likert scale, with the evaluation statements related to comfort, functionality and aesthetics. Each statement had five possible responses: "perfect", "good", "ordinary", "not good" and "bad". The evaluation of the 30 women who tried on the underwear is summarized in Table 1.

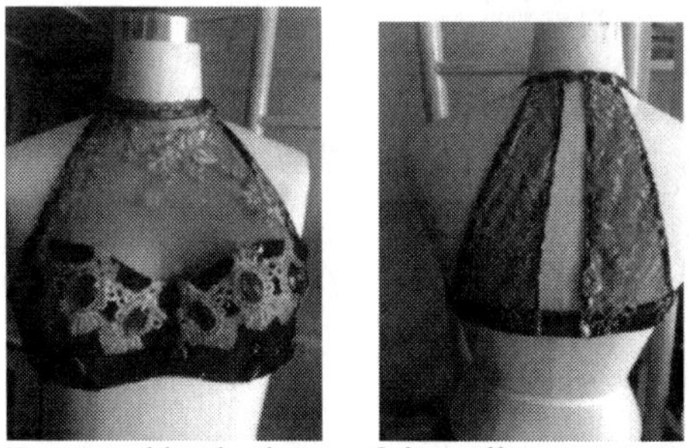

Fig.14 Front and back pieces of the collar structure

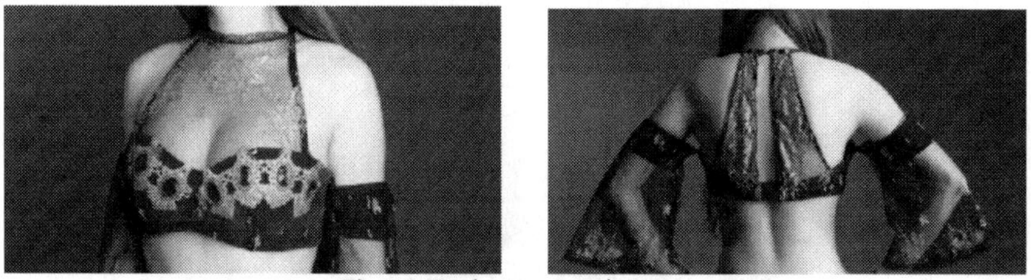

Fig.15 Body try-on pictures

Table.1 The direct try-on feelings based on the five-point scale from five women with a 75B bra size

Sequence	Elements	Perfect	Good	ordinary	Not good	bad
1	Comfort	20%	60%	20%	0%	0%
2	Function	40%	40%	20%	0%	0%
3	Aesthetics	40%	60%	0%	0%	0%

Table.2 Indirect feelings based on the five-point scale from five female bystanders

Sequence	Elements	Perfect	Good	ordinary	Not good	bad
1	Aesthetics	60%	40%	0%	0%	0%
2	Match	40%	40%	20%	0%	0%

When the 30 women with a 75B bra size tried on the underwear, the authors re-selected 10 women and allowed them stand aside and watch, and then make a subjective evaluation of the underwear's aesthetic and external characteristics. The evaluation of 10 indirect bystanders is summarized in Table 2.

4.4 Evaluation

Based on the theoretical literature, the authors analysed the stress caused by lingerie straps and a garment's collar in order achieve the meet this study's research purpose in terms of design and practice. First, we chose a 75B bra with a 1/2 cup, then removed the straps and analysed the stress area. Next, we designed a shoulder collar matching the bra. This design step was limited to some extent because we only chose a typical collar structure to carry out our research. We also chose relevant fabrics and accessories to tailor the lingerie, while taking elements of comfort, function and aesthetics into consideration [15]. However, the limited availability of equipment and style meant that we could only design one type of bra. After producing the bra, we analysed its performance on different bodies, as well as carried out direct and indirect evaluation. On account of the small number of samples, the evaluation of the experimental outcomes are limited in terms of generalizability [16].

Regarding the materials, using plastic red pleuche fabric for the lingerie was advantageous as it covered the surface of the underwear mold cup. The materials used in the underwear collar comprised elastic cord, elastic lace and net yarn. The elastic cord optimized the fitness and comfort, while the elastic lace complemented the overall design and function of the underwear. The micro stretch lace net yarn not only featured a fixing capacity, but also fit the body curve. The final lingerie item was found to be convenient to wear and fix the breast. The stress caused by adding the collar design was smaller than was the case with the original bra.

5. Conclusion

Comfort when wearing a bra is one of the most important factors for the wearer at any time. Based on the theoretical analysis of a garment's collar structure and the analysis of the force of underwear on the shoulder, the authors designed underwear with the characteristics of a shoulder collar structure. This underwear was found to reduce the force on the wearer's shoulder, which was confirmed by a pilot experiment to prove its effectiveness. It is hoped that this study will provide a practical basis for the study of bracelet design in China.

References

[1] http://baike.baidu.com/link?url=T1zC84IcgA0DFz3RMcG91g-O6ajZOh1zqqWEib6Xh_GvCvPyFZdtxuzfHz FGdHYpiaOd9PfbqGHJHLxXuvGcpq.

[2] Ceng Z. The research on the design of women's bra braces. Master's thesis. Donghua University: 2007.

[3] Li Y. Clothing comfort and product development. M. China Textile Publishing House: 2002; 67.

[4] Watkins ER, Richardson AJ. Fire service instructor's undergarment choice to reduce interleukin-6 and minimise physiological and perceptual strain. J Thermal Biology: 2016; 11: 7.

[5] Yu XK. Research on the functional design of a bra shoulder strap. D. Donghua University: 2014; 24.

[6] Meads C, Glover M, Dimmock P, Pokhr S. Parafricta bootees and undergarments to reduce skin breakdown in people with or at risk of pressure ulcers: NICE medical technologies guidance. J Applied Health Economics and Health Policy: 2016; 12: 635-646.

[7] http://baike.baidu.com/link?url=vlcKqHYJAl7p8heqYHI39K51LZCfm5RENO-vtzcVsQsIW8BQ4Wvz-h96t5 nwFhUPp6yVnlO_MjXo_Mj8znbZRs8KLtGzjzBTjgbGka1dggS.

[8] Shi L. Analysis of the garment shoulder framework and relevant factors. J Zhejiang Text Fashion College: 2008; 2: 46-48.

[9] http://baike.baidu.com/link?url=O32Yd1-SUKilWF6TtZVXVAlsHI6-hMggJzOgVmfwUJWBpxpOmk7_4-xF DHHtAtaYePXcMHroy1RWV4gkrcEUmK.

[10] Li LX. The bra straps design research. Knitting Industries: 2008; 05.

[11] Wang LZ, Chen DS, Lin B. Analysis of pressure distribution of a bra's shoulder straps. J Text Res: 2009; 09.

[12] Wang FY. Study on the comfort of a sports bra by the combination of displacement and pressure. Master's thesis. Donghua University: 2013.

[13] Yu XK. Functional analyses and the design of bra straps. PhD thesis. Donghua University: 2014.

[14] He KJ, Lin XQ. The analysis of women's shoulder strap designs. Progress in Textile Science & Technology: 2013; 4: 89-91.

[15] Duan XY, Yu WD. On the factors affecting bra wearing comfort. J Jiangsu teachers University Technol: 2007; 04: 56-62.

[16] Ying SY, Guo FZ. Fabric and Accessories Brochure. Beijing: Chem Ind Press: 2007; 598-602.

The Application of Clothing Fabric Reconstruction

Fu-Hong Wei, Yan-Bo Ji*

School of Apparel & Art Design, Xi'an Polytechnic University, Xi'an, 710048, China

*Corresponding author's email: 785230953@qq.com

Abstract

With the rapid development of science and technology on textiles, thousands of new fabrics have been developed. Fabric reconstruction has become a focus for both designers and consumers. With the definition of fabric reconstruction, a survey questionnaire was derived. In this paper, depth interview and market research methods are adopted. Using the statistical software SPSS17.0, consumers preference to design methods on fabric reconstruction was analyzed. A series of dresses for fabric reconstruction were designed in accordance to the results.

Keywords: Fabric Reconstruction; Fashion Design; Dress; Science and Technology

1. Background

The Chinese fashion industry has developed for decades. Through the trends set by the international and domestic fashion week and large-scale clothing competition, it can be observed that the emphasis of fashion design are on modeling, cutting and that the overall style matches to the later period of 80s and 90s. Since the 21st century, designers have focused on fabric recycling, through combining materials and technology to reflect the theme of design [1]. In general, as a specific classification of clothing, the dress has gradually become a necessity in wardrobe and dress design not only requires changes in color and style, but also on the innovation of texture, structure, material etc. The market demand for the application and diversification of fabric reconstruction in making dresses is growing.

2. Characteristic on the Fabrics Reconstruction

2.1 Definition

Based on the definition set by fashion design, fabric reconstruction is the design and processing methods to recreate existing fabrics, and in producing a visual and tactile reflection of a new texture and pattern of a certain artistic quality [2]. It is an extension of the designer's ideas, with unparalleled innovation. Fabric reconstruction includes two-dimensional graphic design and three-dimensional space design [3]. It broadens the fashion position of fabric in the art design of clothing, and changes the designer's thinking process to inspire a stronger visual impact.

2.2 Characteristics

2.2.1 The Handicraft

The history of fashion and fashion concepts are always changing. fabric reconstruction has the potential to attract consumers for its unique craftsmanship and sewing techniques. For example, Chinese embroidery, inlaid, pick up, patch, junction and other ancient traditional crafts, as well as western costume in tassel, three-dimensional modeling, all belongs to the category of handicraft in clothing fabric reconstruction which occupies a key position in fabric innovation.

2.2.2 Environmental Conservation

The rapid development of the global economy has resulted in the shortage of natural resources and in the rise of environmental problems. The state has advocated the idea of "low carbon" and is actively promoting environmental protection as the new fashion. In order to achieve the balance between fashion and environmental protection, young fashion consumers and white-collar workers are producing fashionable dresses through recycling clothing and reusing them. They make good use of the outdated fabric, digest inventory and reduce cost [4]. For example, natural blended synthetic fiber is used to create more environmentally friendly blankets, yarns and fabrics. The combination of old jeans, tires, postindustrial polyester material and natural materials such as hemp and flax fabrics creates a suitable automobile industry tooling apparel, which not only conforms to the trend of the development on low-carbon economy, it also helps build the foundation for environmental protection. In this way, it not only reduces total carbon emissions, it also recycles clothing.

2.2.3 Artistic Expression

The design of the fabric itself is a form of art. The works of fabric reconstruction involves beautiful colors and a variety of texture. The aesthetics of the design can be reflected from its design style, material composition and color collocation [5]. The art form of fabric reconstruction includes visual style, tactile feeling and hearing effect [6]. The vision emphasizes the innovation of pattern, texture and color on the surface of the fabric and uses new technology to make new arrangement and structure from the original organization. Tactile finger contact is generated when the fabric tactile effect and the stereo effect highlights the characteristics of fabric, namely whether it is soft or stiff, smooth or rough, loose or tight. Combinations of different fabrics made from different materials and ingredients can bring different tactile sensations. Using subtraction or addition design of fabric, the fabric can achieve a three-dimensional effect [7]. Auditory effect refers to the human body through auditory perception. Due to the different characteristics of the fabric itself, different fabrics and materials will produce a different sound when under friction. For instance, silk fabrics will be issued a call with the wire body movement, friction between Eugen yarn will make the wearer feel a sense of air. A large number of silver or metal material are applied on the fabric to create a rhythm to the fabric [8]. These three effects complement each other, to create an interaction and a unique artistic effect that is the basis for the transformation of fabric which is in itself an art of clothing.

2.3 Application of Dress Fabric Reconstruction

The design technique of fabric modification can be controlled from three aspects: structural feature, material addition and deformation method.

2.3.1 Change the Fabric Structure

In order to create a fabric with space order, the beauty and three dimensional sense can be altered by hollow, shearing embroidery, burning flower, tear and wash etc. [9]. For example, by extruding a series of irregular wrinkles on the original smooth surface of the fabric can give the dress a free and elegant character. Conversely, with a thin or slightly thicker fabric of silk or wool through shirring, making a dart and a contraction joint, wave flower forms can be made and applied to the shoulder, chest, skirt hem etc. [10].

2.3.2 Add Material on the Surface of the Fabric

By adding additional decoration with different material, texture and shape of fabric, unique effects with different color and pattern can be produced [11]. Methods such as sewing, embroidery, stitching, weaving, dyeing, adhesive, digital printing methods can be employed to design fabric. Emphasizing modeling was usually accomplished by adding buttons and embroidery on the neck, chest, cuff, waist and other parts [12]. Different combinations of material will have different contrast effects, and greatly enhances the artistic expression of fashion modeling. For example, Elie Saab designed a romantic style of dresses for girls with crystal sequins and feathers.

2.3.3 Deformation of the Original Fabric Morphology

Generally, the shape of the original fabric is relatively smooth. To create more uniqueness consumers will alter the shape and surface of the fabric through techniques such as drawing, shrinking, quilting, embossing to form a three-dimensional texture and a strong sense of touch fabric [13]. For instance, extrusion of irregular lines or shrinkage of different shapes on a smooth fabric can give the clothing a unique texture. In order to create a texture effect on the surface of the sewing process, we can use different colored lines to form other geometric patterns, as well as waving shape with curvature.

3. Questionnaire Investigation

The influencing factors and design techniques of fabric transformation in dress design are summarized above. In order to understand more about the consumers' preferences in color, style, contour, shape, design elements and patterns applied on a dress, a further investigation is needed. Through a study and research, we can design more valuable dresses from examining consumers need. The contents of the questionnaire are shown in Appendix 1.

3.1 Research Procedure

The questionnaire design procedures are as follows:

First, the purpose and theme of the survey was determined. In order to understand the style, color and other factors of fabric design, the title of the questionnaire was made to be "the application of dresses fabric reconstruction". Second, the object of investigation was established. The respondents were mainly young female consumers. There were 300 questionnaires in this study, of which 288 are valid questionnaires and the effective rate of the questionnaire is 96%. Third, the questionnaire questions were created. The topic is divided into single selection and multi selected, with a total of 12 questions. The questions mainly relate to the gender, age and the style of fabric reconstruction, and pays attention to dress design style, color, and whether the application fabric reconstruction on dresses is well liked. The arrangement of the questionnaire was logical and clear, which in turn will bring about a clearer and more accurate result.

The last step is, preliminary investigation and correction. Pre-inviestgation is carried out to identify problems and shortcomings so that the questionnaire can be modified.

3.2 Research Results Analysis

With the increasingly rapid development of society, people have higher and higher requirements for clothing. Traditionally dresses were made for warmth, but nowadays the emphases are more on the pursuit of fashion and aesthetic beauty. Through the analysis of survey results as shown in Fig.1, it can be seen that 40% of women are interested in the artistic expression of the fabric, which suggests that fabric reconstruction is important.

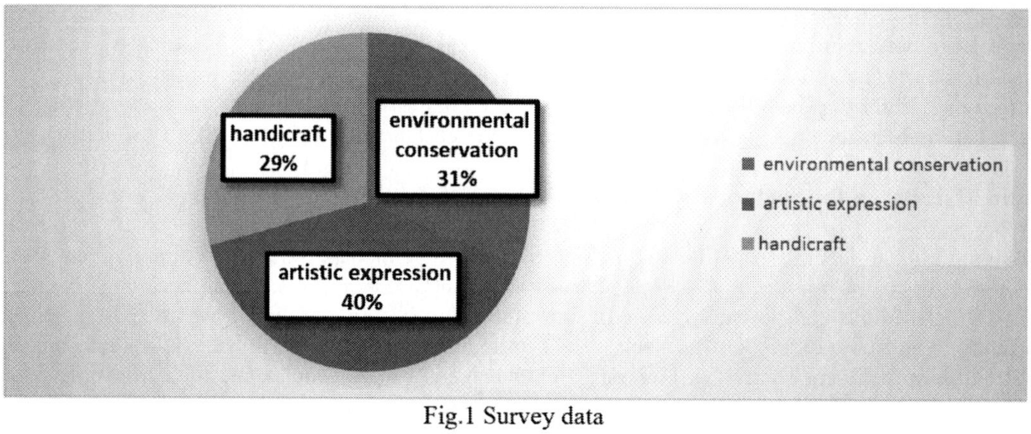

Fig.1 Survey data

The market survey and analysis of the results show that: 63.83% consumers like the dress to be made from fabric reconstruction. 53.9% consumers love printing design elements, 51.77% consumers love hollow design elements, 48.25% consumers love splicing design elements, 43.26% consumers love folding design elements. This can be seen on the survey in Fig.2.

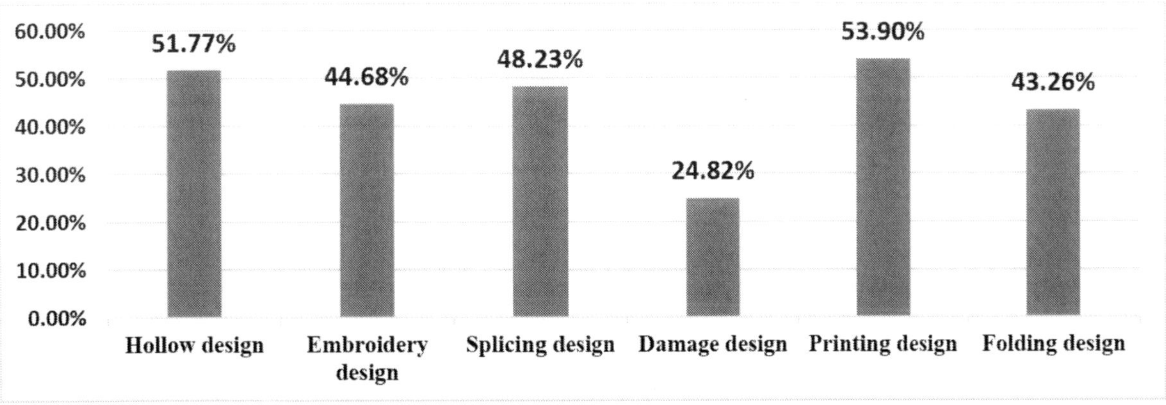

Fig.2 Survey data

There is a high demand for color and dress fabric design for women clothing. 63.12% consumers pay attention to the design of color. Many love the classic colors of black white ash and they are also keen on other colors, such as cold colors. According to the colors that consumers pay attention to, they love the transformation of the fabric in color design, the results of the survey as shown in Fig.3.

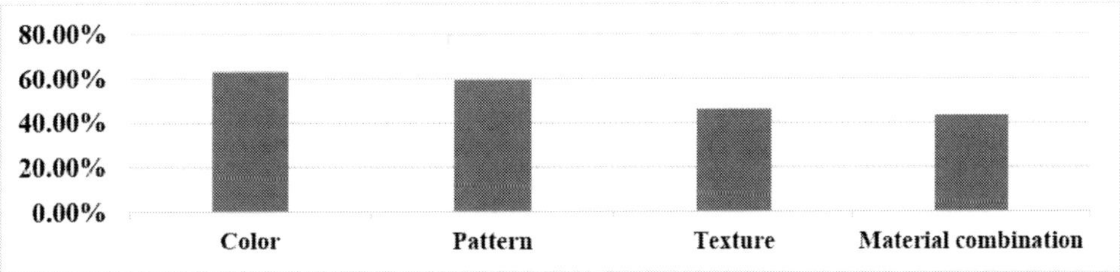

Fig.3 Survey data

The results show that 58.87% of the women like the design of dresses produced by fabric reconstruction techniques. 57.45% women love the folding effect, 54.61% women like application of hollow texture, 48.23% women love split joint effect, the results of the survey as shown in Fig.4.

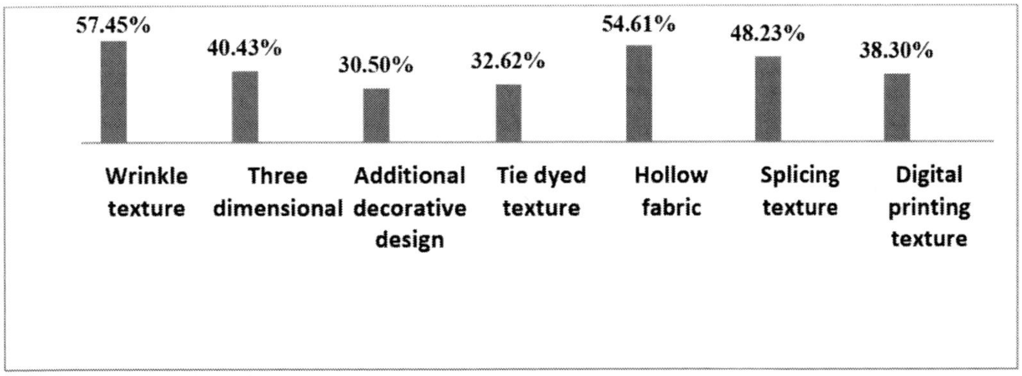

Fig.4 Survey data

In accordance to the material condition of the clothing, the fabric is closely related to the inner and outer structure modeling process with a personal temperament. Everyone's fashion concept is changing, 58.87% consumers pursue a unique temperament, the pursuit of comfort in fabric is 53.19%, the pursuit of fashion is 51.77%. The results of the survey are shown in Fig.5.

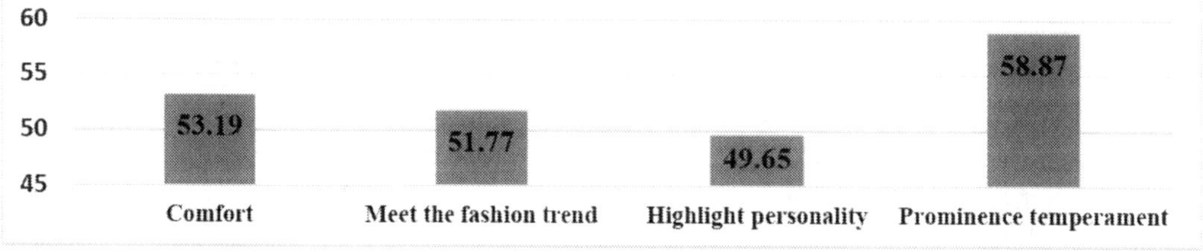

Fig.5 Survey data

For formal dresses, 67.38% consumers pay more attention to cautious and conscientious in style design. 56.03% of consumers love fashion fabrics used in dresses because of its originality. The results of the survey are shown in Fig.6.

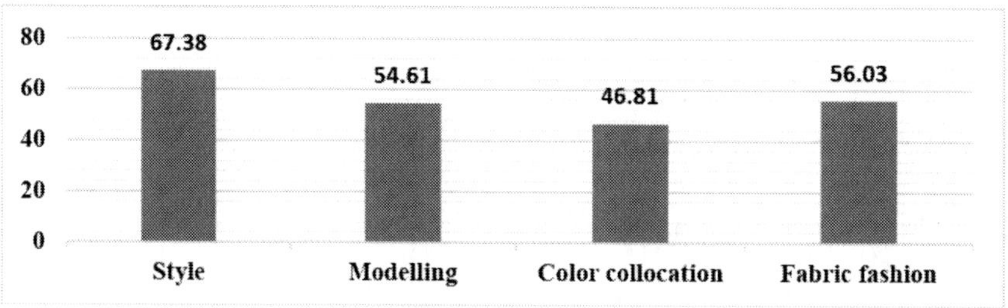

Fig.6 Survey data

Nowadays, women in different ages need different styles of dresses. This is affected by the fashion trend, especially young people's pursuit of fashion and beauty. 58.16% of people like the X type because of more prominent female figure, 45.39% people love the A type, the results of the survey are shown in Fig.7.

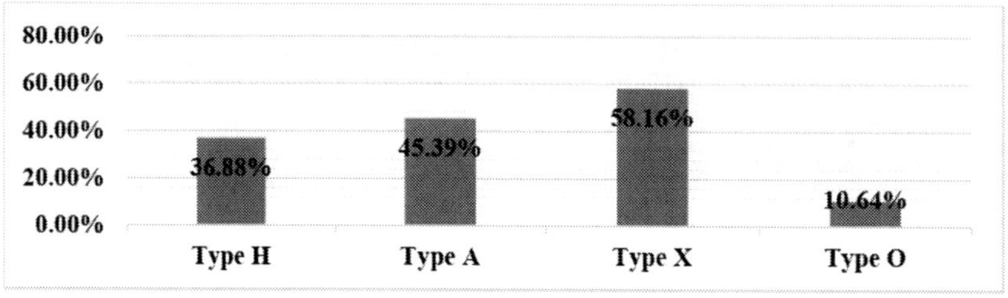

Fig.7 Survey data

The application of fabric reconstruction in dress design has become a new fashion. In the survey 59.57% of women love the transformation in chest fabric design, 54.61% women focus on the design of skirt hem, 53.19% women love waist design. The results of the survey are shown in Fig.8.

Through the above investigation results, conclusion can be drawn that the beauty and artistic expression in fabric reconstruction are loved and recognized by people. It will be applied to the design of fabric redress, not only to increase the overall fashion, but also in highlighting the temperament in line with fashion trends. In accordance to the factors such as design elements, shapes, bright spots, design techniques, profile and dress parts of the consumer, these aspects are analyzed and applied to the later dress design.

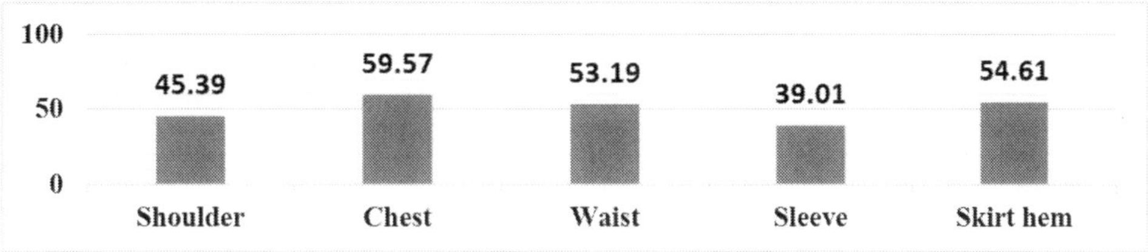

Fig.8 Survey data

4. Fabric Reconstruction in the Dress Design

The design elements were obtained with the results of the survey to design. This chapter analyzes from the aspect of color, fabric, style modeling.

4.1 Color Design Analysis

The series of design were inspired by a fish theme, named "Fish Pleasure". According to the results of the survey, consumers tend to pay attention to the color in dress design. A British psychologist Gregory said: "Color perception is of extreme importance for human beings, it is the core of visual aesthetics, which profoundly affects our emotion state" [14]. This series of dress color uses a mixture of color in gradient blue and light yellow to highlight the visual impact of fabric reconstruction, in view of the overall tone, it creates a fresh and natural look with a sense of emptiness. This kind of digital printing technology formed by the gradual change of the blue is a representation of the sea to create a feminine charm.

4.2 Fabric Design Analysis

A series clothing named "Fish Pleasure" pursues an elegant and crisp look. Wool fabric and twill were selected as the fabrics. S serge wool fabrics were selected in accordance to these characteristics. The texture of fish is made by hollowing out techniques. The white wool fabric is printed with a gradual blue hue by digital printing, the laser cuts out the shape of the scales to make a solid layer, which are the most distinctive feature of this garment. The choice of fabric is an important step in fashion design, the effect of this series of dress design is fully expressed with the design of fabric reconstruction. The process of fabric reconstruction is shown in Fig.9.

Fig.9 Fabric reconstruction process

4.3 Style Design Analysis

In order to exhibit the beauty of the female curve, the series of women clothing named "Fish Pleasure" adopted X type. Bionic design technique is adopted. The bionic design method not only has a unique shape and color, the design was also more creative and challenging [15]. Slim tail skirt highlights the woman's body curves, based on the above factors with the use of aesthetic principles. Clothing painting as shown in Fig.10.

Fig.10 Fashion drawing

4.4 Clothing Production Process

Clothing production can be divided into four main steps. The first is pattern design,using clothing CAD software to draw the clothing structure diagram;then produce the clothing sample;finally modify the pattern according to the result of the fitting. The second step is laser engraving and fabric reconstruction with the use of high-tech technology. The third is digital printing, the position and size are calculated according to the style and pattern. The color of the wool fabric are then transferred with the printing machine. The fourth is clothing production, the pattern of fabric is transformed through cutting and sewing. The basic process of garment production is shown in Fig.11.

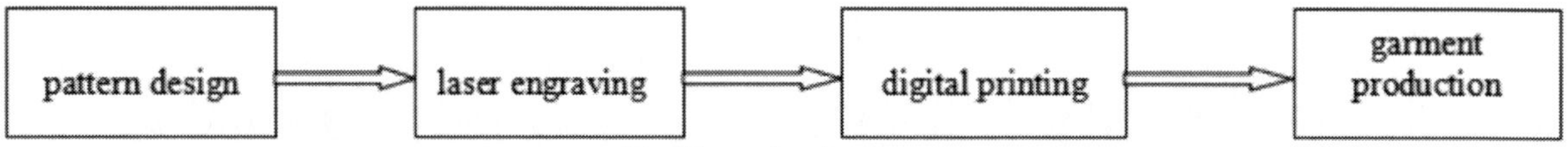

Fig.11 production process

5. Conclusion

With the development of social civilization, sustainable development is becoming more popular. As such fabric reconstruction are emerging with time as it is becoming more favored by fashion designers both at home and abroad. This paper firstly introduces the definition and characteristic of fabric reconstruction. Then based on consumer preferences, a survey was conducted to study the application of design techniques in formal dress. Finally, a conclusion was drawn from the analysis.

With growing environmental awareness, more consumers prefer dresses that are made from reconstruction fabric. Most people like to have a three-dimensional dress that is created through techniques such as folding and hollowing.

With the development of the economy, more consumers are paying attention to the expression of individuality. 40% of consumers hope that the overall design of a dress can reflect the expression of individual personality, at the same time have a strong artistic expression.

Woman's clothing has a great demand for color, 63.12% of consumers pay most attention to color design. In addition to neutral color, most people like warm colors.

In modeling, most women like type X, which highlights the woman's figure. In the survey, 59.57% of consumers pay attention to the chest design, 54.19% focus on the position of hem, 53.19% of consumers pay attention to the waist design.

References

[1] Yang Y. Research on the teaching of the course of fabric redesign [D]. Tianjin University of Technology: 2008.
[2] Wang JR. Process design of fabric texture reconstruction [J]. The textile industry and technology: 2013; 62-66.
[3] Xu DL, Ren S. The two creative design of garment fabrics [J]. Textile Journal: 2016; 37.
[4] Wan L. Application of two design method of fabric in formal dress [J]. Shandong Textile Economy: 2012.
[5] Ma YJ. Application of formal beauty rule in garment fabric redesign [J]. Youth and society: 2014; 17.
[6] Wang CL. Two design and application of clothing material [J]. Intelligence: 2012; 1.
[7] Wang XX. Research on the application of fabric reconstruction and design in modern fashion [D]. Qing dao University: 2014.
[8] Li T. Analyzing the design culture of clothing material originality [J]. Technology vision: 2014; 17.
[9] Zhao M. The expression and application of fiber art in creative design of clothing material [J]. Design: 2016.
[10] Colette W. The Art of Manipulating Fabric [M]. Krause Publ: 1996.
[11] Yao AQ. Discussion on the application of destructive fabrics in fashion design. Art Education Research: 2013.
[12] Li SP. Discussion on the application of fabric reconstruction from the aesthetic of clothing [J]. Humanities Forum: 2012.
[13] Chen JT. Application of fabric embossment in garment design [J]. Journal of LiMing Vocational University: 2014.
[14] Wang JQ. Visual elements and its transmission of clothes shell fabric innovative design [J]. Advanced Materials Research: 2012.
[15] Wang L. Bionic design method of garment modeling bionic design method of garment modeling [J].Hundred Schools In Arts: 2016; 07: 61-64.

Body Mapping as a Method for Design and Engineering of Functional Clothing

Carolina Quintero Rodriguez[1*], Siti Hana Nasir[2], Olga Troynikov[1]

[1]*School of Fashion and Textiles, RMIT University, Brunswick 3056, Australia*
[2]*Faculty of Engineering Technology, Universiti Tun Hussein Onn Malaysia, 81600 Batu Pahat, Johor, Malaysia*

[*]Corresponding author's email: s3188259@student.rmit.edu.au

Abstract

There is an increased demand for innovative-engineered functional clothing for medical use, where not only body morphology and comprising materials are considered but also the responses to the body-garment-function interaction are assessed. The aim of this study is to establish a method for designing and engineering of therapeutic garments - specifically therapeutic gloves and maternity support garments - with enhanced fit, comfort and functionality. 3D body scanning technology and visual analysis of stretch materials were used to examine the locations where the skin strains the most and determine the elongation and contraction of clothing placed on a hand and a pregnant body. The results obtained from the experimental work were then translated into body mapping for development of the segmented therapeutic garment with improved fit, comfort and functionality that ultimately would increase user adherence to the therapy. This study contributes to the existing body of knowledge with a new scientific basis for future design of functional clothing and highlights the importance of future research into body-garment-function interaction for establishing design criteria when developing therapeutic garments.

Keywords: Body Mapping; Functional Clothing; 3D Body Scanning; Therapeutic Garment

1. Introduction

There is an increased demand for innovative-engineered functional clothing for medical use, where not only body morphology and comprising materials are considered but the understanding of biomechanics and body mapping as a response to the body-garment-function interaction is used. Functional clothing for health purposes must fulfil the basic requirements of comfort and the specific demands of the function for which it is created. Mehta and Narrasimham [1], stated that no matter how well clothing is engineered, it cannot be regarded as comfortable if it does not fit properly. At the same time, a garment that does not fit or is uncomfortable will be discarded by the user, regardless of its technical qualities [2].

Therapeutic garments can be defined as garments developed with the purpose of treating or alleviating pain and discomfort. Fit is a key factor that could affect both the therapeutic efficacy as well as wearer comfort and as a result, wearer adherence. The relationship between body-garment-function focus on fit, comfort and therapeutic effects of the garment. Fit in functional clothing is concerned with appearance, comfort and functional ease of the garments [3] and its interaction with the body, taking into consideration body shape, characteristics of the material, functionality of the garment and kinematics. This is to ensure the normal living conditions of the user are restored. Comfort, although difficult to define because of its objective and subjective multidimensionality, is one of the main concerns for wearers. "Indeed, just about every activity a human performs in life involves a process of seeking comfort or relief from environmental and/or mental constrains" [4]. In therapeutic garments, the combination of physical, physiological and psychological responses aim to increase wearer's comfort, quality of life and adherence to the treatment. "When comfort exists, the mind is alert and the body operates at maximum efficiency" [5]. Functionality in therapeutic garments refers to the needs of the specific user while wearing the garment. It is related to the therapy and the movement required when performing activities of daily living (ADL); with the aim to restore normal body functioning on a daily basis.

Despite the increase in developments of functional garments for therapeutic purposes, little has been done to address the design requirements of patients in regards to aspects of sizing, fit, predominant posture, thermal regulation, moisture management, protection and the psychological 'feel good factor' [6]. This paper proposes the use of an objective mix methodology when developing stretch functional clothing with the aim to establish a method for designing and engineering of therapeutic gloves and maternity support garments with enhanced fit, comfort and functionality.

3D scanning technology and visual analysis of stretch materials were used to examine the locations where the skin strains the most and determine the elongation and contraction of clothing placed on the belly area. The results obtained from the experimental work were then translated into body mapping for development of the segmented therapeutic garment with improved fit, comfort and functionality that ultimately would increase user adherence to the therapy.

2. Functional Clothing for Medical Use

Functional clothing for medical use is concern with fit, comfort and functionality for improvement of human mobility and performance, and the effective remedial action of the garment. In this paper, the body mapping methodology was used for developing a framework for therapeutic gloves for the treatment of arthritis and for the development of an improved maternity support garment to improve wear comfort and alleviation of pregnancy-related pains and discomforts.

2.1 Therapeutic Gloves

Therapeutic gloves are widely prescribed by rheumatologists and occupational therapists in routine clinical care to reduce pain, swelling and stiffness, and improve hand function in patients with hand arthritis [7, 8]. Since most hand therapists recommend therapeutic gloves for at least eight hours' wear [7], it is essential that they are comfortable during normal use. The fit of a glove directly influences the hand function of the wearer and resultant wear comfort and users' adherence to use of glove [9]. To optimise the effectiveness of therapeutic gloves and adherence to their wearing, an accurate and efficient measurement of wearers' hand dimensions and characteristics is crucial - especially those that influence hand function and wear comfort.

Although previous studies have considered the influence of hand movements relevant to glove pattern design [9-11], it remains unclear how different hand movements affect hand dimensions and the distribution of skin strain across the hand. Data from the skin strain distribution will provide valuable information for engineers and designers for development of therapeutic gloves with improve fit, comfort, and functionality.

2.2 Maternity Support Garments

Physiological discomforts are common during pregnancy and vary in type and intensity according to the gestational age of the foetus. Some of the discomforts are presented as physical symptoms affecting the mother's quality of life. Of all symptoms reported, the most common discomforts are back pain, Pelvic Girdle Pain (PGP), fatigue and general malaise, which normally arise during the second and third trimesters of the gestational period with the highest intensity between 20 and 36 weeks [12-15].

Non-pharmaceutical relief measures for pregnancy pain such a support garments have been recommended to alleviate discomforts and pain during pregnancy. Although there is no published standard classification of this type of garments, previous researchers described different categories that include four main types: belts or girdles, panties or briefs, cradles, and torso supports, with the belt being the most popular of all [16, 17]. Recently there is an increase in the amount of support garments in the form of shorts and pants that is not considered by previous researchers.

While previous studies show scientific information to confirm the efficacy of the garments in alleviation of pregnancy related pains and discomforts, there is a lack of information about the design principles of these garments. Therefore, the development of this methodology is relevant for future developments.

3. Body Mapping Methods

In order to design and engineer therapeutic functional clothing that enhances body abilities and improves body functions, it is important to find the right balance between fit, comfort and functionality for the specific garment use. For this purpose, multiple methods can be used, including 3D body scanning, analysis of skin strain and analysis of the interaction between materials, body and kinematics of the wearer.

3D body scanning has been used for multiple purposes among the clothing field, such as garment tailoring, sizing surveys, clothing fit etc, as it could provide data for the design and sizing of garments as well as data regarding the interaction between body and garment. 3D body scans capture not only measurements but also the topology of the body reliably [18, 19]. Although 3D body scanners are challenging for capturing textures and colours, it provides fast, accurate and reliable linear and non-linear data that is imperative when engineering therapeutic garments [18, 19, 20]. Through this method, the understanding of the skin strain, body morphology and kinematics of arthritis patients and pregnant women was achieved.

For therapeutic gloves, the skin relaxed-strain ratio was measured and its relationship to skin deformation behaviour during hand movements was quantified. The skin relaxed-strain ratio measures the differences in hand measurements between two landmark points in the static and dynamic postures. Quantification of skin deformation with appropriate wearing ease in garment construction permits unrestricted movement of the wearer's body, thus improving the comfort. The skin relaxed-strain ratio at the dorsal side of the hand was determined as $\lambda_p = [(b-a)/a]\times 100$, where λ_p (%) is the skin relaxed-strain ratio between two landmark points; b (mm) represents the distance between the landmarks in dynamic posture (grip or power grip postures); and a (mm) represents the distance between the landmarks in a static / relaxed posture. 3D INFOOT scanner was used to record the point cloud raw data of the participants in three different hand postures (relax, grip and power grip) as described in Nasir and Troynikov [21]. Data from the skin relaxed-strain ratio were translated into regional hand mapping visualisation with a spectrum of colours representing the range of skin strain (Fig.1). The metacarpal region recorded the greatest change in the skin relaxed–strain ratios in both postures, and this can be observed from the colour representation in the different regions in Fig.1. In the grip posture, the skin relaxed-strain ratios ranged between 38% and 71%, and for power grip posture the values ranged between 55% and 103%. The metacarpal-carpal region recorded minimal changes in skin relaxed-strain ratios; ranging between 5% and 9% in the grip posture and 8 and 13% in the power grip posture. The changes can also be clearly identified based on the colour visualisation in Fig.1. Based on the colour mapping, therapeutic gloves could be developed with regional segmentation to accommodate the skin relaxed–strain ratios of the hand.

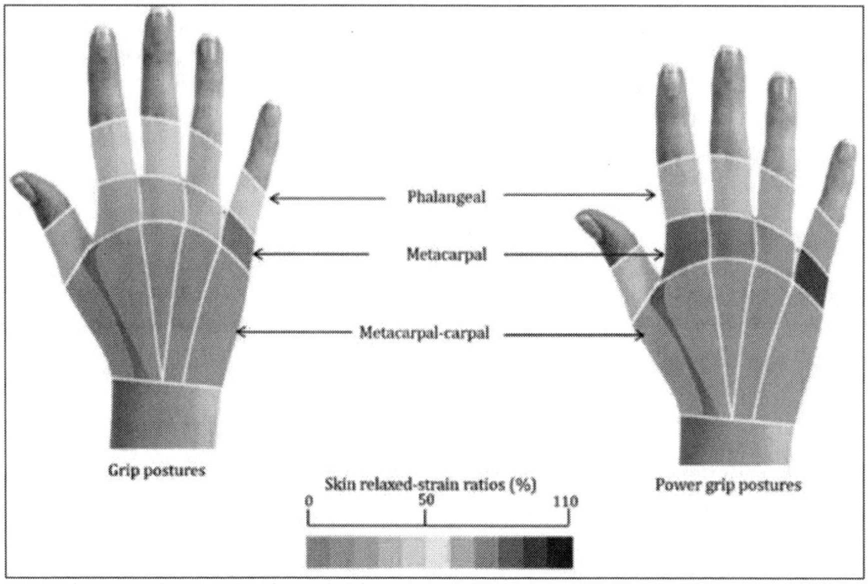

Fig.1 Regional mapping of skin relaxed–strain ratios in different postures.

The elongation and contraction of clothing placed on a pregnant body for the maternity support garment were determined based on the visual analysis of stretch fabrics and visual analysis of garment fit in relevant areas, based on final use [19]. Watkins proposed a visual analysis of stretch materials by studying the curvilinear deformation of a specific grid when placed on a body, which not only influences the fit of the garment but helps to propose an appropriate shape to the garment to ensure functionality and comfort to the wearer [8].

In this study, the visual analysis of stretch fabrics was done through mathematical analysis of expansion of a 2.5cm x 2.5cm grid drawn on a maternity support garment when worn on a pregnant dress form (Fig.2). The garment, in the relaxed state was drawn with the grid and conditioned for 12 hours in standard temperature atmosphere of 20+2 and 65+2% relative humidity as per AS 2001.1-1995 standard [22] and then don on the dress form for measuring. After measuring all cells, the garment was tumble dried at 40°C for 10 minutes to allow for its relaxation for the consecutive measurements. The measurements were repeated five times.

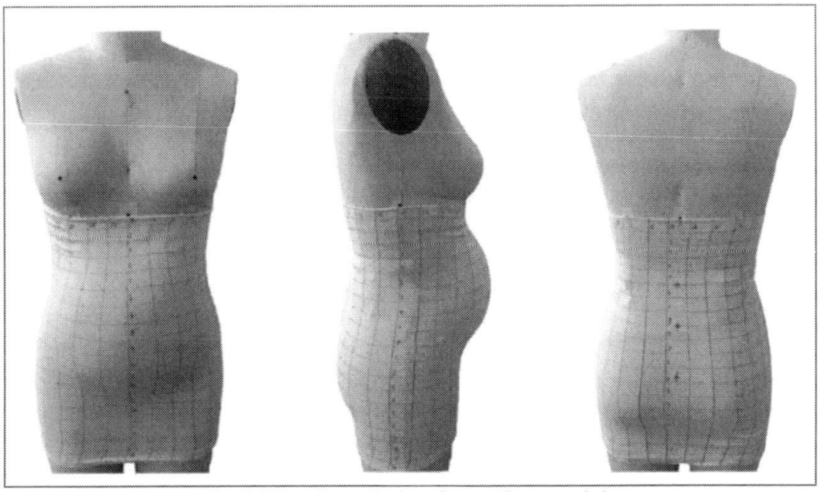

Fig.2 Visual analysis of stretch materials.

The measurements collected were averaged (=AVERAGE) using Microsoft Excel for analysis. The lowest (=MIN) and highest (=MAX) cell values were identified, as to calculate the percentages of stretch of each cell. This data was divided into 5 segments which provided a visual graph of different areas of stretch of the garment when worn in the contours of the body: low extension (between 27% and 39%), medium extension (between 39.1% and 51%), high extension (between 51.1% and 63%), extra high extension (between 63.1% and 75%) and hyper extension (between 75.1% 87.6%) that was represented per 5 different colours. (Fig.3).

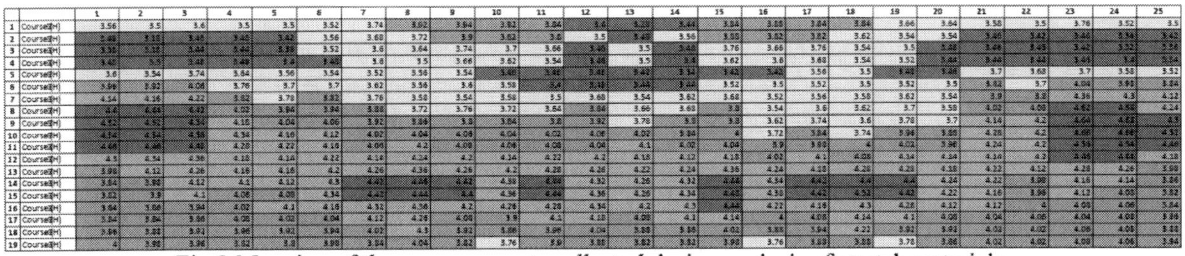

		1	2	3	4	5	6	7	8	9	10	11	12	13	14	15	16	17	18	19	20	21	22	23	24	25
1	Course(H)	3.56	3.5	3.6	3.5	3.5	3.52	3.74	3.82	3.84	3.82	3.84	3.6	3.28	3.44	3.84	3.88	3.84	3.84	3.66	3.64	3.58	3.5	3.76	3.52	3.5
2	Course(H)	3.48	[illegible]	3.48	3.46	3.42	3.56	3.68	3.72	3.9	3.82	3.8	3.5	3.48	3.56	3.88	3.82	3.82	3.62	3.54	3.54	3.46	3.42	3.46	3.34	3.42
3	Course(H)	3.38	[illegible]	3.44	3.44	3.38	3.52	3.6	3.64	3.74	3.7	3.66	3.46	3.5	3.48	3.76	3.66	3.76	3.54	3.5	[illegible]	3.46	[illegible]	3.42	3.32	[illegible]
4	Course(H)	3.42	[illegible]	3.48	3.48	3.4	3.42	3.8	3.5	3.66	3.62	3.54	3.48	3.5	3.4	3.62	3.8	3.68	3.54	3.52	3.44	3.44	3.44	[illegible]	3.4	[illegible]
5	Course(H)	3.6	3.54	3.74	3.64	3.56	3.54	3.52	3.58	3.54	[illegible]	[illegible]	[illegible]	3.42	[illegible]	3.42	3.42	3.56	3.5	[illegible]	[illegible]	3.7	3.68	3.7	3.58	3.52
6	Course(H)	3.96	3.82	4.08	3.78	3.7	3.7	3.62	3.56	3.6	3.58	3.4	[illegible]	3.44	3.44	3.52	3.5	3.52	3.5	3.52	3.5	3.82	3.7	4.04	[illegible]	3.84
7	Course(H)	4.14	4.16	4.22	3.82	3.78	3.82	3.76	3.58	3.54	3.58	3.5	3.68	3.54	3.62	3.68	3.52	3.56	3.58	3.62	3.54	3.9	3.8	4.36	4.3	4.12
8	Course(H)	4.4	4.44	4.42	4.02	3.94	3.94	3.88	3.72	3.76	3.72	3.64	3.84	3.66	3.68	3.8	3.54	3.6	3.62	3.7	3.58	4.02	4.08	4.62	4.58	4.24
9	Course(H)	4.52	4.52	4.34	4.16	4.04	4.06	3.92	[illegible]	3.9	3.84	3.8	3.92	3.78	3.9	3.8	3.62	3.74	3.6	3.78	3.7	4.14	4.2	4.64	[illegible]	[illegible]
10	Course(H)	4.54	4.54	[illegible]	4.34	4.16	4.12	4.07	4.04	4.08	4.04	4.02	4.06	4.02	3.94	4	3.72	3.84	3.74	3.94	3.88	4.28	4.2	[illegible]	[illegible]	[illegible]
11	Course(H)	[illegible]	[illegible]	[illegible]	4.28	4.22	4.18	4.06	4.2	4.08	4.04	4.08	4.04	4.1	4.02	4.04	3.9	3.98	4	4.02	3.96	4.24	4.2	[illegible]	[illegible]	[illegible]
12	Course(H)	4.5	4.54	4.36	4.18	4.14	4.22	4.14	4.24	4.2	4.14	4.22	4.2	4.18	4.12	4.18	4.02	4.1	4.08	4.14	4.14	4.18	4.3	[illegible]	[illegible]	4.18
13	Course(H)	3.98	4.12	4.26	4.16	4.16	4.2	4.26	4.36	4.26	4.2	4.28	4.26	4.22	4.24	4.36	4.24	4.18	4.28	4.28	4.18	4.22	4.12	4.28	4.26	3.98
14	Course(H)	3.84	3.98	4.12	4.1	4.12	4.3	4.42	4.46	4.42	4.38	4.44	4.32	4.26	4.32	4.44	4.34	4.42	4.4	4.4	4.24	4.22	3.98	4.16	4.14	3.86
15	Course(H)	3.82	3.9	4.1	4.06	4.08	4.34	4.42	4.44	4.4	4.36	4.44	4.36	4.36	4.34	[illegible]	4.38	4.42	[illegible]	4.42	4.22	4.16	3.96	4.12	4.08	3.82
16	Course(H)	3.84	3.88	3.94	4.02	4.1	4.18	4.32	4.36	4.2	4.28	4.28	4.34	4.2	4.3	4.44	4.22	4.18	4.3	4.28	4.12	4.12	4	4.08	4.06	3.84
17	Course(H)	3.84	3.84	3.86	4.08	4.02	4.04	4.12	4.28	4.08	3.9	4.1	4.18	4.08	4.1	4.14	4	4.08	4.14	4.1	4.08	4.04	4.06	4.04	4.08	3.88
18	Course(H)	3.86	3.88	[illegible]	3.96	3.92	3.94	4.02	4.1	3.92	[illegible]	3.96	4.04	3.88	3.88	4.02	3.88	3.94	4.22	3.92	3.92	4.02	4.02	4.06	[illegible]	[illegible]
19	Course(H)	4	3.98	3.96	3.82	3.8	3.98	3.84	4.04	3.82	3.76	3.9	3.88	3.82	3.82	3.98	3.76	3.88	3.88	3.78	3.86	4.02	4.02	4.08	4.06	3.94

Fig.3 Mapping of the measurements collected during analysis of stretch materials.

The stretch distribution was then represented in the dress form to visualize the different areas of stretch within the garment (Fig.4).

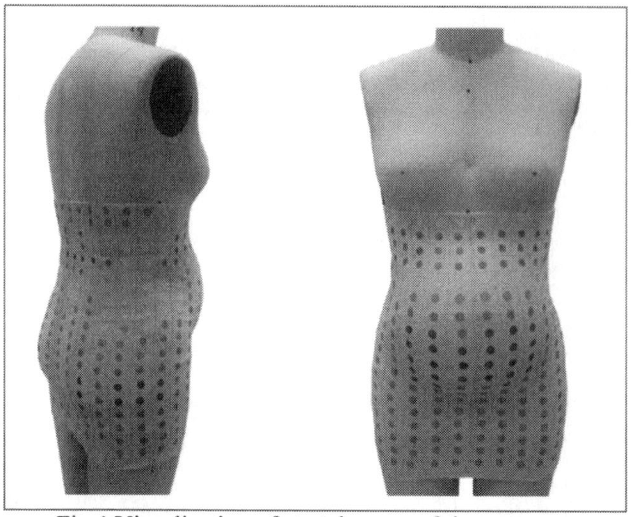

Fig.4 Visualization of stretch areas of the garment.

Finally, the interaction between body-garment was done visually through the analysis of the grid when worn on the body to understand potential movement issues and ease of materials (Fig.5).

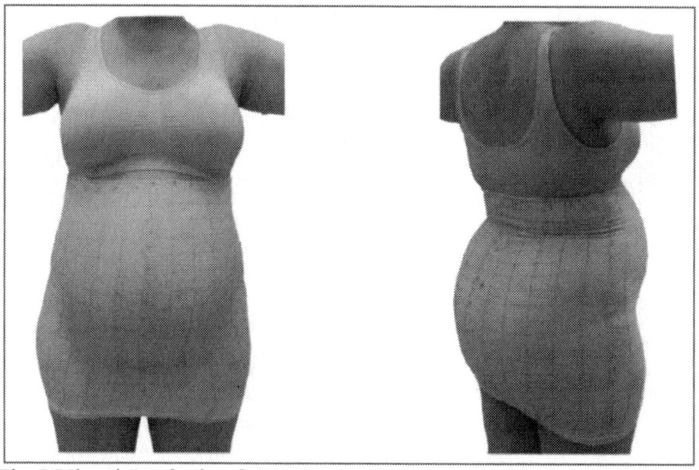

Fig.5 Visual Analysis of stretch garments on a pregnant human body.

Based on the visual analysis and the purpose of the garment, a negative index of the material could be defined to ensure the garment is appropriate for the final use. At the same time, establishing a specific pressure for each part of the garment used in therapeutical applications would increase the balance required for a garment to be effective, comfortable, with good look and fit, for easy user adherence.

4. Conclusion

The use of objective methods for the development of stretch therapeutic garments could bring significant benefits to the wearer. The balance between fit, comfort and functionality will increase adherence to the therapy, alleviation of pain, and user movement that, at the same time, will improve user performance of ADLs.

Multiple methods like 3D body scanning, skin strain analysis and stretch garment analysis can be used to create body mapping for designing segmented therapeutic garments that are able to accommodate the physiological demands of the body. This methodology is expected to facilitate the process of garment design and engineering for therapeutic garment with stretch materials.

References

[1] Mehta R, Narrasimham KV. Clothing Comfort: a review of related properties. Man-made Textiles in India: 1987; p. 327-335.

[2] Gompers M. Sizing up the solution. ISHN: 2004; 38: 75-75.

[3] Boorady LM. Functional clothing- principles of fit. Indian J Fibre Text: 2011; 36: 344-347.

[4] Kilinc-Balci FS. How consumers perceive comfort in apparel. In: Song G, editor. Improving Comfort in Clothing. Woodhead Publishing: 2011; p. 97-113.

[5] Choudhury AKR, Majumdar PK, Datta C. In: Song G, editor. Improving Comfort in Clothing. Woodhead Publishing: 2011; p. 4.

[6] McCann J. Design for ageing well: improving the quality of life for the ageing population using a technology enabled garment system. Adv Sci Tech: 2008; 60: 154-163.

[7] Nasir SH, Troynikov O, Massy-Westropp N. Therapy gloves for patients with rheumatoid arthritis: a review. Ther Adv Musculoskelet Dis: 2014; 6: 226-237.

[8] Hammond A, Jones V, Prior Y. The effects of compression gloves on hand symptoms and hand function in rheumatoid arthritis and hand osteoarthritis: a systematic review. Clin Rehabil: 2016; 30: 213-24.

[9] Torrens G, Campbell I, Tutton W. Design issues in military footwear and handwear. In Spark E, editor. Advances in Military Textiles and Personal Equipment, Woodhead Publishing: 2012; p.139-164.

[10] Bishu RR, Klute G. The effect of extra vehicular activity (Östgaard et al.) gloves on human performance. Int J Ind Ergonom: 1995; 16: 165-174.

[11] Harih G, Dolšak B. Tool-handle design based on a digital human hand model. Int J Ind Ergonom: 2013; 43: 288-295.

[12] Nazik E, Eryilmaz G. Incidence of pregnancy-related discomforts and management approaches to relieve them among pregnant women. J Clin Nurs: 2013; 23: 1736-1750.

[13] Shinkawa H, Shimada M, Hirokane K, Hayase M, Inui T. Development of a scale for pregnancy - related discomforts. J Obstet Gynaecol Res: 2012; 38: 316-323.

[14] Van De Pol G, Van Brummen HJ, Bruinse HW, Heintz AP, Van Der Vaart CH. Pregnancy-related pelvic girdle pain in the Netherlands. Acta Obstet Gynecol Scand: 2007; 86: 416-422.

[15] Wu WH, Meijer OG, Uegaki K, Mens JM, van Dieën JH, Wuisman PI, Ostgaard HC. Pregnancy-related pelvic girdle pain I: terminology, clinical presentation, and prevalence. Eur Spine J: 2004; 13: 575-89.

[16] Ho SSM. Maternity garment treatment for the relief of low back pain. Hong Kong Polytechnic University: Hong Kong: 2008; 244.

[17] Yip J, Yu W. Intimate apparel with special functions, In Yu W, Fan J, Ng S-P, Harlock S, editors. Innovation and Technology of women's intimate apparel. Cambridge:Woodhead Publishing Limited: 2006.

[18] Watkins SM, Dunne LE. Functional clothing design:from sportswear to spacesuits. In Watkins SM, Dunne LE, editors. New York:Fairchild Books: 2005.

[19] Sohn M, Bye E. Visual analysis of body shape changes during pregnancy. Int J Fash Des Tech Edu: 2012; 5: 117-128.

[20] Kim S. Analysis of human body surface shape using parametric design method. Int J Cloth Sci Tech: 2015; 27: 434-446.

[21] Nasir SH, Troynikov O. Influence of hand movement on skin deformation: a therapeutic glove design perspective. Appl Ergon: 2017; 60: 154-162.

[22] Standards Association of Australia. Committee tx, M. O. T. F. T. 1995. Methods of test for textiles. Part 1. Conditioning procedures, Homebush, NSW, Standards Australia.

Research on Constructing a Garment Pattern Design Model for Intelligent Clothing Design

Xiao-Feng Wang*, Xue-Ying Song, Xin Zhang, Bo-An Ying

Apparel & Art Design College, Xi'an Polytechnic University, Xi'an, Shaanxi, 710048, P.R. China

*Corresponding author's email: Mr_wxf@163.com

Abstract

When information became more easily accessible, competition within the clothing industry also became more volatile. To be more competent, mass customization has risen to becomeone of the future trends of the industry. "One person one garment pattern" more suitable with consumer personalized demand. As Made to Measure (MTM) garment production became more significant and popular, there is an urgent growing demand for research on intelligent pattern Computer Aided Design (CAD), making it. a hot research topic in the fashion field in recent years.

In this paper, using tailored suit as an example, the knowledge of intelligent design based on the application of knowledge management was summarized and analyzed, proposing a way to design patterns with Mold. The template model for garment pattern design displayed in this paper was constructed using an AutoPattern platform, in which key components of template design can be easily modified: the module of the design template can be freely combined, superimposed, and/or matched, with the connection between templates easily adjusted to quickly generate a pattern, and eventually creating limitless patterns within a short period of time. This method can promote the application of intelligent clothing CAD, supporting the development of the garment industry.

Keywords: Parametric Clothing Design; Curve Model; Clothing Curve Model.

1. Introduction

With the continuous improvement of living standards, people's demand for clothing is getting higher, especially focusing on satisfying personal needs. Thus, the customized design market share is growing.. Customization design refers to a series of production processes specifically tailored for an individual, which include obtaining accurate measurements of the individual's body parts, using the information to create personalized plate-making and cuttings. To achieve that, clothing enterprises need to pay more attention to the efficiency of plate-making on the basis of the original process. Development of tailored clothing puts forward higher requirements on clothing CAD.

With the development of information technology and artificial intelligence, the idea of intelligent garment pattern design was gaining more attention [1], and can be implemented in the intelligent garment CAD system, The design process of the system is as follows [2]: designer plugs in the required style, body size, shape, data relationship of each part, and design process, etc, to the system, and the system will automatically capture the user's design intention. When the user modifies the parameters in the graphics, the system can also subsequently modify the drawings while maintaining the intended relationship between design objects (such as straight lines and curves) of the original design, generating the required pattern on the computer.

At present, the existing CAD systems, for instance the Gerber system and the OptiTex system and Rick system, replace manual drawings with computerized drawings to realize the clothing pattern design. Several domestic CAD companies are trying to incorporate artificial intelligence functions to the original CAD. For example, domestic BOK intelligence CAD system has integrated various types of prototypes and common garment patterns by parameter processing to establish an automatic correlation between graphics and size specifications based on the intelligent model of automatic recognition technology, which can then quickly generate a new design. Zhi Zun Bao

Fang Intelligent CAD system can automatically design the pattern and entire the whole process of drawing the garment pattern. A garment pattern known as "MTM automatic garment pattern" is automatically generated when the values of the corresponding size table are modified in the record file of the sample version of the same model. The entire MTM pattern-making process imitated the idea of parameterization, and can capture the specific intent of the user during the course of the operation. Even though the above system cited parametric design ideas, it still cannot achieve the free combination of modules, free superposition and automatic matching.

At present, the research area of the relevant institutions are divided into the following major mainstreams:

1) Garment pattern design according to component [2, 3]: for example, Jiao Huiqin of Xi'an Polytechnic University studies the intelligent combination of Business Suit; 2) Parametric design idea [4-6], 3) BP Neural network model, using object-oriented technology to describe garment structure and to establish expert knowledge base and realize automatic generation of pattern based on the idea of parameterization [7-9]; 4) to determine the best matching pattern from the typical pattern library [10, 11] through human-computer interaction and obtaining the models on the basis of the matching correction; 5) Changpeng Hu [12] used Auto CAD 2010 as platform, and utilized Visual LISP and Open DCL as development tools to realize intelligent generation of suit pattern by employing industry expert knowledge and parameterization idea.

Although some achievements have been obtained on the basis of the above research work and extension of intelligent function for CAD system of traditional clothing, they are greatly restricted due to limited data on database system of typical pattern. Since the knowledge of garment pattern design is collected, analyzed, shared and reused without proper knowledge management, the information gathered cannot be effectively incorporated into the clothing CAD system. In order to make the intelligent clothing CAD modules able to freely combine, freely superpose, automatically matching and modifying the relationship between designs to generate patterns quickly, this paper summarized the intelligent design knowledge of patterns for men's suit, and researched on the components and the construction process of intelligent garment pattern model. The model was applied to the design of Auto pattern system, allowing the system to quickly link and modify existing designs to generate as many new patterns as possible within a short period of time.

2. Knowledge Structure and Analysis of Intelligent Fashion Design Model

The design process of intelligent garment CAD is as follows: The designer inputs required style, body size, shape, data relationship for each part of the design process and so on. The system will automatically capture the user's design intention and record all design objects and their relation. When the parameters in the graphics are modified, the system will also automatically modify the drawings, generating the required pattern on the computer while maintaining the intended relationship between the design objects (such as straight lines and curves) in the drawing.

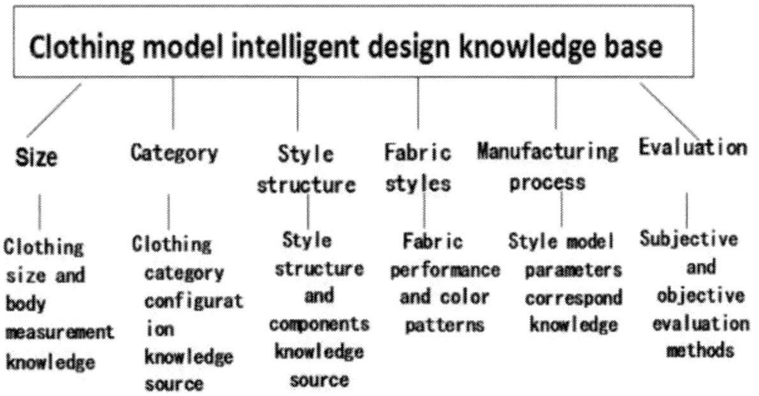

Fig.1 Men's suit intelligent design expert knowledge base framework

By combining design knowledge with design method of garment pattern in the intelligent CAD environment, garment pattern design knowledge can be divided into many categories including dimension knowledge, clothing category knowledge, style structure knowledge, style fabric knowledge, garment pattern-making process knowledge, evaluation, and modification knowledge of garment pattern. Using men's suits as an example, a structure of the clothing model intelligent design knowledge base is illustrated in Fig.1.

Some aspects of knowledge relating to garment pattern intelligent design are as follows:

1) Knowledge of dimensions: including garment size, body size, standard size and dimensions examination;

2) Knowledge of clothing category: according to the different clothing categories such as suits, jackets, and underwear;

3) Knowledge of style structure: covers men's suits of different styles of structural forms and components of knowledge including style part attribute, component change rule and component splicing rule;

4) For example, men's suits can be divided into different designs according to the different styles of clothing profile: four open body, six open body, H profile and V profile. It can also be divided into collar type change and lapel change according to the change of collar, buttons, and other parts of the bottom. Another categorization method is based on bottom change: small arc angle pendulum, slope arc pendulum, and corner pendulum according to the bottom change;

5) Knowledge of pattern-making process: knowledge of pattern- making process includes two aspects: geometric reasoning and rule matching;

6) Knowledge of style fabric: covering fabric performance and compatibility of pattern color and lining material that include physical properties of fabrics, fabric pattern fabric color and compatibility of fabric and lining;

7) Knowledge of process parameters: knowledge of process parameters is aimed at different models, which include sewing process and mechanical properties;

8) Knowledge of model evaluation and revision include the corresponding evaluation of style model, which can be divided into the subjective evaluation and objective evaluation.

3. Research on Construction Method of Intelligent Fashion Design Model

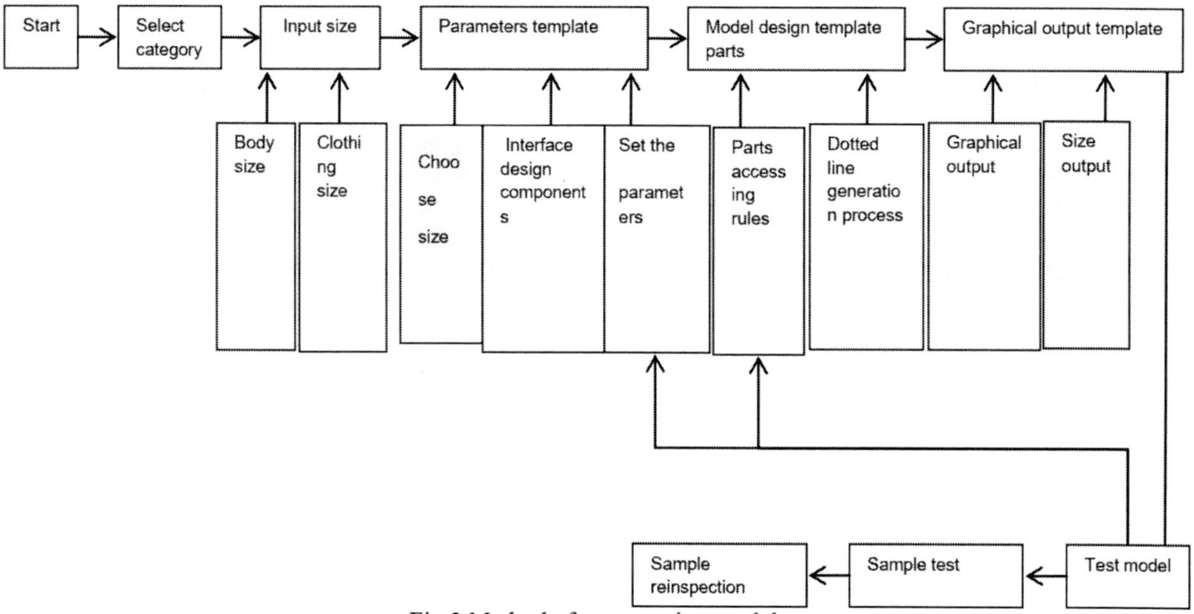

Fig.2 Method of constructing model

A summary of the above information can provide support for the design of garment pattern. The main purpose of the design model is to simplify and make the information more easily accessible and adaptive in the application of the garment pattern design process. In the intelligent CAD system, it is necessary to summarize the knowledge, experience and methods of each step in the design process. In order to form a standard template, the attributes of garment design method and the use of a series of modules are compiled by setting the template constraints.

In the case of men's suit pattern design, the process of building the model is shown in Fig.2:

(1) Category selection: Regardless of the garment type, you need to first choose the garment pattern category. For instance, suits, jackets, shirts or underwear. Then choose the clothing category, and the system will provide a template for this category of clothing input interface;

(2) Size selection: general reference dimensions of pattern design are divided into two categories: human body size and specification size. Regardless of the size, the system will provide the user with a data access interface, in which the user simply has to select the desired size;

(3) Setting of parameter template: Setting of parameter template can be divided into three categories including introduction of human data, interface of style components and parameter settings;

1) Introduction of human body data can be plugged in by introducing the parameters of the human body;

2) Choosing style: according to the different parts of the classification, such as that mentioned in Section 2.3 of this paper, set up different parameters that can be used to control the template design access rules according to the different styles of classification;

3) Parameter template(a template for pattern design using different parameters) settings: parameter template settings can also be used to set up other parameters used in the system. As a section of the garment pattern making process knowledge, fabric knowledge and process knowledge can be used as parameters in the system.

(4) Design module of component template: design module of component template contains two aspects: using template access parameters to determine whether the piece can enter the drawing mode and dot line generation process.

(5) System output: when the system finishes all dot and line-generating processes, the system will produce the template, with the point line graphics generated first, followed by the defined output data.

4. Pattern Checking

The auto-pattern system developed based on the garment pattern design model can be adaptedto draw a garment pattern in the template design. The process is as follows:

1) Information-setting module of user's size

The module does not only allow the user to set his or her personal size, but also choosing the size of the test module, as shown in Fig.3.

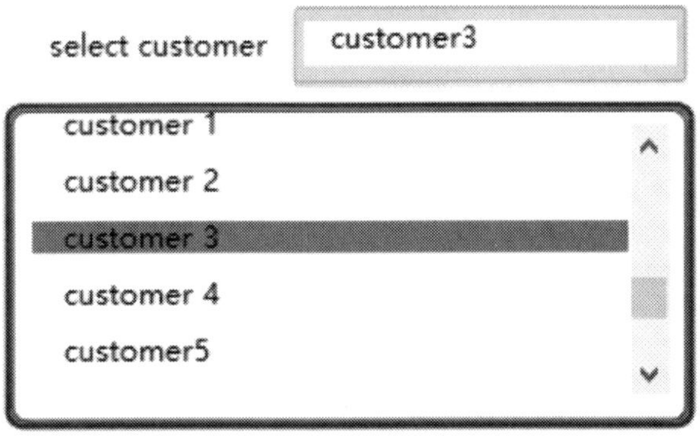

Fig.3 User size information setting module

2) Category-configuring module

The module needs to select or create different clothing categories under pattern configuration. As shown in Fig.4, some of the categories include jackets, suits, dresses, shirts, and bras.

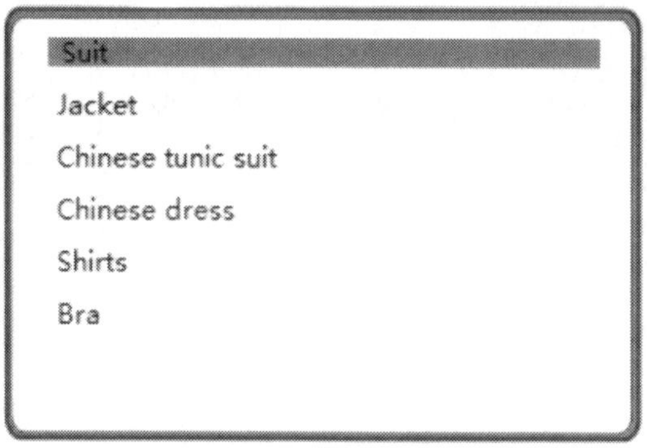

Fig.4 Category configuration module

3) Style-configuring module

This module is to establish a style configuration based on a pre-exisiting clothing configuration and is aimed at clothing styles of different clothing categories, as shown in Fig.5.

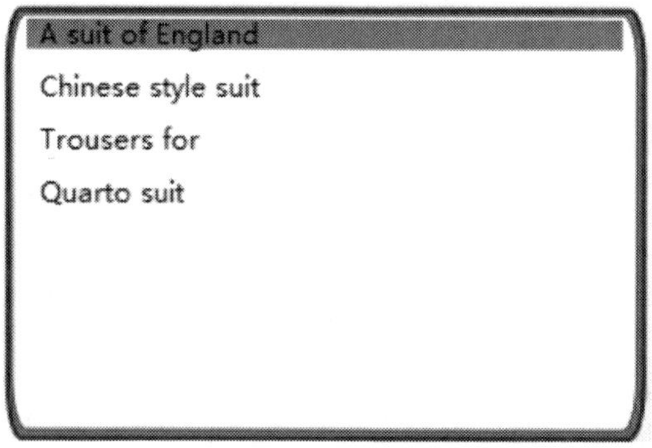

Fig.5 Style configuration module

4) Overall variable-setting module

This module is to add overall variable-controlling parameters before drawing the selected template styles, such as length of Clothes and depth of armpit hole", as shown in Fig.6. The name of each column belongs to a unique part of the clothing with a serial number and a parameter name.

C	035	VS_Chest
C	050	VS_crowtocuff
C	055	VS_depthofscye
C	060	VS_elbow
C	065	VS_frontofshoulde
C	070	VS_fulllength
C	075	VS_halfbackmeasu
C	080	VS_height
C	085	VS_overshoulder
C	090	VS_points
C	095	VS_seat
C	100	VS_seatlength
C	105	VS_waist
C	110	VS_waistlength

Fig.6 Global variable setting module

5) Template-drawing module

The module is based on the knowledge base and designs of the template. It enables users to edit the template in order to meet the actual assembly relationship model according to the geometric topology of the template parts, in which of the location of key points, as shown in the figure, is determined by different rules (formula, interaction, internal grading). Incorporating different rules on lines and curves, the points are then connected, as shown below in Fig.7, by structured lines and parts to create a complete garment pattern.

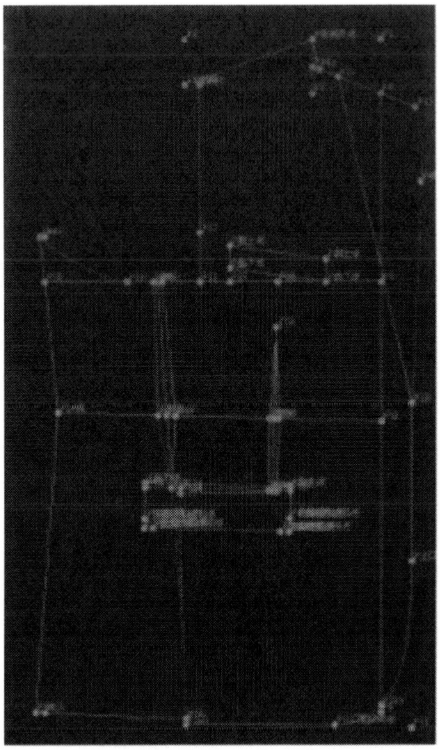

Fig.7 garment pattern drawing module

6) Pattern-generating module

The module is mainly for recording the pattern-drawing process and displaying the results.

7) Garment pattern output module

The module stores and formats the template output. This module can be used for storing documents on the computer and producing the changed format.

Using the above-mentioned model, the template can be modified by adjusting the parameters in the system depending on the individual it is designed for different people.

5. Conclusion

When choosing a different size for the design, the model changes the parameters of template accordingly, while the user can still manually modify the parameter values. However, they can only be modified in the system. Also, the model simplified the processes to designing parametric models. For example, if the user thinks that the intrinsic parameters do not meet the needs of the garment pattern, other parameters can also be added. . To summarize, the model is very flexible, as users can always modify values automatically generated by the computer to suit their individual needs and the modules can be freely combined, superimposed, matched and/or modified to efficiently generate as many new patterns as desired.

Acknowledgement

The name of the project is "Research on Constructing a Garment Pattern Design Model for Garment Intelligent Design" (Project No.J201606), and it is supported by China Textile Industry Federation and Textile Light Technology Education Foundation.

References

[1] Zhang NP, Zhang WY, Zhang HZ. Development and prospects of apparel CAD technology development and prospects [J], Tianjin Institute of Textile Technology: 2006; 9: 120-116.
[2] Hou YL. Research on method of Intelligent garment pattern design [D]. Tianjin University of Technology: 2008.
[3] Cao WL, Xiu Y. Research on aided garment parametric system [D]. Beijing Institute of Clothing: 2007.
[4] Tian ZH. Object-oriented clothing hit / push system [D]. Xiangtan University: 2007.
[5] Ma L, Ying ba, Zhang X. Research on the Methods of the Suit Armholey-Curve Generating in Intelligent CAD System, TBIS: 2015; 04.
[6] Zhou WC, Zhang X, Ying BA. The graph representation of clothing model design knowledge and its Application in Intelligent Design, TBIS: 2016; 03.
[7] Cheng DS, et. Present situation and development trend of garment [J], CAD Journal of Jilin Institute of technology: 2000 (04); 13-15.
[8] Li CZ. research and implementation of personalized clothing CAD based on parametric [D], Shenyang University of Technology, master thesis: 2006
[9] Lu X. The realization of parametric design of garment plate making system based on feature [J]. Chinese Journal of textile science: 2006; 27 (12): 62-65.
[10] Jiang XY L DJ. Garment pattern parameters intelligent generation based on Autolisp [J]. Journal of Beijing Institute of Fashion Technology: 2007; 27 (1): 41-45.
[11] Hu CP, Zhang XL, Liu RP. A basic pattern of deformation of Armhole Curve concave expert knowledge [J]. Journal of Beijing Institute Of Fashion Technology: 2006; 26 (3): 46 50
[12] Hu CP. research on the intelligent design system of suit pattern design [D], Beijing Institute Of Fashion Technology, master thesis: 2010

Research on the Relationship Model of Cheongsam Pattern Based on Personalized Customization

Xin Zhang*, Qian Ma, Bo-An Ying, Bu-Gao Xu

Apparel & Art Design College, Xi'an Polytechnic University, Xi'an, Shaanxi, 710048, P.R. China

*Corresponding author's email: xianzhangxin@sina.com

Abstract

With the rapid development of China's economy, the emergence of personalized costumes has become an inevitable trend. In this paper, the pattern relational model mainly refers to the body structure of the cheongsam is studied, based on the established basic pattern. First, through the analysis of the influencing factors, the interactive parameters are set, and at the same time, find out the relationship between the cheongsam pattern and the basic pattern, to build the cheongsam relationship model; finally, the pattern relationship model is verified by adjusting interaction parameters and trying to evaluate samples.

Keywords: Cheongsam; Interactive Parameters; Pattern Relationship Model; Classification Design.

1. Introduction

With the continuous improvement of people's living standards, personalized clothing customization that possess good quality, strong fitting and their own style, have become the goal of consumers in this modern era [1]. The methodology of mass production has not been able to meet these requirements of the rapid response on the modern market. and the design of garment patterns in mass production is mostly to take the chest circumference, height into the derivation formula, which is difficult to meet the actual needs of customers, but also can not accurately reflect the detailed characteristics of the human body. Therefore, the establishment of garment pattern relationship model based on the characteristics of the human body and the generation technology of the garment pattern will be key to realizing personalized customization. Cheongsam, as the quintessence of China, has a history of more than one hundred years, and its image presented now is the result of social material form and ideology. The cheongsam's style can show off the three-dimensional structure of the human body well. Retaining its original face is understandable. Therefore, the study on cheongsam has a historical significance and I would like to illustrate research conducted by the following domestic scholars:

Chen Liling made a comparative analysis and research on the structural design and technological evolution of the traditional cheongsam, the "improved" cheongsam and the modern cheongsam from horizontal and vertical angles, and summarized the law of structural design and evolutionary process of the three kinds of cheongsam. All of which have an important theoretical value and significance for its study [2]; *Zhang Hao* et al, designed 5 kinds of schemes with different distribution patterns based on the basic pattern of cheongsam, and made sample clothes and corrected them. While, at the same time, he analyzed the relationship between the distribution of the dart and the shape of the side-seam by wrapping the best pattern of the 5 schemes and the development drawing of plaster model, which laid the foundation for the study of cheongsam pattern [3]; *Shi Rui*, through the analysis of the female body regulation, obtained the optimal human dimensions with the help of dynamic and static body measurement, and then studied the loose design and pattern optimization of cheongsam, for different types of women, which provided a scientific and practical method for the quantity and structure design of cheongsam [4]; *Liu Guanbin* et al, used the method of transferring the prototype dart to the cheongsam pattern, and used this principle to change the style of the cheongsam without changing the original style [5]; Liu Jianping developed cheongsam without dart, and made use of the elasticity of the fabric and the common technique of pulling back to meet the needs of the human body curve [6].

Based on the above research, it could be concluded that the research on the style of cheongsam has not yet been realized. Under the background of these researches, this paper will study the automatic generation of cheongsam pattern. Based on the established basic pattern [7], the cheongsam pattern relationship model was studied, which can truly realize the design concept of "one person one pattern, the hierarchical design".

2. Analysis of Influencing Factors of Cheongsam Pattern

2.1 Garment Relaxation

Garment relaxation is also known as the degree of relaxation or volume, the actual meaning is the gap between clothing and the human body. The amount of relaxation has the following two functions: first, to meet the functional requirements of the human body, including the basic breathing, activities and accommodate the needs of underwear level; second, the performance of decorative clothing effect. The loose design is reasonable or not, which can directly determine the wearing comfort of the cheongsam pattern, and can also affect the aesthetic appearance. Therefore, the amount of relaxation is a kind of design, and it is necessary to study the content of the garment pattern relationship model.

Practice has proved that the distribution of clothing relaxation in the chest, back, waist and buttocks have a great influence on the overall movement and shape of the garment. Therefore, this paper takes the body structure of cheongsam as the research subject for relaxation design, from the three aspects of length, circumference and width, for meeting the customer's personalized demand for cheongsam. Table 1 is for the distribution of a loose cheongsam.

Table.1 Relaxation of cheongsam

Part	Relaxation
Dress length	actual measured length+(0.5-1cm)
Bust girth	add 3-5cm
Waist girth	add 3-5cm
Stomach girth	add 3-5cm
Hip width	add 3-5cm
Shoulder width	zero volume
Back width	add 0-1cm
Bust width	zero volume

2.2 Dart Design

The female body is a complex and subtle three-dimensional entity. To design a modern cheongsam in line with the beauty of female body, Intensive study of the dart change technique is essential. Women's darts have chest dart, shoulder dart and waist dart and so on, to adjust any problems with the chest convex, dorsal and waist or buttock.

In the design of a fitted cheongsam, the chest design is important, and the change technique of the chest dart is flexible and easy to use, widely used. It not only lays the foundation for the "semi-stereoscopic" to create three-dimensional clothing, but also develops its principle to "even the dart into the seam "and "dart varies pleat", which would greatly enrich the modern cheongsam structure line and contour.

3. Construction of Cheongsam Pattern Relationship Model

3.1 The Design of Cheongsam Pattern

According to the analysis of the influencing factors of the cheongsam pattern, the cheongsam pattern design, is mainly based on the distribution of loose and dart design. In this paper, the cheongsam pattern is designed from the two aspects, loose distribution and dart design, based on the basis of the established basic pattern. Ease design is mainly reflected in the shoulder, back, chest, waist, buttock and several other detailed sites, which are considered the most critical parts of the cheongsam, with particular focus on the size of the waist relaxation and the change of the waist line, as it can bring a colorful modeling effect.. Furthermore, the length of the cheongsam will vary due to different styles. As for the dart design, the front piece is with the waist dart and underarm dart, and the underarm dart is transferred from the chest dart of the basic pattern; the back is with waist dart design.

3.2 Set up Interactive Parameters

In the design process of the relationship model of the cheongsam pattern, some interactive data is necessary according to the analysis of the special parts, except the size of the human body. The main purpose is to control and adjust the length of the cheongsam pattern, the dart, the distribution of loose in special parts, and hems and slits. These interactive parameters are designed according to the plate making experience. Table 2 lists the interaction values required for the cheongsam pattern relationship model.

Table.2 The interactive information of Cheongsam pattern unit:mm

Serial number	Code	Range
1	X_BH	[10,50]
2	X_BW	[0,50]
3	X_B5	[0,50]
4	X_BB	[0,50]
5	X_FH	[0,50]
6	X_B3	[0,30]
7	X_FB	[0,50]
8	X_FW	[0,50]
9	X1	[-40,20]
10	X2	[0,45]
11	X3	[-100,100]
12	X4	[0,60]
13	X5	[-20,35]
14	X6	[-40,50]
15	X7	[-20,50]
16	X8	[-20,45]
17	X9	[-30,40]
18	X10	[-40,0]

3.3 The Relationship between Cheongsam Pattern and Basic Pattern

In combination with the interactive parameters set in Table 2, the cheongsam pattern is generated on the basic pattern. The characteristic points and characteristic lines are adjusted and transformed on the characteristic points and the characteristic lines of the basic pattern, and the specific changes are shown in Fig.1. The corresponding relations of the characteristic points, the characteristic lines between the two patterns are shown in Table 3 and Table 4.

Table.3 The relationship of the feature points between the cheongsam pattern and the basic pattern

The feature points of the Basic pattern	The feature points of the cheongsam pattern	The change of the cheongsam feature point
FNP	FNP_F	no change
SNP	SNP_F	no change
BNP	BNP_F	no change
SP	SP_F	front shoulder dart turn to breast dart; no change with back shoulder
FA	FA_F	add the interaction value"X_B3"
BA	BA_F	add the interaction value"X_B5"
BP, PP13	BP_F, PP13_F	no change
SS	SS_F	no change
PAP	PAP_F	front sleeves raise "X1+10"
FWP SWP BWP	FWP_F SWP_F BWP_F	waist line up "X2"
FHP	FHP_F	no change
SHP	SHP_F	add the interaction value "X_B"
PP6, PP7 P6, P7	PP6_F, PP7_F P6_F, P7_F	front shoulder dart turn to breast dart; no change with back shoulder
PP8, PP9 PP11, PP12 P8, P9 P11, P12	PP8_F, PP9_F PP11_F, PP12_F P8_F, P9_F P11_F, P12_F	PP8_F raise 15; no change with PP9_F P8_F drop 15; no change with P9_F
——	P14_F, P15_F, PP14_F, PP15_F auxiliary points P16_F P17_F auxiliary points PP16_F、PP17_F	The hem of Cheongsam pattern hem will lengthen; add auxiliary points

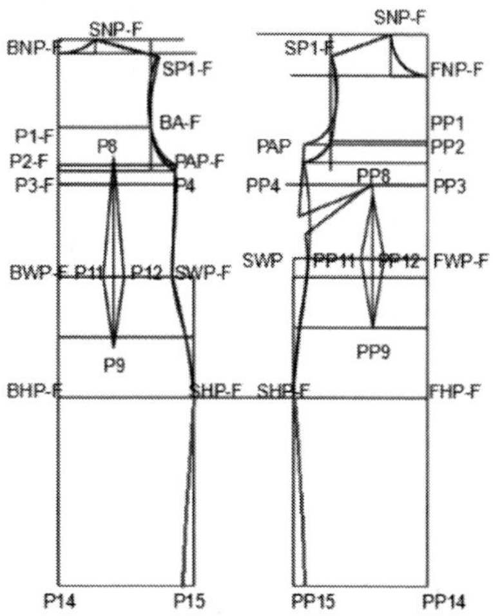

Fig.1 Comparison of cheongsam pattern after adjusting the parameters

Table.4 The relationship of the feature lines between the cheongsam pattern and the basic pattern

The feature lines of the basic pattern	The feature lines of the cheongshen pattern	The change of the cheongsam feature line
FCL	FCL_F	increased by "L_HL-L_AK"
NL1	NL1_F	No change
NL2	NL2_F	No change
SL	SL_F	removing the front shoulder dart, no change with the back shoulder dart
AL1	AL1_F	connected by SP1_F, FA_F, PAP_F, and FA_F, PAP_F have been adjusted
AL2	AL2_F	connected by SP1_F, BA_F, PAP_F, and BA_F, PAP_F have been adjusted
BL1	BL1_F	add the interaction value "X_B3"
BL2	BL_F	add the interaction value "X_B5"
FBL	FBL_F	add the interaction value "X_FB"
BBL	BBL_F	add the interaction value "X_BB"
L1	L1_F	add the interaction value "X_BB"
L2	L2_F	add the interaction value "X_FB"
WL1	WL1_F	add the interaction value "X_FW"
WL2	WL2_F	add the interaction value "X_BW"
HL1	HL1_F	add the interaction value "X_FH"
HL2	HL2_F	add the interaction value "X_BH"
____	LPP14PP15	connected by PP14、PP15
____	LP14P15	connected by P14、P15
SSL	SSL_F	The SSL_F will lengthen
b（BCL）	BCL_F	The BCL_F will lengthen
Lines L1, L2	____	____
L3, L4, L5, L6	L3_F, L4_F, L5_F, L6_F	no change
____	L7_F, L8_F	formed by front shoulder dart turn to breast dart;
l1, l2, l3, l4, l5, l6	l1_F, l2_F, l3_F l4_F、l5_F, l6_F	no change

4. Verification of Cheongsam Pattern Relationship Model

4.1 The Parametric Verification of Cheongsam Pattern Relationship Model

The established cheongsam model is input into the "AutoPattern" plate making software, which is researched by Shaanxi garment engineering technology center, and the cheongsam pattern design model is established; then, substituted with the character size data of the stage. Finally, we adjust the relevant interaction parameters, and automatically generate the contrasted cheongsam model. As shown in Fig.2, the black is on behalf of the pre-adjusted cheongsam pattern, and the red represents the adjusted cheongsam pattern.

The interactive parameters set in the built-in cheongsam model can be adjusted and changed. The change of dart, the waist can be adjusted by these parameteres; at the same time, they can adjust the loose of some parts according to the different cheongsam style and consumer dress habits.

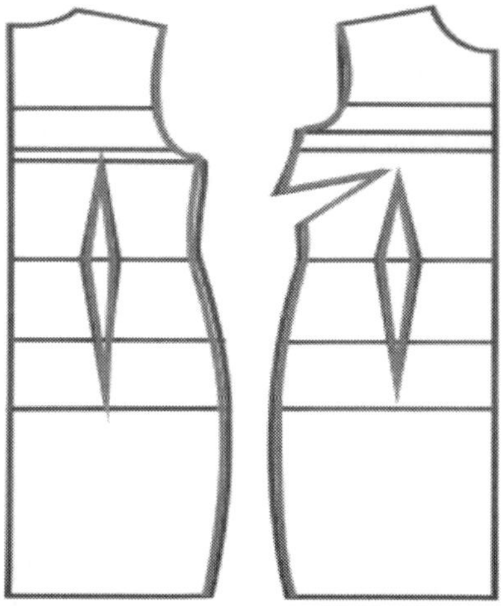

Fig.2 Automatic generation of cheongsam patterns

4.2 The Trial Verification of Cheongsam Pattern Relationship Model

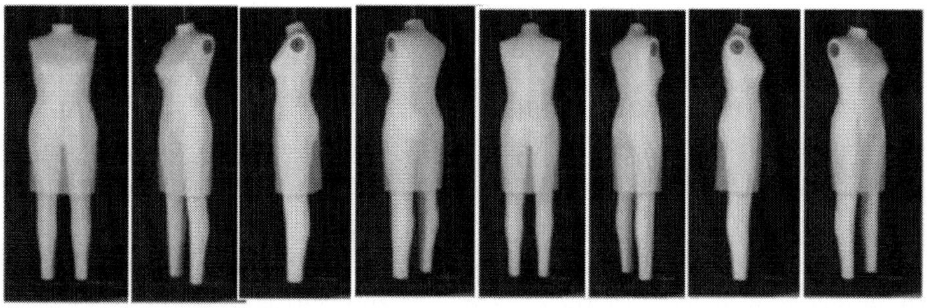

the front 45° the side 135° the back 225° 270° 315°

Fig.3 Eight angles of the sample effect drawing

Choose the shooting system of clothing wearing effect of Shaanxi Garment Engineering Center to try to make the trial verification. The pattern of the cheongsam, which is automatically generated by the pattern design model, is made of white grey fabric, and the model is tested. It is verified by the characteristic line correspondence between the human model and the sample, and the distribution of the ease. The sample fitting effect is shown in Fig.3.

From the horizontal point to view, the collar line of the sample is rounded at the collar of the human body model. And the chest, back, waist, buttock, and knee of the sample are in good agreement with the relative position of the human body model.

From the vertical point to view, the sleeve line of the sample is consistent with the arm circumference of the human body model. The side suture line also cleverly highlights the beauty of the curve of the human body model.

From the perspective of the overall fitted degree of sample clothing, eight angle dress effects are more fit. Only the part that below the hip and above the knee leaves the right loose, which is specially designed to facilitate human walking.

Thus, the establishment of the cheongsam relationship model is reasonable.

5. Conclusion

1) This paper analyzes the influence factors of the cheongsam pattern and then the cheongsam body structure is taken as the study object to carry out the design of cheongsam.

2) Through the analysis of the relationship between the cheongsam pattern and the basic pattern, the pattern relationship model of the cheongsam is built.

3) For the established cheongsam pattern relationship model, this paper verifies its rationality through the interactive parameter adjustment and the sample fitting effect.

6. Acknowledge

China Textile Industry Federation and Textile Light Technology Education Foundation support this study and its project name: Research On The Design Method Of Garment Pattern Design Model For Garment Intelligent Design (project no.J201606).

References

[1] Ning GH. The study of the relationship between the blouses model parametric model based on the characteristics of the female body. [D]. Xi'an Polytechnic University: 2013; 02

[2] Chen LL. Study on the structure design and process evolution of Cheongsam. Jiangnan University: 2010; 7.

[3] Zhang H, Zheng R, Xu F, Wang XX. The relationship between the space distribution of dart and the shape of side seam [J]. JBIFT: 2006; 9 (3): 40-45.

[4] Shi R. Study on the amount of loose cloth in the cheongsam of life and the optimization of pattern. [D]. Donghua University: 2011.

[5] Liu GB, Tang RX. The research and practice of modern cheongsam structure design. [J]. Silk: 2004 (01).

[6] Liu JP, Zhang Y, Jing ML. The exploration and analysis of the shaping method of cheongsam without dart. [J]. Tianjin University of Technology: 2004 (02).

[7] Ma Q, Zhang X*, Xu BG, Ying BA. The Study of the Pattern Design Model for Female Upper Garment based on Mapping Principle, TBIS: 2016 (09); 430-437.

[8] Hu CS, Du HR, Hao RM. Analysis and practice of cheongsam modeling technology from the perspective of ergonomics. [J]. Changshu Institute of Technology: 2016 (08); 43-47.

[9] Zhang QZ. The design and design of silk cheongsam [J]. ITH: 2012 (04); 47.

[10] Li BY. The research of wedding dress design system based on Personalized Customization. [D]. Tianjin University of Technology: 2016.

[11] Li H, Xu JL. Construction of automatic generation model of female upload prototype model based on tailor - made. [J]. FG: 2016; 5 (06): 47.

[12] Hu CS, Du HR, Hao RM. Analysis and practice of cheongsam modeling technology from the perspective of ergonomics [J]. Changshu Institute Of Technology: 2016 (08); 43-47.

[13] Jun ML. The development of an intelligent system for customized clothing making [J]. Expert Systems with Applications: 2010; 37 (1):7 99-803.

[14] Hong J. Detail optimization and virtual display evaluation of men's suits [D], Donghua University: 2010.

[15] Yuan WJ. Research on provincial distribution of tight prototype based on point cloud data [D]. SuZhou University: 2010.

Research on Relational Model for Body Block of Menswear Based on Graph Theory

Xin Zhang[1], Jin Duan[1, 2*], Xue-Lian Kang[1], Bo-An Ying[1]

[1]*Apparel & Art Design College, Xi'an Polytechnic University, Xi'an, Shaanxi, 710048, P.R. China*
[2] *Yangling Vocational & Technical College. Yangling Shaanxi, 712100, P.R. China*

[*]Corresponding author's email: 604951552@qq.com

Abstract

With the consideration of garment construction, point, line and surface are three basic elements in apparel pattern design. The essence of men's apparel pattern design is to investigate the point-line topological relationship based on data collected from parameters of the human body. The process of apparel pattern design includes data between point and line, direction of data flow, basics of pattern design and the suited plane that restricts topological relationship construction which containing menswear styling. In this paper, the design process of the relational model for men's apparel pattern is carried out by applying apparel pattern design classification. Besides, research on pattern design for the first stage of body block of menswear which belonged to the relational model of men's apparel pattern is based on the image processing techniques of graph theory, which establishes the relational model for body block of menswear and enables the knowledge-driven generation of men's apparel pattern automatically in the Auto Pattern software.

Keywords: Men's Apparel Pattern; Graph Theory; Relational Model; Digitized Customization

1. Introduction

The definition of menswear refers to the generic term of apparel that men wear in business, working and leisure occasion. Menswear is the most important sub-sector in the apparel industry, which can be further divided into suit, casual wear, sports wear by consumer demand. The "Analysis Report on the Demand of Production in Chinese Menswear Industry and Marketing Investment Forecast (2015-2020)" pointed out that Chinese menswear industry will enter the period of adjustment. Besides, the report also anticipates a decline of sales growth in China's menswear industry from 16.02% in 2014 to 10.94% in 2018 [1]. The reason underlying the decline is the inability of the current menswear industry meeting each consumer's individual need. In China, studies on men's apparel pattern started relatively late. Professor W.B. Zhang (2002) established the Donghua Basic Block by collecting, extracting and analysing data of some important parts of young men's upper body in East China, South China, and North China, building the unary regression equation among vital detailed measurements of men's upper body, height, and bust [2]. Professor R.F. Liu (2005) obtained the "Standard Basic Body Blocks of Menswear" throng by summarizing and sensibly improving the "Basic Block of Menswear" which was written by Bunka Fashion College and the "Basic lock of Menswear" which was published by Metric Pattern Cutting For Menswear, a book based on the standard body of Chinese adult male [3]. However, men's apparel pattern design by both the Donghua Basic Block and Basic Pattern of Menswear is based on these three designing elements: bust, waist and height in "GB / T1335.1-2008 Standard sizing system for Garment", which all belong to the ready-to-wear pattern design [4]. Since a man's figure incessantly changes during his lifetime and men of the same garment size also tend to differ in accordance with evidently individual differences, consumers have to choose their outfit from the ready-to-wear products available on the current market that approximately suit their figures. As a result, individual customer demand cannot be met.

Based on the idea of apparel pattern design classification, this paper presents a new method for pattern design : the relational model design of menswear pattern, which will ultimately benefit menswear customization. Taking Alva M male form as an example, this paper investigates the first stage of the relational model for men's body block which belong to the retional model of men's apparel pattern. Then, it establishes and analyses the topological relationship for body block of menswear through image processing of the graph. Moreover, it inputs the topological relationship for the body block of men's tops into the Auto pattern to achieve the automatic generation of the relational model of men's apparel pattern. According to the automatically generated pattern, samples can be made and the establishment of the relational model of man's apparel pattern can be justified.

2. Theoretical Analysis

2.1 Definition of the Relational Model for Men's Apparel Pattern

The relational model for apparel pattern of men's tops takes individuals as target customers in the study and customizes the pattern for each individual. To ensure the convenience and precision of apparel pattern in the design process, the design process can be divided into three stages: body block, garment block and styling pattern [5-6].

The relational model for body block of menswear refers to the apparel pattern of men's tops which maximally retains the features of men's upper body without changing the styles nor functions in the basis of relaxing the minimum respiratory volumes. Based on the body block of men's top, the garment block caters to the consumers' individual requirements for styles. Through analysing the special, the garment block for men's top is established in combination with the enterprise's designing elements and processes for these patterns. In addition, the stylistic pattern for men's top is a modification and redesign procedure of special styles according to consumers' demands for styles. This process includes designer's ideas and concepts, and reflects the consumers' individual demands for styles.

The whole process of pattern classification design serves as the basis of the process of relational model for apparel pattern of men's top. In this paper, the design of the relational model of men's apparel pattern is constructed based on the idea of customizing menswear. The classification design of the relational model of men's apparel pattern in the design process ensures a suitable and personalized design of menswear by orderly applying the influencing factors of men's apparel pattern to the relational model of men's apparel pattern. Thus the classification design of the relational model expresses better and stores the pattern generation ideas from the pattern designers. The establishment of the relational model of men's apparel pattern allows for the automatic generation of knowledge-driven menswear patterns. Fig.1 shows an example of procedures of T-shirt pattern design in the relational model for apparel pattern of men's top.

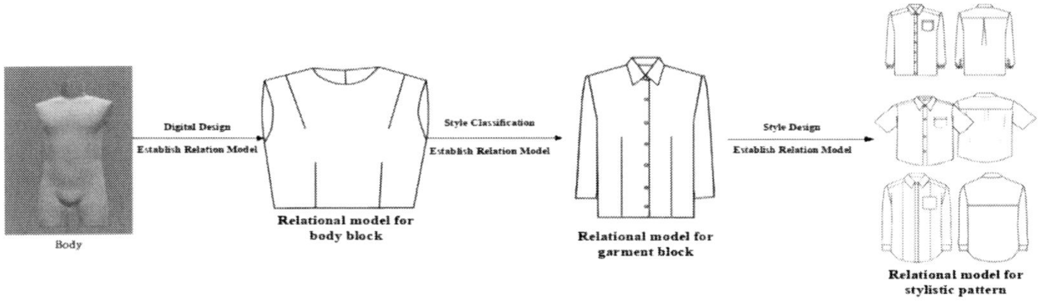

Fig.1 The Classification Design Procedure of the Relational Model for Pattern

2.2 Research on Relational Model for Body Block of Menswear Based on Graph Theoretic Algorithm

J. Duan (2016) presents a method to obtain a two-dimensional planar graph by modelling in the CATIA software and flattening the men's three-dimensional surface morphology and data presented in "Research on Relational Model for Body Block of Menswear based on Digitizing 3D Human Model" [7]. It can be introduced to the graph theoretic algorithm to analyse and design pattern generation. Defining the pattern generation of men's top as graph

G, graph G is composed by a non-empty nodal set $P=\{p_1,p_2,\cdots,p_n\}$ and a frontier set $C=\{c_1,c_2,\cdots,c_m\}$ remarked as $G=(P,C)$ (Fig.2). The node number is rank [8]. Through analysing the most essential contents of graph G, that is, the correspondence between the node and the edge, the relational model for body block of men's top G is an undirected graph. Thus $\psi(c_i)=(p_m,p_k)$ can be introduced among them. In this formula, c_i is the non-directional line with the Wight that links p_m and p_k [9]. The so-called rank refers to the topological relationship from dot to line in apparel pattern.

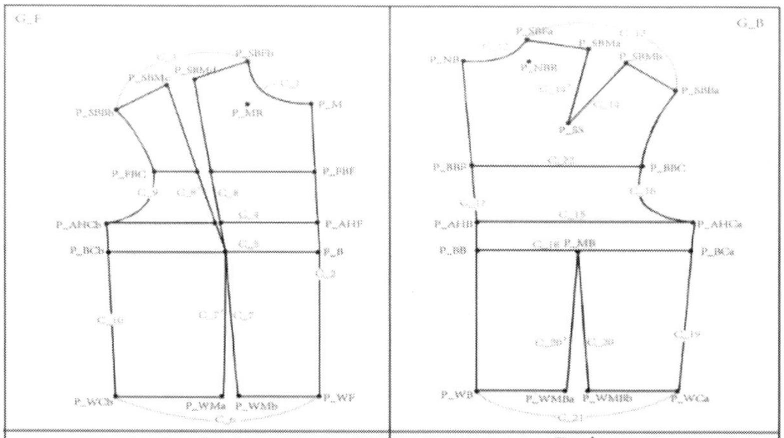

Fig.2 The Relational Model for Body Block of Man's Top Based on the Graph Theory

Therefore, the essence of the relational model establishment of men's apparel pattern is transforming the dot-line topological relations in apparel patterns into a set of geometric data that was derived from the limitations of structural parameters to the graph of the pattern. While constructing the corresponding relationship between graph and data for the structural parameters when the parameter is assigned with a different value, it will plugs in the geometric data to the original pattern, and directly generate a new apparel pattern. From the perspective of apparel pattern designing, parameters in the relational model of apparel pattern can be categorized into the following four types. (Table 1) [10-11]

Table.1 Parameter Type and Parameter Specification

Parameter Type	Parameter Specification
Size parameter	The size of the body's parts that are needed in establishing the relational model of men's apparel pattern
Style parameter	Variables needed in designing the apparel style
Parameter that defines the properties of the graphic object	Variable parameters, that designer's definition of the length, width and other properties of the pattern in designing according to his pattern model knowledge
Composite parameter	A parametric expression formed by combining simple parameters with mathematical or logical operators

2.3 Establishment of the Relational Model for Body Block Based on the Graph Theory

In the relational model for body block of men's top, the size parameter is related to the body measurement. Since there is a corresponding relationship between feature points of the body block of men's top and those of men's upper body (Table 2), the size parameter of the body block of men's top cites the feature points of the body block of men's top, the corresponding feature points of men's upper body and the body measurement data.

Table.2 Size Parameters of the Body Block of Men's Top [7]

Feature lines of body block of men's tops	Feature lines of body block of men's upper body	Feature lines of body block of men's tops	Feature lines of body block of men's upper body
Front neck curve(C_1)	Front neck length (L_FN)	Back neck curve(C_12)	Back neck length (L_BN)
Front centre line(C_2)	Front line length (L_FL)	Back shoulder (C_13)	Side shoulder length (L_SB)
Front shoulder (C_3)	Side shoulder length (L_SB)	Dart line of back shoulder(C_14)	Dart line of back shoulder (L_SS)
Front armhole depth(C_4)	Length of front armhole depth(L_FAD)	Back armhole depth(C_15)	Length of back armhole depth(L_BAD)
Front bust (C_5)	Front bust length (L_FB)	Back armhole curve (C_16)	Back armhole length(L_BAH)
Front waist(C_6)	Front waist line(L_WF)	Centre back (C_17)	Centre back length (L_BC)
Front waist dart(C_7)	Front waist dart length (L_BW)	Back bust(C_18)	Back bust length (L_BB)
Front shoulder dart (C_8)	Front shoulder dart length (L_BS)	Back side seam (C_19)	Side seam line length (L_SSL)
Front armhole curve(C_9)	Front armhole length (L_FAH)	Back waist dart (C_20)	Back waist dart length (L_SW)
Front side seam(C_10)	Side seam line length(L_SSL)	Back waist (C_21)	Back waist length(L_WB)
Front width line(C_11)	Front width line length(L_FB1)	Back width line (C_22)	Back width line length(L_BBW)

In view of the definition, the body block of men's top maximally keeps the body features of men's upper body on relaxing the minimum amount of respiratory allowance, without any changes in style and function [12-13]. Fig.3 is the style parameters that influence the body block of men's top.

Table.3 Types of Style Parameter of the Body Block of Men's Top

Types of Parameter	Parameter Specification
Style Parameter	Front allowance, Back allowance, Front shoulder dart, Back shoulder dart, Front waist dart, Back shoulder dart, Front shoulder slope, Back shoulder slope, Front central dart, Back central dart and Waist side seam in the Body Block of Men's Top

The parameter that defines the properties of the graphic object is the interactive parameter that a pattern designer uses in designing the pattern model. The pattern designer initially chooses the target and obtains the body measurements (size parameter) from the target. Therefore, in the process of designing the body block of men's top, the designer first uses the body measurements of the target, then setting the style parameter through the definition of style data in accordance with styling and modifying the pattern of the relational model for apparel pattern of men's top through setting up interactive parameters (which defines the properties of the graphic object) for the parts that need to be controlled. Table 4 shows the interactive parameter for the body block of men's top based on Alva M Form and the default parameters of pattern measurements. The default interactive parameters come from the body block of Alva M form which was published in 2016 named "Research on Relational Model for Body Block of Menswear based on Digitizing 3D Human Model". The size range is determined based on the knowledge of the pattern designer.

Table.4 Interactive Parameter for the Body Block of Man's Top

Names of the Interactive Data	Defaults	Size range	Specification for the Interactive Value
Front shoulder slope for body block of men's top	21.4°	0-25°	To determine the angle of the front shoulder slash for body block of men's top
Back shoulder slope for body block of men's top	11°	0-25°	To determine the angle of the back shoulder slash for body block of men's top
Front shoulder dart for body block of men's top	8°	0-10°	To determine the intake amount of the front float remainder for body block of men's top
Back shoulder dart for body block of men's top	24°	0-25°	To determine the intake amount for the back float remainder of body block of men's top
Side seam waist for body block of men's top	10MM	0-20MM	To determine the amount of the side seam waist width for body block of men's top
Front waist dart for body block of men's top	5.5°	0-15°	To determine the amount of front waist width for body block of men's top
Back waist dart for body block of men's top	5.5°	0-15°	To determine the amount of back waist width for body block of men's top
Front central dart for body block of men's top	2.5°	0-5°	To determine the amount of the iota chest quantity for body block of men's top
Back central dart for body block of men's top	4°	0-5°	To determine the amount of back upheaval for body block of men's top

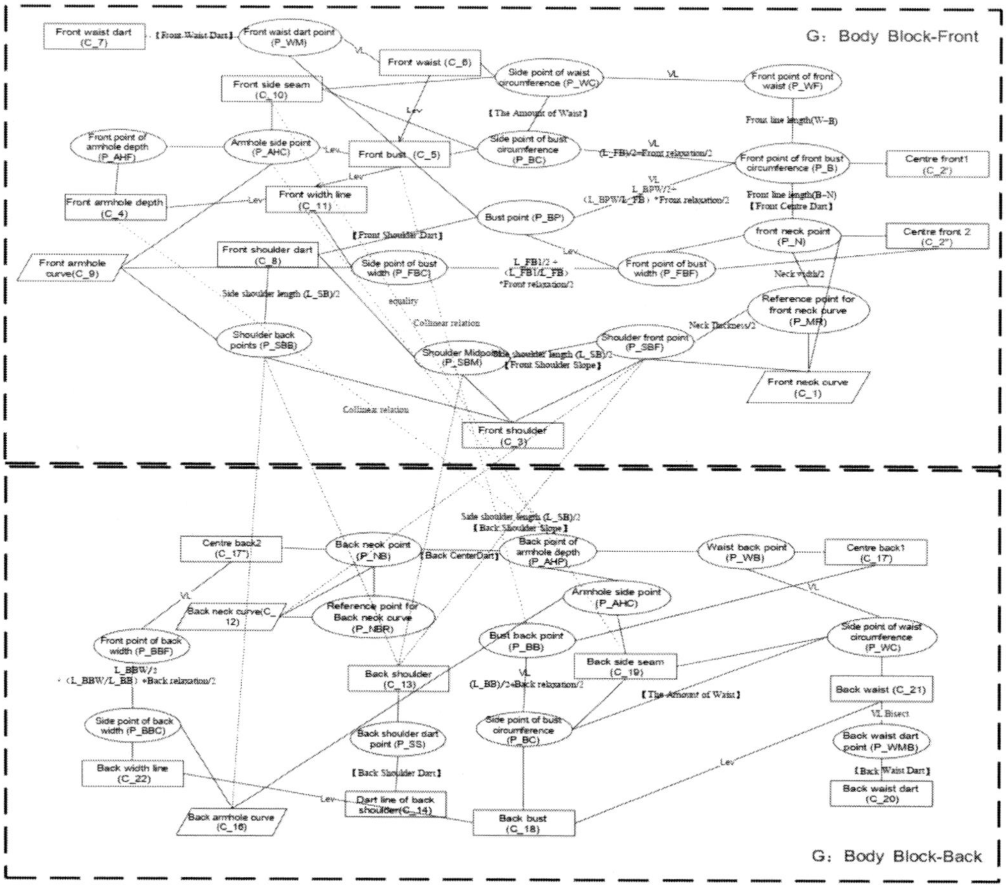

Fig.3 The Relational Model of Body Block Based on the Directed Graph

Fig.3 shows the pattern generation data, procedure and data flow of the rational model for the body block of men's tops in the form of directed graphs. In this graph, ○ represents the dot, □ represents the line, ▱ represents the curve, Lev represents the parallel, VL represents the vertical, → represents the direction of the formation of elements, the signs and formulas on the arrows represent the size of body measurement and the calculation formulas between the size data and the style parameter in the formulation of each element, and 【】 represents the interactive parameter of pattern design.

3. Establishment of the Relational Model for Body Block of Menswear Based on Graph Theory

3.1 Background Introduction of the Relational Model for Body Block Establishment

Auto Pattern is an automatic pattern composition system designed by the Shaanxi Province's Research Center of Apparel Engineering. This technique is based on the knowledge-driven pattern model design software. It automatically generates an apparel pattern by inputting the data name, defining the type of data, and inputting the computing mode for variables. (Fig.4) The pattern generation interface can be divided into four parts from left to right, namely: information configuration area, pattern display area, data input area, and data display area. The relational model for apparel pattern of men's top can be set up as this system can directly configure the individual data, define the type of data, and direct the flow of data.

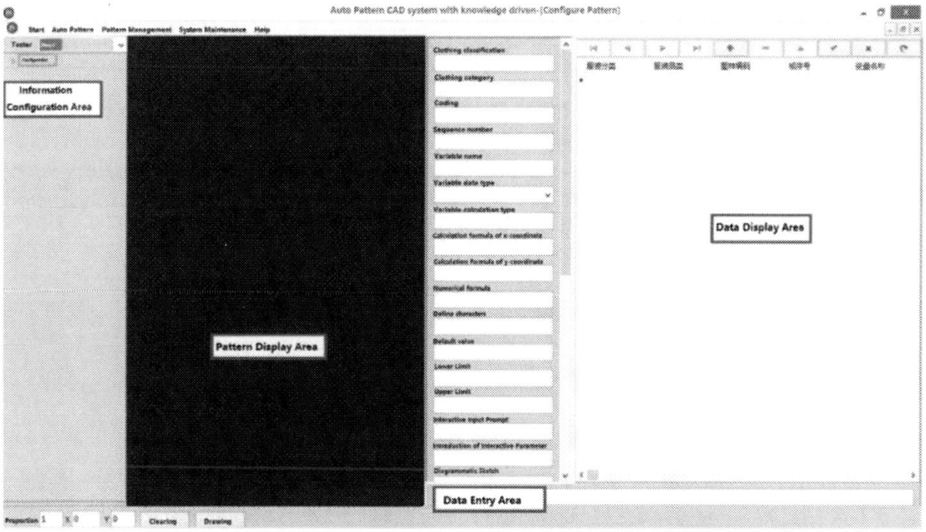

Fig.4 The Interface of Intelligent Pattern Model of Auto Pattern

In the automatic pattern composition procedure of the relational model for apparel pattern of men's top, pattern designers need to select the target, obtain body measurements of target (parameter of size), and quote the body measurements of the target in the design process of the relational model for apparel pattern of men's top. Besides, the designer adjusts the pattern in the rational model of body block of men's apparel pattern through defining the style data (styling parameters) according to the style and setting interactive parameters (which defines the properties of the graphic object) for the parts that need to be controlled. (Fig.4(a)) The relational model for body blocks of men's top has interactive parameter settings which allow pattern designers to adjust the pattern through changing the amount of interactions. For instance, when the interactive parameter for the waist allowance is modified, the relational model for apparel pattern of men's top is automatically adjusted and generated. (Fig.4(b)) The red mark indicates the pattern model which was generated after the modification of interactive parameters and the black mark represents the former pattern of the relational model for apparel pattern of men's top.

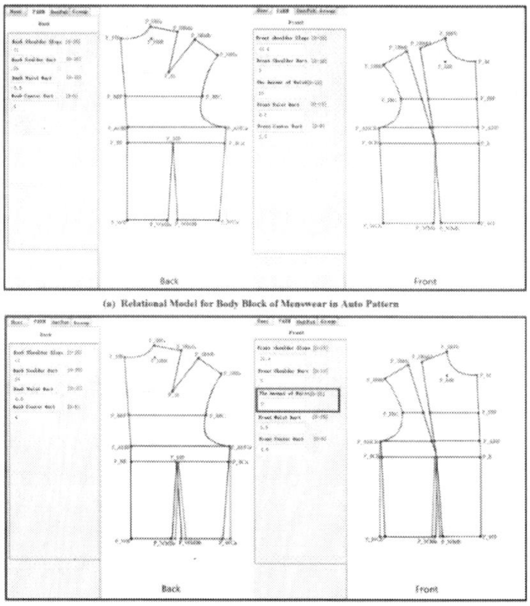

(a) Relational Model for Body Block of Menswear in Auto Pattern

(b) Parameter Adjustment in Auto Pattern

Fig.5 The Design and Variety of the Relational Model for Body Block

3.4 Evaluation of Fitting Sessions for Men's Body Block Samples

Body block samples generated by each stage of men's apparel pattern rational model were put on Model M of Alva Form (Fig.6). Fig.6 shows the fitting effect of each stage of men's apparel pattern rational model when put on Alva M Form. The front central line is vertical and the bust is horizontal. Moreover, the bust allowance is appropriate and the surface of the sample is smooth without any apparent lumps. Thus it is reasonable and scientific to establish the rational model for body block of men's top.

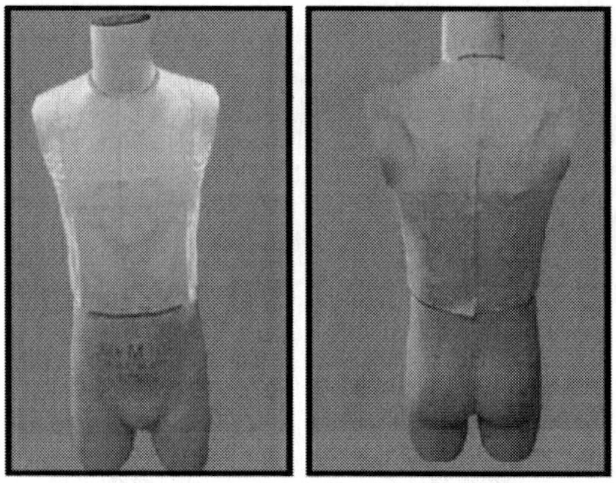

(a) Front (b) Back

Fig.6 Body block samples on Alva M Form

4. Conclusion

Based on the method of classification design for man's apparel pattern and graph theoretic algorithm, this paper explores the correspondence among each stage of pattern model. Meanwhile, in the form of oriented graph, this paper highlights the date flow of apparel pattern as well as their changing rules, directly uncovering the relationship between human body measurements and the relational model of stylistic pattern model for men's tops. Also, it establishes the relational model for body blocks of men's top in Auto Pattern, verifying the validity of the design method of the relational model through fitting sessions and estimation of sample body blocks. Besides, the relational model of men's apparel pattern generated in Auto Pattern automatically creates the men's apparel pattern through adjusting different static and interactive parameters based on the parameterized model of relationship.

5. Acknowledge

The name of the project is "Research on The Design Method of Garment Pattern Design Model for Garment Intelligent Design" (Project No.J201606), and it is supported by China Textile Industry Federation and Textile Light Technology Education Foundation.

References

[1] Prospect Industry Institute. China Men's Clothing Industry Production & Marketing Demand and Investment Forecast Analysis Report: 2015-2020. https://wenku.baidu.com/view/e0b1aa528bd63186bdebbc09.html.
[2] Zhang WB, Zhang WY. Chinese Basic Block-Structure De signing Principle and Technical of Donghua Basic Block. Journal of Donghua University, Natural Sciences: 2002; 28 (3): 44-50.
[3] Liu RB. Pattern Making Theory & Technology -Menswear. China textile press: 2005.
[4] GB/T1335.1-2008 Standard sizing system for Garment
[5] Qi J., Basic Bra Pattern Based on Reflection Theory, Textile Bioengineering and Informatics Symposium Proceedings: 2014; 926-933.
[6] Chen Y. Research on Bra Pattern Design and Evaluation Methods for Digital Made to Measure, Xi'an Polytechnic University: 2014.
[7] Duan J. Research on Relational Model for Body Block of Menswear based on Digitizing 3D Human Model, Textile Bioengineering and Informatics Symposium Proceedings: 2016.
[8] Yu ZG. Discrete mathematics. National Defence Industry Press: 2011.
[9] Xu GJ. Research on Some Problems in Graph Theory. Northwestern Polytechnical University: 2004.
[10] Roller D. An approach to computer-aided parametric. Computer-Aided Design: 1991; 23 (5): 385-391.
[11] Wu JF. Research on Dimension-Driven Parametric Method Based on the Data of Graphics and Building Database. Shanghai Jiao Tong University: 2003.
[12] Wang X. Analysis Study of the Ease Allowance Needed for Garment. Journal of Textile of Research: 2005; 26 (4): 126-128.
[13] Chen Y, Zeng X, Happiette M, et al. A New Method of ease Allowance Generation for Personalization of Garment Design. International Journal of Clothing Science and Technology: 2008; 20 (3): 161-173.

A Study on the Features and Patterns and Colors of Decorative Silk Pictures

Xiao-Sha Zhang [1], Li-Li Mao[1,2]

[1]*Knitted garment design, Xi'an Polytechnic University, School of clothing and art design, Jinhua South Road 19, Xi'an, 710048, China*

[2]*Xi'an Polytechnic University, School of clothing and art design, Jinhua South Road 19,Xi'an,710048,china*

[*]Corresponding author address: 976623712@qq.com

Abstract

As modern aesthetics, environmental awareness, and modern decorations in people's lives occupy an increasingly important position, people are increasingly demanding in their standard of decorations. This gradually formed a "light decoration, heavy decoration" Decoration concept, wool decorative painting in the modern decoration gradually developed, but also occupy an increasingly important proportion. Wool decorative painting can be based on the size of the decorative space, the owner's habits, hobbies and emotional needs. All the above must be coordinated perfectly to show the owner's personality and style.

In the decorative painting, hand painted out of the decorative painting despite the public has been touted and welcome, but more artists in the bold attempt to replace the material, which makes the development of wool decorative painting gradually revealing the use of different yarn to express the decorative screen , Combined with different colors and design, in order to show the picture more tension, so that the viewer has a new mind touch and multi-layer perception of the United States.

The research results of this paper provide theoretical guidance for modern decorative designers, and it is of practical significance to promote the rapid and rapid development of modern decorating industry.

Key words: Wool Decorative Painting; Features; Craftsmanship; Color; Pattern

Introduction

With the continuous development of social means of production and lifestyle changes, modern science and technology develop new materials and adapt according to its technology, inspiring designers' desire to create, they almost use the life of all the materials touched to design and produce works of art to meet aesthetic needs. Material beauty embodies the sensitivity of decorative painters to material selection and the wisdom of material applications. Material properties are the characteristics of sublimation, enriching its artistic qualities. The development of wool decorative paintings has gone through the process of copying paintings from creative paintings. Now gradually develop a mature wool decorative painting.

This paper aims to enrich the modern decorations of the species, improve people's aesthetic horizons, and expand the application of decorative paintings in different fields.

The main point of this thesis is to summarize the techniques of sticking, winding, weaving, weaving, crochet, embroidery, and other techniques of wool decorative painting according to the different types of wool, and the simple analysis of different texture and composition The Influence of Materials on the Visualization of Color and Pattern in Wool Decorative Painting.

1. Wool Decorative Painting

1.1 Decorative Painting Overview

Decorative paintings are considered more pretentious than the general "decorations", decorative paintings are independent pieces of art, and an unique form of painting with its own characteristics.

Modern decorative painting is a special art form which is influenced by modern painting and modern design theory and practice. All the elements of the work for the construction of a visual aesthetic world. Decorative painting should have a "natural" decorative agenda. Like a chair signifying "rest". Independent of the color of the world of black and white decorative painting, which is characterized by black and white as the benchmark, has its own rules, laws and mysteries. For the reason, know the contrast harmony is its special request. [1, 2, 8, 9]

1.2 The Meaning of Wool Decorative Painting

The wool from the decorative painting is simply the incorporation of woolen materials in the creation of decorative paintings according to the contents and emotional elements that decorative painting is trying to convey. The process of creating such a woolen decorative painting involves the selection of appropriate materials for decoration works on the screen, by pasting, winding, weaving, embroidery, crochet and other methods, as well as finding the perfect combination of color; finally selecting the appropriate pattern of yarn for the presentation of the wool decorative painting.

1.3 The Form and Composition of Wool Decorative Painting

The beauty of wool decorative painting is its expression of its aesthetics and methodology. Formal beauty is the service of content. Wool decorative painting is different from ordinary painting, it is often not a form, but is combination of various integrated forms.

The form of wool decorative painting is also important, as even its form is considered content. Because of its decorative vocabulary and the existence of language, the expression of its abstraction determines the endless beauty of the decorative painting, and the pursuit of beauty beyond objectivity. Diversity and unity of the United States; symmetry, balance, rhythm, proportion of its symbol; comparison of the United States of America and its law of harmony, are the main elements of the combination of decorative beauty. We use these principles of beauty to adjust, assign, transform and transition to achieve the unity of opposites. [3, 6]

2. Study on the Technique of Wool Decorative Painting

2.1 The Classification of the Craft of Wool Decorative Painting

In common life woolen decorative paintings mainly consist of paste decorative paintings, decorative paintings that involve winding knitting, woven decorative paintings, decorative painting that involve embroidery, crochets and other forms of decorative painting. These decorative paintings of different styles will give people different visual experiences. [4, 5, 7, 10, 16]

2.1.1 Paste Decorative Painting

Paste decorative paintings involve the use of special beauty wire according to the design to the yarn and pasted a picture, need to use glue or a strong force is fixed, which is characterized by that art highlights the yarn artistic expression, its widespread, abundant sense of color.

The two pictures are made with coral velvet, its soft texture is very strong, this kind of yarn is easy to affect, capable of shaping the vivid image of the object from coral fleece. The texture of the charm adds a touch of warmth to the decorative painting, but also can the paste this approach can better reflect (Fig.1).

(a) (b)

Fig.1 Coral velvet sticker

2.1.2 Winding Decorative Painting

The use of special beauty wire according to the designed pattern, involves pasting the yarn onto the painting, need to use glue or give high tensile fixation pattern according to the picture, choosing the appropriate colored wool materials without any complicated texture accumulation, using only a simple coil winding, without broken back circuitous outline designs, combined with the brightness of objects which reflects the three-dimensional segmentation, creating clear textures and a unique visual experience.

(a) (b)

Fig.2 A yarn wrapped up in decorative painting
(From http://img3.duitang.com/uploads/item/201504/07/20150407H5725_QuxF2.thumb.700_0.jpeg)

These two works are the use of decorative painting techniques and the formation of the wound. In figure a. the knitting winding in the peripheral frame plays a main role in the background painting frame, decorative painting is not a single become dull; figure B is on board a pattern shape, hit nails, wool to back to the screen winding, winding and orderly, filling the painting with the sense of hierarchy and rhythm, as the picture to add some dynamic (Fig.2).

2.1.3 Woven Decorative Painting

Fig.3 A decorative picture of wool piled together

The art of dividing the yarns and cords of different fiber materials into the non - warp and weft lines by the edge of the creation. Can be flat, can also be three-dimensional art; it is characterized by rough mad, fine, and elegant, its broad themes, vivid image, knitting arts art in operation than the interleaving of more flexible materials more widely.

Carefully observe this piece of woven decorative painting, as the whole picture full of rhythm, different color wool weaving techniques in each splicing, without excess gaps, accumulate texture, allowing it to enrich the whole picture's stylish smooth lines (Fig.3). [13, 14]

2.1.4 Interwoven Decorative Painting

According to the warp and weft, mutually intertwined fabric form a regular form of organization. Need to hang the warp and weft and as regularly on woven fiber material for different attributes. These are art characteristics of fine structure texture and fine color composition with a delicate, thick texture and ideal color of the color mixing effect.

The formation of latitude and longitude lines interwoven in the background breaks the quiet pure line, with coral fleece as the pattern texture, showing a sense of accumulation. This breaks the monotony within the screen, with thick coral velvet pile, opening the actual situation and background of the subject, the interleaving the decoration is able to captivate audiences. (Fig.4). [11, 12, 15]

(a) (b)

Fig.4 Coral pile piled up decorative painting

(From http://img4.duitang.com/uploads/item/201608/06/20160806154942_4iPVY.thumb.700_0.jpeg)

2.1.5 Crochet Decorative Painting

Wool Crochet handmade painting, is a new category of art where modern environments and aesthetic interests coincide. The fashionable aesthetic, romantic artistic temperament and elegant inexpensive price of Handmade wool paintings is why they are often the first choice for home decoration. Crochet techniques in decorative paintings, exhibit objects in three dimensions. This method mainly uses the crochet hook decorative painting woven, knitted and crochet techniques according to the requirements of the needle, using Crocheted mostly in the relatively small fabric, The artistic characteristics in the decoration painting involve hollow shapes and are adjusted to the screen space area, allowing this kind of decorative painting to look more delicate.

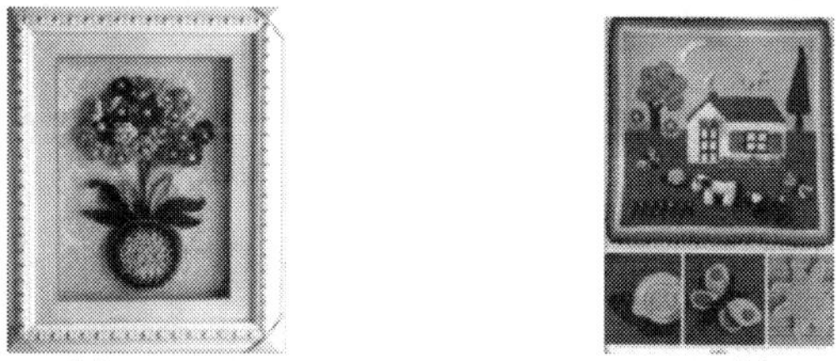

(a) (b)

(c) (d)

Fig.5 Crochet decorative painting

(From http://img3.duitang.com/uploads/item/201608/21/20160821132559_Fc4VM.thumb.700_0.jpeg)

As shown in the fourth diagram, The crochet technique defines the decorative painting. according to the picture's different needs, The length of needle combined with acupuncture, adjusts the tightness of the pattern of the picture, allowing it to appear hollow, so that the screen has a three-dimensional visuality. (Fig.5)

2.1.6 Embroidered Decorative Painting

There are many methods of embroidery on fabric, the most common is the cross stitch and Swiss needle embroidery. These can make embroidery decorative painting that possess a rich texture.

Cross stitch is in knitted fabrics, embroidery method with cross stitch pattern performance. Each coil on the fabric is a grid: involving three-dimensional embroidery on grid paper. Swiss needle embroidery; similarly knitted, which can add new color to knitted fabric with ease. It is used in decorative painting in conjunction with the trend of the coil, according to a certain pattern arrangement to achieve the approximated effect.

Observed from the above diagram, the embroidery decoration painting texture is very strong, and the texture of the arrangement will be neat, at the same time, the selection of wool cotton or lace line usually involves relatively a hard variant of wool. This kind of wool embroidery can stiffen the picture, giving it no extra lines, allowing the picture to exhibit a perfect collocation of cold and warmth; a more vivid Reiki and realistic. (Fig.6).

(a) (b) (c) (d) (e) (f)

Fig.6 A decorative piece of needlework

2.2 A Comprehensive Analysis of Wool Decorative Painting

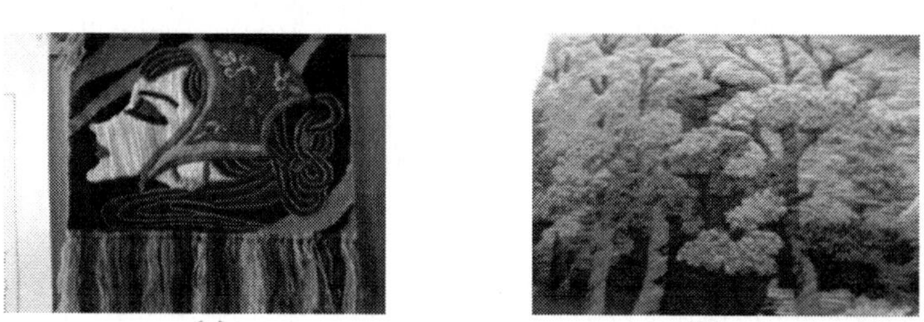

(a) (b)

(From http://img1.imgtn.bdimg.com/it/u=2494007588, 1625219707&fm=214&gp=0.jpg)

(c)

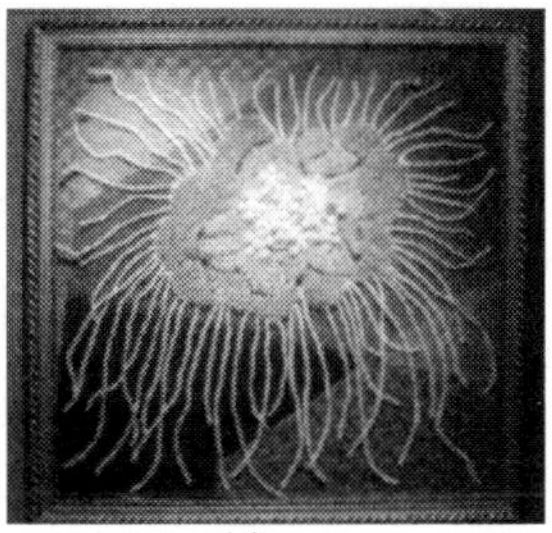

(d)

Fig.7 Synthetic decorative painting (From http://imgsrc.baidu.com/f70cf3ac757b2.jpg)

The four picture is a combination of various techniques of wool decorative painting and the formation of a painting, Fig.7(a) uses paste, winding, embroidery and other techniques, to create a large area block surface, with the cotton lover let people look more soft skin, hair embroidery is the edge, contour hair becomes clearly, the background is used in wool material, can also be three-dimensional characters to the front, the lower part of the picture in the paste some short-term, compact so that the picture can become loose; Fig.7(b) demonstrates two techniques of combining wool decorative painting pasting and winding, creating leaves with dense coral fleece combined with the technique of soft paste, trunk and background by winding techniques. Thus highlighting its strong figure; Fig.7(c) demonstrates the use of embroidery and paste techniques, and some characters in the hands of the umbrella skirt to stick with coral velvet, while highlighting the performance of paste, fluffy, background with embroidery, tight picture; Fig.7(d) demonstrates a combination of pasting, winding and embroidery, background of the subject matter is winding, the loose degree of paste, prioritize (Fig.7).

The integrated use of a complete picture of the wool to convex decorative painting is the needs of a variety of techniques of wool shows unique decorative painting, and according to the characteristics of the different techniques, showing the picture of the actual situation, primary and secondary texture, color and so on, to add the wool decorative painting a sense of fun, not only for decoration space, is a worthy of collection of Art.

3. Study on the Color and Pattern of Wool Decorative Painting

A decorative painting of the complete composition, it needs the right number of elements, the most important is the appropriate collocation of colors and patterns that meet the environment atmosphere to attract audiences, and give people a sense of beauty and appreciate the piece as an addition to modern art. In the wool decorative picture, more consideration is not the so-called painting, but how to use different types of yarn, according to its texture, texture and other different to design a perfect color pattern, and expressive, the theme of this paper is focused on how to use paste, winding technique, using different wool material. The design of a harmonious wool decorative painting.

Color design should fully consider the color tone and its contrast, hue, lightness, and purity, all of which are subject to the main tone. Under the guidance of the main color, to better play the positive role of color, more accurate display of color to bring people's feelings and feelings.

3.1 Study on the Pattern of Wool Decorative Painting

The pattern should be expressed through the design language of the organization, thus giving the aesthetic for the decorative effect. Decorative painting is the main form of the design, people's aesthetic requirements are different. Thus, the variety of decorative patterns become the most direct expression of aesthetics for people of all ages. [17, 18, 19, 20]

3.1.1 Decorative Pattern in a Sweater

Sweater patterns carry certain restrictions on its shape, the decoration painting material for deformation processing, so as to adapt to the basic geometry design has been good, there are specific profile, it has geometric shape or a clear natural shape . Suitable for the shape of the decorative painting to do the basic type of restrictions, not only for the decoration of the restrictions and to provide room for spatial imagination.

Fig.8 Figurative decorative painting
(From http://img3.duitang.com/uploads/item/201506/15/20150615200325_Z5Qjv.thumb.700_0.jpeg)

This piece of wool decorative painting composition is simple: the ship, the sea and the island represent its thematic design, and combined with the color segmentation of images. We can observe this painting, identify the patterns that can be found in life; its form simple with lively rhythm. The lace reinforcing a hard texture, so that the screen looks three-dimensional. (Fig.8)

3.1.2 A decorative pattern in a sweater

With like pattern in the border without limitations, do not show the specific graphics, unable to find specific morphological patterns in life, often appear in the decorative painting, different patterns of graphics based on non-imagery of different reactions; these patterns can be as simple or complex, depends entirely on the mode of thinking.

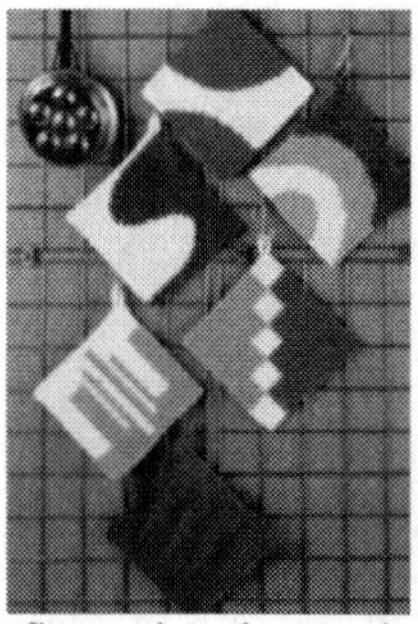

Fig.9 Non figurative decorative painting
(From http://img3.duitang.com/uploads/item/201506/15/20150615095553_dU5zC.thumb.700_0.jpeg)

This is a non-pattern based wool decorative painting, the six picture appears to be simply coherent, but when put together is fun, and gives people a more serene experience, there is no specific painting to reflect the object form, however, so not what picture arrangement is the disharmony, this is not like pattern decoration special.

The unique visual language combined with the pattern's color, so that this trend and the form of fast food into all walks of life. Wool decorative painting is a special art form, it is the pattern of using rules and principle of form beauty, exaggeration, deformation art, that create the beauty of nature. It describes the object, such as figures, landscapes and flowers and birds, such as still like or not like pattern, everything is contained therein. Not only can show the reality of things, but also the performance of the imagination of the reality of things, scenarios, and even the combination of fantasy and reality in the same screen, with a strong decorative meaning. (Fig.9)

4. Conclusion

This paper is mainly about the characteristics of wool decorative painting and the design of pattern design. The paper's intention is to highlight types of modern decorations, to improve people's aesthetic horizons, to provide people with some theoretical knowledge and analysis, from the wool decorative painting Process approach to proceed analysis, obtained the following methods for reference only:

Paste the decorative painting, according to different texture of the wool, with the adhesive tool to stick to the screen to the screen.

Winding the decorative painting, the wool wrapped around the screen.

Knit decorative painting. According to a certain needle method for knitting decorative painting.

Intertwined decorative painting, according to the crossed fabric and latitude each other intertwined, forming a certain form of organization.

Crochet decorative painting, the main use of crochet hook made of decorative painting.

Embroidery decorative painting, in the decorative painting with a color line according to the direction of the coil needle, according to a certain pattern of arrangement, pick embroidery in the fabric, can achieve the effect of similar jacquard.

The innovation of this thesis is that according to the works of different kinds of wool decorative paintings, the techniques of wool decorative painting are put forward and combined with the research of graphic color and graphic composition to perfect the painting of wool decorative painting, and the later researchers Provide some theoretical knowledge.

The main point of this thesis is to summarize the techniques of sticking, winding, weaving, weaving, crochet, embroidery and other techniques of wool decorative painting according to the different types of wool, and the simple analysis of different texture and composition The Influence of Materials on the Visualization of Color and Pattern in Wool Decorative Painting.

The shortcomings of this paper is that some professional analysis is an extension of the existing knowledge system, but also lacks some professional theoretical argument, in some places there are some shortcomings, to be continued to study.

References

[1] Cao XM. "of decorative painting color performance of", "the new campus. On 2014 the 04 Quarterly" period.

[2] Wang YH. "the modeling method of decorative painting" ,"small and medium-sized enterprise management and technology, in 2014 02 Quarterly" period.

[3] Zhai Y. Eagle Decoration performance techniques. China Textile Press: 2004

[4] Zhao AM The "form" of decorative painting, "art grand": 2008

[5] A Ying, crochet basics [M], Beijing: China Textile Press: 2007; 10-31.

[6] Ren H. Discussion on the functional materials and application of modern decorative painting [J], Journal of Harbin Normal University: 2004 (1); 16-17.

[7] Mao LL. Research on the design of functional knitted decorative painting 2009-4-8 Tian Jing: 2006073

[8] Ren CY. Study on the expression of color function in product design [J], author magazine: 2009; 20.

[9] Jing ML, textile color design [M] Beijing: China Textile Press: 2004; 67-69.

[10] Long HR. Knitting Science [M], Beijing, China Textile Press: 2008161 (in Chinese)

[11] Choi MS. Effect of Changes in Knit Structure and Density on Mechanial and Hand Properties of Weft,Knitted Fabrics for Outerwer [M]: 1997.

[12] Unal. Ahmed, Effects of the knit structure on the deformation behavior of weft knitted

[13] Ann Budd. The knitter's Handy book of sweater Parrerns: 2005.

[14] Sandy Blank, The knitter's artistic Patterns: 2005.

[15] Priscilla A. Gibson-Women [J], International Textiles: 2003(12)/2004(1); 32-33.

[16] Zhao N. On the Influence of the Application of Composite Materials on Decorative Painting. [J] Art Education Research]: 2014-1.38-39

[17] Du J. Analysis of decorative art features [J]. Recently Hubei (mid-term): 2015-2.148-148

[18] Yan CY. Analysis of modern decorative painting [J]. Literature and Art, Literature and Art Theory: 2014-11.183-183

[19] Pan YH. On the traditional Chinese decorative painting of the artistic characteristics. [D]. Suzhou. Suzhou University: 2008

[20] Rao S. Artistic design of creative thinking [J]. China Ceramic Industry: 2008-1.52-53

Designing Protective Clothing for Miners to Meet New Challenges

Yi Wang, Hong-FuWang

College of Textiles and Clothing, Jiangnan University, Wuxi 214122, China

*Corresponding author's email: 245297518@qq.com

Abstract

We analyzed the present situation of domestic and foreign miners' protective clothing and summarized their advantages and disadvantages. The second layer of miner protective clothing need to be more personalized. Miner's protective clothing in the new era should be a fusion of safety and fashion, which can enhance miners' sense of belonging and pride for their profession.

Keywords: Miners Protective Clothing; User-Friendly Design; Safety; Fashion

1. Introduction

Miners make a living by doing extremely dangerous work. Faced with harsh working conditions and frequent accidents, their lives are either in their hands or r in a state of uncertainty.. Safety of coal mining in China is problematic [1]. Accidents occur frequently, and miners work under dust, high temperature, high humidity, viruses, bacteria, fungi and other microbial damage which harms them severely. Many develop incurable illnesses such as pneumoconiosis caused by smoking large amounts of dust pollutants, and bursitis caused by long-term kneeling in narrow tunnels, bringing tremendous physical and mental pain to miners and their families [2].

The new era of miner protective clothing requires innovative design for miners to minimize physical and psychological damage from working in harsh environments [3]. The criteria for protective clothing for the human body includes "security protection, functional efficiency, fashion aesthetics." [4]. With the rapid development of the national economy, all kinds of new technology, new equipment, new material and new subject arise. Professional equipment design and development has gradually become a cross-disciplinary new system engineering. The starting point of human vision, fully embodies the humanistic care and the humanitarian spirit [5]. In the labor industry nothing is more important than protecting workers' safety and health. It is important to establish new national standards for miners' protective clothing which embraces technological innovations and aesthetic design.

2. Development of the Fabric of Miners' Protective Clothing

In China, miners' clothing are mostly made of fabrics for traditional work clothes such as cotton or sturdy denim. These are strong and durable, but also bulky, difficult to wash, ugly and clumsy. Consequently, large state-owned coal mines are using blue (black) cotton canvas, which is thin, beautiful, and flexible. As for the applications of special fabrics, in recent years, many coal mines and textile workers have published articles calling for the use of special anti-static protective clothing for releasing static electricity that cannot be eliminated when wearing common cotton clothing. As a result of the special underground working environment, in addition to anti-static requirements, protective clothing should also be dust-proof, anti-oil, fire-retardant, wear-resistant, and so on. Adding conductive wire during the weaving process increases the conductivity of the fabrics, effectively eliminating static electricity within the working environment [6] [7]. Using pure cotton vortex spinning yarn increases breathability. Using the Pulu benzene method gives the fabric a flame-retardant finishing and "three anti" treatment makes the fabric anti-oil, anti-fouling, and waterproof, which means the fabric is pollution-free and stain resistant [8].

Foreign research for dust-proof fabrics started earlier. In 1997, Japanese company Teijin designed "Jijiasitanqi TM", a vacuum-proof ultra-high performance dust clothing based on the dust-proof function required by the conductor industry's plan of increasing memory to "Gigabit" level before 2005. Its performance is 10 times higher than the US dust jacket (Fantanlaersitaiduo 209E) [9].

3. Development of the Structure of Miners' Protective Clothing

Nowadays, miners' work clothes is integrated with cold storage bags [10], controllable airbags and GPS, which can enhance the safety of underground mine operations and provide important assistance during rescues. In ordinary miners' work clothes, cool storage bag of two parts of miners cool suit design can reduce thermal pollution, a common occupational hazard for miners. A life-saving device within the miners' protective clothing is in the fabric and material between join a controlled balloon, the balloon is light weight, comfortable to wear, and does not affect normal work. If a mine floods, the worker can inflate the balloon. The worker's upper body form the airbag cover, effectively preventing the worker from suffering impact damage [11]. Siamese design of miners' protective clothing makes dressing more convenient. Mask and neck baffle offer very good protection [12].

Foreign researchers got involved earlier in research about protective clothing under special work environments [13]. American scholar R F Goldman proposed the "4F" principle for functional clothing design and development process, namely Fashion, Fit, Feel, and Function [14]. In general, the design principles of protective clothing are safety, applicability, beauty, and decency. For protective clothing, foreigners use a non-pocket design: firstly, this avoids accidently being hooked on to machinery, secondly, it prevents a situation where miners cannot better protect themselves during accidents because their hands are in their pockets. This shows the attention to detail in foreign design of protective clothing [15].

In recent years, some scholars in our country also started from the clothes level to study how to reduce injury in miners. But the main focus on fabric development, some existing fabrics with relatively mature, but still exist on structure, function and design a lot of limitations. In general, some foreign countries have advanced fabrics and better design of functions and structures. They have discarded single-color-single-fabric traditional design, and instead use two or more color for sewing of professional clothing.

4. Design Elements of Miners' Protective Clothing

With the continuous upgrade of fabrics and the emergence of new technology, there are higher demands for miners' work clothes. Comprehensive and in-depth research and innovative design in miners' work suits are necessary to improve miners' safety and the efficiency of professional equipment [16] [17].

4.1 Security Miners' Protective Clothing

Underground work is poor, light or dark environment lead to fatigue and lowered mental concentration. Pockets, cuffs and other loose parts of work clothes can get caught with underground mechanical equipment, adding danger to miners. In the structure design of miner's protective clothing, it is possible to alter some conventional designs, such as changing the conventional pocket to the form of a bag cover, plus zipper, invisible pocket [18]. The mouth of the pocket can be tightened to prevent machinery hooking onto trouser cuffs.

Miners often work in dim light. Increasing the area of reflective material on protective clothing will also help to reduce injuries and accidents in dark environments [19].

4.2 Convenience of Miners' Protective Clothing

Most existing miners' work clothes are of a split design because of the hot and humid underground environment in which miners work. Many miners will choose to wear only underwear, leaving their upper body bare – this can lead to preventable injuries and is a violation of safety guidelines. A siamese design prevents this as the whole body is covered to better protect the human body [20].

4.3 Spreadability of Miners' Protective Clothing

The introduction of modern mechanization has changed the traditional downhole operation mode to a large extent, but it is difficult to improve the special working environment. In particular, the safety of miners is threatened by dust pollutants. Reversal and difficult to cure, to numerous miners and miners brought great their family and their body physical and mental pain. The design of miners' protective clothing can be improved by enhancing the fabric's dustproof performance. On the other hand, it is can reduce the skin bare area, tightening sleeve, feet, neck, and dust mask design clothing fusion [21].

4.4 Reactivity of Miners' Protective Clothing

When miners underground work activities are more frequent joints of the limbs, activity too. Can accept PLEAT loose quantity (space) for sport [22].

4.5 Air Permeability of Miners' Protective Clothing

Breathable protective clothing can greatly enhance the comfort of working underground in high heat and humidity environments. Breathable fabric allows better ventilation of areas where the body sweats more, such as the underarm, improving the comfort of the clothing [23].

4.6 Wearability of Miners' Protective Clothing

For underground work, the knees and elbows of work clothes are the most easily worn parts. Elbow and knee pads can be added to increase comfort and durability of miners' protective clothing. (Table 1)

Table.1 Design elements

Design elements	Purpose	Concrete method
Security	Reduced Amount of interdependence hook Enhanced identifiability	Invisible pocket Bright color Anti stitching Reflective strip splicing
Convenience	Wear off convenience Prevention of upper torso exposure of workers	Siamese design
Spreadability	Reduced exposure to skin area	Long sleeves, trousers
Reactivity	Substantial movement requirement	PLEAT
Air permeability	Comfortable heat dissipation	Splicing of breathable fabrics in sweat
Wearability	Prevent abrasion	Elbow pad, Kneepad implantation

Mineworkers can address several dangers. They wear full safety clothing [24]. All mines should stipulate that protective clothing must be worn and the kind of action to be carried out.

Improved visibility: A gear that has high visibility is critical in a mine, particularly when dealing with high traffic. Reflective strips and bright colors always make workers at the mining site visible [25].

Abrasion Resistance: Thick and tough leather offers high abrasion resistance in case of a crash, though fabrics like Cordura is able to provide significant protection too. Fabrics are waterproof, convenient to maintain, cheap and comfortable in hot weather. Thick leather is uncomfortable during hot weather and could result in the loss of control and heat stress. There are fabric gears designed into a mesh-like pattern which is cooling and provides a more stable surface to attach the padding.

Weather Protection: Protection against weather elements is very important in mines. At times, extreme weather makes working conditions rather dangerous or unbearable.

The hands, head, knees, hips, shoulders and elbows are all vulnerable areas.

Full-face mask: Full-face mask is important to protect and is imperative that hardhats be worn [26].

Gloves: Gloves need to be designed specifically for a given mining activity. It is commonly made using Cordura, Kevlar, leather or other given combinations. There are some with carbon fiber knuckle protection and other rigid padding forms. There are gloves that offer wrist protection [27].

Vests: Made using retro reflective materials with high-visibility color. It is possible to wear them over jackets so as to increase visibility. This will be more critical in adverse weather conditions and can also protect the upper limb.

Jackets: Most jackets are made from ballistic nylon, Cordura, leather and other related synthetics. Many jackets feature a special padding on the spine, elbows or shoulders. These padding on the spine, elbows or shoulders could protect the miners' bodies.

Pants: Pants could protect the skin of the lower limbs. They include special protection for hips and knees. Cotton denim jeans that have Kevlar are preferred by some people.

Boots: It's specifically designed for undertaking mining activities and includes plastic caps and reinforcement on the ankles or the toe area [28].

Helmet Visor or Goggles: Something of utmost importance is eye protection. Metal pieces or pebbles that are kicked up towards the eye have sufficient momentum to render significant damage.

Ear plugs: These devices offer protection against any hearing damage and reduce fatigue [29] [30].

5. Developing Trends

Environment analysis: miners' protective clothing is designed for special underground working environments. Based on general work clothes, more concerns should be payed to temperature, humidity, air velocity, lighting, noise and other complex special environmental factors which can bring a lot of negative influences to workers.

Structure and functions: pay more attention to the characteristics of special professions, including common postures, commonly used tools, common occupational diseases to make the design more customized.

Protection functions and high-tech fabrics: Based on the analysis of environmental factors, protective clothing for coal mine operating personnel requires additional considerations for the risks in the mine in order to have certain protective functions, which means miners' work clothing should be upgraded to miners' protective clothing. With dust-proof, anti-static [31], fire-retardant, anti-bacterial, waterproof, anti-oil repellent, and warning and other complex functions, raw fiber materials may turn to high-performance fiber. BErgonomic principles, advanced textile and garment technology will make miners' protective clothing more comfortable, more safe, more efficient and more humane [32].

6. Conclusion

The design of miners' protective clothing for a new era should pay more attention to humanism. The design of miners' protective clothing should enhance workers' safety, comfort and help to improve work efficiency. By introducing fashion elements and improving the original design, work clothes can reflect professional characteristics and enhance special-industry workers' sense of belonging and pride in their profession.

References

[1] Xu L. Research on Humanized Design and Personalized Design in Product Innovation [J]. Art and design: 2010; (9): 199-201.

[2] Yin WT, Li SB, Gao P. Study on division of underground work types in China coal mine [J]. Coal engineering: 2013; (4): 87-93.

[3] Steemson J (1988) Foot work. Occupational Safety and Health: (March) 12-18.

[4] Dong WG. Humanization needs is the fundamental costume design [J]. ART EDUCATION: 2010; (2): 129. Li YG, Zhang LY. Design and evaluation of a new tank cleaning service [J]. Journal of Textile: 2015; 36 (3): 88-91.

[5] Leighton JR. "The Leighton Flexometer and flexibility test", Journal of the Association of Physical and Mental Rehabilitation: Vol. 20; 1966: pp. 86-93.

[6] Gu Y, Dai JM. Brief description of coal mine protective clothing fabric application [J]. Shandong Textile Technology: 2009; (4): 55-56.

[7] Yang JP, Fan WM. Coal mine emergency rescue technology research and supporting equipment [J]. China Conference: 2015; (9): 1-4.

[8] Jin JT. The analysis of clinical and epidemiologic characteristics of low-back pain of coal miners [D]. Sanxi: Shanxi medical university: 2010.
[9] Wang YZ. 21st century dust service [J]. Shanghai Textile Technology: 1997 (5); 50.
[10] Zhang FF, Liang Guo-zhi, Zhou Meng-ying, et al. Application of cooling service in coal mine [J]. Energy Technology and Management, 2011 (2): 114-115.Lu AM. Dust masks - four in one use guide [M]. Beijing: Chemical Industry Press: 2014.
[11] Wang WQ. Special protective clothing collection [J]. Labor protection: 2012 (12); 80-82.
[12] Wang XH, Yu XL, Yu YC. A high temperature protective clothing: CN, CN2198778 [P]: 1995.
[13] Liu HQ, Liu TY, Gao LY, Li Y. Development and Comparison of Protective Clothing at Home and Abroad [J]. Mining Engineering Research: 2016; 31 (3): 71-76.
[14] Li HY, Zhang ZY, Li J. Studies on the study of functional clothing [J]. Silk: 2007 (4); 45-47.
[15] Pringalle C. Service risque chimique et protection individuelle, Efficacité et confortdes équipements de protection individuelle. INRS, Paris, Cahiers de notesdocumentaires-Hygiène et sécurité du travail-N°172, 3e trimester (1998).
[16] McPherson D. Balancing Protection with Comfort, Fit&Style in the PPE SelectionProcess. Kimberly-Clark Professional: March (2008).
[17] Deng HJ. Based on the "people-oriented" concept of modern clothing design [J]. Liaoning silk: 2013; (1): 18-20.
[18] Yao J, Tian DM. Study on special protective clothing for coal mine [J]. JOURNAL OF NORTH CHINA INSTITUTE OF SCIENCE AND TECHNOLOGY: 2014; (1): 69-72.
[19] Zhong CY, Zhong M, Li Juan, et al. Miner uniform with fluorescent strips: CN, CN 202999360 U [P]: 2013.
[20] Yang JP, Fan WM. Coal mine emergency rescue technology research and supporting equipment [J]. China Conference: 2015; (9): 1-4.
[21] Lv AM. Four all-in-one guide for the use of masks [M]. Beijing: Chemical industry press: 2014.
[22] Zhang LN, Wang YY, Li YG. Development and ergonomic evaluation of overall for oil tank cleaners [J]. Journal of Textile Research: 2013; 34 (8): 105-109.
[23] Havenith G. Clothing Protection Against the Natural Environment: The Relevanceof Local Sweating and Local Ventilation for Thermal Comfort.
[24] Keeble VB, Prevatt MB, Mellian SA. "An evaluation of fit of protective overalls manufactured to a proposed revision of ANSI/ISEA 101", in McBriarty, J.P. and Henry, N.W. (Eds), Performance of Protective Clothing: Fourth Volume, ASTM STP 1133, American Society for Testing and Materials, Philadelphia, PA: 1992; pp. 675-97.
[25] Pringalle C. Service risque chimique et protection individuelle, Efficacité et confortdes équipements de protection individuelle. INRS, Paris, Cahiers de notesdocumentaires-Hygiène et sécurité du travail-N°172, 3e trimester (1998).
[26] Mäkinen H (1991) Analysis of problems in the protection of fire fighters by personal protective equipment and clothing - development of a new turnout suit. Tampere University of Technology/Finnish Institute of Occupational Health, Helsinki 1991, 201 pp. Oakley E, Howard N (1984) The design and function of military footwear: a review following experiences in the South Atlantic. Ergonomics: 28 (6) 631-637.
[27] Efficacité et confort des équipements de protection individuelle. Etude réaliséedans 3 entreprises de traitement de surface, INRS: ND 2085 (1998).
[28] De Moya RG (1982). A biomechanical comparison of the running shoe and the combat boot. Military Medicine: (147); 380- 383.
[29] Neeves RE, Barlow DA, Richards JG, Provost-Craig M & Castagno P (1989) Physiological and biomechanical changes in fire fighters due to boot design modifications. Sports Science Center, University of Delaware, Newark, Delaware: August: 1989; 38.
[30] Steemson J (1988) Foot work. Occupational Safety and Health: (March) 12-18.
[31] Huck J, Maganga O, Kim Y. Evaluation of Garment Design and Fit. ProtectiveOveralls: Evaluation of Garment Design and Fit. International Journal of Clothing Science and Technology: 45 (1997).
[32] Zhang YP. Research on co-design of miner's individual protection and distress rescue tools [D]. Shenyang: Shenyang Aerospace University: 2011.

Application of Aran Patterns in the Sweater Design

Fei Li[1], Li-Li Mao[1,2]

[1]Knitted garment design, Xi'an Polytechnic University, School of clothing and art design, Jinhua South Road 19, Xi'an, 710048, China
[2]Xi'an Polytechnic University, School of clothing and art design, Jinhua South Road 19,Xi'an,710048,china

*Conesponding author address: 976623712@qq.com

Abstract

This paper first gives a quick overview of the classification and the meaning of Aran patterns. The application of Aran patterns in sweater is then analyzed in detail, namely the twist needle Aran patterns, diamond Aran patterns, honeycomb Aran patterns and ball of fur Aran patterns. The problems that should be paid attention to in the sweater design when applying Aran patterns are put forward, which is generally to do with the size and layout. In the comprehensive application of Aran patterns, how it is combined with the selection of yarn and the echo between Aran patterns should be considered. This study enriches the application of Aran patterns in sweater design and puts forward a new design idea of how to use Aran patterns in sweater design for its further development.

Key words: Aran Patterns; Sweater Design; Twist Pin; Application

1. Introduction

Sweaters are linked together with clan. The identity of clans is closely linked to the birth of Alan sweaters. There are many stiches when a sweater is knitted. A sweater can convey a lot of information, such as the people who did the knitting and the life of their family. A pattern in a sweater will be used by the same family for generations. There are about 100,000 loops in a carefully knitted sweater which requires approximately 60 days to complete. A sweater can contain any combination of stitches, but the selection of knitting stitch depends on the specific clan pattern. Many of the Aran patterns in a sweater stitch is originated from the Celtic art. Its own unique meaning; this is Aran Island's heritage. The application and research of Aran patterns in the sweater design presented in this paper are divided into two aspects, the first is the application of single Aran patterns in sweater design, and the second is in analysis and study of composite applications of various Aran patterns in sweater design. [1, 2, 3, 4]

Fig.1 The traditional Alan pattern sweater

2. Different Alan Patterns and their Meanings

Table.1 classification and meaning of Alan pattern

Pattern category	Example	Implication
Cable lines		The inspiration of this knitting pattern is a fishing net. The shape is similar to a fishing net, which is more suitable for the design of sport style sweaters. [4, 5, 6]
Honeycomb pattern		A symbol of the fishermen is as busy as a bee. The three-dimensional shape of a circle is similar to a honeycomb. The arrangement of the neat and delicate pattern is suitable for a recreational sweater. [4, 5, 6]
Diamond grain		The shape of the rhombus is designed in the shape of a diamond. The texture is strong, and the moral wealth is many. This pattern can be applied to recreational sweater design. It can also be applied to elegant sweaters when a delicate yarn is chosen to weave a diamond pattern. [4, 5, 6]
Basket pattern		The exterior is more casual and bold mimicking a fish basket and the harvest. The Basket pattern can be used in a sweater with a dilapidated style. [4, 5, 6]
Cliff pattern		The superimposed form is similar to a rock It replicates the impression of fishing in the precipitous rock faces and sheer cliffs. This pattern can be applied to athletic sweaters and make them more recreational. [4, 5, 6]
Twist the needle		Twist the needle pattern is an expression of the fisherman's desire to have a good harvest. There are a wide variety of Twist pin variety. Different sizes of yarn can result in different sizes of Twist the needle pattern. The pattern has a wide application scope, when applied alone, it can be used in conjunction with other tricks. [4, 5, 6]
Zigzag pattern		The zigzag pattern is simple, representing the twisted path of a cliff. It can enrich the tissue texture of a sweater and thereby enhance its appearence. [4, 5, 6]
Tree of life		This pattern is a reflection of the importance of clan unity. It is a symbol to prolong parent's longevity and the desire for children to thrive. This pattern can be applied to the body of the sweater to enhance its appearance. [4, 5, 6]
Diamond pattern	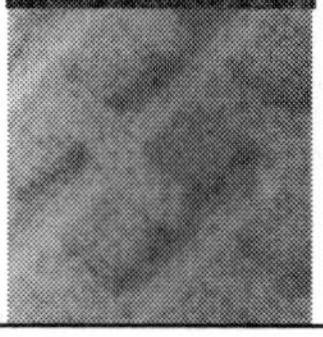	The pattern is connected by the same diamond and is a reflection of an island's geographical characteristics. It expresses the desire for success and wealth. The diamond pattern is simple and generous. [4, 5, 6]

Mao qiu-qiu		People refer to it as a fruit pattern. The more balls there are, the more it symbolizes fruitful. The mao qiu-qiu appearance is simple and can be applied to any part of a garment. The style is more casual and colloquial. [4, 5, 6]

3. Application of Aran Patterns in Sweater Design

A well knitted sweater has many characteristics such as having good drape and is soft etc. These characteristics when combined with the three-dimensional effect of the Aran patterns not only creates an excellent visual effect, it also enhances the quality of the knitted sweater and increases diversity and variation.

3.1 The Application of Twist Pattern in Sweater Design

(a) (b) (c)

(d) (e) (f)

Fig.2 The use of twist needle pattern in sweater

There are various usages for twist of needle pattern in sweater design. It can be distributed throughout the sweater body evenly, as shown in Fig.2(a), the direction and the width of twist needle pattern in each wale are exactly the same. This avoids cluttering and prevents a messy result. The choice of yarn is usually fine for the design of a vest. The application of single Aran patterns should consider the size problem and the layout of the Alan pattern. There are three groups of twist pattern. For the sweater with an evenly distributed pattern as shown in Fig.2(b), because the width and length of the twist pattern is larger, the number of twist pattern cannot be too big otherwise it will destroy the entire style of sweatermaking it slow and clumsy. For this sweater the thinner yarn is chosen. This makes the yarn more contracted. Twist pattern can also create a sweater that has a straightforward style that highlights texture as shown inFig.2(c). The yarn should have a certain roughness for this kind of sweater, and the pattern also cannot too complex. The stitch of this kind of sweater is strong, it highlights the feeling of relief, has a unique style, and the modelling is exaggerated. The application of Twist Pattern in the sweater design is very flexible and its size and pattern direction can be changed. Fig.2(d) is a horizontal design of the twist pattern, it creates a sense of leisure and sports, and it is novel and unique. [7, 8]

Twist pattern can also be used as a decoration on the sweater, its location is not just limited to regular parts like neckline, sleeves and hem. It can also be used in irregular parts like garment body parts, and can be express freely. As shown in Fig.2(e), the design and stitch is a simple twist pattern that appears on the sweater at irregular intervals. It illustrates the diversity in sweater design andmakes the twist pattern rich and interesting..

Aran patterns not only can be used on sweaters in a single manner, it can also be used in a variety of combinations.

The selection of suitable pattern is based according to the sweater style and connotation that the designers want to express. In the application of twist pattern in sweater design, the size problem and the combination and distribution of twist pattern should be considered. Different twist pattern should be designed depending on the sweater style and yarn to create different styles and color.

3.2 The application of diamond in the sweater design

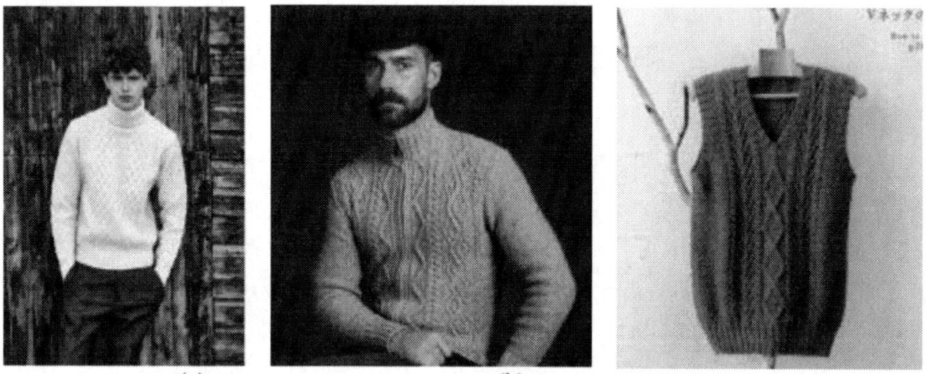

(a) (b) (c)

Fig.3 The use of diamond pattern Sweater

Diamond pattern is an often used pattern of Aran patterns, when the diamond pattern appears in sweater design, it can be applied as a single pattern in the body of the sweater or it can be applied in combination with another pattern. In Fig.3(a), only the diamond pattern is used throughout the entire sweater, the size of each diamond grain are exactly the same, the diamond is spliced in two consecutive ways, the size of the diamond pattern design is very important. [9, 10, 11]

The selection of thickness and color of yarn depends on the style and characteristics of the sweater design. In Fig.3(b), the sweater belongs to tailored style which uses a relatively fine yarn in pattern design. A diamond pattern and twist pattern are selected, it has the style characteristics of retro and elegance. This sweater stereo sense is strong with a design focus on the placket. The sweater sleeve is without any decorative pattern. It is simple and plain which makes it clear what the designer wants to express..

3.3 The Application of Ball of Fur in Sweater Design

Three sweaters in Fig.4 applied the pattern of ball of fur. From its unique appearance, it an be observed that it symbplizes a fruit. The more ball of fur, the greater the symbol of fruits. The size of the ball of fur is the key to the design; there to be a balance in quantity so that the entire style of the sweater will not be ruined. Different sizes of ball of fur are applied in three sweaters reflecting different styles. The appearance and shape of the ball of fur is simple. Because of its simple shape, designers often collocate it with other patterns, and it almost never alone appear independently. In the white knitting coat shown in Fig.4(a), the collocation design of zigzag lines with ball of fur adds a depth and diversity to the sweater design. The ball of fur can be used in different styles of sweaters, as shown in Fig.4(a) and (b). Figure(b) is a variety of Aran patterns combination, easy design and coarser yarns are revealed that the wild wind. Twist pattern on the front piece of the sweater enhances the texture of the sweater. Twist pattern on the sleeve is big, and has a strong stereo sense. Together with the ornament of ball of fur it becomes more straightforward yet does not lose its original loveliness. Ball of fur has a strong adornment effect that is often applied in conjunction with another pattern. When used, the unification of style and coordination with other patterns must be considered.. [12, 13]

(a) (b) (c)

Fig.4 The use of wool ball sweater

3.4 Honeycomb Pattern in the Application of the Sweater Design

(a) (b) (c)

Fig.5 Use of honeycomb pattern sweater

Honeycomb pattern is a typical Aran pattern that is widely used in sweater design. It can appear independently in sweater design or it can be applied in conjunction with another pattern. Honeycomb pattern has a stereoscopic sense and is a symbol of beauty. Through the combination of a variety of patterns across the body of the sweater uniformly, it not only can enrich the visual aesthetics of the sweater, it can also add a lot of fun to this sweater. In Fig.5(a), single honeycomb pattern is applied to the entire body of the sweater and is distributed uniformly. When the honeycomb pattern is combined with other patterns, it can enrich the organization pattern of the sweater. The visual aesthetic feeling of the sweater reflects different styles. In Fig.5(b) not only has the honeycomb pattern, it also has a jacquard pattern echoed with it. The jacquard organization is rich in color and the surface is smooth. Aran patterns have a strong sense of texture pattern and is complementary to many other patterns. Because Aran patterns are the main pattern, jacquard pattern should be smaller in area than Aran patterns. When designing sweaters, the primary and secondary pattern should be considered and made clear. In this sweater, Jacquard pattern plays a role of good foil, it adds a sense of color which creates a rich sweater stitch effect and enhances the visual experience and beauty of the sweater. [14, 15]

4. Conclusion

In this paper, through the application and comprehensive analysis of Aran patterns in sweater design, the different Aran patterns types and its meaning can be further understood. = Aran patterns can be used either independently in sweater design with a uniform appearance or it can be used in conjunction with other patterns. The refined and delicate sweater uses a more delicate pattern and finer yarn. When designing a sweater in a more bold and crude style, a stronger pattern organization is needed with a thicker yarn. Through research, it is necessary to discover the appropriate size of the pattern and how to distribute it in the sweater,The combination and response of different patterns must also be considered,. And it is concluded that wanted to design a sweater not only need to consider the thickness of yarn, sweater style, the size of the pattern and its arrangement, as well as how to choose the pattern style. This article provides a guidance to create innovative sweater designs with Aran patterns.

References

[1] Mao LL. Sweater product design [M]. Beijing: China textile press: 2009; 10: 33-98.
[2] Treasure C. Keitodama. [J]. Henan science and technology press: 2006; 6: 44-47.
[3] Wild ZM, Lang. Ireland island/The legend of Aran Patterns sweater [M]. Woven research news agency: 2002.
[4] Information on http://www.bianzhirensheng.com/a/8904.html
[5] Treasure club. Keitodama. [J]. Henan science and technology press: 2006; 3: 21-55.
[6] Treasure club. World knitting [J]. Henan science and technology press: 2014; 2: 35-45.
[7] Treasure club. Europe weave [J]. Henan science and technology press: 2014; 4: 2-12.
[8] Treasure club. World knitting [J]. Henan science and technology press: 2014; 4: 30-34.
[9] Information on http://fashion.qq.com/a/20160908/030930.htm
[10] Treasure club. World knitting [J]. Henan science and technology press: 2014; 6: 56-61.
[11] Treasure club. Europe weave[J]. Henan science and technology press: 2014; 1: 23-27.
[12] Information on http://www.bianzhirensheng.com/a/18311.html
[13] Treasure club. Keitodama. [J]. Henan science and technology press: 2016; 15: 10-14.
[14] Treasure club. Keitodama. [J]. Henan science and technology press: 2013; 7: 6-17.
[15] Information on http://blog.sina.com.cn/s/blog_626a7ca60102v3ub.html

Visual Analysis of Han Embroidery

Ni Zhang

School of Fashion, Wuhan Textile University, Wuhan, 430073, China

*Corresponding author's email: zhangni333@hotmail.com

Abstract

Han embroidery, a traditional Chinese embroidery technology, has been elected as the national intangible cultural heritage of China. This study aims to explore the way to analyze the visual embellishment value of Han embroidery in fashion. The visual appearance was improved by combining the Chinese culture elements and international fashion trends in fashion design. The purpose of the research is that preserve and develop the traditional handcraft cultural heritage in our country. Especially for the young designers, they has the responsibility to study the Han embroidery used in the contemporary apparel design. Han embroidery diversifies the patterns and textures of Chinese-style clothing which will cater the demands from personalized customers.

Keywords: Han Embroidery; Chu Culture; Visual Value; Fashion Design

1. Introduction

In history, fashion was a good reflection of local culture. The clothing distinguished each other in terms of color, style and technology being used. Modernization has brought quality life to human, but at the same time, it has diminished the culture and resulted in the same fashion all over the world. Traditional clothing has been the selection for some special occasions. How to preserve the almost lost fashion culture has been an important but challenging topic. Being affected by international fashion trend, young generation refuses traditional fashion, which makes the culture preservation even harder. Embedding tradition fashion culture into modern fashion has been proven to be a successful approach, and the so-called Chinese-style clothing favours a variety of customers including the young generation.

Han embroidery is one of the embroidery technologies in Chinese history, it was originated in Chu culture (Geographically in today's Hubei Province). The city of Great Wuhan and its surrounding towns are regarded as the places where the Han embroidery originated and flourished. The emerging of Han embroidery dated back in the Eastern Zhou Dynasty (5th century BC), and it experienced a solid development in Ming Dynasty due to the stabilized social conditions. The late Qing Dynasty had seen the peak of Han embroidery due to the ever-increasing demands. Due to the importance of the embroidery technology in fashion culture, the Han embroidery has been elected as the national intangible cultural heritage of China. In 2008, Han embroidery was included in the expanding program of the first national intangible cultural heritages, which provided a good opportunity in preserving this almost extinct fashion culture.

In this study, the traditional stitch from Han embroidery has been implemented in pattern design to develop a modern fashion work, and the visual value of the clothing have been characterized. Different patterns have been made based on the experimental design and practices. The results demonstrate that the integration of traditional Han embroidery and contemporary fashion trends is an effective way to create innovative and fashionable design.

2. The Visual Characteristics of Han Embroidery in Chu Garment Culture

Embroidery technique dates back to ancient times when wears and garments were developed for human. Embroidery is of great importance in a variety of decorative techniques in textiles. It is created by attaching threads/fibers on clothing or other materials using a needle.

Han embroidery is a regional embroidery which has grown in specific geological conditions, customs and cultures. It has gradually formed its unique visual characteristics of delicacy, brightness and smoothness after a long period of development. With soft clothes and colorful silk threads as the main materials, unique stitches and techniques as the technology, Han embroidery has a strong expressive and aesthetic value.

The Han embroidery is the typical example to illustrate the gorgeousness of the culture and arts of Chu, it enlightens us on the decoration of garments with various figures made for Chu noblemen. Han embroidery enables us to trace back those artworks produced in the Eastern Zhou Dynasty (5th century BC). With the great development of silk and embroidery technique in this period, Chu garments were greatly enriched by a variety of patterns, delicate manufacturing and richness colors. Fig.1 illustrates an early embroidery art treasure excavated from the Ma Shan No.1 Chu Tomb at Jianglin, Hubei province. It is a yellow gown embroidered with phoenix and flower graphics, dedicating for the Warring States Period (Chu country). The size of the gown is about 165 cm in length and 158 cm in sleeve distance. The main characteristics of the style include the right overlapping without buttons, straight fringing, two sleeve opening, decorative fabric with different edges in sleeve and collar. In addition to the basic form of the silk yellow gown, phoenix bird rhombus brocade was embroidered into the straight sleeves, collar and selvage. The selvage and ribbon were created using weft embroideries.

Fig.1 Phoenix and flower graphic embroidery yellow gown.
Cited from Jingzhou Museum, Hubei, China

The needlework of traditional embroidery in Chu country well represents Chinese traditional embroidery technique. Another example is a dragon phoenix tiger embroidered clothing (Fig.2) that was excavated from the Ma Shan No.1 Chu Tomb at Jianglin, Hubei province. In this work, the silk clothing was embroidered with threads in the color of dark red, brown, green, yellow, orange, and black. The chain stitch was utilized to form four kinds of animal patterns as seen. In the embroidery history, the chain stitch was the principal stitch that was used during the Warring States period (475-221B.C.). It can be seen from this work that a decorative sewing stitch in which loops are connected like chains, so called the chain stitch, representing one of the most ancient stitch in Chinese embroidery history [1].

Fig.2 Dragon phoenix tiger embroidered clothing
Cited from Jingzhou Museum, Hubei, China

The adoption of different needling methods resulted in different embroidery styles. In addition to the chain stitch, there are four more basic types stitches for Han embroidery, such as plain stitch, satin stitch, couching stitch and crossed back stitch. Plain stitch is the simplest one used for basic stitch work [1]. The pattern on the plain stitch embroidery is fluent and buoyant, the stitch lines are slim and graceful. The earliest plain stitch is deemed to be the excavated work from Mawangdui Han tomb, Changsha, Hunan, in which two pieces of plain stitch embroideries with respective tree patterns and square patterns were found (Huang Nengfu, 2002). It further indicates that the skills of Chu embroidery had achieved the historical highest level.

Decorative elements created for the purpose of embroidery in Chu garment culture from varied quite a lot from the Warring States Period to Qing Dynasty, due to the geographical and natural characteristics which affected the Chu culture and belief systems. The visual characteristics of Ham embroidery stitches are even stitching (needle spaces) with smooth and flowing curves. In plain stitch, straight lines were embroidered with advancing needling, causing a brief and neat visual effect [1]. Encircle needling was employed in the Chain stitching, resulting in three dimensional and more vivid and attractive thread lines (Wang Libing, Zhong Heng 2007).

Except for the abundant and various stitches, Han embroidery integrates auspicious pictures and good wishes into its products. The widely used symbolic motifs for Han embroidery are phoenixes and dragons, peonies, lotus, people, and they are decorated on costumes, life supplies and Chinese religious products. Pictures for Han embroidery are drawn from popular folk topic, and they often express people's dreams aspirations. For example, the duck couple for romantic love, phoenix and peony for wealth and prosperity (Fig.3), pine and cypress pertaining to the famous "long life embroidery", respectively.

Fig.3 Phoenix and peony pattern of Han embroidery.
Cited from Needlework by Huang Xiaolin, HNCHRC, Hubei, China

3. Examples of contemporary fashion design via Han embroidery

Chu culture has contributed rich patterned art treasures of embroidery and ornamentation techniques. Visual aesthetic characteristics of Han embroidery has historically maintained its richness from its beginning to the modern era. Embroidery keeps on playing an important role in contemporary fashion design collections.

Han embroidery studio of Hubei Province Non-material Cultural Heritage Research Center (HNCHRC) focuses on the applications of traditional embroidery art into fashion design. It aims to determine how embroideries are made and used, how they are used from past until today, and to explore fashion design works utilizing the traditional embroidery techniques.

Pattern, color and technique richness of the Han embroidery continue to be an inspiration for many young designers. HNCHRC has released collections by deriving themes to express Chinese-style aesthetical and commercial values. Inspired by the Han embroidery, a floral motif (orchid flower) was designed and drawn up as a framework on the shining ecru (Chu noble's favorite color) silk fabric, and then the graphics adopted adaptable stitches and form them on the needlework on the chest [2], as shown in Fig.4. Integrating the Chinese culture elements and international fashion trends such as the fashionable silhouette design, this female dress has a reformed Cheongsam shape. This solemn and unsophisticated style fits well the aesthetic features of Chinese females.

Fig.4 Chinese-style dress designed via Han embroidery
Cited from show room by Hongguang Ye, Han embroidery studio of HNCHRC, Hubei, China

Another attempt was the establishment of customized Han embroidery fashion studio (Fig.5). In a fashion show named Chu Style Inspired by Han Culture, there were more than 50 looks in the whole collection, which combine the Han embroidery patterns with fashionable silhouettes. The researchers of HNCHRC dedicate to inherit and protect the Han embroidery cultural heritage, and explore new ways to embed traditional Han embroidery ornamental elements into Chinese-style clothing design. As a result, the aesthetics and commercial value of clothing will be upgraded by interspersing different Han embroidery graphics on apparel. It is a new method that combines Chinese Han embroidery elements with international fashion trends.

Fig.5 Han embroidery fashion studio
Cited from show room by Hongguang Ye. Han embroidery studio of HNCHRC, Hubei, China

4. Conclusion

Chinese Han embroidery renowns for exquisite workmanship, skillful stitches and elegant designs. The original artifacts of these historical wonders in Han embroidery should not be copied but be inherited. Endeavor should be made to develop the traditional embroidery craftsmanship, and explore new ways to embellish apparel using the traditional Han embroidery. The sample dresses decorated by Han embroidery have been developed and presented. However, the possibilities for design are vast. The subjects of the long term research on Han embroidery include the technique used in garment, fabrics, properties of designs, lay out of colors, composition etc. In particular, Han embroidery patterns can be used to decorate the made-by-measure and haute couture, of which the artistic charm will attract increased attention from consumers. It is expected that there is huge potential of value adding in Chinese Han embroidery products.

5. Acknowledgement

The author would like to thank the HNCHRC in Wuhan Textile University for its valuable support. The author is grateful to the foundation of Wuhan Textile University (No.143033) for supporting this research.

References

[1] Sun B. Chinese embroidery: craftsmanship under the needle.Culture Heritage: 2011; 26-29.
[2] Sun PL. Embroidery art to the highest level-a glimpse of splendid Suzhou embroidery. China&The World Culture Exchange: 2011; 38-39.
[3] Wang LB, Zhong H. Discussion on plain stitch embroidery in early China based on new Unearthed materials. Journal of Donghua University. (Eng. Ed.): Vol. 24. No. 3 (2007); 424-426.
[4] Liu GY. Miao people's dragon-patterned embroidery.China&The World Culture Exchange: 2010; 42-43.
[5] Nevbahar G, Nurean K. Decorative elements in Turkish garment culture from past to future: art of embroidery. Journal of Textile and Engineer: Vol. 23. No. 103 (2016); 231-236.

Patterned Fabric Inspection by Image Decomposition based on Sparse Representation

Peng-Fei Li*, Zhuo-Mei Liu, Jun-Feng Jing

College of Electronic and Information, Xi'an Polytechnic University, Xi'an, 710048 China

*Corresponding author's email: Li6208@163.com

Abstract

We present a new approach using image decomposition (ID) of sparse representation to address patterned fabric defect detection. We are inspired by the blind source separation theory, which assumes that the mixed signal is a linear superposition of multiple source signals. Therefore, the fabric defect image is considered to consist of a linear superposition of the defective part, the background part and the noise part. Firstly, histogram equalization pre-processing is used to enhance the image contrast. Then, an ID algorithm, based on sparse representation, is used to decompose the fabric image into a cartoon part and a texture part. Finally, the defective image is segmented by superimposed binarization to show the defective region. The experimental results reveal that the proposed algorithm is effective in patterned fabric defect detection.

Keywords: Patterned Fabric Inspection; Image Decomposition (ID); Sparse Representation

1. Introduction

Competition in the Chinese textile industry is becoming increasingly fierce with the rapid development of science and technology, while the effective detection and control of textile surface defects are central aspects of modern textile enterprises in controlling costs and improving product competitiveness. Traditional fabric defect detection is mainly performed manually, which can lead to false detection, missed detection, low detection rates, and a significant waste of manpower and material resources. Therefore, it is of great significance to develop an efficient, reliable and accurately automatic fabric defect detection algorithm.

Recently, experts and scholars at home and abroad have proposed several fabric defect detection methods, which can be mainly divided into three categories: statistical methods, frequency domain methods and model methods. The objective of defect detection based on statistical methods is to separate the defective area of the inspection image from the background texture. The general algorithms are fractal dimension, grey level co-occurrence matrix, morphological and cross-correlation algorithms. In an earlier work [1], the one-dimensional fractal extraction fabric feature was used to realize the detection of defects, which revealed that the contrast efficiency greatly improves the efficiency of feature extraction. The statistical histogram and the grey level co-occurrence matrix were used in [2] to extract fabric texture features. In the process, 27 types of defects were detected, while the detection success rate was 93.1%. Celik et al. [3] combined morphological and linear filters with offline and online detection of five different types of defects, with an 83% and an 88% success rate, respectively. The autocorrelation function was used in [4] to determine the repeating unit of the fabric surface, in which the size of the repeating unit was calculated as the basis for dividing the fabric image. The aim of spectral methods is to transform the defect image into the frequency domain for image processing and achieve a good processing effect for images with repetitive basic texture primitives. Fabric defect detection, using spectral methods, mainly involves Fourier transform, wavelet transform and Gabor filters. Fourier analysis was used by Millan et al. [5] to extract the structural features of the fabric texture and to design the parameters of the bandpass filter. The defects were extracted from the fabric background by extracting different dimensions of information. In [6], the fabric image was decomposed, reconstructed and separated by warp and weft information by wavelet transform. A mixed Gabor filter and Gaussian model was used to detect the plain weave fabric, for which the success rate was 87% [7]. Multiscale Gabor wavelet was proposed in [8] to detect fabric defects. Model-based methods are primarily suitable for fabric images with stochastic surface variations or for randomly textured fabrics. Model-based methods for defect detection are mainly the Markov random field model and the autoregressive model. Hu et al. [9] proposed the use of a wavelet domain hidden Markov random field to establish the texture model of the defect-free fabric. The model parameters were obtained by training the defect-free images, while the maximum expectation thought was used to determine whether there were any defects. Shape defects and different fabric texture experiments were conducted to verify the reliability of the algorithm. In [11], an autoregressive spectrum estimation model was used to analyse the one-dimensional energy spectral density of the fabric texture, as well as detect the fabric defects from the normal texture. The existing research algorithms are mostly aimed at the simple part of the fabric texture images; only a few can process specific texture images, while there is little research on more complex textiles. Therefore, it is of vital significance to study how to detect textile defects involving a complex texture.

Many scholars have proposed various fabric defect detection methods according to sparse representation theory. Sezer et al. [10] used independent component analysis (ICA) to detect fabric flaws, but the defective identification ability of the method was poor, as it depended on whether the extracted independent part was compatible with the flawed appearance structure. Zhou et al. have carried out a lot of research work on the image sparse representation model, proposing a fabric surface defect detection system based on sparse representation dictionary model reconstruction [11] and an adaptive dictionary [12]. However, these two algorithms only apply to plain and twill fabric surface defect detection. Ye. J et al. [13] proposed the detection of surface defects of pattern fabric based on the Gabor extraction of texture projection features, although the detection of detail defects was very unsatisfactory. A fabric defect detection algorithm based on K-SVD sparse coding dictionary was used in [14], but the algorithm was less robust. In [15], a convolution matching pursuit dual-dictionary (CMPDD) was used for pattern defect detection, but the algorithm was not ideal for the details of the defects and morphological features. This paper proposes a fabric pattern detection method based on sparse representation for ID. This method not only resolves the complex texture issue of fabric image defect detection, but is also suitable for a variety of defects, as well as shows the details of the edge of the defect characteristics wonderfully and with good robustness. Firstly, histogram equalization preprocessing is used to enhance image contrast. Then, the ID algorithm based on sparse representation is used to decompose the fabric image into a cartoon part and a texture part. Finally, the defective image is segmented by superimposed binarization to show the defective region.

2. Patterned Fabric Inspection

The method of ID based on sparse representation for pattern fabric inspection has three main steps: 1) pre-treatment; 2) image decomposition; 3) binary segmentation.

2.1 Pre-treatment

In order to reduce the effect of image distortion, caused by the light being is too bright or too dark during image acquisition, the inaccuracy of the image acquisition components (such as a CCD camera), the various noise of the image transmission process and so on, which will inevitably reduce the quality of the fabric images collected, result in image distortion. Meanwhile, in order to improve image quality, the image needs to be pre-processed by the histogram equalization method known as image enhancement. Histogram equalization [16] involves non-linearly stretching the non-uniform distribution histogram of these images and re-distributing the pixel values of the image, so that the pixels are evenly distributed over the entire grey-scale range of the image in order to enhance the visual effect of the image. The image, equalized by the histogram, can be deduced by formula (1):

$$s_m = T(r_m) = \sum_{i=0}^{m} P(r_i) = \sum_{i=0}^{m} n_i / n \qquad m = 0, 1, 2, \cdots, L-1 \tag{1}$$

where: m is the grey level of the image, n is the total number of pixels in the image, n_i is the number of pixels on the grey level i, $P(r_i)$ represents the probability density on the i grey level, and $T(r_m)$ is a non-linear transform function of the pixels at m grey level. The mapping given by formula (1) is the final image histogram equalization result.

2.2 Image Decomposition

Assuming that the fabric defect image represents the observation vector y, the defect part and the background part represent the two source signal vectors y_1 and y_2, while the noise part in the image represents the additive noise vector v. The mathematical model of the fabric defect image is [17]:

$$y = y_1 + y_2 + v \tag{2}$$

As sparseness is an important feature of defective images, they can be sparsely represented in a particular transformation. Assuming that there are two dictionaries A_1 and A_2, which can mean sparse representation for the defective part and the background part respectively, the model is [18]:

$$\min \|x_1\|_0 \quad s.t. \quad y_1 = A_1 x_1 \tag{3}$$

$$\min \|x_2\|_0 \quad s.t. \quad y_2 = A_2 x_2 \tag{4}$$

Morphological composition analysis, as proposed by Elad et al. [19], is one of the main algorithms for blind source separation. The theory considers the morphological differences of the respective source signals that make up the observed signal, while separating the observed signals that correspond to different representation dictionaries. That is, each source signal has a dictionary, which can represent itself sparsely.

It is assumed that the observed signal y is composed of two different source signals, y_1 and y_2, and a noise v. Assuming that y_1 and y_2 are generated by using the sparse model $M_1 r$ (corresponding to dictionary A_1) and $M_2 r$ (corresponding to dictionary A_2), the morphological analysis model is expressed as follows:

$$\min_{x1,x2} \|x_1\|_0 + \|x_2\|_0 \; s.t. \|y - A_1 x_1 - A_2 x_2\|_2^2 \le \delta \tag{5}$$

As a result, (x_1^δ, x_2^δ) gives a reasonable solution to the problem regarding the separation of observed data: $\hat{y}_1 = Ax_1^\delta, \hat{y}_2 = Ax_2^\delta$. The parameter δ depends on the noise energy, sparse representation and the model error of signals y_1 and y_2. This separation method is called morphological component analysis.

In order to make the above model beneficial to the calculation, the basis pursuit de-noising of the approximation algorithm is performed to transform the non-convex problem into the convex optimization problem (i.e., the l_0 norm is replaced with the l_1 norm), using the error and additive noise as constraints item [20-21]. Thus, equation (5) is transformed into:

$$\min \|x_1\|_1 + \|x_2\|_1 \quad s.t. \quad \|y - A_1 x_1 - A_2 x_2\|_2^2 \le \delta \tag{6}$$

Realizing the separation of the defective part and the background part of the fabric defect image is equivalent to the reconstruction of the two parts of y_1 and y_2. Fig. 1 shows the patterned image separation model of fabric based on sparse representation. Using the above theory, the sparse representation model of fabric defect image separation is obtained as follows [22]:

$$\hat{x}_1, \hat{x}_2 = \arg\min_{x1,x2} \|x_1\|_1 + \|x_2\|_1 + \lambda \|y - A_1 x_1 - A_2 x_2\|_2^2 \tag{7}$$

The function of equation (7) is to replace the constrained optimization in (6) with an unconstrained penalized optimization, as well as replace the l_0 norm with the l_1 norm. This substitution method is called basis tracking de-noising, which is used to convert the non-convex problem into the convex optimization problem; it is an approximation algorithm.

By finding the right pseudo inverse matrix of A_1 and A_2, the substitution $\hat{x}_1 = A_1^{+} y_1, \hat{x}_2 = A_2^{+} y_2$ is used. In turn, the model is converted to:

$$\hat{y}_1, \hat{y}_2 = \arg\min_{y1,y2} \left\|A_1^{-1} y_1\right\|_1 + \left\|A_2^{-1} y_2\right\|_1 + \lambda \left\|y - y_1 - y_2\right\|_2^2 \tag{8}$$

As the defect component has random uncertainty, it is possible to use the curve transform as the defect component dictionary. The texture of the background component is periodic, so the locally discrete cosine transform is used as the background component dictionary.

In this case, since the transformed image is analysed in the direction of the image at different scales, such that it shows a smooth section and smooth edges, the general variation normalization is used to strengthen the quality of the defective component separation and restrain the edge ringing effect. The model is further converted to [23-24]:

$$\hat{y}_1, \hat{y}_2 = \arg\min_{y1,y2} \left\|A_1^{-1} y_1\right\|_1 + \left\|A_2^{-1} y_2\right\|_1 + \lambda \left\|y - y_1 - y_2\right\|_2^2 + \gamma \mathrm{TV}\{y_1\} \tag{9}$$

where: γ denotes the weighting factor of the total variation penalized optimization. Finally, the block coordinate relaxation algorithm [25] is used to solve objective function equation (9), which effectively separates the defect part and the background part and achieves the goal of commendably dividing the defect target.

2.3 Binary Segmentation

The cartoon part I_d is binarized. Using the method of superimposed binarization, a two-threshold segmentation method is proposed, in which the target image is binned twice and then added.

Here, the cartoon part I_d is set as the original image, with the thresholds taken as T_1 and T_2, where $T_1 > T_2$. Thus, the definition of the binarized image is as follows [26]:

$$f(x,y) = \begin{cases} 1, & I_d(x,y) < T_1 \\ 0, & T_1 < I_d(x,y) < T_2 \\ 1, & I_d(x,y) > T_2 \end{cases} \tag{10}$$

where: $f(x,y)$ represents the binary image after segmentation, 0 means black and 1 means white.

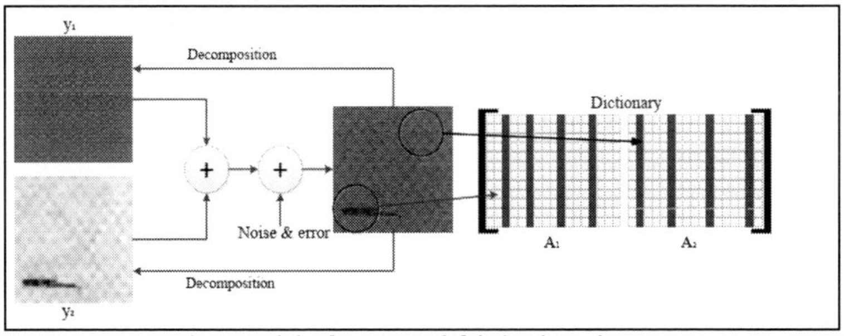

Fig.1 Image separation model of patterned fabrics based on sparse representation

3. Experimental Results and Analysis

3.1 Data Sources and Simulation Environments

This experiment was carried out in the MATLAB R2014a environment, which is commercial mathematics software produced by the American company MathWorks. A printed fabric data set was obtained from the University of

Hong Kong's Industrial Automation Research Laboratory. The total number of samples is 166, including 81 defective images (box: 26, dot: 30, star: 25) and 85 pieces of non-defective images (box: 30, dot: 30, star: 25). The format of the printed fabric images is the BMP format. All of the images have 256 * 256 pixels and a resolution of 600 dpi.

3.2 Inspection Results and Discussion

Fig.2 shows the defective part and background part obtained by the separation method of the square grid, the dot type and the star image. Column A shows the original defect-free image, column B shows the defect image, column C shows the defective component, and column D shows the separation of the background part.

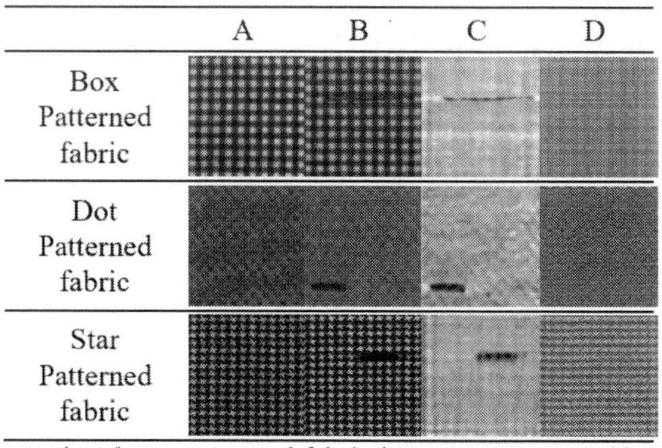

Fig.2 Box-patterned, dot-patterned and star-patterned fabric image as processed by ID: A) defect-free image, B) defect image, C) cartoon part, D) texture part

Fig.3 shows a total of 12 background texture images and defect detection results. The first four columns are of the grid type, the middle four columns are of the dot type, the latter four columns are of the star type, the second row is of the original defect image, the third row is the histogram equalization preprocessing effect chart, the fourth row is the result of the image decomposition algorithm to obtain the cartoon part, and the fifth line is the result of binarization after the binary image is obtained. From the results, we can see that the location and shape of the fabric defects have been well detected.

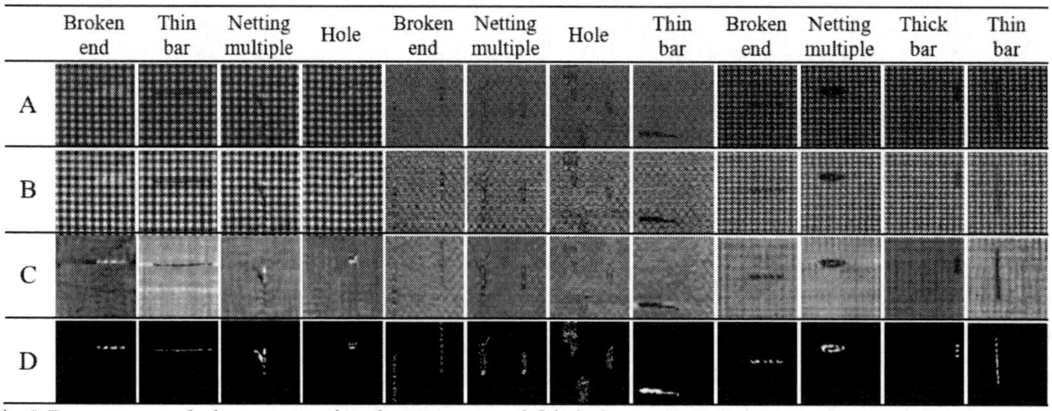

Fig.3 Box-patterned, dot-patterned and star-patterned fabric image Inspection results: A) defect image, B) histogram equalization, C) cartoon part, D) texture part

3.3 Comparison of the Defect Detection of Different Algorithms

In this section, a comparison with the proposed method and the other five advanced defect detection algorithms, including ICA, sparse representation of dictionary model reconstruction (SRDR), Gabor filter, sparse coding dictionary learning (SCDL) and CMPDD. The detection results of the six defect detection algorithms are shown in Table 1.

From the comparison of the test results shown in Table 1, the detection algorithm proposed in this paper has more obvious advantages than other detection algorithms. The main reason is that the method of this paper is to separate the image signal according to the difference of the morphological part, as well as fully maintain the information of each part, with the aim of precise positioning and keeping the details. Under the same conditions, the detection rate of the SRDR algorithm is the lowest, while the SCDL algorithm is higher than SRDR algorithm. The main reason is that the K-SVD algorithm is used for dictionary learning. However, the algorithm has much stricter requirements on fabric texture period. The ICA algorithm only has a good detection rate for the box-patterned fabric, mainly due to the effect of image feature reconstruction. The defect detection algorithm, based on the Gabor filter, has a higher requirement for parameter setting of the filter. The detection rate for the CMPDD algorithm is relatively good, mainly due to the use of CMP and the Gabor-based K-SVD algorithm for sparse coding training and learning to obtain the fusion dictionary, followed by the defective sample image and the image to be detected in the fusion dictionary projection as a feature, as well as to finally determine the defects by distance. The method proposed in this paper not only recognizes the noise, but also uses the sparsity of the defect and the morphological difference of the part of the fabric defect to preserve the details and morphological characteristics of the defect, in terms of offering good robustness and achieving higher detection rates.

Table.2 The comparison of the detection rates of the proposed algorithm and others

Algorithms	Box-patterned	Star-patterned	Dot-patterned	Fine lattice and stripes	Solid colour
Proposed method	97%	97%	95%	97%	98%
ICA	95%	85%	-	-	92%
CMPDD	94%	96%	93%	90%	98%
SRDR	75%	74%	76%	-	-
Gabor filter	92%	90%	-	-	87%
SCDL	95%	92%	-	-	90%

4. Conclusion

Given that sparseness is an important feature of defective images, a patterned fabric defect, based on sparse representation for ID, is proposed. In this method, the part is alternately separated, based on the difference with the morphological part of the defective images. In turn, not only can the information of each part, the precise positioning, the details and the morphological features of the defect be well preserved, but this method can also be applied to various types of defects, especially for the complex texture of the pattern fabric images, with the prospect of good test results.

In this paper, sparse and redundant representations were utilized in the image processing of fabric defects by using the morphological diversity of the observed signals, which can solve the problem of patterned fabric defect detection. Therefore, this study makes sense for patterned fabric defect detection. Further study on this problem is advised, including on the selection of the adaptive strong binary segmentation method to facilitate good visualization of the defective part, and a more effective adaptive threshold update strategy, enabling the decomposition process to be completed and converged quickly.

5. Acknowledgement

The authors gratefully thank the Scientific Research Programme Funded by National Natural Science Foundation of China (61301276) and the Industrial Public Relation Project of Shaanxi Technology Committee (2015GY034).

References

[1] Tian C, Honggang BU, Wang J et al. Fabric defect detection based on fractal feature of time series [J]. Journal of Textile Research: 2010; 31 (5): 44-48.
[2] Selver MA, Avşar V, Özdemir H. Textural fabric defect detection using statistical texture transformations and gradient search [J]. Journal of the Textile Institute: 2014; 105 (9): 998-1007.
[3] Çelik Hİ, Dülger LC. Topalbekiroglu M. Fabric defect detection using linear filtering and morphological operations[J]. Indian Journal of Fibre & Textile Research: 2014; 39 (3): 254-259.
[4] Li SW, Guo L W, Li CH. Fabric defects detecting and rank scoring based on Fisher criterion discrimination [C]. International Conference on Machine Learning and Cybernetics. IEEE: 2009; 2560-2563.
[5] Millán MS, Escofet J. Unsupervised flaw segmentation in textile materials under visible and NIR illumination [J]. Proceedings of SPIE - The International Society for Optical Engineering: 2010; 7723.
[6] Qin Y, Xu F. Analysis and research of the fabric density based on the wavelet transform [C]. Fifth International Symposium on Computational Intelligence and Design. IEEE: 2012; 197-200.
[7] Zhang Y, Lu Z, Li J. Fabric defect detection and classification using gabor filters and gaussian mixture model [Z]. Asian Conference on Computer Vision 2009: 2010; 635-644.
[8] Mak KL, Peng P, Yiu KFC. Fabric defect detection using multi-level tuned-matched Gabor filters [J]. Journal of Industrial & Management Optimization: 2012; 8 (2): 325-341.
[9] Hu G, Zhang G, Wang Q. Automated defect detection in textured materials using wavelet domain hidden Markov models [J]. Optical Engineering: 2014; 53 (9).
[10] Serdaroglu A, Ertuzun A, Ercil A. Defect detection in textile fabric images using wavelet transforms and independent component analysis [J]. Pattern Recognition and Image Analysis: 2006; 16 (1): 61-64.
[11] Zhou J, Semenovich D, Sowmya A et al, Sparse dictionary reconstruction for textile defect detection, International Conference on Machine Learning & Applications: 2012; 1 (4): 21-26.
[12] Zhou J, Wang J. Fabric defect detection using adaptive dictionaries. Textile Research Journal: 2013; 83(17): 1846-1859.
[13] Ye J, Zhu Q, Fan C. Fabric defect detection based on sparse projection features [J]. Sci. Technol. Eng: 2014; 14 (33): 111-115.
[14] Liu SM, Li PF et al. Defect detection based on sparse coding dictionary learning [J]. Journal of Xi'an Polytechnic University: 2015; 29 (05): 594-599.
[15] Jing JF, Fan XT, Li PF. Patterned fabric defect detection via convolutional matching pursuit dual-dictionary [J]. Optical Engineering: 2016; 55 (5): 053109-053120.
[16] Singh K, Kapoor R. Image enhancement using exposure based sub image histogram equalization [J]. Pattern Recognition Letters: 2014; 36 (1): 10-14.
[17] Olsson R K, Hansen L K. Linear state-space models for blind source separation [J]. Journal of Machine Learning Research: 2006; 7 (4): 2585-2602.
[18] Zhou J, Semenovich D, Sowmya A et al. Sparse dictionary reconstruction for textile defect detection [C]. International Conference on Machine Learning and Applications. IEEE: 2012; 21-26.
[19] Starck JL, Murtagh F, Fadili J. Sparse Image and Signal Processing: Wavelets, Curvelets, Morphological Diversity. Cambridge University Press, Cambridge (UK): 2010; 189-200.
[20] Chen S, Saunders MA, Donoho DL. Atomic decomposition by basis pursuit [J]. Siam Review: 2006; 43 (1): 129-159.
[21] Qu L. Study on Underdetermined Image Blind Source Separation Based on Sparse Component Analysis [D]. Tianjin University of Technology and Education: 2014; 14-17
[22] Elad M. Sparse and Redundant Representations: From Theory to Applications in Signal and Image Processing. Springer, New York (NY): 2010; 250-260.
[23] Elad M, Starck JL, Querre P, Donoho DL. Simultaneous cartoon and texture image inpainting using

morphological component analysis (MCA). Journal on Applied and Computational Harmonic Analysis. Vol. 19: 2005 (November); 340-358.
[24] Starck JL, Elad M, Donoho DL. Image decomposition via the combination of sparse representation and a variational approach. IEEE Transaction on Image Processing: 2005; 14, 10: 1570-1582.
[25] Sardy S, Bruce A, Tseng P. Block coordinate relaxation methods for non-parametric signal de-noising [C]. Proceedings of the SPIE - The International Society for Optical Engineering 3391: 1998: 75-86.
[26] Gonzalez RC, Woods RE. Digital Image Processing, Second Edition. Pearson Prentice Hall, New Delhi: 2008.

A Study of Salient Regions in Virtual Fitting Software Interface Design

Yuan Xu, Rong Zheng*, Ling-ling Xu, Yun-chao Zhang

Aimer Human Engineering Research Centre, Beijing Institute of Fashion Technology, Beijing, 100029, China

*Corresponding author's email: yuki.zheng@gmail.com

Abstract

With the increasing popularity of online shopping, virtual fitting software technology is constantly developing and promises to shape a new trend in the clothing industry. This study investigated 8 virtual fitting programs and examined salient regions in their interface designs. Using a SMI eye-tracking device, the eye movements of 19 subjects were recorded while they used the virtual fitting programs. The 8 virtual fitting programs were divided into three categories according to their interface layouts. Using BeGaze Analysis Software, the study conducted a statistical analysis of salient regions for the programs across the three categories. The distribution on salient regions in virtual fitting mobile applications (with horizontal/vertical interfaces) and websites was identified, thus providing a theoretical basis for optimizing interface layout of virtual fitting applicatons and websites.

Keywords: Virtual Fitting; Eye-tracking; Salient Region

1. Introduction

The development of Internet technology has changed people's way of life. As an increasing number of consumers take advantage of the convenience of online shopping, clothing has become one of the retail categories that experience higher volumes of online sales [1]. However, clothes sellers and buyers alike have experienced problems related to returns, refunds and clothing exchanges due to online shopping's inherent shortcoming of buyers being unable to try on apparel in person. As a result, virtual fitting technology came into being and continues to be developed. The term virtual reality (VR), coined by Jaron Lanier in 1987, was defined as a visible and operational three-dimensional environment simulated by a computer [2]. German scientists at the Fraunhofer-Gesellschaft research organisation established the first virtual fitting system, while clothing retailer H&M was the first company to provide an online fitting service [3]. Around 2007, virtual fitting websites such as "face 72", "41go" first appeared in China. Until now, on the whole, domestic virtual fitting software is still in its infancy stage without unified functions and service standards; its usability still needs to be tested and improved.

Salient regions refer to the sections on a web page that attract more of the viewer's attention. Numerous domestic and international studies have been conducted on salient region distribution in web pages, aided by eye-tracking technology. The studies determined that the parts of web pages known as F-shaped region, central region, left region and a golden triangle region attract more attention. Eye-movement tracking technology is used to identify people's interest through an eye-tracking device that records and analyses various eye movement indexes when people browse pictures and read and watch real scenes, trying to visualize what people see [4].

Nielsen asked 232 people to browse thousands of web pages in a study that obtained a huge amount of eye-movement data, from which he mapped out salient regions and visual scanning routes; however, the study concluded that F-shaped pattern is only applicable to searching and browsing pure-text web pages [5]. Through a study of Internet users' eye movement patterns, Schroeder found that users usually concentrate their attention first on the central region of web pages, then shift to the left side and finally to the right side [6]. Following the same method, Sun Linhui et al. determined the focus order when people browse web pages: centre-upper left-upper right/left bottom-right bottom [7]. Buscher et al.'s study of eye movement asked 20 subjects to browse Three hundred and sixty-one web pages, and found that information on the right side of web pages was ignored twice as often as other regions; in terms of information search, analysis and integration, the central left region first attracts more attention than other regions, while in terms of information identification, the upper left region draws attention first [8]. Hotchkiss et al. proposed that there is a "golden triangle" when users browse web pages; the triangle region in the upper left of a webpage attracts most people's attention. However, researchers found that Hotchkiss et al.'s findings do not apply to Chinese website browsing [9].

From the above literature review, it is evident that many researchers used the eye-tracking method to identify salient regions in web page browsing; however, no study has touched upon the functional virtual fitting websites. As the usability of existing virtual fitting software still needs to be proven and improved, this study on salient regions in the virtual fitting software's interface can provide suggestions on optimizing interface design in order to increase users' satisfaction.

2. Experiment Design

Via market surveys, eight 2D/3D virtual fitting programs with different functional characteristics were selected as study objects, including four website-based programs and four applications. Table 1 shows the virtual fitting program applications' and websites' names, types and functions.

Table.1 Information about the selected virtual fitting applications and websites

Type of client gateway	Name	2D / 3D	Type	Function
Application	UNIQLO	3D	Shopping	Provide clothes fitting effect from every angle, realizing online and off-line synchronization
	y.love	3D	Shopping	Clothes matching; combination of multiple brands
	VID Smart iFashion	2D	Entertainment	Match recommendation and tips sharing
	Mallgo	3D	Entertainment	Provide matching options, integrating with its shopping app
Website	Baiyi fitting	2D	Entertainment	Personalized clothes matching ; share experiences with friends
	UNIQLO	2D, 3D	Shopping	Provide clothes fitting effect from every angle, realizing online and off-line synchronization
	My virtual model	3D	Shopping	Provide matching suggestions in simulated real-life situations
	Fitbay	2D	Social networking	Based on users' body shapes, provide purchase and match suggestions in a social interaction format

The study selected undergraduate and graduate students as experiment subjects considering that they are more susceptible to new trends, often shop online and are fairly fluent in English (which is important as the virtual fitting programs under study involve English websites). Because related studies [10, 11, 12] indicate that the subject number should be above a minimum threshold, the study finally selected nineteen subjects with thirteen students majoring in art and six in non-art disciplines. The gender proportion (male:female) was six to thirteen. With their ages ranging from twenty-four to twenty-seven, most of the subjects have Five to six years' experience in online shopping. Generally speaking, the online shopping history of art students was longer than that of non-art students, and female students had a longer history of online shopping than male students.

The experimental devices included the SMI-produced RED distant eye movement-detecting device (as shown in Fig.1) and the HED head-mounted eye movement tracking device. RED was used in the test of virtual fitting websites and HED was used in the test of applications.

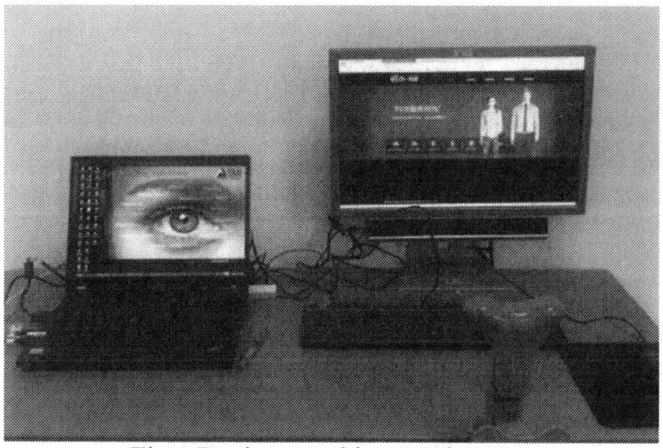

Fig.1 Devices used in experiments.

In the experiment, subjects were instructed to properly wear and use the eye-movement tracking devices, and then tried the eight virtual fitting programs in a randomly arranged order. Since it was the first time that all of them used the eight programs, they needed to follow the director's instructions to use the programs. Instructions were given orally. Based on different functions of the eight programs, special tasks were set, including "enter-set the body shape-search-fitting-rotating model-checking details-purchase-share." Eye-movement tracking devices recorded the test video and eye movement data (mainly about the subjects' attention information).

3. Results and Discussion

By collecting data from the eye-movement tests, the study acquired duration information on task completion of the 19 subjects and their gazes when using the virtual fitting programs. The following section discusses the salient regions in interfaces of virtual fitting programs resulting from the data analysis.

In order to determine the salient region and be conducive to statistical analysis of virtual fitting programs, it was necessary to divide the interface into multiple unit cells. Considering the differences among the three different interface styles of the virtual fitting mobile applications (horizontal version; vertical version) and websites, in order to ensure a scientific statistical study, the study adopted three ways to divide interfaces. In order to ensure content consistency of the unit region, the study adopted the fitting model in the interface as the quantitative parameter, and divided the human body into small grid squares of the same size. Following the method used in the BeGaze software's function of rasterizing interest regions, the study finally divided interfaces of the three versions of virtual fitting programs into 7 rows* 6 columns, 9 rows* 5 columns, and 13 rows*13 columns to investigate salient regions.

With consideration of the virtual fitting programs' functions, the following process was used to define salient regions: 1) adding gaze information at the same grid square for the same task in using different virtual fitting programs-the higher the value, the higher the ranking, thus acquiring Order A; 2) ranking grid squares in light of standard deviation-the lower the value, the higher the ranking, thus obtaining Order B; 3) the sum of their rankings in Order A and Order B generate the overall ranking of each grid square, thus getting Order C; 4) the intersection D between the grid squares occupying the first half of rankings in Order A and the grid squares occupying the first half of rankings in Order C was taken as salient regions; 5) in light of gaze frequencies, grid squares in D were ranked-the higher the value, the higher the ranking. The first half in the ranking of D's grid squares was regarded as salient regions, the following as sub-salient regions.

3.1 Discussion of Salient Regions in the Horizontal Application Interface

The selected applications with a horizontal interface include y.love and VID Smart iFashion. The study mainly analysed eye-movement data in the four tasks of style searching, virtual fitting, 360° rotation of the model to check fitting effect, and purchase.

Based on the eye-movement data (on gaze duration) obtained when subjects completed the four tasks on both programs, the study acquired the gaze duration picture corresponding to the forty-two grid squares after dividing the interface into forty-two grid squares via BeGaze. SPSS was then used for data analysis. The descriptive analysis results about gaze durations are shown in Table 2.

Table.2 Descriptive analysis of gaze durations for grid squares in a horizontal application interface

Grid square unit	Effective sample number(N)	Minimum value (ms)	Maximum value (ms)	Sum (ms)	Average value (ms)	Standard deviation (ms)	Gaze frequencies (N)
1	149	0.00	1397.90	2676.60	17.9638	126.81949	8
2	149	0.00	1560.40	7752.30	52.0289	204.00452	17
3	149	0.00	2756.70	21937.40	147.2309	413.12384	28
4	149	0.00	1817.70	19376.10	130.0409	347.77931	29
5	149	0.00	2517.80	18322.50	122.9698	408.57848	27
6–37				…			
38	149	0.00	4334.70	16664.80	111.8443	437.17465	28
39	149	0.00	3096.60	19804.70	132.9174	399.06584	31
40	149	0.00	4213.90	17838.70	119.7228	428.49579	29
41	149	0.00	2157.60	20608.40	138.3114	382.96177	30
42	149	0.00	1758.30	5673.90	38.0799	171.55812	16

In light of the definition of salient region, the study ranked data of all grid squares. Distribution of the salient regions in the horizontal application interface is shown in Fig.2, with the red squares as the salient and yellow squares as the sub-salient regions. It can be seen that in terms of their location only the salient regions mainly concentrate on the right part of the horizontal application interface.

Fig.2 Salient regions in the horizontal application interface.

3.2 Discussion of Salient Regions in the Vertical Application Interface

The selected applications with a vertical interface include UNIQLO and Mallgo. The study mainly analysed eye-movement data in the six tasks of body shape setting, style searching, virtual fitting, rotating the model, checking details and purchase.

Based on the eye-movement data (on gaze duration) when subjects completed the six tasks on both programs, the study acquired the gaze duration pictures corresponding to the forty-five grid squares after dividing the interface into forty-five grid squares via BeGaze, after which SPSS was used to analyse data. The descriptive analysis results about gaze durations are shown in Table 3.

In light of the definition of salient region, the study ranked data of all grid squares, and finally acquired distribution of salient regions in a vertical application interface as shown in Fig.3. The salient regions mainly cluster in the central part of a vertical application interface. Regions adjacent to the centre in the left also drew a certain amount of attention.

Table.3 Descriptive analyses of gaze durations for grid squares in a vertical application interface

Grid square unit	Effective sample number(N)	Minimum value (ms)	Maximum value (ms)	Sum (ms)	Average value (ms)	Standard deviation (ms)	Gaze frequencies (N)
1	225	0.00	7514.7	75930.5	337.4689	893.22728	73
2	225	0.00	2776.40	30787.4	136.8329	387.11267	51
3	225	0.00	1158.90	10266.5	45.6289	153.58007	30
4	225	0.00	2297.50	14783.6	65.7049	261.00386	30
5	225	0.00	1099.00	12103.1	53.7916	180.97277	31
6–39				…			
40	225	0.00	3894.80	22834.9	101.4884	351.24988	48
41	225	0.00	2496.90	21735.9	95.0040	321.70492	33
42	225	0.00	1618.30	8290.40	36.8462	185.52202	22
43	225	0.00	1000.70	2854.30	12.6858	79.32755	15
44	225	0.00	3176.70	17798.5	79.1044	325.23915	29

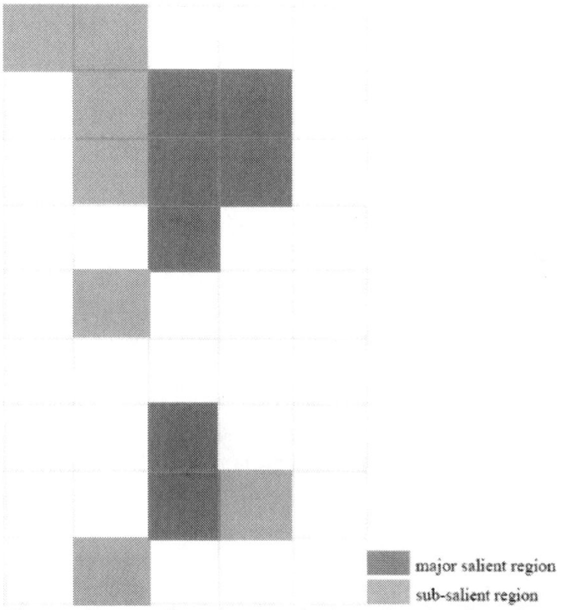

Fig.3 Salient regions in a vertical application interface.

3.3 Discussion of Salient Regions in a Website Interface

The virtual fitting websites selected by the study include UNIQLO, Baiyi fitting, My Virtual Model and Fitbay. The study mainly analysed eye-movement data in tasks of body shape setting, style searching, virtual fitting, rotating the model the whole round, checking details, purchase and sharing matching tips.

Based on the eye-movement data (on gaze duration) when nineteen subjects completed the seven tasks by using the four programs, the study acquired the gaze duration pictures corresponding to the one hundred sixty-nine grid squares of the interface, which was divided via BeGaze, after which SPSS was used to analyse the data. The descriptive analysis results about gaze durations are shown in Table 4.In light of the definition of salient region, the study ranked data of all grid squares, and finally acquired the distribution of salient regions in a website as shown in Fig.4. The salient regions mainly cluster in the central part of a virtual fitting website with the left central part attracting more attention.

Table.4 Descriptive analysis of gaze durations for grid squares in a website interface

Grid square unit	Effective sample number(N)	Minimum value (ms)	Maximum value (ms)	Sum (ms)	Average value (ms)	Standard deviation (ms)	Gaze frequencies (N)
1	390	0.00	718.4	1189.7	3.05051	41.4943	12
2	390	0.00	939.6	1640.4	4.20615	59.2817	11
3	390	0.00	665.8	983.7	2.52231	37.3226	11
4–79				…			
80	390	0.00	1121.1	11837.1	30.3515	133.018	37
81	390	0.00	2714.3	40867.3	104.788	379.607	62
82	390	0.00	5170.7	62999.8	161.538	503.543	88
83	390	0.00	3806.7	61111.9	156.697	429.946	107
84	390	0.00	4539.5	85891.9	220.236	521.953	124
85	390	0.00	4572.6	68102.8	174.623	544.147	93
86–166				…			
167	390	0.00	0.00	0.00	0.0000	0.00000	0
168	390	0.00	0.00	0.00	0.0000	0.00000	0
169	390	0.00	0.00	0.00	0.0000	0.00000	0

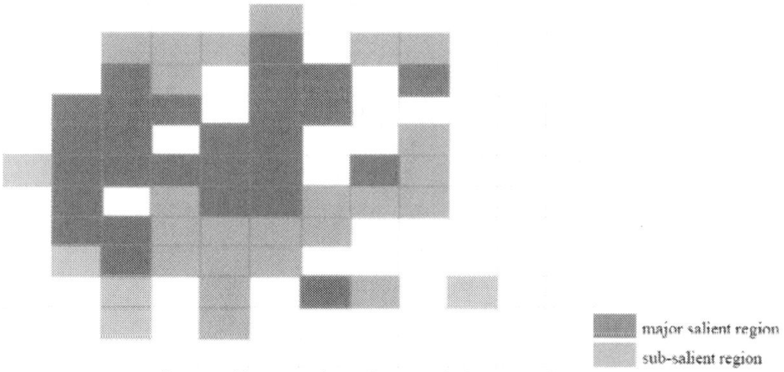

Fig.4 Salient regions in a website interface.

4. Conclusion

The study investigated salient regions in 8 representative virtual fitting programs' interfaces with the aid of SMI eye-movement tracking device. By means of the BeGaze software's function of rasterizing interest regions, the study divided the interface of a specific virtual fitting program into average-sized cells. After statistical analysis of data of the 19 student subjects' first use of the selected programs, salient regions in the programs' interfaces were acquired. After discussion analysing the salient regions, the following conclusions were achieved: salient regions in horizontal applciation interfaces, vertical application interfaces and website interfaces are located in the right part, central part, and left central part, respectively

This study can not only provide a basis for designing the interface of virtual fitting software and even software in other fields, but can also offer guidance to clothing manufacturers to improve their virtual fitting interfaces, thus further promoting the efficiency of customers using the fitting function and improving satisfaction. These features can help promote a garment company's brand influence.

References

[1] China e-business market data survey report. China e-Business Research Centre, 2015. Information on http://www.100ec.cn/zt/2015sndbg/

[2] Chen JX, Cheng YX, Wang WZ. User's need evaluation about virtual fitting function and related interaction interface design. Ming Hsin: 2008; 34 (02): p. 39.
[3] Cordier F, Seo H, Magnenat-Thalmann N. Made-to-Measure technologies for online clothing store. IEEE Computer Graphics and Applications: 2003; 1: p. 38-48.
[4] Qu B. Webtsite browsing-based eye movement study. China: Zhejiang Normal University: 2007.
[5] Nielsen J. F-Shaped pattern for reading web content, 2006. Information on https://www.nngroup.com/articles/f-shaped-pattern-reading-web-content/
[6] Testing web sites with eye tracking. Information on https://articles.uie.com/eye_tracking/
[7] Sun LH, Sun LY, Fu XH et al. College student's eye movement when browsing websites and influence factor analysis. Chinese Ergo: 2010; 16 (2): p.69-71.
[8] Buscher G, Cutrell E, Ringel Morris M. What do you see when you're surfing? using eye tracking to predict salient regions of web pages. Proceedings of the SIGCHI Conference on Human Factors in Computing Systems: 2009; p. 21-30.
[9] Hotchkiss G, Alston S, Edwards G. Eye tracking study: an in depth look at interactions with Google using eye tracking methodology. Kelowna, BC: Enquiro Search Solutions: 2005.
[10] Spryridakis JH. Conducting research in technical communication: the application of true experimental designs. Tech Comm: 1992; 39 (4): p. 607-624.
[11] Nielsen J. Heuristic evaluation. In: Nielsen J, Mack RL, editors. Usability inspection methods. New York: Wiley: 1994; p. 25-62.
[12] Schmitt BH. Experiential marketing: how to get customers to sense, feel, think, act, and relate to your company and brands. New York: The Free Press: 1999.

3D Garment Redressing Based on Virtual Tailoring

Duan Li[1], Ge Wu[1], Peng-Peng Hu[1], Yue-Qi Zhong[1,2*]

[1] *College of Textiles, Donghua University, Shanghai, 201620, China.*
[2] *Key Lab of Textile Science and Technology, Ministry of Education, Shanghai, 201620, China.*

[*]Corresponding author's email: zhyq@dhu.edu.cn

Abstract

In this paper, we presented a new method of redressing virtual garments onto various scanned human models. The "tailor lines" were employed to segment the garment and the human model respectively. The patches of the virtual garment were automatically positioned around the human model by setting up the correspondence via feature matching. Virtual sewing were engaged to obtain the final results of virtual dressing. The collision between cloth model and human model was solved by a geometrical method that untangled the penetrations over iterative position recovery. The experimental results indicate that our method is an efficient approach for redressing various garments onto various human models while maintaining the original geometrical features of clothing.

Keywords: Redressing; Virtual Tailoring; Automatic Positioning; ICP; Laplacian Coordinates

1. Introduction

3D garment dressed on various types of scanned human bodies provided a convenient way to display clothing style. In order to get the dressing effect of 3D garment, Metaaphanon et al [1], Clemenset et al [2] Power et al [3], used CAD to generate 2D garment patterns of clothing, stitched them onto the surface of the human body. Decaudin et al [4],converted the 2D pattern in to an initial 3D surface based on a precomputed distance field around the mannequin, and generated the folds by using procedural modeling of the buckling phenomena observed in the real fabric. Yan et al [5], studied the virtual try-on system for physically disabled people with scoliosis, according to the distortion of the body posture. They deformed corresponding to the 2D pattern Zhang et al [6] and created 3D garments from 2D panels while scaling the 3D garments to fit different 3D bodies.

Many studies [7-10] segmented the 3D human body and flattened the parts to 2D patterns for computer aided design. Others tried to use 3D Garment from scans for virtual dressing. Due to different 3D scanned Garment and human body postures, posture matching is a vital issue that needs to be addressed. Zhong [11], via skeleton-matching algorithm to adjust the posture of the human body for matching the posture of the garment. Huang et al [12], divided the body into different parts, and adjusted the parts position to match the garment posture. Guan et al [13], used the machine learning method to show drape effect. However, they need lots of samples for learning.

Most ICP algorithms [15-16] were used to find the feature points of the human body [17], or compose point clouds. In our approach, ICP algorithm was used in garment parts and human posture matching. In order to sew the 3D garments together, seam lines between each part need to be formed. Zhang et al [6], compared boundary point position to generate seam lines of the 2D parts. In our method, the changes of vertex index were recorded to generate 3D parts seam lines.

The penetration between the human body and garment is a common problem in the virtual try-on system, zhang et al [6], Guan et al [13], fixed the position of the garment vertex by the intersection of the garment vertex and the body mesh. Zhong [11], proposed a normal based method to adjust the vertex position of each garment layer to complete the penetration compensation. In our method, Laplacian-based deformation [18] was used to guarantee the smoothness of the mesh deform in penetration compensation algorithm and garment patches sew algorithm.

In our approach, as shown in Fig.1, 3D garment was decomposed to 3D patches, and scanned body was segmented to 3D parts (Fig.1a), 3D patches and scanned body posture were matched by point cloud feature matching(Fig.1b). After the 3D patches were sewed (Fig.1c), penetration compensation algorithm was used to generate the final wear effect (Fig.1d).

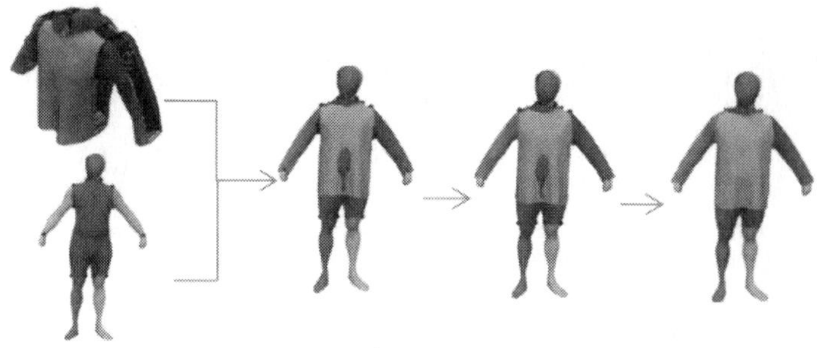

a b c d

Fig.1 (a) Human model and garment decomposition; (b) Automatic positioning garment patches around human model; (c)Virtual sewing; (d) Penetration recovery

2. Methodology

2.1 Sewing path Decomposition

For the convenience of explanation, a sweater was used as an example, as shown in Fig.2a. In order to generate tailoring lines, key points of 3D garment need to be found automatically. AS shown in Fig.2a, the following parts key points are required: armpit, shoulder, neckband, sleeve and coat hem. A set of horizontal planes were employed to slice the sweater along the Y axis from top to bottom, as shown in Fig.2b. In the vicinity of the armpits, due to the separation of the arm and a set of planes were employed to slice the sweater along the Y axis from top to bottom, as shown in Fig.2b. In the vicinity of the armpits, due to the separation of the arm and the body, the slice loop will be changed from one to three. With this feature, the location of the armpits can be determined.

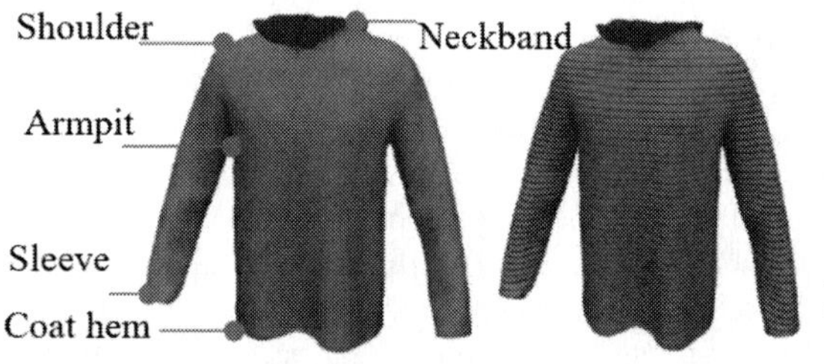

a b

Fig.2 (a) 3D sweater; (b) Sliced sweater.

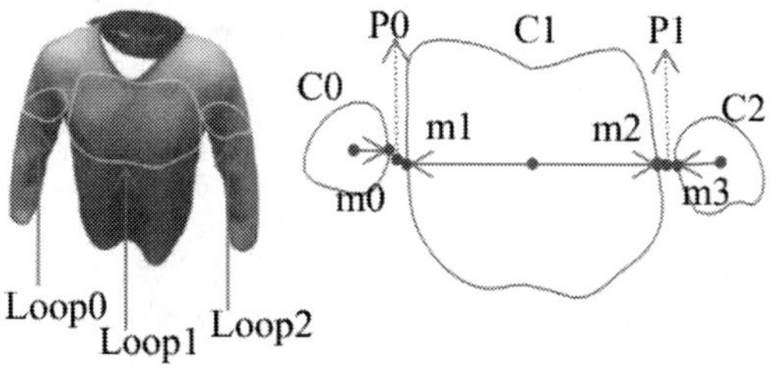

a b

Fig.3 (a) Armpit loops; (b) Armpit points.

As shown in Fig.3a, three separate loops were found corresponding to the right sleeve, the torso, and the left sleeve. The centers of these loops are:

$$c = (\textstyle\sum_{0}^{n} v)/n \tag{1}$$

where c is the center of loop, v is the point in loop, n is the point number in loop.Equation 2 was used to find the intersection of ray through the center and the corresponding loop

$$\begin{cases} m = c + t_0 \times \overrightarrow{cn} \\ m = v + t_1 \times \overrightarrow{vn} \end{cases} \tag{2}$$

where m is the intersection point, c is the center of loop, $\overrightarrow{cn}$ is the direction of ray, v is the point in loops, $\overrightarrow{vn}$ is the direction of two adjacent points, we traverse all points in the loop, four intersection point can be found: m0, m1, m2, m3, as shown in b. Then, right armpit can be found by equation 3, and left armpit point can be found by equation 4;

$$p0 = (m0 + m1)/2 \tag{3}$$

$$p1 = (m2 + m3)/2 \tag{4}$$

The sweater was cut by two vertical planes to obtain s0 and s1, as shown in Fig.4a. After two arm tailoring lines were found, the shoulder points and the armpit points could be calculated by the intersections of the tailoring lines and the rays through the centers of tailoring lines respectively. As shown in Fig.4b, d0 is the right shoulder point, d1 is the left shoulder point, b0 is the right armpit point, b1 is the left armpit point. In order to get tailor points of neckband, cuffs and coat hem, at first, the boundary lines of upper clothing were extracted, as shown in. Fig.5a red lines, there are four boundary lines were extracted.

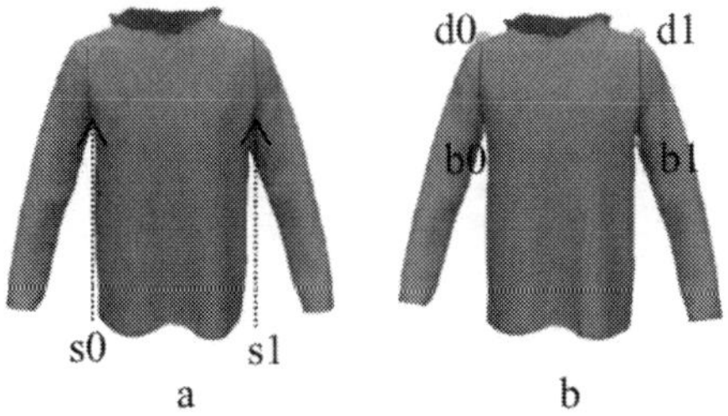

Fig.4 Arm tailoring line.

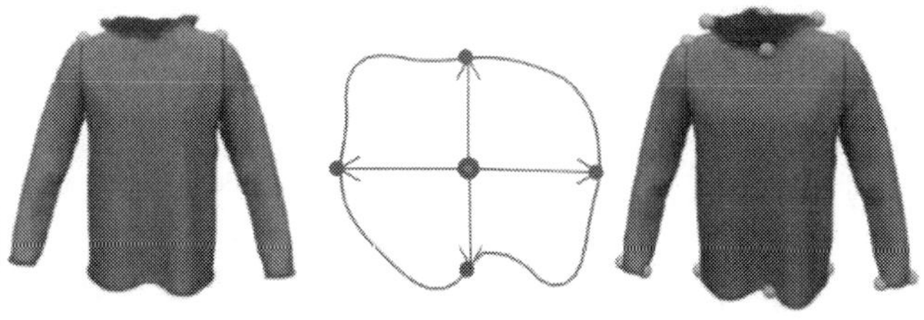

Fig.5 Key points of upper clothes.

Four mutually perpendicular rays were employed to calculate the intersection points of rays and boundaries, as shown in.Fig.5b. The directions of four mutually perpendicular rays are.(1,0,0), (-1,0,0), (0,0,1), (0,0,-1).

Fig.6 Front and back sweater.

Finally, 12 tailor points were found, as shown in Fig.5c, yellow balls. After we got the tailor points, three steps were processed to tailor the sweater:

Fig.7 Tailoring lines.

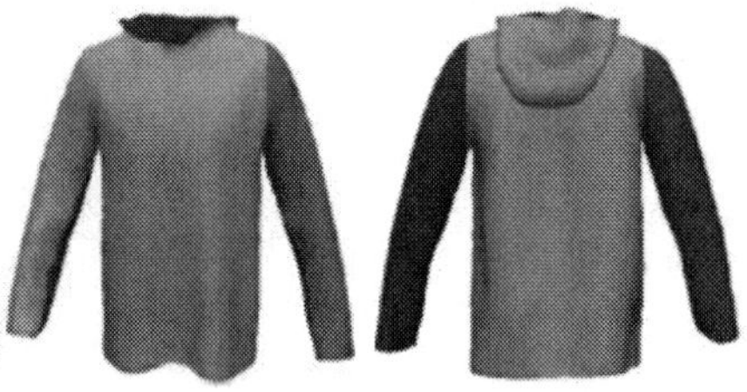

Fig.8 Six patches of the sweater.

1) The sweater was divided into three parts by the right arm and left arm tailor lines, as shown in Fig .6, there are three parts of sweater: left cuffs, right cuffs, and middle clothes.

2) Tailor lines were generated on the three parts of sweater. In order to generated tailor lines, a plane was built by three points to cut sweater, for example ,left armpit point, left shoulder point, and the outer left cuff point were employed to build a plane for cutting left cuff. Use the same method, all the tailoring lines of the sweater could be generated, as shown in Fig.7, blue lines.

3) Sweater was tailored into six patches by 10 tailor lines, as shown in Fig.8, the six patches are: left cuff front patch, left cuff back patch, right cuff front patch, right cuff back patch, middle clothes front patch, middle clothes back patch.

2.2 Seam Lines Generation

In order to sew the patches of the sweater, seam lines between patches should be constructed. As shown in a, two patches were shown as Fig.9a, blue lines and black lines, the red lines were boundaries of two patches. Because the two patches shared the boundary points, the boundary points need to be reproduced, and build seams between them. As shown in Fig.9b, the boundary triangles between two patches were rebuild, and the boundary points were reproduced, we recorded the index of boundary points before triangles rebuild, and after. Two records were compared and the seams of the boundary points could be found. As shown in Fig.9c, the seam lines were generated for six patches of sweater.

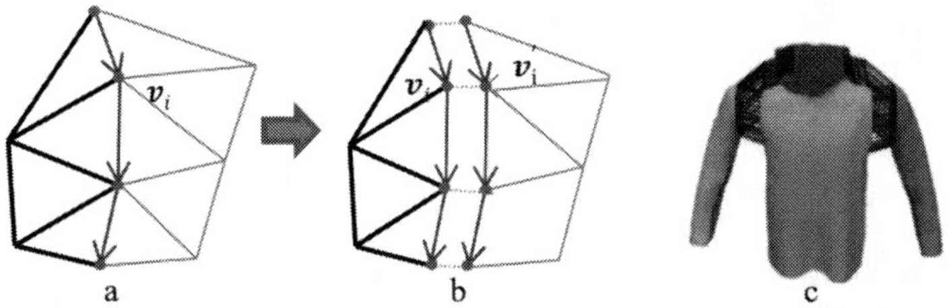

a b c

Fig.9 Seams of sweater.

2.3 Human Body Decomposition

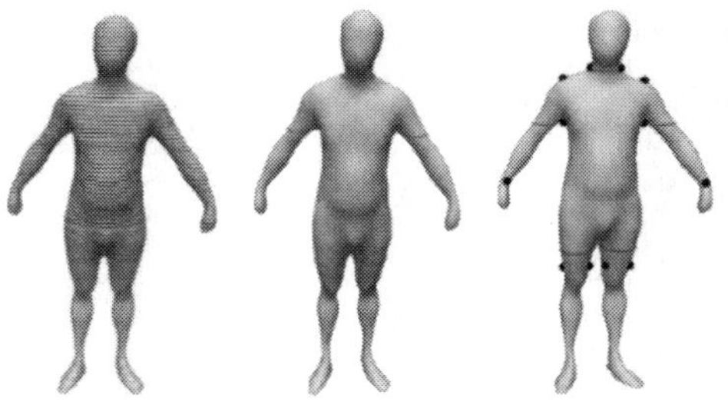

a b c

Fig.10 Found Key points of bod.

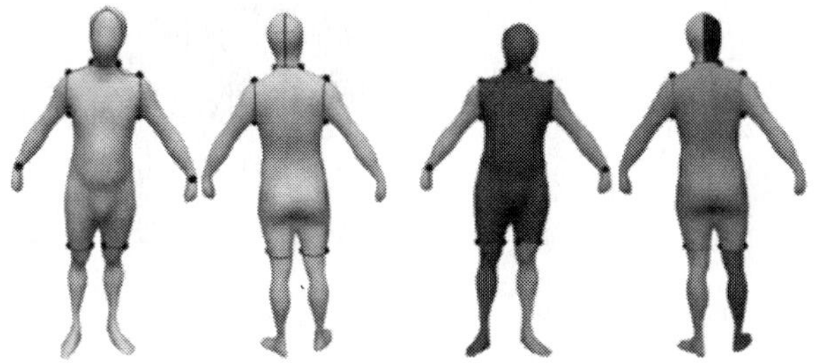

a b

Fig.11 Human body segment parts.

In order to match the 3D garment patches and body parts, human body need to be segmented according to 3D garment patches. The same method as slice 3D garment was used to slice human body. We need a A pose human body, as shown in Fig.10a. the structure of the slice loops was employed to find the segment loops on human body, as shown in Fig.10b. Five segment loops were fixed: neck loop, left arm loop, right arm loop, left leg loop, right leg loop. Body key points could be found based on these segment loops. As shown in Fig.10c, 12 key points of human body were identified. In order to adapt to different sizes of garment, the legs split line was pulled down 10cm from crotch. The two wrist key points are manually identified for sleeve posture match. Segment lines were generated based on human body key points, as shown in Fig.11a, blue lines. By these segment lines, the human body was divided in two steps, as shown in Fig.11b. Firstly, the human body was divided into five parts: head, torso, left arm, right arm, left leg, right leg. Secondly, the head was dived into three parts for hat matching: front, back left, back right.

2.4 Patches Positioning and Patches Sewing

After use our method get divide human body, the sweater shoulder points were used to match human body shoulder points, we can get a result as shown in Fig.12a. From the figure, we can saw that the posture of upper clothes and human body was misplaced, especially the posture of the arm. Due to the different posture of the clothing and the human body, the posture of the clothing and the human body need to be matched. We already have the 3D upper clothes patches in section, and the corresponding human body parts in section 2.4, the corresponding part of garment and human body, has the similar point cloud features, the ICP [14] algorithm was employed to find the matching feature points, and generated Rotation matrix R, and translation matrix T of the sweater. As shown in equation 5.

$$V_{new} = RV_{old} + T \tag{5}$$

where V_{old} is the vertex position of sweater patches after shoulder match with human body as shown in Fig.12a, v_{new} is the vertex position of sweater patches after rotate and translate use ICP algorithm, as shown in Fig.12 b, We can see that the posture of upper clothes and human body is matched. After tailoring the sweater patches matching the human body posture, 3D patches need to be sewed to complete the redressing. We have generated seam lines in section, these seam lines were used to sew the sweater patches. Our sew method based on geometric deformation, and Laplacian-based deformation was employed to smooth the moved points of sweater patches. The sewed result has shown in Fig.13, Fig.13a was the 3D patches before sewed, by our sew algorithm, the gap of the patches was closed, Fig.13b, c, d, e shown the sewed patches.

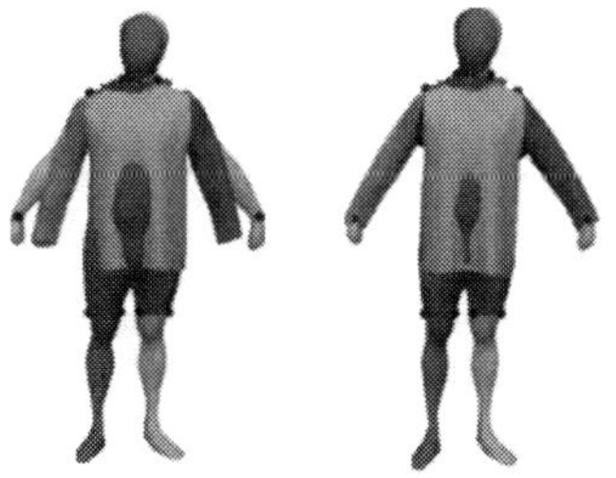

Fig.12 Posture alignment.

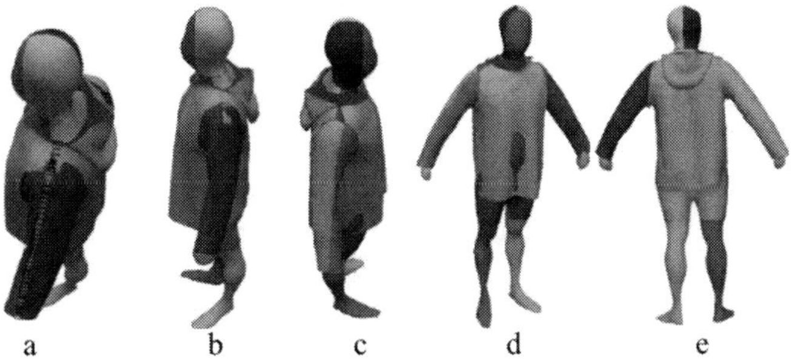

Fig.13 Sewed patches.

2.5 Penetration Recovery

Because the sweater will be redressed on the different types of human body, In order to avoid the phenomenon of human body through clothing, as shown in Fig13b, c, d, e, penetration compensation algorithm was used to solve this problem. Our slice loop penetrating compensation algorithm depicted in flows:

1) The sweater was sliced from top to bottom, human body was sliced in the same area as sweater, as shown in a, Sparse loops for demonstration.

2) As shown in a, on the same level, if slice loop of sweater have intersection points with body and or completely surrounded by the body, as shown in b, the ray $\overrightarrow{OP}$ was used to compute the intersection with human body mesh, $\overrightarrow{OP}$ origin was sweater slice loop point, direction was the normal of triangle which O was located. The intersection P can be got.

3) The triangles which 0 located were moved, the direction was$\overrightarrow{OP}$, and the moved distance was $|OP|$.

4) Laplacian-based deformation algorithms were employed to smooth the neighbor triangles.

The results of the penetration compensation were shown in

Fig.14c.

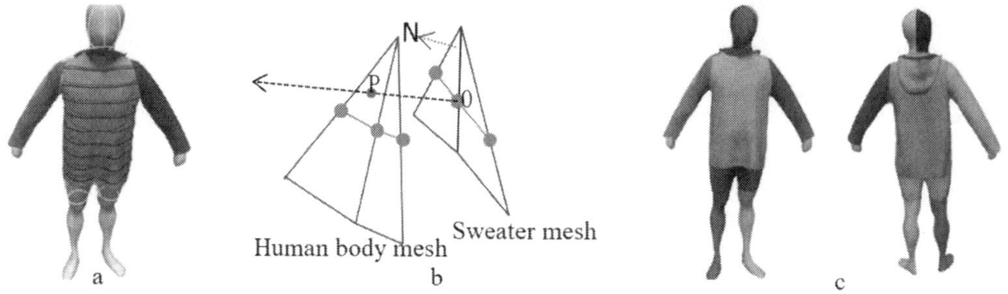

Fig.14 Slice loop algorithm (a) Sparse loops for demonstration; (b)Intersection of ray through sweater slice loop point and human body mesh; (c)The result of sweater penetration compensation.

3. Results and Discussion

We propose a new method for 3D garment redressing, we have the following characteristics:

1) Faster, avoid physical calculation. As shown in Fig.15, variety garments were redressed in different sizes of human body, and the corresponding time cost at each stage were listed in Table 1.

2) The result is more accurate, because the area of the garment was kept. The same garment was redressed in 5 human bodies which have the similar type, the area of the sweater was kept by our method.

3) In our method, arm split method is suitable for a variety of types garment, and human body. Garment tailoring method has a good applicability for kind of garment tailoring and seam lines generating.

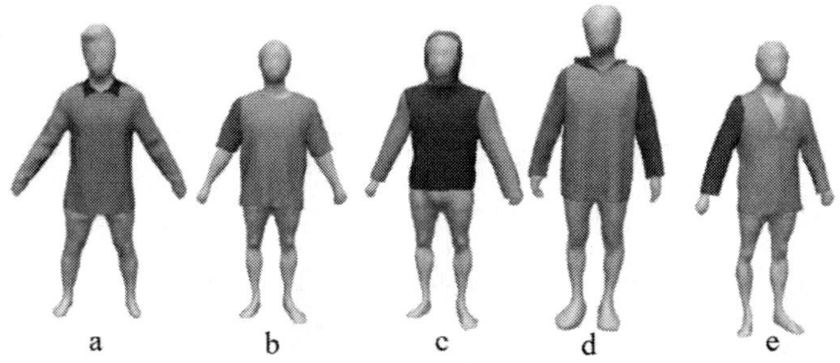

Fig.15 Variety garments redressing.

Table.1 Time of 3D garment redressing.

	No. of Triangles		Time cost (s)
Subject	Garment	Human body	
a	8878	8503	11.04
b	4237	8011	5.1
c	9161	8011	11.87
d	8426	8200	9.4
e	4860	8011	5.58

4. Conclusion

In this paper, a fully automatic garment redressing solution was proposed. The decomposed 3D garment patches were positioned around human body followed by virtual sewing where penetration and over-deformation were well-maintained by Laplacian based representation. Various experimental results indicated that this solution was suitable for tackling the redressing problem when 3D garment model and scanned human bodies were available. It was believed that the proposed method could be used for various types of virtual try-on by minimizing manual intervention while still maintain the original geometrical feature and size stability.

5. Acknowledgement

The authors disclosed receipt of the following financial support for the research, authorship, and/or publication of this article: This work was supported by Natural Science Foundation of China (61572124), Shanghai Natural Science Foundation (14ZR1401100), Postgraduate Innovation Fund Projects of Donghua University (Grant No. CUSF-DH-D-2013025).

References

[1] Metaaphanon N, Kanongchaiyos P. Real-time Cloth Simulation for Garment CAD. International Conference on Computer Graphics: 2005; 83-89.
[2] Clemens, Fuhrmann A, Luckas V. Automatic Pre-Positioning Of Virtual Clothing. Spring Conference on Computer Graphics: 2003; 99-108.
[3] Power J, Apeagyei PR, Jefferson AM. Integrating 3D Scanning Data & Textile Parameters into Virtual Clothing. Proceedings of the 2nd International Conference on 3D Body Scanning Technologies. Hometrica Consulting, Lugano, Switzerland: pp. 213-224.
[4] Decaudin P, Dan J, Wither J, Boissieux L, Sheffer A. Virtual Garments: A Fully Geometric Approach for Clothing Design. Computer Graphics Forum: 2006; 25 (3): 625-634.
[5] Hong Yan, Zeng X, Bruniaux P, et al. Interactive virtual try-on based three-dimensional garment block design for disabled people of scoliosis type [J]. Textile Research Journal: 2016.
[6] Zhang M, Lin L, Pan Z, Xiang N. Topology-independent 3D garment fitting for virtual clothing.Multimedia Tools and Applications: 2015; 74 (9): 3137-3153.
[7] Kim S. Analysis of human body surface shape using parametric design method. International Journal of Clothing Science and Technology: 2015; 27 (3): 434-446.
[8] Sun MP, Choi KM, Lee Y, Yun JN. Multi-purpose three-dimensional body form. International Journal of Clothing Science and Technology: 2011; volume 23 (1): 8-24 (17).
[9] Huang HQ, Mok PY, Kwok YL, Au JS. Block pattern generation: From parameterizing human bodies to fit feature-aligned and flattenable 3D garments. Computers in Industry: 2012; 63 (7): 680-691.
[10] Kim S, Jeong Y, Lee Y, Hong K. 3D Pattern Development of Tight-fitting Dress for an Asymmetrical Female Manikin. Fibers and Polymers: 2010; 11 (1): 142-146.
[11] Zhong Y. Redressing Three-dimensional Garments Based on Pose Duplication. Textile Research Journal: 2010; 80 (10): 904-916.
[12] Huang L, R Yang. Automatic alignment for virtual fitting using 3D garment stretching and human body relocation. The Visual Computer: 2016; 32 (6): 705-715.
[13] Guan P, Reiss L, Hirshberg DA, Weiss A, Black MJ. DRAPE: DRessing Any Person. Acm Transactions on Graphics: 2012; 31 (4): 35.
[14] Besl PJ, McKay ND. Method for registration of 3-D shapes [C]. Robotics-DL tentative.International Society for Optics and Photonics: 1992; 586-606.
[15] Cheng ZQ, Chen Y, Martin RR, Lai YK, Wang A. SuperMatching: Feature Matching using Supersymmetric Geometric Constraints. IEEE Transactions on Visualization & Computer Graphics: 2013; 19 (11): 1885-1894.
[16] Sahillio, Lu Y, Yemez Y. Multiple Shape Correspondence by Dynamic Programming. Computer Graphics Forum: 2014; 33 (33): 121-130.
[17] Sahillioğlu Y, Yemez Y. Multiple Shape Correspondence by Dynamic Programming[J]. Computer Graphics Forum: 2015; 33 (7): 121-130.
[18] Zhang S, Huang J, Metaxas D N. Robust mesh editing using Laplacian coordinates[J]. Graphical Models: 2011; 73 (1): 10-19.

A New Virtual Fitting Method Based on Real Dress

Jing Wang[1], Li-Li Wang[1], Hong-Qin Dai[1,2*]

[1]*College of Textile and Clothing Engineering, Soochow University, Suzhou, Jiangsu 215021, China*
[2]*National Engineering Laboratory for Modern Silk, Soochow University, Suzhou, Jiangsu 215123, China*

[*]Corresponding author's email: daihongqin@suda.edu.cn

Abstract

At present, there has been some achievements on the study of virtual fitting but there are still problems. A new method of virtual fitting is developed in this paper and a new virtual fitting system is established based on real dress that substitutes colors of clothing without changing the original light distribution and clothing physical properties. First the virtual fitting method can automatically save the picture of the real dress. Subsequently the clothing color is substituted using the HSV model. Lastly, the effect of virtual fitting is shown successfully. Experiments show that the authenticity of virtual fitting is improved in the situation where the garment color is bright and the background color is single. Experimental results show the effectiveness of the apparel color substitution and the authenticity of virtual fitting, and provides a reference for further research.

Keywords: Real Dress; Virtual Fitting; Clothing Color Substitution

1. Introduction

For many women, purchasing clothes brings pleasure, but selecting their favorite clothing from a huge array of goods with proper fitting is troublesome [1]. Purchasing online is fast and convenient. You can buy your favorite clothing at home as long as you enter the desired clothing on the internet. With the popularity of purchasing online clothes comes with the demand for virtual fitting. There are some existing difficulties for customers to use virtual fitting. First, there are no three-dimensional character models that have a dynamic 360 degree display [2-5]. Secondly, different garments have different physical properties and different fabrics have different textures. These varied garments have different physical properties such as elasticity and whether its wrinkle free etc. [2-5]. It is important for consumers to have an authentic experience when conducting virtual fitting. In this paper, a virtual fitting method based on real dress is proposed.

Color substitution is widely used in many fields, such as image synthesis, virtual reality and image rendering, etc [6]. It is especially useful for virtual exhibition and internet purchase. In recent years, many scholars at home and abroad have done some research on color substitution [7-9]. Reinhard et al. Proposed a method to render the color appearance of a color image by using the color appearance of another image [8]. Ashikhmin et al. Proposed an algorithm to color the gray image by selecting the target samples from other color images [10]. These methods seem to perform well on natural scene images. However, these methods are not well-preserved light distribution. This paper adopts the HSV model. The real dress effect of the consumer is saved to the constructed system, and the virtual fitting is performed by the HSV model which changes the garment color. It however preserves the light distribution of the original trial effect and the physical characteristics of the garment. Through the use of virtual fitting based on real dress, not only can improve the efficiency of shopping malls, but also to improve the authenticity of virtual fitting.

2. Method for Virtual Fitting

2.1 Method Description

This paper presents a new virtual fitting method based on the real dress that different from the 3D virtual fitting and the traditional fitting. First the picture of the real dress is obtained. Subsequently the clothing color is substituted using the HSV model. Lastly, the effect of virtual fitting is shown successfully without changing the original light distribution and clothing physical properties. The process is shown in Fig.1.

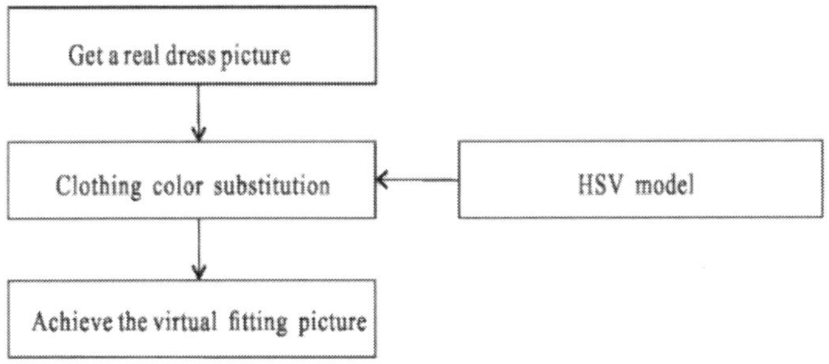

Fig.1 Method Description

2.2 Algorithm of Virtual Fitting

2.2.1 HSV Model

The RGB color space model is not suitable for people's perception of colors. When a person observes a colored object, it is usually described by Hue, Saturation and Value, which generates an HSV color space [11].

HSV color space is a simplified form of the Munsell color space, which is expressed by three elements: Hue (H), Saturation (S) and Value (V). And it is a nonlinear color representation system. Hue is a property that describes solid colors, Saturation is a measure of the degree to which a solid color is diluted by white light, Value is a key parameter that describes the color perception [11-12].

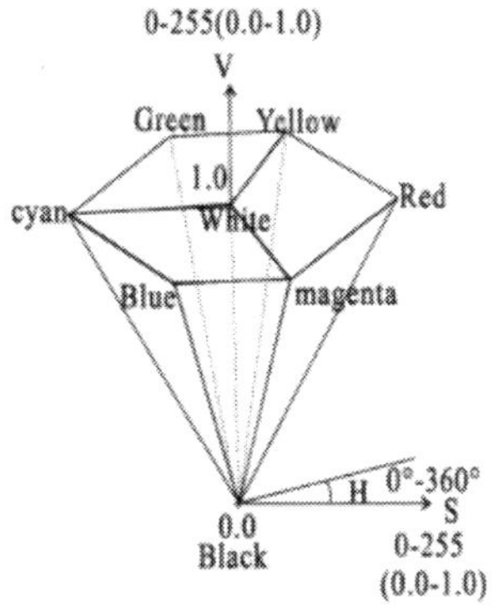

Fig.2 HSV color space model (From http://baike.baidu.com)

HSV color space is consistent with people's perception of colors, The perception of color difference is more uniform in HSV space, and which is also a color space suitable for human visual characteristics. It is beneficial to image processing in the HSV color space, such as edge detection, image segmentation and object recognition. HSV color space model is a cone, as shown in Fig.2.

2.2.2 Substituting Clothing Color

Constructing a system can automatically save the picture based on virtual fitting of real dress. A female researcher stands in the area of the system shooting, and carries out the real dress by using the camera in the system. After the shooting, the system can automatically save the real dress picture, and then using the virtual fitting system to substitute the color of clothing. HSV model is used to substitute clothing color. The system can convert the RGB color space to HSV color space in the real dress picture, and then calculates HSV values of the pixel. When choose to substitute the color, transform the H value and preserve the S and V values of the original image. The algorithm is shown in Fig.3.

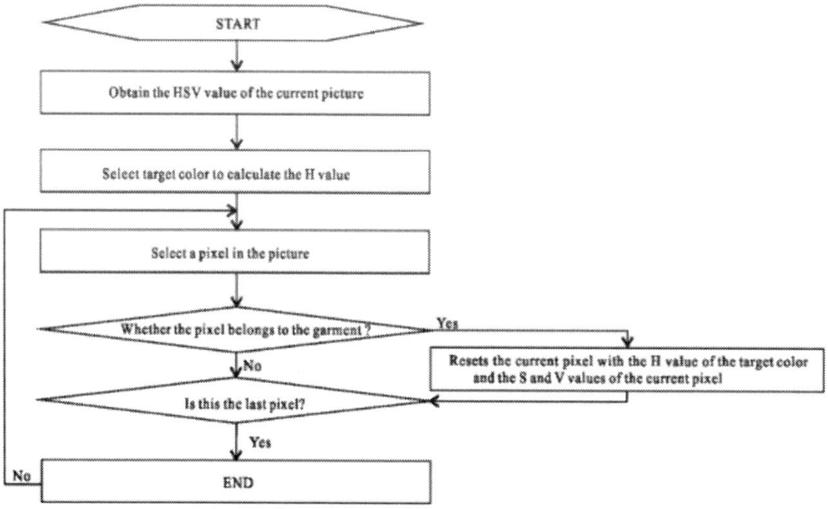

Fig.3 Process of Substituting Clothing Color

3. Case Study

3.1 Experiment

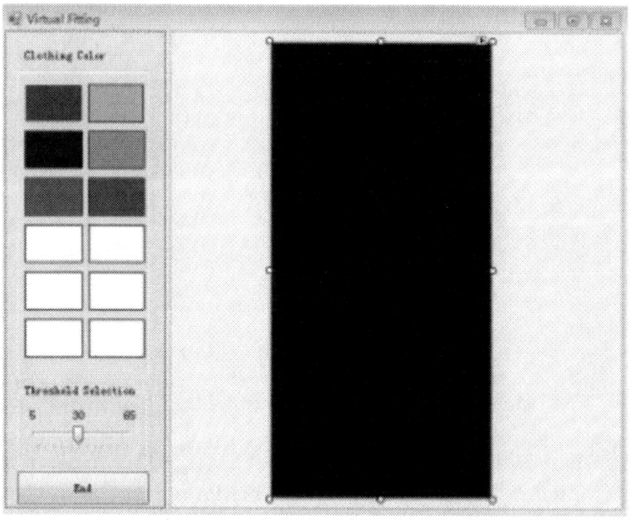
Fig.4 Interface of Virtual Fitting

According to the HSV model, this paper constructs a virtual fitting system based on real dress. The interface of this system is shown in Fig.4. Two women were selected for experiments. One tried on a sweater being shown in Fig.5, and the other tried on a woolen coat being shown in Fig.6. The other colors of sweater and woolen coat can be shown in the virtual fitting system.They need try once which can quickly find their favorite colors.

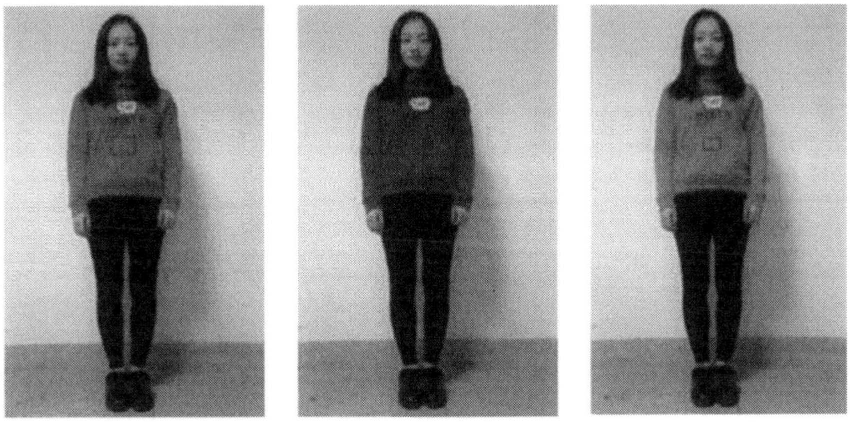

(a) Virtual fitting (b) Real dress (c) Virtual fitting

Fig.5 Sweater of Real Dress and Virtual Fitting

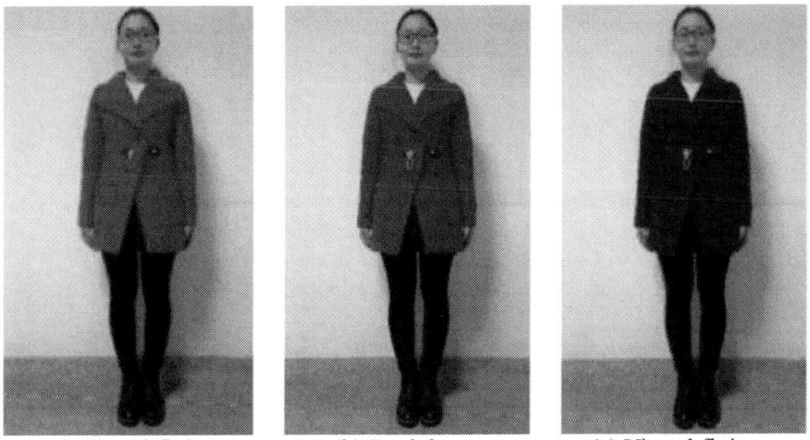

(a) Virtual fitting (b) Real dress (c) Virtual fitting

Fig.6 Woolen coat of Real Dress and Virtual Fitting

3.2 Result and Discussion

Based on the real dress of the virtual fitting successfully demonstrated the substitution of clothing colors using HSV model. It can substitute the color of sweater and woolen coat on the condition of maintaining the original distribution of light and clothing characteristics. A red autumn sweater is shown in Fig.5 (b), and then substitute the color. Change the red to orange as shown in Fig.5 (a) and to green as shown in Fig.5 (c). A red winter woolen coat is shown in Fig.6 (b), and then substitute the color. Change the red to purple as shown in Fig.6 (a) and to blue as shown in Fig.6 (c).

Experiments are very successful using the virtual fitting method based on real dress. This is due to the bright colors of clothing and the single color of background, and keeping the original light distribution and clothing physical properties.

However, the following points need to be discussed:

1)When the color of the garment is similar to the skin color, it cannot substitute the color of the garment. And it need to further study whether the skin color needs to be removed.

2)When the clothing color is black, white or gray, it also cannot change the clothing color. It has not yet succeeded in finding the right solution for the color substitution of black, white and gray clothes.

Therefore, it is our future research direction to explore the color substitution model of separating the color of illumination and garment.

4. Conclusions

In order to make the virtual fitting experience more realistic, and to improve the efficiency of traditional fitting, this paper puts forward a virtual fitting method based on real dress. Through a female researcher try on a garment and save the real fitting picture. Then substitute the color of the garment, and ultimately realize the virtual fitting. Experimental results show the effectiveness of substituting garment color and the realness of virtual fitting, and also provides a reference for further research.

5. Acknowledgment

This work is supported by A Project Funded by the Priority Academic Program Development of Jiangsu Higher Education Institutions, China.

References

[1] Boonbrahm S, Kaewrat C, Sewata L, et al. 3D real time virtual fitting room for women. Design, User Experience, and Usability: Users and Interactions: 2015; 162-171.

[2] Hu J. Research on the theoretical method of three - dimensional Dressing. East China Normal University: 2002.

[3] Liu XY. The value of application on virtual fitting in the clothing brand marketing. Dalian University of Technology: 2013.

[4] Liu Q. Study of the key problems of online shopping based on apparel consumer. North China Electric Power University: 2014.

[5] Xie K. The research on virtual fitting technology based on image. Jilin University: 2016.

[6] Yang JD. The research of virtual fitting system. Beijing Institute of Fashion Technology: 2012.

[7] Levin A, Lischinski D, Weiss Y. Colorization using optimization. Proc. ACM SIGGRAPH: 2004; 689-694.

[8] Reinhard E, Ashikhmin M, Gooch B, Shirley P. Color transfer between images. IEEE Comput. Graphics Appl: 21 (2001); 34-41

[9] Tao S. Object extraction and color replacement of color similarity coefficients [J]. Computer Programming Skills & Maintenance: 2013 (7); 71-72.

[10] Welsh T, Ashikhmin M, Mueller K. Transferring color to grayscale images. ACM SIGGRAPH: 2002; 277-280

[11] Zhao ZS. Research on video vehicle detection and tracking algorithm based on HSV color space. China Ocean University: 2007.

[12] Liu F. Improving the video shot boundary detection using the HSV color space and image subsampling. Lanzhou University: 2015.

Transformation Model of Data Value by Different Measurements of Bust

Tian Yang, Yan-Qun Ning, Xiao-Mei Shang*

College of Textile and Clothing Engineering, Soochow University, Suzhou, Jiangsu 215021, China

*Corresponding author's email: shangxiaomei@suda.edu.cn

Abstract

In anthropometric measurements, the bust size is an important measure of the element. In the design and production of bra, consumers demand a higher degree of fit. Manufacturers want to produce products that make consumers more satisfied with the bra. Merely using the conventional method to measure the size of bust has been unable to meet the needs of manufacturers. A more detailed body bust measurement is needed. This paper first adopts the "conventional measuring method" and the "body-fitted measuring method", two methods of measuring the bust. Two sets of experiments were carried out. the accuracy, stability and practicability of the two methods were then studied. By analyzing the difference and correlation between the two sets of data, a regression model was established and the model was verified. The results were as follows: 1. The discrepancy of the data set obtained by the method 1 was small. the data were more stable than method 2. Method 1's practicability was also better than the method. 2. The data set obtained by method 1 and 2 have a significant difference and correlation. 3. The regression model was established for the two methods, and the optimal regression model was chosen as y = 7.887 + 0.942x.

Keywords: Bust; Conventional Measuring Method; Body-fitted Measuring Method; Transformation Model

1. Introduction

The basic measurement of the human body size is the foundation for fashion design and the pattern design of clothes, while the human body bust size is an important size when it comes to design [1]. In the design of women's bra specifically, more and more people are increasingly seeking comfort and fit. As such in addition to using the conventional method to measure the size of the bust, the bra enterprises also use the body-fitted method to measure the size of the bust in order to improve their competitiveness by producing more comfort and fit bra products [2-3].

Currently, most of the current studies on the human bust are concentrated in the chest curve of the model fit. Liu Hong [4] and others used three-dimensional body measurement to measure 18 young women and in cutting the chest section curve. The quadratic polynomial is used to fit the chest curve. With the drawing method, Lu Chen [5] and others were able to obtain the body bust approximation curve, create an access to bust curve length and chest width as well as chest thickness of the function. The cubic polynomial function was used to fit. Xu Feng [6] and others studied the fitting regression analysis of the bust curve in two-dimensional photographic non-contact measurement. The difficulty of fitting the bust size was analyzed and the formula of bust regression of young women was deduced. Wang Yingying [7] and others studied the Kinect depth data for human body measurement technology. The method could be more accurate to measure the size of the chest. Tian Yannan [8] and others studied the influence of different measurement methods on the accuracy of the measured values. Two methods were used to mesure the body size. The two methods were "control"and "visual" method respectively. Finally found that the data obtained by "control" measurement were more accurate.The measurement of the size of the chest detail was less, it was difficult to fully meet the current users of the bra more and more fit needs [9].

In this paper, the bust of the two measurement methods as the main research object, used two methods to obtained bust size. The accuracy, stability and practicability of the two methods were studied by Excel software. The variability and correlation between the two sets of data were analyzed by SPSS software. The transformation model was established and the model was verified. Which provided a method for establishing the conversion model for different methods of bust.

2. Experiment

2.1 Sample Selection

Participants: 15 female college students were randomly selected from Soochow University as the participants. They are between 18 and 24 years old, and their height is between 155 and 170 cm. They were numbered from NO.1 to NO.15.

Surveyors: 10 women were randomly selected as surveyors. Before the experiment, a unified measurement skills were trained, operational accuracy and proficiency of the basic standards of taking anthropometric measurements. They are between 20 and 25 years old, and their height is between 155 and 170 cm.

Recorders: 10 women were randomly selected as recorders. Before the experiment, a unified record of measurement data was trained. The operation accuracy and proficiency were basically up to standard. They are between 20 and 25 years old. 10 recorders were randomly selected with the 10 surveyors to form measurement partners to complete the measurement experiment.

2.2 Experimental Tools and the Environment

Experimental tools: Tape, pen, record form, record board, laboratory workers dedicated clothes, measuring body tights. Participants were feet closing together, natural standing, hair setting up, took off hers shirt and bra, wearing the measuring body tights. The participants were arm extension, visual front, to avoid shaking when measuring. Surveyors wore laboratory workers clothes, and held the tape. The recorders wore the laboratory workers clothes and hand with the record form and pen.

Experimental environment: The experiment must ensure that at the same time, a unified operating specification environment, and a unified measurement equipment specifications. Environment temperature was 27℃±3℃. Humidity was 60% ± 10%. Ensure that people feel comfortable in laboratory [10]. The light intensity was moderate, to ensure the clear reading.

2.3 Experimental Method

The measuring bust value of the "conventional measuring method" and "body-fitted measuring method" were named method 1 and method 2. The measurement definition and operational requirements of the two methods were as follows:

2.3.1 Measured Value Definition

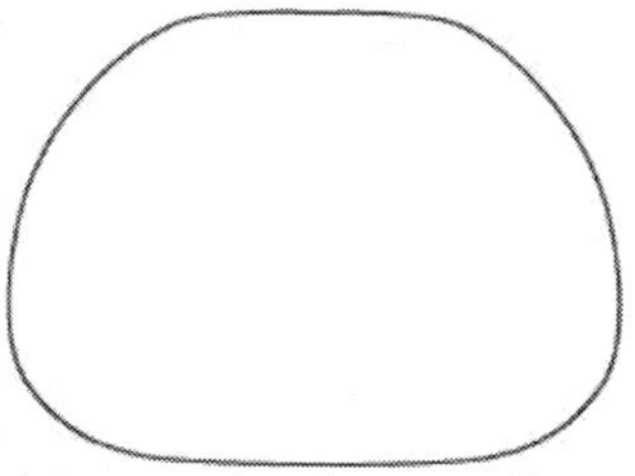

Fig.1 Method 1 measuring the curve diagram

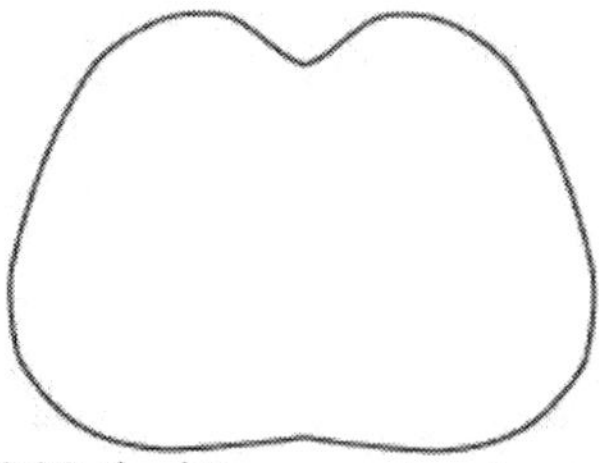

Fig.2 Method 2 measuring the curve diagram

Method 1 for the bust of the "conventional measuring method" is in accordance with GB/T 16160-2008 "Body Parts and Methods of Measurement for Garment" . The definition is the participants upright, normal breathing, with tape measured by the scapula, armpit and nipple the maximum level of girth [11]. This measurement got a rounded bust curve. The curve diagram shown in Fig.1.

Method 2 for the bust of the "body-fitted measuring method" is in the nipple level, from one side of the axillary point to the midpoint of the breast, then to the shoulder blade, eventually return to the side armpit points of the body surface. It is a curve through the body surface [12]. This measurement got a non-circular bust curve. The curve diagram shown in Fig.2.

2.3.2 Experimental Operation Process

Two kinds of measurement methods were used to carry out experimental measurements on the samples. The specific operation methods and precautions were as follows.

10 surveyors measured the bust size of the 15 participants according to the following two different measurement methods. The recorders recorded the data. Operation is as follows:

Method 1: The participants were kept standing. The surveyors stood on the left side of the participants. The left hand caught the tape of the zero-scale line. The right hand controlled the tape movements. First arms surrounded by the participants waist. Left hand held the tape same. The right hand straighten the tape, through the scapula, armpit point and breast point fit the human body forming a circle parallel to the ground. Finally fixed on the left armpit point of the participants. Fixed the tape with your right thumb, the scale was the value of the bust. Took the tape down for an accurate reading [13]. During this measurement, the surveyors didn't moved on the left side of the participants.

Method 2: The participants were kept standing. The surveyors stood on the right side of the participants. The left hand caught the tape of the zero-scale line. The right hand controlled the tape movements. The left hand controlled the zero line fixed on the right armpit point. Right hand controlled the tape through the right breast, and then to the midpoint of the breast. Loosened the left hand and fixed on the tape of the breast midpoint, while the surveyors moved to the front of the participants. Right hand controlled the tape through the left breast point and left armpit point. Loosened the left hand and fixed on the tape of the breast midpoint. The surveyors moved to the left side of the participants. Right hand controlled the tape through the scapula, and then to the right armpit point. While the surveyors moved to the right side of the participants. Fixed the tape with your right thumb, the scale was the value of the bust. Took the tape down for an accurate reading. During this measurement, made sure the measurement of the horizontal plane parallel to the ground.

2.4 Data Processing

This experiment participants were 15 persons. Surveyors were 10 persons. Each surveyor was required to measure 15 participants in turn, each participant was measured 3 times, so there were 450 groups experimental data. Invalid data and abnormal data group were elimimated. Finally there were 420 groups experimental data could be used to study. Then counted the data averaged for each sample, prepared for the data analysis and modeling research.

3. AccuracyAnalysis of Two Methods of Measurements

By using Excel software for the data from two methods of measuring a preliminary statistical analysis, got the mean value, standard deviation and AVEDEV function value (absolute deviation of the function value). The standard deviation indicate the degree of dispersion of a set of data. The greater the standard deviation is, the greater the degree of dispersion. The AVEDEV function value is the average absolute deviation of the data. It reflecte the degree of dispersion of the data. The greater the AVEDEV function value is, the greater the degree of dispersion [14]. Two kinds of measurement method of discrete degree of statistics is shown in Table 1. The bust mean value analysis is shown in Fig.3.

From Table 1 and Fig.3 could be drawn the following conclusions:

1) The mean values of the two methods were different. And the data values measured by the method 2 were all above the method 1. It is indicated that the data values were greater than the method 1. This indicated that the bust size measured by "body-fitted measuring method" is larger than the "conventional measuring method".

2) By comparing the standard deviation and AVEDEV function values, got the method 1's standard deviation and AVEDEV value was less than method 2. The data obtained by method 1 has a smaller degree of dispersion. The data values are more stable. However the data obtained by method 2 has a larger degree of dispersion, not stable. So the measurement accuracy was small. This result is related to the measurement method used by method 2. Because of this method through more reference point, there is a deviation in the base point found by the surveyor. Result in the larger error of the measurement data.

3) Compared the practicability of the two kinds of methods. From the second point above, we can see the data obtained by method 2 has a large degree of dispersion. Explain that method 2's practicability is less than method 1. Method 1 measured bust size by conventional measurement, all clothing enterprises must be done before garment production. Because of developing for many years, the technology is mature. Therefore, in practice, the transformation model could be established by this two methods. Not only reduces the workload, but also can got two sets of data values.

Table.1 Two kinds of measurement methods for the dispersion degree

Participants	Mean value/cm		Standard deviation		AVEDEV function value	
	Method 1	Method 2	Method 1	Method 2	Method 1	Method 2
NO.1	85.77	88.00	0.17	0.50	0.16	0.47
NO.2	83.12	86.33	0.29	0.42	0.27	0.38
NO.3	86.00	89.03	0.41	1.03	0.33	0.96
NO.4	83.57	87.46	0.42	0.33	0.38	0.29
NO.5	85.66	88.10	0.59	0.22	0.56	0.20
NO.6	82.10	85.73	0.37	0.95	0.33	0.84
NO.7	84.47	88.10	0.62	0.83	0.58	0.73
NO.8	87.33	89.50	0.34	0.41	0.31	0.33
NO.9	91.00	93.70	0.82	0.99	0.67	0.87
NO.10	81.10	83.50	0.79	0.98	0.73	0.80
NO.11	84.50	88.77	0.41	0.33	0.33	0.31
NO.12	82.13	84.70	0.66	0.67	0.58	0.60
NO.13	88.00	91.97	0.41	0.86	0.33	0.71
NO.14	80.67	83.97	0.24	0.37	0.22	0.31
NO.15	91.00	93.63	0.57	0.46	0.47	0.42

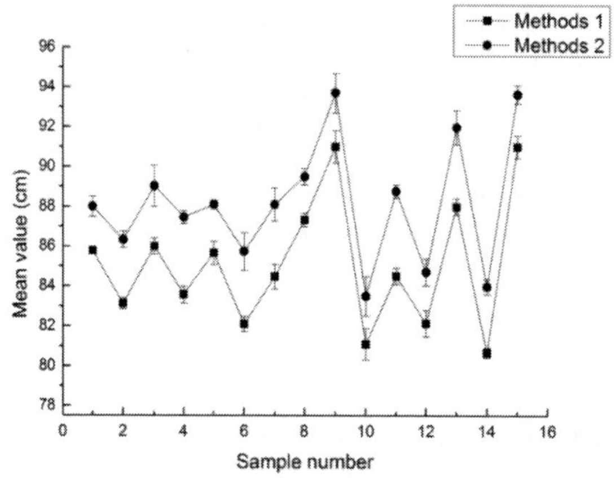

Fig.3 Comparison of bust mean value from two methods

4. Transformation Model Establishment and Validation Analysis

Established the transformation model of the two measurement methods, and verified this transformation model. First of all, performed a paired sample T test of the data for the method 1 and 2. Determined the relevance and differences of the two groups of data. If there is a large relevance and significant differences, it could be established the regression model. Secondly, 13 samples were randomly selected from 15 samples for regression analysis, and established the mathematical model. Thirdly, the remaining two samples were used for model validation and selected the optimal transformation model.

4.1 Significant Analysis

By using the SPSS software system to analyze the significance level of the data sets obtained by the two measurement methods. Set the confidence interval to 95%. Two sets of data were paired sample T test. The results are shown in Table 2 and 3.

Table.2 Method 1 and 2 correlation table

	N	Correlation	Sig.
Method 1 & Method 2	15	0.915	0.000

Table.3 Method 1 and 2 paired samples T test

	Paired Different					t	df	Sig. (2-tailed)
	Mean	Std. Deviation	Std. Error Mean	95% Confidence Interval of the Difference				
				Lower	Upper			
Method 1 & Method 2	-3.27111	1.35839	0.35073	-4.02336	-2.51886	-9.326	14	0.000

From Table 2, we can see that the correlation coefficient of the two methods is 0.915. The significance test of the correlation coefficient shows that the significant value is 0.000. Because Sig. < 0.05, the two sets of data have a significant correlation.

From Table 3, we can see that t is -9.326, the degree of freedom is 14, the differences between the two test significant level is 0.000. Because Sig. <0.05, the two sets of data have a significant differences.

To sum up, the transformation model can be established between the two sets of data for the bust.

4.2 Establish the Transformation Model

By using the SPSS software system to establish the transformation model. The value of the data obtained by method 1 is the independent variable X. The value of the data obtained by method 2 is the dependent variable Y. Linear, Quadratic, Cubic, Compound, and Logistic functions were selected for curve fitting [15, 16]. The fitting results are shown in Table 4. The figure shows the result of the best fit of the selected model. By comparing the square of the correlation coefficient (Rsq) to verify the advantage of the model. The greater the square of the correlation coefficient is, the better the model.

From Table 4, by comparing the Rsq values of all the models. The Rsq values of the linear model, the quadratic polynomial model and the cubic polynomial model are all 0.964, which are higher than the compound model and the logistic model. So the compound model and the logistic model are excluded. In addition, in the cubic polynomial model, b3 value is 0.000. b1 and b2 values are same to quadratic polynomial model. It is indicated that the cubic polynomial model does not exist. So, be excluded.

To sum up, linear model and quadratic polynomial model are the most appropriate to fit the function. In the linear model, the constant term is 7.887 and the coefficient b1 is 0.942. So the model is as follows.

$$y = 7.887 + 0.942x \tag{1}$$

In the quadratic polynomial model, the constant term is -26.081, the coefficient b1 is 1.733 and the coefficient b2 is -0.005. So the model is as follows.

$$y = -26.081 + 1.733x - 0.005x^2 \quad (2)$$

Table.4 Model summary and parameter estimation of two methods

Independent: X

Dependent	Mth	Model summary					Parameter estimation			
		Rsq	F	df1	df2	Sig.	b0	b1	b2	b3
Y	Linear	0.964	296.490	1	11	0.000	7.887	0.942		
Y	Quadratic	0.964	135.782	2	10	0.000	-26.081	1.733	-0.005	
Y	Cubic	0.964	135.782	2	10	0.000	-26.081	1.733	0.005	0.000
Y	Compound	0.962	279.864	1	11	0.000	35.634	1.011		
Y	Logistic	0.962	279.864	1	11	0.000	0.028	0.989		

4.3 Verify the Transformation Model

The remaining two samples (NO.4 and NO.13) are used for the regression model validation analysis. The verification results are shown in Table 5. Among them, the X value is actual measured value for method 1. The Y value is the model validation value for method 2. The Y' value is the actual measured value for Method 2. The Z value is the differential ratio. The formula for Z is as follows. The smaller the differential ratio is, the better the regression model.

$$Z = \frac{|Y - Y'|}{Y'} \quad (3)$$

Table.5 The transformation model of the two sample verification

Sample number	Verify the model category	X	Y	Y'	Z
NO.4	Linear model	83.57	86.61	87.46	0.01
	Quadratic polynomial model		83.83		0.04
NO.13	Linear model	88.00	90.78	91.97	0.01
	Quadratic polynomial model		87.70		0.05

From Table 5, we can see that the linear model's differential ratio is smaller than the quadratic polynomial model. It shows that the linear model is the best fitting result, $y = 7.887 + 0.942x$.

5. Conclusion

In this paper, by studying the data obtained by the two methods of bust measurement, we obtained the following conclusions:

1) The data obtained by method 1 have a smaller degree of dispersion and the data values are stable. This method has a good practicability.

2) The data obtained by method 2 have a larger degree of dispersion and the data values are not stable enough. This method has a poor practicability.

3) Two sets of data have a significant correlation and have a very significant differences. The two groups of data can be established transformation model.

4) Finally, the linear model is defined as the optimal model $y = 7.887 + 0.942x$.

In the conventional human body size measurement, the method 1 is one of the most commonly and important measured size. And compared to the method 2, method 1's reference point less, so the relative error was small. The data measured by method 2 as an essential size of the bra, gradually be needed by garment enterprises. Therefore, it is necessary to establish the transformation model. So that if garment enterprises only get the method 1's bust size, through the transformation model can be get the method 2's bust size. This way not only saves the cost of anthropometric measurement, but also improves the utilization of the measured value. It also provides a reference path for modeling research in the field of human body measurement.

References

[1] Jiang HM. Research on the size setting and measurement method of making outside single up [J]. Textile Industry and Technology: 2016; 3: 63-65.

[2] Liang SZ, Zhang X, Zhang DS. Bra body structure design of the main details of the size of the body based on [J]. Journal of Textile Research: 2008; 12: 69-73.

[3] Chen SS, Hong WJ, Wu Y. Analysis of underwear structure design principle based on bust position change [J].Shandong Textile Science and Technology: 2015; 6: 28-33.

[4] Liu H, Zhang M, Chen DS, et al. Study on bust curve of standard women [J]. Journal of Textile Research: 2011; 2: 117-120.

[5] Lu C, Wang H, Zhang GQ. Fitting method of body bust curve [J]. Journal of Suzhou University (Engineering Science Edition): 2009; 3: 47-50.

[6] Xu F, Zhang H, Zheng R. Regression analysis of chest fitting in photogrammetry [J]. Journal of Textile Research: 2006; 8: 49-52.

[7] Wang YY, Zhong YW, Pan SM, Wang ZW. Human bust measurement method based on Kinect depth data [J]. Computer Engineering and Design: 2015; 8: 2138-2142.

[8] Tian YN, He LW, Shang XM. Analysis of the influence of different measurement methods on the accuracy of measured values [J]. Journal of Beijing Institute of Clothing Technology (Natural Science Edition): 2016; 4: 25-31.

[9] Petrak S. The method of 3D garment construction and cutting pattern transformation models [J]. Hrvatska znanstvena bibliografija i MZOS-Svibor: 2007; 220.

[10] GB/T 16160-2008 "Body parts and methods of measurement of garment" [S].

[11] Li Q, Li PW, Hong ZL. Analysis of the influence of the psychology of surveyor on the measured value[J]. Jiangsu Textile: 2011; 2: 50-51.

[12] Cheng GX. The application of anthropometry in bra design [J]. Private Science and Technology: 2007; 7: 109.

[13] Jia SL, Tian YN, Shang XM. The effect of measurers' view angle deviation on measurement dat [J]. Advanced Materials Research, Materials Science and Engineering Technology: 2014; 936: 2237-2242.

[14] Yang T, Tao DF, Shang XM. Study on the optimization of the acquisition method of the reference point [J]. Journal of Zhongyuan Institute of Technology: 2016; 6: 52-57.

[15] Shang XM, Lu C, Wang H. Curve fitting model based on auto-measure [J]. Journal of Textile Research: 2010; 4: 128-131+138.

[16] Hong L, Dongsheng C, Qufu W. A study of the relationship among clothing pressure and garment bust strain and young's modulus of fabric, based on a finite element model [J]. Textile Research Journal: 2011; 13: 413-6.

Establishment of Three-Dimensional Model of Standard Human Body for Male Youths in Southern China Area

Peng-Peng Cheng*, Dao-Ling Chen

Department of Garment and Art Engineering College, Minjiang University, Fuzhou Fujian, 350108, China

*Corresponding Author's email: cppcdl3344@163.com

Abstract

In the study, 900 young men aged 18-25 years from Southern China area were selected as subjects. The data of 19 body parts was collected by three-dimensional anthropometry, and analyzed by SPSS19.0. The results showed the weighted average coil of standard human body parts of male youths in the area. 3DMAX was used to improve the human body mesh to establish its standard three-dimensional human body model for making up for existing body data in practical application and meeting the body size information requirements for tailored mass production.

Keywords: Southern China Area; Male Youths; Three-Dimensional Anthropometry; 3DMAX; Human Body Model

1. Introduction

According to the "Human Dimensions of Chinese Adults (GB/T 10000-1988)", Southern China area includes Guangdong, Guangxi, Guangdong and Fujian are coastal and textile-garment industry provinces whose economy are relatively developed. With the rise of tailored mass production, the 3D human body model has been widely used in anthropometry, fashion design, garment CAD and virtual fitting. The establishment of 3D standard human body model in each region will make up the existing body shape data in the shortage of the practical application to meet the tailor-made mass production of body information requirements, and ultimately promote the intelligent manufacturing of Chinese garment industry. In this paper, through data extraction, analysis and processing, as well as digitally modeling the human body, three-dimensional model of a standard human is established.

The garment industry is also experiencing a transition from two-dimensional clothing to three-dimensional clothing the establishment of a three-dimensional human body model is imperative. Scholars at home and abroad have different degrees of research and opinions on human modeling, Hinds first proposed the modeling method of garment human body model based on parametric surface expression. It is a significant attempt to model the garment human body. Kayis et al (1999) and Mike Kolich (2003) calculated the required data related to human body parameters which is automatically derived into the dimension of human body model and the morphology of the body layer model based on the structure model of the human body by means of a layer program [1, 2]. Seo et al (2004) presented a multi dimension space radial basis parameters provided by the user based on the value structure of the human body, including using an advanced set of parameters, such as fat percentage ~~to~~ achieve animation effects. This method makes the human body more realistic. The problem however lies in the fact that there are a number of samples in the process of modeling that has different characters, positions that is needed for interpolation [3]. Lv et al. (2008) and Wang et al. (2011) simplified the data derived from European and American human models in Poser and reconstructed a more realistic human model by combining the free deformation algorithm with the variable skeletal control parameters [4-5]. Luo et al. (2009) used polygon modeling technology, triangulation theory and algorithm for segmentation and optimization of point cloud data, using VC++ and OpenGL to build a three-dimensional human model [6]. Cui et al. (2009) used triangles to reconstruct the 3D human body model by using the contour synchronization advance method, the method has the advantages of small amount of data and rapid generation of realistic human body model, which satisfies the requirement of 3D virtual fitting [7]. Tang et al. (2009) used the finite element method to expand and transform the skeletal muscle modules by energy constraints and construct a more realistic human body model by combining the physiological and physiological mechanisms of the human body [8]. Tang et al. (2010) realized the establishment of 3D human body model by referring to the human body two-dimensional template with SolidWorks software [9]; Huang (2010) used the characteristics of the OBJ model file format to express the surface of the general 3D human body model by using the trigonometric primitives, the data of the human body parts are redistributed according to the ratio of the three-dimensional size to the height and combined with the FFD local deformation algorithm for changing body size [10]. Izadi et al. (2011) proposed the use of depth data provided by the camera, the depth of data into the camera coordinate system under the three-dimensional point cloud, the coordinates of the algorithm used to adjust point cloud coordinates with the first frame of the coordinate system, the 3D reconstruction model is based on the point cloud, so it can reflect the convex and concave surface of the real object to a certain extent [11]. Tong et al. (2012) used three Kniect scanners of the Microsoft to perform multi-angle scanning of the human body, and built a personalized human model for virtual fitting based on the global correction algorithm and the human deformed frame sequence [12]. S.Baek et al. (2012) constructed a database of human body models with the distribution of body size to changing the shape of the human body of the user-specified body size based on the existing model [13]. Lin et al. (2012) proposed a method of whole-body reconstruction by using the human body's color frontal and color side facets to reconstruct the 3D human body. Then the contours were measured, the feature points were acquired and the body size was obtained. This method of modeling more realistic, the method for the human body depression (armpits, etc.) to identify the distortion prone to the situation, and a higher background for the photo [14]. D.Alexiadis et al. (2013) also relied on multiple depth cameras to construct the human body model through calibration and synthesis of the data and achieve real-time effects by accelerating, but only a single background of human reconstruction [15]. Yang et al (2015) combined with the characteristics of women's characteristics and national standards of the relevant indicators, using MATLAB technology to obtain the standard intermediate model [16]. Zhang et al. (2015) proposed a design method of virtual product model based on parametric human model. Based on anthropometric data, a standardized anthropometric database was designed and a human body model was designed based on the human measurement database [17].

2. Experimental and Methodology

2.1 Subjects

A sample of young male in South China (Guangdong, Guangxi and Fujian) was screened, and the effective number of samples was 900 male youths aged 18-25 [18]. the determination of the sample size: N= $(1.96\times\sigma^2)/\Delta^2$, N is the number of samples; 1.96 is the standard normal distribution function in the 95% confidence interval value; σ is the standard deviation; Δ is the allowable error. The standard deviation and the allowable error of the sizes of the adult parts are determined according to the standard of GB / T 10000-1988 [19]. Table (refer with Table 1) presents the key points of the youth.

Table.1 Standard deviation and permissible error of each part of the adult

part	standard deviation σ/cm	permissible error Δ/cm	σ/Δ	sample size
Height	6.2	1.0	6.2	148
Bust	5.5	1.5	3.67	52
Waist	6.7	1.0	6.7	173
Front waist length	2.3	0.35	6.57	166
Hip	5.2	1.5	3.47	46

2.2 Experimental Procedure

Anthropometric test in Department of Garment and Art Engineering College (Minjiang University). Figure (refer with Fig.1) presents the measurement was performed by using the HY-3DLS410 non-contact three-dimensional body scanner with an accuracy (±1 mm). The test ambient temperature was (25 ± 2)°C and the humidity was (65 ± 2)%. Figure (refer with Fig.2) presents people should wear a tight-fitting underwear, be in barefoot, wear a white swim cap, fix long hair in the hat, not wear watches, rings and other jewelry, stand in the specified footprints with his or her eyes looking straight ahead, shoulders to maintain natural, breathing uniform. Each person was measured

3 times to ensure the accuracy of measurement, each measurement time was control in 7-10min.

Fig.1 Three-dimensional body measuring instrument

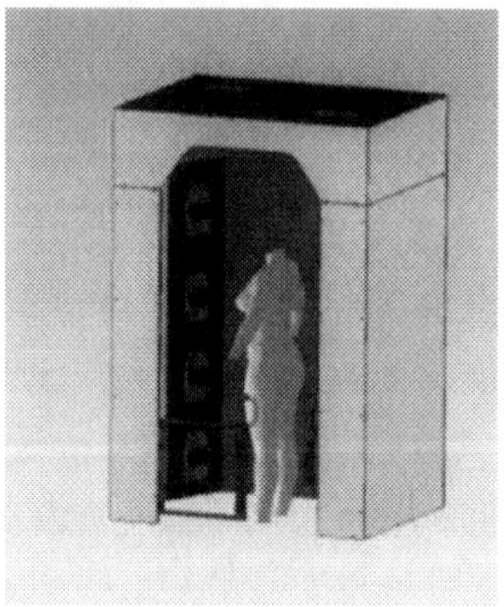

Fig.2 Standing posture

2.3 Data Acquisition

Table.2 Human body measurement items

NO	Measuring position	Measuring method	Schematic diagram of partial measurement
1	Height	the vertical distance from the ground to the top of the head	
2	Cervical point high	the vertical distance from the ground to the cervical vertebra point	
3	Waist high	the vertical distance from the ground to the waist	
4	Chest height	the vertical distance from the ground to the bust	
5	Hip high	the vertical distance from the ground to the hip	
6	Waist	the smallest girth of circumference of the human body waist	
7	Bust	the measurement of horizontal position of human nipples	
8	Hips	the horizontal peak circumference at the hip point	
9	Thigh	thigh maximum girth	
10	Bust-Waist difference	the difference of bust and waist	
11	Hip-Waist difference	the difference of hip and waist	
12	Waist horizontal length	also known as waist width, waist width of the cross section	
13	Bust horizontal length	also known as the width of the chest, chest measured at the torso width	
14	Hip horizontal length	also known as hip width,the widest position on both sides of hips	
15	Back wide	the distance between the left and right axillary points	
16	Shoulder width	the straight line distance between two shoulder points	
17	Bust thickness	the distance of the front and back on chest	
18	Hip thickness	the maximum thickness of the hips	
19	Waist thickness	the maximum thickness of the waist	

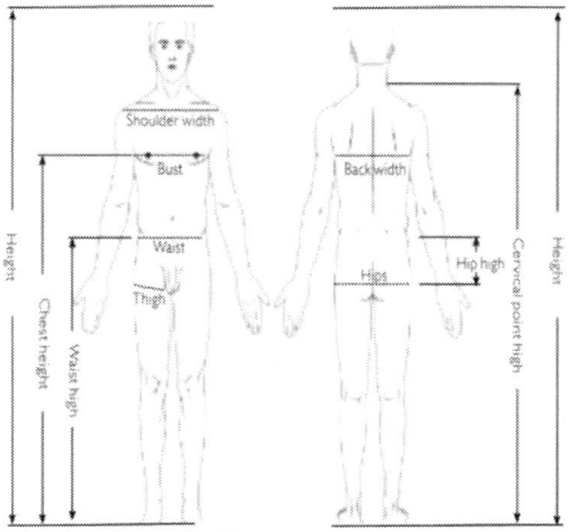

Fig.3 Schematic diagram of the measurement site

In this study, we selected 19 measurement items to measure the human body size based on the relevant research of domestic and foreign scholars [20-22] and GB/T 16160-2008 "Location and method of anthropometric surveys for garments", which is shown in the following Table(refer with Table 2).

Through the HY-3DLS410 non-contact three-dimensional human body scanner to scan the test personnel, access to scan the contour image (refer with Fig.4). These images are transformed into thousands of data points, which correspond to the contours of the object, approaching to an original image called "3D point cloud", which is saved as an OBJ data file, through which accurate measurement data can be obtained.

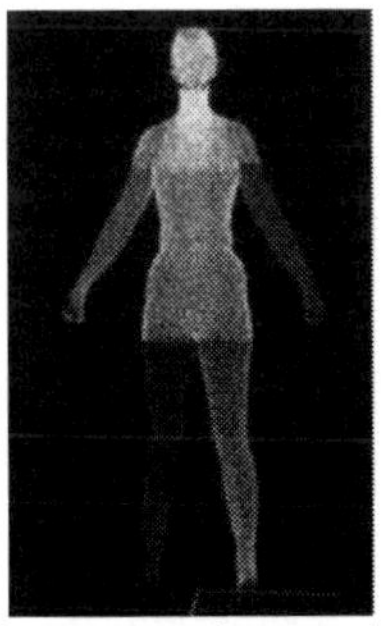

Fig.4 Scanning profile

3. Data Analysis

Table.3 Data comparison of National, Fujian-Guangdong & Guangdong and experimental sample

Part / Sample	Height /cm	Bust /cm	Waist /cm	Hip /cm	(Bust- Waist) /cm	(Hip - Waist) /cm
Nationwide	166.52	88.46	75.12	89.67	13.34	14.55
Fujian-Guangdong & Guangxi	167.43	87.52	74.63	89.23	12.89	14.60
Experimental Sample	173.94	85.54	73.09	83.50	12.45	10.41

Table (refer with Table 3) shows that height and bust-waist difference of young men in Southern China area change significantly. In addition, we also analysis young male body shape based on the width-thickness ratio of body cross-section. Horizontal vector to diameter ratios are mainly used to describe the body parts of the round flat degree (refer with Table 4).

Table.4 Horizontal vector to diameter ratios of adult body BWH measurements in Southern China area

Category	Ratio
Waist width-thickness ratio	1.3186
Hip width-thickness ratio	1.5171
Bust width-thickness ratio	1.6700

There are some analysis methods of human body, such as the Weierweike Index, Roche Index, Body Mass Index (BMI). According to the research report of the Kyoto Women's University in Japan, the Weierweike index is more suitable for the analysis of the somatotype characteristics of Asian human beings than other research methods. Weierweike index (V) is an objective description of the size of the body fat and thin index, the formula is (weight + chest) / height × 100.

Table (refer with Table 5) shows more than half of the young men belong to the intermediate body, that is, the normal body type, and very few are the obese bodies in the study population. This shows that the average body shape of the young men in Southern China area belongs to the standard body.

Table.5 Body fat and thin degree index

Body lassification	Weierweike Index	Number of people / person	Percentage /%
thin	V<81.9	180	20
normal	81.9<V<94.3	702	78
fat	94.3<V	18	2

4. Human Body Modeling

After the analysis of horizontal vector to diameter ratios and body size of the standard body, we can grasp the body characteristics of young men in the region. And then the measured data are imported into 3DMAX, the human body is constructed by NURBS surface. Equation (refer with Eq.1) presents the mathematical model of the rational polynomial NURBS surface with two parameter variables.

$$Q(u,v)=\frac{\sum_{i=0}^{m}\sum_{j=0}^{n}w_{i,j}P_{i,j}R_{i,e}(u)R_{j,f}(v)}{\sum_{i=0}^{m}\sum_{j=0}^{n}w_{i,j}R_{i,e}(u)R_{j,f}(v)} \tag{1}$$

Where, Q(u,v) denotes the surface value point; $w_{i,j}$ denotes the control point weight factor; $P_{i,j}$ is the NURBS surface control network, also known as de Boolean point; $R_{i,e}(u)$ and $R_{i,f}(v)$ denotes respectively B - spline basis of the e-order and f-order on the vector U and V.

Equation (refer with Eq.2, Eq.3) presents U, V node vectors.

$$U=\{0,0,\cdots 0,u_1\cdots u_{n-e+1},1,1\cdots 1\} \tag{2}$$

$$V=\{0,0,\cdots 0,u_1\cdots u_{n-f+1},1,1\cdots 1\} \tag{3}$$

Equation (refer with Eq.4, Eq.5) presents $R_{i,e}(u)$ and $R_{i,f}(v)$ are given, $R_{i,e}(u)$ is taken as an example.

$$R_{i,e}(u)=\begin{cases}1 & u_i\le u\le u_{i+1}\\ 0 & \text{others}\end{cases} \tag{4}$$

$$R_{i,e}(u)=\frac{u-u_i}{u_{i+e}-u_i}R_{i,e-1}(u)+\frac{u_{i+e+1}-u}{u_{i+e+1}-u_{i+1}}R_{i+1,e-1}(u) \tag{5}$$

According to the NURBS curve, the U and V control points are obtained respectively, and then the control points of the whole surface are obtained. The basic model of the standard body is constructed, and the final model is obtained through the later technical processing (refer with Fig.5).

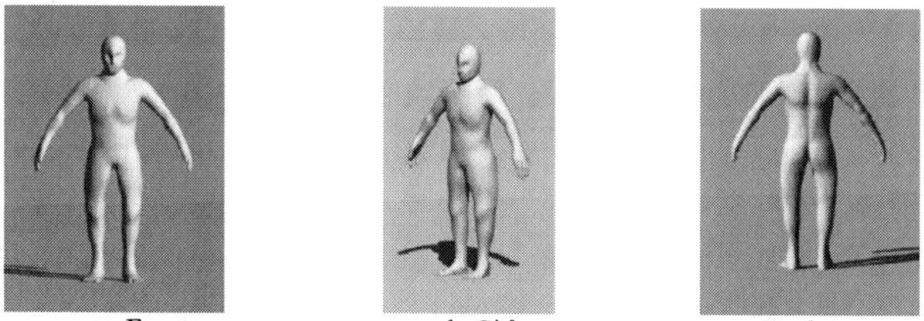

a. Front b. Side c. Back

Fig.5 Standard human body model of male youth in Southern China area

5. Conclusion

To solve the key process of modeling with a few key technologies, the establishment of the apparel industry standards of the human body three-dimensional database is the core competitiveness for the large-scale garment industry. The establishment of the standard volume model will be the key factor to improve and optimize the garment industrialized production and tailor-made industry transformation in Southern China area. It will provide the theoretical reference for the optimization of the garments size standard in this region, and also provide the reference for the young body size. The establishment of the database could provide basis and customized data to support men's garment custom enterprise.

6. Acknowledgement

The authors wish to acknowledgement Fuzhou Science and Technology Project in 2017 (Project number: 2017-G-112) and the education scientific research project of young teachers in Fujian Province in 2015 (Project number: JA15430).

References

[1] Kayis B, Hoang K. Static three-dimensional modeling of prolonged seated posture. Applied Ergonomics: 1999; 30: 255-262.

[2] Mike Kolich. Automobile seat comfort: Occupant preferences vs. anthropometric accommodation. Applied Ergonomics: 2003; 2: 177-84.

[3] Seo H, Magnenat-Thalmann N. An example-based approach to human body manipulation. Graph Models: 2004; 1: 1-23.

[4] Lv ZG, Li Y, He HG. Method of Poser model-based 3D human body modeling. Computer Engineering: 2008; 13: 256-261.

[5] Wang Y, Liu ZD. Human body modeling method based on FFD. In Pro.Int.Conf. Multimedia Technology, Hangzhou, China, Jul: 2011; 352-355.

[6] Luo J, Yang JX, Wang X. Three-dimensional human body modeling technology. Journal of Dalian Polytechnic University: 2009; 5: 378-381.

[7] Cui SQ, Yu SS, Hu XR. The method of individualized human body modeling for 3D-virtual trying on systems. Journal of Huazhong University of Science and Technology (Natural Science Edition): 2009; 10: 26-28.

[8] Tang CY, Zhang G, Tsul CP. A 3d skeletal muscle model coupled with active contraction of muscle fibres and hyperelastic behavior. Biomechanics: 2009; 7: 865-872.

[9] Tang XH, Huang L, Yang Y. Analysis of 3D human body modeling and seat comfort based on Solid Works. Journal of Central South University of Forestry and Technology: 2010; 3: 133-137.

[10] Huang JL. Modeling and implementation of human body based on feature points. Southwest Jiaotong University: 2010.

[11] Izadi S, Kim D, hilliges O, et al. KinectFusion: real-time 3D reconstruction and interaction using a moving depth camera. Proceeding of the 24th annual ACM symposium on User interface software and technology: 2011: 559-568.

[12] Tong J, Zhou J, Liu LG, Pan ZG, Yan H. Scanning 3D full human bodies using kinects. IEEE trans: Visualization and computer graphics: 2012; 4: 643-650.

[13] Baek S, Lee K. Parametric human body shape modeling framework for human-centered product design. Computer-aided design: 2012; 1: 56-67.

[14] YL, Wang MJJ. Constructing 3D human model from front and side images. Expert systems with applications: 2012; 5: 5012-5018.

[15] Alexiadis D, Zarpalas D, Daras P. Real-time full 3D reconstruction of moving foreground objects from multiple consumer depth cameras. IEEE trans on multimedia: 2013; 2: 339-358.

[16] Yang ZT, Yang Y, Shui CC. Method of building standard intermediate people model based on average body model. Journal of Tianjin Polytechnic University: 2015; 2: 28-32.

[17] Zhang SJ, Wang S, Song SJ. Application of parametric manikin in virtual product design. Journal of Xi'an

Polytechnic University: 2015; 4: 471-476.
[18] Zhang Y. Sample size is not "the more the better": talk about sampling investigation to determine sample size. China Statistics: 2008; 5: 45.
[19] GB/T 10000-1988, Human dimensions of Chinese adults. State bureau of technical supervision: 1988.
[20] Guo H, Chi XL. Analyzing body types of undergraduates in Hunan. Journal of Textile Research: 2011; 6: 109-112.
[21] Huang CY. Classification of figure type and size grading of young women of Fujian province. Journal of Textile Research: 2012; 5: 111-115.
[22] Wang AH, Chen MY, Yang ZT, et al. A study on anthropometric specifications of adult men in the north of China. Journal of Donghua University: 2004; 4: 49-55.

Location Change of Siamese Standing Collar Neckline Point

Hong Dai*, Meng Wu

Xi'an, Shaanxi Province, Xi'an, 19 Jinhua South Road, Xi'an Polyechnic University

*Corresponding author's email: ameng990@qq.com

Abstract

Siamese Standing Collar is included in the collar because of its unique style. In modern day clothing, it also occupies an important position, where it is mostly used in business wear, ceremonial clothing, traditional Tang suit and other styles. Therefore, we believe that the research in Standing Collar is necessary.

In this paper, the structure of the standing collar is studied, and related changes in the position on the neckline and the other structural elements of the collar are discussed as well. Mainly through the structure of the collar, human shoulder and neck structure and the collar when the nest point of change in the location of the relevant variables may be speculated, which is analysis and control variables through the method of design experiments to conduct the actual operation. By using the CAD plate to make a physical measurement, and then performing linear regression analysis of the measured data, we obtained a linear regression equation. Its future development prospects can be used to guide the production practice based on some numerical basis.

Keywords: Siamese Standing Collar; Garment Structure; CAD Plate Making; Regression Analysis

1. Introduction

Clothing structural design [1] studies the three-dimensional shape of clothing and the relationship between the development plan, the relationship between the decorative and functional clothing, and the optimal combination of structural decomposition and composition of the course. The basis of modern garment engineering includes style design, structural design and the process of design in three parts. The structural design of garments is an important link in the whole process of garment production. The structural design of garments does not only involve the design and development of clothing styles, but is also the basis of garment design.

Liao Can [2] focused on the fit-type collar, and studied this by analyzing the relationship between the collar structure and the human shoulder and neck structure, and the stand collar structure design. The research also discussed the combined collar design from the amount of control as well as its structural design and precautions.

Li Guilin [3] analysed the human body structure, collar structure, collar parts and clothing production process, and its effects on the relationship with the Chinese standing collar modeling structure.

Because the collar and the body of the collar and the body of the body is connected in whole or in part in the plate when the collar and body in the collar at the nest coincidence, when the collar nest point change in location, Collar of the other styling factors related changes. Most of the existing research is to make its shape first, then to study its plate. This paper uses the reverse research method, using the plate after the study of its modeling variables, to achieve the experimental results. The study of its changing law does not only optimize the garment collar structure so that it is of higher comfort when wearing, but the clothing plate model also improves work efficiency and accuracy. Especially for beginners, they no longer need experience to build a qualified version in plate making, but there is a certain amount of basis for the collar against the structure and shape of the design plate. Therefore, the research on the change of cut-point position of the single-collar collar not only has academic research value but also has a very high social practical value.

2. Probe Into the Influence of the Location of the Neckline Point on the Collar

2.1 The Structure of Stand-up Collar

Stand-up collar refers to the part of the collar where overall or part of the body is connected with the collar. According to its different position in connection with the body, it can be broadly divided as:

1) The front collar is connected to the garment body in its entirety, and the rear collar is integrally connected to the body.

2) The front collar is connected to the garment body and the rear collar is integrally connected to the garment body.

3) The front collar is integrally attached to the body and the rear collar is connected to the body (Fig.1) [4].

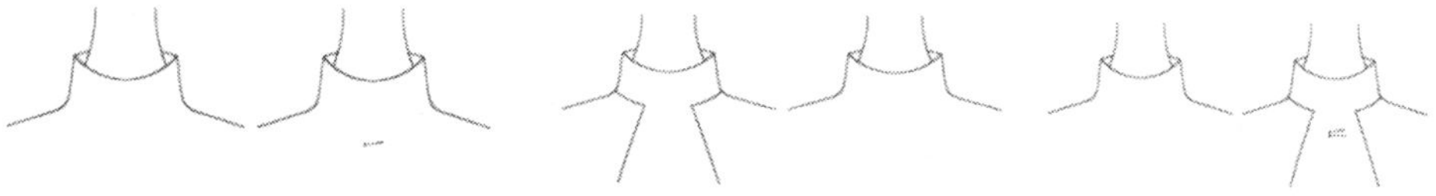

Fig.1 Different Siamese standing collar

2.2 Human Neck and Shoulder Structure

No matter what the kind of collar shape structure design is, we must follow the principle of ergonomics, making clothing suitable for people and the environment to achieve the principle of harmony and comfort. Clothing design should serve the human body, so the analysis of the human body structure and research in the clothing structure design is also very important.

The structural design of the body is also affected by the human neck and the neck structure, and the understanding of the human neck and neck structure can guide the design of the body.

The human neck and shoulder structure can be approximated as a combination of two circular truncated cone (Fig.2) [5-7]. For different people, there are differences in the thickness of the neck, length, ante-version, shoulder oblique degree and so on. While people are breathing, the movement of the neck shape will also change. Therefore, the structural design of the one-piece should pay close attention to the amount of relaxation of the neck, even the fit-type Siamese standing collar should have a certain degree of looseness, and not be designed too close to the human neck.

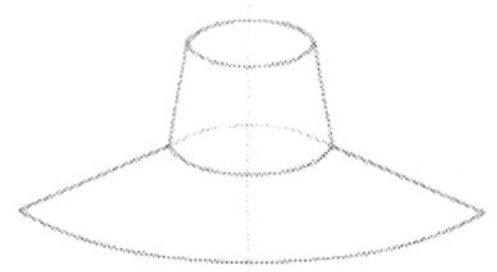

Fig.2 Neck structure

2.3 Analysis of the Effect of the Change of Position of the Nest Point on the Collar

In this chapter, the study of the structure of the collar and the analysis of the neck structure of the human body is combined with existing research [8-10]. When the distance from the tangential point to the front collar is changed, the upper line inclination angle, the distance from the neckline at the centerline of the neckline, and the distance from the neck of the collar side will change, but the specific changes in the law and extent is not known. The design experiment of this paper will further discuss the relationship between the variables produced by the change of the tangential point of the connected body and the analysis of the key points and points for attention.

3. The Mapping Method of Standing Collar

All size specifications in this article are set as follows:

Size: 160/84A;
B=84+10=94cm;
N=33.6+3.4=37cm;
S=39.4+0.6=40cm;
BAL=0.2*G+5=37cm.

3.1 Whole Collar Connected with Collar Stand

The whole Collar connected with collar stand (Fig.3) and the mapping method as shown in Fig.4.

Fig.3 Whole collar connected with collar stand.

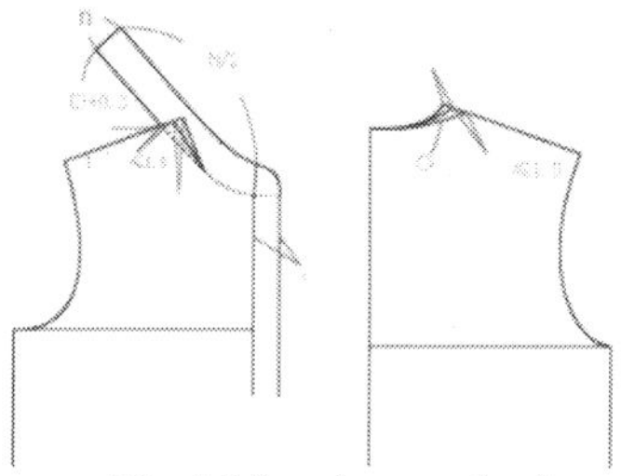

Fig.4 Mapping method

3.2 A Part of the Collar Connected with Collar Stand

3.1.1 Model

Model collar shape as shown in Fig.5, and the mapping steps as shown in Fig.6.

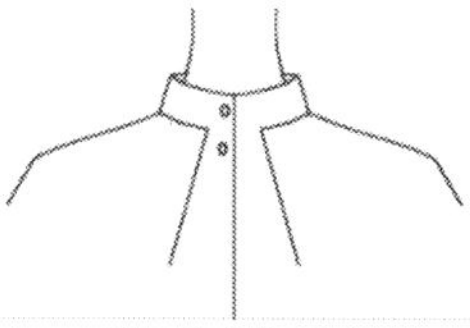

Fig.5 Model

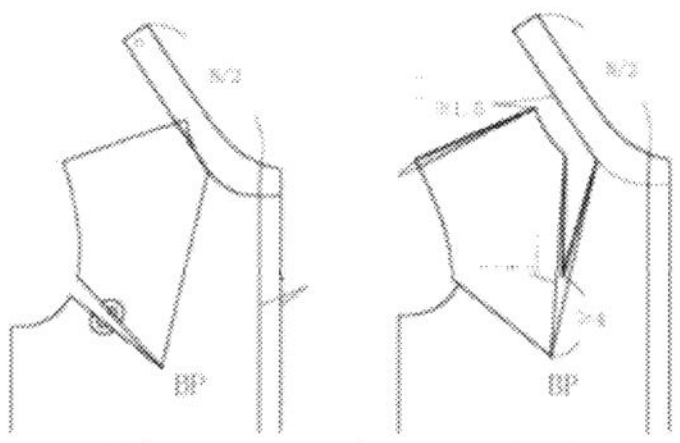

Fig.6 Mapping method

3.1.2 Mode2

Mode2 collar shape as shown in Fig.7, and the mapping steps as shown in Fig.8.

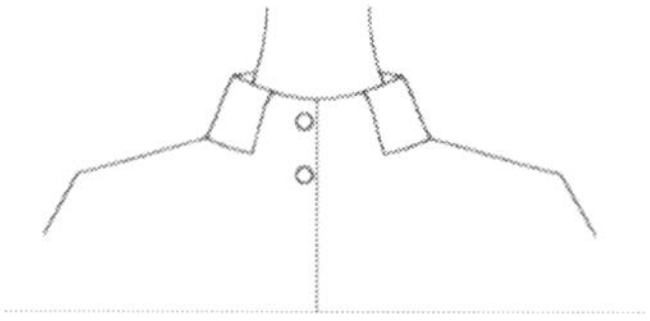

Fig.7 Mode2

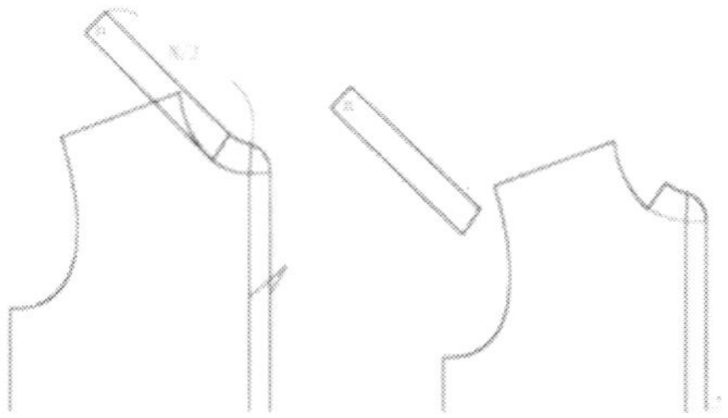

Fig.8 Mapping method

3.2.3 Mode3

Mode3 collar shape as shown in Fig.9, and the mapping steps as shown in Fig.10.

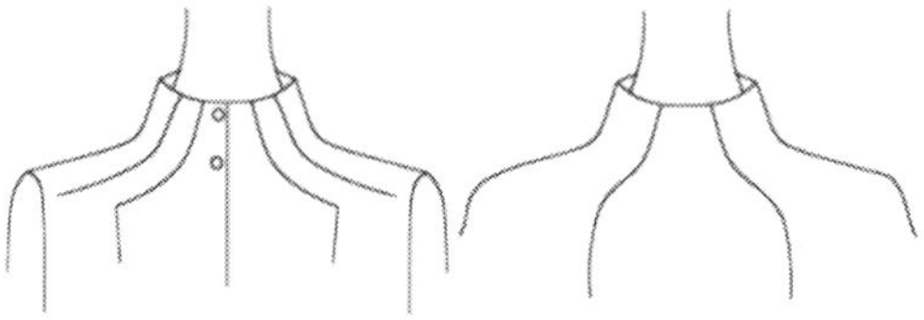

Fig.9 Mode3

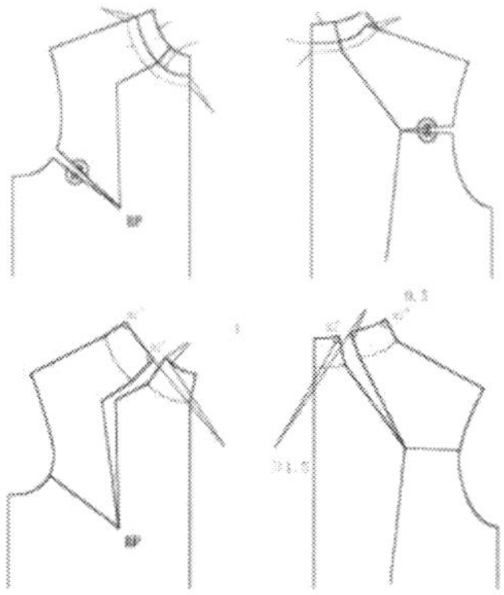

Fig.10 Mapping method

3.3 The Selection of Modeling

This article is the study of the model mentioned chapter two of this paper, the stand-up collar. In this paper, the Japanese culture prototype is used as a template (Fig.11).

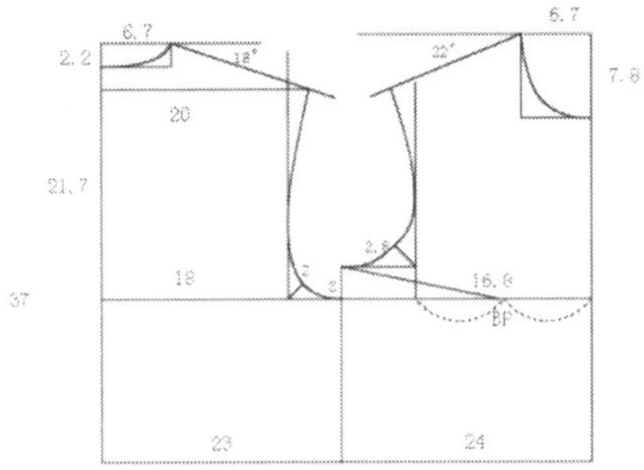

Fig.11 Japanese cultural pattern prototype

4. Experiment Design

4.1 Experimental Ideas and Methods

Through the analysis of the previous three chapters, it was concluded that when the single collar collar nest point changed with changes in the position of the distance from the sides of the collar to the neck, the side obliquity of collar, the tilt angle of roll collar and the back from the collar to the neck (other variables may be temporarily included in the scope of this study). Through the analysis of Li Xiaoyan's paper [11], it was learnt that the front point of the distance between the collar should be between 4cm and 7cm. Specific methods are as follows:

1) According to the analysis and comparison, the distance from the neckline point to the front neck point increases, and is divided into seven gradients of 4.0, 4.5, 5.0, 5.5, 6.0, 6.5 and 7.0 cm [11].

2) The back piece of paper pattern is fixed, the former pattern with the location of the cutting point in turn according to set a good gradient to draw seven different patterns;

3) In order to facilitate the measurement to reduce error, this experiment uses the drawing after the sample with white peeling fabric production, by the same person with the same sewing machine.

4) In this experiment, Japanese culture prototype was used as template. In order to reduce the experimental error, and control uncertainties, the plate in the test are set to a fixed value: collar high value; folded door width of 3.5cm. As the collar width increases; collar depth also increased.

4.2 Experimental Operation

4.2.1 Siamese Standing Collar Shape Description

In order to avoid ambiguity, and also to facilitate the narrative, this paper will introduce the experimental variables and describe its representation. The side obliquity of collar is defined as the angle between the side of the collar and the horizontal line at the neck point (SNP). The tilt angle of rolled collar is between the front collar line and the horizontal line at the collar side. The horizontal distance between the neckline and the neckline is denoted by L_c. The distance from the neck at the rear of the collar is denoted by L_h, as shown in Fig.12.

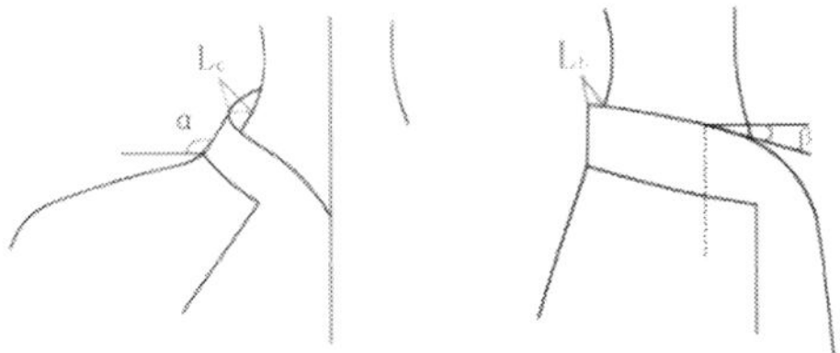
Fig.12 Schematic diagram of measurement site

4.2.2 The Measurement of the Collar

The angle of the collar's roll angle (α) determines the appearance profile of the collar and the three-dimensional shape of the side and rear section of the collar. The camber angle of the collar and the angle of inclination of the neckline are measured by photogrammetry [12]. In order to reduce experimental error, all the data are measured after 3 times, the average as the actual value of the measured value. The measurement results are shown in Table 1.

The reliability of the measured data is analyzed, and the reliability coefficient is 0.92, indicating that the reliability of the scale is very good.

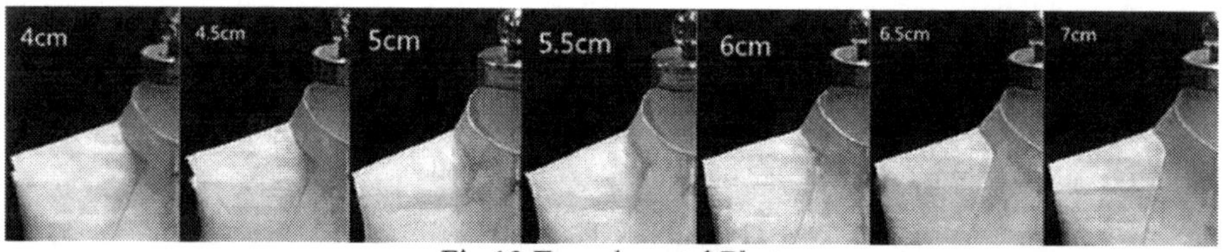

Fig.13 Experimental Photos

Table.1 Measurement data

Distance /cm	4.0	4.5	5.0	5.5	6.0	6.5	7.0
	123.0	129.5	134.7	138.6	142.0	144.4	145.6
α/°	122.8	129.6	134.4	139.3	142.2	144.7	145.8
	123.3	129.4	134.8	139.0	142.1	144.6	145.9
average	123.0	129.5	134.6	139.0	142.1	144.6	145.8
	40.2	43.0	45.7	48.4	49.5	51.0	52.0
β/°	40.0	42.9	46.1	48.3	49.8	51.0	52.1
	39.9	43.2	45.9	48.5	49.7	51.2	51.9
average	40.0	43.0	45.9	48.4	49.7	51.1	52.0
	12.3	11.1	9.9	9.4	7.4	7.3	5.3
L_c/mm	12.6	11.3	10.0	9.4	7.7	7.2	5.1
	12.1	11.4	10.2	9.3	7.8	7.0	5.3
average	12.3	11.3	10.0	9.4	7.6	7.2	5.2
	12.4	11.9	11.1	10.8	9.8	8.3	6.3
L_h/mm	12.3	12.0	11.2	10.7	9.5	8.4	6.5
	12.3	12.2	11.1	10.5	9.7	8.3	6.0
average	12.3	12.0	11.1	10.7	9.7	8.3	6.3

4.3 Correlation Analysis

According to 4.2.3 we can see that the distance between the nest point and the front collar point has some variable relationship with the collar roll angle and the neckline angle of the neckline, which is further analyzed and studied by the formula to get the correlation [13, 14].

$$\rho_{X,Y} = \frac{cov(X,Y)}{\sigma_X \sigma_Y} = \frac{E[(X-\mu_X)(Y-\mu_Y)]}{\sigma_X \sigma_Y} = \frac{E(XY)-E(X)E(Y)}{\sqrt{E(X^2)-E^2(X)}\sqrt{E(Y^2)-E^2(Y)}} \tag{1}$$

$\rho_\alpha = 0.975$; $\rho_{Lc} = 0.991$; $\rho_{Lh} = 0.963$; $\rho_\beta = 0.978$.

When the correlation coefficient approximately equals to 1, it is shown that the two variables are significantly linearly related. Comparison of the size of its relevance: $\rho_{Lc} > \rho_{\beta} > \rho_{\alpha} > \rho_{Lh}$.

4.4 Regression Analysis

4.4.1 Influence of the Side Obliquity of Collar of the Neckline Point

The linear regression analysis was used to establish the regression equation [15, 16]. The distance between the nest point and the front collar point was taken as the independent variable and the side obliquity of collar (α) as the dependent variable. In the residual statistical table of the analysis results, it can be seen that the deviation between the fitted and true values after fitting with a linear regression equation is subject to a normal distribution with a residual value between 0 and 1 (residual In the 2 to -2 between the degree of good model fit), proved to be linear regression analysis is feasible.

Finally, the linear regression equation is:

$Y_1 = 7.564X + 95.354$ ($R^2 = 0.950$, sig = 0.000)

X: The distance between the nest point and the front collar point; Y_1: the side obliquity of collar.

4.4.2 Influence of the Tilt Angle of Roll Collar of the Neckline Point

The linear regression analysis was used to establish the regression equation. The distance between the nest point and the front collar point was taken as the independent variable and the tilt angle of roll collar (β) as the dependent variable. In the residual statistical table of the analysis results, it can be seen that the deviation between the fitted and true values after fitting with a linear regression equation is subject to a normal distribution with a residual value between 0 and 1 (residual In the 2 to -2 between the degree of good model fit), proved to be linear regression analysis is feasible.

Finally, the linear regression equation is:

$Y_2 = 4.000X + 25.157$ ($R^2 = 0.965$, sig = 0.000)

Y_2: The tilt angle of roll collar.

4.4.3 Influence of the Distance from the Collar to the Neck of the Neckline Point

In order to facilitate the observation and comparison, the collar side of the distance from the sides of the collar to the neck (L_c) and the back from the collar to the neck (L_h), the two variables were used together for comparative analysis. In the residual statistical table of the analysis results, it can be seen that the deviation between the fitted and true values after fitting with a linear regression equation is subject to a normal distribution with a residual value between 0 and 1 (residual In the 2 to -2 between the degree of good model fit), proved to be linear regression analysis is feasible.

The resulting linear regression equation is:

$Y_3 = -2.279X + 21.536$ ($R^2 = 0.983$, sig = 0.000)

$Y_4 = -1.914X + 20.586$ ($R^2 = 0.928$, sig = 0.000)

Y_3: The distance from the sides of the collar to the neck.

Y_4: The back from the collar to the neck.

5. Summary

5.1 Conclusion

1) When the distance from the neckline point to the front neck point increases, the increase of modeling factor variables are the side obliquity of collar and the tilt angle of roll collar; while the reduced modeling factor variables are the back from the collar to the neck and the distance from the sides of the collar to the neck.

2) The neckline point to the front neck point distance changes in the different parts of the Siamese standing collar, and the extent of the size of the arrangement are as follows:

The distance from the sides of the collar to the neck>The side obliquity of collar>The tilt angle of roll collar >The back from the collar to the neck.

3) Linear regression model was established by linear regression analysis with SPSS, and linear regression equation was obtained as follows:

$Y_{\alpha}= 7.564X + 95.354$;

$Y_{\beta}= 4.000X + 25.157$;

$Y_{Lc}= -2.279X + 21.536$;

$Y_{Lh} = -1.914X + 20.586$.

It can be seen that the effect of the regression equation is significant, and the effect of the independent variables on the four different dependent variables is also significant.

5.2 Prospect

Despite the author's limited knowledge, in this paper, the use of scientific test methods for test and measurement were as accurate as possible. The process of testing the collar-like clothing production and sewing has a certain degree of uncertainty, and will affect the measurement results. In this paper, we mainly study the four main related variables caused by the change of the location of the tangles, which are not included in the experimental range because of the difficulty of measurement and the low degree of correlation. Therefore, the experimental research may be incomplete. The experiment only selected one kind of single-collar stand-up modeling as an experimental sample, and cannot represent all one-piece stand-up collar models.

The conclusions drawn in this paper also need to be further tested.

References

[1] Zhang WB. pattern making for fashion design. China Textile&Apparel Press: 2012; 146-149.
[2] Liao C. Analysis on the Structure Design of Stand - up Collar. Guangxi Journal of Light Industry: 2009; 2: 98.
[3] Li GL. Discussion on structure design of Chinese collar shape. China Academic Journal Electronic Publishing House.
[4] Wang H. Application of AutoCAD in Garments Design,Shandong Textile Science & Technology: 2004; 2.
[5] Jiang LJ. Research on Collar 3D Construction Digital Technology Based on Neck Characteristics. Donghua University: 2010.
[6] Yi HJ, Liu E. SPSS for Windows 15.0.Social Sciences Academic Press (CHINA): 2003; 96.
[7] Ge Y, Liu GL. Study on Photogrammetric Measurement of Body in Apparel Industry. Journal of Textile Research: 07; 28 (10): 78-81.
[8] Ding YR, Liu C, Wang SH. The Relationship between Lateral Tilt Angle and Neckline of Stand Collar. Progress in Textile Science & Technology: 2013; (4).
[9] Li CM, Zhang WB. The analyze of variable factors on standing collar. Shanghai-Frankfurt am Main: 2008; 1.
[10] Wang LX, Tian BY. An Analysis of Factors Determining the Shape of Collar. Journal of Panzhihua University: 2008; 25 (2): 83-85.
[11] Li XY, Shen Y, Jiang P. Molding Analysis of Grown-on Collar Based on Change of Tangent Point of Collar. Journal of Nantong University: 2014; 13 (1): 55-58.
[12] Huang ZC. The analyze of variable factors on standing collar, Shanghai-Frankfurt am Main: 2010; 3: 66-68.
[13] Jiang LJ. Research on Collar 3D Construction Digital Technology Based on Neck Characteristics. Donghua University: 2010.
[14] Han SJ. Comparison of Undimensionalization in SPSS Cluster Analysis. Science Mosaic: 2008; 3: 229-231.
[15] Zhao L, Chen DS, Gan YJ, Wang JG. Analysis about Male Neck-fitted Stand-collar Based on Relevant Neck Factors. TBIS: 2009.
[16] Charlie CL, Wang Yu Wang, Matthew MF. Yuen. Design automation for Customized apparel Products, Computer-Aided Design: 37 (2005) 675-691.

Back-Curve Fitting Model Based on 3D Anthropometry

Shang Wu, Lin Qi, Xiao-Mei Shang*

College of Textile and Clothing Engineering, Soochow University, Su Zhou 215021, China

*Corresponding author's email: shangxiaomei@suda.edu.cn

Abstract

This paper focuses on the back-curve of adult males measured by 3D anthropometric scanner. Males in group 170/88A were chosen as research samples to make the data collection easier. The back-curve was fitted by a polynomial function, and the cubic polynomial function was determined as a standard reference curve. The fitting function was then turned into a parameter function. The functional relationship between the chest depth, back length and back-curve length was obtained. This is so that the back-curve length can be zoomed through the proportional relationship, and the back-curve length from the back length and chest depth can be calculated when the sample is in the same group. This experiment showed that the error of the back-curve function fitted by our model is smaller than 1cm, which was enough to satisfy the anthropometric measuring for clothing.

Keywords: Anthropometric Measuring; Back-curve; Cubic Polynomial Fitting; Parameter Function

1. Introduction

Since the 21st century, with the high-speed development of computer vision technology, there is an increasing number of research focused on non-contact anthropometric technologies [1]. Due to the non-rigidity of the human body [2], it is very difficult to calculate the curve dimensions of the human body. The general method is to collect contour data of the human body with a 3D scanner. The estimation algorithm is then calculated according to the dimensions which can be measured directly to establish a mathematical model for curve fitting and curve dimensions [3]. Plenty of experiments are needed to prove this model is reliable.

In order to achieve the above measurement technology, the sample curve has to fit and the curve function has to turn into a parametric function. Then the curve dimension can be estimated according to the circumference, width or thickness which can be provided by the profile of the human body. An experiment proceeded by Soochow University collected the curve of human body by manual measurement, and they were fitted with all kinds of function with the least-squares fitting method [4]. The result showed that the fitting goodness of the polynomial function is better than the logarithmic function and exponential function, which meant the polynomial function is more suitable for the model of curve fitting [5].

In this paper, the adult males' back-curve was studied to continue the above experiment, the samples were chosen in the group 170/88A, the back-curve was collecte with a 3D scanner and fitted by the polynomial function.

2. The Collection of Back-curve

In the anthropometry, the human's back-curve is degined as the curve which starts from the 7th cervical vertebra point, through the profile of spine, and ends to the iliocristale point [6].

2.1 The Collection of Formfitting Curve

In this paper, males in the group 170/88A were chosen according to the *Clothing Shape Standard* which published by China. A Size Stream 3D scanner was used to collect the back-curve, and the graphics were imported into the AutoCAD, as shown in Fig.1.

2.2 The Simulation of Tape Measurement

In the anthropometric measurement for garment, people don't take the real formfitting length of the curve as the accurate value of this dimension, but use a tape rule to surround the curve of human body, and measure the length of the external curve [7]. Therefore in this experiment, the trend of tape rule was simulated in AutoCAD, a smooth external curve was built on the back-curve, and its length was taken as the back-curve length dimension measured by tape rule.

As a former experiment has shown the fitting goodness of the polynomial function is better, and the B-spline curve can use arbitrary number of control points to fit the curve [8, 9, 10], they are both in accordance with the characteristics of back-curve. According to the trend of back-curve, the simulative curve of tape measurement must go through the seventh cervical vertebra point, the most salient point, the scapula point and the iliocristale point to constitute the cubic B-spline curve. The effect picture fitted by AutoCAD is shown in Fig.2, the white curve is the formfitting curve, and the red curve is the simulative curve.

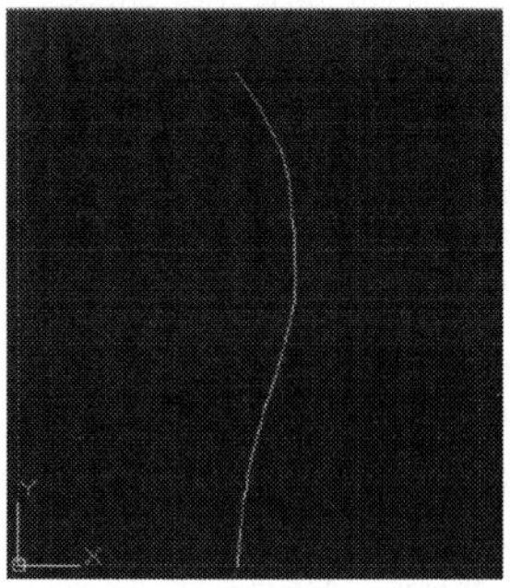

Fig.1 The Back-curve

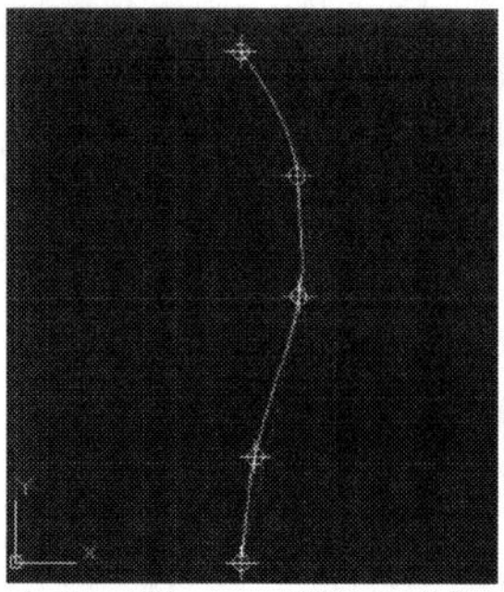

Fig.2 The Simulative Curve of Tape Measuring

The length of simulative curve and the manual measuring were compared in the same sample, and the average error turned out to be about 0.3cm, which was far below the class difference of garment [11]. It shows that the simulation of tape measurement has a good effect and can be in accordance with the need of garment.

3. The Fitting of Back-curve

After the simulative curve of tape measurement was collected, the fitting of curve was proceeded.

In the previous study of Soochow University, the set of equidistance points which cut out from the smooth curve were fitted. The robust regression analysis method [12] of least absolute residuals (LAR) was used to proceed the least squares fitting, and the result shows the fitting effect of the polynomial function is better than the logarithmic function and exponential function.

Therefore the polynomial function was chosen as the fitting function in this experiment, and a deeper discussion about the impact which the power of fitting function has on the goodness of fitting was proceeded .

3.1 Cutting out the Equidistance Points Set

The set of 1cm equidistance points were cut out from the simulative curve of tape measurement in AutoCAD, and the coordinates of all points were recorded in an Excel spreadsheet to be imported into the fitting software.

3.2 Proceeding the Function Fitting

After all the coordinates of the equidistance points were imported into Origin, the abscissa and ordinate were exchanged to adapt to the characteristics of the polynomial function graphics. And the scatter diagram to be fitted is shown in Fig.3.

The scatter diagram was fitted by the polynomial function of different power, and the quadratic polynomial fitting graphics is shown in Fig.4.

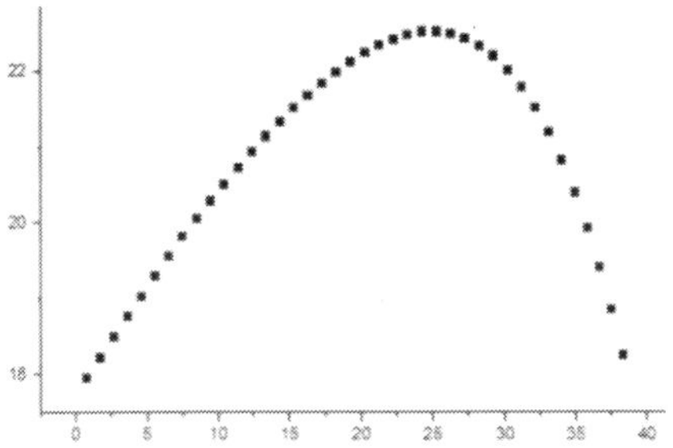

Fig.3 The Scatter Diagram

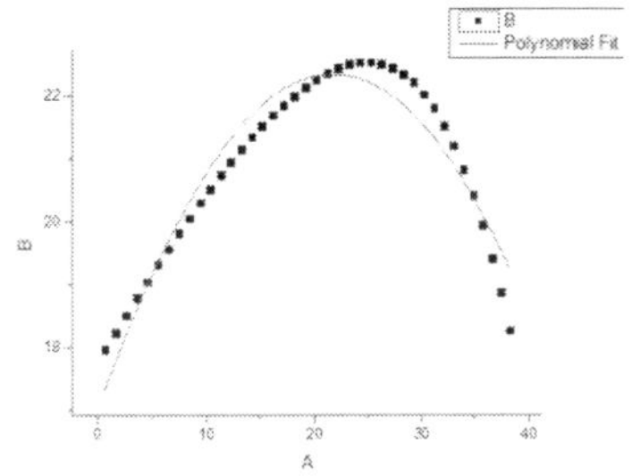

Fig.4 The Quadratic Polynomial Fitting Graphics

4. The Determination of Fitting Function

The fitting result of the quadratic, cubic, four times and five times polynomial function were acquired after the scatter diagram were fitted in Origin. The function expression and the goodness of fitting (R^2) are shown in Table 1. And the fitting graphics are shown in Fig.5-8.

Table.1 The Fitting Result

Power of polynomial function	Function expression	R^2
quadratic	$y=-0.113x^2+0.4917x+16.9932$	0.92931
cubic	$Y=(-3.2962E\text{-}4)x^3+0.0080x^2+0.1888x+18.0010$	0.99562
Four times	$Y=(2.0236E\text{-}7)x^4+(3.3904E\text{-}4)x^3-0.0088x^2+0.3362x+17.7002$	0.99991
Five times	$Y=(-1.0949E\text{-}7)x^5+(2.0875E\text{-}6)x^4+(-3.1225E\text{-}5)x^3-0.0034x^2+0.3052x+17.7437$	0.99998

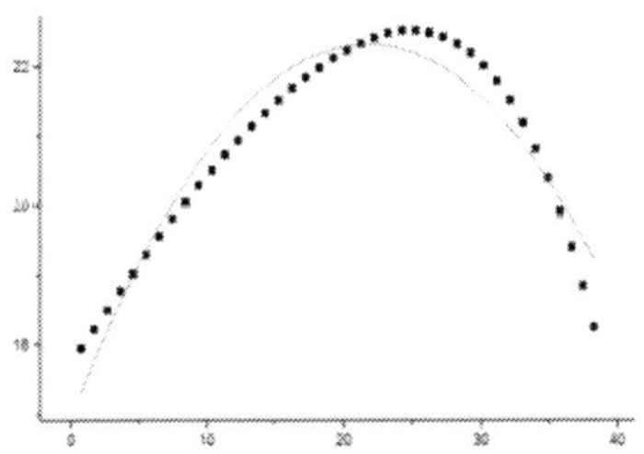

Fig.5 The Fitting Result of Quadratic Polynomial Function

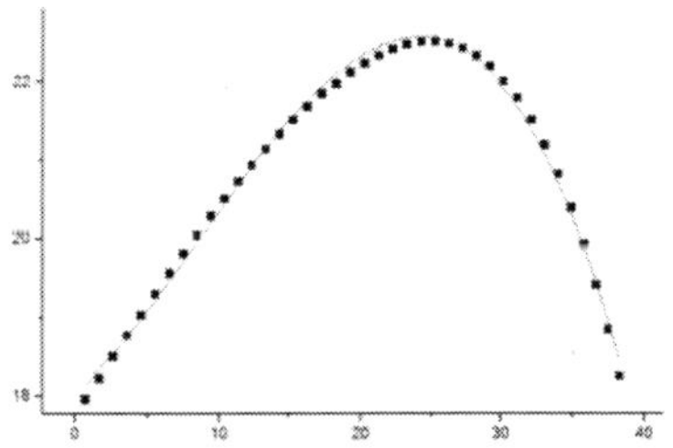

Fig.6 The Fitting Result of Cubic Polynomial Function

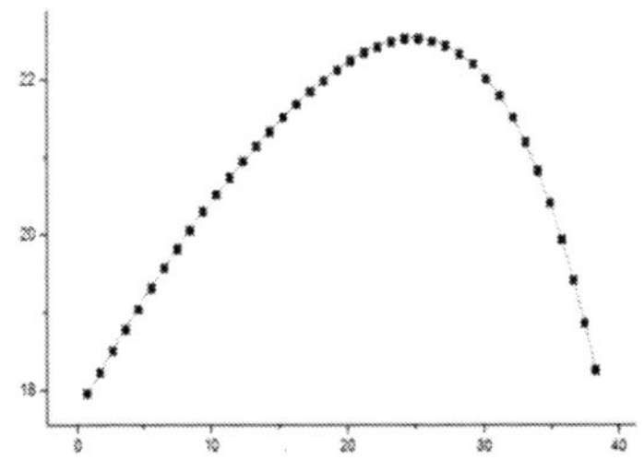

Fig.7 The Fitting Result of Four Times Polynomial Function

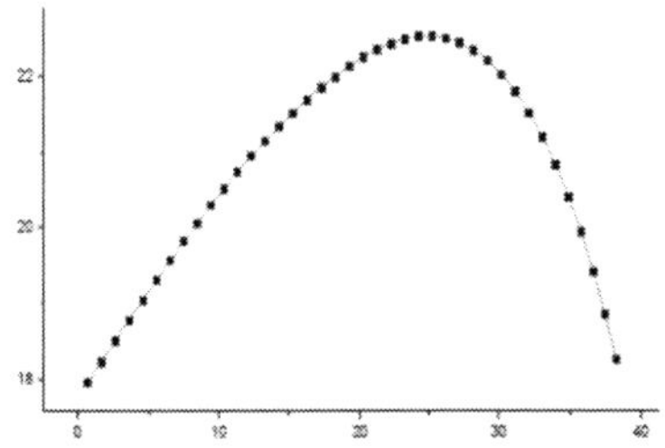

Fig.8 The Fitting Result of Five Times Polynomial Function

According to the result of function fitting, the effect of polynomial function was proved to be satisfactory. As the goodness of fitting and the conciseness of function expression were comprehensively considered, the cubic polynomial function was chosen as the standard fitting function. The length of curve was calculated from the fitting function and compared with the values of manual measuring, and the average error turned out to be about 0.57cm, which was reliable enough for the dimension measuring of garment.

5. The Practical Verification of Fitting Function

After the fitting result of 25 sets of sample data were analyzed, the standard fitting function of males' back-curve in the group 170/88A was determined as:

$$f(x)=(-3.3E\text{-}4)^3 + 0.008x^2 + 0.19x + 18 \quad (1)$$

In fact, this function also determines the relationship between the back-curve length, back length and chest depth [13]. The above function can be converted into a parametric function as:

$$\begin{cases} x = \left(\frac{\alpha}{a}\right)t \\ y = \left(\frac{\beta}{b}\right) f\left(\left(\frac{\alpha}{a}\right)t\right) \end{cases} \quad (2)$$

In the above function, the t is parameter, α and β are the back length and chest depth of the sample, a and b are the back length and chest depth of the standard model.

After the value of x and y were scaled, the definite integral was used to calculate the curve length from the parametric function, and the result was regarded as the back-curve length of the sample [14, 15].

The back-curve of the validation group data were calculated, and the comparison results are shown in Table 2.

Table.2 The Comparison Result of Validation Group (cm)

Back length	Chest depth	Fitting curve length	Calculating curve length	error
38.93	22.16	40.45	40.15	0.3
37.94	20.41	39.89	40.12	0.23
38.73	20.92	39.96	40.25	0.29
39.82	21.90	40.35	40.10	0.25
39.63	21.45	40.23	40.16	0.07

It shows that the error of calculating result is less than 1cm and reliable. The method to use a standard function to calculate the back-curve length in the same group is feasible.

6. Conclusion

In the non-contact anthropometric measurement technology, the dimensions of straight line (as stature, back length, chest depth, etc.) is easy to obtain, and their values are accurate enough. But the curve length and circumference are hard to measure, and the accuracy is not reliable. Soochow University has been studying on these problems for many years, in this paper, the human body curve fitting experiment was continued, the fitting analysis of males' back-curve in group 170/88A was proceeded to test the fitting model, and the conclusions are as follows:

1) The cubic polynomial function is reliable to be the fitting function of back-curve, and the standard function of males' back-curve in group 170/88A is $f(x)=(-3.3E\text{-}4)^3 + 0.008x^2 + 0.19x + 18$.

2) The method to zoom the fitting funciton through the proportional relationship, and calculate the back-curve length from back length and chest depth is proved fasible. It provides a solution to improve the measuring of curve dimensions.

3) In this experiment, the collection scope and number of samples are limited, there are still many dimensions need to be analyzed with more number of samples. Sincerely hoping that the general colleagues can proceed further studies and improve the model.

References

[1] Luo SJ, Zhu SS, Sun SQ. The current situation and trend of development of anthropometric measurement technology [J]. Human ergonomics: 2002; 8 (2): 31-34.

[2] Tian YN, He LW, Shang XM. Influence of measurement methods on the accuracy of measured data [J]. Journal of Beijing Institute of Clothing Technology (Natural Science Edition): 2016; (04): 25-31.

[3] Liu GL, Ji KC, Jin R. Perfection of Girth Fitting in Non-contact Body Measurement System for Young Men [J]. Journal of Silk: 2014; 51 (10): 20-25.

[4] Shang XM, Lu C, Wang H. A fitting model of circumference curve base on auto measurement [J]. Journal of Textile Research: 2010; 4: 128-138.

[5] Lu C, Wang H, Shang XM. A fitting method of human chest curve [J]. Journal of Textile Research: 2009; 6: 38-50.

[6] Wen PZ, Ma C, Hu JR. Measurement of human body's featured imensions based on the national standards [J]. Computer Engineering & Science: 2014; 36 (06): 1114-1119.

[7] Wang YJ. Study of body anthropometrical method in motion state [J]. Journal of Textile Research: 2014; 34 (4): 104-110.

[8] Mukul TN, Venkata R, Anupam S. Constrained Shape Modification of B-Spline Curves [J]. Computer-Aided Design and Applications: 2006; 3: 1-4.

[9] Juan F, Ni XJ. App lication of b-spline fitting surface to measure of body [J]. Application Research of Computers: 2007; (05): 183-185.

[10] Juan F, Ni XJ. App lication of b-spline fitting surface to measure of body [J]. Application Research of Computers: 2007; (05): 183-185.

[11] Pan L, Wang J, Sha S. Study on body typing and garment size grading of young women in Northeast China[J]. Journal of Textile Research: 2014; 34 (4): 131-135.

[12] Roman V, Victor M. Robust Regression Methods for the Analysis of Unreplicated Factorials [J]. Lecture Notes in Engineering and Computer Science: 2009; 2: 17-19.

[13] Sun KK, A research on non-contact anthropometric measurement technology under the natural state of dress [D]. Tianjin university of technology: 2007.

[14] Cai L, Li Q, Li PW. Pattern design method based on reconstruction of parts in blouse [J]. Journal of Textile Research: 2015; (08): 116-120.

[15] Gao PP, Le YM, Shang XM. A review on the research of human station's section model for garment[J]. Modern Silk Science & Technology: 2015; (06): 234-236.

Elliptic Fourier Analysis on Female Chest Contour

Hui-Yao Zhang[1], Duan Li[1], Hao-Yang Xie[1], Yue-Qi Zhong[1,2*]

[1] *College of Textiles, Donghua University, Shanghai, 201620, China.*
[2] *Key Lab of Textile Science and Technology, Ministry of Education, Shanghai, 201620, China.*

*Corresponding author's email: zhyq@dhu.edu.cn

Abstract

In this paper, Elliptic Fourier Analysis (EFA) was used to study the chest shape of female subjects. Three kinds of sampling methods on chest contour analysis were first compared, including curvature sampling, random sampling and equally spaced sampling. According to the experiment, the equally spaced sampling is the best option. Secondly, the appropriate proper sampling density based on the mean distance and the chest circumference was discovered. Finally, the number of harmonics used by EFA was tested for acceptable error. The experimental results indicate that the EFA is an efficient method in female chest shape analysis.

Keywords: Elliptic Fourier Analysis; Sampling Method; Sampling Points; Harmonics

1. Introduction

Fourier transformation has been widely used in many engineering domains, especially, in signal analyzing and processing. Elliptic Fourier has been adopted to present the contour of image and analysis of object shape. In 1982, Kuhl [1] took Elliptic Fourier as the approach to describe closed contour curve. In biology field, Elliptic Fourier was used to describe object silhouette and Elliptic Fourier Descriptors (EFD) was used to measure shape differences [2-4]. For instance, Neto J C used EFDs to describe and analyze the shape of a plant leaf, and the coefficient of EFDs of the plant leaf was taken to identify the type of plant. Martin Friess [5] used Elliptic Fourier to research three dimensional curve forms and extract characteristic curves from human body scans. Then, the analyzing result was used to direct the design of the shape of the protective belt.

The shape and size of the human chest is one of the most important factors whenever a tailor measures a person for suitable clothing [6-8], or when a game engineer rebuilds a virtual body or a costume designer designs virtual [9] or real clothing. In the article, the female chest is selected as target since the male chest slice is easier to fit and reconstruct due to its simple contour.

In this paper, EFA was employed as the core approach to implement the whole study. According to the suggestions from several qualified tailors, a facile hypothesis was proposed that the number of sampling points is based on chest girth and maximum distance error, and mean error range is 1mm and 0.5mm respectively. Different number of sampling points of the same curve was selected to illustrate the huge differences of fitting effect. Next, several different sampling methods and sampling points with certain harmonics was taken to reconstruct the final fitting curve, where eight different models were used to testify the validity of our fitting.

2. Methodology

2.1 Data Collection

Kinnect was used to scan a female bust placed on a rotatable platform and the actual chest circumference was measured manually firstly. To make sure that the obtained data accuracy is sufficient, the environment is set to related standards. The temperature range is from 21 to 25 degree centigrade, and the humidity level is between 60% and 70%. After the human point cloud data was collected, the OBJ file was imported into GeoMagic Studio 12, a well-known professional engineering program brand of 3D system [10]. First, the holes was padded in the scanned model then the number of triangles was reduced for the purpose of computation efficiency. Next, the feature points was chosen manually before scanning, the model was cut horizontally and the chest curve was obtained. Lastly, several different discrete-closed curves was created with various sampling methods.

To testify the scanner accuracy, a cylinder was scanned. The diameter of the cylinder is 25cm and the height is 80cm. The scanned cylinder accuracy is analyzed. According to Fig.1, the scanning error increased from the center to both ends, and the scanning error got the maximum value which is less than 0.5cm, which could satisfy our experiment requirement.

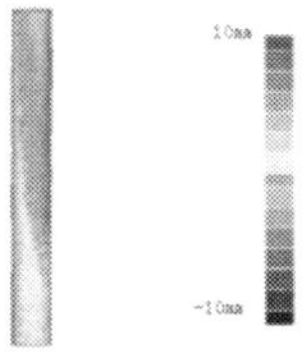

Fig.1 Testify the scanning system error with a cylinder

2.2 Elliptic Fourier Analysis

EFA was one of the most widely used tool for feature extraction and reconstruction of a single closed-curve. A continuous closed curve could be considered as a continuous periodic function. According to the principle of Fourier Series, any periodic functions could be expressed by the sum of a series of sine and cosine functions with increasing frequencies. The coefficients of the Fourier Series were referred to as "Elliptic Fourier Descriptors" (EFDs) [6], which could be implemented to reconstruct the original shape with major features. EFDs was first proposed by [11] and the most significant reason for its popularity was that researchers can use it to represent all kinds of close contour meanwhile retaining the original information with limited number of coefficients. Moreover, it also provided robustness to translation, rotation and scale if the EFDs were normalized [12]. Lots of studies used EFDs to research, for example, the characteristics of animals and plants [13-16], anthropology [17], hand-written recognition [18] and aircraft contour description [19].

Suppose there are *K* discrete sampling points along a continuous-closed contour. This contour can be represented by a sequence of *x* and *y* coordinates which are ordered by counter-clockwise from an arbitrary starting point. The length from the starting point to the *p-th* point is denoted by t_p, and the perimeter of the whole contour is denoted by *T*, and $T = t_K$ where *K* is the total number of the discrete sampling points and Δt_i is the distance between the *(i-1)-th* point and *i-th* point. For each sampling point, the coordinates of *p-th* along x and y directions are:

$$t_p = \sum_{i=1}^{p} \Delta t_i \tag{1}$$

$$x_p = \sum_{i=1}^{p} \Delta x_i \tag{2}$$

$$y_p = \sum_{i=1}^{p} \Delta y_i \tag{3}$$

where Δx_i and Δy_i are the distance from *(i-1)-th* point to the *i-th* point along the *x* and *y* axes respectively. Thus, the Elliptic Fourier expansions of *x* and *y* coordinate along the contour are:

$$x_p = A_0 + \sum_{n=1}^{\infty} (a_n \cos\frac{2n\pi t_p}{T} + b_n \sin\frac{2n\pi t_p}{T}) \tag{4}$$

$$y_p = C_0 + \sum_{n=1}^{\infty}(c_n \cos\frac{2n\pi t_p}{T} + d_n \sin\frac{2n\pi t_p}{T}) \tag{5}$$

The coefficients a_n ,b_n ,c_n and d_n actually have the capacity to describe the outline of the original section and are known as Elliptic Fourier coefficient of the *n-th* harmonic. Notice that each harmonic is denoted by four Elliptic Fourier coefficients and the first harmonic is used to align the contour [19].

$$a_n = \frac{T}{2n^2\pi^2}\sum_{p=1}^{K}\frac{\Delta x_p}{\Delta t_p}\left(\cos\frac{2n\pi t_p}{T} - \cos\frac{2n\pi t_{p-1}}{T}\right) \tag{6}$$

$$b_n = \frac{T}{2n^2\pi^2}\sum_{p=1}^{K}\frac{\Delta x_p}{\Delta t_p}\left(\cos\frac{2n\pi t_p}{T} - \sin\frac{2n\pi t_{p-1}}{T}\right) \tag{7}$$

$$c_n = \frac{T}{2n^2\pi^2}\sum_{p=1}^{K}\frac{\Delta y_p}{\Delta t_p}\left(\cos\frac{2n\pi t_p}{T} - \cos\frac{2n\pi t_{p-1}}{T}\right) \tag{8}$$

$$d_n = \frac{T}{2n^2\pi^2}\sum_{p=1}^{K}\frac{\Delta y_p}{\Delta t_p}\left(\cos\frac{2n\pi t_p}{T} - \sin\frac{2n\pi t_{p-1}}{T}\right) \tag{9}$$

The equations above are the basic principles and tools that were used in this study. First, the Elliptic Fourier coefficient of N harmonics were computed according to sampling points, and then the fitting coordinates were computed based on the original sampling points.

2.3 Error Metrics

Four parameters were selected to measure the similarity of the original contour and fitting contour: 1) perimeter accuracy, 2) mean distance between original sampling points and fitting points, 3) standard deviation of distance between original sampling points and fitting points and 4) maximum distance between sampling points and reconstruction points.

1) Perimeter was treated as one of the error metrics since the number of sampling point is closely related to the contour perimeter, furthermore, the chest circumference is also a significant factor in garment industry.

Taking O as original/actual contour and F denoted as fitting contour. $P(O)$ and $P(F)$ are the perimeters of the original contour and the fitting contour, respectively. Then perimeter accuracy is computed as:

$$A = \frac{P(F)}{P(O)} \tag{10}$$

2) Mean distance ($\bar{D}$) reflects the deviation extent between the original shape and reconstruction shape. Mean distance is computed by the following formula where K is the total number of the sampling points.

$$\bar{D} = \frac{1}{K}\sum_{1}^{K} dist(O_i, F_i) i = 1,2...K \tag{11}$$

3) The standard deviation of distance between original sampling points and fitting points (σ_D) reflects the measure variability which is computed as

$$\sigma_D = std(D_i) \quad (12)$$

4) The maximum distance between sampling points and fitting points is denoted as

$$D_{\max} = Max\{dist(O_i, F_i), i = 1, 2, ..., K\} \quad (13)$$

Now the core algorithm and the methods have been illustrated above as used to measure the errors. During the following sections all of the test results are based on them. Many related works focused on the field of image processing and there is no need to consider this problem because the pixels are naturally discrete [20].

3. Experimental Results and Discussion

3.1 Testing Sampling Methods

Three common sampling methods were included in this study, curvature sampling, random sampling and equally spaced sampling. In this step, these three kinds of sampling methods were used to obtain sampling points and the number of sampling point is equal. So 500 sampling points were selected per curve with different sampling methods and the number of harmonics is 15. Table 1 lists the result of the test. (All three discrete closed curve are created by GeoMagic Studio 12).

Table.1 Comparison of three different sampling methods with 500 discrete points and 15 harmonics

Error Metrics	Curvature Sampling	Random Sampling	Equally spaced Sampling
Mean Distance (mm)	5.56	6.78	0.87
Max Distance (mm)	7.82	8.63	1.72
Standard Deviation	24.39	28.46	0.42
Perimeter Accuracy	92.21%	98.01%	99.53%

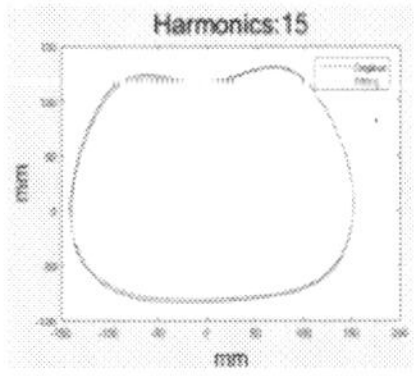

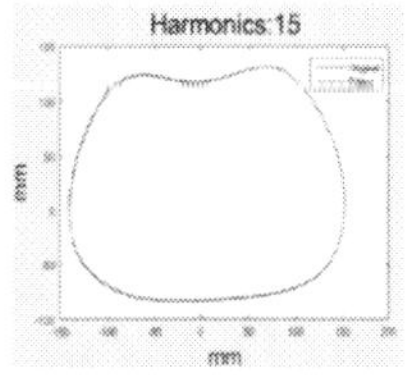

(a) curvature sampling (b) equally spaced sampling

Fig.2 (a) is the result of curvature sampling, at the left bottom corner, the gap between two contour was obvious. (b) is the result of equally spaced sampling. Both of them use 500 sampling points and 15 harmonics. White is the original contour and red is the fitting contour.

As shown in Table 1, the equally spaced sampling has the best performance. Curvature sampling and random sampling have the similar results but are far behind equally spaced sampling. So in the following sections equally spaced sampling was selected by default. Fig.2 (a) and (b) represent the fitting effects of curvature sampling and equally spaced sampling based on 500 sampling points and 15 harmonics.

The fitting effects of curvature sampling and equally spaced sampling

3.2 Sampling Density

The next step was to decide how many discrete points should be taken from the contour. Intuitively, the more points, the more accurate the results. However, the whole procedure would be time consuming if too many discrete points were selected. Conversely, if only a few number of points were employed, the preferable fitting result cannot be achieved even if higher harmonics were adopted. Fig.3 represents an extreme reconstruction example with 120 sampling points and 2000 harmonics. With more harmonics, more details will be kept. However, we cannot achieve the anticipated result with a few number of points in practice. Furthermore, the number of harmonics is constrained by the number of points along a contour, and the number of sampling points is corresponds to the length or size of the object. [21] Sanfilippo used 100 points corresponds with 43 harmonics to describe the shape of human laminas but without the explanation that why 100 sampling points are selected.

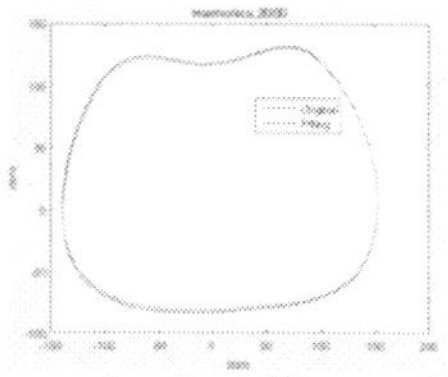

Fig.3 Construction result with 2000 harmonics and only 120 sampling points. The perimeter accuracy is 98.25%.

Intuitively, the number of points is closely related to the perimeter of a shape. This was also the reason why perimeter accuracy was chosen as one of the error metrics above. According to qualified tailors and garment designers' advice, for human measurement and the garment industry, 99% perimeter accuracy, 1mm maximum distance and 0.5mm mean distance can satisfy the practical requirements. Then, we proposed a facile hypothesis and verified the availability in the later section. The distance was used as the sampling interval of equally spaced sampling which means, for example, 0.5mm is the sampling interval so there will be 2000 sampling points if the girth of a closed curve is 100 cm.

3.3 Reconstruction with Limited Harmonics

For now, sampling method has been decided and the number of sampling points was also selected according to the perimeter of a contour. The only variable remaining was the number of harmonics, which was easy to decide according to the different fitting effects and error metrics. The chest circumference of the model was 840mm, so this chest contour could be divided into 1680 sampling points based on the hypothesis proposed in last section.

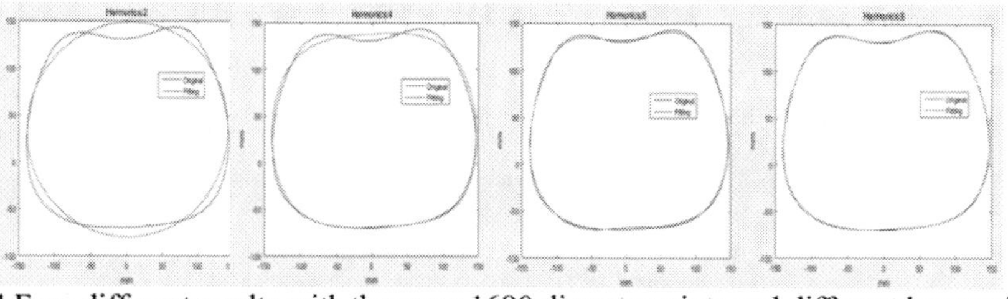

Fig.4 Four different results with the same 1680 discrete points and different harmonics.

Table 2. Errors of four different construction results.

It can be seen that after the number of sampling points was decided, the only thing needed to be done next is to test whether the number of harmonics could satisfy the acceptable errors.

Table.2 demonstrated the different imitations with the same number of sampling points with equally spaced sampling but different harmonics.

Harmonics	Perimeter Accuracy	Mean Distance	Maximum Distance	Standard Deviation
2	96.95%	7.69	17.64	3.88
4	98.55%	3.22	8.13	1.94
6	99.40%	1.52	2.94	0.6
8	99.74%	0.48	0.99	0.18

3.4 Test Validation

In order to validate the hypothesis on sampling points we proposed, other 8 female models are included in this study. For these 8 scanned models, the number of sampling points were decided by the length of chest circumference, and all of acceptable results can be achieved with limited harmonics in practice. Fig.5 demonstrates the imitations and Table 3 represents the error metrics.

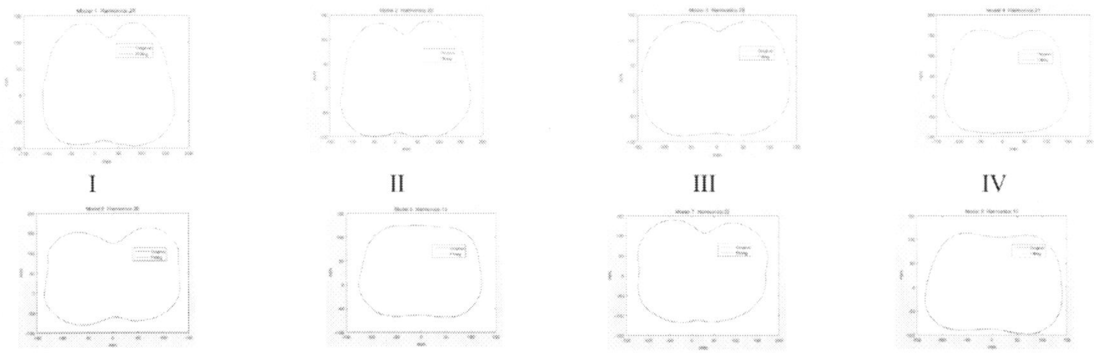

I II III IV

V VI VII VIII

Fig.5 Construction result of other 8 scanned model. The number of sampling points is based on the girth of chest contour and mean error.

Table.3 Error metrics of 8 scanned model. The measure unit of perimeter, mean distance and max distance is millimeter

Model	Perimeter	Sampling Points	Harmonics	Perimeter Accuracy	Mean Distance	Max Distance	Std.
I	880	1760	24	99.70%	0.29	0.99	0.16
II	917	1834	20	99.66%	0.34	0.95	0.2
III	868	1736	23	99.67%	0.3	0.98	0.19
IV	920	1840	21	99.70%	0.3	0.99	0.18
V	891	1782	20	99.67%	0.32	0.91	0.18
VI	738	1476	13	99.70%	0.32	0.93	0.17
VII	950	1900	22	99.70%	0.31	0.96	0.18
VIII	830	1660	14	99.72%	0.31	0.94	0.16

4. Conclusion

In this article, EFA was used as the basic technique to reconstruct the female chest contour. There are three factors which influence the fitting result, which is the sampling method, sampling density and harmonics number. First, in this study, equally spaced sampling is selected as the best one through our experiments. Second, to reduce variables in the fitting process, we proposed a facile hypothesis to decide the sampling interval based on the mean error and the girth of chest, meanwhile validated the availability of this method. After that, the last factor left is the number of harmonics and which was easy to decide according to the fitting result and error metrics. There may be other methods to decide the number of sampling points, however, in practice, the hypothesis we proposed here is effective for the Elliptic Fourier fitting of female chest contour with limited harmonics.

5. Acknowledgement

This work is supported by National Natural Science Foundation of China (Grant No.61572124) and Shanghai Natural Science Foundation (Grant No.14ZR1401100).

References

[1] Kuhi FP, Giardina CR. Elliptic Fourier features of a closed contour [J]. Computer Graphics and Image Processing: 1982; 18 (3): 236-258.

[2] Adebowale A, Nicholas A, Lamb J, et al. Elliptic Fourier analysis of leaf shape in southern African Strychnos section Densiflorae (Loganiaceae) [J]. Botanical Journal of the Linnean Society: 2012; 170 (4): 542-553.

[3] Neto JC, Meyer G , Jones DD, et al. Plant species identification using Elliptic Fourier leaf shape analysis [J]. Computers and Electronics in Agriculture: 2006; 50 (2): 121-134.

[4] Viscosi V, Fortini P. Leaf shape variation and differentiation in three sympatric white oak species revealed by elliptic Fourier analysis [J]. Nordic Journal of Botany: 2011; 29 (5): 632-640.

[5] Friess M, Rohlf FJ, Hsiao HW. Quantitative Assessment of Human Body Shape Using Fourier Analysis [M]//CORNER B D, LI P, PARGAS R P. Three-Dimensional Image Capture and Applications Vi. Bellingham: Spie-Int Soc Optical Engineering: 2004; 117-124.

[6] Liu H, Zhang M, Chen Ds, et al. Study on bust curve of standard female figure [J]. Journal of Textile Research: 2011; 32 (2): 117-120.

[7] Shang Xm, Lu C, Wang H. Human body circumference curve fitting model base on automatic measurement [J]. Journal of Textile Research: 2010; 31 (4): 128-131, 138.

[8] Abels A, Kruusmaa M. (2009). Design of a shape-changing anthropomorphic mannequin for tailoring applications. International Conference on Advanced Robotics (pp.1-6). IEEE.

[9] SlavenkaPetrakDubravkoRogale, (2006), "Systematic representation and application of a 3D computeraided garment construction method", International Journal of Clothing Science and Technology: Vol.18 Iss3 pp. 179 -187.

[10] Hu YF. (2009). The application of geomagic studio software in reverse engineering post-processing. Manufacturing Automation: 5 (5); 501-524.

[11] (1982). Elliptic Fourier Features of a Closed. Contour, Comp. Graphics Image Process.

[12] R N. Bracewell, The Fourier transform, Scientific American: 260 (1989); 62-69.

[13] Iwata H, Niikura S, Matsuura S, Takano Y, Ukai Y. Interaction between genetic effects and soil type in diallel analysis of root shape and size of Japanese radish (Raphaunussativus L), Breeding Science: 54 (2004); 313-385.

[14] Tracey SR, Lyle JM, Duhamelb G. Application of elliptical Fourier analysis of otolith form as a tool for stock identificationi, Fisheries Research: 77 (2006); 138-147.

[15] Rohlf FJ, Archie JW. Acomparison of Fourier methods for the description of wing shape in mosquitoes. (Riteraculicidae), Syst. Zool: 33 (1984); 302-317.

[16] Neto JC, Meyer GE, Jones DD, Samal AK. (2006). Plant species identification using elliptic fourier leaf shape analysis. Computers & Electronics in Agriculture: 50 (2); 121-134.

[17] Daegling DJ, Jungers WL. Elliptical Fourier analysis of symphyseal shape in great ape mandibles, J. Hum. Evol: 39; (1) (2000): 107-122.

[18] Granlund GH. Fourier Preprocessing for Hand Print Character Recognition, IEEE Trans. Comput: 21 (1972) 195-201.

[19] Wallace TP, Wintz PA. An efficient three-dimensional aircraft recognition algorithm using normalized fourier descriptors, Comput. Graph. Image Process: 13 (2) (1980); 99-126.

[20] Freeman H. 1974: Computer processing of line drawing images, Compute Survey: 6; 57-97.

[21] Sanfilippo PG, Grimm JL, Flanagan JG, Lathrop KL, Sigal IA. (2014). Application of elliptic fourier analysis to describe the lamina cribrosa shape with age and intraocular pressure. Experimental Eye Research: 128; 1-7.

Comparison of Male Body and Pattern between Hubei and Shandong Province

Xin Zhao[1], An-Hua Zhong[2] *

[1] *Wuhan Textile University, Mingzu Avenue, Hongshan District, Wuhan City, Hubei Province, 430073, China*
[2] *Wuhan Textile University, Mingzu Avenue, Hongshan District, Wuhan City, Hubei Province, 430073, China*

*Corresponding author's email: zhongah@yeah.net

Abstract

This research conducts static body measurement and data analysis which includes the analysis of means and luster analysis on a random sample of adult men aged between 30-45 years in Hubei and Shandong province. Based on the obtained research data, the center size of the measured samples were collected and compared to. By obtaining the difference of body size and pattern, it may provide a reliable reference for the production and sales of men's clothing in China.

Keywords: Body; Adult Male; Random Sample Survey; Hubei /Shandong Province; Comparsion

1. Introduction

China covers almost fifty degrees of latitude from south to north and over sixty degrees of longitude from east to west, forming different temperature zones and wet & dry areas. As genetic characteristics, economic development, living standards, cultural customs and eating habits vary from region to region, there is an obvious distinction between southerners and northerners in terms of body size. [1] Currently, the size classification of China's garment follow the national standard "garment series" (also called as the standard of GB/T1335.1-97 [2]), which was put forward about two decades ago. Therefore, there is a desperate need to update the body size data. In order to solve this issue, this research conducts a random sample [3] of adult men aged between 30-45 years in Hubei and Shandong Province.Based on the comparative analysis, this thesis reveals the size differences and characteristics of adult males in Hubei and Shandong Province. Additionally, this research obtains the center sizes data of male bodies in these two provinces, then drawing their basic pattern to compare. By analyzing the differences of body and pattern in Hubei and Shandong Province, this research provides a reliable reference for the production and sales of Chinese garment company.

2. Experiment

2.1 Experimental Instruments and Conditions

This paper used a 3-D body laser scanner [3] to measure a male's body size in the two provinces, controlling experimental temperature in the range of (27±3)℃, keeping experimental relative humidity in the range of (60±10)%. Experimental conditions meet the environmental standards of nude measurement. The three dimensions body laser scanner is shown in Fig.1.

2.2 Requirements for Participants

The participant would be naked, wearing a white cap to cover all the hair, and without wearing any jewelry, such as watch and necklace. When being measured, the participant would keep an upright posture which is standing naturally with his feet on the foot mark of the scanning platform, arms on his side, elbows raised slightly to expose the armpits, with a distance from hand to hand of about 8 to 10 cm. The participant should also be breathing naturally, with head facing forwards looking straight ahead, and with minimized shaking [4].

2.3 Sample Size

The sample size (N) refers to the number of adult males in the studied area. The sample size (N) depends on the estimation accuracy. [5] The higher the experimental accuracy required, the larger the sample size. Referring to the formula of physical fitness statistics in China [6], the sample size (N) should be:

$$N = [\ (1.96s/\sigma]^2 \quad (1)$$

Among them, s is standard deviation, and σ is the allowable error. When the coverage rate of sampling is 95%, then the confidence level is 1.96. When there is more variation between samples and a high degree of precision is required for the study, then the sample size needs to be larger. In addition, the sample size will be increased with an increase in the experimental range. The coverage rate is usually at 95% in areas of scientific research and industrial production. In this paper, we consider σ=5%, and therefore the confidence level is 1.96. [7] With reference to the relevant data in GB/T1335 1-97, obtaining the maximum of s/σ is 6.7, and get a sample size (N) of 172. In the study, 710 adult men in Hubei Province and 800 adult men in Shandong Province participated in the measurement sessions, in which the sample size for both areas exceed 172, supporting the validity of the study.

Fig.1 3-D body laser scanner

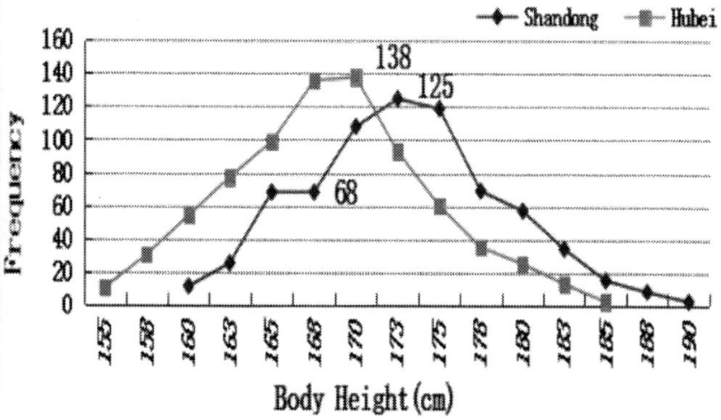

Fig.2 Body Height Distribution

3. Results and Analysis

Removing outliers in the measured data [8], the effective value of Hubei province and Shandong province is 709 and 772, respectively.

3.1 Data Statistics

Grouping the data, a statistical list was obtained, including the sample distribution percentage and the average value for chest and arm length in different height stages. The list is shown in Table.1, in which the data is arranged in ascending order of the participants' body height. From left to right in Table1, the first column is the percentage distribution of samples, the second column is the average chest girth, and the third column is the average arm length. Data from Table 1 shows that height of adult men in Hubei province is in the range of 155 cm to 185 cm, with chest girth in the range of 88.66 cm to 97.04 cm and arm length in the range of 71.30 cm to 84.50 cm. On the other hand, men in Shandong province are around 160 cm to 190 cm tall, with chest girth in the range of 94.00 cm to 101.40 cm, and arm length in the range of 74.00 cm to 87.20 cm. The grey section of Table 1 displays that most of the adult men in Hubei province are around 165 cm to 173 cm tall, whereas those in Shandong province are around 170 cm to 175 cm tall. Summarizing the whole set of data, in general, adult men in Shandong province province have a larger height/chest girth/arm length than those in Hubei province.

Table.1 Body Height vs Chest Girth vs Arm Length (cm)

Body Height	Percent of Samples (%)		Average Chest Girth		Average Arm Length	
	Shandong	Hubei	Shandong	Hubei	Shandong	Hubei
155		1.29		90.06		71.30
158		3.88		87.24		72.40
160	1.55	6.99	94.10	88.09	74.00	73.50
163	3.5	9.97	94.00	88.66	75.10	74.60
165	9.5	12.8	97.20	89.24	76.20	75.70
168	9.59	17.6	95.70	91.17	77.30	76.80
170	15.2	17.8	97.60	91.38	78.40	77.90
173	17.6	12	97.60	91.47	79.50	79.00
175	16.7	7.77	97.30	93.30	80.60	80.10
178	9.7	4.5	96.60	95.21	81.70	81.20
180	8	3.23	99.40	93.77	82.80	82.30
183	4.7	1.68	97.60	97.04	83.90	83.40
185	2.1	0.25	101.40	92.58	85.00	84.50
188	1.1		100.70		86.10	
190	0.28		96.70		87.20	

Fig.2 is a height frequency distribution plot of the two provinces. In Fig.2, the red curve represents Hubei province, and the blue curve represents Shandong province. As a whole, the height distribution of adult males in the two regions shows a normal distribution. In Hubei province, the minimum height of adult males is 155 cm and the maximum height is 185 cm; whereas in Shandong province, the minimum is 160 cm and the maximum is 190 cm. Fig.1 can clearly show the curve's trend. Seen from Fig.2, red curve rises and falls rapidly. The red curve's peak value is 138, with the height in the range of 168 cm to 172 cm at the point. The blue curve has rather less dramatic rise and fall, in which its peak value is 125, with the height in the range of 170 cm to 175 cm. Besides, the blue curve has an obvious plateau in the rising stage. At the plateau, the height is in the range of 163 cm to 168 cm, with a frequency of 68. In general, adult men in Shandong is taller than those in Hubei, while the height distribution in Shandong is more disperse than that in Hubei.

3.2 Correlation Analysis

The human data in Table 1 shows a normal distribution. Using the SPSS software [9], the value of r (Pearson correlation) and Sig.(2-tailed) can be calculated, which can be used to determine whether there is a correlation between the two variables. The formula of r (Pearson correlation) [10] is:

$$r = \frac{\sum_{i=1}^{n}(x_i - \bar{x})(y_i - \bar{y})}{\sqrt{\sum_{i=1}^{n}(x_i - \bar{x})^2 \sum_{i=1}^{n}(y_i - \bar{y})^2}}$$

It can be seen that after the number of sampling points was decided, the only thing needed to be done next is to test whether the number of harmonics could satisfy the acceptable errors.

r is an index used to describe the probability of a correlation. When the value of r is close to 1 or -1, the probability of correlation is higher, but when the value of r is close to 0, the probability of correlation is lower. Generally, when the value of r is larger than 0.05, and the value of Sig. (2-tailed) is less than 0.05, meaning that there is a significant correlation between the two variables. The correlations are shown in Table 2, in which there is a significant correlation between height and chest girth, and between height and arm length. The line of best fit for these two sets of correlations are shown in Fig.3 and Fig.4, respectively.

Table.2 Correlations

Correlations	Height-Chest Girth		Height-Arm Length	
	Shandong	Hubei	Shandong	Shandong
r (Pearson correlation)	0.735**	0.769**	1**	1**
Sig.(2-tailed)	0.005	0.000	0.000	0.000

Fig.3 shows the correlation curve between height and chest girth of adult males in the two provinces. Among Fig.3, the red curve represents Hubei Province, whereas the blue curve represents Shandong province. The two black lines are the lines of best fit of the two provinces.

Among them, the fitting relation-expression of Hubei province is y=0.5925x+87.24, and the correlation coefficient is 0.5925, which shows a high correlation. The fitting relation-expression of Shandong province is y=0.4143x+93.664, and the correlation coefficient is 0.4143, which shows a low correlation. The correlation in the two provinces shows inconsistency. Therefore, chest girth cannot be considered as being correlated to height. In contrast, chest girth increases as height increases, which shows a slight positive correlation.

Fig.4 shows the correlation curve between arm-length and height in Hubei province and Shandong province. In Fig.4, the red curve represents Hubei Province, whereas the blue curve represents Shandong province. The two black lines are the lines of best fit of the two provinces. The fitting relation-expression of Hubei province and Shandong province is y=1.1x+70.2, y=1.1x+70.7, respectively, and the correlation coefficient for both provinces is 1.1, showing a very high correlation. In other words, the correlation of arm length and height in the two provinces shows consistency. There is a significant positive correlation between arm-length and height in Hubei and Shandong Province, and the correlation coefficient is 1.1. In addition, when comparing adult males of the same height in the two provinces, the average arm length of men in Shandong province is higher than that of men in Hubei province. The figures also showed that adult men in Shandong province have longer arms.

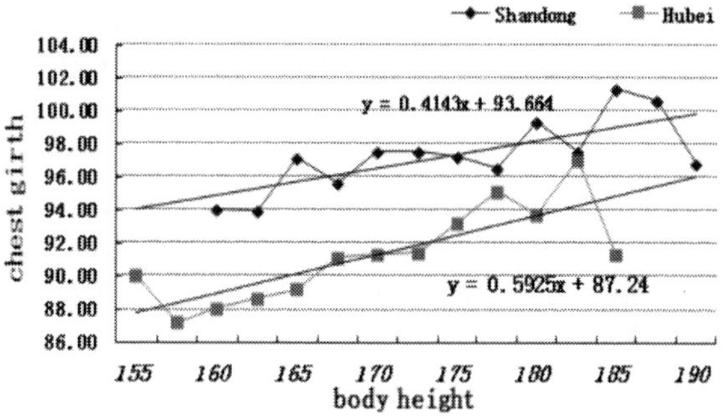

Fig.3 Body Height vs Chest Girth

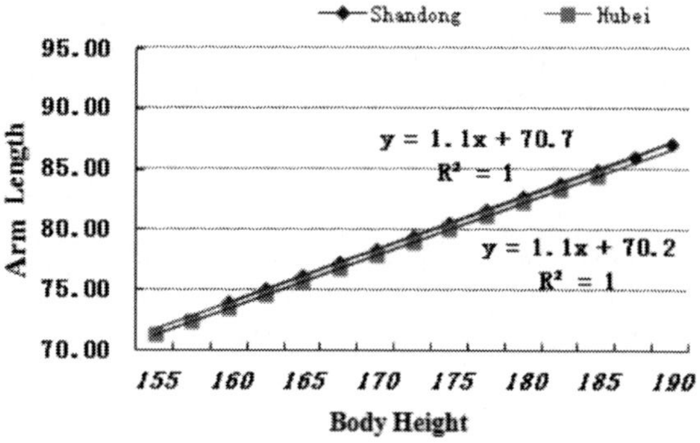

Fig.4 Body Height vs Arm Length

In summary, adult men in Shandong province are taller and stronger in overall. The correlation between height and chest circumference in Hubei and Shandong province is not consistent. It cannot be said that an adult man's chest girth is correlated to height. The correlation between height and arm-length in the two provinces is consistent, which shows a significant positive correlation, and the correlation coefficient is 1.1. That is to say, arm-length is mainly affected by height, whereas geographical and climatic factors have little effect on arm-length.

4. Comparison of the Basic Pattern

An average of the data collected was obtained, which reflects the mean body size of all subjects who participated. In addition, picking up samples that occupied the largest proportion and calculating the average value of these samples, which we refer to here as "the central size" as it can better extrude the body's differences in measured areas [11]. This paper aims to study and analyze the male's body, and compare the body pattern between Hubei province and Shandong province. Thus, this paper uses the central size as the comparative standard. The central size of the two provinces is shown in Table 3. The central size of chest girth in Hubei province is 90.88 cm, whereas height is 168.95 cm and back length is 41.3cm. The central size of chest girth in Shandong province is 97.49 cm, whereas height is 172.75 cm and back length is 42.9 cm. Table 3 also shows that adult male in Shandong province is taller and stronger than men in Hubei.

Table.3 The central size of Hubei/ Shandong province (cm)

Area	Chest	Height	Back length
Hubei	90.88	168.95	41.3
Shandong	97.49	172.75	42.9
D-value	6.61	3.8	1.6

At present, there are three kinds of most popular male patterns: male Donghua prototype [12], Liu Ruipu pattern [13] and cultural prototype [14]. Three parameters of the body need to be measured in order to draw the Donghua prototype, which are height, chest and back length. When comparing the appearances of the three patterns, the Donghua prototype shows an A profile garment, while the other two pattern shows a H profile. The Donghua pattern has a shoulder dart and breast dart [15]. Besides, the neck line of the Donghua prototype is more fitting for human's neck, and it is also more convenient to use it. All in all, Donghua pattern has a higher fitness than the other patterns. Therefore, the Donghua pattern was set as an invariant in the process of comparsion. The Donghua pattern is shown in Fig.5. Using the CAD graphics software [16] to draw their pattern, a comparison of the three patterns is shown in Fig.6.

Fig.6 shows a comparison among the patterns in terms of central size: the red pattern represents Hubei province and the blue pattern represents Shandong province. In overall, the pattern of Hubei area looks like a mini image of that of Shandong province. It is mainly due to the smaller central size of Hubei province. There is an apparent difference in the values of the pattern's length, width, chest girth and neck girth, and meanwhile a smaller difference in the values of armhole's depth and width and shoulder's width.. The drawing formula shows that pattern's length is dependent on back length. The length of front-waste equals to 0.1h+9, and there is an obvious difference in height value between the two province's central size, which is the reason causing the difference of pattern's length. Relatively, armhole's depth and width vary little. The length of armhole curve is 50 cm and 50.89 cm in Hubei and Shandong province, respectively, with a small difference, 0.8 cm, between the two values. Armhole's depth is determined by height vertically, while width is determined by chest loose, chest girth, width of front chest and width of back chest. Among them, their loose is consistent, and the width of front/back chest is concerned with chest girth. All in all, chest girth has little effect on pattern's armhole. The length of shoulder-impulse and chest girth mainly affect the size of the shoulders, Table 3 shows that the difference in values of chest girth reached 6.61 cm, while that of shoulders is only 0.56 cm. It indicates that the difference in value of chest girth has little effect on pattern's shoulders.

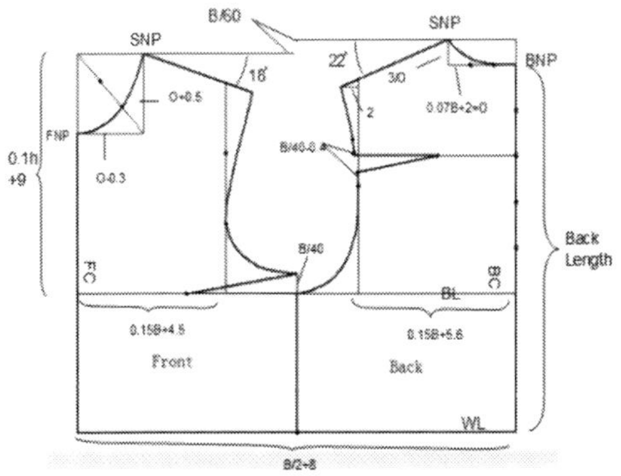

Fig.5 Donghua prototype

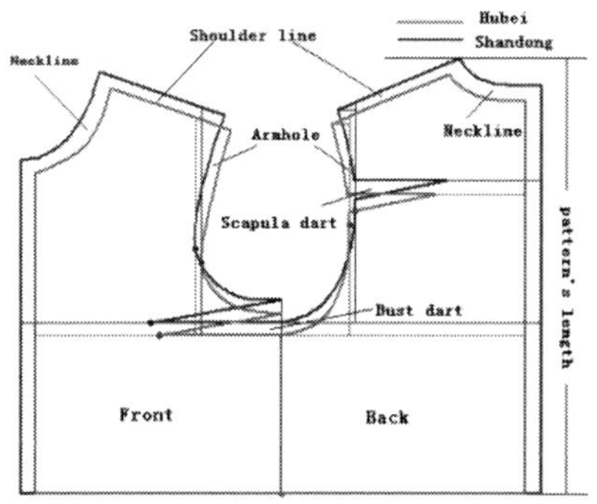

Fig.6 pattern's Comparison in central size

5. Conclusion

Through the above experiments and analyses, the following conclusions are reached:

1) From the data collected, it was demonstrated that men from Shandong province are taller and stronger than those in Hubel province in overall. The height range of adult males in Hubei province is from 155 cm to 185 cm, whereas the chest is in the range of 88.66 cm to 97.04 cm, The range of arm length is from 71.30 cm to 84.50 cm, and most Hubei men's height is in the range of 165 cm to 173 cm. The height range of adult males in Shandong area is from 160 cm to 190 cm, whereas the chest is in the range of 94.00 cm to 101.40 cm. The arm length ranges from 74.00 cm to 87.20 cm, and most Shandong men's height is in the range of 170 cm to 175 cm. However, the height distribution of adult men in Shandong is more disperse, whereas that of Hubei adult men is more compact.

2) The chest-height correlation of the two regions is inconsistent, so it cannot be concluded that adult men's chest girth is correlated with height. In addition, the correlation between height and arm-length of adult males in the two provinces is consistent. The fitting relation-expression of arm length and height in Hubei and Shandong province is y=1.1x+70.2 and y=1.1x+70.7, respectively, and the correlation coefficients of both provinces are 1.1, showing a highly positive correlation. That is to say, arm-length is mainly affected by height, and that geography and climate have little effect on arm length.

3) When comparing patterns, the pattern of Hubei province looks like a minor image of Shandong province in overall. Differences between adult men's body have a larger effect on the pattern's length/width/neckline, while they have little impact on armhole and shoulders.

References

[1] Yang M, Liu GL. Difference analysis of body shape between young and southern males: Journal of Beijing Institute Of Fashion Technology: 2014; 1: 53-57.
[2] Zhong AH. Introduction to modern clothing: Hubei Science and Technology Press: 2016; 5: 256-258.
[3] Hong D. Garment standard and its application: China Textile Press: 2009; 4: 27-29.
[4] Zhu QN, Yang ZT. A comparative study on the somatotype of adult men in East China and Northeast China: Journal of Donghua University: 2004; 3: 24-25.
[5] Sheng Z, Xie SQ, Pan CY. Probability theory and mathematical statistics: Higher Education Press: 2013; 7: 128-317.
[6] Xu DS. Sports statistics: People's Physical Culture Publishing House: 1985; 5: 25-26.
[7] Wang AH. Research on men's body size and setting of male jacket specification in three areas of China based on clothing MTM: Donghua University Press: 2008; 11: 143-148.
[8] Yu SY. National Computer Rank Examination two level tutorial: Higher Education Press: 2013; 5: 231-250.
[9] Teng C, Wang TQ. SPSS statistical analysis: Wuhan university press: 2014; 2: 160-210.
[10] Dong YX, Zhang WB. Prototype's construction of 35-54 year old male in Tianjin area: Tianjin Textile Science & Technology: 2010; 11: 40-47.
[11] Xie L. Research and case of garment structure design: Shanghai science and Technology Press: 2005; 5: 12-36
[12] Zhang WB. Garment structure design: China Textile Press: 2008; 8: 38-54.
[13] Liu RP. Principle and application of pattern design for men's clothing: China Textile Press: 2008; 10: 115-137.
[14] San J. Fashion modeling (Theoretical articles): China Textile Press: 2008; 11: 143-148.
[15] Wang QM, Zhang WB. Research on the prototype and technology of men's clothing based on human body characteristics in Zhejiang Province: Journal of Donghua University Press: 2003; 12: 30-34.
[16] Luo CY. Practical guide to CAD plate making (Second Edition): Posts and Telecommunications Press: 2012; 2: 157-180.

Subdivision of Female Body Types for Recommended Clothes Sizes in Online Shopping

Yi-Di Liu, Rong Zheng*, Ye-Fu Zhang, Yun-Chao Zhang, Hai-Jing Liu

Aimer Human Engineering Research Centre, Beijing Institute of Fashion Technology, Beijing, 100029, China

*Corresponding author's email: yuki.zheng@gmail.com

Abstract

In online clothes shopping, a considerable group of consumers have the problem in selecting a proper clothes size that fits well. Based on 80 measurement items of 2,637 females' scanned anthropometric data, the study, via a principal component analysis, extracted five major factors covering fitness, height, shoulder, angle & protrusion, and real torso length from principal components to describe female body shapes, with a cumulative contribution rate of 69.054%. While retaining such key measurement items as height, bust circumference and waist circumference in China's current clothing sizing system, from the five main factors, four characteristic indexes, including back shoulder width, front body protrusion difference, back body protrusion difference and back length (backing plate), were selected to subdivide female body shapes into 81 types. Finally, the paper described the prospect for size recommendation in online clothes shopping based on the subdivision method.

Keywords: Three-dimensional Anthropometric Data; Body Shape Classification; Female; Size Recommendation

1. Introduction

As the e-business booms, online shopping has become lots of people's choice with their demands in this regard increasing yearly. How to select a proper size in the online clothing shopping website has become a major issue that consumers are facing. [1] Apparel sizing studies [2] show that about 50% of consumers failed to find their formfitting apparel. 30% of returns in online clothes shopping can be ascribed to unsatisfactory fit of the specific clothing item, constituting the top reason for online clothes shopping returns [3]. According to survey data by Zitex & Store-online, currently 77.9% of consumers have the need for apparel fitting. However, due to insufficient knowledge about their own body shapes and various clothes sizing systems, consumers cannot easily select the suitable sizes that fit their bodies well [4]. For clothing manufacturers, garment fitting is always one of the important goals to achieve [5].

Regarding studies on human body shapes, different countries vary in selecting factors to classify body shapes [6]. American ASTM standard takes age, height, weight and girths of different parts as factors to categorize body types. With bust girth difference and hip girth difference of different heights, Japan classifies female body shapes into four types, Y, A, AB and B. Germany takes a similar method with Japan to divide female human body shapes. China issued GB/T1335-91 standard about body types in 1991, which was later revised in 1997 and 2008 [7]. Chinese standard classifies body shapes into four types of Y, A, B and C based on mathematical statistics of the difference between chest and waist circumference. By setting the middle body shape, values of all control parts are ermined. The commonly used sizing and body type collocation is often similar to the following: one size with one body type, one size with multiple body types, and multiple sizes with one body type. Scholars with Yang Yunchu as the representative introduced the torso volume index to vaguely subdivide female body shapes [8]. Regarding size recommendation, after first acquiring two-dimensinal measurement data of such body shape factors as height, neck breath and neck depth, Mao Lianzhong [9] employed the Back Propagation (BP) Neural Network to convert the two-dimensional measurement data into the three-dimensional ones. Based on existing clothes size recommendation theories, He Li [10] introduced Pignet-Vervaeck index evaluating weight fitness, and considering problems in existing studies proposed a size recommendation model after comparing the bust, waist and hip circumferences of various body types with representative garment styles. Compared with domestic size recommendation studies, some developed countries are quite mature in this regard. FitMe.com headquartered in American Silican Valley developed a system by combining 15 measurement items with a multi-tier arrangement system and explanatory notes, the system can select the optimum sizes for customers in light of the principle of choosing the closest grade.

The above literature review leads us to the following conclusions: First, despite a multitude of body shape classifications at home and abroad, considering the complexity of human body shapes, existing classification approaches can't cover all body shapes. Second, overseas studies on clothes size recommendation have reached a high level, with most of them based on relatively scientific body shape data. In contrast, domestic studies in this regard are still at the infancy stage, no effective size recommendation system targeting Chinese people's body shapes is available. With all these in mind, the paper aims to give a relatively comprehensive body shape classification based on Chinese females' 3-D anthropometric data, and give size recommendation for online clothes shoppers.

2. Approach

2.1 Acquisition of Anthropometric Data

The paper selected anthropometric data of 2,637 female objects with their age ranging from 18 to 79 from the basic anthropometric database of Aimer Human Engineering Research Centre affiliated with Beijing Institute of Fashion Technology. By using the two software of Shape Anthropometric System 1.2 [11] and Rapidform XOR3 [12], the study acquired 80 two-dimensional and three-dimensional female body shape measurement items: 16 height items (including body height and waist position height), 14 girth items (including bust girth, waist girth and hip girth), 17 real length items (including length SP2-BP and sleeve crown height SP1), 13 breadth items (neck root breadth and shoulder width), 8 depth items (including BP depth and UBL depth), and 12 angle and protrusion items (right shoulder's slant angle, upper body axis angle and abdominal protrusion).

2.2 Body Shape Data Analysis

A preliminary analysis of the extracted body shape data demonstrated correlation among most measurement items. In order to facilitate the process of selecting characteristic indexes to subdivide body shapes, the paper adopted a dimension reduction method to make principal component analysis of 80 measurement items of the 2,637 female body samples, aiming to acquire fewer measurement factors to categorize female body shapes.

Via SPSS Statistics, the paper made principal component analysis of the selected data. After comparison of different factors, 5 principal components were finally selected to describe body shapes, with a cumulative contribution rate of 69.054% (see table 1)

Table.1 Total variance explained

Factor	1	2	3	4	5
Initial eigenvalue	27.967	14.833	5.644	3.513	3.287
Proportion of measurement	37	19	7	10	5
Rotation squared and loaded	26.218	15.618	4.489	4.488	4.430
Variance interpretation degree%	32.773	19.523	5.611	5.610	5.537
Variance accumulation contribution	32.773	52.296	57.907	63.517	69.054

The first principal component's cumulative contribution rate was 32.773%, covering circumference, breadth and depth items of key body parts including waist circumference (load value 0.950), axillary point circumference (0.950), bust circumference (0.928), under-breast circumference (0.925). As these items are related to a body's weight status, the first principal component was thus named as fitness factor.

The cumulative contribution rate of the second principal component was 19.523%, covering height-related measurement items such as waist position height (0.964), shoulder point height (SP2) (0.954), and shoulder point height (SP1) (0.953). Therefore, it's called height factor.

The third principal component (cumulative contribution rate of 5.611%) includes such shoulder-shape-related measurement items, including shoulder width SP2 (0.875), back shoulder width SP2BNPSP2 (0.871), right chest shoulder width SP2-FNP (0.756), hence called shoulder factor.

The fourth principal component (cumulative contribution rate of 5.61%) contains such factors concerning body angles and protrusions as front abdominal protrusion (0.769), the superior angle of the gluteus protrusion (0.742), t gluteal protrusion (0.703), therefore called angle and protrusion factor.

The fifth principal component (cumulative contribution rate of 5.537%) includes measurement items related to the torso's real length like back length (backing plate) (0.849), Nude back length (0.826), side length WL armpit (0.823), so named as real torso length factor.

3. Results and Discussion

3.1 Classification of Human Body Types

Generally speaking, in selecting characteristic factors to classify body shapes, two aspects should be considered. First, the factor should be able to truly reflect and cover differences of body shapes; second, from the perspective of garment manufacturing, the factor should correlated to parameters used in clothes pattern making. Based on the two principles, in the following section, body shape classification factors would be selected from the results of factor division.

3.1.1 Major Type Division of Body Shapes

Fitness factor (the first principal component) together with height factor (the second principal component) constitutes a cumulative contribution rate of 52.296%, far much higher than the later three principal components. This indicates that fitness and height of key body parts are two important factors influencing body shape. Therefore, corresponding measurement items in the two factors can be selected as characteristic indexes to classify body shapes into major categories. In Female Clothing Size and Type GB/T1335.2-2008, sizing is based on body height, and the type is based on bust girth (waist girth); difference between breast and waist girth was taken as the factor to classify body shapes into type Y/A/B/C. Domestically, garment are mostly designed and manufactured in light of the GB standard. To ensure the recommended size corresponding to the real sizing system, this paper follows the classification approach used in the GB standard in classifying major body shape types.

To take the upper body size and type for example, based on the definition of the upper body size and type in the GB standard, the study gathered the coverage frequency of various size-type combinations among the 2,637 samples according to height and bust girth grades. It was found that into the 160/84A category, the most number of samples fall, at 145.

3.1.2 Subdivision of Body Types

China's GB sizing standard only gives a rough classification of body types, based on which people can specify their suitable size category with their height, breast girth and waist girth values. However human body is an extremely complicated curvy body, not a regular cylinder, and clothes can't only be taken as an exterior wrapper of the human body. From the perspective of clothing fittingness, the sizing system only based on height, breast and waist girth, can not truly reflect states of human bodies. In real life, although a considerable number of people can find their size from the GB sizing system, by wearing the apparel of the specific size, they often find fitness problems involving the shoulder, belly, and buttock parts or even the clothes length. Therefore, it's necessary to subdived body types by taking the relationship between clothes and anthropometric data into account.

In the principal component analysis, the latter three principal components of shoulder factor, angle and protrusion factor and real torso length factor respectively represent morphological characteristics of the body's shoulder, profile protrusion and torso proportion. Taking into account the relationship between clothing and the human body, the three components constitute major consideration factors in garment pattern making. Hence, the paper selected measurement items from the three components as body type subdivision indexes.

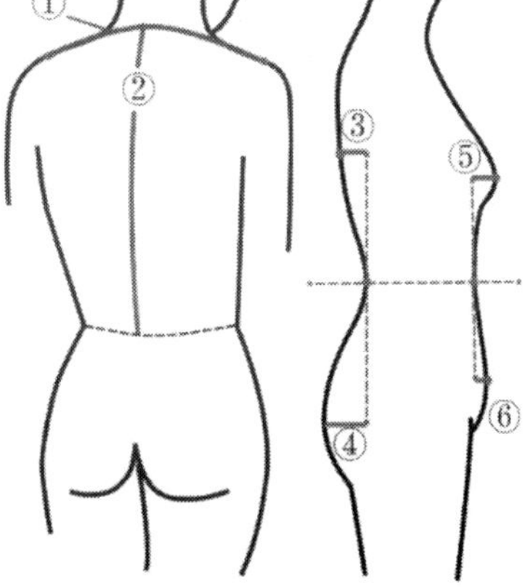

NO.	Measurements	Definition
①	Back shoulder width	The surface curve length between the two SP2 points, passing the back cervical vertebra point
②	Back length (backing plate)	Being nude, with the backing plate on the back, on the torso's back central line, the surface surve length from the back cervical vertebra
③	Thoracodorsal protrusion	The horizontal vector quanitity of the vertical line made from the back protrusion point to the intersection point between the horizental waist line and the back profile curve
④	Gluteal protrusion	The horizontal vector quanitity of the vertical line made from the gluteal protrusion point to the intersection point between the horizental waist line and the back profile curve
⑤	Prothoracic protrusion	The horizontal vector quanitity of the vertical line made from the bust point to the intersection point between the horizental waist line and the profile front curve
⑥	Abdominal protrusion	The horizontal vector quanitity of the vertical line made from the abdominal protrusion point to the intersection point between the horizental waist line and the profile front curve

Fig.1 Names and measurement instruction of characteristic indexes and related diagrammatic drawings

A comprehensive consideration decided the study to select back shoulder width (0.871) from the third principal component as the characteristic index for the shoulder shape subdivision; front abdominal protrusion (0.769), gluteal protrusion (0.703), thoracodorsal protrusion (0.453), and prothoracic protrusion (0.177) from the fourth principal component as indexes reflecting the human body's profile shape; and central back length (backing plate) (0.948) as the characteristic index reflecting torso proportion (as shown in Fig.1)

Back shoulder width (indicated with S) in the third principal component mainly reflects shoulder shape. It's taken as the first characteristic index. Based on its value, shoulder can be subdivided into narrow type S1, standard type S2, and wide type S3.

In the fourth principal component, prothoracic protrusion and abdominal protrusion together give a comprehensive reflection of the front body curve decided by the side body thickness. For simplification, the difference value obtained by using the prothoracic protrusion depth value to subtract the abdominal protrusion depth value is taken as the second characteristic index, named as front body protrusion difference (indicated with PF). In light of its value, the body shape can be classified as abdominal protrusion type PF1, standard type PF2 and chest protrusion type PF3. The third subdivision index is achieved by using the hip protrusion value to subtract the thoracodorsal protrusion value, reflecting the back body curve, therefore named as back body protrusion difference (PB). In light of the PB value, the body shape is subdivided into three types, back protrusion type PB1, standard type PB2 and hip protrusion type PB3.

The back length (backing plate) (L), mainly reflecting length of human torso's upper part, is the fourth subdivision index. Based on its value, the body shape can be subdived into the three types of short upper body L1, standard shape L2 and long upper body L3.

The subdivided types with the classification indexes are shown in Table 2. By permutation and combination of the subdivided types, finally all subdivision types of female body shapes were obtained, $Q=3\times3\times3\times3=81$. That is, with the four classification characteristic indexes, the paper subdivided female body shapes into 81 types.

As for the method about demarcating the critical value that decides the subdivision body types, in order to conform to the principle that in apparel manufacturing the standard size should meet a certain coverage rate, it can be supposed that grade 2 (standard type) has a relatively high coverage rate, and the other two grades have relatively low coverage rates. For example, with $[M-\mu, M+\mu]$ to demarcate the value range for the standard type, according to normal distribution probability graph, the standard grade can cover 68.27% of population with the other two grades covering around 15.87% of people respectively. Manufacturers or clothes brands can specify the critical value for subdivision types in light of their target customer groups.

Table.2 Combination of body shape subdivision indexes

Measurement indexes	Subdivided class 1	Subdivided class 2	Subdivided class 3
Back shoulder width (S)	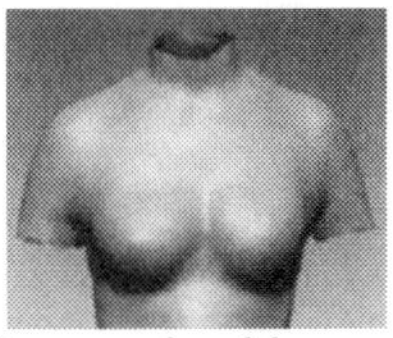 Narrow-shoulder type (S1)	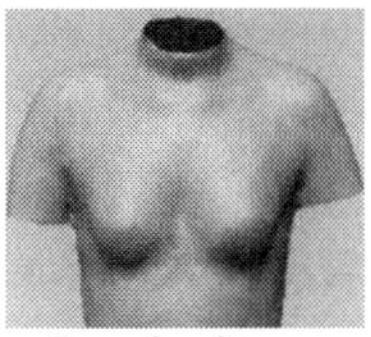 Standard type (S2)	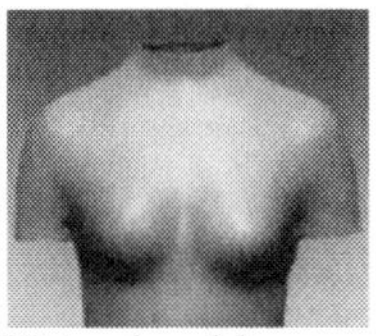 Wide-shoulder type (S3)
Front body protrusion (PF)	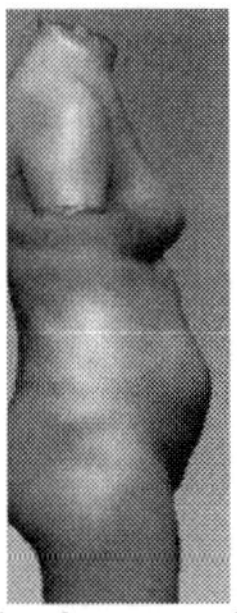 Abdominal protrusion type (PF1)	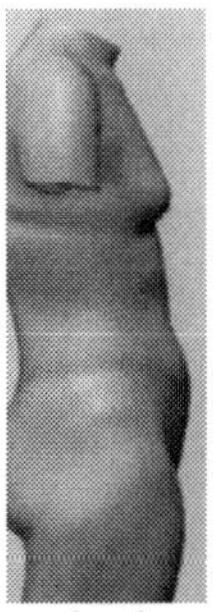 Standard type (PF2)	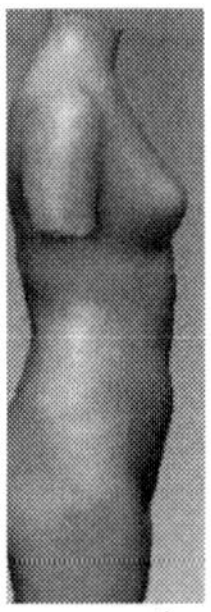 Chest protrusion type (PF3)
Back body protrusion (PB)	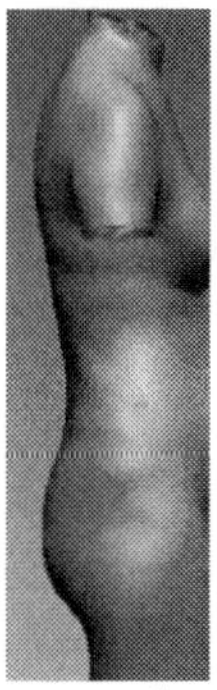 Back protrusion type (PB1)	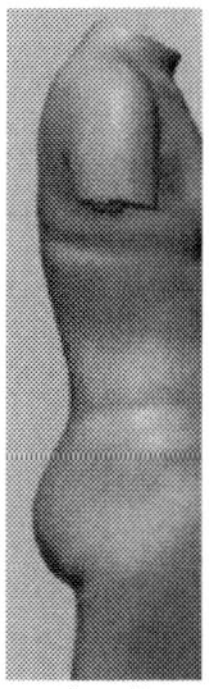 Standard type (PB2)	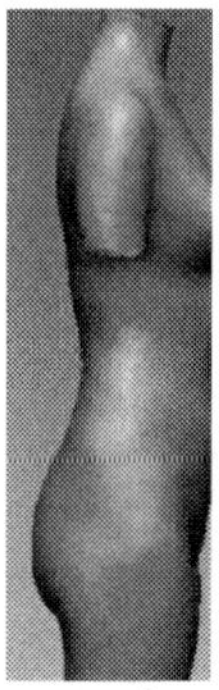 Hip protrusion type (PB3)
Back length (L)	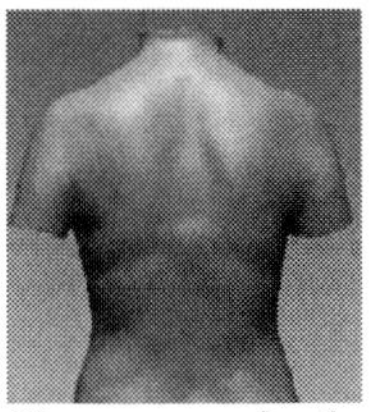 Short upper body (L1)	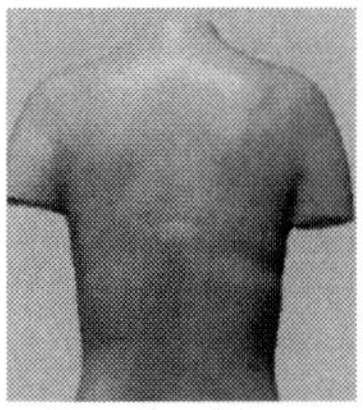 Standard type (L2)	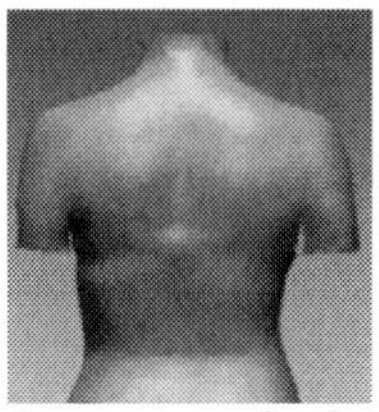 Long upper body (L3)

3.2 Prospects for Apparel Size Recommendation in Online Shopping

With regard to apparels of mass production, for a pattern of apparel, each size only corresponds to a group of fixed measurement specifications. The specific size is formed on the basis of data of relatively standard body shapes falling into the size category. Then the apparel is designed and the pattern is made based on these specifications.

Therefore, many people have found that those clothing items corresponding to their size are not fit, which makes it necessary to try multiple sizes. Then they can find the optimum size. However, in online clothes shopping, there is no way for a customer to really try different sizes, so a relatively scientific size recommendation is needed.

Based on the above-mentioned body type subdivision approach, a more reasonable clothes size recommendation mechanism can be established, targeting the existing sizing system of China's GB standard. To take the female top as an example, the following steps can be taken to provide a customer a proper size recommendation: 1) The system first identifies the customer's anthropometric data; 2) The website service supporter matches the customer with a proper top size based on his or her height, bust and waist girth; 3) service supporters decide the customer's subdivision body type according to his or her four characteristic indexes of back shoulder width, back length (backing plate), front body protrusion difference and back body protrusion difference; 4) by comparing the anthropometric data of the subdivision body type with specifications of the top's different sizes, a matching rule regarding small size, standard size and big size can be established; 5) then the optimum size to the customer's body shape is finally recommended.

4. Conclusion

Prompted by the phenomenon that in online clothes shopping it's hard for consumers to accurately select a fit size, the study made a statistic analysis of 2,637 domestic females' 3-D anthropometric data. In light of the analysis results, on the basis of China's GB clothes standard's major size classification, four characteristic indexes were selected to subdivide body types. After permutation and combination, 81 subdivision body types were obtained. The subdivision approach can finally provide guidance in size recommendation for online clothes shoppers, which will make it possible for online shoppers to more accurately and conveniently select a fit clothes size despite that they can't really try it. In the future, the study plans to establish a complete effective recommendation system by integrating detailed recommendation plans on women's tops, trousers, skirts, dresses and jumpsuits.

References

[1] Chen X. Study on design of visual servo contro system for multi-aerocyst flexible fitting robot. Donghua University: 2014.

[2] Ashdown S P. Review of a Research Program in Apparel Sizing and Fit. Transactions of the Korean Society of Mechanical Engineers A: 2006.

[3] Zhang SY, Zou FY, Ding XJ, Chi JM. Research on the young women's body classification based on SVM.J Journal of Zhejiang Sci-Tech University: 2008; P. 41: 45.

[4] Chen WF. A study on the women's body shaped for clothed fit. Donghua University: 2001.

[5] Ulrich PV, Brannon EL, Anderson-Connell LJ. A consumer-driven model for mass customization in the apparel market. Journal of Fashion Marketing & Management: 2002; p. 240: 258.

[6] Song Q, Zhang YX, A study about chinese female body shape classification approaches. Shaanxi Textile: 2008; p. 21: 22.

[7] GB T 1335.2-1997 Size designation of clothes-Women

[8] Yang YC, Chen MZ, Zou FY. Female body shape parameters analysis based on 3-D scan data. Journal of Textile Research: 2009; p.117: 122.

[9] Mao LZ. The research of recommendation model for clothing size based on image. Donghua University: 2016.

[10] He L. Study on anthropometry and size recommending of clothing marketing online. Jiangnan University: 2008.

[11] Zheng R, Zhang H, Huang HQ, Li J. A kind of anthropometrics analysis method. CHN Patent No. 200410009356.8: 2007.

[12] Wang Y. Development of in minor mannequin based on 3D body shape analysis. Master's degree thesis of Beijing Institute of Fashion Technology: 2010.

Somatotype Factors of Male Actors' Lower Body Parts During Dramatic Movements in Lahunqiang Opera

Bo-Xuan Zhang, Rong Zheng*, Meng-Lin Wang, Wen Wu

Aimer Human Engineering Research Center, Beijing Institute of Fashion Technology, Beijing, 100029, China

*Corresponding author's email: yuki.zheng@gmail.com

Abstract

As a dramatic art form performed locally in the Xuzhou-Huaihai area, *Lahunqiang* opera is characterized as both a tangible and intangible cultural heritage. However, current studies on Lahunqiang opera costumes mostly examines design features (e.g., colour, patterns, accessories) and seldom explore relationships between costumes and actors. This study examined the effect of Lahunqiang opera costumes on five common physical movements of 10 male actors. Through data collection and analysis, the study extracted five factors (crotch, length, left knee, right knee and waist-hip) to describe opera actors' physical movements. The study adds to the scant research on Lahunqiang opera costumes and provides a reference point for further studies on the heritage, protection and development of local drama costumes.

Keywords: Lahunqiang Opera; Xuzhou-Huaihai Area; Heritage; Colourful Pants; Somatotype Factors of Lower Body

1. Introduction

Chinese opera has a long history in China and has developed gradually over thousands of years into a complete drama system. Lahunqiang opera is a local dramatic art form dating back to the reign of Emperor Qianlong and Jiaqing of Qing Dynasty. In its 200-year history, Lahunqiang opera has become known for its special vocal music and is popular in the area of Jiangsu, Shandong, Henan and Anhui Province. It can be classified into Sizhou drama, Liuqin drama and Huaihai drama. UNESCO first listed Sizhou and Liuqin dramas as part of China's national intangible cultural heritage in 2006, while Huaihai drama was listed in 2008 [1]. Old costumes that are preserved and/or used represent tangible heritage [2]. Also, because stage design is closely related to costumes, and ways of dressing and making costumes represent intangible heritage, costumes are an important part of the dramatic arts.

Because Lahunqiang opera is a local drama, there are relatively few studies on the topic that approach it from the perspective of history, including regional artistic features and development. In his master's thesis titled Application of Chinese Traditional Opera Costume Factors in Modern Clothes Design, Chen ZH analyzed a costumes' general characteristics and aesthetic significance [3]. Song JH's doctoral dissertation titled Study on Chinese Ancient Costumes conducted a detailed analysis from the perspective of history of ancient costumes, characteristics of costumes during different periods, relationship among costumes, historical culture and daily clothes [4]. Wan YZ's Analysis of Sizhou Drama is a comprehensive study of the origin, development, artistic characteristics, present situation and future prospects of Sizhou drama [5], while Xue L's Research Review of Lahunqiang Opera summarizes previous studies on the origin, popularity, genre, and vocal and artistic characteristics of Lahunqiang opera [6]. In Innovation and Development of Sizhou Drama, Wei LL also explored the origin and artistic characteristics, development and innovation of Sizhou drama [7], while Zong S classified and summarized four perspectives of traditional drama characteristics of Liuqin drama [8]. In five chapters of the doctoral dissertation From Lahunqiang Opera to Liuqin Drama, Kong PP made detailed and persuasive statements on the historical process of heritage, change and development of Lahunqiang opera's Liuqin drama [9].

From the above studies, we can find that: 1) Several studies have examined Sizhou drama and Liuqin drama but most of them focused on the origin and development, musical characteristics and heritage and protection of intangible cultural aspects of such dramatic forms; 2) Heritage-related factors refer largely to human beings, including dressing and display of costumes; 3) In drama, male characters in martial arts roles have significant movements of lower limbs; and 4) Physical movements result in skin extension changes in different parts of the body. When covering the body, clothes may impede physical movements. Particularly in dramas, costumes not only should be aesthetically pleasing but also allow freedom of body movements, which necessitates further study of the relationship between actors' somatotype and clothes. Based on the above, this study examined male actors who have more than five years of performance experience in Lahunqiang opera in the Xuzhou-Huaihai area as its research subjects and explored the relationship between male actors and clothes in Lahunqiang opera through a study of the Somatotype factors of the actors' lower body parts.

2. Approach

2.1 Acquisition of Dynamic and Static Somatotype Data of Male Lahunqiang Opera Actors' Lower Body Parts

This experiment combined manual measurement, photo taking and 3D scanning to collect data on the lower body parts of 10 male actors who have more than five years of Lahunqiang opera performance experience in the Xuzhou-Huaihai area. The equipment included: weighing machine, Martin measuring instrument, measuring tape, Artec Eva 3D scanner and SLR camera. To ensure correctness of measuring data and 3D scanning, testees all wore measuring shorts, had bare upper bodies and had no footwear. The measuring shorts did not constrain the human body nor were they too loose so as not to produce folds.

2.2 Measuring Method and Approach

This experiment mainly involved two data measurements: static standing condition (abbreviated as static standing A0) and classic movements of Lahunqiang opera static condition (abbreviated as movement static body). Static standing A0 refers to standard standing posture of the testee; movement static body refers to standard classic movements in static condition. Static standing A0 used contact measurement and movement static body used non-contact measurement.

2.2.1 Contact Measurement

We marked the points on the body with pen according to baseline and point of measurement, then used measuring tape and Martin measuring instrument to measure. After measurement, we kept the points unchanged and performed non-contact measurement.

2.2.2 Non-contact Measurement

Non-contact measurement in this experiment involved 3D scanning with the Artec Eva 3D portable scanner. Before scanning, the researchers determined the testee's movement and the testee needed to stay still during scanning.

2.3 Selection and Extraction of Experiment Posture

Static standing A0 was measured according to standing posture of national standard GB5703-1999 [10].

According to features of Lahunqiang opera, static physical data was collected using 3D method with common basic movements, including static standing, horse stance, knee lifting, lunge and backward stretch (see Table 1) and the movements were categorised 3D-scanning postures as follows:

1) Straight: stand with heels close to each other; keep breathing and look straight ahead;

2) Horse stance: legs stand parallel with 3-foot distance between feet; squat, 45-50cm from top of knee to the ground;

3) Knee lifting: left foot on the ground and right foot on a 50cm round chair; shank perpendicular to thigh;

4) Lunge: left leg take a big step forward, bend the knee to make shank perpendicular to the ground; right leg keep straight and sole on the ground;

5) Backward stretch: left leg perpendicular to the ground and right leg lift back to 90°; keep right instep on the table; set a 1-metre tripod 20cm from 45° of left foot and the testee's hands crossed on it.

Table.1 Setting 3D scanning posture

	A1. Straight	B. Horse stance	C. Knee lifting	D. Lunge	E. Backward stretch
Classic movements					
Posture capture and setting				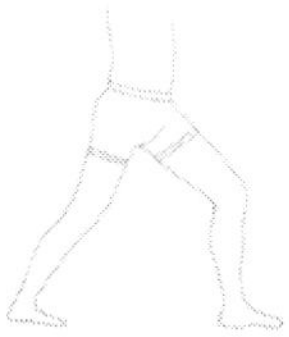	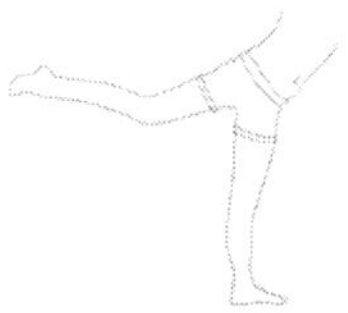

Table.2 Static standing A0 measurement item

	Height measurement item		BWH measurement item		Thickness measurement item
1	Height	15	Horizontal waistline	31	Horizontal waistline
2	Back cervical spine point	16	Abdominal girth	32	Abdominal front point
3	BP point	17	Hipline	33	Buttock back point
4	Horizontal waistline	18	Leg root girth	34	Biggest thigh girth
5	Abdominal front point	19	Biggest thigh girth	35	Biggest shank girth
6	Iliac crest bone front point	20	6cm-above-knee girth		Length measurement item
7	Buttock back point	21	Middle point girth of knee	36	Back
8	Trochanterion point	22	6cm-below-knee girth	37	Front and back upper crotch
9	Navel point	23	Biggest shank girth	38	Front midline
10	Lower crotch	24	Smallest shank girth	39	Back midline
11	Biggest thigh girth		Width measuring item		Other measurement items
12	Knee middle point	25	Horizontal waistline	40	Weight
13	Biggest shank girth	26	Abdominal front point	41	Bust and waist variance
14	Outer ankle	27	Buttock back point	42	BMI
	Age:	28	Trochanterion point		
	Years of practicing opera	29	Biggest thigh girth		
	Years of performance experience	30	Biggest shank girth		

At the same time, for scanning and reducing data error, we normalised the posture of testees' upper part of the body. When doing postures A1, B, C and D, we kept testees' upper part of the body perpendicular to the ground and hands crossed at the chest; when doing posture E, we kept the upper part of the body straight, 130°-135° from left leg.

In the experiment, we first used Martin measuring instrument and measuring tape to measure static standing A0; then the testees adopted postures A1, B, C, D and E and we scanned each posture with the 3D scanner.

2.4 Measuring Items and Definition

2.4.1 Standing Still A0 Measuring Item

Static standing A0 measurement identified 42 parameters closely related to somatotype of the lower part of the body and clothes. Measuring points and baselines were chosen in accordance to national standard GB5703-1999 [10] to reduce measuring error. Table 2 shows body measurement items. Measurement diagrams are shown in Figs.1-4.

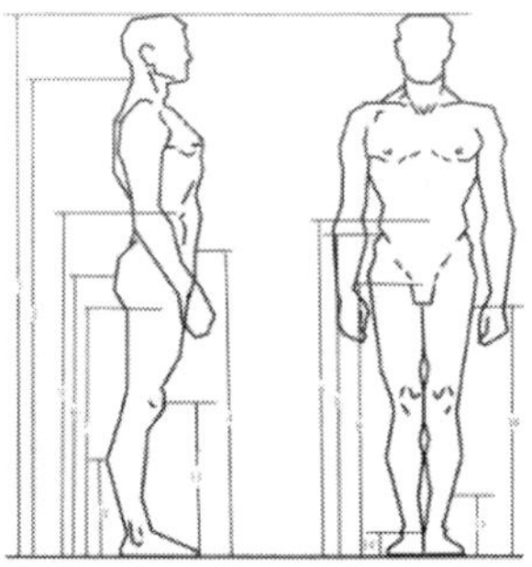

Side view Front view

Fig.1 Height measuring item sketch.

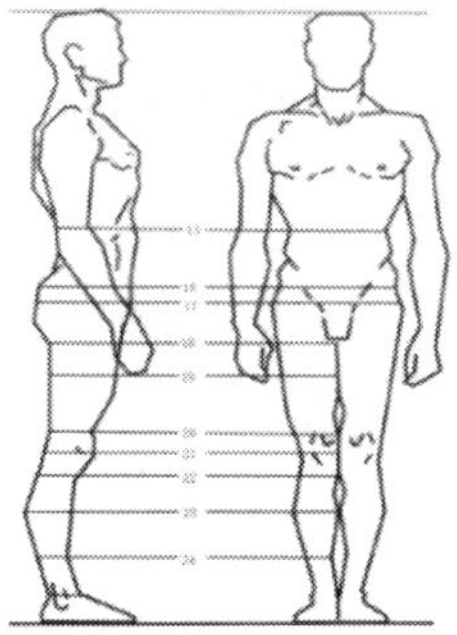

Side view Front view

Fig.2 Girth measuring item sketch.

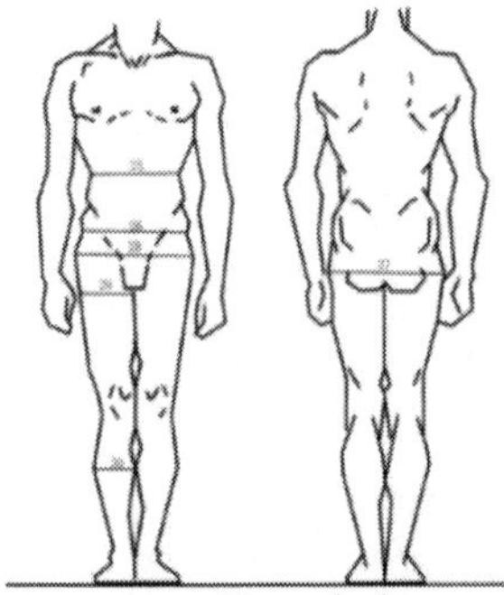

Front view Dorsal view

Fig.3 Width measuring item sketch.

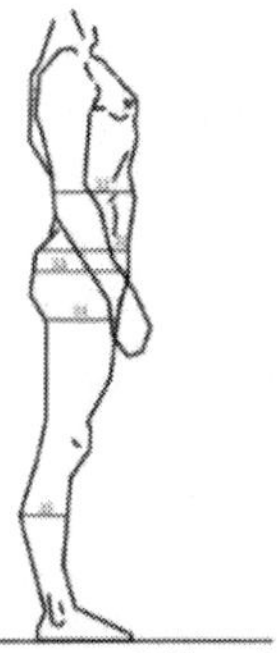

Side view

Fig.4 Thickness measuring item sketch.

2.4.2 Movement Static Body Measurement Items

In order to identify changes in the human body under the condition of static movement, on the basis of the national standard measuring line, the experiment increased some points and lines when in static standing A0 before collection data of movement static body. Points and lines are shown in Fig.5.

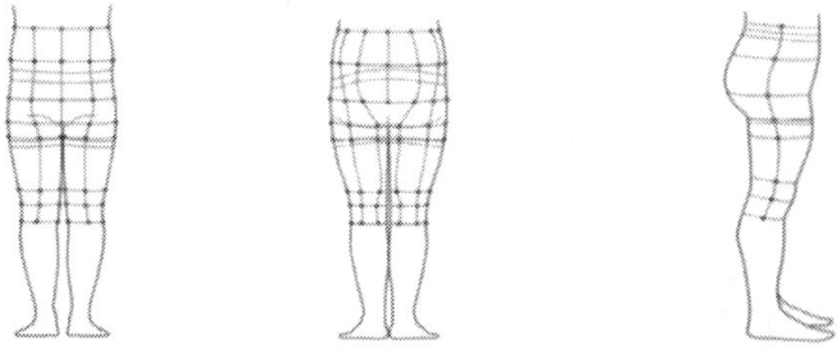

Front view Dorsal view Side view

Fig.5 Experiment gauge point and measurement line.

Girth line: from top to bottom are WL, MHL, HL, FL, FDL, KL, 6cm KUL and 6cm KDL;

Vertical line: front centre line L0, back centre line L5, right line L2, left line L8; the line halves from front waistline, front middle buttock girth and front buttock girth to KDL girth line, right leg L1, left leg L9; the line trisects from back waist girth, back middle buttock girth and back buttock girth to KDL, from right to left are L3, L4, L6, L7; inner side line of right leg is marked LR and that of left leg is marked LL.

3. Results and Discussion

3.1 Static Standing A0 Somatotype Data Statistical Analysis

First, SPSS software was used to analyse extracted data and found that each item is in accordance with normal distribution.

3.1.1 Somatotype Data Descriptive Statistical Analysis

From frequency analysis, we can see that volunteer age distribution is on average 20-29, 30-39, 40-49 and 50-59, accounting for 20%, 20%, 30% and 30%, respectively.

Descriptive analysis results of 42 measurement items is shown in Table 3. Measurement units of items 1-39 and 41 is cm; item 40 is kg.

3.2 Static Movement in Bare State Surface Data Analysis

3.2.1 Experiment Data Preprocessing

As shown in Table 4, The name of body surface measurement line is given. L0 is front centre line; L5 is back centre line. Horizontal and vertical data name and measurement examples (WL and L2) are shown in Fig.6.

At the same time, code standing, horse stance, knee lifting, lunge and backward stretch as A1, B, C, D, E and bare movement A1, B, C, D, E as LA1, LB, LC, LD, LE. Body girth of movement static condition and length data change rate equals value of static movement minus A1 and divides A1. For example, B's change rate=100*(LB-LA1)/LA1.

We used SPSS to do basic tests and get rid of singular value; preprocessed length and girth change rate (part) is shown in Table 5.

Table.3 Experimented somatotype data descriptive statistical analysis (n=10)

No	Measurement item	Distance	Minimum	Maximum	Average	Standard deviation	α value
1	Height	19.50	159.70	179.20	167.90	5.40	0.85
2	Back cervical spine point height	17.00	135.60	152.60	143.27	4.71	0.84
3	Buttock back point height	9.60	80.50	90.10	84.14	2.81	0.84
4	Lower crotch height	13.70	71.20	84.90	76.59	3.83	0.84
5	Biggest thigh girth height	11.50	65.00	76.50	69.36	3.39	0.85
6	Biggest shank girth height	4.00	32.30	36.30	33.71	1.34	0.85
7	Smallest shank girth height	4.70	10.50	15.20	13.12	1.59	0.85
8	Horizontal waistline height	13.20	102.10	115.30	106.96	4.04	0.84
9	Navel point height	15.30	94.40	109.70	99.89	4.02	0.85
10	Abdominal front point height	13.30	91.20	104.50	95.90	3.79	0.84
11	Iliac crest bone front point height	12.80	89.00	101.80	93.01	3.67	0.85
12	Trochanterion point height	11.00	84.00	95.00	87.67	3.30	0.84
13	Knee middle point height	5.80	45.80	51.60	47.19	1.71	0.85
14	Outer ankle height	1.20	6.00	7.20	6.72	0.53	0.85
15	Horizontal waistline	17.20	75.60	92.80	84.02	5.70	0.86
16	Abdominal girth	17.80	76.70	94.50	86.49	5.31	0.85
17	Hipline	8.30	90.20	98.50	95.26	3.01	0.84

18	Leg root girth	6.90	53.40	60.30	55.94	2.41	0.84
19	Biggest thigh girth	6.60	49.90	56.50	53.87	2.17	0.85
20	6cm-above-knee girth	5.00	37.90	42.90	40.29	1.56	0.85
21	Middle point girth of knee	6.90	34.80	41.70	37.84	1.73	0.85
22	6cm-below-knee girth	5.60	31.40	37.00	33.96	1.65	0.85
23	Biggest shank girth	10.30	27.40	37.70	35.18	2.94	0.85
24	Smallest shank girth	2.10	20.80	22.90	21.96	0.69	0.85
25	Horizontal waistline width	5.00	25.50	30.50	28.00	1.51	0.85
26	Abdominal front point width	4.10	28.30	32.40	30.26	1.42	0.85
27	Buttock back point width	2.20	31.60	33.80	32.88	0.76	0.85
28	Trochanterion point width	2.30	30.90	33.20	32.18	0.77	0.85
29	Biggest thigh girth width	2.90	14.40	17.30	15.86	0.89	0.85
30	Biggest shank girth width	1.60	10.20	11.80	11.09	0.54	0.85
31	Horizontal waistline thickness	6.70	19.50	26.20	22.45	2.49	0.85
32	Abdominal front point thickness	7.10	19.50	26.60	22.70	2.46	0.85
33	Buttock back point thickness	5.60	20.20	25.80	22.93	2.14	0.85
34	Biggest thigh girth thickness	11.30	15.50	26.80	17.49	3.35	0.85
35	Biggest shank girth thickness	1.90	10.00	11.90	11.31	0.58	0.85
36	Back length	4.40	37.00	41.40	38.73	1.42	0.85
37	Front and back upper crotch length	15.50	76.00	91.50	86.03	4.70	0.83
38	Front midline length	7.10	8.50	15.60	13.03	1.78	0.85
39	Back midline length	5.30	21.40	26.70	23.72	2.10	0.84
40	Weight	14.50	57.60	72.10	66.54	4.28	0.84
41	Bust and waist variance	13.01	1.51	14.52	7.55	4.58	0.86
42	BMI	7.58	19.71	27.29	23.68	2.21	0.85

Table.4 Horizontal and vertical data measurement items and name

Measurement item	Measurement item name						
Waistline WL	WLL0,WLL1,WLL2,WLL3,WLL4,WLL5,WLL6,WLL7,WLL8,WLL9						
Middle hipline MHL	MHLL0,MHLL1,MHLL2,MHLL3,MHLL4,MHLL5,MHLL6,MHLL7,MHLL8,MHLL9						
Hipline HL	HLL0,HLL1,HLL2,HLL3,HLL4,HLL5,HLL6,HLL7,HLL8,HLL9						
Thigh root girth FL	FLL0,FLL1,FLL2,FLL3,FLL4,FLL5,FLL6,FLL7,FLL8,FLL9						
Biggest thigh girth FDL	FDLL0,FDLL1,FDLL2,FDLL3,FDLL4,FDLL5,FDLL6,FDLL7,FDLL8,FDLL9						
6cm girth from knee KUL	KULL0,KULL1,KULL2,KULL3,KULL4,KULL5,KULL6,KULL7,KULL8, KULL9,KULL						
Knee girth KL	KLL0,KLL1,KLL2,KLL3,KLL4,KLL5,KLL6,KLL7,KLL8,KLL9						
6cm girth below knee KDL	KDLL0,KDLL1,KDLL2,KDLL3,KDLL4,KDLL5,KDLL6,KDLL7,KDLL8, KDLL9						
L0	L0WL	L0MHL	L0HL				
L1	L1WL	L1MHL	L1HL	L1FL	L1FDL	L1KUL	L1KL
L2	L2WL	L2MHL	L2HL	L2FL	L2FDL	L2KUL	L2KL
L3	L3WL	L3MHL	L3HL	L3FL	L3FDL	L3KUL	L3KL
L4	L4WL	L4MHL	L4HL	L4FL	L4FDL	L4KUL	L4KL
L5	L5WL	L5MHL	L5HL	L5			
L6	L6WL	L6MHL	L6HL	L6FL	L6FDL	L6KUL	L6KL
L7	L7WL	L7MHL	L7HL	L7FL	L7FDL	L7KUL	L7KL
L8	L8WL	L8MHL	L8HL	L8FL	L8FDL	L8KUL	L8KL
L9	L9WL	L9MHL	L9HL	L9FL	L9FDL	L9KUL	L9KL

Table.5 Change rate of body length and girth of different movements (parts)

Measurement item	Leg movement B	Leg movement C	Leg movement D	Leg movement E
WLL0	2.27	3.68	0.36	-3.6
WLL1	0.79	-0.27	1.7	8.74
WLL2	5.45	-0.54	1.96	6.57
WLL3	2.99	4.06	1.28	4.13
WLL4	-1.54	-0.44	0.01	6.88
WLL5	-0.8	2.34	-1.49	1.64
WLL6	-1.37	-1.86	3.75	-1.98
WLL7	1.45	-0.61	1.89	1.09
WLL8	1	-0.29	-3.47	-2
WLL9	1.18	2.04	0.98	-5.41
WL	0.46	0.69	0.21	0.3
HLL0	7.85	19.88	7.56	10.96
HLL1	3.08	3.02	4.87	4.08
HLL2	5.83	9.91	9.41	13.45
HLL3	4.78	6.71	-1.67	4.22
HLL4	-1.03	9.66	3.55	2.85
HLL5	4.94	16.1	6.58	0.97
HLL6	1.15	3.13	-4.65	3.27
HLL7	7.64	9.26	4.46	3.46
HLL8	3.75	3.22	4.59	6.3
HLL9	8.53	-1.96	8.78	-1.39
HL	4.62	8.31	4.94	4.05
L1WL	-7.81	-9.97	-6.57	-3.78
L1MHL	-8.71	-49.57	-10.07	-8.5
L1HL	-9.4	-60.09	4.18	6.65
L1FL	11.65	-11.99	8.73	8.31
L1FDL	9.3	13.53	3.71	5.28
L1KUL	11.68	45.44	8.12	9.6
L1KL	8.18	41.03	11.08	18.5
L1	-5.72	-10.07	1.64	3.22
L3WL	-6.55	-12.15	-16.25	-17.44
L3MHL	1.58	3.77	0.78	0.6
L3HL	-3.44	10.69	-22.66	-13.65
L3FL	9.05	21.6	-3.36	-7.34
L3FDL	-6.06	10.19	-3.88	-2.48
L3KUL	-12.7	-30.29	-7.16	-13.28
L3KL	-16	-59.35	-1.72	-4.59
L3	-6.33	-3.33	-6.41	-5.8

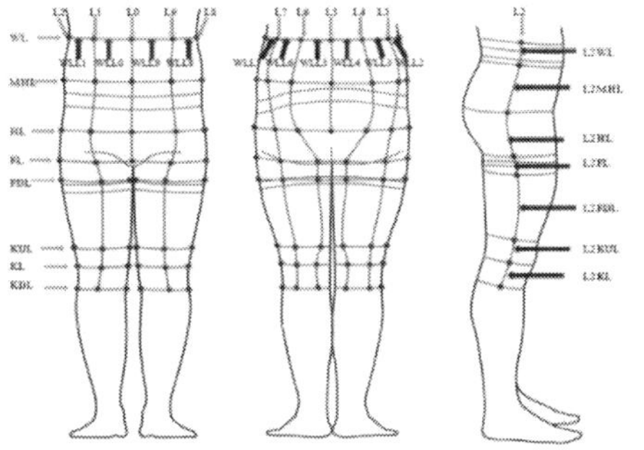

Fig.6 Horizontal and vertical data name and measuring sketch.

3.2.2 Crucial Body Parts

From bivariate analysis of girth and length change rate (see Table 6), we can see that waistline and buttock girth show high correlativity $0.8<|r|$ and L0 and L1 show middle correlativity $0.5\leq|r|\leq0.8$, therefore we can use factor analysis to analyse major elements and extract crucial change in parts of the body during movements.

From the horizontal and vertical variation change rate line charts shown in Figs.7 and 8, we can see the change of every part of the body in static movements. According to change rate distribution and major factors of pants design, we extracted 23 items including waistline girth, middle buttock girth, buttock girth, thigh root girth, biggest thigh girth, knee girth and L0-L1 and used SPSS to undertake major factor analysis of change rate of major measurement items. In these factors, we identified five factors with eigenvalue greater than 1, whose explained variance account for 78.874% of total variance; therefore, we are sure that these five factors can reflect change of body parts. As shown in Table 7, after the maximum variance rotation, we discovered that:

1) In first major element variance, load capacity of FLR, FLL, FDLR, FDLL, LO and L5 is comparatively high, named crotch factor;

2) In second major element variance, high load capacity items are L3, L4, L8 and L9, respectively two trisection lines of right back, side line of left leg and front half line of left leg, named length factor;

3) In third and fourth element variance, high load capacity items include KULR, KLR, KDLR, KLL and KDLL, respectively 6cm up right knee girth, knee girth, 6cm from knee girth, left knee girth and 6cm from knee girth, named knee factor;

4) In fifth major element variance, high load capacity items are waistline girth, middle buttock girth and buttock girth, named waist-hip factor.

It is apparent that waistline girth, buttock girth, thigh root girth, biggest thigh girth and knee girth reflect body girth change and that side line, crotch line and half line of front and back thigh reflect body length change.

Table.6 Correlativity of parts of measuring items

Pearson correlativity	WL	HL	FLR	KLR	L0	L1	L2	L3	L4
WL	1								
HL	0.839	1							
FLR	0.940	0.950	1						
KLR	0.896	0.950	0.993	1					
L0	-0.490	-0.146	-0.194	-0.077	1				
L1	-0.948	-0.806	-0.843	-0.775	0.685	1			
L2	-0.961	-0.740	-0.909	-0.879	0.353	0.830	1		
L3	0.853	0.929	0.976	0.995	0.022	-0.706	-0.857	1	
L4	0.885	0.867	0.968	0.979	-0.029	-0.714	-0.923	0.984	1

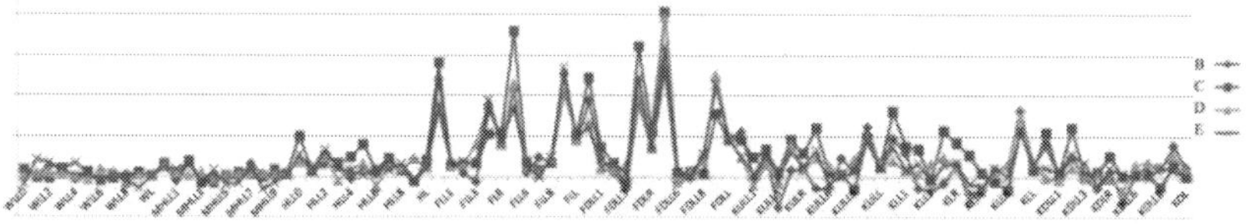

Fig.7 Horizontal variation change rate line chart.

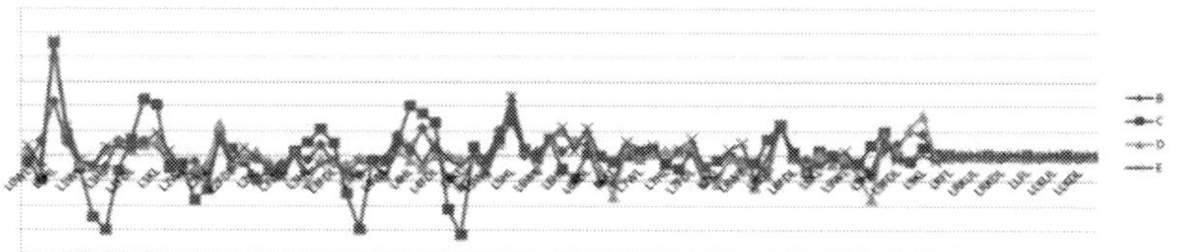

Fig.8 Vertical variation change rate line chart.

Table.7 Major factors and crucial body parts

	Major elements				
	First	Second	Third	Fourth	Fifth
Loaded measurement item	23	23	23	23	23
Value total of initial feature	7.328	4.998	2.217	2.005	1.593
Percentage of variance (%)	31.860	21.731	9.640	8.719	6.924
Accumulated contributing rate (%)	31.860	53.591	63.230	71.950	78.874
High load capacity after maximum rotating (load value)	FLR(0.851), FLL(0.919), FDLR(0.844), FDLL(0.906), L0 (0.703), L5 (0.809)	L3 (0.873), L4 (0.800), L8 (0.901), L9 (0.752)	KULR(0.757), KLR(0.825), KDLR (0.781)	KLL(0.698), KDL (0.780)	WL(0.928), MHL(0.900), HL (0.528)

3.2.3 Extraction of Parts Obviously Influenced by Movement

After determining crucial part size, we did variance analysis of single factor of each part and extracted the parts change most influenced by movements. From single factor variance analysis shown in Table 8, we can see that p value of buttock girth, thigh root girth, biggest thigh girth and front and back crotch line is smaller than 0.05. Lower limb movement has obvious influence on these crucial factors.

Table.8 Single factor variance analysis table

Measurement item	Degree of freedom	Variance	F test	Obviousness
WL	10	36.66	1.84	0.15
MHL	10	34.47	0.75	0.52
HL	10	39.71	3.65	0.02
FLR	10	12.92	1.46	0.44
FLL	10	6.49	5.87	0.02
FDLR	10	7.58	1.25	0.24
FDLL	10	7.73	5.38	0.00
KULR	10	4.22	0.66	0.58
KLR	10	5.49	1.65	0.19
KLL	10	2.72	1.38	0.26
KDLR	10	2.67	0.28	0.83
KDLL	10	1.52	0.63	0.59
L0	10	15.00	2.99	0.04
L3	10	17.11	1.09	0.36
L4	10	17.61	1.01	0.40
L5	10	6.47	2.83	0.05
L8	10	13.79	1.14	0.34
L9	10	11.88	1.17	0.33

4. Conclusion

This study combined qualitative and quantitative research methods, from the perspective of ergonomic 3D somatotype data and 42 static standing somatotype data of 10 male Lahunqiang opera actors aged 20 to 60 with Martin measurement and 3D scanning.

Through data analysis, 1) five factors reflecting body part change were extracted: crotch factor, length factor, right knee factor, left knee factor and waist-hip factor, with a total contribution rate of 78.874%, among which waistline girth, buttock girth, thigh root girth, biggest thigh girth and knee girth reflect girth change of the body and side line, crotch line, front and back thigh half line reflect length change of the body. 2) from single factor variance analysis, we can see buttock girth, thigh root girth, biggest thigh girth and front and back crotch line have obvious influence on lower limb movement of Lahunqiang opera.

Due to the limited number of subjects in our experiment, the study sample is small and cannot be generalised for statistical analysis of all Lahunqiang opera actors. We also used random sampling for our experiment. Future studies could increase the sample size to determine overall somatotype features of Lahunqiang opera male actors and provide more effective data for costume design and development.

References

[1] China Intangible Cultural Heritage: 2016-12-23. Information on http://www.ihchina.cn/54/54_1.html

[2] Song JH. New road to intangible cultural heritage and opera study. Art Study: 2007; 2: 96-104.

[3] Chen ZH. Application of Chinese traditional opera costume factors in modern clothes design (master's thesis). Shanxi Normal University: 2014.

[4] Song JH. Study on Chinese ancient costumes (doctoral dissertation). Zhongshan University: 2002.

[5] Wan YZ. Analysis of Sizhou drama. Anhui, China: Anhui People's Publishing House: 1963.

[6] Xue L. Research review of Lahunqiang opera. J Nanjing U Sci Tech (Soc Sci): 2008; 6: 105-109.

[7] Wei LL. Innovation and development of Sizhou drama. People's Music: 2009; 6: 32-33.

[8] Zong S. On characteristics and tendency of Liuqin drama. Art Genre: 1993; 4: 68-71.

[9] Kong PP. From Lahunqiang opera to Liuqin drama (doctoral dissertation). Chinese Academy of Arts: 2007.

[10] People's Republic of China national standard GB5703-1999. Body measurement basic items for technical design. Beijing: CSBTS: 1999.

Middle-to-Old-Aged Female Lower Body Shape Classification Based on Trouser Structure

Xia-Xin Liu, Rong Zheng*, Shi-Yang Yu, Wen Wu, Ying Wang

Aimer Human Engineering Research Center, Beijing Institute of Fashion Technology, Beijing, 100029, China

*Corresponding author's email: yuki.zheng@gmail.com

Abstract

China's national standard on female clothes size (GB/T 1335.2-2008) categorizes middle-to-old-aged women's body shapes based on chest-waist differences. While suitable for upper body shape classification in designing tops, the standard fails to classify the lower body shape types. This study first made an exploratory analysis of 73 items of anthropometric data on 185 females aged 45-54, and then it extracted factors related to lower body fitness, height and waist-abdomen-hip difference and curvature to describe lower body shapes, with an accumulative contribution rate of 63.942%. A new approach to lower body shape classification was established, creating a theoretical basis for studies on middle-to-old-aged women's clothes structure and a size-designation reference for trousers among clothes manufacturers.

Keywords: Middle-to-old-aged Women; 3D Anthropometry; Body Shape Classification; Trousers Structure

1. Introduction

By 2025, China's population of people aged 65 and over will reach 270 million, accounting for 18.4% of the world total, and 23% of the global old aged population, this is twice the size of that in Europe. Hence, China will have the world's largest market for goods and services catering to elderly people, which creates a significant potential for growth among China's clothing manufacturers [1]. This study thus can meet middle-to-old-aged people's clothing needs better and provide guidance for garment manufacturers seeking to better serve that market segment.

The body shape classification approach used in China's existing national standard on clothes size designation (GB/T 1335.2-2008) does not encompass all body shape features [2]. For example, while young and elderly women may share the same waist and breast circumference, they may differ in hip and abdomen girth, which the national standard does not incorporate into its body shape classification [3]. Therefore, analyzing anthropometric data on middle-to-old-aged women and determining factors related to body shape classification would help establish more accurate size designations for this segment of the population.

Wu Lindi. [4] in 1992 conducted a study on the universal female body shape change law as their age increases. The 2001 study by Li Xinggang and Shen Weiqin found that middle-to-old-aged women's chest-waist and waist-hip ratios decrease as their age increases; subcutaneous fat stored in their breasts decreases but fat accumulated in other parts of the body increases, particularly in the abdomen and hips [5]. Data in the latter studies also reflect greater changes in middle-to-old-aged women's body shape than in younger women.

Western countries took the lead in studying middle-to-old-aged people's body shapes. James F. Annis's 1996 study focused on the influence of age on weight, height and body shape [6]. In 2001, Kothiyal et al. obtained anthropometric data on individuals above age 65, and calculated the average value, standard deviation, mean, minimum value, maximum value and variable coefficient [7]. The U.S. National Textile Center created a subdivision of body shapes of women aged 19-55 based on anthropometric data obtained by 3-D scanning, from which a size standard targeting a specific group of people was developed [8].

2. Data Collection

China classifies people aged 49-59 as being middle-to-old-aged [9]. The study selected women of this age scope as its subjects. Anthropometric data were gathered using a 3D-scanner, camera and Martin measure instrument. Software programs such as Rapidform XOR3 and Shape Anthropometric System 1.2 then were used to process the anthropometric data to acquire data about body surface [10]. All detailed definitions of parameters were based on GB 5703-2010 [11] and existing studies [12].

In light of the purpose and needs of the study, the paper selected 54 direct measurement items and 19 calculated measurement items closely related to the lower body shape and trousers design.

The 54 direct measurement items (as shown in Tables 1 and 2) include 11 on height (e.g., body height), 10 on girth (e.g., breast girth), 10 on depth (e.g., waistline depth), 7 on width (e.g., waistline width) and 7 on length and angle, with measurement methods shown in Figs.1 and 2.

To better describe the lower body shape, the study transformed the direct measurement data on major height and girth items into ratios and difference values, and consequently obtained ratios for height items of the middle-to-old-aged women and calculated items to describe waist, abdomen and hip shapes, as shown in Table 3.

Table.1 Measurement items (I)

No.	Height items	No.	Girth items	No.	Width items
1	Body	12	Breast (with bra on)	22	Waistline
2	Waistline	13	Waist	23	Front ilium
3	Front Iliac crest bone point	14	Front ilium	24	Salient abdominal point position
4	Abdominal convex point	15	Abdominal convex	25	Hip convex point position
5	Hip convex point	16	Hip	26	Side thigh convex point position
6	Side thigh convex point	17	Side thigh convex part	27	Gluteal fold part (right)
7	In-leg	18	Gluteal fold part (right)	28	Knee girth position (right)
8	Gluteal fold	19	Knee (right)	29	Thickest shank part (right)
9	Central knee point	20	Largest shank (right)	30	Thinnest shank part (right)
10	Thickest shank part	21	Smallest shank (right)		
11	Thinnest shank part				

Table.2 Measurement items (II)

No.	Depth items	No.	Length items	No.	Angle items	No.	Convex items
31	Waistline part	41	Between front and back of crotch	45	Axial angle, lower body	52	Front abdominal
32	Front Iliac crest bone position	42	Between front WL to HL (subplate)	46	Front slant angle, lower body	53	Back hip
33	Abdominal convex	43	Between side WL to HL	47	Back slant angle, lower body	54	Side hip
34	Hip convex	44	Between back WL to HL (subplate)	48	Superior angle, abdominal convex		
35	In-leg			49	Inferior angle, abdominal convex		
36	Protruding side thigh part			50	Superior angle, hip convex		
37	Gluteal fold position (right)			51	Inferior angle, hip convex		
38	Knee girth position						
39	Thickest shank part						
40	Thinnest shank part						

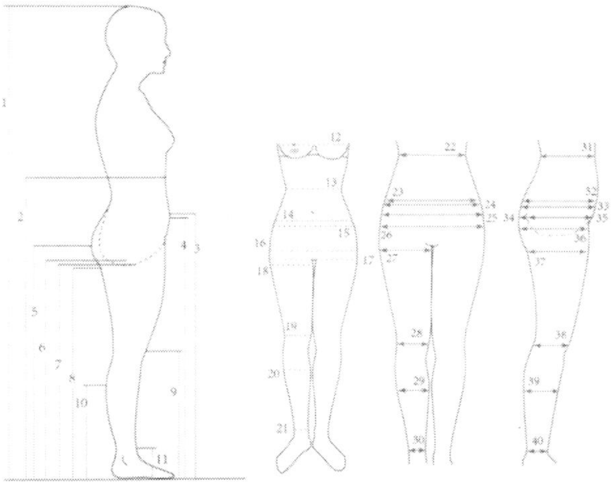

Fig.1 Diagrammatical drawing of measurement items.

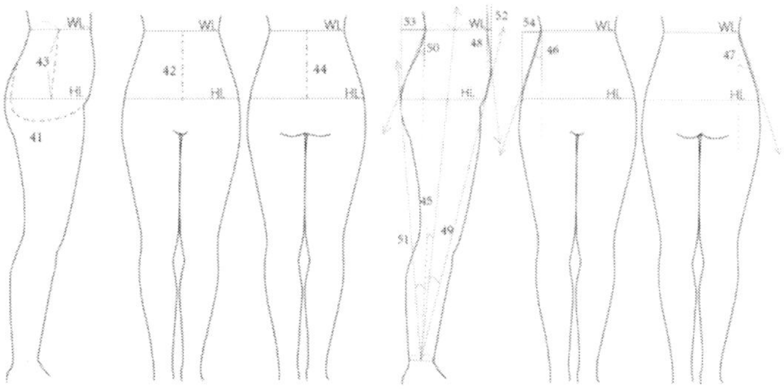

Fig.2 Diagrammatical drawing of measurement items.

Table.3 Calculated items and calculation method

No.	Items	Calculation method
55	Crotch rise	Difference between waistline height and in-leg position height
56	Waistline height/body height	Waistline height to body height ratio
57	Abdomen girth position height/body height	Abdomen girth position height to body height ratio
58	Hip girth position height/body height	Hip girth position height to body height ratio
59	In-leg height/body height	In-leg height to body height ratio
60	Gluteal fold height/body height	Gluteal fold height to body height ratio
61	Knee girth position height/body height	Knee girth position height to body height ratio
62	Thickest shank part height/body height	Thickest shank part height to body height ratio
63	Thinnest shank part height/body height	Thinnest shank part height to body height ratio
64	Waist-hip difference	Difference value between waist and hip girths
65	Waist-abdomen difference	Difference value between waist and abdomen girths
66	Abdomen-hip difference	Difference value between abdomen and hip girths
67	Waist-hip width difference	Difference value between waist and hip widths
68	Waist-abdomen width difference	Difference value between abdomen and waist width
69	Abdomen-hip depth difference	Difference value between hip and abdomen width
70	Waist-hip depth difference	Difference value between hip and waist depth
71	Waist-abdomen depth difference	Difference value between abdomen and waist depth
72	Abdomen-hip depth difference	Difference value between hip and abdomen depth
73	BMI	BMI=body weight kg ÷ body height m^2

3. Data analysis

The study had a tentative analysis of 73 measurement items. The missing value was set in light of the standard of [M-4μ, M+4μ]. After removing singular values, a Q-Q probability graph was used to test whether the samples conform to the normal distribution.

Considering the subjects' age range (from 45-59), significant body shape differences could exist between them. Therefore, the study first used a one-way ANOVA of the key body parts to identify the target group.

The study categorised the subjects into three groups: one group aged 45-49, one group aged 50-54 and one group aged 55-59. Nine major measurement items closely related to lower body shape-such as waist, abdomen and hip girths—were selected to analyse age-related differences. It was found that major measurement items did not show significant difference in subjects aged 45-54. Therefore, 185 women of this age group were identified as the study's subjects.

3.1 Composition of Samples

Frequency analysis of lower body shape data of the 185 female subjects aged 45-54 was conducted. The frequently seen parameters of BMI and body shape were selected for analysis.

First, subjects' age distribution was analysed. As shown in Fig.3, 2% of subjects were slightly lank (with BMI≤18.5), 55% subjects were at a normal level (18.5≤BMI≤23.9), 33% of subjects were plump (24≤BMI≤27.9) and 10% of subjects were classified as obese (BMI≥28).

The study then conducted a statistical analysis of the distribution of the subjects' breast–waist differences. As shown in Fig.4, the distribution of the four body shapes (Y, A, B and C) was nearly even.

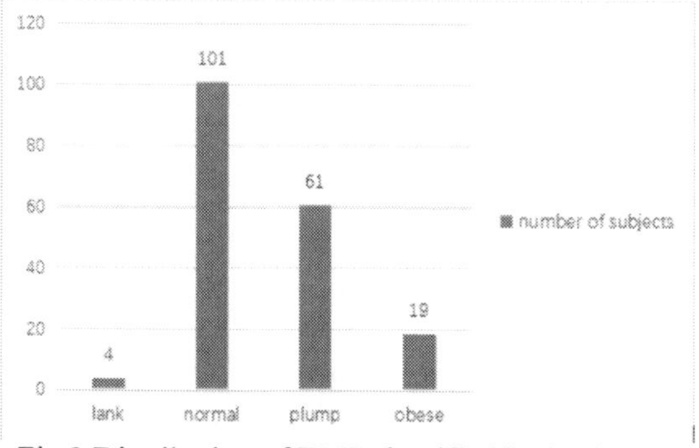

Fig.3 Distribution of BMI-classified body shapes.

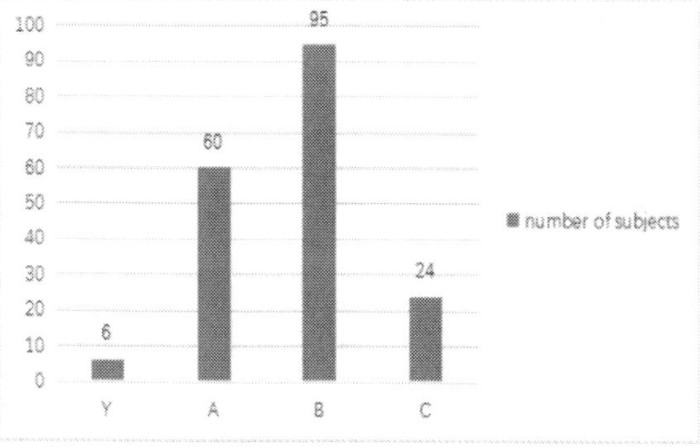

Fig.4 Distribution of breast-waist differences.

GB/T 1335.2-2008 stipulates that the body shapes can be devided by the difference between the chest and waist circumference.The difference between 19 and 24 cm is defined as Y type, between 14 and 28 cm A type,between 9 and 13 cm B type,and between 4 and 8 cm C type.

The analysis of the above three parameters indicated that the age and body shape distribution of the selected subjects was relatively even, which made the further analysis feasible.

3.2 Correlation Analysis

To use as few measurement items as possible to describe body shapes, the study explored the feasibility of dimensionality reduction. First, the Pearson correlation analysis was made with SPSS to determine the linear correlation degree of different measurement items.

Table.4 Correlation analysis of height items (only some measurement items listed)

Pearson correlation	Body height	Waistline height	Front ilium height	Abdominal convex height	Side thigh convex height	Hip convex height	In-leg height	Gluteal fold height	Knee height
Body height	1.00								
Waistline height	0.91	1.00							
Front ilium height	0.86	0.87	1.00						
Abdominal convex height	0.82	0.83	0.80	1.00					
Side thigh convex height	0.80	0.84	0.82	0.73	1.00				
Hip convex height	0.84	0.87	0.85	0.78	0.82	1.00			
In-leg height	0.87	0.91	0.85	0.73	0.83	0.95	1.00		
Gluteal fold height	0.86	0.90	0.88	0.78	0.85	0.94	0.98	1.00	
Knee height	0.86	0.90	0.88	0.83	0.83	0.87	0.90	0.90	1.00

Table.5 Correlation analysis of girth items (only some measurement items listed)

Pearson correlation	Breast girth	Waist girth	Front ilium girth	Abdominal convex girth	Side thigh convex girth	Hip girth	Gluteal fold girth	Knee girth
Breast girth	1.00							
Waist girth	0.90	1.00						
Front ilium girth	0.81	0.86	1.00					
Abdominal convex girth	0.84	0.89	0.95	1.00				
Side thigh convex girth	0.70	0.74	0.85	0.84	1.00			
Hip girth	0.59	0.61	0.75	0.73	0.92	1.00		
Gluteal fold girth	0.61	0.67	0.74	0.73	0.90	0.86	1.00	
Knee girth	0.59	0.66	0.74	0.72	0.79	0.72	0.77	1.00

Tables 4 and 5 show that high and medium-level correlation exists both among girth items (including those for breast, waist, hip and gluteal fold) as well as among height items, including those for waistline, hip, and gluteal fold. Therefore, considering the correlation among these measurement items, a dimensionality reduction approach was adopted to further analyse the data.

3.3 Lower Body Shape Classification

3.3.1 Principal Components Analysis

By means of principal components analysis, the study analysed anthropometric data on the 73 measurement items of the 185 subjects.

First, the study extracted 13 principal components with a respective Eigenvalue of more than 1. From the scree plot of principal components (as shown in Fig.5), it can be observed that characteristics of factors are more obvious when the factor number is 4, as factor distribution is more apparent and it is much easier to explain and name factors; the overall result is superior than when the factor number is not 4. Therefore, the study selected four principal components to describe body shape with the principal factor variance explanation shown in Table 6 and the accumulative contribution rate at 63.942%. After rotation, the factor loading showed a polarisation trend. In light of the factor extraction results after rotation, principal components were named (as shown in Figs.6 to 9).

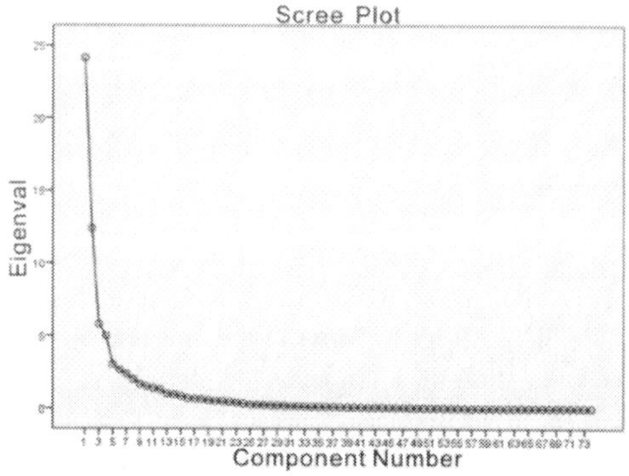

Fig.5 Eigenvalue scree plot.

Table.6 Decomposition diagram of 4 factors' explained variance

Component	Initial Eigenvalues			Extraction of quadratic sum loadings			Rotation of quadratic sum loadings		
	Total	Variance %	Accumulative %	Total	Variance %	Accumulative %	Total	Variance %	Accumulative %
1	24.146	32.63	32.63	24.146	32.63	32.63	22.993	31.071	31.071
2	12.409	16.769	49.398	12.409	16.769	49.398	11.403	15.41	46.481
3	5.779	7.81	57.208	4.779	7.81	57.208	7.697	10.401	56.882
4	4.983	6.733	63.942	4.983	6.733	63.942	63.942	7.059	63.942

The contribution rate of the first principal component was 31.071%, including such factors related to people's fitness degree as hip girth (0.938), body axis (0.910), gluteal fold position girth (0.908), abdominal convex point girth (0.893), the front iliac crest bone point position girth (0.880), side thigh convex girth (0.871) and BMI (0.867). Therefore, the component was named lower body fitness factor.

The contribution rate of the second principal components stood at 15.41%, including factors related to lower body height such as in-leg height (0952), gluteal fold height (0.944), hip height (0.932), crotch rise (0.913), waistline height (0.859) and abdominal convex height (0.788). This component was named lower body height factor.

The contribution rate of the third principal component was 10.401%, covering such measurement items describing the difference of waist, abdomen and hip as waist-hip difference (0.935), difference of waist and hip width (0.901), side hip convex (0.841) and abdomen-hip difference (0.837). The component thus was named waist-hip-abdomen difference factor.

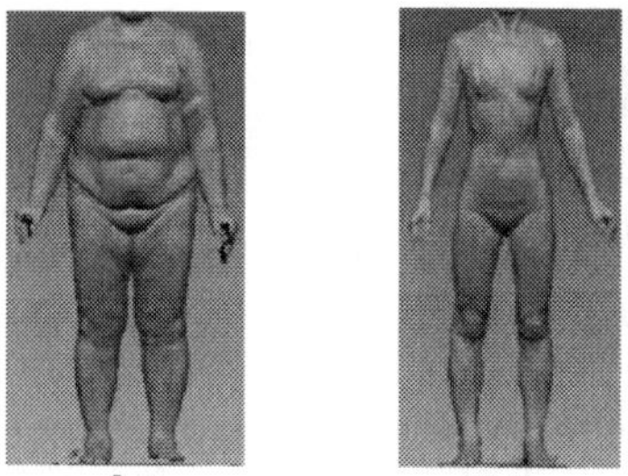

fat thin

Fig.6 Comparison of extreme body shapes in terms of fitness factor.

The contribution rate of the fourth principal component was 7.059%, including such factors reflecting curvature change of waist, abdomen and hip as WL-HL side length (0.741), WL-HL front length (0.704), WL-HL back length (0.666), slant angle of the lower body (0.654), hip convex (-0.644) and abdominal convex (0.557). This component was named waist-abdomen-hip curvature factor.

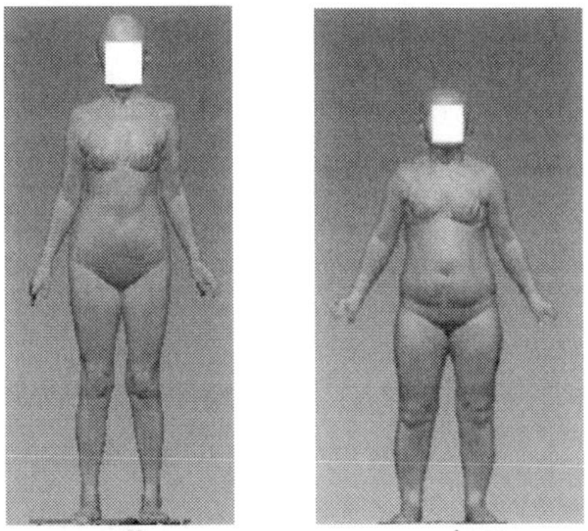

tall short

Fig.7 Comparison of extreme body shape in terms of height factor.

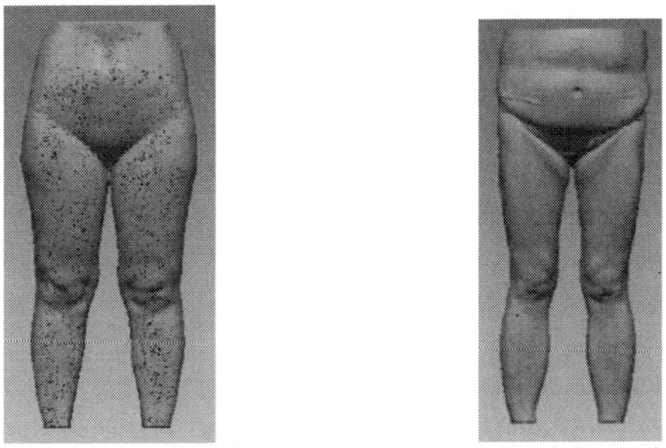

Big waist-abdomen-hip difference Small waist-abdomen-hip difference

Fig.8 Comparison of extreme body shapes in terms of waist-abdomen-hip difference factor.

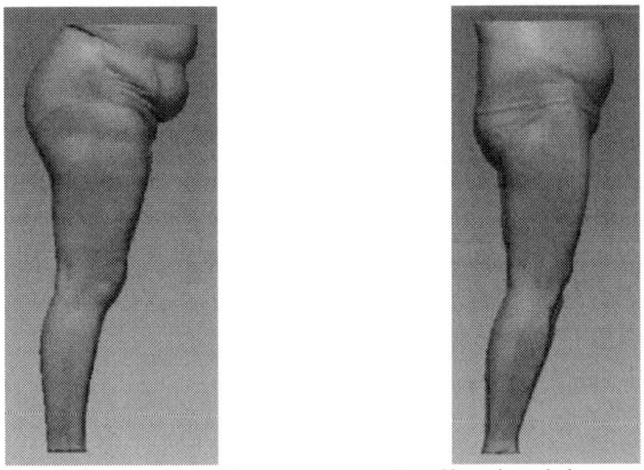

Big waist-abdomen-hip curvature Small waist-abdomen-hip curvature

Fig.9 Comparison of extreme body shapes in terms of waist-abdomen-hip curvature factor.

4. Conclusion

Based on 73 items of anthropometric data on 185 females aged 45-54, the study extracted four principal components-lower body fitness factor, lower body height factor, waist-abdomen-hip difference factor and waist-abdomen-hip curvature factor-via factor analysis to describe lower body shape with the accumulative contribution rate to variance at 63.942%. In comparison with the traditional trousers size designation, our new designation is more inclusive and more accurately describes middle-to-old-aged women's body shapes. This body shape classification approach can serve as a reference for apparel manufacturers and researchers on size designation, and also provide a theoretical basis for apparel structure studies targeting the middle-to-old-aged group.

References

[1] Liang JF, Zhang X, Li YS. Study on the development and strategy of old and middle-aged people clothing market. J Suzhou University: 2002; 22 (1): 86-89.

[2] GB/T 1335.2-2008. Standard sizing systems for garments-Women.

[3] Zhang ZF, Zhang JM. Research of a standard mannequin and the garment industry. J Tex Res: 1994; 15 (5): 45-48.

[4] Wu LD. Research on the body analysis and the laws of the development of Chinese women. J Zhengzhou Institute of Light Industry: 1992; 7: 17-19.

[5] Li XG, Shen WQ. Analysis of the variations of body build of middle-and-old-aged and research on the prototype. J Donghua University: 2001; 27 (5): 121-126.

[6] Anis JF. Aging effects on anthropometric dimensions important to workplace design. Int J Ind Ergonom: 1996; 18 (5): 381-388.

[7] Kothiyal K, Tettey S. Anthropometry for design for the elderly. Int J Occup Saf Ergonom: 2001; 7 (1): 15-34.

[8] Lenda JC. Body scan analysis for fit models based on shape and posture analysis. National Textile Center annual report: 2002; 11.

[9] Information on http://www.who.int/about/zh/34

[10] Zheng R, Zhang H, Huang HJ, Hu XD, Lee J, Duan XJ. A measurement and analysis method of the human body. CN Patent No. ZL200410009356.8: 2007.

[11] GB 5703-2010. Basic human body measurements for technological design.

[12] Wu W, Zheng R. Research on young women's lower body shape and the fit of denim jeans based on 3D scanning data. Beijing Institute of Fashion dissertation database: 2012.

Effect of the Manufacturing Country on Clothing Marketability

KyoungOk Kim [1], Yuta Toyomaru [2], Hong-Wei Li [3, 4], Masayuki Takatera [1*]

[1] *Division of Kansei and Fashion Engineering, Institute for Fiber Engineering (IFES), Interdisciplinary Cluster for Cutting Edge Research (ICCER), Shinshu University, 3-15-1 Tokida, Ueda-shi, Nagano 386-8567, Japan*
[2] *Faculty of Textile Science and Technology, Shinshu University, 3-15-1 Tokida, Ueda-shi, Nagano 386-8567, Japan*
[3] *Interdisciplinary Graduate School of Science and Technology, Shinshu University, 3-15-1 Tokida, Ueda-shi, Nagano 386-8567, Japan*
[4] *Digital Media Course, School of Art Design, Zhejiang Agriculture and Forestry University, Zhejiang, Hangzhou, China*

[*] Corresponding author's email: takatera@shinshu-u.ac.jp

Abstract

The effect of the manufacturing country on garment marketability was investigated by comparing 10 jackets and 10 skirts manufactured by four factories and one sample maker in three countries using five different textiles. We provided the same specifications and sample patterns to each of the factories. The garment marketability was evaluated by 16 Japanese department store buyers using a questionnaire, taking into account garment shape, silhouette, face fabric, sub-materials, anticipated appeal to consumer's claim, and sewing and ironing skills. There was a high correlation between estimated selling price and the evaluation scores in relation to shape, silhouette and face fabric. There were no significant differences between manufacturing countries. The estimated selling prices of garments made with fabric that received a low evaluation were lower than those of other garments. Therefore, it was confirmed that garment quality is not dependent on the manufacturing country because the manufacturing skills of all the selected factories were adequate.

Keywords: Manufacturer; Textile; Clothing Marketability; Department Store Buyer

1. Introduction

Nowadays, garments are both manufactured and sold all over the world [1]. In recent years, at least 80% of the clothing sold in Japan has been manufactured in China [2, 3]. Japanese garment manufacturers have their garments made in China because of the lower labour costs, which are then sold in Japan. These garments are of sufficient quality to sell in Japanese stores as a result of tight quality control by Japanese manufacturers and retailers. Some other developed countries are also employing similar practices. As a result, Japanese apparel-manufacturing companies have been losing their cost competitiveness, and their numbers have decreased by 75% during the period from 1991 to 2013 [2]. The surviving companies are now making either garments with a high level of added value or small runs of garments to enable a quick response to changing demand. Recently, Chinese manufacturers have started to make their own brands of garments, including high-end garments, to sell in China. However, the sewing quality in some of these garments is inadequate for the Japanese market [4]. Marketability of clothing depends on design, textiles used, quality and price. Therefore, the competitiveness of a Japanese company should be clarified for global competition. In this study, we focus on the effect of the manufacturing country on the marketability of garments.

There are many processes undertaken when manufacturing a garment. To make a garment, first, a designer creates an illustrated design and a specification sheet or memo, and then a pattern-maker makes a pattern to represent the design. The pattern is usually checked by making a sample garment. When the sample garment is approved, the specification sheet and the pattern are presented to a factory, where mass production of the garment occurs. Production patterns are created by revising the original pattern so that it is suitable for mass production after taking into consideration the manufacturing efficiency, machining capabilities and workers' skills in the factory in question [5]. Even if the same design illustration is used, the appearance of the final garment can differ depending on the pattern made by the pattern-maker [6]. The combination of patternmaking methods and dress forms' dimensions has an effect on the garment's wearing appearance [7]. Finally, even if the same pattern and specification sheet are provided, the shape and quality of the final garment can differ depending on the manufacturing process used by the factory. The processes involved in making a mass production pattern, selection of material, sewing, and finishing can all differ depending on the factory. In this study, we investigated the effect of the manufacturing country on garment marketability by comparing garments manufactured by different factories in different countries using different materials.

2. Experimental

Two Japanese factories, one Chinese factory, one Polish factory and one individual Japanese sample maker were selected. We provided a sample pattern for a jacket and a skirt based on a designer's illustration to the factories in Japan, China and Poland, and specified the intended target market. Each factory produced both jackets and skirts. We then evaluated the garments produced by each factory.

2.1 Garment Evaluation

The sample jackets and skirts were evaluated by 16 buyers for Japanese department stores. The jackets and skirts were placed on a dress form that matched the specified garment size (French size 38). The subjects evaluated the garments by viewing and touching them and then responding to items in a questionnaire relating to shape, silhouette, face fabric, lining, interlining, button, shoulder pad, anticipated appeal to consumer's claim (below claim), and sewing and ironing skills. Each evaluation item was scored on a five-point scale (0 = poor quality, 4 = very high quality). Information on the place of manufacture and the anticipated selling price was not provided to the evaluators, who were required to provide reasons for each score they gave. Points were accumulated to calculate an average score out of a total score of 100. In addition, we asked the evaluators to estimate the likely selling price and target market among low-price specialty stores, department stores, and luxury brands.

2.2 Experimental Sample

Fig.1 Design illustration [6]

Table 1 shows the specifications of the 10 samples. Samples B–J were assembled using the same pattern. Fig.1 and 2 show the design illustration and the pattern. We added sample A, which was assembled using a pattern made by a different pattern-maker from the same design illustration, as a reference [6]. Samples A, B and C were made from tweed 1. Samples A and B were produced by a Japanese sample maker, individual N, and sample C was made by factory P in Poland, which usually produces French luxury brand clothing. Samples D and F were made by Japanese factory J1, which usually produces women's clothing for a department store brand, and samples E and G were made by Japanese factory J2, which usually produces men's and women's luxury brand clothing. Tweed 2 was used for samples D and E, and worsted fabric was used for samples F and G. Samples H, I and J were made from tweed 3 and tweed 4, assuming different target markets (low-price specialty stores, department stores, and luxury brands).These were made by Chinese factory C, which produces clothing for a range of target markets in China. The wage costs of each factory were assumed based on their usual target market.

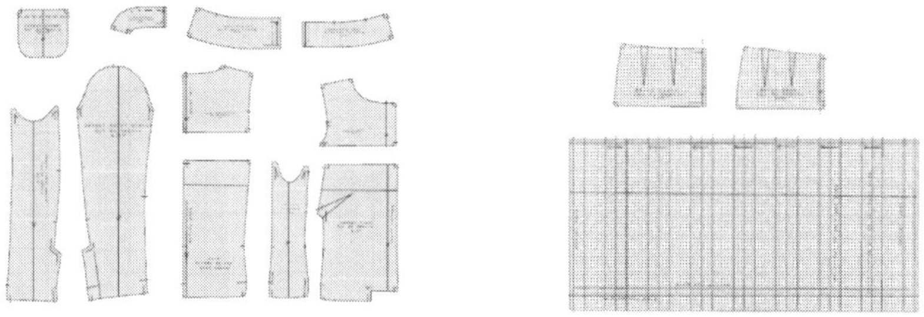

(a) Jacket (b) Skirt

Fig.2 Garment patterns [6]

Table.1 Specifications of sample garments

Sample name	Fabric	Producer	Target market	Nation
A	Tweed 1	Individual N	Luxury brand and department store	Japan
B				
C		Factory P	Luxury brand	Poland
D	Tweed 2	Factory J1	Department store	Japan
E		Factory J2	Luxury brand	
F	Worsted fabric	Factory J1	Department store	
G		Factory J2	Luxury brand	
H	Tweed 3	Factory C	Low-price speciality store	China
I	Tweed 4		Department store	
J			Luxury brand	

3. Results and Discussion

Fig.3 shows the scores for each evaluation item for all the samples. Among samples A, B and C, sample A received a high score for shape, silhouette, claim, and sewing and ironing skills. Sample A was made using a different pattern from that used for the other samples, and because the waist was evaluated as constricted and beautiful, it is considered that this is why the shape and silhouette were evaluated as being better than those of the other samples. The sewing was also rated highly. There were differences between samples B and C in terms of shape, silhouette and interlining. Thus, the difference in manufacturers affected the appearance.

The overall evaluation values for sample E (made by factory J2) were lower than those for sample D (made by factory J1). The target market for clothing made by factory J1 was of a higher quality than that for clothing made by factory J2. Thus, the target market was not positively correlated with the evaluation of the finished garments. This may be related to the type of clothing that each factory usually produces, because although factory J1 mainly produces women's clothing, factory J2 mainly produces men's clothing. It was considered that factory J1 was superior to factory J2 in terms of the technology required to produce women's clothing.

The shape, silhouette and face fabric of samples F and G received lower evaluation scores than those of other samples. Many evaluators expressed negative opinions upon seeing the face fabric. Thus, it is considered that the evaluation of the face fabric affected the evaluation of the shape and silhouette.

Sample H was made for a target market of low-price specialty stores. However, the scores for sample H were higher than those for samples I and J, which were made for department stores and luxury brands. The fabric used for sample H differed from that used for samples I and J, and some evaluators commented that the fabric used for sample H was more suited to the design. Thus, the effect of the face fabric was considered to be the reason for the higher rating of sample H. In addition, sample J was evaluated as being superior to sample I, which was in positive correlation with the target market. Scores for samples made by factory C in China were no lower than those for the other samples.

Multiple comparisons were carried out using the Bonferroni method. There were significant differences between some samples in terms of shape, silhouette and face fabric. Table 2 lists pairs of samples that showed significant differences in these three evaluation items.

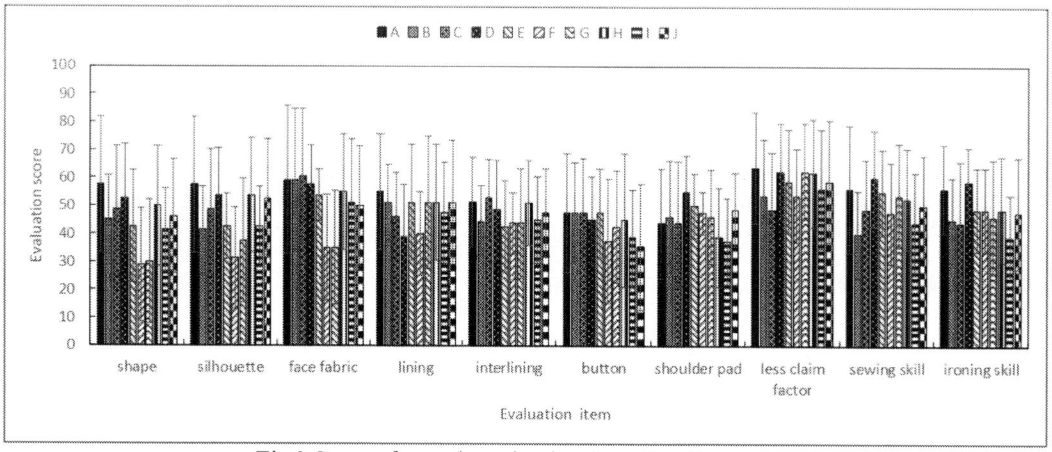

Fig.3 Scores for each evaluation item for all samples

Table.2 Significant differences between pairs for each evaluation item

Evaluation item	Shape	Silhouette	Face fabric
Sample pairs	A–F ** A–G ** D–E *	A–F ** F–D* F–H *	A–F * B–F * C–F * A–G * B–G * C–G *

*: significant at the 5% level
**: significant at the 1% level

Fig.4 shows the estimated average selling price of each sample. The prices for samples F and G (about 20,000 yen for the jacket and about 15,000 yen for the skirt), which were made from worsted fabric, were lower than those for the other samples (about 30,000 yen for the jacket and about 20,000 yen for the skirt). Multiple comparisons were also carried out using the Bonferroni method. There were significant differences between some samples in terms of the estimated selling price. Table 2 lists pairs of samples that showed significant differences. The estimated selling prices of samples F and G were much lower than those of the other samples. The only difference between samples F and G and samples D and E was the textile used.

There was a high correlation between estimated selling price and evaluation scores in terms of shape, silhouette and face fabric, as shown in Fig.5. Face fabric showed the highest correlation with estimated selling price. The estimated selling prices of samples made with fabric that received a low evaluation were lower than those of the other samples. Therefore, it was confirmed that shape and the textiles used have a significant effect on the estimated selling price of garments. Textile value depends on matching with the garment' style rather than the market price. The Chinese factory adjusted its manufacturing processes such as bonding interlining, piping for seam allowance and lining depending on the target market. However, no significant differences were evident despite these changes. The textile that was selected was also suitable for the design.

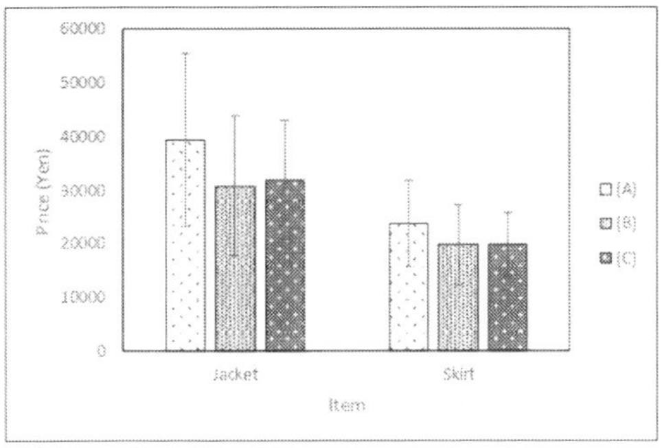

(a) A, B and C

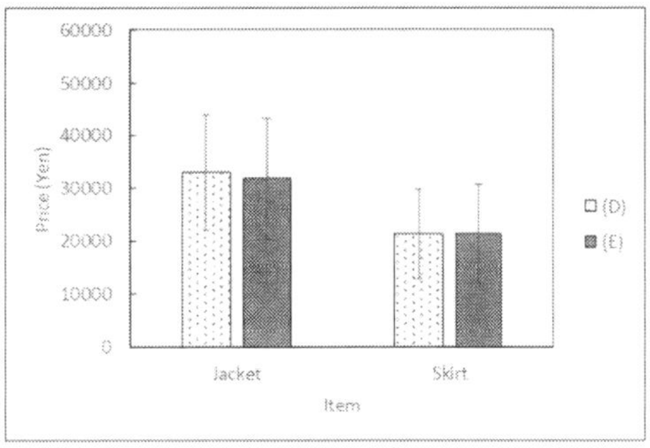

(b) D and F

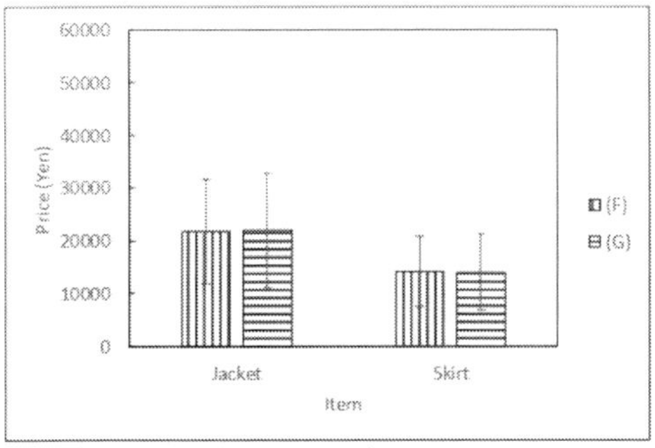

(c) F and G

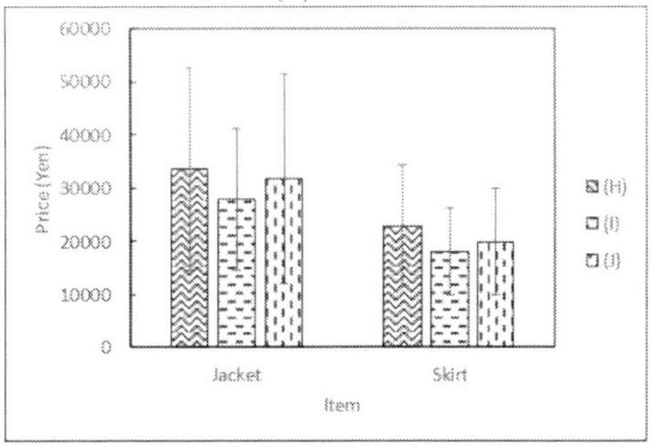

(d) H, I and J

Fig.4 Estimated selling price for each sample

Table.3 Significant differences between pairs of samples in terms of estimated selling price

Evaluation item	Jacket	Skirt
Sample pairs	A–C *	
	A–F **	A–F **
	A–G **	A–G **
	A–I **	D–F*
	D–F*	D–G*
	D–G*	F–H *
	F–H **	G–H *
	G–H **	

*: significant at the 5% level
**: significant at the 1% level

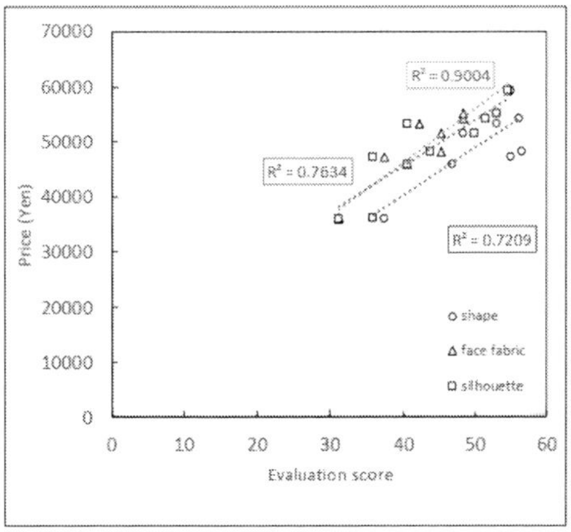

Fig.5 Correlation between estimated selling price and evaluation scores in terms of shape, face fabric and silhouette.

4. Conclusion

To investigate the effect of the manufacturing country on garment marketability, 10 jackets and 10 skirts made by four factories and one sample maker in three countries using five textiles were evaluated by 16 Japanese department store buyers. It was confirmed that the shape of the garment and the textile used have a significant effect on the estimated selling price. There was a high correlation between the estimated selling price and evaluation scores in terms of shape, silhouette and face fabric. Thus, textile selection and garment design are important factors in relation to marketability. However, there were no significant differences among the manufacturing countries. Textile value depends on matching with the garment' style rather than the market price. The garment quality was not dependent on the manufacturing country because the manufacturing skills of all the factories used were adequate. Therefore, Japanese garment-manufacturing companies do not have a competitive advantage in terms of garment quality.

5. Acknowledgment

This work was supported by JSPS (Japan Society for the Promotion of Science) fellows and JSPS KAKENHI under grant numbers JP24220012 and JP16K16255.

References

[1] Tokatli N. Global sourcing: insights from the global clothing industry-the case of Zara, a fast fashion retailer. Journal of Economic Geography: 2008; 8 (1): 21-38.

[2] Ministry of Economy, Trade and Industry of Japan. Report on apparel supply chain research: 2016; http://www.meti.go.jp/committee/kenkyukai/seizou/apparel_supply/pdf/report01_03_00.pdf.

[3] Ministry of Economy, Trade and Industry of Japan. Recent trends in the textile and fashion industry: 2013; http://www.meti.go.jp/committee/materials2/downloadfiles/g9 1225f07j.pdf.

[4] Kim K, Takatera M, Zhu A, Otani T. Comparison of Japanese and Chinese clothing evaluations by experts taking into account marketability. Autex Research Journal: 2015; 15 (1): 67-76.

[5] Keiser SJ, Garner MB. Beyond design: the synergy of apparel product development. New York: Fairchild Books: 2012.

[6] Kim KO, Takatera M, Otani T. Effect of patternmaker's proficiency on the creation of clothing. Autex Research Journal: Sep 2016; DOI: 10.1515/aut-2016-0005, Ahead of print.

[7] Chinami F, Takatera M, Kim K. Effects of combinations of patternmaking methods and dress forms on garment appearance. Autex Research Journal: 2016; DOI: 10.1515/aut-2016-0020, 1-10, Published Online: 2016-09-10.

A Study on Fashion and Fashion Styles through a Survey in Japan

Yu-Tian Long[1], Satoru Hirosawa[2], Tadashi Hayami[2], Saori Kitaguchi[1], Mitsuru Mimura[1], Tetsuya Sato[1*]

[1]*Kyoto Institute of Technology, Matsugasaki, Sakyo-ku, Kyoto, 606-8585, Japan*
[2]*Kyoto Municipal Institute of Industrial Technology and Culture, 91 Chudoji, Aawataguchicho, Shimogyo-ku, Kyoto, 600-8815, Japan*

[*]Corresponding author's email: tsato@kit.ac.jp

Abstract

The purpose of this research is to understand the modern fashion trend of college students through an investigation of fashion awareness and vocabularies, as well as in developing an insight the words that are used in modern fashion. The college student's current awareness of fashion was studied by conducting a questionnaire survey. There were many words in fashion that cannot be found even by referring to fashion literature, and it seems difficult to clearly define the meanings and emotional expressions of those words. It was discovered that fashion style terminology has changed with time, and fashion circumstances are constantly changing.

Keywords: Fashion Style; Fashion Image Word; Fashion Brand; Fashion Magazine; Questionnaire

1. Introduction

Nowadays, the fashion environment is constantly changing. Among young Japanese people, new fashion styles are born one after another, some become fashionable, others disappear quickly. There are also some fashion strategies such as Cool Japan [1] that are proposed by the government or organization for promoting new standards of fashion styles. Under such circumstances, many of us gather information from various media and reflect on them for their own fashion style. In recent years, many popular fashion brands were born from many countries of the world, including many low-priced fashion brands. People who were initially not interested in fashion are gradually becoming interested. It seems that consumers' interest in fashion is increasing these days. In fact, selling stagnation of clothing is often heard by news reports, however in the long term the consumption of textiles is on the rise and the sales of clothing won't be stagnant at all [2, 3].

Fashion is also always changing. As such it is important to study fashion awareness and the behavior of young Japanese people who are sensitive to fashion in order to know the trend of individual fashion. In addition, there are many words related to fashion, such as words that express certain fashion styles. The meaning of the words themselves change with the time. Many new words are born one after another, of which many words are used as Japanese *katakana* words, that is to say these words were brought in from abroad and not born in Japan [4]. To understand what words are used in modern fashion and to understand how people feel about them are important for the fashion business.

Therefore, in this research, a questionnaire survey was conducted in order to understand the current fashion awareness of young Japanese people, like what kinds of fashion style exist among them, and how they feel about the general fashion style.

2. Questionnaire Survey

Nine questions as shown below were included into the questionnaire.

The questionnaire was conducted between July 12 and October 6, 2016. The total number of respondents were 106 including 51 males and 55 females. They were all Japanese who were the undergraduate students or the post-graduate students of Kyoto Institute of Technology, and the age was ranged from 18 to 40 years old.

Question 1: Please tell us your opinion about fashion.
1) Are you interested in fashion?
2) Are you sensitive to new/popular fashions, and intentionally following the fashion?
3) Do you have your own fashion style?
4) Are you often checking fashion information in magazines or on the internet?
Question 2: Please list the fashion brands which you know.
Question 3: Please list the fashion brands which you like.
Question 4: Please list the fashion brands which you had bought their items in the past year.
Question 5: Please list the fashion magazines which you often read.
Question 6: How much do you spend on fashion every month on average?
Question 7: Please write fashion image words expressing fashion styles as many as possible.
Question 8: Please list your fashion styles using the words you answered in question 7.
Question 9: Please list adjectives to describe fashion style as many as possible.

In this questionnaire, the question 1 was the closed-ended questions. The questions 2 to 9 were the open-ended questions.

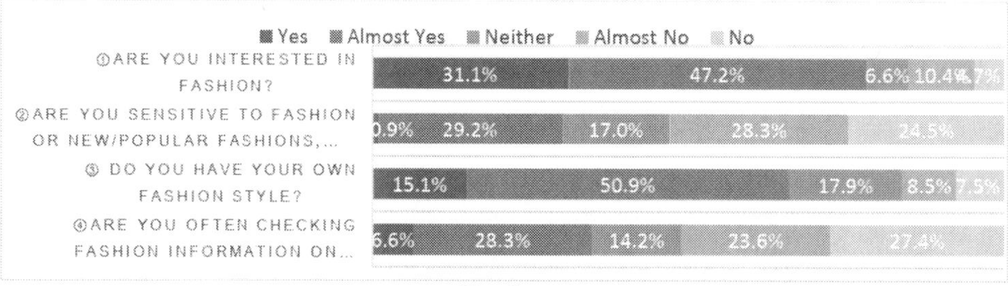

Total

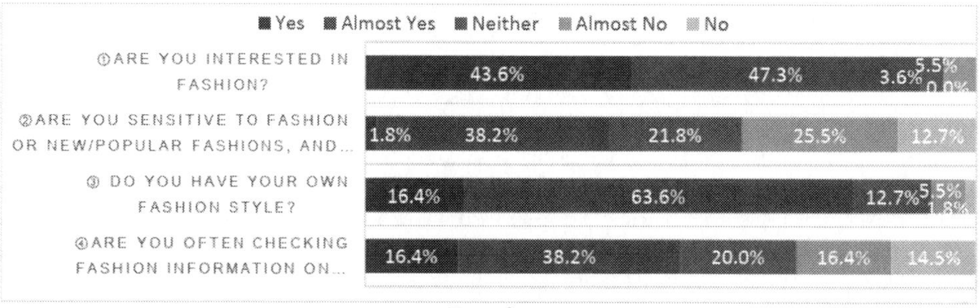

Female

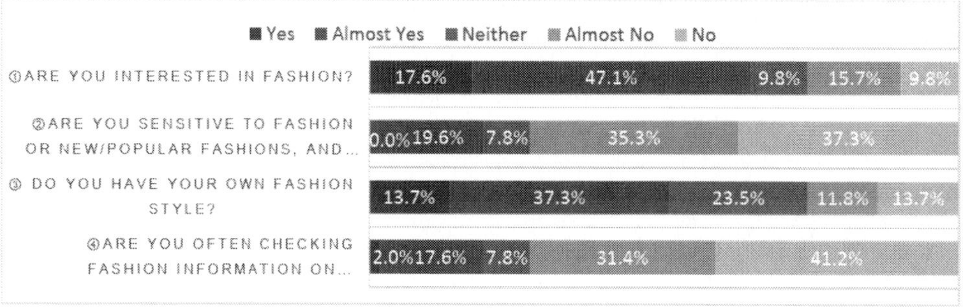

Male

Fig.1 The summary of the question 1.

3. Results and Discussion

3.1 Awareness Survey on Fashion Interest

The results of the question 1 are shown in Fig.1. The result showed that 78.3% and 66.0% were answered "Yes" or "Almost yes" for the questions "Are you interested in fashion?" and "Do you have your own fashion style?", respectively.

The results also showed that the respondents, particularly female respondents, paid close attention to fashion. Over 90.0% of the female respondents were interested in fashion.

30.1% and 34.9% of the respondents answered "Yes" or "Almost yes" for the questions of "Are you sensitive to new/popular fashions, and intentionally following the fashion?" and "Are you often checking fashion information in magazines or on the internet?", respectively. The positive answers of the male respondents were only 19.6% and 19.6% for these questions. However the answer of the female respondents were 40.0% and 50.6%. It indicates that the female respondents paid more attention to fashion than the male respondents.

3.2 Awareness Survey on Fashion Brands

The results of the questions 2 to 4, which were investigated the awareness for fashion brands, are shown as follows.

Table 1 shows the respondents' recognition of fashion brands and their purchase records. It shows the raw number and the ratio of the answers of "Fashion brands you know", "Your favorite fashion brands" and "Fashion brands you've purchased in the past year". Table 2 shows the results of the sex differences. In these three questions, a large number of the respondents listed UNIQLO. This answer was found from the male respondents much more than the female respondents. In addition, the male respondents listed many fast fashion brands, and more sports brands than the female respondents.

The brands recognized by the females and the brands liked by the females were vary widely. The 246 brand names were mentioned by a total number of the 106 respondents as their recognized fashion brands, including the 185 brand names mentioned by the 55 female respondents. These results indicate the diversities of fashion brands and the high concern for fashion of the respondents.

Table.1 Summary of the fashion brand awareness (Total)

	Total (N=106)								
Rank	Known brand	Frequency	%	Favorite brand	Frequency	%	Purchased brand	Frequency	%
1	UNIQLO	65	61.3	UNIQLO	21	19.8	UNIQLO	50	47.2
2	GU	36	34.0	GU	9	8.5	GU	17	16.0
3	H&M	24	22.6	earth music&ecology	6	5.7	H&M	10	9.4
4	ZARA	21	19.8	LOWRYS FARM	6	5.7	BEAMS	7	6.6
5	BEAMS	20	18.9	BEAMS	4	3.8	earth music&ecology	7	6.6
6	GAP	18	17.0	H&M	4	3.8	ZARA	6	5.7
7	earth music&ecology	17	16.0	KBF	4	3.8	LOWRYS FARM	5	4.7
8	LOWRYS FARM	13	12.3	The North Face	4	3.8	GLOBAL WORK	4	3.8
9	UNITED ARROWS	12	11.3	graniph	4	3.8	Paul Smith	4	3.8
10	URBAN RESEARCH	10	9.4	adam et rope	3	2.8	RALPH LAUREN	4	3.8
11	adidas	9	8.5	Bshop	3	2.8	UNITED ARROWS	4	3.8
12	GLOBAL WORK	9	8.5	UNITED ARROWS	3	2.8	graniph	4	3.8

Table.2 Summary of the fashion brand awareness (By gender)

Rank	Female	Frequency	%	Male	Frequency	%
Known brand (F=55, M=51)						
1	UNIQLO	25	45.5	UNIQLO	40	78.4
2	earth music&ecology	15	27.3	GU	21	41.2
3	GU	15	27.3	H&M	17	33.3
4	LOWRYS FARM	12	21.8	GAP	13	25.5
5	ZARA	11	20.0	BEAMS	10	19.6
6	BEAMS	10	18.2	ZARA	10	19.6
7	H&M	7	12.7	Adidas	8	15.7
8	INGNI	6	10.9	UNITED ARROWS	7	13.7
9	GAP	5	9.1	GLOBAL WORK	6	11.8
10	UNITED ARROWS	5	9.1	URBAN RESEARCH	6	11.8
Favorite brand (F=55, M=51)						
Rank	Female	Frequency	%	Male	Frequency	%
1	earth music&ecology	6	10.9	UNIQLO	16	31.4
2	LOWRYS FARM	6	10.9	GU	5	9.8
3	UNIQLO	5	9.1	The North Face	4	7.8
4	GU	4	7.3			
5	KBF	4	7.3			
6	adam et rope	3	5.5			
7	UNITED ARROWS	3	5.5			
Purchased brand (F=55, M=51)						
Rank	Female	Frequency	%	Male	Frequency	%
1	UNIQLO	16	29.1	UNIQLO	34	66.7
2	GU	8	14.5	GU	9	17.6
3	earth music&ecology	7	12.7	H&M	5	9.8
4	H&M	5	9.1	adidas	3	5.9
5	LOWRYS FARM	5	9.1	BEAMS	3	5.9
6	BEAMS	4	7.3	NIKE	3	5.9
7	ZARA	4	7.3	Paul Smith	3	5.9
8	INGNI	3	5.5	RALPH LAUREN	3	5.9
9	Ne-net	3	5.5	URBAN RESEARCH	3	5.9
10	UNITED ARROWS	3	5.5			

3.3 Awareness Survey on Fashion Magazines

Table 3 shows the answers for the question 5, which were the number of fashion magazines frequently read by the respondents. The 14 kinds of magazines were mentioned by the male respondents and the 30 kinds of magazines were mentioned by the 55 female respondents. 58.50% of the respondents (female = 34.5% and male = 84.3%) did not read any fashion magazines in this survey. Comparing with the results of the previous research [5], people who reads any fashion magazines decreased by more than half. Even the female respondents who have a high ratio on reading fashion magazines answered that they "only read fashion magazines at hairdressing salons." This also indicates the slump of paper books since the Internet has diffused across the world.

The magazines PS, JILLE, CHOKiCHOKi and HUgE, which had high ranking in the previous research [5], have already closed down. According to the monthly average circulation number published by the Japan Magazine Publishers Association [6], the circulations of some high ranking magazines - such as MORE, Soen, mina, SPUR, ViVi, JJ, and Ray - decreased by about half from October to December 2008, to April to June 2016. The closing down of fashion magazines and the reduction of the sales could be due to the sudden changes of readers' needs and advertising environments [7]. The results of this study also related to those causes.

Table.3 fashion magazines frequently read by the respondents

Female		
Rank	Magazine	Frequency
1	FUDGE	7
2	MORE	5
2	Soen	5
4	mer	3
4	mina	3
4	Ray	3
4	soup	3
Male		
Rank	Magazine	Frequency
1	MEN'S NON-NO	3
2	ELO	2

3.4 Survey on Purchasing Clothes

The monthly spend on fashion clothing is shown in Fig.2. The overall average spending was 9,063 yen (around 80 US dollar, as present exchange rate 1 US dollar = 110 yen). In addition, the maximum spending of all the answers was 70,000 yen; the minimum spending was 1 yen. The minimum spending of 1 yen probably means that she hardly spends any money on clothes. The average spending of the male respondents was 6,664 yen, and that of the female was 11,283 yen. Comparing with the average of 11,306 yen in the previous research [5], the spending per month was decreased by 2,243 yen. It is thought to be due to the changes in the economic conditions in Japan.

According to the survey of Economic and Financial Data for Japan [8] in 2012, under the terms of "Monthly income and expenditure per household by gender, annual income class, single-family household, Kinki area", the average expenditure on clothing and footwear was 7,499 yen. It decreased by 2,304 yen from 9,803 yen in 2004. According to the survey of National Federation of University Co-operative Associations [9] in 2015, the average of expenses (on clothing, cosmetics, tobacco, bathrooms etc.) of college students was 5540 yen per month. It decreased by 1,850 yen compared with the average of 7,390 yen in 2008 in Japan.

The overall average decreased compared to 7 years ago. Nevertheless, the data of our survey found to be above the average spending across Japan [9]. It can be said that this indicates that respondents have relative high interest in fashion.

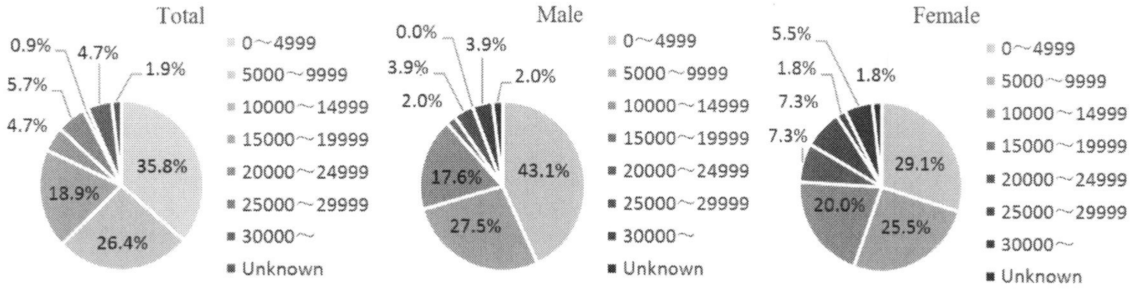

Fig.2 Monthly spend on fashion

3.5 Fashion Image Words Expressing Fashion Style

The results of fashion image words expressing fashion styles are shown as follows.

Table 4 shows the raw numbers of the fashion style listed by the respondents. There were 70 different words from the 106 respondents.

In the data analysis, synonyms in the answers such as "used clothing style "and "vintage" or "mode" and "mode style" were unified. Among the answers, there were many compound words. For example, "office-casual" is a style that combined features of both "formal" and "casual"; "Yurufuwa" expresses the feelings of both "light" and "airy ", and also points out the soft and cute atmosphere. It indicates that nowadays there are further segmentations on fashion styles, and the words expressing fashion styles have also diversified themselves.

Table.4 Words expressing fashion styles

Rank	Fashion style	Frequency	Rank	Fashion style	Frequency
1	Casual	70	27	Military	7
2	Girly	26	28	Elegant	6
2	Simple	26	28	Sexy	6
4	Formal	25	28	Harajuku style	6
5	Mode	22	28	Punk	6
6	Sporty	18	32	Hip-hop	5
6	Feminine	18	32	Conservative	5
8	Chic	14	32	subculture	5
8	Boyish	14	32	Vintage	5
10	American Casual	13	32	Yurufuwa	5
10	Ethnic	13	37	Outdoor	4
10	Natural	13	37	blue letter style	4
10	Mori Girl	13	37	Elegance	4
14	Street	11	37	Colorful	4
15	Cool	10	37	Kawaii	4
15	Normcore	10	37	Gal	4
15	Mannish	10	37	Cute	4
15	Monotone	10	37	flashy	4
15	Rough	10	37	Bohemian	4
20	Stylish	9	37	Retro	4
20	Lolita	9			
22	Neater	8			
22	Gothic lolita	8			
22	Pop	8			
22	Modern	8			
22	Rock	8			

Also, many new words that have not be seen in the literatures of fashion such as "subculture", "a bit bad (Choiwaru)" were in the answers. They are not the formal words and could be considered as a kind of slang, but it is difficult to distinguish them from the official words used in fashion. In this survey, all the answered words were counted.

In addition, the words representing countries and areas such as "Bohemian" and "Ethnic", the words including classifications of the Japanese magazines such as "Blue letter style (Aomoji style)" and " Red letter style (Akamoji style)", and the words coming from the genres of music such as "Rock", "Visual rock", "Punk" were found in the answers. It shows that the fashion styles were named from various fields of cultures.

Some of the high ranked words in the previous research such as "Older sister (Onee style)"," Older brother (Onii style)", "Trad", and "Akiba style" were not answered in this survey. The words that started to become popular in recent years such as "Mori girl", "Normcore", "Pop", and "Subculture" were seen in the high rank. It proves that fashion changes with the times. New fashion styles among Japanese young people are born one after another, some become fashionable, and some go to be eliminated quickly.

3.6 Fashion Styles That the Respondents Think for Themselves

Table 5 shows the rank of the answers of the fashion styles that the respondents think for themselves. "Casual" which also had a large number of the answers in the question 7 occupies the top place. And the other ranks are also similar to the question 7. The 50 different answers were obtained from the 106 respondents. Many respondents thought that they have multiple fashion styles rather than just one style. It indicates that the respondents have diversified choices on fashion styles, and the awareness to "Casual" is overwhelmingly high among them.

Table.5 Fashion styles that the respondents think for themselves

Rank	Fashion style	Frequency	Rank	Fashion style	Frequency
1	Casual	39	9	Outdoor	2
2	Simple	16	9	American Casual	2
3	Natural	7	9	Otaku style	2
3	Rough	7	9	Girly	2
5	Mode	5	9	Neater	2
6	Sporty	3	9	Normcore	2
6	Feminine	3	9	Boyish	2
6	Monotone	3	9	Military	2

3.7 Adjectives to Describe Fashion Styles

Table.6 Adjectives to describe fashion styles

Rank	Adjectives	Frequency	Rank	Adjectives	Frequency	Rank	Adjectives	Frequency
1	Cute	76	10	Simple	13	19	New	6
2	Cool	72	11	Dark	10	19	Calm	6
3	Pretty	41	11	Fit	10	19	Interesting	6
4	Fashionable	25	13	Noisy	8	19	Handsome	6
5	Out of style	22	13	Dirty	8	19	Old	6
6	Flashy	21	13	Strange	8	19	Adult-like	6
7	Sober	16	13	Be one's self	8	19	Painful	6
8	Bright	15	17	Elegant	7			
8	Beautiful	15	17	Stylish	7			

Table 6 shows the number of the words to describe fashion styles. The 205 words were extracted from the 106 respondents. Many of the respondents answered "Cool" and "Cute". It shows that the 2 words are considered to be important.

Among the answers, the similar meaning of the words were listed such as "Snugly (Picchiritoshita)", " Tightly (Kicchiritoshita)"," Tidy (Kicchitoshita)". It suggests the ambiguity and diversity of Japanese. There were also many slang and the words used only among Japanese young people such as" Pretty (Mabui)"," Out of style (Imoi)"," Glitzy (Kebai)" in the answers.

4. Conclusions

In this research, the current fashion awareness of Japanese young people, such as fashion styles exist among them, and opinion of the general fashion style, through the questionnaire survey on fashion.

From the results of the questionnaire survey, it was found that Japanese young people had a wide variety in the favorite fashion brands. The female tends to read fashion magazines more frequently than the male. From other answers we also knew that the students had high interests in fashion, and they had various individual fashion styles and favorite fashion brands. In addition, there were many fashion-style words that could not be found in the literature on fashion. The fashion styles were often described using slang. It could be because of the difficulty to describe and define the meanings of the styles and also because of the change of the fashion style terminology with the times.

Finally, by comparing with the previous research [5], the sales depression of fashion magazines was verified and the reduction of spending for fashion was also observed, it could be because of the increasing of the internet usage. From these results, the constant changes of the fashion circumstances was suggested.

References

[1] Japan Ministry of Economy, Trade and Industry. Cool Japan / Creative Industries Policy. http://www.meti.go.jp/english/policy/mono_info_service/creative_industries/creative_industries.html

[2] Utako H. Present Situation of Consumption of Clothing and Consumer Education. Bulletin of the Faculty of Education, Gunma University: Art, Technology, Physical Education, Life Science Edition: 2003; 38: p. 195-205

[3] Manufacturing Industrial Bureau, Japan Ministry of Economy, Trade and Industry. Supply and Demand of Textiles (1985-2003)

[4] Fashion Dictionary Seventh Edition. Bunka Publishing Bureau: 2008

[5] Tetsuya T. Evaluation of Fashion-styles through Drawing Method. Kyoto Institute of Technology Master's Thesis: 2009

[6] Japan Magazine Publishers Association. Monthly average circulation number. Female Young Magazine / Female Young Adult Magazine http://www.j-magazine.or.jp/magadata/?module=list&action=list

[7] Why? Magazine "AneCan" shut down http://irorio.jp/nagasawamaki/20160810/342633/

[8] Economic and Financial Data for Japan. Table 12 Monthly income and expenditure per household by gender, annual income class, single-family household. Kinki: 2012 http://www.stat.go.jp/data/zensho/2014/kekka.htm/

[9] National Federation of University Co-operative Associations. Summary report of the 51st Student Life Survey: Living Expenses of Boarding Students (2015) http://www.univcoop.or.jp/press/life/report.html

Inspirations from the Industrialization Improvement of Hand-woven Cloth in the Republic of China（1912-1949）to the Industrialization Development of Hand-woven Cloth as an Intangible Cultural Heritage in Nowadays

Ru Zhang*

Institute of safeguarding Intangible cultural heritage, Nanjing Museum, No.321 East Zhongshan Road, Nanjing, 210016, China

*Corresponding author's email: zhangru05@126.com

Abstract

From a professional perspective of clothing and by comparative analyses, we summed up the thoughts and methods from the path of industrialization improvement of the hand-woven cloth in the period of Republic of China (ROC). This may be helpful to the modern development of the Chinese hand-woven cloth, which is an intangible cultural heritage in nowadays. We concluded that to realize the 'living inheritance' of the hand-woven cloth, we can learn from the fashion industry, and find a way of industrialization incorporating with the economy and culture. This can provide some references for the industrialization of the textile based on the traditional intangible cultural heritage in the current society.

Keywords: Hand-woven Cloth in the ROC; Hand-woven Cloth as an Intangible Cultural Heritage; Industrialization Development of the Intangible Cultural Heritage; Living Inheritance; Productive Safeguarding

1. The Role of Hand-woven Cloth in the Clothing Fashion During the Period of the ROC and Nowadays

In recent years, the Chinese economy has been attracting a lot of attention from the world, and the government is paying more and more attention to the 'Cultural and Creative Industry' based on their own tradition. The cultural circle devotes great efforts to promote the productive safeguarding of the intangible cultural heritage. In the fashion field, Chinese style clothing has become a symbol to express the individuality and taste of fashionable people both from China and other countries, due to the advertising made by President Jin-Ping Xi and Mrs. Li-Yuan Peng during important state visits and celebration activities. Under this era background, the hand-woven cloth garments have also returned to today's fashion world. For example, the cheongsam and mandarin jacket made of hand-woven cloth has become the new favourites of some scholars and artists. At the same time, the Hand-woven cloth in China, represented by the Jiangnan hand-woven cloth and the Shandong Lu brocade, has been listed as significant intangible cultural heritage items. Hence some enterprises and individuals have started protective and productive developments on the traditional Hand-woven cloth, and even launched new brands to the market.

In fact, during the period of ROC, Chinese traditional hand-woven cloth had encountered the shock of the foreign Machine-woven cloth in the market, and the national cotton textile handicraft also faced serious crisis. However, with vigorous support by national industrialists, such as Jian Zhang and the ROC government, the great industrialization reform, such as the quality, technological conditions, and the social management, appeared in the industry of the hand-woven cloth. It finally became the popular apparel fabric of governmental officials and fashionable individuals, and played an important role in the history of textile and clothing in modern China (Fig.1). The related historical details can be found in the paper 'The inspirations from hand-woven cloth in the Republic of China (ROC) to modern hand-woven cloth designing' [1].

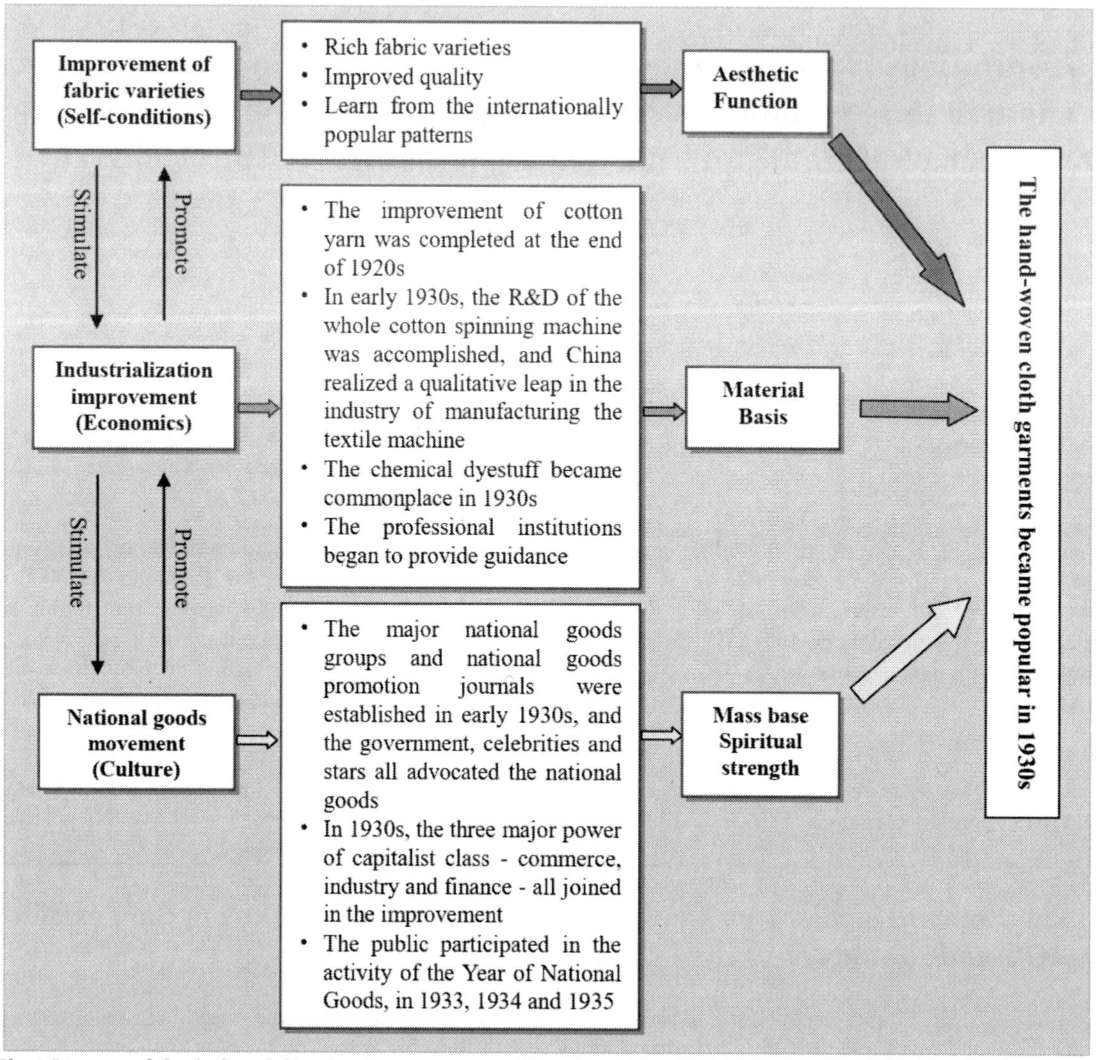

Fig.1 Impact of the industrialization improvement of hand-woven cloth in the ROC on the hand-woven cloth garments

It's not hard to see from the figure that the popularization of the hand-woven clothing in ROC is the result of the mutual promotion and development of the above improvement factors.

2. The Imperfection of the Development of Hand-woven Cloth in Nowadays

Due to the lack of the knowledge about the history of industrialization improvement of hand-woven cloth in the ROC, today's development of traditional hand-woven cloth still remains at a low level. Most developments are simply spontaneous behaviours of individual farmers, and only stay in the field of technology. Today's developing level almost stops at the same stage as the primitive development in late Qing-early Republican era, not to mention the industrialization. The imperfection of today's development of hand-woven cloth is summarized as following:

2.1 Lack of Diversity of the Fabric Varieties

Due to the lack of the knowledge about the fabric varieties of hand-woven cloth in the ROC, people now simply thought that 'the traditional hand-woven cloth has only a few fabric varieties; there are no more than 20 different varieties such as plaid, thinwicker, colour bar and wide stripe etc.' [2]. In fact, thousands of fabric varieties had been created during the period of ROC. And hundreds of them still survive until today.

2.2 The Starting Point of the Technology Development is too Low

The hand-woven cloth industry in the period of ROC already absorbed the most advanced technical achievement in the world at that time. The development of mechanization was almost accomplished and a relatively complete industrial system had been established. In contrast, the development of the hand-woven cloth in nowadays only relies on the practitioners themselves since they have no idea about the level of mechanization, improvement of the hand-woven cloth in ROC. They start from the primitive hand-made stage, only reach or even have not reached the semi-automatization stage, and the manufacture still remains at the level of manual workshop. They have not even noticed how primitive this developmental level is. For example, someone tries to improve the textile technology and efficiency, and achieve a small technical advancement, such as replacing the traditional manual production by the semi-automatic production. Then the exaggerated reports often describe that their contribution is as great as that of Dao-po Huang, one of the most famous revolutionists in the history of Chinese textile technique.

2.3 The Development is not Overall

Firstly, the assortment of the hand-woven cloth garment in the current society still focuses on traditional Chinese costume and is only popularized among a small group of people. It needs much more work to be geared to the international fashion trends and the popular aesthetic standards. In contrast, the hand-woven cloth in ROC was directly used to design the most fashionable styles at that time, such as the cheongsam, new civilized clothes and Chinese tunic suit, etc.

Secondly, the poor understanding of the traditional craft of hand-woven cloth has not resulted in sufficient developments of the hand-woven cloth craft in nowadays, only staying at the level of surface decoration. For example, most practitioners simply print and dye a few hand-woven cloth patterns to express its characteristics, expecting to patternize the traditional craft in a simple way. They have no deep and overall understanding about the characteristics of traditional hand-woven cloth craft such as yarn, mechanism of fabric, the structure and fitness. Therefore, the hand-woven cloth cannot be distinguished from generic modern industrial textiles, and its culture-related characteristics has not been revealed. In contrast, the hand-woven cloth in ROC had been developed in all aspects, such as the cotton types, yarn, fabric structure and weaving process, etc.

2.4 The Advocacy of Related Organizations is not Enough

Today's development usually depends on the practitioners' own efforts, which often lacks the support and guidance from the national and industrial organizations. Hence it has some blindness. In contrast, the industrialization improvement of the hand-woven cloth in ROC was supported by the whole country, such as the governmental organizations (for example, 'the Chinese Goods Maintenance Organization', 'the Shanghai Machine-made National Goods Factory Union', 'the Chinese Goods Production and Distribution Cooperative Association', etc.), industry association (for example, 'the Cotton Guiding Association', 'the Cotton Industry Public Association' (Fig.2) and 'the Improvement Station of Hand-woven Cloth', etc.), specialized industrial educational institutions and the mass media. These organizations provided the step-by-step support and guidance to the hand-woven cloth industry by means of government administration, laws and public opinion. In comparison, although today's government advocates the 'creative culture industry', the concrete measure and the execution are far less considerate and professional.

Fig.2 The restored building the Cotton Industry Public Association (Photographed by Ru Zhang at Nantong Textile Museum)

In short, there are many problems existing in today's development, and the practice in ROC can provide many valuable references and experiences.

3. Inspirations from the Improvement of Hand-woven Cloth in ROC to the Development of Hand-woven Cloth in Nowadays

3.1 The Inspiration to the Developmental Route in Nowadays

To realize the industrialization of the hand-woven cloth, the first step is the industrial mechanization. However, how to improve the hand-worked loom according to the principle of the machine-worked loom is a technical bottleneck that prevents the development in nowadays. In fact, the first step is not even pushed out yet. Therefore, the references from the improvement of hand-woven cloth in ROC can really provide a timely help. It is found by research that the development of the semi-mechanized and mechanized looms in the industry of the hand-woven cloth in ROC has already reached a mature level. There is no need to start the R&D from the most primitive level, or to walk the same way from 'Hand-work to Machine-work' once again.

We think there are several routes, list as follows, about the further development of the hand-woven cloth at present.

3.1.1 Realize the Mechanization by Following the Developmental Route of Post-industrialization

This route must be based on predecessors' achievements and experiences. The improvement of hand-woven cloth in ROC was aimed to industrialization and motorization, and its historical mission was to accomplish the 'Hand-woven to Machine-woven' transformation. This has been realized by the developing of the modern cotton manufacturing technology (Fig.3). We can learn the attitude of the industry in ROC towards the advanced technology, and selectively introduce and absorb the most up-to-date achievements of the cotton manufacturing technology.

Fig.3 The old photo showing the domestic iron machine made by Chinese iron works by modelling on English Henry loom in 1910s (Photographed by Ru Zhang at Nantong Textile Museum)

Currently, the cotton industry is becoming environmental-friendly and individualized. Many developers of the fabric intentionally pursue the effects such as plucked yarns and unevenly dyed and printed patterns, in order to mimic the hand-made effect. It may be great to follow this path in the R&D of hand-woven cloth today. In contrast to following the 'Hand-work to Machine-work' path, one can take the post-industrialization path as 'Machine-work to Hand-work' to realize the mechanization of the hand-woven cloth. It would also be well to enhance the aesthetic features of it, such as plucked yarns. Furthermore, with the increasing awareness of the environmental protection, the natural fabric is becoming more and more important today. We can appropriately take advantage of it in developing the hand-woven cloth. The raw material can be chosen as high-class cotton, like coloured cotton. In addition, we can also build the core competitiveness of the products by improving the quality of the cotton fibre. For example, one can strength the feature of anti-wearing based on the cotton fibre. Since in today's market it is not easy to distinguish the true and fake product, the 'quality' has become the most important factor ever, then the anti-wearing of hand-woven cloth will prove to be a good selling point.

3.1.2 Take the Handmade Development Route

Today, the international fashion world is paying more and more attention to the handmade product, and its price becomes gradually higher than that of the machine-made product. Hence, it is very urgent to create more new varieties by taking the advantage of the hand-work. As to this situation, the development of hand-woven cloth can be classified into high-end and low-end according to different segmentations of the target consumer market.

a. The ordinary low-end development model

This model only demands simple skills of the hand-woven cloth. However, it must be based on the sufficient supply of labour force. During the period of ROC, a westerner living in Shanghai once said: 'when randomly observing a crowd in any town or village in the interior China, from south to north or east to west, 90% are labouring people and more than half of them are working in the fields' [3]. This model still applies to today since the population in the countryside covers more than 40% [4] and can provide enough labour force. If such rich work force can be utilized sufficiently to develop the 'labour-intensive industry' of the hand-woven cloth, many job opportunities could be created and the surplus work force in the current society could be hired. If the quality of product is good enough, it could go into the international market as a high-classed handicraft.

b. High-end and high -class customization development model

This model mainly focuses on the historical, artistic, cultural and emotional values of the hand-woven cloth, all of which can become high added values of the related products. It aims to the consumer groups who extremely pursue the individualized handmade products. They have very high requirements for the handicraft level, the fineness and the patterns of the products. While the heritage of the craftwork of the hand-woven cloth in China is in a bad situation, only old people aged more than 70 living in some remote rural areas still grasp the complicated skills of the traditional hand-woven cloth. These skills have not been handed down from past generations in other regions. The survived skills are also in face of great danger of being lost due to the passing away of those old handicraftsmen, so does the sales business of the clothing made of hand-woven cloth. If the high-class customization development model is adopted, it can fully dig out the potential high-added-value of the craftwork of the hand-woven cloth to create economic benefits. Under the driving of economic benefits, the country can cultivate many inheritors and artists of the intangible culture heritage. Thus, the problem of lacking inheritors can be solved, and the virtuous cycle of creating, learning and studying of the craft of hand-woven cloth may also come to being.

3.1.3 Based on one Particular Element of the Hand-woven Cloth

This route is based on the ready-to-wear for general public. New fashion trend can be created by applying today's industrial dyeing and finishing technology to imitate the pattern, colour and visual quality and combining with today's fashion element. There have been some successful cases for it. For example, the tie-dyeing apparel which is developed based on the traditional craft has already obtained some success in the international fashion world. In nowadays, most of the so-called Chinese style cotton garments that scholars like to wear are no longer made by the real cotton material. Since retaining the visual style of the traditional hand-woven cloth, it is still recognized as hand-woven cloth garment.

To summarize, the three routes above (there are certainly many other routes) can be flexibly used according to various enterprises' design orientation and different target consuming groups.

3.2 The Inspiration to the Operation of Today's Enterprises

3.2.1 The Inspiration to Leaders' Quality

The leading force of the cotton textile industry during the period of ROC was Chinese national industrialists from the beginning to the end. They have devoted themselves into activities of advocating the national goods one after another, and made untiring efforts to improve the quality of national goods and domestic cloth. They fought alone or as a whole in order to promote the production and sales of national goods. For example, Jian Zhang made great contribution to the establishment of infrastructure of cotton and textile industry with his own effort. His impressive story is still widely circulated among Chinese people, and he has been regarded as the spiritual model of the people in Nantong. Most of these industrialists in the ROC are senior intellectuals, even including the number one scholar in the late Qing dynasty. Their personal quality, cultural and moral awareness are worth of being followed by today's business leaders. At present, China is at the critical time that the new national entrepreneurs are replacing the old ones. The born-rich generations have better educational experiences than the elder ones. Most of them have experiences of studying abroad. They have deeper understanding of the world's advanced technology, management, and operation, etc. If they can inherit the essence of the Chinese tradition just like the predecessors in ROC, and take the important responsibility to productive safeguard and industrially develop the hand-woven cloth with the same patriotism and the spirit of dedication, we believe that the hand-woven cloth will revive.

3.2.2 The Inspiration to Operation Mode

During the period of ROC, a lot of advanced management methods from the foreign countries were introduced, such as the most advanced share-holding system at that time (Fig.4), to establish the specialized cotton industry association, sales association and professional education institutions, etc. Besides, the overall business operation mode of the 'expert cultivation-production-operation-distribution' (Fig.5) was established. Therefore, the development of the hand-woven cloth in nowadays could also absorb the most advanced operation mode in the world, such as branding operation, rather than being confined to the small-scaled individual operation.

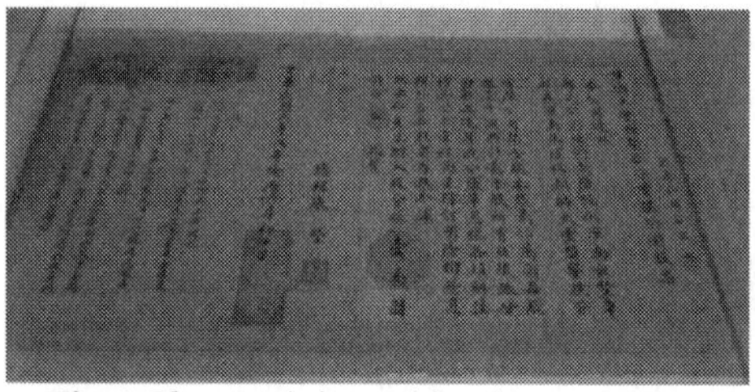

Fig.4 The stock issued by the company Jian Zhang (Collected in Nantong Textile Museum)

Fig.5 Old photo of the classroom of the professional textile college established in the national goods and cotton textile factory (Collected in Zhang Jian Memorial)

In addition, the industry of the hand-woven cloth in nowadays can also fully utilize the advantages of the other industries to develop itself. In the period of ROC, the national power was so weak that the infrastructure of cotton industry had to be established from the very beginning. However, today China has a strong economic strength, and all industries are growing and developing healthily and steadily. The hand-woven cloth industry is powerful enough to exchange fully with other related industries in order to realize self-development.

3.3 The Inspiration to Industrial Spirit and Social Culture

The patriotism presented by Chinese people in ROC when struggling for the industrialization of the national hand-woven cloth and being devoted to the National Goods Movement deserves to be extolled forever by descendants. The spirit of the age, represented by the strong support of the ROC government and the Chinese people's confidence to the national goods, deserves to be rethought and learned by all Chinese people who worship the foreign goods today.

As for the fashion industry, Professor Ze Yuan, who is a famous expert of costume history and fashion design theory, already raised in 2000 in his essay 'Bringing National Goods Consciousness up Again' the point that it is the urgent time to advocate and develop the national goods. He also appeals for protecting the national industry, developing the national brands and strengthening our own advantages. He suggests that CEOs, managers, designers and consumers in the fashion industry must strength their consciousness of the national goods [5].

Today Chinese government has made a firm decision to transform the economic structure, accelerate the speed of cultural industry transition and develop the 'Cultural and Creative Industry' based on Chinese tradition with great effort. This is like a new round of 'the National Goods Movement'. The measures and performance of the people and ROC government during the National Goods Movement and Hand-woven Cloth Movement already provided valuable experience and spiritual wealth to the domestic textile industry.

3.4 Develop the Hand-woven Cloth by Considering the Issue of the Clothing Industry in this New Age

The development of traditional crafts is the historical mission of each era. The historical mission of the development of the hand-woven cloth in ROC is to transform from handmade to machine-made. The industrialization development of the hand-woven cloth today needs not only restorations of previous technical level, but also new creative thoughts to develop by considering the new issues endowed by the age. It has become a new direction of the fashion industry that how to show the historical culture, emotional value and social responsibility contained by the clothing. The historical culture, emotional value and social responsibility contained by the hand-woven cloth cannot be replaced by the machine-made apparel textile which only aims to the commercial exchange without any human emotion. It is the historical mission of the hand-woven cloth to find the balance of these humanistic values and the economic benefits. This needs to be explored by future fashion industry.

4. Conclusion

In conclusion, the Industrialization of Hand-woven cloth in the ROC brings a wealth of valuable experience to the Industry Development of Modern Hand-woven cloth as an Intangible Cultural Heritage. To realize the 'living inheritance' of the hand-woven cloth as a cultural product, we can firstly learn from the fashion industry which focuses on the clothing industry, then find a path of industrialization which combines the economy and culture. This can provide some references for the industrialization of the textile based on the traditional intangible cultural heritage in nowadays.

References

[1] Zhang R. The inspirations from hand-woven cloth in the Republic of China (1912-1949) to modern hand-woven cloth designing. Collection of Papers and Works for 2010 ICCEC. Beijing: China Textile & Apparel Press: 2010; p. 74-77.

[2] Chen JM. The hand-woven cloth paves the way to fortune. Social Spectrum: 2007; 07: 52-53.

[3] Yao XH. Chinese modern history of foreign trade material (1840-1895) the third volume. Beijing: Chung Haw Book Co: 1962; p. 1355.

[4] Information on http://www.stats.gov.cn/english/PressRelease/201701/t20170120_1455922.html

[5] Yuan Z. To Wear Or To Be Worn. Shanghai: China Textile & Apparel Press: 2000; p. 63-64.

The Research on Consumer's Clothing Preference Based on Clothing Sample

Li-Li Wang[1], Jing Wang[1], Hong-Qin Dai[1,2*]

[1] *College of Textile and Clothing Engineering, Soochow University, Suzhou, Jiangsu, 215021, China*
[2] *National Engineering Laboratory for Modern Silk, Soochow University, Suzhou, Jiangsu, 215123, China*

*Corresponding author's email: daihongqin@suda.edu.cn

Abstract

With the explosive growth of electronic commerce, more and more clothing is provided for the consumer. It is very difficult for consumers to find the clothes they need quickly. With the application of a personalized recommendation system, the efficiency of shopping is improved. In order to effectively recommend clothing to customers, it is important to obtain consumer preferences. At present, the main method is through collaborative filtering recommendation, which is based on the history of shopping records. The bottleneck of this method is that it is unable to extract a consumer's interest if it is their first time or if they are not frequent customers. As a result, the method is unable to meet the real demand of the consumer. In this paper, a new method is proposed to obtain the consumer's preference for clothing. Firstly, a clothing sample for a certain type of clothing is constructed. Then, the user evaluation table or the consumer's evaluation of the clothing sample is obtained, the consumer preference in the evaluation table is excavated through the intelligent calculation, and is used for the clothing recommendation. A case study is demonstrated to show that the method is valid.

Keywords: Clothing Samples; Personalized Recommendation; Clothing Attributes; Rough Set

1. Introduction

China is the biggest producer of clothing, there are a lot of clothing that are introduced to the market this year. It has become more difficult for consumers to pick their own clothing because of the extensive network and the rise of electronic commerce. Electronic commerce allows consumers to conduct online transactions more conveniently. The problem however is that the database of information is too large, and the consumer spends too much time filtering the results in search of the clothes they are looking for. [1, 2]. This is a problem as in the growing society, people's pace of life has become faster and they are reluctant to spend a lot of time to find goods. This phenomenon creates a necessity for businesses to create a model that shortens the time spent on purchasing clothing and improve the efficiency of the consumer shopping experience.

Furthermore, with the rapid development of the economy, people are no longer satisfied with the basic function of clothing such as providing warmth. They are paying more attention to personalizing their clothing and such this is how the personalized recommendation system came into being [3]. Personalized recommendation system is to build data model and customer model with data mining technology. After data analysis, the personalized recommendation system will display the clothing which matches with the customer's interest characteristic. It simulates salespeople to provide customers with personalized recommendations to meet the needs of customers to complete the clothing purchase process [4]. At present, the clothing personalized recommendation system is mainly based on the user registration information, browsing history and the last purchase records to achieve personalized recommendation [5-7]. But consumer demands to continue to change with age and social status, and when using their ID, buy clothes for others, the system will default to my needs. So, these data are not a good reflection of consumer preference for dynamic clothing, how to accurately obtain the real demand of consumers is the key to the accuracy of the system recommendation, but also an important direction for the future research of clothing personalized recommendation system [8]. Based on the above problems, a method based on the sample construction is proposed to obtain the dynamic preferences of consumers in this paper. There are two advantages to obtaining consumer preferences through constructed clothing samples: Firstly, the preference information, which is obtained from the evaluation of the clothing samples, belongs to their own. Secondly, clothing samples can be re-evaluated for the consumer and updated. With the change of age and social status of consumers, consumers' demand is also changing. It is a good idea that re-evaluation can grasp current consumers' preferences for certain types of clothing. Sample updates can add the current popular elements into the clothing samples. In this paper, a clothing sample is constructed and the consumers' preferences are obtained through consumers' evaluation of clothing samples. The method can effectively grasp the real demand of the consumer, so you can use it to achieve personalized recommendation of the clothing.

2. Methods to Obtain Consumer's Preferences

2.1 Main Ideas

A method based on sample is proposed to obtain consumers' preferences in order to overcome the shortcomings of existing methods. The main contents of the method are: Firstly, a clothing sample is constructed for a certain type of clothing. Then, the user's evaluation table is obtained through the consumer's evaluation of the clothing sample and the consumer's preference is excavated through the intelligent calculation in the evaluation table. Finally, it is used for the personalized clothing recommendation.

2.2 Construct Clothing Samples

2.2.1 The Composition and Selection of Clothing Samples

We require consumers to evaluate clothing that had been prepared in order to obtain consumers' preferences for certain types of clothing. This set of clothing is called clothing sample. Clothing samples mainly include clothing sample size and attributes as well as the attribute value. The sample is so large that consumers do not want to spend too much time evaluating clothing. Sample size is too small to fully contain the consumers' preferences [9-11]. Therefore, the choice of clothing samples need to follow the principle of small and fine, that is, the sample size as much as possible to cover most of the clothing attributes in the appropriate circumstances, so that the collection of information of consumer's preference is more perfect.

2.2.2 Determination of Clothing Sample Size

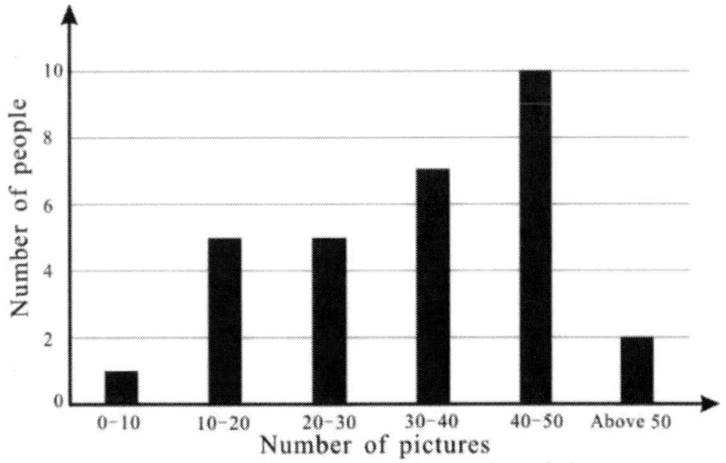

Fig.1 Respondents can accept the number of pictures

This article determine the clothing sample size mainly through the social investigation and experimental optimization. The main purpose of social research is to understand that how large the sample size can be accepted by consumers. In this paper, we mainly discuss the clothing sample size of the lady's winter coat. we asked thirty investigated objects to evaluate the preferences of some of the lady's winter coats that we had dealt with. The investigated object can stop the evaluation when the investigated object does not want to continuously evaluate the next picture. The investigator records the number of pictures evaluated by each investigated object, the investigation results are shown in Fig.1. From the figure we can see the number of pictures evaluated by each investigated object. It can be understood that consumers can accept up to forty to fifty pieces of clothing to evaluate. before purchasing clothing. The purpose of the experimental optimization is to reduce the large sample in a scientific and reasonable way.

2.2.3 Determine the Attributes and Values of Clothing Samples

This article determine the clothing attribute and value through the focus group interviews and expert interviews. The main purpose of focus group interviews is to know that consumer will pay attention to what attributes of clothing before deciding to buy clothing. In this paper, we mainly discuss the attributes of the sample clothing through investigated the lady's winter coat. Organizers asked consumers about attributes of lady's winter coats that they want to know before deciding to buy the clothing and then record several properties. The results was shown in Fig.2. From the figure we can see that the investigated object will focus on the properties of color, material, clothing length, fit when they buy lady's winter coat. Finally, the attributes and values of this kind of clothing are determined by expert interview on the basis of the social survey. The interviewed object directly affects the reliability, comprehensiveness and validity of the research data, so the interviewee is required to be a professional in the field [12, 13].

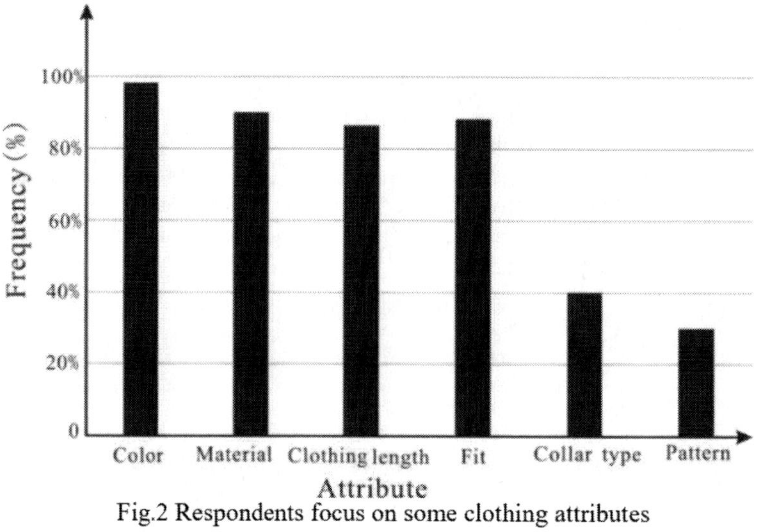

Fig.2 Respondents focus on some clothing attributes

2.3 Consumer's Evaluation of Sample Clothing

The consumer's preference is included in the evaluation of the clothing sample. e is used to represent the consumer evaluation of sample clothing. Its value was - 2, - 1, 0, 1, 2 and indicate that the consumer's evaluation of the sample clothing is "very dislike", "dislike", "neutral", "like" and "very like" respectively. The consumer's decision table is obtained with the evaluation of the consumer's preference for each picture.

2.4 Quantify Consumer's Preference

The sample clothing has the same attribute, but the attributes of each garment may have different values. Consumer's evaluation of every garment can be comprehensive result of all attributes of clothing. In order to quantify the preferences, in this paper, the concept of interest degree of clothing attribute is proposed.

The interest degree of clothing attribute indicates the degree to which a consumer is interested in a particular attribute value. In this paper, we use the lower approximation set of rough set theory to calculate the interest degree [14, 15]. All clothing in lower approximation set have same evaluation for a consumer. Therefore, the interest degree of each attribute value can be expressed by the frequency of the attribute in the lower approximation. s is used to describe single interest degree of the clothing attribute. In addition, in practice, there will be a different evaluation for the same value of clothing, that is to say, the same attribute value may appear in different evaluation values corresponding to the lower approximation set. For a comprehensive consideration of consumer's interest in an attribute value, a weight is given to each evaluation value corresponding to the s. The comprehensive interest degree of clothing, which is called the integrated interest degree of attribute, is obtained and represented by t.

The formula of s is shown in (1).

$$s(X_i)_p^e = \frac{N(X_i)_p^e}{N_e} \tag{1}$$

Where p is attribute; X_i is attribute value ($i=1,2,3\cdots$) of the attribute p; N_e indicates the total number of corresponding attributes p in the lower approximation set when the evaluation value is e ; $N(X_i)_p^e$ is total number of clothing with attribute value X_i in the lower approximation set corresponding p when the evaluation value is e ; $s(X_i)_p^e$ is interest degree of attribute value X_i of attribute p while consumer's evaluation is e .

The formula of t is shown in (2).

$$t(X_i)_p = \sum S(X_i)_p^e \times \delta_e \tag{2}$$

Where δ_e is the weight coefficient of evaluation value e. $t(X_i)_p$ is integrated interest degree of attribute value X_i of attribute p .

The larger the value of e indicates the greater the contribution of the consumer to the overall evaluation. According to this method, the corresponding weights of different e can be obtained: When e =2, δ_e=1, when e = 1, δ_e=0.5, when e =0, δ_e=0, when e =-1, δ_e=-0.5, when e =-2, δ_e=-1[14].

After integrated interest degree of attribute is obtained, consumers' interest in clothing can be computed. $c(i)$ is sum of every integrated interest degree of attribute of garment .

The formula of $c(i)$ is shown in (3).

$$c(i) = \sum t(X_i)_p \tag{3}$$

where i is the clothing number.

Finally, we calculate the consumer's interest for the sale of each garment on the basis of the above formula. A threshold is set before recommending clothing. These garments whose $c(i)$ is bigger than threshold are selected to recommend to consumers.

3. Case Study

3.1 Experimental Clothing Types

In this paper, men's shirts are selected to verify the validity of the method. We choose style, fabric, design and color as the attributes of men's shirts, which are represented by using A, B, C and D respectively [16].

The value of the attribute style is leisure and formal, which are represented by $A1$ and $A2$, respectively. The value of the fabric style is cotton, linen and silk , which are represented by $B1$, $B2$ and $B3$, respectively. The lattice type, pattern, single color and stitching color are selected as the basic attributes of design, respectively represented by $C1$, $C2$, $C3$ and $C4$. The attribute value of the color element includes the dark color and pastel color, which are represented by $D1$ and $D2$, respectively. In this paper, twenty-two different shirts were selected to form a men's shirt sample. That the men's shirt sample was evaluated by the subject formed an evaluation decision table, as shown in Table 1.

Table.1 A subject's evaluation for men's shirt samples

Garment number	Attributes				*e*
1	A1	B1	C1	D1	1
2	A1	B1	C1	D2	2
3	A1	B1	C2	D1	-1
4	A1	B1	C2	D2	1
5	A1	B1	C3	D1	1
6	A1	B1	C3	D2	0
7	A1	B1	C4	D1	-1
8	A1	B1	C4	D2	0
9	A1	B2	C1	D1	1
10	A1	B2	C1	D2	0
11	A1	B2	C3	D1	-2
12	A1	B2	C3	D2	1
13	A1	B2	C4	D1	-1
14	A1	B2	C4	D2	0
15	A1	B3	C1	D1	1
16	A1	B3	C2	D1	-1
17	A1	B3	C3	D1	1
18	A1	B3	C3	D2	2
19	A2	B1	C1	D1	1
20	A2	B1	C1	D2	2
21	A2	B1	C3	D1	1
22	A2	B1	C3	D1	2

3.2 Experimental Objects

Ten male college students from Soochow University were selected as subjects. The age of the subjects were between twenty-three and twenty-six years.

3.3 Experimental Process and Methods

The integrated interest degree for each attribute value in the sample is calculated for each subject by the formulas (1), (2) and (3) according to the subject's evaluation of the men's shirt sample. The integrated interest degree for each attribute value is substituted into the recommended method of Li Qiuyan to recommend the men's shirt to the subject. The recommended clothing are evaluated by the subjects, where full mark is 100 points, 100 points show that subjects like it very much, 50 points show that in general, between 50 and 100 points show that like, and 0 points indicate that don't like very much, between 0 and 50 points is don't like.

3.4 Experimental Results and Discussion

The integrated interest degree for each attribute value in the men's shirt sample is calculated from the above formula, as shown in Table 2. As can be seen from the table, we can quantify the consumer's preference for men's shirts as the integrated interest degree for each attribute value. It shows that we can construct a clothing sample to obtain consumer's preference and can quantify the preferences. The average value of the subjects is shown in Fig.3. Fig.3 shows that the average evaluation values of the clothing recommended by the system for consumer are more than 50 points. The discussion of experimental result is as follows:

1) Based on the integrated interest degree for each attribute value, we can calculate the total interest in each garment for sale. You can recommend a large degree of interest to the consumer, which shows that we can build a clothing sample to obtain consumer preferences, and used for the clothing recommendation.

2) The evaluation values above 50 points show that the method can be used for personalized recommendation by constructing sample to get the preference of consumers, and the overall recommendation effect is good, but the accuracy of the recommendation also needs to be improved.

Table.2 A subject's interest in men's shirts

Attributes	Value	Integrated interest degree of attribute
A	A1	-0.5625
	A2	0.5625
B	B1	0.7500
	B2	-1.0000
	B3	0.2500
C	C1	-0.7500
	C2	0.1875
	C3	0.4375
	C4	0.1250
D	D1	1.1875
	D2	-1.1875

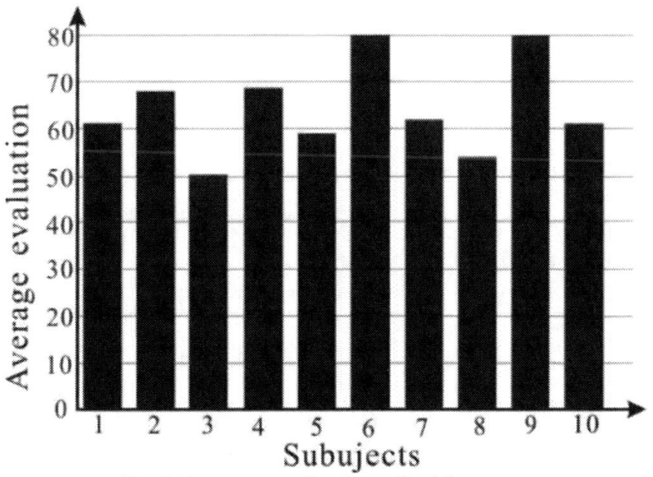

Fig.3 Average evaluation of subjects

4. Conclusions

A personalized clothing recommendation can improve shopping efficiency, it is essential on business to win customers. The key to the accuracy of personalized recommendation is to obtain the real demand of the consumer. In this paper, a new method is proposed to obtain the consumer's preference for clothing in order to obtain the real demand for the consumer and better achieve the personalized recommendation of the clothing. Firstly, we construct a clothing sample. Then, consumer's preference is obtained with the consumer's evaluation of the clothing sample. Finally, the method is used for clothing personalized recommendation. The validity experiment shows that the effect of recommendation obtained by this method is better.

5. Acknowledgment

This work was supported by A Project Funded by the Priority Academic Program Development of Jiangsu Higher Education Institutions, China.

References

[1] Lin YC, Kawakita Y, Suzuki E, Ichikawa H. Personalized clothing-recommendation system based on a modified bayesian network. IEEE/IPSJ International Symposium on Applicatio: 2012; 59 (5): 414-417.

[2] Singh A, Alhadidi B. Knowledge oriented personalized search eEngine: a step towards wisdom web. International Journal of Computer Applications: 2013; 76 (8): 1-9.

[3] Ajmani S, Ghosh H, Mallik A, Chaudhury S. An ontology based personalized garment recommendation system. Wi-iat: 2013; 3: 17-20.

[4] Shan YF, Li BY. Research on apparel recommendation in e - commerce recommendation system. Wool textile journal: 2016; 44 (5): 66-69.

[5] Liu JG, Zhou T, Wang BH. Research progress of personalized recommendation system. Progress in natural science: 2009; 19 (1): 1-15.

[6] Zhu Y, Lin ZN. Review of personalized recommendation methods in electronic commerce. China soft science: 2009; 02: 183-192.

[7] Ding R. E-commerce personalized recommendation development trend in the era of large data. Electronic commerce: 2015; 4: 5-5.

[8] Li QY. Clothing personalized recommendation method. Suzhou University: 2014.

[9] Liu XG. Costume design element theory. Journal of Donghua University Natural Science Edition: 2003; 29 (2): 23-26.

[10] Liu C, Ding GG. Classification algorithm of garment attribute based on vision. Microelectronics and computer: 2016; 01: 28-33.

[11] Jiang L, Zhu Y, Zhou QY. A method for determining the sample size of apparel market research. Journal of Donghua University: 2010; 10: 89-94.

[12] Fang ZZ, Cheng JK. Significance of comparative education research methodology in "Focus Group Interviews". Foreign educational research: 2012; 6: 19-25.

[13] Liu HC. An empirical study on the supplier evaluation index by expert interview. Jiangsu business theory: 2012; 5: 110-111.

[14] Tan ZF, Zhang Y, Wang S. An improved weighting method for rough set theory. Computer Engineering and Applications: 2012; 48 (18): 115-118.

[15] Pawlak Z. Rough set theory and its applications to data analysis. Cybernetics & Systems An International Journal: 2010; 29 (29): 661-688.

[16] Zhou QY, Yang YX, Gan JQ. Analysis on behavior and influencing factors of customer of man’s shirt. Journal of Donghua University (Social Science): 2010; 10 (1): 58-63.

Analysis of High-End Branded Men's Clothing Franchise Stores Shop Display, in Small and Medium-sized Chinese Cities

Wei Tian, Li-Ping He

Xian Polytechnic University, No.19 South Jinhua Road, Xian, Shanxi, 710048, China

Presenting author's email address: 569790315@qq.com; 1137820367@qq.com

Abstract

This article is a case study of Dingbian LILANZ Men's Franchise Store in Yulin City, Shaanxi Province. Using the research methods of literature review, comparative analysis and field investigation, the Dingbian LILANZ store in Yulin City was compared to the Xi'an Bell Tower LILANZ Store in terms of window display, shop display, shop layout and shop environment, etc. The problems of the Dingbian LILANZ franchise store in terms of shop display were summed up. The aim of this research is to summarize the problems in shop display in high-end men's clothing brands stores in small and medium-sized Chinese cities. Finally, the paper puts forward an improvement strategy and planning scheme to improve the long-term and widespread problems in the display areas of high-end branded menswear shops in China.

Keywords: Branded Men's Wear; Store Display; Visual Marketing; Small and Middle-sized Cities; Franchisee

1. Introduction

LL Group, is a high-end Chinese business casual men's brand , founded by three brothers Wang Dongxing, Wang Liangxing and Wang Congxing in 1987. This was when the concept of "business casual men's" [1] first appeared in China. After years of development, nowadays LL is a leading Chinese business men's brand that is responsible for its own product design, product development, production and marketing [2]. LL products are sold in thirty-one provinces, autonomous regions and municipalities in China. In 2008, LL reshaped the brand logo LILANG to LILANZ, this new logo is very international, and also symbolizes LL brand's international strategy. In 2009, LL became the first men's brand from mainland China to be successfully listed in Hong Kong. Until now, despite the impact of e - commerce and inventory pressure LL has been profitable.

At present, there are many scholars in China studying branded clothing from different angles and areas of research. However, there are not many who conduct basic research on shop display, in particular concerning men's brands in small and medium-sized cities. At the same time, according to relevant research, more foreign research has been done in innovating clothing store's spatial design and display [3]. Among them, fast selling fashion brands have a large shopping area, more dressing room space and more features. The shop line planning is clear, in order to minimize a sense of urgency in consumers [4]. Designers and luxury brands place great emphasis in conveying a sense of quality, beauty, and uniqueness in store layout, décor, and how products are displayed.

In this article the authors used the research methods of literature review and the field investigation to compare the Dingbian LL men's franchise store with the Xi'an Bell Tower LILANZ Store on spatial arrangement, window display and shop display, etc. This research points out the existing problems of Ding bian LL men's franchise store and proposes solutions to the problems.

2. Visual Marketing and Shop Display

2.1 The Meaning of Visual Marketing

Visual marketing conveys the product design, structure and overall information to the customer visually, and the ultimate goal is to convince customers to purchase goods [5-6]. Visual marketing has been widely used in the apparel industry, from clothing style positioning, product style, color design, brand packaging, and even advertising design, fashion show, product display, shop and window design, merchandise display, display design, shop design, etc, all of them are a part of visual marketing [7-9]. Visual marketing mainly refers to its role in shop sales. Visual research methods generally mean "use of visual technologies and images in research" [10].

2.2 Understanding of the Shop Display

Effective shop display consists of a reasonable shop layout based on science and has aesthetic appeal, which can effectively lead the customer's line of sight to the expected range, act as an invisible shopping guide and ultimately lead to consumer purchases. [8].

Dingbian LL was chosen as the object of this study, mainly because of LL brand's representativeness in domestic high-end men's wear and profit for LL brand's profitability is also a leading role. But while conducting filedwork the author found that in the small and medium-sized cities to standardize shop displays there are more serious problems, such a franchise will seriously affect the brand image and customer loyalty. I hope that through the study of the brand men in the store shop store management have some help.

3. Analysis on the Display of LL Men 's Franchise Shop in Dingbian County

The LL Group is a high-end business casual menswear brand founded in China in 1987. After years of development, now it sets design, product development, production and marketing of Chinese business men in one of the leading brands [11].

3.1 Dingbian County LL Franchise Stores Display Preparation Process

a. Design the layout plan (as shown in Fig.1).

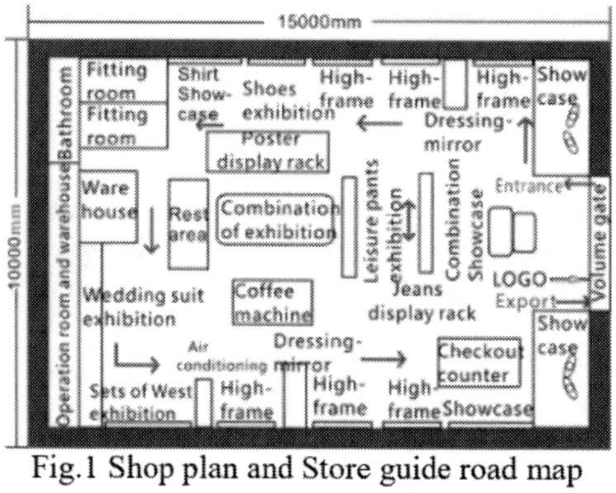

Fig.1 Shop plan and Store guide road map

Table.1 Dingbian County LL brand franchise stores inventory statistics

Year	Inventory	Main Stock Season	The main inventory category	Backlog reasons
2013	328	Winter	Parker, sweaters	The product is too loose
2014	599	Autumn and winter	Cotton, jackets	Product color aging
2015	961	Summer and winter	Jeans, T-shirts	Code number is too large
2016	734	Summer, autumn and winter	Casual pants, jackets, T-shirts	Edition type is too large

b. Commonly used display techniques

Since the opening of LL's franchise stores in Dingbian in the fall of 2012, there has been significant inventory backlog.

In order to reduce the inventory of LL shops the inventory display method used is one which showcases a mix of new products of different styles and colours on the same display. Diversified display is also one of the common display methods used by LL franchised stores (as shown in Fig.2).

Fig.2 Diversified with the display and new products with the display

3.2 Dingbian County LL Franchise Stores Display Specific Analysis

a. LL's Dingbian stores store display

In clothing displays, a reasonable layout of the space for the customer's final choice has a role that cannot be ignored. The general process of buying clothing in a store is as follows: the customer enters the shop, walks around the store, observes and touches the goods before making a purchase. [12] Good space layout can boost sales. LL's Dingbian shop space layout (as shown in Fig.1) is consistent with people's usual habits, to provide customers with a comfortable shopping environment.

Clothing store display and product display, more important is a way of life . The most basic display of clothing shops are with mannequins and exhibition racks. Other props are often used in combination to achieve the purpose of effective display of goods and highlight the brand and brand personality. LL's Dingbian store display a blue and white porcelain plate which attracts the customer's eye, and the painted wooden box seen from the store window makes the shop décor very unique (as shown in Fig.3).

LL shop products,with the display in the middle of the shop to showcase a mix of fashion trends and colors, with customers' needs and acceptance of the store's own business , inventory and sales as the premise. The periphery of the LL shop features clothing with more neutral colors. The color of the operating mode commonly used in front of the light behind the dark method, the interval method and the cross method [13], because the edge LL shop product color is not very rich, so more use of interval method. This way makes the store full of changes, people feel excited and fresh (as shown in Fig.3).

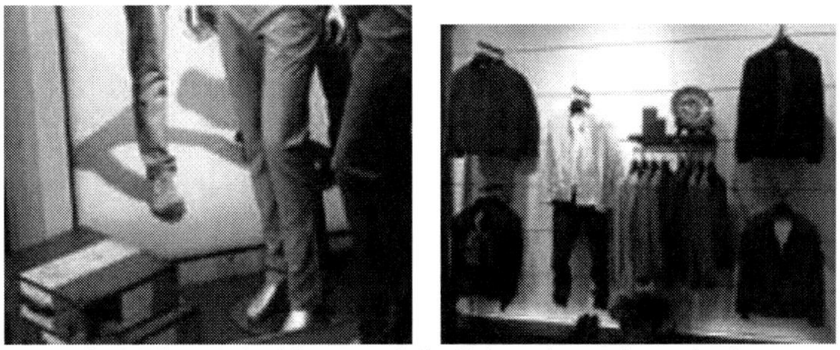

Fig.3 Blue and white porcelain dish and wooden paint box
(Cross method and interval method)

b. LL's Dingbian store window display

Seasonal and festive display methods are the most commonly used window display methods. As shown in Figure 4, big red "blessing" word pendants, and lambs marked with white the "LL" logo convey both brand information and festive blessings to consumers, but also reflects the festive atmosphere of the Spring Festival. The spring shop display showcases new clothes, and thousands of paper cranes in the background brings the entire window display to life (as shown in Fig.5).

c. Dingbian County LL stores display promotional materials

Good promotional material design can effectively attract the attention of customers and guide customers to buy. The most effective promotional material used vy the Dingbian LL shop is POP advertising. A survey conducted by the POPAD on the effects of point-of-purchase advertising on sales show that in-store POP advertising has great impact [10].

The average increase in merchandise sales using ads is 194 percent, making it one of the most effective forms of advertising. POP ads indicating the effect of the sale results in an average increase of 124 percent in sales. POP ads indicating product information only led to a comparatively low 33% increase in sales, while adding POP ads to the product itself increases sales by an average of 18 percent. Thus, the impact of different forms of advertising on sales is clear (as shown in table 2).

Fig.4 Chinese New Year display

Fig.5 Spring on the new

Table.2 POP advertising and sales results

Sun CQ. Sales psychology. M. Tokyo:Corporation Business, 2002; p48.

Name	Average rate of increase in sales
Adding POP ads to the product	Sales increased by an average of 18%
POP ads indicating product information	Sales increased by an average of 33%
POP ads advertising items on sale	Sales increased by an average of 124%
An item in an interstitial ad	Sales increased by an average of 194%

Product features and key information passed to customers through POP advertising can help customers quickly select products most suitable for them [14-15]. The main role of publicity materials is to carry out the most direct and effective information transmission, between the seller, the product and the consumer. Good POP advertising also need to establish the store's image and have its own characteristics.

4. An Analysis of the Comparison between the LL Stores in Dingbian County and Xi'an Bell Tower

4.1 Brief Introduction to Space Planning Map and Basic Information on Xi'an Bell Tower

a. Xi'an Bell Tower LL store space planning map (as shown in Fig.6).

b. LL stores in Xi'an Bell Tower LL and Dingbian County; the difference between the stores' spatial planning.

The total area of Xi'an Bell Tower LL shop is approximately 285 square meters. The layout of the Xi'an Bell Tower store is more reasonable and feels more spacious. The size of the Dingbian LL franchise area is about 190 square meters, but the bad shop display makes the entire space feel cramped.

4.2An Analysis of the Display of LL Direct Shop in Xi'an Bell Tower

a. An Analysis of the Sales of LL Direct Shop in Xi'an Bell Tower

The store plan of Xi'an Bell Tower's LL direct shop is obvious, to the right of the entrance are the suits and formal shirts, followed by new short-sleeves and other products. In the middle of the shop customers can choose from ties, wallets, shoes and other products based on individual needs and checkout – all routes are spacious (as shown in Fig.6).

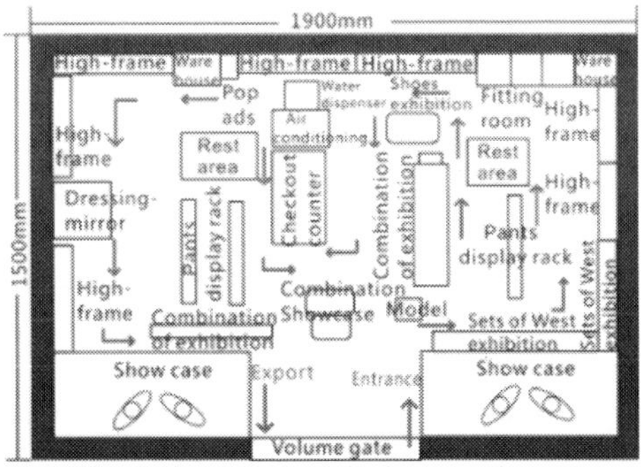

Fig.6 Shop plan and Shop road map

Fig.7 Clothing diversification stacked

Fig.8 Color interval display

Display props used are mainly suitcases, handbags and small potted plants. The shop's products are mainly based on a variety of display methods (as shown in Fig.7), which gives the space a distinct personality. The color of the product is arranged from light to dark,, or displayed using a combination of neutral color products, cool color products and warm color products (as shown in Fig.8).

Compared with the direct shop in Xi'an, the store plan of the Dingbian LL shop is not obvious, the product and product relevance is poor. In terms of theline planning of shop, the shop in Xi'an achieves the purpose of creating an environment that encourages customers to stay in the store for longer. Dingbian's LL store is relatively simple; costumers enter from the right door and leave through the left. In addition, Dingbian LL stores use a wide range of props, resulting in the anti-effect of excessive props looking "piled up".

Fig.9 "Lang's summer" theme display

b. Xi'an Bell Tower: LL store window display

Xi'an LL Direct shop window display commonly uses seasonal and theme displays. 'Lang's summer' (as shown in Fig.9). is an example of a themed window display, it has a rich story and beautiful picture. In contrast, the window display of Dingbian LL shop is relatively weak, lacking in an overall story or theme. Choosing window display props without a clear purpose often only increases the number of props used, but means that many aspects of the display will be incoherent with the theme and style of the display.

c. Xi'an Bell Tower: LL store's display promotional materials

The promotional materials of Xi'an Bell Tower's LL Direct store is designed to be simple and generous. The amount of publicity materials is suitable for the shop by conveying all the vital information to the customer which is very good for publicity. In contrast, Dingbian's LL franchise stores have a large amount of promotional materials. The amount of information is excessive to the point where customers are not willing to read any.

5. Dingbian LL Franchise Stores Display: Problems and Solutions Proposed

5.1 An Analysis of the Problems in the Display of Men's Franchise

5.1.1 Display Methods and Collocation of Specialization is not Enough

According to the analysis at the beginning of this research, we find several issues with the shop displays of Dingbian's LL stores :

a. Products with a sense of the series is not obvious. The store often seems as though only a single product is on display. Product displays lack a central theme, which makes it difficult to achieve joint sales. A large display space using a single display method lacks focus and variety. The goods on display were combined randomly, with no interesting focus or design.

b. "Piling" serious, deliberately create color, special and other marked not clear enough. Sometimes to make the display space seem full, products and unrelated props are put together, leading the theme to be unclear (as shown in Fig.10).

c. Window display lighting imbalance. Window display without consideration to the effects of hyperopia and myopia often loses its appeal, the goods of window display with no simulation and scene effects.

5.1.2 Lack of maintenance in store displays and display props

The damage to the shop's light is serious, props placed neatly, the adjustment of product display was not timely adjustment, the display of the exhibition in a single direction strains the customer's vision. In the walls and props free stickers notice or POP (as shown in Fig.11), leading to space imbalance. In addition, the maintenance of props is not timely, the details of the display is not up to standard.

The field research and analysis by author found that the industry behind store displays lack professional staff and related supervision, especially for small and medium-sized cities franchisedstores lack supervision and management, which is the main reason for these problems.

5.2 Dingbian LL Franchise Stores - the Solution to the Problem

Based on the above research on the display problems of Dingbian LL franchise stores, I propose the following three suggestions:

a. Strengthen management and supervision of store displays. The store must be spacious and always kept clean and orderly. The product display should have focus, standardization, specialization and innovation; a shop display with a good theme and focus is a good silent shopping guide. There is also need to pay attention to how different types of products can be displayed clearly, and to make reasonable use of the gold zone of the shop. In addition, it is necessary to strengthen coordination between departments. Each department directly affects the display, but special attention should be paid to the maintenance of shop display's post-show effects.

b. The theme of window displays should be updated in time. Make use of bright colors to promote the store ad attract customers, put the concept of the brand into the details of each display theme through different theme of market,

c. Increase the level of use color, and to use the diversity and unity of color at the same time. The effective combination of colors will give consumers a different visual experience, simple, refined,and the concept of the window is in line with today's fashion brand personality. In addition, different consumers like different colors, in order to meet the clothing differentiation demand of different consumers, clothing stores will have different styles and colors of clothing, so the use of a variety of colors is very important to the shop.

Fig.10 Deliberately create a sense of fullness of space

Fig.11 Paste materials on the wall

6. Conclusion

Props, use of color and the planning of space is very important in store displays. In this article, the authors compared and analyzed the shop displays of LL stores in Dingbian and the LL direct shop in Xi'an Bell Tower, from window displays, store displays, store space planning to display props, etc. Finally, the authors summed up the problems of Dingbian LL store and proposed solutions for improvement. The aim of this research is to provide solutions to improve shop displays in high-end branded clothing stores in small and medium-sized Chinese cities.

References

[1] LL Retail Terminal Internal Training Manual. M: 2015; p. 23-28.
[2] LL's official website [EB]: http://www.LL.com/investor/briefing.html.
[3] Jin S. A Comparative Study on the Display of Clothing Store at Home and Abroad. J. Text Res J: 2016; 08: p. 97.
[4] Mu YP. Research on the Operation Process of the Shop Display of Clothing Brand. J. Modern business: 2016; 36: p. 190.
[5] Zhao HS. Clothing visual marketing. M. Beijing: China Textile Publishing House: 2008; p. 87-94.
[6] Zhang MN, Xuan Y. Study on the sensory evaluation of underwear display color under different lighting. J. Xi'an: J. Xi'an Polytechnic University: 2016; 02: p. 164-169.
[7] Laila Shin Rohani, May Aung, Khalil Rohani. One step closer to the field: visual methods in marketing and consumer research. J. Qualitative Market Res: 2014; 17: p. 6.
[8] Hao X, Tian W. Research on Display of Chinese Underwear Market Based on Visual Marketing Strategy. D. Xi'an: Xi'an Polytechnic University: 2014; p. 24-32.
[9] LL retail terminal internal training manual. M.
[10] Sun CQ. Sales psychology. M. Tokyo:Corporation Business: 2002; p. 48.
[11] Michel Wedel, Rik Pieters. Visual Marketing: From Attention to Action. J. Taylor & Francis: 2007; 207-217.
[12] Gwenola Guide, Li YJ. Visual marketing practice in Europe. J. Chin clothing newspaper: 2005; 18: p. 4.
[13] Zheng JJ, Ji XF. Consumers' visual perception of clothing display. J. Text Res J: 2016; 03: p. 161-163.
[14] Cai YR, Chen CX, Yang YX. On display helmet eye tracking experiment based on the clothing store window. J. J Donghua University: 2013; 2: p. 81-86.
[15] Li J. Design of display visual marketing silent language. D. Xi'an Polytechnic University: 2011; p. 4-6.

Effect of Brand Logos on Consumer Preference for Clothing

Qian-Yun Chen[1], Xiao-Feng Jiang[1,2*]

[1]*College of Textile and Clothing Engineering, Soochow University, Suzhou, Jiangsu 215021, China*
[2]*National Engineering Laboratory for Modern Silk, Soochow University, Suzhou, Jiangsu 215123, China*

*Corresponding author's email: xfjiangsz@163.com

Abstract

The purpose of this study is to explore the effect of brand logos on consumers' preferences for clothing. First of all, two kinds of logos, one with six well-known brands and the other with six less well-known brands, were rated and selected from clothing brands through questionnaires. Then, each type of logos was further classified into three high-value brands and three low-value brands. Finally, a behavioral experiment based on the pictures with and without logos to evaluate clothing preferences was conducted. The results showed that well-known brand logos have a greater impact on consumer preferences for clothing than the less well-known ones, and high-value brand logos have greater influence on clothing preferences. The current study indicates that there is an effect of superiority for logos of well-known brands or high-value brands on clothing preferences.

Keywords: Brand Logo; Popularity; Brand Value; Clothing Preference

1. Introduction

With the continuous development of the market, more and more enterprises have paid attention to the value and strategy of branding to improve product recognition. Accordingly these companies rely on the brand awareness, brand image, brand value and product quality to satisfy their consumers. As such this topic has intrigued many researchers globally for decades. The influence of branding on consumer behavior has been discussed as early as the 1990s. It was discovered that brand equity had an effect on preference and purchase behavior of consumers if they were familiar with a brand or had positive emotions for it [1]. In fact, brand awareness is constantly viewed as an integrated framework consisting of multi-factors, which can affect the consumers' reactions to various products [2]. Wang adopted the method of theoretical analysis to develop an influence model on consumers' purchasing intention in view of brand awareness and brand image. He suggests that the greater the brand awareness, the greater the consumers' willingness to buy the product. Moreover, the better the brand image, the greater the consumer purchase intention [3].

As a visual symbol, brand logo not only conveys the design concept of products and the culture of enterprises, it also gives products of the brand a certain level of visual aesthetics thereby meeting the psychological demand of consumers [4]. Many previous studies on trademarks only focused on the logo itself instead of its effect on products. Niu discussed the importance of brand logo design, she suggested how to use the pictures and texts to quickly make consumers associate the logo with a brand, and to spread the emotion of products through the concise features of logos [5]. By defining the visual identity theory of a logo, Xu tried to create a logo system to suit clothing style in order to strengthen the brand advantage and enhance the brand's market competitiveness [6].

Existing documents were not focused on the research of consumers, what's more, they consequently disregarded consumers' perception to clothing product. So, how to use the added value of brand trademark to promote the positive association of consumers in the purchase behavior? The study from Lee J and Ko E showed that an individual wearing luxury clothing with brand logo would have an advantage and be more attracted by others in social activities [7]. Park CW examined whether brand names or visual symbols used as signs promote customer' self-identification and generate positive effects of brand perception. The results demonstrated that the positive effects of brand logos on customer brand commitment and firm performance derived from facilitating customer self-identity or expressiveness, representing a brand's functional benefits, and offering aesthetic appeal [8].

The present study explored the clothing preference behavior of consumer based on the stimuli with or without logos and analyzed the influence of different brand logos on consumer preferences from two aspects of brand popularity and brand value, so as to further understand the consumer psychology and to give reference for the brand logo design and brand value maintenance.

2. Experiment

2.1 Participants

30 undergraduates from Soochow University (10 males, 20 females, aged from 20 to 26) participated in the experiment. All subjects were normal or corrected to normal vision, physical health, no neurological or psychological diseases. The subjects were all right-handed and had to be tested individually, as well as had never done a similar experiment.

2.2 Stimuli

1) The determination of clothing brand. Through market investigation, we selected 100 clothing brands, then rated their popularity by questionnaires taken from Wang's measurement table, including four questions, "this brand is very famous", "this brand is widely known", "the advertisements of this brand are often seen" and "the brand's advertising gives very deep impression" [9]. Finally screened out the three most well-known high-value brands (Chanel> Dior> Hermes) and the three least known high-value brands (Alexander McQueen> CÉLINE> Elie Saab), the three most well-known low-value brands (ONLY> Peace Bird> Zara) and the three least known low-value brands (Fairy> Esprit> Innaydn).

2) The choices of the clothing picture. Each brand selected 20-30 clothing pictures from the vogue fashion network or brand official website, and then all pictures were scored based on subjects' preferences by using questionnaires, finally 10 pictures of each brand were selected (the average preference score of these images were 3 to 3.3 points) as the experimental stimuli. The size of each picture is 14 cm × 14cm, and each image must be placed on the same background and all faces of models also must be covered to avoid the effects of extraneous stimuli.

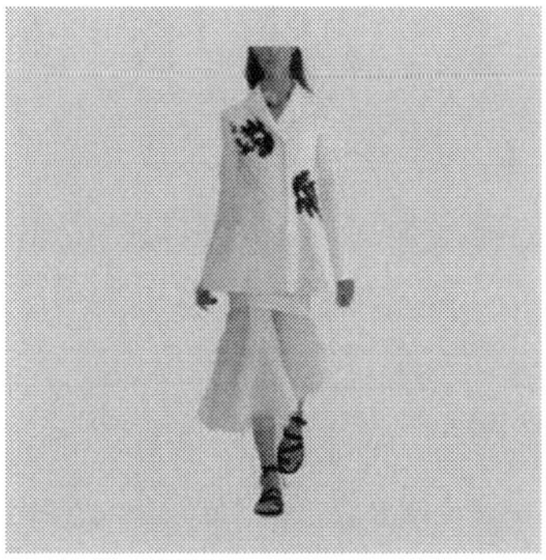

Without Logo

With Logo

Fig.1 Stimulus sample

2.3 Procedure

The experiment is conducted on the computer modulated by the E-Prime2.0 software to measure the preference perception behavior. The experiment was divided into two blocks, the first block required the participants to make preference evaluation on 120 pieces of clothing pictures without the logos, and the other block asked the participants to do it with the brand logos. Before the experiment, subjects sat on the chair about 60cm from the center of the monitor, and read the experimental instruction, according to which subjects pressed keyboard. Prior to the formal experiment, each subject had 6 stimulus pictures used for practice, so that subjects were familiar with the operation and adapted to the rapid stimulus presentation.

Each trail began with the presentation of a fixed cross in the center of the screen for 100 ms, after 400 ms, 120 pieces of clothing pictures appeared in a random way in the middle of the computer screen, and subjects evaluated their preferences based on Likert's five-point scale by pressing keyboard as soon as possible. After the reaction a random interval 500-700ms, and then entered the next test. The entire experiment was about 15 minutes.

2.4 Results

SPSS17 software was used for data statistics and results analysis.

2.4.1 Effect of Logo on Clothing Preference Evaluation

As is shown in Fig.2, comparing the score averages of subjects' preferences under the two situations in which clothing pictures with or without logos. We found that all scores of clothing preference for brands with logos, like Chanel, Dior, Hermes, Only, Peace Bird and Zara, except the last two least known low-value brands, were higher than that of non-logos, especially for the six high-value brands.

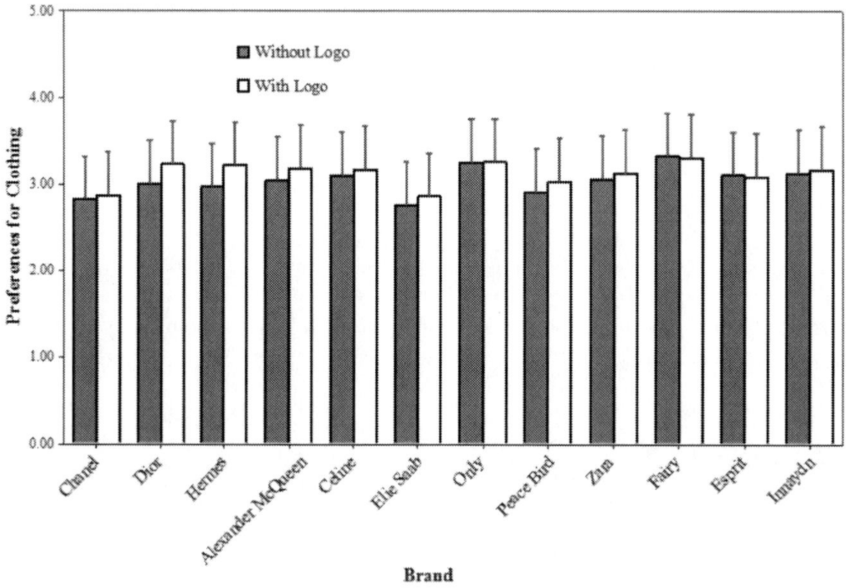

Fig.2 All brands with or without logo clothing preference evaluation

The repeated ANOVA of participants' clothing preferences showed that there is no significant effect for the logo [F = 3.167, P = 0.086> 0.05]. However, the significant effect is notable for the brand [F (11,319) = 3.167, p = 0.045], so we further analyzed the impact of different brand logos on consumers clothing preferences from two aspects of brand value and brand popularity.

1) Effect of different popularity brand logos on clothing preference evaluation

In accordance with the popularity, we divided 12 brands into two groups of well-known brands and less well-known brands, each of them contain three high-value brands and three low-value brands. As shown in Table 1, compared with the logos of less-known brands, the logos of well-known brands effecting on the clothing preferences is more noteworthy [F (1,29) = 5.095, p= 0.032].

Table.1 ANOVA of different popularity brand logos effect on clothing preference evaluation

Source	Well-known brands						Less well-known brands					
	Sum of square	df	Mean square	F	Sig.	ηp^2	Sum of square	df	Mean square	F	Sig.	ηp^2
Logo	1.260	1.000	1.260	5.093	0.032*	0.149	0.240	1.000	0.240	1.156	0.291	0.038
Brand	5.872	2.828	2.076	3.355	0.025	0.104	8.077	1.810	4.461	2.796	0.075	0.088
Logo×Brand	0.750	4.008	0.187	1.792	0.135	0.058	0.333	3.708	0.090	0.922	0.448	0.031

*: The mean difference is significant at the.05 level.

2) Effect of different value brand logos on clothing preference evaluation

According to the pre-experimental questionnaires, we conducted a repeated-measures ANOVA on six high-value brands and six low-value brands respectively. The results indicated that there is a significant effect on consumer preferences for logos of high-value brands [F (1,29) = 4.336, p= 0.046], while it is failed for low-value brands [F (1,29) = 0.361, P = 0.553> 0.05]. As shown in Table 2.

Table.2 ANOVA of different value brand logos effect on clothing preference evaluation

Source	High-value brands						Low-value brands					
	Sum of square	df	Mean square	F	Sig.	ηp^2	Sum of square	df	Mean square	F	Sig.	ηp^2
Logo	1.750	1.000	1.750	4.336	0.046*	0.130	0.084	1.000	0.084	0.361	0.553	0.012
Brand	6.767	3.468	1.951	4.883	0.002	0.509	4.564	2.449	1.864	2.933	0.049	0.092
Logo×Brand	0.506	4.204	0.120	1.539	0.193	0.050	0.244	4.064	0.060	0.905	0.465	0.030

*: The mean difference is significant at the.05 level.

3) The interactive effect among logos, brand popularity and brand value on product preference

we analyzed the interaction effect of four factors, including logo, popularity, brand value and brands. Table 3 reveals the preference average scores with logo or without logo for different popularity and value brands. The statistics suggested that preference averages with logo are higher than that without logo for well-known brands. And preference averages with logo are also higher for less well-known and high-value brands than that of brand without logo, except the lower popularity and lower-value brands, suggesting that subjects with the low-level awareness of a brand have no effect on their clothing preferences.

Table.3 The average data statistics of preference evaluation with or without logo

Logo	Popularity	value	Mean	Std.Error	95% Confidence interval	
					Lower limit	Upper limit
Without logo	Well-known	higher	2.934	0.089	2.751	3.117
		lower	3.079	0.085	2.906	3.252
	Less well-known	higher	2.971	0.103	2.761	3.181
		lower	3.190	0.099	2.987	3.393
With logo	Well-known	higher	3.109	0.113	2.878	3.340
		lower	3.141	0.091	2.956	3.327
	Less well-known	higher	3.076	0.115	2.840	3.311
		lower	3.189	0.116	2.951	3.426

In Table 4, we can see from the ANOVA that brand logos and popularity had a remarkable interactive impact on consumer preferences [F (1,29) = 4.485, p = 0.043].

Table.4 ANOVA of repeated measure for consumer clothing preferences

Sources	Sum of square	df	Mean square	F	Sig.	ηp^2
Logo	1.301	1	1.301	3.167	0.086	0.098
Error (Logo)	11.908	29	0.411			
Logo×Popularity	0.200	1	0.200	4.485	0.043*	0.134
Error (Logo×Popularity)	1.293	29	0.045			
Logo×Value	0.534	1	0.534	2.361	0.135	0.075
Error (Logo×Value)	6.555	29	0.226			
Logo×Popularity×Value	0.001	1	0.001	0.008	0.928	0.000
Error (Logo×Popularity×Value)	1.763	29	0.061			

*: The mean difference is significant at the.05 level.

2.4.2 Reaction Times (RTs) of Preference Evaluation

We also analyzed the reaction time of the preference evaluation. Viewed from Fig.3, the RTs of participants to make preference evaluation with logos is almost all less than that without logos.

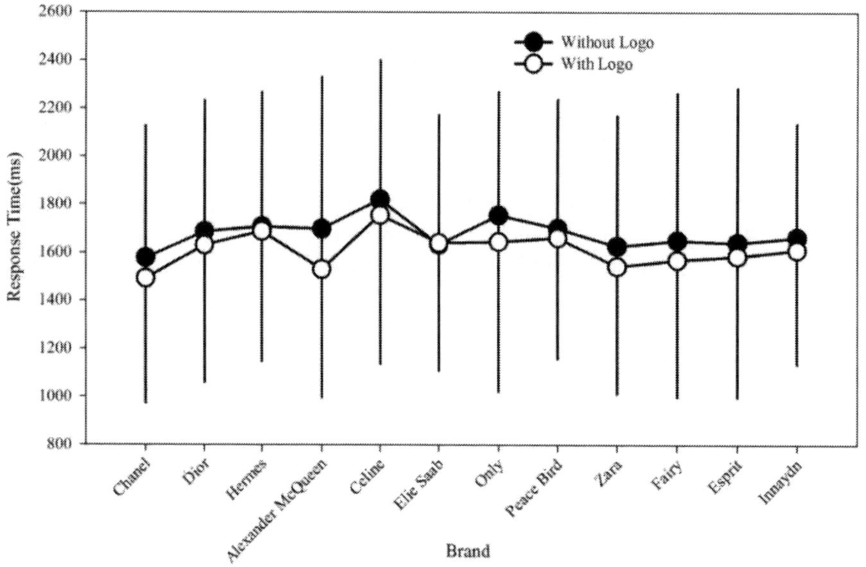

Fig.3 Reaction times (RTs) of clothing preference

We also carried on the analysis of variance to the reaction times of the preference evaluation. The results in Table 5 suggested that: (1) Whether there is a logo or not had no obvious effect on the RTs of preference evaluation, and (2) the interaction effect of the RTs between brand popularity and value was prominent [F (1,29= 4.705, p= 0.038], and (3) the reaction times of participates to make preference evaluation for different popularity brands had apparent differences [F (1,29= 6.682, p = 0.015], and (4) the interactive effect among popularity, value and brands had highly significant differences on the RTs for subjects to make preference evaluation [F (1, 29) = 18.732, p= 0.000].

Table.5 ANOVA of repeated measure for preferences evaluation RTs

Sources	Sum of square	df	Mean square	F	Sig.	ηp^2
Logo	828595.998	1	828594.998	1.025	0.320	0.034
Error (Logo)	2.345	29	808650.082			
Popularity×Value	322612.440	1	322612.440	4.705	0.038*	0.140
Error (Popularity×Value)	1.989	29	68570.589			
Popularity×Brand	570981.841	1	570981.841	6.682	0.015*	0.187
Error (Popularity×Brand)	2.478	29	85451.033			
Value×Brand	33.909	1	33.909	0.000	0.984	0.000
Error (Value×Brand)	2353116.734	29	81141.956			
Popularity×Value×Brand	609881.593	1	609881.593	18.732	0.000*	0.392
Error (Popularity×Value×Brand)	944177.820	29	32557.856			

*: The mean difference is significant at the.05 level.

3. Discussion

In this paper, behavioral experiment was used to investigate the influence of different popularity and value brand logos on consumer preferences. The results illustrated that the popularity and value played important roles in the process of logos affecting consumer preferences. The higher popularity and higher value brands have a conspicuous impact on consumer preferences.

First of all, in the two cases with or without logos, subjects evaluated their product preferences. The experimental results indicated that preference evaluation in the case with logos was higher than that in the case without logo for higher popularity brands, and the same was true for the higher value brands. Brand logo, which is closely linked to the clothing characteristics, will make consumers have a better brand association [10], and produce a positive response to clothing preferences on account of the halo effect of higher popularity and higher value brands.

Secondly, whether there is a brand logo or not does not have a noteworthy influence on the preference perception. However, for popularity brands, the brand logo has a significant impact on the consumer preferences. Similarly, for high-value brands, brand logos also prominently affect the consumer preferences. Brand logo has played a positive role in consumer's attitude towards brand and brand loyalty [11]. Well-known and high-value brands make consumers produce the symbolic benefit association, such as social recognition and self-expression [12]. And these brands often contain high material mental value, which have a competitive advantage in the aspect of brand awareness and product perception [13]. Therefore, a brand logo more obviously impact on consumer product preference for higher popularity and value brands.

Thirdly, in view of the interaction among the brand logo, brand popularity and brand value, popularity is the main factor of brand logos affecting consumer clothing preferences. Regardless of brand value, consumers tend to make a high preference evaluation with logo for well-known brands. The more familiar with the brand the consumers are, the easier for them to make brand evaluation, on the contrary, if consumers are not familiar with the brand, access to brand information is limited [14]. For the lower popularity and value brands, consumers' preference scores with logos was lower than that without logo, it indicated that brand logo does not have a significant effect on consumer preferences when consumers in a low awareness of that brand.

Finally, the RTs of participants during the evaluation of clothing preferences with logos were shorter than that without logos, since brand logos provided a certain information for the subjects, which is related to the description and evaluation of the brand in consumers' memories and is easy for subjects to respond quickly [15].

4. Conclusion

This paper explored the impact of brand logos on consumer product preferences, we obtained the following conclusions:

1) The higher scores were associated with brand logos during the clothing preference evaluation, suggested that subjects had a positive preference reaction owed to the emotional effect of brand logos.

2) There is a superiority effect for the logos of well-known brands or high-value brands on clothing preferences.

3) Subjects can quickly make a preference response because of they getting much information form logos.

References

[1] Keller KL. Conceptualizing, measuring and managing customer based brand equity [J]. JM: 1993; 57 (1): 1-22.

[2] Yang WW, Liu X. Effects of brand cognition on consumer purchasing behavior [J]. Commercial Research: 2010; 03 (395): 158-162.

[3] Wang JB, Qin MY. Effects of brand identity and brand image on consumers' willingness to purchase [J]. Corporation Research: 2011 (2); 50-51.

[4] Xue Q, Zhao YY. The application and value of brand logo in the design of fashion apparel [J]. Science and Technology: 2015 (18); 277.

[5] Niu HC. The importance of logo design in the brand [J]. Modern Decoration (Theory): 2015 (7); 118.

[6] Xu J. Study on visual identity of logo in brand clothing [D]. Qingdao University: 2008.

[7] Lee J, Ko E, Megehee CM. Social benefits of brand logos in presentation of self in cross and same gender influence contexts [J]. JBR: 2015; 68 (6): 1341-1349.

[8] Park CW, Eisingerich AB, Pol G, et al. The role of brand logos in firm performance [J]. JBR: 2013; 66 (2): 180-187.

[9] Wang HZ. Brand measurement and promotion: from measurement model to implementation strategies [M]. Beijing: Tsinghua University: 2007.

[10] Shen Q, Wang FH. Influence of brand identity perception consistency to the brand attitude [J]. Shanghai International Studies University: 2014; 36 (6): 49-54.

[11] Müller B, Kocher B, Crettaz A. The effects of visual rejuvenation through brand logos [J]. JBR: 2013; 66 (1): 82-88.

[12] Wu F. Study on the influence of brand marketing strategy on brand association [J]. Jinan University: 2009.

[13] Feng L. Study on the relationship between luxury package purchase motivation and the conspicuousness of brand logo preference [D]. Nanjing Normal University: 2011.

[14] Zhu YM, Zhou YF. Effects of brand familiarity and sponsorship on consumer response [J]. Journal of Business Economics: 2013; 1 (1): 43-51.

[15] Keller K L. Brand Synthesis: the multidimensionality of brand knowledge [J]. JCR: 2003; 29 (4): 595-600.

Research on the Clothing Consumption Experience Demand Based on the O2O Model

Xi Pu, Jian-Fang Liang*

Apparel & Art Design College, Xi'an Polytechnic University, Xi'an, Shaanxi, 710048, China

*Corresponding authors' email: 342110784@qq.com, 1813339441@qq.com

Abstract

With the rapid development of the Internet and modern e-commerce, the consumer demand for apparel products and services, as well as consumer behaviour, has changed. In turn, the O2O (online-to-offline) business integration model has come into being, while understanding consumer needs with regard to O2O experience in clothing stores has become the top priority.

The purpose of this paper is to explore factors in terms of how to enhance the competitiveness of clothing brands. In this study, a questionnaire survey, SPSS software for statistical analysis and case analysis were used. In the process, this paper puts forward six common attributes of successful O2O experiences in clothing stores.

Keywords: O2O Model; Experience; Clothing Brand; Consumer; Experience Demand

1. Development and Research Status of the O2O Clothing Model

The O2O model is a new type of transaction, which is based on the development of Internet trading platforms, which uses online marketing and purchases to drive offline consumption [1].

For the individual, the O2O experience represents a change in lifestyle. For the market, it is a tempting opportunity. In January 2017, the China Internet Network Information Centre (CNNIC) released the 39th China Internet Development Statistics Report in Beijing. The report shows that, as of December 2016, the number of Chinese Internet users reached 731 million, the Internet penetration rate reached 53.2% and mobile Internet users accounted for 95.1%. The online payment proportion of mobile phone users has increased from 57.7% to 67.5%.

Thus, e-commerce is having a significant impact on the real economy. The O2O model is a new model, which has emanated out of the development trend in e-commerce, while being rooted in real-life service.

The earliest "experience marketing" can be traced back to the book, Future Shock, published by an American scholar, Alvin Toffler, in 1970. In Toffler's view, the thousand-year history of human economic development could be summarized into three stages: product economy, service economy and the era of economic experience [2].

In 1998, Joseph Pine and Jim Jimmy pointed out that economic value has developed from goods and service to experience [3]. Professor Assim Ansari proposed the "experiential strategy model", which divides experience into five branches: feeling, emotion, creative awareness, physical experience and social characteristics.

In 2001, from a strategic point of view, Bernhard Schmidt systematically highlighting the differences between experiential marketing and traditional marketing for the first time, arguing that the tactical experience tool is composed of communication, reputation, product, brand, environment, network and people. He also developed sensory marketing, emotion marketing, thinking marketing, action marketing and related marketing according to five types of customer experience [4].

The domestic research on the O2O experience model started late. The construction of the theoretical system has mainly involved translating foreign viewpoints. In 2000, Xin-gong Gao and Xiang-rong Liu put forward the experience marketing concept, explained its meaning and rationality for existence within the market economy. Then, they analysed five marketing strategies, which embodied "experiential marketing" ideology [5].

In 2002, Professor Feng-jun Liu sought to help out enterprises that were confronted by the marketing dilemma in a new economic era by innovating marketing concepts and methods [6].

In 2004, Qin-xue Deng distinguished the similarities and differences between generalized experience marketing and narrow experience marketing, as well as traditional marketing and experience marketing. He also described the experience marketing situation at home and abroad, as well as proposing seven strategies for experience marketing, which he named "7EP" [7].

In 2005, Xiao-ping Tang concluded that consumer demands have different characteristics in different economic contexts, arguing that consumers are no longer satisfied with the characteristics and effectiveness of products, preferring more personalized satisfaction in the consumption process [8].

In 2013, Xia-zhi Ge also highlighted the importance of the O2O product design [9].

In 2014, Ji-cheng Song discussed the close connection between the product demand changes and the marketing strategies [10]. In addition, some scholars have explored and researched customer experience needs.

In summary, all of these studies focus on consumer experience needs and marketing strategies, but without any specific reference to the O2O model. Based on these previous studies, this paper analyses the experience demands of clothing consumers in the context of the O2O model.

2. Investigation and Analysis of Experience Consumption

Given that college students make copious use of the Internet, they are in a good position to accept new things. Besides, they are a large group of sustainable consumers. As such, the survey mainly concentrated on 19- to 25-year-old college students. At the same time, in consideration of the comprehensiveness of the questionnaire, the survey also includes the grassroots staff, managers and freelancers. In addition, this paper adapted the field and network investigation approach to understand consumers' experience needs in relation to the O2O model.

2.1 The Basic Situation of the Survey

Twenty questions was created for this questionnaire. According to the ratio of 1:6, 120 questionnaires were issued in total. Due to the limitation of time and conditions, this survey only received 108 valid questionnaires. The receipt rate among the sample is 90%. After the reliability and validity analysis of the questionnaire, the reliability coefficient was α=0.83, while KMO=0.87, which proves that the questionnaire had good reliability and was suitable for further study.

2.2 Analysis and Discussion of the Results

2.2.1 Analysis of the Understanding of the O2O Model

Fig.1 presents a cross-analysis of the online shopping experience and the clothing consumption experience regarding the O2O model. As shown in Fig.1, the green-coloured column means that the respondents have a certain understanding of online shopping and the O2O model, while the online shopping experience among survey respondents is proportional to their understanding for the O2O model.

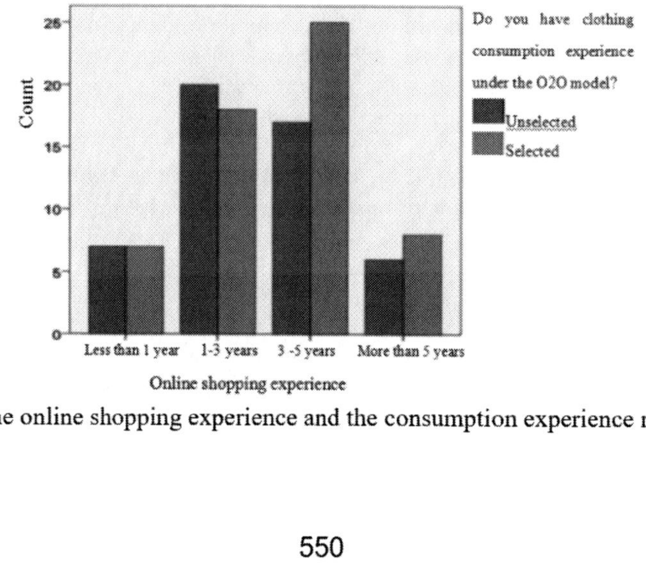

Fig.1 Cross-analysis of the online shopping experience and the consumption experience regarding the O2O model.

Meanwhile. Table 1 presents the subjects' understanding about the clothing brand experience in domestic and foreign shops. Among them, the Metersbonwe brand is a pioneer of the O2O model experience in shops within the domestic apparel industry. The respondents have the highest understanding about it (68.5%). The Nike and the Uniqlo offline experiences have also been developed, with the subjects' understanding of these two brands respectively accounting for 53.7% and 56.5%. Other clothing brands are trying to develop an O2O model, which proves that it represents a future trend in the apparel industry.

Table.1 Clothing experience stores you have heard of or visited

Variable	Item	Frequency	Percentage (%)
The clothing experience store you have heard of or visited (multiple choice)	Metersbonwe	74	68.5
	Uniqlo	58	53.7
	Nike	61	56.5
	Jack Jones	27	25
	Bestseller	19	17.6
	Others	48	44.4

2.2.2 The Analysis of Consumer Experience Needs

Table 2 presents the research findings on consumers' specific requirements about the regional division and functional area in the O2O experience clothing stores. In order to attract consumers, stores should increase the range of functional experience areas, while consumers hope they can receive a more personalized and diversified experience in which to relax. Among the highest expectations are: a recreation experience area (58.3%), a product-related interactive experience area (56.5%) and a personality subject area (54.6%).

Second, expectations about an consultation area and an information browsing area accounted for 52.8% and 42.6%. A consultation area could allow customers to receive professional help and services at any time in order to eliminate their worries while in the store. An information browsing area would be equipped with free Wi-Fi and a variety of Internet devices, so that consumers can look over the brand information and browse fashion information.

51% of consumers want clothing stores that provide a full range display of products, while 56.5% demand virtual fitting walls. As consumers are no longer satisfied with the ordinary shopping experience, their expectations about the product-related interactive experience have increased more and more. 51.9% of consumers want the store to be equipped with tables, chairs, drinks and snacks. In this way, they can enjoy the shopping atmosphere while resting in the store. This proves that human services can more easily satisfy consumers. 48.1% of consumers think that a clothing experience store should have secure payment equipment. Regarding the O2O model, consumers would like the store to eliminate their concerns about payment security and provide a pleasant shopping experience.

Table.2 The general requirements of the O2O experience clothing store

Variable	Item	Frequency	Percentage (%)
The desired division area in a clothing experience store (multiple choice)	Product-related interactive experience area	59	54.6
	Recreation area	63	58.3
	Personality theme area	61	56.5
	Information browsing area	46	42.6
	Cash register area	32	29.6
	Customer information and consultation area	57	52.8
Functional equipment needed in a clothing experience store (multiple choice)	A full range display of products	55	51
	Stereo interactive devices	51	47.2
	Virtual fitting wall	61	56.5
	Tables and chairs, with drinks and snacks	56	51.9
	Free Wi-Fi and a variety of Internet devices	44	40.7
	Safe payment methods and equipment	52	48.1
	Powerful data collection system	43	39.8

Table 3 concerns the survey of consumer demands about product experience and in-store decor in the clothing O2O experience store.

From Table 3, we can see that what consumers most expect are professional guidance and a perfect fitting experience (60.2% and 59.3%, respectively). Currently, almost all those working in China's garment retail sector lack professional knowledge, as there are only a few salespersons who can provide professional guidance to customers, which means that the requirements of consumers are not being met in this regard. Besides, the problem with traditional online shopping exists is that the depictions do not match the actual goods. The O2O experience store could solve these problems more effectively, while consumers can enjoy the perfect fitting experience in store. In addition, 49.1% of consumers hope that the experience store will meet their needs for one-stop shopping, while 41.7% consumers would like to experience diverse clothing styles in store. Consumers also except the store to address their concerns about product quality.

Table.3 Specific expectations about the O2O experience clothing store

Variable	Item	Frequency	Percentage (%)
The desired product experience that should be provided by a clothing experience store (multiple choice)	Good product quality	53	49.1
	Perfect fitting experience	64	59.3
	Professional guidance	65	60.2
	Complete range of product	45	41.7
The desired in-store décor that should be provided by a clothing experience store (multiple choice)	Attractive window displays	61	56.5
	Feeling comfortable in the internal environment	63	58.3
	Diversified product display	58	53.7
	The experience theme in line with the brand culture	65	60.2
	Reasonable combination of products on display	56	51.9

As for the décor in store, consumers are most satisfied with an experience theme that reflects the brand culture, as well as offers a comfortable internal environment (60.2% and 58.3%, respectively). As the theme is the soul of the experience shop, it should "scatter clues" around the shop, which are consistent with the brand culture. This would give consumers a deep and complete impression about the store. However, at present, clothing stores are faced with a serious homogenization problem in their internal environment, which means they struggle to satisfy consumers' needs. At the same time, the window displays and mannequins represent an important "magic wand" to lure consumers into the store. Although 56.5% consumers expect an attractive store window, nowadays, most clothing brand stores lack creative window displays.

Furthermore, 53.7% consumers have high expectations about the variety of products on display and 51.9% would like to see a reasonable mixture of products on display. Diversified product displays and reasonable product portfolios not only highlight the attractive of branded products, but also make the whole store look clean and tidy. But, at present, domestic clothing stores place more emphasis on the number of products.

In light of the above-mentioned investigation and analysis, it can be seen that consumer behaviour has changed a lot. As for consumption structure, the proportion of emotional needs has increased, while consumer content, and personalized and interactive products and services have gradually replaced the demand for popular standard products. Regarding the value of the target, the focus on the product itself has moved towards encouraging feelings of product acceptance. In this regard, consumers are not satisfied with the induction and manipulation of enterprises, but are actively involved in the design and manufacture of products [5].

At the same time, the requirements with regard to the shopping environment have increased more and more. The in-store decor, regional division, functional equipment and other aspects have informed specific expectations among consumers, while expectations about the product extend to the comfortableness of the shopping environment, experience, interactive entertainment, aesthetics and convenience.

2.2.3 Analysis of the Consumer Experience Problems with the O2O Model

Table 4 reflects consumers' intuitive feeling about the O2O experience clothing store nowadays.

As shown in Table 4, consumers are not satisfied with the latest clothing experience stores. Among their most negative issues are service attitude and after-sales service, store distribution and dissatisfaction about product (49.1%, 48.1% and 42.6%, respectively).

Table.4 Feelings about the O2O experience clothing shop

Variable	Item	Frequency	Percentage (%)
Why your experience in current clothing brand shops is unsatisfactory (multiple choice)	Limited store distribution	52	48.1
	Fear of inconsistent goods	46	42.6
	Service attitude and after-sales service	53	49.1
	Logistics is too slow	40	37
	Limited range of goods	24	22.2
Why you are attracted into an O2O experience clothing store (multiple choice)	More promotions	30	27.8
	Have the fitting experience	46	42.6
	Payment is convenient	22	20.4
	Personalized service	54	50
	Relaxed recreation area	49	45.3
	High-tech experience devices	37	34.3
	Free Wi-Fi	42	38.9

Secondly, in terms of attractiveness, consumers express their preference for all items listed in the questionnaire. Among them, personalized service, a relaxing leisure area and try-out experiences are the most attractive factors (50%, 45.3% and 42.6%, respectively).

It can be seen that, although many clothing enterprises are applying the O2O model within their domestic apparel industry, there still exist many obstacles to satisfying consumer experience needs.

1) The atmosphere created by the offline experience store is inadequate

Most of the clothing brand stores offer little in the way of experience, preferring instead to focus on sales rather than exploiting the deep psychological and spiritual needs of consumers. They also fail to realize the importance of in-store decor, displays and colour, which results in a weak experience atmosphere in the store. Besides, customers like to enjoy freedom and a relaxed atmosphere in the store, while companies are concerned that they may turn offline customers into online users. It is also contrary to the self-help experience atmosphere in the store.

2) Lack of offline service value

As good service is a vital element to win favour from consumers, large-scale brand experience stores are built in first-tier cities. However, the physical store coverage rate is very small in second and third tier cities, which means that a considerable number of customers cannot consume with convenience, nor enjoy a complete after-sales service. Moreover, many brands have launched a variety of online marketing activities, attracting many consumers to order products online. Meanwhile, poor-quality offline services can greatly undermine consumer enthusiasm about the experience in stores, which in turn greatly reduces consumers' sense of belonging and loyalty towards brands.

3) Unable to satisfy consumer needs regarding feelings and experience

In an experience built around the O2O model, consumers pay more attention to their own sense of presence and feelings, while hoping to participate in all processes relating to products, such as their design, production and sales. Meanwhile, most brands only provide a sensory and superficial experience, ignoring the scope for interaction with consumers, and not providing them with a personalized, emotional and participatory range of experience activities. Therefore, the stores cannot satisfy consumer demands about feelings and experience.

4) Payment security issues

In the O2O model, online payment often involves consumer information disclosure, property fraud and other issues. At the same time, the offline payment methods in experience stores are also facing security issues with regard to mobile POS machines. The Verizon data leak investigation report said that 2014 was a significant year for retailer data leakage. Retailers were attacked by large-scale POS malicious software, resulting in a large amount of personal bank account and consumption information being leaked. As such, this has become a major concern among consumers.

5) Brand culture is not prominent

In the O2O model experience, consumers look for a brand's products, services and culture to resonate with their psychological needs. Consumers seek an overlap between a brand's culture and their notion of self-worth, which in turn results in a pleasurable shopping experience. Nowadays, however, the apparel industry's products and services are increasingly homogeneous, with brands often pursuing sales, blindly producing hugely popular styles, while ignoring their own culture.

6) Lack of customer loyalty

In the O2O model, the offline experience store relies on the O2O platform to attract more online consumers. But the consumers who are attracted by online promotions cannot build up a real emotional relationship and brand loyalty. Therefore, knowing how to increase customers' brand loyalty by creating the right characteristics shopping for brand experience has become a priority. In addition, the integrity issue has been a problem with O2O consumption patterns. The reason is that consumers do not have enough confidence in clothing brands. As such, clothing brands should improve their service quality and establish greater credibility in order to eliminate consumer concerns.

These problems will be solved in this essay.

3. Suggestions and Countermeasures

3.1 Recommendations on the O2O Experience Clothing Shop

Table 5 presents consumer recommendations about the O2O experience in clothing stores.

According to this table, the recommendations embody the aspects that consumers believe that clothing experience stores should improve. Among the most popular consumer recommendations are that the offline experience store should provide more personalized, diversified, emotional and lively experience activities, should be located in a convenient location and should highlight the brand culture (59.3%, 57.4% and 56.5%, respectively).

Table.5 Recommendations for O2O experience clothing shops

Variable	Item	Frequency	Percentage (%)
Suggestions for offline clothing experience shops (multiple choice)	Located in a convenient location	62	57.4
	Ensure a larger area of experience	41	38
	A clear regional division	45	41.7
	Highlight the brand culture	61	56.5
	Provide more personalized, diversified, emotional and lively experience activities	64	59.3
	Provide access to online information sharing and customization	45	41.7

3.2 Measures to Improve the O2O Consumer Experience

Through the analysis of Metersbonwe and Uniqlo, this paper proposes the following six aspects to help clothing brands improve the consumer experience under the O2O model.

1) Focus on the location and area planning of the experience shop

The two brands' experience shops are located in downtown areas and occupy thousands of square metres of acreage, factors that are not only about offering convenience to consumers, but also about ensuring there is adequate experience space.

The Metersbonwe experience stores are essentially located in the bustling streets of first-tier cities. The first full-brand experience store was built in the most prosperous district of Chongqing in 2014, with a total area of over 4,000 m^2. The product range covers adults to children in order to meet consumer needs regarding a one-stop fashion shopping experience [11]. The Uniqlo experience stores are also located in first-tier cities with convenient transportation. The largest flagship experience store in the world was opened in a bustling commercial street in Shanghai in 2013, occupying more than 8,000 m^2 of business area. The store offers the broadest possible range of goods and the highest level of service to provide consumers with a pleasant shopping experience [12].

2) Sensory marketing - designing an experience store and creating a perfect experience atmosphere

The two brands created a novel experience atmosphere in stores through the in-store décor, which is embodied in the brand concept, the leisure area and high-tech experience devices.

The Metersbonwe Chongqing experience store employs an "art factory" theme, which uses graffiti, illustrations, gears, factory lights, old-fashioned sewing machines and other factory elements to create a retro industry feel. At the same time, the store has set up many high-tech interactive devices, a coffee bar, book bar, a children's play area and an art exhibition to leave consumers with a deep impression. In addition, the new Kunming experience store, built in 2016, splits the original store style into four styles suited to different groups of people, each of which has its own distinct store display design. The Uniqlo store in New York has sought to increase consumer purchasing intentions by providing a Starbucks coffee, as well as iPads, sofas, tables and chairs. The "Magic for All Flagship" experience store opened in Shanghai in 2013. It has not only designed a unique window display for each floor, but also uses Disney elements to divide the store into five differently themed areas. The displays in each area changes in line with the latest popular movies, while electronic sound interactive zones are provided where consumers can interact with the corresponding movie characters. In short, Uniqlo has created a Disney-themed world for consumers [13].

3) Make full use of modern technology

The two brands exploit the advantages offered by modern technology to establish their own new media marketing presence. They have also set up interactive devices in their experience stores as a way of offering novel and convenient services to consumers.

The Metersbonwe Chongqing experience store has installed interactive video projectors, spherical visual magazines and "fashionable collocation" devices. Moreover, the store allocates a tablet PC to their fashion advisers, so that they can show customers specific fashion collocations and share more information of the products. Meanwhile, the tablet PC is connected to all Metersbonwe stores in China, which means that availability of an item in terms of colour and size can be checked nationwide when a customer is in store. This has enhanced the consumer shopping experience, but also reduced the amount of lost sales. In addition, in terms of its future marketing strategy, Metersbonwe plans to observe consumer intentions in store and record purchasing preferences using Wi-Fi data, so that consumer demands can be analysed for the purposes of providing accurate information to its CRM system.

Meanwhile, Uniqlo is utilizing data from Alipay to obtain the regional characteristics and geographical distribution of consumer groups. From this, it can determine the location, publicity and marketing strategy for upcoming experience stores [13]. Uniqlo has also used the WeChat social media platform to interact with consumers and collect their views about products, which are then used to improve physical sales. For their part, consumers can access detailed information, collocation guides and the real-time inventory of products through their smartphone to scan the corresponding code whenever necessary: in short, a convenient shopping experience for consumers. Secondly, customers in the Shanghai experience store can print their self-designed T-shirt instantly, while customers in the Japan Uniqlo store can utilize the 4D virtual fitting system on the official online store and experience the convenient self-checkout points. All of these are possible thanks to modern technology.

Fig.2 Fashion device in Metersbonwe [11]

Fig.3 Virtual fitting system in Uniqlo (http://uniqlo.bigodata.com.cn/u2/mini.php)

4) Demand marketing - to meet the diverse experience demands of consumers

The O2O model emphasizes the fact that consumers enjoy the perfect experience whenever they are online or offline. The two brands both attach great importance to consumers' views and meeting their diverse experience demands. This reflects the fact that the brands pay attention to their consumers.

The four different styles of clothing and displays in the Metersbonwe Kunming experience shop, along with the one-stop shopping service in the Chongqing experience shop, satisfy the demand for the experience of convenience, subdivision and upgrade [14]. The Chongqing experience store is equipped with 3D interactive devices, a leisure bar, a children's play area, which is important to concerned parents, an art exhibition, which reflects the brand culture, fashion advisors offering professional collocation advice, convenient and safe payment services and a fast delivery service. All of these satisfy customers' diversified demands for an interactive experience, a real-life experience, personalized service, acknowledgement of their emotional and other needs [11].

In addition, the Uniqlo brand makes full use of the SNS social networking platform to mobilize users' interactive enthusiasm. Meanwhile, it has carried out a variety of online promotion activities, including online queuing, the Uniqlo alarm clock, Uniqlo recipes and customized Disney T-shirts to offer full-scale interaction with consumers, while making a positive connection between online and offline experiences [13]. The Shanghai flagship store has created five Disney-themed areas and a UT factory custom area to build a Disney shopping paradise for consumers. Besides, the store offers the "My Uniqlo" embroidery service to meet the consumer demand for clothing customization. On the Uniqlooks platform, users can upload photos to share their Uniqlo style, while finding out what others are wearing around the world. In turn, this fulfils consumers' personalized demands for sharing information online, while shortening the distance between brand and consumer. Moreover, in the Japan Uniqlo store, all the goods are labelled with IC tags, which allow consumers to buy products at any time from the self-checkout points.

Fig. 4 Children' play area in Metersbonwe [11]

Fig. 5 UT factory custom area [13]

5) Brand marketing - highlighting the brand's culture and characteristics

The two brands highlight their respective culture's concepts and spiritual values, while enhancing consumers' brand identity through the in-store décor and the cultural atmosphere in the store.

Fig. 6 Leisure bar in Metersbonwe [14]

Fig.7 The LEGO S/S series (http://www.uniqlo.cn/)

The Metersbonwe experience stores emphasize the concept of "one city with a culture, one store with a story", which also reflects the differentiated experiences and cultural connotations [11]. In the Chongqing experience store, the leisure bar allows consumers to enjoy coffee and read books, while the spherical visual magazines lets them access knowledge about film, art, architecture, science and technology easily. In addition, the art exhibition has become an incubation platform of cultural creativity, whereas the fashion collocation devices combine the leisure and fashion concepts of the brand.

The Uniqlo experience store displays different models on each floor to show the all-matching fashion concepts. Besides, the self-checkout points in the Japan Uniqlo store and the online 4D virtual fitting system embody the self-service shopping concept. Furthermore, Uniqlo has implemented the co-designed product ideas of "basic models + famous brands" in recent years: from the Marvel series, to the 2015 F/W (fall and winter) series involving a former Hermes designer, and the LEGO S/S (spring and summer) series, these all reflect the brand concept, while narrowing the distance between famous design and consumers [13].

6) A better integration of the online and offline brand

The two brands offer a good adaptation of the O2O model. Their exclusive official websites, social media accounts and apps are able to attract more consumers to the offline experience shop, while the quality of the in-store services also encourages consumers to go back and shop online. In turn, they have achieve the online-offline integration of their respective brand.

Metersbonwe attracts consumers by creating unique offline experience stores to retain customers. In turn, this extends the length of time that consumers stay in store, which increases offline consumption, while encouraging them to visit the online shopping platform [14]. Meanwhile, the 2D code on a cash receipt acts as an online offer coupon. Moreover, the online shopping platform sends new product coupons, which can be used in offline stores on a regular basis, while in-store advisors use tablet PCs to connect with other stores to check the availability of an item at a nationwide level.

Meanwhile, Uniqlo has standardized the price of its goods available online and offline to facilitate "store pickup". The online shopping platform provides the location of nearby stores, the availability of a product, online ordering and a delivery service to extend the sales radius and improve sales efficiency [12]. The app coupons and the 2D code can also be used in the offline store. In addition, consumers can access detailed product information, collocation guides and a real-time inventory by scanning the 2D code in the offline experience store. Consumers can choose to pay by WeChat, Alipay or cash. As such, Uniqlo attracts customer by a variety of methods, which has enabled it to achieve online and offline connection and integration with great success.

Fig.8 Technical diagram of the Metersbonwe O2O model [12]

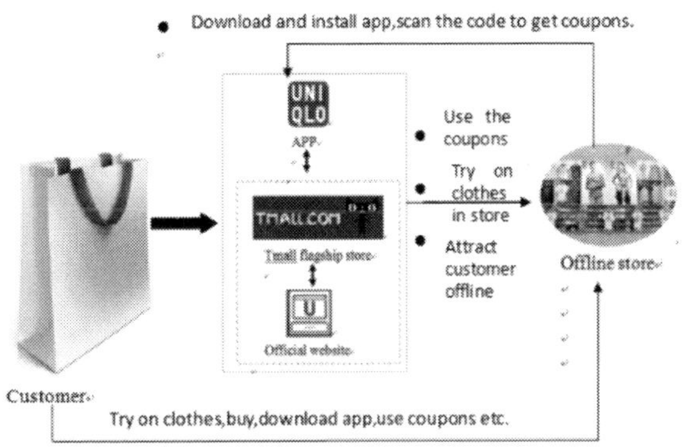

Fig.9 Technical diagram of the Uniqlo O2O [12]

4. Conclusion

Based on the questionnaire survey of the specific consumer groups, this paper summarized the consumer demands according to O2O experience consumption. It was mainly focused on interactive, personalized, emotional, human, cultural and other services. Meanwhile, the survey also showed that consumers are not content with the O2O experience services offered by domestic clothing brands. Among the problems found with domestic offline clothing experience stores are a poor experience atmosphere, poor offline service, consumer experience needs not being satisfied, insecure payment options, unremarkable brand cultures and insufficient customer loyalty.

This paper also analysed the Metersbonwe and Uniqlo clothing experience stores, which identified six successful attributes: focusing on the experience store location and area planning, designing the experience store attentively, creating a perfect experience atmosphere, making full use of modern technology, meeting the consumers' diverse experience demands, highlighting the brand culture and characteristics, and effective online-offline integration.

References

[1] Zhou YL. The O2O model in the apparel industry [J]. Program and Application of Information Technology: 2014; 6: 139-140.

[2] Toffler A. Future Shock [M]. Random House: 1984; 6: 78.

[3] Pine BJ, Gilmore JH. Welcome to the experience economy [J]. Harvard Business Review: 1998; 7: 8.

[4] Schmitt BH. Customer experience management [M]. Wiley Pub.: 2003; 1: 23.

[5] Gao XG, Liu XR. Exploring a new concept in the development of China's retail industry [J]. Journal of Fujian Commercial College: 2000; 6: 21-25.

[6] Liu FJ. Consumer demand and marketing strategy in the experience economic era [M]. China Industrial Economy Press: 2000.

[7] Deng QX, Wang XY. A study of experience marketing [D]. Graduate thesis. Capital Economics University: 2004.

[8] Tang XP. Emotional marketing strategy in the context of the experience economy [J]. The Road to Business: 2005; 7: 39-41.

[9] Ge XZ. User experience study of O2O products [D]. Bachelor's thesis. Wuhan Technology University: 2013; 1-4.

[10] Song JC. Changes in consumer product characteristics, demand and marketing strategy in the context of the O2O trend [J]. Theoretical Studies in Finance and Economics: 2014; 5: 89-94.

[11] Information from: http://fashion.ifeng.com/a/20140331/40002497_0.shtml.

[12] Gu M. Where will O2O go? [J]. Perspective: 2014; 2: 13-15.

[13] Li H. Influence of personalized service on customer loyalty based on the O2O apparel brands [D]. Professional master's thesis. Zhejiang Sci-Tech University: 2015; 20.

[14] Information from: http://corp.metersbonwe.com/Index/ArticleInfo?aid=557.

Exploring the Buyer-supplier Relationships in the Textile and Apparel Industry

Cai-Xia Chen[1], Ying Zhang[1], Yi-Xiong Yang[1,2,3,4*]

[1] *College of Fashion and Design, Donghua University, Shanghai, 200051, China*
[2] *Key Laboratory of Clothing Design and Technology, Ministry of Education, Shanghai, 200051, China*
[3] *Shanghai Style Fashion Design and Value Creation Collaborative Innovation Center, Donghua University, Shanghai, 200051, China*
[4] *Shanghai Institute of Design and Innovation, Tongji University, Shanghai 200092, China*

*Corresponding author's email: yyx@dhu.edu.cn

Abstract

Nowadays, China is the largest textile and apparel exporter in the world. However, low-cost countries, such as Vietnam and Bangladesh, have increasingly become favorite sourcing destination for international buyers. Thus, how to survive in the fierce global competition has become an urgent issue for Chinese textile and apparel suppliers. Furthermore to make matter worse, Chinese manufactures have currently encountered issues regarding the rising cost, high labor turnover and the shortage of skilled workers. This paper aims to find out the relationship between Chinese suppliers and international buyers, throwing new light upon "made in China" subject. Quantitative research method is applied to manifest the relationships between the strategies of suppliers and buyers. The results suggest that the strategies adopted by buyers in the global sourcing supply chain have positive effects on the strategies of Chinese suppliers, which also benefit the suppliers' competitive advantages and business performance.

Keywords: Textile and Apparel; Supply Chain; Buyer-supplier Relationships; Global Competition

1. Introduction

Global sourcing is the trend of international business and has attracted increasing attention from researchers [1-3]. The rise of China's export-oriented apparel industry since the 1990s has been driven largely by global sourcing practices intent on capturing the cost advantages [4].

It has been commonly argued that China has lost its competitive advantage in textile and apparel manufacture [5]. However, trade statistics appears otherwise. Textile and apparel trade volume of China has been increasing since 2004 although Chinese textile and apparel suppliers have faced fierce international competition. Chinese textile and apparel suppliers are still playing important roles in the global fashion supply chain. Therefore, it is important for Chinese suppliers to adjust their strategies in order to adapt to the global competition.

The purpose of this study is to explore the relationships between international buyers' sourcing strategies and operational strategies of Chinese textile and apparel suppliers. In doing so, this study first identified the challenges associated with global sourcing that were perceived by Chinese suppliers. Then it examines how the buyers' sourcing strategies influence suppliers' operation strategies. Finally, the results of Chinese textile and apparel suppliers achieve from their operational strategies were examined in this study.

2. Literature Review

2.1 Buyers' Sourcing Strategy

Textile and apparel industry is a typical buyer-driven commodity chain [6]. Buyers play the pivotal roles in selecting suppliers and set the standard of products in the supply chain. Most of existing studies on the textile and apparel supply chain are from the perspective of buyers [7]. The sourcing strategies of companies with different characteristics (sales volumes, product characteristic and market performance) have been explored in the previous studies [1, 8].

Supplier selecting criteria show the buyers' expectations for supplier performance. The studies on supplier selection and evaluation could date back to 1960s. In the survey of [9], quality, cost and delivery performance were the three most important criteria in supplier selection [9]. However, supplier selection is no longer just the issue of choosing the low-cost products. It deals with the cost covering the whole supply chain activities [10].

2.2 Suppliers' Operation Strategy

Operation strategy is viewed as the effective use of production capability and technology for achieving business and corporate goals [11]. Cost, quality, delivery, and flexibility are commonly used as the important dimensions of operation strategy [12]. Improved quality is thought to reduce cost, as doing things correctly the first time eliminates waste [13]. There are empirical evidences that quality improvement leads to improve business performance [14]. Since apparel is seasonal product, delivery reliability seems very important in the supply chain. Flexibility is a theme in operation management suggested by many scholars. Textile and apparel manufactures even tend to customization, where products are tailored to individual preferences such as Red Collar Group [15]. In addition, volume flexibility is the ability to accelerate or decelerate the rate of production quickly to handle large fluctuations in demand.

2.3 Competitive Advantage

Competitive advantage exists when the firm continuously outperforms its competitors [16]. The empirical literature has been quite consistent with identifying price/cost, quality, delivery and flexibility as important competitive capabilities [17, 18]. Price is no longer the advantage of Chinese textile and apparel suppliers since the labor cost in China increases drastically [4]. In recent years, sourcing from the least developed countries (LDCs) which could offer lower price has become hot issues [19, 20]. Based on the above, the dimensions of competitive advantage constructs used in this study are quality, delivery and flexibility.

2.4 Business Performance

Business performance is a widely debated topic in the literature and various definitions have been proposed. The empirical literatures report a high diversity of performance indicators [21, 22]. Largely, business performance is measured by financial metrics. In previous research, the applied measures include return on asset (ROA), return on investment (ROI), profit margin and market share [23-25].

ROA shows how profitable a company's assets are in generating revenue [26]. If the number is higher, it means that the company could use assets to obtain higher profits. It is useful data for comparing companies in the same industry. ROI is used to evaluate the efficiency of an investment or to compare the efficiency of a number of different investments [27]. Profit margin is a measure of profitability. It is included to measure the business performance comprehensively [28]. Market share is the percentage of a market (defined in term of either units or revenue) accounted for by a specific entity. Considering that the products manufactured in textile and apparel factories are largely in different entities. Many apparel factories produce dress, trousers, shirt and suits with different brand in different aimed markets. It is difficult to figure out the market share of suppliers. Therefore, we obsoleted market share in the survey.

In sum, business performance of Chinese textile and apparel suppliers were measured by the survey respondents' perception of performance. Three performance measures were used, namely ROA, ROI and profit margin.

3. Research Framework

Buyers' behavior and strategies significantly influence the suppliers. In order to analyze the influence of buyers' strategy on the suppliers, we investigated the buyers' supplier selecting strategies from the perspective of suppliers.

Based on the reasoning thus far, we could find that the supplier selecting criteria, reflecting the buyers' sourcing strategy motivate the suppliers to meet the buyers' expectations. Our first hypothesis is therefore:

H1: Buyers' sourcing strategy is positively related to suppliers' operation strategy for global competition.

It has been suggested previously that the firm's strategies have significantly relationship with their competitive advantages and performance [29]. Hence, the following hypotheses are formulated:

H2: Suppliers' operation strategy has positive relationship with their competitive advantages.

H3: Suppliers' operation strategy has positive relationship with their business performance.

Having comparative advantage generally suggested that the suppliers would have one or more of the following capabilities: better quality, shorter delivery time and more flexibility. These capabilities will in turn enhance suppliers' business performance. Therefore, a positive relationship between competitive advantages and business performance can be proposed.

H4: Suppliers' competitive advantages have positive relationship with their business performance.

The above four hypotheses, taken together, support the research framework presented in Fig.1. The corresponding factors and the indicators in the model are shown in Table 1. Measures for the constructs of buyers' sourcing strategy, suppliers' operation strategy, competitive advantages and business performance were developed through a review of the managerial and scholarly literature to establish the content validity of each construct and associated scales.

Table.1 Factors and indicators in the questionnaire

Factors		Source
Buyers' Sourcing Strategy (BSS)		
BSS_1	Our buyers require our products with good price.	[9]
BSS_2	Our buyers require our products with good quality.	
BSS_3	Our buyers require our products with reliable and dependable delivery.	
Suppliers' Operation Strategy (SOS)		
SOS_1	We emphasize the product quality in operation.	[12,13]
SOS_2	We emphasize the reliable delivery in operation.	
SOS_3	We emphasize the product and volume flexibility in operation.	
Competitive Advantage (CA)		
CA_1	We are able to compete on quality.	[17,18]
CA_2	We are able to compete on delivery.	
CA_3	Our company produces diversified products with flexible volumes	
Business Performance (BP)		
BP_1	ROA in our company has increased in the last 3 years.	[23,24]
BP_2	ROI in our company has increased in the last 3 years.	
BP_3	Profit margin in our company has increased in the last 3 years.	

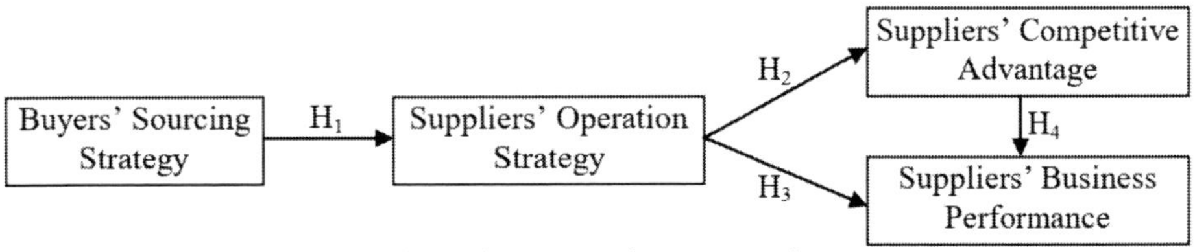

Fig.1 The Research Framework.

4. Research Method

The questionnaire was administered in Chinese, and a thorough review of the literature on each subject of the study was conducted. Questions were designed to be answered from the suppliers' perspective. The items of buyers sourcing strategy, suppliers' operation strategy, competitive advantages and business performance were measured using the instruction such as "Please indicate the extent to which you agree or disagree with the following statements regarding your firm over the last three years" and using five-point Likert scale from 1 (strongly disagree) to 5 (strongly agree).

The research population is based on the membership of China Textile Entrepreneurs Association (CTEA). Only supplier companies were selected from the records. Other companies such as service companies, trade companies and brand companies were discarded from the research population. Thus, 521 textile and apparel manufacturers were selected as the research population from a list of 998 members. Top managers are chosen as respondents because they are directly involved in the management of organizational affairs and have first-hand knowledge on operation strategy and the company performance.

The first step was that an invitation letter was sent to all of the 521 firms comprising the research population. Of the 521 initial letters mailed, a total of 88 were returned undelivered and 46 firms stated that they were not manufactures. Thus, 134 firms were discarded from the research population. Two weeks after the initial letter, survey questionnaires were sent to all 387 potential respondents via e-mail with a covering letter stating the purpose of the research. Approximately 14 days after mailing the questionnaire, reminder letters emailed for those who did not respond. Out of 228 responses received, 4 were excluded due to incomplete information. In order to minimize the effects of the firm size [1, 30], 6 firms employing less than 25 employees were excluded. Therefore, the final sample consisted of 218 Chinese textiles and apparel manufactures, the result response rate was 56.33%. The characteristics of the sample companies are shown in Table 2. Small, medium, and large firms were well represented in the sample.

Table.2 Characteristics of sample companies

Items	Frequency	Percentage (%)	Items	Frequency	Percentage (%)
Years in business			Annual turnover (RMB)		
less than 5 years	25	11.5	less than 20 million	58	26.6
5~9 years	53	24.3	20~100 million	70	32.1
10~19 years	91	41.7	101~500 million	48	22.0
more than 20 years	49	22.5	more than 500 million	42	19.3
Amount of employees			Product Category (multiple choice)		
25~49	43	19.7	yarn	39	17.9
50~249	77	35.3	fabric	82	37.6
250~1000	45	20.6	apparel	147	67.4
1001~5000	38	17.4	home textile	45	20.6
more than 5000	15	6.9	accessories	18	8.3
			anything else	11	5.0

5. Data Analysis

Our model involved a series of correlated variables. A two-stage methodology was engaged in data analysis [31]. First of all, the measurement model was evaluated to determine how given items effect on the constructs in the model. Next, the assessment of structural model hypothesizes were testing by examining whether relationships might or might not exist. We conducted descriptive and correlation analyses by using SPSS 22.0, and conducted a confirmatory factor analysis (CFA) and structural equation modeling (SEM) by using AMOS 17.0.

5.1 Content Validity

To achieve content validity, 6 experts were invited to pilot test the questionnaire, after which minor changes were made to ameliorate both content and clarity of the questionnaire. All the pilot test respondents were excluded from the main samples, which was used for reliability testing, construct validation, and hypothesis testing.

5.2 Reliability

The evaluation of conceptual model includes the estimation of internal consistency for reliability, as well as tests of convergent and discriminant validity for construct validity. Firstly, Cronbach's alpha of this study range from 0.770

to 0.872, exceeding the acceptable threshold of 0.70 for sufficient internal consistency [32]. Compared with Cronbach's alpha, which implicitly assumes that each item carries the same weight, while composite reliability (CR) relied on the actual loadings to construct the factor score, thus CR is a better measure of internal consistency [32]. All CR of constructs have a value higher than 0.70 (between 0.778 and 0.872), indicating adequate internal consistency (see Table 3).

Table.3 Reliability indices for constructs

Items	Loading	SMC	CR	AVE
SOS_1	0.887	0.787	0.872	0.695
SOS_2	0.754	0.569		
SOS_3	0.854	0.729		
BSS_1	0.628	0.394	0.778	0.542
BSS_2	0.825	0.681		
BSS_3	0.743	0.552		
CA_1	0.703	0.494	0.823	0.610
CA_2	0.864	0.746		
CA_3	0.768	0.590		
BP_1	0.832	0.692	0.840	0.638
BP_2	0.870	0.757		
BP_3	0.683	0.466		

5.3 Construct Validity

Theoretically, construct validity consists of two elements: convergent validity and discriminant validity [33]. Convergent validity relates to the degree to which multiple methods of evaluating a variable represent the same results, which is considered acceptable when all factor loadings are greater than 0.5 [34]. As noted in Table 3, item loadings ranges from 0.628 to 0.887, indicating good fit of convergent validity. The value of AVE in our study range from 0.542 to 0.695, exceeding the threshold of 0.5, indicating sufficient discriminant validity between constructs [33].

5.4 Structural Model Assessment

Fig.2 Standardized estimates of model

Table.4 Model fit results

	chi-square (x^2)	Absolute fit measures			Incremental fit			
		df	X^2/df	RMSEA	GFI	AGFI	CFI	NFI
Recommended values	N/A	N/A	<3	<0.08	>0.9	>0.8	>0.9	>0.9
Structural model	90.603	50	1.812	0.061	0.934	0.897	0.968	0.931

Note: Recommended values for samples size >200 (Bagozzi and Yi, 1988; Hair et al., 1998)
The four hypotheses based on the conceptual model (see Fig.1) are tested concurrently using the analysis of moment structures (AMOS) based on the correlations among the construct measures.

According to [35], when the samples size is larger than 200, chi-square/df can be used to determine the fit of the model. The results of the AMOS (see Table 4) reveal that the model fit is acceptable. Fig.2 presents the results of structural model with each factor loading and structural path coefficient.

5.5 Hypothesis Testing

All of the four proposed hypotheses are considered valid. For the confirmative links, H1, predicting a positive relationship between buyers' sourcing strategy to suppliers' operation strategy is supported (β=0.44, t= 5.134, p<0.001). Additionally, H2, examined suppliers' operation strategy positively impact on competitive advantages receives statistical support (β=0.68, t=8.172, p<0.001). H3 examined the effect suppliers' operation strategy on business performance is support by the results (β=0.21, t=2.134, p<0.05). The proposed relationship between competitive advantage and business performance (H4) is also supported (β=0.47, t=4.350, p<0.001) (Table 5).

Table.5 Structure equation modeling results

	Path	β	T-value	P-value	Result
H1	BSS→SOS	0.44	5.134	<0.001	S
H2	SOS→CA	0.68	8.172	<0.001	S
H3	SOS→BP	0.21	2.134	0.033	S
H4	CA→BP	0.47	4.350	<0.001	S

6. Conclusion

This study has examined the influence of buyers' sourcing strategies on the Chinese textile and apparel suppliers. The relationships among the strategy of supply chain partners, competitive advantages, and business performance are explored. With the premise of satisfying the needs of buyers, Chinese textile and apparel suppliers seek to improve their products and service in order to maximize benefits. This study has contributed further understanding of the way in which the suppliers' strategies could be influenced by the buyers, and ultimately have an impact upon suppliers' performance. The research findings reveal that buyers' strategic sourcing has a significant and positive effect on suppliers' operation strategy and suppliers' operation strategy also has a positive effect on the firm's ability to gain competitive advantages and business performance. Therefore, it is concluded that buyers' sourcing strategy is a win-win strategy, which not only good for themselves but also helps to improve their suppliers.

7. Acknowledgement

The authors would like to thank the Shanghai Style Fashion Design & Value Creation Knowledge Service Centre (13S107024), Shanghai Summit Discipline in Design (DB17016) and Donghua University PhD Innovation Project (CUSF-DH-D-2016066) for granting the research and the China Textile Entrepreneurs Association (CTEA) for providing help in the survey.

References

[1] Jin B. Global sourcing versus domestic sourcing: Implementation of technology, competitive advantage, and performance. J Text Inst : 2005; 96 (5): p. 277-286.

[2] Su J, Dyer CL, Gargeya VB. Strategic Sourcing and Supplier Selection in the U.S. Textile- Apparel-Retail Supply Network. Cloth Tex Res J: 2009; 27 (2): p. 83-97.

[3] Kim S. Global sourcing of South Korean apparel industry: sourcing trend and supplier selection. J Text Inst: 2012; 103 (12): p. 1-17.

[4] Zhu S, Pickles J. Bring In, Go Up, Go West, Go Out: Upgrading, Regionalisation and Delocalisation in China's Apparel Production Networks. J Cont Asia: 2014; 44 (1): p. 36-63.
[5] Yang DT, Chen VW, Monarch R. Rising wages: Has China lost its global labor advantage? Pacific Econ. Rev: 2010; 15 (4): p. 482-504.
[6] Gereffi G. International trade and industrial upgrading in the apparel commodity chain. J. Int. Econ: 1999; 48 (1): p. 37-70.
[7] Cho J, Kang J. Benefits and challenges of global sourcing: perceptions of US apparel retail firms. Int Marketing Rev 2001. 18 (5): p. 542-561.
[8] Scarso E, Bolisani E. International manufacturing strategies: experiences from the clothing industry. Int J Oper Prod Manag: 1996; 16 (11): p. 71-84.
[9] Dickson GW. An Analysis of Vendor Selection Systems and Decision. J Purch: 1966; 2 (15): p. 1377-1382.
[10] Teng SG, Jaramillo H. A model for evaluation and selection of suppliers in global textile and apparel supply chains. Int J Physical Distrib Log Manag: 2005; 3046 (7): p. 503-523.
[11] Kim Y, Lee J. Manufacturing strategy and production systems: An integrated framework. J Oper Manag: 1993; 11 (1): p. 3-15.
[12] Badri MA, Davis D, Davis D. Operations strategy, environmental uncertainty and performance: a path analytic model of industries in developing countries. Omega: 2000; 28 (2): p. 155-173.
[13] Ward PT, Duray R. Manufacturing strategy in context: environment, competitive strategy and manufacturing strategy. J Oper Manag: 2000; 18 (2): p. 123-138.
[14] Kaynak H. The relationship between total quality management practices and their effects on firm performance. J Oper Manag: 2003; 21 (4): p. 405-435.
[15] Jia G, Wang J, Zhang R, Zong H, Yao J. Red Collar Group in Qingdao-High-End Clothing Customization Service. 2016: Springer Singapore.
[16] Porter ME. Competitive advantage of nations: creating and sustaining superior performance. 2011, New York: Simon and Schuster.
[17] Li S, Ragu-Nathan B, Ragu-Nathan TS, Rao SS. The impact of supply chain management practices on competitive advantage and organizational performance. Omega: 2006; 34 (2): p. 107-124.
[18] Gotzamani K, Kafetzopoulos D, Psomas E. Quality systems and competitive performance of food companies. Bench: 2013; 20 (4): p. 463-483.
[19] Kathuria LM. Analyzing competitiveness of clothing export sector of India and Bangladesh: Dynamic revealed comparative advantage approach. Compet Rev 2013. 23 (2): p. 131-157.
[20] Lopezacevedo G, Robertson R. The Promise and Peril of Post-MFA Apparel Production, in Economic Premise. 2012: Washington. p. 1-4.
[21] Delaney JT, Huselid MA. The Impact of Human Resource Management Practices on Perceptions of Organizational Performance. Acad Manag J: 1996; 39 (4): p. 949-969.
[22] Roberts N, Grover V. Investigating firm's customer agility and firm performance: The importance of aligning sense and respond capabilities. J. Bus. Res: 2012; 65 (5): p. 579-585.
[23] Goll I, Rasheed AA. The Moderating Effect of Environmental Munificence and Dynamism on the Relationship between Discretionary Social Responsibility and Firm Performance. J Bus Ethics: 2004; 49 (1): p. 41-54.
[24] Hancock JI, Allen DG, Bosco FA, Mcdaniel KR, Pierce CA. Meta-Analytic Review of Employee Turnover as a Predictor of Firm Performance. J Manag: 2013; 39 (3): p. 573-603.
[25] Chi T, Delcampo RG. Business Contingency, Strategy Formation, and Firm Performance: An Empirical Study of Chinese Apparel SMEs. Admin Sci: 2015; 5 (1): p. 27-45.
[26] Needles B. Principles of Accounting, International Edition: 2013, Boston: Cengage Learning.
[27] Phillips JJ. Return on Investment in Training and Performance Improvement Programs. 2nd Edition: 2003, London: Routledge.
[28] Narver JC, Slater SF. The Effect of a Market Orientation on Business Profitability. J. Marketing: 1990; 54 (4): p. 20-35.
[29] Danskin P, Englis BG, Solomon MR, Goldsmith M, Davey J. Knowledge management as competitive advantage: lessons from the textile and apparel value chain. J Knowl Manag: 2005; 12 (2): p. 91-102.
[30] Martín-Tapia I, Aragón-Correa JA, Rueda-Manzanares A. Environmental strategy and exports in medium, small and micro-enterprises. J. World Bus: 2010; 45 (3): p. 266-275.

[31] Gefen D, Straub DW, Boudreau MC. Structural Equation Modeling And Regression: Guidelines For Research Practice. Comm Assoc Inf Syst: 2000; 4 (7): p. 1-76.
[32] Fornell C, Larcker DF. Evaluating structural equation models with unobservable variables and measurement error. J. Marketing Res: 1981; 18 (1): p. 39-50.
[33] Chin WW, Gopal A, Salisbury WD. Advancing the Theory of Adaptive Structuration: The Development of a Scale to Measure Faithfulness of Appropriation. Inf Syst Res: 1997; 8 (4): p. 342-367.
[34] Wixom BH, Watson HJ. An Empirical Investigation of the Factors Affecting Data Warehousing Success. Mis Quart: 2001; 25 (1): p. 17-41.
[35] Bagozzi RP, Yi Y. On the evaluation of structural equation models. J Acad Marketing Sci: 1988; 16 (1): p. 74-94.

Factors Impacting College Students' Online Apparel Shopping Intention

Wen-Jia Li, Rong Zheng*, Hui-Wei Cao, Hai-Jing Liu, Yun-Chao Zhang

Aimer Human Engineering Research Centre, Beijing Institute of Fashion Technology, Beijing, 100029, China

*Corresponding author's email: yuki.zheng@gmail.com

Abstract

With the growing popularity of the Internet and e-commerce in China, online shopping has become an integral part of Chinese people's daily lives, most notably among college students who are adept in online environments. This study seeks to identify factors influencing college students' online shopping. Using college students as its target group, the study collected data via a questionnaire, conducted statistical analysis and extracted 6 factors (accumulative contribution rate of 62.630%) as variables to construct a study model. AMOS software was used to build and revise the research model. Further analysis of data derived from questionnaire samples was undertaken via SPSS descriptive analysis, T-test of independent samples and variance analysis to identify factors influencing college students' online shopping decisions. Findings show that college students may be significantly influenced by factors such as information, norms, online retailers' features, marketing strategies and perception of risks.

Keywords: College Students; Online Shopping; Consumption Behaviour

1. Introduction

Online shopping has gained in popularity due to its geographic acessibility, convenience, rich variety and low-priced goods. It has attracted a growing number of consumers. Data from China Internet Network Information Center [1] indicates that as of June 2016, China's "netizen" population had grown to 710 million, while China's e-commerce transactions had reached 872.41 billion yuan in the second quarter of 2015, increasing by 39.6% year-after-year [2]. The 2015 Investigation Report on Online Shoppers in China showed clothing, shoes and hats, luggage and outdoor goods took the lead amongst all merchandise varieties for online shoppers in 2014, accounting for 67.9% of the total products sold during this period [3].

A host of studies on factors influencing factors consumers' online shopping have been conducted by domestic and international researchers. Moschis (1976) found that consumers' pre-Internet purchasing behaviours were influenced by information searching, source reliability and group pressure [4]. Based on Moschis's theory, Park and Lessig (1977) summarized the three influencing factors of information influence, norm influence and views on expressing personal feelings that impacted consumers' purchasing behaviours [5]. Jarvepnaa and Todd's 1996 study first introduced risk perception into the field, triggering a slew of studies on online shopping behaviours from the perspective of risk perception [6]. Biswas D and Biswas A(2004) pointed out that consumers felt more functional, economic and process-related risks during online shopping than when shopping in a physical store [7]. Ji Hye Park (2004) conducted a study on major approaches to remove risk perception from consumers in online shopping [8]. In his SEM-based study on factors influencing people's online shopping intention, Chinese scholar Wu XX (2008) concluded that risk perception negatively correlated with consumers' online shopping decisions in a significant way [9]. Liu GL and Lu F (2002) held that major factors influencing college students' clothes shopping behaviours included friends' opinions, channels for information acquisition, retailers' promotion activities, prices, comfort of clothes, styles and shopping environment [10].

Surveys show that people aged 20-29 account for 30.4% of the netizen population, the age group with the highest percentage, of which the student group comprises 25.1%. The 2015 Investigation Report on Online Shoppers in China indicated that online shoppers with high academic qualifications outweigh other online shopping groups, with people holding a bachelor degree or above accounting for 77.3% of the total online shopping population [3]. As individuals who are comfortable with high-tech trends and with a capacity for knowledge acquisition, college students constitute a vital demographic for online shopping. Hence, this study selected college students as its subjects and sought to analyse factors influencing their online clothes shopping.

2. Methods

This study on college students' online clothes shopping behaviours was carried out primarily in the form of a questionaire. It adopted the existing approaches and results on consumption behaviours established by numerous previous studies as its theoretical basis. Considering characteristics of modern college students and features of online clothes shopping, the study designed a questionaire consisting of two parts: The first part covered basic personal information, information about Internet usage and online shopping; the second part was mainly about college students' subjective perception and purchase decisions during online clothes shopping.

2.1 Questionnaire Design

The questionnaire encompassed information questions aimed to gather interviewees' basic personal information as well as their Internet usage and online shopping information.

2.2 Design of Scaled Questions

Table.1 Evaluation indexes in the questionaire and their respective sources

Evaluation index	Source
Q1. I would acquire clothes information from expert recommendation via social media like Wechat and microblog.	Yang XK (2013) [11]
Q2. I would share clothes-related information through online shopping platforms like www.taobao.com	
Q3. I would obtain clothes information from website advertisements and celebrity promotion activities.	
Q4. I would ask friends for further information about unfamiliar clothes brands.	Chen Y (2013) [12]
Q5. I would pay attention to other people's clothes brands and clothes collocation.	Wang XF (2013) [13]
Q6. It's important for me that others like the apparel I bought.	Sheng M (2010) [14]
Q7. I want to know what kind of apparel would please or impress people pleasantly.	
Q8. The apparels I bought are usually ones I think people would commend or praise.	
Q9. The online shop's reputation is very important.	Weng SZ (2014) [15]
Q10. The aesthetic effect of the online shop's interface design is very important.	
Q11. It's important for an online shop to provide a huge amount of detailed, adequate and accurate information.	Shi SZ (2011) [16]
Q12. The online store's service attitude and online communication with me would influence my purchase behaviour.	
Q13. I prefer clothes brands with high reputation.	Wang JP (2013) [17]
Q14. Ads often stir my desire for clothes shopping.	Shi SZ (2011)
Q15. Sales and promotion activities would sway my purchase behaviour.	Wang JP (2013)
Q16. Price constitues a lure for me in online clothes shopping.	Shi SZ (2011)
Q17. I would first learn about the online store's credit and reputation information before purchasing apparels there.	Hong CN (2013) [18]
Q18. It's possible that personal information is stolen during online payment, which would spell economic loss.	Wang JP (2013)
Q19. I think it would make me sustain a loss to pay first and then get delivery.	Hong CN (2013)
Q20. If I provided my personal information to online retailors, it would possibly be leaked.	Zhao XJ (2014) [19]

Based on the Stimulus-Organism-Response (SOR) model, the study adapted a basic model framework (as shown in Fig.1). The supposition is that the organism (college student) stimulated by external environment would produce corresponding subjective perception and judgment, upon which they have subjective response (purchase intention). Within the framework, it was supposed that the study subjects would have personal perception information after being stimulated by external environment. College students are exposed constantly to the influence of campus environment and the Internet, receiving information of various kinds. The questionaire aimed to learn about college students' subjective perception of various factors and their online clothes shopping intention after being stimulated by external environment. Therfore, consulting those frequently used evaluation indexes on consumption behaviour studies by scholars at home and abroad, with consideration of current online clothes shopping features, the study finally determined the following 20 evaluation indexes for questions targeting the college student group, so as to describe study subjects' corresponding perception once stimulated by external environment, including information acquisition channels and online marketing. Evaluation indexes and their sources are listed in Table 1.

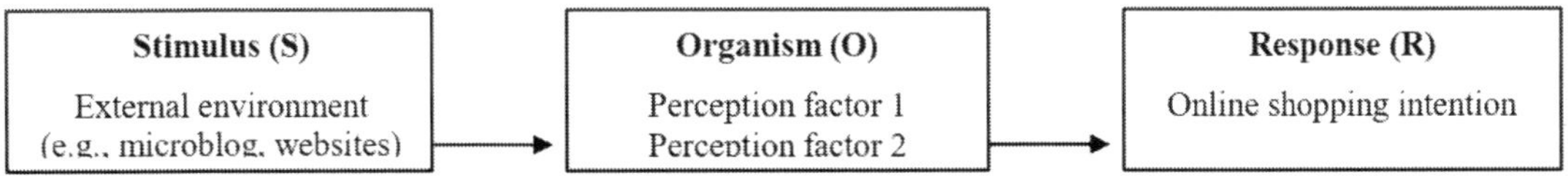

Fig.1 SOR model framework.

3. Results and Discussion

The study collected 308 replied questionaire copies. After screening, 295 effective copies were selected. Then statistical analysis was made of data in the replies.

3.1 Analysis of Basic Sample Information

Table.2 Analysis of samples' online shopping information

Basis Information	Classification	Frequency	Percentage
Average daily time spent Internet surfing	Within 1 hour (including 1 hour)	11	3.73
	1-2 hours (including 2 hours)	34	11.53
	2-3 hours (including 3 hours)	75	25.42
	More than 3 hours	175	59.32
	Total:	295	100.00
Online clothes shopping experience	Below 1 year (including 1 year)	22	7.46
	1-2 years (including 2 years)	38	12.88
	2-3 years (including 3 years)	54	18.31
	More than 3 years	181	61.36
	Total:	295	100.00
Average sum (in yuan) spent for each online clothes purchase	1-200 yuan	110	37.29
	201-400 yuan	111	37.63
	401-600 yuan	51	17.29
	601-800 yuan	8	2.71
	More than 801 yuan	15	5.08
	Total:	295	100.00
Monthly online clothes shopping frequency	0	18	6.10
	1-2 times	188	63.73
	3-4 times	71	24.07
	5-6 times	9	3.05
	More than 6 times	9	3.05
	Total:	295	100.00
Device used in online shopping	Computer	66	22.22
	Tablet computer	11	3.70
	Cellphone	218	74.07
	Total:	295	100.00

Of the 295 respondents, 216 were female college students with online shopping experience and 79 were male college students, representing 73.22% and 26.78% of the sample population, respectively. It is evident that far more female students than male students shop for clothes online. In terms of regions in data collection, North China took the lead with 189 respondents, accounting for 64.07% of the sample population, followed by South China with respondents there accounting for 16.27%. Respondents were mostly senior and postgraduate students, accounting for 61.02% of the population; such students usually have relatively mature views and strong self-awareness, and would be a major force in online shopping. 242 respondents enjoyed a monthly living allowance of more than 1,000 yuan, accounting for 82.04% of the total. The relatively adequate allowance allowed them to shop for clothes online. The detailed online shopping information of the respondents is listed in Table 2.

Of all the interviewees, 175 surfed online more than 3 hours per day, accounting for 59.32% of the sample, indicating that these students are quite familiar with the Internet environment and skilled in using it. There were 181 students with 3 years of online clothes shopping experience, accounting for 61.36 % of the total interviewees. These students had developed online clothes shopping skills, accepted the online clothes shopping model and would continue to use it, demonstrating that the online clothes selling market would grow on a large scale. The average sum spent by the interviewees for an apparel purchase was mostly 1 to 400 yuan (representing 64.92% of the total), which reflected the average living standard and the price range affordable to them. The average monthly online clothes purchase frequency of 188 interviewees was 1-2 times, accounting for 63.73% of the total. 74.07% of the respondents used cellphones for online clothes shopping, signalling that mobile devices have occupied the most half market.

3.2 Extraction of Influencing Factors

A correlation analysis of the 20 evaluation questions found that certain correlations existed among some questions. Therefore dimensionality reduction was undertaken to further summarise the evaluation questions in the questionaire. Factors were extracted through principal component analysis, and Varimax was used to orthogonally rotate factors. Finally 6 principal components were selected to descibe and analyse college students' online clothes shopping behaviours, with variance interpretation of all principal components shown in Table 3 and the accumulative contribution rate of the 6 factors at 62.630%. Factor 1 included four indexes on the initially set online shopping intention, thus named as online shopping intention factor (Q20, Q19, Q18 and Q17) with a contribution rate of 14.067%. Factor 2 described the influence of the community norms and rules on the respondents, named as norm influence factor (Q6, Q5, Q7 and Q4) with a contribution rate of 12.022%. Factor 3 covered various risks in online clothes shopping, named as risk perception factor (Q14, Q15 and Q16) with a contribution rate of 9.536%. Factor 4 touched on features of online retailers from different perspectives, named as retailer feature factor (Q9, Q10 and Q8) with a contribution rate of 9.533%. Factor 5 focused on the influence of various online social platforms and Apps on respondents, so named as information influence factor (Q1, Q3 and Q2) with a contribution rate of 8.836%. Factor 6 was about all marketing and promotion activities and strategies, named as marketing factor (Q12, Q11 and Q13) with a contribution rate of 8.636%.

Table.3 Factors' interpretation degree to the overall project

Factor	Cycling sum of squares loading		
	Total	Variation percentage %	Accumulative contribution rate %
1	2.813	14.067	14.067
2	2.404	12.021	26.089
3	1.907	9.536	35.625
4	1.907	9.533	45.158
5	1.767	8.836	53.994
6	1.727	8.636	62.630

The 6 factors were the evaluation variables in the study: information influence, norm impact, online retailers' features, marketing factors, risk perception and online shopping intention, respectively.

Based on the above analysis, the study made the following hypothesis (as shown in Fig.2):

H1: Information influence plays a role in swaying college students' online clothes shopping intention.
H2: Norm influence plays a role in swaying college students' online clothes shopping intention.
H3: Online retailors' features play a role in swaying college students' online clothes shopping intention.
H4: Marketing factors play a role in swaying college students' online clothes shopping intention.
H5: Risk perception plays a role in swaying college students' online clothes shopping intention.

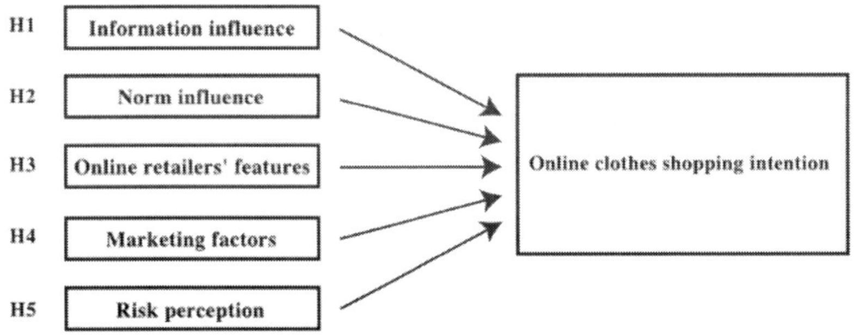

Fig.2 Online shopping intention hypothesis.

3.3 Construction and Verification of Research Model

Table.4 Model Path Factor Table

Path	Path coefficient	Standard error	Value T	Value P
H1 Online clothes shopping intention ← Information influence	.265	.114	2.326	*
H2 Online clothes shopping intention ← Norm influence	.162	.095	1.694	(*)
H3 Online clothes shopping intention ← Retailers' features	.357	.125	2.852	**
H4 Online clothes shopping intention ← Marketing factors	.395	.135	2.929	**
H5 Online clothes shopping intention ← Risk perception	-.192	.074	-2.604	**
(*)P<0.1,*p<0.05, **p<0.01, ***p<0.001				

According to the results compared hypothetical model based on the model path, all the assumptions established.The final validation model is shown in Fig.3.

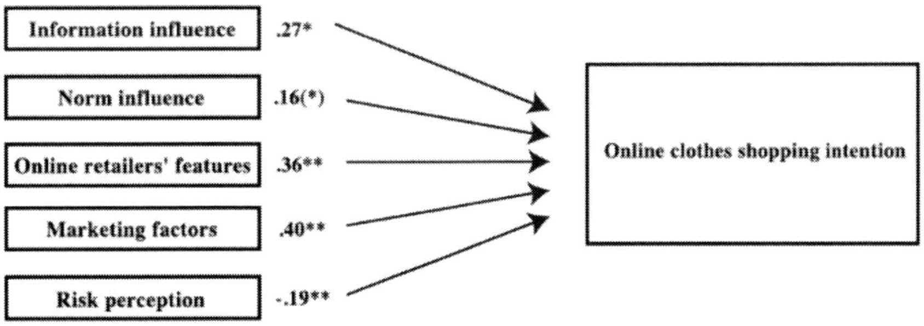

Fig.3 validation model

According to the assumptions in 3.2, Using the AMOS structural equation model software to construct the hypothesis model of this study, using the absolute fitness index (X2, GFI, AGFI, RMR, RMSEA etc.) to the initial model of fitness test and model correction. And using maximum likelihood estimation method was to verify the model path coefficient analysis, the results of the seven direct effects of the path coefficients are shown in Table 4.

The impact of information influence on college students' online shopping intention will vary significantly at the 0.05 level, Thus confirmed the assumption that H1 is established; The influence of norm on college students' online shopping intention is significant at the level of 0.1, which proves that H2 is established; The influence of the retailers' features on college students' online shopping intention is significant at the level of 0.01, then it is assumed that the H3 is established; The influence of marketing factors on college students' online shopping intention is significant at the level of 0.01, and it is verified that the H5 is established.The influence of risk perception on college students' online shopping intention is significant at the level of 0.01, confirming the hypothesis that H6 is established.

3.4 Verification of Influencing Factors

This section discusses the difference analysis of the influence of information, norm, retailers' features, marketing factors and risk perception variables on online shopping intention through single factor's variance analysis.

3.4.1 Information Influence on Online Shopping Intention

Results of variance analysis of data on information influence on online shopping intention are shown in Table 5. As for the online shopping intention, the overall test's F value was 12.658 ($p=0.000<0.5$), showing that the influence on the online shopping intention reaches a significant level. Therefore, the null hypothesis was rejected, indicating that significant difference existed in different degrees of information influence on college students' online shopping intention.

Table.5 Variance analysis of information influence on online shopping intention

ANOVA	Sum of squares	df	Means of squares	F value	Sig.
Online shopping intention among groups	16.876	2	8.438	12.658	.000

Multiple comparisons are shown in Table 6 with the Least Significant Difference (LSD) approach adopted. It can be found that in terms of online shopping intention, the group influenced by information is significantly higher than the group being not sure if they are influenced by information and the group reporting no information influence; the group being not sure if they are influenced by information is significantly higher than the group reporting no information influence.

Table.6 Multiple comparisons

LSD

Dependent variable	Information influence (I)	Information influence (J)	Average difference (I-J)	Standard error	Significance	95% confidence interval Lower limit	95% confidence interval Upper limit
Online shopping intention	No information influence	Not sure	-.310*	.140	.028	-.586	-.034
		Influenced by information	-.690*	.147	.000	-.980	-.400
	Not sure	No information influence	.310*	.140	.028	.034	.586
		Influenced by information	-.380*	.105	.000	-.590	-.174
	Influenced by information	No information influence	.690*	.147	.000	.400	.979
		Not sure	.380*	.105	.000	.173	.586

3.4.2 Influence of Norms on Online Shopping Intention

Results of variance analysis of data on norm influence on online shopping intention are shown in Table 7. The F value of the overall test of norm influence on online shopping intention was 5.817 ($p=0.003<0.05$), showing that the influence on the online shopping intention reached a significant level. Therefore, the null hypothesis was rejected, and the alternative hypothesis was accepted, indicating that significant difference of different degrees of norm influence on college students' online shopping intention existed.

Table.7 Variance analysis of norm influence on online shopping intention

ANOVA

	Sum of squares	df	Means of squares	F value	Sig.
Online shopping intention among groups	8.105	2	4.052	5.817	.003

Multiple comparisons are shown in Table 8 with the LSD approach adopted. It can be found that in terms of the dependent variable (online shopping intention), the group influenced by norms is significantly higher than the group being not sure if they are influenced by norms and the group reporting no norm influence; the group being not sure if they are influenced by norms is significantly higher than the group reporting no norm influence.

Table.8 Multiple comparisons

LSD

Dependent variable	Norm influence (I)	Norm influence (J)	Average difference (I-J)	Standard error	Significance	95% confidence interval	
						Lower limit	Upper limit
Online shopping intention	Influenced by norms	No norm influence	-.401*	.224	.015	.109	.992
		Not sure	.286*	.105	.007	.080	.492
	Not sure	Influenced by norms	-.286*	.104	.007	-.492	-.080
	No norm influence	Not sure	-.551*	.224	.015	-.992	-.110

3.4.3 Influence of Retailers' Features on Online Shopping Intention

Results of variance analysis of data on the influence of retailers' features on online shopping intention are shown in Table 9. The F value of the overall test of the influence of retailers' features on online shopping intention was 9.423 ($p=0.000<0.05$), reaching a significant level. Therefore, significant difference existed in influence of different degrees of college students' attention on retailers' features on online shopping intention.

Table.9 Variance analysis of the influence of retailors' features on online shopping intention

ANOVA

	Sum of squares	df	Means of squares	F value	Sig.
Online shopping intention among groups	12.824	2	6.412	9.423	.000

Results of multiple comparisons with the LSD approach are shown in Table 10. In terms of the online shopping intention variable, the group influenced by retailers' features is significantly higher than the group being not sure if they are influenced by retailers' features and the group reporting no influence of retailers' features.

Table 10. Multiple comparisons

LSD							
Dependent variable	Retailers' features (I)	Retailers' features (J)	Average difference (I-J)	Standard error	Significance	95% confidence interval	
						Lower limit	Upper limit
Online shopping intention	Influenced by retailers' features	No influence of retailers' features	1.258*	.586	.033	.105	2.410
		Not sure	.600*	.157	.000	.292	.908
	Not sure	Influenced by retailers' features	-.600*	.157	.007	-.909	-.292
	No influence of retailers' features	Influenced by retailers' features	-1.258*	.586	.033	-2.410	-.105

3.4.4 Influence of Marketing Factors on Online Shopping Intention

Results of variance analysis of data on the influence of marketing factors on online shopping intention are shown in Table 11. The F value of the overall test of the influence of marketing factors on online shopping intention was 3.539 ($p=0.030<0.05$), reaching a significant level. Therefore, the hypothesis was accepted, indicating that significant difference existed in influence of different marketing factors on online shopping intention.

Table.11 Variance analysis of influence of marketing factors on online shopping intention

ANOVA					
	Sum of Squares	df	Means of squares	F value	Sig.
Online shopping intention among groups	5.005	2	2.503	3.539	.030

Results of multiple comparisons with the LSD approach are shown in Table 12. In terms of the online shopping intention variable, the group influenced by marketing factors is significantly higher than the group reporting no influence of marketing factors, and the group being not sure if they are influenced by marketing factors is significantly higher than the group reporting no influence of marketing factors.

Table.12 Multiple comparisons

LSD							
Dependent variable	Marketing factors (I)	Marketing factors (J)	Average difference (I-J)	Standard error	ignificance	95% confidence interval	
						Lower limit	Upper limit
Online shopping intention	No influence of marketing factors	Not sure	-.401*	.191	.036	-.776	-.026
		Influenced by marketing factors	-.522*	.198	.009	-.911	-.133
	Not sure	No influence of marketing factors	.401*	.191	.036	.026	.776
	Influenced by marketing factors	No influence of marketing factors	.522*	.197	.009	.133	.911

3.4.5 Influence of Risk Perception on Online Shopping Intention

Results of variance analysis of data on risk perception are shown in Table 13. The F value of the overall test of the influence of risk perception on online shopping intention was 1.142 ($p=0.321>0.05$), not reaching a significant level. Therefore, the hypothesis was rejected, indicating that different degrees of risk perception do not have a significant influence on online shopping intention.

Table.13 Variance analysis of influence of risk perception on online shopping intention

ANOVA

	Sum of squares	df	Means of squares	F value	Sig.
Online shopping intention among groups	1.641	2	.821	1.142	.321

4. Conclusion

The study first tried to pinpoint factors influencing college students' online clothes shopping behaviours by means of a questionnaire, with 295 valid replied copies of questionnaire collected. Via statistical analysis of questionnaire responses, the study constructed a model of factors influencing college students' online shopping intention, encompassing information influence, norm influence, retailers' features, marketing factors and risk perception, which all impact students' online shopping intention. The following conclusions can be made from this study:

1) The number of female students who participated in online shopping for apparel was about 3 times that of male students who shopped online for apparel. The market share of women's clothing was larger than that of men's clothing. There is a significant difference in levels of risk perception between the two genders – male subjects show a higher risk perception than female subjects.

2) With the development of smartphones, the potential for mobile shopping platforms continues to grow. The number of subjects who used mobile platforms for online shopping was greater than that who shopped on computers or tablets. There is a higher information influence on mobile users compared to computer or tablet users.

3) The online shopping intention of subjects who were "influenced by information" is significantly higher than those who were "not sure" or with "no information influence"; the online shopping intention of subjects who were "influenced by norms" is significantly higher than those who were "not sure" or with "no norms influence"; the online shopping intention of subjects who were "influenced by retailers' feautres" is significantly higher than those who were "not sure" or with "no influence of retailers' features"; the online shopping intention of subjects who were "influenced by marketing factors" is significantly higher than those who were "not sure" or with "no influence of marketing factors". There is no significant difference between the varying degrees of risk perception on online shopping intention.

5. Limitations

1) There are multiple and complex factors that influence human behaviour. This study involved 5 influencing factors and obtained 295 samples. A larger sample size can be collected in the future for a more meaningful study.

2) This study used a subjective questionnaire method to collect information; the data analysis and results reflect the subjects' subjective views. An objective test method can be incorporated in the future to provide a more comprehensive result.

References

[1] 37th statistical report on Internet development in China. Beijing: China Internet Network Information Center: 2016.

[2] Information on http://www.iresearch.com.cn/report.

[3] 2015 investigation report on online shoppers in China. Beijing, China: China Electronic Commerce Research Center: 2015.

[4] Moschis GP. Social comparison and informal group influence.J Market Res: 1976; 13: 237-244.

[5] Park CW, Lessig VP. Students and housewives: differences in susceptibility to reference group influence. J Consum Res: 1977; 4: 102-110.
[6] Jarvenpaa SL, Todd PA. "Consumer Reactions to Electronic Shopping on the World Wide Web," Journal of Electronic Commerce: 1997; 1 (2): 59-88.
[7] Biswas D, Biswas A. The diagnostic role of signals in the context of perceived risks in online shopping: do signals matter more on the Web? J Inter Market: 2004;18 (3): 30-45.
[8] Park JD, Lee, Ahn J. Risk-focused e-commerce adoption model: A cross-country study. Journal of Global Information Technology Management 7: 2004; 2: 6-30.
[9] Wu XX. A SEM-based study of influencing factors to online shopping intention. Chengdu, China: University of Electronic Science and Technology of China: 2008.
[10] Liu GL, Lu F. Textile and apparel market research and prediction. Beijing: China Textile and Apparel Press: 2003.
[11] Yang XK. Online marketing strategy study based on colleges students' consumption features. Guangxi, China: Guangxi Normal University: 2013.
[12] Chen Y. Perception influence on college students' purchase intention in group buying with consideration of the group role. Nanjing, China: Nanjing University of Aeronautics and Astronautics: 2013.
[13] Wang X. A study of college students' online shopping behaviours based on group influence. Anhui, China: Anhui University of Finance and Economics: 2013.
[14] Sheng M, Lu XX, Qin XM. Influence of online references on college students' purchase decision. J Zhengzhou University of Aeronautics: 2010; 5: 94-98.
[15] Weng SZ. A study of influencing factors to college students' online clothes shopping behaviours. Shanghai: Donghua University: 2014.
[16] Shi SZ. Empirical study of influencing factors to college students' online shopping behaviours. Southwestern University of Finance and Economics: 2011.
[17] Wang JP. An empirical study and analysis of clothes consumption behaviours in digital media. Xi'an, China: Xi'an Polytechnic University: 2013.
[18] Hong CN. A study of college students' online shopping behaviours based on lifestyle theories. Beijing: Beijing University of Posts and Telecommunications: 2013.
[19] Zhao XJ. Analysis of influencing factors to China's online consumption development. Jinan, China: Shandong University: 2014.

Research on Clothing Brand Promotion Based on Network Media

Min Ci[1], Jun Ren[2*]

[1] *School of Apparel & Art Design, Xi'an Polytechnic University, Xi'an, 710048, China*
[2] *Apparel & Art Design, Xi'an Polytechnic University, Xi'an, 710048, China*

*Corresponding author's email: 337006791@qq.com

Abstract

The arrival and long-term development of the information age has cultivated a media culture which revolves around the Internet. Media marketing of clothing has also been flourishing with the emergence of these digitized marketing methods. Internet marketing is a consumer-oriented practice that places heavy emphasis on personalization. Therefore, if any garment enterprise wants to gain a competitive advantage in the market, it is necessary to give full consideration to the needs and psychology of different consumers. Sale across the Internet transcend chronological and spatial constraints; it has become efficient, low cost, wide range and multi-dimensional, interactive and practical. As a result, network marketing has become an important way of clothing brand to promote themselves. This paper mainly studies the current situation of new media, and explains how clothing can be displayed and promoted in the media platform.

Keywords : Network-based Media; Clothing Industry; Promotion; Research

1. Introduction

1.1 Research Background and Current Situation

At present, the new media still exists within a stage of tumultuous development. So, in many ways, further improvements are needed. First of all, the research of this paper is carried out on the basis of previous theoretical research, and the relevant cases of clothing brand are expounded. Secondly, it puts forward new communication strategies from the point of view of consumers, and puts forward constructive suggestions for the dissemination of clothing brands. Therefore, the future development based on the latest brand communication needs further study. [1] The famous American marketing research scholars, Don E. Schultz and Philip J. Kitchen co-authored the "Global Integrated Marketing Communication", expounding the Western enterprises on how to use modern technology to build more efficient marketing systems, as well as listing clear analysis of their cases and putting forward effective marketing guidance. [2] In the domestic new media to promote the continuous development, many researchers have begun to focus on researching new media promotion strategies, and achieved significant results. However, these studies are generally the way to explore new media. Its application to the media is not comprehensive enough due to a lack of systematic, targeted research.

1.2 Goals and Significance of Research

The United Nations Educational, scientific and cultural organization once defined a new media: "new media is network media." [3] According to their article, wired magazines are what defines the new media: "the spread of all people to all." [4] This article combines the characteristics of new media, its potential within the garment industry, and its speculation on potentially appropriate brand communication strategies conducted through in-depth study. It is conducive for China's garment enterprises to provide reference, to guide China's garment enterprises, industry in the dissemination of information and the audience distracted by the dissemination of the environment, fast and accurate to the enterprise brand letter to convey to the audience you want to convey, to increase market share.

1.3 Methodology

1) Literature Articles: Referencing relating materials in animation and costumes.

2) Case Study: Through analysis of domestic and international clothing brands, through new media platforms to summarize the significance of online marketing within clothing brands.

3) Theoretical Analysis: The author puts forward their own understanding of the clothing brand marketing within new media environments, and based on its interactive features and social attributes, analysis of its impact on the clothing brands' proliferation of market share.

4) Applied Theory: According to the new media environment, the issue of marketing, put forward a new fashion brand marketing strategy. We are hoping to apply the aforementioned theories to the future development of clothing brand marketing approaches.

1.4 An Analysis of the Current Situation of New Media

1.4.1 New Media Changes in Lifestyle

The new media brought tremendous change to the social development and people's life has penetrated into all aspects of production, circulation and consumption, driven by information technology and network technology as the representative of the third industrial revolution deepening. It has the characteristics of connection, openness, individuality, intelligence and so on, which makes human society emit the wealth and knowledge bonus which has never been in any previous era. Today, the promotion of new media is a mesh radiation model, which can achieve two-way communication and discussion between users and media, users and users. It has brought great convenience to people's life. [5]

1.4.2 New Media has Covered the World

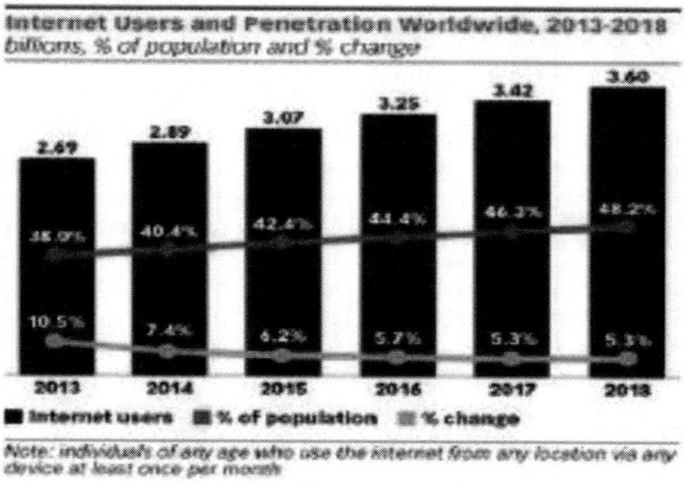

Fig.1 Global Internet users data

With the development of Internet technology, new media applications are becoming more and more popular. [6] The use of new media platform seems to have become an important way of brand marketing. According to market research firm Darkener, the latest report shows that global Internet users are expected to exceed 3 billion by 2016, accounting for nearly half of the global total population. Darkener's new data also pointed out that the Internet user penetration rate in 2016 has, for the first time, exceeded 40% of the total number of Internet users; reaching 2.89 billion, an increase of 6.2% over last year, accounting for the total global population of 42.4%. (Fig.1)

Table.1 Number of active users per social platform

Media	QQ	Webo	QQ Zone	Facebook	WeChat	Google	Twitter
Amount	350M	342M	460M	1026M	780M	300M	690M
NO.	5	6	4	1	2	7	3

According to the research firm – We Are Social, in January 2016 the world's major social applications statistics show that social networking platforms have become an indispensable part of public life. From the above figure, it is not difficult to see that social media in the global market has a wide range of applications with many users, including Facebook, what has the highest rate of users.

2. The Features of New Media Promotion

2.1 Rich and Varied in Manifestations

One of the applications of network media technology for this special brand of information would be to provide technical support, so that the brand can be presented through interactive media such as sound, images, text and other forms to provide consumers with personalized information services, more conducive to shaping the brand's personality. IPTV mainly serves the people through the broadband Internet mode, and has strong interactivity. Mobile media use a wide range, it is very convenient to use, allowing customers to enjoy quality service with ease. The network videos mainly consist of brand based propaganda, video advertisements, fashion shows and other media related to brand imagery. [7] Blogs can support some text, pictures, audio, video and so on. It is characterized its low cost and time effectiveness, allowing for an efficient platform to promote brand awareness. [8]

2.2 Turn Popularization to Specialization

The proliferation of network media, mobile media and outdoor media represents the metamorphosis of media's general properties and points. [9] The audience formed in terms of the popularization model are divided into pieces. Each fragment represents an individual section of its audience. Although the size of each audience is smaller than before, the similarity between individuals in the same group has increased. This is actually more conducive to precise promotion, and the audience and the different groups and benefit from the media personalized development and innovative promotion. [10]

2.3 Make the Interaction Deeper between Promotion and New Media

The emergence of interactive media has greatly changed the quality of marketing promotion. For example, the clothing brand H&M uses "new media" to get a good promotion. Internet and mobile communication technology allow consumers to not only directly engage with marketers and advertisers in two-way dialogue, these forms of communication also allows them to create their own content that they can upload to the Internet, and share with other consumers. On the one hand, this might increase the difficulty of effective promotion; on the other hand, through benign interaction, so that enterprises can extract more accurate market information from their consumers, so that the effectiveness of brand promotion can be greatly increased. [11]

2.4 Make the Promoting Subject Diversification

Table.2 The rate of clothing information recommendation to friends

Will you recommend cloth information to friends	men	women
every time	19.80%	26.1
sometime	72.90%	72.90%
no	7.30%	1%

The source of information: from Chinese TV net data

The general threshold for traditional methods of promotion is high, but in the new media era, any can become the main body of its promotion, we can utilize networking platforms to become their own broadcasting tool that would include tens of millions of visitors to people, with their own words and lifestyles that can affect others. Personal promotion can comprise of an online print-based media, such as the rise of blogs and podcasts. [12] As a result, on one hand, it increases the convenience of word-of-mouth promotion; on the other hand, it could adversely affect brands' ability to conduct damage control in case of a PR crisis.

3. The Impact of New Media on Clothing Promotion

3.1 On the Means of Promotion

Digital technology and media technology changes have spawned new media, in turn, the application of new media is also constantly affecting and promoting changes and updates to these technologies. [13] Among them, a typical representative is the iPad. The iPad is an Apple company Tablet PC (January 27, 2010): its functionalities include browsing the Internet, sending and receiving e-mail, reading e-books, play audio or video amongst other functions. A way of spreading as a new medium. The functional composite terminal based on Ipad represents a direction of the development of new media terminals. Its single channel and terminal compatibility are enhanced. It presents a variety of functions, and provides a carrier and platform for the deep integration and dissemination of clothing brand information. In the digital process, media integration is the trend of media development, including the integration of traditional media as well as the integration of new media. While traditional forms of media have an advantage in allowing a more focused promotion of its content, the new media is highly interactive, personalized and more convenient. Therefore, in the process of media integration, we consider combining traditional media and new media, allowing them to learn from each other to achieve integration.

3.2 On the Main Body of Promotion

Under the new media environment, the communication threshold is reduced, and the communication main body of clothing brand is increasing. The rapid increase in the number of promoters is a direct result from the increase in the absolute amount of information. However, too much information can greatly distract the audience's attention, which is not conducive to the effective and convenient access to useful information. The relationship between the subject of brand communication is becoming more and more complex, and it is very difficult to meet the needs of more and more differentiated audience simply relying on a media. In recent years, with more and more clothing enterprises in China implementing multi brand strategy. Many clothing enterprises at different levels, clothing brand horizontal expansion is more and more common, how to coordinate the relationship between different brands and resource allocation, brand communication work to become a new challenge. A successful case, Metersbonwe with the endorsement of the brand development for Smith, Smith Barney city series. The success of Metersbonwe. The important factor of its success is how to coordinate the relationship between the two brands, and how to allocate resources reasonably. [14]

3.3 On the Promotion Content

In China's clothing brand promotion, the application of new media has shifted from teaching mode to service mode. [15] Clothing promotion through sermons has been unable to meet the need of the audiences, and has since been replaced by a promotion model driven by consumer demand as the core of the service model. [16] Clothing enterprises can use the new media to enrich cultural connotations and personalized characteristics of the brand information to cleverly see to the different needs of the audience to meet their differentiated demands. Clothing enterprises to use new media brand promotion content integration is mainly reflected in the integration of clothing information. Clothing information includes not only the name of clothing, logo, price, quality, design style, but also includes the brand's history and cultural connotation, modes of operation and service concepts. In the new media era, the media integration allows for the process of integrating brand information to be easily achieved, and can reach target audience for a comparatively minimum cost. [17]

3.4 On Consumers

Table.3 IPTV Click Survey

Click	Percentage
VOD	63%
Time shift	7%
On-line TV	30%

The source of information: from Chinese TV net data

Compared with the traditional consumers the new media audience interaction is more interactive, more eager to use in the new media to reflect autonomy and participation. [18] To IPTV, for example, the rate of users on the on-demand, interactive mediums has had a high enthusiasm for the use. According to the Shanghai Media in various places involved in IPTV operations, trial operation of the statistical analysis: IPTV users in the VOD click accounted for 63%, time-shift clicks accounted for 7%, while only 30% of users were watching live TV. Operation of IPTV company hired market research company for professional analysis through their own monthly survey, the final conclusion is: "monthly households gradually deviated from the demand for television programs to demanding more interactive services, and interactive Consumer demand will become more prominent over time." [19] New media users are mainly young, fashionable people. In order to meet their needs, the new media content providers would need to change the content of their product and narrative. Such as interactive TV, to provide users with multiple clues and a variety of optional outcomes, with multiple narrative threads to satisfy its consumers.

4. Clothing Promotion Strategy with New Media

Table.4 What You Think We Could Do to Improve the Impact of New Media on Clothing Promotion

Strategy		Percentage
Clothing Promotion of the Differentiation Strategy	Brand Positioning Differentiation	35%
	To Meet the Needs of Differentiated Audiences	22%

The source of information: from Chinese TV net data

4.1 Clothing Promotion of the Differentiation Strategy

Garment enterprises that implement brand differentiation strategies needs to address two important aspects: First, they should create a differentiated brand positioning. Brand positioning refers to the effective establishment of the corporate brand and the competitor's brand differences between the minds of consumers in shaping a different image and occupying their attention in a separate mental dimension. Secondly, it should meet the increasingly personalized audience demands. As we can see from the success of many of the top clothing brands, only those brands which truly understand consumer needs eventually receive consumer recognition of their brand. In the context of new media, the clothing audience becomes increasingly segmented. Thus, these clothing brands must take the initiative to strengthen their consumers' preferences, by meeting these differentiated demands presented to them by their segmented consumer base.

4.1.1 Brand Positioning Differentiation

Brand positioning refers to the enterprise's role in market positioning and product positioning based on a specific brand in the cultural orientation and personality differences on the commercial decision-making. Brand positioning is a means to establish a target market-related brand image of the process and results. Good brand positioning is the prerequisite for the success of brand management, for the enterprise to forage into the market and expand its market share. If the brand cannot be effectively positioned to establish a unique consumer can agree with the brand personality and image, brand promotion is inevitably obscured by blindness and lack of consistency, will eventually make the product submerged under unnecessary product qualities, unsatisfactory performance and homogenized goods.

4.1.2 To Meet the Needs of Differentiated Audiences

Audience demand is dynamic, it is constantly changing as the environment and conditions change. [20] Changes in the demand for the audience are always accompanied by economic development, as well as scientific and technological progress. Particularly, the rise of economic development has led to an increase in people's income, providing an intrinsic impetus for changes in the needs of the audience. The progress of science and technology has led to the emergence of new products and new technologies and new conditions for its realistic existence.

4.2 Clothing Promotion of the Integration Strategy

Table.5 What You Think We Could Do to Improve the Impact of New Media on Clothing Promotion

Strategy		Percentage
Integration Strategy	Content Integration	53%
	Channel Integration	42%
	Media mode convergence	36%

The source of information: from Chinese TV net data

4.2.1 Content Integration

In the new media era clothing enterprises must incorporate content integration. [21] Content integration is the combination of different media such as mobile phones, television, network, magazines etc. [22] It is unified to promote the design and production of content, rather than a single media as an information terminal. Through the integration of the content, the garment enterprises can establish the connection between different information terminals. This can not only make the product information to be fully shared, but also can reflect the difference, and to achieve complementary advantages. [23] Through content integration, it can enhance the audiences' experience, allow consumers to meet the personalized needs in the new media background, and make the brand clothing achieve unique effects.

4.2.2 Channel Integration

Within the new media-based environment, the promotion of the main body has the potential to infinitely proliferate within massive consumer attention, leading to a high degree of dispersion, but this promotion is not conducive to effectiveness and ease of receiving useful information in regard to the brand. Therefore, the process of promotion channels only operates effectively through integration. To compensate for the shortcomings of a single media to form a complementary advantage. [24] Such as Sina, Sohu and other integrated portal, in the form of cooperative media and other forms of network interaction has been more and more, through the network interaction, to meet the needs of different audiences to obtain information at the same time, the consumer can also enhance their own input towards their clothing style . For the clothing industry, this integration of channels can also be greater development. Not only can we create clothes to become more flexible and diverse, but also allow our clothing to be personalized. Channel integration can obtain high profits at low cost, and lead to the brand image of clothing be better maintained.

4.2.3 Media mode convergence

In today's new media era, the mass media position (based upon traditional media) is fragmented and in its place, new media now presents diversification, leading to the extinction of an absolutely efficient mass promotion, and therefore, a diversified brand communication means becomes more and more important. The development of new media and the use of innovation, in addition to the past commonly used brand promotion methods such as advertising, promotion, public relations, marketing, etc., the Chinese garment enterprises to explore a lot of innovative brand promotion methods, such as e-mail, SMS, virtual Community, experience, word of mouth promotion and club. Ogilvy & Mather Public Relations China General Manager Ke Yingde has put forward the "360-degree integrated marketing promotion" concept, the so-called "360 degrees" refers to a full range means of public relations, which includes corporate image design, advertising, promotion, and media broadcasts, and media interaction. [25] The flexible use of these different methods of brand promotion is now one of the most important challenges and opportunities that China's garment enterprises are facing today. [26]

To integrate the above three ways: First, integration must require a concise goal; secondly, integration is to have a precise strategy research program. Thirdly, integration is about how to allocate and implement more effectively. In the process of integration, we must target, strategy and image, and these three factors are united in one direction. Only in this way can we achieve our goal effectively and quickly.

4.3 The Use of Online Reputation to Enhance the Visibility of Clothing

Table.6 what you think we could do to improve the impact of new media on clothing promotion

strategy	percentage
the use of online reputation to enhance the visibility of clothing	65%

The source of information: from Chinese TV net data

"Word of mouth" refers to the exchange of views and opinions of consumers on specific brands. [27] Word of mouth can have a huge influence on the consumer. The emergence of the Internet has subverted the traditional method of face-to-face interpersonal promotions, instead re-building a kind of indirect media through the Internet to promote, that is, "online reputation." [28] The prerequisite for word of mouth promotion is to let the consumer talk about your product or brand, that is, the product of the enterprise must be conducted in some way into the communication process of word of mouth. [29] There are two reasons why consumers would talk about a product: one is that consumers are talking to others when they use the product or have a positive experience after using the product; The other is the consumer has an unsatisfactory experience with their product and feels the need to tell others. Enterprises use "word of mouth" to promote a positive reputation; it must provide consumers with the notion that their products are high-quality products and provide a good experience. In addition, some of the characteristics of the product itself may also stimulate the consumer to promote their product further, such as innovative products, can show the taste and personality of consumers; the price is very cheap products, consumers will feel affordable; thoughtful service products and convenient use of the product, will let consumers feel warm. To promote real, clear, specific information to consumers is to create a good brand image, to successful implement a positive review through their "word of mouth". At the same time, through the interactive platform, enterprises can also cause the attention of potential customers, but also through the exchange of customers successfully resolve some of the customers for business products, misunderstandings and negative impressions.

4.4 Public Relations Online to Enhance Brand Reputation

Table.7 What You Think We Could Do to Improve the Impact of New Media on Clothing Promotion

Strategy	Percentage
Clothing Online Public Relations to Enhance Brand Reputation	77%

The source of information: from Chinese TV net data

Public relations activities with new media can extend corporate products and brands closer to the target consumer, giving them access to brand information. The interactive features of the new media make the network public relations activities more effective than traditional public relations activities. Through the brand's virtual communities, brand-specific sites and other means, these brands are able to communicate with customers to meet their real demands in regard to information, and provide timely feedback. These online public relation platforms also allow consumers to connected with the brand on a emotional level. Usually the new media public relations activities of the brand promotion, consist of the following three ways: (1) public relations activities with virtual communities. Through the virtual community, these brands carry out new media public relations activities, this is the most commonly used method in brand promotion. Virtual community refers to an organized network for people to communicate online and focus on mutual interest or demand for communication, (namely the clothing brands).

[30] In this community, people with common interests and needs around the world can take the initiative to participate in community activities, as online commutation is free, people would actively take the initiative to communicate with others, get information that they are interested in, and able to freely express their views and feelings. (2) Through the brand-specific website for public relations activities. New media environment establishes the brand through a variety of means of promotion and marketing to convey the brand's emotional appeal, the brand through the exclusive form of the site to display brand positioning and brand culture, to influence the consumer brand association, affecting consumer purchasing decisions. (3) Creating a following for the hot events, online promotion and offline public welfare activities, through public relations activities. Mature branding is mostly people-oriented, based around brand-oriented social and consumer demand. Therefore, we should always use the whole society concerned about the hot events, and combined with the relevant social public service activities as the carrier of brand promotion, to convey a sense of the corporation's social responsibility, with emphasis on creativity and captivation of the majority of target consumers.

5. Conclusion

This paper is based on the relevant literature and cases at home and abroad. First, the article analyzes the current situation of clothing brand communications in the context of new media. Second, to analyze the influence of new media on our clothing brand communication. Third, to explore China's clothing brand communication strategy under the background of new media. Through a series of research, summing up the practical and promotional value of new media in the context of China's clothing brand communication strategy. The new media not only provides new methods for the promotion of the brand, but also brings new challenges to the brand strategy of the enterprise: The question of how to accurately and quickly communicate the brand information of your own business in the fierce market competition to the audience you want to appeal to, and to gain brand awareness in the consumers reputation and loyalty and other aspects of the promotion of good results, it needs careful planning on the new media branding strategy.

References

[1] Xiao F. Study on clothing brand communication strategy based on new media [J]. Enterprise Herald: 2011; 20.

[2] Wang ZL, Xiao J. Sociál media marketing techniques and strategies [M]. Beijing: Publishing House of electronics industry: 2012; 6.

[3] Zhang YF. Relationship marketing [M]. Southwestern University of Finance and Economics press: 2007.

[4] Yu MY, Zhu JD. Brand communication [J]. Shanghai Jiao Tong University press: 2005.

[5] He X. Based on the new media of China's clothing brand communication strategy analysis [J]. Marketing strategy: 2013; 07.

[6] Yang J. New media communication characteristics [D]. Henan University: 2015.

[7] Development and research center of the State Administration of radio, film and television, China audio visual new media development report (2011) [M]. Social sciences literature press: 2011; 2.

[8] The new media features of micro blog [J]. News communication: 2010; 3: 15.

[9] Ban Y. Network new media development status and problem analysis [J]. Enterprise Herald: 2012; 9: 81.

[10] Zhang L. Brand communication in the new media environment [D]. Chongqing Technology and Business University: 2010; 12-14.

[11] Wang XM. New media era of localized brand communication strategy research [D]. Jinan University: 2012; 5-6.

[12] Huang YQ. China clothing advertising communication status and development research [D]. Suzhou University: 2015.

[13] Tom DC. Advertising and integrated marketing communication principle [M]. Beijing: Mechanical Industry Press: 2012.

[14] Holger Rooney, Jean Remy von Matt. Kinetic - the power of today's advertising needs [M]. Beijing: China Light Industry Press: 2012; 76-80.

[15] Yang Y. The new clothing industry competition: the war of channels [N]. Twenty-first Century Business Herald: 2008; 03-15.

[16] News websites enhance the competitiveness of the new media to seize the commanding heights of [N]. China news publishing daily: 2010; 11: 30.
[17] Wang GW. Mobile media advertising research [D]. Beijing University of Posts and Telecommunications: 2016.
[18] Chen PA, Wang J. The reflection of advertising creativity in new media environment [J] .Chinese advertisement: 2005; 4-15.
[19] Zhang L, Wu CW. Demand pull Innovation: the development of new media from the perspective of audience [J]. Chinese journalist: 2006; 11: 01.
[20] Mei M. Clothing popular communication and promotion model of the study [D]. Tianjin University of Technology: 2016.
[21] Li X. The use of new media brand building [D]. East China Normal University: 2014.
[22] Yu X. Analysis of brand communication strategy in the era of social media [J]. Press: 2011; (04): 55-56.
[23] Qiao Z. Advertising and promotion: integrated marketing communication perspective [M]. Dalian: Dongbei University of Finance and Economics press: 2000; 13.
[24] Deng MC. E-imc combat notes [M]. Beijing: Electronic Industry Press: 2012; 1: 74.
[25] Zheng SS. Under the background of new media in China's clothing brand communication strategy research [D]. [Master's Thesis]. East China Normal University: 2011; 03: 01.
[26] Ning J. Clothing brand planning practice [M]. Beijing: China Textile Publishing House: 2012.
[27] Zhang LW. New media integrated marketing communication strategy [J]. Media: 2010; 08
[28] News website to enhance the competitiveness of the Road: to seize the commanding heights of new media [N]. China News published: 2011; 11.
[29] Song SQ. The new media era of clothing marketing communication research [D]. Beijing Institute of Clothing: 2013.
[30] Zhou BB. Experiential terminal construction strategy [D]. Fudan University: 2014.

Design Decision Method for Fashion SME Based On Youth Consumer Behaviour

Yan-Wen Ruan[1], Li-Wen Gu[1,2], Wen Gu[1*], Xiao-Gang Liu[1]

[1] *Donghua University, College of Fashion and Design, West Yan'an Rd, Shanghai 200050, China*
[2] *North Carolina State University, College of Textiles, 1020 Main Campus Drive, Raleigh, NC 27695, USA*

*Corresponding author's email: guwen@dhu.edu.cn

Abstract

Fashion Small and Medium-sized Enterprises (FSME) takes an important role in the fashion industry. However, these enterprises face many problems when making fashion design decisions due to their lack of experience in business operations. To resolve this research gap, a design method for FSME was formed. The method relies on two main research results. One is FSME design decision process and the other is a fashion design decision factor. The decision process was concluded by an interview on twelve FSMEs in China. The decision factors were created based on youth consumer behaviour investigation. As a result, the design decision method for FSME based on youth consumer behaviour was established. It aims to help decision makers select the most appropriate design program and product that will be accepted by the youth consumer market.

Keywords: Design Decision Method; Fashion SME; Youth; Decision Process; Decision Factor; Consumer Behavior

1. Introduction

Small and medium-sized enterprises (SME) usually create more jobs than large companies. The official EU defines SME as enterprises which employ fewer than 250 persons with an annual turnover lower than 50 million USD and/or an annual balance sheet lower than 43 million USD [1]. In China, the officials define SMEs in different industries. SMEs are classified according to three indexes of annual turnovers, the number of employees and general assets [2]. As the data shows, in Europe, SME provide approximately 75 million jobs and represent 99.8% of the 21 million enterprises [3]. They are more and more important to the development of a country. The research on SMEs has been developed for many years [4]. However, in China, there are still some problems when operating a SME, especially in the fashion industry.

The Fashion Small and Medium-sized Enterprises (FSME) plays an important role in the fashion industry despite their growth being limited by their Research and Development (R&D) ability. Most of FSME are in their initial developing stage. They have limited resources and highly depend on the resources and expertise of others [5]. Their small scale and the owner's lack of experience lead to many problems during product development. For instance, when making design decision during the production process, there is a lack of objectivity or standards. These problems may not be initially negative, they will later however become huge obstacles to the growth of these companies.

There are some review studies on the design decision method of industrial design, however there is still a lack of design decision method for FSME. In order to establish the analysis factors system for FSMEs to consider, this study also reviews theories on youth consumer behaviour. The reviews focus on the main factors that influence them to make a consumption decision. The purpose of our research is to establish the design decision method for FSME based on youth consumer behaviour.

2. Problem Formulation

The problems on design decision of FSME mentioned above can easily lead to the products not encountering the needs of consumers and cannot match their branding orientation. Recently, some researchers held the opinion that brand orientation can be a positive force for small business growth [6]. An effective and distinct design decision method takes a guiding role to help FSME offer products to meet their branding orientation and consumer needs. So it is important to build the design decision method to help the growth of FSME.

Problem A: the fashion design decision method based on consumer consuming behavior

This method includes two main parts, the design decision flow part and the analysis factors based on youth consuming behavior. So the problem is divided into two sub-problems as following.

Fashion design decision is an integral process or one step process. This process could solve the problems related to product design development [7]. It is also the process to create and to evaluate multiple programs for making choices [8]. For textile and apparel companies, the design decision process is usually divided into seven steps [9].

From the review above, no fashion design decision method is indicated to FSME. As a special group in fashion industry that takes an important role, how do FSME make their design decisions? Is there any particular detail in their decision making process? A general design decision-making process for FSME is summarized as follows:

Problem A-1: the design decision processing of FSME.

For setting a competitive brand, FSME should understand and respond to multiple internal and external stakeholders [10], especially the consumers who are more relevant to them than to those large scale companies. Additionally, they should make quicker responses to consumers' needs. The speed of decision-making and an eye for market opportunities [11] become the most notable advantages of FSME [12]. So in keeping an understanding of the needs of consumers in time, taking them as the important references in the design process will be the obvious competitive ability for FSME.

Youth consumers are more concerned with their appearance, so they are more interested in purchasing fashionable clothing [13]. They are more concerned with trends than probably any other age groups [14]. Several researchers on consumer behavior focuses on youth and Generation Y [15, 16], because this group of people have the capacity to pay for what they want. They have strong needs on consumer fashion and they are interested in fashion trend. Their consumer behavior is also a distinct influence on the sale and branding of FSME. So putting the fashion consumption decision factors of youth into design decision will benefit FSME to develop their fashion product:

Problem A-2: the analysis factors based on youth consuming behavior.

3. Problem Solution

3.1 Fashion Design Decision Process

3.1.1 Method

Twelve FSME are selected for investigation. Through interviewing the designers and design managers of these companies, we suppose to present and revise the hypothetical decision process of seven steps: 1) determining the design destination, 2) searching the relative information, 3) formulating the standard of design decision, 4) drawing up the probable alternative programs, 5) analysing and choosing the final program, 6) practicing the decision program , 7) evaluating the effect of decision making [9], and to analyse their design decision process.

Comparing the classification standards of EU with the equivalents of China, the two indexes of annual turnover and the numbers of employees are the most used. For the purpose of this research, the classification standard in industrial, retail and wholesale industries in China, which defines SME that employ fewer than 1000 employees and have an annual turnover lower than 400 million RMB [2], will be used.

The name of companies cannot be made public as per the agreement with interviewees, however with their permission, interviews can be recorded for analysis. The interviewing brief was reviewed by five experts in the fashion industry, after deleting and modifying the inappropriate questions, five main questions about garment design formed the final brief. Two terms are defined for clarifying, they are node and decision program. The node is defined as the step which is carried out in the decision making process. The decision program is defined as the plans or schemes that are created for further evaluation and the best one of them will be selected as the final plans or scheme.

3.1.2 Design Decision Process of FSME

According to the investigation, the smaller the scale of the company is, the decision maker will consider the fewer nodes. However, some nodes practiced by FSME are not in the hypothesis. Compared with the theory hypothesis, the following two differences are mostly found.

1) Formulating the decision objective, searching the information, and making the final decision program are three nodes most practiced by FSME. The decision process of some small fashion companies may be without the node of searching the information. They make design decision based on leaders' personal preference without any objective evaluation. 2) Some FSME are more dependent on market feedback. Their design decision process is more detailed and more objective. Some evaluating features are used by them to direct the designers. The designers will give several programs to the decision makers according to these features. Then decision makers will balance the pros and cons of the programs. Sometimes they give scores, then, select the most optimal one.

By comparison with the theory and analysing the results of the survey, a detailed design decision process of FSME is proposed in Fig.1. It is usually divided into two modules. One is to formulate design decision programs and the other is to evaluate design decision programs. The former module includes several nodes as below:

1) Making the decision objective of fashion design. This refers to the aim of activities relative to the fashion design decision. Through these activities, decision makers should define and formulate relevant aims. 2) Screening the decision factors of fashion design while formulating a decision program, it should obey the aim oriented for design decision in FSME. 3) Collecting relevant information that is usually practiced by FSME for thoroughly analysing the relevant influencing factors, because decision programs formulation is based on adequate information resources. 4) Formulating the decision standard of fashion design mainly takes a guidance role towards the decision program formulating. It is the scale that is used to evaluate the programs and to select the final program. It includes two evaluating dimensions, the priority of decision factors and the indexes of decision factors. Through decision factor priority, decision makers can be clear which decision factor should be considered with greater priority during the process of decision making. Through decision factor indexes, decision makers can create the bottom line to each index, in order to realize the decision objective. 5) Formulating the decision programs of fashion design refers to making alternative ones, which meet the decision objective according to clear design effect factors, and all related information.

For the module of evaluating design decision programs, include two nodes as below:

1) Analysing the decision programs of fashion design refers to analysing every alternative one according to the decision standard. The analysis results can be obtained by scoring and calculating. Then the alternatives can be sorted by their scores. 2) Selecting the final design decision scheme refers to selecting and deciding the optimal one by the ranking of scores.

During the decision making process, two modules follow a sequential flow, the module of formulating the decision programs is the first step, while the module of evaluating design decision programs is the following step.

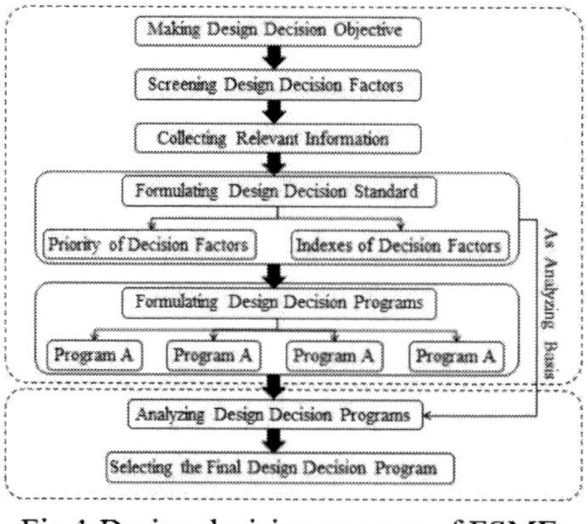

Fig.1 Design decision process of FSMEs

The most detailed design decision process that can be practiced by FSME is put forward in this research. Three main results are concluded. 1) Design decision process of FSME is usually divided into two modules. 2) During the decision making process, two modules follow a sequential flow. 3) Though the design decision process includes many detailed nodes, not all nodes are necessarily practiced by every FSME.

3.2 Youth Consumer Behavior

3.2.1 Method

For the purpose of this research, in order to find the factors related to design on fashion products that will most probably influence the youth consumer behavior, a questionnaire was designed for the youth. The questionnaire includes sixteen questions by using Likert's Method. They are about the fashion design details that consumers may be concerned with, from the silhouette to the fabric. After being reviewed by five experts in the fashion industry, four questions were deleted. Among them, three terms, Garment Plate, Garment Silhouette and Detail Partial Designing, should be defined for clarity. Garment Plate is defined as the garment's integral stereo shape which contributes to the fitness and proportion of a garment. Garment Silhouette is defined as the outline of the 2D shape which a garment shows when it is seen from a certain view. Detail Partial Designing is defined as the design on the components or parts of a garment.

The United Nations defines youth as persons aged between 15 to24. In the meantime, the UN also admits that it awknowledges without prejudice to other definitions set by Member States, such as the age of 18 to 35 [17]. WHO defines youth as persons below the age of 44 [18]. While in China, the National Bureau of Statistic of the People's Republic of China defines youth as persons between the age of 15 to 34 [19]. Because the target consumers of the research are in China, the age of the youth from 15 to 34 will be used. The questionnaire was issued to 160 youths whose age from 15 to 34 randomly through the website 'Sojump' and the shopping districts in Shanghai. The recovery questionnaire number carries on statistics analysis for 141.

Based on the 141 questionnaires, the principal component analysis of factor analysis method is employed through the statistical software SPSS21. Before analyzing the factors, the reliability test was provided. Data shows α is 0.78, which is over 0.6, showing that the reliability analysis test was passed. The table of descriptive statistics is shown in the Table 1.

Table.1 Table of Descriptive Statistics

Code of Factor	Name of Factor	Mean	SD
garm_pl	Garment Plate	4.65	.676
garm_sl	Garment Silhouette	4.71	.567
garm_po	Garment Popularity	3.94	1.208
garm_ps	Garment Individual Style	3.26	1.291
deta_pd	Detail Partial Designing	3.93	1.046
deta_mq	Detail Material Quality	3.86	1.093
deta_mc	Detail Material Color	4.12	1.059
colr_td	Color Tending	4.30	.976
colr_ma	Color Matching	4.34	.917
patt_sz	Pattern Size	4.11	1.094
patt_tp	Pattern Type	3.79	1.168
fabr_pp	Fabric Physical Property	4.54	.722
N Valid		141	

KMO is 0.751(>0.6), the significance is 0.000(<0.05), the data passed the test of KMO and Bartlett, the principal component analysis (PCA) of the statistical method can be used for analysing.

3.2.2 Consumer Analysis Factors

Through the extraction method of PCA, four components that were extracted can explain 62.447% of the total variance shown in Table 2. After rotation and five iterations, four components and their factors are shown in the following Table 3.

The first component includes four factors, which are pattern size, color matching, color trend and pattern type. This component was named as 'Pattern & Color'. The second component includes three factors, which are detail material quality, detail material color and detail partial design. All these factors express the detail design, which is considered most concerned by the youth. So the component was named as 'Detail & Decoration'. The third component also includes three factors, which are garment plate, garment silhouette and fabric physical properties. All these factors describe the look, fit and comfort of a garment that are most perceived by youth consumers. This component was named as 'Modelling & Fit'. The last component includes garment popularity and garment individual style. These two factors are about the spiritual pursuit of youth on costume consumption. The name of 'Taste & Culture' was given to this component.

Table.2 Total Variance Explained

Component	Initial Eigenvalues			Rotation Sums of Squared Loadings		
	Total	Variance[%]	Cumulative[%]	Total	Variance[%]	Cumulative[%]
1	3.679	30.660	30.660	2.590	21.585	21.585
2	1.353	11.271	41.932	1.867	15.559	37.144
3	1.294	10.783	52.715	1.586	13.217	50.361
4	1.168	9.732	62.447	1.450	12.085	62.447

Table.3 Rotated Component Matrix

Code of Factor	Component			
	1	2	3	4
patt_sz	.817	.043	.141	.256
colr_ma	.805	.074	.118	.197
colr_td	.767	.223	.159	-.136
patt_tp	.614	.134	.018	.063
deta_mq	-.025	.853	.130	-.026
deta_mc	.339	.744	.060	.114
deta_pd	.165	.629	.082	.214
garm_pl	.135	.096	.835	.029
garm_sl	.034	.047	.659	.293
fabr_pp	.325	.257	.568	-.341
garm_po	.081	.064	.211	.766
garm_ps	.186	.182	-.048	.689

The weights of four components and their factors should be calculated for the next step. And the result of the weights is the essential of scoring for the fashion design decision method.

$$F_m = \sum_{i=0}^{p=12} a_{im} x_i \tag{1}$$

In Eq.1, F represents the component, m is the number of components, a is the coefficient of factor, x is the factor, p is the number of factors. And as shown in the table (refer with: Table 3), the value range of m and p refers to $\{m|1 \leq m \leq 4, \mathrm{m} \in \mathrm{N}\}$ and $\{p|1 \leq p \leq 12, p \in \mathrm{N}\}$. To calculate the coefficient of factors,

$$a_{pm} = \frac{S_{pm}}{\sqrt{\lambda_m}} \tag{2}$$

In the Eq.2, λ is eigenvector, S is score of eigenvector. The result is,

$$A_F = \begin{bmatrix} a_{11} & a_{12} & \cdots & a_{1m} \\ a_{21} & a_{22} & \cdots & a_{2m} \\ \vdots & \vdots & \ddots & \vdots \\ a_{p1} & a_{p2} & \cdots & a_{pm} \end{bmatrix} = \begin{bmatrix} 0.51 & 0.03 & 0.11 & 0.21 \\ 0.50 & 0.05 & 0.09 & 0.16 \\ 0.48 & 0.16 & 0.13 & -0.11 \\ 0.38 & 0.10 & 0.01 & 0.05 \\ -0.02 & 0.62 & 0.10 & -0.02 \\ 0.21 & 0.54 & 0.05 & 0.09 \\ 0.10 & 0.46 & 0.07 & 0.18 \\ 0.08 & 0.07 & 0.66 & 0.02 \\ 0.02 & 0.03 & 0.52 & 0.24 \\ 0.20 & 0.19 & 0.45 & -0.28 \\ 0.05 & 0.05 & 0.17 & 0.64 \\ 0.12 & 0.13 & -0.04 & 0.57 \end{bmatrix} \tag{3}$$

In order to gain the value of weight, we should calculate the coefficient of integrated score,

$$\kappa_p = \sum\nolimits_{i=0}^{m} |a_{pi}| \cdot D_{\lambda_i}/62.447 \tag{4}$$

$$w_p = \kappa_p / \sum\nolimits_{i=0}^{p=12} \kappa_i \tag{5}$$

In the equation Eq.4 and Eq.5, D is variance contributes, κ is the coefficient of integrated score, and w is the weight after normalization. The result of κ and w shows in the following Table 4.

Table.4 The weight of components and factors

Component No.	Component	Component w	Factor No.	Factor Code	Factor κ	Factor w
F1	Pattern & Color	0.3581	χ1	patt_sz	0.25	0.098
			χ2	colr_ma	0.24	0.093
			χ3	colr_td	0.25	0.1
			χ4	patt_tp	0.17	0.067
F2	Detail & Decoration	0.2450	χ5	deta_mq	0.19	0.073
			χ6	deta_mc	0.24	0.093
			χ7	deta_pd	0.20	0.078
F3	Modelling & Fit	0.2488	χ8	garm_pl	0.19	0.075
			χ9	garm_sl	0.17	0.068
			χ10	fabr_pp	0.27	0.105
F4	Taste & Culture	0.1495	χ11	garm_po	0.19	0.074
			χ12	garm_ps	0.19	0.076

As a sub-conclusion, the analysis factors based on youth consumer behavior, the weight of components and factors were obtained. And the components and factors are the part of design influencing factors based on youth consumer attitude on design decision.

3.3 Fashion Design Decision Method

The fashion design decision method for FSME is composed by the decision process of FSME (top layer) and the decision factors based on youth consumer behavior (bottom layer), which is shown in Fig.3. The decision process helps FSME to fractionize the design decision-making process into steps. And the decision factors help to direct the specific design influencing factors that need to be considered for decision making. The layer of the decision factors is the most important part in the method, because it is specialized designed for decision objective. The objective is to design the garments that are most likely to be accepted by youth consumer.

As shown in Fig.4, the layer of design decision process (LP) is the starting and ending point of the method, it connects to the layer of decision factors (LF). Presupposed information and user-defined information are the two main parts in the LF. The presupposed information includes the components and factors with weights. The user-defined information includes the specific evaluating indexes of every factor.

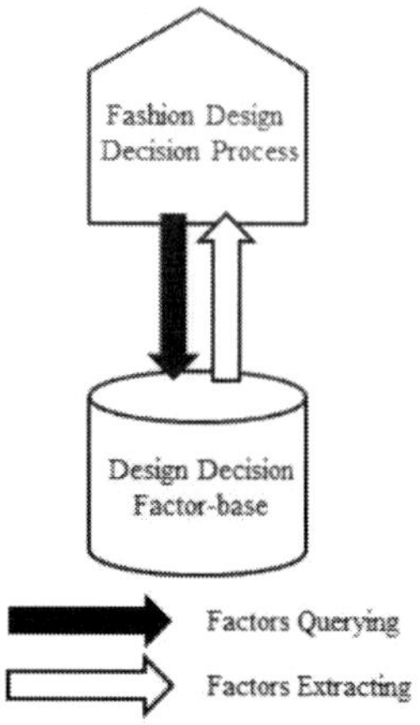

Fig.3 Fashion design decision method.

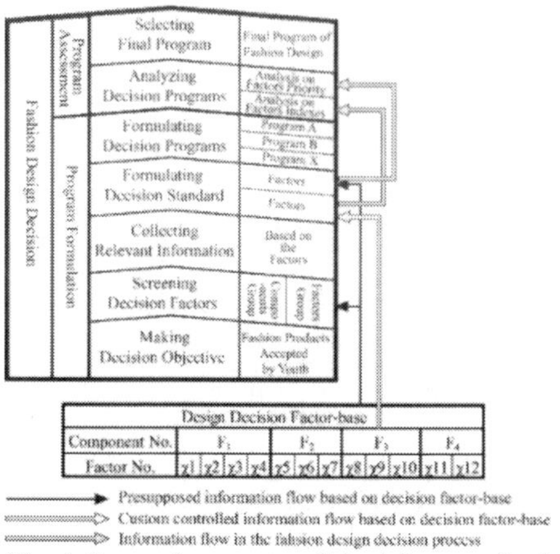

Fig.4 Operating route of decision method

The LF is connected with the nodes of screening the decision factors of fashion design, formulating the decision standard and analysing the decision programs. In the node of screening factors, users can invoke all the factors in LF, or they can use part of them that matches the real situation of their FSME the most to formulate the new factor set. In the node of formulating standard, according to the factor set after screening, formulating the standard by using the priority of factors and indexes of factors. Priority of factors is decided by the ranking of weights. Indexes of factors are defined by the decision makers based on the factors they selected. In the node of analysing programs, according to the standard, users can select the best program by scoring every program.

The aim of establishing the design decision method is to help decision makers select the most appropriate design program in order to lead the product to meet the needs of youth consumers the most. According to the principle of combining subjective and objective, the rule of the operation of the method was formulated. The objective analysing method can limit the entirely subjective judgment during design decision process. However, it can also take a guiding role to the decision makers. On the other hand, for the FSME, the subjective part in the method can also help decision makers play their individual abilities to the most extent on design decision so as to obtain a higher quality decision program based on the consumption behavior of youth.

4. Conclusion

For FSME, designing and producing the fashion product that meets the need of consumers are important for them. While the small scale of the FSME and the lack of experience of their owners can lead to many problems on the product development, especially in design decision. On the other hand, the youth are more likely to follow the current fashion trend and purchase fashion products without minding about the scale of the fashion companies. The youth aged between 15 and 34 can support their fashion consumption by themselves. As they have desire to decorate themselves and dress nicely, they are more concerned with the design and detail of garments. So we think that they are the group that FSME should focus on. Hence, the method of fashion design decision designed for FSME based on the youth consumption behavior is presented in this research.

For the purpose of the research, two problems have been solved. As to the design decision process of FSME, an entirely new decision making process has been formulated based on the survey on 12 FSME in China. Compared with the literature processes, three results were presented. The process is usually divided into two modules, and they are following a sequential flow, and not all nodes in the flow are necessarily practiced by every FSME. As to formulate the factor base, the factors are calculated by using the principal component analysis method based on the survey on 141 youth consumer. In order to investigate their consumption behavior when they decide to pay for a garment, the entirely components and factors AHP list are shown as the result. They composed to the base of design decision factors.

As conclusion, the design decision method for FSME based on the consuming behavior of youth was formulated by combining the two main results. Additionally, the structure and operational route of the method are displayed.

Because of the business privacy of SMEs, the amount of the SME that we interviewed was limited. However, in future research, more SMEs can be interviewed to verify the rationality of the design decision process. Moreover, the empirical analysis on this method should be done to optimize the method. Online FSME are increasing sharply in China, to build a fast and efficient design decision method for them is necessary in the future.

5. Acknowledgement

The authors wish to acknowledge the Shanghai Style Fashion Design & Value Creation Knowledge Service Center (No.ZX201311000031) program, Shanghai Style Fashion Trend Research Based on Knowledge Management of The Fundamental Research Funds for the Central Universities (No.15D110719).

References

[1] European Commission. SME definition: Commission recommendation. Brussels: EUC: 2003.

[2] Information on http://www.gov.cn/zwgk/2011-07/04/content_1898747.htm.

[3] Stawińska A. Key figures on European business with a special feature on SMEs. Publications Office of the European Union, Luxembourg: 2011.

[4] Eggers F, O'Dwyer M, Kraus S, Vallaster C, Güldenberg S. The impact of brand authenticity on brand trust and SME growth: A CEO perspective. J World Bus: 2013; 48 (3): 340-8.

[5] Gilmore A, Carson D, Grant K. SME marketing in practice. Marketing intelligence & planning: 2001; 19 (1): 6-11.

[6] Boyle E. A study of entrepreneurial brand building in the manufacturing sector in the UK. Journal of product & brand management: 2003; 12 (2): 79-93.

[7] Simon HA, Dantzig GB, Hogarth R, Plott CR, Raiffa H, Schelling TC, Shepsle KA, Thaler R, Tversky A, Winter S. Decision making and problem solving. Interfaces: 1987 Oct; 17 (5): 11-31.

[8] Mezher T, Abdul-Malak MA, Maarouf B. Embedding critics in decision-making environments to reduce human errors. Knowledge-Based Systems: 1998 Nov 12; 11 (3): 229-37.

[9] Liu XG, Wang J, Gu W. Methodology of fashion design. China textile press: 2009.

[10] Roper S, Davies G. The corporate brand: Dealing with multiple stakeholders. Journal of Marketing Management: 2007 Feb 1; 23 (1-2): 75-90.

[11] Krake FB. Successful brand management in SMEs: a new theory and practical hints. Journal of Product & Brand Management: 2005 Jun 1; 14 (4): 228-38.

[12] Andersen PH. Relationship marketing and brand involvement of professionals through web-enhanced brand communities: The case of Coloplast. Industrial marketing management: 2005 Jan 31; 34 (1): 39-51.

[13] O'Cass A. Fashion clothing consumption: antecedents and consequences of fashion clothing involvement. European Journal of Marketing: 2004 Jul 1; 38 (7): 869-82.

[14] Martin CA, Bush AJ. Do role models influence teenagers' purchase intentions and behavior?. Journal of consumer marketing: 2000 Sep 1; 17 (5): 441-53.

[15] McCormick H, Livett C. Analysing the influence of the presentation of fashion garments on young consumers' online behaviour. Journal of Fashion Marketing and Management: An International Journal: 2012 Feb 24; 16 (1): 21-41.

[16] Martin CA. From high maintenance to high productivity: What managers need to know about Generation Y. Industrial and commercial training: 2005 Jan 1; 37 (1): 39-44.

[17] Information on_http://undesadspd.org/Youth.aspx.

[18] Information on http://world.huanqiu.com/regions/2013-05/3930101.html.

[19] Information on http://www.stats.gov.cn/tjsj/tjgb/rkpcgb/qgrkpcgb/201104/t20110428_30327.html.

A Survey on Clothing Description for Intelligent Image Understanding

Fei-Fei Sun[1,2], Ping-Hua Xu[3], Tesfaye Mulu Asmamaw[1], Xue-Mei Ding[1*]

[1] *Fashion and design college, Donghua University, No. 1882, West Yan'an Road, Shanghai 200051, PR China,*
[2] *Fine Arts Academy, South-central University For Nationalities, No.708, Minyuan Road, Wuhan, Hubei Province, China,*
[3] *Fashion College, Zhejiang Sci-tech university, 928 Second Avenue, Xiasha Higher Education Zone, Hangzhou, China*

*Corresponding author email: fddingxm@dhu.edu.cn

Abstract

Clothing description is the basis of many other complex clothing vision problems. In this paper, the latest remarkable research works of clothing description in the vision area was reviewed. Firstly apparel visual appearance description based on low-level and middle-level features for Intelligent Image Understanding was reviewed. Next the quantity of images and annotation in the datasets in these studies are summarized. Then recent efforts in clothing recognition and related applications are reviewed. Finally, future research work about clothing image recognition is forecasted. It is hard to understand the clothing images because the detail information of the clothing parts is not easily usable, so we forecasted it to build a novel layered feature mode to achieve a description close to the cognition of human visual system.

Keywords: Clothing Image Description; Clothing Image Recognition; Attribute Recognition; Semantic Recognition

1. Introduction

The growth of online visual social networks such as Facebook, Google+, Instagram, and Pinterest, to name a few, is transforming how people interact with visual media. Fashion has a tremendous impact on our society not only because of its power of reflecting the social status of individuals but also the emergence of online social sites and self-media enabled peoples share photographs very easily and conveniently across the network users. This situation brings enormous growth of online retail sales.

Clothing image data has a very wide range of applications and attractive future prospects, such as clothing intelligent design, clothing brand management, clothing visual recognition, clothing image retrieval, clothing e-commerce data mining. It is becoming a hot spot due to potential benefits in e-commerce application in the huge clothing market.

In face of massive clothing media information, management of this huge data effective description of the clothing image, exploiting garment imaging technology to benefit consumers and societies, are the current opportunities and challenges being faced in the field of clothing computer vision. It is essential to explore the use and management of large number of clothing images among clothing business operators (especially online clothing retailers) and clothing consumers who mainly use computer vision.

The clothing visual image is a poorly researched area and we found very few review articles which lacked the introduction of the latest remarkable developments about clothing description for intelligent image understanding. The rest of this paper is organized as follows. Section 2 reviews apparel visual appearance description based on low-level and middle-level features for Intelligent Image Understanding. The quantity of images and annotation in the datasets in these studies are summarized in Section 3; Section 4 discusses the yearly applications of clothing vision description; finally, future research work about clothing image recognition is forecasted. It is hard to understand the clothing images because the detail information of the clothing parts is easy to use, so we forecast it to build a novel layered feature mode to achieve a description close to the cognition of human visual system.

2. Clothing Description for Visual Content Understanding

Clothing Description for Intelligent Image Understanding is a very challenging high level computer vision problem. It relied on low and mid level cues. Now we will explain the different cues and algorithms which are the base of high level computer vision models.

2.1 Clothing Description Based on Low Level Clue

Typical global features are image color, texture, shape, and so on. Since individual pixels do not contain sufficient information, it is normal to group them into different regions which conceptually correspond to parts of the same object in the image. These are generally compact vector representations of small image patches. The most well-known approach is the SIFT descriptor. As SIFT [1] was originally designed to be used sparsely, there have been several alternatives proposed to be used densely. The most popular one is the Histogram of Oriented Gradients (HOG) [2-4] descriptor.

The current clothing image retrieval technology is mainly based on color, texture, shape and so on. Cheng et al. extract the contours of clothing through edge detection technology, and combine the shape features and color models to express the clothing style information to achieve clothing image retrieval. Hou [5] et al. used the invariant moments and Fourier descriptors to describe the shape of the garment, and combined the color histogram to study garment image retrieval.

Tao Chen identify clothing visual elements: profile, shoulder line, waist line, hem wire, the main colors, color matching based on clothing color and shape features. Ching-I Cheng [6] extracted the color distribution, printing (Sobel), shape (Sobel edge detector) features to characterize clothing, and the clothing vector space were mapped to the sense factor space. Elie [7] located the clothing image through face recognition, and applied video port recognition using histogram comparison, the main color features and texture features as clothing features. Kuramoto [8] established a dress history sharing system called WearReport. The system expressed the clothing by extracting the color and texture features in several divided regions to identify the similar situation. Choi [9] detected the garment area according to the face position, then analyzed the similarity between the clothing by the color histogram information of the garment area in the image. YannisKalantidis [10] quantified the color into 29 in order to optimize the calculation speed and selects the LBPs local texture feature descriptor as the local feature. Wang Hailong [62] use representative color, gray-level co-occurrence matrix and outer contour to express the color, pattern and style of the clothing, and realize garment retrieval by the corresponding matching algorithm.

Low level cues do not encode very prolific information, which is necessary to rely on low and mid level cues.

2.2 Clothing Description Based on Middle Level Semantics Clue

The researchers has attempted to establish the middle semantic concept on the basis of the existing effective image visual characteristic, and divided the "gap" between the low and high level into two smaller gulfs by adding the middle attribute semantic level. The image can be accurately classified into specific semantic categories according to the distribution of semantic concepts. According to the different ways of constructing the concept of semantic concept in the middle of the scene, the existing clothing image represent algorithm based on middle semantic modeling is summarized as follows:

2.2.1 Semantic Segmentation

The main principle of the semantic object method is to form the semantic object region by segmentation and mark these semantic object regions as known semantic object categories using the local classifier. Yamaguchi [11-18] made a series of studies on the visual problem of clothing. The posture estimation proposed by Yang was applied to the overall dress code study. He presents a way to understand the style of clothing based on the retrieval of clothing image/ basing on the large-scale fashion image database. The semantic object area marked in the training images was trained for a series of detection functions. Each detection function was used to find the obvious target for the unlabeled image in the test image. The Maximum a posteriori is used to judge the clothing category. Popularity may be predicted basing on the above description of the clothing. On the basis of this, Simo-Serra, Edgar [19-22] and M. HadiKiapour [23] made similar studies based on massive data. Simo-Serra, Edgar also establish the probability model of the clothing style basing on the random vector field model (56 categories in his model but only 29 in Yamaguchi's model.

Transforming the clothing image classification problem into the recognition of the target is the essence of the semantic object method. However, the target recognition and the image segmentation algorithm are a major obstacle to its application that has to be solved ungently.

2.2.2 Semantic Recognition

The image segmentation in this method is regular sub-regions avoiding the target detection and segmentation process. So middle-level semantic modeling becomes more extensive.

Liu and other established magic closet [3] application system for automatic occasion-oriented clothing recommendation. To narrow the semantic gap between the low-level features of clothing and the high-level occasion categories, middle-level clothing attributes (e.g., clothing category, color and pattern) are adopted as a bridge. Wang and Zhang [24] of the Hewlett-Packard Laboratory approach a model to find the photographs of the same clothing from the life album basing on human skin color detection. There are some attributes used in the clothing search including clothing categories, sleeves, style and so on. Lukas Bossard [25] first detected the upper body, and then extract dense features such as color (LAB), SURF, HOG, LBP, Self-Similarity (SSD). Chen [1] uses four types of features such as SIFT, texture description (maximum filter response), color (LAB), skin predictor in the body's five major regional (torso, upper left arm, right upper arm, lower left arm, lower right arm, left lower limb)to describe the clothing properties. Willimon [26-28] and others designed a LCSH four levels feature model. The low-level visual features were extracted, and 27 kinds of semantic middle-layer clothing were approached by learning. KuanTing Chen [29] Followed the idea of human body posture estimates as Chen and approached the visual bag of features by the color, texture, texture, SIFT and skin characteristics. 72 semantic attributes were classified by SVM. And the relationship between semantic properties are modeled by random vector field.

The high-level descriptions of clothing are complex. Clothing category or style recognition fails basing on local or global statistic features in the clothing.

As the same as the cognition of human visual system, we forecast a novel layered feature model according to the visual process from entirely to locality.

3. Datasets

Though the study of clothing computer vision has just started, there is no open clothing image database with professional annotation [3]. In the past, the database of the related research was formed by the researcher. One part of the picture resource was collected manually [30-32] and was manually annotated. The other part is downloaded from the web via crawler software [33], some of which is annotated with Amazonian robots [3, 16], and the other is extracted from download words as the annotation.

Huizhong Chen [1] collected 1856 images from Sartorialist and Flickr using an upper body detector to select pictures with people, and then got ground truth labels of 26 attributes in total by AMT, including 6 binary-class attributes for pattern, 11 binary-class attributes for color and 6binary-class attributes for miscellaneous and 3 multi-class attributes (sleeve length, neckline shape and clothing category).

The WoW [3] dataset was collected from both shopping websites and sharing websites which has been divided into three subsets: WoW Full (9,469 images), WoW Upper (8,421 images), WoW Lower (6,527 images). Dataset has two types of attributes, one of which is multi-class attribute including 13 clothing categories and the other is detail attribute as same as previous research [1].

Liu constructed two datasets for image retrieval between different scenes, such as Online Shopping (OS) dataset (8293 upper body images and 8343 lower body images) and Daily Photo (DP) dataset (4321upper body images and 4068 lower body images). He defined a set of clothing-specific attributes according to empirical study on clothing catalog. 53 kind of attributes which are manually defined can be summarized into 3 attributes (global, upper-body and lower-body attributes), while lower-body attributes can be further divided into trousers and skirts related attributes.

Yamaguchi [12] introduced a novel dataset(685 photos from fashion blog), and use 2 AMT jobs to gather annotations, one of which collected ground truth pose annotations for the common 14 body parts, and the other got ground truth clothing labels on super-pixel regions. In his study there are 56 different possible clothing labels - more than considered in any previous approach.

Chen [34] collected a large cloth image dataset with fine grained attribute labels which contains 1,108,013 clothing photos with 25 attribute categories (e.g. type, color, pattern, season, occasion).

Liang [35] collect a large dataset composed of cross-scenario image pairs(10000 online merchants product photos and offline customers photos, which are 124 fine-grained semantic attribute types in each image.

DeepFashion [36] contains 800K images, and over 300K image pairs. There are 50 categories, 1, 000 attributes, clothing landmarks (each image has 4-8 landmarks).

4. Application

4.1 Clothing Retrieval

Along with the growing concern of clothing recognition due to the huge clothing e-commerce market, the number of studies in the field of clothing retrieval or recommendation applications has increased [2, 3, 6, 24, 33, 36-47]. Most notably, Liu et al [2] proposes a visual search approach with a sparsely coded transfer matrix to match a fashion street images to shopping images.

In addition to clothing retrieval frequently appearing in the clothing e-commerce, retrieval of similar clothing can help building the datasets for the future study [13, 48].

4.2 Attribute Recognition

Clothing Different clothing attributesare color, shape and pattern and so on. Therefore attribute recognition has been the focus of several clothing recognition literatures [1, 4, 25, 49]. Firstly Borras [50] interpreted five cloth modeling by a graph structure with region features. His approach can automatically describe clothing about color, texture and structural composition. On the basis of a part-based approach Lubomir Bourdev [4] used poselets to recognize attributes, such as the gender, hair style and types of clothes. Chen used conditional random fields to model clothing by considering the relationship between attributes [1]. Di employed a retrieval system basing on fine-grained attribute recognition.

4.3 Person Identification

Identification of person by clothing is another important application. Clothing can be used as a cue to identify the person in a photo and can be applied in surveillance scenario [51-53]. Clothing can identify not only the look of a person, but also the social status, occupation of a person [54-56]. Occupation of the person using clothing clue was recognized in these studies [56, 57]. And social group was identified by fashion style [54, 55].

4.4 Fashion Trend Forecast and Popularity Prediction

Shintami C. Hidayati [58] develop a framework using a novel algorithm automatically that could discover fashion trend elements. Chen [16] firstly established a large dataset to study the similarity between New York Fashion Shows and New York street chic. Chen investigation in fashion trends proved that fashion show has an impact on street fashion. Kota Yamaguchi [16] used clothing image content and text information online to predict popularity of photograph in both online and not online scenarios. Also Simo-Serra [22] approached a CRF model reasoning about the relationships between cloth style, user type, the photograph's setting and fashion score.

4.5 Clothing Parsing

Clothing parsing aim is to assign a semantic object label to each pixel in an image. Human pose is the key clue to recognize garment items. Due to clothing occlusion, deformation and other issues clothing parsing has not been studied much until recently.

Yamaguchi, Kota [12] was the first to develop a novel framework to parse clothing with garment categories, where parsing is formulated as a MAP estimation of image-region labels in conditional random field (CRF) given pose estimation. Jian Dong [56] build a unique DMPM to learn appearance and structure effectively. Nataraj Jammalamadaka [59] proposes a method to segment clothes in images with no assumption on pose of the person, viewing angle, occlusion or clutter. The works [13, 21] build the lager Datasets and parse the clothing by retrieving similar styles. Wei Yang [60]constructed a cloth co-parsing system by two phases of optimization: (a) clothing co-segmentation for extracting coherent clothes regions, and (b) region co-labeling for recognizing various clothes garments. Wang Fan [61] applied mid-level semantic parts to solve the problem of parsing fashion images.

5. Conclusions and Future Work

Costume design relied on human visual cognition and artistic aesthetics. However, too many clothing features caused obstacles to understand clothing images. To provide detailed description of clothing i, There are a small number of research about a multi-level clothing model in computer vision. The following question haven't been yet addressed:

First of all, there is not a professional clothing image database with detailed attributes. This database is the basis for automatic analysis of clothing images.

Secondly in the current works, clothing properties of clothing (such as: clothing length, sleeve length, collar, color, fabric, texture, etc.) were not sufficiently described in the classification of clothing attributes. It is difficult to understand the clothing because that part of the clothing information is not used. So we need to analyze clothing properties and establish semantic model to characterize the appearance of clothing.

Finally, The high-level descriptions of clothing are complex. Clothing category or style recognition is not good basing on single-level local or global statistic features. .

As the same as cognition of human visual system, we forecast a novel layered model according to the visual process from entirely to locality.

6. Acknowledgment

This research is supported by the Fundamental Research Funds for the Central Universities (GrantNo.CSQ13038), Culture research funds for Hubei Ethnic Affairs Commission (Grant No. QSZ13009) and Nantong Science and Technology Project (Grant No.GY12016032).

References

[1] Chen H, Gallagher A, Girod B. Describing Clothing by Semantic Attributes: Springer Berlin Heidelberg: 2012.

[2] Liu S, Song Z, Liu G, Xu C, Lu H, Yan S. "Street-to-shop: Cross-scenario clothing retrieval via parts alignment and auxiliary set," in Computer Vision and Pattern Recognition (CVPR), 2012 IEEE Conference on: 2012; pp. 3330-3337.

[3] Liu S, Feng J, Song Z, Zhang T, Lu H, Xu C, et al. "Hi, magic closet, tell me what to wear!," in Proceedings of the 20th ACM international conference on Multimedia: 2012; pp. 619-628.

[4] Bourdev L, Maji S, Malik J. "Describing people: A poselet-based approach to attribute classification," in Computer Vision (ICCV), 2011 IEEE International Conference on: 2011; pp. 1543-1550.

[5] Hou AL, Z LQ, S DC. "Garment image retrieval based on multi-features," in 2010 International Conference on Computer, Mechatronics, Control and Electronic Engineering: 2010; pp. 194-197.

[6] Cheng CI, Liu DSM, "An intelligent clothes search system based on fashion styles," in 2008 International Conference on Machine Learning and Cybernetics: 2008; pp. 1592-1597.

[7] Khoury EE, Senac C, Joly P. "Face-and-clothing based people clustering in video content," in Proceedings of the international conference on Multimedia information retrieval: 2010; pp. 295-304.

[8] Kuramoto I , Matsuura D, Yamamoto K, Tsujino Y. WearReport: Avoiding clothe conflict against friends based on shared wearing history, in 2013 IEEE 2nd Global Conference on Consumer Electronics (GCCE): 2013; pp. 182-184.

[9] Choi YJ, Kim KJ, Nam Y, Cho WD. Retrieval of identical clothing images based on local color histograms," in

Convergence and Hybrid Information Technology, 2008. ICCIT'08. Third International Conference on: 2008; pp. 818-823.
[10] Kalantidis Y, Kennedy L, Li LJ. "Getting the look: clothing recognition and segmentation for automatic product suggestions in everyday photos," in Proceedings of the 3rd ACM conference on International conference on multimedia retrieval: 2013; pp. 105-112.
[11] Yamaguchi K, Kiapour MH, Ortiz LE, Berg TL. "Retrieving similar styles to parse clothing," IEEE transactions on pattern analysis and machine intelligence: 2015; vol. 37: pp. 1028-1040.
[12] Yamaguchi K, Kiapour MH, Ortiz LE, Berg TL. "Parsing clothing in fashion photographs," in IEEE Conference on Computer Vision & Pattern Recognition: 2012; pp. 3570-3577.
[13] Yamaguchi K, Kiapour MH, Berg TL. "Paper doll parsing: Retrieving similar styles to parse clothing items," in Proceedings of the IEEE International Conference on Computer Vision: 2013; pp. 3519-3526.
[14] Kiapour MH, Yamaguchi K , Berg AC, Berg TL. "Hipster wars: Discovering elements of fashion styles," in European conference on computer vision: 2014; pp. 472-488.
[15] Yamaguchi K. "Understanding Online Fashion Networks," STATE UNIVERSITY OF NEW YORK AT STONY BROOK: 2014.
[16] Yamaguchi K, Berg TL, Ortiz LE. "Chic or social: Visual popularity analysis in online fashion networks," in Proceedings of the 22nd ACM international conference on Multimedia: 2014; pp. 773-776.
[17] Vittayakorn S, Yamaguchi K, Berg AC, Berg TL. "Runway to realway: Visual analysis of fashion," in 2015 IEEE Winter Conference on Applications of Computer Vision: 2015; pp. 951-958.
[18] Yamaguchi K, Okatani T, Sudo K, Murasaki K, Taniguchi Y. "Mix and match: joint model for clothing and attribute recognition," in British Machine Vision Conf: 2015.
[19] Simo-Serra E. "Understanding human-centric images: from geometry to fashion," arXiv preprint arXiv: 1604.08164: 2015.
[20] Simoserra E, Ishikawa H. "Fashion Style in 128 Floats: Joint Ranking and Classification using Weak Data for Feature Extraction."
[21] Simoserra E, Fidler SF. A High Performance CRF Model for Clothes Parsing: Springer International Publishing: 2014.
[22] Simoserra E, Fidler S, Morenonoguer F, Urtasun R. "Neuroaesthetics in fashion: Modeling the perception of fashionability," in Proceedings of the IEEE Conference on Computer Vision and Pattern Recognition: 2015; pp. 869-877.
[23] Kiapour, Mohammadhadi. "LARGE SCALE VISUAL RECOGNITION OF CLOTHING, PEOPLE AND STYLES": 2015.
[24] Wang X, Zhang T, Tretter DR, Lin Q. "Personal clothing retrieval on photo collections by color and attributes," IEEE Transactions on Multimedia: 2013; vol. 15: pp. 2035-2045.
[25] Bossard L, Dantone M, Leistner C, Wengert C, Quack T, Gool LV. "Apparel classification with style," in Asian Conference on Computer Vision: 2012; pp. 321-335.
[26] Willimon B, Birchfield S, Walker I. "Classification of clothing using interactive perception," in Robotics and Automation (ICRA), 2011 IEEE International Conference on: 2011; pp. 1862-1868.
[27] Willimon B, Birchfield S, Walker ID. "Model for unfolding laundry using interactive perception," in IROS: 2011; pp. 4871-4876.
[28] Willimon B, Walker I, Birchfield S. "A new approach to clothing classification using mid-level layers," in Robotics and Automation (ICRA), 2013 IEEE International Conference on: 2013; pp. 4271-4278.
[29] Chen K, Chen K, Cong P, Hsu WH, Luo J. "Who are the Devils Wearing Prada in New York City?," in Proceedings of the 23rd ACM international conference on Multimedia: 2015; pp. 177-180.
[30] Chang YC, Chuang MC, Hung SH, Shen SJC, Chu BS. "A kansei study on the style image of fashion design," in 6. Asian Design Conference: 2003.
[31] Juhee S. "The relationship between design images in clothing style and fabric type in designing blouses," Journal of Home Economics of Japan: 2003; vol. 54: pp. 925-934,.
[32] Zeng X, Koehl L, Wang Y, Chen Y, Chen ZG. "The sensory research on the style of women's overcoats," International Journal of Clothing Science and Technology: 2008; vol. 20: pp. 174-183.
[33] Liu L, Xing J, Liu S, Xu H, Zhou X, Yan S. "Wow! you are so beautiful today!," ACM Transactions on Multimedia Computing, Communications, and Applications (TOMM): 2014; vol. 11: p. 20.
[34] Chen Q, Huang J, Feris R, Brown LM, Dong J, Yan S. "Deep domain adaptation for describing people based

on fine-grained clothing attributes," in Proceedings of the IEEE conference on computer vision and pattern recognition: 2015; pp. 5315-5324.
[35] Liang X, Lin L, Yang W, Luo P, Huang J, Yan S. "Clothes Co-Parsing Via Joint Image Segmentation and Labeling With Application to Clothing Retrieval," IEEE Transactions on Multimedia: 2016; vol. 18: pp. 1175-1186.
[36] Liu Z, Luo P, Qiu S, Wang X, Tang X. "DeepFashion: Powering Robust Clothes Recognition and Retrieval with Rich Annotations," in Proceedings of the IEEE Conference on Computer Vision and Pattern Recognition: 2016; pp. 1096-1104.
[37] Wang X, Zhang T. "Clothes search in consumer photos via color matching and attribute learning," in Proceedings of the 19th ACM international conference on Multimedia: 2011; pp. 1353-1356.
[38] Kiapour MH, Lazebnik XHS, Berg AC, Berg TL. "Where to buy it: Matching street clothing photos in online shops," in Proceedings of the IEEE International Conference on Computer Vision: 2015; pp. 3343-3351.
[39] Huang J, Feris RS, Chen Q, Yan S. "Cross-domain image retrieval with a dual attribute-aware ranking network," in Proceedings of the IEEE International Conference on Computer Vision: 2015; pp. 1062-1070.
[40] Wong WK, Zeng X, Au W. "A decision support tool for apparel coordination through integrating the knowledge-based attribute evaluation expert system and the T-S fuzzy neural network," Expert Systems with Applications: 2009; vol. 36: pp. 2377-2390.
[41] Yonezawa Y, Nakatani Y. "Fashion support from clothes with characteristics," in Symposium on Human Interface: 2009; pp. 323-330.
[42] Dai HQ. "The Research on Intelligent Clothing Recommendation System Based on Ontology," in Advanced Materials Research: 2011; pp. 827-831.
[43] Guo Z, Wong W, Leung S, Li M. "Applications of artificial intelligence in the apparel industry: a review," Textile Research Journal, p. 0040517511411968: 2011.
[44] Iwata T, Wanatabe S, Sawada H. "Fashion coordinates recommender system using photographs from fashion magazines," in IJCAI Proceedings-International Joint Conference on Artificial Intelligence: 2011; p. 2262.
[45] Dong A, Shan D, Ruan Z, Zhou L, Zuo F. "The design and implementation of an intelligent apparel recommend expert system," Mathematical Problems in Engineering: 2013.
[46] Kim D, Kwon YB, Park J. "A scoring model for clothes matching using color harmony and texture analysis," in Graphics Recognition. New Trends and Challenges, ed: Springer: 2013; pp. 218-227.
[47] Miura S, Yamasaki T, Aizawa K. "SNAPPER: Fashion Coordinate Image Retrieval System," in Signal-Image Technology & Internet-Based Systems (SITIS), 2013 International Conference on: 2013; pp. 784-789.
[48] Loni B, Menendez M, Georgescu M, Galli L, Massari C, Altingovde IS, et al. "Fashion-focused creative commons social dataset," in Proceedings of the 4th ACM Multimedia Systems Conference: 2013; pp. 72-77.
[49] Di W, Wah C, Bhardwaj A, Piramuthu R, Sundaresan N. "Style finder: Fine-grained clothing style detection and retrieval," in Proceedings of the IEEE Conference on Computer Vision and Pattern Recognition Workshops: 2013; pp. 8-13.
[50] Borras A, Tous F, Lladós J, Vanrell M, "High-level clothes description based on colour-texture and structural features," in Iberian Conference on Pattern Recognition and Image Analysis: 2003; pp. 108-116.
[51] Anguelov D, Lee K, et al. "Contextual identity recognition in personal photo albums," in Computer Vision and Pattern Recognition, 2007. CVPR'07. IEEE Conference on: 2007; pp. 1-7.
[52] Gallagher AC, Chen T. "Clothing cosegmentation for recognizing people," in Computer Vision and Pattern Recognition, 2008. CVPR 2008. IEEE Conference on: 2008; pp. 1-8.
[53] Wang NandAi H. "Who blocks who: Simultaneous clothing segmentation for grouping images," in 2011 International Conference on Computer Vision: 2011; pp. 1535-1542.
[54] Murillo AC, Kwak IS, et al. "Urban tribes: Analyzing group photos from a social perspective," in 2012 IEEE Computer Society Conference on Computer Vision and Pattern Recognition Workshops: 2012; pp. 28-35.
[55] Kwak IS, Murillo AC, Belhumeur PN, Kriegman DJ, Belongie SJ. "From Bikers to Surfers: Visual Recognition of Urban Tribes," in BMVC: 2013.
[56] Shao M, Li L, Fu Y. "What do you do? occupation recognition in a photo via social context," in Proceedings of the IEEE International Conference on Computer Vision: 2013; pp. 3631-3638.
[57] Song Z, Wang M, Hua XS, Yan S. "Predicting occupation via human clothing and contexts," in Computer Vision (ICCV), 2011 IEEE International Conference on: 2011; pp. 1084-1091.
[58] Hidayati SC, Hua KL, Cheng WH, Sun SW. "What are the Fashion Trends in New York?," in Proceedings of

the 22nd ACM international conference on Multimedia: 2014; pp. 197-200.
[59] Jammalamadaka N, Minocha A, Singh D, Jawahar C. "Parsing Clothes in Unrestricted Images," in BMVC: 2013; p. 2.
[60] Yang W, Luo P, Lin L. "Clothing co-parsing by joint image segmentation and labeling," in Proceedings of the IEEE Conference on Computer Vision and Pattern Recognition: 2014; pp. 3182-3189.
[61] Fan W, Z QY, Tao X, Y BL. "Parsing fashion image into mid-level semantic parts based on chain-conditional random fields," IET Image Processing: vol. 10; pp. 456-463: 2016.
[62] Wang SF,Chen EH et al. Research on Image Emotional Annotation and Retrieval Algorithm Based on Support Vector Machine: 2004; 17(1): 27-33.

Research Application of RFID Technology in the Clothing Industry

Long Wu*

Apparel & Art Design College, Xi'an Polytechnic University, Xi'an 710048, China

*Corresponding author's email: wulong7398@163.com

Abstract

With the rapid changes in the clothing market and the development of the modern manufacturing industry, traditional clothing enterprises are now experiencing bottlenecks, and therefore increasingly unable to meet the requirements for quality management, workshop scheduling and process control. The rapid development of radio frequency identification device (RFID) technology has been, in recent years, providing a strong means of technical support, so as to achieve the synchronization of logistics and information flow in production, and meet the discrete manufacturing requirements of data collection in the production process. The problems that are currently found in information processing within supply chains among clothing enterprises are summarized in this paper. In addition, the application solutions for various departments, such as procurement, production, warehouse, logistics and distribution in the clothing industry via RFID technology, are discussed. RFID technology is considered to eliminate the bullwhip effect caused by the information asymmetry of upstream and downstream segments, thus greatly saving on costs and improving the efficiency of the whole supply chain.

Keywords: RFID; MES; Inventory; Clothing Industry; Warehouse

1. Introduction

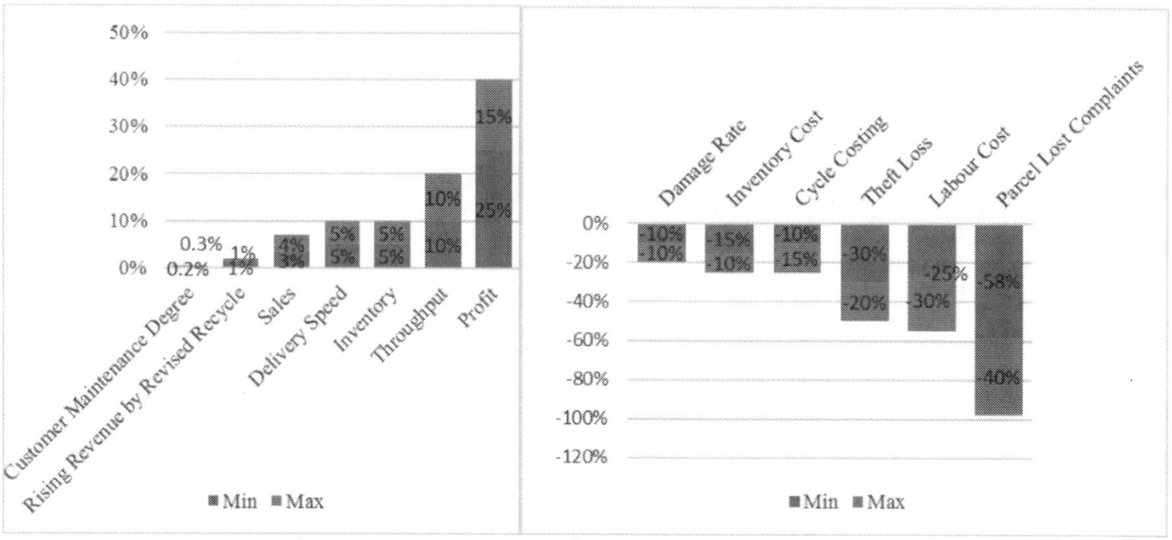

Fig.1 Revenue gained from the use of RFID technology [5]

RFID technology, which is derived from the Internet of Things (physical object network) and used with an electronic product code (EPC), developed by the Massachusetts Institute of Technology in 1999, has revolutionized the supply chain and production process management in the clothing manufacturing industry [1, 2]. The application of RFID technology has significantly improved rapid response capacity and the competitive power of enterprises in both the textile and clothing industries [3] in, for example, Germany, Finland, Japan, the USA, Austria and Portugal

[4]. As is evident from the official statistics of worldwide clothing enterprises (see Fig.1), RFID technology brings about economic benefits for customers through reductions in management cost, lower inventory volumes, increases sales and the circulation speed of commodities, etc. [5].

Neglected for many years by Chinese enterprises, production and logistics departments have continually faced issues in terms of chaotic warehouse management, inventory overstocking, delayed shipments, inaccuracy in stocktaking, delays in treatment and rushing inventories [6]. Logistics and distribution cannot respond quickly enough to meet the ever-changing needs of consumers. With the introduction of enterprise resource planning (ERP) systems in China, Chinese enterprises have gradually accepted the fact that the implementation of these systems will increase competitive capacity and advantage, thus allowing them to increase in strength and size. In the process of implementing ERP systems, however, many Chinese enterprises have discovered that they have many deficiencies and problems, such as online control, role permissions management and a lack of synchronization between funds and information flow. Chinese domestic popular clothing brands, such as Bosideng, Hodo, Baoxiniao, Youngor, K-Boxing, Septwolves, 361 Degrees, X-Tep and Metersbonwe, have started to use RFIDs in their production process to enhance the customer experience in their retail stores and improve management in the warehouse and supply chain [7-9].

In cases where the RFID business process is transplanted into a wireless data collection apparatus, ERP can be used to realize revolutionary changes from post-accounting to monitoring progress, so as to greatly enhance overall operational efficiency and accuracy. IDTechEx indicated that the annual demand for RFID tags (a very common type of ultra-high frequency electronic adhesive label) in the garment industry alone will be close to 20 billion within the next 10 years, while the market share each year could possibly reach USD 1 billion after 2021 [10]. The International Intellectual Property Alliance (IIPA) has even predicted that the scale of the industry market for RFIDs in China will reach RMB 48.5 billion in 2016, and even RMB 62.1 billion in 2017 [11, 12]. The scale of the market for RFIDs in China is anticipated to increase by about 240% from 2013 to 2017, while the annual growth rate is more than 27.88% (see Fig.2) [13]. The latest information from IDTechEx shows that this growth will be consistently over 20% over the next 10 years [14, 15].

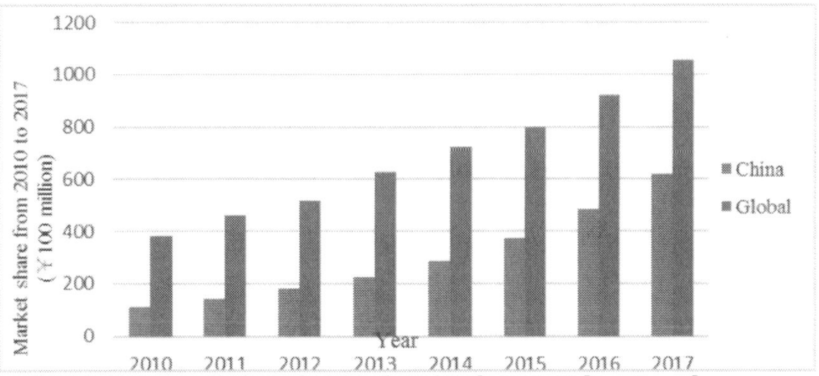

Fig.2 Prediction of scale of market for RFIDs [12, 14, 15].

The aim of this article, therefore, is to provide recommendations and information that could improve the accuracy of production, procurement and delivery through process re-engineering and the integration of internal and external resources, for example, suppliers, distributors, agents and carriers, by using RFID technology in procurement, production, warehouse, logistics, sales and other departments in Chinese enterprises. Consequently, Chinese enterprises can maintain a favourable position in a market environment with strong competition and enhance informatization at the same time to provide prompt responses to the market.

2. Issues

Modern clothing production involves continuous line production; however, it is different from machine manufacturing. Much of the fundamental details, including style, colour, size, fabric and customer identifier, need to be considered from cutting and sewing to ironing to finishing [16]. However, at present, the clothing industry continues to be labour-intensive as it has been traditionally, such that management is still using traditional means. Employees who are working in the clothing industry are accustomed to manual operations and experience-based management, with low levels of quality control. They may resist advanced technology and new styles of management. In terms of production process control, clothing enterprises have continued to implement extensive management practices in productive planning and tasks. Without the development of lean management for each process, enterprise managers do not have a clear grasp of the specific conditions of their production site. Therefore, management decisions will lack structure and efficiency. Other related departments will also have similar problems [17]. The following is an examination of the problems within the traditional garment industry.

2.1 Procurement and Supply

At present, management in general is relatively outdated and chaotic in many of the enterprises, while there are ambiguous roles among the divisions and backwards information technology is used. Therefore, finding ways to coordinate the mechanisms of operations among suppliers has become very important.

Supply chain management (SCM) solutions and modules developed by successful international supply chain companies have emerged in recent years, such as advanced planning and scheduling (APS) and advanced supply chain planning (ASCP), which are both systems from Oracle, demand planning (DP), which is a management process, constraint-based optimization (CBO), which is a minimax strategy, global order promising (GOP), which is a scheduling tool from Oracle, trade matrix scheduler (TMS), trade matrix supply chain execution (TMSCE), supplier relationship management (SRM), value chain management (VCM) and procurement management (PM) strategies, and supply chain optimizer (SCO), which is an operational algorithm [18-26].

The following issues are currently relevant to procurement and supply in the clothing industry in China.

Inadequate management of supply. Suppliers are usually chosen by business executives through subjective impressions or feelings. Even if some of the enterprises have established standards for selecting their suppliers, they have merely considered or focused on the price or quality, while neglecting delivery time and other factors [27, 28]. Most of the garment industries in China are busy simply dealing with market changes, and have therefore neglected the establishment of a relationship with suppliers [29]. However, an open, transparent and collaborative relationship with suppliers should be the driving force for enterprises for improving their business [30].

Procurement business: lack of collaboration with suppliers. Suppliers do not yet understand the implementation of manufacturing systems, so they are unable to prepare in advance [31, 32]. Lack of collaboration often means that the requirements of clothing manufacturing enterprises cannot be met [33, 34].

Lack of collaboration between production and purchasing departments. Since production and purchasing are independent departments, clothing enterprises often rely on experience for purchasing points [35, 36]. After the procurement of the raw materials, the purchase cannot be cancelled [37]. If the sales situation of a product changes, the enterprise cannot respond in a timely manner, meaning that the result is an excessive inventory of raw materials, which would increase the inventory cost [38].

2.2 Production Management

The following issues are currently relevant to production management in the clothing industry in China.

The clothing production process, from the purchase of raw materials to the storage of the finished product, can be divided into three main components: cutting, sewing and ironing. For instance, there are about 120 processes in the sewing work of suits, not including the trousers. In addition, clothing production is obviously seasonal. Taking sewing workshops as an example, the administrator for each line could deal with more than 400 work orders every day. Suppose that the amount of the work-in-process (WIP) inventory is about 3,000 and each work order contains seven packages: if there are 50 workers per line, the total number of items that has to be addressed is about 20,000. Moreover, quality control problems, WIP and schedule management, etc. should also be taken into consideration [39, 40]. Currently, sewing workshops mainly rely on manual records, in which the statistical data are from the work order and quality control records. Only when all of the sewing work is completed and handed over to the next shift, with analysis work carried out without the luxury of real-time statistics [41-43]. Therefore, there is an information gap between production planning and production manufacturing. Due to the greater hysteresis effects in the garment sewing process, work orders and production plans cannot be tracked on time [44, 45]. It is thus more difficult to gain clarity about the progress of an order or plan [46-48].

2.3 Warehouse Management

The following issues are currently relevant to warehouse management in the clothing industry in China.

Low levels of standardization in operation processes. As each enterprise only takes into consideration its own interests, specifications in clothing warehousing hardly follow any unified industrial standards [49, 50]. Consequently, it is not conducive to the development of the globalization of these enterprises [51-53].

Disorganization in layout. Due to lack of planning and disorganization, many Chinese enterprises cannot effectively improve the utilization of their warehouse, given that disorderly product storage is not conducive to the query and control of products [54-56].

Outdated management facility and information system. Inefficiency and high rates of errors found in traditional management procedures result in a heavy workload for goods receipt and delivery, as well as when carrying out a store inventory [57, 58]. Due to the rise in labour costs and the lack of a good information system, enterprises in China cannot conduct real-time supervision of the production process in the warehouse, or have systematic information [59-61]. Therefore, this is not conducive to timely forecasting of market trends or adjusting production strategic plans in real time [62].

2.4 Logistics and Distribution

The following issues are currently relevant to logistics and distribution in the clothing industry in China.

If an enterprise is rapidly expanding, there will be problems in the logistics department. These have not been received enough attention from the industry, such as overstocking during seasonal peak months, chaotic warehouse management, delayed shipments, lack of streamlining, inaccurate inventory, and returns processed in an untimely manner. These could affect logistics management, as they will not be quick enough to respond to rapidly changing consumer demands [63]. The supply cycle is usually between 60 and 90 days for enterprises at present. If 40% of the products cannot be sold at the regular price because they are unable to reach the market in a timely manner, these products then become inventory, meaning that profit margins are reduced and the sustainable development of these enterprises is directly impacted [64, 65]. These could include longer cash conversion cycles, outstanding supplier debt and questionable commercial credit, all of which may result in market shrinkage and loss of competitiveness [66, 67].

Another issue is the procedure involving the use of RFID tags in retail stores. They are placed onto goods when they reach the stores, and then taken off after the sale. Consequently, this could increase the workload of the employees, who also cannot monitor whether every product is actually in place. When RFIDs are used with full container loads (FCLs), then the speed at which products are tagged with RFIDs must be greatly increased [68, 69]. However, this means that the preciseness of identifying products cannot be controlled well. Indeed, sometimes there are mistakes, which require manual verification.

3. Solution to Problems

The following are solutions to the aforementioned problems when using RFID technology.

3.1 Procurement and Supplier

Purchasers may also use different strategies to influence their suppliers with the use of RFID technology. Strategic imperatives and strategies, which make use of research and development subsidies, increase the possibility that suppliers will use RFID technology. The former can reduce the possibility of suppliers waiting, while the latter, in addition to the above, can increase the possibility of using RFID technology in the early stages.

3.2 Production Management

In the clothing industry, there are shortcomings with existing data acquisition equipment for various reasons, such as frequent changes in processing, complex workshop environment, and transmission difficulties of data in production venues. Barcode technology has been used to collect production data in the current discrete manufacturing industry. However, due to its inherent shortcomings, such as being limited to the using environment and less security, the data provided by barcodes have even become the reason for bottlenecks in manufacturing production decisions.

As a production information system, manufacturing execution systems (MES) collect large-scale real-time data and process them during production. The root of the problem with production management in China, as mentioned before, is that data are manually collected. However, this can be addressed by RFIDs, which allow non-contact automatic identification, so that items are automatically identified and monitored by using the transmission characteristics of radio frequency signals and spatial coupling. To control the progress of tasks, each production task is given an electronic tag card for tasks from making to assigning. Each manufacturing process, such as raw material requisition, production, quality inspection, finishing and packing, is equipped with an RFID reader and writer for collecting and writing task schedule information.

For real-time monitoring and scheduling in core departments, such as sewing, the RFID system can be used for real-time monitoring and scheduling of production. Sewn clothing that needs to undergo treatment is classified according to the piece itself. Each piece of clothing is then placed on a hanger with an electronic tag by workers, while each station is equipped with an RFID reader and writer.

MES have allowed industrialization in developed countries and brought about significant economic benefits to many enterprises, applied in discrete and flow manufacturing processes and industries. Many internationally renowned manufacturers use MES, such as Invensys, Siemens, Honeywell, GE Fanuc, Wonderware, Yokogawa, Tecnomatix Tech and SimSci. In spite of the short time that MES have been applied and developed in China and the large gap in their use as compared to other countries, MES have already produced good initial effects.

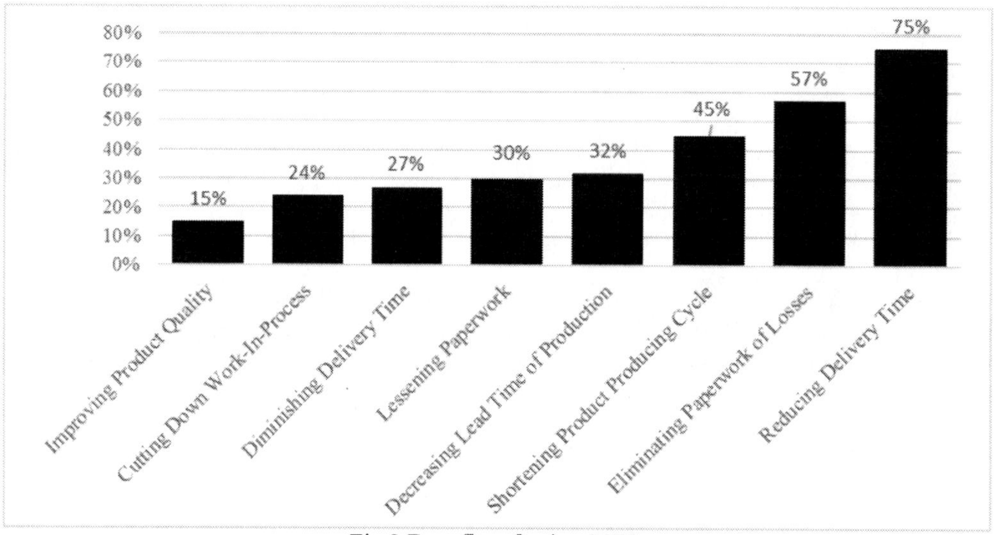

Fig.3 Benefits of using MES

In accordance with an extensive survey conducted by the Manufacturing Execution Systems Association on a worldwide scale in 1996, the general benefits derived from the use of MES are presented in Fig.3 [70].

3.3 Warehouse Management

The modern warehouse management of clothing is not only about the traditional storage management of production materials; there is also the function of providing early warning signals, such as shortages. The tasks of modern warehouse management include the detailed recording of all the information of the materials in storage, including receiving, inventory tracking, goods sorting, packing and shipping, overstocks and shortages. Therefore, the warehouse management system is not only the centre of the supply security and logistics support within the enterprise, but also the core for strategic decision-making and development planning.

Taking warehouse management, based on RFID technology, of a suit enterprise as an example, first, there is A) the receipt of goods. After the last manufacturing procedure, an RFID tag is issued for each suit and hung with the suit. The information includes the article number, style, colour, size, price and fabric composition. When a suit passes under an RFID reader and writer device in the warehouse, the data of the next logistics distribution process are read. Next, there is B) inventory tracking, which uses a portable RFID reader and writer device. The warehouse keeper would walk along the aisles of the warehouse at a normal speed. Since each suit is tagged with an RFID tag and the height of each layer of suits is less than 2 m, the RFID device can read information on all of the RFID tags, thus greatly improving the efficiency of inventory taking. This is followed by C) goods sorting, in accordance with the customer order information. The warehouse keeper looks for the location of the required suit and retrieves the item accordingly. After the portable device reads the information from the RFID tag on this suit from a distance and compares the product attributes with the order request, the corresponding operation is carried out, e.g., reprinting the price tag, so as to improve the accuracy of the delivery. Then, there is D) packing and shipping. After summarizing all the information of the clothing to be delivered, a packing list is generated and written up, while an RFID tag of the packing list is attached to the box. This is the application of RFID technology in third-party logistics and distribution, which provides convenience for warehousing in the exclusive shop. Finally, there is E) overstocking or shortage warnings, that is, when the stock exceeds the storage capacity or when there is less than the minimum required stock. In such cases, inventory warnings will automatically appear on the screen of the system.

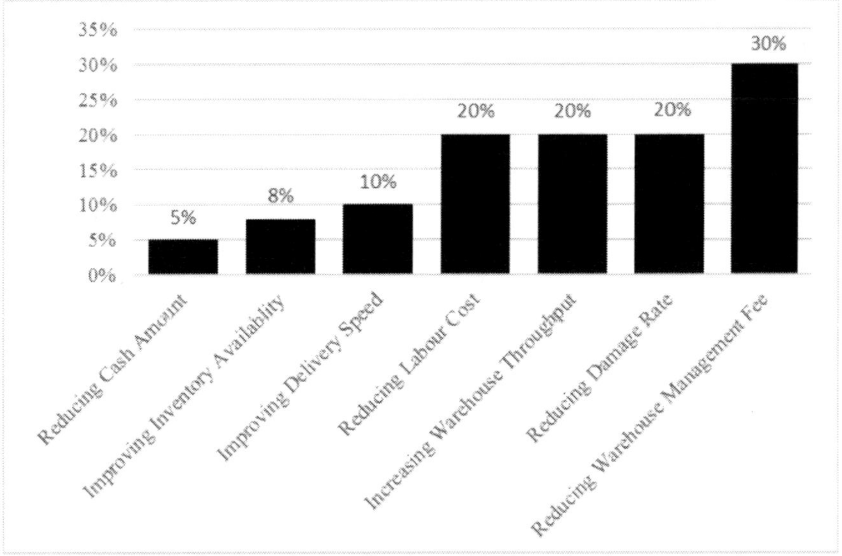

Fig.4 Benefits from the application of RFID technology in logistics

According to statistics, the application of RFID technology in logistics has provided many benefits, which are shown in Fig.4 [70, 71]. RFID technology can ensure the speed of data inputting and the accuracy of each production process in warehouse management, thus allowing enterprises to accurately obtain real-time data in terms of inventory, as well as facilitate reasonable maintenance and control over their inventory.

3.4 Logistics and Distribution

The following is a scenario that exemplifies the use of RFID technology in logistics and distribution.

1) Warehouse entries carried out in stores. When the packing cases arrive at a store, the staff read the RFID tags of the packing list with an RFID device in order to determine whether the delivery is correct. If there is an error, the cases are returned to the head office without the need to open them. If the delivery is correct, the staff then opens the cases and reads the RFID tags for warehouse entries. Automated entries for warehousing can therefore improve

the effectiveness of the store inventory.

2) Clothing sales. When a consumer purchases a suit, the information will be automatically read from the RFID tag on the suit to eliminate the item from the warehouse and account. The store can also give the RFID tag to the customer as a VIP card. When the customer visits the store again, the customer information and purchase record, as well as other relevant information, can be automatically obtained, such that the salesperson can provide targeted recommendations to the customer, which will ultimately improve the quality of customer service. Moreover, because RFID tags are contactless and can be read quickly, the technology can also automatically generate orders and manage the information. In turn, this can greatly improve the efficiency of traditional systems of clothing transactions, while keeping the waiting time of the customer to a minimum.

3) Refunds, exchanges and avoidance of theft. Stores that apply RFID technology not only can carry out refunds and exchanges, but also enjoy a security function from the use of RFIDs. Electronic article surveillance (EAS) systems can be installed for customer identification and theft, while RFID ink-tags, which are a specialized feature in the garment industry, can spray red ink onto clothes when an article is the object of theft.

4) Intelligent fitting function. When a customer is trying on items of clothing, s/he can refer to the displayed graphic information to select suitable items in the store for collocation. At the same time, s/he can also find other styles, colours, sizes etc., which are available simply by looking at the screen. Next, s/he simply informs the staff about his/her requirements without specifically looking for the items. The use of try-on and sales records, in combination with market research, means that enterprises can predict popular trends.

3.5 Solution to Technical Problems with RFID Technology

Although RFIDs have brought about various benefits to clothing manufacturers, they are still not commonly used in China. The primary reasons for the lack of RFID application in clothing enterprises are outlined as follows [72].

1) High cost of one-off investment. Currently, the cost of barcode technology is less than that of RFID technology, which means that the latter may encounter more resistance in use. However, this is changing. In 1996, the cost of an RFID tag was RMB 1.6; over time, the cost of RFID tags has significantly dropped and is now RMB 0.5. This represents a price fall of 70%, while the trend of cheaper RFID tags will probably continue. Therefore, it is now timely to start using RFID tags.

2) Technical difficulties in RFID application. China has started late on conducting research on RFIDs, while research work and breakthroughs are currently lacking on middleware, data processing and key basic technology. Therefore, strengthening related training and professional development means that workers may change their views on this new technology and consequently take part in gaining the related skills that will improve performance management.

3) Lack of alignment of technical standards with universal practices. National standards around RFIDs, which are in any case lacking, may not be compatible with international standards, which means that the processing and distribution of Chinese products worldwide will be problematic. However, the widespread application of RFID tags on clothing production assembly lines means that the cost must first be further reduced through technology and technological innovation, so that they are competitive with the cost of using barcodes.

4) Instability. As RFID technology in China is not yet mature, there are often problems, such as data errors, and reading or writing errors. The user must often ask the background administrator to correct data mistakes. Sometimes, errors that are not corrected in time may result in costly implications. RFID providers in China need to carry out secondary development to meet the personalized needs of the user in time, which would compensate for the shortcomings of the system in use.

5) Security issues. Whether it is the technology itself or encryption needs, RFID systems have already addressed security issues. In terms of current technological development, effective ways to ensure data security are: physical isolation, stopping access to information that is sensitive, joint certification for double tags, authentication, control of data access etc. Therefore, RFID systems are safe to use. Overall, the security of RFIDs should not only rely on the security mechanisms of electronic tags, but addressed as a system as a whole.

4. Conclusion

In this article, by combining the current status of the use of RFID technology in both China and other countries, the existing problems in clothing enterprises have been summarized from procurement, supplying, production, warehouse management, logistics to distribution. Based on the current characteristics of RFID technology and the current practice of the supply chains of the garment enterprises, strategies have been proposed for the corresponding departments. RFID technology can therefore allow managers in clothing enterprises to track each department task from warehouse entry to the production process, thus having access to information on the status of each process and the location in terms of the inventory, semi-finished and finished products, and dispatching and delivery in real time. As such, RFID technology can enhance the visibility of clothing production, improve employee productivity, and strengthen Chinese domestic enterprises.

Acknowledgement

The work was supported by grants from Natural Science Foundation of Shaanxi Provincial Department of Education, China (Project No. 16JK1321), Doctoral Scientific Research Start-up Foundation of Xi'an Polytechnic University (310-0400608) and the Science and Technology Guiding Project Supported by the China National Textile and Apparel Council (2016124).

References

[1] Sarma S, Brock D, Engels D. Radio frequency identification and the electronic product code. IEEE Micro: 2001; 21: 50-54.
[2] Staake T, Thiesse F, Fleisch E. Extending the EPC network: the potential of RFID in anti-counterfeiting. Proceedings of the 2005 ACM symposium on Applied computing: ACM: 2005; 1607-1612.
[3] Legnani E, Cavalieri S, Pinto R, Dotti S. The Potential of RFID Technology in the Textile and Clothing Industry: Opportunities, Requirements and Challenges. Unique Radio Innovation for the 21st Century: Springer, New York (NY): 2011; 309-329.
[4] Moon K, Ngai E. The adoption of RFID in fashion retailing: a business value-added framework. Industrial Management & Data Systems: 2008; 108: 596-612.
[5] Wang K. A study based on increasing costume enterprises' fast reactive capability with RFID technology. Beijing Jiaotong University, Beijing: 2012.
[6] Poon T, Choy KL, Chow HK, Lau HC, Chan FT, Ho K. A RFID case-based logistics resource management system for managing order-picking operations in warehouses. Expert Systems with Applications: 2009; 36: 8277-8301.
[7] Yang Y. Information Technology in Supply Chain Management within the Apparel Industry: 2013.
[8] Wu DL, Ng WW, Yeung DS, Ding HL. A brief survey on current RFID applications. Machine Learning and Cybernetics, 2009 International Conference on: IEEE: 2009; 2330-2335.
[9] Liu L. An Application of Radio-Frequency Identification Systems in Chinese Clothing Supply Chain. Emerging Technologies for Information Systems, Computing, and Management: Springer, New York (NY): 2013; 1325-1333.
[10] Zhu X, Mukhopadhyay SK, Kurata H. A review of RFID technology and its managerial applications in different industries. Journal of Engineering and Technology Management: 2012; 29: 152-167.
[11] Information from: http://www.luxresearchinc.com/news-and-events/press-releases/read/china%E2%80%99s-rfid- card-market-nearly-double-807-million-2017.
[12] IIPA. 2013-2017 China RFID Industry Research Report: 2013.
[13] Information from: http://www.reuters.com/article/ma-lux-research-idUSnBw4G2Fbsa+128+BSW20130312.
[14] Das R, Harrop P. RFID forecasts, players and opportunities 2009-2019. IDTechEx Report: 2009.
[15] Bottani E, Montanari R, Romagnoli G. Improving sales turnover in fashion retailing by means of an RFID-based replenishment policy. International Journal of RF Technologies: 2016; 7: 65-86.
[16] Wang R, Du Y, Mao A, Li Y. Computer engineering database design and application in clothing thermal functional design. Proceedings of TBIS: 2008.
[17] Scott AJ. The changing global geography of low-technology, labor-intensive industry: clothing, footwear, and furniture. World Development: 2006; 34: 1517-1536.
[18] Jonsson P, Holmström J, Ellinger A. Future of supply chain planning: closing the gaps between practice and

promise. International Journal of Physical Distribution & Logistics Management: 2016; 46.
[19] Weng J, Chen L, Lee H, Blanchard R. Demand planning with event-based forecasting. Google Patents: 2005.
[20] Boutilier C, Patrascu R, Poupart P, Schuurmans D. Constraint-based optimization with the minimax decision criterion. Principles and Practice of Constraint Programming-CP 2003: Springer, New York (NY): 2003; 168-182.
[21] Makatsoris H, Chang Y, Richards H. Design of a distributed order promising system and environment for a globally dispersed supply chain. International Journal of Computer Integrated Manufacturing: 2004; 17: 679-691.
[22] Ross DF. Introduction to e-supply chain management: engaging technology to build market-winning business partnerships: CRC Press, Boca Raton (FL): 2002.
[23] Choy KL, Lee W, Lo V. Design of a case based intelligent supplier relationship management system-the integration of supplier rating system and product coding system. Expert Systems with Applications: 2003; 25: 87-100.
[24] Fischer DJ, Squires GM, Sharma R, Ganesan R, Ghodke DM, Rangarajan B. Value chain management. Google Patents: 2008.
[25] Cox A. Relational competence and strategic procurement management: Towards an entrepreneurial and contractual theory of the firm. European Journal of Purchasing & Supply Management: 1996; 2: 57-70.
[26] Ganeshan R, Harrison TP. An introduction to supply chain management. Department of Management Sciences and Information Systems: 1995; 303.
[27] Song H, Fan C. Research of the typical supply chain management models of the garment industry in China. International Journal of Business Administration: 2012; 3: 78.
[28] Zhao Z, Li Y, Xie MB. Silk fibroin-based nanoparticles for drug delivery. International Journal of Molecular Sciences: 2015; 16: 4880-4903.
[29] Barnes L, Lea-Greenwood G, Runfola A, Guercini S. Fast fashion companies coping with internationalization: driving the change or changing the model? Journal of Fashion Marketing and Management: An International Journal: 2013; 17: 190-205.
[30] Taj S, Morosan C. The impact of lean operations on the Chinese manufacturing performance. Journal of Manufacturing Technology Management: 2011; 22: 223-240.
[31] Boyce WS. Supply Chain Relationships in Procurement: Is Collaboration Reality?: University of Missouri-Saint Louis, Saint Louis (MO): 2014.
[32] Hsin Chang H, Tsai YC, Hsu CH. E-procurement and supply chain performance. Supply Chain Management: An International Journal: 2013; 18: 34-51.
[33] Singh R. Practices to compete supply chain performance of organized garment retailing. International Review of Social Sciences & Humanities: 2013; 5.
[34] Zhao Z, Xie M, Li Y, Chen A, Li G, Zhang J et al. Formation of curcumin nanoparticles via solution-enhanced dispersion by supercritical CO2. International Journal of Nanomedicine: 2015; 10: 3171.
[35] Niinimäki K, Hassi L. Emerging design strategies in sustainable production and consumption of textiles and clothing. Journal of Cleaner Production: 2011; 19: 1876-1883.
[36] Niinimäki K, Pedersen ERG, Hvass KK, Svengren-Holm L. Fashion Industry and New Approaches for Sustainability. In: SS Muthu (ed.), Handbook of Sustainable Apparel Production. Handbook of Sustainable Apparel Production: CRC Press, Boca Raton (FL): 2015; 453-474.
[37] Vezzetti E, Alemanni M, Macheda J. Supporting product development in the textile industry through the use of a product lifecycle management approach: a preliminary set of guidelines. The International Journal of Advanced Manufacturing Technology: 2015; 1-12.
[38] Li L, Jingjie L, Ying Z. Research of Chinese garment enterprises collaborative R&D platform architecture. AASRI International Conference on Industrial Electronics and Applications (IEA 2015): Atlantis Press, Amsterdam: 2015.
[39] Xu JX, Li G, Wang JY, Li X. New method for polygynous node automatic creation based on triangulation. Applied Mechanics and Materials: Trans Tech Publ: 2013; 2442-2445.
[40] Olhager J, Cimander B. The impact of manufacturing strategy on the design of production manage. Advances in Production Management Systems: Perspectives and Future Challenges: 2013; 267.
[41] Ulrich K. The role of product architecture in the manufacturing firm. Research Policy: 1995; 24: 419-440.
[42] Deaton DW, Gabriel RG. Method and system for generating incentives based on substantially real-time

product purchase information. Google Patents: 2001.
[43] McFarlane D, Sheffi Y. The impact of automatic identification on supply chain operations. The International Journal of Logistics Management: 2003; 14: 1-17.
[44] Ameri F, Dutta D. Product lifecycle management: closing the knowledge loops. Computer-aided Design and Applications: 2005; 2: 577-590.
[45] Higginson JK, Alam A. Supply chain management techniques in medium-to-small manufacturing firms. The International Journal of Logistics Management: 1997; 8: 19-32.
[46] Ouertani MZ, Gzara L. Tracking product specification dependencies in collaborative design for conflict management. Computer-Aided Design: 2008; 40: 828-837.
[47] Cooper RG. Third-generation new product processes. Journal of Product Innovation Management: 1994; 11: 3-14.
[48] Messer SD. Data processing system for integrated tracking and management of commerce related activities on a public access network. Google Patents: 1999.
[49] Kimball R, Ross M. The Data Warehouse Toolkit: The Complete Guide to Dimensional Modelling. New York (NY): Wiley 2002.
[50] Medina E, Trujillo J. A standard for representing multidimensional properties: The common warehouse metamodel (CWM). Advances in Databases and Information Systems: Springer: 2002; 232-247.
[51] Keung Kwok S, Wu KK. RFID-based intra-supply chain in textile industry. Industrial Management & Data Systems: 2009; 109: 1166-1178.
[52] Lee C, Choy KL, Ho GT, Chin KS, Law K, Tse Y. A hybrid OLAP-association rule mining based quality management system for extracting defect patterns in the garment industry. Expert Systems with Applications: 2013; 40: 2435-2446.
[53] Wang ZG, Tang RZ, Sheng WL, Wang GD. Research on RFID-based production logistics management techniques with application in the garment industry. Technology and Innovation Conference, 2006 ITIC 2006 International: IET: 2006; 383-387.
[54] Chiang DMH, Lin CP, Chen MC. The adaptive approach for storage assignment by mining data of warehouse management system for distribution centres. Enterprise Information Systems: 2011; 5: 219-234.
[55] Roodbergen KJ, Vis IF, Taylor Jr GD. Simultaneous determination of warehouse layout and control policies. International Journal of Production Research: 2015; 53: 3306-3326.
[56] Richards G. Warehouse Management: A Complete Guide to Improving Efficiency and Minimizing Costs in the Modern Warehouse: Kogan Page Publishers: 2014.
[57] Zhu S, Liu J, Zhao L, Wang X. Application of radio frequency identification technology in information system for warehouse management. Proceedings of the 2014 International Conference on Multimedia, Communication and Computing Application (MCCA2014), Xiamen, China, Oct 16-17, 2014: CRC Press: 2015; 163.
[58] Emmett S. Excellence in Warehouse Management: How to Minimise Costs and Maximise Value: John Wiley & Sons: 2011.
[59] Lee C, Choy K, Law K, Ho G. An intelligent system for production resources planning in the Hong Kong garment industry. Industrial Engineering and Engineering Management (IEEM), 2012 IEEE International Conference on: IEEE: 2012; 889-893.
[60] Leung J, Chu SC, Cheung W. Design research guidelines for mindful IT innovations: the case of RFID innovation in supply chain management. 46th Hawaii International Conference on System Sciences (HICSS), 2013: IEEE: 2013; 3727-3736.
[61] Tiwari A, Singhal S, Singh R. To study the impact of referral marketing with one-to-one segmentation in warehousing. International Journal of Management, IT and Engineering: 2013; 3: 54.
[62] Chang YC, Chang KH, Huang CW. Integrate market demand forecast and demand-pull replenishment to improve the inventory management effectiveness of wafer fabrication. Proceedings of the Institution of Mechanical Engineers, Part B: Journal of Engineering Manufacture: 2014; 228: 617-636.
[63] Zhang RX, Yang WX. Discussion on the application of the garment industry's logistics informatization. Applied Mechanics and Materials: Trans Tech Publ: 2014; 2756-2759.
[64] Fernie J, Sparks L. Retail logistics: changes and challenges. Logistics and Retail Management: Emerging Issues and New Challenges in the Retail Supply Chain: 2014; 1.

[65] Cagliano AC, Mustafa MS, Rafele C, Zenezini G. Logistics Performance measurement for sustainability in the fast fashion industry. Next Generation Supply Chains: Trends and Opportunities: 2014; 113.
[66] Gupta A, Dasgupta P. Marketing Innovations in the apparel export manufacturing business: case study of Indian firms. International Journal of Trade & Global Business Perspectives: 2014; 3: 1274.
[67] Jing L. China Textile and Garment Industry to Enhance Competitiveness by Industrial Chain Management. 2015 International Conference on Social Science and Technology Education: Atlantis Press: 2015.
[68] Hao H, Zhao B. Application of RFID Technology in the development of warehouse logistics management system by the Internet of Things. International Conference on Education, Management, Commerce and Society (EMCS-15): Atlantis Press: 2015.
[69] Wang YL, Du JM, Xu SS. Research on logistics warehouse management information system based on RFID. Applied Mechanics and Materials: Trans Tech Publ: 2014; 4906-4909.
[70] Zou C, Jiang CY. The applications of RFID technology in logistics management and constraints. Materials Science Forum: 2006; 1009-1011.
[71] Guo Z, Zhang X. The application of RFID technology in the logistics supply chain. IEEE International Conference on Computer Science and Information Technology: 2010; 518-520.
[72] Li Q. An overview of RFID Technology and Application: 2006; 51-62.

Bohemian Dress Style Analysisof the ETRO Brand

Zhao-Mei Zhang, Wei Tian*

School of Apparel & Art Design, Xian Polytechnic University, No.19 South Jin hua Road, Xian, Shaanxi, 710048, China

*Corresponding author's email: 569790315@qq.com

Abstract

Bohemian style has made a huge impact on the fashion industry. This paper describes Bohemian clothing based on its design. For more details regarding Bo Simi Asian style we used the ETRO brand as an example focusing on its clothing modeling, color composition, fabric selection, design, accessories design and decoration technology research. This was done to gather an in-depth understanding of the ETRO brand clothing that adopts the Bohemi style and why is it so popular in the fashion industry., This paper concludes: the main features that resemble Bohemian style is that it has a rich national flavour and wild free chic. The article provides a different angle for the design of Bohemian style clothing.

Keywords: Bohemian Style; Design Elements; ETRO

1. Introduction

By reviewing the literature research, most of the information studies the characteristics of Bohemian clothing but there is little information regarding the reasons for analyzing brand clothing. The text on Bohemian style clothing are mainly on its origins and clothing characteristics. Popular Bohemian dress in addition of keeping the nature of its exotic nature also adds a fashion elements. [8].

With the development of the society and the improvement of living standards, people are demanding more from clothing. From simply providing cover to the body, it has become a statement of beauty and expression. Bohemia style with its strong color complex design gives off a powerful visual impact. By embodying indigenous patterns on the garment it gives an elegant look.

Bohemian style is an indispensable part of fashion, as it gives a very romantic and wild feel. It not only enriches the concept of clothing design, but also broadens existing forms of clothing and performance techniques.

2. The Bohemian Style Clothes

2.1 An Overview of the Bohemian Dress Style

Bohemian is an original style originated from nomadic ethnic. Tiered full-skirted dress, cascading lace, wax printing, leather tassel, handmade knot, embroidery and beads are all its unique classic elements. Its strong colors gives a strong visual impact, and Bohemian rebel spirit [3].

2.2 The Characteristics of the Bohemian Dress Style Analysis

Although designer styles may be different, there are similar characteristics that all of them have.

Bohemian dress are commonly loose and comfortable. Generally, the top always has a v-neck, U collar and A word. The bottoms are A-line dress. The skirt length go up to knee or even longer, down to the ankle with a wide hem. There are a rich variety of fabric used to make the dress. Natural fabrics are the priority choice such as cotton, linen, wool, suede leather and jean. It always uses tall lightness with ash color, such as white, red, khaki, camel, brown, aureate, light green with partial ash and dark olive green .Furthermore Bohemian dresses uses a combination of strong colors that harmonizes well together such as black, red and white. [1].

Bohemian dress are decorated with beads, sequins, tassels, lace, or flash. The ornaments on the upper half of the body are mainly concentrated on the neckline followed by cuffs and waist line. The ornamental design on the bottom half of the dress are focused along the hip line or skirts. The clothes are often paired with a large number of jewelry including necklace, bracelet or ring. The size of the jewelry is usually large displaying its origins or ethnicity. All the decoration are usually on the belt and bag with long hair, big shawls, boots or flats photograph collocation. The makeup style usually resembles nature giving off an elegant and quite look. Sometimes it is relatively heavy with black, blue and green eye shadow accompanied by long wavy hair or natural straight hair. [7].

3. ETRO Clothing Style Research

3.1 ETRO Briefly

ETRO is a famous Italian brand. It's Chinese name is aiteluo which means "style" in Chinese. Its products cover ladies and men's clothing, handbags, shoes, leather goods and scarves, etc.

ETRO's founder Mr Ji mo loved to travel and he has been all over the world. He is interested in culture and history and loved to collect all kinds of items, especially antique books and clothes. It is the uniqueness and beauty of these discoveries that inspired ETRO fashion brand founded in 1968. ETRO's logo was a flying horse "palin and sotheby's" (Pegasus). The brand of ETRO pursues creativity, kindness and beauty.

ETRO focuses on producing high-grade textile fabrics. They use high quality natural fiber, with elegant design, fashion color and exquisite craft to produce wonderful products.

3.2 Theme

Bohemian has a strong tribal influence. It's design is based around the authenticity of the spiritual feel. [2]. As the population become mixed up together the tribal costumes have begun to adopt several features creating diversity. All of these prompted the evolution and development of the Bohemian dress. Easy going and free nature is an important characteristic of this style. Bohemian style is actually a way to escape reality, an exploration of the future and also the pursuit of individual character. [9]

Fig.3-1 Black and white striped and Turkish skirt

The designer Veronica Etro loves travelling all over the world and learning different historical culture which makes his work wild uninhibited, and free. . It gives the look of fearlessness and unrestrained. Bohemian is especially embodied in the strong contrast between the leather, waist gown and cloak with beads, and silk long skirts to the ankle and pants outfit. Innovation is hidden under each piece of clothing. The design is able to achieve the perfect balance between romantic and practical [2]. Long skirt is a typical Bohemian style design. In order to fully interpret the style of clothing, the designer should pay attention to the adjustment of the waist position as well as the collar and cuffs. [5].

Clan embroidery is filled with bright pink, fluorescent yellow, and green flag colors, or a pattern of black and white stripe (see Fig.3-1). It has the national characteristics of fashionable expressions also highlighted. = Turkish dress either short or long is able to give amorous feelings of Bohemia. Fig.3-1 The Bottom left picture shows the garment with ultra-long sleeves in the arms of brocade coat giving off an ethnic flavor. In Fig.3-1The bottom right clothing bedding has a brown and beige tone with low purity colors such as green, combined with the national printing and a necklace with indigenous tribal pattern. The wide striped long coats are designed to contain the national pattern which is grace and noble.

ETRO uses fashion to interpret his own life, as every design tells its journey and story. The design expresses women's independence and freedom. Some of his designs are like coming from a vast desert, represented by the forces of nature. Some are filled with pure and amorous feelings expressed through free color rendering. Some come from a fantasy garden, with flowers bringing out a women's charm.

3.3 Application Materials

Bohemian style is usually designed manually such as dyeing, cutting and weaving. Using advanced fabrics combined with exquisite handicrafts to achieve the harmonious aesthetic feel will bring new vitality to the clothing.. All of these designs are beautiful and attractive. Paisley print remains popular throughout this series, whether in the long skirt, jumpsuit, or silk mantle long skirt. [4].

Fig.3-2 Tunic and leather

In Fig.3-2 the long coat are made from thick jacquard fabrics, Fig.3-2 the upper left image shows the clothing combination of pants and and ancient long coat, combining nobility and with a wild, free look. In Fig.3-2 the upper right image has a national printed thick coat which contrasts with the silk faille dress. [5]. Fig.3-2 the bottom right image shows the combination of the ancient cloak with popular elegant colors which combines the modern and the traditional. The strip on the black cloak echos with the skirt. Black and dark mauve colors gives an elegant look forming a huge contrast, In Fig.3-2 on the bottom left, Light colors such as yellow are used in greater quantities along with grey shading to make the dress look more artistic. Abstract patterns are printed on the Irregular long skirt combined with the yellow jacket to give a wild energetic visual impact [5].

To imitate the desert atmosphere,, old fabrics are used. After printing and washing, beads, silk and suede frills with soft feathers were sewn on. ,

The garden look on the other hand has hand-embroidered paintings, while introducing rural wind printed skirt. The whole series is rich and interesting. Most of the material used are light in nature to highlight women's elegant posture and slender arms. The front of the V collar structure inadvertently expresses female charm. [4].

3.4 The Beauty of the Accessories

Bohemian style clothing often uses natural materials such as metal and stone to create accessories. It should be noted that raw materials need higher processing technology and collocation methods. [10]

The accessories focuses on adding more complexion to the clothing. Fig.3-3 The saddle bags are able to carry a sketchbook pencil and nickel plating cane hip flask which are basic life necessities Traditional Ian paisley creates strong and dynamic graphics, and redefines the modernism tone. Fig.3-3The middle handbag has a minimalistic look with a yellow and brown tone. This was done to contrast with the colorful and complex clothing. Fig.3-3 The color of the bag is predominantly brown with blue and red decoration to echo the design of the scarf. Fig.3-3 necklace is very striking with different amorous feelings from different regions [5].

Fig.3-3 Functional bag

ETRO women spread out the artistic feel both from the inside and outside. Seasonal accessories have a low profile. Rather than standing out, they are designed to blend in with the clothes. For the accessories, they use warm colors as their primary tone, whether it is asymmetric suede ribbon string stone beads and feathers, delicate feather earrings, leather beltmconcise list of suede sandals or contracted flower pendant, embroidery lace bags and shoes.

3.5 Innovative Design

Bohemian style are seen as free and wild attracting more and more female consumers. Bohemian dress use a lot of innovative elements to create a desire for independence and freedom. Locomotive trousers, jacquard patchwork shorts, elastic movement of wool material wrapped chest and waist silk even pants unlined upper garment, have given this dress a whole new outdoor exploration [6].

Fig.3-4 the left image shows a different form of skirt design. the sporty shorts with long boots gives an athletic feel. The match between the large area strip of upper and small area printing collocation of bottom is very fashionable. Fig.2-4 the garment on the right uses different tones of green to make it seem energetic . The design of the pants makes the strip and print much more fashionable and dynamic!

Fig.3-4 Locomotive installed and wrapped chest

4. Conclusion

Through analyzing the clothing series of clothing designed by ETRO, the biggest character of Bohemian style is its strong ethnic look and wild spirit, emphasizing the art of independence and freedom. Every item has a clothing pattern that displays national characteristics. The pattern displayed is what gives the clothes its uniqueness. New high on both the stage of visual design, and it's life, ethereal light fabrics, handsome appearance shows the little woman's side, both in the clothing design and arrangement of catwalk brings in new different visual feeling, reflecting the enthusiasm of travel culture, not only brings to the clothing the national characteristics, also has a fresh, vigorous vitality. The development process of Etro requires wisdom and curiosity created by exploring exotic culture and art, unique aesthetic pursuit of the east. Bohemian uses traditional techniques to create the garments. Every product is a challenge to craft.

In the design of brand clothing rather than simply looking at color and shapes, it is important to understand the connotations behind Bohemian style. Each series has to have a unified front. Its popularity comes from modern people's passion for spiritual liberation. Bohemian style, not only enriches the concept of fashion design but also expands existing dress performance, form and expression. All of this needs constant innovation.

5. Comment

Fig.3-1~3-4 from the network. (https://www.duitang.com/)

References

[1] Chang S. Bohemia style in clothing design, the application of shallow discussion. Chin: 2016 (10).
[2] Elizabeth Wilson [British] Bohemia, charming exile. Yilin press: 2012; p. 129-150.
[3] Xin JX. Bohemian in Chin. J ten-day times: 2010 (11); 44-46.
[4] Wei P. Garment fabric and accessories (second edition) [M]. Beijing: China textile university press: 2015; 2.
[5] Yang M. Bohemian style in the application of brand clothing. Text Technolo Prog: 2016; (10): 47-49.
[6] Bao QY, Cheng BY. Application of Bohemian Style in Women's Design. Mod Silk Sci Technolo: 2014; (05): 187-189 + 201.
[7] Huang T, Wang YW, Gao J. Bohemian style in the design of clothing applications. Pop Arts: 2015; (23): 126.
[8] Ke TT. Bohemian style in of brand clothing in aesthet embodiment. Shandong Text Technolo: 2016; (02): 45-47.
[9] Zhou K, Jiang K. Tassel in the fashion application of women. Pop arts: 2015; (23): 59-60.
[10] Huang T, Wang YW, Gao J. Bohemian style in the design of clothing applications. Pop Arts: 2015; (23): 126.

Characteristics of West-Road Style Phoenix Costume of Women from She Minority in Xiapu

Yan Yuan

Xiamen academy of arts and design, Fuzhou university, Institute of Road No. 852, Xiamen, Fujian, 361024, China

*Corresponding author's email: 27790467@qq.com

Abstract

Based on the field investigation in Xiapu County, Fujian, this paper studies the west-road style phoenix costume in Xiapu, and explores the origin and development of the Phoenix Costume of women in the She Minority from the perspectives of religious belief, ancestor worship, female worship and ethnic migration. The characteristics of west-road hairstyle - phoenix bun of women in different age groups are further generalized. Also, the features of west-road phoenix costume are summarized, in particular the model and structure size of the basic styling. The main decorative techniques of west-road phoenix costume are also illustrated.

Keywords: Puxia; West-road Style; She Minority; Phoenix Costume

She minority is an ancient ethnic group in Southeast China. At present, She minority mainly lives together with the Han in Eastern China, forming the pattern of "overall scattering and local compact community". She minority in Fujian province mainly concentrates in eastern Fujian (Ningde, Fu'an, Xiapu, and Fuding). Xiapu County in northeast Fujian is one of the major habitations of She minority.

Xiapu County has a history of 1700 years, and was the political, economic and cultural center of eastern Fujian in ancient China. Xiapu County had been the Honshu of Funing before the reign of Emperor Qianlong in Qing Dynasty. In the 12th year of the reign of Emperor Yongzheng (the year of 1734), Funing county was upgraded to Funing Prefecture. The area under the administration of Honshu became Xiapu County which was named after Xiapu Mountain in the territory. Its jurisdiction covered the present Xiapu, Fu'an, Fuding, and Ningde. Demarcated by the gate of Xiapu County, Funing Honshu is divided into the eastern part and the western part. The demarcation of geographical location distinguishes east-road style and west-road style costumes. The current Chongru, She minority Country of Yantian, and Bailukeng in Xinan Town belong to the western part of Xiapu County. The traditional costume of women in this area is called Funing West-Road Costume or West-Road Style. The current Shuimen, Yan Cheng and Sansha of Xiapu County and Fuding City belong to the eastern part of Funing Honshu. Therefore, the traditional costume of women from She minority in this area is called Funing East-Road Costume or East-Road style [1, 2].

1. Origin of "Phoenix Costume" of She Minority

The traditional costume of women from the She minority is called "phoenix costume", which is closely related to the origin of the She Minority and their religious faith. The She have unique beliefs and worship. Their ancestors "Pan Hu and his wife the third princess" are recorded in legends passed down through oral history through successive generations. Emperor Gao song which is the epic of She Minority. According to legend, Panhu, the ancestor of She Minority is mythical colorful dog, and helped Emperor Gaoxin in Shang Dynasty to eradicate foreign tribes. The emperor betroths his third daughter to him, and grants the phoenix coronet and phoenix clothes to her. Later, Panhu and his wife lived in an unfrequented forest and their descendants become the She people. The worship and respect to Panhu and his wife are reflected in all respects of the life of the She. They regard the third princess as the embodiment of the phoenix. They exteriorize their worship to their ancestor through the totem images of the phoenix and dragon dog. She women continually intensify and inherit their culture through costume. The classical bun decorated with red string, and the symbolic phoenix coronet worn in marriage ceremonies is commonly known as "phoenix head". Their costume, aprons, fine embroidery and lace symbolize the waist and neck of the phoenix. The waistband waving with the wind symbolizes the tail. The leggings simulate the phoenix leg, and the silverwares hanging on the costume symbolize twitter of phoenix. To sum up, the main components of the phoenix costume include: the phoenix bun, phoenix outerwear (outerwear decorated with lace), white undercoat, phoenix waist (apron), phoenix belt (mesh belt), dress or pants, and phoenix shoes (Shoes made of cotton fabric).

Because of the worship to their ancestor, the third princess, She traditional culture is endowed with intense women worship. The social status of women in She Minority forms striking contrast to traditional Chinese women. The traditional society of She was dominated by women. They participated in and led daily life and work. This required the Phoenix costume to meet the needs of labor. To this end, the narrow sleeves and leggings were designed. In daily work, the phoenix costume lays stress on simplicity and practicability.

In addition, the She is a migratory group. Before Tang and Song Dynasties, the She was basically a stable nation. After that, they began to migrate. From the late Yuan Dynasty to the Wanli period of Ming Dynasty, they migrated from western Fujian to eastern Fujian and Zhejiang on a large scale. Until Qing Dynasty, the She completed their migration, gradually settled down and lived together with the Han. Consequently, their costume is influenced by the Han to form the current "phoenix costume" through continuous cultural absorption and integration. The She has profound female ancestors and totem worship culture, which gives rise to the unique She female costume – the phoenix costume. The continuous cultural integration in the process of ethnic migration spurred constant change and development in the phoenix costume.

2. Simple and Unsophisticated Xiapu West-road Style Phoenix Bun

Fig.1 Hair style of elderly women, married women, and unmarried girls

Xiapu west-road style bun of woman in She Minority is commonly known as phoenix bun. The hair styles of married women and unmarried girls are different (Fig.1). The phoenix coronet is worn in weddings, scarves and flower hats are worn when working. The traditional style of married women is Yunji and high temples, which are quite complex. Plenty of wigs are utilized, decorated with peach red knitting wool. Another hairstyle feature is the hairpin. The style of hairpin is passed on from generation to generation and is never changed, commonly known as "Hengba", which is 10cm long and 2.5cm wide. The side face of the "Hengba" looks like a bow, and the front looks like two connected leaves decorated with patterns. The hair looks like a dragon from the front, which rises straight up from the temples. The dragon body circles and the dragon head holds high vividly. The hairstyle looks like a phoenix from the side. The upward bun looks like the phoenix head, and the braid decorated with the peach red knitting wool looks like the phoenix body. The tail bun emulates the phoenix tail, scattering on the back side vividly.

Unmarried girls' hairstyle is relatively simple, and commonly known as Pingtouzi. The hair is braided as the flat type, and two hair with the red knitting wool. One beam covers the temples from top to bottom, and the other circles from front to back, looking like a black cap with red edge. The bangs align with the eyebrows. Generally, unmarried girls do not wear ornaments. Elderly women wear "Panlong bun" or "twisted bun" similar to the Han, which is big and flat. They also wear hairnets, hairpins and silver flowers. The forehead is wrapped with a scarf, usually dark fabric or blue calico.

The phoenix coronet, originally called "Hui" or "princess crown", is worn by Han women in weddings or and when they are buried. The top of the Xiapu west-road style phoenix coronet is pointed and the round, topping and gorgeous. The main part of the phoenix coronet is tapered by bamboo skin, covered with black or dark cloth. The middle part of the top is decorated with a delicate small square mirror, mini scissors, ruler, and books, etc. Red silk or beads are fixed on both sides of the cone so as to prevent the phoenix coronet from falling down when worn. The top outside the cone is decorated with an isosceles triangle made of bamboo skin on the right side. The triangle is covered with red cloth, and jade and silver plates are hung on the top of the triangle. Each side is decorated silver plates with the twelve zodiac. Blue beads and butterfly shaped silver are hung on both sides of the triangle, commonly known as butterfly plate, implying many sons and grandsons. There are four beads at the bottom of the triangle, and two pieces of silver plates on the corn below the triangle, commonly known as "Fengpai", meaning the official position. The face cover silver ornament is hung at the bottom of the triangle, commonly known as "Xianxu" which is composed of a quadrate silver plate and nine strings of silver slices, like a curtain hanging down before the face. The big plate is engraved with auspicious patterns, like "two dragons play with a pearl". The nine silver slices are engraved with numerous underlings. The number nine symbol when all is said and done, another meaning of evil spirits. In traditional practice, the new daughter-in-law should also wear the phoenix coronet when working (Fig.2). In the recent 100 years, they wear flower bamboo hats with pointed top.

The Xiapu west-road style flower bamboo hat is commonly known as the flower straw hat, which is characterized by the pointed top (Fig.3), wide brim and its light weight. According to legend, it evolved from the phoenix coronet "princess hat". The diameter is about 38cm, depth is 8cm, and the height of the top is about 3cm. The weight is generally two-thirds or even half of common bamboo hats. The surface is made of 224-240 bamboo skins made in the She Shangshui Village in Xiapu County, the thickness of which is about 0.1 cm [3].

Fig.2 Phoenix coronet

Fig.3 Follower bamboo hat

3. Characteristics of the Xiapu West-road Style Phoenix Costume

The West-style phoenix costume in Xiapu, Fujian mainly emulates the phoenix. The overall style integrates firmness and flexibility, and perfectly combines roughness and delicateness. The She's traditional costume in modern times evolved from the Manchu women's dress with buttons on the right. Combined with the conditions and demands of the She, some elements were accepted, while some have been abandoned. The costume is split into a coat and skirt or pants. The color is generally simple and unsophisticated, with black as the main color. Gorgeous colors and white are match colors. Such collocation is both calm and lively. The material is mainly Chinese grass-ramie, commonly known as ramie cloth. The West-road phoenix costume is divided into informal dress and formal dress. The former is relatively simple, while the latter is more solemn. The patterns of informal dress and formal dress are basically the same. The difference lies in the theme and size of embroidery. Both sides of Xiapu the West-road "phoenix costume" can be worn. The grain side is worn at festivals and the New Year, and the wrong side is worn in daily work (Fig.4). The grain side is solemn, and the wrong side is simple. The patterns of the phoenix costume vary with age, which is mainly reflected in the pattern of embroidery. Younger women's costumes are more elaborate, while the elderly and children's costumes are relatively simple.

Traditionally, women of She minority wear a skirt and a coat with an unchanged pattern. The embroidery of the coat is simple. Influenced by Han culture, the collar of outerwear should be open, and the white shirt should come out. It is said that this is to remember their ancestors. In Tang Dynasty, the ancestors of Zhangpu She minority were suppressed by Chen Yuanguang. The women of She minority were forced to marry soldiers of the Tang army. On the day of wedding, they wore white clothes to remember their sacrificed relatives. This custom continues until today.

Basic garment design and structural size. (The basic size of the garment is the same proportion, and the specific size of each wearer will be different.)

Coat:The coat of Xiapu style (west-road style) phoenix costumes includes a red or blue abdomen wraps up , a white shirt and outerwear. The outerwear is commonly known as "outerwear with lace". Figure 3 shows the classical costume style with narrow sleeves and buttons on the right, stand collar, raglan sleeves, waist, and side slits. The corner has oblique angle, and chest dress has Fudou, this coat no pockets. The right lapel is fixed by Pipa belt made of cloth strip. The front and back of coat is 75cm long, When worn, the bottom edge of the garment is almost flush with the fingers. It is slightly longer than common Han costume. The side silts are about 30cm high, and the sleeves are 15cm wide. The inner margin and cuff are decorated with Gunxiang (repeated sewing) and Tiantiao. Tiantiao and band are mostly blue. The stand collar is narrow, and the middle part is about 2cm. The collarband has button loop which is metal, and made of copper or silver. The surface of the button is decorated with patterns. The lapel is commonly 20cm wider than the central line, with Taojian inside. The inner left Ren of the wrong side is also lapel style and includes Fudou and band [4].

Fig.4 Grain side and wrong side of outerwear with lace

Lower costume: Xiapu style (west-road stlye) phoenix costume is skirt or dress, and an apron which is generally known as "Lanshenqun", "Fenglanyao" or "Heshouji". "Lanshenqun" is worn on the waist ,and tied with a belt. They wear skirt, leggings nasal shoes in wedding.

The apron is an important part of the traditional dress for She women. Composed of skirt and waist , the background color of the skirt is black, and the shell fabric is the same as the pants. The pattern is a symmetrical trapezoid with the height of 33cm, upper width of 33cm, and bottom width of 60cm. There are 5-7 puckers 12cm away from both sides, with the distance of 0.7cm. The waistband is sky blue and 6.5cm wide. There are cloth ears utilized to fasten the belt on both sides (Fig.5). The braid is about 4.5cm to 6cm wide, and 2m long. Both ends are like spiciform. The braid is knitted by She women with white cotton. According to the legend, the braid symbolizes the "Guodai" of official costume. Married women wear two white belts (without pattern) through the cloth ears, and the spike is fastened on both hips. Unmarried women tie a flower belt which is decorated with red or blue character patterns. The belt is first winded backwards and then forward, tying a knot at the front of the waist. The end of the belt falls naturally.

Fig.5 Apron (Fenglanyao)

She women wear skirts or pants under the apron. Pants, particular black pants, are most often worn in daily life, and the style is similar to the Han. The old- fashioned loose pants have been replaced by well-fitted western style pants suitable for modern young people. In addition, they also wear cropped trousers, and the trouser legs stop between the knees and the ankles. Long skirts worn in weddings are black, plain, many skirt. The skirts generally extend to the insteps. Long skirts are divided into straight skirt and surrounded skirt. Leggings, commonly known as "foot warmers", are the proper pattern of traditional costume of She women. Leggings are warm, and is trapezoid, 55cm long and 28cm wide. They are made of white dragon head cloth (purified cotton plain cloth fabric), and worn on the lower part of the leg . The ends are tied with red jade and long lapel. The jade hangs on the shanks naturally.

The traditional shoes worn by She women have thick soles and square toe caps with black welt and black cloth. There is a protruding back in the middle of the vamp, which is commonly known as "single nose shoes". The welt is edged through red, yellow and green wire. In addition, shoes with two protruding backs in the middle of the vamp are commonly known as "double-nose shoes", which are traditional men's shoes for the She.

4. Decorative Techniques of Xiapu West-road Style "Phoenix Costume" - Embroidery and Edge Pressing

Embroidery and edge pressing are the main decorative techniques of the Xiapu west-road style "phoenix costume". The decorative parts mainly include the collar, chest, slit, and apron. The dermatoglyphic pattern of "phoenix costume" make up two parts: "edge" and "flower". The edge includes edge pressing strip and fang pattern. The flower is a design pattern. The edge pressing consists of cloth strips of three to five colors successively. The fang pattern is an embroidery pattern, arranged in zigzag, also known as tiger teeth pattern. The embroidery pattern is rich and colorful, and the theme is bold and unrestrained. The colors are gorgeous, forming striking contrast. The craft is exquisite, and the embroider lines are distinct. The abundant contents have profound implied meaning. The color of embroidery is mainly red (bright red and peach red), matched with green (emerald green and grass green), yellow, white, blue, and sometimes gold. Complex patterns are generally decorated with five or more colors, while relatively simple edging only includes three colors of red, yellow and green. The main outline is embroidered with white wire, so that the whole pattern is clearer on the green background color. The striking contrast with other colors in the embroidery pattern is softened. [5, 6]

The abundant themes of embroidery patterns not only includes the pattern ornamentation with national characteristics, but also integrate Han patterns, including totem worship patterns, abstract geometric patterns, plant patterns, animal patterns, utensil patterns and character patterns. Totem worship patterns include dragons, phoenix and turtles. Commonly used combination patterns include "Two dragons playing with a pearl", "Phoenix and Peony", and "Turtle and Dragon Court". Fang pattern is the most common abstract geometric pattern. In addition, cross pattern, HUI pattern, WAN pattern, floral scrolls (commonly known as nagarjuna pattern), and cloud pattern are also abstract geometric patterns. The most common plant pattern are flower patterns, including peony, lotus, peach blossom, bamboo, orchid, chrysanthemum, and pine, etc. The most common animal patterns include the butterfly, magpie, deer, lion, bat, and crane. The commonly used combination patterns include magpies on plums, bamboo deer, goose paw peony, crane, and two lions playing with a ball, etc. Character patterns are more common in Xiapu east-road style (namely Fuding style), and also occur in west-road style costume. The common patterns include the Eight Immortals, Liu Hai plays with the golden toad, Meng Chong cries at the bamboo, Jiang Ziya fishing, and Emperor Wen, etc. There are also vases and character pattern in She embroidery. The pattern structure includes single pattern, corner pattern and continuous pattern. Most She embroidery craft masters do not prepare a draft, they embroider following their intuition. According to the investigation, the patterns embroidered by craft masters vary even for the same theme.

The embroidery edge of the collar is on the top and bottom collar band lines. The theme of "Flower" is mostly peony, and commonly known as "flower collar", including large-leaf peony, small-leaf peony, lotus and other flowers and plants. "Dragon collar" is embroidered with two dragons playing with a pearl, and "phoenix collar" is embroidered with double phoenixes facing the sun or the peony. The geometric pattern, "fang" patterns form the simple edge pattern [7]. According to the number of edge patterns, collars can be divided into "one-line collars", "two-line collars", and "three-line collars". Such patterns are also mainly used on the oblique angle of the costume.

The embroidery is mainly concentrated on the superior angle of the left front of the right lapel. The left side starts from the middle line of the collar, and the oblique length of the right lapel is about 16cm, the vertical length is about 6.5cm, and the width is 1cm to 10cm, composed of 1-3 groups of embroidery patterns. Each group of embroidery patterns is called "one pond", which can be divided into "One red clothes", "Two red clothes" and "Three red clothes" according to the group number. The flower food is embroidered at the bottom of the flower pond. The embroidery group number of Fudou corresponds to the line number of the collar. That is, "one-line collar" matches "One red clothes", "two-line collar" matches "Two red clothes", and "three-line collar" matches "Three red clothes", which was established by the people through long social practice. "Three red clothes" is the splendid costume of the phoenix costume, and the width of embroidery pattern is up to 10cm, matching the "flower collar", "dragon collar", and "phoenix collar". Such costumes are mainly worn by young women. Elderly women and girls' informal costumes are mainly decorated with narrow embroidery patterns, and the fang pattern is only 1cm wide (Fig.6, 7). Both sides of Lanshenqun and the top are bound with red, yellow, blue, white and green edge pressing strips. The embroidery patterns are within the edge pressing strips, arranged from inside to outside. The patterns are separated by edge pressing strips, varying from one layer to three to five layers. The patterns of traditional Lanshenqun are complex and exquisite, while the modern skirt has become simple, which is mainly two clusters of peonies [8].

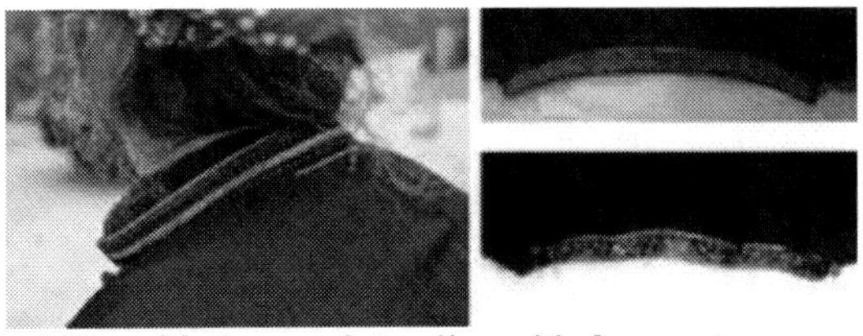

Fig.6 The elderly people' collar with fang pattern,

Fig.7 Fudou embroidery dragon pattern and flower pattern

In addition, She women wear ornaments such as earrings, rings, bracelets, anklets, badges, and necklaces. The big earrings look like a pair of reversed question marks, which are about 3cm long. The rings of the question marks are behind the ear, and the bottoms of the question marks are before the ear (Fig.8). Such earrings are worn by married women, and matched with the "dragon bun". Small earrings are worn by unmarried girls, and the shape is the same as the big one. Rings and bracelets are greatly influenced by the Han and the shape is basically the same.

The patterns of rings include both the common floral scrolls (commonly known as the nagarjuna pattern) and Bagua tactics map of the Han.

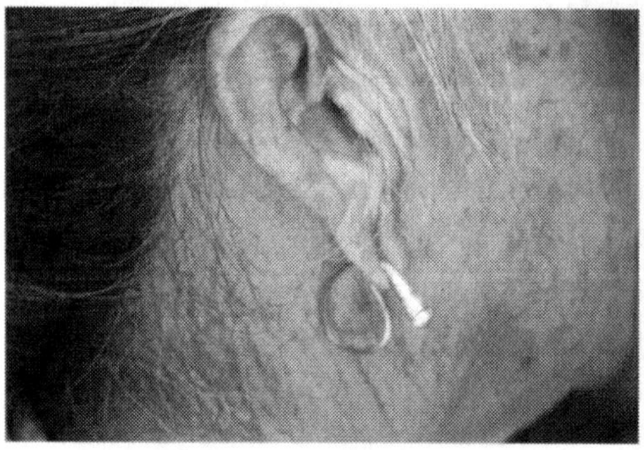

Fig.8 The big earrings

5. Conclusion

The phoenix costume of the She minority in Xiapu, Fujian contains a profound cultural connotation. The phoenix costume is also precious cultural heritage. It reflects not only the belief and habits of the She, but also their evolving culture, ethnic fusion and unique aesthetic tastes [9]. In our research process, it is found that the She people do not have their own written words, and the art and folk customs are passed on by word of mouth from generation to generation. Under the impact of modern civilization, some crafts fail to be handed down from past generations, and the national culture is gradually weakened. Therefore, it is quite urgent to sort out, protect, and excavate its value. These records and summarization of the She's phoenix costume in Xiapu is the basis for further protection and inheritance.

References

[1] Pan HL. Study on the costume type of She Minority in Fujian [J]. Fujian Wenbo: 1987; 2.

[2] Yu YT. Record of She Minority in Xiapu County [M]. Fujian People's Publishing House: 1993; 2.

[3] Chen GQ. She Minority of Chongru Village [M]. Fujian People's Publishing House: 1993; 58: 66.

[4] Great national culture of China. The compilation of the She nationality. Great national culture of China: the She nationality, [M] nationality publishing house: 1999; 16: 21.

[5] Weng YH. Analysis of Embroidery Technique of She Minority [J]. Modern Art: 2014 (01).

[6] Gong RJ. The aesthetic implication of She nationality costume in Xiapu [J] Journa of Fujian institute of socialism No.6: 2012.

[7] Shi Q, Tan ZH. Ethnic Migration and Cultural Inheritance - Research on the Migration Culture of She Nationality in Zhejiang [M] Ethnic Publishing House: 2014; 3.

[8] GWF. Hegel Aesthetics Volume I [M] Commercial Press: 1981.

[9] Dai Ping. Research on Chinese national costume culture [M] Shanghai people's press: 2000; 385.

Modeling and Style Design of Toggery Window Background Wall

Yan Chen[1*], Jian-Ping Shi[2]

[1]*College of Textile and Clothing Engineering, Soochow University, China*
[2]*National Engineering Laboratory for Modern Silk, Soochow University, China*

[*]Corresponding author's email: 1254385643@qq.com

Abstract

The background wall design of clothing stores plays an important role in window design, This paper starts from the design of the toggery window background wall, illustrating the expressing form of transparent, semi-transparent and closed-end background wall design. Then the influence of three different consumer groups on the window background wall style design were analyzed. This paper suggest that different consumer groups will have different preferences for background wall style design. Meanwhile, in the proceeding of design, attention must be paid to the harmonization of the window background wall. The design of the window has to have a primary and secondary order and be consistent with the public aesthetic demands. The purpose of the analysis is to beautify the environment so that the toggery and visual merchandiser can effectively use the information to achieve the ultimate goal of increasing purchasing behaviors.

Keywords: Background Wall Design; Modeling; Style; Window of Toggery

1. Introduction

With the development of the clothing market, more and more international clothing brands have entered into the Chinese market, which forces the domestic clothing brands to be more innovative [1]. The industry is not able to develop without competition. In recent years, people have become attracted to the design of toggery showcase. Window display has become a critical essence to visual merchandising which is akin to the cover of magazines. There are however little studies done on window background design. The background is separated from the store, and it plays a role in stimulating the customer's imagination. An item arranged in a display window stimulates both rational and emotional motivation [2, 3]. Mehrabian and Russell suggested that consumers' inner responses are important mediators that can lead to approach and avoidance behaviors in response to environmental stimuli. Therefore, a good window background wall design not only attracts the attention of consumers, but also enhances the turnover of stores.

2. The Background Wall Design

Window design are part of a broader category of atmospherics called "visual merchandising [4]. The design of the clothing store window originated from the European business industry, which is the product of the industrial age [5]. During the process of development, people have grew to pay more attention to the design of the window. As such the design of the background wall also became popular. There are three types of designs: the closed-end design, the semi-transparent design and the transparent design.

2.1 Closed-end Background Wall Design

Closed-end background wall is the most common design in the terminal stores to attract the consumers eyeball, closed-end wall modeling include plane modeling and stereo modeling. The graphic is composed of point, line and surface elements, which is contribute to the visual effect for brand showcase design. The showcase is combined with plane and stereo, added creative, shape, color, material, lighting and other factors, it is more likely a microcosm of the city [6]. Stereo modeling design is based on plane design, it can more perfectly reflect the stereo backdrop effect. Irregular ups and downs, misalignment changes and free arrangement can also reflect the concave and convex shape of the wall surface, which included the design of wardrobe, external ornament, old wall surface processing and so on.

In the developed capitalist country, such as the United States and Europe, the clothing store background wall design is an essential link, Stereo modeling is not suitable for all of the window background wall design, when the window space is limited, the designers more prefer to choose the plane modeling rather than the stereo modeling. On the basis of plane modeling, if designers have some special needs, they can make full use of the point, line and surface elements to achieve the stereo modeling. Like Fig.1, the background designed with a great deal of curve in one direction, which combined with black and white, it will give the consumers a kind of visual illusion and is different from the plane modeling.

Fig.1 Curve Combination (From http://image.baidu.com/)

2.2 Semi-transparent Background Wall Design

Semi-transparent background wall design is made with translucent material, hollow, or only a block line of sight, like Fig.2. The various design is to separate window from store, then make the showcase independent. There are three representative types: window type, spiral type, vortex type. The semi-transparent background wall is designed to make people feel mystery, then it can stimulate the consumers' interest and attention. This kind of design have many advantages. Firstly, it is convenient and quick. Secondly, it can convey the information to the customers in time. Thirdly, it can balance the overall display between the window and the store, What's more, the use range is wide and the method is flexible. This kind of method not only be used to account the window display, but also be used to display the whole shop

Fig.2 Semi-transparent (From http://www.pop-fashion.com) Fig.3 Transparent (From http://www.eeff.net/)

2.3 Transparent Background Wall Design

Transparent background wall design is a entirely opened form and it is included in the store space, people can see the circumstance directly through the store window glass [7], like Fig.3. In the actual research, this kind of design is rare, the phenomenon can attribute to the good in the window it is easily confused with the store, that is why the showcase clothes can not be seen for the consumers' first sight, and other accessories in the store will not be shown to the consumer.

Transparent background wall design is difficult, demanding the store and the window can form an unified image in color, structure or goods display. But it is also simple, because the entire window design based on the store, and don't need too many modifiers in other items. Thus, it can be seen that a excellent transparent background wall design not only emphasize the design purpose, but also pay more attention to the external conditions.

3. The Style of the Background Wall Design

Style refers to an era, a nation, a genre, or a person's main idea and artistic characteristics [8]. The clothing store background wall design not only to be exquisite, but also to be characteristic.

For example, Chanel, as the incarnation of luxury and elegant, the style of the background wall design always make people feel noble and elegant. Kenzo, using flower pattern to decorate the window,, the purpose is to embody the charming of the flower, using fancy ornament and beautiful color can reflect the brand's affection to the nature. The style of the brand is bound to determine the style of the background wall.

3.1 The Influence of Different Consumer Groups on Setting Wall Style

3.1.1 Children Consumer Groups

Children are the most potential consumer object of the three target consumer groups and the design is not limited. The background wall tend to use wave point element, it can embody the child's lovely. the wave point element is a common and active way to be used in background design. Firstly, there is so many different size and color that the setting wall is very flexible. Secondly, some of the background design often use the scene of the cartoon and the scene of children' activities [9, 10]. Such as: beach, wave, kite, butterfly, boat and so on. Like Fig.4, the cartoon design of Dior, the doll dressed a classic dress made by puppet master, this scene have presented a magic dream scenario. Thirdly, some of the background wall design also use different high purity color or pink color to decorate, the skill will not only attract the eyes of the children but also reflect the child's childlike. Many child's costumes are relatively monotonous, but the flexible design of the setting wall can make up the shortage and achieve the effect that we need.

Fig.4 Cartoon Scene (http://image.baidu.com)

Fig.5 Personality Design (http://www.pop-fashion.com)

3.1.2 Female Consumer Groups

A outstanding clothes brand must have a clear position analysis in the stage of window design, this analysis not only help designers to establish focus, but also provide guarantee for later work, so that it could be carried out smoothly [11]. Women is the most extensive groups,in the consumer market and the style classification is more than other groups, such as: classical style, romantic style, vanguard style, and so on.

Different style design will influence the background wall presented. The style of the classical background wall design is tend to use dark color to highlight women intellectual, dignified. The romantic style is more likely to use flowers and soft color. Vanguard style influenced by pop art, abstract factions, and the modeling characteristics be full of weird and fantasy, Like Fig.5, the design of the Louis Vuitton is so creativity and bizarre that the window looks personality, the designers use the visual language to make people stop and evoke people's emotional resonance. What's more, the species of the women clothing is especially widely, such as business wear, casual wear, home clothes, dress and so on.

3.1.3 Male Consumer Groups

Most of the men background wall design is tend to fainter, dull and monotonous. The clothes reflect the man's personality, heroic bearing, deep taste, the style of design must be concise and capable [12]. So the design is difficult to control.

Masculine type, sport type, vanguard type is the main style of the man's clothing design, each style of the background wall design should be catered to the brand style. For example, the sport type is lively, active, the design of the background wall is generally associated with sports scene, such as football field, basketball court, swimming pool, the runway and so on, All of the scene can stimulate the customer's passion on the sports and the feeling of loving life. In a word, the style of the man's background wall design always make people feel positive and relax. It is rare for Missoni to emphasize individuality by making use of ornate background. In addition to, the window design of Prada is different from the other brands, Like Fig.6, each display and detail in the window has been carefully designed by designers, the design is combined natural texture of wood with perspective sense of geometric, it fully displayed the brand style and culture which is famous for arts.

Fig.6 Perspective Sense Design (http://www.eeff.net/)

Fig.7 Sense of Coordination (http://image.baidu.com)

3.2 Harmonization between the Style of Background Wall and Clothing

The consumers' behavior can be affected by the store environment [13]. From the perspective of arts, harmony is the highest state of aesthetic. The background wall is designed to achieve the balance of physics, during the process of design, we must conform to the aesthetic principle [14]. The successful window design should be combined form with content, it is often counterproductive to pursue pure visual stimulation, So, the key to the window background wall design is to pay more attention to the harmonization between style design and prop, color, light, sample. Armani, a famous designer, who used to work as a clothes display worker in the department store, he said: we will create an exciting and unexpected experience for the customer, then maintain a clear and consistent identification on the whole. Every part of the store is always expressing my ideas about the aesthetic, I hope I can show my design in the space and atmosphere, then provide a profound experience for the customer [15].

The background wall is only a carrier of clothes display. On the one hand, the design have to highlight its own uniqueness to attract consumers, on the other hand, let the consumers pay more attention to the clothing goods instead of background wall, only in this way, it will be the best visual design. Louis Vuitton's clothing window is very big, the display of the product is not sufficient, but each one of them is classic and make consumers feel extremely atmospheric. The window will highlight the clothing goods, the purpose of each auxiliary condition in the window design is to express the subject and get a harmonization between the style of background wall with clothing. During the process of design, the designer have avoid to use too much decoration the purpose is to concentrate the consumer's attention on the clothing. Like Fig.7, the overall design of the window is tend to concise, and accustomed to grasp the unity of background wall design and clothing from the tone, the purpose is to present a scene that can conform to clothing style and temperament. It not only emphasizes the background wall style, but also beautifies the window environment.

4. Conclusion

The design of the toggery window background wall is never isolated from the whole atmosphere, which influenced by the wall shape, the different consumer group, the concept and the style of brand. Firstly, the background wall shape must be designed according to the brand style. Secondly, the three target groups affect brand style, the different consumer groups should have different background wall design. Thirdly, the whole modeling of the background wall design and the style design must be unified. Grasping the various factors, it will be a excellent background wall design.

References

[1] Zhang JF. The study of the men clothing display design. Doctoral Dissertation of Suzhou University: 2010.
[2] Liu CY. Commercial display window design. Chemical Industry Press: 2012; 7.
[3] Kernsom T, Sahachaisaeree N. Strategic Merchandising and Effective Composition Design of Window Display: A Case of Large Scale Department Store in Bangkok. Social and Behavioral Sciences: 2012; 42: 422-428.
[4] Ti C. The Effects of Window Display Setting and Background Music on Consumers' Mental Imagery, Arousal Response, Attitude, and Approach-Avoidance Behaviors. Window Display: 2009.
[5] Zhuang Q. The theme expression of women brand window design. Doctoral Dissertation of Beijing Fashion Institute: 2010; 3.
[6] Zhang L. The research on modern window design. China Science and Technology Expo: 2011; 13: 304-304.
[7] Yin LL. Study on dynamic expression of the window display design. Qingdao Technological University: 2014.
[8] Gao RH. The store display layout and development trend. Education Innovation: 2007; 4: 19-20.
[9] CUITO. Rem Koolhaas: Oma. New York: Amazon Remainders Account: 2002.
[10] M. PRESS, R. COOPER. The Design Experience: The Role of Design and Designers in the Twenty-First Century. Aldershot: Ashgate Publishing: 2003.
[11] Zhao WT, Li Yujie. The window of the brand spirit. Fashion Designer: 2007; 12: 156-159.
[12] Xia XC. The miracle of display design for arts and fashion brand markets. Shopping Mall Modernization: 2008 ; 3: 71.
[13] Somoon K, Sahachaisaree N. Design Elements and Users' Perceptual Response: A Case of Window Display Design for Adolescent Merchandising. Social and Behavioral Sciences: 2012; 50: 685-690.
[14] Qi M. The research form of window design. Xi'an University of Architecture and Technology: 2014.
[15] Peng H. Analysis the color design in clothing store. Charm China: 2010 ; 12: 44-44.

AUTHOR INDEX

AUTHOR INDEX

AUTHOR INDEX

AUTHOR INDEX

AUTHOR INDEX

Textile Bioengineering and Informatics Society (TBIS)
Unit 503, 5/F. Silvercord
Tower 2, 30 Canton Roa
Tsim Sha Tsui
Hong Kong

ISBN 978-1-5108-4951-8
9 781510 849518

CURRAN ASSOCIATES INC.
proceedings.com

10th Textile Bioengineering and Informatics Symposium Proceedings (TBIS 2017)

Wu Han, China
16 – 19 May 2017

Volume 2 of 2

TBIS

10th Textile Bioengineering and Informatics Symposium Proceedings (TBIS 2017)

Wu Han, China
16 – 19 May 2017

Volume 2 of 2

Editors:

Yi Li
Wei Lin Xu

ISBN: 978-1-5108-4951-8

Printed from e-media with permission by:

Curran Associates, Inc.
57 Morehouse Lane
Red Hook, NY 12571

Printed by Curran Associates, Inc. (2018)

Textile Bioengineering and Informatics Society (TBIS)
Unit 503, 5/F. Silvercord
Tower 2, 30 Canton Roa
Tsim Sha Tsui
Hong Kong

Phone: (852) 66382781

tbis@tbisociety.org

Additional copies of this publication are available from:

Curran Associates, Inc.
57 Morehouse Lane
Red Hook, NY 12571 USA
Phone: 845-758-0400
Fax: 845-758-2633
Email: curran@proceedings.com
Web: www.proceedings.com

TBIS

Textile Bioengineering and Informatics Symposium Proceedings

2017

ISSN:19423438

Edited by
Yi Li
Wei Lin Xu

The 10th Textile Bioengineering and Informatics Symposium

May 16-19, 2017 , Wuhan Textile University, Wu Han, China

Welcoming Message

Professor Wei Lin Xu
Chairman, TBIS 2017 Organizing Committee
Wuhan Textile University
Vice President of China Textile Engineering Society
Changjiang Scholar
China

Welcome to Wuhan!

TBIS is an international platform that nurtures the scientific textile community in the exchange and development of innovative and revolutionary ideas in the field of textiles and apparel, and it is a our great pleasure to host the incoming 2017 TBIS.

With the aim of cultivating textile specialists and serving China's textile industry, Wuhan Textile University (WTU) was found in 1958 in central China. Hubei's capital, Wuhan is a metropolis full of universities and talents. WTU is the only college in Central and Southwest China that features studies in textile and apparel, and there are more than 10,000 students studying in the field of textile and apparel. WTU adheres to upgrade its competence by implementing high-level featured strategies for development, promoting its innovative capacity by implementing the strategy for high-level multidisciplinary development, achieving breakthroughs in its faculty team building by introducing and fostering high-end talents. WTU is now extending to different textile universities and forming many collaborations, we look forward to cooperating with you in the near future!

Welcoming Message from TBIS

Professor Yi Li
Chair of Textile Science and Engineering
School of Materials, The University of Manchester
Chairman, TBIS 2017 International Scientific Committee
Chairman, Textile Bioengineering and Informatics Society
Editor-in-Chief, Journal of Fiber Bioengineering and Informatics

With great pleasure, we are now collaborating with Wuhan Textile University to hold TBIS 2017, the 10th Textile Bioengineering and Informatics Symposium in May 16-19, 2017 in Wu Han China.

With continuous success and enthusiastic support from authors and experts around the world, Textile Bioengineering and Informatics Symposiums (TBIS) have been recognized internationally as a brand of high quality conference with vision, creativity, excitement and global network. TBIS proceedings and JFBI (Journal of Fiber Bioengineering and Informatics) have been indexed by worldwide scientific and academic reputable databases such as EI Compendex, Scopus, CPCI/ISI Web of Knowledge. Since 2008, TBIS has published 1686 research papers indexed in the CPCI/ISI Web of Knowledge database, becoming a leading academic conference organizer and publisher.

TBIS has established an effective platform for academic institutions and individual scholars, researchers and students to publish their research works and communicate with peers and experts in the field to get recognitions and increase impact such as citations and H-indexes. The TBIS symposium is a peer-reviewed conference aiming to set up a unique global stage and platform for people working in the whole supply chain from materials, fibers, textiles and apparel and technical textiles to relevant service industries such as biomedical industry, fashion industry and sport industry with common interests in providing high quality of products and services for protection, health, performance and comfort of human populations living in various environments and exposing to various natural and human disasters. To further enhance global networking, this year we have the pleasure to collaborate with Wuhan Textile University to organize TBIS in Wuhan.

TBIS 2017 welcomes you and is looking forward to your active participation and contribution.

Yi Li

TBIS 2017 Organizing and International Scientific Committees

Organizing Committee

Chairman:	Prof. Weilin Xu		
Vice Chairmen:	Prof. Keqin Zhang	Prof. Ruomei Wang	Prof. Xin Zhang
Secretariat:	Ms. Mai Wang Li	Dr. Xin Liu	Dr. Genyang Cao
Members:	Prof. Xue-Mei Ding	Prof. Jun Feng	Prof. Jianxun Gao
	Dr. Junyan Hu	Prof. Jianqiang Li	Prof. Xiangpeng Li
	Prof. Li Lin	Prof. Xiaonan Luo	Dr. Subramanian S.K. Muthu
	Prof. Xiaojun Tong	Prof. Dong Wang	Dr. Zhong Wang
	Prof. Xiongying Wu	Ms. Hui Wu	Prof. Zhaofei Xiong
	Prof. Zhaoyang Xu	Dr. Lei Yao	Prof. Jinbo Yao
	Dr. Boan Ying	Prof. Shangyong Zhang	

International Scientific Committee

Honorary Chairman:	Academician Mu Yao	China		
Chairman:	Prof. Henry Yi Li	UK		
Co-Chairmen:	Prof. Weilin Xu (China)	Prof. Frank Ko (Canada)	Prof. Kanji Kajiwara (Japan)	
Members:	Prof. Seung Kook An (Korea)	Prof. Senem K. Bahadir (Turkey)	Prof. Aizheng Chen (China)	Prof. Guoqiang Chen (China)
	Prof. Weiguo Chen (China)	Dr. Stewart R. Collie (New Zealand)	Prof. ZhaoQun Du (China)	Prof. Emiel Den Hartog (USA)
	Prof. Lubos Hes (Czech Republic)	Dr. Yang Hu (USA)	Prof. Tanveer Hussain (Pakistan)	Prof. Kap Jin Kim (Korea)
	Prof. Gang Li (China)	Dr. Luximon Ameersing (Hong Kong, China)	Prof. Jiri Militky (Czech Republic)	Prof. Qingqing Ni (Japan)
	Prof. Rajiv Padhye (Australia)	Prof. Mishra Rajesh (Czech Republic)	Prof. Tetsuya Sato (Japan)	Prof. Guowen Song (USA)
	Prof. Masayuki Takatera (Japan)	Prof. Kaoru Wakatsuki (Japan)	Prof. Qiang Wang (China)	Prof. Lijing Wang (Australia)
	Dr. Xin Wang (Australia)	Prof. Xungai Wang (Australia)	Prof. Qufu Wei (China)	Prof. Calvin Wong (Hong Kong, China)

Organizing and Supporting Institutions

Hosted by:

Wuhan Textile University
State Key Lab for New Textile
Materials and Advanced Technology

Co-organized by:

Textile Bioengineering and Informatics Society

Supporting Institutions

Aimer Human Engineering Research Centre (HEC)

Soochow University

Shaanxi Province's Research Center of Apparel Engineering and Technology

The University of Manchester

Xian Polytechnic University

Exclusive Sponsor

纤维谷
FIBRE VALLEY
R&D / COMMUNICATION
INCUBATION / CONSULTING
INVESTMENT / COOPERATION / MARKETING
Covering the whole industry chain of Fibre

PROCEEDINGS OF TBIS 2017

TABLE OF CONTENT

Track 1 Textile Biomedical Engineering

Track 2 Textile Biomechanical Engineering

Track 3 Clothing Bioengineering

Track 4 Textile Supply Chain Informatics

Track 5 Textile Thermal Bioengineering

Track 6 Textile Science and Engineering Chemistry

Track 7 Textile Science and Engineering--Physics

Track 8 Textile Material Bioengineering-Chemistry

Track 9 Textile Material Bioengineering-Physics

Factor Study on Wool Felt Fabric Recycling Technology and Application in Clothing

Tian Ping, Xing Zhang*

School of Apparel & Art Design, Xi'an Polytechnic University, Xi'an, 710048, China

*Corresponding author's email: Zhxhg029@126.com

Abstract

In recent years, with the growing development of the human living standard, the application and performance of wool felt fabric is no longer limited to tradition. This paper discusses and studies the factors that influence the development of wool felt fabric and its application in clothing. It sums up a method for wool felt fabric reconstruction, encourages a broader mindset on fabric design, summarizes the application of wool felt transformation in fashion design, provides a reference for the research on the theory of fashion design, and provides a guidance for the transformation of fabric.

Keywords: Wool Felt; Clothing; Technique; Fabric Modification; Factor Study

1. Introduction

In modern fashion design, in order to create a new artistic effect, fabric recycling is often conducted to create new fabrics with a certain style and feature to suit a design. The use of recycled material for design now plays a more and more important role in costume design. In order to create a unique visual effect, designers tend to use various design methods of free collocation, to maximize the excavation potential performance of the materials. Good fabric design is an art itself, and it is a critical factor in determining sale volume of dress in the market.

In the modern world, not only can wool be recycled, it is also biodegradable. As a unique natural material, wool has become the first choice for clothing above synthetic fiber fabric. In the design of wool clothing, its properties such as high durability, soft, elastic etc. gives designers great inspiration and innovation. Through the research on the International Woolmark Prize held by the "THE WOOLMARK COMPANY" the wool material has been re-interpreted. With the recombination of the traditional wool material, and the innovation of many handicrafts, the traditional wool material can be applied many more ways in fashion design. In addition to the excellent performance of wool material, its natural environmental characteristics makes it valuable for theory and also design research. In this paper, techniques such as the three-dimensional relief effect, hollow out, needle felt, wet blankets and other methods are studied for its application on the reconstruction of wool fabric for clothing. This paper aims to change people's impression of traditional wool, increase demand, and to promote further development.

2. The Traditional Technique of Making the Wool Felt

The art of clothing is perpetually changing. The rate of its evolution is proportional to the rate of technological development. The technological influence on clothing centers mainly on the development and use of fabric. Design of modern clothing and the design of clothing materials are integrated together. A new approach has been introduced to clothing design which incorporates the selection and design of good fabric.

The wool felt technique existed for over a thousand years. It is almost as old as the entirety of the human history. Wool felt is produced by wool, it is the oldest fabric known to human and it is otherwise known as "non-woven fabric", which does not require latitude and longitude lines. There are many legends on the origin of wool. After research it was discovered that wool was discovered by ancient nomads dating back to 6500 years BC. It has strong plasticity, and can greatly enrich the scope of modern fabrics. Wool felt is both soft and strong, have good elasticity, is comfortable to be touched, and it also has a good reducibility. When wool products are folded, it can quickly recover and is not easy to be deformed. Fiber structure can be closely intertwined.

2.1 Common Technique of Making Wool Felt

Felting principle: As a traditional handicraft, the wool needle felt can be penetrated very easily. The scales on the surface of the wool is the major cause for the shrinkage of the tapetum. The tapetum will usually open when affected by temperature, pressure, weak alkaline solution conditions, and when rubbed repeatedly by an external force that will make positive and negative scales tie together. According to different principles, wool felting process is divided into two kinds, needle felt and wet felt.

2.1.1 Wet Felt

Wet felt looks like fine wool. It is covered with many invisible squamose tissues. When these scales meet hot water, it will be open and stand up. If subjected to an external force of extrusion, friction and rubbing, the scales will be closely intertwined and shrink together. The basic principle behind this felt state is "wet felt".

Table.1 Advantages and disadvantages of wet felt commonly used tools

	Tools	Thick	Thin	Hard	Encounter water deformation	Recycling use	Non recyclable use
Shuttering	Packing cotton	√		√		√	
	Bubble Wrap	√				√	
	Cardboard		√		√		√
	Folder		√	√			√
Accelerated wool felting tool	eyelet fabric;	√				√	
	Bubble Wrap	√				√	
	Rolling pin	√		√		√	
	Laundry soap powder	√		√		√	

The difficulty lies in spreading the wet felt wool uniformly, the temperature of the liquid and the spraying area etc. The tool is selected based on its performance and its advantage and disadvantages. (as shown in Table 1) Then the mesh cloth is tiled on the table using a 500ml water with a spoon of soap at the combination ratio (500:1) to make bubble water. The bubble film is to a specific shape as needed with scissors. The wool is torn to uniform pieces and kept still. It is then uniformly tiled on bubble paper.. The appropriate soap and water is added to the wool so that it will be easy to rub. In the beginning because the wool is soft, the rubbing intensity should be light otherwise it will be deformed. With increasing speed, the intensity can be increased until the wool becomes half felt. The fabric can then be rolled, and during this process there is no need to worry about it being deformed. After continuous rubbing, the wool will shrink to a predetermined size. [1] The wool then can be cleaned and separated. There are two ways to separate the wools, the first is to separate them with the two flanks (horizontal) and the other way is along the longitudinal direction to make the wool into suitable strips for both sides. In the version of the horizonal and vertical arrangement, the wool is uniformly tiled. After paving, it can be covered with a net cloth before entering the scrubbing stage. When scrubbing the cloth, it needs to cross the network. Scrubbing directly will make the wool wrinkle easily. After a period of time, you can only pull a small amount of wool by hand. The blanket is then over and it is possible to make adjustments according to the designer's own taste. (as shown in Fig.2 and 3)

2.1.2 Needle Felt

In addition to the wet felt, it is also possible to use a special wool needle. The wool can be stabbed so that the scaled tangle with each other in order to creating the effect of felting. This practice is called "needle felt".

Needle felt is less recorded in literature as the ancient Chinese generally used stones, sticks and other auxiliaries to roll the wool into felt. They do not use a specific pin that is more mature, but the principle is the same. The idea is to increase friction to extrude the fiber into felt. Modern needle felt can be used on a large-scale processing method in the industry. Industrial factories makes the felt into all kinds of daily necessities. The problem however is that the signs of wool needle felt is obvious but it has a significant advantage in mass production, low cost and short processing time. Different from the needle felting in the factory, the manual method can form anhydrous felting wool or other wool composition. This kind of handwork has the same principle with the industry felt which uses a specific pin to vertically stab the wool. The main advantage is it can be easily modified. It is possible to design during the making process, the thickness and local shape of the felt is not limited and can be changed which opens up the creativity of the designers [9].

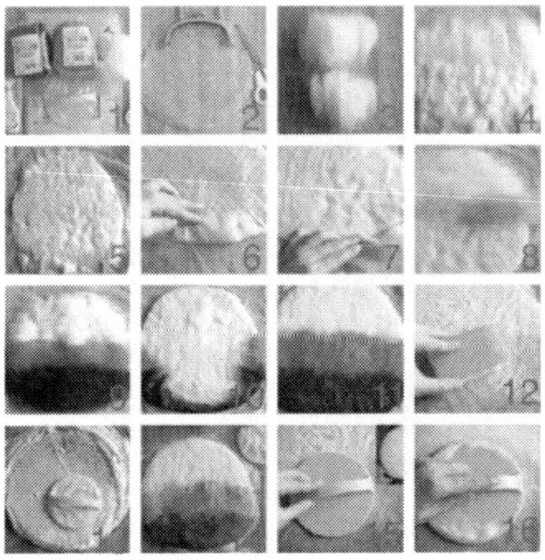

Fig.2 Making process of wet felt

Fig.3 Wet felt works

3. Use the Textile of the Plus Materials to Create a New Clothing of Wool Felt

The first impression of wool felt is that it is white, rough and warm. The plasticity of wool gives designers more space to design. The texture effect should enhance the performance of tension as it is full of fabric layers. When creating the texture wool for clothing, in combining different acupuncture of back, pull and twist etc., it can be carefully carved in the fabric.

3.1 Expressing Method of the Three-dimensional Texture of Felt

Secondary printing and dyeing: To achieve the wanted design results when secondary printing and dyeing, the relationship between wool, garment and patterns must be grasped in order to enrich the decorative beauty of garment fabrics.

Stacking method: A single element or several elements can be overlapped and mutually penetrated to form levels and to achieve a three-dimensional effect based on direct deposition. By repeating the accumulation volume full volume can be achieved. The method is generally applied to felt fabric composite when combined with another wool fabric with needle felt.

Sewing method: It is the organic combination of wool fabric or felt fabric according to the design to create a certain visual affect. The wool fabric collected can be collated according to the inspiration source. The seemingly discordant material and fabric structure can be mixed in a simple profile with the needles. The proper thickness of the needles should be left between inlaying edge of the fabric felt as to achieve the overall rhythmic handmade taste and rich texture. [3].

Weaving: After long wool fibers are entwisted into ropes, they can be directly weaved to keep the wool fiber fluffy. The decorative position of garment is chosen, and the bottom cloth and base material of knots can be completely felt with a needle. After the position is fixed, the needle knot is poked out per share using fou-stare-blankly star sticking. The fiber becomes a stronger form of weaving contour and the hard softness contrast between the felted and none-felted becomes greater. Another method is to weave the long strips after stab felting. The effect is stiffer, tighter and more sculptural, which is suitable for cuffs of decorative clothes [11].

Stereoscopic needling: Stereoscopic needling is type of needle felt. After shaping the wool strip, it is repeatedly stabbed with a special pin. Tools of this process are very simple: a pin, a work bench and the chosen wool. The most critical part is the wool used for for handicraft, which affects the quality of the wool.

Folding: In order to achieve an integrated stitch through felting techniques. The felting fabric is folded in according to the design. The needle is then used to felt together the material at the folds. The needling method is suitable for thin materials. The shaping structure of the fabric can be manually made, such as accordion-fold, syntropy-fold, box-fold etc. Moreover, composite fabric with pleats effect can be made by use of felting properties and friction of wool material. (as shown in Fig.5)

Fig.4 Nail bead design Fig.5 The fold design

Set-folding: The felted fabric is soft, stiff and smooth For instance, this can be achieved through and creating waving lines at the upper body to reflect the felt feeling. The overlap of the needle felt can not only fully highlight the human body curve, it can also enhance the fabric sculpture. With the help of the design concept on sculpture knitting, the felt fabric can be overlapped to add to the feeling of depth and mystery. The cut at the socket can be closed with a needle felt to make the entire garment whole. The method is inspired from garment culture of the early civilization. For the autumn and winter garment, it is can be used for decoration, warmth or safety. [5].

Beading: Bead decoration is a common design in needles fabric. In addition to stapling and sticking, stabling bead can also be used. Before pricking the bottom cloth at early stage, some incompletely punctured thread can be left. The thread can be lead through the bead hole with a needle, and then fixed to the bottom cloth. The ultimate effect of beads is a floating look in the felts. Repeatedly the bead is partly hidden and partly visible with abundant changing (as shown in Fig.4).

3.2 Comprehensive Application of Felt Technique and Other Common Decorative Methods

3.2.1 Comprehensive Application of Felt and Embroidery

Embroidery is a common method that could be done by-hand embroidery, machine embroidery and computation embroidery. As the most common handicraft, it is divided into single-color embroidery, multi-color embroidery, quilting embroidery and cross stitch etc. To achieve large-scale production of garments, most of the enterprises apply the computation embroidery method. This can be achieved by adding embroidery thread on a finished cloth piece so that the fabric is more delicate than before. Besides the embroidery techniques developed on cotton, linen, silk, denim and other traditional fabrics, designers can also apply the delicate decoration method to coarse fabric. Since needle process felting is slightly similar to the wool needle felt process in fabric design and innovative embroidery process, the method is simpler and more suitable for wool fabric reconstruction with specific patterns formed by lines and fibers [4-8]. The embroidery threads used in traditional ethnic embroidery are mostly camel hair, yak line, and even line made from the tendons of an animal, whichdifferent from general ones. They are most natural and the materials coming directly from the living environment. There however is also much more to needle felt design in modern fabrics. Silk color has become more abundant. Embroidery in felt is sometimes visually similar to "quilting" because of the material feeling itself. Because the thickness thread of felt needs be more elastic and solid, every time there is a stitch, a quilted indentation has to be made. In the process, it can be used with the needle so that the embroidery line does not bulge easily. Traditional needle felt embroidery stitch is 0.5 cm, and about three needles per cm, twisted in the shape of S or Z.

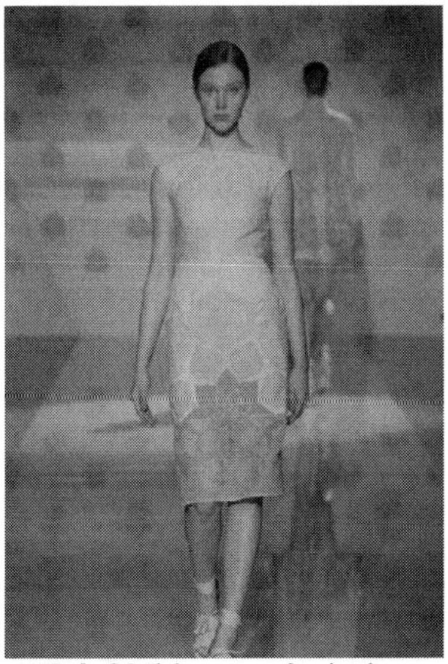

Fig.6 Rahul Mishra award-winning work

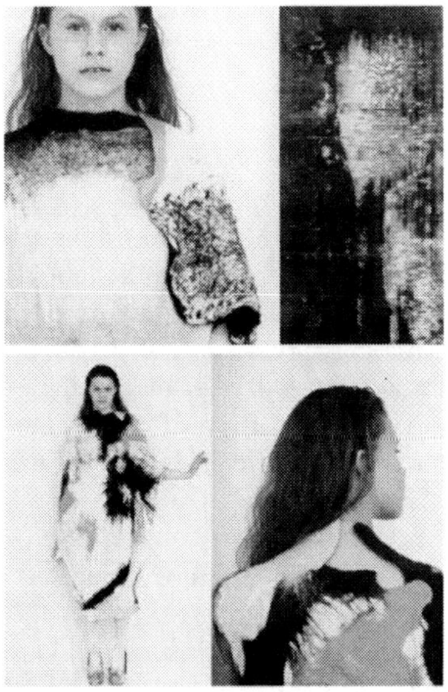

Fig.7 Anita Hirlekar design works

Modern embroidery with CAD, CAM technology are gradually becoming more mature. The production efficiency can be greatly improved when combined with modern machinery of needle felting process. Winner of the 2013 International Wool logo Grand Prix, is Rahul Mishra from India [12]. His series are very feminine made with Indian hand embroidery. Frida Giannini, a Gucci design director and one of the explained the reasons as follows: "We were conquered by the rich embroidery art, the integrity and consistency of the series as well as the designer's passion." Through the unexpected divert of the the diversity of wool fibers, he creatively mixed the embroidery with wool products. The perfect combination of wool and embroidery is itself a work of art. (as shown in Fig.6).

3.2.2 Comprehensive Application of Needle Felting and Wet Felting

Needle felting is generally used when repairing wet felt, which is a common operation method in compliance with the rules for new-learners. In garment design, the processing method for both of them is mostly used in the details. After the wet felt is dried, the details of the needle felt is modfied. Most of the stabbing parts are small. The density relations is compared [5]. The application of combining needle felting and wet felting in garment fabric creates different levels of wool felting to form a new and concave-convex texture with contrast and thickness. The process is similar to the anti-staining process. Tie up where can be without hot water shrinking according to the design to form a regular special texture effect. Usually felt garment can be made by seamless wet felt with creation and imagination. Since the felt is not regular, the effect will be asymmetrical. It is not as neat as those in the industry line, but more arbitrary, especially unmodified edges which is relatively coarse and irregular. When the designer Anita Hirleka produced seamless garment, he splashed paints on different colors of wool. Different textures of felt materials are made because of the different wool shrinking. (as shown in Fig.7)

3.2.3 The Integrated Application of Wool Felt and Needle Felt

The needle felt technique has a positive significance on the application of knitted fabrics. Through certain design methods and techniques, it can be used to resolve defects in fabrics. When dealing with knitted edge, usually a single knit fabric is relatively serious, while a double knitted fabric does not curl. Single knit fabric will show when the package volume status is in a loose state. It has a relationship with the twist of yarn, knitted product itself structure, and the coil density, the main physical reason for producing fabric warp turn to bending deformation from straight, is because that the bend fiber stress will lead the tension and make the edge curling. Curling status, of course, also can be used as one of the design means for personality designer [6]. After the formation of wool fabric and strong toughness, the woven fabric effect becomes very noticeable.

3.2.4 The Integrated Application of Wool Felt and Hollow Out

For hollow decoration, laser hollow is now used more often in addition to manual cutting, the computer typesetting is set in advance and then an automated controlled laser is used to punch in all kinds of fabric. The pattern shape is not restricted and the decorative effect is very strong. The mat will be made from the things cut down from the broken material and then stamped on the fabric to form a hollow d superposition with a more convex stereo effect. The most apparent advantage compared to other materials is, good stability. In the field of jewelry design, there have been many modern style material products. For instance the common felt bag. Except that the hollow also has special edges and it is irregular because of the gunpowder or chemical corrosion, overlay or other reconstruction in the foundation. The composition of wool fiber itself varied according to the thickness of the fiber, such as the selective puncture of the fabric. This is so that the degree of felting is different, and it will form a contrast between the soft and the hard.

3.2.5 The Integrated Application of Wool Felt and Dyeing

Wool is a strong absorbent, it can absorb all kinds of dyes and does not easily fade. The color of the wool products is one of the key advantages in terms of aesthetics. Although there are naturally colored wools, most of them go through a wool dyeing, bleaching and printing process. With the surface of wool scales on different degrees, felt have subtle color change and a halo effect. The gradient can enrich the picture elements, the sense of hierarchy and sense of rhythm. Because wool fiber shrinks in hot water, the dyeing temperature settings is programmed at 30 to 50 degrees Celsius in the mixture. In the full cycle of mordant, the mixture is stirred constantly into dye. The temperature then slowly increases for a minute until the temperature reaches to 90 to 100 degrees Celsius. The fabric in this temperature range will disseminate for at least half an hour.

The dyeing stage can be carried out in three stages, fiber, yarn and cloth. The dyeing effect forms in different stages are different. The Belgian designer Christian Wijnants launched a series of designed wool in 2016, the entire series is tries to maintain the natural texture of the original yarn. It is a simple design concept. In the infinite possibilities of fabric proofing process the designer insists in using the same yarn, and in doing so creates a seamless mosaic Wu node type. The series is inspired by the colors in crystal and ore To prevent the clothes from shrinking, the designers use a unique hand dyeing technology and shadow effect. Each piece of clothing uses the latest method of printing, dyeing and yarn dyeing reweaving after disassembly to create an irregular Dyeing effect.

First dyeing the wool before creating the fabric and creating the fabric before the dyeing process results in a different effect that designers should try more often. In addition to cloth dyeing, dyeing and printing on fabric, it is also common to combine manual tie-dye dyeing. Felt anti-pollution technology manual has a long history. Dyeing technology uses cotton, silk and other materials in Nepal and other minority areas. They still use the technology for dyeing carpet products which is a hand anti pollution technology rarely used in blankets. [10]. Dyeing means is commonly using on the general felt clothing garments, wool clothing dyeing means fabric dyeing wool dyeing is finished in a rotating drum with "mixing", this kind of machine operates softly and the clothing cycle in the dye repetition, such forces can also avoid the wool felting. In addition to the direct dip, dipdye, tie dye, dye injection etc, it also enables the initial dyeing of the wool to be more natural, subtle and have rich changes. [2].

4. The Difference between Needle Felt and Wet Felt

The traditional wool felt process is divided into wet felt method, needle felt method and wet felt needle felt. Wet blanket method is mainly used in the manufacture of large works, rapid prototyping and on the surface of wool works presents smooth compact advantages, disadvantages in the felting process in different materials and colors of the wool can shift easily and make it difficult to achieve the desired designs and patterns, have a short board in the shape of thin section in the works. The surface of needle felt is exquisite, and it is more suitable for the molding of fine details than wet felt. It can be used to make solid patterns and shapes. The combination of wet felt and needle felt will results in a wool felt that has overall and non-local features. The production process however is relatively complicated. People can choose relative methods through their own needs.

5. Sustainable Significance of Wool felt Fabric Design

For decorative fabric, the designer is no longer confined to traditional fabric. Industrial materials and materials in life can be fused according to the designer clothing style after special treatment. From the perspective of the felt process, it is common to find that most designers uses the fabric "No order" recombinant techniques, They use the same or different fabric in certain way to combine the method before changing the fabric appearance. The combination rules are about the same, two different types of contrast, according to a certain law or deliberately irregularly stitching, or other similar places emphasis [8]. The restructuring of the fabric is more interesting for decorative fabrics. The fabric in the modern aesthetic effect, from design concept to performance, the method has a great change, the present technology continues to improve, the recovery effect of handicraft products and markets and also gave the fabric more space to breakthrough, artists, designers are diligently to explore new materials, new technology.

Holland fashion designer Iris Van Herpen is good at using 3D printing and other high-tech technology for the production of advanced custom clothing fabrics (as shown in Fig.10). He designed garments in 2013 to simulate snow [14-15], rime, stalagmite etc, with the help of 3D printing. The material expresses the natural elements accurately and is a major breakthrough on traditional clothing and fabric design. when this kind of clothes is not really popular for customers, the self assembling lab Skylar Tibbits in the MIT shared the concept of 4D printing in 2013 TED speech [13],the fourth dimensions is time, this technology will be a brief program embedded material, and can make the matcrial automatic deformed, and then manipulate the printed parts automatic assembly. The researchers introduced the 4D printing technology and explained how it can be applied for the low efficiency of energy consumption (as shown in Fig.11) and wholescale manufacturing. They also explained how the automatic assembly technology will lead to the revolution of the Royal industries. The master's graduation works of Li Xiao College of art and design focuses on coarse knitting, she joined the silica in

Fig.10 Iris van herpen 3D printing design

Fig.11 Skylar Tibbits the concept of 4D printing

6. Conclusion

From the analysis above, a conclusion can be drawn that the traditional technology and application of highly updated technology has enriched the fabric organization form, broadened the space reconstruction performance of fabric, and created new ideas of fabric reconstruction. Using the traditional handicraft to innovate design has been a popular application in the fashion design for a long time. Wool is an old manual for nonwoven textile, and the main reason why it can be popular in the form of creative handiwork is because the operation process for felt is simple. This paper is based on a great deal of practice and works of designers at home and abroad. It analyzes the feasibility of the felt in fabric design based on the characteristics of felting and develops more methods for decorative fabric design. In terms of the special color mixing effect of wool fiber, different effect for felt, experimental materials found in the process, we can find the potential in needle felt artistic expression. The development of technology has broadened the designer's vision, become an inspiration for new ideas, and promoted modern fabric art reconstruction in a broader context. The conclusion of this paper can provide the reference for clothing design theory and a guidance for the practical application of fabric reconstruction.

References

[1] Ding S, Li JJ, Xu SL. Analysis of wool felt [J]:industrial textiles: 2014; 02.
[2] WOOLMARK on show at Spin Expo[J]: China Textile: 2012; 09: 52-53.
[3] Zhang L. knitted wool fabric the waste felt [D]: Qiingdao University: 2012.
[4] Hao SJ. Traditional art felt products in Inner Mongolia area and its modern design [D]: Jiangnan University: 2013.
[5] Dai JY. Wool in the clothing design innovative use of [J]: art education research: 2014; 3: 90-91.
[6] Zhang JJ. The application of process combination dyeing and wool in knitting in costume design:Beijing Institute Of Fashion Technology: 2015.
[7] Yuan F. Decorative texture design in sweater design function [J]: Xi'an Polytechnic University: 2013; 27 (03): 330-334.
[8] Xin FF, Li H. Design and application of felt material in the field of apparel [C]: Proceedings of the International Conference on art and creative industries: 2012.
[9] Mornu N.500 felt obiects: contemporary explorations of a remarkable material [J]. Lark Crafts: 2011.
[10] Lachman M. felting vintage textiles [J].Quilting Arts Magazine: 2013.
[11] ZhangR, MaXC. HandmadeCharm-Design and manufacture of wool felt products [J]: Beauty and times: 2015 (06)
[12] Information on http://sh.sina.com.cn/fashion/qt/2014-03-06/135184567.html
[13] Information on https://www.ted.com/talks/skylar_tibbits_the_emergence_of_4d_printing
[14] Information on http://www.irisvanherpen.com/
[15] Information on https://www.dezeen.com/tag/iris-van-herpen/

Analysis of Fashion Color in the Tang Dynasty from "Legend of Wu Meiniang"

Xin-Lei Sun

The College of Fine Arts, Chongqing Normal University, Chongqing, 401331

*Corresponding author's email: 812587807@qq.com

Abstract

In Tang Dynasty, much of the textile arts were developed, the color design of the clothes was unique and had a very high aesthetic value. The costumes from the dramas "Legend of Wu Meiniang" has brought an unprecedented visual impact which has caused people's interest to develop on the clothing from the Tang dynasty. In this article, clothes from the Tang Dynasty are taken as an example to study fashion color design of the Tang Dynasty era as well as the "Legend of Wu Meiniang". Through literature investigation, this paper will conduct an in-depth analysis on the characteristics of costume color design of the Tang Dynasty. The intention is to provide a reference for costume piece dress design, and to also play a guiding role on modern fashion color design.

Keywords: Clothing; Color Design; Costume Dramas; The Tang Dynasty; Legend of Wu Meiniang

Color plays a pivotal role on the success of a historical drama [1]. Clothing color is the soul of the costume drama. Therefore the character clothing color performance has a crucial role in the success or failure of many costume dramas. Costume dramas such as the "Legend of Wu Meiniang" is a reduction of prosperity in the Tang Dynasty. There are 260 sets of clothes for the Empress Wu Zetian alone [2]. As the story unfolds, the situation and the status of the characters change unpredictably, as such the costumes that people wear change as well. Adjusting to the appropriate clothing color design based on the situation of the characters is key to a costume drama's success. "Legend of Wu Meiniang" successfully designed more than 3,000 sets of clothes by grasping the characteristics of the role of the characters [2]. Predecessors have conducted related research on designing stage costume, however there is no systematic analysis of the Tang Dynasty clothing color design that is combined with the "Legend of Wu Meiniang". This article will analyze Tang Dynasty clothing color design features by the "legend of Wu Meiniang" in detail.

1. Clothing Color Design in Plays

When fashion designers are given a scenario for a design task, they must first understand the director's intentions and projects so that their clothing designs do not differ from the screenplay. Secondly, they must continue to research on design features of a historical period and the fashion in respect of that particular historical period. Lastly, there must be an in-depth study of the script and the clothes must be designed according to the cultural connotation of the times and the role of the character characteristics, in order to reflect the identity of the characters as well as the situational context that creates the plot atmosphere and enrich the role of the storyline [3].

The three elements of costume design are clothing styles, fabrics and colors [4]. Due to the particularity of costume dramas screen visual effects, clothing color has become the most important aspect about costume in drama fashion design, which not only helps with costumes task, it also expresses the ideas of costume dramas including the feeling of the roles and scenarios.

1.1 Establish the Facts On The Basis of History and Disseminating Historical Culture

Different historical periods have different characteristics of clothing color design. The color concept in Tang Dynasty from conception to development which entails the time, history and culture of the state and the religion are inseparable. For example, red is often used by dignitaries and on festive occasions in the Tang Dynasty, which symbolizes festivity, elegance, warmth, vibrance and so on [5]. On the legitimization ceremony of princess in "Legend of Wu Meiniang", she wore red and white to look dignified and generous, festive and luxurious, as shown in Fig.1; In the coronation for the queen, the color design are red and gold as shown in Fig.2, which fits into the historical and cultural background. In the ancient world, the use of color was quite rigorous, and princesses was not allowed to wear golden costume in the cultural context of the Tang Dynasty. That is guilty and subordinate [6]. The audience can enjoy the different backgrounds of the characters through the screen costumes to understand the different historical and cultural status of the use of color characteristics. In short, the clothing color design must be based on historical facts. At the same time, in order to enhance the visual effect of the historical drama, specific designs can be exaggerated. However it must not distort historical facts in order to avoid giving the audience the wrong idea of indoctrination and which may play a negative impact on the drama. The audiences enjoy both the plot as well as the visuals as well as learning more about the historical and cultural characteristics of the different eras.

Fig.1 Princess clothing [7]

Fig.2 Queen clothing [7]

1.2 The Figures Reflect The Identity Of The Characters And The Situation

Fig.3 Emperor clothing [7]

In costume dramas, clothing color design varies as well as the identity of the characters along with the development of the plot. This is especially true for costume design. Due to the historical and cultural tradition, the color of clothing design should follow the identity of each dynasty level [8]. For instance the color yellow. In the Tang Dynasty it was used as a respect for buddhism which is thought to expel the forces of evil. It also became the Tang Dynasty emperor's special color with a unique grade composition to signify a certain hierarchy. Only the emperor can wear yellow robes and anyone else who wears it will be seen as guilty of conduct [9]. The emperor in the costume drama "legend of the Wu Meiniang" used yellow robes as Wu Meiniang's dress, which is a symbol of identity and status, as well as the sign of the emperor as shown in Fig.3. As such the color of clothing not only reflects the identities of the characters, it also reflects the situational development of the characters.

1.3 Rendering the Story Atmosphere in Context of the Rich Plot

Film and television works can set off the screen characters mood with color to create the plot atmosphere in order to enrich the story [10]. With the help of character costume and scene color, costume dramas jointly present to the audience with a specific visual environment and atmosphere. As such, the color designs occupies an extremely important role in costume dramas, which not only sets people's mood but also creates the whole plot atmosphere so that the story of the plot is better displayed. The dress shape design of Wu Meiniang during the fete ceremony on the " Legend of Wu Meiniang " is ingenuine. The color design is exquisite. Black is used in the background to echo the seriousness of the fete ceremony serious as well as Wu Meiniang's mood of being "aware of the tiger undeterred". The golden dragon embroidery patterns is a symbol of Wu Meiniang's identity and status; and the peacock blue colors of the feathers are used to show the nobility and dignity of Wu Meiniang., The charm is designed so that it does not block the head and the cuffs design is green, bold and creative. On the one hand, it gives a sufficient vjisual impact. On the other hand, it models Wu Meiniang's mind and majestic wisdom. The woman's calm during the murder case renders the visual effects. This creates the story atmosphere and enriches the plot, making it more vivid and attractive as shown in Fig.4.

Fig.4 Fete ceremony clothing [7]

2. Typical Clothing Color Design In "Legend of Wu Meiniang"

As the plot unfolds, the situation of the character and the status of their identity varies in an unpredictable manner, therefore the characters clothes will naturally change. In order to adapt to the situation context of the characters the corresponding clothing color design is key for the success of the costume drama [11].

2.1 Clothing Color Design when Entering the Palace

Wu Meiniang wore an outfit that is pale pink and fruit green when she entered the palace. It was a long yarn which is low key, gentle, serene and soft to reflect a humble status, novelty and the purity of the moment as shown in Fig.5. Pink in gouache, it is composed of red and white colors, whose contrasting colors are blue and bright green [12]. This costume design simply uses the contrast of colors. The clothing is innovative and at the same time suits the identity of the characters as well as the mood of the moment. Pink, an unsaturated red is a symbol of innocence and is a symbol for gentleness and sweetness [13]. As Wu Meiniang's first time to enter the palace, her future is unpredictable. She is only a young girl who is curious, pure and sweet. The design of this clothing suits perfectly with the context of the plot.

Fig.5 clothing color design when first coming to the palace [7]

2.2 Clothing Color Design on the Polo Stadium

When Tang Dynasty was held in marseille, Wu Meiniang wore a purple dress when she rode. She had an irresistible charm as shown in Fig.6. Since ancient times, the color purple represents the emperor and is a symbol of mystery, beauty, elegance and femininity [14]. From lavender to deep purple, purple is worn by many women. Purple can be mixed with white to create a young and energetic atmosphere. This clothing color design is innovative and creative, and portrays Wu Meiniang to be skillful, brave, calm, beautiful and mysterious. The name of "Wu Meiniang" is bestowed from here. Purple color clothing design is also set off by Wu Meiniang at this time by the virtue of their strength to come to the fore.

Fig.6 clothing color design on the polo stadium [7]

2.3 Clothing Color Design on Queen Ceremony

"Red" as the first of seven colors, not only has the physical attributes of nature, it also embodies a rich culture more so than the rest [15]. With a history of five thousand yeras and through the history of precipitation, the red color in China has become a distinct color in the culture [16]. During the coronation ceremony for the queen in "Legend of Wu Meiniang", red is chosen as the background color, with a golden yellow embroidery pattern design. The look is dignified and generous, festive and luxurious, as shown in Fig.7. This further demonstrates Wu Meiniang's status and her identity and nobility. It reflects the peaceful and festive atmosphere which aligns with the development of the story. The ceremonial clothing is also unique in design. With the use of pteris design and the exaggeration of the shoulder, the queen is portrayed is domineering with the status of a mother of the Miriam world. Pteris shape on the diagonal, which is the character position at the moment rising passions and desires thoroughly. In this context, red not only symbolizes blessings and aspiration, it also fully embodies the aesthetic fashion of the Tang Dynasty.

Fig.7 clothing color design on the queen ceremony [7]

2.4 Clothing Color Design on the Emperor Ceremony

Throughout the development of the story there is always a close relationship between the color concept of the Tang Dynasty and the religion. n the chromatogram, the brightness color is yellow. It is the color that represents the life of the sun and embodies the sacredness of Buddhism. Buddhist respects the color yellow and believe it to have the power to expel evil [17]. In "Legend of Wu Meiniang" Wu Meiniang's emperor costumes are yellow robes, which is a symbol of her identity and status as shown in Fig.8. This clothing is designed to not only conform to the historical and cultural features, it also emphasizes Wu Zetian's status and emotions.

Fig.8 clothing color design on the emperor ceremony [7]

3. Conclusion

Regardless of any era, the choice of color is not accidental,, nor is the choice the result of only one factor alone. Whether the color is popular or not depends on many factors such as the political, economic, cultural and religious situation of the time [18]. Therefore, there are profound reasons to study the Tang Dynasty clothing color and color concept.

Tang Dynasty clothing color design are generally based on original colors that are colorful and bright, rich and highly decorated [19]. The Tang Dynasty is the most prosperous dynasty in ancient china. In this era, there were more frequent foreign exchanges which has an influence on Tang Dynasty costume design. The color design in the Tang Dynasty embodies the entire country's essence. "Legend of Wu Meiniang" is a reduction of prosperity in the Tang Dynasty. The careful design of the design of the clothes has a huge impact that not only enhances the effect of the artworks, it also provides a reference for the drama costume design. It also has a direct impact on modern clothing color design through subtle means.

The traditional costume color of Tang Dynasty has a unique artistic charm and a heavy cultural background, which is the source of modern costume design innovation. Now in the process of absorbing the traditional costume culture, how to choose the dress element to reflect the essence and verve of culture in the Tang Dynasty needs to be considered. It is important to absorb and draw lessons from the Tang Dynasty clothing color culture, and to integrate it for more contemporary means. "Fuse tradition with modernity, endow tradition with a new meaning" This quote shows the importance of combining traditional culture to suit modern aesthetics in order to introduce a new vitality to design.

References

[1] Liu EY. Color science and film and television art [M]. Beijing: Beijing Broadcasting Institute Press: 2002.

[2] http://image.so.com/i?q=武媚娘传奇&src=tab_www.

[3] Lv ZQ. The shaping role of fashion design to the film and television characters [J]. Film evaluation: 2007; (14).

[4] Huang YQ. Clothing colorology [M]. Beijing: China Textile Publishing House: 2010.

[5] Shen CW. Chinese dress history. Xi 'an: shaanxi normal university press: 2004.

[6] Zhang FR. On the cultural meaning of "red" and its application in the design of clothing color [J]. Cross-strait traditional color and modern application of academic seminar: 2010.

[7] Liu ZJ. On the "Chinese Red" in the use of modern art design [M]. "Capital Normal University": 2012.

[8] http://image.so.com/i?src=360pic_strong&z=1&cmg=67b04c474e49aba0a92ad5bd7fb60980&q=武媚娘传奇服装

[9] Cui PX. Analysis of clothing color and color concepts in Tang Dynasty [J]. Liaoning Silk: 2006.

[10] Shen CW. Chinese clothing history [M]. Xi'an: Shaanxi Normal University Press: 2004.

[11] Yuan. Chinese clothing history. 1 edition. Beijing: China textile press: 2005.

[12] Cheng YZ. "Color superstition in Chinese culture" [J]. Journal of Chinese culture: 1992; (4).

[13] Zhao QW. The analysis of the rheology of Chinese ancient Chinese studies [J]. Journal of hubei university: 1997; (1).

[14] Na CY. The color view of the official quality and the color of the tang. [J]. Journal of the university of ningxia: 1996.

[15] Du JM. The hierarchy of ancient colors [M].Knowledge of literary history: 1993; (11).

[16] Huang NF, Li DQ. the history of clothing in China and foreign countries. Hubei fine arts press: 2002.

[17] Hu M. Dress folklore [M]. China textile press: 2004; (131).

[18] Zhang ZC. Chinese apparel culture (first volume) [M]. China textile press: 2001.

[19] Tian W. The design of clothing in tang dynasty and its cultural connotation [J]. Costume culture and design: 2007; (3).

Analysis of Manchu Concubine's Accessories in the Early Qing Dynasty

Yun-Tian Liu[1], Jian Li[2*]

Beijing Institute of Fashion Technology, No.2 Yinghua Road, Chaoyang District, Beijing, P.R. China

[*]Corresponding author's email: jean.xiaoyu@163.com

Abstract

Qing dynasty was the last feudal imperial regime in Chinese history, and it was a dynasty in which the Han traditional culture and customs were blended with the Manchu. In this time, the political situation was stable and the economy started to flourish. Thus, the apparels and accessories got highly developed. The accessories that concubines wore were made using the highest techniques, best materials and the most fashionable designs. In early Qing dynasty, Manchu concubines mostly used accessories to ornament their crests, foreheads, ears, necks and hands. Thus, crowns, frontlets, earrings, necklaces and hand jewelry were highly developed. This essay will take the concubines' accessories as the beginning point and use literature research, ancient books research and museum visits as researching methods to briefly analyze the accessories and their characteristics of the early Qing dynasty. The purpose of this essay is to acquaint the readers about the characteristics of early Qing accessories and to make reference to modern accessory design and film costume accessory design.

Keywords: Early Qing Dynasty; Manchu; Concubines; Jewelry; Accessories

Introduction

After Manchu conquered Shanhai Pass, they enacted a dressing policy of shaving hair and changing clothes against the Han Chinese. Although this policy was not extremely strict, it still greatly influenced the traditional Han costumes. Even though Manchu governors claimed that Han and Manchu were considered equal, in reality, Manchu were considered to be nobler. During this time, Manchu harems were one of the noblest people who had the right to wear the most expensive apparel with leading technology and colorful materials. Therefore, this research chose the accessories made for Manchu harems as study objects when exploring accessories of Qing dynasty. The most common jewelry in Qing dynasty were stringing beads, jade accessories and gold or silver jewelry. The technique of inlaying gemstones onto gold and silver started in Yuan dynasty, and in Ming dynasty, gold and silver jewelry had equal importance compared to gemstones' jewelry. But in Qing dynasty, jewelry with inlaid gemstones started to take the leading position [1]. Craftsmen and artists in Qing dynasty often used complicated processes and bright colors to emphasize the designs of the jewelry [2]. In Ming dynasty, people preferred ruby, sapphire, emerald, turquoise and inlaid pearls, but in comparison, Manchu preferred jade, coral, amber, lapis lazuli, etc. [3].

This essay will analyze accessories of early Qing Manchu in several areas, such as crests accessories, frontlets, ear jewelry, neck jewelry and hand jewelry.

1. Crests Accessories

In ancient times, people used to call the ornaments that fastened on the head as "head cloths". There were four sub-categories in crest accessories, which were crowns, tiaras, Bians (a man's cap used in ancient times) and Zes (ancient coif). Crests originally referred to the noble crest accessories, but later, it also included all the head accessories. In early Qing, the crests of Manchu concubines mainly included court crowns, auspicious crowns, hairpins, etc.

According to the official records from Qing, everyone among the royals except odalisques was required to wear court crowns and gowns when grand ceremonies were held. They also needed to wear auspicious crowns and gowns when celebrating jubilant festivals. For daily wearing, they could wear informal crowns and gowns [4]. Respectively, introductions will be given below.

1.1 Court Crowns

The court crowns were the most serious and the most valuable female crest accessories in Qing; they were only used in grand ceremonies. In Qing dynasty, the crowns were divided into two kinds: winter crowns and summer crowns. According to the records from <Regius Qing Collected Statutes Regulations>, the winter court crowns of the queen were made of martens, and the summer crowns were made of grey cashmere. The court crown shown in Fig.1 is a summer crown for the queen. The top of the crown was covered with Zhuwei, which is a filiform ornament for crowns that only appeared in Qing dynasty. The very top part of the crown is called Hualing, which had three layers of golden phoenixes with Tanas (freshwater pearls from northeastern China) in between. Seven phoenix-shaped ornaments were inlaid on the Zhuwei, and every phoenix body was inlaid with nine Tanas, one cat's eye and twenty-one pearls. On the back of the crown, there was a golden Di, which was inlaid with cat's eye and pearls, and the tail of the Di was made of pearl strings which contained three hundred and two pearls total. The collar protector was placed below the golden Di, and there were two bright yellow silk ribbons hanging on each side of the protector. Finally, an azure satin chinstrap was attached to both side of the crown.

The court crowns of concubines were mostly similar to the one mentioned above; meanwhile, court crowns would gradually decrease the materials and the gemstone quantities according to the ranking of the concubines. For example, the imperial concubines of the emperor could put one hundred and ninety-two pearls on their court crowns; low-status concubines had only two layers of Hualings, the golden phoenixes were changed to golden peahens, etc.

Fig.1 Summer Court Crown of the Queen

1.2 Auspicious Crowns

Fig.2.1 Auspicious Crown for the Queen

Fig.2.2 Auspicious Crown for Imperial Noble Consorts

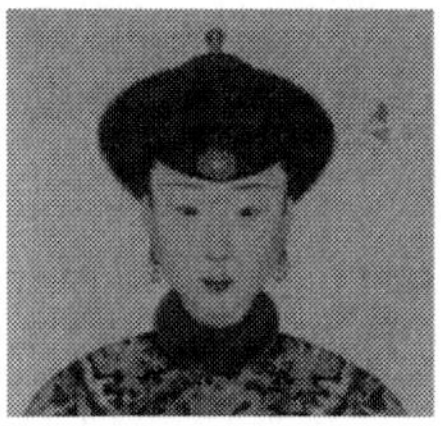

Fig.2.3 Auspicious Crown for Highest-ranking Imperial Concubines

Auspicious crowns were mainly used in officer's official wear in the Qing dynasty, which had to be worn when wearing auspicious gowns. The auspicious crowns for the empress dowager and the queen in Qing were made of martens, on which Tanas were inlaid and Zhuwei was interspersed. The same materials were used for auspicious crowns for imperial noble consorts and highest-ranking imperial concubines [4]. Fig.2.1, Fig.2.2 and Fig.2.3 show the winter auspicious crowns for the queen, the imperial noble consorts and the highest-ranking imperial concubines; the different levels for owner's status could be easily indicated by the differences of peacock feathers of the crowns. According to the different rankings of the concubines, the materials and the ornamenting jewelry gradually decrease. The materials for Hualings from higher classes to lower classes were as follow: corals, ornamental engraving corals, sapphires, lapis lazuli, crystals, tridacnas, gold, etc.

1.3 Tianzis

Tianzis were a kind of ornaments which covered female's hair; their inner tubes were often made with black velvet and satins, and the shells normally used silver wires or copper wires to manufacture with inlaid jade or pearls [5]. The court crowns and auspicious crowns were made for important ceremonies; therefore, they were recorded in official records. But Tianzis never appeared in any known official records. As a result, scholars could only use ancient books to predict which occasions Tinazis could be worn. Based on the information that has been found, Manchu concubines usually wore Tianzis to attend ceremonies. They could also wear Tianzis with auspicious gowns and for daily use. There were three kinds of Tianzis: phoenix-ornamented, fully-ornamented and half-ornamented. Phoenix-ornamented Tianzis were the highest class and were made for the empress dowager, the queen and brides. Ordinary young women usually wore fully-ornamented Tianzis.

Tianzis' decoration is luxuriant. Most Tianzis were inlaid with pearls and gems, so people normally considered Tianzis as a luxury sign, which made Tianzis be used for more occasions and purposes in the mid and late Qing [6]. Tianzi as shown in Fig.3.

Fig.3 Green Tianzi with inlaid pearls

1.4 Bianfangs

In the early Qing, although the up-do styles varied when the Manchu concubines wore the informal crowns, the overall shapes of their hair were the same. The most common hair style in Qing was "Liang-ba Tou", in which Manchu women wrapped their hair around a Bianfang on each end to form a platform-shaped hairstyle [7]. Bianfangs were a rectangular shaped hairstyling tool, as shown in Fig.4. In early Qing, Bianfangs were mainly made from jadeite, mutton fat jade, agalwood, jade, tortoiseshell, etc.; there were also golden Bianfang, which were made from hollowed filigree gold with inlaid gemstones [8].

Fig.4 Filigree Golden Bianfang

1.5 Hairpins & Combs

Combs evolved from hairpins. They served the same purpose and looked very similar. The only difference between them was that combs had two tines while hairpins only had one. In Qing dynasty, Manchu concubines often used Bianfangs to comb their hair and they also used a lot of different types of hairpins, such as bionic hairpins as shown in Fig.5.1, fringes hairpins, Buyaos (big hairpins that have a swaying part attached to the end), etc., to either secure their hair or to decorate their hairstyle. Most of the hairpins were made of precious materials and there were often auspicious patterns carved on them. After the hairstyle was finished, hairpins were often placed at a position where they could be easily seen. Fringes originally referred to the fringes attached to ancient coaches or tents, but later, fringes particularly referred to the hangings attached to hairpins [9], as shown in Fig. 5.2.

In ancient times, phoenix-shaped accessories were used exclusively by the queen and the empress dowager. But in early Qing, pheasant-shaped accessories, which seem very similar to phoenix, were used by concubines [10]. A pheasant is a kind of wild bird that has a long tail, which is very difficult to distinguish from a phoenix. The only way to differentiate pheasant-shaped jewelry from phoenix-shaped jewelry was by the tail. The tails of pheasants always had fringes attached.

Fig.5.1 Bionic Hairpin

Fig.5.2 Fringes Hairpin

2. Frontlets

Manchu concubines also used frontlets in addition to crests accessories. The two main categories of frontlets were Jinyues and Mo'Es. Their main goals were both functional practicability at the very beginning. The function of Jinyues was to secure the hair and the function of Mo'Es was to cover their foreheads from wind. The following part will introduce these two kinds of frontlets respectively.

2.1 Jinyues

As in Fig.6, the main body of the Jinyues were made of gold and they were worn on the forehead before wearing court crowns to keep their hair in place. Jinyues were composed of gold hoops with drooping beads strung behind; their shape looked like a rounded hairpin. In general, people determine grades of Jinyues based on carved-moire gold and quantity of the gems [11]. Jinyues for the queen and the empress dowager had five hanging strings of pearls attached to the back whereas all other concubines had only three strings of pearls.

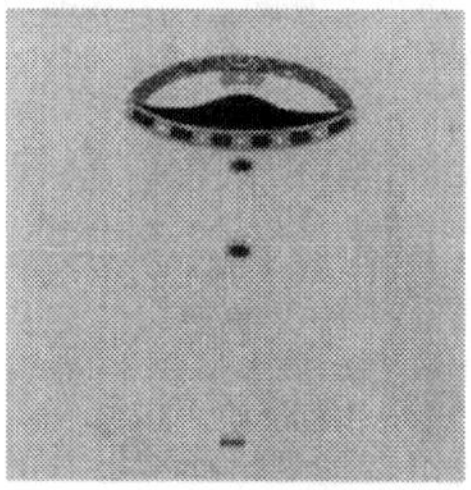

Fig.6 Jinyue

2.2 Mo'Es

Mo'Es as shown in Fig.7, were similar to headbands, hair hoops or eyebrow hoops. It is said that Mo'Es had sedative effects [12]. Mo'Es were originally a kind of forehead ornament for honor guards or dancers, and then they evolved into the headwear of ordinary people. Mo'Es had evolved multifarious names along with the change of dynasties and regions, such as forehead-kerchief and pearls headband. Mo'Es had developed from Ming, and then culminated during Qing. There were a lot of materials could be used to produce Mo'Es. People often used black satins or black brocades in winter and black yarns in summer. Designers could inlay various kinds of jewelry or jades on Mo'Es forming thousands and thousands of Mo'E style.

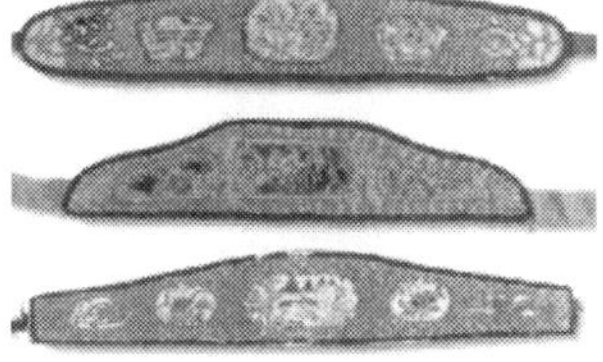
Fig.7 Mo'E

3. Ear Jewelry

Before Qing dynasty, ear jewelry was often worn one piece per ear. However, in Qing dynasty, the number became three. This tradition was passed from Manchu's predecessor, Jurchen. In early and middle Qing dynasty, Manchu nobles usually wore three sets of earrings on each side, but in later Qing dynasty, three per side evolved to one jewelry per side, but each one had three strings attached.

In Qing dynasty, earrings and ear pendants were all called Qian, which meant ear pincers or pincers. Qian referred to those earrings that went through ear holes. The Manchu ear jewelry custom was closely associated with the royal etiquette. During holidays and festivals, complex and elegant jewelry was used; during important ceremonies, golden and Tanas jewelry was used; during funerals, concubines would take off their ear jewelry until the funeral was over. There were rules for different concubine rankings regarding the types of jewelry they could wear. In early Qing, Tanas were the premier gemstone because their place of origin was where Manchus were originally from. Therefore, Tanas were the first choice for concubines' jewelry. Starting from the Tiancong years, the jewelry started to develop from pure gold to inlaid jewelry, and later, Tanas earrings and pearls ear pendants appeared [13].

According to the regulations, ear jewelry for the empress dowager and the queen had two first-class Tanas, imperial noble consorts had two second-class Tanas, consorts had two third-class Tanas, and dames had two forth-class Tanas, as in Fig. 8. Manchu concubines wore three longitudinally aligned sets of ear jewelry on each ear as shown in Fig. 2.1, Fig.2.2, Fig.2.3. Rankings below dame were not allowed to wear Tanas.

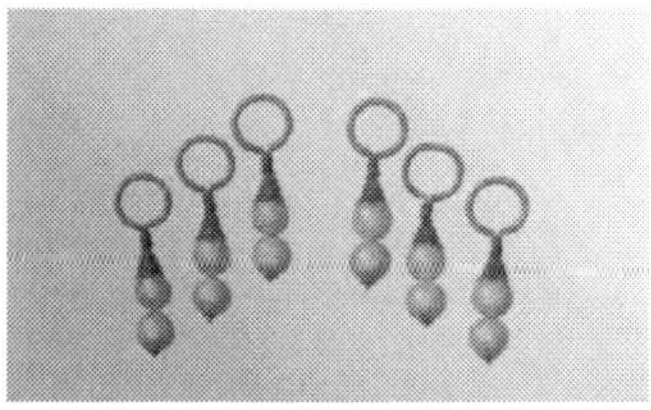
Fig.8 Golden Ear Rings with inlaid Tanas

4. Neck Jewelry

In early Qing dynasty, neck jewelry mainly included beads, necklaces and Lingyues. According to ancient records, concubines had to wear beads and Lingyues on their necks with the proper formal dress during important occasions. Especially during ceremonies, worshipping ancestors or during holidays, beads had to be worn in front of their chests. More information will be given below.

4.1 Beads

The materials of the beads and the color of the ribbons determined the ranking of the concubines. According to the regulations, the queen consort used Tanas, corals and bright yellow ribbons as shown in Fig.9; Imperial noble consorts used ambers, corals and bright yellow ribbons; the materials would downgrade as the ranking decreased

[14]. Beads originated from Buddhism. A set of beads contained one hundred and eight beads and it hung around the neck and drooped in front of the chest. A thin string held the beads together, and the tail string was attached to the back. There were also three sets of small beads attached to the beads; two on one side and one on the other side. Men wore the two sets side to their left and women wore it to the right side.

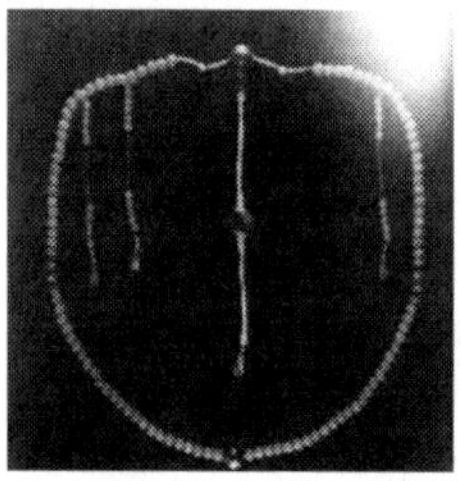

Fig.9 Coral court beads

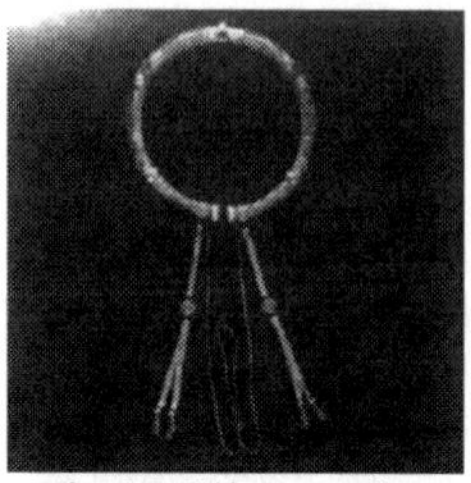

Fig.10 Silver collar with inlaid gemstones

4.2 Lingyues (Collars)

Lingyues were one of the accessories regulated by the Etiquette. When Manchu concubines attended grand ceremonies, they had to obey the rules and dress as required. The complete set of costumes included court dresses, court crowns, Jinyues, three sets of earrings, court beads, Lingyues, etc. The queen and the empress dowager mainly wore Lingyues made of hollowed gold ornamented with eleven Tanas and corals and two bright yellow ribbons on each side, which had corals in the middle and turquoise on the end. Materials varied depending on the concubines ranking, and it was mainly determined by the numbers of Tanas and the color of the ribbons. As in Fig. 10, the collar was made of silver with no Tanas but pearls, which meant this was supposed to be worn by an emperor's concubine [15].

4.3 Necklaces

The heritage collection of Qing dynasty does not have many necklaces. Manchu's neck jewelry was mainly court beads and Lingyues, and concubines from early Qing dynasty did not have the habit of wearing necklaces. Therefore, necklaces were rare, and they were mainly made of gold, silver or beads, and sometimes there was a charm. As shown in Fig.11.

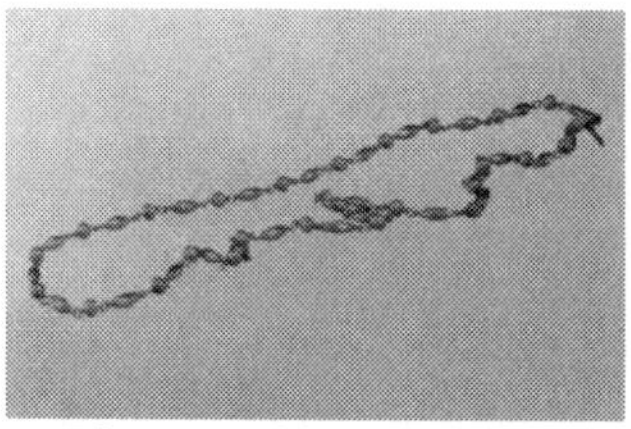

Fig.11 Golden Necklace

5. Hands Jewelry

In Chinese ancient literature, women's hands were often complimented using metaphors of soft weeds, bamboo shoots and spring onions. These metaphors reflect that ancient Chinese aesthetical standard for hands were long, light skin tone and pointed fingertips. Therefore, the main purpose of hands jewelry was to make hands look thinner, lighter and longer. In early Qing Dynasty, other than traditional hands jewelry such as bracelets, bracelet beads and rings, a new kind of hands jewelry, fingernail guards, was invented to make the fingers look even longer. The material differences between rankings were not mentioned in any official records, but there were some minor differences in the design patterns. This essay will mainly focus on bracelets, bracelet beads, rings and fingernail guards.

5.1 Bracelet

Fig.12.1 Mutton fat jade bracelet

Fig.12.2 Coral bracelet with inlaid pearls

The heritage collection in Forbidden City contains a great number of different kinds of bracelets from Qing dynasty. They were often in pairs and made from all kinds of different materials. The design patterns of the bracelets often had auspicious meanings; Bats, 卍 and Buddha hands patterns were the main pattern designs. The bracelets were mainly made of gold, silver, mutton fat jade (Fig.12.1), pearls, jadeite, corals (Fig.12.2), etc. They were made mainly using inlay, filigree and kingfisher craft. The overall purpose was to make the person who wore these bracelets seem more luxurious and splendid.

5.2 Bracelet Beads

Bracelet beads originated from rosary beads in Buddhism, and their purpose also transformed from counting to decorating. Bracelet beads were often made from corals, tourmaline, jadeite, amber, lapis lazuli, agate, etc. Among the Forbidden City museum's collection, there is an odorous bracelet beads that can keep insects away as in Fig.13. This bracelet beads can be put not only on wrists but it can be hung on the buttons of the gowns as well.

Fig.13 18 Counts Bracelet beads

5.3 Rings

Rings are one of the more important hand jewelry and Manchu concubines mainly wore two kinds of rings. The first kind was the old-fashioned, rounded plain rings, which were often made of mutton fat jade seen in Fig.14, corals, jadeite, etc. Sometimes, there were auspicious-meaning patterns carved on them. The other kind are the thin rings that had different kinds of gemstones inlaid on them [16].

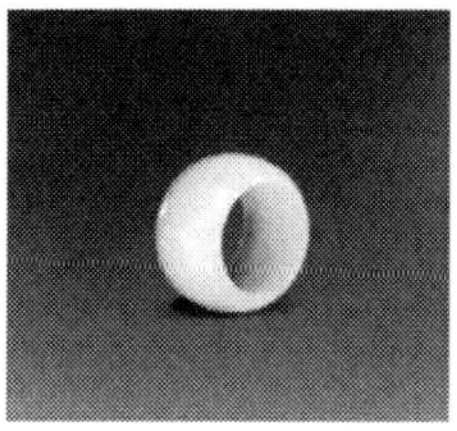

Fig.14 Mutton fat jade plain ring

Fig.15 Filigreed golden fingernail guards

5.4 Fingernail Guards

Among all the early Qing concubine's hand ornaments, fingernail guards had the most Qing characteristics seen in Fig.15 Manchu concubines had the habit of growing their fingernails; thus, the fingernail guards were invented for their nails' protection. Nail guards were often made using inlay, filigree and kingfisher craft. Concubines often wore it on ring fingers and pinkies, and both hands did not have to be symmetrical.

Conclusion

In early Qing, Manchu concubines' main crest accessories includes court crowns, auspicious crowns, Tianzis, Bianfangs, hairpins and combs; The ear jewelry were mainly worn three sets per side. Neck jewelry were mainly court beads, Lingyues, and necklaces. Hand jewelry was mainly bracelets, bracelet beads, rings and fingernail guards. All these accessories were made using the best material and craftsmanship. They not only represented their ornamental purposes but also demonstrated the institutional system and craftsmanship in early Qing dynasty. Moreover, these accessories also set a very solid foundation for the later developments of women's accessories.

References

[1] Yang ZS. Gold and Silver Jewelry of Ancient China (III) [M]. Beijing: Imperial Palace Press: Sept. 2014; 778

[2] Zhou TY. Unique Background of Court Jewelry Design in Qing Dynasty. [J]. Modern Ornament: Theory: 2013 (7); 209

[3] Yang ZS. Gold and Silver Jewelry of Ancient China (III) [M]. Beijing: Imperial Palace Press: Sept. 2014; 779-780

[4] Zhou XB. History of Ancient China Costume [M]. Beijing: China Drama Press: Sept. 1996; 482-493

[5] Chong Y. Miscellanea Since Daoguang and Xianfeng Periods [M]. Beijing: Beijing Ancient Books Press: 1982; 33

[6] Ruan WP. Atlas of Ornaments in Qing Court [M]. Beijing: Imperial Palace Press: March 2015; 19-24

[7] Zhou XB. History of Ancient China Costume [M]. Beijing: China Drama Press: Sept. 1996; 484

[8] Yang ZS. Gold and Silver Jewelry of Ancient China (III) [M]. Beijing: Imperial Palace Press: Sept. 2014; 796

[9] Ruan WP. Atlas of Ornaments in Qing Court [M]. Beijing: Imperial Palace Press: March 2015; 33

[10] Ruan WP. Atlas of Ornaments in Qing Court [M]. Beijing: Imperial Palace Press: March 2015; 23-23

[11] Liu Y, Zhao ZH, Ding Y, et. al. Probe into the Yucui and Fengchi: A Brief Analysis on Headdresses of Women of Manchu in Qing Dynasty [J]. Design: 2015 (11): 95-96

[12] Chen F, Jiang YQ, Zhang Y, et. al, Makeups and Dresses [M]. Beijing: SDX Joint Publishing Company: July 2015, 278-299

[13] Li Y. Time Flows by Ears: Eardrops of Ancient China [M]. Beijing: China Textile Press: Jan. 2015; 175-195

[14] Ruan WP. Atlas of Ornaments in Qing Court [M]. Beijing: Imperial Palace Press: March 2015; 159

[15] Ruan WP. Atlas of Ornaments in Qing Court [M]. Beijing: Imperial Palace Press: March 2015; 159-172

[16] Mo Y. He Cancan. Research on Nail Wrap as Women Ornament [J]. Art Education: 2016; 10: 162-164

Study of Collar-Lapel Structures Used in Burial Pottery Figurines of Northern Wei Dynasty

Jian-Yu Chen, Rong Zheng*, Yue Zhu, Jing Guo

Aimer Human Engineering Research Centre, Beijing Institute of Fashion Technology, Beijing 100029, China

*Corresponding author's email: yuki.zheng@gmail.com

Abstract

This study analyzed collar-lapel structures used in the Northern Wei Dynasty and provides corresponding reasons for their regional distributions. The study selected collar-lapel structures of 683 pottery figurines unearthed from 22 clearly documented tombs of the Northern Wei Dynasty as study objects, classifying collars into two categories (V-shape and round), and lapel structures into three categories-left, right, and non-directional structures. Based on statistical analysis, findings revealed that V-shape collars and left-lapel structures were the dominant fashion in the Northern Wei Dynasty, but differences are found in their regional distributions.

Keywords: Northern Wei Dynasty; Pottery Figurine; Collar-lapel Structure

1. Introduction

The Northern Wei Dynasty (386-557) is an important part of the historical period spanning China's Wei, Jin, Northern and Southern dynasties (220-589). While the Northern Wei Dynasty linked the rich Han, Sui and Tang dynasties in Chinese history, it also witnessed drastic social disruptions and transformations [1]. During its 369-year time frame, the collision and fusion of diversified cultures and ethnic groups finally ushered in the prosperous civilization of the Sui and Tang dynasties. As the starting point of the Northern dynasties (386-581), the Northern Wei Dynasty was the first regime founded by a minority ethnic group in Chinese history, since its ruler was part of the Tuoba family of Xianbei ethnic group [2]. The clothing reforms introduced by Emperor Xiaowen promoted the Hanification of apparel worn by the Northern barbarian tribes at the time. But at the outset, the Han clothing was not well received by the population [3]. According to historical documents, apparel reforms lasted for more than 100 years before they finally exerted their influence on fashion trends in the Northern Wei Dynasty [4].

A review of the literature shows that studies examining clothing of this period tend to collectively refer to the Northern Wei Dynasty, Northern Qi Dynasty and Western Wei Dynasty as the Northern dynasties; therefore, such studies are quite generalized. Moreover, most of the researchers were professionals in fields such as archeology and museology and hence their studies focused mostly on clothing culture, styles and aesthetics, and offer scant details on apparel. For example, there is a lack of information in existing documents related to collar and lapel structures of Northern Wei apparel. Against the background of different ethnic groups' interaction, fusion and integration, the collar-lapel structure seemed to be a particularly important feature to distinguish Han apparel from that used by ethnic minorities.

As traditional Chinese clothing tends to employ silk and linen materials that are not easily preserved, few articles of clothing are available for study by researchers of textile archeology and apparel history. Therefore, three-dimensional pottery figurines used as burial objects are an excellent resource for apparel study. Apparel shown in three-dimensional burial objects provide much greater details than those conveyed by photographs or other two-dimensional illustrations. However, existing clothing and related studies of pottery figurines give only a superficial picture-based description and lack in-depth analysis.

Given the limitations noted above, this study sought to classify and analyse the collar-lapel structures of 683 explicitly documented pottery figurines of the Northern Wei Dynasty from the perspective of garment study, as well the distribution of various structures in regions such as Chang'an and Luoyang, and determine reasons for their popularity.

2. Research

2.1 Scope of the Study

In the collar-lapel structure, the collar is the part that fits around the neck, while the lapel (Jin) refers to the top-front part of a garment [5]. Because it is close to the wearer's face, the collar is the most eye-catching part of a garment, drawing people's attention to the wearer's face and shoulders but also serving as a foil to the face and neck. It also plays a role in beautifying the shoulder and thus constitutes one of the garment's optical centers [6].

In the article Study on Pottery Figurines of the Eastern Wei and Northern Qi Dynasties, Wei Qingli defined pottery figurines as follows: "As burial objects, figurines share a human being's features, trying to simulate human looks" [7]. In a Study of Unearthed Pottery Figurines from Tombs of the Northern Wei Dyansty, Gao Dongyi indicated that ancient Chinese pottery figurines can be classified into categories pertaining to the wearers' civilian and official roles; based on available documents, he speculated that the Northern Wei Dynasty carried forward the tradition of official burial objects, which then extended into the Eastern Wei and Northern Qi dynasties [8].

As a three-dimensional model, pottery-figurine burial objects are an excellent resource for the study of clothing. Because information conveyed by the three-dimensional burial objects is greater than that presented by photos and other two-dimensional illustrations, the study focused on pottery-figurine clothing of the Northern Wei Dynasty. Combing through archeological data, the study selected 88 archeological reports. After comprehensive analysis, the study selected 683 pottery figurines from those unearthed from 22 tombs of the Northern Wei Dynasty, each with well-documented records. These pottery figurines were mainly recorded with photographs, handmade sketches and textual descriptions.

Considering the above-mentioned facts, the study examined archaeological reports on the Northern Wei Dynasty available online or published as major literature sources, as well as numerous periodicals, academic papers and treatises pertaining to clothing of the Northern Wei Dynasty. The study classified collar-lapel structures of the pottery figurines' garments in light of the study purpose, and provided descriptive analyses of pottery figurines' appearance, quantity, distribution and frequency.

Table.1 Number of documented Northern Wei Dynasty tombs according to region

Region	Number
Pingcheng	3
Yecheng	4
Luoyand	9
Chang'an	3
Qingqi	2
Gaoping Town	1
	Total: 22

Table.2 Number of documented Northern Wei Dynasty pottery figurines unearthed in different regions

Region	Number
Pingcheng	115
Yecheng	84
Luoyand	293
Chang'an	135
Qingqi	11
Gaoping Town	45
	Total: 683

The figurines under study are all burial objects unearthed from explicitly documented tombs of the Northern Wei Dynasty in the following six regions: Pingcheng, Yecheng, Luoyang, Chang'an, Qingqi and Gaoping Town. The tombs were mostly distributed in economically and politically developed regions and their adjacent areas such as Luoyang and Pingcheng (as shown in Table 1). (Note: Pingcheng and Luoyang were both the capital of the Northern Wei Dynasty during different periods of time.) The number of figurines unearthed in Luoyang and

Pingcheng is also relatively high. Although the number of tombs in Pingcheng is lower than that in Yecheng, the number of unearthed pottery figurines in Pingcheng is much higher than that in Yecheng (see Table 2). Archaeological studies show that those pottery figurines were mostly subordinates of various ranks under the direction of magnates and dignitaries. The greater number of tombs found in Luoyang and Pingcheng is congruous with their having been capitals of the Northern Wei Dynasty during different periods, resulting in a greater number of powerful public figures.

2.1.1 Classification of Pottery Figurines

Classification is one of the major foci of archaeological studies. The study's literature review shows that archeological reports have adopted numerous approaches in the classification of pottery figurines, as do research documents such as academic papers and treatises. This study thus adopted a relatively mainstream and simple approach to classify pottery figurines, namely placing them into four major categories: guardian figurines, honour guard figurines, daily-life figurines and figurines of other types. The study directly adopted names of figurines cited in archaeological report. The study focused on one part of the figurines' garments: collar-lapel structure. Because the examination of the collar-lapel structure alone does not involve pottery figurine classification, the study classified pottery figurines only generally (as shown in Table 3). The classification was used in the analysis of reasons for distributions of pottery figurines' collar-lapel structures.

Table.3 Variety quantity of documented Northern Wei Dynasty pottery figurines in different regions

Region	Number
Pingcheng	19
Yecheng	14
Luoyand	65
Chang'an	23
Qingqi	4
Gaoping Town	8
	Total: 133

2.2 Methodology

1) Checking and verification through literature review: The study examined archaeological reports on the Northern Wei Dynasty available online or published as major literature sources, as well as numerous related periodicals, academic papers and treatises pertaining to clothing of the Northern Wei Dynasty.

2) Statistical analysis: The study classified collar-lapel structures of pottery figurines' garments in light of the study purpose, and provided descriptive analyses of quantity, distribution and frequency of pottery figurines.

3. Results and Analysis

3.1 Neckline Shapes of Pottery Figurines

A B

Fig.1 Crossed neckline with (A) right lapel and (B) left lapel. Feng tomb cluster, Jing County, Hebei Province.

Preliminary research shows that the most frequently seen collar-lapel structures within the scope of the study are as follows: crossed neckline with the left and right lapel (Fig.1); turn-down collar with the left and right lapel (Fig.2); and round neckline with the left and right lapel (Fig.3). Apart from the above six types, there are also other collar-lapel structures such as the V neckline, criss-cross neckline and round neckline without identified wrapping of two front pieces (Fig.4).

A B

Fig.2 Turn-down collar with (A) right lapel and (B) left lapel. Yangji tomb, Luoyang, Henan Province.

A B

Fig.3 Round neckline with (A) right lapel and (B) left lapel. Northern Wei tomb, Luoyang, Henan Province.

A B C

Fig.4 (A) round neckline without identified wrapping of two front pieces, mural tomb, Chencun Village, Mengjin County, Luoyang, Henan Province; (B) V neckline, Nanchen tomb, Mengjin County, Luoyang; (C) criss-cross neckline.

Fig.5 V-shape necklines.

Fig.6 Round necklines.

As noted above, the study's examination of pottery figurines' collar-lapel structures focuses mainly on neckline shape and the wrapping direction of two front pieces. This section describes neckline shapes of the Northern Wei Dynasty's pottery figurines.

In order to clearly analyse distributions and changes of neckline shapes once popular in the Northern Wei Dynasty, the study classified pottery figurines' neckline shapes into two major categories: V-shape neckline (including those close to a V shape; see Fig.5) and round-shape neckline, including all round necklines of different styles.

Based on the definition, the paper summarises the distribution of two types of necklines, as shown in Table 3.

Table.3 Quantity and varieties of V-shaped and round necklines in the Northern Wei Dynasty

Item	V-shaped neckline	Round neckline	Total
Variety of pottery figurines	112	21	133
Percentage of total varieties	84.22%	15.78%	100%
Number of pottery figurines	597	86	683
Percentage of total number of pottery figurines	87.41%	12.59%	100%

The Northern Wei Dynasty was a transitional period from the Han-style garment (mainly featuring crossed neckline and right lapel) to the round-neck gown of the Sui and Tang dynasties that marked the beginning of the Northern dynasties. As the Emperor Xiaowen introduced the clothing reform, the Han-style garment was most popular and the trend towards round-necked gowns had just begun. The distribution of the figurines' collar types supports this point, as shown in Fig.3. Whether in terms of variety or quantity, the V-shaped neckline outweighed the round neckline.

The quantity and variety of the Northern Wei Dynasty pottery figurines with V-shaped and round necklines in different regions are shown in Tables 4 and 5.

Table.4 Number of Northern Wei Dynasty figurines with V-shaped and round necklines per region

Pottery figurines	Pingcheng	Yecheng	Luoyang	Chang'an	Qingqi	Gaoping Town
No. with V necklines	79	69	284	133	6	26
No. with round necklines	36	15	9	2	5	19

Table.5 Number of varieties of Northern Wei Dynasty figurines with V-shaped and round necklines per region

Pottery figurines	Pingcheng	Yecheng	Luoyang	Chang'an	Qingqi	Gaoping Town
No. of varieties with V necklines	15	12	60	21	3	1
No. of varieties with round necklines	4	2	5	2	1	7

As shown in Tables 4 and 5, apart from Gaoping Town, the other five regions all saw the dominance of the V-shaped neckline whether in terms of varieties or quantity; although round-neckline figurine varieties are most prominent in Gaoping Town in terms of quantity, V-neckline figurines are by far the most common across the other five regions. This supports the study's conclusion that V-neckline apparel was more popular during the Northern Wei Dynasty.

3.3 Wrapping Direction of Top Front Pieces

Traditional Chinese garments usually have a uniform two-dimensional cross-shaped structure with the collar and front integrated into one component [8]. Hence the wrapping direction between the two front pieces can directly influence the visual effect of the collar and front. The wrapping direction of the two front pieces had been a focus of Chinese clothes designs in ancient times.

During the period of Wei, Jin, Northern and Southern Dynasties, China experienced considerable social disruptions and turmoil. As a variety of cultures exchanged and blended with one another, the dominant crossed neckline and right lapel fashion suffered impingement of other ethnic groups' fashion during the Qin (221 to 207 BC) and Han (202 BC to 220 AD) dynasties. During the Northern Wei Dynasty, the prudent attitude towards the wrapping direction of two front pieces changed somewhat, which led to the commonly seen random dressing.

This study sought to describe the specific wrapping methods of garment front in the Northern Wei Dynasty. Prior to the data analysis, it is necessary to define and clarify concepts regarding front wrapping:

Right Ren (also called right lapel; *Ren* literally means front): refers to the left front of the garment being stretched to the right underarm and tied with a belt, wrapped over the right front (see Fig.1A). The reverse is called left Ren (or left lapel).

Left Ren (or left lapel): refers to the right front of the garment being stretched to the left underarm and tied with a belt, wrapped over the left front (see Fig.1B). The reverse is called right Jin (or right lapel).

Right Jin (or right lapel): refers to the left front of the garment being stretched to the right underarm and tied with a belt, wrapped over the right front. The reverse is called left Jin (see Fig.2A and Fig.3A).

Left Jin (or left lapel): refers to the right front of the garment being stretched to the left underarm and tied with a belt, wrapped over the left front. The reverse is called right Jin (see Fig.2B and Fig.3B).

Non-directional lapel: a term used when there is no way to judge the wrapping direction of Jin or Ren (as shown in Fig.4).

In order to have a clear presentation of the regional distribution and changes about the wrapping direction of garment front in the Northern Wei Dynasty, the study classified various front wrapping directions into three major categories: right, left and non-directional. A preliminary data collection and analysis resulted in findings shown in Table 6 on the distribution of wrapping direction of pottery figurines' collar-lapel structure in the Northern Wei Dynasty:

Table.6 Quantity and varieties of pottery figurines with right, left, and non-directional lapel

Item	Right lapel	Left lapel	Non-directional lapel	Total
Number of pottery figurine varieties	44	61	28	133
Percentage of total pottery figurine variety	33.08%	45.87%	21.05%	100%
Number of pottery figurines	225	331	127	683
Percentage of total number of pottery figurines	32.94%	48.46%	18.60%	100%

As early as 770-476 BC, Confucius made it clear that the left lapel represented barbarian dress features, while the right lapel represented orthodox Han Chinese clothing characteristics, a view that is still accepted by mainstream Chinese to this day. It is an evident manifestation of the dual nature (spirituality, function) of the collar-lapel structure on the front wrapping of the garment. As mentioned above, the Northern Wei Dynasty was the first regime established by a minority ethnic group in China's history. Its rulers carried out the clothing reform in a bid to adopt the Han Chinese clothing system, though such reform was not embraced at the outset. Study results show that varieties and quantity distribution of pottery figurines' garment wrapping directions in the Northern Wei Dynasty are as follows: left lapel>right lapel>non-directional structure (as shown in Table 6). This is consistent with the historical data indicating that the left lapel was more favoured by people and widely used.

Pottery figurine varieties and quantity with the three garment wrapping directions in different regions of the Northern Wei Dynasty are shown in Tables 7 and 8.

Table.7 Number of varieties of Northern Wei Dynasty pottery figurines with right, left, and non-directional lapel structure per region

Type of pottery figurine	Pingcheng	Yecheng	Luoyang	Chang'an	Pingqi	Gaoping Town
With right lapel	0	1	37	4	2	0
With left lapel	12	11	18	19	1	0
With non-directional lapel structure	7	2	10	0	1	8

Table.8 Number of Northern Wei Dynasty pottery figurines with right, left, and non-directional lapel structure per region

Type of pottery figurine	Pingcheng	Yecheng	Luoyang	Chang'an	Pingqi	Gaoping Town
With right lapel	0	13	196	12	4	0
With left lapel	95	68	43	123	2	0
With non-directional lapel structure	20	3	54	0	5	45

A horizontal comparison shows that in terms of the number of varieties of pottery figurines with a right lapel, Luoyang>Chang'an>Qingqi>Yecheng; in terms of the number of figurines with a right lapel, Luoyang>Yecheng> Chang'an>Qingqi; in Pingcheng and Gaoping Town, no pottery figurine with a right lapel has been found. In terms of the number of varieties of figurines with a left lapel, Chang'an>Luoyang> Pingcheng>Yecheng>Qingqi; in terms of the number of figurines with a left lapel, Chang'an>Pingcheng> Yecheng>Luoyang>Qingqi; in Gaoping Town, no pottery figurine with a left lapel has been discovered. In terms of the number of varieties of pottery figurines with a non-directional lapel, Luoyang>Gaoping Town>Pingcheng>Yecheng>Qingqi; in terms of the number of figurines with a non-directional lapel, Luoyang>Gaoping Town>Pingcheng>Qingqi>Yecheng; in Chang'an, no pottery figurine with a non-directional lapel has been discovered.

A vertical comparison shows that in Pingcheng, in terms of both variety and quantity, the left lapel outweighed the non-directional lapel structure; no pottery figurine with a right lapel has been found in Pingcheng. In Yecheng, in terms of variety, left lapel>non-directional lapel>right lapel; in terms of quantity, left lapel>right lapel> non-directional lapel. In Luoyang, in terms of variety, right lapel>left lapel>non-directional lapel; in terms of quantity, right lapel>non-directional lapel>left lapel. In Chang'an in terms of both variety and quantity, left lapel>right lapel; non-directional lapel has not been found among unearthed pottery figurines in Chang'an. In Qingqi, in terms of variety, right lapel>left lapel=non-directional lapel; in terms of quantity, non-directional lapel>right lapel>left lapel. In Gaoping Town, only pottery figurines with non-directional lapel have been found.

From the above analysis, it can be concluded that among unearthed pottery figurines in Luoyang, the right lapel was the dominant fashion; in Chang'an, Pingcheng and Yecheng, the left lapel prevailed; in Gaoping Town, only the non-directional lapel was found.

According to historical records, Emperor Xiaowen moved the capital from Pingcheng to Luoyang in 493, which led to Xianbei people's massive southward migration. Thirty years after settling in Luoyang, the Xianbei group had been basically integrated, which corresponds to the above-mentioned conclusion [9]. Back then, Luoyang as the capital was the key city for implementation of the dynasty's Hanification policy; in contrast, other cities had a relatively slow pace in this regard. Therefore, in Luoyang, the right lapel was probably more popular. However, in places like Chang'an and Pingcheng, people favoured the left-lapel garment.

4. Conclusion and limitations

4.1 Conclusion

1) The paper classifies the collar structures of pottery figurines in the Northern Wei Dynasty into two categories: V-neckline shape and round-neckline shape. Based on a descriptive analysis, it can be concluded that in the Northern Wei Dynasty, the V-shape collar was more popular, and favoured by the public in daily wear. The conclusion is consistent with the historical fact that in that period Hanfu had been inherited, which was embodied in people's clothing shape; Hanfu features crossed collar (V-shape collar) and right lapel.

2) In its analysis, the study classified the lapel structure of pottery figurines in the Northern Wei Dynasty into three categories: right lapel, left lapel and non-directional lapel. An analysis of the selected 683 pottery figurines of the Northern Wei Dynasty shows that the left-lapel structure prevailed. It is speculated that back then the left-lapel garment was the leading fashion and favoured by people in places such as Chang'an, Pingcheng and Yecheng. However, among unearthed pottery figurines from Luoyang, it is found that the right-lapel structure was the dominant fashion, consistent with the historical fact that the Emperor Xiaowen introduced clothing reform there after moving the capital to Luoyang. It is speculated that so influenced by the new policy, people in Luoyang accepted the Hanfu trend better than people in other places; therefore, right-lapel garments prevailed. It is curious that among pottery figurines unearthed in Gaoping Town, the non-directional lapel structure was the main fashion;

further studies are warranted to identify if a historical reason can explain the trend.

3) At the same time, I hope to provide some theoretical basis for the study of clothing history.

4.2 Limitations

There are some limitations in using pottery figurines for the study of clothing. First, most pottery figurines have sustained different degrees of damage, which led to the increased efforts to identify broken or missing parts. Second, pottery figurines mostly represented figures of lower social classes, so their fashion cannot be used exclusively to study overall clothing fashion of this specific period. Third, artisans produced pottery figurines, and it was highly possible that their understanding of clothing was inaccurate, and they may even have fabricated some clothing styles. Therefore, conclusions obtained by taking pottery figurines as clothing study objects are not fully convincing and remain open to questioning. Fourth, although the basic principle for creating pottery figurines was to make them look like real persons, considering that they were taken as burial objects and there was no unified production standard, it is difficult to determine whether their clothing was an authentic representation of real life, or perhaps represented special burial costume patterns [10].

References

[1] Song BL. A study of non-religion garments in northern dynasties. Jinan, China: Shandong University: 2008.

[2] Zhang J, Wang MF, Yang K. The historical origins of women clothing in Northern Wei Dynasty. Lantai World: 2012; 3: 71-72.

[3] Li J. The clothing changes during Xiaowen reform in Northern Wei Dynasty. J Shanxi Datong University (Soc Sci ed.): 2013; 27: 110-112.

[4] Zhu XE. Analysis on Hu and Han people's culture relations from clothing changes during Northern Wei Dynasty. J Hexi College: 2006; 22: 27-30.

[5] Xinhua Zidian [new Chinese dictionary]. Beijing: Commercial Press: 1993.

[6] Xu XH. Caoyong and Song Shaohua. Introduction to clothing. Beijing: China Textile Press: 2013.

[7] Wei QL. Study on pottery figures of Eastern Wei Dynasty and Northern Han Dynasty. Zhengzhou, China: Zhengzhou University: 2007.

[8] Gao YD. Study on the unearthed pottery figures in the tombs of the Northern Wei Dynasty. Guangzhou, China: Jinan University: 2014.

[9] Information at https://zhidao.baidu.com/question/571510197.html

[10] Song BL. A brief analysis of the limitations of image materials in ancient costumes research-Exampling by costumes in Northern Wei Dynasty. J Nanjing Arts Institute Art and Design: 2009; 4: 51-54.

An Analysis of the Communication Strategy of Tai Chi Costume Culture

Yu Jin, Jian-Fang Liang*

Apparel & Art Design College, Xi'an Polytechnic University, Xi'an, Shaanxi, 710048, China

*Corresponding author's email: 342110784@qq.com; 412942044@qq.com

Abstract

In recent years, Tai Chi sports has received increasing amount of attention. It offers a platform for the spread of Tai Chi costume culture. Therefore, it is important to consider how to effectively promote Tai Chi Costume culture. This study uses literature review, personal interview and questionnaire method to analyze the communication strategy of Tai Chi costume culture. SPSS software is used to analyze the attitude of the audience in a data analysis form, which draws a strategy for the spread of Tai Chi clothing (a kind of Chinese traditional sportswear) culture. The study suggests that taking advantage of the new media to create an international brand of Tai Chi costume is the most effective.

Keywords: Tai Chi Costume; Culture; Media Choice; Communication Strategy

1. Research Background

Clothing is not only an essential part of human life but also conveys culture. The clothing design, fabric selection and colors expresses different cultural characteristics. Chinese dress culture has existed since ancient times. In China, clothing culture is often associated with religious, ideological and martial arts culture together [1]. Clothing culture is more than the combination of material and spiritual, it's the transcendence of material. It is also the combination of human thought and spiritual creativity. The clothing is an important part of Tai Chi and its design reflects upon the culture.

There is growing concern regarding Tai Chi culture in recent years, largely because of its rich connotations and health maintenance. Currently in china and all over the world, there seems to be a movement focused around Tai Chi. Tai Chi clothing as an important part of the culture is also developing rapidly [2]. But at the present, the promotion of Tai Chi clothing are mediocre with only the people involved in it buying the products. Modern society relies heavily on advertising which is the direction in which Tai Chi Clothing should be heading. Media selection is the key factor for the success of the advertising strategy because of the characteristics of Tai Chi products. This article is to provide a viable approach for the Tai Chi clothing culture based on an analysis of the spreading strategy of Tai Chi costume culture in this environment.

2. Literature Review

2.1 Research Status of Communication Strategies

In the 1940s, communication technology was advancing. In fact, the development of communication in China has not exceeded 20 years [3]. There are basically no studies regarding communication strategies. Scholars focus on the study of all kinds of brand enterprises to improve the brand communication and marketing strategy. However, advertising media research has developed for a long time as a part of communication studies. As early as 1997, Junliang Chen published a book called "Advertising Media Research" which became known as the first Chinese media. This book has introduced scientific media strategy, purchasing method and also describes the combination of advertising media and advertising media planning [4]. Jinsong Zhao's book "Advertising Media Practice: Advertising Media Research, Planning and Purchase Guide" was published in 2001 [5]. It explores the choice of media advertisement. In 2002, Xiaoyun Hu, Liping Zheng translated the "Advanced Media Program" by John R.

Rossiter. This book makes the concept of media planning closer to the essential goal of communication advertisement [6]. At present, more and more scholars have turned their attention to media such as Yi Sun's article. This paper analyzes the common problems of brand marketing and the importance of brand marketing in the new media age. It put forward several strategies for brand markéting [7].

2.2 Research Status of Costume Culture

People's attention is mainly focused on the following aspects when it comes to the culture of costumes [8]. First, from a historical point of view, as shown by Mr. Congwen Shen's "Ancient Chinese Clothing Research" [9]. Secondly, from the perspective of cultural interpretation as shown "The Deep Implication of Chinese Clothing Culture" [10], "Human Clothing Culture" [11] and a series of related works. The third is from the perspective of literary phenomenon, such as "Luo Yi and its Cultural Implication in the Tang Poetry " [12]. A small number of scholars have taken the media representation of clothing culture as a research point. However, there are little research on the communication strategy of clothing culture of which one is Le Xing's study. This article combines the current situation of traditional clothing culture communication in the centeral plains as well as put forward modern communication strategy for promoting traditional costume culture [13].

2.3 Research Status of Tai Chi Costume Culture

At present there are only a few, domestic and foreign research on Tai Chi clothing culture. The main literature article is Li Ye's journal "Talking about the New School of Women's Tai Chi Competition Routine Clothing Cultural Meaning". The study pointed out that after the requirements of the clothing were relaxed in the "Wushu routine competition rules", women's Tai Chi clothing had greatly changed from the production method, fabric and color [14]. The article gives a detailed description of the cultural connotation of Tai Chi costume from the process, fabric, style, color and some other aspects. It hopes to provide a broader space for the development of women's Tai Chi clothing.

3. Investigation Method

In fact, there are still many problems when spreading Tai Chi clothing culture. To resolve these issues, the key is to understand the audience's views on the spread of Tai Chi clothing culture. Through questionnaires and personal interview method, we can understand the audience attitude and perspective.

The methodology are as follows: first, Investigators are engaged in a one-on-one discussion with some volunteers in Xi'an where the Tai Chi enthusiasts are gathered. This allows the understanding of their basic views and regarding Tai Chi clothing culture. The second step is to use a questionnaire survey. The questionnaire takes the form of open and closed questions with multiple choices. 105 questionnaires were randomly distributed.

4. The Basic Situation of the Investigation

A total of 105 questionnaires were sent out in this survey. Among it, 78 questionnaires were distributed on the spot and 27 questionnaires were distributed on the internet. 103 questionnaires were collected with the recovery rate being 98.1%. After screening, three invalid questionnaires were removed and 100 valid questionnaires were obtained. The effective questionnaire rate was 95.2%.

Descriptive statistics and Chi square test in the SPSS software were used to analyze the data. The SPSS software was used in the beginning to test the reliability and effectiveness of the questionnaire. After analysis, the questionnaire was according to Cronbach's Alpha 0.839 "Table 1". Cronbach's Alpha > 0.8, indicating that it is reliable.

In the survey, 56% were male and 44% were female. The sex ratio is relatively balanced. From the age distribution of the respondents, 75% respondents were aged 31 years or older while 34 people were over the age of 45 as shown on"Table 2". Middle and old people's interest to Tai Chi culture were the strongest. 76% of the respondents were educated for more than 12 years.

Table.1 Reliability Statistics

Cronbach's Alpha	Number of items
0.839	33

Table.2 Age Distribution of Respondents

	Frequency /per	Percentage /%
Under 16 years old	2	2.0
16～25 years old	12	12.0
26～30 years old	10	10.0
31～35 years old	21	21.0
36～45 years old	23	23.0
46～55 years old	7	7.0
56～65 years old	15	15.0
Over 65 years old	10	10.0
Total	100	100.0

The respondents are living in all sectors of society (Fig.1). The distribution of occupation is helpful to improve the reliability of the results. Looking at the individual's monthly income from the respondents, 38% of respondents income ranges between 3001~4000 yuan (Fig.2). This part of the population belongs to average income groups and may represent the majority of people's consumption attitude. 61 respondents had bought Tai Chi clothes before.

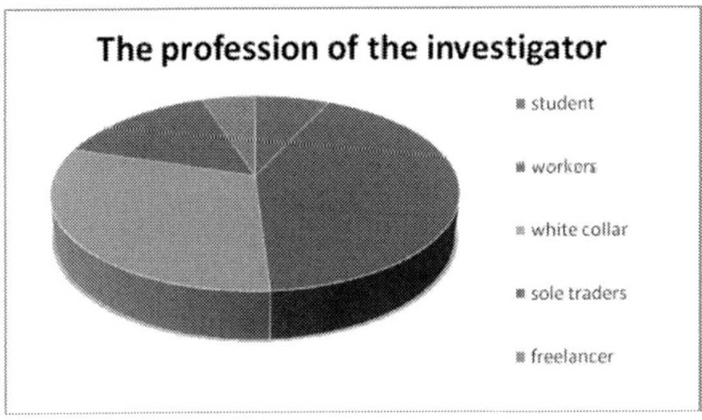

Fig.1 The profession of the investigator

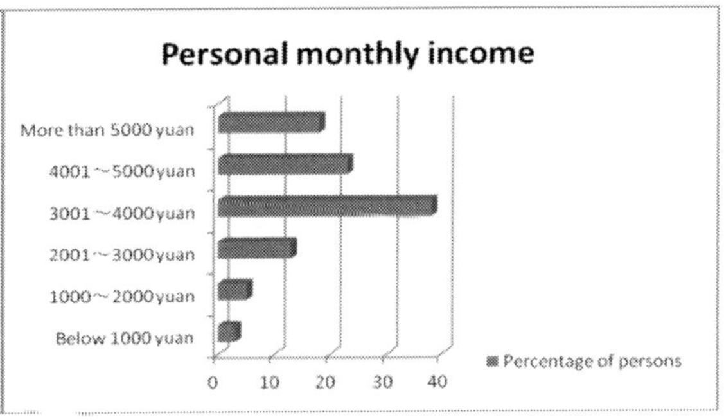

Fig.2 Personal monthly income

5. Analysis and Discussion of the Results

5.1 Media Contacts and Credibility

Looking at the survey, traditional media contact rate is the highest with better credibility than other media. The contact rate of TV and mobile phone rate is the highest. 70% of the respondents watch TV every day, and 88% of respondents use their phones. Middle-aged people prefer newspaper. If you want to improve the effectiveness of advertising, television, mobile media may be the better choice according to the data. 63% of respondents believe that television has more credibility. On the contrary, only 12% of respondents believe that the internet is trusthworthy (Fig.3). Although the credibility of the media is not as good as the past, the traditional media is still more reliable according to users opinion.

People's age has a greater impact on how often the media is used. Through the Chi-square test ($P<0.05$), it has significant statistical difference as shown on "Table 4". In the survey, middle aged and old people are highly relying on TV, newspapers, magazines and radio. Young people prefer mobile phones and the internet. We also discovered that a small number of elderly people began to use the internet as a source of media.

Table.3 Chi-square test

	numerical value	df	Gradual Sig. (bilateral)
Pearson chi-square	7.134a	2	0.028
likelihood ratio	7.519	2	0.023
Linear and linear combination	5.920	1	0.015
N in effective cases	50		

When asked what kind of advertising media they prefer, most young people choose new media. The new media is a relative concept, is that we usually see on television, newspapers, radio and other traditional media, new media forms beyond. The reason for the young people is that the new media is interactive and participatory (Fig.4). This suggests that to expand the Ta Chi apparel consumer group it is important to consider the use of micro-blog. WeChat as an example of new media is a good choice. Whether it is an official micro-blog, or WeChat public platform, the cost of operation and maintenance are quite low. Many clothing brands have begun to use these means to maintain customers.

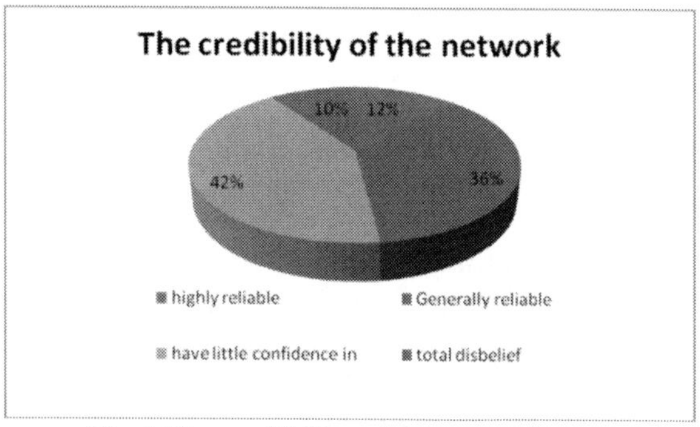

Fig.3 The credibility of the network

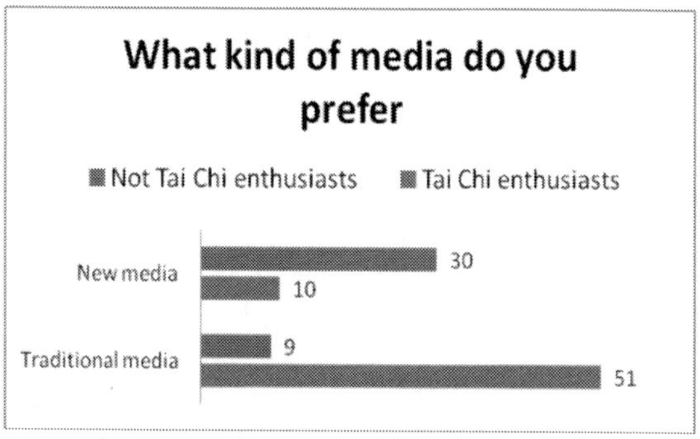

Fig.4 What kind of media do you prefer

5.2 The Display of Tai Chi Clothing

Tai Chi clothing consumers wants to fully understand the brand and details regarding the clothing.

When buying Tai Chi clothes (Fig.5), 38% of the respondents will give priority to the brand. Meanwhile, 23% of the respondents will give priority to the fabric (Fig.6). Younger generation's first look at brand followed by style. Therefore, to successfully promote Tai Chi Clothing it is important to consider the media that can best highlight the brand and display the fabrics and workmanship. Television and network are able to provide audio visual on the clothing products. Print media has the ability to describe the products in all aspects. For display of Tai Chi clothing, 37% of the respondents chose the details of the clothing, such as collar and cuffs. 34% of the respondents believed that the model display can show Tai Chi clothing in all aspects.

Fig.5 Example images of Tai Chi clothing [15]

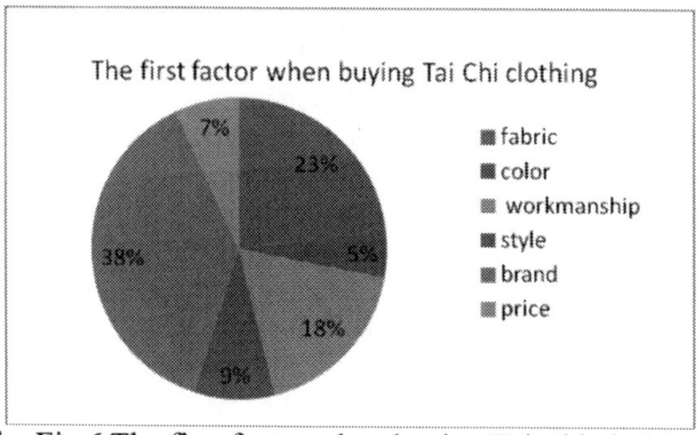

Fig.6 The first factor when buying Tai Chi clothing

Unlike ordinary consumers, Tai Chi enthusiasts will give priority to the brand and fabric. Through the Chi-square test ($P<0.05$), two groups have significant statistical differences. Tai Chi hobbyists know more about Tai Chi suits, and they also have a high brand awareness. At the same time, they pay more attention to the fabric as it determines the comfort of the clothes when worn. When considering the display of Tai Chi Clothing, 80% stresses the importance of detail as they reflect the quality of clothing.

5.3 The Dissemination Status

Tai Chi clothing ads are mainly concentrated on traditional media as promotion method. How do consumers obtain information of Tai Chi dress? In the survey results Tai Chi clothing are promoted through friends' recommendations, newspapers, flyers, outdoor advertising and other traditional media (Fig.7).

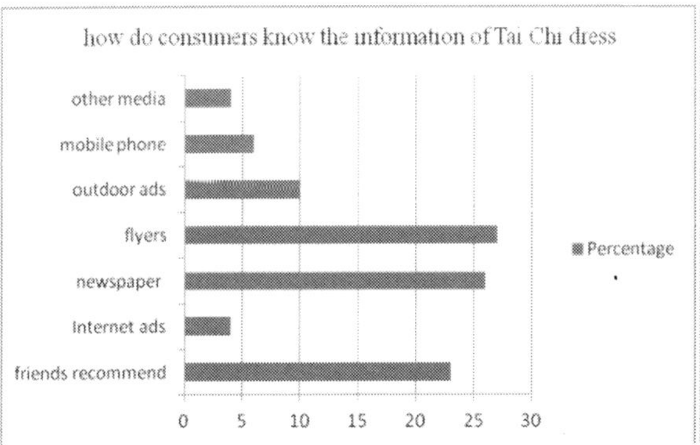

Fig.7 How to know the information of Tai Chi dress

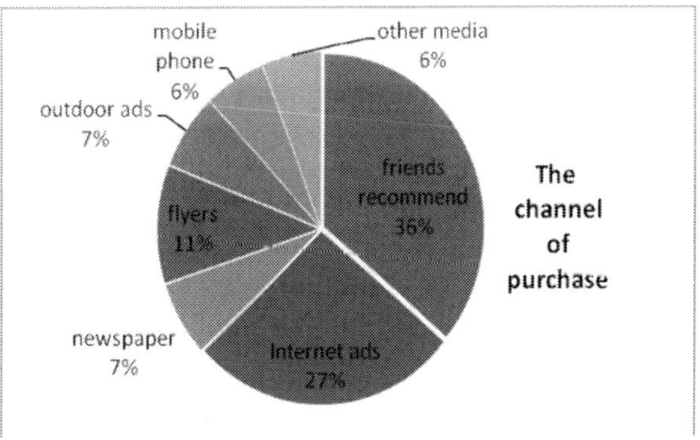

Fig.8 The channel of purchase

Secondly, the reputation of Tai Chi clothing will largely influence the sales. The network has also become an important channel of purchase. 36% of the respondents have been recommended to buy Tai Chi clothing (Fig.8) from a friend or family which highlights the importance of reputation. 27% of the respondents have bought Tai Chi clothes because of online advertising. After communication with the respondents, we found that most of the choice of online advertising is from network channels, such as Taobao.

Thirdly, consumers favor mobile phones and paper media. Network video, film implantation is concerned too. 72% of respondents believed that newspapers and magazines reveal most comprehensive information of Tai Chi clothing. Respondents also mentioned that online video, film are always ways to promote Tai Chi clothing.

Fourth, there is a lack of innovation when it comes to advertising strategies of Tai Chi Clothing. 97% of respondents said they have not seen the much advertisements regarding Tai Chi clothing. With the development of network technology, data mining and analysis has become mature. The internet can automatically match consumer preferences. For example, if the same IP address in the network have searched Tai Chi several times, the internet will automatically display advertisement regarding Tai Chi clothing.

5.4 The Dissemination Strategy

The importance of brand name can be seen from the survey. High-income groups pay attention to the brand name. 68% of respondents said that they prefer to choose brand. 75% of respondents believe that brand is more important in an advertisement (Fig.9). Most of the respondents whose income are more than 4000 yuan think that the added value of Tai Chi clothing is important (Fig.10). This is related to the concept of consumption. High-income groups pay more attention to show off their social status and spiritual enjoyment. Brand apparel generally control the quality of the product more stringently. Customers can achieve social superiority through brand value. Brand apparel not only meet their material requirements, but also their spiritual requirements. Tai Chi as a fitness exercise, has gradually infiltrated into the upper class. To open up the high end markets it is important for people to understand the attitudes and habits of high income consumers.

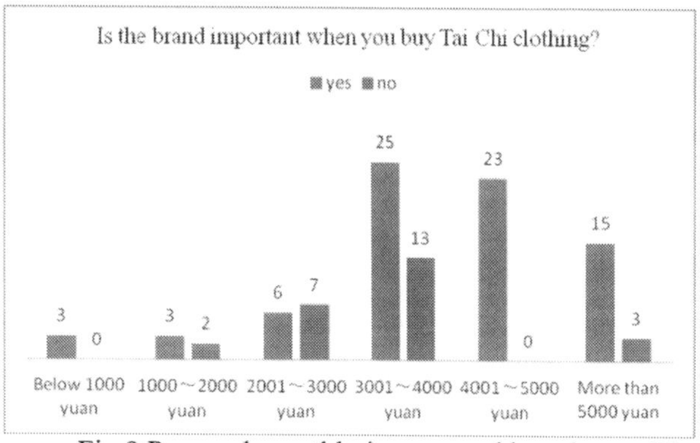

Fig.9 Personal monthly income and brand

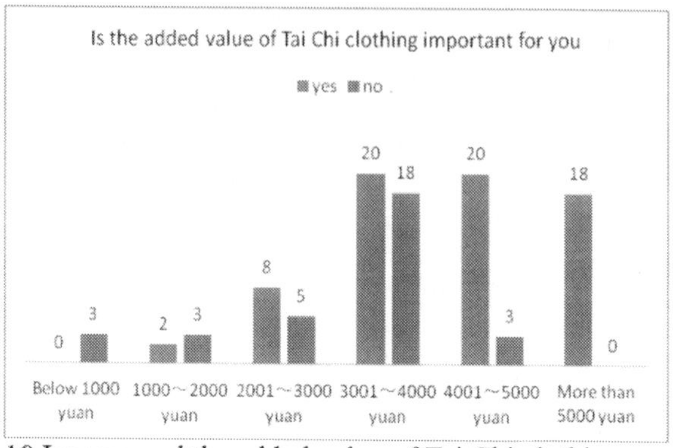

Fig.10 Income and the added value of Tai Chi clothing

Public relations activities, represented by sporting events, are favored by targeted consumers. Almost all of the respondents believed that public relations activities have a certain role in the spread of Tai Chi clothing culture. 56% of the respondents believes that relevant sporting events can draw attention to Tai Chi dress culture (Fig.11)

Advertisement plays a huge role when promoting Tai Chi clothing culture. When asked which way is the best way to stimulate the desire to buy Tai Chi clothing, 52% of the respondents chose advertisement. 20% of the respondents selected the sales promotion simultaneously "Table4". In the survey, the majority of respondents said they rarely see advertisements of Tai Chi clothing.

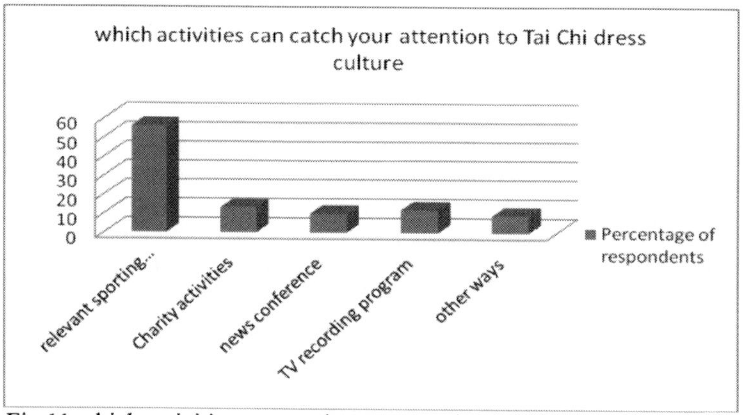

Fig.11 which activities can catch your attention to Tai Chi dress culture

Table.4 Which methods can best stimulate your desire

	Frequency / per	percentage/%
Sales promotion	20	20.0
Public relations activities	17	17.0
Interpersonalcommunication (word of mouth)	11	11.0
Advertisement	52	52.0
Total	100	100.0

In this survey, several respondents believe that Tai Chi clothing culture can be spread from preserving Chinese traditional culture and supporting domestic products. If we can integrate the essence of Tai Chi clothing culture and enlarge the product benefit of Chinese national brand, it may promote the spreading of Tai Chi Clothing.

6. Summary and Recommendations

If Tai Chi clothing wants to have a presence in the world, it has to represent the national Chinese culture [16]. To develop a Tai Chi clothing culture, specific strategies have been recommended.

The first is market segmentation. Pricing and developing style according to the audience's income. Choosing the right media according to age characteristics of the audience. Selecting a method of communication based on the education status of the audience When promoting Tai Chi clothing, it is important to pay more attention to the dissemination of cultural meaning, but at the same time comfort and performance.

Secondly it is important to improve brand awareness and establish a brand image. With the development of the market economy, the brand image has become vita. For Tai Chi clothing, the use of brand strategy will enable more people to become aware of Tai Chi. Cooperating with sportswear brand, provides reassurance to consumers.

The third is to invest in advertisement by fully taking advantage of the media. Such as inviting celebrities to endorse Tai Chi clothing. In addition, we can make full use of the efficiency of the internet.

Fourth, using public relations activities to expand the awareness of Tai Chi clothing. At present, Tai Chi clothing is generally used in public sports activities and live performances. The advantage of such activities is that Tai Chi can be brought into a deeper cultural atmosphere. The disadvantage is that the scope of communication is limited. In fact, you can also introduce a variety of activities that promotes Tai Chi. Such as Tai Chi clothing T-stage shows and Tai Chi cultural exchange week.

Fifth, marketing Tai Chi clothing verbally from mouth to mouth. Word of mouth marketing is the way people communicate with each other [17]. It varies from high credibility of media advertising, brand recommendation, or public welfare activities.

References

[1] Zhang ZP. Clothing Brand Communication and Advertising Strategy. Chin Text Lead: 2010; (8).

[2] Zhang XG. A Study on the Strategy and Revitalization of Tai Chi Culture in China. University of International Business and Economics: 2013; 03.

[3] Tian ZY. The Basis of Communication. Yuelu Press: 2009; 002.

[4] Chen JL. Advertising Media Research. China Market Press: 2007; 08.

[5] Zhao JS. Advertising Media Practice: Advertising Media Research, Planning and Purchase Guide. World Affairs Press: 2001; 11.

[6] Rossiter J, Danaher P. Advanced Media Program. Zhejiang University Press: 2002; 12.

[7] Xing L. A study of Han Dynasty's costume culture evolution and modern communication in Central Plains. Jiangnan University: 2017; 03.

[8] Zeng YH. Clothing: a Carrier of Culture and Media. Journal of Silk: 2013; (1): 58.

[9] Shen CW. The Chinese Ancient Clothing Study. The Commercial Press: 2011; 03.

[10] Lin SX. The Deep Meaning of Chinese Clothing Culture. Fudan J (Soc Sci Ed): 1997; (3): 62-68.

[11] Hua M. Human Clothing&Ornament culturology. Tianjin People Press: 1995.

[12] Zeng YH. "Luo Yi" in the Tang Dynasty and its Cultural Implication. Journal of Chaohu College: 1997; (3): 56-60.

[13] Sun Y. Communication strategy of brand marketing in New Media Age. New Media Research: 2017; 03: 63.

[14] Li Y. Talking About the New School of Women's Tai Chi Competition Routine Clothing Cultural Meaning. Journal of Military Physical Education and Sports: 2008; (4): 27.

[15] https://image.baidu.com.

[16] Qian XN, Zhang N. Cultural Management of Clothing Brand Personality. Journal of Liaoning Normal University(Soc Sci Ed): 2009; 6.

[17] Ardent J. Role of Product-Related Conversations in the Diffusion of a New Product. Journal of Marketing Research: 1967; 4.

A Study on Career Suits of Female College Students Based on Kansei Design

Jia Lv[1*], Yue Sui[2]

[1]*College of Clothing and Artisitc Engineering, Minjiang University, Fuzhou,350108, China*
[2] *Department of Physics and Electronic Information Engineering, Minjiang University, Fuzhou, 350108, China*

[*]Corresponding author's email: 61012023@qq.com

Abstract

In order to analyze Kansei evaluation of respondents on career suits, we established the corresponding relationship between the intention of consumers and specific design elements. T study regards career suits as the research carrier and considers female college students as experimental objects. The author applies principal component analysis as the initial Kansei words to do dimension reduction process and extracts two factors with the higher contribution rate as representative factors, and names it as elegant factor and simple factor. When designing it, the author conducted style planning from these two aspects. By calculating scores of every sample on these two factors, the author analyzes the interactive relationship between samples and Kansei factors, and analyzes attribute features of high-grade samples in two factors, so as to design career suits particularly.

Keywords: Career Suits; Kansei Design; Factor Analysis; Kansei Evaluation

1. Introduction

Career suits differ from vocational smock and uniforms. Except that it should be presentable in the working environment. It must integrate with personalized and fashionable features to shape the overall image of wearers [1, 2]. However, the design of career suits is stuck at a primary level with single styles which restrains the development of career suits in China. For double requirements of service and spiritual functions, as well as revolution of design level and consumer demands, the design of career suits develops towards Kansei design direction and emphasizes on detail perception and emotional experience of consumers [3, 4].

In this paper, the author regards career suits as the research carrier, considers female college students as experimental objects and conducts quantitative analysis on Kansei information of experimental samples, studies correlation between Kansei information and design elements, hoping to provide some idea and reference for designing career suits based on Kansei design.

2. Experiment

2.1 Materials

The author collects 140 styles of female career suits from network, magazines and other channels. With primary screening, the author eliminates slur styles and obtains 65 basic samples. By screening again, the author eliminates samples with the same shape or style to get 40 stimulus materials for the experiment. In order to eliminate influences of other factors, all stimulus pictures conduct uniform treatment and are presented in the form of black and white. The background is white. Luminance and contrast remains the same. The size is 800x1200 pixels, including 20 dark color samples and 20 light color samples, named from 1-40, respectively.

2.2 Subjects

86 healthy female college students who major in clothing design and garment engineering are selected as respondents to do professional evaluation. Their age ranges from 19 years old to 22 years old. The mean age is 21.4 years old.

2.3 Produce

By collecting Kansei adjectives from network, magazines, and books, the author selects 18 pairs of adjectives with high frequency and opposite meaning as Kansei words. 20 workers who engage in studying clothing are invited to collect adjectives to classify and conclude. 9 pairs of adjectives are used as Kansei words to stand for extraction of emotional factors, marked as X_1—X_9, respectively. The specific Kansei words are shown in Table 1.

Table.1 Kansei Adjective Pairs and Codes

Adjective pairs	codes
Neutral—feminine	X1
Contracted—complicated	X2
Fashionable—classical	X3
Lively—serious	X4
Vocational—casual	X5
Mature—childish	X6
Novel—ordinary	X7
Elegant—robust	X8
Individual—popular	X9

Stimulus pictures are installed with an E-prime2.0 computer (17 inch screen, resolution ratio 1280x 1024, 60Hz) to present at random. The experiment of emotional factors applies stannic scale. For example, for Kansei evaluation: neutral—feminine, the score closes to 1, indicating that it tends to be neutral. The score closes to 9, indicating that it tends to be feminine. The number 5 stands for the middle value. Every picture presents a clue for 5s, reminding that respondents should prepare for the next picture to score. Every picture is presented for 20s. During the period, respondents should mark it. In principle, it should give priority to first impression and don't reflect too much.

2.4 Results

SPSS is used to extract information of Kansei factors and corresponding design elements are analyzed.

2.4.1 Correlation Coefficient Matrix of Kansei Words

Every respondent's 9 pairs of SD scale scores for 40 pictures are conducted correlation analysis, checking whether it is suitable apply factor analysis to extract factors between original variables. Results show that correlation coefficient between most of variables is higher and every variable presents the linear relationship. As shown in the Table 2, KMO metric given by Kaiser shows that the original variables are suitable for factor analysis.

2.4.2 Information Extraction of Kansei Factors

According to correlation coefficient matrix of every variable, Kansei evaluation of different samples is based on the principal component analysis. Orthogonal rotation conversion is applied. The eigenvalue factor of which eigenvalue is selected, obtaining the larger principal component and variables with high load in common factors.

Table.2 Correlation Coefficient Check of Kansei Adjective Pairs

Test of KMO and Bartlett		
Sample sufficient degree of Kaiser-Meyer-Olkin for measurement		.776
Bartlett's test of sphericity	Approximate chi-square	366.435
	df	36
	Sig.	.000

As shown in the total variance of explanation, eigenvalue the first factor and the second factor is 4.737 and 2.36, respectively. Contribution rate of the accumulated variance for 2 factors is 78.85%, closing to 80%. As a result, top 2 factors are extracted, marked as factor 1 and factor 2.

Varimax is applied to implement orthogonal rotation on initial factor loading matrix. After 3 times of iteration, rotation is restrained, obtaining the rotated component matrix. Variables with more than 0.5 of loads are selected to extract variables with high loads in factor 1. According to the meaning, it can be summarized as the elegant factor. Variables with high loads in factor 2 are extracted. According to the meaning, it can be summarized as the simple factor (Table 3).

Table.3 Kansei Evaluation Factor Analysis

	Factor 1	Factor load	Factor 2	Factor load
Kansei words	X8 elegant—robust	.953		
	X7 novel—ordinary	.823	X4 lively—serious	.881
	X9 individual—popular	.783	X5 vocational—casual	.879
	X6 mature—childish	.766	X2 contracted—complicated	.731
	X3 fashionable—classical	.721		
Eigenvalue	4.737		2.360	
Contribution rate of variance	52.629		26.221	
Accumulated contribution rate	52.629		78.850	
Factor naming	Elegant		Simple	

2.4.3 Calculation of Kansei Factor Scores

According to the component score coefficient matrix, factor score function is written. Score of every sample on each factor is calculated. Thus, when calculating variables of two factor score variables, weight of X1, X6 and X6 variables is higher with the opposite direction. This is identical to the actual meaning of factors. Expression of FAC is expressed as:

$$\text{Factor }1=-0.202X_1+0.048\,X_2+0.148\,X_3-0.025\,X_4+0.076\,X_5+0.271\,X_6+0.186\,X_7+0.273\,X_8+0.171\,X_9$$
$$\text{Factor }2=0.038\,X_1-0.248\,X_2+0.123\,X_3+0.287\,X_4-0.304\,X_5-0.266\,X_6+0.082\,X_7-0.106\,X_8+0.098\,X_9$$

2.4.4 Correlation Analysis of Kansei Factors and Samples

By calculating scores of factor 1 and factor 2 on each sample, ranking situation of each sample is obtained to do correlation analysis with two factors, respectively and study interactive relationship between samples and Kansei factors. As shown in Fig.1, factor 1 of sample 23 gets the highest score, indicating that its Kansei intention in elegance is much stronger than other samples. The score of sample 15, 27, 38, 37 and 32 is also higher. The factor 2 of sample 33 gets the higher score, indicating that Kansei intention in the simple aspect is much stronger than other samples, followed sample are 30, 6, 31, 37 and 34, respectively.

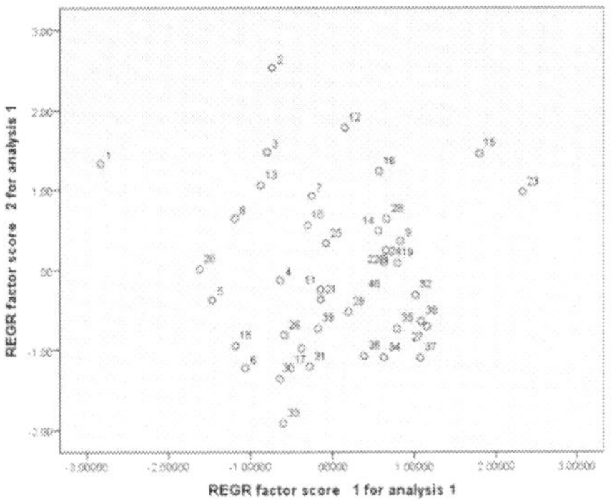

Fig.1 Scatter Diagram of Two Factor Score Variables of Each Sample

3. Discussion

The study regards career suits as the research carrier, considers female college students as experimental objects, extracts Kansei information of career suits and studies it:

1) By conducting dimension reduction process on original variables, the factor with the higher contribution rate is obtained. The results show that contribution rate of accumulated variance for 2 factors is close to 80%, indicating that these two factors can stand for 9 initial Kansei words to analyze Kansei respondents' intention level of career suits. As a result, top 2 factors can be extracted to use as representative factors to analyze corresponding design elements. According to variable load, they are named as the elegant factor and simple factor, respectively.

2) It can be observed by further analyzing variable loads in the factor 1that a set of career suit with elegant intention also is novel, fashionable, mature and individual. It can be observed by further analyzing variable loads in the factor 2 that a set of career suit with vocational intention is contracted and serious. This also conforms to results of traditional experience in Kansei evaluation of career suits.

3) By calculating factor 1 and factor 3 of each sample and scores of comprehensive factors, ranking situation from high to low scores of each sample is acquired. According to factor, top 6 samples are extracted typical design elements. Representative elements of the elegant factor include X silhouette, waist, clothes length around hipline, decoration on the waistline, linear collarless or mandarin collar, decoration of hachure or satin face on the edge of the neckline, symmetrical form on front fly, simple fastener or hook connection, natural line on shoulders, skirt length over the knees, and straight skirt. The overall style emphasizes on waistline and lines, auxiliary detail decoration.

According to the factor 2, top 6 samples are extracted for classical design elements. Representatives of the simple factor include type H, no waist, clothes length above hipline, lapel, no decoration on the top collar, symmetric front fly, single-breasted form, stand out official sense, clothes length above knees, straight or type A skirt. Overall style presents sense of lines, but it doesn't emphasize. Block surface treatment is used as a primary decorative technique. It is natural as a whole. In the future, career suits for female college students are designed from these two aspects and it is matched with Kansei intention.

Fig.2 Representatives of the elegant factor

Fig.3 Representatives of the simple factor

4. Conclusion

For female college students, elegance and contraction are representative factor variables of Kansei intention for career suits. In the future, the design can conduct style planning from these two aspects, so as to realize reduction of dimension and simplification. According to high scores of samples in two factors, classical design elements are extracted as the representatives of Kansei factors. As designing, it can be used as reference, so as to do Kansei design of career suits with targets. This paper only presented data for the women's career suits styles as a case. Future work should be conduct on analysing other types of clothing, in order to construct a more precise model to measure consumers' emotional activities.

5. Acknowledgement

This study has been supported by the talent introduced fund from Minjiang University (MJY15002) and the science and technology program of Minjiang University (MYK15006).

References

[1] Chuan G. The study on emotional design and Kansei engineering system's development of office suit [D]. Shanghai university of engineering science: 2010.
[2] Wei Y. Analysis of modern of fice suit design styles and development tendency [J]. Journal of Tianjin Polytechnic University: 2002; 21 (5): 53-59.
[3] Wei G, Jun X. Individual differences in perceptual evaluation for fashion design: taking female students as research subjects [J]. Journal of Textile Research: 2014; 35 (5): 137-141.
[4] Jia L, Chen DS. Yue S. Event-related potentials technique using in affective fashion design [J]. International Journal of Clothing Science and Technology: 2016; 28(1): 77-91.

Research on Sensory Marketing Experience in Clothing Retail Terminal

Bo Ji, Zhao-Hui Gu*

Apparel and Art Design College of Xi'an Polytechnic University, Jinhua Road, Xi'an, Shan Xi, 710048, China

*Corresponding author's email: lightgu2008@126.com

Abstract

According to reports on the current economic environment and problems of domestic retail terminal, the experience marketing strategy contains the following, themes, situational experience, interactive experience, product experience and effect management. As a result of the difficult application of the marketing strategy in clothing retail terminal, five senses are brought in the experience marketing strategy in this paper. Meanwhile, according to the analysis of consumers' market survey, the method to use five senses in marketing experience is put forward and a marketing plan is made. The purpose is to enrich the clothing experience marketing strategy system and better guide the clothing retail terminal to carry out experience centered marketing.

Keywords: Sensory Marketing Experience; Clothing Retail Terminal; Five Senses; Marketing Strategy

1. Introduction

Extensive concern was raised when the concept of experience marketing was brought into China. Hence, in order to adapt to the marketing system, not only does the product design need to be promoted, the retail terminal also needs to be promoted. Issues related to marketing experience has become a hot topic of the apparel industry, for instance, how to use experience in the clothing retail terminal marketing to incorporate "five feeling senses" into the marketing strategy etc.

In 1998, the concept of experience was launched by Joseph Pine and James Gillmor. [1] In 2009, the evaluation system of five senses in the marketing of the clothing store is established, which is based on three major theories (the theory of apparel stores marketing, experiential marketing and human perception). [2] In 2013, in order to enhance the consumer's experience, Wang Nan pointed out that the product, environmental facilities, interactive advertising, service promotion, branding and other means should be considered in experience marketing. [3] In 2015, Zhang Fan on the basis of experiential marketing (the sensory marketing, emotion marketing, thinking, action marketing, relation marketing, etc.) took into consideration of real marketing cases and developed a study on them. [4] In 2017, Zong Tingting and Guo Ruiliang analyzed the present situation of customized men's clothing development and the related theories of experiential marketing. They introduced the existing customized men's marketing strategy from the aspects of sense, emotions, thinking and action respectively. [5] Accordingly, the marketing experience based on five sensors have been gradually applied in the field of clothing .

2. Experience Marketing Strategy of Clothing Retail Terminal

2.1 Problems of Clothing Retail Terminal Experience Marketing

With the development of experience marketing, more and more clothing retailers have come to realize the importance of enhancing clothing experience. [6] There are however misunderstandings of retail terminal, for instance the lack of theme and activities management, and the difficulty of attracting customers to participate in marketing activities. As such it is difficult to meet the demand of the consumers' multi-level consumption.

2.2 The Clothing Experience Marketing Strategy Based on Customer Contact Points

With the development of the experience economy, the demand of consumers are also changing. Therefore, the feeling and psychological reaction of consumers should be considered in clothing marketing strategy. In the paper of "Clothing Display Design Based on the Theory of the Experience Marketing Research", the deep connotation of clothing experience marketing strategy was discussed and a set of clothing experience marketing strategy was developed that is focused on the consumer's feelings.

2.3 The Content of the Experience Marketing Strategy

In accordance to the contact between individuals and the given situation, products and other individuals, in order to build a complete experience strategy pattern, the experience marketing strategy needs to incorporate three things namely, situationajl atmosphere, type of products and interaction with consumers. [7] The main contents includes themed experience, situational experience, interactive experience, product experience and effect management respectively (Fig.1). [8]

2.4 The Characteristics of the Experience Marketing Strategy

1) Resolve the problems of experience marketing strategy, such as the lack of the theme and activities management and the low rate participation of marketing activities of customers.

2) Attract consumers to participate in the clothing marketing activities and in creating a unique experience.

3) Increase brand awareness through interacting with customers.

2.5 The Improvement of the Experience Marketing Strategy

Although this marketing system puts emphasis on contact with customers, it does not stress the important elements (the five senses) through which the customers experience the world. [9-12] The five senses "sight, hearing, touch, smell, taste" are the ways in which the consumers communicate with their reality. [13-15] As such the five senses were brought in the mid of this experience marketing strategy in order to enrich the strategy system (Fig.2).

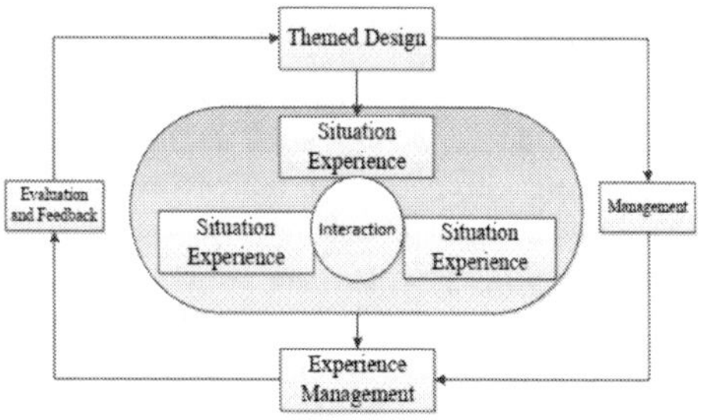

Fig.1 Model Architecture Diagram

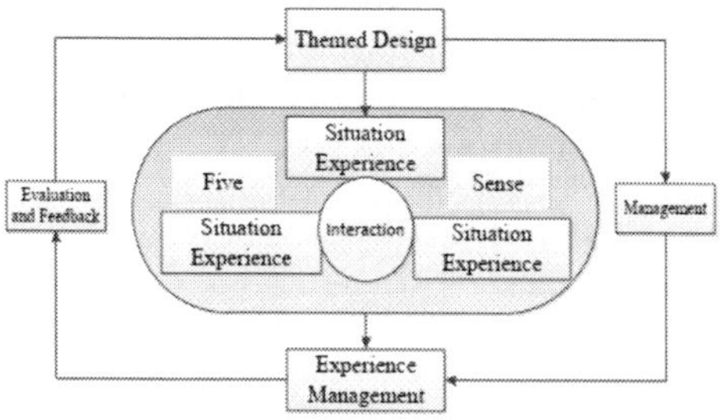

Fig.2 The Improvement of the Experience Marketing Strategy

3. The Analysis of Market Research

In order to analyze the influence of the five senses on consumers, a questionnaire was conducted (including 80 online and 120 paper-printed questionnaires). The object of the investigation were college students, government official, enterprise staff, self-employed and freelancer. The survey was conducted in Xi 'an Kai Yuan Shopping Mall and Xi 'an Sega Shopping Center respectively.

3.1 The Analysis of Reliability and Validity Hence

1) The reliability analysis of the questionnaires

Generally speaking, the reliability of questionnaire refers to the consistency and reliability of the results. The methods of reliability measurements are the following, retest reliability method, split-half reliability, split-half method and Cronbach α. Most of the survey questionnaires were based on attitude. [10], it is reasonable to use split-half method to analyze the reliability of the questionnaires. The SPSS 13.0 for windows was applied in this paper. If the Cronbach α is more than 0.9, then it indicates that the reliability is high. Between 0.8-0.9, reliability is considered as perfect, if less than 0.8, the questionnaire needs to be modified. Table 1 shows the reliability results for the different objects.

2) The validity analysis of the questionnaires.

Validity refers to the level by which the measurement methods can accurately measure the desired matters. The methods used in the questionnaire validity analysis are content validity, structural validity and criterion validity. This kind of questionnaire was conducted for the first time, as such only the content of structural validity was analyzed. If KMO is more than 0.9, the questionnaire is perfect, a digit between 0.8-0.9 indicates that the questionnaire is suitable for the investigation. If KMO is less than 0.7, then it does not meet to the requirement. Table 2 shows the validity results of the different objects.

Table.1 Results of the Reliability

Occupation	Students	Government Official	Enterprise Staff	Self-Employed	Other
Reliability	0.931	0.834	0.806	0.829	0.826

According to Table 1, the reliability of students, government official, enterprise stuff, self-employed and other individuals are 0.931, 0.834, 0.806, 0.829 and 0.826 respectively, all the results are between .8~0.9 which shows that the reliability of this questionnaire is high.

Table.2 Results of the Validity

Occupation	Students	Government Official	Enterprise Staff	Self-Employed	Other
Reliability	0.923	0.861	0.909	0.916	0.901

According to the Table 2, the validity of students, government official, enterprise' stuff, self-employed and other individuals are 0.923, 0.861, 0.909, 0.916 and 0.901 respectively, thereby demonstrating that the validity of this questionnaire is good.

3.2 Statistical Analysis

Table 3 shows that 200 individuals were investigated and 190 out of 200 questionnaires are valid, including 80 out of 120 online and 110 out of 120 paper-printed questionnaires.

Table.3 Valid Rate of the Survey

	Amount (Out)	Amount (Back)	Response Rate
Xi 'an Kai Yuan Shopping Mall	52	48	92%
Xi 'an Sega Shopping Center	68	62	91%
Online	80	80	100%
Total	200	190	95%

Table.4 Gender Distribution

Variable	Option	Number	Percentage
Gender	Men	82	43.16%
	Women	108	56.84%
Total	——	190	100%

According to Table 4, the percentage of female in the survey is 58.64%, the age distribution is mainly between 18 to 28 and 29 to 39 (64.21% and 23.68% respectively). It covers a very wide range of subjects (students account for approximately 51.06%, government officials take up around10%, enterprise staff share 20%, the percentage of self-employed is 7%). (Table 5- 6)

Table.5 Age Distribution

Variable	Option	Number	Percentage
Age	18—28	122	64.21%
	29—39	45	23.68%
	40—50	20	10.53%
	>50	3	1.58%
Total	——	190	100%

Table.6 Occupation of the Object

Variable	Option	Number	Percentage
Occupation	Student	97	51.06%
	Government Official	20	10.53%
	Enterprise Staff	38	20.00%
	Self-Employed	7	3.68%
	Freelancer	20	10.53%
	Other	8	4.21%
Total	——	190	100%

1) The Effect of Sight

Fig.3 and Fig.4 shows the atmosphere of a store account for the major influence on the consumers' buying appetite. The table suggests that it is important to create a good atmosphere that will promote the consuming behavior of the customers. Fig.5 shows that the right products exhibition is the most vital part of the experience marketing strategy.

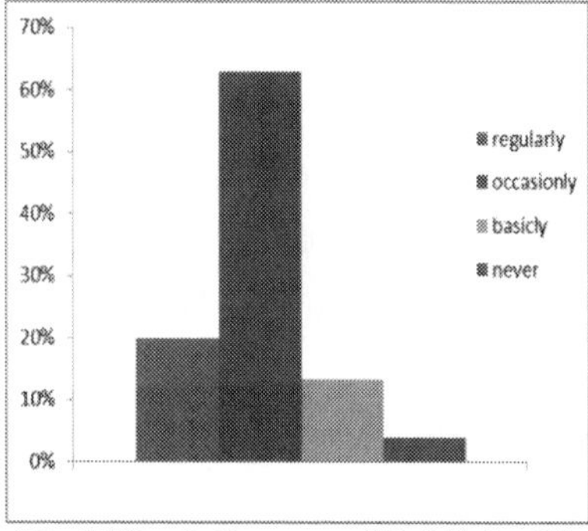

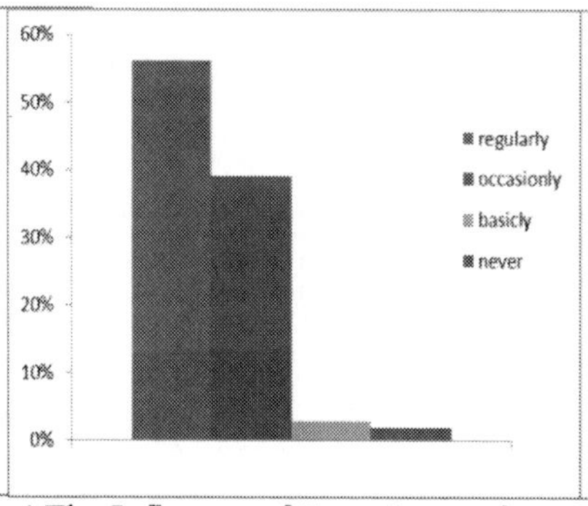

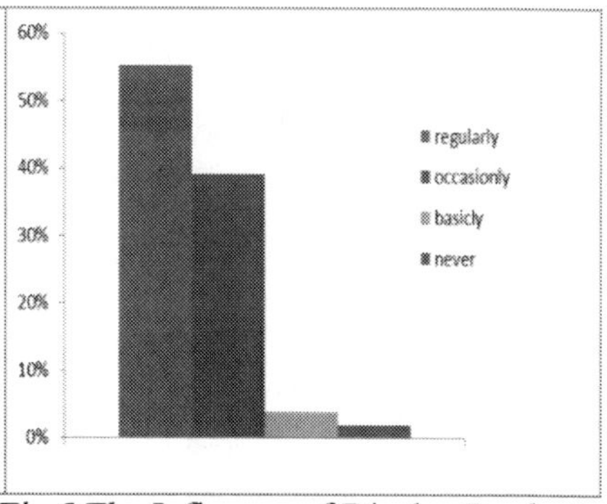

Fig.3 The Influence of Publicity Fig.4 The Influence of Store Atmosphere Fig.5 The Influence of Display Products

2) The Effect of Hearing

Fig.6 and Fig.7 illustrates that background music and the altitude of sales staff profoundly affects customers' purchasing behavior (around 45% and 70% respectively).

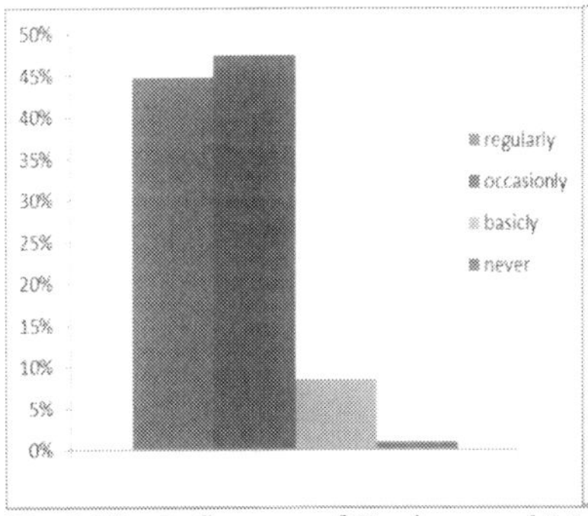

Fig.6 The Influence of Background Music

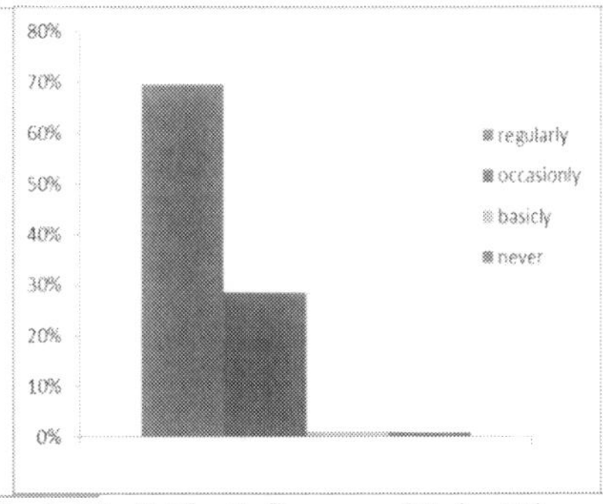

Fig.7 The Influence of Sales Discourse

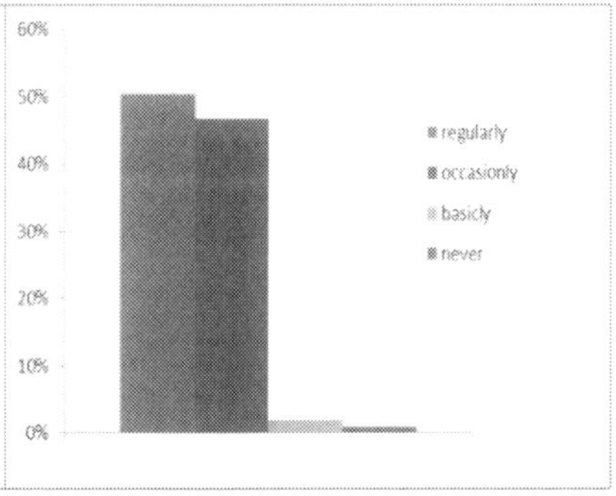

Fig.8 The Influence of Environmental Odor

3) The Effect of Smell

Fig.8 shows that the sense of smell in the experience marketing strategy can bring consumers a better overall shopping atmosphere. A good smelling environment will make consumers feel joy abd extend their shopping time thus influencing their shopping behavior.

4) The Effect of Taste

As an experience in marketing, taste is a strategy that aims to achieve customer's loyalty. This is done through differentiation of service and it is beyond things like color to attract the attention, smell and odor to bring calm, sound to stimulate staying in a place and taste for surprising the taste. This issue leads the customer to understand the unique value of the experience and creates a memorable experience for them that stimulates the replication and extension of this experience, regardless of the price (Fig.9).

5) The Effect of Touch

Touch is the direct way consumers feel the products. It is the most vital part in the five senses, which enables customers a comprehensive understanding of the function of both the product and quality. They feel the charm of the product, which will affect their psychology and behavior. (Fig.10).

Touch is the largest sensory organ of the body, by touching the products, customer behavior and shopping attitude is positively affected. Touching the product encourages shoppers to interact with it. Material, temperature, weight and shape postively impacts on the touch experience and enhances customer loyalty.

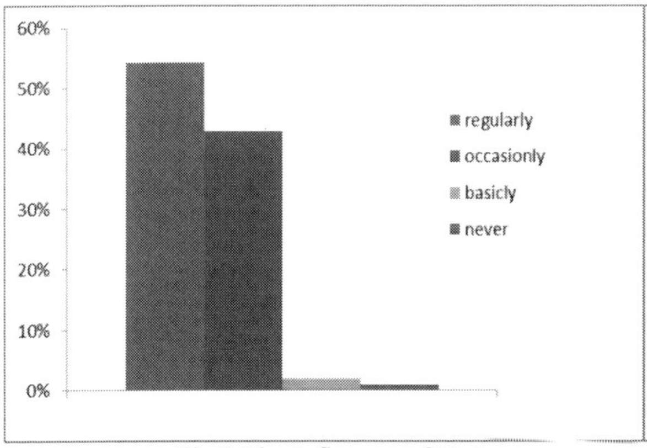

Fig.9 The Influence of Taste

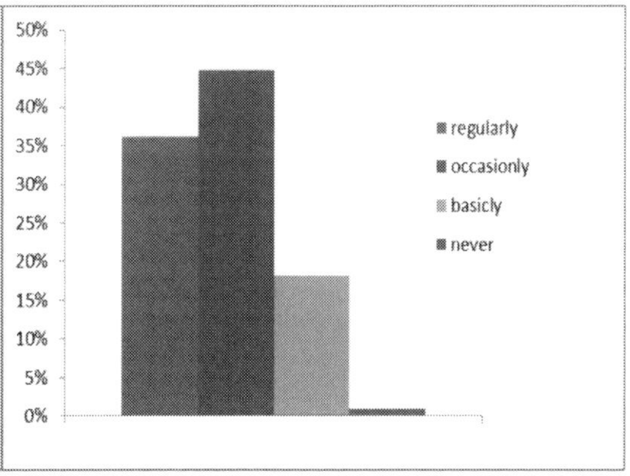

Fig.10 The Influence of Product Tactile

3.3 The Survey Conclusions

Through the analysis of the survey, different influence of the five senses on consumers can be found. Store atmosphere, good product display and store propaganda has a strong influence on consumers as sight elements. The background music and sales staff language profoundly affects customers hearing. A good smelling environment will make consumers feel comfortable. It can extend the consumer's shopping time and affect shopping behaviors.

3.4 The Usage of Five Senses in Experience Marketing Strategy

According to the results of the investigation, some tips are put forward to incorporate the five senses in experience marketing strategy.

1) Sight: the propaganda and the overall decoration of the store play a significant role in affecting the buying behavior of the customers.

2) Hearing: the duration time of the customers is a result from the background music and the attitudes of the salesman. It affects the buying behavior indirectly.

3) Smell: the smell of the store creates a relaxed environment for the consumers which could extend the shopping time.

4) Taste: the role of the taste plays gently in affecting the emotion of the customers, however, details of the brand determine the duration of consuming.

5) Touching plays a key role in affecting the consumers' behavior, and it directly shapes shopping behavior.

4. The Application of Five Senses in Experience Marketing

In regards to the new experience marketing strategy and survey conclusion, a marketing planning for clothing brand that uses the five senses in experience marketing is made in this paper.

4.1 Themed Experience Design

The poster points out the theme of the activity from the earlier propaganda, at the same time, with the help of the promotional video on the LED screen, an impression were given to the consumers from sound and sight. Fig.11 is the poster of the theme.

4.2 Situation Experience Design

1) Sight

A dark tonal model of the indoor sand bar is made, and the overall store is designed with a tropical beach theme set in southern California. The dim light, electronic screen display with the scenery of the west bank of California, the tropical totem element wall and the slogans written in California style are all elements that help build the feeling of freedom, comfort and happyiness

Fig.11 The Poster

Fig.12 Experience Activity Process

2) Hearing

Mixed with the electronic trance music, the background music gives the customer a hearing feast.

3) Smell

Perfume is the direct resource of smell, So Cal is mixed with wood and fruit incense. They are the mark of ae beach lifestyle. As such it is clear that So Cal uses smell in the store. The scent makes customers feel better in the music festival.

4) Taste

Store offers cocktails and whiskey and dessert for the consumers to enjoy

5) Touch

The store is decorated with dark gray wood which aligns with the theme of retro California.

4.3 Product Experience Design

The store model was made for the music festival specifically. All customers can try on the clothing, and they enable the customers to see, touch, feel and smell.

4.4 Interactive Experience Design

An activity is launched to allow consumers to take photos with our clothes. All the photos will be shown on WeChat, and consumers who participated in this activity can win a big prize. Fig. 12 shows the experience activity flow chart.

4.5 Effect of Management Design

In order to rapidly adjust experience strategy, to adapt to the market, the effect of this plan will be evaluated in case of resource waste of enterprises.

5. Conclusions

The conclusions are as follows:

1) According to analysis of the current economic environment and problems of domestic retail terminals, theme experience, situationional experience, interactive experience, product experience and effect management are quoted in the experience marketing strategy. As a result of the difficulty in implementing an effective marketing strategy in clothing retail terminal, five senses are brought in the experience marketing strategy.

2) The survey shows the different influence of five senses on consumers. Store atmosphere, good product display and store propaganda has a key influence on consumers and are part of the sight element. The background music and the attitudes of the staff affects customers as hearing element. The smell of the store makes the consumers feel comfortable. It can increase the consumer's shopping time and affect shopping behaviors. The other senses have a more subtle influence on the consumers.

3) According to the improved experience marketing strategy, a marketing plan is made.

References

[1] Erving G. The Presentation of Self in Everyday Life. Beijing: Peking University Press: 2008; 78.

[2] Li B. Research on the Five Feeling Experience in Apparel Stores Marketing. Dong Hua University Master's Degree Thesis: 2009.

[3] Wang N. The Elements of Experience Marketing Implementation in Clothing Retail Research. Science times: 2013; 22.

[4] Zhang F. Experiential Marketing Analysis. Chinese and Foreign Entrepreneurs: 2015; 103-103.

[5] Zong TT, Guo RL. Analysis of the Experience Marketing Strategy for Male Customization. Progress in Textile Science & Technology: 2017; 1.

[6] Sun YH. Experiential Marketing Application in the Clothing Brand. Journal of Textile Industry and

Technology: 2011; 40: 89-91.
[7] Li P. Clothing Display Design Based on the Theory of the Experience Marketing Research. Journal of Commerce of China: 2012; 2: 54-55.
[8] Zhao M. Martin L. Brand Sense. Tianjin Education Publishing House: 2011.
[9] Li MY. Five Senses Comprehensive Experience in the Application of Environmental Art Study, Shan Dong Normal University Master's Degree Thesis: 2011; 4.
[10] Costa MF. Sensory marketing: consumption experience of the Brazilian in the Restaurant Industry. Int J Bus Str: 2012; 12 (4): 165-171.
[11] Achrol R. S, Kotler P. Frontiers of the Marketing Paradigm in the Third Millennium, J Acad Mar Sci: 2012; 40 (1): 35-52.
[12] Farias S. A Conceptual Frame Work and Research Propositions for An Extraordinary Customer Experience . Int Bus Res: 2014; 7 (2): 87-99.
[13] Krishna A. Sensory Marketing, Embodiment, and Grounded Cognition: A Review and Introduction. J Consum Psychol: 2014; 24 (2): 159-168.
[14] Krishna A. Product Scent and Memory. J Con Res: 2010; 37 (1): 57-67.
[15] Krishna A. An integrative Review of Sensory marketing: Engaging the Senses to Affect Perception. Judgment and Behavior. J Consum Psychol: 2012; 22 (3): 332-351.

Application of Traditional Apparel Pattern of She Nationality in Fujian Province to the Adolescents' Design

Xiu-Mian Chen*, Jin-Lian Bai

Department of Commercial Art, Fujian Business College, Wufeng Street Gulou District, FuZhou, FuJian,350012,China

*Corresponding author's email: chenxiumian558@126.com

Abstract

Fujian She nationality apparel with its magnificent patterns and ribbon technology is well known, but now it is undergoing chinesization, and at a risk of becoming lost. The traditional apparel characteristics of She nationality in Fujian were analyzed based on a survey of traditional apparel and apparel pattern of She nationality in Fujian province. Through the questionnaire survey of the adolescents subjects, we found the characteristics of their costume consumption and the aesthetic preference of apparel patterns. Fujian She traditional apparel pattern and its application to adolescent design are discussed based on layout, color, craft and other several aspects.

Keywords: Fujian She Nationality; Traditional Apparel Pattern; Adolescent Costume; Costume Consumption Characteristics; Aesthetic Preference

1. Introduction

She traditional apparels are unique and are mostly made of self-woven ramie cloth. The color of She costumes are black and blue. Woman apparels are mainly the right length of the top jacket, cardigan jacket, skirt, apron and Phoenix clothing. Clothes and apron are embroidered with a variety of flowers, birds and geometric colorful patterns. Accessories are mainly headdress, necklace, earrings, bracelets and rings. Men's apparel type includes right short jacket, right long coat, jacket, trousers, leggings and hats. Usually in the collar, skirts and cuffs have a variety of decorative patterns. During their ancestor's wedding they wear a long shirt that are embroidered with dragon-shaped pattern. Children apparel type includes right long coat, jacket, pants, hat and tiger-head shoes. Industrious women of the She nationality are knitting and embroidery craftsmen, because they produce a wide range of handicrafts with colorful, unique styles, such as weaving ribbons. The patterns are diverse with beautiful colors, which have significant national and regional characteristics.

With the prosperity of the economy, and the mixing with the Han nationality, fewer people are wearing She traditional apparel. Furthermore it has become difficult to differentiate the difference between the decoration of the She people and the Han. Such as some of the contemporary young people will not choose to embroider She traditional apparel pattern and use the ribbon technology to weave which is a huge threat to the She traditional apparel inheritance and development. To protect the traditional apparel culture of She Nationality it is important to promote it to the society.There has not been much data gathered that shows She traditional apparel pattern being used by foreigners, while domestically, She National apparel patterns have be searched, they are limited to the study of the art characteristics of the She Nationality apparel, cultural connotation and its use in handicrafts [1-11]. The study that She traditional apparel patterns are applied to the costume design is very little, especially in the use of the adolescents' design is still blank. The traditional costumes of the She nationality were applied to the adolescents' design based on a survey of apparel patterns of Fujian She nationality and adolescents' questionnaire survey. So the adolescents' understanding of the clothing culture of the She nationality was strengthened. It is a way to promote the inheritance and protection of the She nationality apparel culture.

2. The Characteristics of Traditional Apparel Patterns of Fujian She Nationality

2.1 The Classification of Traditional Apparel Patterns of She Nationality

The themes of traditional apparel patterns of She nationality can mainly be divided into two types: figurative and abstract patterns. The former are vivid manifestations of concrete things, such as plants, animals and characters, Plum blossom, orchid, bamboo trees, chrysanthemum, peach blossom, peony, curly grass are very common plants in plant patterns, as shown in Fig.1.while butterflies, dragonflies, magpies, cranes, tiger leopards, lions can be found in animal patterns. As for character patterns, boys are the most favorable decorative elements on curtains and hats. Sometimes, plant patterns and animal patterns will appear together. For example, the combination of phoenix and peony; the combination of lotus and boy; and the combination of dragon and phoenix. These combinations are widely known as auspicious patterns.

Abstract patterns are highly generalized symbols which can not only express people's longing for a better life and also reflect the traditional culture of the She ethic group. Take Phoenix costumes in She culture for example. The design of this costume, including head-wear and shoes, is a complete imitation of the bird phoenix, which symbolizes that the female is the royal princess. Totem patterns, abstract geometric patterns and text patterns are the three common themes in abstract patterns. Dragon squama and Phoenix are the most representative totem patterns in She culture.

The abstract geometric patterns, such as canine, million words, cloud hook, coins, willow stripes, especially those continuous triangular canine patterns, are the most common designs featuring for strong sense of modeling. Some text patterns express auspicious message, such as Fortune text pattern, as shown in Fig.2. while other text patterns can tell locations, surnames and so on, which are generally woven on ribbons.

Fig.1. The plant pattern on Fu'an women apron

Fig.2. The text pattern on Luoyuan women costume

2.2 Characteristics of Pattern Composition on Traditional Apparel of She Nationality

There are four ways of pattern composition on traditional apparel of She nationality, separate patterns, suitable patterns, corner patterns and two sides continual patterns. These patterns are partially applied to the costume collar, cardigan, cuffs, hem and other parts. Different decorative patterns are used in different parts of the costume, among which separate patterns and two sides continual patterns are more widely used. Separate pattern composition of plant and animal is usually applied on the chest part of the costume and aprons. Corner pattern composition with the combination of animal and plant patterns are usually applied on aprons and hats. The most classic pattern composition is to choose several cloth strips embroidered with continuous pattern and abstract geometric pattern, and then arrange them orderly on cardigan parts of She costume. Thereby, it will form a decorative design made of three to more than 10 trims, as shown in Fig.3. It shows the unique artistic charm of the She nationality apparel pattern, and also demonstrate the unique ethnic customs of the She Nationality.

Fig.3. The pattern on Luoyuan Woman cardigan

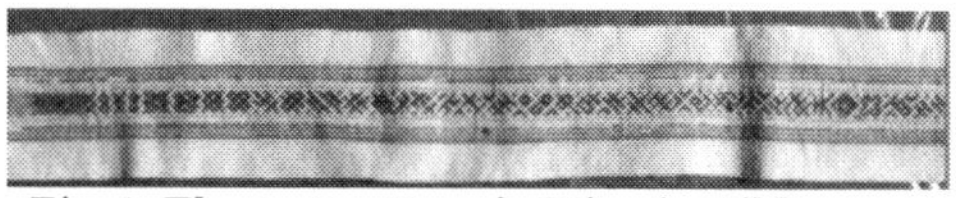

Fig.4. The text pattern in Ningde Ribbon

2.3 Color Expression of Traditional Apparel Patterns of She Nationality

In our country, different nationalities have their own views on color and aesthetic taste [12]. People of She nationality love black and blue, because they think these two colors are solemn and simple. The most distinctive feature of She nationality apparel is colorful patterns. Colorful ribbons, cloth pieces and embroidery patterns are added to costume collar, cardigan, cuffs, hem and other parts. Those "rainbow colors", including red, orange, yellow, green, blue, indigo and purple, have strong brightness and great purity, as shown in Fig.3, and they can enhance brightness of overall effect of the costume. A wide variety of hues with a small area of uniformity are usually chosen, and different patterns are orderly arranged on collar, placket, sleeves and other parts of the costume. In that way, these color patterns with unexpected changes and strong brightness can be naturally assimilated into the dark background color of the costume. Therefore, a variety of colors can be mixed harmoniously on the apparel of She nationality so as to express the beauty of nature.

2.4 Technological Process of Traditional Apparel Patterns of She Nationality

Weaving and embroidery are the two main technologies in making traditional apparel pattern of She nationality [13]. People of She nationality weave different apparel patterns mainly on ribbons, which is usually decorated with text patterns, as well as coins, words and other patterns. The handmade woven ribbons can show craftsmaships of She people. When weaving the ribbon, they have to fix one end of the ribbon, and the other end is fixed around the waist of the weaver. Patterns are not weaved along the two sides of the ribbon. Instead, the main pattern is along the center line, as shown in Fig.4. Both sides of the ribbon are white without patterns, while the middle ribbon is weaved with black lines with those red, green and blue lines close on both sides.

The embroidery craft of She nationality has a long history, and it still retains its unique folk craftsmaships. Embroidery techniques mainly include plain embroidery, cross stitch and patch embroidery. Embroidery themes are free creation out of She people. Those exquisite embroidered patterns are both attractive and very practical. Take embroidery pattern on "Phoenix" costume for example. It takes at least one week or even one month to embroider flowers on a Phoenix costume simply on hands. These embroidery techniques, with a high decorative value, is an important vehicle to carry the traditional cultural heritage and national spirit of She nationality.

3. The Significance of the Application of the Traditional Apparel Pattern of Fujian She Nationality in Adolescents' Costume

Through the study of questionnaire, the characteristics of the adolescents' costume consumption and the aesthetic preference of apparel pattern are analyzed. This study will guide the rational application of traditional apparel pattern of She nationality in the design of adolescent costume. In this survey, a total number of 400 questionnaires were distributed, with 400 copies recovered, and 50 invalid. The questionnaire rate was 100% and the effective rate was 87.5%. The questionaire is made of three parts: The first part is the basic information, including school, grade, place of birth, parental occupation; The second part is on features of clothing consumption, including life attitudes, purchase motives, purchase sites, purchase frequency; The third part is on aesthetic preferences, including apparel pattern selection preferences, composition characteristics preferences, color expression preferences, process performance preferences and other related issues. The results of this survey provides a theoretical basis for the rational use of the traditional apparel patterns of the She nationality in the design of the adolescent costume.

3.1 Adolescent Costume Consumption Characteristics

The adolescent population, at the age of 13-18, was mainly born at the year of 2000, and then the birth rate fell to a record low. It was a time when China's per capita GDP exceeded 10,000 Geary-Khamis Dollar and was surfing towards 20000 Geary-Khamis Dollar, and might even break through the 30000 Geary-Khamis Dollar, into the cultural industry boom. It was a time of prosperous development for the generation born after 2000 [14].

People born after 2000 live in a time when the market economy is more mature. The development of market economy to a civilized and legal society also help people born after 2000 to realize that the real happiness of life lies in the satisfaction of personal values. Social and economic environment has impact on adolescent's costume purchase motives, purchase sites, purchase frequency. The survey results show that contemporary adolescent clothing consumption has a distinctive characteristic of the times, as shown in Table 1. From Table 1, it can be seen that the consumption characteristics of adolescent groups of the people born after 2000 are perceptual and impulsive. Their are in the rapid development both physically and mentally. Meanwhile in the Internet era, they are more passionate, self-centered, and rebellious. They like online shopping without considering the price and they pursue "non-mainstream" personality.

Table.1Characteristics of Adolescents' Costume Consumption

Questionnaires	Survey results	consumption characteristics
Life attitude	They are good at expressing their own ideas, and they like life full of passion and new things.	Sensuality, passion, self
Purchase motive	They will buy them as long as they see the favorite costume. 58.3%of adolescent swill not consider the price.	Impulsive, rich life
Purchase place	64.9% of adolescents choose online shopping and micro-business, 35.1% of them choose to buy costume at speciality stores and shopping malls.	Internet era, convenient
Purchase frequency	51.2% of adolescents buy costumes at least once every two months, and they would buy multiple costumes One-time.	Like consumption, impulsive
Knowing ways	39.6% of adolescents know information through the network, 23.1%of adolescents know it through magazines, advertising, television, 11.2% of adolescents know it through classmates, friends recommended.	Internet subculture
Purchase preferences	They will not pay attention to the brand and the style.They like leisure style. They like black and blue color.	Non-mainstream, publicity personality

3.2 Adolescents' Aesthetic Preference for Apparel Pattern

People born after 2000 have more resources and broader perspectives to experience the real world and the Internet. Internet pop culture such as animation culture, game culture, the popular culture with its strong inclusive and "flowing" aesthetic taste deeply attracted the adolescent [15]. Adolescent have a good knowledge of the popular network culture, but they know little about the national culture. The survey results show that 78.2% of adolescent do not understand She Nationality. Through the survey of aesthetic preferences of adolescent apparel patterns, it is hoped that the She traditional apparel pattern will be applied to the design of adolescent costume more properly.

The questionnaire covers the apparel pattern, composition, color, technology and other aspects of the relevant issues.

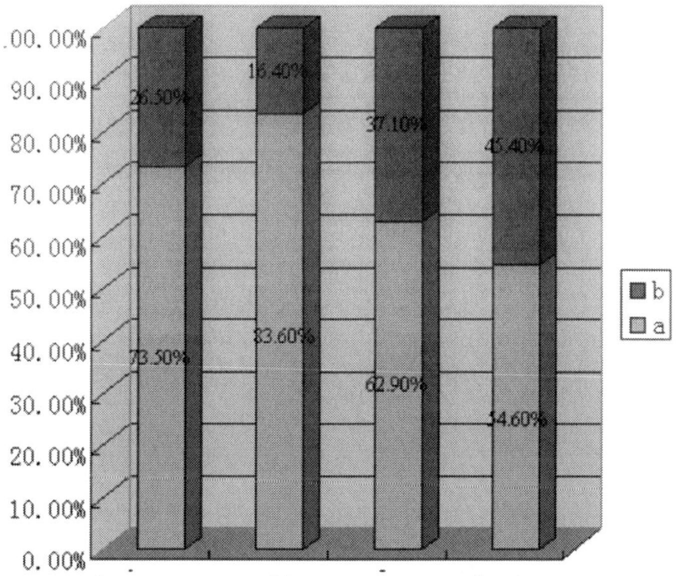

Fig.5. Adolescents' aesthetic preference for apparel pattern

Fig.5 is an analysis on the theme of apparel pattern: a for the concrete matter; b for the abstract theme; 73.5% of adolescent love plants, animals and other subject matter of the apparel pattern. Figure a stands for the local application; b stands for the overall application; 83.6% of adolescent prefer the pattern which is partly applied to the costume design so as to highlight the characteristics of costume style. Only 16.4% of the adolescent like the costume to be full of pattern, but most adolescents said it was too fancy. Adolescents want to grow up quickly, and get the recognition from parents and teachers. They like to wear clothes of dark colors, such as black and gray because they think it would show that they are more mature. And as for the color of patterns 62.9% of adolescent like blue, red, yellow and other bright colors (a). As for the apparel pattern processing, they tend to choose simple and convenient process (a); 45.4% of adolescent like the classic pattern process, but they think it is too complicated. From the survey results we can see that network culture can subtly influence values and aesthetic preferences of the adolescent.

3.3 The Significance of the Combination of She Traditional Apparel Pattern Elements and Adolescent Costumes

Adolescents are growing up rapidly, they need to constantly change their clothes. With the improvement of people's life, the costume consumption of adolescents has become a huge market which can produce as much as 30 billion annual sales profits [16]. At present, China's adolescent costume market, especially in the area of casual costume market, the sale volume is still dominated by foreign brands [17].

In recent years, with the development of Chinese economy, China's culture is gradually open up to the world. Chinese traditional culture and works of art have attracted worldwide attention. In the most fashionable costume field, China's traditional costumes, with its distinctive national characteristics, have become the source of inspiration to the world's fashion design , and have been widely used by costume designers around the world. They extract design elements from the style, structure, craftsmanship, fabric, color, pattern from traditional ethnic costumes. If the apparel pattern of the traditional She Nationality is applied to adolescent costume design, it will boost up the domestic brand of adolescent costumes. On the one hand, the costume of domestic brand has the advantages of price and information, because it will reduce costs and improve profits. It will have a bright market prospect, and can speed up marketization of the domestic brands of adolescent costume. On the other hand, it will make the She traditional apparel elements assimilate into the national culture of the adolescent, which can not only highlight the national spirit, but also help the adolescents understand the spiritual and cultural connotation of She nationality people, which has a very important practical significance for the inheritance and protection of the traditional costumes of the She nationality.

4. The Application of Traditional Apparel Pattern of She Nationality to the Design of Adolescent Costume

Adolescents are in rapid physical development, so adolescents are more suitable for casual costume. She traditional apparel patterns are applied to the design of adolescent costume, so that adolescent costume has national cultural characteristics of content. Fujian She nationality apparel pattern is one of the traditional Chinese apparel culture. It should follow a certain principle when She apparel pattern is applied to the design of adolescent costume, so that the She nationality apparel culture is effectively integrated of adolescents' costume.

4.1 A rational selection of Traditional Apparel Pattern of She Nationality based on the Adolescents' Costume Consumption Psychology and Aesthetic Preferences

The survey results show that adolescents are passionate, self-centered, and have strong personality. Their costume consumption is emotional, impulsive. Combined with the survey results of adolescents' aesthetic preference for apparel patterns, when the designer select pattern theme of adolescent costume, the concrete matter of the She traditional apparel pattern should be considered first, such as plant matter like plum and chrysanthemum and animal matter like butterflies and dragonflies. At the same time the most representative totem pattern, such as Long Qi, Phoenix and other patterns can be properly selected. She traditional apparel pattern is applied to the design of adolescent costume. It is not a slavish copy of the pattern theme in the design. The pattern theme should be abstracted, summarized, refined and re-created. Based on the adolescent costume consumption psychology and aesthetic preference, the most typical features of the traditional pattern, such as the shape, lines will be combined with the beauty modern design. And then the new traditional patterns are applied to the design of adolescent costume, so that adolescent costumes have national characteristics and are also very trendy.

4.2 She's Traditional Apparel Pattern Should be Partly Applied to the Design of Adolescent Costume

When the traditional apparel pattern of the She nationality is applied to the design of adolescent costume, it should take into account the application form of the pattern. The survey results show that adolescents do not pay attention to brand. They love leisure style and prefer costumes which are partly decorated with traditional patterns. Therefore, She's traditional apparel pattern should be partly applied to the design of adolescent costume, and composition form of the pattern should be reasonably selected. Two sides continual pattern can be used on the collar, cuffs, placket, hem of the costume. The chest, sleeves, back, shoulders and other parts can choose a separate pattern or corner pattern. The size of the pattern should be adjusted in order to coordinate with the style and size of the costume. As shown in Fig.6, the text pattern in the ribbon of Fujian She nationality is decorated with the placket and the collar of male adolescent costume, which make the style of adolescent costume fashionable, and the special artistic aesthetic effect also highlights the ethnic characteristics of the She nationality

4.3 A Reasonal Combination of Color and Pattern Between She's Traditional Apparel and Adolescent Costume

The color match should consider the preferences of the adolescent, the design concept of adolescent costume and the environment. The color of She's traditional apparel pattern itself has a strong national characteristics. People of She nationality advocate black and blue, but the color of apparel pattern is "rainbow color", including red, orange, yellow, green, blue, indigo and purple. According the survey, adolescents are fond of black, gray, white-based color, because they think these colors seem more mature to them. The colors of decorative pattern are blue, yellow, red and other bright color embellishment. We can see that adolescent and people of She nationality both like dark colors, decorative patterns for the bright color. When design the adolescents' costume, they should make clever use of this common ground. As shown in Fig.7, the color of the costume is black; the color of apparel pattern is red; and a little blue and purple can enhance the brightness of the overall effect of the costume. This costume is decorative and interesting with characteristics of She's apparels both in pattern and color.

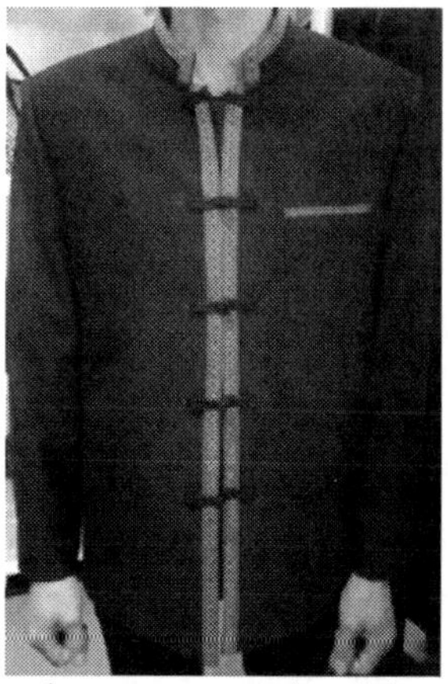

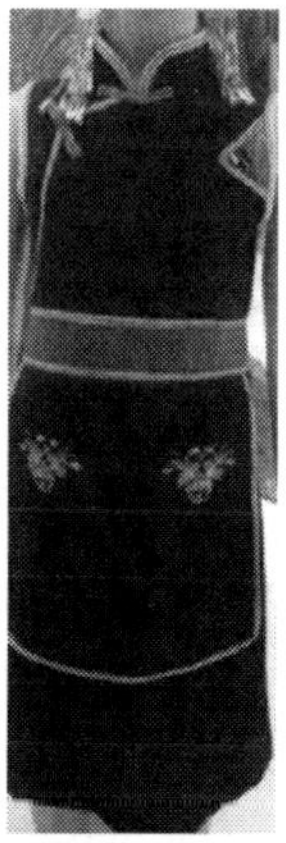

Fig.6 Two sides continual pattern on male adolescent costume Fig.7. The color application of She costume Pattern

4.4 A rational selection of Processing Technology of She Traditional Apparel Pattern according to the Fabrics of Adolescent Costume

Today, there are a wide variety of processes of costume pattern, including organic printing, weaving, hand printing and dyeing, embroidery, knitting. Different means of technology of the pattern processing have their own characteristics [18]. In the survey of the adolescent, the results show that adolescents love leisure style, simple, convenient pattern making technology. According to a comparison of the most commonly used processes, the machine printed technology is the most convenient, efficient. The pattern on the costume is vivid and it is easy to wash, and this technology is suitable for a variety of fabric processing. Adolescents like casual costume which is generally made of knitted fabrics, so it is suitable for using machine processing technology. For the use of dragon and phoenix pattern, you can choose embroidery processing technology, and embroidery pattern has a high artistic aesthetic value. Now the cost of hand embroidery is very high, so machine embroidery is more widely used. So the design of adolescent costume of She traditional apparel pattern is a combination of machine printing and embroidery of the integrated technology. Adolescent costume tend to more creative, trendy, and it also show the characteristics of She costume culture.

5. Conclusion

Fujian She traditional apparel pattern, with its rich subject matter, smart composition, bright colors, outstanding craft, is well known around the world, and it shows traditional culture of the She people and their aesthetic taste. Adolescents are the future of the country, the hope of the country, and they shoulder the responsibility of protecting and inheriting Chinese traditional culture. The apparel patterns, which are the distillation of the essence of the She's art, should be applied to the design of adolescent costume. When the traditional elements are integrated into the modern fashion design, those elements are also integrated into the study and life of the adolescent, which would deepen the understanding of traditional costume culture for the adolescent. It also would awaken the protection consciousness of the traditional costumes of the She nationality, so that She nationality apparel culture will be well protected and inherited.

6. Acknowledgement

This research was supported by the Young Teacher's Education and Research Projects in Fujian Provincial Department of Education (No.JAS160665).

References

[1] Zheng J. Implied meanings and expressions of Shes' finery decorative patterns. Art and design: 2010; 10: 275-276.
[2] Chen X, Chen DS. Shes female costumes pattern in Ningde and Xiapu of Fujian. Journal of textile research: 2009; 3: 88-92.
[3] Ma HR. Artistic research on the silverware pattern of She Nationality in the east Fujian. Journal of Fujian Agriculture and Forestry University: 2014; 5: 109-112.
[4] Zhang KY. The beauty of She Nationality dress in the east area Fujian. Big stage: 2008; 1: 43-45.
[5] Hu ZJ. Discussion on the traditional costume of She Nationality. Journal of Guangxi arts college: 2006; 4: 32-34.
[6] Shangguan ZQ. Research on costume art and cultural connotation of the She Nationality in Fujian province. Shanxi University the thesis for master degree: 2008; 6: 33-36.
[7] Qiu HL. She dress culture symbol of application design.Zhejiang Sci-tech University the thesis for master degree: 2014; 2: 16-23.
[8] Lin Q. The expression of She Nationality's traditional cultural elements in the design of modern mublic space landscape. Artistic life: 2013; 5: 68-70.
[9] Lei HZ. The traditional pattern elements of She Nationality in decoration of Longquan celadon. Art Review: 2013; 5: 122-15.
[10] Chen XH, Liao HJ, Sen W. The creative application of dress patterns of She minority in the design of brands and packaging. National art research: 2015; 5: 139-142.
[11] Zhou JR. In the ceramic decoration embodies the She nationality dress pattern. Fine art: 2012; 12: 58.
[12] Han HJ. Research on clothing color. Southwest Normal University Press, 2002. p.33-36.
[13] Zhou Y. Chinese minority dress handicrafts. China Textile Publishing Press, 2014.p.47-50.
[14] Information on http://www.360doc.com/content/14/0831/22/7542574_406130581.shtml
[15] Liu S. Animation, games and other popular network culture on the aesthetic impact of young people. Art Education Research: 2016; 4: 80-81.
[16] Information on http://news.ef360.com/Articles/2014-7-30/316375.html
[17] Bai XL, Li X. The Development Prospect of Youth clothing market in China. New campus Journal: 2013; 6: 21-23.
[18] Wang F. Pattern Design for Fashion and Accessory. Shanghai People's Fine Arts Publishing Press: 2007; p.155-160.

Application of Tang Dynasty's Baoxiang Patterns in Modern Men's Clothing Design

Yuan-Zhuo Cai, Zhao Lv*

Apparel & Art Design College, Xi'an Polytechnic University, Jinhua road, Xi'an, Shaanxi, 710048, P.R. China

*Corresponding author's email: 1183369029@qq.com

Abstract

Being a traditional floral motif in ancient China, the Baoxiang pattern was one of the main decorative costume patterns in Tang Dynasty. This paper analyzes the art form of Baoxiang pattern from the Tang dynasty and discusses the significance of combining the pattern with Men's Clothing Design. Research shows that the Baoxiang pattern has great aesthetic value and is a Chinese cultural heritage. When combining Baoxiang pattern with modern men's clothing, the pattern will be modified to meet the modern people's aesthetic and psychological demands.

Keywords: Key words: Baoxiang Pattern in Tang Dynasty; Modern; Men's Clothing Design; Applied Research

1. Introduction

In the context of economic globalization, the facilitation of cultural transmission is becoming greater. The cultural influence of countries overseas has had a huge impact on traditional Chinese culture, as such the traditional Chinese culture has started to fade in light of globalization [1, 2, 3]. Although there are many new design concepts and techniques in contemporary fashion, it is critical for every designer to learn, inherit and preserve the essence of the traditional national costume culture, while at the same time create new forms and changes [4]. It is only through this method can traditional Chinese art form be preserved.. The Baoxiang pattern, a type of traditional Chinese classical pattern has high aesthetic and cultural values in which is an integral part of the traditional Chinese culture. The life of design lies in innovation, and the soul of innovation rests with cultural inheritance. Both of which have a dynamic connection [5]. Based on the achievements made by contemporary designers and through personal experience, further explorations are conducted and some suggestions are put forward on the application of the Baoxiang pattern in the design of men's clothing.

2. Baoxiang Pattern in Tang Dynasty

2.1 The Definition of Baoxiang Pattern in Tang Dynasty

The term "Baoxiang" is originally derived from Buddhism. It is an ancient saying that "the bearing of the emperor is inward, while Baoxiang gives off its charm." This illustrates the noble status of the Buddha [6]. As such the Baoxiang flower in Tang Dynasty was a combination of patterns based on lotus petals and other plants and flowers. The Baoxiang flower embodies an intense religious and symbolic color(such as: gold and yellow). With time the flower has evolved into a model pattern and reached its peak in Tang Dynasty when Buddhism gained unprecedented development. This phenomenon indicated that the Baoxiang pattern and religious culture are closely interconnected [7]. During the Kaiyuan period of the Tang Dynasty, the Baoxiang pattern developed into a purely auspicious pattern and its stylized image symbols were established and perfected. The symbols includes a lotus pattern, acanthus design (scroll design), pomegranate design, cloudy design and peony design [8]. The patterns were symmetrical with a circumference and were full flower shapes rich in petals. The Baoxiang pattern absorbed these design complexities and appealing features to form a grand plant decoration motif. The colors of the

Baoxiang flower are mainly blue, green, red and gold as they were contrastive of one another, thus they were able to provide a rich and honorable visual perception.

2.2 The Cultural Connotation of Baoxiang Pattern in Tang Dynasty

The Baoxiang pattern stems from the imagination of Buddhists. The pattern resembles a dragon, phoenix and other similar things from traditional Chinese culture. The Baoxiang pattern characteristics determine its completeness and reunion. Its patterns are naturally dense, which has certain relations with the traditional aesthetics in China. From the perspective of the pattern's image, its overall contour and partial patterns are both characterized by circles that symbolize people's hope of a more happy and successful life.As such it is possible to perceive the shadow of traditional culture in China through the research on the Baoxiang pattern. The research is conducive to the study of ancient culture [9].

3. The Manner of Formation and Expression of Baoxiang Patterns in Tang Dynasty Costumes

3.1 The Formation Characteristics of the Baoxiang Patterns in the Tang Dynasty

3.1.1 Symmetry

Symmetry indicates that the elements of the Baoxiang Pattern repeat frequently and but do not clutter. The overall composition of the pattern adopts the structure of Chinese character "十". The rudiment of Baoxiang pattern is Chinese character "米" based on Chinese character "十"; the foundation of the pattern is circular radiation (as Fig.1) or a positive diamond-shaped radiation (as Fig.2). The symmetrical distribution of four ends (upper, lower, left, and right) is a structure followed by simple successive and extensive patterns. This variety of symmetrical concept has a certain relationship with China's traditional culture. For instance, Chinese buildings and decorations are tied to symmetry. Based on the above structure and foundation, a pattern of 8 petals shaping like character "米" will be formed after 45 degrees of rotation. A pattern of 8 petals enriches the details of the overall pattern. In addition to the 8-petals style pattern, there are other patterns of 10 petals, 12 petals and 14 petals patterns. These are all based on structure of character "十".

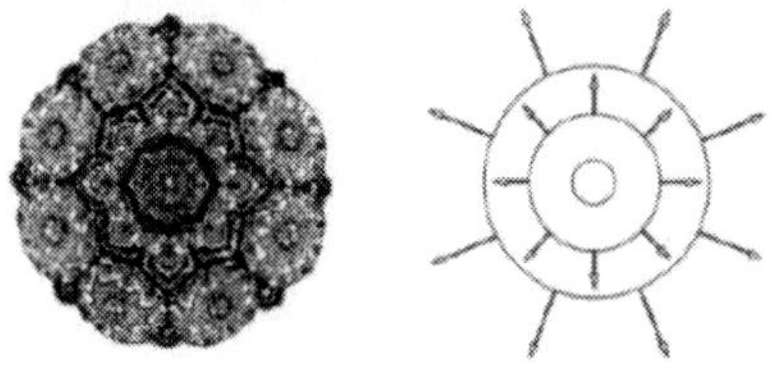

Fig.1 the Baoxiang pattern of circular radiation

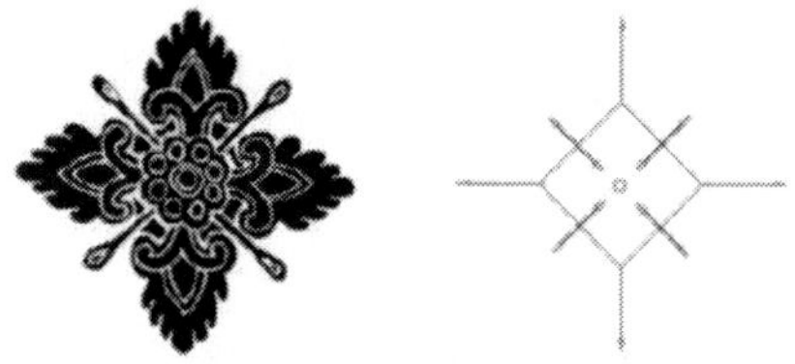

Fig.2 the Baoxiang pattern of diamond-shaped radiation

3.1.2 Rhythm

The Baoxiang pattern is composed of a variety of visual symbols with rhythmical charms created by a series of rich-layer patterns. Changing parts of the symbol adds to the rhythmic elements of the pattern. Although the pattern is static, there is still a dynamic feeling.

3.2 Expression form of the Baoxiang Pattern in Clothing of Tang Dynasty

The Baoxiang patterns are applied as decorative arts on fabric through dyeing, weaving cloth and embroidery etc. The pattern of brocade commonly known as "Tang Jin" is the most beautiful pattern (Fig.3). Distinguished from the traditional wrap-knitting method prior to Tang Dynasty, the weaving method of the brocade adopts a flat-knitting method that is capable of weaving complex and colourful Baoxiang patterns. Compared to traditional wrap-knitting, the flat-knitting fabric is rather conducive to colour change and illustrates the pattern very well.

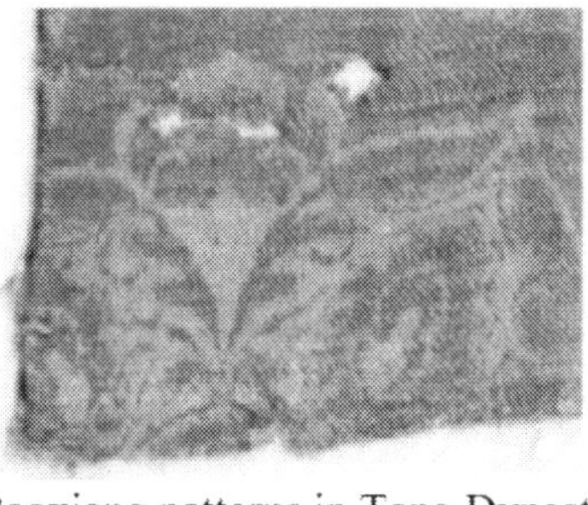

Fig.3 Brocade with Baoxiang patterns in Tang Dynasty(www.baidu.com)

4. The Application of Tang Dynasty's Baoxiang Pattern in Modern Men' s Clothing Design

4.1 Make Adjustments to Color Scheme of Patterns to Adapt it to Modern Aesthetic Standard

There is a disparity in aesthetics between traditional and modern people, and colors being their main difference. For example, traditional people use monotonous colors of green, red, yellow, white and black. This is because at the time there were limited. On the other hand in modern society there are hundreds of color types that allows people to wear clothing with a wide range of colors. It also presents a developmental trend using combinations of a variety of colors in modern men's clothing. This is a contrast to the monotonous colors used in the past. The younger generation has a particularly strong desire for color, so charming and polychromatic clothes can capture their attention. Therefore, it is favorable to increase color types and their combination modes to enhance the beauty of Baoxiang patterns. For instance, designers can enhance the brightness and saturation of colors while also increasing fashion elements of the Baoxiang pattern to render it as a modern art feature. The fashion traits of the Baoxiang pattern will be revealed and stay in accordance with the characteristics of modern people and then further conform to the modern aesthetic standards (as Fig.4). China's clothing designers should be bold in their innovations in the field of Baoxiang patterns by using more colors when designing to achieve better results [10].

Fig.4 Multicolor Baoxiang pattern

4.2 Transformation Processing

It is a common practice to conduct an appropriate transformation of original patterns and it is widely applied in the field of modern fashion design. This approach can also be adopted with the design of Baoxiang patterns. For instance, simplistic styles in modern fashion design draw more attention. Therefore, it is advisable to transform and change complex Baoxiang patterns while retaining maximum morphological features and main decorative patterns. The transformation purifies and flattens the traditional patterns, while maintaining the original characteristics of the

Baoxiang pattern. A simple and decent design effect will be gained and a win-win outcome can be delivered. The Baoxiang pattern consists of multi-layer flowers and each layer are decorated with smaller flower patterns. The superimposed layers increase the intricacy of the pattern. The fashion sensation may be reduced if the patterns are completely reflected in modern clothing designs. However, concise designs have easier access to meet the aesthetic needs of modern people (as Fig.5). Therefore, the simplification process of the Baoxiang pattern is a feasible means to incorporate into the modern design of men's clothing. In addition, simplification is a favorable practice to mirror male disposition of simplicity and in being capable and experienced [11].

Fig.5 Simple transformation processing of the Baoxiang pattern in Modern Men's Clothing Design

4.3 Exaggerated Transformation Processing

Exaggerated transformation treatment is a relatively disparate means when compared to the above transformation processing. Changes in exaggerated transformation are even greater. It re-innovates the original Baoxiang pattern while preserving the fundamental elements. The decorative features of the floral pattern are rather distinct. They can be boldly pronounced during theexaggerate transformation process. This presents a more obvious decorative effect. For example, the size of a flower layer can be enlarged so that they appear more conspicuous. The shape of the flowers can be distorted to bring about a magical result. This kind of audacious transformation treatment is even more suitable when applied towards designing clothes of a younger generation because it aligns with the finger popping trend, while also conforming to the characteristics of the youth.

4.4 Appropriate Decomposition and Reorganization of the Pattern

The decomposition and reorganization of the Baoxiang pattern is a daring innovation. In the process of decomposition, the colour, image, and the performance techniques are decomposed. There are lots of skills for decomposition. In the process of reorganization, all the components or part of the elements can be reorganized into a new pattern. The Baoxiang pattern, through decomposition and reorganization, generate even more enjoyment and modernization. Of course, the basic elements of the pattern have to be retained and it is not permitted to rip the pattern apart. The traditional cultural elements should be reserved.

4.5 Integration of Baoxiang Pattern and Geometric Pattern

An effect of combining the old and new can be attained through the integration of modern geometric patterns and the Baoxiang pattern. For example, designers can make distributed diagonal designs of the Baoxiang pattern as well as the right-angle trim designs. The gap can then be filled with eye catching red geometric shapes. This variation of design will create a visual attention on the chest. The application of geometric graphics will increase the effect of flexibility and relaxation. The Baoxiang pattern can also enhance the beauty of clothing. They both play different roles in the market, and the result of clothing design will be effectively improved with the integration of these two factors. It will be favoured by youth. (as Fig.6).

Fig.6 Integration of Baoxiang pattern and geometric pattern in Modern Men's Clothing Design

4.6 Combining Baoxiang Pattern with the New Technology and Materials

The Baoxiang pattern can be manifested by means of embroidery such as pearl embroidery, machine embroidery, and even knitting. Regarding the materials, metal, precious stones and other new materials can be used to deliver better visual effects [12, 13, 14]. The combination of the traditional Baoxiang pattern with new technology enriches clothing design and it represents artistic structure of the Baoxiang pattern well.

5. Conclusion

As part of China's traditional culture, Baoxiang pattern should be protected and inherited by the current generation. Because of its extremely rich aesthetic value in design, the Baoxiang pattern can be applied in fashion design, especially in men's clothing line. It not only can enhance the artistic perception of the male apparel, it can also amplify the elegance and internal attractiveness of the male in such apparels. Therefore, when male apparel designers adopt Baoxiang pattern into their designs, they should integrate this art form with the addition of modern pop elements. In the meantime, the pattern location and the spatial arrangement of the patterns in the apparel require more attention [15]. Furthermore, the patterns should suit the styles of the apparel and meet the requirements of the apparel craftmanship. Only when the charm of the national tradition is well manifested in the modern apparel meanwhile conforming to the aesthetic preferences of modern society can such Chinese element be favourably applied in modern men's Clothing Design.

6. Acknowledgement

We would like to thank Shaanxi Province Social Science Fund Project support, "The sorting and research of various ethnic minorities along the Silk Road in the Tang Dynasty "project, item number: 13SC040 and innovation Fund for postgraduate of Xi'an Polytechnic University support, "Study on Uighur crowns in Tang Dynasty "project, item number: CX201703.

References

[1] Li XL, Zhang Y. the Integration of Local Fashion Innovative Design and Traditional Culture [J]. Art and design: 2010; 2: 224.
[2] Li YF, Hu Y. Fashion Art Design [M]. China Textile & Apparel Press: 2006; 235.
[3] Liu XG, Feng L. Fashion Design Introduction [M]. Donghua University Press: 2010; 316.
[4] Sun Y. The Application of Chinese Traditional Clothing Elements in Modern Dress Designing [J]. Art and Design: 2008; 5: 197.
[5] Zheng QX, Liang HE. Application of Floral Medallion Pattern on Costumes in Tang Dynasty in Modern

Women's Wear Design [J]. Journal of Silk: 2014; 8: 51.
[6] Yang SL. Study on Baoxiang Pattern [J]. Commercial culture: 2011; 9: 327.
[7] Shen CW. Chinese ancient costume research [M]. Beijing: The Commercial Press: 2011; 433-436.
[8] Zhu XY, Shi J. Splendid Blooming Baoxing Flowers-a Study on Patterns of Baoxing Flowers in Tang Dynasty and Its Application in Modern Accessory Design [J]. Journal of Gems and Gemmology: 2012; 6: 59.
[9] Zhou ZH. Evolution of the Dermatoglyphic Pattern from Lotus Flower to Baoxiang Flower [J].Journal of Shanghai Business School: 2007; 6: 42-43.
[10] Hu LX. Baoxing flower pattern and its evolution in the Sui and Tang Dynasties [J]. World of Antiquity: 2009; 7: 34-38.
[11] Liu Z, Xu C. Analytics of Tang Dynasty Treasure Appearance Floral Lines Pattern Construction Form [J]. Packaging Journal: 2010; 4: 74-76.
[12] Wei YL. Study on innovative design and application of embroideryin fashion design [J]. Wool Textile Journal: 2014; 7: 32.
[13] Zhou LL. Huang CC.Ye P. On the protection and inheritance of embroidery in china [J]. Wuhan university journal: 2016; 3: 108.
[14] Kuang YH, Liu H. Study on Layout of Embroidered Patterns in Han and Tang Dynasties [J]. Journal of silk: 2014; 8: 29.
[15] Cao YQ, Qian XN, Zhang N, Zhang SK. Application of Huaiyang clay dog art form in pattern design of men's shirt [J]. Journal of Textile research: 2016; 9: 116.

Interview-Based Status Analysis of Colorful Trousers Used in Traditional Lahunqiang Opera

Hui-Jun Cao, Rong Zheng*, Meng-Lin Wang, Wen Wu

Aimer Human Engineering Research Centre, Beijing Institute of Fashion Technology, Beijing, 100029, China

*Corresponding author's email: yuki.zheng@gmail.com

Abstract

This study examined colourful trousers used for costumes in the traditional Lahunqiang opera that have undergone little change for decades. Chi-square testing was used to determine correlation between selection of trouser styles for 20 Lahunqiang plays and in 111 major roles whose outfits include gowns, coat, jacket, shoes and boots. Based on interviews with 10 actors, the study analyzed trouser usage and factors influencing their selection. The study found that formality influences selection of trouser colour and style, subject to factors such as type of role and costuming, including shoes and boots. Trousers' unique design and stable pattern and structure are closely associated with their principal feature of allowing a wide range of movement. While it remains inconvenient to don the trousers and difficult to preserve them, their pattern features and coordination with costumes have been passed down and preserved. From the perspective of clothes patterns and people–clothes relations, the study identified reasons for the colourful trousers' continued use that contributes to the development and protection of intangible cultural heritage.

Keywords: Traditional Lahunqiang Opera; Colourful Trousers; Design Pattern; Interview; Influence Factor

1. Overview

The Lahunqiang Opera, also known as Zhouguzi [1], is now popular mainly in Xuzhou in Jiangsu Province, Linyi in Shandong Province and some places in Anhui Province. Tune prolongation is used for singing during the finale with assisting chorus; male characters' ending sound remains at the same octave while female characters employ coloratura portamento, in a higher octave, which has been regarded to have the effect of drawing in people's soul. Lahunqiang has evolved from its beginnings as talking and singing performances by street artists to the current local opera with professional actors. The opera's performance groups have also developed from small theatrical troupes to amateur opera troupes and finally to professional troupes. The effects of modern media and people's changing attitudes towards traditional operas have lessened the popularity of Lahunqiang. As the traditional Lahunqiang plays and dramatic forms change gradually, the corresponding costume culture also has been affected. Therefore, targeted studies can help record and preserve such traditional cultural treasures.

Since the Republic of China (1912-1949) period, costumes used by Lahunqiang theatrical troupes have been purchased mostly from other parts of the country. All troupes have strict rules on costume management [2] and use similar terminology. The articles and costumes used in theatrical performances are stored in special "theatrical chests": the Dayi-literally "first clothes," referring to a wide range of gowns, capes and dresses for specific types of roles-chest is used to store gowns for singing and dancing characters; the Eryi-"second clothes" which, depending on the Dayi for a certain type of role, include various tops and some accessories-chest is for costumes of martial performance and all tops; and the Sanyi ("third clothes") chest is used to store auxiliary articles including colourful trousers and sweat-absorbent jackets. Therefore, it can be seen that the colourful trousers have subordinate position in costuming. The trousers' colour selection hinges on a character's gown and other exterior costumes, and their style is decided by the character's shoes and boots [3]. The Chinese Opera and Folk Art Dictionary defines the colourful trousers as "A kind of theatrical costume usually in red, white and black colour, embroidered. Its draped fabric texture make the colourful trousers flexible and lightweight, suitable for dance movements" [4].

A review of a variety of dictionaries summarises common characteristics of the colourful trousers: 1) red and black are the most frequently used colours; 2) patterns can be embroidered or non-embroidered; 3) style features loose pattern and heightened crotch; 4) fabric is usually silk or cotton cloth; and 5) functional design to facilitate dance and other movements.

The development of various traditional Chinese operas and exchanges among them has brought constant changes to their costumes. Lahunqiang's theatrical costumes have also changed. While absorbing and introducing costumes of operas from other areas, Lahunqiang has abolished and changed some of its costumes; still, according to the study's investigation, some colourful trousers purchased in the 1950s are still in use, which means that the colourful trousers are the bearers of theatrical costume traditions. With that in mind, the study first analysed the correlation between the trousers' shape and structure classification in various plays and the type of roles and other costumes. Then, based on interviews and from the perspective of cultural heritage, the paper analysed humanistic and social reasons for both the evolution and preservation of the traditional colourful trousers [5], which together can be the basis for further studies in the field.

2. Basic Information about the Colourful Trousers

2.1 Frequently used Colours and Styles

2.1.1 Classification of Colourful Trousers

The colourful trousers can be classified in terms of colour, texture and style. Although the gender for a specific type of role is fixed, for practical performance purposes men can also act as women. The main differences between men's and women's trousers are in their colours and embroidery, but the style and structure is basically the same. Fig.1 shows the classification of colourful trousers.

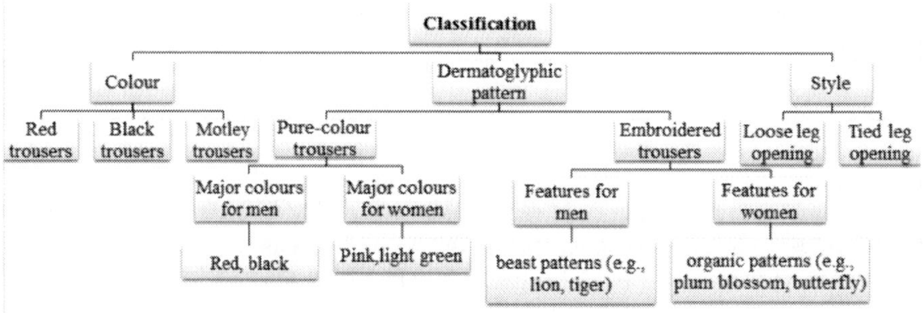

Fig.1 Classification of colourful trousers

2.1.2 Colour

Red, black and white are the major colours for traditional Lahunqiang opera colourful trousers. Male characters are often dressed in red trousers while female characters wear light-coloured trousers such as white, pink or light green. A certain rule governs the use of colour for trousers. One principle is that red represents nobility and black signifies humility. The principle conforms to the class concept embodied in the use of colour for official gowns during the Ming Dynasty; for example, officials above the role of magistrate of a prefecture wore red trousers while those below magistrate wore black trousers. A second principle pertains to colour match between the trousers and Dayi/Eryi. Dayi and Eryi constitute the main body of the theatrical costume outfit, covering most parts of the actor's body, thus, the trousers' colour should match them. Finally, a third principle is that the colourful trousers should correspond to the type of role in a play. For example, young female characters (e.g., an unmarried young lady of a noble family) should wear trousers of bright colours such as pink and light green, while trousers for young and middle-aged female characters in main female roles should have soft colours such as muted green and pale pinkish purple. [6].

2.1.3 Patterns and Texture

The colourful trousers' style is similar to the traditional Chinese Miandang trousers, featuring long and loose waists, wide legs and deep crotch [7]. Comparison of the two types of trousers is shown in Fig.2. The colourful trousers have two styles, loose legged and tied leg. Loose-legged colourful trousers should be worn when the character is also wearing a pair of low shoes; usually, the trousers' legs should cover the instep. Trousers with tied legs should be worn when the character wears boots; after being tied, the trousers' legs should be placed inside the boots. Silk and crepe-back satin is often used in making colourful trousers, and cotton cloth is used occasionally. Nowadays, imitation crepe fabric is also used in making the trousers, but accompanied by greige cloth for the waist part, which can be fastened to prevent deformation and the trouser from dropping. Usually the fabric used in colourful trousers is soft and fluid, which can enhance the character's graceful movements and make them more lively. Compared with everyday trousers, the colourful trousers are much looser from the waist to the bottom and have a deeper crotch, which can be conducive to the character's movement by reducing the clothes' constraint.

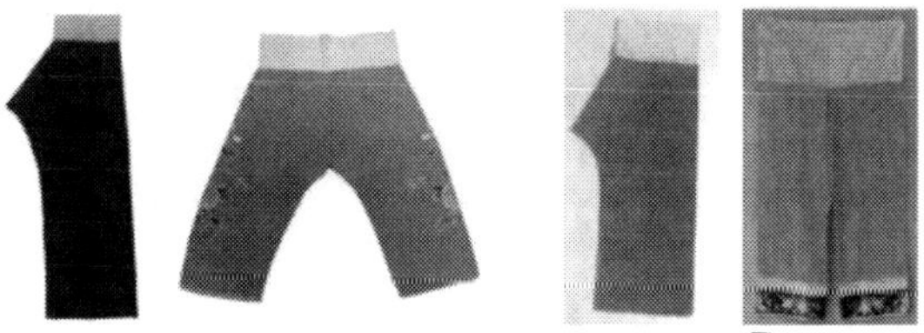

A B

Fig.2 Comparison of colourful trousers (A) and traditional Chinese Miandang trousers (B).

Note: Miandang Ku is a type of traditional Chinese trouser with a large waist band that is folded and fastened with a belt.

2.2 Colourful Trousers in Repertoire Environment

The study collected and sorted data on colourful trousers' patterns and colours used by 111 major characters in 20 traditional plays of the Lahunqiang opera. Chi-square testing was used to find the correlation among the selection of colourful trousers, the type of roles and other costume items including Dayi and Eryi and footwear.

As shown in Fig.3, it can be observed that red, white and black are most frequently used in colourful trousers. To coordinate trouser colour with Dayi and Eryi colour, yellow Dayi/Eryi would be matched with yellow and purple colourful trousers; blue Dayi/Eyri would be matched with white, blue and black trousers; red Dayi/Eryi would be matched with red and white trousers; black Dayi/Eryi would be matched with red and black trousers; and white Dayi/Eryi would be matched with white or black trousers.

Based on our interviews and investigation of the 111 plays, seven types of colourful trousers can be summarized in terms of pattern, numbered as 1 to 7, with their respective features shown in Table 1. It can be seen that the numbers 1, 5, 6 and 7 type of colourful trousers are most frequently used. Frequencies of colourful trousers of different patterns or colours used to match different Dayi/Eryi, type of role and footwear are shown in Figs.3-6. The value of the Chi-square test of the correlation between type of roles and colourful trouser patterns was 165.120 with the freedom degree of 132 and P value of the two-sided test at 0.027. At the significance level 0.05, the original hypothesis of the two variables being independent of each other was rejected. Therefore, with the probability of 95%, the two variables are correlated with each other. With the Cramer V coefficient at 0.498, it can be proved that the type of roles influences the selection of colourful trousers' patterns. From Fig.4, it can be observed that in the 20 traditional plays, in terms of type of roles, the number of dan (a type of female character) and sheng (a type of male character) outweigh other characters. The dan characters mostly selected No. 5 and No. 7 colourful trousers, and sheng characters mostly used No.1, No.6 and No.7 colourful trousers.

The value of the Chi-square test of the correlation between Dayi/Eryi and colourful trouser patterns was 229.168 with the freedom degree of 132 and P value of the two-sided test at 0.00. At the significance level 0.01, the original hypothesis of the two variables being independent of each other was rejected. Therefore, with the probability of 99%, the two variables are correlated with each other. With the Cramer V coefficient at 0.587, it can be concluded that Dayi/Eryi influence the selection of colourful trousers' patterns. From Fig.5, it can be seen that in the 20 traditional plays, the most frequently used costumes are short embroidered capes, zhezi (a kind of robe under the overcoat), culotte coat, jianyi (a kind of martial gown) and tea gown. Short embroidered capes are often matched with No. 5 and No. 6 colourful trousers; zhezi often matches No. 1 colourful trousers; culotte coat is often matched with No. 7 colourful trousers; and jianyi and tea gown often match No.1 and No.7 colourful trousers.

The value of the Chi-square test of the correlation between shoes/boots and patterns of colourful trousers was 76.264 with the freedom degree of 30 and P value of the two-sided test at 0.00. Therefore, at the significance level 0.01, the original hypothesis of the two variables being independent of each other was rejected. Hence, with the probability of 99%, the two variables are correlated with each other. With the Cramer V coefficient at 0.371, it can be concluded that shoes/boots influence the selection of colourful trousers' patterns. From Fig.6, it can be seen that in Lahunqiang opera, thin-sole shoes, colourful shoes, thick-sole boots and cloud-shaped shoes are often used.

The correlation coefficients obtained via Chi-square testing proved that selection of colorful trousers in plays is influenced by specific types of roles, Dayi/Eryi and shoes/boots. The test results also provide theoretical basis for the following interview questions and selection of nodes in data coding.

Table.1 Frequencies of colourful trousers of different patterns used in the 111 traditional Lahunqiang plays

No.	Pattern	Frequency
1	Bound legs + Inserted front & back pieces* + Loose large waist	33
2	Bound legs + Inserted back pieces + Loose large waist	1
3	Loose legs + Inserted front & back pieces* + Loose large waist	7
4	Loose legs + Inserted front piece + Loose large waist	1
5	Bound legs + Bound waist	20
6	Bound legs + Loose large waist	13
7	Loose legs + Loose large waist	36

*"Inserted front & back pieces" means pieces are inserted into front and back of trousers' crotch part; unless otherwise indicated, the crotch part has no inserted piece.

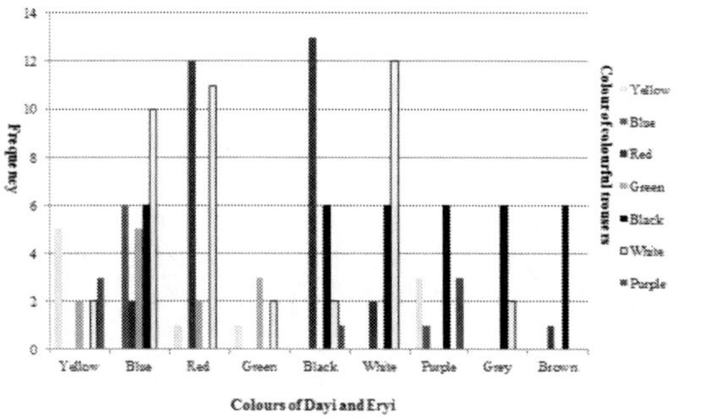

Fig.3 Colour match between Dayi/Eryi and colourful trousers.

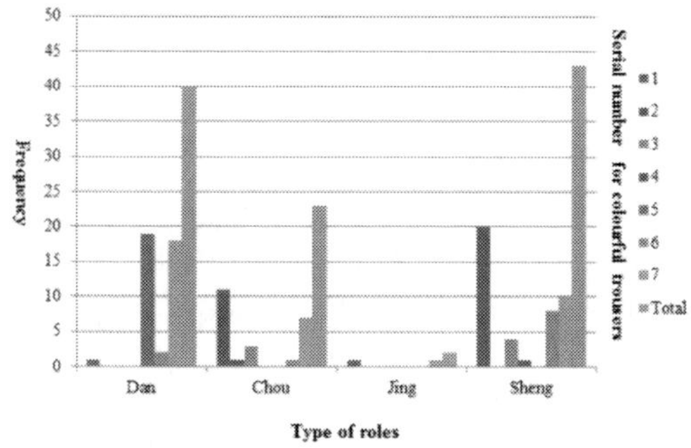

Fig.4 Relation between type of roles and selection of different styles of colourful trousers.

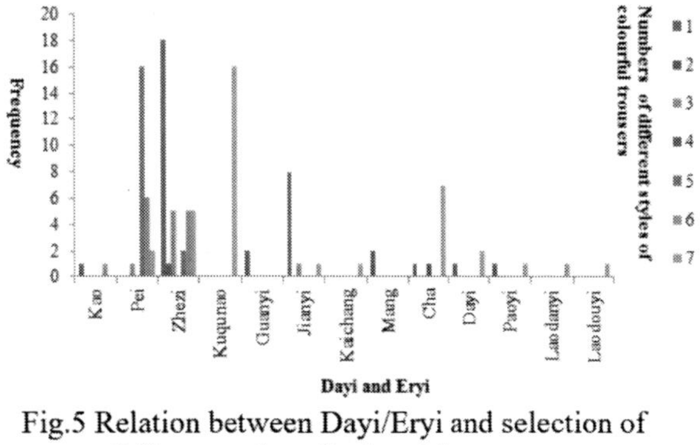

Fig.5 Relation between Dayi/Eryi and selection of different styles of colourful trousers.

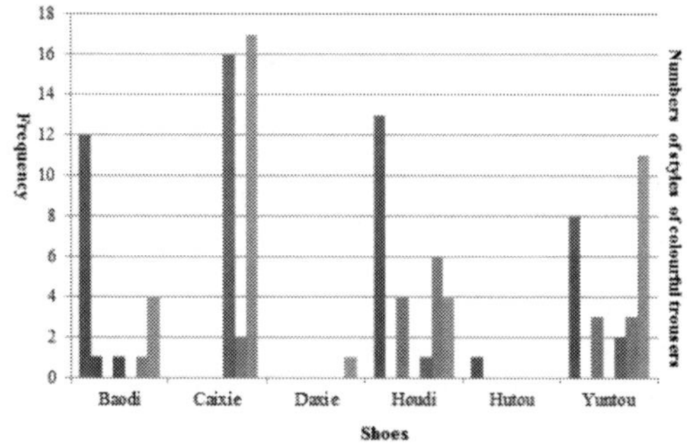

Fig.6 Relation between shoes/boots and selection of different styles of colourful trousers.

Notes to Figs.5-6: Kao, an armour-like costume item for a martial character in a play, usually has four flags inserted at its back. Pei, a short embroidered cape, is often worn by officials and their wives in such colours as red, yellow and blue. Zhezi refers to gowns worn by civilian characters in a play. Kuqunao, a dress outfit worn by a daughter of a humble family or a servant girl in a play, consists of a short and tight padded top and a pair of embroidered pants. Guanyi are modified robes are usually worn by civilian officials in a play. Jianyi refers to a type of gown for riding and shooting, originally worn by northern barbarian tribes, with horse-hoof-shaped cuffs, a round collar, large front and body-long length. Kaichang literally means open cloak, with a skew collar, large cuffs and a bottom hem. Mang, the official dress for emperors, kings, generals and high-ranking civilian officials, features a round collar, large front and cuffs, and embroidered patterns of dragons and sea water. Chayi refers to the costume for clowns, a type of short blue cotton-cloth upper garment. Dayi refers to the costume for acrobatic fighting in black or white. Paoyi refers to a type of grey-blue gown for government runners, in the same shape as Zhezi but without long sleeves. Laodan Yi refers to the costume for senior female characters in a play. Laodou Yi refers to the costume worn by old civilian characters in a play, made of tussah silk of the original colour with the same shape as Zhezi. Baodixie are thin-soled shoes, while Caixie are cloth-made hard-bottom shoes for young female characters with embroidery, 1cm-thick sole and a toe cap decorated with silk fringe. Daxie, cloth-made shoes for martial characters, have an upper with cross patterns created by alternate black and white cloth stripes. Hutou Xie, which literally means tiger-head shoes, are thick-soled boots made from satin, with high boot legs just below the knee, with the toe cap patterned like a tiger head, thus accounting for its name. Yuntou Xie refers to shoes with their toe caps decorated with cloud patterns, often used by celestial beings and Taoist priests.

3. Colourful Trousers Selection and use Status Based on Interviews

Costumes, including colourful trousers, have both material and spiritual properties that should meet the needs of the character on stage and also satisfy the spiritual needs of the actor [8]. Based on interviews, this section explores the colourful trousers selection principle from the subjective perspective of actors with consideration of the relationship between people and costumes. The interview objectives were follows:

From the perspective of Lahunqiang opera practitioners, the study tried to learn about heritage and evolution of traditional colourful trousers;

From actors' physiological and psychological perspective, the study sought to analyse humanistic and social reasons for the traditional colourful trousers' heritage and evolution, and extracted actors' needs in such aspects as comfort, movement and flexibility.

3.1 Interview Content and Objects

The research team used a semi-structured interview format in which the researchers prepared interview outlines beforehand and posed questions to the interviewees based on the outlines, giving interviewees a great degree of freedom in their replies. The interviews were adjusted or extended based on interviewees' replies [9].

Interviews were conducted at participating Lahunqiang opera practitioners' workplaces. The familiar environment made it more conducive for interviewees to provide comprehensive and thoughtful replies to the questions. Duration of interviews ranged from 30-60 minutes and all interviews were transcribed into text format for further analysis. The interviews focused mainly on the following topics:

1) What factors and reasons underpin the heritage and evolution of the Lahunqiang opera's colourful trousers?

2) What are the factors that determine sizes of the colourful trousers' different parts?

3) What are participants' physiological and psychological feelings when dressed in costume on stage?

Interview participants were 10 Lahunqiang opera practitioners including actors, directors and clothing managers (3 females, 7 males). Interviewees' ages ranged from 35 to 75, and average interview length was 33 minutes. Interviewees were numbered as CF001–CF010.

3.2 Coding and Analysis of Interview Contents

The study used NVivo11 software for data analysis of interview contents. NVivo is a set of quantitative analysis software (developed by QSR) that enable coding, modeling and classification of text, video and picture content. It is an ideal tool for quantitative analysis covering longitudinal, dialogue, content, behaviour, and anthropological analyses, and can also quantify qualitative data as much as possible [10].

The operation and analysis steps are as follows:

1) Import interview contents into NVivo via external data option;

2) After repeated reading of the imported documents, combine and classify the freely coded documents into three-grade nodes, potential and high-order topics;

3) Analyse logical relationship among various topics;

4) Summarise and export the analysis results.

After combing through the imported interview content, two high-order topics were summarised, including four potential topics, with the overall hierarchical relations shown in Table 2. "Number of document sources" means how many interviews cover the specific node; "Quantity of the reference node" indicates the total number of a specific node; "Node coverage rate" refers to the percentage of words under the node vs. total words in the document.

Table.2 Coding of interview information

High-order topics	Potential topics	Third-grade nodes	Number of document sources	Quantity of the reference node	Node coverage rate
Costume development	Heritage	Purchase and maintenance	7	14	7.78%
		Classification in plays	3	4	2.33%
		Size rule	8	17	7.43%
		Match rule	9	17	10.69%
		Shape and structure features	10	32	23.97%
	Development	Style change	9	14	10.08%
		Influence of social development	9	16	11.13%
People–clothes relationship	Physiological level	Facilitating movement	7	13	5%
		Comfort needs	7	15	5.78%
	Psychological level	Aesthetic needs	8	20	9.15%
		Sense of shame	3	10	2.03%
		Sense of confidence	3	5	1.65%

Part of third-grade nodes are explained as follows:

Size rule is about the frequently used colourful trousers' sizes and making in Lahunqiang's traditional plays;

Match rule is mainly about the match among colourful trousers, Dayi/Eryi and shoes/boots;

Shape and pattern characteristics are mainly about features of sizes, structures and ways to wear colourful trousers.

Sense of shame refers to actors' mindset on stage when trying to prevent colourful trousers from falling, which requires reinforcement of the waist part prior to performance.

3.3 Analysis Results and Discussion

From the node number and their coverage rates shown in Table 3, the following can be concluded: 1) With regard to colourful trousers, heritage outweighs evolution as factors related to matching, shape/structure, size and maintenance basically remain unchanged; however, style has evolved somewhat due to changing social environments. 2) From the perspective of human–costume relations, physiological and psychological factors have equal importance, implying that actors' needs in both regards exert a similar degree of influence on costume and trousers' development.

Table.3 Node correlation analysis

Node A	Node B	Pearson correlation coefficient
Node\\physiology level	Node\\heritage	0.747817
Node\\psychology level	Node\\heritage	0.731375

Using NVivo11's "exploration" toolbar and selecting "cluster with node" based on similar terms, correlations among various nodes were analysed. Among correlations of the third-grade nodes, the Pearson correlation coefficient between costume shape and structure characteristics and size rule was 0.833, 0.773 between costume shape and structure characteristics and match rule; in addition the Pearson correlation coefficient between costume shape and structure characteristics and two other items under the "heritage" factor were both above 0.6. Considering that "shape and structure characteristics," "size rule" and "match rule" account for 80% of the "heritage" part, the third-grade nodes in the heritage part can be combined and analysed. Correlation analysis was also made respectively between the heritage node and psychology/physiology nodes, with results shown in Table 4. The Pearson correlation coefficients were both above 0.7, which indicated that both physiological and psychological factors have great influence on costume heritage, with the psychological being more significant.

The study further analysed correlation between heritage and third-grade nodes under physiology and psychology levels with correlation coefficients, as shown in Table4.

Table.4 Correlation coefficients between heritage and the third-grade nodes

Node A	Node B	Pearson correlation coefficient
Heritage	Facilitating movement	0.818
	Comfort needs	0.739
	Aesthetic needs	0.703
	Sense of shame	0.493
	Sense of confidence	0.373

In light of values of the Pearson correlation coefficient, it can be concluded that facilitating movement is most closely correlated with heritage; both comfort needs and aesthetic needs have close correlation with heritage; sense of shame and sense of confidence have a low degree of correlation with heritage.

Next, correlations among third-grade nodes under "physiology" and "psychology" were analysed, and modeled by means of NVivo's "exploration" tool bar. The model presents a clear picture of relationship among various factors, as shown in Fig.7 (black bold solid lines signify close relationship; grey solid lines indicate apparent correlation; dotted lines indicate distant correlation).

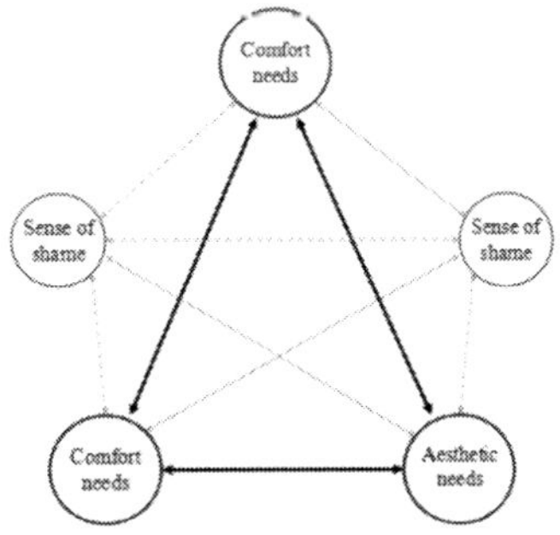

Fig.7 Correlation model of heritage factors for colourful trousers used in Lahunqiang opera.

Fig.7 shows clearly the correlations among the five factors. The three major factors of "facilitating movement," "comfort needs" and "aesthetic needs" have mutual influence on each other; "sense of confidence" and "sense of shame" show low degree of correlation with other factors, but influence the heritage factor of Lahunqiang opera colourful trousers to a certain extent.

4. Conclusion

Via literature review and field investigation, the study used SPSS and NVivo software to analyse and summarise characteristics of and classification methods for Lahunqiang opera colourful trousers. The study also analysed the correlation between selection of colourful trousers and factors such as Dayi/Eryi, type of roles and footwear, and discussed the humanistic reasons for inheritance of colorful trousers' shape and structure. Major conclusions are summarised as follows:

1) In terms of colour, style and fabric texture, colourful trousers can be classified in three ways. In terms of style, colourful trousers can be divided into two types: loose leg and tied leg. The trousers' colour selection follows a certain rule, influenced by type of roles, Dayi/Eryi and footwear.

2) There are seven styles of colourful trousers frequently used in Lahunqiang plays, with pattern changes found mainly in waist, crotch and leg bottom. The most frequently used colourful trouser styles are as follows: trousers with bound leg, large waist and inserted piece at the crotch; trousers with bound legs and bound waist; trousers with bound legs and loose large waist; and trousers with loose legs and large waist.

3)Through discussion about heritage of colorful trousers from a perspective of the people–clothes relationship, it was found that "facilitating movement," "comfort needs" and "aesthetic needs" are major factors influencing heritage. Considering correlation coefficients between heritage and the third-grade nodes, it can be concluded that the stability of the colourful trousers' patterns has a high correlation with the "facilitating movement" factor.

The unique pattern features of colourful trousers are conducive to substantive movements that often are a part of theatrical performances. The major requirements for colourful trousers among opera practitioners are a fastened waist and a crotch part not likely to be ripped; as for practitioners' comfort needs, the trouser fabric should not stick to the body when sweating. Against the broad background of Lahunqiang opera, the colourful trousers' shape and structure characteristics conform to the aesthetic needs of the opera field, and can enhance wearers' confidence and satisfaction. Therefore, despite the colourful trousers' shortcomings (e.g. difficult to don; loose and large compared with people's everyday wear; difficult to be preserve), their shape and structure characteristics and match rules remain constant and have been passed down to the present generation. In light of colourful trousers' basic information and results of its interview analyses, the study identified and discussed colourful trousers' selection rules in specific Lahunqiang opera plays in order to contribute to the development of theatrical costumes. From the perspective of a people–clothes relationship, the study discussed reasons for the heritage aspect of colourful trousers in an attempt to promote the continued development and protection of this important facet of intangible cultural heritage. As time is limited, I only use 10 people to complete the test,and I will increase the sample size in the future experiments.

References

[1] Zong S. On characteristics of Liuqin opera plays and future orientation. J Hundred Schools in Arts: 1993; 4: 68-71.

[2] Yang Y. Folklore interpretation of Sizhou drama. Hefei, China: Anhui University: 2012.

[3] Wu TB. Dictionary on Peking opera knowledge. Tianjin, China: Tianjin People's Press: 2007.

[4] Shanghai Art Research Institute. Chinese opera and folk art dictionary. Shanghai: Shanghai Dictionary Press: 1985.

[5] Li DQ. Introduction to clothology. Beijing: Higher Education Press: 1998.

[6] Kong PP. From Langhunqiang to Liuqin opera. Hangzhou, China: Chinese National Academy of Arts: 2007.

[7] Wei LL. Innovation and development of Sizhou opera. J People's Music: 2009; 6: 32-33.

[8] Song JH. New orientation of intangible cultural heritage and drama studies. J Lit Art Studies: 2007; 2: 96-104.

[9] Li PP. A study of Baiku Yao costumes based on the relationship between people and clothes. Beijing: Beijing Institute of Fashion Technology: 2016.

[10] An YF. Application analysis of qualitative data computer analysis software NVivo. J China Sci Tech Info: 2012; 5: 66-67.

Exposure of Bangladeshi Handicraft in International Haute Couture

Tawsif-Al-Mehran[1*], Hui Tao[1,2], Tamanna Hayder[2], Abdullah-Jaman Jony[2], Tanzila Tabassum[2]

[1] *Wuhan Textile University, No.1 FangZhi Road, Wuhan city, Hubei, Post code:430074, China*

*Corresponding author's email: tawsifalmehran@yahoo.com

Abstract

The most significant industries in Bangladesh in early and medieval times are handicraft industries. Bangladeshi handicrafts are unique in their richness, sustainability and beauty. The traditional fabrics made of handloom, wooden beads, wood-carvings, sun-dried coconut shell for carvings and fibres from discarded husks are used to make rope laces and accessories as well as bags and footwear. The crafts and cultural traditions are deeply rooted and interconnected. Here, the crafts reflect in equal measure its history, landscape and way of life. Famous international brands like Chanel, Dior are promoting handicrafts through haute couture globally. By creating a connection between International brands and Bangladeshi handicrafts, rich handicraft can be enabled to make a global identity and as such are able to survive in this automated machine era. This paper is an attempt to study specific Bangladeshi handicraft and haute couture brand and to discover the process of exposure of the former on the latter.

Keywords: Bangladeshi Handicrafts; Haute Couture; Going Local to Global

1. Introduction

Handicrafts are distinctive expressions and representation of a culture, tradition and heritage of a country. Bangladesh handicrafts sector is a cottage industry, which requires minimum expenditure, infrastructure or training to set up. It uses existing skills and locally available materials with or without electricity, involves minimum capital investment and local available raw materials and it also provides part-time or full-time employment. Handicrafts can be an important productive sector in Bangladesh. Products of handicrafts bear marks of the artisans'. Artisans underwrite their knowledge of local materials and give form to human experience. The artisan sector is primary, small scale, village based manufacturing sector that provides the essential manufactures, functions and services catering to local communities. Now by default, it has become an ever increasing export market. It is also a critical and essential element of the village economy. The traditional crafts of Bangladesh have been a focal point of interest for visitors from abroad for centuries. Per United Nations Educational, Scientific and Cultural Organization/Information Technology Community (UNESCO/ITC) International Symposium on "Crafts and the International Market: Trade and Customs Codification", Manila, Philippines, October 1997: "Products which are produced either completely by hand or with the help of tools. Mechanical tools may be used if the direct manual contribution of the artisan remains the most substantial component of the finished product. Handicrafts are made from raw materials and can be produced in unlimited numbers. Such products can be utilitarian, aesthetic, artistic, creative, culturally attached, decorative, functional, traditional, religiously and socially symbolic and significant. The importance of the Handicrafts sector in Bangladesh cannot be traced back to any point of time. Ingrained deeply in the long legacy of the land, it gradually grew over centuries representing the amusing, rich and unique traditions of the soil, people and their distinctive culture. The love of the craftsperson for their craft coupled with their untiring effort to survive against odds helped the sector flourish. Today, the handicraft industry of Bangladesh looks much well organized than it was a decade back. The growth of international markets for handcrafted products and an increased interest in global goods have opened new-market opportunities for artisans. [3] On the other hand, for more than a century, couture has been emblematic of the triumph of costume and fashion. It represents the fusion of fashion—the modern entity that combines novelty and synergy with personal and social art—and costume—the arts of dressmaking, tailoring, and crafts component to apparel and accessories.

Haute Couture is the prestigious front for French creative fashion (recently a number of fashion houses outside Paris have been afforded membership) and original design and refers to "wildly expensive garments made to measure for an elite cadre of the world's ultra-rich...where designers unleash their creativity, using the finest materials and techniques".[10] Haute couture creates the illusion through brand association that accessories, (bags, scarves, sunglasses etc.) are as exclusive as the couture products at the premium end of the range. Couture generates high levels of publicity, which funnel down into sales in the ready to wear and mass production stages. [6] The couture house is customarily composed of two parts, one devoted to dressmaking, the other devoted to tailoring. Skilled workers in each area practice the arts suitable to the area. Embellishments and accessories are added incrementally as applied decoration, often from sources outside the couture house. Nevertheless, haute couture persists in providing us with a paragon of the most beautiful clothing that can be envisioned and made in any time. [14]

2. Objectives

To study the strength of Bangladeshi handicrafts.
Reasoning the comparative specific handicraft product and international haute couture brand.
To study the possibilities of establishing Bangladeshi handicrafts through international haute couture brands.
To elaborate the possible threats to the Bangladeshi handicrafts.

3. Research Methodology

The data was collected from Bangladeshi Village handicrafts house to know the exact nature and the quantum of the various parameters under the research study to authenticate & check its accuracy and relevance. The information was collected through a face to face situation from the root level artisans, Bangladeshi garments factory, local handicrafts boutique. The data was also obtained from archives of BHMEA (Bangladesh Handicrafts Manufacturers' & Exporters Association), The Story of Chanel's Métiers d'Art, Social media, news and journal. Considering Bangladeshi Handicraft research and current practices and uses of handcrafted materials and procedures in international haute couture market, the purpose of our research was to examine the exposure of Bangladeshi handicraft in international haute couture from the perspective of foreign executives from six different countries. foreign executives were selected as the sample because these respondents were identified as acquainted with handicraft and as highly involved with haute couture and shopping, making them important potential consumers of haute couture associated with Bangladesh handicraft. We used an instrument suitable for quantitative analysis that consisted primarily of groups of items related to Bangladesh Handicraft, international haute couture and added creative marketing strategy to create exposure. Product included 7 different types of products such as bridal wear, outer wear, suit, shawl, accessory item like bag, jewelry and footwear. Respondents were asked to choose two product types that they would most prefer to buy. For each of these two product types, respondents used our list of 6 features to identify product aspects that they would like to buy. These products contain medium price range because of their production origin, high quality and contemporary design aspect. Products were chosen from Bangladeshi brand- Aarong. To aid the representation of Bangladeshi handicraft in haute couture, creative marketing approach for instance story behind Bangladeshi handicraft, sustainable aspect of these products is introduced to respondents before they answered questions. Questions included the willingness to spend time and money for products. Respondents were asked to indicate their agreement or disagreement with each item using a 7-point Likert-type scale (-4 = strongly disagree, 4= strongly agree). Scores on these six items were summed together for data analysis. Preference for what to buy was also measured using a nominal scale (yes/no). The instrument was administered to a convenience sample of 131 foreign executives from six different countries (United Kingdom, Italy, Portugal, China, Sri Lanka and Germany) hired by textiles and clothing manufacturing and Export companies. Most respondents were females (79%). Mean age of both males and females was 27 years. The respondents were European (80.5%) & Asian (19.5%). Most respondents shopped for clothes for themselves every month (51%) and over one-fourth of respondents shopped for clothes for themselves every week (27%). Respondents were most likely to shop for clothes in customized store and boutique house. Over 65% of respondents had no experience with sewing or designing their garments. Due to the limited number of male subjects (N = 25), the statistical analysis was more extensive for female subjects. The data on male subjects were used only to examine gender differences in preferred product.

Table.1 Seven Product Types and Six Features That Respondents Preferred to Buy (N = 131)

Product type (%)

Product Feature %	Bridal wear (Featuring Jamdani)	Suit (Featuring Khadi)	Outer wear (Featuring Monipuri)	Shawl (Featuring Nakshikantha)	Bag (Featuring Jute)	Footwear (Featuring Jute)	Jewelry (Featuring Jute and wood)
Price (medium)	97%	85%	95%	93%	98%	97%	83%
Quality(high)	94	93	86	43	95	95	67
Design (contemporary)	71	63	69	53	70	63	96
Eco product	78	81	71	59	44	73	79
Tell a story (Marketing strategy)	98	89	90	71	89	88	95
Availability in online shop (Marketing strategy)	91	93	93	89	95	90	99

Among 7 different product types included in the instrument, subjects were most likely to buy product depends on price followed by quality and marketing strategies. We examined the features of the seven product types subjects wanted to buy. For all seven product types, quality was the most important feature respondents wanted to buy (88%). Because of proper marketing strategies, respondents wanted to buy products over 90%. At the same time, less than 70% of respondents preferred sustainability and design as a feature for all seven product types.

4. Bangladeshi Handicrafts

4.1 Nakshi Kantha

Nakshi Kantha, a type of embroidered quilt, is an old traditional art in Bangladesh. The basic material used is thread and old cloth. The most basic stitch found in kantha is the running stitch which is called kantha stitch. In the rural area of Bangladesh "Nakshi kantha" is a common handicraft. It is an art of neglected village women. Sometimes, some fashionable women of rural area pass their leisure period by designing Nakshi kantha. Nakshi kantha is a tradition of village. [15]

Fig.1 Bangladeshi Nakshi Kantha

Hand stitching on Nakshi Kantha (https://www.trustedclothes.com/blog/2016/08/17/nakshi-kantha/), Cushion cover made of Nakshi Kantha (http://www.jatrabd.com) (from left to right)

4.2 Jute Goods

Diversified jute products are a preferred choice of global consumers due to their environment-friendly nature. In market crowded with handicraft items, jute handicrafts and novelties easily stand out, for obvious reasons. Jute handicrafts are very trendy. Jute products are appreciated due to their sustainability, innovative designs, vivid colors, contemporary patterns and shapes. Jute made products are bags, shoes, sacks, jewelry etc. Obviously, a jute handicraft's distinctive look takes a little help from accessorizing or creative jazzing up in the form of appliques, embroideries, patches, gems, beads and woods.

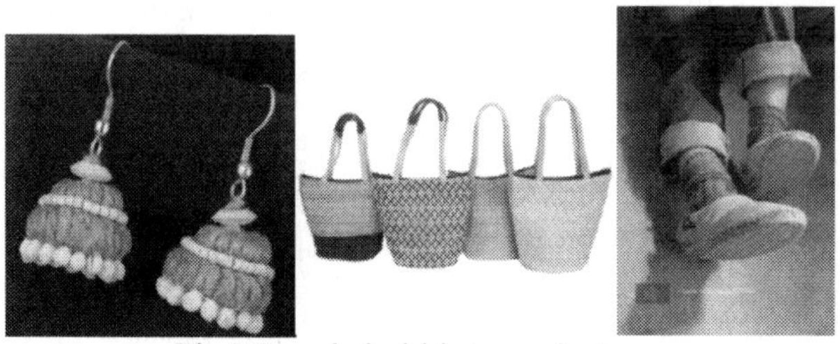

Fig.2 Bangladeshi jute products

Jute ear ring (www.aarong.com), Jute bag (http://global.rakuten.com/en/store/anuenue/item/mb-1/), Jute shoes (http://global.rakuten.com/en/store/tompa/item/10008667/)(from left to right).

4.3 Jewelry

Art jewelry emphasizes creative expression and design, and is characterized using a variety of materials. it forms a counterbalance to the use of "sustainable materials" such as wood, plastic, clay, beads, thread etc.

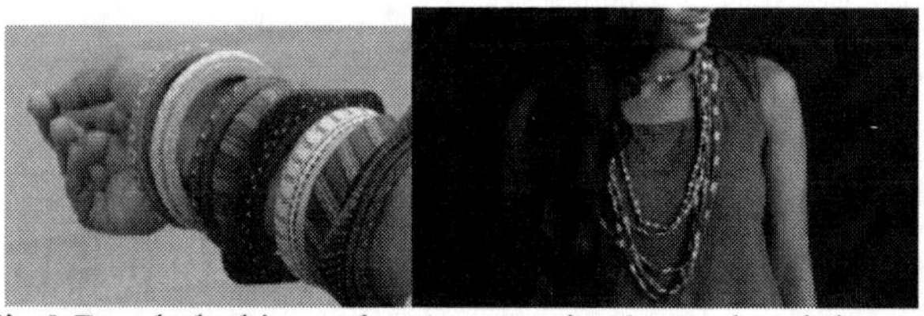

Fig.3 Bangladeshi wooden Accessories (www.jatrabd.com)

Wooden ear ring, Wooden bangles, Wooden beads necklace (from left to right)

4.4 Jamdani, Khadi, Monipuri

Jamdani is one of the finest muslin textiles of Bangladesh, produced in Dhaka District, Bangladesh for centuries. UNESCO has declared the traditional art of weaving jamdani as an Intangible Cultural Heritage of Humanity. [12] Khadi or Khaddar is a term for handspun and hand-woven cloth from Bangladesh mainly made out of cotton. The cloth is usually woven from cotton and may also include silk, or wool, which are all spun into yarn on a spinning wheel called a charkha. It is a flexible fabric, cool in summer and warm in winter. [7]

Manipur is famous for its special fabrics like moiraingphee, leirum, lasingphee and phanek. Tribal textiles are woven on a loin-loom, a type of back-strap loom commonly used by hill people. The warp of manageable length and breadth is prepared and fastened at one end generally to the wall of a house or to two fixed poles while the other is tied to the waist of the weaver with a cotton or leather belt. Sometimes the belt is woven out of cane or bamboo.

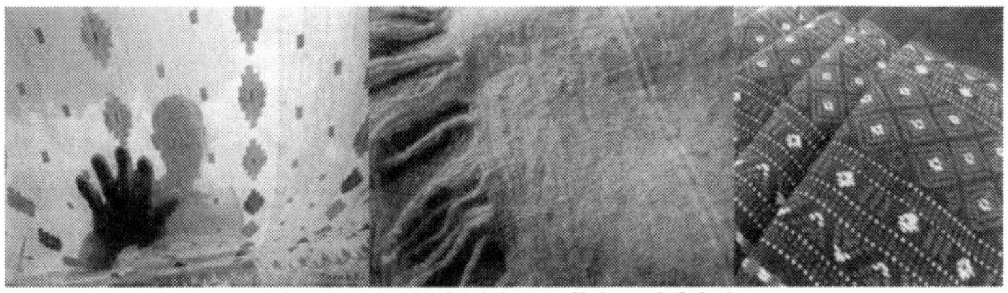

Fig.4 Bangladeshi Traditional fabric

Jamdani (https://skydancingblog.com/tag/jamdani-weaving/),Khadi(http://www.einesaite.com/khadi/). Monipuri (http://www.manipuronline.in/About/Tourism/Shopping-in-Manipur.html) (from left to right).

5. Case Study

International haute couture brands like Chanel, Dior, Yves Saint Laurent are already occupied with handicraft houses around the world.

Table.2 Handicraft houses involved working for International haute couture brands [8]

Haute Couture Brand	Hired Handicraft House	Handcrafted Goods
Chanel	Desrues, Costume Jeweler and Accessory Maker	Button
Balenciaga, Nina Ricci, Christian Dior and Chanel	Lemarié, Feather and Flower Maker	Flower made from peacock, swan and ostrich feathers
Marlène Dietrich, Barbara Hutton, Elizabeth Taylor, Romy Schneider and Chanel.	Massaro, Handmade Bootmaker	Boot
Pierre Balmain, Cristobal Balenciaga, Yves Saint Laurent, Christian Dior, Hubert de Givenchy, Christian Lacroix, Chanel in 2002.	Lesage, Embroiderer	Fabric embroidery
Dior, Givenchy, Yves Saint Laurent and Chanel.	Goossens, Jeweler	Silver and gold-plated bronze, presenting semi-precious stones, quartz and cultured pearls
Chanel	Guillet, Corsage and Floral Decoration maker	Floral accessories
Chanel	Montex, Embroiderer, a specialist in tambour or Lunéville beading	Beads, sequins, the house also sews, piques and embroiders.
Chanel	Barrie Knitwear, Cashmere producer The Scottish knitwear house	Cashmere
Chanel	Lognon, Pleater	Fabric pleating
Chanel	Causse, Glove maker	Glove

Not only the International handicraft houses, Bangladeshi local Handicraft boutiques are also exporting handmade products individually. However, they don't have access to haute couture brands like Chanel & Dior. [5] For instance, Aarong is one of the 18 social enterprises of BRAC – the world's largest development organization. Aarong's export business started in 1986. It started exporting gradually in Europe, North America, Australia and New Zealand. In 1991 and 1992, Aarong opened its own retail outlets in Vancouver and London. [4]

Thanapara Swallows another handicraft producer exports its garments made of handloom fabrics to Japan, Denmark, Italy, Spain, the Netherlands, Finland and Sweden. It employs 250 people, per M.D Raihan Ali, executive director of the non-governmental organization. [2]

These Bangladeshi Organizations are exporting globally. So, there is a demand of Bangladeshi Handicrafts in international market. There is a huge possibility to sell Bangladeshi rich handicrafts to international haute couture brands as they are already purchasing handmade products from different handicraft producers.

6. Problem Formulation

There are a lot of artisans and they are making many handicrafts but there is no link to distribute it all over the world through brands. There are talented and creative people in different parts of Bangladesh. They are scattered, unheard of and underappreciated. While they possess skills which not all of us are blessed with to create beautiful products, they have very limited means to make a living out of it.

Bangladesh's handicraft product faced a period of decline because the handicrafts are mainly regulated by non-technical persons who possess very little technical knowledge and there is no involvement of engineers or well-trained people (thus product development and quality product is quite impossible). The workers are mainly un-skilled women and children.

Young designers are interested in industrial designs. They don't take steps or interest for this sector's improvement.

Today, handicrafts represent less than 1 percent of exports from Bangladesh. In contrast, the global market for handicrafts is estimated to be close to $100 billion. Surely this represents a huge opportunity for this country in terms of rural employment creation. [16]

Despite investment, mainly from international donors, the growth of the industry has been starkly different from its cousin the garment industry, which now dominates more than 70 percent of Bangladesh's GDP. Clearly something isn't working. [13]

The one of the major constraints of handicraft exports is poor planning for marketing the products. Bangladesh is lacking in export marketing plans for handicraft sector.

Shortage of access to capital and financial credits is a major constraint in developing and expanding handicraft exports.

Absence of centralization is one of the biggest problems in this sector. Inadequate infrastructural facilities in terms of marketing facilities (particularly storing), good communication, transport bottlenecks, port facilities, and the biggest of all lack of product development research centers are important hindrances in the growth and development of handicraft products.

Though some training is provided to manufacturers by Bangladeshi Cottage Industries Corporation (BSCIC) and Banglacraft, these are basically theoretical in nature and in most cases, are provided to urban based handicraft manufacturers, who have better linkages. Bangladesh export promotion bureau (EPB) is also trying to pick up good craftsmen through handicraft marketing organizations to send them abroad for higher training. These training programs are not sufficient to cover even a sizeable portion of craftsmen. A few NGO (Non-Governmental Organization) also provide training to the producers. [1]

Reluctance at Government level to find out new and potential markets for handicraft products is another constraint for the development and export of handicraft products. Lack of proper incentives and facilities are responsible for the failure in exploiting demands of handicraft products in the world markets. Although the Government has undertaken some steps to enhance the export earnings from this sector, it is still insufficient to develop the interest among the manufacturers and exporters. Technology has a tremendous impact on production. Bangladesh has been suffering from quality products and less competitive price due to inadequate modern technology and chemical treatments. The price of the handicraft products is often so high due to high price; handicraft products find difficulties in entering the world market. The exporters have been facing tremendous price competition mainly from China, Korea, Philippines and neighboring country India. [11]

There's no opportunity to participate in international trade fair with this rich culture. No necessary steps have taken from Government or private organizations till now to improve this sector through participating in trade fairs or arranging exhibitions.

On the other hand, International haute couture brands are only working with selective handicraft houses. They are not exploring the diversified and rich handicrafts of the world which should be their main priority. [9]

There is no proper access between these two extensively rich fashion ends, one is fully developed haute couture brands and another is underdeveloped but rich Bangladeshi handicrafts culture.

7. Problem Solution

Without a doubt, the global market for handicrafts and all things hand-made is expanding. When customers are faced with two products of similar price, quality and design, they will invariably select hand-made over mass produced.

It is important to understand that pricing is one of the key factors to success in expanding this industry. Since success in this area will only be achieved by meeting the needs of the customer, we must first consider what the customer is looking for. International buyers of handmade products are clear in their requirements. They need large volumes, consistently high quality, timely delivery and effective communication. If the Bangladesh textile handicraft sector can provide all this, then the likelihood is that the buyers will supply the designs. Projects bringing in designers to work with artisans for short periods of time equally bringing small-scale success for only a short period will no longer be necessary.

Large volumes cannot be produced by a single small group of artisans. Timely delivery is a challenge to us all every day and so much more so to artisans with little or no understanding of the countries that their products are destined for. Effective communication requires literacy, language skills and IT skills and equipment in a world where customers want questions answered immediately. They also need good quality raw materials, effective sampling facilities and centralized quality control, finishing, packing and dispatch. The major production work for handicrafts can be done in various rural parts of the country in small-scale groups of artisans, both living and working locally.

To expand the rural production of handicrafts specifically for export we need to combine the craft skills of the artisans with the business skills of SMEs, who could provide all the centralized services along with bulk purchasing of raw materials, sampling, marketing and communication. Setting up design development center is urgently needed for producing new product and product diversification. Crafts men to be trained for better performance, innovation and technology to be provided for overall development of the sector in a design development institute. Participation in the international general trade fairs, single country exhibitions, specialized trade fairs and sending business delegations abroad are necessary for expansion and consolidation of existing markets and creation of new markets for handicraft products.

Such initiatives have neither been taken properly at the Government level nor in the private sectors. Bangladesh Government should organize various trade fairs and provide necessary cooperation to the exporters, so that they can participate in trade fairs both at home and abroad. But again, it should be a both way relation, International haute couture brands should also spread their interest to work with Bangladeshi handicrafts in proper terms and regulations. Bangladeshi rich indigenous culture will help to enhance haute couture brands' unique handmade sophistication at the same time these international brands can help to survive Bangladeshi handicrafts in this machine age.

The good news is now under the one belt one road initiative, Bangladesh and China will be able to open the new age for cooperation in more business area.

8. Conclusion

The Handicrafts Sector can play a significant and important role in the country's economy. It provides employment to a vast segment of craft persons in rural and semi urban areas and generates substantial foreign exchange for the country. The Handicraft sector has, however, suffered due to its being unorganized, with the additional constraints of lack of education, low capital, and poor exposure to new technologies, absence of market intelligence, and a poor institutional framework. However Bangladeshi Handicraft has great growth potential in the changing scenario with its basic strength being the abundant and cheap availability of manpower and being a traditional profession of millions still requires very low investment. We're living in a new golden era of good design, and collaboration between designers & craftspeople is largely the reason to be thankful for. As consumers, we're more aware of materials now - not just their visual and sensual appeal, but also where they come from, why they matter and their sustainability. We're more curious about how things are made and who made them. Collaborative innovation between international designer and craftsperson is a means of expanding the handicraft culture and tapping contemporary markets. It is also argued that a link between the apparently conflicting tenets of sustainability can be achieved through responsible and strategic design innovation which integrates the social, economic, ecological and cultural aspects. Bangladesh can develop this sector as a wide earning sector to contribute to their national economy by involving experts in this sector and by developing manpower skill. Finally, it should be helpful to introduce their tradition or culture to the globe. Bangladesh has improved on handicraft business. Some local brands are also exporting handicrafts to foreign countries. Through this business they are being able to create a relationship with different countries. But this is happening in a very small scale. Finally, we can say that this sector has potential to raise, and there are people willing to be the part of this sector, all they need is the help from the government, which can effectively contribute to the establishment of business, and make this sector work as other organized sectors are working.

References

[1] Chowdhury AH. Handicrafts - Export Marketing Strategies. Souv on Nat Craft Show. Die EPB and B craft, Dhk, BD: 1995; 4: 17.

[2] Bhuiyan MZH. Export Performance of Handicraft Products: A Review. Dhk Uni Jnl of Biz Studies: 1992; 13: 66.

[3] Pye, Elwood A. Artisans in Economic Development - Evidence from Asia. Int Dev Res Cen, Ottawa: 2006; 2: 25-30.

[4] Mondol AR. Export of Bangladesh handicrafts present performance and prospects. Souv, Jatio Karuponno Prodorshoni: 2016; 4: 25-27.

[5] Shah Jalal MD. Concept Note on Development of Handicraft manufactured by the Small and Medium Enterprises of Bangladesh seeking Switch Asia 2008 Project, help of European Commission in promoting Sustainable consumption and Production in Bangladesh. B craft (BHMEA):2010; 2: 21-24.

[6] Catry B. The great pretenders: the magic of luxury goods. Biz Strat Rev: 2003; 14 (3): 10-17.

[7] Jain LC. A Heritage to Keep: The Handicrafts Industry, 1955-85. Econ and Pol Wkly: 1988; 21 (20): 873-887.

[8] Issa C. Chanel Métiers d'Art: for the love of luxury. The Telegraph: 2013; 13: 22.

[9] Phau I, Leng YS. Attitudes toward domestic and foreign luxury brand apparel: A comparison between status and non-status seeking teenagers. Jnl of Fash Mktg and Mgmt: 2008; 12 (1): 68-89.

[10] Barchfield J. Couture in crisis. The West Australian Jnl of Consr Res: 2009; 15: 139 - 168.

[11] Park HJ, Rabolt NJ, Jeon KS. Purchasing global luxury brands among young Korean consumers. Jnl of Fash Mktg and Mgmt: 2008; 12 (2): 244-259.

[12] Alam S. Bangladeshi handicrafts get global recognition for certified fair trade. The daily Star/Asia News Network: 2016; 4-7.

[13] Holbourne D. Bangladesh-Consultancy on Handicrafts. Report on a Mission, Proj No. BD/91/004, Int'l Trade Cen UNCTAD/WTO: 1996; 4: 2-13.

[14] Latter C, Phau I, Marchegiani C. The Roles of Consumers Need for Uniqueness and Status Consumption in Haute Couture Luxury Brands. Jnl of Global Fash Mktg: 2010; 1: 1-5.

[15] Khan MWM. The Traditional Epic Art of Bangladesh (Intangible Cultural Heritage of Humanity). Int'l Jnl of Asian His, Ce and Trad: 2016; 3 (2): 31-36

[16] Anwar, Syed F. Handicrafts: Production - Marketing Interaction. BMET Proj, Uni Grants Commn of BD: 1988; 1: 2-11.

Investigation and Analysis on the Fashionability of Modern Women's Professional Dress

Yan-Xia Wang, Zhao-Hui Gu*

College of Apparel & Art Design, Xi'an Polytechnic University, NO.19 Jinhua Road, Xi'an, Shannxi, 710048, China

*Corresponding author's email: lightgu2008@126.com

Abstract

At present, most of the domestic professional women's clothing market is mostly designed under the influence of traditional principles. It still cannot meet the requirements of modern fashion women's dress code, emphasizing on the need to optimize the fashionability of modern professional women's dress. The purpose of this paper is to master more detailed and accurate information of the demand and ideas for modern women's professional dress. Research methods such as documentation, questionnaire survey and data analysis are used in this paper. The questionnaire mainly covers four aspects, namely basic information of the respondents, consumers' demand and preferences for fashion styles of professional women's dress, consumers' accessories preferences for fashionable professional women's dress, and channels for consumers to obtain fashion information of professional women's dress. The data was obtained and analyzed by SPSS (Statistical Package for the Social Science) statistical software. Finally, conclusions were derived based on market research data, which provided first-hand market research data for fashionable designs of modern professional women's dress.

Keywords: Professional Women's Dress; Fashionability; Market Research

1. Introduction

1.1 Present Situation of Modern Professional Women's Dress

At present, professional women's dress in our country still lingers at a conservative level [1]. In order to boost development and gain an advantage in the market, it is crucial to abandon the conservative thinking. The market needs fashionable, artistic, and open professional women's attire. If it does not meet the needs of the terminal consumers' demand, it cannot survive in the market [2]. Especially for modern young professional women, looking fashionable is of superior importance to practicality of the dress. In order to gain a larger market share in China, the design of modern professional women's dresses needs to be optimized and evolved.

In the west, the emergence and development of western professional women's dress has started over hundreds of years ago. Their longer history and experience in research on designs have given them fruitful results [3]. The book "Fashion in Costume" written by Joan Nun gave detailed analyses on the evolution of western clothing over 800 years. It shows vividly dressing habits of Europeans combining with different periods of productivity level and fashion elements [4, 5]. After the First World War, the difference between uniforms and frocks is that professional women still like to wear these fashionable professional dresses to visit friends or go shopping. History has proven that the focus on professional women's clothing has always been a mixture of fashion and practicality, with distinctive fashion features [6]. Nowadays, clothing is designed in developed countries, and almost all well-known fashion designers are from Europe, America, or Japan. In contrast to Chinese designers, overseas designers have demonstrated a harmonious relationship between appearance and workability. The designs of foreign professional women's dresses are more diverse and decent than the domestic dresses, and they certainly look more fashionable [7].

1.2 Problems of the Professional Women's Dress Market in China

There is an apparent problem in professional Chinese women's attire, which is monotonous and lacks aesthetic sense. Specifically, the designs are always from one of the following dull colors: gray, white and black [8]. At present, the design of professional women has not been able to meet the needs of young women in the workplace.

The positioning of professional women's clothing brands is vague. Since most of the companies that design and sell professional women's clothing lack the expertise to conduct thorough market research, they stand at a blind spot where their designs cannot satisfy the consumers' needs [9]. This phenomenon leads to the brand characteristics of the modern professional women's attire are not prominent. In fact, companies must do a good job in market positioning if they want to seize the attention of the consumers. Not only do they have to segregate the market, but also to identify the psychological needs of consumers. This reflects the necessity and significance of this paper to conduct detailed market research on consumers' perception of fashionable, modern professional women's attire.

This phenomenon led to the current professional women's brand, there is homogenization

1.3 Fashion Trends in the Development of Professional Women's Dress

With the development of economic globalization, people pay more attention to the wearing of feelings. With domestic and foreign exchange more and more frequently, young women in the workplace affected by international fashion and began to seek a better interpretation of their personality, publicity and steady fashion occupation dress [10]. Fashion personalized is an important part of modern professional women's clothing. The progress of the times in the performance of clothing is fashionable progress. The modern professional women's requirements for fashion is very high because they have a high degree of attention to fashion. Therefore, the brand enterprises of the professional women's dress need to be ideologically consistent with these young women and keep pace with the times. Striving to make the product feel fashionable to meet the needs of different industries of professional women whether its shape, color or decoration by increasing popular trend. Research on fashionability of the professional women's attire design can better meet the requirements for beauty of contemporary women.

2. Questionnaire Design and Implementation

2.1 Research Objectives

To comprehend the lifestyle of target consumers and their perception, needs and preferences of the professional women's fashion. To provide original data for up-to-date fashion analyses and design of professional women's attire.

2.2 Research Targets

The project is mainly about the fashionability of professional women's attire, therefore, the research targets are college graduates about to enter the workplace and women who are already working. They can be divided into the following categories with consideration to various factors .Consumers surveyed specific samples are listed below:

1) age composition (college graduates aged from 18 to 25 who are about to enter the workplace, women aged from 25 to 40 who are already working).

2) education background (high school, college and undergraduate, graduate and above).

3) occupational hierarchy (company staff, executive class).

2.3 Research Content

The main part of the survey is divided into four major modules: basic information of the surveyed targets, consumer demand and preferences for women's fashion, consumers' preferences for professional accessories, main channel access to fashion information.

2.4 Research Methodology

Surveys were randomized and distributed using the anonymous method according to the proportion of the above research object classification). Data was obtained and analyzed by SPSS Statistics V17.0 (Statistical Package for the Social Science) statistical software developed by IBM corporation of America [11].

2.5 Research Execution Process

After preparing the survey questions, I sent the survey to 20 female graduate students who major in fashion design and engineering, asking for their opinion and evaluation to optimize the questions before sending the final version to the target study groups. Each student was required to complete four tasks: 1) carefully fill out the questionnaire; 2) modify the ambiguous questions; 3) add important options that are missing / delete redundant options; 4) adjust the order of certain items.

The 20 graduate students formed a group of experts. I conducted an in-depth interview with each expert after collecting the questionnaires. I observed and noted their attitude and evaluation of each group member, and further discussed with them. More members were involved when serious comments are not required. With the valuable input from the experts, I modified the wordings used in the questions, options available after each question, and order of the questions. In short, the design of the questionnaire followed the following principles: 1) related to the purpose and theme of the investigation; 2) clear and accurate use of language; 3) answer options are not redundant nor restrictive; 4) demonstrate respect and understand the target groups.

With the finalized questionnaire, they were issued to the respondents. Taking into account the effectiveness of the questionnaire, our group decided to take one-on-one supervision on respondents filling out the questionnaire, and actively participated in the whole process. We gave each of the respondents patient guidance in order to fully grasp the real situation. Questionnaire subjects were picked in a way to cover a wide range of industries and consumers of all ages. The locations of the self-filling survey were mainly in shopping malls, pedestrian streets, schools, and enterprises.

2.6 Collection and Improvement of the Questionnaires

Our purpose was relatively strong in the whole process of distribution and collection of the questionnaire. There are relatively clear target groups. At the same time we conducted an one-on-one supervision and guidance in the whole process. As a result, 192 questionnaires were effectively collected. The remaining eight were not filled out seriously, which were unrealistic response; thus these questionnaires were removed from the analysis. Then we continued the investigation until the number of final valid questionnaires reached 200.

3. Data Analysis of Research Results

3.1 The Reliability Analysis of the Questionnaires

The reliability index is an indicator of the consistency, reliability and stability of the test results [12]. Reliability analysis is the stability and consistency of the measurement results of a particular type of questionnaire, rather than the broad general consistency, whose size is expressed by the reliability coefficient [13]. The SPSS 17.0 for windows was applied in this paper to test Cronbach α to analyze the reliability of the questionnaires. Usually the value of Cronbach coefficient is between 0 and 1. If the X coefficient does not exceed 0.6, it is generally believed that the internal consistency lacks confidence. If the X coefficient reaches 0.7-0.8, the scale has a considerable degree of reliability. If it is up to 0.8-0.9, the scale reliability is very good [14]. Table 1 shows the reliability results for different subjects. According to Table 1, the Cronbach coefficient is 0.875, meaning that the reliability of this questionnaire is higher.

Table.1 Reliability Statistics

Cronbach's Alpha	Cronbach's Alpha Based on Standardized Items	N of Items
0.875	0.881	32

3.2 Basic Information of the Research Subject

Basic information of the survey targets is shown in Table 2.

Analysis results of the target consumer segmentation show that the potential consumers of fashion professional women's dress can be divided into two categories according to age composition. The first category is college graduates aged from 18 to 25, who are about to enter the workplace; the second category is professional women aged from 25 to 40. Since the participants grew up in different backgrounds, these two groups of consumers have very different consumer attitudes, values, and taste. Therefore, the brand design of fashionable professional women's dress must reflect the characteristics of different ages and focus on personalized fashion needs.

Table.2 Basic information of the research object

Items		Proportions
Age composition (years)	18-25 years old	41.10%
	26-40 years old	58.90%
Academic distribution	High school and below	3.70%
	College and undergraduate	70.37%
	Graduate and above	25.93%
The disposable expenses used to purchase professional dress (yuan)	600 yuan or less	15.48%
	600-1000 yuan	31.47%
	1001-3000 yuan	40.56%
	more than 3000 yuan	12.59%

3.3 Consumers' demand and Preferences for Fashionability of Professional Women's Dress

3.3.1 Investigation on Fashion Style Preferences for Professional Women's Attire

Professional women were affected by a variety of design ideas and trends in the development process, and formed a variety of design styles. The leisure trend of professional women's attire appeared in the early 90s, which formed a "casual style" of professional women's attire. In the late 1990s, the prevalence of neutral minimalist design created "neutralization" professional women's style. Nowadays, stable styles of professional women's attires such as "masculine style", "national style", "sweet style", and "casual style" are influenced by fashion and form in different time periods. Fashion trends play an important guiding role for brand style positioning of professional women's attire.

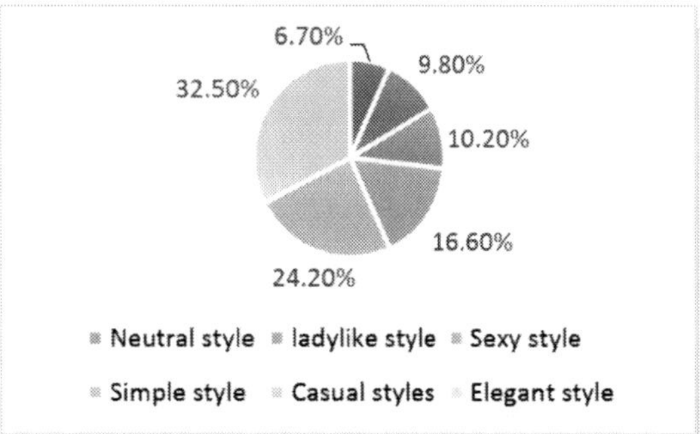

Fig.1 Preferences for fashion style

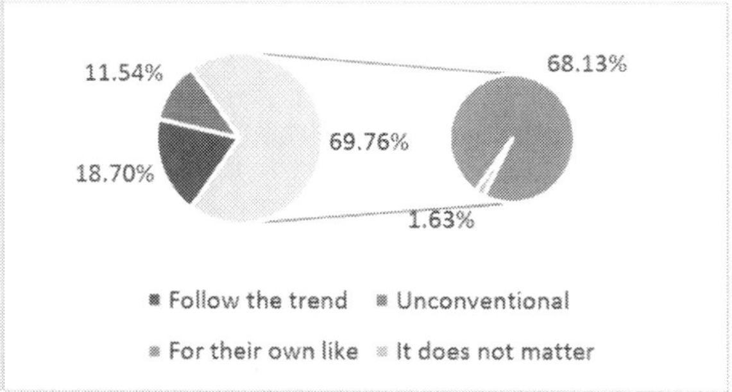

Fig.2 Target consumers' views on the fashion trend

Research results are shown in Fig.1. It can be found that an elegant style is the most popular choice, followed by a leisure style and a contracted style. Thus it can be seen that fashion style design of professional women's clothing should be rather simple. The details should be designed to be simple and add elements of elegant temperament to cater to the needs and preferences of most target consumers.

The results of the survey on fashion trends shows that regardless of the age of the consumers, suitability for themselves is the top choice (Fig.2). It can be seen that the logic behind consumers; purchases of professional clothing is rather rational: they choose products suitable for themselves rather than blindly following the trend. Therefore, fashion design for professional women's clothing should be appropriately combined with popular elements of the season while maintaining its own characteristics. Characteristics of simplicity should be maintained at all times. On this basis, the popular elements suited to the characteristics of the brand could be assimilated into the designs, so that the brand will not lose its loyal customers and its brand image.

Results were sorted according to different factors which reflect what consumers pay attention to when they purchase professional dress (Fig.3). It is shown that style is the most important factor for consumers, followed by price and brands. Therefore, style should be as rich as possible to meet the preferences of most consumers.

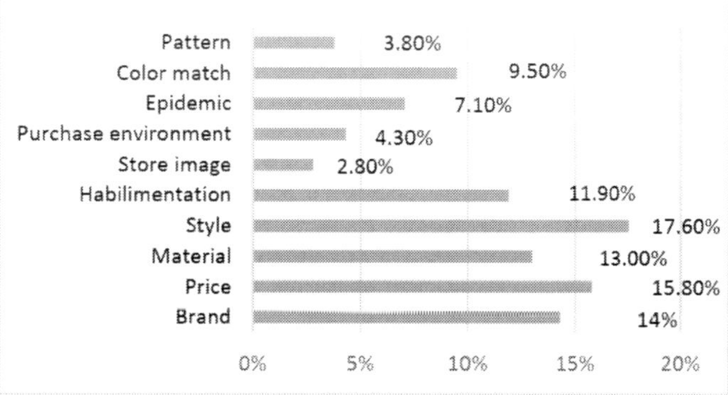

Fig.3 Factors sorting results

3.3.2 Investigation on the Fashion Color Preferences for Professional Women's Attire

Color is a vital component in any part of a piece of clothing, as it reflects the characters of the clothing and the design. Different colors carry a different impact on the clothing style [15]. Since there is a large variety of colors, it is impossible to list all specific colors; therefore, the questionnaire primarily focuses on two color schemes: achromatic colors which include black, white and gray; and chromatic colors which include red, orange, yellow, green, blue, blue, and purple. The detailed statistical results are shown in Fig.4.

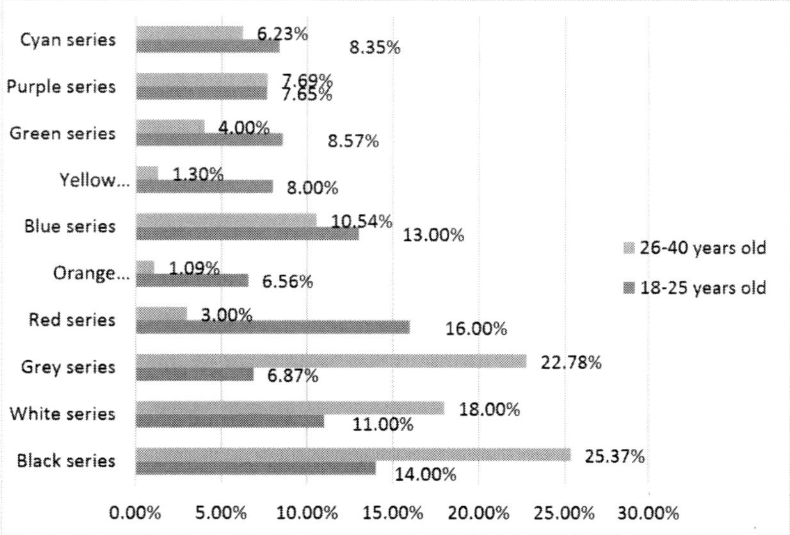

Fig.4 Fashion color preferences for professional women's clothing

Research results show that women from different age groups have a different preference for the colors used in professional clothing (Fig.4). Young women of 18 to 25 years old prefer more colorful schemes as to the traditional achromatic colors. As a result, we can mix black and white, and turn them into a spiral for young professional dress fashion color design. Excessive colors or auxiliary colors can be used to make the young professional women's clothing more youthful and bright. Also, such designs will be more in line with their age characteristics. Women among 25 to 40 years old are considered as relatively senior professionals. They want to convey an efficient and friendly tone, in which they tend to choose the professional wear which are professional but not lack of affinity. They are more inclined to achromatic and cool colors, but the color scheme used should not be too dull either.

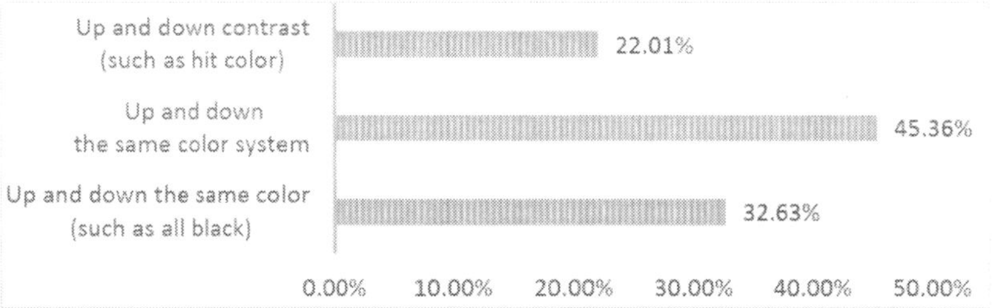

Fig.5 Respondents' preference for color collocation preference

The results of the survey on professional women's color collocation shows that 45.36% of the respondents would like to choose a fluctuation outfit from the same color system (Fig.5). 32.63% chose a fluctuation outfit in the same color (such as pure black blouse and skirt). 22.01% of the respondents chose contrast color collocation. Thus, consumers are more conservative in terms of color collocation. Contrasting colors can be added to the design. Generally, it is used in a single color clothing, such as a black suit with a silver glittering corsage, which makes the clothing more lively. Another example is to wrap the pocket edge of a suit with contrasting colors, as to make the outfit look more alive.

1.4 Consumers' preference for Accessories of Fashion Professional Women's Dress

1) Dress collocation for professional attire

The overall design of professional clothing should reflect the consumer's taste and quality. In France, Italy and other foreign countries, other accessories such as a matching belt and a handbag are designed alongside the clothing to create an overall outfit [16]. Survey results on dress collocation for professional attire(Fig.6) show that 81.2% of the consumers will consider the problem of the whole collocation ,such as the color factor, style factor and fabric factor .

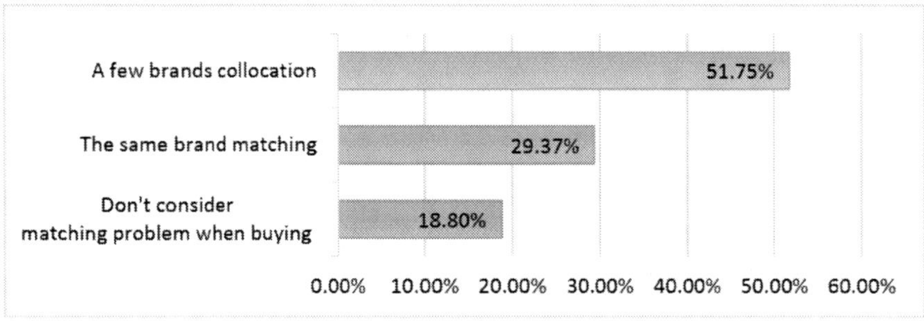

Fig.6 Purchase way for professional attire

As to cater to all the concerns of consumers regarding professional clothing, companies can provide an overall package design, which does not only give them a competitive edge in the market but also to guide the aesthetics of consumers. Currently, such full package outfits for professional women are not seen in China. Given the complexity of the number of factors involved in designing an entire matching outfit, designers must master related products design, and acquire the ability to harmonize various products in an outfit.

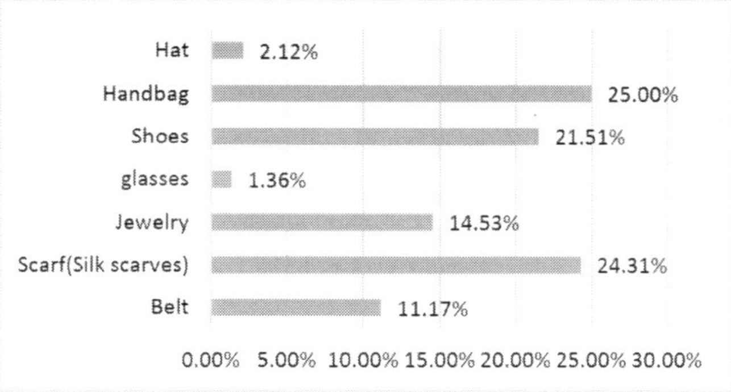

Fig.7 The survey results of fashion accessories

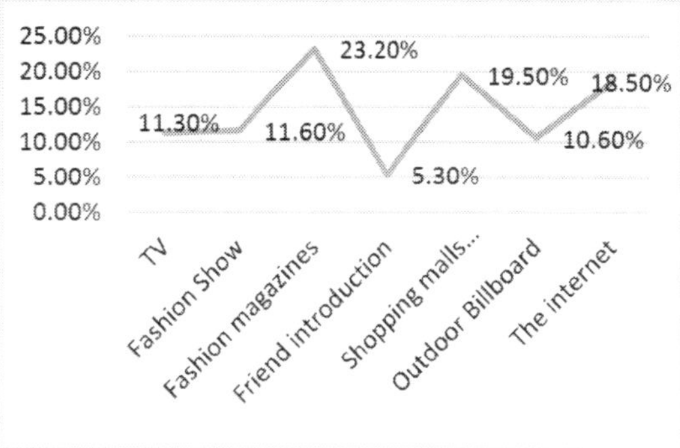

Fig.8 The information channel

2) Survey results on consumers' preference for accessories of professional dress

The above chart (Fig.7) displayed that respondents pay the most attention to bags, scarves and shoes when considering clothing accessories. To attract consumers and guide their aesthetics, companies should then design the overall outfit which includes accessories like bags, scarves and shoes.

3.5 The Channels for Consumers to Obtain Fashion Information of Professional Women's Dress

The surveys clearly revealed that consumers most frequently obtain fashion information by reading fashion magazines, followed by visiting shopping malls and reading on the internet (Fig.8). Thus, fashion brands should promote their clothing series and design concept via fashion magazines, so to increase their chance of getting the consumers' attention. However, with the rapid development of the internet, most companies prefer to advertise and promote their brands online. Only when a design captures both fresh fashion trends and practicality can it satisfy consumers' demand.

4. Conclusions

1) Since the two groups of consumers come from different socio-economical background, they have demonstrated different attitudes, values and taste towards the designs of professional dresses. Therefore, the fashionable design of these dresses should reflect the characteristics of women of different ages and focus on personalized fashion needs.

2) The survey results show that only six percent of the consumers interviewed were satisfied with the designs of professional women's attire on the market. They hope the attire can reflect their taste, interests, status and personality, which may help boosting their career. This can be achieved through more fashionable and individualistic designs. All consumers interviewed hope that they can pick a design that is most suitable for them, in terms of the personality portrayed while wearing the attire, on the market. Therefore, designs should be simple, but can be enlightened by elements of elegant temperament to cater to the needs and preferences of most target consumers. Young women aged between 18 to 25 like to use some chromatic colors to the traditional achromatic color of professional women's attire as the ornament color. They think that bright and warm colors look more fashionable and can relax them in a stressful work environment. Women aged between 25 to 40 tend to choose designs with neutral colors and cold colors. 82.20% of the consumers consider dress collocation when buying professional attires. Consumers most frequently obtain fashion information through reading fashion magazines.

3) Analysis of the survey results regarding designs of professional women's dress suggests that designers should fully consider a variety of factors during the design process. To be specific, designers should grasp the overall feeling they wish to portray in the design, and combine their ideas with current fashion trends in order to produce coordinated and fashionable professional women's dresses.

5. Acknowledgement

The author would like to express their gratitude towards the teacher and the Apparel & Design College of Xi'an

Polytechnic University. Thanks to all the teachers and classmates who have contributed to the study in any way. I would like to thank my mentor for patiently giving me valuable comments and suggestions. Due to my limited knowledge, there is room for improvement in either the text or the study design, and I sincerely welcome all teachers and colleagues to give valuable advice.

References

[1] Wei N. The Analysis of the Fashion of Young Women's Clothes. Northeast Normal University: 2015.
[2] Li GS. The Development Process of Women's Professional Dress. Inform Eng Res I,USA: 2012; 4.
[3] Shi J. The Theory Of Feminism Research On The Influence Of Women's Professional Style In The 20th Century. Wuhan Textile University: 2015; 3-5.
[4] Neil H, Karen J, Natassia C. Unbuttoned: The Interaction Between Provocativeness of Female Work Attire and Occupational Status. Sex Roles: 2015; 72 (3-4): 105-116.
[5] Thinada P, Rujirutana M. Psychological Antecedents of career Women's Fashion Clothing Conformity. J of Fashion Marketing Manag: 2016; 28: 2.
[6] Sun Z. Study of Dress Requirements and Trends of the Urban Career Women. Shanxi University of Science and Technology: 2014.
[7] He JH.A Review of the Current Situation of Female Career Development in Foreign Counties. Foreign Econ Manag: 2006; (01): 23-29.
[8] Xiang L, Yang XL. The Current Situation and Countermeasure of Chinese Women 's Brand Development. Manag 'J: 2010; (02): 212.
[9] Wan L, Liang HE. Research on the Style of the Vocational-Dress. J Wuhan Univ Sci Eng: 2006; (11): 10-12.
[10] Lin Y, Jiang XW. On the Development Trend of Fashionable Fashion. New West: 2010; (12): 123+108.
[11] He K, Zhan LL, Li XM. Discussion on Teaching Methods of Market Investigation and Analysis. J Liaoning Univ Technol: 2016; (01): 108-109+138.
[12] Meng HL, Li JW. The Application of SPSS in Education Questionnaire Analysis. Intelligent Computer and Applications: 2013; (1): 81-83.
[13] Luo Z, Zha YF, Gu CJ. Analysis on the Reliability and Validity of KAP Questionnaire on AIDS Prevention for Female Entertainment Workers. J Environ Occup Med: 2011; (11): 710-712.
[14] Dong Z. Use SPSS to Analyze the Quality of Test Papers. China Sci Technol Inform: 2009; (15): 100-102.
[15] Zhang J, Li N. Discussing on the design of fashion professional women's dress. Int Text Leader: 2013; (04): 79-80+82.
[16] Ye WY. On the Focus of Designing Modern Vocational Women' s Dress. J Zhejiang Text Fashion Coll: 2005; (03): 26-27.

The Somatotype of Middle-aged and Older Women in Xi'an

Jing Qi*, Long Wu, Xin Zhang, Bo-An Ying

Apparel & Art Design College, Xi'an Polytechnic University, No. 19 Jinhua South Road, Xi'an, Shaanxi, 710048, China

*Corresponding author's email: Jing Qi qijing0928@126.com

Abstract

There is a big difference between the body shape of the young, the middle-aged and the elderly. The current standard sizing system in China for GB1335 garment is not divided according to age groups. It is necessary to understand the size of the middle-aged and elderly, as well as develop sizing systems for the garments they wear. In this study, 198 female subjects aged from 35 to 75 years were randomly selected from the body measurements database. We then measured them using their 3D images and obtained nine body measurements. Their body shape was evaluated using several body shape indices, while the samples were classified according to the GB1335 method. When comparing the results, we found that middle-aged and older women are fatter. Six circumference difference measurements were calculated and used as classification indices, with the samples divided into five somatotypes. The characteristics of each type were analysed by comparing the mean of the body measurements. The study's findings can be used in the development of sizing systems for garments and apparel pattern designs for the middle-aged and elderly.

Keywords: Middle-aged and Older Women; Body Shape; Somatotype

1. Introduction

The ageing of the population has become a major social problem. As we get older, our body's metabolic capacity weaken and the storing of fat becomes easier. This is why the middle-aged and elderly tend to be more overweight than young people. The current standard sizing system in China for GB1335 garments is not divided according to age groups, which means it is unable to reflect the body shape characteristics of different age groups. Thus, a series of problems arises, such as the fact that some middle-aged and elderly people cannot buy fitting clothes, while many garment manufacturing enterprises have to face the problem of excess inventory and stock holding.

Some countries, such as the US and Japan, have begun to consider age factors in the study of body shape characteristics among the middle-aged and elderly, as well as continuously improving their standard somatotype system [1, 2]. Many Chinese scholars have also carried out research on body shape characteristics, somatotype, and sizing systems for garments for the middle-aged and elderly [3-11]. However, most of their work is limited to their respective regions, meaning that there are some limitations in the application of their research results. The China National Institute of Standardization organized activities related to adult human body measurement in 2015. Adults aged from 18 to 75 years were measured using three-dimensional (3D) human body scanners and manual techniques. The subjects were divided into four groups by age: 18-24, 25-34, 35-59 and 60-75 years [12]. The data will be used in the development of various national and industry standards, including the apparel industry.

The Research Center of Clothing Engineering and Technology of Shaanxi Province in Xi'an Technology University, as the sampling point of this activity, organized and measured adults living in Xi'an City, and the surrounding towns and rural areas. The sampling scheme adopted a hierarchical random cluster sampling method. The occupations of the subjects included menial and light workers working, for example, in universities and government institutions, and heavy manual workers, such as those involved in construction, catering and housekeeping.

In the course of this measurement work, it was found that two age groups (35-59 and 60-75 years) accounted for 67% of the sample [12]. Among them, 198 women were randomly selected as the sample for this paper, whose 3D images were measured in the process. Using the statistical analysis method, this paper studies the body shape characteristics and somatotype.

2. Measurement Items and Methods

The sample comprised 198 female subjects aged from 35 to 75 years, who were randomly selected from the database. Nine body measurements were measured on 3D images (see Fig. 1), while two body measurements (body height and weight) were measured by manual methods. In order to compare our measurements with the GB1335 data, the same measuring items and methods from GB/T 16160 were adopted [13].

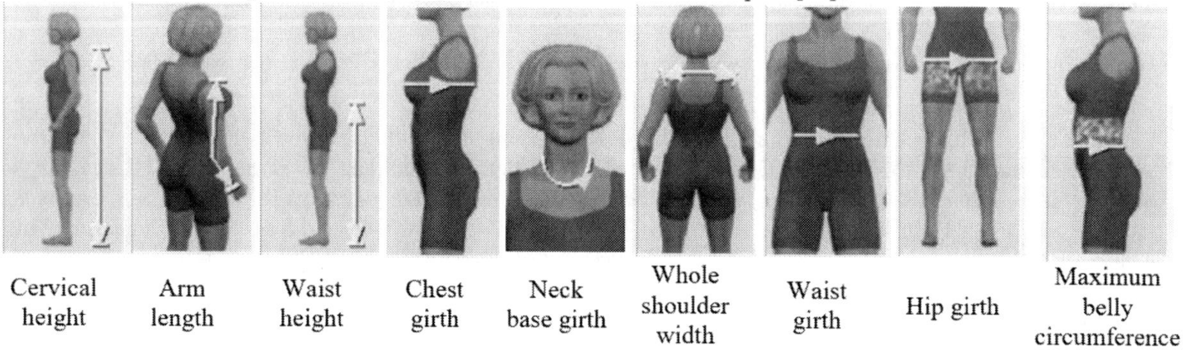

Fig. 1 Measurement items and methods

3. Body Shape Index

As one of the most important methods for the quantitative evaluation of body shape, the body shape index is of great significance in the study of physique. It involves the use of human body data, which is subject to a mathematical formula to obtain a new value that indicates the body characteristics. The commonly used body shape index includes the chest-height radio index (the ratio of the chest circumference and body height), the Rohrer index, the Pignet index, the Pignet-Vervaeck index [3] and the body mass index (BMI).

According to the Rohrer index, the Pignet index and the Pignet-Vervaeck index, which are calculated based on body height, weight and chest circumference, the somatotype is divided into slender, moderate and obese.

The chest-height ratio index is mainly used to evaluate the chest shape of the human body. Specifically, it refers to the aspect ratio of the body. Using the ratio of chest circumference and body height, the somatotype is divided into narrow chest type, normal chest type and wide chest type.

Calculated based on human height and weight values, the BMI is now commonly used around the world to measure body fat and healthy levels.

The formulas for the above indices are as follows:

$$\text{Rohrer index} = \left(\frac{\text{Weight}}{\text{Body height}^3}\right) \times 10^7 \tag{1}$$

$$\text{Pignet index} = \text{Body height} - (\text{Chest girth} + \text{Weight}) \tag{2}$$

$$\text{Pignet - Vervaeck index} = \left(\frac{\text{Weight} + \text{Chest girth}}{\text{Body height}}\right) \times 100 \tag{3}$$

$$\text{Chest - height ratio} = \left(\frac{\text{Chest girth}}{\text{Body height}}\right) \times 100 \tag{4}$$

$$\mathrm{BMI} = \frac{\mathrm{Weight}}{\mathrm{Body\ height}^2} \tag{5}$$

Table 1 shows the somatotype judgement standard of the above indices and the distribution of the samples.

As can be seen from the classification results of the somatotype, as shown in Table 2, most people are of the obese type according to the Rohrer index, the Pignet index and the Pignet-Vervaeck index. When classification is made using the chest-height ratio, most people belong to the wide chest type. Meanwhile, according to the classification results of the BMI index, most people are moderate and overweight, with obesity an obvious trend.

Table.1 Body shape index

Body shape index	Body shape	Evaluation of criteria	Mean of the samples	N of the samples
	Slender	<129	121.3	26
Rohrer index	Moderate	130-149	142.7	49
	Obese	>150	167.6	123
	Slender	>26	30.7	8
Pignet index	Moderate	25.9-20.1	22.5	8
	Obese	<20	-2.3	182
	Slender	<81.9	80.3	7
Pignet-Vervaeck index	Moderate	82.0-94.2	89.7	42
	Obese	>94.3	103.8	149
	Narrow chest	<50	49.0	3
Chest-height ratio	Normal chest	50-55	52.9	15
	Wide chest	>55	62.5	180
	Light	<19	18.1	6
	Moderate	19-24	22.1	88
BMI	Overweight	24-29	25.8	86
	Obese	29-34	30.6	15
	Very fat	>34	35.3	3

4. Somatotype Classification

Many research studies on somatotype are currently using different classification indices.

1) Circumference difference. The main body part involves chest circumference, waist circumference and hip circumference. These are not necessarily synchronized. For example, the same chest circumference but a different waist circumference will show a different somatotype. Therefore, some circumference difference can be used as the basis for distinguishing somatotype.

This method is used in Chinese National Standard GB1335. According to the difference between the chest and waist circumference, the somatotype is divided into Y, A, B and C types [14, 15].

According to the difference between the chest and hip circumference, Japanese adult women have been divided into the Y, A and B types [16].

2) The difference between the front and the back of the neck to the waist. This is the best index to distinguish the difference between normal, big chest, convex belly and bent back types. Similarly, some body measurements or their differences, which show a body bend, may also be used as a basis for somatotype classification.

3) Indices of body size. Examples include the fullness index, which is the ratio of weight and body height, the ratio of a particular circumference and body height, and the ratio of a particular circumference and so on.

4) The difference between and the ratio of the chest and waist circumference. This is a combination of the first and third methods, which can avoid the drawbacks when using only one method.

5) Specific variable. Generally, this is a key measurement of the human body.

6) Factors extracted from factor analysis. This classification method is widely used because of the extracted factor, which contains most of the information about the original variables involved in the data analysis.

In this paper, according to the GB1335 somatotype classification method, the difference between the chest and waist circumference is used as the classification index to inspect the distribution of the samples.

4.1 Somatotype Classification by the Difference between the Chest and Waist Girth

Based on the difference value of the chest and waist, the Chinese body shape has been divided into four types: Y, A, B and C [14, 15]. Y represents a thin shape, A represents a normal shape, B represents a slightly fat shape and C represents a fat shape. On this basis, we added two extra kinds of somatotype, named X and D. If someone's breast-waist difference is bigger than type A, the somatotype is X. If it is smaller than type C, the somatotype is D. The proportion of each somatotype among the samples was compared with the GB1335 data for the whole country and middle western areas (see Table 2).

Table.2 Comparison of somatotype frequencies

Somatotype	Breast-waist difference (cm)		Proportion			
			GB1335 [15]		Samples	
	Difference range [15]	Calculated value	Whole country	middle western areas	N	%
X	—	>=24.5	—	—	0	0.00%
Y	19~24	18.5~24.4	14.82%	17.50%	15	7.60%
A	14~28	13.5~18.4	44.13%	46.79%	88	44.40%
B	9~13	8.5~13.4	33.72%	30.34%	78	39.40%
C	4~8	3.5~8.4	6.45%	4.52%	16	8.10%
D	—	<=3.4	—	—	1	0.50%
Total			99.12%	99.15%	198	100%

As shown in Table 2, most people belong to type A and type B, accounting for 83.8% of the total number. The proportion of type C is greater than in GB1335, which shows that the middle-aged and older women have a tendency to gain a fatter shape.

Table 3 presents a comparison of the mean of body measurements between the samples and GB1335. The height of everyone in the samples is shorter than the GB1335 mean, but the circumference in the former case is much larger. This also shows that people are fatter than before.

4.2 Somatotype Classification by Circumference Difference

From the above analysis, it can be seen that the main body shape feature of middle-aged and older women involves obesity. The circumference difference of the body is the best index to reflect the degree of obesity.

Six circumference difference measurements were calculated in respect of chest and waist, chest and belly, chest and hip, hip and waist, hip and belly, and belly and waist. These measurement were then used as classification indices in the K-means cluster analysis of the somatotype classification of the samples.

In K-means cluster analysis, the number of clusters must be given first, while the biggest difficulty is to determine the most appropriate number of clusters. Lehmann (1979) suggested that the number of clusters should be between n/30 and n/60. N represents the number of samples. Therefore, as the sample number is 198, the number of clusters should be between three and seven. As the current national standards (GB1335) are divided into four somatotypes, the number of clusters should be greater than four in order to achieve the purpose of subdivision. Therefore, the number of clusters separately used was five, six and seven, which were compared with the classification results.

After cluster analysis, all the results were found to have a value of significance equal to 0, which shows that the three classification results were all meaningful. Table 4 shows the results relating to the number of people in every cluster are relatively balanced. Their final cluster centre also shows an obvious difference.

Table.3 Comparison of the mean of the main body measurements (cm)

Somatotype		Body height	Cervical height	Arm length	Waist height	Chest girth	Neck base girth	Whole shoulder width	Waist girth	Hip girth
Y	Samples	157.6	134.1	51.2	98.2	93.3	37.3	39.1	73.8	92.9
	GB1335	160.0	136.2	50.4	98.2	84.0	33.4	39.9	63.6	89.2
	Difference	-2.4	-2.1	0.8	0.0	9.3	3.9	-0.8	10.2	3.7
A	Samples	157.2	133.7	50.9	97.5	95.3	37.5	39.6	79.7	93.8
	GB1335	160.0	136.0	50.4	98.1	84.0	33.7	39.9	68.2	90.9
	Difference	-2.8	-2.3	0.5	-0.6	11.3	3.8	-0.3	11.5	2.9
B	Samples	157.6	134.1	51.2	97.9	97.8	38.3	40.1	85.9	96.3
	GB1335	160.0	136.3	50.5	98.0	88.0	34.7	40.3	76.6	94.8
	Difference	-2.4	-2.2	0.7	-0.1	9.8	3.6	-0.2	9.3	1.5
C	Samples	157.2	134.1	51.5	97.4	104.0	40.0	40.4	97.2	105.5
	GB1335	160.0	136.5	50.5	98.2	88.0	34.9	40.5	81.9	96.0
	Difference	-2.8	-2.4	1.0	-0.8	16.0	5.1	-0.1	15.3	9.5

Table.4 Comparison of the classification results

Number of cluster		5	6	7
	N	58 / 55 / 20 / 27 / 38	20 / 24 / 35 / 14 / 41 / 64	18 / 25 / 46 / 23 / 52 / 14 / 20
Final cluster centre (cm)	Chest-waist difference	14.33 / 14.14 / 16.25 / 11.19 / 12.58	11.09 / 11.71 / 10.97 / 17.16 / 16.35 / 14.32	11.02 / 12.1 / 15.7 / 17.35 / 12.61 / 17.16 / 9.82
	Chest-belly difference	8.89 / 4.82 / 5.99 / 1.2 / 7.52	0.27 / 6.85 / 5.06 / 5.87 / 5.48 / 8.83	-0.07 / 7.19 / 5.62 / 11.36 / 6.88 / 5.87 / 4.03
	Chest-hip difference	-2.86 / 2.03 / 6.56 / -1.23 / -7.21	-1.23 / -7.77 / 3.28 / 7.14 / 0.55 / -4.00	-1.29 / -7.99 / 1.1 / -5.49 / -2.52 / 7.14 / 4.99
	Hip-waist difference	11.48 / 16.17 / 22.81 / 9.96 / 5.01	9.86 / 3.36 / 14.25 / 24.31 / 16.9 / 10.32	9.73 / 3.56 / 16.8 / 11.86 / 10.09 / 24.31 / 14.8
	Hip-belly difference	6.04 / 6.85 / 12.55 / -0.03 / 0.31	-0.97 / -0.92 / 8.34 / 13.01 / 6.04 / 4.82	-1.36 / -0.8 / 6.73 / 5.87 / 4.36 / 13.01 / 9.01
	Belly-waist difference	5.44 / 9.32 / 10.26 / 9.99 / 4.69	10.82 / 4.27 / 5.91 / 11.29 / 10.87 / 5.5	11.09 / 4.36 / 10.07 / 5.99 / 5.73 / 11.29 / 5.79

The results of clustering should make the difference between the groups as large as possible, while the difference within the group should be as little as possible. This can be determined by the size of the F value. The larger the F value, the better the clustering results.

Table 5 shows the F value for every cluster index when the number of clusters is respectively five, six and seven. When the number of clusters was seven, the F value of the chest-waist difference was the largest. But the F value of the chest-belly difference was the largest when the number of clusters was five. As the size of the F value of each index was not consistent in the results, it is impossible to determine which classification was the best.

Each index was ranked according to the size of the F value, then the number of the ranking of each clustering index in each classification and the total score of each classification were calculated. For example, the F value of chest-waist difference was the largest when the number of clusters was seven, which means that a cluster number of seven is first and receives a score of '3'. The F value of the chest-belly difference was in the middle when the number of clusters was seven, which means the cluster number of seven is second and receives a score of '2' (and so on; see Table 6).

Table.5 Comparison of the F value

Number of clusters / Index	5	6	7
Chest-waist difference	10.4	31.5	41.0
Chest-belly difference	41.3	30.9	37.1
Chest-hip difference	121.5	100.2	119.6
Hip-waist difference	224.3	193.1	170.3
Hip-belly difference	114.8	89.3	81.3
Belly-waist difference	34.2	54.6	31.6

Table.6 Ranking and score

Number of clusters / Number of each index ranking	5	6	7
1st	4	1	1
2nd	1	3	2
3rd	1	2	3
Score	15	11	10

Scoring method: 1st=3, 2nd=2, 3rd=1

A cluster number of five received the highest score, which means that the classification effect in this respect was the best. Therefore, the sample somatotype can be divided into five categories and the mean of the body measurements of each type can be calculated (see Table 7).

Table.7 The mean of the body measurements of five somatotypes (cm)

Somato type	Body height	Cervical height	Arm length	Waist height	Chest girth	Neck base girth	Whole shoulder width	Waist girth	Hip girth	Maximum belly circumference
X1	155.9	132.5	50.2	96.3	98.8	38.9	39.8	87.6	97.6	97.6
X2	156.3	133.0	50.8	97.0	98.5	37.9	40.1	84.2	95.7	89.6
X3	157.2	133.9	51.2	97.9	102.5	39.7	40.8	90.3	95.3	95.0
X4	158.7	135.1	51.6	98.5	93.8	37.2	39.5	79.7	95.9	89.0
X5	159.0	135.6	51.6	99.4	86.8	36.3	38.7	70.6	93.4	80.8

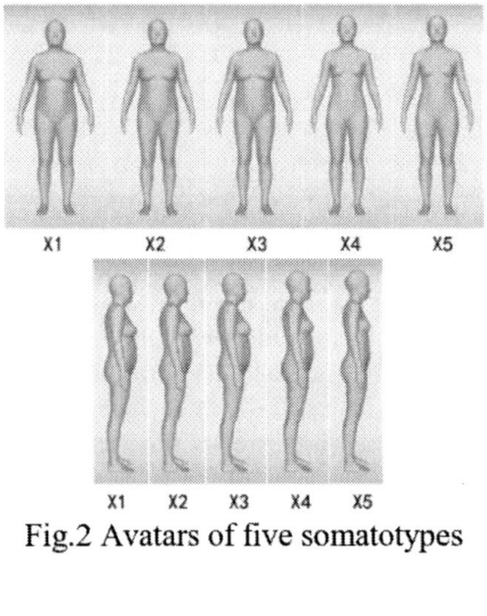

Fig.2 Avatars of five somatotypes

Ranking from low to high by the value of body height, Table 7 highlights the fact that the difference between somatotypes is mainly reflected in the circumferences. X1 has similar value of chest girth to X2, but other values of circumferences are much larger. X3 has obvious characteristics of an enormous breast, a thick waist and a small hip. X4 has a similar value concerning the hip and belly girth with X2, but the value for the chest and waist girth is much smaller. X5 has the tallest and the thinnest across the five types. Using these data, five avatars were created with the aid of Vidya Avatar software (see Fig.2)

5. Conclusion

A random selection of 198 female subjects aged from 35 to 75 years was made from the database. The subjects were then measured using their 3D images to obtain nine body measurements. Their body shape was evaluated using several body shape indices. Most subjects belonged to the obese type according to the Rohrer index, the Pignet index and the Pignet-Vervaeck index. According to the classification by chest-height ratio, most people belong to the wide chest type, while, according to the classification results of the BMI index, most people are moderate and overweight, with obesity an obvious trend. According to the GB1335 classification method, which was used to classify the samples, the proportion of each somatotype among the samples was compared with the GB1335 data. Most subjects belonged to type A and type B, accounting for 83.8% of the total number. The height of every woman in the sample was shorter than the corresponding height in GB1335. But their circumference was much larger than it should be according to GB1335. This shows that people are fatter than before. Six circumference difference measurements were calculated for the chest and waist, the chest and belly, the chest and hip, the hip and waist, the hip and belly, and the belly and waist. These measurements were then used as classification indices in the K-means cluster analysis of the somatotype classification of the samples. By separately determining the number of clusters as five, six and seven, and comparing the classification results of the number of people involved, a final cluster centre and the size of the F value led us to the conclusion that the best number of clusters is five. Therefore, the sample somatotypes were divided into five categories, while the mean of the body measurements of each type were calculated and compared.

6. Acknowledgement

The authors wish to acknowledge the "Research on the digital design model of the female breast with the continuously changed surface under controllable constraints" programme (Grant No. 61303120, NSFC).

References

[1] ASTM Committee. Standard Terminology Related to Body Dimensions for Apparel Sizing. USD521999: 1999.
[2] Gupta D, Zaka R. Anthropometry, Apparel Sizing and Design. Woodhead Publishing: 2014.
[3] Yu XK, Hu F, Zhu DH, Luo Y, Guan JJ. Research on the body shape of middle- and old-aged women in the Shanghai area. Journal of Beijing Institute of Clothing Technology: 2016; 36: 9-17.
[4] Qian LL, Wang HF, Hu CJ. Research on the body size classification of 50~75-year-old middle-aged and old women from Jiangsu and Zhejiang based on 3D anthropometry. Shandong Textile Science & Technology: 2014; 2: 40-44.
[5] Zhou P, Wang SZ. Construction of a middle- and old-aged clothing database in Henan Province, Shanghai Textile Science & Technology: 2012; 40: 5-7.
[6] Liu YM, Dai H. Research on the body build of middle- and old-aged women in the Chengdu area. Journal of Textile Research: 2010; 31: 110-115.
[7] Wand BH. Study of the body shapes of the middle- and old-aged women in East Liaoning Province. Knitting Industries: 2008; 12: 33-35.
[8] Song X, Zhang YX. Probe into the classification of the Chinese female body build. China Fiber Inspection: 2008; 7: 71-73.
[9] He L, Zhang HQ. Comparative study on the classification of women's shape. Progress in Textile Science & Technology: 2007; 5: 95-97.

[10] Liu Y, Zhang ZF. Study of the size designation of clothes based on the body feature of middle- and old-aged men. Journal of Donghua University: Social Sciences Edition: 2004; 4: 20-23.
[11] Li YH. Research on the body build of middle- and old-aged men in Shanghai. Shanghai: Donghua University: 2002.
[12] China National Institute of Standardization. Sampling Scheme for Body Measuring of Adults: 2015.
[13] GB/T 16160-2008. Location and Method of Anthropometric Surveys for Garments: 2008.
[14] GB/T 1335.1-2008. Standard Sizing Systems for Garments - Men: 2008.
[15] GB/T 1335.2-2008. Standard Sizing Systems for Garments Women: 2008.
[16] Size Designation of Clothes; Standard Editor Groups; Instruction and Application of Chinese Standards. Beijing: Chinese Standard Publications: 1992.

Research on Chinese Enterprises' Carbon Disclosure's Status Quo and Amelioration

Jia-Yu Li, An Lu*, You-Fang Chen*

Business School of Beijing Institute of Fashion Technology, Beijing 100029, China

Corresponding author's email: sxylan@bift.edu.cn

Abstract

In recent years, under the influence of low carbon economic environment and corporate events, the development of the productive forces and the climate change aspect on daily lives cannot be ignored. However, the carbon disclosure in China is still in its infancy. This paper will illustrate the status quos of foreign countries' carbon disclosure and the current problems of carbon disclosure in China. The goal of this paper is to lay a solid foundation for future disclosure.

Keywords: Low-carbon; Climate Change; Carbon Information; Carbon Disclosure

1. Overview of Carbon Disclosure

Under the circumstance of low-carbon constraint, carbon intensity of enterprises determines the size of the regulatory risk they are facing. That means carbon disclosure is the bridge conjoined enterprises and other stakeholders. The main goal of low-carbon economy is: to reduce greenhouse gas emissions, to relieve the pressure faced by human society, and to evantually achieve sustainable development [1]. With the signing of the Kyoto Protocol and the convening of a series of international conferences like Copenhagen climate summit, the fact that the measurement method is been changing by the climate change is raising a widespread concern in academic circles. As carbon emissions-related activities of enterprises are still in progress, =the traditional measurement methods cannot fulfill all the requirements of carbon accounting and carbon disclosure which leads to the rise of carbon disclosure.

1.1 Definition of Carbon Disclosure

Internationally, carbon disclosure is often used to measure and monitor the amount of greenhouse gas (GHG) released by a tissue or product. Carbon disclosure is based on the financial combination that is disclosed in the form of a text description, to promote green production with the main objective of saving energy and reducing emissions during production. The aim is to create a new requirement, measurement methods, recording methods and forms of reporting [2].

1.2 Objects of Carbon Disclosure

The target objectives of carbon disclosure are: low-carbon investment activities, low-carbon business activities and low-carbon financial activities.

Low-carbon investment includes the introduction of advanced energy-saving equipment, low-carbon technology investment in research and development, and the purchase of low-carbon funds.

Low-carbon business activities include the procurement of raw materials, raw materials for further processing, recycling and transportation process. In the procurement of raw materials, companies can buy clean energy to reduce greenhouse gas emissions. In the material processing and production stage carbon dioxide emissions can be reduced during the production process. In the cycle and transit phase, companies can choose cleaner, more environmentally friendly fuels to reduce carbon emissions [12].

Low carbon finance activities include low-carbon special funds established by the government, reputable green credit projects of banks, and financial assistance from developed countries through the Clean Development Mechanism (CDM) project.

2. Present Situation of Enterprise Carbon Disclosure

What standards are followed by foreign companies for carbon disclosure, and what information about carbon-related activities is available. This section mainly describes the relevant forms of disclosure in Table 1, and content descriptions of several major international organizations.

Table.1 International Group on Carbon Disclosure [3]

Name	The background and reasons for the establishment of the organization
CRDI: Climate Risk Disclosure Initiative	A group of leading institutional investors from around the world released the Global Framework for Climate Risk Disclosure-a new statement on disclosure that investors expect from companies-in October 2006. Investors require this information in order to analyze a company's business risks and opportunities resulting from climate change, as well as the company's efforts to address those risks and opportunities. The Framework encourages standardized climate risk disclosure to make it easy for companies to provide and for investors to analyze and compare companies.
CDSB: Climate Disclosure Standards Board	The Climate Disclosure Standards Board (CDSB) is a non-profit organization working to provide material information for investors and financial markets through the integration of climate change-related information into mainstream financial reporting. CDSB operates on the premise that investors and financial institutions can make better and informed decisions if companies are open, transparent and analyses the risks and opportunities associated with climate change-related information.
PwC: Price Waterhouse Coopers	Price Waterhouse Coopers (trading as PwC) is a multinational professional services network headquartered. Dedicated to the future of business development needs, providing the world's first service in the enterprise carbon disclosure example.
GRI: "Sustainability Reporting Guidelines" issued by Global Reporting Initiative	The Global Reporting Initiative (known as GRI) is an international independent standards organization that helps businesses, governments and other organizations understand and communicate their impacts on issues such as climate change, human rights and corruption. As of 2015, 7500 organizations used GRI Guidelines for the sustainability reports. GRI Guidelines apply to multinational organizations, public agencies, smaller and medium enterprises, NGOs, industry groups and others. For municipal governments, they have generally been subsumed by similar guidelines from the UN ICLEI.
CDP: Carbon Disclosure Project	The CDP (formerly the "Carbon Disclosure Project") is an organization based in the United Kingdom which works with shareholders and corporations to disclose greenhouse gas emissions of major corporations. As disclosure of emission related data as CDP's primary activity, the quality of the data reported to CDP is the key. In 2014, nearly 2000 businesses reported climate change data to CDP. The value of CDP's reports for investors and NGOs is contested. Furthermore, the quality of the data on which CDP's reports are premised is questionable.

2.1 The Development of Corporate Carbon Information Disclosure in Foreign Countries

China has played a role to actively carry out energy-saving emission reduction activities through supporting carbon disclosure, Enterprises are gradually beginning to promote the relevant carbon measurement laws and regulations.

It will start from the basic situation and development of carbon disclosure of domestic and foreign enterprises, and lay the foundation for exploring the methods of carbon disclosure research and improvement with detailed contents and concrete introduction [8].

In developed countries, some industry regulatory bodies requires companies to disclose carbon emissions information, and the development of carbon emissions auditing standards. For example, in 2003, the American Institute of Certified Public Accountants issued a "greenhouse gas emissions information certification business." The same year, the Institute of Chartered Accountants of Canada issued the "Practice Guidance: Audit of Greenhouse Gas Emissions Information". In 2008, Australia's Climate Change Division developed the National Guidelines for Emissions and Energy Consumption of Greenhouse Gases, which provided guidance on the conduct of carbon emissions audits.

The Climate Risk Disclosure Initiative (CRDI) defines the content of carbon disclosures: First, the general situation of carbon emissions in the past, present and future; Second, the strategic analysis of climate and emissions, corporate governance; Third, measuring the direct consequence of rising temperatures; Fourth, the analysis of risks associated with carbon emissions, events and their uncertainties, projections of carbon emissions regulations and estimates of carbon costs.

The Climate Disclosure Standards Board (CDSB) is primarily about regulatory standards: first, climate change strategy analysis; second, climate change regulatory risk; and third, the tangible risk of climate change; and finally, The scope of carbon information is defined, including carbon information within the organization's boundaries during the reporting period.

"PricewaterhouseCoopers" (PWC) first described management evaluation: the purpose of corporate reporting on carbon disclosure, strategic deployment and measures to address environmental change, governance arrangements, environmental change and carbon trading. Operating and financial impact of carbon emission reduction performance; Second, the enterprise's carbon emissions statements; again, the enterprise's carbon emissions statements Note: the emission reporting policy, the sub-source carbon emissions and the regional emissions of the business data, combined or separation of the impact of carbon emissions; the final is the third-party verification report [10].

The Sustainability Reporting Guidelines developed by the Global Reporting Initiative (GRI) provide: First, the strategy and profile: corporate profile, strategy and analysis, governance mechanisms, management commitment, stakeholder engagement Management, etc.; Second, the management approach: to provide information on how to deal with specific issues of various types of business background, in order to facilitate understanding of business and other aspects of the efforts made; Third, the performance indicators: specific sub-economic, environmental and social aspects , Disclosure of corporate direct and indirect carbon emissions, carbon reduction measures and results.

"Carbon Disclosure Project" (CDP: Carbon Disclosure Project) is a voluntary number of international companies to carry out a typical example of carbon information disclosure. CDP has developed a relatively complete corporate carbon information disclosure framework, which includes four aspects: First, the risks and opportunities arising from climate change, including the risk of laws and regulations, natural risks, competition risks and reputation risks; Regulatory opportunities, visible opportunities and other opportunities. Second, the accounting of carbon emissions, including the choice of carbon accounting methods, the preparation of carbon reduction accounting reports, external verification and audit, direct carbon emissions and indirect carbon emissions tons, Third, carbon reduction management, including emission reduction projects, emission intensity, energy costs, emissions trading, emission reduction planning and so on. Fourth, climate change governance, including emissions reduction responsibilities and their respective contributions.

2.2 The Development of Carbon Disclosure in China

Because of the lack of relevant mandatory disclosure regulations in China, carbon information disclosure of listed companies is of voluntary disclosure. Relevant reports, disclosure of the structure is not standardized.

Climate change may affect the operating performance of listed companies, including energy, petrochemical, steel and cement industries such as high energy consumption, including banking, insurance, hotel tourism and other low-power industries. In the past three years, China's listed companies have become more and more aware of climate change hence has adopted a positive attitude, with increasing information disclosure. However, compared with the international market, China's listed companies regarding the levels of carbon information disclosure is still low. The quality of information disclosure needs to be improved. While the Chinese government is actively promoting the Copenhagen climate change summit, Chinese enterprises have also shown positive attitude towards climate change at the summit. The Chinese business delegation issued "Our Commitment and Hope: The Copenhagen Declaration of the Chinese Business Community", solemnly committed to establishing a climate change strategy and guiding the development of enterprises; to reduce the carbon footprint in production and business activities to actively promote the establishment of specific emission reduction targets. "On the promotion of China's low-carbon economy development proposal" became the 2010 national "two sessions" on the 1st proposal, the development of low-carbon economy and low-carbon life increasingly strong atmosphere. Chinese enterprises actively explore the development model of low-carbon economy, low-carbon strategy, new energy investment, carbon management and carbon trading, low-carbon technology, low-carbon finance and other fields proactive and innovative efforts. IThe emergence of a group of low-carbon pioneer, explore the development of low-carbon enterprises, for instance: Changsha Yuanda, Dezhou Huangming, Baoding Yingli and Wuxi Shangde, etc. [11].

Under the constraints of building a low-carbon environment, the disclosure of corporate carbon information reflects the size of the enterprise's regulatory risk, the degree of social responsibility and the value of the enterprise. Due to the limited knowledge in this field, only a few enterprises can complete, comprehensive and detailed disclosure, thus the data on China's carbon disclosure related research is still inadequate.

3. The Problems in the Disclosure of Carbon Information in Chinese Enterprises

The structure is disorganized, and relevant information regarding carbon disclosure is scattered in the annual report. The information disclosed is mainly qualitative. Carbon emissions are mainly described by brief introductions. Few companies report carbon emission reductions directly. The industry differences in carbon disclosure are obvious. Since carbon information is not included in the audit and appraisal of third parties, the information on carbon disclosure is more subjective and random [5].

3.1 The Unbalance of Voluntary Disclosure of Carbon Information in Enterprises

Table 2 shows the number and percentage of specific disclosures in the six disclosure parts of the carbon information. It can be seen that the highest number of disclosure is "carbon emission reduction initiatives and performance", the main reason may be the promulgation of the "Energy Conservation Law", which provides that listed companies must have carbon emission reduction initiatives and performance in the annual report [6].

Table.2 Voluntary disclosure of carbon information in listed companies

Items	Numbers of the companies who disclose (%)
Opportunities, risks and strategies	918 (58.96%)
Carbon emission	100 (6.42%)
Carbon reduction measures and performance	1271 (81.63%)
Carbon trading participation information	36 (2.31%)
Third party audit and assurance	0 (0%)
Others	958 (61.53%)

As seen from Table 2, China's corporate carbon information disclosure "carbon information third-party audit forensics" project disclosure percentage is 0, indicating that China's listed companies have not yet carried out the audit of carbon disclosure.

3.2 Comparison of Carbon Disclosure of Chinese Enterprises and Foreign Enterprises

As the development of carbon information disclosure in China is still not perfect, different industries are faced with a completely different market environment and government management system. The impact on the environment during production and operation is very different [9]. The environmental problems caused by heavily polluting industries are more disruptive and more likely than those in non-polluting industries. Moreover, environmental information disclosure regulations and public environmental awareness are under greater pressure. So they bear the responsibility of society to maintain and improve the environment. In order for enterprises to obtain resources they must comply with the requirements of the system. Therefore, with the public's concern of environmental change and the introduction of carbon emissions system, heavily polluting industries are under increasing pressure to consciously respond to the pressure of carbon emissions and to fully disclose relevant carbon Information. Under the pressure of the government's regulatory system, the degree of carbon information disclosure was significantly higher when comparing heavily polluting industries to non-polluting industries. However, there is no significant relationship between the high-carbon industry and carbon disclosure, and it is inferred that high energy consumption and heavy industry companies are reluctant to disclose too much carbon information [7][13].

4. Suggestions on Improving Carbon Disclosure of Chinese Enterprises

To improve carbon disclosure of Chinese Enterprises it is important to, popularize the concept of carbon emissions information disclosure. Improve the quantitative criteria for carbon disclosure, and formulate a set of carbon emission information disclosure index system and data statistics caliber suitable for China's market. In addition, the regulatory body needs to increase relevant training, so that more companies understand and master basic quantitative methods of the disclosure of carbon emissions information.

Furthermore it is important to introduce the regulatory bodies to listed companies to disclose carbon information rules. In the future, regulators should be listed as soon as possible,, gradually converting voluntary disclosure to mandatory disclosure while ensuring the requirements are clear [4].

Finally to improve carbon disclosure, there is a need of developing relevant products, and guide the whole society on carbon efficiency concerns. Based on the development of carbon efficiency index, financial institutions can try to introduce a variety of financial products.

5. Conclusion

In this paper, the main problems and suggestions for improvement of carbon disclosure in Chinese enterprises are discussed, which will lay a foundation for future carbon disclosure research. The development of carbon information disclosure is still in the initial stage. The form of evaluation system and the content of carbon disclosure are not unified. However, we are confident of the future development of China's carbon information disclosure, and will establish a unified standard gradually improving the carbon information disclosure system, and continuously pushing China's carbon disclosure to a more mature direction.

References

[1] Chen H, Wang HY, Chen Z. Corporate Characteristics and Voluntary Disclosure of Carbon Information-Based on the Analysis of Legitimacy Theory, Accounting and Economic Research: 2014; 4.

[2] Chen H, Wang HY, Jing X. Chinese company's carbon disclosure, Content definition, measurement methods and the status quo: 2013; 12.

[3] He Y, Tang QL, Wang KT. Carbon Disclosure, Carbon Performance and Capital Costs, Accounting Research: 2014; 1.

[4] Tan DM, Zou SL. International Development of Carbon Disclosure and Construction of China's Carbon Disclosure Framework, Statistics and Decision Making: 2010; 11.

[5] Wang ZB, Jin XC. Correlation Study of Carbon Disclosure and Corporation Value, Macroeconomic Research: 2013; 1.

[6] Zhang QL. Study on Carbon Accounting and Information Disclosure, Contemporary Finance and Economics: 2010; 4.

[7] Zhang QL, Song WB, Tan J. Carbon Emission, Carbon Disclosure Quality and Enterprise Value, Nanjing Institute of Audit Journal: 2013; 2.

[8] Patten DM. The relation between environmental performance and environmental disclosure: a research note. Accounting, Organizations and Society: 2002; 8.

[9] Qiang DY, Wen GY. Build a enterprise low carbon accounting system thinking. Friends of Accounting: 2010;

2.
[10] Chen LB. The development status of the Western carbon accounting and the development of China's carbon accounting lag reasons Exploration. Era of economic and trade: 2010; 9.
[11] Tang QL, Luo L. Carbon Management Systems and Carbon Mitigation. Australian Accounting Review: 2014; 1.
[12] Lambert R, Leuz C, Verrecchia RE. Accounting Information, Disclosure, and the Cost of Capital. Journal of Accounting Research: 2007; 2.
[13] Bewley K, Li Y. Disclosure of Environmental Information by Canadian Manufacturing Companies: A Voluntary Disclosure Perspective. Environmental Accounting & Management: 2000; 1.

Reducing Environmental Impacts of Swimsuits

Rui-Qiong Wen, You-Fang Chen*

Business school, Beijing Institute of Fashion Technology, Beijing, 100029, China

*Corresponding author's email: sxychenyf@bift.edu.cn

Abstract

With the development of the leisure tourism industry and the enhancement of young people's swimming consciousness, the consumption of swimwear has been increasing in recent years. Meanwhile, people have paid more and more attention to environmental protection. The production of swimwear causes impacts on the environment, and the expansion of swimwear consumption increases the impacts. It is necessary to study how to reduce the swimsuits' environmental impacts. This paper discusses the environmental impacts of swimsuits in two aspects, namely, raw materials and production process. It also puts forward corresponding measures to reduce environmental footprints. In addition, it is acknowledged that one-time consumption of swimwear is the key impact of swimsuits on environment, therefore, this paper makes corresponding measures for it.

Keywords: Swimsuit; Environmental Impact; Production Process; Consumption Attitude

1. Introduction

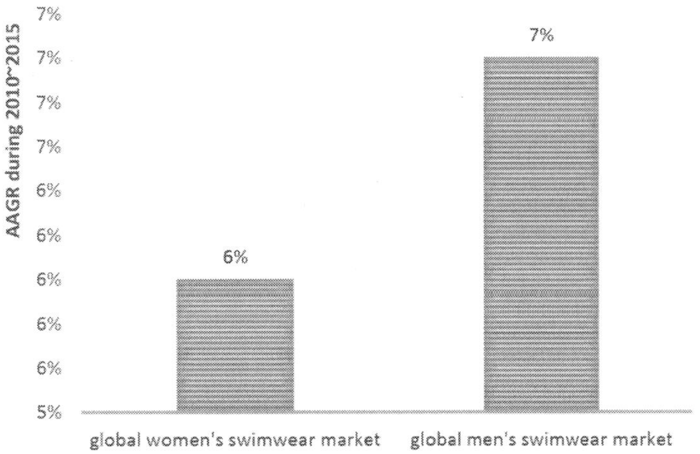

Fig.1 The AAGR of global swimwear market during 2010-2015.
Cited from Global Swimwear Market to Witness Growth Through 2020 by Technavio, Business Wire (English)

Swimsuits refer to the special garments people wear when they swim in water [1]. On the one hand, in the international arena, swimming is essential for many people during the summer, as such swimwear has become a fast fashion product. There is a strong worldwide demand for swimsuit, with the average consumption reaching about 1 billion annually [2]. The consumption of swimsuit has become greater, which drives the market to rise year after year. The AAGR of global swimwear market during 2010-2015 is shown in Fig.1 [3]. Technavio's analysts forecasts the global swimwear market to grow at a CAGR of 5.97% during the period 2017-2020 [4]. In China, with the development of the leisure tourism industry and the enhancement of young people's sense of swimming, swimwear consumption has great increased in recent years. It is becoming a big part of China's apparel industry. On the other hand, according to experts, people's attitudes to swimwear consumption in Europe, the United States and other developed areas are different from that in China. In these developed areas, almost all women would use swimsuits during their vocations, with each owning 3 to 5 sets per year [2]. However, in China, most customers change their swimsuits once a year, and a few consumers change once every two years. In addition, the data of market research firm NPD shows that the US market swimsuit sales reached 4.4 billion USD in 2013, with an increase of 6% over the previous year. Behind the $ 4.4 billion market size, there are only 300 million people on the scale. While China's population is several times as much as that of the United States, swimsuit sales are only a fraction of the United States [5]. The annual consumption in developed areas, including Europe, United States etc., is higher than China, in which consumers' one-time consumption attitudes to swimsuits play a direct role.

Although industrialization has improved the living conditions of human beings, it also has brought great environmental problems, such as the manufacture of swimsuit. The swimsuit production process has many adverse effects on environment, and swimwear consumption is so great. From the international perspective, swimsuit waste is very serious because of one-time consumption concept. Therefore, the environmental impacts caused by swimwear become more and more serious. How to reduce swimsuits' impacts on environment should be taken into account, and it is worth conducting a more comprehensive discussion. This paper mainly explores swimwear in saving resources, reducing emissions on the environment impacts, and how to reduce the impacts. The swimsuits talked about here mainly refer to those people wear in the water or beach activities.

2. Main Materials of Swimwear and the Environmental Impacts in Production Process

2.1 Main Materials of Swimsuits

Raw material is critical for textiles and clothes, and has become the center of textile industry's sustainable development [6]. With the analysis of swimsuits' materials, a very good foundation will be laid for the subsequent analysis. Analyzing the materials of swimsuits provides a more profound, more comprehensive understanding of sustainable development of swimsuits.

Materials used to make swimwear need to have some special features. In order to meet the users' needs in water, the fabrics used in swimwear should be different from ordinary ones. For example, they are supposed to be non-absorbent, quick-drying, flexible and so on. Therefore, the majority of swimsuits' fabrics are polyester, nylon and a small amount of spandex [7]. The polyester fabric is less flexible, so it is mainly used in swimming trunks or split women swimsuits instead of one-piece. Nylon fabric is currently the most commonly used swimsuit fabric. Since its elastic softness is better, it is usually used for medium-priced products. Spandex is the best elastic fabric, and it can be extended to 5 times as long as the original. It is usually mixed with nylon or polyester as a fabric to enhance the swimwear stretch, and accounts for only a small part of swimwear. Table 1 shows the functions and usages of these fibers.

Table.1 The main functions and uses of swimwear

name	function	category
polyester	High strength, moderate elasticity, high wear resistance, poor hygroscopicity	Chemical fiber
nylon	High wear resistance, high strength, good elasticity, light weight	Chemical fiber
spandex	Excellent elasticity, high seawater resistance	Chemical fiber

Source: sorted by the author from the website information

2.2 Swimsuits' Environmental Impacts in Production Process

Swimsuits' production process includes fiber production, printing and dyeing, cutting and sewing, etc. In these links, fiber production and printing and dyeing produce the most environmental impacts.

The first part is swimsuits' fiber production process. Since polyester and nylon are the major materials used in the production process, with a few spandex to improve elasticity, polyester and nylon are mainly discussed in this paper.

Polyester fiber is a petrochemical product, and its impacts are mainly about petrochemical industry, including extraction cost and transportation costs. Petroleum is scarce and non-renewable, and it consumes large amounts of energy to convert raw oil to usable fiber. It is estimated that 1 kilogram of polyester fibers need 109 MJ energy consumption, 49 MJ to generate raw materials and 63 MJ to process raw materials into fiber [8]. Chemical fiber needs less water than natural fiber, with little or no water, but the production of chemical fiber causes a lot of environmental pollution. It produces a series of harmful substances to environment such as heavy metals and greenhouse gas.

Nylon is similar to polyester, and the main impacts on resource and environment are heavy metal salts, other toxic substances pollution and greenhouse gas emissions. Producing nylon also emits nitrous oxide, which is a powerful greenhouse gas. It is estimated that greenhouse effects created by nitrous oxide emissions in the UK are equivalent to more than 3% of the UK's total carbon dioxide emissions [9].

Both polyester fiber production process and nylon production process cause a lot of waste water and gas emissions, and need to consume a lot of energy.

The second part is dyeing and printing process. At present, swimwear in the market has many colors and patterns, and swimwear fabric's printing and dyeing need to be considered.

The dye bath contains processing (auxiliary) chemicals as well as dye. The dyeing process could be influenced by several factors, such as dye materials, machine, liquor ratio, time and temperature. This process is resources intensive, including water, energy and chemicals, and could produce highly-colored effluents. Dyeing process is the most likely source of major metal pollutants, such as zinc, copper and chromium, and it is hard to reduce this pollution. No single color could be produced without any pollution. For example, green and yellow use one harmful metal substance as materials. Generally, the darker the shade is, the greater the amount of dye lost to effluent is.

The chemicals required for printing are complex and it is stained in specific areas. There are many printing technologies, and they have different influence on environment. In printing, the main impact on environment is printing paste. There are many harmful chemicals in printing paste, so reducing the waste of printing paste can achieve the effect of environmental protection.

3. Measures to Reduce the Environmental Impacts of Swimwear Production Process

Based on the analysis mentioned above, the corresponding measures to reduce environmental impacts will be proposed.

3.1 Measures to Reduce the Environmental Impacts of Swimwear in the Fiber Production Process

As mentioned above, it can be learned that the main environmental impacts of the fiber production process of swimwear are mainly waste water and gas emissions, and energy consumption. For these, efforts can be made from the following three aspects.

Firstly, waste water and gas from the production process should be processed to reduce harmful emissions.

Currently, the main waste water treatment technologies are biological treatment, adsorption and oxidation treatment and some other forms. Continuous reactor technology is widely adopted internationally, and it is a relatively advanced technology. Waste water is pretreated by chemical oxidation method, and then the part that is not fully treated is reprocessed to ensure that the waste water does not cause great pollution to environment.

The main treatment technologies of waste gas are spray treatment and combustion treatment, but the effect of spray treatment is not good, with much many residual harm. The method of combustion treatment is easy to explode. At present, most companies adopt the approach which is the combination of spray and combustion treatment. Catalytic combustion technology is also a better way.

Secondly, the production process of polyester and nylon should be improved to reduce the emissions and to save energy by technology innovations. However, this is difficult to come true.

Thirdly, substitute of polyester and nylon could be used, to achieve the goals of reducing pollutant emissions and saving energy by finding the proper alternative fibers. These alternative fibers need to be environmentally friendly and have the special functions that polyester and nylon have. People in polyester industry believe that the development of renewable energy technology will be their savior [10]. Textiles made of bio-polymer will reach rapid growth in the coming year, which will reduce the dependence on polyester.

Considering the specific properties required by swimsuits, among the preexisting fibers, an alternative fiber that is currently very environmentally friendly is poly-lactic acid. It is a material currently synthesized with a renewable material called corn. It is different from the traditional synthetic materials, which comes from gas, but it is a kind of crop that is annual harvest and a renewable resource. Poly-lactic acid could be converted to fiber or yarn, which is similar to polyester with a certain degree of ductility and strength. However, the current cost of the production of poly-lactic acid is relatively high, with a low melting point, and the dyeing process could cause material waste. The application of poly-lactic acid is limited, but the technological advances may overcome these restrictions.

In addition, DuPont is promoting a renewable resource of bio-polymers, which is an intermediate substance in the Sorona production process. As a high-performance breakthrough, Sorona is derived from bio-based rather than petrochemical products, and it has multiple applications. DuPont claims that 37% of the raw materials are from natural renewable resources rather than petrochemical raw materials due to the use of renewable raw materials, thereby reducing dependence on oil resources [11]. Sorona's production process reduces energy use and greenhouse gas emissions, compared with equivalent nylon production. Filament or staple fibers made of Sorona are appropriate materials for high-quality products and have various applications. It has many advantages, such as flexibility, comfortable stretch, resilience, anti-chlorination, and anti-ultraviolet performance, colorful and it is easy to be taken care of. It is the ideal choice for swimwear. However, due to high production cost, the current application is not very broad. The environmental value of the Sorona polymer produced by the biochemical method is shown in the figure below.

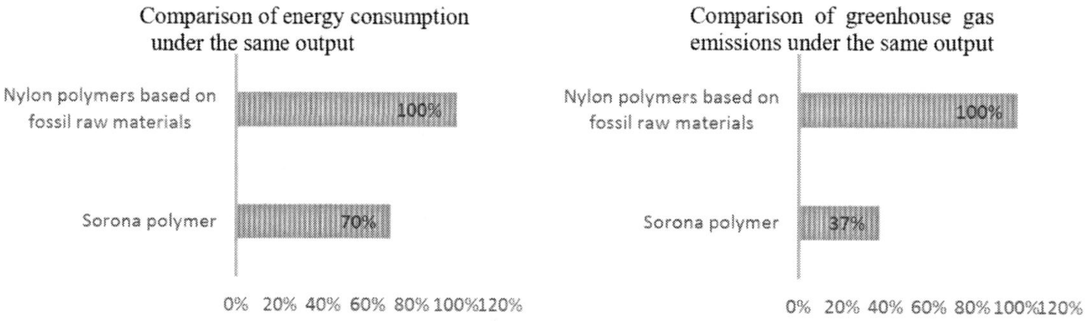

Fig.2 The environmental value of the Sorona polymer Cited from the official website of DuPont

3.2 Measures to Reduce the Environmental Impacts of Swimwear in Dyeing and Printing Process

According to the above analysis, in the dyeing process of swimwear, effective measure to reduce pollutant emissions is more advocacy of light-colored swimsuit consumption. In the printing process of swimwear, effective measure is to use some advanced printing technologies.

According to the explorations mentioned above, in dyeing, the darker the shade is, the greater the pollution to the environment is. Although the environmental benefit of avoiding dyeing to dark shades may be offset in laundering, as light shades tend to show the dirt more easily and need frequent laundering, causing more impacts during the use phase. However, for swimwear, laundering process is very simple, and the water consumption is very small, almost no consumption of detergent. Therefore, light-colored swimsuits produce less impact on environment than dark ones. In terms of dyeing process, what can be done is to encourage consumers' preference to light-colored swimwear. Thus, manufacturers will produce more light-colored swimwear. To a certain extent, the effect of reducing pollutant emissions in the dyeing process can be achieved.

In printing, there are many printing technologies. Some of them are environmentally friendly, such as silk screen printing which can reduce the waste of printing paste. There are many harmful chemicals in printing paste, and reducing the waste of printing paste could protect the environment.

For swimsuits, it is worth mentioning an advanced technology called transfer printing technology, which has many advantages in comparison with other methods of printing. This technology is beneficial for polyester, which is the most commonly used fabric in swimwear.

Transfer printing technology uses volatile disperse dyes. First of all, images are printed on the paper with volatile-disperse dyes. Then, the paper is heated with the fabric in a reactor. In this process, the dyestuff is transferred from the paper to the textile material by sublimation. In transfer printing, only the dyestuff, no other chemical, is deposited on the fabric, so no washing off is required and no effluent is generated. This method consumes only 2 kilogram water to print 1 kilogram fabric, compared with 250 kilogram water in conventional method [12]. However, this technology is immature, and it is not suitable for massive industrial production. There are certain limitations, but it could be a direction for the development of printing in the future.

4. Reducing the Environmental Impacts of Swimwear Consumption

According to the discussion in the introduction, it can be seen that the gradual increase in swimwear consumption is the main aspect of the environmental impacts of swimwear, while the one-time concept of swimwear is one of the main reasons for the increase in swimwear consumption.

Consumers' attitudes toward swimsuits are very critical but not easy to achieve. However, it is worthy considering. When swimsuit reaches the hands of consumers after sale, it will be washed for repeated use or discarded as disposable item. This is related to the attitude of consumers. In China, people often re-use swimwear. After swimming, they wash the swimwear in clean water and hang it to dry.

However, western consumers' attitudes toward swimwear are very different from those of Chinese consumers. As mentioned above, in most developed areas, swimwear is considered as a one-time item and will be discarded after the first use. The consumption phrase produces very little impacts on environment, while the production phrase produces the most part of environmental impacts. Swimsuits' main materials are polyester, nylon and other man-made fibers, so the suitable temperature for washing is relatively low, in the range of 30 to 40 Degrees Celsius [13]. Repeated use can save around 10 to 20 times energy, so re-use of swimsuit is much more energy-efficient and environmentally friendly. This requires a global change in the attitude toward swimwear consumption. The global consumption attitudes toward swimsuits should be changed, and specific ways are listed below.

People's concepts are formed in a long term, and they are difficult to change. Once a change occurs, it has a very significant impact. In order to change the consumer's one-time consumption of swimsuit, the core is to trigger the emotional contact between consumers and swimwear. People wouldn't discard things that have emotional ties with themselves. So, the disposable swimsuit which has emotional contact with the consumer into reusable can be transferred, and then sustainable development results could be achieved. To trigger this emotional connection, there are several ways (Fig.3).

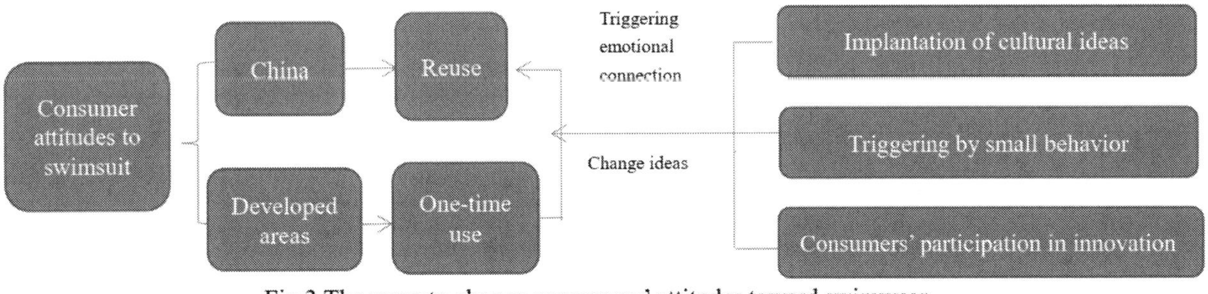

Fig.3 The ways to change consumers' attitudes toward swimwear
Source: sorted by the author

4.1 Implanting Cultural Ideas into Swimwear

Fashion and clothing are two different categories, and they satisfy peoples' needs in different ways, in terms of conception and function [14]. Clothing is a material product but fashion is a symbolic product. Although their usages and appearances sometimes look similar, they are different when they connect with people. Fashion is about time and space, and its purpose is to meet the psychological needs of individuals, but clothing is to meet the functional needs of the body from the role of shelter and protection. To change consumers' attitudes toward swimwear, swimwear should be given the meaning of the cultural heritage of fashion, not just clothing. Once the swimsuit has a cultural connotation, it will become valuable to consumers. Consumers will become more cautious when buying swimsuits and they are probably to use them multiple times because of their cultural connotations. In this way, consumers' attitude to their own swimsuit is no longer a disposable one, but a symbol of their own ideological content, a symbol of their culture and fashion accomplishment. In this way, consumers will not let so meaningful swimwear be discarded.

4.2 Triggering Emotional Connection by Small Behavior

Although the value and meaning of clothes are influenced by complex factors, it seems that the life of clothes can be prolonged by triggering people's feelings about the clothes in a simple way. For example, Otto von Busch has done a much deeper experiment [15]. Instead of buying clothes, customers were required to exchange the new clothes with their own clothes they wore that day. The customers were required to write down their feelings about their own clothes, to think about their relationships with their clothes. Eventually, many people chose to keep the piece they wore. Similarly, for swimwear, some small things can be done to trigger the emotional connection between consumers and swimwear. Here, a relatively simple example by designing a simple scenario will be given. For example, there is a woman who has just lost weight, and her body is very charming. She has a strong self-confidence, and she wants to show her body to other people eagerly. She takes her swimsuit to the beach where a lot of people are swimming. She was not in good shape before, and ashamed to show her body; this swimsuit is a one-piece one. Then, she could transform this one-piece swimsuit into another shape with a simple cut, into a two-section swimsuit. This old swimsuit not only has not been abandoned by her, but has been transformed into a new style. It brings her new life experience, too. This new life experience gives emotional contact between the consumer and the swimsuit, extending life cycle of the swimwear, so as to achieve the effect of saving resources.

4.3 Introducing Consumers to the Swimsuits' Innovation Process

In the 1970s, the cultural critic Ivan Illich put forward a concept like this, which has a significant impact on the relationship between customer and textile now [16]. People are recast in roles other than simply that of consumers, they are also competent individuals who are potential producers of their clothes, or suppliers of skills and resources enabling them to create as well as consume. Customers could participate in the designing or making process, and they could produce their own unique swimsuits. This process brings them a sense of pleasure and enhances their love for swimsuits. Some modern information tools, such as Micro-blogging, We-Chat and Facebook can be used to provide consumers platforms, so that they can communicate with each other on these platforms to enhance their swimwear making and bring them self-recognition value, as well as to further strengthen their emotional ties with swimwear and achieve the goal of re-use.

5. Conclusion

The development of leisure tourism industry and the enhancement of young people's swimming consciousness make the consumption of swimwear increase year by year, coupled with the shortage of resources. Environmental protection has drawn more and more attention, and it is necessary to study how to reduce the swimsuits' environmental impacts. This paper discusses and puts forward the main materials of swimwear, the main environmental impacts of the production process and the corresponding measures to reduce the environmental impacts of swimwear production. The impacts in production process include the production of swimwear fibers and the dyeing and printing process of swimwear. The production process has many negative impacts on environment, including energy consumption, greenhouse gas emissions, and so on. In these two terms, some measures like taking technical innovations, adopting alternative materials and so on should be taken. Besides, worldwide one-time consumption attitude is the key to reduce the environmental influences, and this paper also makes corresponding measures in response to this problem. It is believed that consumers' attitudes should be changed by the implantation of cultural concept, consumers' participation in innovation and other emotional contacts with swimwear. Changing the attitudes could contribute to reducing the environmental influences of swimsuits.

6. Acknowledgment

This work was supported by Study on the Present Recycling Condition and Recycling Model Discarded Garment in Beijing Based on the Classification of Renewable Resources, 15JDJGB010, Study on Recycling and Recycling System of Waste Textile in the View of Urban Waste Reduction, SZ20161001208, Study on the Recycling of the Discarded Garment and the Resource and Benefit System in Beijing - Tianjin - Hebei Cooperation, PXM2016_014216_000022, and Study on the Cultivation and Education of Green Consumption in Beijing Residents' Clothing, SZ20171001209.

References

[1] Information on http://baike.baidu.com.
[2] Information on http://news.sohu.com
[3] Technavio R. Global swimwear market to witness growth through 2020. Business Wire (English): 2016; 5.
[4] Newswire PR. Global swimwear market 2016-2020. Newswire US: 2016; 3.
[5] Yicong G. Research on the upgrading strategy of Xingcheng swimsuit industrial cluster. Master's Degree Essays: 2016; 6.
[6] Simpson P. Global trends in fibre prices, production and consumption. Textiles Outlook International: 1999; 82-106.
[7] Information on http://news.sina.com
[8] Laursen E, Hansen B. Environmental assessment of textiles. Danish Environmental Protection Agency: 1995; 84.
[9] Pierre SD. Du Pont outlines plans to curb nylon emissions. ENDS Report: 1996; 7-8.
[10] Allwood JM, Malvido DR. Advanced technology to save energy. Institute of Manufacturing: 2003; 29.
[11] Information on http://dupont.com
[12] Union NE. The textile industry and the environment. Industry and the Environment Technical Report: 1993; 56.
[13] Information on http://www.marksandspencer.com/gp/node/n/50713031
[14] Song G. On the difference of clothing design in our country from the traditional. Intelligence: 2009; 3.
[15] Information on http://www.kulturservern.se/wronsov/italyan
[16] Illich I. Tools for conviviality, London: 1975; 26.

Quantification of Water Footprint on China's Textiles and Clothing Exports

Wan-Wen He[1], Yi Li[2], Lai-Li Wang[1, 2,3*]

[1] *School of Fashion Design & Engineering, Zhejiang Sci-Tech University, Hangzhou, 310018, China*
[2] *Zhejiang Provincial Key Research Institute of Philosophy and Social Sciences for Ecological Civilization, Hangzhou, 310018, China*
[3] *Zhejiang Sci-Tech University (Engineering Research Center of Clothing of Zhejiang Province), Hangzhou, 310018, China*

*Corresponding author's email: wangll@zstu.edu.cn

Abstract

China is the world's largest producer and exporter of textiles and clothing. Water resources are 'virtually' transferred from China to trade partners through the international trade network of textiles and clothing. This paper provides a systematic analysis of virtual water trade flows related to China's textiles and clothing exports in the period 2001-2014 with the water footprint (WF) methodology. The results indicate that both WF of textiles exports ($WF_{exports}(T)$) and WF of clothing exports ($WF_{exports}(C)$) showed continuous increase from 2001 to 2007, and fluctuation changes from 2008 to 2014. The total $WF_{exports}(C)$ was 40.71 Gm^3, a little larger than the total $WF_{exports}(T)$ with a volume of 39.96 Gm^3. Water pollution caused more severe impacts on the environment as grey $WF_{exports}$ was larger than blue $WF_{exports}$. Policies that encouraged the development and popularization of freshwater saving and wastewater pollutants reduction technologies in China's textiles industry promoted the reduction of water footprint intensity of textiles and clothing industrial production. This effectively inhibited the increase of $WF_{exports}$, especially after 2008.

Keywords: Water Footprint; Virtual Water Flow; Textiles and Clothing Exports; Water Resources Transfer

1. Introduction

Water plays a vital role in the ecosystem and is necessary for life. Global water consumption continues to increase due to the growing population. It is estimated that 47% of the world's population will be living under severe water stress by 2030 [1]. As an essential component of national and local economies, the consumption of water is generally categorized by its three major uses: agriculture, industry, and domestic use [2]. The projections indicate that global water demand for the manufacturing industry is expected to increase by 400% from 2000 to 2050, leading all other sectors [3]. Generally, there is no general positive relation between water demand and availability. A water-scarce country can thus aim at importing products that require a lot of water in their production and exporting products or services that require less water [4]. This is called 'virtual water trade' and will relieve the pressure on the country's own water resources [5].

The virtual water trade has been investigated by a remarkable variety of scientific contributions from different spatial scales (global, national, regional and river basin, etc.) and products with different methods. For example, there are studies on virtual water trade of China, Malaysia, Spain, EU 27, Mid-Atlantic states, Nile Basin and EU river basins [6-12]. It is pointed out that China has been the largest net virtual water exporter [13]. Besides, many literatures focused on the virtual water trade of agricultural products, industrial products, building, forest and biofuel [14-17]. The methods adopted in virtual water trade research included HJ-CCD remote-sensing, multi-regional input-output, input-output model, life-cycle assessment, and etc.

China is the world's largest producer and exporter of textiles and clothing. Textile industry is one of the most water intensive and polluting industries. Large quantities of freshwater are used in the industrial production of textiles and clothing. Wastewater contains a complex mixture of organic and inorganic chemicals is discharged to the natural water body, causing severe impacts on the water environment and serious environmental damage. In China, more than 2700 million tons of freshwater is consumed and 2500 million tons of textile wastewater is discharged on a yearly basis [18]. According to the concept of virtual water trade, China exports freshwater and imports wastewater pollutants in virtual forms. However, there has no literature published that focus on the virtual water trade in relation to China's international trade of textiles and clothing. In this context, we aim to quantify the volume of virtual water trade flows related to textiles and clothing exports in the period 2001-2014 with the water footprint methodology. It will give useful insight and orientation for potential nation-wide policy implications for water security. Furthermore, this paper will also provide information contributes to the sustainable management of national water resources.

2. Methodology and Data

2.1 WF Methodology

WF is a multi-dimensional indicator measuring water resources appropriation by source and polluted volumes by type of pollution. It was first introduced by Hoekstra and Hung that referred to the concept of virtual water [19]. WF has three components: green WF (WF_{green}), blue WF (WF_{blue}) and grey WF (WF_{grey}). WF_{green} refers to consumption of green water resources (rainwater insofar as it does not become run-off). WF_{blue} refers to consumption of blue water resources (surface and groundwater) along the supply chain of a product. WF_{grey} refers to pollution and is defined as the volume of freshwater that is required to assimilate the load of pollutants given natural background concentrations and existing ambient water quality standards [20].

In this paper, only the industrial production stage of exported textiles and clothing is considered. WF_{blue} and WF_{grey} can be calculated as:

$$WF_{blue} = BlueWaterEvaporation + BlueWaterIncorporation + LostReturnFlow \quad (1)$$

$$WF_{grey} = \max\left(\frac{L^k}{c^k_{max} - c^k_{nat}}\right) \quad (2)$$

where L^k is the load of pollutant k generated in industrial production processes. c^k_{max} is the maximum acceptable concentration of pollutant k that can be discharged. c^k_{nat} is the natural concentration of pollutant k in the receiving water body. The 'max' outside standard brackets means that WF_{grey} is determined by the most critical pollutant that is associated with the largest pollutant-specific WF_{grey}.

2.2 Virtual Water Trade Flows

Water 'embedded' in textiles and clothing may also be considered to be 'traded' when the products are traded. Virtual water trade flows are thus calculated as:

$$WF_{exports} = WFI_{in\text{-}China} \times V_{exports} \quad (3)$$

where $WF_{exports}$ is the virtual water flows of China's textiles and clothing exports. $WFI_{in\text{-}China}$ is the water footprint intensity of textiles and clothing industrial production. It equals to the result of WF divided by the total industrial output value (i.e., WF/V_{TIOV}). $V_{exports}$ is China's export value of textiles and clothing.

2.3 Data Sources

Data used in this paper cover the period from 2001 to 2014. China's export values of textiles and clothing were collected from International Trade Statistics (2002-2015) published by World Trade Organization (see Table 1). Data of freshwater consumption, water pollutants discharge and the total industrial output value were collected from Annual Statistic Report on Environment in China (2001-2014). COD_{Cr} was the most critical pollutant according to former literature [21]. The maximum acceptable concentrations of COD_{Cr} in textile effluents used for WF_{grey} calculations refer to GB 4287 Discharge standard of water pollutants for dyeing and finishing of textile industry (120 mg/L). Natural concentrations of COD_{Cr} in many regains is 15 mg/L according to China's Water Resources Bulletins. Therefore, we select 120 mg/L and 15 mg/L for simplify in this paper.

Table.1 Data of water and economic values from 2001 to 2014

Year	Export values (billion dollars)		Freshwater consumption (Gm^3)		COD_{Cr} discharge (Mt)		Total industrial output value (billion dollars)	
	Textiles	Clothing	Textiles	Clothing	Textiles	Clothing	Textiles	Clothing
2001	16.83	36.65	1.54	0.07	0.24	0.01	35.92	3.58
2005	41.05	74.16	2.02	0.11	0.30	0.02	58.41	5.90
2010	77.00	130.00	2.90	0.15	0.30	0.01	158.83	12.04
2014	112.00	187.00	2.38	0.21	0.24	0.02	189.89	21.23

3. Results and Discussion

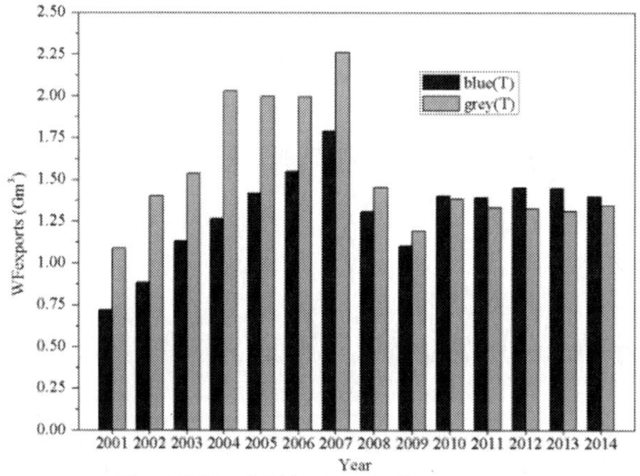

Fig.1 WF of China's textiles exports.

Fig.1 shows the calculated results of WF of China's textiles exports ($WF_{exports}(T)$). It can be seen that grey $WF_{exports}(T)$ ($WF_{exports,grey}(T)$) was a larger than blue $WF_{exports}(T)$ ($WF_{exports,blue}(T)$) before 2009. $WF_{exports,blue}(T)$ increased from 0.72 Gm^3 in 2001 to 1.79 Gm^3 in 2007 with an average increase rate of 16%. $WF_{exports,grey}(T)$ increased from 1.09 Gm^3 in 2001 to 2.26 Gm^3 in 2007 with an average increase rate of 13% though there were slight decreases in 2005 and 2006. Large decreases, approximately 27% and 35% respectively, appeared in 2008 for the two kinds of $WF_{exports}(T)$ and they continued decreasing in 2009. $WF_{exports,blue}(T)$ surpassed $WF_{exports,grey}(T)$ from 2010 and they changed slightly with a rate less than 5% in the period of 2010 and 2014. The total $WF_{exports,grey}(T)$ and the total $WF_{exports,blue}(T)$ was 21.67 Gm^3 and 18.29 Gm^3 respectively in the research period.

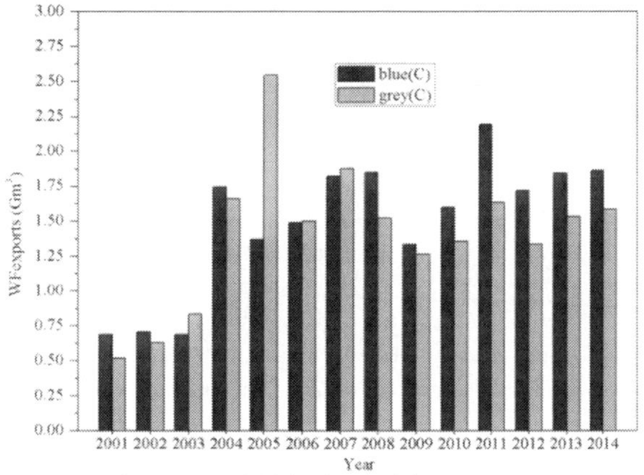

Fig.2 WF of China's clothing exports.

WF of China's clothing exports ($WF_{exports}(C)$) is shown in Fig.2. The two kinds of $WF_{exports}(C)$ changed largely and the total blue $WF_{exports}(C)$ ($WF_{exports,blue}(C)$) was little larger than the total grey $WF_{exports}(C)$ ($WF_{exports,grey}(C)$) from 2001 to 2014. $WF_{exports,blue}(C)$ and $WF_{exports,grey}(C)$ increased steadily from 2001 to 2003. Then, they jumped up to 1.75 Gm^3 and 1.66 Gm^3 with increase rates of more than 155% and 100% respectively in 2004. From 2004 to 2014, $WF_{exports,blue}(C)$ and $WF_{exports,grey}(C)$ showed fluctuation changes mainly between 1.26 Gm^3 and 1.85 Gm^3.

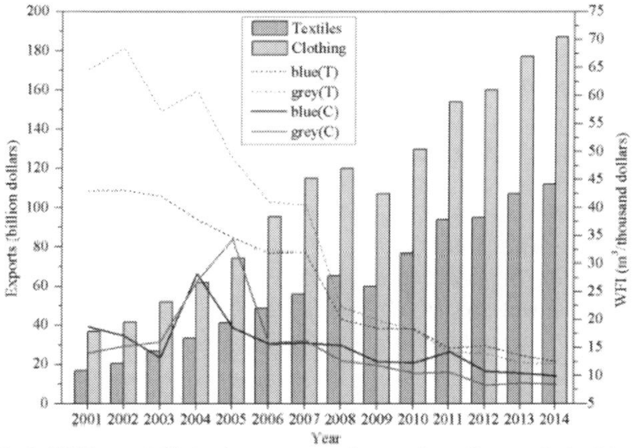

Fig.3 WFIs and China's export values of textiles and clothing.

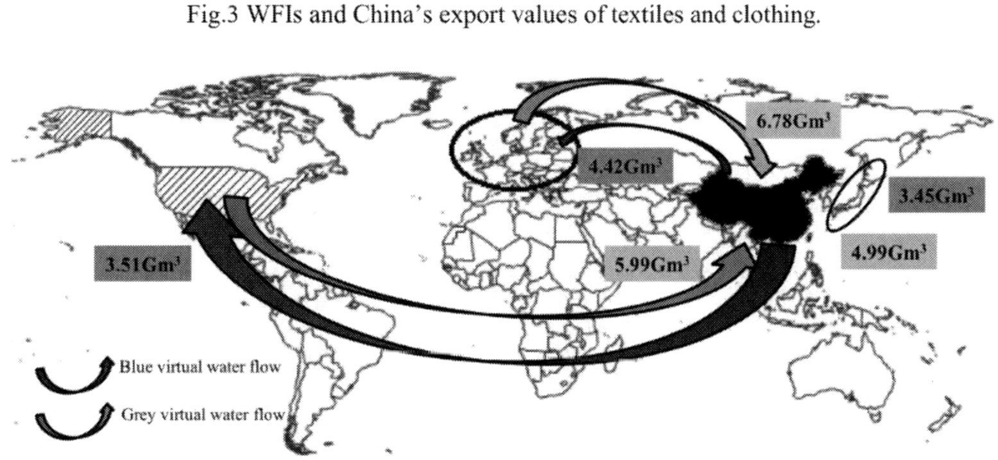

Fig.4 Virtual water flows of China's textiles and clothing exports.

China's international trade of textiles and clothing gained a further and faster development since China officially became the 143rd member of WTO on 11 November 2001. The exports of textiles and clothing increased more than six times and five times respectively, from 2001 to 2014 as showed in Fig.3. Larger quantities of freshwater were consumed for the industrial production of fiber, yarns, fabrics and clothing for exports. This directly generated quantities of textile effluents that discharged into the environment. The pressure on freshwater resources shortage and wastewater pollution in China was increased due to China's increasing exports of textiles and clothing referring to the virtual water trade theory. For example, more than 4.42 Gm^3, 3.51 Gm^3 and 3.45 Gm^3 freshwater that embedded in the exported textiles and clothing were transferred into EU, the USA and Japan from 2001 to 2014, respectively (see Fig.4). This $WF_{exports,blue}$ was more than two times of the total yearly water storage of Taihu Lake. More severally, quantities of wastewater pollutants were discharged in China. Another more 17.76 Gm^3 (the sum of 6.78 Gm^3, 5.99 Gm^3 and 4.99 Gm^3) were needed to assimilate the load of discharged wastewater pollutants in order to alleviate the impacts on the environment and human health. Though this $WF_{exports,grey}$ was only an indicator of the degree of freshwater pollution as we actually treated wastewater to reduce pollutants emission instead of dilute pollutants, the impacts of wastewater pollutants caused by the industrial production of exported textiles and clothing were real.

Fortunately, Chinese government has issued many policies (e.g., 'Entry criteria on printing and dyeing industry 2010', 'Norm of water intake standards for textiles and clothing', 'Action Plan for Prevention and Control of Water Pollution') on restrictions of freshwater consumption, wastewater generation and pollutants discharge in the industrial production of textiles and clothing. These policies promoted the development and popularization of freshwater saving concepts and wastewater treatment technologies in China's textiles industry. Therefore, the WFIs decreased from 2001 to 2014 (see Figure 3) inhibiting the increase of $WF_{exports}$, especially after 2008.

4. Conclusion

This paper presents an investigation of the virtual water trade flows of China's textiles and clothing exports in the period 2001-2014 according to the water footprint methodology. As the world's largest producer and exporter of textiles and clothing, large quantities of freshwater were transferred into other countries (especially developed countries) from China with textiles and clothing exports. From 2001 to 2007, both $WF_{exports}(T)$ and $WF_{exports}(C)$ showed increase trend though there were slight decreases in 2005 and 2006. They showed fluctuation changes from 2008 to 2014 and the fluctuation range of $WF_{exports}(T)$ was not so fierce compared to that of $WF_{exports}(C)$. The total $WF_{exports}(C)$ was a little larger than the total $WF_{exports}(T)$ though the clothing exports was much larger than textiles exports. This was because the industrial production of textiles was more water intensive (see Figure 3) than that of clothing as it included most of the wet processing processes (e.g., filaturing, retting, starching, desizing, scouring, bleaching, mercerizing, dyeing). Water pollution caused more severe impacts on the environment as $WF_{exports,grey}$ was larger than $WF_{exports,blue}$.

The government issued many policies and standards in order to alleviate freshwater scarcity and wastewater pollution in China's textile industry. This also promoted the decrease of $WF_{exports}$. However, there is still much work to do in order to reduce freshwater resources shortage pressure and wastewater pollution impacts. Water intensity reduction must be continued taken up on a priority basis. Freshwater saving and wastewater treatment technologies can directly reduce $WF_{exports}$. Moreover, industrial structure adjustment is also important. The industrial production of water intensive textiles and clothing should be decreased and promote exports of high value-added textiles and clothing with less water intensity.

5. Acknowledgment

The authors are grateful to the National Natural Science Foundation of China for providing funding supports to this research through project 71503233, to the Zhejiang Provincial Natural Science Foundation for providing funding supports to this research through project LQ16G030012 and LY17G030035, to Zhejiang Sci-Tech University for providing funding supports to this research through project 'Zhejiang Sci-Tech University Scientific Research Foundation (15072022-Y)'.

References

[1] World Water Council. Water and Green Growth: beyond the Theory for Sustainable Future. Marseille: 2015.

[2] Cardone R. Wet business risks. Corporate Knights Mag: 2004; 2: 16-17.

[3] United Nations World Water Assessment Programme. The United Nations World Water Development Report 2015: Water for a Sustainable World. Paris, UNESCO: 2015.

[4] Hoekstra AY, Hung PQ. A quantification of virtual water flows between nations in relation to international crop trade. Water Res: 2002; 49: 203-209.

[5] Allan JA. Virtual water-the water, food, and trade nexus. Useful concept or misleading metaphor? Water Int: 2003; 28: 106-113.

[6] Zhang Z, Yang H, Shi M. Spatial and sectoral characteristics of China's international and interregional virtual water flows e based on multi-regional input-output model. Econ Syst Res: 2016; 28:1-21.

[7] Hassan A, Saari MY, Tengku Ismail TH. Virtual water trade in industrial products: evidence from Malaysia. Environ Dev Sustain; in press.

[8] Duarte R, Pinilla V, Serrano A. The effect of globalisation on water consumption: a case study of the Spanish virtual water trade. Ecol Econ: 2014; 100: 96-105.

[9] Serrano A, Guan D, Duarte R, Paavola J. Virtual water flows in the EU27: a consumption-based approach. J Ind Ecol: 2016; 20: 547-558.

[10] Wang YD, Lee JS, Agbemabiese L, Zame K, Kang SG. Virtual water management and the water-energy nexus: a case study of three Mid-Atlantic States. Resour Conserv Recycl: 2015; 98: 76-84.

[11] Zeitoun M, Allan JA, Mohieldeen Y. Virtual water 'flows' of the Nile Basin, 1998e2004: a first approximation and implications for water security. Glob Environ Chang: 2010; 20: 229-242.

[12] Vanham D. An assessment of the virtual water balance for agricultural products in EU river basins. Water Resour Ind: 2013; 1-2: 49-59.

[13] Liu X, Klemes JJ, Varbanov PS, Cucek L, Qian Y. Virtual carbon and water flows embodied in international trade: a review on consumption-based analysis. J Clean Prod: 2017; 146: 20-28.

[14] Zhang Y, Zhang JH, Wang C, Cao JJ, Liu ZH, Wang LC. China and Trans-Pacific Partnership Agreement countries: Estimation of the virtual water trade of agricultural products. J Clean Prod: 2017; 140: 1493-1503.

[15] Chapagain AK, Hoekstra AY. The global component of freshwater demand and supply: an assessment of virtual water flows between nations as a result of trade in agricultural and industrial products. Water Int: 2008; 33: 19-32.

[16] Han MY, Chen GQ, Meng J, Wu XD, Alsaedi A, Ahmad B. Virtual water accounting for a building construction engineering project with nine sub-projects: a case in E-town, Beijing. J Clean Prod: 2016; 112: 4691-4700.

[17] Luo K, Tao F. Monitoring of forest virtual water in Hunan Province, China, based on HJ-CCD remote-sensing images and pattern analysis. Int J Remote Sens: 2016; 37: 2376-2393.

[18] Ministry of Environmental Protection of the People's Republic of China. Annual Statistic Report on Environment in China. China Environmental Science Press, Beijing: 2015.

[19] Elena G, Esther V. From water to energy: the virtual water content and water footprint of biofuel consumption in Spain. Energ Policy: 2010; 38: 1345-1352.

[20] Hoekstra AY. Virtual Water Trade: Proceedings of the International Expert Meeting on Virtual Water Trade. Value of Water Research Report Series No. 12. UNESCO-IHE, Delft, The Netherlands: 2003.

[21] Wang LL, Ding XM, Wu XY. Blue and grey water footprint of textile industry in China. Water Sci Technol: 2013; 68: 2485-2491.

Aerogel Embedded Electrospun Nanofiber Layers for Thermal Insulation

Mohanapriya Venkataraman*[1], Rajesh Mishra[1], Jaromir Marek[2], Klara Kucerova[2], Jiri Militky[1]

[1]*Department of Materials Engineering, Technical University of Liberec, Czech Republic*
[2]*Department of Nanotechnology and Informatics, Technical University of Liberec, Czech Republic*

*Corresponding author's email: mohanapriya.venkataraman@tul.cz

Abstract

Human's quest for survival against harsh climatic conditions is assisted by the clothes they wear. It is imperative that the efficacy of such clothes is high under difficult conditions. Experimentation on protective clothing using different fibers, fabrics and coating materials is a continuous process. In this regard, the thermal properties of nanofibers and their potential protection against cold environments have to be explored further. Thermal insulation battings incorporating nanofibers could possibly decrease the weight and bulk of current thermal protective clothing, and increase mobility for wearers. In this research, the mechanisms of heat transfer was studied through fibrous insulation where the fiber diameter is sub micrometer (μm). Electrospinning process was used to produce flexible nanofibers. Electrospun PUR and PVDF nanofibers were embedded with silica (SiO_2) aerogel. The thermal properties of the samples were evaluated and statistically analyzed. The microscopic examination confirmed the presence of aerogel particles. The results showed enhancement of thermal insulation by increasing the number and the weight per unit area of both nanofibrous layers. The results confirmed that embedding silica aerogel in nanofibrous layers leads to increased thermal insulation. From the research of nanofibers embedded with aerogel, it could be concluded that nanofibers can provide efficient thermal insulation.

Keywords: Nanofiber; Aerogel; Thermal Insulation; Conduction; Heat Transfer

1. Introduction

Electrospinning is a simple and low-cost method for making polymer and ceramic fibers with superfine diameters [1-3]. It is gaining increasing acceptance due to the promising properties of structured and assembled nanofibers developed via electrospinning. Understanding heat transfer through nanofiber structures will enable the exploitation of the unique properties of polymer nanofibers for applications such as improved cold weather clothing and hand wear, sleeping bags, and tent liners, as well as applications for food service refrigeration and storage equipment [4]. Literature searches on the subject of submicron fibers in thermal insulation reveal very limited fundamental or applied work using polymer nanofibers for thermal insulation applications. Silica aerogel is a highly porous material with pore diameters in the range of 2–50nm [5, 6]. The nanoporous structure of the silica aerogel having a high porosity above 90% makes the aerogel a highly thermal insulating materials with a super-low thermal conductivity as low as $0.013 W \cdot m^{-1} \cdot K^{-1}$. It has wide applications in aircrafts and aerospace, chemical engineering, building constructions, and so forth [7-9].

In this paper, the mechanism of heat transfer was studied through fibrous insulation where the fiber diameter was less than 1 micrometer (μm). The thermal insulating efficiency of fiber-based insulation is known to increase as the fiber size is reduced. Flexible electrospun nanofibrous layers embedded with silica aerogel was produced via electrospinning process. The electrospun PUR and PVDF nanofibrous microstructures were fabricated and then used to reinforce the SiO_2 aerogel. The effects of thermal properties of the electrospun nanofibrous layers embedded with SiO_2 aerogel were evaluated.

2. Materials and Methods

2.1 Materials

Silica aerogel powder and granules were purchased from Cabot aerogel Corp. Polyurethane (PUR) & PVDF was used from the CXI lab (nanocenter, TUL, Czech Republic).

2.2 Methods

2.2.1 Electrospinning of PVDF& PUR Nanofibrous Layers

The PUR nanofiber was dissolved in Dimethylformamide (DMF) at room temperature at a concentration of 18 wt.% (g/mL) and at the same concentration PVDF also was first stirred for 2 hours and then with SiO_2 aerogel in both powder and granule forms were added. These mixtures were stirred for 1-2 hrs at room temperature prior to electrospinning and were then electrospun at room temperature. The prepared solution was placed in a cylinder containing active electrode parallel to collecting electrode. The fibers were collected on a spunbond polypropylene fabric.

2.2.2 Instruments and Characterizations

The electrospun PUR and PVDF nanofibrous layer embedded with silica aerogel have been investigated on their morphology and microstructure by using a SEM (VEGA TESCAN Inc. USA) at 30 kV. The densities of the samples were determined by measuring the weight and volume. To investigate the comparative thermal properties of insulating materials, alambeta instrument developed at the technical university of Liberec, Czech Republic was used. Electrospinning was carried out using Nanospider technology as a modified electrospinning technique, Nanospider laboratory machine NS LAB 500S from Elmarco s.r.o. Alambeta measuring device was used for measuring of transient and steady state thermo-physical properties (thermal insulation and thermal contact properties) and also the sample thickness.

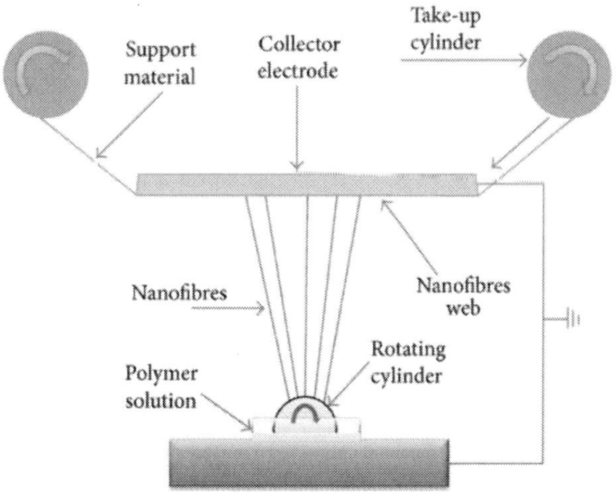

Fig.1 Schematic of electrospinning setup -"Nanospider."

Electrospinning is widely accepted as a technique to fabricate submicron polymer fibers. It is a fiber-forming process, where high voltage is used to create an electrically charged jet of polymer solution or melt from the needle. The polymer solidifies as it travels toward the collecting plate, often producing nanometer scale fibers [10-13] Nanospider is a modified electrospinning method which requires the use of a high-voltage electrostatic field to create an electrically charged stream of polymer solution or melt. The innovatory idea of Nanospider is based on the possibility of producing nanofiber from a thin layer of liquid polymer. In this case, Taylor cones (the source of nanofiber) are created on the surface of a rotating roller, immersed in a polymer solution. Because the Taylor streams are formed next to each other, throughout the entire length of the roller, this revolutionary idea produced many advantages, such as high productive ability. This commercial method for production of polymeric nanofiber is used in industrial range. This is a simple and versatile method for production of ultrathin fibers from a variety of materials that include polymers. In addition, Nanospider has the ability to process a wide range of polymers in diameters of 50–300 nm into nonwoven webs [14].

Table.1 Sample details of electrospun PVDF nanofibrous layer embedded with silica aerogel.

Samples	Type	Sample description (Spun PP +NFA)	Areal density [g/m^2]	Thickness [mm]
SPUR1	Aerogel nanofibrous layer with Spunbond PP back up	only PUR	34.01 (± 1.71)	0.290 (± 0.014)
SPUR2		PUR + aerogel (Powder)	33.11 (± 1.65)	0.254 (± 0.013)
SPUR3		PUR + aerogel (Granular)	35.28 (± 1.76)	0.300 (± 0.015)
SPUR4		PUR + aerogel (Powder)	35.29 (± 1.65)	0.390 (± 0.020)
SPUR5		PUR + aerogel (Granular)	38.58 (± 1.93)	0.370 (± 0.019)
PUR1	Aerogel nanofibrous layer	Only PUR	6.01 (± 0.31)	0.083 (± 0.004)
PUR2		PUR + aerogel (Powder)	5.11 (± 0.26)	0.089 (± 0.005)
PUR3		PUR + aerogel (Granular)	7.28 (± 0.03)	0.112 (± 0.006)
PUR4		PUR + aerogel (Powder)	7.29 (± 0.07)	0.206 (± 0.010)
PUR5		PUR + aerogel (Granular)	10.58 (± 0.21)	0.248 (± 0.012)

Note: "±" is the upper and lower 95% confidence interval of the mean.

Table.2 Sample details of electrospun PVDF nanofibrous layer embedded with silica aerogel.

Samples	Type	Sample description (Spun PP +NFA)	Areal density [g/m^2]	Thickness[mm]
SPVDF1	Aerogel nanofibrous layer with Spunbond PP back up	only PVDF	34.89 (± .79)	0.48 (± 0.045)
SPVDF2		only PVDF	38.62 (± 0.93)	0.32 (± 0.036)
SPVDF3		PVDF + aerogel (Powder)	34.58 (± 2.73)	0.46 (± 0.048)
SPVDF4		PVDF + aerogel (Powder)	44.00 (± 2.38)	0.38 (± 0.091)
SPVDF5		PVDF + aerogel (Granular)	37.18 (± 1.26)	0.41 (± 0.025)
SPVDF6		PVDF + aerogel (Granular)	39.89 (± 1.95)	0.40 (± 0.102)
PVDF1	Aerogel nanofibrous layer	only PVDF	6.80 (± 0.34)	0.11 (± 0.005)
PVDF2		only PVDF	10.62 (± 0.43)	0.20 (± 0.012)
PVDF3		PVDF + aerogel (Powder)	6.58 (± 0.13)	0.17 (± 0.065)
PVDF4		PVDF + aerogel (Powder)	16.00 (± 0.23)	0.05 (± 0.025)
PVDF5		PVDF + aerogel (Granular)	9.18 (± 0.60)	0.19 (± 0.033)
PVDF6		PVDF + aerogel (Granular)	11.89 (± 0.55)	0.23 (± 0.015)

Note: "±" is the upper and lower 95% confidence interval of the mean.

3. Results and Discussion

3.1 Microstructures of Nanofibrous Layer with Aerogel

Fig.2 and 3 shows the morphologies and microstructures of electrospun PUR & PVDF nanofibrous layers. The electrospun nanofibrous layers have better integrity and flexibility. The different microstructures could be observed with and without aerogel particles present which were electrospun from the solutions with the concentration of 18wt.%.

PUR 1 PUR 2 PUR 3 PUR 4

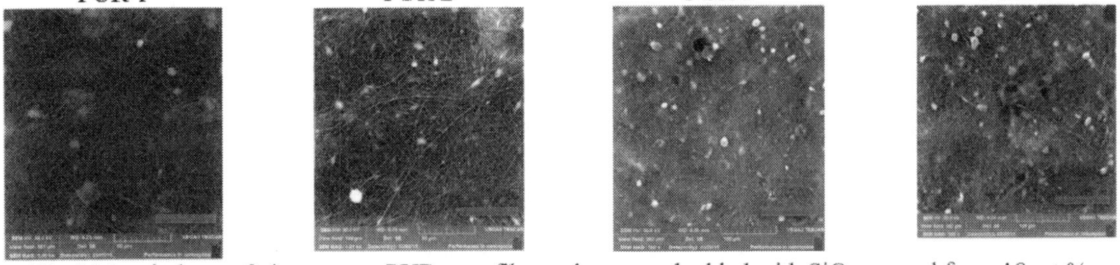

Fig.2 Morphology of electrospun PUR nanofibrous layers embedded with SiO_2 aerogel from 18 wt.%.

PVDF 1 PVDF 2 PVDF 3 PVDF 4

Fig.3 Morphology of electrospun PVDF nanofibrous layers embedded with SiO_2 aerogel from 18 wt.%.

3.2 Thermal Properties of Electrospun Nanofibrous PUR and PVDF Layer with SiO_2 Aerogel

Thermal conductivity as a function of areal density for PUR and PVDF electrospun nanofibrous layer embedded with silica aerogel is shown in Fig.4.As shown in Fig.4, thermal conductivity of the electrospun nanofibrous layer decreased with increase in density. This can be explained by the fact that as the density increases; it makes the fibrous structure more packed. This causes the mean free path (distance travelled by a photon before it collides with another fiber surface for a photon movement to decrease thus causing a decrease in the heat transfer because of radiative conduction. When the density comes to a critical point, the increase in conduction through solid phase (fibers) and decrease in radiation conductivity results in an increase in total thermal conductivity. In fact, in fibrous structures the small size of the pores and the tortuous nature of the air channels present prevents any heat transfer by convection. Moreover, in fibrous insulation materials because of low fiber volume fraction, heat conduction through the solid phase (the fibers) is not significant and conduction through air is usually considered to be the conductivity of still air that is poor at room temperature. Thus, radiative conductivity is the prevalent mechanism of conductivity since it has a high porosity percentage of fibrous structures. By adding a nanofiber web, thermal conductivity was enhanced noticeably that is believed to be because of their extremely fine fiber and very high porosity of web. The superfine fibers in the web have better radiation absorption and extinction since their higher surface-area-to-volume ratio leads to decrease in the thermal conductivity. Moreover, smaller pore size between nanofibers decreases the mean free path for photon movement resulting in lower radiative energy transfer. This improvement becomes more significant when bulk density is increased. In high densities, increase in the thermal conductivity of the sample containing web was diminished which may be attributed to the presence of nanofiber and their natural compact structure that could compensate for increased thermal conductivity. According to thermal conductivity curves which is apparent in figures 4, decrease in the average nanofiber diameter leads to lower limit of conductivity. Higher specific surface of thinner fibers [15] means more surface area for radiative absorption that results in lower thermal conductivity. Further, higher porosity of the web with a nanofiber diameter around 150 nm could be the other reason for their lowest thermal conductivity. Another explanation for reduction in conductivity can be smaller pore size in the web containing thinner nanofibers leading to lower radiative conductivity. For better understanding of how reduction in fiber diameter affected porosity of the nanofibers, it is clearly shown in SEM images in Fig.2 and 3. In this context, it could be understood that using thinner nanofibers leads to noticeable performance and helps in achieving very low limit of thermal conductivity. Of particular interest are the results for the two nanofiber insulation materials (electrospun PUR and PVDF nanofibrous layer). Both materials showed excellent reduction in overall heat transfer compared to standard low-density fibrous insulating materials (at areal densities above 40 g/m^2).

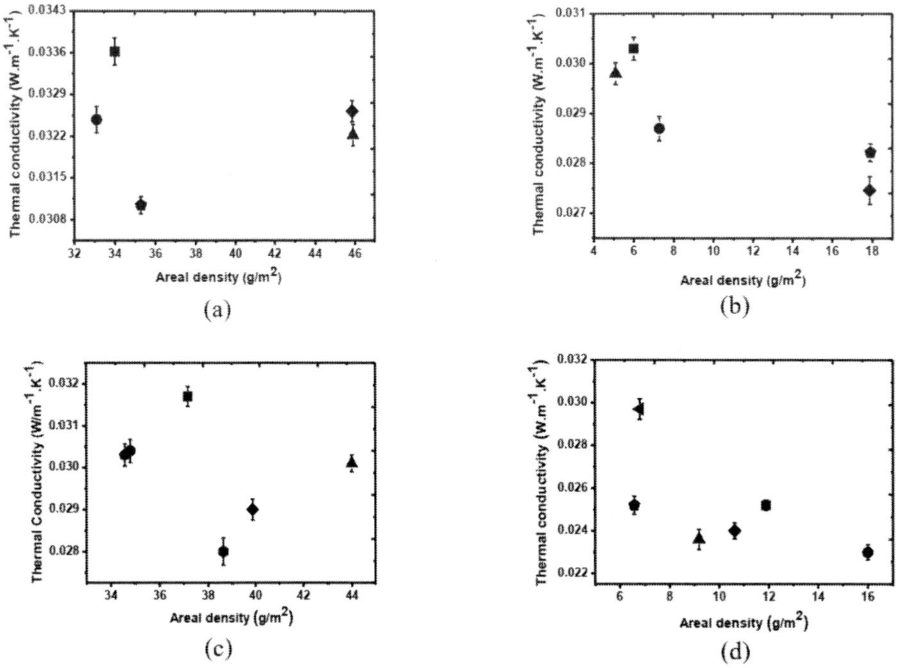

Fig.4 Thermal conductivity Vs GSM for (a) PUR samples with spunbond PP; (b) Electrospun PUR nanofibrous layers embedded with silica aerogel; (c) Electrospun PVDF nanofibrous layers embedded with silica aerogel backed up with spun bond PP and (d) Electrospun PVDF nanofibrous layers embedded with silica aerogel

The PVDF nanofibrous layer, in particular, showed superior insulation at higher areal density values. Thermal conductivity testing confirmed that decreasing fiber diameter tends to increase the thermal resistance of fibrous insulation materials. However, the nanofiber/aerogel becomes an effective insulator since the aerogel structure suppresses conduction and convection, and the fibers reduce radiation heat transfer while increasing the strength of the brittle and weak aerogel structure. Although the aerogel/nanofiber combination has good thermal properties, the volume fraction of fiber must be fairly high to support and protect the aerogel matrix. Thus the aerogel materials can't achieve the same thermal conductivity at densities as fibrous insulation, but they do achieve better thermal resistance for an equivalent thickness of material. High porosity of electrospun fibrous mesh is able to trap air which potentially gives it a good thermal insulation property. This is confirmed using thermal conductivity tests which show that decreasing fiber diameter leads to an increase in thermal resistance [4]. As mentioned previously, nanofiber/aerogel have shown superior insulation properties for applications where thickness is of concern. With respect to high porosity fibrous insulation materials, aerogel/nanofiber has excellent insulation per unit thickness properties, as shown in Fig.5. From the figures, it can be seen that the thermal resistance increases with the increases in thickness. Also, the electrospun PVDF nanofibrous layer embedded with silica aerogel is higher than electrospun PUR nanofibrous layer embedded with silica aerogel. The data were examined by one-way analysis of variance (ANOVA) with 95 % confidence level. A significant difference ($p<0.05$) has been observed between the samples

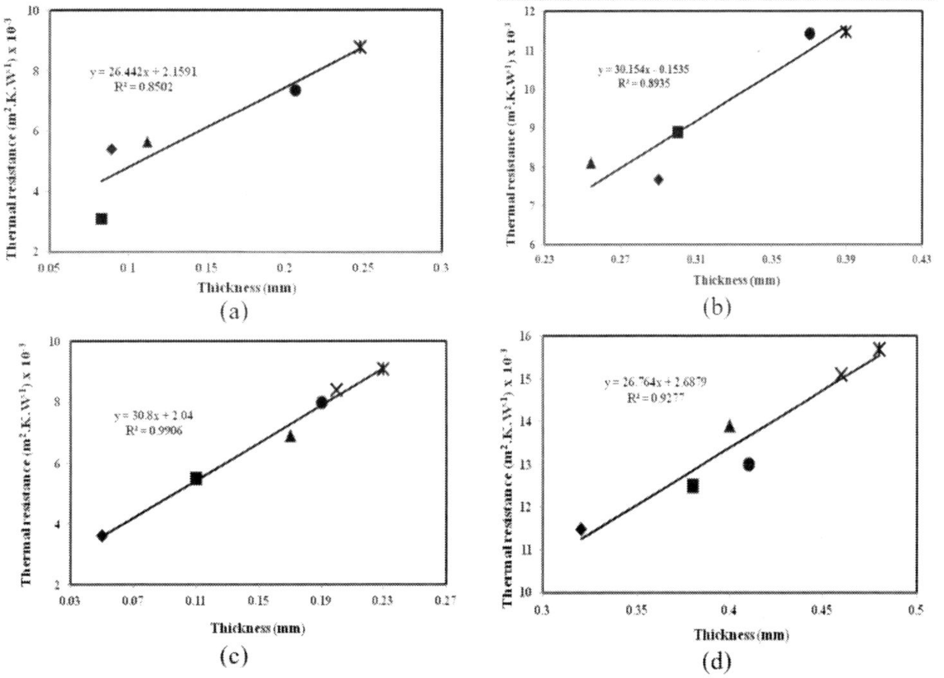

Fig.5 Thermal resistance Vs Thickness (linear function) (a) Electrospun PUR nanofibrous layers embedded with silica aerogel; (b) Electrospun PUR nanofibrous layers embedded with silica aerogel backed up with spun bond PP; (c) Electrospun PVDF nanofibrous layers embedded with silica aerogel and (d) Electrospun PVDF nanofibrous layers embedded with silica aerogel backed up with spun bond PP.

3.3 Air Permeability of Nanofibrous Layer Embedded with SiO_2 Aerogel

Air permeability is a very important parameter for thermal insulation of electrospun nanofibrous layer. Lower air permeability causes lower air flow; consequently, more thermal insulation. The air permeability of electrospun nanofibrous layers are shown in Fig.6. According to the figures; samples containing PUR nanofiber with double layer showed less air permeability. This behavior can be attributed to the finer diameter of PUR nanofiber compared to PVDF nanofiber diameter.

As can be seen from the figures, by increasing the number of nanofibrous layers, lower air permeability was achieved, confirming the relation of this important parameter with thermal insulation ability. Fig.6 shows that sample PUR4, PUR5 and PVDF5 were impermeable at 100 and 200 Pa.

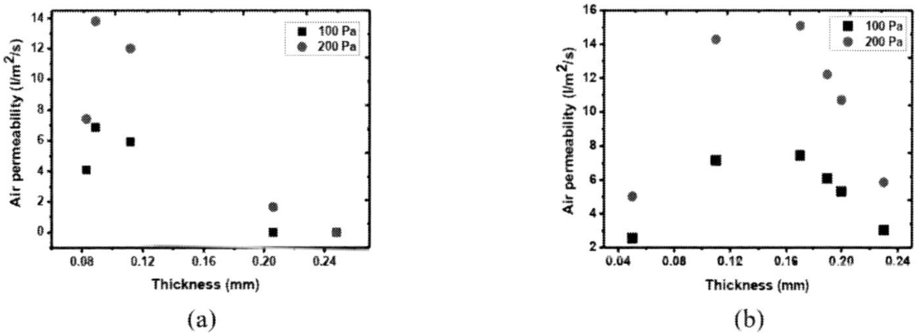

Fig.6 Air permeability (a) Electrospun PUR nanofibrous layer embedded with silica aerogel and (b) Electrospun PVDF nanofibrous layer embedded with silica aerogel.

4. Conclusions

Literature searches on the subject of nanofibrous layers in thermal insulation reveal no fundamental or applied work using polymer nanofibers for thermal insulation applications. The effects of electrospun PUR and PVDF nanofibrous layers embedded with silica aerogel on thermal behavior were studied. The results show enhancement in thermal insulation by increasing the number and the weight per unit area of both nanofibrous layers. Higher thermal resistance was observed in the case of samples containing PUR and PVDF nanofibrous layers, which can be attributed to the low air permeability and fiber diameter. Moreover, thermal measurements show that embedding silica aerogel in nanofibrous layers leads increased thermal insulation. We have shown in this study that our experimental insulation material can enhance thermal resistance. Furthermore, weight and thickness can be reduced by means of nanofiber layers. The work presented in this paper did not show nanofibers to be useful for highloft thermal insulation. However, they may be useful as components in hybrid battings with high bulk densities. Fibers below 1 μm in diameter are not thermally efficient at low fiber volume fractions; this corresponds with previous research on fiberglass insulation. Performance gains in existing thermal insulation materials may be possible by incorporating a proportion of nanofibers into the structure, but large diameter fibers would still be necessary for durability and compression recovery. Although the electrospun nanofibrous layers have been proposed to strengthen the aerogel, the preparation technique of the electrospun nanofibrous layers embedded with aerogel with larger size, and lower thermal conductivity has to be further developed.

References

[1] Ukponmwan JO. The Thermal-Insulation Properties of Fabrics. Text. Prog: 1993; 24: 5.

[2] Sang P. Heat and Mass Transfer Analysis of Fabric in The Tenter Frame. Text. Res. J: 1997; 67 (5): 311-316.

[3] Ammar AS, El-Okeily M. Heat Transfer Through Textile Fabrics: Mathematical Model. Mathematical and Computer Modelling: 1998; 12 (9): 1187.

[4] Frydrych I, Dziworska G, Bilska J. Comparative analysis of the thermal insulation properties of fabrics made of natural and man-made cellulose fibres. Fibre Textiles Eastern Europe: 2002; 39: 40-44.

[5] Farnworth B. Mechanism of heat flow through clothing insulation. Text. Res. J: 1983; 53: 717-725.

[6] Martin JR, Lamb GER. Measurement of thermal conductivity of nonwovens using a dynamic method. Text. Res. J: 1987; 57: 721-727.

[7] Mao N, Russell SJ. The thermal insulation properties of spacer fabrics with a mechanically integrated wool fiber surface. Text. Res. J: 2007; 77: 914-922.

[8] Li Y, Plante M, Holcombe BV. Fibre hygroscopicity and perceptions of dampness part-II: Physical mechanisms. Text. Res. J: 1995; 65.

[9] Hes L, Araujo M, Djulay V, Effect of Mutual Bonding of Textile Layers on Thermal Insulation and Thermal Contact Properties of Fabric Assemblies. Text. Res. J: 1996; 66: 245.

[10] Cohen E. Thermal properties of advanced aerogel insulation, in Dept. of Mechanical Engineering. 2011, Massachusetts Institute of Technology: Massachusetts.

[11] Gun AD. Dimensional, Physical and Thermal Comfort Properties of Plain Knitted Fabrics Made from Modal Viscose Yarns Having Microfibers and Conventional Fibers. Fibr. Polym.: 2011; 12 (2): 258-267.

[12] Arlon AJ. Jantzen CA. Cao WQ. Aerogel - a high performance insulating material at 0.1 bar. Insulation materials: testing and applications, ed. W.D. Graves RS. Vol. 2. 1991, Philadelphia: American Society for Testing and Materials.

[13] Kistler SS. Coherent expanded-aerogel. Journal of Physical Chemistry: 1931; 36: 52-64.

[14] Varkiyani S, Rahimzadeh H, Bafekrpoor H. Influence of punch density and fibre blends on thermal conductivity on nonwoven. Open Tex. J.: 2011; 4: 1-6.

[15] Venkataraman M, Mishra R, Jasikova D, Kotresh TM, Militky J. Thermodynamics of aerogel treated nonwoven fabrics at subzero temperatures. J. Ind. Tex.: 2014.

Thermal Protection of Fireproof Fabrics with Shape Memory Alloy Springs Under Hot Surface Contact

Ni-Ni Ma[2], Li-Jun Wang[2], Ye-Hu Lu[1, 2*], Hong-Qin Dai[1, 2]

[1] *National Engineering Laboratory for Modern Silk, Soochow University, Suzhou 215123, China*
[2] *College of Textile and Clothing Engineering, Soochow University, Suzhou 215021, China*

*Corresponding author's email: yhlu@suda.edu.cn

Abstract

Thermal protective clothing is widely used to ensure the health and safety of firefighters and emergency rescuers. In this study, shape memory alloy (SMA) springs were fixed between moisture barrier and thermal liner to develop novel thermal protective materials with a dynamic adaptive structure. The changes in temperatures at different layers were measured under exposure to hot surface contact, and the effect of SMA springs on the thermal protective performance was assessed. The results indicated that SMA springs could greatly decrease the heat transfer to the human body, and thereby improve thermal protection. It was concluded that the arrangement mode of SMA springs has impact on the protective performance. The research findings may be applied to design optimized thermal protective clothing with the proper arrangement of SMA springs.

Keywords: Thermal Protective Clothing; Shape Memory Alloy; Hot Surface Contact; Arrangement Mode

1. Introduction

Fire-fighters and emergency rescuers often suffer a lot of hazards, including thermal (fire, radiation and convection), biological (blood borne and pathogens), chemical (skin contact), physical (impact, debris and rough surface) and environmental (extreme temperature and high humidity) disasters [1]. Ideally necessary protection against flame and heat should be provided by a clothing system. The main focus of thermal protection and comfort properties of protective clothing is to control the factors affecting rescue time and efficiency.

The quality of traditional protective clothing is improved by increasing the thickness of the fabrics or the number of layers [2, 3]. These approaches however have an adverse effect of thermal physiological load on the wearer, resulting in heat stress issues, and thus impairing the wearer's ability to perform their task effectively and efficiently [1, 4, 5]. A reported method of overcoming the shortcomings is by using aerogel or phase change materials [2, 6, 7]. Thermal protective clothing embedded with aerogel materials is cumbersome and has poor air permeability. Meanwhile, it does not have dynamic adjustment ability [2, 7]. The phase change materials easily leak from the substrate in extreme hot environments, resulting in droplet formation that may damage the human skin [6]. Furthermore, SMA or shape memory polyurethane (SMPU), trained to return to a given shape when an actuation temperature is reached, is incorporated into protective clothing systems [8, 9]. The spring made by SMA or SMPU shows a flat shape at room temperature, which does not affect the daily wear. Once the clothing layer is actuated, the thickness will increase, resulting in an increased air gap [10-12]. Nevertheless, previous studies have not systematically considered the effect of arrangement mode of SMA springs, and the performance of SMA springs under hot surface contact is not evaluated as well.

In this study, SMA springs were incorporated between moisture barrier and thermal liner to develop novel thermal protective materials with a dynamic adaptive structure [9, 11]. The changes in temperatures at different layers were characterized under exposure to hot surface contact, and the impact of SMA springs on the thermal protective performance was evaluated.

2. Design of Fabrics for Thermal Protective Clothing

2.1 Shape Memory Alloy Springs

The SMA springs must be as flat as possible between the fabric layers to ensure minimum thermo-physiological load when the thermal insulation is not required. Consequently, a 1.5 mm diameter wire, flat conical spring with an actuation temperature of about 45°C that extends when exposed to heat was designed, as shown in Fig.1(a). The maximum diameter of the spirals is 28 mm and minimum diameter is 14 mm. The copper based SMA spring will be deformed in conical shape with the maximum height of 32 mm when the temperature exceeds the deformation temperature, as shown in Fig.1(b). It is possible to achieve thermal protection by introducing an expanding air gap between two layers of fabrics when the spring is actuated, and thus improving wearing comfort without constraint of movement when springs are in original state.

(a) Original state (b) Actuated state

Fig.1 Shape memory alloy springs in original state (a) and actuated state (b).

2.2 Design of Fabric Combination

A typical three-layer fabrics system was used, i.e., fireproof outer layer, moisture barrier and thermal liner. The basic properties of flame resistant fabrics are shown in Table 1. The dimensions of the fabric layers were 50 mm*50 mm. The SMA springs were incorporated between moisture barrier and thermal liner in five different arrangements. A total of six scenarios were designed, i.e., One, Two diag, Two para, Three diag, Three tria and CON (control group), as shown in Table 2. The fabrics were fixed in two diagonals to prevent the fabric slipping during the spring's deformation, which simulated the interaction between two fabrics in reality.

Table.1 The basic properties of flame resistant fabrics

Fabric code	Content	Structure	Weight (g/m^2)	Thickness (mm)
Outer layer	98% of aramid1313, 2% of aramid 1414	twill	193.7	0.49
Moisture barrier	100% of aramid / PTFE film	lamination	108.3	0.85
Thermal liner	100% of aramid mats and base fabric	needle-punched	200.0	0.72

3. Experimental

In this study, the hot plate at 400°C was used to simulate the hot surface contact during fire-fighting and rescuing, as shown in Fig.2. The exposure time was set at 20 s. During the hot surface contact test, the real-time temperature of T_1 (temperature between the moisture barrier and the outer surface), T_2 (temperature between thermal liner and moisture barrier) and T_3 (inner surface of thermal liner) were recorded by NI Data logger (NI 9231, USA). The NI Data logger includes 32 gauge T-type thermocouples and thermocouple modules. The sensors for measuring T_1, T_2 and T_3 were fixed at the central position of the outer surface of moisture barrier, the outer and the inner surface of thermal liner. Each experiment was repeated at least three times.

Table.2 Arrangements of SMA springs

Scenario	Description	Diagram
CON	No spring	
One	A spring was placed in the middle	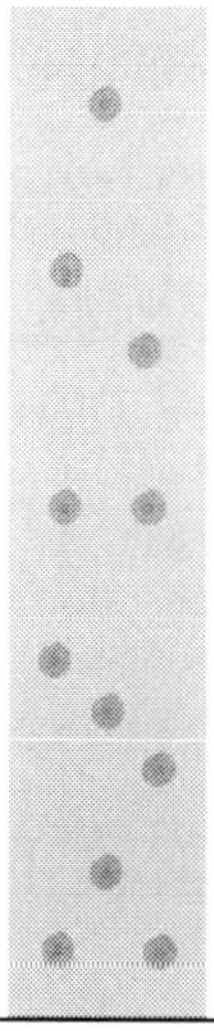
Two diag	Two springs were placed diagonally	
Two para	Two springs were placed in parallel	
Three diag	Three springs were placed diagonally	
Three tria	Three springs were placed in equilateral triangles	

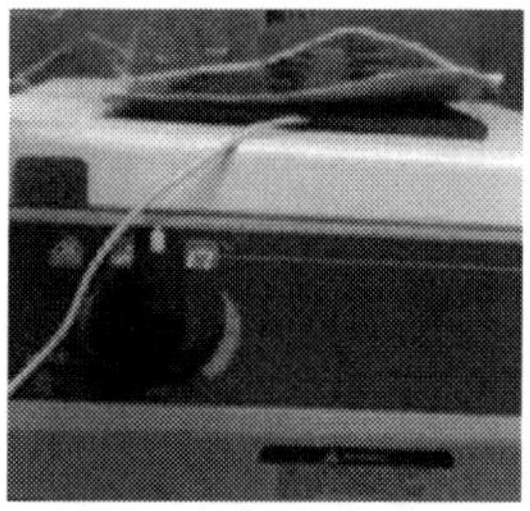

Fig.2 Hot surface contact test.

4. Results and Discussion

4.1 Temperature between the Moisture Barrier and the Outer Surface T_1

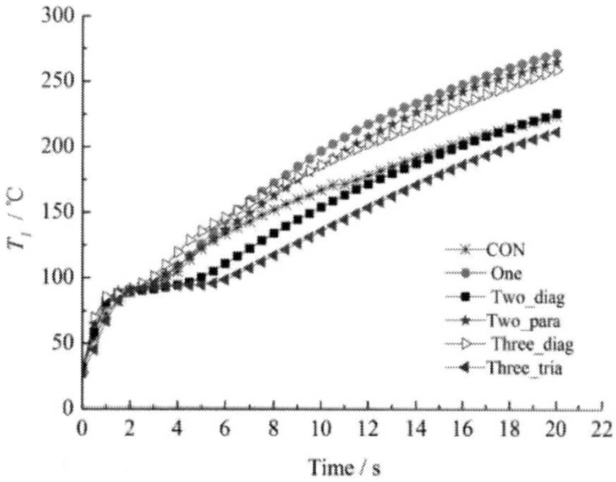

Fig.3 The change curves of T_1 in different arrangements of SMA springs.

Fig.3 shows the change curves of T_1 in different arrangements of SMA springs. The observed T_1 quickly increased to about 100°C at 0 ~ 2 s and then rose slowly. Later, it continued to increase rapidly after 3 s, and finally reached 210 ~ 270°C. The temperature curves of all samples showed similar change trends. T_1 in scenario Two diag and Three tria increased slowly at initial 2 ~ 5 s and final temperatures were relatively lower. Furthermore, the final temperature of T_1 in scenario CON was close to the Two diag, whereas the others were higher than CON.

The SMA springs in scenario One and Three diag had a higher temperature because the sensor fixed in center was opposite to the springs sewn in the inner center of moisture barrier. With the springs deformed by heating, the moisture barrier and the outer surface would separate resulting in an increased air gap, which was one of the major determining factors for insulation, thus the outer surface was closer to hot plate. The reason why T_1 in scenario Two para was higher maybe that the narrower space made by two closer SMA springs led to faster heat transfer.

4.2 Temperature between Thermal Liner and Moisture Barrier T_2

Fig.4 shows the change curves of T_2 in different arrangements of SMA springs. After a rapid increase to about 65°C within 0 ~ 2 s, T_2 exhibited a slight decline and maintained relatively stable later. Subsequently, the temperature T_2 gradually increased until the end of the test and reached 76 ~ 115°C. The temperature in scenario Three tria was the lowest. All samples developed a similar trend, however, presented different ratios.

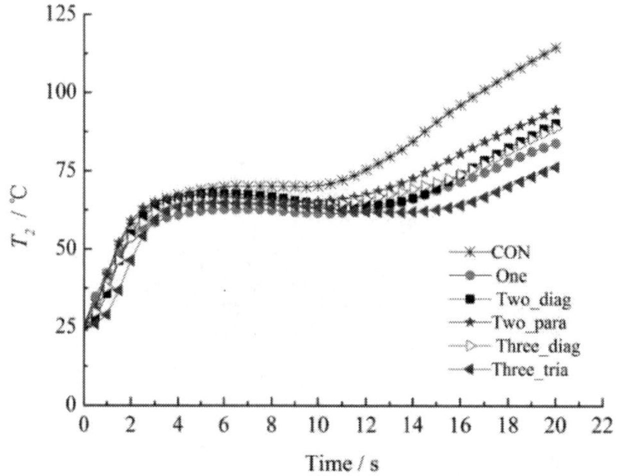

Fig.4 The change curves of T_2 in different arrangements of SMA springs.

Table.3 Temperature changes of T_2

Scenario	CON	One	Two diag	Two para	Three diag	Three tria
Initial growth interval	0~3 s	0~2.5 s	0~2.5 s	0~2.5 s	0~3 s	0~3.5 s
Initial reached temperature	65 ± 4.8°C	55 ± 11.6°C	60 ± 3.8°C	64 ± 4.7°C	62 ± 2.6°C	63 ± 1.6°C
Initial change rate	12.8 ± 1.9°C /s	11.7 ± 1.0°C /s	12.0 ± 1.4°C /s	14.3 ± 1.4°C /s	11.9 ± 0.4°C /s	13.4 ± 0.7°C /s
Plateau interval	3~10 s	2.5~11 s	2.5~14 s	2.5~10 s	3~10 s	3.5~16 s
Final change rate	4.5 ± 1.1 °C /s	2.8 ± 0.8°C /s	4.0 ± 0.2°C /s	3.2 ± 0.7°C /s	3.8 ± 1.5°C/s	3.0 ± 0.1°C/s
Temperature at 20s	115 ± 12.4°C	84 ± 7.6°C	90 ± 4°C	95 ± 9.2°C	8 9± 19.4°C	76 ± 3.9°C

In addition, the change ratios presented in Fig.4 is listed in Table 3. It was obvious that T_2 rapidly increased within 0 ~ 3.5 s because a significant temperature difference between hot surface and fabrics led to fast heat transfer at the beginning contact. As the springs deforming, the air layer gradually increased, and the heat transfer slowed down until the springs reached the maximum height, i.e., the air layer thickness was stable. That was why T_2 had a slight decline and maintained relatively stable later. The temperature in scenario CON was relatively stable within 4 ~ 10 s when heat absorption was equal to heat dissipation, and then increased rapidly. Compared with CON, there was a slightly decrease for T_2 of samples with various arrangements of the SMA springs, moreover, the arrangement affected the thermal protective performance. Temperature in scenario of Three tria was 39°C lower than CON at 20 s, depending on the higher insulation property of fabrics with SMA springs and their ability to hold thicker dead air. The insulation provided a barrier against flame and intense heat, therefore the increase of T_2 in scenario of Three tria was tardily (3.0°C/s). However, owing to the thin air layer where sensor was fixed, T_2 in scenario of Two para was higher than others.

4.3 Inner Surface of Thermal Liner T_3

Fig.5 shows change curves of T_3 in different arrangements of SMA springs. T_3 of all samples developed a similar trend, exhibited a slight increase at the first 2 seconds, and then presented different change trends. In scenario One and Three diag, T_3 gradually increased to about 45°C. After slow growth of T_3 in scenario of CON, it increased rapidly within 1 ~ 6 s and rose slowly until the maximum temperature reached about 58°C. Later, it had a slight decline and finally reached around 54°C. Temperature in scenario Two diag showed similar change trend with the CON with the peak temperature of 57 °C and the final temperature of 52°C at 20 s. Similarly, the maximum T_3 in scenario Two para was about 53°C and final temperature was 47°C. Temperature T_3 in arrangement of Three tria increased sharply and then continued to gradually increase to 53°C. It declined to 48°C and kept relatively stable till the end of the test. Different from the above scenarios, the scenario One kept increasing to 45°C during the 20 s. Owing to the thermal insulation of outer fabric and moisture barrier, the increment in inner surface temperature T_3 of all samples was slow. The temperature in scenario CON, Two diag, Two para and Three tria had a rapidly increase later. This was attributed mainly to no SMA spring in the center of fabrics. The placement of SMA springs was farther from the center, then the increase of air gaps induced by SMA springs deformation was slower, hence the temperature rise was higher. There was a slight drop in temperature when the outward heat dissipation was faster than inward heat absorption with air gaps. At a certain moment, T_3 tended to be relatively stable when the outward heat dissipation was equal to the inward heat absorption. Unlike T_1 and T_2 in the scenario Three tria, inner surface temperature T_3 was higher. It should be noted that two diagonals of fabric was sewed as mentioned above to prevent the fabrics slipping and simulated the interaction between fabrics in reality. Due to the limited internal layer space, the deformation of springs was not completed, therefore the air gap did not reach the maximum height.

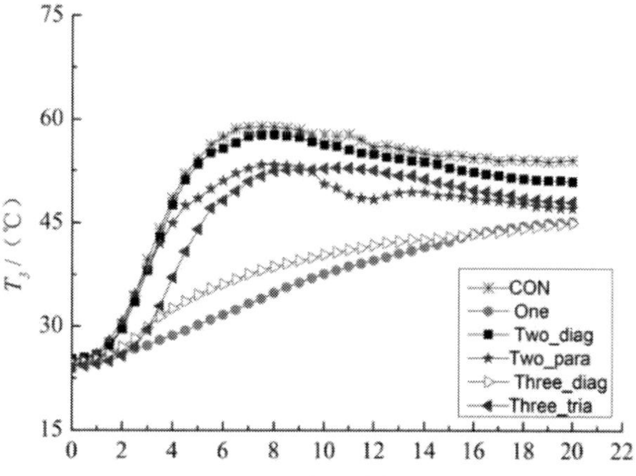

Fig.5 The change curves of T_3 in different arrangements of shape memory alloy springs.

In scenario One and Three diag, the inner surface temperature T_3 gradually increased to about 45°C. Compared with CON, it was about 10°C lower, which was about 18.5% of the temperature in CON. It indicated that the insulation material–a stable and thick air layer in fireproof fabrics combination with the SMA springs fixed in center could effectively reduce the temperature rising rate, improving the protective performance.

Human skin damage occurs when the skin temperature is more than 44°C (temperature rise is at least 12°C) and second-degree burns will appear when the skin temperature exceeds the temperature of 56°C (temperature rise is at least 24°C). Analyzing the time showed in Table 4, the effect of SMA springs on the thermal protective performance could be assessed. On the basis of time required to reach 44°C, the scenario One and Three diag could effectively increase the fireman's rescue time to 17 s, which was 386% higher than the CON, and according to occurrence of second degree burn, rescue time increased to more than 20 s, which was at least 300% higher than the traditional protective clothing. The results showed that the SMA springs could greatly decrease the heat transfer to human body, and thus improved the thermal protection, which was consistent with previous studies [9, 10, 12]. Overall, it indicated that intelligent protective clothing with dynamic adaptive structure was able to improve insulating performance, rescue efficiency and reduce property loss by spontaneous dynamic structure adjustment.

Table.4 The thermal insulating properties of samples in different arrangements

Scenario	Time to reach 44°C	Time to reach 56°C
CON	3.5 ± 0.7 s	5.5 ± 0.6 s
One	17.1 ± 2.8 s	\
Two_diag	3.7 ± 0.2 s	6.1 ± 0.2 s
Two_para	3.9 ± 1.4 s	\
Three_diag	17.7 ± 1.4 s	\
Three_tria	5.0 ± 0.3 s	\

5. Conclusion

This study aimed at demonstrating the application of shape-memory springs in thermal protection of fireproof fabrics under hot surface contact. Under exposure to high temperature of 400°C, the changes in temperatures at different layers were measured, and the protective performance was evaluated. The conclusions could be drawn as follows:

1) Outer surface temperature T_1 increased quickly, then rose slowly and continued to increase rapidly later. The final temperature in scenario Three tria was the lowest.

2) The temperature T_2 between thermal liner and moisture barrier, increased slowly at the first 2 s, then had a slight decline and maintained relatively stable later. Subsequently, the temperature T_2 kept increasing until the end of the test, and T_2 in scenario Three tria was the lowest.

3) Inner surface temperature T_3 in scenario One and Three diag kept increasing to 45°C within 0~20 s and was lower 8~14°C. For other scenarios, the temperature had a slow decline after rapid increase, and gradually kept stable.

4) Before the skin damage occurred, the scenario One and Three diag could effectively increase the fireman's rescue time to 17 s, which was 386% higher than the traditional protective clothing. On the basis of no second degree burn of the body, rescue time increased to more than 20s, which was four times longer than the traditional protective clothing.

5) The SMA springs could greatly decrease the heat transfer to human body, and thus improved the insulating performance. The arrangement of SMA springs had impact on the average thickness of the air layer in fireproof fabrics, and thus affected the thermal protective performance.

6. Acknowledgment

This study was financially supported by the National Natural Science Foundation (No. 51506137), the Natural Science Foundation of Jiangsu Province (No. BK20161255), and Open Projects of Advanced Textile Engineering and Technology Center of Jiangsu Province (No. XJFZ-2016-5).

References

[1] Wang Y. Textile fabrics used for the manufacture of firefighter turnout garment. Fire Technology Products Information: 2006; 6: 53-54.

[2] Zhang XJ, Wu HF, Kong XM. Analysis of thermal protective performance of aerogel-based new combined firefighters' clothing. China Personal Protective Equipment: 2013; 6: 20-24.

[3] Nocker W, Seibert J. Fire fighter garment with non textile insulation. Paper presented at the RTO HFM Symposium on Blowing Hot and Cold: Protecting Against Climatic Extremes: 2001; Dresden, Germany.

[4] Lu Y, Song G, Li J. A novel approach for fit analysis of thermal protective clothing using three-dimensional body scanning. Applied Ergonomics: 2014; 45 (6): 1439-1446.

[5] Song G. Clothing air gap layers and thermal protective performance in single layer garment. Journal of Industrial Textiles: 2007; 36 (3): 193-205.

[6] Rossi R, Bolli W. Phase Change materials for the improvement of heat protection. Advanced Engineering Materials: 2005; 7 (7): 368-373.

[7] Qi Z, Huang D, He S, Yang H, Hu Y, Li L, et al. Thermal protective performance of aerogel embedded firefighter's protective Clothing. Journal of Engineered Fibers & Fabrics: 2013; 8 (2): 134-139.

[8] Mondal S, Hu J, Yang Z, Liu Y, Sento YS. Shape memory polyurethane for smart garment. Research Journal of Textile and Apparel: 1999; 6 (2): 75-80.

[9] Chen Y, Lin LT, Ren QQ. Investigation on the instant fireproof composite fabric made of shape memory alloy. Shanghai Textile Science & Technology: 2002; 40 (2): 43-46.

[10] Yoo S, Yeo J, Hwang S, Kim YH, Hur SG, Kim E. Application of a NiTi alloy two-way shape memory helical coil for a versatile insulating jacket . Materials Science & Engineering A: 2008; 481-482 (1): 662-667.

[11] Yaters DA. Design and evaluation of a thermally responsive firefighter turnout coat. Master Dissertation: 2012; University of Worcester.

[12] Conaltion D. Shape memory alloys for use in thermally activated clothing, protection against flame and heat. Fire & Materials: 1999; 23 (5): 223-226.

Application of Manikin: Characterizing Effect of Fabric Properties on Protective Performance Against Hot Water Spray

Si Chen[2], Ye-Hu Lu[1, 2*], Xiao-Qun Dai[1, 2]

[1]*National Engineering Laboratory for Modern Silk, Soochow University, Suzhou, China*
[2]*College of Textile and Clothing Engineering, Soochow University, Suzhou, China*

[*]Corresponding author's email: yhlu@suda.edu.cn

Abstract

In the area of protective clothing, much attention was paid on the protection against heat and flame, however, the relatively unexplored hot liquid hazard is a common risk in workplace safety from an industrial scale to home scale. In this study, a hot water spray manikin was used to investigate the effects of fabric properties on thermal protection. The results indicated that the water repellence finish significantly affected the protective performance, while the fabric weight and air permeability had slight effect on the protection. The weft flexural rigidity of the fabric affected the performance. The overall performance increased with the increase of air gap size for permeable garments. The findings are helpful for improving the design of the protective clothing.

Keywords: Manikin Test System; Thermal Protective Performance; Protective Clothing; Hot Water Spray

1. Introduction

Protective clothing is widely used to protect the wearer's health and safety against radiant heat, flame, molten metal substances, steam and hot liquid splashes [1, 2]. The relatively unexplored hot liquid hazards present a life-threatening environment to end users. The statistical report shows that the highest proportion of burns results from contact and scalds [3]. The proportion of scald injuries is about 33.7% between 2004 and 2013 in America, which is approximate to that of flame injuries (National Burn Repository, 2014). Based on the analysis of collected burn injury data, hot liquid hazard is a considerable risk in many industries. Personal protective equipment is required to minimize the energy transferred to skin and protect workers' safety and health. The investigation of factors influencing the performance of protective clothing against hot liquid splashes is necessary.

Using the modified test apparatus described in ASTM F2701, researchers have investigated the protective performance of fabrics against hot liquid splashes. The protective performance was primarily determined by the fabric structure and physical properties [4, 5]. Among the physical properties, air permeability was a critical factor in thermal protection against hot liquid. Impermeable fabric systems could reduce mass transfer and provide high performance. For impermeable and semi-permeable fabrics, the thickness of fabric appeared to be the dominant factor in thermal protection; for permeable fabrics, it was shown that the increase of liquid penetration resistance significantly improved thermal protection [2]. Fabric surface property also affected protection against splashes from hot liquid [2]. The position of the membrane in fabric systems influenced thermal performance [4]. This was because the introduction of a semipermeable or impermeable membrane into the fabric system could minimize heat and mass transfer by blocking hot liquid flow further from the skin. Generally, minimizing mass transfer during hot liquid splash was the most important mechanism to provide high level protective performance [2]. Furthermore, the quantitative investigation of mass transfer characteristics through protective materials under the same exposure condition has been conducted [2]. The effects of fabric surface property, thickness, fabric combination and membrane on liquid absorption and penetration were explored. In addition, the liquid properties such as viscosity and surface tension had impact on the protective performance of fabric system. Neal [6] measured the heat transfer of flame resistant rainwear materials under exposure to canola oil. Lu et al. analysed the effect of different exposure liquids on protection of fabrics and found the thermal diffusivity of liquid was the key factor to protective performance. Mass transfer rate and amount had great influence on the heat transfer to skin. When the fabric was located in horizontal, the 2nd degree burn time decreased. Lu et al. studied the effect of air gap on thermal protection of fabrics. The results showed that the heat transfer to the sensor depended on the fabric properties, liquid properties and test configurations. The air gap significantly improved the thermal performance of fabric. However, vapour transfer through hydrophobic permeable fabric and its condensation on sensor attenuated the positive effect of air gap on thermal protection.

Instrumented manikins are very helpful to evaluate the protective performance of full scale garments. The air gap between clothing and the body is one of the most important factors influencing heat transfer to skin [7]. The effects of garment style and fit, and body geometry on flame protection have been investigated using flame manikin system [8, 9]. Furthermore, the protective performance of garments against hot steam environment was evaluated by using a copper manikin [10].The test results obtained from fabrics and garments showed good agreement. The impermeable fabrics and garments provided greater protection. Moreover, the thicker fabrics and garments provided better performance [10]. The performance of industrial thermal protective clothing against hot liquid hazard was furthered evaluated using a hot water spray manikin [2]. The results demonstrated that impermeable clothing provided better protection. Minimizing mass transfer was recognized as the critical factor for protection from hot water, which was consistent with findings obtained in bench scale tests. The design features showed effects on the clothing performance. Adding fabric layers could improve thermal protective performance of garments. The size of garment showed effects on thermal protection of permeable garments. Burn injury mainly occurred at the areas of compression upon water spray and heavy water flow.

The existing testing standard of thermal protection against hot liquid splashes is used to characterize the thermal performance of fabrics. The influencing factors of thermal protection have been explored in bench scale tests. However, few studies on protection of full scale garment have been carried out. This study aimed to investigate the impact of fabric properties on protective performance of whole garments using an instrumented hot water spray manikin system. The overall objective was to give suggestions for designers to engineer protective materials and develop high performance protective clothing.

2. Experimental

2.1 Testing Garments

To explore the effect of fabric properties on the thermal protection against hot water spray, several typical flame resistant (FR) cotton protective garments for industrial workers were used in this study. Garment C1~C4 were permeable. C5 and C6 were semipermeable due to polymer finishing. The configurations of test samples were listed in Table 1. The fabric thickness was measured in accordance with ASTM D1777-96 under pressure of 1kPa. The flexural rigidity of the fabrics was measured using cantilever method according to ASTM D1388-08. The fabric air permeability was tested according to ASTM D 737-04. A non-contact VITUS Smart 3D whole body laser scanner by Human Solutions was used to characterize the size of air gap between the clothing and manikin. All the garments have similar pattern with size of 42R in US clothing size. Pockets and reflective tapes were also designed at specific regions.

Table.1 Configuration of the test samples

	Fibre content	Weight (g/m^2)	Thickness (mm)	flexural rigidity (uN·m)		Air permeability ($cm^3/cm^2\cdot s$)	Average air gap (mm)
				warp	weft		
C1	cotton	360	0.67	102.7	106.4	2.97	33.7
C2	88/12 cotton/nylon	305	0.69	49.6	19.5	5.62	28.0
C3	88/12 cotton/nylon	237	0.62	98.8	9.3	26.9	26.1
C4	88/12 cotton/nylon	305	0.65	151.4	30.6	18.3	27.1
C5	C4 with finishing	322	0.66	177.2	52.5	0	29.7
C6	88/12 cotton/nylon with finishing	412	0.67	207.6	45.8	0	27.8

2.2 Testing Apparatus

In this study, a duplicate of flame manikin model was employed for the 3-D body scanning by a VITUS Smart 3-D whole body laser scanner from Human Solutions. Pictures of the dressed manikin were taken and compared across garments of the same style to ensure similar wear condition. The raw data obtained from nude and clothed scan were imported in Rapidform XOR. The scan data was meshed, rewrapped and the holes were filled. Subsequently, healing wizard and optimized mesh was applied to smooth the model. Finally, the refined model was exported for air gap measurement. The nude and clothed scans were required to be overlapped and aligned as accurate as possible in Rapidform XOV. The overall air gap distribution over the body surface was calculated.

The hot water spray manikin test system, shown in Fig.1, was employed in this study. The male manikin of size 40R, equipped with 110 skin simulant sensors, was the same with flame manikin [8]. The manikin was hung and fastened to stand upright. Four sets of cylinder spray jets were used to spray the manikin torso. Each group consisted of three bottom-up nozzles. Hot water was simultaneously sprayed by twelve nozzles. The hot water was heated up by super-heater and pumped from a 20L tank by a motor. In this study, pressure of 250kPa was used to simulate the hot water hazards occurred in industrial working accidents [11]. The heat flux at the manikin skin under clothing was recorded by data acquisition system and input to predict the skin burn distribution over the body and the total absorbed energy (TAE) during test. The absorbed energy during hot water spray (EAE) and the total energy transmitted to skin when the 2nd degree burn occurred (BAE) were calculated as well. The three-layer Pennes model and Henriques Burn Integral were applied to predict the skin burn injury. The procedure and parameters were described in ISO 13506.2008.

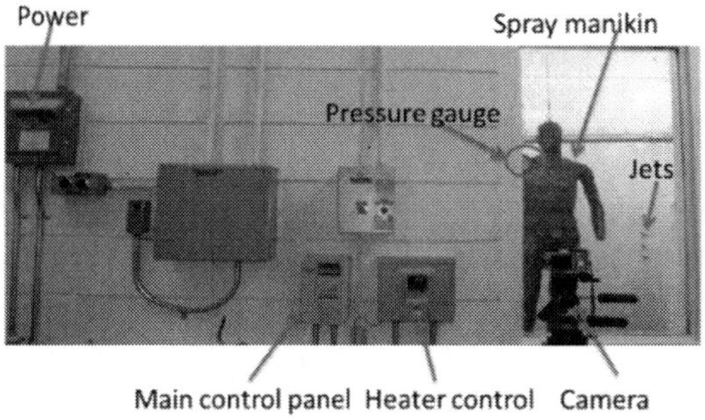

Fig.1 Hot water spray manikin test system.

2.3 Protocol

The garments were preconditioned in a standard climate for at least 24 hours prior to testing. The air gap between protective clothing and manikin surface was determined using 3D body scanning. Protective clothing was dressed on the spray manikin according to pictures taken during 3D body scanning to ensure a similar wear condition. The hot water was heated to 85°C and the garment was exposed for 10s. The data acquisition system was set for 60s. The percentage of second and third skin burn, total absorbed energy and skin burn distribution were predicted. Three replicates of each garment were measured.

3. Results and Discussion

The percentage of 2nd, 3rd degree burn and total skin burn area (TSBA), total absorbed energy (TAE) during test, were shown in Table 2. The different garments provided different skin burn injury and TAE while exposed to hot water spray. It was shown that little 3rd degree burn was predicted during the test. There was less than 10% burn area observed while wearing garments C5 and C6. Approximate 50% skin generated burn injury while wearing C3 or C4. The garment C1 exhibited the best performance among the selected permeable samples in this study. The TAE showed the same trend with TSBA.

Table.2 Test results of selected garments

	2nd degree burn [%] (SD)	3rd degree burn [%] (SD)	TSBA [%] (SD)	TAE [kJ/m^2] (SD)
C1	16.77 (5.50)	0.00 (0)	16.77 (5.50)	87.90 (4.22)
C2	38.02 (2.12)	1.17 (1.02)	39.18 (1.65)	130.82 (3.62)
C3	49.63 (1.30)	2.17 (1.93)	51.80 (0.83)	154.23 (4.88)
C4	46.87 (0.28)	2.40 (2.26)	49.27 (2.18)	147.86 (6.83)
C5	8.92 (2.01)	0.00 (0)	8.92 (2.01)	51.21 (4.30)
C6	6.35 (1.92)	0.00 (0)	6.35 (1.92)	46.98 (1.92)

Note: -SD is the abbreviation of standard deviation.

It was reported that if the percentage of second degree burn exceeded 20%, or that of third degree burn was higher than 5% of surface area, the victim should take medical care [6]. The burn injury below 20% of the whole body is one of the protective performance criteria for a military flame/thermal protective clothing system [12]. In this study, TSBA less than 20%, with third degree burn lower than 5%, was considered as the criterion to assess the overall performance of a garment. Therefore, garments C1, C5 and C6 provided sufficient protection; while the other garments failed to protect wearer's health.

Fig.2. showed the relationships between fabric weight and TSBA ($y=-0.286x+121.22$, $R^2=0.675$, $p=0.045$) and TAE ($y=-0.650x+315.53$, $R^2=0.635$, $p=0.05$). It was clear that the TSBA decreased with the increasing of fabric weight, and TAE showed a similar trend to TSBA. The linear regression indicated that both the TSBA and TAE presented a negative linear correlation with weight. The effect of fabric density on overall protective performance was shown in Fig.3. A similar negative linear relationship between density and TSBA ($y=-0.479x+339.82$, $R^2=0.642$, $p=0.05$) or TAE ($y=-0.209x+131.01$, $R^2=0.671$, $p=0.046$) was found.

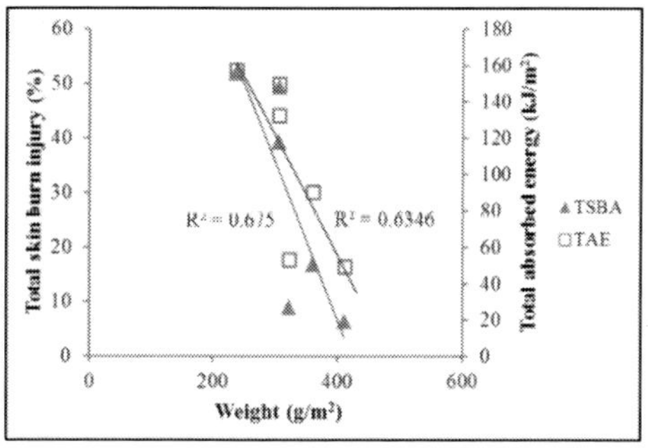

Fig.2 Relationships between fabric weight and TSBA and TAE.

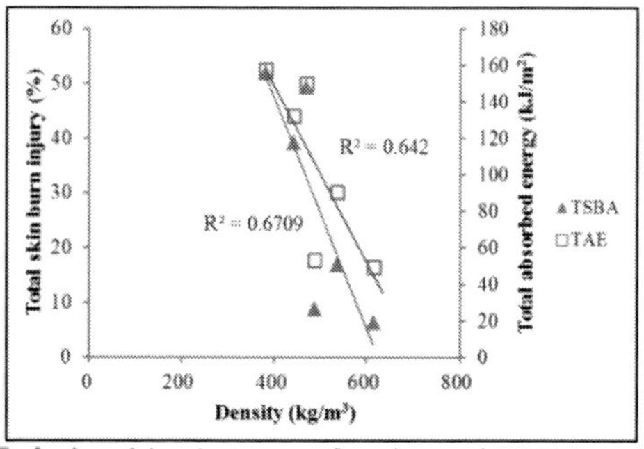

Fig.3 Relationships between density and TSBA and TAE.

The impact of air permeability on TSBA ($y=1.655x+13.88$, $R^2=0.804$) and TAE ($y=3.764x+71.68$, $R^2=0.758$) were presented in Fig.4. It indicated that both TSBA and TAE increased with the increasing of air permeability of fabric. The positive linear empirical equation demonstrated that decreasing fabric air permeability can effectively reduce total skin burn injury and improve overall thermal protective performance.

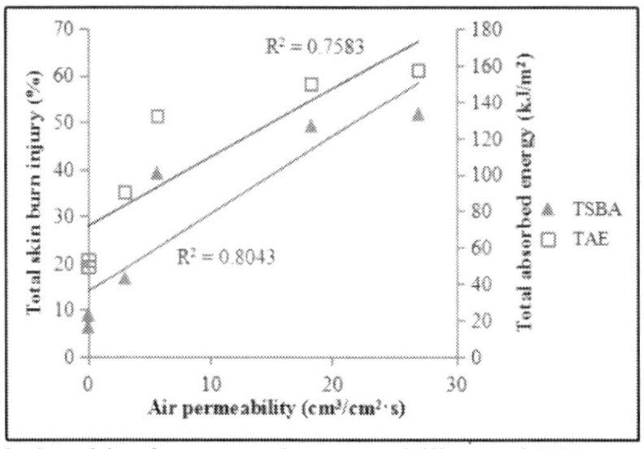

Fig.4 Relationships between air permeability and TSBA and TAE.

The correlation analysis was conducted to understand the effects of fabric weight, density and air permeability on protective performance. The Pearson correlation coefficient between fabric weight or air permeability and TBSA or TAE was listed in Table 3. Fabric weight and air permeability showed significant correlation with TSBA and TAE; air permeability showed significant correlation with TAE, whereas fabric weight didn't. The partial correlation was further analysed and found that the correlation between fabric weight or air permeability was not significant ($p = 0.556$, $p = 0.198$). The results indicated that the effect of fabric weight and air permeability was interfered.

Table.3 Pearson correlation results

Variables		TSBA	TAE
Fabric weight	Correlation	-0.822	-0.797
	Sig. (2-tailed)	0.045	0.058
Fabric density	Correlation	-0.819	-0.801
	Sig. (2-tailed)	0.046	0.055
Air permeability	Correlation	0.897	0.871
	Sig. (2-tailed)	0.015	0.024

There was no significant difference in thermal protection provided by C3 and C4. The two garments were made of fabrics with different weight, which demonstrated that the weight did not show significant effect on protective performance of permeable garments. Garment G5 and G6 were constructed of fabrics with different weight and both were treated by polymer finishing, resulting in no air permeability. There was no significant difference in TSBA and TAE, demonstrating the fabric weight did not have significant effect on the protection of semipermeable garments.

Semipermeable garments (C5 and C6) provided better protection than permeable garments (C1~C4), which was consistent with results in previous study [11]. This was due to the blocking or minimizing of mass transfer. Comparison of C5 with C4 demonstrated the importance of water repellence. The garment C5 was treated by EPIC polymer finishing on C4. The water repellent polymer finishing showed a significant effect on thermal protection.

The fabric flexural rigidity might affect the air gap size of the garment, and thus influenced thermal protective performance. The cantilever was used to determine the flexural rigidity of fabric. The relationship between fabric flexural rigidity and air gap size was analysed in Fig.5. The average air gap size showed significant linear correlation with weft flexural rigidity, whereas the warp stiffness did not show significant relationship with air gap size.

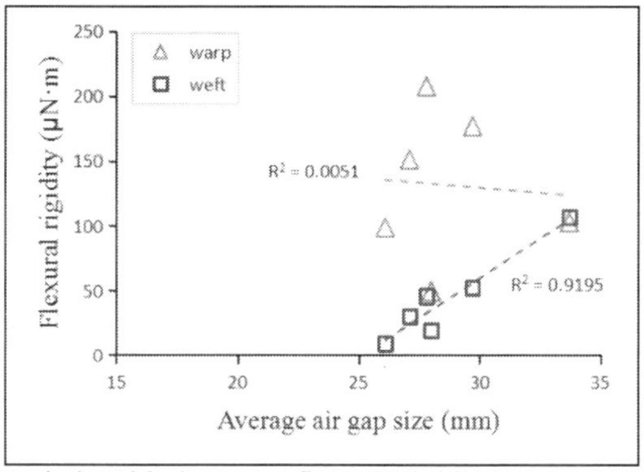

Fig.5 Relationship between flexural rigidity and air gap size.

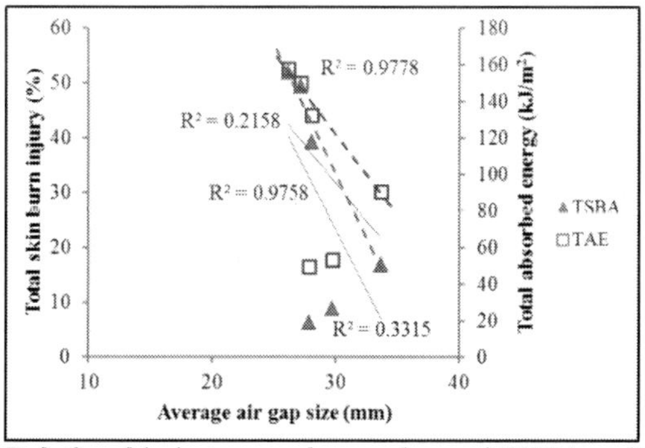

Fig.6 Relationship between air gap size and TSBA and TAE.

The correlations of air gap size with TSBA and TAE were presented in Fig.6. There was no obvious relationship found in the selected garments. However, the air gap size of permeable garments showed strong linear correlation with TSBA ($y=-4.6258x+172.13$, $R^2=0.9758$, $p=0.012$) and TAE ($y=-8.6946x+382.29$, $R^2=0.9778$, $p=0.011$). It indicated that the effect of air gap on thermal performance against hot liquid splashes was associated with the permeability of the garment. It has been confirmed that the semipermeable garments show better protection than permeable ones whatever the air gap size is.

4. Conclusion

The thermal protective performance of protective clothing under exposed to hot water spray was investigated using an instrumented manikin in this study. The results showed that semipermeable garments provided better performance than permeable ones among the selected coveralls. The water repellence finish significantly affected the protective performance. Partial correlation analysis showed that fabric weight and air permeability did not have significant effect on the TSBA and TAE. The effect of air gap size on thermal protection was interfered by the garment permeability. The protective performance increased with the increasing of air gap size for permeable garments. The weft flexural rigidity presented strongly positive correlation with air gap size, and thus affected thermal protective performance. The findings provided useful basis and guidance to design high-performance protective clothing.

5. Acknowledgement

The experiment was conducted in University of Alberta and the authors gratefully appreciated the technical support, especially Dr. Guowen Song in Iowa State University at present. This study was financially supported by the National Natural Science Foundation (No. 51506137), the Natural Science Foundation of Jiangsu Province (No. BK20161255), and Jiangsu Postdoctoral Science Foundation (1402108C).

References

[1] Sati R, Crown EM, Ackerman M, et al. Protection from steam at high pressures: Development of a test device and protocol. INT J OCCUP SAF ERGO: 2008; 14: 29-41.

[2] Lu YH, Song GW, Ackerman M, et al. A new protocol to characterize thermal protective performance of fabrics against hot liquid splash. EXP THERM FLUID SCI: 2013; 46: 37-45.

[3] Taylor AJ, McGwin G, Cross JM, et al. Serious occupational burn injuries treated at a regional burn center. J BURN CARE REHABIL: 2002; 23: 244-248.

[4] Jalbani SH, Ackerman MY, Crown BM, et al. Modification of ASTM F 2701-08 apparatus for use in evaluating protection from low pressure hot water jets. In: "9th symposium on performance of protective clothing and equipment: emerging issues and technologies". California: Anaheim: 2011.

[5] Ackerman MY, Song GW, Gholamreza F, et al. Analyzing thermal protective clothing performance against the impact of small splashes of hot liquid. In: "9th symposium on performance of protective clothing and equipment: emerging issues and technologies". California: Anaheim: 2011.

[6] Berry MF, Holmes JH, Schwab CW. Burn Management. In: Atluri P, Karakousis GC, Porrett PM and Kaiser LR, editors. The Surgical Review: An Integrated Basic and Clinical Science Study Guide. Philadelphia: 2006; p. 289-295.

[7] Torvi DA, Dale JD, Faulkner B. Influence of air gaps on bench-top test results of flame resistant fabrics. J FIRE PROT ENG: 1999; 10: 1-12.

[8] Mah T, Song G. Investigation of the contribution of garment design to thermal protection. Part 2: Instrumented female mannequin flash-fire evaluation system. TEXT RES J: 2010; 80: 1473-1487.

[9] Crown EM, Ackerman MY, Dale JD, et al. Design and evaluation of thermal protective flightsuits: Part II: Instrumented mannequin evaluation. CTRJ: 1998; 16: 79-87.

[10] Desruelle AV, Schmid B. The steam laboratory of the Institut de Medecine Navale du Service de Sante des Armees: a set of tools in the service of the French Navy. EUR J APPL PHYSIOL: 2004; 92: 630-635.

[11] Ackerman MY, Crown EM, Dale JD, et al. Project update: Protection from steam and hot water hazards. In: "Protective Clothing Systems for Safety '11". Canada: Edmonton: 2011.

[12] Kim IY, Lee C, Li P, et al. Investigation of air gaps entrapped in protective clothing systems. FIRE MATER: 2002; 26: 121-126.

The Study of Thermal Transfer of Laminated Knitted Fabrics

Bao-Fang Wei[1], Hui Zhu[1], Ji-Hong Wu[1*]

[1] *Wuhan Textile University, Jiangxia District Sunshine Road No.1, Wuhan, Hubei, 430200, China*

[*]Corresponding author's email: jihong_wu@wtu.edu.cn

Abstract

Laminating with hot melt adhesive is a commonly used method. The laminated fabrics which are made in this way are better than the single fabrics. In this experiment, four knitted fabrics (cotton knitted fabric, polyester/cotton knitted fabric, ramie/polyester knitted fabric and polyester knitted fabric) were used as raw materials, which were then laminated into knitted fabrics. The thermal transfer properties of single-layer knitted fabrics and laminated knitted fabrics were determined to study what factors affected it. The contact surface, materials, thickness and mass per unit area of the fabric have effects on the thermal transfer properties of laminated knitted fabrics. Various materials and different contact surfaces make a significant difference on thermal transfer. In a certain range, thermal transfer properties like heat retention ratio can be improved by increasing the thickness and mass per unit area of the fabrics.

Keywords: Laminated Knitted Fabric; Thermal Transfer Properties; Influence Factors

1. Introduction

1.1 Thermal Transfer Properties

The heat and water vapor transfer properties are the fundamental criterion when measuring the comfort of textiles. At present, the main researches on thermal transfer properties are studied from two aspects: one is about the mechanism of fabrics' thermal transfer properties, including the thermal transfer models, its influence factors, and the test analysis by simulating the wearing conditions and the comprehensive evaluation methods; the other is about the principles and methods of the tests of the thermal transfer properties [1].

1.1.1 Mechanism of Thermal Transfer

At a macroscopic level, the fabric is an inhomogenous assembly which is composed of air and textile materials. The methods of thermal transfer include thermal conduction, thermal convection, thermal radiation and latent heat transfer with the water vapor transmission. Generally speaking, thermal conduction plays a major role in the thermal transfer process of the fabric. Under certain circumstances, the impact of thermal convection and thermal radiation counts for a lot in thermal transfer [2]. From a microscopic mechanism point of view, there are multiple mechanisms in thermal transfer of textiles. It is attributed to fibers and textiles' characteristics, i.e. the assemblies possess a complex structure with many pores, and the transmission of light can be available. Therefore, the mechanism of thermal transfer is very complex.

At present, the researches on the model of fabrics' thermal transfer properties have yielded a great number of positive results. For example, to compare the insulation properties of different textiles, the CLO value model has been established. And some parameters were taken into account, such as the human physiological parameters, psychological senses, temperature and humidity of the environment and wind speed conditions [3]; Farnworth, et al. established a model of thermal conduction with radiation to analyze the thermal conductivity of the fabrics and the radiation between the layers of fiber and garment [4]; furthermore, the combined model of thermal and wet resistance was established by Lotens and his team. The model gave an evaluation on the heat insulation and moisture transfer of the overall garments in a certain circumstance [5].

1.1.2 Test Method of Thermal Transfer

Many researchers have done a lot of studies on the test method of thermal transfer. For the moment, the methods for testing fabric thermal transfer properties are as follows: The principle of constant-temperature method is that placing the test fabric on one side of the isothermal heating element, and other sides of the heating elements are good insulation materials. With the energy supply, the heating elements are kept at a constant temperature, and the fabric's conductivity coefficient of thermal transfer, thermal resistance and heat retention ratio, the fabric's heat retention properties can be determined by testing the energy. Cooling Rate Method is placing the test fabric on one side of the heating element, or wrapping the heating element fully with the fabric, then making it cool down naturally after it is heated to a certain temperature. The fabric's insulation performance can be indicated by the cooling rate of the heating element. Heat-flowmeter Method is putting the fabric in the center of a hot plate and a cold plate with the constant temperature. The fabric's heat flow can be determined by a circular plate heat flow sensor [6]. Microclimate which was put forward by Harada Takashi concentrates on temperature, oxygen consumption, sweat and other physiological indexes when people wear clothing in various environmental conditions [7]. Based on this theory, the THERMOLABO-I and THERMOLABO-II type device was developed by Kawabata to measure the heat flow in the dry state, non-contact dry state, wet contact and non-contact wet state [8]. The thermal manikin can be used to simulate the heat exchange process between the simulated human body, garments and environment in a specific environment. Now, a lot of apparatuses have been developed, such as dry thermal manikin, sweating thermal manikin, breathing thermal manikin, immersion thermal manikin, numerical thermal manikin, thermal artificial limb and head [9].

1.2 Laminated Fabric

The laminated fabric is a kind of shell fabric which is cohered by two kind of fabrics or by one kind of fabric and a polymer film. Through the finishing process, the advantages of different fabrics can be concentrated on the one fabric according to the different demands. By means of such a technology, clothes could be windproof, waterproof, breathable, and warm. At present, the researches of laminated fabric mainly are concentrated on the field of polymer materials and multifunctional materials. And the general studies are about the influence of laminating and finishing process to laminated fabrics. For example, Zhao-hui Li [10] has studied the laminating process of the three-layer fabrics which were waterproof and hygroscopic. Hind Abdellaoui [11] has studied the fabrication, characterization and modelling of laminated composites based on woven jute fibers reinforced epoxy resin. However, as to the thermal transfer properties, there is a large space remains to be improved in the field of laminated knitted fabric. Therefore, this experiment used some knitted fabrics with different specifications and parameters, then, the laminated knitted fabrics were made with the four knitted fabrics and hot melt adhesive. As a result, how its contact surface, fabric materials, fabric weight and thickness affected the thermal transfer performance of laminated fabric were studied.

2. Experiments and Methods

2.1 Materials and Apparatuses

In order to study the thermal transfer properties of the laminated knitted fabrics, a series of samples were produced. In this experiment, four single-layer knitted fabrics (cotton knitted fabric, polyester/cotton knitted fabric, ramie/polyester knitted fabric and polyester knitted fabric) were used. With the help of hot melt adhesive, the laminated knitted fabrics were made by the above-mentioned four single-layer knitted fabrics.

As primary parameters, thickness and mass per unit area of the fabrics were determined. The thickness of knitted fabric was determined by using an apparatus for measuring thickness with a pressure. The mass per unit area of the knitted fabric was determined by weighing the sample with an area of 1 m^2. So, the apparatuses used in this study were as follows, electronic balance, fabric thickness testers, sweating guarded hot plate. The basic specifications of the fabrics were shown in this paper (Table 1).

Table.1 Specifications of the fabrics

Sample code	Fabric description	stitch	Fabric weight(g/m^2)	Thickness(mm)
a	cotton knitted fabric	Jersey	236.61	0.59
b	polyester/cotton knitted fabric	Jersey	122.09	0.58
c	ramie/polyester knitted fabric	Jersey	156.24	0.56
d	polyester knitted fabric	tuck	166.01	0.58
e	a laminated with d	—	402.69	1.18
f	a laminated with b	—	359.52	1.19
g	a laminated with c	—	393.47	1.17
h	d laminated with b	—	288.55	1.16
i	d laminated with c	—	322.76	1.15

2.2 Method

The sweating guarded hotplate was used to test the thermal transfer properties of the laminated knitted fabrics. It simulated the transfer of heat and water vapor processes between the human skin, the textile layers and the environment. In order to achieve the semi-turbulent airflow across the test plate, airflow hood was set at the top of the hotplate assembly. And the heating was provided by resistance wire on the inner surface of each part. Before the beginning of this experiment, all the samples were placed in a chamber for 24 h at 20℃ and 65% relative humidity of the air. During the tests, the sweating guarded hotplate was also placed in an environmental chamber where the following conditions were established: 20 ℃ and 65% relative humidity of the air [12], and the power supply kept the plate at constant temperature of 35℃ which was near to the temperature of human body.

The four kinds of single knitted fabrics, cotton knitted fabric, polyester/cotton knitted fabric, ramie/polyester blended knitted fabric and polyester knitted fabric, were cut into smaller piece as a scale of 30 cm × 30 cm. The laminated fabrics were made by hot melt adhesive. When the temperature of the apparatus was in a steady state, the thermal transfer properties of the single-layer knitted fabrics and the laminated knitted fabrics were determined.

3. Results and Discussion

Table.2 The thermal transfer properties of samples

Fabric	Heat retention ratio(%)(STD)	Conductivity coefficient [$W/(m^2 \cdot K)$] (STD)	CLO value (STD)	Thermal resistance ($m^2 \cdot K/W$) (STD)
a	12.71(0.274)	54.08(1.552)	0.12(0.006)	0.02(0.000)
b	18.41(0.204)	33.07(0.443)	0.20(0.006)	0.03(0.000)
c	9.50(0.130)	70.54(1.085)	0.09(0.000)	0.01(0.000)
d	8.77(0.204)	77.22(1.954)	0.08(0.006)	0.01(0.000)
e(a)	21.89(0.460)	28.54(0.780)	0.23(0.000)	0.04(0.006)
e(d)	14.36(0.806)	47.20(2.807)	0.14(0.010)	0.02(0.000)
f(a)	22.30(0.271)	25.90(0.405)	0.25(0.000)	0.04(0.000)
f(b)	23.07(0.282)	24.23(0.380)	0.26(0.010)	0.04(0.000)
g(a)	21.09(0.862)	29.77(0.589)	0.22(0.006)	0.03(0.000)
g(c)	17.19(0.462)	35.77(1.165)	0.18(0.010)	0.03(0.000)
h(d)	21.84(0.623)	27.58(0.374)	0.23(0.000)	0.04(0.000)
h(b)	24.97(0.386)	23.16(0.325)	0.28(0.000)	0.04(0.000)
i(d)	24.52(0.323)	24.46(0.491)	0.26(0.006)	0.04(0.000)
i(c)	30.09(0.377)	17.27(0.601)	0.37(0.015)	0.06(0.000)

The abbreviation of e(a) means that the contact surface of the fabric which was laminated with cotton knitted fabric and polyester knitted fabric was cotton when testing the heat retention ratio, and the abbreviation of e(d) means that the contact surface of the laminated fabric was polyester. The same principle can be applied to other laminated fabrics.

The performance indexes of thermal transfer properties were measured, such as the single-layer knitted fabrics and the laminated knitted fabrics' heat retention ratio(%), conductivity coefficient[W/($m^2 \cdot K$)], CLO value and thermal resistance($m^2 \cdot K/W$). When testing the thermal transfer properties, the thickness of the fabric is a critical parameter. As shown in Table 1, the thicknesses of single-layer knitted fabrics are very similar (0.58±2mm), this conclusion also can be applied to the laminated knitted fabrics (1.17±2mm). Therefore, when discussing the influence of raw materials, the impact of thickness can be ignored. As indicated in Table 2, the results of thermal transfer properties are shown. The heat retention ratio of the laminated knitted fabric is higher than that of the single-layer knitted fabric. As for the laminated knitted fabric, different contact surface and material have effects on the thermal transfer properties.

3.1 Influence of the Contact Surface

When the different side of the fabric was being contacted with the heating element, the heat retention ratio of the laminated fabrics were different as indicated in Fig.1.

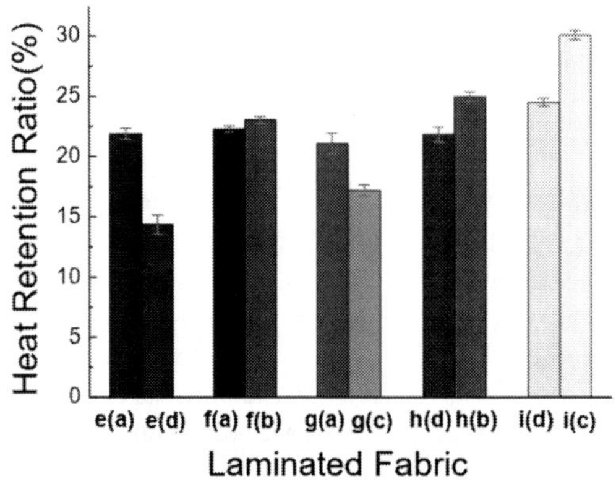

Fig.1 The heat retention ratio of different contact surface

Comparing the Table 2 and Fig.1, for the sample labelled e, the heat retention ratio of the laminated knitted fabric whose contact surface was cotton (21.89%) was better than that of the polyester (14.36%). Similarly, for the sample labelled f, the heat retention ratio of the laminated knitted fabric whose contact surface was polyester/cotton (23.07%) was better than that of the cotton (22.30%). And for the sample labelled g, the heat retention ratio of the laminated knitted fabric whose contact surface was cotton (21.09%) was better than that of the ramie/polyester (17.19%). Besides, the heat retention ratio of the laminated knitted fabric whose contact surface was polyester/cotton (24.97%) was better than that of the polyester (21.84%). In the same way, for the sample labelled i, the heat retention ratio of the laminated knitted fabric whose contact surface was ramie/polyester (30.09%) was better than that of the polyester (24.52%). For the samples labelled e(d), h(d) and i(d), the contact surface was the same polyester knitted fabric, but the heat retention ratio of the polyester was low. For the samples labelled e(a), f(a) and g(a), the heat retention ratio of the cotton was high. So, when testing the same laminated knitted fabric, the thermal transfer properties were different in the fabric face and fabric back. Due to the difference of single-layer knitted fabric, the transfer of thermal flux is also different from the hotplate to the fabric. Because of the dissimilar resistance, it has formed a gradient of thermal transfer. Then, the heat retention ratios are changed. Therefore, this conclusion can provide a reference for the designing of knitted sportswear with different contact surface, especially the various functions like warming and cooling in the fabric face and fabric back.

3.2 Influence of the Materials

The relationship of single-layer knitted fabric material and its heat retention ratio is shown in Fig.2. In addition, the relationship of laminated knitted fabric material and its heat retention ratio is shown in Fig.3.

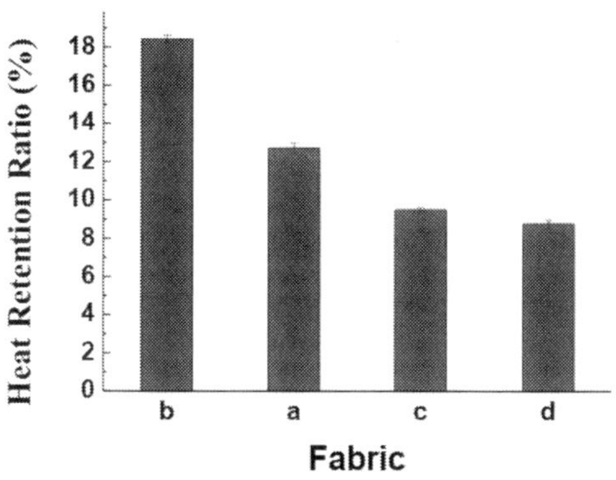

Fig.2 Relationship of single-layer knitted fabric material and heat retention ratio.

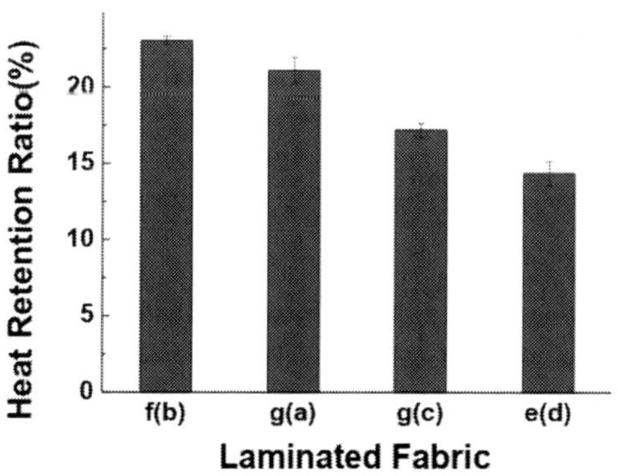

Fig.3 Relationship of laminated knitted fabric material and heat retention ratio.

The heat retention ratios of single-layer fabrics and laminated knitted fabrics with different materials have the following relationship: polyester/cotton is the best, cotton is the second best, ramie/polyester is poor than cotton, and the polyester is the worst, i.e. different materials have different thermal transfer properties. And the heat retention of the natural cellulose fiber and its blended fabrics is better than that of the polyester fabrics. This is owing to the structure of fiber. There is a lumen in cotton fiber and ramie fiber. And the lumen allows the fibers to contain more still air. In addition, the thermal conductivity coefficient of cotton fiber is about 0.071 $W \cdot m^{-1} \cdot °C^{-1}$, and the coefficient of polyester fiber is 0.084 $W \cdot m^{-1} \cdot °C^{-1}$. Therefore, the heat retention of the cotton fabric is better. Similarly, when companies are designing the warm clothing, by selecting the natural cellulose fibers and its blended materials as raw materials, the good heat retention properties can be achieved.

3.3 Influence of the Fabric Thickness and Weight

To find the relationship of laminated knitted fabric thickness, weight and heat retention ratio, laminated knitted fabrics were made by using the same size polyester/cotton knitted fabric with hot melt adhesive. And the thermal transfer properties of the laminated polyester/cotton knitted fabrics are shown in Table 3, Fig.4 and Fig.5.

A conclusion can be drawn from the Fig.4 and Fig.5. When laminating the polyester/cotton with different layers, the heat retention ratios of laminated polyester/cotton knitted fabrics are different. In a certain range, the heat retention ratio of laminated polyester/cotton knitted fabric increases with the thickness of knitted fabric assemblies. When the thickness ranges from 0 to 2 millimeters, the heat retention ratio increases sharply, when the thickness is more than 2 millimeters, the heat retention ratio increases slowly. Similar conclusions can be concluded from the fabric weight. In a certain range, the heat retention ratio of laminated polyester/cotton knitted fabric increases with the weight of knitted fabric assemblies. When the fabric weight ranges from 0 to 400 $g \cdot m^{-2}$, the heat retention ratio increases at a high rate, when the fabric weight is more than 400 $g \cdot m^{-2}$, the heat retention ratio increases at a low rate. Therefore, by increasing the fabric thickness and fabric weight, the thermal transfer properties like heat retention ratio can be improved.

Table.3 Thermal transfer properties of laminated polyester/cotton knitted fabrics

Fabric layer	Thickness (mm)	Fabric weight(g/m^2)	Heat retention ratio(%)(STD)	Conductivity coefficient [$W/(m^2*K)$] (STD)	CLO value(STD)	Thermal resistance ($m^2 \cdot K/W$) (STD)
1	0.58	122.09	18.41(0.204)	33.07(0.443)	0.20(0.006)	0.03(0.000)
2	1.17	250.56	32.16(0.210)	16.93(0.567)	0.38(0.000)	0.06(0.000)
3	1.75	375.78	40.15(0.671)	11.59(0.305)	0.56(0.021)	0.08(0.006)
4	2.34	505.11	44.23(0.653)	9.02(0.141)	0.70(0.012)	0.10(0.000)
5	3.01	630.45	47.11(0.263)	8.02(0.140)	0.81(0.015)	0.13(0.006)

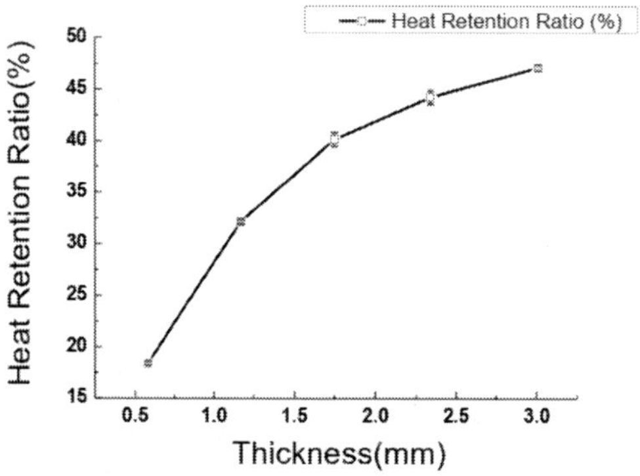

Fig.4 Relationship of laminated knitted fabric thickness and heat retention ratio.

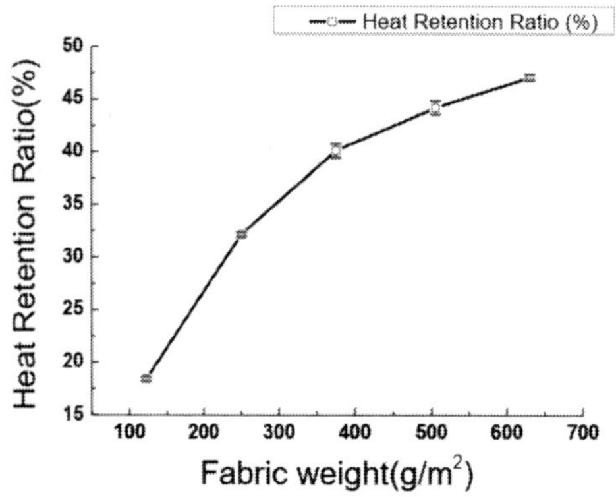

Fig.5 Relationship of laminated knitted fabric weight and heat retention ratio.

4. Conclusion

The purpose of this experiment is to analyze and determine the relationship between the knitted fabric parameters and the thermal transfer properties. Thus, some laminated knitted fabrics have been made with four single-layer knitted fabrics (cotton knitted fabric, polyester/cotton knitted fabric, ramie/polyester knitted fabric and polyester knitted fabric) and hot melt adhesive. The stitches of the single-layer knitted fabrics are jersey and tuck. By testing the basic parameters and thermal transfer properties, it was found that the thermal transfer properties in steady state were mainly influenced by the contact surface, materials, fabric thickness and fabric weight.

In summary, the thermal properties of the laminated fabric are better than that of the single-layer knitted fabrics. When the laminated fabric are determined with the different contact surface, the different heat retention properties are shown in the fabric face and fabric back. By making the laminated knitted fabric with hot melt adhesive and different single-layer knitted fabric, the transfer of thermal flux is different from the hotplate to the fabric when testing the different side of the laminated knitted fabric. Owing to the dissimilar resistance, it has formed a gradient of thermal transfer. What's more, when the materials of contact surface are the natural cellulose fiber and its blended fabrics, the heat retention ratio is the best, and the heat retention ratio of polyester is the worst. It's due to the structure of fiber and the thermal conductivity coefficient of textile materials. So, by using the different inner and outer materials like natural cellulose fiber and its blended fabrics, the different thermal transfer properties can be designed.

For the laminated knitted fabrics with the same materials, the fabric thickness and fabric weight also have an effect on the thermal transfer properties. The heat retention ratio increases sharply with the thickness and weight of knitted fabric assemblies in a certain range. When it goes beyond a certain range, the heat retention ratio increases slowly. Therefore, it is an effective way to enhance the heat retention by changing the fabric thickness and weight. In order to have a better understanding of the mechanism of heat and water vapor transfer through the laminated fabrics, other aspects are in need of research in the future, such as stitches, structures, the way of lamination, the thickness of air-layer, etc.

5. Acknowledgement

The authors would like to express the sincere thanks to all those who have helped the experiment in the course of writing this paper. First of all, it's an honor to take this opportunity to show the sincere gratitude to the supervisor, Mr. Wu, who has given so much useful advices on writing. Secondly, by providing the testing equipment of this experiment, the authors gratefully acknowledge the technical assistance of Department of Textile Science and Engineering. The authors wish to acknowledge the deep processing and product development of ramie fibers program. And also thanks for supporting the scientific and adequate literature review of this study.

References

[1] Wu HJ, Qian K, Cao HJ. Effect of fabric structure parameters on its heat transfer property. Text. Res. J: 2007; 28: 21-23.
[2] Shi HL. The review of heat transmission mechanism of the textiles. Journal of Nantong Institute of Technology: 2001; 4: 44-47.
[3] Gagge AP, Burton AC, Bazett HC. A practical system of units for the description of the heat exchange of man with his environment. Sci: 1941; 94: 428-430.
[4] Farnworth B. Mechanisms of heat flow through clothing insulation. Text. Res. J: 1983; 53: 717-725.
[5] Lotens WA, Havenith G. Calculation of clothing insulation and vapor resistance. Ergonomics: 1991 34: 233-254.
[6] Yu WD, Chu CY. Textile Physics. Shanghai: Donghua University: 2009; 401-408.
[7] Duan HQ. Research on testing method of fabric heat transmitting property. Shanghai: Donghua University: 2004; 1-9.
[8] Jiang PQ. The dynamic heat-moisture comfort properties of fabrics. Shanghai: Donghua University: 1999; 6-12.
[9] Huang JH. Research on thermal manikin at home and abroad.Building Energy & Environment: 2006; 25: 24-29.
[10] Li ZH. Three-lay moisture management composite fabrics with Softshell elastic fiber. Yinran: 2013; 21: 29-31.
[11] Hind A, Hala B, Jamal E. Fabrication, characterization and modelling of laminated composites based on woven jute fibers reinforced epoxy resin.Materials & Design: 2015; 68: 104-113.
[12] Cubric IS, Skenderi Z, Mihelic A. Experimental study of thermal resistance of knitted fabrics. Exp. Therm. Fluid Sci: 2012; 38: 223-228.

Relationship between Clothing Local Ventilation and Air Gap Thickness

Ying Ke[1*], Xiang-Hui Zhang[2], Shi-Ying Zhang[2], Hong-Fu Wang[1]

[1] *Jiangsu Non-material Culture Heritage Laboratory, College of Textiles and Clothing, Jiangnan University, Wuxi 214122, China*
[2] *Fashion & Art Design Institute, Donghua University, Shanghai 200051, China*

[*]Corresponding author's email: keying0312@163.com

Abstract

The purpose of this study is to investigate the relationship between clothing local ventilation and the air gap thickness. Three working jackets with the same construction but different fabrics were chosen. Local ventilation and air gap thickness of the arm, chest and back were measured in standing still conditions. The results showed that the back had the highest ventilation, followed by the arm and the chest. While garments with impermeable fabric had the biggest local air gap thickness. The local air gap thickness of the back was greatest, and the chest was smallest. For the three garment locations, clothing local ventilation had negative liner correlation with the average air gap thickness.

Keywords: Clothing Local Ventilation; Air Gap Thickness; Tracer Gas; Air Permeability

1. Introduction

Clothing ventilation is a key factor affecting human thermal comfort. It directly determines human body sweat evaporation rate and heat loss. According to measurement objects, clothing ventilation can be divided into whole ventilation and local ventilation. Clothing ventilation differs from various garment locations, therefore there are lots of studies focused on clothing local ventilation [1-4].

Methods and setups for measuring clothing local ventilation have already been developed. Satsumoto et al. developed a local ventilation measuring system for diapers based on the steady state method [5]. Ueda et al. used Crocford's unsteady-state method to investigate the regional ventilation rates of several working jackets [6]. And Ke et al. set a ventilation measuring system for working jackets, based on the steady state method [1].

Air gap thickness affects the air exchanges between clothing microclimate and the environment. Clothing thermal insulation is directly affected by air gap thickness. Bouskill et al. investigated the relationship between clothing ventilation and thermal insulation. They found that clothing ventilation and thermal insulation had a negative correlation [7]. This study only discussed whole ventilation. For local ventilation, the results may be different. The reason is that: Clothing local ventilation (local air exchange) has three pathways: directly through fabric, through garment apertures and the air exchanges between garment locations. The problem is that only the first two pathways affect human heat loss, which is called effective ventilation. Effective ventilation is affected by air gap thickness. Air exchange between garment locations had impact on the heat loss from human body. As such the relationship between clothing local ventilation and the air gap thickness may be more complicated.

The purpose of this study is to discuss the relationship between clothing local ventilation and the air gap thickness. Three experimental garments with the same construction but different fabrics were chosen. Local ventilation and air gap thickness of the arm, the chest and the back were measured separately. Relationship between clothing local ventilation and the average air gap thickness of the arm, the chest and the back were discussed. The study could give some suggestions on clothing comfort evaluation and functional garment design.

2. Experimental

2.1 Experimental Garments and Manikin

2.1.1 Experimental Garments and Fabrics

Fig.1 Experimental garments

Three working jackets with the same size (170/88A) and construction but different fabrics were chosen as the experimental garments. The fabrics were named as F1, F2 and F3. And the experimental garments were named according to the fabric, G1, G2 and G3. Fig.1 showed the drawing of the experimental garments. And Table 1 illustrated the basic properties of the three experimental fabrics. The main difference of these fabrics was the air permeability. F1 was impermeable. And F3 was most permeable relatively. In addition, F1 was the F2 laminated with an impermeable thin coating.

Table.1 Basic properties of the experimental fabrics

Type	Thickness (mm)	Material composition	Density warp/weft (inch)	Weight (g/m²)	Thermal insulation (℃·m²/W)	Vapor resistance (Pa·m²/W)	Air permeability (mm/s)
F1	0.48	100% cotton coated with TPU	78/55	248.90±3.64	0.0071±0.0004	--	0.00
F2	0.48	100% cotton	101/56	233.14±5.01	0.0096±0.0008	3.15±0.35	59.00
F3	0.48	100% cotton	101/56	186.30±1.10	0.0162±0.003	2.80±0.06	135.18

2.1.2 Manikin

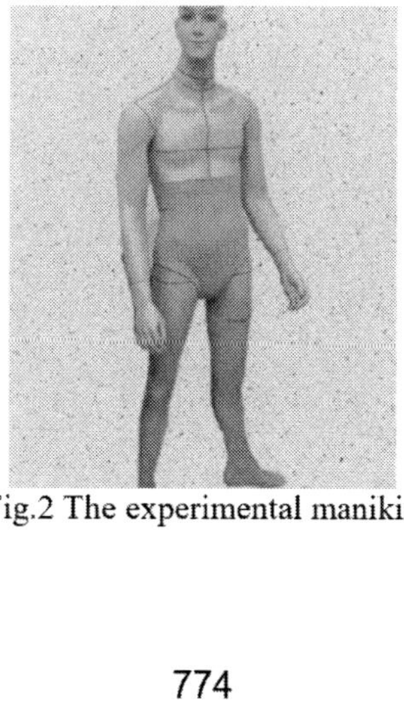

Fig.2 The experimental manikin

A standard shopping manikin of 170/88A was chosen as the experimental subject, as was shown in Fig.2. The height of the manikin was 170 cm, and the chest circumference was 88 cm.

2.2 Measurement of the Local Ventilation

Local ventilation of the arm, the chest and the back were measured separately using the setup developed by Ke et. al [1]. N_2 was chosen as the tracer gas. Clothing local ventilation was computed according to Eq.1.

$$Vent=R*(C_{in}-C_{out})/(C_{out}-C_{air}) \quad (1)$$

Where R is the flow rate of the local circulating system, L/min; C_{in} is the tracer gas concentration of the inlet flow, %; C_{out} means the tracer gas concentration of the microclimate, %; C_{air} is the tracer gas concentration of the environment, %.

2.3 Measurement of the Air Gap Thickness

Method proposed by Lu et al. was used to get the average air gap thickness of the experimental garments [8]. The $[TC]^2$ body scanner was used to capture the 3D body shape and the clothing profile. And Geomagic qualify 12.0 software was used to deal with the scanner data and get the average air gap thickness.

3. Results and Discussion

Table 2 showed the local ventilation and average air gap thickness of the experimental garments. Back had the highest local ventilation, followed by the arm and the chest. The local ventilation of G3 was higher than these of G1 and G2. In addition, back had the biggest air gap thickness, followed by the arm and the chest. The air gap thickness of G1 was biggest, and G3 was smallest for the same garment location.

Table.2 Local ventilation and the average air gap thickness of the experimental garments

Garments	Local ventilation (L/min)			Air gap thickness (mm)		
	Arm(SD)	Chest(SD)	Back(SD)	Arm(SD)	Chest(SD)	Back(SD)
G1	7.4(0.9)	7.3(0.8)	10.9(1.0)	13.1(1.3)	12.3(1.0)	15.1(1.3)
G2	13.8(1.6)	13.3(1.0)	24.2(2.2)	12.3(0.9)	11.8(0.7)	14.3(0.9)
G3	21.9(1.7)	19.4(1.2)	27.9(2.4)	11.1(1.0)	10.7(1.1)	13.5(1.2)

3.1 Local Ventilation of the Experimental Garments

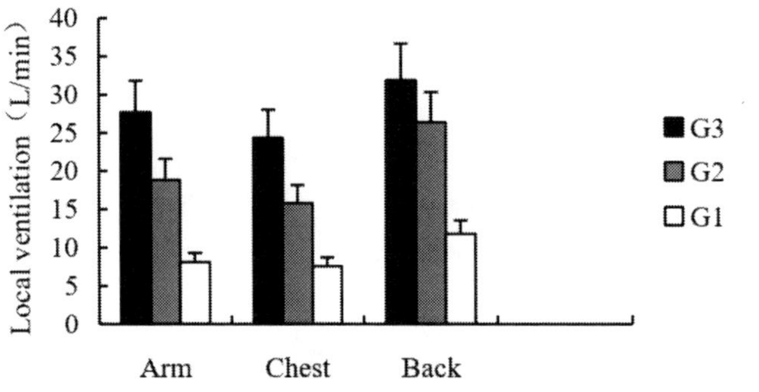

Fig.3 Local ventilation of the arm, the chest and the back of the experimental garments

Fig.3 showed the local ventilation of the arm, the chest and the back of the experimental garments. G1 had the lowest ventilation and G3 had the highest, as the fabric of G3 has better air permeability than that of G1. Back had the highest ventilation and chest had the lowest. This differed from the result of Ke et al. The reason was that clothing local ventilation was affected by garment fit. The manikin of the two studies was different. This caused the various microclimate conditions of the experimental garments.

3.2 Local Air Gap Thickness of the Experimental Garments

Fig.4 presented the average air gap thickness of the arm, the chest and the back. The air gap thickness of the back was highest, followed by the arm and the chest. In addition, G1 had the biggest air gap thickness. And the local air gap thickness of G3 was smallest.

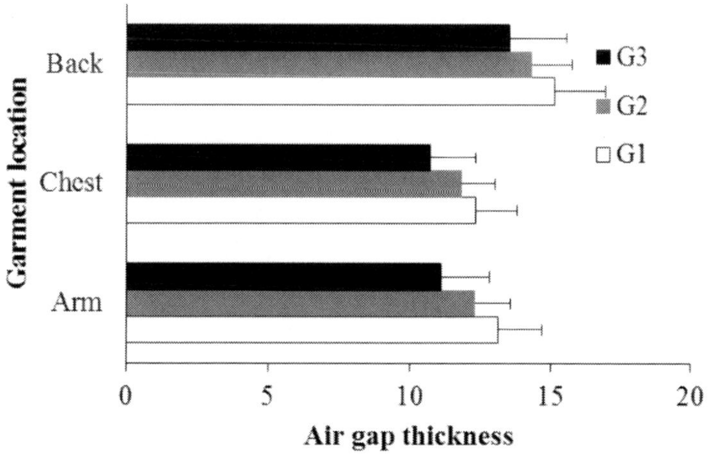

Fig.4 The average air gap thickness of the arm, the chest and the back of the three experimental garments

3.3 Relationship between Clothing Local Ventilation and Air Gap Thickness

Fig.5 illustrated the relationship between clothing local ventilation and air gap thickness. A good liner correlation (R^2=0.998, 0.957, 0.904 for the arm, the chest and the back) was found between clothing local ventilation and air gap thickness. This meant that the bigger the air gap thickness, the lower the local ventilation.

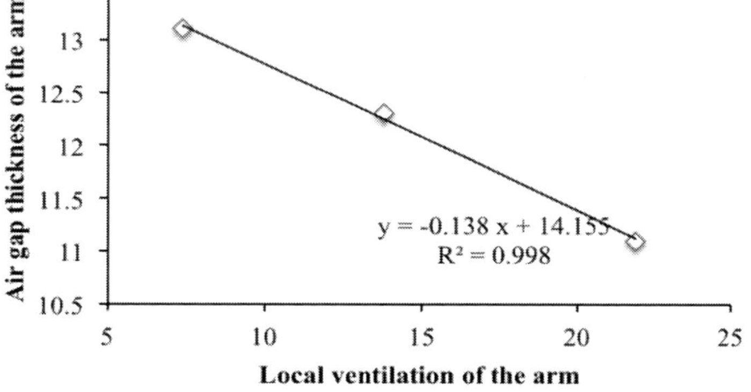

(a)

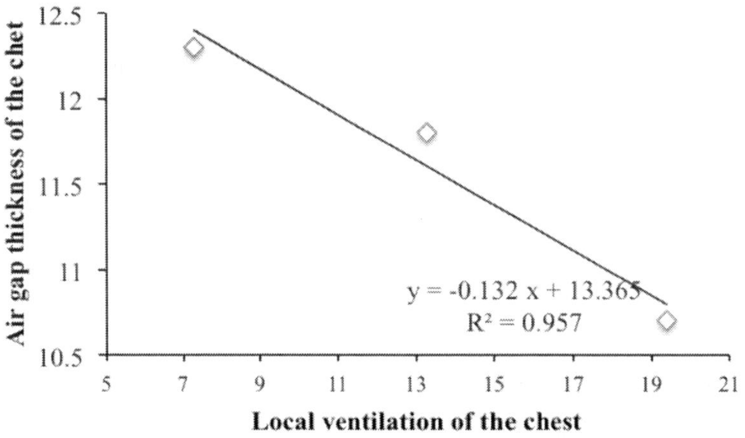

(b)

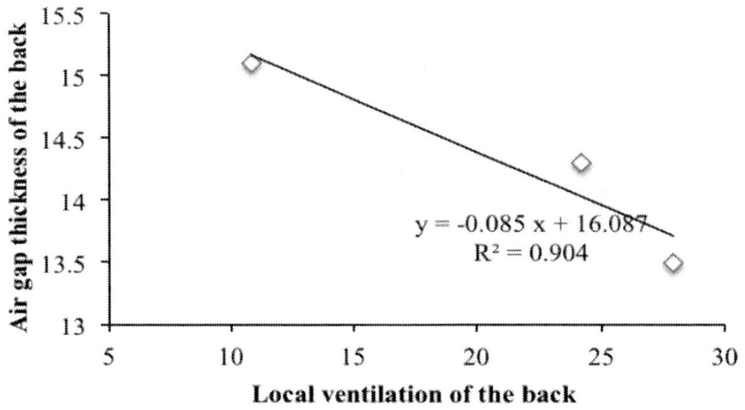

(c)

Fig.5 Relationship between clothing local ventilation and the air gap thickness. (a) arm; (b) chest; (c) back.

4. Conclusion

An experimental study of the relationship between clothing local ventilation and air gap thickness was conducted. For the tested samples, back had the highest local ventilation and chest had the lowest. For garments with the same construction, garment of air permeable fabric had higher local ventilation than that of the impermeable fabric. Back had the biggest local air gap thickness and chest had the smallest. For garment locations, local ventilation had liner negative correlation with air gap thickness. The results presented that air gap thickness also affected local microclimate air exchange directly. Location differences should be considered when conducting garment design and evaluation.

5. Acknowledgement

The authors wish to acknowledge the National Natural Science Foundation (51506076), the China Postdoctoral Science Foundation funded project (2016M591763) and the Fundamental Research Funds for the Central Universities (JUSRP51735B).

References

[1] Ke Y, Li J, Havenith G. Local ventilation and wear response of working jackets with different fabric permeability. Int J Cloth Sci Tech: 2015; 27: 835-851.

[2] Ke Y, Havenith G, Zhang XH, Li XH, Li J. Effects of wind and clothing apertures on local clothing ventilation rates and thermal insulation, Text Res J: 2014; 84: 941-952.

[3] Ke Y, Li J, Havenith G. An improved experimental method for local clothing ventilation measurement, Int J Ind Ergonom: 2014; 44: 75-81.

[4] Ke Y, Havenith G, Li J, Li XH. A New Experimental Study of Influence of Fabric Permeability, Clothing Sizes, Openings and Wind on Regional Ventilation Rates, Fibers Polym: 2013; 14: 1906-1911.

[5] Satsumoto Y, Havenith G. Evaluation of overall and local ventilation in diapers. Text Res J: 2010; 80: 1859-1871.

[6] Ueda H, Inoue Y, Matsudaira M, et al. Regional micro- climate humidity of clothing during light work as a result of the interaction between local sweat production andventilation. Int J Cloth Sci Tech: 2006; 18: 225234.

[7] Bouskill L, Havenith G, Kuklane K, et al. Relationship between clothing ventilation and thermal insulation. AIHA J: 2002; 63: 262-268.

[8] Lu YH, Song GW, Li J. A novel approach for fit analysis of thermal protective clothing using three-dimensional body scanning. Appl. Ergon: 2014; 45: 1439-1446.

Thermal Protective Performance of Protective Fabrics on Steam Hazard

Yun Su[1,2], Rui Li[2], Guo-Wen Song[2*], Jun Li[1,3]

[1] *College of Fashion and Design, Donghua University, Shanghai 200051, China*
[2] *Iowa State University, Ames 50010, Iowa, USA*
[3] *Key Laboratory of Clothing Design and Technology, Donghua University, Ministry of Education, Shanghai 200051, China*

*Corresponding author's email: gwsong@iastate.edu

Abstract

An experimental investigation was carried out in order to examine thermal protective performance of fabric system when exposed to steam. The skin bio-heat transfer model and Henriques burn integral model were employed to predict the required times to reach 2nd and 3rd degree skin burn. For studying the effect of moisture barrier on thermal protective performance, the moisture barrier with different orientations was inserted in fabric system. The results obtained from this research demonstrated that the thickness, mass and moisture regain of fabric could affect protective levels provided by fabric system, while air permeability was the dominant influential factor of thermal protective performance. Even though the moisture barrier provided excellent protective performance for steam exposure, the orientation of moisture barrier presented a decisive influence on the role of moisture barrier. The findings obtained in this study can be used to design the reasonable collocation of multilayer fabric system that provides better protection for hot steam exposure.

Keywords: Thermal Protective Performance; Hot Steam; Protective Fabric; Skin Burn

1. Introduction

Thermal protective clothing made of functional textile materials is an effective equipment to protect the body from thermal hazards, such as flash fire, high-intensity thermal radiation, hot liquid, hot gases and pressurized steam [1-4]. As stipulated in standard NFPA 1971 and EN 469, a typical thermal protective clothing can be a one-layer coverall or composed of three different fabric layers, i.e., an outer shell, a moisture barrier, and a thermal liner. The role of outer shell is to provide flame resistance, protection against heat radiation, resistance to water, and certain levels of abrasion. As a middle layer, the moisture barrier is designed to not allow liquid water to penetrate from outside but promote water vapor transport outward from the wearer [5, 6]. In particular, the thermal liner acts as an insulating medium against heat conduction and heat flow.

Different kinds of thermal protective clothing can be used to protect workers from steam leakage that is a severe hazard for workers in oil sector and other heavy industries [7]. Hot steam has a considerable capacity of storing thermal energy and rapidly penetrates through protective clothing system, thus producing steam burns or scalds and even fatalities [8]. Therefore, some preliminary explorations were carried out to characterize the performance of current thermal protective clothing exposed to a pressurized steam. For simulating hot steam in fire ground, some researchers developed the bench scale tests [9-11] as well as the full-scale thermal manikin tests [7, 10] to measure the protective capacities of fabrics and clothing under hot steam conditions. The horizontal bench top tests developed by Liu et al. [9] and Murtaza [11] were used to investigate the performance of fabrics under hot steam exposures. Liu et al. selected different steam pressures (50.6kPa and 152kPa) to expose the specimen for a fixed period of 20s while Murtaza evaluated the performance of fabrics under steam pressure of 210 kPa at 150 °C for 10s exposure. The main difference between two test apparatuses is the steam flow directions: upward and downward, respectively. Because different steam flow directions can affect the rate of heat and moisture transfer in fabric system, horizontal steam flow more conforms to the actual fire ground according to the body standing state. Therefore, a vertical test device was developed by Derscuell and Schimid [10]. The device can change the steam splash distance and steam pressure to simulate different experimental conditions, but the major limitation of

Derscuell and Schimid's work is that the heat flux sensors have relatively large error for higher skin temperature. Sati et al. [7] established the test device of a cylindrical shape to measure heat transfer through a fabric while exposing it to moderately high-pressure steam (69 and 207kPa). The fabric could be mounted with or without a space to provide an air gap between the cylinder and the fabric that is closer to the practical wearing state. Derscuell and Schimid [10] employed a thermal manikin and a steam climatic chamber to evaluate the performance of the whole clothing. The results demonstrated that steam penetration and heat transfer in protective clothing depends on fabric properties, such as resistance to water vapor diffusion, air permeability, thermal insulation and total heat loss, as well as fabric coating/laminate.

Additionally, Su and Li [12] evaluated the combined effect of steam and radiant heat exposure on thermal protective performance of protective clothing, indicating that the fabric assemblies entrapping moisture barrier can better resist the penetration of steam through the fabric system [13]. However, few studies have been done on investigating the effect of outer shell and moisture barrier on thermal protective performance of protective clothing under a pressurized steam. Therefore, the aim of this paper was to examine the thermal protective performance of outer shell and moisture barrier used for steamy condition. The times to cause skin burn and heat transfer in various fabrics were analyzed in order to provide proper suggestions to improve the thermal protective performance of clothing in a steamy condition.

2. Experimental Part

2.1 Fabric Materials

Four types of fabrics currently used in thermal protective clothing were selected as samples. Two kinds of composite fabric with different air permeability were used for outer shell. Two kinds of waterproof and breathable fabric were selected for moisture barrier in this study. The basic specifications of testing samples were listed in Table 1. The thickness of test specimens was measured in accordance with standard ASTM D 1777-96. The fabric's air permeability was tested in a pressure drop of 2000 Pa according to ASTM D 737. The oven can be used to measure moisture regain of all samples in a constant atmosphere (20 ℃ temperature, 65% relative humidity). Single layer fabrics can be assembled into double layer fabric system. Different faces of moisture barrier can be inserted into double layer fabric system. For example, the OS1+MB1-frontside fabric system means that outer shell is exposed to steamy condition and the substrate face of moisture barrier is contacted to skin-simulant sensor, while the membrane face of moisture barrier is directed to skin-simulant sensor for the OS1+MB1-backside fabric system.

Table.1 Basic physical properties of protective fabric

Fabric code	Fiber content	Fabric structure	Thickness (mm)	Mass (g/m^2)	Air permeability ($cm^3/s/cm^2$)	Moisture regain
OS1	Nomex/Kevlar/P-140	Twill	0.66	260.07	262.20	2.98%
OS2	Nomex/Kevlar	Plain	0.54	248.03	125.20	1.65%
MB1	PTFE+PBI and Kevlar mixture	Twill	0.27	187.48	0.48	0.95%
MB2	PTFE+Polyester	Plain	0.18	123.13	0.33	0.07%

2.2 Testing Apparatus

In order to evaluate the thermal protective performance of protective fabric system, an improved bench-scale test apparatus was developed by Labs for Functional Textiles and Protective Clothing (Iowa State University, USA) [14], as shown in Fig.1. This apparatus can control the required steam exposure conditions, such as steam pressure and steam flow. It includes a steam generator, a delivery spout, heat exposure cabinet, specimen fixed component and data acquisition system. The testing specimen was exposed to a saturated steam at a set pressure produced by a small 3 kW boiler with an added super heater. The steam temperature ranged from 100 to 150℃ by controlling the electrically heated super heater. The pressurized steam was introduced into a steam box by a nozzle having an inner diameter of 4.6 mm. The T-type thermocouple with a diameter of 0.254 mm (OMEGA: TC-GG-T-30) was fixed near the steam nozzle to monitor the steam temperature. The air gap sizes between the fabric and skin surface can be adjusted by using spacer blocks with different thicknesses. The skin-simulant sensor that has the similar thermal properties with human skin was used to measure the temperature of skin surface [15].

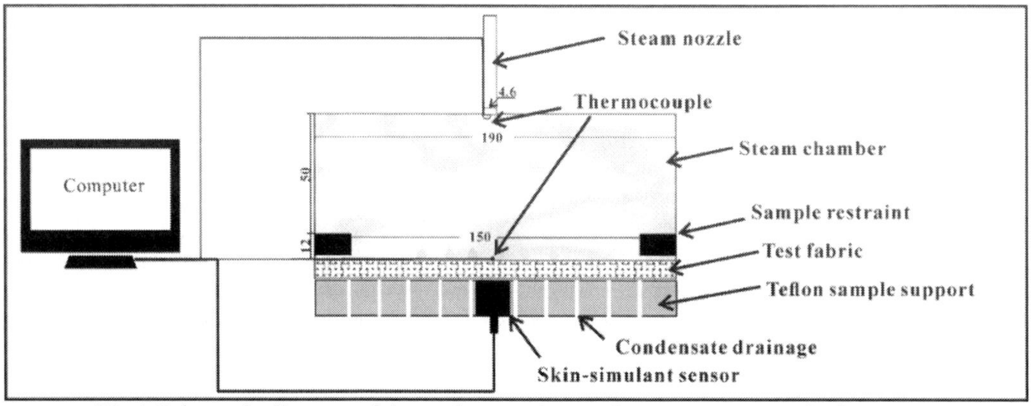

Fig.1 Schematic diagram of thermal protective performance tester under hot steam

2.3 Testing Method

Before mounting the testing sample, it is essential to calibrate the exposure conditions. The steam pressure was set at 40 psi. These specimens were conditioned in a standard atmosphere (20℃ and 65% RH) for at least 24 hours before the experiments. The fabric specimens were exposed to the preset steamy condition for 20 s. After the exposure, these sensors continue to collect the thermal data for the next 40 s as the cool down period. The thermal response of skin surface could be employed to calculate times to 2nd and 3rd degree burn based upon Pennes bio-heat transfer model [16] and Henriques burn integral model [17]. Each specimen was tested three times and the average was obtained.

3. Results and Discussion

3.1 Thermal Protective Performance of Fabrics

The predicted times to 2nd and 3rd degree skin burn under a pressurized steamy conditions are described in Table 2, and the standard errors of three tests for each fabric system are given. The protective levels provided by the selected fabric system ranged from 0.19 s to 7.43 s in steam heat exposure of 20 s. Most skin burns were caused by thermal energy from steamy condition, not by thermal energy stored in protective fabric system, except for time to 3rd degree burn with OS2+MB1-frontside fabric system. With regard to single-layer fabric, the lowest protective level was observed for the OS1 fabric with the highest air permeability. The protective levels of moisture barrier were obviously greater than that of outer shell. The maximum differences between outer shell and moisture barrier were around 31.86 times and 3.66 times, respectively for 2nd and 3rd degree skin burn. It was said that moisture barrier provided a resistance to liquid water, but allowed the penetration of water vapor due to the pore size of moisture barrier [18]. For two kinds of moisture barrier, the time to skin burn when MB1 fabric was exposed to steamy condition was larger than that of MB2 exposed to steamy condition. This reason might be that the fabric's thickness and mass affected the protective level of fabric system. In addition, the MB2 fabric with higher moisture regain could absorb more steam penetrating through membrane, thus increasing the amount of stored thermal energy in the fabric. Therefore, despite the fact that the thickness and mass of protective fabric system could influence the protective performance of clothing, the air permeability of protective fabric played a dominant role in protective level under steam heat exposure.

Comparing to single-layer fabric, the protective performance provided by double layer fabric system did not present a significant increase. In contrast, some double layer fabric systems had a worse protective level under steamy condition. When the substrate face (backside) of moisture barrier was directed to outer shell, the times to 2nd and 3rd degree skin burn were far lesser than that of the substrate face of moisture barrier contacted with sensor, including single and double layer fabric. It was indicated that the position of membrane and substrate faces of moisture barrier exerted an important impact on protective level afforded by moisture barrier. The resistance to steam transfer in moisture barrier was mainly provided by the membrane made of PTFE, while the substrate of moisture barrier could accumulate some condense water. If the membrane was connected with the skin-simulant sensor, the hot steam penetrated firstly through the substrate and following through the membrane. This could result in the production of more condensed water on the skin-simulant sensor and more severe skin burn. The substrate of moisture barrier did not absorb steam through membrane but accumulated a larger amount of thermal energy. In addition, the OS2 fabric with better protective level provided the greater protective performance for double-layer fabric system when OS1+MB1-frontside and OS2+MB1-frontside fabric systems were compared.

Table.2 Times to cause 2nd and 3rd degree skin burn with various protective fabrics

Fabric code	Time to 2nd degree skin burn (s) (SD)	Time to 3rd degree skin burn (s) (SD)
OS1	0.19(0.02)	5.10(0.20)
OS2	0.49(0.03)	5.84(0.09)
MB1	5.95(0.24)	18.67(0.56)
MB2	3.50(0.07)	11.68(0.11)
OS1+MB1-frontside	6.29(0.50)	19.61(1.67)
OS1+MB1-backside	0.94(0.06)	7.58(0.24)
OS2+MB1-frontside	7.43(0.48)	21.26(2.17)
OS2+MB1-backside	0.79(0.07)	7.58(0.34)

3.2 Effect of Moisture Barrier on Heat Transfer

In order to further analyze the thermal protective performance of fabric systems, the variation in the temperature of skin surface over time for single- and double-layer fabric systems were respectively illustrated in Figs.2 and 3. It was clear that the maximum temperatures for two kinds of outer shell during the exposure were respectively 103.39 ℃ and 101.13℃, which were both close to the temperature of steam flowed from steam nozzle. This indicated that hot steam was easy to penetrate through outer shell. There was a jolt in the thermal energy absorbed by the outer shell during steam transmission. With the increase of exposure time, fabrics OS1 and OS2 took 3.9 s and 3.4 s to reach the peak temperature, respectively. Thus, the dominant heat transfer mode in outer shell was steam heat transfer that could rapidly increase the skin temperature. However, there was a decreasing trend in the temperature of skin surface for outer shell after the peak temperature. The condensation of steam occurred on the skin surface during the exposure. The accumulated liquid water could evaporate to loss a portion of thermal energy, which resulted in the decrease of skin temperature.

For the moisture barrier, it was found that skin temperature witnessed a continuous increase during steam exposure. The peak skin temperature was when the steam exposure was ended. There had no sharp change of skin temperature for moisture barrier. The increasing rate of temperature decreased gradually over the exposure time. It could be inferred that conductive heat transfer played an important role in heat transportation in moisture barrier. Due to the resistance to steam penetration, the thermal energy carried by steam was firstly transmitted to moisture barrier and then transferred to skin tissue in conductive heat transfer. The temperature gradient between outside and inside of moisture barrier was driving force to affect heat transfer rate. Thus, steam heat transfer has a higher heat transfer rate than conductive heat transfer.

After the exposure, the temperature drop on the skin surface was determined by the ambient temperature and the amount of condensed water. The decrease of ambient temperature quickened up the conductive heat transfer due to the increased temperature difference between the outer and inner of fabric system. For the decreasing rate of skin temperature, the difference between two kinds of moisture barrier was caused by the thickness of moisture barrier that affected conductive heat transfer. Also, the liquid water on the skin surface could be evaporated to absorbed thermal energy from skin tissue. It was for these reasons that the outer shell contained more liquid water presented a higher decreasing rate comparing with the moisture barrier. In addition, the heat transfer after the exposure was dependent on the stored thermal energy within fabric system. The discharge of stored thermal energy could exaggerate the skin burn as well [19].

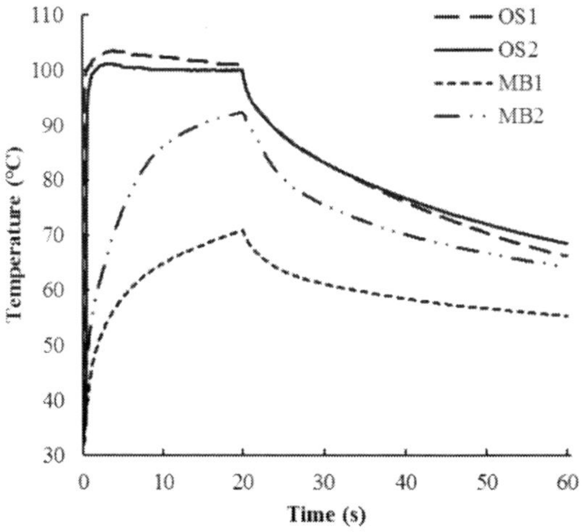

Fig.2 Changes in the skin temperature for single-layer fabrics versus time

The changes in the skin temperature for double-layer fabric system were described in Fig.3. The change trend in all temperature curves showed a similar with MB1 and MB2 (see Fig.2). According to the above conclusion, the steam penetration though double-layer fabric system was mostly resisted due to the existence of moisture barrier. However, there was still significant difference when the moisture barrier was placed in different orientations. With regard to OS+MB1-frontside fabric system, hot steam could penetrate through outer shell. A large amount of steam was resisted in the outer shell due to the waterproof membrane. Another small portion of steam transferred through membrane could be absorbed by the substrate of moisture barrier. Therefore, steam heat transfer just existed in the outer shell, while conductive heat transfer occurred in the membrane and substrate of moisture barrier. For OS+MB1-backside fabric system, hot steam could penetrate through outer shell and the substrate of moisture barrier so that hot steam was condensed within outer shell and the substrate. Since the membrane of moisture barrier was near to skin surface, a small portion of steam could be condensed on the skin surface. In addition, the membrane of moisture barrier was a thin PTFE membrane so that the released thermal energy from hot steam in the outer shell and the substrate was quickly transmitted to the skin surface in conductive way.

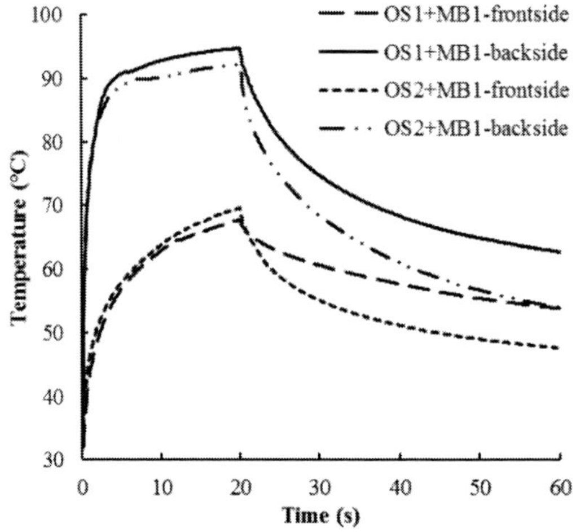

Fig.3 Changes in the skin temperature for double-layer fabrics versus time

When moisture barrier was entrapped between the outer shell and skin-simulant sensor in the same orientation, the skin temperature for double-layer fabric system experienced a similar change trend over the exposure time. The marginal difference was determined by the basic properties of outer shell. For example, the OS2 fabric with lesser air permeability reduced the amount of steam penetration while the OS1 fabric with better moisture regain provided the larger capacity to storing steam condensation. After the exposure, it was obvious that the double-layer fabric system containing OS2 presented the greater decreasing rate. The OS1 fabric could store more thermal energy and condensed water, thus increasing the heat transfer rate due to conduction from the fabric system to skin tissue.

4. Conclusions

In this study, the thermal protective performance of single- and double-layer fabrics was examined under a pressurized steam exposure. The results showed that moisture barrier played an important role in thermal protective performance provided by fabric system. The maximum difference between outer shell and moisture barrier was around 31.86 times and 3.66 times, respectively for 2nd and 3rd degree skin burn. Comparing with thickness, mass and moisture regain of fabric, the air permeability was more important influencing factor. This was because steam heat transfer that existed in outer shell or substrate of moisture barrier could be a quicker heat transfer mode than conductive heat transfer.

The position of membrane and substrate faces of moisture barrier presented an important impact on protective level afforded by moisture barrier. With regard to OS+MB1-frontside fabric system, hot steam could penetrate through outer shell while a large amount of steam was resisted in the outer shell due to the waterproof membrane. For OS+MB1-backside fabric system, hot steam could penetrate through outer shell and the substrate of moisture barrier so that hot steam was condensed within outer shell and the substrate. Decreasing the amount of steam entering into fabric system including the steam condensation and penetration reduced heat transfer rate during the exposure and stored thermal energy that could affect the cooling effect after the exposure. Therefore, moisture barrier is critical in determining thermal protective performance of fabric system under hot steam, especially for the orientation of moisture barrier. Further study should be conducted to examine the relationship between protective level provide by moisture barrier and transport properties in hot steam and even hot liquid.

5. Acknowledgement

The authors would like to acknowledge the financial support from the National Nature Science Foundation (Grant NO. 51576038), Donghua University PhD Thesis Innovation Funding (NO.16D310701), and the Open Funding Project of National Key Laboratory of Human Factors Engineering (Grant NO. SYFD150051812K)

References

[1] Barr D, Gregson W, Reilly T. The thermal ergonomics of firefighting reviewed. Applied ergonomics: 2010; 41: 161-72.
[2] Mäkinen H, Smolander J, Vuorinen H. Simulation of the effect of moisture content in underwear and on the skin surface on steam burns of fire fighters. Performance of Protective Clothing: Second Symposium, ASTM STP: 1988; p. 415-21.
[3] Barker RL, Stamper SK, Shalev I. Measuring the protective insulation of fabrics in hot surface contact. Performance of protective clothing: second symposium, ASTM STP: 1988; p. 87-100.
[4] Song G, Lu Y, Gholamreza F. Clothing for protection against hot-liquid splash and steam hazards. Protective Clothing: Managing Thermal Stress: 2014; p. 90-111.
[5] Aldridge D. Firefighter garment with combination facecloth and moisture barrier. US Patent No. 5640718: 1997.
[6] Reischl U, Stransky A. Comparative assessment of Goretex™ and Neoprene™ vapor barriers in a firefighter turn-out coat. Textile Research Journal: 1980; 50: 643-7.
[7] Sati R, Crown EM, Ackerman M, et al. Protection from steam at high pressures: development of a test device and protocol. Int J Occup Saf Ergo: 2008; 14: 29-41.
[8] Moritz AR, Henriques Jr FC, McLean R. The effects of inhaled heat on the air passages and lungs: an experimental investigation. The American journal of pathology: 1945; 21: 311.

[9] Liu H, Cui Z, Zhang W. Investigation of protective performance of heat resistant fabrics under external hot steam exposure. Proceedings of the third international conference on modelling and simulation: 4th Volume Modelling and simulation in biology, ecology & environment: 2010; p. 435-8.

[10] Desruelle AV, Schmid B. The steam laboratory of the institut de medecine navale du service de sante des armees: a set of tools in the service of the french navy. European journal of applied physiology: 2004; 92: 630-5.

[11] Murtaza G. Development of fabrics for steam and hot water protection. University of Alberta: 2012.

[12] Su Y, Li J. Development of a test device to characterize thermal protective performance of fabrics against hot steam and thermal radiation. Measurement Science and Technology: 2016; 27: 1-9.

[13] Su Y, Li J. Analyzing steam transfer though various flame-retardant fabric assemblies in radiant heat exposure. Journal of Industrial Textiles: 2016: 1528083716674907.

[14] Ackerman M, Crown E, Dale J, et al. Development of a test apparatus/method and material specifications for protection from steam under pressure. Performance of Protective Clothing and Equipment: 9 th Volume, Emerging Issues and Technologies. ASTM International: 2012.

[15] Mandal S, Song G. Thermal sensors for performance evaluation of protective clothing against heat and fire: a review. Textile Research Journal: 2015; 85: 101-12.

[16] Pennes HH. Analysis of tissue and arterial blood temperatures in the resting human forearm. Journal of applied physiology: 1948; 1: 93-122.

[17] Henriques FC. Jr. Studies of thermal injury; the predictability and the significance of thermally induced rate processes leading to irreversible epidermal injury. Archives of pathology: 1947; 43: 489-502.

[18] Gohlke DJ, Tanner JC. Gore-Tex® Waterproof breathable laminates. Journal of Industrial Textiles: 1976; 6: 28-38.

[19] Song G, Gholamreza F, Cao W. Analyzing thermal stored energy and effect on protective performance. Textile Research Journal: 2011; 81: 1124-38.

The Thermal Protective Performance of Wetted Fabrics-A Review

Meng Chen, Fang-Long Zhu*, Qian-Qian Feng, Qiu-Yue Yang

College of Clothing, Zhongyuan University of Technology, 41 Zhongyuan Middle Road, Zhengzhou, 450007, PR.China

*Corresponding author's email: zhufanglong_168@163.com

Abstract

Moisture accumulation in firefighter protective clothing has complex effects on its thermal protective performance. The moisture transfer through wetted protective clothing may increase or decrease the heat transfer depending on various factors like the type, intensity and duration of exposure, the amount and location of moisture as well as the type of materials, etc. Significant amounts of research have been conducted on the effects of moisture on the heat protection of thermal protective clothing. This paper presented a brief review of the research and development in the field of thermal protective performance of wetted fabrics. The main factors affecting thermal protective performance were discussed and the models coupled with heat and moisture were highlighted. It was found that there is still a lack of systematic understanding of moisture on the thermal protective performance of fabrics for heat exposures of different types and intensity. Further systematic research adopting new methods and facilities is required to understand the influence parameters of the thermal protective performance for wetted fabrics.

Keywords: Thermal Protective Performance; Moisture Content; Thermal Exposure; Moisture Location; Heat and Moisture Transfer Model

1. Introduction

Firefighters are surrounded by a hot but also moist environment during firefighting. They can produce 1200 to 1800 g/h of sweat during an assignment causing moisture to accumulate in their turnout garments [1]. Furthermore, the extinguishing water may also be absorbed by the textile layers and it may cause a high-water vapor pressure in the environment.

The absorbed moisture present in the fabric have two forms: bound water and free water. The moisture absorbed by solid fibers is known as bound water. The remaining water would then occupy the void space between fibers after reaching saturation state, which is known as free water. Heat transfer through fabrics in dry state involves heat transfer through solid fibers and air present in the voids. Similarly, in the presence of moisture heat transfer through fabrics involves heat transfer through solid fibers and water [2]. It is a well-known fact that the moisture present in the garments can alter the thermal properties of the protective clothing system such as thermal conductivity and heat capacity. Besides, the water vapor at the fire ground caused by the water extinguisher would be heated by the thermal radiation. Then the hot steam from outside or being evaporated from the wetted textile layers can move towards the body and lead to steam burns. Therefore, the moisture present in fabrics have a complex influence on the thermal protective performance. The study on the influence of humidity on heat protection in heat protective clothing assemblies were already conducted by some researchers but they obtained contradictory results due to different conditions. This research aims to discuss the main factors affecting the thermal protective performance and highlight the model that couples heat and moisture transfer.

2. Influence Factors

The thermal protective performance of moist fabrics is affected by a variety of different factors such as the type of heat transfer (i.e. convection, radiation and conduction), the duration and intensity of thermal exposure, the moisture content and its source, the location in clothing system, the type and conditioning of the materials [3-5].

2.1 Effects of Moisture Content

The different amount of moisture may have different degrees of impact since the thermal conductivity and heat capacity of fabrics can be altered by moisture. Barker et al. [6] found in their study that at low-level thermal radiation of 6.3 kW/m^2, the addition of moisture negatively impacted the predicted burn protection to the greatest degree when the moisture was added at a comparatively low level of approximately 15% of turnout composite weight. Song et al. [7] studied the effects of moisture on the thermal protective performance for higher radiant heat (21 kW/m^2) and convection/radiant heat (42 kW/m^2 and 84 kW/m^2) exposures. The thermal protective performance was found to be lower for wet fabrics with 15% add-on as compared to dry fabrics. Schacher CCG [8] also analyzed the effects of moisture on the thermal protective performance of multi-layer clothing assembly in low level (6.3 kW/m^2 to 21 kW/m^2) radiant heat exposures. It was observed that the thermal protective performance of moist clothing decreased, in which lower amount (12-50%) of water was added. However, the thermal protective performance of moist clothing increased when increasing the moisture content sequentially (near saturation state) and may become comparable to that under dry condition. This suggested that there is not a general linear correlation between the thermal protection performance and moisture content. Thus further systematic research is required to understand the relation.

2.2 Effects of Thermal Exposure Intensity

Heat transfer through fabrics is affected by the type and intensity of heat exposure [9]. The effects of intensity on the heat transfer of wetted fabrics is more complex because that the water may change phase by absorbing heat and the hot steam may cause skin burns. Lee and Barker [10] analyzed the effects of moisture on the woven, knitted and nonwoven fabrics subjected to heat exposure. They found that the moisture present in the fabric increased or decreased the heat transfer through the fabric depending upon the intensity of the thermal radiation. The moisture improved the thermal protection at 20 kW/m^2, while it negatively impacted the thermal protection at 84 kW/m^2. A similar result was obtained by Zhu and Li [11]. They found that the moisture increased second-degree burns time in a low intensity heat exposure of 21 kW/m^2, but it tended to increase heat transfer through the thermal protective fabric system under high heat flux exposures, such as 42 kW/m^2, and the tolerance time of the same fabrics would reduce. Fu et al. [12] analyzed the combined effects of moisture and radiation on the thermal protective performance of clothing by using a sweating thermal manikin. A typical thermal protective clothing was exposed to low level (2 kW/m^2, 2.5 kW/m^2 and 3 kW/m^2) radiant heat fluxes under different sweating rates (100, 200 and 300 g/m^2h).They observed that the wet skin temperature increased and the heat loss from the manikin surface decreased as intensity of radiant heat exposure increased. Therefore, increasing the intensity of heat exposure has a negative influence on the thermal protective performance of wetted fabrics, which is identical to the situation of dry fabrics.

2.3 Effects of Fabrics' Type

Several researchers have analyzed the thermal protective performance of protective fabrics for different types. Hirschler MM. [13] conducted experiments on two fabrics applied in protective clothing: an aramid and a modified viscose cellulosic fabric. Their thermal protective performance was assessed by the temperature transmitted through the fabric when exposed to high heat, both in their dry state and moist state. They found that the thermal insulation properties of modified viscose fabric improved in moist state, while the aramid fabric remained unaffected (or perhaps was negatively affected) by the water spray. Fu et al. [14] conducted bench scale thermal radiation tests to investigate the effects of low thermal radiation on the heat transfer of multilayer thermal protective clothing. They found in their study that the thermal protective performance of the inner layer could be improved with the moisture transfer, while the outer layers, especially the moisture barrier, will be heated by the external radiation and the released heat by the moisture transfer. Hence, it could be found that the fabrics' type plays an important role in thermal resistance for wetted fabrics.

The type of fabric may include these factors: thickness, structure and weave, thermo-physical properties, porosity, etc. In general, as fabric thickness increases, its thermal resistance increases and therefore the second degree burn time and TPP increases [2]. Structure of knitted fabrics and weave type of woven fabrics play crucial role in the thermal resistance. Majumdar et al. [15] compared three fabric structures and found that the thermal resistances are least and air permeability are highest for plain knitted fabrics followed by rib and interlock knitted fabrics. The thermo-physical properties include thermal conductivity, surface reflectivity and heat capacity, which all have important influences on the thermal protective performance of fabrics. Porosity is also one aspect of the thermo-physical porosity. It is affected by the structure and weave of the fabrics and also affect the air permeability and therefore the thermal resistance.

2.4 Effects of Moisture Location

Lawson et al. [3] studied effects of both internal and external moisture on a two-layer fabric system exposed to a high-heat-flux flame exposure (83kw/m^2) and a low-heat-flux radiant exposure (10kw/m^2) respectively. They found that not only the intensity and type of exposure but the location of moisture affects the protective performance of fabric system. Corinne Keiser et al. [16] quantified moisture evaporation and moisture transfer in a multilayer firefighter protective clothing system with defined wetted layers exposed to low thermal radiation by using X-ray radiography. It was observed that the evaporation rate per surface area was independent of the initial amount of moisture supplied to the clothing assemble, but depended on the location of the moisture. Moisture evaporated faster from the outer layers of the assembly than from the inner ones. Thus an additional (moisture-assisted) heat flux takes place by evaporation of moisture in the outer layers and recondensation near the body.

3. Test Apparatus

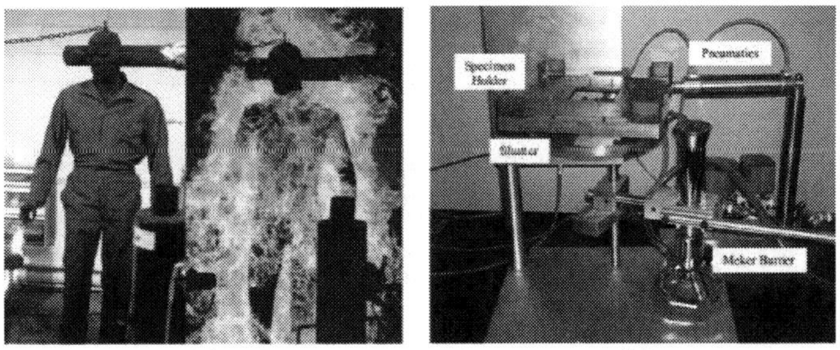

(a) (b)

Fig.1 (a) the full scale manikin testing and (b) the bench scale testing (b) [2].

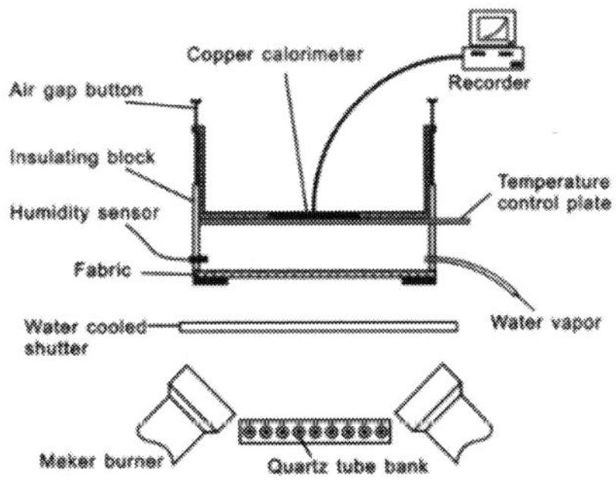

Fig.2 the new testing apparatus was CSI-206 (Custom Scientific Instrument Corporation, USA) with addition of microclimate regulation chamber [19].

The performance of thermal protective clothing can be evaluated by two ways basically: Full scale manikin testing and bench scale testing, which were shown in Fig.1. Various full scale manikins have been developed to analyze the thermal protective performance of firefighters' garments over the past few decades. The manikin was exposed to averaged heat flux of 84 kW/m^2 provided by eight propane burners for up to 20 s to simulate actual conditions encountered by firefighters [17]. Then second degree burns were predicted based on the data obtained by heat flux sensors placed on the surface of the manikins. In the recent years, many such manikins have been developed including some sweating manikins. Bench scale, developed by Behnke [18], provides a relatively easy way to analyze the performance of the thermal protective fabric samples. The point of intersection of temperature rise curve of the sensor and Stoll curve gives the tolerance time of human skin. The manikin test is more applicable than manikin test because the manikin test is very complicated, costly, time consuming and requires years to build the facility. However, the evaluation on the performance of wetted fabric involves the effects of various microclimate relative humidity for different air gaps, which the current bench top tests can't evaluate well. Thus some modified approaches have been developed. For example, Li et al. [19] developed a new modified TPP testing apparatus (shown schematically in Fig.2), which was consisted of a typical TPP tester and RH adjustable microclimate chamber. And they verified the newly developed testing apparatus could be well used to evaluate the effects of moisture on the thermal protective performance of flame-resistant fabrics.

4. Models

Numerical modeling of the heat and mass transfer through thermal protective clothing can be useful in understanding the complex heat and mass transfer phenomenon. Textile fabrics are composed of fiber or yarns woven in a particular manner (plain, twill or satin) with void spaces in between and it can be treated as a kind of porous material [2]. Hence moisture can be absorbed into the void space in the form of liquid or water vapor. The numerical model that couples heat and moisture transport have been studied extensively due to the developed research on moisture transport in fabric and its effects on protective performance of the garment. An easier study that consider both heat and moisture transmission simultaneously into account was by Nordon and David [20], which was shown in Eq.1.

$$\rho C_P(C_f)\frac{\partial T}{\partial t} = k\frac{\partial^2 T}{\partial x^2} + h_{sd}(C_f)\frac{\partial C_f}{\partial t} \tag{1}$$

Where ρC_P and h_{sd} are volumetric heat capacity and heat of sorption/desorption of water vapor by solid phase, respectively. Both of the two parameters were considered to be the function of water vapor concentration in fibers (C_f). Concentration of water vapor in fiber, C_f was determined using law of conservation of mass, which was shown in Eq.2.

$$D\frac{\partial C_a}{\partial x^2} = \frac{\partial C_f}{\partial t} + \varepsilon\frac{\partial C_a}{\partial t} \tag{2}$$

Where D is diffusion coefficient, C_a is concentration of water vapor in air and ε is porosity of fabric.

Whereafter, more general models coupled heat and moisture transfer in the porous media were developed by several researchers [21-24]. But some important phenomena were neglected in their models like absorption of water and mass transfer through solid, evaporation of bound water in the solid, the effects of wicking, condensation etc. These phenomena may play significant roles on the moisture transmission. Later, coupled heat and moisture transfer through fabrics considering various phenomenon like condensation [25, 26], wicking [27], absorption [28,29], phase change [30,31] and radiation attenuation[32-34] were modeled based on the theory of porous medium, conservation of mass and energy [2]. However, these models were seldom applied to fabrics exposed to flash fire. Prasad et al. [35] firstly developed a one-dimensional heat and mass transfer model through multi-layered firefighter's clothing under flashover conditions. Conservation of energy over a small control volume of homogeneous fabric is given as Eq.3.

$$\rho_T C_T \frac{\partial T}{\partial t} = \frac{\partial}{\partial x}\left(k_T \frac{\partial T}{\partial x}\right) - \rho_c \Delta h \frac{\partial M}{\partial t} - \frac{\partial q_R}{\partial x} \quad (3)$$

Where ρ_T, C_T and k_T are total density, total heat capacity and total thermal conductivity, respectively. All of these were considered to be function of the moisture. Δh is heat of absorption per unit mass, ρ_c is dry density of fabric and M is local moisture content. q_R is radiative flux obtained by solving the radiative transfer equation (RTE) using two-flux method. Above equation was solved simultaneously with equation of conservation of mass shown in Eq.4.

$$\frac{\rho_c}{\varepsilon} \frac{\partial M}{\partial t} = \frac{\partial}{\partial x}\left(D \frac{\partial p}{\partial x}\right) - \frac{\rho_g W_1}{P W_m} \frac{\partial p}{\partial t} \quad (4)$$

Where p is vapor pressure, ρ_g is gas density, W_1 and W_m are molecular weight of water vapor and mean molecular weight of system, respectively. While the developed model was able to account the effects of evaporation, condensation and sorption in fabric, it has completely ignored the effects of capillary action and attenuation of radiation by the moisture. This model was recently extended to incorporate the absorption of thermal radiation by the moisture [36].

Chitrphiromsri and Kuznetsov [37] also developed a numerical model of heat and moisture transport in firefighter protective clothing during a flash fire exposure. They found that the moisture was transported from the outside to the inside of the fabric and into the air gap due to the temperature gradient. Liquid water and water vapor transfer were considered in the model developed by V.A.F. Costa et al. [38], including the capillary effects for the liquid phase, as well as the air transfer inside the porous medium. Later, Zhu and Li [11] analyzed the heat and moisture transfer through moist cotton fabric. They incorporated the heat-induced changes in fabric thermo physical properties and the drying process was described by a one-step chemical reaction in the model.

In general, the researchers took more factors into account when modeling moisture and heat transfer in recent years, which gave a deeper understanding of the complex phenomenon during heat and moisture transfer to new scholars in this area.

5. Conclusion

This paper elucidated the advancement in the field of studying on the thermal protective performance of wetted fabrics. The important investigations have given people a deeper understanding about the thermal protective performance of wetted fabrics. But there is still a lack of systematic understanding of moisture on thermal protective performance of fabrics for heat exposures of different types and intensity. Besides, the effects of moisture amount on the heat protection of protective clothing system hasn't been fully comprehended. The researchers need to deeply explore the thermal protective performance of the wetted clothing system by adopting new methods and developing new experimental facilities in the further.

6. Acknowledgement

This study was financially supported by the National Natural Science Foundation of China (51576215), and a joint project between the Henan-Province and the National Natural Science Foundation of China (U1304513).

References

[1] Lawson JR. Fire fighters' protective clothing and thermal environments of structural firefighting, in "Performance of Protective Clothing," J. O. Stull and A. D. Schwope (Ed.), ASTM STP, American Society for Testing and Materials, West Conshohocken, PA, USA: 1997; 334-352.

[2] Udayraj, Prabal T, Das A, Alagirusamy R. Heat and mass transfer through thermal protective clothing- A review. International Journal of Thermal Sciences: 2016; 106: 32-56.
[3] Lawson LK, Crown EM, Ackerman MY, Dale JD. Moisture effects in heat transfer through clothing systems for wildland firefighters. International Journal of Occupational Safety and Ergonomics: 2004; 10: 227-238.
[4] Keiser C, Rossi RM. Temperature analysis for the prediction of steam formation and transfer in multilayer thermal protective clothing at low level thermal radiation. Textile Research Journal: 2008; 78: 1025-1035.
[5] Stull JO. The effect of moisture on firefighter protective clothing thermal insulation: A Review of Industry Research, in "Performance of Protective Clothing," C. N. Nelson and N.W. Henry (Ed.), ASTM STP, American Society for Testing and Materials, West Conshohocken, PA, USA: 2000; 557-576.
[6] Barker RL, Schacher CG, Grimes RV, Hamouda H. Effects of moisture on the thermal protective performance of firefighter protective clothing in low-level radiant heat exposures. Textile Research Journal: 2006; 76: 27-31.
[7] Song G, Cao W, Gholamreza F. Analyzing stored thermal energy and thermal protective performance of clothing. Textile Research Journal: 2011; 81: 1124-1138.
[8] Schacher CCG. Evaluating the effects of moisture on the thermal protective performance of firefighter protective clothing in low level heat exposures (PhD thesis). Raleigh, North Carolina: Fiber and Polymer Science, North Carolina State University: 2007.
[9] Udayraj P, Talukdar R, Alagirusamy AD. Heat transfer analysis and second degree burn prediction in human skin exposed to flame and radiant heat using dual phase lag phenomenon. International Journal of Heat and Mass Transfer: 2014; 78: 1068-1079.
[10] Lee YM, Barker RL. Effect of moisture on the thermal protective performance of heat-resistant fabrics. Journal of Fire Sciences: 1986; 4: 315-331.
[11] Zhu FL, Li KJ. Numerical modeling of heat and moisture through wet cotton fabric using the method of chemical thermodynamic law under simulated fire. Fire Technology: 2011; 47: 801-819.
[12] Fu M, Weng WG, Yuan HY. Combined effects of moisture and radiation on thermal performance of protective clothing: experiments by a sweating manikin exposed to low level radiation. International Journal of Clothing Science Technology: 2015; 27: 412-425.
[13] Hirschler MM. Analysis of thermal performance of two fabrics intended to use as protective clothing. Fire and Materials: 1997; 21: 115-121.
[14] Fu M, Weng WG, Yuan HY. Quantitative assessment of the relationship between radiant heat exposure and the protective performance of multilayer thermal protective clothing during dry and wet conditions. Journal of Hazardous Materials: 2014; 276: 383-392.
[15] Majumdar A, Mukhopadhyay S, Yadav R. Thermal properties of knitted fabrics made from cotton and regenerated bamboo cellulosic fibres. International Journal of Thermal Sciences: 2010; 49: 2042-2048.
[16] Keiser C, Wyss P, Rossi RM. Analysis of steam formation and migration in firefighters' protective clothing using X-Ray radiography. International journal of occupational safety and ergonomics (JOSE): 2010; 16: 217-229.
[17] ASTM F1930-13. Standard test method for evaluation of flame resistant clothing for protection against fire simulations using an instrumented manikin: 2013.
[18] Behnke WP. Thermal protective performance test for clothing. Fire Technology: 1997; 13; 6-12.
[19] Li X, Lu Y, Li J, Wang Y, Zhou L. A new approach to evaluate the effect of moisture on heat transfer of thermal protective clothing under flashover. Fibers and Polymers: 2012; 13: 549-554.
[20] Nordon P, David HG. Coupled diffusion of moisture and heat in hygroscopic textile materials. International Journal of Heat and Mass Transfer: 1967; 10: 853-866.
[21] Whitaker S. Simultaneous heat, mass, and momentum transfer in porous media: a theory of drying. Advances in Heat Transfer: 1977; 13: 119-200. New York: Academic Press.
[22] Sahota MS, Pagni PJ. Heat and mass transfer in porous media subject to fires. International Journal of Heat and Mass Transfer: 1979; 22: 1069-1081.
[23] Farnworth B. A numerical model of the combined diffusion of heat and water vapor through clothing. Textile Research Journal: 1986; 56: 653-665.
[24] Ghali K, Jones B, Tracy J. Modeling moisture transfer in fabrics. Experimental Thermal and Fluid Science: 1994; 9: 330-336.
[25] Murata K. Heat and mass transfer with condensation in a fibrous insulation slab bounded on one side by a cold

surface. International Journal of Heat and Mass Transfer: 1995; 38: 3253-3262.
[26] Li Y, Zhu Q. Simultaneous heat and moisture transfer with moisture sorption, condensation, and capillary liquid diffusion in porous textile. Textile Research Journal: 2003; 73: 515-524.
[27] Ghali K, Jones B, Tracy J. Modeling heat and mass transfer in fabrics. International Journal of Heat and Mass Transfer: 1995; 38: 13-21.
[28] Fohr JP, Couton D, Treguier G. Dynamic heat and water transfer through layered fabrics. Textile Research Journal: 2002; 72: 1-12.
[29] Wu H, Fan J. Study of heat and moisture transfer within multi-layer clothing assemblies consisting of different types of battings. International Journal of Thermal Sciences: 2008; 47: 641-647.
[30] Ye C, Huang H, Fan J, Sun W. Numerical study of heat and moisture transfer in textile materials by a finite volume method. Computer Physics Communications: 2008; 4: 929-948.
[31] Li Y, Zhu Q. A model of heat and moisture transfer in porous textiles with phase change materials. Textile Research Journal: 2004; 74: 447-457.
[32] Fan J, Wen X. Modeling heat and moisture transfer through fibrous insulation with phase change and mobile condensates. International Journal of Heat and Mass Transfer: 2002; 45: 4045-4055.
[33] Fan J, Cheng X, Wen X, Sun W. An improved model of heat and moisture transfer with phase change and mobile condensates in fibrous insulation and comparison with experimental results. International Journal of Heat and Mass Transfer: 2004; 47: 2343-2352.
[34] Cheng X, Fan J. Simulation of heat and moisture transfer with phase change and mobile condensates in fibrous insulation. International Journal of Thermal Sciences: 2004; 43: 665-676.
[35] Prasad K, Twilley W, Lawson JR. Thermal performance of fire fighter's protective clothing. 1. Numerical study of transient heat and water vapor transfer. Technical report NISTIR 6881. Gaithersburg, MD: National Institute of Standards and Technology: 2002.
[36] Fu M, Yuan MQ, Weng WG. Modeling of heat and moisture transfer within firefighter protective clothing with the moisture absorption of thermal radiation. International Journal of Thermal Sciences: 2015; 96: 201-210.
[37] Chitrphiromsri, P, Kuznetsov AV. Modelling heat and moisture transport in firefighter protective clothing during flash fire exposure. International Journal of Heat and Mass Transfer: 2005; 4: 206-215.
[38] Costa VAF, Mendonc ML, Figueiredo AR. Modeling and simulation of wetted porous thermal barriers operating under high temperature or high heat flux. International Journal of Heat and Mass Transfer: 2008; 51: 3342-3354.

Effect of Sportswear Type on Thermoregulation During Exercise

Ze-Jing Qi[1], Guo-Wen Song[2], Xiao-Qun Dai[1,2*]

[1]*Soochow University, College of Textile and Clothing Engineering, Suzhou 215006, China*
[2]*Iowa State University, College of Human Sciences, Ames IA50011, USA*

*Corresponding author's email: daixqsz@suda.edu.cn

Abstract

In this paper, three typical types of sport ensembles representing a difference cover area were used to investigate the effect of garment cover area on human thermoregulation in a mild environment. The thermoregulation of a male body during exercise at two different intensities was simulated by using a sweating thermal manikin. It was found that small cover area helped to maintain low skin temperature. For low intensity exercise with low sweat rate, small cover area helped to maintain a lower core temperature and less sweat amount. For high intensity exercise with high sweat rate, large cover area helped to evaporate more sweat and thus keep a lower core temperature and less sweat amount.

Keywords: Thermoregulation; Core Temperature; Skin Temperate; Sweat Rate; Sportswear

1. Introduction

During exercise, body heat production increases and sweat rate also increases. To maintain a constant body temperature, additional heat needs to be diffused quickly. To keep skin dry and comfortable, sweat needs to be taken away from skin surface and sweat evaporation is a primary means of cooling the body during exercise [1]. For such apparent and latent heat transfer, clothing plays a complicated role. On one hand, garments covering the body are big barriers for heat and moisture diffusion. On the other hand, wearing garments might enhance the efficiency of evaporative cooling. Not all the amount of sweat is effective in cooling down, excess sweat dripping from the body has no contribution to cooling [2]. When the sweat rate is very high, garments can absorb the sweat before it flows down the skin. The sweat held in the garment can then evaporate from its outer surface and diffuse to the surrounding environment, thus helping to cool down the body. This implicates that the area of garments covering the body can make a difference in thermoregulation. It is therefore necessary to investigate such effect to understand clothing physiology and provide useful advice for athletes and exercisers in choosing sportswear.

Although the sweat absorbed by the garment can still contribute to evaporative cooling of the body, its cooling efficiency is reduced because of the increased distance between the evaporation locations and the skin [3]. The air layer under garments acts as a big barrier for the heat and moisture diffusion from the body. In recent years, tight-fit sportswear has become very popular among consumers and the manufacturers claim such tight-fit sportswear can assist exercisers in thermoregulation during exercise in hot environment [4]. Gavin and Babington investigated whether thermoregulation is improved during exercise in moderate environment by the use of clothing made of synthetic material compared with cotton using human trials [5]. However, there is few report about the effect of garment cover area.

Due to the thermo-physiological difference among individuals and the variation of an individual's thermo-state across time, it is difficult to investigate the effect of garment covering area by conducting real human wear trials with limited number of subjects. Recent developments on thermoregulatory model controlled sweating manikins have shown that they have the potential to replace human subjects to simulate human physiological responses over a range of test scenarios [6, 7]. Instead of real human wear trials, a sweating thermal manikin was used to simulate thermal regulation of a human body in this study. Three typical types of sports ensembles were chosen to represent difference cover areas of garments. Their effect on human thermoregulation during moderate exercises in mild environment was investigated.

2. Materials and Methods

Three types of sports ensembles were used for tests: Long, a long sleeve T-shirts and a pair of long pants made of 100% polyester; Short, a short sleeve T-shirts made of 100% polyester and a pair of short pants made of 100% cotton; Brief: a brief made of 100% nylon. All of them were of medium size and they are shown in Fig.1. A 34-segment Newton type sweating manikin (MTNW, Seattle, USA) of body height 175cm was used for the tests, and the physiological control mode of the manikin was used to simulate the thermal comfort of a human body wearing the sports ensembles respectively. Under this mode, the manikin surface temperature and sweating rate are controlled in analogy to the prediction of the Fiala human thermoregulatory model [7, 8].

Comfort tests were carried out in a moderate environment (Ta=21°C, RH=50%, Va=0.4m/s). To investigate the effect of different exercise intensity, two protocols were used. Protocol A: a 30-min rest (1.8 MET, manikin standing still), and then a 30-min mild exercise (6 METs, manikin walking at 55 double steps per minute), followed by a 20-min rest (1.8 MET, manikin standing still). Protocol B: a 40-min vigorous exercise (10 METs, manikin walking at 55 double steps per minute), followed by a 20-min rest (1.8 MET, manikin standing still). Before starting the physiological control mode, the manikin was preheated to a defined thermal neutral condition to make sure that the manikin started from a same state for all tests. The skin temperature, the body core temperature and the whole body sweat rate Sw over time were recorded every 30 seconds. Mean skin temperature (MST) was obtained by weight averaging skin temperature over surface areas of the manikin zones.

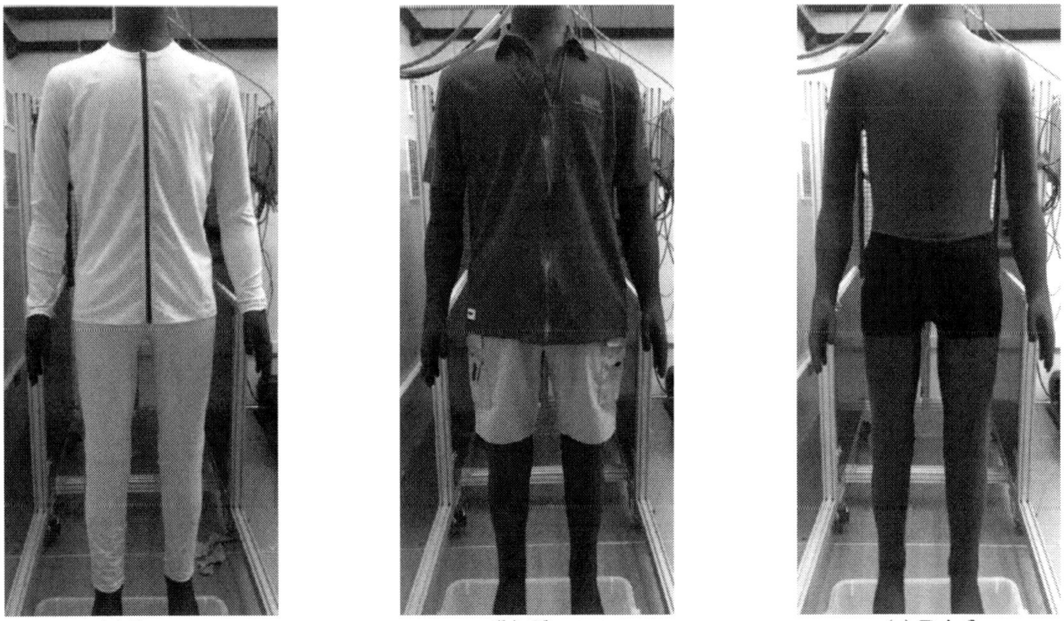

(a) Long (b) Short (c) Brief

Fig.1 Manikin tests

3. Results and Discussion

3.1 Mean Skin Temperature

As shown in Fig.2(a), during the first 30-min of rest, mean skin temperature (MST) for Long is the highest while the MST for Brief is the lowest. Because of the smallest cover area of Brief, most of the skin surface was exposed to the environment, resulting in the lowest MST. This also confirmed the difference of thermal insulation among three ensembles: Brief the lowest, and Long the highest. During the 30-min vigorous exercise, MST for the first about 15-min kept stable, and then started to drop greatly. During the 20-min of rest recover, MST still kept dropping for all three ensembles. During the whole 80-min period, MST for Brief kept as the lowest and MST for Long was the highest.

In Fig.2(b), MST shows similar trend as that in Fig.2(a). For the first 5-min, because of the ventilation caused by the manikin walking, MST for three ensembles dropped greatly, and then kept stable for another about 5-min, and then again began to drop greatly, till the end of exercise. At 40-min when the manikin stopped to walk, MST increased a little because of the absence of ventilation caused by walking. The MST difference among three ensembles in Fig.2(b) was bigger than the respective ones in Fig.2(a).

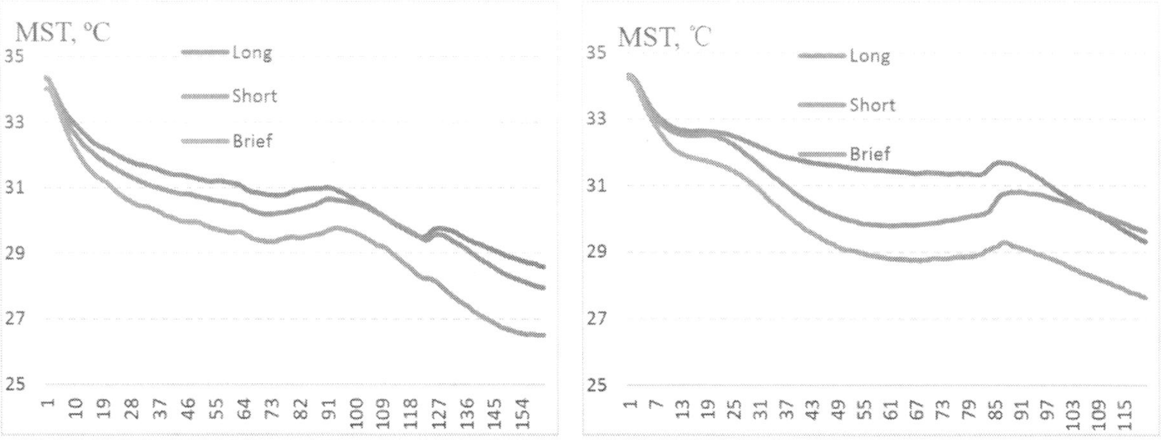

(a) Protocol A (b) Protocol B

Fig.2 Mean skin temperature (MST)

3.2 Body Core Temperature

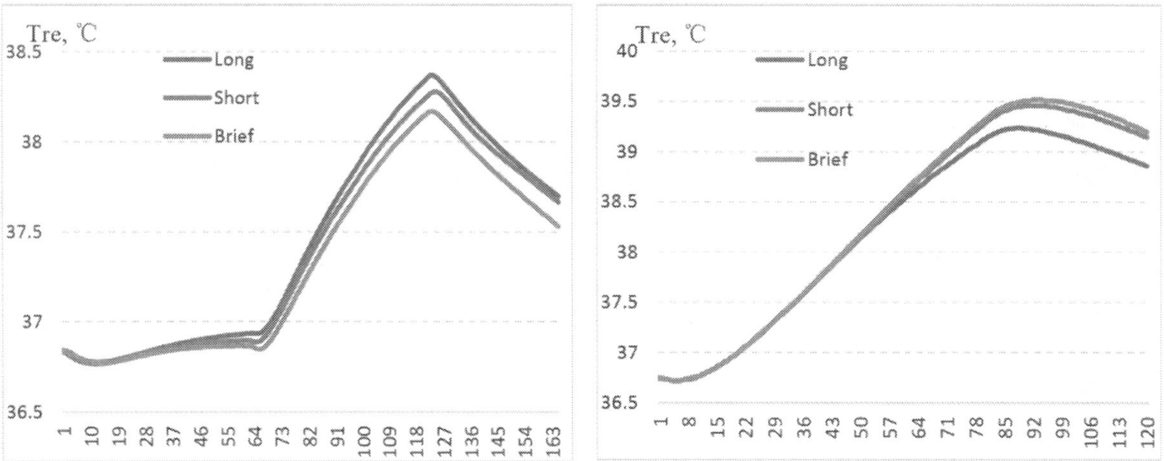

(a) Protocol A (b) Protocol B

Fig.3 Body core temperature (Tre)

As shown in Fig.3(a), during the first 30-min of rest, the core temperature (Tre) for three ensembles was close to each other and kept almost constant. When the manikin began exercise at 30-min, Tre increased greatly till 60-min when it stopped to move, and then Tre turned to decrease. During the whole period, Tre for Long kept to be the highest and Tre for Brief was the lowest although the difference was not big as that for MST.

As shown in Fig.3(b), Tre for three ensembles kept increasing for the 40-min exercise period, reached peaks a few minutes after exercise stopped, and then turned to drop. During the first 25-min of exercise, Tre for three ensembles was almost the same. After 25-min exercise, Tre for Long began to be obviously lower than Tre for the other two ensembles. Tre for Short and Brief kept close to each other. The difference was contrary to that shown in Fig.3(a).

3.3 Whole Body Sweat Rate

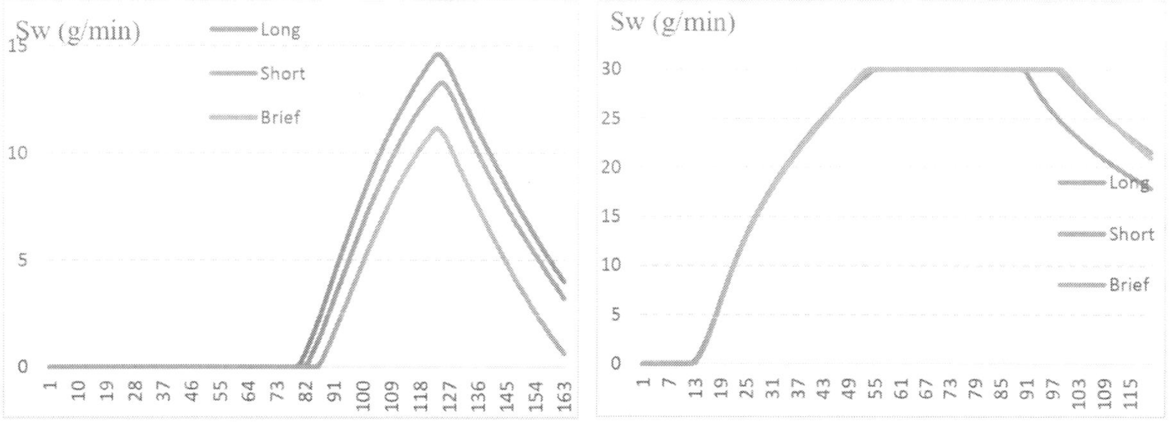

(a) Protocol A (b) Protocol B

Fig.4 Whole body sweat rate (Sw)

Fig.4 shows the whole body sweat rate (Sw) during two simulated exercise periods. In Fig.4(a), Sw for Long is the first to begin to increase (10-min after exercise started), about 1-min later Sw for Short began to increase and another 2-min later Sw for Brief began to increase. Sw for three ensembles reached peaks at almost the same time a couple of minutes after the exercise stopped. The peak Sw for Long reached was the highest while that for Brief was the lowest.

As shown in Fig.4(b), Sw for three ensembles began to increase 6-min after exercise started, and reached same peak of 30g/min at a same time after exercise started for 26-min. Since 30g/min is the maximum sweat rate for an average man [9], it is set as the maximum sweat rate for the manikin. At 5-min after exercise stopped, Sw for Long began the decrease, 4-min later, Sw for both Short and Brief began to decrease. Refer to Fig.3, it is clear that it is the core temperature of body (Tre) controlling the onset of sweating.

3.4 Effect of Cover Area

The smaller cover area of garments resulted in lower mean skin temperature (MST) during both low and high intensity exercises. However, the core temperature (Tre) showed difference between two exercise intensities. The increase of Tre resulted from heat storage (Sto). Eq.(1) shows the human heat balance: (1) The metabolism (Met) and the external work (Wext) were the same for different garment types. The conductive and respiration heat loss were low and were neglected. The difference caused by different garment types were the dry heat loss consisting of the convective heat loss (Conv) and the radiant heat loss (Rad), and the evaporative heat loss (Evap). For the exercise of low intensity (protocol A), since the sweat rate (Sw) was low, the dry heat loss was the primary way to cool down the body. For Brief of which the cover area was the smallest, Conv, Rad and Evap were the highest. Thus Sto for Brief was the lowest and resulted in the lowest Tre. On the contrary, Long with most of the body surface covered, acted as a big barrier for heat and moisture diffusion, resulted in the highest Sto and Tre.

For the exercise of high intensity, Tre reached higher values, and Sw was also much higher, and reached the maximum. For such case, only the dry heat loss which is often limited is not efficient for cooling down, evaporation become the primary way for heat loss. Compared to Evap, the difference of Conv and Rad was lower and could be neglected. The major difference contributing to Sto was Evap. Long, which covered the most of the skin surface, helped most part of the whole body sweat amount to evaporate and resulted in the highest Evap and thus the lowest Tre. On the contrary, more part of the whole body sweat amount for Brief seemed to drop down quickly before it evaporated, and lost in vain. Consequently, wearing Long, helped to keep lower body temperature while sweated less. It was confirmed that the cover area of garments affected on thermoregulation differently while the body was doing exercise of different intensity.

4. Conclusion

In this paper, three typical types of sports ensembles representing difference cover area were used to investigate the effect of garment cover area on human thermoregulation in mild environment. The thermoregulation of a male body during doing exercise of two different intensities was simulated by using a sweating thermal manikin. It was found that small cover area helped to keep low skin temperature. The effect on body core temperature and sweating was more complicated, and depended on the exercise intensity. For low intensity exercise with low sweat rate, small cover area helped to keep lower core temperature and less whole body sweat rate. For high intensity exercise with high sweat rate, large cover area helped to evaporated more sweat and thus keep lower core temperature and less whole body sweat rate. It was concluded large cover area of garment would help human body in high sweat rate state to evaporate effectively, and keep lower body core temperature while lose less body water. However, real human wear trial needs to be carried out to confirm this finding in future.

5. Acknowledgement

The corresponding author would like to acknowledge the China Scholarship Council for the financial support to study overseas.

References

[1] Watkins SM. Clothing: The Portable Environment, Ames, IA, Iowa State University Publishers: 1984; p: 1-24.
[2] Kerslake D. The stress of hot environment, Cambridge, Cambrige University Press: 1972.
[3] Burton AC, Edholm OG. Man in a cold environment. London: E. A. Arnold Ltd: 1955; p: 53-55.
[4] Kicklighter TH, Edsall JR, Martin M. Effect of moisture-wicking garments on temperature regulation during exercise. Int J Athl Ther Trai: 2011; 16: 9-13.
[5] Gavin TP, Babington JP. Clothing fabric does not affect thermoregulation during exercise in moderate heat. Med Sci Sports & Exerc: 2001; 2124-2130.
[6] Foda E, Sirén K. A thermal manikin with human thermoregulatory control implementation and validation. Int J Biometeorol: 2012; 56: 959-971.
[7] Psikuta A, Richards M, Fiala D. Single-sector thermophysiological human simulator. Physiol Meas: 2008; 29: 181-192.
[8] Psikuta A, Wang LC, Rossi RM. Prediction of the physiological response of humans wearing protective clothing using a thermophysiological human simulator. J Occup Environ Hyg: 2013; 10: 222-232.
[9] American Society of Heating, Refrigerating and Air-Conditioning Engineers: Physiological principles and thermal comfort. ASHRAE Handbook Fundamentals. ASHRAE, Atlanta: 1993; 8.1-8.32.

Distance Measurement Based on Three-dimensional Motion Capture Apparatus and Influence Factor Analysis

Qin He[1,4], Yi-Dan Wang[1], Yi Tang[1], Zheng Liu[1,2,3*]

[1]*Fashion College, Zhejiang Sci-Tech University, Hangzhou, Zhejiang, 310018, China*
[2]*Key Laboratory of Silk Culture Inheriting and Products Design Digital Technology, Ministry of Culture, Hangzhou, Zhejiang, 310018, China*
[3]*Zhejiang Engineering Laboratory of Clothing Digital Technology, Hangzhou, Zhejiang, 310018, China*
[4] *Engineering Research Center of Clothing of Zhejiang Province, Hangzhou, Zhejiang, 310018, China*

*Corresponding author's email: Koala@zstu.edu.cn; 1940125600@qq.com

Abstract

Measuring space distance between clothing and wearer's skin with 3D motion capture system is an effective way to obtain three-dimensional human body dress form. The purpose of this paper is to test the accuracy of the distance measurement of 3D motion capture system (QUALISYS, Sweden). A device which simulates a dressed human cross section, set device height, angle of measuring distance, gasket height, simulated human body movements, marker ball diameter as variable parameters was designed. The aim is to analyze the influence of the parameters during 3D motion capture measuring experiment. The experimental result shows that the passive marker ball diameter and active marker ball's gasket height have significant effect on 3-D dynamic capture meter space distance measurements.

Keywords: 3D Motion Capture System; Space Between Clothing and Wearer's Skin; Distance Measurement; Influence Factor

Introduction

Air gap between clothing and wearer's skin represents the spatial shape between human and garment model. It is an important factor that influences the garment shape [1]. Acquiring spatial information under the dressed clothing plays an important role in influencing clothing fitness evaluation [2], performance of the protective clothing [3-4], the intelligence of 3D garment design process and the rationality of the design results.

3D body scanning was used to measure air gap in the former study [5-8] .With the progress of computer image technology, many new instruments have been available for 3D measuring. 3D motion capture device was used to collect dressed clothing spatial information. The device (QUALISYS, Sweden) has two types of marker ball (passive maker ball receive infrared light from lens and be captured by camera, active marker ball depend on itself shine and be captured by camera) that stick on the body skin and clothing respectively. Through capturing the distance between two type balls in real time, the distance of air gap was measured.

Before 3D motion capture system was applied, the test of result accuracy should be conducted. Therefore, a 3D motion capture system of QUALISYS was taken as the equipment and the device was designed to simulate a dressed human cross section. The device height, angle of measuring distance, gasket height, simulated human body movements, marker ball diameter as variable parameters was set and the influence of the parameters in 3D motion capture measuring experiment was analyzed.

1. Experiment

1.1 Experimental Design

The device simulated the human body dress section was pasted with active and passive marker balls. And manual measuring the distance between active and passive marker balls, which was regarded as the actual measuring distance D1.We used 3D dynamic capture instrument and acquired the active and passive 3D coordinate data, defined as the distance D2. Comparing the distance difference of D1 and D2 as d(ΔD). Set the device's height, the angle of measuring distance as variable parameters, and variance analysis was carried out on the ΔD under the different parameters. Finally we examined the significant effect of various parameters on the result of the experiment.

Experimental variables parameters are shown in the following figures. Fig.1 was device height variable, which tested the influence of the space distance measurement in different height cases. Fig.2 was the angle of measuring distance variable, which tested the effect of space distance measurement under the clothing fluctuation. Fig.3 was the parameter of gasket height variable, which tested the measurement effect of marker in different body bulges. Fig.4 is the female（160/84A）hold device at the chest and executed actions of whirling, walking, running motions respectively, which tested the variety of space distance of 3D dynamic capture apparatus in dynamic condition

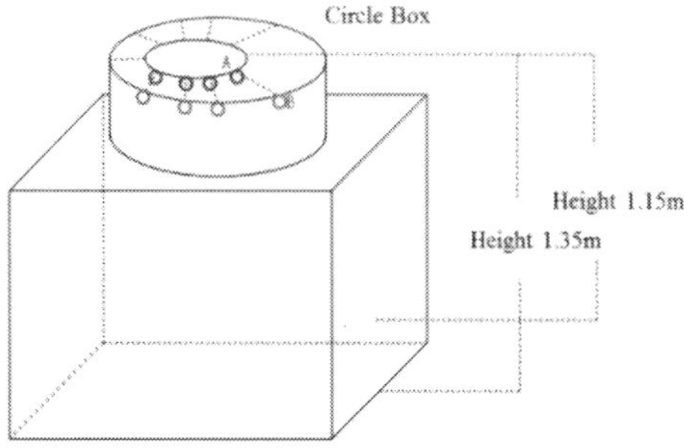

Fig.1 Height

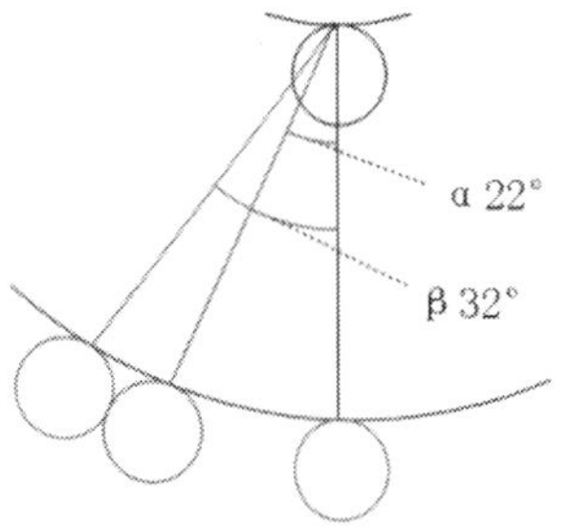

Fig.2 Angle

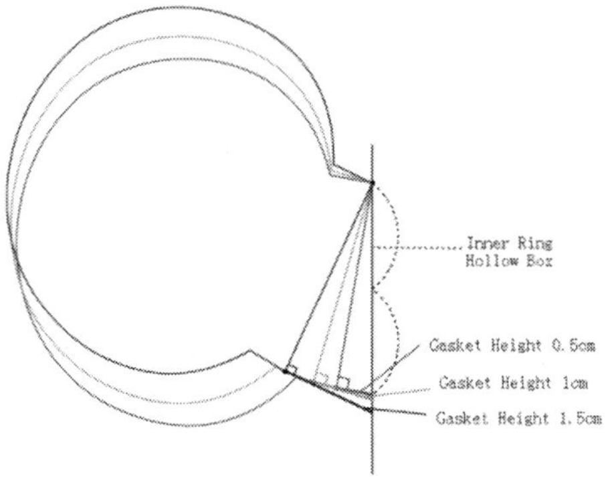

Fig.3 Gasket Height

Fig.4 Dynamic motion

1.2 Experimental Equipment

Experimental equipment with three-dimensional motion capture system, the system contains capture camera, connection cable, power supply and data exchange with the hub hardware. Calibration systems and capture the reflective balls. It is based on Computer Graphics (CG) theory. By more than one video capture devices arranged in space, the moving object (tracker) velocity and image are recorded, then the computer processing of the image data obtains different objects in the space coordinates (X, Y, Z) with different time units. Under the 112fp gathering frequency, the accuracy of system scanning is 0.01 mm. QUALISYS supports two types of marker balls. We used the active marker ball with 16 mm width and 10 mm length, the passive marker balls with 6 mm and the 16 mm diameter.

Experiment device was made of two different radius of the hollow boxes, which imitating a dressed human cross section. Experimental device was shown in Fig. 5.We Stick on an active marker ball every 5 cm inner circular box. Along the active marker ball's extension cord of radius, the position of eight passive marker balls was confirmed. Fig.6 was the simulated height of the gasket.

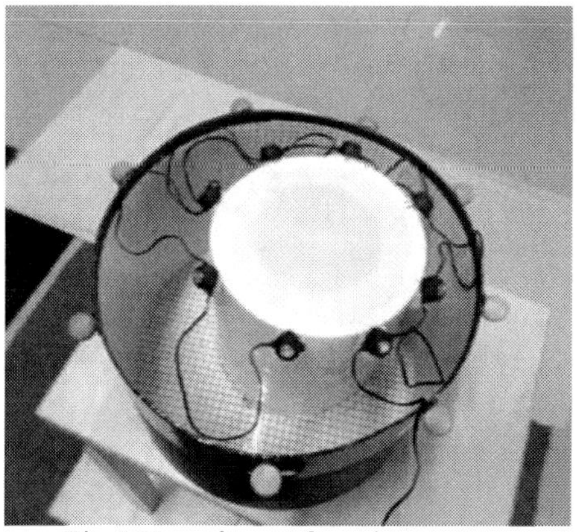

Fig.5 Experimental equipment

Fig.6 Gasket.

1.3 Experimental Steps

First, we manual measured eight section labeled distances 20 times, took average data as the D1.Second, captured the active and passive points three times by dynamic capture device, select all active and passive spot's data at the same time. Computed the D2 and the ΔD. Last, repeated step 1 and step 2 to get the ΔD under different parameters according to the five parameters of experimental design level.

2. Experimental Results and Analysis

2.1 Static Experiment

The difference between passive marker balls of 16mm diameter and passive marker balls of 6mm diameter in static experiment was significant. As shown in Fig.7, when a passive marker ball was 16mm, ΔD was very small. However, when a passive marker ball was 6mm, ΔD became bigger. Which the distance difference was 5-7mm.

The difference ΔD of two passive marker balls had a significant difference (significance level of 0.05). Meanwhile, when the marker ball's diameter is 6mm, ΔD had some difference .The difference range was 5 to 7 mm.

The set parameters of tilt angle (A3) and raised height (A2) had a significant effect on ΔD as shown in table1.

Fig.8 showed the variable value of ΔD when stuck on passive marker balls which diameter is 6mm.The effect of tilt angle on ΔD as shown in table 2.

Table 3 showed highly significant effect (significance level of 0.05) of different raised height of ΔD.

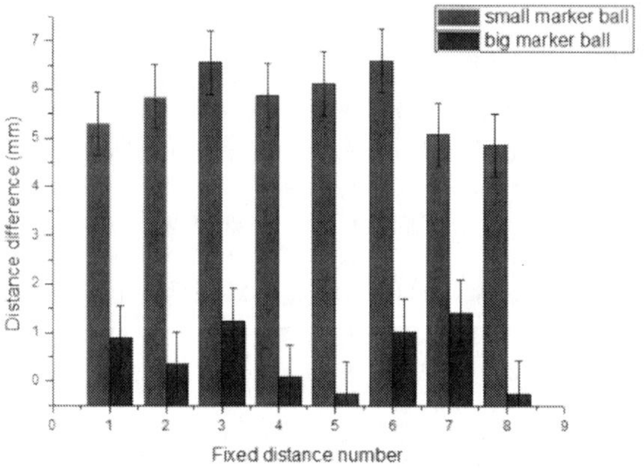

Fig.7 Distance difference

Table.1 Inter - subjective test

Origin	df	MS	F	Sig.
Calibration Model	6	101.826	31.433	.000
Intercept	1	25.809	7.967	.006
A2	3	125.466	38.731	.000
A1	1	.259	.080	.778
A3	2	96.119	29.671	.000
Error	73	3.239		
Sum	80			

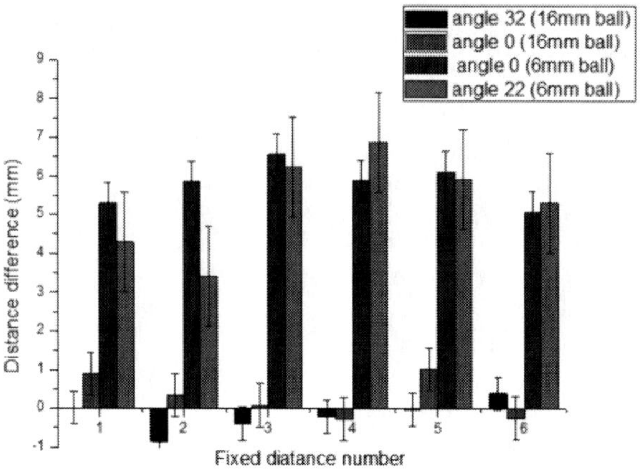

Fig.8 Distance difference

Table.2 Comparison of angle parameter

(I) angle	(J) angle	(I-J)	Standard error	Sig.	95% confidence interval	
					lower limit	upper limit
0°angle	22°angle	-5.35080*	.680279	.000	-6.70659	-3.99500
	32°angle	-.71020	.510209	.168	-1.72705	.30664
22°angle	32°angle	4.64059*	.779358	.000	3.08734	6.19385

Table.3 Gasket height parameters

(I) Gasket Height	(J) Gasket Height	(I-J)	Standard error	Sig.	95% confidence interval	
					lower limit	upper limit
0cm	0.5cm	2.44584*	.551089	.000	1.34753	3.54416
	1cm	4.12406*	.551089	.000	3.02574	5.22238
	1.5cm	5.81895*	.551089	.000	4.72064	6.91727
0.5cm	1cm	1.67821*	.636343	.010	.40998	2.94644
	1.5cm	3.37311*	.636343	.000	2.10488	4.64134
1cm	1.5cm	1.69490*	.636343	.010	.42667	2.96313

2.2 Dynamic motion Experiment

In the dynamic motion experiment, we set the height (A1), the raised height (A2), dynamic motions (A3) as three parameters. From the results of analysis of variance, the raised height had a significant impact (significance level of 0.05) on the ΔD as shown in table4.

Table.4 Inter - subjective test

origin	df	MS	F	Sig.
Correction model	44	21.094	2.926	.000
intercept	1	242.309	33.608	.000
A2	2	447.467	62.063	.000
A3	2	.096	.013	.987
A4	4	.779	.108	.980
A2 * A3	4	.912	.126	.973
A2 * A4	8	.751	.104	.999
A3 * A4	8	.523	.073	1.000
A2 * A3 * A4	16	.640	.089	1.000
error	435	7.210		
sum	480			

2.3 Results and Discussion

There was no significant difference for ΔD, when used of 16mm active and passive marker balls. While used of 6mm passive marker balls, the ΔD have obvious difference. The gap (ΔD) was 5-7mm.It can be found that the passive marker ball's diameter variable had significant influence on the ΔD. Which can be interpreted as, supposed 16mm active marker balls sent out the center light from light emitting area, 16mm passive marker ball receives light point was the center of the ball. The measurement of the distance D2 had no difference with the D1. The gap 5mm was the difference of the radius of the two passive marker balls as shown in Fig.9. The variable of tilt angle had significant impact on ΔD because the change of marker ball diameter had a significant impact on ΔD. It also can be explained that the gasket height parameter had significant influence on the ΔD. Passive marker ball center did not change while the active marker luminous point changed. So even used a 16mm passive marker ball, the result still has a significant effect on ΔD. And the higher of the gasket height, the greater difference of ΔD as shown in Fig.10.

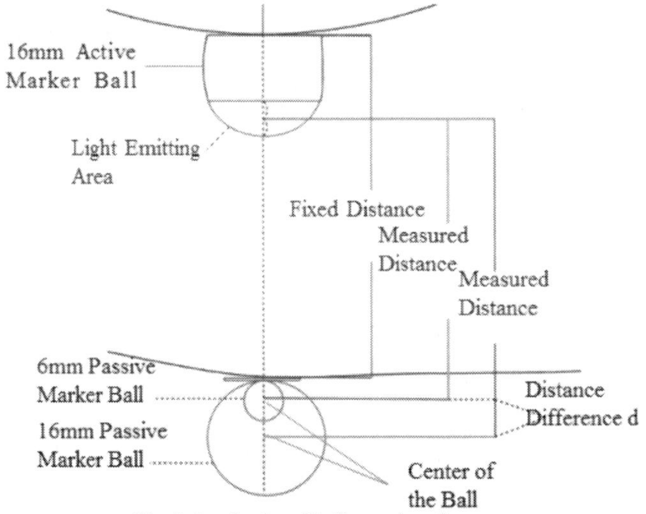

Fig.9 Analysis of influencing factors

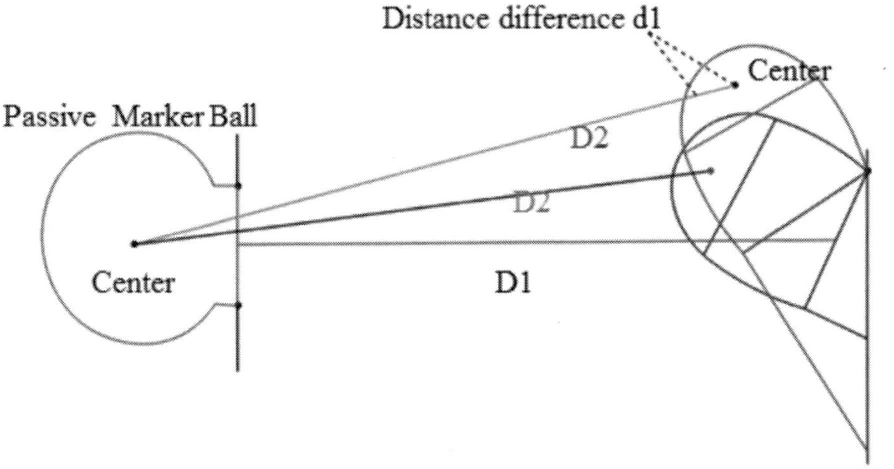

Fig.10 Gasket height impact analysis

3. Conclusion

From the theoretical analysis and experimental results, both the diameter of the passive marker ball and the gasket height of the active marker ball had significant effect on space measurement in 3D motion capture system. The active marker ball sends out the light source from the center of the ball, passive marker ball received light point is the center of the ball, so the distance between them is the difference of the radius of the passive marker ball. When gasket height changed, the center point of active marker ball altered correspondingly. The distance difference between spatial distance and fixed distance had changed.

4. Acknowledgement

This research was supported by Zhejiang Provincial Natural Science Foundation of China under Grant No. LY15E050028, the Science Foundation of Zhejiang Sci-Tech University under Grant No.13072177-Y.

References

[1] Liu Z, Li JT, Lu GD. Review on 3D Allocation and Ease Allowance and its Applicationsin Garment Design. Journal of Computer-Aided Design & Computer Graphics: 2012; 10: 1294-1296

[2] Paquette S. 3D scanning in apparel design and human engineering. IEEE Computer Graphics and Applications: 1996; 5: 11-15.

[3] Kim Y, Lee C, LI P, et al. Investigation of air gaps entrapped in protective clothing systems. Fire and Materials: 2002; 3: 121-126.

[4] Song G. Clothing air gap layers and thermal protective performance in single layer garment .Journal of Industrial Textiles: 2007; 3: 193-205.

[5] Xu JH, Zhang WB. Influential factors of distance ease on typical cross sections. Journal of Texitile Research: 2009; 5: 104-107.

[6] Yu M, Li J, Wang YP. Correlation of fabric performances and space between garment and wearer's skin. Journal of Texitile Research: 2012; 4: 101-105.

[7] Zhang ZH, Wang YY, Li J. Effect of thickness of air layer under clothing on heat transmission of wearer. Journal of Texitile Researc: 2010; 2: 103-107.

[8] Wang YY, P JH, L SY. Study on the relation between the air gap size under clothing and the comfort sensations. Journal of Donghua University:English Edition: 2006; 5: 101-107.

Relationship between Human Body Anthropometric Measurements and Basal Metabolic Rate

Hui Zou, Li Liu*, Ya-Huan Li

Beijing Institute of Fashion Technology, Beijing, 100029, China

*Corresponding author's email: fzyll@bift.edu.cn

Abstract

Through the use of 3D body measurement technology and cardiopulmonary function test equipment, the body size data and basal metabolic rate of 116 healthy young subjects were obtained. This study aims to find the relationship between the size of the human body and basal metabolic rate. Factor analysis, univariate analysis, linear regression analysis was performed on 13 observed items (selected from 152 human data) by SPSS data analysis software. The 13 observed items include the largest abdominal circumference, waist circumference, chest circumference (horizontal), thigh circumference, hip circumference, weight, total shoulder width, neck circumference, height, waist height, high cervical point, hip height, and chest height. The results indicate that girth and height factors are correlated with the predicted basal metabolic rate as well as the measured basal metabolic rate. The predicted basal metabolic rate is significantly correlated with weight, height, hip circumference, and neck circumference. The measured basal metabolic rate is significantly correlated with the neck circumference as well as height.

Keywords: 3D Body Measurement; Basal Metabolic Rate; Factor Analysis; Univariate Analysis; Linear Regression Analysis

1. Introduction

Many studies reveal that the size of the human body not only can indicate the beauty of the human body, but also can predict the physiological parameters of the disease [1, 2, 12]. Several studies show that neck girth has a high correlation with obstructive sleep apnea-hypopnea syndrome, metabolic syndrome and female hyperandrogenism [3, 4]. Onat et al found that neck circumference was associated with metabolic syndrome even more than waist circumference [4]. In 1996,Seidell et al found that deaths from cardiovascular disease increased three times in 48287 Dutch with body mass index greater than 25 kg / m^2 [5]. In 2012, Kai Liang et al found that waist-height ratio is the best predictor of abnormal glucose metabolism in both men and women. 3011 residents subjects were recruited over 20 years old, in 4 cities, by comparing the predictive value of different genders, body mass index, waist circumference, waist hip ratio and waist height ratio for abnormal glucose metabolism [6].

Basal metabolism is the minimum energy required for vital movement. Basal metabolic rate can assess the state of the human body [7, 8]. In medicine, the determination of basal metabolic rate is a method of diagnosing disease [9, 10, 11]. Based on the relevant investigations of using size of the human body as physiological prediction parameters to diseases, this paper will elaborate the relationships between body size and basal metabolic rate.

2. Methodology

2.1 Subjects

116 healthy young men and women at 18-26 years old were invited to participate in the study, embodying 95 females and 21 males. Through the relevant literature, 13 human data were selected for observation. Table 1 shows the 13 basic dimensions and body mass index (BMI) of the 116 subjects, while Table 2 demonstrates the experimental data on metabolic rate of 116 subjects.

Table.1 Body size and BMI index of 116 subjects

Testing items	Max	Min	Mean
Body height (cm)	176.0	152.0	162.2
Weight (kg)	82.2	40.3	54.6
Bust girth (horizontal) (cm)	119.0	78.0	87.9
Waist girth (cm)	115.9	60.5	71.7
Buttock girth (cm)	118.1	81.1	92.8
Maximum belly circumference (cm)	123.3	64.9	79.2
Mid neck girth (cm)	38.6	28.3	32.8
Thigh girth (right) (cm)	64.5	43.8	52.3
Across shoulder (cm)	49.3	37.0	41.8
Bust height (cm)	128.4	102.2	117.0
Waist height (cm)	109.8	87.5	101.0
Neck height (cm)	150.8	122.1	138.2
Buttock height (cm)	89.2	69.4	80.5
BMI (kg/m^2)	30.2	16.2	20.2

Table.2 Experimental data on metabolic rate of 116 subjects

Parameter	Max	Min	Means
Oxygen uptake (L/min)	0.326	0.127	0.228
Carbon dioxide output (L/min)	0.269	0.107	0.191
Respiratory quotient (%)	1.070	0.62	0.838
Heart rate (bpm)	111.000	57.000	77.731
Measured basal metabolic rate (%)	2260.000	902.000	1595.999
Predicted basal metabolic rate (%)	2070.000	1359.000	1549.148
Lipid oxidation rate (%)	84.000	0.000	43.426

2.2 Testing Equipment

We obtain relevant body size data through Voxelan (HAMANO ENGINEERING CO.LTD.JPN) 8CCD 3D scanner, which can obtain regional images by the camera and use the model software to convert to the spatial point. CORTEX series MetaMax 3B cardiopulmonary exercise telemetry function test instrument from Germany can help us obtain relevant metabolic rate. It can be employed to collect parameters of gas metabolism during the course of breathing exercise, such as oxygen consumption (VO^2), carbon dioxide output (VCO^2), respiratory rate, heart rate, respiratory exchange rate, ventilation and environmental temperature, pressure and so forth.

2.3 Experimental Protocol

This experiment was conducted in an artificial climate chamber (temperature: 25 ± 2℃, humidity: 50 ± 2%). All subjects were required to wear sportswear, of whom the female shall put on the special underwear with gym short while the male shall wear the sport short only. Before starting the experiment, subjects were expected to fill in the basic personal information, such as age, gender, weight and so on. When taking 3D body scanning, the subjects erectly stood on the scanner platform and spread feet 15 cm with arms lifted outward and smooth breathing. Scan will be completed within 5-10s. After that the body fat composition of the subjects would be measured. Besides, while taking the metabolic test, subjects sat in a quiet room with physical relaxation. The testing time varies from person to person and about 15 min is required for most subjects.

3. Results and Discussion

3.1 Factor

The KMO value is greater than 0.05 and close to 1, sig.=0.000<0.05, so the 13 observed items are suitable for factor analysis(Table 3). For Table 4 reveals high communality of each factor, the extracted components can be well described by these variables. Meanwhile, in the light of Table 5, the eigenvalues of the first two factors are 6.985 and 3.833 respectively, accounting for 83.596% of the total variance. The first two factors explain the variance of 83.596% of the original 13 factors, hence, we will confirm to extract the two principal components.

Table.3 Kmo and bartlett's test

Kaiser-Meyer-Olkin measure of sampling adequacy		.843
Bartlett's test of sphericity	Approx. Chi-Square	2403.025
	df	78.000
	Sig.	.000

Table.4 Communalities

	Initial	Extraction
Maximum belly circumference	1.000	.849
Waist girth	1.000	.867
Bust girth (horizontal)	1.000	.830
Thigh girth(right)	1.000	.750
Buttock girth	1.000	.768
Weight	1.000	.919
Across shoulder	1.000	.544
Mid neck girth	1.000	.629
Body height	1.000	.930
Waist height	1.000	.954
Neck height	1.000	.974
Buttock height	1.000	.901
Bust height	1.000	.952

Table.5 Analysis on all variances

Factor	Initial eigenvalues			Extraction sums of squared loadings			Rotation sums of squared loadings		
	Total	% of variance	Cumulative %	Total	% of variance	Cumulative %	Total	% of variance	Cumulative %
1	6.985	53.729	53.729	6.985	53.729	53.729	5.741	44.165	44.165
2	3.883	29.867	83.596	3.883	29.867	83.596	5.126	39.431	83.596
3	.853	6.560	90.156						
4	.383	2.948	93.103						
5	.247	1.899	95.003						
6	.214	1.650	96.652						
7	.135	1.040	97.693						
8	.111	.855	98.548						
9	.067	.517	99.065						
10	.051	.390	99.455						
11	.041	.315	99.770						
12	.018	.140	99.910						
13	.012	.090	100.000						

In order to name these factors, we rotated the factors so that the coefficients were polarized to 0 and 1. By rotating the factor matrix, the factor can be named and interpreted (Table 6). Factor 1 is named the girth factor since it can represent waist girth, maximum belly circumference, bust girth (horizontal), thigh girth (right), buttock girth, weight, across shoulder and mid neck girth. Factor 2 is named the height factor as it can represent waist height, neck height, bust height, buttock height and body height.

The coefficient of principal component score is shown in Table 7.

Standardized first factor =0.178×maximum belly circumference + 0.171×waist girth + 0.171×bust girth (horizontal) + 0.166×thigh girth (right) + 0.154×buttock girth + 0.135×weight + 0.112× across shoulder + 0.104 ×mid neck girth - 0.017×body height - 0.043×waist height - 0.022×neck height - 0.044×buttock height - 0.037× bust height.

Standardized second factor = -0.066×maximum belly circumference - 0.036×waist girth - 0.047×bust girth (horizontal) - 0.059×thigh girth (right) - 0.015×buttock girth + 0.047×weight + 0.023×across shoulder + 0.051× mid neck girth + 0.189×body height + 0.203×waist height + 0.196×neck height + 0.198×buttock height + 0.200 ×bust height.

Table.6 Rotation component matrix

	Component	
	1	2
Waist girth	.928	.077
Maximum belly circumference	.919	-.070
Bust girth (horizontal)	.911	.017
Thigh girth (rihgt)	.865	-.049
Buttock girth	.862	.156
Weight	.848	.448
Across shoulder	.679	.287
Mid neck girth	.674	.418
Waist height	.062	.975
Neck height	.171	.972
Bust height	.094	.971
Buttock height	.051	.948
Body height	.192	.945

Table.7 Component score coefficient matrix

	Component	
	1	2
Maximum belly circumference	.178	-.066
Waist girth	.171	-.036
Bust girth (horizontal)	.171	-.047
Thigh girth (rihgt)	.166	-.059
Buttock girth	.154	-.015
Weight	.135	.047
Across shoulder	.112	.023
Mid neck girth	.104	.051
Body height	-.017	.189
Waist height	-.043	.203
Neck height	-.022	.196
Buttock height	-.044	.198
Bust height	-.037	.200

3.2 Factors and Predicted Basal Metabolic Rate

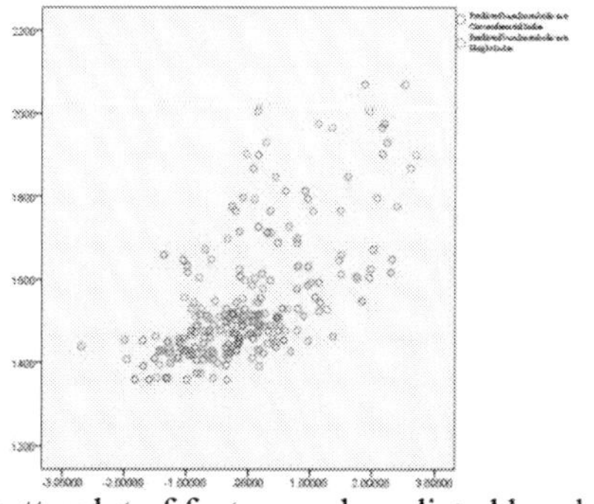

Fig.1 Simple scatterplot of factors and predicted basal metabolic rate

According to Fig.1, the girth and height are highly related to the predicted basal metabolic rate with linear correlation. The correlation coefficients between predicted basal metabolic rate and girth index, predicted basal metabolic rate and height index are 0.627 (sig.=0.000<0.01, reject null hypothesis) and 0.634 (sig.=0.000<0.01, reject null hypothesis) respectively. The results unveil there is a significant correlation between predicted basal metabolic rate and girth index, predicted basal metabolic rate and height index respectively. (Table 8)

Table 9 displays the sources of variation, degrees of freedom, mean square, F value and the significance of F.

The inter-group mean square of model 2 is much larger than the internal mean square of the group, and the F value is 216.155 (sig. =0.000<0.05). Therefore, it can be deduced that the established regression equation is effective.

According to Table 10, in model 2, dependent variable Y regression on the two independent variables X1 and X2 of the non-standardized regression coefficients are 98.698 and 97.650 respectively while T values of the corresponding saliency detection are 14.780 and 14.624 respectively, and the significant level of their regression coefficient (sig.) is .000, which is less than 0.05. Hence, it can be deduced that there is a definite linear relationship between the two factors and measured basal metabolic rate.

Table.8 Correlation analysis between two factors and predicted basal metabolic rate

		Predicted basal metabolic rate	Circumferential index	Height index
Predicted basal metabolic rate	Pearson's correlation coefficient	1	.627**	.634**
	Sig.(2-sided)		.000	.000
	N	114	114	114
Circumferential index	Pearson's correlation coefficient	.627**	1	.000
	Sig.(2-sided)	.000		1.000
	N	114	114	114
Height index	Pearson's correlation coefficient	.634**	.000	1
	Sig.(2-sided)	.000	1.000	
	N	114	114	114

Table.9 Anova

Model		Sum of squares	df	Mean square	F	Sig.
1	Regression	1100764.555	1	1100764.555	75.320	.000[a]
	Residual	1636821.445	112	14614.477		
	Total	2737586.000	113			
2	Regression	2178287.472	2	1089143.736	216.155	.000[b]
	Residual	559298.528	111	5038.725		
	Total	2737586.000	113			

a. Predictor variables: height indicator; b. Predictor variables: height indicator, girth indicator;
c. Dependent variable: predicted basal metabolic rate

3.3 Factors and Measured Basal Metabolic Rate

In the light of Fig.2, to some extent, the girth and height index are related to the measured basal metabolic rate, but the linear relationship is not significant. The consequence indicates that the correlation coefficient between the measured basal metabolic rate and the height index, the measured basal metabolic rate and the girth index are 0.303(sig. =0.001<0.01, reject null hypothesis) and 0.349(sig. =0.000<0.01, reject null hypothesis) respectively. Accordingly, we can conclude there is an insignificant correlation between the measured basal metabolic rate and the height index, measured basal metabolic rate and the girth index respectively (Table 11).

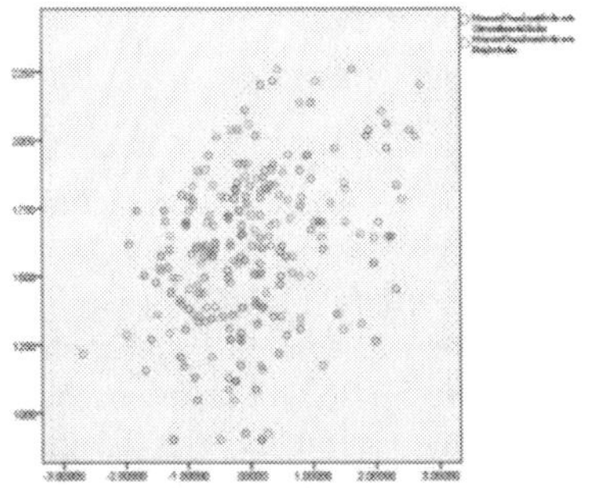

Fig.2 Simple scatterplot of factors and measured basal metabolic rate

Table.10 Regression coefficient

Model		Non-standardized coefficients B	Std. Error	Standardized coefficients Beta	t	Sig.
1	(Constant)	1551.000	11.322		136.985	.000
	Height index	98.698	11.372	.634	8.679	.000
2	(Constant)	1551.000	6.648		233.294	.000
	Height index	98.698	6.678	.634	14.780	.000
	Girth index	97.650	6.678	.627	14.624	.000

Coefficients [a]

Dependent variable: predicted basal metabolic rate

Table.11 Correlation analysis between two factors and measured basal metabolic rate

		Measured basal metabolic rate	Circumferential index	Height index
Measured basal metabolic rate	Pearson's correlation coefficient	1	.303**	.349**
	Sig.(2-sided)		.001	.000
	N	114	114	114
Circumferential index	Pearson's correlation coefficient	.303**	1	.000
	Sig.(2-sided)	.001		1.000
	N	114	114	114
Height index	Pearson's correlation coefficient	.349**	.000	1
	Sig.(2-sided)	.000	1.000	
	N	114	114	114

3.4 Univariate and Predicted Basal Metabolic Rate

According to Table 12, except for the right thigh variable, weight, height, hip circumference and neck circumference variables are embedded into regression model. Besides, there is high R square value on model 1,2,3,4,5 and 6, which unveils the dependent and independent variable are highly correlated (Table 13). The F value in model 6 is 358.808, and sig in all models is 0.000, less than 0.01, which has a strong significance. Meanwhile, all the regression variances are greater than residuals, indicating the established regression equation is effective (Table 14)

Table.12 Modelling

Variables entered / removed [a]

Model	Variables entered	Variables removed	Method
1	Weight	.	Stepwise(criteria:Probability-of-F-to-enter<=.005,Probability-of-F-to-enter>=.100)
2	Thigh girth(right)	.	Stepwise(criteria:Probability-of-F-to-enter<=.005,Probability-of-F-to-enter>=.100)
3	Body height	.	Stepwise(criteria:Probability-of-F-to-enter<=.005,Probability-of-F-to-enter>=.100)
4	Buttock girth	.	Stepwise(criteria:Probability-of-F-to-enter<=.005,Probability-of-F-to-enter>=.100)
5	Mid neck girth	.	Stepwise(criteria:Probability-of-F-to-enter<=.005,Probability-of-F-to-enter>=.100)
6	.	Thigh girth(right)	Stepwise(criteria:Probability-of-F-to-enter<=.005,Probability-of-F-to-enter>=.100)

Table.13 Model summary [a]

Model	R	R square	Adjusted R square	Std. error of the estimate
1	.898[a]	.806	.804	68.926
2	.943[b]	.890	.888	52.127
3	.952[c]	.906	.904	48.245
4	.959[d]	.919	.917	44.965
5	.964[e]	.930	.927	42.170
6	.964[f]	.929	.927	42.104

a. Predictive variable: (constant), weight.
b. Predictive variable: (constant), weight, thigh girth (right).
c. Predictive variable: (constant), weight, thigh girth (right), body height.
d. Predictive variable: (constant), weight, thigh girth (right), body height, buttock girth.
e. Predictive variable: (constant), weight, thigh girth (right), body height, buttock girth, mid neck girth.
f. Predictive variable: (constant), weight, body height, buttock girth, mid neck girth.
g. Dependent variable: predicted basal metabolic rate.

Table.14 Anova

Model		Sum of squares	df	Mean square	F	Sig.
1	Regression	2205499.952	1	2205499.952	464.241	.000[a]
	Residual	532086.048	112	4750.768		
	total	2737586.000	113			
2	Regression	2435975.842	2	1217987.921	448.250	.000[b]
	Residual	301610.158	111	2717.209		
	Total	2737586.000	113			
3	Regression	2481549.406	3	827183.135	355.379	.000[c]
	Residual	256036.594	110	2327.605		
	Total	2737586.000	113			
4	Regression	2517200.603	4	629300.151	311.244	.000[d]
	Residual	220385.397	109	2021.884		
	Total	2737586.000	113			
5	Regression	2545528.093	5	509105.619	286.286	.000[e]
	Residual	192057.907	108	1778.314		
	Total	2737586.000	113			
6	Regression	2544352.653	4	636088.163	358.808	.000[f]
	Residual	193233.347	109	1772.783		
	total	2737586.000	113			

a. Predictor variables: (constant), weight. b. Predictor variables: (constant), weight, thigh girth (right).
c. Predictor variables: (constant), weight, thigh girth (right), body height.
d. Predictor variables: (constant), weight, thigh girth (right), body height, buttock girth.
e. Predictor variables: (constant), weight, thigh girth (right), body height, buttock girth, mid neck girth.
f. Predictor variables: (constant), weight, body height, buttock girth, mid neck girth.
g. Dependent variable: predicted basal metabolic rate.

In the light of Table 15, the standard equation of multiple linear regression equation about independent variable and dependent variable is: Y=0.771 weight + 0.268 body height - 0.243 buttock girth + 0.197 mid neck girth.

Table.15 Regression coefficient

Model		Non-standardized coefficients B	Std. error	Standardized coefficients Beta	t	Sig.
1	(Constant)	591.996	44.975		13.163	.000
	Weight	17.615	.818	.898	21.546	.000
2	(Constant)	1164.971	70.904		16.430	.000
	Weight	23.644	.900	1.205	26.258	.000
	Thigh girth (right)	-17.221	1.870	-.423	-9.210	.000
3	(Constant)	297.992	206.631		1.442	.152
	Weight	19.243	1.298	.981	14.829	.000
	Thigh girth (right)	-11.486	2.162	-.282	-5.313	.000
	Body height	4.909	1.109	.202	4.425	.000
4	(Constant)	588.554	204.637		2.876	.005
	Weight	20.533	1.248	1.046	16.455	.000
	Thigh girth (right)	-5.100	2.525	-.125	-2.020	.046
	Body height	5.343	1.039	.220	5.141	.000
	Buttock girth	-8.278	1.971	-.243	-4.199	.000
5	(Constant)	142.712	222.059		.643	.522
	Weight	15.874	1.653	.809	9.603	.000
	Thigh girth (right)	-2.024	2.490	-.050	-.813	.418
	Body height	6.055	.991	.249	6.111	.000
	Buttock girth	-7.457	1.860	-.219	-4.009	.000
	Mid neck girth	10.517	2.635	.186	3.991	.000
6	(Constant)	56.615	194.877		.291	.772
	Weight	15.131	1.375	.771	11.001	.000
	Body height	6.504	.822	.268	7.916	.000
	Buttock girth	-8.266	1.569	-.243	-5.269	.000
	Mid neck girth	11.180	2.502	.197	4.469	.000

Coefficients [a]

a. Dependent variable: predicted basal metabolic rate

3.5 Univariate and Measured Basal Metabolic Rate

Table.16 Modelling

Variables entered / removed [a]

Model	Variables entered	Variables removed	Method
1	Mid neck girth	.	Stepwise (criteria:Probability-of-F-to-enter<=.005,Probability-of-F-to-enter>=.100)
2	Body height	.	Stepwise (criteria:Probability-of-F-to-enter<=.005,Probability-of-F-to-enter>=.100)

a. Dependent variable: measured basal metabolic rate.

According to Table 16, the mid neck girth and body height variables can be embedded into the model. The goodness of fit of model 2 is better than model 1, but the R value of the model 2 is 0.252, which indicates that independent variable can explain the change of dependent variable 25.2% (Table 17). In the model regression analysis, the goodness of fit is general. In addition, for the probability of F value greater than F critical value (sig.) is about 0.000, we can deduce that there are correlations between measured basal metabolic rate and the mid neck girth, measured basal metabolic rate and body height respectively (Table 18).

Table.17 Model summary

Model	R	R square	Adjusted R square	Std. error of the estimate
1	.473[a]	.223	.216	254.233
2	.502[b]	.252	.239	250.619

a. Measure variables: (constant), mid neck girth.
b. Measure variables: (constant), mid neck girth, body height.

Table.18 Anova

Model		Sum of squares	df	Mean square	F	Sig.
1	Regression	2081624.638	1	2081624.638	32.206	.000[a]
	Residual	7239032.599	112	64634.220		
	Total	9320657.237	113			
2	Regression	2348739.947	2	1174369.974	18.697	.000[b]
	Residual	6971917.290	111	62810.066		
	Total	9320657.237	113			

a. Measure variables: (constant), mid neck girth.
b. Measure variables: (constant), mid neck girth, body height.
c. Dependent variable: measured basal metabolic rate.

Table.19 Regression coefficient

Coefficients [a]

Model		Non-standardized coefficients		Standardized coefficients	t	Sig.
		B	Std. error	Beta		
1	(Constant)	-22.175	287.115		-.077	.939
	Mid neck girth	49.413	8.707	.473	5.675	.000
2	(Constant)	-1128.222	606.438		-1.860	.065
	Mid neck girth	38.379	10.114	.367	3.795	.000
	Body height	8.940	4.335	.199	2.062	.042

a. Dependent variable: measured basal metabolic rate.

In the light of Table 19, since P values of the two independent variables are 0.000 and 0.042 respectively, the mid neck girth and body height are related to the basal metabolic rate. Meanwhile, after considering all the factors of the independent variable, we can deduce the final regression equation: Y= -1128.222 + 38.379× mid neck girth + 8.940× body height.

4. Conclusion

In this study, after undertaking the factor analysis, linear regression analysis, univariate analysis and other analysis methods, we can draw the following conclusions:

1) There is commonality among three dimensional body measurement data, embracing maximum belly circumference, waist girth, bust girth (horizontal), thigh girth (right), buttock girth, weight, across shoulder, mid neck girth, waist height, neck height, bust height, buttock height and body height, which can be well divided into girth and height factor, with the waist girth, maximum belly circumference, bust girth (horizontal), thigh girth (right), buttock girth, weight, across shoulder, mid neck girth included in the girth factor while waist height, neck height, bust height, buttock height, body height contained in the height factor.

2) Girth and height factors are correlated with the predicted basal metabolic rate as well as the measured basal metabolic rate. They have a significant linear relationship with the predicted basal metabolic rate whereas there is no significant linear relation between the two factors and the measured basal metabolic rate.

3) There are several variables linearly related to the predicted basal metabolic rate and basal metabolic rate, embracing waist girth, maximum belly circumference, bust girth (horizontal) , thigh girth (right), buttock girth,

weight, across shoulder, mid neck girth and body height. The predicted basal metabolic rate is in a significant correlation with weight, body height, buttock girth and mid neck girth while the basal metabolic rate is correlated with mid neck girth and body height.

5. Acknowledgement

The authors gratefully acknowledge the BIFT-BIEM.L.FDLKK Golf Clothing Ergonomics Centre

References

[1] Piers L. Cornelissen, Martin J. Toveé, Melissa Bateson. Patterns of subcutaneous fat deposition and the relationship between body mass index and waist-to-hip ratio: implications for models of physical attractiveness. J Theor Biol: 2008; 256 (3): 343-252.

[2] Adrian Furnham, K.V. Petrides, Androulla Constantinides. The effects of body mass index and waist-to-hip ratio on ratings of female attractiveness, fecundity, and healthy. Pers Indiv Differ: 2005; 38 (6): 1823-1834.

[3] Caffo B, Diener-West M, Punjabi NM, et al. A novel approach to prediction of mild obstructive sleep disordered breathing in a population-based sample. Sleep: 2010; 33 (12): 1641-1648.

[4] Onat A, Hergenc G, Yaksel H, et al. Neck circumference as a measure of central obesity: associations with metabolic syndrome and obstructive sleep apnea syndrome beyond waist circumference. AM J Clin Nutr: 2009; 28 (1): 46-51.

[5] Seidell JC, Verschuren WM, Van leer E et al. Over weight underweight and mortality. A prospective study of 48, 287 men and women. Arch Intern Med: 1996; 156 (6): 958-963.

[6] Liang K, Hou XG, Wang XH, et al. The predictive value of human obesity measurement parameters on the risk of abnormal glucose metabolism. Journal of Shandong University: 2012; 50 (9): 59-63.

[7] Chu YS, Wang YL, Mao DQ, et al. Study on the basis of rural health adult metabolic rate. J Med Pest Co: 2011; 27 (10): 889-892.

[8] Sun R, Zhang YH. Study on basal metabolic rate of young female. J Clin Exp Med: 2006; 40 (6): 450-451.

[9] Purnell JQ, Kahn SE, Albers JJ, et al. Effect of weight loss with reduction of intro abdominal fat on lipid metabolism in older men. J Clin Endocrinol Metab: 2000; 85 (3): 977-982.

[10] Wing RR, Shoe maker M, Marcus MD, et al. Variables associated with weight loss and improvements in glycemic control in type II diabetic patients in behavioral weight control programs. Int J Obes: 1990; 14 (6): 495-503.

[11] Jason Weeden, John Sabini. Physical attractiveness and health in western societies: A Review. Psychol Bull: 2005; 131 (5): 635-653.

[12] Sybil A. Streeter, Donald H. McBurney. Waist–hip ratio and attractiveness new evidence and a critique of ''a critical test''. Evol Hum Behav: 2003; 24: 88-98.

Design and Research of Intelligent Temperature Monitoring Clothing for Children

Yu Wang[1], Yong-Mei Deng[1*], Xiao-Yu Yang[1,2]

[1] *School of Apparel & Art Design, Xi'an Polytechnic University, No.19 Jinhua Rd (S), Xi'an, Shaanxi, 710048,China*

[2] *Chinese academy of sciences Computer network information center, No. 4 Zhongguancun Street, Beijing, Haidian ,100190, China*

[*]Corresponding author's email: dengymxpu@sina.com

Abstract

At present, the research on smart clothing is mainly on the theoretical side and technical trial stage and there are a few research on the intelligent design of children's wear. In this paper, the design of intelligent temperature monitoring clothing for children is explored. It introduces the principles, concepts, elements and implementation strategies of design for intelligent temperature monitoring clothing. Based on the characteristics of children's psychology, physiology, behavioral cognition, ergonomics, costume design and other interdisciplinary styles, structures and temperature monitoring equipment, the design elements of intelligent temperature monitoring clothing are analyzed in detail. On this basis, this paper summarizs that the design principles of the children' s temperature monitoring clothing.

Keywords: Intelligent; Design; Control Theory; Temperature; Sensors

1. Introduction

In recent years, researchers payed more and more attention to electronic intelligent clothing. Developed countries have invested a lot on human and material resources, and focused on the research of development functional [1, 2] and industrial products. This field of study is lead by United States, Japan and some European countries. In China, the development of intelligent electronic clothing is still in the experimental stage [3, 4], and will have an important impact on China's textile industry.

At present, the integration of electronics, information, semiconductors, wireless communication technology and the development of textile and garment technology has opened up new fields for the development of electronic textiles and Smart clothing [5, 6]. Electronic clothing will be a combination of fashion and high tech functions that are,an important component of smart clothing. Electronic clothing not only detects the external pressure, temperature, charge and stimulation, but can also expand the functionality of clothing in response to the stimuli by using high-tech fabrics, clothing style, structural design perfect integration, expand the functionality of clothing [7].

In the 20th century, the American-led research team began exploring smart clothing. The earliest appearance of intelligent clothing is an experiment done in the 1970s by Mann.Sin " 《Smart Clothing: The Shift to Wearable Computer》 .Writing on the computer requires a lot of space for installation ,and there are wireless networks with a variety of modern equipment [8]. Steve had wanted the equipment to be wearable but it was only introduced in the 20th country after 90 years of research and development. A representative study at the time, the agency had MIT's Media Lab and the United States Oregon University, where smart clothing was first applied to electrocardiogram monitoring, such as smart clothing [9].

On the basis of green environmental protection, the awareness of security is gradually supported by the public. Children's clothing is to be developed from meeting basic attributes such as durability and aesthetic to security, comfort and property changing functions. Therefore the development of smart clothing for children emphasizes health, comfort and security. The behavioral cognitive ability and self-protection ability for 3-7 year-Old Children [10] in preschool and early childhood are relatively weak therefore poorly fit ordinary clothing are not very suitable.

Intelligent temperature monitor clothing for children will have the necessity of research and design.

2. Children' s Physical and Behavioral Cognitive Characteristics

2.1 Children Body Feature

Children and adult body size was significantly different, generally compared with adults. The thickness of the waist of children [11]. The waist circumference is close the circle, sometimes has the very long front and rear thickness situation. Boys and girls at the age of 8 do not have much physical differences. As shown in Fig.1.

Fig.1 Body shape of children

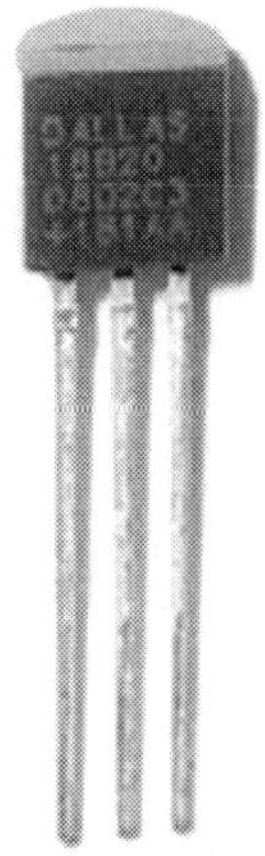

Fig.2 DS18B20 physical map

2.2 Children's Cognitive Behavioral Characteristics

Young children's cognitive behavior is relatively weak during early childhood but gradually develop with age. [12]; Children's control of their behavior are susceptible to external factors Self-regulation mechanism of children are relatively weak as they cannot consciously organize activities and need adult guidance .As they grow, children's self-awareness and control are developed .

In short, children within the ages of 3-7 have low behavioral cognitive ability, self-regulation ability hence rely heavily on parents and other guardians, Clothes that have self-adjusting temperature feature is beneficial to children's physical health [13].

3. Key Technology

3.1 Development Program

The temperature measuring instrument is a new type of sensor which measures the nervous system of intelligent clothing [14], Different kinds of miniature sensors are embedded in the clothing, which can detect changes in the micro climate and physiological parameters such as body temperature, heart rhythm and blood pressure [15].

In this project, we used DS18B20 temperature sensor [16, 17] which is small in size, has reliable performance, high precision, strong anti-interference, low operating voltage and is also harmless to the human body. At present, many new integrated digital sensor circuits are simple, small in size, have high precision, therefore more suitable for clothing. As shown in Fig.2.

When designing children's intelligent temperature monitoring clothing, not only is it important to consider its aesthetic look and practicality, but also the position of the sensor and control unit =The temperature monitoring [18] section includes both hardware system design and software programming. Therefore part of the design is to minimize the space accommodated by the control functions so that it is portable and lightweight.

3.2 Overall Description of Control System

The goal of the control system for children' s intelligent temperature clothing is to monitor the ambient temperature of the garment in real time through the temperature sensor embedded inside the garment. Then transmit the information to the central processing unit, which is used to compare, analyze, process, store and display the data. By programming the heating components and the heat dissipation components of the switch, the clothing can adjust the ambient temperature of the garment within the preset range of 28-32 degrees. The work frame of the central processor is shown in Fig.3.

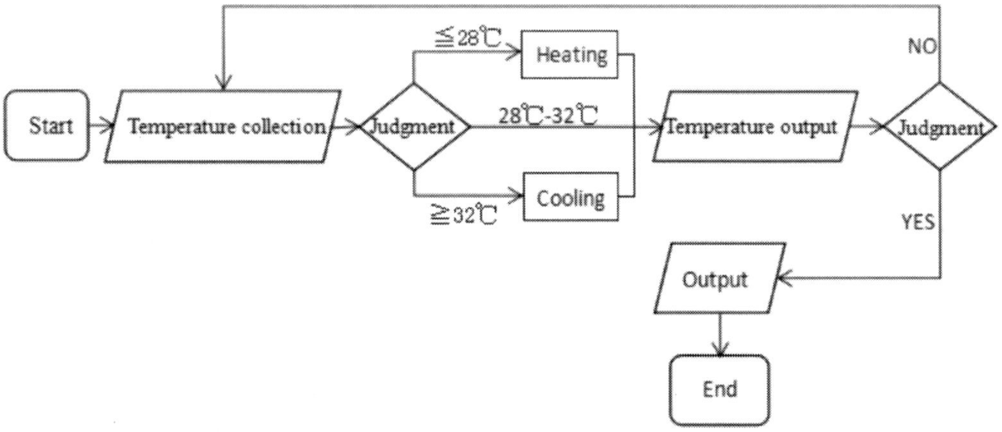

Fig.3 Central processor work frame

Processor Temperature indexs and output Formula:

$$Y=\begin{cases} x+b & x<28℃ \\ x & 28℃\le x\le 32℃ \\ x-b & x>32℃ \end{cases}$$

Y-Temperature value of final output;
x-Displayed temperature value;
b-Add heat or Bulk Heat.

There are three temperature sensors in this clothes. In order to prevent other factors interference, the system control heating and cooling devices are based on two or three temperature sensors that collected data. Because the temperature range required for control in this subject is 28-32 ℃, the maximum temperature setting of the central processor is 32 ℃ and the minimum temperature is 28 ℃.The central controller can adjust the temperature based on the differences between the temperature of the individual and the interior of the garment to the user's need.

4. Garment Prototype Design

Due to special characteristics of children's body, they wear the H-type, A-type and O-type. Before child's body features were taken in consideration, the clothing type were based on chest convex abdomen, chest waist and hip measurements size gap. The development of the children's smart clothing were limited to H-type. H-type clothing have long, simple, relaxed, comfortable features. As children's clothing need to be embedded with the equipment, it is designed to have three layers: the inner layer (lining), the middle layer (functional layer), the outer layer (surface). The design is shown in Fig.4.

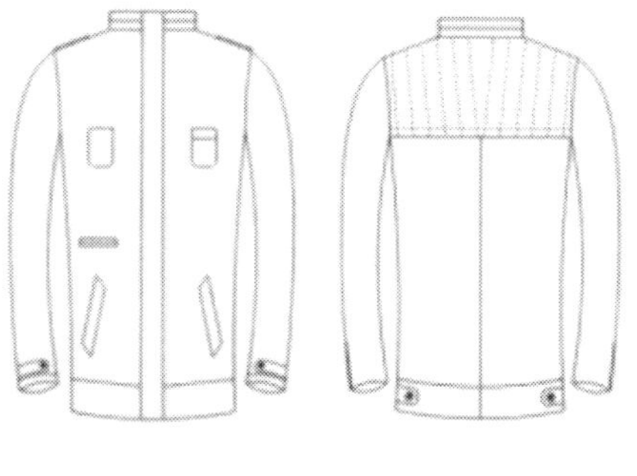

a. Front b. Back

Fig.4 Children' s clothing prototype

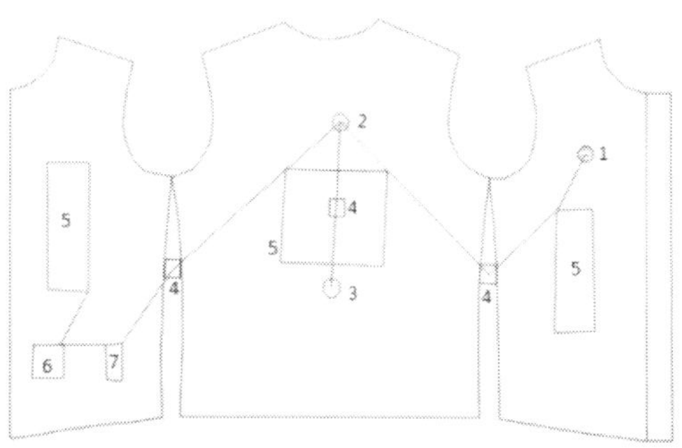

1-1sensor, 2-2sensor, 3-3sensor, 4-heat sink
5-Heater, 6-Central processing unit, 7-power supply

Fig.5 Clothing internal structure design

5. Internal Structure Layout

There are three presets to intelligent clothing temperature sensor: First, the temperature sensor location is set in front of the f left chest. The second temperature sensor is located at the back towards the center position. The third temperature sensor is set at the center of the waist.

There are different ambient temperatures when underwear, shirt and cardigan sweater are worn. The air layer temperature in the different areas of the clothing is lower than that of the armpit and the abdomen particularly the three parts which are chest, back and waist [19]. We placed three temperature sensors in the chest, back and waist to insure the air layer temperature is within the minimum value of comfort range. Radiator Layout Design: Number First radiator position is set at the center of the back, second and third radiator is set in the left and right armpit 6cm, respectively.

Central processor, Power layout design: The central processing unit and the power supply are set on the left front of the belly of the garment to ensure maximum wearer comfort as it is the largest part while. For the convenience, all equipment can be disassembled. The layout and connection diagram of each component in the garment are shown in Fig.5.

6. Clothing Performance Test Analysis

6.1 Experimental Measurements

Test method: Air-conditioning was used to adjust the indoor ambient temperature to the experimental temperature. The subject wore a smart temperature suit on top of a long sleeved shirt. The precision of the temperature sensor is at 0.1 degrees

Table.1 Temperature data

Experimental temperature (℃)	Balance time (s)	Clothing temperature(℃)						Mean variance(℃)		
		Sensor 1		Sensor 2		Sensor 3		Sensor1	sensor2	sensor3
		max	min	max	min	max	min			
16	1010	34	20	34	19	33	18	26.1	25.4	24.3
21	652	34	24	34	24	33	23	28.5	28.5	27.5
25	530	35	27	35	27	34	26	30.7	30.7	29.7
30	308	35	32	34	31	34	31	33.4	32.4	32.4

Experimental scheme: After adjusting the laboratory temperature, the subject enters the room and within one minute, the system starts to work and begins to record the experimental data. The experimental data are shown in table 1.

Experiment Objective: To test the time and the maximum and minimum temperature of the controlled clothes from start to the preset temperature range in different environment, and to calculate the temperature rate.

6.2 Experimental Data Analysis

Through the human body wearing test and statistical analysis of the relevant data, the development of children's intelligent temperature monitoring clothing are able to achieve the initial design goals. When the experimental temperature is at 16 degrees, the average heating rate is 0.015 degrees/Sec, the time achieved a temperature balance is 1010 seconds, and the heating effect is significant when the maximum heating rate is 0.042 degrees/sec. In this experiment, the temperature control can increase the maximum temperature of the human body by 14 degrees and greatly improve the wearing comfort. When the test temperature is between 21 to 25 degrees, the clothing heat rate increases at 0.02 degrees/sec at the average temperature of 21 degrees Celsius and 25 degrees/sec at 0.19 degrees Celsius. Under high temperatures of 30 to 35 degrees, the clothing can regulate the temperature to 32 to 33 degrees to ensure comfort. The images of data processing results are shown in Fig.6, Fig.7.

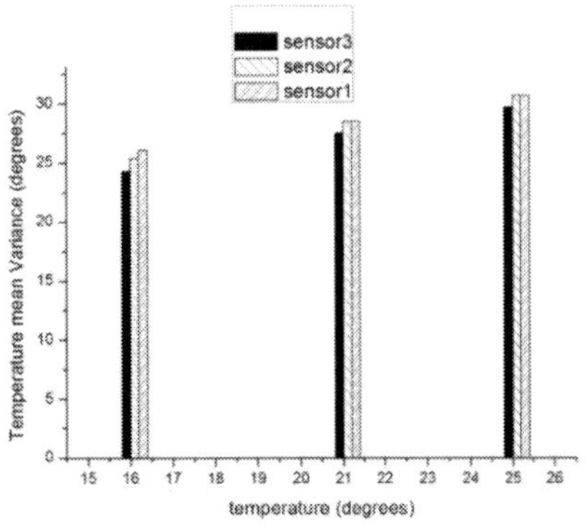

Fig.6 Three sensor temperature mean variance bar chart

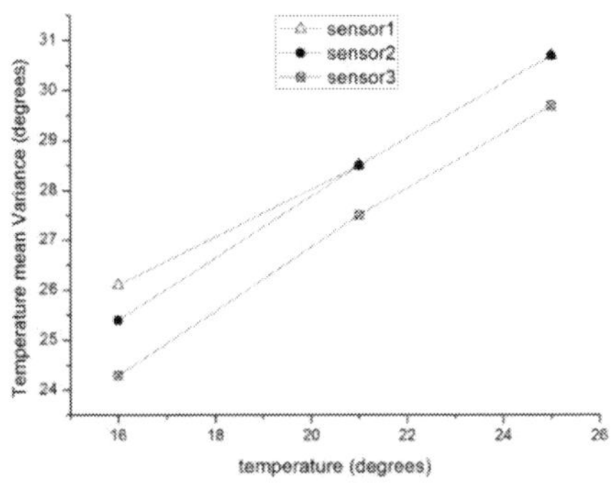

Fig.7 Three sensor temperature mean variance line rate graph

7. Conclusion

The follow up development on the research of intelligent clothing will be on clothing fabrics to make it more responsive to the needs of children. Furthermore to add on equipment fastening accessories to ensure the garment is solid and more convenient. With the progress of science and technology, intelligent systems will be smaller and more convenient [20, 21]. The effectiveness of intelligent clothing will be more comprehensive, efficient and flexible [22]. The gradual improvement of the intelligent system will expand the consumer market not only to children's wear, but also to other industries, so that everyone can enjoy the safety, convenience and comfort brought by technology.

8. Acknowledgement

This paper is supported by HaoYun Lai Innovation Fund(2014KJ-048)- Knitted woolen Dress Design Study(HYL201608); Science and Technology Planning Project of Shan xi Province-Natural Science Foundation Research Program (Grant No:2015JM5167); Special Research Program of Education Department of Shan xi Province (Grant No:12JK0571).

References

[1] Ngai EWT, Peng S, Alexander Paul. Decision support and intelligent systems in the textile and apparel supply chain: An academic review of research articles. Expert Syst Appl: 2014; 1: 81-91.

[2] Guo R. Electronic intelligent clothing development status and market analysis. Text & Tecnol: 2010; 3: 55-58.

[3] Yan NN, Zhang H, Deng YM. Wearable research situation and development trend of medical care clothing. J ind Text: 2015; 6: 162-168.

[4] Honarvar, Mozhdeh Ghahremani, Latifi, Masoud. overview of wearable electronics and smart textiles. J Text I: 2017; 4: 631-652.

[5] Guan CY, Qin SF, Ling Wessle. Apparel recommendation system evolution: an empirical review. Int J Cloth Sci Tech: 2016; 6: 854-879.

[6] Wei BL, Xu WL, Duan JH. Advances in the application of UWB in biomedical electronics. Tech Coloproctol: 2014; 10: 132-136.

[7] Stoppa M, Chiolerio A. Wearable Electronics and Smart Textiles. Sensor Rev: 2014; 11957-11992.

[8] Sun B. Embedded system in the application of smart clothing. Int J Cloth Sci Tech: 2009; 01: 34-39.

[9] Gulrajani ML, Gupta Deepti. Emerging techniques for functional finishing of textiles.Indian J Fibre Text: 2011; 4: 388-397.

[10] Xu ZY, Wu NY, Min RF. 3-year-old children control their behavioral development of his point and influence the development of factors. Biol Psychol: 1987; 1: 18-24.

[11] Zhu SW, Liu JW. Clothing Materials. China Textile Publishing House: 2011; 4: 79-100.

[12] Marisa C, Michael C.The development of children's ability to track and predict turn structure in conversation.J Mem Lang: 2017; 92: 234-253.

[13] Xiao H. Clothing health and comfort application. J Donghua University: 2009; 5: 24. 25.

[14] Cui YM. children's clothing design. Donghua University: 2009; 72-81.

[15] Zhao HF. Intelligent heating clothing and thermal properties. Journal of Tianjin University (Sci & Tecnol): 2014; 1-2.

[16] Li L. Design of intelligent garment with transcutaneous electrical nerve stimulation function based on the intarsia knitting technique. Text Res J: 2010; 3: 279-286.

[17] Li HQ, Yang HJ, Li EB. Wearable sensors in intelligent clothing for measuring human body temperature based on optical fiber Bragg grating. Opt Express: 2012; 11: 11740-11752.

[18] Lesnikowski, Jacek. Research on poppers used as electrical connectors in hign speed textile transmision lines. Autex Res J: 2016; 4: 228-235.

[19] Tae-gyu Lee, Seong-Hoon Le. Dynamic bio-sensing process design in mobile wellness information system for smart healthcare. Wireless Pers Commun: 2016; 86: 201-215.

[20] Zhu QH, Yi J, Sheng SY. A Computer-Aided modeling and measurement system for environmental thermal comfort sensing.Ieee T Instrum Meas: 2015; 2: 478-486.

[22] Yu XG, Miao CY, Li HQ. Research on demodulation system for human body temperature measurement of intelligent clothing based on arrayed waveguide grating. Spectrosc Spect Anal: 2012; 8: 2032-2036.

Thermal and Compression Performance of Aerogel Incorporated Textiles

Xiao-Man Xiong[1*], Tao Yang[1], Rajesh Mishra[1], Jiri Militky[1], Hiroyuki Kanai[2]

[1] *Department of Material Engineering, Faculty of Textile Engineering, Technical University of Liberec, Liberec 46117, Czech*
[2] *Department of Advanced Textile and Kansei Engineering, Faculty of Textile Science and Technology, Shinshu University, 3-15-1 Tokida, Ueda Nagano 386-8567, Japan*

[*]Corresponding author's email: xiaoman.xiong@tul.cz

Abstract

A new approach to apply silica aerogel into textile by using laser treatment and laminating technique was proposed. Alambeta instrument and FLIR ThermaCAM TVS300 thermal camera were employed to measure thermal insulation performance of the prepared composites. ORIENTEC STA-1225 Universal Testing Machine was used for testing compression properties. Results showed that the aerogel-incorporated structure has significant effect on thermal insulation enhancement while the air-filled voids give small increment to thermal resistance value. A temperature gap of 1°C to 1.5°C was observed between aerogel-filled samples and regular samples. Nonwoven fabrics with air-filled voids or aerogel-filled voids were observed to have lower compression resistance and compression resilience than regular nonwovens. Results also indicated that the decrease in compression resilience induced by air-filled voids and aerogel-filled voids are respectively ranging from 4.50 % to 12.76 % and 7.49 % to 11.38 %.

Keywords: Silica Aerogel; Nonwoven Fabrics; Laser Treatment; Thermal Insulation; Compression Properties

1. Introduction

Silica aerogel has across-linked internal structure of silica dioxide with a large number of air-filled nano-scale pores. Nowadays, silica aerogel is well acknowledged as one of the most attractive thermal insulating materials for applications in air crafts, building constructions, and so forth. Recently, the successful and cost-effective production of silica aerogels by use of inexpensive precursors and ambient pressure drying method has been achieved [1]. This raises the possibility of continuous production with lower capital and operating costs for industrial application. However, due to the weak strength resulted from high porosity and cracking along the drying process, aerogels cannot be easily applied to conventional applications. They are usually filled into hollow spaces to provide high thermal resistance. Thus, incorporating aerogels in textile structure, such as woven fabric, spacer fabric and nonwoven is quite effective since silica aerogels can act as a medium to fill the interstitial space among fibers in these structures.

A lot of effort has been made by scientific researchers to explore the combination of silica aerogel with available supporting materials as thermal insulator, but only a few investigations regarding the incorporation of aerogel with textile structure for thermal insulation enhancement can be found in existing literature. Aspen Aerogels Inc., USA, produced an "Aerogel blanket" for winter apparel products application by using a fiber matrix impregnated with an aerogel forming precursor and supercritical drying method [2]. G. Rosace, E. etc. [3] selected three cotton woven fabrics to apply silica aerogel on the surface by coating and found that pique showed the lowest values in thermal resistance and satin showed the highest values while plain weave lied in between. It was also stated that the thermal properties of treated high-density cotton plain weave fabric were strongly influenced by finishing agent concentration. Zhengkun Qi etc. [4] embedded aerogel into the thermal barrier layer of firefighters' protective clothing (FPC) and concluded that the backside temperature of the FPC samples with aerogel was about 100 °C lower than that of the samples without aerogel when exposed to radiant heat. Abu Shaid etc. [5] incorporated super hydrophobic silica aerogel nanoparticles in 65/35 wool-Aramid blended fabric and analyzed the thermo-physiological comfort of the composite. Their results showed that only 2% coating of aerogel nanoparticle increases thermal resistance by up to 68.64% and can reduce air permeability by up to 45.46% whereas 4% aerogel coating can reduce 61.76% air permeability. These studies all confirmed that the present aerogel in textile structure significantly improve the thermal performance of the composite, however, the application of aerogel granules has so far been limited in a few methods such as coating, padding and impregnating, with the assistant of additive agent. Since aerogel granules are exposed or filled into the void space in textile structure, and the porous space of the loose textile structure is partly filled by additive agent, these will reduce the thermal performance because the porous space is essential and useful to entrap air pockets to enhance thermal insulation properties. Furthermore, the prepared composite will be lack of compressibility and compression resilience, which influences the final use and the durability of thermal-insulating function.

In this work, a new approach to apply silica aerogel into textile for thermal insulation was proposed. High porous nonwoven fabrics as well as sponge foam were selected as support layers to produce aerogel-encapsulated composites. The support layers were pretreated by laser treatment to form go-through holes along the thickness direction as designed, a thin fabric layer was bonded on one side of the support layer, aerogel granules were subsequently filled into these holes followed by the laminating of the same fabric layer on the other surface to prevent against aerogel loss. Since both sides of the support layer were covered by soft fabrics to achieve a closed fabric system and the adhesion of aerogel with the support structure was not involved, these multilayer fabrics will have light weight, excellent thermal resistance and good compression property simultaneously.

2. Methodology

2.1 Materials

Nonwoven and sponge foam are both high porous and flexible thermal insulators widely used in clothing or thermal insulating field. In this work, three types of nonwoven fabrics made by 70 % PET and 30 % bi-component PET as well as a PU sponge foam purchased from Shijiazhuang Chuanghong Technology Co., Ltd were selected as the support layer to hold silica aerogels, their characterizations are listed in Table 1. Fabric weights per unit area were determined according to ASTM D1059 standard using electronic weighing scales, fabric thicknesses were measured according to ASTM D1777-96 standard with the SDL digital thickness gauge at a pressure of 100 Pa. A thermo-bonded nonwoven fabric with thickness 0.16 mm and GSM 49.55 g/m^2 was used as cover layer to laminate onto both sides of the support layers. Silica aerogel granules were purchased from Cabot aerogel Corp., the specifications of the aerogel granule are shown in Table 2. In order to bond the covering with support layer and maintain the flexibility of the multilayer composite, an extra soft textile adhesive film was used for laminating, which can perfectly adapt to the fabric surface without increasing thickness.

Table.1 Structural parameters of samples used as support layers

Sample codes	Structure	Bulk density kg/m^3	Thickness mm	GSM g/m^2
A	High-loft nonwoven	25.56	12.42	317.51
B	High-loft nonwoven	26.79	9.68	259.28
C	High-loft nonwoven	16.48	12.05	198.64
D	PU sponge	29.54	6.67	197.01

Table.2 Specifications of aerogel granules

Properties	Value range
Particle size (mm)	0.1-0.7
Pore diameter (nm)	~ 20
Particle density (kg/m^3)	About 120
Thermal conductivity (W/m·K)	0.012 (at 25℃)

2.2 Methods

2.2.1 Preparation of Aerogel Incorporated Composite

The aerogel incorporated fabric system was designed with a composite structure. This composite has three layers: a support layer, a base layer on the bottom and a face layer on the surface. The support layer was pretreated by laser treatment to make uniformly distributed go-through holes along the thickness direction, using a commercial pulsed CO2 laser system GFK Marcatex FLEXI-150 under atmospheric condition in air. The generated wavelength of laser beam was set at 10.6 μm and the input voltage was set at 100V. Before the laser treatment, the holes pattern file was designed in grey scale by Photoshop CS4 graphic design software. Then the samples were placed in the laser treatment cabinet for testing with different parameters, followed by the laser treatment on nonwovens or foam with optimized parameters. The specifications of parameters are listed in Table 3.

Table.3 Specifications of parameters for laser treatment

Parameter	A	B	C	D
Duty circle %	50	50	45	45
Pixel time μs	100	50	50	80
Resolution dpi	96	96	96	96

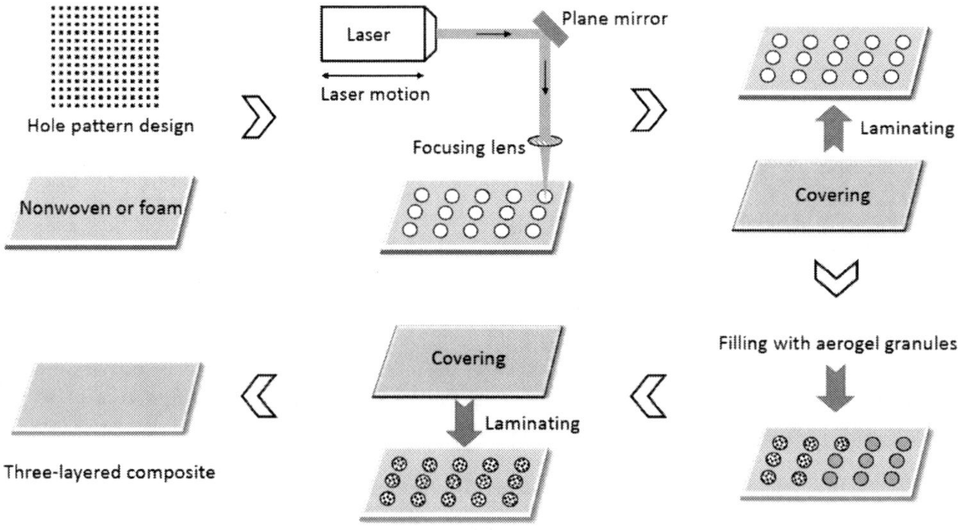

Fig.1 Fabrication process of aerogel incorporated composite.

After laser treatment, the covering fabric was bonded onto one side of the nonwoven or foam using textile adhesive film. Aerogel granules were uniformly filled into these holes, and another covering fabric was subsequently bonded on the upper surface to form a closed fabric system. The three-layered system with aerogel-filled voids were prepared. The fabrication process was illustrated in Fig.1. Different cases of samples including untreated support layer and support layer with air-filled voids were prepared as control samples. In the case of aerogel-incorporated composites, the aerogel content differs with the support layer, with values of 31.89 g/m^2, 42.35 g/m^2, 34.68 g/m^2 and 18.76 g/m^2 for support material A, B, C and D respectively.

2.2.2 Measurements

Thermal insulation performance like thermal conductivity, thermal resistance were examined by Alambeta Instrument according to EN 31092 Standard. Tests were carried out in the standard atmospheric condition of 65 ±

4% and 23 ± 2°C. All samples were conditioned under the standard condition at 23°C and 65% relative humidity for 24 h before testing. The mean value of six measurements for each sample was calculated.

The thermal imaging measurement was conducted by using FLIR ThermaCAM TVS300 thermal camera. Tests were carried out in a thermography chamber with dimension 45 cm × 32 cm × 12 cm, a guard hot plate with constant temperature 33°C was located in the thermography chamber to provide uniform thermal radiation, the thermal camera was placed in the air space with a distance of 40 cm from the hot plate. When the specimen was placed on the hot plate, pictures were taken every 5 second up to when the heat transfer reaches steady state. The room temperature was kept at 23± 2°C. The temperature of the surface was calculated using Avio Thermography Studio 2007 software in which each pixel of the picture was allocated to one temperature value.

The ORIENTEC STA-1225 Universal Testing Machine was used for testing the compression property of the composites. The sample size used in the compression test was 20 cm × 20 cm and the loading speed was set at 2 mm/min. The maximum pressure used was 54.92 kPa, the pressure foot area was 36.3 cm^2 with diameter 68 mm. In each case, 5 specimens were tested and the average values were reported.

3. Results and Discussion

3.1 Thermal Insulation Performance

Nonwoven fabrics possess a large amount of void space, which can entrap large volumes of still air. Resulted from the much lower conductivity of still air in comparison to textile fibers, the thermal insulation performance of nonwoven textiles is determined by the trapped air in the inter-fiber spaces. The measured thermal conductivity and thermal resistance of prepared composites are shown in Fig.2 and Fig.3. M, N and W refer to regular composite without laser treatment, composite with air-filled voids and with aerogel-filled voids respectively.

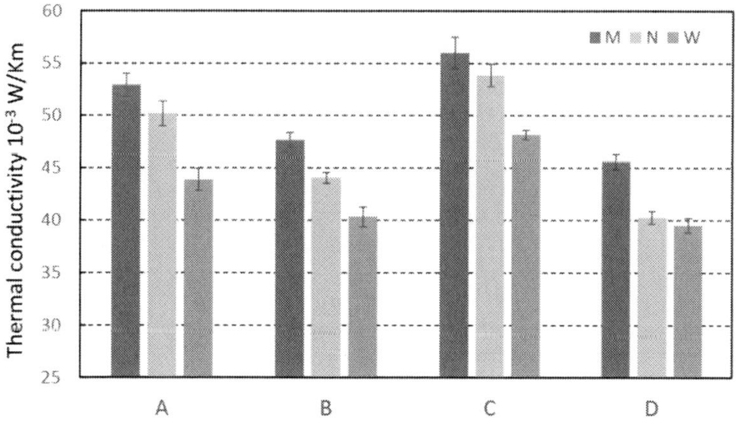

Fig.2 Thermal conductivity of different composites.

It is clear that for all sample groups the air-filled voids give slightly decrease to thermal conductivity value and small rise to thermal resistance as compared to regular samples. This is because the large voids formed by laser treatment will enlarge the amount of still air in the composites structure, causing less heat transfer through the fabrics by conduction. Heat transfer by free convection in nonwoven materials has been reported to be negligible because the fibers subdivide the gas into sufficiently small pores and the tortuous nature of air channels present prevent any heat transfer by convection [6]. Thus, the regular samples can be assumed to be free of convection. These samples gained lots of huge air-filled voids with a size about 5mm × 5mm after laser treatment, which may allow air flows through and cause heat transfer by convection. However, the laminating of covering fabrics on the surfaces could minimize this heat convection.

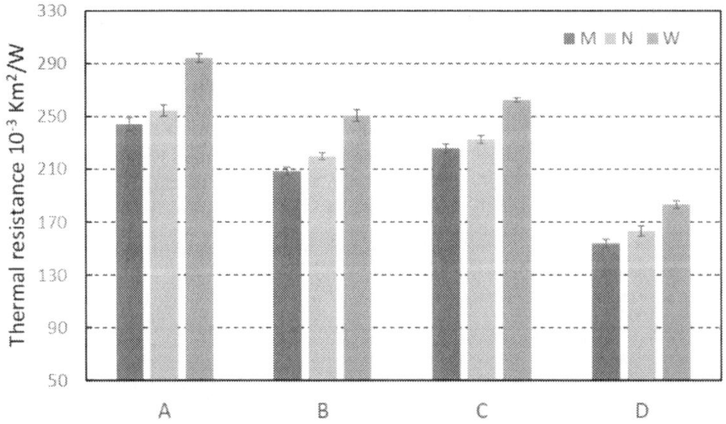

Fig.3 Thermal resistance of different composites.

It is also observed that the aerogel-filled voids give decrease to thermal conductivity while the thermal resistance significantly increases. For the aerogel-filled composite, the large open voids are filled by nanoporous structural aerogel, this will further prevent the convection current transfer through the composites because aerogel granule can be approximately considered as air-proof material. Furthermore, since the pore size in the aerogel granules is lower than the mean free path of air molecules, these extremely small pore size causes a very low gaseous thermal conductivity due to Knudsen effect [7], resulting in less heat transfer through the composite and significant improvement of thermal insulation performance.

3.2 Infrared Thermography

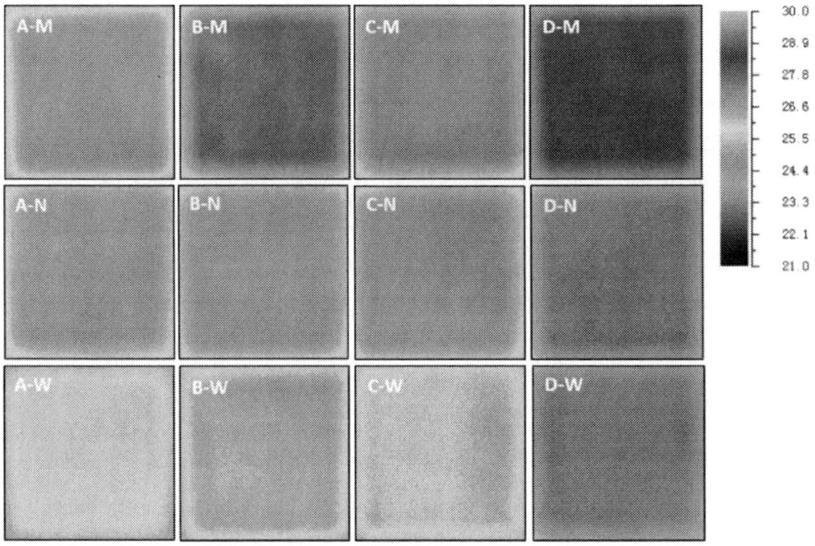

Fig.4 Infrared thermography images.

Infrared thermography images are shown in Fig.4. In general, thermal images are actually visual displays of the amount of infrared energy emitted, transmitted, and reflected by an object. The amount of radiation emitted by the heat plate through the fabric was detected, therefore the thermography allows to see temperature variations. For each group, the detected temperatures from samples with air-filled holes are found to be lower than that from regular samples, and the lowest values are observed from aerogel-filled samples, indicating that aerogel-filled samples exhibit best thermal insulation ability. The maximum temperature detected from regular samples A-M, B-

M, C-M and D-M are 27.3°C, 27.6°C, 27.5°C and 28.3°C respectively, whereas those for samples with air-filled holes are only 26.8°C, 27.0°C, 27.0°C and 27.5°C while these values for aerogel-filled samples are 25.8°C, 26.5°C, 26.2°C and 27.1°C respectively. Generally, a temperature gap of 1°C to 1.5°C was observed between aerogel-filled samples and regular samples under the condition of a temperature difference 10 °C between hot plate and environment. Normally, this temperature gap will be enlarged under greater temperature gradient condition.

3.3 Compression Property

Compression property plays an important role in textile insulators because compressibility has a linear relationship with thermal conductivity. Compression can be defined as a decrease of initial thickness that occurs with an appropriate increase of force. Compression behavior of a fabric is generally described by the relationship between the applied force per unit area and the resulting fabric thickness, in the form of compression load-displacement curve. The compression resistance and compression resilience of all the fabrics were calculated from compression hysteresis curves, using the following Equations:

$$\text{Compression resistance (\%)} = [(T_0-T_1)/T_0]\times 100$$
$$\text{Compression resilience (\%)} = [W_R/W_C]\times 100$$

where T_0 is the initial thickness, T_1 is the thickness at maximum pressure, W_C is the work done during compression and W_R is the work done during recovery process.

The compression resistance of different fabrics under 560 gf/cm^2 pressure in terms of the percentage change in thickness, is shown in Fig.5.

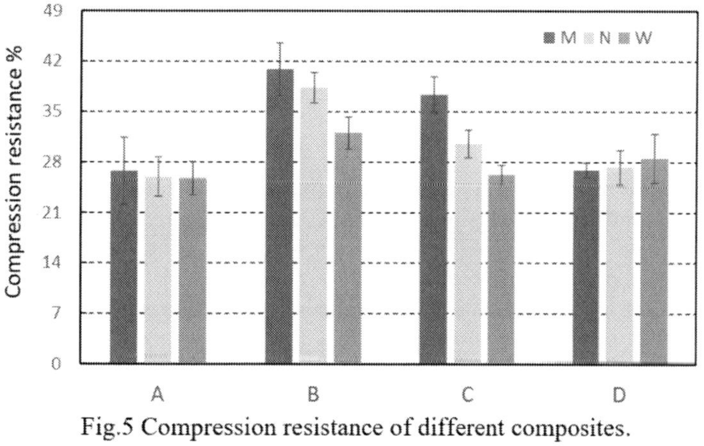

Fig.5 Compression resistance of different composites.

It is apparent that nonwoven fabrics with air-filled voids (A-N, B-N, C-N) or aerogel-filled voids (A-W, B-W, C-W) generally have lower compression resistance than those regular nonwovens (A-M, B-M, C-M), indicating that nonwoven fabrics consisting of air-filled voids and aerogel-filled voids can be compressed easily under the same pressure range to a much higher extent. During the process of fabric compression, the inter-fiber spaces decrease continuously, the resistance force necessary to compress a fabric has to overcome the internal stresses of the fibers and the inter-fiber frictional force [7]. Thus, the compression resistance of a sample is closely related to fiber quantity, samples with less fibers have a lower compression resistance if the fiber arrangement of the fabrics are assumed to be the same. However, it is found that in comparison to sample D-M the air-filled voids (D-N) show insignificant effect on compression resistance while the aerogel-filled voids (D-W) give slightly increase to compression resistance. This is because in sponge foam there is no fiber-to-fiber slippage and fiber entanglement happened during the process of compression, the deformed network is easy to return back to their earlier position.

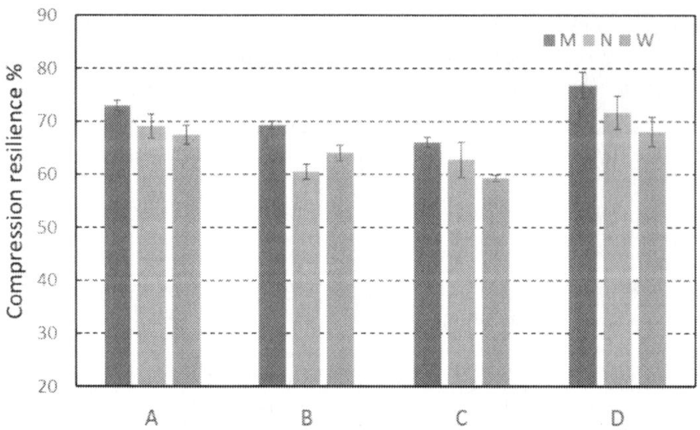

Fig.6 Compression resilience of different composites.

Fig.6 shows the compressive resilience of different composites. The compression resilience values of regular samples (A-M, B-M, C-M and D-M) are found to lie in the range of 66.11 % - 76.82 %, samples with air-filled voids (A-N, B-N, C-N and D-N) and aerogel-filled voids (A-W, B-W, C-W and D-W) are observed to recover slightly less than regular samples. This may be attributed to the fact that the amount of fibers per unit area of the laser-treated nonwovens is much lower than those of regular samples, the compressive load will be shared by a fewer number of fibers, resulting in a stronger fiber-to-fiber slippage, fiber bending and fiber entanglement during the compression phase, this may lead to higher energy loss and reduced percentage compression resilience. Hence, the recovery properties after compression greatly depend on the composite structure, samples containing less fibers have low compression resilience than those containing more fibers in their structures. Results also indicates that the decrease in compression resilience induced by air-filled voids and aerogel-filled voids are respectively ranging from 4.50 % to 12.76 % and 7.49 % to 11.38 %.

4. Conclusion

Three kinds of nonwoven fabrics and one type of sponge foam were selected as support layers to carry out laser treatment, silica aerogel granules were applied into these support layers by using laminating technique. The prepared aerogel-incorporated textiles were examined in terms of thermal insulation performance, Infrared thermography images and compression properties.

Results showed that the aerogel-filled voids give remarkable rise to thermal resistance while the air-filled voids slightly improve the thermal insulation performance. A temperature gap of 1°C to 1.5°C is observed between aerogel-filled samples and regular samples under the condition of a temperature difference 10 °C between hot plate and environment. It is also found that nonwoven fabrics with air-filled voids or aerogel-filled voids generally have lower compression resistance than those regular nonwovens, the air-filled voids show insignificant effect on compression resistance of sponge foam while the aerogel-filled voids give slightly increase to compression resistance. The compression resilience values of regular samples are found to lie in the range of 66.11 % - 76.82 %, samples with air-filled voids and aerogel-filled voids are observed to recover slightly less than regular samples. Results also indicates that the decrease in compression resilience induced by air-filled voids and aerogel-filled voids are respectively ranging from 4.50 % to 12.76 % and 7.49 % to 11.38 %.

5. Acknowledgement

This work was supported by the research project of Student Grant Competition of Technical University of Liberec no. 21195/2017granted by Ministry of Education Youth and Sports of Czech Republic.

References

[1] Bargozin H, Amirkhani L, Moghaddas JS, Ahadian MM. Synthesis and application of silica aerogel-MWCNT nanocomposites for adsorption of organic pollutants. Sci. Iran: 2010; 17 (2F): 122.

[2] Ryu J. Flexible aerogel superinsulation and its manufacture, United State Patent No. 6068882: 2000.

[3] Rosace G, Guido E, Colleoni C, Barigozzi G. Influence of Textile Structure and Silica Based Finishing on Thermal Insulation Properties of Cotton Fabrics. International Journal of Polymer Science: 2016.

[4] Chemical Rubber Company, Lide DR. CRC handbook of chemistry and physics: a ready-reference book of chemical and physical data (86. ed). Boca Raton: CRC Press: 2005.

[5] Shaid A, Fergusson M, Wang L. Thermo-physiological comfort analysis of aerogel nanoparticle incorporated fabric for fire fighter's protective clothing. Chemical and Materials Engineering: 2014; 2: 37-43.

[6] Bhattacharyya RK. Heat-Transfer Model for Fibrous Insulations, Thermal Insulation Performance, edited by D. L. McElroy and R. P. Tye, ASTM Special Technical Publication 718, American Society for Testing and Materials, Philadelphia: 1980; p. 272-286.

[7] Pierre, Alain C, Gérard M. Pajonk. Chemistry of Aerogels and Their Applications. Chemical Reviews: 2002; 102 (11): 4243-66.

Preparation of Magnetic Immobilized Glucose Oxidase and Bleaching of Cotton Fabrics

Fang Qiu*, Feng-Yan Li, Qing-Yu Yang

School of Textile, Tianjin Polytechnic University, Tianjin, 300387, China

*Corresponding author's email: 1046915591@qq.com

Abstract

To improve the stability and reproducibility of glucose oxidase, the poly diallyl dimethyl ammonium chloride-iron oxide (PDDA-Fe_3O_4) magnetic carrier is bound to glucose oxidase by electrostatic attraction to prepare immobilized glucose oxidase. The immobilization conditions of immobilized enzyme are then further studied. The result shows that the optimal immobilization temperature is 5 °C, the pH value is 7, and the mass ratio of magnetic carrier to enzyme is 30: 1. The thermal stability of immobilized glucose oxidase is improved, its range of pH value is enlarged, and the optimum pH value is 8. After 5 times of reuse, the glucose oxidase still retain a higher activity. After being bleached by immobilized glucose oxidase, the maximum whiteness of the cotton fabric is close to 90.

Keywords: Glucose Oxidase; Magnetic Carrier; Immobilized Enzyme; Stability; Refuse; Bleaching

1. Introduction

People are gradually paying attention to the sustainable development of the textile industry,. Traditional cotton fabric bleaching is mainly conducted by oxidation [1] and reduction [2]. Both of them use a large number of chemical reagents which have significant damage to the fabric, and produce environmental problems when consuming energy. The high efficiency, specificity, mildness and biodegradability of biological enzymes make them more and more important in textile bleaching [3]. Glucose oxidase can catalyze the decomposition of glucose into hydrogen peroxide and gluconic acid for bleaching, and gluconic acid can chelate metal ions, so the bleaching process don't need any other chemical reagents [4]. Compared with traditional methods, glucose oxidase bleaching have relatively small damages to the fabric and environment. However, due to the existence of free enzyme inactivation, difficulty of recovery, high cost and other issues, it is difficult to apply on a large-scale [5].

Immobilized enzyme has the advantages of high stability, easy recovery, reusability, etc. In recent years, its application research in the field of textile has been paid more and more attention [6]. Enzyme immobilization methods mainly includes adsorption method [7], cross-linking method [8], covalent bond method [9] and embedding method [10]. In this paper, adsorption method was used and the operation is simple and easy. Zong Wei, Liu Yanfang et al [11] optimized the immobilization conditions, the lipase was immobilized on the magnetic chitosan carrier by physical adsorption. Chen Shanshan et al [12] used anion exchange resin pectinase immobilization through the ion adsorption method. The combination of the magnetic carrier and enzyme make the separation and recovery simple, low cost, and the use of external magnetic field can control the magnetic material immobilized enzyme movement direction to improve the catalytic efficiency of immobilized enzyme [13]. However, the immobilized conditions need to be further explored. In this paper, glucose oxidase and self-made poly diallyl dimethyl ammonium chloride-iron oxide (Fe_3O_4-PDDA) magnetic nanoparticles are combined to obtain good stability, high relative activity and easy recovery. The optimum conditions for immobilization of glucose oxidase and the changes in enzyme performance were studied. The bleaching effect of immobilized glucose oxidase on cotton fabric was also discussed. Fig.1, it is the preparation and application of immobilized glucose oxidase.

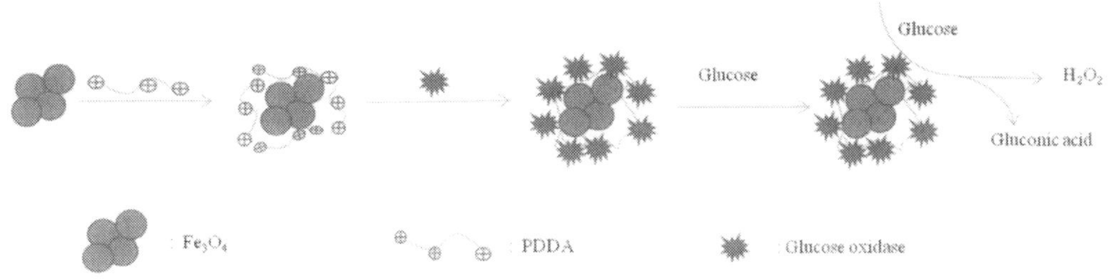

Fig.1 The preparation and application of immobilization glucose oxidase

2. Experimental

2.1 Materials and Instruments

Poly diallyl dimethyl ammonium chloride (PDDA) was purchased from Shandong. Glucose oxidase was puchased from Shanghai. Iron oxide was prepared by our research group. Indigo carmine, sodium hydroxide, glucose, crystalline sodium acetate, glacial acetic acid, ferrous chloride, and ferric chloride sodium pyrophosphate were puchased from Tianjin.

PHS-2F digital pH meter, Electronic analytical balance, WSD-3U fluorescent whiteness Meter, V-1200 spectrophotometer, HH-2K4 two rows of four-hole water bath, KQ2200DB CNC Ultrasonic Cleaner, FD-1-50 freeze dryer.

2.2 Immobilization of Glucose Oxidase

The preparation method of the experimental PDDA-Fe_3O_4 magnetic composite particles refer to published literature [14]. 0.15 g of PDDA-Fe_3O_4 nanoparticles and a certain amount of glucose oxidase were added to 3 mL 0.2 mol/L soum acetate buffer with different pH. After 12 h at a certain temperature f, washed it by magnetic separation technology, to remove unfixed immobilized glucose oxidase. The final substance was immobilized glucose oxidase, freeze-dried and stored it at 5 °C.

2.3 Determination of Glucose Oxidase Activity

At 37 °C, when catalytic glucose, the amount of enzyme needed to produce 1ug hydrogen peroxide in a minute can be defined as an enzyme activity unit, i.e. 1 U. Glucose oxidase catalyzes the oxidation of glucose to produce hydrogen peroxide, it can make indigo carmine fade at pH = 4 and 100°C, and the reaction rate is proportional to the concentration of hydrogen peroxide in a certain range [15]. Therefore, the activity of glucose oxidase was measured by indigo carmine fading method.

, Placed appropriate amount of enzyme into a color tube with 2 mL PH = 8 buffer. And 2 mL 0.2 mol/L glucose oxidase solution was added, reacted at 37°C for 10 min, then killed the enzyme at high temperature, chilled-down it for 5 min. added 3 mL pH = 4 sodium acetate buffer and 1.3 mL 1.0×10^{-3} mol/L indigo carmine solution to the tube, and diluted the mixture to 25 mL, then bathed at 100°C for 13 min, finally, chilled-down it for 5 min. The absorbance was measured with a spectrophotometer. The concentration of H_2O_2 was calculated by compared with the H_2O_2 standard curve, thereby calculated enzyme activity.

2.4 PH Stability of Free Enzyme and Immobilized Enzyme

Under the condition of 37°C, a certain amount of free glucose oxidase and immobilized glucose oxidase were added to the color tubes, then 2 mL acetic acid-sodium acetate buffer at pH = 3, 4, 5, 6, 7, 8, 9 were added to the tubes respectively, after 30 min, 0.2 mol/L glucose solution was added, andreacted for 10 min. After that, the reaction was terminated, and the activity of enzyme was measured by indigo carmine fading method.

2.5 Thermal Stability of Free and Immobilized Enzymes

Appropriate amount of enzyme was dissolved in 2 mL pH=7 buffer . Several samples were prepared and kept in colorimetric tubes for 30 min at temperature of 30°C, 40°C, 50°C, 60°C, 70 °C and 80°C, respectively. Then at 37°C, 2 mL 0.2 mol/L glucose solution was added , after 10 min, the reaction was terminated at high temperature, and the activity of enzyme was measured by indigo carmine fading method.

2.6 Effects of Immobilized Enzyme Concentration on Glucose Decomposition Rate

Under the condition of 60°C, 52 mg, 156 mg, 260 mg, 364 mg, 468 mg immobilized glucose oxidase were separately dispersed in 1 mL pH = 8 buffer solution . 30 min later, 2 mL 0.2 mol/L glucose solution was added into it and reacted for 10 min at 37°C. Subsequently, the reaction was terminated andobtain an enzymatic solution, the activity of enzyme was measured by indigo carmine fading method.

2.7 Bleaching of Cotton Fabrics by Immobilized Glucose Oxidase

At pH = 8 and 37 °C, an appropriate amount of immobilized glucose oxidase was placed in a beaker, and an amount of 0.2 mol/L glucose solution was added, then reacted for a period of time to obtain bleach. Under the condition that the bath ratio was 1: 50, pH = 10.8, temperature was 97°C, took 2 g of cotton fabric, infiltration, then bleaching for 85 min, finally, the fabric was rinsed and dried naturally.

3. Results and Discussion

3.1 Standard Curve

0, 1, 2, 3, 4, 5, 6 mL 12 mg/L hydrogen peroxide standard solution were added into 25 mL colorimetric tubes respectively. 1.3 mL 1.0×10^{-3} mol/L indigo carmine solution and 3.0 mL sodium acetate buffer were also added to the tubes, then diluted them to 25 mL with distilled water. After reacted in boiling water for 13 min, chilled-down the mixtures for 5 min. With distilled water as a reference(A_0), the absorbance (A, A_0) was measured at a wavelength of 615 nm. Table 1, it shows different absorbance values correspond to the concentration of hydrogen peroxide, according to the results, we can draw the standard curve line. Fig.2, it is the linear regression equation of standard curve, by the formula we can calculate the corresponding concentration of H_2O_2, then measure the activity of enzyme.

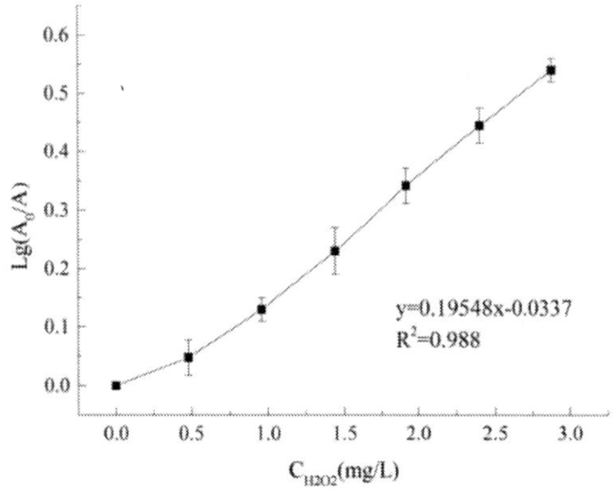

Fig.2 Standard curve line

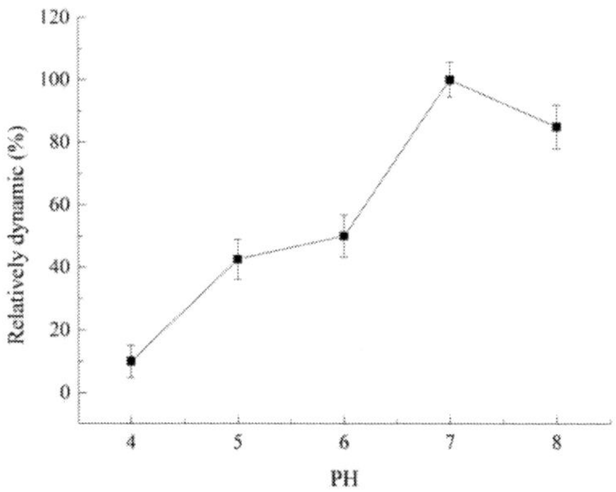

Fig.3 Effect of pH on immobilized enzyme activity

Table.1 different absorbance values correspond to the concentration of hydrogen peroxide

Volume (mL)	0	1	2	3	4	5	6
Abs(A)	0.635	0.569	0.471	0.374	0.289	0.228	0.183
$\lg(A_0/A)$	0	0.0477	0.1298	0.2299	0.3419	0.4448	0.5403
C_{H2O2} (mg/L)	0	0.48	0.96	1.44	1.92	2.4	2.88

3.2 Optimization of Immobilization Conditions of Glucose Oxidase

3.2.1 Effect of pH Value

The isoelectric point of glucose oxidase is 4.2, PDDA/Fe_3O_4 particles are positively charged, when the pH of the medium is greater than the isoelectric point, the glucose oxidase surface is negatively charged, so the pH of the buffer has a great effect on the immobilization of the enzyme. At 5 °C, 5 mg glucose oxidase was immobilized with 0.15 g PDDA-Fe_3O_4 magnetic carrier in different pH buffer. Fig.3, it is the effect of pH on immobilized enzyme activity, with the increase of PH value, the activity of immobilized glucose oxidase enzyme increase first and then decrease, and the optimum pH value is about 7.

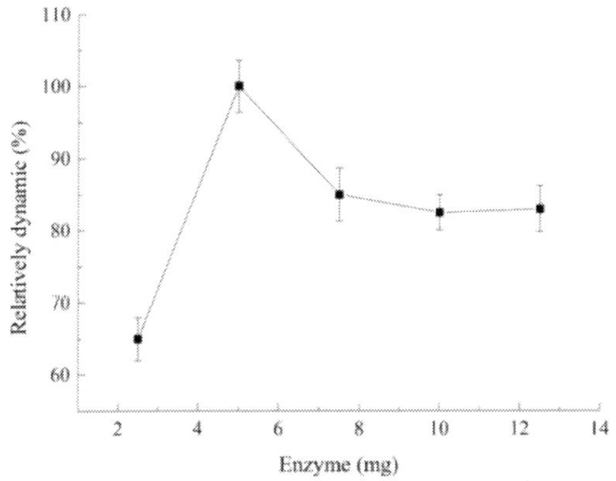

Fig.4 Effect of enzyme on immobilized

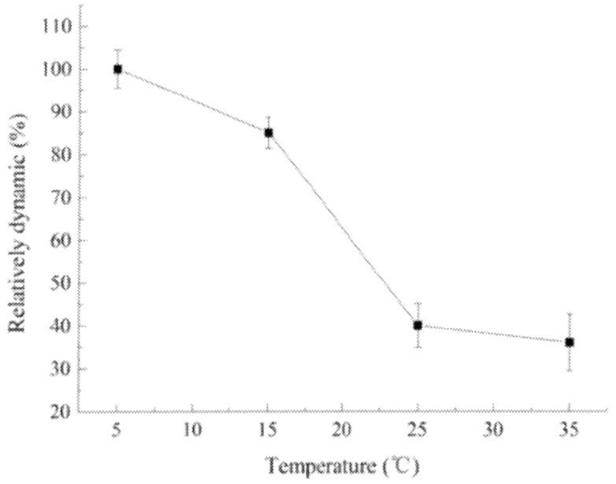

Fig.5 Effect of temperature on immobilized enzyme activity

3.2.2 Effect on the Amount of Enzyme

Immobilized glucose oxidase enzyme was prepared with different amounts of glucose oxidase, then measured its activity. The results are shown in Fig.4, when the amount of enzyme is less than 5 mg, the activity of immobilized enzyme increase with the increase of the amount of glucose oxidase. However, when the amount of enzyme is more than 5 mg, the activity of immobilized enzyme decrease. Maybe due to excessive enzyme, the binding sites for the effective immobilization of glucose oxidase on the vector are reduced. And the carrier on the enzyme molecular density increases, resulting in increasing steric effect , so the immobilized glucose oxidase enzyme activity decrease. This indicates that under certain conditions, the amount of enzyme protein on per gram of carrier load has a limit.

3.2.3 The Effect of Immobilization Temperature

At pH=7, the immobilized enzyme was prepared with 5 mg glucose oxidase at different temperature, then the activity of immobilized enzyme was measured. Fig.5, with the increase of temperature, the enzyme activity of immobilized glucose oxidase decrease significantly. This is due to the high temperature, the protein structure of the glucose oxidase changs, and the enzyme deteriotate and inactivate. So it should be immobilized at low temperature.

3.3 The stability of Immobilized Glucose Oxidase

3.3.1 Thermal Stability

Fig.6, it is the effect of temperature on free enzyme and immobilized activity, the maximum enzyme activity of the free enzyme and immobilized enzyme are defined as 100%. The figures show that the activity of free enzyme and immobilized glucose oxidase increase with the increase of temperature between 40°C and 60°C, and decrease after 60°C. But the activity of immobilized enzyme decrease slowly, it explains that the oxidative stability of glucose oxidase is increased after immobilization.

3.3.2 PH Stability

The pH stability of free glucose oxidase and immobilized glucose oxidase are shown in Fig.7, the maximum enzyme activity of the free and immobilized enzymes are defined as 100%. The figure shows that the activity of immobilized enzyme is higher in the range of pH 4-8, and the range of free enzyme is limit to 5-7, which indicate that the stability of immobilized glucose oxidase is better than that of free enzyme. Compared with free enzyme, it has a wider pH value, and the alkali resistance has been improved. The optimum pH for immobilized glucose oxidase is 8.

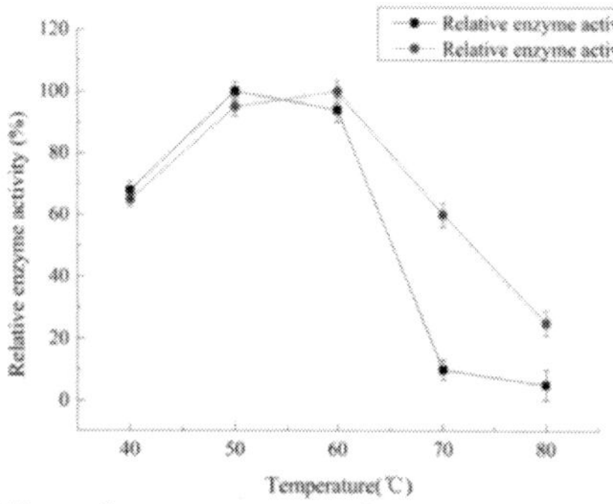

Fig.6 Effect of temperature on free enzyme and immobilized enzyme

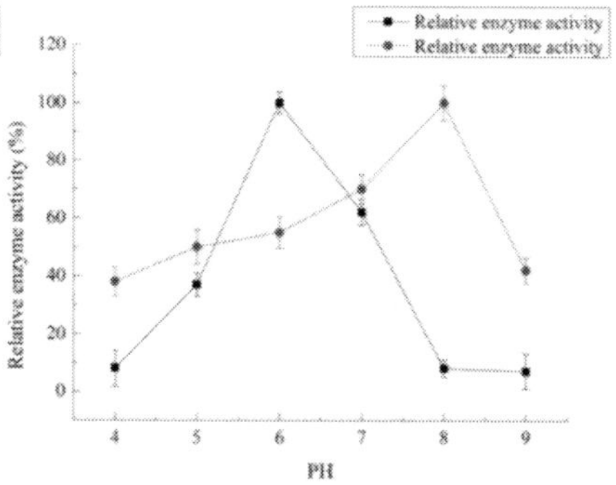

Fig.7 pH stability of free enzyme and immobilized enzyme

3.3.3 Application Stability

The immobilized enzyme was recovered by external magnetic field, and 1 mg immobilized glucose oxidase was used to continuously catalyze the substrate for 5 times, measured their enzyme activity respectively. The results are shown in Fig.8, with an initial enzyme activity of 100%. The activity of immobilized glucose oxidase decrease gradually with the increase of cycles, and 68% of the original enzyme activity remain after 5 cycles, indicating that immobilized glucose oxidase has good reproducibility.

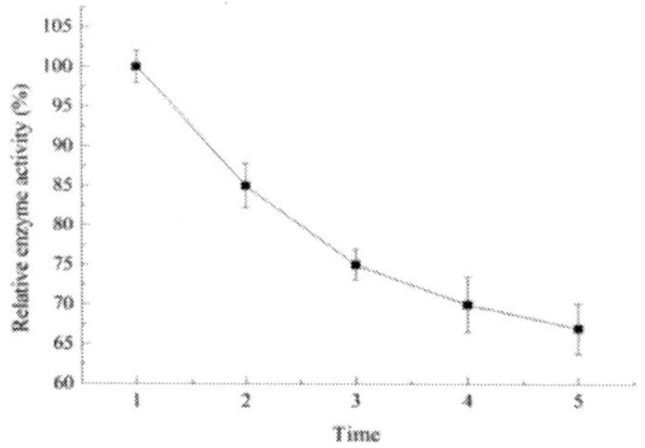

Fig.8 Repeated utilization of immobilized enzyme

3.3.4 Comparison of the Immobilized Glucose Oxidase Degradation of Glucose at Different Concentrations

Under the condition of 37°C and pH=8, different concentrations of immobilized enzyme solution were reacted with 2 mL 0.2 mol/L glucose oxidase. After 10 min, the reaction was terminated and the concentration of hydrogen peroxide was measured. Table 2, with the increase of hydrogen peroxide concentration, the amount of hydrogen peroxide show a trend of increase and then decrease, indicating that when immobilized glucose oxidase decompose the glucose, the rate of decomposition increases with the amount of immobilized enzyme increase in a certain range. When the amount of immobilized enzyme out of the range, the steric hindrance will hinder the decomposition of glucose due to excessive dosage. The optimum concentration is 260 mg/mL.

Table.2 Different concentrations of immobilized glucose oxidase to decompose glucose to produce hydrogen peroxide

Concertration (mg/mL)	52	156	260	364	468
Abs	0.406	0.342	0.260	0.295	0.272
$C_{H_2O_2}$ (mg/L)	1.17	1.56	2.16	1.88	2.06

3.4 Fabric Bleaching Effect

The hunt whiteness values of the original and bleached fabrics were measured by the fluorescence whiteness meter. 5 groups of samples weremeasured, the average value of the whiteness value was89.22, the original whiteness value was75.56. This shows that immobilized glucose oxidase can catalyze glucose production to produce enough hydrogen peroxide to bleach, and it has reached the actual production requirements.

4. Conclusion

In this paper, glucose oxidase was immobilized on self-made PDDA-Fe_3O_4 magnetic carrier by electrostatic attraction. The immobilization conditions were optimized, the properties of immobilized glucose oxidase and its bleaching effect on cotton fabric were discussed. The main conclusions are as follows.

1) The optimal conditions for immobilization of glucose oxidase areas follows: temperature is 5°C, pH=7, mass ratio of enzyme to magnetic carrier is 30: 1.

2) The stability of the immobilized glucose oxidase is higher than that of free enzyme, and the scope of PH value is expand. The optimum pH value is 8. After 5 times reused, the relative enzyme activity still keep at 68 %

3) It is very effective for bleaching of fabrics with immobilized glucose oxidase, and the whiteness value is up to 89.22.

References

[1] Wang YY. Cotton fabric bleaching technologies. Progress in Textile Science and Technology: 2009; 4: 12-15.

[2] Jin Y, Li W. Decolourisation and bleaching of grey down. Textile Auxiliaries: 2003; 4: 39-41.

[3] Chen SS, Qiu NX. Studies on the immobilization of pectinase by anion-exchange resin. Journal of Shaanxi Normal University: 2008; 2: 102-105.

[4] Zhao Z, Chen TC, Zhu Q, et al. Activated low temperature bleaching of cotton fabric with glucose. Printing and Dyeing: 2011; 10: 6-9.

[5] Sun XM, Pan RN, et al, The preparation and property evaluation of a new immobilized xanthine oxidase.Fine Chemicals: 2016; 4: 11-15.

[6] Roessl U, Nahálka J, Nidetzky B. Carrier-free immobilized enzymes for biocatalysis. Biotechnology Letters: 2010; 3: 341.

[7] Wang Y, Zheng XH. Study of immobilization of laccase on mesoporous molecular sieve MCM-41. Journal of Chemical Engineering of Chinese Universities: 2008; 1: 83-87.

[8] LI Q. Study of application and development of immobilization technique in food industry. Sichuan Food and Fermentation: 2006; 44: 16-19.

[9] Wang BK. Effect of preparation of porous glass microsphere carrier on immobilized α-amylase. Anhui Chemical Industry: 2007; 6: 24-27.
[10] Zhang L, Zhang Y, Hou HP. Developments in cells immobilization technique. Sichuan Food and Fermentation: 2006; 2: 17-18.
[11] Wei Z. Investigation of immobilization of lipase in magnetic chitosan microspheres. Food and Machinery: 2008; 1: 13-15.
[12] Chen SS, Qiu NX. Studies on the immobilization of pectinase by anion-exchange resin. Journal of Shaanxi Normal University: 2008; 2: 102-105.
[13] Bai G, Liu Y, Ling FY, et al. Preparation of immobilized glucose oxidase on magnetic composite particles. Journal of Textile Research: 2014 ;3: 62-0.
[14] Li FY, Yang QY, Qiu F. Modification of superparamagnetic iron oxide nanoparticles with poly (diallyldimethylammonium chloride) at air atmosphere. Polymers for Advanced Technologies: 2016; 11: 1530-1534.
[15] Zhou JQ, Chen SH, Wang JW. A simple and convenient method to determine the activity of glucose oxidase. Experimental Technology and Management: 2008; 12: 58-60.

Durable Hydrophobic Silk Fabric Coating with Nanoscale TiO_2 Film Via Atomic Layer Deposition

Si-Qi Zhang[1,2], Xing-Fang Xiao[1,2], Wei-Lin Xu[1,2*]

[1]*Key Laboratory of Green Processing and Functional Textiles of New Textile Materials, Ministry of Science and Technology, Wuhan Textile University, Wuhan 430073, P. R. China*
[2]*College of Material Science and Engineering, Wuhan Institute of Technology, Wuhan 430073, P. R. China*

Corresponding author's email: Weilin-xu@hotmail.com

Abstract

Hydrophobic surface was successfully prepared by depositing various thickness of TiO_2 film on fabric via atomic layer deposition (ALD). The surface morphology of silk fabric was measured through scanning electron microscopy (SEM), the results revealed that uniform TiO_2 layer were deposited on the surface of silk fabric. The thermal and mechanical properties of the TiO_2 silk fabric were then measured. The thermal stability was superior to those of the uncoated substance while the mechanical properties were slightly lower than the uncoated substrate. The static water contact angles of ALD coated silk fabrics raised from 0° to around 133°, and the thickness of TiO_2 layer had no effect on hydrophobic property. Furthermore, the wear resistance test showed that ALD coated silk fabrics had durable hydrophobic surface.

Keywords: Atomic Layer Deposition; Titanium Dioxide; Silk Fabrics; Hydrophobic

1. Introduction

Silk fiber as a natural filament has been widely applied for a long time in human history, as it has many advantages, such as elegant appearance, soft and smooth texture, warmth, and biodegradability [1]. However, the large bunch of hydrophilic groups on silk protein molecules make the silk products easily stainable by dirt. Hence, fabrication of functional silk fabrics with self-cleaning properties is a pressing matter in the textile industry [2]. The superhydrophobic surface with a static water contact angle over 150° have appealed to rich investigative researches for the self-cleaning and anti-snow sticking properties, which would increase the potential application value of materials. At present, a variety of techniques have been rapidly proposed to prepare hydrophobic silk fabrics. Cao et al. fabricate multifunctional silk fabrics with electrical conductive, anti-ultraviolet and water repellent properties via graphene oxide repeatedly coating and chemical reduction method [3]. Huang et al. combine magnetron sputter methods to prepare PTFE coating on silk fabrics, which make the surface of silk fabric is hydrophobic [4]. Satreerat et al. prepare hydrophobic silk fabrics by SF6 plasma [5]. Zulfqar produce low-formaldehyde hydrophobic and cum crease resistant silk fabrics through cross-linker [6].

In addition, many techniques have been used to prepare superhydrophobic surfaces [7-11], such as sol-gel [8], bio-inspired engineering [9]. However, these reported methods are not suitable in many cases, especially for natural protein fibers. Thermal degradation happens on natural protein fiber when the treatment temperature is high when it yellows and become fragile. Moreover, the binding strength between superhydrophobic layer and natural protein fibers is weak. An alternative approach is to apply the chemical reaction on the nature protein fabrics to fabricate superhydrophobic surface at a relative low temperature. Hyde et al. formed a layer of Al2O3 on cotton via atomic layer deposition (ALD) at 100℃. The result showed that the static contact angle raised from 0° to 127° [12]. Although cotton and silk fiber have different structure and chemical groups, they have the same wetting ability, and both are hydrophilic. Therefore, atomic layer deposition technique might be an approach to change the hydrophilic surface of silk fiber to hydrophobic. As far as I can see, there are few investigative report about using ALD technology obtained hydrophobic silk fabric.

An ALD process is usually combined with two time-separated half-reactions. As two different precursor transport into reactor separately, the first precursor absorbs on the surface of the substrate, and the second precursor reacts with the first one, each of the surface reaction is self-limiting while the two reactions proceed sequential, thus completely deposit an atomic level uniform film [13]. The binary reaction for TiO_2ALD is [14]

$$Ti(OCH(CH_3)_2)_4+2H_2O \rightarrow TiO_2+4(CH_3)_2CHOH$$

Due to the strong formation among Ti-O bond during TiO_2 ALD, the surface reactions are very efficient and self-limiting, so that the TiO_2 ALD can be deposited at relative low temperature available for silk fabrics. In this study, we deposited TiO_2 onto silk fabrics by ALD at 120 ℃ to fabricate a durable hydrophobic silk fabric. The morphologies, thermal performance, mechanical properties and hydrophobic properties of samples were characterized.

2. Experimental Section

2.1 Materials

Silk fabrics (twill-weave, 200g/m^2) were purchased from a local fabric store, and titanium tetraisopropoxide (TIP) was acquired from Strem Chemicals, Inc., Deionized water (I degree, specific resistance 18ΩM·cm) was purified with by UP system.

2.2 Preparation of Nano TiO_2 on Silk Fabric via Atomic Layer Deposition

The formation of TiO_2 films performed on silk fabrics and silicon wafer via atomic layer deposition, and the color change of silicon wafer was acted as a reference whether the TiO_2 was deposited on the surface. ALD experiment was conducted in a hot-wall closed chamber-type ALD reactor. Titanium tetraisopropoxide (TIP) and deionized water were used as the titanium and oxygen source to deposite TiO_2. Prior to deposition, silk fabrics were placed in the ALD reactor and dried at 120℃ for 30min in vacuum (20 Pa) with a steady N_2 gas stream (20 sccm). The temperature of the reactant lines for TIP and water was 80℃ and room temperature, respectively. The reactor was purified by the nitrogen before ALD process, then TIP and water were alternately brought into the reactor separated by purging the nitrogen, which would remove byproduct and remaining reactant. In the first half of the cycle, TIP reacted with the silk fabric for 10 s following the step of nitrogen purge for 25 s. In the second half of the cycle, water reacted with TIP that had been absorbed on the surface of silk fabric for 10 s following the step of nitrogen purge for 25 s. In this study, deposition cycle of TiO_2 on silk fabrics were 5, 25, 50, 100, 200, 400, 800. Pulse time for water and TIP were both 0.5s.

2.3 Characterization

Scanning electron microscopy (SEM): the morphologies of control silk fabric, TiO_2-coated silk fabric were investigated on a scanning electron microscope (JSM-7001, JEOL Co. Ltd, Japan) at 20 kV after gold coating.

Thermogravimetry analysis (TGA): the thermal performance of control silk fabrics and ALD coated silk fabrics were carried out using a TG209 F1 (NETZSCH Instruments, Germany) thermogravimetric analyzer with a heating rate of 10℃/min in N_2.

Mechanical test: the mechanical properties of control silk fabric and TiO_2-coated silk fabric were measured with a universal material testing machine (YG208, Beijing Tongde Co., Ltd.) at 22℃ and relative humidity of 65% at a gauge length 20 mm and tensile speed of 100mm/min. The breaking strength and breaking elongation are the average of 3 measurements.

Static water contact angles: the static water contact angles on the control silk fabric and TiO_2-coated silk fabric were measured using a Kruss contact angle goniometer (DKSH(China) Co., Ltd.). All reported contact angle values are obtained within 5 s after the dropping on the silk fabrics.

Wear resistance test: the control silk fabric and TiO_2-coated silk fabric were rubbed with YG522 fabric wear tester for 5 times. And the rubbed silk fabrics were measured using a Kruss contact angle goniometer (DKSH (China) Co., Ltd.) to evaluate the binding strength between TiO_2 coating and silk fabric.

3. Results and Discussion

3.1 Morphology of Control Silk Fabric and TiO_2-coated Silk Fabrics

Typical SEM images of control silk fabric and TiO_2-coated silk fabric are shown in Fig.1. It was clear that without ALD coating, the fiber had a smooth surface morphology although several attached particles could be visualized (Fig.a1), which might be the silk gum and impurity. The SEM images of TiO_2-coated silk fabric were identical with the control silk fabric (From Fig.b1 to h1). In the high-magnification image of the fabric, the fiber without ALD coating had a smooth surface morphology whose diameter was around 10μm, although several vertical stripes could be observed (Fig.a2), which might be the truth that the fine structure of silk fiber was combined with fibril and microfibril. From Fig.b2 to Fig.h2, the fine structure of silk fiber could also be observed, suggesting that conformal TiO_2 nanolayer was successfully deposited onto the fiber surface through the ALD technique.

Fig.1 SEM images of (a) control silk fabric (b) 5-cycle (c)25-cycle (d)50-cycle (e) 100-cycle (f) 200-cycle (g) 400-cycle (h) 800-cycle TiO_2-coated silk fabrics

3.2 TG Analysis of Control and TiO_2-coated Silk Fabrics

Typical TG curves of the control silk fabric and TiO_2-coated silk fabric are shown in Fig.2, and the thermal analysis data from Fig.2 are listed in Table 1. The TG curves of the control and TiO_2-coated samples are divided into three mass-loss processed with similar temperature ranges. The initial stage mass lose is from 30-300°C, which is attributed to the loss of silk gum and absorbed moisture on the silk fabric; the major decomposition stage is from 300-330°C, which is caused by the decomposition of silk fabric; the third stage is from 330-700°C, which is primarily created by char decomposition of the silk fabric. However, control and TiO_2-coated silk fabric are of different mass loss, the value of mass loss decreases as the TiO_2 ALD cycle increases, except for 50-cycle TiO_2-coated silk fabric, as shown in Figure.2 and Table 1. Especially, the mass loss of 800-cycle TiO_2-coated silk fabric is 9.3% greater than that of control samples, which indicates that TiO_2 is deposited on the silk fabric successfully.

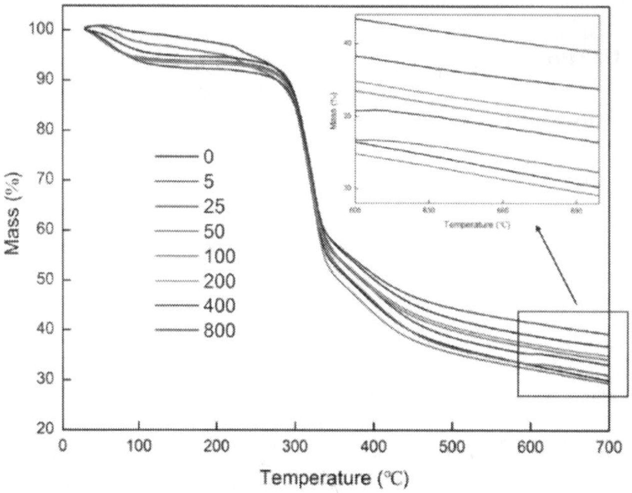

Fig.2 TG curves of control (0-cycle) and TiO_2-coated silk fabric

Table.1 Thermal Analysis Data from TG Curves of control and TiO_2 Coated Silk Fabric

Silk fabric (cycle)	0	5	25	50	100	200	400	800
Mass loss (%)	69.9	68.8	66.8	70.4	65.7	65.0	63.1	60.6

3.3 Mechanical Properties of Control and TiO_2-coated Silk Fabrics

To our best knowledge, mechanical properties is critical for estimating the property of the silk fabric under actual application. Average values for the breaking strength, breaking elongation of silk fabrics are shown in Fig.3. From Fig.3(a), the control silk fabric exhibits a breaking strength of 217.1N, whereas these values are slightly lower than the TiO_2-coated silk fabric. The slight decrease of breaking strength for the TiO_2-coated silk fabric might be caused by the treatment temperature. The high treatment temperature which was 120 ℃ would damage mechanical properties of silk fabric. In Fig.3(b), the uncoated silk fabric exhibits a breaking elongation of 4.20mm, whereas some values are similar or even greater for the TiO_2-coated silk fabric, especially for the cycle number below 200. However, other values are lower for the TiO_2-coated silk fabric whose cycle over 200. The decrease of breaking elongation might be the silk fabric had been kept in high temperature for a long time, resulting in a decline in its deformation ability. In general, the mechanical properties of the silk fibers are slightly decreased by the ALD process.

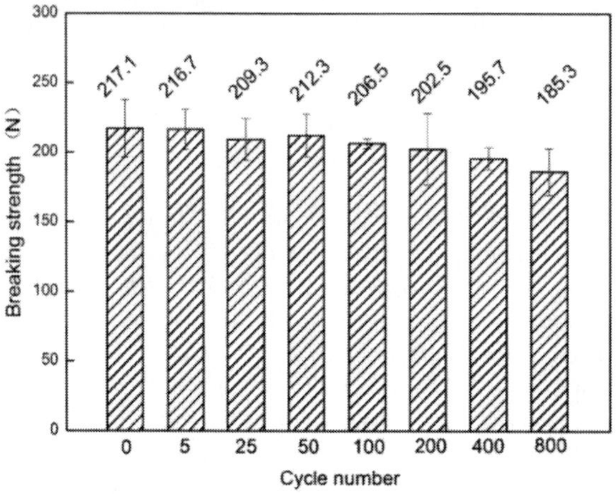

Fig.3(a) Breaking strength of control and TiO_2-coated silk fabric

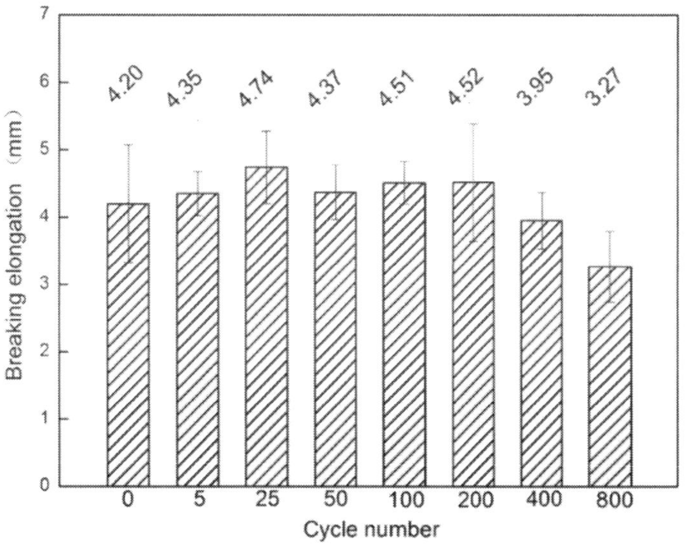

Fig.3(b) Breaking elongation of control and TiO_2-coated silk fabric

3.4 Hydrophobic Performance of Silk Fabrics

Fig.4 shows the static water contact angles of silk fabrics. The static water contact angle of control wool fabrics is about 0°, which is hydrophilic. Interestingly, the static water contact angle of TiO_2 coated silk fabrics with different cycle increased to around 133°, which suggested that TiO_2 coated silk fabrics had a hydrophobic surface. Besides, the thickness of TiO_2 film had no effect on the wetting property of silk fabrics. Fujishima [15] reported that titanium dioxide nano film materials would produce electronic-hole pair under the irradiation of ultraviolet. Then the electronic reacted with Ti^{4+}, while the hole reacted with super-oxygen ion, thus formed Ti^{3+} and oxygen vacancies, respectively. The formation of Ti^{3+} could be quickly oxidized by oxygen, and the oxygen vacancies reacted with H_2O to generate -OH, which led to strong interaction with the polar water molecules on the surface. Therefore, the nanometer titanium dioxide thin film is hydrophilic by the irradiation of ultraviolet. However, when to stop UV irradiation, oxygen on the air would replace hydroxyl, which make the surface was hydrophobic. After the formation of one cycle of TiO_2 film, the surface would generate fresh hydroxyl groups, and combined with the oxygen in the air, which made the silk fabric was hydrophobic after the deposition of TiO_2.

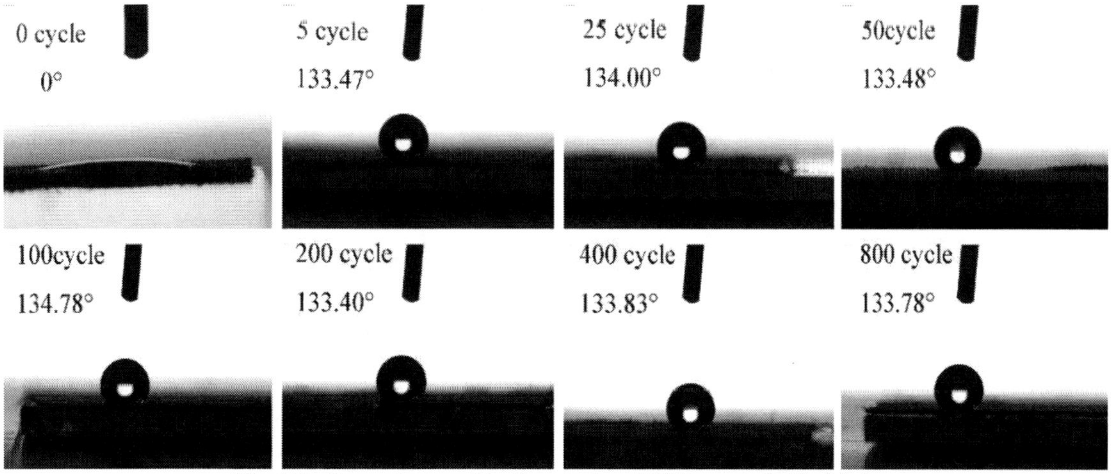

Fig.4 Static contact angle of control and TiO_2-coated silk fabric

3.4 Durable Hydrophobic Performance of Silk Fabrics

Fig.5 shows the static contact angle of the silk fabrics with and without ALD coating after the friction of fabric wear tester. The silk fabric coated with TiO_2 remained hydrophobic with the static contact angle was around 130°. However, these values were slightly lower compared with unrubbed silk fabric, the specific change of degree was shown on Table 2. Interestingly, static contact angle of silk fabric whose cycle number less than 100 dropped greater than silk fabric whose cycle number over 100, this might because the content of TiO_2 on the silk fabric surface was different. In a word, the binding strength between ALD coating and silk fabric was strong and the hydrophobic property of silk fabric with ALD coating was durable.

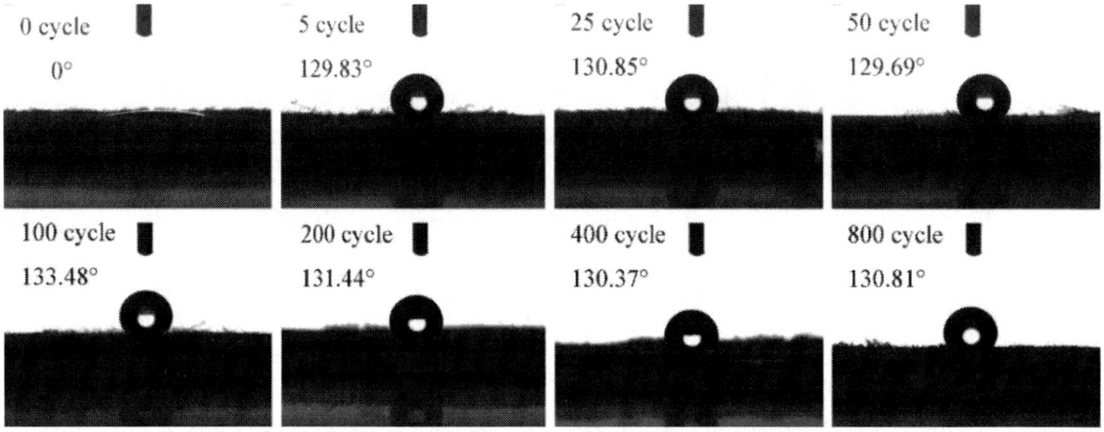

Fig.5 Durable hydrophobic performance of control and TiO_2-coated silk fabric

Table.2 Static contact angle change of control and TiO_2-coated silk fabric

Silk fabric（cycle）	0	5	25	50	100	200	400	800
Contact angle change（deg）	0	3.64	3.15	3.79	1.30	1.96	3.46	2.97

4. Conclusion

The results revealed that the silk fabrics were successfully coated with uniform TiO_2 film via the ALD process. The TG measurements results suggest that TiO_2-coated silk fabric had superior thermal stability compared to the control one, and the ALD process yields typical ALD growth behavior. The ALD TiO_2 coating on silk fabrics changed surface wetting ability, resulting into the increasing static water contact angle from 0° to 133°. Interestingly, the results showed that ALD coated silk fabrics could keep the hydrophobic surface after rubbed by the fabric wear tester, indicating the higher durability of hydrophobic of ALD coated silk fabrics. It was confirmed that ALD coating method could be applied in preparing various kinds of hydrophobic products.

5. Acknowledgments

We greatly acknowledge the support by the National Funds for Distinguished Young Scientists (Project No. 51325306), the Graduate student innovation fund of Wuhan Textile University.

References

[1] Liu XS, Xing TL, Xu DM, Chen GQ. Study on novel eco-friendly anti-creasing agents for natural silk fabric. Chin. Chem. Lett: 2012; 23: 665-668.

[2] Xu B, Cai ZS. Fabrication of a superhydrophobic ZnO nanorod array film on cotton fabrics via a wet chemical route and hydrophobic modification. Appl. Surf. Sci: 2008; 254: 5899-5904.

[3] Cao JL, Wang CX. Multifunctional surface modification of silk fabric via graphene oxide repeatedly coating and chemical reduction method. Appl. Surf. Sci: 2017; 405: 380-388.
[4] Huang FL, Wei QF, Liu Y, Gao WD, Huang YB. Surface functionalization of silk fabric by PTFE sputter coating. J Mater Sci: 2007; 42: 8025-8028.
[5] Hodak SK, Supasai T, Paosawatyanyong B, Kamlangkla K, Pavarajarn V. Enhancement of the hydrophobicity of silk fabrics by SF6 plasma. Appl. Surf. Sci: 2008; 254: 4744-4749.
[6] Raza ZL, Anwar F. Low-formaldehyde hydrophobic cum crease resistant finishing of woven silk fabric. Fibers Text in East Eur: 2015; 23: 116-119.
[7] Jeyasubramanian K, Hikku GS, Preethi AVM, Benitha VS, Selvakumar N. Fabrication of water repellent cotton fabric by coating nano particle impregnated hydrophobic additives and its characterization. J. Ind. Eng. Chem: 2016; 37: 180-189.
[8] Xue CH, Jia ST, Chen HZ, Wang M. Superhydrophobic cotton fabrics prepared by sol–gel coating of TiO2 and surface hydrophobization. Sci. Technol. Adv. Mater: 2008; 9: 035001.
[9] Ahmad I, Kan CW. A review on development and applications of bio-Inspired superhydrophobic textiles. Materials: 2016; 9: 892.
[10] Xu B, Cai ZS, Wang WM, Ge FY. Preparation of superhydrophobic cotton fabrics based on SiO2 nanoparticles and ZnO nanorod arrays with subsequent hydrophobic modification. Surf. & Coat. Tech: 2010; 204: 1556-1561.
[11] Wang N, Xiong DS, Deng YL, Shi Y, Wang K. Mechanically robust superhydrophobic steel surface with AntiIcing ,uv-durability, and corrosion resistance properties. ACS Appl. Mater. Interfaces: 2015; 7: 6260-6272.
[12] Hyde GK, Park KJ, Stewart SM, Hinestroza JP, Parsons GN. Atomic layer deposition of conformal inorganic nanoscale coatings on three-dimensional natural fiber systems: effect of surface topology on film growth characteristics. Langmuir: 2007; 23: 9844-9849.
[13] George SM. Atomic layer deposition: an overview. Chem. Rev: 2009; 110: 111-131.
[14] Xiao XF, Liu X, Chen FX, Fang D, Zhang CH, Xia LJ, Xu WL. Highly anti-uv properties of silk fiber with uniform and conformal nanoscale TiO2 coatings via atomic layer deposition. ACS Appl. Mater. Interfaces: 2015; 7: 21326-21333.
[15] Fujishima A, Zhang XT. Titanium dioxide photocatalysis: present situation and future approaches. C. R. Chimie: 2006; 9: 750-760.

Preparation and Characterization of TiO_2 Nanoparticle Multilayer Films on Cotton Fabric

Qin Zhou[1], Jia-Yi Wu[1], Xin-Wang Cao[1], Ji-Hong Wu[1], De-Shan Cheng[1 *]

[1] *School of Textile Science and Engineering, Wuhan Textile University, Wuhan, China*

[*]Corresponding author's email: chengcds@163.com

Abstract

The TiO_2 nanoparticles which were oppositely charged have been assembled in multilayer films layer-by-layer (LbL). Self-assembly technique was used on the pre-treated cotton fabric to enhance the UV protection. The multilayer films were characterized by UV-vis absorption spectroscopy and Scanning Electron Microscope (SEM). The results confirmed that the presence of TiO_2 nanoparticles films and the nanoparticles distributed uniformly on the surface of the cotton fiber. After the assembly of multilayer films, the UV protection property of cotton fabric was measured by Ultraviolet Protection Factor (UPF), and the laundering experiments were carried out to determine the durability of TiO_2 nanoparticles on cotton fabric. The results showed that the UV protection property of cotton fabrics assembled by TiO_2 nanoparticles still maintained at a high level after 5 launderings.

Keywords: TiO_2 Nanoparticles; Cotton Fabric; LbL Self-assembly; Multilayer Nanocomposite Films; UV Protection Property

1. Introduction

Cotton is one of the most versatile natural fibers. Due to its comfort and durability, it is extensively used in upholstery, slipcovers and curtains. It is known that UV irradiation can cause a severe damage to cotton fabric such as color, mechanical strength, and human physiological comfort [1]. More importantly, it has been known that the wavelength of UVA (315-400 nm) and UVB (280–315 nm) rays have negative effect on the skin such as sunburn, allergies, and even skin cancer [2, 3]. However, the UPF value of cotton fabrics was too low to provide protection against UV irradiation [4]. Consequently, improving the UV protecting activity of cotton fabric is a major challenge to extend it's application areas.

Several past years have witnessed growing research on the preparation of UV protecting cotton fabrics through the immobilization of TiO_2 nanoparticles on fabric surface. Several techniques such as sol-gel [5], atomic layer deposition [6], mini-emulsion polymerization [7], hydrothermal method [8] and layer-by-layer self-assembly [9] have been explored to generate protection coatings on fabric substrates. Among these approaches, the LbL self-assembly technique is environmentally friendly, relatively fast and simple, and there is no need for expensive equipment or organic solvents. The methods of LbL self-assembly was initially reported by Decher et al. [10, 11] These methods are based on the alternating adsorption of charged cationic and anionic species on the solid surfaces through electrostatic interactions. [12] The technology of LbL self-assembly has been widely used in surface coating, which attributes to its easy process, low-cost, precise control of film composition and thickness, and suitable to a more variety of solid substrates with different shapes. [13, 14]

In this paper, we utilized the technology of LbL self-assembly to impart the UV protection on the surface of cationic cotton fabrics by depositing alternating layers with oppositely charged TiO_2 nanoparticles. The cotton fabric being assembled by TiO_2 nanoparticles multilayer films was characterized by UV-vis and SEM. Additionally, the UV protective ability of cotton fabric being assembled by multilayer films was measured by UPF and the laundering experiments were carried out to determine the durability of TiO_2 nanoparticles on cotton fabric.

2. Experimental

2.1 Materials

The cotton fabric has a plain woven structure with 135 g/m^2 mass, the warp density is 570 per 10cm and the warp density is 280 per 10cm. 3-chloro-2-hydroxypropyl tri-methyl ammonium chloride was used as a cationising agent (CHTAC, CR2000, TCI Co) to create cationic surface charge on the cotton fabric. Nano-powder of titanium dioxide (TiO_2, Degussa P25 grade, 20% rutile and 80% anatase) was used as a UV absorber. All the chemicals were analyzed and were used without further purification.

2.2 Preparation of TiO_2 Nanoparticle Suspension

TiO_2 Nano-powders were dispersed in deionized (DI) water, and the stirring of rotate speed is 1500 rpm for 20 min, then sonicated for 10 min under ultrasonic treatment. The concentration of the suspension was adjusted to 0.1 wt%. The pH of the suspension was adjusted with dilute 0.1 mol/L HCl or 0.1 mol/L NaOH.

2.3 Pretreatment of Cotton Fabrics

Cationic cotton fabric was prepared by 3-chloro-2-hydroxypropyl tri-methyl ammonium chloride,which introduce the activity sites onto the surface of fiber. CHTAC was activated by reacting it with NaOH at a 1:2 mol ratio with the solution pH maintained below 11. Cotton fabric was padded through this mixture solution and squeezed it untill the wet pick up to 100%. The padded fabric was put in plastic bag at room temperature overnight, and then rinsed several times with water and finally neutralized with a dilute acetic acid (1 g/l).

2.4 Multilayer Assembly on Cotton Fabrics

The alternating assembly of the multilayer on cotton fabric was prepared as follows: At the first stage, the cat ionized cotton fabric was immersed into the negatively charged TiO_2 suspension for 10 min and rinsed with water, and the modified substrate was dried with a gentle stream of nitrogen. As a result, the surface charge of the cotton fabric was reversed to be negative. Then the negatively charged cotton fabric was immerged into the positively charged TiO_2 suspension for 10 min and rinsed with water. After each layer deposition, samples were rinsed with water and dried with gentle stream of nitrogen. The adsorption and rinsing steps were repeated until the desired number of bilayers was obtained.

2.5 Characterization of the Assembled Cotton Fabrics

The FTIR spectra were recorded on Nicolet 5700 FTIR spectrophotometer (Nicolet Co., US) using a transmission model. The spectra were taken over a wavenumber range of 4000–400 cm^{-1} with a resolution of 2 cm^{-1} at room conditions. The particle size distribution and surface zeta-potential were assessed by a ZetaSizer 3000HSA (Malvern Instruments Ltd, Worcestershire, UK). Ultraviolet-visible spectroscopy (UV-vis) (U-4100, Hitachi, Japan) was used to characterize the TiO_2 nanoparticle suspension and cotton fabric after assembled TiO_2 nanoparticles with different layer. The surface morphologies and elemental composition of cotton fabrics were observed with scanning electron microscopy (SEM) (JSM-5600LV, JEOL, Japan). The Ultraviolet Protection Factor (UPF) was measured by a UV-1000F (Ultraviolet Transmittance Fabric Analyzer) from US Lab sphere. The cotton samples were washed in a Haier automatic washing machine according to the AATCC Test Method 135-2000.

3. Results and Discussion

3.1 FT-IR Analysis

FT-IR spectra were used to identify the presence of functional groups on the cotton fabric. Fig.1 shows FT-IR spectra of pristine and cat ionized cotton fabrics. Fig.1a shows FT-IR spectrum of pristine cotton. The spectra of characteristic bands corresponding to their chemical structures, Fig.1 shows the hydrogen bonded OH stretching observed at 3440 cm^{-1} and the CH_2 stretching centered at 2990 cm^{-1}. A strong adsorption band with a maximum at 1030 cm^{-1}, it is a result of the overlapping bands attributed to the C-C, C-O, and C-O-C stretching vibrations of cellulose. The CH wagging bands were obtained at 1430 cm^{-1} and 1320 cm^{-1} which are characteristics for cellulose structure in cotton fibers. Fig.1b shows the FT-IR spectra of cat ionized cotton fabric. Compared with pristine cellulose, we can see that the cat ionized cotton fabric has an obvious new absorption peak at 1,480 cm^{-1}, which was attributed to the quaternary ammonium groups, the result was in accordance with the literatures report [15, 16]. This result indicates that the reaction has successfully to insert cationic groups on the surface of cotton fibers.

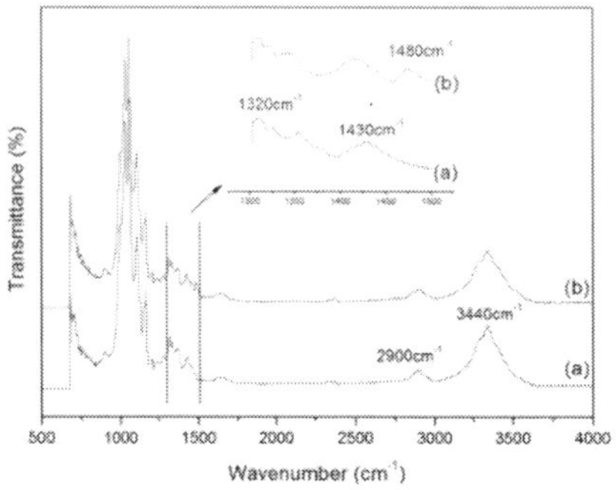

Fig.1 FT-IR spectra of pristine cotton fabric (a) and cat ionized cotton fabric (b)

3.2 Zeta Potential Measurement

The size dispersion of the TiO_2 nanoparticles in the suspension is shown in Fig.2a. The size distribution of these particles is narrow ranging from 50 to 120 nm, and the average diameter is 92 nm which is automatically calculated. In addition, the extinction peaks are symmetrical on both sides, indicating that a stable suspension of TiO_2 nanoparticles against agglomeration. The isoelectric points (IEP) of TiO_2 appears when the pH is 4.7-6.2 [17]. The zeta potential of the TiO_2 nanoparticles in the suspension can be controlled by adjusting the PH of the suspension to a desired level. The pH value of the TiO_2 nanoparticles suspension was adjusted to 2.0 by adding HCl, resulting in a relatively lower pH value than that of TiO_2 IEP, the average zeta potential is 38.3mv (Fig.2b). Conversely, the pH value of the nanoparticle suspension was adjusted to 11.0 by adding NaOH to achieve a larger pH, the average zeta potential is -30.2 mv. It is thus evident that the charge of the TiO_2 nanoparticles suspension can be altered by adjusting pH, the oppositely charged TiO_2 nanoparticles will help to form the multilayer films on cotton fabric surface.

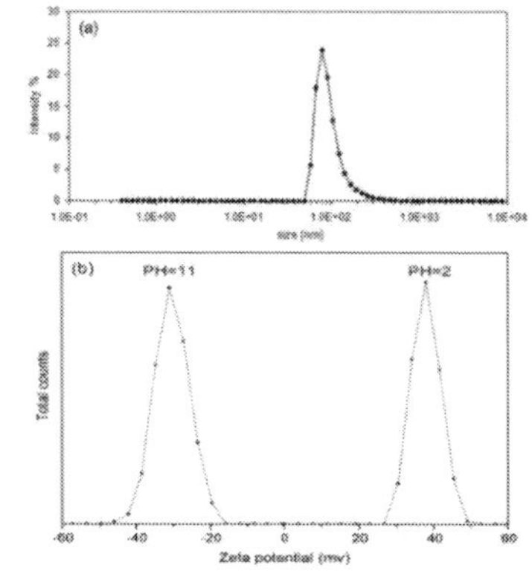

Fig.2 The size distribution of the TiO_2 nanoparticles suspension (a), zeta potential of the TiO_2 nanoparticles suspension with different PH values (b)

3.3 UV-vis Spectroscopy

Fig.3a shows the UV-vis absorption spectra of the TiO_2 nanoparticles suspension. The strong adsorption peak was discovered at 294 nm. Fig.1b shows the UV-vis absorption spectra of the pristine cotton fabric with 1, 3, 5, 7, and 10 bilayers of TiO_2 nanoparticles assembled on cotton fabric. The pristine cotton does not absorb in the spectral region monitored. The adsorption peak of cotton fabric was observed at broad maximum wavelengths from 285 nm to 325 nm after assembled TiO_2 nanoparticles multilayer films. Moreover, it is clearly seen that the absorbance increases with the increase of the number of layer pairs. We can assume that the absorption peak at the UV region is attributed to the absorption of TiO_2 nanoparticles on cotton fabrics surface. UV-vis adsorption spectroscopy confirmed the consecutive growth of the multilayer films.

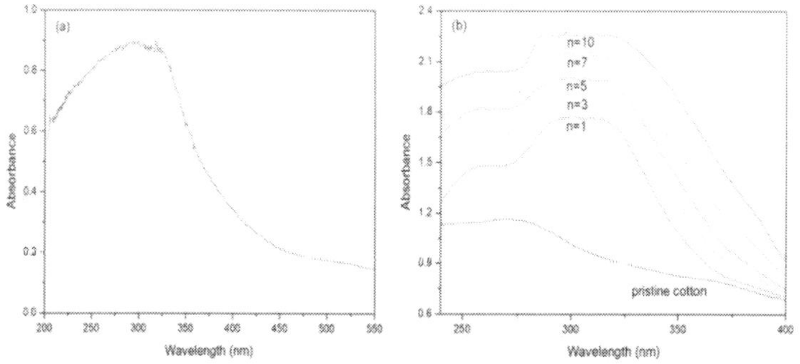

Fig.3 UV-vis absorption spectra of the TiO_2 nanoparticles suspension (a), pristine cotton fabric and after assembled TiO_2 nanoparticles multilayer films with different numbers of bilayers (b)

3.4 Surface Morphology of Cotton Fabrics

The SEM images of the cotton fabric before and after assembled by 10 bilayers TiO_2 nanoparticles multilayer films are shown in Fig.4a-c. Fig.4a shows a typical SEM image of the pristine cotton fabric. The surface of these fibers is smooth and clean with many tiny striations. Fig.4b illustrates that the nanoparticles could be clearly seen on the fiber surfaces after assembled TiO_2 nanoparticles. And the higher magnification image of the single fiber in Fig.4c shows that the surface of cotton fiber was covered with dense films of nanoparticles after assembled 10-bilayers TiO_2 nanoparticles multilayer films. SEM micrograph showing the presence of nanoparticles on cotton fiber after assembled multilayer films. It is evident that there is no significant aggregation of TiO_2 nanoparticles taken place during assembling process, which shows that the similarly charged colloidal particles cannot come close enough together to agglomerate.

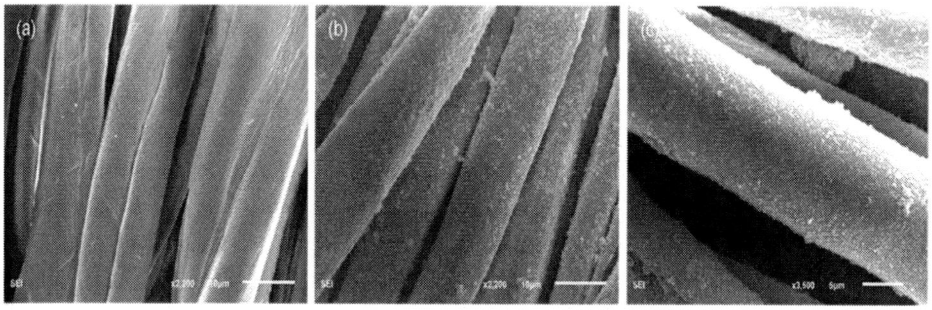

Fig.4 SEM images of the pristine cotton fabric (a) and after assembled 10-bilayers TiO_2 nanoparticles multilayer films with different magnification (b, c)

3.5 UV Protective Property

The UPF values of the pristine cotton fabric and after assembled TiO_2 nanoparticles in different assembled layers were listed in Table 1. As shown in Table 1, the pristine cotton fabric showed a low UPF of 6.5, while the cotton fabrics showed increase in UPF value after assembled TiO_2 nanoparticles. The UPF value of sample increased as the number of assembled layers increase. When the assembling layer was 10, the UPF value was increased to 93.3. Generally, a UPF value of 40-50+ indicates excellent protection yielded by clothing against damage to the human skin from UV radiation [18]. The cotton fabrics could obtain excellent UV protection ratings of UPF > 40 after assembled several bilayers of TiO_2 nanoparticles. The increased UPF value of samples confirms the remarkable UV shielding property of TiO_2 nanoparticles coating, which shows that the TiO_2 nanoparticles adhere well onto the surface of cotton fabrics.

To determine the durability of the UV protective property, cotton fabric samples were washed 5 times at 40 °C for 30 min with a washing machine. By the end of 5 washing cycles, the UPF values of cotton fabrics was decreased after assembled TiO_2 nanoparticles, but still imparted to the cotton fabric a very good protection against UV radiation. The good laundering durability of cotton fabrics loaded with TiO_2 nanoparticles can be attributed to the electrostatic attraction between TiO_2 nanoparticles and cotton fabrics surfaces. These results showed that the cotton fabric had excellent UV protection and good adhesion of the nanoparticles onto the fabric surface even after the launderings.

Table.1 the Effect of TiO_2 nanoparticles treatment and washing cycles on the UV protection properties of cotton fabrics.

Sample	UPF value	After 5th wash
pristine cotton fabric	6.5	6.5
assembled 5 bilayers TiO_2 nanoparticles	57.8	41.7
assembled 10 bilayers TiO_2 nanoparticles	93.3	66.5

4. Conclusion

The sequential adsorption of oppositely charged TiO_2 nanoparticles by LBL self-assembly is an efficient method for obtaining Nano composites films on cotton fabric. The multilayer assembly process can be monitored by UV-vis. UV-vis adsorption spectroscopy confirmed the consecutive growth of the multilayer films. The morphology of the cotton fabric after assembled TiO_2 nanoparticles multilayer films have been studied by SEM. It is observed that the TiO_2 nanoparticles were uniformly dispersed on the surface of cotton fiber. The UPF values confirm that the cotton fabrics exhibited excellent UV protection after assembled TiO_2 nanoparticles multilayer films. The laundering experiment shows that the good adhesion of the TiO_2 nanoparticles onto the surface of cotton fabrics.

5. Acknowledgments

This research was supported by the National Natural Science Foundation of China (51503164). Also, this research was supported by Wuhan Textiles University Research Fund, 2014 and Scientific Research Project of Hubei Provincial Department of Education, 2016.

References

[1] Xiao X, Liu X, Cao G, Zhang CH, Xia LG, Xu WL, Xiao SL. Atomic layer deposition TiO2/Al2O3 nanolayer of dyed polyamide/aramid blend fabric for high intensity UV light protection. Polym. Eng. Sci: 2015; 55: 1296-1302.

[2] Gorenšek M, Recelj P. Nanosilver functionalized cotton fabric. Textile Res. J: 2007; 77: 138-141.

[3] Zhang Y, Yu L, Ke S. TiO2/SiO2 hybrid nanomaterials: synthesis and variable UV-blocking properties. J. Sol-Gel. Sci. Techn: 2011; 58: 326-329.

[4] Fakin D, Veronovski N, Ojstršek A. Synthesis of TiO2–SiO2 colloid and its performance in reactive dyeing of

cotton fabrics. Carbohyd. Polym: 2012; 88: 992-1001.
[5] Abidi N, Cabrales L, Hequet E. Functionalization of a cotton fabric surface with titania nanosols: applications for self-cleaning and UV-protection properties. ACS. Appl. Mater. Inter: 2009; 1: 2141-2146.
[6] Hyde GK, McCullen SD, Jeon S. Atomic layer deposition and biocompatibility of titanium nitride nano-coatings on cellulose fiber substrates. Biomed. Mater: 2009; 4: 025001.
[7] Lu H, Fei B, Xin JH. Fabrication of UV-blocking nanohybrid coating via miniemulsion polymerization. J. Colloid. Interf. Sci: 2006; 300: 111-116.
[8] Daoud WA, Xin JH. Nucleation and growth of anatase crystallites on cotton fabrics at low temperatures. J. Am.Ceram. Soc: 2004; 87: 953-955.
[9] Ugur ŞS, Sariişik M, Aktaş AH. The fabrication of nanocomposite thin films with TiO2 nanoparticles by the layer-by-layer deposition method for multifunctional cotton fabrics. Nanotechnology: 2010; 21: 325603.
[10] Decher G. Fuzzy nanoassemblies: toward layered polymeric multicomposites. Science: 1997; 277: 1232-1237.
[11] Decher G, Eckle M, Chmitt J. Layer-by-layer assembled multicomposite films. Curr. Opin. Colloid. In: 1998; 3: 32-39.
[12] Zhou, CH, Shen: ZF, Liu LH. Preparation and functionality of clay-containing films. J. Mater. Chem: 2011; 21: 15132-15153.
[13] Sun J, Wu T, Liu F. Covalently attached multilayer assemblies by sequential adsorption of polycationic diazo-resins and polyanionic poly (acrylic acid). Langmuir: 2000; 16: 4620-4624.
[14] Yan D, Lu J, Wei M. Heterogeneous Transparent Ultrathin Films with Tunable-Color Luminescence Based on the Assembly of Photoactive Organic Molecules and Layered Double Hydroxides. Adv. Funct. Mater: 2011; 21: 2497-2505.
[15] Fan L, Cao M, Gao S. Preparation and characterization of a quaternary ammonium derivative of pectin. Carbohyd. Polym: 2012; 88: 707-712.
[16] Khalil-Abad MS, Yazdanshenas ME, Nateghi MR. Effect of cationization on adsorption of silver nanoparticles on cotton surfaces and its antibacterial activity. Cellulose: 2009; 16: 1147-1157.
[17] Chadwick MD, Goodwin JW, Lawson EJ. Surface charge properties of colloidal titanium dioxide in ethylene glycol and water. Colloid. Surface. A: 2002; 203: 229-236.
[18] Nateghi MR, Shateri-Khalilabad M. Silver nanowire-functionalized cotton fabric. Carbohyd. Polym: 2015; 117: 160-168.

GSM (gm/m^2) Change of Grey Knitted Fabric from Pretreatment to Dyeing and Finishing

K. M. Faridul Hasan[1, 2], Md. Abdul Mueeid Alam[1, 2], Gen-Yang Cao [1*]

[1]*Wuhan Textile University, Sunshine Road, Wuhan, Hubei, 430200, China*
[2]*Southeast University, Tejgaon Road, Tejgaon, Dhaka, 1208, Bangladesh*

Corresponding authors email: 249747014@qq.com

Abstract

GSM (gram per square meter) changes of grey knitted fabrics from pretreatment to finishing were investigated and analyzed through this research. GSM change from Pretreatment to dyeing and finishing depends on the properties of fabric such as the structure of fabric, wales per inch, courses per inch, stitch length, yarn count and fabric spirality etc. GSM of fabric also changes with the types of shade, which needs to be dyed for instance dark, medium or light shades in different processing stages such as scouring, bleaching, enzyme treatment, dyeing, soap wash and fixation. GSM is changed to a greater extent in colouration stage. During finishing processes, different machines like stenter, compacting, brushing, sueding are used where the GSM is changed to a significant level. If there are sufficient information about the processing procedure, then GSM change can easily be controlled. Fabric quality is totally dependent on the GSM. As such this paper studies the change of GSM in every stage and take note on how much GSM is changed during the processing stages for different types of fabric. It is observed that; an average 10 to 25% GSM is changed from grey stage to finishing stage in a knit dyeing production unit.

Keywords: Fabric GSM; Pretreatment; Dyeing; Finishing

1. Introduction

GSM means gm per square meter (gm/m^2). It is an essential property for knitted fabric. Due to GSM, the knitted fabric can become heavier or lighter. GSM varies from fabric to fabric especially with the count and stitch length of the fabric [1]. During pretreatment, dyeing and finishing GSM is changed to a greater extent. During these processes, GSM change depends on the types of chemicals and dyes used and also the processes used. GSM depends on the fabric stitch length, course per unit length, wales per unit length and yarn count. GSM indicates the properties of the fabric such as fabric weight. Fabric GSM calculations depend on the yarn count system as well [2, 3].

Fabric quality is an important issue for colored fabrics. If the quality is not not upto standard the buyer will not take the ordered fabric or garments. Therefore, it is extremely necessary to ensure the quality of the dyed fabric. The quality of the fabric is mostly controlled during dyeing and finishing. GSM is one of the most important property of fabric that determines its quality which depends on various factors. If the GSM of the grey fabric is not perfect then the coloration would also not be perfect [4, 5]. During dyeing, different processing chemicals and dyes are responsible for the changing of GSM. In scouring and bleaching GSM is extremely reduced. During the coloration stage, GSM is increased to a considerable level depending on the types of shade dyed. During soap washing as the excess colors are removed from the fabric, the GSM is again changed to small amount. However, during the finishing stages the change of GSM is excessive [6, 7]. If the GSM is not properly controlled in this stage fabric quality would be strongly hampered. The tone of the colored shade would also change. Through controlling the overfeed and underfeed mechanism, GSM change is controlled in the stenter machine. Here, the diameter of the fabric is reduced for 2-3% to control the GSM as well as other qualities of the fabric such as shrinkage, spirality, twisting bowing etc. Still now, there are not so many research on this process. To ensure better quality of the fabric; more research is needed [8, 9].

2. Experimental

2.1 Materials

Knitted fabric, chemicals and dyes available in the factory of Esquire knit composite limited situated in Bangladesh.

2.2 Instruments

Dyeing machines (Sclavos, Athena, Fongs) available in Esquire knit composite limited, Bangladesh. GSM cutter was used to cut the GSM of fabric collected after different processing stages. Electric balance was used to measure the samples GSM. The size of the GSM cutter is equal to the area of 1 cm x 1 Cm. The radius of the GSM cutter is 6cm X 6cm. Dewatering, stenter and compacting machine were used in finishing operations.

2.3 Methods

Firstly, different fabrics such as single jersey, interlock and lacoste fabrics were selected for this research. Then the samples were collected after the different processing stages of dyeing such as as scouring, bleaching, enzyme, dyeing, soap wash, and fixation. Samples were also collected after the different finishing operations like as stenter, compacting and after the moisture absorption from nature. The GSM of each samples were measured and the deviations were calculated. For the accuracy of the findings, every samples were checked for four times by taking the GSM measurement.

3. Results and Discussion

Table.1 GSM change of single jersey fabric in different processing stages.

Fabric name	Buyer and Colour	M/c no.	Lot no.	Count	Brand of yarn	Treatment observation stage	Obtained GSM
Single Jersy	TEMA; Pink	Athena - 1/1	03	40/1 Count	GTNIL-YRN	Grey stage	120
						Pre-treatment	108
						Enzyme	106
						Dyeing stage	119
						Fixing and Softening	120
						Stentering stage	109
						Compacting stage	138
						Final/finished stage	140

During overall wet processing and finishing processes of single jersey fabric GSM was changed to a greater extent (Table 1). GSM was mainly changed in pretreatment, dyeing, stentering and in compacting stages. During pretreatment overall impurities were removed by using scouring and bleaching agent for why GSM was extremely reduced than the grey fabric (Table 2). During antipilling protruding fibres were removed by using enzyme for why GSM was reduced. During dyeing, lots of dyes are used for the combination shade for why GSM increased. During soaping extra unfixed color is removed for why GSM is reduced. During stentering GSM is reduced. During compacting GSM is increased to a greater extent due to increase of the Overfeed%. GSM is increased by absorbing moisture (3-5%) from nature after compacting as well. After the overall procedure the GSM changed from 120 to 140.

Table.2 Difference of GSM between processes to final stage of single jersey fabric.

GSM Change in processing stages	Changing Value	Difference between Process to final stage GSM	Required value of GSM for finishing from process
Grey to Pretreatment	120-108=12 ↓	Grey to Finishing	140-120=20
Pretreatment to Antipilling/Enzyme	108-106=2 ↓	Pretreatment to Finishing	140-108=32
Antipilling to Dyeing	119 - 106=13 ↑	Antipilling to Finishing	140-106=34
Dyeing to Soaping	119-118=1 ↓	Dyeing to Finishing	140-119=21
Soaping to Fixing and Softening	120-118=2↑	Soaping to Finishing Softening and Fixing to Finishing	140-118=22 140-120=20
Fixing and Softening to Stentering	120-108 =1 2↓	Stentering to Finishing	140-109=31
Stentering to Compacting	138-109 = 29↑	Compacting to Finishing	140-138=2
Compacting to Finished stage	138 -140 =2 ↑	Finishing (Required) to Finishing (Actual)	140-140 =0

During overall wet processing and finishing process of double lacost fabric, GSM was changed to a greater extent (Table 3). GSM was mainly changed in pretreatment, dyeing, stentering and in compacting stages (Table 4). During pretreatment, overall impurities are removed by using scouring and bleaching agent for why GSM is extremely reduced than grey fabric. During antipilling protruding fibres are removed by using enzyme for why GSM is reduced. During dyeing lots of dyes are used for the combination shade; for why GSM is increased. During soaping unfixed color is removed for why GSM is reduced. During stentering GSM is reduced. During compacting GSM is increased to a greater extent due to the increasing Overfeed%. GSM is increased by absorbing moisture (3-5%) from the nature after compacting.

Table 3. GSM change of double lacoste fabric in different processing stages.

Fabric name	Buyer and Colour	M/c no.	Lot no.	Count	Brand of yarn	Treatment observation stage	Obtained GSM
Double Lacost	ZARA; Lavenda	HT-6	28	26/1	KAMYA-YRN	Grey Fabric	230
						Pre-treatment	221
						Dyeing	236
						Soaping	234
						Fixing and Softening	236
						Stentering	225
						Compacting	258
						Final/finished	260

During overall wet processing and finishing process of 1×1 Rib fabric, GSM was changed to a greater extent (Table 5). GSM was mainly changed in pretreatment, dyeing, stentering and in compacting stages. During pretreatment overall impurities are removed by using scouring and bleaching agent for why GSM is extremely reduced than grey fabric. During antipilling protruding fibres are removed by using enzyme for why GSM is reduced (Table 6). During dyeing lots of dyes are used for combination shade; for why GSM is increased. During soaping extra unfixed color is removed for why GSM is reduced. During Stentering GSM is reduced. During compacting GSM is increased to a greater extent due for increasing Overfeed%. GSM is increased by absorbing moisture (3-5%) from the nature after compacting. GSM was changed from 140 to 169.

Table.4 Difference of GSM between different dyeing and finishing processes to final stage of double lacoste fabric.

GSM Change in processing stages	Changing Value	Difference between Process to final stage GSM	Required value of GSM for finishing from process
Grey to Pretreatment	230-221=9 ↓	Grey to Finishing	260-230=30
Pretreatment to Dyeing	236-221=15 ↓	Pretreatment to Finishing	260-221=39
Dyeing to Soaping	236-234=2 ↓	Dyeing to Finishing	260-236=24
Soaping to Fixing and Softening	236-234=2↑	Soaping to Finishing Fixing and Softening to Finishing	260-234=26 260-236=24
Finishing to Stentering	236-225= 11↓	Stentering to Finishing	140-108=32
Stentering to Compacting	258-225 = 33↑	Compacting to Finishing (Required)	140-140=0
Compacting to Finished stage	260-258 =2 ↑	Finishing (Required) to Finishing (Actual)	140-141 =1

Table.5 GSM change of double lacoste fabric in different processing stages.

Fabric name	Buyer and Colour	M/c no.	Lot no.	Count	Brand of yarn	Treatment observation stage	Obtained GSM
1×1 Rib	C and A; Mock Orange	Athen a -4/1	36	28/1	Paradis e	Grey fabric	140
						Pre-treatment	133
						Enzyme	132
						Dyeing	146
						Soaping	145
						Fixing and Softening	147
						Stentering	139
						Compacting	166
						Final/finished fabric	169

Table.6 Difference of GSM between different dyeing and finishing processes to final stage of rib fabric.

GSM Change in processing stages	Changing Value	Difference between Process to final stage GSM	Required value of GSM for finishing from process
Grey to Pretreatment	140-133=7 ↓	Grey to Finishing	170-140=30
Pretreatment to Antipilling/Enzyme	133-132=1 ↓	Pretreatment to Finishing	170-133=37
Antipilling to Dyeing	146 - 132=14 ↑	Antipilling to Finishing	170-132=38
Dyeing to Soaping	146 – 145=1 ↑↓	Dyeing to Finishing	170-146=24
Soaping to Fixing and Softening	145-147=2↑	Soaping to Finishing Fixing and Softening to Finishing	170-145=25 170-147=23
Fixing and Softening to Stentering	147-140 = 7↓	Stentering to Finishing	170-140=30
Stentering to Compacting	166-140 = 26↑	Compacting to Finishing (Required)	170-166=04
Compacting to Finished stage	166-169=3 ↑	Finishing (Required) to Finishing (Actual)	170-69 =1

3.1 Effect of GSM on finished fabric

1) 10-25% GSM of grey knit fabric is increased after overall dyeing and finishing process.
2) If GSM is increased than diameter is decreased.
3) With the change of fabric GSM shrinkage and spirality of fabric is changed.
4) Fabric remains heavier or lighter with the change of GSM.
5) If GSM is increased in uncontrolled way than the fabric may be damaged due to extra overfeed which may hamper the fabric quality.

4. Conclusion

It is strongly tried, to find out the Changes of GSM in different fabrics from pretreatment to finishing in whole fabric dyeing and finishing section. For this purpose, I had to collect the samples of each and every processes. For some fabrics, more than one samples were taken to ensure the accuracy of the result. All most all the GSM values found are near about to acceptable limit. When more than one GSM values observed for same fabric and same processes it was seen that the differences were not so high. This research would provide a guideline to take the technical assumptions about the fluctuations of GSM in different processing stages of dying, in order to control the GSM properly. More research and development is required in future to monitor the Change of GSM in different processes.

During dyeing significant amount of dyes were used depending on light, medium and dark shade; for why GSM were increased. During stentering 10-15% GSM were increased by increasing overfeed % (up) and controlling overfeed % (down). During compacting 10-15% GSM were increased by increasing overfeed % (up). By absorbing the humidity from the atmosphere 2-3% GSM were increased after compacting and buyer allowed ±5% deviation regarding the expected GSM.

References

[1] Moin CJ, Mahabubuzzaman A. Process for level dyeing of 100% cotton knit fabrics with reactive dye. J Innov Dev Strategy: 2009; 3: p. 1-8.

[2] Raja D, et al. Study on dyeing behavior of cotton/organic cotton knitted fabrics. Indian Journal of Science and Technology: 2010; 3 (7): p. 746-751.

[3] Amin T. Overview of a Knit-Dyeing Factory with Necessary Production Formulas. Global Journal of Research In Engineering: 2014; 14 (1).

[4] Rahman M, et al. Effect of shade percentage on various properties of cotton knitted fabric dyed with reactive dyes. International Journal of Research in Engineering and Technology: 2014; 3 (2): p. 339-343.

[5] Hossain I, Hossain A, Choudhury IA. Dyeing process parameters optimisation and colour strength prediction for viscose/lycra blended knitted fabrics using Taguchi method. The Journal of The Textile Institute: 2016; 107 (2): p. 154-164.

[6] Mia MS, et al. Effective Processing Time & Cost Management of Dyes, Chemicals & Utilities Used in the Combed & Compact Knit Fabric Colouration in the Dyeing Textile Industries. American Journal of Energy and Environment: 2016; 1 (1): p. 1-16.

[7] Hasan KF. Study on the changes of gsm (gm/m 2) of grey knitted fabric from pretreatment to finishing. International Journal of Textile Science: 2015; 4 (6): p. 119-136.

[8] Shore J. Cellulosics dyeing: 1995: Society of Dyers and Colourists.

[9] Baker JR. Principles of biological microtechnique. A study of fixation and dyeing. Principles of biological microtechnique. A study of fixation and dyeing: 1958.

Dyeing of Meta-aramid Fabric with DMAc-$CaCl_2$ System

Bing-Qian Liu, Dan Sheng, Yun-Yi Wang, Gen-Yang Cao*

School of Textile Science and Engineering, Wuhan Textile University, Wuhan, China

*Corresponding author's email: 249747014@qq.com

Abstract

Meta-aramid fabrics are difficult to dye owing to their high glass transition temperatures and crystallinity. In this study, meta-aramid fabrics were dyed with the *N,N*-dimethylacetamide (DMAc)-$CaCl_2$ system by applying cationic dyes. The effect of DMAc-$CaCl_2$ system on the microstructure, surface morphology, and dyeing properties of meta-aramid fibers was studied by X-ray diffraction (XRD), thermogravimetric analysis, scanning electron microscopy, and colorimetry. XRD results showed that the crystallinity of meta-aramid fiber was only slightly influenced by the DMAc-$CaCl_2$ system when the concentration of DMAc was 40%. The thermal stability of meta-aramid fiber also decreased and the surface became rougher. These results indicated that the DMAc-$CaCl_2$ system had a swelling effect on the amorphous region of the meta-aramid fiber, which proved beneficial as it enhanced the uptake of cationic dyes. Furthermore, the washing and rubbing fastness of the dyed meta-aramid fabrics were generally good. The optimum dyeing conditions determined according to the results of the characterization were: 3% o.w.f. (on weight fabric) of the cationic dye, 40% DMAc, 40 g L^{-1} $CaCl_2$, bath ratio of 1:30, dyeing temperature of 95 °C, and a dyeing time of 1 h.

Keywords: Meta-aramid Fabric; Cationic Dyes; DMAc-$CaCl_2$ System; K/S Value

1. Introduction

Meta-aramid (poly isophthaloyl meta-phenylenediamine) also known as aramid fiber 1313, is a high-performance fiber which has been widely used for the preparation of fire retardant clothing, racing suits, military uniforms, spacesuits etc. These fibers possess extraordinary properties such as outstanding thermal stability, flame retardant ability, stable electric insulation, chemical stability, and corrosion resistance. Additionally, the strength and elongation of meta-aramid fibers are similar to ordinary polyester which makes them easier to be spun into fabric [1].

The meta-aramid fibers have the advantage of good regularity in their macromolecular chains, which results in high crystallinity and a high glass transition temperature (T_g, 270 °C). Moreover, the surface of meta-aramid fibers is relatively smooth and lacks active functional groups. The latter attribute especially makes it difficult to dye these fibers using conventional dyeing methods [2]. The usual dyeing methods of meta-aramid fabrics mainly included dope dyeing and swelling dyeing. Colored aramid fibers are mostly produced by dope dyeing, which has the advantages of being a quick and low-cost process that results in good color fastness. It has been observed that when dyes were added into the spinning solution, it causes problems in the spinning process because of the incompatibility between the dyestuff and spinning solution. Furthermore, it is difficult to control the color [3]. The swelling dyeing method, with the assistance of special treatment techniques, auxiliaries, and solvents, promotes the swelling of the meta-aramid fibers. Such swelling is favorable for the dyes to enter the internal space of the fibers. Related research on effective dyeing methods of meta-aramid fibers includes supercritical carbon dioxide dyeing [4], low temperature plasma treatment, liquid ammonia pretreatment, ultraviolet/ozone treatment [5], and dyeing with a carrier such as *N,N*-diethyl-*m*-toluamide, *N*-methylformanilide, and 2-phenoxyethanol [6-8]. However, the abovementioned methods require high temperature and pressure conditions, which consumes a lot of energy and the concept becomes antithetical to the principles of green chemistry. Kim et al have reported that the use of cationic dyes on meta-aramid fibers results in an improvement of color depth and fastness [9]. Other studies have utilized the *N,N*-dimethylacetamide (DMAc) salt system in meta-aramid spinning to destroy their hydrogen bonding network structure [10]. Apart from the hydrogen bonds, the crystal structure and morphology of meta-aramid fibers treated with DMAc are also changed greatly, and according to a published report, this intensely promotes the dye uptake [11]. Therefore, it has been proposed that the low temperature dyeing of aramid fibers may be achieved by using a known concentration of DMAc and $CaCl_2$. In this study, meta-aramid fabrics were dyed or treated with cationic dyes with the DMAc-$CaCl_2$ system. The effect of DMAc-$CaCl_2$ system on the microstructure, surface morphology, and dyeing properties of meta-aramid fibers were also investigated and the optimum dyeing conditions were determined.

2. Experimental

2.1 Materials

Meta-aramid fabrics (weight 185.2 g m^{-2}) were supplied by Taihe New Materials Co. Ltd (China). CI Basic Red 46 and CI Basic Yellow 24 were purchased from Desida Dyestuff Co. Ltd (China). CI Basic Blue 3 was purchased from Shanghai Jiaye Dyestuff Industry Co. Ltd (China). The structures of the three dyes are shown in Fig.1. Analytical grade DMAc and $CaCl_2$ were purchased from Sinopharm Chemical Reagent Co. Ltd (China). Cinscour PS, used as soaping agent before dyeing and Cinfast EF, used as soaping agent after dyeing were provided by Bozzetto Chemical Co. Ltd, Ningbo (China) (industrial grade).

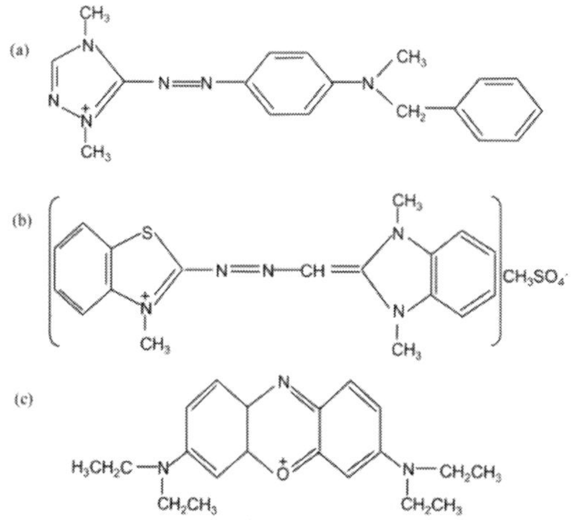

Fig.1 Dye structures
(a) CI Basic Red 46, (b) CI Basic Yellow 24, (c) CI Basic Blue 3

2.2 Pretreatment

Before the dyeing process, the meta-aramid fabric was soaked in water containing 2 g L^{-1} of Cinscour PS with a bath ratio of 1:20 at a temperature of 75°C for 15 min to remove oil stains and chemical reagents on the fabric. Subsequently, the fabric was rinsed thoroughly with water and dried at 50°C in preparation for the following process.

2.3 Dyeing

The meta-aramid fabric was dyed or treated in an Infrared Universal Dyeing Machine (HIF-24; Huaxia Sci&Tec Co. Ltd, China) according to the dyeing process depicted in Fig.2. The dyebath was prepared with CI Basic Red 46 (CI Basic Yellow 24 and CI Basic Blue 45 were also employed for fastness test), 3% o.w.f., DMAc (from 0 to 50%), and $CaCl_2$ (from 0 to 70 g L^{-1}), by applying a bath ratio of 1:30. The dyebath temperature was increased at a rate of 3°C min^{-1} from room temperature to 95°C, at which it was kept consistent for 1 h. Subsequently, the dyebath temperature was dropped to 50°C at a rate of 4°C min^{-1}. After dyeing, the meta-aramid fabric was first washed with running water and then with a soap solution containing 2 g L^{-1} Cinfast EF with a bath ratio of 1:20 at 75°C for 15 min in order to remove the unfixed dyestuff. Finally, a large amount of flowing water was used to rinse the meta-aramid fabric until it no longer faded.

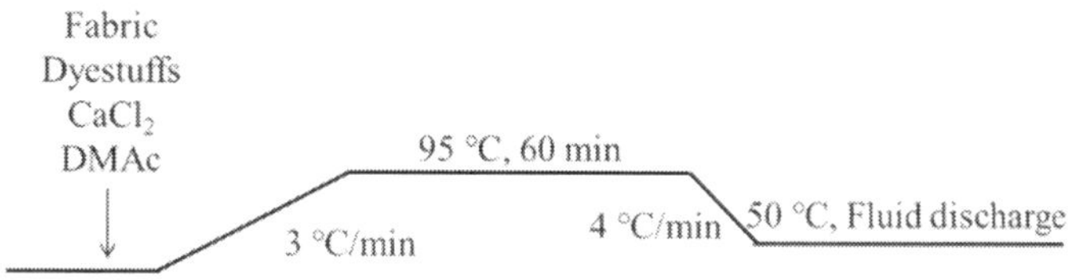

Fig.2 Dyeing process of meta-aramid fabric with the DMAc-$CaCl_2$ system.

2.4 X-ray Diffraction Analysis

X-ray diffraction analysis was used to analyze the crystalline states of the samples. An X-ray diffraction instrument (D8 ADVANCE X, Bruker, Germany) was employed to determine the crystallinity of original sample and meta-aramid fabrics treated with DMAc-$CaCl_2$ system. The scanning range was from 0 to 40°.

2.5 Thermogravimetric Analysis

Thermal degradation of original sample and meta-aramid fibers treated with DMAc-$CaCl_2$ was measured using a TG analyzer (209F1, Netzsch, Germany). The temperature range of the test was from 25 to 800°C at a heating rate of 10°C min^{-1}. The balance was protected by nitrogen throughout the heating process.

2.6 Surface morphology of meta-aramid fiber

The surface morphologies of original and DMAc-$CaCl_2$ treated samples were examined with a scanning electron microscope (SU1510, Hitachi, Japan). The meta-aramid fiber was fixed on the test bench with the conductive adhesive and prepared to be sprayed with gold. The surface morphology of the meta-aramid fiber was observed under 5000x magnification, and the scanning voltage was 5 kV.

2.7 Colorimetric Measurement

The color yield of dyed meta-aramid fabric was evaluated by the L^* and K/S values by employing a Computer Color Matching System (Color i7, X-Rite, USA) under the illuminant D_{65} and using a 10° standard observer. The L^* and K/S values of the dyed fabrics were measured at the maximum absorption wavelength. The instrument was calibrated before the measurement. The samples were folded into four layers.

The L^* value indicated the lightness of the dyed sample, and the sample was found to display a greater depth of shade (as evidenced by the lower L^* value). The K/S value derived from the Kubelka-Munk equation was used for evaluating the color depth of dyed sample [12]:

$$K/S = (1-R)^2/2R \qquad (1)$$

where K and S are the absorption and scattering coefficients of the dyed sample, respectively. R is the reflectance of the fabric corresponding to each wavelength.

2.8 Fastness Test

Washing and rubbing fastness of the three cationic dyes on meta-aramid fabrics with recommended DMAc-$CaCl_2$ system and control samples was evaluated according to GB/T 3921-2008 standard. A washing fastness tester (SW-12A; Dahua Electronic Instrument Co. Ltd, China) was employed to test the dyed meta-aramid fabrics.

Similarly, both dry and wet rubbing fastness were also tested (Y571N; Hongda Experimental Instrument Co. Ltd, China).

3. Results and Discussion

3.1 X-ray Diffraction Analysis

The influence of DMAc-$CaCl_2$ system on the crystallinity of meta-aramid fiber is illustrated in Fig.3. It is found that the crystallinity of the meta-aramid fiber treated with 40% DMAc is slightly lower than that of the original sample, which confirms that the DMAc-$CaCl_2$ system (when the concentration of DMAc = 40%) had little influence on the crystallinity of meta-aramid fiber. This may be explained by the limited swelling effect on the amorphous region of the meta-aramid fibers by the DMAc-$CaCl_2$ system. In contrast, the crystallinity of the meta-aramid fiber treated with 50% DMAc was significantly decreased, suggesting that it was possible to reduce the crystallinity of the meta-aramid fiber by using the DMAc-$CaCl_2$ system when a specific amount of DMAc was employed. This phenomenon might be attributed to the excessive swelling of the meta-aramid fiber caused by the DMAc-$CaCl_2$ system. The surface structure of the fiber was changed after treating with it with this system, which was also evident in the surface morphology analysis. The DMAc molecule penetrated the inner space of the fiber, and the crystalline region was partially broken. This morphological change proved beneficial as the dye molecules could enter deeper into the fiber, resulting in increased dye uptake. However, the significant decrease in the crystallinity of fiber probably led to some deterioration of mechanical properties.

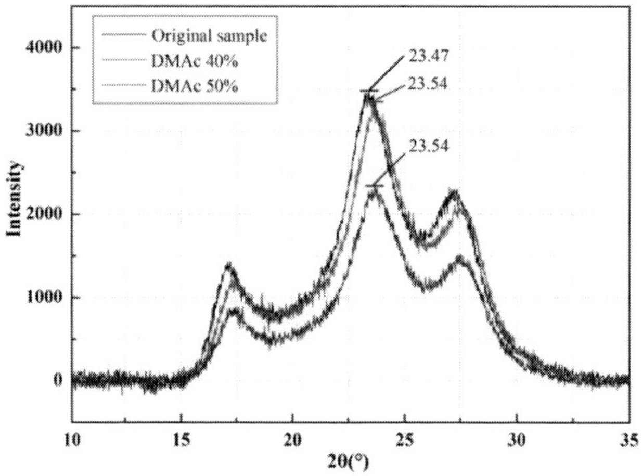

Fig.3 X-ray diffraction curves of the original and DMAc-$CaCl_2$ treated meta-aramid fibers.(Treatment conditions: 40%/50% DMAc, 40 g L^{-1} $CaCl_2$, bath ratio 1:30, temperature 95°C, 1 h).

3.2 Thermal Analysis

The TGA curves are shown in Fig.4, and depict the thermal degradation of meta-aramid fibers. The temperatures corresponding to the residual ratios (95, 90, 75, and 55%) are shown in Table 1. It can be seen that the treated meta-aramid fibers at the residual ratio of 95% have little higher decomposition temperature values than the original sample. All fibers have almost same decomposition temperature when the residual ratios up to 90%. However, the treated meta-aramid fiber with DMAc 50% decomposed more faster after 443°C. Its decomposition temperature at the residual ratio of 75% was 502°C, which was decreased by 12°C in comparison with the other two. In addition, when the residual ratio of the fibers came to 55%, the temperature of the treated fiber with DMAc 50% was 655°C, DMAc 40% was 703°C and the original sample was 729°C. The decomposition temperature of treated mete-aramid fibers was significantly lower than that of the original sample, which suggested that the former had much worse thermal stabilities. The reduction of the thermal stability is likely caused by the swelling effect on the amorphous region of meta-aramid fibers in the presence of DMAc and $CaCl_2$, which was also evident from

X-ray diffraction analysis. Therefore, less heat energy was required for breaking down the structure of the fibers.

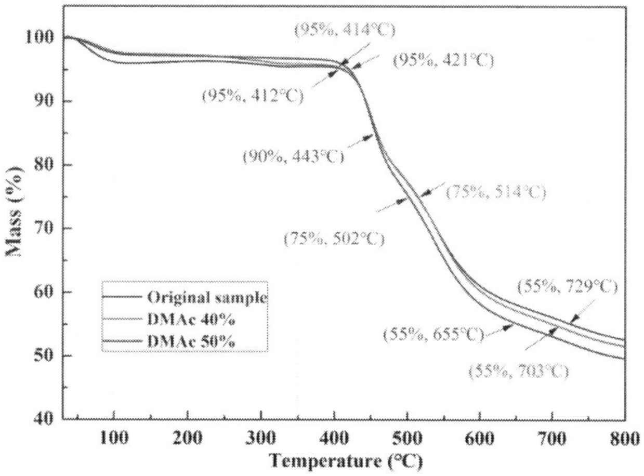

Fig.4 The TGA curves of the original sample and meta-aramid fibers treated with DMAc-$CaCl_2$ system. (Treatment conditions: 40%/50% DMAc, 40 g L−1 CaCl2, bath ratio 1:30, 95°C, 1 h).

Table.1 Temperatures for the two samples at different mass levels (°C).

Samples	95%	90%	75%	55%
Original sample	412	443	514	729
DMAc 40%	414	443	514	703
DMAc 50%	421	443	502	655

3.3 Surface Morphology Analysis

The surface morphologies of the original and treated meta-aramid fibers are shown in Fig.5. The surface of the original sample was found to be rough, as there were a small amount of dents that formed during the production of the meta-aramid fibers. However, the surface of the original sample was relatively smooth compared with the fiber treated with DMAc-$CaCl_2$ system. It was found that the meta-aramid fiber became more roughened, and the grooves on the fiber evidently became deeper on treatment with DMAc-$CaCl_2$. The change in the surface morphology of the meta-aramid fibers could be attributed to the swelling effect caused by this system. The rougher surface structure of the fiber was favorable for the adsorption of more dye molecules. With increasing amount and depth of the grooves, dyes had more space to adhere to the fibers and were more easily able to penetrate the internal locations of the fibers. In this manner, more dyestuffs could be fixed on the fabric and the DMAc-$CaCl_2$ system proved beneficial for improving the dyeing property of meta-aramid fibers.

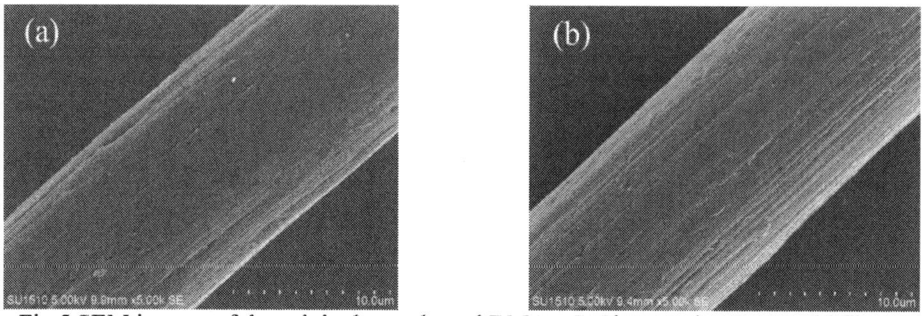

Fig.5 SEM images of the original sample and DMAc-$CaCl_2$ treated meta-aramid fibers.
(Treatment conditions: 40% DMAc, 40 g L^{-1} $CaCl_2$, bath ratio 1:30, temperature 95°C, 1h).
(a) Original sample; (b) Meta-aramid fiber treated with DMAc-$CaCl_2$ system.

3.4 Effect of DMAc-$CaCl_2$ System on the Color Yield of Meta-aramid Fabric

The colorimetric data including L^* and K/S values obtained for the fabric samples that were treated with varying amounts of DMAc (0 to 50%) and $CaCl_2$ (0 to 70 g L^{-1}) is shown in Fig.6.

Fig.6(a) shows that with an increase in the DMAc content, the L^* value decreased to a varying extent at a specific CaCl2 concentration. A greater amount of DMAc caused a greater decline of L* value. This value was also decreased with growing content of CaCl2 when the DMAc content was fixed, especially when the concentration of CaCl2 increased from 0 to 10 g L−1. Subsequently, the L* value decreased slightly with a growing concentration of CaCl2. These results revealed that the depth of shade of the meta-aramid fabrics increased with increasing amounts of DMAc and CaCl2.

Fig.6(b) shows the K/S value of meta-aramid fabrics dyed with DMAc-$CaCl_2$ system. The K/S value increased with the increasing amount of DMAc when the concentration of $CaCl_2$ was fixed. These values for the meta-aramid fabrics dyed with 50% DMAc were larger than those determined for the fabric when 40% of DMAc was used, which might be attributed to the evidently decreasing crystallinity of the meta-aramid fibers. As the content of DMAc exceeded 40%, the thermal stability and mechanical properties of the fabric possibly changed significantly. Consequently, the optimum content of DMAc was chosen to be 40%. Furthermore, the K/S value increased with increasing concentrations of $CaCl_2$ when the content of DMAc was fixed. It increased remarkably when the amount of $CaCl_2$ increased from 0 to 10 g L^{-1}, and the degree of increase was greater when the content of DMAc was also increased. These results implied that the effect of DMAc and $CaCl_2$ on meta-aramid fabrics dyeing is synergistic. The K/S value increased slightly when the concentration of $CaCl_2$ exceeded 40 g L^{-1}. Accordingly, the recommended amount of $CaCl_2$ was chosen as 40 g L^{-1} in order to achieve a satisfactory dyeing property.

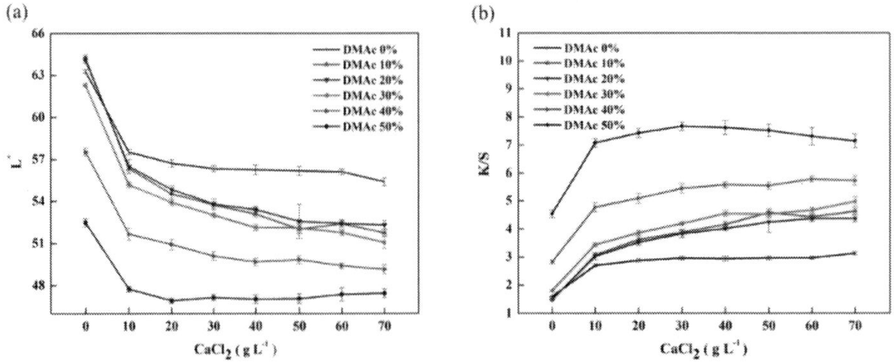

Fig.6 L^* and K/S value of meta-aramid fabrics dyed with DMAc-$CaCl_2$ system. (Dyeing conditions: CI Basic Red 46 3% o.w.f., DMAc content from 0% to 50%, $CaCl_2$ concentration from 0 to 70 g L^{-1}, bath ratio 1:30, temperature 95°C, 1 h). (a) L^* value (b) K/S value.

3.5 Fastness Properties

Table.2 Washing and rubbing fastness of the dyed meta-aramid fabrics (Dyeing conditions: CI Basic Red 46 CI/ Basic Yellow 24/ CI Basic Blue 3, 3% o.w.f., 40% DMAc, 40 g L^{-1} $CaCl_2$, bath ratio 1:30, temperature 95°C, 1 h) according to GB/T 3921-2008 standard.

Dyes	Sample	Washing fastness			Rubbing fastness	
		Color change	Staining		Dry	Wet
			Cotton	Meta-aramid		
CI Basic Red 46	Control	4	4-5	4-5	4-5	4
	DMAc-$CaCl_2$ system	4-5	4-5	4	4	3-4
CI Basic Yellow 24	Control	4-5	4-5	4-5	4	4-5
	DMAc-$CaCl_2$ system	4-5	4-5	4-5	3-4	4
CI Basic Blue 3	Control	4	4-5	4	4-5	4
	DMAc-$CaCl_2$ system	4-5	4-5	4	4	4

The results of washing and rubbing fastness of the three cationic dyes on meta-aramid fabrics with and without the DMAc-$CaCl_2$ system are shown in Table 2. The grades of color change were found to have improved only slightly when meta-aramid fabrics were dyed with the recommended DMAc-$CaCl_2$ system. The three cationic dyes showed excellent washing fastness on the cotton fabric, with a staining rating of 4-5. While the CI Basic Red 46 dye with DMAc-$CaCl_2$ system showed a marginally lower washing fastness on meta-aramid fabric staining, its rubbing fastness (both dry and wet) was moderately good. The rubbing fastness of meta-aramid fabrics dyed with DMAc-$CaCl_2$ system was marginally lower compared with the control sample. Overall, the DMAc-$CaCl_2$ system was found to have little effect on the color fastness of meta-aramid fabric.

4. Conclusion

In this paper, the effect of different concentrations of DMAc-$CaCl_2$ system on the microstructure, surface morphology, and dyeing properties of meta-aramid fiber were investigated. The results showed that the crystallinity of the meta-aramid fibers treated with the DMAc-$CaCl_2$ system decreased in varying degrees. When the DMAc content was 40%, the crystallinity was slightly reduced. When the DMAc content was increased to 50%, there was a stark decrease in the crystallinity of the meta-aramid fibers. Such fabrics dyed with the DMAc-$CaCl_2$ system achieved considerable color depth, which increased with the increasing concentration of DMAc and $CaCl_2$. These results can be explained by the swelling effect of the DMAc-$CaCl_2$ system on the fibers and the increasing roughness of their surface. Furthermore, the color fastness was also good, with a value above 3. The optimal dyeing conditions were: 3% o.w.f. of the cationic dye, 40% DMAc, 40g L^{-1} of $CaCl_2$, bath ratio of 1:30, dyeing temperature of 95°C, and the total dyeing time of 1 h.

References

[1] Liu ZX, Zheng JX, Chen GJ. Carrier dyeing of aramid fiber. Dyeing & Finishing: 2008; 3: 19-30.
[2] Wang JM, Wang HX, Xie FF. Dyeing properties of Aramid fiber 1313. Dyeing & Finishing: 2009; 18: 17-19.
[3] Zhang YJ, Liu LQ, Chen L, et al. Progress on Dyeing Technology of Meta-Aromatic Polyamide Fiber. Hi-Tech Fiber & Application: 2009; 34: 45-51.
[4] Zheng HD, Zheng LJ. Dyeing of Meta-aramid Fibers with Disperse Dyes in Supercritical Carbon Dioxide. Fibers and Polymers: 2014; 15: 1627-1634.
[5] Dong YY, Jang JH. The enhanced cationic dyeability of ultraviolet/ozone-treated meta-aramid fabrics. Coloration Technology: 2011; 3: 173-178.
[6] Lei LY, Mao YH, Xu XF, et al. Effect of N,N-diethyl-m-toluamide on the structure and dyeing properties of meta-aramid and para-aramid fibre. Coloration Technology: 2014; 5: 349-356.
[7] Mohammad TI, Francesco A, Ada F, et al. Use of N-methylformanilide as swelling agent for meta-aramid fibers dyeing: Kinetics and equilibrium adsorption of Basic Blue 41. Dyes and Pigments: 2015; 113: 554-561.
[8] Cao GY, Sheng D, Xu WL, et al. Structural and dyeing properties of aramid treated with 2-phenoxyethanol. Coloration Technology: 2015; 5: 384-388.
[9] Kim EM, Choi JH. Synthesis of cationized anthraquinone dyes and their dyeing properties for meta-aramid fiber. Fibers and Polymers: 2013; 12: 2054-2060.
[10] Ren J, Li XJ, Xu QS, et al. Study on the dissolution and coagulation properties of aramid fiber. Synthetic Fiber in China: 2016; 3: 17-23.
[11] Teng ZY, Shan CH, Zhang HP. Influence of polar solvent on structure and properties of m-aramid fiber. Advanced Textile Technology: 2016; 2: 1-3.
[12] Kubelka P, Munk J. Ein Beitrag zur Optik der Farbanstriche. Z Tech Phys: 1931; 12: 593-601.

Ultrasonic Dyeing of Cotton Fabric with Nature Plant Dye from Perilla Leaf

Kun-Di Zhu, Hui-Yu Yang, Wen-Fang Xu, G M Nazmul Islam, Meng-Mei Zhou, Gui-Zhen Ke*

School of Textile Science and Engineering, Wuhan Textile University, Wuhan, 430200, China

*Corresponding author's email: kgz66@126.com

Abstract

The authors extracted the natural plant dyestuff from perilla leaf and dyed cotton fabric with ultrasonic process. The optimum extraction process of perilla leaf was investigated through single factor and orthogonal experiments. The effects of dyeing time, temperature, pH value of dyeing solution, ultrasonic frequency and salt concentration on K/S value of the dyed fabric were discussed. The extraction results showed that the optimum extraction process of Perilla was at 60℃, solid-liquid ratio 1:60, time 90min, NaOH concentration 30g/L. Dyeing results showed that ultrasonic improved K/S value of the dyed cotton fabric. The optimum ultrasonic dyeing process: ultrasonic frequency 60 kHz, dye concentration 0.5C, dyeing time 60min, dyeing temperature 75℃, pH 6, salt concentration 20g/L.

Keywords: Natural Dye; Perilla Leaf; Ultrasonic Dyeing; Cotton

1. Introduction

Currently, a lot of researches about natural dyes have been done [1-4]. Natural dyes are generally more environmental friendly. At present, natural dyes are mostly used in the dyeing of natural fiber such as wool, silk and cotton fiber [5-8].

Perilla is a Lamiaceae annual herb and used as traditional medicine and edible plants [9]. As a multi-purpose economic crop, it has 2000 years of cultivation history in China, mainly used for medicine, oil, spice, food and so on. As the first batch of medicine promulgated by Ministry of Health, it has potential physiological function and antimicrobial effect on escherichia coli, shigella, staphylococcus [10]. Perilla leaf contains abundant Anthocyanins, including 9 anthocyanins and cis-isomers, the highest levels are malonylshisonin and perrilylAlcohols [11]. The pigment extracted from the basil leaves is anthocyanin pigment. It has better stability to light and heat and can be used as a raw material of natural pigment [12]. Nowadays, environmental protection and health awareness become more and more important, therefore the natural green plant dye is worth exploring [13].

Ultrasonic cavitation effect can make the dye disperse in the dye solution evenly in a single molecule state, reduce the energy consumption in the process of dye diffusion, improve dye diffusion velocity and dyeing efficiency [14]. The literature survey indicates that there have been relatively few works focusing on ultrasonic dyeing of natural plant dye [15]. Therefore, to make full use of the natural plant dyes, and to develop some high value-added multi-functional textiles, Perilla was extracted and cotton fabric was dyed with perilla extraction using ultrasonic method.

2. Experimental

2.1 Materials

Cotton poplin (bleached fabric) and perilla were procured commercially. NaOH, CH_3COOH and NaCl were of analytical grade. Distilled water was used in all the experiments.

2.2 Extraction of Perilla Dye Solution

Extraction experiment was carried out according to $L_9(3^4)$ orthogonal experiment. The factors and levels of orthogonal experiment are presented in Table 1.

Table.1 Factors and levels of orthogonal experiment

Level	Bath ratio	Extraction time (min)	Extraction temperature (℃)	NaOH concentration (g/L)
1	1:100	40	40	10
2	1:80	60	60	20
3	1:60	90	90	30

1g perilla was put into NaOH solution and kept at constant temperature in the thermostatic heater (water bath) for the specified time. The extraction was filtrated and the filtration was concentrated to a fixed volume. The absorbance of perilla extraction was measured on a Shimadzu 2550 UV/visible spectrophotometer. Optimal extraction process was selected by orthogonal experimental and the extraction was taken as the original extract.

2.3 Dyeing of Cotton Fabric with Perilla Extraction

The conventional and ultrasonic dyeing conditions were used to explore the dyeing properties of Perilla on cotton fabrics. The dye was extracted by the optimum extraction process. The original dye concentration is C. Cotton fabric is 1g and the bath ratio is 1:40. Perilla dye concentration, dyeing temperature, dyeing time and dyeing pH value and salt content were studied by single factor experiments.

2.4 K/S Value Test

K/S value of the dyed fiber was measured with UV/Vis spectrophotometer plus an integrating sphere attachment and color measuring software, using illuminant D65 and 10° standard observer. K/S value can be determined using Kubelka–Munk equation:

$$K/S=(1-R)^2/2R \tag{1}$$

Where K is coefficient of absorption, S is coefficient of scatter, and R is fractional reflectance.

2.5 Fastness Testing

The color fastness to washing and rub was determined according to GB/T3921-2008 and GB/T 3920-2008 respectively.

3. Results and Discussion

3.1 UV-Vis Absorption Spectrum of Perilla Extraction

The visible spectrum of Perilla dye extraction with different NaOH concentration are shown in Fig.1.

The figure shows Perilla dye has a maximum absorption at 651.5nm. NaOH concentration has not changed the maximum absorption wavelength. In the following test, all absorbance was measured at λ_{max} 651.5nm.

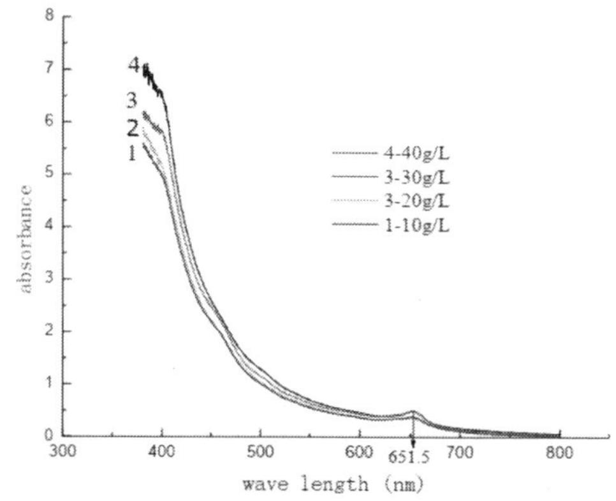

Fig.1 Different concentration of NaOH (10 to 40 g/L) of Perilla dye solution of the visible spectrum

3.2 The Extraction of Perilla Dye

Extraction experiment was carried out according to $L_9(3^4)$ orthogonal experiment.

Bath ratio, dyeing temperature, dyeing time and NaOH concentration were chosen as four factors to design $L_9(3^4)$ orthogonal experiment to explore the impact of these factors on extraction of Perilla dye solution.

Table.2 Orthogonal trial results for the extraction of perilla

Test number	Bath ratio	Temperature (℃)	Time (min)	NaOH concentration (g/L)	Absorbance
1	1:100	60	40	10	0.103
2	1:100	80	60	20	0.305
3	1:100	100	90	30	0.488
4	1:80	60	60	30	0.300
5	1:80	80	90	10	0.575
6	1:80	100	40	20	0.638
7	1:60	60	90	20	0.415
8	1:60	80	40	30	0.718
9	1:60	100	60	10	0.730
average1	0.298	0.273	0.486	0.469	
average2	0.504	0.533	0.445	0.453	
average3	0.621	0.618	0.493	0.502	
range	0.323	0.345	0.048	0.033	

According to the literature and some of the previous single factor experiments, the main factors affecting the extraction of perilla are the bath ratio, temperature, time and alkali concentration, so these four factors are selected to design $L_9(3^4)$ orthogonal experiment.

From the above table, it's seen that the temperature is the most important factor that influences the Perilla extraction. Secondly, it's bath ratio, the influence of extract time and NaOH concentration is relatively small. In the process of extraction experiment, the greater bath ratio, more fully Perilla pigment is made. Besides, the effect of extraction temperature on the extraction is also quite important. The higher the temperature is, the big the absorbance of Perilla extraction, it's consistent with the experimental results. Although extracting time has a certain influence on pigment extracting, when extracting time exceeds 90 min, the concentration of the solution reaches saturated state, and the pigment is difficult to be extracted even if the extraction process continues.

Based on full consideration of orthogonal analysis, the optimum extract conditions are that 1g perilla is dipped in

30g/L NaOH solution with bath ratio 1:60 at 100 °C for 90min.

3.3 The Effects of Dye Concentration on K/S Value

The influence of concentration of dyeing liquid (0.1C-0-9C) on K/S value is shown in Fig.2.

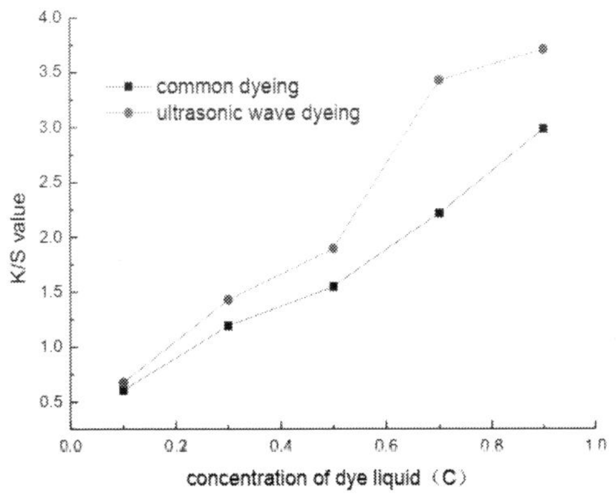

Fig.2 The influence of concentration of dye liquid on K/S value

Fig.2 shows that K/S value and concentration of dyeing liquid are positively correlated. When concentration of dyeing liquid is low, the K/S value of common dyeing and ultrasonic wave dyeing are similar, but when the concentration of dyeing liquid exceeds 0.5C, the effect of ultrasonic dyeing is significantly superior to the common dyeing. Taking into the consideration of cost saving and the dyeing effect, the optimal concentration of dyeing liquid is chosen as 0.5C.

3.4 The Effects of Dying Time on K/S Value

The influence of time (10-90min) on K/S value is shown in Fig.3.

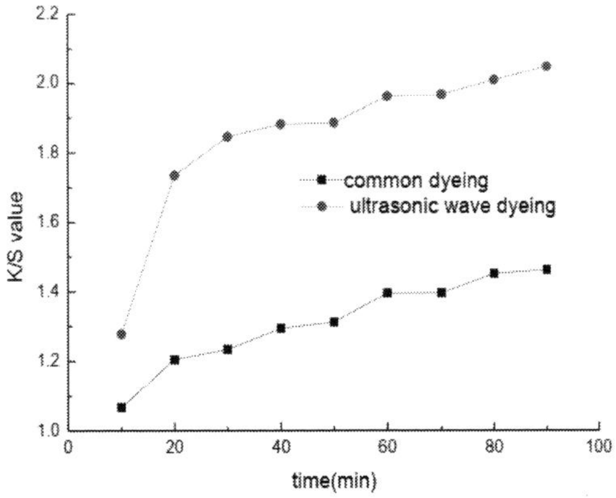

Fig.3 The influence of time on K/S value

It's observed that K/S value and dyeing time also are positively correlated, and the dyeing rate is fastest in the first 20 minutes, after 20 minutes, the dyeing rate becomes slower and reaches stable. At the beginning, the dye concentration in the dyeing liquor is high, the dye concentration in fibers is low, thus dye adsorption rate is faster during first 20min. At the same time, the effect of ultrasonic dyeing is significant. Both K/S value and dyeing rate increases obviously when dyed with ultrasonic. Undoubtly, in the process of dyeing, increasing dyeing time is helpful to raise the utilization ratio of the dye, but in consideration of the efficiency, 60min was chosen as dyeing time.

3.5 The Effects of Dyeing Temperature on K/S Value

The influence of dyeing temperature on K/S value is shown in Fig.4.

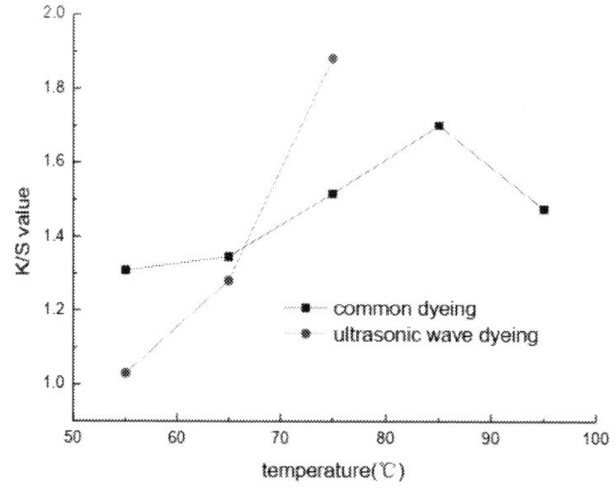

Fig.4 The influence of temperature on K/S value

Fig.4 shows that K/S value gradually increases with the increase of dyeing temperature. It's due to the increase of the kinetic energy of dye molecules and diffusion rate at higher dyeing temperature. At low temperature, the conventional dyeing is superior to ultrasound wave dyeing, but At high temperature (75℃), the ultrasound wave dyeing overtakes conventional dyeing. Ultrasonic wave dyeing is in its best performance at 75℃ and the K/S vale gained at this temperature is even higher than that gained at 95℃ with conventional dyeing. Finally, the best ultrasonic dyeing temperature is chosen as 75℃.

3.6 The Effects of pH Value on K/S Value

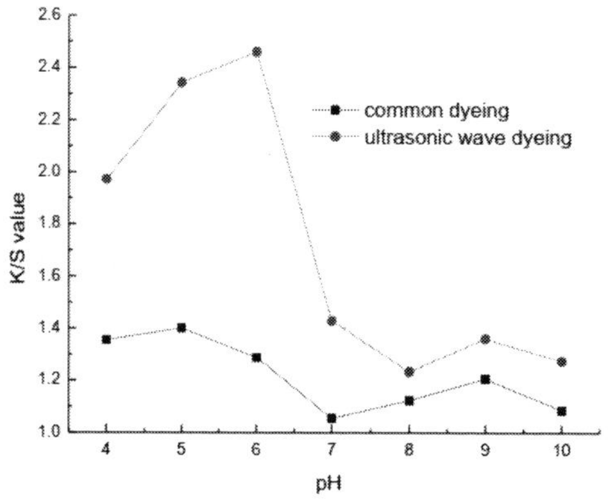

Fig.5 The influence of pH value on K/S value

The influence of pH value on K/S value is shown in Fig.5.

It's observed that when the fabric is dyed in acidic solution, the K/S value of the dyed cotton is significantly better than that of alkaline conditions. Considering the strength loss of cotton fiber in acid solution. The pH value is chosen as 6 during following dyeing.

3.7 The Effects of Salt Concentration on K/S Value

The influence of salt concentration (10 to 50 g/L) on K/S value is shown in Fig.6.

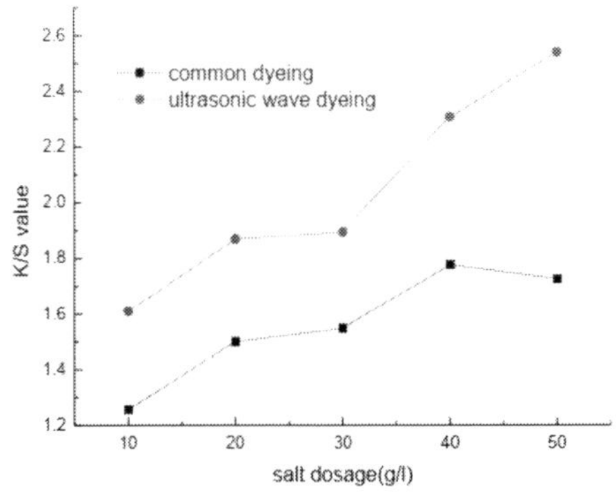

Fig.6 The influence of salt dosage on K/S value

Fig.6 shows K/S value of the dyed cotton fabric is positively proportional to the salt dosage. Adding salt to dye solution will affect coulomb force between dye molecule and the fiber. To reduce the use of dyeing auxiliary, the optimal salt concentration is chosen as 20g/L.

Comparing the K/S value from the Fig.2 to Fig.6, ultrasonic dyeing is superior to the conventional dyeing, it's due to the cavitation effect of ultrasonic wave. Ultrasonic wave caused the mixing, dispersion and degassing of dye solution and made the dye evenly disperse in the dye solution as single molecule state, and thus promoted the dyeing effect.

3.8 The Effects of Ultrasonic Intensity on K/S Value

The influence of ultrasonic intensity (40-80KHZ) on K/S value are shown in the figure below:

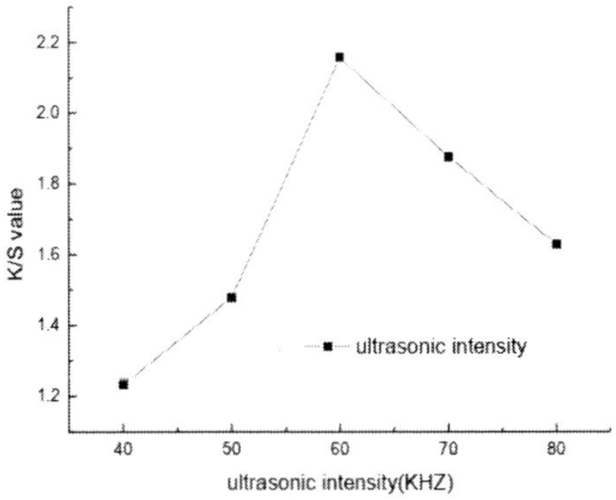

Fig.7 The influence of ultrasonic intensity on K/S value

From the Fig.7, it is observed that firstly K/S value increases with the increase of ultrasonic intensity, and then decrease, the best ultrasonic intensity is 60KHZ. When the ultrasonic frequency is moderate, ultrasound can accelerate the movement rate of dye molecules, when ultrasonic frequency is too high, excessive vibration makes the adsorption dye on the fabric fall off, and thus the dyeing depth is reduced.

3.9 Color Fastness Evaluation

The color fastness of the dyed cotton fabrics were measured according to corresponding standard. The results of washing and rubbing fastness are shown in Table 3.

Table.3 Color fastness of the cotton fabric dyed with perilla extraction

Dying type	Wash-fastness		Dry rub-fastness		Wet rub-fastness	
	Fading	Staining	Fading	Staining	Fading	Staining
convention	2-3	3-4	3-4	4-5	3	4
ultrasonic wave	3	3-4	4	4-5	3-4	4

It's noted that color fastness of perilla dyed cotton fabric to rubbing is better than washing. The washing fastness is acceptable. Ultrasonic dyeing improved half grade of washing fastness to fading.

4. Conclusion

The optimized process for extraction of Perilla: extraction temperature was 100℃, bath ratio was 1:60, extraction time was 90min, concentration of NaOH was 30g/L. The optimized ultrasonic dyeing method: ultrasonic intensity was 60KHZ, concentration of dye liquor was 0.5C, dyeing time was 60min, dyeing temperature was 95℃, pH=6, concentration of salt was 20g/L. All the dyed fabrics had better dry rubbing fastness than wet rubbing fastness. And ultrasonic dyeing improved half grade of the color fastness to washing. The limitation and shortcoming of this study are color fastness and color reproduction, and the next research plan is to find a new fixation method in the future work.

5. Acknowledgment

The authors greatly appreciate for the financial support by Education Department of Hubei Modern Textile Engineering and Technology Research Center of Wuhan Textile University (2016), China.

References

[1] Ratnapandian S, Fergusson SM, Wang LJ. Application of acacia natural dyes on cotton by pad dyeing. Fibers and Polymers: 2012; 13 (2): 206-211.

[2] Zhang RP, Cai ZS. Study on the Natural Dyeing of Wool Modified with Enzyme. Fibers and Polymers: 2011; 12 (4): 478-483.

[3] Acquaviva S, D'Anna E, De Giorgi ML, Della Patria A, Baraldi P. Physical and chemical investigations on natural dyes. Applied Physics. Section A: Materials Science and Processing: 2010; 100 (3): 823-828.

[4] Cristea D, Vilarem G. Improving light fastness of natural dyes on cotton yarn. Dyes and Pigments: 2006; 71: 39-46.

[5] Geelani SM, Ara S, Mir NA, Bhat SJA, Mishra PK. Dyeing and fastness properties of Quercus robur with natural mordants on natural fibre. Textiles and Clothing Sustainability: 2017; 2 (1): 1-10.

[6] Dehnavi E, Shams-Nateri A, Khalili H. Wool dyeing with binary mixture of natural dyes. Pigment & Resin Technology: 2017; 45 (1): 52-61.

[7] Güneş E, Atav R. The use of nutshell firstly as a natural dye for cotton and wool and then as a natural adsorbent for colour removal of basic dye effluent. Coloration Technology: 2017; 13 (1): 88-93.

[8] Ammayappan L, Shakyawar DBB. Dyeing of Carpet Woolen Yarn using Natural Dye from Cochineal. Journal

of Natural Fibers: 2016; 13 (1): 42-53.
[9] Jiao L, Liu S. Study on mordant-free dyeing of flax fabrics with pigment from purple perilla [J]. Textile Auxiliaries: 2011; 7: 014.
[10] Zhang J. 1, Wang SX, Wang B, Zhu P. Mordant dyeing of cotton fabric with natural dyestuffs [J]. Dyeing and Finishing: 2005; 16.
[11] El-Shishtawy RM, Shokry GM, Ahmed NSE, Kamel MM. Dyeing of modified acrylic fibers with curcumin and madder natural dyes Fiber. Polym: 2009; 617-624
[12] Ping LIN. Study on Preparation and Application of Natural Dyes [J]. Silk: 2009; 06.
[13] Sivakumar V, Anna JL, Vijayeeswarri J, et al. Ultrasound assisted enhancement in natural dye extraction from beetroot for industrial applications and natural dyeing of leather [J]. Ultrasonics Sonochemistry: 2009; 16 (6): 782-789.
[14] Zhao H, Wang L, Lin J. Mechanism of ultrasonic dyeing on cotton fabric with reactive dye [J]. Journal of Textile Research: 2009; 3: 016.
[15] Sivakumar V, Vijaeeswarri J, Anna J L. Effective natural dye extraction from different plant materials using ult rasound [J]. Industrial Crops & Products: 2011; 33(1):116-122.

Dissolution of Cotton Linters in Novel Alkali Solution

Sheng Li[1], Yang Liu[1], Yu-Miao Gong[1], Jian-Wu Dai[1], Yi-Xuan Mei[1], Xin-Wang Cao[1,2,*]

[1] *College of Textile Science and Engineering, Wuhan 430200, China*
[2] *National engineering laboratory for advanced yarn and fabric formation and clean production, Wuhan Textile University, Wuhan 430200, China*

[*]Coresponding author's email: aswang1984@163.com

Abstract

The solubility and swelling condition of cotton linters in the novel alkaline complex solvent NaOH/thiourea/urea solutions are studied. The FTIR results show that the crystals precipitated during the precooling process of solvent is thiourea and contains no urea and sodium hydroxide. In addition, the viscosity cellulose solution differs according to cellulose mass fraction. The structural viscosity index of cellulose solution increases with the increase of cellulose mass fraction. The solubility, swelling condition differs according to the cellulose mass fraction and how the cellulose is delivered.

Keywords: Cellulose; Sodium Hydroxide/ Urea/ Thiourea; Solubility; Viscosity; Swelling Condition

1. Introduction

Reviews of the early work on direct dissolution have been produced by Turbak [1, 2]. He records the efforts to dissolve cellulose directly as a base using sulfuric, phosphoric and nitric 'protonic' acids, or using, thiocyanateszinc chloride, iodides and bromides as Lewis acids. With regard to cellulose acting as an acid, he lists hydrazine, sodium zincate, and sodium hydroxide as inorganic solvents, and amines, quaternary ammonium hydroxides, dimethylamine/dimethylsulfoxide mixtures and amine oxides as organic bases. He obtained 14% solutions in dinitrogen tetroxide/dimethylformamide systems [3] and 16% cellulose solutions in lithium chloride/ dimethylacetamide systems [4]. Cuculo and Hudson [5] obtained solutions of up to 14% cellulose in an extensive study of the ammonia/ammonium thiocyanate solvent system. The dimethylsulfoxide/paraformaldehyde system, discovered by Johnson and Nicholson [6], proved capable of dissolving cellulose with a wide range of degrees of polymerization without causing degradation. However, despite early promises, the problems of developing fibre production routes using these systems have so far proved insurmountable, with the single exception of the amine oxide route.

However in this paper, dissolution of a number of cellulose samples in these aqueous solvent were investigated with respect to the influence of solvent environment (weight ratio of NaOH), precooled temperature, dissolving method and stirring rate. The process of direct dissolution of cellulose was observed using a polarized light microscope (PLM) because of the high crystallinity of cotton linters. The solvents were investigated by Fourier Transform Infrared Spectroscopy (FTIR). The solutions were investigated by solubility and viscosity.

2. Experimental

2.1 Dissolution of Cellulose

NaOH/urea/thirourea/H_2O with the weight ratio of 8:8:6.5:77.5(wt%) were blended and stirred until transparent. They were then pre-cooled in the refrigerator till needle-like crystal appeared under the bottom of the breaker. Cellulose with different weight were delivered in the precooled solutions and stirred under ice-water.

The dissolving property of the same solvents effected by the de livery methods was check by change the order of precooling and dissolving which were showed in Table 1.

Table.1 Preparation of several cellulose solutions

Cellulose (g)	NaOH (g)	Thiourea(g)	Urea (g)	water (g)	delivery method
3	8	8	6.5	77.5	Pre-cold and then dissolution
4	8	8	6.5	77.5	Pre-cold and then dissolution
6	8	8	6.5	77.5	Pre-cold and then dissolution
3	8	8	6.5	77.5	dissolution and freeze
4	8	8	6.5	77.5	dissolution and freeze
6	8	8	6.5	77.5	dissolution and freeze

2.2 Fourier Transform Infrared Spectroscopy (FTIR)

The precipitate crystals, NaOH, Urea and thiourea were investigated by Nicolet Model Impact 410 Fourier Transform Infrared Spectroscopy (FTIR).

2.3 Solubility Test

The cellulose solubility were test by weight lost. The solutions were centrifuged for 3 min with the speed of 10000 rad/min, which were calculated by,

$$Sa=(W_1-W_2) / (W_1-W_0) \quad \text{Formula 1}$$

Where Sa is solubility,W_0 is the net weight of centrifuge tube (g),W_1: total weight of centrifuge tube and solution(g),W_2 is the total weight of centrifuge tube and solution subtracting the upper pure water(g).

2.4 Viscosity Test

The viscosities of solution with diiferent weight percent were tested by NDJ-79 Digital Viscometer with the speed of 30rpm/min.

2.5 Polarized light microscope (PLM)

The cellulose solutions were screened by Polarized light microscope (PLM).

3. Results and Discussions

3.1 FTIR

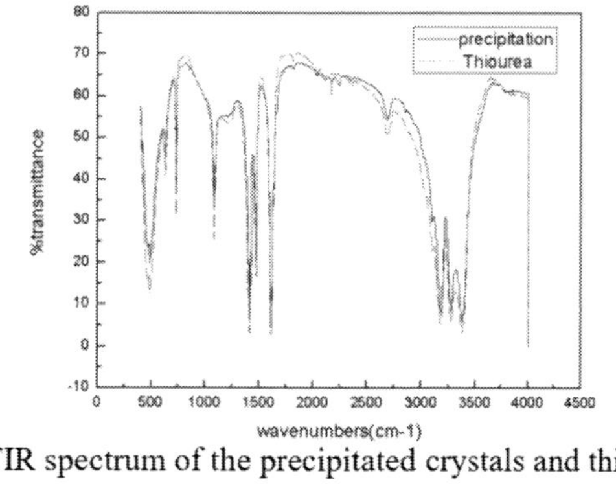

Fig.1 FTIR spectrum of the precipitated crystals and thiourea

During our experiment, it showed stronger solubility with the precooling solution with precipitated crystals.

The precipitated crystals were filtered by Buchner funnel and vacuum dried under 70℃ for 24h. The FTIR spectrum of the precipitated crystals, NaOH, urea and thiourea were showed in Fig.1. The FTIR spectrum of the precipitated crystals and thiourea looks the same, where showed a little difference under 2170cm^{-1} which was the absorption peak attributed by the impurity (thiocyanate ion). This means that The precipitated crystals during precooling was thiourea but not NaOH or urea.

3.2 Study on the Swelling Condition

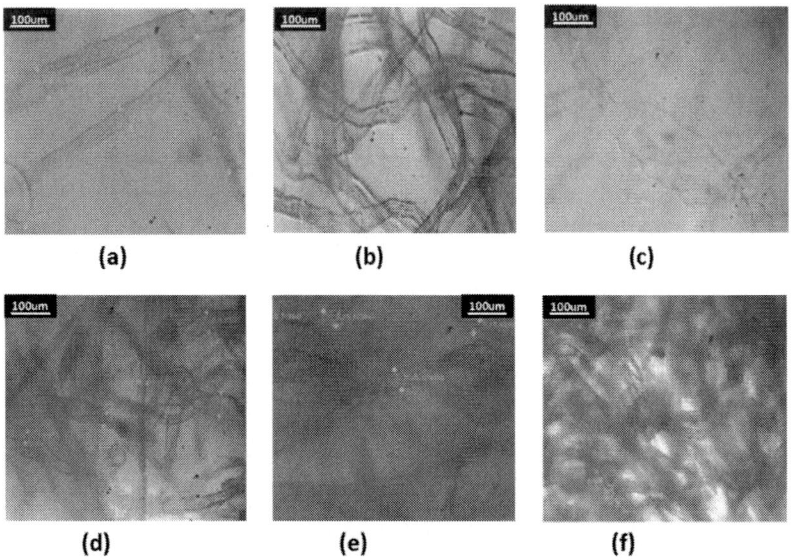

Fig.2 Microscope photographs of solutions 1-6

Microscope photographs with the magnification of 400 of cellulose solutions were showed in Fig.2, where the light points were not homogeneous, but with shadows and uneven thickness. It means that the cellulose were swelled and dissolved by the action of NaOH. Swelling were showed the homogeneous light points and dissolving were corresponding to uneven thickness. The diameters of cellulose from the cellulose solutions with same weight percentage resulted by different delivery method and solutions with different weight percentages resulted by same delivery method showed different change which were showed in Table 2.

Table.2 the diameter distributions of cellulose in different solutions

NO.	Delivery method	d (mm)	d_{max} (mm)
1	3 g precool and then dissolve	0.036	0.045
2	3g dissolve and then unfreeze	0.036	0.043
3	4 g precool and then dissolve	0.018	0.025
4	4 g dissolve and then unfreeze	0.018	0.025
5	6g precool and then dissolve	0.017	0.023
6	6 g dissolve and then unfreeze	0.017	0.019

From the investigation of the diameter cellulose fibers change, we could see the swelling property. The perfect swelling properties were showed in the greater diameter change with the effect by solvent. Table 2 showed the swelling properties of solutions with different weight percentages by same delivery method were 3g>4g > 6g, and the precooling method showed better swelling property of solutions with same weight percentage. That is to say, the 3 constituents did not only swell the cellulose but show the complementarity and reconstruction. Because the NaOH/urea/thiourea/water system, cellulose intrammolecular were destroyed, and the polar function groups in urea and thiourea such as C=O, C=S and $-NH_2$ would form hydrogen bonded with cellulose macro-molecule, which were good for the cellulose dissolving in the complex solvents.

3.3 Effect of Cellulose Weight Percentages on the Viscosity

Table.3 Viscosity tests of cellulose solutions with different weight percentages

Cellulose weight	Percent torque (%)	Viscosity (mPa.S)
3g	24	1250
4g	36	3050
6g	78	3300

In the new solvent system, the viscosities were increased with the increase of the cellulose weight percentages, and the increasing speed up with the same trend. That could be interpreted by the reeological properties effected with the concentration change. When the solution with low concentration the cellulose molecule keep in a clew shape, which could not be oriented during spinning.

3.4 Effect of Delivery Method on the Solubility

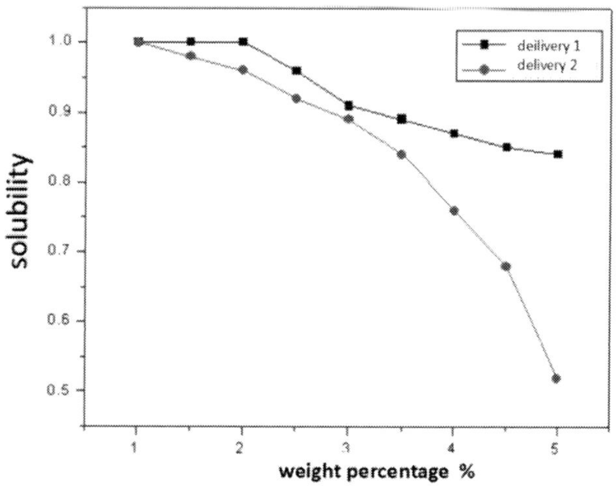

Fig.4 Effect of delivery method on the solubility

Effect of delivery method on the solubility were studied by dissolving cellulose with stirring for 3min in -10℃ precooled solvents and continuous stirring for certain times in salt-ice water bath with the temperate of -2-0℃ and putting cellulose in the new solvent stirring for 5 min and then cool for 12 h under the temperature of -8℃, unfreezing with strong stirring.

To check the solubility, cellulose solutions were gained by two different delivery methods with weight percentage from 1 to 5 wt%. Fig.4 showed the effect of delivery method on the solubility. Where it showed that when the weight percentage was greater than 3%, the solubility of delivery method 1 was higher than the method 2, but with little difference when the weight percentage was less than 3%. This could be interpreted by that to prepared high weight percentage solution with method 2, the freezing time of 12h, the uncompleted dissolute cellulose was as the crystal nucleus, which made the thiourea precipitate, change the constituents of the solvent, and decrease its solubility. In addition, the uncompleted dissolve made the cellulose gelling and effect the further dissolve. In the solutions with lower weight percentage, the cellulose were dispersed in the solvent, there were no so much thiourea precipitate. The constituents of the solvent did not change too heavily. Even with short time stirring, the cellulose could dissolve completely. With method 1 nothing were precipitate, the solvent did not change.

4. Conclusions

In a summary, we have gained new insight into the dissolution process, dissolving mechanism of cellulose in NaOH/thiourea/urea aqueous solution. It not only had academics significance and prospect of exploration application, but also provided imports and application of the cellulosic products. It may offer an alternative route to replace more hazardous existing methods for the production of regenerated cellulose fibers.

5. Acknowledgment

The authors wish to acknowledge the National Natural Science Foundation of China (51503162) and Natural Science Foundation of Hubei Province (2016CFB459).

References

[1] Turbak AF, et al. A Critical Review of Cellulose Solvent Systems, ACS Symposium Series 58, American Chemical Society: 1977

[2] Turbak AF. Proceedings of the 1983 International Dissolving and Speciality Pulps Conference: 1983; p. 105

[3] Turbak AF. Hammer RB. US Patent 4 056 675 (1 November 1977), to International Telephone and Telegraph Corp.

[4] Turbak AF, El-Kafrawy A, Snyder FWJr, Auerbach AB. US Patent 4302252 (24 November 1981), to International Telephone and Telegraph Corp.

[5] Cuculo JA, Hudson SM. US Patent 4367 191 (4 January 1983), to Research Corporation.

[6] Johnson DC, Nicholson MD. US Patent 4 097 666 (27 June 1978), to the Institute of Paper Chemistry.

Microbial Pigments and Application in Textile

Yin-Li Xu[1, 2], Zheng Li[1, 2*], Ji-Xian Gong[1, 2], Jian-Fei Zhang[1, 2]

[1]*School of Textiles, Tianjin Polytechnic University, Tianjin 300387, P.R. China*
[2]*Key Laboratory of Advanced Textile Composites (Tianjin Polytechnic University), Ministry of Education, Tianjin 300387, P.R. China*

*Corresponding author's email: Lizheng_nx @163.com

Abstract

A growing interest in the use of naturally produced pigments for textile dyeing has led to increased attentions towards the microbial pigments. The microbial pigments are used to enhance the appearance of textile and in making them preferable or attractive, and some fabrics were endowed with antibacterial properties. This review intends to provide and describe a comprehensive overview on the source of the strains, pigment properties, especially dyeing properties and effects. Meanwhile, poor light fastness has become a critical problem. This paper pointed out that more work should be focused on the improvement of production conditions, high efficiency of pigment extraction and screening of high yield strains.

Keywords: Microbial Pigment; Textile; Dyeing Property; Natural Pigment

Introduction

With the increasing application of synthetic dyes or colorants in dyeing industries and inefficiencies in dyeing resulted in a large amount of dyestuff, an estimation of 10,000,000 tons per annum is directly lost in the wastewater [1]. The waste dyestuff has caused serious environmental pollution and there is a report on the biological toxicity in some of synthetic dyes, which can be potentially dangerous to the human body [2]. Recently, there is a reviving interest for natural colorants in dyeing textiles among consumers because of their biodegradability, renewability with the environment and low allergic reaction [3, 4].

Natural colorants which were produced by plants had been applied in production before synthetic dyes were invented, but in very low yields. Furthermore, the use of plants in producing colorants is not environmentally friendly and sustainable due to the large amount of biomasses produced [5, 6]. Natural colorants from microorganisms are a promising alternative in relation to other additives extracted from animals and plants [7, 8]. Microbial pigments have production advantages compared with the similar products extracted from plants or animals, since they do not exhibit the problem of seasonal availability and are often more stable as well as being potent to produce pigments of different colors and shades, easier to scale-up production and growth on inexpensive substrates, for example, agro-industrial residues [7, 9, 10].

For the textile industry, the dyes produced by microorganisms are relevant. It is possible to guarantee that the metabolite of interest will be produced under controlled conditions, regardless of external factors and seasonal supply of raw materials. Also the batch-to-batch variability can be minimized. Some examples of the pigment producing bacterial species include Chromobacterium sp (violet pigment, violacein) [11], Serratia marcescens (red pigment, prodigiosin) [12], fungus Dermocybe sanguinea (yellow pigment, emodin and dermocybin) (Fig.1) [13]. Besides, some natural colorants, especially anthraquinone and flavonoid type compounds, have shown remarkable antibacterial activity in addition to providing bright colors, which could serve as functional dyes in producing colored antimicrobial textiles [14]. In this review, the species and structures, color, properties, and applications in textile of microbial pigment will be reviewed.

1. Microbial Pigments

1.1 Prodigiosin

Prodigiosins (PG) are tripyrrole red-colored compounds produced as secondary metabolites by many terrestrial and marine bacterial strains, including species of Serratia, Streptomyces, Vibrio and so on [12]. The color of prodigiosin is pH-sensitive. It presents purplish red under acidic and neutral conditions, while orange-yellow under alkaline condition [15].

Prodigiosin as a textile dye seems to be renewed as few publications, and suggesting that biosynthesized prodigiosin could be used as a functional dye for different textile materials. Alihosseini et al. [6, 16] characterized the bright red pigment prodigiosin from Vibrio spp. and suggested that it could be used to dye many fibers including wool, nylon, acrylics and silk. However, the dyeing performances are different, depending on the types of fiber. From the colorfastness testing, the dyed fabrics also have the ability to maintain its color under several external conditions such as perspiration, washing, and rubbing.

Ryazantseva et al. [17] found a natural bacterial pigment from Serratia marcescens strain 9986 was used as a dye of polyolefines (polyethylene, ultratene). The pigment suspension was introduced gradually up to rolled polymer sheet for the equilibrium coloration of the dyed-stuff. The pigment preparation prodigiosin was used in the technology dyeing as a thick suspension-concentrate without spraying in the air.

Prodigiosins are almost insoluble in water at room temperature, while they have certain solubility at high temperature because of the -NH groups in their molecules [15]. Yanfei Ren [18] et al got prodigiosin nanomicelles by the fermentation of Serratia marcescens through adding the nonionic surfactant to the culture media. The average pigment diameter was 184.3nm. Then dyed cotton fabric with the nanosuspension of prodigiosin micelles and discussed the optimum dyeing process. In addition, the antibacterial activity of dyed cotton against Escherichia coli and Staphylococcus aureus was investigated. Meanwhile, polyester fabrics were dyed with the nanosuspension of prodigiosin micelless as a type of cleaner disperse dye and the optimum dyeing condition was obtained in terms of the dyeing result, antibacterial activity and property of dyed polyester [15].

1.2 Violacein

Violacein is an indole derivative which produced mainly by bacteria of the genus Chromobacterium. It is a purplish-black needle prism. It is insoluble in water, slightly soluble in ethanol, moderately soluble in dioxane and acetone, and soluble in DMSO, methanol and ethyl acetate. Violacein exhibits important antiparasitic, antitumoral and antimicrobial properties.

In textile industry, it is possible to use natural fibers, such as silk, cotton, and wool and also chemical fibers, such as nylon and polyester. Violacein was manufactured by culturing Chromobacterium and Janthiobacterium in media containing beast hair degradation products as C supplements and was used for dyeing fiber materials and nylon cloth [19].

Chidambaram K. Venil [20] et al evaluated the dyeing potential of violet pigment from Chromobacterium violaceum UTM5. Dye-ability of violet pigment on different fabrics, i.e. pure cotton, pure silk, pure rayon, jacquard rayon, silk satin, cotton and polyester were tested using mordant and color fastness tests were performed. Results from this study showed that violet pigment is capable of dyeing both natural and synthetic fibers.

"Batik" is a popular gown like dress worn mostly by woman in South East Asian region and the most popular motifs include leaves, flowers and geometrical design [16]. Ahmad [21] et al. observed the applications of violacein in batik making for the first time.

1.3 Monascus Pigment

Monascus pigments are a group of fungal secondary metabolites called polyketides, which have similar molecular structures as well as similar chemical properties [22]. Monascus can produce yellow, orange, and red pigments. The six main monascus pigments are insoluble in water, but dissolved in ethanol, acetic acid, and so on. Usually, most monascus pigments are stable under 80°C and pH 6.0-8.0 [23]. The monascus pigments are sensitive to lights, especially to sunlight and ultraviolet lights, and the yellow pigment constituents are more stable than the red ones

[24].

Liu Yanchun et al.[25] used it on the dyeing of silk, the effect of pH value, mordant concentration and mordant method on the dye uptake and K/S value was investigated, and the soaping fastness and light fastness of the dyed samples were also researched. The results indicated that the rare earth praseodymium chloride has strong complex effect, can form a complex structure between monascus pigment with fiber, and thus improve the dye uptake and K /S value. The soaping fastness of the dyed samples was up to rating 4 grades and the light fastness up to rating 3-4 grades.

1.4 Phenazine Pigment

Phenazines, a series of heterocyclic compounds are produced naturally and substituted at different points around their rings by different bacterial species. Small modifications of the core phenazine structure give rise to a full spectrum of colors [26]. Pseudomonas is known to secrete different shades of phenazine derivatives, such as 1-hydroxy-5-methyl phenazinium betaine (pyocyanin), phenazine-1-carboxylic acid, phenazine-1-carboxamid (oxychlororaphin) and 7-imino-5methyl phenazine-1 carboxylate (pyorubin) [27, 28].

Rajasekar Saranya et al [27] extracted phenazine colorants of various shades from different Pseudomonas sp. Respectively, pyorubin was extracted from 2-day-old Pseudomonas aeruginosa MTCC 2474 culture supernatants, oxychlororaphin was extracted in ethyl acetate (1:1 v/v)from the culture supernatants of Pseudomonas aeruginosa (B).The dyeing potential of pyorubin and oxychlororaphin as colorants for silk dyeing were examined, the result showed that the optimum condition for dyeing was 90℃ at pH 3 and dyeing time 90min for oxychlororaphin, and 70℃ a t pH 3 and dyeing time 60 min for pyorubin.

Moreover, there is an increasing interest in adding value to textiles with the use of natural dyes [9]. Antimicrobial textiles with improved functionality find wider applications in health and hygiene products [16, 27]. Phenazines are heterocyclic nitrogen compounds with carboxy- and hydroxyl-substituted derivatives and are known to possess antimicrobial activity [29]. It was observed that the silk fabric dyed with pyorubin showed highly effective antimicrobial activity against all tested microorganisms when compared with oxychlororaphin [27].

1.5Anthraquinones Pigment

A number of anthraquinone derivatives, have been identified from various species of fungi and Lichens [30]. These metabolites are of interest because many of them possess significant antibiotic activity, primarily against Gram-positive bacteria and Pseudomonas aeruginosa [31]. Anthraquinones are also reported to have antiprotozoal and cytotoxic activities [14].

F.A. Nagia et al [14] isolated a strain Fusarium oxysporum from the roots of citrus trees. Two kinds of anthraquinone compounds were extracted from the culture of F.oxysporum, 2-acetyl-3, 8-dihydroxy-6-methoxy anthraquinone and 3-acetyl-2, 8-dihydroxy-6-methoxy anthraquinone, respectively. Cultures were filtered after 4-6 weeks, and the filtrate was adjusted to pH 3 with HCl. The acidified filtrate was then passed through a column of Amberlite XAD-7 for absorption of the pigments. When dye-ability of anthraquinone pigment on wool fabric, the values of dyeing rate constant, half-time of dyeing and standard affinity were calculated and discussed. The result shown that dye bath pH was 3, dyeing time 60min and temperature was 100℃.

Anthraquinone derivatives were previously isolated from the fungus Dermocybe sanguinea [13, 32]. Hynninen et al isolated anthraquinones from the fungus Dermocybe sanguinea by the enzymatic method. 2-acetyl-3, 8-dihydroxy-6-methoxy anthraquinone and dermocybin were separated from the anthraquinone mixture by a multiple liquid-liquid partition method. And the results from the UV/Vis and mass spectrometric measurement showed that the purity grade of emodin and dermocybin was 99% [13].

Then Raisanen R [13, 32] et al used emodin and dermocybin as disperse dyes for polyester and polyamide fabrics with a high-temperature dyeing method for the first time and used the mordant dyeing technique for wool and polyamide. The results showed that emodin dyes polyester bright yellow and dermocybin bright reddish-orange, and emodin dyes wool and polyamide yellow and red, whereas dermocybin confers purple and violet color with color-fastness properties from good to moderate, respectively. The fabrics showed excellent color-fastness results. This study showed that pure natural anthraquinone compounds can produce bright hues and color-fastness properties equivalent to those of synthetic dyes, and thus providing useful alternatives to the synthetics.

1.6 Other Microbial Pigments

In addition to the above pigments, which are mentioned in the literature could be used in textile dyeing also include the following: Velmurugan [33, 34] et al. studied the dyeing potential of five different water-soluble colorants extracted from M.purpureus, Isaria spp., Emericella spp., Fusarium spp., and Penicillium spp. The authors optimized the dyeing process in leather and cotton yarn resulting in uniform dyeing and intense and bright shade. The results indicated that the fungal colorants could be an alternate for leather and cotton yarn dyeing as a natural dye.

Genevieve Weber [35] et al. used various concentrations of green, red, and yellow pigments extracted with dichloromethane from the wood-staining fungi Chlorociboria aeruginosa, Scytalidium cuboideum, and S. ganodermophthorum, respectively, for the effective dyeing of bleached cotton, spun polyamide (nylon 6.6), spun polyester (Dacron 54), spun polyacrylic, and worsted wool.

Wan Azlina Ahmad [21] et al. got a kind of yellow pigment from chryseobacterium sp. by a via the boiling method. Several tapes of fabrics namely Natural silk, Dubai silk, Linen, Japanese cotton and Indian cotton were dyed by immersing the fabrics in the pigment solution at 80-90℃ and time at 60min.

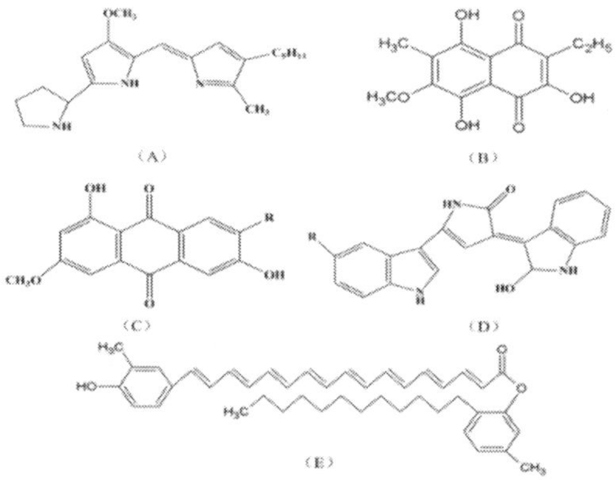

Fig.1 Chemical structural formula of prodigiosin (A), naphthoquinone (B), emodin (C), violacein (D), flexirubin (E)

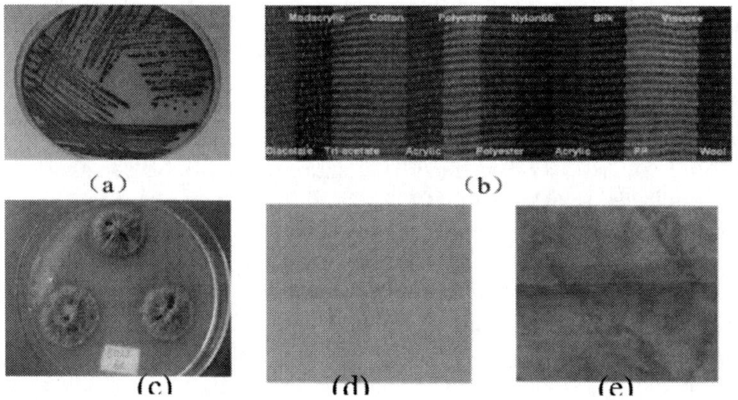

Fig.2 Compound fabrics dyed with Prodigiosin (a) extracted from Vibrio gazogenes (b) [36], silk fabrics dyed with Monascus (c, e) [37] and cotton fabrics (d) dyed with violacein [20]

2. Current Challenges

Colorfastness to light and washing are most important parameters to evaluate the performance of a textile and decide about its end use. Although colorfastness to rubbing and perspiration are also important especially if it is to be used as apparel [38]. Most natural dyes have poor to moderate light fastness, while synthetic dyes represent the full range of light fastness properties from poor to excellent [39]. Shahitha et al. [40] applied purified prodigiosin extracted from S. marcescens on cotton. Bringing the solution to boiling, the cotton cloth to be dyed was dipped in it and the bath was allowed to simmer for about 20 min. The cotton was then washed with water and dried in the shade, the result showed good color tone and did not change the color in washing, but it was sensitive to sunlight drying.

3. Future Perspectives

Currently, the potential of microbial pigment sources for textile applications are promising considering the ever rising demand by the consumers to replace their synthetic dyes. But only limited research studies are available on exploration of microorganisms for pigment production which really points towards exploring microbial pigments in more details [8].The research on microorganism pigments signify that genetic engineering for strain improvement, optimization of bioprocess modeling and utilizing cheap agro-industrial residues as substrates are key developmental strategies to maximize pigment production from microorganisms. And engineering of microbial pathway enzyme can also produce high amount of pigments in industrial process [10].

References

[1] Torres FAE, Zaccarim BR, Jozala AF, Santos CAD, Teixeira MFS, Santos-Ebinuma VC. Natural colorants from filamentous fungi. Appl. Microbiol. Biot: 2016; 100: 2511.

[2] Kiumarsi A, Gashti MP, Salehi P, Dayeni M. Extraction of dyes from Delphinium Zalil flowers and dyeing silk yarns. J Text I: 2016; 1.

[3] Hou X, Chen X, Cheng Y, Xu H, Chen L, Yang Y. Dyeing and UV-protection properties of water extracts from orange peel. J Clean Prod: 2013; 52: 410.

[4] Mapari SA, Meyer AS, Thrane U. Colorimetric characterization for comparative analysis of fungal pigments and natural food colorants. J. Agric. Food Chem: 2006; 54: 7027.

[5] Durán N, Teixeira MFS, De Conti R, Esposito E. Ecological-Friendly Pigments From Fungi. Crit. Rev. Food Sci. Nutr: 2002; 42: 53.

[6] Alihosseini F, Ju KS, Lango J, Hammock BD, Sun G. Antibacterial Colorants: Characterization of Prodiginines and Their Applications on Textile Materials. Biotechnol. Prog: 2008; 24: 742.

[7] Lopes FC, Tichota DM, Pereira JQ, Segalin J, de Oliveira Rios A, Brandelli A. Pigment Production by Filamentous Fungi on Agro-Industrial Byproducts: an Eco-Friendly Alternative. Appl. Biochem. Biotechnol: 2013; 171: 616.

[8] Tuli HS, Chaudhary P, Beniwal V, Sharma AK. Microbial pigments as natural color sources: current trends and future perspectives. J. Food Sci. Technol: 2015; 52: 4669.

[9] Torres FAE, Zaccarim BR, de Lencastre Novaes LC, Jozala AF, Santos CAD, Teixeira MFS, Santos-Ebinuma VC. Natural colorants from filamentous fungi. Appl. Microbiol Biot: 2016; 100: 2511.

[10] Venil CK, Aruldass CA, Dufosse L, Zakaria ZA, Ahmad WA. ChemInform Abstract: Current Perspective on Bacterial Pigments: Emerging Sustainable Compounds with Coloring and Biological Properties for the Industry -An Incisive. Evaluation. Cheminform: 2014; 4: 39523.

[11] Venil CK, Aruldass CA, Halim MHA, Khasim AR, Zakaria ZA, Wan AA. Spray drying of violet pigment from Chromobacterium violaceum UTM 5 and its application in food model systems. Int. Biodeterior. Biodegrad: 2015; 102: 324.

[12] Stankovic N, Senerovic L, Ilic-Tomic T, Vasiljevic B, Nikodinovic-Runic J. Properties and applications of undecylprodigiosin and other bacterial prodigiosins. Appl. Microbiol. Biotechnol: 2014; 98: 3841.

[13] Raisanen R, Nousiainen P, Hynninen PH. Emodin and dermocybin natural anthraquinones as mordant dyes for wool and polyamide. Text. Res. J: 2001; 71: 1016.

[14] Nagia FA, EL-Mohamedy RSR. Dyeing of wool with natural anthraquinone dyes from Fusarium oxysporum. Dyes & Pigments: 2007; 75: 550.

[15] Ren Y, Gong J, Fu R, Li Z, Yu Z, Lou J, Wang F, Zhang J. Dyeing and functional properties of polyester fabric dyed with prodigiosins nanomicelles produced by microbial fermentation. J Clean Prod: 2017; 375.

[16] Venil CK, Zakaria ZA, Ahmad WA. Bacterial pigments and their applications. Process. Biochem: 2013; 48: 1065.
[17] Ryazantseva I, Andreyeva I. Application of prodigiosin as a colorant for polyolefines. Adv. Biol. Chem: 2014; 04: 20.
[18] Ren Y, Gong J, Fu R, Li Z, Li Q, Zhang J, Yu Z, Cheng X. Dyeing and antibacterial properties of cotton dyed with prodigiosins nanomicelles produced by microbial fermentation. Dyes & Pigments: 2017; 138: 147.
[19] Durán M, Ponezi AN, Faljoni-Alario A, Teixeira MFS, Justo GZ, Durán N. Potential applications of violacein: a microbial pigment. Med. Chem. Res: 2012; 21: 1524.
[20] Venil, Yusof CK, Aruldass NZ, Ahmad CA, Wan-Azlina. Application of violet pigment from Chromobacterium violaceum UTM5 in textile dyeing. Biologia: 2016; 71: 121.
[21] Wan AA, Wan YWA, Zakaria ZA, Yusof NZ. Application of Bacterial Pigments as Colorant: Springer Berlin Heidelberg: 2012.
[22] Vendruscolo F, Bühler RM, de Carvalho JC, De OD, Moritz DE, Schmidell W, Ninow JL. Monascus: a Reality on the Production and Application of Microbial Pigments. Appl. Biochem. Biotech: 2016; 178: 211.
[23] Feng Y, Shao Y, Chen F. Monascus pigments. Appl. Microbiol. Biot: 2012; 96: 1421.
[24] Mapari SAS, Meyer AS, Thrane U. Photostability of natural orange-red and yellow fungal pigments in liquid food model systems. Journal of Agricultural & Food Chemistry: 2009; 57: 6253.
[25] Liu YC, Bai G. Bioecological mordant dyeing of silk with monascorubrin [J]. Text. Res. J.2012; 33 (4): 78-81.
[26] Pricewhelan A, Dietrich LEP, Newman DK. Rethinking 'secondary' metabolism: physiological roles for phenazine antibiotics. Nat Chem Biol: 2006; 2: 71.
[27] Saranya R, Jayapriya J, Tamilselvi A. Dyeing of silk fabric with phenazine fromPseudomonas species. Color. Technol: 2012; 128: 440.
[28] Hernandez ME, Kappler A, Newman DK. Phenazines and other redox-active antibiotics promote microbial mineral reduction. Appl. Environ. Microbiol: 2004; 70: 921.
[29] Dwivedi D, Johri BN. Antifungals from fluorescent pseudomonads: Biosynthesis and regulation. Currentence: 2003; 85: 1693.
[30] Yagi A, Okamura N, Haraguchi H, Abot T, Hashimoto K. Antimicrobial tetrahydroanthraquinones from a strain of Alternaria solani. Phytochemistry: 1993; 33: 87.
[31] Agarwal SK, Singh SS, Verma S, Kumar S. Antifungal activity of anthraquinone derivatives from Rheum emodi. J Ethnopharmacol: 2000; 72: 43.
[32] Räisänen R, Nousiainen P, Hynninen PH. Emodin and Dermocybin Natural Anthraquinones as High-Temperature Disperse Dyes for Polyester and Polyamide. Text res J: 2001; 71: 922.
[33] Velmurugan P, Kamala-Kannan S, Balachandar V, Lakshmanaperumalsamy P, Chae J, Oh B. Natural pigment extraction from five filamentous fungi for industrial applications and dyeing of leather. Carbohyd Polym: 2010; 79: 262.
[34] Velmurugan P, Kim MJ, Park JS, Karthikeyan K, Lakshmanaperumalsamy P, Lee KJ, Park YJ, Oh BT. Dyeing of cotton yarn with five water soluble fungal pigments obtained from five fungi. Fiber Polym: 2010; 11: 598.
[35] Weber G, Chen H, Hinsch E, Freitas S, Robinson S. Pigments extracted from the wood-staining fungiChlorociboria aeruginosa, Scytalidium cuboideum, and S. ganodermophthorum show potential for use as textile dyes. Color. Technol: 2014; 130: 445.
[36] Darshan N, Manonmani HK. Prodigiosin and its potential applications. J. Food Sci. Technol: 2015; 52: 5393.
[37] Yu YY. Study on microbial dyeing of silk and silk fabric. [D] Soochow university: 2015; 79.
[38] Saxena S, Raja ASM. Natural Dyes: Sources, Chemistry, Application and Sustainability Issues: Springer Singapore: 2014.
[39] Cristea D, Vilarem G. Improving light fastness of natural dyes on cotton yarn. Dyes & Pigments: 2006; 70: 238.
[40] Shahitha. Enhanced Production of Prodigiosin Production in Serratia Marcescens. J. Appl. Pharm. Sci: 2012; 2.

Processing Properties of PLA Fiber Fabric

Zhi-Lin Gao[1], Xing Zhang[2*]

Xi'an Polytechnic University, Apparel & Art Design college No.19 South Jinhua Road, Xian, Shanxi, 710048, China

[*] Corresponding author's email: Zhxhg029@126.com

Abstract

PLA polylactic acid fiber, made from renewable materials such as corn with fermentation techniques, is a type of eco-friendly new material that is biodegradable under suitable conditions.

In this paper, the basic properties of fabrics were discussed by comparing PLA / C (50/50) fabric, cotton fabric, and PLA / C (40/60) fabric. Many aspects have been studied including PLA fiber morphology, strength and other physical properties and thermal stability. It is shown that PLA / C (50/50) fabric has excellent performance in all aspects.

Studies have shown that polylactic acid has high fabric physical strength, high moisture permeability and good drape. However, its advantages are likely compromised by static electricity and an anti-pilling weak problem. The blended polylactic acid fiber PLA / C (50/50) has the advantages of being breathable,moisture-proof and effectively reduce static and pilling, especially for summer apparel fabric development.

With the strengthening of people's health and environmental awareness, the future of polylactic acid fabric PLA / C (50/50) development research will have significant social and economic benefits.

Keywords: PLA Fiber; Green Environmental Protection; Fabric Properties; Fabric Development

1. Introduction of PLA Polylactic Acid Fabrics

PLA polylactic acid fiber mainly comes from a new and environmental polymer fiber materials developed by Japan and USA . The main raw material of PLA is polylactic acid. It is made from recycled plant resources (corn, wheat, grain, sugar beet, etc.) as a starting material and then processed to produce polylactic acid fibers. PLA fiber can be discarded as it is biodegradable and can be degraded with water, soil and microbes. [1]

In the development of modern green polymer, the greatest development has been undoubtedly the work on biodegradable polymer materials. The abundance of PLA polylactic acid fiber's makes it a "21st century environmental recycling materials". Ecological fiber in the future has great potential for development.

1.1 PLA Polylactic Acid Fiber Structure

PLA polylactic acid fiber has a relatively high strength and elongation at break as well as a curl and smooth surface. It also has a good resilience and its performance is more suitable for textile fabrics. In addition, PLA polylactic acid fiber has a high stability in high temperatures and a weak alkali resistance. As such, it needs to be protected by an appropriate processing method. PLA polylactic acid fiber is a type of polymer which can be completely biodegraded. Under suitable conditions, it can decompose into water and carbon dioxide. Therefore, PLA polylactic acid fiber is a good green textile product. [2]

The PLA polylactic acid fiber chemical structure are made of simple chiral molecules. This is an important feature as it is the enantiomer of the polymerized monomer, so PLA polylactic acid fiber has three isomers namely, L-PDA, D-PDA and PDLLA. Among them, L-PLA and D-PLA are crystalline structures with a degree of 60%, thermoplastic polymer, PDLLA is a non-crystalline structure. Therefore, it is precisely because PDLLA has no shape structure which results in the relatively fast to the degradation rate. [3] The mechanical strength levels of D-PLA and L-PLA are relatively high and can be used as a material for doctors to use for surgical sutures. The lactic acid obtained by fermentation is 99.5% isomer, L-PLA is a high concentration of lactic acid polymer (crystal), especially for the production of textile fabrics and other products. [4] L-PLA can be degraded to L-lactic acid, in the process of human metabolism without tissue reaction.

Fig.1 Fiber longitudinal

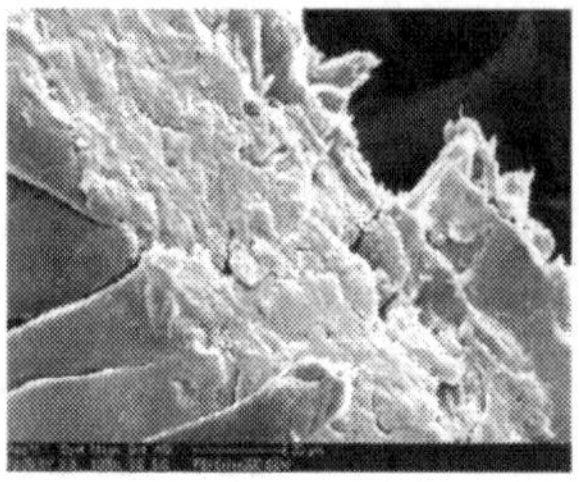

Fig.2 Fiber end

Different PLA polylactic acid fiber molecular structure will also change the performance of the fiber. With a different level of content of PLA polylactic acid in different fabrics, PLA polylactic acid polymer will have a different performance.

The lattice structure parameters of the L-PLA crystals cultured in dilute solution were determined by electron microscopy, X-ray diffraction and atomic force microscopy. [5] The lattice has the following parameters a = 1.078 nm, b = 0.604 nm, and c = 2.87 nm. [6]

PLA polylactic acid fiber formation is derived from the melting of the fibrous polymer melt, this is done from the nozzle through exerting air pressure into a filament. The PLA polylactic acid fiber showed signs of curl with a very good processing performance. The short fibers show a curl in a certain degree, and the cross section is a circular shape. There are no regularities in the longitudinal section of the fiber and the presence of streaks is discontinuous. The cross section is also a circle with spots. [7] The formation of these spots and stripes is a part of the non-crystalline section of the PLA. In the presence of water, oxygen and bacteria, it can quickly break down.

1.2 The Basic Properties of Polylactic Acid Fiber

Table.1 Basic properties of PLA

Density	Softening temperature	Melting temperature	Tensile stress	Stretch	Impact strength
1.24 g/mL	54°C -58°C	140°C -175°C	53 MPa	6%	12.8 J/m
Proportion	Regain	Elongation	Strength	Fineness	Crimp
1.25 g/cm3	0.4%-0.6%	30%-40%	30-50 cN/tex	1.5 dtex	30%-50%

PLA polylactic acid fiber combines the characteristics and advantages of natural fibers and synthetic fibers. It has many excellent properties such as a good biodegradable performance, excellent mechanical properties, great dyeing and wet performance, good to touch, high resilience and UV stability as well as being flame retardant and anti-fouling.

From the table 2 it can be seen that the physical properties of polylactic acid lies between polyester and nylon in terms of strength, hygroscopicity, elongation and dyeing. In detail:

1) Polylactic acid fiber specific gravity is less than polyester but greater than nylon. The manufactured products of polylactic acid fiber are relatively light.

2) Due to the elastic modulus of polylactic acid fiber between the polyester and nylon, the style is dry.

3) Polylactic acid fiber melting point is relatively low.

4) Polylactic acid fiber has high strength, medium flexibility. I t has enough strength, which is suitable for making all kinds of clothes. It is also a general fiber material and high practicality; it has a lower modulus, which allows the fiber fabric to have a good processing performance; Polylactic acid fiber breaking strength and elongation at break are close to that of polyester; This gives the fabric high strength, good elongation, soft, drape and good resilience.

5) Polylactic acid fiber moisture absorption is relatively small. It is close to polyester but has a better wicking. As such, during the water wetting process, the water diffusion is good with good comfort.

6) Polylactic acid has a good elastic recovery rate, the appropriate glass transition temperature gives it a good shape and anti-wrinkle properties.

Table.2 Comparison of properties of polylactic acid fiber and other fibers

	Nylon	Polyester fiber	Acrylic fibres	polylactic acid	Viscose	Cotton	Hair
Proportion	1.14	1.39	1.18	1.25	1.52	1.52	1.31
Young modulus	300 kg·mm-2	1200 kg·mm-2	-	400-600 kg·mm-2	-	-	-
Strength	42-48 CN/tex	45-55 CN/tex	30-35 CN/tex	30-35 CN/tex	20-25 CN/tex	20-40 CN/tex	12-20 CN/tex
Regain	4%-4.5%	0.2%-0.4%	1.0%-2.0%	0.4%-0.6%	11%-13%	7%-8%	14%-18%
Melting point	215°C	255°C	320°C	170°C	-	-	-

1.3 Special Properties of Polylactic Acid Fiber

1.3.1 Green Environmental Performance

PLA polylactic acid fiber is a veritably organic polymer "green product". The most prominent advantage of PLA polylactic acid fiber is that it is green. First of all, the raw materials for the production can be recycled and the source of raw materials is in line with the requirements of sustainable development. The raw material used in the synthesis of PLA is maize and it can be fermented to obtain lactic acid. The lactic acid further reacts to form a cyclic dimer, which is polymerized to obtain PLA polylactic acid. In addition, the raw materials can also be other types of agricultural products. [8]

The green environmental performance of PLA fiber has an advantage when compared to the limitations of chemical fiber fabrics. For the availability of raw materials, PLA polylactic acid is easier to obtain, healthier and less costly. In the production process, the whole synthesis process of PLA fiber is non-toxic. In regards to the energy consumption, the melting point of PLA polylactic acid is 170℃ which is lower than ester, nylon and polypropylene. As a result, the energy consumption of PLA fiber fabric is the lowest in the production process of various fabrics. In reference to the treatment of polluted waste water, the biological enzyme in PLA fiber is easier to be degraded compared to soil. [9] Therefore, it is more environmentally friendly and in line with the requirements of green production.

1.3.2 High Temperature Resistance

The temperature and melting point of PLA polylactic acid fiber glass are very low: at only 57℃ and 175℃ which is significantly lower than polyester. The heat setting temperature should be between the glass transition temperature and melting point. It will shrink when heated to 140℃, so the temperature of polylactic acid fiber products can not be too high during the process. Since PLA fiber is not thermostable, it is not suitable for some applications and processing. The higher the content of fiber D-isomer, the lower the crystallinity of the fiber, and therefore the lower the melting point. If the ratio of the two isomers in the PLA fiber changes, the melting point of the fiber may vary between 120℃ and 170℃. The temperature requirements are high during the dyeing and finishing process of the PLA polylactic acid fiber: heat setting temperature can not exceed 135℃. Cargill Dowpolyme recommends a pre-cure temperature of 120-130℃ and a heat setting temperature of 135℃ and a duration of 30 seconds. The ironing temperature can not be too high during the nursing process. [10]

1.3.3 The Absorbability by Human Body

PLA polylactic acid fiber and its copolymer have good biocompatibility and biodegradability. It can be safely implanted into the body and can be gradually degraded to carbon dioxide and water in the human body with no accumulation and toxic side effect. Poly-L-lactic acid (PLA) in the human body can be degraded and absorbed so it was used as a special surgical device, graft and plastic material, such as for surgical sutures, bone clips, bone nails in the 1970s and the encapsulation materials of delayed-released drug. [11] At present, polylactic acid has been widely used in medical bandages, disposable surgical clothing, anti-adhesion film, medical equipment and so on.

1.3.4 Anti-ultraviolet Performance

PLA polylactic acid fiber also has UV resistance. The UV absorption value of Polylactic acid is low. It can prevent UV intrusion and is suitable for outdoor applications and interior decoration with the fabric. [12]

1.4 PLA Research Status of Fiber Fabrics

There is a long history of product development and performance research on PLA polylactic acid fiber, and the development of PLA fiber fabrics emerged in the late 1990s. [13] Shanghai Watson Company has made corn fiber polylactic acid for weft, and the cotton thread as a warp since 2003, the "corn shuttle" and other fabrics developed in the imported shuttle loom have been in the market. [14] According to the survey, some companies in Shanghai, Zhejiang and Shandong used PLA short fiber to twist yarn products. Their fabrics were studied preliminarily and was tried in the production process\. The product variety however was single and the production scale was small. The companies concentrated on the spinning process and on non-woven fabrics based on the theory of the polylactic acid fiber applications and fiber properties. [15] The research is focused on how to use polylactic acid fiber to develop products as well as the spinning, weaving and dyeing process. It is to determine the processing parameters and there is less input on the configuration.

In general, research on PLA fiber and its textiles have attracted more and more attention both domestically and abroad, however the research on fiber is much more than PLA fabric research. Thus PLA fiber in the textile field, in particular to do with fabrics have a long way to go.

2. Study on Properties of Polylactic Acid Blended Fabric

In order to study the performance of PLA / C blended fabric, the performance of PLA raw materials was further analyzed and studied. The tensile strength, permeability, pilling and abrasion resistance of PLA / C (50/50), PLA / C (40/60) and cotton products were tested and compared.

2.1 Sample Specifications

The samples of PLA / C (50/50), PLA / C (40/60) (Yizheng Xinghai Chemical Fiber Co Ltd.production) and cotton samples (Hebei Junxing Textile Co Ltd. production) were collected, and the brief specifications are shown in Table 3.

Table.3 Samples and specifications

Item number	sample	Specifications			
		The warp	weft	Warp density /5cm	Weft packing /5cm
1	PLA/C (50/50)	29	18.2×2	223	120
2	Pure cotton	29	18.2×2	220	118
3	PLA/C (40/60)	29	18.2×2	220	118

2.2 PLA Method for Testing Performance of Fabric Samples

2.2.1Test Instrument

Model HD026N electronic fabric strength tester, show in Fig.3

Beijing Xin Wei Austria Technology Development Co., Ltd.

Applicable standards: GB/T3923.1.2; ISO13934.1.2; ISO13935.1.2; ISO11857;
Main technical parameters:
Type: constant speed elongation (CRE)
Range of force measurement: (four specifications)
0~1000N resolution: 0.1N 0~2000N resolution: 0.1N
0~3000N resolution: 0.5N 0~5000N resolution: 1.0N
Measuring accuracy: + 0.5%
Spacer length: 25-350mm digital setting
Maximum elongation: 600mm (clamping distance 25mm)
Stretching speed: 20~400mm/min (numerical setting)
Pre tension method: 1, external tension clamp 2, built-in intelligent settings
Test methods: 1. Constant speed tension 2, peel test 3, tear test

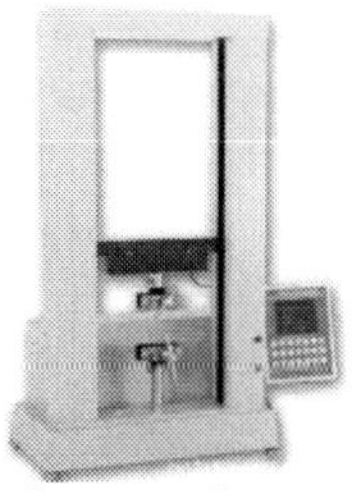

Fig.3 Model HD026N electronic fabric strength tester (http://www.texindex.com.cn/)

Y522 type disc type fabric flat grinding instrument, show in Fig .4
Changzhou Depp Textile Technology Co., Ltd.
Applicable standard: GB/T5453-85
Main technical parameters
Working disc: dia 125mm, speed 70r1pm
Wheel: (diameter * thickness) Phi 56 * 12mm
Weights: weights (1) 750g, weights (2) 250g, weights (3) 125g

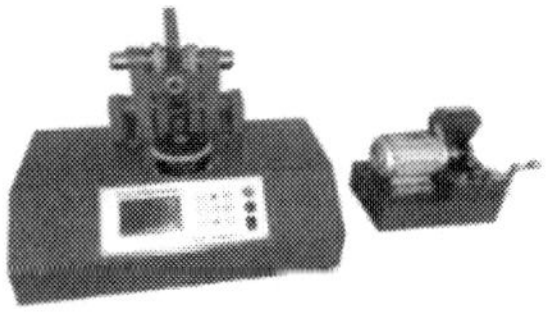

Fig.4 Y522 type disc type fabric flat grinding instrument (http://www.texindex.com.cn/)

YG461 medium pressure air permeability gauge, show in Fig.5
Ningbo Textile Instrument Factory
Applicable standard: GB/T5453-97
Main technical parameters
Pressure range 0~4000Pa
The measured fabric thickness is less than 8mm
The test error is less than 2%
Measurable gas permeability 3.5-11932L/m2.s

Fig.5 YG461 medium pressure air permeability gauge (http://www.texindex.com.cn/)

YG502 fabric pilling tester, show in Fig.6
Technology center of Wenzhou Jigao Testing Instrument Co. Ltd
Applicable standard: GB/T4802.1-1997 DIN 53863.2
Main technical parameters
Pilling trajectory: a circular trajectory of 40mm in diameter.
Adjustable brush: nylon brush with adjustable height.
Specimen specification: diameter 112.8mm round, area 100 square centimeter.
Sample reciprocating speed: 60 /min
Motor: model JX5022, power 60W, speed 2800r/min

Fig.6 YG502 fabric pilling tester (http://www.jigaotest.com/)

2.2.2 Test Conditions

Standard atmospheric pressure; temperature: 20°C soil 2°C; relative humidity: 65% ± 3%.

2.3 Fabric Performance Test Results and Analysis

2.3.1 Fabric Tensile Test

The breaking strength and elongation at break of the fabric were measured by stretching the edge yarns. The specified size of the sample was stretched at a constant elongation until fracture. The breaking strength and elongation at break were recorded. The tensile and fracture properties of PLA / C (50/50), cotton and PLA / C (40/60) were tested. The procedures are as follows: Preparation of the sample -The sample cloth is cut on a variety of full-size fabrics with a length of about 70 cm. The test cloth should be tested once in the test. Since the tensile strength of the fabric is determined by the breaking load and elongation of the 6 cm wide strip on the strength machine, the sample should be cut to a width of 6 cm and a length of 30-33 cm. The longitude test strip, the meridian of each sample and the weft test number of the provisions of five fabric should be at least 1/10 wide fabric from the edge of the fabric using a variety of different physical, mechanical, chemical and other types of effects and gradually destroyed. In general, mechanical force is the main power. The strength of the fabric is one of the main indicators to measure the durability of the fabric. The test results are shown in Table 4 below.

Table.4 Fracture strength of different fabrics

	Warp		
	1 PLA / C (50/50)	2 Pure cotton	3 PLA / C (40/60)
experimental data	1305.02N±13.28N	1326.41N±10.22N	1311.23N±11.81N
	Weft		
	1 PLA / C (50/50)	2 Pure cotton	3 PLA / C(40/60)
experimental data	693.28N±5.29N	724.16N±6.24N	704.32N±4.33N

Table 5 shows that because the sample warp is greater than the weft density, the warp fracture strength test is greater than the zonal. The breaking strength of cotton fabric is slightly larger than PLA-containing fabric, which is mainly due to the large difference in elongation between cotton and PLA polylactic acid fiber. The tensile strength of cotton fabric is greater than PLA, and the ratio of blended yarn strength is the lowest at the lowest value.

Table.5 Elongation at break of different fabrics

	Warp		
	1 PLA / C (50/50)	2 Pure cotton	3 PLA / C (40/60)
experimental data	19.02%±0.28%	12.41%±0.22%	17.23%+0.81%
	Weft		
	1 PLA / C (50/50)	2 Pure cotton	3 PLA / C (40/60)
experimental data	20.28%±0.29%	12.16%±0.24%	18.32%±0.33%

2.3.2 Fabric Abrasion Resistance Test

The round fabric sample is fixed on a working plate which rotates at a constant speed The grinding wheel has a frictional effect on the sample so that the sample becomes worn. The weight loss rate of the sample was used as the evaluation index of abrasion resistance. The greater the weight loss rate of the sample, the less the wear of the fabric. Test steps:

The fabric was cut into round samples with a diameter of 125 mm. A small hole is cut in the center of the sample and it was weighed with a balance. Then the sample is screwed tight on the working plate, the appropriate pressure was selected as well as the weight of the grinding wheel. The model of grinding wheel is A-150 and weighs 50 gram. Adjust the vacuum tube height and perform the experiment. The experiment on each sample was repeated twice, and then the sample was weighed corrected to three decimal places.

The weight loss rate of the sample was used as the evaluation index of abrasion resistance. The greater the weight loss rate of the sample, the worse the abrasion resistance of the fabric. The weight was tested before and after friction; the weight loss rate is calculated to an accuracy of 2 decimal places.

Table 6 shows the weight loss rate of the PLA sample at the same plane grinding speed, which is less than cotton fabric. This indicates that the difference between the PLA fiber and the cotton fiber is comparable to that of the PLA fiber difference.

2.3.3 Fabric Permeability Test

Based on PLA fiber fabrics, we select test tube of 50 mm caliber, use standard air pressure, 50 liter of the volume of air ,measuring the pressure difference of two chambers to determine the volume of the flow of the air. With a steady gas flow on both sides of the sample fabric, The greater the difference between the measured value of the sample, the better the permeability of the fabric.

Test procedure:

The sample is spread directly on the circle of the sample diameter. It is then placed in the sample to stretch the pressure ring; the level of the indenter is adjusted so that the pressure handle and the vertical line is at an angle of 15-20°. If the pressure gauge is 100 Pa, the current pressure P can be directly read out otherwise another nozzle can be chosen.

According to the differential pressure gauge readings and the size of the selected nozzle diameter, the table can be checked to obtain the breathability value of the sample. Each sample is carried out 10 times. The experimental data is measured up to 2 decimal places. The volume and the ability of air can penetrate through the fabric is defined as the breathability of the fabric, which has a significant effect on the comfort of the garment.

The permeability of the fabric is generally expressed as the amount of air flowing through the fabric per unit time on the side of the fabric that is under a specified pressure difference. The greater the breathability is, the better the penetrability of the fabric will be. As can be seen from the table, the permeability of the three fabrics are not very good. The springs and autumn fabrics require a certain degree of resistance to wind, greater fabric density, tighter fabric and less breathability. The permeability of PLA / C fabric however is higher than cotton fabric. Under the same tissue density condition, PLA fiber elasticity and curl are better, the yarn is relatively fluffy, and PLA fiber fabric is relatively porous making it slightly better.

Table.6 different fabric friction weight loss rate

Serial number		Average weight	loss rate
1 PLA / C (50/50)	Before wear	2.876g	0.52%
	After wear	2.861g	
2 Pure cotton	Before wear	2.243g	0.98%
	After wear	2.221g	
3 PLA / C (40/60)	Before wear	2.566g	0.77%
	After wear	2.546g	

Table.7 Ventilation of different fabrics L/m2·s

Serial number	1 PLA / C (50/50)	2 Pure cotton	3 PLA / C (40/60)
No.1	70.45 L/m2·s	65.89 L/m2·s	71.49 L/m2·s
No.2	71.68 L/m2·s	64.35 L/m2·s	72.59 L/m2·s
No.3	70.98 L/m2·s	69.24 L/m2·s	69.33 L/m2·s
No.4	72.59 L/m2·s	63.29 L/m2·s	70.45 L/m2·s
No.5	77.40 L/m2·s	64.53 L/m2·s	69.82 L/m2·s
No.6	73.48 L/m2·s	66.12 L/m2·s	71.92 L/m2·s
No.7	71.44 L/m2·s	67.00 L/m2·s	68.47 L/m2·s
No.8	72.52 L/m2·s	68.71 L/m2·s	72.00 L/m2·s
No.9	72.86 L/m2·s	66.89 L/m2·s	69.70 L/m2·s
No.10	73.30 L/m2·s	68.78 L/m2·s	69.73 L/m2·s
Average value	72.67L/m2·s	66.48L/m2·s	70.55L/m2·s

2.3.4 Fabric Anti - tee Test

The rotating tray is gently lifted, the temperature is set at 180°C, and the fabric is placed smoothly on the working position. Then, the sample chuck is added which makes the sample contact with the abrasive surface. The instrument automatically opens the default number of friction. The finished rolling mill is brought out in a classification box. The density assessment is used to assess the level of pilling of each sample according to the size of the sample.

The fabric continues to withstand friction in the actual wear and washing process. The fabric surface is left exposed. When the fabric shows a lot of plush, it is kneaded into many small spherical particles which will affect the fabric appearance and even fabric durability.

The sample will be graded in degree classification:

5 - slightly hairy, no pilling;
4 - slightly hairy, little pilling;
3 - medium number of pilling;
2 - slightly serious pilling;
1 - serious pilling.

Calculate the arithmetic mean of the five sample levels.

The test results are as follows:
1 PLA / C (50/50)-the sample level is 4, the surface has a little pilling and fluff;
2 Pure cotton -sample level is 4.5, no pilling is slightly raised;
3 PLA / C (40/60)-sample level is 4.5;

The fluff of the fabric is related to the strength, elongation, elasticity and cross-sectional shape of the fiber. Fibers with high strength, elongation and good elasticity are more likely to show pilling. Besides, the fibers with a circular cross section which are fluffy makes the fabric fluffy, napped and pilled. High elongation and good elasticity leads to good wear resistance of the fabric. Once the fabric surface fluffs, hair tangles to the ball and the phenomenon of pilling begins to show. Although PLA fiber has high strength, high elongation and good elasticity, the fiber surface viscosity increases the friction among the fibers; the fabric surface is improved, and anti-pilling performance becomes slightly better. Cotton fiber has worse elasticity, elongation and wear resistance, so the surface becomes slightly fluffy and there is no pilling phenomenon.

2.3.5 Material Degradation Experiment

Four samples of each fabric were embedded in a self-made degradation tank 1 at a height of 3 cm from the bottom. The other four samples were placed in a self-made constant temperature chamber 2 at a height of 3 cm. The degradation tank 1 was placed in a constant temperature and humidity chamber. The degradation tank 2 was placed on a thermostatic heating plate and the thermostat plate temperature controller was commissioned to ensure that the soil temperature was constant at 40°C. The samples were removed from the proposed time, specifically for 1 week, 2 weeks, 3 weeks and 4 weeks. The samples were washed with deionized water and dried. The sample was placed under standard atmospheric conditions for 24 hours to prepare for testing.

Four samples of each fabric were taken out at the time of preparation and after cleaning and drying the results were shown in the table. 8.

With the growth of knitted fabrics in the soil, the degradation of the three knitted fabrics occurs, and the degradation rate, appearance and mechanical properties of the samples change with degradation. PLA / C (50/50) knitted fabric has the greatest degree of degradation.

Table.8 20°C sample quality changes

Test number	initial	1week	2week	3week	4week
1 PLA / C (50/50)	0.60g	0.50g	0.40g	0.30g	0.20g
2 Pure cotton	0.60g	0.59g	0.59g	0.58g	0.57g
3 PLA / C (40/60)	0.60g	0.52g	0.44g	0.36g	0.28g

3. Research Conclusion

The comparative samples were prepared on a PLA / C (50/50), cotton sample, PLA / C (40/60) market as a PLA / C fabric.

A comparison was made on the basic properties of PLA / C (50/50), cotton samples, PLA / C (40/60) such as tensile strength, abrasion resistance, breathability, pilling, biodegradability.

and the following conclusions were drawn:

PLA / C (50/50) fabric, the tensile elongation of PLA / C (50/50) fabric is better than the other two fabrics. PLA / C (50/50)) fabric's wear resistance is better than cotton fabric. Thus, in general, the fastness of PLA / C fabrics is generally stronger than of cotton fabrics. PLA / C fabric permeability is better than cotton fabric. PLA / C fabric's fluff resistance is stronger than cotton fabric. PLA / C have better degradability than cotton fabric. [16]

Through the above comparative analysis, it can be seen that PLA / C fabric is soft and that the pilling degree, the fabric light weight and breathable performance are all better than cotton fabrics. In particular, the characteristics of degraded green fibers can be designed into more environmentally friendly fabrics.

References

[1] Liao XJ, Huang XJ, Liu XL, PLA multi-component knitting with blended yarn and the development of knitted fabrics. J. Knitting Industry: 2007; 02: 11-17.
[2] Wu Q, Wu JH, Shen J, Development of PLA / cotton wet and dry knitted fabrics. J. Shanghai Textile Science and Technology: 2009; 01: 36-38.
[3] Wang T, Green environment-friendly PLA fiber fabric processing technology and performance research. D. Suzhou University: 2007.
[4] Liu Y, Meng JG, Green environmental fiber PLA knitted product development. J. Shanghai Textile Technology: 2006; 01: 55-56.
[5] Liu J, Zhang YP, PLA cotton knitted fabric design and development J. Knitting industry: 2006; 08: 37-39.
[6] Gao B, Development of PLA weft knitted products J. Knitting Industry: 2006; 08: 40-42.
[7] Wu QY, Wang C, Fang HT, Zhu JW, Study on desizing process of corn filament fabric (PLA) J. Modern Silk Science and Technology: 2010; 04: 1-2 + 9.
[8] Fu ZJ, Yu LS, Sun YF, PLA fiber and its fabric dyeing and finishing technology progress. J. Dye and dyeing: 2013; 03: 30-37.
[9] Zhang RY, Polylactic acid fiber core yarn performance and product development research. D. Donghua University: 2013.
[10] Guo YL, Cotton men's shirt fabric feel style and comfort performance test and evaluation. D. Suzhou University: 2009.
[11] Feng XW, Pan ZJ, Wang JG, Zhao RZ, Structure and Properties of PHA / PLA Composite Fibers J. Modern Silk Science and Technology: 2016; 01: 1-3 + 6.
[12] Shi HP, Li XM, Effects of lipase treatment on the hydrophilicity of polylactic acid nonwovens. J. Shanghai Textile Science and Technology: 2016; 06: 26-27 + 36.
[13] Zeng LP, Guo S, Fu ZJ, Yu LS, Preliminary Study on Thermodynamics of PLA Fiber Dyeing. J. Auxiliaries: 2008; 09: 24-28.
[14] Yu LL, Milk. bamboo pulp, Tencel and other multi-component fabric performance research D. Zhejiang University of Science and Technology: 2010.
[15] Liu P, Performance of PLA filament and interlaced fabric D. Zhejiang Polytechnic University: 2011.
[16] Wang LZ, Chen DS, Lin B, Analysis of pressure distribution of bra's shoulder straps. J. TEXT RES: 2009; 09.

Preparation and Photocatalytic Activity of Photocatalyst Ag-X(Br, I) /$BiVO_4$ for Acid Orange 7 under Visible Light

Jian-Hua Ran[1,2], Cong Zhang[1], Felix Telegin[1*]

[1]*Hubei Key Laboratory of Biomass Fibers and Eco-dyeing & Finishing, Wuhan Textile University,No.1 Fangzhi Road, Wuhan, Hubei, 430073, China*
[2]*Ivanovo State University of Chemistry and Technology, Sheremetevsky Ave 7, Ivanovo, 153000,Russia*

[*]Corresponding author's email: felix.telegin@gmail.com

Abstract

Ag-X (Br, I) modified $BiVO_4$ composite photocatalysts were prepared by hydrothermal synthesis and chemical deposition method. The prepared photocatalysts were characterized by X-ray diffraction (XRD), scanning electron microscopy (SEM), UV-vis diffuse reflectance spectra techniques (UV-vis). The photocatalytic performance of Ag-X (Br, I)/$BiVO_4$ photocatalysts was evalutated by the degradation of Acid Orange 7 under visible-light irradiation (λ>420nm). It was found that the doping of Ag-X (Br, I) could greatly improve the catalytic degradation performance for Acid Orange 7. The AgBr/$BiVO_4$ composite catalyst presented a high photocatalytic activity with degradation efficiency of 92.96% for acid orange 7 after 80 min under visible-light irradiation.

Keywords: Ag-X (Br, I)/$BiVO_4$; Chemical Deposition; Photocatalytic; Acid Orange 7

1. Introduction

With the development of economy and society, environmental pollution is increasing seriously, especially in water pollution. Dye pollution is a major source of enviromental pollution in waste fields which is difficult to eliminate with conventional water treatment technologies [1]. Recently, photocatalytic technology is a very effective method for degrading organic water pollutants. It not only degrades different dyes, but also causes their complete mineralization to carbon dioxide (CO_2), water (H_2O) and mineral acids [2]. Monoclinic bismuth vandate ($BiVO_4$) has attracted much attention due to its relatively narrow band gap(Eg=2.4eV), high chemical stability, non-toxic [3-5]. However the photocatalytic activity of pure $BiVO_4$ is relatively low due to its poor adsorptive performance and difficult migration of electron-hole pairs [6]. Currently, there are two major methods to improve its photocatalytic activity. The first method is to increase the specific suface area by producing $BiVO_4$ with different morphologies and smaller size, such as spheres, sheets and tubes [7-9]. On the other hand, researchers found that some noble metals(Ag,Pt,Pd) loaded on the $BiVO_4$ surface or coupling with other semiconductors could efficiently improve its photocatalytic activity, such as CuO/$BiVO_4$ [10, 11], Co/$BiVO_4$ [12], TiO_2/$BiVO_4$ [13], $BiVO_4$/ZnO [14] and Cu_2O/$BiVO_4$ [15].

In recent years, the use of co-catalysts (metals or metal oxides) [16-18] has been an important alternative method to improve the visible-light photocatalytic activity of pure semiconductor materials. In comparison with a single compent conventional semiconductor photocatalysts, the coupling semiconductor has been proved to be effective way to develop high performance photocatalysts for the decomposition of various organic contaminants, attributing to improving the charge separation efficiency and multifunctional spectral reponse. Until now, various coupled photocatalysts based on constituted semiconductors with matched band potentials, such as TiO_2-SnO_2 [19, 20] and $CuBi_2O_4$/$BiVO_4$ [21], have been designed and investigated. The silver/silver halides (denoted as Ag/AgX,X=Cl,Br) is one type of plasmonic photocatalysts which can work under visible light for polutant degradation, water disinfection, and carbo dioxide reduciton [22, 23]. For example, a plasmonic Ag-AgBr/TiO_2 [24] compositephotocatalysts exhibited obviously enhanced photocatalytic activity for the degradation of MB dye in the visible light. AgI/TiO_2 [25, 26], Ag/$BiVO_4$ [27-29], AgBr/BiOBr [30] and other composit photocatalysts are synthesized, these photocatalyst containing silver halide maitained a good stability and photocatalytic ability. In

Chen's paper [31], $AgI/BiVO_4$ was sythesized by an in situ precipitation, which exhibited the higher degradation efficiency (94.91%) within 1 h of visible light irradation.

In this paper, Ag-X (Br, I) modified $BiVO_4$ composite photocatalysts were prepared by hydrothermal synthesis method and chemical deposition method. The prepared photocatalysts were characterized by X-ray diffraction (XRD), scanning electron microscopy (SEM), UV-vis diffuse reflectance spectra techniques (UV-vis). The photocatalytic performance of Ag-X (Br, I)/$BiVO_4$ was evalutated by the photodegradation of Acid Orange 7 undervisible light irradiation (λ>420nm).

2. Experimental

2.1 Materials

All of the reagents are analytical grade and used without further purification. Solutions were freshly prepared with distilled water. Bismuth nitrate pentahydrate ($Bi(NO_3)_3 \cdot 5H_2O$), ammonium vanadate (NH_4VO_3), nitric acid (HNO_3), ammonia ($NH_3 \cdot H_2O$), sodium hydroxide(NaOH), silver nitrate($AgNO_3$),potassium iodide (KI), potassium bromide (KBr) and ethanol (C_2H_5OH) were purchased from standard sources (China National Medicines Corporation Ltd.)

2.2 Preparation of Composite Photocatalyst

$BiVO_4$ photocatalysts were synthesized through a hydrothermal process. 5 mmol $Bi(NO_3)_3 \cdot 5H_2O$ was initially dissolved in 40mL of nitric acid, followed by addition of 5mmol NH_4VO_3 and stirred at room temperature for 30 minutes. The pH value was adjusted to 7 with aqueous ammonia during continuous stirring for 30 minutes. Then this solution is transferred to a 100 mL Teflon-lined stainless-steel autoclave, and incubated at 130°C for 24 h. After this autoclave is cooled down to room temperature, the vividly yellow powder was filtered by vacuum filtration and washed with distilled water for more than 3 times. Finally, the prepared catalyst was dried under vacuum drying box at 80°C for 4 h.

The Ag-X (Br, I)/$BiVO_4$ photocatalysts were prepared by chemical deposition method. Typically, 0.2g of prepared $BiVO_4$catalyst was dispersed in 20ml distilled water by ultrasonic for 10min, followed by the addition of 0.0147g of KBr (0.0205 KI), and then ultrasonic treated for 10 min (marked solution A). 0.021g of $AgNO_3$was dispersed in 10ml distilled water, followed by addition of 1ml (25%) ammonia to form $[Ag(NH_3)_2]^+$ solution which was rapidly added to the solution A. The mass fraction of Ag is 5% (note $AgBr/BiVO_4$, $AgI/BiVO_4$). Then the mixed solution was magnetically stirred in the dark for 4h, and then washed by distilled water and ethanol for several time respectively. Finally, the prepared catalyst Ag-X (Br, I)/$BiVO_4$was dried under vacuum drying box at 80°C for 4 h.

2.3 Characterization of Catalysts

Phase analysis of the catalysts was performed by using powder X-ray diffraction technique, the Dutch X'Pert Pro X-ray diffractometer; morphology of the samples was received from scanning electron microscopy analysis (SEM), Phenom desktop scanning electron microscope; UV-vis spectrophotometry of dye solutions was performed by SPECORD@210 Plus, Analytikjena; CEL-HXUV300 Xenon lamp was used as visual light source; the elemental analysis was carried out energy dispersive X-ray fluorescence spectrometer (EDX).

2.4 Photocatalytic Performance

The photocatalytic activities of the as-obtained Ag-X (Br, I)/$BiVO_4$photocatalysts were evalutated through the degradation of Acid Orange 7 dyes under the visible light irradiation. Photocatalytic reactions were carried out in a 100 mL reactor. In all experiments, the 0.02 g of catalyst was added into 50 mL dye solution in glass reaction vessel with a circular condensate water system. The initial concentration of the dye solution was 20 mg/L. No more pH adjustment was used during the entire course of the degradation process. Dye solution with the catalyst was stirred in the dark for 30 min in order to achieve equilibrium of dye adsorption on the catalyst, then dye solution was exposed to light source which induces photocatalytic reaction of dye destruction. Residual dye concentration was achieved by taking 4 ml solution from reaction vessel with a certain time interval, centrifugal separation of the liquid and spectrophotometric analysis the solution. Extent of dye degradation for a certain timer was evaluated by the following formula:

$$D = \frac{C_0 - C}{C_0} \times 100\% = \frac{A_0 - A}{A_0} \times 100\% \quad (1)$$

Where C_0 and C - initial and current concentrations of dye solution, respectively; A_0 and A - correspondent initial and current absorbance of dye solution at the wave length of 485 nm, respectively.

3. Results and Discussion

3.1 XRD Analysis

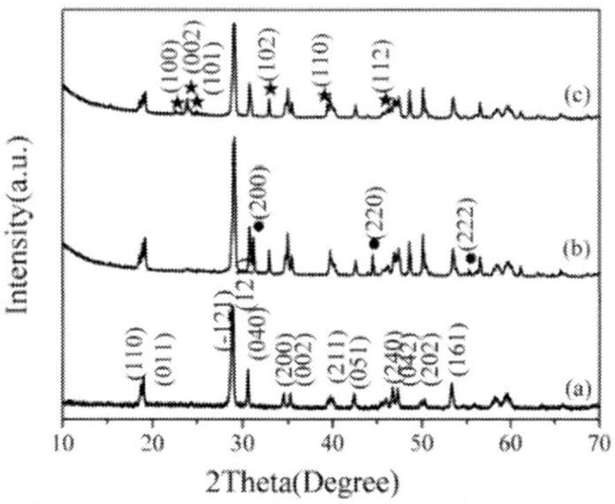

Fig.1XRD patterns of photocatalysts: (a) $BiVO_4$; (b) $AgBr/BiVO_4$; (c) $AgI/BiVO_4$

Phase analysis of the prepared catalysts was performed by using powder X-ray diffraction technique, the Dutch X'PertPro X-ray diffractometer. Fig.1 showed the crystal structure and phase purity of the Ag-X (Br, I) /$BiVO_4$samples and pure $BiVO_4$ samples. The majors diffraction peaks of pure $BiVO_4$ (Fig.1a) were observed clearly with 2θ at 19.0°, 28.9°, 30.5°, 39.9° and 53.3°, which consisted with the basis of monoclinic $BiVO_4$ JCPDS No:14-0688(cell parameter a=0.5195nm, b=0.1701nm, c=0.5092nm), so the pure $BiVO_4$was monoclinic. The diffraction peaks of pure $BiVO_4$ catalyst had high strength, so the catalyst had good crystallinity. Compared with pure $BiVO_4$, diffraction peaks (200), (220) and (222) appeared in $AgBr/BiVO_4$ (Fig.1b) composite photocatalysts, which consisted with the basis of AgBr standard pattern JCPDS No:06-0438 (cell parameter a = b = c = 0.5774 nm). This result showed that AgBr was successfully introduced into the composite photocatalyst $AgBr/BiVO_4$. In the Fig.1c there were (100), (002), (101), (102), (110) and (112) diffraction peaks appeared in $AgI/BiVO_4$ composite photocatalysts, which consisted with AgI standard pattern JCPDS No:09-0374 card (cell parameters a=b=0.4592nm, c=0.7510nm). This result showed that AgI was successfully introduced into the composite photocatalyst $AgI/BiVO_4$. The results above showed that the composite photocatalysts Ag-X (Br, I) /$BiVO_4$ maintained the pure monoclinic structure of $BiVO_4$, so AgBr and AgI were complex with $BiVO_4$ on the surface of $BiVO_4$ crystal to from heterojunction structure.

3.2 SEM Analysis

Scanning electron microscope (SEM) was used to study the morphology and dispersion of the prepared photocatalysts. Fig.2 showed the SEM picture and EDX energy spectrum of pure $BiVO_4$ and Ag-X (Br, I)/$BiVO_4$ composite photocatalysts. As can be seen from the Fig.2a, the pure $BiVO_4$photocatalyst was composed of a large number of small particles to form adendritic structure, and there was a longitudinal growth trend. The composite photocatalyst $AgBr/BiVO_4$ (Fig.2b) obviously appeared irregular solid particles with smooth surface, and the composite photocatalyst $AgI/BiVO_4$ (Fig.2c) had oval structure with holes. So the AgBr and AgI didn't change the original shape of pure $BiVO_4$, and only dispersed in the $BiVO_4$ samples. The energy spectra of Ag and Br elements appeared in the composite photocatalyst $AgBr/BiVO_4$ samples (Fig.2d), and the composite photocatalyst $AgI/BiVO_4$ (Fig.2e) also contained the spectra of Ag and I elements. This result consisted with XRD result, which proved that the prepared composite catalysts contained AgBr and AgI.

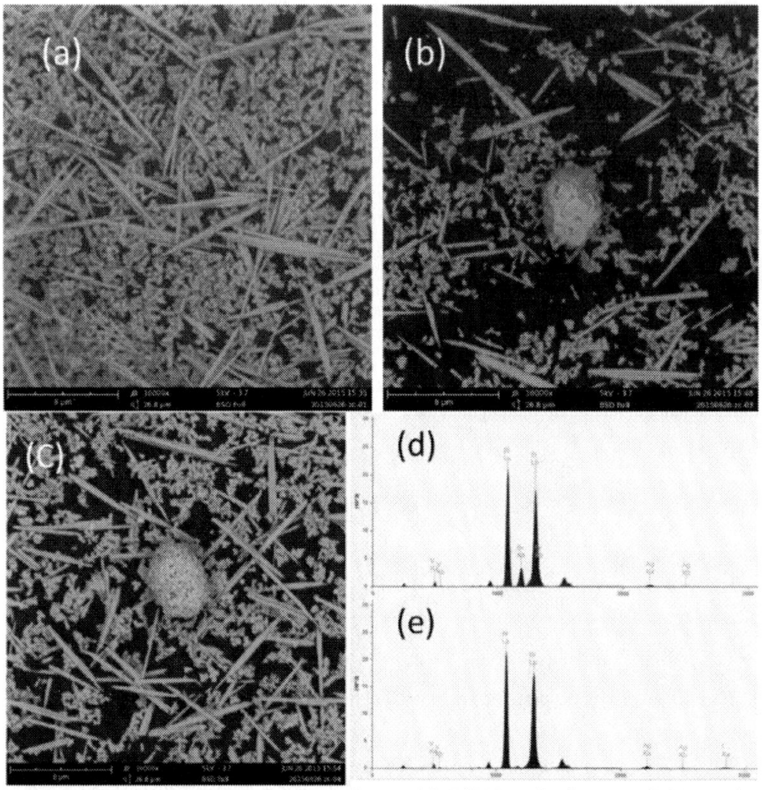

Fig.2The SEM and EDX of composite photocatalysts: (a) $BiVO_4$; (b,d) $AgBr/BiVO_4$; (c,e) $AgI/BiVO_4$

3.3 UV-vis Diffuse Reflection Spectra

Optical absorption property of catalyst has been recognized as the key factor affecting its photocatalytic performance. Fig.3 showed the UV-vis diffuse reflectance spectra of pure $BiVO_4$ and Ag-X (Br, I) /$BiVO_4$ composite photocatalyst. In Fig.3a, it can be seen that pure $BiVO_4$ and Ag-X (Br, I)/$BiVO_4$ composite photocatalysts had great absorption in the ultraviolet region, and also had a different utilization ratio of visible light. The difference between observed absorption of photocatalyst samples was due to the different band gap transition between each sample. UV-vis diffuse absorption of $AgBr/BiVO_4$ composite photocatalyst was the highest in the visible light range. It meant that the catalyst $AgBr/BiVO_4$ had the best visual light utilization ability among all the samples obtained in this research.

Moreover, the band energy of catalyst is important in determining its photocatalytic activity. The relationship of absorbance and incident photon energy hv can be described as:

$$Ah\nu = A_0 (h\nu - Eg)^{1/2} \quad (2)$$

Where A represent the absorption coefficient; Eg represent the band gap energy; h represent Planck's constant;ν represent the incident light frequency; A_0 denotes a constant.

Calculated band gap energy of the samples was depicted in Fig.3b. The band gap energy (Eg) of the obtained catalysts can be estimated from a plot depicting $(Ahv)^2$ versus hv. Estimated band gap energies for $BiVO_4$, $AgBr/BiVO_4$ and $AgI/BiVO_4$ were 2.47eV, 2.45 eV and 2.44eV respectively. The band gap level reflected the bond strength degree of the valence electrons. The results showed that the introduction of Ag-X (Br, I) narrowed the band gap of the photocatalysts.

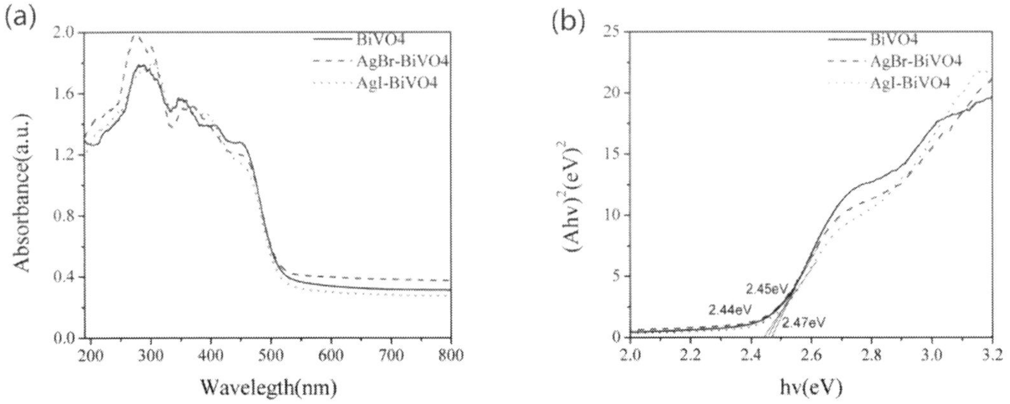

Fig.3UV-vis diffuse reflectance spectrum of $BiVO_4$ and Ag-X(Br、I)/$BiVO_4$

3.4 Photocatalytic Mechanism

Table.1 Absolute electronegativity, estimated band gap, conduction band edge, and valence band

Semiconductor	Absolute electronegativity (X)	Estimated energy band gap E_g(eV)	Conduction band edge (eV)	Valence band (eV)
AgBr	5.81	2.6	0.01	2.61
AgI	5.38	2.77	-0.51	2.26
$BiVO_4$	6.16	2.47	0.43	2.9

Coupling with two semiconductors could efficiently improve photocatalytic activity by reducing the recombination of eletrons and holes. In the Ag-X (Br, I) /$BiVO_4$ composite photocatalysts system, $BiVO_4$ and AgBr were combined, the heterojunction would be produced. The conduction band edge (E_{CB}) and value band edge (E_{VB}) of a semiconductor were calculated by the following formulas:

$$E_{CB} = X - E_C - \frac{1}{2}E_g \tag{3}$$

$$E_{VB} = E_{CB} + E_g \tag{4}$$

Where X is the absolute electronegativity of the atom semiconductor; E_C is the energy of free electrons of the hydrogen scale (4.5eV); E_g is the band gap of semiconductor; E_{VB} is the VB edge potential and the E_{CB} is the CB edge potential. The energy band of semiconductor was illustrated in Table 1. In the Ag-X (Br, I)/$BiVO_4$ composite photocatalysts system, nano silver was generated by light reduction reaction in visible light. Then silver nanoparticles were excited by visible light to generate electrons and holes, which could be transmitted to the electronic conduction band of silver halide (CB). Then part of the electronic conduction band could continue transfer to $BiVO_4$ (CB), thus leaded to the photogenerated electrons and holes effectively separated, thereby improving its photocatalytic degradation ability. In the process of reaction, the metal ions were reduced from the dye molecules, thus maintaining the balance of the whole system.

3.5 Photocatalytic Activity

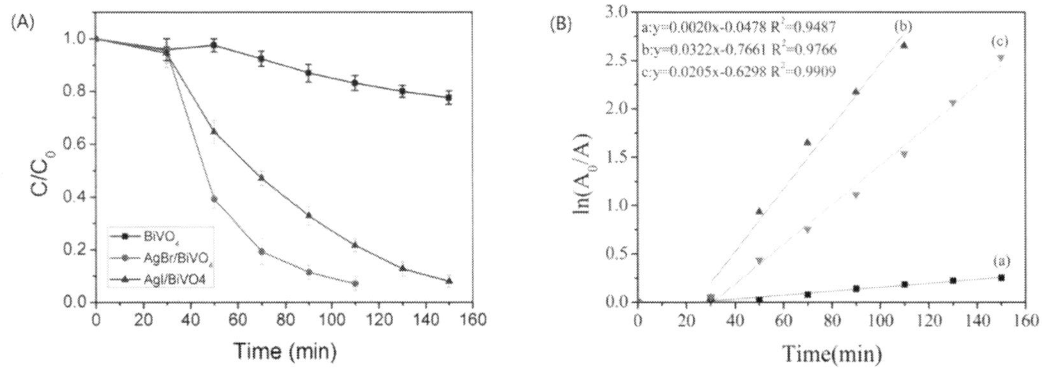

Fig.4Degradation of Acid Orange 7 by photocatalysts: (a) $BiVO_4$; (b) $AgBr/BiVO_4$; (c) $AgI/BiVO_4$

Fig.4 shows the photocatalytic degradation of Acid Orange 7 with different prepared photocatalysts in the visible light. It was found that pure $BiVO_4$ and Ag-X (Br, I)/$BiVO_4$ composite photocatalyst samples had different effects on the degradation of Acid Orange 7. As can be seen in Fig.4a, The degradation ratio of Acid Orange 7 for pure $BiVO_4$ and $AgI/BiVO_4$ samples were 22.43%, 92.04% respectively after 120 min visible light irradiation. For composite catalyst $AgBr/BiVO_4$, the degradation ratio of Acid Orange 7 was 92.96% only after 80 min visible light irradiation. So the photocatalytic activity of composite catalyst Ag-X (Br, I)/$BiVO_4$was successfully improved.

In order to further evaluate the photocatalytic activity of $BiVO_4$ and Ag-X (Br, I)/$BiVO_4$, the first-order kinetics reaction of catalyst for Acid Orange 7 was shown in Fig.4B. The first order kinetic reaction equation follows:

$$kt = \ln\frac{C_0}{C} = \ln\frac{A_0}{A} \tag{5}$$

As shown in Fig.4B, the catalytic degradation of Acid Orange 7 was in accordance with the first-order kinetic reaction. The reaction rate of pure $BiVO_4$, $AgBr/BiVO_4$ and $AgI/BiVO_4$photocatalyst were 0.002 min^{-1}, 0.032 min^{-1}, and 0.020 min^{-1} respectively. It can be seen that the photocatalytic activity of $AgBr/BiVO_4$ was the best. It can be explained that after loading Ag-X (Br, I), part of Ag-X(Br, I) was reduced to nano silver in the light, which was adsorbed on the surface of photocatalyst. The Ag-X (Br, I)/$BiVO_4$ light response range extended to the visible light region due to the plasma resonance effect produced by Ag nanoparticles. Under visible light irradiation, the photogenerated electron hole pairs were formed on the Ag nanoparticles, which firstly transfered to the conduction band of silver halide (CB), and then continued to conduction band of $BiVO_4$ (CB), leading to separate electrons and holes effectively. At the same time, the hole combined with Br^- presented in the Ag-X (Br) catalysts to form Br^0 radicals which is a strong oxidant to decompose Acid Orange 7. Therefore, the photocatalytic activity of $AgBr/BiVO_4$ was better than $AgI/BiVO_4$.

4. Conclusion

Ag-X (Br, I) modified $BiVO_4$ composite photocatalyst were prepared by hydrothermal synthesis and chemical deposition method. The results of XRD and SEM showed that doping of Ag-X (Br, I) didn't change the crystal structure of $BiVO_4$, and AgBr (AgI) were complex with $BiVO_4$ on the surface of $BiVO_4$ crystal to from heterojunction structure. UV-vis diffuse reflectance (UV-vis) reflected that Ag-X (Br, I)/$BiVO_4$ composite photocatalyst extended the light response region and improved the visual light utilization ability, because these catalysts can effectively separate the photoelectrons and holes. Under visible light, it was found that the doping of Ag-X (Br, I) could greatly improve the degradation of Acid Orange 7. The photocatalytic activity of $AgBr/BiVO_4$ composite catalyst was the best, which yielded 92.96% degradation of Acid Orange 7 after 80min of visible light irradiation. In the visible light, Ag-X (Br, I) was partially reduced to Ag^0 formed resonance effect between them, which greatly promoted its degradation of Acid Orange 7 by separating photoelectrons and holes.

Acknowledgements

Authors acknowledge the financial support from Hubei Key Laboratory of Biomass Fibers and Eco-dyeing & Finishing of Wuhan Textile University. We also would like to thank the cooperation of Ivanovo State University of Chemistry and Technology in Russia.

References

[1] Yagub MT, Sen TK, Afroze S, Ang HM. Dye and its removal from aqueous solution by adsorption: a review. Adv Colloid Interface Sci: 2014; 209: 172-184.
[2] Forgacs E, Cserhati T, Oros G. Removal of synthetic dyes from wastewaters: a review. Environ Int: 2004; 30: 953-971.
[3] Zeng J, Zhong J, Li J, Xiang Z, Liu X, Chen J. Improvement of photocatalytic activity under solar light of BiVO4 microcrystals synthesized by surfactant-assisted hydrothermal method. Mater Sci Semicond Process: 2014; 27: 41-46.
[4] Zhang A, Zhang J, Cui N, Tie X, An Y, Li L. Effects of pH on hydrothermal synthesis and characterization of visible-light-driven BiVO4 photocatalyst. J Mol Catal A-Chem: 2009; 304: 28-32.
[5] Zhang X, Ai Z, Jia F, Zhang L, Fan X, Zou Z. Selective synthesis and visible-light photocatalytic activities of BiVO4 with different crystalline phases. Mater Chem Phys: 2007; 103: 162-167.
[6] Xue Y, Wang X. The effects of Ag doping on crystalline structure and photocatalytic properties of BiVO4. Int J Hydrogen Energy: 2015; 40: 5878-5888.
[7] Abdullah AH, Moey HJM, Yusof NA. Response surface methodology analysis of the photocatalytic removal of Methylene Blue using bismuth vanadate prepared via polyol route. J Environ Sci: 2012; 24: 1694-1701.
[8] García Pérez UM, Sepúlveda-Guzmán S, La Martínez-de Cruz A, Ortiz Méndez U. Photocatalytic activity of BiVO4 nanospheres obtained by solution combustion synthesis using sodium carboxymethylcellulose. J Mol Catal A-Chem: 2011; 335: 169-175.
[9] Li H, Liu G, Duan X. Monoclinic BiVO4 with regular morphologies. Mater Chem and Phys: 2009; 115: 9-13.
[10] Li J, Cui M, Guo Z, Liu Z, Zhu Z. Synthesis of dumbbell-like CuO–BiVO4 heterogeneous nanostructures with enhanced visible-light photocatalytic activity. Mater Lett: 2014; 130: 36-39.
[11] Yu C, Yang K, Yu JC, Cao F, Li X, Zhou X. Fast fabrication of Co3O4 and CuO/BiVO4 composite photocatalysts with high crystallinity and enhanced photocatalytic activity via ultrasound irradiation. J Alloys Compd: 2011; 509: 4547-4552.
[12] Geng Y, Zhang P, Li N, Sun Z. Synthesis of Co doped BiVO4 with enhanced visible-light photocatalytic activities. J Alloys Compd: 2015; 651: 744-748.
[13] Tong R, Wang X, Zhou X, Liu Q, Wang H, Peng X, Liu X, Zhang Z, Wang H, Lund PD. Cobalt-Phosphate modified TiO2/BiVO4 nanoarrays photoanode for efficient water splitting. Int J Hydrogen Energy: 2016; 1-9.
[14] Yang JS, Wu JJ. Low-potential driven fully-depleted BiVO4/ZnO heterojunction nanodendrite array photoanodes for photoelectrochemical water splitting. Nano Energy: 2017; 32: 232-240.
[15] Min S, Wang F, Jin Z, Xu J. Cu2O nanoparticles decorated BiVO4 as an effective visible-light-driven p-n heterojunction photocatalyst for methylene blue degradation. Superlattices Microstruct: 2014; 74: 294-307.
[16] Qiu P, Park B, Choi J, Cui M, Kim J, Khim J. BiVO4/Bi2O3 heterojunction deposited on graphene for an enhanced visible-light photocatalytic activity. J Alloys Compd: 2017; 706: 7-15.
[17] Mao M, Chen F, Zheng C, Ning J, Zhong Y, Hu Y. Facile synthesis of porous Bi2O3-BiVO4 p-n heterojunction composite microrods with highly efficient photocatalytic degradation of phenol. J Alloys Compd: 2016; 688: 1080-1087.
[18] Xu J, Wang W, Wang J, Liang Y. Controlled fabrication and enhanced photocatalystic performance of BiVO4@CeO2 hollow microspheres for the visible-light-driven degradation of rhodamine B. Appl Surf Sci: 2015; 349: 529-537.
[19] Farhadi A, Mohammadi MR, Ghorbani M. On the assessment of photocatalytic activity and charge carrier mechanism of TiO2@SnO2 core-shell nanoparticles for water decontamination. J Photochem and Photobiol A:

2017; 338: 171-177.
[20] Lei C-X, Huang X, Liu X, Wang LS, Zhang GS, Peng DL. Photoelectrochemical performances of the SnO2-TiO2 bilayer composite films prepared by a facile liquid phase deposition method. J Alloys and Compd: 2017; 692: 227-235.
[21] Nishikawa M, Hiura S, Mitani Y, Nosaka Y. Enhanced photocatalytic activity of BiVO4 by co-grafting of metal ions and combining with CuBi2O4. J Photochem and Photobiol A: 2013; 262: 52-56.
[22] Liu L, Xu H, Li H, Xu Y, Xia J, Yin S. Synthesis and characterization of the efficient visible-light-induced photocatalyst AgBr and its photodegradation activity. J PhysChem of Solids: 2012; 73: 523-529.
[23] Amornpitoksuk P, Suwanboon S. Photocatalytic decolorization of methylene blue dye by Ag3PO4-AgX (X=Cl−, Br− and I−) under visible light. Adv Powder Technol: 2014; 25: 1026-1030.
[24] Sui Y, Su C, Yang X, Hu J, Lin X. Ag-AgBr nanoparticles loaded on TiO2 nanofibers as an efficient heterostructured photocatalyst driven by visible light. J Mol Catal A-Chem: 2015; 410: 226-234.
[25] Wang Q, Chen M, Zhu N, Shi X, Jin H, Zhang Y, Cong Y. Preparation of AgI sensitized amorphous TiO2 as novel high-performance photocatalyst for environmental applications. J Colloid Interface Sci: 2015; 448: 407-416.
[26] Xue B, Sun T, Wu J-K, Mao F, Yang W. AgI/TiO2 nanocomposites: ultrasound-assisted preparation, visible-light induced photocatalytic degradation of methyl orange and antibacterial activity. Ultrason Sonochem: 2015; 22: 1-6.
[27] Gao XM, Fu F, Zhang LP, Li WH. The preparation of Ag-BiVO4 metal composite oxides and its application in efficient photocatalytic oxidative thiophene. Physica B: Condensed Matter: 2013; 419: 80-85.
[28] Zhu S, Li Q, Li F, Cao W, Li T. One-pot synthesis of Ag+ doped BiVO4 microspheres with enhanced photocatalytic activity via a facile hydrothermal method. J Phys Chem Solids: 2016; 92: 11-18.
[29] Fan H, Wang D, Xie T, Lin Y. The preparation of high photocatalytic activity nano-spindly Ag-BiVO4 and photoinduced carriers transfer properties. Chem Phys Lett: 2015; 640: 188-193.
[30] Gupta G, Kaur A, Sinha A, Kansal SK. Photocatalytic degradation of levofloxacin in aqueous phase using Ag/AgBr/BiOBr microplates under visible light. Mater Res Bull: 2017; 88: 148-155.
[31] Chen F, Yang Q, Sun J, Yao FB, Wang SN, Wang YL, Wang XL, Li XM, Niu CG, Wang DB, Zeng GM. Enhanced photocatalytic degradation of tetracycline by AgI/BiVO4heterojunction under visible-light irradiation: mineralization efficiency and mechanism. ACS Appl Mat Interfaces: 2016; 8: 32887-32900.

Study on Granulation Performance of Waste Automobile Sound-absorbing Cotton

Qian Yang[1], Qin Shi[2], Ru-Quan Zhang[1*]

[1]College of Textile Science and Engineering, Wuhan Textile University, Wuhan, 430200, China;
[2]Guangzhou Sanmu Automobile Parts Co. Ltd, Huadu District Dongfeng Motor City Avenue North Field, 510800, Guangzhou

*Corresponding author's email: zhangruquan@wtu.edu.com

Abstract

This paper studied the granulation performance of waste automobile sound-absorbing cotton. Waste recycling is beneficial to alleviate the shortage of resources in the automotive interior decoration industry and reduce environmental pollution. It has certain guiding significance for the breakthrough recycling technology bottlenecks, and is conducive to high-performance polyester recycling. This paper adopted the method of reactive extrusion melt chain extension, which belongs to the chemical recovery method. The waste material is crushed by a pulverizer and then dried by a vacuum drying oven. Then the melted extrusion reaction is followed by the addition of different chain extender. And the extruded strip is subjected to traction, cooling and artificial shearing to obtain samples. In this paper, infrared spectrometer, melt flow rate meter, thermogravimetric analyzer and differential scanning calorimeter were used to test the properties of the sample. The results shows that: Analyzed by infrared spectrometer, the component of sound-absorbing cotton is 100%PET. Proved by experiments, the waste automobile sound-absorbing cotton can be recycled by melt reaction extrusion chain method, and the granulation performance is good. Two kinds of epoxy chain extender were used to reduce the melt index, and the CX-2 chain extender is more effective.

Keywords: Sound-absorbing Cotton; Granulation Performance; Recycling; Polyester Fiber

1. Introduction

1.1 The Purpose and Significance of the Study

In recent years, the study on waste polyester recycling has been developing rapidly in our country. The factories for recycling polyester bottles are mostly in Jiangsu, Zhejiang and Shanghai. For example, Jiangsu Huayang light cashmere products Co.Ltd., produced coarse denier hollow staple fiber with polyester flakes and recycled materials on VD405 spinning winding machine. In order to recycle waste polyester products, Jiangsu Xiake environmental protection color spinning limited by shares built the web site to produce differentiated fibers and colored polyester yarn. Annualrecycling capacity of the company could be 60 thousand tons, and the utilization rate is 100% [1]. At present, most of polyester industry in China generally adopts the phase polycondensation equipment and technology of Buler company. There are many advantages applied to it, such as simple process, unique design, easy operation and maintenance, high control accuracy, etc.

Recycling of the worn-out car sound-absorbing cotton is a new type of industry which is rich in resources, less investment and significant benefits. It not only alleviates the shortage of resources in the automotive interior decoration industry, but also reduce the environmental pollution caused by the industry. It is of enormous economic and social benefits. At present, China is vigorously promoting the construction of a resource-saving and environment-friendly society to develop a recycling economy. The main reason why the recycling and utilization of the waste automotive interior materials in China is not at a large-scale is due to the fact that the comprehensive utilization technology of materials is not mature enough. Many key technologies have not been solved. Therefore, the urgent need of advanced scientific and technological means to support the study on the key technologies and the equipment of recycling the waste textiles. Break through the technical bottleneck of the recycling process and improv the comprehensive utilization level of waste automobile interior decoration materials. Currently, there is a new polyester recycling technology which is called reactive extrusion chain method [2]. It is a new kind of polyester viscosity increasing method in the world. It not only can be used to prepare high molecular weight polyester, but also can be used for the recovery , utilization and high performance of condensation polymer. Consequently, the related research is necessary and of practical significance. Especially high performance polyester is consistent with the current social trends. While the polyester fiber has been increasingly used for variety of applications, the waste in the form of the apparel textiles or nonwovens are increased, which has been exerted great environment and resource pressures to the society. Not only can the process of recycling PET reduce the environmental pollution and save resource, but also turn waste to treasure and create good economic and social benefits. Recycling PET waste is the trend of social development, it will be conducive to the sustainable development of social economy. It is the topic that grabs the widespread attention all over the world, and there are many domestic and foreign scholars have devoted themselves to the study.

1.2 Introduction of the Chain Extending Method for Melt Reaction Extrusion

There are 3 main methods for the reuse of waste polyester. They are energy recovery, physical recovery and chemical recovery [3-6]. Firstly, energy recovery of waste polyester. The polymer with higher heat value of waste polyester is converted into heat by incineration, the energy is used for power generation. It is more suitable to adopt energy recovery method for the waste polyester which cannot be recycled. Secondly, physical recovery. Physical recycling refers to the simple processing of waste polyester, then it can be recycled, or made some primary polyester raw materials recycling methods. Thirdly, Chemical recovery. The chemical recovery is the degradation of high molecular polymer in waste polyester into monomer, and the monomer is polymerized and spun to form a new chemical fiber. This paper mainly dealed with the reaction extrusion chain method which belongs to chemical recovery.

The reaction extrusion chain is a new method of PET recycling in recent years. This method is used to increase the molecular weight of the product by using the reaction extruder to melt the material after adding the chain extender. It has many advantages, such as short technological process, less equipment investment, fast reaction speed, high efficiency, strong applicability, convenient operation, etc.

The extruder is a continuous reactor, and the mixture is subjected to a particular chemical reaction during melt extrusion. Specifically, it is the unique function of the extruder to handle high viscosity polymers. On each section of the screw, we need to solely control the temperature, material retention time and shear strength, and in each section of the transmission process, make the material finish a series of basic chemical unit operation which are solid conveying, pressurized fusion, material mixing, melt pressure, chemical reaction, removal of by-products and unreacted monomers, melt conveying and pumping molding. Accordingly, it is an ideal melting reaction pressure vessel for high viscosity material. Compared to solid phase polycondensation, Chemical coupling chain extension has many advantages, therefore, the corresponding research is more [7-12]. The research prospects of relevant reaction extrusion method is better. The so-called reaction extrusion chain is the use of extruders as a reactor for chemical coupling chain extension, chain extension reaction is often associated with chain extender. Chain extender refers to certain low molecular weight compounds used to increase the relative molecular mass of the polymer in a short period of time. The chain extender should be able to speed up the reaction and can be carried out on an extruder. The chain extender of PET is mainly composed of condensation type chain extender, hydroxyl addition chain extender, carboxyl addition chain extender, carboxyl hydroxyl and addition chain extender. The chain extender used in this paper belongs to hydroxyl addition chain extender. The mechanism of epoxy chain extender: The electrophilic addition reaction exist between epoxy group and the carboxyl end group of PET, whose activity is stronger than that of the PET chain hydroxyl group. Therefore, the PET epoxy chain reaction is mainly the addition reaction between epoxy group and carboxyl group. The chain reaction of epoxy compounds can be divided into two steps, which are sealing and coupling. When a PET carboxyl group reacts with an epoxy group, it generates capped PET molecule. When the two or two more carboxyl groups of PET react with a plurality of epoxy groups, they generates the PET coupling products, then, the molecular weight of PET increases.

2. Experimental

2.1 Materials

The sound-absorbing cotton was provided by Mori Memi Automation Co. Ltd., Hubei, China. The CX-2, a kind of epoxy chain extenders, was provided by Shanghai Xiuyuan chemical Co. Ltd., Shanghai, China. The SAG-008, a kind of chain extenders, was provided by Shanghai Xiuyuan chemical Co. Ltd., Shanghai, China.

2.2 Experimental Installation

The name, type and origin of the equipment involved in the experiment are shown as table 1.

Table.1 Experimental Installation

Laboratory equipment	Equipment model	Equipment origin
grinder	miniature	Tianjin Tai Site Instrument Co. Ltd.
melt flow index rheometer	XNR-400AM	Chengde Kecheng Testing Machine Co. Ltd.
vacuum drying oven	DZ-3BC	
co-rotating twin screw extruder	SHJ-20	Nanjing giant Machinery Co. Ltd.
infrared spectrometer	TENSOR-27TGA-IR	AXS
powder compressing machine	769YP-15A	Tianjin keqi high-tech company
DSC	DSC 204F1	Germany NETZSCH
TGA	TG 209F1	Germany NETZSCH
electronic balance	JM-A10002	Zhuji chaoze Equipment company
microbalance	METTLER TOLEDO XP6	Hangzhou remit instrument

2.3 Experimental Method and Procedures

Firstly, this paper determined the composition of sound-absorbing cotton by infrared spectrometer. After analyzing we concluded that the composition of sound-absorbing cotton is pure PET. Secondly, crushed the raw materials with the pulverizer.Thirdly, determined the melt index of the raw material by melt indexer. Melt index is used to set the process parameters of melt extruder. Melt flow rate instrument is for pellets, but the material we studied is flocculent, After melting and cooling, some of the raw materials are inclined to be hard and brittle. In order to determine the temperature of the melt and flow of the material, different temperature values are required to set. Fourthly, the raw materials need to be dehydrated by vacuum drying oven. This step is extraordinary critical. PET is a saturated linear polymer, containing hydrophilic groups in the molecular structure of -COOR, which is very easy to absorb moisture. If the raw material contains moisture, it will cause hydrolysis in the melt extrusion. The viscosity and molecular weight of the melt decreased significantly, which will lead to the abnormal phenomena such as foaming, embrittlement and yellowing of extruded sheets, etc. Then, the material was developed by melt bl -ending and extrusion graining with twin-screw extruder. In the experiment, there is no melt pressure to form strip. At length, the molecular weight of PET was increased by adding chain extender to deal with this problem. Finally, added two kinds of chain extender to the raw material. During melt extrusion, CX-2 chain extender was added to the material to obtain the sample 2 and SAG-008 chain extender was added to the material to obtain the sample 3. And there was no chain extender in sample 1.

3. Results and Discussion

3.1 Melt Index

Under certain temperature and pressure, the melt is expressed that in ten minutes the weight of melt passed the standard capillary (g/10min). In general, for the chemical structure of a certain resin, the smaller the melt index, the greater the molecular weight, and the fracture strength, hardness, toughness, aging resistance and other properties will be improved. the larger the melt index, the lower the molecular weight, and the formation of liquidity on the corresponding will be better.

Adding the chain extenders is to increase the molecular weight of polymer. Testing the melt indexes of three samples is to compare the CX-2 with the SAG-008. The "θ" stands for melt temperature. The "monm" stands for applied pressure. The "t" represents cutting interval. "m" says the average mass of two times of cutting. "MFI" indicates the melt index.

Table.2 Melt indexes of different samples

Experimental sample	Sample 1	Sample 2	Sample 3
parameters θ (℃)	265	265	265
monm(Kg)	1.2	1.2	1.2
time(s)	5	5	5
mass(g)	1.030	0.6581	0.8100
MFI(g/10min)	123.600	77.988	97.200

The CX-2 and the SAG-008 belong to epoxy chain extender, which is composed of functional monomers with epoxy functional groups. Chain extension mechanism is the fact that the chain extender belongs to multi monomer polymerization.

There are epoxy groups on side chains. The chemical reaction which plays a role of coupling between epoxy groups and carboxyl groups in PET molecules. It makes the PET molecular chain longer and improves the relative molecular mass, thereby, it increases the intrinsic viscosity of the system. At the same time, the carboxyl group existing in the PET system will decrease due to the chain reaction which consumes the carboxyl groups in the molecule.

Through analysis of the data, after adding the chain extender, the melt index of the samples decreased, the flow properties of the samples became worse, and the melt viscosity increased. The comparative analysis between sample 2 and sample 3 shows that the effect of CX-2 is more obvious for decreasing the melt index.

3.2 Infrared Data

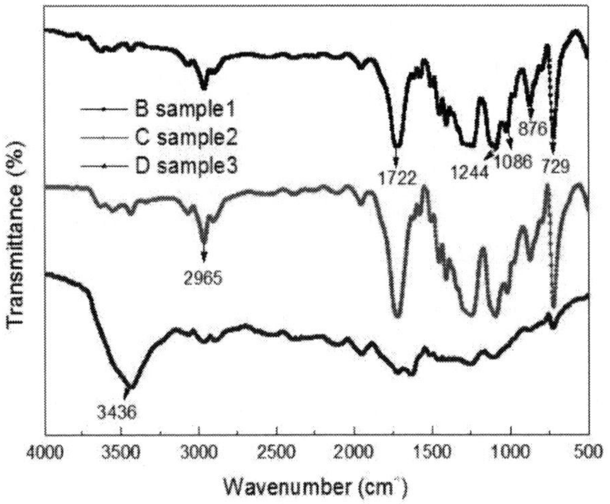

Fig.1 Infrared spectrogram

Fourier transform infrared spectroscopy (FTIR) has been widely used to detect the molecular reaction and determine the composition of polymer. Fig.1 shows that there is a strong absorption peak around 1700cm^{-1}, so we can know that which molecule contains carbonyl. There are two peaks close to each other around 1500-1300cm^{-1}, then we can determine the presence of methylene and methyl in the molecule. There are two small absorption peaks around 1600-1400cm^{-1}, and an absorption peak appears around 3100-3000cm^{-1}, it is possible to determine the benzene ring structure is in the molecule. Thus, it can be concluded that the corner waste is polyester fiber. The reason that the strong peak appears around 1700cm^{-1} is the conjugation of benzene rings. Affected by the carbonyl group, the vibration absorption peak of C-H bond on the benzene ring shifts to723cm^{-1}. The reason that the double peaks appears between 1128cm^{-1}and 1099cm^{-1} is the asymmetric stretching vibration of O-CH2-CH2- bond.

Table.3 Wave number to different groups

Wave number	Corresponding group
723	C-H
1099	O-CH2-CH2-
1128	O-CH2-CH2-
1722	C=O
2965	-CH2-
3436	O-H
910	-C-O-C-

The curves of sample1 and sample2 almost coincide, so we can know that the addition of CX-2 had no effect on the raw material groups. Through the comparison between sample3 and sample1, it can be concluded that the addition of SAG-008 has some effects on the raw material groups.

3.3 Thermo Gravimetric Data

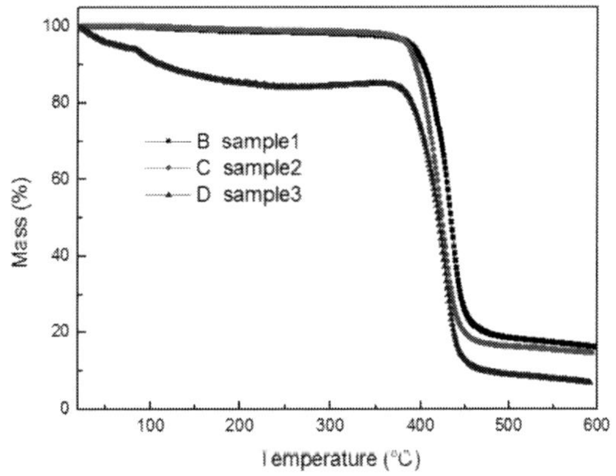

Fig.2 TG of different samples at different temperatures

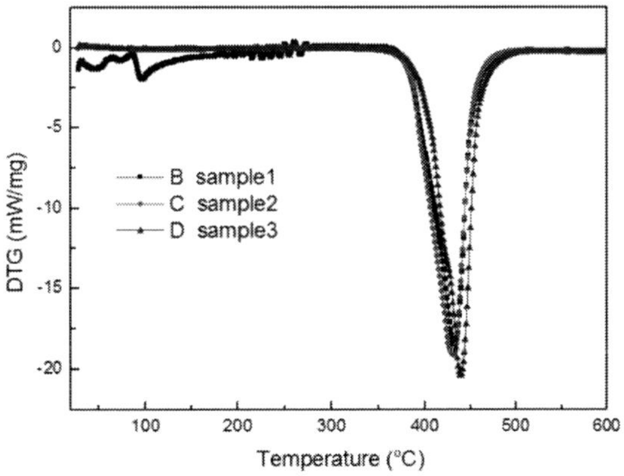

Fig.3 DTG at different temperatures

Thermo gravimetric analysis shows that there are mass changes in the process of heating or cooling. In addition, the size of the change and the temperature are closely related to the chemical composition and structure of the material. Consequently, in the process of heating and cooling, the quality of material can be distinguished and identified.

According to the residues of the samples at different temperatures in Fig.2, it can be seen that the thermal stability of the sample 2 is the best, while the sample 1 is the worst.

The Fig.2 and Fig.3 represent thermo-gravimetric curve and derivative thermo-gravimetric curve respectively. The peaks of the starting and ending points correspond to the steps of the starting and ending points, and the number of steps is equal to the number of peaks. As shown in the thermo-gravimetric curve, there are two platforms whose qualities are basically unchanged. The steeper the downhill, the greater the mass loss, and the faster the mass loss rate, the faster the decomposition rate of PET.

Table.4 TG data of different samples

sample	starting point	midpoint	inflexion	termination	remnants
1	404.0	424.5	429.4	445.6	6.9%
2	410.4	429.4	434.2	449.5	16.4%
3	408.7	427.0	432.9	448.1	14.5%

3.4 DSC Data

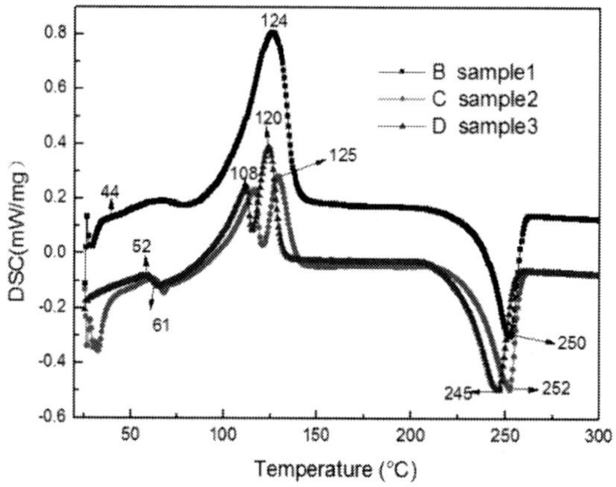

Fig.4 DSC of different samples

Table.5 DSC data of different samples

sample	Tg(℃)	Tcc(℃)	Tm(℃)
1	42	121	248
2	56	112	250
3	61	108	243

In Fig.4, the glass transition temperatures of the samples increase gradually. With chain extender, the crystallization peak of sample 1 raises to two which even is one without chain extender. The melting point of sample 2 is the highest, and the peak temperature of sample 2 is higher than that of sample 3.

For the same crystal component, the higher the general melting point, the narrower the melting peak and the larger the area, which shows the more perfect crystallization, crystallization ability and the higher the degree of crystallization, generally. If there are multiple peaks in the melt, the different crystal forms, crystalline perfection and thermal history can be explained. Furthermore, if cold crystallization is found in the heating curve, the crystallization ability of polymer can be determined according to the location and area of cold crystallization peak.

4. Conclusion

According to the study on the structure and performance characteristics of the samples, the main conclusions achieved from the researches in this paper were summarized as follows:

1) The component of sound-absorbing cotton of unknown waste vehicle was 100% PET by infrared spectrometer, a strong absorption peak around 1700cm^{-1} indicates that which molecule contains carbonyl. two peaks close to each other around 1500-1300cm^{-1} can determine that methylene and methyl exist in molecules. Two small absorption peaks around 1600-1400cm^{-1} and an absorption peak appears around 3100-3000cm^{-1}, shows the benzene ring structure in the molecule. Thus, it can be concluded that the corner waste is polyester fiber.

2) It is proved that the worn-out automobile sound-absorbing cotton can be recycled and expanded by melting, reaction and extrusion.

3) The CX-2 and the SAG-008 belong to epoxy chain extender, which is composed of functional monomers with epoxy functional groups. After adding the chain extender, the melt index of the samples decreased, the flow properties of the samples became worse, and the melt viscosity increased. Through comparative analysis, the effect of CX-2 is more obvious for decreasing the melt index.

4) The thermal stability of sample 2 was the best, while the thermal stability of the sample1 was the worst.

5) Through comparative analysis, the glass transition temperature of the samples increases gradually, after adding chain extender, a crystallization peak is changed into two peaks, and the melting point of sample 2 is the largest. Through comparison between sample 2 and sample 3, the crystallization peak temperature of sample 2 was higher than that of sample 3.

5. Acknowledgements

This study was supported by Hubei Province Natural Science Fund Project of (2013CFA090).

References

[1] Shuo W. Study on the synthesis of chain extender and chain extender of PET [D]. Dalian University of Technology: 2005.

[2] Tai YL, Zhao YJ, Song H. A new recycling technology of waste PET bottle [J]. plastics technology: 2004; 04 (1): 61-64.

[3] Shi YY, Chen YX, she, she, she Xiao Hong. Rerecycling of polyester recycled materials and development of textile products [J]. progress in textile science and technology: 2008; 01 (7): 29-31.

[4] Zhang Z. Current situation of domestic and international textile recycling industry [J]. foreign textile technology: 2001; 02 (2): 37-39.

[5] Yang Y, Lu Y, Xiang H, et al. Study on methanolytic depolymerization of PET with supercritical methanol for chemical recycling [J]. Polymer Degradation and Stability: 2002; 75 (1): 185-191.

[6] Mutlu A, Lee BK, Lee CH, et al. Sludge minimization and production improvement by changing filtration and reaction techniques in vinyl acetate-ethylene copolymerization [J]. Journal of Cleaner Production: 2012; 24 (5): 109-116.

[7] Sako T, Okajima I, Sugeta T, et al. Recovery of constituent monomers from polyethylene terephthalate with supercritical methanol [J]. Polymer Journal: 2000; 32 (2): 178-181.

[8] Wang DC. New technology for polyester waste recycle. [J]. Synthetic Fiber Industry: 2006; 02 (2): 46-49.

[9] Wang ZY, Zhao YM. Applications of Chain-Extending Reactions in the Synthesis of Polymeric Mate [J]. Chemical Propellants and Polymeric Materials: 2003; 06 (3): 23-26.

[10] Liu F. Study on viscosity of recycled PET modified engineering plastics and processed [D]. Xihua University: 2008.

[11] Wu T, Li Y, Cai FL, Tan HM. Effects of chain extender technology on chain reaction of PET.[J]. Polyester Industry: 2002; 02 (1): 20-24.

[12] Chen CJ. Polyester containers for packaging and their progress. [J]. Shanghai Plastics: 2001; 04 (4): 13-16.

Dyeing Property Research on Photoluminescent Polyamide Fiber

Hai-Yan Ni[1,3*], Shi-Wei Fu[2], Yong-Gui Li[1]

[1]*Minjiang University, Wenxian Road No.1, 350108, Fuzhou, China*
[2]*Fujian Fiber Inspection Bureau, Zhaoyu Road No.17, 350026, Fuzhou, China*
[3]*Fuzhou Technology Innovation Center for Textile and Clothing Industry,350108, Fuzhou, China*

[*]Corresponding author's email: mjuhaiyan@163.com

Abstract

In order to broaden the application field of photo-luminescent polyamide fiber, the dyeing properties of nylon filament were studied. A series of experiments on accumulating luminous fiber were done. The fibers were dyed with red, yellow and blue dyes of different consistency and their brightness were tested after dyeing. Effects of dye concentration and dye type on the luminescent properties were analyzed. The factors affecting the dye uptake of the luminescent polyamide were selected, such as concentration, temperature and time. The optimal dyeing process was determined and the soaping and rubbing fastness tests were carried out on the dyed fibers.

Keywords: Luminescent Polyamide; Dyeing Process; Dyeing Concentration; Performance Testing; Dyeing Fastness

1. Introduction

The photoluminescent polyamide fiber is a type of photo-induced light-storage luminescent fiber which is based on polyamide raw material and manufactured with the special spinning process by adding the rare earth luminescent activating material and nanometer functional auxiliary agent. This fiber is recognized by people for the advantages of high initial brightness, long afterglow time and no radioactive hazard [1, 2]. With the expansion of light-emitting fiber applications, issues of color compatibility occur in the product development combining light-emitting nylon and other fiber materials [3, 4]. Therefore, light-emitting fibers with a rich variety of colors are needed to be studied. To make light-emitting nylon fibers achieve the color diversification similar to ordinary textile fibers [5], this experiment begins with the dyeing effect of light-emitting nylon, determines the main two influencing factors of dye concentration and dyestuff color and gains their influencing rules on the luminescent properties of luminescent dyed nylon through experiments and mechanism analysis [6]. Optimization experiments on the dyeing process parameters are done to get the optimal dyeing process of the photoluminescent polyamide fiber.

2. The Experimental Part

2.1 Dyeing Materials

The main experimental materials are as follows photo-luminescent polyamide fiber (150D/36F), three kinds of dyes (red, yellow and blue) of weak acid red GR, weak acid yellow and weak acid brilliant blue RAWL; dyeing auxiliaries (leveling agent and acetic acid, etc.).

2.2 Experimental Methods

2.2.1 Dyeing Method of Photoluminescence Nylon Fiber

Use weak acid dyes for high temperature dyeing. The dyeing bath ratio is of 50: 1 with the single sample (weight of 5 g), and the amounts of dye used are respectively 0.2%, 0.5%, 0.7%, 1%, 3%, and 5% (owf). Use the pure water as the solvent and prepare the three dyes into a series of different concentrations of dyes. Place the dye solution and the sample into the steel tube for sealing, place them in automatic high-temperature dyeing test machine, heat up to 100℃, insulate and dye for 30min, and then cool, wash and dry them.

2.2.2 Experimental Method of the Luminescent Performance for the Luminescent Fiber Dyed

Use the PR-305 long afterglow phosphor tester, with the ambient temperature of (22±3)℃, the relative humidity RH≤70%, the illumination value of 1000lx, the excitation time 15min and the test time of 10h [8].

2.2.3 Experimental Method of the Dyeing Performance

Use the UV/V-1000 spectrophotometer to test the dyeing the percentage of the dye after dyeing.

2.2.4 Experimental Method of the Color Fastness

Take the GB / T3921-2008 test standard for the soaping performance test and GB / T3920-2008 test standard for the rubbing resistance test.

3. Results and Analysis

3.1 Analysis on the Luminescent Properties of the Dyed Luminescent Polyamide

3.1.1 Comparison of Different Colors under the same Concentration

For a convenient observation on the luminescent property changes of the luminescent polyamide before and after the dyeing, firstly, the luminescent property test is taken on the undyed luminescent polyamide, with the testing result shown in Fig.1. In order to compare colors' influence on the luminescent properties of the luminescent polyamide under the same concentration, three kinds of weak acid dyes are used to dye the photoluminescent fibers, the dyeing test is done with the dye dosages of 0.2%, 0.5%, 0.7%, 1%, 3% and 5% respective in different concentrations of dyes, six pairs of luminescence performance data of the dyed nylon are gained, as for the similar laws and the limited space, here three concentrations are selected by intervals, with the test results shown in from Fig.2 to 4.

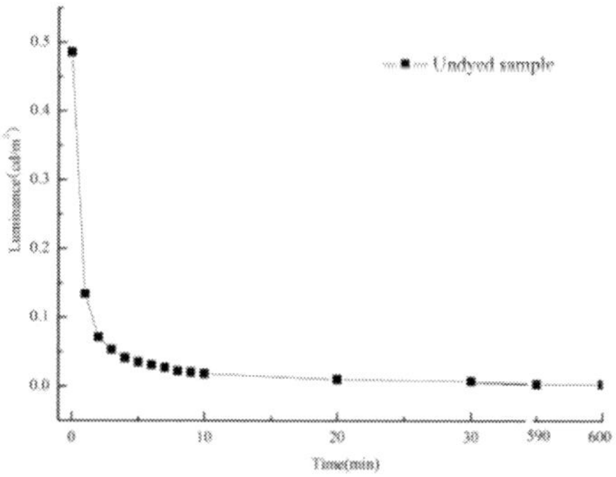

Fig.1 the luminous performance for the undyed luminous nylon

It can be seen from the above figure that, the luminous intensity of the dyed luminescent polyamide has the similar rule as the changes of time and the un-dyed fiber, in the first ten minutes, the fiber brightness decreases significantly and gets gentle and close after that. However, after dyeing in different colors, the luminescent properties of the fibers have been declined to different degrees. As the weak acid dye molecules would go into the fibers and cover the luminescent material in the fiber, the luminescent polyamide's light absorbing, storing and emitting effects would get weak. The luminescent properties of luminescent polyamide under the same concentration and in different colors are separately: yellow> blue> red. This is because of the deep red, moderate blue and lighter yellow colors, which further explain that the colors of the dye would influence the luminescent fibers' luminescent property.

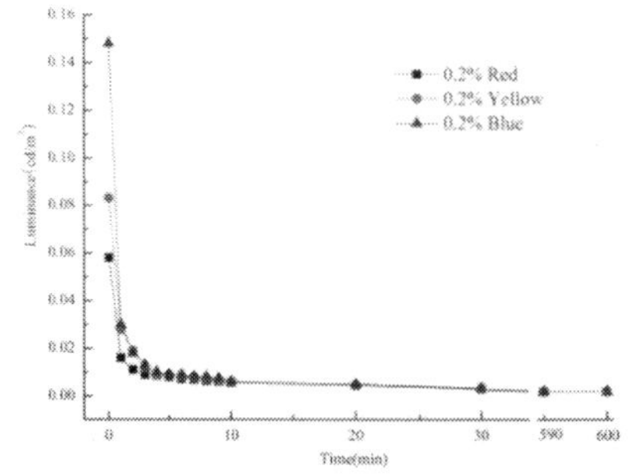

Fig.2 The luminous performance for the luminous nylon dye with the concentration of 0.2%

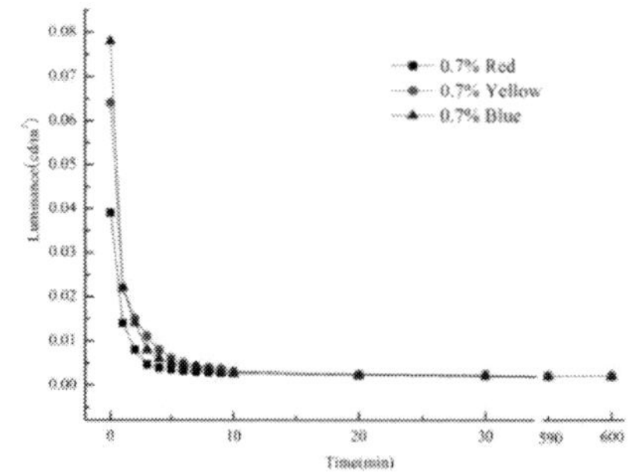

Fig.3 The luminous performance for the luminous nylon dye with the concentration of 0.7%

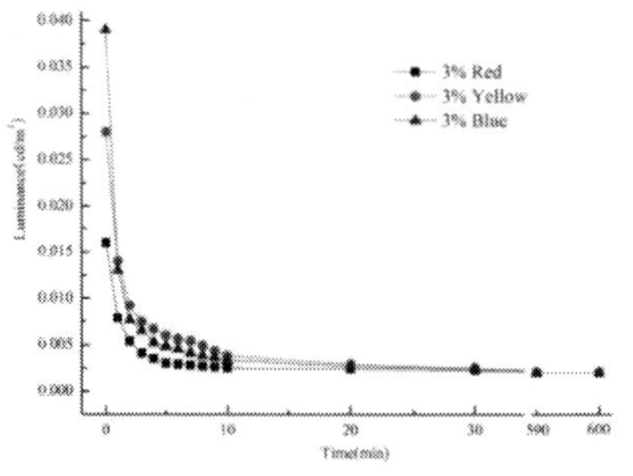

Fig.4 The luminous performance for the luminous nylon dye with the concentration of 3%

3.1.2 Comparison of Different Concentrations in the same Color

Three pairs of test data for the luminescent polyamide fiber under different concentrations in the same dye are gained, as shown from Fig.5 to 7.

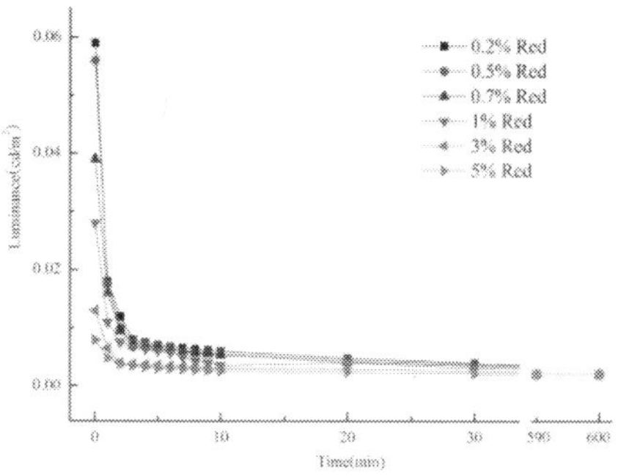

Fig.5 The luminous performance for the luminous nylon in the red dye

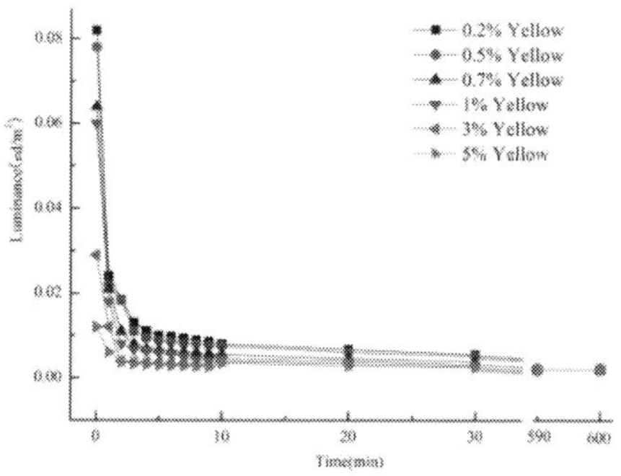

Fig.6 The luminous performance for the luminous nylon in the yellow dye

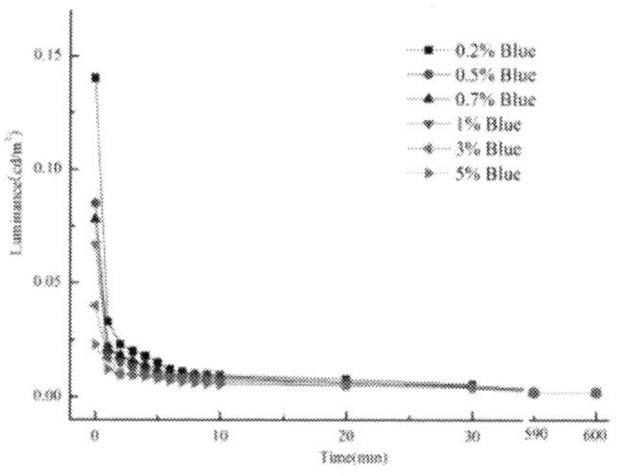

Fig.7 The luminous performance for the luminous nylon in the blue dye

We can find from Fig.5 to 7 that after dyeing, the luminescent brightness of the dyed luminescent polyamide decrease as the dye concentration increases, the lower the dye concentration is, the better the luminescence and the glowing are, conversely, the luminescence effect is poorer. The reason is that as the dye concentration increases, fiber molecules attached to the surface of the dye increase to reducing the absorption and releasing functions of the luminescent material within the fiber [9]. The test result shows that the dye concentration of 0.2%, 0.5% and 0.7% have good effects for the dyeing of luminescent polyamide fiber.

3.2 The Optimum on the Dyeing Process of the Luminescent Polyamide

Based on the analysis on the luminescent property of the luminescent polyamide under the above-mentioned dyeing concentrations, the reasonable dyeing concentration range is obtained. Combining the factors of temperature and time in the dyeing process, the optimization analysis on the dyeing process of the luminescent polyamide is analyzed and the simple analysis method of the orthogonal experiment is applied, as shown from Table 1 to 4, and the optimal dyeing process is obtained after the analysis on the dyeing percentage of the luminescent polyamide fiber.

Table.1 Level and factor table of orthogonal experiment

Factor / Level	Concentration/%	Time/min	Temperature/℃
	A	B	C
1	0.5	20	90
2	0.7	30	100
3	1	40	110

Table.2 The dyeing percentage of the first test

Item / Dye color	Dyeing percentage/%		
	$A_1B_1C_1$	$A_1B_1C_2$	$A_1B_1C_3$
Weak acid red/GR	94.9	96.1	90.8
weak acid yellow/G	88.4	89.3	80.8
weak acid blue/RAWL	91.9	95.0	87.7

Table.3 The dyeing percentage of the second test

Item / Dye color	Dyeing percentage/%	
	$A_1B_2C_2$	$A_1B_3C_2$
Weak acid red/GR	96.1	95
weak acid yellow/G	92.3	92.2
weak acid blue/RAWL	96.3	94.8

Table.4 The dyeing percentage of the third test

Item / Dye color	Dyeing percentage/%	
	$A_2B_2C_2$	$A_3B_2C_2$
Weak acid red/GR	98.2	94.6
weak acid yellow/G	94	91.5
weak acid blue/RAWL	97.2	94.4

As Table 4 shows, the dyeing rate decreases as the dye concentration increases, which means that the extra high dye concentration wouldn't be absorbed as the luminescent polyamide gets saturated in the dye. According to the experimental results, the dyeing percentage is the highest when the dye concentration is 0.7%, therefore, the optimal dyeing process parameters for the luminescent polyamide are: the concentration is 0.7%, the temperature is 100 ℃, the time is 30mins and the pH value is 5.

3.3 The Dyeing Fastness Analysis on the Dyed Luminescent Polyamide

Prepare the test samples by the gained optimal dyeing process of the photoluminescent polyamide [7], and make the dyeing fastness test by the standards of fiber soaping fastness and rubbing fastness test, with the test result shown in Table 5.

Table.5 Color resistance of luminescent polyamide fiber

Dye color \ Item	Washable resistance		Friction resistance
	Discoloration / level	Staining / level	Friction staining / level
Weak acid red/GR	4	4-5	4-5
weak acid yellow/G	4-5	4-5	4
weak acid blue/RAWL	4	4-5	4-5

4. Conclusions

1) As the dye concentration increases, the luminescent intensity of the dyed luminescent polyamide fiber decreases. The appropriate dyeing proportion should be selected in the actual production, avoiding that the luminance of the luminous fibers fails to reach the requirement for being dyed too much.

2) The light-colored luminescent performance is better than that of deep colors under the same concentration, and dyes with higher brightness and light colors after be dyed should be selected in the dyeing and finishing process of luminous fiber products in actual production, thus to reduce the dyeing depth and improve the luminous effects.

3) After the analysis on the main factors affecting the dyeing of the photoluminescence dyestuffs, the optimal dyeing parameters are obtained as follows: the concentration is 0.7%, the temperature is 100℃, the time is 30min and the pH value is 5.

4) For the luminescent polyamide fiber dyed under the optimal dyeing conditions, the dyeing performance is good, both the soaping and rubbing fastness are above 4 levels.

5. Acknowledgement

This research was financially supported by the Fuzhou Science and Technology Project [No.2016-G-56], the Education Department of Fujian Province [No.JAT160376] and Minjiang University Science and Technology Project [No. MYK14007].

References

[1] Xu J, Liu JH, Yu ZJ. The application and development situation of luminous textile. China Fiber Inspection: 2013; 10: 82-84.
[2] Yu Y, Ge MQ. Types and application status of rare earth luminescence textile. Modern Textile Technology: 2013; 5: 62-64.
[3] Tang XF, Du HB. The analyses of dyeing technology development of polyamide fiber. China Textile Leader: 2014; 5: 69-73.
[4] Xu RX, Ge MQ, Li YG. Dyeing properties of accumulating Luminous fibers. Printing and Dyeing: 2005; 8: 9-11.
[5] Guo XT, Zhao YM, Ling P. The research and development of Nylon 6 fiber dyed with dye. Synthetic Fiber: 2006; 1: 27-34.
[6] Ji HX. Study on dyes suitable for fine denier nylon fabrics and dyeing depth. Donghua University: 2007.
[7] Huang L. Comparison among testing standards for color fastness and analysis of common fastness problem. Journal of Printing and Dyeing: 2010; 7: 36- 40.
[8] Chen Y. Experimental course of dyeing and finishing technology. China Textile Press: 2004; 94-97.
[9] Li YH, Zhang ZT. Study on the preparation and luminescent mechanism of aluminates long afterglow photoluminescence materials. Materials Review: 2000; 14:23-25.

Study on Water Repellent and Flame-Retardant Finishing of Polyester/cotton Blended Fabric

Yun-Li Wang[1, 2, *], Xiao-Jian Lv[1], Yang Liu[3], Qiao-Lin Xu[4], Gen-Yang Cao[3], Liang-Jun Xia[3], Cong Luo[1], Lu Zhou[1]

[1]*College of Chemistry and Chemical Engineering, Wuhan Textile University, Wuhan* 430073, *P. R. China*
[2]*Hubei Key Laboratory of Biomass Fibers and Eco-Dyeing & Finishing, Wuhan Textile University, Wuhan* 430073, *P. R. China*
[3]*Key Laboratory of Green Processing and Functional Textiles of New Textile Materials, Ministry of Education, Wuhan Textile University, Wuhan* 430073, *P. R. China*
[4]*Hubei Province Fiber Inspection Bureau, Wuhan* 430061, *P. R. China*

[*]Corresponding author's email: ylwang@wtu.edu.cn

Abstract

In this study, the water repellent finishing agent FK-518 and the flame-retardant finishing agent FK-103 were used to treat polyester/cotton blended fabric so that the fabric can simultaneously have excellent water repellent and flame-retardant finishing properties. The results show that the one bath finishing process can result in a better water repellent and flame-retardant effect, whereas in the two-step finishing process, it is difficult to obtain good flame-retardant performance.

Keywords: Water Repellent Finishing; Flame-retardant Finishing; Polyester/cotton Blended Fabric; One Bath Finishing; Two Step Finishing.

1. Introduction

In order to meet people's demand on apparel, decoration, industry and national defence, the performance of fabric needs to be improved. It is necessary to carry out anti-crease, water- and oil- repellent and flame-retardant finishing on textiles [1-5]. With the development of textile materials, multifunctional textiles have attracted a lot of attention. The key to the development of multifunctional textiles is how to solve the problem of compound functionality. It however often leads to the weakening or disappearance of another function for the increase in a particular function [6].

Polyester/cotton blended fabric is widely used in the textiles field. This is especially true for protective clothes and interior decoration and other environmental situations which are required to have good flame-retardant performance. In addition, some cases also need to have water repellence properties. The flammability of the polyester/cotton blended fabric is not equal to the simple sum of the polyester and cotton component. The "skeleton" effect formed in the combustion of polyester/cotton blended fabric results in the combustibility to be far much stronger than expected [7]. Researchers have done some work on the pure polyester fabric or cotton fabric to create a separate water repellent or flame-retardant finishing [8-9]. However, due to the choice of finishing agent and the problem on the one bath finishing process, the development of water repellent and flame-retardant composite finishing is slow on polyester/cotton blended fabric. It is difficult to take into account the fabric's water repellent and flame-retardant effect simultaneously [10]. There is a very important theoretical and practical significance in promoting the progress of the dyeing and finishing technology, which requires the realization of the functional combination in the polyester/cotton blended fabric, and in meeting the multifunctional requirements of the people.

2. Problem Formulation

2.1 Instrument

EL-400 type vertical squeeze roll (Shanghai Longcolor Textile Equipment Co., Ltd., Shanghai, China); DHG-9246A type electric heating constant-temperature drying oven (Shanghai Jing Hong Laboratory Instrument Co., Ltd., Shanghai, China); JY-PHb type contact angle tester (Chengde Testing Equipment Manufacturing Co., Ltd., Hebei, China); YG(B)815D-II type fabric flame-retardant performance tester (Wenzhou Jigao Testing Instrument Co., Ltd., Zhejiang, China).

2.2 Chemicals

Polyester/cotton blended fabric (the blending ratio of polyester and cotton is 65/35, and the surface density is 166.8g/m^2, purchased by Wuhan Hanzheng Street).

Water repellent finishing agent FK-518 (Organic Fluorine Type Chemicals), flame-retardant finishing agent FK-103 (Phosphorus Type Chemicals), and resin 062 were all provided by Beijing CTA-Tex Chemicals Co., Ltd.

85% wt. phosphoric acid (H_3PO_4), magnesium chloride hexahydrate ($MgCl_2 \cdot 6H_2O$) were all supplied by China National Medicines Co., Ltd.

2.3 Experimental Method

1) One bath finishing process of water repellent and flame-retardant: Preparing the finishing solution (containing FK-518, FK-103, resin 062, $MgCl_2 \cdot 6H_2O$ and 85% H_3PO_4) → Padding (Two dip two rolling) → Drying (80°C, 3min) → Curing → Washing → Drying → Measurement.

2) Two step finishing process (Flame-retardant finishing after water repellent finishing): Water repellent finishing → Flame-retardant finishing → Washing → Drying → Measurement.

3) Two step finishing process (Water repellent finishing after flame-retardant finishing): Flame-retardant finishing → Water repellent finishing → Washing → Drying → Measurement.

2.4 Measurement

1) Flame-retardant performance was measured according to FZ/T 01028-93 standard "textile fabric combustion performance test".

2) The water repellent property of the fabric was evaluated by testing its contact angle.

3. Problem Solution

3.1 One Bath Finishing Process of Water Repellent and Flame-retardant

3.1.1 Effect of Water Repellent Finishing Agent Concentration on Water Repellent and Flame-retardant Properties

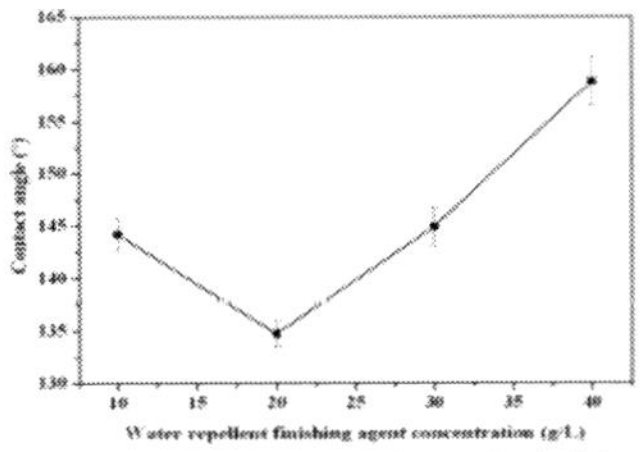

Fig.1 Effect of water repellent finishing agent concentration on water repellency of the fabric.

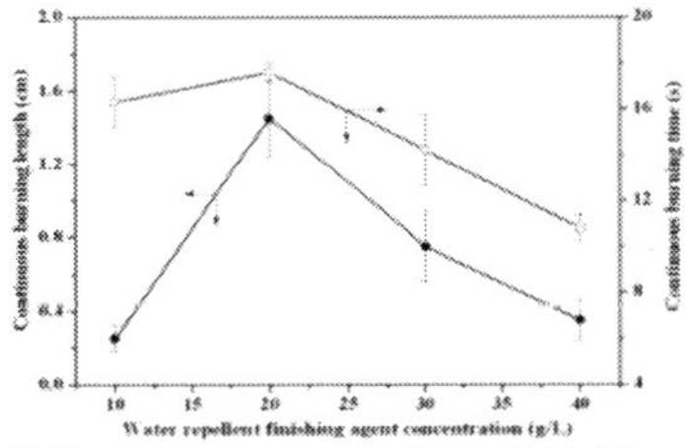

Fig.2 Effect of water repellent finishing agent concentration on flame-retardant properties of the fabric.

It can be seen from Figs.1 and 2 that the water repellency of the fabric becomes better in the process of treatment with the increase of the concentration of FK-518. The water repellency effect of the fabric is the best when the concentration of water repellent finishing agent reaches 40 g/L. At this time, the continuous burning length and time have smaller values, attaining the good flame-retardant effect. On the other hand, Flammability fluctuates greatly with the change of water repellent finishing agent concentration. This is due to the synergistic effect of the two water repellency and flame-retardant finishing agents.

The concentration of water repellent finishing agent was determined to 40 g/L through overall consideration of water repellency and flame-retardant finishing effects.

3.1.2 Effect of Flame-retardant Finishing Agent Concentration on Water Repellent and Flame-retardant Properties

Figs.3 and 4 indicate that the flame-retardant effect and water repellency are the best at the concentration of flame-retardant finishing agent FK-103 reaching 400 g/L. The flame-retardant effect of the fabric becomes worse when the flame-retardant finishing agent concentration decreases. In addition, the contact angle of the fabric fluctuates notablely. This may be related to the uneven treating process.

Combined with water repellency and flame-retardant effect, one bath process for water repellent and flame-retardant finishing was determined as: water repellent finishing agent FK-518 concentration was 40 g/L, flame-retardant finishing agent FK-103 concentration was 400 g/L, resin 062 concentration was 70 g/L, magnesium chloride hexahydrate was 4 g/L, 85% phosphoric acid was 14 g/L, pre-drying temperature was 80 °C and time was 3 min, curing temperature was 160 °C and time was 3 min. Under this process, the composite finishing can achieve better water repellency and flame-retardant effect.

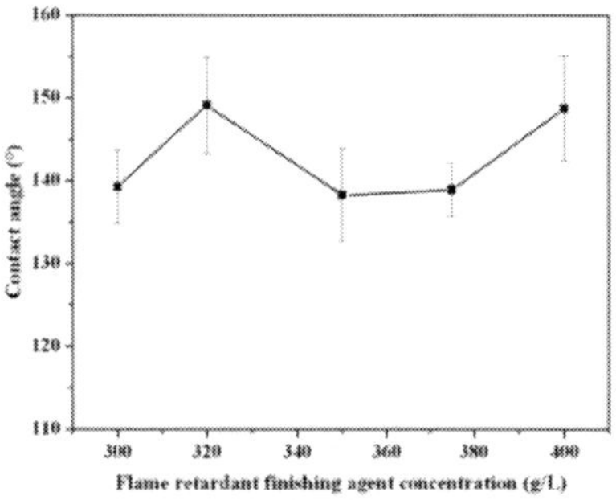

Fig.3 Effect of flame-retardant finishing agent concentration on water repellency.

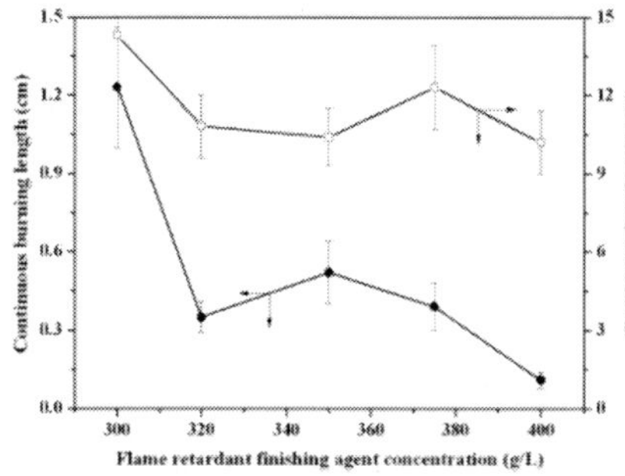

Fig.4 Effect of flame-retardant finishing agent concentration on flame-retardant properties.

3.2 Two Step Process of Water Repellent and Flame-retardant Finishing

Table.1 The effect of two step process of water repellent and flame-retardant finishing on the polyester/cotton blended fabric

FK-103 concentration (g/L)	Flame-retardant finishing after water repellent finishing			Water repellent finishing after flame-retardant finishing		
	Water repellency	Flame-retardant properties		Water repellency	Flame-retardant properties	
	Contact angle (°)	Continuous burning length (cm)	Continuous burning time (s)	Contact angle (°)	Continuous burning length (cm)	Continuous burning time (s)
300	154.1[±1.7]	9.8[±1.3]	42.2[±4.1]	158.7[±1.2]	20.2[±1.1]	90.5[±3.9]
350	148.3[±0.6]	12.0[±2.1]	66.5[±3.7]	153.0[±2.6]	19.1[±2.8]	97.8[±2.5]
375	146.7[±1.5]	10.3[±1.7]	41.8[±2.5]	153.3[±1.5]	20.3[±1.3]	111.5[±4.6]
400	163.2[±2.6]	8.6[±1.4]	34.6[±2.1]	147.1[±1.6]	20.4[±2.2]	136.9[±3.4]

The results of two step finishing process were listed in Table 1. It can be seen from Table 1, fabric flame-retardant properties become very poor whether it is flame-retardant finishing before or after on the polyester/cotton blended fabric adopting two step finishing process. In addition, the two step finishing process is also more complex compared with one bath finishing process.

It may be because that the continual water-repellent network film formed on the surface of the fiber, restricting in the fixation reaction of flame-retardant finishing agent with fiber when water repellent finishing is carried out on the fabric firstly. Moreover, the thin film formed by water repellent finishing agent covered on the surface of the flame-retardant layer also affected the flame-retardant finishing effect of the fabric when flame-retardant finishing is performed on the fabric firstly.

4. Conclusion

1) The better water repellency and flame-retardant finishing effects of the polyester/cotton blended fabric can be got through one bath finishing process. The suitable finishing process was obtained through experimental results.

2) It is difficult to simultaneously acquire excellent water repellency and flame-retardant finishing effects on the polyester/cotton blended fabric whether it is flame-retardant finishing before or after using two step finishing process.

5. Acknowledgement

We are very grateful for the financial support from the Natural Science Foundation of Hubei Province of China (Project Code 2014CFB753), and the Foundation of Wuhan Textile University (Project Code 143058).

References

[1] Ji B, Qi H, Yan K, Sun G. Catalytic actions of alkaline salts in reactions between 1,2,3,4-butanetetracarboxylic acid and cellulose: I. Anhydride formation. Cellulose: 2016; 23: 259-267.

[2] El-Hady MMA, Farouk A, Sharaf S. Flame retardancy and UV protection of cotton based fabrics using nano ZnO and polycarboxylic acids. Carbohyd Polym: 2013; 92: 400-406.

[3] Mohsin M, Carr CM, Rigout M. Novel one bath application of oil and water repellent finish with environment friendly cross-linker for cotton. Fiber Polym: 2013; 14: 724-728.

[4] Shao H, Sun JY, Meng WD, Qing FL. Water and oil repellent and durable press finishes for cotton based on a perfluoroalkyl-containing multi-epoxy compound and citric acid. Text Res J: 2004; 74: 851-855.

[5] Vasiljević J, Tomšič B, Jerman I, Orel B, Jakša G, Simončič B. Novel multifunctional water- and oil-repellent, antibacterial, and flame-retardant cellulose fibres created by the sol–gel process. Cellulose: 2014; 21: 2611-2623.

[6] Vasiljević J, Zorko M, Štular D, Tomšič B, Jerman I, Orel B, Medved J, Kovač J, Simončič B. Structural optimisation of a multifunctional water- and oil-repellent, antibacterial, and flame-retardant sol–gel coating on cellulose fibres. Cellulose: 2017; 24: 1511-1528.

[7] Rahimi MH, Parvinzadeh M, Navid MY, Ahmadi S. Thermal characterization and flammability of polyester fiber coated with nonionic and cationic softeners. J Surfact Deterg: 2011; 14: 595-603.

[8] Kuan C-F, Kuan H-C, Ma C-CM, Chen C-H. Flame retardancy and nondripping properties of ammonium polyphosphate/poly(butylene succinate) composites enhanced by water crosslinking. J Appl Polym Sci: 2006; 102: 2935-2945.

[9] Weil ED, Levchik SV. Flame retardants in commercial use or development for textiles. J Fire Sci: 2008; 26: 243-281.

[10] Fonseca VM, Fernandes Jr. VJ, Araujo AS, Carvalho LH, Souza AG. Effect of halogenated flame-retardant additives in the pyrolysis and thermal degradation of polyester/sisal composites. J Therm Anal Calorim: 2005; 79:429-43.

Ozone/UV Collaborative Treatment on the Colour Fading of Dyed Cotton

Zheng-Lei He[1], Xin-Yi Zhou[1], Dan-Ying Zuo[2], Chang-Hai Yi[1*]

[1]*School of Textile Science and Engineering, Wuhan Textile University, Wuhan, 430200, China;*
[2]*School of Materials Science and Engineering, Wuhan Textile University, Wuhan, 430200, China;*

*Corresponding author's email: 751067874@qq.com

Abstract:

Ozone (O_3) has been used to fade dyed textiles to achieve worn style because of its eco-friendly characteristic. Ozone/UV collaborative decolourisation on wastewater is a hot topic as UV irradiation was regarded as a catalyst in the treatment. In this study, the ozone/UV collaborative treatment on colour fading reactive dyed cotton was investigated in comparison with conventional fading processes of hydrogen peroxide (H_2O_2) and chlorine (NaClO) in terms of the effect on colour fading, mechanical properties (the tensile strength and elongation, tearing strength and abrasion resistance) and surface microstructure in SEM. UV clearly illustrated that the ozone fading effect on dyed cotton and the decolourisation efficiency of it is much better than conventional treatments with less negative impact on all of the mechanical property related items above. Overall, it possessed an excellent potential to alternate the conventional processes as it not only performs better in the evaluations but is also capable to save a huge amount of water and power.

Keywords: Ozone; UV; Reactive dye; Colour Fading

1. Introduction

Ozone (O_3) is seen as an environmentally friendly oxidant because of its strong oxidability without emitting secondary pollution and can be decomposed rapidly to O_2 after processing. It is a perfect alternative for traditional oxidizing agents and bleaching agents such as hydrogen peroxide and chlorine bleach which have been used for decades to decolour wastewater [1]. Researches of the last few decades had accumulated a wealth of knowledge on wastewater decolourisation, especially on the area of textile, like investigating the ozone decolourisation and degradation mechanism of different textiles dyes [2-5]. Ozone is also intensively used in recent years for bleaching, dyeing,colouring industry and paper-making [6] as it needs no aqueous bath and is able to save a considerable amount of water.

Colour faded textile and garment products, such as denims with worn and vintage looks, are increasingly popular among young customers [7]. However, these looks generally are achieved by chemical methods--denim washing, which has a negative influence on the environment. Denim washing which aims at getting distinctive and desirable fashionable looks is not only resource-consuming and inefficient but also pollution-causing and very harmful to workers' skin [8]. Instead, exploiting ozone to decolourise or fade the dyed fabric for vintage style may make a difference.

Researches on the bleaching effect of ozone on textile material range from natural fibers such as cotton, wool and silk to manmade fibers have been reported [9, 10]. Ultraviolet (UV) irradiation is regarded as a catalyst when applied in combination with ozone in water treatment [11]. In this study the effect of UV collaborated ozone treatment on the colour fading of reactive dyed cotton fabric is investigated. It was compared with conventional treatments of hydrogen peroxide and chlorine in terms of the effect on fabrics mechanical properties and fabric surface microstructure.

2. Experimental

2.1 Materials

The undyed and desized 3/1 twill woven cotton fabric (325.7g/m^2) was used in this study. Reactive Blue FL-RN (RB-RN), Reactive Red FL-2BL (RR-2BL) and Reactive Yellow FL-2RN (RY-2RN) supplied by Colour Root, Hubei are bifunctional reactive dyes containing fluorotriazine (commercial quality). The chemicals of NaOH, HCl, $Na_2SiO_3 \cdot 9H_2O$, 30% H_2O_2, Na_2SO_4, Na_2CO_3, NaClO and $Na_2S_2O_3 \cdot 5H_2O$ supplied by Sinopharm Limited, China were used in this study and were of analytical grades. OP-20 (nonionic surfactant) was of chemically pure and supplied by Tianjin Guangfu, China.

2.2 Apparatus

The equipment used for colour fading is mainly composed by CF-G50 ozone generator (Guolin, China. a corona discharge type generator fed by pure and compressed dry oxygen from oxygen cylinder) and a reactor made of glass. Four UV lamps (maximum emission at 253.7nm; Philips, TUV 15W) were mounted on the interior top of the reactor. The exhaust from reactor was destroyed by heater before venting to the atmosphere. The setup of reactor is shown in Fig.1.

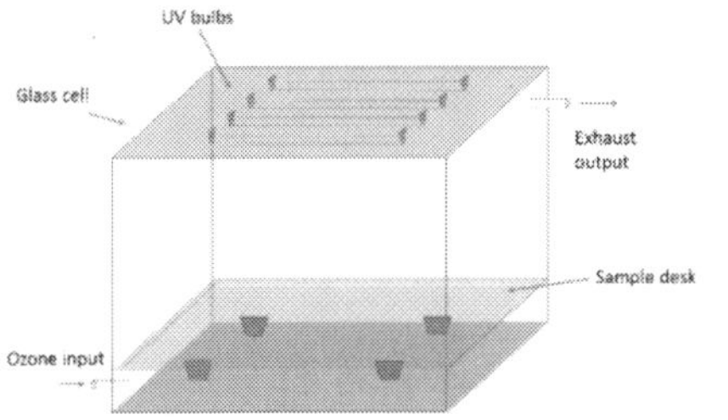

Fig.1 Diagram of the ozone/UV reactor setup

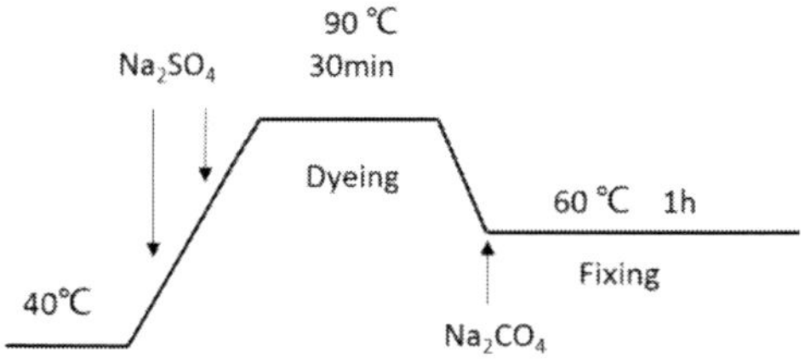

Fig.2 Dyeing profile of the reactive dyes.

2.3 Dyeing Procedure

Pretreatment of dying on cotton fabrics in this study based on the followed parameters: 100℃, 8g/L NaOH, 3g/L OP-20, 5g/L $Na_2SiO_3 \cdot 9H_2O$, liquor ratio of 15:1, 15min. The cotton fabric was dyed with 3% (on weight of fabric) reactive dyes of RB-RN, RR-2BL, and RY-2BN respectively according the profile illustrated in Fig.2 under the same conditions as follows: 70g/L Na_2SO_4, 20g/L Na_2CO_4, and liquor ratio of 15:1.

2.4 UV Collaborated Ozone Fading Treatment

In this treatment, dyed fabrics with different colour were cut respectively to pieces and then were putted evenly on a net sample desk into the reactor (Fig.1), the ozone gas flow was 2 L/min, and the ozone concentration was about 70g /m^3. A pretreatment was employed before ozonation that the fabrics were wetted (moisture is about 75%) by aqueous HCl solution (pH=4). The treatment was under room temperature, and lasted for 1h. Fabric samples (one piece for each colour, dyed by RB-RN, RR-2BL, and RY-2BN respectively) were picked out from the reactor sequentially in per 5min before 30min, per 10min after 30min. For comparing the difference between the treatments with and without UV, an experiment was done under the same conditions without UV.

2.5 Hydrogen Peroxide Fading Treatment

A liquor ratio of 30:1 was used in hydrogen peroxide treatment with the concentration of 30% H_2O_2 (on weight of fabric), the temperature of 100 and the pH of 12 which was adjusted by NaOH aqueous solution. The treatment was last for 1h for each colour of fabric.

2.6 Chlorine Fading Treatment

Concentration of 22ml/L for NaClO and 16g/L for $NaCO_4$ were combined in aqueous solution which was used with the liquor ratio of 20:1 and the temperature of 50℃ in fading treatment. After treating for 20min by chlorine, fabrics of each colour were washed in the aqueous solution of $Na_2S_2O_3 \cdot 5H_2O$ (10g/L) with a liquor ratio of 20:1 for 8 min.

2.7 Colour Measurement

Depend on Kubela-Munk function that $K/S = \frac{(1-R)^2}{2R} = k\ C$, it is known that K/S_{sum} value is linear to the colour depth and is able to evaluate the effect of fading treatments on colour yield of cotton fabrics. In this study, K/S_{sum} of fabric samples treated by different fading processes were measured by using a spectrophotometer Datacolour 110 (Datacolour, USA). Four measurements that the relative error within 0.3% were obtained from each sample, after conditioning at 20±2℃ temperature and 65±2% humidity.

2.8 Mechanical Properties and Microstructure Measurement

Tensile strength and elongation, tearing strength and abrasion performance were obtained by certain testers to evaluate the loss of mechanical properties of the treated fabric samples in this study. A scanning electron microscope (SEM) JSM-6510 (JEOL, Japan) was used to observe the surface microstructure of samples in this study. Each sample were conditioned at 20±2℃ temperature and 65±2% humidity for 24hours.

3. Result and Discussion

3.1 Effect of Ozone/UV on Colour Fading of Dyed Cotton

A comparative investigation was carried out to study the effect of UV's collaboration on the colour yield of cotton fabric in ozone treatment. Fig.3 shows the relation between K/S and Time in both of the ozone fading treatment with UV and without UV. It is easy to find out that ozone/UV generally had a higher efficiency than ozone treatment in each single fading experiment of different colour fabrics.

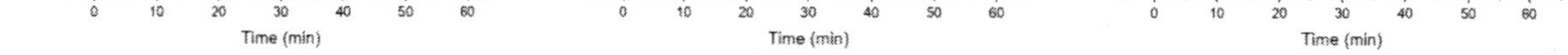

(a) (b) (c)

Fig.3 The relation between K/S and Time in ozone and ozone/UV treatment for fading fabrics dyed by (a)RB-RN, (b)RR-2BL, and (c)RY-2BN respectively.

To analyse the kinetic grates of the fading effect of both ozone and ozone/UV treatments on different colour fabrics, the linear fitting on ln $(\frac{K/S_0}{K/S})$, $\frac{1}{k/s} - \frac{1}{k/s_0}$ and $\frac{1}{k/s^2} - \frac{1}{k/{s_0}^2}$ *vs* time depend on the result in Fig.3 were studied respectively, and the result of the R^2, i.e. correlation coefficient of linear fitting, were demonstrated in Table. 3. As the R^2 in all of the reactions with different dyed cotton have similar tendencies in both of ozone and ozone/UV treatments and certain reactions generally follow the same kinetic grates, it is supposed that the ozone have played a more significant role than UV in the reactions and in affecting the kinetic grates. On the other hand, the difference on kinetic grates between different colour samples probably reveals that the reaction between ozone (or ozone/UV) and RR-2BL dyed cotton are more complicated than others.

Table.1 Comparison of correlation coefficient in different kinetic grates

		1st order	2nd order	3rd order
RB-RN	Ozone	0.82584	0.97523	0.91701
	Ozone/UV	0.78418	0.96653	0.76211
RR-2BL	Ozone	0.47821	0.90323	0.94784
	Ozone/UV	0.57176	0.9656	0.97803
RY-2RN	Ozone	0.85725	0.98715	0.93467
	Ozone/UV	0.79069	0.97239	0.84391

Fig.4 shows the linear fitting results of ozone and ozone/UV fading on cotton dyed by RB-RN, RR-2BL, and RY-2BN respectively. Corresponding k (min^{-1} or d $k/s \cdot min^{-1}$, constant of fading rate) values of the reactions are illustrated in Table 2. It is easy to find out that UV promoted the fading efficiency in both ozone and ozone/UV treatments for all of the cotton samples, and this effect were more apparent in the samples dyed by RB-RN and RY-2RN (k is approximately 2.7 and 2.5 times respectively to the one without UV). Instead, samples dyed by RR-2RN only demonstrated a slight improvement for colour fading effect (0.20518 from 0.11077 d $k/s \cdot min^{-1}$). The explanation for the gap of UV collabrated fading effect between fabrics dyed by different dyes involve a consideration of that RR-2BL reacting in certain treatment was more complicated than others, UV can barely impact the fading rate in such ozone fading treatment while relatively remarkable in others.

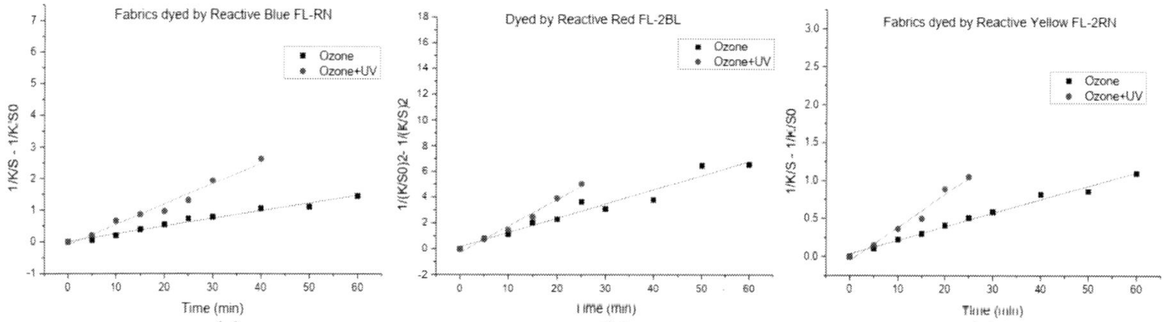

(a) (b) (c)

Fig.4 The kinetic fitting results of ozone and ozone/UV faded fabrics dyed by (a)RB-RN, (b)RR-2BL, and (c)RY-2BN respectively.

Table.2 Related parameters of k and R^2 in kinetic fitting.

		k_{ap} / min^{-1}(d $k/s \cdot min^{-1}$ for RR-2BL)	R^2
RB-RN	Ozone	0.02435	0.97523
	Ozone/UV	0.06461	0.96653
RR-2BL	Ozone	0.11077	0.94784
	Ozone/UV	0.20518	0.97803
RY-2BN	Ozone	0.01787	0.98715
	Ozone/UV	0.04349	0.97239

The probable reason for the collaborative effect of UV in ozone treatment for colour fading on dyed cotton may involve the photocatalysis of UV. Mcgrath and Norrish have studied the photolysis mechanism of ozone in gas phase and revealed that the ozone reaction under UV irradiation follows the chain as Eq.1 and Eq.2 bellow [12]:

$$O3 + hv \rightarrow O_2 + O\,(^1D) \quad (1)$$
$$O\,(^1D) + H_2O \rightarrow 2\cdot OH \quad (2)$$

UV acts as a catalyst, helps in forming hydroxyl radical which is a powerful and non-selective chemical oxidant, and consequently, the oxidation efficiency in ozone/UV is better than that in the fading treatment of only ozone.

3.2 Comparative Colour Fading Effect on Different Treatments

An investigation in order to contrast the fading effect on ozone/UV and conventional methods such as hydrogen peroxide (H_2O_2) and chlorine (NaClO) was done in this work. In Fig 4, the fabric samples of each colour treated by ozone/UV for 5min were used to be compared with the samples treated by hydrogen peroxide (H_2O_2) and chlorine (NaClO) in terms of fading efficiency ($\frac{K/S_0-K/S}{K/S_0}$), as shown in Fig.5 (a))and the appearances of samples (Fig.5 (b)). By contrast, the ozone/UV process could be carried out at room temperature in a shorter period (5min *vs* 1h and 28min) and the water bath could be avoided. Which means that the water consumption and wastewater productivity in ozone/UV fading process would be far lower than conventional treatments.

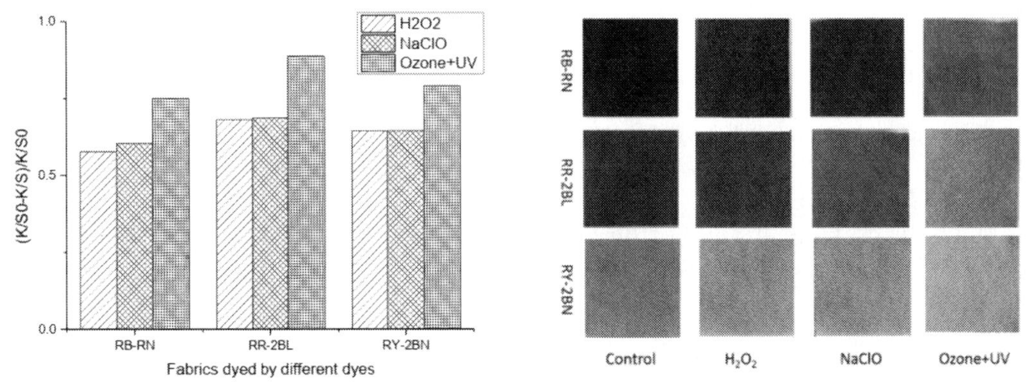

(a) (b)

Fig.5 Contrast between the processes of H_2O_2, NaClO and Ozone/UV on colour fading cotton fabrics dyed by RB-RN, RR-2BL, and RY-2BN respectively in (a) $\frac{K/S_0-K/S}{K/S_0}$ and (b) sample appearance.

Table.3 Comparison on related consumptions between the fading processes of H_2O_2, NaClO and ozone/UV.

	H_2O_2		NaClO		Ozone/UV	
Duration(min)	60		28		5	
Temperature (℃)	100		50		Room temperature	
Water Consumption for 1kg fabrics (kg)	30		40		A bit	
Chemicals Consumption for 1kg fabrics	30% H_2O_2 NaOH	0.3kg a bit	NaClO Na_2CO_4 $Na_2S_2O_3\cdot 5H_2O$	440ml 320g 200g	HCl	a bit

The bar chart (Fig.5 (a)) indicated that ozone/UV leads the colour fading effect on every single colour of fabrics, though hydrogen peroxide and chlorine are also effective (and chlorine is slightly better than hydrogen peroxide) in this issue. Different from the mechanism of ozone/UV on dyed cotton which has been mentioned above, hydrogen peroxide mostly bleaching or fading colour base on the ion of HO_2^- which may influence the reaction in two ways [13]. It may react directly with the double bonds in the dyes or induce the hydrogen peroxide decomposed to be free radical species of $\cdot HO_2$ and$\cdot OH$ to react with dyes as shown in Eq.3 and Eq.4:

$$HOOH + OH^- \rightarrow H_2O + [O] + O_2 \quad (3)$$

Or $$HOOH + OH^- \rightarrow HOO^- + H_2O \qquad HOO^- + H_2O_2 \rightarrow \cdot HOO + \cdot HO + OH^- \quad (4)$$

On the other hand, NaClO mostly destroy and decompose the dyes by [O] which is produced by Eq.5[14]:

$$NaClO + H_2O \rightarrow HClO + NaOH \qquad HClO \rightarrow HCl + [O] \quad (5)$$

In sequence, the better process for fading colour on reactive dyed cotton fabrics in this study ranged from H_2O_2, NaClO to ozone/UV. But the treatment of ozone/UV only took 5min and consumed a little amount of water, power and chemicals, the detailed comparisons are demonstrated in Table 3.

3.3 Effect of Colour Fading Treatments on Fabric Mechanical Properties

As the fact that all of H_2O_2, NaClO and ozone/UV have a decomposed effect on both of the chromophore on dyes and the fiber in fabrics, cotton fabrics treated by different process may perform variedly in the respect of fabric mechanical properties. The tensile strength and elongation, tearing strength and abrasion resistance were used in this study to generally evaluate the variation of fading treatments on fabrics mechanical properties. Specifically, the abrasion resistance of fabric was characterized by the number of turns of the friction wheel when the first visible hole arise on the fabric. It means that the larger the number, the better the abrasion resistance fabric sample possess. The result of tests are demonstrated in Table.4. In addition, for illustrating the variation directly, the loss ratio was employed according to the equation as shown as Eq.6.

$$\text{Loss ratio } (\%) = (\text{control value} - \text{tested value}) \div \text{control value} \times 100\% \quad (6)$$

It can be seen that with increasing the colour fading effect, there is an increase in mechanical loss in the treatment. It is easy to found that the ozone/UV treatment has the minimum influence on fabric mechanical properties, and this effect generally was followed by chlorine (NaClO) and hydrogen peroxide in sequence. Tensile strength shows the most apparent decrease for all the fading treatments which is 27% for ozone/UV, but 29.5% and 32% for hydrogen peroxide and chlorine (NaClO) respectively. However, this result is not evident in elongation loss (about 5% for all of the three treatments), though it is also related to the tensile property. Loss ratio from ozone/UV on fabrics' tearing strength and abrasion resistance are approximate (13% and 14% respectively), such phenomenon is resemble in hydrogen peroxide treatment as well (17% and 16% accordingly). It is worth to note that the chlorine (NaClO) has a comparatively rapid impact on fabric's abrasion resistance with 20% loss ratio and 19% on tearing strength, which is just in accordance to the performance in fading colour. Overall, the impact on fabrics chemical properties of each treatment in this work matches their ability for fading colour, and this relation is about linear.

Table.4 Mechanical properties of fabrics treated by H_2O_2, NaClO and ozone/UV respectively.

	Tensile strength (CN)	Loss ratio (%)	Elongation (%)	Loss ratio (%)	Tearing strength (CN)	Loss ratio (%)	Abrasion resistance (r)	Loss ratio (%)
Control	1331.4	-	26.8	-	55.01	-	981.2	-
H_2O_2	940.8	29.5	25.5	5	45.88	17	819.5	16
NaClO	907.6	32	25.9	3	44.42	19	785	20
Ozone/UV	969.8	27	25.7	4	47.86	13	840.1	14

3.3 Analysis of Surface Microstructure

Fig.6 shows the 150, 1000 and 5000 times SEM of the fabrics treated by H_2O_2, NaClO and ozone/UV respectively.

It was found that the compact structure of the yarn in control sample was damaged inordinately in different treatments. This was particularly evident in the samples treated by chlorine, which possesses a wide range of tassel type of wound and cracks over the cotton fiber. Such tassel type of wound probably is the outcome from the decomposition of the lignin and xylan by NaClO [15]. Moreover, it seems that there are more of the fibers on the surface of yarn that were broken by NaClO than that by other treatments. The microstructure of samples treated by H_2O_2 shows a lot of corrosion-similar damage, rather than tassels, around the surface of the fibers which may reveal that the damage from H_2O_2 only stay on the fiber surface and influence little of the core of cotton and so that the mechanical properties loss of fabrics in H_2O_2 are less than in NaClO. As well as the ozone/UV treatment, which mostly maintains the surface microstructure of control, only has a little impact on the cotton. The observations on cotton and fabrics treated by different fading treatments match their mechanical performances, that the influence on cotton and fabrics from NaClO are harder than that from H_2O_2 and ozone/UV.

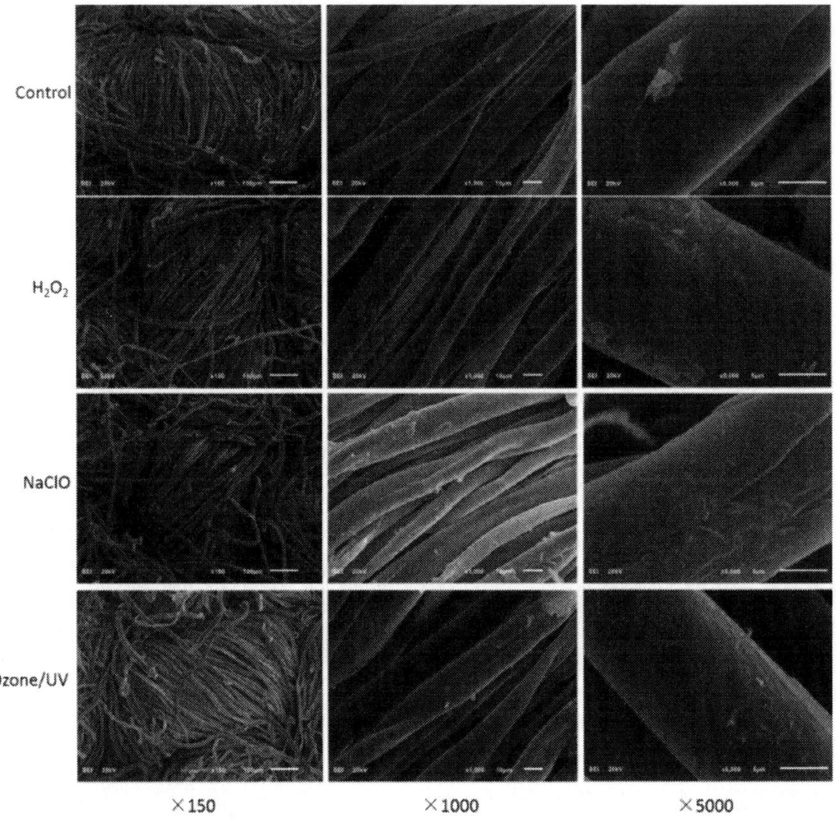

Fig.6 The 150, 1000 and 5000 times SEM of the fabrics treated by H_2O_2, NaClO and ozone/UV respectively

4. Conclusion

The ozone/UV collaborative treatment on colour fading reactive dyed cotton and the comparison with conventional fading processes such as hydrogen peroxide (H_2O_2) and chlorine (NaClO) in terms of the effect on colour fading, mechanical properties and surface microstructure were investigated. It was found that ozone/UV has an excellent potential to alternate the conventional processes as it performs better in the colour fading evaluations and less negative influences on fabric mechanical properties, more importantly, it is capable to save a huge amount of water and power. This work also has clearly shown that the treating time can be decreased dramatically if treating dyed cotton by ozone/UV instead of hydrogen peroxide (H_2O_2) and chlorine (NaClO).

5. Acknowledgments

The authors wish to express their gratitude to the Denim Garment Institute of Hubei for financial support (Project no. 52300101202) for this work.

References

[1] Gottschalk C, Libra JA, Saupe A, Gottschalk C, Libra JA, Saupe A. "Ozonation of water and waste water: a practical guide to understanding ozone and its applications," International Journal of Environmental Studies: 2010; pp. 795-796.
[2] Tehranibagha AR, Mahmoodi NM, Menger FM. "Degradation of a persistent organic dye from colored textile wastewater by ozonation," Desalination: vol. 260; pp. 34-38: 2010.
[3] Turhan K, Ozturkcan SA. "Decolorization and Degradation of Reactive Dye in Aqueous Solution by Ozonation in a Semi-batch Bubble Column Reactor," Water, Air, & Soil Pollution: vol. 224; pp. 1-13: 2013.
[4] Venkatesh S, Quaff AR, Pandey ND, Venkatesh K. "Decolorization and mineralization of C.I. direct red 28 azo dye by ozonation," Desalination & Water Treatment: vol. 57; pp. 4135-4145: 2016.
[5] Wu J, Eiteman MA, Law SE. "Evaluation of Membrane Filtration and Ozonation Processes for Treatment of Reactive-Dye Wastewater," Journal of Environmental Engineering: vol. 124; pp. 272-277: 1998.
[6] Deng Q. "The Current Situation and Developing Tendency of Ozone Technology," Water Purification Technology: 2001.
[7] Card A, Moore MA, Ankeny M. "Performance of garment washed denim blue jeans," Aatcc Review: vol. 5; pp. 23-27: 2005.
[8] Khalil E, Sarkar J. "Effect of Industrial Bleach Wash and Softening on the Physical, Mechanical and Color Properties of Denim Garments,": vol. 1; pp. 46-49: 2014.
[9] Eren S, Gümüs B, Eren HA. "Colour stripping of reactive-dyed cotton by ozone treatment," Coloration Technology: vol. 132; pp. 466-471: 2016.
[10] Shao J, Carr CM, Rowlands CP, Walton J. "XPS, SIMS, and ESR Studies of UV/Ozone-irradiated Silk and Wool," Journal of the Textile Institute: vol. 90; pp. 459-468: 1999.
[11] Tezcanligüyer G, Ince NH. "Individual and combined effects of ultrasound, ozone and UV irradiation: a case study with textile dyes," Ultrasonics: vol. 42; pp. 603-609: 2004.
[12] Mcgrath WD, Norrish RGW. "Studies of the Reactions of Excited Oxygen Atoms and Molecules Produced in the Flash Photolysis of Ozone," Proceedings of the Royal Society A Mathematical Physical & Engineering Sciences: vol. 254; pp. 317-326: 1960.
[13] Zeronian SH, Inglesby MK. "Bleaching of cellulose by hydrogen peroxide," Cellulose: vol. 2; pp. 265-272: 1995.
[14] Estrela C, Estrela CR, Barbin EL, Spanó JC, Marchesan MA, Pécora JD. "Mechanism of action of sodium hypochlorite," Braz Dent J: vol. 13; p. 113: 2002.
[15] Dianyuan C. "Study on application of chlorine dioxide in cotton fabric pretreatment ", Donghua University: 2013.

Studies on the Dyeing Properties of Wool Fibers treated with Gardenia Pigments

Wen-Fang Xu[1], Kun-Di Zhu[1], Chun-Hua Zhang[2, 3], Yun-Li Wang[3*]

[1]*College of Textile Science and Engineering, Wuhan Textile University, Wuhan, 430200, China;*
[2] *College of Textiles & Garments, Southwest University, Chongqing, 400715, P. R. China*
[3]*Key Laboratory of Green Processing and Functional Textiles of New Textile Materials, Ministry of Education, Wuhan Textile University, Wuhan, 430200, China*

*Corresponding author's email: ylwang@wtu.edu.cn

Abstract

Natural dyes are often perceived as an environment-friendly product and has been receiving increasing attention, especially amongst urban populations. In this paper, gardenia pigments were employed to explore the dyeing mechanism of wool fibers. Moreover, the effects of mordant agents on the dyeing efficiency of gardenia pigments were also investigated. The results show that mordant agents played an essential role in dyeing wool with gardenia pigments and therefore the dyes absorbed on wool fibers fixed well. Furthermore, the dyeing efficiency of gardenia pigments was reduced significantly with the increasing of dye concentration.

Keywords: Wool Fiber; Natural Dyes; Dyeing; Property

1. Introduction

The disposal of textile dyeing waste water has become a serious issue in recent years due to the discharge of inorganic salts and colorful chemical dye-stuffs [1]. In order to protect the environment, eco-friendly dyeing methods for fibers using natural dyes should be advocated and expanded. In comparison with chemical dye-stuffs, natural dyes are more likely to exhibit better biodegradability and are generally environmentally-friendly [2]. However, the main differences between natural and chemical dye-stuffs are stability and cost: chemical dye-stuffs are generally more stable and less expensive than natural ones [3, 4]. The difference in stability and cost leads to the enduring popularity of chemical dye-stuffs directly, and hinder the development of natural dyes. Hence, the dyeing proficiency of natural dyes should be enhanced.

Many studies on the application of natural dyes have been investigated by researchers recently. The study of the dyeing properties of natural dyes are mainly concentrated in Asia, especially in India and Japan. Gardenia is a kind of rubiaceae gardenia plant, which had been extracted into a traditional dye in ancient China. The gardenia yellow pigment, a kind of water-soluble pigment, is extracted from the fruit of the gardenia. The colorings of gardenia yellow is blended primarily of crocetin, pectin, tannic and crocin of terpenoids. The gardenia yellow pigment has many excellent performance, such as strong tinctorial power, vivid color, soft hue, favorable stability, strong solubility, little off-flavor and little toxicity [5]. Because of its bright color, the clothes dyed with these natural dyes are more likely to used for royal families, however, the residents seem to have no access to them. Gardenia blue is derived from the gardenia yellow pigment processed by the food enzyme. People have been using the seeds of gardenia in which carotenoid-based yellow pigment crocin and base material of the source of blue pigment geniposide existed to extract gardenia yellow pigment since ancient time. Due to the coexistence with the yellow pigment, gardenia blue pigment produced by geniposide forms green and have a tone of dark. Therefore, domestic and foreign research scholars are committed to investigated the mechanism of the formation of gardenia blue pigment, and separate the crocin and geniposide to produce high quality pigment respectively and effectively. Gardenia pigment was chosen to dye wool fiber because it is a kind of natural and "green" food additive which has a long history.

The study on gardenia pigment is mainly about the dyeing of leather, ramie, cotton fabric, silk and bamboo fiber, and the dyeing of wool fibers is rarely reported [6]. In this study, gardenia pigment was applied to the dyeing of wool fibers to explore the dyeing mechanism of this natural dye.

2. Experimental

2.1 Materials and Methods

2.1.1 Materials

Gardenia pigments (Jiangsu Hans chemical agent limited company); Wool fibers (Shandong RUYI Group); Sodium chloride; Alum (AR); Ethanol (AR); Natural soap; Oscillating dyeing machine HBC-24 (Ronggui Hui Bao dyeing and finishing plant , Shunde, Foshan)

2.1.2 Dyeing Experiments

Mordant agent could play an important role in the improvement of dye uptake and color fastness due to the inadequate substantivity of natural pigments. Both of the Gardenia blue and gardenia yellow can be used for dyeing directly, and also be used for dyeing with mordant. In order to discuss the effects of dyes and mordant on the dyeing properties of wool fibers separately, the experimental variables of this study are shown in Table 1. Gardenia blue and gardenia yellow were fixed to the wool fibers by mordant agent after dyeing in this scheme.

Table.1 Experimental variables

Sample	Dye color	Dye concentration/%(owf)	Mordant concentration/%(owf)
GB1	gardenia blue	6	4
GB2	gardenia blue	3	4
GB3	gardenia blue	6	8
GB4	gardenia blue	3	8
GY1	gardenia yellow	6	8
GY2	gardenia yellow	6	4
GY3	gardenia yellow	3	4
GY4	gardenia yellow	3	8

By consulting relevant material and combining the characteristics of wool fibers itself, using single factor analysis, under the conditions of changing dyeing temperature, dyeing time, pH value and other process, the wool fibers were dyed. Therefore, after a large number of experiments, the optimal dyeing process was summarized. Here is the specific process curve:

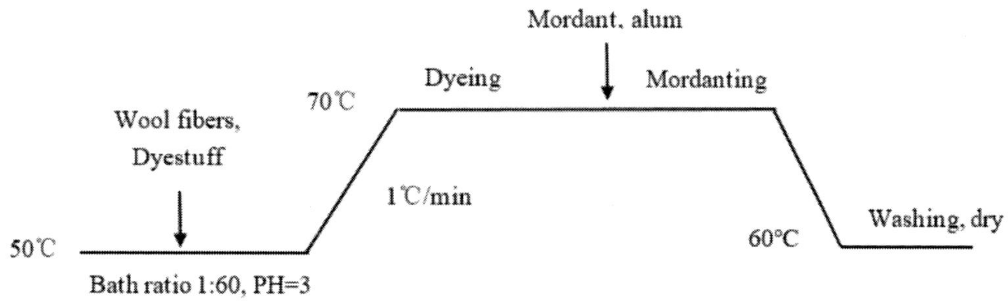

Fig.1 The process diagram of dyeing of wool fibers.

2.2 Characterization

2.2.1 Scanning Electron Microscope (SEM)

Surface images of wool fibers were observed using Scanning Electron Microscope (SEM) (JSM-6510LV) after gold palladium coating at a voltage of 20 kV with a magnification of 1000×.

2.2.2 K/S

The color depth of wool fibers before and after soaping were measured by X-Rite Color i7 Spectrophotometer. Each sample was tested 15 times.

2.2.3 Fourier Transform Infrared Spectra (FTIR)

FTIR spectra were recorded using a spectrometer (TENSOR37) with the reflection absorption spectroscopy technology in the range of 4000 - 600 cm^{-1}. The data were collected over 128 scans with a resolution of 4 cm^{-1}.

3. Results and Discussion

3.1 Morphology of Wool Fibers Before and After Dyeing

The surface microscopy of wool fibers before and after dyeing with gardenia pigments are shown in Fig.2. The wool fiber without any treatment is presented in Fig.2a. It can be seen from the figure that scales on the surface of the wool fiber are distributed evenly, and the distance between scales of the wool fiber is 7.6×10^{-3}-13.4×10^{-3} mm. The scales are smooth and the structure is clear and complete. Fig.2b-e shows the wool fibers dyed with different concentration of gardenia pigment and mordant. In comparison with the wool fibers without any treatment, the scale layers on the surface of fibers after dyeing are relatively smooth, and the apparent morphology of the wool fibers is almost the same as the unstained wool fiber. There is also a small amount of powdered particles on the surface of the dyed wool fibers, which are considered to be the residual salts. It can be observed that negligible changes have taken placed in the morphology of fibers. Therefore, we can draw a conclusion that different concentrations of gardenia pigment and mordant seem to have little effect on fiber morphology, and this dyeing method of gardenia pigment could lead to less damage to wool fibers.

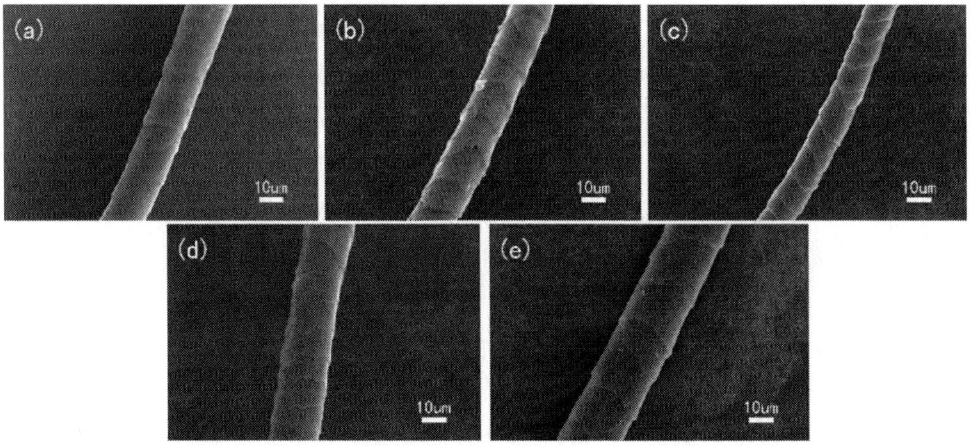

Fig.2 SEM images of wool fibers: (a) SEM images of wool fibers untreated; SEM images of wool fibers dyed by 6% gardenia blue and treated with (b) 4% mordant agent, (d) 8% mordant agent; SEM images of wool fibers dyed by 3% gardenia blue and treated with (c) 4% mordant agent, (e) 8% mordant agent.

Wool fibers dyed with different concentrations of gardenia yellow pigment and mordant are presented in Fig.3. Compared with the wool fibers without treatment, the scales on the surface of fibers showed in Fig.3a-d are distributed evenly, and the structure of it is basically integrated. However, subtle changes are still be found. It is observed that the scale layer on the surface of wool fiber showed in Fig.3a is thin, and part of the surface of them is rough and accompanied by small cracks. The structure of the scales showed in Fig.3b is blurred and damaged seriously. The scales showed in Fig.3c-d are smooth and the structure is clear and complete, the apparent morphology of them is almost the same as the unstained wool fiber. The results show that the high concentration of gardenia yellow pigment has a certain influence on the surface structure of wool fiber. In comparison with gardenia blue pigment, gardenia yellow has a great influence on the apparent morphology of wool fibers, although gardenia pigment is a kind natural dyes which would not destroy the surface structure of wool fibers basically.

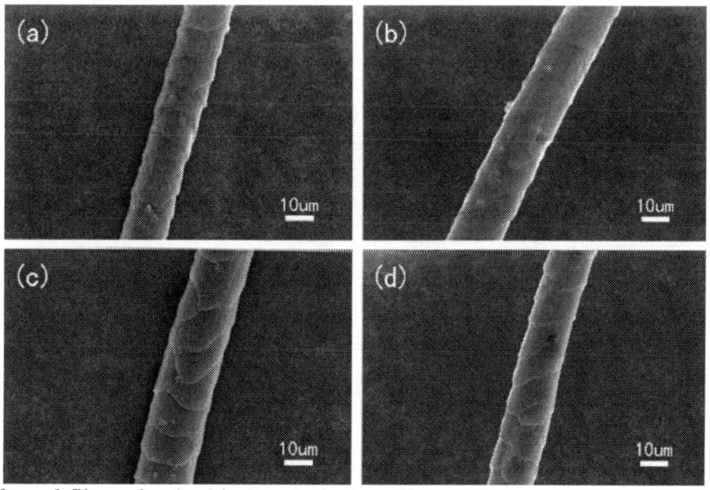

Fig.3 SEM images of wool fibers dyed with 6% gardenia yellow and treated with (a) 8% mordant agent, (b) 4% mordant agent; SEM images of wool fibers dyed with 3% gardenia yellow and treated with (c) 4% mordant agent, (d) 8% mordant agent.

3.2 K/S Analysis

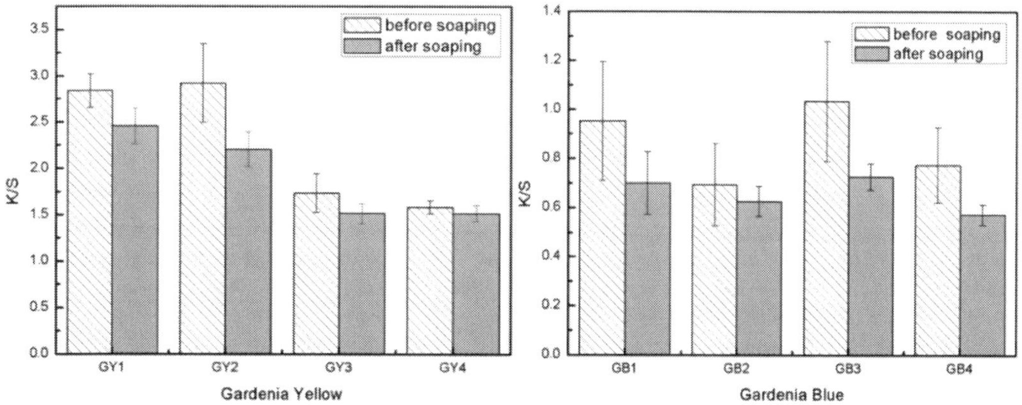

Fig.4 Effects of the concentration of gardenia dyes and mordant on K/S of wool fibers.

To evaluate the effect of the concentration of dyes and mordant on staining results, the K/S statistical test was performed. Fig.4 shows the K/S that reflects the color depth of wool fibers dyed with gardenia blue and gardenia yellow, respectively. Because the color on the surface of wool fibers is deeper, the K/S value of wool fibers dyed with high concentration of dyes is relatively larger. However, the distribution of dyes is not uniform on fibers.

When the concentration of gardenia pigments is low, the K/S value of the wool fiber will be reduced, and the unevenness of the color will be reduced at the same time. Thus it can be seen that the dye concentration is not proportional to the dyeing uniformity.

The utilization ratio of the high concentrations of dyes seems to be lower than that of the low concentrations of dyes, which may because the K/S before and after washing of wool fibers dyed with high concentrations of dyes decreased more significantly than that of fibers dyed with low concentrations of dyes. Therefore, the fastness of the fibers dyed with high concentrations of dyes seems to be relatively low.

When treating with the same concentration of mordant, the K/S value of wool fibers dyed with gardenia yellow increased dramatically than that of fibers dyed with gardenia blue. Accordingly, with the increasing of the concentrations of dyes, the dyeing efficiency of gardenia yellow was improved significantly, and it was better than that of gardenia blue after the concentration of dyes increased to a certain extent.

The experimental results show that mordant has a certain effect of deepening color on wool fibers dyed with gardenia pigments, but with the increasing of the concentrations of mordant, the color of the dyed fibers is not deepening much more. It is thus clear that the increase of the amount of dyes appropriately could deepen the color of wool fibers. However, surplus dyes dosage will lead to exorbitant cost. Hence, we must consider the cost of production, and choose the right amount of dye.

3.3 FTIR Analysis

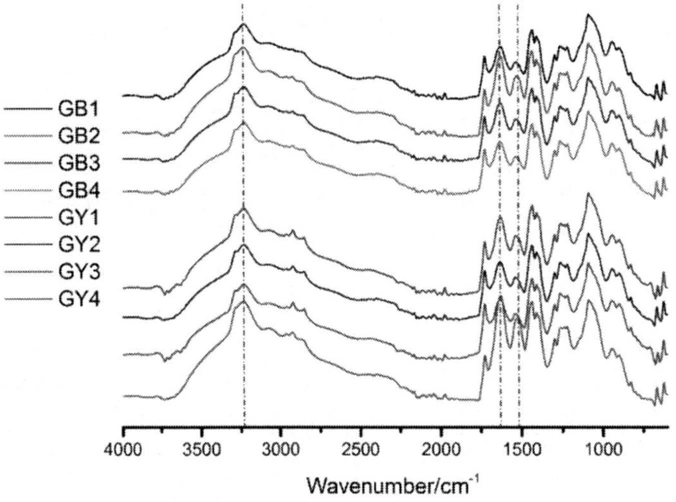

Fig.5 FT-IR spectra of dyed wool fibers.

Wool fibers dyed with different concentration of gardenia pigment and mordant were measured by FTIR (Fig.5). The dyed wool fibers contained typical wool protein characteristic absorption bands at 1626 cm^{-1} (amide I), 1524 cm^{-1} (amide II) and 1238 cm^{-1}(amide III). Comparably, the wool fiber dyed with gardenia yellow showed a little shift to low wave numbers at 3229 cm^{-1} comparing with those of gardenia blue. The little shift is because of the formation of hydrogen bonds, which could be created between the wool fibers and the gardenia yellow. Moreover, these results showed that the different gardenia pigment did not change the chemical structure, and also proved that the main chemical structure of wool fibers remains unchanged.

4. Conclusion

The study revealed that the principal structure of wool fibers dyed with gardenia blue and gardenia yellow would not be destroyed obviously, and the addition of mordant could contribute to the improvement of the dyeing properties of gardenia dyes. The gardenia dyes and mordant may play an essential role in enhancing the bright color and the fluffy feeling of wool fibers. By comparing the dyeing efficiency of gardenia blue and gardenia yellow on wool fibers, it is obvious that the surface of the wool fibers dyed with gardenia yellow damaged even worse than the fibers dyed with gardenia blue, which indicates that the dyeing process of gardenia blue seemed to be more moderate. With the increasing of the concentrations of dyes, the dyeing efficiency of gardenia yellow was improved dramatically, and it was better than that of gardenia blue after the concentration of dyes increased to a certain extent. Mordant has a certain effect of deepening color on wool fibers dyed with gardenia pigments, however, with the increasing of the concentrations of mordant, the color of the dyed fibers was not deepening much more. The application of natural dyes is of great significance to environmental protection and resource conservation. Moreover, the exploration of the dyeing theories of natural pigments and the optimization of the dyeing process may provide a basic information for a wider applications of natural pigments.

References

[1] Wu XB, et al. Effective Removal of Reactive Blue 19 from Aqueous Solution by Catalytic Ozonation Method. The 89th Textile Institute Word Conference: 2014; 50 (13): 2815-2818.

[2] McGregor BA, Butler KL. Indices for the identification of biologically productive cashmere goats within farms. Small Ruminant Research: 2015; 129: 11-17.

[3] Brouta-Agnésa M, Balsells S, Paul R. Synergistic effect of hydrolyzed collagen in the dyeing of wool. Dyes & Pigments: 2013; 99 (1): 116-119.

[4] Sérgio RK Velho, Luis FW Brum, Carlos O Petter, João Henrique Z dos Santos, Štefanija Šimunić, Wilfried HK. Development of structured natural dyes for use into plastics. Dyes and Pigments: 2017; 136: 248-254.

[5] Ozaki A, Kitano M, Furusawa N, Yamaguchi H, Kuroda K, Endo G. Genotoxicity of gardenia yellow and its components. Food and Chemical Toxicology: 2002; 40 (11): 1603-1610.

[6] Zhou Y, Gao D, Ma J. Dying Effect on Wool and Collagen Fiber by Natural Dyes. China Leather: 2014; 7.

PH-induced Fixation of Tea-based Biocolorant for Fixation in dyeing of Cotton Fabric

Ji-Xian Gong[1,2,*], Jiang-Fei Lou[1,2], Yan-Fei Ren[1,2], Kun Liu[1,2], Qiu-Ya Zhang[1,2], Zheng Li[1,2], Qiu-Jin Li[1,2], Jian-Fei Zhang[1,2,**]

[1]*School of Textiles, Tianjin Polytechnic University, No. 399 Binshui Xi Road, Xi Qing District, Tianjin 300387, P. R. of China.*
[2]*Key Laboratory of Advanced Textile Composites, Ministry of Education of China, Tianjin 300387, China*

[*]Corresponding author's email: gongjixian@126.com

Abstract

Clean production has become a popular dyeing and finishing technology in recent years. Biological materials provide a more environment-friendly and greener alternative to produce dyes and pigments. Therefore, the application of biomass dyes in the dyeing process has attracted more and more attention. However, fastness properties are still the main problems in the application of bio-colorants derived from natural compounds. In this study, a novel dyeing method of pH-induced fixation was established with natural pigment derived from tea stalk. Fastness and color strength were measured to evaluate the effect of dyeing process. The mechanism of pH-induced fixation was deduced from detecting the oxidation reduction potential of dyebath. The colorant derived from tea stalk can be fixed with alkaloid and alkali in the dyeing process and meanwhile, the fastness can be increased. Polychrome colors can be obtained by using alkaloids such as berberine, caffeine and theobromine as fixers. The natural pigment in tea stalk can be oxidized in the fixation process with alkaloid or alkali, and this might be the reason of pH-induced fixation.

Keywords: Natural Pigment; Fixation; PH-induced; Oxidation; Tea

1. Introduction

Clean production has become the goal of endeavor in improving the dyeing and finishing technologies in recent years. Biological materials were considered as more environment-friendly and greener alternative to produce dyes and pigments [1-3]. Therefore, the application of biomass dyes has become more important in the dyeing process [4-6]. It is characterized by the sustainability of the source of materials, the cleanness of the preparation process and the ecology of the product [7-9]. In the last decade, studies about the use of natural dyes in textile dyeing process had been conducted by various research groups [3-5]. Some of the natural dyes were successfully introduced into a technical dyeing process. However, fastness properties were still the main problems in the application of natural dyes [1, 5]. The fixation mechanism of natural dyes in fibers should be further studied and the dyeing technology of natural dyes was expected to be improved. It was noticed that the existence of natural pigments in vivo largely depended on the acidic or alkali niche in biomass, which can give people some inspiration for dyeing with natural pigments. The strategy of textile dyeing and finishing can be drawn up from the existence and distribution of pigment in biomass [10, 11].

In this study, a novel dyeing method of pH-induced fixation was established with natural pigment derived from tea stalk. Fastness and color strength were measured to evaluate the effect of the dyeing process. The mechanism of pH-induced fixation was introduced by detecting the oxidation reduction potential of dyebath.

2. Experiment

2.1 Materials

The scoured cotton knitted fabric was purchased from its retailer. Tea stalk was obtained from one of the retailers of Tie Guanyin tea. All the other chemicals used were laboratory-scale.

2.2 Methods

2.2.1 Extraction of Natural Compounds

Plant materials such as tea-stalk of Tie Guanyin tea were selected and applied in this study. The obtained plant samples were dried and grinded by an electrically operated grinder into powder.

The obtained powder materials (30g) were dispersed into water (1000ml) in a soxhlet extractor and then heated for 60 minutes to extract the natural products.

2.2.2 Dyed Methods

The cotton fabric samples were padded twice-rolling fluid rate of 80% with tea extract solution (30g/L), followed by drying at 70℃/7 min and then curing at 100℃/10 min in a steamer. The cured fabric samples were then padded once-rolling fluid rate of 80% with fixer solution, followed by curing at 100℃/7 min in a steamer. The cured fabric samples were then washed for 10 minutes at 90℃ having the use of 2g/L soap flakes, rinsed thoroughly; and dried finally at 70℃/5 min. Typical formulations used in this study are given explicitly in the text.

2.3 Measurements

2.3.1 Color Measurement

The color strength and color depth of both dyed and mordant samples were determined colorimetrical (color space: CIE Lab (1976) / D65) by light reflectance technique using Datacolor 600.

2.3.2 Color Fastness Tests

The color fastness of the dyed cotton fabric samples was assessed by means of rubbing (dry and wet), washing and light exposure. The measurements of rub fastness, wash fastness and light fastness of the fabrics were carried out on the principle of ISO 105-C01, ISO 105-C06 and ISO 105-B02 respectively.

3. Results and Discussion

Natural compounds have experienced a long process of evolution. The existence of pigments in organism depends greatly on the development status of cellular fluid or intercellular fluid, such as pH, concentration of electrolyte or free radical. In some plants, alkaloids play a significant role in maintaining the structure and function of the pigment. In point of tea, theophylline is contained in the part of tea leaf.

In this study, the role of alkaloids was studied in the process of fixation of tea pigment with fiber. The alkaloids, such as berberine, caffeine and theobromine, were used in dyeing process as fixing agents. Color values and color strength of the dyed cotton fabric were measured to evaluate the effect of alkaloids on tea pigment fixation (Table 1).

The result shows that shades can be obtained with extract solution of tea in cotton fabric. It can also be seen that the color of fabrics fixed with different alkaloids varied greatly. This phenomenon may be caused by the color of each alkaloid itself. The color of fabrics processed with alkaloids overlap both the color of the tea pigment and the color of alkaloid. Thus, bright yellow can be obtained with the berberine treatment, light brown can be generated with the caffeine and camel can be produced with the theobromine.

Table.1 Color Values and Color Strength Data of the Dyed Cotton Fabric

Dye-fixing agent	*K/S*	L^*	a^*	b^*
Control	7.3	83.1	2.1	10.7
Berberine	15.1	71.1	2.5	14.4
Caffeine	14.1	71.3	3.6	12.4
Theobromine	11.7	72.9	2.5	9.1

However, color fastness is a noticeable problem in dyeing of natural pigment [5, 6]. In this study, fastness properties of fabrics fixed with different alkaloids was measured and the data was given in Table 2.

It was obviously shown that washing fastness and rubbing fastness were improved by fixation process with alkaloids. Fixed with alkaloids, light fastness was still low in fabrics though the value of it was increased than the controlled one. This result indicated that the existence of alkaloids played a positive role in fixation of tea pigment in fabric.

Table.2 Fastness Properties of Fabrics Fixed with Different Alkaloids

Dye-fixing agent	Washing fastness	Rubbing fastness		Light fastness
	Color change	Dry	Wet	
Control	3	3	3/4	2
Berberine	3/4	4	4	3
Caffeine	3	4	4	3
Theobromine	3/4	4	4-5	3

Considering that alkaloid can act as a kind of effective fixative of tea pigment, alkali was used in the subsequent research to make a comparison of the effects caused by alkaloids and alkali respectively. Two common inorganic bases, NaOH and $Ca(OH)_2$, were chosen in the next experiment, and color values and color strength of the dyed cotton fabric were measured (Table 3).

Table.3 Color Values and Color Strength of Fabric Fixed with Inorganic Alkali

Dye-fixing agent	*K/S*	L^*	a^*	b^*
Control	7.3	83.1	2.1	10.7
NaOH	10.2	75.4	2.2	12.9
$Ca(OH)_2$	12.4	73.0	2.6	16.9

It can be seen that color strength (*K/S*) of fabrics fixed with inorganic bases was lower than that of alkaloid on the grounds that the alkaloids have its own color. However, the color of dyed fabric only demonstrated the hue of tea pigment since the inorganic alkali was colorless.

Furthermore, it can be seen that the *K/S* values of fabrics fixed with inorganic alkali were increased compared with the control (Table 4), which may be caused by the transformation of the structure of tea pigment in fixation process.

Table.4 Fastness Properties of Fabric Fixed with Inorganic Alkali

Dye-fixing agent	Washing fastness	Rubbing fastness		Light fastness
	Color change	Dry	Wet	
Control	3	4	3/4	2
NaOH	4/5	5	4/5	3
$Ca(OH)_2$	4/5	5	4/5	4

The result in Table 4 shows that the values of wash and rub fastness of the samples treated with inorganic alkali was lower than that of alkaloid. Nevertheless, the fastness was improved 0.5-1 grade after fixation with alkalis.

Due to the improved color fastness of fabrics fixed with high pH condition, it is necessary to explore the fixation mechanism of pigments from tea stalk in alkaline condition.

Dyeing process with tea stalk extracting solution with different pH was carried out at 70℃ and 100℃ respectively for 7 minutes. Then, the oxidation reduction potential was measured. The results were shown in Table 5.

Table.5 Influence of Temperature on Tea Polyphenols Solution Oxidation Reduction Potential

No	pH=1	pH=3	pH=7	pH=11
1	347	238	0	-222
2	334	105	-14	-166
3	344	106	48	25
4	342	97	63	51

(1 Aqueous Solution, 2 Tea Polyphenols Solution, 3 Tea Polyphenols Solution at 70℃, 4 Tea Polyphenols Solution at 100℃)

The result illustrated that, along with the increase of temperature, which indicated the increased oxygen of tea polyphenols, the oxidation reduction potential has been gone up accordingly. When pH reached 7, the changes of the oxidation reduction potential of tea polyphenols were gentle, the changes were greater in case of pH amounted to 11. The results demonstrated that tea polyphenols have a higher oxidized degree under alkaline condition. So the fact that tea polyphenols can be oxidized at high pH and turn into big macromolecule substance may be the reason of pH-induced fixation.

4. Conclusions

1) The colorant derived from tea stalk can be fixed with alkaloid and alkali in dyeing process with increased fastness.

2) Polychrome colors can be obtained by using alkaloids, such as berberine, caffeine and theobromine, as fixers.

3) The natural pigment in tea stalk could be oxidized in the fixation process with alkaloid or alkali, and this might be the reason of pH-induced fixation.

5. Acknowledgements

This research was financially supported by the National Key Research Project (2016YFC0400503-02) and Important Special Research Project of Xinjiang Province (2016A03006-3).

References

[1] Colombini MP, Andreotti A, Baraldi C, Degano I, Tucejko JJ. Colour fading in textiles: A model study on the decomposition of natural dyes. Microchem J: 2007; 85: 174-182.

[2] Paul R, Solans C, Erra P. Study of a natural dye solubilisation in o/w microemulsions and its dyeing behaviour. Colloid Sureface A: 2005; 253: 175-181.

[3] Ibrahim NA, El-Zairy WM, El-Zairy MR, Eid BM, Ghazal HA. A smart approach for enhancing dyeing and functional finishing properties of cotton cellulose/polyamide-6 fabric blend. Carbohyd Polym: 2011; 83: 1068-1074.

[4] Diaz Blanco C, Gonzulez MD, Monmany JMD, Tzanov T. Dyeing properties, synthesis, isolation and characterization of an in situ generated phenolic pigment, covalently bound to cotton. Enzyme Microb Tech: 2009; 44: 380-385.

[5] Zarkogianni M, Mikropoulou E, Varella E, Tzanov T. Colour and fastness of natural dyes: revival of traditional dyeing techniques. Color Technol: 2011; 127: 18-27.

[6] Guimarães C, Kim S, Silva C, Cavaco-Paulo A. In situ laccase-assisted overdyeing of denim using flavonoids. Biotech J: 2011; 6: 1272-1279.

[7] Grifoni D, Bacci L, Zipoli G, Albanese L, Sabatini F. The role of natural dyes in the UV protection of fabrics made of vegetable fibres. Dyes Pigments: 2011; 91: 279-285.

[8] Punrattanasin N, Nakpathom M, Somboon B, Narumol N, Rungruangkitkrai N. Silk fabric dyeing with natural dye from mangrove bark (Rhizophora apiculata Blume) extract. Ind Crop Prod: 2013; 49: 122-129.

[9] El Ksibi I, Ben Slama R, Faidi K, Ben Ticha M, M'Henni MF. Mixture approach for optimizing the recovery of colored phenolics from red pepper (Capsicum annum L.) by-products as potential source of natural dye and assessment of its antimicrobial activity. Ind Crop Prod: 2015; 70: 34-40.

[10] Naik RR, Brott LL, Rodriguez F, Agarwal G, Kirkpatrick SM, Stone MO. Bio-inspired approaches and biologically derived materials for coatings. Prog Org Coat: 2003; 47: 249-255.

[11] Singh R, Jain A, Panwar S, Gupta D, Khare SK. Antimicrobial activity of some natural dyes. Dyes Pigments: 2005; 66: 99-102.

The Effect of Wet Textile Parameters on Washing Efficiency in a Domestic Front-loading Washer

Zhi-Feng Xu[1], Hong Liu[1], Xiong-Ying Wu[1,2*], Xue-Mei Ding[1,3]

[1]*College of Fashion and Design, Donghua University, Shanghai 200051, China*
[2]*Shanghai Entry-Exit Inspection and Quarantine Bureau, Shanghai 200135, China*
[3]*Key Laboratory of Clothing Design & Technology, Ministry of Education, Donghua University, Shanghai 200051, China*

[*]Corresponding author's email: wuxiongying@vip.sina.com

Abstract

Textile properties is one of the main factors affecting mechanical action during washing, and to some extent mechanical force affects textile movement. This paper imitated the actual washing condition and investigated wet textile properties, which include friction coefficient, draping and other properties, in terms of textile properties and washing performance. Ten different textiles were selected in the experiment to compare the differences of washing performance. Factors, which have a significant influence on washing performance, were selected and a multiple regression model was established to analyze the relationship of textiles' wet properties and washing performance. The result shows that wet friction coefficient, weight, volume and draping have significant influences on textile washing performance and mechanical action, especially the wet friction coefficient. It is beneficial to find a suitable washing condition among various washing factors, and achieve higher washing performance and lower fabric damage.

Keywords: Front Loading Washer; Wet Textile Properties; Washing Performance; Textile Movement; Mechanical Action

1. Instruction

Mechanical action is one of the most important factors affecting washing performance, which could make textiles rotate, fall, and rub. There was a direct correlation between textile movement and washing performance [1, 18]. The major factors were found to be the parameters of textile and washer, which would affect mechanical action and washing performance. At present, a great deal of attention has been dedicated to studying washer parameters and geometric parameters [1-15]. The results show that washing performance increases with washing time. Also, there is a positive correlation between washing speed and washing performance. So it was an appropriate method to obtain higher washing performance and smoothness by increasing washing speed, rinsing speed and washing time [2, 3, 6, 8-10]. Relative studies on geometry parameters on the washer was established earlier, and they found that geometric parameters of drum had a significant effect on washing performance. Drum diameter needed to match rotated speed, and the number and angle of lifter should be appropriate. Otherwise, drum depth and drum diameter also affect washing capacity and washing performance [2].

However, previous studies paid little attention to the influence of textile properties to washing performance. Lee et al. found that water absorption of textiles affects "folding" of textile movement and washing efficiency [4]. Most of the previous studies studied [5, 7] the relationship between mechanical action and washing efficiency. Park, Yun et al. explored the relationship between textile properties and textile movement [5], as well as the relationship between textile movement and washing performance [1, 5, 19-20]. They discovered that textile drape, weight, and water content have great influence on textile movement in the front-loading washing machine. Subsequently, they studied the relationship between textile movement and washing performance, and it indicated that washing efficiency has a direct correlation with textile movement in a front-loading washer.

Most research of textile property and washing performance focused on the relationship between textiles' dry properties and washing performance, which could not reflect the actual washing conditions. Textiles are in wet condition during washing. While there are significant differences between the properties of dry textile and wet textile. According to the actual washing condition, it is necessary to study how do wet-textile properties influence washing performance. However, it is difficult to test the wet textile properties [16] because of the unstable water content during the testing process [15]. In order to solve these problems, the existing instruments were modified, and textile humidity was controlled in a stable state. This study aims to examine wet textile properties and their effect on washing performance. In order to achieve the objectives, ten types of cotton textiles were selected, and textiles' properties have been explored in wet condition. High-speed camera was used to record the trajectory of textile movement under the same washing condition [17]. Then correlation analysis and multiple regression analysis were used to establish the regression equation between wet textile properties and washing performance. Finally, the washing mechanism of textile was interpreted by the trajectory of textile movement, which had a direct correlation with textile properties.

2. Experimental Details

2.1 Materials

Ten kinds of cotton textiles were selected and cut into the same size (80cm×40cm). The samples were placed at room temperature of (22±2) ℃ and relative humidity of (65±5) % for 24 hours. The samples were immersed in water for 4 hours and hung on the hanging rod until no more water dropping in a time of 30 seconds, which could be considered that the wet textile is at the saturated state. Then wet textile characteristics were tested. Detergent A* [21] was used for the whole experiment, except for the measurement of textile movement. The fabric soiled by carbon black and mineral oil was used to evaluate washing performance, in that it is more sensitive to mechanical force. Thread removal material was used to evaluate mechanical force and fabric damage. Washing performance and fabric damage was respectively measured by the GB/T 4288-2008 [22] and IEC 62473 [23] methods. Textile characteristics are shown in Table 1.

Table.1 Textile characteristics

Textile	Thickness (mm)	Weight (g/m^2)	Structure
1#	0.52	180.50	Plain
2#	1.71	287.10	Towel
3#	0.68	172.70	Twill
4#	0.56	165.80	Plain
5#	1.00	224.15	Plain
6#	0.34	85.23	Plain
7#	0.47	149.38	Knit
8#	0.37	137.88	Twill
9#	0.70	245.68	Plain
10#	0.73	261.38	Knit

2.2 Instruments and Methods

In order to ensure measurement accuracy of wet textile properties, the whole testing process was performed at room temperature of (22±2) ℃ and relative humidity of (65±5) %.

2.2.1 Test of Wet Textile Properties

1) Friction coefficient

Friction is an important factor, which can affect mechanical during washing. It includes the friction between the textile, and the friction between textile and drum. According to the actual washing condition, the textile movement is also influenced by the friction between the textile and the friction between textile and drum. So it is necessary to test friction coefficient between textile and drum, and friction coefficient between the textile.

The INSTRON instrument was modified to test friction coefficients, as shown in Fig.1. The principle of uniform linear motion is used to evaluate friction, so the tension of the rope is equal to the friction. Friction coefficients could be calculated according to the formula (1)

$$\mu = \frac{f}{F_N} \tag{1}$$

Where f is friction or traction (N), F_N is pressure (N), and μ is friction coefficient.

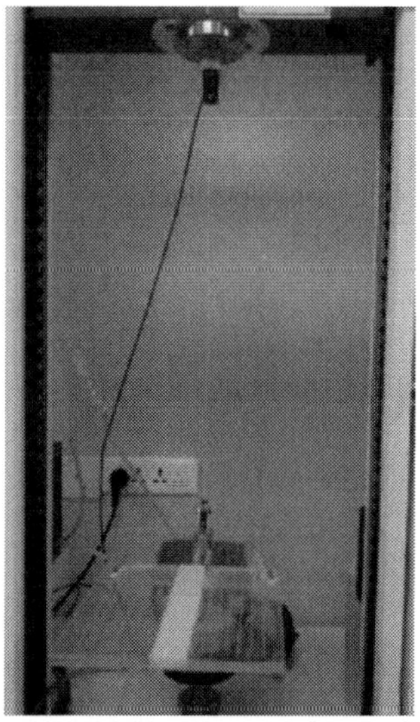

Fig.1 Modified INSTRON instrument for testing textile friction

2) Bending stiffness [24]

It is a factor that can affect bending deformation of textile and washing performance. Therefore, it needs to test the bending stiffness of textiles. When the moisture content of fabric was controlled at a saturated state, the textile softness tester was used to measure bending length, and bending stiffness are calculated according to equation (2)

$$B = G \times 10^{-2} \times C^3 \tag{2}$$

Where B is bending stiffness (mN/cm), G is quality (g/m2), and C is bending length (cm).

3) Drape coefficient

Drape coefficient is closely related to fabric deformation, which can affect washing performance during washing. In the study, drape tester was used to test textile drape coefficient in wet condition.

4) Tensile strength [25]

The INSTRON tester was also used to test textile warp/weft tensile strength. Tensile strength can reflect the tensile intensity, which has a close connection with textile deformation. The above instruments and methods were used to test textiles' wet properties, and the results are shown in Table 2.

Table.2 Wet textile properties

textiles	Weight (g/m^2)	Volume (cm^3)	Friction coefficient		Drape coefficient (%)	Bending stiffness (mN.cm)		Tensile strength (N)	
			1	2		weft	warp	weft	warp
1#	611.47	786.47	0.04	0.01	6.50	2.39	2.34	303.40	204.75
2#	1061.10	1016.36	0.01	0.01	15.05	3.62	3.75	720.00	446.95
3#	526.57	556.58	0.07	0.02	24.65	2.81	6.45	722.35	1194.67
4#	618.67	677.57	0.05	0.02	26.00	4.94	6.66	1020.83	880.30
5#	606.57	629.17	0.07	0.02	31.60	2.30	3.83	364.99	1122.82
6#	289.00	483.98	0.12	0.05	42.75	6.39	6.06	348.91	658.99
7#	648.77	750.17	0.02	0.01	11.30	10.57	5.08	83.86	51.57
8#	376.40	483.98	0.10	0.05	24.75	6.43	6.42	204.57	901.46
9#	234.70	362.98	0.17	0.06	11.40	3.83	4.07	155.14	408.33
10#	781.60	883.26	0.03	0.01	12.50	5.99	5.63	180.34	196.99

2.2.2 Test of Washing Performance

The fabric soiled by carbon black and mineral oil was used to evaluate washing performance, and five pieces of the soiled fabric were attached to the base fabric for each experiment. After washing, the whiteness meter was used to test the reflection of the soiled fabric, as shown in Fig.2. Washing performance is calculated according to equation (3).

$$\mathrm{Dr} = \frac{Rw - Rs}{Ro - Rs} \times 100\% \tag{3}$$

Where Dr is washing performance, Rw is the reflection of the soiled fabrics after washing, Rs is the reflection before washing, and Ro is the reflection of raw fabrics.

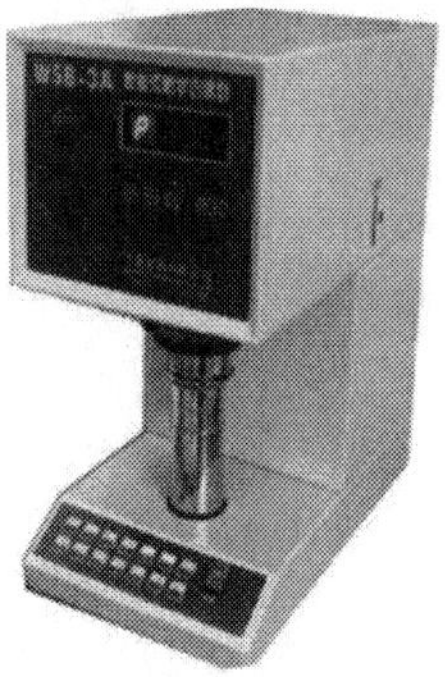

Fig.2 Whiteness meter

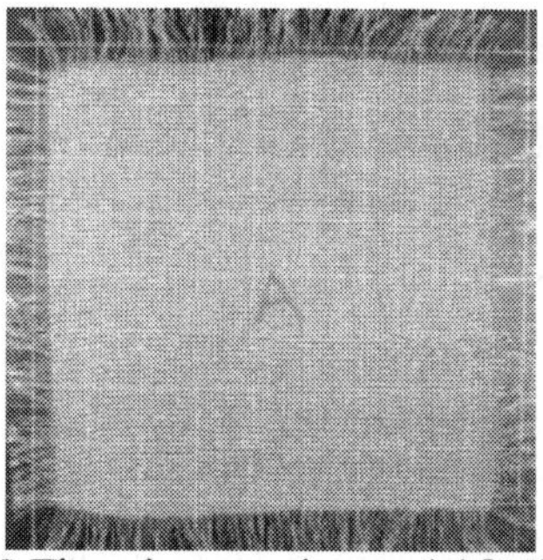

Fig.3 Thread removal material [23]

2.2.3 Test of Mechanical Action

Thread removal materials were selected to evaluate mechanical action and fabric damage, as shown in Fig.3. Five pieces of thread removal material (10cm×10cm) were attached to the base fabric for each experiment. Then the number of warp and weft yarn was counted to calculate fabric damage after washing. The formula is as shown in below.

$$A = \frac{\text{Not remaining threads}}{\text{Total threads warp and filling of original}} \times 100\% \quad (4)$$

Where A is fabric damage (%).

2.2.4 Test of Fabric Movement

The trajectory of fabric movement was recorded by a high-speed camera (OLYMPUS PEN Lite E-PL5). The fabric movement diagrams of anti-clockwise rotating are symmetrical to those obtained from clockwise rotation [18], so anti-clockwise rotation diagrams were chosen to analyze fabric movement in this study. After washing for 3 minutes, the anti-clockwise rotation of the first washing cycle was observed. In order to qualitative analysis of fabric movement, the outline and centroid of textile movement were traced and recorded by using the image processing software.

2.3 Washing Conditions

A front-loading washer (MINIJ 6-P, maximum capacity 2.8 kg) was used to study the washing performance and fabric damage, and washing parameters remained consistent. Specific washing parameters are shown in Table 3.

Table.3 Washing parameter setting

Washing speed (rpm)	Washing time (min)	Rotating time (s)	Off-time (s)	Rinsing speed (rpm)	Dehydration speed (rpm)	Washing load (g)
35	20	10	10	35	600	500

3. Results and Discussion

Washing performance and fabric damage have been measured, as shown in Fig.4 and Fig.5 respectively. Then the relationship between wet textile parameters and washing performance is established, as well as fabric damage.

3.1 Evaluation of Washing Performance

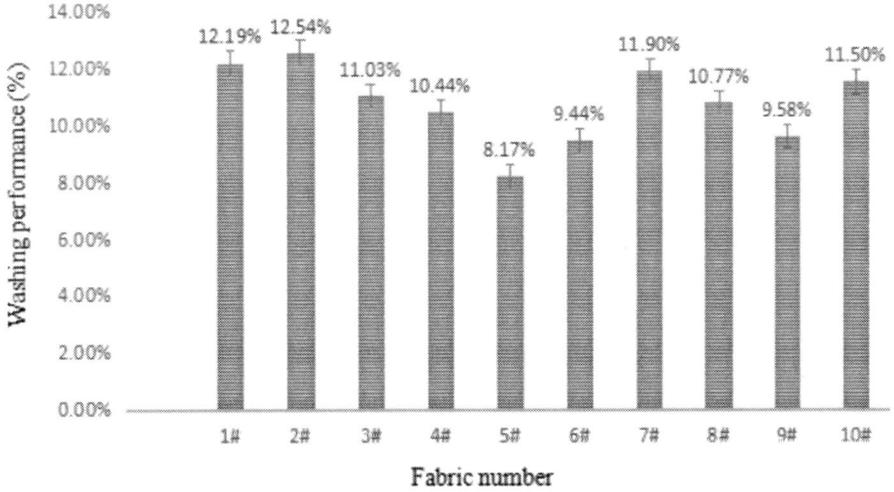

Fig.4 Washing performance of different textiles

There are significant differences between washing performances of ten kinds textiles according to Fig.4, then the

Pearson correlation analysis was used to analyze the correlation between wet textile properties and washing performance. The analysis results show that the absolute value of the correlation coefficient is more than 0.5 and sig value is less than 0.05 between washing performances and some textile properties, which are found to be drape coefficient, friction coefficient, volume and weight.

A regression model is established for drape coefficient, friction coefficient, volume, weight and washing performance by multiple-regression analysis, which is used to predict washing performance during washing. As shown in below,

$$Dr = 12.88 + 75.46 \times X_1 - 42.35 \times X_2 - 0.07 \times X_3 + 0.002 \times X_4 - 0.002 \times X_5 \qquad (5)$$

Where Dr is washing performance, X_1 is friction coefficient between textiles, X_2 is friction coefficient between textile and drum, X_3 is drape coefficient, X_4 is volume, and X_5 is weight. The adjusted R^2 is 0.796, indicating a good fit for this model.

It shows that friction coefficient has the greatest influence on washing performance, followed by drape coefficient, weight and volume. At the same washing condition, drape property of textiles increases with the decrease of drape coefficient. Drape property of textiles is closely related to fabric deformation. Besides, fabric deformation can promote mass transport of textile and washing solution, and improve washing performance. Friction coefficient is another important factor affecting washing performance, especially the friction coefficient between textiles. Also, the friction increases as friction coefficient of textiles. It is beneficial to remove stains. Similarly, textiles' volume and weight also have a significant influence on the impact force of washing, which eventually improve washing performance. In a word, textile properties affect washing performance by influencing mechanical force during washing, such as friction, impact force and free-fall gravity.

3.2 Evaluation of Fabric Damage

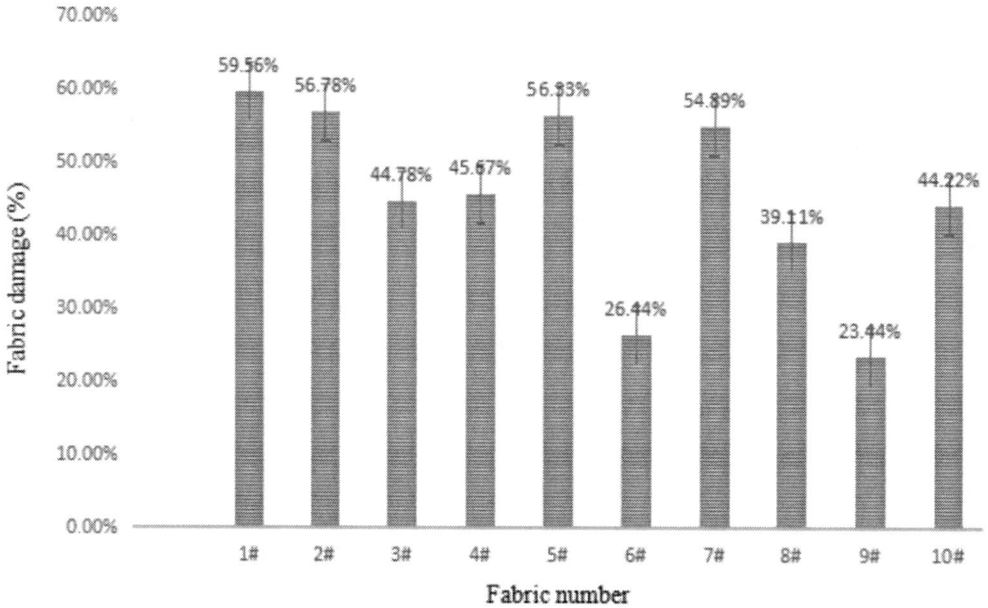

Fig.5 Fabric damage of specimens

There are significant differences between fabric damages of ten cotton fabrics according to Fig.5. Then the Pearson correlation analysis was used to analyze the correlation between textile property and fabric damage, and the result shows that the absolute value of the correlation coefficient is greater than 0.5 and sig value is less than 0.05 between fabric damage and some textile properties that found to be friction coefficient, volume and weight.

A multiple regression equation is established for the above textile properties and fabric damage by multiple regression analysis. Fabric damage decreases with the increase of the friction coefficient and volume, while it increases with wet textile weight. Therefore, washing capacity and filling degree of the washer should be controlled at a suitable level. The prediction model is used to evaluate the effect of mechanical force on textiles, and fabric damage can be used to characterize mechanical force during washing.

$$A = 75.50 - 396.96 \times X_1 - 116.90 \times X_2 - 0.03 \times X_3 + 0.02 \times X_4 \quad (6)$$

Where A is textile abrasion, X_1 is friction coefficient between textile and drum, X_2 is friction coefficient between textiles, X_3 is volume, and X_4 is weight. The adjusted R^2 is 0.807, indicating a good fit for this model.

It indicates that friction coefficient has the largest influence on fabric damage, followed by volume and weight. Friction coefficient directly affects friction, and it has the greatest influence on fabric damage. The mechanical force increases with wet textile weight in the process of the free-fall movement. Finally, it leads to the higher fabric damage. Also, friction coefficient between textile and drum has greater effect on fabric damage than friction coefficient between textiles, which suggests that friction between textile and drum influence on fabric damage is greater than the friction between textiles.

The results show that friction coefficient, drape coefficient, weight and volume have a more significant influence on washing performance and fabric damage. Compared with the friction coefficient between textile and drum, the friction coefficient between textiles has a larger influence on washing performance. While the influence of friction coefficient to fabric damage is different, friction coefficient between textile and the drum has a larger influence on fabric damage. Visibly, friction between textiles is beneficial to improve washing performance, and friction between textile and the drum has a larger influence on fabric damage. Also, washing performance and fabric damage of different textiles is different at the same washing condition. The study of the relationship between wet textile properties and washing performance is helpful to obtain higher washing performance and lower fabric damage.

3.3 Analysis of Textile Movement

In order to analyze washing mechanism and the relationship between textile property and washing performance, trajectories of textile movement were recorded during washing. Furthermore, the centroid and outline of trajectories were recorded to explain the relationship among textile properties, textile movement and washing efficiency. The trajectories of textile movement are shown in Fig.6.

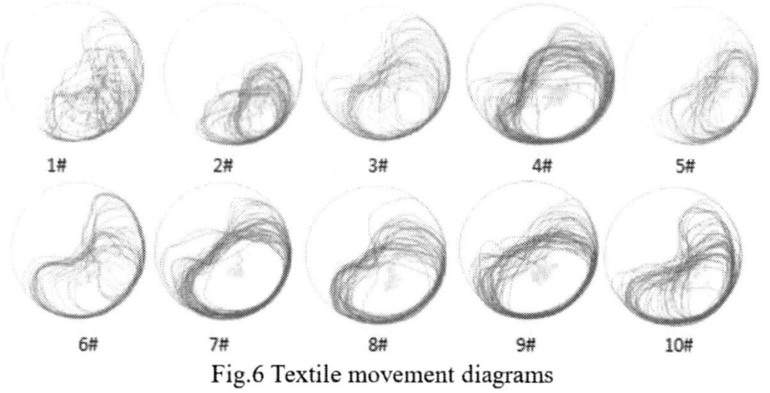

Fig.6 Textile movement diagrams

Textile properties affect mechanical action during washing, and it also has a close connection with textile movement and washing performance. The images of textile movement show that the outline of textile movement happened to the quadrants in the absence of the second, and folding and sliding are the main textile movement. Among all of the textile movement diagrams, the textile movement of No.2 only has relatively rubbing and folding movement, which caused by heavy weight and big volume. Also, the impact force and friction imposed on the No.2 textile can accelerate stains to detach from the textiles' surface and obtain higher washing efficiency. Wet weight and volume are the main factors affecting washing performance of rubbing and folding. For the same reasons, the textile of No.1, No.3 and No.10 also have higher washing efficiency. Conversely, the textile of No.5 has lower washing performance. There are only sliding and less textile deformation that can lower mechanical force and washing performance. The centroid and the outline of textile movement are important to evaluate washing performance, in that it can reflect the mechanical force and textile deformation. Wet weight, volume and bending deformation of textiles are the main factors affecting textile movement. And textile properties influence washing efficiency by affecting mechanical force and textile movement.

In addition to study the effects of wet textile properties on textile movements and washing properties, further researches should be carried out to analyze interaction effects of textile properties and washer parameters on textile movement and washing performance.

4. Conclusion

According to actual washing condition, wet textile properties were measured to investigate the relationship between textile properties and washing efficiency. The results of correlation and regression analysis show that main textile properties affecting washing performance are wet weight, volume, friction coefficient between textiles, friction coefficient between textile and drum, and drape coefficient. The regression model of washing performance consists of the five independent variables of textile properties, which can be used to evaluate and improve washing efficiency.

Except for wet textile properties, washer parameters including washing parameters and geometric parameters are other factors affecting washing efficiency. It is recommended to further investigate the interaction effects of textile properties and washer parameters on washing efficiency, and build a prediction model to improve washing performance and reduce fabric damage.

Acknowledgements

The authors are grateful to the Fundamental Research Funds of the Central Universities for providing support to this research. We also want to thank the reviewers and editors of this paper.

References

[1] Yun C, Park CH. The effect of textile movement on washing performance in a front-loading washer II: under various physical washing conditions. Textile Research Journal: 2015; 85: 251-261.

[2] Liu DX. Parameters design of pulsator washer. Journal of household electrical appliances of science and technology: 1985; 05: 4-7.

[3] Wu YC, Fan ZW. Distance measurement of inner and outer in front-loading washing machine. Electrical appliances: 2011; (S1): 272-273.

[4] Lee A, Seo MH, Yang S, et al. The effects of mechanical actions on washing efficiency. Fiber Polymeter: 2008; 9: 101-106.

[5] Park S, Yun C, Kim J, et al. The effects of the textile properties on textile movement and the prediction of the textile movements in a front-loading washer. Textile Research Journal: 2013; 83: 1201-1212.

[6] Wang FX, Zeng JF, Ding XM. The influence of washing parameters on washing performance of cotton shirts in the pulsator washing machine. Journal of Beijing institute of fashion technology (natural science edition): 2014; 01: 32-37.

[7] Kissa E. Kinetics and mechanisms of detergency part I: liquid hydrophobic (oily) soils. Textile Research Journal: 1975; 45: 736-741.

[8] Yuan JR. Wrinkling mechanism of woven textile washing in the front-loading washing machine. Donghua University: 2014.

[9] Hu B. Washing parameter optimization of silk textile in the household washing machines. Donghua university: 2015.

[10] Wang YY. Wrinkling mechanism of cotton woven textile research and prediction in the pulsator washer. Master degree theses of Donghua University: 2015.

[11] You YH. Design of structure parameters in the front-loading washing machine—the ways to improve washing performance. Journal of household electrical appliances of science and technology: 1986; 01:12-13.
[12] Wang B, Zhou CL. Design of washing speed and application of washing methods in the front-loading washer [A]. Proceedings of China household appliances technology conference in 2014. Chinese household electrical appliances association: 2014; 4.
[13] Xu ZG, Lu JG, Zhao WB, Chen ST. Washing performance analysis in household electric machine. Journal of electronics: 2012; S1: 346-350.
[14] Wei L, Sun RJ, Liang GY, Yu L. Friction performance of honeycomb tissue textile under the condition of different wet [J]. Journal of Xi 'an engineering University: 2011; (5): 626-630.
[15] Wang X. Friction performance study between textile and skin. Donghua University: 2011.
[16] Liu YF. Contact comfort research of wet textiles. Suzhou University: 2010.
[17] Wang YY, Ding XM. Measurement of textile movement in domestic washing machine. Journal of electronic science and technology: 2014; 07: 42-44.
[18] MacNamara C, Gabriele A, Amador C, Bakalis S. Dynamics of textile motion in a front - loading domestic washing machine. Chemical Engineering Science: 2012; 75: 14-27.
[19] Yun CS, Chung Hee Park. The effect of textile movement on washing performance in a front-loading washer III: Focus on the optimized movement algorithm. Textile Research Journal: 2016; 86 (6): 563-572.
[20] Yun CS, Yoonyung Cho, Chung Hee Park. Washing efficiency and fabric damage by beating and rubbing movements in comparison with a front-loading washer. Textile Research Journal: 2016; 0 (00) 1-7.
[21] KS C IEC 60456. Clothes washing machines for household use-methods for measuring the performance.
[22] GB/T 4288-2008. Electric washing machines for household and similar purposes.
[23] IEC PAS 62473. Clothes washing machines for household use-methods for measuring the mechanical action in household washing machines.
[24] GB/T 18318.1-2009: Determination of flexural properties of textiles first parts: inclined plane method.
[25] GB/T 3923.1-2013: Tensile properties Part 1: determination of breaking strength and breaking elongation (strip method).

Evaluation of Acoustic Characteristics of Woven Fabrics in Relation to Fiber and Weave

Hafsa Jamshaid[1,2], Rajesh Mishra [2*], Jiri Militky[2]

[1] *National Textile University, Faculty of Textile Engineering Faisalabad, Pakistan.*
[2] *Technical University of Liberec, Faculty of Textile Engineering, Liberec, Czech Republic*

Corresponding author's email: hrntu@hotmail.com

Abstract

The amount of research conducted on sound absorption properties of woven fabrics in contrast with nonwoven and spacer textile fabrics is extremely limited. This work deals with the investigation of the acoustic characteristics of woven fabrics in relation to fabric structural parameters and air flow resistivity. This work deals with the study of the acoustic characteristics of woven fabrics in relation to fiber and weave .In order to accomplish the goals of the research, sound absorption coefficient of woven fabric samples was determined via an impedance tube method. Based on results, the effect of fabric structural parameters on the sound absorption coefficient was established. In this work NRC of the samples were also obtained. Correlation between acoustic impedance and modulus was also calculated. The results of measurements were statistically analyzed.

Keywords: Sound Absorption Coefficient; Noise Reduction Coefficient; Basalt Hybrid Woven Structures; Acoustic Impedance

1. Introduction

Presently noise control is given much importance as noise pollution is ranked 3rd after air and water pollution in the world. The acoustic phenomenon and its consequences in environments, such as work place and residential homes, have gained vital importance. The acoustic characteristics of a room or an auditorium can be controlled by covering the walls, ceiling and floor with materials having suitable acoustic absorptive characteristics. Noise control and its principles play an important role in creating an acoustically pleasant environment. This can be achieved when the intensity of sound is brought down to a level that is not harmful to human ears. Achievement of a pleasant environment can be obtained by using various techniques that employ different materials. One such technique is by absorbing the sound. Researchers always seek new materials and arrangements to enhance the sound absorption efficiency. Fibrous, porous and other kinds of materials have been widely accepted as sound absorptive materials. These porous textile materials with air inside the pores act as absorbing or damping medium. Damping refers to the capacity of the material to dissipate energy; current, sound absorption materials commercially available for acoustic treatment consist of glass, mineral or synthetic fibres. A study on sound absorption characteristics of rock wool and glass fibre indicated that these fibres behave in a similar fashion [1]. However, when the issue of safety, insulation, environment protection and health are considered, they pose interference to human health and surroundings. Thus, environmentally friendly materials of high flame resistance and high insulation as well as good sound absorption properties are required. Public awareness and concern about the negative effects of pollution has driven the consumers towards the favour of environmentally friendly materials, less contaminating processes and recycled products. Therefore, it is important to increase research on acoustical materials based on renewable resources that can lead to viable alternatives against conventional materials for current and future application [2-5].

Prior to the present research, other researchers, [5-8] studied the acoustic properties of different materials and their relationship between the flow resistance and the absorption characteristics of fabrics .Normally nonwoven materials are preferred for noise reduction. Despite this fact, woven fabrics have less sound absorption properties, woven fabrics are preferred in end-uses, where nonwoven textiles cannot be used due to both technical and economic reasons. Shaoshi [9] studied sound absorption properties by combining multi-layers of non-woven fabric and single layer of woven fabric. It is found that when the single layer of woven fabric was put on the front of multi-layers of non-woven fabric, the best performance was achieved, and the single layer of fabric contributed about 20-40% to the sound absorption of this combination.

This research intends to study the acoustic properties of woven hybrid and non-hybrid fabrics made from Basalt(B), Polypropylene (PP), Polyester (PET) and jute (J) jute fabric ,for noise reduction in house hold appliances, automotive and architectural applications. It is foreseen that the result can provide scientific guidance for the development and the design of sound absorption composite structures.

2. Material and Methods

2.1 Materials

Raw materials involved in this study were woven hybrid and non-hybrid fabrics made from basalt, Polypropylene (PP), Polyester (PET) and jute (J). A total number of 27 samples were produced by keeping same thread density i.e 12ends/cm and 7 picks/cm.

2.2 Materials

All Structural parameters (Thickness, GSM) are measured as per standards.

2.2.1 Sound Absorption Measurement

In order to achieve the objectives of the research, sound absorption properties of hybrid and non-hybrid woven fabric samples was determined. It was measured by using two-microphone impedance tube according to ASTM E1050-08. It uses plane sound waves that strike the material straight and so the sound absorption coefficient (SAC) is determined. The device is used to determine the sound absorption coefficient, SAC (α) of circular samples with a diameter of 100 mm for the frequency range from 50-1600 Hz and 29mm for the determination of sound absorption coefficient, SAC (α) of circular samples for the frequency range of 500-6400Hz. 20 measurements were repeated for each sample. The variability was restricted to 95% confidence interval. For comparison among fabrics mid frequency range is used i.e. 250-2000 Hz. Typically, perforated materials only absorb the mid-frequency range unless special care is taken in designing the facing to be as acoustically transparent as possible.

In this work, the sound absorption properties of all fabrics samples were calculated using two terms, namely sound absorption coefficient (SAC) and noise reduction coefficient (NRC) .SAC is the measure of sound absorbed by the samples for each and every frequency and overall sound absorption of different materials can be compared through NRC. It is defined as the arithmetic average of SAC (α) at four different frequencies, namely 250 Hz, 500 Hz, 1000 Hz and 2000 Hz [1]. SAC varies from zero (0) to one (1).

$$NRC = \frac{\alpha_{250\text{Hz}} + \alpha_{500\text{Hz}} + \alpha_{1000\text{Hz}} + \alpha_{2000\text{Hz}}}{4} \quad (1)$$

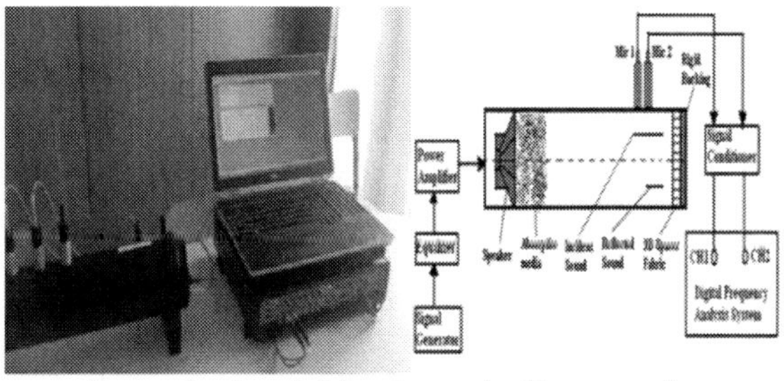

Fig.1(a) Impedance tube method and (b) schematic diagram of an experimental set-up.

2.2.2 Air Flow Resistance

In this work acoustical characteristics of hybrid woven fabrics including sound absorption coefficient and factors affecting sound absorption coefficient via the impedance tube method are determined. The factors that chiefly determine a materials sound absorption property are its fiber type, fiber size, material thickness, density, porosity and air flow resistance. Among these factors, the air flow resistance is the major contributing parameter in various materials. Flow resistance has been used as an important parameter in theoretical equations by many researchers. It is therefore important to consider the flow resistance of an acoustic sample, which is calculated by Equation 2. The air flow resistance can yield good absorption regardless of the thickness of the fabrics. Air Flow Resistivity, is calculated as follows:

$$r = \frac{\Delta p}{v\ h} \tag{2}$$

Where, r - air flow resistivity (Pa.s.m-2)
Δp - Pressure difference in Pascal,
v -linear air flow velocity in m/s
h- Thickness in meters

The influence of the structural characteristics on acoustical and sound resistance of the developed fabric samples are measured and interpreted.

2.2.3 Relationship between Tensile Moduli and Sound Propagation.

Porous materials are usually good sound absorbers but bad sound reflectors. Specific acoustic impedance serves as an index to explain how efficiently energy sound can be transferred from one medium to another. The specific impedance of a medium is “opposing the propagation of a plane sound wave. The SI unit of acoustic impedance is the pascal second per cubic meter (Pa•s/m^3).

Correlation between specific acoustic impedance and modulus was calculated by Equation 3

$$Z = \sqrt{E\rho} \tag{3}$$

Where E is moduli in N/m2 and p is density of the medium in kg/m3

3. Results & Discussion

From Fig.2, it can be observed that with the increase of frequency, the sound absorption coefficient (SAC) of all the samples increased. It is known that with the increase of frequencies sound energy becomes larger which accelerates the air vibration inside fabrics and creates the opportunities of friction between air and pore walls. This process causes sound energies to be converted into heat energies and gradually diminishes the sound effect. This is indicated as an increase in SAC.

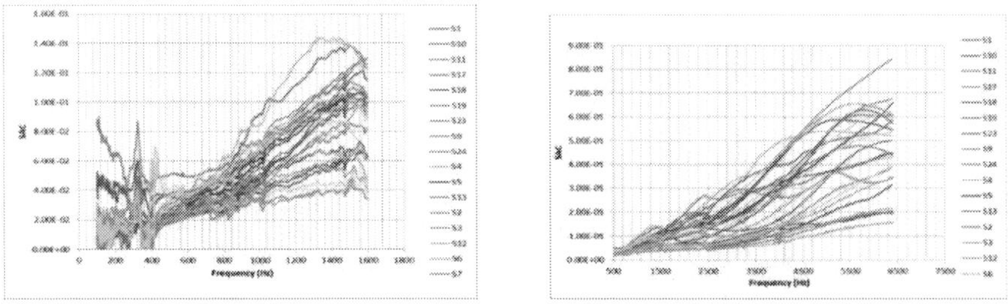

(a)At low frequency (b) At high frequency

Fig.2 Sound absorption coefficient of all samples.

It can be seen that fabrics with polyester fiber have better SAC. The reason for better SAC of polyester fabrics may be finer fibers which result in higher number of fibers per unit area of the material. This leads to higher total fiber surface area, and greater possibilities for a sound wave to interact with the fibers and ultimately dissipate inside the structure. 100% Jute and Basalt fabric show lower SAC values in both frequency ranges. The use of basalt fiber for hybrid structures improved SAC for all fabrics.

Porosity is one of the most important parameters which affect sound absorption. This effect can be explained in terms of the friction between the air inside the material (within pores, between fibers) and the solid walls of the material. Moreover when sound propagates within a porous material, it is reflected many times because there are several air-solid-air interfaces. Due to friction and to the multiple reflections within the pores, part of the energy associated with sound is transferred to the solid skeleton of the material, by making its particles vibrate, and therefore, it is dissipated.

As stated earlier, the viscous resistance of air in the porous material has an important influence on the sound absorption mechanism. Flow resistance has been used as an important parameter in theoretical estimations by many researchers. It is therefore important to consider the flow resistance of an acoustic sample.

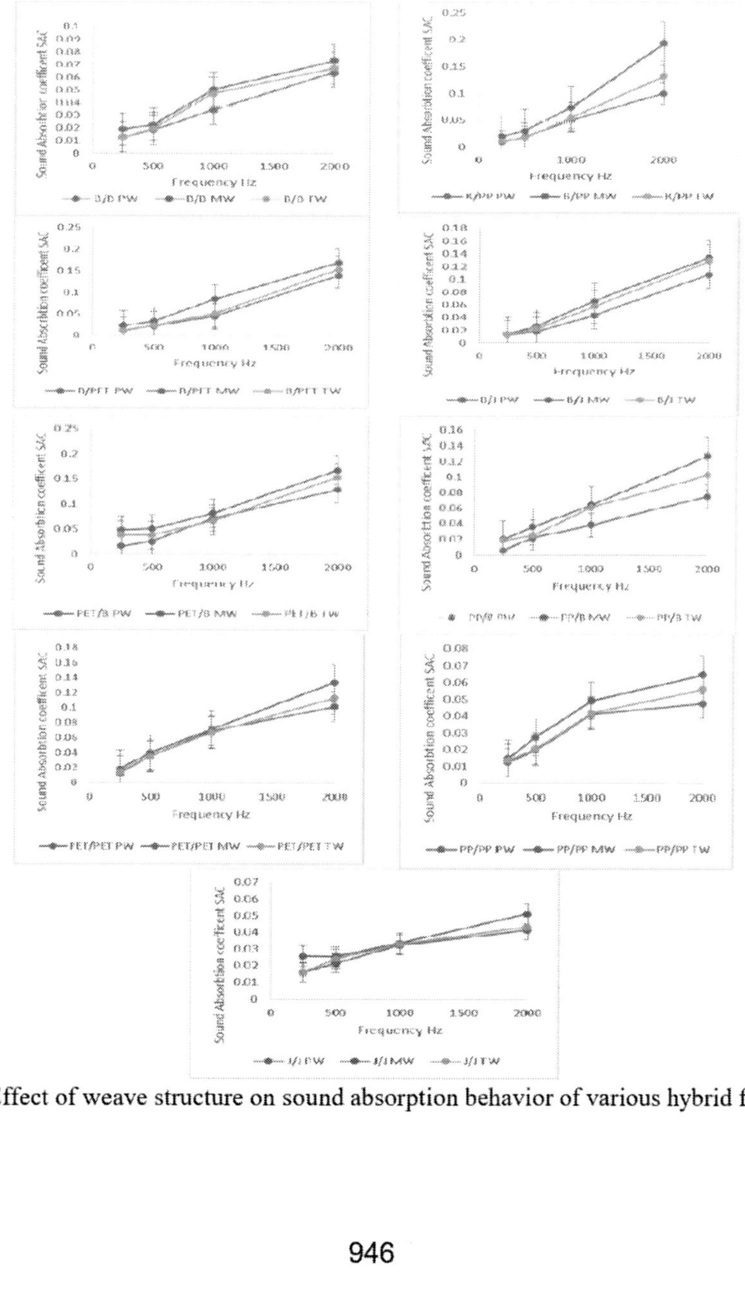

Fig.3 Effect of weave structure on sound absorption behavior of various hybrid fabrics.

Typically, perforated materials only absorb the mid-frequency range unless special care is taken in designing the facing to be as acoustically transparent as possible. Smaller denier yields more fibers per unit mass of the material; when the total fiber surface area increases, the possibility of frictional flow resistance of sound wave increases. Normal sound absorption coefficient was measured experimentally using the impedance tube for all samples having different grades of porosity, pore size and pore opening. Static flow resistance could also measure acoustical performance.

The static flow resistance determines that, as the pore size opening is increased, the static flow resistance decreases. Elastic materials and steel frames can transmit vibrations throughout the building and are not good for sound insulation. The maximum sound absorption for all samples occurs at a frequency range of 1000 -2000 Hz. Fig.3 depicts the effect of weave structure on sound absorption behavior of various hybrid fabric

A relationship between fabric porosity, airflow resistivity and NRC is shown in Fig.4.

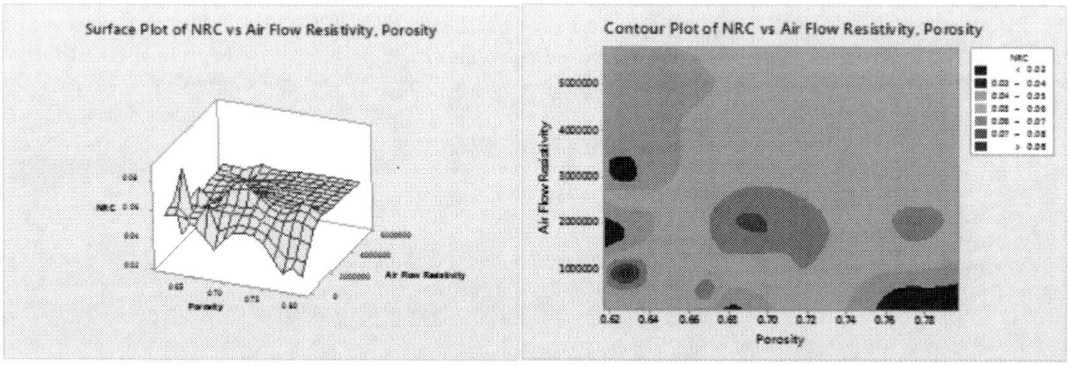

Fig.4 Relationship between fabric porosity, airflow resistivity and NRC.

This phenomenon can be justified by considering that, upon reaching the fabric surface depending on fabric characteristics, the incident sound wave is partially absorbed, transmitted or reflected. Yarn intersection points in the fabric act as frictional elements that resist sound wave propagation through the fabric. Fabric internal tortuosity causes sound wave amplitude to decrease. This in turn leads to conversion of sound energy into heat. Plain weave has maximum number of intersection points. The maximum and minimum NRC values were obtained for plain and non-plain (matt and twill) fabrics. This is due to different interlacement of yarn in these weave types. The plain weave has the greatest number of yarn inter-sections, which results in increases in weight and density of this fabric. Additionally, shorter free float length causes less air to be entrapped within the thread spaces in the fabric. Thus, the NAC of the fabric against sound waves is decreased. Therefore, absorption occurs due to energy loss as the sound wave passes through the fabric and the frictional resistances offered by the fibers and entrapped air in the fabric is overcomed. Accordingly, the sheer crimping of the yarns in the plain weave offers more resistance to sound waves. The length of free float of the yarns in twill and matt fabric is an additional contributing factor in absorption of large proportion of the incident sound wave by the fabric. The increase in yarn twist results in increases in yarn compactness. This in turn reduces both fabric cover and the amount of voids in the yarns, thus the observed reduction in fabric NRC/SAC.

It can be seen that fabrics with polyester fiber have better NRC. The reason for better NRC of polyester fabrics may be finer fibers which results in higher number of fibers per unit weight of the material. This leads to higher total fiber surface area, and greater possibilities for a sound wave to interact with the fibers and ultimately dissipate inside the structure. Finer fibers yields more fibers per unit mass of the material; when the total fiber surface area increases, the possibilities of sound wave frictional flow resistance increases. Additionally, PET yarn is bulkier yarn and also it has less twist. The increase in yarn twist results in increases in yarn compactness. This in turn reduces both fabric cover and the amount of voids in the yarns, thus the observed reduction in fabric NRC/SAC. 100% Basalt fabric shows lowest NRC values in both frequencies. By hybridization with basalt, the noise reduction coefficient increases as the combination of dissimilar yarns results in higher crimp of individual yarns and more porosity in the structure.

Matt weave structures have highest value of NRC due to more porosity. This phenomenon can be justified by considering that, upon reaching the fabric surface depending on fabric characteristics, the incident sound wave is partially absorbed, transmitted or reflected. Yarn intersection points in the fabric act as frictional elements that resist sound wave propagation through the fabric. Fabric internal tortuosity causes sound wave amplitude to decrease. This in turn leads to conversion of sound energy into heat. Matt weave has more number of pores as compared to plain weave. Thus, the NRC of the fabric against sound waves is increased. Therefore, absorption occurs due to energy loss as the sound wave passes through the fabric and the frictional resistances offered by the fibers and entrapped air in the fabric is overcome.

Specific acoustic impedance

As elastic modulus of material increases, impedance value increases i.e. material can be good sound reflector, but not good sound absorber. As it is obvious from one axis of graph, density has an indirect relation with acoustic impedance. As density increases acoustic impedance starts decreasing, but this effect is not pounced as compared to elastic moduli, which have significant effect on acoustic impedance .As elastic moduli increases, acoustic impedance increase. Dependence of specific acoustic impedance on elastic moduli and density of woven fabrics is shown in Fig.5.

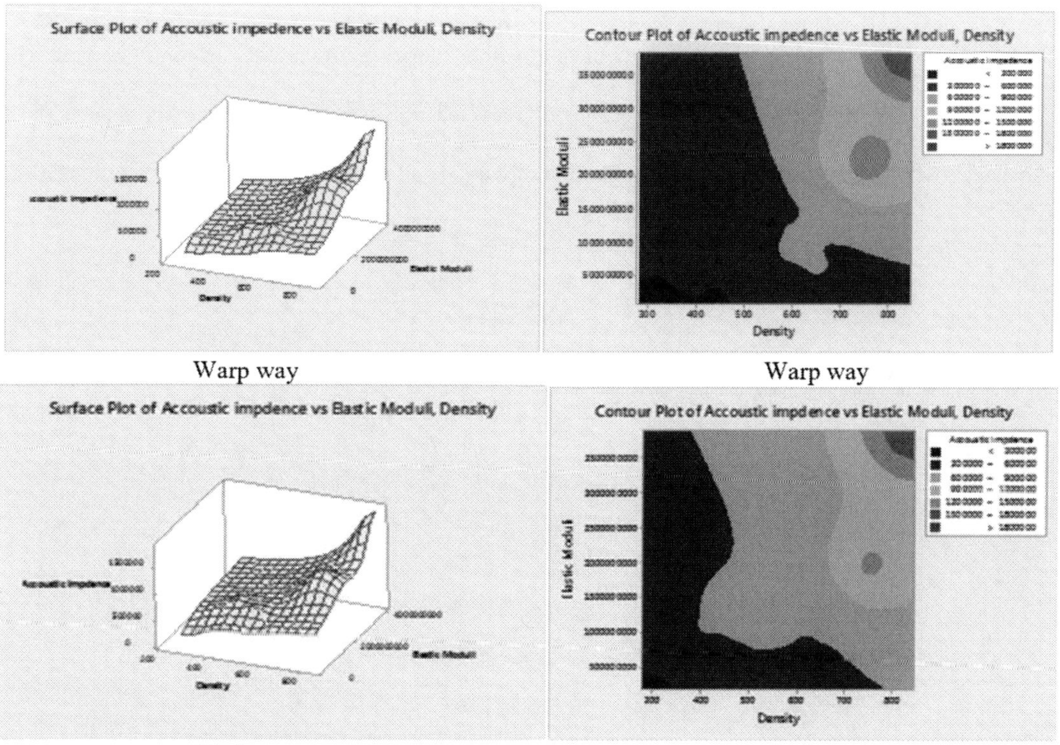

Fig.5 Dependence of specific acoustic impedance on elastic moduli and density of woven fabrics.

This effect can be explained in terms of the friction between the air inside the material (within pores, between fibers) and the solid walls of the material. Moreover when sound propagates within a porous material, it is reflected many times because there are several air-solid-air interfaces. Due to friction and to the multiple reflections within the pores, part of the energy associated with sound is transferred to the solid skeleton of the material, by making its particles vibrate, and therefore, it is dissipated.

4. Conclusion

It can be seen that fabrics with polyester fiber have better Noise Reduction Coefficient. The reason for better NRC of polyester fabrics is attributed finer fibers which results in higher number of fibers per unit weight of the material. This leads to higher total fiber surface area, and greater possibilities for a sound wave to interact with the fibers and ultimately dissipate inside the structure. By hybridization with basalt, the noise reduction coefficient increases as the combination of dissimilar yarns results in higher crimp of individual yarns and more porosity in the structure. Matt weave structures have highest value of NRC due to more porosity.

References

[1] Aso K, Kinoshita R. Absorption of sound wave by fabrics, Part 1: Absorption mechanism. TMSJ: 1963; 8: 32-39.

[2] Fatima Sm Mohanty AR. Acoustical and fire-retardant properties of jute composite materials, Appl. Acoust.: 2011; 72: 108-114.

[3] Nick A, Becker U, Thoma W. Improved Acoustic Behavior of Interior Parts of Renewable Resources in the Automotive Industry. J Polym. Environ: 2002; 10: 115-118.

[4] Asdrubali F. "Survey on the Acoustical Properties of New Sustainable Materials for Noise Control," Proceedings of Euronoise, Tampere, Finland: 2006.

[5] Aso K, Kinoshita R. Absorption of sound wave by fabrics, Part 3: Flow resistance. TMSJ: 1964; 10: 236-241.

[6] Aso K, Kinoshita R. Sound absorption characteristics of fiber assemblies. TMSJ: 1964; 10: 209-217.

[7] Na YJ, et al. Sound absorption coefficients of micro-fiber fabrics by reverberation room method. TRJ: 2007; 77: 330-335.

[8] Lee YE, Joo CW. Sound absorption properties of recycled polyester fibrous assembly absorbers. AUTEX: 2003; 3: 77-84.

[9] Shoshani YK. "Noise absorption by a combination of woven and nonwoven fabrics. JTI: 1991; 82: 500-503.

Relationship Between Sound Absorption and Transfer Properties of Struto Nonwoven Fabrics

Tao Yang[1], Xiao-Man Xiong [1], Rajesh Mishra[1], Jan Novák[2], Jiri Militky [1]

[1] *Department of Materials Engineering, Faculty of Textile Engineering, Technical University of Liberec, Czech Republic*
[2]*Department of Vehicles and Engines, Faculty of Mechanical Engineering, Technical University of Liberec, Liberec 46117, Czech Republic*

*Corresponding author's email: tao.yang@tul.cz

Abstract

This paper presents an investigation on sound absorption behavior, air permeability and thermal properties of Struto nonwoven fabrics and their relationships. Seven Struto nonwoven fabrics were selected for the examination of sound adsorption through obtaining average values of sound absorption coefficients ($\bar{\alpha}$) as well as air permeability and thermal insulation properties. It was also observed that sound absorption has an insignificant correlation with thermal conductivity while it strongly correlates with thermal resistance. The correlation coefficient of $\bar{\alpha}$ and thermal resistance was 0.94853, which is an indication that that $\bar{\alpha}$ is directly proportional to thermal resistance of lofty nonwovens. It was also observed that $\bar{\alpha}$ was inversely proportional to air permeability, with a correlation coefficient of 0.95. It was concluded that a higher thermal resistance and lower air permeability indicates a better sound absorption performance for a Struto nonwoven fabric.

Keywords: Struto Nonwoven; Sound Absorption; Thermal Conductivity; Thermal Resistance; Air Permeability

1. Introduction

Nonwoven fabrics have very high porosity, high specific surface area, economical price, light weight, good elasticity and a great flexibility in air permeability. Given their properties, nonwoven fabrics have a large number of fiber assemblies which is widely applied in the thermal insulation and sound adsorption field [1].

Thermal properties of nonwoven fabrics are extensively studied. Generally, thermal insulation properties of nonwoven fabric have a strong correlation with its dimensional and structural parameters [2]. Arambakam numerically investigated the thermal performance of fibrous insulation materials and concluded that heat conduction through fibrous structures can be increased by increasing the solid volume fraction, fiber diameter, and fibers' through-plane orientations [3]. Researchers summarized that for a fixed weight, thermal insulation increases with thickness [4]. Vallabh confirmed that fabric density is a significant factor in influencing the radiation component of effective thermal conductivity and that radiative thermal conductivity decreases with an increase in fabric density [5]. Soltani and Zarrebini reported that the acoustic characteristics of woven fabrics are related to fabric structural parameters such as weave type, yarn linear density, yarn twist, and fabric thickness of woven fabric [6]. Moreover, they stated that air permeability of woven fabrics, which is strongly dependent on fabric cover factor, can be used as a criterion of sound absorption behavior. Yang and Yu conducted experimental study and found that nonwoven fabrics with the highest value of air permeability exhibit inferior acoustic absorbency [7]. Kucuk M and Korkmaz Y measured sound absorption properties of eight different nonwoven composites including different fiber types mixed with different ratios, they stated that the increase in thickness and the decrease in air permeability results in an increase in sound absorption properties of the material [8]. Struto nonwovens exhibit excellent compression recovery, softness and very good sound absorption performance since Struto nonwovens have highloft structure where all the fibers are orientated in the vertical plane [9-11]. Tascan and Vaughn investigated the acoustical insulation of Struto nonwoven fabrics made from different polyester fibers, and stated that fabrics made from 3 denier fibers have better sound insulation than those made from 15 denier fibers. It was also found that Struto nonwoven fabrics made from 4DG and trilobal fibers showed better sound insulation than those made from round fibers [12-13]. Klara Kalinova reported that Struto nonwovens (perpendicularly laid fiber web) have better sound absorption performance than longitudinally laid fiber web [14].

Since the transmission behavior and acoustic property of nonwoven both strongly depend on its structure parameters, for a specified nonwoven fabric there should be some relation between these performances. However, there is very limited research devoted to understanding this relationship. In this paper, seven Struto nonwovens were selected to measure sound absorption properties, air permeability and thermal performance. The relationship between sound absorption and thermal properties were studied as well.

2. Methodology

2.1 Materials

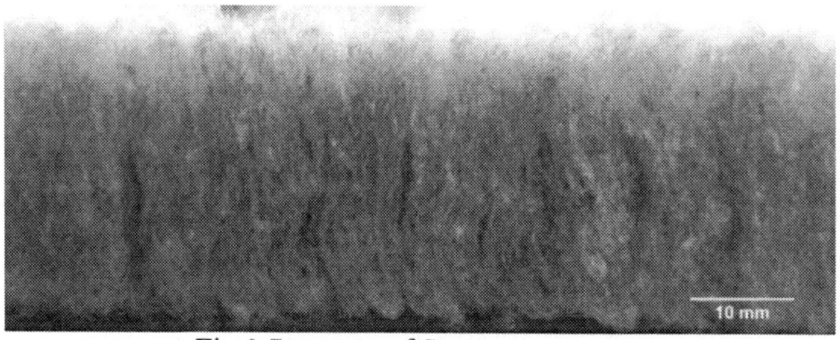

Fig.1 Structure of Struto nonwoven.

Table.1 Fiber specifications

Fiber Code	Types of PET	Diameter (μm)	Fineness (dtex)	Staple Length (mm)	Ratio of Core and Sheath
i	Hollow PET	28.64	4.92	70.00	----
ii	PET	15.25	1.76	50.00	----
iii	Bicomponent PET	22.02	3.24	50.00	3:1
iv	PET	26.91	6.70	57.00	----
v	Bicomponent PET	14.58	2.20	38.00	3:1

Table.2 Characteristics of Struto nonwovens

Samples code	Content	GSM (g/m^2)	Thickness (mm)	Porosity (%)	Bulk density (kg/m^3)
A	30%-i 45%-ii 25%-iii	507.51	24.09	97.57	20.07
B		478.32	28.36	98.06	16.87
C		465.22	27.48	98.05	16.93
D	70%- iv 30%-v	335.68	20.82	98.67	16.12
E		317.51	19.85	98.69	15.99
F		198.64	20.12	99.19	9.87
G		259.28	20.66	98.97	12.55

Seven Struto nonwoven fabrics made by Technical University of Liberec were selected in this study. Struto nonwoven is developed on a Struto line which consists of a carding machine, a Struto vertical lapper and a through-air thermobonding chamber [13]. The carded web is fed onto conveyor belt and a reciprocating forming comb pulls the carded web towards the hold back roller to form a fold. The fold is pulled off the comb by a system of needles placed on a reciprocating compressing bar and pushed to the fiber layer which is created and moved between the conveyor belt and a wire grid. The fiber layer is bonded by melting bonding fibers present in the fiber blend when it passes through thermo-bonding chamber [15]. The Struto nonwovens samples were made by different types of polyester fibers, the fiber specifications are shown in Table 1. The characteristics of the specimens are shown in Table 2. In Struto nonwoven structure, fibers are orientated in the vertical plane as shown in Fig.1.

2.2 Methods

2.2.1 Evaluation of Sound Absorption Performance

Sound absorption of Struto nonwovens were measured according to ASTM E 1050: Standard Test Method for Impedance and Absorption of Acoustical Properties Using a Tube, Two Microphones and a Digital Frequency Analysis System. A Brüel and Kjær measuring instrument containing Type 4206 Impedance Tube, PULSE Analyzer Type 3560, and Type 7758 Material Test Software was used for testing within the frequency range 50Hz–6.4 kHz. A large tube (100 mm in diameter) and a small tube (29 mm in diameter) were set up for measuring the sound absorption in low-frequency range from 50-1600Hz and high-frequency range 500-6400Hz respectively. The curves from both measurements were merged. The testing principle is illustrated in Fig.2. Sound absorption of five different samples from each material was measured.

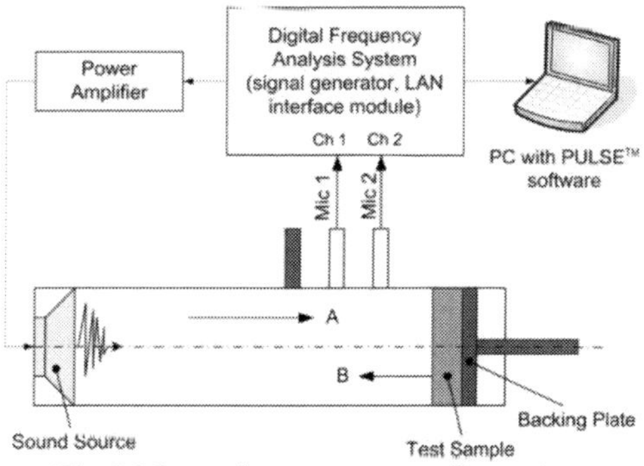

Fig.2 Measuring system configuration.

2.2.2 Measurement of Air Permeability

Air permeability measures the ability of a porous medium to transmit fluids. It depends on the porous geometrical structure. The air permeability of Struto nonwovens was measured using FX3300 Textech Air Permeability Tester. The fabric sample is fixed as an obstacle in a flow of air by the clamping holder. A pressure difference Δp between both sides of the fabric sample develops as a consequence of hydraulic losses. The pressure difference is recorded by using of the manometer. The measured value is a speed of air in meter per second or a volume rate of the flow in liter per hour.

2.2.3 Measurement of Thermal Properties

Alambeta instrument was used to measure thermal conductivity and thermal resistance, according to EN 31092 standard. Each specimen was tested five times and the results were averaged.

Thermal conductivity

Thermal conductivity, λ, measures the rate at which heat is transferred through unit area of the fabric across unit thickness under a specified temperature gradient and thus is defined by the relation [16]

$$\lambda(W \cdot m^{-1} \cdot K^{-1}) = \frac{Q}{F\tau \frac{\Delta T}{h}} \quad (1)$$

where Q is the amount of conducted heat, F is the area through which heat is conducted, τ is the time of heat conduction, ΔT is the gradient of the temperature and h is the fabric thickness.

Thermal resistance

Thermal resistance expresses the ability of material to prevent heat flow through the thickness over unit surface area. Fabric thickness and thermal conductivity are important factors governing thermal insulation of textiles. Usually, the higher the thermal resistance, the lower the heat loss. The thermal resistance, R, is connected with the thermal conductivity, λ, and the fabric thickness, h, as follows

$$R(m^2 \cdot K \cdot W^{-1}) = \frac{h}{\lambda} \tag{2}$$

3. Results and Discussion

3.1 Sound Absorption Coefficient

Sound absorption coefficient indicates how much of the sound is absorbed by the testing material. The absorption coefficient can be expressed as

$$\alpha = \frac{I_a}{I_i} \tag{3}$$

where I_a is sound intensity absorbed in W/m^2 and I_i is incident sound intensity in W/m^2.

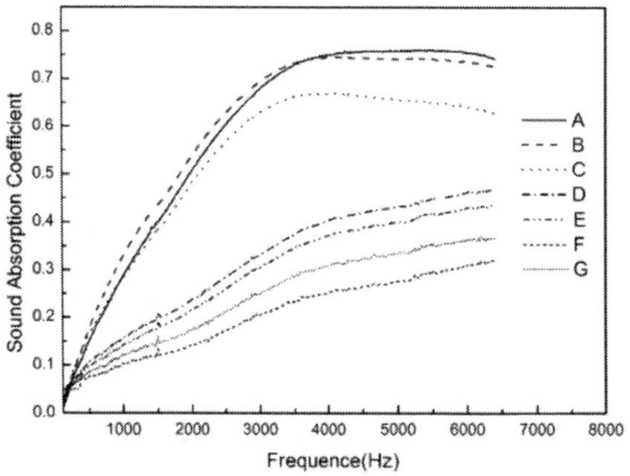

Fig.3 Sound absorption coefficient of Struto nonwovens

The normal incidence sound absorption coefficient (α) of Struto nonwovens were determined as a function of the sound frequency (f), as shown in Fig.3. It is observed that sound absorption coefficient of the test samples lies in the range of 0.017-0.76. Apparently, the value of absorption coefficient for samples A, B and C sharply increases at frequency bands 50 Hz-3500 Hz and the maximum value of absorption coefficient occurs at frequency bands 4000 Hz-5400 Hz. However, samples D, E, F and G show lower absorption coefficient value in comparison to samples A, B and C, and the value of absorption coefficient increases with the increasing of frequency at the whole measurable frequency bands (50 Hz-6400 Hz). Results indicate that Struto nonwoven exhibits much better sound absorption ability at frequency bands 3000 Hz-6400 Hz.

Average value of sound absorption coefficient ($\overline{\alpha}$) is the average of the absorption coefficients for a specific frequency bands. The higher the $\overline{\alpha}$, the better the material absorbs sound. The $\overline{\alpha}$ of Struto nonwovens is calculated using the following equation:

$$\bar{\alpha}=\frac{\int_{F_1}^{F_2}\alpha(f)df}{F_2\text{-}F_1} \tag{4}$$

where f is sound frequency, F_1 (50 Hz) is lower bound of sound frequency in testing and F_2 (6400 Hz) is upper bound of sound frequency in measuremen.

The computed $\bar{\alpha}$ values between 50 Hz and 6400 Hz for Struto nonwovens are shown in Table 3. It is observed that the $\bar{\alpha}$ values of Struto nonwovens are ranging from 0.201 to 0.588.

Table.3 Average value of sound absorption coefficients

Samples code	$\bar{\alpha}$	CV
A	0.580±0.008	0.015
B	0.588±0.007	0.012
C	0.524±0.006	0.011
D	0.315±0.028	0.089
E	0.291±0.017	0.060
F	0.201±0.007	0.035
G	0.243±0.014	0.056

3.2 Air Permeability

In order to examine the effect of fiber deformation on the measured air permeability of Struto nonwovens, different pressure gradients 50Pa, 100Pa, 150Pa and 200Pa were chosen to carry out air permeability testing. The measured air permeability is listed in Table 4. Each value is the average of five test results.

Table.4 Measured air permeability of samples

Samples Code	Air Permeability (mm/s)			
	ΔP=50Pa	ΔP=100Pa	ΔP=150Pa	ΔP=200Pa
A	320.1±18.418	597.6±33.781	848.7±47.877	1070.5±63.177
B	320.1±58.586	652.1±74.994	923.3±102.373	1177.6±136.177
C	370.6±15.407	687.3±27.713	968.3±36.003	1227±46.440
D	908.8±33.417	1598±58.052	2166±74.027	2712±95.237
E	929.8±59.667	1628±96.281	2220±130.192	2760±159.844
F	1548±42.071	2608±95.237	3508±167.093	4336±308.432
G	1234±30.496	2104±42.779	2824±58.992	3468±40.866

Note: Samples were held by clamping arm under the load 50N

3.3 Thermal Properties

Thermal conductivity and thermal resistance of Struto nonwovens are presented in Table 5. For nonwovens with approximately the same thickness (samples D, E, F and G), thermal resistance increases with the increasing of fabric areal density and thermal conductivity shows an adverse trend. This is because increase in GSM causes increase in fibre to fibre contact and packing density as well as tortuosity, so less heat flows through the channels in nonwoven, and thermal resistance therefore increases correspondingly. Moreover, for samples A, B and C, since they have slightly different GSM and varying thickness, the increase of thermal resistance with the increase in sample thickness indicated that fabric thickness plays a major role in deciding thermal resistance of Struto nonwovens.

Table.5 Thermal properties of Struto nonwovens.

Sample codes	Thermal conductivity			Thermal resistance		
	Mean value 10^{-3} W·m^{-1}·K^{-1}	SD	CV	Mean value 10^{-3} K·m^{2}·W^{-1}	SD	CV
A	55.84	1.86	0.03	431.80	15.48	0.04
B	61.70	3.28	0.05	460.60	22.78	0.05
C	60.04	1.95	0.03	458.20	18.71	0.04
D	63.30	1.35	0.02	329.20	6.61	0.02
E	63.42	0.59	0.01	313.00	13.17	0.04
F	71.10	0.70	0.01	283.00	6.36	0.02
G	67.00	1.49	0.02	308.40	16.04	0.05

3.4 Relation Between Sound Absorption Performance and Air Permeability

Estimated correlation between air permeability and sound absorption is presented in Fig.4. It is observed that $\bar{\alpha}$ has significant correlation with air permeability, the correlation coefficient is about 0.95, indicating the existence of an inverse relation between air permeability and sound absorption of Struto nonwovens. That means Struto nonwoven with lower air permeability usually exhibits better sound absorption performance. This may provide a new method to evaluate the sound absorption property of Struto nonwovens by means of air permeability testing.

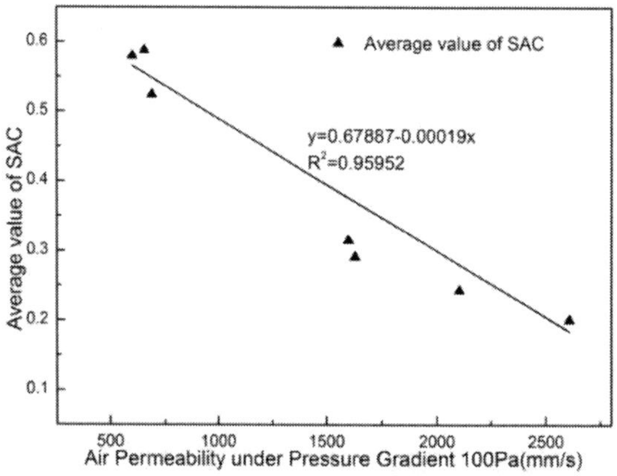

Fig.4 Estimated correlation between air permeability and sound absorption.

3.5 The Relationship Between Acoustic and Thermal Properties

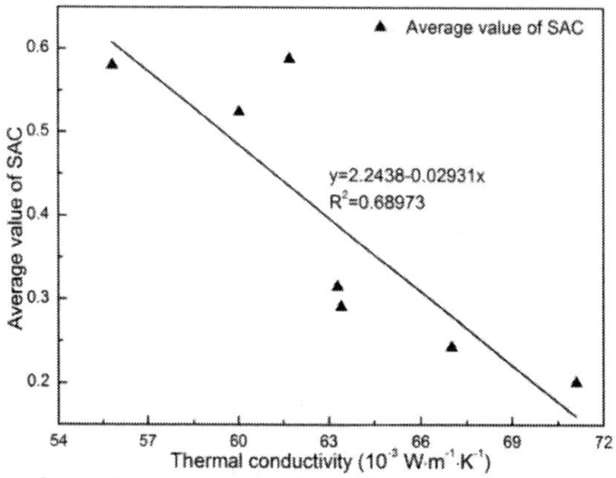

Fig.5 Estimation of correlation between thermal conductivity and sound absorption.

Fig.5 illustrates the estimation of correlation between thermal conductivity of Struto nonwovens and sound absorption. It is observed that $\bar{\alpha}$ has insignificant correlation with thermal conductivity, the correlation coefficient is 0.68973. The reason could be that thermal conductivity is not so strongly dependent on fabric thickness while fabric thickness is a determining factor for sound absorption.

Estimation of correlation between thermal resistance and sound absorption is presented in Fig.6. The correlation coefficient between $\bar{\alpha}$ and thermal resistance is 0.94853, indicating that the sound absorption performance of Struto nonwovens has a very strong correlation with thermal resistance. It can be concluded that $\bar{\alpha}$ are directly proportional to thermal resistance of Struto nonwovens. That means for different Struto nonwovens, a higher thermal resistance suggests a better sound absorption performance.

The relationship between sound absorption performance and thermal resistance can be explained by the strong dependence of sound absorption and thermal insulation on pore characteristics of nonwoven fabrics. The thermal resistance of porous materials is directly proportional to fabric porosity. Moreover, the sound absorption performance of porous materials depends on their specific airflow resistance, and this specific airflow resistance is also proportional to porosity. Generally, for highly porous Struto nonwovens, when the pressure wave penetrates into the material through the open pores of the surface and gets scattered by a high number of internal, these reflections transfer energy to the solid structure through frictional losses and efficiently absorb sound. As the porosity decreases, less energy is transferred into the solid structure and more is reflected from the surface. Thus, the sound absorption performance of Struto nonwovens is indirectly correlated to their thermal resistance.

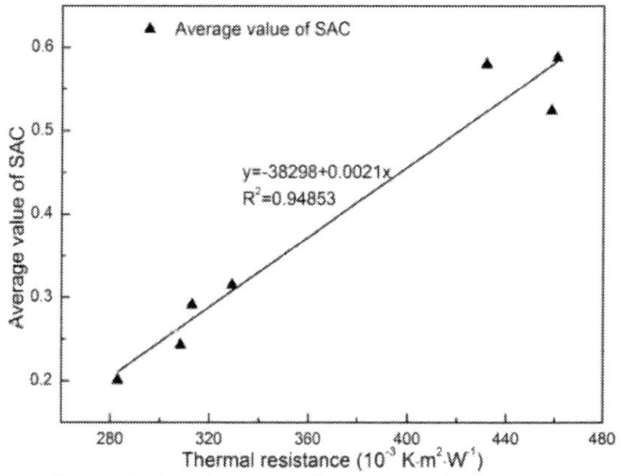

Fig.6 Estimation of correlation between thermal resistance and sound absorption.

4. Conclusion

Seven Struto nonwovens were selected to investigate their acoustic property as well as air permeability and thermal properties. Results showed that Struto nonwoven is ineffective for sound absorption at frequency lower than 3000 Hz while it exhibits much better sound absorption ability at frequency bands 3000 Hz-6400 Hz. Results indicated that there exists an inverse relation between air permeability and sound absorption of Struto nonwovens. It can be stated that a lower air permeability suggests a better sound absorption performance for a Struto nonwoven fabric. It is also observed that sound absorption has insignificant correlation with thermal conductivity while it is strongly correlated with thermal resistance. The correlation coefficient $\bar{\alpha}$ and thermal resistance is 0.94853, which indicates that $\bar{\alpha}$ is directly proportional to thermal resistance of Struto nonwovens. Thus, a higher thermal resistance suggests a better sound absorption performance for a Struto nonwoven fabric.

Acknowledge

This work was supported by the research project of Student Grant Competition of Technical university of Liberec no. 21197 granted by Ministry of Education Youth and Sports of Czech Republic.

References

[1] Thilagavathi G, Pradeep E, Kannaian T, Sasikala L. Development of Natural Fiber Nonwovens for Application as Car Interiors for Noise Control. J IND TEXT: 2010; 39: 267-278.

[2] Xiong X, Yang T, Huang J, et al. Heat transfer through thermal insulation materials Part I- Nonwoven fabrics. Recent developments in fibrous material science: 2015; 2: 85-104.

[3] Arambakam R, Vahedi Tafreshi H, Pourdeyhimi B. A simple simulation method for designing fibrous insulation materials. Mater Design: 2013; 44: 99-106.

[4] Behera BK, Mishra R. Artificial neural network-based prediction of aesthetic and functional properties of worsted suiting fabrics. Int J Cloth Sci Tech: 2007; 19: 259-276.

[5] Vallabh R, Banks-Lee P, Mohammadi M. Determination of radiative thermal conductivity in needle punched nonwovens. J Eng Fiber Fabr: 2008; 3: 46-52.

[6] Soltani P, Zarrebini M. Acoustic performance of woven fabrics in relation to structural parameters and air permeability. J Text I: 2013; 104 (9): 1011-1016.

[7] Yang S, Yu W. Air permeability and acoustic absorbing behaviour of nonwovens. J Fiber Bioeng Infor: 2011; 3: 203-207.

[8] Kucuk M, Korkmaz Y. The effect of physical parameters on sound absorption properties of natural fiber mixed nonwoven composites. Text Res J: 2012; 82 (20): 2043-2053.

[9] Vasile S, Van-Langenhove L. Automotive industry a high potential market for nonwovens sound insulation. J Text Appl Tech Mgmt: 2004; 3: 1-5.

[10] Parikh DV, Calamari TA, Goynes WR, Chen Y, Jirsak O. Compressibility of cotton blend perpendicular-laid nonwovens. Text Res J: 2004; 74: 7-12.

[11] Jirsak O, Burian T, Sasková P. Improvements in Compressional Properties of Highlofts. Fibred Text East Eur: 2003; 11 (3): 80-83.

[12] Tascan M, Vaughn EA. Effects of fiber denier, fiber cross-sectional shape and fabric density on acoustical behavior of vertically lapped nonwoven fabrics. J Eng Fiber Fabr: 2008; 3 (2): 32-38.

[13] Tascan M, Vaughn EA. Effects of total surface area and fabric density on the acoustical behavior of needle punched nonwoven fabrics. Text Res J: 2008; 78 (4): 289-296.

[14] Kalinova K. Theoretical assessment of sound absorption coefficient for anisotropic nonwovens. QScience Connect: 2013; 3: 3-10.

[15] Jirsak O, Wadsworth L. Nonwoven Textiles. Durham, NC: Carolina Academic Pr: 1998; 57-58.

[16] Gun AD. Dimensional, physical and thermal comfort properties of plain knitted fabrics made from modal viscose yarns having microfibers and conventional fibers. Fiber Polym: 2011; 12: 258-267.

Compression Behavior and Energy Absorption of 3d Weft Knitted Spacer Fabrics

Veerakumar Arumugam, Rajesh Mishra, Jiri Militky, Blanka Tomkova, Dana Kremenakova, Mohanapriya Venkataraman

Dept. of Materials Engineering, Faculty of Textile. Technical University of Liberec, Liberec, Czech Republic 461 17

*Corresponding author's email: veerakumar27@gmail.com

Abstract

Spacer fabric is a three-dimensional network which has two outer layer structure connected with spacer yarn. Spacer yarn acts as a linear spring which offers more energy absorption during compression and excellent recovery at the time of deceleration. This paper studies the influence of different structural characteristics on the compressive behavior and energy absorption capability of a weft knitted spacer fabric. The potential compression mechanism of the fabric was identified with support of a compression stress-strain curve, work done and efficiency at different compression stages. Advance statistical evaluation and two-way analysis of variance is used to analyze the significance of various factors such as thickness, spacer yarn diameter and surface structures on energy absorption at maximum compression load and deformation.

Keywords: Weft Knit Spacer Fabrics; Compression Stress; Compressive Energy Absorption; Efficiency

1. Introduction

The properties of spacer fabrics such as 3D fiber disposition permits the possibility for it to be used as different materials and for a single step production system. This enables them to be used in different application areas. The major application areas are acoustics and automotive, aerospace, civil engineering, medical field, geotextiles, protective textiles, sportswear and composites [1]. Spacer fabrics are a class of material with unique properties and applications. They are lightweight and designed to undergo very large deformations. They are typically used in packing and to absorb energy. Due to complex micro structure and three dimensional networks with low density, the spacer fabrics undergo large deformation and its compression characteristic is also better than conventional textile structures. Compression resilience is an important attribute of spacer fabrics, which is related to the sensation of mechanical comfort. Modern consumers consider compression as one of the most important attributes to comfort sensation such as the hand of clothing material. Fabric thickness and compressibility have a linear relationship with thermal conductivity. Compression characteristic of knitted fabrics has been studied by various researchers [2, 3]. Postle (1974) indicated that bulk density or compression property of knitted structures is related to the effective diameter of the yarn inside of the fabric and also fabric thickness [4]. Xu-hong et. al., analysed the stress-strain behaviour of warp knitted spacer fabrics when compressed [5]. Mecit Armakan et. al., investigated the compression characteristic of warp knitted spacer fabrics on the basis of spacer yarns in their structure [6]. They noted that, the material, pattern and thread of spacer yarns have a significant effect on the compression characteristic of the fabrics. It was also observed that the location angle and the amount of the spacer yarns influence the compression behaviour of fabrics.

The lack of comprehensive studies on the compressive behaviour of knitted 3D spacer fabrics are sound basis for this research. In the present study, the paper reports the compression behaviour of knitted spacer fabrics specially developed for car seats and back support applications. With an attempt to increase the energy absorption during compression and efficiency of the cushioning material, various structural parameters such as structure, thickness and spacer yarn liner density are varied during development of spacer fabrics. At the end, the effect of all these factors on compressibility and energy absorption of different samples were carefully compared and analyzed using advance techniques. It is expected that a clear picture for tailoring a knitted spacer fabric with promising cushioning property for automobile upholsteries

2. Materials and Methods

Six different types of spacer fabrics were developed using computerized Mayer & Cie, OVJA 1.6 E 3 WT knitting machine. These fabric samples were classified into two groups for convenient analysis of results, the first group has been developed using Polyester/Polypropylene blend with three different proportions and second group with Polyester/Polypropylene/Lycra blend having another 3 different compositions As a spacer yarn, three different types of 88 dtex Polyester monofilament yarn and Polyester multifilament yarns (167 dtex and 14.5 tex) were used [9]. 14.5 tex Polypropylene yarn was used on both the surfaces in group 1 samples. In group 2 samples polypropylene (14.5 tex), without and with lycra (44 dtex) were used for the top and bottom surface of the spacer fabrics (Table 1). The loop length of the fabric without lycra (WES 1) was 2.46 mm and samples (WES 2 and WES 3) were 2.78. The loop length weft knit spacer fabrics with Lycra on the surface (WES 4) was 1. 28 mm and for samples (WES 5 and WES 6) were 1.52 mm.

Table.1 Fabric particulars.

Fabric sample No.	Fabric layers	Technical face	Spacer yarn	Technical back	Fiber composition (%)
Group 1 - Without Lycra					
WES1	Type of yarns and linear density	Polypropylene (PP) - 14.5 tex	Polyester monofilament (PES monofil - 88 dtex	Polypropylene (PP) -14.5 tex	58% PP, 42% PES monofilament
WES2		Polypropylene (PP)- 14.5 tex	Polyester (PES) - 14.5 tex	Polypropylene (PP) -14.5 tex	45% PP, 55% PES
WES3		Polypropylene (PP) - 14.5 tex	Polyester (PES) - 167 dtex	Polypropylene (PP) -14.5 tex	41%PP, 59% PES
Group 2- With Lycra					
WES4	Type of yarns and linear density	Polypropylene (PP)- 14.5 tex Lycra - 44dtex	Polyester monofilament (PES monofil) - 88 dtex	Polypropylene (PP) -14.5 tex	55%PP, 39%PES monofilament, 6% Lycra
WES5		Polypropylene (PP)- 14.5 tex Lycra - 44dtex	Polyester (PES) - 14.5 tex	Polypropylene (PP) -14.5 tex	42% PP, 52% PES, 6% Lycra
WES6		Polypropylene (PP)- 14.5 tex and Lycra - 44dtex	Polyester (PES) - 167 dtex	Polypropylene (PP) -14.5 tex	39% PP, 55% PES, 6% Lycra

The fabric characteristics such as areal density, stitch density, structure, thickness, are presented in Table 2.

2.1 Compression Tests

All the compression tests have been carried out on a universal testing machine (TIRA) fitted with 5kN load cell. The speed of compression had been chosen at 12mm/min in accordance to the ASTM d 575 (Test methods for rubber properties). The test performed is a compression one and the machine is equipped by 2 strictly parallel plates with a diameter of 150mm and a smooth surface and the samples were cut with dimensions of 100 mm x 100 mm. All the spacer fabric specimens are compressed up to 80 % of the initial thickness in an atmospheric condition of 20°C and 65% relative humidity. Five tests were carried out for each sample under each testing condition and the average compression stress-strain cure are presented throughout this paper.

Table.2 Characteristics of spacer fabrics.

Weft Spacer Samples	Areal Density (g.m^{-2})				Thickness (mm)				Density (kg.m^{-3})	Stitch Density (Stitches/cm^{2})			
	Mean	ME	LL	UL	Mean	ME	LL	UL		Mean	ME	LL	UL
WES 1	493	0.16	492.84	493.16	4.4	0.88	3.52	5.28	112	200	0.1	199.9	200.1
WES 2	443	0.12	442.88	443.12	2.62	1.1	1.52	3.72	169.1	150	0.04	149.96	150.04
WES 3	477	0.2	476.8	477.2	2.74	0.61	2.13	3.35	174.1	150	0.12	149.88	150.12
WES 4	632	0.1	631.9	632.1	4.4	0.55	3.85	4.95	144.8	350	0.06	349.94	350.06
WES 5	657	0.12	656.88	657.12	3.5	0.86	2.64	4.36	187.7	280	0.1	279.9	280.1
WES 6	695	0.22	694.78	695.22	3.4	0.45	2.95	3.85	205.4	280	0.1	279.9	280.1

*ME-Margin of Errors; LL - Lower Limit; UP- Upper Limit

Overall compressive stress-strain trend of the spacer fabric samples are presented in the Fig.1. Normally the compressive behaviours of spacer fabrics are classified in to four stages with respect to changes in the slope. The four stages are initial (1), elastic (2), plateau (3) and densification (4). In the first stage, the surface layer of spacer fabric undergoes compression, the lower slope is observed for loose structure and slope increases with increase in stitch density. Here the spacer yarn has low contribution in constraint during initial compression. Further compression (2nd stage) leads to rapid increase in stress, it might be due to jamming of surface yarns allow monofilament to buckle at larger extend [6]. In spacer fabrics, third stage is quite complex behavior to analyze because the compressive stress and strain have been affected by buckling, shearing and inter-contacting of spacer yarns [7]. It is also noticed that faster increase in stress in 4th stage because of fabric achieves high density.

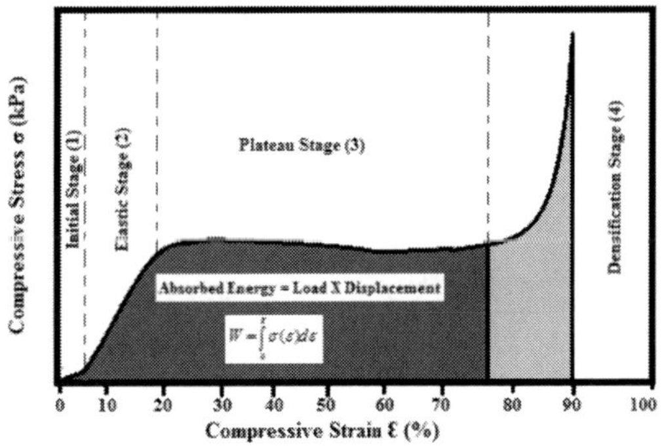

Fig.1 Compressive behavior of 3D spacer fabrics.

2.2 Energy Absorption During Compression

It is necessary to evaluate and analyze the spacer fabrics energy absorbing ability during compression. it would be more useful to get a better understanding on the cushioning behavior of the spacer fabric. The compression curves reveal long deformation plateaus, suggesting that all spacer fabrics samples may potentially be good energy-absorbing materials [8]. The area under the load-displacement curve represents the total energy absorbed and it can be calculated by multiplying the area under the stress-strain curve by the volume of the sample. The energy absorption capacity per unit volume, W, can be calculated by integrating the compression stress-strain curve, as given by Eq.1:

$$W = \int_0^{\varepsilon} \sigma(\varepsilon) d\varepsilon \tag{1}$$

Where, σ is the compression stress, $\mathcal{E}$ is the compression strains where $\mathcal{E}$ is the strain at the end/beginning of densification stage. In order to better understand the energy-absorption capacity of a cushioning material, the energy-absorption efficiency E can be used to analyze its energy-absorption process. The efficiency E is defined as the ratio of the energy absorbed by a real cushioning material compressed to a given strain and energy absorbed by an ideal cushioning material that transmits a constant stress of the same value at the same given strain. It is useful to plot the efficiency as a function of the stress to obtain the indication for optimum usage. The efficiency E is expressed by Eq.2)

$$E = \frac{Ah\int_{0}^{\varepsilon} \sigma(\varepsilon)d\varepsilon}{Ah\sigma} \tag{2}$$

Where A - area, h - thickness, σ - stress at the strain $\mathcal{E}$. The energy-absorption and efficiency of all the spacer fabrics are compared and analyzed to find the suitable material for automobile seat applications.

3. Results and Discussions

3.1 Effect of Fabric Characteristics on Compression Behavior of Weft Knit Spacer Fabrics

In this section, the effect of thickness, types of spacer yarn and surface characteristics on compressive behavior of weft knitted spacer fabrics have been carefully evaluated and analyzed. Six weft knit spacer fabrics (WES 1 -WES 6) produced with same outer layer structure but with different stitch density, thickness and types of spacer yarn. As shown in Figs.2a &b, the compressive stress-strain curve reveals that the compressive resistance of spacer fabric without lycra made up of monofilament yarn is low in linear and elastic stage, but sudden increase in compressive stress observed in plateau and densification stage. From Fig.2b, spacer fabrics with lycra made up of monofilament spacer yarn constantly offers high compression resistance in all four stages.

Among the fabrics made up of multifilament spacer yarn, the compressive stress - strain curve shows that, compressive resistance has indirectly proportional to thickness of the spacer fabrics. It is also observed that the denser fabrics require higher compressive stress to undergo same compressive deformation than the fabric with low density. It is also observed that the thicker fabric has ability to undergo larger deformation under low loading condition. The thickness of the fabric should be selected according to the amount of the energy to be absorbed and the allowed stress level.

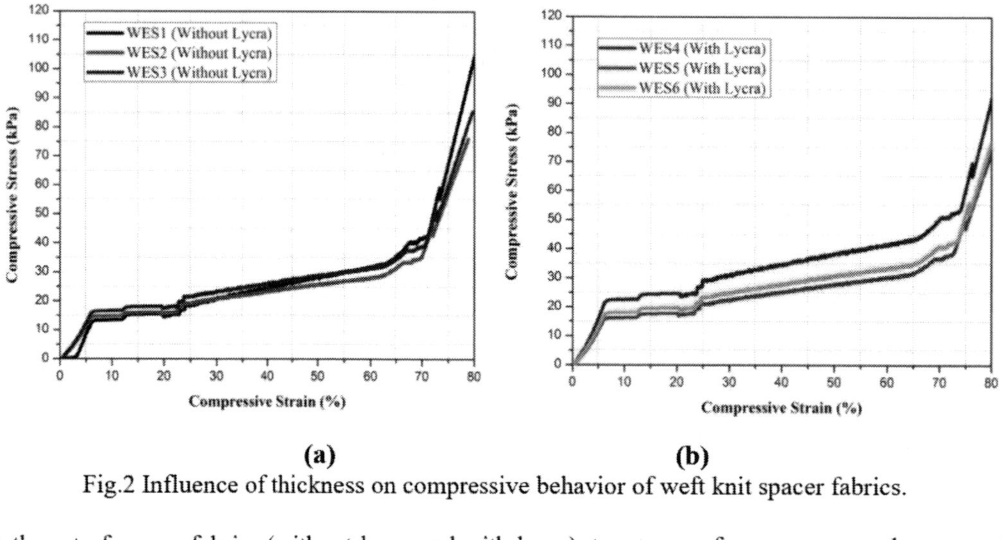

(a) **(b)**

Fig.2 Influence of thickness on compressive behavior of weft knit spacer fabrics.

In both the set of spacer fabrics (without lycra and with lycra), two types of spacer yarn such as monofilament and multifilament spacer yarns were used for convenient analysis of its effect on compression. Normally the monofilament spacer yarns act as a linear spring which offers more resistance towards compression as compared to other type of materials in cushioning applications. From the Fig.3, it is observed that, the compressive resistance is high for the fabrics with monofilament yarn for both the groups than that of fabric with multifilament spacer yarn. In plateau region (3rd stage), the marginal differences has been observed between the fabrics made up of monofilament spacer yarn (WES1 & WES4) in both the groups. It might be due to the fact that the large differences in density between these two samples. But the fabrics with multifilament spacer yarn (WES2, 3 & WES5, 6) don't show significant differences in compressive strength because the densities of these samples have almost closer to each other. It has also been found that the outer layer structures could affects the stitch density of the fabrics. The stitch density on the surface layer directly affects the compressive strength of the spacer fabrics. The compressive resistance increases with increase in stitch density. The lower stitch density on the surface of the fabric results in large surface deformation. In stage 4, the lower deformation was observed in the fabrics with monofilament spacer yarn than the fabrics made up of multifilament spacer yarn. It might be the fact that the spacer yarns have comes in to contact with each other and also makes locking effect with surface structure as quick as possible.

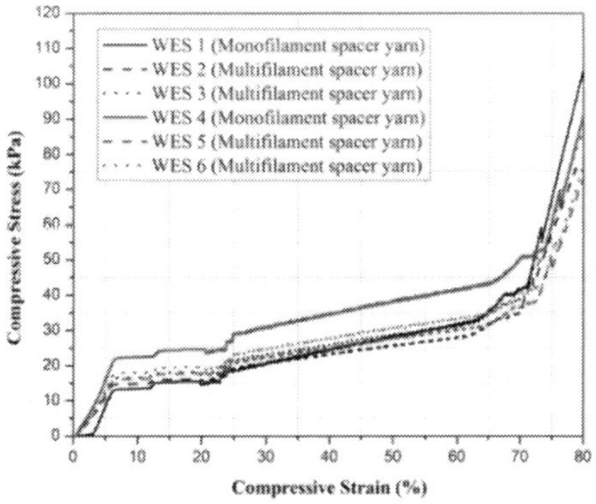

Fig.3 Influence of spacer yarn on compressive behavior of weft knit spacer fabrics.

3.2 Compressive Energy Absorption of Weft knit Spacer Fabrics

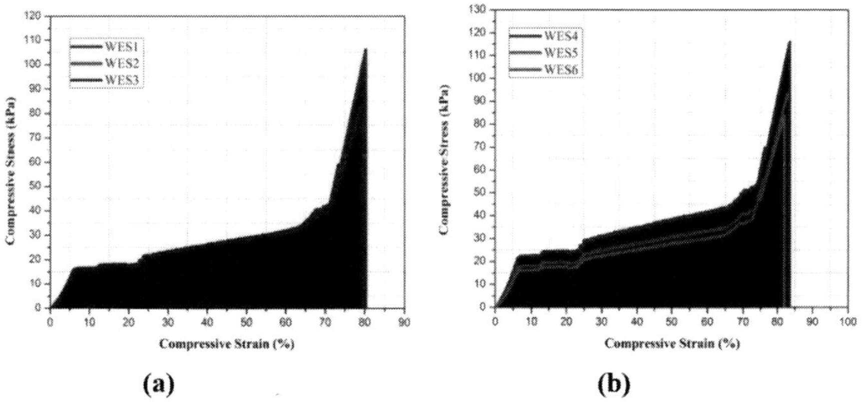

Fig.4 Compressive Energy absorption during compression of weft knit spacer fabrics.

It is convenient to evaluate the fabrics energy-absorbing ability under the out-of-plane compression. The compression curves reveal long deformation plateaus, suggesting that three-dimensional spacer fabrics may potentially be good compressive energy absorbing materials. Fig.4 (a & b) presents the work done of all spacer fabrics under compression load and also it compares the response compressive stress with effect of deformation and structural characteristics. The figures reveal that thicker spacer fabrics with monofilament spacer yarn have higher work done than that of thin fabrics with multifilament spacer yarn when it undergoes compression. Irrespective of the structural variations, the compressive work done shows same trend for all the samples. In both the groups, over all work done values are higher for the spacer fabrics made (WES1 & WES2) up of monofilament yarn. The density of spacer fabric is also plays a vital role in compressive behavior, negative linear correlation was observed between density and compressive work done.

Fig.5 presents the graphical analysis of compressive stress - absorbed energy - efficiency. It is observed from the figure that the absorbed energy of weft knit spacer fabrics (WES1 -WES6) linearly increases with the stress in the initial stage of compression. The marginal differences in energy absorption between the samples can be seen when the compressive stress reaches towards the third stage for both the groups. At the start of densification stage, the rapid increases in stress results in small deformation and energy absorption. From the energy absorption graph, it is easy to find the stress associated with the required amount of energy to be absorbed. So, it is more convenient to select the suitable spacer fabrics for car seat and back support application with optimum compressive performance [9].

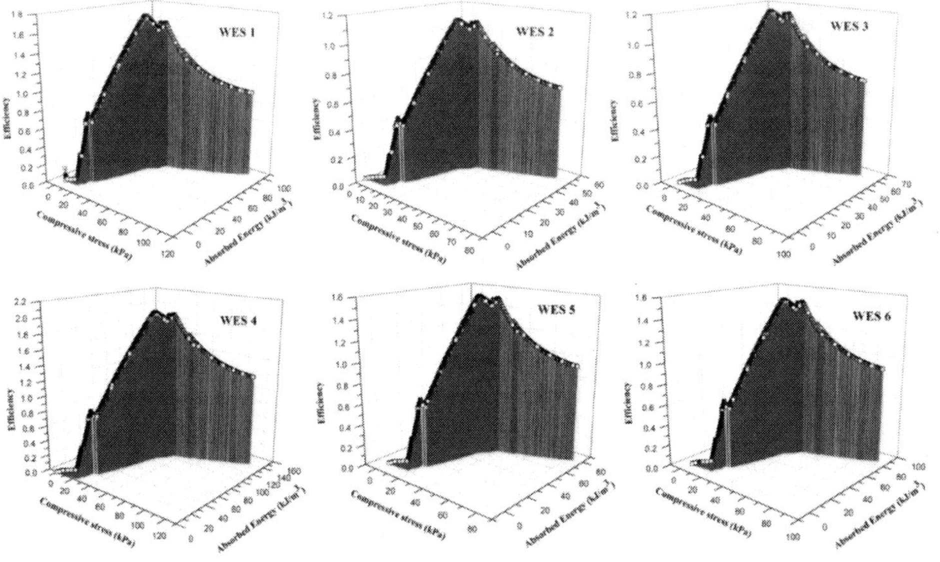

Fig.5 Energy absorption and efficiency of weft knit spacer fabrics.

The maximum energy-absorption efficiency is obtained at the end of the plateau stage. As noticed from the Fig.5, in the densification stage, efficiency decreases with rapid increase in stress level. It is also because of dramatic increase in volume density of the spacer fabrics. The point at the maximum energy-absorption efficiency can also be considered a critical point between the plateau zone and the densification zone. Overall it is observed that the compressive energy and efficiency is higher for the thicker fabrics made up of monofilament spacer yarn with low density. Also it is found that fabrics with finer spacer yarns undergoes large amount of work done as well as high efficiency during compression mechanism. The spacer fabric with large compressive deformation, high efficiency and energy absorption until plateau stage is the suitable finding for automobile upholsteries especially for seats and back supports.

3.3 Statistical Evaluation for Compressive Behavior Response of Weft Knit Spacer Fabrics

The above-mentioned results are confirmed by analysis of variance (ANOVA), and the result is significant influence of the surface structure, types of spacer yarn and thickness on fabric properties. In this section, one-way ANOVA is analyzed and the selected value of significance for all statistical tests in the study is $\alpha = 0.05$ levels.

The degree of freedom is 1, 8, the F_{critic}al is 5.318, and degree of freedom 3, 16, the $F_{critical}$ is 3.239. If the statistic is smaller than the critical value, we retain the null hypothesis because the p-value must be bigger than α , and if the statistic is equal to or bigger than the critical value, we reject the null hypothesis because the p-value must be equal to or smaller than α . Also pair wise comparison using Scheffé's method and Z score was calculated and presented in Table 3. The results of the ANOVA are listed in Table 3, which analyses the effect of groups of thickness and surface characteristics and types of spacer yarn of spacer fabric samples with compressive stress. The value of $F_{critical}<F_{actual}$ proves that the changes in the thickness, types of spacer yarn and surface layer structure (stitch density) of -knitted spacer fabric have significant influence on the above-mentioned fabric compressive stress. The insignificant difference in compressive stress is obtained between the pair, sample made up of multifilament spacer yarn without lycra on surface and with lycra. But the quite significant values are obtained in compressive stress between the other samples with multifilament spacer yarn.

Table.3 Statistical evaluation for compressive behavior of weft knit spacer fabrics.

							Pairwise comparis (Scheffé's method)		
One Way Anova - Influence of various factors on compression stress of weft knit spacer fabrics									
Test of factor influence : Influence of monofilament spacer yarn between both the groups									
		F_{cri}	F_{cal}	Prob.	Conclusion	Z-score (95% interval)	Compared Pair	Prob.	Significance
a	Without Lycra	5.318	30.520	0.0006	Significant	2.859	a-b	5.57E-04	Significant
b	With Lycra					3.126			
Test of factor influence : Influence of multifilament spacer yarn between both the groups									
a	Without Lycra (Thickness - 2.62mm)	3.239	92.238	2.03E-08	Significant	1.974	a-b	4.98E-05	Significant
b	Without Lycra (Thickness - 2.74mm)					2.217	a-c	0.000561	Significant
							a-d	2.22E-08	Significant
c	With Lycra (Thickness - 3.5mm)					2.169	b-c	0.614266	Insignificant
							b-d	0.000637	Significant
d	With Lycra (Thickness - 3.4mm)					2.410	c-d	5.59E-05	Significant
Test of factor influence : Influence of types of spacer yarn within the group - Without Lycra									
a	Monofilament Spacer Yarn (Thickness - 4.4 mm)	5.318	409.016	3.73E-08	Significant	3.138	a-b	3.73E-08	Significant
b	Multifilament Spacer Yarn (Thickness - 2.62 mm)					-2.248			
Test of factor influence : Influence of types of spacer yarn within the group - With Lycra									
a	Monofilament Spacer Yarn (Thickness - 4.4 mm)	5.318	187.617	7.78E-07	Significant	3.215	a-b	7.78E-07	Significant
b	Multifilament Spacer Yarn (Thickness - 3.4 mm)					-2.572			

4. Conclusion

Compressive behavior, energy absorption and efficiency of weft -knitted spacer fabrics were studied. The compression deformation mechanism of the fabric was identified based on the analyses of the load-displacement curve. The results showed that the 3D spacer fabrics are more resilient towards compression stress. Indeed, this structure enables a vertical alignment of spacer yarn; this z axis yarn has the effect to provide maximum recovery to the material. On the other hand, it has been shown that the spacer fabrics dissipate more energy in compression than other foam and nonwoven materials. Overall it is observed that the compressive energy and efficiency is higher for the thicker spacer fabrics with low density. Also it is found that fabrics with monofilament spacer yarns undergoes large amount of work done as well as high efficiency during compression mechanism. The spacer fabric with large compressive deformation, high efficiency and energy absorption until plateau stage is the suitable finding for automobile upholsteries especially for seats and back supports.

5. Acknowledgment

This work was supported by Department of Materials Engineering, Faculty of Textile, Technical University of Liberec, Czech republic and project No. 21195 by students' grant scheme, Czech Republic

References

[1] Arumugam V, Mishra R, Militky J, Tunak M. In-plane shear behavior of 3d knitted spacer fabrics. J. Ind. Text: 2016; 46: 868-886.
[2] Ajayi JO, Elder HM. Fabric friction, handle and compression", J. Text. I: 1997; 88 (3): 232-241.
[3] Taylor PM, Pollet DM. Static low-load lateral compression of fabrics", Text Res J: 2002; 72 (11): 983- 990.
[4] Postle R. A geometrical assessment of the thickness and bulk density of weft knitted fabrics", J. Text. I: 1974; 65 (4): 155-163.
[5] Mecit Armakan D, Roye A. A study on the compression behaviour of spacer fabrics designed for concrete applications, Fibers and Polym: 2009; 10 (1): 116-123.
[6] Xu HM, Ming QG. The compression behaviour of warp knitted spacer fabric, Fibers Text East Eur: 2009; 16; 1 (66): 90-92.
[7] Liu YP, Hu H, Long H, Zhao L. Impact compressive behavior of warp-knitted spacer fabrics for protective applications, Text. Res. J: 2012; 82: 773 - 288.
[8] Avalle M, Belingardi G, Montanini R. Characterization of polymeric structural foams under compressive impact loading by means of energy-absorption diagram, Int. J. of Impact Eng: 2001; 25: 455-472.
[9] Arumugam V, Mishra R, Militky J, Salcova J. Investigation on Thermo-physiological and Compression Characteristics of Weft Knitted Spacer Fabrics. J. Text. I., Epub ahead of print 12 August 2016. DOI: 10.1080/00405000.2016.1220035.

Preparation of and Study on Thermo-mechanical Properties of Carbon Fibre-reinforced Polyamide1012 Composites

Rui Yang[1], Fu-Jie Chen[2*], Mohammad Irfan Iqbal[3], Lei Wang[1], Cheng-Qi Zhang[1], Hai-Feng Bao[1]

[1] *School of Materials Science and Engineering, Wuhan, Textile University, Hubei 430200, China.*
[2] *School of Chemistry and Chemical Engineering, Wuhan, Textile University, Hubei 430200, China.*
[3] *School of Textile Science and Engineering, Wuhan, Textile University, Hubei 430200, China.*

[*]Corresponding author's email: whcfj@163.com

Abstract

The development of surface modified carbon fibre-reinforced polyamide1012 (PA1012) composites with excellent thermo-mechanical properties were studied thoroughly in this research. Carbon fibre was used to enhance the toughness and heat-resisting properties of polyamide1012. During the manufacturing process of the composite, carbon fibre was used to enhance the property of PA1012 by varying fibre content, for instance, by 5%, 10%, 15%, 20%, 30% and 40%. In this paper, the composite of carbon fibre and polyamide1012 was prepared, while their thermo-mechanical properties and fracture morphology were characterized by means of thermogravimetric analysis (TGA), a microcomputer-controlled electronic universal tester, and Scanning Electron Microscopy (SEM), respectively. The results show that the initial decomposition temperature of the composite was above 400°C. The tensile strength and elastic modulus were improved with an increase in carbon fibre content. It was found that the maximum values of tensile strength and elastic modulus were 87.01 MPa and 438.31 MPa, respectively, when the carbon fibre content was 15%. It was found that the surface modified carbon fibre is superior to original carbon fibre. It was also observed that the surface of modified carbon fibre had some particles that may have contained the ester group, which could have improved the interfacial bonding strength.

Keywords: Carbon Fibre; PA1012; Composite; Thermo-mechanical Properties.

1. Introduction

PA1012, which has a high-carbon alkyl group, was used to replace the metal material because of its numerous advantages, such as lower water absorption, stable size, high strength and good toughness. It is widely used in machinery, automobiles, the military, aerospace and other fields [1]. Thus, improving its mechanical properties was the main focus of current research. Wang et al. [2] used phenylethylene-(ethylene-butene)-phenylethylene multipolymer, grafted by maleic anhydride (SEBS-g-MAH), to improve the toughness of PA1012. But the effect was not good. Chen et al. [3] used an acrylic co-polymer, grafted by the maleic anhydride co-polymer (ACR-MAH), to enhance the performance of PA1012. However, the percentage grafting was rather low and the consistency of the interface was not so good. As such, other materials should be attempted to enhance to PA1012.

Carbon fibre is a new type of material, which has high strength and modulus. It also has other excellent properties, such as like low density, low coefficient of thermal expansion and good heat resistance. Carbon fibre is usually used as reinforced material because of its outstanding performance in aircraft manufacturing, national defence, military industry, automobile, medical equipment, sport equipment and so on [4]. The surface of original carbon fibre was non-polar because of the copious presence of inert graphite crystallite [5, 6]. This structure resulted in bad wetting properties and interfacial bonding properties with the resin. There were three ways to modify the surface of carbon fibre. One was to introduce active functional groups, such as carbonyl, carboxyl and hydroxy, to enhance the wetting properties and bonding capacity [7]. The second was to sculpture some groove on the surface of the carbon fibre to increase the specific surface area for the sake of forming a mechanically interlocked structure with the resin [8]. The third was to remove the weak interface layer on the surface of the fibre. Yang et al. [9] used carbon fibre, modified by a silane coupling agent, to reinforce the epoxy and found that the interlaminar shear strength increased by about 42%. The wetting properties were also improved. Yu et al. [10] used potassium peroxydisulphate (K2S2O8)/silver nitrate (AgNO3) to treat the surface of the carbon fibre. The results showed that the interlaminar shear strength increased by 62.5%, while the surface was not destroyed. This was a pleasing outcome.

The thermoplastic resin reinforced by carbon fibre is one of the main materials of interest to current researchers all over the world [11-14]. The composite can also be recycled because of the thermoplasticity of the resin. Ge et al. [15] used carbon fibre to reinforce polyamide1010 (PA1010) and found that carbon fibre reinforcements not only enhance the tensile strength and hardness, but also reduce the friction coefficients and wear rates of the nylon composites, compared with neat nylon materials.

Fibres used to reinforce materials have been significantly applied in many fields [16-18]. In our study, we used carbon fibre, whose surface was modified to reinforce PA1012, with a view to obtaining some composite materials (C/PA1012). We wanted to find the influence factors of the thermal and mechanical properties of these reinforced materials.

2. Experimental Section

2.1 Materials

Carbon fibre (T700) was purchased from the Toray Company in Japan. PA1012 was procured from Shandong Guangyin New Material Company. Aluminium foil was obtained from Tianjin Angesi Steel Trade Company.

2.2 Preparation and Characterization

The composite materials of carbon fibre and PA1012 were prepared by a two-step process. First, carbon fibre and PA1012 were mixed in the Haake Rotational Rheometer (RheoDrive 7), which was purchased from ThermoFisher Scientific Co., Ltd., for about 10 min by varying the percentage of carbon fibre content by 5%, 10%, 15%, 20%, 30% and 40% respectively. The temperature of the Banbury mixer was set at 210°C and the rotate speed was set at 30 r/min. In subsequent steps, the mixture was put into the tablet press, which was obtained from Shanghai Xima Weili Rubber and Machinery Co., Ltd., to cut it into a tablet under 30 MPa of pressure. Then, the composite tablet was cut into a standard spline whose effective length, width and thickness were 20 mm, 5 mm and 1 mm, respectively.

The mechanical property of the composite materials was tested with a stretching rate of 50 mm/min using the WDW-20E Microcomputer-controlled Electronic Universal Testing Machine, manufactured by Jinan Shidai Shijin Testing Machine Group Co., Ltd. The TGA of the PA1012 and composite materials was performed by using a TG 209F1 from NetzschScientific Instruments Trading (Shanghai) Ltd. The samples were run at 10 K/min for the heating rate and 20 K/min for the cooling rate in an ambient temperature up to 600°C, under the atmosphere of which the shielding gas and sweeping gas were respectively 20 mL/min and 50 mL/min. The original and modified carbon fibre were used to perform the Fourier transform infrared spectroscopy (FTIR) test. The FTIR spectra were recorded on a Tensor-27 instrument produced by Buker Optics Company in Germany. The sample was tested by reflective infrared. The microcosmic surface and cross section of the carbon fibre and composites were observed using the JSM-6510LV SEM, manufactured produced by Questar China Ltd.

3. Results and Discussion

3.1 Mechanical Properties of Composites

The mechanical properties of C/PA1012 composite materials containing various percentages of carbon fibre are shown in Table 1. From the table, it can be seen that the tensile strength and elastic modulus of the composite materials were increased along with an increase in carbon fibre content at the beginning. When the percentage rose beyond 15%, there was no remarkable improvement. The tensile strength and elastic modulus were increased 30% and 40%, compared with pure PA1012. Thus, the optimum result was found with regard to 15% of carbon fibre in C/PA1012. However, when the content of the carbon fibre rose beyond 15%, the tensile strength and elastic modulus were decreased sharply. Thus, we could say the tensile strength and elastic modulus were poorer than for pure PA1012 when the content of the carbon fibre was over 15%. The dispersion uniformity and the saturability of the bonding became poor when the resin was filled with lots of carbon fibre. The space between the carbon fibres lacked the resin for sufficient bonding. This caused micro crack defects in the composites. When under a tensile force, the extension of the micro crack led to decreasing affordability. The result observed at the macro level was a decrease in tensile strength and elastic modulus. As for elongation at the break, the value of this physical quantity also decreased with the increase in carbon fibre content. The toughness of the composite material declined indirectly.

Table.1 The mechanical properties of C/PA1012 composite materials

Serial number	Sample	Tensile strength/MPa	Elastic modulus/MPa	Elongation at break/%
1	Pure PA1012	62.61±0.97	339.01±3.99	288.00±63.57
2	5% C/PA1012	72.42±4.39	362.57±10.53	34.25±2.13
3	10% C/PA1012	82.89±8.26	405.42±30.16	35.91±2.80
4	15% C/PA1012	87.01±3.19	438.31±14.03	34.28±1.52
5	20% C/PA1012	57.38±0.96	337.75±23.08	28.80±1.28
6	30% C/PA1012	47.89±9.37	316.14±48.89	28.30±2.26
7	40% C/PA1012	46.07±3.29	262.48±30.10	26.88±1.54

3.2 TGA

The temperature of the initial decomposition in pure PA1012 and 15%C/PA1012 was about 400 °C. In Fig.1, the maximum temperature of decomposition was 450.2 °C, with the mass loss was 96.57%, compared with the original sample. However, when we used 15% carbon fibre in the pure PA1012, the maximum temperature of decomposition was raised to 467.1°C, while the mass loss was 87.93%. Hence, the maximum temperature of decomposition was improved a little by adding carbon fibre. Thus, carbon fibre improved the heat resistance of the composite material.

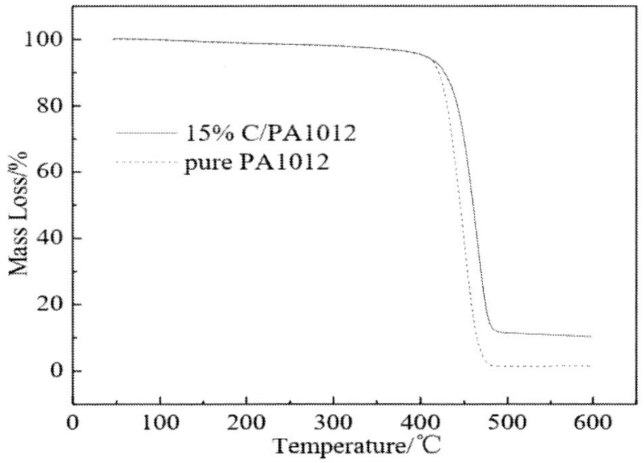

Fig.1 The TGA of pure PA1012 and 15%C/PA1012

3.3 FTIR Analysis

The FTIR spectrum of the original and modified carbon fibre is shown in Fig.2. A strong absorption peak was observed at 1,662.34 cm^{-1}, which refers to the strong stretching vibrations of the carbonyl (C=O) functional group. The band at 1,230.36 cm^{-1} was identified as stretch vibrations of the C-O-C bond, while the bands at 2,917.77 cm^{-1} and 2850.27 cm^{-1} were respectively identified as stretch vibrations of methyl ($-CH_3$) and methylene ($-CH_2$). As for the absorption peak at 3,357.46 cm^{-1}, this represented the characteristic absorption band of hydroxy (-OH) in the water contained in the fibre. The original carbon fibre was modified by the polyvinyl acetate (PVAc). IR spectroscopic study confirmed the existence of PVAc in the modified fibre.

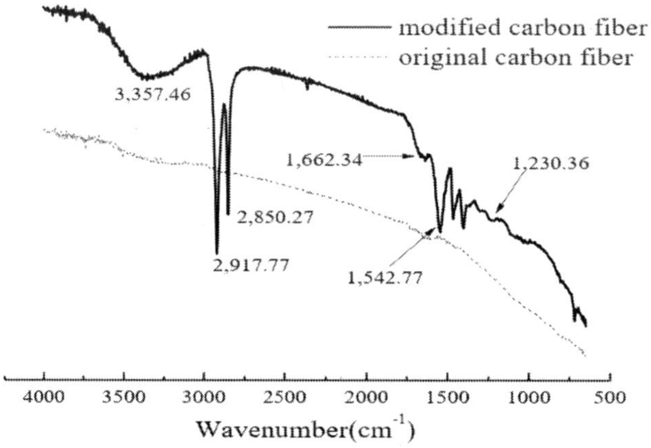

Fig.2 FTIR spectrum for original carbon fibre and carbon fibre modified with polyvinyl acetate

3.4 Surface Morphology

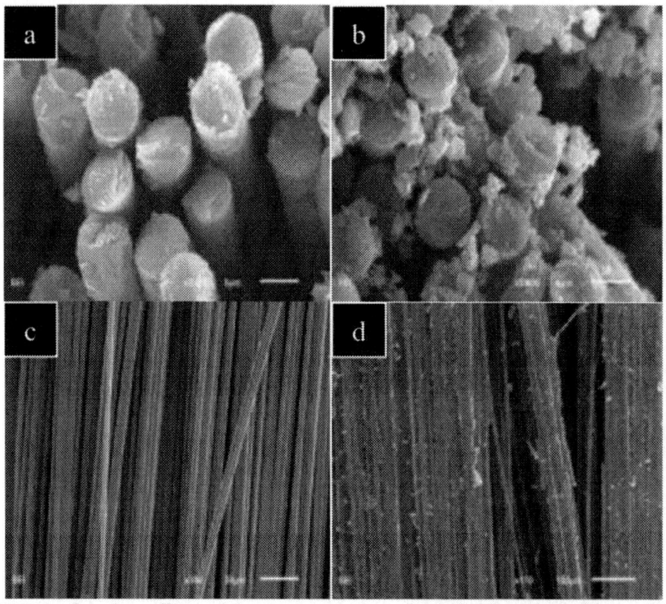

Fig.3 SEM of carbon fibre: (a) cross section of original carbon fibre, (b) cross section of carbon fibre modified by PVAc, (c) surface of original carbon fibre, (d) surface of carbon fibre modified by PVAc

The surface morphology of pure carbon fibre and surface-modified carbon fibre is shown in Fig.3. It was already known that carbon fibre must be modified by a surface treating agent before being added to the resin. In Fig.3, there were many small particles in the modified fibre's surface, compared to the carbon fibre that was without modification by PVAc. The particles were adhered to the surface of the carbon fibre, which contained PVAc with an ester group. This was performed in order improve the interfacial bonding strength and relieve the internal stress. This different structure can be observed in Fig.3, which shows that many modified fibres were bonded together.

Fig.4 demonstrates the SEM of surface and tensile fracture morphology for pure PA1012 and 15% C/PA1012. We observe from Fig.4(a) that the carbon fibre was tightly packed by the pure PA1012. This explains the smooth surface of the composite material. It was also found that the carbon fibres were arranged irregularly to make a mesh structure within the composites, thus improving the tensile strength and elastic modulus of the composite material. The fracture behaviour of pure PA1012 during the tensile test in shown in Fig.4(b), which is explained by the orientation of the fibre within the composites being stretched during the tensile test. In Fig.4(c), the carbon fibre was packed by PA1012 in order to improve the performance of the composite. Given the high strength and modulus of the carbon fibre, it can bear higher pressure than pure PA1012. When the composite material was under pressure, the force was transmitted from the matrix to the reinforcement in order to disperse the pressure.

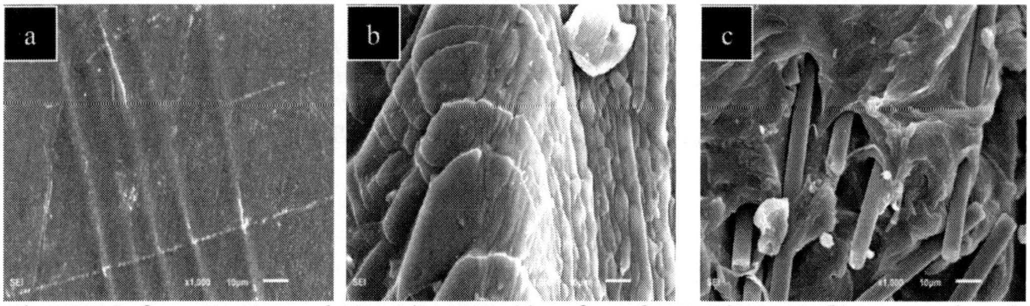

Fig.4 SEM of pure PA1012 and 15% C/PA1012: (a) surface of 15% C/PA1012, (b) cross section of pure PA1012, (c) cross section of 15% C/PA1012

4. Conclusion

The thermal and mechanical properties of C/PA1012 composites were improved by mixing carbon fibre with PA1012 using melt blending method. The experimental data obtained in this work proved that the fastest decomposition temperature of the composite was above 467.1°C, which was higher than for pure PA1012. It was also found that, with the increase in carbon fibre content, the tensile strength and elastic modulus of the composites increased up to a certain limit. For a 15% carbon fibre content, the maximum values of tensile strength and elastic modulus were found to be 87.01 MPa and 438.31 MPa, respectively, which 30% and 40% higher than for pure PA1012, also respectively. Based on the performances of surface-modified carbon fibre, we can say that PVAc in modified carbon fibre not only improves the interfacial bonding strength, but also relieves internal stress.

References

[1] Li FM. Development and applications of long carbon chain nylon. New Chemical Materials: 2006; 34: 6-9.
[2] Wang YW, Wang YD, Zhao QX, Liu MY, Guo FF, Fu P. Study on toughening of nylon1012 by SEBS-g-MAH. China Plastics Industry: 2008; 36: 57-59.
[3] Chen XJ, Xu YB, Bai Y, Wu GF. Study on ACR-MAH toughened PA1012 resin. China Plastics Industry: 2016; 44: 23-26.
[4] Chand S. Review Carbon fibers for composites. Journal of Materials Science: 2000; 35: 1303-1313.
[5] Guo H, Huang YD, Meng LH et al. Interface property of carbon fibers/epoxy resin composite improved by hydrogen peroxide in supercritical water. Materials Letters: 2009; 63: 1531-1534.
[6] Fitzer E, Geigl KH, Hüttner W et al. Chemical interactions between the carbon fiber surface and epoxy resins. Carbon: 1980; 18: 389-393.

[7] Rich MJ. Adhesion of graphite fibers to epoxy matrices: I. The role of fiber surface treatment. Journal of Adhesion: 1983; 16: 1-30.

[8] Liu B, Wang X, Yang J et al. Research progress of surface modification of carbon fiber. Chemical Research: 2015; 26: 111-120.

[9] Yang J, Xiao J, Zeng J et al. Matrix modification with silane coupling agent for carbon fiber reinforced epoxy composites. Fibers and Polymers: 2013; 14: 759-766.

[10] Yu J, Meng L, Fan D et al. The oxidation of carbon fibers through $K_2S_2O_8/AgNO_3$ system that preserves fiber tensile strength. Composites Part B Engineering: 2014; 60: 261-267.

[11] Li L, Cao CL. Effect of carbon fiber content on mechanical properties of short carbon fiber reinforced PE composites. Foundry Technology: 2012; 1: 26-28.

[12] Peng SW. Study on PA66 reinforced with carbon fiber. Engineering Plastics Application: 1998; 9: 5-7.

[13] Sun YP, Wang HQ, Li L, Gao R. Carbon fiber-reinforced polymer composites. Journal of Shandong University of Technology: 1997; 27: 244-249.

[14] Kumar S, Doshi H, Srinivasarao M et al. Fibers from polypropylene/nano carbon fiber composites. Polymer: 2002; 43: 1701-1703.

[15] Ge SR, Zhang DK, Zhu H, et al. Mechanical properties and their influence on the friction and wear of the carbon fibers reinforced polyamide 1010. Acta Materiae Compositae Sinica: 2004; 21: 99-104.

[16] Wu SH, Wang FY, Ma CCM. Mechanical, thermal and morphological properties of glass fiber and carbon fiber reinforced polyamide-6 and polyamide-6/clay nanocomposites. Materials Letters: 2001; 49: 327-333.

[17] Licea-Claverie A, Carrillo FJU, Alvarez-Castillo A et al. Characterization of mixed fiber nylon composites incorporating composite scrap. Polymer Composites: 2010; 20: 314-320.

[18] Ishak ZAM, Berry JP. Impact properties of short carbon fiber reinforced nylon 6.6. Polymer Engineering & Science: 1993; 33: 1483-1488.

Laundering Performance of TBCC-activated Peroxide System

Dong-Yan Shao, Chang Sun, Chang-Hai Xu*

College of Textile and Clothing, Jiangnan University, Wuxi, Jiangsu 214122, China

Corresponding author's email: changhai_xu@jiangnan.edu.cn

Abstract

In this study, a cationic bleach activator, N-[4-(triethylammoniomethyl) benzoyl] caprolactam chloride (TBCC), was combined respectively with equimolar amount of sodium perborate (SPB), sodium percarbonate (SPC) and carbamide peroxide (CP) to create a novel activated peroxide system. Cotton fabric can be bleached in a near-neutral condition at low temperature through TBCC-activated peroxide systems. Meanwhile, the colored stains on the bleached cotton fabric can be removed by these systems. Experimental results showed that TBCC-SPB, TBCC-SPC and TBCC-CP activated peroxide systems exhibited different bleaching performance, which was mainly caused by dissociation of these peroxides to peracids in alkaline condition with different levels. The activated peroxide system could perform better and did less damage to fabrics in the near-neutral condition. Hence, by incorporating with TBCC, the beaching performance of the peroxides was highly enhanced in a mild condition, which improved the application of peroxides in Chinese domestic laundering.

Key words: TBCC; Peroxide; Cotton Fabric; Bleaching; Laundering

1. Introduction

Laundry detergent is an important commodity in daily life which makes cloths clean and bright. Normally, a detergent is consisted of plentiful surfactant, alkali, essence and others like peroxides, enzyme which can impart detergent with different applications. Detergent with peroxides is mainly used the bleaching of fabrics and removal of colored stains [1, 2]. Although peroxides are economic, safe and environmental, they work in strong alkali condition (pH=10-11) with high temperature (90-100 °C). However, in China, laundry is generally carried out at room temperature, unless with a special machine, besides strong alkali condition is harmful for human beings and also the washing machines.

The most common peroxide is hydrogen peroxide (H_2O_2), but it is a solution and difficult to store and use. Sodium perborate (SPB), sodium percarbonate (SPC) and carbamide peroxide (CP) are solid peroxides and can be used in detergent, as shows in scheme 1. SPB ($2NaBO_3 \cdot H_2O$) was first included in detergent with a simple manufacture technology, low price, great water solubility and stability at room temperature [3]. SPC ($Na_2CO_3 \cdot H_2O_2$) is an adduct of sodium carbonate and hydrogen peroxide and can release hydrogen peroxide in water quickly. With additional stabilizing coatings, SPC can be stable at room temperature [3, 4]. CP($CO(NH_2)_2 \cdot H_2O_2$) also has simple manufacture technology, widely uses raw materials, low cost and great water solubility without side-effect and harm which is a substitution for SPB and SPC. However, CP is not stable at room temperature, and it should be stored at low temperatures (2-8 °C) [1].

In order to overcome the shortcomings of peroxides, bleaching activators were developed and used, which can activate peroxides forming much more activated peracids [5]. Peracids can bleach in a near neutral condition at low temperature which is less harmful for fabrics [6, 7]. Etraacetylethylenediamine (TAED), nonanoyloxybenzene sulphonate (NOBS) and N-[4-(triethylammoniomethyl) benzoyl] caprolactam chloride (TBCC) are three kinds of activators which are investigated intensively. Though TAED is widely used in domestic laundry, poor water solubility still limits its application [7-9]. While NOBS tends to generate unwanted diacylperoxide byproduct when NOBS is excess or applies at low pH [7]. TBCC is one kind of cationic bleaching activators which has an excellent water solubility and inherent substantivity for cellulosic fibers [10-13]. Many works have revealed that TBCC performs better when the molar ratio of TBCC, H_2O_2 and $NaHCO_3$ is 1:1.2:1.4 [14, 15]. Scheme 2 shows the chemical structure of TBCC and the perhydrolysis reaction of TBCC and H_2O_2.

SPB 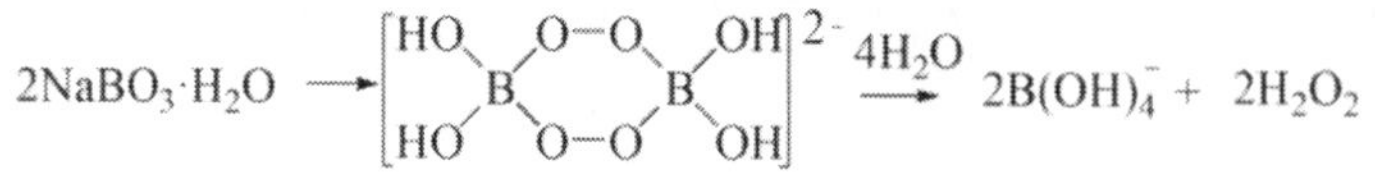

SPC $Na_2CO_3 \cdot 1.5H_2O_2 \longrightarrow Na_2CO_3 + 1.5H_2O_2$

CP $CO(NH_2)_2 \cdot H_2O_2 \longrightarrow CO(NH_2)_2 + H_2O_2$

Scheme 1 Dissolution of SPB, SPC and CP

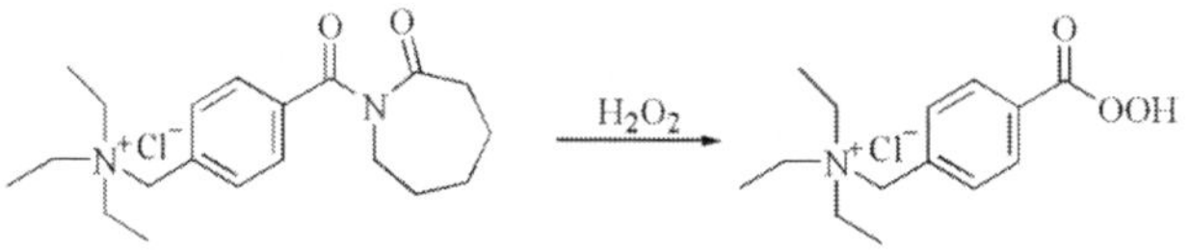

TBCC TBA

Scheme 2 Chemical structure and perhydrolysis of TBCC

In this study, new TBCC-activated bleaching systems were developed by combining TBCC with three kinds of peroxides (SPB SPC and CP), and their bleaching ability and laundering performance are investigated.

2. Experiment

2.1 Material

Cotton greige knitted fabric and cotton bleached printing fabric were kindly provided by Hongdou Group Textiles Co., Ltd. (China). TBCC was synthesized and purified to ≥ 97% according to the previously reported method [13]. Sodium perborate monohydrate, Sodium percarbonate, Carbamide peroxide and Bis (ethylenediamine) copper (II) hydroxide solution were purchased from Sigma-Aldrich (China). $NaHCO_3$ and NaOH were purchased from Sinopharm Group (China). Scouring agent, wetting agent (Penetrant JFC) and peroxide stabilizer (DM 1403) were kindly provided by Dynamic Chemicals (China). Red wine, black coffee and tea were purchased online.

2.2 Scouring of Cotton Greige Knitted Fabric

Cotton greige knitted fabric was scoured in a solution contained NaOH (20 g/L) and scouring agent (1 g/L) for 2 hours at 100 °C, and the bath ratio was 1:50. After scoured, the fabric was rinsed thoroughly with copious amounts of water, and dried under ambient conditions.

2.3 Preparation of Stained Fabrics

Cotton bleached printing fabric was stained by tea, black coffee and red wine as follows:

1) Put a bag of tea into 300 mL of 80°C drink water for 5 min, then take out the tea bag and put 4 g of fabric in for 5 min, and dry the fabric under ambient conditions.

2) Dissolve 13 g of black coffee in 300 ml of 80°C drink water totally, then put 4 g of fabric in for 5 min, and dry the fabric under ambient conditions.

3) Put 4 g of fabric to the solution which contained 300 mL of red wine for 5 min, and dry it under ambient conditions.

2.4 Establishment of TBCC-activated Peroxide System

Table 1 shows the prescription for process of scoured fabrics and stained fabric. Unless other stated, these fabrics were processed for 30 min at 25 °C and the bath ratio was 1:50. At the meantime, use a pH value detector to detect the pH value of bleaching system before and after bleaching.

Table.1 The activated peroxide system process prescription and comparative experiment

Chemicals	Non-activated system			TBCC-activated system		
SPB(g/L)	1.50	-	-	1.50	-	-
SPC(g/L)	-	1.57	-	-	1.57	-
CP(g/L)	-	-	1.41	-	-	1.41
TBCC(g/L)	-	-	-	5.50	5.50	5.50
JFC(g/L)	1	1	1	1	1	1
EDTA(g/L)	1	1	1	1	1	1

2.5 Measurement of Cotton Fabrics

2.5.1 Degree of Whiteness

The processed cotton fabric was folded into a four-layer sample and measured on a Datacolor 650 spectrophotometer (Datacolor International, USA). The instrument was set with the CIE Illuminant D65, CIE 1964 Supplemental Standard Observer, large area view (30 mm), and specular included. CIE whiteness index (WI) was calculated according to the AATCC Test Method 110-2010. Each sample was measured four times with 90° rotation between measurements to give an average value of CIE WI.

2.5.2 Degree of Polymerization

Cellulose dispersions were prepared from processed cotton fabrics using Bis (ethylenediamine) copper (II) hydroxide solution as a solvent, and the fluidity (F) of cellulose dispersions was determined according to the AATCC Test Method 82-2007. The degree of polymerization (DP) of cotton fabrics was calculated from the fluidity by Eq.(1). The DP value is correlated with chemical damage of cotton fibers resulting from oxidative bleaching. A higher DP value indicates a lower fiber damage.

$$DP=2032\log_{10}\left(\frac{74.35+F}{F}\right)-573 \quad (1)$$

3. Results and Discussion

3.1 Bleaching Effect of TBCC-activated Peroxide Systems on Scoured Fabrics

The scoured fabrics were treated with 6 different bleaching systems, the degree of whiteness of bleached cotton fabric is shown in Fig.1a, and the pH value of bleaching system is shown in Fig.1b. It can be seen that cotton fabrics can be bleached by peroxide system at room temperature, and the whiteness of cotton fabrics bleached by non-activated peroxide system was lower than TBCC-activated system with SPB and SPC which means that TBCC can improve the performance of SPB and SPC. However, TBCC-CP bleaching system did not get a satisfied result. As shows in Fig.1b, TBCC-SPB and TBCC-SPC system can maintain pH in a near-neutral conditions before and after bleaching without extra alkali. But TBCC-CP system had an acidic pH value before and after bleaching which was not positive for perhydrolysis of TBCC and peroxide. In this situation, TBCC-CP system was ineffective for bleaching which lead to the lower whiteness of bleached cotton fabric.

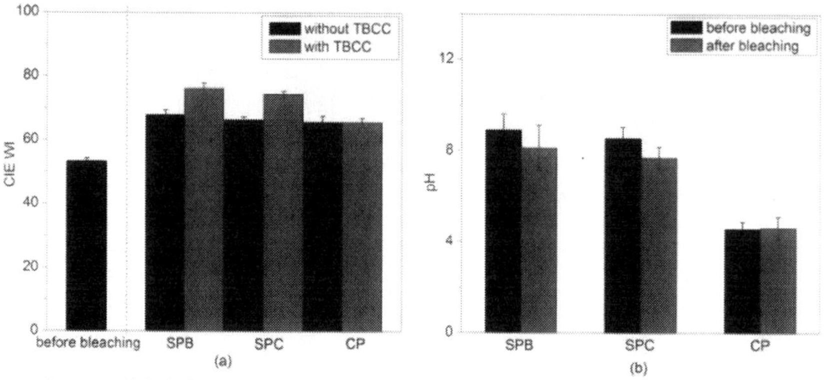

Fig.1 Whiteness of cotton fabric bleached with different systems (a) and pH value of TBCC-activated peroxide systems before and after bleaching (b)

3.2 Bleaching Effect of TBCC-activated Peroxide Systems on Stained Fabrics

Fig.2a shows the whiteness of cotton fabrics stained with tea, coffee and red wine, separately. As can be seen from Fig.2a, at room temperature, stained cotton fabric processed by non-activated bleaching system did not get a high whiteness compared to TBCC-activated bleaching system except TBCC-CP system, which means TBCC can enhance bleaching performance of SPB and SPC at room temperature. Whether CP bleaching system or TBCC-CP bleaching system cannot work well because of its low pH value of CP solution. The reason is same as that in section 3.1, that is, TBCC-SPB and TBCC-SPC system can maintain pH in a near-neutral condition before and after bleaching, therefore they performed better than TBCC-CP system in acidic pH value, as shown in Fig.3b.

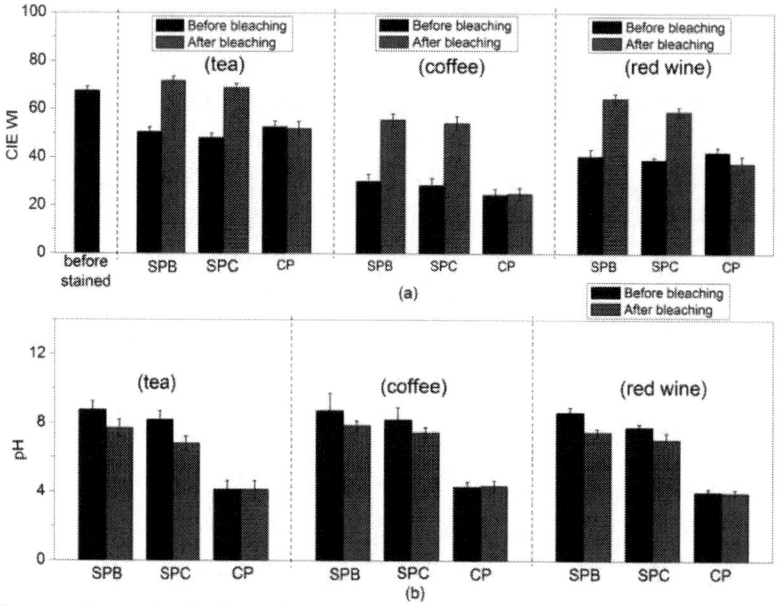

Fig.2 Whiteness of tea, coffee and red wine stained cotton fabric processed with different systems (a) and pH value of TBCC-activated peroxide systems before and after bleaching (b)

In Fig.2a, the whiteness of tea stained cotton fabric is higher than red wine stained fabric, and the whiteness of coffee stained fabric is the lowest, which indicates that bleaching system performed better for tea stain and red wine stain. The reason is that the degree of contamination of fabrics with these three different stains is different, and the removal or bleaching for higher degree of contamination is much more difficult in a bleaching and laundry system.

In this study, the degree of contamination of coffee is the highest, while tea is the lowest. For another reason, the pH valve of tea, coffee and red wine were all lower than 7 in which the performance of TBCC-activated bleaching system was inhibited, especially coffee and red wine.

3.3 Improvement of TBCC-CP System

In order to overcome the defect of TBCC-CP bleaching system that TBCC-CP system is acidic before and after bleaching, Na_2CO_3 (0.8-2.3 g/L) was added as alkali to adjust the pH value of bleaching solution to near-neutral condition (pH=7-8) for a better performance of TBCC-CP bleaching system. After adjustment, red wine stained cotton fabrics was bleached with the improved TBCC-CP system to process, and then the change of whiteness and pH value was observed.

As shown in Fig.3, with increase of the concentration of Na_2CO_3, the whiteness of cotton fabric got a great promotion, especially when pH value was near 8. As the pH value of TBCC-CP-Na_2CO_3 bleaching system over 9, the whiteness of cotton fabric decreased, which is caused by alkali-catalyzed bimolecular decomposition of peracids in alkali condition [15, 16] as well as the rapid hydrolysis of TBCC. The result in Fig.3 is conform to the relationship of pH value and whiteness of cotton fabric bleached with TBCC-SPB and TBCC-SPC system.

The improved TBCC-CP-Na_2CO_3 system (Na_2CO_3=1.70 g/L) was used to process scoured cotton fabric and stained cotton fabrics, as shown in Fig.4. The degree of whiteness of scoured fabric was near 80 after bleaching, which was near even over the whiteness of the un-stained one. The improvement of TBCC-CP system successfully overcame the shortcoming and increased the bleaching effect of this system.

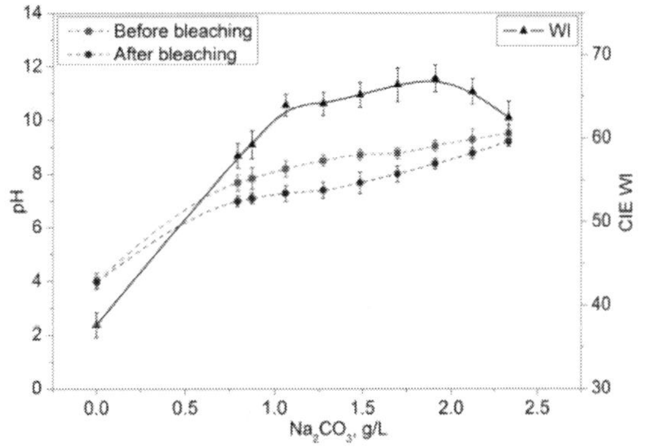

Fig.3 Whiteness of cotton fabric bleached with TBCC-CP system and pH value of TBCC-CP system before and after bleaching by adding different concentration of Na_2CO_3

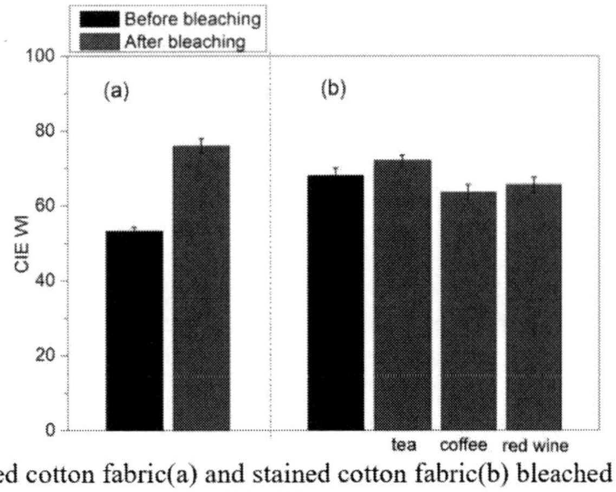

Fig.4 Whiteness of scoured cotton fabric(a) and stained cotton fabric(b) bleached with improved TBCC-CP-Na_2CO_3 system

3.4 Chemical Damage of Cotton Fibers

The degree of polymerization (DP) of cotton fabrics before and after bleaching was showed in Fig.5, as can be seen, the DP value of cotton fabrics before bleaching was almost the same as after bleaching. The TBCC-SPC did nearly no damage to cotton fabrics. Bleaching or cleaning by TBCC-activated bleaching system was protective for cotton fibre.

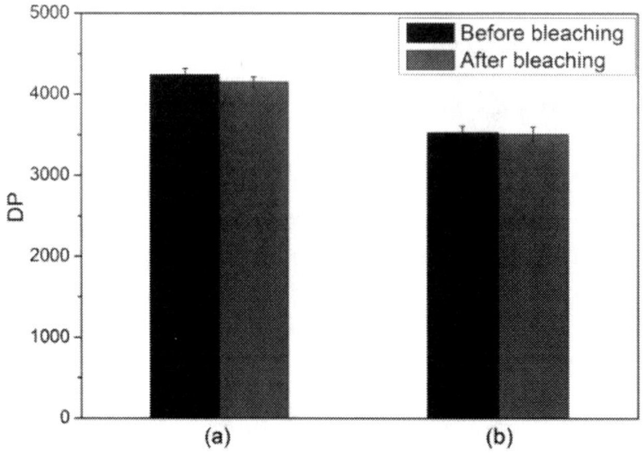

Fig.5 DP Value of scoured cotton fabric (a) and red wine stained cotton fabric (b) before and after processed with TBCC-SPC system

4. Conclusion

Three kinds of peroxide systems were established by using TBCC as a bleach activator and SPB, SPC and CP as peroxides for cotton bleaching and stain cleaning. At 25°C, TBCC-SPB and TBCC-SPC bleaching systems can achieve satisfied effects for cotton bleaching and stain cleaning, and the degree of whiteness of bleached cotton fabric can reach to 75. However the bleaching performance of TBCC-CP bleaching system was not as great as the other two systems due to the pH value of bleaching system which can be overcame by adding extra alkali-Na_2CO_3. When the pH value of TBCC-CP-Na_2CO_3 bleaching system was around 8, the whiteness of bleached cotton fabric was the highest. Compared to conventional bleaching system, TBCC-activated systems provided cotton fabric with the approximate whiteness but higher degree of polymerization. The combination of TBCC and peroxides is a foundation for TBCC-activated bleaching using in domestic laundry which may provide laundry detergent with high effect and environmental protection.

5. Acknowledgments

The work was supported by the National Natural Science Foundation of China (Grant No. 21276106), the Major Research and Development Program of Jiangsu Province (Grant No. BE201596).

References

[1] Giulia O. Bianchetti, Carolyn L. Devlin, Kenneth R. Seddon. Bleaching systems in domestic laundry detergents: a review [J]. The Royal Society of Chemistry: 2015; 5: 65365-65384.

[2] Neal J. Milne. Oxygen Bleaching Systems in Domestic Laundry [J]. Journal of Surfactants and Detergents: 1998; 1 (2): 253-261.

[3] Mohammadreza K, Nahid M, Masumeh R, et al. Quantitative determination of sodium perborate and sodium percarbonate in detergent powders by infrared spectrometry [J]. Journal of Analytical Chemistry: 2012; 67 (4):

330-334.
[4] Takeshi W, Nobuyoshi K. Chemical Composition of Sodium Percarbonate: An Inquiry-Based Laboratory Exercise [J]. Journal of Chemical Education: 2013; 90 (8): 1048-1052.
[5] Hofmann J, Just G, Pritzkow W, et al. Bleaching activators and the mechanism of bleaching activation [J]. Journal für Praktische Chemie/Chemiker-Zeitung: 1992; 334 (4): 293-297.
[6] Zeronian SH, Inglesby MK. Bleaching of cellulose by hydrogen peroxide [J]. Cellulose: 1995; 2 (4): 265-272.
[7] Cai JY, Evans DJ, Smith SM. Bleaching of natural fibers with TAED and NOBS activated peroxide systems [J]. AATCC Review: 2001; 1 (12): 31-34.
[8] Long X, Xu C, Du J, et al. The TAED/H_2O_2/$NaHCO_3$ system as an approach to low-temperature and near-neutral pH bleaching of cotton [J]. Carbohydrate Polymers: 2013; 95: 107-113.
[9] Xu C, Long X, Du J, et al. A critical reinvestigation of the TAED-activated peroxide system for low-temperature bleaching of cotton [J]. Carbohydrate Polymers: 2013; 92 (1): 249-253.
[10] Lim SH, Gursoy NC, Hauser P, et al. Performance of a new cationic bleach activator on a hydrogen peroxide bleaching system [J]. Coloration Technology: 2004; 120 (3): 114-118.
[11] Lim SH, Lee JJ, Hinks D, et al. Bleaching of cotton with activated peroxide systems [J]. Coloration Technology: 2005; 121 (2): 89-95.
[12] Xu CH, Hinks D, Shamey R. Development of a novel bleaching process for cotton [J]. AATCC Review: 2011; 11 (6): 73-77.
[13] Wei D, Sun C, Wang MZ, et al. Synthesis of N-[4-(dimethylalkylammoniomethyl)benzoyl] caprolactam chlorides as cationic bleach activators for low-temperature bleaching of cotton fabric under near-neutral pH conditions [J]. Coloration Technology: 2014; 130: 432-436.
[14] Xu C, Hinks D, Sun C, et al. Establishment of an activated peroxide system for low-temperature cotton bleaching using N-[4-(triethylammoniomethyl)benzoyl] butyrolactam chloride [J]. Carbohydrate Polymers: 2015; 119: 71-77.
[15] Chen W, Wang L, Wang D, et al. Recognizing a limitation of the TBLC-activated peroxide system on low-temperature cotton bleaching [J]. Carbohydrate Polymers: 2016; 140: 1-5.
[16] Koubek E, Haggett ML, Battaglia CJ, Ibne-Rasa KM, Pyun HY, Edwards JO. Kinetics and mechanism of the spontaneous decompositions of some peroxoacids, hydrogen peroxide and t-butyl hydroperoxide. Journal of the American Chemical Society: 1963; 85 (15): 2263-2268.

Selection of Cosmetic Cotton Based on Absorbency and Pipetting Performance

Shi-Yu Zhu[1], Hua Ji[2], Jian-Qiang Li[1*]

[1]*Wuhan Textile University, Hubei,Wuhan, 430200, China*
[2]*Winner Medical (Jingmen) Co., Ltd, Hubei, Jingmen, 448121, China*

[*]Corresponding author's email: ljq@wtu.edu.cn

Abstract

Based on the market and the development of cosmetic cotton, the relationship between suction shift, fluid volume and weight of cotton can be observed by comparing the specifications of cosmetic cotton. This paper tested the liquid absorbability of 5 kinds of cosmetic cotton, with different amount of make-up water (substituted by the distilled water during the experiment), 1ml, 2ml, 3ml, 5ml and saturation state respectively. They were tested for pipetting performance after a simulation experiment. This paper aims to find the basic principles of cosmetic cotton, and provide new ideas for developing new products.

Keywords: Cosmetic Cotton; Nonwovens; Absorbent Properties; Pipetting Performance

1. Introduction

Makeup has a long history in female beauty technology. With the development of the society more and more women like makeup. There are a wide range of cosmetic products in the market of which when used all requires an auxiliary product - cosmetic cotton products.

As an indispensable tool in makeup, cosmetic cotton replaced the hand or cotton swabs when putting on and taking off makeup. Among the varieties of cosmetic cotton, spunlace cotton pad has many advantages like clean, soft, high liquid absorbability, no wrinkle and no hair scraps. In the past ten years, the cosmetics market have developed rapidly and it has become a hot new area in the medical and healthy materials market.

With the rapid development of the cosmetics market, the cosmetic cotton market also has a high-speed growth. Along with this trend, new products continue to emerge, such as color cosmetic cotton for girls as well as men's cosmetic cotton, etc. The development of cosmetic cotton products relies on the market. as a new research area, there are few relevant scientific and technological articles about cosmetic cotton, most of them relate to papers about finishing. [1]. Due to the many incorrect concepts for cosmetic cotton design, this paper will discuss the selection of cosmetic cotton based on the absorbency and pipetting performance.

2. Raw Materials of Cosmetic Cotton and Processing Technology

The choice of cosmetic cotton raw materials and processing technology characteristics, determines the final product performance.

2.1 Raw Material

As a sanitary material, safety requirements must be met. Cosmetic cotton is an external product, must also have good biological adaptability and environmental harmlessness. From the perspective of biological adaptability, cotton, viscose, polyester, etc. undoubtedly have its requirements; and from the perspective of natural degradation and absorbency liquid of materials, cotton and viscose is a better choice; but raw material production from the point of view of pollution-free, cotton has almost become the only choice.

Cotton is the best raw material that possessed the best performance. Its characteristic are as follow: 1) the cotton fiber has the highest proportion of cellulose nature fiber , it have better performance in physical and chemical properties than other natural fibers; 2) Cotton fiber is a high absorbent fiber, the absorption property is good, which conforms to the cosmetic cotton functional requirements; 3) Cotton fiber is also a natural "breathing" fiber, which can effectively prevent the passage of liquid, but allows the passage of gas and water vapor; 4) In the wet condition, the cotton fiber has a higher breaking force (which, contrary to viscose fiber, its wet breaking force is lower than the dry breaking force), and does not have sliding touch as synthetic fiber [2].

People on the material comfort, health and safety requirements, resulting in the 100% absorbent cotton in the present stage of cosmetic cotton market is the most used. The best raw material for cosmetic cotton is absorbent cotton.

2.2 Processing Technology

Absorbent cosmetic cotton processing method is a kind of non-woven processing. The absorbent cosmetic cotton processing: the use of absorbent cotton orderly or disorderly superimposed together to form a cotton with a certain thickness and strength.

The use of non-woven fabric processing methods: web + reinforcement [3]. The fibers are formed by different ways, such as carded web or air laid web, fiber network by superimposing reach the design thickness. In order to improve the strength of the product and reduce the surface hairiness, usually after the superposition of the fiber mesh for a slight surface spunlace treatment

2.3 Flow Chart

Process for purified cotton spunlace production: raw fiber material→mixture and opening→carding→pre wetting→hydroentanglement→dip→mangling→drying→coiling.

Flowing chart shown in Fig.1.

Flow chart: cotton opening and carding, then cross lapping to achieve the required thickness. Degreasing and bleaching the cotton roll under the condition of high temperature and high pressure, and then feeding it in spunlaced machine. The use of spunlace reinforcement to increase cosmetic cotton breaking force, while reducing the plush surface of the cotton pad, after the completion of the spunlace process, drying and winding into a roll [4].

For large package products according to the requirements of cutting, packaging, that is a complete cotton production. The market has a variety of different weight of absorbent cotton's cosmetic cotton, processed in this way.

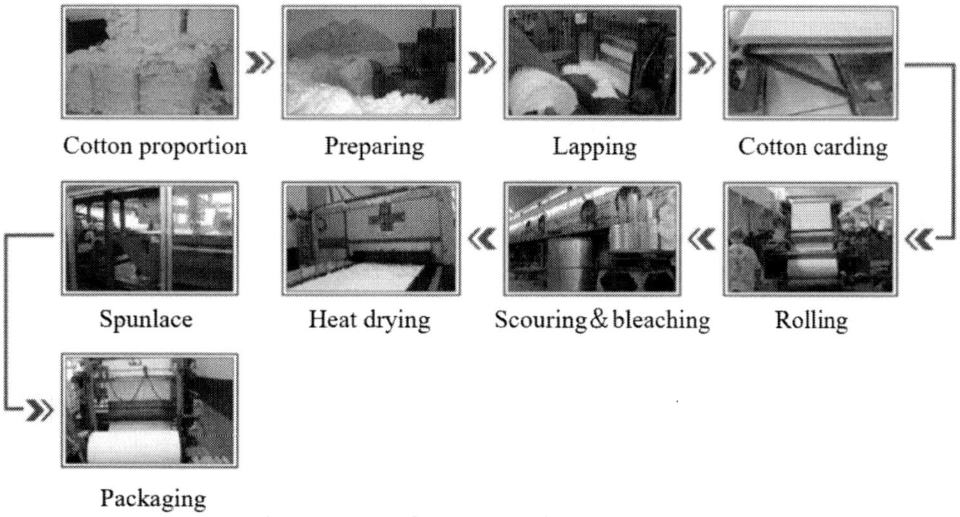

Fig.1 Process of cotton spunlace production.

3. Sample Size

Select the market 5 kinds of different thickness of cosmetic cotton. Thickness from small to large numbered a, b, c, d, e.

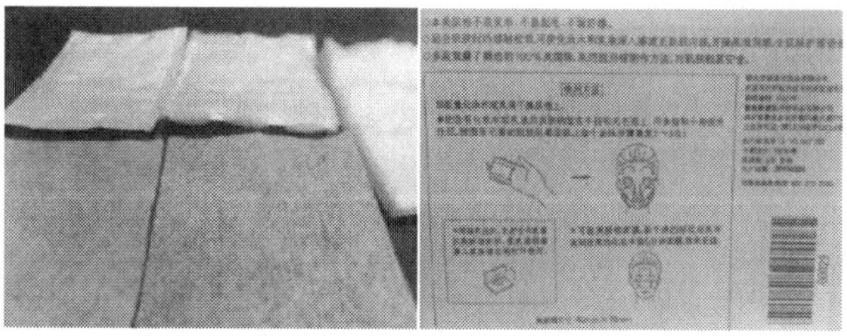

Fig.2 Different cosmetic cotton products

Table.1 Test of cosmetic cotton

	a	b	c	d	e
Thickness (mm)	0.40	0.43	1.60	2.00	3.18
Weight (g/piece)	0.131	0.194	0.359	0.450	0.860
gram weight (g/m²)	40	50	120	150	200

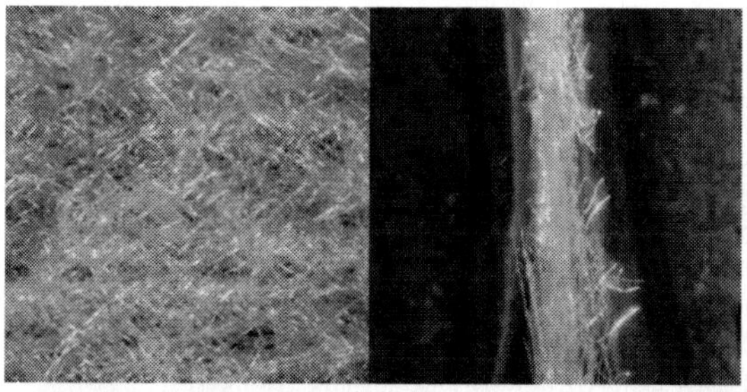

(A)

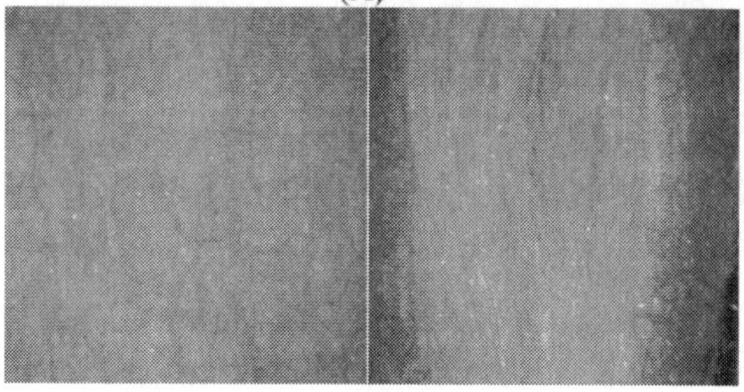

(B)

Fig.3 (A), (B) are the horizontal and vertical photos of the thinnest and thickest cosmetic cotton respectively

4. Cosmetic Cotton Sample Analysis

For show more clearly the absorbent ability of a cosmetic cotton, a experiments were carried out on the condition that the cosmetic cotton was not cropped. In the distilled water by adding the blue ink, put 5 kinds of specifications cosmetic cotton on the white paper, white paper is the following waterproof material, carried out the test of the four groups. In cotton center instillation of 1ml, 2ml, 3ml and 5ml reagent (blue water), rest for 5 seconds, the degree of imbibition of cotton (left), and then take off the cotton pad, observe their degree of penetration in the white paper of the cotton (right).

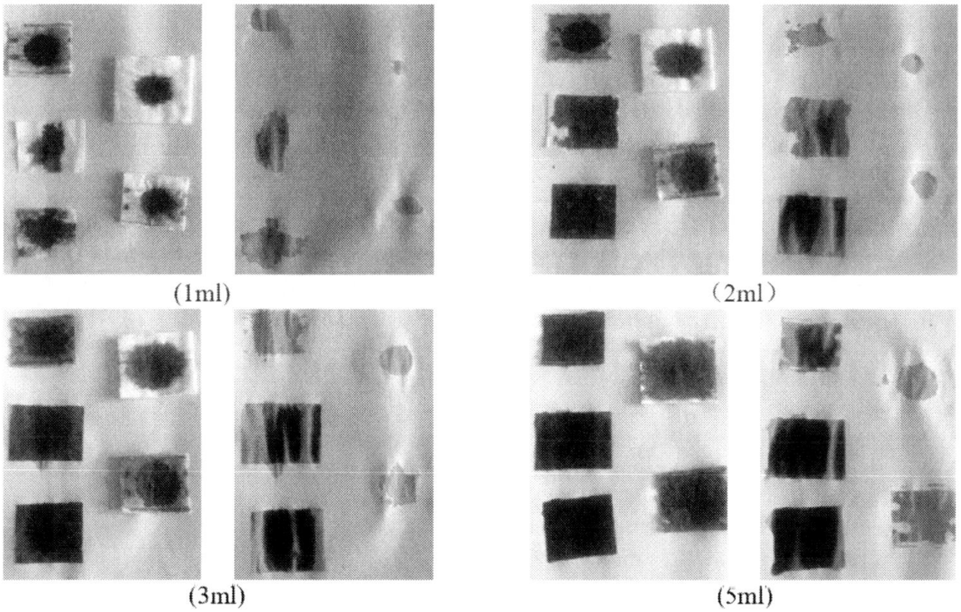

Fig.4 Imbibition experiment

It can be seen from Fig.4 that the use of distilled water 1ml, 5 kinds of cotton pad on the white paper residue less, can effectively absorb. In the use of 2ml and above, sample a, b was saturated, liquid absorption capacity is weak. And of 2ml and 3ml sample c, d can better absorb the distilled water, of 5ml distilled water will produce some leakage. Thick cotton sample e of liquid capacity is very strong, very little distilled water penetration out.

In the following aspiration capacity and pipetting ability test, will be in the five kinds of cotton cut into uniform size of the case of its liquid absorption and the amount of fluid to do the analysis and comparison.

4.1 Cosmetic Cotton Absorbency Test

As for the product fuction, the absorbent property of the cosmetic cotton is one of the most important factors, and also the primary basis for consumer choice. Good cotton need excellent liquid absorbability, when pouring the makeup water on the surface of the cotton pad, it can absorb quickly. And there is no waste phenomenon like liquid leakage [5].

Relevant data shows that the use of cosmetic cotton, a standard amount of make-up water is 2ml [6]. According to this information, this experiment on 5 kinds of cosmetic cotton, with different amounts of 1ml, 2ml, 3ml, 5ml , and saturation state distilled water absorption test respectively [7].

Absorption test method:

1) Sample cutting with of cosmetic cotton:5cm×6cm;

2) Weighting the dry weight of each samples;

3) Pouring distilled water (1ml, 2ml, 3ml, 5ml) into the center of the 5 kinds of cosmetic cotton, and the last group of saturated absorption test, a total of five groups experiments;

4) The samples standing for a few seconds;

5) When without distilled water dripping, weigh the weight of the cosmetic cotton after moisture absorption;

6) Calculate the ability of different cosmetic cotton to absorb liquid, X(g). In the same figure to make 5 sets of line graph, (such as the use of 1ml distilled water, 5 kinds of cosmetic absorbent volume data express in a group of line chart), and comparative analysis.

4.2 Cosmetic Cotton Pipetting Performance Test

As for the product function and cost, the cosmetic cotton pipetting performance is undoubtedly one of the most important performance, but also the factor consumers often ignore. Good cosmetic cotton need excellent ability to release water and also need to transfer the cosmetic water to human skin perfectly, which is called pipetting performance. After using the cosmetic cotton, the less makeup water left on it, the better pipetting performance it has.

This experiment was carried out on a simulated use of a cosmetic cotton. After 5 groups of water absorption tests, the cosmetic cotton was subjected to a cover absorption test with a fixed weight of the absorbent material under the same pressure. Since the liquid absorbability of the absorbent material is greater than the human skin, so test of the pipetting capacity value for the cosmetic cotton is usually larger. But can be compared under this one standard analysis of different grams of cotton effect [8].

Pipetting test method:

1) After the water absorption test, the X values were obtained;

2) Prepare absorbent material, and weigh;

3) After five groups of experiments, 5 kinds cosmetic cotton absorbed by using absorbent materials, under the same pressure operation, unified absorption of the 5 times;

4) Test the weight of the absorbent material, minus the own weight of the absorbent material to obtain five sets of experimental pipetting volume Y (g);

5) The weight of the residual water in the cosmetic cotton, $Z=X-Y(g)$;

5. Test Results and Analysis

The absorption data can be used as an evaluation of the absorbent capacity of the cosmetic cotton; figure 5 shows the result of different weight cotton samples.

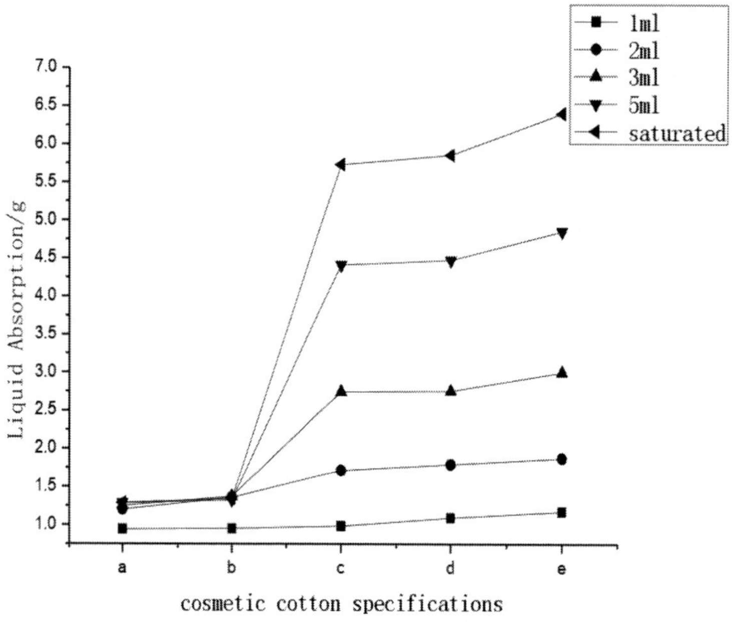

Fig.5 Different weight samples absorption values

The pipetting liquid data can be used as the evaluation of the effectiveness of cosmetic cotton; figure 6 shows the result of different weight cotton samples.

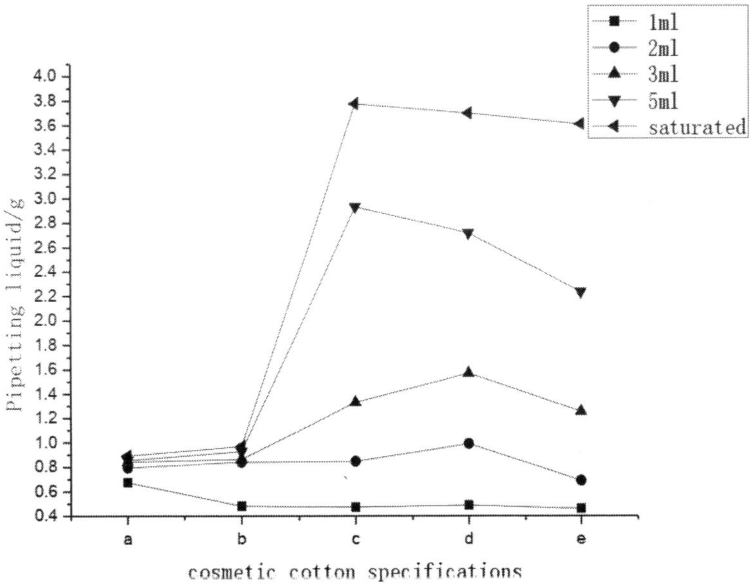

Fig.6 Different weight samples pipetting liquid

The amount of residual weight of the cotton pad after the pipetting test is shown in Fig.7.

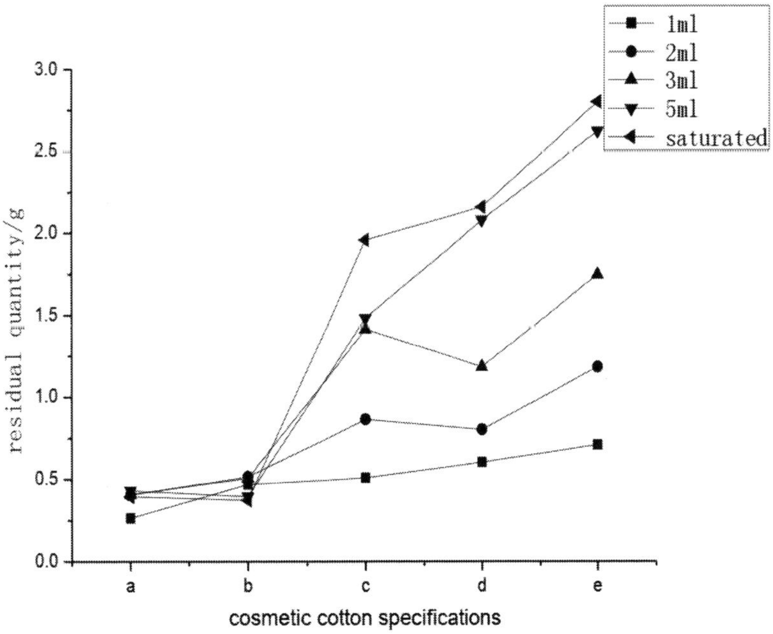

Fig.7 Different weight samples residual quantity

It can be seen from Fig.5 line: different specifications of the cotton pad in 1ml distilled water test liquid absorption little difference, from the data essentially complete absorption 1g of distilled water.

The cosmetic cotton samples a, b, in the use of 2ml, 3ml, 5ml and saturated absorption, the liquid absorption capacity is less than 2g. Samples c, d, e of the cotton pad each absorption values of the numerical difference is not significant, and with the weight increase in the number of liquid absorption gradually increased. So to infer 40g/㎡ and 50g/㎡ of cosmetic cotton saturation of about 1.5g, 120g/㎡ and 150g/㎡ of cosmetic cotton is the 5g or so, and 200g/㎡ cosmetic cotton is the 6g or so.

In Fig.6 we can evaluate which specifications of cosmetic cotton are better transfer the make-up water when the use of the same amount of make-up water. And through can be seen in Fig.7, which specifications themselves absorb (or preserve) more make-up water.

For the amount of liquid transfer or pipetting, according to Fig.6 data, each specification has the ability to transfer liquid as a carrier. However, due to the amount of absorption, the pipetting volume values are difference. In general, the line of Fig.6 pipetting liquid has a tendency to first increase and then decrease. In the use of 1ml distilled water test, it can be seen from Fig.5 in their absorption of liquid under the same conditions, as the cotton pad specifications weight increases, due to the autoabsorption, the amount of liquid is a gradually decreased, the ability to transfer the liquid gradually weakened.

In the use of 2ml, 3ml of distilled water test, the liquid absorption of samples c, d, e in Fig.5 gradually increased Fig.6, the value of the quantity of the pipetting liquid is increased first and then decreased, the best transfer ability is 150g/㎡ of cotton pad.

Samples a, b of cosmetic cotton, because of its absorption capacity is limited, so the volume of liquid is less than 1g. While samples c, d of cosmetic cotton in the use of 5ml and above distilled water, the pipetting liquid volume increased significantly, from the data perspective, are about half the liquid absorption. Sample e of cotton absorbing fluid volume is very large, but due to its autoabsorption of distilled water, so the pipetting liquid amount was less than the samples c, d.

In Fig.7, overall, as the weight increases, due to its own absorption, the transfer capacity of distilled water weakening, so in the use of make-up water, the residue is increased.

Note: the above tests use distilled water for testing, aimed at imitate make-up water absorption and transfer states, the actual situation may be difference.

6. Conclusion

Conclusion 1: Cotton cosmetic in market can achieve the daily use.

From the absorbent properties and pipetting performance tests, It is clear that the products with low weight (such as 40g/㎡ or 50g/㎡) are more economical under the premise of satisfying the basic function of absorption and transfer,. However, their liquid absorption absorbability is relatively weak. For those products with higher weight (such as 200g/㎡), they usually has better liquid absorbability but waste more make-up water.

Conclusion 2: the choice of cosmetic cotton should consider the convenience and cost.

To grasp some information about a product, in addition to using test data for comparison, It is necessary to understand the product development process and market conditions. While the higher price cosmetic cotton is the heavier weight cotton pad, the cheaper cosmetic cotton is the lower weight cotton pad. But in use, may need to be repeated use, and each time make-up water should not exceed 1g, otherwise waste more serious. But every time the use of make-up water, the big heavy weight of the product is residual too much, shortening the use of make-up water period, that is not recommended. While the middle weight of the product has a very good liquid absorption capacity and pipetting performance, can achieve the effectiveness of the use of cosmetic cotton.

Conclusion 3: the choice of cosmetic cotton should consider other properties

Through the test data, type of thick cotton relative to thin cotton will waste more make-up water, usually may shorten the use of the cycle about 30%-50%. However, particularly sensitive skin or especially high dressing sense and flexibility requirements of users can chose 120 g/㎡ of the cotton. As for those 200 g/㎡ above the cosmetic cotton, will retain make-up water, and increase the cost.

Overall, the ideal cosmetic cotton can not only absorb the appropriate amount of make-up water, but also transfer well and release the make-up water, consumer can use make-up water more efficiently, less make-up water waste. At the same time it should have some advantages such as softness and comfort. However, the physiological factors and psychological factors of each consumer are different; And it may lead different choices to the cosmetic cotton products.

References

[1] Ding S, Xu H. Development of Cosmetic Cotton by Living Enzyme Scouring Process. Shandong textile science & technology: 2007; 48(5):10-12.
[2] Yu WD. Textile Materials Science. China textile press: 2006.
[3] Ke QF, Jin XY. Nonwovens. Shanghai: Donghua University press: 2010; 1-140.
[4] Wang YF. Elementary Research on Spun-Laced Technique and Non-Woven Fabrics. Shandong textile science & technology: 2007; 04: 43-45.
[5] Gao J. How to properly use different materials of cosmetic cotton. Every one is Well: 2015 (12)
[6] Ming Y. Buy cosmetic cotton tips. China Anti-Counterfeiting Report: 2016 (4); 122-122.
[7] Ma XH. Study and Analysis on Moisture Absorption Properties of Nonwoven Materials. Manager: 2011 (14); 387-387.
[8] Yu XF. Experimental technology of textile materials. Beijing: China Textile Press: 2004; 3: 230-242.

Factors of Cotton Fabrics Strength Loose and Crease Angle in BTCA Anti-wrinkle Finishing

Qing-Wei Chen[1], Heng Pan[2], Gen-Yang Cao[2*]

[1]*General Design Institute of Hubei Aerospace Technology Research Institute*
[2]*Ministry of Education Key Laboratory of New Textile Materials Green Processing and Functioning, Wuhan Textile University, Wuhan, Hubei, 430073, China*

[*]Corresponding author's email: genyang.cao@wtu.edu.cn

Abstract

In this study, 1, 2, 3, 4-butane tetracarboxylic acid (BTCA) was applied to analyze the factors contributing to the strength loss of cotton fabrics, which were identified to be acid degradation and crosslinking. When pH ranged from 1 to 3, the contribution of acid degradation to the strength loss of the fiber was greater, and when the pH was from 4 to 7, the contribution of crosslinking was greater. At pH 4 and a baking temperature of 140°C, the desired anti-wrinkle effect could be achieved and high strength was observed.

Keywords: 1, 2, 3, 4- Butane Tetracarboxylic Acid; Anti-wrinkle; PH; Mechanical Properties; Wrinkle Recovery Angle

1. Introduction

Cotton fiber, which is one of the centuries-old textile raw materials, still occupies an important place in the textile industry [1]. Cotton is favorable because of its irreplaceable comfort and biodegradability [2]. Pure cotton fabric has several advantageous characteristic of natural fibers, including breathability and softness, however there is wrinkle formation upon washing. Anti-crease finishing of cotton fabric has an important strategy for improving the fabric's performance, it however also results in the loss of strength [3].

The purpose of crease resistant finishing is to identify the equilibrium point between the increase in the fabric crease angle and the decrease in strength, which has become the intrinsic motivation for new process development. Traditional anti-crease finishing involves the addition of hydroxyethyl methylacrylate reagent or formaldehyde; all these methods increase the crease angle but do not address the problem of strength degradation or fabric toxicity after finishing [4-5]. In this study, 1, 2, 3, 4-butane tetracarboxylic acid (BTCA) was applied to solve the aforesaid problems, and to analyze the effect of baking temperature and pH on the crease angle and strength [6-8].

2. Experimental

2.1 Material and Reagents

Cotton bleached fabric was purchased from Anhui Huamao Co., Ltd. The fabric weight, warp density, and weft density were 105g/m^2, 710/10cm, and density 560/10cm, respectively. Sodium carbonate was procured from Tianjin Hengxing Chemical Reagent Co., Ltd. Glacial acetic acid was obtained from Shanghai Shenbo Chemical Co., Ltd. Oxalic acid was supplied by Qingzhou Xinsheng Chemical Co., Ltd. BTCA was purchased from Hubei Youmao Co., Ltd.

2.2 Experimental Procedure

The fabrics were dipped into a solution of BTCA or oxalic acid, rolled (the rolling residual rate was 60%-70%), and then dried at 85℃ (moisture rate: 9%-15%). Next, the fabrics were baked at different temperatures (130℃, 140℃,

150℃, 160℃) for 6 min. Then, the fabrics were washed and neutralized using 5g/L of sodium carbonate and 3g/L of acetic acid solution. Finally, the fabrics were dried at 110℃ for 5 min.

2.3 Analysis of Crease Recovery Angle

An SDL - M003A crease recovery angle tester was used for measuring the fabric wrinkle recovery angle(WRA). The sample size was 40 mm×15 mm, and each group was subjected to 10 rounds of tests.

2.4 Whiteness Index Analysis

The whiteness of the cotton fabrics was measured by a WSB-2 white degree instrument, in the standard D_{65} light source, and the fabrics were combed to the same thickness and opaqueness. The test area was 5cm×5cm, and each sample was tested 5 times over different areas.

2.5 Fabric Strength Analysis

The breaking strength of the cotton fabrics was tested by using an Instron 5566 universal material tester with the standard of GB/T3923.1-1997, The test area was 20cm×5cm, The gauge was 10 cm, and the speed was 10 mm/min. Each sample was subjected to 10 measurements, and the average of the results was calculated.

3. Results and Discussion

3.1 Effect of Baking Temperature and pH on WRA and Whiteness

Before the experiment measured average untreated original sample fabric, the crease recovery Angle of untreated original sample fabric is 41.25° and the whiteness index average is 54.8.

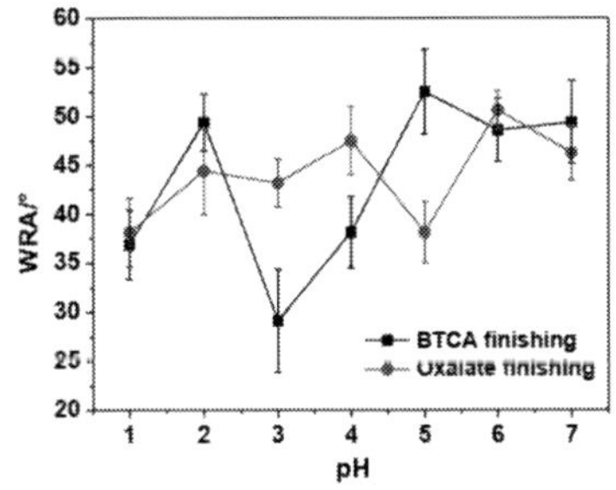

Fig.1 WRA under 85℃ baking with the pH change

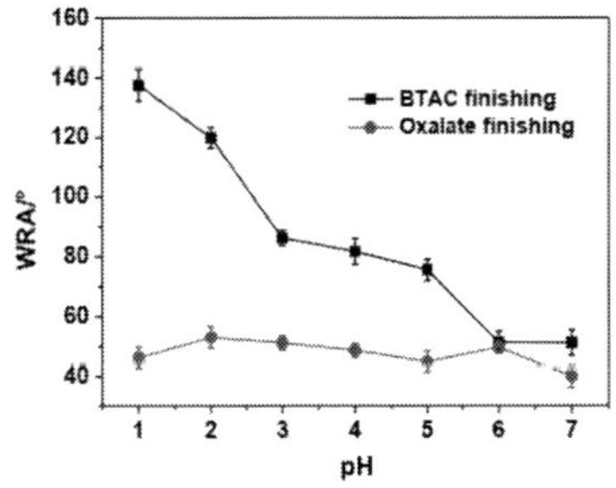

Fig.2 WRA under 130℃ baking with the pH change

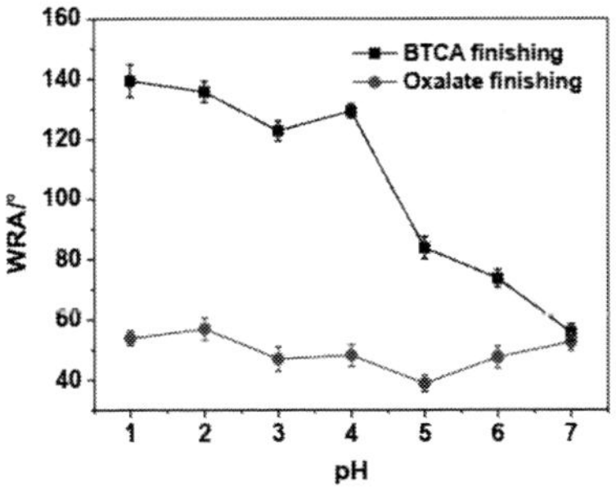

Fig.3 WRA under 140℃ baking with the pH change

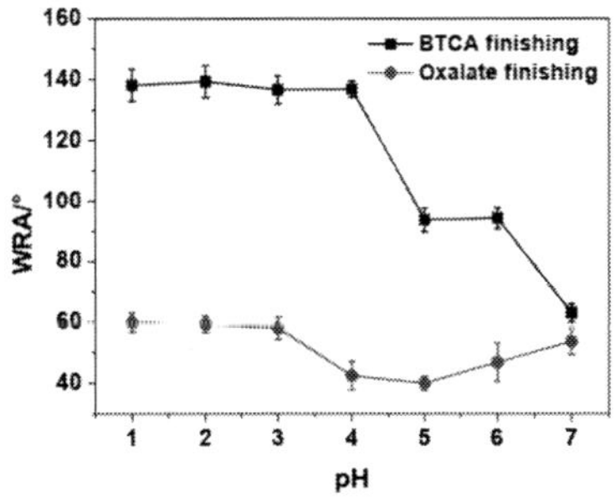

Fig.4 WRA under 150℃ baking with the pH change

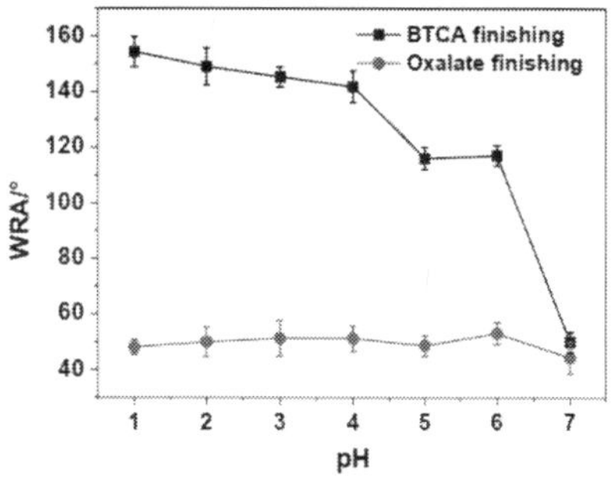

Fig.5 WRA under 160℃ baking with the pH change

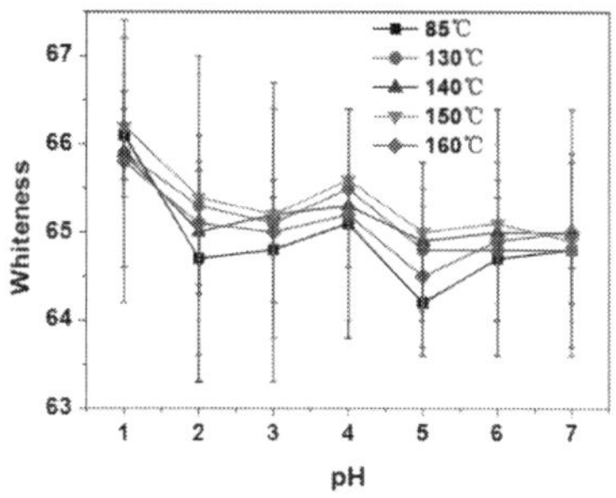

Fig.6 Effect of baking temperature and pH value to whiteness

As seen in Figs.1 to 6, if only oxalic acid was used to adjust the pH and the fabrics were treated at 85℃, 130℃, 140℃, 150℃, 160℃, the crease angle showed no obvious change with increasing pH and temperature and remained stable at 40° to 60°. However, when BTCA was applied for pH adjustment and the fabric treatment temperatures were 85℃, 130℃, 140℃, and 150℃, the recovery angle increased with temperature; when the temperature was 160℃, the wrinkle recovery angle reached the maximum value.

At pH 1 to 4, the WRA was stable against increasing pH, with the maximum WRA being 154.38°. When the pH was more than 4, the change in the crease recovery angle was more pronounced. At pH 7, the WRA decreased with increasing pH, and there was no significant improvement in the crease recovery angle. The baking temperature had little effect on the whiteness of the cotton fabric, and the original whiteness was almost unaltered. Hence, upon treatment with BTCA, the wrinkle resistance of the fabrics increased. When the pH ranged from 1 to 4 and the baking temperature ranged from 140℃ to 160℃, the crease angle increased by approximately 140% and the whiteness remained constant.

3.2 Effect of Baking Temperature and pH on Strength and Strength Loss Rate

Table.1 Effect of baking temperature and pH value to strength

Strength	Strength of cotton fabric (N)									
Baking temperature	85℃		130℃		140℃		150℃		160℃	
Conditioning fluid PH value	Oxalic acid	BTCA	Oxalic acid	BTCA	Oxalic acid	BTCA	Oxalic acid	BTCA	Oxalic acid	BTCA
1	194.9±4.5	185.1±5.6	100.0±3.8	120.9±3.4	78.6±3.1	108.9±6.8	54.5±6.0	104.0±4.9	44.0±8.4	94.89±5.3
2	201.4±3.8	203.4±6.9	114.9±7.8	138.1±10.2	139.0±3.9	123.5±3.8	82.5±3.7	107.7±3.5	58.99±6.4	105.9±6.2
3	207.1±2.6	200.5±8.6	164.3±4.9	141.1±6.7	156.6±3.8	125.4±7.9	100.1±8.4	115.6±8.5	108.1±7.8	109.8±3.5
4	190.5±6.7	195.2±3.9	188.8±8.0	150.4±6.7	184.4±4.9	121.8±5.8	160.5±6.9	114.7±7.2	165.4±7.9	103.0±5.8
5	196.4±6.4	188.7±5.9	201.9±8.2	177.2±4.1	199.6±1.3	164.1±3.8	194.0±4.5	138.4±3.7	194.3±8.5	130.8±6.9
6	205.3±3.7	178.4±3.8	195.1±1.5	188.9±3.4	199.3±5.6	164.8±4.8	187.1±5.9	135.1±6.2	190.0±5.0	125.0±3.1
7	203.2±5.6	188.9±4.6	198.7±3.8	194.5±3.9	202.6±4.1	171.4±8.9	208.1±7.5	138.9±3.4	188.9±4.9	146.9±3.7

The average strength is 211.0N before treatment, Table 1 shows that as temperatures rise, with BTCA and SHP mixture treatment with oxalic acid samples, as temperatures rise, puissant drop. It is believed that during curing of cotton fabrics at high temperatures, the crosslinking reaction between BTCA and cellulose ester proceeds to completion, thus increasing the crosslinking degree and stress concentration. At the same time, extensive acid hydrolysis of the cotton fiber occurs, resulting in damage to the fiber and decline in the fabric strength.

In order to analyze the main reason for the decrease in cotton fabric strength loss, the temperature, pH of the contribution to the fabric of the weight loss, the data are treated as follows:

$$\text{Strength loss rate} = \frac{a-c}{a} \times 100 \Big/ 100 \tag{1}$$

Where a and c refer to the strength of the cotton fabric before and after treatment, respectively.

Table.2 Effect of baking temperature to strength loss rate

Strength loss rate	Strength loss rate of cotton bleached fabric (%)									
Baking temperature	85℃		130℃		140℃		150℃		160℃	
Conditioning fluid PH value	Oxalic acid	BTCA	Oxalic acid	BTCA	Oxalic acid	BTCA	Oxalic acid	BTCA	Oxalic acid	BTCA
1	7.6	12.3	52.6	42.7	62.7	48.4	74.2	50.7	79.1	55
2	4.5	3.6	45.5	34.5	34.1	41.5	60.9	49.0	72.0	50.2
3	1.8	5.0	22.1	33.1	25.8	40.6	52.6	45.2	48.8	48.8
4	9.7	7.5	10.5	28.7	12.6	42.3	23.9	45.6	21.6	51.2
5	6.9	10.6	4.3	16.0	5.4	22.2	8.0	34.4	7.9	38.0
6	2.7	15.5	7.5	10.5	5.5	21.9	11.3	36.0	10.0	40.8
7	3.7	10.5	5.8	7.8	4.0	18.8	1.4	34.2	10.5	30.4

As seen from Table 2, the strength loss of the samples treated with BTCA increased with an increase in temperature and decreased with an increase in pH value. The loss rate of the sample treated with oxalic acid increased with an increase in temperature and decreased with an increase in pH. When the temperature was 85℃, there was no obvious regularity between the strength loss of BTCA or oxalic acid, and the reaction solution reacted weakly with the cotton fiber at 85℃. When the fabrics were treated with BTCA or oxalic acid, the strength decreased as the baking temperature increased; this could be due to the increased degree of crosslinking between BTCA and cellulose, and the drastic acid hydrolysis of the cotton fiber [6]

When it was at the same temperature, as the pH value increased, the effect of acid on the strength weakened. Table 1 shows that when the temperature was 130℃-160℃ and the pH was 1 to 3, the strength loss of fabric subjected to finishing with BTCA was small but the strength loss of the fabric treated by oxalic acid was large, and WRA was larger. This indicated that the crosslinking effect was very prominent in this range, and the strength loss of the crosslinking was small. It is possible that the extent of reaction was very high, in the finishing liquid depth degree, the cross-linking occurred mainly at the surface of a single fiber chain. This would increase the strength of the fiber. When the pH was more than 4, with the effective reduction in BTCA in the finishing fluid, the degree of this reaction was reduced, and acid hydrolysis also weakened; thus, the proportion of the crosslinked fabric was increased. When pH was 4, the loss ratio was the largest, while the overall loss was moderate. It could be inferred that at pH 4, under the conditions of crosslinking, the desired anti-wrinkle effect could be achieved [9].

4. Conclusions

1) When the baking temperature was fixed, the WRA decreased with increasing pH. However, when pH ranged from 1 to 4, the WRA, fabric whiteness, and breaking strength remained stable and unchanged. When pH was changed from 4 to 7, the WRA showed an obvious decline, the breaking strength increased rapidly, and the

whiteness index of fabric did not change.

2) When the pH was fixed, the WRA increased with the baking temperature, the breaking strength decreased rapidly, and the fabric whiteness was unchanged. When the temperature is 140℃, the desired anti-wrinkle effect could be achieved and high strength was observed.

3) When pH ranged from 1 to 4, the effect of acid on the fabric strength was more pronounced than that of crosslinking, and when the pH ranged from 4 to 7, the contribution of the pH was greater than that of the acid to the hydrolysis of the cotton fiber. When pH was 4, the strength of the crosslinking had the maximum loss ratio, the WRA was small, and the total strength loss was moderate. It is concluded that at pH 4, the desired anti-wrinkle effect and high strength could be achieved.

5. Acknowledgment

The authors would like to acknowledge Yunli Wang for their assistance in this work.

References

[1] Wang HF, Chen YY, Lin H. The research history and development trend of cotton fiber [J]. Jounal Of Soochow University Engineering Science Enition: 2003; (4): 13-19.

[2] Chen L. Application status of modification of cotton fiber [J]. Hebei Textile: 2006; (1): 55-58.

[3] Qi H, Huang YE, Ji BL, Sun G. Anti-crease finishing of cotton fabrics based on crosslinking of cellulose with acryloyl malic acid [J]. Carbohydrate Polymers: 2016; 135: 86-93.

[4] Cheng HN. Cotton fabric grafted HEMA anti wrinkle effects on fabric properties [J]. Chemical Fiber & Textile Technology: 2014; 43 (1): 9-12.

[5] Huang L, Lu YP. Advances increase-resist agents for cotton fabric [J]. Printing and dyeing: 2003; 9: 43-45.

[6] Montazer M, Shamei A, Alimohammadi F. Stabilized nanosilver loaded nylon knitted fabric using BTCA without yellowing [J]. Progress in Organic Coatings: 2012; 74 (1): 270-276.

[7] Alimohammadi F, Gashti MP, Shamei A. A novel method for coating of carbon nanotube on cellulose fiber using 1,2,3,4-butanetetracarboxylic acid as a cross-linking agent[J]. Progress in Organic Coatings: 2012; 74 (3): 470-478.

[8] Kono H, Fujita S. Biodegradable superabsorbent hydrogels derived from cellulose by esterification crosslinking with 1, 2, 3, 4-butanetetracarboxylic dianhydride [J]. Carbohydrate Polymers: 2012; 87 (4): 2582-2588.

[9] Xu WL, Cui WG, Li WB, Guo WQ. Two-step durable press treatment of cotton fabric [J]. Coloration Technology: 2001; 117: 352-355.

Cross-section of Flame Retardant Multi-fiber Blended Yarn

Xin-Ping Lu, Ying-Qi Xu*

Fashion School, Shanghai University of Engineering Science, Long Teng Road, Shanghai 201620, China

*Corresponding author's email: xyqq911@hotmail.com

Abstract

In order to study the fiber distribution of multi fiber blended yarn, the preparation of cross-section image are discussed in this paper. According to the performance of flame retardant blended yarn, which is composed of meta-aramid, poly (phenylene-1,3,4-oxadiazole) (POD), flame retardant viscose and flame retardant vinylon, dyes Argazol Red NF-3B and Cationic Blue SD-GSL are applied to dyeing the flame retardant blended yarn by two baths via a two-stage process. After surface treatment by plasma, the yarn was made into continuous slices by using the technique of resin embedding. Then the cross section image of the blended yarn was obtained by CX41RF Microscope (×100). Finally, by means of Photoshop image processing tool box the fiber distribution of the blended yarn were analyzed according to the different geometric morphological characters. Results show that each component of the flame retardant blended yarn can be gained quickly and simply and recognized easily with this method.

Keywords: Dye; Plasma Treatment; Embed; Slice; Cross Section

1. Introduction

Flame retardant yarn is a type of yarn that is difficult to be ignited and is beneficial to reduce fire hazard [1]. After many inspections [2-3] by the national fire equipment quality supervision and inspection center, the limit oxygen index of flame retardant fiber (LOI) is more than 32%. Flame retardant fabric made of flame retardant fibre in flame combustion can decrease the flammability of fabrics as well as slow down the spread of flame speed. Large areas of combustion is not formed. The fabric will self-extinguish fast leaving the burning flame. Application of flame retardant yarn is very broad: decorative textiles in aviation, aircraft, railway, bus, ship, automobile and the interior decorative textile in bedding, high-rise buildings, theatres, radio and television, dance halls, restaurants, hotels and so on. The flame retardant yarn used for military protective clothing needs to meet the requirements of durability, high strength, good wear resistance, good flame resistance and good wear comfort.

The flame retardant yarn in this paper is a kind of special blend yarn composed of meta-aramid, poly (phenylene-1,3,4-oxadiazole) (POD), flame retardant viscose and flame retardant vinylon with permanent flame-retardant effect contains no halogen elements and contains only carbon, hydrogen, oxygen and a small amount of phosphorus. Compared with the traditional halogen flame retardant agent, the flame retardant fiber does not produce asphyxiating poisonous gas when burning, has low smoke, good flame retardant effect, good thermal stability, non-toxic etc. Although this kind of yarn has excellent flame retardant performance, it has high strength irregularity and is easy to break. In order to evaluate the relationship between yarn structure and strength, the cross-section of the blended yarn were studied in this paper.

2. Experimental

2.1 Materials, Reagents and Instruments

2.1.1 Materials

Meta-aramid	1670 dtex/1000f	Yantai Tayho Advanced materials Co.,LTD

POD	1.67 dtex*70cm	Jiangsu POD New Materials CO., LTD
Flame retardant viscose	200 dtex/50f	Jilin Chemical Fiber CO ., LTD
Flame retardant vinylon	1.5 dtex*70cm	Shanghai All Cosmos Bio-Tech Co ., LTD
Flame retardant blended yarn	36 s	Shandong Woyuan new fabric co., LTD

2.1.2 Reagents

Argazol Red NF-3B	Argus(Shanghai) Enterprise Group
Cationic Blue SD-GSL	Shanghai Luojing Dyestuff Chemical Co.
Citric Acid	Yixing second chemical reagent factory
Sodium Dihydrogen Phosphate	Sinopharm Chemical Reagent Co.,LTD
JFC	Shanghai Auxiliary Factory Co., LTD.
E44(6101) Epoxy Resin	
650 Low Molecular Polyamide Resin	Hangzhou Wuhuigang adhesive Co., LTD.
DBP	

2.1.3 Instruments

AL104 Electronic Balance	Shanghai MEITO Metrical Scientific & Technical Co., LTD.
JT601N Electronic Balance	Shanghai Jing Tian Electronics Instrument Co., LTD.
Linkup Series Infrared Dyeing Machine of High Temperature and High Pressure	Wuxi Lipu Technology Co .,LTD.
STR/I Plasma Processor	Shanghai Textile Research Institute Co., LTD
Y802N 8-basket Conditioning Oven	Changzhou Textile Instruments Factory Co.,LTD
KD-202 Hand-style Slicer	Zhejiang Jinhua Kedi Instrumental Equipment Co., LTD
CX41RF Microscope	Olmpus(China) Co., LTD
KQ-700B Ultrasonic Cleaner	Kunshan Ultrasonic Instruments Co., LTD

2.2 Methods

The main procedure of preparation of cross section image of the flame retardant multi-fiber blended yarn includes: Dye yarns→ Modify the surface of fibers→ Embed yarns with resin→ Use hand-style slicer to make embedded objects into sections→ Obtain the cross-section images of the blended yarn for processing.

2.2.1 Yarn Dyeing

1) Processing parameters of the first step

Bath ratio	1: 20
Argazol Red NF-3B/% (owf)	x
JFC/g•L-1	0.5
PH	7

2) Processing parameters of the second step

Bath ratio	1:20
Cationic Blue SD-GSL/% (owf_	y

3) Processing route

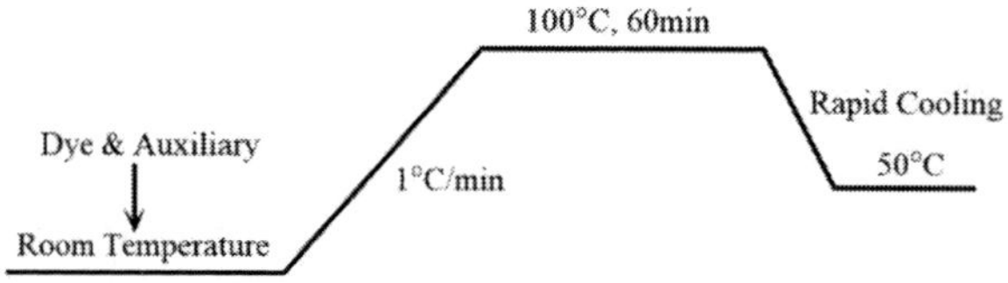

First of all, all of the dyes and auxiliaries should be put into the distilled water at the room temperature. Then the temperature would increase to 100 °C by 1 °C/min. Yarns should be dyed for 60 minutes under this condition. After that, temperature should be cooled rapidly to 50 °C. The dyeing processing routes of the two steps of two stage process are the same. After each step, yarns should be washed repeatedly with distilled water in KQ-700B Ultrasonic Cleaner and dried by Y802N 8-basket Conditioning Oven.

2.2.2 Fiber Surface Modification

Vacuum degree: 0.7

Processing time: x s

Fiber can be easy to combine with the resin after modifying by STR/I Plasma Processor and is not easy to slip and lodge in slicing. In order to avoid the plasma treatment failure, the surface of treated yarns should be covered with resin as soon as possible.

2.2.3 Resin Embedding

According to product specification, process parameters are as follows: E44(6101) Epoxy Resin: DBP: 650 Low Molecular Polyamide Resin= 100: 15: 90.

Important points:

1) Mix BDP with E44(6101) Epoxy Resin at first.
2) Use KQ-700B Ultrasonic Cleaner to eliminate bubbles.
3) Do not mix water with embedding medium.

The two ends of a dried yarn should be fixed at the diagonal position of a polyethylene pipette with 5 mm diameter and 5-10 cm in length. The yarn should be straight, and the middle part of it should be in the middle of the pipette, not close to the wall of the pipette, and resin would be filled in the pipette. Then seal the one end of the pipette with a double-sided adhesive, and let the pipette upright in a small hole in the thick plastic foam board. At last, use Y802N 8-basket Conditioning Oven to dry it.

2.2.4 Slicing Technology

Set the initial scale of the KD-202 Hand-style Slicer to 25 μm, and slice the embedded object into sections until a complete section has been achieved. Then set the initial scale to 10 μm to get the final section. At last, a 10 μm complete section with both sides being smeared with E44(6101) Epoxy Resin should be put on the glass slide and covered by a cover slip. The photo should be taken after eliminating bubbles.

2.2.5 Micrographs of Cross-sections Preparation

Use CX41RF Microscope (×100) to observe the sections, and take photos. Then use the quick selection tool of Photoshop software to separate fibers from complex background.

3. Results and Discussion

3.1 Dyeing Process

3.1.1 Dyeing Material

According to the performance of the compositions of the flame retardant blended yarn dyeing conditions and effects of the four kinds of fibers are shown in Table 1.

Both meta-aramid and POD have similar circular sections, but their dye-uptakes are different due to their crystalline structure [4-6], so they can only be distinguished by color. Since meta-aramid and Flame retardant viscose will be harmed by acid reactive and cationic dyes are good choice [7]. The zigzag cross section of flame retardant viscose is so obvious that it can be recognized evidently by using reactive dyes to dye it [8-10]. Although all the dye uptake of flame retardant vinylon is very low, the kidney-shaped section of this kind of fiber can be identified obviously [11]. In this paper, in order to get different dyeing characteristics so as to have a clear distinction, Argazol Red NF-3B and Cationic Blue SD-GSL are selected to dye the flame retardant blended yarns.

Table.1 Dyeing conditions and effects of the four kinds of fibers

Fiber	Shape of cross section	Performance of resistance to acid and alkali	Suitable dyes	Dyeing effect
Meta-aramid		Good alkali resistance and bad strong acid resistance	Cationic dyes Disperse dyes	High crystallinity lead to the low dye uptake
POD		Good alkali resistance and good acid resistance	Cationic dyes Disperse dyes Acid dyes	High dye uptake
Flame retardant viscose		Good alkali resistance and bad acid resistance	Reactive dyes	High dye uptake
Flame retardant vinylon		Good alkali resistance and good acid resistance	Reactive dyes Cationic dyes Disperse dyes	Low dye uptake

3.1.2 Dyeing Method

Generally, there are several kinds of process for dyeing multi-fiber blended yarn. For example one bath one stage and two bath two stage processes. The dyeing effects of them are shown in Fig.1.

(a) One bath one stage (b) Two bath two stage

Fig.1 Dyeing effects of the two different dyeing methods

As can be seen from the Fig.1, fibers can't be recognized from one bath one stage process, but can be recognized easily from two bath two stage process. It can be seen in Fig.1.(b) that meta-aramid is blue rotundity with white core, POD is homogeneous stain blue rotundity, flame retardant viscose appears red and flame retardant vinylon is in purple. It's because that different dyes have different dyeing conditions, such as temperature, PH, leveling agent, electrolyte and so on. They may influence each other so as to change the shade, reduce the dye-uptake, colour fastness and discrimination of different, and it will be more difficulty to recognize fibers [12]. So in this paper, two bath two stage process is select to dye yarns.

3.1.3 Dye Concentration

3.1.3.1 Concentration of Argazol Red NF-3B on dyeing Flame Retardant Viscose

In the dyeing process with Argazol Red NF-3B, only flame retardant viscose is dyed. So the effects of concentration of Argazol Red NF-3B on dyed flame retardant viscose are compared, which is shown in Fig.2.

Among the four concentrations of Argazol Red NF-3B, the outline of flame retardant viscose is not clear in 1% (owf). When concentration of Argazol Red NF-3B is changed to 2% (owf) and 3% (owf), flame retardant viscose is homogeneously dyed with clear outline. When concentration of Argazol Red NF-3B is increased to 4% (owf), flame retardant viscose is tend to be dyed unevenly and its outline begins to blur. In this paper, Argazol Red NF-3B with 2% (owf)concentration is finally selected to dye the flame retardant blended yarns.

(a) 1% (owf) Argazol Red NF-3B (b) 2% (owf) Argazol Red NF-3B (c) 3% (owf) Argazol Red NF-3B (d) 4% (owf) Argazol Red NF-3B

Fig.2 Effect of different concentration of Argazol Red NF-3B on dyeing flame retardant viscose

3.1.3.2 Concentration of Cationic Blue SD-GSL on dyeing Meta-aramid and POD

In this paper, the dyeing effect of meta-aramid and POD in three different concentrations of Cationic Blue SD-GSL with 2% (owf) Argazol Red NF-3B are compared. See the below details shown in Table 2 and Fig.3.

Table.2 Formulation of Cationic Blue SD-GSL and Argazol Red NF-3B

Formulation	Cationic Blue SD-GSL (g): Argazol Red NF-3B (g)
A	1:25
B	1:50
C	1:75

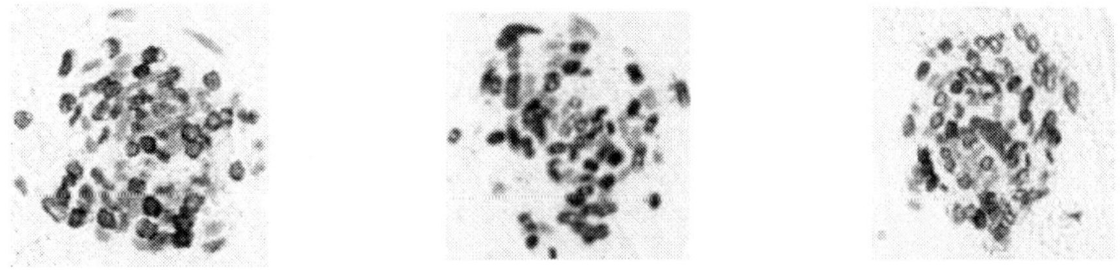

(a) Formulation A (b) Formulation B © Formulation C

Fig.3 Cross sections of dyed yarns of different formulation of Cationic Blue SD-GSL and Argazol Red NF-3B

As shown as Fig.3 four kinds of fibers can be distinguished quite clearly with Formulation C in which Cationic Blue SD-GSL (g): Argazol Red NF-3B (g) is 1:75. The cross section of meta-aramid is blue rotundity with white core. The cross section of POD is homogeneous stain blue rotundity. Flame retardant viscose appears red zigzag cross section. Flame retardant vinylon will present a cross section of a purple kidney-shape. In this paper, formulation C is selected to dye the flame retardant blended yarns.

3.2 Time of Plasma Treatment

The flame retardant yarn is hydrophobic, in order to improve the quality of the cross-sections, fiber's surface needs to be modified before being embedded [13]. Plasma treatment is a surface modification technique that readily primes any surface for better acceptance of secondary manufacturing applications. More and more conventional industrial pretreatment methods are being replaced by plasma technology in order to make processes more effective and environmentally friendly. It can be used to improve the hydrophilicity of the material [14-16] while other performance won't be affected [17-19].

In the condition of that vacuum degree is 0.7, yarns are treated with plasmas for different time, and results are shown in Fig.4.

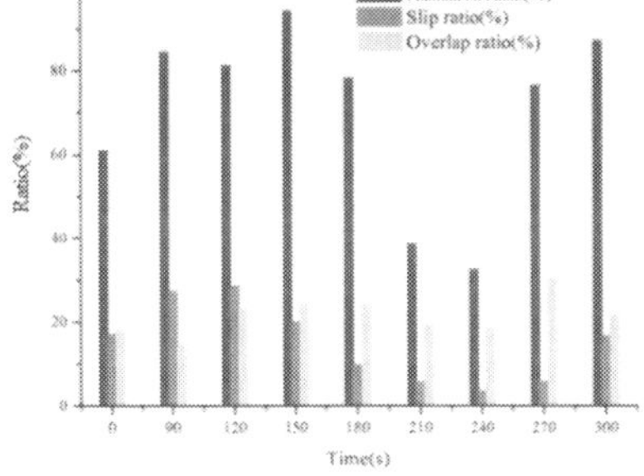

Fig.4 The effect of different plasma treatment time on fiber surface modification

Adhesion ratio, Slip ratio and Overlap ratioare used to describe the effect of different plasma treatment time on fiber surface modification. They can be calculated by below formulas:

$$\text{Adhesion ratio} = \frac{\text{The number of adherent fibers}}{\text{The number of totle fibers}} \times 100\% \quad (1)$$

$$\text{Slip ratio} = \frac{\text{The number of slip fibers}}{\text{The number of totle fibers}} \times 100\% \quad (2)$$

$$\text{Overlap ratio} = \frac{\text{The number of overlap fibers}}{\text{The number of totle fibers}} \times 100\% \quad (3)$$

As show in the bar graph, when the duration of plasma treatment is in the range of 0 s to 300 s, only 210 s and 240 s are better than the untreated fiber surface based on comparison of the quality of the cross section image. When overlap ratio is compared, the difference between these two periods is not significant, but slip ratio and overlap ratio of 240 s is slightly lower than that of 210 s'.

Cross sections of dyed yarns before and after plasma treatment for 240 s are compared in Fig.5. It can be seen that after the plasma treatment for 240 s, the fibers are completely filled with epoxy resin E44 in blended yarn. Contact between the fiber and the resin are more closely, fibers are not easy to distort by squeezing during slicing [20]. In addition, the fibers are not easy to lodge and pull out [21], so it is much easier to judge the distribution of yarn components.In this paper, time of 240 s is selected to treat the blended yarns with plasma in order to modify the surface of fibers.

(a) Before plasma treatment (b) cross section after plasma treatment

Fig.5 Cross sections of dyed yarns before and after plasma treatment

3.3 Analysis of Cross-section Images of the Flame Retardant Multi-fiber Blended Yarn

After blended yarns have been dyed by dyes Argazol Red NF-3B and Cationic Blue SD-GSL as the formulation of C (see Table5) with two bath two stage process, modified by STR/I Plasma Processor for 240 s, embedded by resin in accordance with a certain proportion(E446101 Epoxy Resin: DBP: 650 Low Molecular Polyamide Resin= 100: 15: 90), sliced by KD-202 Hand-style Slicer, taken photo by CX41RF Microscope (×100) and processed by 'Photoshop', the final cross-section images of fibers and flame retardant blended yarn will be obtained as shown in Fig.6.

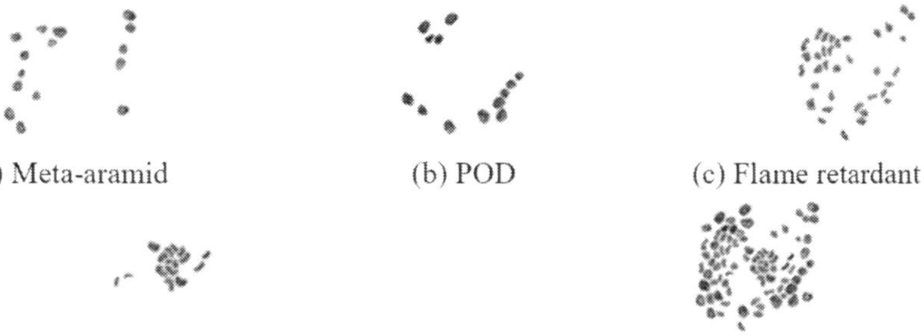

(a) Meta-aramid (b) POD (c) Flame retardant viscose

(d) Flame retardant vinylon (e) Flame retardant blended yarn

Fig.6 The cross-sections of fibers and flame retardant blended yarn after image processing

According to a quantity of images, it can be concluded that meta-aramid and POD are distributed in outer blended yarn, and flame retardant viscose and flame retardant vinylon are distributed in inner blended yarn.

4. Conclusion

In this paper, a study on the cross-section of flame retardant multi-fiber blended yarn is investigated by two bath two stage process with different dye concentration and plasma treatment. Finally, the following conclusions can be obtained:

1) After dyeing blend yarns by two bath two stage process, the color of four component fibers are different generally. According to the features of different color and shapes of the cross sections of each fiber, the distribution of fibers can be estimated.

2) When the mass ratio of Cationic Blue SD-GSL to Argazol Red NF-3B is 1:75, meta-aramid and POD can be identified easily.

3) With plasma treatment for 240s, the yarns can be made into continuous slices easily after embedded by epoxy resin. This process can insure fibers not easy to lodge and pull out so as to make it easier to judge the distribution of each component of blended yarn.

4) By image processing techniques it can be concluded that meta-aramid and POD are distributed in outer blended yarn, and flame retardant viscose and flame retardant vinylon are distributed in inner blended yarn.

References

[1] Feng JB. Research on flame retardant fiber, flame retardant yarn and flame retardant fabric[J]. Light and Textile Industry and Technology: 2004; 33 (3): 35-39. (in Chinese)

[2] Oguz Demiryürek, Erdem Koc. Predicting the Unevenness of Polyester/Viscose Blended Open-end Rotor Spun Yarns Using Artificial Neural Network and Statistical Models,Fibers and Polymers. J: 2009; 10 (2): 237-245.

[3] Oguz Demiryürek, Erdem Koc. The Mechanism and/or Prediction of the Breaking Elongation of Polyester/Viscose Blended Open-end Rotor Spun Yarns, Fibers and Polymers. J.: 2009; 10 (5): 694-702.

[4] Wang JM, Wang HX, Xie FF, Li Y. Dyeing properties of Aramid fiber 1313, Dyeing & Finishing: 2009; (18): 17-19. (in Chinese)

[5] Liu H. Research on the structure and property of Newstar® [D]. Dong Hua University: 2010. (in Chinese)

[6] Liu LQ, Chen L, Hu ZM, Yu JR, Jing Z. Structure and Properties of Dope-Dyed Poly (m-phenylene isophthalamide) Fibers By Wet Spinning [J]. Asian Journal of Chemistry: 2013; 25 (7): 3952-3956.

[7] Li WT, Wu M, Zhang CB, Yao SH, Sun J. Properties and applications of aromatic polyoxadiazoles fiber [J]. Hi-Tech Fiber & Application: 2013; 38 (1): 31-33,71. (in Chinese)

[8] Shi MW, Ren PF. The Structure and Properties of High Tenacity PVA Fibers [J]. China Fiber Inspection: 2011; (07): 76-77. (in Chinese)

[9] Ge WB, Fu XW, Jiang MJ, Xu JJ, Ye GD. Structures and properties of flame retardant vinylons and development trend thereof [J]. Journal of Textile Research1: 2012; (06): 136-143. (in Chinese)

[10] Ru XW, Yan YJ, Shi MW, Jin Z, Zhou JG. Wet processing of vinylon blended fabric with high tenacity and flame retardancy[J]. Dyeing & Finishing: 2013 (5); 26-28. (in Chinese)
[11] Chen Y, Li C, He YF, Zheng QK. Dyeing of flame retardant viscose fiber blending with polyphosphazene[J]. Dyeing & Finishing: 2015; (02); 27-29: 38.(in Chinese)
[12] Song XY. One bath dyeing of multi-component textile with reactive dyes and the assistants [J]. Textile Auxiliaries: 2009; 26(9): 1-8.(in Chinese)
[13] Dong WG. Surface treatment technology for high performance fiber [J]. Shandong Textile Science & Technology: 2001; 11 (6): 44-46. (in Chinese)
[14] Sun SY, Sun J, Yao L. Wettability and sizing property improvement of raw cotton yarnstreated with He/O2 atmospheric pressure plasmajet [J].Applied Surface Science: 2011; 257: 2377-2385.
[15] PANDIYARAJ Navaneetha K, SELVARAJANV. Non Thermal plasmatreatment for hydrophilicity improvement of grey cotton fabrics [J]. Journal of Materials Processing Technology: 2008 (199); 130-139.
[16] TEMMERMAN T Eef, LEYS Christophe. Surface Modification of cotton yarn with a DC glow dischargein ambientair [J]. Surface & Coatings Technology: 2005 (200); 686-689.
[17] Huang SJ, Shi ZG, Gu YP. Influence of Plasma Processing on Mechanical Properties of PSA Fiber [J]. Journal of Shanghai University of Engineering Science: 2009; 23 (4): 321-324 (in Chinese)
[18] Dai JJ. Application of plasma technology in textile processing [J]. Journal of Textile Research: 1996; 17 (6): 20-23. (in Chinese)
[19] Mao ZY. Introduction of low temperature plasma surface treatment technology for textiles [J]. Nonwovens: 2003; 11 (4): 20-22.(in Chinese)
[20] Shao JD. Application of low temperature plasma technology in surface modification of textile materials [J]. Sichuan Silk: 2005 (3); 20-22. (in Chinese)
[21] Niu Q, Luo JJ, Xia YX, Sun SQ, Chen Q. Surface modification of bio-char by dielectric barrier discharge plasma for Hg-0 removal [J]. Fuel Processing Technology: 2017; 156: 310-316.

Manufacturing Three Dimensional Tubular Fabric Reinforced Composite with a Unibody Molding Method

Li-Li Wang[1], Shang-Yong Zhang[2], An-Chang Xu[3*]

[1]*College of Textile Science and Engineering, Wuhan Textile University, Sunshine Avenue, Wuhan, Graduate student, 735887129@qq.com, China*
[2]*College of Textile Science and Engineering, Wuhan Textile University, Professor, 313598106@qq.com*

*corresponding author's email: xuanch@live.cn

Abstract

The study of fiber reinforced plastic composites are attractive to researchers because of their advantage in being lightweight and strong. Three dimensional fabric reinforced composites are now the most favored materials for they not only possess the advantages of normal composites they are also unibody moldable which can reduce processing time and cut costs. In 3D tubular composite the force condition is unlike plant composites, the stress concentration in 3D tubular is not as serious as free edge of plant composites. As such it is reasonable to believe that the 3D tubular composite will be more widely used in the aerospace vehicles industry as structural parts. The purpose of this study is to develop a unibody molding method to manufacture three dimensional tubular fabric reinforced plastic composite. Angle interlock is used as the basic fabric weave to fabricate a tubular fabric, by changing the interweaving between warp and weft thread the thickness of the tubular fabric can be designed.

Keywords: Three Dimensional Tubular Fabric; Tubular Composite; Unibody Molding; Mechanical Performance

1. Introduction

The 3D tubular fabric is woven by combining tubular weaving process and 3D fabric weaving process, when the high performance fibers such as aramid fibers are employed as the reinforcing material in tubular composite, they are with many advantages such as lightweight and strong which make them are suit for widely using in industry, transportation, military, aerospace and some other fields [1]. For tubular fabrics and other unconventional geometric fabric reinforced composite materials, the design of the mold is very important in composite fabrication, because it is concerned to if it is possible for unibody molding.

A lot of scholars have studied 3D tubular fabric composite material on the topic of mechanical properties of tubular composite materials, use various experimental and theoretical methods to investigate the relationship between performance and fabric structure. However, there is insufficient research on the fabric in 3D tubular composites along the wall thickness direction of the superposition effect and the corresponding mechanical properties, this paper will explore the relationship between two kinds of composite pipes.

2. Experiment

2.1 The Design of Tubular Fabrics

Integral forming dimensional tubular fabric weaving process is when double fabric weaving process is combined with 3D fabric weaving process. This paper selects the following 3D fabric structure, 2 , 3, 4 layer weft angle interlock structure, as shown in the layer pattern design. As comparison the twill fabric is fabricated and winded into tube as tubular reinforcement.

In the weaving, the raw material used is Kevlar K29, 1500D, Produced by the United States DuPont.

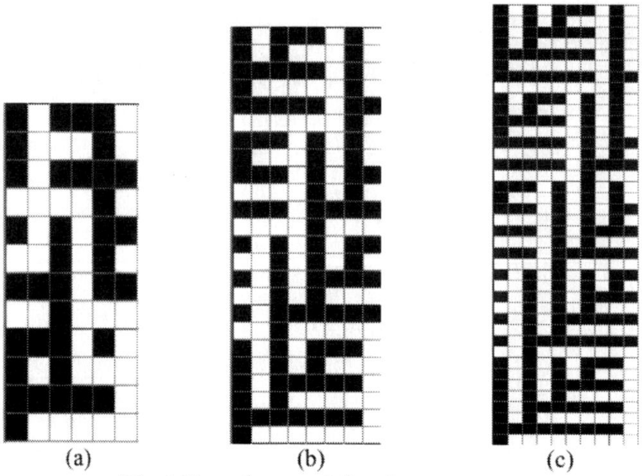

Fig.1 Two, three and four layer pattern

2.2 Design of Lining Mold

This paper uses vacuum assisted resin transfer molding composite technology. This technology have developed very fast in recent years, it is widely used in the aircraft industry, automobile industry, shipbuilding industry and other fields while meeting the needs of different applications [2].

In the VaRTM process, to ensure that the molding material has a good geometric shape and that the resin infiltration is done evenly, the mould and the preform requires good adhesion and sealing. During the resin curing process, the tubular composites formed will have a certain degree of contraction, resulting in the preform being tightly attached to the middle of the cylinder liner on the outer wall of the cylinder lining mold. When removed by external force, friction between cylinder lining and composite cylindrical inner wall will hinder the mould take off. It has caused great difficulties in taking off the mould. So it is very important to design a new kind of mould, by using which the tubular composite can be released from the mould through simple operation.

The vacuum assisted resin transfer molding process was employed in tubular composite fabrication and the newly designed mould is shown in Fig 2. This mould includes a cylindrical lining body with a set of inner casing and an outer casing placed in the two ends of the cylindrical lining body. The cylindrical lining body can be divided into some segments along the axial direction. The inner casing is convex shaped cylinder, and its small diameter is the same as the inner diameter of the cylindrical lining body, its large diameter is the same as the outer diameter of the cylindrical lining body, this design helps hold the separated cylindrical lining body in cylinder. The outer casing is a hollow tube and its inner diameter is the same as the outer diameter of the cylindrical lining body, which helps the inner casing hold the cylindrical in position. During compounding process vacuum will provide pressure on the cylindrical lining mould, which will help the mold hold the cylindrical shape. The detail of composite fabrication setup will be described later.

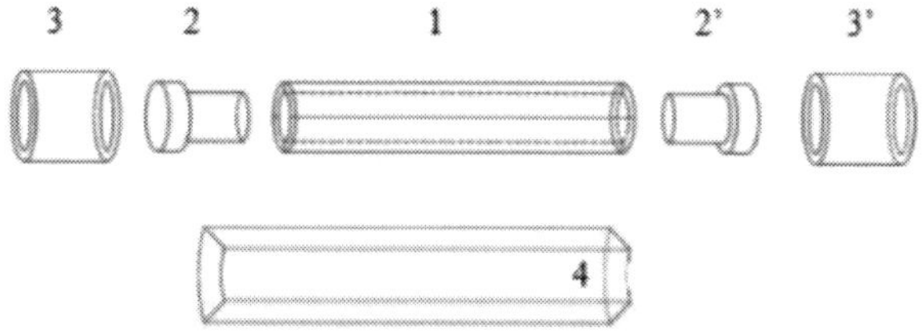

Fig.2 Schematic diagram of the mould, 1: cylindrical liner; 2and 2': inner casing; 3 and 3': outer casing; 4: separation part

2.3 Mould Principle

When taking off the mould after the composite is finished, because the resin has a certain degree of contraction during the curing process, is the preform is tightly attached to the middle cylinder lining on the outer wall with a resistance of:

$$F= \mu PS \tag{1}$$

μ is the friction coefficient between the inner surface of the cylinder liner and the inner wall of the tubular composite material; P is the pressure of cylinder liner; S is the contact area between the cylinder liner and the inner wall of the tubular composite material.

Because of the geometry of the tubular inner lining mould with a regular shape and uniform stress, the outer wall of cylinder liner body is always under a certain pressure. By separating the cylindrical liner into several segments, the contact area S between the cylinder liner and the inner wall of the tubular composite material can be reduced, which will effectively decrease the pulling force F required during demold process.

The outer surface of the cylindrical body lining smearing release agent, can not only effectively reduce the friction coefficient between the cylindrical lining surface and the inner wall of the tubular composite material, it also can avoid resin adhesion with lining mould. The cylinder lining body is segmented to effectively reduce the contact area S and reduce resistance F during the removal of the mould. The greater the segmentation of the body, the smaller the resistance, operand the easier it is to take the cylinder lining off.

2.4 Compound Process

The composite design is shown in the following figure:

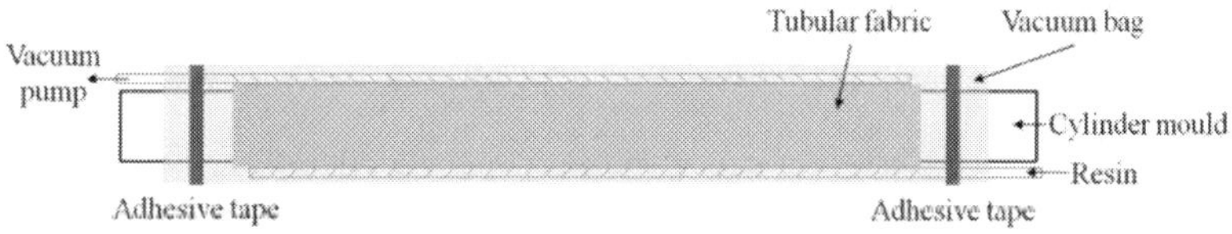

Fig.3 Schematic diagram of the VaRTM setup

The entire composite process:

1) Sample preparation: After the assembly of the cylinder liner body 1, the release agent is applied on the outer surface of cylindrical lining body, the inner casing 2,2' and outer casing 3,3' were placed in the ends of the cylindrical lining body which help the cylindrical lining hold the cylinder shape from both inside and outside of the cylindrical lining. Then the whole setup of part 1 and 2,2' and 3, 3' shown in Fig 2 forming a cylinder was inserted into tubular fabric. As shown in Fig 3 the VaRTM setup is made, under the vacuum the tubular fabric will cling to thr cylinder lining body which gives the final composite a perfect cylinder shape.

2) The stripping cloth and the diversion network were wrapped around on the tubular fabric, which makes the resin can flow smoothly in the fabric and fully infiltrated the entire tubular fabric.

3) After debugging the device, the vacuum pump is opened. The resin gradually infiltrates the entire fabric along the diversion net. When the infiltration is complete, the vacuum pump is closed, and the two sides of the diversion pipe is sealed by clamping them. The composite process is completed and left static for 24 hours for resin curing.

4) When the resin curing is completed, the outside wrapping auxiliary material is first removed, and then take off outer casing 3,3 ' and inner casing 2,2', without the supporting of inner and outer casing, the segments of cylindrical lining can be easily take off from the inner face of the tubular composite, finally after demolding there will be a perfect cylinder shape composite for further testing.

2.4 Mechanical Performance Testing and Analysis

The mechanical performance of tubular composite includes tensile, bending, compression, impact, shearing and more. In consideration of the text fixture and form that the weft was forced into during the test form, the compression test and bending test was chosen. The compression test examines stress and strain while the bending test examines force and displacement.

Testing machine is Electronic Universal Testing Machine, RGM-4100Model, The range is 100KM, product by Shenzhen Rui-Geer Instrument Company.

3. Result and Discussion

3.1 Compression Property Test

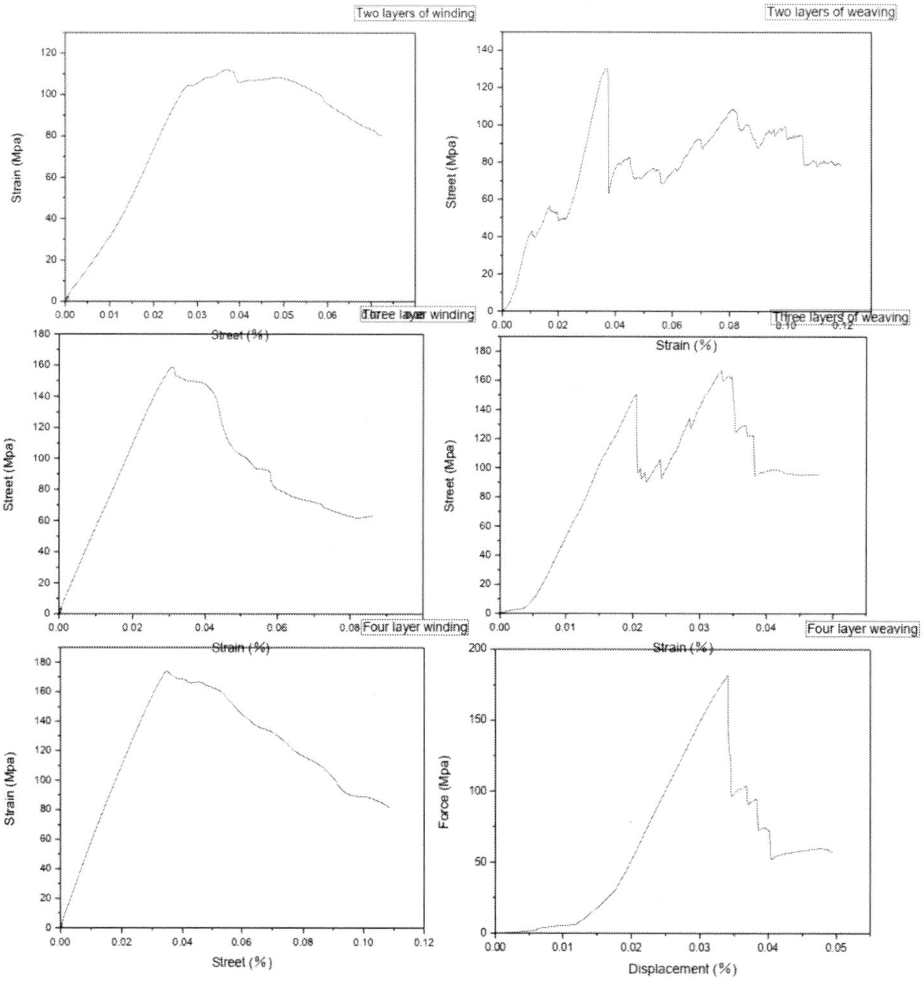

Fig.4 Compressive stress-strain curves of composite pipe

1) It can be seen from the longitudinal comparison of stress and strain that the maximum stress increases with the increase of layers both in weaving and winding. The horizontal contrast shows that the maximum stress of each weaving tube layer is slightly larger than the winding tube.

2) In the compression experiment, the main force is along the axial. The number of wrap in fabric has a great

influence on the mechanical performance of tubular composite, and this experiment uses the same number of warp in weaving to create a difference. Due to the interweave way of weft in which the weaving tube is very complex, it can bear a certain level of axial pressure when forced., As such, the weaving tube compression performance is slightly better than the winding tube.

3) The stress-strain curve of winding tube is relatively uniform, stress initially increased but then decreased when the maximum stress was achieved. The stress-strain curve of the weaving tube is irregular with the curve rising and falling continuously. This indicates that the weft plays a certain role in the binding process during warp breaking. The weft can bear a part of the pressure which increases stress. Therefore, the whole stress-strain curve shown as jagged.

3.3 Bending Performance Test

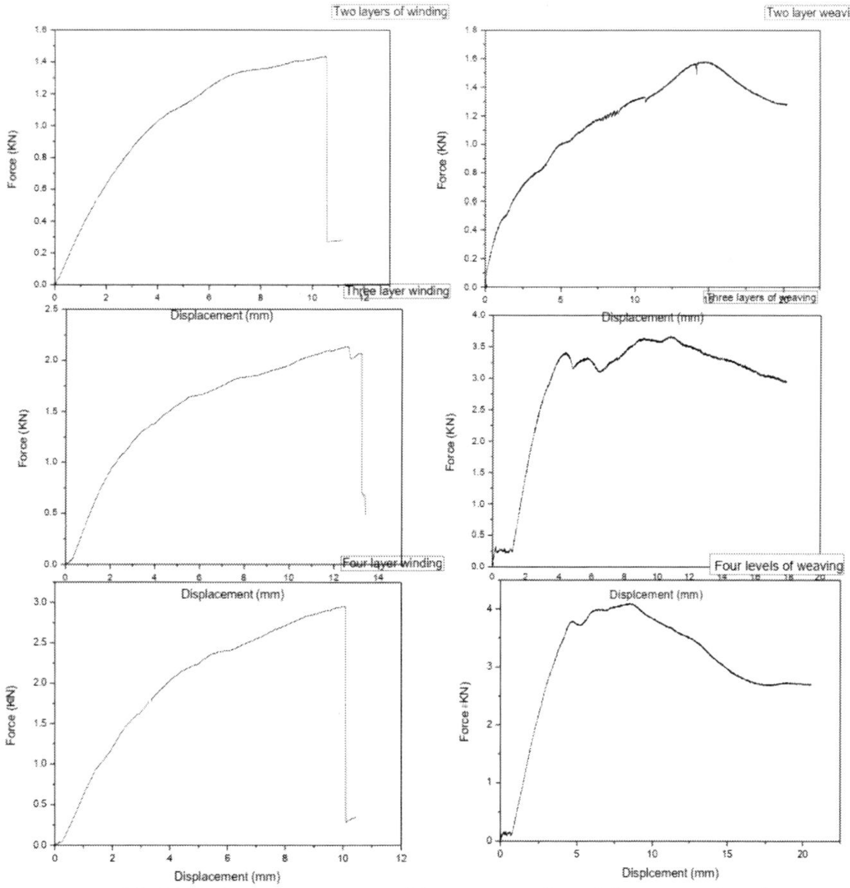

Fig.5 Displacement curves of the compression of the composite pipe

1) From the longitudinal comparison of the force-displacement curve, it can be seen that the maximum bearing capacity increases with the increase layers in both weaving and winding. The horizontal comparison shows that the maximum force of each weaving tube layer is slightly larger than the winding tube.

2) The curves above all show the force of winding tube . After the force peaked it decreased rapidly while the tube was breaking The force of the weaving tube however decreased slowly when peaked. The main reason is that there are no intertwining fiber between the layer and layer winding tube. Instead it under the influence of a simple adhesion with resin. When the force reaches a certain value, there will be a separation between the layers, resulting in no more yarn to withstand the pressure when the tube is damaged. k When the force of weaving tube is peaked while breaking, intertwined weft will continue to support pressure by drive yarn through other layers, resulting in

the rest of the forces to decrease slowly.

4. Conclusion

In this work a newly designed cylinder mould was proposed for fabrication of 3D tubular fabric reinforced composite. The 3D fabrics weaving process was combined with the double layer fabrics weaving process. The process parameters were set reasonably so that weaving 3D tubular fabrics was successful.

3D tubular composite materials were made successfully by the vacuum assisted resin molding process and newly designed lining mold. The design philosophy of the mould is based on reducing the friction between the outer surface of the cylinder lining and the inner surface of the tubular composite material.

Each layer of weaving tube is slightly better than the winding tube both in the compression and bending performance test. In the process of force, the weft can connect the warp yarns into an integrated component, reduce the deformation of the warp in the force, and bear a part of the force thereby improving the composite materials mechanical performance.

5. Acknowledgment

This work was supported by project 'Application of Aramid Fiber Angle Interlock Fabric Reinforced Composites in Anti-riot Helmets' (2015KF25).

References

[1] Shen LP, Bai Y. Fabric weave and design [M]. Beijing: Chemical Industry Press: 2014; 174-175.
[2] Bai G, Zhang M, Guo Y. The test standard of advanced composite material mechanical properties [M]. Beijing: Chemical Industry Press: 2015; 1.

The Properties Characterization of Cotton Fabric Treated by Etherifying 2D Resin

Yong Wu[1], Heng Pan[2], Gen-Yang Cao[2*]

[1] *Wuhan Second Ship Design and Research Institute, Wuhan 430025, China*
[2] *Ministry of Education Key Laboratory of New Textile Materials Green Processing and Functioning, Wuhan Textile University, Wuhan, Hubei, 430073, China*

[*] Corresponding author's email: genyang.cao@wtu.edu.cn

Abstract

In this study, an etherifying 2D resin was applied to identify the factors causing strength loss of cotton fabric. The strength, wrinkle recovery angle, and whiteness of the cotton fabrics were tested at different pH levels and baking temperatures. The results showed that the strength of the fabric decreased with increasing baking temperature, while WRA increased; however, the whiteness changed to a small extent. When the temperature was kept constant as the pH was increased, the fabric strength increased and the WRA decreased, but the whiteness remained unchanged. When the pH was 3 and the baking temperature was 140℃, the desired anti-wrinkle effect and high strength was achieved.

Keywords: Etherifying 2D Resin; PH; Mechanical Properties;, Wrinkle Recovery Angle

1. Introduction

Cotton, one of the most popular textile materials has many applications because of its irreplaceable comfort and excellent wearability [1]. However, cotton fabrics wrinkle easily after laundering which makes it inconvenient for users. The lack of resilience in cotton is caused by the structure of the biopolymer cellulose in its fibers, as it has hydrogen bonds as the major intermolecular interactions [2].

Anti-wrinkle finishing is a possible solution to this problem. Urea formaldehyde or phenolic has been used in the treatment of cotton for enhancing wrinkle resistance properties. However, when urea formaldehyde or phenolic treated cotton fabric is subjected to multiple laundering cycles, the fabric produces a large amount of formaldehyde, which is considered hazardous to public health [3]. Subsequently, dimethylol dihydroxy ethylene urea (DMDHEU) has been used for improving wrinkle resistance; in this process, DMDHEU gradually hydrolyzes to form N-methylol groups [4-6] that is good for anti-wrinkle properties, However, fabric finishing with DMDHEU also has health related problems. In this study, etherifying the aldehyde groups in 2D resin is applied to solve those problems. The effects of baking temperature and pH on the crease angle and strength are also analyzed.

2. Experimental

2.1 Material and Reagents

Cotton bleached fabric was purchased from Anhui Huamao Co., Ltd. The fabric weight was 105g/m^2, warp density was 710/10cm, and weft density was 560/10cm. Sodium carbonate was obtained from Tianjin Hengxing Chemical Reagent Co., Ltd. Glacial acetic acid was purchased from Shanghai Shenbo Chemical Co., Ltd. Oxalic acid was obtained from Qingzhou Xinsheng Chemical Co., Ltd. The etherifying 2D resin was purchased from Guangdong De Mei Fine Chemical Co., Ltd. $MgCl_2 \cdot 6H_2O$ was procured from Hubei Youmao Co., Ltd.

2.2 Solution Preparation

Formulation 1: The solution was composed of the etherified 2D resin (200g/L) and $MgCl_2 \cdot 6H_2O$ (24%), and the pH was adjusted from 1 to 7 with oxalic acid or anhydrous sodium carbonate.

Formulation 2 (control group): The pH of solution was adjusted from 1 to 7 by using oxalic acid.

2.3 Experiment Procedure

The fabrics were dipped into solution (composed of etherified 2D resin or oxalic acid), rolled (rolling residual rate: 60%-70%), and dried at 85℃ (moisture rate: 9%-15%). Next, fabrics were baked at different temperatures (130℃, 140℃, 150℃, 160℃,) for 6 min. Then, the fabrics were washed and neutralized using 5g/L of sodium carbonate and 3g/L of acetic acid solution, and finally dried at 110℃ for 5 min.

2.4 Analysis of Crease Recovery Angle

An SDL-M003A crease recovery angle tester was used for determining the fabric wrinkle angle. The sample size was 40 mm×15 mm, and each group was tested 10 times [7].

2.5 Analysis of Whiteness Index

The whiteness of the cotton fabric was measured by means of a WSB-2 white degree instrument, in the standard D_{65} light source, and the fabrics were combed to the same thickness and opaqueness. The test area was 5cm×5cm, and each sample was tested 5 times over different areas.

2.6 Analysis of Fabric Strength

The breaking strength of the cotton fabrics was tested using an Instron 5566 universal material tester with the standard of GB/T3923.1-1997, The test area was 20cm×5cm. The gauge was 10 cm, and speed was 10 mm/min. The samples were subjected to 10 rounds of measurements, and the average of the results was obtained.

3. Results and Discussion

3.1 Effect of Baking Temperature and pH Value to WRA and Whiteness

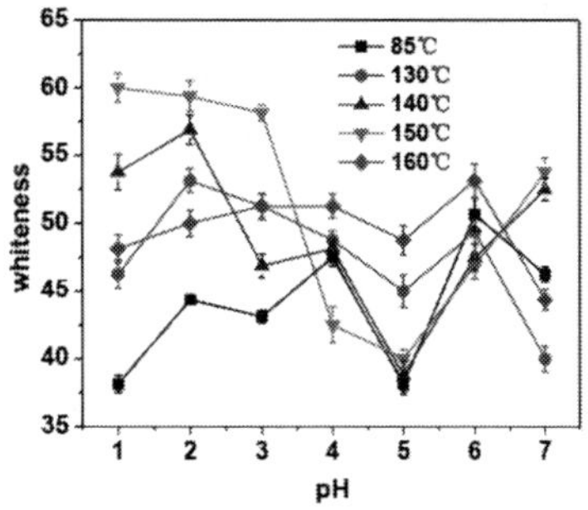

Fig.1 Effect of baking temperature and pH to whiteness

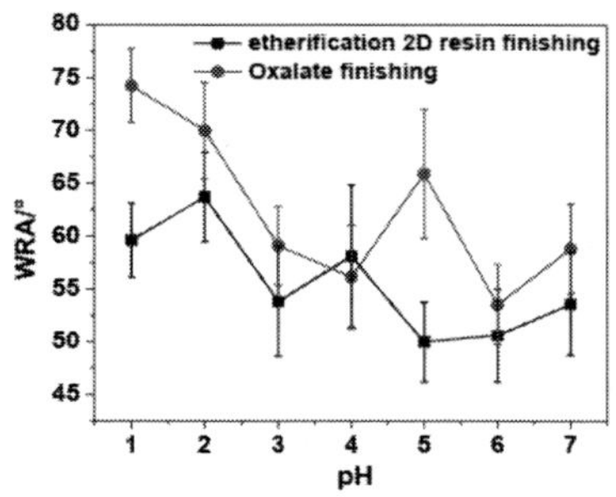

Fig.2 WRA under 85℃ baking with the pH change

Before the experiments, the average crease recovery angle of the untreated sample fabric was 54°, and the average whiteness was 64.84. As seen in Figs.1 to 3, if only oxalic acid was used to adjust the pH and if the fabrics were treated at 85℃, 130℃, 140℃, 150℃, and 160℃, the crease angle showed no obvious change with increasing pH and temperature and remained between 40° and 80°. However, if the etherified 2D resin was used to adjust the pH of the treatment fluid and if the fabrics were treated at 85℃, 130℃, 140℃, 150℃, and 160℃, the recovery angle increased with increasing temperature; however, as the pH increased, the WRA decreased. The WRA of the fabrics after baking without neutralization and washing was greater than that of the fabrics after neutralization and washing. According to the experimental results, the baking temperature should be controlled at 140℃-150℃. This showed that oxalic acid may not occur and etherification crosslinking, 2D resin cross-linking was obvious.

Acidic conditions have a better effect on the fabric wrinkle resistance. As the pH was increased, the anti-wrinkle resistance of the fabric declined, because the high temperature conditions had a negative effect on the crystallization of the crosslinked acid body, and the crystallinity decreased to a certain degree. When the temperature was increased, the reaction degree increased, so that the WRA showed an increasing trend. However, at the same temperature, the WRA would decrease after washing. This is because of the formation of a film on the surface of the cotton fabrics after finishing using the etherified 2D resin, which enhanced the resilience of cotton, so that the WRA increased. The membrane surface after neutralization and water washing was destroyed, so the WRA of the fabric WRA declined. However, the WRA of the finished fabric was larger than that of the original. The reason for this is that the low formaldehyde resin underwent polycondensation, leading to a mesh structure in the cotton fibers, and the amorphous region was deposited on the fiber, which changed the molecules or the basic structural unit of the cellulose phase through physical and mechanical functions, thus improving the crease resistance [8-10].

From Fig.1, it could be seen that the whiteness of the samples treated with oxalic acid was almost the same as that of the original sample, and that changes with pH, temperature, and process conditions were very small.

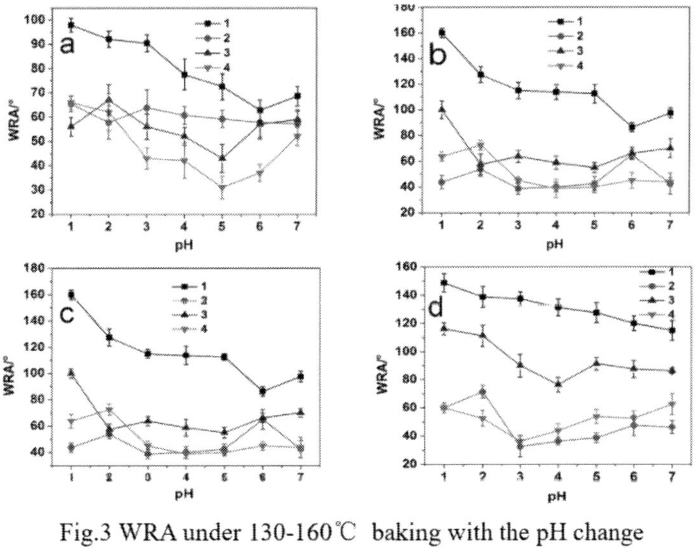

Fig.3 WRA under 130-160℃ baking with the pH change

1. The cotton fabric treated with etherified 2D resin; 2. The cotton fabric treated with oxalic acid; 3. The cotton fabric treated with neutralization and water washing after etherified 2D resin finishing; 4. The cotton fabric treated with neutralization and water washing after oxalic acid finishing.

a. baking temperature is 130℃; b. baking temperature is 140℃; c. baking temperature is 150℃; d. baking temperature is 160℃.

3.2 Effect of Baking Temperature and pH on Strength (N) and Strength Loss Rate

The average strength is 211.0cN before treatment, Fig.4 shows that the strength of the fabric sample treated with oxalic acid increased first and then decreased with the increase in pII. When the baking temperature was increased, the strength of the oxalic acid treated samples decreased.

The changes in the samples after neutralization and washing were not obvious. However, the strength of the fabric after baking without neutralization and washing was similar to that of the fabric after neutralization and washing; the strength of the fabric sample treated with etherified 2D resin increased with an increase in pH. As the baking temperature increased, the tensile strength of the sample decreased. The fabric strength was lower than that of the original after finishing with the etherified 2D resin.

In order to analyze the main reason for the decrease in the strength loss of the cotton fabric, the temperature, pH of the contribution to the fabric of the weight loss, the data are treated as follows:

$$\text{Strength loss rate} = \frac{a-c}{a} \times 100\% \tag{1}$$

where a and c refer to the strength of the cotton fabric before and after treatment, respectively.

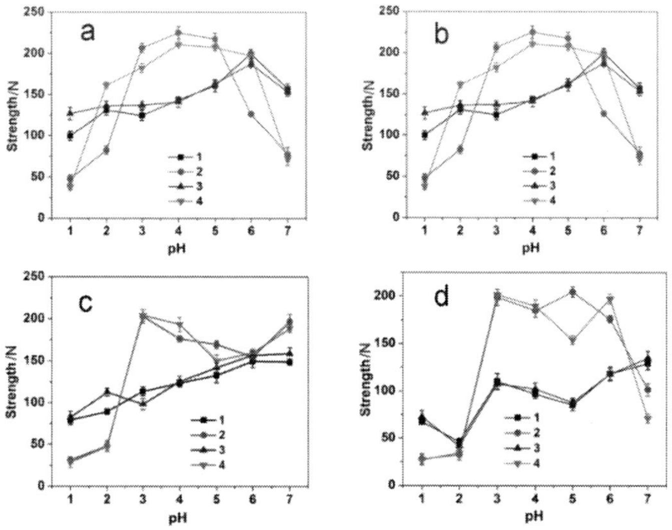

Fig.4 Strength under 130-160℃ baking with the pH change

1. The cotton fabric treated with etherified 2D resin; 2. The cotton fabric treated with oxalic acid; 3. The cotton fabric treated with neutralization and water washing after etherified 2D resin finishing; 4. The cotton fabric treated with neutralization and water washing after oxalic acid finishing.

a. baking temperature is 130℃; b. baking temperature is 140℃; c. baking temperature is 150℃; d. baking temperature is 160℃.

Table.1 Effect of baking temperature to strength loss rate

Strength loss rate	Strength loss rate of cotton fabric (%)									
Baking temperature	85℃		130℃		140℃		150℃		160℃	
conditioning fluid pH value	oxalic acid	etherified 2D resin	oxalic acid	etherified 2D resin	oxalic acid	etherified 2D resin	oxalic acid	etherified 2D resin	oxalic acid	etherified 2D resin
1	31.71	21.04	60.61	41.18	77.27	52.7	85.12	62.54	86.35	61.11
2	17.63	-0.66	6.64	25.73	60.92	37.77	77.18	57.8	77.53	46.59
3	13.46	-2.32	7.44	7.82	2.04	40.76	3.89	46.16	3.22	53.48
4	13.79	-2.18	9.67	10.57	-6.68	31.9	16.45	41.42	8.34	40.62
5	7.54	1.42	9.05	14.21	-3.13	23.65	19.67	37.11	28.86	32.75
6	11.47	4.79	10.95	6.87	39.95	5.4	26.54	29.15	24.36	25.78
7	8.44	2.13	15.4	-0.71	63.32	25.5	6.78	29.53	10.66	24.83

Table 1 shows that for the fabric finished with etherified 2D resin at a temperature was 85℃, the strength loss rate was less than zero, the WRA was about 55°, and the whiteness was around 62, because of the limited or no crosslinking reaction.

At the same temperature, the strength loss rate with the increase of pH in general was showing a trend of first increase and then decrease, while baking under the condition of the same pH value, WRA increased significantly along with the increase of temperature, the whiteness decreased, the strength decreased, the strength loss rate increases, to obtain the good anti-wrinkle effect and high strength, when pH is 3, the comprehensive effect is good, the largest tensile strength could reach 166.3N, accounted for the largest percentage of crosslinking strength loss and the loss of the whole is moderate, the crease recovery angle could get from 56° to 90°, the whiteness of the appropriate could get 65.

4. Conclusions

1) The fabric treated with the etherified 2D resin showed increased anti-wrinkle elasticity but decreased strength and whiteness. Because of the strong interaction between the factors, it is necessary to control the process conditions stringently, especially the baking temperature and pH.

2) When the fabric treated with the etherified 2D resin and the pH was kept fixed, the strength of the fabric decreased with an increase in the baking temperature, while WRA increased and the whiteness hardly changed. When the temperature was constant, the strength of the fabric increased, WRA decreased, and whiteness changed very little with an increase in pH.

3) Based on the above results, it could be concluded that the best baking temperature was 140℃, and when the pH of the etherified 2D resin was 3, the desired anti-wrinkle effect could be achieved and reserve high strength. The tensile strength of the finished fabric was 137.3N, the loss rate of crosslinking strength was 34.93%, and the strength loss rate was about 13.51%.

5. Acknowledgment

The authors would like to acknowledge Yunli Wang for their assistance in this work.

References

[1] Lewin M. Handbook of Fiber Chemistry, Third Edition [M]. Crc Press: 2006.

[2] Huan Q, Huang YG, Ji BL, Sun G. Anti-crease finishing of cotton fabrics based on crosslinking of cellulose with acryloyl malic acid [J]. Carbohydrate Polymers: 2016; 135: 86-93.

[3] Qi H, Pan J, Qing FL, et al. Anti-wrinkle and UV protective performance of cotton fabrics finished with 5-(carbonyloxy succinic)-benzene-1,2,4-tricarboxylic acid. [J]. Carbohydrate Polymers: 2016; 154: 313.

[4] Petersen H. The chemistry of crease-resist crosslinking agent [J]. Coloration Technology: 2008; 17 (1): 7-22.

[5] Harifi T, Montazer M. Past, present and future prospects of cotton cross-linking: New insight into nano particles [J]. Carbohydrate Polymers: 2012; 88(4): 1125-1140.

[6] Peng H, Yang CQ, Wang X, et al. The Combination of Itaconic Acid and Sodium Hypophosphite as a New Cross-Linking System for Cotton [J]. Industrial & Engineering Chemistry Research: 2012; 51 (35): 11301-11311.

[7] Tang P, Ji B, Gang S. Whiteness improvement of citric acid crosslinked cotton fabrics: H_2O_2, bleaching under alkaline condition [J]. Carbohydrate Polymers: 2016; 147: 139.

[8] Xu W, Wang X. 4-Durable press treatments to improve the durability and wrinkle resistance of cotton and other fabrics[J]. Understanding & Improving the Durability of Textiles: 2012; 120 (25): 70-81.

[9] Wang CM, Miao QH. Moisture crosslinking process of etherified 2D resin [J]. Dyeing & Finishing: 2007.

[10] Hua ZH, Sun WG, Yang D. Study on Hydrophilic Finishing of Polyester Fabric with PEG/Etherified 2D Resin [J]. Synthetic Fiber in China: 2010.

Preparation and Mechanical Properties of Polybutylene Terephthalate/Carbon Fiber Composites

Cheng-Qi Zhang, Zi-Qing Cai*, Rui Yang, Lei Wang, Hai-Feng Bao

School of Materials Science and Engineering, Wuhan Textile University, Wuhan 430200, China.

*Corresponding author's email: zqcaicup@gmail.com

Abstract

This paper studied the mechanical properties of polybutylene terephthalate (PBT) /carbon fiber composites. Some carbon fibers (CFs) were modified with polyvinyl acetate (PVAc), and they were named p-C. Some other CFs were modified by ethylene acrylic acid (EAA), and they were named e-C. After modification, the groups of carboxyl were introduced into the surface of p-C. For e-C, the groups of C=C were introduced. For PBT/p-C composites, the maximum tensile strength reached 68.22 MPa and elastic modulus reached 3263.18MPa, while the content of CFs is 15 wt.%. For PBT/e-C composites, with the addition of the same content of CFs, the tensile strength and elastic modulus could reach 90.76 MPa and 4334.76 MPa respectively. Compared with pure PBT, the mechanical properties of the composites were improved significantly, and the reinforcement effect was more obvious in PBT/e-C composites. The decomposition temperature of the composites was between 378 °C and 417 °C, and they presented better heat resistance property.

Keywords: PBT; Carbon Fibers; Composites; Mechanical Properties

1. Introduction

Carbon fibers (CFs) with high tensile strength, high temperature, friction and other advantages, are used to improve the mechanical properties of polymer materials [1-4]. They are mixed with thermoplastic resin and present high strength, good rigidity, light quality, and are recyclable [5-7]. These kinds of composites have excellent abrasion resistance, self-lubricity, anti-electrostatic, electromagnetic interference, small thermal expansion, coefficient fatigue resistance and creep resistance properties and were used in aerospace equipment, aircraft parts, electronic instrument, weaponry and automobile [8-11]. In the case of CFs without any surface treatment, the interfacial adhesion is poor in the composites and it is difficult to make full use of the advantages of CFs [12]. Thus, the surface modification of carbon fibers is very important to improve the interfacial adhesion between carbon fibers and resin matrix [13-15]. There are some surface treatments, such as gas phase oxidation, acid oxidation treatment, plasma treatment and so on [16-17]. In this article, I select surface sizing, it is simple and effective.

2. Experimental Section

2.1 Materials

CFs (T700) were purchased from TORAY Company in Japan. PBT pellets were purchased from Xin jiang Blue Mountain station River Polyester Co. Ltd.

2.2 Preparation of Samples

Before melt blending, PBT pellets and CFs were dried at 80°C for 24 h in a vacuum oven. Composites with varying CFs contents were prepared by melt blending in a Haake mixer (RheoDrive 7) at 230°C for 5 min with 30rpm.

Plate specimens were prepared by compression molding at 235°C for 10 min. The drawing dumbbell specimens were cut to 1.2 mm thickness, 5 mm width, and 55 mm length.

Table.2-1 Mass fraction of CFs in PBT

Carbon fiber contents	5%	10%	15%	20%	30%	40%
CFs (g)	3	6	9	12	18	24
PBT (g)	57	54	51	48	42	36

2.3 Characterization

Fourier Transform infrared spectroscopy (FTIR, TENSOR-27) was used to characterize the functional groups on CFs. The spectra were measured between 400 and 4000 cm^{-1}, and the resolution was set at 4 cm^{-1}.

Thermogravimetric Analysis (TGA, TG-209F1) was employed to characterize the stability of composites under argon atmosphere. TGA curves were obtained at temperatures ranging from 50°C to 600°C at 10°C/min heating rate.

The mechanical properties of the samples were determined using an electronic universal tensile testing machine (WDW-20). The tensile strength, elastic modulus and elongation at breakage point were measured at room temperature at 50 mm/min.

Scanning Electron Microscope (SEM, JSM-IT300) was used to obverse the surface of CFs and the cross-section of composites.

3. Experimental Results and Discussion

3.1 Analysis of CFs by FTIR

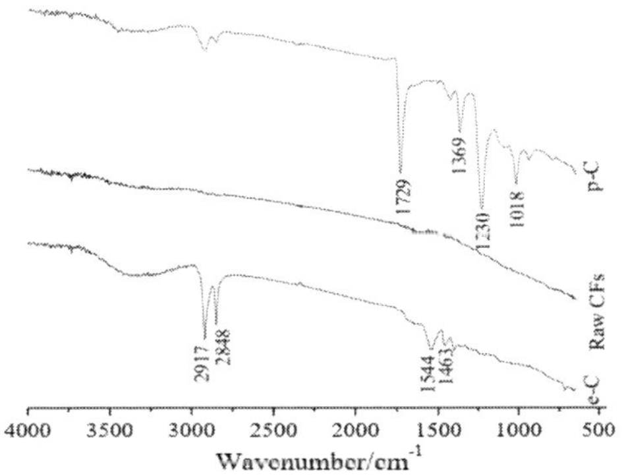

Fig.3-1 FTIR spectra of raw CFs and surface treated CFs

The absorption peaks of functional groups in CFs is shown in Fig.3-1. Compared with raw CFs, for p-C, the strong absorption peaks at 1729cm^{-1},1230cm^{-1} and 1018cm^{-1} indicating that the groups of carboxyl are introduced into the surface of p-C, and a peak at 1369cm^{-1} is the signal that the surface of CFs have some methyl groups. For e-C, the strong absorption peaks at 2917cm^{-1} and 2848cm^{-1}swhich suggest that the surface of CFs have some C-H bonds, and the strong absorption peaks at 1544cm^{-1} and 1463cm^{-1} state that the C=C bonds are introduced into the surface of e-C.

3.2 TGA Curves of PBT/CFs Composites

The quality of material changed in the heating process is shown in Fig.3-2. The decomposition temperature of the composites were between 378°C and 417°C. After adding CFs, the maximum thermal decomposition temperature of the material increased from 402°C to 407.7°C. The decomposition temperatures of three kinds of materials were comparatively analyzed. It is found that the thermal stability of PBT/CFs composites are improved.

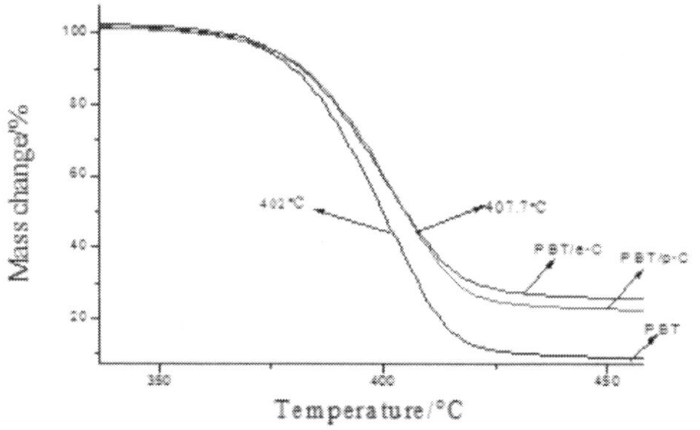

Fig.3-2 TGA curves of PBT resin, PBT /p-C and PBT /e-C

3.3 Mechanical Properties of PBT / CFs Composites

For Table 3-1, with the increase of carbon fiber contents to 15%, tensile strength and elastic modulus reached a maximum. For table 3-2, with the increase of carbon fiber content, the tensile strength first decreases before increasing, the elastic modulus increases gradually. By comparing the p-C and e-C, the enhancement effect of e-C has gotten better. The elongation at break of composite has decreased and the brittleness of the material has increased.

When the carbon fiber content is less, the reinforced effect of CFs is not obvious in the composite material, but when the carbon fiber content is too high, in the matrix resin, the distribution of CFs is uneven leading to stress concentration. Because the carbon fiber has high strength, high modulus characteristics, the elastic modulus of composites has obvious enhancement by adding CFs. Meanwhile. Carbon fiber has a certain degree of brittleness which will lower the elongation at break of the material.

Table.3-1 Mechanical properties of PBT/p-C composites

	Samples	Tensile Strength (MPa)	Yield strength (MPa)	Elastic modulus (MPa)	Elongation at break (%)
1	PurePBT	51.45±2.17	54.02±2.32	1662.32±260.59	4.06±0.36
2	5% p-C/PBT	45.25±8.70	-	1871.04±194.35	3.02±0.62
3	10% p-C/PBT	68.22±5.29	-	2969.44±375.12	2.83±0.22
4	15% p-C/PBT	64.02±6.98	-	3263.18±161.21	3.12±0.62
5	20% p-C/PBT	50.38±8.60	-	2279.25±145.96	2.97±0.55
6	30% p-C/PBT	33.14±3.63	-	2279.30±153.47	1.88±0.15
7	40% p-C/PBT	51.95±6.22	-	3150.90±483.11	2.84±0.51

Table.3-2 Mechanical properties of PBT/e-C composites

	Samples	Tensile Strength (MPa)	Yield strength (MPa)	Elastic modulus (MPa)	Elongation at break (%)
1	Pure PBT	51.45±2.17	54.02±2.32	1662.32±260.59	4.06±0.36
2	5% e-C/PBT	59.80±8.70	-	2162.25±33.59	3.57±2.13
3	10% e-C/PBT	90.16±9.78	-	2695.86±221.99	4.09±0.43
4	15% e-C/PBT	90.76±0.83	-	3157.28±179.13	3.6±0.17
5	20% e-C/PBT	79.80±4.74	-	3366.40±156.78	3.14±0.42
6	30% e-C/PBT	74.18±3.99	-	4261.19±409.07	2.16±0.37
7	40% e-C/PBT	76.52±8.09	-	4334.76±535.47	1.70±0.60

3.4 SEM Analysis of PBT /CFs Composites

The binding force of fibers and matrix are determined by the properties of fiber's surface. From Fig.3-3a, it can be seen that the surface of CFs are very smooth. For Fig.3-3b and Fig.3-3c, the surface area is increased, when the CFs are modified, it is better to be combined with matrix resin.

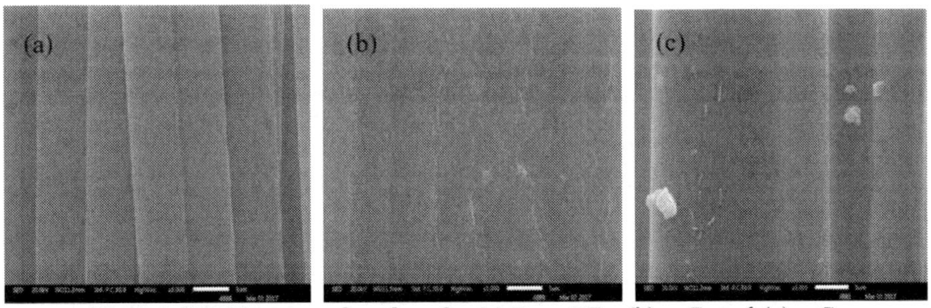

Fig.3-3 SEM images of surface for (a) Raw CFs, (b) p-C, and (c) e-C.

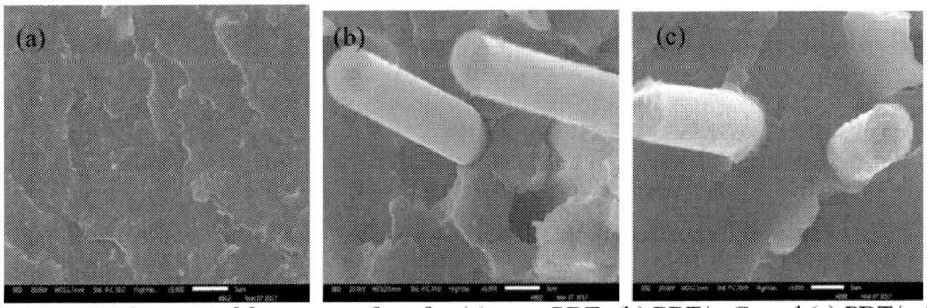

Fig.3-4 SEM images of fracture surface for (a) pure PBT, (b) PBT/p-C, and (c) PBT/e-C.

Due to the external force, fibers were pulled out from the matrix and left with some holes. In Fig.3-5a, the wall of the hole is relatively smooth, whereas in Fig.3-5b, it is more rough. In Fig.3-4a, the fracture surface of PBT has stratified indicating that there is a brittle fracture. In Fig.3-4b, there is a gap between the fiber and matrix. The fiber is pulled out by an external force. The surface of fiber is smoother. These phenomena shows s that there is a poor bonding performance between the fiber and matrix. On the contrary, there is almost no gap between fibers and PBT. In Fig.3-4c, the surface of fibers has a layer of matrix material indicating that the e-C is modified better. The adhesion of CFs and PBT resin has been greatly improved, the resin matrix can be better to transfer the stress to the CFs, the mechanical properties of the material are improved.

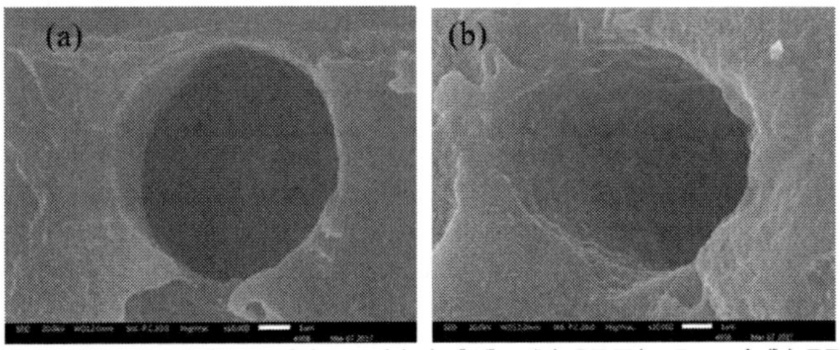

Fig.3-5 SEM imagines of the inner surface of the voids left for (a) PBT/p-C, and (b) PBT/e-C after fibers were pulled out.

4. Conclutions

According to the mechanical properties of the two kinds of composites, when the content of CFs was 15 wt.%, the mechanical properties of the composites reached the optimum level. The reinforcing effect of e-C in PBT was more obvious, the maximum of tensile strength could reach 90.76 MPa, and the elastic modulus reached 3157.28 MPa, the mechanical properties of the composites were improved significantly. Meanwhile, through the SEM images of CFs in the composites, they indicated that the surface of CFs were inert, and they were difficult to be dispersed well in resin. The adhesion of interface between the matrix and fibers need to be strengthened. When CFs were modified by EAA, the compatibility of the CFs with matrix was improved, which was beneficial when mixing the CFs with PBT resin. Meanwhile, the thermal stability of PBT/CFs composites was improved.

References

[1] Li XP, Zhang XP, Wang HW. Preparation and Application of Carbon Fiber, Journal of High-tech Fiber and Application: 2005; 5: 28-34 + 44.

[2] Chen XM. Property, development and application research of carbon fiber, Textile Auxiliaries: 2015; 07: 1-4.

[3] Donent JB, Park SJ. Surface characteristics of pitch-based carbon fibers by inverse gas chromatography method, Carbon: 1991; 29: 955-961.

[4] Qian BZ. Domestic and international application of carbon fiber applications, market demand and carbon fiber production capacity,High-teach fiber: 2010; 01: 43-46.

[5] La Rosa AD, Banatao DR, Pastine SJ, Latteri A, Cicala G. Recycling treatment of carbon fibre/epoxy composites: Materials recovery and characterization and environmental impacts through life cycle assessment, Composites Part B: Engineering: 2016; 104: 17-25.

[6] Li X, Bai RB, McKechnie J. Environmental and financial performance of mechanical recycling of carbon fibre reinforced polymers and comparison with conventional disposal routes, Journal of Cleaner Production: 2016; 127: 451-460.

[7] Longana ML, Ong Natalie, Yu HN, Potter KD. Multiple closed loop recycling of carbon fibre composites with the HiPerDiF (High Performance Discontinuous Fibre) method, Composite Structures: 2016; 153: 271-277.

[8] Ning FD, Cong WL, Qiu JJ, Wei JH, Wang SR. Additive manufacturing of carbon fiber reinforced thermoplastic composites using fused deposition modeling, Composites Part B: Engineering: 2015; 60: 369-378.

[9] Naito K, Oguma H. Tensile properties of novel carbon/glass hybrid thermoplastic composite rods, Composite Structures: 2017; 161: 23-31.

[10] Lai Y, Liu Y, Ma D. Automatically melting snow on airport cement concrete pavement with carbon fiber grille, Cold Regions Science & Technology: 2014; 103: 57-62.

[11] Zhang FY, Wang HQ, Zhang XP. Application prospect of carbon fiber / thermoplastic resin composites in China's civil industry, New chemical materials: 1994; 01: 5-7.

[12] Cai ZQ, Meng XY, Ye HM, Cong CB, Wang YF, Cui LH, Zhou Q. Reinforcing polyamide 1212 nanocomposites with aligned carbon nanofibers, Materials & Design: 2014; 63: 691-698

[13] Jiang B, Zhang T, Zhao LW, Huang YD. Interfacially reinforced carbon fiber composites by grafting modified methylsilicone resin, Composites Science and Technology: 2017; 140: 39-45.

[14] Gao B, Zhang RL, He MS, Wang CG, Liu L, Zhao LF, Wen ZJ, Ding ZP. Interfacial microstructure and mechanical properties of carbon fiber composites by fiber surface modification with poly (amidoamine)/ polyhedral oligomeric silsesquioxane,Composites Part A: Applied Science and Manufacturing: 2016; 90: 653-661.

[15] Zhang RL, Gao B, Du WT, Zhang J, Cui HZ, Liu L, Ma QH, Wang CG, Li FH. Enhanced mechanical properties of multiscale carbon fiber/epoxy composites by fiber surface treatment with graphene oxide/ polyhedral oligomeric silsesquioxane, Composites Part A: Applied Science and Manufacturing: 2016; 84: 455-463.

[16] Zhang YY, Zhang YZ, Liu Y, Wang XL, Yang B. A novel surface modification of carbon fiber for high-performance thermoplastic polyurethane composites, Applied Surface Science: 2016; 382: 144-154.

[17] Lee H, Ohsawa I, Takahashi J. Effect of plasma surface treatment of recycled carbon fiber on carbon fiber-reinforced plastics (CFRP) interfacial properties, Applied Surface Science: 2015; 328: 241-246.

Influence on the Physical Properties of Different ZnO-content Coated Cotton Fabrics

Chun-Hong Zhu[1*], Minori Ishimori[2], Jian Shi[3], Hideaki Morikawa[1]

[1]*Faculty of Textile Science and Technology, Shinshu University, 3-15-1 Tokida, Ueda, Nagano 386-8567, Japan*
[2]*Graduate School of Science and Technology, Shinshu University, 3-15-1 Tokida, Ueda, Nagano 386-8567, Japan*
[3]*Faculty of Systems Science and Technology, Akita Prefectural University, 84-4 Aza Ebinokuchi Tsuchiya, Yurihonjo, Akita 015-0055, Japan*

[*]Corresponding author's email: zhu@shinshu-u.ac.jp

Abstract

In this study, we proposed to investigate the suitable weight percentage of zinc oxide nanoparticles (ZnO NPs) coated on cotton fabrics while taking the physical properties into account. Different weight percent of ZnO NPs solution was coated on cotton fabrics, by the traditional dip- pad- dry- cure process. Self-cleaning property was evaluated by the color change of ZnO-content coated fabrics which were stained by coffee qualitatively and quantitatively. Moreover, we discussed the influence of ZnO-content on the thickness and air permeability of fabrics. The results showed that the 9 wt.% coated fabric had the best self-cleaning property than the others. For the physical properties, such as thickness and air permeability, the 9 wt.% fabric showed little difference compared to the lower weight percent ZnO NPs solutions coated fabrics. These results can be used for the functional fabric development taking self-cleaning performance and comfort into account in the future.

Keywords: Cotton Fabric; ZnO Nanoparticles; Coating; Self-cleaning; Air Permeability

1. Introduction

Photocatalysis is a sustainable technology which utilizes the renewable solar energy to activate chemical reactions by oxidation and reduction, in order to provide solution for environmental problems [1]. Photocatalysis involves a photochemical reaction at the surface of a metal oxide semiconductor. During this process, at least two reactions must be occurring simultaneously. The first one involves oxidation coming from photo-induced positive holes and the other one involves reduction, coming from photo-induced negative electrons [2]. Several kinds of metal oxide can be applied as photocatalyst, such as titanium dioxide (TiO_2), zinc oxide (ZnO), iron (III) oxide (Fe_2O_3), zirconia (ZrO_2), niobium pentoxide (Nb_2O_5), tungsten trioxide (WO_3) and vanadium (V) oxide (V_2O_5) and so on [1, 3].

ZnO is a wide band-gap semiconductor with exciton binding energy of 60 meV at room temperature [4, 5]. It is compatible with living organisms, leaving no risk to human health, therefore it is an environmental friendly material [6]. Moreover, ZnO has nearly the same band gap energy with TiO_2, therefore its photocatalytic capability is expected to be similar to that of TiO_2 [7]. Moreover, ZnO is relatively cheaper compared with TiO_2. The greatest advantage of ZnO is the ability to absorb a wide range of solar spectrum and more light quanta than some semiconducting metal oxides [8]. Therefore, investigations were focused on the photodegradation of ZnO, especially on the replacement of TiO_2 [9-12]. Zhang X et al [13] demonstrated the fabrication of different aspect ratios of ZnO nanorods with surface defects by mechanical-assisted thermal decomposition method. The results showed that ZnO nanorods with higher aspect ratio and surface defects have higher photocatalytic performance. Moore JC et al [14] measured the photocatalytic activity and stability of thin, polycrystalline ZnO films fabricated on sapphire substrate via direct current sputter deposition of Zn-metal films and described the growth parameters resulting in porous. However, during these previous studies, there was little research on the relationship between fabrics coated by different weight percent of ZnO nanoparticles (ZnO NPs) solutions and their physical properties, such as thickness and air permeability.

In this study, we purposed to investigate the suitable weight percent of ZnO-content for coating on cotton fabrics which took physical properties into account. Therefore, different weight percent of ZnO NP-solutions were coated on cotton fabrics, by the traditional dip- pad- dry- cure process. Self-cleaning performance was investigated using the as- prepared cotton fabric to evaluate the color change of stained coffee qualitatively and quantitatively. Moreover, influence of ZnO-content on the thickness and air permeability of cotton fabrics were discussed.

2. Experimental

Fabric samples used in this experiment is 100% cotton plain fabric, and the warp and weft density is 72 and 44 $inch^{-1}$ separately, with the weight per unit area of 125 g/ m^2. Samples were pretreated to remove the grease, wax and some others using the same method according to the literatures [15-16] and then air dried under environmental conditions for 24 h.

Fabric coating process was carried out by the dip- pad- dry- cure process as introduced in literature [17]. During this process, different weight percent of 3 wt. %, 5 wt. % 9 wt. % ZnO nanoparticles (Sigma-Aldrich Corporation, USA) solution used as the photocatalyst solutions. Furthermore, uncoated cotton fabrics were also used for comparisons.

Surface morphology and element analyze of all the samples were firstly investigated by scanning electron microscope (SEM) (JSM-6010LA, JEOL, Japan), including the energy dispersive X-ray spectroscopy (EDS). Then the samples were used for evaluating the self-cleaning performance. A 2 cm square shape sample was cut from cotton fabrics, and dropped by a 10μL of coffee stain. A solor simulator (XES-40S2-CE, San-ei Electric, Japan) was used to simulating the sunlight and stained fabrics were irradiated under the solor simulator for 24 h. In order to investigate the effect of weight percent on the self-cleaning quantitatively, colormeter (CM-2600d, Konica Minolta, Japan) was applied to measure the color difference $\Delta E^{*}ab$ by measuring coffee colors before and after 24 h irradiation for each sample. $\Delta E^{*}ab$ can be calculated from Eq.1 as follows:

$$\Delta E^{*}ab = \sqrt{(\Delta L^{*})^2 + (\Delta a^{*})^2 + (\Delta b^{*})^2} \quad (1)$$

where ΔL^{*}, Δa^{*}, and Δb^{*} can be obtained based on Eq.2 to 4.

$$\Delta L^{*} = L_a^{*} - L_b^{*} \quad (2)$$

$$\Delta a^{*} = a_a^{*} - a_b^{*} \quad (3)$$

$$\Delta b^{*} = b_a^{*} - b_b^{*} \quad (4)$$

where L_b^{*} and L_a^{*} show the lightness of color before and after 24 h irridiation respectively. L^{*}=0 yields black and L^{*}=100 indicates diffuse white. a_b^{*} and a_a^{*} are the position of color in a three- dimensional color space between green and red/ magenta before and after 24 h irradiation. Negative value of a^{*} is green, while positive value means red/ magenta. Finally, b_b^{*} and b_a^{*} indicate the position of color in color space between blue and yellow before and after 24 h sunlight exposure, respectively. Here negative value of b^{*} indicateds blue, while positive value means yellow.

Simutaneouly images were taken for the qualitative comparision. Moreover, in order to find out the relationship between weight percent of coatings and the basic physical charactetistics of fabrics, Kawabata Evalustion System (KES) was applied for the measurement of compressional property (thickness) and permeability. Five times of samples were mearsured and the average was obtained.

3. Results and Discussion

3.1 Surface Morphology

SEM images of uncoated, 3 wt.%, 5 wt. % and 9 wt.% ZnO NPs coated fabrics are shown in Fig.1 respectively. Surface of uncoated cotton fiber is seen to be comprised of ridges and groves; where in some places the continuity of this structure is shown unevenness. These bumps and valleys are in the micrometer level. Fig.1(b) displays the lowest fraction of ZnO NPs, showing some deposition of ZnO NPs on the fiber surface, which makes fiber to be smooth. With the increase of weight percent, deposition can be found more and more clear. Simultaneously, for 9 wt.% coated fiber, uneven can be found in some part of fiber because the much more deposition of ZnO NPs due to the highest weight percent.

Apart from the SEM images, element analyze of samples was carried out, with the uncoated and 5 wt.% coated samples shown in Fig.2. Compared with the EDS results of two samples, it demonstrates that after coating process, element Zn appeared on the surface of cotton fiber. Moreover, according to the result shown in Fig.2(a), it can be seen that pretreatment of cotton fabric has been carried out successfully.

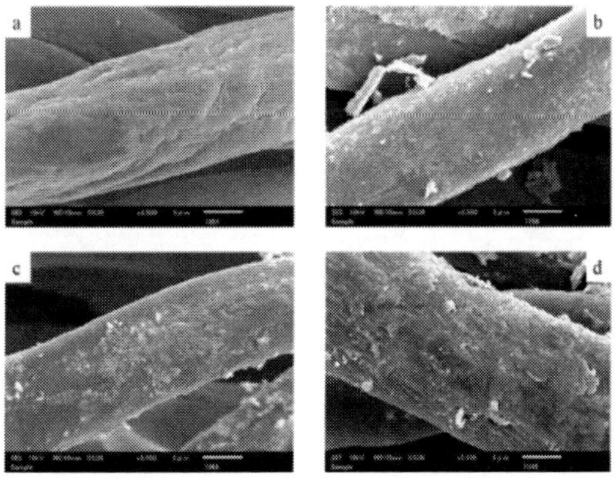

Fig.1 SEM images of (a) uncoated, (b) 3 wt.%, (c) 5 wt. % and (d) 9 wt.% ZnO NPs coated fabrics.

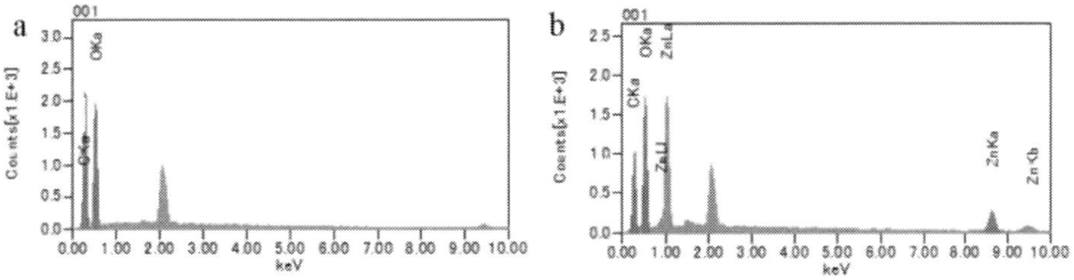

Fig.2 EDS spectra of (a) uncoated and (b) 5 wt.% ZnO NPs coated fabrics.

3.2 Coffee Stain Degradation for Self-cleaning Property

The self-cleaning property is evaluated by the level of stain removal from fabrics under sunlight irradiation. In this experiment, quantitative and qualitative evaluation approaches were employed. Firstly decoloration of coffee stained fabrics was evaluated before and after 24 h sunlight irradiation with the results shown in Fig.3. For all of the samples, the color degradation can be found after 24 h irradiation obviously. The color difference ΔE^*ab, used for qualitative evaluation, along with the standard deviation is shown in Fig.4. This figure showed that with the increase of weight percent, color difference also increased. For ZnO NPs coated fabrics, the ΔE^*ab was larger than 10, which meaned that colors were more similar than oppostie after sunlight exposure [18]. The ΔE^*ab of uncoated fabric was around 5.7, showed a weaker self-cleaning function than the coated fabrics, although the color difference can be perceptible at a glance. The mechanism for the color degradation can be explained by the photocatalyst. ZnO particles absorb photons from the UV light which can initiate the photocatalytic reactions. The electrons in the valence band of ZnO can be promoted to the conduction band, and a positive hole can be produced in the valence band. The generated holes and electrons induces oxidation- reduction reaction, which can be used for the oxidative degradation of organic compounds. Therefore, increasing the weight percent was to increase the electron- hole for catalyze the oxidative degradation of stained coffee. Thus, the color difference was increased according to the mass franction, and the self-cleaning performance was also followed to this rule.

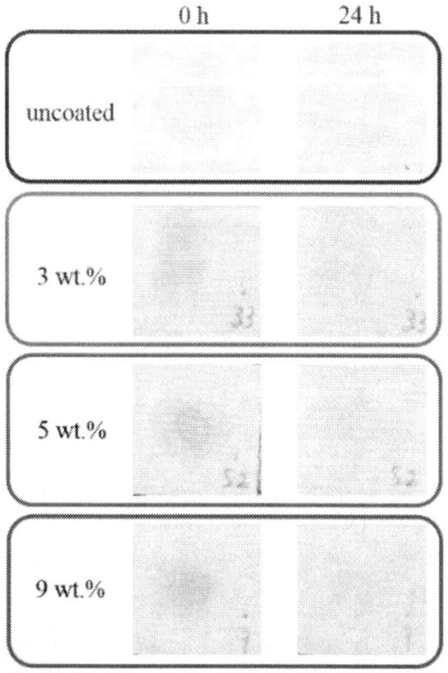

Fig.3 Decoloration of coffee before and after 24 h sunlight irradiation.

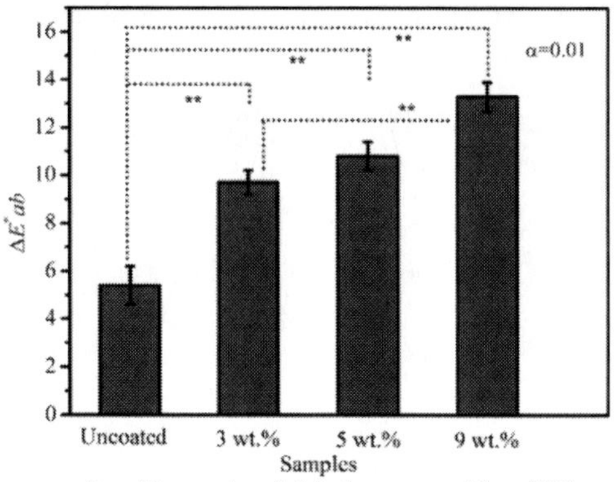

Fig.4 Color difference of coffee stained fabrics coated by different weight percent.

3.3 Physical Characteristics of Coated Fabrics

The thickness and air permeability are very important physical properties which influences the comfort of fabric. Therefore, these two parameters were evaluated in this study to investigate whether there was an effect of coating on the fabric comfort. Results on the effect of ZnO NPs-coating on fabric weight and thickness are shown in Fig.5.With the increase of weight percent, the weight per unit area was increased. This demonstrated that ZnO NPs were coated on fabrics for certain. However, fabric thickness showed different trends. Compared with uncoated samples, the coated ones had decreased thickness. Moreover, the thickness changed little when weight percent of

ZnO NPs changed. This phenomenon can be explained by the coating process. During the coating process, fabrics were dipped into ZnO NPs solutions for a while, which made all directions around cotton fibers adsorb the nanoparticles, not only on the surface of fabric. Therefore, it can be seen that there is little relationship between weight percent of coating solution and the fabric thickness.

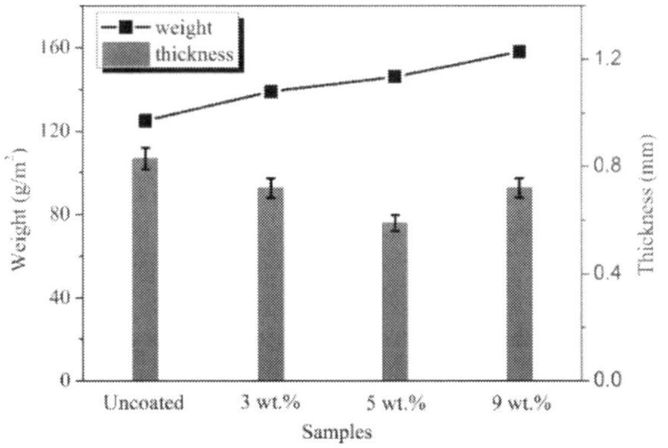

Fig.5 Weight and thickness of different weight percent ZnO coated cotton fabrics.

Fig.6 shows the air permeability of different weight percent of ZnO NPs solutions coated on cotton fabrics. It demonstrated that ZnO NPs coated on cotton fabrics caused decrease of air permeability. This may be due to the coating on fibers which made porosity of fiber smaller and then decrease the air permeability. Moreover, the results showed little influence of weight percent on the air permeability as by the 3, 5 and 9 wt.% ZnO NPs coating, the air-flow resistant did not changed obviously.

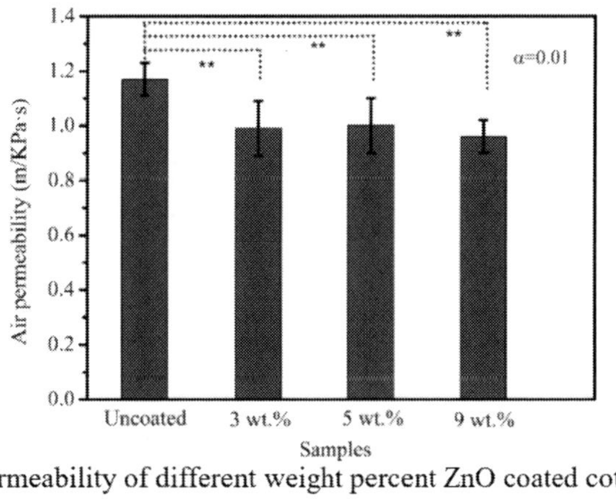

Fig.6 Air permeability of different weight percent ZnO coated cotton fabrics.

According to the self-cleaning property and the thickness and air permeability, the 9 wt.% ZnO NPs coated cotton fabrics showed the best photodegradation. Meanwhile, the thickness and air permeability showed almost the same to the cotton fabrics coated by lower weight percent of ZnO NPs solution.

4. Conclusion

In this study, we discussed the effect of the weight percent of ZnO NPs solutions on the self-cleaning performance of coffee stained cotton fabrics by quantitative and qualitative evaluation approaches. Moreover, the physical properties, focused on the thickness and air permeability of coated fabrics were also discussed. The results showed that for the image observation and color difference analyzing, the 9 wt.% coated fabric owned the best photodegradation than others. For the physical properties, such as thickness and air permeability, the 9 wt.% fabric showed little difference compared to the lower weight percent ZnO NPs solutions coated fabrics. However, compared to the uncoated one, the air permeability was lower regardless of the weight percent coated on fabrics. These results can be used for the functional fabric development taking functions and comfort into account in the future.

5. Acknowledgement

This work was supported by the JSPS KAKENHI, grant number JP16K16256.

References

[1] Kudo A, Miseki Y. Heterogeneous photocatalyst materials for water splitting. Chem. Soc. Rev: 2009; 38: 253-278.

[2] Fujishima A, Zhang XT, Tryk DA. TiO_2 photocatalysis and related surface phenomena. Surf. Sci. Rep: 2008; 63: 515-582.

[3] Vinu R, Madras G. Environmental remediation by photocatalysis. J. Indian Inst. Sci: 2010; 90: 189-229.

[4] Janotti A, Walle VD. Fundamentals of zinc oxide as a semiconductor. Rep. Prog. Phys: 2009; 72: 126501.

[5] Peng WQ, Qu SC, Cong GW, Wang ZG. Synthesis and temperature-dependent near-band-edge emission of chain-like Mg-doped ZnO nanoparticles. Appl. Phys. Lett: 2006; 88: 101902.

[6] Mende LS, Driscoll JLM. ZnO- nanostructures, defects and devices. Mater. Today: 2007; 5: 40-48.

[7] Lee KM, Lai CW, Ngai KS, Juan JC. Recent developments of zinc oxide based photocatalyst in water treatment technology: A review. Water Res: 2016; 88: 428-448.

[8] Behnajady MA, Modirshahla N, Hamzavi R. Kinetic study on photocatalytic degradation of C.I. Acid Yellow 23 by ZnO photocatalyst. J. Hazard. Mater: 2006; 133: 226-232.

[9] Cakir BA, Budama L, Topel O, Hoda N. Synthesis of ZnO nanoparticles using PS-b-PAA reverse micelle cores for UV protective, self-cleaning and antibacterial textile applications. Colloid Surface A: 2012; 414: 132-139.

[10] Chandrappa KG, Venkatesha TV. Light-induced self-cleaning properties of ZnO nanowires grown at low temperature. Appl. Catal. A- Gen: 2012; 411-412: 7-14.

[11] Ghayempour S, Montazer M. Ultrasound irradiation based in-situ synthesis of star-like Tragacanth gum/zinc oxide nanoparticles on cotton fabric. Ultrason Sonochem: 2017; 34: 458-465.

[12] Hatamie A, Khan A, Golabi M, et al. Zinc oxide nanostructure-modified textile and its application to biosensing, photocatalysis and as antibacterial material. Langmuir: 2015; 31: 10913-13921.

[13] Zhang X, Qin J, Xue Y, Yu P, Zhang B, Wang L, Liu R. Effect of aspect ratio and surface defects on the photocatalytic activity of ZnO nanorods. Sci Rep: 2014; 4: doi: 10.1038/srep04596.

[14] Moore JC, Louder R, Thompson CV. Photocatalytic activity and stability of porous polycrystalline ZnO thin-film grown via a two-step thermal oxidation process. Coatings: 2014; 4: 651-669.

[15] Zhu C, Takatera M. Effect of fabric structure and yarn on capillary liquid flow within fabrics. J Fiber Bioeng Inform: 2013; 6: 205-215.

[16] Zhu C, Takatera M. Effects of hydrophobic yarns on liquid migration in woven fabrics. Tex Res J: 2015; 85: 479-486.

[17] Zhu C, Shi J. Self-cleaning property of cotton fabrics coated with nano-sized zinc oxide. Textile Bioengineering and Informatics Symposium Proceedings. Melbonurne, Australia: 2016.

[18] http://zschuessler.github.io/DeltaE/learn/

Geometrical Modelling of Jacquard Warp Knitted Fabric

Junaid Khalid, Kei Wei, Zhong-Min Deng*, Abdul Rehman Akbar, Md. Saiful Islam Bhuiyan, Md. Abdul Mueeid Alam

College of Textile Science and Engineering, Wuhan Textile University, Sunshine Road, Wuhan, Hubei, 430200, China

*Corresponding author's email: hzcad@163.com

Abstract

This research revolves around the geometrical modelling of jacquard warp knitted fabric for the prediction of warp knitted structures and improvement of warp knitting simulation software i.e HZCAD. Geometric modelling of warp knitted fabric has become the topic of interest for scientists and research studies for the last few years. The yarn is a basic fundamental raw material for all kinds of fabric industry. After considering yarn's performance and appearance the major attention is to resolve the cost. The objective of this project is to develop a geometrical model for simple jacquard warp knitted fabric, to predict actual required yarn quantity and its actual yarn cost and to develop and improve a warp knitting simulation software for the simulation of these warp knitting design. The variable that usually influence yarn consumption are gauge of the warp knitting machine, yarn thickness, yarn composition, number of stitches in one repeat of the fabric, total number of needles in one repeat of the fabric and number of stitches per inch. It was concluded that the derived geometrical formula predicted the yarn consumption accurately up to 89.84% and using these geometries to improve the HZCAD software was effective. In the industry this geometrical formula can save time and can be helpful in calculating the cost of yarn. It will expand profit margin and save time.

Keywords: Geometrical Modelling; Jacquard Warp Knitted Fabric; Yarn Consumption Estimation; Formula Of Yarn Consumption

1. Introduction

Geometric modelling is known to be a branch of applied mathematics and computational geometry that studies ways and algorithms for the mathematical description of shapes [1]. Now-a-days most geometric modelling are performed with computers and for computer-based applications. Two-dimensional models are necessary in processors typography and technical drawing [1, 2].

The yarn is considered as a basic and important material of every fabric material. Yarn significantly effect the appearance, feel, texture, balance, proportion and quality of the warp knitted fabric. After yarn selection cost of the yarn is resolved. Yarn cost can be defined as the cost of actual yarn that is used in warp knitted fabric production, yarn wastages during winding, rewinding, warping, beaming, unused yarn in stock etc. [1, 2].

The objective of this project is to develop a geometrical model for simple jacquard warp knitted fabric to predict actual required yarn quantity and its actual yarn cost. There are large research available on yarn properties which effect yarn consumption [1, 3]. This paper will talk about the geometry of the warp knitting jacquard fabric, make a mathematical formula for it and check the accuracy of the formula.

2. Problem Formulation

Geometric models are typically distinguished from procedural and object-oriented models that outline the form implicitly by an opaque algorithmic rule that generates its appearance [4]. They're additionally contrasted with digital pictures and geometrical models that represent the shape as a set of a fine regular partition of space; and with shape models that offer an infinitely definition of the shape [6, 7]. However, these distinctions are usually blurred: as an example, a digital image can be taken as a group of squares; and geometric shapes like circles are outlined by implicit mathematical equations. Also, a shape model yields a constant quantity or implicit model once its definition is truncated to a finite depth [8, 9].

A stitch is a configuration made from the intra-looping, inter-looping or interlacing of yarns(s) in a specific repeated unit on the material, in a material, through the material or without the material. [10] The difference between interlacing, interlooping and intralooping is the configuration in which yarn(s) is/are configured to form the stitch. Interlacing is passing of yarns over and around yarn or a loop of another yarn; Interlooping involves one yarn loop passing through another loop which is made by another yarn; and in intralooping a loop is passed through another loop formed by same yarn [11]. Interlacing is often seen in the weaving of the fabric, interlooping is seen in the warp knitted fabric which can't be unravel whereas the intralooping can be seen in ravelable knitted fabric and some stitches [3].

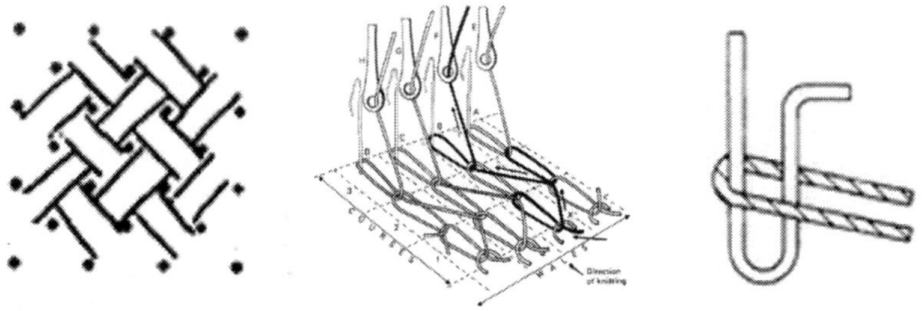

Fig.1 Interlacing, interlooping and intralooping (left to right) [11]

Different Textile structure's geometrical modelling are been examined over the several ages. It involves different software, models and methods available to predict or calculate the yarn axes of one of the simplest textile structure i.e the woven fabrics. In many researches, the yarn axis is fixed in one plane (2D space) [5, 12]. However, the yarn axis of knitted structures is fixed in the 3D space and is worked only with sections of the fabrics. There exists a lot of models for loop form calculation but most of them focus on the single loop geometry of simple structures, mostly weft knitted [13, 18]. The warp knitted structures have been examined less than other textile structures, for the fact of its high complexity - the structure of the warp knitting fabrics depends on the lapping movement, threading and yarn material, where small variations in the threading or the lapping movement cause in very substantial changes in the topology of the structures [12, 18].

The research on warp knitting loop, earlier was done by G.L Allison who put forward a simplified model shown in the Fig. below [15]. With radius of 2d semicircle arc, with two straight lines and the third line as the circle column and extend line respectively. But the model does not consider the stretch of coil, there is no points before and after guide bar [15]. Because the model and the real structure has a lot of difference, so future generations on the basis of this model is constantly improving with more realistic and more accurate models, thus promotes the research development of the textile industry. In 1964 Grossberg (P. Grossberg) proposed the loop as an elastic rod with mathematical analysis [15, 16].

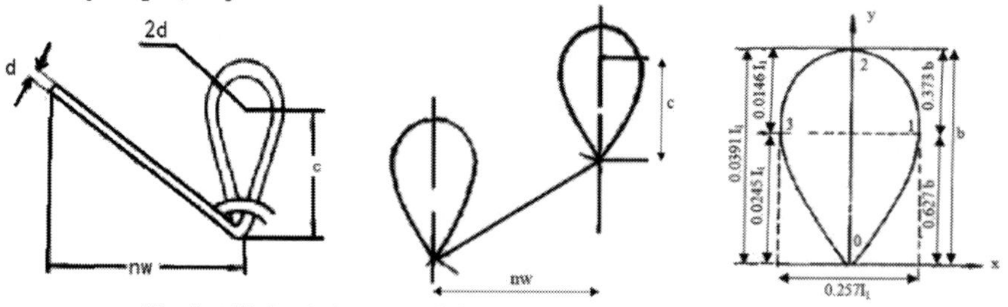

Fig.2 Allision's loop model and grossberg's first model [15]

According to the grossberg's model the loop was broken down in small segment. The equation for these loops are written as following [21].:

The length from 0 to 1 was.

$$0\ to\ 1 = 0.627b \tag{1}$$

The length from 1 to 2 was given by the following equation.

$$1\ to\ 2 = 0.373b \tag{2}$$

The length from point 2 to 3 was given by the following equation.

$$2\ to\ 3 = 0.0146\ l_1 \tag{3}$$

The length from point 3 to 4 was given by the following equation.

$$3\ to\ 4 = 0.0245 l_1 \tag{4}$$

For the full length of the loop the equation proposed was.

$$b = 0.0391 l_1 \tag{5}$$

The research objective of this research was to formulate a Geometrical model for simple jacquard warp knitted fabric for predicting yarn consumption and using these geometrical shapes to improve our warp knitted fabric simulation software HZCAD. The flowchart of the experimental plan includes the formulation of mathematical formula for the fabric by using geometrical modelling technique, sampling of the fabric for the calculation of the actual yarn consumption. The estimation of the yarn consumption for manufacturing of the fabric by using the derived formula, comparison and analysis of the accuracy and reliability of the derived formulas. After the confirmation of the accuracy of the formulae the geometries and the mathematical calculation were used to improve the HZCAD warp knitting simulation software.

3. Problem Solution

I have used the HZCAD warp knitting simulation software for the making of the fabric computer designs in this research. The technical route of the prediction the yarn consumption of the simple jacquard warp knitted fabric is deriving the formula of the four basic color of the jacquard i.e Black (/or white), Red, Green and Blue. Then selecting the given design of the fabric and calculating the total number of black, red green and blue color and multiply it with the derived formula to get the total number of yarn consumption for one repeat of the fabric [14, 22].

Fig.3 Black color of jacquard warp knitted fabric; left to right (a) red (b) green and (c) blue color of jacquard warp knitted fabric and the indication of major semi axis (a) and minor semi axis (b) of the ellipse [14, 22]

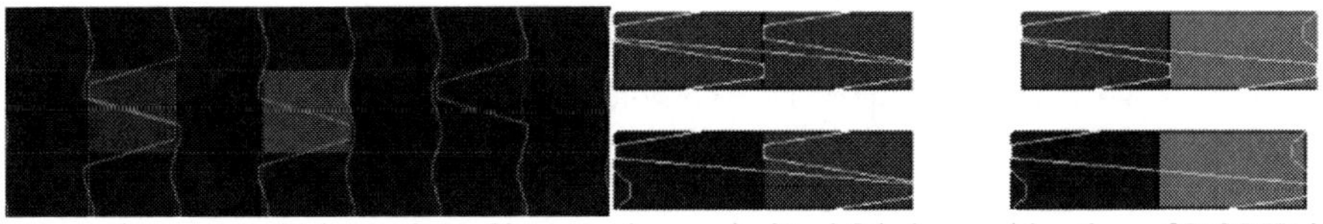

Fig.4 Left to right red green and blue color of jacquard warp knitted fabric combination of RGBW jacquard colors [14, 22]

3.1 Geometrical Model

The geometrical modelling of the simple warp knitting fabric is done by making the mathematical formula of the basic four colors red, green, blue and black of the warp knitted jacquard fabric and estimating the yarn consumption of the fabric. The detail derivation of the mathematical formula is discussed in this segment. As seen in the Fig.4, the black color of the warp knitted jacquard fabric consists of one yarn only so the mathematical formula for it would be as following

$$B = AB + DF = \frac{1}{2}\left\{\pi\left(\frac{D_1}{2}\right)^2\right\} + \frac{1}{2}\left\{\pi\left(\frac{D_1}{2}\right)^2\right\} = \pi\left(\frac{D_1}{2}\right)^2 \quad (6)$$

Where

B = the length of the yarn in black color of the warp knitted jacquard fabric.

D_1= the half of the length of one course

Length of one course $D_1 = \frac{Total\ no\ of\ wales\ in\ a\ repeat}{total\ length\ of\ repeat\ (cm)}$

As it can be seen in fig. 5 (a), the shape of the yarn in the black color jacquard structure is almost same like a circle so the mathematical formula of the circumference of the circle is used here for the estimation of the yarn consumption. Where, D_1 is the diameter of that circle. The mathematical model for the red color jacquard warp knitted fabric sturcture was calculated by using the formula for the estimated circumference of the ellipse $\{2\pi\sqrt{\left(\frac{a^2+b^2}{2}\right)}\}$ [23]. Where, a & b are the major and minor diameters of the ellipse as shown is Fig.4.

$$R = AO + OF = \frac{1}{2}\left\{2\pi\sqrt{\left(\frac{a^2+b^2}{2}\right)}\right\} + \frac{1}{2}\left\{2\pi\sqrt{\left(\frac{a^2+b^2}{2}\right)}\right\} = 2\pi\sqrt{\left(\frac{a^2+b^2}{2}\right)} \quad (7)$$

Where

R = length of the yarn used in red color of the warp knitted jacquard fabric

a, b = major and minor semi axis of the ellipse, respectively as shown is Fig.4

a = W = 25.4mm/E(guage) where W is the distance between two needles.

$b = \frac{1}{2}\left(\frac{Total\ no\ of\ wales\ in\ a\ repeat}{total\ length\ of\ repeat\ (cm)}\right)$ as shown is Fig.4, $H = \frac{10mm}{Density}$ where $b = \frac{H}{4}$

The mathematical formula for green and blue color of the warp knitted jacquard fabric will be the same as it is the equal but opposite in direction. The yarn consumption in the green and blue color is almost same like the red color yarn.

$$G = AD + DG = \frac{1}{2}\left\{2\pi\sqrt{\left(\frac{a^2+b^2}{2}\right)}\right\} + \frac{1}{2}\left\{\pi\left(\frac{D_1}{2}\right)^2\right\} \quad (8)$$

$$D = AD + DG = \frac{1}{2}\left\{\pi\left(\frac{D_1}{2}\right)^2\right\} + \frac{1}{2}\left\{2\pi\sqrt{\left(\frac{a^2+b^2}{2}\right)}\right\} \quad (9)$$

Where

G = length of the yarn used in green color of the warp knitted jacquard fabric

D = length of the yarn used in blue color of the warp knitted jacquard fabric

a, b = major and minor semi axis respectively as shown is Fig.4

a = Guage of the machine/2.45 as shown is Fig.4

$b = \frac{1}{2}\left(\frac{Total\ no\ of\ wales\ in\ a\ repeat}{total\ length\ of\ repeat\ (cm)}\right)$ as shown is Fig.4

But as discussed, getting the length of the actual consumption of the fabric is really difficult as the length of the yarn may be increased due to the stresses of pulling during the process of unravelling of the yarn so the weight of the yarn is calculated. For this reason the estimated yarn was also taken in the weight formula. Hence the result formula for prediction of weight of yarn in the segment of black, red, green and blue color in grams is shown is Equation (6), (7), (8) and (9) respectively.

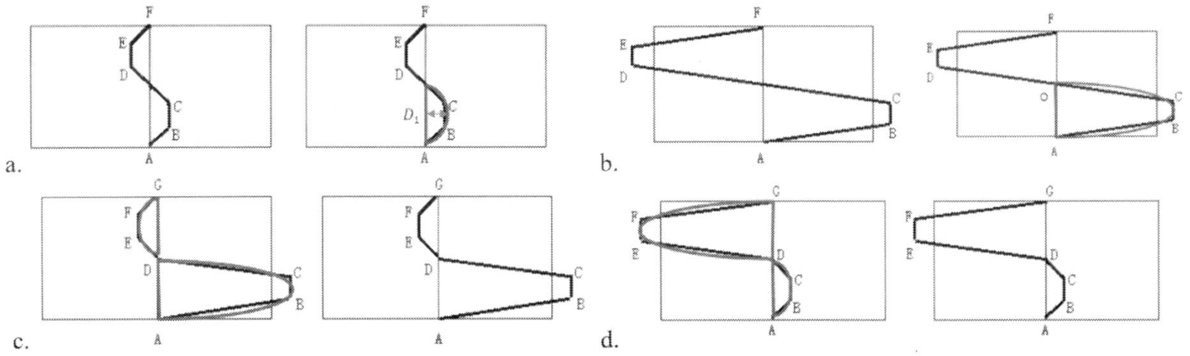

Fig.5 Geometrical model of a. black, b. red, c. green and d. blue color jacquard

$$W.B = \left\{\pi\left(\frac{D_1}{2}\right)^2\right\} x\ W_b \quad (10)$$

$$W.R = 2\pi\sqrt{\left(\frac{a^2+b^2}{2}\right)}\ x\ W_r \quad (11)$$

$$W.G = [\frac{1}{2}\left\{2\pi\sqrt{\left(\frac{a^2+b^2}{2}\right)}\right\} + \frac{1}{2}\left\{\pi\left(\frac{D_1}{2}\right)^2\right\}]\ x\ W_g \quad (12)$$

$$W.D = \left[\frac{1}{2}\left\{2\pi\sqrt{\left(\frac{a^2+b^2}{2}\right)}\right\}\ \frac{1}{2}\left\{\pi\left(\frac{D_1}{2}\right)^2\right\}\right] x\ W_d \quad (13)$$

Where
W_b = the weight (grams) of 1cm of the yarn used for the black color
W_r = the weight (grams) of 1cm of the yarn used for the red color
W_g = the weight (grams) of 1cm of the yarn used for the green color
W_d = the weight (grams) of 1cm of the yarn used for the blue color

3.2 Factor Affecting Yarn Consumption

The factor affecting the yarn consumption of the warp knitted fabric is type of warp knitted fabric, yarn tension, type of machine, stitch density, yarn composition. Every structure have different yarn consumption. As Yarn tension is increased consumption of yarn decreases and vice versa. Tricot, raschel and jacquard has different yarn consumption as the rpm of the machine is different, More stitches for an inch requires short stitching cycles and fabric become more bulky and very compact. GSM of fabric increases with yarn consumption. It is observed that if 100% polyester (PES) yarn is utilized rather than 100% cotton yarn, consumption of yarn is comparatively less. [19, 28]

3.3 Feasibility Analysis of the Experimental Plan

3.3.1 Sampling

Sampling was the most initial process of the procedure. The design of the sample was designed in HZCAD warp knitted fabric simulation software, then the samples was manufactured by warp knitting machine.

3.3.2 Unraveling

After manufacturing of the samples, the obtained samples were separated with rest of their stitch types and weighted. Then samples was conditioned at standard temperature and pressure for 24 hours at the temperature of 25±2°C and humidity of 65±5 to remove the extra moisture according to the ASTM standard ASTM D1776 - 08e1[20, 24].

3.3.3 Actual Yarn Calculation

We can use two methods of actual yarn calculation. It can be done by length measurement in which the fabric is manufactured, unraveled and yarns are collected in the polyvinyl bag. Then the length of each yarn is measured in such a way that all the crimps of the yarn are removed [25].

The second method is by weight measurement in which the weight of the sample is obtained after getting the yarns by weighing it on sensitive digital weighing balance. It is more precise method than length measurement method [17, 26]. So, in this research the actual yarn consumption is calculated by weighing sampled obtained. The weight of each sample was noted and recorded in a Microsoft Excel sheet for analysis.

3.3.4 Analysis

Formula for the yarn consumption was formulated via geometrical modelling. Afterwards, the prediction of length and weight of the yarn with respect to the formula was done for each and every sample of the design of experiment and the percentage relative error was calculated for model validation. [17].

4. Conclusion

After the formulating of formula for all the 4 color of stitches in warp knitting jacquard fabric. The model was validated for all the 27 sample. The predicted length of yarns were measured for each sample by the putting the input factors in the formula. Then the weight of 1 cm of the respective yarn was calculated by multiplying it with the predicted length to calculate the predicted weight of the yarn and compared it with the actual of the yarn calculated. Table 1 shows the comparison between actual and predicted yarn consumption along with its absolute relative error.

The trend line of the scatter diagram plotted between the predicted weights of the yarn vs the actual weight of shows that the modal was acceptable and near to the actual measurement with R2 value of 0.8984 as shown in Fig.6. The accuracy of the model is indicated by R2 value. High value to R2 shows that the geometric model proposed for the simple RGB jacquard warp knitted fabric has taken into account all the variables on which the thread consumption depends and 89.84% accurate.

After the confirmation of the accuracy of the geometric model they were used in the HZCAD software for the improvement of the simulation algorithm design in the coding of the software. It helped in the better simulation and more accurate design of the simple warp knitting jacquard designs of the fabric.

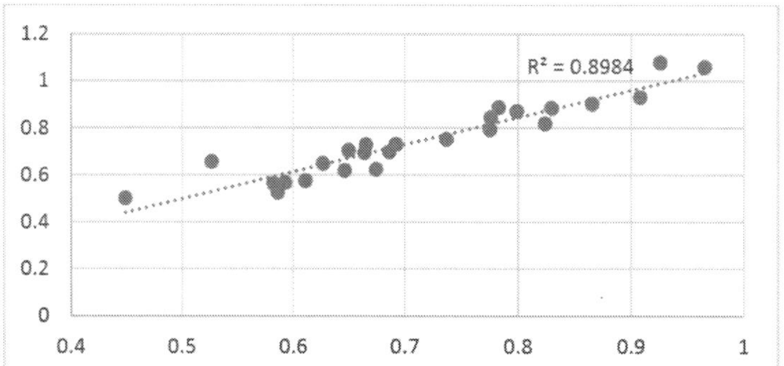

Fig.6 Scatter Diagram between Actual Weight and Predicted Weight of Yarn

This paper studied the geometry of the four basic fundamental stitch making unit of the jacquard warp knitting fabric regarded as the black/white, red, green and blue color in the literature review of the warp knitting CAD/CAM simulation systems. We formulated a mathematical formula for the geometrical representation of the structure with the help of the technique acknowledged as geometrical modelling. After making of the formula it was tested for prediction of the yarn consumption to check where it was accurate or not. Through experimentation, it was concluded that the formula proposed from the geometric modelling of simple warp knitted jacquard fabric was 89.84% accurate. It proves that this mathematical formulae can be used for the prediction yarn calculation in research and industries and for the costing calculations of the fabric order in terms of the yarn consumption. This model was also used for the improvement of the HZCAD warp knitted fabric simulation software for the algorithmic coding and make it more precise and user-friendly.

Table.1 Comparison of the predicted and actual weight with the %age absolute relative error

No. of Experiments	Predicted Weight (gm)	Actual Weight (gm)	%age Absolute Relative error
1	0.448	0.5	10.236
2	0.586	0.524	11.757
3	0.582	0.562	3.452
4	0.592	0.566	4.621
5	0.664	0.693	4.224
6	0.728	0. 730	0.300
7	0.649	0.702	7.535
8	0.824	0.817	0.750
9	0.798	0.869	8.129
10	0.521	0.535	2.755
11	0.611	0.574	6.400
12	0.674	0.623	8.246
13	0.626	0.647	3.250
14	0.736	0.749	1.674
15	0.775	0.791	2.001
16	0.692	0.727	4.890
17	0.865	0.901	3.912
18	0.908	0.929	2.245
19	0.526	0.655	19.648
20	0.646	0.618	4.530
21	0.686	0.697	1.527
22	0.665	0.727	8.495
23	0.775	0.843	8.009
24	0.829	0.883	6.056
25	0.783	0.886	11.643
26	0.965	1.058	8.823
27	0.926	1.077	14.034

References

[1] Mao A, Luo J, Li Y, Lin Y, Han Y. Knitted fabrics design and manufacture: A novel CAD system for qualifying bagging performance based on geometric-mechanical models. CAD: 2016; 75: 61-75.

[2] Li Y, Yang L, Chen S, Xu L. Three dimensional simulation of weft knitted fabric based on surface mode. Comput Model Technol: 2014; 18: 52-7.

[3] Rasheed A, Ahmad S, Mohsin M, Ahmad F, Afzal A. Geometrical model to calculate the consumption of sewing thread for 301 lockstitch. J TEXT I: 2014; 105 (12): 1259-64.

[4] Kyosev Y, Renkens W. About the 3D Modelling of Jacquard Warp Knitted Structures. FATE: 2016; 2015.

[5] Zhang LZ, Jiang GM, Miao XH, Cong HL. Three-dimensional computer simulation of warp knitted spacer fabric. FIBRE SCI TECHNOL: 2012.

[6] Renkens W, Kyosev Y. Geometry modelling of warp knitted fabrics with 3D form. Text Res J: 2011; 81 (4): 437-43.
[7] Renkens WKY. editors, editor 3D simulation of warp knitted structures–new chance for researchers and educators. 7th International Conference-TEXSCI: 2010.
[8] Renkens W, Kyosev Y, Renkens W, Kyosev Y. Geometrical modelling of warp knitted fabrics. Finite element modelling of textiles and textile composites St Petersburg: 2007; 26-8.
[9] Kyosev Y, Renkens W, editors. Virtual warp knitted fabrics-a toolkit for engineers and designers. Proceedings, Aachen-Dresden International Textile Conference, Aachen: 2007.
[10] Raz S. Warp knitting production: Melliand: 1987.
[11] Paling DF. Warp knitting technology: Columbine Press Manchester: 1965.
[12] Leaf G, Glaskin A. 43-The geometry of a plain knitted loop. Journal of The Textile Institute Transactions: 1955; 46 (9): T587-T605.
[13] Spillane RT, Kowalski MS. Warp knitted textile spacer fabric, method of producing same, and products produced therefrom. Google Patents: 1995.
[14] Goktepe O, Harlock S. Three-dimensional computer modelling of warp knitted structures. Text Res J: 2002; 72 (3): 266-72.
[15] Wu WL, Hamada H, Maekawa ZI. Computer simulation of the deformation of weft-knitted fabrics for composite materials. J Text I: 1994; 85 (2): 198-214.
[16] Mingzhen ZRC. MWK (Multiaxial-Warp-Knitting) Technology and Its Application in Compsite [J]. JWTSI: 1999; 3.
[17] Kyosev Y, Angelova Y, Kovar R. 3D modelling of plain weft knitted structures of compressible yarn. RJTA: 2005; 9 (1): 88-97.
[18] Rock M, Lohmueller K. Three-dimensional knit spacer fabric for footwear and backpacks. Google Patents: 1999.
[19] Robitaille F, Clayton B, Long A, Souter J, Rudd C. Geometrical modelling of industrial preforms: warp-knitted textiles. Proc Instn Mech Eng: 2000; 2 (1): 71-90.
[20] Standard A. Standard Practice for Conditioning and Testing Textiles. ASTM International, West Conshohocken, PA: 2008.
[21] Spencer DJ. Knitting technology: a comprehensive handbook and practical guide: CRC Press: 2001.
[22] Goktepe O, Harlock S. A 3D loop model for visual simulation of warp-knitted structures. J Text I: 2002; 93 (1): 11-28.
[23] Hodgman CD. Mathematical tables from handbook of chemistry and physics: 1946.
[24] Moesen M, Lomov S, Verpoest I, editors. Modelling of the geometry of weft-knit fabrics. Tech Texti symposium, Frankfurt: 2003.
[25] Knobil H. Warp Knitting Technology: 1954.
[26] Kyosev Y, Renkens W. Geometrical modelling tool for biaxial reinforced weft knitted structures. DWI Reports, Ed B KuØppers, DWI an der RWTH Aachen e V, Aachen: 2005; 129.

Research on Yarn-dyed Fabric Defect Detection Based on Regression Using Deep Learning

Jun-Feng Jing*, Ming Li, Xun Li, Peng-Fei Li

Xi'an Polytechnic University, College of Electronic and Information, 19 Jinhua South Road, Xi'an City, Shaanxi, 710048, China

*Corresponding author's email: jingjunfeng0718@sina.com

Abstract

Due to the poor stability and low processing speed of traditional artificial fabric defect detection, a yarn-dyed fabric defect detection method is proposed. The method, combined with the deep learning algorithm, is based on regression analysis. YOLOv2 (You Only Look Once) model is adopted to detect fabric defects. Its design philosophy is to consider the object detection problem as a regression problem in order to predict the target area and the category. Meanwhile, the YOLOv2 model is superior to the YOLO model in terms of accuracy and processing speed. Experimental results show that it is feasible and effective to detect the defect of the dyed fabric, which includes stripes and grids, by using the proposed detection model based on regression. The accuracy of regression and classification can reach 98.96% and 99.88%, respectively.

Keywords: Yarn-dyed Fabric; Defect Detection; Regression; Deep Learning; YOLOv2

1. Introduction

With the improvement in the quality of life, considerable attention has been paid to the integration of clothing and life. Yarn-dyed fabric is popular among consumers because of the generous solemn style. The production of the yarn-dyed fabric places more emphasis on small batches and diversification. As a result, the defect detection of the yarn-dyed fabric is made more difficult [1].

The human visual system is able to inspect fabric defects quickly with high accuracy. As soon as skilled workers in the textile factory inspect the defective fabric, they can grasp the type and location of the defects immediately. But the human visual system is prone to be influenced by the working environment as well as labour intensity. In other words, speed and precision will be greatly decreased in situations involving high-intensity working. With the rapid development of modern technology, automatic fabric defect detection and classification systems are becoming valued production models in the textile and garment industry. Considerable effort has been given to automated yarn-dyed fabric defect detection, for example, the ICA algorithm based on PCA [2], patterned fabric defect detection via combining the optimal Gabor filter and golden image subtraction [3], defect classification based on LBP and GLCM [4], and printed fabric defect detection via a Gabor filter and a regular band [5].

Deep learning is a discipline that specializes in computers simulating or implementing human learning behaviour. A computer using deep learning acquires new knowledge so as to enhance its accuracy by ceaselessly relearning existing knowledge models [6]. In recent years, deep learning methods have been developed rapidly. Therefore, it is in a position to meet the demand for industrial automation, both in detection accuracy and in efficiency. In addition, Deep leaning has been applied by Google to real-time automated machine translation systems.

In this paper, the method of object detection, based on regression, using the deep learning algorithm, is proposed to realize the automatic recognition and detection of fabric defects. The paper is organized as follows. Section 2 mainly describes the structure of YolOv2, while Section 3 presents the experimental results of defect detection and evaluates the performance of the proposed method. A detailed conclusion is given in Section 4.

2. Yarn-dyed Fabric Defect Detection

Deep learning is a method based on machine learning to train deep neural networks. The computer learns large amounts of data using deep learning to generate a stable knowledge model [7, 8]. In the research, it predicts the location and category of the yarn-dyed fabric detect automatically by learning the given yarn-dyed fabric defect images. The main acquisition process of the knowledge model is shown in Fig.1, which is divided into three steps: preprocessing the image training samples, running convolutional neural network [9, 10] (CNN), and non-maximal suppression (NMS) optimization [11]. YOLOv2 uses unified detection to unite the separate components of object detection into a single neural network.

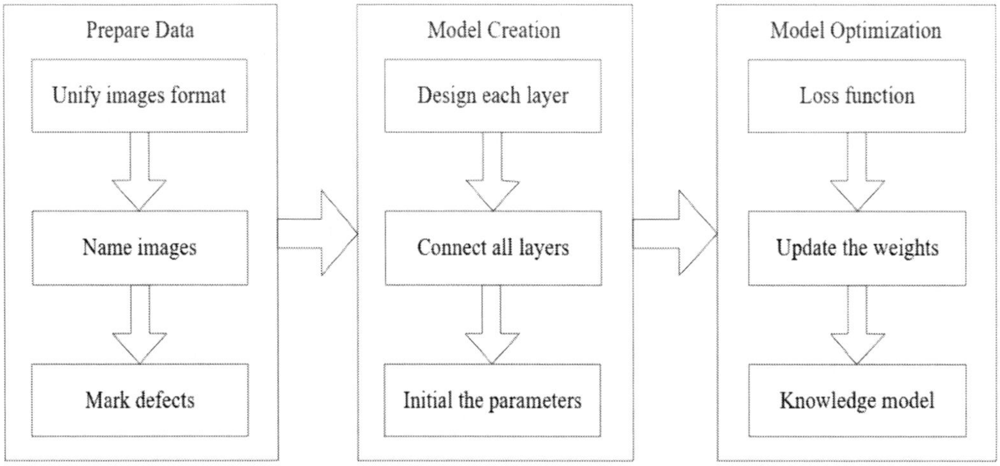

Fig.1 The training process of the knowledge model

2.1 Unified Detection

YOLO [12] follows the design conception of end-to-end training and real-time detection. The realization of this algorithm is to utilize the integrated detection scheme. In other words, a single neural network is used to predict the probability of the defect location and the defect category directly from the defective image. The input images are divided into S * S grids. Each grid cell predicts n bounding boxes while training. Each bounding box corresponds to five prediction parameters: x, y, w, h and confidence. The centre, width and height of the bounding box are represented by (x, y), (w, h) coordinates. The confidence score indicates the probabilities that the bounding box contains an object, as well as how accurate the box is predicted compared with the ground truth in the model. The confidence score is calculated as:

$$confidence\ score = \Pr(Object) * \mathrm{IOU}_{pred}^{truth} \tag{1}$$

If there is no defect in the bounding box, $\Pr(Object)=0$; otherwise, $\Pr(Object)=1$. The confidence score is equal to the intersection over union (IOU). The IOU is calculated by the predicted box and ground truth. At the same time, the posterior probability, $\Pr(Class_i \mid Object)$, indicates which class the defect belongs to in case there is defect in the bounding box. Regardless of the number of boxes B, each grid cell shares one set of class probabilities: $\Pr(Class_i \mid Object), i = 1, 2, \cdots, C$. At test time, the confidence score of the bounding box is as follows:

$$\Pr(Class_i \mid Object) * \Pr(Object) * \mathrm{IOU}_{pred}^{truth} = \Pr(Class_i) * \mathrm{IOU}_{pred}^{truth} \tag{2}$$

2.2 Optimization and Improvement

YOLOv2 [13] is optimized, based on YOLO, by increasing batch normalization [14], a high-resolution classifier [13], convolution with anchor boxes [8], dimension cluster algorithms [13] and so on. Besides, the input images are divided into $W_0 * H_0$ in order to balance the speed of processing and prediction accuracy, instead of applying the previous $S * S$. It is calculated by $W_i * H_i$ and n, corresponding to the size of the input of CNN and the number of pooling layers, respectively.

2.3 Implementation

As shown in Fig.2, the architecture of YOLOv2 consists of 30 layers, including 22 convolutional layers, five max-pooling layers and three regression prediction layers. The input image of the proposed detection model is divided into grids and predicts bounding boxes that represent confidence and class probability. These prediction results are encoded into the $W_0 * H_0$ filter dimensional tensor. It is calculated as:

$$filter = (classes + coords + 1) * num \tag{3}$$

where $classes$ represents the number of defect classes, $coords$ has a value of 4, corresponding to x, y, w, h, 1 represents confidence, and num represents the number of defects predicted in each bounding box.

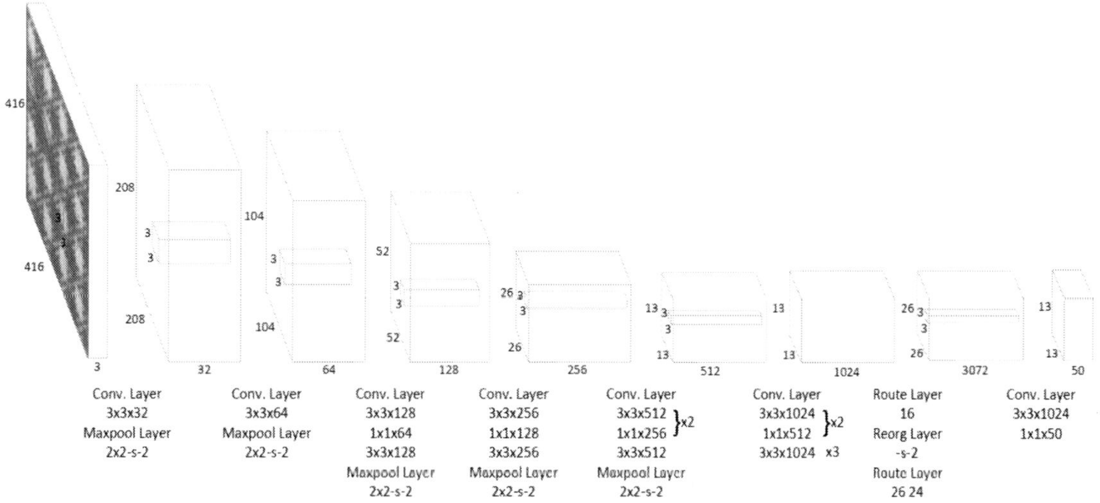

Fig.2 The architecture of YOLOv2

3. Experiment and Result Analysis

3.1 Preparation Data

All the images are converted, via uniform format conversion, to the JPG format for the subsequent operation with reference to the VOC data set [15]. Each type of defect image is named sequentially. The specific naming rule and its significance are shown in Table 1.

Table.1 The naming rules and significance of the defective images

Defect type	Images	Name	Significance
Hole		######.jpg	(1) The first two "#" represent the type of fabric defect. For example, grid fabric belongs to the category C3R1, which is expressed as 31, while striped fabric C3R3 is expressed as 33. (2) The third "#" indicates the defect type in this paper: namely, the hole, greasyDirt, scratch, linellae and wrinkle, which are represented by 1, 2, 3, 4 and 5, respectively. (3) The last three "#" are used to represent the index of the images under that category. For example, 315042.jpg represents the 42nd image under the category of wrinkle in the grid fabric.
GreasyDirt	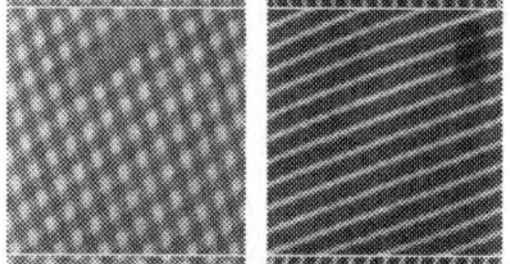		
Scratch			
Linellae			
Wrinkle	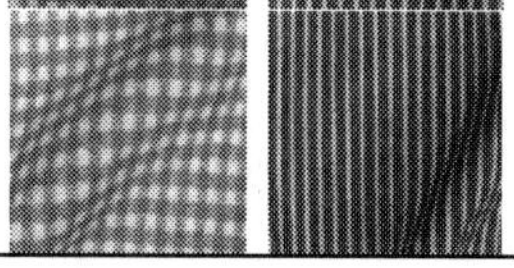		

3.2 Experimental Setups

All our experiments were conducted on a computer with a 3.0 GHz Intel Core^{TM} i7-5930K CPU and a 64 GB memory. The GPU card used in training the deep CNN network is Nvidia TITAN X. We evaluated the model performance over the two data sets on Ubuntu14.04. Each data set was split into 150 training images, 50 validation images and 50 test images.

3.3 Model Training

a b

Fig.3 The validation accuracy of the knowledge trained on grid fabric and striped fabric

With the data prepared in Section 3.1, the YOLOv2 model was iterated 3,000 times for about 4 h. It learned 64 sample images in each iteration. Each validation was made after training eight images. We calculated the validation accuracy rate every 100 times. The validation accuracy rate curve generated after 3,000 times of iteration is shown in Fig.3. From Fig.3a, we can see that, after the model was iterated 2,900 times, its regression position reached the highest accuracy. The accuracy of regression and classification was found to reach 98.96% and 99.88%, respectively. In Fig.3a, we can see that, after the model was iterated 2,900 times, its regression position reached the highest validation accuracy. The accuracy of regression and classification reached 98.44% and 99.93%, respectively.

3.4 Results Analysis

After training the knowledge model, the recognition effect of the model required testing. Table 2 shows the test results of 50 striped fabric images and 50 grid fabric images. The test results show that the knowledge model has a good application in terms of yarn-dyed fabric defect detection. It can also be seen that the knowledge model obtained by training given fabric images using YOLOv2 is pleasingly efficacious. Table 3 shows the results of five samples selected randomly from each class of grid defect fabric and striped defect fabric, respectively.

As for the test time, it took about 223 ms and 224 ms to test 256 * 256 images and 768 * 576 images, respectively, which shows that the detection rate improved greatly. According to the experimental results, we know that the test time was mainly influenced by the input size defined in the model. It was largely unaffected by the size of the measured image.

However, the proposed method still needs to be improved in several aspects. On the one hand, the accuracy rate needs to be improved with regard to defect detection, given that some defects may have leaked in the detection test. The accuracy rate decreased with regard to the detection of smaller defects, such as the hole and linellae, which indicates that the universal results of YOLOv2 model are unavailable in the detection of small defects. On the other hand, the proposed knowledge model is not perfect enough, due to the lack of training samples. A good object detector should improve when supplied with more training data [3]. Its accuracy rate will be further improved after expanding the training samples.

In this paper, accuracy rate, loss rate and false rate were utilized to evaluate the performance of the YOLOv2 on grid fabric and striped fabric. The definitions of the three indices are expressed as follows:

$$R_{accuracy} = (D_{right} \div N_{all}) \times 100\% \tag{4}$$

$$R_{false} = (D_{false} \div N_{all}) \times 100\% \tag{5}$$

$$R_{loss} = (D_{loss} \div N_{all}) \times 100\% \tag{6}$$

where $R_{accuracy}$ is the accuracy rate, R_{false} is the false rate, R_{loss} is the loss detecting rate, D_{right} is the number of defective fabric images detected correctly, D_{false} is the number of defective fabric images detected mistakenly, and N_{all} is the number of all the fabric images, including defective and defect-free images.

Table.2 The test results on 50 images of striped fabric and 50 images of grid fabric

Defect categories		Hole	GreasyDirt	Scratch	Linellae	Wrinkle	Average value	STD
Grid fabric	Accuracy rate	80%	90%	90%	80%	90%	86%	5.5%
	Loss rate	20%	10%	10%	10%	10%	12%	4.0%
	False rate	0	0	0	10%	0	2%	4.5%
Striped fabric	Accuracy rate	80%	100%	90%	80%	90%	88%	8.4%
	Loss rate	20%	0	10%	20%	10%	12%	8.4%
	False rate	0	0	0	0	0	0	0

Table.3 Test results on different classes of striped fabric and grid fabric

	Hole	GreasyDirt	Scratch	Linellae	Wrinkle
Striped fabric				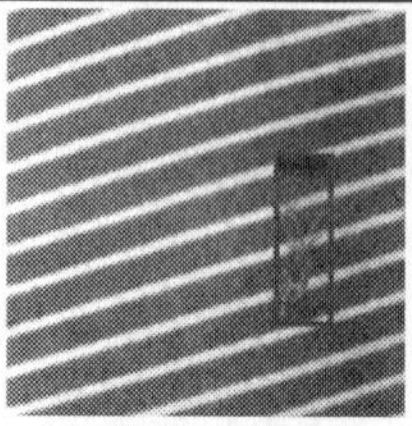	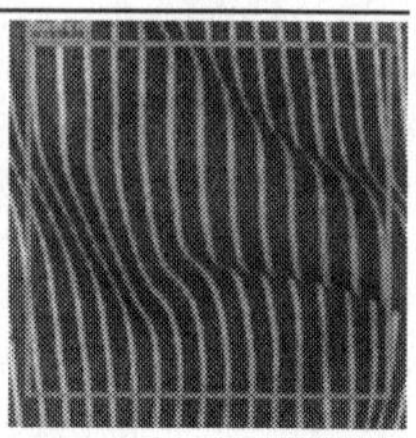
Grid fabric					

4. Conclusion

In this paper, a yarn-dyed fabric defect detection method is proposed. The method is combined with the YOLOv2 model, which is based on regression analysis. According to the experimental results, within the limited value of the premise, the more iterations there are, the higher the validation accuracy of classification and regression. In the proposed method, the model can automatically predict the location and category of striped and grid fabric defects by learning the training yarn-dyed fabric defect images. The performance of the proposed defect detection model has been evaluated through extensive experiments with various types of real fabric samples. The proposed detection method has proven to be effective and robust, and in a position to meet the demands for industrial automation both in detection accuracy and in efficiency. Future work may include expanding the data sets so that the model can handle a wider variety of image types, such as printed fabrics.

5. Acknowledgement

The authors gratefully acknowledge the financial support from the Scientific Research Program Funded by National Natural Science Foundation of China (61301276) and the Industrial Public Relations Project of Shaanxi Technology Committee (2015GY034).

References

[1] Li W, Xue W, Cheng L. Intelligent detection of defects of yarn-dyed fabrics by energy-based local binary patterns. Textile Research Journal: 2012; 82 (19): 1960-1972.

[2] Jing J, Zhao J, Li P et al. The algorithm of ICA based on PCA for fabric defect detection [J]. Journal of Fiber Bioengineering and Informatics: 2015; 8 (4): 687-696.

[3] Jing J, Chen S, Li P. Automatic defect detection of patterned fabric via combining the optimal Gabor filter and golden image subtraction [J]. Journal of Fiber Bioengineering and Informatics: 2015; 8 (2): 229-239.

[4] Zhang L, Jing J, Zhang H. Fabric defect classification based on LBP and GLCM [J]. Journal of Fiber Bioengineering and Informatics: 2015; 8 (1): 81-89.

[5] Kang X, Yang P Jing J. Defect detection on printed fabrics via Gabor filter and regular band [J]. Journal of Fiber Bioengineering and Informatics: 2015; 8 (1): 195-206.

[6] Goodfellow I, Bengio Y, Courville A. Deep Learning [M]. MIT Press: 2016.

[7] Girshick R. Fast r-cnn [C]. Proceedings of the IEEE International Conference on Computer Vision: 2015; 1440-1448.

[8] Ren S, He K, Girshick R et al. Faster r-cnn: Towards real-time object detection with region proposal networks [C]. Advances in Neural Information Processing Systems: 2015: 91-99.

[9] Krizhevsky A, Sutskever I, Hinton GE. ImageNet classification with deep convolutional neural networks [C]. International Conference on Neural Information Processing Systems. Curran Associates Inc: 2012; 1097-1105.
[10] Turaga SC, Murray JF, Jain V et al. Convolutional networks can learn to generate affinity graphs for image segmentation [J]. Neural Computation: 2010; 22 (2): 511-538.
[11] Neubeck A, Gool LV. Efficient non-maximum suppression [C]. International Conference on Pattern Recognition. IEEE Computer Society: 2006; 850-855.
[12] Redmon J, Divvala S, Girshick R et al. You only look once: unified, real-time object detection [J]. Computer Science: 2016; 779-788.
[13] Redmon J, Farhadi A. Yolo9000: better, faster, stronger. arXiv preprint arXiv: 1612.08242: 2016.
[14] Ioffe S, Szegedy C. Batch normalization: Accelerating deep network training by reducing internal covariate shift [J]. Computer Science: 2015.
[15] Everingham M, Gool L, Williams CK et al. The pascal visual object classes (voc) challenge [J]. International Journal of Computer Vision: 2010; 88 (2): 303-338.

Research on the Development of Weft Knitted Hosiery Fabrics Made of Hemp, Cotton and Nylon 66 Blended Yarns

Hui Zhu[1], Qin Zhou[1], Ji-Hong Wu[1*]

[1]*School of textile Science and Engineering Wuhan Textile University, Wuhan, Hubei, 430200, China*

[*]Corresponding author's email: jihong_wu@wtu.edu.cn

Abstract

The hemp/cotton and Nylon66 blended yarns in this paper was chosen as raw materials to design and develop weft knitted hosiery fabrics, whose functions are antibacterial, moisture permeability and so on. We achieved blended yarns by changing the twist and fineness of the nylon. When changing the fineness of the nylon, the proportion of blended yarns also changed. The knitting fabrics were obtained from the flat knitting machine. The hairiness index, uniformity of the fineness of the yarns, abrasion resistance, pilling resistance were tested and analyzed to find the effects of twisting and the proportion of blended yarns on the properties of the fabric. The results show that the hairiness index, unevenness of the blended yarns, and the abrasion resistance all improved. At present, there are a few scholars who are researching hemp. By changing the content and twist, we have obtained a little achievement.

Keywords: Hemp/Cotton and Nylon66 Blended Yarns; Hosiery Fabric; Property

1. Introduction

Marijuana is one of the annual herbaceous plants with an upright stem,branches, longitudinal groove and cortical rich in fiber. As the leaves of the marijuana flowering plants contain THC, restrictions have been made on the cultivation of cannabis. It is banned in the majority of places in the world. However Due to its superior biological characteristics that are relevant to scientific and technological circles, lower levels of tetra hydro cannabis (THC) were chosen to be planted. Internationally, products with less than 0.3% tetra hydro cannabis (THC) content are called Industrial Hemp, any higher than 0.3% content are called medicinal and drug marijuana (marijuana and hashish). For the sake of reunification, different varieties of industrial cannabis are collectively referred to as "China Hemp". It can't be used as a drug directly, but only as raw materials for industrial purposes, beneficial at a large scale [1].

We know that hemp fiber not only contains a variety of trace elements and is naturally antibacterial, but is also good for the human body, comfortable and safe to use. Moreover, the surface of the hemp has many cracks, and the medium cavity is larger, which makes the moisture perspiration, breathable thermal conductivity particularly good [1-3]. But when the hemp fiber is bending and has torsional stiffness, the holding force is small. It is easy for the fiber head end to slide out of the yarn which makes it hard to spin [2-5].

Nylon, also known as polyamide fiber contains amide-based synthetic fibers in each chain. Nylon-66 is made of adipic acid and hexamethylene diamine. Firstly, the equimolar hexamethylene diamine and hexagonal are twisted to produce nylon-66 salt, and then it in the molten state is condensation in the nitrogen poly whose melting point is about 260℃. The heating, melting, spinning stretching are in the nitrogen stream [6].

The wear resistance, breaking strength, large stretch, elasticity and fatigue resistance of nylon fiber are very good, but the biggest shortcomings is nylon fiber's moisture absorption.

The hosiery yarn of Hemp/Cotton and Nylon 66 developed in this paper not only solves the problem of poor hemp linen, but also combines the advantages of hemp/cotton and nylon 66 fibers, and does not affect each other's performance. Simultaneously, it was developed with antibacterial, moisture absorbent and fast drying and other functions.

2. Experimental

2.1 Material and Equipment

The main experimental materials used in this study are shown in Table 1, and the main equipment used is shown in Table 2.

Table.1 The yarn used in this study

Experimental materials	Fineness
Hemp/cotton blended yarns	27.76tex
Nylon66	4.44tex(34F/Z)
Nylon66	7.78tex(24F/Z)
Nylon66	11.11tex(24F-Z)
Spandex	4.44tex

Table.2 The machinery used in the experiment

Machinery	Model
Doubling and twisting machine	HFX-06
Hairiness tester	YG173A
Evenness tester	YG133B/M
Flat knitting machines	12G,914.4mm
Pilling - box tester	YG502
Fiber abrasion tester	YG522

2.2 Preparation of Twisted Yarns

The twisted yarns required in the experiment were made on a small doubling and twisting machine, model HFX-06 [7]. The simple model of the twisting machine is shown in Fig.1. The materials are 21 S the hemp/cotton blended yarns with nylon 66 whose fineness are 40D, 70D and100D. The yarns took shape by twisting the hemp/cotton blended yarns and three kinds of nylon yarns. We changed the twist of the hemp/cotton and Nylon66 blended yarns when they were twisting with the same denier of nylon, so that we can get 15 kinds of yarns. The twisted yarns of the 15 kinds of yarns are sequentially labeled, as shown in Table 3.

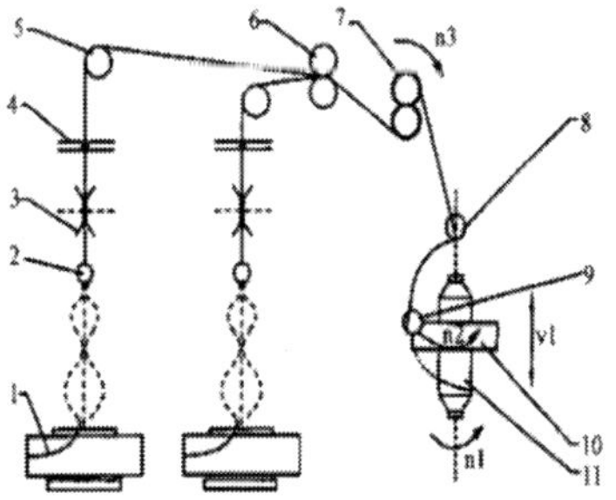

Fig.1 simple model of the twisting machine

1-Yarn bobbin 2-Guide yarn 3-Tensioner 4-Yam cleaner 5-Glass rods 6-double yarn roll 7-Traction roll 8-Guide hook 9-Bead ring 10-Steel collar 11-Spindle

Table.3 Twisted yarns

Mark	Materials	Ratio of the yarn blending	Twist
1	27.76tex Hemp/cotton yarns and 4.44tex Nylon66	Hemp/cotton/Nylon66(47/39/14)	40
2	27.76tex Hemp/cotton yarns and 4.44tex Nylon66	Hemp/cotton/Nylon66(47/39/14)	60
3	27.76tex Hemp/cotton yarns and 4.44tex Nylon66	Hemp/cotton/Nylon66(47/39/14)	80
4	27.76tex Hemp/cotton yarns and 4.44tex Nylon66	Hemp/cotton/Nylon66(47/39/14)	100
5	27.76tex Hemp/cotton yarns and 4.44tex Nylon66	Hemp/cotton/Nylon66(47/39/14)	120
6	27.76tex Hemp/cotton yarns and 7.78tex Nylon66	Hemp/cotton/Nylon66(43/35/22)	40
7	27.76tex Hemp/cotton yarns and 7.78tex Nylon66	Hemp/cotton/Nylon66(43/35/22)	60
8	27.76tex Hemp/cotton yarns and 7.78tex Nylon66	Hemp/cotton/Nylon66(43/35/22)	80
9	27.76tex Hemp/cotton yarns and 7.78tex Nylon66	Hemp/cotton/Nylon66(43/35/22)	100
10	27.76tex Hemp/cotton yarns and 7.78tex Nylon66	Hemp/cotton/Nylon66(43/35/22)	120
11	27.76texHemp/cotton yarns and 11.11tex Nylon66	Hemp/cotton/Nylon66(39/32/29)	40
12	27.76texHemp/cotton yarns and 11.11tex Nylon66	Hemp/cotton/Nylon66(39/32/29)	60
13	27.76texHemp/cotton yarns and 11.11tex Nylon66	Hemp/cotton/Nylon66(39/32/29)	80
14	27.76texHemp/cotton yarns and 11.11tex Nylon66	Hemp/cotton/Nylon66(39/32/29)	100
15	27.76texHemp/cotton yarns and 11.11tex Nylon66	Hemp/cotton/Nylon66(39/32/29)	120

2.3 Fabric

Plating stitch means that all the coils or portions of the coil on the fabric are formed of two yarns. Plating stitch is a color organization which is formed by the two yarns in accordance with the requirements of the fabric in front and back. Plating stitch can be divided into two categories. The one is that all the coils are made up of two yarns in the fabric. The one side of the fabric is exposed by one yarn and the other is exposed by another yarn. When knitting fabric with two different colors and yarns of different nature, the fabric has different colors and suitability. The other is that some coils are made up of two yarns, and the rest of yarns are made up of one yarn. the another is part of the coil Tim organization [8-9].

The treated yarn is processed into a fabric by plating stitch which is a flattening of all the coils based on the plain stitch on a knitting machine. The loop pattern draft of plating stitch shown in Fig.2. The knitting structures of the plating stitch is shown in Fig.3. The veil yarns are two identical yarns that have been treated, and the underground yarn is spandex. A yarn angle diagram is shown in Fig.4.

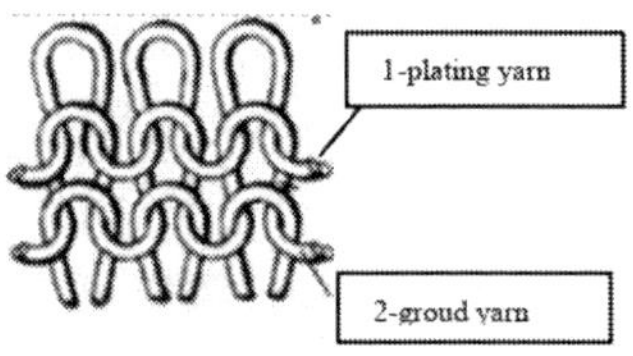

Fig.2 plating stitch

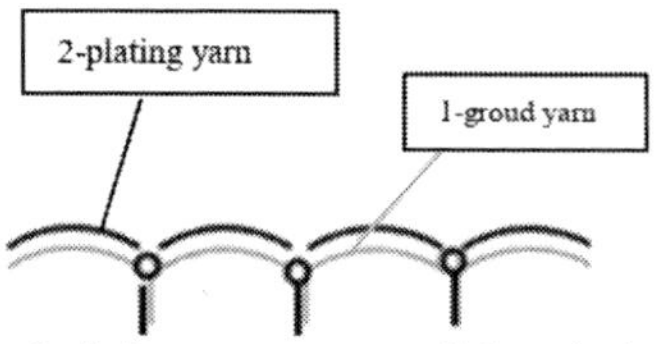

Fig.3 The knitting structures of the plating stitch

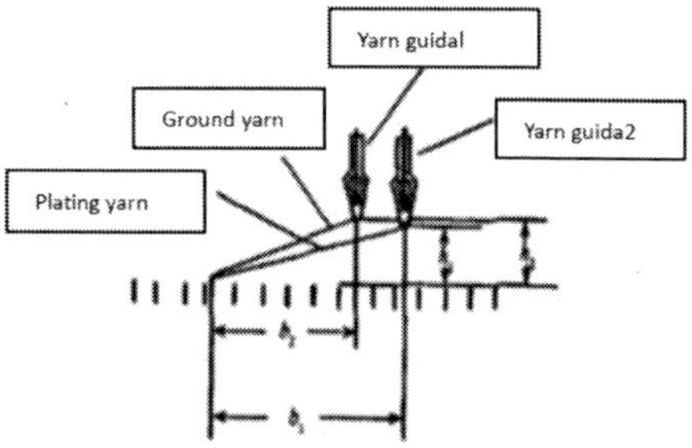

Fig.4 Yarn angle diagram

3. Results and Discussion

3.1 Yarn Hairiness Test

There are mainly two waysabout form of the hairiness. One refers to the hairiness formed during the twisting.

The other is the tail of the twisted fiber is not required to be wrapped around the inside that let the surface of the yarn easy to extend out of the surface to form a backward hairiness without external force. There will be some fibers pulled out of the yarn surface if those are not close with the yarn body that will make the original hairiness pull out, elongated to the surface. The process will form the end feathers, hairiness and floating hairiness [10].

The method of yarn hairiness test in this experiment is a photoelectric measurement which uses an optical amplification system. The working principle of this instrument is to test the length of the hairiness measured by projection or count, calculate the hairiness characteristics of the indicators, and then automatically detect the length of the yarn within the hairiness. The improvement rate of yarn hairiness index at 3mm length is calculated as shown in Fig.5.

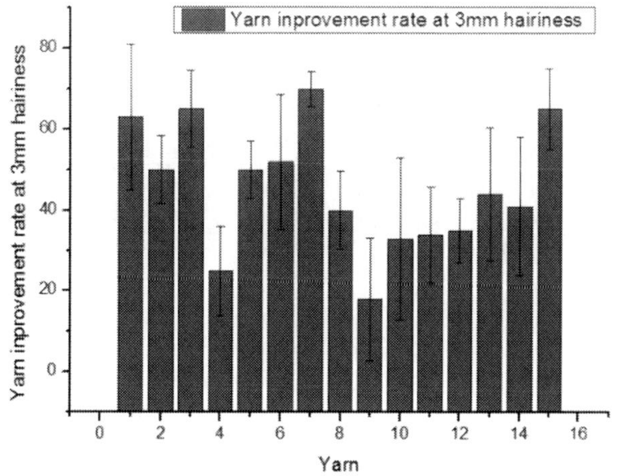

Fig.5 Yarn improvement rate at 3mm hairiness

From the above figure, when the ratio of the yarn is 47/39/14 hemp/cotton/nylon 66, yarn hairiness in the 3mm improvement rate also appears to correspond to fluctuations with the increase of the twist. When the twist is 80, the improvement rate of hairiness is the best; when the twist is 100, the improvement rate of hairiness is the smallest. When the yarn ratio of hemp/cotton/nylon 66 is 43/35/22, the yarn hairiness in the 3mm improvement rate generally increases first, but decreases when the twist is in higher than 60. The improvement rate of hairiness is the biggest of 15 kinds of yarn when the twist is 60. In contrast, the improvement rate of hairiness is the smallest of 15 kinds of yarn when the twist is 60. When the ratio of yarn is 39/32/29 hemp/cotton/nylon 66, the improvement rate of hairiness increased with the increase of twist. The hairiness improvement rate of the treated yarn is better than untreated hemp/cotton yarn.

3.2 Testing Yarn Uniformity

In this experiment, the method of electronic bar is used to test the uniformity of the yarn (the dryness). At present, the most common use of the electronic strip dryness uniform instrument is a capacitive strip uniformity meter.. The coefficient of variation of the yarn (the dryness) at thickness is an important parameter that characterizes the uneven thickness of the yarn (unevenness of the bar). The unevenness of the yarn will not only produce yarn defects, affect the appearance and strength of the yarn, but also cause weaving to break and stop. Therefore, uniformity of yarn fineness has become one of the important indexes of yarn quality evaluation [11]. Fig.6 shows unevenness CV of the yarn.

As shown in Fig.6, when the ratio of the yarn is 39/32/29 hemp/cotton/nylon 66, i.e. when the content of nylon 66 is the highest, the yarn unevenness (CV) value is lower than the other two kinds of yarn. The yarn uneven CV value increases at the first, and then reduces with an increased twist. When the yarn twist is 40, the yarn uneven CV value is the least of 15 kinds of yarn. However, when the twist of the yarn is 100, the unevenness of the yarn is the largest. When the ratio of the yarn is 43/35/22 hemp/cotton/nylon 66 the variation in uniform CV value of the yarn is small as the twist increases. When the ratio of the yarn is 47/39/14 hemp/cotton/nylon 66, that is to say the content of the hemp yarn is the highest, the unevenness of the yarn is larger than the other two proportions of the yarn, indicating that the proportion of the hemp yarn is not conducive to the uniformity of the yarn. When the yarn twist is 80, the unevenness of the yarn is the largest of the 15 yarns. Using the other 4 kinds of twist, the yarn unevenness CV value is almost equal.

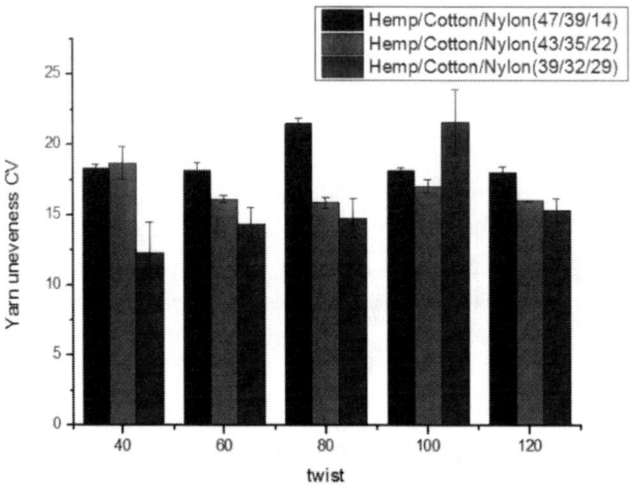

Fig.6 Yarn unevenness CV.

3.3 Fabric Abrasion Resistance Test

In this experiment, the test machine used a rotary whose method is flat grinding. The specimen was fixed on a working disc rotated at 3000r/min for constant speed with a diameter of 120 mm. There are two tents on the disc. In the two brackets, there are two grinding wheel discs on their own axis. During the test, the test piece on the working disc was in contact with the two grinding wheels and they were moved in relation to each other so that the sample is subject to many aspects of wear and tear. In this paper, the percentage drop in fabric weight was measured to calculate the fabric's wear resistance after withstanding a certain amount of friction [12]. Fig.7 shows the results of the test.

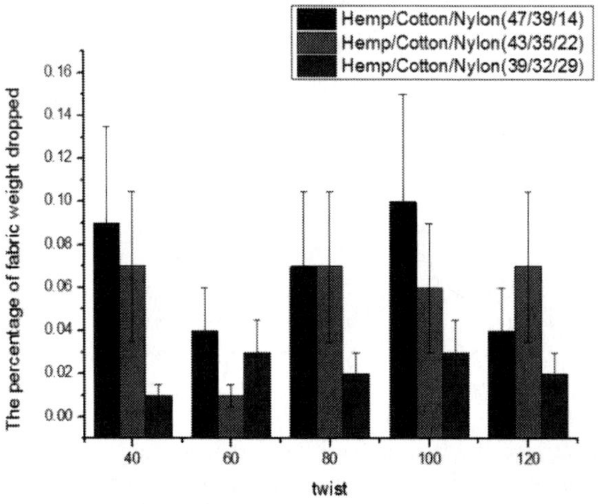

Fig.7 The percentage of fabric weight dropped

As shown in Fig.10, when the proportion of yarn is 39/32/29 hemp/cotton/nylon 66; the highest content of nylon

66 yarn, the quality loss of the fabric is lower than the other two proportions of yarn, indicating that the increase in the content of nylon 66 yarn is conducive to the wear resistance of the fabric. At this yarn proportions, the quality loss rate of the fabric decreases with the increase of the twist.

When the proportion of the yarn is 43/35/22 hemp/cotton/nylon 66, the mass loss rate of the yarn initially increases, but then decreases with the increase of the twist. When the yarn twist is 60, the mass loss rate of the fabric is smallest in the 15 yarns; when the ratio of the yarn is 47/39/14 hemp/cotton/nylon 66; with the highest content of the hemp yarn, the mass loss rate of the fabric fluctuates greatly, and the quality of the fabric loss rate is greater than the other two yarns, indicating that the increase in the content of the hemp yarn is not conducive to the abrasion resistance of the fabric. When the yarn twist is 100, the mass loss rate of the fabric is the greatest of the 15 yarns.

3.4 Fabric Pilling Performance Test

The pilling is an important parameter that characterizes the uneven thickness of the yarn. The unevenness of the yarn will not only produce yarn defects, affect the appearance and strength of the yarn, but also cause weaving to break and stop, form the piliing. Therefore, uniformity of yarn fineness has become one of the important indexes of yarn quality evaluation [13].

In this experiment, the circular trajectory method is used to measure the pilling performance of the fabric. That is, in a certain pressure, let the fabric and the nylon brush for the relative friction as the circular motion of the track. After several times, under the specified lighting conditions, compare the standard sample to assess the degree number of the fabric pilling. The sample was divided into 1 to 5 grades. The best grade is 5 that is not easy to form the pilling; the worst grade is 1, applied when there is serious pilling of the fabric. In conclusion, the higher the grade, the better anti-pilling performance of the fabric [14-15]. The pilling grades of 15 fabrics are shown in Table 4.

Table.4 The pilling grades of 15 fabrics

	1	2	3	4	5	6	7	8	9	10	11	12	13	14	15
Pilling grades	1	2	1	2	1	3	3	4	2	3	4	5	5	4	4

From Table 4, the fabric pilling grades of 15 kinds of fabrics generally showed an increasing trend. Indicating a fabric's pilling grade increased along with the proportion of nylon 66. This suggests that as the content of nylon 66 increases, the anti-pilling of the fabric improves. When the yarn ratio is 39/32/29 hemp/cotton/nylon 66, and the twist of yarn is at 60 or 80, the pilling grade of the fabric is the best in the 15 fabrics.

4. Conclusion

1) Yarn made of a combination of hemp, cotton and nylon 66 were tested on hairiness, uniformity, and pilling by changing the proportion of the different yarns. Overall, the properties of the yarn had improved. Better performing fabrics are those with a higher nylon 66 content and lower hemp content.

2) When the content of yarns are the same, yarn hairiness, uniformity of the fabric, wear resistance of the fabric, and the pulsating performance of the fabric are different with the change in yarn twist, but in general, the yarn and fabric performance is better when the yarn twist is 60.

3) Blended yarns (containing hemp and cotton with nylon 66 as the main raw materials) are antibacterial, hygroscopic and breathable, and also solves the problems. Moreover, the hairiness, uniformity of the yarn, fabric wear resistance and pulsating performance have improved.

We can take full advantage of the excellent performance of these three yarns to make the fabrics comfortable, antibacterial, and so on. At present, there are a few people who are researching yarns combining hemp, cotton and nylon. So we will research about this study.

References

[1] Nancollas H. In vitro studies of calcium phosphate crystallisation. In: Mann S, Webb J, Williams RJP, editors. Biomineralization. Chemical and biochemical perspectives. New York: VCH: 1989; p.157-182.

[2] Brown W, Chow LC. Combinations of sparingly soluble calcium phosphates in slurries and paste as mineralizers and cements. US Patent No.4612053: 1986.

[3] Wang BH. China Textiles Finishing and Progress Volume. China Light Industry Press: 2013; 09: 14.

[4] Zhou MJ. China marijuana textile status and development of my opinion. Beijing textile: 2001; 4: 9.

[5] Huang CR, Yu WD. The development of linen denim fabric. Journal of Textile Science: 2005; 6: 84-85.

[6] Zhou JH, Yang RN, Zhang GH. Development and utilization of fiber with cannabis. Chinese paper making: 2001; 5: 63.

[7] Chen YH, Wang GH. Study on Sexual Bear of Hemp/Cotton Blended Yarn with Different Blending Ratio. Modern Silk Science and Technology: 2013; 28: 11.

[8] Luo YB editor. Insulation and lubrication materials chemistry. Wuhan University Press: 2005; 1: 134.

[9] Gao ZY, Zhang WH. Research on the Biological Characteristics and Application of Hemp. Textile Science and Technology: 2006; 6: 37.

[10] Jiang GM editor. Knitting. China Textile Press: 2012; 9: 86.

[11] Shi JX, SGL, Liu GH. Discussion on the Technology of Bending Tissue of Double - Shaped Saw: 2012; 5: 13.

[12] Huo H, edited by Chen HF. Textile Inspection (2nd Edition). China Fortune Publishing House: 2014; 06: p. 209.

[13] Yang LF, Zhang HT, Li JP editor, Chen YN trial. Textile materials and testing. Dong hua University Press: 2013; 03: page149.

[14] Wu PY, Geng QY editor. Textile fiber and products (below) test technology. Suzhou University Press: 2007; 9: page 181.

[15] Zhu SK, Gao WD editor. Machine science second edition. China Textile Press: 2015; 01: page 148.

Development of Electrically Conductive Activated Carbon Fabric from Kevlar Fabric for Effective Emi Shielding Applications

Daniel Karthik[1*], Vijay Baheti[2], Veronika Tunakova, Jiri Militky, Sundaramurthy Palanisamy

Department of Material Engineering, Faculty of Textile Engineering, Technical University of Liberec, Studentska 2, Liberec 46117, Czech Republic

*Corresponding author Email: danielkarthiktex10@gmail.com

Abstract

In the present work, porous and electrically conductive activated carbon fabric was produced by heating Kevlar fabric wastes using novel single stage carbonization. The influence of carbonization temperature of 800°C, 1000°C and 1200°C on physical and morphological properties of activated carbon fabric was studied from EDX and SEM analysis. Additionally, the electrical conductivity was also measured. At the end, the utility of prepared activated carbon web was investigated for electromagnetic shielding ability in high frequency (i.e. 2.45 GHz) and low frequency regions (i.e. below 1.5 GHz) using waveguide method and coaxial transition line method respectively. The activated carbon fabric produced at 1200 °C showed maximum shielding effectiveness in both high and low frequency regions. For single layers of 1200 °C carbon fabric, the electromagnetic shielding effectiveness of 40.5 dB, 41.8 dB, 45.1 db and 50.9 dB was found for respective frequencies of 30 MHz, 100 MHz, 1GHz and 1.5 GHz. This behaviour was attributed to increased absorption of electromagnetic radiations due to its higher porosity and also to increased reflections of electromagnetic radiations due to its higher electrical conductivity.

Keywords: Kevlar Fabric Wastes; Physical Activation; Activated Carbon; Specific Surface Area; Electrical Conductivity; Electromagnetic Shielding

1. Introduction

In recent years, research on electromagnetic interference (EMI) shielding materials has attracted significant attention due to an increase in electromagnetic population from widespread applications of computer and telecommunication technologies [1, 2]. Electromagnetic interference refers to the radiant electromagnetic signals emitted by electrical instruments during their operation. The emitted electromagnetic radiations are a concern since they interfere with the working of other appliances as well as causing serious health risks to the consumers [3].

The EMI shielding is related to reflection or absorption of electromagnetic radiations by shielding material. Reflection is a commonly used shielding mechanism by high electrical conductivity materials such as metals and their nanoparticles. However, high density, lack of flexibility, easy corrosion, costly processing and weak microwave absorption are main drawbacks of metals [4]. In addition, the reflection of radiations from their surface interferes with working of other devices. Recently, carbon nanostructures and graphene have been reported as promising alternatives to metal-based shielding materials [3, 5].

Among the continuous fibers, carbon fibers are dominant, due to their low density, high modulus, high strength, wide availability and are especially attractive in their electrical conductivity, which relates to EMI shielding effectiveness. The carbon-based shielding materials are expected to be predominant in effective shielding mechanism due to the synergetic effect of electrical conductivity and multiple reflections. [6, 7]

Activated carbon fibers, because of their characteristics, have been used for a wide range of applications such as: Operation of air conditioning systems, creation of personal protection, in the chemical industry, for the protection of wounds by semi-permeable band aids that do not allow micro-organisms access and for removing VOCs (benzene, acetone, dichloromethane, n-hexane, acrylonitrile, etc.)

Feedstock materials for Activated carbon fibers are usually of low crystallinity, such as viscose rayon, polyacrylonitrile, or coal tar pitch. Alternatively, aramid fibers have been proposed as high crystallinity precursors and some studies have reported using Nomex and Kevlar as precursors [8]. For most high-volume applications, however, it is necessary to consider their production costs due to the lengthy, multistep process for the conversion of starting precursor fibers into high-performance carbon fibers. During the processing, the precursor fibers require oxidative stabilization prior to carbonization heat-treatment. This oxidative stabilization step is slow and affects the final mechanical properties of the carbon fibers [9].

A variety of organic precursors have been exploited for the preparation of carbonaceous materials. Aramids, particularly, Kevlar fibres have attracted much attention of researchers as a precursor for high modulus carbon fibres and high efficiency active carbon fibres because they are composed of linear single aromatic rings and no stabilization reaction in the oxygen atmosphere that is often required for the carbonization of low melting organic precursors, is necessary. Moreover, if the Kevlar flocks are used as the precursor, the cheaper price may be another merit since they are the wastes from the Kevlar production process [10, 11].

Kevlar is the polycondensation product of 1,4-diaminobenzene and terephthalic acid and thus the polymeric molecules are composed of linear chains of single aromatic rings. An extensive inter-chain hydrogen bonding leads to molecular sheets that are radially arranged along the fiber axis. This structural feature imparts high modulus and high thermal stability to the fibers [10].

It appeared in one report that the carbon yield of Kevlar fibres range from 25% to about 40% [11]. This low carbon yield likely results from the linear nature of polymeric chains in Kevlar fibers. In general, when linear polymers are subjected to a heat-treatment, bond cleavage leads to a formation of small molecular fragments that are eventually lost during the pyrolysis process. One strategy to improve the carbon yield in Kevlar pyrolysis may thus be the conversion of linear aromatic chains to more condensed/ cross-linked polyaromatic intermediates before the serious thermal decomposition takes place. It is intuitive that once crosslinked networks are generated, fragmented molecules are not so small that volatile loss could be minimized. [12].

Seoung Ho et al [10], in their work have demonstrated that Carbon yield after 1000^{0}C heat treatment of PPTA fibres surprisingly increased up to 60wt% exhibiting more than 20% increase than conventional one step carbonization.

The graphitizing properties of these high-performance aromatic polymeric fibers have not been fully studied so far although they may be useful for the preparation of not only the carbon fibers but also activated carbon fibers (ACF), electrodes of secondary batteries, etc. For the production of these carbon goods from aromatic polymers, simple direct carbonization process is attractive to save the production cost. Although it is not easy to produce carbon fibers with excellent tensile properties, carbon fibers with specific properties, e.g., heat resistance, absorption properties, electrical conductivity [2], chemical resistance, could be producible [9].

To improve the properties, a detailed understanding of the physical and chemical structure of the fibres, the evolution of structure with heat treatment, and the relationship of structure to mechanical properties [13].

The present study deals with the simple and novel method of preparation of porous and electrical conductive activated carbon fabric from Kevlar fabric wastes. The prepared activated carbon is advantageous over carbon made from other materials because of low cost, high density and better purity. The activated carbon fabric was prepared by physical activation of Kevlar fabric wastes. The carbonization was performed under the layer of charcoal at 800°C, 1000°C and 1200°C with the heating rate of 300°C h^{-1} and without any holding time. Further, electrical conductivity, EDX, SEM analysis was carried out to study the effect of carbonization temperature on physical and morphological properties of activated carbon fabric. At the end, the electromagnetic shielding ability of the produced three activated carbon fabrics was investigated with respect to change in carbonization temperature using two different measurement approaches (i.e. waveguide method and coaxial transition line method). In this way, the present study served to utilize large quantity of Kevlar fabric wastes for development of effective electromagnetic shielding materials.

2. Experimental Methods

2.1 Materials

The Kevlar woven fabric wastes were obtained from VEBA textiles, Czech Republic.

2.2 Preparation of Activated Carbon Fabric

The Kevlar fabric waste was dipped in acetone for 24 hr to remove surface finish and impurities. The cleaned fabric was then transferred to high temperature furnace (Elektrické Pece Svoboda, Czech Republic) for direct carbonization without any stabilization step. The single stage carbonization and physical activation in presence of air was performed at 800 °C, 1000 °C and 1200 °C with heating rate of 300 °C h^{-1} and without any holding time. This was achieved by controlled carbonization under the layer of charcoal.

2.3 Characterization of Activated Carbon Web

Energy dispersive x-ray (EDX) analysis. It was performed on Oxford Instruments, LZ 5 EDX detector, UK to know the change in relative proportion of different elements with respect to change in carbonization temperature.

Electrical conductivity. Hewlett Packard 4339 B high resistance meter was used to measure the electrical resistance of activated carbon web samples. It was measured at temperature 22°C and relative humidity 40 %. The specific voltage potential was applied across opposite ends of each activated carbon web and resultant current flowing across the sample was measured.

SEM analysis. The field emission scanning electron microscope Sigma, Zeiss, Germany was employed to investigate the morphology of Kevlar fabric and prepared activated carbon fabric of 800 °C, 1000 °C and 1200 °C carbonization temperature. This helped to understand the development of porosity characteristics of activated carbon web with change in carbonization temperature. The micrographs were taken at 2 kV accelerated voltage and 1000 x magnification. As the prepared activated carbon web was electrically conductive, it was not metalized before test. However, the Kevlar waste woven fabric was metalized by sputter coating.

2.4 Electromagnetic Shielding Effectiveness of Activated Carbon Web

The electromagnetic shielding effectiveness of prepared activated carbon web was determined from two different measurement principles (i.e. waveguide method and coaxial transition line method).

Waveguide method. This method was used to determine the shielding effectiveness of samples for microwave frequency range of 2.45 GHz [14]. Fig.1(a) shows the measurement set up of waveguide method. The device consisted of a rectangular hollow waveguide having electrically conductive walls. A receiving antenna was placed inside of this waveguide, while a sample was placed at the entrance to the waveguide. The end of the waveguide was filled with foam saturated with carbon particles to absorb the electromagnetic field passed through the sample. Transmitting antenna was placed in front of the waveguide input at 16 cm distance apart. Anetwork analyzer Agilent E 4991A was used to generate the electromagnetic waves, and a high frequency analyzer HF-38B (Gigahertz Solutions) was used to receive the electromagnetic signals. The electromagnetic shielding effectiveness SE [dB] was calculated based on following formula:

$$SE = 10\log\left(\frac{P_t}{P_i}\right) \quad \text{(Eq. 1)}$$

Where P_t and P_i are electromagnetic field density [W/m2] measured in presence of sample, and without the sample respectively.

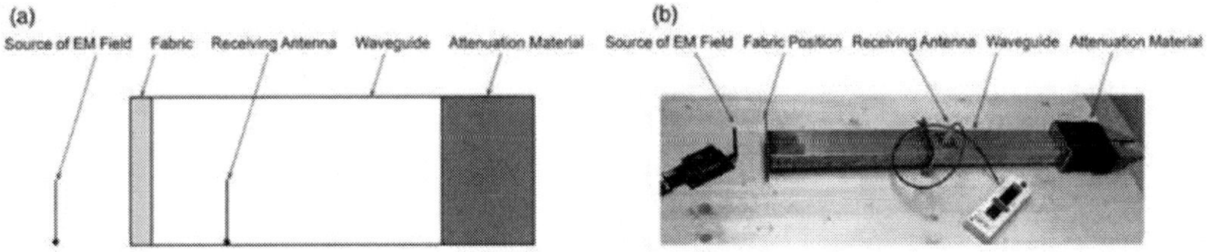

Fig.1(a) Measurement set up of waveguide method

Coaxial transition line method. For more detailed electromagnetic shielding analysis, coaxial transition line method was used in frequency range of 30 MHz -1.5 GHz according to ASTM D 4935-10. This device determined electromagnetic shielding effectiveness using the insertion-loss method. Fig.1(b) shows the measurement set up of coaxial transition line method. The set-up consisted of a sample holder with its input and output connected to the network analyzer. A shielding effectiveness test fixture (Electro-Metrics, Inc., model EM-2107A) was used to hold the sample. The network analyzer (Rohde & Schwarz ZN3) was used to generate and receive the electromagnetic signals.

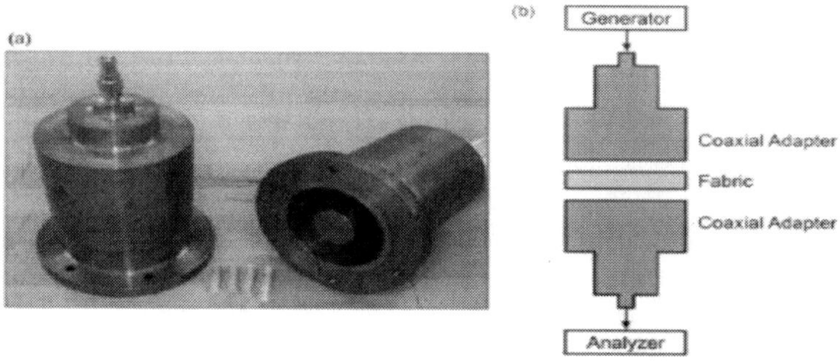

Fig.1(b) Measurement set up of coaxial transition line method

3. Results and Discussions

3.1 Characterization of Activated Carbon Web

3.1.1 EDX Analysis

Energy disperse x-ray spectroscopy was performed to know the relative proportion of different elements present in the activated carbon fabrics. From Table 4, the increase in carbon content and reduction in oxygen content was found with increase in carbonization temperature from 800°C to 1200 °C. The activated carbon web produced at 1200 °C exhibited 92.35% increase in carbon content and 4.56% reduction in oxygen content. This behavior was attributed to removal of hydrogen, sulfur, nitrogen and other elements due to decomposition at higher temperature.

Table.1 Effect of carbonization temperature on elemental composition of activated carbon fabric

AT. (%)	C	N	O	Na	S	Cl	K	Ca	Total
1200(°C)	92.35	2.27	4.56	0.09	0.23	0.05	0.39	0.05	100.00
1000(°C)	87.60	5.11	6.15	0.25	0.27	0.01	0.54	0.06	100.00
800(°C)	73.88	9.19	12.22	3.39	0.96	0.07	0.21	0.08	100.00
Untreated	69.45	14.07	15.39	0.68	0.34	0.04	0.00	0.03	100.00

3.1.2 Electrical Conductivity

For electromagnetic shielding effectiveness to be achieved through reflection of radiations, the knowledge of electrical conductivity of samples is necessary. Fig.2 shows the results of electrical resistivity for different activated carbon web samples. The electrical volume resistivity was found to decrease with increase in carbonization temperature. The 1200°C activated carbon sample exhibited 1000 times reduction in electrical resistivity over 800°C activated carbon sample. The higher electrical conductivity of 1200°C activated carbon sample was attributed to more graphitization.

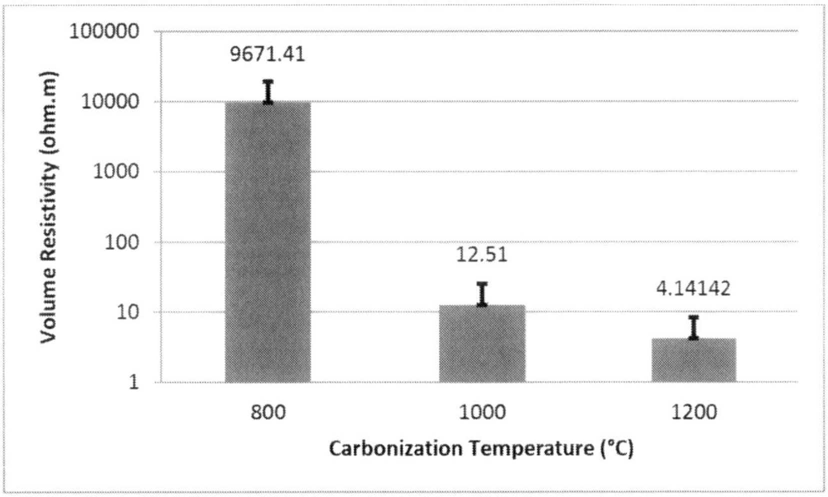

Fig.2 Effect of carbonization temperature on electrical conductivity of activated carbon web (Error bars added)

3.1.3 SEM Morphology

In order to know the development of porosity after carbonization of Kevlar fabric wastes, the surface morphology of Kevlar fabric wastes before and after carbonization was studied from SEM images. Figs.7 (a) and 7(b) show the SEM images of Kevlar fabric and activated carbon fabric produced at temperature of 1200°C. The activated carbon fabric showed noticeable rough surface as compared to Kevlar fabric web. The surface roughness was found to increase with increase in carbonization temperature, which indicated the development of more porous structure after physical activation of Kevlar fabric wastes.

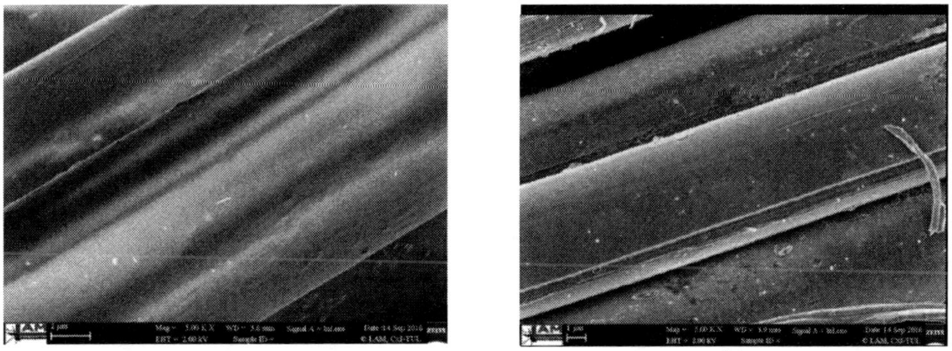

(a) (b)

Fig.3 SEM image of (a) Untreated Kevlar fabric (b) 1200°C activated carbon fabric

3.2 Electromagnetic Shielding Ability

3.2.1 Waveguide Method

Fig.4 shows the average values in 95 % confidence interval for electromagnetic shielding effectiveness of prepared activated carbon web samples in single and double layers measured at 2.45 GHz frequency. The electromagnetic shielding effectiveness was found to increase with increase in number of layers and increase in carbonization temperature.

Then, the shielding effectiveness was found to increase dramatically over a very narrow range of carbonization temperature, which was connected to the amount of carbon phase present in the structure. This point is called the percolation threshold (similar to electrical percolation threshold). It is an important phenomenon showing minimum carbonization temperature required for maximum increase in conductivity for higher shielding effectiveness. In present study, the percolation threshold was found between the range of 800 °C and 900 °C carbonization temperature. The maximum shielding effectiveness in this range was attributed to increased absorption of electromagnetic radiations due to higher porosity of activated carbon and also to increased reflections of electromagnetic radiations due to higher electrical conductivity of activated carbon [15]. The dramatic increase of shielding ability could not be expected with further increase of carbonization temperature (T>1100 °C). Therefore, the usage of carbonization temperature at 1000 °C was considered optimal with regard to its relatively high electromagnetic shielding ability and satisfactory mechanical properties. At 2.45 GHz, the electromagnetic shielding effectiveness of 40.17 dB, 32.04 dB and 11.93 dB was exhibited by single layers of activated carbon fabric produced at 1200°C, 1000°C and 800°C, respectively.

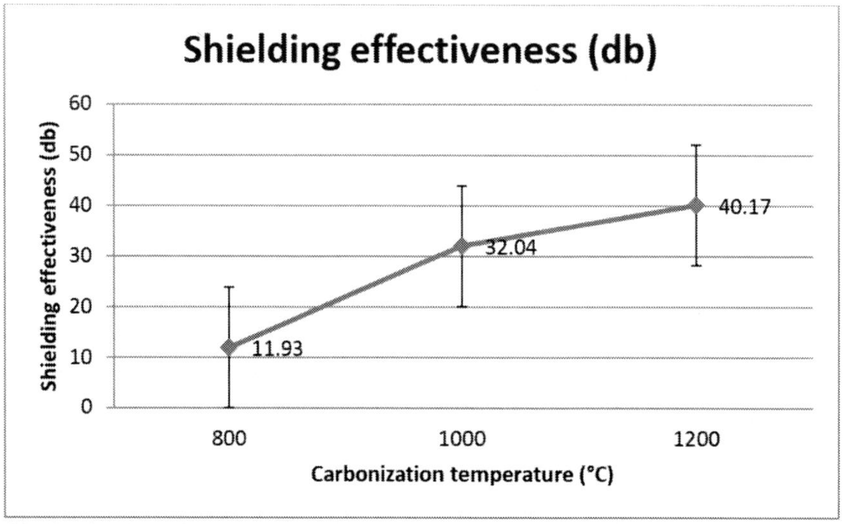

Fig.4 Effect of carbonization temperature on electromagnetic shielding effectiveness in high frequency region (Error bars added)

3.2.2 Coaxial Transition Line Method

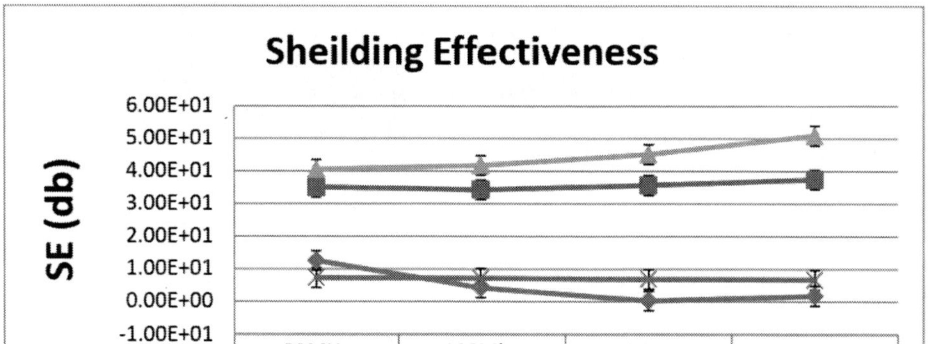

	30MHz	100Mhz	1GHz	1.5GHz
untreated	1.25E+01	4.21E+00	2.48E-01	1.76E+00
800	7.26E+00	7.14E+00	6.83E+00	6.70E+00
1000	3.51E+01	3.43E+01	3.57E+01	3.75E+01
1200	4.05E+01	4.18E+01	4.51E+01	5.09E+01

Fig.5 Effect of frequency on electromagnetic shielding effectiveness (Error bars added)

This method was used for estimation of shielding effectiveness according to ASTM 4935-10 standard in low frequency region from 30 MHz to 1.5 GHz. Fig.12 (a-b) show the mean values of electromagnetic shielding effectiveness for single layers of different activated carbon samples in frequencies of 30 MHz, 100MHz, 1 GHz and 1.5 GHz. The increase in shielding effectiveness with increase in carbonization temperature was observed. The 1200 °C activated carbon web exhibited the shielding ability of 40.51 dB, 41.75 dB, 45.13 dB and 50.90 dB for respective frequencies of 30 MHz, 100 MHz, 1 GHz and 1.5 GHz. This behaviour was attributed to increased absorption of EM radiations due to its higher porosity and also to increased reflections of EM radiations due to its higher electrical conductivity.

4. Conclusion

The present study was focused on development of porous and electrically conductive carbon based electromagnetic shielding materials. A simple and novel approach was employed to introduce the both properties of absorption and reflection of electromagnetic radiations into the shielding materials. This was achieved by physical activation of woven Kevlar fabric wastes into activated carbon fabric. The carbonization was performed under the layer of charcoal at 800°C, 1000°C and 1200°C with the heating rate of 300°C h^{-1} and without any holding time. Further, the influence of carbonization temperature on physical and morphological properties of activated carbon web was studied from EDX, SEM analysis and electrical conductivity. In the end, the utility of prepared activated carbon web was investigated for electromagnetic shielding ability in high frequency (i.e. 2.45 GHz) and low frequency regions (i.e. below 1.5 GHz) using waveguide method and coaxial transition line method respectively. At 2.45 GHz, the electromagnetic shielding effectiveness of 40.17 dB, 32.04 dB and 11.93 dB was exhibited by single layers of activated carbon fabric produced at 1200°C, 1000°C and 800°C, respectively. On the other hand, for low frequency regions, the 1200 °C activated carbon web exhibited the shielding ability of 40.51 dB, 41.75 dB, 45.13 dB and 50.90 dB for respective frequencies of 30 MHz, 100 MHz, 1 GHz and 1.5 GHz. This behaviour of 1200 °C activated carbon web was attributed to increased absorption of EM radiations due to its higher porosity and also to increased reflections of EM radiations due to its higher electrical conductivity.

5. Acknowledgement

This work was supported by the research project of Student grant competition of Technical University of Liberec no. 21197 granted by Ministry of Education Youth and Sports of Czech Republic.

References

[1] Šafářová V, Tunák M, Militký J. Prediction of hybrid woven fabric electromagnetic shielding effectiveness. Text Res J: 2015; 85: 673-86. doi: 10.1177/0040517514555802.
[2] Šafářová V, Militký J. Electromagnetic shielding properties of woven fabrics made from high-performance fibers. Text Res J: 2014; 84: 1255-67. doi: 10.1177/0040517514521118.
[3] Chung DD. Electromagnetic interference shielding effectiveness of carbon materials. Carbon N Y: 2001; 39: 279-85. doi: 10.1016/S0008-6223(00)00184-6.
[4] Rubežienė V, Baltušnikaitė J, Varnaitė-Žuravliova S, Sankauskaitė A, Abraitienė A, Matuzas J. Development and investigation of electromagnetic shielding fabrics with different electrically conductive additives. J Electrostat: 2015; 75: 90-8. doi: 10.1016/j.elstat.2015.03.009.
[5] Wang LL, Tay BK, See KY, Sun Z, Tan LK, Lua D. Electromagnetic interference shielding effectiveness of carbon-based materials prepared by screen printing. Carbon N Y: 2009; 47: 1905-10. doi: 10.1016/j.carbon: 2009.03.033.
[6] Salman N, Baheti V, Tunakova V, Militky J, Daniel K, Tomkova B. Development of porous and electrically conductive activated carbon web for effective EMI shielding applications. Carbon: 2017; 111: 439-447.
[7] Junhua W, Chung DD. Increasing the electromagnetic interference shielding effectiveness of carbon fiber polymer–matrix composite by using activated carbon fibers, Carbon N Y: 2002; 40: 445-467.
[8] Seong-Ho Y, et al. Multi-Staged Carbonization of Aramid Fibres

[9] Tomizuka I, Isoda Y, Amamiya Y. Tanso: 1981; 106: 93-101.
[10] Ki SK, Chul WP, Seong-Ho Y, Seung MO. Preparation of Kevlar-derived carbon fibers and their anodic performances in Li secondary batteries, Carbon: 2001; 39: 1619-1625.
[11] Giraldoa L, Ladinob Y, Pirajánc JCM, Rodríguezb MP. Synthesis and characterization of activated carbon fibers from Kevlar. Ecl. Quím: 2007; 32 (4): 55-62.
[12] Kawahara Y, Otoyama S, Yamamoto K, Wakizaka H, Shinahara Y, et al. Direct Carbonization of High-performance Aromatic Polymers and the Production of Activated Carbon Fibers. J Textile Sci Eng: 2015; 5: 219. doi: 10.4172/2165-8064.1000219.
[13] Quichen Z, Ying L, Steven BW, Partial carbonization of aramid fibers. Journal of Polymer Science: 1994; 32 (13): 2207-2220.
[14] Safarova V, Tunak M, Truhlar M, Militky J. A new method and apparatus for evaluating the electromagnetic shielding effectiveness of textiles. Text Res J: 2016; 86: 44-56. doi: 10.1177/0040517515581587.
[15] Chen Z, Ren W, Gao L, Liu B, Pei S, Cheng H-M. Three-dimensional flexible and conductive interconnected graphene networks grown by chemical vapour deposition. Nat Mater: 2011; 10: 424-8. doi: 10.1038/nmat3001.

The Influence of Noil Gauge on the Evenness of Cotton-web in Comber

Yu-Heng Chen[1], Jia-Zhi Ren[1*], Qing-Guo Feng[1], Guo-Xin Jia[2]

[1] *College of Textile, Zhongyong University of Technology, Zhengzhou, Henan 450007, China;*
[2] *Department of Textile, Henan University of Engineering, Zhengzhou, Henan 450007, China*

*Corresponding author's email: rjzhi@163.com

Abstract

According to the length and the weight percentage curve of the cotton bundle clamped by nipper jaw, the influence of noil gauge in comber on the evenness of cotton-web has been analyzed. When noil gauge was 8mm, 9mm, 10mm, 11mm, using the method of cutting and weighing, the weight percentage curves of separation bundle and cotton-web were obtained. The 2mm weight CV value of the separation bundle and the cotton-web are calculated based on the separation-joint principle on comber. The results show that the smaller the noil gauge, the better the evenness of the cotton-web.

Keywords: Comber; Noil Gauge; Separation-joint; Separation Bundle; The Evenness of Cotton-web

Introduction

The distance of the nipper leading edge to the surface of the separation roller is called noil gauge when the nipper swings to the most forward position on comber. It is an important process parameter to adjust the combing quality [1, 2, 3]. The selection range of the noil gauge is 8-11mm. If the noil gauge is too large or small it will not meet the combing quality [4, 5, 6]. Some researchers have studied the effect of noil gauge when short fibers and neps are eliminated [7, 8]. Still, some scholars analyzed the effect of noil gauge on the process performance of the comber [9, 10, 11] through computer programming. However, there is still a lack of systematic research and analysis on the effect of noil gauge on the evenness of separation bundle and cotton-web outputted by separation rollers. Therefore, researching the influence of the noil gauge on the evenness of cotton-web has an important significance to design new comber and optimize process parameters.

1. Theoretical Analysis

As shown in Fig.1, a single separating bundle produced by separation rollers in a work cycle, the serial cotton-web produced by these fiber bundles head tail connected at an interval of a certain length. Therefore, the evenness of the cotton-web outputted by separating roller is closely related to the shape and weight distribution of the separating bundle. The separation bundle is formed by fibers which are pulled out quickly from nipper jaw by the detaching rollers. The shape of separation bundle is determined from cotton bundle morphology clamped by the nipper jaw. The noil gauge is an important factor affecting the length and weight distribution of the cotton bundle clamped by nipper jaw. So it also has a great influence on the shape of separation bundle and the evenness of cotton-web evenness on comber.

1.1 The Length of Cotton Bundle Clamped by Nipper Jaw

B indicates that separation gauge (mm), K is feeding coefficient (0<K<1), A is feeding length (mm), In the process of forward feeding on comber [12, 13]. The equation of cotton bundle length clamped by nipper jaw is:

$$L = B + (1 - K)A \quad (1)$$

It can be seen from formula (1): The length of the cotton bundle clamped by nipper jaw increases with the increasing of separation gauge B. When other process parameters are constant on comber, the weight distribution of cotton bundle clamped by nipper jaw changes along with the increase of noil gauge. Thus, the number of fibers outputted by detaching roller is also affected by it (as shown in Fig.1) [14].

1.2 The Weight Distribution of Cotton Bundle Clamped by Nipper Jaw

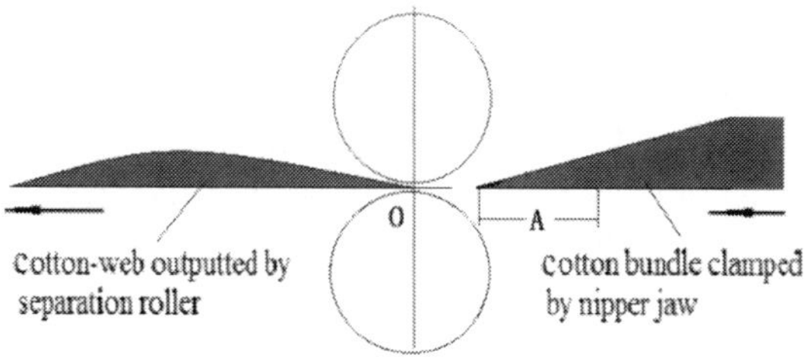

Fig.1 The model of separation-joint on comber

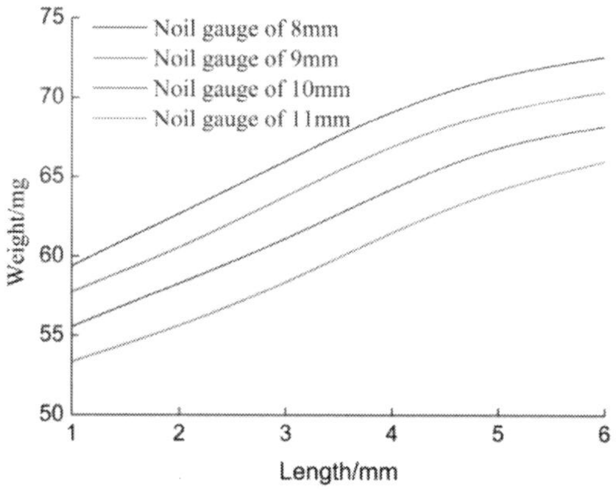

Fig.2 The weight distribution of cotton bundle clamped by nipper jaw at different noil gauge

The length of cotton bundle clamped by nipper jaw is equal to the length of feeding length A in a work cycle. When noil gauges is 8mm, 9mm, 10mm, 11mm, the weight distribution of the 0-6mm section of cotton bundle clamped by nipper jaw is shown in Fig.2. The head end of the separation bundle as the origin, the length of the separation bundle as X-axis , the weight percentage as Y-axis. It can be seen that the weight of the fiber in the cross section of the head of the cotton bundle 0-6mm decreases as the noil gauge increases. The number of fibers outputted by separation roller at the same time is reduced.

In summary, when other process parameters are constant on comber, as the increases of noil gauge, the length of the cotton bundle increases and the number of fibers in the cross section of the head of the cotton bundle decreases. The number of fibers outputted by the separation roller during the same period of time is reduced. It results in a change of the separation bundle morphology and the evenness of cotton-web outputted by separation roller.

2. Experiment

2.1 Condition of Experiment

100% fine cotton, staple length: 29.6mm, short fiber rate below 16mm: 13.8%, the effective output length of comber: 26.48mm, lap weight: 70g/m, feeding length: 5.2mm, cylinder location: 37 index, lab scale of detaching roller: 0 scale, top combing depth: 0mm, output weight: 22g/5m.

2.2 Scheme of Experiment

When other process parameters are constant on comber, changing noil gauge, the separation bundles of different noil gauges are obtained by using the method of manual rotation of comber a cycle. The weight percentage curves of different separation bundles are obtained by using the method of 2mm cutting and weighing. The characteristic parameters of different separation bundle are calculated. Cotton-web is made by multiple separation bundles of the head distance is equal to the effective length on comber [15]. Based on above principle, the weight percentage curve of cotton-web is drawn according to the weight percentage curve of single separation bundle. And then, the characteristic parameters of cotton-web are calculated.

3. Results and Discussion

3.1 The Weight Percentage Curve of Separation Bundle

On comber, when noil gauges is 8mm, 9mm, 10mm, 11mm, the separation bundles of different noil gauges are obtained by using the method of manual rotation of comber a cycle. Weight percentage of the separation bundle each 2mm fragment is calculated by using the method of 2mm cutting and weighing. The head end of the separation bundle as the origin, the length of the separation bundle as X-axis , the weight percentage as Y-axis, the weight percentage curves of different separation bundles are drawn (as shown in Fig.3). The peak weight percentage, peak position and weight percentage of 2mm fragment and other characteristic parameters are shown in Table 1.

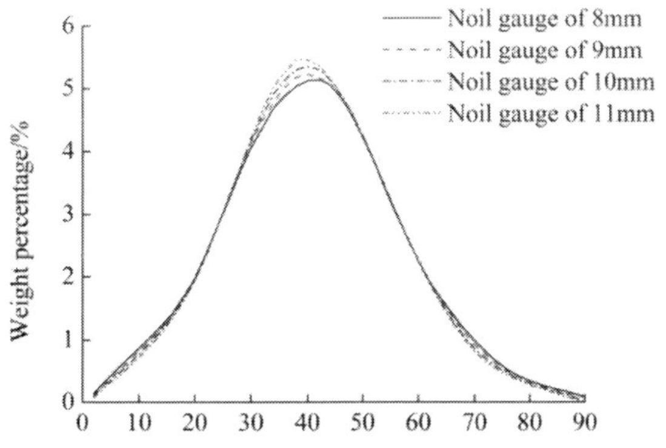

Fig.3 Weight percentage curves of the separation bundles of different noil gauges.

Table.1 Characteristic parameters of the separation bundles of different noil gauges.

Noil gauge/mm	Weight percentage at 10mm /%	Weight percentage at 70mm /%	Peak weight percentage /%	Peak position /mm	The weight CV of separation bundle /%	The length of separation bundle /mm
8	0.86	0.98	5.14	42	78.2	90.8
9	0.81	0.92	5.23	40	79.9	90.2
10	0.76	0.85	5.34	40	79.87	89.4
11	0.72	0.81	5.46	38	81.7	88.5

As shown in Fig.3 and Table 1: (1) The weight percentage curves of the separation bundles are close to the normal distribution. (2) With the increasing of the noil gauge, the peak of the weight percentage curve of the separation bundle increases and the peak of the separation bundle is shifted to the left. (3) When the noil gauge increases, the weight percentage at 10mm and 70mm of the separation bundle decreases and The weight CV of the separation bundle increases.

3.2 The Weight Percentage Curve of Cotton-web

The effective output length of comber is 26.48mm. According to the basic principle of separation-joint on comber, the cotton-web outputted by separation roller is made by multiple separation bundles of the head distance is equal to 26.48mm on comber. When the noil gauge is 8mm, 9mm, 10mm, 11mm, the weight percentage curve of cotton-web is drawn according to the 2mm weight percentage curve of single separation bundle (as shown in Fig.4 (a-d) and Fig.5. The characteristic parameters of cotton-web are calculated, as shown in Table 2.

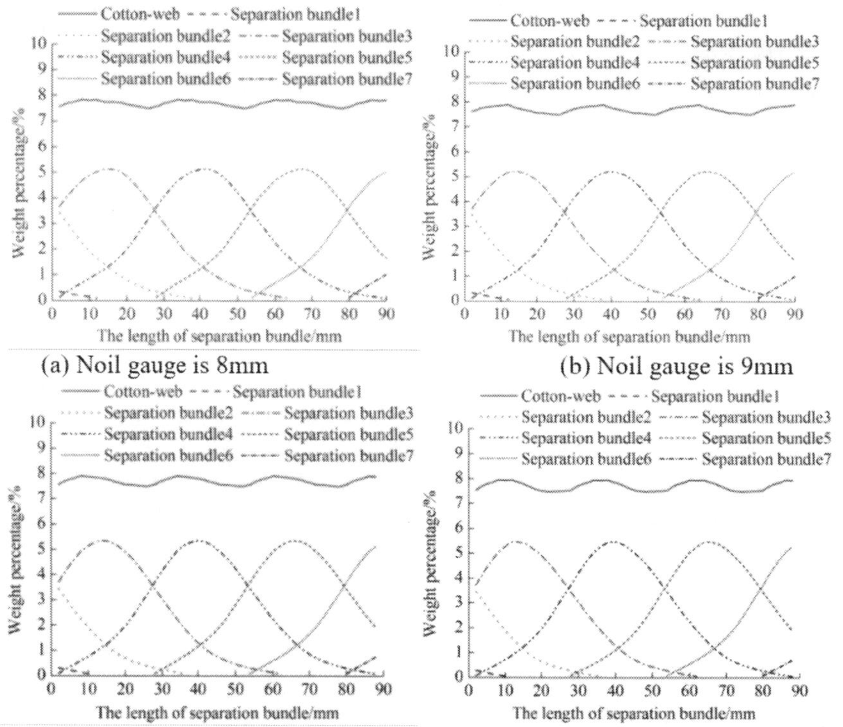

(a) Noil gauge is 8mm (b) Noil gauge is 9mm

(c) Noil gauge is 10mm (d) Noil gauge is 11mm

Fig.4 The curves of the separation bundles and the cotton-webs under the different noil gauges.

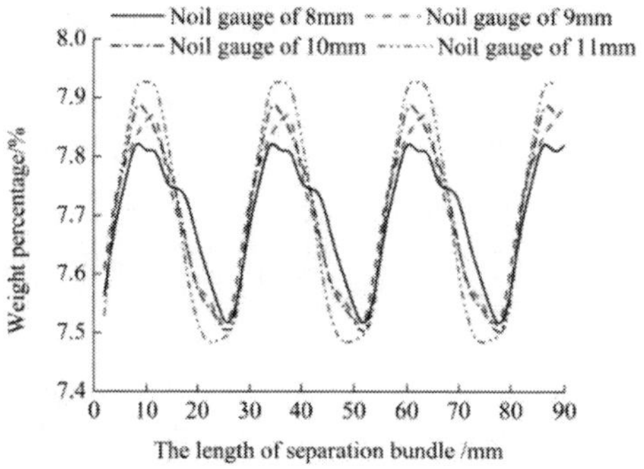

Fig.5 The contrast of the cotton-web curves under the different noil gauges

Table.2 The characteristic parameters of the cotton-webs under the different noil gauges

Noil gauge/mm	8	9	10	11
Minimum weight percentage of 2mm fragment /%	7.5	7.492	7.482	7.478
Maximum weight percentage of 2mm fragment /%	7.83	7.88	7.88	7.93
Average weight percentage of cotton-web/%	7.7	7.7	7.7	7.7
Weight percentage CV of cotton-web /%	3.45	3.68	3.84	4.23

As shown in Fig.4, Fig.5 and Table 2:

1) The weight percentage curve of cotton-web has obvious peak and trough with periodic. The distance of two adjacent peaks or troughs is equal to the effective output length of comber.

2) The difference between peak and trough and weight CV of cotton-web increase as the noil gauge increases. That is, the increasing of noil gauge aggravate the evenness of cotton-web.

4. Conclusions

1) When other process parameters are constant on comber, the length and the weight of the cotton bundle clamped by nipper jaw change with the change of noil gauge. Furthermore, the morphology of separation bundle has also changed.

2) The weight percentage curve of the separation bundle makes a periodic change with the change of noil gauge. With the increase of noil gauge, the length of separation bundle decreases, the peak of the separation bundle is shifted to the left, the peak weight percentage and the weight percentage CV of separation bundle gradually increase.

3) The curve morphology of cotton-web outputted by separation roller depends on the weight distribution of the single separation bundle and the effective output length on comber. The decreases of noil gauge and effective output length is beneficial to reduce cotton-web weight CV and improve the evenness of cotton-web. When noil gauge increases, adjusting lap scale of separation roller is helpful to reduce the weight CV of separation bundle and improve the quality of the cotton-web.

References

[1] Zhou JG, Cao YC. Separation-joint of comber and the quality of sliver [J]. Cotton Textile Technology: 1994; 22 (12): 38-40.
[2] Ren XX, Xue SL, Zheng L, et al. The analysis of separation-joint process of FA261 comber [J]. Cotton Textile Technology: 1997; 25 (4): 17-20.
[3] Ren JZ, Ma HQ, Feng QG. Technology characteristics and spinning property of E66 comber [J]. Cotton Textile Technology: 2011; 39 (9): 46-49.
[4] Ren JZ, Yu CW. Influence of dropping scales of E7/6 comber on process performance [J]. Journal of Donghua University (Eng. Ed): 2005; 22 (2): 23-25.
[5] Ren JZ, Zhang YF. Processing technology and quality analyses of comber [J]. Cotton Textile Technology: 2011; 39 (10): 19-22.
[6] Ren JZ. Influence of cylinder combing radian on process of E7/6 comber [J]. Cotton Textile Technology: 2004; 32 (8): 20-23.
[7] Qian YS. Discussion on advancing carding degree of comber [J]. Cotton Textile Technology: 2005; 33 (4): 13-16.
[8] Ren JZ, Ma C, Zhang YF. Discussion of efficient combing technology [J]. Cotton Textile Technology: 2012; 40 (11): 20-23.
[9] Ren JZ, Jia GX. Influence of noil gauge of E7/6 comber on process performance [J]. Journal of Textile Research: 2006; 27 (6): 78-80.
[10] Zhang FN, Zhu H. Theoretical analysis of separation-joint process on comber [J]. Journal of Textile Research: 1981; 2 (4): 49-54.
[11] Ren JZ, Zhang YF. Comber processing technology and quality analyses [J]. Cotton Textile Technology: 2011; 39 (10): 19-22.
[12] Zhang FN, Chen ZX. The analysis of noil and feeding process on comber [J]. Cotton Textile Technology: 1985; (5): 11-15.
[13] Ren JZ, Yang YG, Jia GX, et al. Feed processing research of E65 comber [J]. Cotton Textile Technology: 2009; 37 (6): 14-16.
[14] Ren JZ, Zhang HY, Chen YH, et al. Influence of comber feed length on output web uniformity [J]. Cotton Textile Technology: 2015; 43 (5): 17-20.
[15] Ren JZ, Gao WD, Xie CP, et al. Influence of forward motion timing of comber detaching roller on evenness of cotton web [J]. Journal of Textile Research: 2014; 35 (3): 127-131.

A Study on the Influence of Structural Changes of Corefil Spun Yarn Properties

Chi-Yu Fu[1], Min-Jie Xin[1], Jian-Jian Zhang[2], Zhi-Gang Xia[1,3*]

[1] *College of Textile Science and Engineering, Wuhan Textile University, Wuhan, 430200, China;*
[2] *Hubei Province Fibre Inspection Bureau, Wuhan 430000, China;*
[3] *Xiangyang Jihua 3542 Textile Company Limited, Xiangyang, Hubei 441002, China*

*Corresponding authors' email: zhigang_xia1983@hotmail.com

Abstract

The study purpose was to reveal the mechanism and regulations on the influence of filaments location, configuration and amount of changes in yarn core structure on the core-fil composite yarn properties. This paper produced corefil yarns with various filament counts and locations; then their properties were studied comparatively. Surface hairiness results showed that the increased filament amount in yarn core induced very slight yarn surface hairiness change; the filament partly located in yarn surface could largely reduce yarn hairiness. However the increased filaments could reduce yarn irregularity due to the uniform elements improvement; the changed filament configuration in yarn structure caused little variation in yarn irregularities. Tensile properties results indicated yarn strength improved as the filament amount increased and partly surface exposure. In specialty, the initial part of tensile curves became evenly compact, while the yield part got server fluctuated during the fracture process as the filament amount increased. This might be due to the increased filaments changing the overlap style with the staple strand.

Keywords: Filament/staple Fiber Composite Yarn; Yarn Structure; Spinning Process; Yarn Properties

1. Introduction

1.1 Development of Core-fil Spinning

The history of corefil spinning can be traced back to the mid 60s of the last century. At first, in order to enhance cotton canvas, using cotton fiber as leather, polyester staple fiber as the core, staple fiber core-fil yarn was developed. The main purpose was to keep the water repellency of cotton fiber when it absorbs water. Using the advantages of wet polyester, such as anti-stretch, tear resistance and anti-shrinkage, etc [1]. In the mid-1970s, the United States Rockwell company launched an electrostatic spinning equipment which can produce core spun yarn. The device uses a static electric field to attach a staple fiber as an outer layer to the core. With this device they produced yarn with polyester filament as the core and cotton fiber as the cladding layer [2]. Due to the advantages in the strength, evenness and hand appearance performance of the hollow spindle wrapped spinning [3], this technology was rapidly developed in a short span of 20 years from the 1970s [4]. However, the hollow spindle wrapping yarn has its own defects which are mainly the unevenness of the fiber after drawing that cannot be compensated by the untwisted core yarn, as it affects the mechanical properties of the yarn, restricting its application in the fields of worsted.

In the early 80s of the last century, the core-fil spinning technology has been greatly developed. In 1981, Z. J. Grosicki developed an elastic air vortex spinning [5], before that, the most common method of production of elastic core-spun yarn is to coat a stretched elastic core with one or several "rigid" yarns on an ordinary twister with a hollow spindle. At the same time, Swiss Patent (CH2PS) No. 395819 announced a new type of core-spun yarn spinning method [6]: two or more core yarns are drawn in the same drafting device, and two or more core yarns are fed into the front roller, the outer wrapping yarn is joined with the core yarn and then twisting and winding on the bobbin.

1.2 Development Status and Problems of Core-fil Spinning

In recent years, the related research scholars and technicians have mainly studied and discussed the factors that affect the covering properties of core-spun yarn, put forward some effective measures to improve the yarn quality. However these measures are mostly based on reforming the ring spinning machine equipment and process parameters, such as rational allocation of distance, increase the roving twist, and installation of roving yarn guide hook etc. Dating back to the mid 1960s, core spun yarn has more than 40 years of history. According to statistics, there are about 1,000 million spindles in the production of core-spun yarn, the expected annual growth is 20-30 million spindles [7].

At present, there is still a difficulty that the bad coating effect in the performance of staple fiber/filament core-spun yarn. Poorly covered yarn defects with hollow core, exposed core, empty sheath, tangled yarn and elastic in-homogeneity etc, especially the exposed core [8]. Core-spun yarn fabric often appear to a large number of spots and defects, this phenomenon is due to the poor coating effect of core-spun yarn. Because of the limitations of traditional ring spinning itself, it is impossible to fundamentally solve this problem.

The purpose of this study is to improve the quality of yarn, increase the mechanical properties of core-spun yarn, reduce hairiness and improve evenness. This study provides a deeper understanding of core-spinning system. It is of great significance to the study of structure and properties of the yarn and improve the research of the ring spinning system, so as to guide the production practice more effectively, at the same time, it can reduce the consumption and cost of production, and increase the added value of the products.

2. Experimental

2.1 Experimental Materials and Instruments

The raw materials used in this experiment include quantitative 6.49g/10m cotton/hemp roving, and fineness of 5tex nylon filament.

The equipment used in the experiment include a ring spinning-frame which has a installation of a set of guide wheel, as shown in the Fig.1.

Fig.1 A yarn guiding wheel is installed on a spinning-frame

Yarn guiding wheel can induce the passage of nylon filaments as well as adjust the position of the filament. In order to study the influence of the position of filament, it can be achieved by adjusting the position of the guide wheel.

2.2 Design the Experimental Scheme

Two groups of experiments are set up in this paper. The first group in order to study the component ratio of different effects on yarn quality, relative position of the filament and staple fiber is kept constant, in which the filament is located at the center of the strip, only change the amount of filaments. Respectively, do not feed the filament, add a filament, add two filaments, add three filaments for core-spun yarn.

In the second groups, in order to study the different structure of component effects on yarn quality, each sample is added with two filaments while keeping the twist factor constant. The discrepancy between these samples is that the filament's position is different. They are respectively designed as follows: a filament located in the center of the fiber, while the other filament on the right side; a filament located in the center of the fiber, while the other one on the left side; two filaments are located on the left and right sides of the fiber bundle.

This paper requires seven experimental programs, numbered ①-⑦, respectively. The design of the core-spun yarn is shown in Table 1.

Table.1 The experimental scheme

Scheme	Variables
①	No filament
②	A filament
③	Two filaments
④	Three filaments
⑤	A filament located in the center while the other one is on the left
⑥	A filament located in the center while the other one is on the right
⑦	Two filaments are located on the left and right sides respectively

2.3 Spinning Parameters

A cotton and hemp blended sliver (649tex fiber linear density) was used as the staple fiber and a nylon filament (5tex) as the filament yarn feed into the front roller. Composite yarn twist factor remain unchanged.

Table.2 Parameters of each experimental scheme

Scheme	Spindle speed m/min	Twist T/m	Total draft	Twist factor
①	9500	1187	29.62	556
②	8572	1071	29.62	556
③	7872	984	29.62	556
④	7319	914	29.62	556

In order to study the effect of the composition ratio and the structure on the yarn performance, we control the twist factor constant. Z-twist was used throughout the whole experiment. The spinning parameters of composite yarn are shown in Table 2 (The parameters of scheme ③⑤⑥⑦ is the same).

The experimental and schematic diagrams are shown in Fig.2 and Fig.3.

3. Performance Testing

All produced composite yarns were conditioned for more than 24 hours under standard atmospheric conditions (20±2℃ and 65±2% relative humidity [RH]) before measuring yarn physical properties including tensile strength, elongation, unevenness, and hairiness. The yarn tensile strength was measured using an YG068C single yarn strength tester according to GB/T 3916-1997 single yarn break strength and elongation testing standard. Yarn test speed was 500mm/min and test gauge was 500mm. Then the tensile results were averaged after 20 tests for each composite yarn. The yarn irregularity was obtained on YG139BA capacitive evenness tester according to the CN GB/T 3292.1-2008 capacitance testing standard; the testing speed was 400m/min, and the testing time was 0.5 minute for each composite yarn. The yarn hairiness was measured using a YG173A hairiness tester according to the CN textile industry standard FZ/T 01086-2000. The test speed was 30m/min and the testing length for each yarn was 100m. All of the tests were conducted under the standard atmospheric conditions. [9] Every testing totally meets the standard of textile industry.

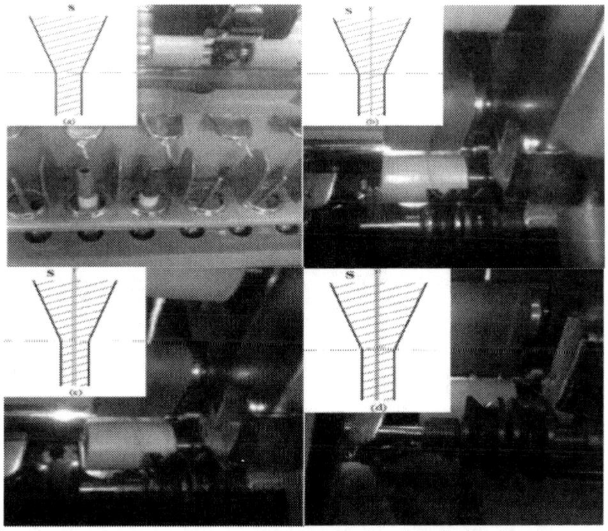

Fig.2 The design of the different components core-spun methods; (a): Spinning process without filament; (b): Add a filament and located in the centre; (c): Add two filaments and located in the centre; (d): Add three filaments and located in the centre.

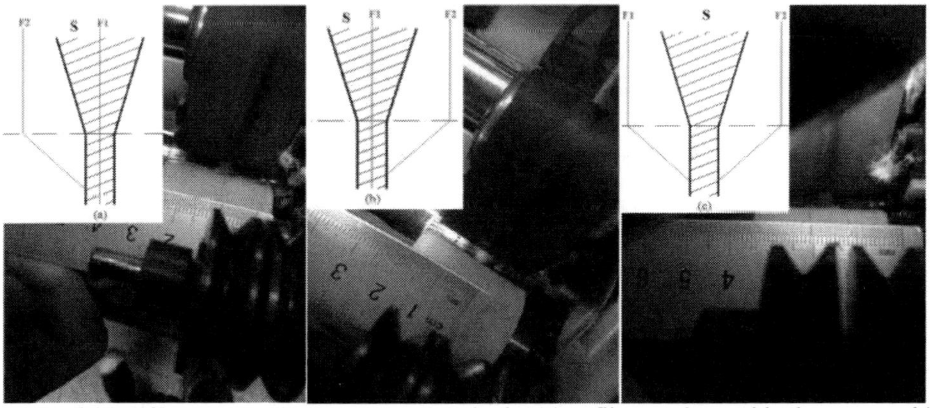

Fig.3 The design of the different structures core-spun methods; (a): a filament located in the center while the other one is on the left; (b): a filament located in the centre while the other one is on the right; (c): two filaments are located on the left and right sides respectively.

4. Results and Discussion

4.1 Effect of Different Component and Structural on Yarn Tensile Strength

In this research experiment, the proportion of the nylon filament have a dramatic influence on the core-fil composite yarn properties, which is indicated in Table 3.

Table.3 Tensile properties of different composite yarns

Scheme	Strength (cN)	Tenacity (cN/Tex)	Elongation (%)	Time(s)	Work (cNcm)
①	205.17	9.37	3.25	1.95	181.54
②	287.12	10.67	5.36	3.22	381.34
③	486.76	15.25	10.83	6.50	1488.27
④	820.28	22.22	33.56	20.14	9649.39
⑤	502.45	15.75	9.50	5.70	1088.08
⑥	504.43	15.81	9.96	5.98	1192.89
⑦	523.41	16.68	10.07	6.04	1200.78

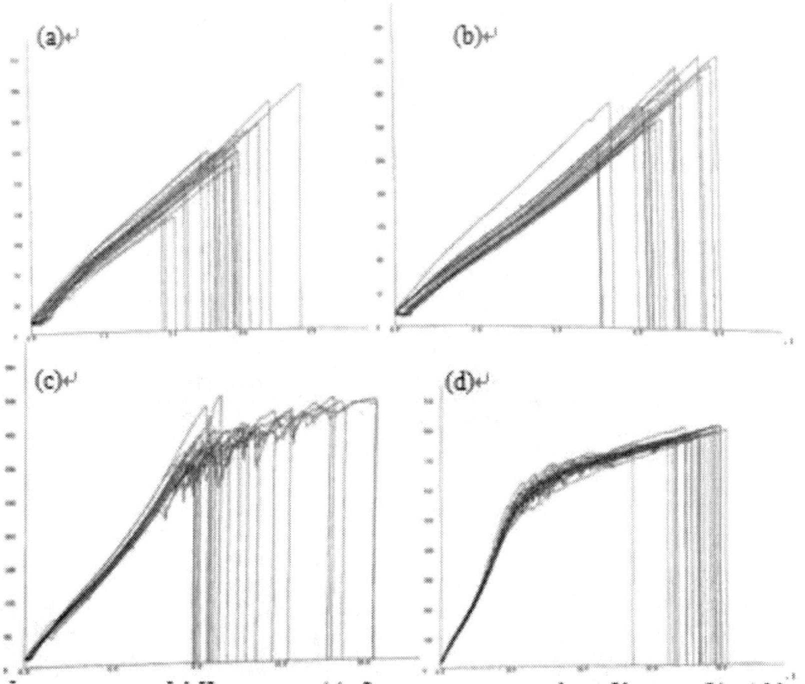

Fig.4 Tensile fracture curves of different yarn; (a): Spinning process without filament; (b): Add a filament and located in the center; (c): Add two filaments and located in the center; (d): Add three filaments and located in the center.

The data in it illustrate that with the increase of filament proportion, the yarn strength significantly increased. However, as for the sample ③, ⑤, ⑥, and ⑦, there induced very slight change in strength, this may due to the same proportion of these composite yarns.

Through the analysis of the tensile fracture curve, it is found that there is a significant difference between them. Different tensile fracture curves are provided in Fig.4, as shown in the Fig.4(a), (b), (c) and (d). Tensile properties results indicated yarn strength improved as the filament amount increased. Having compared the curves in Fig.4, it can be found that (a) and (b) are similar, in other words, their fracture characteristics are the same. This is due to only a filament is joined into the yarn, yarn in the process of tensile fracture mainly shows the mechanical properties of staple yarn, so two tensile fracture curves of (a) and (b) are roughly the same. However, with the increasing number of filaments, they showed significant differences in the back-end curves. In detail, the initial part of tensile curves became evenly compact, while the yield part got server fluctuated during the fracture process as the filament amount increased. In this case, Fig (c) and (d) appears more obvious. This might be due to the increased filaments changing the overlap style with the staple strand, more specifically, as the filaments increase, the yarn exhibits the fracture mechanical properties of the filaments. Curve violent fluctuations due to fracture formation of these filaments.

4.2 Effect of Different Component and Structural on Hairiness of Yarn

The amount of yarn hairiness with different length decreases exponentially as their length increases [10], which is indicated in Table 4. The date shows that the increased filament amount in yarn core induced very slight yarn surface hairiness change. Yarns spun by the siro core-spun have much less hairs than normal core-spun yarn. That means the filament partly located in yarn surface could largely reduced yarn hairiness. This may due to the nylon filament that wrapped around the yarn can capture the surface staple fiber, and tie them on to the composite spun yarn body. More specifically, with the addition of center filaments increase, the coating rate of the short fibers is reduced, resulting in increased hairiness.

Table.4 Surface hairiness of different composite yarns

Scheme	1mm	2mm	3mm	4mm	5mm	7mm	10mm
①	2170.9	615.2	214.1	101.6	46.5	9.8	0.8
②	2797.5	808.2	300.9	134.0	57.5	8.5	0.5
③	2665.7	807.8	311.2	152.8	73.6	16.3	0.8
④	2619.3	726.2	279.8	133.0	65.9	13.7	1.6
⑤	1738.5	327.3	93.3	34.8	14.8	2.8	0.2
⑥	1842.0	343.7	100.7	39.8	16.8	3.7	0.8
⑦	2364.0	529.1	157.4	60.7	23.3	4.5	0.4

4.3 Effect of Different Component and Structural on Unevenness

Table.5 Unevenness of different composite yarns

Scheme	CV%	Thin places -50%	Thick place +50%	Neps +200%
①	39.945	6500	5207.5	4662.5
②	30.815	2360	3915	3055
③	25.82	405	2637.5	1885
④	22.905	52.5	1830	1267.5
⑤	27.285	697.5	3135	2222.5
⑥	27.66	745	3120	2372.5
⑦	26.715	577.5	2840	2282.5

The coefficient of mass variation per unit length (CVm) is widely used to access yarn unevenness [11, 12]. The unevenness of different composite yarns are showed in Table 5. It is obvious that in the composite yarn, the effect of component ratio on yarn unevenness is more significant than that of yarn structure. The important factor could be that the filament component is more uniform than the staple ingredients. With the amount of nylon filaments increasing, the CVm value becomes lower than other composite yarns. Therefore, the reasonable proportion of the components is beneficial to improve the evenness of the yarns.

5. Conclusion

In this paper, the different processes of core-spun yarns were introduced including the different the filaments location, configuration and amount changes in the yarn core structure.

1) It was concluded that surface hairiness results showed the increased filament amount of yarn core induced very slight yarn surface hairiness change. However, the filament partly located in yarn surface could largely reduce yarn hairiness. This is due to the surface of wrapped filaments has an effect on hairs. In detail, hairs are wrapped onto the yarn body to decrease hairiness.

2) Yarn unevenness results showed that the increased amount of filaments could reduce yarn irregularity due to the uniform elements improvement. The changed filament distribution of yarn structures caused little variation in yarn irregularities.

3) As for the tensile properties, yarn strength improved as the filament amount increased and partly surface exposure; different tensile fracture curves showed that the proportion of filament has a great influence on the yarns fracture process. The repeated peaks of the curve indicate that each component of the composite core yarn is not broken at the same time; this phenomenon is more obvious with the increase of filament.

6. Acknowledgement

We greatly acknowledge the support from the National Natural Science Foundation of China (Project No. 51403161).

References

[1] Sheng X. Core-spun yarn [J]. Beijing textile: 1993; 5 (2): 1-3.
[2] Huang X. Spinning core spun yarn with electrostatic spinning machine [J]. Shanghai Textile Science and technology: 1976; 23 (3): 1-2.
[3] Audivert R. Advantages of staple-fibre yarns covered with a filament [J]. Textile Institute and Industry: 1974; 12 (9): 271 - 272.
[4] Xie Y, Oxenham W, Grosberg P. A study of strength of Wrapped Yarns, Part Ⅰ: The Theoretical Model [J]. J. Text. Inst: 1986; 77(3): 295 - 304.
[5] Grosicki ZJ, Zhun W. Elastic air vortex core spun yarn [J]. Foreign textile technology: 1981; 32 (3): 54-55.
[6] Xu H. Spinning method of core-spun yarn [J]. Foreign textile technology: 1983; 07 (2): 33-34.
[7] Rong L, Wei M. Properties and spinning of core-spun yarn [J]. Chinese yarn quality and new product development technology forum: 2009; 5 (12): 71-73.
[8] Xin H. Development and problems of core spun spinning technology [J]. Textile Science and technology: 2008; 6 (6): 9-10.
[9] Xia Z, Wang X, Wei X, et al. Fiber trapping comparison of embeddable and locatable spinning with sirofil and siro core-spinning with flute pipe air suction. Textile Res J: 2012; 82 (13): 1225-1262.
[10] Viswanathan G, Munshi VG, Ukidve AV, Chandran K. Comparative Evaluation of Yarn Hairiness by Different Methods. Text Res J: 1988; 58 (4): 477-479.
[11] Hasler A, Honegger E. Yarn evenness and its determination. Textile Res J: 1954; 24 (4): 73-85.
[12] Carvalho V, Monteiro JL, Soares FO, Vasconcelos RM. Yarn evenness parameters evaluation: a new approach. Textile Res J: 2008; 78 (34): 119-127.

The Frequency Response Characteristics of the Frequency Selective Fabrics with Periodical Structure Made of U-shaped Velvet and Looped Pile

Ya-Jing Wang [1,2], Hong Xiao [2*], Mei-Wu Shi [2], Qun Wang [3], Zhang-Hong Tang [3]

[1] *College of Textiles, Donghua University, Shanghai 201620, China*
[2] *The Quartermaster Equipment Research Institute of Logistical Support Department, Beijing 100082, China*
[3] *College of material, Beijing University of Technology, Beijing 100124, China*

[*] Corresponding author's email: 76echo@vip.sina.com

Abstract

In this paper, two kinds of frequency selective fabrics (FSFs) with the periodic structure separately made of U-shaped velvet and looped pile, the stainless steel blended yarn as the conductive unit material, were prepared. The transmission coefficient and reflection coefficient of specimens were tested and the experiment results were analyzed to explore the differences in frequency response characteristics of FSFs. With the same size, the resonance frequencies of the U-shaped velvet and looped pile samples are close, but the resonance peak value of the latter is smaller than the former. An increase of the unit length leads to resonance frequencies of two kinds of specimens moving to the lower and then being stable. When the basic structure spacing b increases, the resonance peak value of looped pile sample becomes smaller, and that of U-shaped velvet sample becomes larger. The looped pile sample with cube unit has more abundant and stable frequency response characteristics because of its regular unit shape and good connectivity between looped piles. Although a lot of researches need to be done on looped pile FSFs, this paper contributes to future research.

Keywords: U-shaped Velvet Fabric; Looped Pile Fabric; Frequency Selective Surface; Frequency Response Characteristics

Frequency selective surfaces (FSSs) are composed of the periodic structure with the same elements in one or two dimensions [1]. Conventional FSSs are prepared by metal material. With the application of FSSs in the field of wearable technology [2] and shielding technology [3], the higher requirements on flexibility and quality are put forward [4]. FSFs are soft and have light weight. Meanwhile, FSFs have the characteristic of selective frequency for electromagnetic waves. Electromagnetic waves in a particular frequency band can be prevented passing through (band-stop FSS) or allowed to pass through (band-pass FSS). The FSF is a cross-field product of electromagnetics and textile science. It has not only the electromagnetic characteristics of FSSs, but also the advantages of fabrics, such as light weight, good flexibility and so on.

Plane FSFs were made by many scholars via the method of electroplating [5], ink jet printing [6], screen printing [7], computer embroidery [8, 9], weaving [10], knitting [5], etc. Compared with planar FSFs, three-dimensional FSFs have more advantages including multiple resonance points and angle stability and so on. Currently, there are a few researches on the three-dimensional FSFs at home and abroad. The literatures [11, 12] investigated the influence factors on frequency response characteristics of three-dimensional velvet fabric with FSS, in which the silver yarn was used as unit material. Many conclusions have been drawn before. For example, with the increase of cell size or spacing, the resonance frequency becomes lower. In order to further optimize the frequency response characteristics, FSFs fabricated by the stainless steel blended yarn as unit material, which is more available and economy than the silver yarn, were prepared and tested. This paper focuses mainly on the differences between U-shaped velvet and looped pile FSFs, especially the differences on the effects of the unit size, the unit inner structure distance and other factors. Specimens with the cubic unit structure were manufactured to provide data reference for the future product.

1. Experimental Design

A kind of conductive yarn was used in this paper. Yarn parameters and sample parameters are shown in Table 1 and 2, respectively.

Table.1 Yarn parameters for the FSF unit

Name	The stainless steel/cotton blending ratio	Strand conductivity (S/m)	Strand linear density (tex)	Plied yarn linear density×number of filaments (tex)
the stainless steel/ cotton blended yarn	30/70	6666.7	46.5	93×8

1.1 FSF Samples Based on U-shaped Velvet

A number of U-shaped velvets or looped piles constitute the FSF unit. As shown in Fig.1, the FSF of unit length L was prepared by two U-shaped velvet structures and the unit material is the stainless steel blended yarn. In order to explore the influence of the number of U-shaped structures on frequency response characteristics, FSF samples with periodic elements of 2, 3 and 4 U-shaped velvets were made and tested. The sample of 2 U-shaped structures is shown in Fig.2. The substrate layers are polyester fabric and cellular PE plates supporting conductive yarns to prevent collapsing.

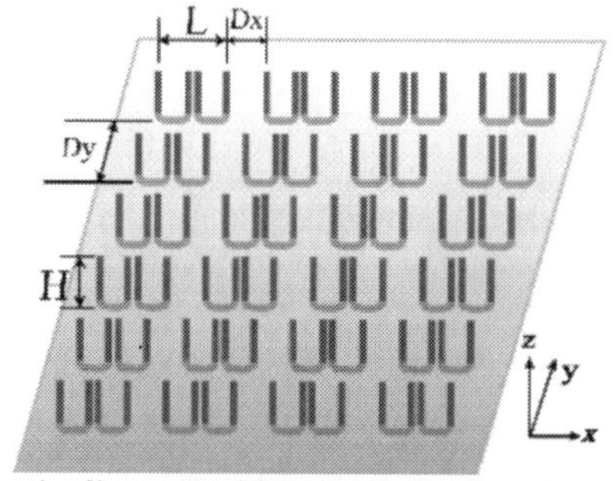

Fig.1 The schematic diagram of FSF samples based on U-shaped velvet

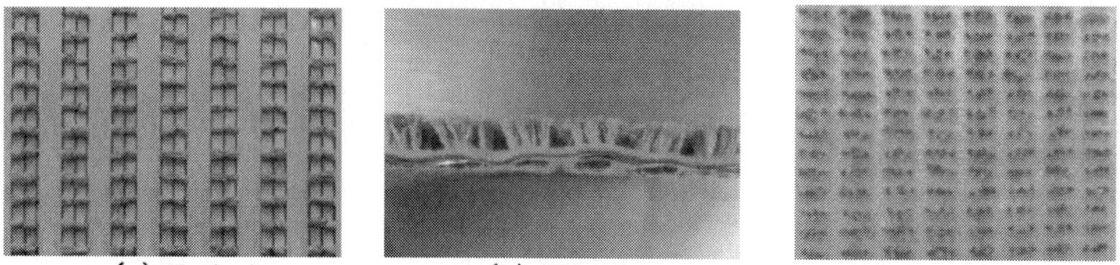

(a) Back view (b) Side view (c) Front view

Fig.2 Photographs of samples with unit length L=6mm

1.2 FSF Samples Based on Looped Pile

In comparison with the U-shaped unit, the looped pile unit (shown in the right sketch of Fig.3) has better connectivity. In this paper, the length of single U-shaped velvet or looped pile is 3mm for the reason that the length of single U-shaped velvet or looped pile of the tufted fabric product is about 3mm. With the manufacturing method of looped pile, the FSF samples, the same size as U-shaped velvet samples in the 1.1 chapter, were fabricated to investigate the differences in frequency response characteristics. Looped pile samples have the similar images with U-shaped velvet samples, as depicted in Fig.2(a).

1.3 FSF Samples with Different Basic Structure Spacing

Each looped pile is connected and different from the disconnected U-shaped velvet in Fig.3. The increase of the basic structure spacing b leads to the fact that the connectivity of U-shaped velvet is changed, and that the connectivity of the looped pile has no effect. U-shaped velvet and looped pile samples with the basic structure spacing b of 3mm, 6mm and 9mm (shown in Fig.4) were fabricated and tested. The impact of b on the resonance point was analyzed.

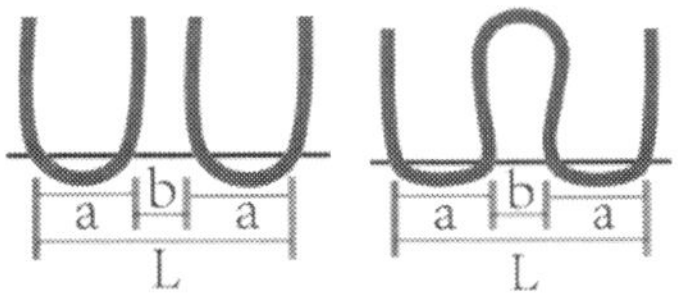

Fig.3 The unit of U-shaped velvet (left) and looped pile (right)

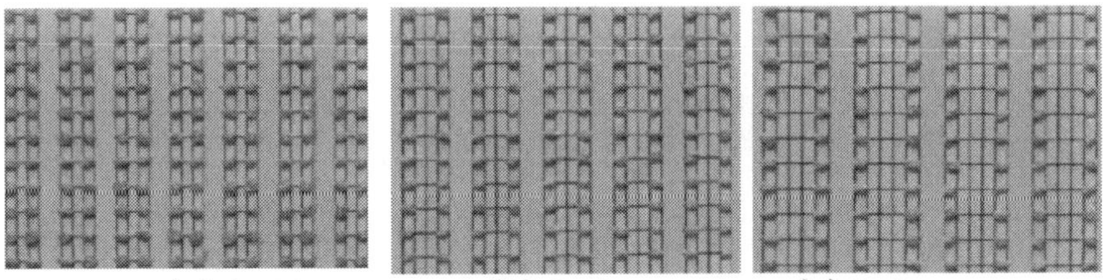
(a) b=3mm (b) b=6mm (c) b=9mm

Fig.4 Photographs of samples with different basic structure spacing length b

Table.2 Sample parameters based on the stainless steel blended yarn

No.	Unit parameters			Spacing Dy=Dx (mm)	Villus height H (mm)	The basic structure of the unit	The number of the basic structure in the unit
	Unit length L (mm)	The basic structure length a (mm)	The basic structure spacing b (mm)				
U-L1-#	6	3	0	6	6	U-shaped	2
U-L2-#	9	3	0	6	6	U-shaped	3
U-L3-#	12	3	0	6	6	U-shaped	4
Q-L1-#	6	3	0	6	6	looped	2
Q-L2-#	9	3	0	6	6	looped	3
Q-L3-#	12	3	0	6	6	looped	4
U-b1-#	9	3	3	6	6	U-shaped	2
U-b2-#	12	3	6	6	6	U-shaped	2
U-b3-#	15	3	9	6	6	U-shaped	2
Q-b1-#	9	3	3	6	6	looped	2
Q-b2-#	12	3	6	6	6	looped	2
Q-b3-#	15	3	9	6	6	looped	2

1.4 FSF Samples with the Cube Structure

The connectivity of the yarn in unit affects frequency response characteristics. U-shaped sample UL-12 (shown in Fig.5 (a)) and looped pile sample QL-12 (shown in Fig.5 (b)) were prepared to demonstrate the influence of the yarn connectivity on frequency response characteristics. It was worth noting that the periodic structure in this part was cube rather than a single row unit used in all samples above. Cubic length and width were set as 12mm, villus height was 9mm, and the basic structure spacing and unit spacing were 0mm and 6mm, respectively. Compared with FSFs based on U-shaped structure, looped pile FSFs have better connectivity and more regular unit shape. Experimental results will verify whether the sample with looped pile has more excellent frequency characteristics or not.

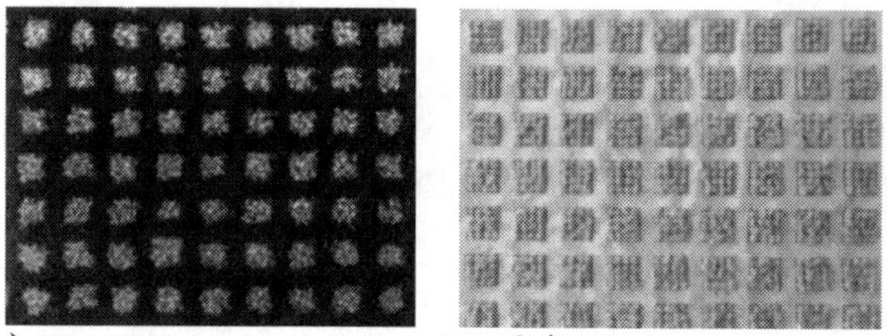

(a) The U-shaped velvet sample (b) The looped pile sample
Fig.5 Photographs of three-dimension FSFs with cube unit

2. Experimental Measurement

Testing system included an Agilent E8257D signal generator (250KHz-40GHz), an E7405AEMC spectrum analyzer (100Hz-26.5GHz), two horn antennas (1GHz-18GHz), an absorbing screen and connecting cables etc. The system devices were set up as Fig.6 [13, 14]. Signal generator and spectrum analyzer were connected by a computer to output data. Transmitting antenna, receiving antenna and sample on the test holder were on the same horizontal line in the testing system for transmission coefficient. The incident angle of electromagnetic wave was 15° during testing reflection coefficient. The sample holder was 18cm×18cm. Transmission and reflection coefficients were measured and these parameters reflected the frequency response characteristics in the experiments. Due to the test equipment, the data in the 2-15GHz band is valid.

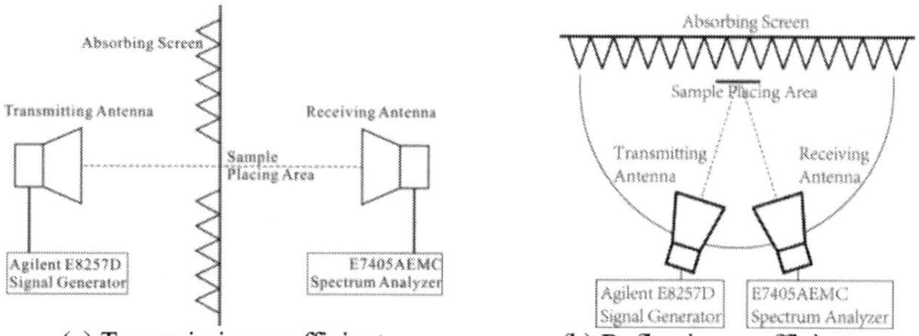

(a) Transmission coefficient (b) Reflection coefficient
Fig.6 The testing system

The formula for calculating transmission coefficient S_{21}:

$$S_{21} = 10\lg\frac{10^{\frac{p_1}{10}}}{10^{\frac{p_2}{10}}} = p_1 - p_2$$

In the formula, S_{21} is the transmission coefficient (unit: dB). p_1 and p_2 are the receiving power with or without samples respectively.

The formula for calculating reflection coefficient:

$$S_{11} = 10\lg\frac{\left(\frac{10^{\frac{p_1}{10}}}{10^{\frac{p_2}{10}}}\right)}{\left(\frac{10^{\frac{p_3}{10}}}{10^{\frac{p_2}{10}}}\right)} = (p_1 - p_2) - (p_3 - p_2) = p_1 - p_3$$

In the formula, S_{11} is the reflection coefficient (unit: dB). p_1 and p_3 are the reception power with the sample and the aluminum plate placed on the sample holder respectively. p_2 is the transmitting power.

3. Results and Discussion

When the incident wave is perpendicular to samples, electrons in the direction of electric field (generated by the decomposition of electromagnetic waves) produce oscillation. When the electromagnetic wave passes through the sample, part of the energy is transmitted or reflected and the rest is absorbed by electrons. At a certain frequency, all transmission energy of electromagnetic wave is changed to electrons. In other words, the additional scattering produced by electrons can offset the electromagnetic wave scattering field. The resonance phenomenon will occur. This frequency point is called resonance frequency [15].

3.1 Frequency Response Characteristics of FSFs with U-shaped Velvet

The unit length of U-shaped sample U-L1-#, U-L2-# and U-L3-# are 6mm, 9mm, 12mm separately. There are two resonance points in the test curves of the three samples in Fig.7. The two resonance frequencies of U-L1-# are 14.26GHz and 5.76GHz, that of U-L2-# are 14.09GHz and 5.08GHz, and that of U-L3-# are 14.09GHz and 4.91GHz. The FSFs are composite periodic structures, namely, a large periodic structure with unit length L and a small periodic structure consisting of some U-shaped structures. Although both two kinds of structures have influence on the frequency response, the former mainly affects the first resonance frequency, and the latter is related to the second resonance frequency. With the increase of unit length, the resonance frequency moves to lower frequency and then tends to be stable, owing to the effect of equivalent capacitance and unit coupling. The frequency of the peak in the reflection coefficient curve is in 9-12GHz, which corresponds to the frequency of the transmission coefficient curves close to 0dB. For example, the frequency of Sample U-L1-# in the peak of reflection coefficient curve is around 11.37GHz, which is close to 11.20GHz in transmission coefficient curve.

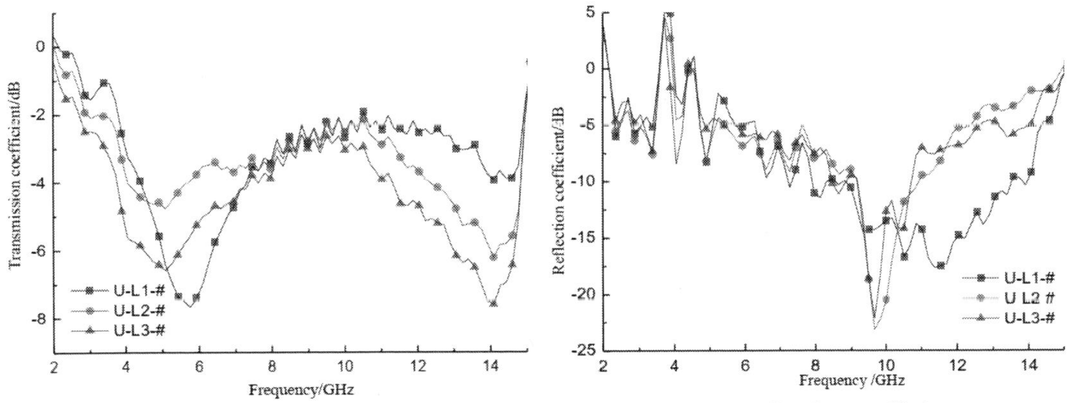

(a) Transmission coefficient (b) Reflection coefficient

Fig.7 Frequency response characteristics of samples with different unit length

3.2 Comparison of Frequency Response Characteristics of U-shaped Velvet and Looped Pile FSFs

For exploring the differences in frequency response characteristics of U-shaped velvet and looped pile samples, looped pile sample Q-L1-# (L=6mm), Q-L2-# (L=9mm) and Q-L3-# (L=12mm) with the same size as U-shaped velvet samples were prepared. Fig.8 shows that frequency response characteristics of looped pile samples are similar with U-shaped velvet samples, from which we can find the changes of two resonance points in 2-15GHz. The resonance parameters of the samples are shown in Table 3. There is a wide band in the resonance phenomenon of Q-L2-# and Q-L3-# for 3-7GHz range, which may be related to the looped pile structure. In spite of the need for a lot of experiments and theoretical support, this is also a significant discovery. With the increase of unit length L, the resonance frequency moves to lower frequency, especially the first resonance frequency, and the half peak width of resonance increases gradually. The discontinuous U-shaped structure results in the poorer connectivity of U-shaped velvet unit. Therefore, the resonance peak value of looped pile samples is smaller than that of U-shaped velvet sample with the same size.

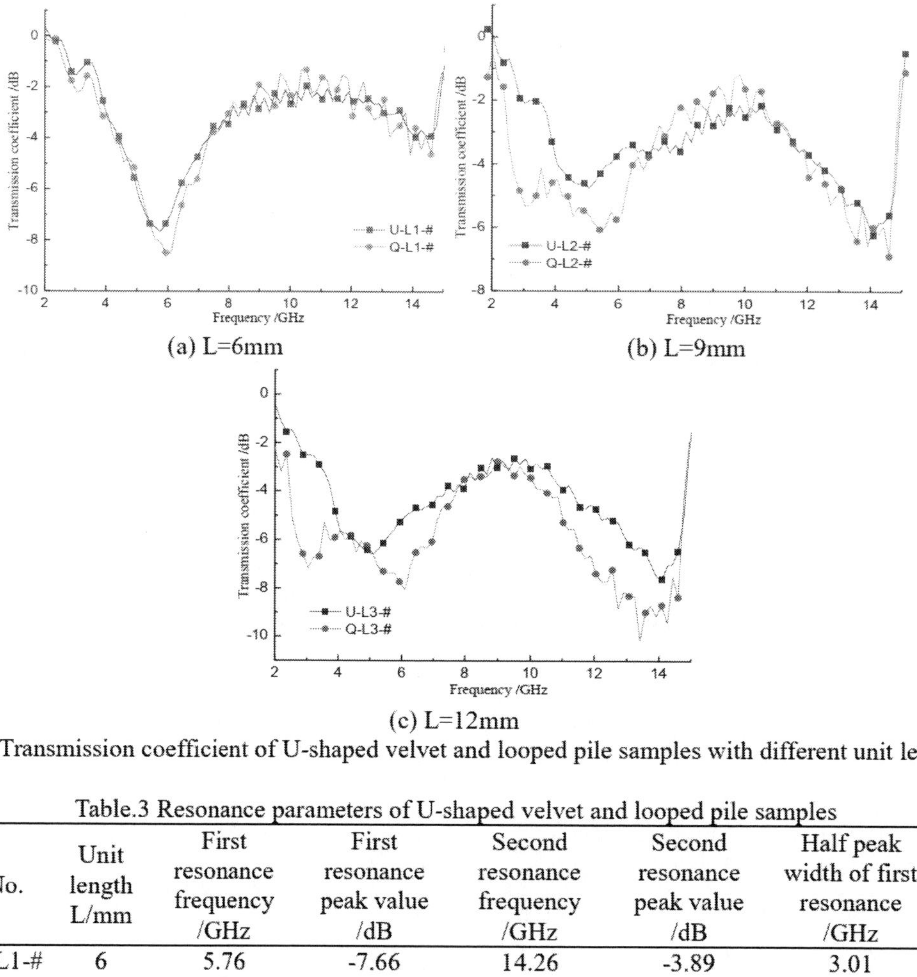

(a) L=6mm (b) L=9mm

(c) L=12mm

Fig.8 Transmission coefficient of U-shaped velvet and looped pile samples with different unit lengths

Table.3 Resonance parameters of U-shaped velvet and looped pile samples

No.	Unit length L/mm	First resonance frequency /GHz	First resonance peak value /dB	Second resonance frequency /GHz	Second resonance peak value /dB	Half peak width of first resonance /GHz
U-L1-#	6	5.76	-7.66	14.26	-3.89	3.01
U-L2-#	9	5.08	-4.76	14.09	-5.99	3.52
U-L3-#	12	4.91	-6.59	14.09	-7.60	4.56
Q-L1-#	6	6.10	-8.50	14.60	-4.60	3.10
Q-L2-#	9	4.57	-5.99	14.26	-6.20	4.73
Q-L3-#	12	4.40	-8.01	14.25	-10.13	5.27

3.3 Frequency Response Characteristics of FSFs with Different Basic Structure Spacing

In above experiments, the basic structure spacing b is equal to 0mm. This section discusses frequency response characteristics of samples with different values of b. When the basic structure length a is a known constant and b is increased (such as 3mm, 6mm, 9mm), the corresponding unit length L becomes larger (corresponding to 9mm, 12mm and 15mm). Resonance parameters of samples are shown in Table 4.

In Fig.7, two resonance frequencies are caused by the large periodic structure with unit length L and the small periodic structure composed of U-shaped structures. When the basic structure spacing b is large enough and increases in the small periodic structure, mainly affecting the second resonance peak, Fig. 9(b) indicates that the second resonance phenomenon disappears in 2-15GHz. With the same size, the U-shaped velvet and looped pile within the periodic structure have little influence on the frequency response characteristics in the 3.2 chapter, where b is equal to 0mm. That is to say, factors that affect the frequency response characteristics are the conductivity and the size of the whole structure. However, the frequency response characteristics of U-shaped velvet samples are different with that of looped pile samples accompanied by the increase of b. With the increase of b, the second resonance frequency shifts to lower one due to the continuous looped pile structure and the increase of unit length L (shown in Fig.9(a)).

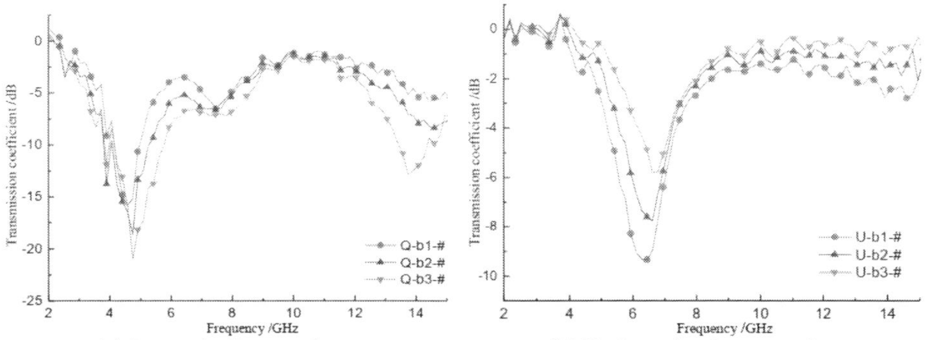

(a) Looped pile samples (b) U-shaped velvet samples

Fig.9 Transmission coefficient of samples with different basic structure spacing b

Table.4 Resonance parameters of samples with different basic structure spacing b

No.	the basic structure spacing b/mm	First resonance frequency /GHz	Second resonance frequency /GHz
U-b1-#	3	6.44	-
U-b2-#	6	6.61	-
U-b3-#	9	6.61	-
Q-b1-#	3	4.57	14.77
Q-b2-#	6	4.74	14.60
Q-b3-#	9	4.74	13.75

3.4 Frequency Response Characteristics of FSFs with Cube Unit Shape

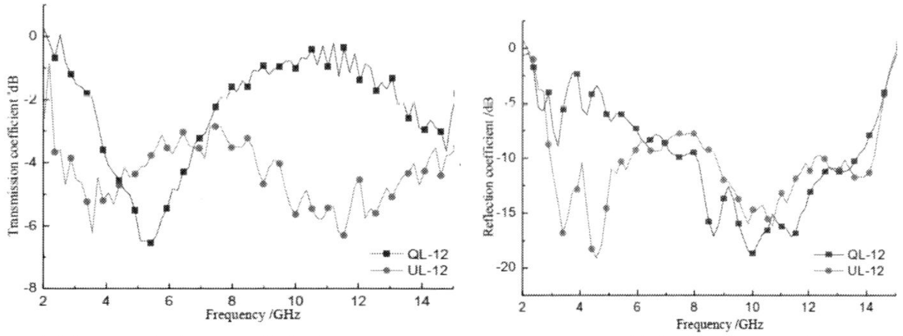

(a) Transmission coefficient (b) Reflection coefficient

Fig.10 Frequency response characteristics of samples with cube unit shape

The unit shape of the U-shaped velvet sample UL-12 and the looped pile sample QL-12 is cube, whose side length is 12mm and villus height is 9mm. The unit shape of sample QL-12 made by hand is regular. The transmission coefficient is smooth. But the unit shape of the U-shaped velvet sample prepared by tufted carpet weaving machine is irregular and the villi are dispersed. Test results are that the resonance phenomenon is not obvious in Fig.10 (a) and the reflection coefficient curve in Fig.10(b) is unstable. The first resonance frequency of sample QL-12 is 5.25GHz and that of sample UL-12 is 3.55GHz. Due to dispersed U-shaped villi leading to larger unit length, the first resonance frequency of sample UL-12 is less than that of sample QL-12. Experimental results show that looped pile samples have more abundant and stable response. Further research can be done to improve the regularity of the unit and to optimize the frequency selective characteristic of tufted products.

4. Conclusions

This paper introduces two kinds of FSFs and mainly to explore the differences in frequency response characteristics of U-shaped velvet and looped pile FSFs. For the sizes mentioned in this paper, the conclusions are as follows. 1) The frequency response characteristics of periodic structure made of either U-shaped velvet or looped pile samples are similar. There are two resonance points in the 2-15GHz. 2) With the increase of the unit length, the resonance frequency of the two kinds of samples moves to lower frequency, particularly the first resonance frequency. 3) With the increase of the basic structure spacing b, the second resonance frequency of looped pile sample moves toward lower frequency, and that of U-shaped velvet sample may vanish. 4) With the cube unit, the looped pile sample fabricated by manual has stabler and more abundant frequency response characteristics than the U-shaped velvet sample manufactured by tufted machine. Researches in this work reveal the complexity and variability of the periodic structure, which is worth to study. The further research is continuing.

5. Acknowledgement

The project is supported by Natural Science Foundation of China No. 51673211.

References

[1] Munk BA. Frequency selective surfaces theory and design. Beijing: Science Press: 2009; 1-3.
[2] Guan FW, Xiao H, Shi MW, Wang FM. Fabrication and application of fabric based frequency selective surface. Journal of Textile Research: 2016; 02: 141-148.
[3] Seager RD, Chauraya A, Bowman J, Broughton M, Philpott R, Nimkulrat N. Fabric based frequency selective surfaces using weaving and screen printing. Electronics Letters: 2013; 49: 1507-1509.
[4] Shi MW, Xiao H, Wang Q. Research on textile electromagnetism and development of electromagnetic textile. Journal of Textile Research: 2013; 34: 73-81.
[5] Tennant A, Hurley W, Dias T. Experimental knitted, textile frequency selective surfaces. Electronics letters: 2012; 48: 1386-1388.
[6] Chauraya A, Whittow WG, Vardaxoglou JC, Li Y, Torah R, Yang K, Beeby K, Tudor J. Inkjet printed dipole antennas on textiles for wearable communications. Microwaves, Antennas & Propagation: 2013; 7: 760-767.
[7] Guan FW, Xiao H, Shi MW, Wang Q. The novel frequency selective fabric and application research. Journal of Industrial Textiles: 2015; 11: 1-17.
[8] Lv ZR, Tang ZH, Xu XX, Wang Q, Xiao H, Shi MW. Preparation and transmission characteristics of flexible frequency selective surfaces. Safety and Electromagnetic: 2014; 4: 79-81.
[9] Wu LJ, Shi MW, Wang Q, Tang ZH, Guan FW. Transmission method for transmission characteristics of frequency selective surface by different preparation methods. Environmental Technology: 2014 (S1); 4-8.
[10] Tanaka M, Jang JH. Wearable microstrip antenna. Antennas and Propagation Society International Symposium: 2003; 2: 704-707.
[11] Wang YJ, Xiao H, Cheng HH, Shi MW, Wang Q, Tang ZH. Influence factors of frequency response characteristics for new velvet fabric with frequency selective surface. Journal of Textile Research: 2017; 38: 90-98.
[12] Xiao H, Cheng HH, Shi MW. Exploration of three- dimension periodic structure frequency selective fabric based on U-taped velvet. Science & Technology Review: 2016; 34: 1-7.
[13] 2008 G J B. Measuring method for shielding effectiveness of electromagnetic shielding materials: 2008; 1-4.
[14] 2004 G J B. Test methods for radio-frequency absorber performance: 2004; 2-15.
[15] Wu LJ. The influence of the periodic structure unit and physical parameters on the frequency response. Beijing University of Technology: 2013; 31-44.

A Design Method of Hollow Structure Woven Fabric

Chun-Hong Zhu[1*], Kasumi Hayashi[2], Akio Sakaguchi[1], Qing-Qing Ni[1], Hideaki Morikawa[1]

[1]*Faculty of Textile Science and Technology, Shinshu University, 3-15-1 Tokida, Ueda, Nagano 386-8567, Japan*
[2]*Graduate School of Science and Technology, Shinshu University, 3-15-1 Tokida, Ueda, Nagano 386-8567, Japan*

[*]Corresponding author's email: zhu@shinshu-u.ac.jp

Abstract

In this study, a kind of hollow woven fabric named [(N+1)-N layer] structure was proposed, and the equations for number of warp and weft yarns of this structure were derived using layer number and interweaving times as parameters. After that, the structure was simulated by the Digital Fabric System (DFS) to see the hollow structure and also to find out whether the derived equations can work. Moreover, a Jacquard loom was applied for fabricating. During both structure simulation and fabrication, layer number and the interweaving times were set differently. The results showed that the proposed structure can be used for hollow structure and the number of warp and weft yarns can be obtained by the derived equations.

Keywords: Hollow Structure; Fabric; Design; Simulation; Weaving

1. Introduction

In order to reduce the weight, textile- reinforced composites made of high- performance fibers possess a great potential to supplant metallic components in different areas, such as aeronautic, transportation, wind energy, building/ construction, and sports and so on. In order to develop these kinds of composites, the textile technology, such as weaving, knitting, and sewing as well as multi-axial technology are used [1-2]. 3D fabrics is one of these textile materials as well as fabricated weaving, knitting, sewing and some other textile technologies. It is defined as fabrics which have either an overall 3D shape or a more complex internal 3D structure or both [3]. Although all textiles have a 3D internal structure, macroscopically most can be regarded as thin 2D sheets. The following forms are one kind of 3D fabrics, such as single-layer materials with an overall shape, multilayer hollow materials, solid planar materials with multiple layers, solid multilayer materials with an overall 3D shape and so on.

3D woven fabrics, have been used in many fields, such as medical science for vascular vents, prosthetic applications and artificial arteries, structures for air-frame and self-supporting [4]. It can be classified into two methods [5]: the first is fabrics woven by a 2D weaving system, the second is a 3D weaving system and non-interlacing weaving method (noobing), classified based on type of weaving process. The other is based on types of 3D structure, such as 3D solid structure, 3D hollow structure, 3D shell structure and 3D nodal and so on. Weaving technology can be used to fabricate different shapes or features of fabrics.

In 3D woven fabrics, the most widely used weave structure are generally multilayer, angle interlock and orthogonal weave architectures [5]. Orthogonal weave architecture needs a special designed 3D weaving machine to be produced, as the orthogonal weave structure consists of three sets of yarns that are perpendicular to each other. Multilayer weave structures consist of multiple layers, and each of these layers has its own sets of warp and weft yarns. Self- stitching (existing yarns) or central stitching (external sets of yarns) are applied for the connection of the layers. On the other hand, for the angle interlock weave structures, at least two sets of yarns are needed, such as warp and weft. Stuffer yarns will be added in order to increase fiber volume fraction and in- plane strength. Therefore, multilayer and angle interlock weave structures can be produced with conventional 2D weaving machines, such as shuttle loom, dobby loom and Jacquard loom. In the case of 2D weaving, two sets of perpendicular yarns are interlaced, irrespective of whether it is woven as single layer or multilayer. Another set of yarns, knowns as pile or binder yarns, can be introduced in the direction of fabric thickness. Fabrics can be produced by 2D techniques with different sets of warp yarns in the following methods [6]: such as by effective utilization of warp and weft in a single layer, by the use of multilayer warp and weft or multilayer ground warp, binder warp and weft, and by utilizing three sets of yarns (pile fabric).

In this study, we proposed a kind of multilayer hollow woven fabric, from which the shape of the hollow can be changed on the decision of users conveniently. The minimum repeat for the structure was proposed, and the number of warp and weft yarns needed was also derived using layers and interweaving times as parameters. Then the designed structure was simulated by DFS (Vision Japan) and woven by a Jacquard loom to find out whether the derived equation can be worked, by changing layers and interweaving times.

2. Proposal of a New Kind of Hollow Structure

2.1 Design of the Hollow Structure

In this study, we proposed a new kind of hollow structure, with the minimum repeat shown in Fig.1. As this woven repeat can be divided into two parts, part A composes of three layers, and part B is two layers, we defined this kind of structure as [3-2 layer] structure. In this minimum full woven repeat, 6 warp yarns and 15 weft yarns were utilized to form several hollow shapes, shown as the red marks. Plain weave is used for this structure. In order to form hollow structure, some of the weft yarns in part B should be interwoven twice by different warp yarns from part A.

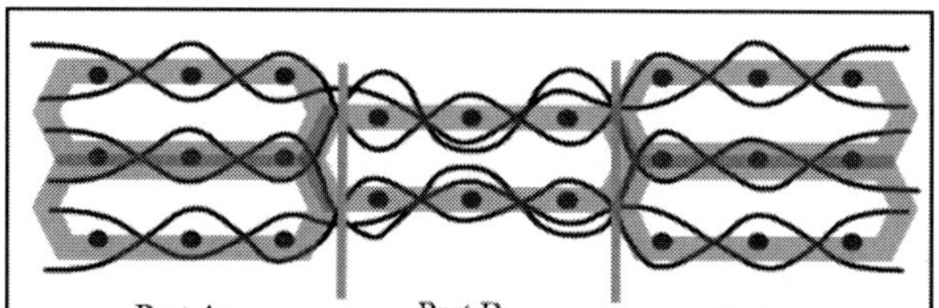

Fig.1 The minimum woven repeat of a kind of hollow fabric.

In order to design several different sized structures, firstly we proposed the general design method for this kind of structure based on the [3-2 layer] structure. Fig.2 shows the representative of the general design.

In one repeat, structure can also be divided into two parts, the N layer structure and the N+1 layer structure. Therefore, the general design was named as [(N+1)-N layer] struture. As a multilayer fabric, the general design was created by connecting the top and bottom layer using the additional yarns comming from the second and the of the N+1 layer. In the [(N+1)-N layer] structure, the number of warp and weft yarns can be calculated by the following equations:

$$P_j = 2 \times (N + 1) \tag{1}$$

$$P_w = (2n + 1) \times (2N + 1) \tag{2}$$

where, P_j, P_w represent the number of warp and weft yarns in a full repeat, respectively. n is the natural number, and it should be larger than 1. That is to say, in every layer, in order to form a hollow, interwoven of the warp and weft yarn should be odd number and larger than 3. The example in Fig.2 shows the interweave time is 5.

Here, we can easily change the layers of the structure by changing the N. In order to change the size of hollow, it is necessary to change the number of n. Thus, for hollow shape structure, two parameters, N and n are used for making changes on the layers and the hollow size.

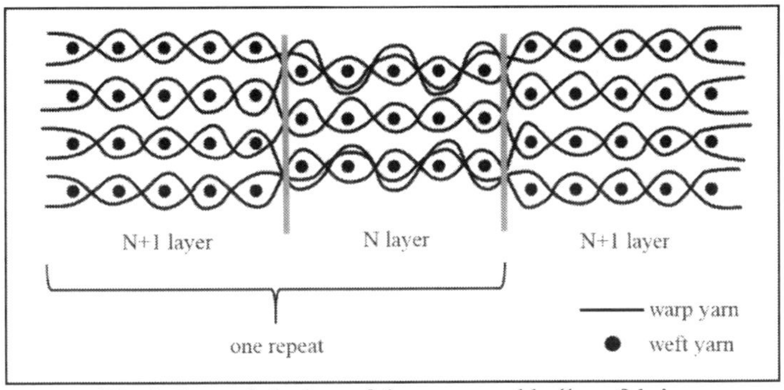

Fig.2 General design of the proposed hollow fabric.

Fig.3 shows the warp cross section and weave diagram of a minimun repeat of [4-3 layer] structure, with n as 1 (the interweaving in every layer of 3 times). From Fig.3(b), the weave diagram can be divided into six parts, during which part I, II and part III refered to the [N+1] layer, while part IV, V and part VI refered to the [N] layer. Here part I and III, part IV and VI are the same weave because the interwoven of warp and weft yarns in both parts are the same. Therefore, if n is changed, the weave diagram can be drawn by repeating the part II and III, and part V and VI corresponding to n.

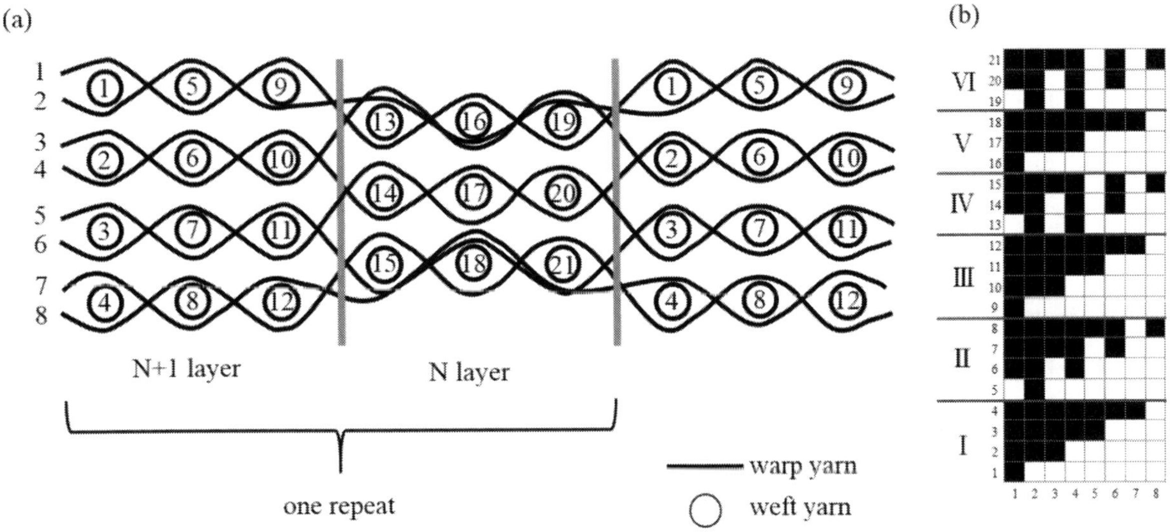

Fig.3 Cross section and weave diagram of a [4-3 layer] structure interwoven for 3 times in every part.

3. Structure Simulation and Weaving

Before weaving it is necessary to do some simulation on the proposed structure to see if there is some hollow in it and also to find out whether the derived equations can work. Therefore, DFS was used for fabric structure simulation. The simulation process is shown as follows: firstly set the weaving density in both warp and weft directions, then input the weave diagram of the structure, after that set the yarn, including yarn kind, yarn density and so on, finally simulate. The overall procedure is shown in Fig.4. The weaving density was set as the same as the ones on the real Jacquard loom, shown in Table 1. In the yarn selecting process, there is a library from which we can choose the suitable yarn to simulate. In our study, yarns similar to the ones in the Jacquard loom were selected.

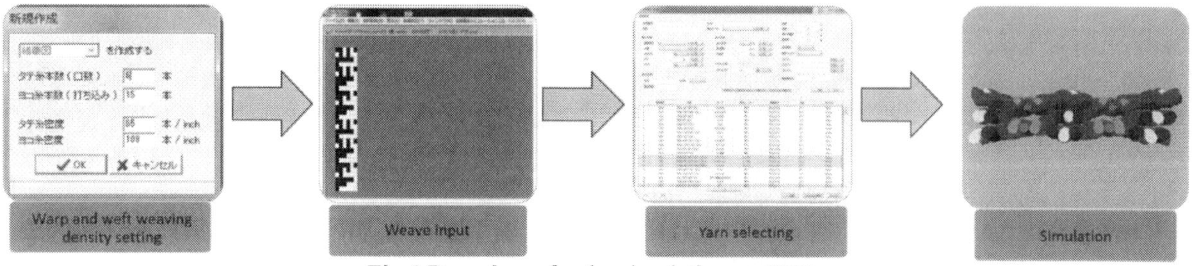

Fig.4 Procedures for the simulation on DFS.

Besides the simulation, weaving on a Jacquard loom was also carried out. The loom setting was shown in Table. 1. Both the simulation and weaving, the parameter N was set to 2 and 3, and parameter n was set to 1, 3, 5, and 7. When N was set to 2, with the increase of the hollow size n, the results of the simulation and weaving are shown in Fig.5. From this figure, it can be found that with the increase of n, interweaving length for each layer and the hollow size became larger. When n was 1, interweaving time was 3, causing the hollow structure of the real fabric difficult to be marked. When increased to 3, it became a little larger, but also difficult to be marked. However, when n reached to 5 or more than 5, as shown in Fig.5, in the real fabric, the hollow structure can be found clearly. Moreover, in Fig.5, n defined as 7 showed larger hollow space in the real fabric than n was 5. For the results of structure simulation, as shown in Fig.5, the hollow structure can be found difficulty. For the [(N+1)-N layer] structure, the hollow shape can be found only in the N-layer. Certainly this phenomenon is determined by the setting of warp and weft density and yarn count.

Table.1 Specification of the Jacquard loom

Parameters	Warp yarn	Weft yarn	Warp weave density	Weft weave density
Specifications	2/40s polyester	135D polyester	115/ inch	66/ inch

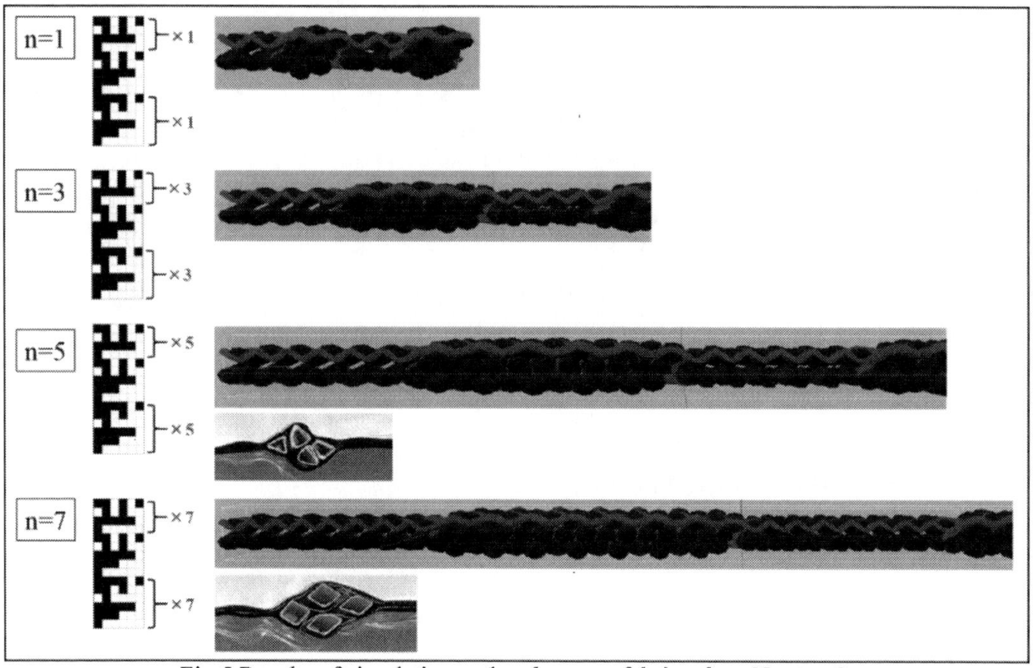

Fig.5 Results of simulation and real woven fabric when N set to 2.

Fig.6 shows the results of simulation and real woven fabrics when the layers were set to 2 and 3 respectively. In this figure, the parameter n was set to 5, meaning that the interweaving times at each part of each layer was 11.

When N was set to 3, layers in each part of fabric were increased. However, the hollow structure in the simulation results showed difficult to certification than N was 2. In this figure, different layers of simulation and real object demonstrated that the proposed method can be used to design hollow structure fabrics easily.

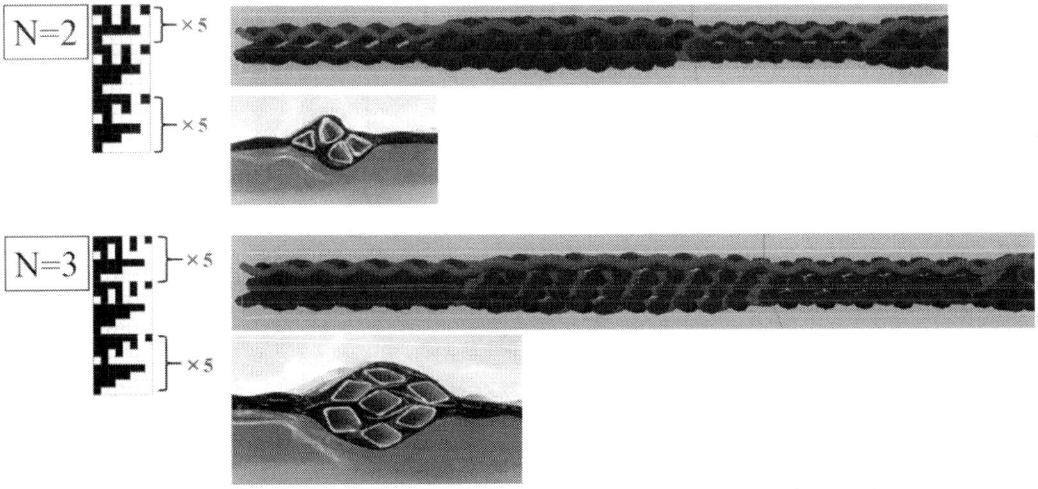

Fig.6 Results of simulation and real objects when N set to 2 and 3 (n=5)

4. Conclusion

In this study, a design method for hollow structure fabric was proposed. Number of warp and weft yarn were derived using layers and interweaving times as parameters. Different layers and hollow sizes of fabrics can be designed by changing the mentioned two parameters. Moreover, DFS was utilized for hollow structure fabric simulation to verify the derived equation, and a Jacquard loom was used for fabricating real fabric. The result showed that the hollow structure can be changed by changing the layer numbers and the interweave times shown in the derived equation, which was verified to be used for hollow structure design.

5. Acknowledgement

This work was supported by the JSPS KAKENHI, grant number JP16K16256.

References

[1] Adil M, Michael L, Gerald H, Chokri C, Knut G. 3D woven near-net-shape preforms for composite structures. Adv. Eng. Mater: 2016; 18: 391-396.

[2] Pelin GU. 3D woven fabrics. In: Han-Yong Jeon, editors. Woven fabrics. Intech: 2012; 91-120.

[3] Chen XG. Advances in 3D textiles. Woodhead publishing: 2015.

[4] Briggs GA, Townsend K. Textile design: principles, advances and applications. Woodhead publishing in textiles: 2011.

[5] Behera BK, Mishra R. 3-dimensional weaving. Indian J Fibre Text: 2008; 33: 274-287.

[6] Gokarneshan N, Alagirusamy R. Weaving of 3D fabrics: A critical appreciation of the developments. Textile Progress: 2009; 41: 1-58.

Automated Fabric Defect Detection Based on NSCT and Twin SVM

Qiao-Min Dai[2], Shun-Lin Fu[2], Yu-Chen Yan[3], Li Yuan[1*]

[1] *School of Electronic and Electrical Engineering, Wuhan Textile University, Wuhan, Hubei 430200, China*
[2] *School of Textile Science and Engineering, Wuhan Textile University, Wuhan, Hubei 430200, China*
[3] *Electronic Information School, Wuhan University, Wuhan, Hubei 430072*

*Corresponding author's email: yuanli@wtu.edu.cn

Abstract

A novel detection approach is proposed to detect various uniform and structured fabric defects based on Non-Subsampled Contourlet Transform (NSCT) and Twin Support Vector Machine (TSVM) model. The inspection algorithm is composed of two phases. First of all, the NSCT and the mixture of the generalized Gaussian model are used to obtain compact and accurate signatures for fabric texture description, and TSVM results to learn signatures of defected and non-defected classes. In the second phase, defects are detected on new images via the trained TSVM and an appropriate decomposition of images into blocks. The performance of the proposed detection method is evaluated off-line through extensive experiments based on various types of fabric. Experimental results reveal that the proposed method is effective and robust, and superior to recent state-of-the-art methods in terms of the high detection rate and low false alarm rate.

Keywords: Fabric Defect Detection; Non-subsampled Contourlet Transform (NSCT); Generalized Gaussian Mixture Model; Twin Support Vector Machine (TSVM)

1. Introduction

The fabric defect detection plays a crucial role in the manufacturing process for textiles. According to some studies, the defects in fabrics may reduce the price of a product by 45-65% [1]. However, the level of fabric defect detection in some countries is traditional quality detection which is done manually, which could be exhausting, neglectful and error-prone. It has been revealed that human eyesight system can recognize only 50-70% of the overall defects in fabric textile [2]. With the rapid development of computer and image processing technology, computer vision has been widely used in textile industrial production. Therefore, the development of automated inspection systems at hand is a major requirement for improving reliability and speeding up the quality control, which may increase the productivity. Nevertheless, the surfaces of fabric usually exhibit complex patterns that show high similarity all over the image, while in a local region, the textile may vary in their gray level distribution, pattern shape and size, as shown in Fig.1. These variations make defect detection more challenging when compared with those non-textured surfaces such as glass panel and plastic film [3, 4].

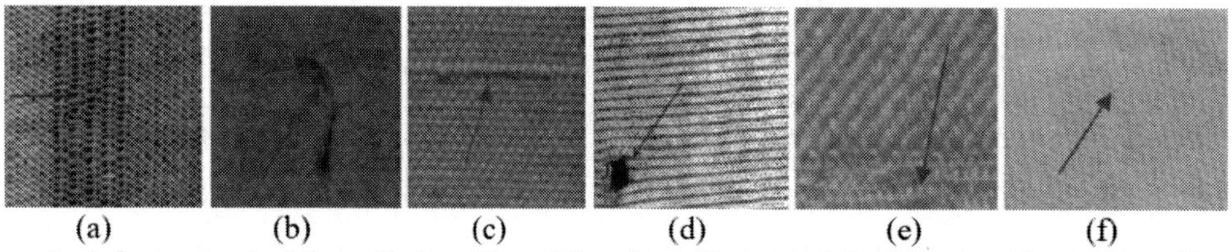

Fig.1 Example defects namely (a) needle breaking, (b) weft curling, (c) slub, (d) hole, (e) stitching, (f) rust stains.

In recent years, the methods for fabric defect detection have been concerned on many researchers in the world, and different algorithms have been proposed to reduce the cost. In [5], different signatures of fabrics, such as contrast, homogeneity, energy and correlation properties of co-occurrence matrix were obtained and online fabric defect detection system was performed. Sakhare et al. [6] developed spectral domain models for fabric defect detection. In order to identify fabric defects, various block processing algorithms of Fourier transform, Wavelet and Gabor filter, and these filters were used, which can be customized with different scale and angle values according to the texture structure. Tong et al. [7] developed a defect detection model via optimized Gabor filters. The proposed method uses composite differential evolution (CoDE) to optimize the parameters of Gabor filters. Zhou and Yang [8] developed the criteria of choosing the optimal Gabor filter. Then, the optimal Gabor filter could be used to detect defections for glass fiber clothes. Another approach based on the multiple Gabor filters and Kernel Principal Component Analysis (KPCA) was proposed to detect uniform and structured fabric defects [9]. In the proposed method, fabric images are filtered by various Gabor filters with four orientations and six scales to compute feature vectors firstly. Then, the high-dimension feature vectors were reduced by using KPCA. The similarity matrix was computed by Euclidean norm and segmented with OTSU method. This method obtained a high turn detection rate for online fabric inspection successfully. However, it is not suitable to use spectral approaches for fabrics containing random texture, and the choice of optimal filter parameters is quite difficult.

Nowadays, learning-based approaches have demonstrated excellent performance on texture classification. Tsan et al. [10] developed a novel Elo rating (ER) approach to obtain defect detection in the spirit of sportsmanship. The ER based method was carried out well in the dot-and star-patterned fabric. The proposed method achieved 96.89% accuracy for dot-patterned fabrics finally. Eldessouki and Hassan [11] developed the objective pilling classification system of the fabric images using an adaptive neuro-fuzzy system. They use basic textural features extracted from the fabric images to obtain better distinctive features of the fabric surface. Furthermore, there are fabric defect detection studies made through such classifiers as the Support Vector Machines (SVM) [12]. Thanks to the parallel operation, Artificial Neural Network (ANN) is considered to be heavily used in the future fabric defect detection systems. Convolutional Neural Networks (CNNs) have attracted much attention in many fields such as object detection and object classification.

Recently, non-subsampled contourlet transform (NSCT) and twin support vector machine (TSVM) have achieved significant success in image processing and patterns recognition, but applying less in fabric defect detection. Due to the comparative mature theoretical research and fitness for any form of distribution, NSCT and TSVM have a great development prospects in fabric defect detection. In this paper, a novel fabric inspection approach, which has the capability to cope with different types of defects in textile textures, is proposed. The algorithm is based on two phases. In the first phases, for each textile category, a set of training images of that category and label them at the block level as defected or defects-free. Signatures for fabric texture description are obtained from NSCT and mixture of generalized Gaussian model. In a nutshell, these signatures describe the distribution of contourlets coefficients at different scales and orientations of the image. In the training set, distances are computed between signatures and defected and defect-free classes, and discriminated by TSVM. In the inspection phase, new images are analyzed and classified based on trained TSVM. Experimental results reveal that the proposed detection model is effective and robust, and is superior to recent state-of-the-art methods in terms of the high detection rate and low false alarm rate.

The rest of the sections are organized as follows. Section 2 presents a brief background theory for NSCT and TSVM. In Section 3, the scheme of proposed model is described in detail. Section 4 demonstrates the experimental results for a variety of textured surfaces. Finally, the discussion and conclusions in this research are provided in Section 5.

2. Statistical Textured Surfaces Modeling

2.1 The Non-Subsampled Contourlet Transform (NSCT)

An efficient multiscale image representation forms the foundation of many image processing tasks. By allowing the redundancy, the shift-invariance which means less sensitivity to the image shift can be achieved in the NSCT [13]. Compared with the contourlet transform (CT), which is one of the state-of-the-art multiscale analysis techniques, the Gibbs phenomenon due to the coefficient modification is suppressed to a great extent, since the interpolation of filters replaces the image decimation. The multiscale property of the NSCT is achieved by the nonsubsampled pyramid (NSP). At the same time, the multi-directional property is obtained by the nonsubsampled directional filters (NSDFs), which further integrate high-frequency coefficients into wedge-shaped directional subbands [14].

Filters in the NSP are denoted as $H_0(z^2), H_1(z^2)$ for the first stage, $H_0(z)$, $H_1(z)$ for the second stage, while the NSDFs are written as $U_0(z^2)$, $U_1(z^2)$ for the first stage, $U_0(z)-U_1(z)$ for the second stage. The inverse transform achieves as the inverse process of the decomposition, resorting to the reconstruction filter bank. The number of the direction sets in a scale is the exponentiation 2^n , where n is a nonnegative integer, such as two or four in this example.

The filter design is the core of the NSCT structure. These 2D filters should satisfy the Bezout identity or the perfect reconstruction condition as follows:

$$\sum_{i=0}^{P-1} D_i(z^{2(L-1)})R_i(z^{2(L-1)}) = 1, (l = 1 \sim L) \tag{1}$$

where $D_0 - D_{P-1}$ mark the decomposition filters; $R_0 - R_{P-1}$ mark the corresponding reconstruction filters; l is the index of L stages. A side from (1), the frequency responses also constrain filters. Filters in the NSP should be low-pass and high-pass, while NSDFs should be directional. The 2D frequency plane could be divided into the wedge shaped subbands by NSDFs. It proves that the fan filter and the quadrant filter can be resampled to generate the bank of NSDFs which has 2^n directions, while the former can convert into the latter by rotation. Note that the diamond-shaped filter can be modulated to the fan filter by π in either the ω_0 or ω_1 frequency variable. Therefore, the key of the NSDFs is to design the diamond-shaped filter. Fig.2A illustrates the idealized frequency partitioning of a typical two-stage NSCT. The corresponding structure is displayed in Fig.2B.

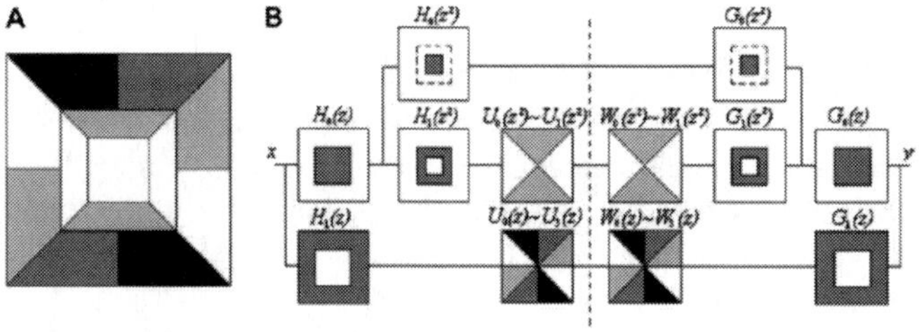

Fig.2 A typical two-stage NSCT: (A) the idealized frequency partitioning and (B) the corresponding structure of the NSCT.

2.2 Modeling of NSCT Coefficients

The general Gaussian distribution for a univariate random variable $x \in R$ is defined as follows:

$$p(x|\mu,\sigma,\beta) = \frac{\beta\sqrt{\frac{\tau(3/\beta)}{\tau(1/\beta)}}}{2\sigma\tau(1/\beta)} \exp(-A(\beta)\left|\frac{x-\mu}{\sigma}\right|^{\beta}) \tag{2}$$

where $A(\beta) = \left[\frac{\tau(3/\beta)}{\tau(1/\beta)}\right]^{\frac{\beta}{2}}$, $\tau(\bullet)$ denotes the gamma function. μ and σ are respectively the distribution mean and standard deviation parameters. The β is the flexible shape parameter which fits the kurtosis of the probability density function (pdf) and determines whether the distribution is peaked or flat. As $\beta \to \infty$ the distribution becomes uniform; whereas when $\beta \to 0$, the distribution becomes a delta function with center at center at μ. As $\beta = 2$ the function becomes a Gaussian. For multi-modal data, the marginal distribution of a random variable $x \in R$ is the resultant of a mixture of generalized Gaussian distributions (MoGG). Given a MoGG with K components, this is given by:

$$p(x|\Theta) = \sum_{i=1}^{K} \pi_i p(x|\mu_i, \sigma_i, \beta_i) \tag{3}$$

where $0 < \pi_i \leq 1$, $\sum_{i=1}^{K} \pi_i = 1$ and Θ denotes the set of model parameters $\{\pi_i, \mu_i, \sigma_i, \beta_i, i = 1, \ldots, K\}$. The model selection and parameter estimation of the MoGG is achieved in an unsupervised fashion using the minimum message length (MML) principle. The MML provides a natural tradeoff between model complexity and goodness of fit for a given sample of data $\chi = \{x_1, x_x, \ldots, x_n\}$. The details of parameter estimation of the MoGG are given in [15].

2.3 Similarity Measurement of NSCT-MoGG Signatures

Do and Vetterli [16] have used a closed-form Kullback Leibler divergence (KLD) to measure similarity between two statistical distributions of wavelet subband coefficients. When these distributions are multi-modal, a closed form solution is intractable. Given two MoGG models $P(x) = \sum_{i=1}^{K} \pi_i p(x|\theta_i)$ et $Q(x) = \sum_{j=1}^{M} w_j q(x|\theta_j)$, the KLD between these models is defined as follows:

$$KLD(P \| Q) = \int P(x) \log(\frac{P(x)}{Q(x)}) dx \tag{4}$$

The modified KLD by Monte Carlo integration is given by:

$$KLD_{mc}(P \| Q) = \frac{1}{n} \sum_{i=1}^{n} \log \frac{P(x_i)}{Q(x_i)} \approx_{n \to \infty} KLD(P \| Q) \tag{5}$$

The approximation method by sampling aims to generate a sufficiently large sample $\chi = \{x_1, x_x, \ldots, x_n\}$ drawn independently from P in order to approximate the KLD using the Monte Carlo integration (5).

2.4 Twin Support Vector Machine

Assume a binary classification problem in the n-dimensional real space R^n. The set of training data points is represented by $T = \{(x_i, y_j) | i = 1, 2, \ldots, m\}$, where x_i is input and $y_i \in \{+1, -1\}$ is the corresponding output. Let matrix A represent the samples belonging to class +1 and matrix B represent the samples belonging to class -1. Linear TSVM solves the binary classification problem by seeking two nonparallel separating hyperplanes.

$$x^T w_1 + b_1 = 0, x^T w_2 + b_2 = 0 \tag{6}$$

TSVM generates the hyperplanes by solving the following quadratic programming problems (QPPs):

$$\min_{w1,b1,\xi} \frac{1}{2} \|Aw_1 + e_1 b_1\|^2 + c_1 e_2^T \xi \text{, s.t } -(Bw_1 + e_2 b_1) + \xi \geq e_2, \xi \geq 0 \tag{7}$$

$$\min_{w2,b2,\eta} \frac{1}{2} \|Bw_2 + e_2 b_2\|^2 + c_2 e_1^T \eta \text{, s.t } -(Aw_2 + e_1 b_2) + \eta \geq e_1, \eta \geq 0 \tag{8}$$

where e_1 and e_2 are column vectors of ones of appropriate dimensions, ξ and η are slack variables, and c_1 and c_2 are penalty parameters with positive values. A new test sample is assigned to class $r \in \{+1, -1\}$ depending on which of the two hyperplanes given by (8) it lies closer to.

3. Proposed Algorithm for Defect Detection

As mentioned previously, the defect detection process could be divided into two phases: a *learning phase* and an *inspection phase*. The algorithm operates at the block level of images which aims at localizing defects at fine resolutions allowed by the block decomposition.

The training set R is composed by defect-free subsets $B=\{B_1,B_2,...,B_n\}$ and defective subsets $\tilde{B}=\left\{\tilde{B}_1,\tilde{B}_2,...,\tilde{B}_m\right\}$ respectively. For each block in the training set R, NSCT-MoGG signature, which combines several orientations and scales of the texture, is calculated. The algorithm trains a TSVM classifier and results into a set of reference blocks RB that will be used to inspect testing images. The training process will be iterating on the sets B and $\tilde{B}$ by selecting a new reference block to add to the set RB at each time. Finally, the training blocks could be classified via TSVM classifier, then the parameters of classification error ε and the set C of false defect detection will be obtained. This process is repeated until the classification error ε is null or higher than ε_p, the classification error at previous iteration. The training process of the defect detection system is given by the script of Alg. 1.The inspection phase is performed on testing images of the same type as the one used in the learning phase. The steps of the inspection process are given in the script of Alg. 2. First, the input textile image is decomposed into blocks with the same dimension as the training ones. Each block is then classified as containing detects or defect-free using the reference set RB and the trained TSVM classifier.

Algorithm 1 The learning phase

Data: $B=\{B_1,B_2,...,B_n\}$ and $\tilde{B}=\left\{\tilde{B}_1,\tilde{B}_2,...,\tilde{B}_m\right\}$

Result: Set of reference blocks: RB, TSVM classifier.

Generate NSCT-MoGG signature for each block in B and $\tilde{B}$

$RB \leftarrow B_1; \varepsilon \leftarrow \infty; C \leftarrow \varnothing; N \leftarrow 0;$

Repeat

1- $N \leftarrow N+1$;

2- $\varepsilon_p \leftarrow \varepsilon$;

3- Update the set RB from C;

4- Calculate the $KLDs$ $\{d_1,...,d_n\}$ and $\{\tilde{d}_1,...,\tilde{d}_m\}$;

5- Train a TSVM classifier on the calculated $KLDs$;

6- Classify the $KLDs$ and calculate the error ε;

7- Update C the set of false defect detections;

Until ($\varepsilon = 0$ OR $\varepsilon \geq \varepsilon_p$)

Algorithm 2 The inspection phase

Data: Input image I, $RB=\{B_{r1},...,B_{rN}\}$ and the TSVM classifier.

Result: Image with blocks classified.

Decompose the image into blocks;

Generate NSCT-MoGG signature for each block;

For (each block B_k) do

1- Calculate the $KLDs\{KLD_{r1},...,KLD_{rN}\}$;

2- Choose $d_k = \min\{KLD_{r1},...,KLD_{rN}\}$;

3- Use TSVM to classify B_k as containing defects or defect-free

End for

4. Experimental Results

In the section, the performance of the proposed fabric defect detection model is evaluated by using a database which consists of 65 defect-free and 60 defective fabric samples acquired by using a CMOS camera satisfy the definition of fabric defects from TILDA database [17] in the textile industry. These fabric samples contain common types of defects that always appear in the textile industry. All the samples are 512×512 pixels with 256 gray levels, and subdivided into overlapping blocks with size of 16×16 pixels. Detection models used in the experiments are realized under the image processing toolbox of MATLAB prototyping environment. Experiments are conducted on a workstation under a Win7 operating system running on an Intel(R) Xeon(R) CPU E3-1230 processor.

In this paper, the four measurements, which are precision (PR), sensitivity (SE), specificity (SP) and accuracy (AC) defined in [7], are employed to judging the performance of defect detection models. In order to further verify the effectiveness of the proposed model for fabric defect detection, the detection performance is compared with other two existing approaches. The evaluations are under the same environment. Between the two selected models, the first one is based on independent component analysis (ICA) reference method [18], which is widely considered as a classic method for fabric defects detection. The second one is based on the redundant contourlet transform (RCT) and Bayesian classification (BC) proposed in [19]. Table 1 shows that the detection method proposed in this paper outperforms the other two existing models in all four aspects. Precision indicates the percentage of correct alarm during detection. Sensitivity indicates the percentage of defective samples that are correctly detected. Specificity manifests the percentage of defect-free samples that are correctly classified as normal. Accuracy indicates the percentage of correct classification of all testing samples.

Table.1 Performance comparison of defect detection models

Detection method	Precision (%)	Sensitivity (%)	Specificity (%)	Accuracy (%)
ICA reference method	79.3	75.8	80.2	78.3
RCT based method	83.5	88.4	79.5	84.9
The proposed method	95.7	93.2	94.8	96.3

The comparison results of time performance are summarized in Table 2 in terms of average operation time (AOT) consumed in processing each fabric image sample. It is noticed that the AOT of the proposed method is not the most efficient one, but due to the two hyperplanes introduced, the overall AOT is still comparable with the other two methods.

Table.2 Time cost comparison of defect detection models

Detection model	AOT(s)
ICA reference method	0.12
RCT based method	0.38
The proposed method	0.72

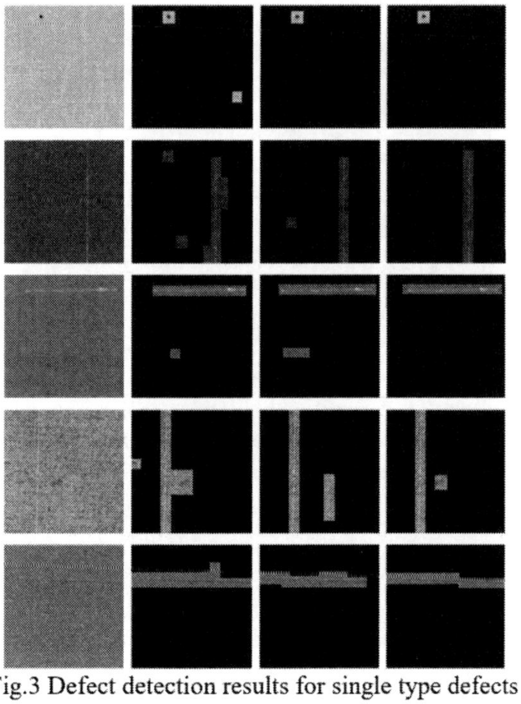

Fig.3 Defect detection results for single type defects

Some typical detection results, which on single type and complex type defects, are shown in Fig.3 and Fig.4. The images in the first column are the original defective images, and the second to the third columns are the detection results obtained by methods based on ICA and RCT, respectively. Figures in the last column are the detection results obtained by the proposed model. In Fig.3, the defect detection results for single type defects show that the proposed detection method could distinguish all defect blocks and overall is the best one among all methods. Firstly, the ICA reference method and RCT-based method could detect almost defects, but generate false positive results (i.e., it detects defect free blocks as defective) as well. Secondly, the texture description algorithm used in RCT-based method is relatively close to that of the proposed model. However, the inspection results are incomparable, because only Bayes classifier utilized in detection phase. Finally, it is noticed that there are lots of false detections in the results obtained by the detection method based on ICA and RCT for complex type defects, as shown in Fig.4. The reason is that ICA and RCT based methods are not able to accurately distinguish the textural components between normal fabric texture and defective texture, which results in poor defect segmentation and misclassification of the normal area. But the proposed method keeps the same high degree of reliability in detecting defects on both textile types. Thanks to the signatures obtained from NSCT and TSVM, the proposed algorithm has been able to classify very precisely image blocks containing defects with single and complex models.

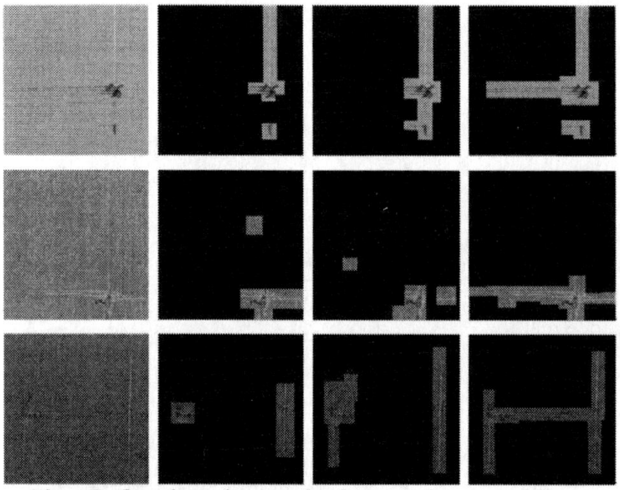

Fig.4 Defect detection results for complex type defects

5. Conclusion

In this paper, an automated and efficient fabric defect detection method has been proposed. This method has two phases: a learning phase and defects detection. Firstly, by taking advantages of the NSCT and mixture of generalized Gaussian distributions, compact and accurate signatures for fabric texture description could be obtained, and training TSVM operation is implement consequently. Secondly, defects are detected on new images using the trained TSVM and an appropriate decomposition of images into blocks. Extensive experiments have been carried out on different fabric samples acquired by using a CMOS camera satisfy the definition of fabric defects from TILDA database in the textile industry. A series of results demonstrates its effectiveness on the detection of defects of various shapes, sizes and locations. The overall comparative experiment results reveal that the proposed defects detection method achieves not only high successful detection rate but also low false alarm rate. In the future, more attention will be devoted to the developing multi-channel signatures for colored fabric defects inspection.

Acknowledgement

The authors sincerely thank the Natural Science Foundation of Hubei Province (2014CFB754), Hubei Provincial Education Department (Q20141607) and Undergraduate Training Programs for Innovation and Entrepreneurship of Wuhan Textile University.

References

[1] Hanbay K, Talu MF, Ömer Faruk Özgüven. Fabric defect detection systems and methods-A systematic literature review. Optik - International Journal for Light and Electron Optics: 2016; 127 (24): 11960-11973.

[2] Mahajan PM, Kolhe SR, Patil PM. A review of automatic fabric defect detection techniques. Advances in Computational Research: 2009; 1 (2): 18-29.

[3] Zhang WY, Zhang J, Hou Y, et al. MWGR: A new method for real-time detection of cord fabric defects. International Journal of Advanced Mechatronic Systems: 2012; 458-461.

[4] Jing JF, Zhang HH, Wang J, et al. Fabric defect detection using Gabor filters and defect classification based on LBP and Tamura method. Journal of the Textile Institute: 2013; 104 (1): 18-27.

[5] Raheja JL, Ajay B, Chaudhary A. Real time fabric defect detection system on an embedded DSP platform. Optik - International Journal for Light and Electron Optics: 2014; 124 (21): 5280-5284.

[6] Sakhare K, Kulkarni A, Kumbhakarn M, et al. Spectral and spatial domain approach for fabric defect detection and classification. International Conference on Industrial Instrumentation and Control: 2015; 640-644.

[7] Tong L, Wong WK, Kwong CK. Differential evolution-based optimal Gabor filter model for fabric inspection. Neurocomputing: 2016; 173: 1386-1401.

[8] Zhou J, Yang K. The Criteria of Choosing the Optimal Gabor Filter and Defect Detection Using the Optimal Gabor Filter. International Conference on Intelligent Human-Machine Systems and Cybernetics: 2015; 224-227.

[9] Jing J, Fan X, Li P. Automated Fabric Defect Detection Based on Multiple Gabor Filters and KPCA. International Journal of Multimedia & Ubiquitous Engineering: 2016; 11 (6): 93-106.

[10] Tsang CSC, Ngan HYT, Pang GKH, Fabric inspection based on the Elo rating method, Pattern Recognition 51: (2016); 378-394.

[11] Eldessouki M, Hassan M. Adaptive neuro-fuzzy system for quantitative evaluation of woven fabrics' pilling resistance. Expert Systems with Applications: 2015; 42 (4): 2098-2113.

[12] Basu A, Chandra JK, Banerjee PK, et al. Sub image based eigen fabrics method using multi-class SVM classifier for the detection and classification of defects in woven fabric. Third International Conference on Computing Communication & NETWORKING Technologies: 2012; 1-6.

[13] Do MN, Vetterli M. The contourlet transform: an efficient directional multiresolution image representation. IEEE Transactions on Image Processing: 2005; 14 (12): 2091-106.

[14] Cunha ALD, Zhou J, Do MN. The nonsubsampled contourlet transform: theory, design, and applications. IEEE Transactions on Image Processing: 2006; 15 (10): 3089-101.

[15] Allili MS, Baaziz N, Mejri M. Texture Modeling Using Contourlets and Finite Mixtures of Generalized Gaussian Distributions and Applications. Multimedia IEEE Transactions on: 2014; 16(3):772-784.

[16] Do MN, Vetterli M. The contourlet transform: an efficient directional multiresolution image representation. IEEE Transactions on Image Processing: 2005; 14(12):2091-106

[17] Workgroup on texture analysis of DFG. TILDA textile texture database, http://lmb.informatik.uni-freiburg.de/research/dfg-texture/tilda (accessed August 1996).

[18] Sezer OG, Ercil A, Ertuzun A. Using perceptual relation of regularity and anisotropy in the texture with independent component model for defect detection. Pattern Recognition: 2007; 40 (1): 121-133.

[19] Yapi D, Mejri M, Allili MS, et al. A Learning-Based Approach for Automatic Defect Detection in Textile Images. Ifac Papersonline: 2015; 48 (3): 2423-2428.

Study on Testing Mmethod of PTT Fabric's Shape Memory Function

Cui-Ting Chen[1,2,3], Si-Xia Chen[1], Jian-Hua Sui[1,2,3,*]

[1] *College of Textile and Clothing Engineering, Soochow University, China, Suzhou, 215021*
[2] *Modern Textile and Silk Science &Industrial Technology Research Center, China, Nantong 226018*
[3] *National Engineering Laboratory for Modern Silk (Suzhou), China, Suzhou, 215123*

[*]Corresponding author's email: suijianhua@suda.edu.cn

Abstract

In this paper, a rectangular groove forming test device and a sliding imitating stroking test device were developed to study the objective evaluation methods for PTT fabric's shape memory function. Based on the two devices, a test method for evaluating the plasticity was designed. PTT, PTT/PET and some conventional material fabrics, such as cotton, silk, viscose, were selected as the test samples, and the plastic forming and shape recovery ability of these fabrics were tested. The results showed that PTT, PET with high compact structure and PTT/PET fabrics all have plasticity. Among the three key factors, material, structure and compactness, the fiber material is the most important. Compared with other fibers, PTT fibers, used as warp and weft yarn, could improve shape plasticity and recovery performance of fabrics.

Keywords: Shape Memory Fabric; Plastic Forming Performance; Shape Recovery Performance; Test and Evaluation Method

PTT fiber is a new type of polyester fiber obtained through polycondensation of PTA and PDO [1]. It has excellent shape and recovery plasticity, and has gained consumer acceptance while catching the attention of many companies when the concept of "shape memory" was proposed.

The development upsurge of shape memory fabric initiated by PTT fiber has been about 10 years. There are many academic reports on evaluation of fabric's shape memory function [2-6], but it seems that there are no specific methods with easy operation and evaluation. Some information shows that the fabric's shape memory function includes easy plastic forming and shape recovery function. Plastic forming refers to the fabric's ability to change from an original shape into a specified shape under external effects, and shape recovery refers to the fabric's ability to return to the original shape under external effects [7-13]. Therefore, a rectangular groove forming test device and a sliding imitating stroking test device were developed to evaluate objectively fabric's shape memory function.

1. Design of Test Equipment and Test Method

1.1 Groove Plastic Forming Test Device

1.1.1 Device Design

The schematic of a rectangular groove forming device is shown in Fig.1. The device consists of a groove seat and a plastic strip. The groove seat consists of three rectangular grooves with the same size and spacing. The groove width b is 10mm, the groove height h is 15mm, the groove length l is 100mm, and the groove spacing a varies from 2mm to 10mm. The number of strips is equal to that of grooves in the groove mould. The strip width b' is 2mm to 3mm less than b, the strips height h' is 5mm to 10mm higher than h, and the strips length l' is 5mm to 10mm more than l. The tests for different thickness of the samples will be adjusted by using different layers of 0.1mm thickness films.

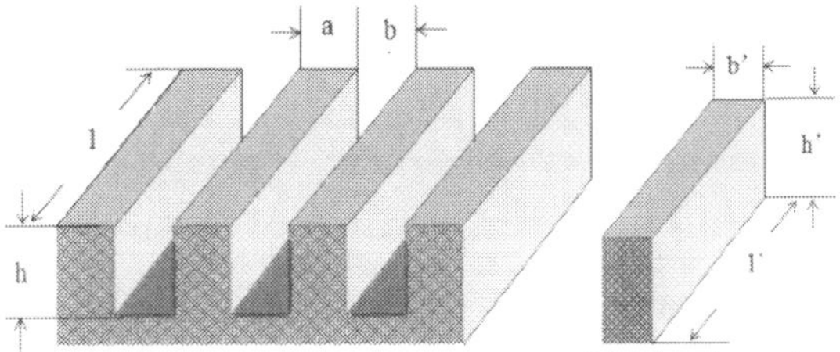

(a) Plastic Mould (b) Plastic Strip

Fig.1 Rectangular Groove Forming Device

1.1.2 Test Method

1) Sample preparation: All samples were 10mm×10mm rectangle fabrics which were cut along warp-wise, welf-wise and oblique direction with an angle of 45° respectively. After ironing these samples were put in a container with constant temperature (20 ± 2 ℃) and humidity (65 ± 2%) for 24 hours. The thicknesses of samples were measured, and films were used to adjust the value of strip width *b'*.

2) Plastic forming: As shown in Fig.2(a) and (b), the samples were placed on the rectangular groove mould in alignment direction, and then 5kg pressure was applied on the strips embedded in groove for 5 minutes to produce a concave-convex shape on fabric.

3) Samples hanging: As shown in Fig.2(c), (d) and (e), pressure was removed and this state was kept for 5 minutes. Then the samples were gently taken out and hung for 30 minutes.

4) Measurement and evaluation:

a. As shown in Fig.(f), the samples were gently placed on the experimental bench. The shapes of the samples were observed, and the shape plastic functions of them were researched according to the fabric arch and creases.

b. The highest value of each arch of each sample was measured by sliding caliper. The average highest arch values of each sample were recorded as H_{ji}, H_{wi} and H_{xi} respectively. There are five samples in every direction. And the average highest arch values of five samples were recorded as H_j, H_w, H_x, see the by the formula 1. The unit is millimeter, and the precision is 0.1mm. The larger the H value, the better the plastic formability.

$$H = \frac{\sum_{i=1}^{5} H_i}{5} \tag{1}$$

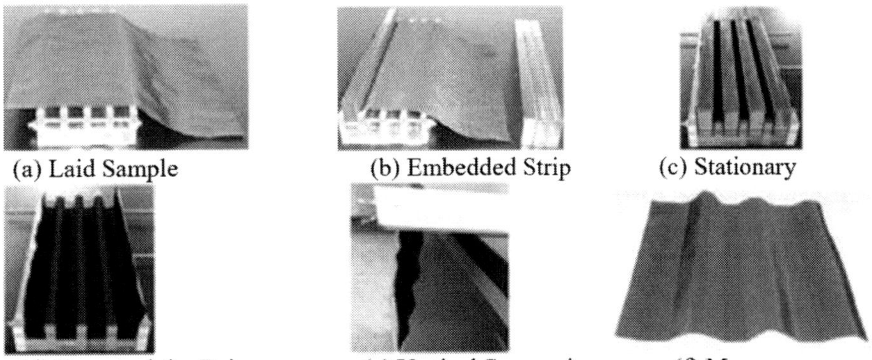

(a) Laid Sample (b) Embedded Strip (c) Stationary

(d) Removed the Strip (e) Vertical Suspension (f) Measure

Fig.2 Design of Fabric Shape Plasticity Test

1.2 Sliding Shape Recovery Test Device

1.2.1 Device Design

The principle of the sliding imitating stroking test device is to simulate the process of hands stroking fabric quickly (the temperature of hands was ignored). As shown in Fig.3, this device is mainly composed of horizontal platen, sample clip, slider, slider rail, traction line, guiding roller, pulley and gravity weight. The length of the horizontal platen is 500mm, the width of that is 270mm. A scale is provided along the length direction, and the unit is millimeter. The weight of slider varies from 200g to 300g, and the slider with 120mm length and 30mm width can slide from left to right along the guide rail in the traction of the gravity weight.

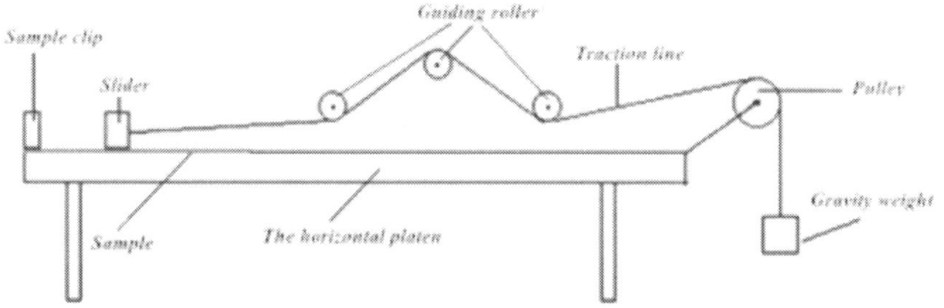

Fig.3 Sliding Block Stroking Device

1.2.2 Test Method

1) Sample preparation: Five samples were 210mm×100mm rectangle fabrics which were cut along warp-wise and welf-wise respectively. After ironing these samples were put in a container with constant temperature (20 ± 2 ℃) and humidity (65 ± 2%) for 24 hours.

2) Folding and forming: The samples were folded back and forth along the length direction for 10 times. Each folding length is 20mm, and the total crease length is 200mm. Then 5kg pressure was applied on the folded samples for 24 hours. After removing pressure, the samples were gently placed on the experimental bench for five minutes. The sample lengths were measured by scale. The average value of lengths was recorded as L_0. The unit is millimeter, and the precision is 0.1mm.

3) Sliding imitating stroking test: The test process was shown in Fig.4(a) and (b). The unfold part of sample with about 10mm length was clamped by clips, and the slider which was connected with traction line was placed on the sample adjoin the clips. The other side of traction line was 100g gravity weight. After releasing gravity weight, the slider was slid through the sample quickly.

4) Measurement and evaluation: The slider was removed, and the sample was left on the platen for 5 minutes. Then the sample lengths was measured by scale. The average value of lengths was recorded as L_i. The unit is millimeter and the precision is 0.1mm. Crease recovery rate ε_i of each sample was calculated by the formula 2. The average value of ε_i was recorded as ε. The larger the value ε, the better the shape recovery performance.

$$\varepsilon_i = \frac{L_i - L_0}{180} \times 100 \tag{2}$$

$$\varepsilon = \frac{\sum_{i=1}^{5} \varepsilon_i}{5} \tag{3}$$

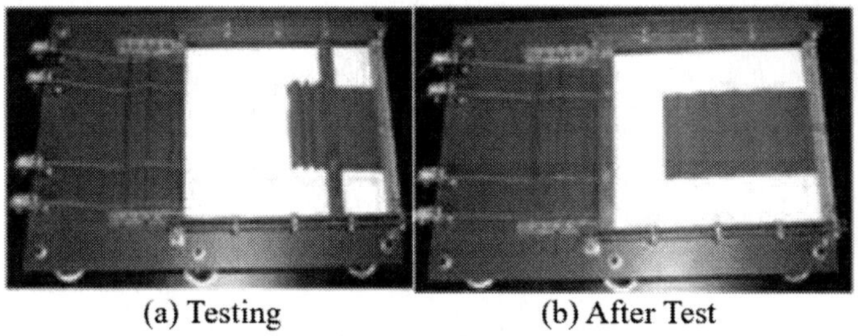
(a) Testing (b) After Test

Fig.4 Design of Fabric Shape Recovery Test

2. Verification Testing

2.1 Test Samples

Material, structure and weaving density of test samples were shown in Table 1.

Table.1 The Samples Size

Number	Structure	Warp Combination × Weft Combination	Warp Density × Weft Density number*10cm^{-1}	Grammes Per Square Metre g*m^{-2}
1#	plain weave	40Ne/2cotton yarn×21Ne/2cotton yarn	470×310	120
2#		80Nm/2silk yarn×32Nm/2silk yarn	430×270	125
3#		2/120D viscose filament×2/120D viscose filament	430×270	118
4#		2/75DPETfilament×150DPET filament	550×330	105
5#		40Ne/2 Milk/Tencel blended yarn×21Ne Milk/Tencel blended yarn	430×320	135
6#		75D PTT filament×150DPTT filament	820×410	112
7#		75D PTT filament×150DPET/PTT conjugated yarn	650×450	100
8#		75DPET filament×150PET filament	900×390	118
9#	$\frac{2}{2}$ twill	75D PTT filament×150DPTT filament	830×430	133
10#		75D PTT filament×75DPTT filament	830×620	133
11#	$\frac{31}{11}$ twill	75D PTT filament×150DPTT filament	870×430	138

2.2 Verification of the Testing Method

1) Samples 1# to 8#, which have similar structure but different materials, were tested by the method introduced in 1.1.2 with a value of 2mm. The shape forming effects of these samples were observed, and the shape plastic functions of them were studied.

2) Samples with plastic forming properties were tested by groove plastic forming test with a value of 6mm and 10mm. Combined the results with a value of 2mm, the results of H_j, H_w and H_x were analyzed to choose the best a value. Then the CV value of H was calculated by formula 4 to analyze the confidence level of this test method.

$$CV_H = \frac{\sum_{i=1}^{5}(H_i - H)}{5} \tag{4}$$

3) Samples 9# to 11# were tested by using the best a value. Then combined the results of samples 1# to 8# with the best a, the relationship between plastic forming properties and fabric structure were analyzed.

4) In order to analyze the relationship between the fabric shape recovery performance and fabric materials and structure, all the samples with plastic forming function were tested by the method introduced in 1.2.2. Then the CV value of ε were calculated by formula 5 to analyze the confidence level of this test method.

$$CV_\varepsilon = \frac{\sum_{i=1}^{5}(\varepsilon_i - \varepsilon)}{5} \tag{5}$$

3. Results and Discussion

3.1 Fabric Shape Plastic Forming Performance Test

The results of plastic forming tests were shown in Table.2.

Table.2 The “H”Value of Fabric Shape Plastic Performance Test(Unit: mm)

Number	Plastic Forming Visual Discrimination			a=2mm			a=6mm			a=10mm		
	Arch	Crease	Forming Property	H_j	H_w	H_x	H_j	H_w	H_x	H_j	H_w	H_x
1#	faint	faint	poor	—	—	—	—	—	—	—	—	—
2#	faint	faint	poor	—	—	—	—	—	—	—	—	—
3#	faint	faint	poor	—	—	—	—	—	—	—	—	—
4#	none	none	poor	—	—	—	—	—	—	—	—	—
5#	faint	none	poor	—	—	—	—	—	—	—	—	—
6#	strong	strong	outstanding	7.27	7.91	9.28	9.45	10.65	8.21	6.80	9.30	6.74
7#	obvious	obvious	good	6.88	5.56	7.40	5.51	6.11	6.04	5.79	6.68	5.21
8#	obvious	obvious	good	5.58	5.29	5.64	4.92	5.63	6.92	5.56	5.81	5.09
9#	—	—	—	—	—	—	7.99	7.77	6.56	—	—	—
10#	—	—	—	—	—	—	5.89	5.98	5.35	—	—	—
11#	—	—	—	—	—	—	8.41	7.75	7.86	—	—	—

1) As shown in Table.2, samples 1# to 5# of non PTT fabric had no obvious plastic shape effect, sample 6# of pure PTT fabric had most obvious plastic effect, and sample 7# of PTT and PET mixed fabric had certain plasticity. Although sample 8# was a pure PET fabric, the fabric had certain plasticity because of its high density.

2) The results of correlation analysis of *H* with different *a* values were shown in Table.3. It could be seen that the correlation coefficient of the *H* value was more than 0.95 when *a* was 6mm, and the test result was the most stable compared with the results of *a* of 2mm and 10mm. Therefore, the optimal *a* value is 6mm.

Table.3 Correlation Coefficient of “H” Value Under Different a Values

a/mm	H_j with H_w	H_j with H_x	H_w with H_x
2	0.696	0.804	0.949
6	0.973	0.9644	0.949
10	0.896	0.899	0.942

3) The yarn specification and warp/weft density of sample 9# were similar to sample 6#, but the *H* value of sample 9# was obviously less than that of sample 6#. The analysis revealed that the average float of sample 6# with plain weave structure was 1, and this made the fabric become most closely interweave. However, the average float of sample 9# with twill weave structure was 2, and this made the structure of fabric become loose. The compactedness of the fabric significantly affected the plastic formability of the fabric.

4) The test result of sample 10# was significantly smaller than that of sample 9#.The analysis revealed that the small linear density of filling yarn of sample 10# led to the decline of plastic forming performance.

5) Compared with sample 9#, the test results of sample 11# showed that the *H* value increased. The analysis

revealed that the yarn linear density and the warp/weft density of the two samples were basically the same, and they all used a twill weave with an average float of 2. But the shape plastic forming performance increased because of the partial plain structure of sample 11#.

6) Although the structure, yarn material and yarn line density of sample 8# and sample 6# were the same, and even the warp/weft density of sample 8# was higher than sample 6#, the H value obtained from the experiment was slightly smaller than that of sample 10#. It indicated that the shape memory function of the material was more prominent.

7) When a was 6mm, the CV values of H values of sample 6# to 11# were shown in Table.4. It could be seen that the CV values of H values were all less than 5%. This meant the test equipment and test method could be adopted.

Table.4 The Value of CV When a is 6mm (Unit: %)

Number	CV_{H_j}	CV_{H_w}	CV_{H_x}
6#	1.00	0.76	0.12
7#	4. 22	2. 33	2.93
8#	1.28	2.02	0.82
9#	3.99	2.73	3.92
10#	1.03	1.16	2.86
11#	0.11	3.13	0.37

3.2 Fabric Shape Recovery Performance Test

Table.5 Results of Fabric Shape Recovery Test

Number	ε_j/%	$CV_{\varepsilon j}$/%	ε_w/%	$CV_{\varepsilon w}$/%
6#	92.78	1.32	94.44	2.77
7#	86.44	5.33	80.89	4.33
8#	82.39	4.61	84.06	4.62
9#	82.22	2.11	82.67	4.59
10#	84.44	3.94	80.67	2.30
11#	85.63	4.10	87.78	3.57

The results of samples shape recovery performance test were shown in Table.5, and included the following information:

1) Sample 6# of PTT "shape memory" fabric had the best plastic recovery performance. The plastic recovery performances of samples 9# to 11# were slightly lower than that of sample 6#. Sample 7# of PTT*PET/PTT interwoven "semi-memory" fabric and sample 8# of pure PET "imitating memory" fabric all had a certain plastic recovery performance.

2) The yarn specification and warp/weft density of sample 9# were similar to those of sample 6#, but the ε value of sample 9# was obviously less than that of sample 6#. The analysis revealed that the structure of sample 6# was plain weave, while the structure of sample 9# was twill. It meant the compactedness of fabric significantly affected the shape recovery performance of fabric.

3) Compared with sample 9#, weft line density of sample 10# decreased, but the weft density of that increased. Therefore, it had little effect on the test results.

4) The ε value of sample 11# was higher than that of sample 9#. The analysis revealed that the yarn linear density and the warp/weft density of the two samples were basically the same, and they all used a twill weave with an average float of 2. But the shape recovery performance increased because of the partial plain structure of sample 11#.

5) The CV values of crease recovery rate of samples 6# to 11# were basically less than 5%. This meant the test equipment and test method could be adopted.

4. Conclusion

In this paper, a rectangular groove forming test device and a sliding imitating stroking test device were developed to study PTT fabric's shape memory function. Based on the two devices, a testing method for evaluating the plastic forming and shape recovery ability of fabrics was designed. The results were as follows:

1) When a rectangular groove forming test device with groove spacing a of 6mm was used in the test, the CV value of plastic forming performance of PTT shape memory fabric was less than 5%. When a sliding imitating stroking test device was used in the test, the CV value of shape recovery performance of PTT shape memory fabric was also less than 5%. Therefore, the designed fabric shape memory function test method was feasible.

2) The test results showed that PTT fibers used as warp/weft materials, compact structure such as plain weave and the appropriate warp/weft density could improve the plastic forming performance and shape recovery performance of fabric. Among the three key factors, material, structure and compactness, the fiber material was the most important, and the PTT fiber was obviously superior to other fibers.

References

[1] Hong JH, Han X. Analysis of shape memory performance of PTT fabric [J]. Higher Vocational Education Forum: 2009; (9): 18-20.

[2] Xiao H, Shi MW, Liu J. Impact of tension on the structure and properties of PTT/PET filaments [J]. Journal of Textile Research: 2008; 29 (7): 1-5.

[3] Liu XX, Dai ZY. Development and research of shape memory alloy textiles [J]. Technical Textiles: 2007; (12): 5-9.

[4] Hu JL, Liu XX. Research and development of shape memory polymers for textiles [J]. Journal of Textile Research: 2006; 27 (1): 114-120.

[5] Zhao LH, Qin L, Wang FM. Mechanism of "shape memory" PTT fabric returning to its initial shape [J]. Journal of Textile Research: 2009; 30 (3): 34-38.

[6] Hu JL. Shape Memory Material for Textile [M]. China Textile & Apparel Press: 2006.

[7] GB/T 3819-83《Textile fabrics-Determination of the recovery from creasing of a folded specimen by measuring the angle of recovery》.

[8] Qin L, Zhao LH, Wang FM. Exploration for testing shape memory effects of PTT fabrics [J]. Journal of Textile Research: 2009; 30 (6): 39-43.

[9] Xiao H, Liu J, Shi MW. Influence of duration of treatment and media on the properties of PET/PTT filaments [J]. Journal of Textile Research: 2008; 29 (9): 16-19.

[10] Wu HY. Property Test and Analyses of PTT Fibre [J]. Cotton Textile Technology: 2008; 36 (2): 28-31.

[11] Liu Y, Hu JL. Shape Memory Textile and its Development [J]. Progress in Textile Science & Technology: 2005; (5): 7-9.

[12] Zhao LH, Wang FM. Development and Outlook of Shape Memory Fabrics [J]. China Textile Leader: 2008; (2): 50-52.

[13] Zhou X, Zhu M, Wang SQ. A kind of lining with shape memory ability [P]. China: 99236045.5: 2001-02-21.

Estimation of Facial Feature Points Depth Based on a Single Photo

Le Song[1], Lin Gu[1*], Hu Dong[2], Zheng-Lei Jia[1], Qiao-Qiao Du[1]

[1] *School of Computer Science, Xi'an Polytechnic University, Xi'an 710048, China*
[2] *School of Electronics and Information, Xi'an Polytechnic University, Xi'an 710048, China*

[*]Corresponding author's email: 396500021@qq.com

Abstract

In this paper, we use neighbourhood weighting and a back-propagation (BP) neural network to estimate the depth value of the feature points from a single face photograph. First, the K-value is selected by using the cross-validation method and the neighbour weighting method, so that the error is minimized after recombination. Then, in a BP neural network, the Levenberg-Marquardt algorithm is used to optimize the weights and thresholds of the neural network to avoid them falling into the local minimum. The results of MATLAB simulation show that the error between the true value and the estimated value, when using the neighbourhood weighted method, is higher than the error when using the BP neural network. Therefore, the depth information of the feature points obtained by using the optimized BP neural network can be used to determine the position of the feature points in three-dimensional (3D) face modelling. This paper offers a quantitative guarantee about the location of the feature points in 3D face modelling, and also a certain application value.

Keywords: Depth Estimation; Neighbourhood Weighting; Back-propagation Neural Network; Cross-validation; Levenberg-Marquardt

1. Introduction

With the rapid development of e-commerce, more and more scholars are committed to studying technology related to virtual fitting system [1-3]. One of the important aspects in this field is 3D face reconstruction from a single image [4-6]. Feature modelling based on feature points is one of the more mature face modelling methods, in which the extraction of feature points is both a precondition and a key attribute [7-9]. On a face picture, the plane information (x, y coordinates) of each feature point is relatively clear, but the depth value (z coordinates) is missing. Thus, the key to feature points extraction is to obtain depth information from facial feature points.

In recent years, many scholars have studied the depth information of facial feature points based on a single two-dimensional (2D) face photograph. Tu Yi, using an improved ASM method, was able to automatically obtain 2D face feature points, while, via the fitting sparse morphable model, it was possible to obtain depth information of the given image [10]. Zheng Yin used statistical learning to map the texture and depth of images, as well as used mapping to estimate depth from texture in 3D face modelling. In turn, face alignment techniques were incorporated into the learning process to dramatically promote depth estimation accuracy [11]. He LX used a shape-from-shading (SFS) method to exploit the change in an image's grey values in order to reconstruct the relative height and normal surface [12]. However, at present, most of SFS methods are based on the Lambertain model. The Lambertain model cannot describe the characteristics of speculate reflectance, as reflection characteristics of the face surface are more complex in the real imaging process. In recent years, the neural network estimation method has been recognized and adopted by scholars in terms of its high efficiency and large-scale processing ability, as well as its superior robustness. In particular, some background knowledge and reasoning rules are unclear, while better performance has been reported with regard to the partial loss of input mode and other treatment issues [13-15].

In this paper, two kinds of estimation method, based on learning, are used to estimate the depth values of the human face's feature points from a single face picture. The first method is based on a 3D facial database, which is used to estimate the depth information of facial feature points by adopting an improved nearest neighbour approach. The second depth estimation method uses the neural network method to estimate the feature point depth value from a single face picture on the basis of the 3D face database. At the same time, five different optimization functions are used to optimize the neural network weights and thresholds. Through the analysis and comparison of the training results of the neural network, a better optimization algorithm is selected for the neural network in order to estimate the depth of the face feature points. Finally, the results of these two methods are compared and analysed.

2. Analysis of Algorithms

2.1 K-nearest Neighbour Weighting

When used to estimate the depth of the feature points, the nearest neighbour method only focuses on individual differences, while ignoring the connection between feature points in the sample library [16, 17]. Moreover, in the process of calculating the minimum error between two feature points, sometimes because a number of points can have the same error value, such that the feature point depth value has multiple reference points and cannot be determined. Thus, the K-neighbour weighting method is used to estimate the depth of the feature points in view of the shortcomings of the nearest neighbour method.

First of all, in the set of 2D sample feature points S^{2d} , as projected by the 3D sample feature points set S^{3d} , it is necessary to look for the K projection feature points that are closest to the unknown feature point. Secondly, based on the K projection feature points, their coordinates (x,y) should then be re-linearly combined with the unknown feature point, after which the weighted value of the K feature points' linear combination is calculated when the feature point error is minimized. Finally, in the 3D face sample library, the depth value (z) of the K closest feature points is linearly weighted according to the recombination weights, so that the depth value of the unknown feature point is obtained. Thus, this problem can be equivalent to that when calculating the reconstructed coefficient w_i of the K-nearest neighbours when the reconstruction error ε is the minimum value:

$$\varepsilon(w) = \min \| s^{2d} - \sum_{i=1}^{k} w_i s_i^{2d} \|^2 \tag{1}$$

2.2 BP Neural Network

BP is a multilayer feed-forward neural network, in which the neural element transfer function is the S function. It implements nonlinear mapping from the input to the output. The structure of a BP neural network includes an input layer, a hidden layer and an output layer. Although the principle of a BP neural network is relatively simple, the network is often subject to unreasonable weight and threshold settings, resulting in slow network convergence and a local optimal solution.

In order to solve the above problem, improve the convergence speed of the network and avoid the local optimal situation for the network, in this paper, five algorithms are used to optimize weights and thresholds of the neural network. Through experimental analysis and comparison, we select a better algorithm to estimate the depth of the feature points. In addition, we adjust the training parameters in the network according to the results from network training, so that the BP neural network can achieve a better estimation state.

3. Experimental Process and Results Analysis

In this experiment, 320 face photographs are selected, and 300 of them are selected as the training sample set, with 20 selected for the test sample set.

First, we collect and normalize 3D data about the top of the nose feature point on each image, and save them in a 3D database, using the data normalization function mapminmax (A) in MATLAB, Version: 8.3.0.532 (R2014a). Fig.1(a) shows the 3D coordinates of the feature point at the tip of the nose prior to normalization, while Fig.1(b) shows them after normalization of the feature point at the tip of the nose.

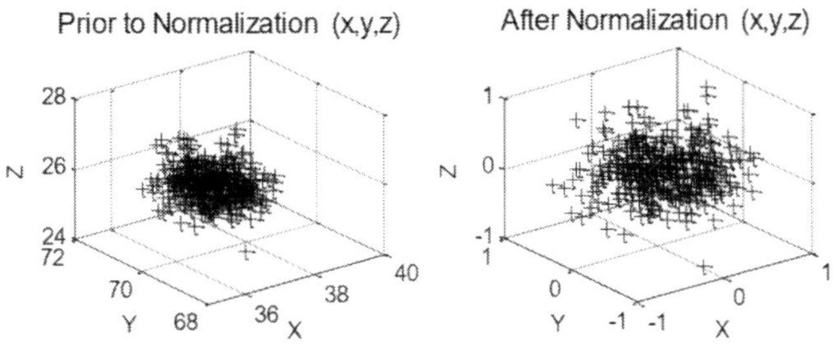

(a) Prior to normalization (b) After normalization

Fig.1 Data normalization

3.1 K-nearest Neighbour Weighting

Step 1: Determine the K-nearest neighbour of the target point. We use Euclidean distance to find the K feature points that are nearest to the unknown point s_i^{2d}. Next, we choose a value of K that will have a great impact on the result of the algorithm, that, is, if the K-value is too small, over-fitting will occur, while, if it is too large, the possibility of learning the approximate error will increase. Table 1 shows the K-value for some of the test samples involving the minimum reconstruction error. When the sample reconstruction error is minimized, we use cross-validation to determine the K-value.

Table.1 The K-value for the minimum reconstruction error

No.	Error	K	No.	Error	K
1	-0.0131	6	5	-0.0065	19
2	-0.0113	1	6	-0.0751	2
3	0.0031	11	7	0.4508	3
4	0.0374	9	8	-0.0245	7

Step 2: Calculate the reconstruction coefficient. The distance between the sample feature point and the feature point to be measured is denoted by L. First of all, the K feature points nearest to the feature point to be measured are arranged by distance from near to far, after which the percentage of the distance L_i (i = 1,2, ..., k) of the K-nearest neighbour feature points and the maximum value L_{max-k} of K distances are calculated. The closer the distance, the greater the impact, which means that the reconstruction coefficients of the K-nearest neighbour feature points can be expressed as:

$$w_i = 1 - \frac{L_i}{L_{max-k}} \tag{2}$$

Step 3: Estimate the depth value of the target point. This step is based on the reconstructed coefficients of the K-nearest neighbour feature points as calculated in Step 2. As the depth values of the K feature points are linearly combined according to the reconstruction coefficients, t the depth estimate of the unknown feature point in the test samples can be expressed as:

$$\hat{s}_i^{3d}(z) = \sum_{i=1}^{k} w_i s_i^{3d} \tag{3}$$

The above represents the entire process for estimating the depth value of facial feature points by using the adjacency-weight method. It is mainly based on the improved nearest neighbour method, in which the K-nearest neighbour feature points are re-linearly combined to obtain the unknown point depth value.

3.2 BP Neural Network

In this paper, the BP neural network structure is as follows: the input layer has two neurons, which can be expressed as (x, y); the hidden layer has 10 neurons; the number of hidden layer neurons can be changed during the experiment; the output layer contains one neuron and can be represented by z. The learning rate parameter and the momentum factor of the neural network model are $\eta = 0.7$, $\alpha = 0.9$, respectively. The error goal in network training is $Goal = 0.001$.

After the structure of the network is set up, the weights and thresholds of each layer need to be considered. In the BP neural network, five different training functions are used to optimize the weights and thresholds in the network. The training results are shown in Table 2. According to the comparison of the data, the Levenberg-Marquardt training function is used to train the neural network. The algorithm is a combination of the gradient descent method and the Gauss-Newton method. The basic idea is to allow the error in the iterative process to search along the deterioration direction and, through the adaptive adjustment between the gradient descent method and the Gauss-Newton method, realize the purpose of optimizing the network weights and thresholds.

Table.2 Comparison of different training function results

Training function	Accuracy	Error	Iterations	Time	Saturation value
Traingd	0.00565	0.109	10,000	8 s	Accuracy
Traingdm	0.00421	0.128	10,000	7 s	Accuracy
Trainrp	0.00299	0.047	325	16 s	Gradient
Trainlm	0.00201	0.046	1,182	7 s	Gradient
Trainbr	0.00270	0.043	1,165	6 s	mu

3.3 Analysis of Results

Through the preparation of the above experimental data and parameters, the nearest neighbour, the neighbour weighting and the BP neural network methods are used to estimate the depth value (z coordinate) of the feature points. Meanwhile, the error between the estimated depth value $\hat{s}^{3d}$ and the true depth value s^{3d} of the feature point is compared thus:

$$err(z) = \frac{1}{k}\sum_{i=1}^{k} | s_i^{3d} - \hat{s}_i^{3d} | \tag{4}$$

The results of the three methods are simulated in MATLAB. Fig.2(a) shows that different methods can be used to predict the depth value, with the dotted line representing the true value of the depth of the feature point, the asterisked solid line representing the depth value, as estimated by the nearest neighbour method, the circle solid line representing the depth value, as estimated by the neighbourhood weighting method, and the square solid line representing the result of the BP neural network estimation of the depth value. From the degree of the curve, we can see that the value estimated by the neighbourhood weighting method is close to the real depth value. Fig.2(b) shows the percentage error of the depth value as predicted by the different methods. The star-dotted line represents the percentage of the predicted error in the nearest neighbour, while its variation range is [-0.03756 0.03739]. Solid circular and square lines represent the distribution of the percentage of the depth value error estimated by the neighbourhood weighting and BP neural network methods, respectively. The error percentages range are [-0.02377 0.01782] and [-0.01542 0.0122]. Based on the statistical analysis of the errors of the three estimation methods, the average estimation error of the feature points is, respectively, $\mathrm{err}(\bar{z}\_nn) = 0.3603$, $\mathrm{err}(\bar{z}\_knn) = 0.1174$, and $err(\bar{z}\_BP) = 0.1438$. Therefore, although the range of error percentage for the BP neural network is small, the accuracy is greater when using the neighbourhood weighting method.

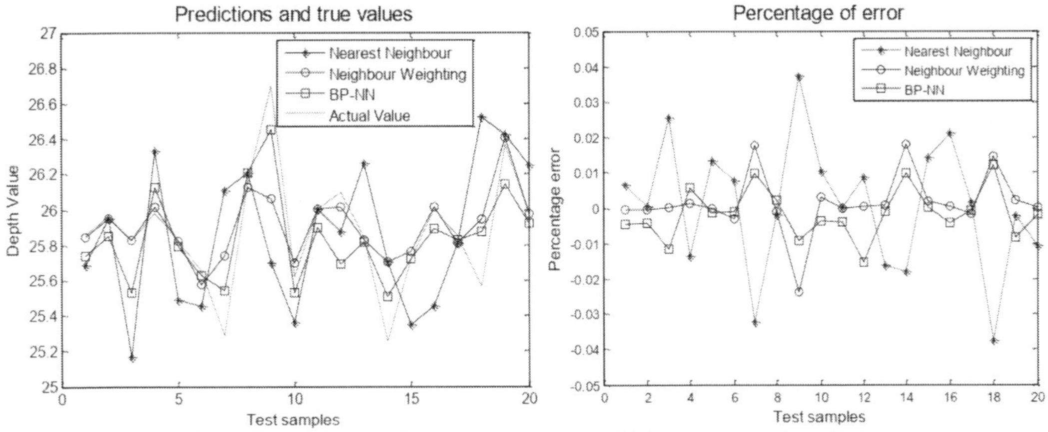

(a) Predicted depth value (b) Percentage of prediction error

Fig.2 Predictive value comparison

4. Conclusion

This paper mainly solves the problem of modelling distortion caused by missing 3D data on facial feature points in the process of reconstructing a 3D model of a face based on a single face picture. In the process, the nearest neighbour, the K-nearest neighbour weighting and the BP neural network methods have been used to estimate the depth of the feature points. In the case of the K-nearest neighbour weighting method, the K-value is chosen via cross-validation, which can avoid over-fitting the results and error increase, after which the K points of the nearest neighbour re-linearly combine in order to estimate the depth of the unknown feature point. In the BP neural network algorithm, the Levenberg-Marquardt algorithm is used to optimize weights and thresholds of the network by comparing the optimization function, so as to avoid the local optimal solution. By analysing and comparing the simulation in MATLAB, the error between the true depth and the predicted value is the least when using the K-nearest neighbour weighting. Therefore, the method that is based on the feature point of the weighted neighbourhood, in order to estimate depth, provides 3D coordinates for feature points in 3D face modelling, thereby offering significant practical value.

References

[1] Gao F, Dong LF. Research on technique of Net-based 3D fitting room. Computer Simulation: 2006; 06: 209-212.

[2] Wang HB, Huang X, Zeng GP et al. Design of intelligent 3D virtual clothing model simulation system. Application Research of Computers: 2009; 04: 1405-1408.

[3] Niu XX, Wang CR, Gu GP. 3D shape reconstruction based on single face image. Computer Simulation: 2011; 04: 230-233.

[4] Kemelmacher-Shlizerman I, Basri R. 3D face reconstruction from a single image using a single reference face shape. IEEE Transactions on Pattern Analysis and Machine Intelligence: 2011; 33 (2): 394-405.

[5] Deng QP, Zhao YM. 3D face reconstruction method based on single frontal photo. Computer Engineering: 2010; 20: 176-178.

[6] Xiong P, Lu Y. Modified 3D facial reconstruction algorithm based on a single photo. Journal of Computer Applications: 2013; 08: 2359-2361, 2369.

[7] Li YL, Jin Y, Wang JM et al. Face feature point extraction: a review. Chinese Journal of Computers: 2016; 07: 1356-1374.

[8] Zhang XB, Pan G, Ren HY et al. Real-time localization of 3D facial landmarks. Journal of Computer-aided Design & Computer Graphics: 2013; 09: 1325-1337.

[9] Han YF, Wang XL. Method of human facial feature points positioning based on improved ASM. Computer Science: 2013; 04: 271-274.

[10] Tu Y, Zhao QF, Lu FF. 3D face reconstruction based on generic model and single image. Application Research of Computers: 2009; 10: 3998-4000.
[11] Zheng Y. Image-based 3D face modeling. University of Science and Technology of China: 2009; 69-75.
[12] He LX. Research on 3D reconstruction of a single image based on SFS methods. Shenyang: Shenyang University of Technology: 2009; 1-87.
[13] Zhang XL. Optimized character recognition algorithm based on BP artificial neural network. Computer Engineering and Applications: 2012; 35: 182-185.
[14] Yang XL, Zhou JZ, Ding JH, Zhang YC. Flood disaster evaluation based on improved BP neural network model. Applied Mechanics and Materials: 2012; 2025 (220): 2462-2465.
[15] Zhao H. Character image recognition based on the improved BP neural network. Applied Mechanics and Materials: 2011; 1287 (58): 2655-2658.
[16] Wang XY, Wang ZO. An improved K-nearest neighbor Algorithm. Journal of Electronics and Information Technology: 2005; 03: 487-491.
[17] Zeng Y, Shu H, Hu JP et al. Adaptive pseudo nearest neighbor classification based on BP neural network. Journal of Electronics & Information Technology: 2016; 11: 2774-2779.

Theory on the Super High Count Yarn Production using Vulnerable Fibers

Cai-Ling Ding[1, 2], Jian-Yong Yu[1*], Chao Chen[2], Hui Li[3], Guang Qin[3], Zhi-Gang Xia [4, 5**]

[1]*College of Textiles, Donghua University, Shanghai 201620, China*
[2]*National Engineering Research Center for Spinning Technology, Jining 272073, Shandong Province China*
[3]*Shandong Ruyi Textile Company Limited, Jining 272073, Shandong Province China*
[4]*College of Textile Science and Engineering, Wuhan Textile University, Wuhan 430200, China*
[5]*Xiangyang Jihua 3542 Textile Company Limited, Xiangyang, Hubei 441002, China;*

[*]Corresponding authors' email: yujy@dhu.edu.cn; 117462379@qq.com

Abstract

This study purpose was to establish and analyze a protective systematic process. Theoretical analysis indicted that strict hydrothermal conditions could keep fibers in an optimal state during opening; gentle carding effectively reduced the fiber damage by reducing flats of normal carding; the false twisting method could enhance fiber control and decrease fiber loss in a loose assemblage during the combing; false twisting roving could achieve compact roving structure with low twist density to eliminate unexpected drawing; in particular, the embeddable and locatable spinning model was developed to avoid end breakage and fiber loss during the super high count yarn using tender fibers. Guided by theoretical modeling analysis, super high count yarn of vicuna fibers was produced successfully by adopting systematic processing procedures.

Keywords: Super High Count; Vulnerable Fiber; Low Damage; Spinning

1. Introduction

Softness is the key factor endowing the high qualified fabric with delicate and comfortable handle. The ultra soft fabric can be produced by two methods: the first one is the post-treating of fabric with soft active agent; the second one is to produce woven or knitted fabric using super-high count yarn of super fine fibers. The post treatment of fabric consumes lots of waters and does harm to environments; therefore it is very important to seek a green and effective method for super-high count yarn production. These fine fibers including vicuna fiber, cashmere, superfine wool and red deer velvet are tender, warm and delicate. Among these natural fibers, vicuna fibers are extremely rare and precious as their annual usable output is only about 5000 kg in total [1]. Under this condition, vicuna fine fibers await an extremely effective utilization. Related studies have validated that each fiber processing stage can cause fiber damage, which ultimately deteriorates the spinning performance and quality of the products [2]. Unfortunately, vicuna fibers are so tender that they are very easy to be damaged and even broken apart; then vicuna fiber shortness and strength become extremely deteriorated, failing in producing qualified yarn. Therefore it seems impossible for the vulnerable tender vicuna fibers converting into a super-high count yarn with high quality and efficiency.

With the aim to produce super-high count yarn using vulnerable vicuna fibers, a protective systematic process was theoretically modeled to minimize fiber damage before the fiber strand twisting to form yarn, to avoid end-breakage and fiber loss during the fiber strand twisting to form yarn. Then low damage carding, false twist roving and embeddable and locatable spinning were employed to produce super high-count yarn from vicuna fibers under the guidance of above theory. The corresponding spun yarn properties were also comparatively studied.

2. Protective Systematic Process Modeling

2.1 Vulnerable Fiber Damage Minimization Model

Protecting vulnerable fibers from fatal damages is very essential to the successful production of high qualified staple yarn. To minimize the vulnerable staple fiber damages, improved systematic processing is developed (Fig.1). Firstly the low temperature and humidity can cause the protein fiber fragile during opening [3] and electrostatic [4]; very high temperature and humidity encourage wool fibers felting easily by external mechanical forces [5]. Therefore the opening of vulnerable fibers should be conditioned strictly. To minimize vicuna fiber damage, which tends to happen with a Scutcher opener [6], the opening room temperature and humidity should be strictly controlled at 25±1°C and 68±1% respectively on basis of practical summary. Secondly, fibers also incur cracks and end-breakages during the carding process [7], especially for the tender fibers [8]. Therefore gentle carding should be also applied by reducing flats, apart from adopting other optimum carding parameters similarly in normal carding system [9]. After carding, the increased shortened fibers [10] should be removed in time; thereafter, combing process is necessary for the high quality yarn production [11]. Because of very short fibers exhibiting weak cohesive forces, normal untwisted loosen sliver of vicuna fibers easily suffers unexpected drafting during its usage on a roving frame. Under this situation, false twisting combing should be applied to increase the sliver compactness and fiber cohesion, enhancing fiber control and avoiding unexpected draft during the roving process [12]. Similarly the loose roving with a low twist density suffers unexpected draft problems; however, the compact roving with a high twist density is also hard to get a uniform draft, even failing to be draft apart. To overcome this conflict, false twisting is also applied to produce a relatively strong roving with low twist density.

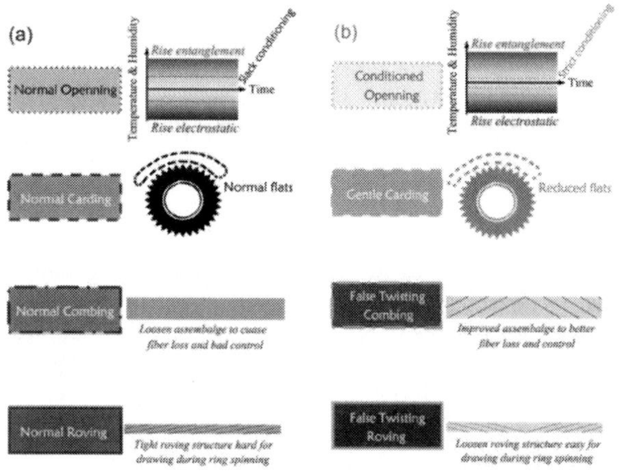

Fig.1 Processing model of fiber damage minimization

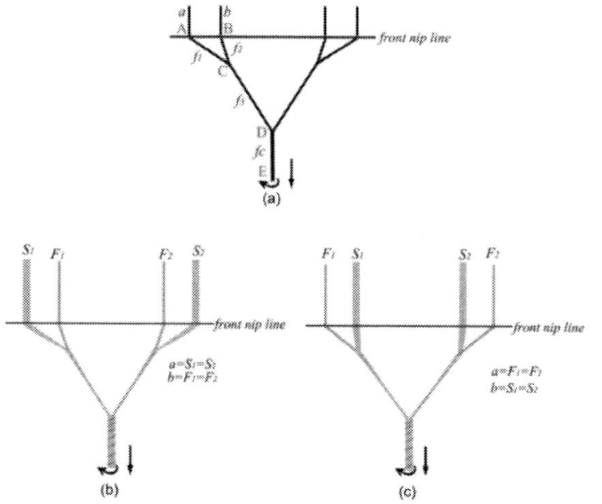

Fig.2 Tension optimization to avoid staple breakage.

2.3 Protective Model for Vulnerable Fiber Strand Spinning

To protect super high-count spinning strand from end-breakage, embeddable and locatable spinning (denoted as ELS, also known as "RUYI SPUN") system is developed by employing two water-soluble multi-filaments as supporters [13]. This spinning system resembles a combination of Sirofil [14] and Sirospun [15] processes in a single spinning step (Fig.2). The classical ELS system has two types: the first one is featured as the two staple fiber strands (S_1 and S_2) locating at outer position and the two filaments (F_1 and F_2) feeding between them (Fig.2b); the second one is featured as the two filaments (F_1 and F_2) locating at outer position and the two staple fiber strands (S_1 and S_2) feeding between them (Fig.2c).

Assuming f_1, f_2, f_3, and f_c are spinning tensions of section AC, BC, CD and DE respectively; v_1 and v_2 are the velocities of the component a and b respectively; ρ_1 and ρ_2 are the densities of the components a and b respectively;

R_1 and R_2 are the radius of the components a and b respectively. According to previous theoretical analysis, we can get:

$$f_1 \propto f_3 \propto f_c \tag{1}$$

$$f_2 \propto f_3 \propto f_c \tag{2}$$

$$\frac{f_1}{f_2} = \frac{R_1^2 \rho_1 v_1^2}{R_2^2 \rho_2 v_2^2} = \frac{R_1^2 \rho_1}{R_2^2 \rho_2} \tag{3}$$

Equations (1) and (2) indicate that end-breakage of staple fiber strand can be decreased by reducing main composite spinning tension f_c (i. e., using light traveler, low spinning speed). Equation (3) indicates that spinning tension distribution on components a or b is directly proportional to the component dynamic linear mass. When the composite spinning tension is constant, both the staple fiber strand count reduction and the filament count increase can further decrease the spinning tension on the staple fiber strand. Therefore, it seems to be irrelative with symmetric component location changes for the spinning tension optimization to protect staple strand from breakage.

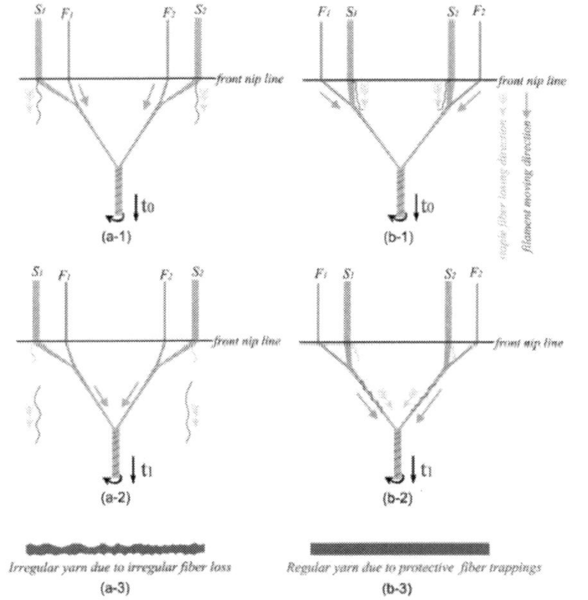

Fig.3 Systematic trapping ability to avoid fiber loss

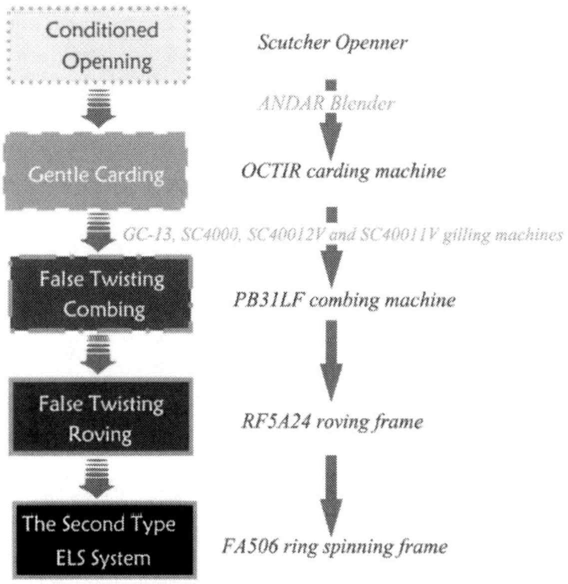

Fig.4 Practical systematic processing procedures

Fig.3 illustrates the comparison of systematic trapping abilities in the first (Fig.2b) and second (Fig.2b) type ELS systems. During the super high count yarn production, ELS system should have an excellent staple fiber trapping capacity. However, the first type ELS system cannot converge the staple fiber losing direction and the filament moving direction, failing to trap losing fibers by the filaments in the yarn formation zone, shown in Fig.a1 and a2. This irregular fiber loss in first type ELS system will cause an uneven fiber mass distribution in yarn lengthwise, increasing composite spun yarn unevenness (Fig.a3). In contrast, the second type ELS system can combine the staple losing fibers with the filaments in time (Fig.b1), twisting the losing fibers with the composite spinning strand again (Fig.b2) in the convergent section of composite spinning strands to avoid fibers distracting from the ELS yarn formation zone. The avoidance of fiber loss secures all the strand fibers spun into yarn with a high regularity (Fig.b3). Therefore, the second type ELS system shows improved performance in super-high count yarn production than the first type one.

3. Practical Systematic Production Solution

Under the guidance of aforementioned theoretical analysis, systematic processing procedures including strict conditioning opening, low damage carding, false twist combing, false twist roving, and the second type ELS were employed to collaboratively producing super-high count yarn of Vicuna fibers, shown in Fig.4. The used vicuna fibers had parameters below: average diameter 13.1 μm, tenacity 1.17 cN/dtex, and breaking elongation 9.23%. After scouring, Vicuna fiber was entangled, and contains a small amount of grass clippings. Therefore opening was applied to separate the entangled fibers. To minimize fiber damage, which tends to happen with a Scutcher openner, the temperature and humidity in the opening room was controlled at 25±1°C and 68±1%, respectively. The low damage carding technology was also used to minimize damage to both fiber surface and length, and to reduce the potential for fiber damage in subsequent processing. In the combing process, a false twist device was used to improve the strength and handling of the combed slivers. This could help ensure a high quality vicuna top was produced for roving and spinning. Finally, the second type ELS system was established on a FA506 ring frame: two outer filaments to support the two inner vinuca strands during spinning. According to above theoretical analysis, the vicuna strand ends-down can be significantly reduced to secure the spinning continuity. In addition, with the filament supporter, yarn twist can be reduced during spinning, endowing the resultant yarn with improved softness [16].

During the processing, vinuca fiber properties were monitored by the following instruments: Laserscan (CSIRO) for fiber diameter measurement, SEM (Germany Zeiss) for fiber surface characterization, and LLY- 06E electricity for vicuna fiber evenness and strength. Resultant spun yarns were stored for at least 24 hours under a standard atmospheric condition (20±2% RH and 65±2 ℃). Then, each bobbin of single cotton yarns was tested repeatedly in terms of unevenness CV, tensile properties and yarn count. YG029-I Automatic Single-yarn Tensile Tester was employed to test the tensile properties of the yarns; each sample was tested for 20 times with a test speed of 500 mm/min and a gauge length of 500 mm. YG136 evenness tester was used to test yarn unevenness; the testing speed was 400 m/min, and the testing length for each samples was 400 meters at each designated testing positions. All the tests were under the standard condition.

4. Results and Discussion

4.1 Vicuna Fiber Properties after Carding

Gentle fiber opening and carding were used by reducing the flat number from 108 to 54 on the carding machine to minimize fiber damage. By this way, the entangled vicuna fiber tuft was individualized and separated out gently (Fig.5). Thus short fibers showed a very slight increase after the gentle carding process (Table 1).

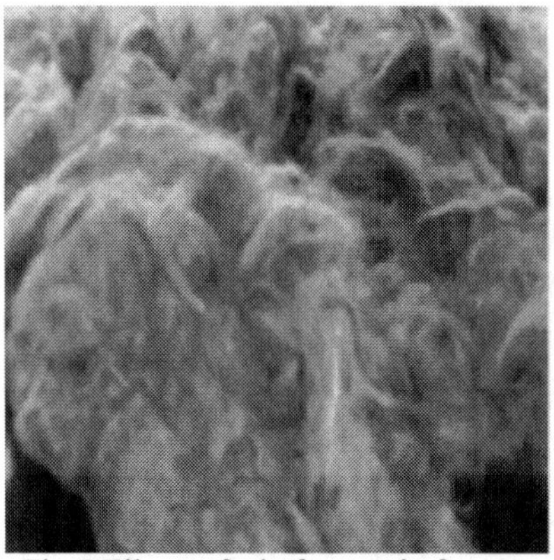
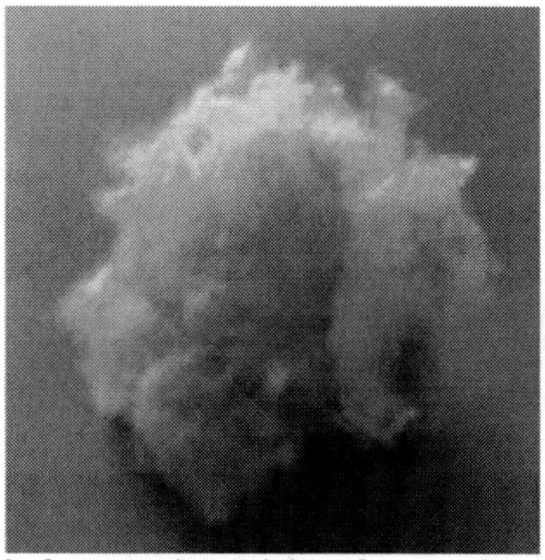

Fig.5 Fiber tufts before and after opening; Left: before opening; Right: after opening

Table.1 Vicuna fiber properties before and after carding

items	Before carding	After carding
Ave. diameter /um	13.0	14.5
Diameter CV/%	20.2	19.2
Barbe Length /mm	30.7	25.7
Length CV/%	41.2	38.2
≤30mm/%	55.1	65.1
≤20mm/%	14.2	30.2
Impurities/%	0.03	0.00
Coarse fibre percentage/%	0.4	0.1

In specialty, vicuna fiber surface damage were largely reduced during the gentle carding processing than that during the normal carding process (Fig. 6), achieving the fiber damage minimization effect.

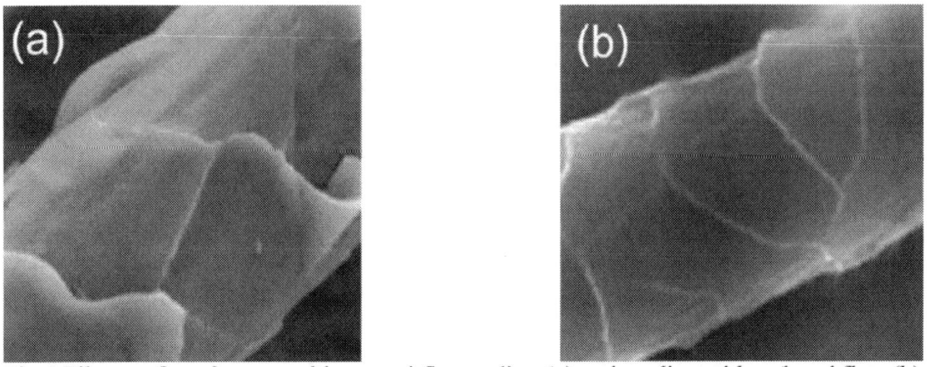

Fig.6 Fiber surface damage with normal flat carding (a) and carding with reduced flats (b)

4.2 Influence of False Twisting on Vicuna Roving Properties

Compared with the conventional roving, the false twisting technology could better the fiber control for roving process; therefore a high draft were allowed to get a high count roving for the processing with false twisting (shown in Table 2).

Table.2 Roving parameters

Experimental scheme	Roving draft	Roving weight /g.m-1	Spinning draft
Processing with false twisting	11.8	0.22	18.48
Processing without false twisting	10.68	0.25	21

The high draft of roving could reduce the draft ratio of spinning to get an improved spun yarn irregularity (Table 3).

Table.3 Yarn unevenness parameters

Experimental scheme	CV /%	Thin -50% / km	Thick +50% / km	Nep +200% / km	CVb /%
Processing with false twisting	20.9	538	171	152	0.95
Processing without false twisting	21.8	600	210	140	1.31

4.3 Vicuna Super High Count Yarn Properties

The qualified high count roving could produce the essential material for high count yarn production in the same ratio-drafted conventional ring frame (Table 4). However the produced yarn extreme counts were limited by the end-breakage and fiber loss due to the weak adherence and low tenacity of the high count spinning strand.

Then the second type ELS system was adopted to break the limitation: water-soluble PVA filaments were used as F1 and F2; two qualified high count roving used as S1 and S2. The spun yarn results were listed in table 4. Obviously, the second ELS system largely reduced the limitation of the spun yarn count; the second ELS spun yarn also showed a much higher strength than the conventional spun ones. It should be noted that, in table 4, the second ELS spun yarns were vicuna/PVA composite yarns, since the PVA filaments were not dissolved before yarn testing. The composite yarns were sufficiently strong for subsequent processing. Typically dissolving was conducted to remove the PVA filaments off the vicuna/PVA composite yarns after fabric formation. The disappearance of PVA filaments components increased the yarn looseness and count in the resultant fabric. Finally, the super high count vicuna fiber yarn was formed into the resultant ultra tender fabric structure.

Table.4 Vicuna fibre yarn count and strength comparison

Material (lot.)	Conventional ring spinning			The Second ELS System		
	Yarn count /tex	Twist level	Yarn strength /cN	Yarn count /tex	Twist Level	Yarn strength /cN
Roving produced with false twisting technology	16.6×2	740	132	10.0×2	900	178
Roving produced without false twisting technology	20.0×2	650	178	12.5×2	820	228

5. Conclusion

To overcome processing and spinning difficulties of vulnerable fibers, fiber damage minimization model was built to protect fibers during opening, carding, roving processes; the fiber trapping and protective model was also theoretically developed to avoid end-breakage and fiber loss during the strand twisting to form yarn. According to theoretical model guidance, super high count yarn of vicuna fibers could be produced by employing strict condition opening, low damage carding, false twist roving and embeddable and locatable spinning technologies.

6. Acknowledgement

We greatly acknowledge the support from the National Natural Science Foundation of China (Project No. 51403161).

References

[1] Mahapatra NN. Processing of vicuna fibres in textile industries, Colourage: 2014; 61: 63-66.

[2] Brack N, Lamb RN, Pham D, Phillips T, Turner P. Effect of physical processing on the wool fiber surface, Textile Res. J: 2001; 71: 911-915.

[3] Xia Z. Mechanism of moisture and temperature influence on cellulose textile fibers' tensile properties and its application in smooth yarn production, Doctoral Thesis of Donghua University: 2012; 48-49.

[4] Gomez N, Juliá R, Erra P, Naik A. Effect of quaternized amphiphilic peptides on pilling and electrostatic diacharge of wool fabrics, Textile Res. J: 1994; 64: 648-652.

[5] Chaudri MA, Whiteley KJ. The influence of natural variations in fiber properties on the felting characteristics of loose wool, Textile Res. J: 1970; 40: 297-303.

[6] Happey F. 1982, Contemporary Textile Engineering, The Greystone Press, New York: p.108-109.

[7] Carnaby GA. Fiber breakage during carding: Part I: theory, Textile Res. J: 1984; 54: 366-369.

[8] Carnaby GA, Burling-Claridge RG. Carding of tender wool: part I: theory, Textile Res. J: 1996; 66: 90-98.
[9] Simpson J, DeLuca LB, Fiori LA. Effect of Carding Rate and Cylinder Speed on Fiber Hooks and Spinning Performance for an Irrigated Acala Cotton, Textile Res. J: 1967; 37: 504-510.
[10] Atkinson K; Stannard W; Bagshaw K. A Method for the Continuous Measurement of Combing Noil and Its Application in Wool Topmaking, Textile Res. J: 2010; 80: 439-447.
[11] Erdumlu N; Oxenham W; Ozipek B. The impact of combing and processing parameters on the structure and properties of fine count vortex yarns, Textile Res. J: 2013; 83: 396-405.
[12] Jeddi AAA, Oxenham W. The Potential of False Twist to Assist in Handling Low Strength Cotton Slivers, Textile Res. J: 1995; 65: 423-426.
[13] Xu W, Xia Z, Wang X, Chen J, Cui W, Ye W, Ding C, Wang X. Embeddable and locatable spinning, Textile Res. J: 2011; 81: 223-229.
[14] Xia Z, Liu H, Huang J, Gu S, Xu W. A study on the influence of a surface-contacting spinning strand on yarn appearance during sirofil spinning, Textile Res. J: 2015; 85: 128-139.
[15] Xia Z, Feng Y, Guo Q, Ye W, Xu W. A comparative study of hair trapping by a short grooved surface during conventional and siro-spinning, Textile Res. J: 2015; DOI: 10.1177/0040517515619355.
[16] Hearle JWS, Gupta BS, Merchant VB. Migration of fibres in yarns, Part 1: characterization and idealization of migration behavior, Textile Res. J: 1956; 35: 329-334.

Influence of Different Measuring Tools' Width on Accuracy of Measurement Data

Li-Wen He[1], Zhan-Wei Bao[2], Meng-Jie Guo[1], Xiao-Mei Shang[1*]

[1]*College of Textile and Clothing Engineering, Soochow University, Suzhou, Jiangsu 215021, China*
[2]*Research and Development Center of Jiangsu Hongdou Industry Co., Ltd., Wuxi, Jiangsu 214000, China*

[*]Corresponding author's email: shangxiaomei@suda.edu.cn

Abstract

The paper is based on the method of manual measurement to conduct the experiment,. It studies the influence of different measuring tools' width on the accuracy of measurement data, which were the neck base girth and the armscye girth. In the research, the measurement of the neck base girth and the armscye girth were obtainedt by using measuring tools of different width that included 2cm, 1cm and 0.5cm. The influence of the experimental results based on three methods were studied and analyzed. To obtain experimental data by repeating the experiments many times, the average, the standard deviation and the AVEDEV function value of the measurement data was calculated using a mathematical statistic's software. ANOVA test and statistical hypothesis test was applied to the measurement data. The research found that different measuring tools' width had remarkable effect on the measurement data of the neck base girth and the armscye girth, the measuring tool is also wider or thinner, the dispersion degree of the measurement data is larger, the accuracy of the measurement data is smaller.

Keywords: Anthropometry; Measuring Tool; Width; The neck Base Girth; The Armscye Girth

1. Introduction

Anthropometry is necessary to design and to make clothing in the garment industry, and it is the primary foundation for making apparel [1]. Measuring the human body manually in the present garment industry involves using measuring tools that directly touch the human body in order to obtain body data. Even if normative methods of measurement are taken, the influence of the objective factors of measuring tools' on the accuracy of the body measurement data is inevitable [2, 3]. As people improve their standard of living, their requirement for coats grow. A standard coat needs to be both beautiful and of good fitting. As such the accuracy of the human body measurement data needs to be improved constantly [4]. The neck base girth and the armscye girth have an important effect on fitness and pattern of the collar and sleeve for the coat, so they are important measurement data [5]. This paper mainly studied the influence of using different wide measuring tools to obtain data of the neck base girth and the armscye girth. In the manual measurement experiment, different wide measuring tools were used namely 2cm, 1cm and 0.5cm to obtain the measurement data of the neck base girth and the armscye girth. of the influence of the three methodson the experimental result was studied, then analyzed for their degree of influence. To obtain experimental data by repeating the experiments many times, the average, standard deviation and AVEDEV function value of the measurement data was calculated using a mathematical statistic's software. ANOVA test and statistical hypothesis test of the measurement data was applied and used to evaluate the influencing degree of the measuring tools' width on the accuracy of the measurement data. It can provide a reference for constantly improving the accuracy of the human body measurement data and measuring methods [6].

2. Anthropometric Experiment

2.1 Experimental Samples

The participants: 10 women were selected randomly from Soochow University, their age was between 19 and 25, the height was between 155cm and 170cm, the number was from No.1 to No.10.

The measurers: 10 measuring people were selected randomly that were trained uniformly to measure human body from the university, had the measuring experience, measured body expertly and the accuracy of the measurement data obtained up to the standard.

The recorders: 10 recording people were selected randomly that were trained uniformly to record the measurement data, had the anthropometric knowledge and recorded data expertly. They had randomly a combination with 10 measurers to accomplish the work of measuring and recording.

2.2 Experimental Environment and Measuring Tools

Experimental environment: the experimental site was in the anthropometric laboratory of Soochow University, the temperature was 27℃±3℃ and the humidity was 60%±10% of indoor. Besides the light intensity was moderate in order to ensure legible reading [7]. In addition to the experiment request variable, the experiment should be carried under the same time, the same environment arrangement and the same operation. During the environment processing, in order to obtain accurate results should guarantee the measuring equipment consistent.

Measuring tools: the experimental tools included the tapes that their width were 2cm, 1cm and 0.5cm, the rulers, the coordinate papers, the special measurement forms, the recording boards, the recording pens, the special measurement garments and so on. The measurers and recorders wore the special garments and gauze masks, the participants wore the special measurement tights. The measurers took the measurement tools. The recorders took the special measurement forms, the recording board and the recording pen.

2.3 Dimension Definition and Measuring Methods

According to GB/T16160-2008 [8] that is Location and method of anthropometric surveys for garments. The neck base girth is measured throughthe seventh cervical vertebra point surface, the neck base girth surface and the anterior neck landmarks with a tape. The armscye girth is measured with a tape, the right acromion landmark is a starting point, passing over the anterior and the back armpit landmark, then up to the starting point.

1) The method of measuring the neck base girth

The participant stood erect, looking straight ahead. The heels were together with the weight distributed equally on both feet. The shoulders and upper extremities were relaxing with the palms facing the thighs [9]. In the whole measurement process, the measurement was at the maximum point of quiet respiration, and the participant was in the same standing posture.

The participant kept the standing posture. The measurer with a 2cm tape stood in front of the participant, whose left hand put the zero graduation line of the tape in the anterior neck landmarks of the participant and right hand controlled the tape. Making the tape pass over the right lateral neck landmark, the seventh cervical vertebra and the left lateral neck landmark, then fixed in the anterior neck landmarks. This scale of the tape's top edge was the neck base girth [10], as shown in Fig.1. The participant kept the standing posture and was measured repeatedly 3 times. Then used 1cm and 0.5cm tape to replace it with the same measuring method. The reading must be the top edge number of different measuring tools closed to the neck base dimension.

2) The method of measuring the armscye girth

The participant stood erect in side, looking straight ahead. The heels were together with the weight distributed equally on both feet. The shoulders and upper extremities were relaxing with the palms facing the thighs. The measurement was at the maximum point of quiet respiration. In the whole measurement process, the participant was in the same standing posture.

The participant kept the standing posture. The measurer with the 2cm tape stood in side of the participant, whose hand took the starting point of the tape in the right acromion landmark, the other hand made the tape pass over the anterior and back armpit landmark, then up to the starting point of the acromion landmark. This scale of the tape's top edge was the armscye girth [11], as shown in Fig.2. The participant kept the standing posture and was measured repeatedly 3 times. Then made 1cm and 0.5cm tape replace it with the same measuring method. The reading must be the top edge number of different measuring tools closed to the armscye dimension.

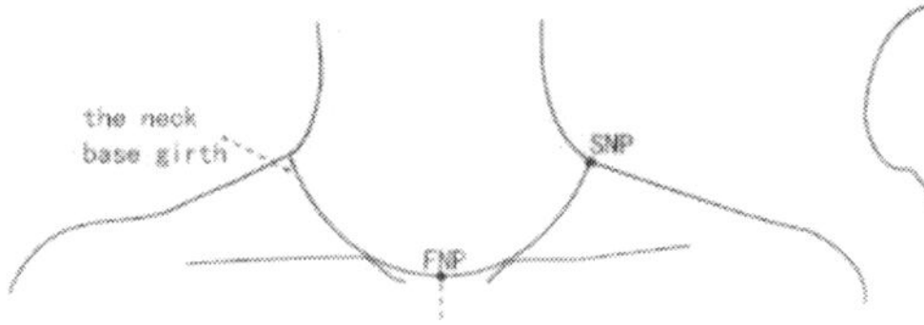

Fig.1 The sketch of the neck base girth

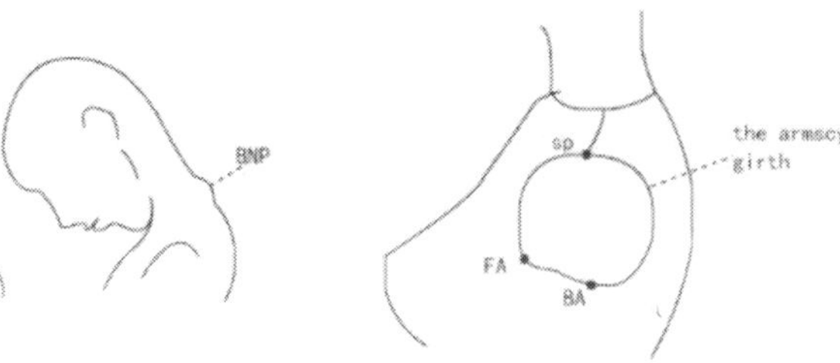

Fig.2 The sketch of the armscye girth

2.4 Analysis of Measurement Data

The measurement data of the neck base girth and the armscye girth obtained by using different wide measuring tools that were 2cm, 1cm and 0.5cm. In the experiment, the measurers were 10, every measurer measured the same participant 3 times. So the data sets of a sample were 30, there were 10 participants, the accumulated data sets were 300. The void data and the abnormal data were took out, finally the efficient measurement data sets of doing the experiment were 292. Then the average of the measurement data were calculated, which were obtained by using different methods to measure the neck base girth and the armscye girth.

3. Accuracy of Measurement Data

According to the measuring experience and analysis of the measurement data, it found that the width of the measuring tools had little influence on the measurement data of horizontal dimensions. So the chest girth was selected as the reference object, the neck base girth and the armscye girth were selected as the research objects. the measurement data got by three methods were dealt with and analyzed, the result of stability and volatility of the measurement data was used to judge whether beyond the range of allowable error. Then the average, the standard deviation and the AVEDEV function value of the measurement data in different measuring methods were calculated and analyzed by Excel software. If the standard deviation is big, the difference between the measurement data sets and the average is big. The AVEDEV function value is the average of the absolute deviations between the data sets and its average. If the AVEDEV function value is big, the dispersion of the measurement data sets is big.

3.1 Comparison of Stability on Measurement Data

Firstly, the 10 participants were signed from 1 to 10,which were noted as No.1~No.10. Three methods that used three measuring tools of the 2cm wide tape, the 1cm wide tape and the 0.5cm wide thin rope were noted as A1, A2, A3. Through the clustering analysis of body shape, it found that the 10 participants could be divided into three classes. The participant No.1, No.3, No.7 belonged to the shape 155/82A, the participant No.4, No.5, No.8, No.9 belonged to the shape 160/84A, the participant No.2, No.6, No.10 belonged to the shape 165/88A. A participant chosen as the research object from each shape class, which were No.1, No.5, No.2. According to GB/T1335.2-2008 [12] that is The Standard sizing systems for garments and the requirements of the accuracy of the human body measurement data for garments, it transformed the grading values of all dimensions into the scope of the measurement data as a condition in terms of judging stability. So the scope of chest girth, the neck base girth and the armscye girth were 2cm, 1cm and 0.5cm. In the garment industry, the measuring tool of 1cm wide tape is used generally to measure human body. So the paper described the wave curve of the measurement data obtained by the method A2, as shown in Fig.3 and Fig.4.

As can be seen from the Fig.3 and Fig.4, the chest girth measurement data of 3 samples were fluctuating in the error range of less than 2cm. But the measurement data of the neck base girth and the armscye girth were volatility in the range of 4cm, it was more than the allowable error range. So the stability of the measurement data of the neck base girth and the armscye girth is poor, and the data volatility is large relatively.

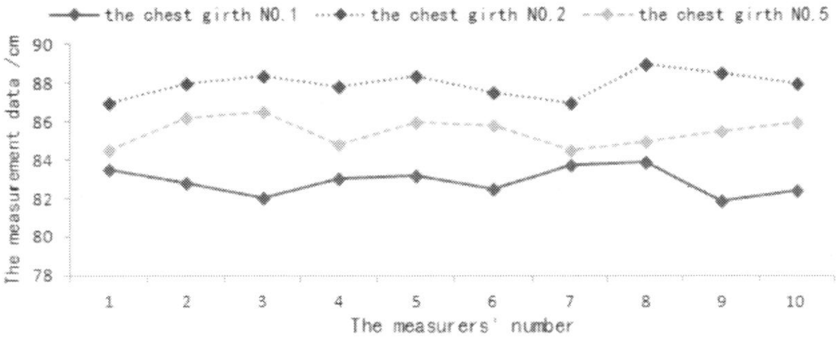

Fig.3 The wave curve of the chest girth by the way of A2

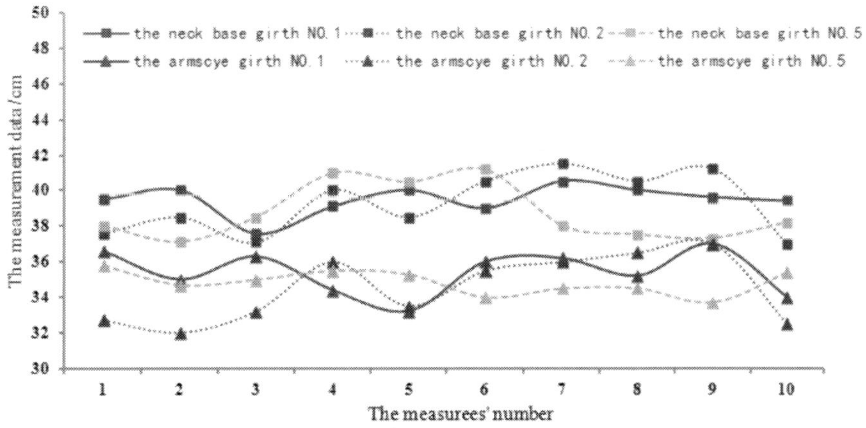

Fig.4 The wave curve of the neck base girth and the armscye girth by the way of A2

3.2 Comparison of Discreteness on Measurement Data

The measurement data got by A1, A2, A3 had quantitative analysis of the discreteness degree with Excel software. The average, the standard deviation and the AVEDEV function value of the data sets were calculated, then compared the difference of the discreteness degree among three methods, the statistical result such as Table 1.

The following conclusions can be gotten from Table 1,

1) The standard deviation of the chest girth was less than 1cm, the AVEDEV function value was less than $0.5cm^2$. According to the wave curve of Fig.3, the discreteness degree of the chest girth was small. The data sets obtained by different measurers were fluctuating in the allowable error range, the measurement data was accurate relatively. So the standard deviation and the AVEDEV function value of the chest girth's measurement data could be a standard, which to judge the discreteness degree of the data sets. That provided a reference for the measurement data of the neck base girth and the armscye girth.

2) The standard deviations of the neck base girth and the armscye girth were both more than 1cm, and the AVEDEV function value were both more than $0.5cm^2$. According to the discreteness degree of the chest girth's measurement data could speculate that the discreteness degree of the neck base girth and the armscye girth was large, and the measurement data was not accurate.

3) For 3 measuring samples, the standard deviation and the AVEDEV function value of the measurement data abtained by the method A2 were less than the method A1 and the method A3. The same data sets obtained by three methods had the difference, which shown that the 1cm wide tape was the right relatively measuring tool, and wider or thinner measurement tools had an influence on the stability and the accuracy of the measurement data on the neck base girth and the armscye girth.

Table.1 The discreteness degree of measurement data got by different methods

The Dimension	The Samples	Measuring Methods	Average	Standard Deviation	AVEDEV
The chest girth	NO.1	A2	82.89	0.69	0.47
	NO.2	A2	87.96	0.65	0.31
	NO.5	A2	85.48	0.73	0.42
The neck base girth	NO.1	A1	39.15	0.90	0.68
		A2	39.47	0.80	0.55
		A3	39.34	1.24	1.09
	NO.5	A1	39.75	1.73	1.45
		A2	39.24	1.62	1.20
		A3	40.50	1.69	1.50
	NO.2	A1	39.31	1.90	1.51
		A2	38.73	1.56	1.30
		A3	40.52	1.84	1.58
The armscye girth	NO.1	A1	36.29	1.17	0.97
		A2	35.54	1.11	0.92
		A3	36.09	1.28	1.03
	NO.5	A1	35.81	2.54	2.06
		A2	34.71	1.86	1.65
		A3	36.39	2.25	1.81
	NO.2	A1	36.08	0.69	0.58
		A2	34.78	0.67	0.55
		A3	36.12	0.86	0.60

3.3 ANOVA Test and Statistical Hypothesis Test of Measurement Data

1) In the paper, the two factors individual analysis of variance was used to analyze the measurement data. The experiment had two factors which were A and B. The factor A was the measuring method and divided into 3 levels, the factor B was the participant sample and divided into 10 levels. Every level had one observed value for the same sample. The average of every level measured repeatedly was obtained, every observed value was affected by the factors A, B and the random error. So every experimental group added up to 30. The participant samples were 3. The measurement dimensions were 2. There were 180 the observed value in the experiment.

2) Verification principle about the two factors individual analysis of variance for the experimental results,

$$F = MS_t^2/MS_e^2 \tag{1}$$

$$MS_t = SS_t/df_e \tag{2}$$

$$MS_e = SS_e/df_e \tag{3}$$

$$SS_t = \sum_{i=1}^{k} n_i(\bar{x}_{in} - \bar{x}_{...})2 \tag{4}$$

$$SS_e = \sum_{i=1}^{k}\sum_{j=1}^{ni}(\bar{x}_{in} - \bar{x}_{...})2 \tag{5}$$

$$df_t = k\text{-}1 \qquad df_e = N\text{-}k \tag{6}$$

3) Statistical hypothesis test, in the two factors individual analysis of variance for the experimental results about the model of F-test, there were two hypotheses that included the null hypothesis H_0 and the alternative hypothesis H_A, that ensured the Sig. was $\alpha=0.05$ and $\alpha=0.01$. According to the null hypothesis H_0, the F test statistic of the experimental data could be calculated, then compared the F value with the $F_{0.05}$ value and the $F_{0.01}$ value, deduced if the null hypothesis H_0 is right, and decided if the null hypothesis H_0 could be accepted [13]. By designing the above-mentioned 6 experiments, it had separately the two factors individual analysis of variance and the statistical hypothesis test, the statistical result was shown Table 2. Then it could compare Sig. with the defined α and judge the significant degree of the data sets' difference further.

Table.2 Two factors individual analysis of variance for the measurement data

The Dimensions	The sample	The factor	Sum of squares	Df	Mean square	F	Sig.
The neck base girth	NO.1	A	16.82	2	1.87	2.89	0.003
		B	0.52	9	0.26	0.40	0.684
	NO.2	A	70.46	2	7.83	12.39	0.001
		B	8.03	9	4.02	6.35	0.732
	NO.5	A	77.37	2	8.34	19.81	0.000
		B	16.68	9	5.60	10.42	0.571
The armscye girth	NO.1	A	35.17	2	3.91	6.43	0.021
		B	1.91	9	0.96	1.57	0.421
	NO.2	A	43.76	2	4.86	3.79	0.010
		B	0.96	9	0.48	0.38	0.690
	NO.5	A	9.21	2	4.60	27.06	0.001
		B	7.49	9	0.94	5.54	0.346

The following conclusions can be gotten from Table 2,

1) In the 6 designing experiments, the corresponding Sig. value of the factor B was more than 0.05. It turned out that the difference of different measurers was not significant. That was to say, the measuring technology of 10 measurers didn't have the significant difference.

2) For the neck base girth, the corresponding Sig. value of the participants No.1, No.2, and No.5 was less than 0.01. The results showed that the measuring methods of using different wide measuring tools had highly significant influence on the measurement data.

3) For the armscye girth, the corresponding Sig. value of the participants No.2 and No.5 was less than or equal to 0.01. The results showed that the measuring methods of using different wide measuring tools had very significant influence on the measurement data. The corresponding Sig. value of the participants NO.1 was more than 0.01 but less than 0.05. The results indicated that the measuring methods of using different wide measuring tools had significant influence on the measurement data.

In conclusion, three measuring methods of three different wide measurement tools had the significant influence on the measurement data of the neck base girth and the armscye girth. The discreteness degree of the measurement data sets obtained by the method A2 which was the 1cm wide measuring tool was smaller, the accuracy of the measurement data was higher.

4. Conclusion

The paper mainly made the human body measurement data of the neck base girth and the armscye girth act as the research objects by the manual measurement experiment. That used three measuring methods of different wide measuring tools that were 2cm, 1cm and 0.5cm to get the measurement data, took and designed the experiment. Making use of the Excel software, it could calculate and analyze the average, the standard deviation and AVEDEV function value, and had the two factors individual analysis of variance and statistical hypothesis test by the SPSS software. The study found that measuring methods of using three different wide measurement tools had the significant influence on the measurement data of the neck base girth and the armscye girth, and the discreteness degree of the measurement data sets obtained by the method A2 which was the 1cm wide measuring tool was smaller. The accuracy of the measurement data was highest. The conclusion of this experiment provides a reference for improving the accuracy of the human body measurement data in the manual measurement for garments.

References

[1] Guo PP, Liu GL. Reliability Analysis of the Data Based on the Measurement of Young Female Body [J]. Journal of Beijing Institute of Clothing Technology: 2015; 35 (2): 35-39.

[2] Jia SL, Tian YN, Shang XM. The effect of measurers' view angle deviation on measurement data [J]. Advanced Materials Research, Materials Science and Engineering Technology: 2014; 936: 2237-2242.

[3] He LW, Bao ZW, Guo MJ, Shang XM. Research on the Conversion Model of Perineum Height in Different Measurement Methods [J]. Journal of Zhejiang Fashion Institute of Technology: 2016 (4); 18-22.

[4] Cao B. The Study on Generating and Proving of Intelligent Pants Pattern for East China Domain Youth [D]. Suzhou: Soochow University: 2014.

[5] Tian YN, Le YM, He LW, Shang XM. The measure method's optimization for neck base girth through controlling the benchmark [J]. Textile Reports: 2016 (9); 35-39.

[6] Hong ZL, Li PW, Shang XM. Psychological Factors Affect Measured Data of Human Body [J]. Textile Bioengineering and Informatics Symposium Proceedings: 2015; TBIS2015: 122-127.

[7] He YJ, Shi XQ, Wang HF. Camber features of women's shoulder and influence thereof on fitted clothes' structure [J]. Journal of Textile Research: 2016; 37(10): 94-99.

[8] GB/T16160-2008. Location and method of anthropometric surveys for garments [S].

[9] Zhang JH, Wang HF. Study of womens' shoulder based on 3-D body measurement [J]. Journal of Textile Research: 2011; 32 (5): 95-97.

[10] Hong ZL, Le YM, Shang XM, et al. Human Shoulder Morphology Influences the Acquisition of Acromion Points [J]. Zhejiang Fashion Institute of Technology: 2016; (3): 25-30.

[11] Tian YN, He LW, Shang XM. Influence of Measurement Methods on the Accuracy of Measured Data [J]. Journal of Beijing Institute of Clothing Technology: 2016; 32 (4): 25-31.

[12] GB/T1335.2-2008. The standard sizing systems for garments-Women [S].

[13] Jin AY, Gao PP, Shang XM.Optimization design of cutting path based on characteristics of garment pattern [J]. Journal of Textile Research: 2015; 36(9): 89-93.

Preparation of PANI/Cu/carbon Nanofibers for the Detection of Hydroquinone

Da-Wei Li, Qing-Qing Wang, Feng-Lin Huang, Qu-Fu Wei*

Key Laboratory of Eco-Textiles of Ministry of Education, Jiangnan University, Wuxi 214122, P. R. China

*Corresponding author's email: qfwei@jiangnan.edu.cn

Abstract

PANI/Cu/carbon nanofibers (PANI/Cu/CNFs) were prepared by a facile and efficient method, which were further applied to the laccase based biosensor for the detection of hydroquinone. The morphology and composition of the fabricated nanocomposite were characterized by scanning electron microscopy (SEM), transmission electron microscopy (TEM), X-ray diffraction (XRD), and Fourier transform infrared spectroscopy (FTIR). The electrochemical behavior of the novel biosensor was studied using cyclic voltammetry (CV) and chronoamperometry. The results showed that the constructed biosensor possessed good selectivity, reproducibility, repeatability, high sensitivity (41.65 μA/mM) and a low detection limit (2.37 μM).

Keywords: PANI; Copper; Electrospinning; Carbon Nanofiber; Biosensor; Hydroquinone

1. Introduction

In recent years, as a highly efficient facility with high sensitivity, simplicity, fast response and excellent selectivity, electrochemical biosensors have become of interest in various areas such as viruses, environmental pollution monitoring, food, and so on [1-2]. Laccase (Lac) is a blue oxidase with the ability to catalyze the oxidation of hydroquinone, a promising candidate for the detection of hydroquinone. The most vital element in constructing a biosensor with excellent biosensing performance is the selection of materials used for the immobilization of Lac. The materials applied to biosensors include metals, metal oxides, polymers, and carbon-based materials [3-5].

Because of its good electrical conductivity, high surface area, porosity, uniformity and mechanical strength, carbon nanofibers (CNFs) were used in this research. Carbon nanotubes (CNTs) and CNFs attract the highest attention among carbon materials. While compared to CNTs, CNFs can be synthesized easier and has many more edge sites which is far superior to CNTs. It has been proven that CNFs has a high surface area and is much more suitable for the development of biosensors [6].

Nowadays, metal nanoparticles especially copper nanoparticles have been found with all kinds of inspiring properties such as good biocompatibility, catalytic activity, and electrical conductivity [7]. The copper nanoparticles are known to be relatively cheap and can also increase response current. CNFs have been proved to be an ideal material for supporting nanosized metallic particles in electrodes [8]. The synthesis of CuCNFs can combine the advantages of Cu and CNFs together. In this work, CuCNFs composite nanofibers were chosen as the immobilization matrix of enzymatic biosensor.

Polyaniline (PANI), a conducting polymer, is one of the most important conducting polymers. PANI is known for its facial preparation at low cost, high polymerization yield, high electrical conductivity, good redox reversibility and environmental stability [9]. In this work, novel PANI/CuCNFs nanocomposite was prepared by in situ polymerization. The application of conducting polymer in the matrix of CuCNFs for the fabrication of electrode was expected to improve the analytical performance of biosensor. Besides, it was beneficial for the immobilization of enzyme, thus promoting electrical conductivity and elecrocatalytic property of enzymatic electrodes.

2. Experimental

2.1 Materials

Aniline monomer (An), ammonium persulfate (APS), Hydrochloric acid, $(CH3COO)_2Cu{\bullet}H_2O$, CH3COOH, CH3COONa, potassium ferricyanide ($K_3[Fe(CN)_6]$), potassium hexacyanoferrate ($K_4[Fe(CN)_6]{\bullet}3H_2O$), N,N-Dimethylformamide (DMF) with a purity of 99.5%, hydroquinone, guaiacol, vanillin, 3,5-dinitrosalicyclic acid, phenol ethanol and catechol were purchased from Sinopharm Chemical Reagent Co., Ltd. (Shanghai, China). Polyacrylonitrile (PAN, average molecular weight = 79100) powder and laccase (from Rhusvernicifera, crude acetone powder, activity ≧50 units/mg) were supplied by Sigma Aldrich Chemical Co., Ltd. (St. Louis, MO, USA). Nafion (5% w/w) was obtained from E. I. DuPont Company. All the reagents were used as received without further purification. The supporting electrolyte consisted of 0.1 M acetate buffer solution was prepared by mixing stock solutions of CH3COOH and CH3COONa. Hydroquinone solutions were freshly prepared before use. Deionized water (DIW) was used in all experiments.

2.2 Preparation of CuCNFs

First, 10 wt% PAN and $(CH3COO)_2Cu{\bullet}H_2O$ powder were added into DMF under stirring with the weight ratio of 2:1. Being stirred for 12 h successively, the solution was poured into a syringe for electrospinning. The membrane of PAN/(CH3COO)2Cu•H2O composite nanofibers was prepared through electrospinning. Then, CuCNFs were obtained by pre-oxidation and carbonization. The membrane was pre-oxidized at 280 °C for 1 h in the air and carbonized at 800 °C for 2 h in nitrogen atmosphere with a heating rate of 2 °C/min in Muffle furnace. Then it recovered to room temperature, CuCNFs were well prepared.

2.3 Preparation of PANI/CuCNFs

PANI/CuCNFs were synthesized by in situ polymerization of aniline on the surface of CuCNFs. HCl and APS were used in the reaction as a catalyst and an oxidant, respectively. The polymerization of aniline was carried out in an ice/water bath of aniline solution at the temperature of 0 - 5 °C. First, 0.05 g of CuCNFs were added into 230 mL of 1.4 M HCl solution with 0.05 g of aniline in it. It took 30 minutes for CuCNFs to absorb the aniline monomer. Next, 20 mL of 1.4 M HCl solution containing APS was dripped into the solution mentioned above drop by drop. The mole ratio of aniline to APS was 1:1.The successive polymerization lasted for 5 h. Then the mixed solution was centrifuged for 10 minutes. Being washed with DIW two times and dried in a drying oven at 50 °C for 24 h, the composite nanofibers were successfully obtained finally.

2.4 Preparation of Biosensor

The glassy carbon electrode (GCE) was carefully polished with alumina slurry until a mirror-like surface appeared, thoroughly rinsed by DIW and washed with ethanol in an ultrasonic bath prior to modification. The CuCNFs and PANI/CuCNFs were ground in an agate mortar before utilization. The PANI/CuCNFs modified by laccase were glued on the surface of the pretreated GCE by the following method: the mixture of 0.4 mg/mL PANI/CuCNFs, 0.6 mg/mL laccase and 15 mg/mL Nafion was added into 1 mL 0.1 M acetate buffer solution (pH 4.0) and then stirred for 2 h. 10 μL of the solution was dropped onto the surface of GCE and the GCE was dried at 4 °C in refrigerator. Finally, the electrode named as Nafion-PANI/CuCNFs-Lac-GCE was freshly fabricated. The remaining control sample electrodes were fabricated using the same procedures as Nafion-PANI/CuCNFs-Lac-GCE. All the electrodes were stored at 4 °C in refrigerator before use to keep the activity of laccase.

2.5 Electrochemical Measurement

Electrochemical experiments were recorded on the CHI 660E electrochemical workstation (CH Instruments, Shanghai, China) with a conventional three-electrodes system. A platinum wire electrode and an Ag/AgCl electrode filled with saturated KCl solution were used as a counter and reference electrode, respectively. The modified glassy carbon electrode (GCE) was used as a working electrode. All the electrodes were purchased from Gaoss Union Technology Co., Ltd, (Wuhan, China).

2.6 Characterization

In order to observe the surface of PANI/CuCNFs, Hitachi SU1510 scanning electron microscope (SEM) and JEOL/JEM-2100 transmission electron microscope (TEM) from Japan were employed. The chemical components of PANI/CuCNFs were analyzed using energy dispersive X-ray spectroscopy (EDX, EDAX-TSL, AMETEK USA) and Powder D8 Advance X-ray diffraction (XRD, Bruker AXS D8). Fourier transform infrared spectroscopy (FTIR) spectra were obtained by a Nicolet iS10 FT-IR spectrometer (Thermo Fisher Scientific).

3. Results and Discussions

3.1 Characterizations

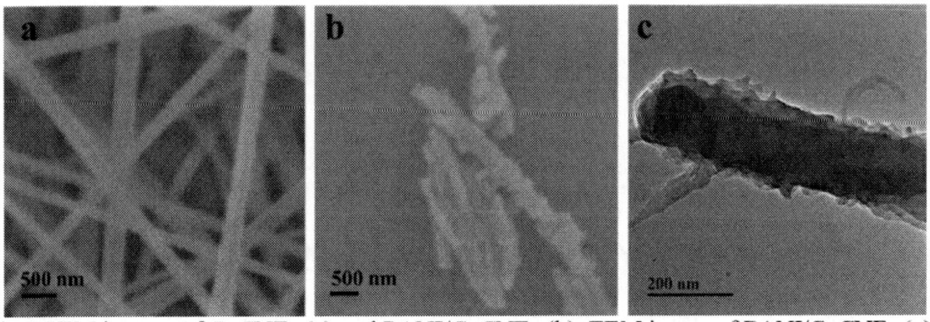

Fig.1 SEM images of CuCNFs (a) and PANI/CuCNFs (b); TEM image of PANI/CuCNFs (c).

Fig.1a presents a SEM image of the CuCNFs. Figs.1b and 1c present a SEM and a TEM image of the PANI/CuCNFs, respectively, which were prepared by the in situ polymerization. As shown in Fig.1a, the CuCNFs were straight cylindrical fibers with smooth surfaces. A significant change in the morphology of the nanofibers can be observed from Fig.1b that some nanoparticles were randomly distributed on the surface of the CuCNFs. From Fig.1c, it can be seen much more clearly that the PANI was coated on the CuCNFs. The coating of PANI onto CuCNFs greatly enlarged the active surface area and formed an effective access for the electron transfer, which was expected to improve the sensitivity of the biosensor.

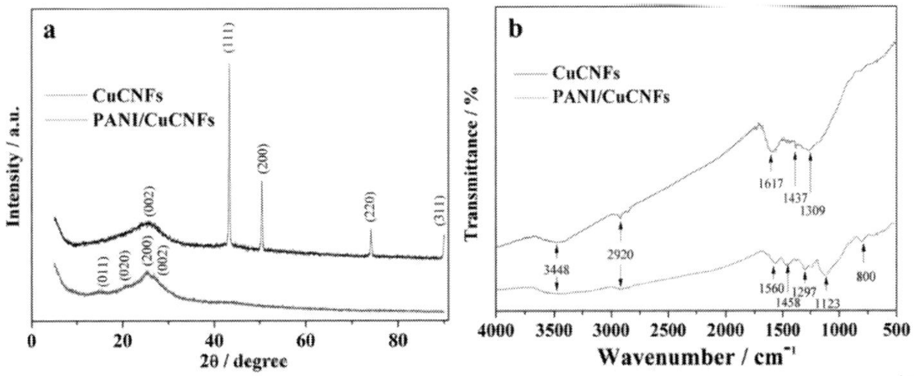

Fig.2 (a) XRD patterns of CuCNFs and PANI/CuCNFs; (b) FTIR spectra of CuCNFs and PANI/CuCNFs.

XRD was employed to ascertain the composition of the synthesized material. XRD patterns of CuCNFs and PANI/CuCNFs are displayed in Fig.2a. For comparison, three peaks, (011), (020), (200) appeared at ca. 15.24°,

20.82°, 25.22° respectively, showing on the XRD pattern of PANI/CuCNFs and none of them appeared on the XRD pattern of CuCNFs. The three diffraction peaks were originated from PANI [10]. The four diffraction peaks at ca.43.36°, 50.50°, 74.12° and 89.92° respectively corresponding to (111), (200), (220) and (311) appeared on the XRD pattern of CuCNFs indexed with the copper can hardly be seen on the pattern of PANI/CuCNFs, which may result from the densely coating of PANI. Both of the nanofiber composites showed a broad and weak diffraction peak, (002), in the range of ca. 20° - 30°, which was assigned to the planes of disordered carbon structure [11]. The above results demonstrated that CuCNFs were obtained and PANI was well coated on the CuCNFs.

The FTIR spectra of CuCNFs and PANI/CuCNFs are shown in Fig. 2b. It gives five peaks at 3448, 2920, 1617, 1437 and 1309 cm^{-1} on the spectrum of CuCNFs. The PANI/CuCNFs spectrum consists of seven distinct peaks at 3448, 2920, 1560, 1458, 1297, 1123 and 800 cm^{-1}. The peaks at 3448 and 2920 cm^{-1} were caused by O-H and C-H stretching vibration, respectively [12]. The peak at 1617 cm^{-1} indicated the stretching vibration band of C=O groups [13]. The distinct peaks at 1437 and 1309 cm^{-1} were ascribed to C-H vibrations of different modes [12]. The characteristic bands located at 1560and 1458 cm^{-1} resulted from the C=C stretching vibration of the quinoid ring and the benzene ring, respectively, illustrating the presence of oxidation state of PANI. And the bands at 1297, 1123 and 800 cm^{-1} were attributed to the C-N stretching vibration of the secondary aromatic amine, C=N stretching vibration and the C-H bending vibrations in the aromatic ring [14-15]. These characterizations indicated that the PANI was successfully doped in the composite materials which was in accordance with the results mentioned above, further demonstrated the successful synthesis of PANI/CuCNFs.

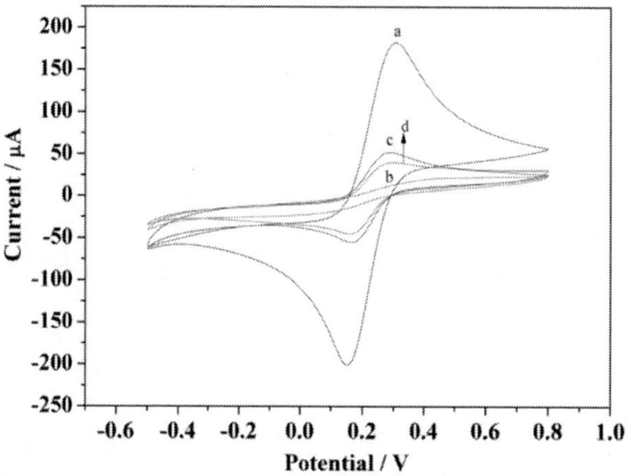

Fig.3 Cyclic voltammograms of bare GCE (a), Lac-Nafion-GCE (b), CuCNFs-Lac-Nafion-GCE (c) and PANI/CuCNFs-Lac-Nafion-GCE (d) in 0.1 M KCl and 1 mM $Fe(CN)_6^{3-/4-}$ solution at scan rate of 100 mV/s.

Fig.3 shows the cyclic voltammetric curves of bare GCE (a), Nafion-Lac-GCE (b), Nafion-CuCNFs-Lac-GCE (c) and Nafion-PANI/CuCNFs-Lac-GCE (d). From the results, it can be seen that the redox peaks of all the fabricated electrodes were well-defined and the electrodes modified with the materials containing laccase showed rather low current values in comparison with that of bare GCE. This can be explained by that laccase is a kind of protein which is nonconductive and it restrained the electron transfer. Both Cu and CNFs are good conductors of electricity while PANI is a semiconductor. As a result, the oxidation peak current for PANI/CuCNFs-Lac-Nafion-GCE slightly decreased in comparison with that of CuCNFs-Lac-Nafion-GCE. Despite this, it was still much higher than that of Lac-Nafion-GCE, suggesting that the synthesized material, PANI/CuCNFs, facilitated electron transfer and improved the electrocatalytic performance of the biosensor.

Fig.4 shows the cyclic voltammograms of bare GCE, Lac-Nafion-GCE, CuCNFs-Lac-Nafion-GCE and PANI/CuCNFs-Lac-Nafion-GCE in 0.1 M acetate buffer solution with hydroquinone in it. On both curve a and b in Fig.4, there were a pair of well-defined redox peaks (0.40 V and 0.10 V) and (0.43 V and 0.20 V), which were attributed to the redox reaction of hydroquinone. Two pairs of redox peaks appeared on curve c and d, namely, (0.447 V and 0.11 V) and (0.4 V and 0.07 V), which were ascribed to the high efficient catalysis of laccase toward hydroquinone, and (0.04 V and-0.12 V), (0.03 V and -0.11 V), which were caused by the redox reaction of copper. For PANI/CuCNFs-Lac-Nafion-GCE, the oxidation peak current (26.2 μA) resulting from the catalysis of laccase toward hydroquinone increased strikingly in comparison with the other three current values. It means that the PANI coated on CuCNFs greatly expanded the surface area for the immobilization of laccase, thus improved the electrocatalysis of the biosensor. The reaction mechanism of the biosensor in the process is described below. Hydroquinone was oxidized into 1,4-benzoquinone as it got into contact with Lac during the presence of molecular oxygen. After that the 1,4-benzoquinone was reduced into hydroquinonr electrochemically on the surface of GCE.

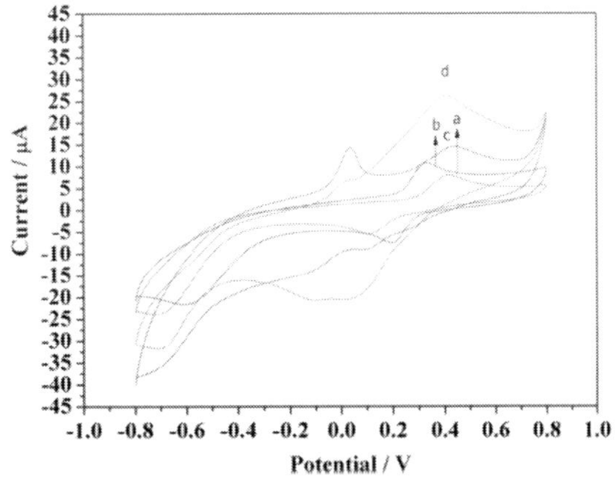

Fig.4 CVs of bare GCE (a), Lac-Nafion-GCE (b), CuCNFs-Lac-Nafion-GCE (c) and PANI/CuCNFs-Lac-Nafion-GCE (d) in 0.1 M acetate buffer solution (pH 4.0) containing 9.95 mM hydroquinone at a scan rate of 100 mV/s.

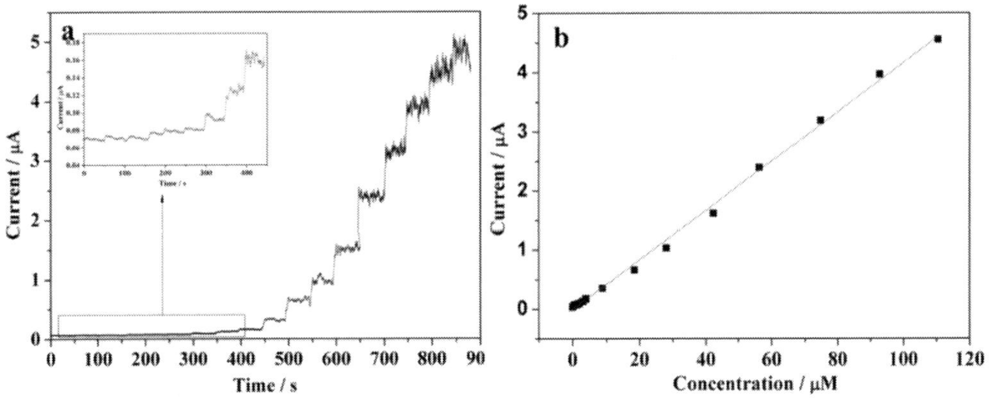

Fig.5 (a) The typical amperometric responses of PANI/CuCNFs-Lac-Nafion-GCE on successive injection of different amounts of hydroquinone into stirring 0.1 M acetate buffer solution (pH 6.0) at 0.4 V. (b) Calibration curve of the current vs. the concentration of hydroquinone.

Through comparison experiment, pH value of 6.0 and the applied potential of 0.40 V were used in the following experiments. The PANI/CuCNFs-Lac-Nafion-GCE for the determination of hydroquinone was developed. And the performance of the biosensor was evaluated by amperometry. Fig.5a exhibits a typical current-time curve of PANI/CuCNFs-Lac-Nafion-GCE for successive addition of hydroquinone into stirred acetate buffer solution. Once the hydroquinone was injected into the buffer solution, the response current rose sharply and achieved a steady-state value within 5s which indicated that the electrode responded rapidly to the change of hydroquinone concentration. As shown in Fig.5b, the response currents displayed a linear increase with increasing concentration of hydroquinone. The linear range of the constructed biosensor was found at 500 nM to 110 μM. The linear regression equation was Current (μA) = 0.00171 + 0.04165 Concentration (μM), with a correlation coefficient of 0.99768, a sensitivity of 41.65 μA/mM and a detection limit of 2.37 μM. The comparison of the biosensors applied to hydroquinone detection reported in recent years was summarized in Table 1. As shown in Table 1, the prepared biosensor showed satisfactory biosensing performance. It exhibits wider linear response range and higher sensitivity, which suggested that the biosensor can detect the concentration of hydroquinone within a shorter time. The excellent performance can be attributed to that the attachment of nanostructured PANI on the surface of CuCNFs improved the interfacial area for the immobilization of enzyme. Besides, the high conductivity of CuCNFs also made a great contribution to the favorable biosensing performance.

Table.1 Sensing performance comparison of different sensors toward hydroquinone.

Electrode description	Detection limit (μM)	Linear range (μM)	Sensitivity (μA/mM)	Reference
N-GCE	0.2	5 - 260	27.31	[16]
CPE-tyrosinase-Nafion	1.6	9 - 120	11.3	[17]
HRP/MMPs/Au electrode	0.4	0.5 - 4.5	1.298	[18]
Lac/AP-rGOs/Chit/GCE	7	3 - 2000	14.16	[19]
Cu/CNFs/Lac/Nafion/GCE	1.18	9.95 - 9760	33.1	[20]
Nafion-PANI/CuCNFs-Lac-GCE	2.37	0.50 - 110.44	41.65	this work

3,5-dinitrosalicylic acid, vanillin, guaiacol, phenol and catechol were used in the experiment to investigate the anti-interference performance of the manufactured biosensor, a crucial characteristic of biosensors. The experiment was conducted by adding the interferents into 0.1 M acetate buffer solution (pH 6.0) containing 1 μM hydroquinone by the same amount. Before and after adding each interferent, the current value was recorded, respectively. No obvious increase occurred in current response during the process of experiment, indicating negligible interference. The biosensor possessed a good selectivity which can be attributed to the strong anti-interference performance of Nafion and the high catalytic specificity of laccase for hydroquinone.

The prepared electrode was tested six times successively and the relative standard deviation (RSD) was 1.74% which demonstrated that the prepared biosensor possessed a superb repeatability. In addition, the reproducibility was investigated by testing three separate electrodes prepared under the same conditions. The RSD was 3.75%, confirming that the biosensor was highly reproducible. Furthermore, the storage stability of the biosensor was also investigated. The biosensor used in the experiment was stored in buffer at 4 °C in refrigerator. The response current of the biosensor decreased to 98.6% after being stored in refrigerator for a week. However, a month later, it still remained 92.7% of its initial response, which suggested that the stability of the biosensor was satisfactory.

4. Conclusion

In summary, electrospinning, carbonization and in situ polymerization formed the fabrication process of the new hybrid nanomaterial, PANI/CuCNFs. A novel biosensor was manufactured using the PANI/CuCNFs, laccase, Nafion and GCE. It showed excellent electrochemical properties and highly efficient catalysis toward hydroquinone. Moreover, it also exhibited good selectivity, repeatability, reproducibility and stability, attributing to the favorable biocompatible microenvironment of the composite matrix. The biosensor also performed well in the practical application of hydroquinone detection. The outstanding performance for hydroquinone detection of the biosensor indicated that the PANI/CuCNFs would be a promising substrate material for the construction of enzymatic biosensors.

Acknowledgements

This work was financially supported by the Priority Academic Program Development of Jiangsu Higher Education Institutions, the Natural Science Foundation of Jiangsu Province (BK20150155), Six talent peaks project in Jiangsu Province (2014-XCL001), the Fundamental Research Funds for the Central Universities (JUSRP115A04, JUSRP51505 and JUSRP11701).

References

[1] Dong SB, Zhao RT, Zhu JG, Lu X, Li Y, Qiu SF, Jia LL, Jiao X, Song SP, Fan CH, Hao RZ, Song HB. Electrochemical DNA Biosensor Based on a Tetrahedral Nanostructure Probe for the Detection of Avian Influenza A (H7N9) Virus. ACS Appl Mater Interfaces: 2015; 7: 8834-8842.

[2] Guo Y, Chen JH, Chen GN. A label-free electrochemical biosensor for detection of HIV related gene based on interaction between DNA and protein. Sensors Actuators B Chem: 2013; 184: 113-117.

[3] Mao CJ, Chen XB, Niu HL, Song JM, Zhang SY, Cui RJ. A novel enzymatic hydrogen peroxide biosensor based on Ag/C nanocables. Biosens Bioelectron: 2012; 31: 544-547.

[4] Odaci D, Telefoncu A, Timur S. Pyranose oxidase biosensor based on carbon nanotube (CNT)-modified carbon paste electrodes. Sensors Actuators B Chem: 2008; 132: 159-165.

[5] Rahmanian R, Mozaffari SA. Electrochemical fabrication of ZnO-polyvinyl alcohol nanostructured hybrid film for application to urea biosensor. Sensors Actuators B Chem: 2015; 207: 772-781.

[6] Adabi M, Saber R, Faridi-Majidi R, Faridbod F. Performance of electrodes synthesized with polyacrylonitrile-based carbon nanofibers for application in electrochemical sensors and biosensors. Mat Sci Eng C-Mater: 2015; 48: 673-678.

[7] Batra B, Lata S, Sharma M, Pundir CS. An acrylamide biosensor based on immobilization of hemoglobin onto multiwalled carbon nanotube/copper nanoparticles/polyaniline hybrid film. Anal Biochem: 2013; 433: 210-217.

[8] Vamvakaki V, Tsagaraki K, Chaniotakis N. Carbon nanofiber-based glucose biosensor. Anal Chem: 2006; 78: 5538-5542.

[9] Zhang S, Zhang DW, Sheng QL, Zheng JB. PANI-TiC nanocomposite film for the direct electron transfer of hemoglobin and its application for biosensing. J Solid State Electr: 2014; 18: 2193-2200.

[10] Yan J, Wei T, Fan ZJ, Qian WZ, Zhang ML, Shen XD, Wei F. Preparation of graphene nanosheet/carbon nanotube/polyaniline composite as electrode material for supercapacitors. J Power Sources: 2010; 195: 3041-3045.

[11] Saranya K, Subramania A, Sivasankar N, Mallick S. Electrospun TiC embedded CNFs as a low cost platinum-free counter electrode for dye-sensitized solar cell. Mater Res Bull: 2016; 75: 83-90.

[12] Tian CH, Yi JA, Wu YQ, Wu QL, Qing Y, Wang LJ. Preparation of highly charged cellulose nanofibrils using high-pressure homogenization coupled with strong acid hydrolysis pretreatments. Carbohyd Polym: 2016; 136: 485-492.

[13] Zhao X, Su Y, Chen W, Peng J, Jiang Z. Grafting perfluoroalkyl groups onto polyacrylonitrile membrane surface for improved fouling release property. J Membr Sci: 2012; 415: 824-834.

[14] Pang ZY, Fu JP, Luo L, Huang FL, Wei QF. Fabrication of PA6/TiO2/PANI composite nanofibers by electrospinning-electrospraying for ammonia sensor. Colloid Surface A: 2014; 461: 113-118.

[15] Xu GH, Xu DD, Zhang JN, Wang KX, Chen ZM, Chen JF, Xu Q. Controlled fabrication of PANI/CNF hybrid films: Molecular interaction induced various micromorphologies and electrochemical properties. J Colloid Interface Sci: 2013; 411: 204-212.

[16] Wang XY, Xi M, Guo MM, Sheng FM, Xiao G, Wu S, Uchiyama S, Matsuura H. An electrochemically aminated glassy carbon electrode for simultaneous determination of hydroquinone and catechol. Analyst: 2015; 141: 1077-1082.

[17] Sys M, Pekec B, Kalcher K, Vytras K. Amperometric Enzyme Carbon Paste-Based Biosensor for Quantification of Hydroquinone and Polyphenolic Antioxidant Capacity. Int J Electrochem Sci: 2013; 8: 9030-9040.

[18] Elyacoubi A, Zayed SIM, Blankert B, Kauffmann JM. Development of an amperometric enzymatic biosensor based on gold modified magnetic nanoporous microparticles. Electroanal: 2006; 18: 345-350.

[19] Zhou XH, Liu LH, Bai X, Shi HC. A reduced graphene oxide based biosensor for high-sensitive detection of phenols in water samples. Sens Actuators B: 2013; 181: 661-667.

[20] Fu JP, Qiao H, Li DW, Luo L, Chen K, Wei QF. Laccase Biosensor Based on Electrospun Copper/Carbon Composite Nanofibers for Catechol Detection. Sensors: 2014; 14: 3543-3556.

Fabrication and Characterization of Electroresponsive Hydrogel Fiber Based on Xanthan Gum and Gelatin

Ru-Quan Zhang[1], Yong-Zhen Tao[2,3*], Xiao-Juan Wang[3], Quan He[3], Ren Cheng[3], Zheng-Ming Wang[3], Cong Zhang[3], Wei-Lin Xu[2]

[1] *College of Mathematics and Computer Science, Wuhan Textile University, Wuhan, 430073, China*
[2] *Key Laboratory of Green Processing and Functional Textiles of New Textile Materials, Ministry of Education, Wuhan Textile University, Wuhan 430073, China*
[3] *School of Material Science and Engineering, Wuhan Textile University, Wuhan 430073, China*

[*]Corresponding author's email: yongzhentao@126.com

Abstract

Xanthan gum and gelatin methacrylate were synthesized to prepare photocrosslinked hybrid hydrogels. Time sweep and creep-recovery tests were performed to monitor the photocrosslinking process, and to evaluate the mechanical deformation and recovery for xanthan gum and gelatin methacrylate hydrogels. The results indicated that the crosslinking reaction was triggered by the UV light with an initiator to form network structures and achieved reaction equilibrium rapidly (less than 380s). The photocrosslinked hydrogels represented less deformation and higher ratio of recovery than those of the native hydrogels. The thixotropy behaviour was studied, and the results suggested that the native hydrogels exhibited a pseudo-plastic behaviour with positive values of hysteresis loop and relative high viscosity. Furthermore, photocrosslinked hydrogel fibers were fabricated and characterized. The relatively high concentration of negative charges (COO^-) was contributed to relatively high bending speed and total bending angle for the hydrogel fibers. This work provides some valuable and fundamental information of the photocrosslinked hydrogels based on nature polymers for further application in functional biomaterials and tissue engineering.

Keywords: Xanthan Gum; Gelatin; Methacrylate Modification; Rheological Properties; Photocrosslinked Hydrogel Fiber; Electroresponsive Behavior

1. Introduction

Smart hydrogels with the response to environmental stimulus such as temperature, pH, light, and electric signal have attracted wide attention in application of controlled release, sensors, and actuators [1]. Gelatin is obtained from hydrolysis of collagen, and the gelatin based hydrogels have been widely applied for various biomedical and tissue engineering fields [2]. However, it is not easy to form hydrogel fiber or other designed patterns with perfect appearance and self-support properties. Xanthan gum (XG), an anionic polysaccharide, consists of a (1→4)-β-D-glucose backbone substituted on every second unit with a charged trisaccharide side chain [3]. The unique chemical structure of xanthan gum confers it high viscosity and shear thinning property. Furthermore, both the gelatin and xanthan gum are polyelectrolyte, which entitles them a characteristic of response under electric stimuli. Therefore, the hybridization of the gelatin with xanthan gum can result in hydrogels with biocompatibility, fiber forming, and electroresponsive properties. Electroresponsive hydrogels are promising materials for the applications of mechanical hands, artificial fish, soft robotic, and various artificial muscle and nervous constructs [4]. Polyelectrolyte hydrogels with insoluble and swellable network structures carry cations or anions in the main chain, leading to electroresponsive behavior. A typical function of the polyelectrolyte hydrogels is to bend under electric field stimulation. The fixed charged groups on the chains of hydrogels and the mobile ions of surrounding and inner hydrogel networks play the important roles on the bending speed and direction [5].

In this work, xanthan gum and gelatin were modified with methacrylic anhydride to obtain photocrosslinkable xanthan gum and gelatin methacrylate. Rheological measurements were performed to investigate the forming process and the mechanical deformation and recovery properties of the photocrosslinked hydrogel. The thixotropic behaviour was carried out to evaluate the structure destruction and recovery and flow behaviour. Furthermore, the hybrid hydrogel fibers were fabricated based on xanthan gum and gelatin methacrylate, and the electroresponsive behaviour and mechanical properties were investigated.

2. Experimental Methods

2.1 Materials

Xanthan gum was purchased from Shanghai Ryon Biological Technology Co. Led (Shanghai, China). Gelatin, Methacrylic anhydride (MA) and 2-hydroxy-4′-(2-hydroxyethoxy)-2-methylpropiophenone (PI) were purchased from Sigma-Aldrich Co. (USA). Analytically pure NaOH, NaCl, KCl, KH_2PO_4, and Na_2HPO_4 were obtained from Shanghai Chemical Co. in China.

2.2 Synthesis of Xanthan Gum and Gelatin Methacrylate

XG and gelatin methacrylate were synthesized by reacting XG or gelatin with methacrylic anhydride. Briefly, XG was dissolved in deionized water (pH=8.0) at room temperature, and kept stirring at 50 ℃ for 2 h. MA was added to this solution and the reaction was continued for 6 h at 50 ℃. Periodically, pH (8.0) was adjusted with 0.1 M NaOH solution. The modified XG (MXG) solution was purified by dialysis for at least 7 days against deionized water to remove the excess of MA. Lyophilize the purified MXG and stored at a dry place protected from light. Gelatin was dissolved in phosphate buffer saline (PBS, pH=7.4) at room temperature, and kept stirring at 60 ℃ for 1 h. MA was added to this solution and the reaction was continued for 3 h at 60 ℃. The resultant gelatin methacrylate (GM) solution was purified by dialysis for at least 7 days against deionized water to remove the excess of MA. Lyophilize the purified GM and stored at a dry place protected from light.

2.3 Fabrication of Photocrosslinked Hydrogel Fibers

MXG and GM were dissolved in PBS containing 1% (w/v) photoinitiator, 2-hydroxy-1-[4-(2-hydroxyethoxy) phenyl]-2-methyl-1-propanone to make solutions with varying MXG concentrations (2%, 3%, (w/v)) and 5% (w/v) of GM. The resultant solutions were transferred into a syringe to prepare hydrogel fibers with a flat needle, and then the hydrogel fibers were exposed in UV light with the power of 8.4 mW/cm^2 for 6 min to obtain the photocrosslinked hydrogel fibers. The photocrosslinked hydrogel fibers were coded as MXG2-GM5 and MXG3-GM5 in accordance with the MXG and GM content.

2.4 Rheological Measurements

Rheological data were measured on a Haake MARS rheometer (Thermo Electron Co., Germany) in the parallel plate mode with a plate diameter of 20 mm and a gap of 1 mm. The samples were stayed for 180 s at 25 ℃ before testing to allow stress relaxation and temperature equilibration. Time sweep was used to monitor *in situ* the photocrosslinking process of the MXG3, GM5, and MXG-GM5 hydrogels at 25 ℃. Creep-recovery tests of the samples were performed under controlled stress mode at 25 ℃, and the shear strain (γ) was recorded with respect to time. The thixotropic curves for the photocrosslinked hydrogels were examined at 25 ℃. In the tests, the shear rate increased from 0.001 s^{-1} to 450 s^{-1}, and then stayed at 450 s^{-1} for 180 s, finally decreased from 450 s^{-1} to 0.001 s^{-1}.

2.5 Electroresponsive Test

The photocrosslinked MXG2-GM5 and MXG3-GM5 hydrogel fibers were dried by vacuum at room temperature, and then were cut to prepare fibers with length of 34 mm. The resultant hydrogel fibers were placed in a 35 mm Petri dish. Aluminium wires used as electrodes were placed 35 mm apart and connected to a DC power source. The hydrogel fibers were fixed at the centre using double sided sticky tape. About 15 mL of 0.2 M NaCl aqueous solution was added into the Petri dish, and immediately, the power of 30 V was turn on. Imaging was performed with a digital camera.

2.6 Measurement of Mechanical Properties

Single fibre strength tester (YG001A, China) with 20 mm gauge length and 2 mm/min displacement rate was used to measure the mechanical properties of the hydrogel fibers. The tensile strength (σ_b), extent of elongation at failure (ε_b), as well as the work of fracture of the fibers were computed from the tensile stress-strain graphs. The average obtained from a minimum of five test runs per sample was used for computing the mechanical properties of the samples.

3. Results and Discussion

3.1 Synthesis of Xanthan Gum and Gelatin Methacrylate

In this work, XG and gelatin methacrylate were synthesized to prepare photocrosslinked hybrid hydrogels, which was shown in Fig.1. The rheological behavior of hydrogels was performed to investigate the photocrosslinking process, creep-recovery and flow behavior. Furthermore, the hydrogel fibers were fabricated and the electroresponsive behavior and mechanical properties of the hydrogel fibers were evaluated.

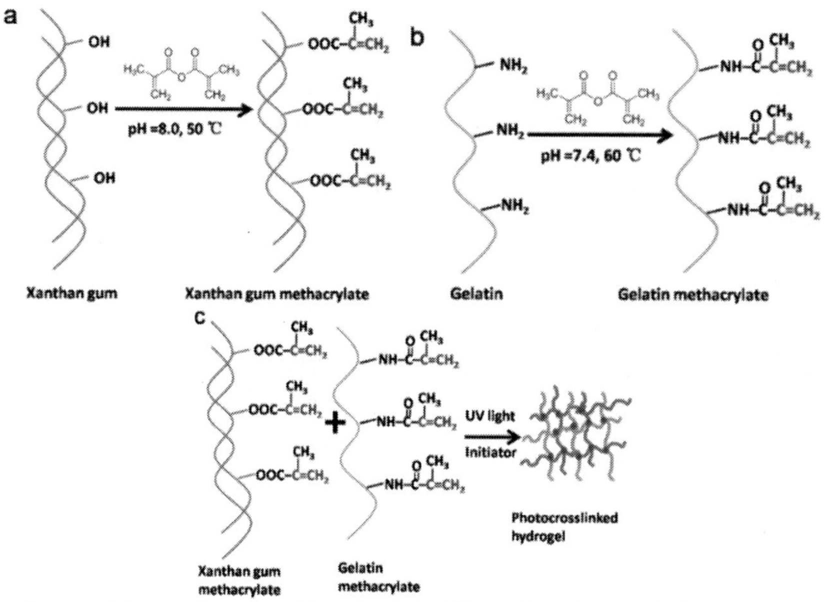

Fig.1 Synthesis scheme of the xanthan gum (a) and gelatin (b) methacrylate, and photocrosslinked hydrogel (c).

3.2 Rheological Behavior

3.2.1 Time Sweep

Dynamic rheology is an effective method for monitoring the crosslinking process of the polymers. To examine the effect of time on the properties of MXG3, GM5, and MXG-GM5 hydrogels, time sweep was performed at 1 Hz at 25 ℃ over an extended period of time when the samples were exposed in the UV light. The changes of G' and G'' with the exposed time for the MXG3, GM5, and MXG-GM5 samples are shown in Fig.2a. The elastic response is dominating ($G'>G''$), and the storage modulus (G') increase rapidly from the beginning of the test, which is attributed to the photocrosslinking process. The G' values reach a plateau region up to 380 s, 250 s, 190 s, and 180 s, respectively, for the GM5, MXG2-GM5, MXG3-GM5, and MXG5-GM5. It indicates that photocrosslinking network structures have formed in the MXG, GM, and MXG-GM samples. Furthermore, the changes of G' and G'' with the UV exposed time for the MXG3 were shown in the magnified picture (Fig.2b). The result shows that G' values increase slowly over the exposed time, which indicates the crosslinking process can develop as time increasing.

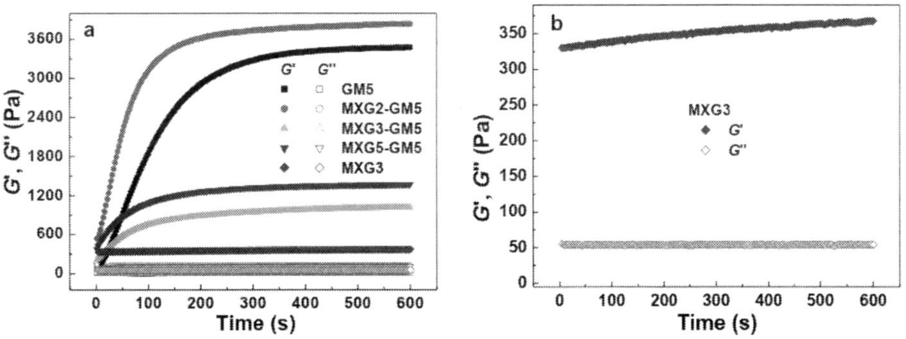

Fig.2 Time dependence of G' and G'' for the MXG3, GM5, and MXG-GM5 samples (a), and magnified picture of the MXG3 sample (b).

3.2.2 Creep-recovery Test

Creep-recovery tests were performed to investigate the viscoelastic properties of the native and photocrosslinked MXG, GM, MXG-GM samples. Fig.3 depicts the creep-recovery curves of deformation as a function of time, which is composited of the creep phase in the period from 0 to 300 s, and the subsequent recovery phase from 300 to 600 s. The deformation curves for the native GM5 show a nearly linear response with a relative high maximum deformation of 1.03×10^{7}% in the creep phase and almost no recovery during the relaxation period (Fig.3a). It indicates that the native GM exhibits a viscous behaviour. However, for the photocrosslinked GM, the deformation is relative small and reaches a plateau value of 0.58% immediately in the creep phase, and the deformation can recover promptly after unloading (Fig.3b). It suggests that the photocrosslinked GM hudrogel shows an elastic solid-like behaviour. Furthermore, the deformation of the native and photocrosslinked MXG3 and MXG-GM5 samples increases steadily with time when the constant stress of 20 Pa applied, and then a delayed recovery takes place after the stress removed (Fig.3c and d). It indicates that the native and photocrosslinked MXG3 and MXG-GM5 samples exhibit the typical deformation behaviour of viscoelastic polymers. The maximum deformation values are represented in Fig.3 for the samples before and after being exposed in the UV light, respectively. The photocrosslinked hydrogels exhibit smaller deformation than the native samples due to the formation of the covalently crosslinking network structure triggered by UV. The deformation of the photocrosslinked MXG2-GM5 is smallest among these tested samples, indicating a highest crosslinking density. A smaller deformation means the sample can resist the given shear stress more effectively. In addition, the compliance (J), which is defined as the deformation per unit stress, can be estimated at any given time based on the corresponding deformation and the constant stress applied. The recovery responses of samples were quantified according to the following equation [6]:

$$R_{ex}(\%) = \frac{J_{300} - J_{600}}{J_{300}} \times 100 \tag{1}$$

where $R_{ex}(\%)$ is the recovery percentage in this work, and J_{300} and J_{600} represent the compliance values at 300 s and 600 s, respectively. The values of the recovery percentage for the GM5, MXG2-GM5, MXG3-GM5, MXG5-GM5, and MXG3 samples were calculated to be 0, 69.3, 57.8, 63.4, and 62.7% before exposed in the UV light, and 96.7, 98.5, 92.8, 92.5, and 91.6% after exposed in the UV light, respectively. The recovery percentage values of the photocrosslinked hydrogels are more than 90%, and far higher than those of the native samples. It suggests that the photocrosslinked hydrogels exhibit the majority of elastic properties. It is notable that the photocrosslinked MXG2-

GM5 displays the smallest deformation and the highest value of the recovery percentage in the tests, which is contributed to the relative high elastic property of the dense crosslinking network structures.

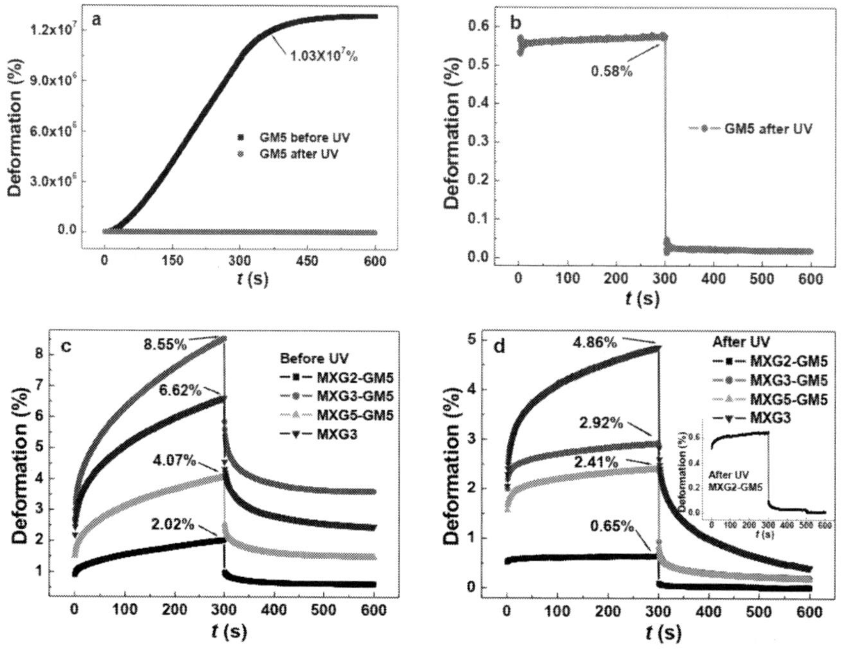

Fig.3 Creep-recovery curves for the GM5, MXG-GM5, and MXG3 samples before and after UV exposure. Insert of d is the magnified graph of MXG2-GM5 after UV exposure.

3.2.3 Thixotropic Behavior

Fig.4a shows the steady shear flow curves of shear viscosity (η) vs. shear rate ($\dot{\gamma}$) for the native GM5, MXG-GM5, and MXG3 samples in the shear rate range from 0.001 to 450 s at 25 ℃ before exposed in the UV light. The values of shear viscosity decrease with increasing shear rate, indicating a shear-thinning behaviour of the samples. The initial values of shear viscosity are relative high due to the hydrogen bonding between the polymer chains. Thus, the GM5, MXG-GM5, and MXG3 samples can be easily injected into desired styles, and the shape of the injected hydrogel can be fixed perfectly. Therefore, the viscous and shear-thinning GM5, MXG-GM5, and MXG3 samples are suitable to apply as injectable materials for fiber forming or 3D bioprinting.

Generally, shear-thinning materials exhibit thixotropic properties as it takes time to recover the microstructures. The degree of thixotropy is evaluated by the area of hysteresis hoop, which is measured from the upwards and downwards flow curves of shear stress vs. shear rate for the samples. Fig.4b shows the hysteresis loop of the GM5, MXG-GM5, and MXG3 samples. The area of hysteresis loop for the GM5, MXG2-GM5, MXG3-GM5, MXG5-GM5, and MXG3 samples is -95, 15839.3, 10126.8, 4944.1, and 5128.9 pa/s, respectively. The thixotropic behavior can be attributed to the disentanglement and alignment of the polymer chains by shearing and re-entanglement when shearing is removed gradually [7]. The entanglement of the polymer chains is broken progressively with increasing shear rate, and reformed as diminishing the shear rate. The negative value of the hysteresis loop for GM5 suggests that the structure of GM5 was destroyed completely, whereas the positive values of the MXG2-GM5, MXG3-GM5, MXG5-GM5, and MXG3 samples represented structure recovery. A larger area of the hysteresis loop for the MXG2-GM5sample suggests a stronger thixotropy behavior. Therefore, the MXG2-GM5 sample is an ideal candidate for fiber forming or 3D bioprinting due to the strong pseudoplasticity with relatively high viscosity and large value of hysteresis loop.

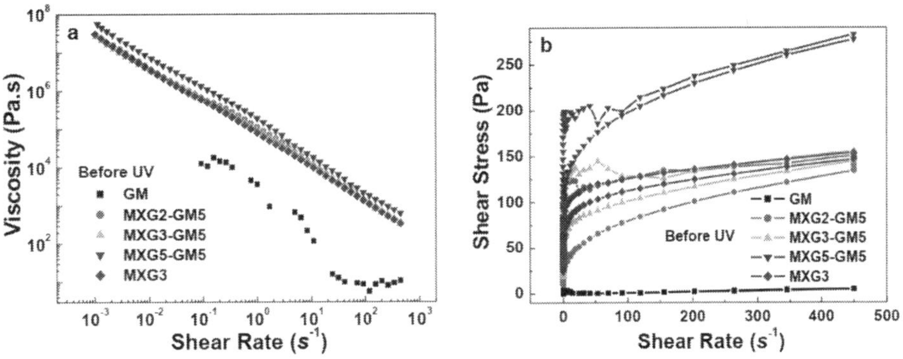

Fig.4 Dependence of the steady shear viscosity on the shear rate (a), and hysteresis loops (b) of the native GM5, MXG-GM5, and MXG3 samples

3.3 Electroresponsive Behavior

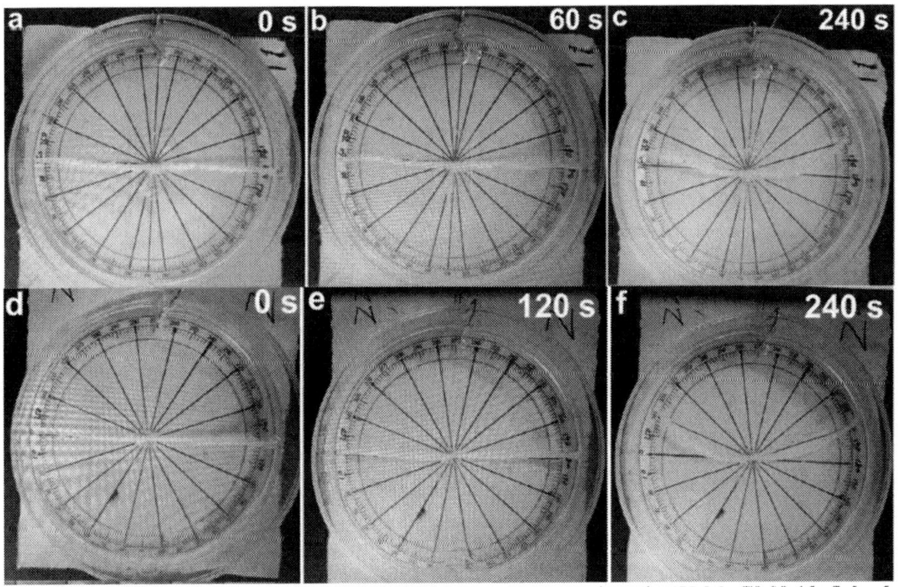

Fig.5 The bending trajectory of the photocrosslinked MXG2-GM5 (a–c) and MXG3-GM5 (d–f) hydrogel fibers at given time.

The MXG2-GM5 and MXG3-GM5 solutions were injected to form hydrogel fibers, and then were exposed in the UV light to prepare photocrosslinked hydrogel fibers. The dried photocrosslinked MXG2-GM5 and MXG3-GM5hydrogel fibers were immersed in 0.2 M NaCl aqueous solution with 30 V of electric field, and bending angle was measured to evaluate the electroresponsive behavior. Fig.5 illustrates the bending trajectory of the hydrogel fibers at given time. The bending occurs from 60 s for the MXG2-GM5 hydrogel fiber and 120 s for the MXG3-GM5 hydrogel fiber. Based on the rheological results, the MXG2-GM5 hydrogel fiber possesses higher density of crosslinking and lower MXG concentration than that of the MXG3-GM5 hydrogel fiber. Thus, it took relative short time to achieve equilibrium of swelling and to start bending for the MXG2-GM5 hydrogel fiber. It is noting that the maximum bending angle was achieved in 30 V of electric field applied for 4 min for the MXG2-GM5 and MXG3-GM5 hydrogel fibers. The bending angles of 5°and 20°are observed for each terminal of the MXG2-GM5 hydrogel fiber. However, both terminals achieve maximum bending angle of 20°for the MXG3-GM5 hydrogel fiber. The relatively high concentration of MXG and relatively low crosslinking density resulted in much more negative charges (COO⁻), which is contributed to relatively high bending speed and total bending angle for the MXG3-GM5 hydrogel fiber [8].

3.4 Mechanical Properties

Uniaxial testing of the single fibre was performed at room temperature to investigate the mechanical behaviour of the photocrosslinked MXG2-GM5 and MXG3-GM5 hydrogel fibers dried by vacuum. The strength-strain curves of the fibers are shown in Fig. 6. From the curves, we were able to determine the values of tensile strength, extent of elongation at failure, and the work of fracture are 289.6 cN, 2.55%, and 82.98 cN*mm for the MXG2-GM5 fiber, and are 290.1 cN, 2.80%, and 88.67 cN*mm for the MXG3-GM5 fiber, respectively. The results show that the photocrosslinked MXG2-GM5 and MXG3-GM5 hydrogel fibers exhibit almost the same mechanical properties. A linear relationship of strength on strain was observed for the photocrosslinked MXG2-GM5 and MXG3-GM5 hydrogel fibers, and elongation at failure is relatively small, which is characteristic of fragile materials.

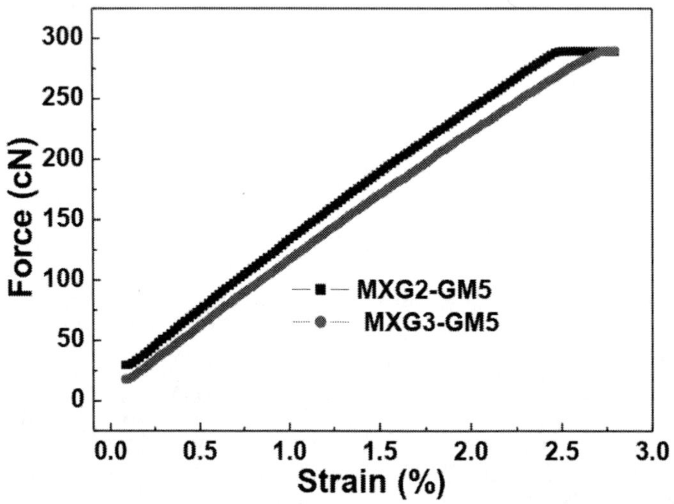

Fig.6 The force-strain curves of the photocrosslinked MXG2-GM5 and MXG3-GM5 hydrogel fibers.

4. Conclusion

The xanthan gum and gelatin were modified with methacrylic anhydride to prepare xanthan gum and gelatin methacrylate, which were applied to form photocrosslinked hydrogels and hydrogel fibers. Time evolution measurement during the *in situ* photocrosslinking process showed that the photocrosslinked hydrogels formed rapidly, and the exposed UV time required to reaction equilibrium was less than 380s. The photocrosslinked hydrogels exhibited smaller deformation on given constant shear stress, and the values of recovery percentage were higher than those of the native samples through the creep-recovery tests. The thixotropy tests indicated that the hydrid composites of xanthan gum and gelatin methacrylate samples were suitable to apply as injectable materials for fiber forming or 3D bioprinting due to the strong pseudoplasticity with relatively high viscosity and positive hysteresis loop. The photocrosslinked hydrogel fibers exhibited electricresponsive behaviour, which were tunable by changing xanthan gum methacrylate concentration. The photocrosslinked hydrogel fibers fabricated from the xanthan gum and gelatin methacrylate exhibit the potential for functional biomaterials and tissue engineering applications.

5. Acknowledgement

The authors wish to acknowledge the National Natural Science Foundation of China (21344006), and the

Foundation of Hubei Educational Committee of China (D20161603).

References

[1] Yu L, Gu L. Effects of microstructure, crosslinking density, temperature and exterior load on dynamic pH-response of hydrolyzed polyacrylonitrile- blend -gelatin hydrogel fibers. Eur Polym J: 2009; 45: 1706-1715.

[2] Yue K, Santiago GT, Alvarez MM, Tamayol A, Annabi N, Khademhosseini A. Synthesis, properties, and biomedical applications of gelatin methacryloyl (GelMA) hydrogels. Biomaterials: 2015; 73: 254-271.

[3] Roy A, Comesse S, Grisel M, Hucher N, Souguir Z, Renou F. Hydrophobically modified xanthan: an amphiphilic but not associative polymer. Biomacromolecules: 2014; 15: 1160-1170.

[4] Prudnikova K, Utz M. Polyelectrolyte hydrogels as electromechanical transducers. Frontiers in Sensing: From Biology to Engineering. Vienna: Springer Vienna: 2012; 351-361.

[5] Sun S, Wong YW, Yao K, Mak AFT. A study on mechano-electro-chemical behavior of chitosan/poly (propylene glycol) composite fibers. J Appl Polym Sci: 2000; 76: 542-551.

[6] Tian Z, Duan L, Wu L, Shen L, Li G. Rheological properties of glutaraldehyde-crosslinked collagen solutions analyzed quantitatively using mechanical models. Mat Sci Eng: C 2016; 63: 10-17.

[7] Benchabane A, Bekkour K. Rheological properties of carboxymethyl cellulose (CMC) solutions. Colloid Polym Sci: 2008; 286: 1173-1180.

[8] De Barros GG, Huang MW, Frisch HL. Interpenetrating polymer networks: High molecular weight natural rubber and poly(2,6-dimethyl-1,4 phenylene oxide). J Appl Polym Sci: 1992; 44: 255-261.

Effect of Silver Content on Electrospinning of SA/PVA/Ag

Yan Feng[1*], Run-Jun Sun[2]

[1] *State Key Laboratory for Mechanical Behavior of Materials, Xi'an Jiaotong University, Xi'an, 710049, China.*
[2] *School of Textiles and Materials, Xi'an Polytechnic University, Xi'an, 710048, China*

[*] Corresponding author's email: fengyan861124@163.com

Abstract

Sodium alginate (SA) / Poly (vinyl alcohol) (PVA) / silver fibers were prepared by electrospinning with different silver content. The properties of spinning solution were measured. The morphology of the fibers and the diameter of fibers were researched, and the antibacterial activity of the fiber felts was tested. The results show that the electrospun process and fiber morphology are affected by the increase of silver content. The measurements show that the surface tension and electrical conductivity of the solutions increase with the increase of silver content, which are two important parameters to the electrospun process and fiber morphology. When the silver content is 0.1%, the average diameter of the fiber is smallest, the diameter distribution is relatively concentrated, and the electrospinning felt has a certain antibacterial activity.

Keywords: Electrospinning; Sodium Alginate; Silver Content; Solution Properties; Diameter Distribution; Antibacterial Activity

1. Introduction

Alginate fibers have many superior properties, such as high moisture absorption, easy removal from wound, high oxygen penetration, biodegradation absorption and biocompatibility [1, 2]. It has been widely used in medical textiles, and is especially suitable for wound dressing [3-5]. Alginate fibers can be become gelatinous when they react to sodium ions which occurs when there is a leakage from wound. The gelatin can reduce the pain of the sick because its moisture makes it difficult for the wound to conglutinate, and it can accelerate the wound healing [6]. Poly (vinyl alcohol) (PVA) has good chemical stability, biocompatibility and degradability [7]. It can also improve the mechanical properties of alginate fiber [8]. Silver which can be used for bactericidal in vitro and for drinking water purification, possesses broad spectrum antibacterial activity and its effect is lasting [9, 10]. The electrospun fibers are characterized by a large specific surface area, high porosity, thin fiber diameter and lightweight [11]. The electrospinning of SA/PVA/Ag fibers have many advantages and are suitable for wound dressings [12-14].

There are many different methods to add silver in to the polymer, such as chemical reduction, photo reduction, heat process, and radiolytic method. The silver prepared by the chemical reduction is added in the spinning solution in the form of silver particle or silver sol. The silvers prepared by the other three methods are added in the solution in the form of silver ion (Ag^+), and a different treatment is needed to reduce Ag^+ to Ag after electrospinning, however the SA will be degraded by the reduction methods like photo, heat and radiolysis. As such adding the silver particle or silver sol in the spinning solution is an appropriate method to SA.

This paper prepared ultrafine SA/PVA/Ag fibers by electrospinning, and the effect of different silver content on electrospinning solution properties and the diameter distribution were researched. The antibacterial activity of PVA felt, SA/PVA felt and SA/PVA/Ag with 0.1% silver felt were tested. To our knowledge, it has not been reported in the literature the effect of silver sol on PVA/SA/Ag solution properties. The antibacterial activity of these electrospun felts were also not reported.

2. Experimental

2.1 Materials and Apparatus

Sodium alginate (SA) was supplied by Qingdao Bright Moon Seaweed Group Co., Ltd. The viscosity was 150 mPa·s. Poly (vinyl alcohol) (PVA) was purchased from Shanghai Mengjia chemical technology Co., Ltd. The polymerization degree was 2400 and saponification degree was 88%. Silver nitrate ($AgNO_3$), 88% formic acid, sodium hypophosphite, sodium hexametaphosphate and Poly (N-vinylpyrrolidone) (PVP) were all analytical reagent. All chemicals were used without any further purification.

2.2 Preparation of Spinning Solutions

The silver sol was used as the source of silver in the SA/PVA/Ag fibers. The preparation of the silver sol needed to prepare oxidized solution and reduction solution respectively. The oxidized solution was $AgNO_3$ / deionized water solution. The sodium hypophosphite, sodium hexametaphosphate and PVP were dissolved in deionized water, and formic acid was added to adjust the PH, which composed the reduction solution. Then the oxidized solution was instilled into the reduction solution by stirring using magnetic stirrer at 50℃, and the solution immediately became brownish red. The silver sol was prepared successfully after continuing to stirring for 30 min.

SA was dissolved in deionized water by stirring at room temperature with the concentration of 4 wt%. PVA was dissolved in deionized water by stirring at 70℃ with the concentration of 10 wt%. Then the blending solution of silver sol, SA and PVA were stirred for 4 hours (silver sol addition was the mass fraction of the silver sol in the solute). SA and PVA solution were mixed at blended ratio of 2: 8 (mass ratio), and the addition of the silver sol was 0, 0.1%, 0.2% and 0.3%, respectively. The concentration of SA and PVA and the blending ratio were based on the previous researched.

2.3 Electrospinning Process

The electrospinning setup consist of a direct-current high voltage generator (ZGF, Chengdu Chuangao electrical technology Co., Ltd., China), micro-injection pump (WZ-50C6, Zhejiang University medical instrument Co., Ltd., China), collector (aluminium foil), syringe and needle (internal diameter = 1.4 mm). The spinning solution was placed in a 20-mL syringe having a stainless steel needle, which mounted on a syringe pump. The needle was connected to the cathode and the collector was connected to the anode of the high voltage generator. In the process of the electrospinning, applied voltage was 20 kV, flow rate was 0.1 mL / h, needle tip-to-collector distance was 12 cm, temperature was about 25 ℃, and relative humidity was about 40 RH%.

2.4 Characterization

The silver sol particle size was analyzed by laser diffraction particle size analyzer (Mastersizer 2000, Malvern Instruments Ltd., UK). The surface tension of solution was measured by contact angle measurement (JC2000C3, Shanghai zhongchen Digital Technic Apparatus Co., Ltd., China). Each sample was measured three times, and calculated the average value. The electrical conductivity was measured by conductivity meter (DDS-11D, Shanghai INESA Scientific Instrument Co., Ltd., China). Each sample was measured three times, and calculated the average value.

The morphology of the fibers were observed by scanning electron microscopy (SEM) (JMC-5000 Neoscope, Japan) after gold coating. The diameters of fibers were measured by Image J soft. Every sample was measured at 100 points, and the average diameter was calculated.

Shaking flask method was used and the GB15979-2002 (Hygienic Standard for Disposable Sanitary Products) was the main reference in the antibacterial activity test [15]. PVA felt, SA/PVA felt and Ag/SA/PVA felt were tested. All of the felts were prepared by electrospinning. The strain was Escherichia coli bacteria (ATCC25922).
The index was bacteriostasis rate, and the calculation method was showed in equation 1.

$$Xs = \frac{A\text{-}B}{A} \times 100 \qquad (1)$$

where Xs was bacteriostasis rate, %. A was the average numbers of bacteria colony before oscillation. B was the average numbers of bacteria colony after oscillation.

The effective condition of the test: the average numbers of bacteria colony of no sample was 1×10^4 - 9×10^4 cfu / mL, and The average numbers of bacteria colony difference was 10% between before and after oscillation.

The evaluation standard of antibacterial activity: when $Xs \leq 0$, the bacteriostasis rate was 0. When bacteriostasis rate difference between samples with and without antibacterial constituent was greater than 26%, the sample had the antibacterial activity.

3. Results and Discussion

Silver particle size distribution D (v, 0.5) was 0.15μm which meant 50% particles size were less than 0.15μm. The particle size was enough small for the experimental requirement. The silver sol was still well dispersed after one month.

As shown in Fig.1(a), the electrical conductivity increased with the increase of silver content, and the electrical conductivity in the silver content of 0-0.1% was growing faster than the rate in the silver content 0.1%-0.3%, which was attribute to the high electrical conductivity of the silver. Fig.1(b) shows that the surface tension increased with the increase of silver content. In the silver content of 0-0.1%, the growth rate of the surface tension was smaller than the rate in the silver content of 0.1%-0.3%. It was due to the existence of the silver sol in the solution. The silver existed in the form of particles in the solution, and it can not dissolve in the solution or react with the solution.

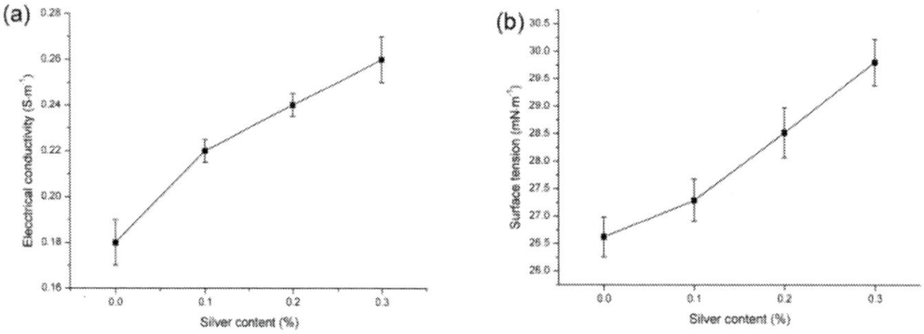

Fig.1 Electrical conductivity (a) and surface tension (b) of solution with different silver content

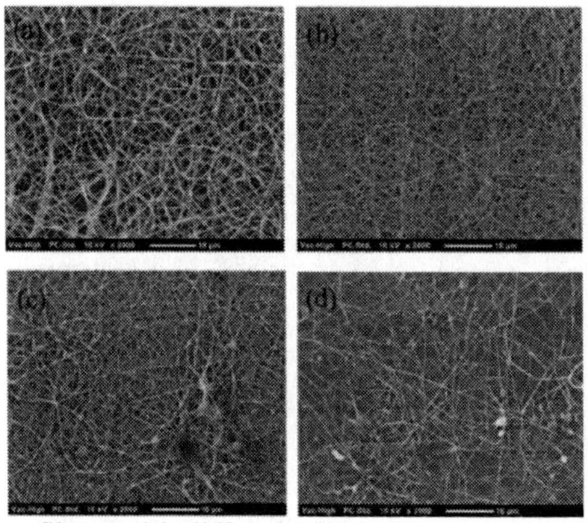

Fig.2 SEM images of electrospun fibers with different silver content: (a) 0, (b) 0.1%, (c) 0.2% and (d) 0.3%

Table.1 Fiber diameter distribution with different silver content

Silver content (%)	Average diameter (nm)	CV (%)
0	315	18.74
0.1	200	11.21
0.2	216	14.30
0.3	280	18.89

Fig.2 shows the scanning electron micrographs of the electrospun fibers with different silver content. It showed that when the silver content was 0.1%, the fiber morphology was better than others and there was no bead. Table 1 shows that the fiber average diameter was smallest and the diameter distribution was relatively concentrated with 0.1% silver content. When there was no silver, solution electrical conductivity was low and jet flow didn't get enough stretch, so the fiber diameter was larger and formed beads easily. After adding a small mount of silver, solution electrical conductivity increased and the quantity of electric charge of jet increased, leading to enough stretch, so the fiber diameter was decreased. But when the silver content continued to increase, electrical conductivity increased more, leading to the quantity of solution which flowed from the nozzle increased, and the surface tension increased more, which needed more electrical force to overcome the surface tension, so the fiber diameter increased and more beads deposited on the felt. Ag/SA/PVA felt with 0.1% silver content was used in antibacterial activity test because the solution with 0.1% silver can keep spinning continuous, and the felt did not have obvious beads.

Table 2 shows the average numbers of Escherichia coli bacteria of PVA felt, SA/PVA felt and SA/PVA/Ag felt with different diluted concentration after incubating for 48 hour at 37℃. The bacteria liquid at time 0 was not oscillated and other samples were all oscillated.

Table.2 Result of antibacterial activity in different samples

Sample	Dilution and The average numbers of Escherichia coli bacteria colony				Report system (cfu/mL)	Bacteriostasis rate (%)
	10	10^2	10^3	10^4		
Bacteria liquid without oscillated at time 0	uncountable	224	153	12	22400	—
Bacteria liquid oscillated for 1 hour	uncountable	241	164	18	24100	—
PVA	uncountable	248	98	13	24800	0
PVA/SA	uncountable	201	54	7	20100	10.27
Ag/PVA/SA	uncountable	122	32	4	12200	45.55

PVA felt didn't have the bacteriostasis and antibacterial activity because the bacteriostasis rate was 0 (Table 2). Then PVA felt can be viewed as the sample of no antibacterial constituent. SA/PVA felt had a certain bacteriostasis and the bacteriostasis rate was 10.27%. The difference between SA/PVA felt and PVA felt was 10.27%, which was smaller than 26%, so it didn't have the antibacterial activity. The PVA felt didn't have the bacteriostasis, so SA had the bacteriostasis, but not the antibacterial activity.

The difference between SA/PVA/Ag felt and PVA felt was 45.55% which was larger than 26%, so it had the antibacterial activity, it also illustrated Ag had the antibacterial activity.

4. Conclusion

The effect of silver content on the properties of SA/PVA/Ag blended solution, the morphology of the fibers and the antibacterial activity were researched. It showed that surface tension and electrical conductivity were increased with the increase of silver content. The fiber average diameter was smallest and the diameter distribution was relatively concentrated with 0.1% silver content. SA/PVA/Ag felt prepared by electrospinning has a certain antibacterial activity through antibacterial activity test, combining with the advantages of SA, PVA, Ag and electrospun felts, so the SA/PVA/Ag felt prepared by electrospinning is suitable for the wound dressings.

References

[1] Zhan YZ, Zhu P. The preparation and development of alginate fiber in medical dressings. Technical Textiles: 2007; 8: 39-43.

[2] QIN YM. The characterization of alginate wound dressings with different fiber and textile structures. J Appl Polym Sci: 2006; 3: 2516-2520.

[3] Xie H, Lu H, Huang H, Lu L, Guan H. Study of modification of polyvinyl alcohol/sodium alginate electrospun fibers and its drug release ability in vitro. J Third Mil Med Univ: 2011; 33: 2485-2489.

[4] Kaassis AYA, Young N, Sano N, et al. Pulsatile drug release from electrospun poly(ethylene oxide)-sodium alginate blend nanofibers. J Mater Chem: 2014; 2: 1400-1407.

[5] Ma G, Fang D, Liu Y, et al. Electrospun sodium alginate/poly(ethylene oxide) core–shell nanofibers scaffolds potential for tissue engineering applications. Carbohyd Polym: 2012; 87: 737-743.

[6] QIN YM. Absorption characteristics of alginate wound dressings. J Appl Polym Sci: 2004; 2: 953-957.

[7] Zhu XS, Pang J, Zhou ZH, et al. Electrospinnability of poly (vinyl alcohol) and its biopolymer blending solutions. Synthetic fiber in China: 2006; 10: 4-8.

[8] Safi S, Morshed M, Hosseini Ravandi SA, et al. Study of electrospinning of sodium alginate, blended solutions of sodium alginate/poly(vinyl alcohol) and sodium alginate/poly(ethylene oxide). J Appl Polym Sci: 2007; 5: 3245-3225.

[9] Ma WS, Cui Y, Zhao YY, et al. Progress of antibacterial mechanisms study on nanoparticles. Acta Biophys Sin: 2010; 8: 638-648.

[10] Jin WJ, Lee HK, et al. Preparation of polymer nanofibers containing silver nanoparticles by using poly (N-vinylpyrrolidone). Macromol Rapid Commun: 2005; 26: 1903-1907.

[11] Ramakrishna S, Fujihara K, Teo WE, Lim TC, et al. An introduction to electrospinning and nanofibers. Singapore: World Scientific Pub Co Inc: 2005; p 10-12.

[12] Alborzi S, Lim LT, Kakuda Y. Electrospinning of sodium alginate-pectin ultrafine fibers. J Food Sci: 2010; 75: 101-107.

[13] Moon SC, Ryu B, Choi JK, et al. The morphology and mechanical properites of sodium alginate based electrospun poly(ethylene oxide) nanofibers. Polym Eng Sci: 2010; 49: 52-59.

[14] Lee YJ, Lyoo WS. Preparation and characterization of high-molecular-weight atactic poly(vinyl alcohol)/sodium alginate/silver nanocomposite by electrospinning. J. Polym. Sci., Part B: Polym. Phys: 2009; 47: 1916-1926.

[15] Shen W, Lu M, Yang HP, et al. GB15979-2002, Hygienic Standard for Disposable Sanitary Products.

Fabrication of Size-controllable Nanoparticles from Feather Waste

Pei Lyu[1], Xin Liu[2*]

[1]*College of Textile Science and Engineering, Wuhan Textile University, Wuhan, 430200, China;*
[2]*Key Laboratory of Green Processing and Functional Textiles of New Textile Materials, Ministry of Education, Wuhan Textile University, Wuhan, 430200, China*

Corresponding author's email: 94026047@qq.com

Abstract

Feather is a waste product generated in large quantities from industrial poultry processing, which has distinctive properties such as high protein content, biodegradability and biocompatibility [1-5]. In this study, the feasibility of producing nanoparticles from feather waste by conversion of alkaline and acid was investigated. The effect of NaOH concentration on the size of particles was evaluated to optimize the best condition. The results show that the smallest size of particles around 50 nm were obtained in the presence of 8% wt sodium hydroxide. SEM indicated the fiber fibrillation and degradation as it was progressively converted into particles form. FTIR showed no remarkable changes in the chemical composition of prepared samples.

Keywords: Size; Feather; Nanopaticles; Alkali; Acid

1. Introduction

Environmental concerns continue to prompt research into the replacement of synthetic materials with an increasing variety of natural materials [6-8]. Poultry feathers are a significant waste material produced by the meat processing industry [9-11]. Fibers from feathers have several distinctive features such as: surface toughness, flexibility, high length to diameter ratio, hydrophobicity, and a highly organized morphology characterized by its complex hierarchical structure [12, 13]. However, in most cases, the feathers are disposed of by burial, whereas an improved, more effective, and hopefully profitable utilization of feather waste is desirable [14, 15]. Therefore, recycling of this renewable source of biopolymers has been the objective of many researches. Since the protein particles could keep the original properties of the material without destroying the microstructure, it has been widely applied in modern industries. In the investigation of Barone et al. [16], while polyethylene-based composites were prepared using keratin feather fiber obtained from chicken feathers, the best composite properties were found at processing temperature of 205℃, which seems to be relatively high. Bullions et al. concluded that polypropylene matrix composites could be fabricated using feather fiber, recycled kraft pulp fiber, recycled news pulp fiber, retted kenaf bast fiber [17].

However, it seems difficult to obtain different sizes of particles required for specific fields. In this paper, size-controllable nanoparticles were obtained from feather fibers by the treatment with sodium hydroxide. The effect of alkaline on the particles morphology were characterized. Results showed that most spherical particles were around 50 nm treated by 3 wt% sodium hydroxide. The influence of alkaline concentration on the size of feather nanoparticles were also studied, which is believed to be one of the most promising approachs due to the milder process conditions leaving no harmful byproducts.

2. Experimental

2.1 Materials

Feather fibers were kindly supplied by Zhejiang Natural Group. Sodium hydroxide (AR) and sulfuric acid (CP) were pruchased from Sinopharm Chemical Reagent Co.,Ltd (China). Distilled water was employed in all experiments.

Instruments: HH-2 Digital-Thermostatic Water Bath, SHZ-D (III) Circulating Water Vacuum Pump, JY-1002 Electronic Scales, FD-1000 Lyophilizer, ZEN-3600 Laser Diffract Instrument, 769YP-15A Powder Compressing Machine, TENSOR 27 FTIR Spectrometer, JSM-6510LV Scanning Electron Microscope.

2.2 Preparation of Nanoparticles

One gram of feather fibers were weighed and dissolved in the various concentrions of sodium hydroxide solutions, which are 1% wt, 3% wt, 5% wt, 8% wt and 10% wt respectively. After the feather fibers were completely wet, the samples were placed in the water bath and kept at 30℃ for 4h. For the solution of sodium hydroxide solution with feather fiber, filter was used for vacuum filtration. Then dilute sulfuric acid with a mass fraction of 3% was applied to adjust the pH value around 7 to 8. Finally, the resulting solution was placed in the dialysis bag and kept in deionized water for 3 days, to remove the residual inorganic salt ions in the solution and to get the ideal pure liquid.

3. Characterization

3.1 Dynamic Light Scattering

The height of the solution in the polystyrene cuvette is approximately 1 cm. The sample was irradiated with a laser at the laboratory temperature and humidity (25 ± 1℃, 65 ± 5%).

3.2 Fourier Transform Infrared Spectroscopy

FTIR spectra of the samples were recorded using a spectrometer (TENSOR 27) with the reflection absorption spectroscopy technology in the range of 4000 - 400 cm^{-1}. The data were collected over 128 scans with a resolution of 4 cm^{-1}. The solid powder was obtained by freeze-drying before testing.

3.3 Scanning Election Microscopy

The surface morphology and size of the particles were observed by scanning electron microscopy (JSM-6510LV), all samples were coated with gold for 7 min before analysis.

4. Results and Discussion

4.1 Size Control

The solution of nanoparticles obtained by the dissolution of feather fibers in the 1%, 3%, 5%, 8%, 10% sodium hydroxide solution and the Tyndall phenomenon were shown in Fig.1.

Fig.1 Tyndall phenomenon of nanoparticles prepared by the dissolution of feather fibers in the 1%, 3%, 5%, 8%, 10% of sodium hydroxide

As it can be seen from the figure above, when the liquid was irradiated with a laser, the liquid can form a bright passage with a dindal effect, indicating that the liquid obtained by dissolving the fiber with the lye was colloid.

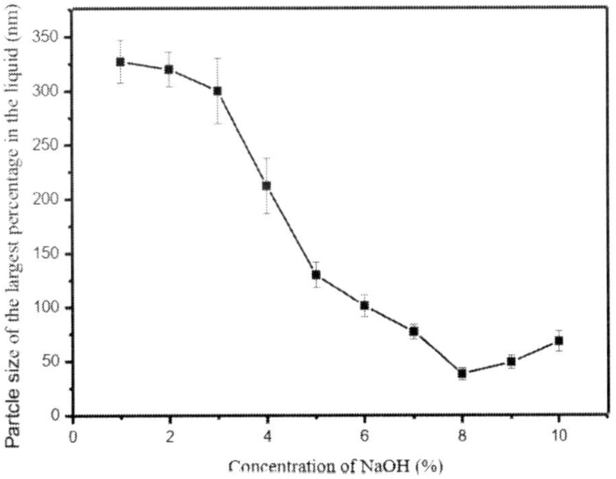

Fig.2 Particles size of the largest percentage gained from feather fibers dissolved by different concentrations of lye

The particle size range in the solution can be obtained directly by the dynamic light scattering test. As one can see, when the lye concentration was 1%, nearly 44.9% of the particles had a diameter of 341.9 nm. When the concentration of sodium hydroxide was 3%, the particle size distribution range was narrowed, and the particle diameter was reduced. About 35.6% of the particle diameter was 295.3 nm. When the mass fraction was 5%, 8%, 10%, data could be available from the test results that 16.8% of the particle sized 141.8nm, 27.9% of the particle owned the diameter of 37.8nm, 23.9% of the particle sized 58.7nm respectively.

The results indicate that different sizes of particles could be well fabricated using different concentrations of lye. According to this method, we can prepare size-controllable nanoparticles which can be used in the bio-medicine, food industry and other fields to achieve the purpose of recycling utilization and environmental protection.

4.2 FTIR Analysis of Nanoparticles

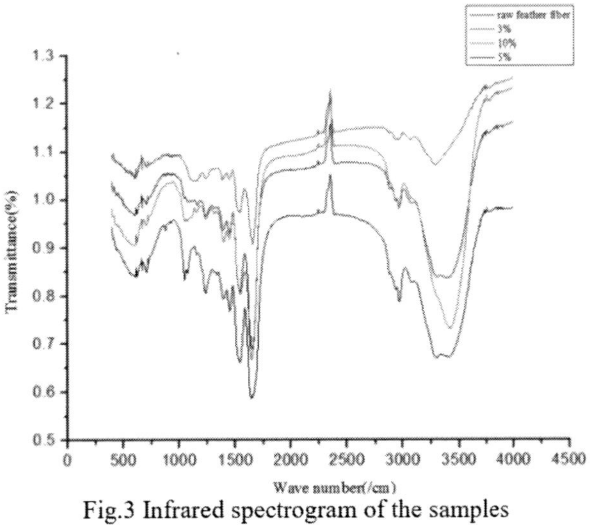

Fig.3 Infrared spectrogram of the samples

Fourier transform infrared spectroscopy was used to analyze the effect of alkali concentration and pH value on the chemical structure of feather fiber particles. As can be seen from Fig.3, the protein characteristic group peaks centered at 3257 cm^{-1}, 2969cm^{-1}, 1625 cm^{-1}, 1513 cm^{-1}, 1229 cm^{-1} and 1046 cm^{-1} corresponding to amide A, C-H stretching vibration, amide Ⅰ, amide Ⅱ, amide Ⅲ, C-N stretching vibration.

For the comparison of the infrared spectra, the 3% lye-treated feather fibers, and the original feather fibers, it can be seen that the peak position of the infrared spectrum was almost the same as that of the original feather, and the stretching vibration peak of C-H disappears. Additionally, there was an increase of 1132 cm^{-1}: C-O stretching vibration peak. This showed that it had less impact for the chemical groups when treated with 3% concentarion of the alkali. As for the 5% mass fraction of the alkali, there was an increase of 3407 cm^{-1}: O-H stretching vibration peak; and 701 cm^{-1}: N-H out-of-plane bending vibration peak disappeared. Therefore, the effect of 5% lye on feathering protein was also slight. Combined with infrared spectroscopy analysis, when using alkali solution to dissolve the feather protein, the peak position of the spectrum would be a certain change, indicating that when destructing the protein's molecular chain, it may generate other groups correspondingly, and may also cause the coupling peak to become a single peak. Comprehensively, under the experimental design of the lye concentration, the amide I band, C = O stretching vibration peak, had not been the destruction of lye, showing that the alkali resistance of the fiber was great.

4.3 SEM analysis of nanoparticles

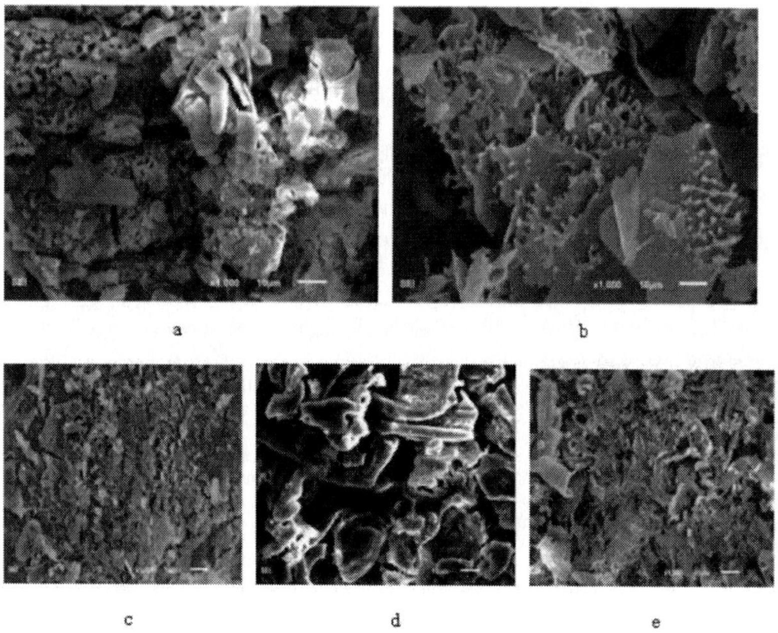

Fig.4 SEM of samples

Fig.4, from a to e, shows the size of particles obtained by dissolving feather fibers with mass fraction of 1%, 3%, 5%, 8%, 10% alkali respectively. By the combination of scanning electron microscopy and dynamic light scattering, the surface morphology and size of the obtained particles were analyzed as following: When the mass fraction of sodium hydroxide was 1%, the resulting particle size was large. As it can be seen from Fig.4-A, the particles were aggregated together to form a block structure; When the lye concentration was 3%, the size of the particles decreased and the surface was smooth. The morphology of the particles could be clearly seen from the photographs of the scanning electron microscope. With the increase of the alkali solution's pH value, the particle size was still decreased and it had the agglomeration tendency when the mass fraction of sodium hydroxide was 5%. When the alkali concentration was 8%, the size of the particles was smaller and the particle surface was smooth. When the mass fraction came to 10%, from Fig.4-E, it formed a flocculent structure.

When the concentration of lye is small, the alkali powder destroys the feather fibers to form nanoparticles. As the concentration of lye increases, nanofibers will form. The conclusion is still to be further validated.

5. Conclusions

Feather fibers were dissolved in sodium hydroxide solution with the mass fraction of 1%, 3%, 5% and 10%. The obtained particles could show the Tyndall effect in the solution, indicating that the liquid obtained by dissolving the feather fiber was colloid under experimental conditions. At the same time, the size of the particles in the solution and the surface morphology of the solution could be obtained by the dynamic light scattering test (DLS) and the scanning electron microscopy (SEM).

When providing 1% sodium hydroxide for dissolving feather fiber, the size of the particles in the liquid was distributed in the range from 295.3nm to 458.7nm, and the resulting particles were larger in size, gathering to form a block structure. When the mass fraction of sodium hydroxide was 3%, nearly 70% of the particle diameter was distributed in the range from 255nm to 295nm, of which nearly 35.6% of the particle size was 295.3nm, showing that the particle size was reduced but with smooth surface. When the concentration of Alkaline was 5%, there were nearly 58% of the particle size distributing from 122nm to 190nm, which was further reduced, having the tendency to reunite. When the mass fraction of alkali was 8%, the size of the particles ranged from 28.2nm to 615.1nm, and 27.9% of the particles had a size of 37.8nm, which was still further reduced.

The electron micrographs indicated that the particles' surface was smooth. When the mass fraction of lye was 10%, the size of the particles in the liquid ranged from 37.8nm to 824.9nm, and about 60% of the particle diameter was between 50nm to 68nm. The results of the electron microscopy showed that there were flocculent structure generation.

In combination with infrared spectroscopy (IR) analysis, the peak position of the spectrum changed when dissolving the feather protein, which indicated that as destroying a group on the molecular chain, it may produce other groups, meanwhile, the coupling peak might change into a single peak. In conclusion, the protein characteristic group peaks centered at 1625.8 cm^{-1} corresponding to amide A, which manifesting that $C = O$ stretching vibration peak has not been destroyed by lye. At the same time, when the mass fraction of sodium hydroxide is 8%, the damage to feather fibrin is the greatest and the resulting particle size is the smallest.

The experimental results show that different concentrations of lye can get different size of particles, in which way we can use to prepare size controllable nanoparticles of feather fibers, applying in medicine, food industry and other fields to achieve recycling utilization and environmental protection.

References

[1] Shannahan JH, Lai X, Ke PC, et al. Silver Nanoparticle Protein Corona Composition in Cell Culture Media [J]. Plos One: 2013; 8(9): e74001.

[2] Kohane DS. Microparticles and nanoparticles for drug delivery [J]. Biotechnology & Bioengineering: 2007; 96 (2): 203.

[3] Huda S, Yang Y. Feather Fiber Reinforced Light-Weight Composites with Good Acoustic Properties [J]. Journal of Polymers and the Environment: 2009; 17 (2): 131-142.

[4] Evans MR, Karcher D. Properties of plastic, peat, and processed poultry feather fiber growing containers [J]. Hortscience A Publication of the American Society for Horticultural Science: 2004; 39 (5): 1008-1011.

[5] Ullah A, Wu J. Feather Fiber-Based Thermoplastics: Effects of Different Plasticizers on Material Properties [J]. Macromolecular Materials & Engineering: 2013; 298 (2): 153-162.

[6] Verbeek CJR, Berg LEVD. Extrusion Processing and Properties of Protein-Based Thermoplastics [J]. Macromolecular Materials & Engineering: 2010; 295 (1): 10-21.

[7] Khalil HA, Tehrani M, Davoudpour Y, et al. Natural fiber reinforced poly (vinyl chloride) composites: A review [J]. Journal of Reinforced Plastics & Composites: 2013; 32 (5): 330-356.

[8] Laohakunjit N, Noomhorm A. Effect of Plasticizers on Mechanical and Barrier Properties of Rice Starch Film [J]. Starch - Stärke: 2004; 56 (8): 348-356.

[9] Laohakunjit N, Noomhorm A. Effect of Plasticizers on Mechanical and Barrier Properties of Rice Starch Film [J]. Starch- Stärke: 2004; 56 (8): 348-356.

[10] Song NB, Jo WS, Song HY, et al. Effects of plasticizers and nano-clay content on the physical properties of chicken feather protein composite films [J]. Food Hydrocolloids: 2013; 31 (2): 340-345.
[11] Jackson M, Mantsch HH. The use and misuse of FTIR spectroscopy in the determination of protein structure. [J]. Critical Reviews in Biochemistry and Molecular Biology: 1995; 30 (2): 95.
[12] And MNA, Dufresne A. Plasticized Starch/Tunicin Whiskers Nanocomposites. 1. Structural Analysis[J]. Macromolecules: 2000; 33 (22): 8344-8353.
[13] Gupta VB, Drzal LT, Lee YC, et al. The temperature-dependence of some mechanical properties of a cured epoxy resin system [J]. Polymer Engineering & Science: 1985; 25 (13): 812-823.
[14] Pommet M, Redl A, Guilbert S, et al. Intrinsic influence of various plasticizers on functional properties and reactivity of wheat gluten thermoplastic materials. [J]. Journal of Cereal Science: 2005; 42 (1): 81-91.
[15] Bamford CH, Hanby WE, Happey F. The Structure of Synthetic Polypeptides. I. X-Ray Investigation [J]. Proceedings of the Royal Society A: 1951; 205 (1080): 30-47.
[16] Barone JR, Schmidt WF, Liebner CFE. Compounding and molding of polyethylene composites reinforced with keratin feather fiber [J]. Composites Science & Technology: 2005; 65 (3-4): 683-692.
[17] Bullions TA, Hoffman D, Gillespie RA, et al. Contributions of feather fibers and various cellulose fibers to the mechanical properties of polypropylene matrix composites [J]. Composites Science & Technology: 2006; 66 (1): 102-114.

Preparation and Properties of Silk Fibroin Microsopheres

Shu Chen, Chen Lu, Jie Zhang, Shu-Qin Yan, Ren-Chuan You, Qiang Zhang*

School of Textiles Science and Engineering, Wuhan Textile University, Wuhan, Hubei, 430073, P.R.China

*Corresponding author's email: qiang.zhang@wtu.edu.cn

Abstract

In this study, a quickly and effective method to prepare well dispersivity microspheres has been reported. The isopropyl alcohol has been used as denaturant to process the fresh silk fibroin solution which can induce the formation of β-sheets from random coil structure quickly, then the process of freezing had been used to induce the self-assembly of the β-sheets silk fibroin macromolecules. The size and distribution of silk fibroin microspheres can be controlled by changing the molecular weight of silk fibroin macromolecules. The surface features were observed by scanning electron microscopy (SEM). The results indicated that the silk fibroin microspheres are spherical, the size and dispersvity of silk fibroin microspheres diameter were determined by article size analysis. The result showed that the diameter of silk fibroin microspheres prepared from LiBr solution at 90 °C decreased by 45.1% compared to 60 °C, with the distribution of the diameter ranging from 50 nm to100 nm, and the average value is 86.5 nm; the diameter of silk fibroin microspheres prepared from $Cacl_2/H_2O$/Et-OH solution at 100 °C decreased by 43.2% compared to 72 °C, and the diameter is 60-120 nm, the average value is 101.7 nm; the second structure of silk fibroin microsphere were investigated by Fourier Transform Infrared Spectroscopy (FTIR), the result suggested that the structure of silk fibrion molecule of microspheres has changed to β-sheet or α-helix structure. In this way, the micro and nanospheres can be prepared quickly and effectively, which will largely broaden the applications of silk fibroin microspheres.

Keywords: Silk Fibroin; Microsphere; Self-assembly; Molecular Weight

1. Introduction

As a kind of natural protein fiber, silk has excellent mechanical properties, good biocompatibility and biodegradability. Compared with other polymeric materials, silk fibroin, a natural polymer material, derived from living organisms, and some of the biological information containing its amino acid sequence can be easy identified by cells; Silk fibroin is a high purity protein secreted by the inner wall of the endothelial cells on silkworm silk gland, containing no organelles, with no risk of cross infection between human pathogenic microorganisms, biological safety and reliable security; silk fibroin can be biodegraded, the final product is amino acids or oligopeptides, easily absorbed by the body and dissolved, purified and processed into a variety of forms of materials. Based on the above advantages, silk fibroin has been used in cosmetics and food additives for about 30 years, in the field of surgical sutures nearly a hundred years [1]. In drug delivery and regenerative medicine as micro or nano-carriers, silk fibroin microspheres have special affinity with some cells and tissues [2], and can concentrate the drugs contained in the targets [3], thus achieving targeted therapy for the body lesions [4], showing a great potential for application and market prospects. Therefore, the preparation, structure, property and application of silk fibroin sustained-release microspheres have become the research focus and it mainly applied in the field of pharmaceutical preparation and regenerative medicine.

Self-assembly refers to the process that micro and nanometer particles spontaneously form regular structures. The self-assembly of proteins and peptides is mainly based on the interaction of hydrogen bonds, electrostatic interactions and hydrophobic interactions. With an average diameter of 980 nm, silk fibroin microspheres were prepared by using self-assembly method [5]. Nanoparticles were first induced by ethanol and then added to polyvinyl alcohol (PVA) solution, frozen for 24 h, and then the drug is loaded by adsorption. Drug-loaded silk fibroin microspheres were prepared by self-assembly via dissolving the model drug PTX in ethanol, then mixing with the silk fibroin solution, and then freezing for 24 h at -20 °C [6]. And also, a method jointed ethanol, freezing and self-assembly were employed to fabricate nanoscale microspheres [7], which was more high-efficiency than the particles derived form frozen shear self-assembly method although without adding any toxic chemical crosslinking agent [8]. In order to get SF microsphere by green, mild ways, another electric field was adopted silk fibroin nano microspheres as template, using the self-assembly way to obtain the calcium carbonate microspheres in diameter of 10-2000 nm, no toxic chemical reagents added in the whole process of microsphere preparation, fully meet the makeup, paint, drug loaded materials and other requirements [9]. Cheng et al. [10] also obtained the silk fibroin microspheres by self-assembly method, and the microspheres were prepared with the chloroform solution, the micro-capsule size controllable, good permeability, is a potential drug carrier. Wang et al. [11] prepared silk fibroin (SF)/recombinant human bone morphogenetic protein-2 (rhBMP-2) sustained-release microparticles in a round shape with a particle size of 300-900 nm and a drug loading rate of 4.5% by freeze-drying. The results showed that the ratio of SF to the crosslinking agent, the mass ratio of rhBMP-2 to SF had some influence on the particle size and drug loading rate. Liu YC, Qian JC, et al. [12] prepared the curcumin-containing silk fibroin microspheres in a round shape with a particle size of 210-320 nm and the dispersion index was about 0.17 and the entrapment efficiency and encapsulation efficiency were 40.27% and 1.22% respectively, and the study showed that the particle size of drug-loaded microspheres increased with the adding of the mass fraction of silk fibroin, and the initial mass concentration of curcumin increased, the particle size of microspheres decreased slightly.

The method of preparing microspheres by self-assembly is simple, the reaction condition is mild, the yield is high, the harmful solvent is not introduced, and the hydrophobic or hydrophobic drug loading can be realized. However, the size and dispersibility of the microspheres has yet to be optimized. This study focused on the advantages of using self-assembly method to prepare silk fibroin microspheres, at the same time, using different methods to prepare silk fibroin solution to obtain different molecular weight of silk fibroin macromolecules corresponding average particle size is small and narrow distribution and silk fibroin microspheres ~~with~~ has better dispersibility.

2. Materials and Methods

2.1 Preparation of Silk Fibroin Microspheres

The silkworm was degummed in a Na_2CO_3 of 0.06% at a bath ratio of 1:50, at 98-100°C, treated with deionized water for 30 minutes twice, then followed by degumming for 45 min. the degummed silk fibers were washed thoroughly ~~cleanly~~ with deionized water every time. The silk fibers were dried at the 55 °C in oven to obtain the refined silk fibroin. In order to dissolve the silk fibroin fibers, 9 M LiBr solutions were used as solvent at a bath ratio of 1: 5. The solutions were stirred at 60±2°C, 72±2 °C, 90±2 °C, 100±2 °C for 2 hours to obtain a silk fibroin solution. And also the silk fibroin fiber was dissolved in a solvent system of $CaCl_2/CH_3CH_2OH/H_2O$ (1:2:8 in molar ratio) at 72±2 °C, 90±2 °C, 100±2 °C for 2 h. The obtained solution was dialyzed with deionized water for 4 days (Spectrum, Shanghai, China, molecular weight cutoff 8000-14000 Da). The water was changed every 2 hours. The silk fibroin solution was obtained by filtration and centrifugation in a refrigerator at 4 °C for 15 min.

The concentration of the silk fibroin was determined by equation (1). Here, The three aluminum plates were put into the 105 °C oven for 1h, its weight was recorded as W (g). 1mL of silk fibroin solution was injected into the plates using a pipette and its weight was W_1 (g), After 4 h drying, its weight was recorded as W_2 (g). The concentration of the silk fibroin solution was calculated by the equation (1). The concentration of fresh silk fibroin solution was about 4 wt%.

$$C=\frac{W_2 - W}{W_1 - W} * 100\% \tag{1}$$

The fresh silk fibroin solution prepared by different methods was diluted into 1wt%. The mixed solution of isopropyl alcohol and freshly prepared silk fibroin was mixed at a volume of 1: 5, stirred at 37 °C for 5 min, freezed at -20 °C for 24 h, followed by thawing at room temperature to obtain silk fibroin microspheres.

2.2 Scanning Electron Microscope

The morphology of the silk fibroin and porous SF/HA materials was observed by Hitachi. S-570 Scanning Electron Microscope (SEM).

2.3 Fourier Transform Iinfrared (FTIR)

The blend porous scaffolds were cut into micro-particles with radius less than 40 μm, and then samples were prepared in KBr pellets. The spectra were obtained with a Nicolet Avatar-IR360.

3. Results and Discussion

3.1 Microspheres Morphology

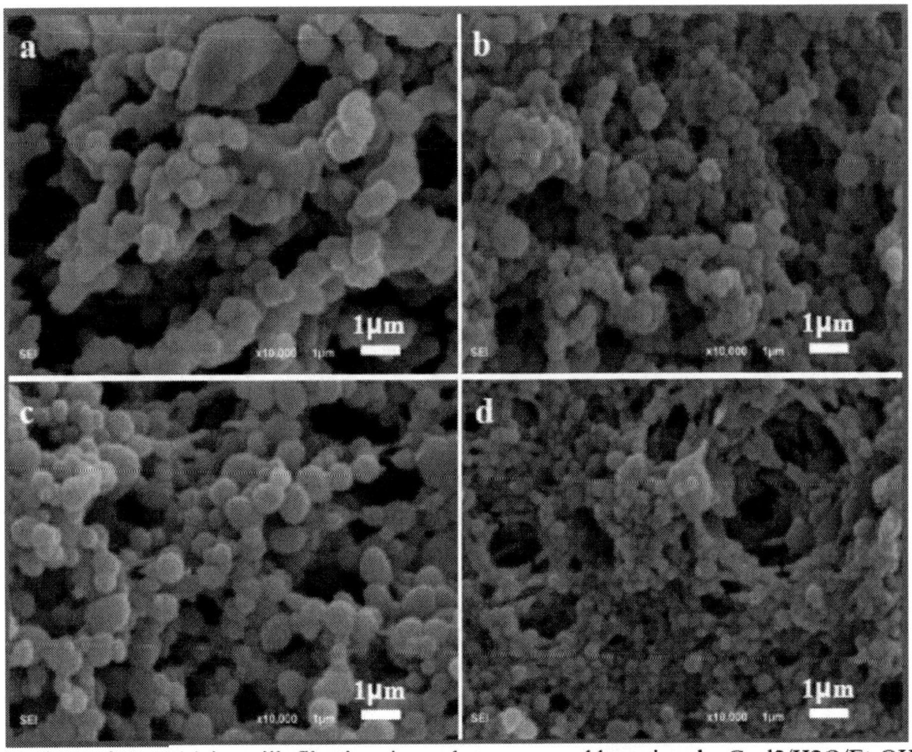

Fig.1 Fibroin microspheres; (a) is a silk fibroin microsphere prepared by using the Cacl2/H2O/Et-OH solution at the concentration of 1wt% under the temperature of 72 °C; (b) is a silk fibroin microsphere prepared by using the Cacl2/H2O/Et-OH solution at the concentration of 1wt% under the temperature of 100 °C; (c) is a silk fibroin microsphere prepared by using a solution of LiBr at a concentration of 1wt% under the temperature of 60 °C; (d) is a silk fibroin microsphere prepared by using a solution of LiBr at a concentration of 1wt% under the temperature of 90 °C.

The morphology of the microspheres was observed by scanning electron microscopy (SEM), and the diameter of the microspheres was analyzed preliminarily. The results showed that the morphology of silk fibroin microspheres was spherical when the concentration of silk fibroin solution was the same and the same preparation method of silk fibroin. At the same time, the experimental results showed that the particle size of silk fibroin and the preparation of silk fibroin solution method have a clear relationship. Under the same conditions, the silk fibroin microspheres prepared by dissolving the silk fibroin solution with Cacl2/H2O/Et-OH solution under the temperature of 100°C were smaller than that of the silk fibroin microspheres prepared by dissolving the silk fibroin solution with Cacl2/H2O/Et-OH solution under the temperature of 72 °C; At the same time, the silk fibroin microspheres prepared by the solution of LiBr solution at 90 °C were smaller than the silk fibroin microspheres prepared by the solution of LiBr solution at 60 °C. The results showed that the condition of preparing silk fibroin solution was relatively strong, the breaking degree of the obtained silk fibroin macromolecule would be relatively higher, the relative molecular weight of the obtained silk fibroin macromolecule may be smaller, thus further affecting the diameter of the microsphere and diameter distribution.

3.2 Particle Size Analysis

The average particle diameter and size distribution of silk fibroin microspheres prepared by different methods were analyzed by SmileView software. The size of the silk fibroin microspheres prepared by using the Cacl2/H2O/Et-OH solution at 72 °C was 100-250 nm and the average particle size was 179.3 nm; The size of the silk fibroin microspheres prepared by using the Cacl2/H2O/Et-OH solution at 100 °C was 60-120 nm and the microspheres with the diameter of about 80 nm accounted for the majority, the average diameter was 101.7 nm. The size of the silk fibroin microspheres prepared by the LiBr solution at 60 °C was 100-300nm and most of the particles were 150-200 nm. The mean particle size was 157.6 nm. The size of the silk fibroin microspheres prepared by the LiBr solution at 90 °C was 50-100 nm, and most of the particles size was at 70 nm and the average diameter was 86.5 nm. The results showed that silk fibroin solution prepared by different methods had obvious relation to the diameter and distribution of the final silk fibroin microspheres. The width and average size of the silk fibroin microspheres prepared at 100 °C in Cacl2/H2O/Et-OH solution was significantly smaller than those at 72 °C. Similarly, the particle size distribution and average size of the silk fibroin microspheres prepared at 90 °C in LiBr solution were significantly smaller than those at 60 °C. It is conjectured that the condition of the preparation of silk fibroin solution has a certain influence on the molecular weight and distribution width of the obtained silk fibroin molecules, which can further control the particle size and distribution width of the prepared silk fibroin microspheres.

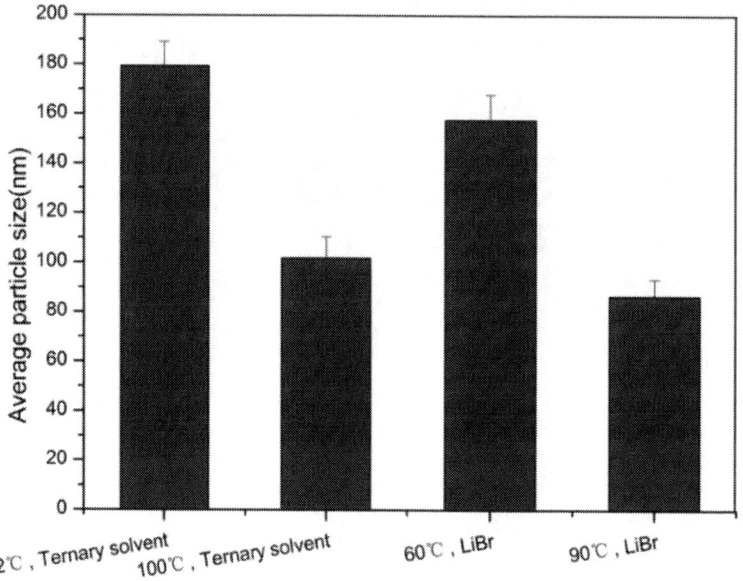

Fig.2 Average particle size of silk fibroin solution prepared under different conditions

3.3 FTIR

The second structure of the microspheres was analyzed by fourier transform infrared spectroscopy (FTIR). The infrared absorption spectrum of silk fibroin showed the absorption peak of Silk I are band of 1650-1655 cm^{-1} (amide I), 1525-1540 cm^{-1} (amide II), 1266 cm^{-1} (amide III), 669 cm^{-1} (amide V). The absorption peaks of Silk II are band of 1620-1635 cm^{-1} (amide I), 1531 cm^{-1} (amide II), 1230-1235 cm^{-1} (amide III) and 695 cm^{-1} (amide V). The absorption peak of random coil appeared in the range of 1655-1660 cm^{-1}, 1535-1545 cm^{-1}, 1235 cm^{-1}, 650 cm^{-1}, while the freshly prepared silk fibroin was unstable random coil structure [12]. The results showed that the infrared absorption peak of silk fibroin microspheres prepared by silk fibroin solution was basically the same. The infrared absorption peaks at 1650 cm^{-1} and 1535 cm^{-1} appear obviously. And 1650 cm^{-1} is the characteristic absorption peak of α-helix structure, and 1535 cm^{-1} is the characteristic absorption peak of β-sheet structure. The results indicated that the structure of silk fibroin was partially β-sheet, forming a stable β-sheet structure. Meanwhile, there was also a relatively metastable α-helix structure.

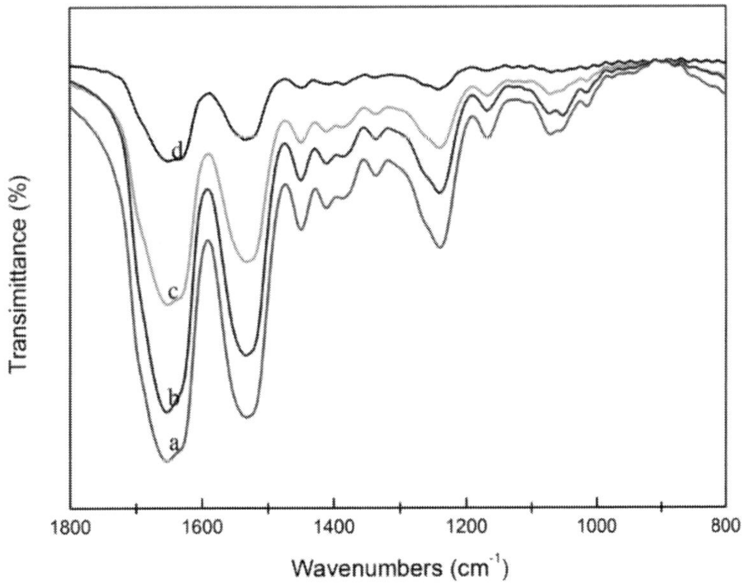

Fig.3 FTIR of silk fibroin microspheres: (a) is silk fibroin microsphere prepared by using a solution of LiBr under the temperature of 90 ℃; (b) is silk fibroin microsphere prepared by using the Cacl2/H2O/Et-OH solvent under the temperature of 100 ℃; (c) is silk fibroin microsphere prepared by using the Cacl2/H2O/Et-OH solution under the temperature of 72 ℃; (d) is silk fibroin microsphere prepared by using a solution of LiBr under the temperature of 60 ℃.

4. Conclusions

In this paper, silk fibroin microspheres were prepared by using different solutions, and the effects of various silk fibroin solutions on the size and distribution of silk fibroin microspheres were discussed. It was concluded that the silk fibroin microspheres prepared by silk fibroin solution obtained by relatively severe way had smaller particle size and particle size distribution width. The results also showed that the silk fibroin microspheres can be successfully prepared by the solution of the silk fibroin dissolved at 100℃ with the Cacl2/H2O/Et-OH solution and the silk fibroin solution dissolved in the LiBr solution at 90℃, and the distribution width of the microspheres is narrow. This method can be quickly and effectively prepared micro and nano-silk fibroin microspheres, which makes silk fibroin microspheres more promising in the drug-loaded aspects of the application.

5. Acknowledgments

This work was supported by the National Nature Science Foundation of China (51403163, 51303141 and 31600774), Educational Commission of Hubei Province, China (154004).

References

[1] Moyer L, Lee A, Zalka A. Commonly used suture materials in skin surgery. Am Fam Physican: 1991; 44: 2123-2128.
[2] Veparia C, Kaplan DL. Silk as a biomaterial. Prog.Polym.Sci: 2007; 32: 991-1007.
[3] Han F, Liu S, Liu X, et al. Woven silk fabric-reinforced silk nanofibrous scaffolds for regenerating load-bearing soft tissues. Acta Biomater: 2014; 10: 921-930.
[4] Cheng XM, Li YB, Zuo Y, Zhang L, Li JD, Wang HN. Properties and in vitro biological evaluation of nano-hydroxyapatite/chitosan membranes for bone guided regeneration. Materials science & engineering: 2009; 29.
[5] Shi P, Gohjch. Release and cellular acceptance of multiple drugs loaded silk fibroin particles. International Journal of Pharmaceutics: 2011; 420 (2): 282-289.
[6] Chen M, Shao Z, Chen X. Paclitaxel-loaded silk fibroin nanospheres.Journal of Biomedical Materials Research Part A: 2012; 100 (1): 203-210.
[7] Cao ZB, Chen X , Shao ZZ. The preparation of regenerated silk fibroin microspheres. Soft Matter: 2007; 3 (7): 910-915.
[8] Yaowalak Srisuwan, Prasong Srihanam, Yodthong Baimark. Preparation of Silk Fibroin Microspheres and Its Application to Protein Adsorption. Journal of Macromolecular Science. Pure and Applied Chemistry: 2009; 46.
[9] Dmitry V. Volodkin, Regine von Klitzing, Helmuth Mohwald. Pure Protein Microspheres by Calcium Carbonate Templating. Angewandte Chemie: 2010; 49.
[10] Cheng C, Ian T, Oliver B. Stimuli-Responsive Capsules Prepared from Regenerated Silk Fibroin Microsphere. Macromolecular Bioscience: 2014; 14 (6): 807-816.
[11] Wang P, Pi B, Wang JN, Zhu XS, Yang HL. Preparation and properties of calcium sulfate bone cement incorporated with silk fibroin and Sema3A-loaded chitosan microspheres. Frontiers of materials science: 2015; 9.
[12] Liu YC, Qian JC, et al. Blend membrane of regenerated silk fibroin, poly (vinylalcohol), and peroxidase and its application to a ferrocene mediating hydrogen peroxide secsor. Journal of Applied Polymer Science: 1996; 61: 641-647.

A Review on Fabricating Graphene Fiber Wet Spinning

Heng Zhai[1], Yi Li[2*], Yang-Yang Fan[2]

[1,2]*School of Materials, University of Manchester, Oxford road, Manchester, M13 9PL, UK*

[*]Corresponding author's email: henry.yili@manchester.ac.uk

Abstract

Since the single-layer flakes of graphene were first isolated by Geim and his co-workers at the University of Manchester in 2004, the research on graphene has made a great breakthrough and the applications of graphene have extended to various aspects in devices, sensors, composites and so on. In this paper, a design of graphene fiber via wet spinning was introduced. Firstly, the basic structure and properties of graphene are introduced, followed by the production process of graphene. To change the poor dispersibility of graphene, the most accessible precursor of graphene, chemically oxidized graphene or graphene oxide was introduced. In this part, the most popular method to produce GO by Hummers is explained in detail. A subsequent study was carried out to examine the different forms of graphene oxide liquid crystals (GO LC) and its complicated behaviors to fabricate graphene fiber via wet spinning. In the last part of the paper, a new attempt of graphene/viscose composite fiber was given to compensate the disadvantages of monotonous graphene fibers.

Keywords: Graphene; Graphene Oxide; Wet Spinning; Dispersibility; Graphene Fiber.

1. Introduction

1.1 The Graphene

Graphene is a flat two-dimension (2D) monolayer of sp^2-hybridized carbon atoms tightly piled into a honeycomb lattice, it is the elementary component of all the graphene-based materials [1]. Graphene is a fast becoming a rising star in the field of material science and condensed-matter physics [2]. It has attracted much attention due to its extraordinary electronic, mechanical and thermal properties which mainly arises from its unique atomic thickness and 2D structure [3]. It was only in 2004 when Geim and his co-workers at the University of Manchester managed to isolate the single layer samples from graphite did the development of graphene turn a new page [4]. Since then, the promising applications of graphene in electronic devices, sensors and composites have lead to an explosion interest.

1.2 The Structure of Graphene

As referred above, graphene is a single layer of carbon atoms tightly packed into a honeycomb crystal lattice. Each atom has four bonds, one σ bond with each of its three neighbors and one π bond that is oriented out of plane. On one hand, graphene can be seen as a strictly two-dimensional material, which gives it a promising crystal quality that allows electrons to travel without scattering. On the other hand, on the basis of both theory and experiment, the perfect two-dimensional crystals cannot exist in free state [5]. In fact, studies by transmission electron microscopy (TEM) show that suspended graphene sheets are not completely flat, they exhibit intrinsic microscopic waving, and the corrugations in the third dimension may offer the subtle possibilities for the stability of two-dimensional crystals [6].

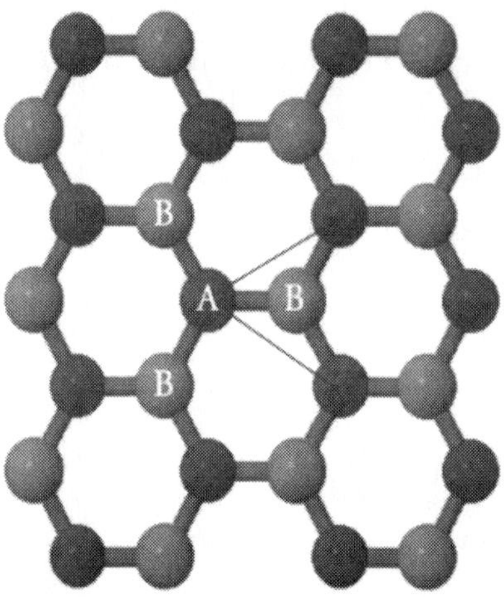

(A) (B)

Fig.1 Picture (A) is the schematic diagram of idea single graphene sheet. In Picture (B), the blue ball stands for the sp^2-hybridized carbon atom (pictures are cited from Wikipedia).

1.3 The Properties of Graphene

The electronic property is one of the most important properties of graphene when it is used in smart conductive materials. It has been ascertained that graphene is a zero-gap two-dimensional semimetal with valence and conductance bands being slightly overlapped. Besides, the charge carriers in it only has litter scattering when moving under ambient conditions. When applying gate voltage to graphene, the concentration of charge carriers can go up to 10^{13} cm^{-2} and room-temperature mobility of 10000 cm^2 V^{-1} s^{-1}, which exhibits strong ambipolar electric field effect [7]. However, for suspended graphene, this amount could be approaching 20000 cm^2 V^{-1} s^{-1} for carrier densities below 5* 10^9 cm^{-2} at low temperature. The elastic properties and intrinsic breaking strength of free-standing monolayer graphene membranes were measured by nanoindentation using an atomic force microscope (AFM) [8]. The measurement of breaking strength is 42N m^{-1} and the Young's modulus of E is 1.0 TPa, which indicates its promising potential to be the strongest material ever measured. The elastic deformation of chemically reduced graphene oxide has been studied by AFM, which shows that the reduced graphene oxide can be reversibly folded and unfolded several times without breakage. The bent point appeared to be the same location [9]. The thermal conductivity of single-layer suspended graphene at room temperature ranges from 3000 to 5000$Wm^{-1}K^{-1}$, which is depended on the size of the graphene sheet [10]. While the measured thermal conductivity of graphene is about 600$Wm^{-1}K^{-1}$, which is an order of magnitude lower than suspended graphene roughly, but it is still two times higher than copper and fifty times higher than silicon which are most commonly used for electronics today [11].

1.4 The Production of Graphene

In the discovery on the original nature of graphene and the exploration on the innovative use of graphene, there is an inevitable problem on how to produce high-quality graphene to satisfy different occasions. Basically, the method for the production of graphene can be divided into wet chemical approaches and non-wet chemical methods, each of which has its own advantages and limitations [12].

Non-wet methods for the production of graphene.

The first way to produce graphene was by the mechanical exfoliation of highly oriented pyrolytic graphite using an adhesive tape, which was the original method that graphene was obtained [7]. As the physical and chemical process was not sophisticated, the graphene crystal produced has less defects, but they were hard to produce in bulk amounts.

The second non-wet method is to grow graphene on an insulating silicon carbide (SiC) surfaces with a high temperature in a vacuum. An advantage of this method is that the grown graphene can be used directly without transferring when using the insulating SiC substrates. However, there is a limitation of large-scale structure quality because of deficiency in uniformity aand continuity of the grown film [13].

The third non-wet approach for the production of graphene is built on chemical vapor deposition (CVD), which is the most widely used and promising approach. During the CVD process, the chemical reaction occurs when the reactor is full of gas species and the hydrocarbon precursors in the hot zone decompose to carbon radicals at the metal substrate surface to form graphene with single-layer and few-layers [14]. Generally, polycrystalline Ni films are used as mental substrate for its relatively high carbon solubility at an increased temperature compared with copper. However, copper is used for surface of synthesis to make more continues films of graphene. Produced with the method of CVD, the graphene can be obtained with high quality, but needs to be transferred to a more desirable surface.

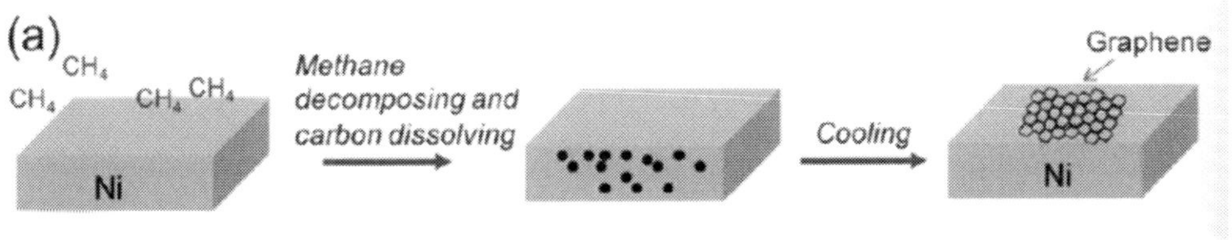

Fig.3 (a)Schematic diagram of graphene formation on Ni [14].

Wet methods for production of graphene

In addition to those non-wet approaches based on solid-phase deposition, graphene can also be obtained by chemical reaction in wet condition. It is demonstrated that single-layer graphene can be synthesized by the chemical reaction of ethanol and sodium followed by low-temperature flash pyrolysis, then provided with a gentle sonication of the carbon product, finally the product is washed in water and dried to obtain pure graphene [15]. The other method is via the organic synthesis with polyaromatic hydrocarbons. We can also gain the graphene though reduction of graphene oxide which will be introduced in the following page.

1.5 Composite of Graphene

Combined with such remarkable properties, graphene is an ideal material to be integrated into the macroscopic assembled materials to translate the splendid properties of individual graphene into macroscopic graphene-based materials. However, before further exploiting these properties in certain applications, reliable synthetic routes need to be developed for the preparation of graphene and its derivatives, arising from the bottom-up epitaxial growth to the top-down exfoliation of graphene through oxidation, intercalation and/or sonication [16]. There are two main graphene-based composites in the last few decades, the first one is graphene-inorganic nanostructure composite, which includes metal like Au, Ag, Pt, Ni, Cu, Ru and Rh [17], oxides like TiO_2, ZnO, SnO_2, MnO_2, Fe_3O_4 and SiO_2 [18]. The second one is the graphene-polymer composite which is depended on the 3D arrangement of graphene and the interaction between graphene sheet and the polymer. There are also other kinds of graphene-based composites like organic crystals, metal-organic frameworks(MOF), biomaterials and carbon nanotubes(CNTs).

2. Fabrication, Structure and Properties of Graphene Oxide (GO)

Although graphene has exhibited so much splendid properties as introduced above, it still has some instinctive deficiencies which hinder the development of the direct scalable assembly of graphene. For example, in most cases, graphene has limited dispersibility in common solvent. However the most accessible precursor of graphene, which is named as chemically oxidized graphene or graphene oxide (GO), with an abundant oxygen functional groups on its basal planes and edges is able to be completely exfoliated and colloidal dispersions of single sheet are formed in water and other polar organic solvents [19]. The superiority of GO, such as long-term solubility, easy handling and scalable production with low cost makes it reliable in assembling 2D sheets into macroscopic graphene-based architectures [1].

2.1 The Structure of GO

Due to the intriguing properties of graphene oxide, interest in the structure of GO has increased. Although there is widely acceptance that epoxy and hydroxyl groups are its main chemical functional groups, the accurate structure of GO has remained elusive [20]. During the early period, there are the Hofmann Model which only considers C-O structure, Ruess Model which contains both C-O and O-H structure, and the Scholz-Boehm & Nakajima-Matsuo Model which has regular period structure [21]. However, the most acceptable structural model is given by the Leaf-Klinowski Model and Decany Model. The two model differs considerately for the different chemical oxidation process used. In the Leaf-Klinowski Model, the containing of five- and six-membered-ring lactols and an occasional 2-hydroxynaphthalic anhydride or 1,3-dihydroxyx- anthone has been proved by experimental support [20].

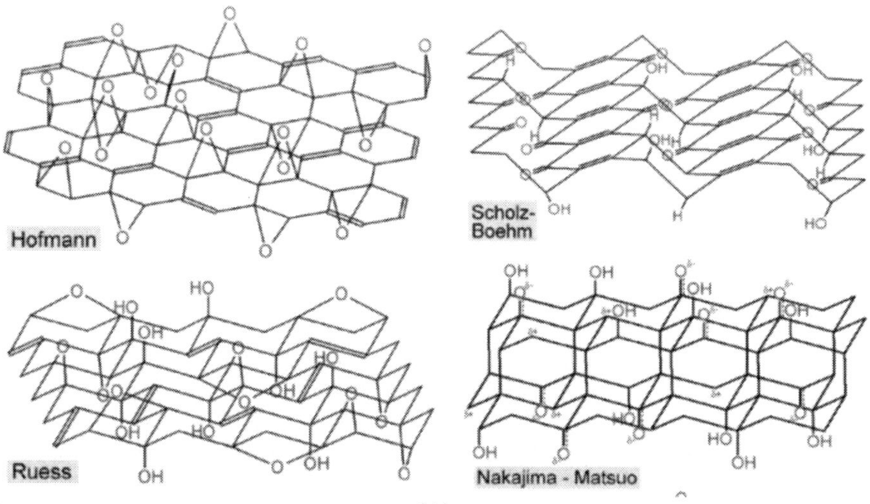

(A)

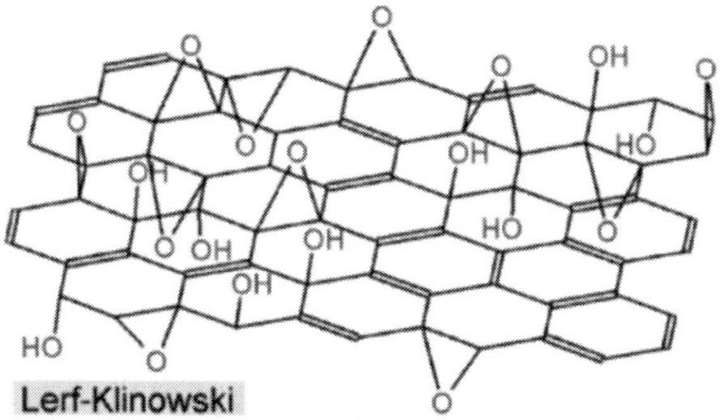

(B)

Fig.4 (A) The four structural models of GO in the early period. (B) The improved Leaf model in recent years.

2.2 The Properties of GO

2.2.1 The Physical Properties of GO

Although during the oxidative process of graphene, quantities of functional groups are introduced into the graphene which causes some defects, the GO still maintains almost the same two-dimensional topology structure and excellent mechanical properties as the original graphene. GO is also accompanied with a series of physical properties that the original graphene does not have, such as good dispersibility.

Mechanical property

The original Young's Module and breaking strength of graphene are tested to be 1.0 TPa and 130 GPa relatively through tip-induced deformation experiment. By the same testing method, Gomez-Navarro found that despite their defect content, the single reduced GO sheets exhibit an unexceptionable stiffness (E= 0.25 TPa), together with a high flexibility which enables it to bend easily in their elastic regime [22]. Schniepp tested the folding and bending properties of reduced GO through atomic force microscope (AFM). he observed that the single sheets can be reversibly folded and unfolded several times without breakage, at the meantime, the folding almost appeared in the same location [9].

Dispersibility

After oxidation of graphene, the surface of GO contains abundant functional groups which weaken the interactions of Van der Waals force between conjugated structure of graphene, meanwhile, there are strong solvation function between the small solvent molecules. Hence, GO exhibits good dispersibility in water and many other organic solvents. In the experiment of Paredes, dispersion of GO produced by Hummers methods has been tested in different organic solvents. According to Figure.6, GO shows long-term stability in water, DMF, NMP and THF but short-term stability in acetone, ethanol, DMSO and pyridine, however, it cannot disperse in dichloromethane and n-hexane [23].

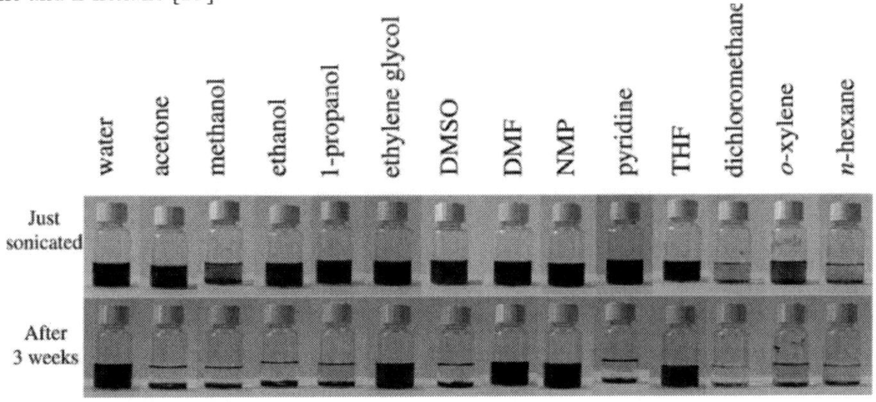

Fig.5 Digital pictures of as-prepared graphite oxide dispersed in water and 12 solvents through bath ultra-sonication(1h). Top: dispersion immediately after sonicated. Bottom: 3 weeks after sonicated [23].

Electrical conductivity of reduced GO

According to Leaf Model, the conjugated region of GO is isolated in the non-conjugated network, which means GO is an electrical insulator. The reduction process of GO is the elimination of oxygen functional groups and recovery of conjugated structure. The reduced GO is not produced until the linked conjugated network is formed, and the electrical conductivity in some certain degree is recovered.

There are many methods for the reduction of GO, different methods determine the various reduced degree of GO. Currently, hydrazine hydrate, hydroiodic acid and chemical heat-reduce are the best way to gain higher electrical conductivity of rGO. For example, the electrical conductivity of rGO can reach between 300 and 7200 S/m when using hydrazine hydrate [24], $3.0*10^4$ S/m when adopting hydroiodic acid [25], $1.0*10^5$ or when using the method of chemical heat-reduced in vacuum condition [26].

2.2.2 Chemical Properties

If we focus on the chemical structure models of GO, the chemical properties of GO can be divided into two main parts. One is relating to the undamaged conjugated structure of GO which is similar to the original graphene, the other one relates to the oxygen functional groups in GO.

Conjugated structure

In the Leaf Model, the conjugated regions isolated in the sp3-network have some certain chemical activity, even though there exist highly-active double bonds. Hence, GO exhibits abundant double-bond chemical properties. Zhong explored the application of modified reduced GO through aryne cycloaddition, which allows it to disperse better in more organic solvents [27].

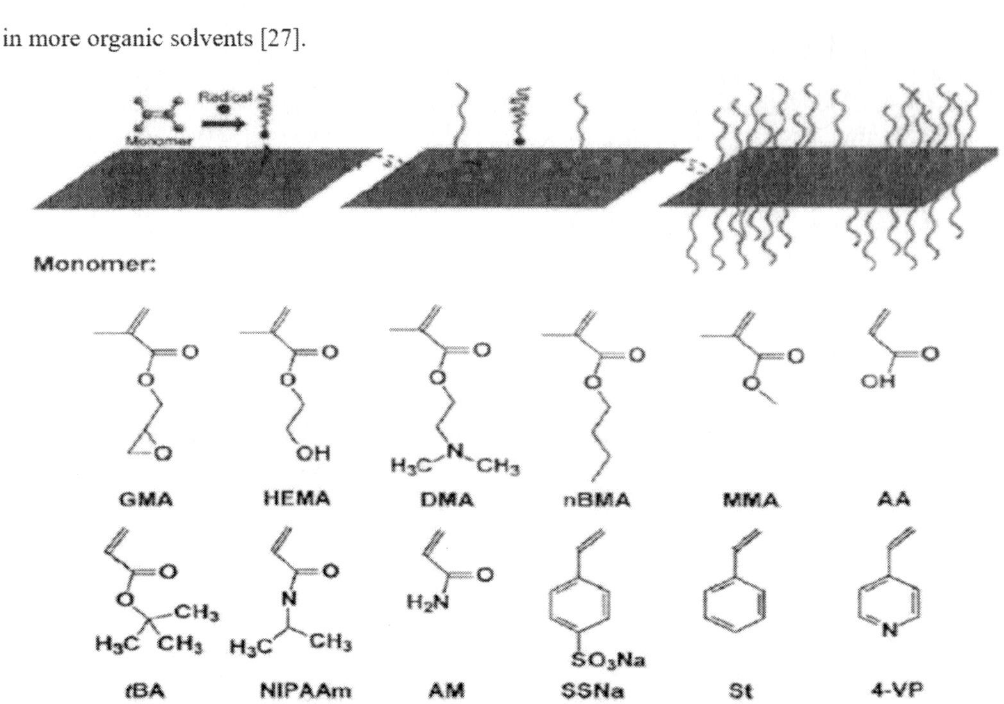

Fig.6 Preparation of vinyl polymers-grafted graphene oxide by in-sity free radical polymerization and the list of adopted monomers[28].

Oxygen functional groups

There are three main oxygen functional groups in graphene oxide: carboxyl group, hydroxyl group and epoxide group. This ample oxygen functional groups support an excellent platform for the modification of GO, thus the regulation of the surface properties of GO are realized, resulting in advancing the dispersibility of GO and optimizing the properties of graphene-based materials. The solubility of modified graphene in the THF solution can be 0.5mg/ml. Furthermore, it can be dissolved in non-polar solvents, such as carbon tetrachloride and methylene dichloride as well. It has been confirmed by AFM that modified graphene presents a single-layer dispersion in non-polar solvents [29].

Reduction of GO

The scale production of graphene can be realized by the reduced process of GO. However, how to synthetize high-quality graphene through chemical reduction approach is also a crucial research project. At present, the common reduction methods are chemical reduction and heat-reduction. The other methods contain photo-reduction, electrochemical reduction, microwave reduction and so on [30].

Chemical reduction

Hydrazine hydrate is used for the effective reductive agent of GO in the earlier period, which can reduce the GO in both dispersed solution and steamy form. The chemical mechanism of hydrazine hydrate is to open the loop of epoxy to eliminate the oxygen atom and single nitrogen atom to form double bond through dehydration.

Different from using the hydrazine hydrate, the new method with hydroiodic acid will not produce plenty of gases that destroys the materials. The graphene-based material by reduction of hydroiodic acid has high electrical conductivity and becomes firmer which will help to enhance its structural integrity and mechanical properties [25].

Heat-reduction

Under elevated temperature, the functional groups in GO decompose, and conjugated structure at the same time recovers to some extent. During the process of heat-reduction, gases with small molecular exhausted, which will have a good reduced effect on very thin assembly such as thin films. However, in regards to the large thick GO assembly, the gases cannot be released properly, resulting in the damage of the integral structure. To this point, the chemical reduction can be firstly used followed by the heat-reduction to optimize the reduced degree.

2.3 The Production of GO

The theory on chemical synthesis of GO is based on addition of strong oxidant, such as potassium perchlorate and potassium permanganate, and in destroying the conjugated structure and surficial oxygen functional groups of GO within the condition of strong acid. On one hand, this oxygen functional groups recedes the Van de Waal's force among the graphene sheets. On the other hand, they enhance the solvation among the dispersed solvent molecules. In the end, within the combination of these two functions, the graphite oxides dissolve into single-layer graphite oxides, which is graphene oxide. The methodology to produce GO are Brodie's method, Staudenmaier's method and Hummers' method.

2.3.1 Brodie's Method

Brodie was the forerunner of creating the abundant chemical properties of graphite and graphene, his methody to obtain high-oxidation GO needs several repetitive oxidative processes. When adopting the potassium chlorate and nitrosonitric acid as the oxidizing agent and reactive solvent, the carbon oxygen ration of graphite oxide created was 2.2:1. According to the knowledge at that time, Brodie confirmed the molecular formula of graphite oxide as $C_{2.19}H_{0.8}O_{1.00}$. At the same time, he found graphite oxide had a good dispersion in water and alkaline solution, but could not be dispersed in the acid solution. So graphite oxide was named as graphite acid [31]. When it has been heated up to 220°C, the carbon oxygen ration rose to 5.5:1, which is the experimental rudiment of heat-reduction of GO at present.

2.3.2 Staudenmaier's Method

Forty years after graphite oxide had been synthesized by Brodie, L. Staudenmaier for the first time through overcoming the disadvantage of repetitive oxidation by dividing the addition of potassium perchlorate into graphite several times instead of the one-step addition by Brodie. The shortcoming of Staudenmaier's method is that there are strong corrosive gases created during the synthesis of GO, and particular safety devices are needed for mass production [32].

2.3.3 Hummers' Method

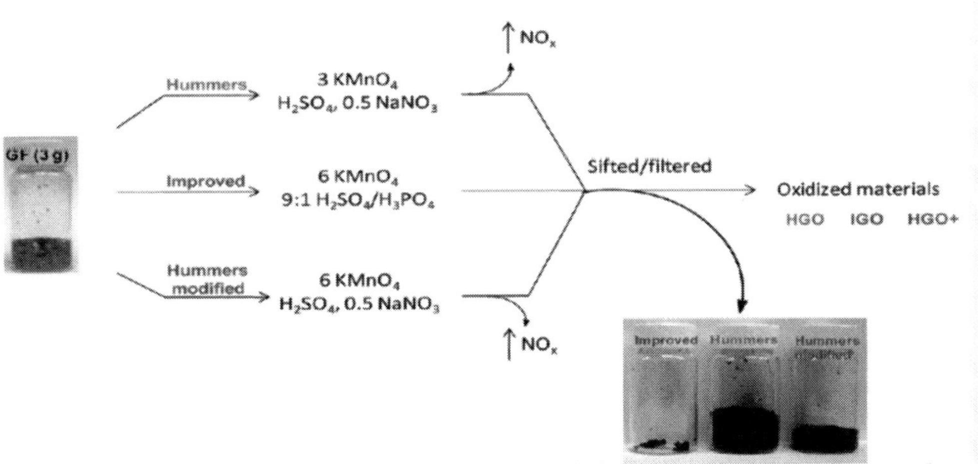

Fig.7 The contrastive experiments of Hummers, Improved Hummers and Hummers modified. The procedure starts with graphite flakes(GF). Through purification of IGO, HGO and HGO+, the increased efficiency of IGO method is accompanied with the small amount of oxidized materials produced [35].

To avoid the exhaustion of toxic gases, Hummers and Offeman employed the oxidative system of potassium permanganate and concentrated sulfuric acid to produce highly oxygenated graphite oxide [30]. Without providing special safety devices, Hummers' method has been widely used. In this system, Mn_2O_7 is the intermediate that makes valid oxidative functions, which dissolves in the concentrated sulfuric solution and presents atrovirens observed during the experiment [33]. Because of the strong protonation by the concentrated sulfuric, the oxidative agents enter the gaps between the graphene sheets accompanied with concentrated sulfuric at the same time. The oxidative process starts from the crystal edge of graphite, with the continuity of reaction, the defects of conjugated regions enlarge gradually. Finally, it turns into highly-oxidative graphene oxide.

Based on the Hummers' method, a variety of modified methodologies have come up, for example, synthesizing ultra-large GO sheets through modifying technology [34]. Another instance is optimizing the post treatment technology by adopting phosphoric/sulfuric acid compound system [35].

3. Fabrication of Graphene fibers via Wet Spinning

With the help of Web of Science, plenty of research results related to our project Design of graphene fiber via wet spinning can be shown which can help us to know this field deeply.

3.1 Research on Web of Science

Searching topic: Graphene fiber wet spinning

Result found: 70

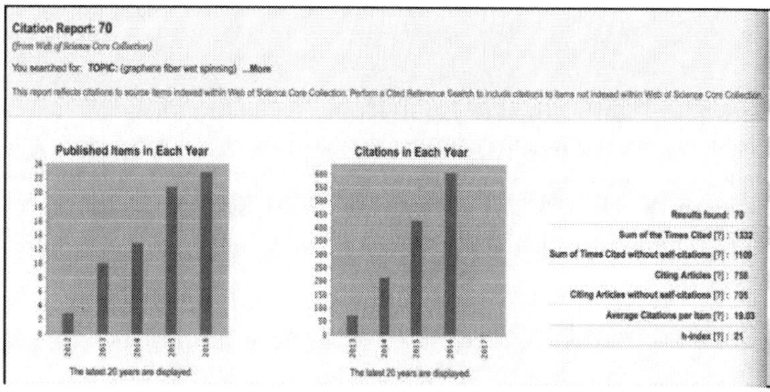

(A)

Field: Authors	Record Count	% of 70	Bar Chart
GAO C	15	21.429 %	
WALLACE GG	10	14.286 %	
XU Z	9	12.857 %	
KOU L	6	8.571 %	
RAZAL JM	6	8.571 %	
SUN JK	5	7.143 %	
WANG ZH	5	7.143 %	
ZHAO XL	5	7.143 %	
ZHENG BN	5	7.143 %	
ZHU MF	5	7.143 %	
Field: Authors	Record Count	% of 70	Bar Chart

(B)

Field: Organizations	Record Count	% of 70	Bar Chart
ZHEJIANG UNIV	15	21.429 %	
UNIV WOLLONGONG	12	17.143 %	
DONGHUA UNIV	9	12.857 %	
QINGDAO UNIV	7	10.000 %	
CHINESE ACAD SCI	4	5.714 %	
DEAKIN UNIV	4	5.714 %	
SICHUAN UNIV	2	2.857 %	
UNIV DELAWARE	2	2.857 %	
UNIV MELBOURNE	2	2.857 %	
UNIV SYDNEY	2	2.857 %	
Field: Organizations	Record Count	% of 70	Bar Chart

(C)

Field: Publication Years	Record Count	% of 70	Bar Chart
2016	23	32.857 %	
2015	21	30.000 %	
2014	13	18.571 %	
2013	10	14.286 %	
2012	3	4.286 %	
Field: Publication Years	Record Count	% of 70	Bar Chart

(D)

Fig.7 (A),(B),(C),(D) are the results from Web of Science. (A) is the Citation Report, (B) is field of Authors, (C) is the field of Organizations and (D) is the field of Publication Years.

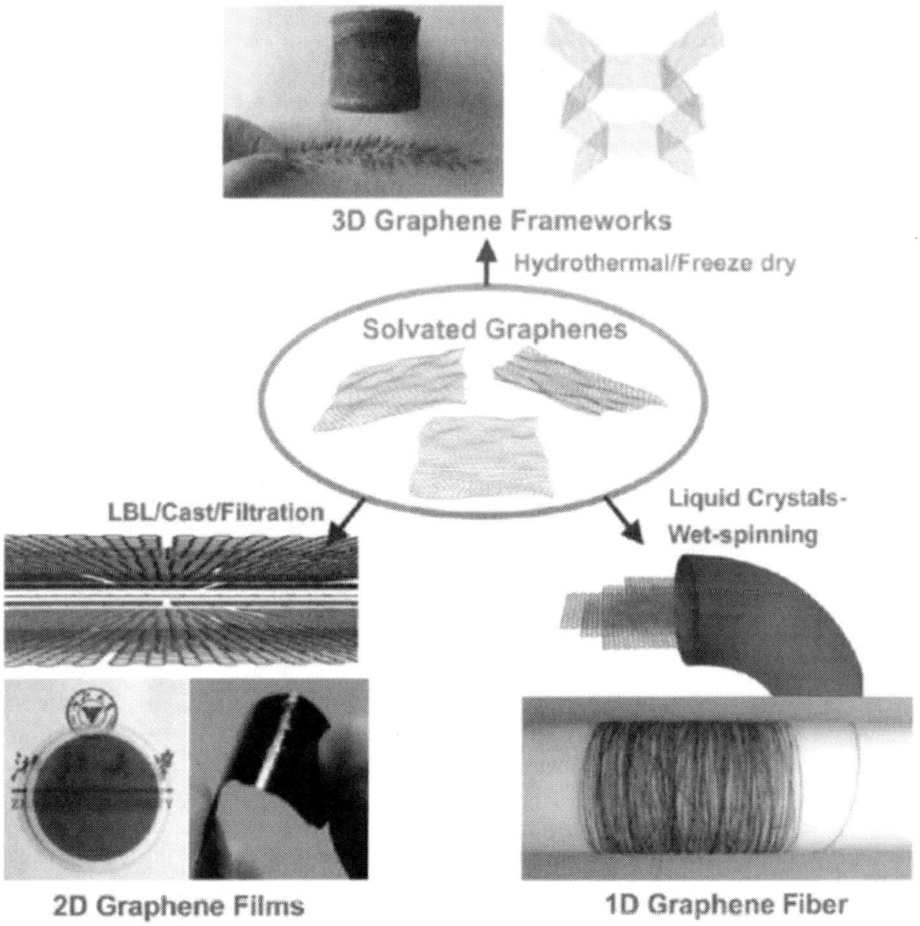

Fig.8 Macroscopic assembly of 1D graphene fiber, 2D graphene films and 3D graphene frameworks [36].

From the research results, it is possible to deduce that research on graphene fiber via wet spinning started roughly from 2012. A total of 70 papers has been published till now. Since the first discovery of graphene in 2004, there has been a rapid development in this field, including research on the structure and properties of graphene and its derivations of graphene oxide. It was until 2012 that the research about graphene-based macroscopic material, graphene fiber, has been broken through and published. The dominated figures in this field of graphene fiber are Gao Chao, Wallace GG and Xu Zhen. Among them, Gao Chao and Xu Zhen are in the same research group from Zhe Jiang University who published the most papers among all the organizations. The main contribution of Gao and Xu is that they found new graphene chiral liquid crystals which can better translate the excellent properties of graphene sheets to macroscopic assembled fibers.

As an example, Zhen Xu and Chao Gao introduced a way to turn graphene to graphene oxide liquid crystals, and finally fabricate graphene fiber via wet-spinning after the reduction of GO fiber. Besides the whole fabrication process, they conducted most research on the mesophase of graphene oxide liquid crystals. Except for the ordinary nematic and lamellar mesophase of GO LC, they found a new form of mesopahase which is chiral liquid crystals (CLC) as GO concentration further increases to 0.38 vol %(nematic phase at 0.23 vol %). They also concluded three important characteristics of this new mesophase: (1) fingerprint-like texture along with aligned bands; (2) strong optical activity; (3) quasi-long-range lamellar ordering. The positive effect of this discovery not only compensates the poor dispersibility of graphene, but also offers an effective possibility to control the surface properties.

Through the analogy of GFs' structure with well-organized polymeric fibers, they found a way to enhance the mechanical strength of graphene fiber. There are three important points to achieve it: (1) introduce high energy bonds to increase the interlayer interaction; (2) reduce the structural defects; (3) advance the regular alignment of graphene paralleled in the fiber axis. The final GFs product will gain enhanced mechanical strength with this method.

The unexceptionable physical properties of graphene have attracted the interest of material scientists all over the world. The main research purpose is to assemble macroscopic material based on the graphene unit and promote the wide applications in real life. Due to the instinct properties, graphene cannot be assembled under the molten state, so the most effective and valid way is to assemble the graphene into macroscopic materials in dispersed solution. The previous researches focus on two-dimensional graphene films and three-dimensional aerogel/aquogel. However, in recent years, the one-dimensional graphene fiber has exhibited superiority in graphene-based smart materials and shown the potentials in scale productions.

However, except for expanding multiple macroscopic assembled forms of graphene-based materials, how to design and fabricate the materials to optimize the properties of graphene is an important part as well. The most outstanding properties of graphene are its mechanical and electrical properties. To apply these two properties in graphene fiber, especially the excellent electrical property which is undoubtedly necessary for conductive graphene fiber, the first inevitable work that needs to be completed is realizing regular ordering of graphene in graphene fiber.

As referred above, the only viable method to achieve macroscopic assembled fibers is to assemble them in dispersed solution. Combining these two necessary requirements, which are liquid-phase assembly and regular ordering of graphene, the core task is to form graphene liquid crystals. The previous research and theory on two dimensional liquid crystalline collision has proved that graphene liquid crystal can be formed.

3.2 Two-dimensional Liquid Crystalline Colloids

Liquid crystal is the phase state between crystal and liquid, which possesses both the viscosity of liquid and the anisotropy of crystals. Since its discovery in 1888, liquid crystal has been widely used in displayer and high-performance fiber. According to the formation conditions of liquid crystal, it can be divided into thermotropic liquid crystal and lyotropic liquid crystal. Because of the extremely high melting point of graphene colloid, the graphene liquid crystal can only be formed by a method of lyotropic liquid crystal. The methodology is to disperse graphene colloid in appropriate solvents, it then proceeds to ordering phase when the solution comes to a certain concentration [37].

With the discovery of more and more two-dimensional liquid crystalline colloids, the species of their liquid crystals have become richer as well. Nematic, columnar and smectic liquid crystalline phase of two-dimensional colloids have been already proved, other novel phase of liquid crystals like chiral liquid crystal have been found and gradually accepted recently.

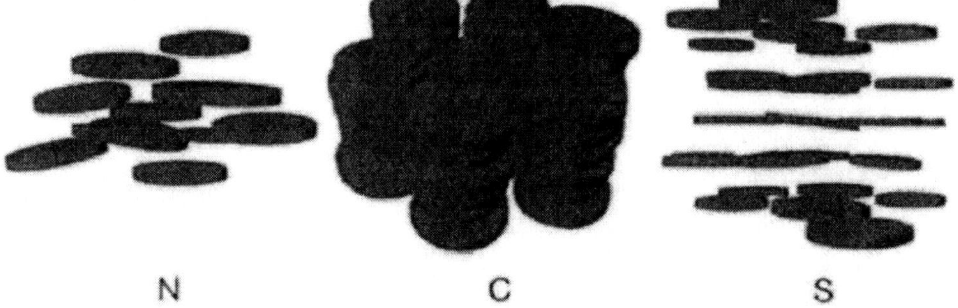

Fig.9 The structural model for nematic (N), columnar (C), and smectic liquid crystalline phases of two-dimensional plate-shape colloids [38].

3.3 Graphene Oxide Liquid Crystals

From the experimental rule of thumb and analogy analysis of small-molecular liquid crystals, it can be concluded that the basic elements for the formation of liquid crystal are asymmetric structures (large width-to thickness ration) and good enough dispersibility. As a two-dimensional nano material, graphene meets the former element of an asymmetric structure. Even while restricted with the instinct poor dispersibility, graphene can be oxidized into graphene oxide which has good and long-term dispersibility. Having met all these two requirements, graphene oxide liquid crystal is proven to exist.

GO Nematic and Lamellar Liquid Crystals

One of the convincing evidence of nematic LC is that GO dispersion began to exhibit Schlieren textures and formed stable mesophases at 5-8 mg/cm3 in POM (polarized optical microscopy) inspection. Through quantitative small-angle X-ray scattering (SAXS), visual SEM and TEM, the orientation ordering along with disclination defects of GO nematic could be confirmed as well [39].

In 2013, Xiang C.S found that the GO nematic family has expanded to contain GO sheets with lateral width between sub-micrometers and several micrometers, and giant GO with scores of micrometers [40]. In Dan's theory, the lateral size plus its size distribution of GO were proven as crucial factors to their liquid crystalline behavior, which was larger in size and higher in aspect ratio which results in a lower concentration at which phase transition begins [41]. According to this theory, the greater size distribution is accompanied with more narrow concentration range of phase transition [42]. So the conclusion can be deduced that preparing GGO and their LCs with low concentration and low viscosity will be benefit for fluid processing to fabricate the graphene fiber eventually.

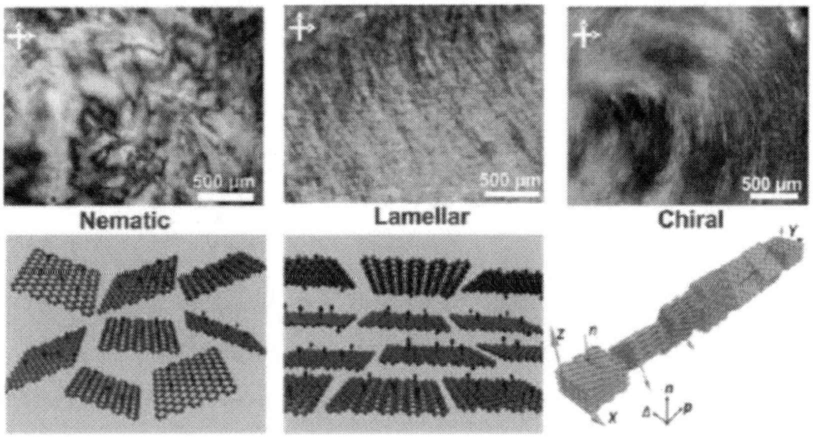

Fig.10 Typical GO LCs optical textures of Nematic, Lamellar and Chiral and their corresponding structure models [36].

GO New Chiral Liquid Crystal of 2D Colloids

Except for the ordinary nematic and lamellar mesophase, there might be other forms of mesophase. Xu and Gao discovered a new GO chiral liquid crystal (GO CLC), there are three fundamental characteristics of this new mesophase: (1) fingerprint-like texture with aligned bands, (2) strong optical activity, (3) quasi-long-range lamellar ordering [19]. For this discovery, chiral liquid crystals can be formed in a twist-gain-boundary phase-like model with simultaneous lamellar ordering from graphene oxide. Finally, aqueous graphene oxide liquid crystals can be continuously spun into macroscopic graphene oxide fibers.

3.4 Wet-spinning of Graphene Fibers

From graphene LC to graphene fibers

Graphene fiber is one of the most ideal materials of graphene in one-dimensional form. To continuously assemble graphene fiber, the graphene should be aligned regularly and individually. Through fluid assembly, the continuous aligment in the fluid state of graphene can transfer to graphene fibres in the solid state, which means graphene LCs promise to syntheze continuously regular aligned graphene fibers. By wet spinning, the advantageous properties of graphene can be translated into the graphene fibers and other graphene-based composite fibers.

Neat graphene fibers

Applying the parallel principle of wet-spinning for polymeric fibers, the continuous injection of GO LC dispersions into the coagulation bath, produces continuous graphene oxide fibers. After the reduction process, graphene oxide fibers turn to be graphene fibers (GFs). The GOF and GFs with wet-spinning keep the orientation ordering of GO LC. Hence, they inherit the excellent mechanical performance. It has been measured that the mechanical strengths of the first made GFs is 140 MPa at 5.8% elongation[36]. What is more, they are also flexible enough to be fastened to knots with no damages. These properties ensure GFs to apply in the conductive fabrics with other threads.

To enhance the mechanical properties of the GFs, the regular alignment of graphene sheets should be enhanced along the fiber axis. This is so that the solution employs the giant GO sheets. The reason for this is, because there are alignment of wrinkles on giant GO fiber surfaces and intensively impacted laminates on fiber cross-sections. With the employment of giant GO, the mechanical strengths of reduced giant GOF are enhanced markedly, measuring up to 303 MPa at 6.4% elongation [36].

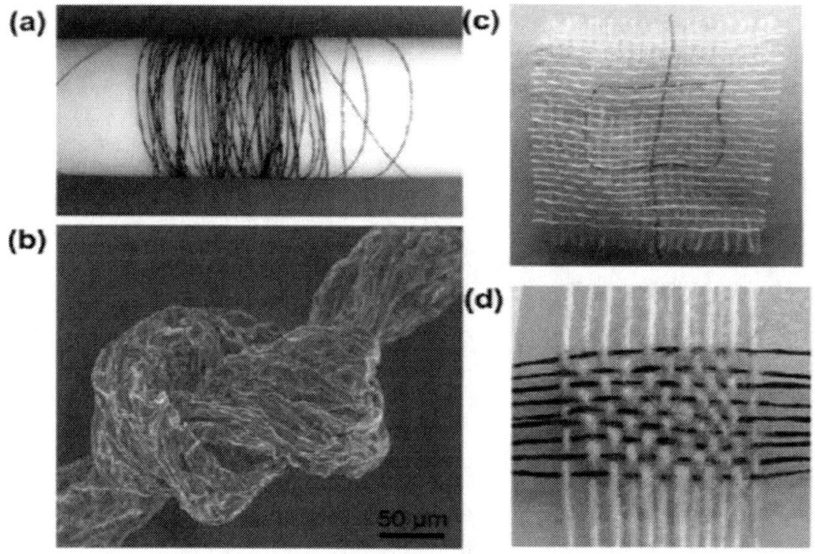

Fig.11 (a) Continuous macroscopic GFs. (b) Graphene fiber knot image by SEM. (c)&(d) Patterns of GFs[19].

Control of wet-spinning

Wet-spinning process is the final and vital process for the fabrication of graphene fibers. It is the process based on the phase transformation of graphene from ordered fluids to ordered solids. There are three steps comprised in this phase transformation process.

1) Step one: GO sheets tend to be flow-induced unidirectional alignment and homogenize orientation orderings paralleled in the flow direction.

2) Step two: solvents exchanging starts between coagulation bath and dopes. Under this function, the solvated GO interconnect to each other and self-standing gel fibers are formed. The alignment of GO tends to be more regular after the wet-drawing.

3) Step three: the final step is the drying procedure. Resulting in the shrinkage of gel fibers and typical wrinkled surface [36].

In wet-spinning, two important control factors have been investigated to affect their structure and macroscopic performances, which are coagulation and bath stretching. The dreamed coagulation bath should meet the basic requirement solvents of dopes which needs to be miscible and also be able to maintain a balanced solvent exchanging rate. Coagulation like Ethanol/water mixture with divalent ions has been tested to be effective. As the other important factor, stretching not only perfects the compact structure, more importantly, it promotes the regular alignment of graphene, which could be proven by the higher Young's modulus (up to 47 GPa) [43] and tensile strength (up to 0.5 GPa) [42] of GFs after stretching process.

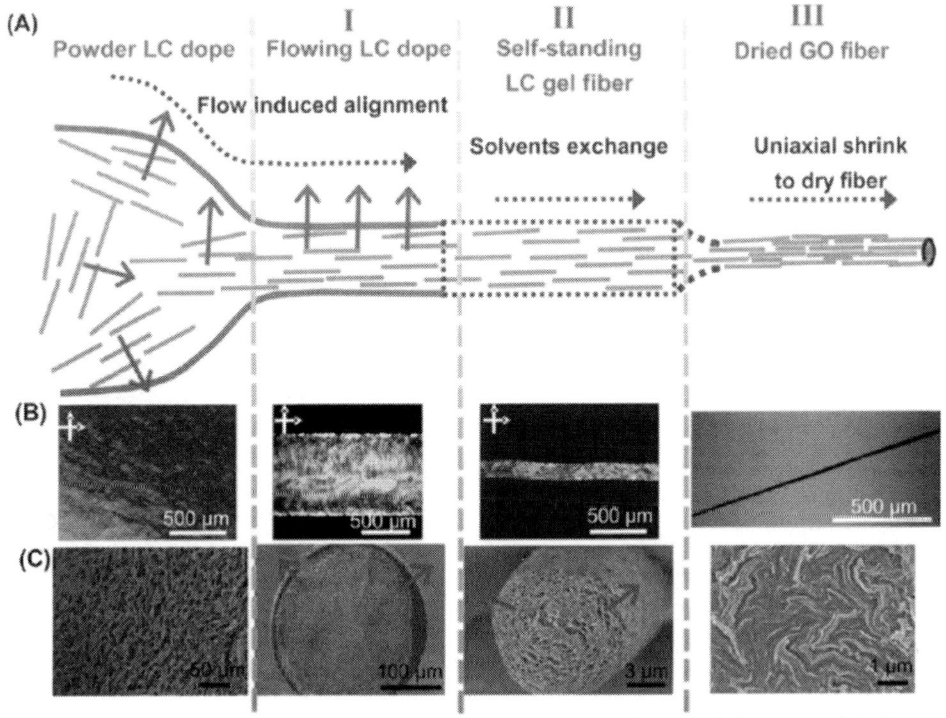

Fig.12 (A)structure evolution in the wet-spinning process. (B) the tracking result by POM. (C) the tracking result by visual SEM on the fiber section. (I)(II)(III) are the enhanced alignment during wet-spinning [36].

3.5 Fabrication of Composite Graphene Fibers

The fabrication of graphene fiber makes graphene-based materials popular in the family of smart materials. However, the direct application of graphene fiber in smart textiles have some disadvantages, such as it being too expensive, having not good enough hygroscopicity, softness and dyeability, which are the important properties as smart textile in garments. Even though there are many appearances of graphene/polymer composite fibers in recent years, such as graphene/polyurethane composite fibers [44], graphene/polyimide composite films [45] and graphene/polyaniline composite films [46], these defects seem not to be solved very well because of the intrinsic drawbacks of polymers or the poor integration of polymer and graphene. Here, a new attempt of graphene/viscose composite fibers will be introduced to solve these drawbacks.

Viscose is cheap and are easily produced. It is a fundamental material in the field of textile for its excellent performance in hygroscopicity, breathability, softness, drapability and dyeability and has been fully applied in apparel, fabric, home textiles and so on. Most importantly, the produced process of viscose fiber is via wet-spinning as well, which shows compatibility with produced process of graphene fiber. With the integration of graphene, the viscose materials are endowed with outstanding electronic, thermal and mechanical properties, which makes the graphene/viscose composite fiber a spectacular performance in smart textile material.

However, there are rarely reported articles about graphene/viscose composite fiber. The obstacles are through to be (1) how to create homogeneous-dispersed graphene oxide suspension, (2) in what proportion of graphene oxide dispersion and viscose dispersion mix are needed to optimize the properties of composite fiber. (3) how to control the coagulation bath (4) which way are possible to reduce the graphene oxide fiber.

To solve this, Patent CN 201110354746 is given as an example. Modified Hummers' method is used to produce graphene in this patent. Then, a strong base, urea, sulfur urea and deionized water were mixed by weight (3 to 10) :

(4 to 8) : (1 to 10) : (72 to93) in -10~0 ° C freezer about 12 hours, then cellusose was added under the room temperature and keep stirring for 12 to 24 hours. Finally, in the mixing process, the following concentration gradient of cellulose by 0.3%, 0.5%, 1.0%, 2.0% and 10.0% of mass fraction are used to test the best proportion of cellulose content, which can be realized by testing the properties of the final composite fiber.

4. Conclusion

Through many research and findings of graphene abundant liquid crystalline behaviors, the brand-new graphene fibers accompanied with promising performances and multi-functionalities will be established. As such, the big challenges to translate the excellent properties of graphene into the macroscopic assembled graphene fibers are related to phase transformation of GO LC during the wet-spinning process. To achieve a more regular alignment of GO sheets, more novel liquid crystalline behaviors of graphene are expected to be discovered in the recent future. Through the fabrication process of wet spinning, plenty of other materials with splendid properties can be introduced to fabricate graphene-based smart materials.

Acknowledgement

We would like to thank the support of the EU Horizon 2020 and University of Manchester through projects with project codes 644268 - ETEXWELD- H2020-MSCA-RISE-2014, AA14512 (UMRI project "Graphene-Smart Textiles E-Healthcare Network") and R1119938.

References

[1] Cong HP, Ren XC, Wang P, Yu SH. "Wet-spinning assembly of continuous, neat, and macroscopic graphene fibers," Scientific reports, vol. 2: 2012.
[2] Geim AK, Novoselov KS. "The rise of graphene," Nature materials, vol. 6: pp. 183-191; 2007.
[3] Geim AK. "Graphene: status and prospects," science, vol. 324: pp. 1530-1534; 2009.
[4] Allen MJ, Tung VC, Kaner RB. "Honeycomb carbon: a review of graphene," Chemical reviews, vol. 110: pp. 132-145; 2009.
[5] Novoselov K, Jiang D, Schedin F, Booth T, Khotkevich V, Morozov S, et al. "Two-dimensional atomic crystals," Proceedings of the National Academy of Sciences of the United States of America, vol. 102: pp. 10451-10453; 2005.
[6] Meyer JC, Geim AK, Katsnelson MI, Novoselov KS, Booth TJ, Roth S. "The structure of suspended graphene sheets," Nature, vol. 446: pp. 60-63, 2007.
[7] Novoselov KS, Geim AK, Morozov SV, Jiang D, Zhang Y, Dubonos SV, et al. "Electric field effect in atomically thin carbon films," science, vol. 306: pp. 666-669; 2004.
[8] Lee C, Wei X, Kysar JW, Hone J. "Measurement of the elastic properties and intrinsic strength of monolayer graphene," science, vol. 321: pp. 385-388; 2008.
[9] Schniepp HC, Kudin KN, Li JL, Prud'homme RK, Car R, Saville DA, et al. "Bending properties of single functionalized graphene sheets probed by atomic force microscopy," ACS nano, vol. 2: pp. 2577-2584; 2008.
[10] Balandin AA, Ghosh S, Bao W, Calizo I, Teweldebrhan D, Miao F, et al. "Superior thermal conductivity of single-layer graphene," Nano letters, vol. 8: pp. 902-907; 2008.
[11] Jevšnik S, Yi L, Hu J, Xiao H, Xinxing W, Primentas A. "Thermal-Mechanical Sensory Properties of Hot-Air Welded Textile Transmission Lines," Tekstilec, vol. 59: 2016.
[12] Eigler S, Hirsch A. "Chemistry with graphene and graphene oxide—challenges for synthetic chemists," Angewandte Chemie International Edition, vol. 53: pp. 7720-7738; 2014.
[13] Emtsev KV, Bostwick A, Horn K, Jobst J, Kellogg GL, Ley L, et al. "Towards wafer-size graphene layers by atmospheric pressure graphitization of silicon carbide," Nature materials, vol. 8: pp. 203-207; 2009.
[14] Zhang Y, Zhang L, Zhou C. "Review of chemical vapor deposition of graphene and related applications," Accounts of chemical research, vol. 46: pp. 2329-2339; 2013.
[15] Choucair M, Thordarson P, Stride JA. "Gram-scale production of graphene based on solvothermal synthesis and sonication," Nature Nanotechnology, vol. 4: pp. 30-33; 2009.

[16] Huang X, Qi X, Boey F, Zhang H. "Graphene-based composites," Chemical Society Reviews, vol. 41: pp. 666-686; 2012.
[17] Zhou X, Huang X, Qi X, Wu S, Xue C, Boey FY, et al. "In situ synthesis of metal nanoparticles on single-layer graphene oxide and reduced graphene oxide surfaces," The Journal of Physical Chemistry C, vol. 113: pp. 10842-10846; 2009.
[18] Chen S, Zhu J, Wu X, Han Q, Wang X. "Graphene oxide− MnO2 nanocomposites for supercapacitors," ACS nano, vol. 4: pp. 2822-2830; 2010.
[19] Xu Z, Gao C. "Graphene chiral liquid crystals and macroscopic assembled fibres," Nature communications, vol. 2: p. 571; 2011.
[20] Gao W, Alemany LB, Ci L, Ajayan PM. "New insights into the structure and reduction of graphite oxide," Nature chemistry, vol. 1: pp. 403-408; 2009.
[21] Szabó T, Berkesi O, Forgó P, Josepovits K, Sanakis Y, Petridis D, et al. "Evolution of surface functional groups in a series of progressively oxidized graphite oxides," Chemistry of materials, vol. 18: pp. 2740-2749; 2006.
[22] Gómez-Navarro C, Burghard M, Kern K. "Elastic properties of chemically derived single graphene sheets," Nano letters, vol. 8: pp. 2045-2049; 2008.
[23] Paredes J, Villar-Rodil S, Martínez-Alonso A, Tascon J. "Graphene oxide dispersions in organic solvents," Langmuir, vol. 24: pp. 10560-10564; 2008.
[24] Moon IK, Lee J, Lee H. "Highly qualified reduced graphene oxides: the best chemical reduction," Chemical Communications, vol. 47: pp. 9681-9683; 2011.
[25] Pei S, Zhao J, Du J, Ren W, Cheng HM. "Direct reduction of graphene oxide films into highly conductive and flexible graphene films by hydrohalic acids," Carbon, vol. 48: pp. 4466-4474; 2010.
[26] Becerril HA, Mao J, Liu Z, Stoltenberg RM, Bao Z, Chen Y. "Evaluation of solution-processed reduced graphene oxide films as transparent conductors," ACS nano, vol. 2: pp. 463-470; 2008.
[27] Zhong X, Jin J, Li S, Niu Z, Hu W, Li R, et al. "Aryne cycloaddition: highly efficient chemical modification of graphene," Chemical Communications, vol. 46: pp. 7340-7342; 2010.
[28] He H, Gao C. "General approach to individually dispersed, highly soluble, and conductive graphene nanosheets functionalized by nitrene chemistry," Chemistry of Materials, vol. 22: pp. 5054-5064; 2010.
[29] Niyogi S, Bekyarova E, Itkis ME, McWilliams JL, Hamon MA, Haddon RC. "Solution properties of graphite and graphene," Journal of the American Chemical Society, vol. 128: pp. 7720-7721; 2006.
[30] Pei S, Cheng HM. "The reduction of graphene oxide," Carbon, vol. 50: pp. 3210-3228; 2012.
[31] Laveran A. "Note sur un nouveau parasite trouvé dans le sang de plusieurs malades atteints de fièvre palustre," Bull Acad Med, vol. 9: pp. 1235-6; 1880.
[32] Staudenmaier L. "Verfahren zur darstellung der graphitsäure," Berichte der deutschen chemischen Gesellschaft, vol. 31: pp. 1481-1487; 1898.
[33] Simon A, Dronskowski R, Krebs B, Hettich B. "The crystal structure of Mn2O7," Angewandte Chemie International Edition in English, vol. 26: pp. 139-140; 1987.
[34] Zheng Q, Ip WH, Lin X, Yousefi N, Yeung KK, Li Z, et al. "Transparent conductive films consisting of ultralarge graphene sheets produced by Langmuir–Blodgett assembly," Acs Nano, vol. 5: pp. 6039-6051; 2011.
[35] Marcano DC, Kosynkin DV, Berlin JM, Sinitskii A, Sun Z, Slesarev A, et al. "Improved synthesis of graphene oxide," ACS nano, vol. 4: pp. 4806-4814; 2010.
[36] Xu Z, Gao C. "Graphene in macroscopic order: liquid crystals and wet-spun fibers," Accounts of chemical research, vol. 47: pp. 1267-1276; 2014.
[37] Anderson VJ, Lekkerkerker HN. "Insights into phase transition kinetics from colloid science," Nature, vol. 416: pp. 811-815; 2002.
[38] Davidson P, Gabriel JCP. "Mineral liquid crystals," Current Opinion in Colloid & Interface Science, vol. 9: pp. 377-383; 2005.
[39] Xu Z, Gao C. "Aqueous liquid crystals of graphene oxide," ACS nano, vol. 5: pp. 2908-2915; 2011.
[40] Xiang C, Behabtu N, Liu Y, Chae HG, Young CC, Genorio B, et al. "Graphene nanoribbons as an advanced precursor for making carbon fiber," ACS nano, vol. 7: pp. 1628-1637; 2013.
[41] Dan B, Behabtu N, Martinez A, Evans JS, Kosynkin DV, Tour JM, et al. "Liquid crystals of aqueous, giant graphene oxide flakes," Soft Matter, vol. 7: pp. 11154-11159; 2011.
[42] Xu Z, Sun H, Zhao X, Gao C. "Ultrastrong fibers assembled from giant graphene oxide sheets," Advanced

Materials, vol. 25: pp. 188-193; 2013.
[43] Xiang C, Young CC, Wang X, Yan Z, Hwang CC, Cerioti G, et al. "Large flake graphene oxide fibers with unconventional 100% knot efficiency and highly aligned small flake graphene oxide fibers," Advanced Materials, vol. 25: pp. 4592-4597; 2013.
[44] Hsiao ST, Ma CCM, Tien HW, Liao WH, Wang YS, Li SM, et al. "Preparation and characterization of silver nanoparticle-reduced graphene oxide decorated electrospun polyurethane fiber composites with an improved electrical property," Composites Science and Technology, vol. 118: pp. 171-177; 2015.
[45] Liao WH, Yang SY, Wang JY, Tien HW, Hsiao ST, Wang YS, et al. "Effect of molecular chain length on the mechanical and thermal properties of amine-functionalized graphene oxide/polyimide composite films prepared by in situ polymerization," ACS applied materials & interfaces, vol. 5: pp. 869-877; 2013.
[46] Wu Q, Xu Y, Yao Z, Liu A, Shi G. "Supercapacitors based on flexible graphene/polyaniline nanofiber composite films," ACS nano, vol. 4: pp. 1963-1970; 2010.

Preparation and Characterization of Chitosan/silk Fibroin Based Membranes with Different Blend Ratios

Kanza Hina[1], Han-Tao Zou[1*], Meng Zhu[1], Wang Bing[1], Dan-Ying Zuo[1], Chang-Hai Yi[1]

[1] *Key* Laboratory *of Green Processing and Functional Textile of New Textile Materials (Wuhan Textile University), Ministry of Education, Wuhan, Hubei 430073, PR China*

[*]Corresponding author's email: hantao.zou@etu.edu.com

Abstract

The silk fibroin is extracted from mulberry silk cocoons and then chitosan/ silk fibroin membranes are prepared using different silk fibroin weight percentage (0wt%, 5wt%, 10wt%, 15wt%, 20wt% and 25wt%) by the solution-cast technique. Scanning electron microscope (SEM) and Fourier transform infrared spectroscopy (FTIR) were used for the characterization of prepared Chitosan/Silk fibroin membranes. Moisture regain, swelling, mechanical property, air permeability and in vitro degradation rate are evaluated. SEM reveals that all blended membranes are compactly packed and morphology or surface depends on blend ratio. The FTIR analysis shows that chitosan and silk fibroin are physically blended and compatible with each other. It is found that with increasing silk fibroin content, the moisture regains and swelling decreases due to the lack of hydrophilic groups of silk fibroin. The introduction of silk fibroin leads to weak intermolecular forces between the two components cause the decrease in mechanical properties of blend membranes such as the strength of membranes is reduced from 5.04 N/mm to 3.04 N/mm and the elongation of membranes decrease from 17.8% to 8.23%. In comparison to the pure chitosan membrane, the blend membranes have higher air permeability i.e. 52%-68% and lesser in vitro degradation rate i.e. 89-60%, which is more suitable for medical use.

Keywords: Chitosan; Silk Fibroin; Blend Membrane; Air Permeability; In Vitro Degradation

1. Introduction

Silk has been commercially used as surgical sutures for decades and as a source of textile-grade fibres for centuries. Bombyx mori silk consists of two kinds of proteins i.e. sericin and fibroin. Recently, interest on silk fibroin has been growing resulting in a dramatic increase on the use of several processed forms of the solubilized silk fibroin in (bio) technological materials and biomedical applications [1-6]. Similar to human skin, silk fibroin film has good oxygen permeability in its wet state, which suggests the promising applications of silk fibroin as a wound dressing. However, the SF film is too brittle to be used in biomedical applications. Due to this reason, it is usually used by blending with other natural polymers such as chitosan [7-11].

Chitosan (CS) is also the most abundant biomaterial and has a repeating structure of 1,4-linked 2-amino-2deoxy-b-D-glucan. It has useful functional groups such as hydroxyl and amino groups that can be modified to various derivatives. CS is well known to have biocompatible and biodegradable properties, particularly as a biomaterial for wound dressings and artificial skin due to its unique properties such as the acceleration of wound healing [12-15]. It can be easily shaped into fibres and films, hollow fibres, membranes, medical gauze and wound dressings [16, 17].

It is also an important route for improving SF film by blending with natural macromolecular structures. Noishiki prepared the SF/cellulose whisker composite films with varied compositions. It was found that a conformational change of SF from random coil to β-sheet structure induced by the contact of SF molecules with a highly ordered surface of cellulose whisker. This conformational change was also observed in the SF/chitosan blend films by Kweon et al. It indicated that the mechanical properties of SF films enhanced by blending 10 40% chitosan and the two compositions were miscible [10]. Similarly, some blend films such as SF/sodium alginate, SF/gelatin, and SF/S-carboxymethyl kerateine were prepared using solution blend method. However, there is still an absence of a detailed description on the mechanism of conformational transition of SF in these blend films [18, 19].

In this paper, silk fibroin is extracted from mulberry silk cocoons and then blends with chitosan. The chitosan/silk fibroin membranes were prepared with different blend ratios. SEM (Scanning electron microscope) and FTIR (Fourier transform infrared spectroscopy) were used for characterization of prepared membranes. Moisture absorbency, mechanical properties, air permeability and in vitro degradation are also evaluated. The effect on the properties of chitosan/silk fibroin membranes because of different blend ratio of silk fibroin and chitosan are analyzed.

2. Experiment

2.1 Material

Mulberry silk was purchased from the market. Chitosan, anhydrous sodium carbonate, lithium bromide in powder form, glacial acetic acid and polyethylene glycol were obtained from Sinopharm Chemical Reagent. Phosphate buffer solution was provided by Thermo Fiser Scientific limited.

2.2 Preparation of Chitosan/silk Fibroin Membranes

Chitosan solution of 2% weight concentration solution of chitosan was prepared in 1% acetic acid aqueous solution by stirring for 3 hours at 70^0C and then cooled in the refrigerator at 4^0C.

The mulberry cocoons were degummed with 0.25% Na_2CO_3 solution for 30 minutes at 98^0C -100^0C and washed in deionized boiling water for 30 minutes. After washing, the silk was dried in an incubator at 50^0C. This dried silk was cut and stirred in 9.3ml LiBr solution at 60^0C until the dissolution was complete. After cooling, blending solution was put in dialysis bags with molecular weight cut-off 14000Da. The dialysis bags were immersed in deionized water for three days. The deionized water was changed three times in a day. Silk fibroin solution was shrunk to increase silk fibroin concentration in solution by using 20% polyethylene glycol and dialysis bags of molecular weight cut-off 3500.

Finally, chitosan solution mixed with silk fibroin solution with five different blend ratios (silk fibroin 5wt%, 10wt%, 15wt%, 20wt% and 25wt%). The blending chitosan/silk fibroin solution was stirred and deaerated. The chitosan/silk fibroin based membranes were prepared by the solution-cast technique onto PTEE board. In order to compare with the property, the pure chitosan membrane was also prepared by the same method.

2.3 Characterization

The chitosan/silk fibroin membranes were weighted after placing at 20^0C and 65% relative humidity conditioned atmosphere until moisture equilibrium and then compared with the weight that obtains after drying sample at 120^0C in DHG-9241A thermostatic dryer.

Swelling of membranes was measured by measuring weight at moisture equilibrium in conditioned atmosphere i.e. G_1 and weight when membranes were immersed in water at 20^0C for 24 hours and drain surface water i.e. G_2.

$$\text{Swelling} = [(G_2\text{-}G_1)/G_1]\times 100\% \tag{1}$$

Air permeability was measured using GB T12704 fabric dish standard method. In this method, membranes were encapsulated in a glass with a cup mounted with 1 ml distilled water, and used an air circulation drying oven ZJ-841Y-1 at 38^0C as the environment for samples. Samples were weighted after every 1 hour to find air permeability The 5556 Instron Universal Strength Tester was used to determine mechanical properties such as tensile strength, and elongation. The sample with 80mmx5mm was used at 10mm/min speed to find tensile strength and elongation.

The morphology micrographs of the membranes were examined after sputter-coating the sample using an FEI Quanta 200 scanning electron microscope (FEI Company, Eindhoven, Netherlands), with an accelerating voltage of 20 kV.

The chemical groups of the membranes were examined using Fourier transform infrared (FTIR) spectrum analysis with a spectrometer (Tensor 27, Bruker Optics, Germany), in the scanning range of 4000 to 400 cm^{-1}.

The biodegradability of the chitosan/silk fibroin blended membranes was measured using in vitro degradation rate. The absolute dry sample with weight W_1 put in beaker with 10ml buffer solution for six day. After six days the sample was washed, absolute dried and then weighted to get W_2. The degradation rate then measured by following equation.

$$\text{Degradation rate (\%)} = (W_1-W_2)/W_1*100\% \quad (2)$$

3. Result and Discussion

Fig.1 shows moisture content of chitosan/silk fibroin membranes with different content of silk fibroin. It is evident from the figure that as the content ratio of silk fibroin increased up to 25wt%, moisture regain decreased from 22.06% to 16.18%. Chitosan has hydroxyl and hydroxymethyl groups that help in adsorbing moisture and hydroscopic nature of membranes. Silk moisture regain (11%) is less than chitosan moisture regain (22% in Fig.1). Thus, decrease in moisture regain is because of the drop of chitosan content ratio. The swelling percentage of chitosan/silk fibroin membranes with different content ratios of silk fibroin is also shown in Fig.1. It is shown that the introduction of silk fibroin in membranes can cause a rapid drop of swelling as compared to the pure chitosan membranes. The swelling percentage drops to almost half from 51.44% of pure chitosan to 27.78% of the blend membrane with only 5% silk fibroin. The drop in swelling is attributed to the lack of hydrophilic groups in silk. The increasing content of silk fibroin cause decrease in hydroxyl groups of the blend membranes. Due to this, less water entrance in the membrane structure and eventually drop in swelling percentage of membranes.

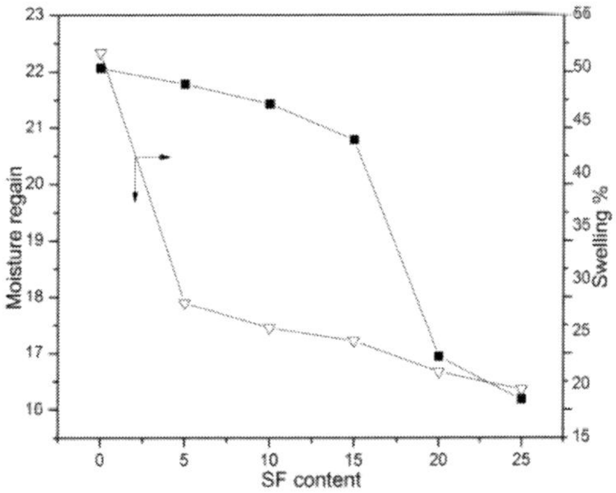

Fig.1 Water adsorption properties of chitosan/silk fibroin membrane with different silk fibroin content ratio (a) moisture regain (%) (b) swelling (%)

The air permeability of chitosan/silk fibroin membranes is shown in Fig.2. The air permeability of all membranes is more than 50%. With the increase of silk fibroin content, the air permeability factor of the blend membranes also increases due to excellent breathability and porous structure for silk structure. The increasing factor of air permeability is very significant because it increased from 52% to 68% but it is beneficial because membranes have sufficient pores that allow the exchange of nutrients water vapours and gas.

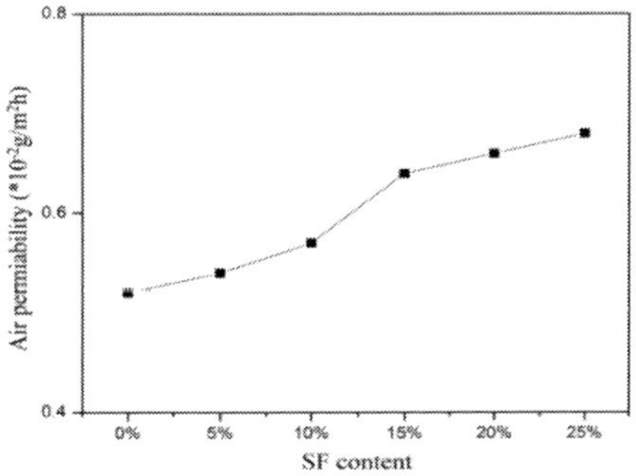

Fig.2 Relation between silk fibroin content ratio and air permeability in chitosan/silk fibroin membranes

Mechanical properties such as strength and elongation of chitosan/silk fibroin membranes are shown in Fig.3. Silk fibroin has poor mechanical properties as compared to chemical fibre chitosan due to week intermolecular forces. So the addition of silk fibroin in the membranes causes the decrease in mechanical properties such as strength and elongation of membranes. Silk fibroin also not form any chemical bond with chitosan so intermolecular forces of membranes are also not strong as pure chitosan membranes and cause the further decrease in mechanical properties. But in spite, the fact the mechanical properties are dropping from 5.04 N/mm to 3.04 N/mm and elongation reduce from 17.8% to 8.23% but they are still sufficient for medical use.

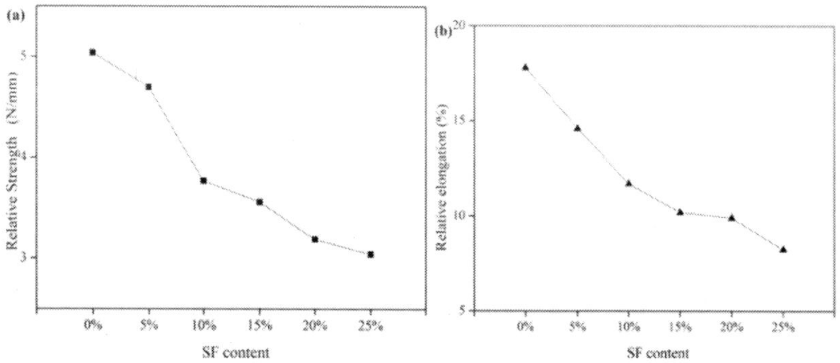

Fig.3 Mechanical properties of the prepared chitosan/silk fibroin membranes

The Fourier Transform Infrared (FTIR) spectra of membranes to evaluate chemical structure are shown in Fig.4. The broad peak of 3100-3600 cm^{-1} corresponds to the stretching and vibration mode of -OH. Peaks at 1626 cm^{-1}, 1527 cm^{-1}, and 1360 cm^{-1} correspond to the characteristic peak of silk β sheet conformation for amide I, amide II, and amide III respectively [7]. On the other hand the peaks at 1010 cm^{-1} is because of skeleton vibration in chitosan involving C-O stretching and 892 cm^{-1} attributed to saccharide structure of chitosan. Chitosan -COOH (carboxyl) stretching bond vibration is shown by the peak at 1400 cm^{-1} [17]. It is also obvious from the figure that as the silk fibroin concentration increase the peaks attributed to silk amide I, amide II and amide III become more prominent and clear and FTIR analysis also shows that chitosan and silk fibroin are physically blended and compatible with each other.

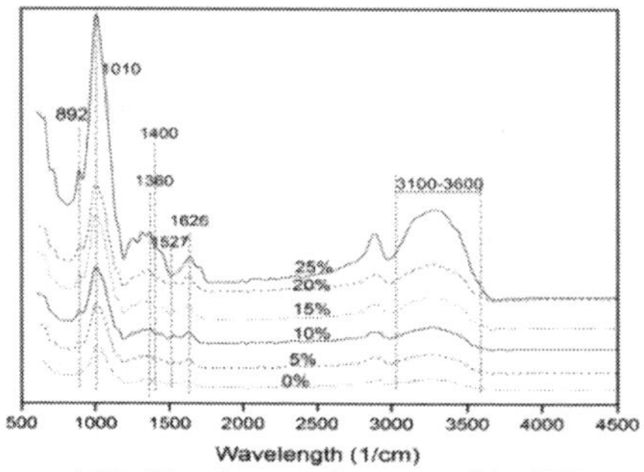

Fig.4 FTIR spectra of chitosan/silk fibroin membranes with different silk fibroin content ratios

The SEM images in Fig.5 reveals the surface morphology of chitosan/silk fibroin based membranes, indicates pure chitosan membrane has a uniform surface, but morphology of blended membranes depends on the blend ratio. 5wt% and 20wt% chitosan/silk fibroin membranes show compact membranes with uneven and rough surface attributed to the contraction of silk fibroin. 25wt% chitosan/silk fibroin membranes have strips on the surface with a rough surface whereas 10wt% and 15wt% chitosan/silk fibroin membranes have smooth surface almost alike pure chitosan membranes. So, from the results, it can be concluded that all membranes are densely packed and showing that blend ratio of chitosan and silk fibroin also have the significant effect on the morphology of membranes. As all membranes are not homogeneously blended but more hydrophobic fibroin molecules tend to shift towards surface (air-liquid interface).

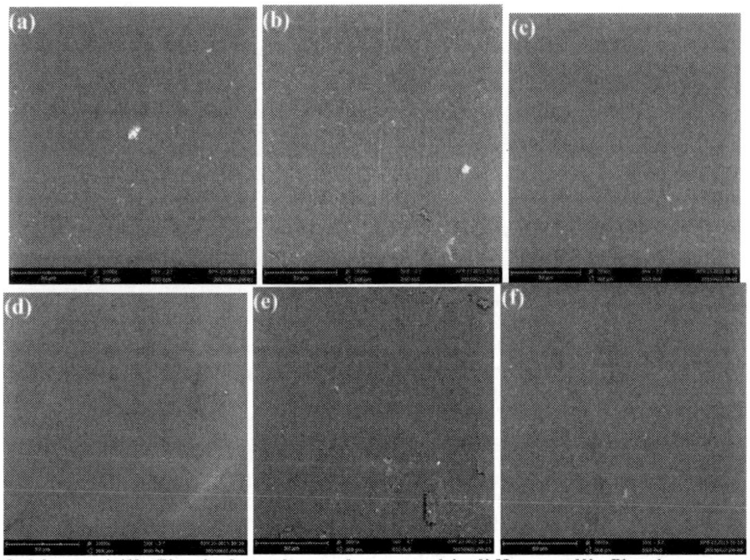

Fig.5 SEM images of chitosan/silk fibroin based membranes with different silk fibroin content ratios as (a) 0% (b) 5% (c) 10% (d) 15% (d) 20% (e) 25%

Fig.6 shows in vitro degradation rate of chitosan/silk fibroin membranes and results reveal that as the amount of silk fibroin increase, the degradation rate of membranes drops from 89.45% to 60.43%. The drop in degradation rate is because chitosan content ratios decrease and increase silk content cause the decrease in polar groups in the blend, that leads to decline in hydrophilic degree and all this eventually leads to the less degradations. Also, bipolarity slow down the intrusion of the low molecular weight compound as degradation rate is manily depends on these low molecular weight compounds. The drop in in vitro degradation rate with increasing silk fibroin weight content means less enzymes attack and healthier membranes for bandages, and scaffolds etc.

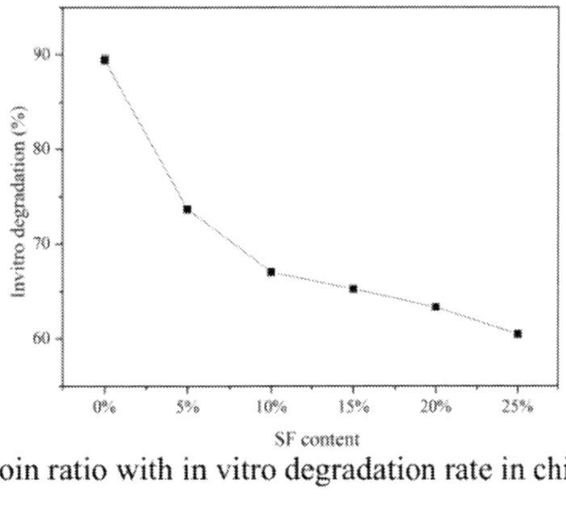

Fig.6 Relationship of silk fibroin ratio with in vitro degradation rate in chitosan/silk fibroin membranes

4. Conclusion

Chitosan/Silk fibroin membranes are prepared by different blend ratios by the solution-cast technique. The effect of blending ratio on properties of membranes is investigated. SEM shows that all membranes are compactly packed and surface morphology depends on blend ratio of silk fibroin in membranes as hydrophobic fibroin molecules tend to shift towards the surface. FTIR analysis reveals that chitosan and silk fibroin are compatible with each other in physically blended membranes. Moisture regain decreases from 22.06 to 16.18 and swelling decrease from 51.44% to 19.56% because chitosan blend ratio starts decreasing because chitosan has a high number of hydroxyl and hydroxymethyl groups and these groups are responsible for water absorbency and retain ability. Relative strength and elongation also decrease with the increase in silk fibroin blend ratio because silk fibroin is brittle in nature and have week intermolecular forces as compare to chitosan. The air permeability of blended membrane shows an increment in the result from 52% to 68% as silk fibroin content starts to increase and it is because of good breathability of silk helps in exchange of nutrient, gas or vapours. In vitro degradation rate greatly drops from 89.45% to 60.43% because chitosan blend ratio drops showing that silk is less susceptible to enzymes and membranes with high blend ratio of silk is healthier for medical use.

5. Acknowledgement

The financial support from the National Nature Science Fund of China (51303139) and the Scientific Research Foundation of Hubei Provincial Education Department (No: Q20121710) is greatly appreciated.

References

[1] Kundu B, et al. Silk proteins for biomedical applications: Bioengineering perspectives. Prog. Polym. Sci: 2014; 39: 251-267.

[2] He J, Wang Y, Cui S, Gao Y, Wang S. Structure and properties of silk fibroin/carboxymethyl chitosan blend films. Polym. Bull: 2010; 65: 395-409.

[3] Jin HJ, Park J, Valluzzi R, Cebe P, Kaplan DL. Biomaterial films of Bombyx mori silk fibroin with poly (ethylene oxide). Biomacromolecules: 2004; 5: 711-717

[4] Bhardwaj N, et al. Potential of 3-D tissue constructs engineered from bovine chondrocytes/silk fibroin-chitosan for in vitro cartilage tissue engineering. Biomaterials: 2011; 32: 5773-5781.

[5] Delezuk JAM, et al. Silk fibroin organization induced by chitosan in layer-by-layer films: Application as a matrix in a biosensor. Carbohydr. Polym: 2017; 155: 146-151.

[6] Bhardwaj N, Kundu SC. Chondrogenic differentiation of rat MSCs on porous scaffolds of silk fibroin/chitosan blends. Biomaterials: 2012; 33: 2848-2857.

[7] Bhardwaj N, Kundu SC. Silk fibroin protein and chitosan polyelectrolyte complex porous scaffolds for tissue engineering applications. Carbohydr. Polym: 2011; 85: 325-333.

[8] Kaushik S, Sarma MK, Thungon PD, Santhosh M, Goswami P. Thin films of silk fibroin and its blend with chitosan strongly promote biofilm growth of Synechococcus sp. BDU 140432. J. Colloid Interface Sci: 2016; 479: 251-259.

[9] Basal G, Altiok D, Bayraktar O. Antibacterial properties of silk fibroin/chitosan blend films loaded with plant extract. Fibres Polym: 2010; 11: 21-27.

[10] Kweon H, Um IC, Park YH. Structural and thermal characteristics of Antheraea pernyi silk ® broin / chitosan blend ® lm: 2001; 42: 6651-6656.

[11] Luangbudnark W, Viyoch J, Laupattarakasem W, Surakunprapha P, Laupattarakasem P. Properties and biocompatibility of chitosan and silk fibroin blend films for application in skin tissue engineering. ScientificWorld Journal: 2012; 697201: 10.

[12] Park SJIN, Lee KY, Ha WANS, Park SOOY. Structural Changes and Their Effect on Mechanical Properties of Silk Fibroin / Chitosan Blends. J. Appl. Polym. Sci: 1999; 74: 2571-2575.

[13] Park WH, Jeong L, Yoo DI, Hudson S. Effect of chitosan on morphology and conformation of electrospun silk fibroin nanofibres. Polymer (Guildf): 2004; 45: 7151-7157.

[14] Rujiravanit R, Kruaykitanon S, Jamieson AM, Tokura S. Preparation of Crosslinked Chitosan/Silk Fibroin Blend Films for Drug Delivery System. Macromol. Biosci: 2003; 3: 604-611.

[15] Sionkowska A, et al. Biopolymer blends as potential biomaterials and cosmetic materials. Key Eng. Mater: 2013; 583: 95-100.

[16] Sionkowska A, Lewandowska K, Płanecka A. Miscibility and physical properties of chitosan and silk fibroin mixtures. Journal of Molecular Liquids: 198; 2014: 354-357.

[17] Sionkowska A, Płanecka A. Preparation and characterization of silk fibroin/chitosan composite sponges for tissue engineering. J. Mol. Liq: 2013; 178: 5-14.

[18] Sionkowska A, Płanecka A. Surface properties of thin films based on the mixtures of chitosan and silk fibroin. J. Mol. Liq: 2013; 186: 157-162.

[19] Srivastava CM, Purwar R. Chitosan-finished Antheraea mylitta silk fibroin nonwoven composite films for wound dressing. J. Appl. Polym. Sci: 2017; 134: 1-16.

Effective Removal of Colorants from Aqueous Suspension through Cellulose Nano whisker Templated Titanium Dioxide/Cellulose Nanocomposite

Mohammad Neaz Morshed[1], Xiao-Lin Shen[1*], Xing-Ya Zhang[1], Hridam Deb[1], Shamim Al Azad[2]

[1]*School of Textile Science and Engineering, Wuhan Textile University, Wuhan, 430073, People's Republic of China*

[2]*College of Chemistry and Chemical Engineering, Wuhan Textile University, Wuhan, 430073, People's Republic of China*

*Corresponding author's email: xiaolin.shen@wtu.edu.cn

Abstract

A facile approach to synthesize one-dimensional TiO_2 nanomaterials by using natural cellulose whiskers as a carrier was taken, and the potentiality of TiO_2/cellulose nano-composite as an effective eliminator of commercial colorants from waste water is reported in this study. TiO_2/cellulose nanocomposite has been fabricated via a sol–gel process commencing the hydrolysis of $Ti(OC_2H_5)_4$ by using native cellulose nanowhiskers as a green template. The structures and properties of the composite nanoparticles has been characterized with Fourier Transform Infrared Spectroscopy (FTIR), X-Ray Diffraction (XRD) analysis, Field Emission Scanning Electron Microscope (FESEM) and Energy Dispersive Spectroscopy (EDS) Analysis. The resulting nanocomposite exhibit superior competency to the catalytic decoloration of contaminated dyeing water with commercial anthraquinone reactive textile dye solution. Catalytic decoloration was observed by using UV-visible spectrophotometer with quantitative measuring method. The outcome shows that, the superior decoloration efficacy of simulated dye solution in presence of UV light irradiation approaches to more than 99.50% within a time frame of 40 mins. By replacing toxic and highly reactive organics with natural cellulose nanowhiskers, this study suggests a significant prospect of broad range on environmental remediation.

Keywords: Cellulose-nanowhisker; Nano-composite; Commercial Dye; Catalytic Decoloration;

1. Introduction

The conventional approaches to complete the degradation of the dyes from wastewater treatment systems through precipitation, adsorption, flocculation, flotation, oxidation, reduction, electrochemical, aerobic, anaerobic, and biological treatment methods have a contrary back draw. These technologies with inherent limitations of less effectiveness and in producing secondary sludge is a costly issue for subsequent disposal.[1] Adsorption of dyes on many adsorbents (e.g., SiO_2, Al_2O_3, TiO_2, CaO, MgO, Fe_2O_3, Na_2O, K_2O, bentonite, and montmorillonite) is a simple and effective method/design. It is also easy to use and can be implemented for dye treatment even in small plants. Still, it typically produces a large extent of sludge, particularly in the wastewater with high dye concentrations [2, 3]. However, Titanium dioxide (TiO_2) is one of the most favourable semiconductors that have great photo-catalytic properties [4]. Due to its high photo-catalytic activity, non-toxicity, and physicochemical stability, its application has been increased [5, 6]. In the last decades of study, TiO_2 has grown to have a versatility of application in material science and engineering involved in UV resistance, anti-fogging and self-cleaning properties [7]. Hydrophilic TiO_2 coated surfaces with photocatalytic properties are able to degrade various organic pollutants [8, 9] based on advanced oxidation processes (AOPs), dye sensitization, [10] doping, [11] coupling, [12] and capping, [13] pesticide degradation and water splitting of TiO_2 [14]. Goldie Oza and co-workers [15] developed a micro-filter or more specifically a nano-filter which is able to remove microorganisms from water. Mahmoodi, N.M. and M. Arami [16, 17] showed photocatalytic treatment of wastewater through immobilization of TiO_2 have outstanding de-coloration property along with significant amount of toxicity reduction. Photo catalysis of TiO_2 nanoparticles can be categories into following two types of reaction, suspension and immobilization. The suspension type photo catalysis the catalyst particles needs to be separate and recycle before discharging the treated wastewater so can be defined as time killing costly procedure. Moreover, The catalyst particles possess a high amount of absorption capability causing the poor intensity of penetration of UV light [16, 17]. Anodic aluminum oxide, polyelectrolyte membranes, electrospun mats [8, 9] or diblock polymers are also novel candidates for the carrier of TiO_2 nanomaterials, but the templates used thus far have not been cost effective, and they have often been prepared in a multistep approach. Therefore, from a green chemistry perspective, it is highly desirable to develop a simple and low-cost method for carry TiO_2 nanomaterials. It is interesting that native cellulose whiskers also can be used as carrier for the preparation of functional inorganic nanomaterials. Compared with the conventional inorganic or organic templates that are used for the preparation of nanomaterials, cellulose whiskers are easier to get and cheaper, and their most important virtue is their regenerative and environmental friendly nature and easy removability [18, 19]. This hypothesis aimed at fabrication of nanocrystaline titanium-dioxide on cellulose nanocrystal as a carrier and investigate the catalytic performance for effective removal of commercial model colorant as a green approach. For this objective, cellulose nannocrystal has been synthesized from cotton linter by acid hydrolysis method [18, 20, 21]. Then as synthesized cellulose nanocrystal has been used as carrier for in-situ sonosynthesis of nanocrystaline titanium dioxide. Catalytic colorants decoloration from aqueous solution of nano-composite has been investigated. This will expected to provide a new and ''green'' trail for the tedious strategy and assembly of other serviceable nanomaterials and open novel use assortments for cellulose

2. Experimental

2.1 Materials

For this hypothesis, Cotton powder was procured from Hubei Chemical Fiber Co., Ltd. (Xiangfan, Hubei, China) and Tetrabutyl Titanate (Ti $(OC_4H_9)_4$, TBT) was chosen as the titanate source. All the raw materials such as, tetrabutyl titanate ($Ti(OC_4H_9)_4$, TBT, 98%), sodium hydroxyl (NaOH, powdered), Hydrochloric Acid (HCl 98%) and ethanol (C_2H_5OH, 99.7%) were obtained from Sinopharma Chemical Reagent Co. limited (China). Analytical grade chemical used in this experiment and did not experienced any further purification before use in the laboratory. Water used in all experiments was purified using water purification system from Chengdu Ultra Technology Co., ltd. China (ULUPURE Model: UPH-II-20T) with resistivity higher than 18 MΩ cm.

2.2 Methodology

2.2.1 Preparation of Cellulose Nanoparticles

In this work, cellulose nanocrystal has been synthesized from cotton powder by acid hydrolysis method followed by the method reported in literature [19, 22]. Nevertheless described in the literature, hydrolysis parameters for synthesis of cellulose nanocrystal several trial experiment has been studied to find suitable molar concentration of acid and corresponding actions. Here, 5.0 g cotton powder has been dispersed into 100 ml of 40% (w/w) sulfuric acid solution under vigorous stirring at a constant temperature of 45°C for 120 minutes. The as obtained suspension was further subjected to dilution afterwards by using distilled water to halt and suspend ongoing reaction of hydrolysis between cotton powders and H_2SO_4. Hydrolyzed solution of cellulose be there introduced a mild handling with 30% NaOH suspension to maintain the pH in a neutral condition. Then suspension were centrifuged with speed of 6000 rpm and repeated the action for three times tailed by watering down with distilled water and set in an ultrasonic bench top bath. The solution were subjected to ultrasonic irradiation for 120 minutes at constant sonic power of 100W for better dispersion of nanocrystals. As obtained suspension was centrifuged at 6000 rpm for 4 times and freeze dried for 48 hours to obtain cellulose nanocrystal. As obtained nanocrystal was preserved, characterized and used for further application.

2.2.2 In Situ Sonosynthesis of TiO_2/cellulose Nano-composite

The application of ultrasound in nanocrystalline synthesis has recently gained attention of researchers. It is well-known that physical and chemical effects of ultrasound arise from acoustic cavitation phenomena, the rapid formation, growth, and the collapse of bubbles in liquid [23]. In this study, nanocrystaline titanium dioxide has been synthesized by using cellulose nanocrystals as template under ultrasonic irradiation through insitu solgel method. In a typical manner, two solution (solution A and B) for this sonosynthesis has been prepared. Solution A contains 3 ml of tetreabutyl titinate dispersed in 30 ml ethanol solution through magnetic stirring having pH 4.0 (HCl was used to maintain the medium acidic). On the other hand, required amount of as synthesized cellulose nanocrystal were isolated into 60 ml of distilled water and under vigorous stirring (250 rpm) in 250 mL glass beaker for 5 min with 10 ml ethanol to prepare solution B. Solution B was placed in an ultrasonic bechtop bath and subjected to ultrasonic irradiation at 100 W sonic power. Solution A were added descent wise into solution B under ultrasonic radiation at constant sonic power, temperature. The ultrasonic bath was well-appointed with ordinal heating regulator method. To find proper initial and final temperature of process as well as appropriate sonication time for formation of desired TiO_2/Cellulose nano-composite were carried. Based on the preliminary results, the temperature was adjusted 50°C to obtain gel of titanium dioxide. The formed gel was freeze dried for 48 hours and obtained Cellulose nanocrystal template cellulose/Titanium dioxide nano-composite. Prepared nano-composite were characterized to validate this output. Synthesis of nanocrstaline titanium dioxide using cellulose nanocrystal template is a proved concept. However, sonosynthesis of nanocrystaline titanium dioxide in cellulose nanocryatal is rather a novel approach for superior applicability.

2.2.3 Catalytic Decoloration of Model Colorants

Cellulose /titanium dioxide nano-composite has been applied to remediation of common industrial colorants under light intensity of the UV light. Dyes solution at concentration of 50 mg/l of commercial dyestuffs was prepared in the laboratory as a model solution to examine catalytic decoloration efficiency of nanoreactor. 40 ml of as prepared solution was taken and 0.05 g of nano-composite was dispersed into the solution. The set up for photo catalysis was performed as described by Nandanwar, R., and coworkers [24]. Each reactor was mechanically agitated vigorously under a self-made UV light source. Within several time interval 1ml of fluidic sample was carried out and reserved in a transparent tube. Prior to quantitative measurement by a UV-visible spectrophotometer from Shimadzu Corporation, Japan (UV-2550), the sample was diluted to 3 mL with demineralized water. The final solution was filtered by using whatman filter paper for further reuse of nano-composite.

2.3 Characterizations

Surface morphology, existence of particular functional groups, decoloration of model dyes solution and photosensitive characteristics of as synthesized TiO_2/Cellulose nano-composite were analyzed by STOE STADI P transmission X-ray powder diffractometric arrangement by observing the deflection angle from 20 to 70 (2h) using monochromatized Cu Ka (k=1.54051*A) radiation, Series 100 Thermo Nicolet FTIR (Nicolate, USA) spectrophotometer to examine the existence of particular functional groups where the spectra were noted at a resolution of 1cm^{-1} where the skimming was 500-4000 cm-1 and an average of 20 scans was recorded, FESEM and EDS of synthesized TiO2-NPs and TiO2 NPs assembled cotton fibers by using Horiba EMAX X-ray Detector (EDS) followed by quantitative measurement by a UV-visible spectrophotometer (UV-2550, Shimadzu Corporation, Japan). The amount of Remaining residual dyes after decoloration of model dyes solution was determined by an equation based on absorbance calibration curve of known standard solutions % (Eqtn.1) [25]

$$\text{Decoloration } \% = [(C_0-C_t)/C_0] \quad (1)$$

Here,
C_0= Concentration of model colorants at 0 min,
C_t = Concentration of model colorants at different time intervals (i.e. 5, 10,20,30,40, 60 mins)

3. Results and Discussions

3.1 XRD Analysis of TiO_2/Cellulose Nanocomposite

The structure and crystallinity of nanocrystaline titanium dioxide, cellulose nanocrystals and cellulose templated titanium dioxide nanocomposite has been analyzed by STOE STADI P transmission X-ray powder diffractometric system by monitoring the diffraction angle from 10 to 70 (2θ) using monochromatized Cu Ka (k=1.54051*A) radiation as shown in Fig.1. It reveals that all diffraction peaks are indexed as typical tetragonal crystal structure of nano-TiO_2. And the diffraction peak at $2\theta = 31.25°$ corresponding to the (121) planes of rutile TiO_2, while the peaks at $2\theta = 25.23°$, 37.98°, 48.04°, 54.74°, and 63.35° are attributed to the (101), (004), (200), (105), and (204) planes of anatase TiO_2, respectively. The characteristic diffraction peak of the anatase nano-TiO2 at $2\theta = 25.23°$ was stronger than other peaks; this is due to the (101) planes of anatase TiO_2 has lower surface energy and is the steadier structure phase in the mixed crystal structure. All the prepared TiO_2 were composed as crystal structure instead of amorphous structure.

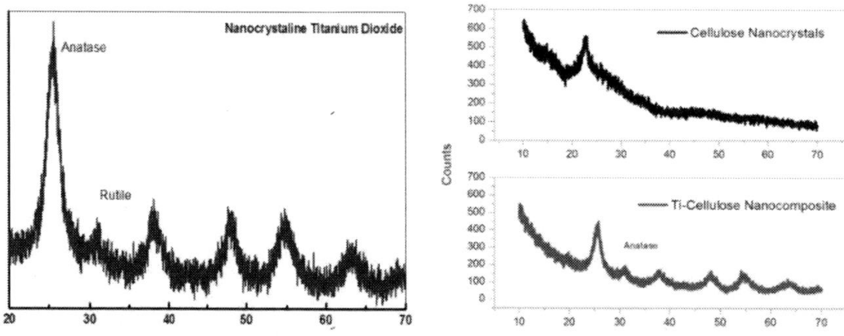

Fig.1 XRD analysis of the TiO_2/cellulose Nanocomposite (a) nanocrystaline titanium dioxide, (b) cellulose nanocrystal, (c) Cellulose nanocrystals template titanium dioxide nanocomposite

3.2 FESEM and EDS Analysis of TiO_2/Cellulose Nanocomposite

FE-SEM images of TiO_2/Cellulose Nano composite to investigate their morphology and structure has been shown in Fig.2-a, and 2-b.

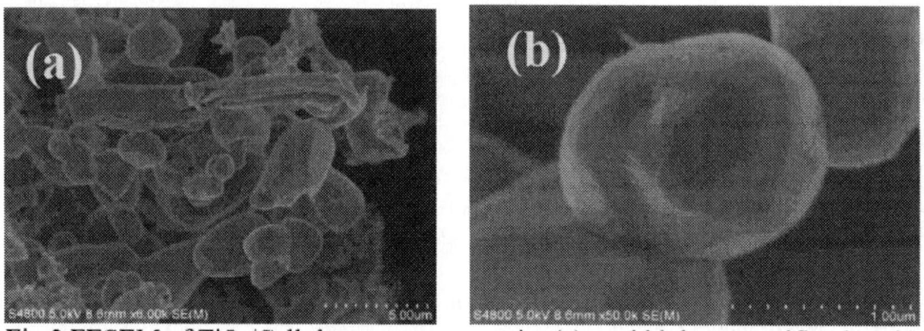

Fig.2 FESEM of TiO_2/Cellulose nano-composite (a), and higher magnification (b)

It has been analyzed that, the average particle size of loaded TiO_2 on cellulose Nano crystals is approximately 100 nm although these nanoparticles could be involved several aggregated particles. It can be obviously seen from Fig.2-a, that, the surfaces of cellulose nanocrystals were completely covered by titania nanoparticles. The nanoparticles were highly aggregated; nevertheless, some part of the same sample was observed with low aggregated TiO_2 particles.

Energy-dispersive X-ray spectrometry (EDS) is an analytical technique used for the elemental analysis or chemical characterization of a sample and detection of the elements using X-rays. The EDX spectrum shown in Fig.3-b corresponding to Fig.3-a is EDX spectrum of TiO_2/Cellulose Nano composite where Fig.3-d corresponding to Fig.3-c is EDX spectrum of pristine cotton. From the EDX spectra of Fig.3-b is has been seen that 23.50% of weight percentage of sample was found as Titina which confirmed the presence of TiO_2 NPs. Additionally, weight percentage (W%) and atomic percentage (A%) of Ti element was not found for the pristine cotton showed in Fig.3-d in comparison with sample of TiO_2/Cellulose Nano composite, indicating TiO_2 NPs content is missing. These results were consistent with corresponding FE-SEM images shown in left of all EDX spectrum.

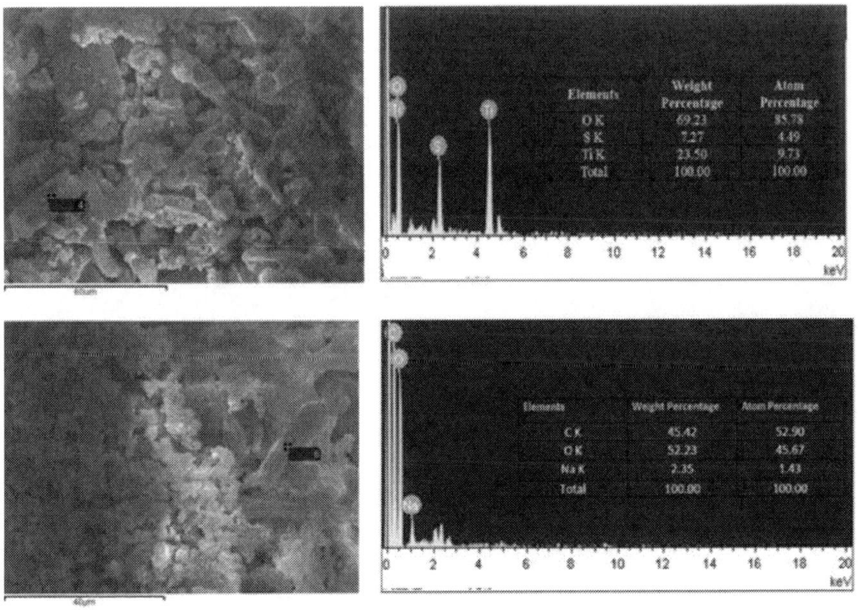

Fig.3 (a, b) FESEM and (c, d) EDS Analysis of TiO_2/Cellulose nano-composite and (e.f) EDS Analysis blank (pristine cotton) by using Horiba EMAX X-ray Detector (EDS)

3.3 FTIR Chemical Analysis of TiO_2/Cellulose Nano-Composite

Fig.4 displays the FTIR spectra of the pristine cotton, cellulose nanocrystal, cellulose/titanium dioxide nano-composite at a resolution of $1cm^{-1}$ and the scanning was 500-4000 cm^{-1}.

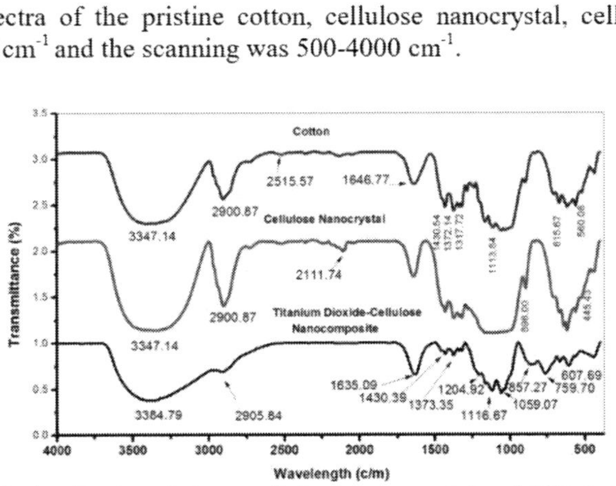

Fig.4 FTIR Chemical Analysis of cotton linters, cellulose nanocrystal and TiO_2/cellulose nano-composite at a resolution of $1cm^{-1}$ and the scanning was 500-4000 cm^{-1}

The FTIR spectrum of the cotton shows typical characteristic peaks of cellulose a broad band occurred at 3200-3500 cm^{-1}, which is assigned to O–H stretching. Two peaks at 2900.5 and 2515.57 cm^{-1} are attributed to the asymmetric and the symmetric stretching of ethylene (−CH2−) groups in long alkyl chains for both cotton and cellulose nanocrystal where it was nearly missing for cellulose nanocrystal and titanium composite sample, which indicated the presence of waxes on the cotton fibers and where for nano-composite it can be presence of metal oxide in alkayl chains. The band centered around 1646.77 cm^{-1} corresponds to deformation vibration of absorbed water molecules. The bands in the range of 1500−800cm^{-1} appeared as a result of C−H, O−H, C−O, and C−O−C vibrations, representing the fingerprint of cellulose. Some essential difference in the FTIR spectra of cotton, cellulose nanocrystal and titanium dioxide cellulose nano-composite were found in nearly 1000−500cm^{-1} which indicated that the assembly of nanocrystaline titanium dioxide has visibly influence the chemical structures of prepared nano-composite.

3.4 Catalytic Decoloration of Model Dyes by TiO_2/Cellulose Nano-Composite

With a view to investigate the outcome on environmental application, the TiO_2/cellulose nanocomposite were used to decolorize the simulated aqueous solution of textile colorants.

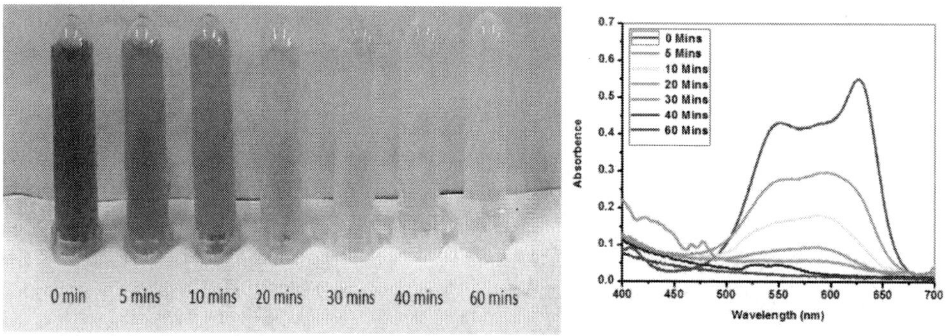

Fig.5 Photographs of the dyes solution treated with cellulose nanocrystal template titanium dioxide nano-composite and UV-vis spectra of a solution (50 mg/l, 40 ml) at a time interval of 0 min, 5 mins, 10 mins, 20 mins, 30 mins, 40 mins, and 60 mins

Commonly used colorant in textile processing industries which is an organic dye with containing benzene rings was used for this experiments. A method of high temperature decomposition method was found by Sungmin Chin et.al. [26] Which is less beneficial approach to decolorize large volume of waste water for environmental management. TiO_2/cellulose nano-composite were able to remove dyes at around more than 96% with a time frame of 40 min and 99% within a time frame of 60 mins (see Fig.5). The decoloration experiment was carried out with dyes concentration of 50 mg/L under a laboratory scale UV radiation light at ambient room temperature. With 5 mins interval the blue color of similated aquous solution was disappeared to a substantial amount and further gradual decline of color from the solution was occurred. UV-visible spectrophotometer analysis was also performed to determine. We observe that the typical absorption peak of simulater aquous solution of commercial textile dyes at 626 nm was decreased after 5 min interaction of TiO_2/cellulose nano-composite. With the extension of time to 60 min decoloration was continued. The main reason of the decoloration phenomena of the TiO_2/cellulose nano-composite is due to the reactive nature of TiO_2 nanoparticles in presence of UV light.

3.5 Absorbance Calibration Curve Analysis

The as treated TiO_2/cellulose nanocomposite was recovered by filtration of that solution after treatment under UV irradiation for 60 mins and dried at ambient condition for re-use. To illustrate the remediation capability of fresh

$$Decoloration\% = \% = [(C_0 - C_t)/C_0] \quad (2)$$

Here, C^0= Concentration of simulated aqueous solution at at 0 min,
C^t = Concentration of simulated aqueous solution at different time intervals (i.e. 5, 10,20,30,40, 60 mins)

TiO_2/cellulose composite and reuse for 1st, 2nd and 3rd time was demonstrated by plotting the percentage of decoloration against contact time. The calculation relate to decoloration % is as above equation (2).

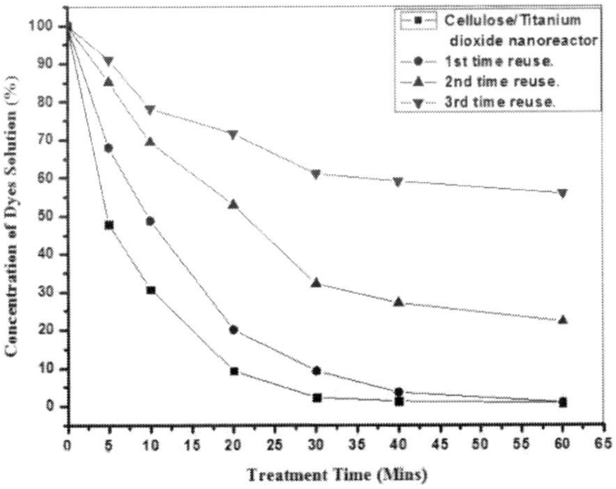

Fig.6 Remaining dyes fraction percentage of prepared solution treated using freshly prepared Cellulose/TiO_2 nanoreactor, 1st, 2nd and 3rd time reuse of nano-composite.

In this method Cellulose/TiO_2 nanoreactor prepared to accomplish the study and proper assessment. From the Fig.6 it can be seen that 1st, 2nd and 3rd time reused Cellulose/TiO_2 nanoreactor were able to remove the blue color dye solution within the exposure time of 60 min at a considerable level. It may be due to further association of TiO_2 on Cellulose/TiO_2 nanoreactor causing easily reusable and recyclable.

4. Conclusion

TiO_2 gained extra attention due to its superior photocatalytic application. A facile approach has been introduces by synthesis of nanocrystaline TiO_2 by using cellulose nanocrystal as a template. XRD, FESEM, EDS, FTIR and catalytic decoloration efficiency of the synthesized nanoreactor proves the ability of formation of crystal TiO_2 and the synthesis method by using ultrasonic irradiation has some impact on the chemical properties of cellulose and titian. The phenomena shows superior reusability by easy recovering. By catalytic decolotaion rate analysis it has been proved that, about 99.50% of commercial colorants has been removed from the model aqueous solution by means of 40 mins of treatment. So this methods can brings a new strategy to an eco-friendly treatment of waste water and minimizing the pollution from waste contents.

5. Acknowledgement

The authors wish to express their appreciation to the Hubei Provincial Department of Education for providing support through key project (No.: 20121701) and the stuffs of analysis and measurement center as well as technological institute of Wuhan Textile University for their help of handling laboratory equipment.

References

[1] Julkapli Muhd, Bagheri NS, Bee S. Abd Hamid, Recent advances in heterogeneous photocatalytic decolorization of synthetic dyes. The Scientific World Journal: 2014; 2014.

[2] Gong JL, et al., Removal of cationic dyes from aqueous solution using magnetic multi-wall carbon nanotube nanocomposite as adsorbent. Journal of Hazardous Materials: 2009; 164 (2): p. 1517-1522.
[3] Peralta-Hernández J, et al, Comparison of hydrogen peroxide-based processes for treating dye-containing wastewater: decolorization and destruction of Orange II azo dye in dilute solution. Dyes and Pigments: 2008; 76 (3): p. 656-662.
[4] Kääriäinen ML, Kääriäinen T, Cameron D. Titanium dioxide thin films, their structure and its effect on their photoactivity and photocatalytic properties. Thin Solid Films: 2009; 517 (24): p. 6666-6670.
[5] Fujishima A. Electrochemical photolysis of water at a semiconductor electrode. nature: 1972; 238: p. 37-38.
[6] Zhang X, et al. Double-layered TiO_2-SiO_2 nanostructured films with self-cleaning and antireflective properties. The Journal of Physical Chemistry B: 2006; 110 (50): p. 25142-25148.
[7] Wang R, et al. Light-induced amphiphilic surfaces. Nature: 1997; 388: p. 431-432.
[8] Soutsas K, et al. Decolorization and degradation of reactive azo dyes via heterogeneous photocatalytic processes. Desalination: 2010; 250 (1): p. 345-350.
[9] Rauf M, Ashraf SS. Fundamental principles and application of heterogeneous photocatalytic degradation of dyes in solution. Chemical Engineering Journal: 2009; 151 (1): p. 10-18.
[10] Miller RD, et al. Surface electron transfer processes: 1995: VCH New York.
[11] Carp O, Huisman CL, Reller A. Photoinduced reactivity of titanium dioxide. Progress in solid state chemistry: 2004; 32 (1): p. 33-177.
[12] Zhang X, et al. A novel CdS/S-TiO_2 nanotubes photocatalyst with high visible light activity. Separation and Purification Technology: 2009; 66 (2): p. 417-421.
[13] Bedja I, Kamat PV. Capped semiconductor colloids. Synthesis and photoelectrochemical behavior of TiO_2 capped SnO_2 nanocrystallites. The Journal of Physical Chemistry: 1995; 99 (22): p. 9182-9188.
[14] Ali R, Hassan SH. Degradation studies on paraquat and malathion using TiO_2/ZnO based photocatalyst. The Malaysian Journal of Analytical Sciences: 2008; 12 (1): p. 77-87.
[15] Oza G, et al. Photocatalysis-assisted water filtration: using TiO_2-coated vertically aligned multi-walled carbon nanotube array for removal of Escherichia coli O157: H7. Materials Science and Engineering: C: 2013; 33 (7): p. 4392-4400.
[16] Mahmoodi NM, Arami M. Degradation and toxicity reduction of textile wastewater using immobilized titania nanophotocatalysis. Journal of Photochemistry and Photobiology B: Biology: 2009; 94 (1): p. 20-24.
[17] Ray AK, Beenackers AA. Novel photocatalytic reactor for water purification. AIChE Journal: 1998; 44 (2): p. 477-483.
[18] Eichhorn SJ, et al. Review: current international research into cellulose nanofibres and nanocomposites. Journal of materials science: 2010; 45 (1): p. 1.
[19] Habibi Y, Lucia LA, Rojas OJ. Cellulose nanocrystals: chemistry, self-assembly, and applications. Chemical reviews: 2010; 110 (6): p. 3479-3500.
[20] Shopsowitz KE, et al. Free-standing mesoporous silica films with tunable chiral nematic structures. Nature: 2010; 468 (7322): p. 422-425.
[21] Cranston ED, Gray DG. Morphological and optical characterization of polyelectrolyte multilayers incorporating nanocrystalline cellulose. Biomacromolecules: 2006; 7 (9): p. 2522-2530.
[22] Rahman MM, et al. Preparation and characterization of jute cellulose crystals-reinforced poly (l-lactic acid) biocomposite for biomedical applications. International Journal of Chemical Engineering: 2014; 2014.
[23] Sadr FA, Montazer M. In situ sonosynthesis of nano TiO_2 on cotton fabric. Ultrasonics sonochemistry: 2014; 21 (2): p. 681-691.
[24] Nandanwar R, et al. Preparation of TiO_2/SiO_2 nanocomposite with non-ionic surfactants via sol-gel process and their photocatalytic study. Oriental Journal of Chemistry: 2014; 30 (4): p. 1577-1584.
[25] Xiao S, et al. Size-tunable Ag nanoparticles immobilized in electrospun nanofibers: synthesis, characterization, and application for catalytic reduction of 4-nitrophenol. RSC Advances: 2012; 2 (1): p. 319-327.
[26] Chin S, et al. Photocatalytic degradation of methylene blue with TiO_2 nanoparticles prepared by a thermal decomposition process. Powder Technology: 2010; 201 (2): p. 171-176.

Novel Method for Synthesis of N-[4-(triethylammoniomethyl) benzoyl]Caprolactam Chloride as a Cationic Bleach Activator for Low-temperature Bleaching of Cotton Fabric

Shuang-Shuang Cui, Chang Sun, Chang-Hai Xu*

College of Textile and Clothing, Jiangnan University, Wuxi, Jiangsu 214122, China

Corresponding author's email: changhai_xu@jiangnan.edu.cn

Abstract

Low-temperature bleaching of cotton fabric can be achieved by N-[4-(triethylammoniomethyl) benzoyl]caprolactam chloride (TBCC) as a cationic bleach activator which allows cotton fabric to bleach under near-neutral pH conditions. In this study, TBCC was synthesized by one-pot method which reduces material consumption and increases the final yield of TBCC by 21.1% compared to conventional synthesis method, the TBCC product was confirmed by ^{1}H nuclear magnetic resonance spectroscopy and mass spectroscopy. Bleaching experiments revealed that TBCC prepared by the one-pot synthesis method had the same bleaching effect as those synthesized by the conventional synthesis methods. Low-temperature bleaching can give the same white value degree to cotton fabric compared with the conventional peroxide bleaching. The one-pot synthesis of TBCC has advantages such as simplified operation, shortened process, reduced energy consumption, increased product yield, and lowered cost.

Keywords: Low-temperature Bleaching; Peroxide Bleach Activator; Cotton Fabric; One-pot Synthesis

1. Introduction

Cotton fabric needs to be bleached before printing and dyeing due to the presence of natural pigment impurities in cotton fibers. Hydrogen peroxide are widely used in cotton fabric bleaching with its excellent bleaching properties [1]. The conventional hydrogen peroxide bleaching of cotton fabric is often carried out under alkaline conditions (pH > 11) at temperatures greater than 95°C. This bleaching process causes not only extensive energy consumption but also severe fiber damage [2-5].

N-[4-(triethylammoniomethyl)benzoyl]caprolactam chloride (TBCC) is a new type of bleach activator, based on which an activated peroxide system can be established for low-temperature bleaching of cotton fabric under near-neutral pH conditions. Compared with the conventional hydrogen peroxide bleaching system, the TBCC-activated peroxide system has the advantage of being energy saving and is able to protect fibers from chemical damage [6-11]. However, TBCC is usually synthesized by a two-bath and two-step process which is a long and complex processes, consequently leading to a low product yield as well as a high cost [11-13]. This thesis explores the one-pot synthesis of TBCC in order to simplify the synthetic steps and reduce synthetic cost. TBCC was synthesized using a new method confirmed by ^{1}H NMR and HPLC-MS, and was applied for low-temperature bleaching of cotton fabric.

2. Experiment

2.1 Material

4-Chloromethylbenzoyl chloride (98%) was purchased from J&K. 4-(Chloromethyl)benzoic acid were purchased from TCI. Triethylamine, Thionyl chloride, caprolactam (99%) and hydrogen peroxide (H_2O_2, 30% w/w) were purchased from Sinopharm (China). 100% cotton single jersey circular-knitted fabric (178 g/m^2) was obtained from the Hongdou Textile Group (China). A wetting agent (DM-1230) and peroxide stabilizer (DM-1403) were purchased from Dymatic Chemicals (China).

2.2 Synthesis of TBCC

TBCC was synthesized by the one-pot method as shown in Scheme 1. In comparison, TBCC was also synthesized using a two-bath and two-step reaction procedure as shown in Scheme 2.

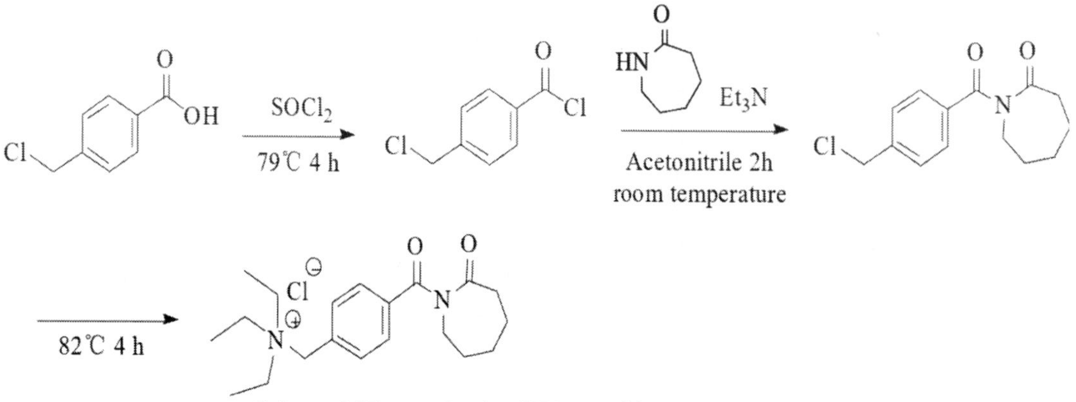

Scheme.1 The synthesis of TBCC with one-pot method

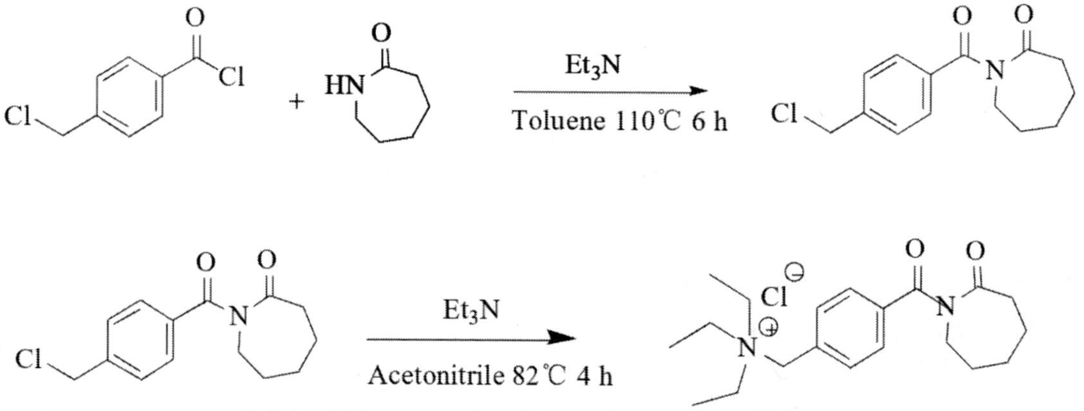

Scheme.2 The conventional synthesis procedure of TBCC

Step1: Synthesis of 4-Chloromethylbenzoyl chloride

Quantities of 0.02 mol of 4-(Chloromethyl)benzoic acid were added in 30 ml of thionyl chloride. The mixture was heated to reflux and stirred at reflux for 4 h. The solvent of thionyl chloride was evaporated to afford 4-chloromethylbenzoylchloride.

Step2: Synthesis of 4-Chloromethylbenzoylcaprolactam(CMBC)

Quantities of 0.024 mol of caprolactam and 0.06 mol of triethylamine were added to 50 ml of acetonitrile under nitrogen purge gas. A quantity of 0.02 mol of 4-chloromethylbenzoylchloride was dissolved in 10 ml of acetonitrile and then the solution was slowly added to the mixture solution of caprolactam and triethylamine. The reaction was stirred at room temperature for 1-4 h.

Step3: Synthesis of N-[4-(triethylammoniomethyl)benzoyl]caprolactam chloride

The reaction mixture was heated to reflux and stirred at reflux for 4 h under nitrogen purge gas. The reaction progress was monitored by TLC using a mixture of hexane and ethylacetate (1:1) as a mobile phase. After completion of the reaction, the mixture was cooled to room temperature. A quantity of 0.024 mol of $NaHCO_3$ was added and the mixture was heated to 45°C and stirred for 1 h. The reaction mixture was filtered while the solution was not cooled. The solvent of filtrate was evaporated. 100 ml of acetone was added, and the mixture was heated briefly and cooled to room temperature. The white precipitate was filtered, washed with acetone three times, and dried under vacuum conditions to afford TBCC.

2.3 Instrumentations

^{1}H nuclear magnetic resonance (NMR) spectra were recorded in $CDCl_3$ solution using an Avance III 400 MHz digital NMR spectrometer (Bruker, Switzerland). Chemical shifts (d) were relative to tetramethylsilane (TMS, d = 0.00) as an internal standard and expressed in ppm. Mass spectra (MS) were recorded in positive electric spray ionization (ESI+) mode using a Waters (USA) MALDI Synapt Q-Tof mass spectrometer.

2.4 Bleaching of Cotton Fabric

Bleaching method

A bleaching bath was prepared as Table 1. A sample of 5 g of cotton fabric was added to 100 ml of the prepared bleaching bath. Bleaching was carried out in a Datacolor (USA) Ahiba Nuance IR laboratory dyeing machine at 50°C for 1 h. After completion of the bleaching, the bleaching bath was cooled to room temperature, and the bleached cotton fabric was rinsed thoroughly in tap water and dried under ambient conditions.

Table.1 Process prescription of bleaching

	Low-tempcrature bleaching	Conventional bleaching
TBCC ($g \cdot L^{-1}$)	9.2	-
H_2O_2 ($g \cdot L^{-1}$)	3.4	6.0
$NaHCO_3$ ($g \cdot L^{-1}$)	2.9	-
NaOH ($g \cdot L^{-1}$)	-	3.0
wetting agent ($g \cdot L^{-1}$)	1	1
peroxide stabilizer ($g \cdot L^{-1}$)	1	1
temperature (°C)	50	95
time (min)	60	60
Liquor ratio	1:20	1:20

Whiteness measurement

Cotton fabric was measured on a Datacolor 650 spectrophotometer with illuminant D65 and a CIE 1964 standard observer. Each sample was folded twice to give an opaque sample with four plies and the whiteness was measured four times at different surfaces [12]. The CIE whiteness index (WI) values were calculated using AATCC Test Method 110-2005.

3. Results and Discussion

3.1 Synthesis of TBCC

Fig.1 shows the ^{1}H NMR spectroscopy of TBCC prepared by one-pot synthesis method. The data is given as follows: ^{1}H NMR (400 MHz,CDCl3):δ7.66 (d, J = 8.2 Hz, 2H), 7.51 (d, J = 8.1 Hz, 2H), 4.88 (s, 2H), 3.95 (s, 2H), 3.39 (q, J = 7.1 Hz, 6H), 3.08 (qd, J = 7.3, 4.9 Hz, 1H), 2.66 (d, J = 6.7 Hz, 2H), 2.14 (s, 1H), 1.82 (s, 6H), 1.39 (t, J = 7.2 Hz, 9H). It is seen from the ^{1}H NMR spectroscopy that TBCC products synthesized by one-pot method contained a very small amount of triethylamine hydrochloride.

An amount of 150 ml of deionized water was added to end the reaction after CMBC was synthesized. The mixture was stirred thoroughly and filtered to give white precipitate product. The white precipitate was washed by 5% $NaHCO_3$ aqueous solution and deionized water. Then the product was dried to afford pure CMBC product. Table 2 shows the effect of the yield of CMBC and reacting time.

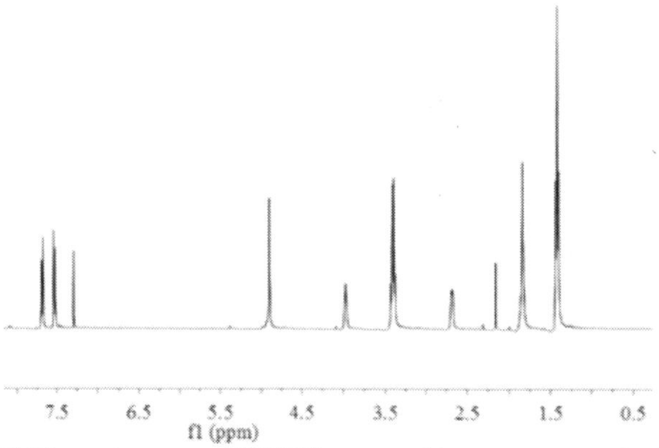

Fig.1 ^{1}H NMR spectroscopy of TBCC prepared by one-pot synthesis method

Table.2 Effect of the yield of CMBC and reacting time

Time(h)	Yield of CMBC(%)	STD
1	87.18	0.9786
1.5	84.02	0.8225
2	84.92	0.9464
4	81.44	0.7368

It is seen from Table 2 that the yield of CMBC decreased as the reaction time increased. The reason is that CMBC reacted with triethylamine to form TBCC while CMBC was generated. It means that the step3 started in the course of step2 with a slightly lower speed on account of low temperature (room temperature). The reaction time of step2 of one-pot synthesis method was set to 2 h in order to ensure the reaction of step2 proceeded thoroughly.

After completion of the third-step reaction, $NaHCO_3$ was added to remove the triethylamine hydrochloride which is generated in step2 to make sure the TBCC product is pure. Compared with the conventional synthesis method, the one-pot synthesis method used acetonitrile as the solvent instead of toluene to make it available that step2 would be conducted under room temperature and the whole synthesis reaction can be carried out in one bath which simplified the synthetic process significantly. And the reaction time of step2 was reduced from 6 h to 2 h with the yield of TBCC increasing from 57.1% to 78.2%. Table 3 shows the TBCC synthetic consumption comparison of per mole without the consideration of time and energy consumption. It is seen that one-pot synthesis make less material cost than conventional synthesis. Therefore, one-pot synthesis method of TBCC has the advantages of low energy and material consumption, high efficiency and high yield. The most important is that one-pot synthesis method enables the reaction to be carried out continuously to make it is easy to operate which has greatly shortened the synthetic cycle of TBCC.

Table.3 TBCC synthetic consumption comparison of per mole

	One-pot synthesis	Conventional synthesis
4-(Chloromethyl)benzoic acid(g)	218.36	-
Thionyl chloride(L)	1	-
4-Chloromethylbenzoyl chloride(g)	-	330.82
Caprolactam(g)	173.81	237.64
Triethylamine(g)	388.57	531.25
Acetonitrile(L)	3.84	3.84
Toluene(L)	-	3.84
Acetone(L)	7.50	7.50
$NaHCO_3$(g)	129.04	1250
Water(L)	-	15

3.2 Bleaching Performance

TBCC can react with hydrogen peroxide in aqueous solution with the caprolactam leaving group was cleaved to generate 4-(triethylammoniomethyl) peroxy benzoic acid (TPA) which is more kinetically active than hydrogen peroxide and allow cotton fabric to be bleached at lower temperatures. Scheme 3 shows the mechanism of TBCC-activated peroxide system.

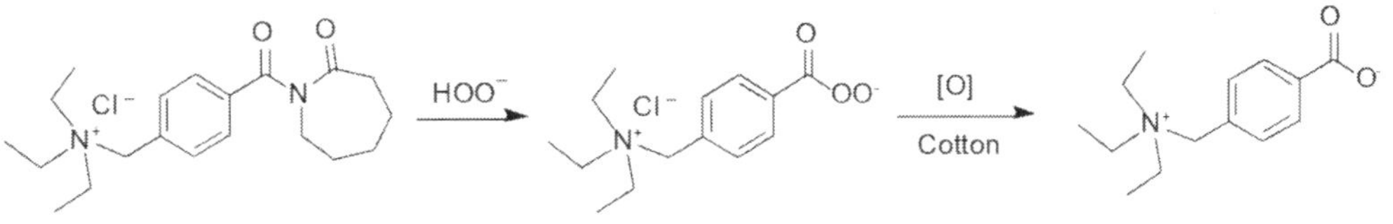

TBCC TPA TBA

Scheme.3 The mechanism of TBCC-activated peroxide system

Table 4 shows the performance comparison of low-temperature bleaching and conventional hydrogen peroxide bleaching. It is seen that The TBCC product synthesized by one-pot method had the same bleaching effect as that synthesized by the conventional synthesis method. $H_2O_2/NaHCO_3/$ TBCC bleaching activated system can provide the same white value to cotton fabric as conventional hydrogen peroxide bleaching.

Table.4 Performance comparison of low-temperature bleaching and conventional hydrogen peroxide bleaching

Bleaching method	White value (CIE)	STD
Blank	7.65	0.8212
Conventional bleaching	64.37	0.7258
Bleaching with TBCC (conventional synthesis)	63.86	0.6975
Bleaching with TBCC (one-pot synthesis)	64.15	0.8365

4. Conclusion

Compared with the conventional synthesis method, the one-pot synthesis method of TBCC can provide the same quality product and less energy consumption, short synthesis period, high yield (21.1% increased). TBCC-activated peroxide system for low-temperature bleaching based on TBCC which is synthesized by one-pot method and conventional method had the same bleaching effect on cotton fabric. Compared with the conventional bleaching peroxide system, TBCC-activated peroxide system can provide cotton fabric with equivalent degree of whiteness.

References

[1] Wei D, Sun C, Wang MZ, Du JM, Xu CH. Systhesis of N-[4-(triethylammoniomethyl)benzoyl]caprolactam chlorides as cationic bleach activators for low-temperature bleaching of cotton fabric under near-neutral pH conditions [J]. Coloration Technology: 2014; 130 (6): 432-436.
[2] X CH, X L, et al. A critical reinvestigation of the TAED-activated peroxide system for low-temperature bleaching of cotton. Carbohydrate Polymers: 2013; 92 (1): 249-253.
[3] X L, X CH, et al. The TAED/H2O2/NaHCO3 system as an approach to low-temperature and near-neutral pH bleaching of cotton [J]. Carbohydrate Polymers: 2013; 95 (1): 107-113.
[4] Wang MZ, Long XX, Du JM, Sun C, Fu SH, Xu CH. X-ray photoelectron spectroscopy analysis of cotton treated with the TBCC/H2O2/NaHCO3 system [J]. Textile Research Journal: 2014; 84 (20): 2149-2156.
[5] X CH, Hinks D, Sun C, et a. Establishment of an activated peroxide system for low-temperature cotton bleaching using N-(4-(triethylammoniomethyl) benzoyl) butyrolactam chloride[J]. Carbohydrate Polymers: 2015; 119: 71-77.

[6] X CH, Hinks D, et al. Bleaching cellulosic fibers via pre-sorption of N-[4- (triethylammoniomethyl) benzoyl]-butyrolactam chloride. Cellulose: 2010; 17 (4): 849-857.
[7] Hashem M, El-Bisi M, Sharaf S, Refaie R. Pre-cationization of cotton fabrics: An effective alternative tool for activation of hydrogen peroxide bleaching process. Carbohydrate Polymers: 2010; 79 (3): 533-540.
[8] Xu C, Shamey R, Hinks D. Activated peroxide bleaching of regenerated bamboo fiber using a butyrolactam-based cationic bleach activator. Cellulose: 2010; 17 (2), 339-347.
[9] Zeronian SH, Inglesby MK. Bleaching of cellulose by hydrogen peroxide [J]. Cellulose: 1995; 2 (4): 265-272.
[10] Liu Y, Tao J, et al. Removing polysaccharides-and saccharides-related coloring impurities in alkyl polyglycosides by bleaching with the H2O2/TAED/NaHCO3 system [J]. Carbohydrate Polymers: 2014; 112: 416-421.
[11] Qin X, Song M, et al. Low-temperature bleaching of cotton fabric with a binuclear manganese complex of 1,4,7-trimethyl-1,4,7-triazacyclononane as catalyst for hydrogen peroxide [J]. Coloration Technology: 2012; 128 (5): 410-415.
[12] Lee JJ, Hinks D, Lim SH, Hauser P. Hydrolytic stability of a series of lactam-based cationic bleach activators and their impact on cellulose peroxide bleaching Cellulose: 2010; 17: 671-678.
[13] Jun Jin Lee, Sang-HoonLim, Peterltauser, et al. Stability of a Novel Cationic Bleach Activator in Aqueo Solution [J]. Color Technol: 2005; 121 (1): 37-46.

The Effect of Dissolution Conditions on the Structure and Properties of Regenerated Antheraea Pernyi Silk Fibroin

Lu-Ping Wang, Jie Zhang, Yu-Hang Guo, Yan-Fei Feng, Qiang Zhang, Ren-Chuan You*

State Key Laboratory of New Textile Materials and Advanced Processing Technologies, College of Textile Science and Engineering, Wuhan Textile University, No.1 Yangguang Avenue, Wuhan ,Hubei,430200, China

*Corresponding author's e-mail: yourenchuan1182@126.com

Abstract

In this study, the *Antheraea pernyi* silk fibroin (ASF) was dissolved in $Ca(NO_3)_2$ solution with different dissolution temperature and times, and the effects of the dissolution conditions on the solubility, molecular weight, and mechanical properties of regenerated ASF film were investigated. The solubility was significantly dependent on the dissolution temperature and time. The extending dissolution time and increasing temperature induced the degradation of ASF chain, making molecular weight descend. Consequently, the strength of regenerated ASF films significantly decreased with the extending dissolution time and increasing dissolution temperature, indicating that a mild dissolution process that induces less degradation of the ASF chain is indispensable to achieve the ASF-based materials with a high mechanical performance. The optimal dissolution temperature and time are 90 °C and 4 h respectively.

Keywords: *Antheraea Pernyi* Silk Fibroin; Regeneration; Dissolution; Molecular Weight; Mechanical Properties

1. Introduction

Biomaterials containing inherent biological signals that can promote the cell attachment and proliferation which possess great potential for tissue engineering. *Antheraea pernyi* silk is a protein fiber spun by wild silkworms, which is one of the most extensively applied silk species in textile industry. *Antheraea pernyi* silk fibroin (ASF) contain inherent marginal-glycyl-aspartic (RGD) tripeptide sequence [1, 2], which is known as the receptor of cell integrin and may function as a biological recognition signal and promote cell adhesion [3]. It was found that ASF promoted the cell adhesion compared to Bombyx mori silk fibroin and collagen [4, 5]. Recently, ASF has been exploited for the fabrication of tissue engineering scaffolds due to its outstanding biocompatibility and sustainable degradability.

It is difficult to collect enough aqueous silk fibroin from the silkworm's silk gland. Additionally, the biodegradation of natural *Antheraea pernyi* silk fibers is very slow due to its high crystallinity. This limited the use of *Antheraea pernyi* silk as a biodegradable biomaterial [6]. Therefore, natural silk fibers must be dissolved into aqueous solution for designing desired material forms, such as films and porous scaffolds. However, the structure of ASF is remarkably different from that of Bombyx mori silk fibroin, the poly(–Ala–)$_n$ β-sheets make *Antheraea pernyi* silk fibers difficult to dissolve in salt solvents [4, 7]. Although many salt solvents have been utilized for the dissolution of Bombyx mori silk fibroin, few solvents have been reported for *Antheraea pernyi* silk fibers, namely $Ca(NO_3)_2$ and LiSCN [8, 9].

Though it was reported that *Antheraea pernyi* silk fibers could be completely dissolved in heated $Ca(NO_3)_2$ and LiSCN solution [8, 9], the effect of various dissolution conditions on the molecular weight of regenerated ASF including dissolution temperature and time has not been studied in detail. As we know, the molecular weight is a key factor influencing the structure and properties of regenerated SF materials. In this study, the *Antheraea pernyi* silk fibers were dissolved in $Ca(NO_3)_2$ solution with different dissolution temperature and time. The effects of dissolution temperature and time on the solubility, molecular weight, and mechanical properties of regenerated ASF were investigated.

2. Materials and Methods

2.1 Extraction and Dissolution of Antheraea pernyi Silk Fibers

Antheraea pernyi silk fibers were degummed by the established procedure described previously [1]. In brief, *Antheraea pernyi* silk fibers (Dandong, Liaoning, China) were boiled for 30 min in an aqueous solution of 0.25 wt% Na_2CO_3 and rinsed thoroughly with deionized water to extract the sericin proteins, which was repeated for three times. Then, the sericin-free silks were dried in an oven. The extracted *Antheraea pernyi* silk fiber were dissolved in molten $Ca(NO_3)_2$ solution (bath ratio=1:10) at 90 °C for 4 h, 90 °C for 5 h and 100 °C for 4 h, respectively. The dissolved solutions were dialyzed against deionized water for 3 days with a 9-12 kDa molecular weight cut off dialysis tube. The purified ASF solution was centrifuged at 10,000 rpm for 3 min to remove aggregates and debris and stored at 4 °C before using it in all experiments.

2.2 Molecular Weight Distribution of Regenerated ASF

To measure the molecular weight distribution (MWD) of regenerated ASF, the sodium dodecyl sulfate polyacrylamide gel electrophoresis (SDS-PAGE) test was performed. The concentration of stacking gel is 5%, and the concentration of separating gel is 10~12%. Pre-stained protein served as the MW markers (10-200 kDa), and the gels were stained with an Easy Stain Coomassie Blue Kit (Invitrogen, Carlsbad, CA). Furthermore, the MWD of the ASF were measured by gel permeation chromatography (GPC; Waters 1525) and a GL column (TSK G3000 swxl 6.5 * 300mm + TSK G2000 swxl 6.5 * 300mm) was used as the solid phase. For sample preparation, the ASF samples were diluted into 5 mg/ml. The MWs were determined after a calibration curve was obtained by measuring with standard protein sample (12.4-200 kDa).

2.3 Preparation of Regenerated ASF Films

The regenerated ASF films were prepared using regenerated solutions cast upon polystyrene plate. After casting, the films were allowed to dry for 24 h at 25 °C in a fume hood. After drying, the films were peeled off and immersed in 75% ethanol for 2 h for mechanical tests in wet state.

2.4 Mechanical Tests of ASF Films

To study the effect of dissolution conditions on mechanical properties of the regenerated ASF films, the tensile strength of ASF films was tested using an Instron 5967 mechanical testing instrument at 25 °C and 50 ± 5% RH for dry strength and wet strength. The films were cut for uniform strips with the size of 50 mm × 10 mm (n=5 for each sample). The gauge length and the drawing speed were preset as 10 mm and 10 mm/min, severally.

3. Result and Discussion

The dissolution conditions, including salt concentration, heating temperature and dissolution time, showed significant effect on the solubility of ASF fibers [8, 9]. Table 1 shows the solubility of ASF fibers in $Ca(NO_3)_2$ solution under varied dissolution conditions. During the dissolution of ASF fibers, the solubility is strongly dependent of the heating temperature and time. ASF fibers were hardly dissolved after 5 h heating at 80 °C. At 90 °C, the most of ASF fibers have been dissolved after 3 h heating, and a more clear solution was achieved after 4 h heating. ASF fibers have been almost dissolved after 3 h heating when the temperature was increased to 100 °C. The results indicated that the increased heating temperature promoted the dissolution of ASF fibers, and the heating temperature at least 90 °C is suggested .

To visualize the MWD of regenerated ASF after dissolution, SDS-PAGE was performed and the images obtained from the gels were analyzed. As shown in Fig.1, the SDS-PAGE results showed that the ASF solutions produced by the different dissolution conditions all have sequential bands from 20 to 200 kDa, indicating that the regenerated ASF was a mixture of polypeptides with a wide MWD. The clear bands at 20-25, 40-70 and 150-200 kDa could be observed after dissolving at 90 °C (Fig.1(line a and b)). However, the band at 40-70 kDa disappeared after dissolving 4 h at 100 °C, and the bands at 20-25 and 150-200 kDa weaken (Fig.1(line c)), suggesting that increased dissolution temperature enhanced the degradation of ASF molecules. Furthermore, the GPC was used to measure the MW of the different ASF samples. Although the regenerated ASF showed similar average molecular weight (Mw: 32.2 kDa and 33.2 kDa) after dissolving for different time at 90 °C, the position of main peak is significantly different. As shown in Figure 2, the sample obtained by dissolving ASF at 90 °C for 4 h exhibited a main peak at about 67 kDa (Fig.2(A)), whereas the sample obtained by dissolving ASF at 90 °C for 5 h exhibited a main peak at about 47 kDa (Fig.2(B)), indicating prolonged dissolution time enhanced the degradation of ASF molecules. Our results showed that the dissolution conditions had a greater impact on the degradation of molecular chain, and we conclude that the preferable dissolution conditions of ASF fibers in $Ca(NO_3)_2$ solution are at 90 °C for 4 h.

Table.1 The solubility of *Antheraea pernyi* silk fibroin fibers in $Ca(NO_3)_2$ solution.

Samples	Temperature (℃)	Solubility
1	80	Insoluble
2	90	Partially insoluble
3	90	Completely soluble
4	100	Completely soluble

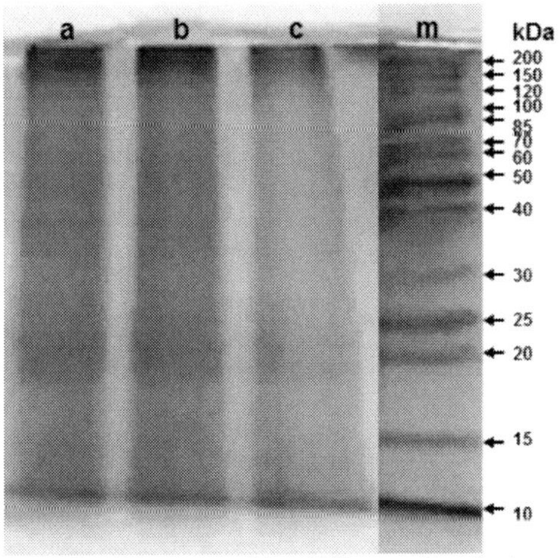

Fig.1 The SDS-PAGE of the regenerated ASF prepared by different dissolution conditions. (a) 90 °C-4 h; (b) 90 °C -5 h; (c) 100 °C-4 h, where M is the molecular weight marker.

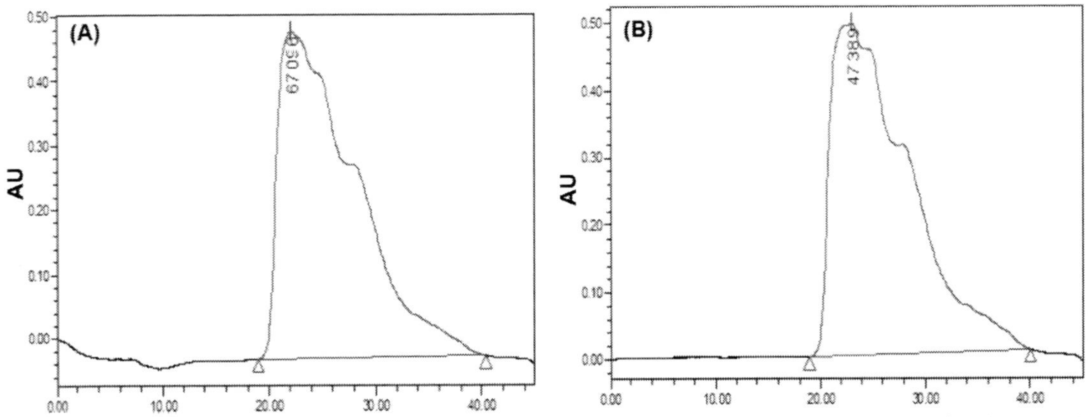

Fig.2 GPC analysis of regenerated ASF prepared by different dissolution conditions: (A) 90 °C-4 h (B) 90 °C-5 h

The MW is a key factor influencing the structure and properties of regenerated SF materials [10, 11]. It can strongly affect material forming ability and mechanical properties, and it is difficult to form ASF film if the MW is excessive low [11, 12]. For the different ASF solutions prepared by dissolution conditions, the ASF films could be fabricated (Fig.3). Furthermore, the tensile test was carried out to evaluate the mechanical properties of regenerated ASF films. The stress-strain curves are presented in Fig.4 and the values were listed in Table 2. It can be found that the breaking stress and breaking elongation of ASF films showed significant reduction with prolonged dissolution time and elevated heating temperature. Severe dissolution conditions resulted in the chain scission of ASF molecules, which lead to negative influences on mechanical performance of regenerated ASF films, especially in wet state. Therefore, a mild dissolution process that induces less degradation of the ASF chain is indispensable to achieve the ASF-based materials with high mechanical performance.

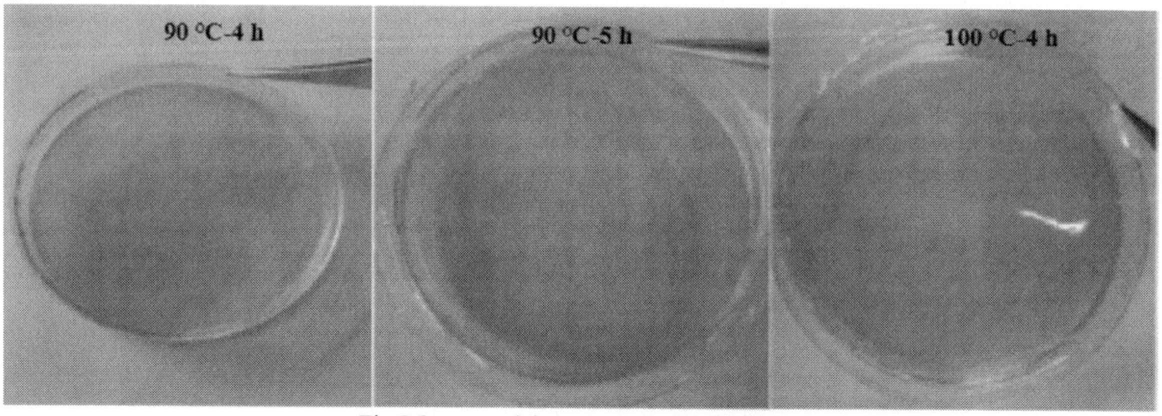

Fig.3 Images of the regenerated ASF films.

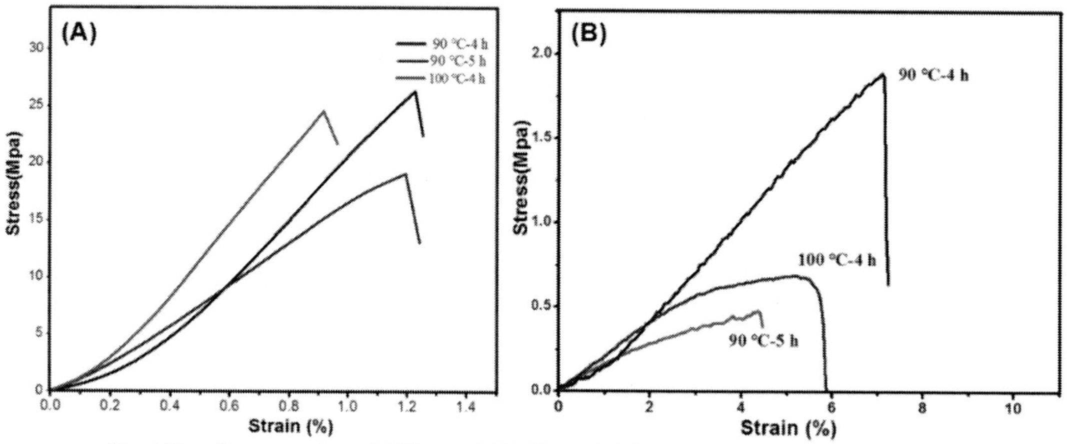

Fig.4 Tensile properties of different ASF films: (A) in dry state, and (B) in wet state.

Table.2 Tensile test of ASF films in dry and wet state (n=5).

Groups		Breaking stress (MPa)	Breaking strain (%)
Dry films	90 ℃-4 h	28.1±2.4	1.3±0.1
	90 ℃-5 h	20.5±1.4	1.2±0.1
	100 ℃-4 h	26.0±2.1	0.8±0.1
Wet films	90 ℃-4 h	2.2±0.3	10.2±2.1
	90 ℃-5 h	0.5±0.1	4.3±0.6
	100 ℃-4 h	0.6±0.1	5.1±0.9

4. Conclusion

In this study, we demonstrated that dissolution conditions have a great influence on the degree of degradation of ASF. The increasing heating temperature and dissolution time would lead to a considerable degradation of ASF molecules, and we conclude that the preferable dissolution conditions of ASF fibers are at 90 °C for 4 h. Moreover, our results showed that dissolution conditions affected the mechanical performance. Severe dissolution conditions resulted in degradation of ASF molecules, which lead to negative influences on mechanical performance of regenerated ASF films, especially in wet state, indicating that a mild dissolution process is indispensable to achieve the ASF-based materials with high mechanical performance.

5. Acknowledgement

This work was supported by the Natural Science Foundation of Hubei Province (2016CFB260), and the National Nature Science Foundation of China (31600774).

References

[1] Li XF, Li MZ, Zhang Q, Yan SQ, You RC. Aqueous-based electrospinning of regenerated *Antheraea pernyi* silk fibroin. Fiber Polym: 2016; 17: 1421-1427.

[2] Shao WL, He JX, Sang F, Ding B, Chen L, Cui SZ, Li KJ, Han QM, Tan WL. Coaxial electrospun aligned tussah silk fibroin nanostructured fiber scaffolds embedded with hydroxyapatite–tussah silk fibroin nanoparticles for bone tissue engineering. Mater Sci Eng: C 2016; 58: 342-351.

[3] Hersel U, Dahmen C, Kessler H. RGD modified polymers: biomaterials for stimulated cell adhesion and beyond. Biomaterials: 2003; 24: 4385-4415.

[4] You RC, Xu YM, Liu Y, Li XF, Li MZ. Comparison of the in vitro and in vivo degradations of silk fibroin scaffolds from mulberry and nonmulberry silkworms. Biomed Mater: 2014; 10: 015003.

[5] Li XF, Zhang Q, Ye DZ, Zhang J, Guo YH, You RC, Yan SQ, Li MZ, Qu J. Fabrication and characterization of electrospun PCL/*Antheraea pernyi* silk fibroin nanofibrous scaffolds. Polym Eng Sci: 2017; 57: 206-213.

[6] Zhao CX, Wu XF, Zhang Q, Yan SQ, Li MZ. Enzymatic degradation of *Antheraea pernyi* silk fibroin 3D scaffolds and fibers. Int J Biol Macromol: 2011; 48: 249-255.

[7] Kundu SC, Kundu B, Talukdar S, Bano S, Nayak S, Kundu J, Mandal BB, Bhardwaj N, Botlagunta M, Dash BC, Acharya C, Ghosh AK. Nonmulberry silk biopolymers. Biopolymers: 2012; 97: 455-467.

[8] Kweon H, Park YH. Dissolution and characterization of regenerated *Antheraea pernyi* silk fibroin. J Appl Polym Sci 2001; 82: 750-758.

[9] Tao W, Li MZ, Zhao CX. Structure and properties of regenerated *Antheraea pernyi* silk fibroin in aqueous solution. Int J Biol Macromol: 2007; 40: 472-478.

[10] Aznar-Cervantes SD, Vicente-Cervantes D, Meseguer-Olmo L, Cenis JL, Lozano-Perez AA. Influence of the protocol used for fibroin extraction on the mechanical properties and fiber sizes of electrospun silk mats. Mater Sci Eng: C, 2013; 33: 1945-1950.

[11] Cho HJ, Ki CS, Oh H, Lee KH, Um IC. Molecular weight distribution and solution properties of silk fibroins with different dissolution conditions. Int J Biol Macromol: 2012; 51: 336-341.

[12] Wang Q, Chen Q, Yang YH, Shao ZZ. Effect of various dissolution systems on the molecular weight of regenerated silk fibroin. Biomacromolecules: 2013; 14: 285-289.

Novel Elastic Fiber (SUSTANS®) With Different Spinning Parameters

Tao Cai[1,2,3*], Yong-Sheng Feng[2], Xi Zheng[2], Hong-Yan Liu[2], Zha-Xian Zheng[2], Wen-Xian Ouyang[2]

[1]*Quanzhou Normal University, Quanzhou, Fujian, 362000, China*
[2]*Haixing Material Technology Co., ltd, Shishi, Fujian, 362700, China*
[3]*Hosa International Limited, Jinjiang, Fujian, 362000, China*

[*]Corresponding author's email: robincaitao@163.com

Abstract

A novel elastic fiber (SUSTANS®) was prepared with different spinning parameters. The different viscosity, draw ratio and temperature of drawing roll were changed during the experimental process of the elastic fiber. The effect of these parameters on breaking strength, original modules, breaking elongation and roll elastic were studied. It showed that with the decrease viscosity difference of the elastic fiber, the roll and elastic properties of the composite fiber were increased and the breaking strength was decreased. In addition, with the increase of draw ratio and roll temperature, the breaking strength and elasticity were increased and the breaking elongation was decreased.

Keywords: Elastic Fiber; Spinning; Parameter

1. Introduction

Elastic fiber has high application value at present which could provide good fitting property for clothing [1]. There are two kinds of elastic fiber in the market: spandex and polyolefin. Spandex ages easily and has poor resistant to oxidation while costing a lot to obtain. What's more, it cannot be used individually which limited the application of it. Polyolefin has some application defects such as small elastic modulus and poor elastic recovery, which is not preferable for clothing [2, 3]. SUSTANS® was a new kind of elastic fiber with good elastic property, curl property and has good resistance to fatigue and chemicals. SUSTANS® could be prepared by PET and PTT with different melt pipe by complicate composite spinning technology. SUSTANS® fiber was greatly influenced by viscosity difference, draw ratio and stretch roll temperature. There has been very few research done on it. In this paper, the effect of spinning parameters on the elastic property of SUSTANS® was investigated. In addition, the mechanical properties of SUSTANS® fibers were studied.

2. Experiment

The dried PET and PTT section were transferred and heated at 260℃ and 285℃ through a melt pipe respectively by a screw conveyor. The melted solutions were distributed by compound spinning. After spinning solutions were controlled with reasonable ratio of the metering pump, the initial composite fiber would finally join together by the extrusion from the same spinneret holes.

Table.1 characteristics of material

Type	Intrinsic viscosity (dl/g)	Melt point (℃)
PTT	1.01	227
PET①	0.89	252
PET②	0.69	253
PET③	0.47	249

The DSC measurements were made on DSC2910 (TA, USA). SUSTANS® fibers were cut to less than 2mm and the test sample weight was 10mg. The heating rate was set at 10k/min and the final temperature was 300℃. The crystallization melting curve of the samples would be record. The mechanical properties were performed by a XD-2 vibration type fiber fineness instrument and XQ-2 fiber strength tester (Donghua University, China). The clamping length was 20mm and the draw ratio was 5mm/min. Mean data was collected of 30 samples. The cross-sections of the fibers were observed by a Ha chip tester (Hongda, China).

3. Result and Discussion

3.1 Morphology of the SUSTANS® Fiber

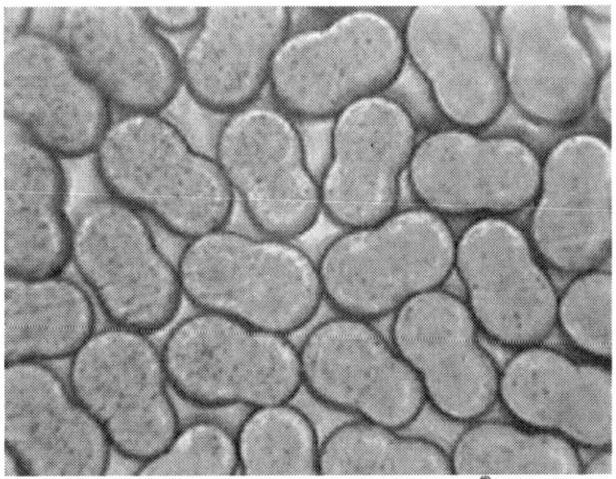

Fig.1 Cross-section of SUSTANS® fiber

Note: PTT/PET 60:40; viscosity difference: 0.32pa.s; draw ratio: 3.5; stretch roll temperature: 120℃

Fig.1 showed the cross-section of SUSTANS® fiber. It could be seen clearly that the cross-section was like peanut and the edge was smooth, which mean PET and PTT were composited well. No trend of separation could be found. In addition, good interface bonging of PET and PTT could provide SUSTANS® stable mechanical properties. There were some black spots in the figure which were the flatting agent of PET and PTT.

3.2 Thermal Property of SUSTANS® Fiber

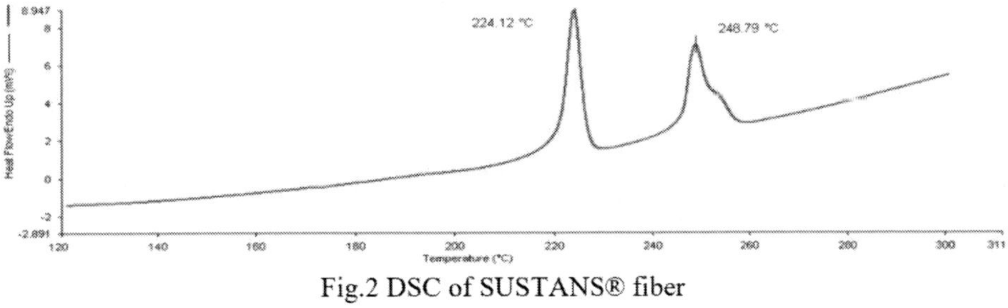

Fig.2 DSC of SUSTANS® fiber

Note: PTT/PET 60:4; viscosity difference: 0.32pa.s; draw ratio: 3.5; stretch roll temperature: 120℃

Fig.2 showed that there were double peaks in the DSC graph of SUSTANS® fiber. The peak of 224.12℃ belonged to PTT melt peak and the 248.29℃ belonged to PET. The melt points of PET and PTT chip were 227℃ and 253℃, which mean there were difference due to the interaction change of PET and PTT during the spinning process. What's more, there was a shoulder peak near 248.29℃ would be the interaction crystal peak of PET and PTT which appeared in many research [4].

3.3 The Influence of Viscosity Difference on SUSTANS® Fiber

Table.2 Properties of SUSTANS® fiber

Viscosity difference (Pa.s)	Breaking strength (CN/dtex)	Initial modulus (cN/dtex)	Breaking elongation (%)
0.12	3.64	35	42
0.32	3.37	28	53
0.54	3.21	22	64

Note: PTT/PET: 60:40; draw ratio: 3.5; stretch roll temperature: 120℃

Viscosity difference was a very important parameter of SUSTANS®. Three different viscosity of PET were prepared with a viscosity reducing equipment. Table 2 showed that with the increasing of viscosity difference, the breaking strength was decreased and the breaking elongation was increased. The main stress of PET/PTT composite fiber was concentrate on the PET agent, which mean with the decreasing of viscosity of PET the mechanical properties of SUSTANS® fiber was reducing.

3.4 The Influence of Draw Ratio on SUSTANS® Fiber

Table.3 Properties of SUSTANS® fiber

Draw ratio	Breaking strength (CN/dtex)	Initial modulus (cN/dtex)	Breaking elongation (%)	Number of crimp (/25mm)
2.5	3.27	28	60	4.5
3.5	3.37	28	53	5.0
4.5	3.56	32	45	6.5

Note: PTT/PET: 60:40; viscosity difference: 0.32pa.s; stretch roll temperature: 120℃

Spinning solution would be stretched from the spinneret orifice to form an oriented structure. With the increasing of draw raito, the breaking strength and the initial modulus were increased and the breaking elongation was decreased, which showed a regular pattern of chemical fiber process. What's more, the number of crimp of SUSTANS® fiber was on the rise with the increase of draw ratio.

3.5 The Influence of Stretch Roll Temperature on SUSTANS® Fiber

Table.4 Properties of SUSTANS® fiber

Temperature (℃)	Breaking strength (CN/dtex)	Initial modulus (cN/dtex)	Breaking elongation (%)	Number of crimp (/25mm)
120	3.37	28	53	5.0
130	3.42	30	50	6.5
150	3.61	41	45	7.5

Note: PTT/PET: 60:40; viscosity difference: 0.32pa.s; draw ratio: 3.5

The stretch roll temperature was a very importance spinning parameter which promoted the structure of fiber. With the increasing of temperature, the breaking strength and initial modulus were on the rise, so was the number of crimp. The difference of two material was the base of elastic of SUSTANS® fiber and with the increase of roll temperature, fiber showed a uptrend elastic property which gave fiber more elastic potential.

4. Conclusion

The different viscosity, draw ratio and temperature of drawing roll were change during the process of the elastic fiber of SUSTANS®. It showed that with the decrease of viscosity difference of the elastic fiber, the roll and elastic of the composite fiber were increased and the breaking strength was decreased. In addition, with the increasing of draw ratio and roll temperature, the breaking strength and elastic were increased and the breaking elongation was decreased.

References

[1] Guo J, You CF, Mu SY, Zhang L, Guan FC, Liu YF, Zhang S. Properties of PET/PTT and PET/PBT bicomponent filaments. Journal of Dalian Polytechnic University: 2015; 34 (6): 463-466.

[2] Yang XH, Yang K. The Production Technology of PET-PTT Composite Elastic Fiber. Synthetic Fiber in China: 2014; 43 (4): 18-20.

[3] Yu HF, Yang MG. Effect of drawing process on properties of PET/PTT composite fiber. China Synthetic Fiber Industry: 2012; 35 (3): 58-60.

[4] Yu HF, Yang MG. Production process of side-by-side PET/PTT composite fiber. China Synthetic Fiber Industry: 2011; 34 (6): 53-55.

Research of Polysulfone Amide Micro-nanofiber Forming Technology

Yao-Wei Chen, Qian-Wen Yu, Chen Chen, Wei-Wei Zhang, Chen Yu, Cheng-Hong Yang, Shi-Xin Jin, Bin-Jie Xin*

School of Fashion Technology, Shanghai University of Engineering Science, Shanghai 201620, P. R. China

Corresponding author's email: xinbj@sues.edu.cn

Abstract

Polysulfone amide has excellent heat resistance, thermal stability, high dimensional stability, flame retardancy, good electric insulation and resistance to electromagnetic radiation; it also has good mechanical properties, chemical stability and dyeability. There are two methods used for the preparation of polysulfone amide fibers: wet-spinning and electrospinning. In this paper, the theory of the two preparation methods, research status, major problems and development trend were studied respectively, laying the technical foundation for progress research.

Keywords: Polysulfone Amide; Wet-spinning; Electrospinning; Micro-nano Molding

Polysulphone Aramid (PSA), belongs to the para-aramid fiber series and is scientifically named as polybenzene sulfone terephthalamide fiber. It has excellent heat resistance, thermal stability, high dimensional stability, flame retardancy, good electric insulation and resistance to electromagnetic radiation, it also has good physical properties, chemical stability and dyeing properties [1].

China has the independent intellectual property rights of PSA [2] and has full control of the industrialized production of the organic fiber, which is an important basic material for advanced protection and the synthesis of structural materials. PSA is a new textile material, and has high strength, light weight and high temperature resistance, which is often used as protection materials in China's aerospace field and other high temperature working environments, as well as in some civil fields for fire retardant requirements, functioning in an important role in China's national defense military and modern industry. Wet-spinning and electrospinning are the two main forming processes of this kind of material.

Although PSA has many excellent properties, it also has certain weakneses. For example, PSA cannot withstand ultraviolet radiation [3-5] because of the -NHCO- group in the fiber macromolecule chains. The volume of the fiber is higher than resistance, which makes it easy to generate static electricity. This leads to many problems in subsequent processing. As such the mechanical properties of PSA needs to be improved as it limits its application.

General structure of PSA monomer:

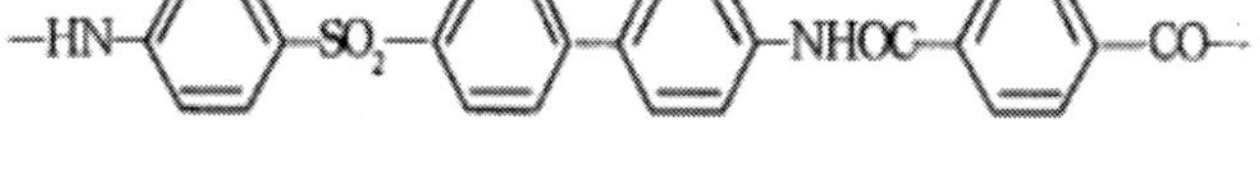

This paper mainly introduces the basic principles, current research status, main problems and development trends in wet-spinning and electrospinning.

1. Wet-spinning

Wet-spinning method is one of the main spinning methods of chemical fibers. Wet-spinning has a variety of forming methods, and spinning machine also has many different structures. For example, there are single bath or double bath methods, deep bath or shallow bath methods, funnel shaped or tube forming methods, etc. Wet-spinning requires not only bulk solutions and many large-scale machines for preparations, but also the coagulation bath, recycling and recovery equipment. Due to the complex process, high cost of factory constructions and low production efficiency, the cost and expense is high.

Actually, the wet-spinning process as the traditional method of preparing fiber appeared much earlier than the electrospinning process, but now we still use the wet-spinning process frequently in the fields of medicine, environment, biology and so on. In 2010, Zhang Hong [6] used wet-spinning process to studied the chitosan of different concentration of deacetylation, and now chitosan has been widely used in biomedical field. We can foresee that the wet-spinning technology as a traditional process will play an important role in studying medical absorbable materials, tissue-engineering vector materials. In 2012, Zhou Huan [7] researched on regenerated silk fibroin by wet-spinning process. This research improved the effect of the subsequent water vapor tension of silk protein fiber, and improved the morphology of regenerated silk protein fiber. In 2016, Zhang Xuchen et al [8] prepared BSCF hollow fiber membrane by wet-spinning process. The hollow structure prepared by this test could increase the surface area of the hollow membrane and increase the oxygen adsorption rate. We have reasons to believe that with the further study of more and more scholars, the traditional wet-spinning process is bound to make more progress based on the original achievements.

1.1 The Basic Principles of Wet-spinning

The basic processes of wet-spinning includes the following steps: 1) preparation of spinning solution; 2) extruding the solution from the spinneret hole forming filament; 3) filament solidifying into spun fibers; 4) packaging or reprocessing the spun fibers.Diffreent devices are currently for wet spining (Fig.1.1), but their principles are basically the same.

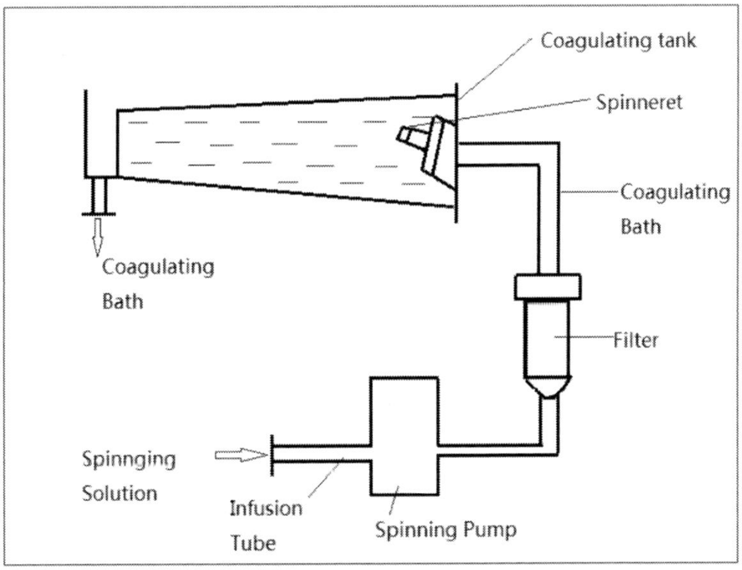

Fig.1.1 The diagram of wet spinning device

The basic principle of wet-spinning is dissolving the fiber in a suitable solvent in order to form the spinning pregnant solution which has good viscosity, compositions, and spinning performance. The forces between the large molecules decrease when the macromolecules of polymer are under the action of the solvent molecules, and then mix evenly with each other until the polymer macromolecule and solvent molecules dispersed homogeneous system, which is called the dissolution process of polymer. The polymer dissolution process is divided into two steps: Swelling (Fig.1.2) and Dissolution (Fig.1.3). Then the spinning pregnant solution is carried to the spinning machine [9] by the circulation pipe and forming fiber.

In 2011, Chen Zhuoming et al [10] used the spinning solution of polysulfone amide content of about 13% after filtration and deaeration to prepare the PSA fiber in a ternay solidifying solution system that composed of CaCl2-DMAC-H2O by wet-spinning process.

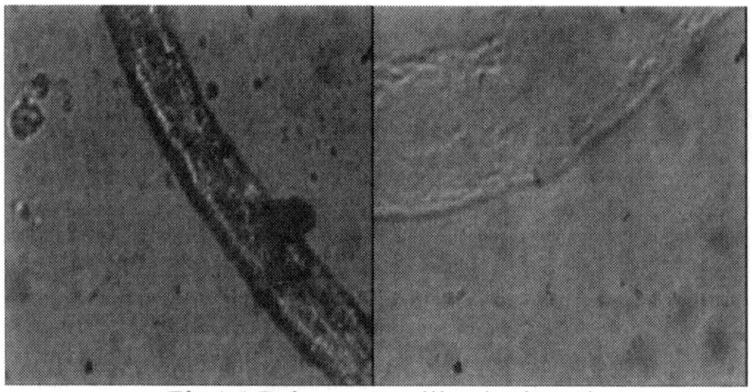

Fig.1.2 Polymer Swelling in the solvent

Fig.1.3 Polymer dissolving in the solvent

1.2 Research Status of Wet Spinning Process for Preparing Polysulfonamides Fibre

Overall, at present as the preparation of polysulfone amide fiber method, the wet-spinning is more mature than the electrospinning process. Polysulfonamide fibers prepared by wet-spinning are with higher crystallinity, better mechanical properties and thermal stability than the fibers by electrospun. Deng Jie [12-14] and many other domestic scholars were involved in the study on improvement of the polysulfonamide fiber characteristics such as anti ultraviolet, conductive, anti friction performance etc.

In recent years, the polysulfonamide fiber as the composite material to functional fibers has been attracting the attention of more and more scholars gradually. In 2007, Liu Jiexia [15] et al synthesized the polysulfone amide / inorganic nano powder composites and prepared the fiber by wet-spinning, the test improved the mechanical properties of the polysulfone amide fiber being prepared by wet-spinning successfully. Most of the domestic scholars studied on how to improve the polysulfoneamide fiber's disadvantages which was prepared by wet-spinning and only less scholars reaearched on how to improve the comprehensive performance. In 2010, Xin Binjie [16] et al mixed PSA with nano-TiO2 to prepare the PSA/nano-TiO2 composite spinning solution (Fig.1.4). and the composite fibers and composite films were prepared by wet-spinning process and digital display technology, which improved the crystallinity, mechanical properties, thermal stability and anti ultraviolet properties of the composite fibers successfully. In2012, anonymous [17] prepared the PSA hollow fiber membrane by wet-spinning, and the experiment put forward a feasible scheme for the realization of controllable PSA hollow fiber size and good separability.)

However, the mainly direction these studies discussed was how to improve a single performance of polysulfonamide fiber. It was not effective to solve the comprehensive problem of polysulfone amide. Most of them just improved the problem that polysulfonamide fiber produce static electricity easily and its low anti UV properties. Furthermore, most of the fibers were prepared in micron order of magnitude, and we still need further study on blending modification with wet-spinning system.

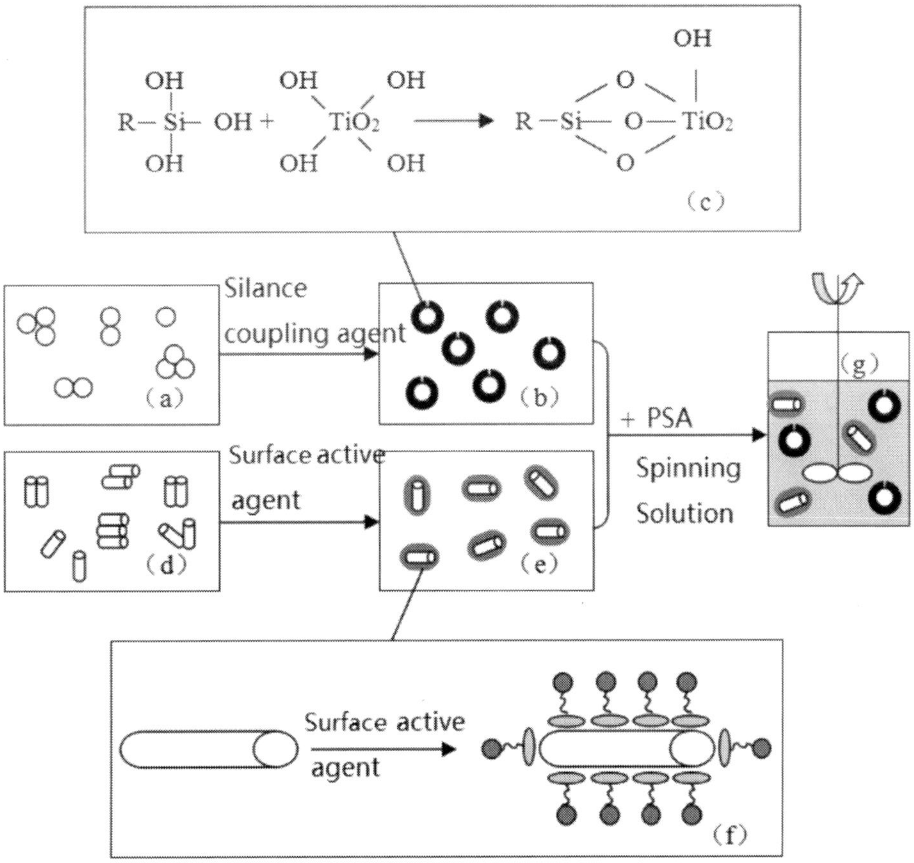

Fig.1.4 The physical cross blend mechanism model of carbon nanotubes and nano-TiO2 and PSA

2. Electrospinning

Electrospinning is a special process of polymer fluid electrostatic atomization. At the time of atomization, not micro drops but micro jets are split out, which can run a long distance and solidify into fibers last.

Although the electrospinning has been developed in 1930s. But it was not widely concerned, until the United States paid attention on the technology and explored in depth in 1990s. The research on electrospinning technology in Chian startded later but the development was so fast. Now the electrospinning technique has been used in many fields such as environmental protection, medicine, cleaning, filtering etc. because of its unique characteristics such as the ability of prepare nano scale fiber, the lower cost than wet-spinning technique's, simple equipment, controllable operation and other advantages.

In 2010, Chu Xue-hui [18] et al prepared chitosan nano fiber by electrospinning technique. The experiment showed that the surface of electrospun nanofibrous membrane was conducive to liver cell for adhesion and growth. The environmental problems have concerned gradually in recent years and the electrospinning as a simple and practical technology has great significance in the environmental problems. In 2012, Yao Chun-mei [19] et al prepared poly lactic acid fiber with a diameter of about 620nm and then they obtained the composite filter material by electrospinning technique. The test results of this material showed that the filtration effect of impurities in different diameter was improved. In 2014, Wang Yan [20] was prepared the mesoporous alumina fiber, silica - alumina core shell fiber membrane and flexible alumina fiber membrane by electrospinning technique. The nano fiber membrane worked very well in the many fields such as wastewater treatment, industrial dust filtration.

In general, the electrospinning technology has developed rapidly in recent years and its characteristics gave its unique advantages, but there is still many problems has not been resolved. Now there are hundreds of polymer processed by the electrospinning technology, but only the technology to prepare among the polyvinyl alcohol, polyurethane and poly lactic acid, chitosan and polyacrylonitrile are mature.How to use the electrostatic spinning technology to prepare other fibersn, how to prepare the continuous and ordered nano fibers and how to manufacture on a large scale, these questions are the main directions of the further research.

2.1 The Basic Theory of Electrospun Nanofibers

The electrostatic spinning device mainly comprises the following four parts (Fig.2.1): a high voltage direct current power supply, a solution supplying device, a spraying device and a receiving device. High-voltage DC power supply device provides DC power to split and stretch the electrospinning solution or melt; spraying device connects with a supply device which is full of polymer solution and use it to control the flow rate of the liquid and the receiving device is usually made of proll or frame .

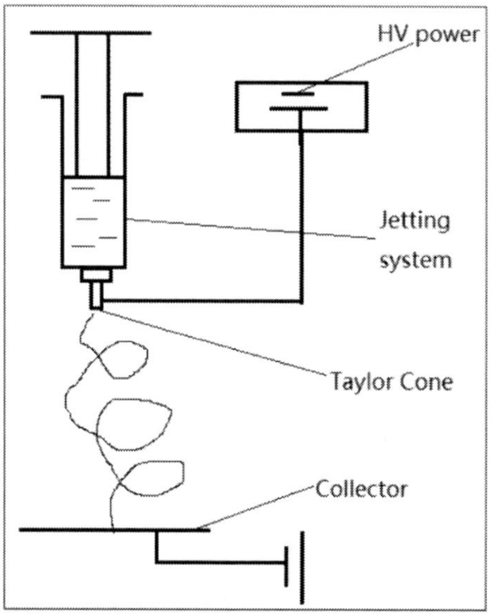

Fig.2.1 The diagram of electrostatic spinning device

The basic principle of electrostatic spinning can be explained as follows: thousands or tens of thousands of volts of high voltage electrostatic is applied in polymer solution or polymer melt. the "Taylor cone [21-23]" will be formed when the electriferous droplets on the needle under the synergic forces of electrostatic repulsive, surface tension and adhesion. When the enough electrostatic field force is applied, the polymer solution can overcome the surface tension and form the tine and long jet.Finally it evaporates in the spray process and form the fiber felt which deposit at accept device similar to nonwoven fabric. This is the electric field in a simulator (Fig.2.2, parameter: spinneret hole 0.05mm, voltage 28kv, spinning distance 15cm. Because the positive voltage is applied to the needle so the electric field in the needle is the most intensive, and more far away from the needle, the electric field distribution is more loose. The electric field lines extend from the needle to the receiving device [24]. This simulation model shows that the solution spouted from the spinneret hole in that moment the solution has the largest electric field force. By then, with the scope of the electric field gradually expands the electric dield force decreases. The solution deposits at accept device under the force of the electric field and the fiber is nanometer generally [25].

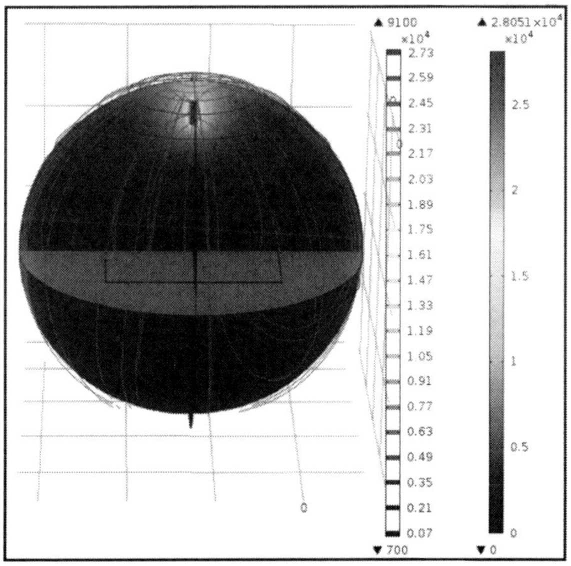

Fig.2.2 The diagram of the electric field in a simulator

2.2 The Research Status of Electrospinning Process for Preparing PSA Fibre

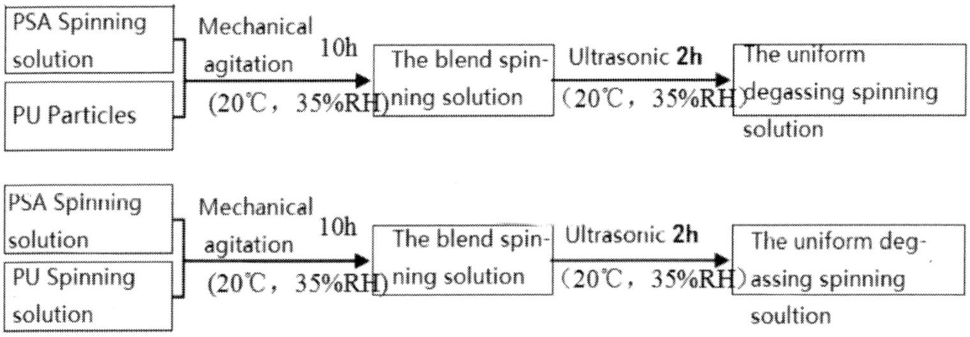

Fig.2.4 Two methods of preparing PSA/PU spinning solution

The wet-spinning technology can prepare micron fibers, and the electrostatic spinning technology can prepare nano fibers. The researches in China of how to prepare polysulfonamide fiber by electrospinning technique have began to increase and deep gradually in recent years. In 2005, Yao Yong-yi [26] et al designed a new air electrostatic spinning device which was adding the jet component in the nozzle. The result showed that the polysulfone amide nano fiber prepared by gas jet electrostatic spinning was fine, uniform and higher yield. In 2013, Liu Li [27] invented a preparation method of polysulfonamide nano fiber mats. The device made polysulfonamide spouting from the top under the action of electric field and formed the electrospinning membrane in the aluminum foil. The polysulfone amide electrospinning membrane obtained by this device had excellent adsorption properties and thermal stability. In the same year, Chen Wen-jie [28] used electrospinning technology for preparing various polysulfoneamide fiber samples of different soil content of polysulfonamide spinning solution, different spinning distance and different electrostatic voltage. The results showed the higher solid content of the polysulfonamide spinning solution, the higher the heat resistance of the sample and the heat resistance decreased when the voltage rised. This study provided important reference for the preparation of polysulfoncamide fibers with specific heat resistance. In 2014, Xin Binjie [29] prepared natural cellulose /LiCl/DMAC spinning solution and polysulfonamide after ultrasonic defoaming. Then putted them into DMAc solution methylacetamide for preparing polysulfonamide spinning solution by ultrasonic-soild mixing. Finally parpared natural cellulose / polysulfone amide nano fiber by electrospinning process. The dual component nano fiber with this mothed which could be used for many fields such as high temperature resistance, flame retardant requirements of filtration, medical and other fields had the characteristics of stable thermal performance, good moisture absorption and pore size. In 2015, Xi Tong [30] prepared PSA/PU spinning solution which blended of polysulfonamides (PSA) and polyurethane (PU) (Fig.2.4). Do not use electrospinning technology and the number of significant technology made of composite fiber and composite film. The increase of PSA retarded the thermal decomposition behaviors of PU and the PU enhanced the intensity and break elongation of PSA.This experiment provided the reference for how to improve the low strength of polysulfoneamide fiber which was prepared by electrospinning.)

Generally, most of the researches of how to prepare the polysulfone amide fiber by electrospinning are making use of polysulfonamide unique advantages such as flame retardant, high temperature. A some part of the studies are compounding PSA with other materials to improve its shortcomings such as strength and anti UV etc. But the greatest problem of preparting polysulfone amide fiber by electrostatic spinning is industrialization. So far, only a small number of scholars have been studied on this problem and this problem will greatly limits the promotion of polysulfone in the practical application. How to efficiently produce polysulfone amide composite fiber with better comprehensive performance remains to be further studied.

3. Conclusion

Whether it is wet-spinning or electrostatic spinning technology, how to improve and optimize the performance of polysulfone amide is one of the most important problems in the field of high temperature resistant materials now. Of course, In China the cognition and the research of polysulfone amide is still in the initial stage. The future development direction of polysulfonamide fiber should be in the multifunctional composite fiber material without destroying the polysulfonamide fiber itself and improving ins anti UV, conductive and mechanical properties. Therefore, it still need to make a deep study of blending moification of PSA by the technology of nano blending and modified mechanism in the further.)

References

[1] Wang XF, Zhang YH. Properties and Application of PSA. China Textile Leader: 2005; 1: 18-20.

[2] Anonymous. High temperature "Chinese fiber" into the industrialization stage. China Chemical Industry News: 2006; 1: 49-49.

[3] Liu XY, Yu WF. Study on degradation of thermo - photo - aging of PSA. Hi-tech Fiber & Application: 2005; 30 (4): 31-36.

[4] Lipika G, Mohammad HF, Hiroshi, et al. Synergistic effet of hyperthernal atomic oxygen beam and vacuum ultraviolet radiation exposures on the mechanical degradation of high-modulus aramid fibers [J]. Polymer: 2006; 47 (19): 6836-6842.

[5] He MJ, Chen WX. Polymer Physics. Shanghai Fudan University Press: 1990; 10: 284.

[6] Zhang H. Study on the Properties of Chitosan Spinning Solution and Its Wet Spinning Process. Beijing Institute of Clothing Technology: 2010.

[7] Zhou H. Study on wet spinning of regenerated silk protein fiber. Fudan University: 2012.

[8] Zhang XC, Cheng JG, Li SS, et al. Preparation of Ba0.5Sr0.5Co0.8Fe0.2O3-lk protein fibernd Its Wet Spinning Procession exposures on the mechanical degradation. Journal of the Chinese Ceramic Society: 2016; 44 (3): 392-396.

[9] Chen GP. Wet spinning device. CN104342768A [P]: 2015.

[10] Chen ZM, Xin BJ, Wu XJ, et al. Research Status and Analysis of Functional Polysulfone Amide Nanocomposites. Technical Textiles: 2011; 29 (6): 1-8.

[11] Xi T, Xin BJ. Comparison and Analysis of the Forming Process of Polysulfone. Shanghai Textile Science & Technology: 2016; 1: 19-23.

[12] Deng J. Preparation of Polysulfone Amide Composites. Shanghai University: 2005.

[13] Zou SG, Dong ZM, Qian XL, et al. Study on Hybrid Reinforced Friction Materials of PSA. National Conference on Tribology: 1997; 4: 35-40.

[14] Tang ZY, Zhang DeR, Sun JL, et al. Polysulfone amide / ZnO nanocomposite and preparation method. China,

CN200710041151.1 [P]: 2007-11-07.
[15] Liu JX, Tang ZY, Zhang DR, et al. Preparation and Characterization of Polysulfone Amide Nanocomposites. Technical Textiles: 2007; 25 (2): 14-20.
[16] Chen ZM, Xin BJ, Wu XJ, et al. Preparation and Characterization of Polysulfone Amide / Nanometer Titanium Dioxide Composites. Synthetic Fiber in China: 2011; 40 (12): 1-5.
[17] Anonymous. Preparation of Polysulfone Amide Hollow Fiber Membrane by Wet Spinning. High - Technology Fiber and Application: 2013 (6); 69-69.
[18] Chu XH, Shi XL, Gu JY, et al. Study on the effect of chitosan nanofiber membrane on hepatocyte in vitro. Chinese Journal of Biomedical Engineering | Chin J Biomed Eng: 2010; 29 (1): 144-149.
[19] Yao CM, Huang FL, Wei FF, et al. Study on Filtration Performance of Electrospun Polylactic Acid Nanofiber Composite Filter Media. New Chemical Materials: 2012; 40 (4): 122-124.
[20] Wang Y. Preparation of Alumina Fiber by Electrospinning and Its Application in Environmental Field. Shan Dong University: 2015.
[21] Fong H, I Chun, DH. Reneker, Beaded nanofibers formed during electrospinning. Polymer: 1999; 40 (16): 4585-4592.
[22] Reneker DH, I Chun. Nanometre diameter fibres of polymer, produced by electrospinning. Nanotechnology: 1996; 7 (3): 216-293.
[23] Reneker DH, AL Yarin. Electrospinning jets and polymer nanofibers. Polymer: 2008; 49 (10): 227-231.
[24] Teo WE, Ramakrishna S. [J]. Nanotechnology: 2006; 17(14): 89-106.
[25] Srinivasan G, Reneker DH. Structure and morphology of small-diameter electrospun aramid fibers. Polymer International: 1995; 36 (2): 195-201.
[26] Yao YY, Zhu PX, Hai Y, et al. Preparation of Polysulfone Nanofibers by Electrospinning and Airflow - electrospinning. Acta Polymerica Sinica: 2005 (5); 687-692.
[27] Liu L, Dong SG, Yu Y, et al. Preparation method of polysulfonamide nanofiber mat and electrospinning device. CN104032483A [P]: 2014.
[28] Chen WJ, Xin BJ, Wu XJ, et al. Preparation and Characterization of Electrospun Polysulfonamide Nanofibers [J]. Technical Textiles: 2013 (7); 7-15.
[29] Xin BJ, Cao YJ, Wu XJ, et al. A preparation method of natural cellulose / polysulfonamide nanofibers. CN105002590A [P]: 2015.
[30] Xi T, Xin BJ. Preparation and Characterization of Sulfonamide / Polyurethane Nanocomposite Fibers. Materials review: 2015; 29 (s2): 63-67.

Dyeing of Acrylic Fibers with Natural dye from Phellodendron Amurense

Kun-Di Zhu, Gui-Zhen Ke*

Department of Textile Science and Engineering, Wuhan Textile University, Wuhan 430200, China

*Corresponding author's email: kgz66@126.com

Abstract

Natural plant dye phellodendron amurense was extracted using a conventional water extract process and optimal extract process was selected by orthogonal scheme. The orthogonal analysis results showed extract temperature was the most important factor. The stability investigation showed phellodendron amurense extract was stable in a weak acid and alkaloid environment and common metal ion such as Na^+, K^+, Ca^{2+}. Acrylic fiber had been dyed with phellodendron amurense aqueous solution. The dye uptake rate of acrylic fibers increased with the increase of dyeing temperature. The optimal dyeing process was selected by orthogonal experimental as follows: dyeing temperature 90 °C, pH value 3, dyeing concentration X, dyeing time 80min and bath ratio 80:1. Color fastness to washing and rubbing of the dyed acrylic were both above 4 grade.

Keywords: Natural Plant Dye; Phellodendron Amurense; Acrylic Fiber; Dyeing; Extraction

1. Introduction

Currently, there is a movement to reuse natural dyes, and considerable research on natural dyes has been reported [1-5]. To some extent, it is a result of some stringent environmental standards imposed on the toxic and allergic reactions associated with synthetic dyes [6]. Natural dyes are environmentally friendly, as such the interest in the use of natural dyes have been revived. At present, natural dyes are mostly used in the dyeing of natural fiber such as wool, silk and cotton fiber [7-10], occasionally there are some reports about the use of natural dyes on the dyeing of synthetic fibers [11, 12].

Natural dyes are obtained from plants, minerals and animals; most of them are extracted from plants which are herbs themselves. According to the chemical structure, the natural plant dye can be divided into indigo type, anthraquinone type, flavone type, alkaloids type, poly-hydroxybenzene type, diketone type, benzopyran type, and carotenoid type [13, 14]. Berberine dye is a widely distributed berberidaceaen alkaloids that has been employed in traditional medicine as an antiprotozoal and antidiarrheal agent [15, 16]. It is found mostly in the roots of rhzioma coptidis and stems of phellodendron amurense. The Latin name of phellodendron amurense is Cortex Phellodendri Chinensis. Phellodendron amurense is the dried bark of Rutaceae Clausena lansium tree or Amur corktree, which can be found near mountain forests or in the valley streams of north, west to central China. In addition to the wild resources, at present phellodendron amurense can also be artificially cultivated.

As a yellow natural plant dye, phellodendron amurense extraction is usually used for the dyeing of natural fiber. To develop its dyeing function, in this work, phellodendron amurense dye was used to dye synthetic fiber-acrylic fiber. Acrylic fiber was dyed with phellodendron amurense aqueous solution and the dyeability was studied in terms of the thermodynamic and kinetic properties and dyeing process.

2. Experimental

2.1 Materials

Mid-length type acrylic fiber (51~7 6mm, about 1.1 dtex) was procured commercially, which was made by copolymerization of acrylic (the first monomer, content of more than 85%), methyl acrylate (the second monomer, content 3~12%) and sodium allylsulfonate (the third monomer, content 1~3%). Phellodendron amurense was obtained from medicine market. All other reagents were of analytical grade.

2.2 Extraction of Phellodendron Amurense Dye

Extraction experiment was carried out according to $L_9(3^4)$ orthogonal experiment. The factors were water addition, extraction temperature and extraction time, and each factor adopted three levels. Phellodendron amurense (2.5 g) was minced and dipped in distilled water at constant temperature for some min, then the content was cooled at room temperature and filtrated. The filtrate was concentrated to 500 mL and the absorbance of extraction was measured. Optimal extraction process was selected by orthogonal experimental and the extraction was taken as the original extract.

2.3 UV-Vis Spectrum of Phellodendron Amurense Extract

UV/visible absorption spectrum of extraction solution of phellodendron amurense were measured in the range of 200-800 nm on a Shimadzu 2550 UV/visible spectrophotometer.

2.4 Measurement of Extraction Stability

The stability of phellodendron amurense extract under different conditions was investigated by spectrometry analysis and color change observation.

Different ion solution (10 mg/mL) was added into the corresponding extract respectively, and the absorbance of each solution was determined after 24 h.

The pH value of the original extract was adjusted with sodium hydroxide and hydrochloric acid. The absorbance of the extract under different pH value was measured and the color was observed.

2.5 Dyeing Rate

Phellodendron amurense was extracted with above optimum extraction conditions. The filtrate was concentrated to 500 mL and the extraction was taken as the original extract and used for dyeing.

For the study of dyeing rate, 0.5 g fiber was dyed with phellodendron amurense extraction at the temperatures of 70 °C, 80 °C and 90 °C, keeping the liquor-to-material ratio at 80:1. The extent of dye uptake(%)was determined using equation (1) by the absorbance spectroscopic analysis of the dyebath at different time intervals, and the exhaustion curves were obtained.

$$\text{dye uptake}(\%)=\frac{A_0-A_1}{A_0}\times 100(\%) \tag{1}$$

where, A_0 is the absorbance of the dyebath before dyeing, A_1 is the absorbance of the dyebath after dyeing.

2.6 Dyeing of Acrylic Fiber with Phellodendron Amurense

Acrylic fiber was dyed directly with solution of phellodendron amurense extraction, which was called "direct dyeing". Direct dyeing experiment was carried out according to $L_9(3^4)$ orthogonal experiment with bath ratio 80:1. The four factors were temperature, pH value, dyeing concentration and dyeing time, and each factor adopted three levels. Color difference(ΔE) of the dyed fiber was measured with UV/Vis spectrophotometer plus an integrating sphere attachment and color measuring software, using illuminant D65 and 10 ° standard observer. The effects of dyeing conditions such as dye concentration, pH value of dyebath, dyeing temperature and dyeing time on color difference of the dyed were discussed, and optimal dyeing process was selected by orthogonal experimental.

2.7 Fastness Testing

The colour fastness to washing and rubbing was determined according to GB/T3921-2008 and GB/T 3920-2008 respectively.

3. Results and Discussion

3.1 UV-Vis Absorption Spectrum of Phellodendron Amurense Extraction

The UV-visible absorption spectrum curve of phellodendron amurense aqueous extract is shown in Fig.1.

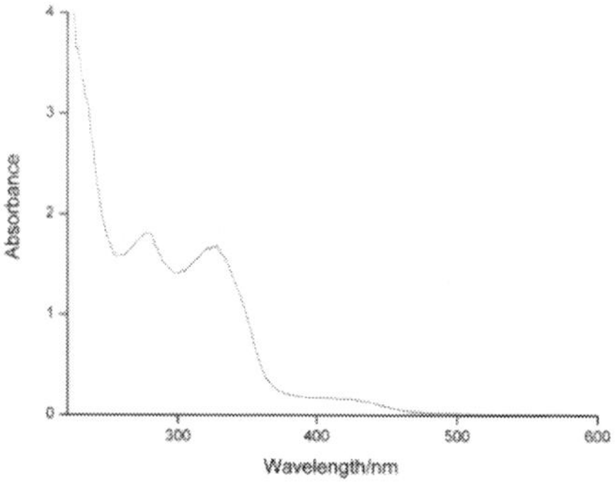

Fig.1 UV-visible absorption spectrum of phellodendron amurense aqueous extract

It is observed that there are two strong absorption peaks at λmax 325 nm and 278 nm, and a relatively weak absorption peaks near λmax 410 nm. These suggest that except for dyeing function, phellodendron amurense may have some anti-ultraviolet effects. In following test, all absorbance was measured at λmax 325 nm.

3.2 The Extraction of Phellodendron Amurense Dye

Water extraction process of Chinese medicine transfers water into herb inside by solvent diffusion and osmosis and finally causes internal components to dissolve out. Keep adding new solvent, the effective composition dissolve out continuously through concentration difference between internal and external medicinal plant cell until saturation. Phellodendron amurense was extracted using conventional water extract process and optimal extract process was selected by orthogonal scheme. The results of orthogonal experiment simple analysis are showed in Table 1.

Table.1 Orthogonal trial results for the extraction of phellodendron amurense

Number	Water addition /mL	Temperature/ °C	Time /min	Abs.
1	150	60	50	0.304
2	150	70	60	0.547
3	150	80	70	0.802
4	200	60	60	0.565
5	200	70	70	0.723
6	200	80	50	0.592
7	250	60	70	0.569
8	250	70	50	0.546
9	250	80	60	0.783
$\overline{K_1}$	0.501	0.479	0.481	
$\overline{K_2}$	0.627	0.605	0.632	
$\overline{K_3}$	0.633	0.726	0.698	
R	0.132	0.247	0.217	

The orthogonal analysis results show extract temperature is the most important factor, extract time and water addition followed. The optimum extract conditions are that 2.5 g phellodendron amurense is dipped in 250mL distilled water at 80 °C for 70 min.

3.3 Stability of Phellodendron Amurense Extraction

The stability of phellodendron amurense extract under different conditions was investigated by spectrometry analysis and color change observation. The absorbance of phellodendron amurense extract with different metal ion is presented in Table 2. It is observed that the extract is stable to common metal ion such as Na^+, K^+, Ca^{2+}, while the addition of Fe^{2+} and Fe^{3+} makes absorbency value and color value of the extract increase. When chelation reactions happen between the iron and copper ion and the chromophore, level spacing of electronic transitions in natural dye system will change to some extent, and thus changes the color of dye.

Table.2 Effects of metal ion on absorbance of phellodendron amurense extraction

ion	without	Cu^{2+}	Fe^{3+}	Fe^{2+}	Na^+	K^+	Mg^{2+}	Ca^{2+}
Abs.	0.665	0.732	1.164	0.810	0.667	0.657	0.669	0.702

The absorbance of phellodendron amurense extract under different pH value is shown in Fig.2. It is observed that the absorbance is relatively stable when pH value varies from 2 to 8, but when pH value continues to increase, it begins to decrease gradually. It indicates that phellodendron amurense extract is stable to mild acid and alkaloid environment, and strong alkaloid environment results in the change of absorbance value.

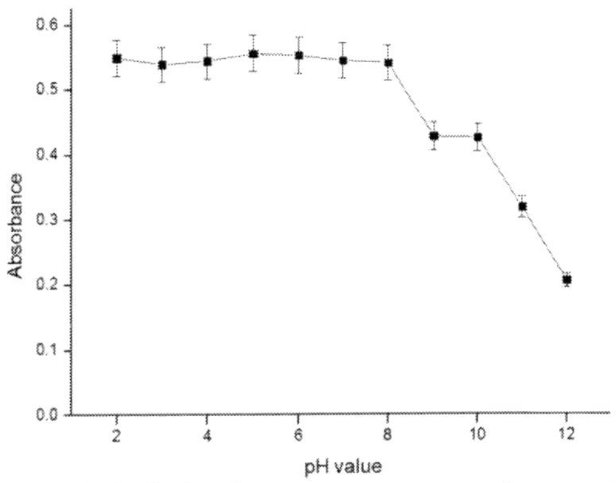

Fig.2 The stability of phellodendron amurense extraction to acid and alkaline

3.4 Dyeing Rate Curves

Dyeing rate curve is an important method to study dyeing kinetics. The dyeing rate curves of phellodendron amurense on acrylic fibers are presented in Fig.3, which provides complete description of dye adsorption rate under different dyeing temperature.

As Fig.3 shows, the adsorption of phellodendron amurense on acrylic fiber is rapid in the initial dyeing phase, i.e. initial dyeing rate is high. And the equilibrium dye uptake amount increases with the increase of dyeing temperature. When dyeing at 90°C, the adsorption of phellodendron amurense dye is much higher than that at 70 °C.

Dye uptake rate depends on the dye diffusion rate in the fiber. The dye diffusion performances in the fiber are affected by the characters of dyes and fibers, and dyeing conditions such as dyeing concentration, temperature, pH etc. When acrylic fiber is dyed near glass transition temperature, molecular chain segment movement becomes quite fierce, and the gap between macromolecular is broadened, so more phellodendron amurense dyes are absorbed and the time towards adsorption equilibrium is shortened.

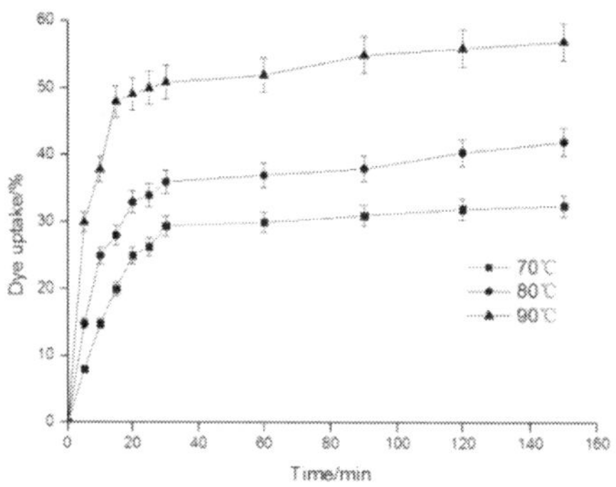

Fig.3 Dyeing rate curves of phellodendron amurense on acrylic fiber at different temperature

3.5 Dyeing Process of Acrylic Fiber with Phellodendron Amurense Extraction

The results of orthogonal experiment for direct dyeing and simple analysis are shown in Table 3.
The results indicate the pH value of dyeing solution is the most important factor in direct dyeing of acrylic fiber with phellodendron amurense, then follows dyeing concentration, dyeing temperature and dyeing time in sequence. As shown in Table 3, Higher ΔE can be obtained when dyeing in acid environment. The higher the dyeing temperature and dyeing concentration, the greater the color difference(ΔE) is. The effect of dyeing time on color difference is relatively small. According to ΔE, the optimum dyeing conditions for direct dyeing are selected as follows: dyeing temperature 90 °C, pH value 3, dyeing concentration X, dyeing time 80min and bath ratio 80:1.

Table.3 Orthogonal trial results for direct dyeing of acrylic fiber

Number	Temperature/ °C	pH value	Dyeing concentration	Dyeing time/min	ΔE
1	70	3	X	60	13.26
2	70	5	(1/2) X	70	6.34
3	70	7	(1/4)X	80	4.40
4	80	3	(1/2) X	80	15.44
5	80	5	(1/4)X	60	6.99
6	80	7	X	70	9.99
7	90	3	(1/4)X	70	10.56
8	90	5	X	80	14.04
9	90	7	(1/2) X	60	8.52
$\overline{K_1}$	8.00	13.09	12.43	9.59	
$\overline{K_2}$	10. 81	9.12	10.10	8.96	
$\overline{K_3}$	11.04	7.63	7.32	11.29	
R	3.04	5.46	5.11	2.33	

X represents initial dyeing concentration.

3.6 Color Fastness

Color fastness measurement results are presented in Table 4. It shows that both wash-fastness and rub-fastness are above 4 grade, which indicate that natural plant dye phellodendron amurense is suitable for acrylic fiber.

Table.4 Color fastness of the dyed acrylic fiber with phellodendron amurense extraction

Fastness type	Wash-fastness		Dry rub-fastness		Wet rub-fastness	
	Fading	Staining	Fading	Staining	Fading	Staining
Fastness grade	4~5	4~5	5	5	4~5	4~5

4. Conclusions

Natural plant dye phellodendron amurense was extracted with water. The optimum extract conditions were that 2.5 g phellodendron amurense was dipped in 250 mL distilled water at 80 °C for 70 min. Phellodendron amurense extract was stable to weak acid and alkaloid environment and common metal ion such as Na^+, K^+, Ca^{2+}. Strong acid environment and addition of Fe^{2+} and Fe^{3+} changed the absorbance of the extract. Dye uptake rate curves showed that the dyeing rate at the early stage increased with the increase of dyeing temperature. When dyeing temperature reached 90 °C, the equilibrium dye uptake was much higher than that at 70 °C. pH value of dyebath was the most important influencing factor during direct dyeing, and color difference value increased clearly with the decrease of pH value. The color fastness measurement showed that the dyed acrylic fiber had good wash-fastness and rub-fastness.

5. Acknowledgement

The authors greatly appreciate for the financial support by Education Department of Hubei Modern Textile Engineering and Technology Research Center of Wuhan Textile University (2016), China.

References

[1] Ratnapandian S, Fergusson SM, Wang LJ. Application of acacia natural dyes on cotton by pad dyeing. Fiber Polym: 2012; 13 (2): 206-211.

[2] Zhang RP, Cai ZS. Study on the Natural Dyeing of Wool Modified with Enzyme. Fiber Polym: 2011; 12 (4): 478-483.

[3] Acquaviva S, D'Anna E, De Giorgi ML, Della Patria A, Baraldi P. Physical and chemical investigations on natural dyes. Appl Phys A: Mater Sci Process: 2010; 100 (3): 823-828.

[4] Mirjalili M, Nazarpoor K, Karimi L. Eco-friendly dyeing of wool using natural dye from weld as co-partner with synthetic dye. J Clean Prod: 2011; 19 (9-10): 1045-1051.

[5] Cristea D, Vilarem G. Improving light fastness of natural dyes on cotton yarn. Dyes Pigments: 2006; 71: 39-46.

[6] Kamel MM, El-Shishtawy RM, Yussef BM, Mashaly H. Ultrasonic assisted dyeing: III. Dyeing of wool with lac as a natural dye. Dyes Pigments: 2005; 65 (2): 103-110.

[7] Geelani SM, Ara S, Mir NA, Bhat SJA, Mishra PK. Dyeing and fastness properties of Quercus robur with natural mordants on natural fibre. Text Cloth Sustain: 2017; 2 (1): 1-10.

[8] Dehnavi E, Shams-Nateri A, Khalili H. Wool dyeing with binary mixture of natural dyes. Pigm Resin Technol: 2017; 45(1): 52-61.

[9] Güneş E, Atav R. The use of nutshell firstly as a natural dye for cotton and wool and then as a natural adsorbent for colour removal of basic dye effluent. Color Technol: 2017; 13 (1): 88-93.

[10] Ammayappan L, Shakyawar DBB. Dyeing of Carpet Woolen Yarn using Natural Dye from Cochineal. J Nat Fibers: 2016; 13 (1): 42-53.

[11] Gupta D. Mechanism of dyeing synthetic fibres with nature dyes. Colourage: 2000; (3): 23.

[12] Shams Nateri A, Dehnavi E, Hajipour A, Ekrami E. Dyeing of polyamide fibre with cochineal natural dye. Pigm Resin Technol: 2016; 45 (4): 252-258.

[13] Teli MD, Roshan P, Pardeshi PD. Natural dyes: classification, chemistry and extraction methods. Colourage: 2000; (12): 43-48.

[14] Gilbert KG, Cooke DT. Dyes from plants: Past usage, present understanding and potential. Plant Growth Regul: 2001; 34: 57-69.

[15] Zhang HJ. Manual for Identifying of Chinese Native Medicinal Materials. Guangdong: Guangdong Tour Publishing Company: 2000; p. 283-286.

[16] Chen AH, Lee YT, Chen CY. Microwave accelerated extraction and capillary electrophoresis analysis of Berberine from the cortices of Phellodendron wilsonii and Phellodendron amurense. J Chin Chem Soc: 2005; 52 (4): 781-784.

Preparation and Properties of Nano-structured TiO_2 Film Deposited by Magnetron Sputtering

Xiao-Hong Yuan*, Li Zhao, Dong-Sheng Chen, Ying-Jin Gan

Faculty of Clothing and Design, Minjiang University, Fuzhou, Fujian, 350121, China

*Corresponding author's email: yxhong1981_2004@126.com

Abstract

Titanium dioxide (TiO_2) films on polyester fabrics were fabricated by magnetron sputtering. The surface morphologies and microstructure of thin films deposited on polyester fabrics were analyzed by scanning electron microscopy (SEM) and x-ray diffraction (XRD). The photocatalytic activities and anti-ultraviolet properties of polyester fabric coated with TiO_2 films were investigated. The results indicated that TiO_2 films deposited on polyester fabric substrates were relatively smooth, dense and uniform. XRD analysis revealed that the structure of TiO_2 film was non-crystalline structure. Compared with the original fabric samples, the polyester fabrics coated with TiO_2 films have better photocatalytic activities and anti-ultraviolet property.

Keywords: Titanium Dioxide Film; Magnetron Sputtering; Polyester Fabric; Photocatalytic Activities; Anti-ultraviolet Properties

1. Introduction

As time goes on and with the development of science and technology, people gradually have higher requirements for the quality of textiles. Modern people's pursuit of textile strays further away from the original modesty and cold proof period, but is filled with curiosity on functional with properties that are high-quality, comfort, health care and protection.

Functional textiles having different properties from general textiles [1]. Preparation methods of functional textiles include physical methods, chemical methods and a combination of physical and chemical methods. Magnetron sputtering technology is one of the physical preparation methods [2-5].Magnetron sputtering technique have been considered as a very effective deposition method of thin films which can prepare super-hard films, corrosion and friction resistant films, superconducting films, magnetic films, optical films, and other special films. [6].

In this paper, the author fabricated the Titanium dioxide (TiO_2) film on polyester fabric through magnetron sputtering and analyzed the surface morphologies and microstructure of thin films deposited on polyester fabrics by scanning electron microscopy (SEM) and x-ray diffraction (XRD). We also investigated Photocatalytic activities and anti-ultraviolet properties of polyester fabric coated with TiO_2 films.

2. Experimental Section

2.1 Materials

The author cut polyester fabrics into 5 cm diameter circular samples, and immersed them in an acetone solution, then sonicated them for 30min by the KQ – 50B type ultrasonic cleaners to remove impurities on the surface of the fabrics. After been repeatedly washed with deionized water, the fabrics were dried in an oven at 60 °C.

99.99% pure titanium (Ti) as the sputtering target material was been used.

2.2 Deposition of TiO2 Film

There are two methods to prepare titanium dioxide thin films. The first one is titanium dioxide films deposition on the fabric's substrate using the target titanium by radio frequency (RF) magnetron sputtering. The second one is titanium films deposition on the substrate using the target titanium by direct current (DC) magnetron sputtering and then placing the fabric substrate in the air for a period of time, to completely oxidize Ti to TiO_2 because it is easily oxidized in the air. The first method was chosen in this paper, and the sputtering unit (JPG-450 type) was used in the experiment.

The author also used Argon (Ar) as sputtering gas with a gas flow rate of 50mL/min, oxygen (O_2) as reaction gas with a gas flow rate of 50mL/min, a base pressure of 1.5×10^{-3}Pa, a rotating speed of 10r/min, a working gas pressure of 0.8Pa, a sputtering power of 300w and for a sputtering time of 60min to deposit the titanium film

2.3 Characterizations

The author examined the microstructures of TiO_2 film on the textile substrate by scanning electron microscopy (SEM, JSM-5610LV, Japan), and the coated fabrics by X-ray diffraction (XRD) measurements on a Bruker-AXS X-ray diffractometer system with Cu Kα radiation. Scanning range was 2°- 90°(2θ).

According to GB/T18830-2009, anti-ultraviolet properties of the coated fabrics were tested by ultraviolet transmittance analyzer (UV-1000F, Lapsphere, America). The evaluation indexes of anti-ultraviolet properties included solar UV-A spectral transmittance [T(UVA)], solar UV-B spectral transmittance [T(UVB)] and ultraviolet protection factor (UPF). Each sample was tested five times, and the average values were reported.

Photocatalytic performance test of nano-meter TiO_2 film is as follows. 50mL of rhodamine B solution with a concentration of 15mg / L was used as the degradation target, UV light source for 20w, UV germicidal lamps, stirrer for the magnetic stirrer were need in the test, and there was an irradiation time of 2h. The absorbance of the rhodamine B solution at 554 nm was measured by using a 721 visible spectrophotometer.

The solution degradation rate η is calculated as follows:

$$\eta = \frac{A_0 - A}{A_0} \times 100\% \qquad (1)$$

Where A_0 and A are the absorbance of the solution before and after illumination [7, 8].

3. Results and Discussion

3.1 Surface Morphologies and Microstructure of TiO_2 Film

The surface morphology of TiO_2 film was examined by SEM, and the SEM image is shown in Fig.1.

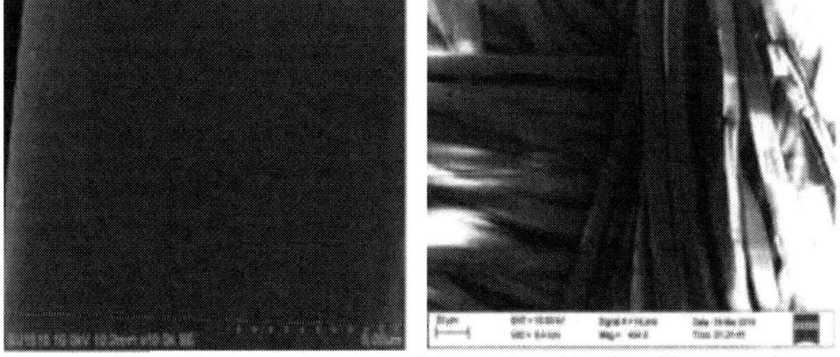

(a) original sample (b) sample with TiO_2 film coated

Fig.1 SEM of polyester fabric with and without nanoscale TiO_2 film coated.

Fig.1a was the SEM image of polyester fiber, the surface of which was relatively smooth, and almost with no particles. From Fig.1b, it can be seen that the surface of the TiO_2 film deposited on polyester fiber was relatively smooth, dense and uniform. It indicated that TiO_2 film was deposited on polyester fiber densely and evenly.

Fig.2 illustrates the X-ray diffraction (XRD) patterns of original sample and sample with TiO_2 film coated.

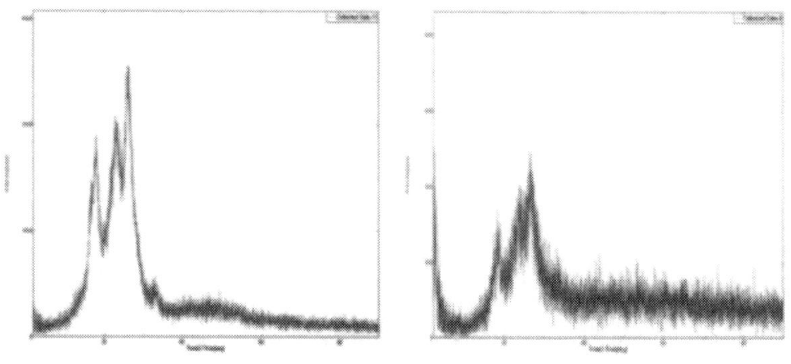

(a) original sample (b) sample with TiO_2 film coated

Fig.2 XRD patterns of polyester fabric with and without nanoscale TiO_2 film coated

In Fig.2, it can be seen that the XRD pattern of original fabric is similar to that of the sample with TiO_2 film coated. Compared with the XRD pattern of original fabric, the TiO_2 diffraction peak was not obvious. Therefore, the structure of TiO_2 film is the non-crystalline structure.

3.2 Anti-ultraviolet Property

Table 1 shows the anti-ultraviolet properties of original sample and sample with TiO_2 film coated. T (UVA) and T (UVB) express solar UV-A spectral transmittance and solar UV-B spectral transmittance. The smaller the transmittance value is, the higher the anti-ultraviolet properties are. Conversely, the greater the ultraviolet protection factor (UPF) is, the higher the anti-ultraviolet properties are.

Table.1 Experimental result of anti-ultraviolet property

Samples	T(UVA) /%		T(UVB) /%		UPF	
	Average value/%	Standard deviation/%	Average value/%	Standard deviation/%	Average value	Standard deviation
Original fabric	38.62	0.42	5.80	0.15	9.72	0.88
Fabric coated with titanium dioxide film	10.12	0.18	4.56	0.25	24.68	1.02

From Table 1, it can be seen that the anti-ultraviolet properties of the original sample were lower than that of the sample coated with titanium dioxide film. The values of T (UVB) the sample coated with titanium dioxide film was less than 5%, and the value of UPF was close to 25. Nevertheless, compared with the original samples, the anti-ultraviolet properties of the fabric deposited with TiO_2 film were greatly improved.

3.3 Photocatalytic Activities

Fig.3 illustrates the photocatalytic decomposition rate of formaldehyde after 2 hours decomposed by nanometer TiO_2 film.

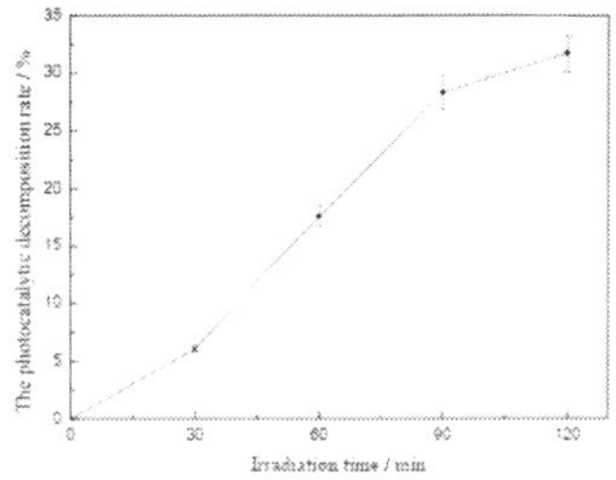

Fig.3 The photocatalytic decomposition rate of nano titanium dioxide films

From Fig.3, it could be seen that with the extension of time, the degradation rate gradually increased, and the photocatalytic decomposition rate of formaldehyde after 2 hours decomposed by nanometer TiO_2 film was 31.7%. It indicated that the polyester fabric coated by TiO_2 film could decompose formaldehyde, and have better photocatalytic performance.

4. Conclusion

TiO_2 film was successfully deposited on polyester fabric by RF magnetron reaction sputtering technique with pure Ti targets. TiO_2 film deposited on textile substrates was even and dense, and the structure of TiO_2 film was non-crystalline structure. Compared with the original fabric, anti-ultraviolet properties of the fabric coated with TiO_2 film were improved greatly. The polyester fabric coated by TiO_2 film has better photocatalytic performance.

5. Acknowledgement

This work is supported by The Education Department of Fujian Province (No. JK2014042), Textile Science and Engineering, key discipline of Fujian province, Ministry of education of Fujian (No.2012-136); Clothing design and engineering, teaching trial reform of universities in Fujian, Department of Finance and Ministry of education of Fujian (No.2012-41-130), and the Technical Plan Project of Fuzhou City (2016-G-77).

References

[1] Sun F, Xu HY. Application of nano material to multifunctional finish of textile. Dyeing & Finishing: 2005; 22:1-4.

[2] Waykar GR., Pawbake AS, Kulkarni RR, et al. Influence of RF power on structural, morphology, electrical, composition and optical properties of Al-doped ZnO films deposited by RF magnetron sputtering. Journal of Materials Science: Materials in Electronics: 2016; 27: 1134-1143.

[3] Li LX, Xu D, Zhang N, et al. Effect of film orientation on the dielectric properties of bismuth magnesium niobate thin films prepared by RF magnetron sputtering. Journal of Materials Science: Materials in Electronics: 2015; 26: 288-293.

[4] Konischev ME, Kuzmin OS, Pustovalova AA, et al. Structure and Properties of Ti–O–N Coatings Produced by Reactive Magnetron. Sputtering Russian Physics Journal: 2014; 56: 1144-1149.

[5] Jean MD. Optimisation of tribological performance of sputtering Zr/ZrC coatings. Surface Engineering: 2015; 31: 103-113.

[6] Ramazanov SM, Kurbanov MK, Safaraliev GK, et al. Structural properties of the epitaxial (SiC)1-x(AlN)x solid solution films fabricated by magnetron sputtering of SiC-Al composite targets. Technical Physics Letter: 2014; 40: 300-302.

[7] Yi G, Zhang BW. The effects of the concentration of individual reactants on Fe-Ni-P-B nanoparticles fabricated by chemical reduction. Joumal of Materials Processing Technology: 2001; 117: 37-42.

[8] Yi G, Guo Y, Zhang BW, et al. Preparation and thermal properties of amorphous Fe-W-B alloy nanopowders. Journal of Materials Processing Technology: 1998; 74: 10-13.

Application of Transparent Nano-fibrous Membrane onto Fiberglass Mesh for PM2.5 Removal

Guo-Cheng Zhu[1,2,3], Lian-Ying Zhao[1,2], Lv-Tao Zhu[2]

[1] *Key Laboratory of Advanced Textile Materials and Manufacturing Technology, Ministry of Education, Zhejiang Sci-Tech University,Xiasha, Hangzhou, 310018, China*
[2] *College of Materials and Textiles, Zhejiang Sci-Tech University, Xiasha, Hangzhou, 310018, China*
[3] *Hubei Key Laboratory of Advanced Textile Materials & Application, Wuhan Textile University, Wuhan, 430200, China*

Corresponding author's email: zgc100100@hotmail.com

Abstract

In order to capture PM2.5 and to improve the indoor air quality, one kind of window screen was developed by applying transparent nano-fibrous membrane onto fiberglass mesh. The most important two properties, pressure drop and filtration efficiency, of nano-fibrous window screen (NFWS) were evaluated. The results revealed that the pressure drop of NFWS at an airflow speed of 5m/min was very low (<25 Pa) compared with the values from other reported work, and the filtration efficiency of sample 620 was more than 83% for PM1.88. Besides this, the NFWS showed very good transparency.

Keywords: Nanofiber; Electrospinning; Pressure Drop; Filtration Efficiency; PM2.5

1. Introduction

Owing to rapid industrialization, urbanization and modernization in many developing countries, air pollution has become a very serious problem in affecting environment, public health and people's living quality. Smog is the most direct form of air pollution which contains massive suspended particulate matters including sulfate, nitrates, ammonia, sodium chloride, black carbon, mineral dust and water [1]. Particulate matter with a diameter of 2.5 microns is called PM2.5. PM2.5 in air can penetrate and lodge deep inside the lungs. Chronic exposure to particles contributes to the risk of developing cardiovascular and respiratory diseases, as well as lung cancer [2-6]. According to WHO record in 2012, around 3.7 million premature deaths worldwide are caused by air pollution and the greatest number was in the Western Pacific and South-East Asia regions [1].

The guideline value of PM2.5 annual mean is 10μg/m^3, and 24-hour mean is 25μg/m3 according to WHO 2005 guideline [1]. But in many cities of China, the concentrations of PM2.5 were over 100μg/m^3 especially in northern China. The concentration of PM2.5 in some cities can reach to hundreds μg/m^3. Therefore, PM2.5 has attracted a lot of attention from governments and researchers. The Chinese government unveiled its action plan for air pollution control (2013-2017), which will dedicate 1.7 trillion RMB towards pollution control. Some researchers had applied Polyacrylonitrile (PAN), Poly (vinyl pyridine) (PVP), Poly (methyl methacrylate) (PMMA), and Nylon-66 nano-fibrous membrane onto metallic mesh and nylon mesh to capture PM2.5 [7, 8]. Their samples had good transparency and very high filtration efficiency. However, the pressure drop of sample was very high which strongly influence the air exchange rate between indoor and outdoor. Besides, the protection of nanofibers onto the mesh was not taken into consideration.

Poly (vinylidene fluoride) (PVDF) has been widely used as a membrane material thanks to its good chemical resistance, great mechanical properties and high thermal stability [9-11]. It is broadly applied in ultrafiltration, microfiltration, membrane distillation and other membrane processes [12, 13]. Therefore, in this work the PVDF nano-fibrous membrane was prepared by electrospinning technology and applied onto fiberglass mesh. The pressure drop and filtration efficiency of samples were evaluated.

2. Materials and Methods

2.1 Materials

PVDF powder provided by Arkema Group, Dimethylacetamide (DMAc) supplied by PENTA Chemicals, Tetraethylammonium bromide (TEAB) obtained from Sigma Aldrich, fiberglass mesh provided by Jinwu Glass Fiber CO., LTD., were used in this work.

PVDF was dissolved in DMAc with concentration of 12% in w/w. Before electrospinning process, 2% TEAB was added into the PVDF solution. Nanofibers were electrospun from PVDF solution by Nanospider™ NS-500. The temperature and humidity during electrospinning process were controlled at 23±0.5℃ and 18±1%. The specification of processing parameters is given in table 1.

Table.1 Specifications of processing parameters of electrospinning

Samples	Voltage (kV)	Electrodes distance (mm)	Slit diameter (mm)	Rotating wire speed (mm/min)	Cartridge speed (mm/s)	Winding speed (mm/min)
620	-10/+50	180	0.5	80	300	620
720	-10/+50	180	0.5	80	300	720
830	-10/+50	180	0.5	80	300	830
920	-10/+50	180	0.5	80	300	920
1000	-10/+50	180	0.5	80	300	1000
1150	-10/+50	180	0.5	80	300	1150

2.2 Methods

The conductivity of solution was measured by SCHOTT instruments (Model: k00090-L), the solution viscosity was measured by HAAKE Viscotester E from Thermos Scientific Company. The fiber morphology and nano-fibrous window screen was observed under scanning electron microscope (SEM). The pressure drop of samples was evaluated according to ASTM D737-1996 by air permeability tester FX3300, which measures the airflow rate under constant air pressure drop, or measures the pressure drop under constant airflow rate. Six measurements for each sample were carried out, the average value and the standard deviation was determined. The filtration efficiency was evaluated according to DIN EN 1822-3 and ISO 29463-3 standards by PALAS (AGK 2000). Three measurements for each sample were carried out, and then the average value was determined.

For better understanding the process parameters, the main part of electrospinning is shown in Fig.1. The polymer solution was contained in a cartridge which has a slit on it; a steel wire acted as an electrode passed the cartridge through the slit and contacted with polymer solution; the cartridge was able to do reciprocating motions onto the steel wire, and left solution droplets on it. Therefore, the solution droplets were exposed to a strong electrical field which provided by high voltage power supply through two electrodes. Once the applied voltage exceeded a critical value, jetting of polymer solution was initiated, and collected by the supporting material. As the charged jets travel toward the supporting material, the solvent evaporates, resulting in solid fibers.

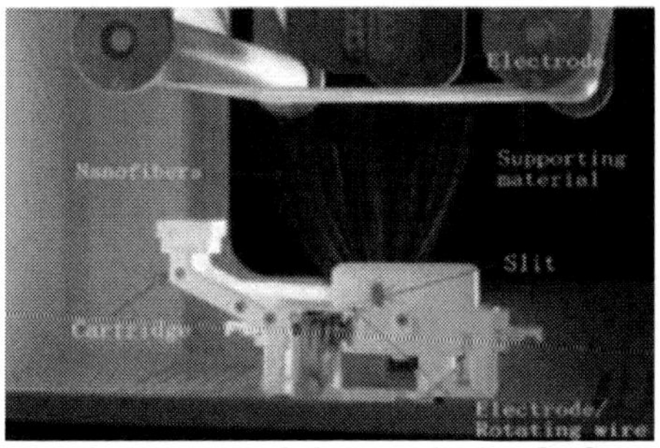

Fig.1 Image of electrospinning [14]

3. Results and Discussions

3.1 Surface Morphology of Samples

It's well known that both viscosity and conductivity are important factors for electrospinning, which are given in table 2. The solution had good electric conductivity, which tends to a high charge density at the surface of the jets, thus decreasing the tendency of droplet formation during electrospinning [15].

Table.2 Viscosity and conductivity of PVDF solution

Solution Concentration (%)	Viscosity (mPa·s at 23°C)	Conductivity (μS/cm at 23°C)
12	1317.1	90.4

Many researchers have found that the solution concentration had a most direct influence on nanofiber diameter and morphology [16, 17]. In our previous work, the beads and flake-like nanofibers were observed when the solution concentrations were 8 wt% and 10 wt%. Therefore, 12 wt% solution was prepared and the fiber-like structure nanofibers were obtained (Fig.2A). Meanwhile, the diameter of nanofibers was evaluated, which was 54.5±12.7 nm (Fig.2A). The lateral profile of nano-fibrous window screen is given in figure 2B, the nano-fibrous membrane layer was strongly bonded onto fiberglass mesh. Besides, the nano-fibrous window screen got very high uniformity and good transparency (Fig.2C).

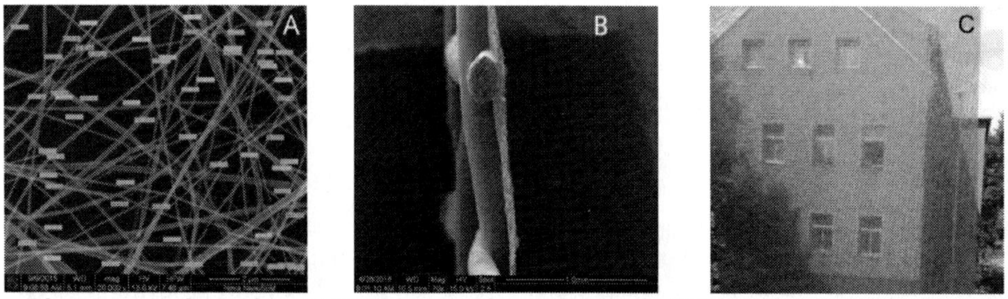

Fig.2 Image of samples (A) SEM image of nanofibers; (B) SEM images of nano-fibrous window screen; (C) Image of nano-fibrous window screen from camera

3.2 Pressure Drop

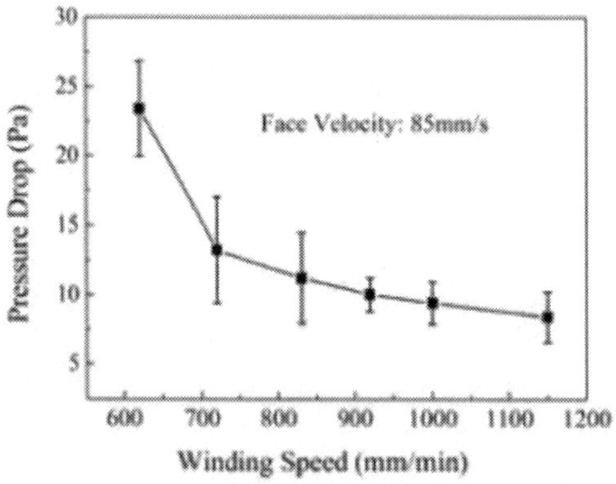

Fig.3 Relationship between pressure drop and winding speed

Pressure drop represents the pressure loss when airflow goes through the nano-fibrous window screen from one side to another. Since there hasn't a standard for evaluating the pressure drop of window screen, the airflow at a speed of 5 m/min was assumed for a normal condition. Under the tests, the airflow velocity was controlled at 85±1.3 mm/s. The results revealed that the pressure drop of all the samples was less than 25 Pa (Fig.3), and the lowest pressure drop was 8.4 Pa for sample 1150, which was much smaller than the values from other reports [7, 8]. The low pressure drop indicated that high air exchange rate could be expected. Besides, the pressure drop declined as the increase of winding speeding, which it's easy to be understood. The higher winding speed was, the less nanofiber was deposited onto fiberglass mesh, thereby, the less airflow resistance was.

3.3 Filtration Efficiency

The filtration efficiency of samples was evaluated with different particle size from 0.35 μm to 1.88 μm, and the filtration efficiency increased as the increase of particle size (as shown in Fig.4). Besides, the filtration efficiency declined as the increase of winding speed. Specifically, the filtration efficiencies of samples (620, 720, 830, 920, 1000, and 1150) for PM1.88 were 83%, 76.9%, 71.5%, 62.5%, 54.1%, and 49.3%, respectively. According to the principle of filtration efficiency, the values of samples can be expected higher for PM2.5.

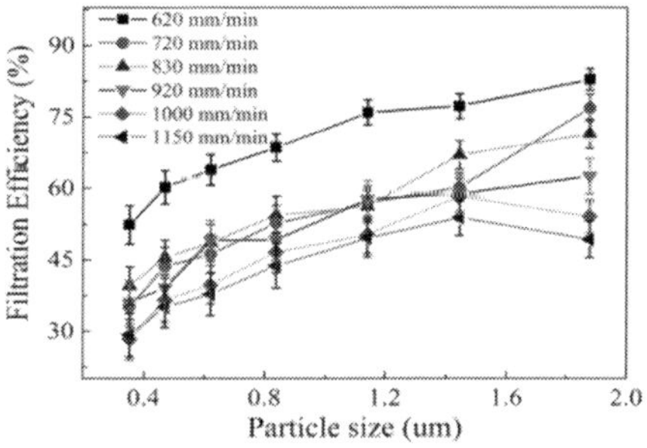

Fig.4 Filtration efficiency of samples for different particle size

4. Conclusions

Transparent PVDF nano-fibrous membrane was successfully applied onto fiberglass mesh in this work. The transparency of nano-fibrous window screen was good. The pressure drop of all samples was less than 25 Pa at an airflow speed of 5 m/min, which a high air exchange rate can be expected. Meanwhile, the pressure drop declined as the increase of winding speed, the lowest pressure drop was 8.4 Pa for sample 1150. The filtration efficiency increased as the increase of particle size, and decreased as the increase of winding speed. The highest filtration efficiency for PM1.88 was 83% obtained from sample 620.

5. Acknowledgment

This work was supported by Zhejiang Sci-Tech University (Project Number: 16012168-Y), The Young Researchers Foundation of Key Laboratory of Advanced Textile Materials and Manufacturing Technology, Ministry of Education, Zhejiang Sci-Tech University (Project Number: 2016QN05), and Hubei Key Laboratory of Advanced Textile Materials & Application, Wuhan Textile University (Project Number: Fzxcl2017006).

References

[1] Information on http://www.who.int/mediacentre/news/releases/2016/air-pollution-rising/en/.

[2] Darrow L, Klein M, Flanders W, Mulholland J, and Tolbert P. Air pollution and acute respiratory infections among children 0-4 years of age: an 18-year time-series study. American Journal of Epidemiology: 2014; 10: 968-977.

[3] Wang K, and T Chau. An association between air pollution anddaily outpatient visits for respiratory disease in

a heavy industry area. Plos One: 2013; 10: e75220
[4] Hoek G, Krishnan RM, Beelen R, Peters A, Ostro B, Brunekreef B, and Kaufman JD. Long-term air pollution exposure and cardio-respiratory mortality: a review. Environmental Health: 2013; 12: 1-15.
[5] Jung KH, Hsu SI, Yan BZ, Moors K, Chillrud SN, Ross J, Wang S, Perzanowski MS, Kinney PL, Whyatt RM, Perera FP, and Miller RL. Childhood exposure to fine particulate matter and black carbon and the development of new wheeze between ages 5 and 7 in an urban prospective cohort. Environment International: 2012; 45: 44-50.
[6] Artola A, Kamal A, Ramakers G, Biessels G, and Gispen W. Cardiovascular effects of fine and ultrafine particles. Journal of Aerosol Medicine the Official Journal of the International Society for Aerosols in Medicine: 2005; 1: 1-22.
[7] Khalid B, Bai XP, Wei HH, Huang Y, Wu H, and Cui Y. Direct blow-spinning of nanofibers on a window screen for highly efficient PM2.5 removal. Nano Letters: 2017; DOI: 10.1021/acs.nanolett.6b04771.
[8] Liu C, Hsu PC, Lee HW, Ye M, Zheng GY, Liu NA, Li WY, and Cui Y. Transparent air filter for high-efficiency PM2.5 capture. Nature Communications: 2015; 6: 1-9.
[9] Wang DL, Li K, and Teo WK. Preparation and characterization of polyvinylidene fluoride (PVDF) hollow fiber membranes. J. Membr. Sci.: 1999; 163: 211-220.
[10] Yeow ML, Liu YT, and Li K. Morphological study of poly(vinylidene fluoride) asymmetric membranes: effects of the solvent, additive, and dope temperature. J. Appl. Polym. Sci.: 2004; 92: 1782-1789.
[11] Tan XY, Tan SP, Teo WK, and Li K. Polyvinylidene fluoride (PVDF) hollow fibre membranes for ammonia removal from water. J. Membr. Sci: 2006; 271: 59-68.
[12] Liu F, Hashim NA, Liu YT, Abed MRM, and Li K. Progress in the production and modification of PVDF membranes. J. Membr. Sci.: 2011; 375: 1-27.
[13] Cui ZL, Drioli E, and Lee YM. Recent progress in fluoropolymers for membranes. Prog. Polym. Sci.: 2014; 39: 164-198.
[14] Information on http://www.elmarco.com/photogallery/technology/.
[15] Joachim H, Wendorff, Agarwal S, and Greiner A, Eds. "Electrospinning: materials, processing, and applications", 1st ed, Wiley-VCH Verlag GmbH & Co.KGaA, Weinheim, Germany: 2012.
[16] Wang X, Niu H, Wang XG, and Lin T. Needleless electrospinning of uniform nanofibers using spiral coil spinnerets. Journal of Nanomaterials: 2012; 10: 1875-1890.
[17] Beachley V, and Wen XJ. Effect of electrospinning parameters on the nanofiber diameter and length. Materials Science & Engineering C-Biomimetic And Supramolecular Systems: 2009; 3: 663-668.

Preparation and Characterization of CuO-deposited Cellulose Acetate Nanofibers

Shi-Li Xiao[1,2*], Peng-Jun Xu[3], Yan Chen[2], Qing-Yan Peng[2]

[1]*State Key Laboratory of New textile Materials and Advanced Processing Technologies, Wuhan Textile University, Wuhan 430200, China*
[2]*School of Textile Science and Engineering, Wuhan Textile University, Wuhan, Hubei 430200, China*
[3]*Faculty of Clothing and Design, Minjiang University, Fuzhou, Fujian, 350108, China*

[*]Corresponding author's email: xshili@yahoo.com

Abstract

Advanced antibacterial wound dressing material is quite important in the biomedical field. This study reports a facile method to prepare antibacterial CuO nanoparticle-based nanofibrous mats. In this process, cellulose acetate (CA) nanofibers were fabricated by using electrospinning technology. Then, CA nanofibers acted as a substrate to load CuO nanoparticles synthesized using copper chloride and sodium hydroxide via solution process. The synthesized CuO-deposited CA nanofibrous mats (CuO-CA mat) are thoroughly characterized by scanning electronic microscopy, Fourier transform infrared spectroscopy, X-ray diffraction and thermogravimetric analysis. It showed that CuO nanoparticles are uniformly deposited onto the surface of CA nanofibers with a layered, monoclinic structure, and the content of CuO is 30.2%.

Keywords: Cellulose Acetate; Electrospinning; Nanofiber; Antibacterial

1. Introduction

Metal-based nanoparticles with biocidal properties have received much interest as promising antimicrobial agents because of their effective inhibition of bacterial growth. Several kinds of metal or metal oxide (e.g., silver, zinc, Cu and CuO, etc.) have been widely studied [1-3]. Among these, copper has been drawn special attention to due to its advantages such as showing effective growth inhibition of bacteria, fungus, and algae, serving a number of important roles in human health. Most importantly, it displays no acute copper toxicity at high concentration (2 mg/L) [4, 5]. Various forms of copper-based nanoparticles including zero-valent copper, copper oxide, ionic copper and copper-containing molecular complexes have been studied. For instance, N. Cady coated zero-valent copper nanoparticles are placed onto cotton fabric through layer-by-layer electrostatic self-assembly technology and in situ chelation and reduction process [6]. A layer of sub-5 nm copper nanoparticles have been produced on the surface of cotton substrate. The copper-coated cotton exhibited rapid and efficient killing of a multidrug resistant bacterial wound pathogen. Recently, Xu and co-workers developed a simple method for deposition CuO nanoparticles on a cellulose paper [7]. In that study, cellulose paper was soaked in the $Cu(NO_3)_2$ for 24 h, and then transferred to NaOH solution and kept inside for 6 hours. Finally, the CuO nanoparticles formed on cellulose paper after washing and subsequent drying in oven at 60 ^{0}C. The CuO nanoparticles were stably anchored on cellulose paper and CuO paper showed good and stable antibacterial activities against gram positive and negative bacteria. However, these substrates are in micro-scale, which is not beneficial for the chelation of Cu ions onto substrate due to limited active functional group, and the pretreatment is needed before complexing Cu ions for the formation of Cu-based antibacterial agents. Thus, there is still room for preparing Cu-based composite for antibacterial application.

It is known that electrospun fibers have been widely used in the biomedical field as kind of advanced materials [8, 9]. The electrospun nanofibers have a nanoscale diameter and large specific surface area that is benefit for the location of functional cargos. Moreover, the porous structure and high porosity of electrospun nanofibrous mats make them quite suitable for mimicking the extracellular matrix structure, which can allow cell attachment and growth [10]. Taken advantage of electrospun nanofibers, we hypothesis that in situ coating CuO nanoparticles onto electrospun cellulose acetate nanofibers to produce antibacterial mats for wound dressing would be effective.

2. Materials and Methods

2.1 Materials

Cellulose acetate (powder, viscosity 300-500 mPa s), acetone, N, N-dimethylformamide (DMF), $CuCl_2.3H_2O$, and NaOH were purchased from Sinopharm Chemical Reagen Co., Ltd. Water used in all experiments was purified using a Milli-Q Plus 185 water purification system (Millipore, Bedford, MA) with resistivity higher than 18 MΩ cm.

2.2 Preparation of CuO-deposited Cellulose Acetate Nanofibers

Before depositing CuO nanoparticles onto cellulose acetate (CA) nanofibers, CA nanofibers were fabricated by electrospinning a 120 g/L CA solution according to a procedure in our previous study [11]. The CA solution was prepared by dissolving a measured amount of CA powder in a mixed solvent of acetone and DMF (2/1 v/v) for 12 h with vigorous stirring. The homogenized CA solution (10 mL) was loaded into a syringe controlled by a syringe pump with a feed rate of 1.0 mL/h. The applied voltage was kept at 20 kV, and the collection distance was set at 20 cm. Pieces of fresh CA nanofibrous mats were then soaked in prepared $CuCl_2$ (0.5 M) aqueous solution overnight. The soaked pieces were then transferred to the aqueous solution of NaOH (0.25 M) for the formation of CuO-deposited CA nanofibrous mats (CuO-CA mat). 6 h later, the final pieces (CuO-CA mat) were taken out, washed thoroughly with deionized water to remove loose particle on the CA nanofibrous mats and dried in oven at 60 ^{0}C. CuO-CA mat showed dark brown color.

2.3 Characterization of CuO-deposited Cellulose Acetate Nanofibers

The morphologies of CA and CuO-deposited CA nanofibrous mats were observed using scanning electron microscope (SEM) (JSM-5600LV, JEOL Ltd., Japan) with an operating voltage of 15 kV. Prior to SEM measurements, samples were sputter-coated with 10 nm thick Pt films. Fourier transform infrared (FTIR) spectra were recorded using a Nicolet 5700 FTIR spectrometer (Thermo Nicolet Corporation, United States) in a wavenumber range of 4000-500 cm^{-1} at ambient conditions. The X-ray diffraction (XRD) pattern was recorded in reflection mode using Cu Kα radiation at 25 ^{0}C (D/Max-2550 PC, RIGAKU, Japan). The CuO content in the nanofibrous mats was determined by thermogravimetric analysis (TGA) using a TG 209 F1 (NETZSCH Instruments Co., Ltd., Germany) thermogravimetric analyzer with a heating rate of 10 ^{0}C min^{-1} in air.

3. Results and Discussion

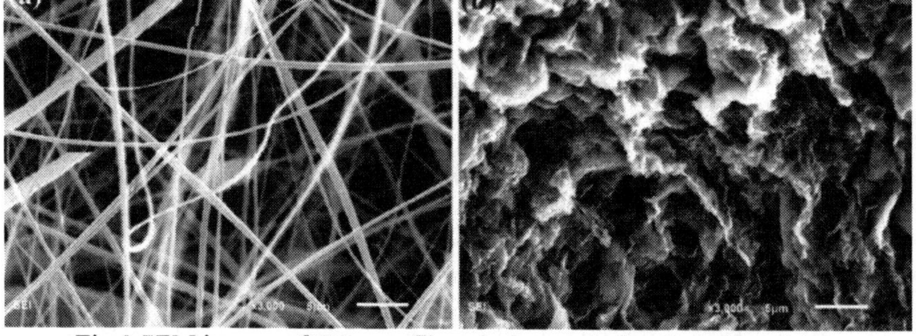

Fig.1 SEM images of CA nanofibers before and after CuO deposition.

Surface morphologies of CA nanofiber before and after CuO deposition were shown in Fig.1. It is clear that smooth and continuous CA nanofibers with random orientation were produced under the selected electrospinning parameters. The electrospun CA nanofibrous mats exhibit a 3D structure with pores in micrometers, which is a typical feature for extracellular matrix, although the diameter of CA nanofibers was not uniform. However, the fibrous structure of CA nanofibers became indistinct after depositing dense CuO nanoparticles, only some pores and clusters with layered structure could be observed.

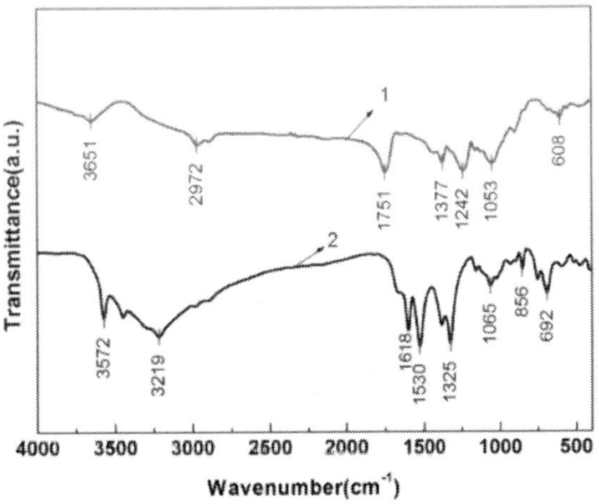

Fig.2 FTIR of CA nanofiber before (curve 1) and after (curve 2) CuO deposition.

Fig.2 shows FTIR spectra of the pristine CA and CuO-CA nanofibrous mats. We can see that there are some differences on the adsorption bands of cellulose acetate nanofibers before and after CuO deposition. For CA nanfibers' spectrum, the absorption band at 1053 cm^{-1} is attributed to the asymmetric stretching of C-O-C in cellulose. The bands at 1242 cm^{-1} and 1375 cm^{-1} are attributed to deformation, and bending of CH, respectively. The band at 1751 cm-1 is ascribed to the vibration of C=O. Absorption band at 2972 cm^{-1} is attributed to bending of CH and band at 3651 cm^{-1} can be assigned to adsorbed water in cellulose fibers. After depositing CuO nanoparticles onto CA nanofiber, adsorption band shift occurred and the absorption intensity changed due to the interaction between CuO nanoparticles and CA nanofibers. The asymmetric stretching of C-O-C and adsorption band due to adsorbed water in cellulose shifted to 1065 cm^{-1} and 3572 cm^{-1}, respectively. In addition, three new bands at 1618 cm^{-1}, 1530 cm^{-1}, and 1325 cm^{-1} ascribed to C=O vibration, CH_2 bending and C-O stretching vibration were collected, respectively.

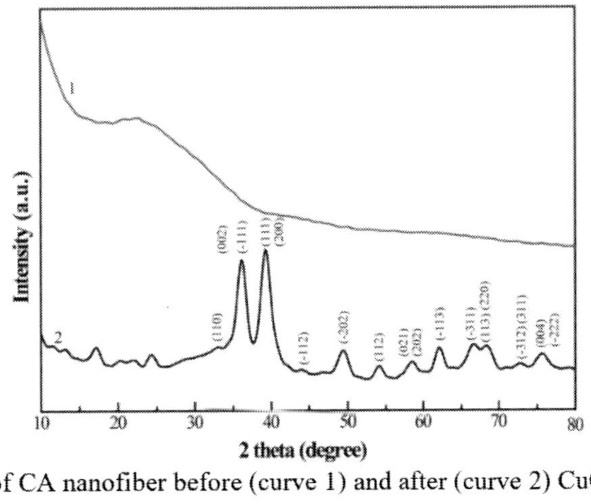

Fig.3 XRD of CA nanofiber before (curve 1) and after (curve 2) CuO deposition.

The formation of CuO nanoparticles was further confirmed using XRD. Fig.3 shows that no significant characteristic peak was observed on the XRD spectrum of CA nanofibers, except for a broad around 22.8^0 indexed to cellulose crystal (curve 1). Whereas after deposition of CuO nanoparticles onto CA nanofibers, some characteristic peaks ascribed to diffraction of copper oxide with a momoclinic structure (PDF 01-080-1916)[7] appeared (curve 2), such as the strong characteristic peaks at around 35.4^0 and 38.5^0 corresponding to the (002), (-111), (111), and (200) reflections, respectively, confirming the existence of crystalline CuO nanoparticles in the CA nanofibrous mats.

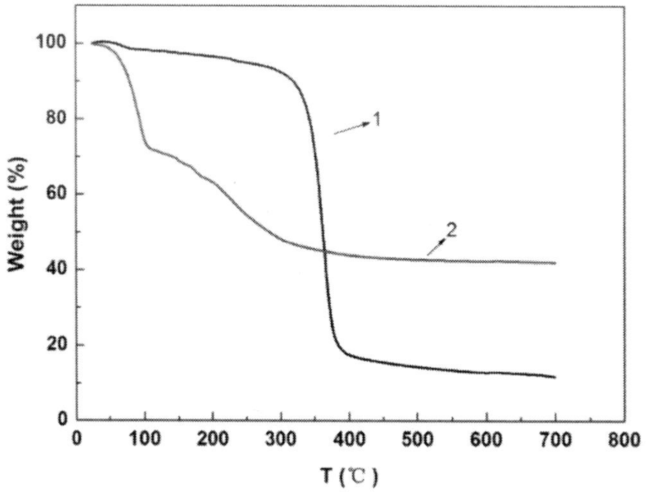

Fig.4 TG curves of CA nanofiFbrous mats before (curve 1) and after (curve 2) CuO deposition.

Fig.4 shows TGA measurement of pristine CA nanofibrous mat and CuO-CA nanofibrous mat. It showed that CA nanofibrous mat undergone a two-step decomposition process. Within 81 °C, a small weight loss was obtained, which was probably due to the evaporation of residual water in the CA nanofibers. CA nanofibrous mat starts to decompose at around 239 °C, and stop decomposition at around 412.2 °C. At 700 °C, the residual weight of CA nanofibrous mat was 11.9%. Whereas CuO-CA nanofibrous mat showed a sharp weight loss within 100 °C and then gradually decomposed with the temperature increasing until reaching a plateau at 405 °C. The rapid weight loss might be caused by the water evaporation adsorbed onto mats during storage. In addition, the presence of transition metal oxide such as CuO can accelerate the decomposition process of the organic matrix due to the catalytic effect [12]. The stable residual weight percentage of CuO-CA nanofibrous mat at 700 °C is 42.1%. According to these data, the CuO loading on the CA nanofibrous mat could be calculated to be 30.2%.

4. Conclusion

In summary, CuO nanoparticles have been successfully immobilized onto electrospun cellulose acetate nanofibers with in situ chelation and a simple chemical reaction. The synthesized CuO-CA nanofibrous mats showed a layered, porous structure and CuO nanoparticles deposited on the CA naonfibrous mats are typical monoclinic structure. The CuO loading on the composite nanofibrous mats was 30.2%. The developed CuO-CA nanofibrous mats would find a potential application in wound dressing because of the excellent antibacterial capability of CuO nanoparticles. Moreover, the method using electrospun nanofibrous mats as substrates to anchor functional nanoparticles can also be extended to prepare antibacterial nanofiltration for purifying drinking water.

5. Acknowledgment

This research was financially supported by the National Natural Science Foundation of China (51403165 and 51403098), the Education Commission of Hubei Province (D20161602) and the Science Foundation of Wuhan Textile University (2014CFB161).

References

[1] Gordon T, Perlstein B, Houbara O, Felner I, Banin E, Margel S. Synthesis and characterization of zinc/iron oxide composite nanoparticles and their antibacterial properties. Colloids Surf A Physicochem Eng Asp: 2011; 374: 1-8.

[2] Raffi M, Mehrwan S, Bhatti TM, Akhter JI, Hameed A, Yawar W, ul Hasan MM. Investigations into the antibacterial behavior of copper nanoparticles against Escherichia coli. Ann Microbiol: 2010; 60: 75-80.

[3] Son WK, Youk JH, Park WH. Antimicrobial cellulose acetate nanofibers containing silver nanoparticles. Carbohyd Polym: 2006; 65: 430-434.

[4] Borkow G, Gabbay J. Copper, an ancient remedy returning to fight microbial, fungal and viral infections. Curr Chem Biol: 2009; 3: 272-278.

[5] Ibrahim M, Wang F, Lou M, Xie G, Li B, Bo Z, Zhang G, Liu H, Wareth A. Copper as an antibacterial agent for human pathogenic multidrug resistant Burkholderia cepacia complex bacteria. J Biosci Bioeng: 2011; 112: 570-576.

[6] Cady NC, Behnke JL, Strickland AD. Copper-based nanostructured coatings on natural cellulose: nanocomposites exhibiting rapid and efficient inhibition of a multi-drug resistant wound pathogen A. baumannii, and mammalian cell biocompatibility in vitro. Adv Funct Mater: 2011; 21: 2506-2514.

[7] Booshehri AY, Wang R, Xu R. Simple method of deposition of CuO nanoparticles on a cellulose paper and its antibacterial activity. Chem Eng J: 2015; 262: 999-1008.

[8] Wang S, Zheng F, Huang Y, Fang Y, Shen M, Zhu M, Shi X. Encapsulation of amoxicillin within Laponite-doped poly(lactic-co-glycolic acid) nanofibers: preparation, characterization, and antibacterial activity. ACS Appl Mater Interfaces: 2012; 4: 6393-6401.

[9] Xu X, Zhong W, Zhou S, Trajtman A, Alfa M. Electrospun PEG-PLA nanofibrous membrane for sustained release of hydrophilic antibiotics. J Appl Polym Sci: 2010; 118: 588-595.

[10] Xu C, Inai R, Kotaki M, Ramakrishna S. Electrospun nanofiber fabrication as synthetic extracellular matrix and its potential for vascular tissue engineering. Tissue Eng: 2004; 10: 1160-1168.

[11] Xiao S, Wu S, Shen M, Guo R, Huang Q, Wang S, Shi X. Polyelectrolyte multilayer-assisted immobilization of zero-valent iron nanoparticles onto polymer nanofibers for potential environmental applicaitons. ACS Appl Mater Interfaces: 2009; 1: 2848-2855.

[12] Nakayama J, Miyake A. Catalytic effect of copper(II) oxide on oxidation of cellulosic biomass. J Therm Anal Calorim: 2012; 110: 321-327.

Study on Silica Gel Finishing and Properties of Vertical Cotton

Bo-Ya Zhang, Heng-Tong Li, Ru-Quan Zhang*

Wuhan Textile University, Sunshine avenue 1st, Jiangxia district Wuhan, 430200, China

*Corresponding author's email: zhangruquan@wtu.edu.cn

Abstract

Liquid silica sol is used to further improve the compress and resilient behavior of vertical cotton. Five groups of experimental samples were taken from the same vertical cotton, one of which was not treated as a reference. With the dipping method, the amount of dipping liquid used were: 6g, 12g, 18g, 24g which then decided the most appropriate group. In order to explore the effect of curing agent and silicone oil on the experiment, the amount usage of curing agent were: 1%, 2%; silicone oil: 0, 1%, 2%. After the treatment for each group of dipping liquid, we then tested the compression elasticity performance change of the vertical cotton. The experimental results show that the compression elasticity of the sample is improved after being dipped. When the amount of liquid silica gelled was 12g and curing agent is 1%, the compression elasticity of the impregnated samples were the best.

Keywords: Vertical Cotton; Silica Gel; Compression Elasticity; Vertical Spread

1. Introduction

1.1 The Purpose of Study and Significance

With the rapid development of the national economy, people's material and cultural living standards continue to improve. The elastic fiber material quality and quantity requirements continue to increase. Liquid silica gelled [1] as a textile finishing agent, can make textiles soft and straight, smooth and elastic, smooth and plump, and can give the fabric water repellent properties after finishing. This experiment involves the use of liquid silica gel to improve the compression elasticity of vertical cotton to get the maximum flexibility and compression resistance [2-3].

1.2 Vertical Cotton Introduction

Erect polyester fiber cotton is made from 100% polyester fiber. It is is composed of cotton with vertical netting which is a nonwoven technology. The fibers in the main body are not in a horizontal arrangement, but in a vertical direction [4]. The material has a cavernous three-dimensional vertical web structure or honeycomb web structure which is different from the traditional parallel web structure of non-woven materials. The vertical alignment of the web and the use of heat-shrinkable fibers gives the 3D upright cotton material a better and softer touch. The three-dimensional structure gives the product more than 90 percent recovery rate, which can save up to 20 percent of the weight compared to a normal product. After compression, the 3D upright cotton can be 30-60 per cent thicker and insulated than the non-vertical network material [5]. Observe from a longitudinal profile, non-woven high elastic solid material to erect web structure or the web of honeycomb structure, different from other parallel web structure of non-woven material. The different ways of laying the net form a completely different structure, making the material more elastic and compressive. The most elastic and compressive properties were obtained through verifying multiple independent experiments again and again.. After the long-term or repeated to load can still keep good resilience, vertical into mesh material can maintain its function in the use of process performance. In this regard, its compressibility and resilience are close to the sponge. Its environmental performance is its greatest strength, direct contact with the human body, it is the nearest natural and harmless to the human body to the environment pollution-free new environmental protection material. Adopt new process to make fiber products, green environmental protection, no additives. Flexible, comparable to a sponge, is a good substitute for a sponge

[6]. The vertical cotton have the characteristics of high resilience, high compression resistance, high air permeability, non-toxic environmental protection, good folding and discharging performance, and can be used in bra cup mold filling, mattress furniture, seat interior, filter material, sound insulation Materials, medical and health supplies [7].

2. Preparation of Vertical Cotton

Production process: Ingredients calculation→open mixing →mixing box→first carding laying→reciprocating laying→roll-card carding laying→vertical laying→drying oven →heating and reinforcement→edge cutting →winding →cutting→finished products.

2.1 Fibrous Materials

Fiber type and proportion: Polyester fiber HCF (old H-3): Polyester fiber HCF (H-7): Polyest-er fiber (Yuan Fang2028)=25%:35%:40%

2.2 Mixing and Opening

The production process uses batch mixing in which each fiber component of the batch is weighed at a desired rate and then the entire batch of all fiber components is fed into the slant feed Such as the stage. The raw material is opened with the first mixing chamber after being opened, and the horizontally laid fiber layer is picked up vertically by the reclaiming device. This method is called "vertical spreading" to ensure uniforms mixing of raw materials. The loosening machine is opened and then sent to the second mixing chamber, mixed again, and then in the fine open by the fine machine processing, into the post-processing into the net [8].

The process is as follows: the entire batch of raw materials by mixing the proportion of fiber weighing→bale opener→the first mixed warehouse→opening→the second mixed warehouse→ fine open. Production is the use of multi-warehouse mixers and fine open-loosening.

2.3 Carding

A roller-cylinder-type card machine is used in the production, which consists of a feed roller such as a roller, spine, scylin, working roller, doffer and chopping knife, or peeling roller, to the final output fiber net. In which the card is mainly produced between the needle rolls of the working rollers and the cylinders. The stripping roller serves to strip the fibers which have been agglomerated on the working rollers during the card process and to transfer them back to the cylinders for the next card unit Combing. The arrangement of the cloths on each of the working elements is similar to that of the opening and closing device, and the density of the front and rear clothing is also "front and back", and the thickness of the cloth is " In order to satisfy the complete comb and the requirement of reducing fiber injury. In the original single-cylinder double-doff on the basis of the increase in a cylinder, making the card work area doubled in area, that is, in the cylinder unit area of fiber load unchanged, the increase in area to improve efficiency, and single cylinder comparing the same to achieve the effect of increasing production, and combing the quality of more easily controlled.

2.4 vertical Laying

Vertical laying is a single layer of fiber mesh folded up and down, so that fibers in the network to approximate the vertical arrangement, the thickness also increased significantly. Card machine output of the web under the guidance of the guide plate steel, With the swinging of the runner to fold, paved or so thick mesh, the layers of fabric on the pressure at the same time, the fibers form a certain degree of mechanical entanglement. This form of thick network of network, the width is also determined by the width of the card machine.

2.5 Thermal Bonding

The principle is [9]: most of the polymer materials with thermoplastic, that is heated to a certain temperature will soften and melt into a certain flow of viscous fluid, cool and then re-curing, into a solid, thermal bonding non-woven technology is the use of Thermoplastic polymer materials, this feature, so that part of the fiber web or hot melt softening melting powder, resulting in adhesion between fibers, When cooled, the fibers are strengthened and become hot bonding non-woven materials The hot air penetration bonding line used in this project is a flat mesh type hot air penetration oven [10]. This equipment can be based on the needs of the process, the entire length of the work is divided into several different temperature zones, the effective drying of the network of hot air, flat-type drying room in addition to single-layer, there are multi-layer, so that can save the area, in maintaining a certain production rate, can increase the heating time of the fiber network, so as to ensure that the bonding material to get adequate melting and bonding quality.

3. Experimental Program

3.1 Experimental Materials and Equipment

Table.1 Experimental materials

Material	Type	Manufacturer
Polyester fiber HCF(old H-3)	3D×64MM	Long Ta Trading Co
Polyester fiber HCF(H-7)	7D×64MM	Long Ta Trading Co
Polyester fiber(2028)	2D×51MM	Yuan Fang Trading Co
Liquid mold silica gel	Additive silica gel	Tai Xing Trading Co
Hardener	Chemical reagents	Tai Xing Trading Co
Hydroxy silicone oil	Chemical reagents	Xing Fei Long Chemical Co

Table.2 Experimental instrument

Instrument	Type	Manufacturer
electronic balance	FA2004	Da Rong textile equipment
Fabric thickness gauge	YG141D	Da Rong textile equipment

3.2 Dipping Finishing

3.2.1 Sampling

On the same piece of vertical cotton, the experimental sample was 8 cuboids with 8cm ×8cm, and the thickness was T=4cm and the mass was 6.4g with thickness tester. Two of them recorded as A, B do not do finishing. The rest are numbered 1-16. Prepare a glassware of appropriate size of impregnation.

3.2.2 Preparation of Silica Gel Solution

First take a certain amount of silica gelled into the glass container, and then in accordance with the experimental program by adding the appropriate amount of curing agent and hydroxyl silicone oil dripping into the glass container in turn, stirring. Silicone and curing agent must be mixed evenly, if not evenly mixed, the mold will appear one has been cured, one is not cured. The silicone oil was added in order to improve the softness of the sample after immersion, and the viscosity of the silica gel, and can not add too much [11].

3.2.3 Finishing

The silica gel solution was prepared according to the experimental protocol, and the prepared sample was placed in the impregnation vessel at one time. The pad is padded after completion of impregnation so that the sample can be completely impregnated and uniformly impregnated and the excess silica gel solution to the sample can be removed.

After immersion, place at room temperature for 24 hours. The experimental sample components are shown in Table 3.

Table.3 Dipping scheme

Number	Silica gel dosage/g	Proportion of Hardener/%	Proportion of silicone oil/%
1	6	2	0
2	12	2	0
3	18	2	0
4	24	2	0
5	6	1	0
6	12	1	0
7	18	1	0
8	24	1	0
9	6	2	1
10	12	2	1
11	18	2	1
12	24	2	1
13	6	2	2
14	12	2	2
15	18	2	2
16	24	2	2
A	0	0	0
B	0	0	0

3.3 Compression Elasticity Test and Characterization

Compression elasticity testing principle [12]: measuring the sample of the light and heavy pressure load, and remove the load after the test thickness changes, obtained fabric thickness, compression elasticity and bulkiness and other indicators.

Method of operation [13]: First, the sample placed in the middle of a square plate, and then use the same piece of pressure plate placed on the sample. First, place a weight of 1.50kg (the test pressure is $0.23N/cm^2$) above the platen. After one minute, measure the height of the platen with a ruler and take the average value was taken as T_0. Next, a weight of 4.50kg (at this time, a test pressure of $0.69N/cm^2$) was placed over the platen. After one minute, the height of the platen was measured with a ruler and the average value were taken as T_m.

Finally, remove the weight, making the sample rebounded on zero load two minutes later, with a ruler to measure the height around the plate, take the average, recorded as T_1.

Commonly used quantitative characterization of sample compression performance indicators are as follows [14]:

1) Constant pressured thickness (mm): T_0 and T_m is the compression process for a given pressure of the thickness. T_0 is the thickness of the sample of a pressure on $0.23N/cm^2$, that is, the thickness in the initial state. T_m is the thickness of the test specimen under the maximum test pressure P_m=$0.69N/cm^2$.

2) Compression ratio of work W_c ($N\times cm/cm^2$): W_c for the compression process of external force on the unit area of the work done by the sample, the greater the general compression work, the more fluffy sample. Correspondingly, the compressive recovery functioned W_c' is used to characterize the work done by the pressure on the sample in the process of recovery.

3) Compression worked recovery rate R_C (%): R_C is the percentage of compression work, which means the compression elastic recovery performance of the sample. Generally, the bigger the value is, the better the compression recovery performance.

$$R_C=W_c'/W_c \tag{1}$$

4) Compression curve linearity L_c: L_c said that the compression work and that the sample compression curve of

the degree of buckling in the value of pressure equal to the curve under the curve of the area under the curve of the ratio of the area under the two ends of the connection, the characterization of textile materials in the entire compression The degree of difficulty in the process of being compressed

$$L_c=2W_c/ P_m(T_0- T_m) \quad (2)$$

5) Compression rate EMC (%) [6]:can was characterized by the deformation of the flakes ability, that is, anti-compression ability, the greater the value of that flakes more easily deformed, that is, poor compression resistance.

$$EMC=(T_0- T_m)/T_0 \quad (3)$$

6) Compressible thickness T' (mm): T' is the thickness of the specimen compressed during the compression test. The larger the value of T', the greater the void loss in the specimen during compression.

$$T'=T_0- T_m \quad (4)$$

4. Test Results

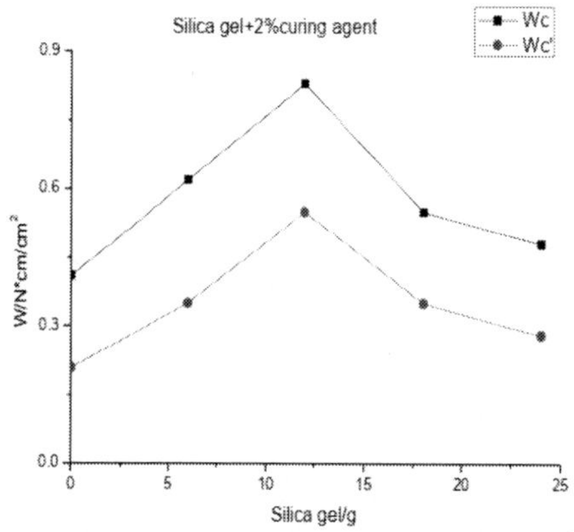

Fig.1 The relationship between the amount of silica gel and Wc, Wc '

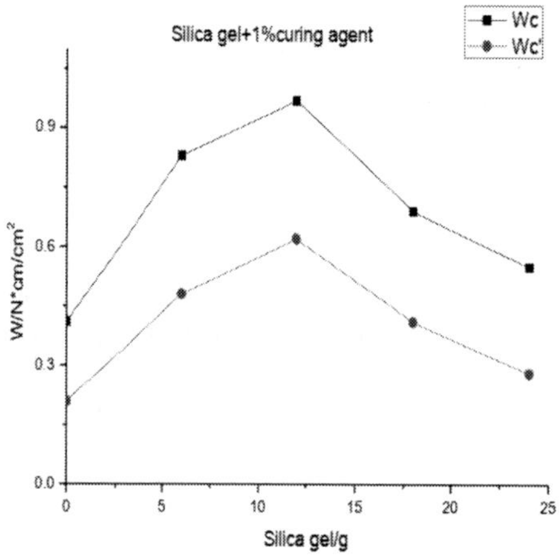

Fig.2 The relationship between the amount of silica gel and Wc, Wc '

Table.4 Compressive thickness test results

Number	T_0(mm)	T_m(mm)	T_1(mm)	T-T_1(mm)
1	31.00	21.00	34.00	6.00
2	29.00	17.00	32.00	8.00
3	29.00	19.00	34.00	6.00
4	29.00	22.00	36.00	4.00
5	29.00	17.00	33.00	7.00
6	29.00	15.00	31.00	9.00
7	29.00	19.00	37.00	6.00
8	30.00	22.00	36.00	4.00
9	29.00	20.00	35.00	5.00
10	28.00	16..00	32.00	8.00
11	30.00	22.00	35.00	5.00
12	30.00	23.00	36.00	4.00
13	25.00	13.00	32.00	8.00
14	26.00	15.00	33.00	7.00
15	28.00	18.00	34.00	6.00
16	29.00	19.00	35.00	5.00
A	34.00	28.00	37.00	3.00
B	35.00	29.00	37.00	3.00

Table.5 Sample compression performance of the results

Number	Wc(N×cm/cm^2)	W_c'(N×cm/cm^2)	R_C(%)	EMC(%)	T'(mm)
1	0.69	0.41	60.00	32.26	10.00
2	0.83	0.55	66.67	41.38	12.00
3	0.62	0.41	62.00	31.03	9.00
4	0.48	0.28	57.00	24.14	7.00
5	0.83	0.48	58.33	41.38	12.00
6	0.97	0.62	64.29	48.28	14.00
7	0.69	0.41	61.00	34.48	10.00
8	0.55	0.28	51.00	26.67	8.00
9	0.62	0.35	55.60	31.03	9.00
10	0.83	0.55	66.67	42.86	12.00
11	0.55	0.35	62.50	26.67	8.00
12	0.48	0.28	57.14	23.33	7.00
13	0.83	0.55	66.67	48.00	12.00
14	0.76	0.48	63.64	42.31	11.00
15	0.69	0.41	60.00	35.71	10.00
16	0.69	0.35	50.00	34.48	10.00
A	0.41	0.21	50.00	17.65	6.00
B	0.41	0.21	50.00	17.14	6.00

Using Table 4 data, according to the formula obtained in Table 5 characterization values.

Analysis of Figures 1 to 5 [15]:

1) after liquid silicone impregnation, vertical cotton W_c, W_c', R_C have improved. This is because when the liquid silicone rubber in the vertical cotton internal curing, formed into an elastic body, increased its compression recovery elasticity.

2) When adding 2% of the curing agent in liquid silicone, W_c is the first increase and then decrease, The maximum value is 0.83N×cm/cm^2 when liquid silicone is used in 12g. W_c' first increased and then decreased, the maximum is when the amount of liquid silica gelled 12g, 0.55N×cm/cm^2. R_C increased first and then decreased, the maximum is when the amount of liquid silicone 12g, 66.67%. Indicating the amount of silica gel 12g, the best impregnation effect. With the increase in the amount of silica gel into the vertical cotton internal formation of the

more elastic body, too much elastomer makes vertical cotton harden, but reduce its compression resilience.

3) When adding 1% of curing agent in liquid silicone, W_c first increased and then decreased, the maximum value is 0.97N×cm/cm^2 when liquid silicone is used in 12g. W_c' first increased and then decreased, the maximum is when the amount of liquid silicone 12g, 0.62N ×cm/cm^2. R_C increased and then decreased, the maximum is when the amount of liquid silicone 12g, 64.29%. Indicating the amount of silica gelled 12g, the best impregnation effect. With the increase in the amount of silica gel into the vertical cotton internal formation of the more elastic body, too much elastomer makes vertical cotton harden, but reduce its compression resilience. However, the amount of curing is reduced, so that the curing effect is lowered, so that the maximum value of compression work and the value of compression recovery increased.

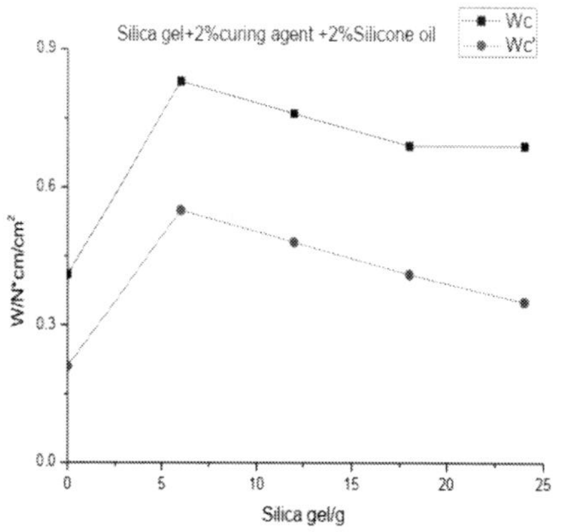

Fig.4 The relationship between the amount of silica gel and Wc, Wc '

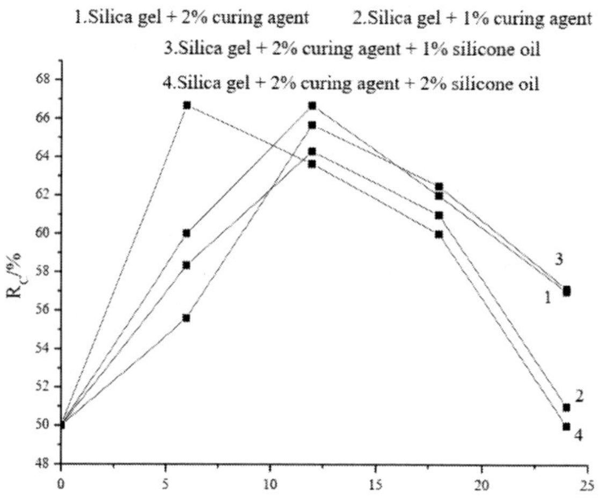

Fig.5 The amount of silica gel and RC (%) relationship

4) When adding 2% solidifying agent and 1% hydroxyl silicone oil to liquid silicone, W_c first increases and then decreases, the maximum value is 0.83N×cm/cm^2 when the amount of liquid silica gel is 12g. W_c' first increased and then decreased, the maximum is when the amount of liquid silica gelled 12g, 0.55N×cm/cm^2. R_C increased and then decreased, the maximum is when the amount of liquid silicone 12g, 65.67%. Indicating the amount of silica gelled 12g, the best impregnation effect. With the increase in the amount of silica gel into the vertical cotton internal formation of the more elastic body, too much elastomer makes vertical cotton harden, but reduce its compression

resilience. But the addition of 1% hydroxyl siliconed oil, after curing to reduce the hardness of the formation of elastomers and to improve the softness of vertical cotton, affecting its compression recovery elasticity.

5) When adding 2% curing agent and 2% hydroxyl silicone oil into liquid silicone, W_c first increased and then decreased, the maximum value is $0.83N \times cm/cm^2$ when the amount of liquid silica is 6g. Wc' first increased and then decreased, the maximum is when the amount of liquid silicone 6g, $0.55N \times cm/cm^2$. R_C increased and then decreased, the maximum is when the amount of liquid silicone 6g, 66.67%. Indicating that the amount of silica gel 6g, the impregnation effect is the best. With the increase in the amount of silica gel in the upright cotton internal formation of the more elastic body, too much elastomer makes upright cotton harden, but reduce its compression resilience. But the addition of 2% silicone oil, with the addition of 1% silicone oil and silicone oil is not added compared to have greatly reduced after curing in the formation of upright cotton elastomer hardness, and make upright cotton is more flexible. Thus affecting the compression elasticity of upright cotton.

5. Conclusion

1) The compressive elasticity of the upright cotton samples increased after immersion in the silica gel solution. When the amount of liquid silicone molded 12g, 1% curing agent, the sample after the best resilience impregnation.

2) When preparing liquid silicone, the amount of silica gelled used, the amount of curing agent added, and whether or not to add silicone oil, and the amount of silicon oil added to the silicone oil, have an effect on the elastic properties and other properties of the impregnation.

3) Compression elasticity of upright cotton samples increased first and then decreased to the increase in liquid silica gelled content. This is because after impregnation, liquid silicone gel in the vertical cotton to form elastomers. When the amount of liquid silica gelled increases, the solid elastomer formed in the vertical cotton sample increases, Make it softer and less flexible. So that the compression elasticity of the sample is lowered.

4) When the amount of curing agent is 1% and 2%, respectively, the former after impregnation of cotton compression elasticity is better than the latter. The results show that when the amount of the liquid mold is the same, the compressive elasticity of the sample after impregnation decreases to the decrease of the amount of curing agent.

5) When the amount of curing agent is 2%, compared whether to use hydroxyl silicone oil. When the amount of silicone oil is 1%, the compressive elasticity of the corresponding sample does not change too much. When the amount of silicone oil is 2%, the compressive elasticity of the corresponding sample is slightly reduced, But the compression thickness has increased. This is because the addition of silicone oil reduces the hardness of the cured elastomer and increases the sample's softness.

6. Acknowledgements

This project was sponsored by Huibei Province Natural Science Fund Project (Project number. 2013CFA090. The authors gratefully acknowledge this support.

References

[1] Chen GL. Development of Elastic Polyester Lacey 9.7 tex Blended Yarn [J]. Cotton Textile Technology: 2014; 42 (9): 18-27.

[2] Zhang SY. Preparation of three different fine denier polyester composite yarn [J]. Polyester Industry: 2010; 23 (6): 26-28.

[3] Xu FJ.11 dtex / 6 f polyester ultrafine low elastic yarn development [J]. Chemical fiber and textile technology: 2015; 23 (1): 4-6.

[4] Teijin, fiber ELK as a bra pad attention [J]. Textile and Apparel Weekly: 2011; 23 (11): 101-102.

[5] Yao X. Research on Improving the Back-elastic Production Technology of Hollow Polyester Staple Fiber [D]. Suzhou University: 2005; 25 (3): 16-18.

[6] Wang L. Three dimensional Curved Hollow Polyester Staple Silicone Oil Finishing Production Process [J]. Synthetic Fiber Industry: 2002; 25 (1): 53-54.

[7] Wang X. Thermal properties of composite hot melt fiber and preparation of special paper [J]. China Papermaking: 2009; 28 (11): 71-73.

[8] Zhou JP. Study on Compressive Properties of Kapok Fiber Aggregate [D]. Donghua University: 2011; 32 (23): 15-17.
[9] Zhou YX. Review on the Application Field of Liquid Silicone Rubber [J]. Power Technology Monthly: 2013; 26 (2): 195-196.
[10] Lin LH, KM. Surface activity water repellency properties of cleavable modified silicone surfactants. Colloids and Surfaces A: Physics, Chem. Eng. Aspects: 2007; 16 (1): 393-340.
[11] Zhou JP, Chen S. KES fabric style instrument test index analysis and application [J]. Modern Textile Technology: 2005; 14 (6): 37-40.
[12] Fang C. Kapok fiber aggregation of the compression properties of [D]. Donghua University: 2011; 32 (23): 15-17.
[13] Lou Y, Wang FM, Liu W, et al. Analysis of compressive properties of kapok [J]. Textiles: 2007; 28 (1): 10-13.
[14] Chama A, Reddy BD. New Stable Mixed Finite Element Approxi-Mations for Problem in Linear Elasticity. Computer Methods in Applied Mechanics and Engineer: 2013; 23 (256): 211-233.
[15] Chen XL, Nie CY, Liu C, et al. Fabric elasticity test and its classification [J] Guangxi Textile Science and Technology: 2009; 38 (1): 20-22.

Mechanical Properties and Structural Changes of Raw Silk Fibers Induced by Pre-wetting Twisting

Shun-Lin Fu[1], Chi-Yu Fu[1], Jing Wang[1], Zhi-Gang Xia[1,2*], Wei-Lin Xu[3*]

[1]*College of Textile Science and Engineering, Wuhan Textile University, Wuhan, 430200, China;*
[2]*Xiangyang Jihua 3542 Textile Company Limited, Xiangyang, Hubei 441002, China;*
[3]*Key Laboratory of Green Processing and Functional Textiles of New Textile Materials, Ministry of Education, Wuhan Textile University, Wuhan, 430200, China*

*Corresponding authors' email: zhigang_xia1983@hotmail.com

Abstract

The study purpose was to regulate the structure of molecular chains in the fiber and to endow helical conformation characteristics by using pre-wet tensional drawing. It is predicted that pre-wet tensional drawing could reset the internal molecular structure in the fiber, reduce the initial modulus and then improve its flexibility. Guided by prediction, silk fibers were selected to conduct structure resetting by drafting and twisting in wet and dry conditions respectively. The structural and mechanical property results indicated that unlike the dry drafted twisting treatment, the pre-wet one induced helical structural formation with little crystal destruction and molecule rupture. The silk softness also improved although the breaking strength and elongation decreased after pre-wet twisting treatment.

Keywords: Pre-wet Drawing; Silk Fiber; Structural Readjustment; Mechanical Properties

1. Introduction

In nature, spiders and several worms including mites, butterflies and moths, can produce silks. Silk is a synthetic fibrin synthesized in specialized epithelial cells that line glands in these organisms [1]. In the textile field, silk fibers from silkworms are a kind of natural biopolymer produced by domestic (Bombyx mori) and wild silkworms. Silk has been used as a source of textile material for nearly 5000 years. Rightly called "the queen of textiles", could be used for a long time because of its unique luster, tactile properties, durability, and dye-ability [2]. In non-textile field, silk fibroin has been used to make cosmetic and food additive [3-5]. Moreover, the surgical suture made by silk has been largely applied in clinics [6]. Silk fibroin which is non-toxic and non-irritating, has good bio compatibility and is beneficial for cells of the human body and many kinds of mammals to adhere and proliferate [7-11]. So in recent years, it also has been widely studied as biomedical materials, and used for wounds, controlled drug release carriers, tissue engineering scaffolds and repair materials for skins, bones, ligaments, etc. Silk fibers have outstanding natural attributes that rival the most advanced synthetic polymers, yet the production of silk does not require harsh processing conditions hence there are lots of ongoing research[12]. Recently, considerable attention has been paid to silks by a range of scientists from textile engineers to polymer chemists and bio-medical researchers. Compared with the most advanced synthetic polymers, silk fibers have excellent natural attributes, such as excellent mechanical properties. It hasthe combination of strength and toughness not found in today's man-made materials together with its excellent bio-compatibility [13], high resistance to chemicals [14], eco-friendly properties and so on.

These appealing physical properties originate from the structure of the silk and therefore, structural analysis will be the key to enhance the usage of the silk. Pe´rez-Rigueiroa.J. et al [15-16] performed the tensile tests on silkworm silk fibers submerged in liquid environments (water, acetone, ethanol and isopropanol). The result indicates that water disrupts hydrogen bonds before present in the amorphous phase, while the other solvents contribute to the formation of new hydrogen bonds in the amorphous phase of silk. They also performed mechanical tests on single brins of Bombyx mori silk and get conclusion that the presence of sericin in bave, contributing to the integrity of the sample cross-section but adding little to the fiber's ability to resist tensile deformation. Magoshi J et al [17] investigated the thermal behavior of films of amorphous silk fibroin in the random-coil conformation in a temperature range 25-220℃ by differential scanning calorimetry.

It is predicted that pre-wet tensional drawing could reset the internal molecular structure in the fiber and reduce the fiber initial modulus and then improve the fiber flexibility. Thus the aim of the present study was to regulate the structure of molecular chains in the fiber and to endow helical conformation characteristics by using pre-wet tensional drawing. A broader objective was to reset the internal molecular structure in the fiber and reduce the fiber initial modulus and then improve the fiber flexibility. To accomplish this goal, we chose the precious silk as the experimental object, preparing a series of processes to resetting structure by drafting and twisting, such as giving the raw yarn different draft and twist in wet and dry conditions differently. The mechanical properties of both raw and treated silk fibers were carried out by single fiber tensile strength tester. Moreover, the surface and inner structures of silk fibers were examined by scanning electron microscopy and fourier transform infrared spectrometer.

2. Experimental

2.1 Materials

Mulberry raw silk with about 20.0tex fineness was provided by the Huguangshanse silk Co. Ltd., Jiangsu province, China. The silk samples with the length of 20.0 cm were glued on a black flannel board, and each black flannel board was affixed with twenty silk samples, nine black flannel boards were prepared totally. All the water used in the experiment was distilled water.

2.2 Sample Preparation

2.2.1 Preparation of Silk Sample Twisted in Dry Condition

Using two clips to clamp the end of the silk (20.0 cm), one of the clips is fixed on the black flannel board by magnets, and then hold another clip to twist the silk sample to the target twist. The silk samples were inserted with 0, 250, 500, 750 and 1000 twists/meter respectively. Twist distribution remained in steady state. The twisted silk samples were smoothly put on a black flannel board. After the completion of the twisting process, silk samples (20.0 cm) were gently sprayed by distilled water until completely wet. Let the silk samples maintain wet state for five minutes at room temperature, then using a blower to dry them. Thereafter a collection of every 20 silk samples were fixed on the holder with two magnets, and two clips were hanged in the center of the silk collection for 2 days. This treatment was in order to give tension treatment (the applied tension was 2.55×10^{-3} cN). Then the samples with the treatment of twistingin in dry condition were ready for testing.

2.2.2 Preparation of Silk Sample Twisted in Wet Condition

The silk samples were neatly put on the black flannel board and moisturize it completely with distilled water. The silk samples were soaked in water. Two clips were applied to clamp the end of the silk (20.0cm) which is similar to the twisting procedure introduction in dry condition. The silk samples were finally twisted to 0, 250, 500, 750, and 1000 twists/meter respectively. Twist distribution remained in steady state. After the completion of the twisting process, drying process and pre-tension process are consistent with the process of twisting silk samples in dry condition. Then the samples with the treatment of twisting in wet condition were ready for test.

2.3 Measurement

The mechanical properties of silk samples were tested by a multifunction automatic single fiber tensile strength tester (FAVIMAT+BOBOT2 Type, Germany) using standard technique at 22 ℃ and 65% RH with the clamping length of 20 mm and elongation rate of 20 mm/min. The morphologies of surface of silk samples were examined using a scanning electron microscopy (SEM) instrument (S4800, Japan) at 15 kV of acceleration voltage. Before placing the samples in the SEM chamber, the samples were mounted onto an aluminum stud and sputter-coated with gold/palladium for 180 s (E-1010 ION SPUTTER, Hitachi, Japan) to prevent charging. The spectra of samples were obtained by a Fourier transform infrared spectroscopy (FTIR) instrument (Nicolet 5700, America) in the range of 6000-4500 cm^{-1}.The samples were prepared in KBr pellets.

3. Results and Discussion

3.1 SEM Observation

SEM was used to analyze the surface morphology of silk fibers produced by the twist treatment in different conditions. The longitudinal micro graphs of original and treated sample are shown in Fig.1. It is observed that silk fibers arranged in parallel with each other and the sericin can be seen clearly at some junctions. The glue materials of silkworm play an important role to join some filaments together. Longitudinal micro graphs showed that the original silk fibers were parallel, smooth, and intact.

The longitudinal micro graphs showed the silk fibers got larger radical deformations as the twists increased. However, it could be observed that the twist of wet twisted silk was more uniform with less damage than the dry ones.This also indicated that the inner structure of wet twisted samples was relatively uniform, with same endurance to external force. In the contrary, the dry twisted silks showed fractional morphology which was a sign that the inner structure of the treated sample is not uniform, with low capability to endure external force.

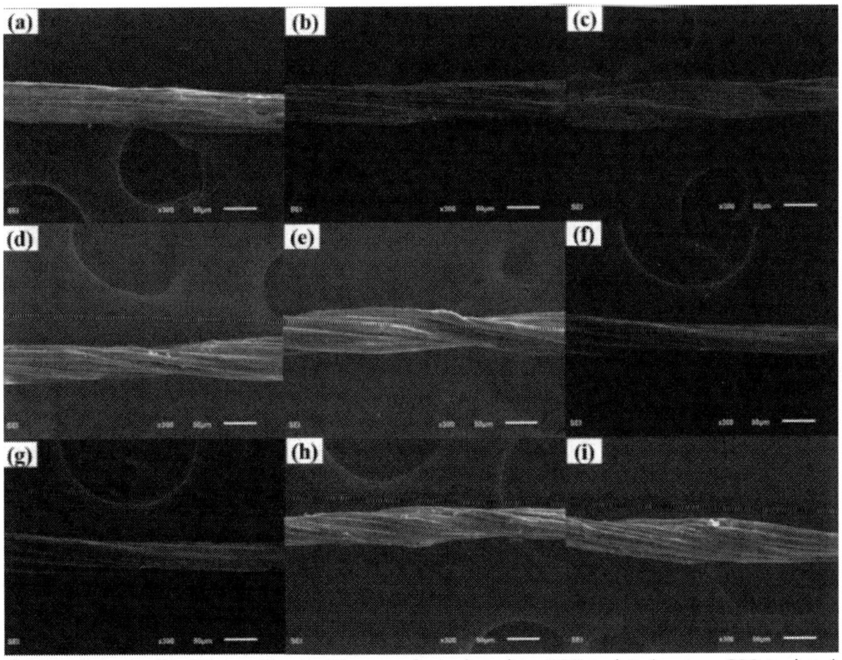

Fig.1 SEM of (a) original silk fibres; (b-e) silk sample twisted to 250 twists/meter, 500 twists/meter, 750 twists/meter, 1000 twists/meter in dry condition;(f-i) silk sample twisted to 250 twists/meter, 500 twists/meter, 750 twists/meter, 1000 twists/meter in wet condition.

3.2 Physical Properties

Fig.2 showed the initial modulus of the silk fibers with different twists under different processing conditions. It is obvious that initial modulus of the silk fibers twisted in wet condition decreases significantly as the twist ratio increases, while this of another sample is about slightly up at first and then down. But the initial modulus of silk fibers twisted in dry condition is always higher than the contrast.

The reason for this phenomenon is the effect of distilled water and pre-tension process. Immersion in distilled water for a period of time disrupts the hydrogen bonds between chain segments in the amorphous phase, leaving van der Waals bonds to dominate [15].This would lead to a reduction in the initial stiffness of silk fibers. As a result, the distilled water can help produce the compliant fibers. However, the silk fibers immersed in water would promote the substitution of protein-protein hydrogen bonds by water-protein hydrogen bonds. This substitution should affect a decrease in the initial modulus of the silk, consistent with experimental observation of others. For silk fibers twisted in dry condition, the destruction of external forces gradually stronger than the own cohesion of itself with the degree of twist increasing.

Fig.3 compares the breaking force of two silk fibers that treated in different condition. No significant differences are apparent before silk fibers twisted less than 500 twists/meter. After that, the breaking force of wet-twisted silk reaches a lowest point at 750 twists/meter and then rises in the reverse direction, while the breaking force of the dry-twisted silk always maintain the declining trend.

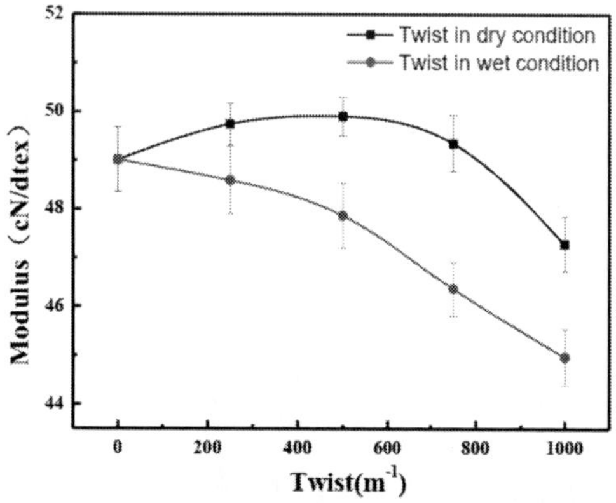

Fig.2 The silk initial modulus comparison

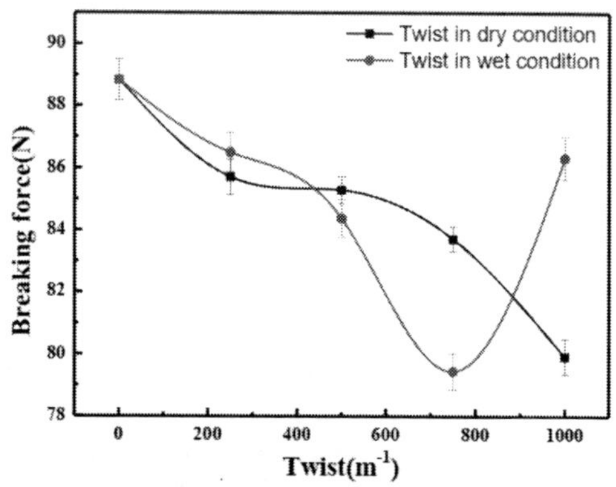

Fig.3 The silk breaking force comparison

Fig.4 compares the elongation at break of two silk fibers treated under different condition. It is obvious shown that breaking length of the silk fibers twisted in wet condition increases steadily with the inserted twist ratio increases. This can be attributed to the shrinkage of the twisted silk fibers. When silk fibers twisted in the wetting environment, the role of water can make the fiber molecules becoming loose than before and thus can move easily. For dry-twisted silk, the beginning of the decline is due to the inter molecular bonding more closely, and later rising can be attributed to the increased fiber shrinkage of the twisting.

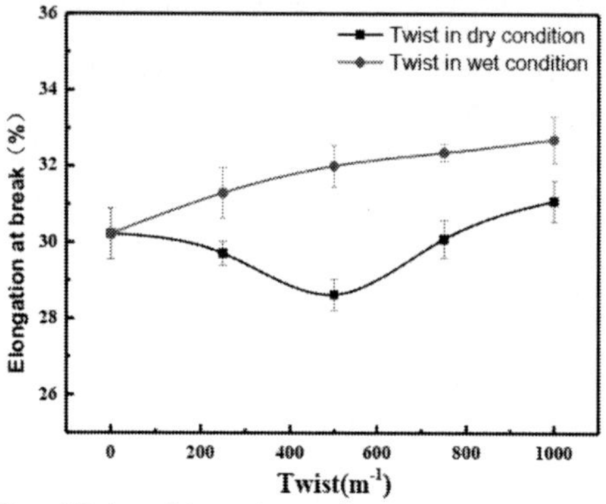

Fig.4 The silk breaking elongation comparison

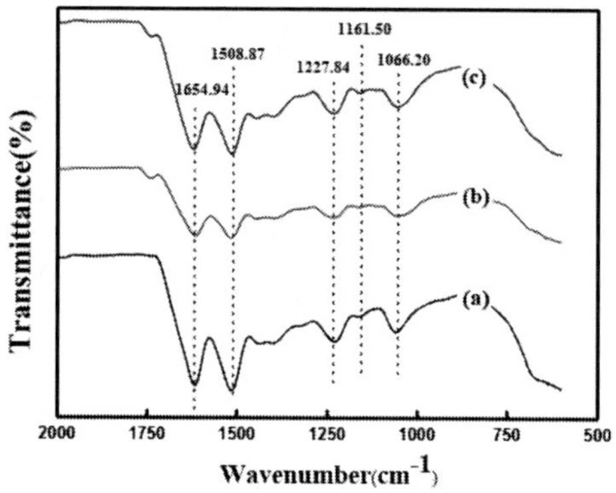

Fig.5 The silk FT-IR spectra comparison

3.3 Infrared Analysis

Fourier transform infrared spectroscopy (FTIR) has been widely used to detect the molecular reaction and conformation transition of silk fibroin, the position and intensity of amide peaks can well indicate the molecular conformation [18-20]. The studies on the molecular conformation of silk fibroin by Tsukada et al. [21-22] show that silk fibroin is characterized by α-helix absorption peaks around 1655 cm^{-1} (amide I), 1546 cm^{-1} (amide II), 1270 cm^{-1} (amide III), and β-sheet absorption around 1630 cm^{-1} (amide I), 1520 cm^{-1} (amide II) and 1240 cm^{-1} (amide III).

The FTIR spectrum of different silk fibers is shown in Fig.5. The absorption peaks of three different silk fibers appear at 1654 cm^{-1} (amide I), 1508 cm^{-1}(amide II), 1227 cm^{-1} (amide III), 1161 cm^{-1} (amide III), 1066cm^{-1} (amide Ⅳ). It indicating that the main molecular conformations inside silk fiber through the twisting process are still β-helix structure. And the characteristic peaks of the infrared spectra of silk before and after treatment were almost unchanged. Herein, it could be concluded that there is no new functional group or chemical structure in the treated sample, and the twisting treatment has no obvious effect on the chemical structure of the silk fibers. No difference is shown between the peaks. It seems that the crystallinity degree of the fabrics before and after treatment was not changed; or even if it changed, the change could not be tested by FTIR. This indicates that the treatment did not have much influence on the conformation of β-sheet in the silk fibers, which means that the treatment did no harm to the inner structure of the silk fibers. However, we can see that the characteristic peaks of (c) are more pronounced than (b), which is due to the fact that silk has excellent hydrophilic. In the process of wet treatment, under the influence of local stress, sufficient water and time, the chain segments of silk fibroin molecules can stretch repeatedly and rotate around the main axis. That causes the rearrangement of hydrogen bonds and the growth of β-sheet nuclei.

4. Conclusion

According to the structure and performance characteristics of silk fibers, this paper purpose was to regulate the structure of molecular chains in the fiber and to endow helical conformation characteristics by using pre-wet tensional drawing. The main conclusions achieved from the researches in this paper were summarized as follows:

1) Compared with the untreated silk, those treated silk fibers produced helical grains in their surface and the grains under the wet preparation, without any fracture, were better twisted than those under the dry preparation.

2) After the treatment of twisting, there is no new functional group or chemical structures generate in the treated sample, and the peaks have differences. The twisting treatment has no obvious effect on the chemical structure of the silk fibers and the crystallinity degree of the fabrics.

3) The initial modulus of the silk fibers after the pre-wet drawing and twisting treatment are obviously reduced, that is, the degree of softness is increased. In addition, the initial modulus of wet-twisted silk fibers is lower than those treated in the dry state. The breaking strength of that prepared silk, compared with unprepared silk fibers, decreased while the softness increased. In a certain range, the breaking strength of the dry-twisted silk is higher than that of the wet-twisted one. But with the increasing of the twist, the silk twisted in the wet condition has a greater breaking strength than that in the dry condition. And elongation of the wet-twisted become longer while this of dry-twisted descends first and then rise, but always lower than the former.

5. Acknowledgement

We greatly acknowledge the support from the National Natural Science Foundation of China (Project No. 51403161).

References

[1] Kaplan D, Mcgrath K, Kaplan D, et al. Protein-Based Materials. Q Rev Biol: 1998.
[2] Reddy RM. Innovative and Multidirectional Applications of Natural Fibre, Silk - A Review. Acad J Entomol: 2009.

[3] Hoppen HJ, Leenslag JW, Pennings AJ, et al. Two-ply biodegradable nerve guide: basic aspects of design, construction and biological performance. Biomaterials: 1990; 11 (4): 286-90.
[4] Tamapa A, et al. Application of the Bacterial Milk-Clotting Enzyme Preparation "Sirenin" in the Production of Soft and Solid Cheese Prepared with Sheep and Cow Milk. Int J Bio Macromol: 1987; 2 (4): 35-40.
[5] Chen K, Iura K, Takano R, et al. Effect of fibroin administration on the blood cholesterol level of rats loaded with cholesterol. Journal of Insect Biotechnology & Sericology: 2016; 62.
[6] Moy RL, Lee A, Zalka A. Commonly used suture materials in skin surgery. Am Fam Physician: 1991; 44 (6): 2123-2128.
[7] Chiarini A, Petrini P, Bozzini S, et al. Silk fibroin/poly(carbonate)-urethane as a substrate for cell growth: in vitro interactions with human cells. Biomaterials: 2003; 24 (5): 789.
[8] Armato U, Migliaresi C, Motta A. Bio-artificial substrate for the production of animal and, in particular, human tissse and organs: wo, ep1218490: 2002.
[9] Sofia S, Mccarthy MB, Gronowicz G, et al. Functionalized silk-based biomaterials for bone formation. J BIOMED MATER RES B: 2001; 54 (1): 139.
[10] Gomes S, Leonor IB, Mano JF, et al. Silk-Based Biomaterials. Biomimetic Approaches for Biomaterials Development: 2012; 75-92.
[11] Li M, Ogiso M, Minoura N. Enzymatic degradation behavior of porous silk fibroin sheets. BIOMATERIALS: 2003; 24 (2): 357-365.
[12] Chen Z, Kimura M, Suzuki M, et al. Synthesis and Characterization of New Acrylic Polymer Containing Silk Protein. FIBER: 2003; 59 (5): 168-172.
[13] Asakura T, Suzuki Y, Nakazawa Y, et al. Silk structure studied with nuclear magnetic resonance [J]. PROG NUCL MAG RES SP: 2013; 69 (2): 23.
[14] Raghu K, Raghu K, Noorunnisa Khanam P, Venkata Naidu S. Chemical Resistance Studies of Silk/Sisal Fiber-Reinforced Unsaturated Polyester-Based Hybrid Composites.J REINF PLAST COMP: 2010; 29 (3): 343-345.
[15] Pérez-Rigueiro J, Viney C, Llorca J, et al. Mechanical properties of silkworm silk in liquid media. POLYMER: 2000; 41 (23): 8433-8439.
[16] Pérez-Rigueiro J, Viney C, Llorca J, et al. Mechanical properties of single-brin silkworm silk. J APPL POLYM SCI: 2000; 75 (10): 1270-1277.
[17] Magoshi J, Magoshi Y, Nakamura S, et al. Physical properties and structure of silk. V. Thermal behavior of silk fibroin in the random-coil conformation. J POLYM SCI POL PHYS: 1977; 15 (9): 1675-1683.
[18] Asakura T, Iwadate M, Demura M, et al. Structural analysis of silk with ~ (13) C NMR chemical shift contour plots. INT J BIOL MACROMOL: 1999; 24 (2-3): 167-171.
[19] Odian G, O'Callaghan MP, Chien CK, et al. Zwitterion polymerization of tetrahydro - 1 - [4 - hydroxy - 3 - (2 - hydroxyethoxy)phenyl]thiophenium hydroxide inner salt. MACROMOLECULES: 1990; 23 (4): 918-925.
[20] Gotoh Y, Tsukada M, Minoura N. Chemical modification of silk fibroin with cyanuric chloride-activated poly (ethylene glycol): analyses of reaction site by 1H-NMR spectroscopy and conformation of the conjugates. BIOCONJUGATE CHEM: 1993; 4 (6): 554.
[21] Li M, Tao W, Lu S, et al. Compliant film of regenerated Antheraea pernyi silk fibroin by chemical crosslinking. INT J BIOL MACROMOL: 2003; 32 (3-5): 159-163.
[22] Tao W, Li M, Zhao C. Structure and properties of regenerated Antheraea pernyi silk fibroin in aqueous solution. INT J BIOL MACROMOL: 2007; 40 (5): 472.

Determination of the Content of Calcium Oxalate Crystals in Antheraea Pernyi Cocoon Layers Based on the Inductively Coupled Plasma-Atomic Emission Spectrometry

Meng-Ru Li[1], Zheng-Lei He[1], Tian-Pei-Jin Cui[2], Zhi-Yue Wu[2], Ji-Song Wu[1*]

[1] *School of Textile Science and Engineering, Wuhan Textile University, Wuhan, China;*
[2] *College of Advanced Textile, Wuhan Textile University, Wuhan, China*

[*]Corresponding author's email: 1345270619@qq.com

Abstract

It is known that calcium oxalate crystal is a very important component of a cocoon shell, which is thought to be responsible for providing the initial defence to any predator attack and other outside physical attacks. In particular, the content of calcium oxalate crystals is related to the thermal insulation property of cocoon. In this study, we chose Anthraea pernyi cocoon (A. pernyi cocoon) as a model to quantitatively study the content of calcium oxalate crystals in the shell of cocoon using inductively coupled plasma-atomic emission spectrometry (ICP-AES). Multi-factor experiments have been conducted to optimize the treatment temperature, the type of crystal dissolving agents (EDTA, EDTA-2Na and concentrated HNO_3) and the concentration of dissolving agents. The results indictated that the content of calcium oxalate is independent on the type of crystal dessolving agents but varies with the producing places. The A. pernyi cocoon from Liaoning Dandong shows a much higher crystal content (96.1%) than that from Henan Nanyang (77.4%).

Keywords: ICP-AES; Antheraea Pernyi; Cocoon Layers; Calcium Oxalate Crystals; Calcium Ion

1. Introduction

As we know, silkworm cocoon can be categorized into three types: domestic, semi-domestic and wild. For each type there are, some variations in terms of composition and structure. For example, there are great differences between A. pernyi cocoon layers and the domestic one in terms of composition, structure and properties. The domestic silkworm cocoon is composed of silk fibroin and sericin. There are a number of crystals, which are loosely attached to the silk and the gap between the silk with different size and shape (mostly in brick shape), in semi-domestic and wild silkworm cocoon [1-3]. These crystals are reported to be calcium oxalate crystals, which were secreted during the silkworm spinning process. These hard crystals are thought to be responsible for providing the first point of defense to any predator attack and other physical attacks from the outside, and hence is essential for survival [4-6]. However, there is limited information when calculating the quantity of calcium oxalate crystals in the cocoon shell and the effect of the content of calcium oxalate crystal on the structure and property (e.g. thermal insulation) of the cocoon, and how it may affect the fabric and textile composition [7]. Therefore, it is of vital importance to systematically study the content of calcium oxalate crystals in the cocoon shell.

The chemical formula of achromatic calcium oxalate, which has a molecular weight of 128.10 g/mol , is CaC_2O_4. In general, there are three types of calcium oxalates: calcium oxalate monohydrate (COM, monoclinic system), calcium oxalate dehydrate (COD, tetragonal system), calcium oxalate trihydrate (COT, triclinic system) [8, 9]. There are several kinds of morphologies of calcium oxalate cyrstals, including the single crystal, clustered crystal and needle crystal [10]. Before measuring the content of calcium oxalate crystal in cocoon shell, we detached the crystal from the cocoon shell and dissolved it. However, the crystals have stable chemical properties and only start to lose the crystal water when the temperature is around 200℃.In addition, the crystals have a hardly solubility in water and acetic acid, only can dissolve in the nitric acid and hydrochloric acid. The crystals should be dissolved by some chemical reagents.

In this study, we aim to quantitatively analysis the content of calcium ion in the cocoon shell using inductively coupled plasma-atomic emission spectrometry (ICP-AES) because of its excellent features, such as easy operation, high sensitivity, fast analysis and high reproducibility [11-13]. To this end, Anthraea pernyi cocoon (A. pernyi cocoon) was selected as a model. Different agents including EDTA, EDTA-2Na and concentrated HNO3 were used to dissolve calcium oxalate crystals, and the concentration of dissolving agents was optimized. A. pernyi cocoons before and after treatment were characterized using scanning electron microscopy (SEM) and FTIR spectroscopy.

2. Experimental Design

2.1 Materials and Preparations

Ethylenediaminetetraacetic acid (EDTA), EDTA disodium salt, concentrated nitric acid, sodium hydroxide and hydrochloride were bought from Sinopharm, China. A. pernyi cocoons were purchased from Dandong, Liaoning Province and Nanyang, Henan Province. The cocoons were cut into pieces, then rinsed by deionized water before use.

2.2 Experimental Methods

2.2.1 Physical Demineralization

Cocoons pieces (1 g) were collected in the beakers and then physically demineralized in deionized water at 45℃, which were placed in numerous control ultrasonic wave washing machines (KH-250DB, China). The rate of power of washing machine was 40 KHz, with a power of 250W and a heating power of 600W. At desired time (30 min, 60 min and 120 min), cocoon pieces were taken out and dried in the oven before use. The remaining sediment of three beakers was dried in an oven and the Fourier transform infrared spectra (FTIR) of the crystals were obtained using FTIR spectrometer (NEXUS470, Nicolet, America).

2.2.2 Chemical Demineralization

EDTA and EDTA-2Na solutions were prepared by dissolving desired amount of EDTA and EDTA-2Na powder into 1.2 mol/L sodium hydroxide solution, respectively. The pH of the EDTA and EDTA-2Na solutions were adjusted to 7.4 by 1mol/L hydrochloride solution. Series of EDTA solutions with different concentrations, namely 0.01M, 0.02M, 0.04M, 0.06M, 0.08M were prepared; and similarly, EDTA-2Na solutions with various concentrations of 0.01M, 0.03M, 0.06M, 0.12M, 0.25M were also prepared. Cocoons pieces were added into each of the EDTA and EDTA-2Na solution, under thermostatically heated magnetic stirrer (DF-101S, Wuhan Cole Instruments Equipment Co., Ltd), at 40 -60 ^{0}C for 2 hours, in which the amount of cocoon pieces in a 100 mL solution was 1 g.

The cocoons were also treated with 63% concentrated nitric acid for 2 hours in the dark. The filtrate should be diluted with a certain concentration of sodium hydroxide solution until its pH value goes up to 6-7. The filtrate and the wash water were fixed volume in the 1L volumetric for the measurement of calcium ion.

2.2.3 Morphological Characterization

The morphologies of the cocoons treated with different agents were observed using a scanning electron microscopy (SEM) (JSM-6510, JEOL, Japan) after the chemical demineralization.

2.2.4 Measurement of Calcium Ion

In this study, the content of calcium ion in A. pernyi cocoon shell was determined by inductively coupled plasma-atomic emission spectrometry (ICP-AES, ICP2060T, Jiangsu, China) with a power of 1.2 KW, a frequency of 50 Hz, a voltage of 220 V, and a measurement range from 190 nm to 50 nm, which is helpful to quantitatively analyze the content of calcium oxalate in the cocoon layers. The standard solution used was 10mg/L calcium ion solution.

3. Results and Discussion

3.1 FITR of the Crystals

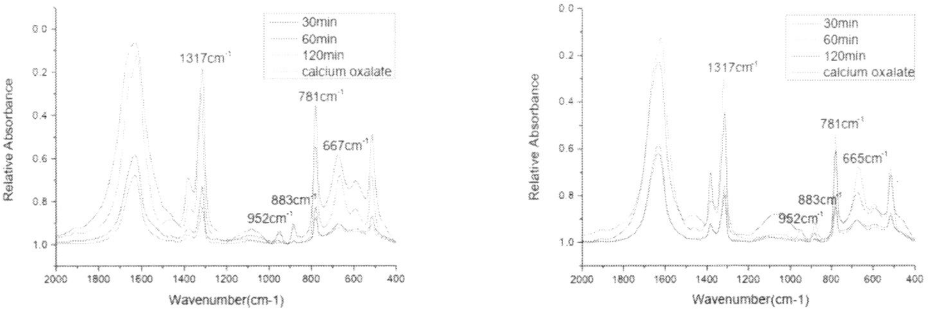

(a) (b)

Fig.1. FTIR spectra of the commercial purchased calcium oxalate crystals and the crystals obtained by ultrasonic method from the A. pernyi cocoons at different times: (a) Liaoning; (b) Henan.

Fig.1 shows the FTIR spectra of the commercial purchased calcium oxalate crystals and the crystals obtained by ultrasonic method from the A. pernyi cocoons at different times. The FTIR spectra of materials is a reflection of the molecular structure. The absorption peaks in spectra correspond with the molecular vibration of each group. The absorption peaks of both the commercial purchased calcium oxalate crystals and the crystals obtained by ultrasonic method were similar. In the spectra, the peak at 1680-1640 cm^{-1} represents C=C bonds stretching vibration. The analysis of the further fingerprint region, namely, the frequency range from the 1000 cm^{-1} to 675cm^{-1} were olefin C-H out-plane flexural vibration. Comparing FTIR spectra of the crystals with that of the commercial calcium oxalate crystals, we could easily identify the type of calcium oxalate crystals. In the spectra of pure COM crystals, the absorption peaks at 1313, 945, 880, 780, 664 cm^{-1} were the main features of COD crystals. According to this, the calcium oxalate crystals in the A. pernyi cocoons was ascribed to calcium oxalate monohydrate (COM) crystals.

3.2 Optimization of Chemical Method using EDTA to Remove Crystals from A. Pernyi Cocoons

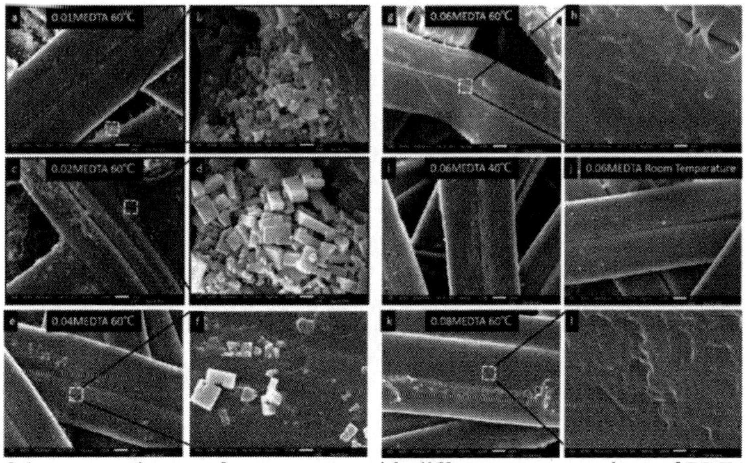

Fig.2. SEM images of the cocoon layers after treatment with different concentrations of EDTA solution at different temperatures.

Fig.2 shows the SEM images of the A. pernyi cocoons fiber from Liaoning treated with different concentrations of EDTA solutions at 60℃. It showed that there were plenty of brick crystals of different sizes embedded in the cocoon layers treated with 0.01M EDTA solution (Fig.2a). When the concentration of EDTA solution increased to 0.02 M, no significant changes was observed in the morphology of calcium oxalate (Fig.2c). Most of the calcium oxalate crystals remained in the cocoon layer with intact shape and were not destroyed at all. When the concentration of EDTA solution increased to 0.04 M, the shape of the calcium oxalate crystals became round and the remaining number of crystals decreased (Fig.2f). When the concentration of EDTA solution increased to 0.06 M, the crystals were almost dissolved and the silk fiber was also undamaged (Fig.2g and 2h). It can be interpreted that 0.06M EDTA solution is sufficient to dissolve the calcium oxalate crystals in the cocoon layer. Therefore, it can be considered that the optimum conditions for the treatment of the cocoon layer are using EDTA solution of 0.06 M concentration, with a reaction temperature of 60℃ for 2 hours at pH 6-7.

Analysis of the SEM images also shows the process of calcium oxalate crystals being destroyed during the chemical demineralization. The morphological changes of crystals can be recorded when the crystals were transferred from EDTA solutions of lower to higher concentrations. When the cocoon layer was treated with 0.01 M EDTA solution, the crystals began to crack (Fig.2b). Most of the crystals became fragmented when incubated with a higher concentration of 0.04 M EDTA solution, and the crystals completely dissolved, leaving only silk mark on the silk (Fig.2h), at the higher concentration of 0.06M treatment.

3.3 Optimization of Chemical Reaction with EDTA-2Na to Remove the Crystals in A. Pernyi Cocoons

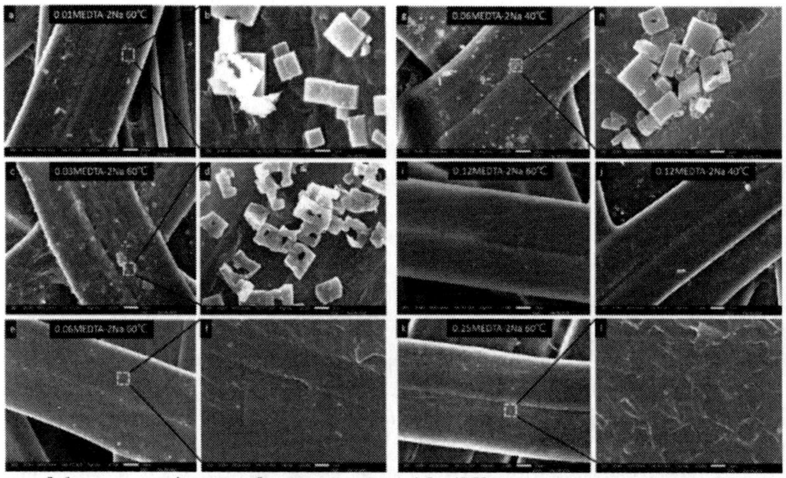

Fig.3 SEM images of the cocoon layers after treatment with different concentrations of EDTA-2Na solution at different temperatures.

Fig.3 shows the SEM images of the A. pernyi cocoons fiber from Liaoning treated with different concentrations of EDTA-2Na solutions at 60℃. It showed that most of the calcium oxalate crystals remained in the cocoon layer when treated with 0.01M EDTA-2Na solution (Fig.3a). When treated with 0.03M EDTA-2Na solution, most of the calcium oxalate crystals remained in the cocoon layer, but most of the crystals were corroded in the middle, leaving round holes with different diameters (Fig.3d). The shape of the calcium oxalate crystals became round and the number of crystals remaining in the cocoon layer abruptly decreased when the cocoon layer was treated with 0.06 M EDTA-2Na solution (Fig.3f). When the cocoon layer was treated with 0.25 M EDTA-2Na solution, the crystals completely dissolved (Fig.3k).

The results showed that treatment of cocoon layers with various concentrations of EDTA-2Na solution could lead to the same trend as the EDTA solution treatment. However, the EDTA-2Na solution is not as effective as the EDTA solution in terms of removing the crystals from the cocoons. The optimum conditions were EDTA-2Na solution at the concentration of 0.25M at a reaction temperature of 60℃with a pH of 6-7 reaction for 2 hours.

This conclusion is in line with the conclusions drawn by Kaur J [14], who reported the SEM images of cocoons subjected to chemical demineralization using EDTA solutions of different concentrations and various stages of crystals breakage. He remarked, "When EDTA solution at a concentration of 0.01 M was used, the crystals started to disintegrate. At a slightly higher concentration of 0.03 M, the crystals were fragmented, and when the concentration was further increased to 0.06 M, the crystals were completely removed, leaving brick shaped marks on the sericin layer over the silk fibers", which is in line with the results observed in this study.

3.4 Treatment with Concentrated Nitric Acid

Both crystals and A. pernyi cocoons were dissolved completely in concentrated nitric acid. There is no particle in the solution, since the major components of silk is protein, and the sulfur bonds of silk was oxidized by concentrated nitric acid. The silk then turned yellow, due to the fact that the tyrosine on the surface of the silk experiences an oxidation reaction under the concentrated nitric acid, which is consistent with the literature review [15].

3.5 Analysis of Ca^{2+} Content

The concentration of calcium ions determined using ICP-AES was shown in Table 1. For A. pernyi cocoons from Liaoning Dandong province, the Ca^{2+} concentration was slightly higher (96.115%) than those treated with EDTA and EDTA-2Na solutions. It might be the reason that the reaction between the cocoons and the concentrated EDTA solution and EDTA-2Na solutions is not as drastic as that of concentrated nitric acid.

Table.1 Determination of calcium ion in A. pernyi cocoons

A. pernyi cocoons	Treatment program	calcium ion measured value/(mg/L)
Liaoning Dandong	0.06M EDTA solution	95.545
	0.25M EDTA-2Na solution	94.603
	Concentrated nitric acid	96.115
Henan Nanyang.	Concentrated nitric acid	77.383

The results also showed that the content of calcium oxalate crystals in Dandong A. pernyi cocoons was about 1.3 times that of Henan corresponding part.

4. Conclusion

This study investigated the type of calcium oxalate crystals and the content of calcium oxalate crystals in A. pernyi cocoons from two different areas by FTIR and ICP-AES, respectively. Physical method was used to obtain the crystals from A. pernyi cocoons in order to identify the crystals. Chemical methods were used to remove the crystals from the cocoons using three reagents. The content of calcium ion in A. pernyi cocoons from two different areas was different, which should be used to study the influence of calcium oxalate crystals on the thermal insulation of the cocoons quantitatively. In this paper, the calcium oxalate crystals were removed without damaging the silk, which is helpful to reel the silk easily, rather than destroying the strength of it. Once the type and content of the crystals were confirmed, the data and results can be used in the further study of the simulation of cocoons and contribute to the development of new composites. The ICP-AES can accurately detect the content of calcium ion in the cocoons and has many advantages, such as high precision, good accuracy, fast and simple.

5. Acknowledgement

We acknowledge the support from the College of Chemistry and Chemical Engineering (CCCE) in Wuhan Textile University for this work. We would also like to thank Mr. Furong Guo for his help with the inductively coupled plasma-atomic emission spectrometry.

References

[1] Kaur J, Rajkhowa R, Tsuzuki T, et al. Crystals in Antheraea assamensis silkworm cocoon: Their removal, recovery and roles [J]. Mater. Des.: 2015; 88: 236-244.
[2] Chen F, Porter D, Vollrath F. Morphology and structure of silkworm cocoons, Mater. Sci. Eng. C: 2012; 32 (4) 772-778.
[3] Chen F, Porter D, Vollrath F. Silk cocoon (Bombyx mori): multi-layer structure and mechanical properties [J]. Acta Biomater: 2012; 8 (7): 2620-2627.
[4] Chen RX, He JH, Kong HY. Waterproof and dustproof of wild silk: a theoretical explanation. J. Nano Res.: 2013; 22: 61-63.
[5] Zhang J, Kaur J, Rajkhowa R, et al. Mechanical properties and structure of silkworm cocoons: a comparative study of Bombyx mori, Antheraea assamensis, Antheraea pernyi and Antheraea mylitta silkworm cocoons [J]. Mater. Sci. Eng. C: 2013; 33(6): 3206-3213.
[6] Danks HV. The roles of insect cocoons in cold conditions [J]. Eur J. Entomol: 2004; 101 (3): 433-438.
[7] Zhang J, Du S, Kafi A, et al. Surface energy of silk fibroin and mechanical properties of silk cocoon composites [J]. RSC Adv: 2015; 5(2): 1640-1647.
[8] Ouyang JM, Yao XQ. Chemical Research and Application: 2001; 13: 353.
[9] Chen L, Man RL, Li ZM. Analysis of Calcium Oxalate Crystalline Phases by Infrared Spectrophotometry [J]. J of Instru Ana: 2007; 26 (4): 582-584.
[10] Han QQ. Calcium oxalate crystals in the application and analysis of traditional Chinese medicine identification [J]. China practical medicine: 2007; 2 (26): 66-67.
[11] Ma AZ, Qi WX, Yang YF. Determination of Pb in cosmetics by inductively coupled plasma emmision spectrometer [J]. Analytical Instrumrntation: 2008; 2008 (3): 39-41.
[12] Davidowski L, Grosser Z, Thompson L. Heavy Metal Analysis for the Toy Industry by Inductively Coupled Plasma-Optical Emission Spectrometry [J]. Rock & Mineral Analysis: 2009; 28 (2): 543-548.
[13] Xin RX. Inductively coupled plasma atomic emission spectroscopy, the second edition, Che. Industry Publishing House: 2011.
[14] Kaur J, Rajkhowa R, Tsuzuki T, et al. Crystals in Antheraea assamensis silkworm cocoon: Their removal, recovery and roles[J]. Mater. Des.: 2015; 88: 236-244.
[15] Li Y, Cai N, Wang M, et al. Fluorometric detection of tyrosine and cysteine using graphene quantum dots [J]. Rsc Advances: 2016; 6 (39): 33197-33204.

Preparation and Characterization of Copper Coated Bamboo Rayon Fabric Pretreated with 3-aminopropyltrimethoxysilane by Magnetron Sputtering

Ling-Hui Peng[1], Rong-Hui Guo[1*], Jian-Wu Lan[1], Zhou-Yi Zhang[2], Jun-Liang Zhang[2]

[1]*College of Light Industry, Textile and Food Engineering, Sichuan University, Chengdu, China,*
[2]*Ningbo Grand Home Furnishing Products Co. Ltd., Ningbo, China*

*Corresponding author's email: ronghuiguo214@126.com.

Abstract

Bamboo rayon fabric was pretreated with 3-aminopropyltrimethoxysilane (silane) and copper nanoparticles were then coated by magnetron sputtering. The copper nanoparticle coated fabrics were characterized by fourier transform infrared spectroscopy (FTIR), scanning electron microscope (SEM), and ultraviolet spectroscopy. Contact angle and color characteristics of the copper coated bamboo rayon fabric were tested. The results suggest that the copper coated bamboo rayon fabric modified with the silane has excellent hydrophobicity and UV blocking properties.

Keywords: Bamboo Rayon Fabric; Magnetron Sputtering; Copper; 3-aminopropyltrimethoxysilane; Hydrophobicity

1. Introduction

Bamboo rayon fibers have attracted a lot of attention because they possess unique properties including excellent moisture absorption and permeability, wicking properties, dyeing and finishing abilities and abrasion resistance [1, 2]. Bamboo rayon fibers have been applied in the textile industry in the form of undergarments, sports textiles, T-shirts and socks in some Asian countries [3].

Recently, more and more research is devoted to coating fibers with copper nanoparticle films to obtain multifunctional textiles. Berendjchi A [4] et al coated copper nanoparticles on cotton fabrics for superhydrophobic and antibacterial applications. Kim [5] et al synthesized copper-plated nylon/spandex fabric as electrotherapeutic pad electrode. Jiang [6] et al investigated adhesion and durability of copper film on polyester fabric through finishing. There has been growing use of organosilanes as coupling agents before coating metal thin films on the surface of fibers [7]. Generally, organosilanes form a layer on substrates and then react with metal nanoparticles on the fibers via chemical bonds [8].

Magnetron sputtering technique is a kind of efficient and environmental friendly method for fabrication of high quality thin films [9, 10]. Magnetron sputtering technique enables coating of uniform copper nanoparticle films on various substrates [11, 12]. Coating copper nanoparticle films by magnetron sputtering on fabrics improves UV blocking, superhydrophobic and electromagnetic shielding properties of fabrics [13]. The copper coated fabrics can be used as many materials such as bullet-proof jacket, helmets, aerospace materials, electromagnetic shielding military clothing and home wearing.

In this study, copper nanoparticle films were coated on the surface of bamboo rayon fabrics modified with the silane by magnetron sputtering. Copper coated bamboo rayon fabrics were characterized by FTIR and SEM. Contact angle, UV protection and color characteristics of copper coated bamboo rayon fabrics were evaluated.

2. Experimental

2.1 Materials and Preparation

100% twill bamboo rayon fabrics (148 g/m^2) and 99.995% copper target (ZhongNuo New Material Technology Co., Ltd.) were used. 3-aminopropyltrimethoxysilane (silane) was purchased from Xiya Co., Ltd.

Prior to the coating of copper nanoparticles, the bamboo rayon fabrics were initially cleaned in ultrasonic baths with acetone, ethanol, and deionized water for 15 min, respectively. The pretreatment experiment was carried out using 1 vol.% of 3-aminopropyltrimethoxysilane in ethanol (40ml). The bamboo rayon fabrics were immersed in the solution at 40℃ for 1h. The treated fabrics were dried in an oven at 50 ℃ for 24 h.

Copper films were coated on bamboo rayon fabric at room temperature by magnetron sputtering technology. Base pressure is 5×10^{-4}Pa. Electrical current, flow rate of Ar and working pressure are 0.2A, 105 sccm, and 0.26 Pa, respectively. Target-substrate distance is 10 cm. The sputtering time was 60 min.

2.2 Characterization

FTIR was used to characterize the presence of chemical groups on the surface of the silane modified bamboo rayon fibers. The spectra were recorded on Nicolet 6700 for each measurement over the spectral range of 400–4000 cm^{-1} at resolution of 4 cm^{-1}.

Surface morphology of the copper coated bamboo rayon fiber was examined by a field emission scanning electron microscope (JSM-5900LV) with an acceleration voltage of 5 kV and 8mm of working distance.

Water contact angles were measured by a contact angle analyzer to determine the hydrophobicity of the original and copper coated bamboo rayon fabrics. Five points were chosen for each sample with 5μL droplet. The average value was calculated by using the baseline of the droplet and the circle method with the HARKE-SPCA×1.

Color measurements of the original and copper coated bamboo rayon fabrics were conducted by a spectrophotometer (COLOR I5) under Pulsed Xenon D65 and processed using software with the small aperture.

UV transmittance was used to evaluate UV protection of copper coated bamboo rayon fabric modified with the silane. Ultraviolet transmittance spectra of copper coated bamboo rayon fabrics were measured by a spectrophotometer (Shimadzu UV-2700) with wavelength changing from 200 nm to 400 nm.

3. Results and Discussion

3.1 FTIR Spectra

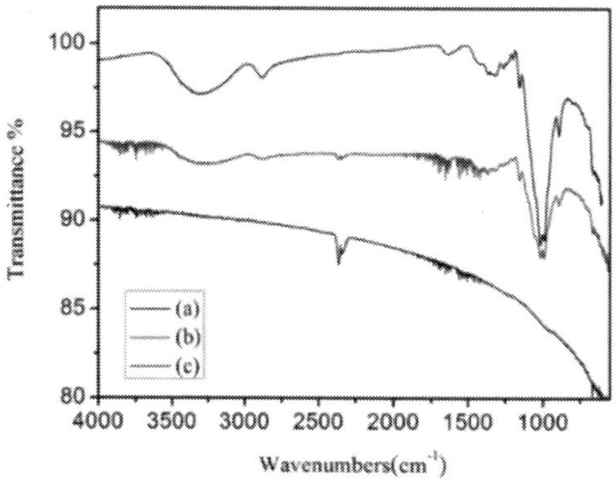

Fig.1 FTIR spectra of (a) original and (b) silane modified bamboo rayon fabrics and (c) copper coated bamboo rayon fabric modified with silane

FTIR spectra of the original fabric, the silane modified fabric and copper coated bamboo rayon fabrics modified with the silane are shown in Fig.1. A broad peak corresponding to the hydroxyl group of bamboo rayon fabric is detected at 3328 cm^{-1} and a peak of the carbonyl group is observed at 1635 cm^{-1} as shown in Fig.1(a). Peak of symmetric C-H stretching is obtained at 2901.8 cm^{-1}. Characteristic peaks at 1423 cm^{-1}, 1161cm^{-1}, 1056 cm^{-1}, 1032cm^{-1} and 995 cm^{-1} can be identified as cellulose. Compared with the FTIR spectrum of original bamboo rayon fabric, characteristic peaks of cellulose also can be detected on the silane modified bamboo rayon fabric (Fig.1(b)). The characteristic peaks of silane can not be detected due to relative less amount of the silane on the surface of fabric. Fig.1(c) shows FTIR spectrum of copper coated bamboo rayon fabric modified with the silane. Both the characteristic peaks of bamboo rayon fabric and silane can not be detected due to thick copper film on the surface of the copper coated bamboo rayon fabric.

3.2 Morphology

SEM images of original, silane modified and copper coated bamboo rayon fiber are shown in Fig.2. It can be seen from Fig.2(a) that the bamboo rayon fiber consists of fibrils that is arranged along the length of the fiber. There are full of grooves between fibrils in the longitudinal fiber surface and apertures along the length of the fibrils. A few granules are observed on the surface of bamboo rayon fiber due to impurities such as hemicelluloses, lignin, waxes and oils. However, the surface of silane modified fiber is covered with layer of silane which makes the surface of fiber smooth (Fig.2(b)). It can be seen from Fig.2(c) that the fibers are completely coated with dense copper nanoparticles. The copper nanoparticles are small and even. There is a layer of uniform copper film with some bigger aggregations on the surface of copper coated fibers. The phenomenon can be explained by that surface of fiber is covered by silane film before coating, which can be chelated with copper nanoparticles. More copper nanoparticles are coated on the surface of the silane modified fiber, therefore, a layer of copper film is evenly formed on the surface of the fiber.

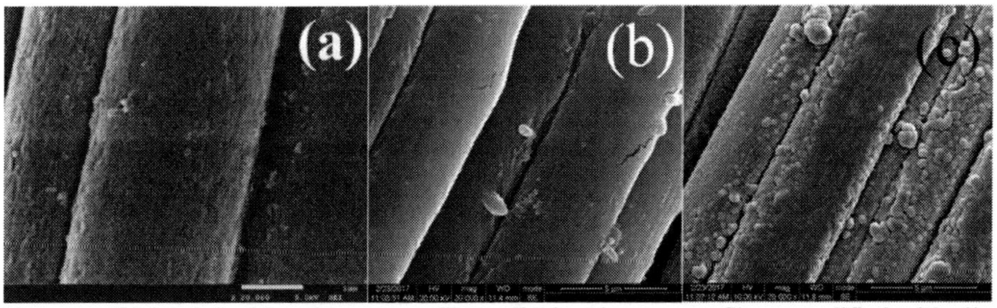

Fig.2 SEM images of (a) original bamboo rayon fibers, (b) bamboo rayon fibers modified with silane, and (c) copper coated bamboo rayon fibers modified with the silane

3.3 Contact Angle

Contact angles of the original fabric, the silane modified fabric and copper coated bamboo rayon fabric modified with the silane are shown in the Fig.3. The contact angle of the original bamboo rayon fabric (Fig.3(a)) is 0°, indicating that the original bamboo rayon fabric is hydrophilic due to abundant of hydroxyl groups. However, the contact angle of silane modified bamboo rayon fabric (Fig.3(b)) is higher than that of original fabric and reaches 123.3°. The result shows that the silane modified bamboo rayon fabric has better hydrophobic property than the original bamboo rayon fabric. This is because the fibers are covered with silane films which are hydrophobic, accordingly, the surface of the bamboo rayon fiber changes from hydrophilic to hydrophobic. The contact angle of the copper coated bamboo rayon fabric is 139.6°which is higher than that of silane modified bamboo rayon fabric as shown in Fig.3(c), indicating that the copper coated bamboo rayon fabric has excellent hydrophobic property. The phenomenon can be explained by that silane films are hydrophobic and the copper films on the surface of fibers are rough, which increase the contact angle of the copper coated fabric. In addition, copper nanoparticles are not only deposited on the surface of the fibers, but also deposited in the spaces between the fibers, therefore water penetration of the copper coated bamboo rayon fabric decreases.

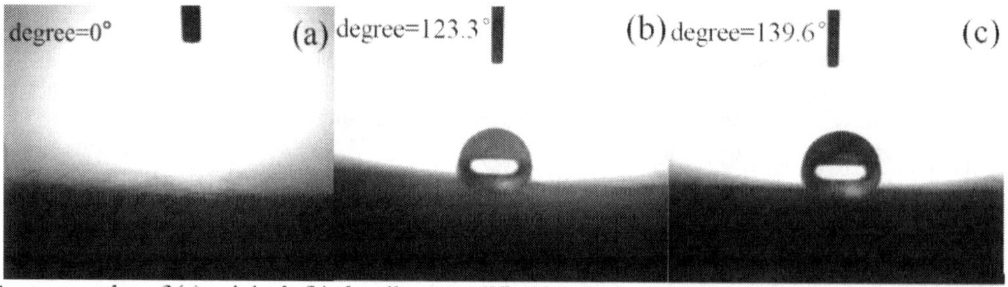

Fig.3 Contact angles of (a) original, (b) the silane modified bamboo rayon fabric and (c) copper coated bamboo rayon fabric modified with the silane

3.4 UV Blocking

UV blocking properties of original and copper coated bamboo rayon fabrics modified with the silane are illustrated in Fig.4. It is obvious that the UV transmittance of original bamboo rayon fabric is from 80% to 20% when wavelength ranges from 200 nm to 400nm, which is far higher than that of copper coated bamboo rayon fabric. The UV transmittance of copper coated fabric is almost 0% at wavelength ranging from 200nm to 400nm, indicating that the copper coated fabric shows excellent UV blocking property. The phenomenon can be explained by that copper nanoparticles coating on the fabric enhances the light reflection and scattering. Moreover, copper nanoparticles are deposited not only on the fiber surface but also in the spaces between the fibers and yarns during the copper sputtering. The coverage of the spaces can prevent the penetration of the UV radiation through the fabric and reach the skin. Thus, the UV transmittance of copper coated bamboo rayon fabric modified with the silane is much higher than that of original bamboo rayon fabric. The result indicates that copper coated bamboo rayon fabric modified with the silane has excellent ultraviolet radiation protection property.

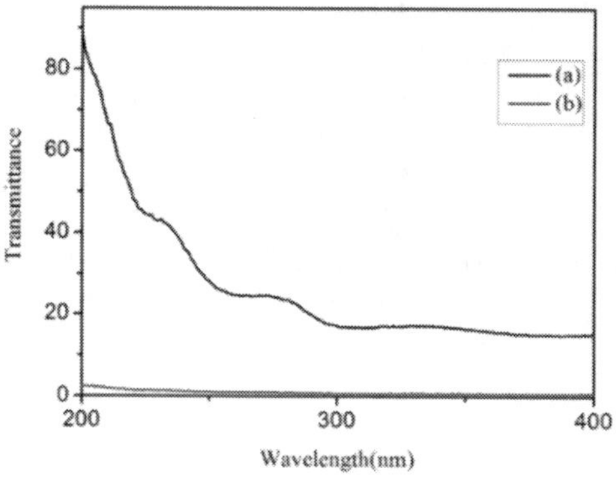

Fig.4 UV transmittance of (a) original and (b) copper coated bamboo rayon fabric modified with the silane

3.5 Color Characteristics

L*, a* and b* values of original fabric, silane modified and copper coated bamboo rayon fabric modified with the silane are presented in Table 1. The lightness (L*) represents the darkest black and brightest white when L* is 0 and 100, respectively. The color channels (a* and b*) are indicators of red-green (redder if positive and greener if negative) and the yellow-blue colors (yellower if positive and bluer if negative), respectively. The result shows that lightness (L*) for original bamboo rayon fabric and bamboo rayon fabric modified with silane are very high. Lightness (L*) of copper coated bamboo rayon fabric modified with silane is the lowest. The values of a* and b* of the copper coated bamboo rayon fabric modified with the silane are higher than that of original bamboo rayon fabric. The result indicates that the color of copper coated bamboo rayon fabric is redder and yellower due to red color of copper metal. The result indicates that copper coated bamboo rayon fabric modified with silane is redder than original fabric because copper nanoparticles are coated on the bamboo rayon fabric modified with the silane.

Table.1 L*, a* and b* values of original and copper coated bamboo rayon fabrics modified with the silane

Sample	L*	a*	b*
Original bamboo rayon fabric	86.36	3.77	-13.89
Silane modified bamboo rayon fabric	86.22	3.05	-12.03
Copper coated bamboo rayon fabric modified with the silane	49.38	9.03	9.86

4. Conclusion

Copper nanoparticle films were successfully coated on bamboo rayon fabric modified with 3-aminopropyltrimethoxysilane by magnetron sputtering. The silane modification enables copper nanoparticles to deposit on the surface of the bamboo rayon fiber evenly and densely. The contact angle and UV transmittance of copper coated bamboo rayon fabric modified with the silane is 139.6°and 0%, respectively. The color of copper coated bamboo rayon fabric is red. The results suggest that the copper coated bamboo rayon fabric modified with the silane possesses excellent hydrophobicity and UV blocking properties. There is a potential application of copper coated bamboo rayon fabrics modified with the silane in superhydrophobic and UV protection textiles.

Acknowledgement

This work was financially supported by Ningbo Municipal Science and Technology Bureau (15H0640) and Chengdu Science and Technology Bureau (2015-HM01-00380-SF).

References

[1] Correia VC, dos Santos V, Rodier LB, et al. Bamboo fiber at macro-, micro-and nanoscale for application as reinforcement. Green Materials: 2016; 4 (1): 41-52.

[2] Yusoff RB, Takagi H, Nakagaito AN. Tensile and flexural properties of polylactic acid-based hybrid green composites reinforced by kenaf, bamboo and coir fibers. Industrial Crops and Products: 2016; 94: 562-573.

[3] Peng L, Guo R, Lan J, et al. Microwave-assisted coating of silver nanoparticles on bamboo rayon fabrics modified with poly (diallyldimethylammonium chloride). Cellulose: 2016; 23 (4): 2677-2688.

[4] Berendjchi A, Khajavi R, Yazdanshenas ME. Fabrication of superhydrophobic and antibacterial surface on cotton fabric by doped silica-based sols with nanoparticles of copper. Nanoscale research letters: 2011; 6 (1): 594.

[5] Kim SH, Oh KW, Bahk JH. Electrochemically synthesized polypyrrole and Cu-plated nylon/spandex for electrotherapeutic pad electrode. Journal of Applied Polymer Science: 2004; 91 (6): 4064-4071.

[6] Jiang S, Miao D, Li A, et al. Adhesion and durability of Cu film on polyester fabric prepared by finishing treatment with polyester-polyurethane and aqueous acrylate. Fibers and Polymers: 2016; 17 (9): 1397-1402.

[7] Peng L, Guo R, Lan J, et al. Silver nanoparticles coating on silk fabric with pretreatment of 3-aminopropyltrimethoxysilane in supercritical carbon dioxide. Journal of Industrial Textiles: 2016; 4-14.

[8] Khimi SR, Pickering KL. The effect of silane coupling agent on the dynamic mechanical properties of iron sand/natural rubber magnetorheological elastomers. Composites Part B: Engineering: 2016; 90: 115-125.

[9] Pierson JF, Wiederkehr D, Billard A. Reactive magnetron sputtering of copper, silver, and gold. Thin Solid Films: 2005; 478 (1): 196-205.

[10] Hussain S, Erikson H, Kongi N, et al. Heat-treatment effects on the ORR activity of Pt nanoparticles deposited on multi-walled carbon nanotubes using magnetron sputtering technique. International Journal of Hydrogen Energy: 2016.

[11] Stranak V, Wulff H, Rebl H, et al. Deposition of thin titanium–copper films with antimicrobial effect by advanced magnetron sputtering methods. Materials Science and Engineering: C: 2011; 31 (7): 1512-1519.

[12] Tengdelius L, Broitman E, Lu J, et al. Hard and elastic epitaxial ZrB_2 thin films on Al_2O_3 (0001) substrates deposited by magnetron sputtering from a ZrB_2 compound target. Acta Materialia: 2016; 111: 166-172.

[13] Suryaprabha T, Sethuraman MG. Fabrication of copper-based superhydrophobic self-cleaning antibacterial coating over cotton fabric. Cellulose: 2017; 24 (1): 395-407.

Microwave Irradiation Time Effect on Expandable Graphite for Graphene Oxide Synthesis

Yang-Yang Fan[1], Heng Zhai[2], Yi Li[2*]

[1,2]*School of Materials, University of Manchester, Oxford road, Manchester, M13 9PL, UK*

[*]Corresponding author's email: henry.yili@manchester.ac.uk

Abstract

Graphene oxide was nowadays wildly investigated by scientists due to extraordinary properties of graphene and its derivatives. With rapid development of synthesising graphene oxide, there are still few issues to be discussed. This paper investigated the time effect of microwave irradiation on expanded graphite for further oxidation reaction. Expandable graphite was expanded by microwave irradiation. Microwave irradiation time of 0s, 10s, 20s, 30s and 40s were compared confirming that with the time increasing, expansion ratio also increased. After 30s, the rising volume became inconspicuous. The graphene oxide was then oxidized by mixing 20s, 3 0s and 40s expanded graphite with potassium permanganate in concentrated sulfuric acid. No obvious change in chemistry structure under different irradiation time, while mean size of which grew from 20s to 30s and dropped slightly after another 10s. This paper investigated the time effect of microwave irradiation on expanded graphite for further oxidation reaction. The results gave the inspiration of industry producing graphene oxide with size and other properties controlled.

Keywords: Expanded Graphite; Microwave; Graphene Oxide; Size Distribution

1. Introduction

Graphene firstly discovered by A.K. Geim and K.S. Novoselov [1, 2] has been reported to have many attractive properties, with an ideal tensile strength of 130 GPa, elastic modulus of 1.0 TPa, remarkable flexibility, electrical conductivity of 10^8 S/m, thermal conductivity of 5000 $Wm^{-1}K^{-1}$, transmittance of almost 97% and high temperature, chemical resistance. Due to its extraordinary properties, lots of efforts have been made in preparation of graphene and its derivatives. For pure graphene, many methods have already been developed, i.e. micro-mechanical exfoliation, liquid phase exfoliation, explosion, chemical vapor deposition, physical vapor deposition, electrochemical exfoliation and microwave irradiation. However, considering the cost and processibility of monolayer graphene, the application of graphene is restricted in industry application. To acquire better product with lower cost, graphene oxide as an oxidation product of graphene has great processibility, mechanical properties and potentially electrical conductivity when reduced [3]. The most commonly used method of graphene oxide preparation is based on Hummers' method. 10g graphite was added with 5g sodium nitrate and 230ml concentrated sulfuric acid to form intercalated graphite. After being intercalated, 30g potassium permanganate was added to the suspension to oxidize graphite [4]. Full exfoliation from graphite oxide material into individual, single-layer graphene oxide sheets was then achieved by sonication. It was then improved by Marcano, et al, with a higher extend of oxidation of graphene oxide and a more regular structure [5]. Another process of producing single sheet graphene oxide was firstly reported by Schniepp, et al [6, 7]. The first step is similar to Hummers' method, mixing graphite, sulfuric acid, nitric acid, and potassium chlorate for 96h. The intercalated graphite was then put into 1050°C tube furnace for 30s. The graphite oxide was then expanded into graphene oxide. This method brought in further benefits, not only efficient but also better structural property.

As another way to expand graphite, microwave irradiation is fast and low cost. It is also much easier to set up the experiments compared with tube furnace. Xin, et al [8], utilized microwave oven to exfoliate expandable graphite. The reactant was then oxidized using Hummers' method. The stable monolayer graphene oxide was generated after oxidization and gentle shaking. Differ from traditional methods, graphene oxide generated in this method has large flake size as it is not sonicated. In this method, graphene oxide showed good physical structure and electrical, thermal conductivity after reduction. However, potentially similar to tube furnace, properties of graphene oxide synthesised by expanded graphite can be affected by heat irradiation time. How the microwave irradiation time affects synthesis of graphene oxide is still not very clear. In this paper, effects of irradiation time were investigated systematically. The aim of this research is to get proper graphene oxide size and quality after oxidation reaction with microwave irradiation time controlled for industry application.

2. Method

0.5g Expandable graphite flakes (>300μ, Sigma Aldrich) was irradiated by commercial microwave (700w) for 20s, 30s, 40s. The expanded graphite acquired was then put into a 3 neck flask with 100ml concentrated sulfuric acid (95%, Fisher Scientific). 5g potassium permanganate was slowly added into flask with a magnetic stirrer at 200rpm. The reaction was run for 24h in water bath. Deionized water was slowly added into the flask in an ice bath, followed by 25ml H_2O_2 (30%, Fisher Scientific). After the flask was continuously stirred for another 30min, 100ml HCl (10%, Fisher Scientific) was mixed into solution. The dispersion was washed by centrifuge for 5 times at a speed of 14500rpm until PH is over 6 and undergoes a 3000rpm to remove unreacted graphite or multiple layers graphite oxide.

In order to investigate size distribution of graphene oxide, SEM and ImageJ were applied to analyse graphene flakes spin coated on silicon wafer. Atomic force microscope (AFM) (Dimensional 3100) was used for graphene oxide thickness test. Graphene oxide dispersion (50μg/ml) was coated on a pre-cleaned silicon wafer with 300nm silicon dioxide coating using spinning coater at a spin rate of 3000rpm. Fourier transform infrared spectroscopy (FTIR, NICOLET 5700) was tested with graphene oxide and potassium bromide well mixed, then pressed into tablet. Raman spectroscope with 514.5nm laser was used to analysis chemical structure of graphene oxide on silicon wafer.

3. Results and Discussion

Fig.1 SEM image of expandable graphite treated by microwave irradiation for (a) 0s, (b) 10s, (c) 20s, (d) 30s, (e) 40s (scale bar 100μm), digital image of (f) expanded graphite treated by microwave irradiation for 0s, 10s, 20s, 30s and 40s in vials

In Fig.1 (a) to (e), expanded graphite of 0s, 10s, 20s, 30s and 40s irradiation time was compared. It is obvious that when irradiation time increase at early age, the size of expanded graphite worm increased slightly from 0s to 10s, as graphite flakes were partially expanded. From 10s to 20s, most intercalated layers were expanded, which dramatically increased the size of graphite worm. Further irradiation time from 20s to 40s may cause an over sintering problem, which tend to further decrease the size of graphite. Moreover, the over heat generated may cause damage to microwave oven. In this case, the maximum time of irradiation is limited to be 40s. Fig.1 (f) showed the trend of expansion ratio during irradiation. The expandable graphite was not expanded much at first 10s. The volume of graphite then sharply increased due to release of gas intercalated from 10s to 20s. As most graphite had been expanded after 20s, the effect forming thermal cracking might dominate the volume of expanded graphite worm.

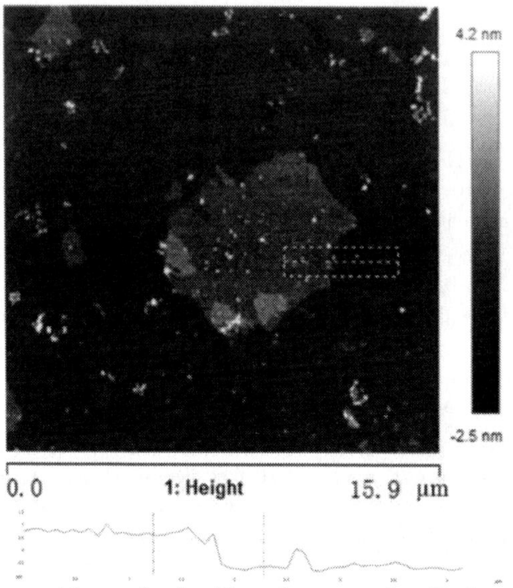

Fig.2 (a) AFM image of graphene oxide synthesized by microwave-irradiation then chemical-oxidation method

It is commonly agreed that monolayer of graphene is about 0.34nm [1], while as introducing of oxygen-containing groups, the thickness of graphene oxide is around 0.8nm-1.3nm. AFM image of graphene oxide indicated that the graphene oxide flakes synthesized are mostly monolayer graphene oxide (Fig.2a).

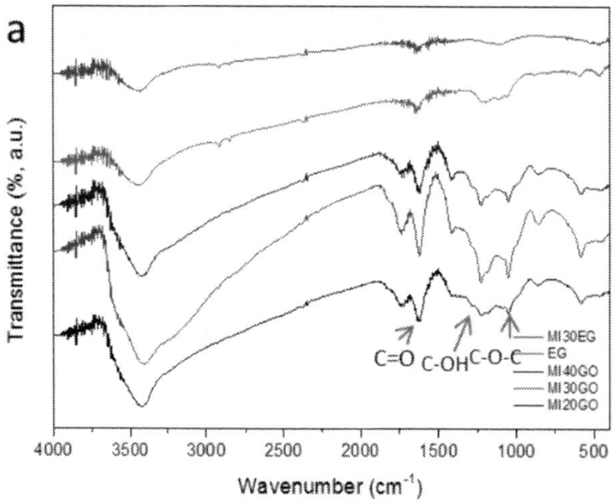

Fig.3 Fourier transform infrared (FTIR) spectrum of (a) MI20GO, MI30GO and MI40GO

The FTIR spectroscopy results in Fig.3 a, b confirmed the basic chemical structure of microwave-irradiated-20s expanded graphite (MI20EG), microwave-irradiated-20s graphene oxide (MI20GO), microwave-irradiated-30s graphene oxide (MI30GO), and microwave-irradiated-40s graphene oxide (MI40GO). The peak at 1720 cm^{-1} confirmed the existence of C=O stretching bond. This indicated after microwave irradiation, the general amount of C=O bond did not change a lot, while during oxidation, it increased sharply. The peaks at around 1100 cm^{-1}, 1250 cm^{-1} were provided by C-OH and C-O-C stretching. The existence of C=O, C-OH and C-O-C bonds supported that all the three samples have similar chemical structure with epoxide, hydroxyl and carboxyl groups.

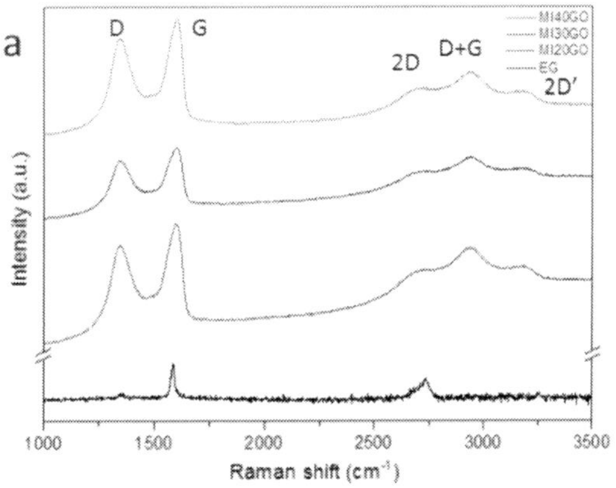

Fig.3 (a) Raman spectroscopy of MI20GO, MI30GO and MI40GO

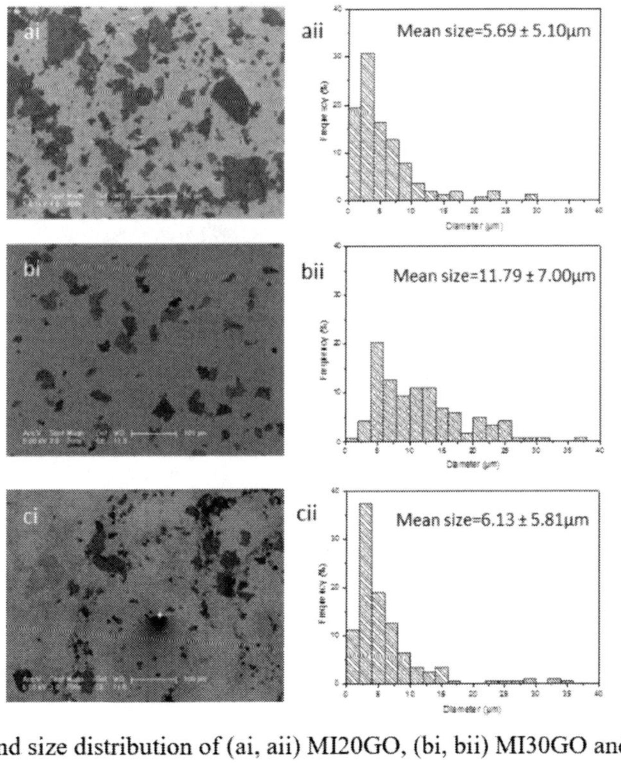

Fig.4 SEM image and size distribution of (ai, aii) MI20GO, (bi, bii) MI30GO and (ci, cii) MI40GO, scale bar seen in image

Raman spectroscopy was utilized to characterize chemical structure difference between MI20GO, MI30GO, MI40GO and EG. The two most intense features of Raman spectra of graphene and bulk graphite are the G peak at 1580 cm^{-1} and a band at 2700 cm^{-1}. G peak is normally called defect-free graphite peak. Other peak at around 1350 cm^{-1} is defected graphite, called D band peak. According to Kudin [9], the G band of graphite is 1580 cm-1, with the increasing of oxidation extend, the G band will be located at a higher frequency. In our Raman spectroscopy, G band peak in MI20GO, MI30GO, MI40GO are at 1598.6, 1601.4 and 1604.1 cm^{-1} respectively. It can be predicted that the order of oxidation extend is MI40GO > MI30GO > MI20GO. ID/IG represented a disorder extend, the higher it is, the material is more disordered. According to the results analysis, ID/IG of MI20GO, MI30GO, MI40GO are 0.84, 0.83 and 0.85, i.e. the MI30GO has best chemical structure of carbon ring contained.

All the 3 specimens showed a trend that small graphene oxide flakes dominated the size distribution. According to Fig. 4 aii, bii, cii, with increasing of irradiation time, mean size of graphene oxide increased from 5.69 ± 5.10 μm to 11.79 ± 7.00 μm, and then decreased to 6.13 ± 5.81 μm. The distribution of MI20GO size showed an asymmetrical peak, where the frequency of small size was quite high. MI30GO showed a broader shoulder with more large size graphene oxide flakes resulting in a mean size of 11.79 ± 7.00 μm. It is also obvious that MI30GO has much more flakes with an area larger than 10 μm, as the frequency of flakes higher than 10 μm is quite high compared with the other two samples. MI40GO, however, showed a similar trend to MI20GO with more small size graphene oxide flakes (especially from 2 μm to 4 μm). The results of size distribution showed that despite of inhomogeneity of the GO size distribution, the average aspect ratio of MI20GO, and MI30GO and MI40GO are calculated to be 7113, 14750 and 7663.

This can be explained that low irradiation time gives more small size few-layer graphite flakes exfoliated from graphite and less large-size graphite exfoliated, i.e. insufficient expanding towards expandable graphite. In one hand, this resulted in more graphite oxide and unreacted graphite formed after oxidation reaction, which will be removed during centrifugation. In the other hand, with short irradiation time, only small size graphite was exfoliated from expandable graphite, while with 30-secend-irradiation more large size few-layer graphite was exfoliated and was oxidized into graphene. It is assumed that when irradiation time is too long, no more expansion between graphite layers will happen. The heat that is remaining and generated tends to contribute to thermal cracking on graphite. This will decrease its size into small flakes dramatically.

4. Conclusion

The time effect of microwave irradiation was investigated in this research. Expandable graphite started to expand from 0s to10s. Most graphite was expanded from 10s to 20s, while after 20s the effect of thermal cracking starts to dominate the irradiation. With 20s, 30s, 40s irradiated expanded graphite, the reactant graphene oxide has similar chemical structure. However, the size of graphene oxide tended to increase and then decrease as time going. Among all, MI30GO has largest graphene oxide mean size and a broader size distribution. This research may help with controlling size of synthesis graphene oxide flakes, which can be applied in some certain area, e.g. preparation of graphene oxide fibres, graphene ink and graphene films.

5. Acknowledgement

I would like to thank the support of the EU Horizon 2020 and University of Manchester through projects with project codes 644268-ETEXWELD-H2020-MSCA-RISE-2014, AA14512 (UMRI project "Graphene-Smart Textiles E-Healthcare Network"), R1119938 and AA01906.

References

[1] Novoselov KS, Geim AK, Morozov SV, Jiang D, Zhang Y, Dubonos SV, Grigorieva IV, Firsov AA. Electric field effect in atomically thin carbon films. Science: 2004; 306: 666-669.

[2] Novoselov KS, Geim AK, Morozov SV, Jiang D, Katsnelson MI, Grigorieva IV, Dubonos SV,Firsov AA. Two-dimensional gas of massless dirac fermions in graphene. Nature: 2005; 438: 197-200.

[3] Zhu Y, Murali S, Cai W, Li X, Suk JW, Potts JR, Ruoff RS. Graphene and graphene oxide: Synthesis, properties, and applications. Advanced materials: 2010; 22: 3906-3924.

[4] Hummers WS, Offeman RE. Preparation of graphitic oxide. J Am Chem Soc: 1958; 80: 1339-1339.
[5] Marcano DC, Kosynkin DV, Berlin JM, Sinitskii A, Sun ZZ, Slesarev A, Alemany LB, Lu W,Tour JM. Improved synthesis of graphene oxide. Acs Nano: 2010; 4: 4806-4814.
[6] Schniepp HC, Li JL, McAllister MJ, Sai H, Herrera-Alonso M, Adamson DH, Prud'homme RK, Car R, Saville DA,Aksay IA. Functionalized single graphene sheets derived from splitting graphite oxide. J Phys Chem B: 2006; 110: 8535-8539.
[7] McAllister MJ, Li JL, Adamson DH, Schniepp HC, Abdala AA, Liu J, Herrera-Alonso M, Milius DL, Car R, Prud'homme RK,Aksay IA. Single sheet functionalized graphene by oxidation and thermal expansion of graphite. Chem Mater: 2007; 19: 4396-4404.
[8] Xin GQ, Yao TK, Sun HT, Scott SM, Shao DL, Wang GK, Lian J. Highly thermally conductive and mechanically strong graphene fibers. Science: 2015; 349: 1083-1087.
[9] Kudin KN, Ozbas B, Schniepp HC, Prud'homme RK, Aksay IA,Car R. Raman spectra of graphite oxide and functionalized graphene sheets. Nano Letters: 2008; 8: 36-41.

Preparation and Properties of Magnetic Collecting Umbrella for Annulus Logging

Wen-Jun Dou, Ru-Quan Zhang*

School of Textile Science and Engineering, Wuhan Textile University, Wuhan, China 430200

Corresponding author's email: zhangruquan@wtu.edu.cn

Abstract

A kind of magnetic collecting umbrella fabric with water-proof function is prepared by magnetic finishing. Nd2Fe14B micro-particle with waterborne-polyurethane is attached to the base fabric surface, and magnetic property of Nd2Fe14B micro-particle and effect of times, the concentration of $Nd_2Fe_{14}B$ micro-particle, the magnetizing voltage on magnetic properties are studied. The result showed that the $Nd_2Fe_{14}B$ micro-particle is the permanent magnet, whose residual flux Mr is 44.659emu/g, and saturation flux Ms is 79.554emu/g. At the condition of 4 times of spraying, 35wt% of Nd2Fe14B concentration and 1200v magnetizing voltage, the final residual flux is 7.1 mT. The contact angle is all above 110°after spraying. Three is the best time of spraying.

Keyword: $Nd_2Fe_{14}B$ Particles; Magnetic Property; Finishing; Residual Flux; Contact Angle

1. Introduction

Annulus logging is type of oil well production logging method which directly acquires a set of dynamic parameters by winch pulling the cables and test equipment from the annulus into the well, in the cases of non-stop production [1]. As an important part of the annulus logging instrument, the collecting umbrella will carry out the process of opening umbrella, testing, closing umbrella, obtaining the number of turbine impulses and water holding response frequency [2]. Collecting umbrella mainly is made from umbrella fabric and umbrella ribs. The role of the umbrella rib is to support the collecting umbrella when it is opening or folding. Due to the column-like structure and the annular space, in the process of opening umbrella and closing umbrella, collecting umbrella often collides with the tubing, which makes steel burr easy to wear off the banding line, scrape off the umbrella pieces and destroy the umbrella, which will ultimately cause the annulus logging to fail [3-4].

In this article, the traditional method of fixing the collecting umbrella fabric with the umbrella ribs by mechanical friction is abandoned. With the new coating method, the micro-particle is attached to the base fabric surface, enabling the base fabric to be waterproof. In the external magnetic field, the atomic magnetic moments in the particles rearrange their direction along the external magnetic fields, after removing the external magnetic field, the magnetic particles can still maintain a strong magnetic, the umbrella fabric closely fits in the collecting umbrella ribs [5].

2. Experimental Methods and Procedures

2.1 Materials and Apparatus of Experiment

In this experiment, a plain weave fabric made from polyamide (nylon 66) was used as base fabric. With the 50D liner density of warp yarns and the 40D linear density of weft yarns respectively. The weight per square meter of fabric is $58g/m^2$.

The $N_2dFe_{14}B$ magnetic micro-particle was purchased from Guangzhou new noble transmission components co., LTD, with particle size from 38μm to 78μm and a purity of more than 99%. Waterborne-polyurethane (WPU), model TF667, was purchased from Zhejiang Transfar CO., LTD. The basic performance indicators was shown in table 1.

The PM-3633 waterproof finishing agent was purchased from 3M China co., LTD.

The air spray gun, model W-71-3G, with a standard 400 mL spray gun cans, was purchased from Japan iwata co., LTD. Instrument used in this study included JB25-C precision power electric blender, R-3 automatic stereotypes dryer and OTS-750-30L air compressor.

Table.1 basic performance indicators of WPU

Project	Solid content	Viscosity	100%modulus
Index	35%±1%	>80CPS	30kg/cm^2

2.2 Fabrics Magnetic Spraying

Table.2 materials rates

No.	Materials	Mass ratio
1	$Nd_2Fe_{14}B$:WPU	35:65
2	$Nd_2Fe_{14}B$: PM-3633:WPU	33:2:65
3	$Nd_2Fe_{14}B$: PM-3633:WPU	31:4:65
4	$Nd_2Fe_{14}B$: PM-3633:WPU	29:6:65

The spray nozzle diameter is 1.5mm, the spraying distance is 200mm, the spraying air pressure 3.0Mpa, the amount of air 230L / min, the amount of paint sprayed 190mL / min, spray width 185mm, the time of spraying is 30s.

At the end of first spraying, the basic fabric was sent to dry at the temperature of 100℃ for 3 minutes, bake at the temperature of 150℃ for 2 minutes, which means the first spraying and ready for the second spraying. The same as first spraying, the second spraying last for 30s, followed by drying and baking. The third and fourth spraying were exactly same with the first and second spraying, and also followed by drying and baking [6].

2.4 Magnetization

The magnetization step was carried out in a magnetizing apparatus (MZM-1530, Hangzhou Ming Zhe magnetic Technology Co., Ltd), by aligning the base fabric with the magnetizing fixture and adjusting magnetization voltage from 600V to 1200V to make the basic fabric different magnetic flux.

2.5 Characterization

The $Nd_2Fe_{14}B$ magnetic micro-particle was characterized by vibrating sample magnetometer (VSM, Lakeshore-EM4-HVA) with an applied filed of 15KOe at room temperature.

For the studies of magnetic property, a Tesla meter was used (TM-801, KANETEC co.) to test the remanent magnetism at the room temperature.

Waterproof performance was carried by hydro tester (FX 3000-Ⅳ,TEXTEST) and fabric contact angle device at the room temperature.

3. Results and Analysis

3.1 $Nd_2Fe_{14}B$ Magnetic Micro-particle Hysteresis Loop Analysis

As Fig.1 showed the $Nd_2Fe_{14}B$ magnetic micro-particle hysteresis loop. The magnetic moments of each different magnetic domain were in the statement of mixed and disordered without any magnetic treatment. The magnetic moments neutralized each other, the vector sum was zero, presenting the statement of demagnetization. From Fig.1, we can see that the magnetic moments of the $Nd_2Fe_{14}B$ magnetic micro-particle increased with the increasing of the external magnetic field. In a small external magnetic field, direction of the internal magnetic domain tent to be consistent with the increase of external magnetic field, so that the magnetic domain of each different magnetic moments would no longer neutralize each other, showing a weak magnetic property. [7] External magnetic field continued to increasing, the external magnetic field-oriented magnetic domain volume constantly increased, external magnetic field-disoriented magnetic domain volume constantly decreased. As a result, the total magnetic moments increased with the external magnetic field in the amount consistent with the external magnetic field in the direction. At the point A, the external magnetic field was 13591.4G, and the mass ratio of the magnetic micro-particle was saturated to 72.4195emu/g. Displacement of the domain wall accelerated with the increase of the external magnetic field, and finally the magnetic domain external magnetic field-disoriented disappeared completely, reaching the saturation point Ms where the saturation mass ratio of the $Nd_2Fe_{14}B$ magnetic micro-particle was 79.554emu/g. Reducing external magnetic field, magnetic moments was not along the original magnetization curve, showing the hysteresis phenomenon. The external magnetic field reduced to zero, residual flux Mr was 44.659emu/g. Magnetic moments continued to reducing with the increase of the reverse magnetic field. When magnetic moments reduced to zero, reaching the point -Hc, the external magnetic field intensity was 6334.5 G, coercive force Hc was 6334.5 G. Continue to increasing reverse magnetic field, located in the third quadrant of the hysteresis loop was the same as the hysteresis loop located in the first quadrant, merely, the opposite directions. Eventually the saturation mass ratio of $Nd_2Fe_{14}B$ magnetic micro-particle magnetic moments was 79.554emu/g, coercive force was 6334.5 g, remanence 44.659emu/g. [8]

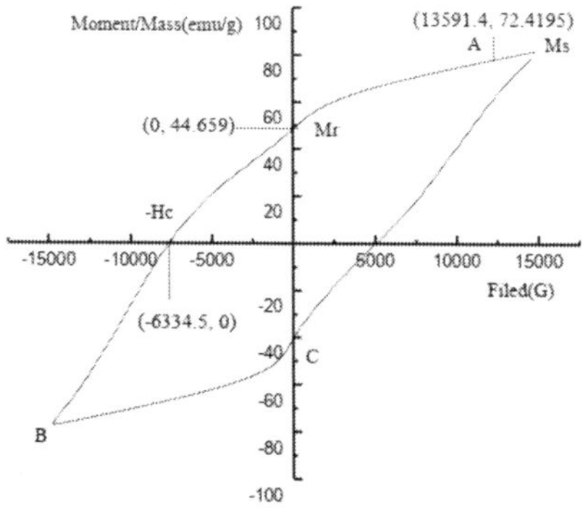

Fig.1 $Nd_2Fe_{14}B$ magnetic micro-particle hysteresis loop

3.2 Effect of Magnetizing Voltage on Magnetic Properties

The magnetization process was actually the change of the direction of the magnetic domain. The current pulse generated by the magnetizer generates a strong magnetic field in the coil making the samples exposed in external magnetic field, whose spontaneous magnetization direction in line with the direction of external magnetic field or a small angle. The volume of the magnetic domain increases with the external magnetic field, so that magnetization direction of the magnetic domain was further turned to the direction of the external magnetic field; the other spontaneous magnetization whose direction opposite to external magnetic field or a large angle, volume of magnetic domain gradually reduced, showing external magnetism.

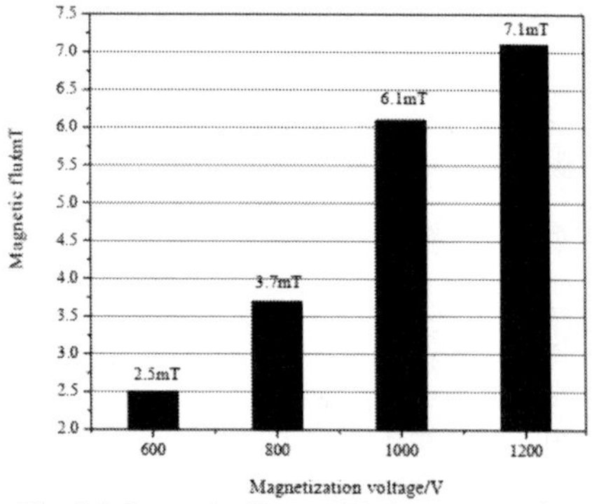

Fig.2 Magnetic flux of 4 times spraying

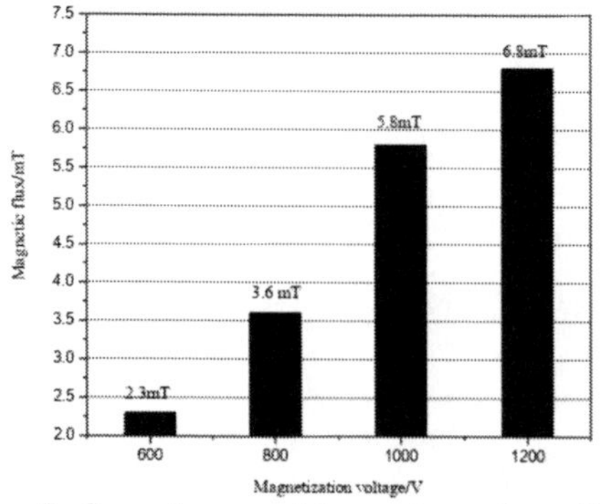

Fig.3 Magnetic flux of magnetizing voltage from 600v to 1200v

Fig.2 shows the remanent magnetism characterization of magnetization voltage in the situation of four times of spraying, magnetic micro-particle concentration of 35wt %. Fig.3 shows the remanent magnetism characterization of magnetization voltage in the situation of four times of spraying, magnetic micro-particle concentration of 33wt %.

Residual magnetic flux increased with the increasing of magnetizing voltage, through a large number of early experiments. It was found that at 600v of magnetizing voltage, the magnetization had preliminary effect, showing a weak external magnetism, then residual magnetic flux had increased significantly, for the 2.5mT. The magnetic flux increased with increasing of magnetizing voltage, which was in the magnetic field generated by the magnetizing apparatus, the magnetic moments in $Nd_2Fe_{14}B$ magnetic micro-particles had broken the original balance in the magnetic field and gradually inclined to the external magnetic field direction. When magnetizing voltage increased to 800V, the residual magnetic flux gradually increased to 3.7mT, with an increase of 1.2mT, which the increase rate was 50%. When magnetizing voltage was 1000V, the residual magnetic flux was 6.7mT, with a significant increase of 4.2mT, which the increase rate was 168%. When magnetizing voltage was 1200V, the residual magnetic flux was 7.1mT, with residual magnetic flux increased by 4.6mT, which the rate of increase was 184%. However, for 6.2mT of the residual magnetic flux for 1000V, residual magnetic flux increased only by 1mT, with an increase rate of only 16%. This was due to the low magnetizing voltage, strength of magnetic field generated by the magnetizing apparatus was low, so the magnetic domain walls were subject to internal stress and other obstacles in the pinning effect, with domain wall movement blocked, vector of the magnetic moments was small. The magnetic domain walls try to overcome the role of pinning to move, due to weak magnetic field generated by the low magnetizing voltage such as 800V where domain was unable to get out of the obstacles. The effect of pinning was smaller than that of the external magnetic field, the magnetic domain wall was separated from the pinning action and got out of the obstacle and moved in accordance with the direction of the external magnetic field when the magnetic field was further enhanced at higher voltages such as 1000v and 1200v. The volume of magnetic domain whose direction in line with the direction of external magnetic field increased until the maximum. The volume of magnetic domain whose direction opposite to external magnetic field decreases until disappear to the magnetic saturation point. Removing the external magnetic field, due to the demagnetizing field which was inevitable caused by material itself, whose direction was opposite to the whole total magnetic moments, residual magnetic flux was less than the saturation flux.

3.3 Effect of Times of Spraying on Magnetic Properties

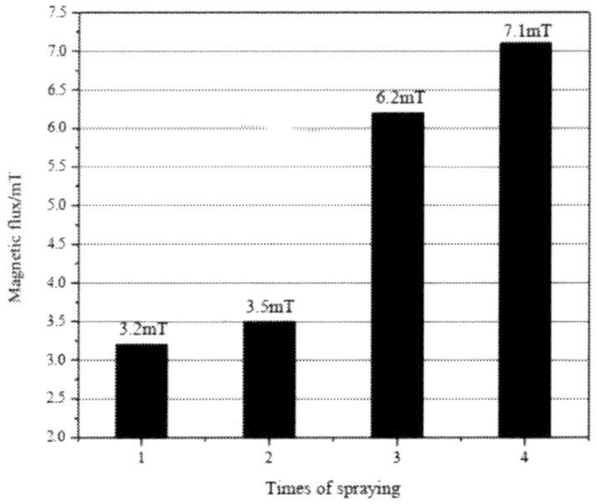

Fig.4 1200v voltage, 35wt % magnetic concentration

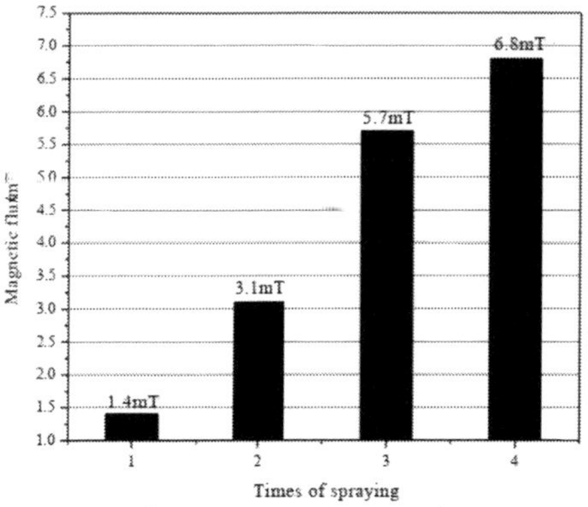

Fig.5 1200v voltage, 33wt % magnetic concentration

Fig.4 shows the remanent magnetism characterization of times of spraying in the situation of 1200v magnetizing voltage, magnetic micro-particle concentration of 35wt %. Fig.5 shows the remanent magnetism characterization of times of spraying in the situation of 1200v magnetizing voltage, magnetic micro-particle concentration of 33wt %.

In the case of only one times of spraying, 1200v magnetizing voltage, the residual magnetic flux was 3.2mt. Although $Nd_2Fe_{14}B$ magnetic micro-particle was exposed in the strong magnetizing voltage and external magnetic field, due to few content of magnetic particle on the base fabric, total magnetic moments vector was small, residual magnetic flux was small too. 3 times of spraying, the residual magnetic flux was 6.2mT, with an increase of 3mT, which an increase rate was 93%, the most obvious increase of all.

Compared with only one time of spraying, content of three layers of magnetic particle increased by 3 times, in 1200v magnetization voltage, total magnetic moments vector within the magnetic particle greatly increased, along the external magnetic field direction. 4 times of spraying, the residual magnetic flux was 7.1mT. The increase rate was only 14% for that of 3 times of spraying, and the increase had become gentle. Therefore, 3 times was the best spraying time.

3.4 Effect of Concentration of PM-3633 on the Water-proof Properties

Water droplets spread out completely in the fabric surface, forming a layer of water film, fully infiltrating the fabric without any finishing treatment. The contact angle was 0. The contact angles of sprayed fabric with different times of spraying were shown in the Fig.7.

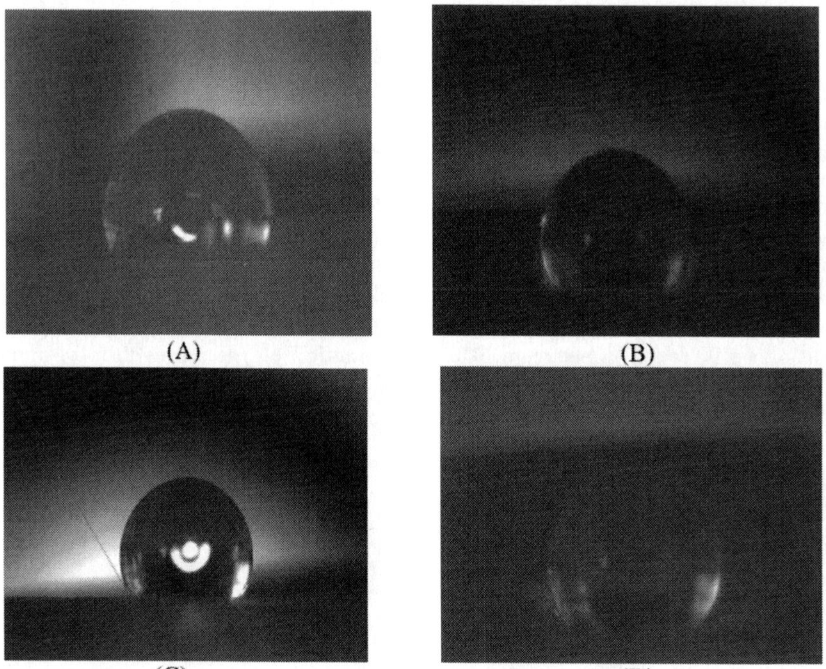

Fig.7 A (1 time) B (2 time) C (3time) D (4time)

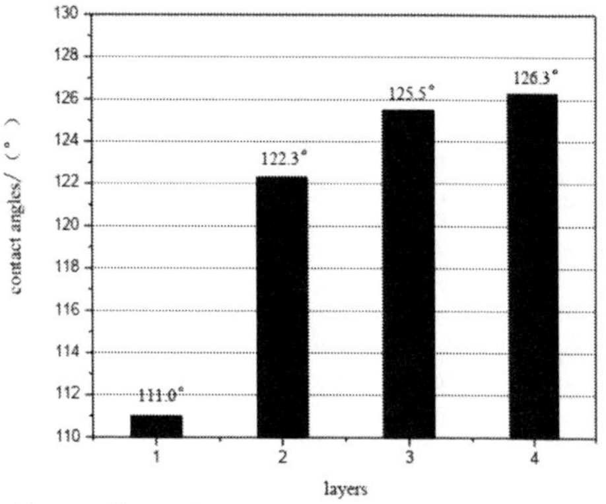

Fig.8 effect of spraying times on contact angle

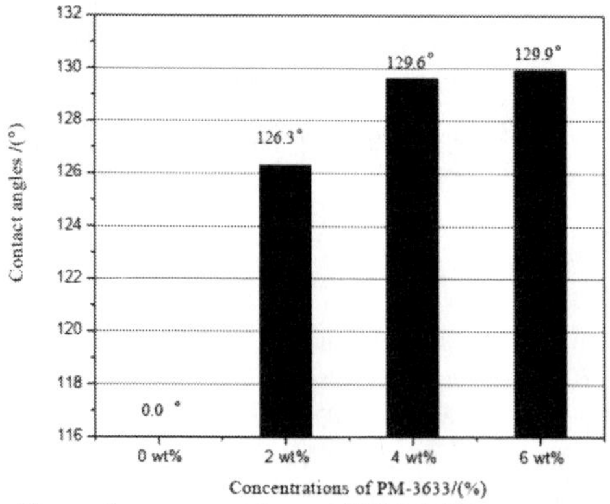

Fig.9 effect of concentrations of PM-3633 on contact angle

As can be seen from Fig.8, the contact angle was 110° when only one time spraying. It was a great ascension of waterproof property, the change from hydrophilic to water repellent. The times of spraying increased from one to two, the contact angle had a relatively large increase, fabric had a contact angle of 122.3°, with an increase of 12.3°. Increasing the times of spraying, from two to three times, the contact angle increased to 125.5°, at this time the increase of the contact angle had become flat, only increase of 3.2°. The times of spraying increased from 3 to 4, and the contact angle increased by only 0.8 °. Therefore, the three times was the best spraying time.

The contact angle of different concentration of PM-3633 in the situation of 2 times spraying was shown in Fig.2. The contact angle of original fabric was 0, meaning no waterproof function at all. When the concentration of PM-3633 was 2wt%, the contact angle increased sharply to 126.3°. When the concentration increased to 4wt %, the increase of the contact angle was 3.3°. Continue to increase the concentration of PM-3633, reaching 6wt %, the change of contact angle of fabric was not obvious, with only increased by 0.3°.

4. Conclusions

1) After magnetic finishing, the fabric gained the magnetic property, at the situation of 4 times of spraying, under 35wt% of Nd2Fe14B concentration and 1200v magnetizing voltage, and the final residual flux was 7.1 mT.

2) Water-proof properties was raised, contact angle was all above 110° after finishing. At the situation of 6wt% PM-3633 and 4times of spraying, the contact angle reached 129.9°.

5. Acknowledgements

The support of the project was sponsored by Hubei Province Natural Science Fund Project (Project number: 2013CFA090). The authors gratefully acknowledge this support.

References

[1] Wang YL. Research and application of annulus logging technology [J]. Science and Technology Innovation Review: 2013; (33): 31-32.

[2] Dong YL, Yang Y, Song R, et al. Application Effect Analysis of Annulus Logging Technology. Journal of Daqing Petroleum Geology and Exploitation: 2005; 24 (5): 96-98.

[3] Duan YL, Guo HM, Yuan W, et al. Application of Centralized Annulus Logging Technology in Jilin Oilfield [J]. Journal of Petroleum and Natural Gas: 2014; (9): 74-77.

[4] Meng QW, He W. Technology and Application of Annular Well Logging [J]. Chinese Oil and Chemical Industry Standards and Quality: 2012; 33 (8): 120-120.

[5] Hao XX, Feng QC. Magnetic paper's development [J] Paper and Make Paper: 2003; 26 (4): 44-49.

[6] Chen YR, Yu JY, Chu CY. Effect of Water Repellent Finishing on the Hydrostatic Pressure Resistance of Cotton Fabrics [J]. Journal of Donghua University: 2009; 35 (1): 35-39.

[7] Jiang RL, Liu YJ, Liu YC, et al. Study on dysprosium ferrite magnetic magnetic [J]. Journal of China University of mining: 2003; 32 (6): 698-700.

[8] Mao MY, Zhou Y, Xu F, et al. The research progress of spinel ferrite catalytic reduction of CO_2 [J]. Journal of shantou university (natural science edition): 2013 (2); 20-29.

Study on the Bending Fatigue Fracture Characteristic of High Performance Fibers

Meng-Yun Yang, De-Shan Chen, Wen-Bin Li, Guang-Ming Cai*

College of textile, Wuhan Textile University, YangGuang Avenue, WuHan, 430073, Hubei, China

*Corresponding author's email: guangmingcai2006@163.com

Abstract

The damage of high performance fibers occurs easily in bending fatigue. In this paper, we discussed the effect of pre-tension and bending angle on the bending fatigue fracture characteristic of high performance fibers through a new fatigue testing method. The experimental results indicated that the fracture characteristic of fibers are different at different re-tension and bending angle. The morphologies characteristic obtained from the optical apparatus can be used to explain the bending failure mechanisms of high performance fibers.

Keywords: High Performance Fibers; Bending Fatigue; Fracture Characteristic

1. Introduction

High performance flexible fibers such as aramid, PBO, and UHMWPE fibers have high strength, high modulus, and high temperature resistant properties. They are widely used in various industries, such as the military, aeronautics, and aerospace. However, the bending strength, shearing strength and shearing modulus of high performance fibers are comparatively low [1], which could greatly limit applications in various fields [2]. Previous studies show that high performance fibers exhibit excellent tensile strengths and high modulus, however, the compressive and shearing strength and modulus are comparatively low [3-5]. An important method to illustrate the bending property of fiber is bending fatigue test [6], which test the major deformation caused by the bending force. In this paper, a new equipment developed by authers [7], which can perform the bending fatigue test under given bending angle, pre-tension and frequency. The bending fatigue fracture characteristic of single high performance fiber is measured and observed by a microscope. The purpose of this study was to investigate the effects of pre-tension and bending angle on the bending fatigue fracture characteristic of high performance fibers using this new apparatus.

2. Experimental

2.1 Testing Apparatus

A photo of a new bending fatigue test apparatus used in this study is shown in Fig.1. Fig.1 displays the details of the bending system, which includes the following parts: the 1) upper jaw, 2) positioning pin, 3) lower jaw, 4) running plane, 5) microscope, 6) CCD camera, 7) temperature sensor, 8) heater and 9) force sensor. The jaws are used to clamp fiber samples. The positioning pin is used to hold the fiber and fix the bending point under bending process. The running plane can bend the fiber by rotating different angles. The microscope and CCD camera provide the optical system from which the fiber can be observed. The heater and temperature sensor are used to control the temperatures of the bending point. The temperature can be regulated electrically from room temperature up to 400℃. The force sensor is used to record and set the pre-tension, the maximum of which is set at 4000 mN and has accuracy of 0.2 mN.

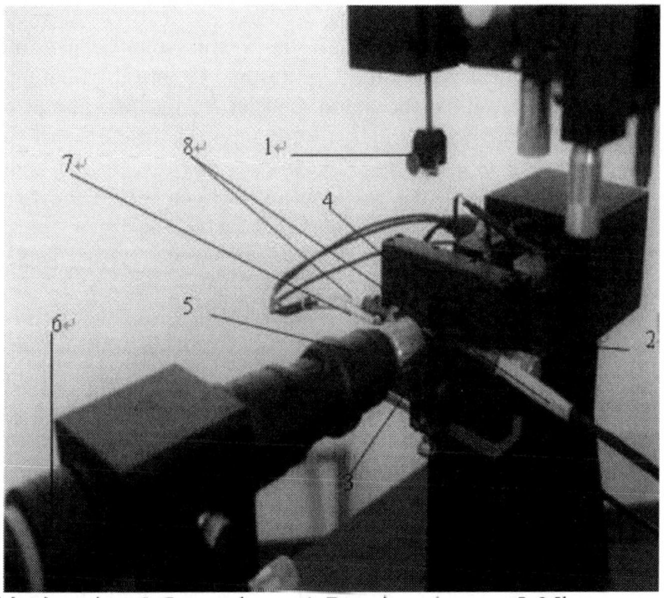

1. Upper jaw 2. Positioning pin 3. Lower jaw 4. Running plane 5. Microscope 6. CCD camera 7. Temperature Sensor 8. Heater 9. Force sensor

Fig.1 The photo of bending fatigue test apparatus

2.2 Experiment Materials

The Kevlar 49, PBO, Nomex, UHMW-PE fibers were chosen for the single-fiber bending fatigue test. The linear densities of the four kinds of fibers are 2.2 dtex, 4 dtex, 2.2 dtex and 6 dtex respectively. The length of the fibers used in the experiment is 6 cm.

3. Results and Discussion

3.1 The Bending Fatigue Fracture Characteristic of Kevlar 49 fiber

Fig.2 The fracture morphology of Kevlar 49 fiber (a) the pretension is 7 cN/dtex; bending angle 450 and cyclic number is 7145 (b) the pretension is 7 cN/dtex; bending angle 450 and cyclic number is 8385 (c)the pretension is 4 cN/dtex; bending angle 450 and cyclic number is 35254

The photics micrographs of single Kevlar 49 fiber fractured ends, which have been carefully selected from a very large number of photographs, were presented in this section. Fig.2(a)-(f) show the bending fatigue fracture morphology of Kevlar 49 fiber. As can be seen in Fig.2, the fracture-ends seem to be irregular, but most of the fracture-ends of Kevlar 49 fiber are fibrillar splitting. Fig.2(a) and (b) are the fracture morphology of fiber under the same pretension and bending angle. It can be found that the fracture morphology can be different under the same fatigue conditions. Fig.2(a) has obvious fibrillar splitting, but the difference of fibrillar length is no obvious, which show a typical "brush" fracture morphology. Fig.2(b) has obvious fibrillar splitting, but the difference of fibrillar length is very obvious, which show the axial pull type fracture and the fatigue lifetime is longer than Fig.2(a). The axial pull type fracture indicate that there maybe is resistance bending fatigue. Fig.2(c) is the fractured ends of Kevlar 49 fiber under small pretension, which show a typical friction fracture and no obvious fibrillar splitting. As can be seen that the bending fatigue farcture due to the friction fracture under small peretension. Fig.2(d)-(f) are typical fibrillar splitting fracture ends. From these fracture ends, we can find the bending fatigue fracture of high performance Kevlar fiber is the main feature of fibrillation splitting failure pattern.

3.2 The Bending Fatigue Fracture Characteristic of PBO Fiber

Fig.3 The fracture morphology of PBO fiber (a)the pretension is 8cN/dtex; bending angle 450 and cyclic number is 7046 (b) the pretension is 8cN/dtex; bending angle 450 and cyclic number is 8787 (c)the pretension is 8 cN/dtex; bending angle 450 and cyclic number is 9145

The optical microscopy pictures of the bending fatigue fracture ends of the PBO fiber are illustrated in Fig.3. It can be seen that its similar to the fracture end of Kevlar 49 fiber, which show the greater fibrillation splitting characteristics. As can be seen in Fig.2, the fracture-ends seem to be irregular, some of them are orderly while others are untidy, but most of the fracture-ends of PBO fiber are fibrillar splitting fracture. Fig.3(a) shows the fracture end of PBO fiber is orderly and the difference of fibrillar length is unconspicuous. The fracture end of Fig.3(b) and (c) are typical fibrillation splitting fracture. Compared with fracture-ends of Fig.3(a), the difference of fibrillar length are more obvious than Fig.3(b) and (c), and the bending fatigue life is longer, which indicate the reistance fatigue is better. The morphology of the PBO fiber rupture ends displays greater fibrillation splitting characteristics. When fiber is bent, the outer layer is pulled and the fiber inside layer is pressed. The fractured ribbons of the fiber surface begin to peel off. It is evident that the PBO fibers are broken due to the fibrillation or axial split, which can be explained by the fibrillar structure, the high orientation and the low intermolecular force.

3.3 The Bending Fatigue Fracture Characteristic of UHMW-PE Fiber

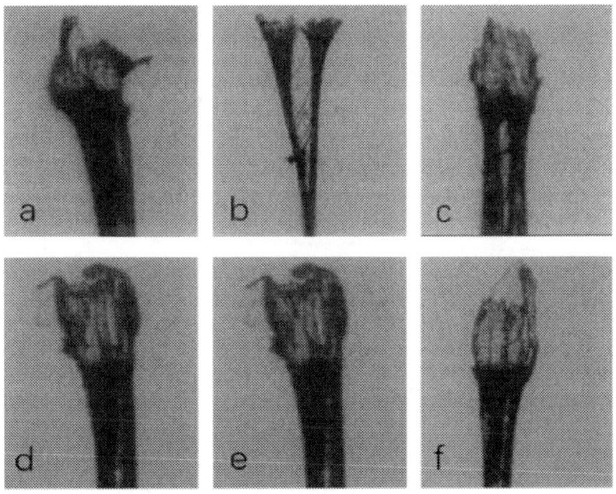

Fig.4 The fracture morphology of UHMWPE fiber (a)the pretension is 8cN/dtex; bending angle 600 and cyclic number is 9426 (b) the pretension is 8cN/dtex; bending angle 600 and cyclic number is10487 (c)the pretension is 8 cN/dtex; bending angle 600 and cyclic number is 11413

The fatigue fracture ends are illustrated in Fig.4. As can be seen from the picture that the fracture ends of the UHMW-PE fiber have mild fibrillation and mild fibrous fracture and fatigue fracture caused by obvious accumulation of plastic deformation. The difference in the morphology between the aramid fibers and the UHMW-PE fibers is evident, the former is a typical fibrillated end and the latter shows the breakage of plastic deformation. Due to the different molecular, supermolecular and fiber structure, the UHMW-PE is more flexible and plastic than the Kevlar and PBO fibers. The fatigue fracture-end of the fiber morphology shows obviously plastic deformation damage rather than the fibrillating damage. At the same time, it can be found that the UHMW-PE fiber has a lower glass transition temperature and melting point, the temperature of the bending point increased when repeatedly bending the fiber, which takes on the performance of plastic and a significant accumulation of the fatigue characteristics of plastic.

3.4 TheBbending Fatigue Fracture Characteristic of Nomex Fiber

Fig.5 shows the fracture ends of the Nomex fiber. It can be found that the fracture ends have no apparent separation and splitting of the micro-fiber, which shows a long tongue fracture end (Fig.5(b)) and tearing-type fracture end (Fig.5(c)). The main reasons are the Nomex fiber has greater breaking elongation, flexibility, and the elongation of internal and external layers are different. A small part of micro fibers may lose during the repeated fiber bending period.

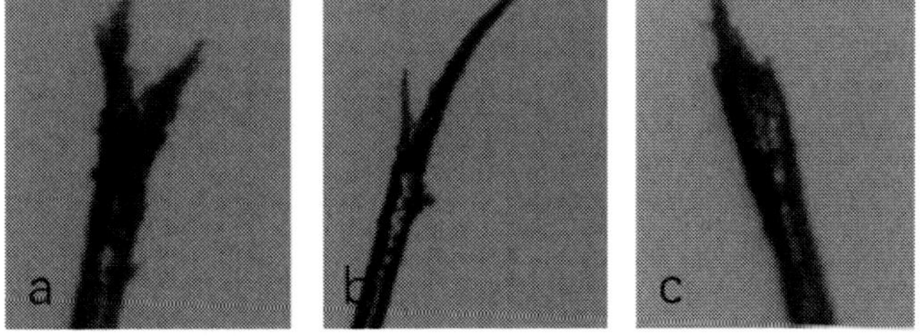

Fig.5 The fracture morphology of Nomex fiber (a) the pretension is 1.5cN/dtex; bending angle 600 and cyclic number is 12016 (b) the pretension is 1.5cN/dtex; bending angle 600 and cyclic number is11435 (c) the pretension is 2 cN/dtex; bending angle 450 and cyclic number is 14145

4. Conclusion

In this paper, we discussed the effect of pre-tension and bending angle on the bending fatigue fracture characteristic of high performance fibers by a new fatigue testing method. The fatigue fracture ends morphologies of fibers show that the failure mechanisms of Kevlar 49 and the PBO fiber are broken due to the fibrillation or axial split, which can be explained by the fibrillar structure and the high orientation and low intermolecular force. The fibril was suffered from tension and compression during the repeated fiber bending. When the suffered stress of fibril reaches its breaking stress, the fibril ruptures. Nomex fibers showed a torn tongue damage. The fracture ends show that the fatigue failure mechanism of the UHMW-PE fiber is obviously the accumulation of plastic deformation. At the same time, the fracture characteristic of fibers are different at different re-tension and bending angle.

5. Acknowledgement

This research was supported by the National Natural Science Foundation of China (NSFC 51503164), and the Natural Science Foundation of Hubei Province, China (No.2014CFB760).

References

[1] Ward IM, Cansfield DL. High performance fibers. In: Mukhopadhyay SK (ed.) Advance in fiber science. Chap. 1. Manchester: The Textile Institute: 1992; p. 10-12.
[2] Batra SK, Davis HA. Fiber microstructure and fatiue, Annual Report National Textile Center: 1997; 11: 225-228.
[3] Hearle JWS, Wong BS. Flexural fatigue and surface abrasion of Kevlar-29 and other high-modulus fibers. J Mater Sci: 1977; 12: 2447-2455.
[4] Hobbs RE, Burgoyne CJ. Bending fatigue in high-strength fiber ropes. Int J Fatigue: 1991; 13: 174-180.
[5] Liu XY, Yu WD. Bending fatigue properties of single aramid fibers. Chem Fibers Int: 2004; 54: 173-175.
[6] Gu BH. Bending fatigue properties of single wool fibers. J Text Res: 1992; 14: 157-159.
[7] Cai GM, Wang XG, Shi XJ, Yu WD. Evaluation of fatigue properties of high performance fibers based on fixed-point bending fatigue test. J Compos Mater: 2012; 46: 831-840.

Analysis on Van der Waals and Electrostatic Interaction of Crystalline Cellulose Iβ Based on Molecular Dynamics

Xue-Wei Jiang[1*], Hong-Hui Zhang[1], An-Hua Zhong[1]

[1]*College of Apparel Engineering, Wuhan Textile University, 1 Textile Road, Wuhan, Hubei, 430073, China*

[*]Corresponding author's email: xwjiang@wtu.edu.cn

Abstract

Cellulose Iβ is a main component in natural cotton and ramie fibers. The analysis on the interaction of cellulose Iβ is critical in understanding the dissolution and modification mechanisms of cellulose fiber. In this paper, we analyzed Van der Waals and electrostatic interactions of cellulose Iβ using the molecular dynamics simulations. We found that the Van der Waals and electrostatic interaction energy per chain are -43.91 kcal/mol and -26.24 kcal/mol at 298 K respectively. In order to understand these interaction mechanisms, the energy distribution of each residue was analyzed and divided into three types -- the intrachain, interchain and intersheet. The results show that Van der Waals interaction is important for stacking sheets, and the electrostatic interaction plays a certain role in the stability of intrachain and interchain.

Keywords: Van Der Waals Interaction; Electrostatic Interaction; Cellulose Iβ; Molecular Dynamics

1. Introduction

Cellulose fibers, in their natural and regenerated form, are an important part of textiles and other materials. All types of studies on natural crystalline cellulose and synthetic crystalline cellulose have attracted great attention in recent years [1-6]. Native crystalline cellulose consists of two crystal forms Iα and Iβ. Phase Iα exists widely in cell walls of some algae and bacteria, whereas cotton and ramie fibers are rich in Iβ phase [7]. Cellulose Iβ is found to be more thermally stable than Iα [8]. The research conducted by Nishiyama and coworkers describes the basic atomic structures and hydrogen bonding networks of native cellulose Iβ and Iα using a synchrotron X-ray and neutron diffraction analyses [4]. Understanding of the interaction of cellulose Iβ can help us understand the fundamental mechanisms of cotton and ramie fibers. It is pivotal in the modification of materials. In order to collect more information about the interaction of cotton and ramie fibers, the cellulose Iβ is discussed in this work.

Cellulose Iβ is a monoclinic $P2_1$ structure with two cellobiose chains in each unit cell, while a = 7.784 Å, b = 8.201 Å, c = 10.38 Å, α = β = 90°, γ = 96.5°, and c is the chain direction [4]. Crystalline cellulose Iβ is formed by stacking cellulose sheets, which are composed of linear chains. It is well known that the hydrogen-bonding interaction generates the main binding force for the maintenance of the molecular crystals. Each sheet is stabilized by the interchain hydrogen-bonding (O6H6...O3′) between the O6H6 (donor) in one chain and O3 (acceptor) in the neighboring chain. Each chain is stabilized by intrachain hydrogen bonds O3H3...O5 and O2H2...O6 [2, 4]. The binding forces between the sheets are thought to be Van der Waals forces and weak intersheet C-H...O hydrogen bonds [4, 9]. This assumption indicates the hydrogen bond is significant for thermal stability and solubility of cellulose. As such most studies on cellulose are mainly focused on the hydrogen bonding interaction [10-12]. Though the hydrogen bonding is the main binding force among chains, it is not all for stacking sheets. Van der Waals forces and electrostatic interaction should also be taken into consideration when studying the interaction of cellulose Iβ.

Molecular dynamics approach is an effective computer modeling technology in the study of theoretical mechanism at the atomic level. It can help us to understand the molecular interactions of cellulose. As the most direct method of theoretically investigating the behaviors of complex molecular systems, molecular dynamics method is well fit for studying the molecular interactions of cellulose Iβ. Molecular dynamics simulation has been widely used to comprehend the hydrogen-bonding interactions of cellulose in former studies [1, 2, 6, 10]. The simulation results are in good accordance with experimental results. These works are not accounting for Van der Waals force and electrostatic interaction. The interactions between cellulose sheets are not referred to within those references. In this paper, molecular dynamic methodology was adopted to investigate the Van der Waals and electrostatic interactions.

In the process of molecular dynamics simulation, the selection of force field is very important. In the work of Stortz and his co-worker, eighteen empirical force fields were compared with experiments and the semi-empirical quantum methods for studying β-cellobiose, α-maltose, and α-galabiose [13]. The results show that the empirical force fields GLYCAM-06, GROMOS, and MM3 are in accord with experiment [14]. They are also closely followed by MM4, CSFF, and OPLS-2005. Although the explicit hydroxyl hydrogens has been considered in GROMOS, united atom is used to represent aliphatic groups. The Van der Waals forces and electrostatic interactions are not considered in this force field. This leads to the relatively large discrepancy in the predicted unit cell parameters compared to the crystallographic data. GLYCAM-06 is an all atom carbohydrate force field. It is consistent and transferable for modeling carbohydrates and glycoconjugates. The work of Spiwork indicates that this force field was remarkably correspond to quantum metadynamics studies for free energy surface calculation [15]. This indicates that the GLYCAM-06 is a reasonable force field to study the molecular interactions of cellulose. Amber is introduced to simulate the Van der Waals and electrostatic interaction of cellulose Iβ, and this force field is compatible with Amber. In light of these works, we deployed GLYCAM-06 force field to explore some unsolved problem regarding interaction of cellulose Iβ.

2. Simulation and Interaction Analysis Methodology

In this section, a cellulose Iβ was built using the program cellulose-builder, and molecular dynamics simulations were carried out at first. The interactions were analyzed by MMPBSA.py of AmberTools.

2.1 Modeling and Simulation

Fig.1 The model of cellulose projected onto the ab base plane.

The initial structure of cellulose Iβ was built up according to cellulose elementary fibrils proposed by Ding and Himmel [9]. Cellulose Iβ was composed of 36 parallel chains, each chain was consisting of eight glucoses units (Fig.1). Cellulose Iβ was a monoclinic P21 structure, while unit cell parameters were set as a = 7.784 Å, b = 8.201 Å, c = 10.38 Å, α = β = 90°, γ = 96.5°, and c was the chain direction. The initial atom coordinates of cellulose Iβ were automatically generated by a user-friendly program cellulose-builder [16]. Except for the crystal direction c, no translational symmetry operations were done along other directions. No special action was set to endow crystallites with translational symmetry along the crystal direction. This meant there was no further translational symmetric operation after initial model was built. In order to set periodic covalent bonding along crystal direction, a hydrogen atom (H) and a hydroxyl group (OH) were not added to both ends of every chain respectively. In this model, Crystallographic surfaces (100) and (010) were exposed.

The molecular dynamics simulations were adopted to investigate the crystallization in water solution under the conditions of various temperatures (298 K, 350 K, 400 K, 450 K, 500 K and 550 K). As mentioned in introduction, the force field GLYCAM-06 was selected in the simulation process. In the force field GLYCAM-06, the 1-4 electrostatic and non-bonded interaction scaling factors were set to unity. All simulations were performed with Amber software and explicit water. We put a 12Å buffer of TIP3P water around the cellulose Iβ in each direction. In this way, all atoms in the cellulose Iβ initial structure were no less than 12 Å from the edge of the periodic solvent box. The next step was to choose a value for the non-bonded cut off. The larger the cut off value is, the less possibility the error is likely to emerge for the non-bonded force evaluation. With the value of threshold increasing, the computation becomes more complex and the calculation tends to be more time consuming. Taken together these two factors, 12 Å is a reasonable trade-off for all minimization and molecular dynamics simulations of this model. Here, non-bonded interactions were calculated with cutoff radii 12 Å in all molecular dynamics simulations.

The minimization procedure for solvated cellulose was integral and it was the best way to run a dual stage minimization. The first stage used position restraints on cellulose so that only the water was minimized. Then in the second stage, it turned to the whole system. The steepest descent minimization was performed 1000 steps firstly and then 1000 steps of conjugate gradient minimization. In the second stage, the steepest descent minimization was performed 2000 steps firstly and then 2000 steps of conjugate gradient minimization. The equilibration protocol allowed the whole system to heat up from 0 K to the target temperature. In order to avoid problems with wild fluctuations in the solute, the Langevin temperature equilibration scheme was used. Since the calculation of pressure was imprecise in the first few ps at low temperatures while using constant pressure periodic boundaries, we executed 1 ns of molecular dynamics at constant volume with weak restraints initially. The second stage of molecular dynamics was to run production dynamics at constant temperature and pressure (NPT) and get it closer to laboratory conditions. The motion balance was integrated with the time step of 2 fs. The bonds involving hydrogen atoms were kept rigid with the SHAKE algorithm. The Langevin dynamics were used to control the temperature corresponding to the canonical (constant T) ensemble. The pressure was controlled to be 1 bar through the Brendsen barostat. At each temperature, the procedure ran 100 ns and trajectory were saved every 10 ps.

2.2 Interaction Analysis Method

The topology files of cellulose Iβ and the simulation trajectory were used to calculate the interaction energy by program MMPBSA.py of AmberTools. This was a post-processing method with which representative snapshots from trajectory were used to calculate the interaction energy change between two glucose residues. This procedure performed Molecular Mechanics and Generalized Born Surface Area (MM/GBSA) calculations. Pairwise energy decomposition with 1-4 electrostatic interaction increased electrostatic potential and 1-4 Van der Waals interaction increased Van der Waals potential terms. It is generally known that the system can reach equilibrium within 50ns for cellulose Iβ of target temperature, so the snapshots of last 50ns from the whole trajectory were extracted to calculate the interaction between glucose residue pairs.

The Van der Waals and electrostatic interaction energy of each glucose residue i were reckoned by the formulas $VDW_i = \sum_j VDW_{ij}$ and $EEL_i = \sum_j EEL_{ij}$ respectively, where VDW_{ij} and EEL_{ij} represent the Van der Waals and electrostatic energy between glucose i and j respectively, j belongs to set A. The set A incudes the glucose residues belonging to the sheet or chain which we are interested in. The distribution of the Van der Waals and electrostatic interaction energy are described by the profile of (VDW, P(VDW)) and (EEL, P(EEL)) respectively, where P(VDW) = N(VDW)/N and P(EEL) = N(EEL)/N, N(VDW) and N(EEL) are the number of residues belonging to the value of free energy equals to VDW and EEL, N is the total glucose residues number.

3. Results and Discussions

As discussed the intersheet hydrogen bonding is very weak in the interior and surface regions, although the structure of cellulose is stable. This suggests that intersheet hydrogen bonding does not play the main role for the cohesion of sheets. Therefore, Van der Waals and electrostatic interactions of cellulose Iβ deserve further discussion.

3.1 Average Interaction Per Chain

Fig.2 displays the average Van der Waals and electrostatic intermolecular energy per chain at different temperatures. At 298 K, the Van der Waals and electrostatic intermolecular energy per chain are -43.91 kcal/mol and -26.24 kcal/mol respectively. Their standard deviations are 2.72 kcal/mol and 1.83 kcal/mol respectively. The low standard deviation indicates that the energy of each chain tends to be very close to the mean. This result is nearly consistent with the average nonbonded interaction energy (39.19 kcal/mol and 19.60 kcal/mol) determined by the earlier molecular dynamics simulation [6]. Attention should be paid to Van der Waals and electrostatic intermolecular energy especially when the stability of cellulose Iβ is mentioned. The Van der Waals energy gradually reduces when the temperature rises from 298 K to 400 K, meanwhile it remains almost unchanged as the temperature goes beyond 400 K. The electrostatic energy increases initially and then decreases. Results show that the Van der Waals energy per chain of different temperatures is higher than electrostatic energy at different temperatures. This indicates that the contribution of Van der Waals is greater than that of electrostatic to stability of cellulose. In order to analyze the correlations of Van der Waals and electrostatic energy to temperature furtherly, the details were discussed in the following. The following discussion focuses on the differences of the intermolecular energy distribution between intrachain, interchain and intersheet.

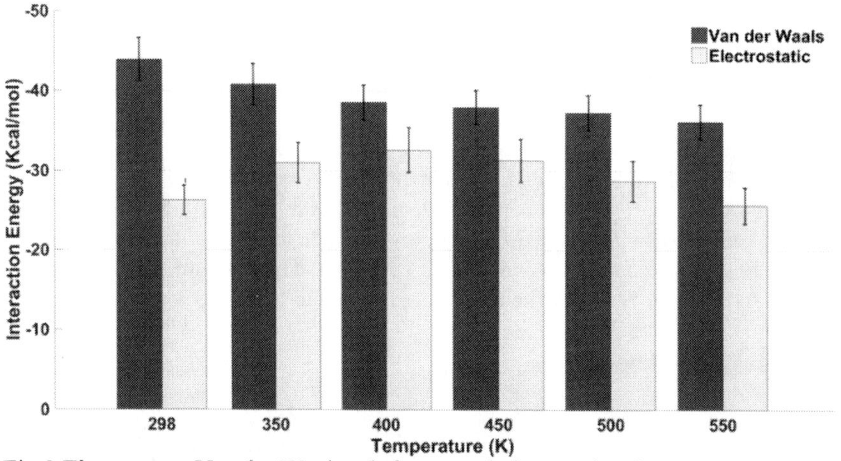

Fig.2 The average Van der Waal and electrostatic intermolecular energy per chain

3.2 Interaction Details

The Van der Waals interaction and electrostatic energy distributions of each glucose residue for the intrachain, interchain and intersheet of cellulose Iβ are demonstrated in Fig.3, Fig.4 and Fig.5 respectively. The absolute values could be discussed since interaction energy is negative. Fig.3 and Fig.4 show that the Van der Waals energy of all glucose residues for intrachain and interchain is less than 2 kcal/mol at 298 K in the main. The electrostatic energy of most glucose residues for intrachain lies on 11-14 kcal/mol and 5-8 kcal/mol (Fig.3), while located on 7-10 kcal/mol and 2-5 kcal/mol at 298 K (Fig.4) for interchain. These results indicate that the electrostatic interaction of intrachain is stronger than that of interchain, and the Van der Waals of intra-sheet is lower than electrostatic at room temperature. At 298 K, the electrostatic interaction energy of all glucose residues for intersheet is almost less than 2kcal/mol and weakest among different temperatures (Fig.5). Van der Waals interaction energy of most glucose residues between sheets is mainly concentrated in 8-10kcal/mol and 12-14 kcal /mol (Fig.5). The intersheet Van der Waals interaction is higher than electrostatic and thus Van der Waals is a significant binding forces between the sheets.

The Van der Waals and electrostatic interaction energy of residues pairs are almost less than 0.1 kcal/mol. This leads to the fact that they are quite small and do not contribute to the stability of cellulose Iβ. These interactions are often overlooked in the past works. According to the discussion above, we find that the Van der Waals and electrostatic interaction energy of many glucose residues are fairly strong at room temperature and they should not be ignored. Therefore, combining with the analysis of hydrogen bonding, we can conclude that the intrachain and interchain hydrogen bonding are main interactions among chains. This result is consistent with the views of most scholars [2, 4]. The electrostatic interaction energy in intrachain and interchain also plays a role for the stability of cellulose Iβ at room temperature. Based on the preceding discussion, intersheet hydrogen bonding energy is weak and the Van der Waals is relatively strong. The stacking sheets are stabilized by strong Van der Waals interaction energy and weak hydrogen bonding at room temperature. This result is in accordance with the findings of Nishiyama and other researchers [2, 6].

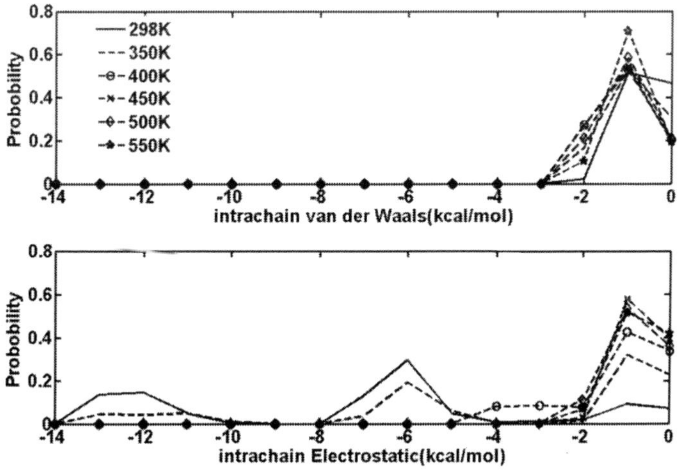

Fig.3 The distribution of intrachain Van der Waals and electrostatic interaction energy.

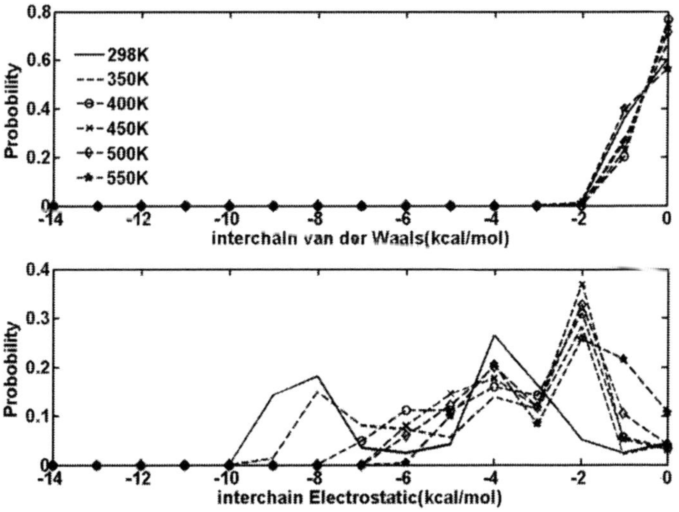

Fig.4 The distribution of interchain Van der Waals and electrostatic interaction energy.

The Van der Waals interaction energy in intrachain increases slightly as the temperature rises (Fig.3) and there is almost no growth for interchain (Fig.4). These results reveal that the intrachain and interchain Van der Waals interactions have no devotion to the thermal stability of cellulose Iβ. Compared to 298 K, the intrachain electrostatic energy decreases to some degree at 350 K (Fig.3). The electrostatic energy of some glucose residues for intrachain are still located on 11-14 kcal/mol and 5-7 kcal/mol, but probabilities are less than that of 298 K. As temperature increases to 400 K, the peak of electrostatic energy shifts to 3-4 kcal/mol. The electrostatic interaction energy of the most glucose residues reduces to the scope of 0-2 kcal/mol when temperature is higher than 450 K (Fig.3). The intrachain electrostatic energy is higher than Van der Waals at temperature 298-400 K and there is no significant difference between them at temperature higher than 400 K. These results suggest the intrachain electrostatic interaction contributes to the thermal stability of cellulose Iβ at temperature below 400 K. For the interchain, even though the electrostatic interaction decreases while the temperature increases, the electrostatic energy of some glucose residues is lying on the range of 3-6 kcal/mol at 550 K (Fig.4). The interchain electrostatic energy is higher than Van der Waals at temperature 298-550 K. This indicates that the interchain electrostatic interaction has more contribution to the thermal stability than that of intrachain

The intersheet Van der Waals interaction decreases slightly with the temperature increases, the Van der Waals energy of most glucose residues is located on the range of 6-12 kcal/mol at 550 K (Fig.5). This indicates that the Van der Waals interaction is important for the stacking sheets at different temperatures, and likewise plays a role to the thermal stability of cellulose Iβ. The electrostatic interaction for intersheet increases gradually with the temperature increasing from 298 K to 450 K, the electrostatic energy of most glucose residues is located on the range of 4-6 kcal/mol at 450 K (Fig.5). When the temperature rises from 450 K to 550 K, this interaction begins to decline. At 550 K, it is mainly distributed in 2-4 kcal/mol. These results show that the intersheet electrostatic energy are much less than Van der Waals energy at different temperatures. In spite of the electrostatic energy is much lower than Van der Waals, it should not be ignored and play a minor role to stack sheets at different temperatures. These results are consistent with the case of interaction energy per chain.

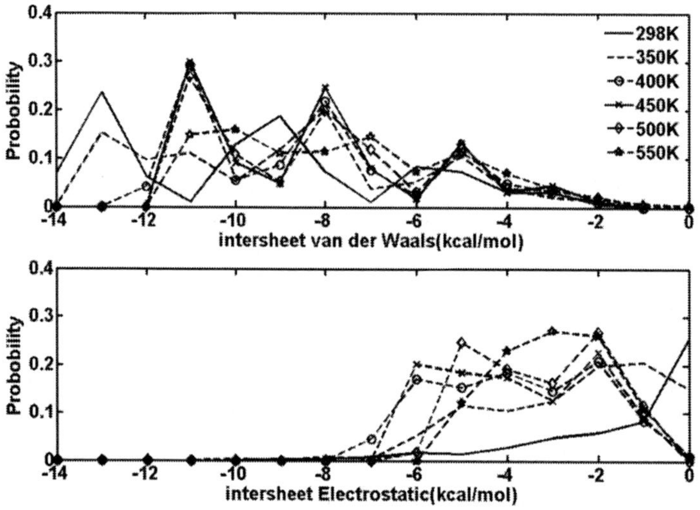

Fig.5 The distribution of intersheet Van der Waals and electrostatic interaction energy.

4. Conclusion

In summary, Van der Waals and electrostatic interactions of cellulose Iβ were studied through molecular dynamics simulations with the GLYCAM06 force field. The Van der Waals and electrostatic interactions for each glucose residue were analyzed combined with the interaction energy per chain. The results show that the Van der Waals and electrostatic interactions should not be ignored, although they are often regarded to be negligible. The stacking sheets are stabilized by strong Van der Waals interactions, minor electrostatic interaction and weak intersheet hydrogen bonding. The hydrogen bonding is the main factor for stabilizing sheets, but the intersheet electrostatic interaction should also be considered. These findings can shed some light on the mechanism of cellulose Iβ dissolution and modification. It should be noted, however, that the interactions were discussed only in water, and no comparisons were made to other solvents. Molecular interaction in different solvent environments will be examined in future.

5. Acknowledgement

This work is supported by the Scientific Research Project of Hubei Provincial Department of Education under Grants No. Q20141604. We would like to thank the Laboratory of Biomolecular Dynamics, Wuhan Institute of Physics and Mathematics, Chinese Academy of Sciences for computing resources.

References

[1] Liu H, Sale KL, Simmons, BA, Singh S. Molecular dynamics study of polysaccharides in binary solvent mixtures of an ionic liquid and water. J Phys Chem B: 2011; 115: 10251-8.

[2] Nishiyama Y, Johnson GP, French AD, Forsyth VT, Langan P. Neutron crystallography, molecular dynamics, and quantum mechanics studies of the nature of hydrogen bonding in cellulose Iβ. Biomacromolecules: 2008; 9: 3133-3140.

[3] Yui T, Okayama N, Hayashi S. Structure conversions of cellulose IIII crystal models in solution state: a molecular dynamics study. Cellulose: 2010; 17: 679-691.

[4] Nishiyama Y, Langan P, Chanzy H. Crystal structure and hydrogen-bonding system in cellulose Iβ from synchrotron X-ray and neutron fiber diffraction. J Am Chem Soc: 2002; 124: 9074-9082.

[5] Wada M, Kim UJ, Sasaki S. X-ray diffraction study on the thermal expansion behavior of cellulose Iβ and its high-temperature phase. Polym Degrad Stabil: 2010; 95: 1330-1334.

[6] Heiner AP, Sugiyama J, Teleman O. Crystalline cellulose Iα and Iβ studied by molecular dynamics simulation. Carbohyd Res: 1995; 273: 207-223.

[7] Atalla RH, Vanderhart DL. Native cellulose: a composite of two distinct crystalline forms. Science: 1984; 223: 283-285.

[8] Debzi EM, Chanzy H, Sugiyama J, Tekely P, Excoffier G. The Iα-Iβ transformation of highly crystalline cellulose by annealing in various media. Macromolecules: 1991; 24: 6816-6822.

[9] Qian X, Ding SY, Nimlos MR, Johnson DK, Himmel ME. Atomic and Electronic Structures of Molecular Crystalline Cellulose Iβ: A First-Principles Investigation. Macromolecules: 2005; 38: 10580-10589.

[10] Cai L, Liu Y, Liang H. Impact of hydrogen bonding on inclusion layer of urea to cellulose: Study of molecular dynamics simulation. Polymer: 2012; 53: 1124-1130.

[11] Rabideau BD, Agarwal A, Ismail AE. Observed Mechanism for the Breakup of Small Bundles of Cellulose Iα and Iβ in Ionic Liquids from Molecular Dynamics Simulations. J Phys Chem B: 2013; 117: 3469-3479.

[12] Watanabe A, Morita S, Ozaki Y. Study on temperature-dependent changes in hydrogen bonds in cellulose Iβ by infrared spectroscopy with perturbation-correlation moving-window two-dimensional correlation spectroscopy. Biomacromolecules: 2006; 7: 3164-3170.

[13] Stortz CA, Johnson GP, French AD, Csonka GI. Comparison of different force fields for the study of disaccharides. Carbohydr Res: 2009; 344. 2217-2228.

[14] Kirschner KN, Yongye AB, Tschampel SM, Gonzalez-outeirtno J, Daniels CR, Lachele Foley B, Woods RJ. GLYCAM06: a generalizable biomolecular force field. J Comput Chem: 2008; 29: 622-655.

[15] Spiwok V, Kralova B, Tvaroska I. Modelling of β-D-glucopyranose ring distortion in different force fields: a metadynamics study. Carbohydr Res: 2010; 345: 530-537.

[16] Gomes TC, Skaf MS. Cellulose-Builder: A toolkit for building crystalline structures of cellulose. J Comput Chem: 2012; 33: 1338-1346.

AUTHOR INDEX

AUTHOR INDEX

AUTHOR INDEX

AUTHOR INDEX

AUTHOR INDEX

Textile Bioengineering and Informatics Society (TBIS)
Unit 503, 5/F. Silvercord
Tower 2, 30 Canton Roa
Tsim Sha Tsui
Hong Kong

ISBN 978-1-5108-4951-8

9 781510 849518